Constitution of Binary Alloys

Metallurgy and Metallurgical Engineering Series

ROBERT F. MEHL, *Consulting Editor*

MICHAEL B. BEVER, *Associate Consulting Editor*

BARRETT · *Structure of Metals*

BRICK AND PHILLIPS · *Structure and Properties of Alloys*

BRIDGMAN · *Studies in Large Plastic Flow and Fracture*

BRIGGS · *The Metallurgy of Steel Castings*

BUTTS · *Metallurgical Problems*

DARKEN AND GURRY · *Physical Chemistry of Metals*

HANSEN · *Constitution of Binary Alloys*

HEINDLHOFER · *Evaluation of Residual Stress*

KEHL · *The Principles of Metallographic Laboratory Practice*

RHINES · *Phase Diagrams in Metallurgy*

SEITZ · *The Physics of Metals*

SMITH · *Properties of Metals at Elevated Temperatures*

WILLIAMS AND HOMERBERG · *Principles of Metallography*

CONSTITUTION OF BINARY ALLOYS

Dr. phil. **Max Hansen**
Managing Director, Metallgesellschaft A.G.
Frankfurt-Main, Germany. Formerly Manager, Metals Research Department
Armour Research Foundation, Chicago, Illinois

Second Edition

Prepared with the Cooperation of

Dr. rer. nat. **Kurt Anderko**
Research Metallurgist, Metallgesellschaft, A.G.
Frankfurt-Main, Germany. Formerly Research Metallurgist,
Metals Research Department
Armour Research Foundation, Chicago, Illinois

McGRAW-HILL BOOK COMPANY, INC.

NEW YORK TORONTO LONDON

1958

Constitution of Binary Alloys

Library of Congress Catalog Card Number 56-8862

THE MAPLE PRESS COMPANY, YORK, PA.

Preface

This book is a complete revision of "Der Aufbau der Zweistofflegierungen," published by Julius Springer, Berlin, in 1936. Since that time an enormous amount of information on the constitution of binary alloy systems and the crystal structure of intermediate phases has been published, the papers being widely scattered throughout the world's metallurgical, physical, chemical, and crystallographical literature. A large number of systems have been more accurately established, and phase diagrams of many additional systems have been determined, such as those of the lesser-known elements. Whereas "Der Aufbau der Zweistofflegierunegn" covered 828 systems and contained 456 diagrams, the present work includes 1,334 systems and 717 diagrams (with the addition of 48 systems and 33 diagrams in proof). The number of references has increased from about 5,500 to about 9,800.

As regards the deadline up to which the literature could be covered, it varies from the end of 1955 to the autumn of 1957 depending on the position of the respective system in the alphabetical order. However, the literature of 1956 and 1957 had in most cases to be considered in the form of Supplements or Notes Added in Proof; wherever necessary, additional diagrams were inserted.

Beside including these new results, the present book differs from the older German text in several respects:

1. Because of the many new papers to be considered, the mode of presentation had to be altered. Otherwise the book would have grown into a cumbersome volume and a prohibitive price would have had to be set. No longer could the results from different authors be critically discussed at great length and compared as to accuracy, reliability, etc. Instead, the text had to be confined to a more condensed presentation of the data. The authors have analyzed and evaluated all information as compiled in an extensive search. However, they have decided not to report all their deliberations and conclusions or to give in all cases the reasons why certain data were given preference. Their final conclusions are incorporated in the diagrams, which are, for the most part, composite diagrams. Text and diagram must be considered as a whole, and the reader is urged not merely to make use of the diagram but to read the related text as well. Uncertainties still existing and portions not yet established in the phase diagrams are indicated by dashed lines and/or question marks.

2. The diagrams differ from those in the German edition in a significant point: they are presented linearly in atomic per cent rather than in weight per cent. The authors themselves had no doubt that at this state of development of physical metallurgy the phase diagrams should be presented in atomic per cent. However, they have consulted in this matter some 60 leading metallurgists affiliated with research institutions, industry, and government agencies. A sizable majority have decided in favor of atomic per cent. Somewhat surprisingly, more than 50 per cent of the metallurgists working in industry or industrial laboratories preferred atomic per cent over weight per cent. Even more important than this numerical evaluation were

some of the comments made. Indeed, these were stronger in the answers favoring atomic percentage. After all, the interconversion of atomic and weight per cent is much less time-consuming than is widely assumed, if use is made of the conversion factors and tables included in the book.

3. The presentation of data concerning the constitution of binary alloy systems would be incomplete if crystallographic data were not considered systematically. As compared with the older work, not only is the symmetry of the intermediate phases given but their lattice spacings as well. The literature dealing with the variation of lattice parameters of primary solid solutions has also been considered.

For further details regarding these three points, reference is made to the Introduction.

The preparation of the manuscript was sponsored by the United States Air Force through the Wright Air Development Center, Air Research and Development Command, under Contract No. AF33(616)-193 with Armour Research Foundation of Illinois Institute of Technology, Chicago. The project was initiated by Mr. J. B. Johnson, Technical Director, Aeronautical Research Laboratory, and administered by Mr. E. J. Hassell, Chief, Metallurgy Research Branch, Aeronautical Research Laboratory, Wright Air Development Center. The authors are greatly indebted to both gentlemen for their initiative and continuous interest and advice during the course of the project. Likewise, they wish to express their sincere appreciation to the Armour Research Foundation, which provided encouragement and additional financial assistance to start and complete this undertaking. They are particularly grateful to Dr. C. E. Swartz and Mr. R. A. Lubker.

The authors were ably assisted by Miss Violet Johnson, who did, besides other helpful work, the laborious literature search, organized the card index, procured the foreign publications not available in Chicago libraries, and prepared excerpts of papers. Without her unselfish devotion to the work and her unfailing vigilance and enthusiasm, the project would not have progressed so well as it did. We owe our deepest thanks to Miss Johnson for her valuable contribution.

Also, the authors gratefully acknowledge the help given by a number of gentlemen who have aided in translating orally many of those papers published in foreign languages: Professor L. F. Mondolfo (Italian), Mr. A. Yamamoto (Japanese), and Mr. N. T. Bredzs and Dr. F. Jucaitis (Russian). Members of the Max Planck-Institut für Metallforschung, Stuttgart; the Institut für Metallkunde of the Bergakademie Clausthal; and Metall-Laboratorium of Metallgesellschaft A.G., Frankfurt, have assisted in preparing some 40 systems to expedite the completion of the manuscript. Valuable literature was kindly furnished, in particular, by Professor I. Obinata, Professor W. Köster, Dr. K. Schubert, and Dr. J. D. Summers-Smith.

The publication of this book would not have been possible without the relinquishing of publishing rights held by the Springer-Verlag, Berlin, for the revision of "Der Aufbau der Zweistofflegierungen." The senior author wishes to express his gratitude to Dr. Julius Springer for his liberal decision.

M. Hansen
K. Anderko

Acknowledgment

Preparation of this book was made possible through the cooperation of the Armour Research Foundation and the Aeronautical Research Laboratory, Wright Air Development Center.

The United States Air Force in its appreciation of scientific achievements felt that this extensive compilation and evaluation of technical data would aid metallurgists and engineers in their constant search for new knowledge and new metal products.

Contents

Systems are arranged in alphabetical order of the chemical symbols of the elements. This order is identical with that of the systems in the text. For an index containing the systems in alphabetical order of the names of the elements, see page 1287.

Systems with a diagram are indicated by an asterisk (*).

NOTE: For systems with columbium, see the respective system of niobium (Nb).

Contents

Introduction

Phase Designation. The nomenclature of phases in metallic systems has ever been a point of controversy. Some time ago Subcommittee III of Committee E-4 of the American Society for Testing Materials, under the chairmanship of Professor P. A. Beck, again took up this subject. The Subcommittee agreed that phase designation should be based on crystal structure and advanced proposals for a new nomenclature (based on the Bravais lattice concept) which is, however, still in the discussion stage. The authors have therefore decided to retain Greek letters for phases of variable composition and, wherever feasible, to couple certain crystal structures with certain letters, for example, β for b.c.c. and γ for the γ-brass type of structures. Chemical formulas are used for phases of fixed and nearly fixed composition. Primary solid solutions are designated by the chemical symbol of the base element in parentheses; in some cases a Greek letter is given additionally. It is realized that the designation used is neither consistent nor free from arbitrary exceptions.

Phase Diagrams. For the sake of greater clarity only the one-phase fields are lettered. Characteristic compositions in the diagrams (eutectics, solid solubilities, etc.) are given in both atomic and weight per cent, the latter in parentheses. A dashed curve indicates that the phase boundary, or portions thereof, has either not been reliably established or not been investigated as yet. Data points have been inserted in numerous diagrams to illustrate (*a*) discrepancies in experimental results or (*b*) the amount of work on which the results are based. Ticks at the upper composition abscissa indicating weight percentage, together with the compositions given in both percentages, in the diagrams as well as in the text, allow the reader to draw on graph paper a weight-percentage-vs.-atomic-percentage curve. This curve will serve as a means of interconverting with rough approximation the atomic and weight percentages.*

Interconversion of Atomic and Weight Percentages.† The conversion (for binary alloys) is done by means of the following formulas:

$$Y = \frac{100\dfrac{X}{A}}{\dfrac{X}{A} + \dfrac{100 - X}{B}} \qquad X = \frac{100YA}{YA + (100 - Y)B}$$

where X and Y are the percentages by weight and by atoms, respectively, of the element that has atomic weight A in a binary alloy with another element that has atomic weight B. The following rearrangement of these equations is easier to use:

* It is little known that if, for example, there are available 12 compositions as coordinates to draw this curve, actually 12 additional coordinates can be used without additional computation: if 30 wt. % A corresponds to about 42.5 at. % A, then 57.5 (100 − 42.5) wt. % A corresponds to 70 (100 − 30) at. % A.

† From C. S. Smith, *AIME Contribs.*, no. 60, 1933.

$$Y = \frac{100X}{X + \frac{A}{B}(100 - X)} \qquad X = \frac{100Y}{Y + \frac{B}{A}(100 - Y)}$$

The relation can be expressed also as a function of the ratio of the amounts of the two elements, because the ratio by weight is merely the ratio by atoms multiplied by the ratio of atomic weights, that is,

$$\frac{X}{100 - X} = \frac{A}{B}\,\frac{Y}{100 - Y}$$

On both sides of this equation, expressions of the form $x/(100 - x)$ occur. By the use of a table of $f(x) = \log [x/(100 - x)] + 10$ and a table of $\log (A/B)$, the multiplication becomes simple addition just as in the use of ordinary logarithmic tables.

In this book, a table of $f(x) = \log [x/(100 - x)] + 10$, with x varying from 0 to 4.99, 5.0 to 98.9, and 99.0 to 99.99, is given on page 1282. Values of $\log (A/B)$ are not tabulated but, for the convenience of the user, are listed in the heading of each alloy system, for example (see page 13):

$\bar{1}.9821$
0.0179

Ag-Cd Silver-Cadmium

Of these two figures, the upper represents

$$\log \frac{\text{atomic weight of Ag}}{\text{atomic weight of Cd}}$$

the lower,

$$\log \frac{\text{atomic weight of Cd}}{\text{atomic weight of Ag}}$$

In general, in a binary system A-B, the upper figure represents

$$\log \frac{\text{atomic weight of } A}{\text{atomic weight of } B}$$

the lower,

$$\log \frac{\text{atomic weight of } B}{\text{atomic weight of } A}$$

Use of the Conversion Factors. 1. Find from Table E (page 1282) the value of $f(x)$ corresponding to the percentage figure (either atomic or weight per cent) that is to be converted.

2. Look up the value of log (at. wt. A/at. wt. B) for the two components concerned. A is the element whose percentage is to be converted; B is the other, usually major, component.

3. If conversion is from *atomic* to weight per cent, *add* the two values. The resulting value of $f(x)$ corresponds to the desired weight percentage, which is found by consulting Table E again.

4. If conversion is from *weight* to atomic per cent, *subtract* $\log (A/B)$ to obtain a new value of $f(x)$ corresponding to the desired atomic percentage.

Examples. (*a*) What weight per cent corresponds to 20.0 at. % Cd in the system Ag-Cd?

$$\log \frac{x}{100 - x} + 10, \text{ for } x = 20.0 \ldots\ldots\ldots\ 9.3979$$

$$\text{Add: } \log \frac{\text{at. wt. Cd}}{\text{at. wt. Ag}} \ldots\ldots\ldots\ldots\ldots\ldots\ldots\ \underline{0.0179}$$

$$9.4158$$

The result, as can be seen from Table E, is 20.67 wt. % Cd.

(*b*) What atomic per cent corresponds to 2.41 wt. % Ag in the system Ag-Cd?

$$\log \frac{x}{100 - x} + 10, \text{ for } x = 2.41 \ldots\ldots\ldots\ 8.3926$$

$$\text{Subtract: } \log \frac{\text{at. wt. Ag}}{\text{at. wt. Cd}} \ldots\ldots\ldots\ldots\ldots\ \bar{1}.9821$$

$$8.4105$$

The result, as can be seen from Table E, is 2.51 at. % Ag.

Conversion of Small Percentages. Small percentages may be converted by using the approximate formula $Y = (B/A)X$. Sometimes it is simpler to use a correction factor $X(B/A - 1)$ to add to the weight percentage in order to obtain the atomic equivalent directly. Neither method should be used for percentages greater than about 0.3% except when the value of A/B is close to 1.

Temperature Conversions. A table of interconversion of degrees centigrade and degrees Fahrenheit will be found on page 1276.

Crystallographic Data. For crystallographic data, reference is also made to Table B, Structural Data of the Elements (page 1265), and Table C, Crystal-structure Types According to "Strukturbericht" (page 1270).

Efforts have been made to present all lattice spacings in "true" angstrom units (1 A = 10^{-8} cm). Lattice spacings given in angstroms (A or Å) in the literature up to 1946, when a new value for Avogadro's number was adopted at the International Conference in London sponsored by the British Institute of Physics, are actually in kX units (Siegbahn). Converting these to "true" angstrom units requires multiplication by the factor 1.00202.

Since 1946, unfortunately, some investigators continued to call the Siegbahn units "angstroms." Therefore, if the characteristic wavelength on which the calculations are based is not stated in publications, results expressed in A units may still really be in kX units.

0.6019
$\bar{1}$.3981

Ag-Al Silver-Aluminum

The Composition Range 0–42 At. (15.3 Wt.) % Al. Phase boundaries in this region have undergone considerable changes since the first diagram was published [1, 2]; those presented in Fig. 1 are based on the work of [3–7].

For the liquidus, the temperatures reported by [3] were used. Liquidus temperatures covering the whole composition range [8, 9] and the regions 17.4–26.6 at. % [10] and 25–40 at. % Al [11] are in substantial agreement with these data. The solidus curves of the α_{Ag} [4] and ζ [5] phases were established micrographically, whereas that of the β phase was obtained from heating curves of previously homogenized alloys [5]. The temperature of the peritectic reaction, melt + $\alpha_{Ag} \rightleftarrows \beta$, was reported as 770 [1], 772 [8], 771 [9], 779 [3], 778 [10], and 779.8°C [5]. The peritectic reaction, melt + $\beta \rightleftarrows \zeta$, was found to occur at 721 [1], 727 [8], 722 [9, 11], 729 [3], and 726.9°C [5].

The solid solubility of Al in Ag was the subject of numerous studies using micrographic [3, 12, 4, 5, 13, 14] and lattice-spacing methods [6, 7]. There is very good agreement in the results of [4–7] for the temperature range 780–450°C. Solubilities differ by maximal 0.6 at. % at 450°C and less at higher temperatures. At temperatures below 450°C, however, the divergence between the results of the micrographic analyses [4, 5] and the findings of the lattice-parametric work [6, 7] is considerable, particularly for the range 200–400°C [15]. The tabulated data of [6] in at. (wt.) % Al, used to plot the solubility boundary in Fig. 1, are as follows: 17.84 (5.15) at 780°C, 18.18 (5.27) at 750°C, 18.85 (5.50) at 700°C, 19.62 (5.76) at 650°C, 20.34 (6.0) at 610°C, 20.19 (5.95) at 550°C, 20.06 (5.90) at 500°C, 19.92 (5.86) at 450°C, 17.02 (4.88) at 400°C, 12.52 (3.46) at 300°C, and 8.75 (2.34) at 200°C [18]. Solubilities according to [7] are slightly less but do not differ at any temperature by more than 0.4 at. %.

Phase relationships and transformations below the solidus have been investigated by [12, 13, 16, 14, 17, 10, 11, 5]; however, only [12, 13, 14, 17, 5] have determined the solubility changes with temperature. Prior to the work of [5], it was believed that (*a*) the β phase decomposes eutectoidally into α_{Ag} and ζ and (*b*) the phase Ag_3Al, being of fixed composition, is formed by a peritectoid reaction of α_{Ag} and ζ. The temperatures of the three-phase equilibria were reported as about 610 [12], 615 [13], 600 [14, 17], 606 [10], and 610°C [5] for the eutectoid and about 390 [12], 420 [13, 17], 430 [14], 456 [10], and 448°C [5] for the peritectoid transformation.

High-temperature X-ray data [16] indicated that the alloy with 25 at. (7.70 wt.) % Al (Ag_3Al) lay in the ζ field at 570 and 465°C, in contrast to the diagrams of [12, 13, 14, 11] which would require the structure at these temperatures to consist of $\alpha_{Ag} + \zeta$. The diagrams of [17, 10] show the composition Ag_3Al nearly to coincide with the $\zeta/(\alpha + \zeta)$ boundary at temperatures between the two three-phase equilibria.

A thorough reinvestigation of the range 15–42.5 at. (4.2–15.6 wt.) % Al by means of micrographic analysis [5] resulted in the phase boundaries shown in Fig. 1. They are to be regarded as having a very high degree of accuracy. This study evidenced that (*a*) the ζ phase extends considerably further toward the Ag-rich side of the diagram than had previously been supposed; (*b*) the phase μ, formed peritectoidally at 448°C, is of variable, rather than fixed, composition and may just include the composition Ag_3Al at lower temperatures; (*c*) the $\zeta/(\mu + \zeta)$ boundary given by [13, 14, 17] was substantially correct. The minimum of the β field lies at about 24.5 at. (7.5 wt.) % Al, 603°C, and the three phases at 448°C have the compositions 19.9 [6], 23.2, and 23.8 at. % Al (5.86, 7.03, 7.28 wt. %).

The Composition Range 42 (15.3 Wt.) to 100 At. % Al. Liquidus temperatures [1, 8, 19, 20] agree quite well for the region above about 75 at. % Al; however, they diverge somewhat between 42 and 75 at. % Al. Therefore, the eutectic composition is still uncertain, as indicated by the approximate values reported in wt. % Al:

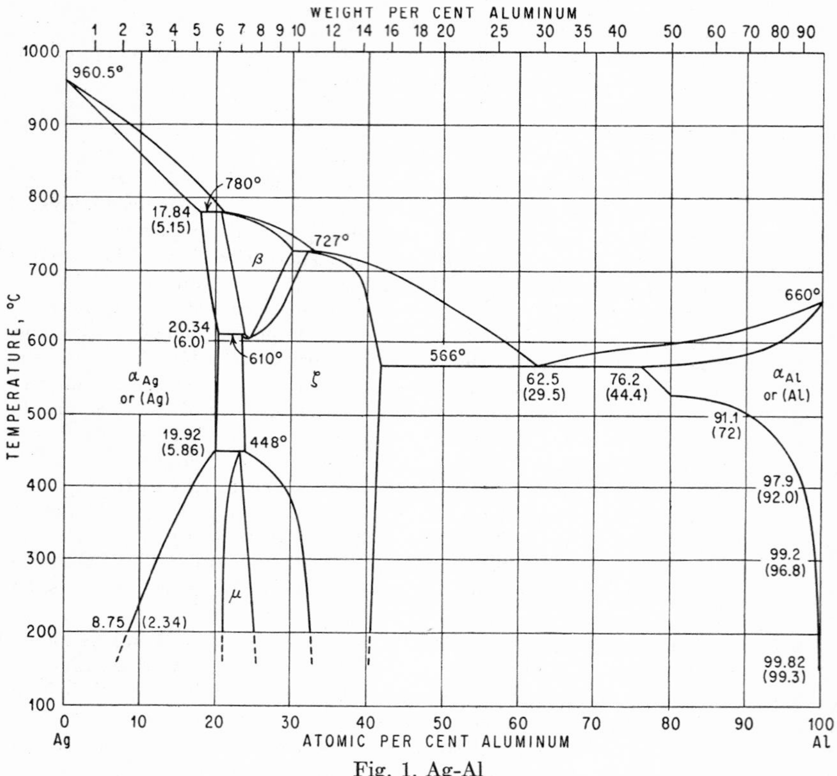

Fig. 1. Ag-Al

30 [1], 35 [21], 25.5 [8], and 33.3 [20]. The eutectic lies probably between 26 and 30 wt. (about 58.5–63 at.) % Al. The alloy with 29.5 wt. (62.5 at.) % Al was reported to consist entirely of eutectic [31]. The eutectic temperature was conclusively established as 566°C [22]; earlier values were 567 [1], 565 [21], 568 [8], and 558°C [19, 20].

The extensive solid solubility of Ag in Al was first recognized by [19]. It was found to increase from about 0.75 wt. (0.2 at.) % at 200°C to about 48 wt. (18.75 at.) % Ag at the eutectic temperature, on the basis of micrographic work. Results reported by [8] and [23] are in substantial agreement with this. Recently, [22] redetermined the solubility boundary for temperatures above 420°C and the solidus of the α_{Al} phase by micrographic analysis of alloys prepared by using a higher-grade Al than that of previous investigators. They found the maximal solubility to be even higher, 55.6 wt. (23.8 at.) % Ag at 566°C, and the solubility curve to have a marked change in direction near 49.4 wt. (19.6 at.) % Ag, 526°C (Fig. 1) [24]. The portion of the α_{Al} phase boundary between 150 and 420°C is based on electrical-resistivity measurements and microexamination [25]. According to the findings of [22, 25], the solubility of Ag in Al in wt. (at.) % Ag is 0.7 (0.18) at 150°C, 1.1 (0.29) at 200°C, 3.2 (0.8) at 300°C, 8.0 (2.15) at 400°C, 16 (4.5) at 450°C, 28 (8.9) at 500°C, 49.4 (19.6) at 526°C, and 55.6 (23.8) at 566°C.

Crystal Structures. For lattice parameters of the α_{Ag} phase, see [26–28, 12] and especially [6]. As shown by high-temperature X-ray work, the β phase (3:2 electron compound) is b.c.c. of the A2 type [13, 16]. The μ phase is isotypic with β-Mn (A13 type) [26, 29, 12, 13, 16, 14], a = 6.92–6.93 A [26, 13, 14]; see also [30]. The ζ phase, whose composition includes the electron concentrations of 3:2 and 7:4, is h.c.p. [26, 13, 16]. Its lattice dimensions change from a = 2.871 A, c = 4.662 A, c/a = 1.624 at 27 at. % Al to a = 2.885 A, c = 4.582 A, c/a = 1.588 at the Al-rich saturation limit [26].

The lattice spacing of Al was believed [26, 32, 33] to remain the same when Ag is taken into solid solution, owing to the near identity of the atomic radii of the elements. Indeed, [34] has shown that the parameter is practically unchanged up to 6 at. % Ag; between 6 and 14 at. % there is a small but definite increase in spacing (from about 4.0495 to 4.0504 A), while further additions up to 27 at. % Ag cause no further increase. Contradictory results were reported by [35].

1. G. I. Petrenko, *Z. anorg. Chem.*, **46,** 1905, 49–59.
2. Previously, H. Gautier (*Compt. rend.*, **123,** 1896, 109) had outlined the liquidus curve and C. T. Heycock and F. H. Neville [*Phil. Trans. Roy. Soc.* (*London*), **A189,** 1897, 69] had determined the effect of small Al additions on the melting point of Ag.
3. T. P. Hoar and R. K. Rowntree, *J. Inst. Metals,* **45,** 1931, 119–124.
4. W. Hume-Rothery, G. W. Mabbott, and K. M. Channel-Evans, *Trans. Roy. Soc.* (*London*), **A233,** 1934, 66–70.
5. W. Hume-Rothery, G. V. Raynor, P. W. Reynolds, and H. K. Packer, *J. Inst. Metals,* **66,** 1940, 209–239.
6. F. Foote and E. R. Jette, *Trans. AIME,* **143,** 1941, 151–157.
7. E. A. Owen and E. A. O. Roberts; see E. A. Owen and D. P. Morris, *J. Inst. Metals,* **76,** 1949-1950, 157–158.
8. E. Crepaz, *Atti congr. nazl. chim. pura ed appl. 3d Congr.*, **1929,** 371–379.
9. F. E. Tishchenko, *Zhur. Obshchei Khim.*, **3,** 1933, 549–557.
10. F. E. Tishchenko, *Zhur. Obshchei Khim.*, **9,** 1939, 729–731.
11. E. E. Cherkashin and G. I. Petrenko, *Zhur. Obshchei Khim.*, **10,** 1940, 1526–1530.
12. N. Ageew and D. Shoyket, *J. Inst. Metals,* **52,** 1933, 119–129.
13. I. Obinata and M. Hagiya, *Kinzoku-no-Kenkyu,* **12,** 1935, 419–429; *Science Repts. Tôhoku Univ., K. Honda Anniv. Vol.*, **1936,** 715–726; *Mem. Ryojun Coll. Eng.*, **10,** 1937, 1–10.
14. H. Kato and S. Nakamura, *Nippon Kwagaku Kwaishi,* **58,** 1937, 694–705.
15. See discussion by E. A. Owen and D. P. Morris, *J. Inst. Metals,* **76,** 1949-1950, 157–158; and W. Hume-Rothery, *J. Inst. Metals,* **76,** 1949-1950, 680–682.
16. W. Hofmann and K. E. Volk, *Metallwirtschaft,* **15,** 1936, 699–701.
17. S. Ishida, H. Tajiri, and M. Karasawa, *Rept. Aeronaut. Research Inst. Tokyo Imp. Univ.*, **13,** 1937, 694–705; H. Tajiri, *Tetsu-to-Hagane,* **24,** 1938, 357–370.
18. The corresponding wt. % values given by [6] are not correct.
19. M. Hansen, *Z. Metallkunde,* **20,** 1928, 217–222; *Naturwissenschaften,* **16,** 1928, 417–419.

20. F. E. Tishchenko and I. K. Lukash, *Zhur. Fiz. Khim.*, **9**, 1937, 440–448; Al used contained 0.9% Fe, 0.19% Si.
21. M. Tazaki, paper abstracted in *Kinzoku-no-Kenkyu*, **4**, 1927, 34.
22. G. V. Raynor and D. W. Wakeman, *Phil. Mag.*, **40**, 1949, 404–417; preliminary results in *J. Inst. Metals*, **75**, 1948-1949, 143–144.
23. G. Rassmann; see W. Guertler, *Metallwirtschaft*, **19**, 1940, 435–444.
24. K. Hirano and Y. Takagi (*J. Phys. Soc. Japan*, **9**, 1954, 730–735) redetermined the solubility boundary and reported it to coincide with that due to [22].
25. L. Rotherham and L. W. Larke, *J. Inst. Metals*, **81**, 1952-1953, 67–71.
26. A. Westgren and A. J. Bradley, *Phil. Mag.*, **6**, 1928, 280–288.
27. R. T. Phelps and W. P. Davey, *Trans. AIME*, **99**, 1932, 234–245; discussion, pp. 245–263.
28. C. S. Barrett, *Metals & Alloys*, **4**, 1933, 63–64, 74.
29. S. Fagerberg and A. Westgren, *Metallwirtschaft*, **14**, 1935, 265–267.
30. B. W. Roberts, *Phys. Rev.*, **75**, 1949, 1629.
31. E. C. Ellwood and K. Q. Bagley, *J. Inst. Metals*, **76**, 1949-1950, 631–642.
32. R. F. Mehl and C. S. Barrett, *Trans. AIME*, **93**, 1931, 90.
33. H. J. Axon and W. Hume-Rothery, *Proc. Roy. Soc.* (*London*), **A193**, 1948, 1–24.
34. E. C. Ellwood, *J. Inst. Metals*, **80**, 1951-1952, 605–608.
35. A. P. Guljaev and E. F. Trusova, *Zhur. Tekh. Fiz.*, **20**, 1950, 66–78; *Structure Repts.*, **13**, 1950, 6.

0.1584
$\bar{1}.8416$

Ag-As Silver-Arsenic

The diagram given in Fig. 2 was determined by thermal, microscopic [1], and lattice-parametric [2, 3] investigations. Earlier information is obsolete [4–6]. The liquidus curve of the As-rich melts is realizable only under increased pressure. The composition of the intermediate phase ϵ, which forms by a peritectic reaction at 595°C and decomposes eutectoidally at 374°C [1] or a temperature above 390°C [7], has not been determined precisely as yet; 7.5 wt. (10.5 at.) % As was given as a likely composition [1, 7]. Apparently it has a very small range of homogeneity [1, 7]. The solid solubility of As in Ag, determined by lattice-parameter measurements [2, 3], was found as follows: 595°C, 8.3–8.8 at. (5.9–6.3 wt.) % (extrapolated); 545°C, 8.0–8.5 at. (5.7–6.1 wt.) %; 500°C, 7.9 at. (5.6 wt.) %; 400°C, 5.2 at. (3.7 wt.) %; and 300°C, 4.3 at. (3.0 wt.) % As [8]. Ag is practically insoluble in solid As [7].

The ϵ phase is h.c.p. of the Mg (A3) type, $a = 2.897$ A, $c = 4.731$ A, $c/a = 1.633$ [7], or $a = 2.98_6$ A, $c/a = 1.63$ [9]. Lattice parameters of the Ag-rich phase were reported by [7] and especially [2].

1. W. Heike and A. Leroux, *Z. anorg. Chem.*, **92**, 1915, 119–126.
2. E. A. Owen and V. W. Rowlands, *J. Inst. Metals*, **66**, 1940, 361–378.
3. E. A. Owen and E. A. O'D. Roberts; see E. A. Owen and D. P. Morris, *J. Inst. Metals*, **76**, 1949-1950, 145–168, 701.
4. K. Friedrich and A. Leroux, *Metallurgie*, **3**, 1906, 192–195.
5. A. Descamps, *Compt. rend.*, **86**, 1878, 1023.
6. W. Guertler, "Handbuch der Metallographie," vol. 1, pp. 853–855, Verlagsbuchhandlung Gebrüder Borntraeger, Berlin, 1912.
7. S. J. Broderick and W. F. Ehret, *J. Phys. Chem.*, **35**, 1931, 3322–3329.
8. See also W. Hume-Rothery, *J. Inst. Metals*, **76**, 1949-1950, 682.
9. M. Miwa and Y. Watanabe, *J. Phys. Soc. Japan*, **3**, 1948, 52–56.

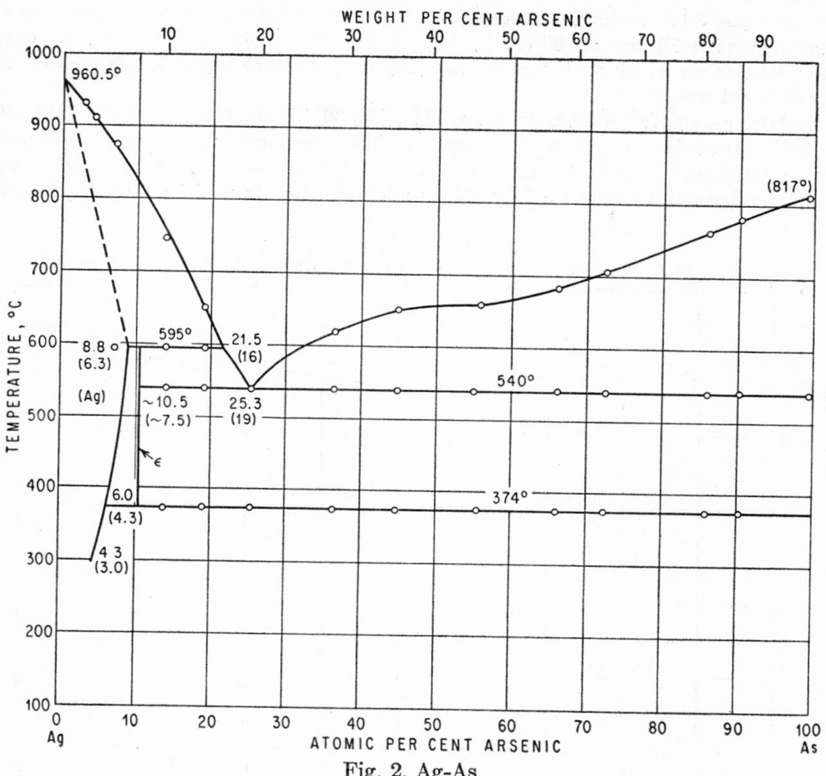

Fig. 2. Ag-As

1.7380
0.2620

Ag-Au Silver-Gold

The liquidus and solidus curves (Fig. 3) are due to [1, 2]. Results agree very well. Additional thermal-analysis data were reported earlier [3–6]. As the solidus curve was obtained by thermal analysis only, it cannot be regarded as an equilibrium curve. Indeed, [43] calculated the width of the gap between the liquidus and the solidus at 50 at. % to be only 1.3°C in contrast to an experimental value of approximately 10°C.

The results of studies of numerous physical properties are compatible with the assumption that the system consists of a continuous series of solid solutions, with no formation of ordered structures at temperatures between the solidus and room temperature [7]. The properties investigated are electrical conductivity [8–13], electrical resistivity [14, 15], temperature coefficient of electrical resistance [16, 10, 11, 13], thermal conductivity [12], thermoelectric force [10, 17, 18, 12, 13], magnetic susceptibility [19, 14, 15], Hall effect [20, 21], coefficient of linear expansion [22, 13], and density [23–25]. Measurements of the electrochemical potential at room temperature and elevated temperatures (200–745°C) have been carried out by [26, 27] and [27–30], respectively. Determinations of the rate of solution are due to [31–33].

Certain anomalies of thermodynamic properties (e.g., activity coefficient) have been considered compatible with the existence of intermediate phases or ordered structures [28, 30]. A diagram suggested by [28] shows two new phases below 800°C in the regions 0–18 and 50–80 at. % Au. It has been pointed out, however, that measurements on which such conclusions [28] are based are not accurate enough to be considered reliable data [29].

Lattice parameters have been reported by [34–39, 32, 40, 41]; those by [38, 32, 40] are to be regarded as the most accurate. The parameter-vs.-composition curve shows a minimum close to 70 at. % [38, 41] or close to 50 at. % Au [32, 40]. No superstructure could be detected by [42] in alloys of the compositions Ag_3Au, AgAu,

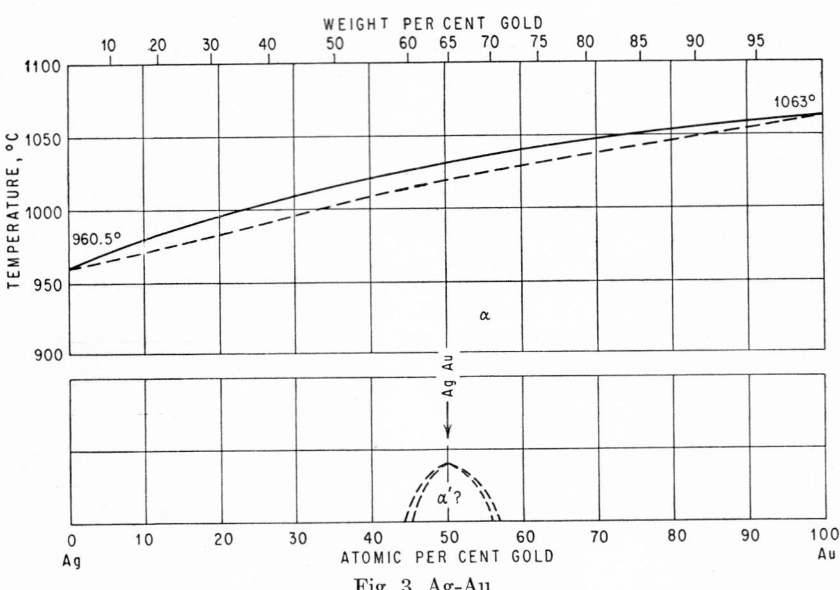

Fig. 3. Ag-Au

and $AgAu_3$. However, [42] did find evidence for a partial ordering of the atoms in the 50 at. % alloy. Also, [44] has shown that considerable short-range order is present in this alloy.

1. E. Jänecke, *Metallurgie*, **8,** 1911, 599–600.
2. U. Raydt, *Z. anorg. Chem.*, **75,** 1912, 58–62.
3. T. Erhard and A. Schertel, *Jb. Berg-u. Hüttenwes. Sachsen,* **1879,** 164.
4. H. Gautier, *Bull. soc. encour. ind. natl.*, **1,** 1896, 1318; Société d'encouragement pour l'industrie nationale, Paris, Commission des alliages, "Contribution à l'études des alliages," p. 93, Typ. Chamerot et Renouard, Paris, 1901.
5. C. T. Heycock and F. H. Neville, *Phil. Trans. Roy. Soc. (London)*, **A189,** 1897, 69. (Effect of small additions of gold on the melting point of silver.)
6. W. C. Roberts-Austen and T. Kirke-Rose, *Proc. Roy. Soc. (London)*, **71,** 1903, 161–163; *Chem. News,* **87,** 1903, 1–2.
7. J. A. M. van Liempt, *Rec. trav. chim.*, **45,** 1926, 203–206.
8. A. Matthiessen, *Pogg. Ann.*, **110,** 1860, 219–220.
9. W. C. Roberts-Austen, *Phil. Mag.*, (5)**8,** 1879, 58.
10. Strouhal and C. Barus, *Abh. kgl.-böhm. Ges. Wiss.*, **12,** 1883-1884.
11. B. Beckmann, Dissertation, Uppsala, 1911.

12. E. Sedström, *Ann. Physik,* **59,** 1919, 137–138; Dissertation, Stockholm, 1924.
13. W. Broniewski and K. Wesolowski, *Compt. rend.,* **194,** 1932, 2047–2049.
14. Y. Shimizu, *Science Repts. Tôhoku Univ.,* **21,** 1932, 829–834.
15. H. Auer, E. Riedl, and H. J. Seemann, *Z. Physik,* **92,** 1934, 291–302.
16. A. Matthiessen and C. Vogt, *Pogg. Ann.,* **122,** 1864, 42, 45–46, 53.
17. E. Rudolfi, *Z. anorg. Chem.,* **67,** 1910, 85–88.
18. G. Borelius, *Ann. Physik,* **53,** 1917, 615–628.
19. E. Vogt, *Ann. Physik,* **14,** 1932, 8–10.
20. L. S. Ornstein and W. C. van Geel, *Z. Physik,* **72,** 1931, 488–491.
21. E. van Aubel, *Z. Physik,* **75,** 1932, 119.
22. C. H. Johansson, *Ann. Physik,* **76,** 1925, 448–449.
23. A. Matthiessen, *Pogg. Ann.,* **110,** 1860, 36–37.
24. C. Hoitsema, *Z. anorg. Chem.,* **41,** 1904, 66–67.
25. E. Gebhardt and S. Dorner, *Z. Metallkunde,* **42,** 1951, 353–358.
26. A. P. Laurie, *J. Chem. Soc.,* **65,** 1894, 1031–1039.
27. G. Tammann, *Z. anorg. Chem.,* **107,** 1919, 144–152.
28. A. Ölander, *J. Am. Chem. Soc.,* **53,** 1931, 3577–3588.
29. C. Wagner and E. Engelhardt, *Z. physik. Chem.,* **A159,** 1932, 241–267.
30. A. Wachter, *J. Am. Chem. Soc.,* **54,** 1392, 4609–4617.
31. H. Borchers, *Metall u. Erz,* **29,** 1932, 392–398.
32. M. LeBlanc and W. Erler, *Ann. Physik,* **16,** 1933, 321–336.
33. I. N. Plaksin and S. V. Shibaev, *Izvest. Sektora Fiz.-Khim. Anal.,* **9,** 1936, 159–182.
34. L. W. McKeehan, *Phys. Rev.,* **20,** 1922, 424–432.
35. H. Weiss, *Proc. Roy. Soc. (London),* **108,** 1925, 652–654.
36. H. Jung, *Z. Krist.,* **64,** 1926, 425–429.
37. S. Holgersson, *Ann. Physik,* **79,** 1926, 42–46.
38. G. Sachs and J. Weerts, *Z. Physik,* **60,** 1930, 481–490.
39. P. Wiest, *Z. Physik,* **81,** 1933, 121–128.
40. O. Nygaard and L. Vegard, *Skrifter Norske Videnskaps-Akad. Oslo, I. Mat. Naturv. Kl.,* no. 2, 1947, 37–40; see *Structure Repts.,* **11,** 1947-1948, 126.
41. F. Hund and E. Trägner, *Naturwissenschaften,* **39,** 1952, 63.
42. A. Guinier and R. Griffoul, *Compt. rend.,* **221,** 1945, 555–557; A. Guinier, *Proc. Phys. Soc.,* **57,** 1945, 310–324.
43. C. Wagner, *Acta Met.,* **2,** 1954, 242–249.
44. J. W. Fitzwilliam, unpublished work; see S. Siegel and B. L. Averbach in "Phase Transformations in Solids," pp. 370, 384, John Wiley & Sons, Inc., New York, 1951.

0.9987
$\bar{1}$.0013

Ag-B Silver-Boron

Amorphous B does not dissolve in molten Ag at 1500–1600°C (in a hydrogen atmosphere) [1]. Cementation experiments of Ag with B had negative results [2].

1. H. Giebelhausen, *Z. anorg. Chem.,* **91,** 1915, 261–262.
2. W. Loskiewicz, *Przeglad Górniczo-Hutniczy,* **21,** 1929, 583–611; abstract in *J. Inst. Metals,* **47,** 1931, 516–517.

$\bar{1}$.8951
0.1049

Ag-Ba Silver-Barium

The diagram of this system (Fig. 4) is based exclusively on thermal-analysis data [1]. Because of experimental difficulties it was impossible to position the phase boundaries accurately in the range 50–100 at. % Ba. According to the investigator, the existence of the compounds $BaAg_4$ (24.15 wt. % Ba), Ba_3Ag_5 (43.31 wt. % Ba), and Ba_2Ag_3 (45.91 wt. % Ba) is fairly well established. This may hold for the compounds $BaAg_4$ and Ba_2Ag_3, both having maximum melting points. However, the composition of the phase formed peritectically at 797°C is somewhat doubtful although a corresponding compound was also found in the system Ag-Sr (page 54).

The constitution of the alloys with more than 50 at. % Ba is open to speculation. Certain evidence was found which could indicate the existence of two additional

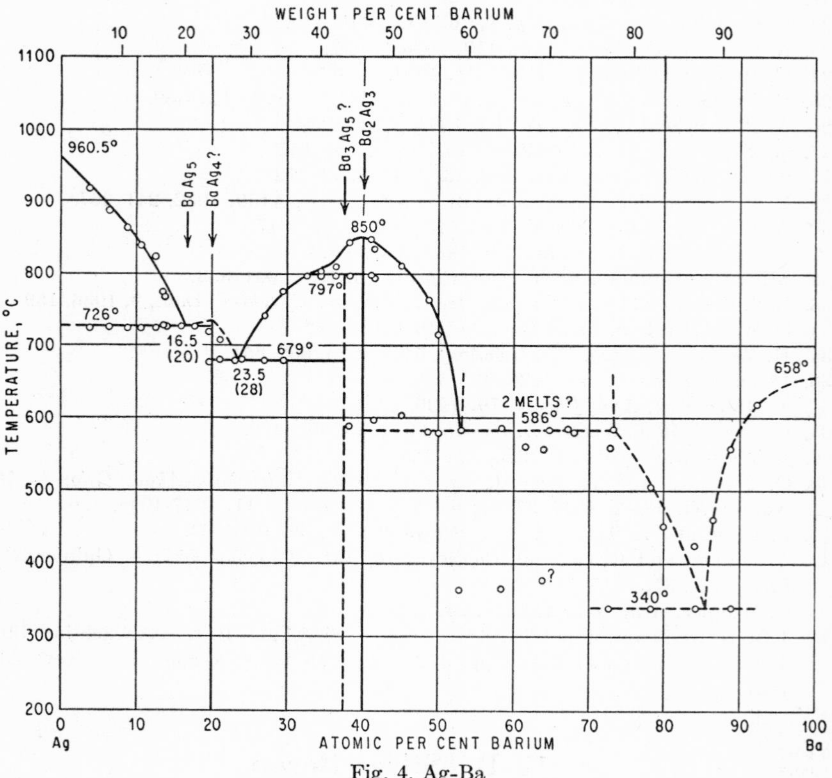

Fig. 4. Ag-Ba

compounds, Ba_4Ag_3 [57.14 at. (62.93 wt.) % Ba] and Ba_3Ag (79.25 wt. % Ba); however, if one assumes that the nearly horizontal portion of the liquidus (?) curve between about 53 and 73 at. % Ba indicates the occurrence of a miscibility gap in the liquid state [2], this does not appear too likely. Concerning the liquidus temperatures of the Ba-rich melts, it must be mentioned that the barium used contained 1.9% Sr and 0.15% N and solidified at 629°C rather than 658°C.

Recently, the phase $BaAg_5$ [16.67 at. (20.30 wt.) % Ba] was identified by determining its crystal structure. It is of the $CaCu_5$ type, with a = 5.708 A, c = 4.636 A, c/a = 0.812 [3]. Evidently the phase $BaAg_5$ would replace the compound $BaAg_4$ indicated in Fig. 4.

1. F. Weibke, *Z. anorg. Chem.*, **193,** 1930, 297–310.
2. Some evidence of the formation of two layers was reported.
3. T. Heumann, *Nachr. Akad. Wiss. Göttingen Math.-physik Kl.*, **2,** 1950, 1–6.

1.0781
$\bar{2}$.9219

Ag-Be Silver-Beryllium

Thermal and microscopic investigations of the entire system [1, 2] arrived at phase diagrams which are characterized by a eutectic—at 1.5 wt. (15.4 at.) % Be, 878°C [1], and 0.97 wt. (10.5 at.) % Be, 881°C [2]—between Ag-rich and Be-rich terminal solid solutions and a transformation at 750°C interpreted as an allotropic transformation of the beryllium-rich phase [3]. The detection of an intermediate

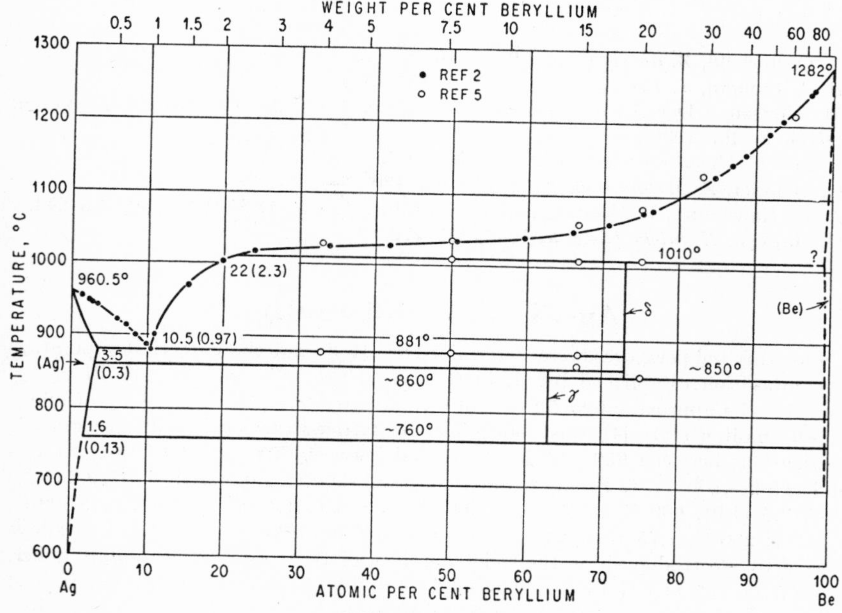

Fig. 5. Ag-Be

phase having a crystal structure of the $MgCu_2$ (C15) type [4], however, indicated that the phase relationship cannot be of the simple eutectic type. Indeed, a micrographic analysis of heat-treated alloys with 25, 33.6, 50, 66.7, 75, and 83 at. % Be (2.7, 4.0, 7.7, 14.3, 20, and 29 wt. %), corroborated by thermal analysis and X-ray studies [5], revealed the existence of two intermediate phases, stable only within a certain temperature range, with approximately 63 and 73 at. % Be (12.5 and 18.5 wt. %) (Fig. 5). The large discrepancy between the data might, at least in part, be due to the slow rate of the peritectic reaction at 1010°C and especially of the transformations in the solid state at approximately 860, 850, and 760°C. Although the existence of the γ and δ phases appears to be well established, additional work is warranted to fix their composition and temperature range of stability more precisely.

The liquidus curve given in Fig. 5 and the eutectic temperature (881°C) and composition [10.5 at. (0.97 wt.) % Be] are based on accurate data [2]. The solid solubility of Be in Ag was found by micrographic tests to be 3.5 at. (0.3 wt.) % Be at

881°C and about 1.6 at. (0.13 wt.) % Be at 750°C [2]. It is doubtful, however, that these figures represent equilibrium conditions. The transformation horizontal at 750°C, found by thermal analysis earlier [1, 2], cannot be identical with the 760°C eutectoid horizontal of Fig. 5 since the decomposition of the γ phase is very sluggish and would not be detected as a thermal effect on cooling curves. The solid solubility of Ag in Be was given as approximately 10 at. % [2] and "considerably under 5 at. % Ag" [5] (57 and 38 wt. %). An alloy with 90 wt. (99.1 at.) % Be was reported to consist largely of one phase, with small amounts of a white compound in the form of globules [6].

The δ phase has an $MgCu_2$ (C15) structure, with $a = 6.299_7$ A [4] or 6.34 A [5], although its Be content is considerably higher than that of Be_2Ag.

Supplement. The eutectic point was located at 16.5 at. (1.63 wt.) % Be, 878°C [7].

1. G. Oesterheld, *Z. anorg. Chem.*, **97**, 1916, 27–32.
2. H. A. Sloman, *J. Inst. Metals*, **54**, 1934, 161–176.
3. M. Hansen, "Der Aufbau der Zweistofflegierungen," pp. 12–13, Springer-Verlag OHG, Berlin, 1936.
4. L. Misch, *Metallwirtschaft*, **15**, 1936, 163–166.
5. O. Winkler, *Z. Metallkunde*, **30**, 1938, 162–170.
6. A. R. Kaufmann, P. Gordon, and D. W. Lillie, *Trans. ASM*, **42**, 1950, 785–844.
7. O. Hájiček, *Hutnické Listy*, **3**, 1948, 265–270.

$\bar{1}.7128$
0.2872

Ag-Bi Silver-Bismuth

Liquidus temperatures were determined by [1–3] and the eutectic temperature and composition established by [1, 3] (Fig. 6). The solid solubility of Bi in Ag, reported to be approximately 5.5 wt. (3 at.) % after "very slow" cooling of alloys from the molten state [4], was determined by lattice-parametric and micrographic investigations between 900 and 266°C [5] and between 259 and 200°C [6]. It was found as follows in wt. % Bi (at. % in parentheses): 900°C, 1.0 (0.5_2); 800°C, 2.3 (1.2); 700°C, 3.6 (1.9); 600°C, 5.05 (2.6_7); 500°C, 4.95 (2.6_3); 400°C, 3.85 (2.0_3); 300°C, 2.4 (1.2_5); 266°C, 1.75 (0.9_1); 226°C, 0.85 (0.4_4); and 200°C, 0.6 (0.3_3). In alloys prepared by electrodeposition the solid solubility of Bi in Ag is considerably higher, about 2.5 wt. (1.3 at.) % Bi [7]. The solid solubility of Ag in Bi, not yet determined in alloys prepared by melting [8], was found to be negligibly small in electrodeposited alloys [7]. For lattice parameters of Ag-rich alloys, see [5].

Supplement. [9] reported the eutectic point to lie at 94.95 at. (97.3 wt.) % Bi, 262°C.

1. C. T. Heycock and F. H. Neville, *J. Chem. Soc.*, **61**, 1892, 895; **65**, 1894, 73.
2. C. T. Heycock and F. H. Neville, *Phil. Trans. Roy. Soc. (London)*, **A189**, 1897, 67–68.
3. G. I. Petrenko, *Z. anorg. Chem.*, **50**, 1906, 136–139.
4. S. T. Broderick and W. F. Ehret, *J. Phys. Chem.*, **35**, 1931, 2627–2636.
5. E. Raub and A. Engel, *Z. Metallkunde*, **37**, 1946, 76–81; see also E. Raub and A. v. Polaczek-Wittek, *Z. Metallkunde*, **34**, 1942, 93–96.
6. H. H. Chiswik and R. Hultgren, *Trans. AIME*, **137**, 1940, 442–446.
7. E. Raub and A. Engel, *Z. Metallkunde*, **41**, 1950, 485–491.
8. See A. Goetz and A. B. Focke, *Phys. Rev.*, **45**, 1934, 170–199.
9. O. Hájiček, *Hutnické Listy*, **3**, 1948, 265–270.

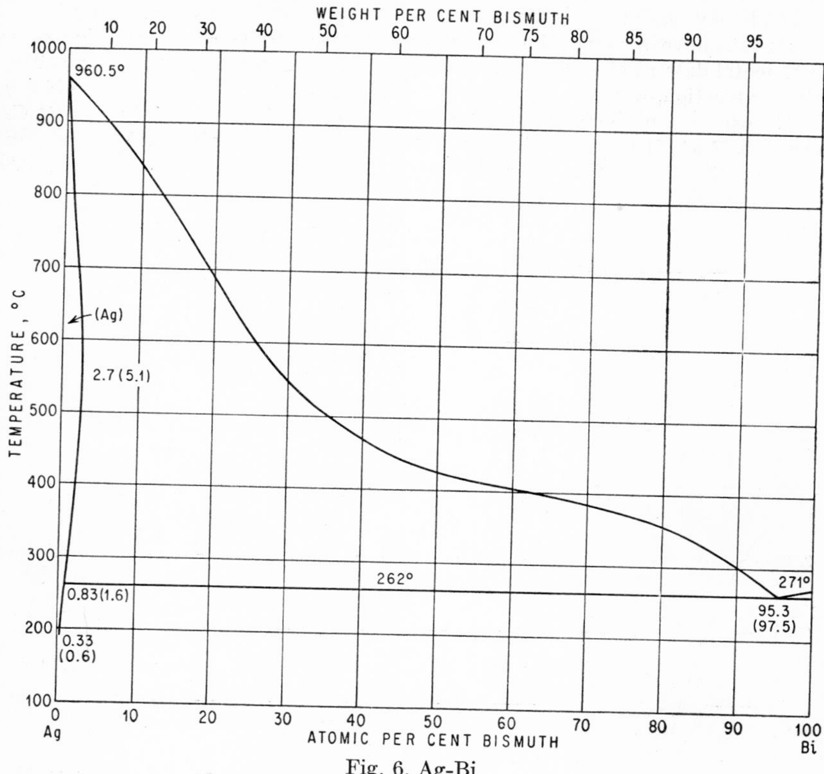

Fig. 6. Ag-Bi

0.9534
$\bar{1}$.0466

Ag-C Silver-Carbon

According to [1], the solubility of C in molten Ag at 1660, 1735, and 1940°C (assumed to be the boiling point) is 0.0012, 0.0025, and 0.0022 wt. %, respectively. [2] found a solubility of 0.026–0.04 wt. % C, at an unreported temperature. On cooling, the entire carbon content crystallizes in the form of graphite.

1. O. Ruff and B. Bergdahl, *Z. anorg. Chem.*, **106,** 1919, 91.
2. W. Hempel, *Z. angew. Chem.*, **17,** 1904, 324.

0.4300
$\bar{1}$.5700

Ag-Ca Silver-Calcium

The main diagram of Fig. 7 was determined by thermal analysis exclusively [1]. The constitution in the range 50 at. (27 wt.) % to 70 at. (46.5 wt.) % Ca is still uncertain. The author [1] suggested that solid solutions crystallizing from the melts with 50 at. % Ca (CaAg) to 66.6 at. % Ca (Ca_2Ag?) decompose at temperatures between 533 and 557°C into the compounds CaAg and Ca_2Ag(?). His presentation contradicts the laws of heterogeneous equilibria.

The diagram suggested by [2] shows the composition Ca_2Ag as the Ag-rich end of a solid solution extending from about 67 to 73 at. % Ca. The latter alloy was indicated to have a maximal melting point of 556°C. The interpretation of the experimental data as given in the inset of Fig. 7, showing that Ca_2Ag is formed by a peritectic reaction, would be more likely.

The existence of the compounds $CaAg_4$ (8.50 wt. % Ca), $CaAg_3$ (11.02 wt. % Ca), $CaAg_2$ (15.67 wt. % Ca), and CaAg (27.09 wt. % Ca) was also indicated by potential measurements [3]. Additional evidence for CaAg was given by [4]. On the other

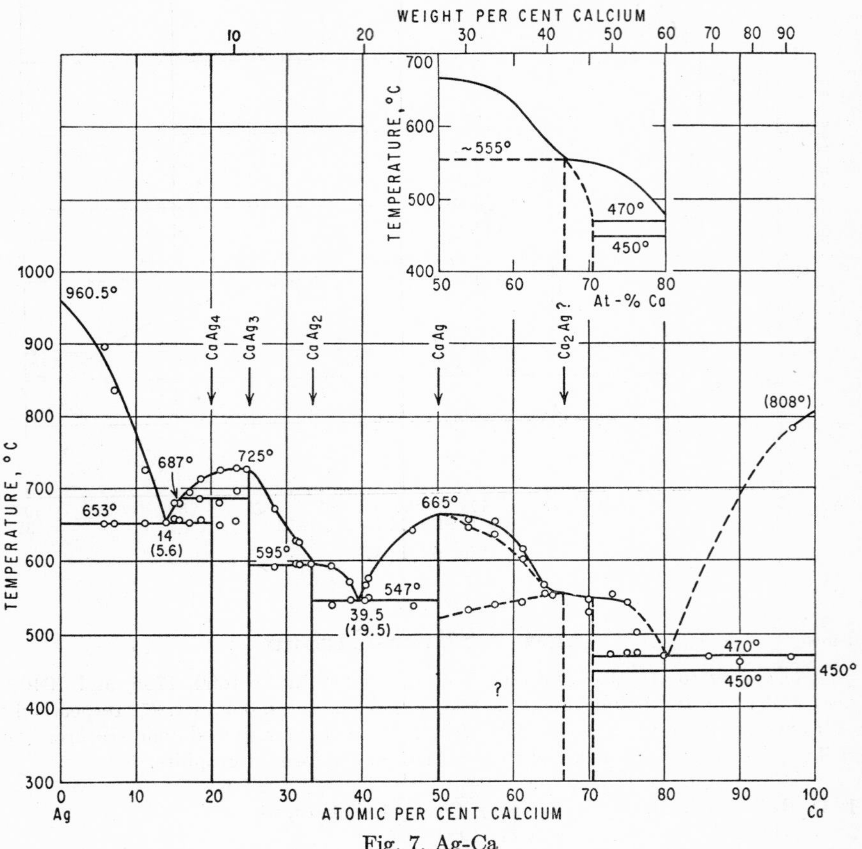

Fig. 7. Ag-Ca

hand, X-ray analysis [5] of alloys having compositions of the intermediate phases and eutectics according to Fig. 7 only confirmed the presence of $CaAg_3$ and CaAg. The remaining alloys were reported to consist of mixtures of Ag with $CaAg_3$, $CaAg_3$ with CaAg, and CaAg with Ca.

CaAg was reported to have a f.c.c. structure, $a = 9.09$ A [5]. However, the structure is definitely of lower symmetry [6–8]. By substituting 4 at. % Mg for Ag in $CaAg_2$ the hexagonal structure of the $MgZn_2$ (C14) type is formed [7]. The structure of $CaAg_3$ was given as tetragonal, $c/a = 0.88$ [5]. [8] confirmed that it is not cubic.

1. N. Baar, *Z. anorg. Chem.*, **70**, 1911, 383–392.
2. J. L. Haughton, "International Critical Tables," vol. II, p. 421, McGraw-Hill Book Company, Inc., New York, 1927.
3. R. H. Kremann, H. Wostall, and H. Schöpfer, *Forschungsarb. zur Metallkunde*, **1922**, no. 5.
4. C. A. Kraus and H. F. Kurtz, *J. Am. Chem. Soc.*, **47**, 1925, 43–69.
5. C. Degard, *Z. Krist.*, **90**, 1935, 399–407.
6. E. Zintl and G. Brauer, *Z. physik. Chem.*, **B20**, 1933, 245–271.
7. H. Nowotny, *Z. Metallkunde*, **37**, 1946, 34.
8. A. Iandelli, *Rend. seminar. fac. sci. univ. Cagliari*, **19**, 1949, 133–139.

$\bar{1}.9821$
0.0179

Ag-Cd Silver-Cadmium

Solidification. The liquidus curves determined by [1, 2] are, within the limits of experimental error, in agreement and also comport substantially with that presented in Fig. 8. Earlier work [3, 4] on the solidification of Ag-Cd alloys is obsolete. [2] were the first to recognize that the constitution of the system is characterized by the existence of two terminal solid solutions and three intermediate phases of variable composition [5] which form peritectic equilibria with the adjacent phases at 722, 630, 578, and 337°C.

The liquidus of Fig. 8 is based on the thermal data by [6], [7], [8], and [9], which cover the approximate composition ranges 0–16, 0–39, 39–67, and 59–100 at. % Cd, respectively. The peritectic temperatures are due to [8, 9].

Solid State. *The α Phase.* The solidus curve was determined micrographically [6], and the solubility boundary by micrographic [6] and lattice-parametric work [10]. Results as to the latter boundary agree at temperatures around 700°C but deviate by 0.5, 0.7, 0.75, and 0.8 at. % at 600, 500, 400, and 300°C, respectively, the solubilities based on X-ray data being higher; see [11]. Mean values were used to plot the curve in Fig. 8. There is a slight decrease in solubility between 400 and 250°C [10].

The Composition Range 42–60 *At.* % Cd. The constitution of this region was the subject of numerous investigations using thermal [2, 12, 8], micrographic [12, 8], and X-ray methods [13–18], as well as emf measurements at temperatures between 330 and 555°C [19]. Additional contributions were made by [20, 21]. The crystal structures will be reviewed below.

Phase boundaries in this range [12, 19, 8, 17] agree in principle but deviate as far as the location of the phase boundaries and the three-phase equilibria is concerned. Results by [8], established by thermal and micrographic analysis, are to be regarded as most accurate (Fig. 8). The partial diagram given by [12] is based almost totally on thermal data. The boundaries above 250°C at the Ag side of the β and ζ fields, determined by the lattice-parameter method [17], are in accordance with those reported by [8]; however, there is considerable divergence at the Cd side of these phase fields. The X-ray method admittedly is likely to give less reliable results because the β phase may not always be retained, even on rapid quenching [17]. The boundaries based on emf measurements [19] are in remarkable agreement with those shown in Fig. 8. On the whole, differences are not greater than one would expect with phase relations of this kind.

The existence of the three phases β, ζ, and β' in the vicinity of 50 at. % has been conclusively established by determination of the crystal structures (see below). By means of emf measurements, [19] found a transformation in the γ phase at temperatures varying between 436°C at the Ag side and 470°C at the Cd side of the homogeneity range (Fig. 8). [20] confirmed a transformation by means of measurements of modulus of elasticity vs. temperature; it starts at a much lower temperature, about 330°C, and probably ends at 460°C.

The Composition Range 60–100 *At.* % Cd. The boundaries of the phases γ, ϵ, and η were established by means of micrographic [8, 9] and X-ray work [17]. The latter authors found the ($\gamma + \epsilon$) field to be approximately 1.5 at. % wider than [9]. Its boundaries, shown in Fig. 8, were obtained by graphical interpolation of the data

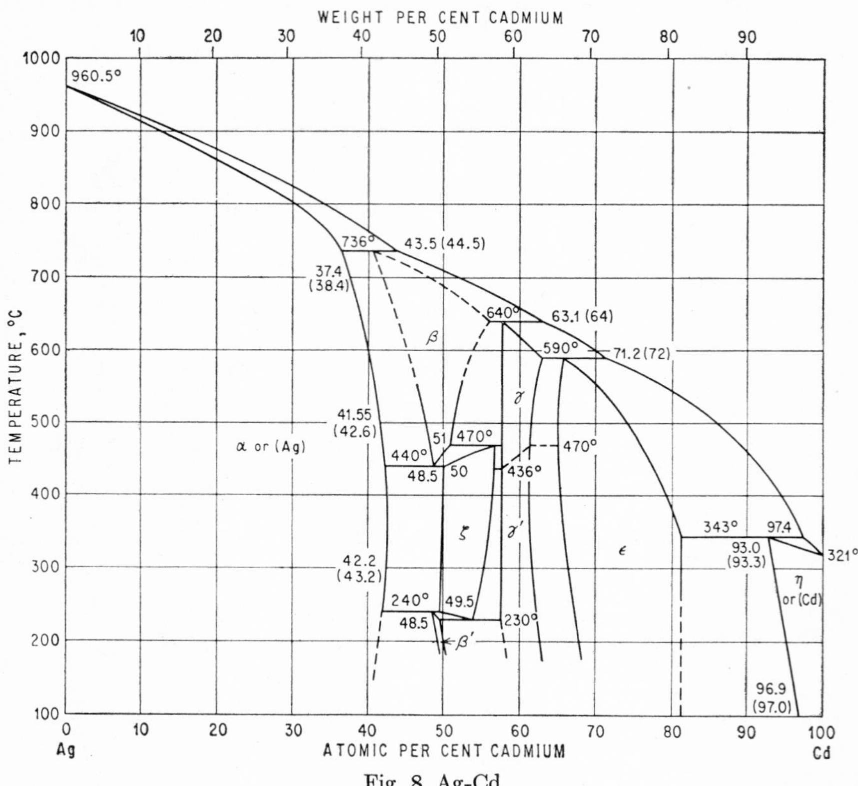

Fig. 8. Ag-Cd

of [8, 17]. The $\epsilon/(\epsilon + \eta)$ boundary is practically vertical at 81.4 [9] or 82.1 at. % Cd [17].

The solid solubility of Ag in Cd was found parametrically as 6.35, 5.55, 4.70, 3.9, and 3.1 at. (6.1, 5.3, 4.5, 3.75, and 3.0 wt.) % Ag at 300, 250, 200, 150, and 100°C [17]. According to [9], the solubility is 6.25 at. (6.0 wt.) % at 343°C and 5.4 at. (5.2 wt.) % Ag at 250°C.

Crystal Structures. Lattice parameters covering the whole α range are available [13, 14, 22, 10]. At 42.8 at. % Cd (saturation limit at 440°C) the parameter is $a = 4.1856$ A [10].

The β phase, a 3:2 electron compound [23], is b.c.c. (A2 type) [13, 24, 15, 17, 18]. [14, 16] believed this phase to be h.c.p.; however, it has been conclusively established, especially by high-temperature X-ray work [18], that the phase stable above 440–470°C is b.c.c., whereas ζ is h.c.p., with $c/a = 1.61$ [13, 15, 17, 25, 18]. Results reported by [16] are incomprehensible, since these authors claimed that the ζ phase was b.c.c., on the basis of X-ray patterns obtained at about 270°C. Lattice parameters of β were given by [13, 24, 14, 17]: a = 3.207 A at 47.4 at. % and a = 3.332 A at 49.9 at. % Cd [17]. The axial ratio of ζ changes from 1.619 at 46.8 at. % to 1.610 at 55.4 at. % Cd [14]. The β' phase was reported to be b.c.c. and very likely ordered [15, 16, 18]. [17] were unable to confirm this phase; however, there is no doubt that ζ undergoes a transformation with fall in temperature [2, 12, 8, 20, 18].

The γ phase is isotypic with γ-brass ($D8_2$ type) [13, 14, 24, 17, 26], with a = 9.955 A at 55.4 at. % and a = 10.002 A at 67.1 at. % Cd [14]. γ is a 21:13 electron compound [27]; its characteristic composition Ag_5Cd_8 (61.54 at. % Cd) lies close to the Cd side of the homogeneity range. Either the transformation $\gamma' \rightleftarrows \gamma$ is an order-disorder change or γ' has a slightly distorted γ lattice.

The ϵ phase, a 7:4 electron compound, is h.c.p. [13, 14, 24, 17, 28]. The axial ratio varies between 1.582 at 67.1 at. % and 1.555 at the Cd side of the homogeneity range, and the a axis between 3.046 and 3.099 A [14]. The lattice constants of the saturated η phase are $a = 2.99_6$ A, c/a = 1.816, as compared with a = 2.979 A, c/a = 1.886 for pure Cd [14].

1. G. Bruni and E. Quercigh, *Z. anorg. Chem.*, **68,** 1910, 198–206.
2. G. I. Petrenko and A. S. Fedorow, *Z. anorg. Chem.*, **70,** 1911, 157–168.
3. H. Gautier, *Bull. soc. encour. ind. natl.*, **1,** 1896, 1315; *Compt. rend.*, **123,** 1896, 173.
4. T. K. Rose, *Proc. Roy. Soc. (London)*, **A74,** 1905, 218–230.
5. Originally, [2] reported the existence of four intermediate phases. This was corrected, however, in a later paper published in *Z. anorg. Chem.*, **71,** 1911, 215–218.
6. W. Hume-Rothery, G. W. Mabbott, and K. M. Channel-Evans, *Phil. Trans. Roy. Soc. (London)*, **A233,** 1934, 1–97.
7. W. Hume-Rothery and P. W. Reynolds, *Proc. Roy. Soc. (London)*, **A160,** 1937, 282–303.
8. P. J. Durrant, *J. Inst. Metals*, **56,** 1935, 155–164.
9. P. J. Durrant, *J. Inst. Metals*, **45,** 1931, 99–113.
10. E. A. Owen and E. W. Roberts, *Phil. Mag.*, **27,** 1939, 294–327.
11. E. A. Owen, *J. Inst. Metals*, **76,** 1949-1950, 707.
12. W. Fraenkel and A. Wolf, *Z. anorg. Chem.*, **189,** 1930, 145–167.
13. G. Natta and M. Freri, *Rend. reale accad. Lincei*, **6,** 1927, 422–428, 505–511; **7,** 1928, 406–410.
14. H. Astrand and A. Westgren, *Z. anorg. Chem.*, **175,** 1928, 90–96.
15. A. Wolf, *Z. anorg. Chem.*, **189,** 1930, footnote on p. 152; *Z. Metallkunde*, **24,** 1932, 270; see also *Z. Metallkunde*, **22,** 1930, 369.
16. G. F. Kosolapov and A. K. Trapeznikov, *Zhur. Tekh. Fiz.*, **6,** 1936, 1131–1134.
17. E. A. Owens, J. Rogers, and J. C. Guthrie, *J. Inst. Metals*, **65,** 1939, 457–472.
18. L. Muldawer, M. Amsterdam, and F. Rothwarf, *Trans. AIME*, **197,** 1953, 1458–1459.
19. A. Ölander, *Z. physik. Chem.*, **A163,** 1933, 107–121.
20. W. Köster, *Z. Metallkunde*, **32,** 1940, 154–155.
21. G. R. Speich and D. J. Mack, *Trans. AIME*, **197,** 1953, 549–553; see also discussion, **200,** 1954, 675–676.
22. W. Hume-Rothery, G. F. Lewin, and P. W. Reynolds, *Proc. Roy. Soc. (London)*, **A157,** 1936, 167–183.
23. W. Hume-Rothery, P. W. Reynolds, and G. V. Raynor, *J. Inst. Metals*, **66,** 1940, 191–207.
24. V. M. Goldschmidt, *Z. physik. Chem.*, **133,** 1928, 397–419.
25. K. Moeller, *Z. Metallkunde*, **34,** 1942, 171–172.

26. H. Perlitz and R. Aavakivi, *Acta Comm. Univ. Tartu.*, **A35**, 1939, no. 2.
27. W. Hume-Rothery, J. O. Betterton, and J. Reynolds, *J. Inst. Metals*, **80**, 1952, 609–616.
28. E. O. Wollan, *Phys. Rev.*, **53**, 1938, 203.

$\bar{1}.8864$
0.1136

Ag-Ce Silver-Cerium

Thermal-analysis data and microstructures of seven alloys, prepared with Ce of 99.8+% purity, were used to outline the phase diagram given in Fig. 9 [1] and to

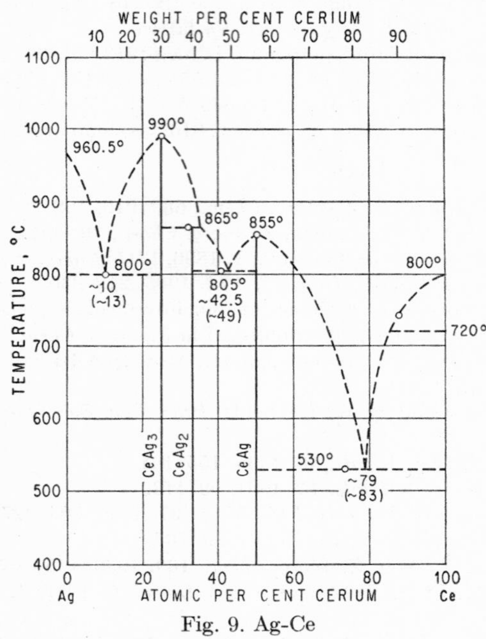

Fig. 9. Ag-Ce

prove that Ag forms the same compounds with Ce as with La and Pr: $CeAg_3$ (30.22 wt. % Ce), $CeAg_2$ (39.37 wt. % Ce), and CeAg (56.50 wt. % Ce). CeAg is b.c.c. of the CsCl (B1) type, $a = 3.75_3$ A [2], $a = 3.73_8$ A [3].

1. R. Vogel and T. Heumann, *Z. Metallkunde*, **35**, 1943, 29–42; see also R. Vogel and H. Klose, *Z. Metallkunde*, **45**, 1954, 670–671.
2. A. Iandelli, *Atti congr. intern. chim.*, *10th Congr. Rome*, **2**, 1938, 688.
3. H. Bommer and E. Krose, *Z. anorg. Chem.*, **252**, 1943, 62–64.

0.2625
$\bar{1}.7375$

Ag-Co Silver-Cobalt

Ag and Co have been reported to be completely insoluble in each other in the liquid state at 1600°C. The solidified mixtures consisted of two layers of the pure metals [1]. Potential measurements proved to be at least not contradictory to these results [2]. The solid solubility of Co in Ag, as determined by magnetic measurements of alloys prepared by adding very small amounts of Co to liquid Ag heated to 1000 and 1200°C, was found to be 0.0007 and 0.0004 wt. %, respectively [3].

[4] investigated the effect of Ag on the temperatures of the magnetic and polymorphic transformation of Co and concluded that up to 2.8 wt. % Ag was soluble in Co at temperatures above about 460°C. This is highly doubtful, however, considering the data given above [5]. In addition, an alloy with 0.2 wt. % Ag proved to be heterogeneous [5].

1. G. I. Petrenko, *Z. anorg. Chem.*, **53**, 1907, 215.
2. F. Ducelliez, *Bull. soc. chim. France*, **7**, 1910, 506–507.
3. G. Tammann and W. Oelsen, *Z. anorg. Chem.*, **186**, 1930, 279–280.
4. U. Haschimoto, *Nippon Kinzoku Gakkai-Shi*, **1**, 1937, 177–190.
5. W. Köster and E. Horn, *Z. Metallkunde*, **43**, 1952, 333.

0.3168
$\bar{1}$.6832

Ag-Cr Silver-Chromium

The phase diagram presented in Fig. 10 is based on the thermal-analysis data given and the fact that two layers were observed in the alloys with 15, 50, and 92 wt. (27, 67.5, and 96 at.) % Cr [1]. Alloys were prepared using Cr of 99.9% purity. At

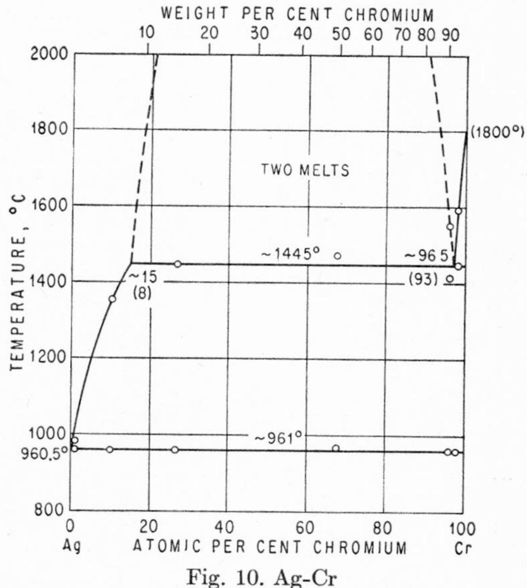

Fig. 10. Ag-Cr

the monotectic temperature, found as 1444, 1472, 1414, and 1488°C, the miscibility gap was assumed to extend from about 8 wt. (15 at.) % to about 93 wt. (96.5 at.) % Cr. The diagram suggested earlier [2] was similar, with the monotectic temperature at approximately 1465°C. However, the alloys were prepared with a low-grade Cr. In addition, they were heavily contaminated on melting, as clearly indicated by the low melting point of "chromium" observed (1550°C).

1. A. T. Grigoriev, E. M. Sokolovskaia, and M. I. Kruglova, *Moscow Univ. Vestnik, Ser. Fiz.-Mat. Estest. Nauk*, **9**, 1954, 77–81.
2. G. Hindrichs, *Z. anorg. Chem.*, **59**, 1908, 423–427.

0.2299
$\bar{1}$.7701

Ag-Cu Silver-Copper

Solidification. The liquidus curve was first published as early as 1875 [1]; and the existence of a eutectic (28.1 wt. % Cu, 748°C), formerly assumed to be the compound Cu_2Ag_3 (28.19 wt. % Cu) [2], was recognized by the same author [3]. The whole liquidus was determined by [4–7] and in the range 88–100 wt. % Cu by [8]. The curve shown in Fig. 11 was taken from the diagram proposed by [9]; it is based on the data of [4, 8]. The eutectic temperature was conformably given as 778–779°C; careful determinations by [10] yielded 779.4 ± 0.1°C, equivalent to 779.8°C on the

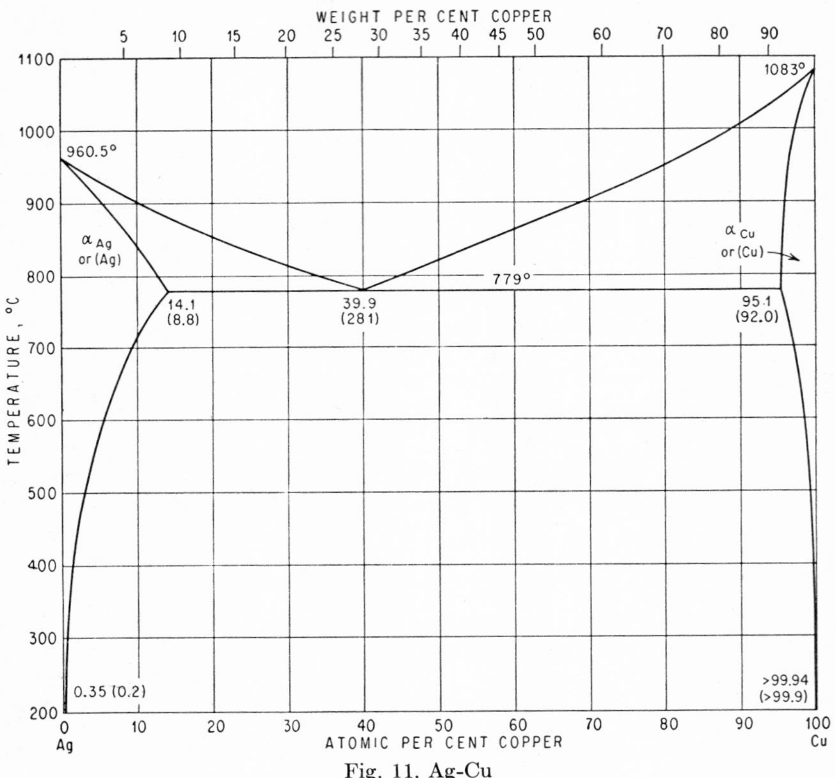

Fig. 11. Ag-Cu

1948 International Temperature Scale [9]. The eutectic composition was reported to be 28.0 [4], 28.5 [6], 28.06 [11], and 29.0 wt. % Cu [7]. Metallographic determination of the eutectic point proved to be impossible, as primary α_{Ag} or α_{Cu} or both appear in the microstructure of the same alloy, in the region 28–28.5 wt. % Cu, according to the rate of solidification [12]. The figure selected, 28.1 wt. (39.9 at.) % Cu, is probably not more than ±0.1 wt. % in error [9].

For the solidus of the Ag-rich phase [13–16], the curve determined micrographically by [15] was adopted. Determinations of the solidus of the Cu-rich phase [17, 15, 8] are in good agreement between 0 and 5 wt. % Ag; above this composition the curve by [15] probably is somewhat more accurate than that by [8]. Data by [16] do not represent equilibrium conditions.

Solubility of Cu in Ag. The change of the solubility with temperature has been determined by micrographic [13, 15], electrical-conductivity [15], and X-ray methods [18, 19] (Table 1). Possible reasons for the deviations were discussed by [20, 21]. The solubility of 1.3 wt. % Cu at 450°C, based on X-ray work [22], is in excellent agreement with 1.2 wt. % determined by [19] and 1.4 wt. % Cu by [15]. Determinations of the solubility by other authors using various methods [5–7, 13, 14, 23–33] need not be considered here [34], since they either do not represent equilibrium conditions or are only partial determinations. For a thermodynamic treatment, see [35].

Table 1. Solid Solubility of Cu in Ag in Wt. % Cu (At. % in Parentheses)

	779°C	750°C	700°C	600°C	500°C	400°C	300°C	200°C	100°C	0°C
Ref. 18	8.9*	7.0	5.2	3.1	1.7	1.0	0.6	0.4	0.2*	0.1*
Ref. 15	8.8	7.4	5.8	3.5	1.9	1.1	...	<0.8		
Ref. 19	8.5*	7.5	6.1	3.3	1.8	0.7	0.4	0.2	...	0.1*
Fig. 11..	8.8	7.4	5.8	3.3	1.8	0.7	0.4	0.2		
Fig. 11..	(14.1)	(11.9)	(9.5)	(5.5)	(3.0)	(1.2)	(0.7)	(0.35)		

* Extrapolated.

Solubility of Ag in Cu. Observations and partial determinations by [26, 5, 27, 28, 36, 32, 7] do not permit drawing quantitative conclusions. Also, the data reported by [30, 17, 6, 37, 38] need not be considered for various reasons. As regards the results of micrographic [15, 8], electrical-conductivity [15], and X-ray determinations [39, 40, 41, 19], see Table 2. At 450°C the solubility was found as 0.83 wt. % Ag [40], in very good agreement with [39, 19].

Table 2. Solid Solubility of Ag in Cu in Wt. % Ag (At. % in Parentheses)

	779°C	700°C	600°C	500°C	400°C	300°C	200°C
Ref. 39........	7.0*	4.4	2.4	1.3	0.6	0.25	<0.1
Ref. 15........	8.2	4.7	2.5	1.3	0.7		
Ref. 8.........	7.9	4.4	2.1	0.9	0.4		
Ref. 41........	9.1*	4.8	2.7	1.5	0.9		
Ref. 19........	8.4	5.5	2.9	1.4	0.5	0.2	0.1
Fig. 11........	8.0	5.2	2.6	1.3	0.5	0.2	<0.1
Fig. 11........	(4.9)	(3.1)	(1.55)	(0.8)	(0.3)	(0.12)	(<0.06)

* Extrapolated.

Crystal Structures. For lattice parameters of the Ag-rich and Cu-rich solid solutions, see [18, 22, 19] and [39, 40, 41, 19], respectively.

1. W. C. Roberts-Austen, *Proc. Roy. Soc. (London)*, **23,** 1875, 481–495.
2. A. Levol, *Ann. chim. et phys.*, **36,** 1852, 193; **39,** 1853, 163.
3. W. C. Roberts-Austen, *Engineering*, **52,** 1891, 579–580.
4. C. T. Heycock and F. H. Neville, *Phil. Trans. Roy. Soc. (London)*, **A189,** 1897, 32–36, 57–58.
5. K. Friedrich and A. Leroux, *Metallurgie*, **4,** 1907, 297–299.
6. T. Hirose, *Mem. Imp. Mint, Osaka No.* 1, 1927, pp. 1–74; *Proc. World Eng. Congr. Tokyo* (1929), **36,** 1931, 37.

7. W. Broniewski and S. Kostacz, *Compt. rend.*, **194**, 1932, 973–975.
8. C. S. Smith and W. E. Lindlief, *Trans. AIME*, **99**, 1932, 101–114.
9. J. C. Chaston, Annotated Equilibrium Diagram Series, no. 10, The Institute of Metals, London, 1953.
10. W. F. Roeser, *Bur. Standards J. Research*, **3**, 1929, 343–358.
11. D. Stockdale, *J. Inst. Metals*, **43**, 1930, 193–211.
12. J. A. A. Leroux and E. Raub, *Z. anorg. Chem.*, **178**, 1929, 257–271.
13. M. Hansen, *Z. Metallkunde*, **21**, 1929, 181–184.
14. M. Haas and D. Uno, *Z. Metallkunde*, **22**, 1930, 154–157.
15. D. Stockdale, *J. Inst. Metals*, **45**, 1931, 127–140.
16. G. Tammann and H. J. Rocha, *Z. Metallkunde*, **25**, 1933, 133–134; H. Nipper and E. Lips, *Z. Metallkunde*, **27**, 1935, 242–243.
17. F. Möller, *Metallwirtschaft*, **9**, 1930, 879–885.
18. N. Ageew and G. Sachs, *Z. Physik*, **63**, 1930, 293–303.
19. E. A. Owen and J. Rogers, *J. Inst. Metals*, **57**, 1935, 257–266; see also E. A. Owen, *J. Inst. Metals*, **73**, 1947, 471–489.
20. W. Hume-Rothery, *J. Inst. Metals*, **45**, 1931, 142–145.
21. N. Ageew, *J. Inst. Metals*, **45**, 1931, 147–148.
22. J. D. Bernal and H. D. Megaw, *J. Inst. Metals*, **45**, 1931, 149–152; H. D. Megaw, *Phil. Mag.*, **14**, 1932, 130–142.
23. A. Matthiessen, *Pogg. Ann.*, **110**, 1860, 190.
24. A. Matthiessen and C. Vogt, *Pogg. Ann.*, **116**, 1869, 369.
25. V. Barus and C. Strouhal, *U.S. Geol. Survey Bull.* 14, 1885, p. 85; *Ann. Physik Beibl.*, **9**, 1895, 353.
26. F. Osmond, *Compt. rend.*, **124**, 1897, 1094–1097, 1234–1237; *Bull. soc. encour. ind. natl.*, **2**, 1897, 837.
27. W. v. Lepkowski, *Z. anorg. Chem.*, **59**, 1908, 289–291.
28. N. Kurnakow, N. Puschin, and N. Senkowski, *Z. anorg. Chem.*, **68**, 1910, 123–140.
29. W. Fraenkel and P. Schaller, *Z. Metallkunde*, **20**, 1928, 237–243.
30. C. H. Johansson and J. O. Linde, *Z. Metallkunde*, **20**, 1928, 443–444.
31. A. L. Norbury, *J. Inst. Metals*, **39**, 1928, 149–150.
32. O. Weinbaum, *Z. Metallkunde*, **21**, 1929, 397–405.
33. M. Hansen, *Z. anorg. Chem.*, **186**, 1930, 41–48.
34. Some of the earlier results are discussed by [13, 14, 15, 8].
35. M. K. Arafa, *Proc. Phys. Soc. (London)*, **B62**, 1949, 238–241.
36. W. Fraenkel, *Z. anorg. Chem.*, **154**, 1926, 388.
37. P. Wiest, *Z. Physik*, **74**, 1932, 225–253; **94**, 1935, 176–183.
38. R. W. Drier, *Ind. Eng. Chem.*, **23**, 1931, 404–405, 970; see also C. S. Smith, *Ind. Eng. Chem.*, **23**, 1931, 969–970.
39. N. Ageew, M. Hansen, and G. Sachs, *Z. Physik*, **66**, 1930, 350–376.
40. H. D. Megaw, *Phil. Mag.*, **14**, 1932, 130–142.
41. E. Schmid and G. Siebel, *Z. Physik*, **85**, 1933, 41–55.

0.2859
$\bar{1}$.7141

Ag-Fe Silver-Iron

Data concerning the mutual solubility of Ag and Fe are controversial, and there appears to be some confusion in the discussion of the subject as to whether liquid solubility (alloyability) or solid solubility is meant.

Cooling curves of melts heated up to 1600°C showed thermal effects at the melting points of the pure metals, indicating that there is no appreciable solubility in the liquid state [1]. Silver added to molten steel is exuded in the form of globules on solidification [2]. Liquid solubilities as high as 0.7 wt. (0.37 at.) % Ag [3] and as low as 2 to 7 $\times 10^{-3}$% [4] have been reported to be retained in electrolytic iron and Armco iron, respectively.

Using a very sensitive magnetic method in studying the solid solubility of Fe in Ag, solubilities of 0.0006 and 0.0004% Fe were reported. The melts were heated to 1600 and 1000°C, respectively, prior to slow solidification [5].

Resistance measurements of Fe-rich alloys, prepared by sintering powder compacts at 950°C, were claimed to indicate a solid solubility of 0.5–1.0 wt. % Ag in Fe [6], although photomicrographs showed the presence of Ag as a separate constituent in an alloy containing 0.25 wt. (0.13 at.) % Ag. It has been stated [7], without giving data, that Ag is insoluble in solid Fe. However, dilatation curves showed that the temperature of the $\alpha \rightarrow \gamma$ transformation of Fe is lowered, indicating a certain solid solubility [8]. No diffusion of liquid Ag into Fe was observed at 1000°C [9].

1. G. I. Petrenko, *Z. anorg. Chem.*, **53**, 1907, 215.
2. J. Stodart and M. Faraday, *Quart. J. Science*, **9**, 1820, 319–330.
3. C. F. Burgess and J. Aston, *Trans. Electrochem. Soc.*, **22**, 1912, 241–250.
4. A. J. Dornblatt, *Trans. Electrochem. Soc.*, **74**, 1938, 280–283.
5. G. Tammann and W. Oelsen, *Z. anorg. Chem.*, **186**, 1930, 277–279.
6. C. G. Fink and V. S. de Marchi, *Trans. Electrochem. Soc.*, **74**, 1938, 271–280.
7. F. Wever, *Arch. Eisenhüttenw.*, **2**, 1928-1929, 739–746; *Naturwissenschaften*, **17**, 1929, 304–309.
8. E. Raub and W. Plate, *Z. Metallkunde*, **40**, 1949, 206–214.
9. N. W. Ageew and M. I. Zamotorin, *Izvest. Leningrad Politech. Inst., Otdel. Mat. Fiz. Nauk*, **31**, 1928, 15–29.

0.1896
$\bar{1}$.8104

Ag-Ga Silver-Gallium

The constitution of this system (Fig. 12) was established by thermal, micrographic, and X-ray investigations and is mainly due to [1]. More detailed studies were devoted to portions of the liquidus curve (up to 29.5 at. % Ga) [2, 3] and the solidus curve of the α phase [3, 2] and the limit of solid solubility of Ga in Ag [1–6]. The α-phase boundary was the subject of considerable controversy as regards its exact position. The reader is referred to this interesting discussion [6–8].

In Fig. 12 the α solidus curve by [3] and the solubility curve by the same investigators [3], both based on micrographic analysis, have been adopted, the latter essentially because of the long times of annealing used to ensure equilibrium. X-ray analysis [4, 5] yielded a very similar curve, shifted about 0.7 to 0.4 at. % to lower Ga concentrations [6].

The temperature of the peritectic reaction $\alpha + L \rightleftarrows \zeta$ was given as 619 [1] and 611°C [3]. The temperature of the peritectoid reaction $\alpha + \zeta \rightleftarrows \zeta'$ was found to lie at 440 [1] and 375–380°C [3, 6]. The former value is based on thermal analysis [1] (see data points) and the latter on micrographic and high-temperature X-ray work [3]; see also [5, 6]. No obvious reason can be given for this great temperature difference; however, the lower transformation temperature appears to be well founded. In accepting 380°C as the peritectoid temperature in the $\alpha + \zeta(\zeta')$ field, the transformation temperature in the $\zeta(\zeta')$ + liquid field was arbitrarily located at a temperature 5°C lower (Fig. 12). By thermal analysis the temperature difference was found to be about 2°C (440 vs. 438°C). The boundaries of the ζ, ζ', and δ phases [1] have not been positioned accurately as yet. [11] reported that he was unable to find the δ phase, in spite of annealing for 3 months at 300°C.

Crystal Structures. Lattice spacings of the α phase were determined by [4, 3]. The ζ and ζ' phases, also designated as β and γ, respectively [1], are 3:2 electron compounds [9]. $\zeta(\beta)$ has a h.c.p. structure [10, 3, 11], with a = 2.936 A, c = 4.757 A, c/a = 1.620 [11]. $\zeta'(\gamma)$ is isomorphous with the hexagonal ζ phase of the Ag-Zn system [10, 3], a = 7.81₆ A, c = 2.88₆ A [11].

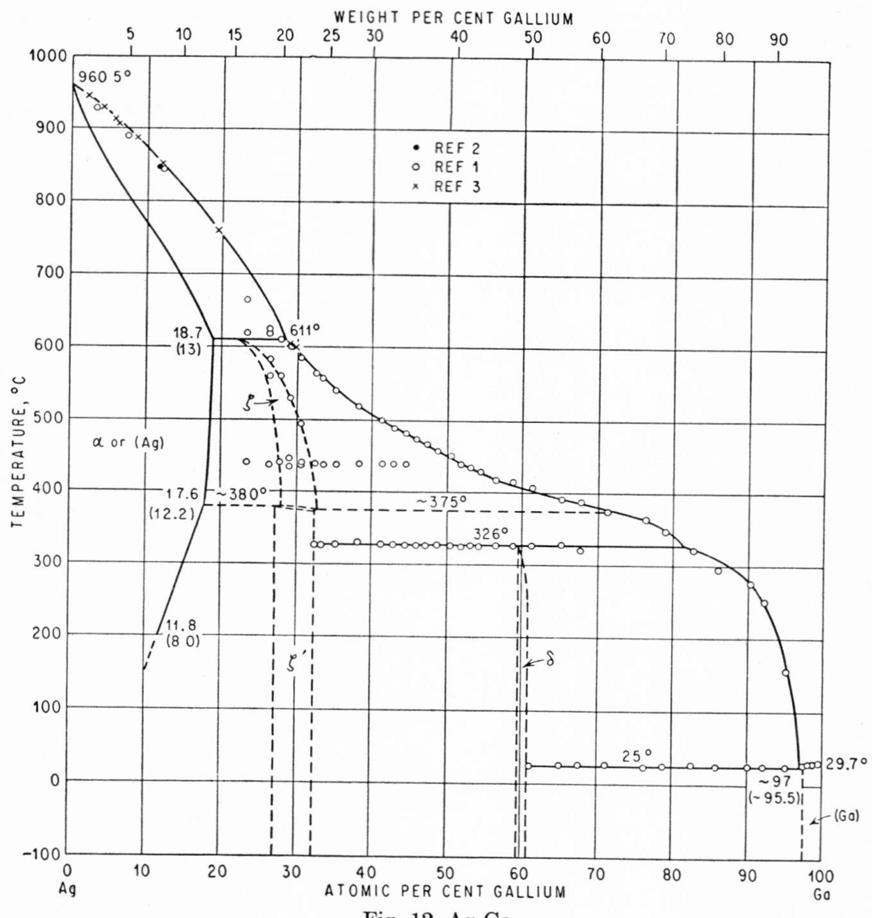

Fig. 12. Ag-Ga

1. F. Weibke, K. Meisel, and L. Wiegels, *Z. anorg. Chem.*, **226,** 1936, 201–208.
2. W. Hume-Rothery, G. W. Mabbot, and K. M. Channel-Evans, *Phil. Trans. Roy. Soc. (London)*, **A233,** 1934, 1–97.
3. W. Hume-Rothery and K. W. Andrews, *J. Inst. Metals*, **68,** 1942, 133–143.
4. E. A. Owen and V. W. Rowlands, *J. Inst. Metals*, **66,** 1940, 361–378.
5. E. A. Owen and D. P. Morris, *J. Inst. Metals*, **76,** 1949-1950, 145–168; especially pp. 160–162.
6. E. A. Owen and D. P. Morris, *J. Inst. Metals*, **76,** 1949-1950, 677–679, 699–700.
7. W. Hume-Rothery, *J. Inst. Metals*, **76,** 1949-1950, 679–680.
8. K. W. Andrews, *J. Inst. Metals*, **76,** 1949-1950, 690–691.

9. W. Hume-Rothery, P. W. Reynolds, and G. V. Raynor, *J. Inst. Metals*, **66**, 1940, 191–207.
10. K. Moeller, *Z. Metallkunde*, **31**, 1939, 19–20.
11. E. Hellner, *Fortschr. Mineral.*, **29–30**, 1950-1951, 58–61.

0.1720
$\bar{1}$.8280

Ag-Ge Silver-Germanium

The whole system was investigated twice by thermal and microscopic analysis [1, 2] (Fig. 13). The liquidus curves agree fairly well (see data points); the eutectic was given as 26.0 at. (19.1 wt.) % Ge, 650°C [1], and 24.1 at. (17.6 wt.) % Ge, 649°C

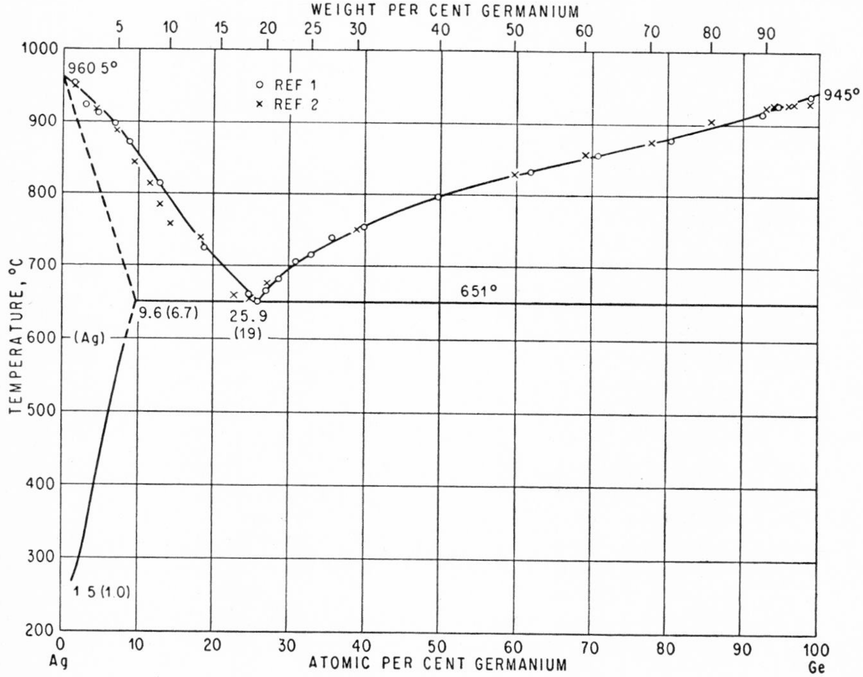

Fig. 13. Ag-Ge

[2]. The solid solubility of Ge in Ag at the eutectic temperature was reported as about 6.5 at. (4.5 wt.) % Ge [1], 7.9 at. (5.5 wt.) % Ge [2], and 8.1 at. (5.6 wt.) % Ge [3]. More careful work [4], using lattice-spacing studies of well-annealed alloys, however, showed the solubility to be 9.6 at. (6.7 wt.) % Ge (extrapolated), 7.9 (5.5), 6.7 (3.9), 3.8 (2.6), 2.7 (1.8), and 1.5 at. (1.0 wt.) % Ge at 651, 575, 470, 375, 308, and 270°C, respectively (Fig. 13). The solid solubility of Ag in Ge is negligible [1, 2]. Results of microscopic work [5] on alloys with 10–20 at. % Ge are in accordance with Fig. 13; the eutectic temperature was very accurately determined as 651 ± 0.5°C. Lattice parameters of the Ag-rich phase with up to 7.16 at. % Ge were reported by [4]; the Ag lattice is expanded by about 0.05%.

1. T. R. Briggs, R. O. McDuffie, and S. H. Willisford, *J. Phys. Chem.*, **33**, 1929, 1080–1096.

2. H. Maucher, *Forschungsarb. Metallkunde u. Röntgenmetallog.*, no. 20, 1936.
3. H. Nowotny and K. Bachmayer, *Monatsh. Chem.*, **81**, 1950, 669–678.
4. E. A. Owen and V. W. Rowlands, *J. Inst. Metals*, **66**, 1940, 361–378.
5. W. Hume-Rothery, G. V. Raynor, P. W. Reynolds, and H. K. Packer, *J. Inst. Metals*, **66**, 1940, 237–238.

2.0295
$\bar{3}$.9705

Ag-H Silver-Hydrogen

The solubility of hydrogen in solid Ag in the range of 200–900°C and at pressures from 50 to 800 mm was investigated by [1]. At a pressure of 800 mm the solubility, in cc H_2 per cc Ag, was found to be 0.019 at 600°C, 0.025 at 700°C, 0.036 at 800°C, and 0.046 at 900°C. At atmospheric pressure, the lattice parameter of Ag, in the range 15–140°C, is not affected by hydrogen additions [2]. For a review of the literature, see [3].

1. E. W. R. Steacie and F. M. G. Johnson, *Proc. Roy. Soc. (London)*, **A117**, 1928, 663–679; see also *Z. Metallkunde*, **21**, 1929, 44.
2. E. A. Owen and J. I. Jones, *Proc. Phys. Soc. (London)*, **49**, 1937, 590.
3. D. P. Smith, "Hydrogen in Metals," pp. 80–83, 268–269, The University of Chicago Press, Chicago, 1948.

$\bar{1}$.7306
0.2694

Ag-Hg Silver-Mercury

The phase diagram (Fig. 14) is essentially based on the thermal and microscopic investigation by [1] and the roentgenographic studies by [2–4]. The liquidus curve between 200 and 6°C was obtained by determining the solubility of Ag in liquid Hg (see below). In the alloys with more than about 35 at. % Ag, there is a strong tendency for the formation of metastable conditions on cooling after fusion. This is indicated by the fact that the peritectic reactions at 276 and 127°C are entirely or partially suppressed, according to the rate of cooling [1, 5].

The Intermediate Phases. Because of the work of [1–3] and others it now appears to be well established that there are two intermediate phases, ϵ (generally designated as β) and γ, with compositions around 60 wt. (44.65 at.) % Hg and about 70–71 wt. (55.7–57 at.) % Hg, the latter formerly often designated as Ag_3Hg_4 [6–9] [57.14 at. (71.26 wt.) % Hg]. In the earlier literature [6, 7, 10–12] the existence of numerous different compounds was suggested; however, most of the conclusions were based on inconclusive evidence, more so since equilibrium conditions could not have been attained in these studies. Also, most of the methods used would give unreliable results [12].

The Solid Solubility of Hg in Ag. According to [1], the solubility limit—formerly reported to lie at only 2 wt. % [13] and about 17 wt. (10 at.) % Hg [8]—is 44–45 wt. (29.7–30.6 at.) % Hg at 276°C and approximately 50 wt. (35 at.) % Hg at lower temperatures. [2] found 46 ± 2 wt. (about 30–33 at.) % Hg at 100°C. Lattice-parametric work [4] showed the solubility to be 37.3, 36.7, and 36.5 at. (52.4, 51.8, and 51.6 wt.) % Hg at 276, 200, and 100°C, respectively (Fig. 14). The solidus curve, first outlined by [1], was also determined parametrically between 560 and 416°C [4] (Fig. 14).

The Solubility of Ag in Liquid Hg. The composition of the liquid phase in equilibrium with the γ and ϵ phases was determined by [14] (14–163°C), [15] (80–200°C), [16] (9–81°C), and [17] (6–19°C); see also [18]. According to these investigations, the solubility in wt. % Ag (at. % in parentheses) at 20, 50, 100, 150, and

200°C is 0.035 (0.066), 0.08 (0.145), 0.23 (0.41), 0.48 (0.92), and 0.93 (1.75) % Ag, respectively.

Crystal Structures. There are many publications dealing with this subject [19–21, 2, 3, 9, 22, 4, 23–28]. The existence of a b.c.c. phase with the structure of γ-brass ($D8_{1-3}$ type) was reported by [19–21, 2, 3, 22–24, 28]; it has a lattice parameter of about 10.0 A [2, 3, 22–24]. According to [3], the parameter of the γ phase changes

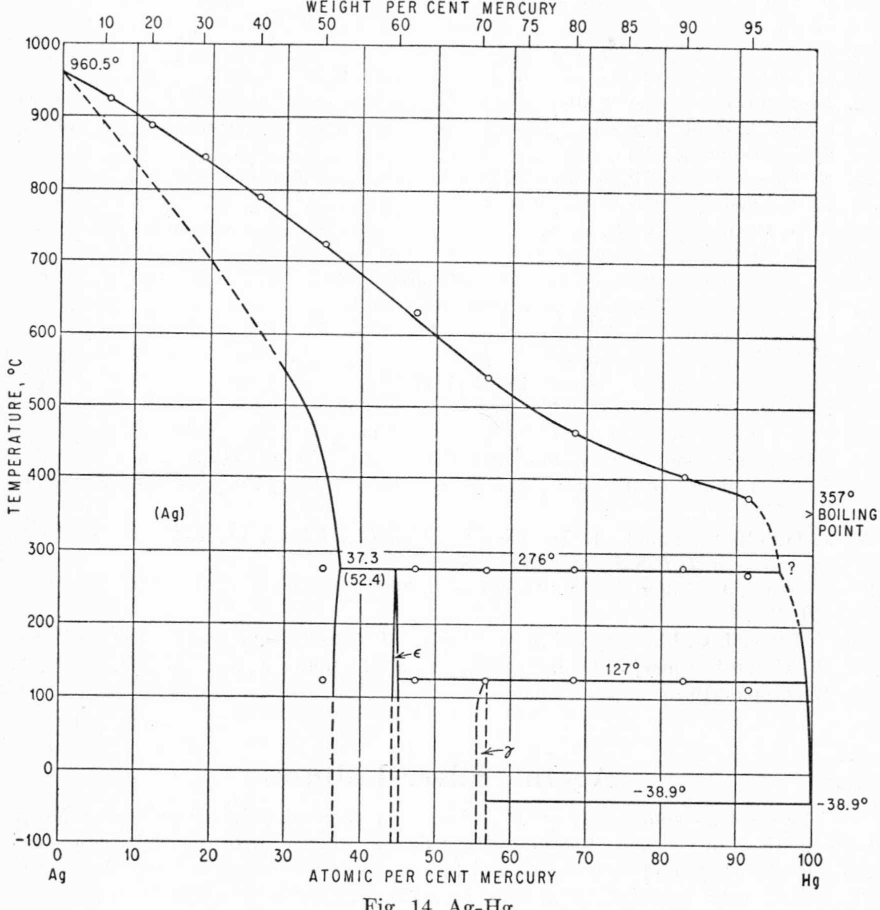

Fig. 14. Ag-Hg

from 10.033 to 10.051 A within the narrow range of homogeneity. The ideal composition of γ would be Ag_5Hg_8 [61.54 at. (74.85 wt.) % Hg]; however, this composition lies in the (γ + Hg) field [1]; see also [29].

The ε phase (mostly designated as β) has a h.c.p. lattice (A3 type) with parameters reported by [19, 2, 3]; $a = 2.970$ A, $c = 4.840$ A, $c/a = 1.631$ if saturated with Ag, and $a = 2.993$ A, $c = 4.840$ A, $c/a = 1.617$ if saturated with Hg.

Lattice parameters of the α or (Ag) phase were reported by [2, 3, 22, 4, 27].

1. A. J. Murphy, *J. Inst. Metals*, **46**, 1931, 507–522; see also discussion, pp. 528–535.
2. G. D. Preston, *J. Inst. Metals*, **46**, 1931, 522–527.

3. S. Stenbeck, *Z. anorg. Chem.*, **214**, 1933, 16–18.
4. H. M. Day and C. H. Mathewson, *Trans. AIME*, **128**, 1938, 261–280.
5. For further details, reference must be made to the original papers.
6. A. Ogg, *Z. physik. Chem.*, **27**, 1898, 290–311.
7. W. Reinders, *Z. physik. Chem.*, **54**, 1906, 609–627.
8. G. Tammann and T. Stassfurth, *Z. anorg. Chem.*, **143**, 1925, 369–376.
9. A. Weryha, *Z. Krist.*, **86**, 1933, 335–339; *Compt. rend. soc. polon. phys.*, **7**, 1926, 57–63.
10. A. Fedorow, *Chem.-Ztg.*, **36**, 1912, 220.
11. R. Müller and R. Hönig, *Z. anorg. Chem.*, **121**, 1922, 344–346.
12. No mention is made of numerous other publications on the subject. For a brief review, see M. Hansen, "Der Aufbau der Zweistofflegierungen," pp. 30–34, Springer-Verlag OHG, Berlin, 1936.
13. N. Parravano and P. Jovanovich, *Gazz. chim. ital.*, **49**(1), 1919, 6–9.
14. R. A. Joyner, *J. Chem. Soc.*, **99**, 1911, 195–208.
15. A. A. Sunier and C. B. Hess, *J. Am. Chem. Soc.*, **50**, 1928, 662–668.
16. R. E. DeRight, *J. Phys. Chem.*, **37**, 1933, 405–415.
17. R. J. Maurer, *J. Phys. Chem.*, **42**, 1938, 515–519.
18. D. R. Hudson, *Metallurgia*, **28**, 1943, 203–206.
19. V. M. Goldschmidt, *Z. physik. Chem.*, **133**, 1928, 397–419.
20. A. Westgren, *Metallwirtschaft*, **7**, 1928, 701.
21. A. Westgren, *J. Inst. Metals*, **46**, 1931, 533–534.
22. F. Heide, *Naturwissenschaften*, **25**, 1937, 651–652.
23. N. A. Shishakov, *Bull. acad. sci. U.R.S.S., Classe sci. chim.*, **1941**, 683–689.
24. S. R. Swamy and T. G. Shammana, *Current Sci. (India)*, **21**, 1952, 7–8.
25. A. E. Aylmer, G. I. Finch, and S. Fordham, *Trans. Faraday Soc.*, **32**, 1936, 864–871.
26. Z. G. Pinkser and L. I. Tatarinova, *Zhur. Fiz. Khim.*, **15**, 1941, 96–100; *Acta Physicochim. U.R.S.S.*, **14**, 1941, 193–200.
27. F. Hund, J. Müller, *Naturwissenschaften*, **38**, 1951, 303; *Z. Elektrochem.*, **57**, 1953, 131–138.
28. G. Ryge, J. C. Moffett, and A. G. Barkow, *J. Dental Research*, **32**, 1953, 152–167.
29. W. Hume-Rothery, J. O. Betterton, and J. Reynolds, *J. Inst. Metals*, **80**, 1951-1952, 609–616.

$\bar{1}.9732$
0.0268

Ag-In Silver-Indium

The diagram shown in Fig. 15 is based on thermal, micrographic, and X-ray investigations; it is essentially due to [1]. The liquidus curve of the α phase between 2.9 and 15.1 at. % and between 2.6 and 11.1 at. % In was also determined by [2] and [3], respectively (see data points in Fig. 15). The micrographically determined solidus curve of the α phase [2], shown in Fig. 15, lies at temperatures slightly higher than that obtained by thermal analysis [1]. The boundaries of the solid solubility of In in Ag, as fixed micrographically [2] and by lattice-parameter measurements [4], agree very well. The phase relationships involving the high-temperature β phase are shown in greater detail in the inset of Fig. 15; they are entirely the result of thermal analysis. The boundaries of the ζ phase, or γ phase according to [1, 5], have not yet been positioned accurately [1].

It has been suggested [1] that the ζ phase transforms with fall in temperature between 187 and 303°C into a solid solution called δ (not shown in Fig. 15) which is homogeneous in the range 25.6 to 31.8 at. % In. This assumption was based on the thermal-data points shown in Fig. 15 and lattice-parameter values of alloys with 21.0, 23.7, 26.8, and 28.7 at. % In. The alloy of the composition Ag_3In (26.18 wt. % In) was assumed to be duplex in structure, $\alpha + \delta$. It was reported later [5], however, that (*a*) the composition Ag_3In is homogeneous at 600 as well as 150°C; (*b*) the ζ-phase field, extending between about 24.9 and 32.7 at. % In at 400°C, narrows down to the range 24.9 to 26.2 at. % In at temperatures below 200°C (Fig. 15), thus giving rise to a two-phase field ($\zeta' + \gamma'$), between about 26.2 and 33.2 at. % In, instead of a

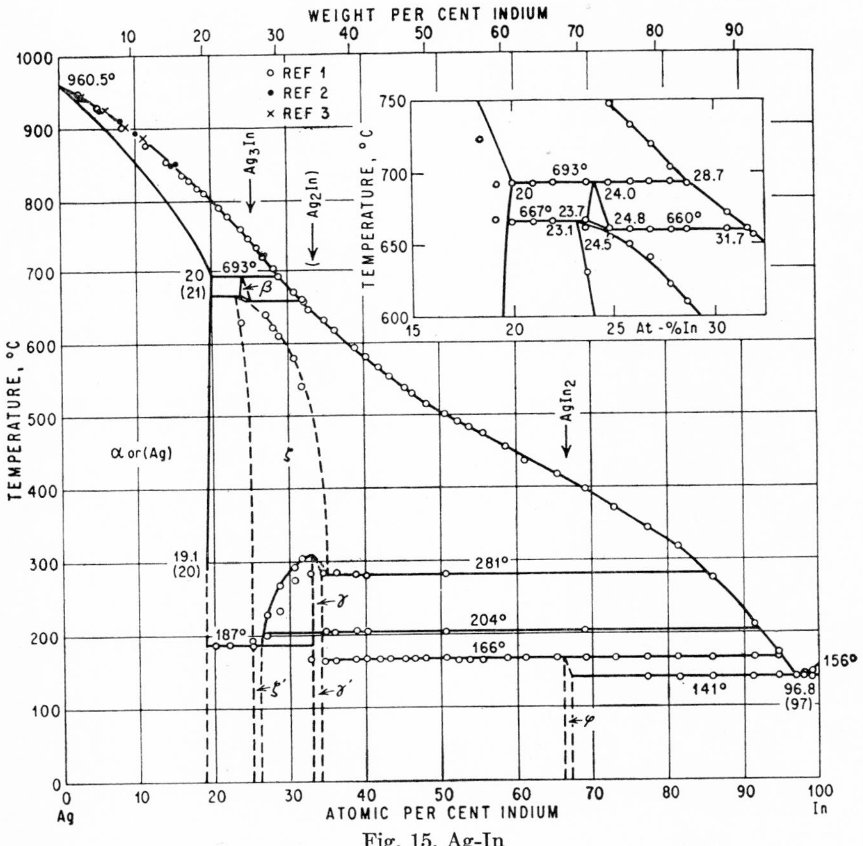

Fig. 15. Ag-In

one-phase field δ, according to [1]; and (*c*) the γ and γ' phases are stable within 33.2 and 34.2 at. % In (including the composition Ag_2In with 34.72 wt. % In) instead of 31.9 to 34.2 at. % In [1]. The findings of the two studies [1, 5] contradict each other to a considerable degree, and it appears somewhat doubtful that the phase relations in this composition range (Fig. 15) are entirely correct. Specifically, the change of the lattice parameters of alloys with 21 to 29 at. % In and the thermal data (Fig. 15) are not compatible with a rather broad two-phase field ($\zeta' + \gamma'$) and with the assumption that the γ phase extends from 33.2 to 34.2 at. % In and includes the composition Ag_2In. It should be noted, however, that the diagram given by [1] conflicts with the phase rule in that no two-phase field between ζ' and γ' is shown.

Assuming the existence of the ($\zeta' + \gamma'$) field, as shown in Fig. 15, the reaction at 187°C had to be interpreted as an order → disorder transformation, $\zeta' \rightarrow \zeta$ [5]. Similarly, the transformation in the γ phase must be ascribed to an ordering process on cooling, $\gamma \rightarrow \gamma'$ [5].

The composition of the φ phase could not be accurately fixed by [1]; it was assumed to lie at about 73 at. % In and to have a very narrow range of homogeneity. [5] showed that this phase is the compound $AgIn_2$ (68.04 wt. % In), by determining its crystal structure.

Crystal Structures. Concerning the lattice parameters of the α phase, see [1, 6, 4]. The β phase presumably is a 3:2 electron compound with a b.c.c. lattice [7]. ζ is h.c.p. [8, 1, 9, 5], with $a = 2.95_6$ A, $c = 4.80_6$ A, $c/a = 1.626$ [8]; $a = 2.954$ A, $c = 4.804$ A, $c/a = 1.626$ for the alloy with 25 at. % In above 300°C [5]. Assuming ζ' (Ag_3In) to have an ordered structure, it would be of the Mg_3Cd ($D0_{19}$) type. The γ phase (Ag_2In or more probably Ag_9In_4), formed by transformation of the ζ phase, is isotypic with γ-brass [5, 10], $a = 9.905$ A [5]; and the φ phase is isotypic with $CuAl_2$ (C16 type), $a = 6.883$ A, $c = 5.615$ A, $c/a = 0.816$ [5]. No difference in structure was found between γ and γ' [5].

1. F. Weibke and H. Eggers, *Z. anorg. Chem.*, **222,** 1935, 145–160.
2. W. Hume-Rothery, G. W. Mabbott, and K. M. Channel-Evans, *Phil. Trans. Roy. Soc. (London)*, **A233,** 1934, 1–97.
3. W. Hume-Rothery and P. W. Reynolds, *Proc. Roy. Soc. (London)*, **A160,** 1937, 282–303.
4. E. A. Owen and E. W. Roberts, *Phil. Mag.*, **27,** 1939, 294–327.
5. E. Hellner, *Z. Metallkunde*, **42,** 1951, 17–19. This paper, by mistake, does not give the corrected diagram. It is presented in the paper by E. Hellner and F. Laves, *Z. Naturforsch.*, **2a,** 1947, 180–181.
6. W. Hume-Rothery, G. F. Lewin, and P. W. Reynolds, *Proc. Roy. Soc. (London)*, **A157,** 1936, 167–183.
7. W. Hume-Rothery, P. W. Reynolds, and G. V. Raynor, *J. Inst. Metals*, **66,** 1940, 191–207.
8. V. M. Goldschmidt, *Z. physik. Chem.*, **133,** 1928, 397–419.
9. L. L. Frevel and E. Ott, *J. Am. Chem. Soc.*, **57,** 1935, 228.
10. W. Hume-Rothery, J. O. Betterton, and J. Reynolds, *J. Inst. Metals*, **80,** 1951-1952, 609–616.

$\bar{1}.7472$
0.2528

Ag-Ir Silver-Iridium

Iridium was reported to be insoluble in molten silver [1].

1. H. Rössler, *Chem.-Ztg.*, **24,** 1900, 733–735.

$\bar{1}.8902$
0.1098

Ag-La Silver-Lanthanum

The phase diagram presented in Fig. 16 is based on thermal-analysis data, corroborated by metallographic examination of alloys cooled from the molten state [1]. The existence of the three intermediate phases $LaAg_3$ (30.03 wt. % La), $LaAg_2$ (39.16 wt. % La), and LaAg (56.29 wt. % La) is well established; see also [2]. According to [2], the melting points of $LaAg_3$ and LaAg are 990 and 835°C, respectively, and that of the eutectic Ag-$LaAg_3$, 790°C. The composition of the eutectic LaAg-La was reported as 82 wt. (78 at.) % La [2]. The solubilities in the solid state have not been investigated.

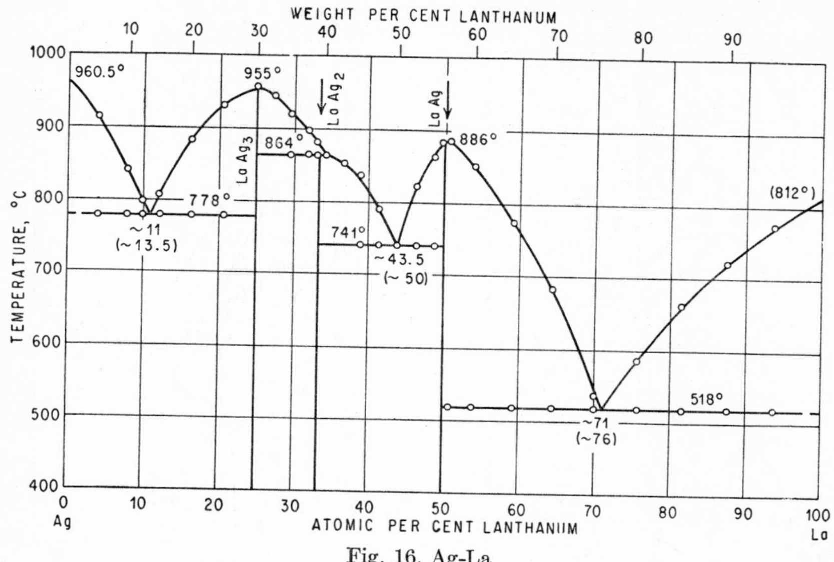

Fig. 16. Ag-La

LaAg is b.c.c. of the CsCl (B2) type, a = 3.79 A [3]; a = 3.77 A [4].

1. G. Canneri, *Metallurgia ital.*, **23,** 1931, 815–819.
2. R. Vogel and H. Klose, *Z. Metallkunde*, **45,** 1954, 670.
3. A. Iandelli, *Atti congr. intern. chim.*, *10th Congr. Rome*, **2,** 1938, 688–694.
4. H. Bommer and E. Krose, *Z. anorg. Chem.*, **252,** 1943, 62–64.

1.1916
$\bar{2}$.8084

Ag-Li Silver-Lithium

The first study of the system [1] resulted in a phase diagram characterized by the existence of two phases of fixed composition, LiAg and Li_2Ag, melting congruently at 955 and 450°C, respectively, and forming eutectics at about 30 at. % Li, 610°C; 70 at. % Li, 410°C; and 100% Li. A careful investigation [2], using thermal, micrographic, and X-ray diffraction methods, has shown that the constitution is strikingly different (Fig. 17). This equilibrium diagram can be considered as having a high degree of accuracy. The solidus was established by micrographic analysis and the constitution in the solid state by the same method, as well as by thermal and X-ray work.

Crystal Structures. An alloy of the composition LiAg was reported to have the b.c.c. structure of the CsCl (B2) type, a = 3.23 A [3], a = 3.17 A [4]. A γ-brass type of structure was found in alloys containing 76–80 at. % Li, a = 9.96 A [5]. Later this phase was more properly denoted $Li_{10}Ag_3$ (76.92 at. % Li) [6].

The more comprehensive X-ray analysis by [2] had the following results: The β phase was confirmed to be isotypic with CsCl, a = 3.169 A at 49.88 at. % Li. The γ_3, γ_2, and γ_1 phases appear to be closely related both to each other and to the γ-brass structure, and it is probable that they are based on the compositions Li_9Ag_4 (69.23 at. % Li), $Li_{10}Ag_3$ (76.92 at. % Li), and $Li_{12}Ag$ (92.31 at. % Li), respectively. Differences probably exist in the distributions of the Ag and Li atoms on the various atomic sites available. The lattice spacing of the γ_3 phase was found as a = 9.51 A at the

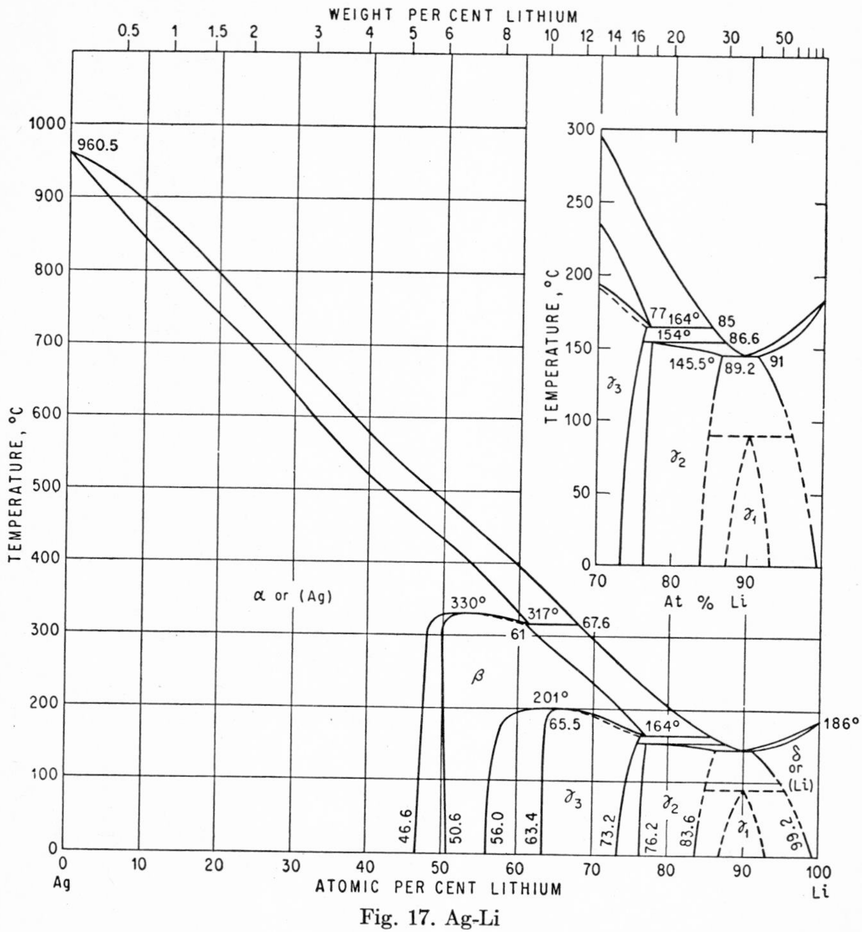

Fig. 17. Ag-Li

Ag-rich composition limit, that of the γ_2 phase as a = 9.70 A at 79.56 at. % Li, and that of the γ_1 phase as a = 9.82 A at 91.65 at. % Li.

1. S. Pastorello, *Gazz. chim. ital.*, **61,** 1931, 47–51.
2. W. E. Freeth and G. V. Raynor, *J. Inst. Metals*, **82,** 1953-1954, 569–574.
3. S. Pastorello, *Gazz. chim. ital.*, **60,** 1930, 493–501.
4. E. Zintl and G. Brauer, *Z. physik. Chem.*, **B20,** 1933, 245–271.
5. H. Perlitz, *Z. Krist.*, **86,** 1933, 155–158.
6. W. Hume-Rothery, J. D. Betterton, and J. Reynolds, *J. Inst. Metals*, **80,** 1951-1952, 609–616.

0.6470
$\bar{1}$.3530

Ag-Mg Silver-Magnesium

The general form of the diagram was first established by [1]; however, the phase boundaries had not been fixed with certainty, and the presence of Mg-rich solid solutions had not been detected. Electrical-conductivity measurements of alloys annealed at 400°C were in good agreement with this early diagram and also indicated the existence of solid solutions of Mg in Mg_3Ag in the range of approximately 75–77.3 at. % Mg, as well as of Ag in Mg above about 97.5 at. % Mg [2]. With the exception of the solidus curve of the β' phase and the upper end of the $\beta'/(\beta' + \epsilon)$ boundary, the diagram (Fig. 18) is now accurately known, based on thermal, micrographic, and X-ray analysis data [3–6].

The entire liquidus curve was determined by [1] and in the regions 0–39.7 and 81.6–100 at. % Mg by [6] and [5], respectively. The solidus curve of the α phase is based on micrographic analysis [6]. The $\alpha/(\alpha + \beta')$ and $\beta'/(\alpha + \beta')$ boundaries

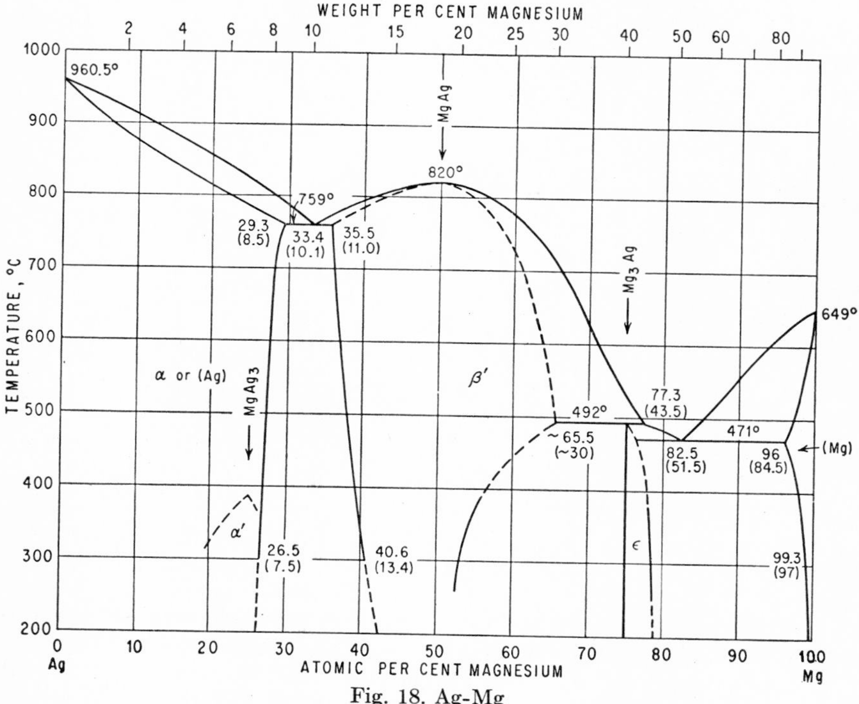

Fig. 18. Ag-Mg

were determined by lattice-parameter measurements [3] and micrographic work [6]. The two sets of data differ considerably, as shown in Table 3. In Fig. 18 the data by [6] have been used [17]. Both papers definitely proved an earlier conclusion [7], that the Ag-rich primary solid solution extends to about 65 at. % Mg, to be incorrect. The $\beta'/(\beta' + \epsilon)$ boundary, between 390 and 260°C, is based on lattice-parametric work [6]. The ϵ phase extends from 75 at. (40.3 wt.) % to 78.7 at. (45.5 wt.) % Mg [3].

Differences in the solidus curve of the (Mg) phase, determined micrographically, are slight [4, 5]. The solubility boundary of this phase was determined by micrographic [4, 5] and X-ray work [3]. Results differ substantially for temperatures below 450°C, as shown in Table 4, indicating that the solubility found by the lattice-spacing method [3] is lower than that obtained by metallographic work [4, 5].

The existence of an order-disorder transformation in the α phase at 25 at. (6.99

wt.) % Mg, already suspected by [6], was definitely established by X-ray and electrical-resistance studies [8]. The transformation temperature at the composition $MgAg_3$ is about 387°C. (See Supplement.)

Table 3. $\alpha/(\alpha + \beta')$ and $\beta'/(\alpha + \beta')$ Boundaries in At. % Mg (Wt. % in Parentheses)

°C	$\alpha/(\alpha + \beta')$		$\beta'/(\alpha + \beta')$	
	Ref. 3	Ref. 6	Ref. 3	Ref. 6
700	32.1 (9.6)	28.0 (8.1)	40.9 (13.5)	36.0 (11.3)
600	32.1 (9.6)	27.6 (7.9)	41.5 (13.8)	36.9 (11.7)
500	31.6 (9.4)	27.2 (7.8)	42.3 (14.2)	37.9 (12.1)
400	30.2 (8.9)	26.9 (7.7)	43.4 (14.7)	39.2 (12.7)
300	28.4 (8.2)	26.5 (7.5)	44.5 (15.3)	40.6 (13.4)

Table 4. Solid Solubility of Ag in Mg in At. % Ag (Wt. % in Parentheses)

	471°C	400°C	300°C	260°C	200°C
Ref. 3...........	~3.7 (~14.5)	1.6 (6.8)	0.45 (2.0)	0.2 (0.9)	
Ref. 4...........	3.9 (15.3)	2.2 (9.0)	0.9 (3.9)		
Ref. 5...........	4.0 (15.5)	1.6 (6.8)	0.7 (3.0)	....	<0.2 (<1.0)
Fig. 18..........	4.0 (15.5)	1.9 (8.0)	0.7 (3.0)	0.45 (2.0)	<0.2 (<1.0)

Crystal Structures. Lattice parameters of the α phase were determined by [3, 6, 9, 10]; see especially [3, 6, 9]. The β' phase has a structure of the CsCl (B2) type, with lattice parameters given by [11, 12, 3, 9, 10], especially [3, 9]. The degree of ordering is highest at 50 at. (18.4 wt.) % Mg [3, 9]. Because of its ordered structure, stable up to the melting point [13], the designation β' (instead of β) for this 3:2 electron compound was introduced by [14]. The ϵ phase, based on the composition Mg_3Ag (40.34 wt. % Mg), was reported to have a hexagonal lattice with 8 atoms per unit cell and $a = 4.94$ A, $c = 7.82_6$ A, $c/a = 1.58$ [3]. However, [9] were unable to verify this structure; see also [10]. Concerning the lattice parameters of the (Mg) phase, see [3, 10, 15, 16], especially [3, 10].

Supplement. The unit cell proposed by [18] for the $MgAg_3$ superstructure phase, cubic with 256 atoms [19], was not corroborated by [20], who found a structure derived from the $L1_2$ (Cu_3Au) type by the regular disposition of atomic ordering faults (*Verwerfungsebenen*).

1. S. F. Zemczuzny, *Z. anorg. Chem.*, **49**, 1906, 400–414.
2. W. J. Smirnow and N. S. Kurnakow, *Z. anorg. Chem.*, **72**, 1911, 31–54.

3. N. V. Ageew and V. G. Kuznezow, *Izvest. Akad. Nauk S.S.S.R., Otdel. Khim. Nauk,* **1937,** 289–309; see also *J. Inst. Metals,* **60,** 1937, 361–363.
4. W. Hume-Rothery and E. Butchers, *J. Inst. Metals,* **60,** 1937, 345–350.
5. R. J. M. Payne and J. L. Haughton, *J. Inst. Metals,* **60,** 1937, 351–356.
6. K. W. Andrews and W. Hume-Rothery, *J. Inst. Metals,* **69,** 1943, 485–493.
7. F. Saeftel and G. Sachs, *Z. Metallkunde,* **17,** 1925, 258–264.
8. L. M. Clarebrough and J. F. Nicholas, *Australian J. Sci. Research,* **A3,** 1950, 284–289.
9. H. R. Letner and S. S. Sidhu, *J. Appl. Phys.,* **18,** 1947, 833–837.
10. S. Goldsztaub and P. Michel, *Compt. rend.,* **232,** 1951, 1843–1845.
11. E. A. Owen and G. D. Preston, *Phil. Mag.,* **2,** 1926, 1266–1270.
12. A Westgren and G. Phragmén, *Metallwirtschaft,* **7,** 1928, 700–703.
13. W. L. Bragg and E. J. Williams, *Proc. Roy. Soc. (London),* **A151,** 1935, 540.
14. W. Hume-Rothery, P. W. Reynolds, and G. V. Raynor, *J. Inst. Metals,* **66,** 1940, 191–207.
15. R. S. Busk, *Trans. AIME,* **188,** 1950, 1460–1464.
16. R. S. Busk, *Trans. AIME,* **194,** 1952, 207–209.
17. B. R. T. Frost and G. V. Raynor, *Proc. Roy. Soc. (London),* **A203,** 1950, 133–134.
18. A. M. Mathieson, quoted in [8] as private communication.
19. J. F. Nicholas, *Proc. Phys. Soc. (London),* **A66,** 1953, 201–208.
20. K. Schubert, B. Kiefer, M. Wilkens, and R. Haufler, *Z. Metallkunde,* **46,** 1955, 692–715, especially 705.

0.2931
$\bar{1}$.7069

Ag-Mn Silver-Manganese

The existence of a wide miscibility gap in the liquid state was found by [1]. By means of thermal and microscopic investigations [2], a phase diagram was obtained which shows the essential features of that presented in Fig. 19, with the exception of the effect of Ag on the transformation points of Mn which were not known at that time. It was found (*a*) that the miscibility gap extends from 31 to 93 wt. (46.8–96.2 at.) % Mn at the monotectic temperature, 1180°C, and (*b*) that the peritectic reaction, melt [about 33 at. (20 wt.) % Mn] + Mn $\rightleftarrows$ α, occurs at 980°C.

In a reinvestigation of the system [3] using thermal, micrographic, resistometric, and X-ray analysis, special attention was paid to the solid solubility of Mn in Ag and the phase relationships in the Mn-rich alloys (Fig. 19) [4]. The temperatures of the peritectic and eutectoid equilibria involving the γ_{Mn} phase practically coincide. The saturation curve of the α phase is well established, especially below 750°C. The solubility of Ag in γ-Mn was determined parametrically after quenching from 1150 and 990°C and given as 3.1 at. (5.8 wt.) % Ag and 2.6 at. (4.9 wt.) % Ag, respectively. In view of the fact that the γ_{Mn} phase is not completely retained on quenching [3, 5] and transforms to the metastable tetragonal γ_{Mn} phase [5], these solubility values appear somewhat doubtful. The solubility of Ag in β-Mn and α-Mn was found to be smaller than **0.2** at. (0.4 wt.) % Ag and practically nil, respectively [3].

Lattice parameters of alloys up to about 60 at. % Mn and between about 88 and 100 at. % Mn (the latter after quenching from 990 and 1150°C) were given by [3].

1. G. Hindrichs, *Z. anorg. Chem.,* **59,** 1908, 437–441.
2. G. Arrivaut, *Z. anorg. Chem.,* **83,** 1913, 193–199; *Compt. rend.,* **156,** 1913, 1539–1541.
3. E. Raub and A. Engel, *Z. Metallkunde,* **37,** 1946, 62–64; see also E. Raub, *Z. Metallkunde,* **40,** 1949, 359–360.

4. The alloys were prepared using distilled Mn containing about 0.1% Si in a hydrogen atmosphere.
5. U. Zwicker, *Z. Metallkunde,* **42,** 1951, 246–252.

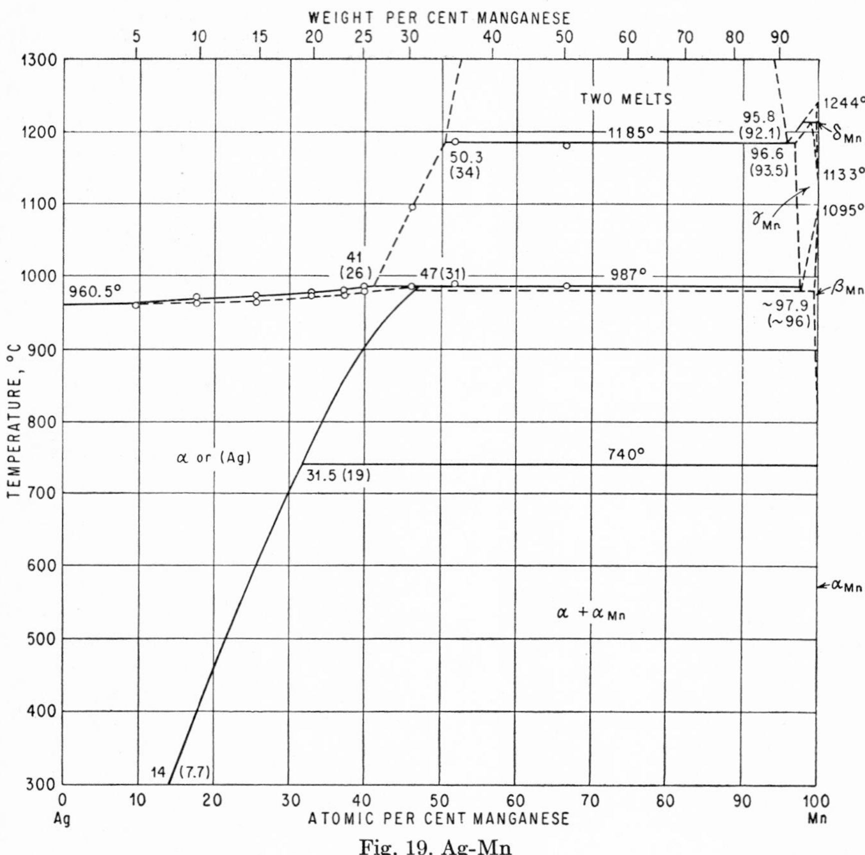

Fig. 19. Ag-Mn

0.0509
1̄.9491

Ag-Mo Silver-Molybdenum

At 1600°C at least 5 wt. (5.6 at.) % Mo is soluble in Ag. The microstructure of the 5% Mo alloy consisted of primary Mo crystals in a matrix of pure Ag [1].

1. Dreibholz, *Z. physik. Chem.,* **108,** 1924, 4.

0.8866
1̄.1134

Ag-N Silver-Nitrogen

Nitrogen is insoluble in solid and liquid silver (investigated up to 1300°C) [1].

1. A. Sieverts and W. Krumbhaar, *Ber. deut. chem. Ges.,* **43,** 1910, 894; E. W. R. Steacie and F. M. G. Johnson, *Proc. Roy. Soc.* (*London*), **A112,** 1926, 542–558.

0.6713
$\bar{1}.3287$

Ag-Na Silver-Sodium

Results of thermal analyses (Fig. 20) by [1] and [2] agree fairly well. Between 10 and 100 wt. % Na the liquidus curve according to [2] runs 10–30°C below that due to [1]. The melting point of Na is lowered 0.09°C by 0.3 wt. % Ag [3]; see also [4]. The fact that the liquidus curve in the range of approximately 40–75 at. (12–40 wt.) % Na is only slightly inclined toward the composition axis indicates a tendency

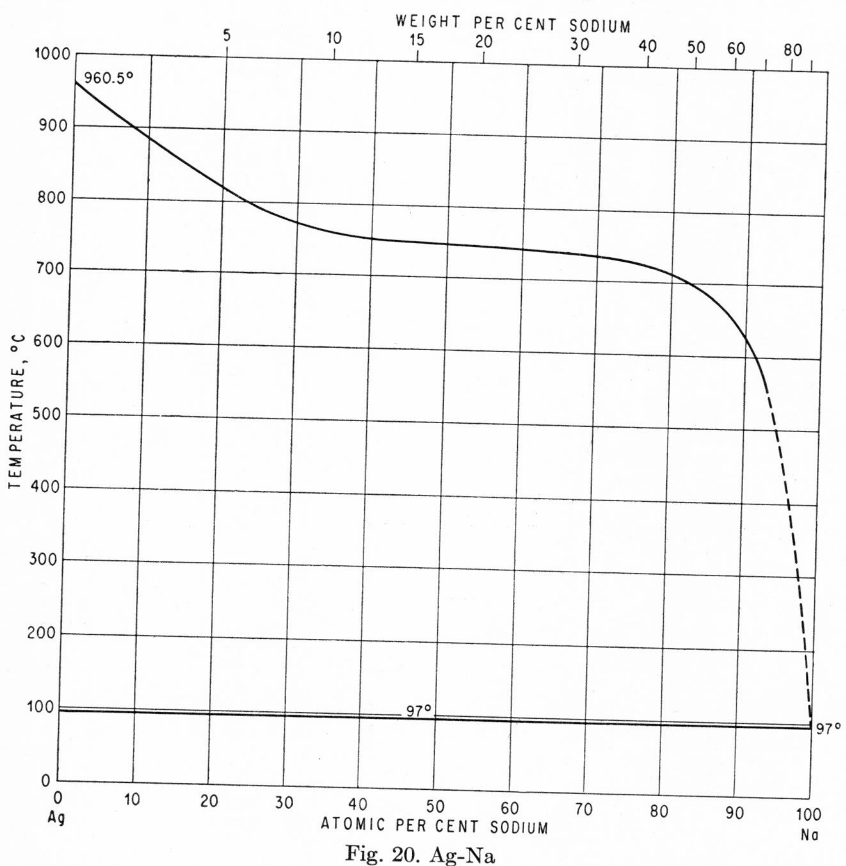

Fig. 20. Ag-Na

to the formation of a miscibility gap in the liquid state. However, layer formation has not been observed [1, 2]. The solid solubility of Na in Ag is small, 0.22 wt. (1 at.) % Na at the most [2]; that of Ag in Na is negligible. The absence of intermediate phases was confirmed by [5–7].

1. E. Quercigh, *Z. anorg. Chem.*, **68**, 1910, 301–306.
2. C. H. Mathewson, *Intern. Z. Metallog.*, **1**, 1911, 51–63.
3. G. Tammann, *Z. physik. Chem.*, **3**, 1889, 447.
4. The very slight decrease of the melting point of Na is consistent with the observa-

tion that Ag is not noticeably soluble in Na at its melting point: C. T. Heycock and F. H. Neville, *J. Chem. Soc.*, **55,** 1889, 674.
5. C. A. Kraus and H. F. Kurtz, *J. Am. Chem. Soc.*, **47,** 1925, 43–60.
6. W. M. Burgess and E. H. Smoker, *J. Am. Chem. Soc.*, **52,** 1930, 3573–3575.
7. E. Zintl, J. Goubeau, and W. Dullenkopf, *Z. physik. Chem.*, **A154,** 1931, 1–46.

0.2644
$\bar{1}$.7356

Ag-Ni Silver-Nickel

The phase diagram of Fig. 21 is based on data by [1, 2, 3]. The extent of the liquid-miscibility gap at the monotectic temperature of 1435°C has not been determined accurately. Analyses of the solidified layers gave the following results: for the

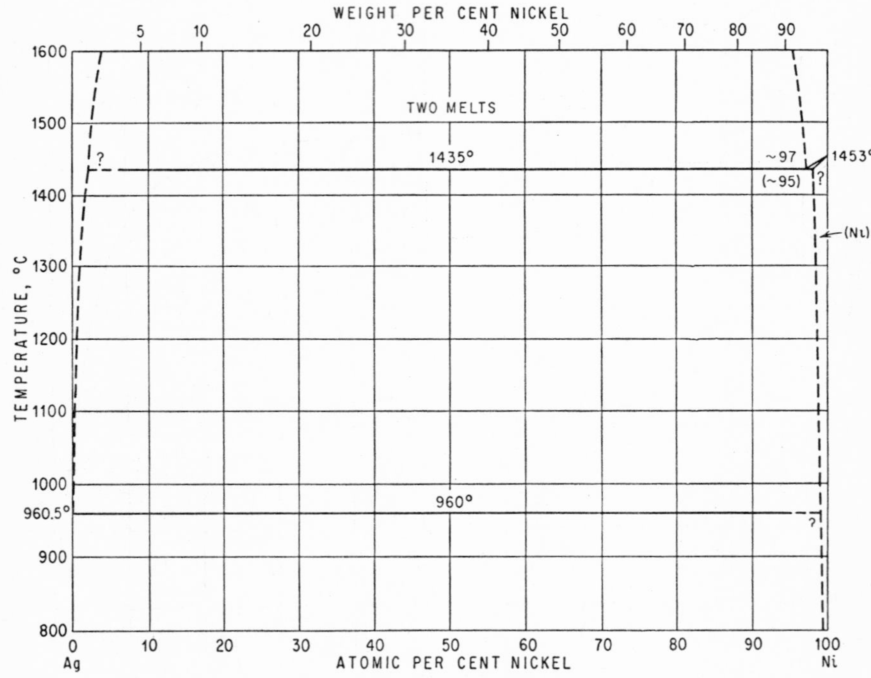

Fig. 21. Ag-Ni

Ag-rich layer, 1.5 wt. (2.7 at.) % Ni [2, 3] and 0.4 wt. (0.7 at.) % Ni [1]; and for the Ni-rich layer, 3.65 wt. (2.0 at.) % Ag [1] and 3.9 wt. (2.15 at.) % Ag [2]. The limit of the solid solubility of Ni in Ag was determined by means of measurements of the specific magnetization of samples annealed for 4 hr at 940°C, slowly cooled to various temperatures, and quenched after annealing for 2–5 hr: 922°C, 0.102%; 860°C, 0.084%; 785°C, 0.066%; 702°C, 0.044%; 640°C, 0.032%; 600°C, 0.026%; 510°C, 0.018%; and 400°C, 0.012 wt. % Ni [4]. The solid solubility of Ag in Ni decreases with fall in temperature [3]; no quantitative data are available, however.

1. G. I. Petrenko, *Z. anorg. Chem.*, **53,** 1907, 212–215.
2. P. de Cesaris, *Gazz. chim. ital.*, **43**(2), 1913, 365–379.

3. W. Guertler and A. Bergmann, *Z. Metallkunde*, **25**, 1933, 56.
4. G. Tammann and W. Oelsen, *Z. anorg. Chem.*, **186**, 1930, 264–266.

0.8288
$\bar{1}$.1712

Ag-O Silver-Oxygen

A phase diagram for a constant pressure of 1 atm (760 mm Hg) was suggested by [1], using data on (*a*) the solubility of O in Ag at temperatures between 200 and 800°C and pressures of 50–800 mm Hg [2]; (*b*) the solubility of O in liquid Ag at temperatures between 973 and 1125°C [3]; (*c*) the solubility in molten Ag at 1075°C in dependence upon pressure [4]; (*d*) the dissociation pressure of Ag_2O at 173–191°C [5] and 374–500°C [6]; and (*e*) the effect of pressure on the lowering of the melting point of Ag

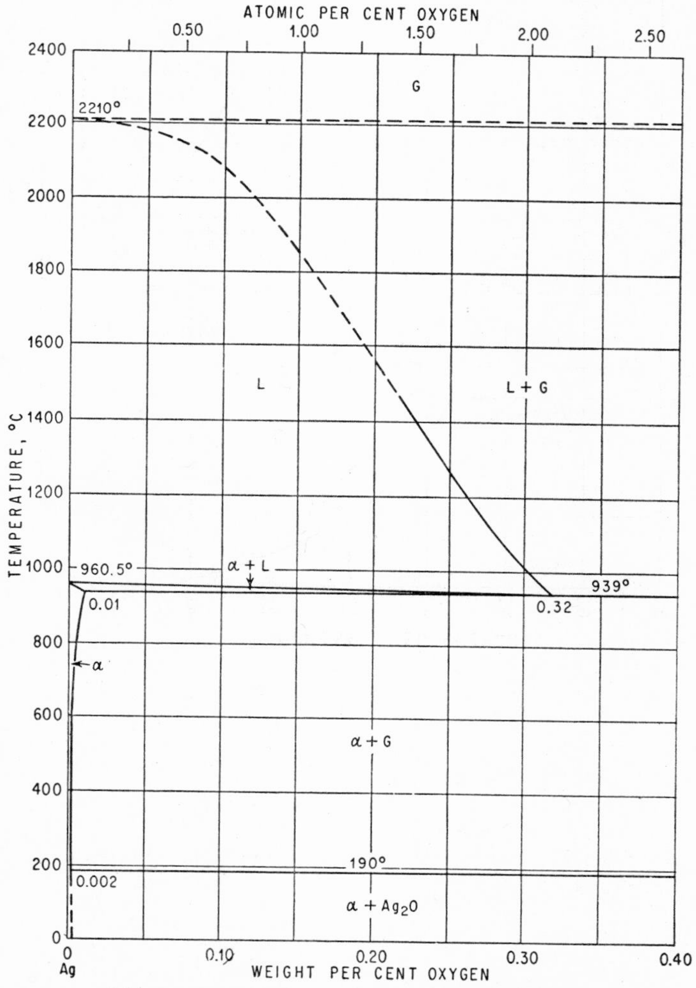

Fig. 22. Ag-O. (According to H. C. Vacher.)

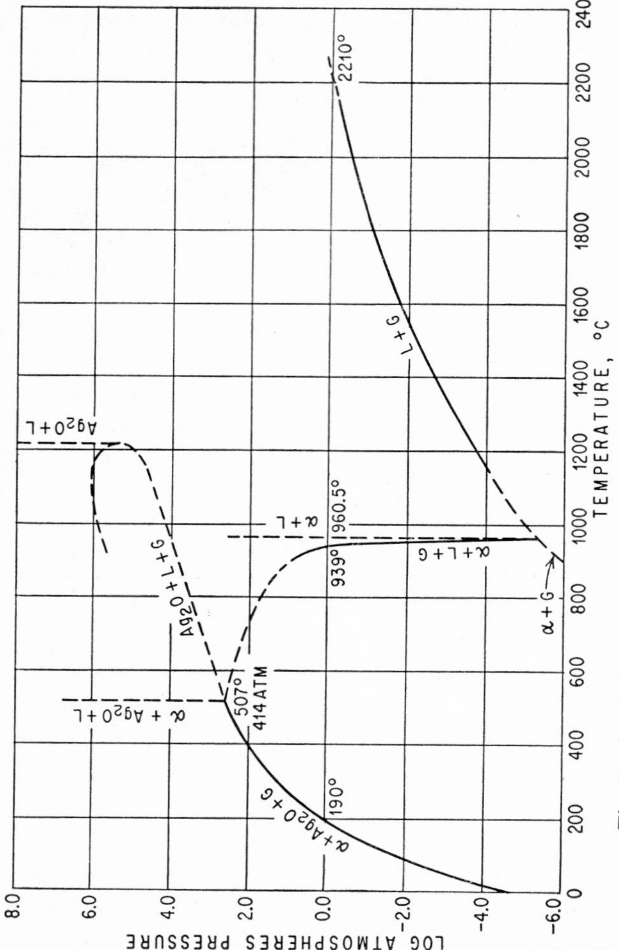

Fig. 23. Ag-O. Pressure vs. temperature. (According to H. C. Vacher.)

by O [7]. This diagram, as taken from the "Metals Handbook" [1], is presented in Fig. 22. As regards the fact that the solubility was found to be lowest at 400°C and probably greater at room temperature than at 200°C, [2] suggested that up to about 400°C the dissociation of Ag_2O, possibly existing in the melt, may be effective.

[1] also designed the pressure-vs.-temperature diagram given in Fig. 23. It indicates that the curve of the dissociation pressure of Ag_2O [6] and the melting-point-vs.-pressure curve [7] intersect at 507°C and 414 atm. One can assume that close to this temperature the eutectic Ag-Ag_2O exists provided that the dissociation of Ag_2O can be suppressed by a sufficiently high pressure. Indeed, an alloy prepared in a steel bomb at 600°C showed primary Ag in a matrix of the Ag-Ag_2O eutectic [7]. Ag_2O is isotypic with Cu_2O (C3 type), a = 4.727 A [8].

1. H. C. Vacher, in "Metals Handbook," 1948 ed., pp. 1151–1152, The American Society for Metals, Cleveland, Ohio.
2. E. W. R. Steacie and F. M. G. Johnson, *Proc. Roy. Soc. (London)*, **112**, 1926, 542–548; *Z. Metallkunde*, **21**, 1929, 43.
3. A. Sieverts and J. Hagenacker, *Z. physik. Chem.*, **68**, 1909, 115–128.
4. F. G. Donnan and T. W. A. Shaw, *J. Soc. Chem. Ind.*, **29**, 1910, 987–989.
5. A. F. Benton and L. C. Drake, *J. Am. Chem. Soc.*, **54**, 1932, 2186–2194.
6. F. G. Keyes and H. Hara, *J. Am. Chem. Soc.*, **44**, 1932, 479–485.
7. N. P. Allen, *J. Inst. Metals*, **49**, 1932, 317–340.
8. P. Niggli, *Z. Krist.*, **57**, 1922, 253–299.

0.5419
$\bar{1}$.4581

Ag-P Silver-Phosphorus

It has long been known [1–3] that P is soluble in liquid Ag but is largely ejected on solidification in the form of elementary phosphorus. Claims that preparations of the compositions AgP [4], Ag_2P_3 [2], and AgP_2 [5] were compounds appeared questionable, as the homogeneity of these products was not proved. By thermal decomposition of products rich in P, taking pressure-concentration isotherms at 400–500°C, it was found that, under these conditions, only AgP_2 (36.48 wt. % P) and AgP_3 (46.28 wt. % P) exist [6]. This was corroborated by X-ray studies.

The ejection of P from Ag-P melts on solidification takes place violently. Even on rapid quenching of the saturated melt, not more than 0.4–0.5 wt. (1.3–1.7 at.) % P is retained in the silver in the form of AgP_2 [7]. The solubility of P in liquid Ag at 960°C and atmospheric pressure was found as 1.45 wt. (5.0 at.) % P. The solid solubility of P in Ag is less than 0.026 wt. % P [7].

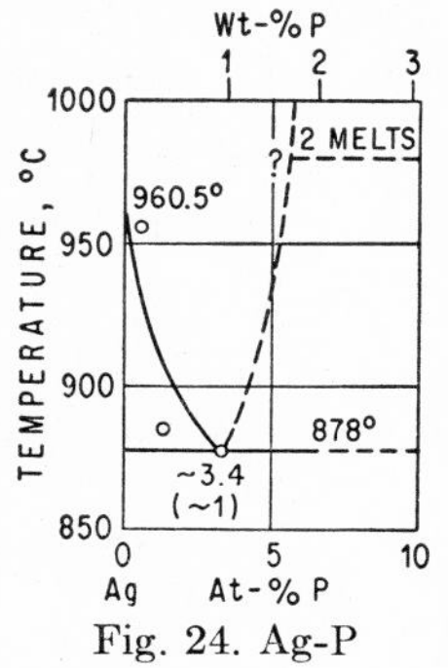

Fig. 24. Ag-P

Figure 24 shows a partial diagram obtained by thermal analysis of a few melts in evacuated sealed porcelain tubes [7]. A eutectic lies at 0.97 wt. (about 3.25 at.) % P and 877–879°C. Alloys with P contents of 2 wt. (6.7 at.) % and higher consisted of two layers. The P content of the upper (friable) layer was 18–20 wt. (about 43.5–46.5 at.) %, corresponding to about 52% AgP_2. An alloy with 39.7 wt. % P, sintered at 470–480°C, proved to be one phase, indicating that AgP_2 and AgP_3 form solid solutions [7].

1. Pelletier (1788, 1792); see Gmelin-Kraut, "Handbuch der anorganischen Chemie," vol. V(2), p. 127, Carl Winter's Universitätsbuchhandlung, Heidelberg, 1914.
2. A. Schrötter, *Ber. Wien. Akad.*, **2**, 1849, 301.

3. P. Hautefeuille and A. Perry, *Compt. rend.*, **98,** 1884, 1378.
4. O. Emmerling, *Ber. deut. chem. Ges.*, **12,** 1879, 152.
5. A. Granger, *Compt. rend.*, **124,** 1897, 896–898.
6. H. Haraldsen and W. Biltz, *Z. Elektrochem.*, **37,** 1931, 504–506; H. Haraldsen, *Skrifter Norske Videnskaps-Akad. Oslo*, **9,** 1932, 1–63.
7. H. Moser, K. W. Fröhlich, and E. Raub, *Z. anorg. Chem.*, **208,** 1932, 227–230.

$\bar{1}.7165$
0.2835

Ag-Pb Silver-Lead

The entire liquidus curve was determined by [1–4], with good agreement of the data. The eutectic composition was found as 95.9 (97.8) [5], 95.3 (97.5) [2, 6], about 94.4 (97) [3], and 95.7 at. (97.7 wt.) % Pb [4], and the eutectic temperature as 304 [1, 2, 3, 5] and 301°C [6].

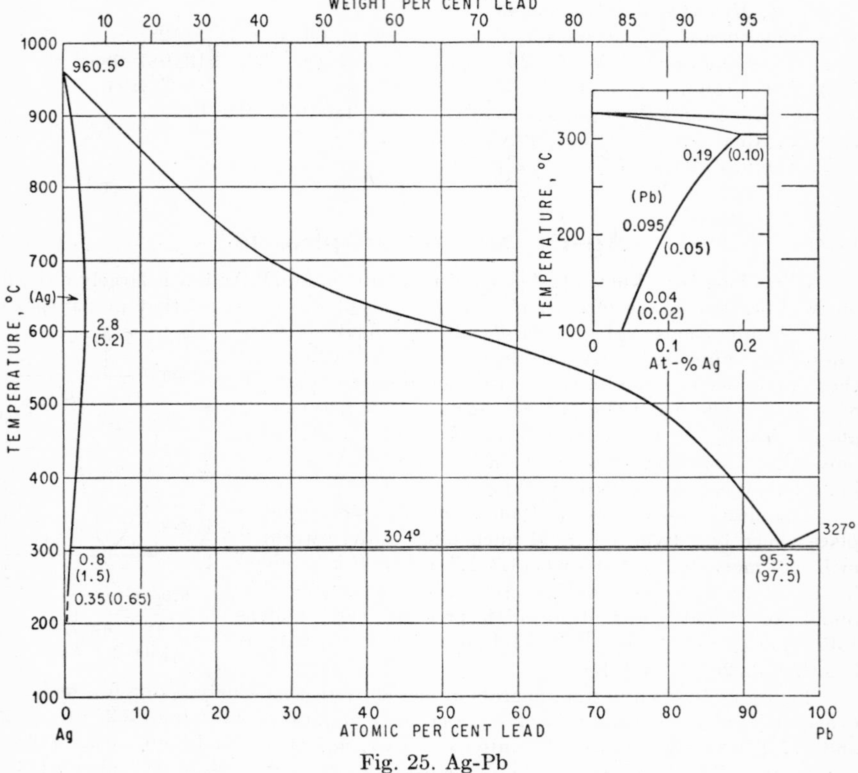

Fig. 25. Ag-Pb

The solubility of Pb in Ag was determined between 900 and 300°C by means of micrographic and lattice-parametric investigations [7], and between 300 and 250°C by means of lattice-parametric measurements [8], after the existence of Ag-rich solid solutions had already been assumed [9]. The solubility in at. % Pb (wt. % in parentheses) was found as follows: 900°C, 1.1 (2.0); 800°C, 2.1 (3.95); 700°C, 2.6 (4.9); 600°C, 2.8 (5.2); 500°C, 2.1 (3.95); 400°C, 1.5 (2.8); 300°C, 0.75 ± 0.15 (1.4 ± 0.3);

280°C, 0.6 (1.1); and 250°C, 0.35 (0.65) % Pb. In electrodeposited alloys, more than 10 wt. (5.5 at.) % Pb can be held in supersaturated solid solution [10].

The solid solubility of Ag in Pb is very restricted; see inset of Fig. 25. According to [11] and [12], it is 0.055–0.065 wt. % at 250–270°C and 0.065–0.07 wt. % at 285°C, based on diffusion experiments. By means of thermoresistometric investigations slightly higher solubilities were found: about 0.1 wt. % Ag at the eutectic temperature and about 0.05 wt. % Ag at 200°C [11]. A considerably lower value was reported by [13]: 0.039 wt. (0.075 at.) % at 232°C. [14], using various methods, accepted as the most probable values 0.02 wt. (0.04 at.) % at 100°C, 0.05 wt. (0.095 at.) % at 200°C, 0.09 wt. (0.17_5 at.) % at 280°C, and 0.10 wt. (0.19 at.) % Ag at 300–304°C (Fig. 25).

Lattice parameters of the Ag-rich and Pb-rich alloys were reported by [7, 8] and [14], respectively.

1. C. T. Heycock and F. H. Neville, *Phil. Trans. Roy. Soc. (London)*, **A189,** 1897, 37, 39, 58–60.
2. K. Friedrich, *Metallurgie*, **3,** 1906, 396–406.
3. G. I. Petrenko, *Z. anorg. Chem.*, **53,** 1907, 201–204.
4. F. Yoldi and D. L. de A. Himinez, *Anal. soc. españ. fís. y quím.*, **28,** 1930, 1055–1065; abstract in *J. Inst. Metals*, **47,** 1931, 76.
5. C. T. Heycock and F. H. Neville, *J. Chem. Soc.*, **61,** 1892, 907; **65,** 1894, 72–73.
6. O. Hájiček, *Hutnické Listy*, **3,** 1948, 265–270.
7. E. Raub and A. v. Polaczek-Wittek, *Z. Metallkunde*, **34,** 1942, 93–96, also published by E. Raub and A. Engel, *Z. Metallkunde*, **37,** 1946, 77.
8. H. H. Chiswick and R. Hultgren, *Trans. AIME*, **137,** 1940, 442–446. The data were slightly corrected by [7].
9. M. Hansen, "Der Aufbau der Zweistofflegierungen," pp. 46–47, Springer-Verlag OHG, Berlin, 1936.
10. E. Raub and M. Engel, *Z. Elektrochem.*, **49,** 1943, 89–97.
11. W. Seith and A. Keil, *Z. physik. Chem.*, **B22,** 1933, 350–358.
12. G. v. Hevesy and W. Seith, *Z. Elektrochem.*, **37,** 1931, 530–531; see also W. Seith and J. G. Laird, *Z. Metallkunde*, **24,** 1932, 195.
13. A. Pasternak, *Bull. intern. acad. polon. sci., Classe sci. math. nat., sér. A*, 1951, pp. 177–192.
14. Y. Fuke and A. Kondo, *Nippon Kinzoku Gakkai-Shi*, **16,** 1952, 611–614.

0.0048
$\bar{1}$.9952

Ag-Pd Silver-Palladium

The liquidus temperatures and approximate solidus temperatures determined thermally by [1] indicate the crystallization of a continuous series of solid solutions (Fig. 26). This was confirmed by [2]; no data were given, however. Also, measurements of the lattice parameter [3–8, 18], electrical conductivity (or resistivity) [9–15, 6], temperature coefficient of electrical resistance [9, 12, 15], thermal conductivity [10, 11], magnetic susceptibility [13], and coefficient of linear expansion [16] confirm the conclusion that the system is devoid of transformations in the solid state.

On the other hand, [7] concluded from the occurrence of discontinuities in the thermal expansion and the change of electrical resistivity with temperature that new phases are formed in the solid state. Likewise, [17] claimed, from studies of the ternary system Ag-Cu-Pd, that in the system Ag-Pd a miscibility gap forms below 400°C.

1. R. Ruer, *Z. anorg. Chem.*, **51,** 1906, 315–319.
2. E. Ya. Rode, *Izvest. Sektora Platiny*, **13,** 1936, 167–175.

3. L. W. McKeehan, *Phys. Rev.*, **20**, 1922, 424–432.
4. F. Krüger and A. Sacklowski, *Ann. Physik*, **78**, 1925, 72–82.
5. W. Stenzel and J. Weerts, "Festschrift zum 50-jährigen Bestehen der Platinschmelze," pp. 288–299, G. Siebert G.m.b.H., Hanau, 1931; J. Weerts, *Z. Metallkunde*, **24**, 1932, 138–141.
6. F. Krüger and G. Gehm, *Ann. Physik*, **16**, 1933, 190–193.
7. Ch. Thierer, Dissertation, Stuttgart T.H., 1935; see "Gmelins Handbuch der anorganischen Chemie," System No. 68A (5), pp. 653–660, Verlag Chemie G.m.b.H., Weinheim/Bergstrasse, 1951.
8. G. Rosenhall, *Ann. Physik*, **24**, 1935, 297–325. (Ag-Pd-H alloys.)
9. W. Geibel, *Z. anorg. Chem.*, **70**, 1911, 240–242.
10. F. A. Schulze, *Physik. Z.*, **12**, 1911, 1028–1031.

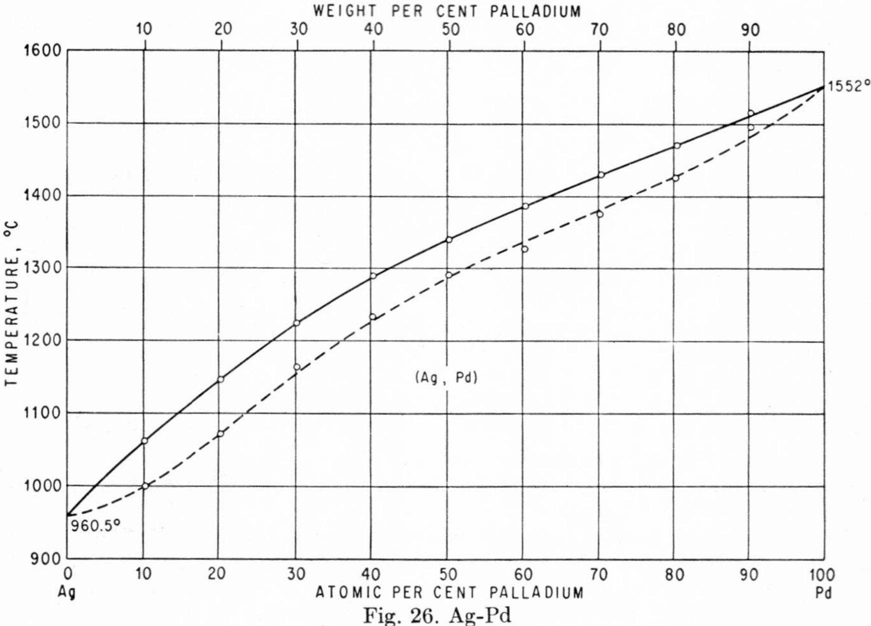

Fig. 26. Ag-Pd

11. E. Sedström, Dissertation, Stockholm, 1924; Akademie Abhandlung Lund, 1924.
12. F. E. Carter, *Proc. AIME, Inst. Metals Div.*, 1928, pp. 759–782.
13. B. Swensson, *Ann. Physik*, **14**, 1932, 699–711.
14. J. Wortmann, *Ann. Physik*, **18**, 1933, 233–250. (Ag-Pd-H alloys.)
15. R. F. Vines, "The Platinum Metals and Their Alloys," pp. 125–126, The International Nickel Company, Inc., New York, 1941.
16. C. H. Johansson, *Ann. Physik*, **76**, 1925, 445–454.
17. F. Glander, *Metallwirtschaft*, **18**, 1939, 357–361.
18. V. G. Kuznecov, *Izvest. Sektora Platiny*, **20**, 1946, 5–20.

$\bar{1}.8840$
0.1160

Ag-Pr Silver-Praseodymium

The phase diagram of Fig. 27 is based on thermal analysis and metallographic examination of alloys cooled from the molten state. Solubilities in the solid state were not investigated. There are three intermediate phases: $PrAg_3$ (30.33 wt. % Pr), $PrAg_2$ (39.51 wt. % Pr), and PrAg (56.64 wt. % Pr) [1]. PrAg was reported to be isotypic with CsCl (B2 type), a = 3.74 A [2].

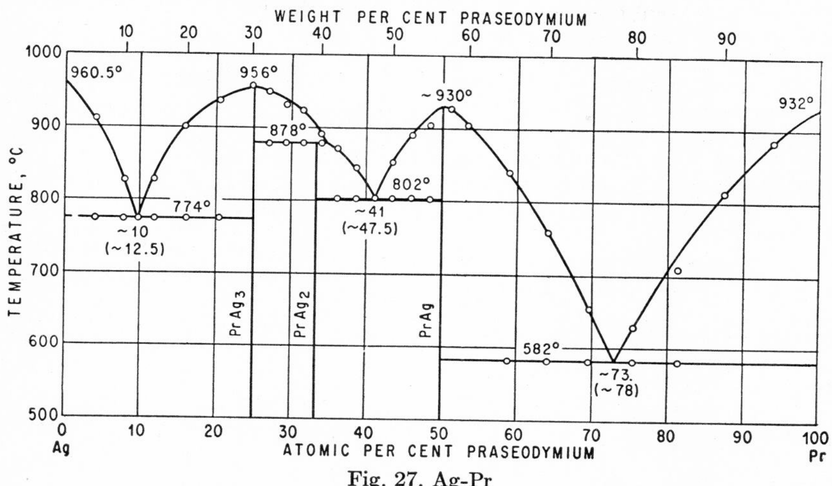

Fig. 27. Ag-Pr

1. G. Canneri, *Metallurgia ital.*, **26**, 1934, 794–796.
2. Quoted from C. J. Smithells, "Metals Reference Book," p. 183, Butterworth & Co. (Publishers) Ltd., London, 1949; the original reference could not be located.

Ī.7424
0.2576

Ag-Pt Silver-Platinum

The liquidus and solidus temperatures given in Fig. 28 were determined by [1], indicating that there is a peritectic equilibrium between Ag-rich and Pt-rich solid solutions at 1185°C. Previously, [2] had shown that 3.6 wt. (2.0 at.) % Pt raises the melting point of Ag to 990°C. Liquidus points of five alloys with 6–42 at. % Pt were reported by [3]; however, the data were inaccurate.

The phase diagram by [1] was substantially confirmed by metallographic examinations, with the solubility limits of the two solid solutions located between 23.5 and 27 at. % and 77.5 and 86.5 at. % Pt, respectively [4]. The boundaries of the miscibility gap were more accurately determined by means of electrical-resistance-vs.-composition curves, resistance-temperature measurements, and lattice-parametric studies, using carefully annealed specimens [5]. The saturation limits obtained are indicated in Fig. 28 by data points. Also, it was found that intermediate homogeneity ranges of phases having ordered structures exist at temperatures below 750°C. The compositions PtAg (64.41 wt. % Pt) = β and Pt_3Ag (84.45 wt. % Pt) = γ' appeared to be the ideal compositions of these phases, both of which have ordered f.c.c. structures. γ' was found to undergo a transformation into another ordered structure, γ, at approximately 600°C.

A thorough reinvestigation of the phase boundaries and transformations in the solid state yielded the diagram presented in Fig. 28 [6]. Lattice-parameter determinations after annealing at various temperatures between 590 and 1130°C and quenching, besides measurements of resistance vs. temperature, were the chief method of investigation (see data points). As the rate of transformation is extremely

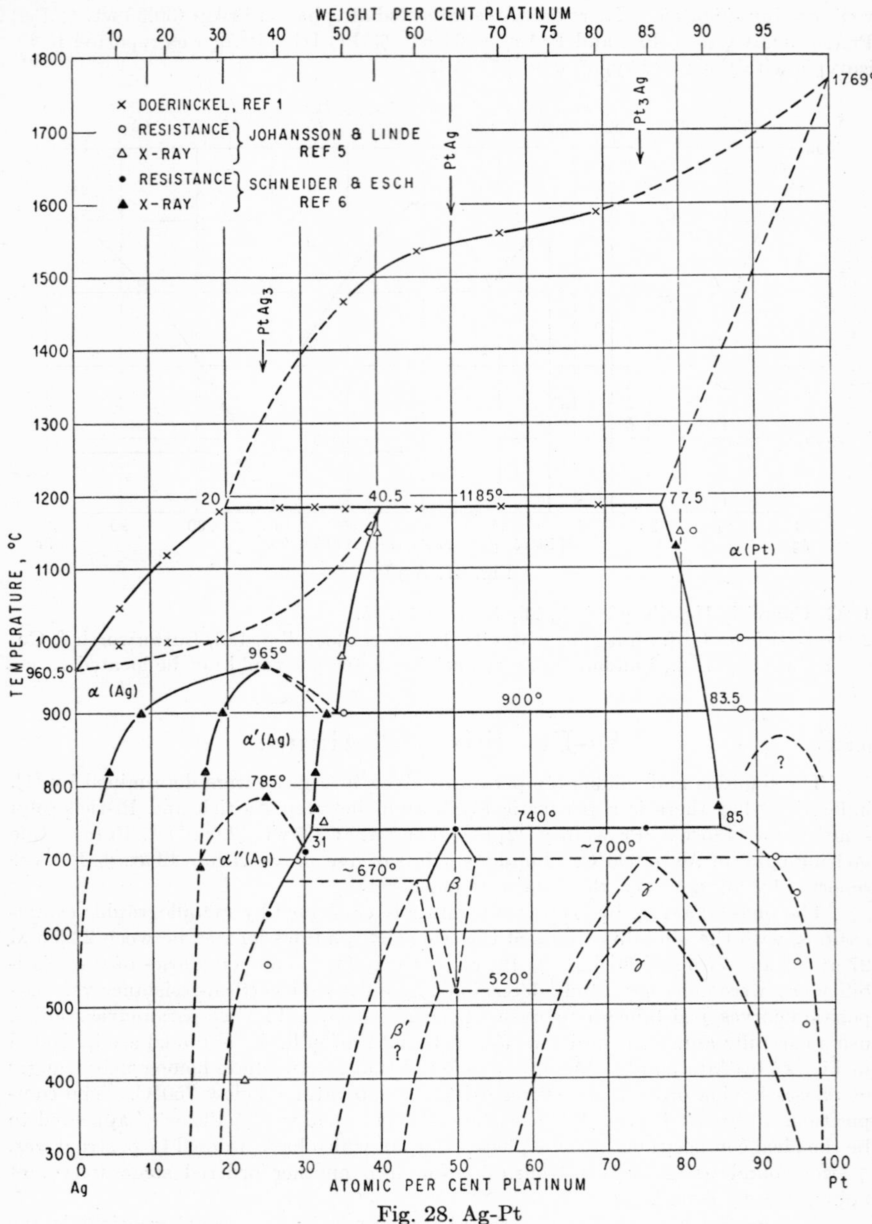
WEIGHT PER CENT PLATINUM
× DOERINCKEL, REF 1
○ RESISTANCE
△ X-RAY
JOHANSSON & LINDE REF 5
• RESISTANCE
▲ X-RAY
SCHNEIDER & ESCH REF 6
Pt Ag
Pt_3Ag
$PtAg_3$
1769°
1185°
20
40.5
77.5
α(Pt)
960.5°
965°
α (Ag)
900°
83.5
α′(Ag)
785°
740°
85
31
α″(Ag)
~700°
~670°
β
γ′
γ
520°
β′
?
TEMPERATURE, °C
ATOMIC PER CENT PLATINUM
Ag
Pt

Fig. 28. Ag-Pt

sluggish, the alloys had to be deformed and annealed for very long times, e.g., 108, 31, 90, 78, and 14 days at 685, 770, 820, 900, and 1130°C, respectively. Nevertheless, some of the transformations, especially α(Ag) → α'(Ag) → α''(Ag) and α'(Ag) + α(Pt) → β, were far from being complete. The dashed phase boundaries in Fig. 28 are largely hypothetical.

Crystal Structures [7]. Tabulated lattice parameters and parameter-vs.-composition curves are given in the paper by [6]. The α'(Ag) phase ($PtAg_3$) has a f.c.c. structure with random atomic distribution and a parameter smaller than that of α; e.g., a = 4.025 A for α and 3.902 A for α' in the 25 at. % Pt alloy, after annealing at 820°C. α'' has an ordered f.c.c. structure of the Cu_3Au ($L1_2$) type with a = 3.887 A at 25 at. % Pt. The β phase probably has an ordered rhombohedrically distorted cubic structure. The β phase according to [5] probably is identical with the γ' phase, since the latter is formed after relatively short times of annealing. The structure of β' appears to be similar to that of β. γ' is ordered f.c.c. formed from the f.c.c. lattice of the components by doubling the unit cell, with a = 3.925 A for the simple cell. γ is ordered f.c.c. with a simple cubic cell, a = 3.885 A at 85 at. % Pt. There are some indications of a transformation in the α(Pt) solid solution.

1. F. Doerinckel, *Z. anorg. Chem.*, **54,** 1907, 338–344.
2. C. T. Heycock and F. H. Neville, *Phil. Trans. Roy. Soc. (London)*, **189,** 1897, 69.
3. J. F. Thompson and E. H. Miller, *J. Am. Chem. Soc.*, **28,** 1906, 1115–1132.
4. N. S. Kurnakow and W. A. Nemilow, *Z. anorg. Chem.*, **168,** 1928, 339–348.
5. C. H. Johansson and J. O. Linde, *Ann. Physik*, **6,** 1930, 458–486; **7,** 1930, 408.
6. A. Schneider and U. Esch, *Z. Elektrochem.*, **49,** 1943, 72–89.
7. Results given in this paragraph are based on the work of [6]. Lattice parameters were given by the authors in angstroms (10^{-8} cm) and, therefore, were taken as true angstroms. For additional information on lattice structures, see [5].

$\bar{1}$.7627
0.2373

Ag-Re Silver-Rhenium

Re is insoluble in both liquid and solid Ag [1].

1. U. Holland-Nell and F. Sauerwald, *Z. anorg. Chem.*, **276,** 1954, 155–158.

0.0205
$\bar{1}$.9795

Ag-Rh Silver-Rhodium

According to [1], Rh appears to be insoluble in molten Ag. However, [2] have prepared alloys in all proportions, which were shown by X-ray studies to consist of practically pure Ag and a solid solution of at most 0.1 at. % Ag.

1. H. Rössler, *Chem.-Ztg.*, **24,** 1900, 733–735.
2. R. W. Drier and H. L. Walker, *Phil. Mag.*, **16,** 1933, 294–298.

0.0256
$\bar{1}$.9744

Ag-Ru Silver-Ruthenium

X-ray investigation of an alloy with 5 wt. (5.3 at.) % Ru indicated slight solid solubility and the absence of an intermediate phase [1].

1. J. L. Byers, AIME Preprint 10, p. 16, 1932; not published otherwise.

0.5269
$\bar{1}$.4731

Ag-S Silver-Sulfur

Investigations of the partial system Ag-Ag_2S by thermal analysis were carried out by [1–6] after [7] had already established its general features. The work of [1] is obsolete because it does not reveal the existence of a miscibility gap in the liquid

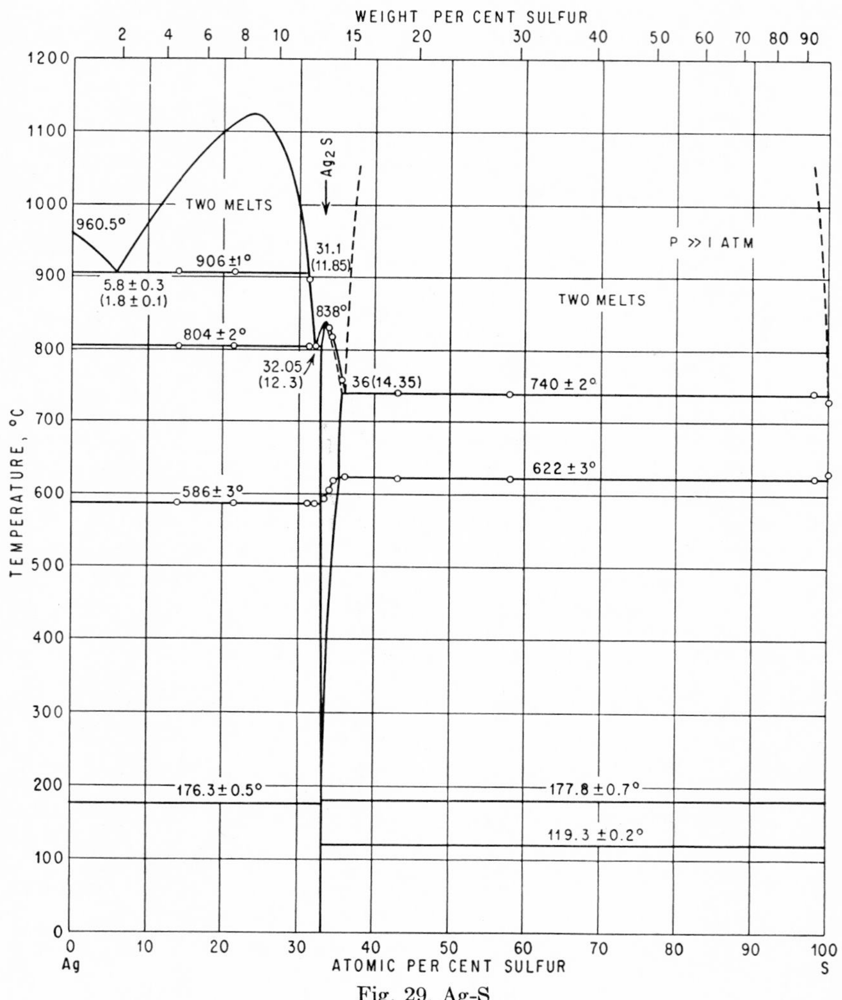

Fig. 29. Ag-S

state. Data by [2–6] are presented in Table 5. Figure 29 is substantially based on the accurate work of [6].

The melting point of Ag_2S and its lower transformation point were also determined by other investigators: 840–845 [8], 825 [1], 834 [9], 812 [10], and 837°C [11]; and 175 [12, 9, 11], 178 [10], and 180°C [13]. The upper transformation point was given as 591.5°C [11]. The lower values of the melting point are due to insufficient protection against oxidation and the fact that the melting point is strongly lowered by a small excess of both Ag and S (Fig. 29). The variability of the transformation points of Ag_2S with composition [6] clearly indicates that this phase is of variable composition, especially at higher temperatures. According to [14], the composition of the Ag_2S phase coexisting with Ag at 200°C is 33.31 at. % S ($Ag_{2.002}S$), whereas Ag_2S in equilibrium with liquid S has very nearly ideal stoichiometric composition. The difference in the transformation temperature of Ag_2S in equilibrium with Ag or S, respectively, was calculated as 1.7°C, as compared with 1.6 ± 1.1°C according to [6].

Table 5. Characteristic Temperatures in °C and Compositions in At. % S (Wt. % in Parentheses) of the System Ag-Ag_2S

	Ref. 2	Ref. 3	Ref. 4	Ref. 5	Ref. 6
Monotectic temperature	905°C	905°C	904°C	~900°C	906 ± 1°C
Eutectic temperature	807°C	806°C	804°C	~800°C	804 ± 2°C
Ag-rich melt at monotectic temperature	5.8–6.2 (1.8–1.95)	6.4 (~2.0)	6.7–7.0 (2.1–2.2)	6.1 (1.9)	5.8 (1.8)
S-rich melt at monotectic temperature	>30.8 (>11.7)	31.8–32.0 (12.2–12.3)	31.6 (12.1)	...... ?	31.1 (11.85)
Eutectic composition	....	32.0 (~12.3)	33.0 (12.8)	32.0 (12.3)	32.0 (12.3)
Melting point of Ag_2S	....	842°C	815°C	842°C	838 ± 2°C
Transformation points of Ag_2S	175°C	175°C	....	576°C 175°C	See Fig. 29

The boundary of the miscibility gap in the liquid state between Ag and Ag_2S was determined by studying the reaction $Ag_2S + H_2 \rightarrow 2Ag + H_2S$ and measuring the ratio $H_2S:H_2$ in dependence upon temperature and composition [15]. At 1000 and 1100°C the two liquids in equilibrium contain about 12 and 30 at. (3.9 and 11.3 wt.) % S and about 20 and 26.5 at. (6.9 and 9.7 wt.) % S, respectively. The critical point lies at approximately 24 at. (8.6 wt.) % S and 1125 ± 25°C.

The solid solubility of S in Ag is extremely small and probably less than 0.05 wt. (about 0.17 at.) % S [2].

Phase relations in the partial system Ag_2S-S, as shown in Fig. 29, are due to [6].

Crystal Structures. The modification of Ag_2S stable between about 180 and 625°C, Ag_2S II, was reported to have a b.c.c. structure, with 2 molecules per unit cell (the Ag ions being statistically distributed) and a = 4.84–4.90 A [13, 16–18], a = 4.865 A [19]. The low-temperature modification Ag_2S III is monoclinic with 8 molecules per unit cell and a = 9.49 A, b = 6.93 A, c = 8.30 A, β = 124°C [18, 20]. [19] reported that, on heating at 350–400°C for 1–2 min, the lattice was transformed irreversibly to f.c.c., a = 5.412 A.

1. H. Pélabon, *Compt. rend.*, **143**, 1906, 294–296; *Ann. chim. et phys.*, **17**, 1909, 526–566.
2. K. Friedrich and A. Leroux, *Metallurgie*, **3**, 1906, 361–367.
3. F. M. Jaeger and H. S. van Klooster, *Z. anorg. Chem.*, **78**, 1912, 248–252.
4. C. C. Bissett, *J. Chem. Soc.*, **105**, 1914, 1223–1228.
5. G. G. Urasow, *Ann. inst. polytech., Petrograd*, **23**, 1915, 593–627; abstract in *J. Inst. Metals*. **14**, 1915, 234–235.

6. F. C. Kracek, *Trans. Am. Geophys. Union*, **27**, 1946, 364–374.
7. F. Roessler, *Z. anorg. Chem.*, **9**, 1895, 34–39.
8. H. Pélabon, *Compt. rend.*, **137**, 1903, 920.
9. W. Truthe, *Z. anorg. Chem.*, **76**, 1912, 168.
10. C. Sandonnini, *Rend. accad. nazl. Lincei*, **21**, 1912, 480.
11. E. Jensen, *Skrifter Norske Videnskaps-Akad. Oslo, I. Mat. Naturv. Kl.*, **1947**, no. 2; abstract in *Chem. Abstr.*, **43**, 1949, 3274.
12. M. Bellati and S. Lussana, *Z. physik. Chem.*, **5**, 1890, 282.
13. R. C. Emmons, C. H. Stockwell, and R. H. B. Jones, *Am. Mineralogist*, **11**, 1926, 326–328.
14. C. Wagner, *J. Chem. Phys.*, **21**, 1953, 1819–1827.
15. T. Rosenquist, *Trans. AIME*, **185**, 1949, 451–460.
16. J. Palacios and R. Salvia, *Anales fís. y quím.* (*Madrid*), **29**, 1931, 269–279.
17. P. Rahlfs, *Z. physik. Chem.*, **B31**, 1936, 157–194.
18. L. S. Ramsdell, *Am. Mineralogist*, **28**, 1943, 401–425.
19. H. Wilman and A. P. B. Sinha, *Acta Cryst.*, **7**, 1954, 682–683.
20. L. S. Ramsdell, *Am. Mineralogist*, **10**, 1925, 281–304; **12**, 1927, **25**; see also [16].

$\bar{1}.9474$
0.0526

Ag-Sb Silver-Antimony

Liquidus-curve data covering the whole system and portions of it have been reported by various authors [1–7]. Whereas the first diagram published [3] showed only one intermediate phase, Ag_3Sb (27.34 wt. % Sb), to exist and to be formed peritectically at 559°C, determination of the hydrogen overvoltage of the alloys [8] indicated that there is an additional phase between 0 and 25 at. % Sb. This was first confirmed by X-ray studies [9] and verified by later investigators [10, 11, 12, 6, 13, 7, 14, 15]. Detailed or partial determinations of the solid-state phase limits have been reported: $\alpha/(\alpha + \zeta)$ [10, 6, 16, 7]; $\zeta/(\alpha + \zeta)$ [10, 6, 7]; $\zeta/(\zeta + \epsilon)$ [10, 6, 7]; $\epsilon/(\zeta + \epsilon)$ [7]; $\epsilon/(\epsilon + \text{Sb})$ [13, 7].

The partial diagram up to 18 at. (20 wt.) % Sb presented in Fig. 30 is based on the careful micrographic analysis by [6]. Their data on the $\alpha/(\alpha + \zeta)$ boundary agree within a few tenths of a per cent with those of [16], based on lattice-parameter measurements. The less completely determined $\zeta/(\alpha + \zeta)$ and $\zeta/(\zeta + \epsilon)$ boundaries by [10, 7] are in substantial agreement with the findings of [6]. The $\zeta/(\zeta + \epsilon)$ boundary of Fig. 30 was determined by [7], and the ϵ solidus curve and $\epsilon/(\epsilon + \text{Sb})$ boundary accepted in Fig. 30 are due to [13]. The ϵ phase undergoes a transformation at 440–449°C which appears to be of the order-disorder type [7].

Crystal Structures. For lattice parameters of the α phase see [17, 16]. The ζ phase—also designated as ϵ [9, 10, 7]—is a 3:2 electron compound [18] having a h.c.p. structure with lattice parameters given by [9, 11, 7]. The ϵ and ϵ' phases are 7:4 electron compounds [19] having an orthorhombic structure with lattice parameters reported by [9, 11, 7]. The solid solubility of Ag in Sb appears to be negligibly small [11, 7].

1. H. Gautier, *Compt. rend.*, **123**, 1896, 173; *Bull. soc. encour. ind. natl.*, **1901**, 93–111.
2. C. T. Heycock and F. H. Neville, *Phil. Trans. Roy. Soc.* (*London*), **A189**, 1897, 52–54.
3. G. I. Petrenko, *Z. anorg. Chem.*, **50**, 1906, 139–144.
4. D. Stockdale, *J. Inst. Metals*, **43**, 1930, 193–211.
5. W. Hume-Rothery and P. W. Reynolds, *Proc. Roy. Soc.* (*London*), **A160**, 1937, 282–303.

6. P. W. Reynolds and W. Hume-Rothery, *J. Inst. Metals*, **60**, 1937, 365–374.
7. F. Weibke and I. Efinger, *Z. Elektrochem.*, **46**, 1940, 52–60; see also F. Weibke and I. Efinger, *Z. Elektrochem.*, **46**, 1940, 61–69.
8. M. G. Raeder and J. Brun, *Z. physik. Chem.*, **133**, 1928, 28–29; M. G. Raeder, *Z. physik. Chem.*, **B6**, 1929, 40–42.
9. A. Westgren, G. Hägg, and S. Eriksson, *Z. physik. Chem.*, **B4**, 1929, 461–468.
10. N. Ageew and M. Hansen, unpublished work (1930); see M. Hansen, "Der Aufbau der Zweistofflegierungen," p. 57, Springer-Verlag OHG, Berlin, 1936.
11. S. J. Broderick and W. J. Ehret, *J. Phys. Chem.*, **35**, 1931, 2627–2636.
12. W. Guertler and W. Rosenthal, *Z. Metallkunde*, **24**, 1932, 8.
13. D. Stockdale, *J. Inst. Metals*, **60**, 1937, 375–376.

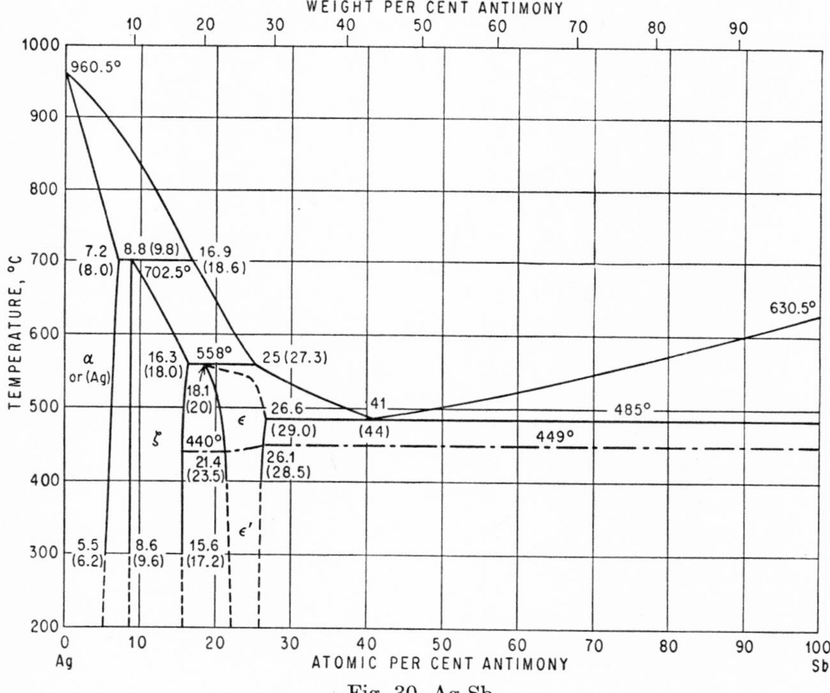

Fig. 30. Ag-Sb

14. W. G. John and E. J. Evans, *Phil. Mag.*, **22**, 1936, 417–435.
15. G. O. Stephens and E. J. Evans, *Phil. Mag.*, **22**, 1936, 435–445.
16. E. A. Owen and E. W. Roberts, *Phil. Mag.*, **27**, 1939, 294–327.
17. W. Hume-Rothery, G. F. Lewin, and P. W. Reynolds, *Proc. Roy. Soc.* (*London*), **A157**, 1936, 167–183.
18. W. Hume-Rothery, P. W. Reynolds, and G. V. Raynor, *J. Inst. Metals*, **66**, 1940, 191–207.
19. B. R. T. Frost and G. V. Raynor, *Proc. Roy. Soc.* (*London*), **A203**, 1950, 134–135.

0.1355
$\bar{1}$.8645

Ag-Se Silver-Selenium

Thermal and microscopic investigations of the partial system Ag-Ag_2Se (26.79 wt. % Se) [1, 2] showed that there is a miscibility gap in the liquid state as reported earlier [3]. The monotectic point is located at approximately 9 wt. (12 at.) % Se, 890°C, and the composition of the eutectic Ag-Ag_2Se, which was not accurately determined, was found to lie close to the composition of the selenide (Fig. 31). The findings of both papers agree well but are not in accord with those of another thermal study [4], which failed to detect a monotectic equilibrium and, instead, showed a eutectic at 19.5 wt. (24.9 at.) % Se, 830°C. An alloy with 0.2 wt. (0.27 at.) % Se proved to be heterogeneous, after cooling from the molten state [1].

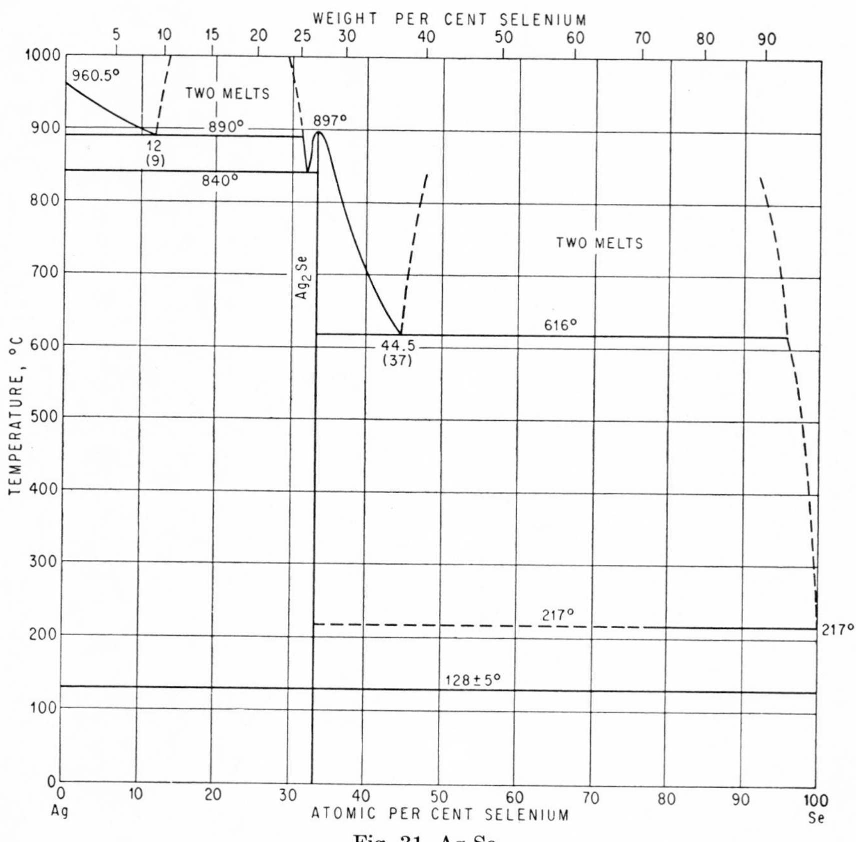

Fig. 31. Ag-Se

The partial system Ag_2Se-Se (Fig. 31) is based on the conformable thermal-analysis data of two investigations [4, 2]. Ag_2Se undergoes an allotropic transformation at 133 [5] or 122°C [2]. The low-temperature α modification has been described as either orthorhombic or monoclinic [6]. The β modification, stable above 128 ± 5°C, was reported to have the structure of the CaF_2 (C1) type with a = 4.993 A at 170–190°C [7]; see, however, [6].

1. K. Friedrich and A. Leroux, *Metallurgie*, **5**, 1908, 357–358.
2. G. Pellini, *Gazz. chim. ital.*, **45**, 1915, 533–539.
3. F. Roessler, *Z. anorg. Chem.*, **9**, 1895, 39–41.

4. H. Pélabon, *Compt. rend.*, **143,** 1906, 294–295; *Ann. chim. et phys.*, **17,** 1909, 558–560. No numerical data are given.
5. M. Bellati and S. Lussana, *Z. physik. Chem.*, **5,** 1890, 282.
6. B. W. G. Wyckoff, "Crystal Structures," vol. I, chap. IV, p. 34, Interscience Publishers, Inc., New York, 1948.
7. P. Rahlfs, *Z. physik. Chem.*, **B31,** 1936, 157–194.

0.5844
$\bar{1}$.4156

Ag-Si Silver-Silicon

The phase diagram presented in Fig. 32 was established by thermal analysis [1], which confirmed earlier observations [2–4] that Ag and Si do not form a silicide. An

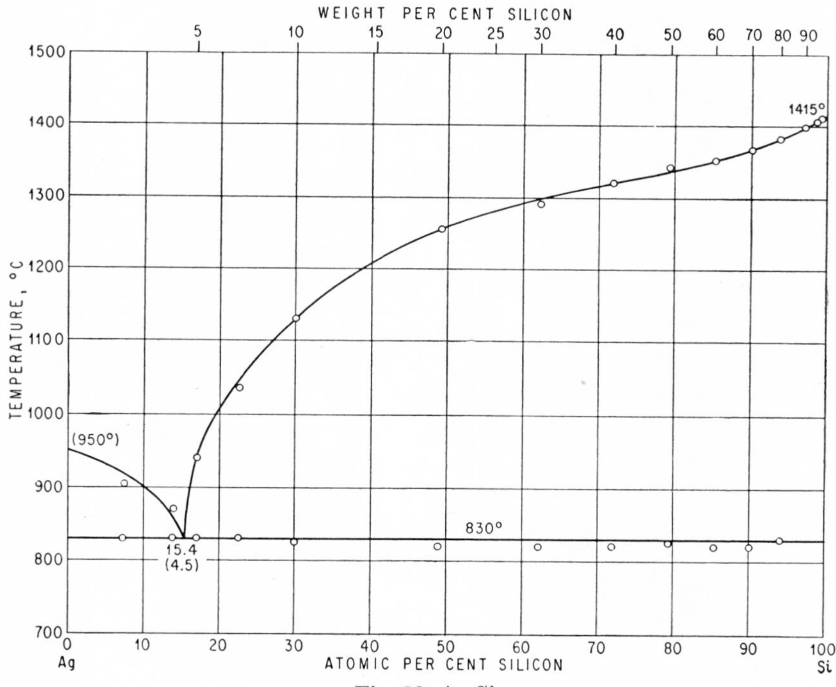

Fig. 32. Ag-Si

arc-melted alloy with 4.5 wt. (15.4 at.) % Si was found to consist entirely of the eutectic [5], in fair agreement with the thermal data. According to [6], the eutectic lies at 14 at. (4.05 wt.) % Si.

From positive deviations of the activity coefficient of liquid Ag-rich alloys it was concluded that there is an appreciable solid solubility of Si in Ag [7], confirming diffusion tests of Si in Ag [8]. On the other hand, determinations of the lattice constants indicated that the mutual solid solubility is practically nil [9].

1. G. Arrivaut, *Z. anorg. Chem.*, **60,** 1908, 436–440; *Compt. rend.*, **147,** 1908, 859; *Rev. mét.*, **5,** 1908, 932. The silicon used was 99% pure.
2. H. Moissan, *Compt. rend.*, **121,** 1895, 625–626.
3. H. Moissan and F. Siemens, *Compt. rend.*, **138,** 1904, 1299–1303.

4. E. Vigouroux, *Compt. rend.*, **144**, 1907, 1214.
5. W. R. Johnson and M. Hansen, AF Technical Report 6383, June, 1951.
6. O. Hájiček, *Hutnické Listy*, **3**, 1948, 265–270.
7. H. M. Schadel, G. Derge, and C. E. Birchenall, *Trans. AIME*, **188**, 1950, 1282–1283.
8. W. Loskiewicz, *Przeglad Górniczo-Hutniczy*, **21**, 1929, 583–611; Congrès International des Mines, etc., in Liége, 1930, Section de Métallurgie.
9. E. R. Jette and E. B. Gebert, *J. Chem. Phys.*, **1**, 1933, 753–755.

$\bar{1}.9585$
0.0415

Ag-Sn Silver-Tin

The diagram presented in Fig. 33 is the result of studies of numerous investigators and is essentially based on the work by [1–10]. The liquidus curve was determined, in part and in full, by [1, 11, 7, 12] and [13, 2, 3, 4], respectively. Results agree very well. The eutectic point was repeatedly found to be located at 96.5 wt. (96.2 at.) % Sn, 221°C [1, 3, 4, 7]. [21] reported 97.15 at. % Sn, 221°C. The existence of the ζ phase, first detected by potential measurements [14] and definitely established by thermal and micrographic studies [4], was confirmed by X-ray investi-

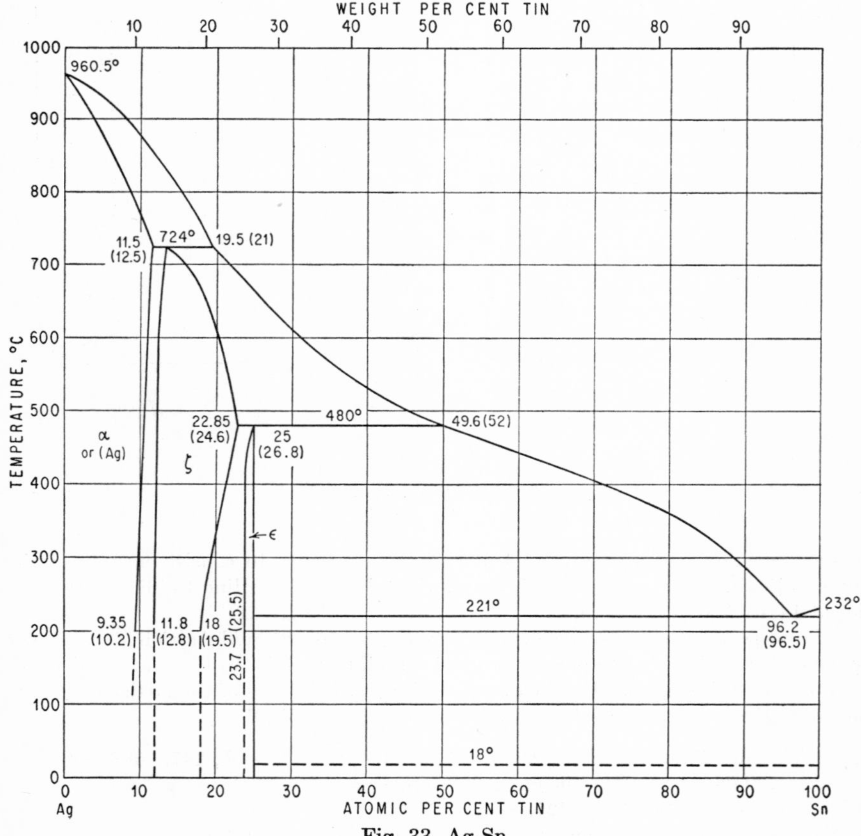

Fig. 33. Ag-Sn

gations [15, 5, 10]. The existence of the ε phase was detected by [3] and confirmed by (4, 16, 5, 10, 17].

The solidus curves of the α and ζ solid solutions were determined micrographically by [4, 6] and [4], respectively. The solid solubility of Sn in Ag, as found by micrographic analysis [4, 6], is slightly higher at temperatures above 650°C and lower at temperatures below 600°C than that determined by X-ray parametric work [8, 10]. Results obtained by the latter differ by only 0.2–0.3%. The maximum solubility at 724°C is about 11.3 at. (12.2 wt.) % Sn according to [8] and about 12.2 at. (13.3 wt.) % Sn according to [4]. The former value is more consistent with the solubility limits at lower temperatures, which lie very close to 11.0 (12.0), 10.7 (11.7), 10.25 (11.2), 9.9 (10.8), and 9.35 at. (10.2 wt.) % Sn at 600, 500, 400, 300, and 218°C [8, 10], respectively. The $\zeta/(\alpha + \zeta)$ curve was found to lie at 12.5 (13.6), 12.15 (13.2), and 11.75 at. (12.8 wt.) % Sn at 600, 500, and 218°C [10]. The $\zeta/(\zeta + \epsilon)$ curve, determined by micrographic investigations [4], is practically identical with that obtained by X-ray analysis [10]. The ε phase is homogeneous between 23.7 at. (25.5 wt.) % and 25.0 at. (26.8 wt.) % Sn [4]; the latter composition corresponds to the formula Ag_3Sn.

The solid solubility of Ag in Sn was reported to be as follows: less than 0.1 wt. (0.09 at.) % [4]; 0.06 wt. (0.055 at.) % at 210°C and 0.02 wt. (0.18 at.) % at "room temperature" [7]; approximately 0.1 at. % at 200°C [18]; and 0.04 wt. (0.036 at.) % Sn at 220°C [9]. The solidus curve of the Sn-rich solid solution was determined by [9]. The ε phase (Ag_3Sn) was found to undergo a transformation at about 60°C [4], the nature of which is unknown. X-ray investigation showed that it is not accompanied by a lattice change [16]. For a thermodynamic study of the system, see [22].

Crystal Structures. Lattice spacings of the α phase were determined by [5, 19, 8, 10]; see also [20]. The ζ phase has a h.c.p. lattice of the Mg (A3) type [15, 5, 10]; lattice parameters of various compositions are given by [5, 10]. The ε phase, first reported to be h.c.p. [16], was shown to have a slightly rhombically deformed h.c.p. superlattice with 4 atoms per unit cell [5, 10, 17], $a = 2.995$ A, $c = 4.780$ A, $c/a = 1.596$ at 25 at. % Sn [5].

Supplement. [23] measured the heats of formation of liquid and solid Ag-Sn alloys at 450°C and suggested a reinvestigation of the crystal structures of the intermediate phases.

1. C. T. Heycock and F. H. Neville, *J. Chem. Soc.*, **57**, 1890, 376–393.
2. C. T. Heycock and F. H. Neville, *Phil. Trans. Roy. Soc.* (*London*), **A189**, 1897, 40–41, 58–60.
3. G. I. Petrenko, *Z. anorg. Chem.*, **53**, 1907, 204–211.
4. A. J. Murphy, *J. Inst. Metals*, **35**, 1926, 107–124.
5. O. Nial, A. Almin, and A. Westgren, *Z. physik. Chem.*, **B14**, 1931, 83–90.
6. W. Hume-Rothery, G. W. Mabbott, and K. M. Channel-Evans, *Phil. Trans. Roy. Soc.* (*London*), **A233**, 1934, 1–97.
7. D. Hanson, E. J. Sandford, and H. Stevens, *J. Inst. Metals*, **55**, 1934, 115–131.
8. E. A. Owen and E. W. Roberts, *Phil. Mag.*, **27**, 1939, 294–327.
9. C. E. Homer and H. Plummer, *J. Inst. Metals*, **64**, 1939, 169–200.
10. M. M. Umanskiy, *Zhur. Fiz. Khim.*, **14**, 1940, 846–849.
11. C. T. Heycock and F. H. Neville, *J. Chem. Soc.*, **65**, 1894, 65–76.
12. W. Hume-Rothery and P. W. Reynolds, *Proc. Roy. Soc.* (*London*), **A160**, 1937, 282–303.
13. H. Gautier, *Compt. rend.*, **123**, 1896, 173; Société d'encouragement pour l'industrie nationale, Paris, Commission des alliages, "Contributions á l'étude des alliages," pp. 93–118, Typ. Chamerot et Renouard, Paris, 1901.

14. N. Puschin, *Z. anorg. Chem.*, **56,** 1908, 20–22.
15. A. Westgren and G. Phragmén, *Z. Metallkunde,* **18,** 1926, 279–284.
16. G. D. Preston, *J. Inst. Metals,* **35,** 1926, 118–119, 129.
17. P. Michel, *Compt. rend.*, **235,** 1952, 377–379.
18. E. Jenckel and L. Roth, *Z. Metallkunde,* **30,** 1938, 135–144.
19. W. Hume-Rothery, G. F. Lewin, and P. W. Reynolds, *Proc. Roy. Soc.* (*London*), **A157,** 1936, 167–183.
20. E. A. Owen, *J. Inst. Metals,* **73,** 1947, 471–489.
21. O. Hájiček, *Hutnické Listy,* **3,** 1948, 265–270.
22. R. O. Frantik and H. J. McDonald, *Trans. Electrochem. Soc.,* **88,** 1946, 253–260.
23. O. J. Kleppa, *Acta Met.,* **3,** 1955, 255–259.

0.0903
$\bar{1}$.9097

Ag-Sr Silver-Strontium

The diagram of Fig. 34 is based merely on the thermal-analysis data given by [1]. The composition of the four intermediate phases $SrAg_4$ (16.88 wt. % Sr), Sr_3Ag_5

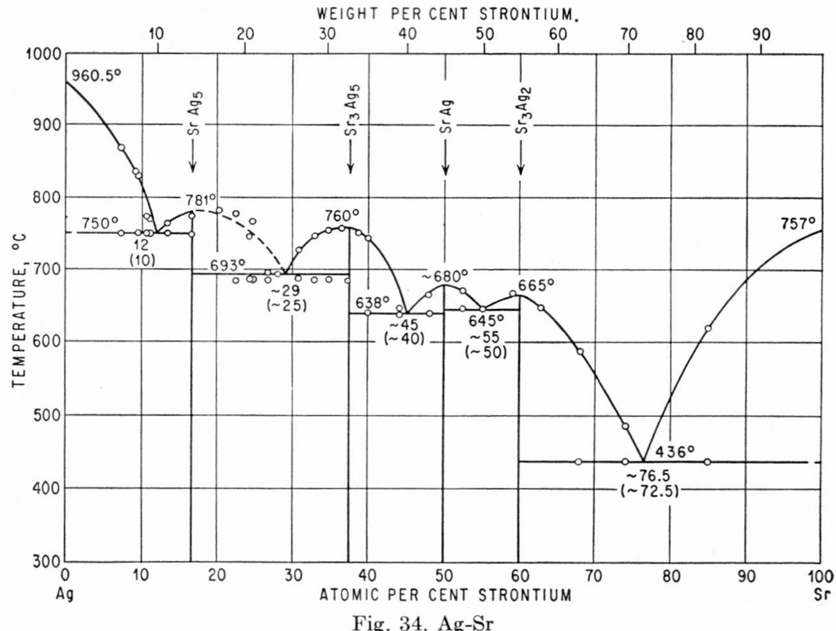

Fig. 34. Ag-Sr

(32.77 wt. % Sr), SrAg (44.82 wt. % Sr), and Sr_3Ag_2 (54.92 wt. % Sr) was ascertained from the position of the maxima of the liquidus curve and the lengths of the eutectic crystallization times. Instead of the composition Sr_3Ag_5 (37.5 at. % Sr), perhaps the simpler atomic ratio Sr_2Ag_3 (40 at. % Sr) could be taken into consideration too. Recently the existence of the phase $SrAg_5$ [16.67 at. (13.98 wt.) % Sr], identified by its crystal structure, has been reported [2]. In fact, the analysis of the specimen designated by [1] as $SrAg_4$ showed the actual composition of $SrAg_5$.

$SrAg_5$ is isotypic with $CaCu_5$; a = 5.675 A, c = 4.619 A, c/a = 0.814 [2].

1. F. Weibke, *Z. anorg. Chem.*, **193**, 1930, 297–310.
2. T. Heumann, *Nachr. Ges. Wiss. Göttingen, Math.-physik. Kl.*, **2**, 1950, 1–6.

$\bar{1}.7756$
0.2244

Ag-Ta Silver-Tantalum

The two metals do not form alloys [1].

1. H. Moissan, *Compt. rend.*, **134**, 1902, 411.

$\bar{1}.9271$
0.0729

Ag-Te Silver-Tellurium

Five papers deal with the thermal and microscopic investigation of this system [1–5]. The existence of the compound Ag_2Te (37.16 wt. % Te), reported earlier [6], was confirmed by all these studies. However, the composition given for the second intermediate phase, formed peritectically, differs widely among Ag_7Te_4 [36.33 at. (40.33 wt.) % Te] [3], $Ag_{12}Te_7$ [36.84 at. (40.83 wt.) % Te] [4], Ag_3Te_2 (44.09 wt. % Te) [5], and AgTe (54.17 wt. % Te) [2]. Table 6 contains the compositions and temperatures of the characteristic points as found by the various authors.

Table 6. Characteristic Temperatures in °C and Compositions in At. % Te (Wt. % in Parentheses) in the System Ag-Te

	Ref. 1	Ref. 2	Ref. 3	Ref. 4	Ref. 5
Ag-Ag_2Te eutectic	825°	872°	870°	860°	870°
	~20	13–14	~11	13	8
	(~23)	(15–16)	(~12.5)	(15)	(9.3)
Melting point of Ag_2Te	955°	959°	957°	958°	959°
Peritectic temperature	......	~443°	446–447°	443°	465°
Peritectic melt	......	~53	~53	51	51
		(~57)	(~57)	(55)	(55)
Te-rich eutectic	345°	351°	347°	348°	351°
	~63	66.5	~64.5	~64.5	67
	(~67)	(70)	(~68)	(~68)	(70.6)

Another point of discrepancy is whether a miscibility gap in the liquid state exists in the partial system Ag-Ag_2Te. The facts (*a*) that several authors were either unable [2] to detect primary thermal effects between about 12 at. % Te and Ag_2Te or found a very peculiarly shaped liquidus curve in this range [3, 4] and (*b*) that there is a range of liquid immiscibility in the partial systems Ag-Ag_2S and Ag-Ag_2Se seemed to be in favor of a miscibility gap. Indeed, its occurrence was established by the work of [5]. The diagram of the partial system Ag-Ag_2Te, shown in Fig. 35, was drawn on the basis of the information given in an advance abstract of a paper yet to be published [5].

The scanty data available make it difficult to decide between the various compositions suggested for the second intermediate phase. Although the compositions corresponding to the formulas Ag_7Te_4 [3] and $Ag_{12}Te_7$ [4] differ by only 0.5 at. % or wt. % Te, seemingly indicating good agreement between two investigators, it is felt that Ag_3Te_2 is more likely. This is suggested because of the simpler atomic ratio as well as the fact that the X-ray data [4], claimed to be in favor of $Ag_{12}Te_7$, are not entirely convincing. Also, the composition Ag_7Te_4 is not based on reliable data [3]. The experimental evidence found for the composition Ag_3Te_2 is unknown since no information as to this point is given in the abstract [5].

The solid solubility of Te in Ag and Ag in Te is negligible [4]. Ag_2Te undergoes a transformation, the temperature of which was given as 135°C on cooling [7], 141°C

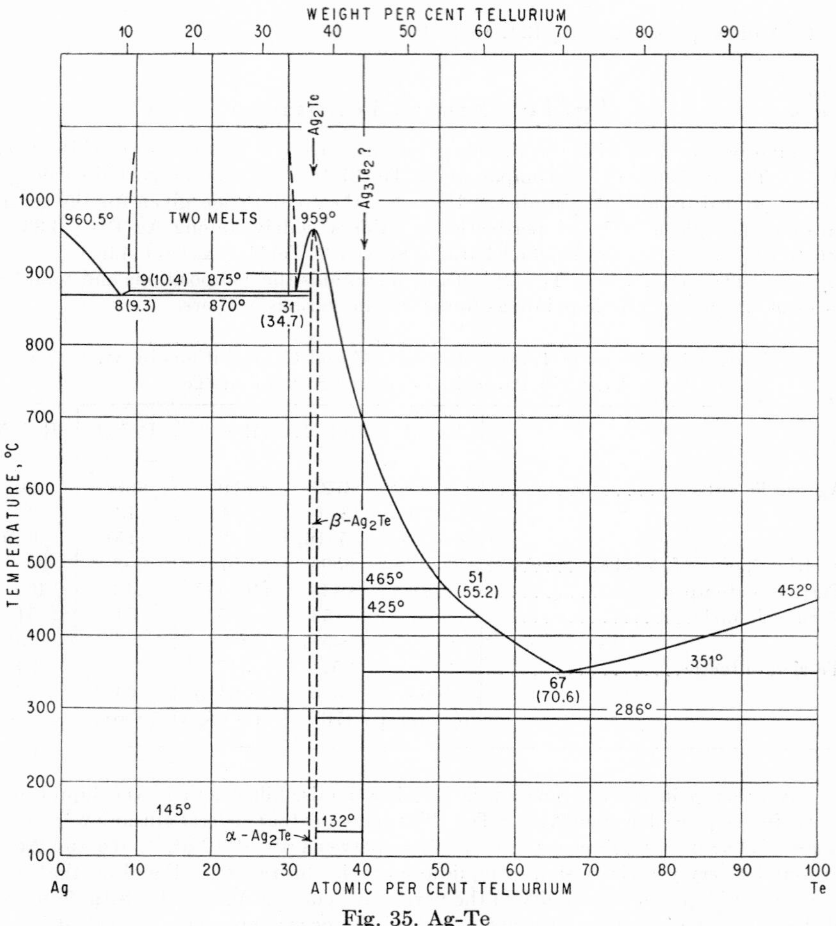

Fig. 35. Ag-Te

on heating [7], 149.5°C [8], and 140°C [4]. [5] reported that the inversion was found on heating at 145°C in the pure compound and in the presence of excess Ag, and at 132°C in the presence of excess Te. This would indicate a certain range of homogeneity. Ag_3Te_2 (?) can exist in at least two, probably three, modifications, with transformation temperatures reported as 412 [2, 3], 408 [4], 425 [5], and 286°C [5].

Crystal Structures. The structure of the low-temperature modification of Ag_2Te (in Fig. 35 designated as α-Ag_2Te in contrast to [9], who called the high-temperature form α-Ag_2Te) and the mineral hessite has been the subject of several studies, with contradictory results [10, 11, 12, 4, 13]. It is probably orthorhombic [9, 4, 13]. The dimensions of the unit cell were reported as follows: $a = 13.0$ A, $b = 12.7$ A, $c = 12.2$ A [4], and $a = 16.27$ A, $b = 26.68$ A, $c = 7.55$ A, with 48 [Ag_2Te] per unit cell [13]. The high-temperature form of Ag_2Te was found to be f.c.c., with 4 [Ag_2Te] per unit cell and $a = 6.585$ A at about 250°C [9], $a = 6.64$ A at 155°C [13], and $a = 6.59$ A at about 150°C [14]; see also [15].

The low-temperature form of $Ag_{12}Te_7$ (?) was reported to be hexagonal, with 57 atoms per unit cell and $a = 13.456$ A, $c = 8.468$ A, $c/a = 0.6293$ [4]. For the structure of the minerals stuetzite and empressite, see [5, 16].

1. H. Pélabon, *Compt. rend.*, **143**, 1906, 295–296.
2. G. Pellini and E. Quercigh, *Atti accad. nazl. Lincei*, (2)**19**, 1910, 415–421.
3. M. Chikashige and I. Saito, *Mem. Coll. Sci. Univ. Kyoto*, **1**, 1916, 361–368.
4. V. Koern, *Naturwissenschaften*, **27**, 1939, 432; *Acta et Commentationes Univ. Tartuensis (Dorpatensis)*, **A35**, 1940, no. 4.
5. F. C. Kracek and C. J. Ksanda, *Trans. Am. Geophys. Union*, **1940**, 363; abstract of a paper still unpublished.
6. J. Margottet, *Ann. sci. école norm. supér.*, (2)**8**, 1879, 253; B. Brauer, *Monatsh. Chem.*, **10**, 1889, 421; J. B. Senderens, *Compt. rend.*, **104**, 1887, 175; R. D. Hall and V. Lenher, *J. Am. Chem. Soc.*, **24**, 1902, 918; C. A. Tibbals, *J. Am. Chem. Soc.*, **31**, 1909, 909.
7. C. Tubandt, M. Reinhold, and A. Neumann, *Z. Elektrochem.*, **39**, 1933, 227–244.
8. H. Borchert, *Neues Jahrb. Mineral. Geol., Beilage Bd.*, **A69**, 1935, 465 (natural hessite).
9. P. Rahlfs, *Z. physik. Chem.*, **B31**, 1936, 157–194.
10. M. L. Huggins, *Phys. Rev.*, **21**, 1923, 211.
11. L. Ramsdell, *Am. Mineral.*, **10**, 1925, 289.
12. L. Tokody, *Z. Krist.*, **82**, 1932, 154–157; **89**, 1934, 416.
13. J. F. Rowland and L. G. Berry, *Am. Mineral.*, **36**, 1951, 471–479.
14. U. Zorll, *Ann. Physik*, **14**, 1954, 385–390.
15. H. Nowotny, *Z. Metallkunde*, **37**, 1946, 38–40; and *Structure Repts.*, **10**, 1945-1946, 25, 82.
16. R. M. Thompson, M. A. Peacock, J. F. Rowland, and L. G. Berry, *Am. Mineral.*, **36**, 1951, 458–470.

$\bar{1}.6672$
0.3328

Ag-Th Silver-Thorium

The diagram in Fig. 36 is chiefly based on the thermal analysis of alloys prepared from pressed pellets of powder mixtures melted under argon [1]. Above about 40 at. % Th, results are uncertain because of contamination of the melts by reaction with the crucible materials [2]. The Th-rich alloys contained an unstated amount of ThO_2. All alloys were analyzed, and the oxide content was considered in computing the metal ratio.

The solid solubility of Th in Ag, as determined by X-ray and micrographic studies, was reported to be 0.2–0.3 wt. (about 0.1–0.14 at.) % Th at the eutectic temperature. It is likely that besides $ThAg_3$ (41.77 wt. % Th) and Th_3Ag_5 (56.35 wt. % Th) intermediate phases richer in Th exist.

1. E. Raub and M. Engel, *Z. Elektrochem.*, **49**, 1943, 487–493.
2. Alloys low in Th were melted in Pythagoras crucibles, those richer in Th in corundum crucibles.

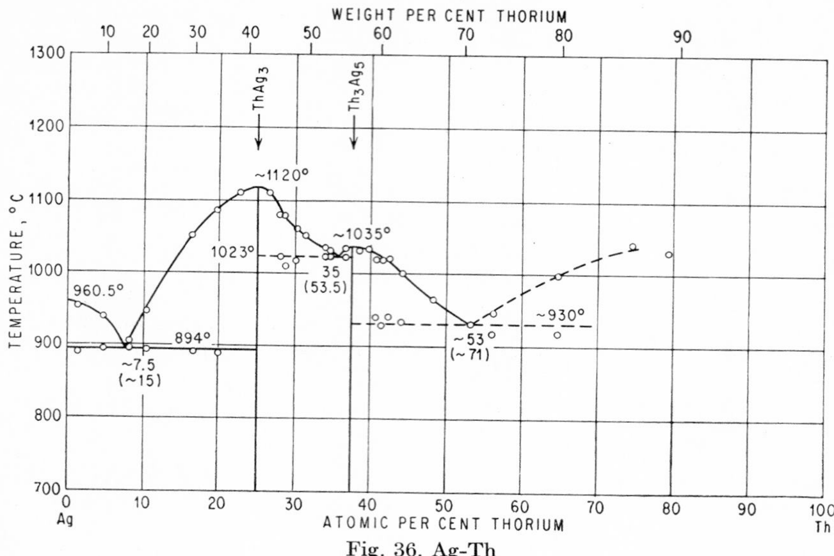

Fig. 36. Ag-Th

0.3526
$\bar{1}.6474$

Ag-Ti Silver-Titanium

In contrast to previous reports that Ag and Ti do not form an intermediate phase [1] and are completely immiscible in the liquid state [2], it has been found that the two metals alloy in all proportions [3, 4] and form at least one, possibly two intermediate phases, TiAg (69.25 wt. % Ag) [3, 4, 5] and another phase based on Ti_3Ag (42.88 wt. % Ag) [6].

Micrographic studies [3, 4] yielded two tentative phase diagrams which agree as to the general phase relationships. They are characterized by the existence of only one intermediate phase, TiAg (Fig. 37). There is also agreement as regards the maximal solubility of Ag in β-Ti—about 16 at. (30 wt.) % Ag [3] and 16.3 ± 0.6 at. (30.5 ± 1 wt.) % Ag [4]—and the temperature of the eutectoid decomposition of β, about 850 [3] and 849 ± 4°C [4]. The eutectoid composition was given as about 7.2 at. (15 wt.) % Ag [3] and 8.4 at. (17 wt.) % Ag [4], and the temperature of the peritectic formation of TiAg as approximately 1025 ± 20 [3] and 1039 ± 3°C [4]. There is still insufficient evidence to establish whether a peritectic or eutectic phase relation exists at the Ag-rich end of the system.

[6] studied the solid-state equilibria in the 0–30 at. % Ag range with Mg-reduced Ti and the 0–13.5 at. % range with iodide Ti-base alloys, using micrographic and X-ray methods. As indicated by the partial diagram in the inset of Fig. 37, results disagree with those reported by [3, 4] in that evidence was found for the occurrence of an intermediate phase γ, with about 30 at. (51 wt.) % Ag. Neither of the other investigators recognized this single-phase field, although alloys in its reported region of homogeneity were examined. It is interesting that this phase was claimed to be isomorphous with Ti_3Cu, Zr_3Cu, and Zr_3Ag, which were reported by [7, 8] but denied by other workers [9–12].

The major differences evident in these investigations make it impossible to draw a final, or even compromise, diagram at this time. Apart from the existence of the γ

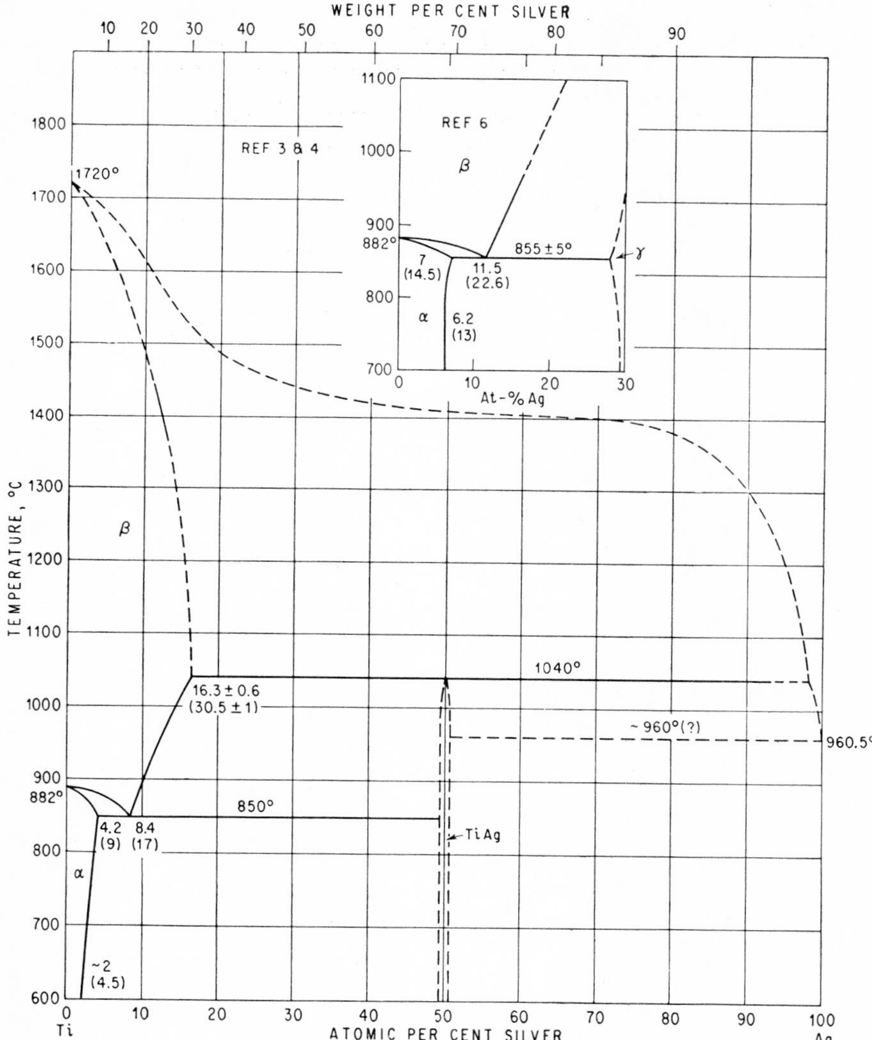

Fig. 37. Ag-Ti. (A tentative diagram combining the results of [3, 6] was suggested by A. D. McQuillan and M. K. McQuillan in their book "Titanium," pp. 257–258, Butterworths Scientific Publications, London, 1956.)

phase, the diagrams of [3, 4] and that of [6] differ in the following points: (*a*) the extent of the β field [13], (*b*) the composition of the eutectoid [14], and (*c*) the extent of the α field [15]. The shape of the liquidus curve in Fig. 37 was suggested on the basis of observations on melting [4].

Crystal Structures. The structure of the phase TiAg is ordered f.c. tetragonal (CuAu, $L1_0$ type), with a = 4.104 A, c = 4.077 A, c/a = 0.993 [5]. According to [6], the γ phase is ordered f.c. tetragonal (Cu_3Au, $L1_2$ type), with a = 4.187 A, c = 3.950 A, c/a = 0.943. Lattice parameters of the α phase were given by [6].

1. F. Laves and H. J. Wallbaum, *Naturwissenschaften*, **27**, 1939, 674; H. J. Wallbaum, *Naturwissenschaften*, **31**, 1943, 91–92.
2. E. Raub, P. Walter, and M. Engel, *Z. Metallkunde*, **43**, 1952, 112–118.
3. R. J. Van Thyne and H. D. Kessler, unpublished research, using iodide Ti-base alloys. The diagram was reproduced in the paper of N. A. De Cecco and M. J. Parks, *Welding J. (N.Y.)*, **32**, 1953, 1071–1081.
4. H. K. Adenstedt and W. R. Freeman, *WADC Tech. Rept.* 53–109, part I, 1953; the 0–13 at. % Ag region was studied with iodide Ti-base alloys; sponge Ti was used for the balance of the alloys.
5. R. J. Van Thyne, W. Rostoker, and H. D. Kessler, *Trans. AIME*, **197**, 1953, 670–671.
6. H. W. Worner, *J. Inst. Metals*, **82**, 1953-1954, 222–227.
7. N. Karlsson, *J. Inst. Metals*, **79**, 1951, 391–405.
8. N. Karlsson, *Acta Chem. Scand.*, **6**, 1952, 1424.
9. A. Joukainen, N. J. Grant, and C. F. Floe, *Trans. AIME*, **194**, 1952, 766–770.
10. W. Rostoker, *Trans. AIME*, **194**, 1952, 209–210.
11. C. E. Lundin, D. J. McPherson, and M. Hansen, *Trans. AIME*, **197**, 1953, 273–278.
12. R. W. Kemper, Master's thesis, Oregon State College, Corvallis, Ore., 1952.
13. Alloys with 20–30 at. % Ag would still be solid at 1100°C according to [6], while they would contain liquid according to [3, 4].
14. The composition of the eutectoid was found by [6] to be about 11.5 and 11 at. % Ag with alloys based on iodide Ti and sponge Ti, respectively.
15. The solubility of Ag in α-Ti was not investigated in detail by [3]. The α-phase boundary reported by [4] was closely bracketed by microexamination; that of [6] is based on lattice-parameter determinations.

$\bar{1}.7225$
0.2775

Ag-Tl Silver-Thallium

Liquidus temperatures were reported by [1–4]. The entire liquidus curve was determined by [3, 4]; that shown in Fig. 38 is based on the work of [4]. The eutectic temperature was found as 291 [4], 289 [1], and 287°C [3, 6], and the eutectic composition as 97.2 at. (98.5 wt.) % [4], 97.6 at. (98.7 wt.) % [1], 95.4 at. (97.5 wt.) % [3], and 94.75 at. (97.2 wt.) % Tl [6].

The curve of retrograde solubility of Tl in Ag between 900 and 300°C is based on X-ray and micrographic data [7], between 287 and 150°C on X-ray data [8]. The solubility [9] of Tl in at. % (wt. % in parentheses) was found as follows: 900°C, 1.1 (2.0); 800°C, 4.2 (7.7); 700°C, 7.0 (12.5); 600°C, 7.5 (13.2); 500°C, 7.5 (13.2); 400°C, 6.6 (11.8); 300°C, 5.3 (9.5); 287°C, 5.0 (9.0); 200°C, 3.6 (6.6); and 150°C, 2.3 (4.2).

The solubility of Ag in Tl, not determined as yet, can only be slight, since the temperature of the $\alpha \rightleftarrows \beta$ transformation of Tl is not appreciably affected by Ag [3, 5]. For lattice parameters of the Ag-rich alloys up to 8 at. % Tl, see [7].

1. C. T. Heycock and F. H. Neville, *J. Chem. Soc.*, **65**, 1894, 33.
2. C. T. Heycock and F. H. Neville, *Phil. Trans. Roy. Soc. (London)*, **A189**, 1897, 55.
3. G. I. Petrenko, *Z. anorg. Chem.*, **50**, 1906, 133–136.
4. E. Raub, unpublished work, see [5].
5. E. Raub, *Z. Metallkunde*, **40**, 1949, 432.
6. O. Hájiček, *Hutnické Listy*, **3**, 1948, 265–270.

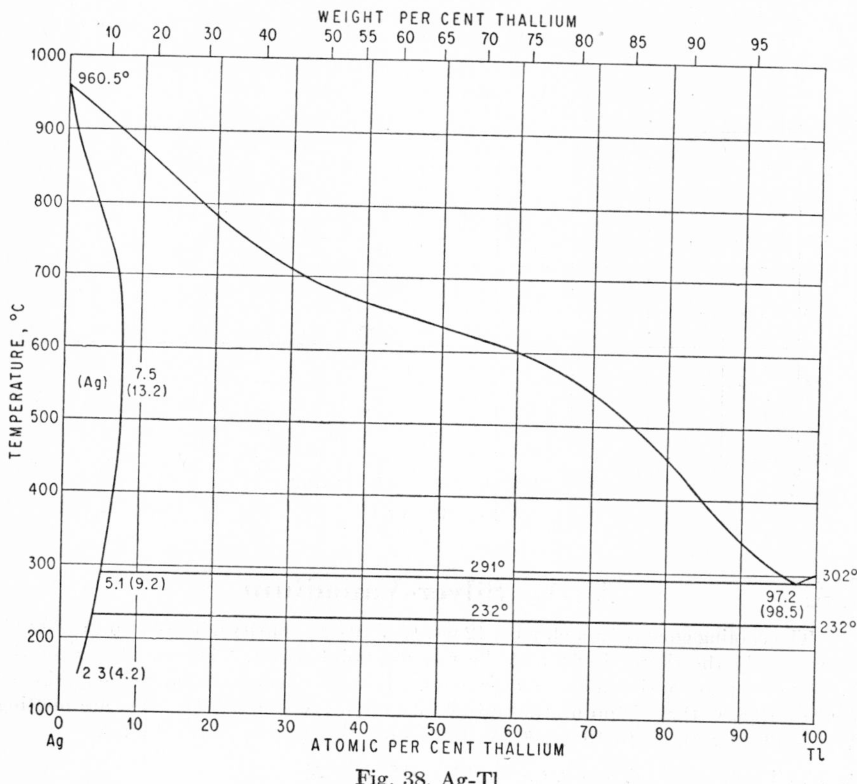

Fig. 38. Ag-Tl

7. E. Raub and A. Engel, *Z. Metallkunde*, **37**, 1946, 76–81.
8. H. H. Chiswik and R. Hultgren, *Trans. AIME*, **137**, 1940, 442–446.
9. Alloys investigated for solubility determinations by F. R. Hensel (*Trans. AIME*, **166**, 1946, 399–412) were not in equilibrium.

$\bar{1}.6562$
0.3438

Ag-U Silver-Uranium

Thermal-analysis, microscopic, and X-ray data [1] yielded the phase diagram presented in Fig. 39. The solid solubility of U in Ag at the eutectic temperature was estimated as less than 0.4 wt. (about 0.19 at.) % U; that of Ag in U is practically nil.

1. R. W. Buzzard, D. P. Fickle, and J. J. Park, *J. Research Natl. Bur. Standards*, **52**, 1954, 149–152.

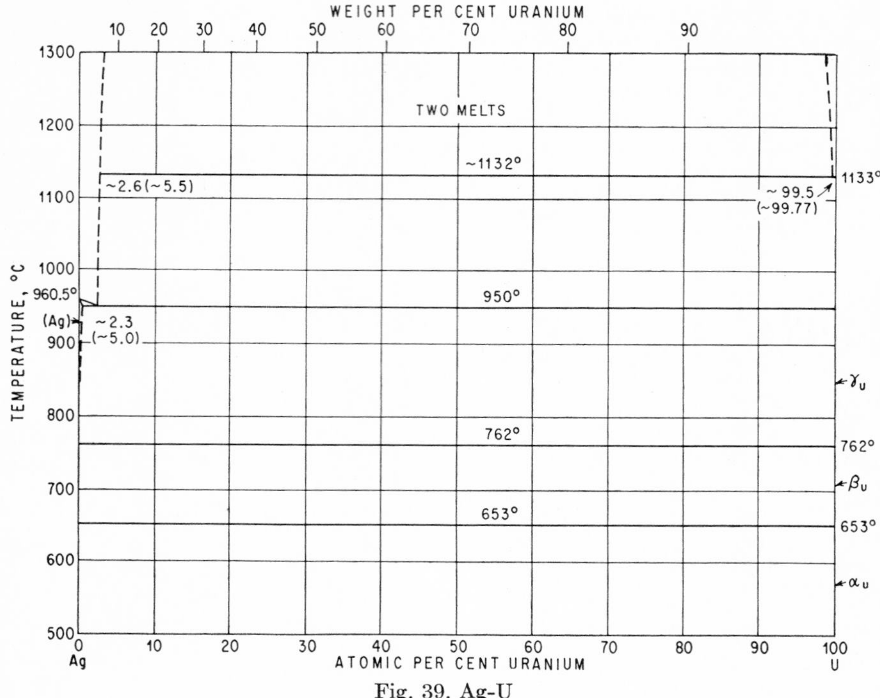

Fig. 39. Ag-U

0.3258
$\bar{1}$.6742

Ag-V Silver-Vanadium

The cooling curve of a melt with 12 wt. % V (94.2% purity), heated up to 1800°C, showed only the thermal effect at the melting point of Ag. The two layers of the ingot proved to consist of the components. It was concluded from this experiment that at 1800°C there is no mutual miscibility of the two metals [1]. On arc-melting V-rich alloys with up to 10 wt. % Ag the entire Ag addition distilled off [2].

1. H. Giebelhausen, *Z. anorg. Chem.*, **91,** 1915, 256–257.
2. A. Yamamoto and W. Rostoker, *Trans. ASM,* **46,** 1954, 1136–1163.

$\bar{1}$.7683
0.2317

Ag-W Silver-Wolfram

The two metals do not form alloys; i.e., they are practically insoluble in each other in the molten state [1].

1. F. A. Bernoulli, *Pogg. Ann.*, **111,** 1860, 587–588; M. v. Schwarz, "Metall-u. Legierungskunde," p. 73, Ferd. Enke Verlag, Stuttgart, 1929.

0.2175
$\bar{1}$.7825

Ag-Zn Silver-Zinc

The constitution of this system has attracted great attention since the liquidus curve was first determined [1]. The main features of the phase diagram were established by [2, 3], and the $\beta \rightleftarrows \zeta$ transformation was clarified by [4]. The solid-state phase boundaries were determined by [5, 6], using lattice-parametric and micrographic methods, respectively. For reasons given below, the findings of [6], particularly in the range 0–60 at. % Zn, appear to be more reliable. The whole diagram of Fig. 40, including the solidification temperatures, is due to [6]; it is to be regarded as established to a high degree of accuracy. The crystal structures of the phases β, γ, and ϵ were first determined by [7].

Solidification. The liquidus curve, covering the entire composition range, was established by [1, 2, 8, 6], with the data of [1] and [6] being in excellent agreement;

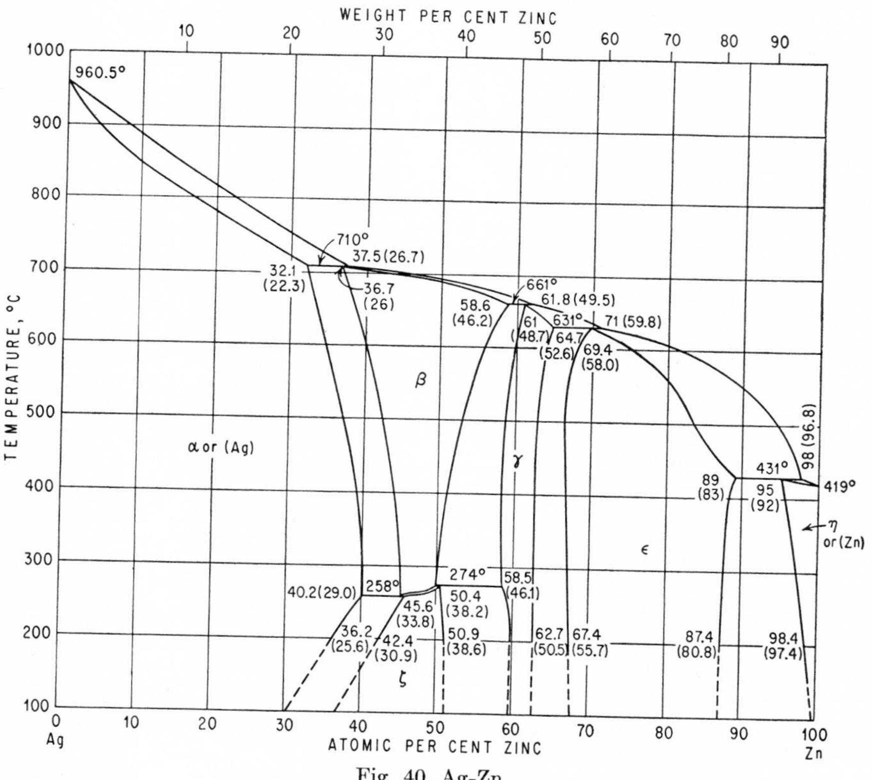

Fig. 40. Ag-Zn

see [6]. The temperatures of the four peritectic equilibria and the compositions of the peritectic melts in at. % Zn were accepted to be 710°C, 39.0%; 665°C, 61.3%; 636°C, 71.2%; and 430°C, 97.8%, on the basis of the work of [1, 2, 8, 9]. Those indicated in Fig. 40 are due to [6].

The solidus curve of the α phase was determined by means of microexamination and heating curves [6]. It is to be regarded as more accurate than that of [10], the latter, determined micrographically, being as much as 10°C higher. The solidus curves of γ, ϵ, and η are based on micrographic work, that of β on heating-curve data [6].

The α Phase. The solubility of Zn in Ag was determined using lattice-spacing [11] and micrographic [6] methods. Between 500 and 258°C both curves are in very good agreement; however, in the temperature range 500–700°C solubilities due to [6] are somewhat lower. At 710°C the solubility was found as 32.1 at. (22.3 wt.) % Zn [6] as compared with 33.2 at. (23.15 wt.) % Zn [11]. For reasons given by [6], the results of [6] are to be considered as more correct. At temperatures below 258°C, where α is in equilibrium with ζ, the solubility decreases considerably [12, 13, 6], the decrease being more pronounced according to [12] and especially [6] (Fig. 40). The specimens examined by [13] were not annealed long enough to attain equilibrium. [12] reported that an alloy with 33 at. % Zn, heavily cold-worked and annealed at 100°C, was duplex in structure; see also [22].

The β Phase. The boundaries of this phase shown in Fig. 40 are based on the micrographic data of [6]. X-ray analysis gave completely unreliable results since the powder specimens decomposed on quenching from the β field, resulting in a much narrower β range [11]; see discussion by [6, 15].

The $\beta \rightleftarrows \zeta$ Transformation. Earlier interpretations of this phase change are obsolete [2, 3, 8, 16–18]. The dependence of the transformation temperature upon composition was already observed by [3, 16–18]. The transformation temperatures indicated in Fig. 40 are based on high-temperature X-ray work after equilibration at proper temperatures [6]. [13] gave the transformation temperature as varying from about 270°C in the Ag-rich alloys to 289°C in the Zn-rich alloys, also based on high-temperature X-ray studies. As to the change in crystal structure see below.

The ζ Phase. The boundaries of this phase were determined by [13, 6]. According to the latter investigators (Fig. 40) the ζ field, under equilibrium conditions, widens much more with fall in temperature than found by [13]. This was confirmed by [12], who reported that a heavily cold-worked alloy with 36.1 at. % Zn, after annealing at 100°C, appeared to consist entirely of ζ; see also [22].

The γ-phase boundaries were determined by [11, 13, 19] as well as [6]. The boundaries on the Ag side of the homogeneity region agree only between 500 and 600°C. They intersect the peritectic horizontal at 661°C at 61.4 at. % [6] and 59.5 at. % Zn [11], the latter value being incompatible with the accurate $\beta/(\beta + \gamma)$ boundary according to [6]. At the 274°C horizontal the difference is even greater: 58.5 at. % [6] vs. about 61.0 at. % Zn [13]. The $\gamma/(\gamma + \epsilon)$ boundaries given by [19] and [6] run nearly parallel between 200 and 500°C at 63.6 at. % and 62.7 at. % Zn, respectively, and intersect the 631°C horizontal at 64.7–64.8 at. % Zn [19, 6].

The ϵ Phase. The $\epsilon/(\gamma + \epsilon)$ boundaries, based on the data of [19] and [6], run nearly parallel at a distance of only approximately 0.5 at. %. The limiting solubility value at 631°C was given as 68.95 [19] and 69.4 at. % Zn [6]. The boundaries on the Zn side, at temperatures below 400°C, run also nearly parallel at a distance of about 1.5 at. %, that of [6] lying at higher Zn contents. The limiting values at 431°C are 86.7 [19] and 89.0 at. % Zn [6].

The η Phase. The solid solubility of Ag in Zn was determined by lattice-spacing measurements in the temperature ranges 408–343°C [19] and 408–145°C [20], and micrographically between 420 and 200°C [6]. Results of [6, 20] are in nearly quantitative agreement, whereas [19] reported a slightly higher solubility above 350°C (6.2 at. % Ag at 431°C). The solubility decreases from 5.0 at. (8.0 wt.) % Ag at 431°C to 4.3, 2.9, and 1.0 at. (6.9, 4.7, and 1.6 wt.) % Ag at 400, 300, and 150°C, respectively [6, 20].

Crystal Structures. Parameter values of the α phase were reported by [7, 21, 11, 13, 22, 23]. The β phase is b.c.c. (A2 type) [4, 24, 25]. On quenching from the β field the structure transforms into the metastable phase β', which is b.c.c. ordered of the CsCl (B2) type [7, 4, 11, 24, 25, 26, 27], with parameters reported by [7, 11]. The transformations $\beta' \rightarrow \zeta$ and $\zeta \rightleftarrows \beta$ were studied by [28, 29].

The ζ phase was first suggested to be h.c.p. disordered with 9 atoms per unit cell and $c/a = 0.367$ [28]. A hexagonal structure was also reported by [30]. [13, 14] claimed the structure to be h.c.p. with 54 atoms per unit cell and $a = 7.631$ A, $c = 5.660$ A, $c/a = 0.7417$ for 48.15 at. % Zn. Recently, [31] confirmed the assumption of [28] that ζ was hexagonal with 9 atoms per unit cell and $a = 7.6360$ A, $c = 2.8197$ A, $c/a = 0.3693$ for the 50 at. % alloy. The structure is closely related to that of β.

The γ phase is isotypic with γ-brass ($D8_2$ type); parameters were given by [7, 21, 19, 11, 32]; see also [33]. ϵ is h.c.p. of the Mg type [7, 21, 19]; the axial ratio changes from about 1.59 on the Ag side to about 1.55 on the Zn side of the homogeneity region. For parameters of the terminal solid solution η, see [21, 19, 20].

Supplement. The structure of the ζ phase determined by [31], using powder data only, was corroborated by the single-crystal work of [34].

1. C. T. Heycock and F. H. Neville, *J. Chem. Soc.*, **71**, 1897, 407–418.
2. G. I. Petrenko, *Z. anorg. Chem.*, **48**, 1906, 347–363.
3. H. C. H. Carpenter and W. Whiteley, *Intern. Z. Metallog.*, **3**, 1913, 145–167.
4. M. Straumanis and J. Weerts, *Metallwirtschaft*, **10**, 1931, 919–922; see also [28].
5. E. A. Owen and I. G. Edmunds, *J. Inst. Metals*, **57**, 1935, 297–306; **63**, 1938, 265–278, 279–290, 291–301.
6. K. W. Andrews, H. E. Davies, W. Hume-Rothery, and C. R. Oswin, *Proc. Roy. Soc. (London)*, **A177**, 1940-1941, 149–167.
7. A. Westgren and G. Phragmén, *Phil. Mag.*, **50**, 1925, 311–341.
8. S. Ueno, *Mem. Coll. Sci. Kyoto Univ.*, **A12**, 1929, 347–348.
9. G. I. Petrenko and B. G. Petrenko, *Z. anorg. Chem.*, **185**, 1930, 96–100.
10. W. Hume-Rothery, G. W. Mabbott, and K. M. Channel-Evans, *Phil. Trans. Roy. Soc. (London)*, **A233**, 1934, 1–97.
11. E. A. Owen and I. G. Edmunds, *J. Inst. Metals*, **63**, 1938, 265–278; see also [14].
12. C. Haase and F. Pawlek, Vorträge der Hauptversammlung 1938 der deutschen Gesellschaft für Metallkunde, *Z. Metallkunde*, **30**, 1938, 57HV–60HV.
13. E. A. Owen and I. G. Edmunds, *J. Inst. Metals*, **63**, 1938, 279–290; see also [14].
14. E. A. Owen and I. G. Edmunds, *J. Inst. Metals*, **63**, 1938, 291–301.
15. W. Hume-Rothery, K. W. Andrews, and E. A. Owen, *J. Inst. Metals*, **76**, 1949-1950, 683, 690, 705, 712.
16. G. I. Petrenko, *Z. anorg. Chem.*, **165**, 1927, 297–304.
17. B. G. Petrenko, *Z. anorg. Chem.*, **184**, 1929, 369–375.
18. G. I. Petrenko, *Z. anorg. Chem.*, **149**, 1925, 395–400; **184**, 1929, 376–384.
19. E. A. Owen and I. G. Edmunds, *J. Inst. Metals*, **57**, 1935, 297–306; see also [14].
20. A. v. Wiedebach-Nostiz, *Z. Metallkunde*, **37**, 1946, 56–60.
21. E. A. Owen and L. Pickup, *Proc. Roy. Soc. (London)*, **A140**, 1933, 344–358.
22. H. Lipson, N. J. Petch, and D. Stockdale, *J. Inst. Metals*, **67**, 1941, 79–85.
23. M. E. Straumanis, *Acta Cryst.*, **2**, 1949, 82–84; see also *Structure Repts.*, **12**, 1949, 127.
24. E. A. Owen and I. G. Edmunds, *Proc. Phys. Soc. (London)*, **A50**, 1938, 389–397.
25. P. Gruzin and E. Kaminsky, *Zhur. Tekh. Fiz.*, **8**, 1938, 2069–2072.
26. S. T. Shavlo and G. A. Alaverdov, *Zhur. Eksptl. i Teoret. Fiz.*, **9**, 1939, 59; *Met. Abstr.*, **7**, 1940, 62.
27. L. Muldawer, *J. Appl. Phys.*, **22**, 1951, 663–665.
28. J. Weerts, *Z. Metallkunde*, **24**, 1932, 265–270.
29. W. Köster, *Z. Metallkunde*, **32**, 1940, 151–152.
30. A. Westgren, *Metallwirtschaft*, **10**, 1931, 921, footnote.
31. I. G. Edmunds and M. M. Qurashi, *Acta Cryst.*, **4**, 1951, 417–425.
32. R. E. Marsh, *Acta Cryst.*, **7**, 1954, 379.

33. W. Hume-Rothery, J. O. Betterton, and J. Reynolds, *J. Inst. Metals*, **80**, 1951-1952, 609–616.
34. G. Bergman and R. W. Jaross, *Acta Cryst.*, **8**, 1955, 232–235.

0.0728
$\bar{1}$.9272

Ag-Zr Silver-Zirconium

Phase relations in the composition range 0–50 at. % Zr (Fig. 41) are due to [1]. X-ray analysis [1, 2] confirmed the existence of ZrAg (45.82 wt. % Zr) and showed [2] that there is only one additional phase richer in Zr [3]. The composition of the latter phase, which was also observed by [1, 4], is still uncertain. [2] reported that in sintered compacts the maximum intensity of reflection of this phase was obtained between 70 and 85 at. % Zr, indicating the composition Zr_3Ag (71.73 wt. % Zr).

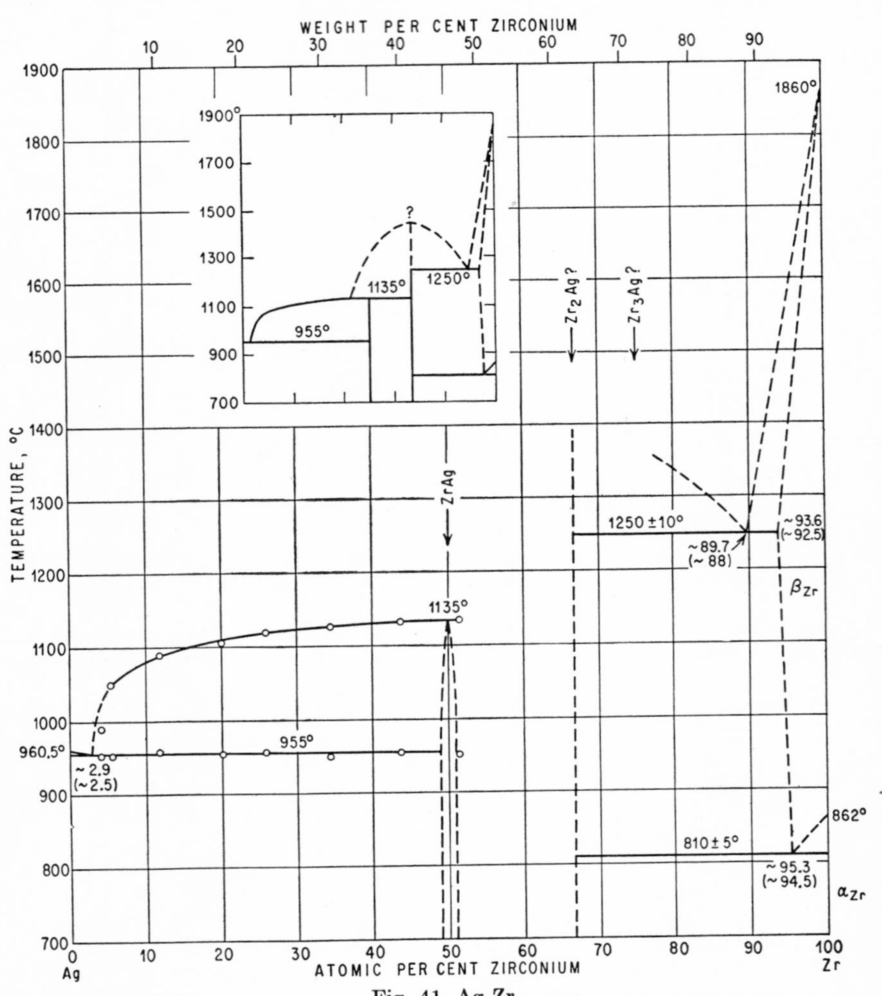

Fig. 41. Ag-Zr

However, it was not possible to reach equilibrium and to obtain the phase in pure form. According to metallographic examination of arc-melted alloys, the composition would lie between 58 and 69 at. % Zr and possibly correspond to the formula Zr_2Ag (62.84 wt. % Zr) [4]. The structure was claimed to be isotypic with that of Zr_3Cu [2]. This would be in favor of Zr_3Ag.

Also, additional information is quite conflicting: (*a*) Some evidence was reported [4] to indicate a peritectic formation of ZrAg. If this is true, the liquidus curve in the range of about 40–50 at. % Zr shown in Fig. 41 would not be correct. However, the fact that the liquidus is very flat in this region would be compatible with a peritectic equilibrium at about 1135°C. (*b*) According to [4], the phase Zr_2Ag (or Zr_3Ag) forms a eutectic with the β-Zr solid solution at 1250°C, indicating that the melting point of the compound is still higher. On the other hand, [2] reported that Zr_3Ag (?) was stable only below 800°C.

The constitution in the range 80–100 at. % Zr (Fig. 41) is based on thermal and metallographic work of arc-melted alloys prepared by using magnesium-reduced Zr. It is to be considered as tentative [4].

At present it is not possible to combine the two partial diagrams proposed by [1, 4] and to consider the findings of [2]. The inset of Fig. 41 shows a possible construction of the phase diagram, essentially based on the results of [1, 4].

The solid solubility of Zr in Ag is negligibly small [1, 2]. The same holds for the solubility of Ag in α-Zr [2, 4].

Crystal Structures. ZrAg has a tetragonal structure [1, 2] with 4 atoms per unit cell and constants changing from $a = 3.348$ A, $c = 6.603$ A, $c/a = 1.904$ to $a = 3.476$ A, $c = 6.629$ A, $c/a = 1.907$ [2]. The phase Zr_3Ag was claimed to be isotypic with Zr_3Cu; the structure may be considered as a tetragonally deformed Cu_3Au ($L1_2$) type. The smallest and largest dimensions observed were $a = 4.566$ A, $c = 3.986$ A, $c/a = 0.873$ and $a = 4.582$ A, $c = 3.989$ A, $c/a = 0.871$ [2].

1. E. Raub and M. Engel, *Z. Metallkunde*, **39**, 1948, 172–177.
2. N. Karlsson, *Acta Chem. Scand.*, **6**, 1952, 1424–1430.
3. Earlier, C. Sykes (*J. Inst. Metals*, **41**, 1929, 179–190) had examined the microstructure of alloys with up to 30 wt. (33.6 at.) % Zr. Results indicated that the intermediate phase coexisting with Ag was richer in Zr.
4. R. S. Kemper, unpublished M.S. thesis, Oregon State College, Corvallis, Ore., 1952

$\bar{1}.5565$
0.4435

Al-As Aluminum-Arsenic

By melting aluminum and arsenic together at about 800°C and atmospheric pressure, or by reaction of arsenic vapor with aluminum powder at about 500°C, a compound is formed which has definitely been identified as AlAs (73.52 wt. % As) [1–3] with a melting point of >1200 [2] or >1600°C [4]. Earlier, [4] had erroneously concluded from thermal-analysis data that the compound formed was Al_3As_2 (64.92 wt. % As). AlAs is insoluble in both molten Al and As [4]. According to the data available, the phase diagram of this system for increased pressure will consist of a horizontal liquidus curve lying at the melting point of AlAs (>1600°C) and extending from 0 to 100% As (corresponding to a complete miscibility gap in the liquid state) and two eutectic horizontals at the melting points of the elements extending up to the composition AlAs.

AlAs has a f.c.c. structure of the zincblende (B3) type, $a = 5.63$ A [1, 2].

Supplement. [5] corroborated by microscopic and thermal work that AlAs is the only intermediate phase in the Al-As system. They suggest a liquidus which rises smoothly from (practically) the melting points of the elements to that of the compound (>1600°C); however, the experimental evidence given is insufficient to warrant the nonexistence of a miscibility gap in the liquid state.

1. V. M. Goldschmidt, *Skrifter Norske Videnskaps-Akad. Oslo, Math. naturv. Kl.*, 1927, no. 8.
2. G. Natta and L. Passerini, *Gazz. chim. ital.*, **58,** 1928, 458–460.
3. E. Montignie, *Bull. soc. chim. France*, **9,** 1942, 739–740.
4. Q. A. Mansuri, *J. Chem. Soc.*, **121,** 1922, 2272–2277.
5. W. Köster and B. Thoma, *Z. Metallkunde*, **46,** 1955, 291–293.

$\bar{1}.1361$
0.8639

Al-Au Aluminum-Gold

In spite of quite extensive investigation by thermal [1] and some micrographic analysis [2] as well as X-ray studies [3–9], the composition of some of the intermediate phases and the phase relations in the range 70–90 at. (94.5–98.5 wt.) % Au are still not fully known.

The liquidus curve and temperatures of the three-phase equilibria given in Fig. 42 are based on the work by [1]. Only the Al-$AuAl_2$ eutectic was redetermined later [10] and found to lie at 0.7 at. (5.0 wt.) % Au, 642°C (melting point of Al: 660°C) as compared with 1.1 at. (7.5 wt.) % Au, 648°C (melting point of Al: 655°C) according to [1].

There is no doubt as to the existence of the compounds $AuAl_2$ (78.52 wt. % Au) [11, 1, 5–8] and Au_2Al (93.60 wt. % Au) [1, 7, 8], both having maximum melting points. The phase formed peritectically at 625°C was suggested by [1] to be the compound AuAl (87.97 wt. % Au). Although this is quite likely, no clear evidence of this composition was found [7, 8].

In the phase diagram by [1], there is no indication of the formation of Au_5Al_3 [62.5 at. (92.42 wt.) % Au] by a peritectic reaction, although this phase has been mentioned in the literature to be a 7:4 electron compound with a h.c.p. structure [12]. Neither did [7, 8] find any evidence of its existence.

The phase formed peritectically at 575°C was suggested by [1, 2] to be either Au_5Al_2 [71.43 at. (94.82 wt.) % Au] or Au_8Al_3 [72.73 at. (95.12 wt.) % Au]. [7] confirmed the existence of a phase in this range with a composition closer to Au_8Al_3; but, according to the thermal-analysis data (Fig. 42), the composition Au_5Al_2 is more likely. This phase probably has a structure similar to that of γ-brass, however, apparently distorted into a hexagonal cell [7]. The ideal composition of a 21:13 electron compound would be Au_9Al_4 [69.23 at. (94.27 wt.) % Au].

On the basis of X-ray work, [8] asserted that in the neighborhood of 75 at. % Au a phase, not found by others [1, 7], is formed peritectically; but there is no indication of a peritectic reaction in this range (Fig. 42). On the other hand, the occurrence of Au_3Al (95.64 wt. % Au) has been reported on the basis of X-ray studies [3, 13], which revealed the existence of a phase with the structure of β-Mn. This was interpreted as indicating that Au_3Al is a 3:2 electron compound. However, it was shown that a phase of the composition Au_3Al does not exist [9] and that the phase with the structure of β-Mn has the composition Au_4Al rather than Au_3Al [7, 9]. Also, [1] and [8] have claimed the compound Au_4Al to exist. Its structure was found to be similar to, but not completely identical with, that of β-Mn [8, 9].

As to the constitution in the range 70 to 90 at. % Au, [2] suggested the diagram in the inset of Fig. 42, on the basis of thermal and micrographic studies. It shows a high-temperature phase, designated by [2] as β. From the limited data reported [2], it is not possible to explain fully the phenomena observed or to draw the phase relationships for equilibrium conditions. It was suggested [14] that the inset diagram represents, at least partially, metastable states. Indeed, [2] found that the compound Au_4Al is formed under equilibrium conditions at about 520°C. On the other hand, [8] reported that Au_4Al is not stable below 400–500°C, whereas [7] found Au_4Al to be stable at room temperature. Neither [7] nor [8] found any indication of the existence of the high-temperature β phase, although [8] made X-ray studies at 500°C. This

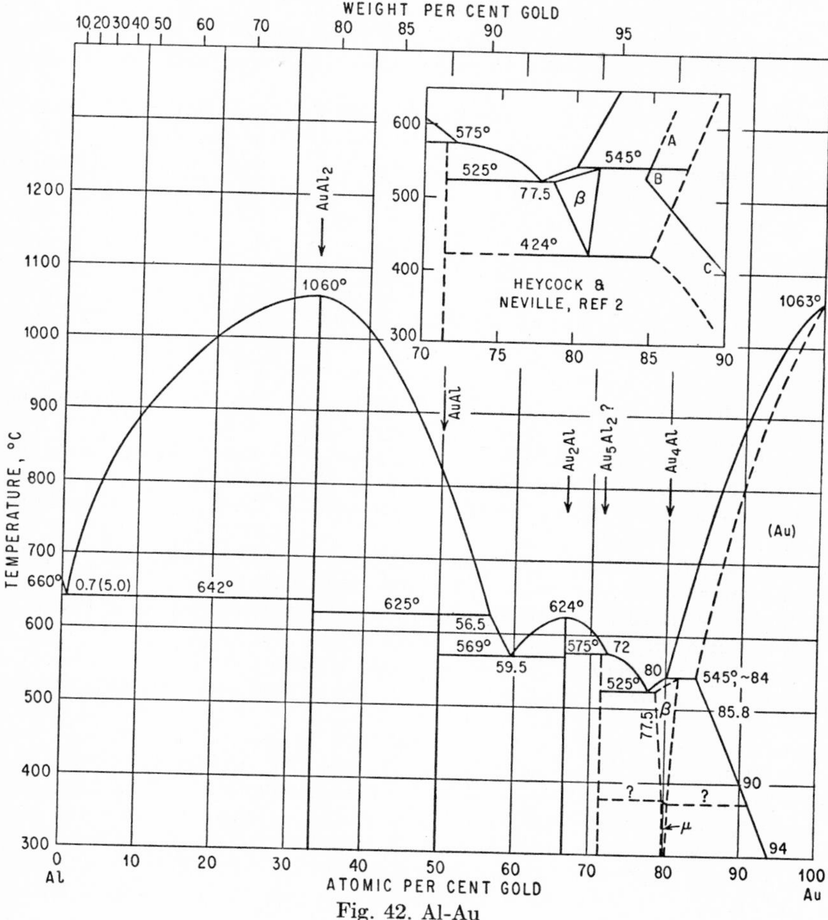

Fig. 42. Al-Au

problem has apparently been solved recently by [9]. It was found that the phase of the composition Au_4Al undergoes a polymorphic transformation at a temperature between 300 and 400°C, but that the low-temperature form μ can be metastable up to about 500°C (see Crystal Structures).

As to the range of homogeneity of the various phases, [8] concluded from lattice-parameter measurements that all of them are very narrow. However, [7], using the same method, estimated that Au_2Al, Au_5Al_2 (or Au_8Al_3), and Au_4Al have homogeneity ranges of approximately 0.5–1.0 at. %.

The solid solubility of Au in Al has not been determined yet. The lattice-parametric method is evidently not sensitive enough [7, 8]. [10] stated that it is below 1 wt. (<0.2 at.) % Au at the eutectic temperature.

The solid solubility of Al in Au was determined parametrically between 200 and 500°C [7], at 510°C (12.4 at. % Al) [8], and between 400 and 570°C [15]. Results by [7] and [15] agree very well. The solubility was found to be 14.7, 15.5, 14.2, 12.0, 10.0, 8.0, and 6.0 at. % Al (2.3, 2.45, 2.2, 1.85, 1.5, 1.15, and 0.85 wt. %) at, respectively, 570, 530, 500, 450, 400, 350, and 300°C [7, 15]. The saturation curve (curve *ABC* in the inset of Fig. 42) changes direction at about 530°C. As the accuracy is not higher than ±0.5 at. % [15], 530°C can be considered to correspond to the peritectic temperature, 545°C.

Crystal Structures. $AuAl_2$ has the structure of the CaF_2 (C1) type, a = 6.01 A [5, 6] or a = 5.999 A [7]. The phase presumably of the composition AuAl was reported to have a very complex powder pattern [7, 8]. A thin film of the composition AuAl, prepared by vapor deposition in high vacuum and annealed at 400°C, showed the structure of the zincblende (B3) type, a = 6.05 A. Perhaps this structure is due to the method of preparation used [16]. Au_5Al_2 (or Au_8Al_3) is probably similar in structure to γ-brass, but apparently of hexagonal distortion [7]. The low-temperature modification μ, of the phase having the composition Au_4Al (formerly considered to be Au_3Al), has a structure very similar to that of β-Mn [3, 4, 13, 7, 8, 9; see especially 8], with a = 6.919–6.923 A [4, 13, 7, 9]. The high-temperature modification of Au_4Al, β, is b.c.c. (A2 type) with a = 3.24 A at 500°C [9].

1. C. T. Heycock and F. H. Neville, *Phil. Trans. Roy. Soc. (London)*, **A194,** 1900, 201–232.
2. C. T. Heycock and F. H. Neville, *Phil. Trans. Roy. Soc. (London)*, **A214,** 1914, 267–276; *Proc. Roy. Soc. (London)*, **90,** 1914, 560–562.
3. A. Westgren and G. Phragmén, *Metallwirtschaft*, **7,** 1928, 70; *Trans. Faraday Soc.*, **25,** 1929, 379.
4. N. Katoh, *J. Chem. Soc. Japan*, **52,** 1931, 851–854; *Chem. Abstr.*, **26,** 1932, 4991.
5. C. D. West and A. W. Peterson, *Z. Krist.*, **A88,** 1934, 93–94.
6. E. Zintl, A. Harder, and W. Haucke, *Z. physik. Chem.*, **B35,** 1937, 354–362.
7. A. S. Coffinberry and R. Hultgren, *Trans. AIME*, **128,** 1938, 249–258.
8. O. E. Ullner, *Arkiv Kemi, Mineral. Geol.*, **14A,** 1940, 1–20.
9. V. G. Kuznetsov and V. I. Rabezova, *Doklady Akad. Nauk S.S.S.R.*, **81,** 1951, 51–54.
10. N. Ageew and V. Ageewa, *Trans. AIME*, **128,** 1938, 259–260.
11. W. C. Roberts-Austen, *Proc. Roy. Soc. (London)*, **50,** 1891-1892, 367–368.
12. W. Hume-Rothery, "The Structure of Metals and Alloys," p. 113, Institute of Metals, London, 1950; see, however, G. V. Raynor in "Progress in Metal Physics," vol. 1, p. 22, Pergamon Press, New York, 1949.
13. S. Fagerberg and A. Westgren, *Metallwirtschaft*, **14,** 1935, 265–267.
14. M. Hansen, "Der Aufbau der Zweistofflegierungen," pp. 77–81, Springer-Verlag OHG, Berlin, 1936.
15. E. A. Owen and E. A. O'Donnell Roberts, *J. Inst. Metals*, **71,** 1945, 213–254.
16. O. Eisenhut and E. Kaupp, *Z. Elektrochem.*, **37,** 1931, 472.

0.3968
$\bar{1}$.6032

Al-B Aluminum-Boron

Prior to the first attempt to determine phase relations in the Al-rich alloys, the existence of two compounds, AlB_2 (44.51 wt. % B) and AlB_{12} [92.13 at. (82.78 wt.) % B], had been reported in the literature [1]; see also the review by [2]. [3] found that "amorphous boron" dissolves noticeably in molten Al at 1000°C.

According to the results of thermal studies in the range 0–8.5 wt. (18.8 at.) % B, an Al-AlB_2 eutectic would have to be assumed at 15–18 wt. (about 31–36 at.) % B and about 565°C [2]. However, [4] gave conclusive evidence for the type of phase diagram presented in Fig. 43, which indicates that the solidus temperature practically coincides with the melting point of Al. The solubility of B in liquid Al was determined to be as follows in at. % B (wt. % in parentheses): 4.8 (2.0) at 1300°C, 1.7 (0.7) at 1100°C, 0.85 (0.35) at 950°C, 0.41 (0.17) at 785°C, and 0.22 (0.09) at 730°C [5]. This is in good agreement with the statement [4] that primary crystals of AlB_{12} occur in an alloy with about 4 wt. % B. The temperature of the peritectic reaction liq. + $AlB_{12} \rightleftarrows AlB_2$, given as about 1350°C [4] in Fig. 43, is only an approximate value and may lie above 1400°C [6]. The solubility of B in solid Al is negligibly small, according to lattice-parameter measurements [4].

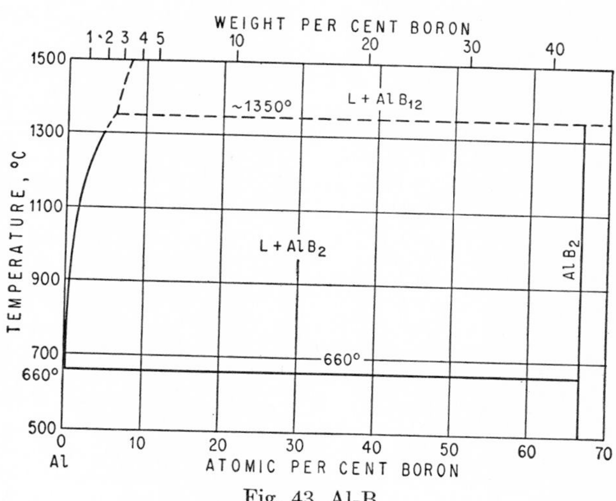

Fig. 43. Al-B

AlB_2 is hexagonal, with a = 3.01 A, c = 3.25 A, and 1 formula weight per unit cell (C32 type) [7, 4]. The structure of AlB_{12}, long known to be "crystallized boron," was determined by [8] and [9, 10]. It exists in two modifications, a monoclinic and a tetragonal [10]. The assumption that this phase has the composition AlB_{13} [9] rather than AlB_{12} was abandoned in the final paper [10]. Concerning observations of the decomposition of AlB_2, see [6].

1. F. Wöhler, *Liebigs Ann.*, **141**, 1867, 268; W. Hampe, *Liebigs Ann.*, **183**, 1876, 75; H. Biltz, *Ber. deut. chem. Ges.*, **41**, 1908, 2634; **43**, 1910, 300; H. Funk, *Z. anorg. Chem.*, **142**, 1925, 277.
2. P. Haenni, *Rev. mét.*, **23**, 1926, 342–352; *Compt. rend.*, **181**, 1925, 864–866.
3. H. Giebelhausen, *Z. anorg. Chem.*, **91**, 1915, 262.
4. W. Hofmann and W. Jäniche, *Z. Metallkunde*, **28**, 1936, 1–5.
5. Unpublished work at Aluminum Research Laboratories (Alcoa), reported in "Metals Handbook," 1948 ed., p. 1155, The American Society for Metals, Cleveland, Ohio.
6. F. Lihl and P. Jenitschek, *Z. Metallkunde*, **44**, 1953, 414–417.
7. W. Hofmann and W. Jäniche, *Z. physik. Chem.*, **B31**, 1936, 214–222; *Naturwissenschaften*, **23**, 1935, 851; see also [4].

8. St. v. Naráy-Szabó, *Z. Krist.*, **A94,** 1936, 367–374.
9. F. Halla and R. Weil, *Naturwissenschaften*, **27,** 1939, 96.
10. F. Halla and R. Weil, *Z. Krist.*, **A101,** 1939, 435–450.

$\bar{1}$.2932
0.7068

Al-Ba Aluminum-Barium

By thermal analysis of alloys containing between 64 wt. (90 at.) % and 100% Al [1], the Al-rich part of the phase diagram of Fig. 44 was obtained [2]. The phase coexisting with Al was identified as $BaAl_4$ (44.00 wt. % Al) [3]. $BaAl_4$ has a b.c. tetragonal structure with a = 4.540 A, c = 11.16 A, c/a = 2.46, and 2 $BaAl_4$ per unit cell [3].

More recently the system was studied between 0 and 20 wt. (56 at.) % Al by [4], using thermal and X-ray diffraction analysis. The thermal as well as X-ray data

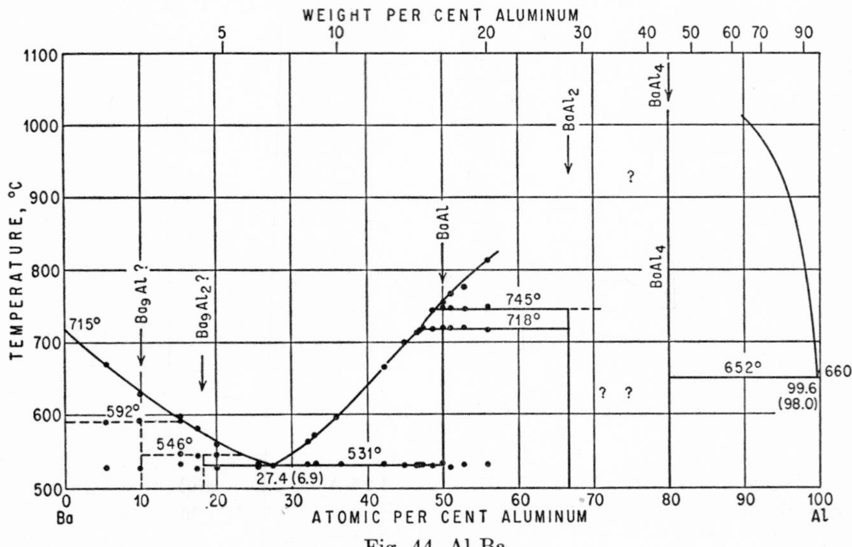

Fig. 44. Al-Ba

indicated the presence of two peritectic compounds in this range of composition, BaAl (16.42 wt. % Al) and $BaAl_2$ (28.20 wt. % Al). For two additional peritectic compounds, Ba_9Al and Ba_9Al_2, only some thermal evidence was obtained (see Fig. 44); "additional data are necessary to confirm their formation, since X-ray diffraction analysis has thus far failed to give evidence of their existence." Single-crystal work showed that BaAl is hexagonal, a = 6.01 A, c = 17.78 A, c/a = 2.96, with 16 atoms per unit cell. The solubility of Al in solid Ba was estimated to be less than 0.1 wt. (0.5 at.) %.

1. E. Alberti, *Z. Metallkunde*, **26,** 1934, 6–9.
2. The alloys were prepared by reduction of BaO in molten aluminum at 1000°C under argon; they contained about 0.24% Fe, 0.24% Si.
3. K. R. Andress and E. Alberti, *Z. Metallkunde*, **27,** 1935, 126–128.
4. E. M. Flanigen, thesis, Syracuse University, Syracuse, N.Y., 1952.

0.4762
$\bar{1}.5238$

Al-Be Aluminum-Beryllium

The whole liquidus curve was determined by [1, 2]. In the range 35 to 95 at. % Be temperatures differ by 30–60°C, those reported by [2] being higher. The curve given in Fig. 45 represents average values for this composition range; below 35 at. % Be the data by [1] have been adopted.

The eutectic temperature was determined to be 644 [1], 645 [3, 2], and 647°C [4], and the eutectic composition, 1.4 (4.1) [1, 5, 6], 1.1 (3.2) [2], 0.87 (2.5) [3], and 0.5 wt.

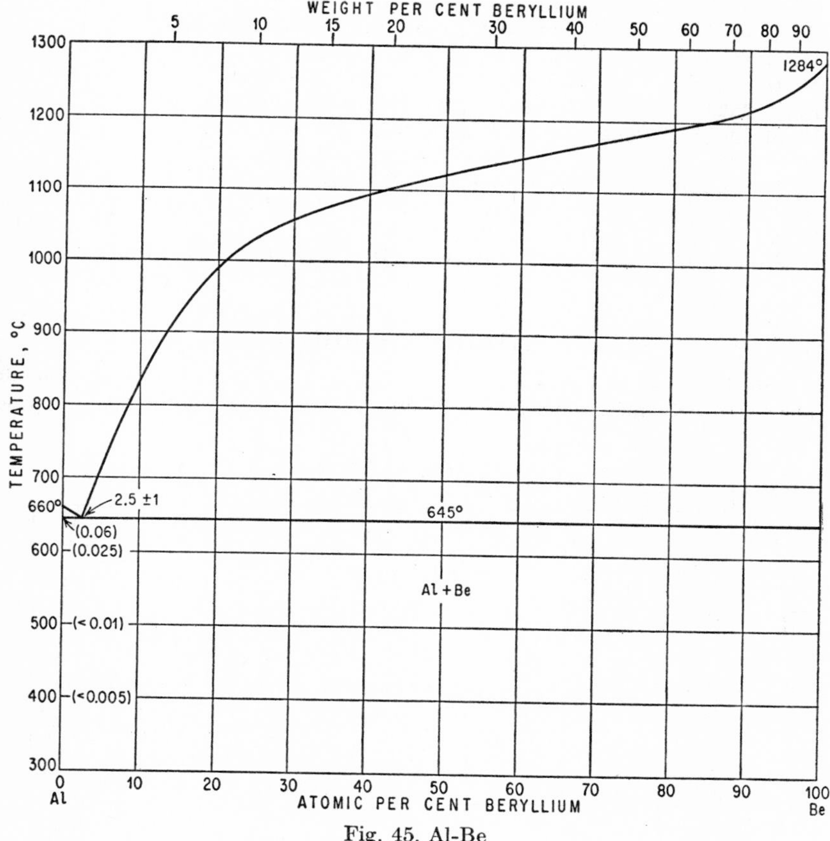

Fig. 45. Al-Be

(1.5 at.) % Be [4]. In all likelihood, the eutectic composition lies below 2.5 at. %, probably close to 1.5 at. % Be.

The solid solubility of Be in Al reported by various investigators differs by orders of magnitude. There is no doubt that the values given by [5, 6, 7, 2] are by far too high. Those found by [3] (hardness), [4] (lattice parameter), and [8, 9] (microhardness) agree remarkably well. It must be assumed that the solubility in wt. % is about 0.05–0.06 at 645°C, 0.02–0.03 at 600°C, and 0.005–0.01 at 500°C, and practically zero at lower temperatures [10]; 0.05 wt. % corresponds to 0.17 at. % Be.

Thermal-analysis data, corroborated by microscopic examination and hardness measurements, have been asserted to indicate the formation of Be-rich solid solutions estimated to be as high as 4–5 wt. (1.4–1.7 at.) % Al [2]. However, on the basis of parametric and microscopic studies, [11] reported that the solubility is small or negligible, definitely under 1 wt. (0.35 at.) % Al [12].

Supplement. [13] reported 644°C, 3.67 at. % Be as eutectic data.

1. G. Oesterheld, *Z. anorg. Chem.*, **97,** 1916, 9–14.
2. L. Losana, *Alluminio*, **9,** 1940, 8–13.
3. R. S. Archer and W. L. Fink, *Trans. AIME*, **78,** 1928, 616–643; especially pp. 625–633; *Z. Metallkunde*, **20,** 1928, 446–447.
4. W. I. Micheeva, *Izvest. Akad. Nauk S.S.S.R.* (*Khim.*), **1940,** 775–782.
5. W. Kroll, *Metall u. Erz*, **23,** 1926, 613–616.
6. M. Haas and D. Uno, *Z. Metallkunde*, **22,** 1930, 277–278.
7. E. S. Makarov and L. Tarschisch, *Zhur. Fiz. Khim.*, **9,** 1937, 350–358.
8. H. Bückle, *Z. Metallkunde*, **37,** 1946, 43–47.
9. H. Bückle, *Compt. rend.*, **230,** 1950, 752–754.
10. G. Masing and O. Dahl, *Wiss. Veröffentl. Siemens-Konzern*, **8,** 1929, no. 1, 249–250.
11. A. R. Kaufmann, P. Gordon, and D. W. Lillie, *Trans. ASM*, **42,** 1950, 785–844.
12. H. A. Sloman, *J. Inst. Metals*, **49,** 1932, 369.
13. O. Hájiček, *Hutnické Listy*, **3,** 1948, 265–270; *Chem. Abstr.*, **43,** 1949, 4935.

$\bar{1}.1109$
0.8891

Al-Bi Aluminum-Bismuth

Thermal analysis and microscopic examination [1] showed that the two metals are almost entirely immiscible in the liquid state [1]. Layer formation was also

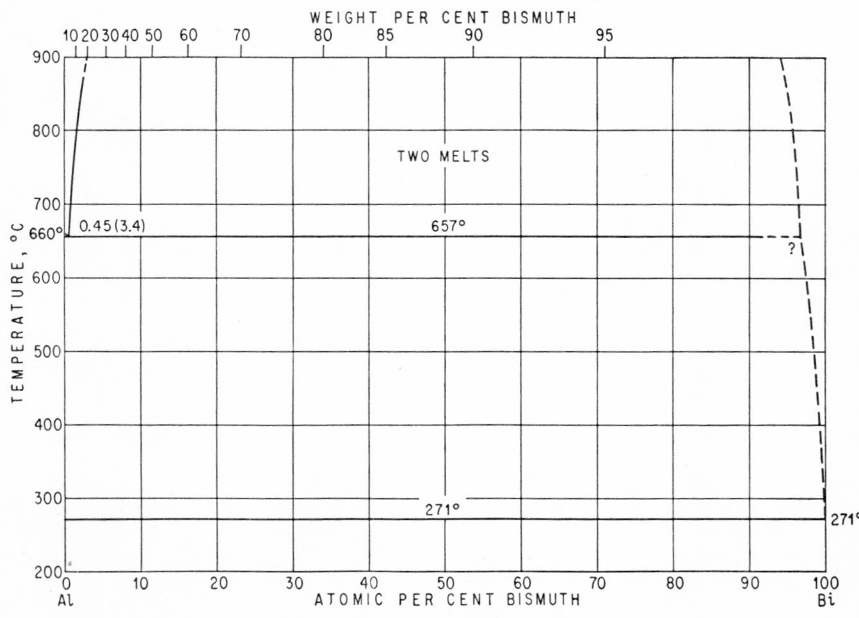

Fig. 46. Al-Bi

observed by [2–4] and by [5] (above 30 wt. % Bi). The monotectic temperature was found to lie about 5 [1], 3.5 ± 0.5 [6], or 3°C [7] below the melting point of Al. At the monotectic temperature the miscibility gap extends from 3.4 wt. (0.45 at.) % Bi [7] to at least 98.5 wt. (90 at.) % Bi, probably higher [1] (Fig. 46). At 700, 800, and 880°C, respectively, about 5.5, 9.9, and 15.2 wt. (0.75, 1.4, and 2.3 at.) % Bi are soluble in liquid Al [7]. [8] reported that the melting point of Bi is lowered 0.25°C by 0.13 wt. (1 at.) % Al. The solid solubility of Bi in Al at 657°C is less than 0.2 wt. ($<0.02_6$ at.) % [7].

1. A. G. C. Gwyer, *Z. anorg. Chem.*, **49**, 1906, 316–319; Al of 99.4% purity was used.
2. C. R. A. Wright, *J. Soc. Chem. Ind.*, **11**, 1892, 492–494; **13**, 1894, 1014–1017.
3. W. Campbell and J. A. Mathews, *J. Am. Chem. Soc.*, **24**, 1902, 255–256.
4. W. Guillet, *Bull. soc. encour. ind. natl.*, **101** 2, 1902, 259; *Rev. mét.*, **18**, 1921, 461.
5. H. Pécheux, *Compt. rend.*, **138**, 1904, 1501–1502; see also **143**, 1906, 397–398.
6. M. Hansen and B. Blumenthal, *Metallwirtschaft*, **10**, 1931, 925–927; Al of 99.9% purity was used.
7. L. W. Kempf and K. R. Van Horn, *Trans. AIME*, **133**, 1939, 81–92; electrolytically refined Al was used.
8. C. T. Heycock and F. H. Neville, *J. Chem. Soc.*, **61**, 1892, 893.

0.3515
$\bar{1}$.6485

Al-C Aluminum-Carbon

The solubility of carbon in molten Al is small—estimated to be <0.05 wt. % at 1300–1500°C [1]—but there is a great tendency to form Al carbide, Al_4C_3 [42.86 at. (25.03 wt.) % C] [2]. According to an abstract of a Russian paper [3], the solubility of C in Al at 1000–1100°C is practically nil. Small particles of Al found in technical Al_4C_3 indicate that at high temperatures Al dissolves in the carbide and separates again on cooling. On fusing Al_4C_3 with Al above 2000°C, the system separates into two layers; and on cooling, pure Al separates from the carbide layer and the carbide from the Al layer.

[4] determined the melting points of Al-C reaction products with about 6.3–17 wt. % C (containing 3–6.4% Al_2O_3 as a contaminant) and concluded that the carbide Al_3C (12.91 wt. % C) exists (?).

Al_4C_3 has a rhombohedral structure, a = 8.55 A, α = 22°28′, 1 molecule per unit cell ($D7_1$ type) [5].

1. J. Czochralski, *Z. Metallkunde*, **15**, 1923, **276**.
2. O. Ruff, *Z. Elektrochem.*, **24**, 1918, 159.
3. M. P. Slavinsky, I. A. Nazirov, and L. R. Edelson, *Metallurg*, **1934**, no. 9, 12–22; abstract in *Met. Abstr.*, **2**, 1935, 461–462.
4. E. Baur and R. Brunner, *Z. Electrochem.*, **40**, 1934, 156–157.
5. M. v. Stackelberg and E. Schnorrenberg, *Z. physik. Chem.*, **B27**, 1934, 37–49.

$\bar{1}$.8281
0.1719

Al-Ca Aluminum-Calcium

On the basis of thermal studies [1] the phase $CaAl_3$ (33.12 wt. % Ca) [2] was erroneously assumed to decompose on melting at 692°C into two melts with approximately 15 and 43 wt. % Ca [3]. A more comprehensive thermal and thermoresistometric analysis and microscopic examination [4] resulted in the phase diagram presented in Fig. 47, with the exception that instead of $CaAl_4$ (27.08 wt. % Ca), as shown, the phase $CaAl_3$ was assumed to exist. The temperature of the Al-rich eutectic was found to be 610 [1], 616 [4], and 613°C [5] and its composition as approximately 8 [1] and 7.6 wt. % Ca [4] (5.5 and 5.3 at. % Ca). The solid solubility of Ca in Al was given as about 0.6 wt. (0.4 at.) % at 616°C and about 0.3 wt. (0.2 at.) % at "room temperature" [4]; however, these values can be considered only qualitative. On the other hand, parametrically determined solubilities of about 2.8, 2.2, and 1.7 wt. (1.9, 1.5, and 1.2 at.) % Ca at 616, 400, and 300°C [6], respectively, appear to be far too high, since measurements of the electrical resistivity do not indicate a noticeable solubility [7, 5].

The phase coexisting with Al was established by X-ray work to be $CaAl_4$, analogous to $SrAl_4$ and $BaAl_4$. It has a b.c. tetragonal structure of the $BaAl_4$ ($D1_3$) type,

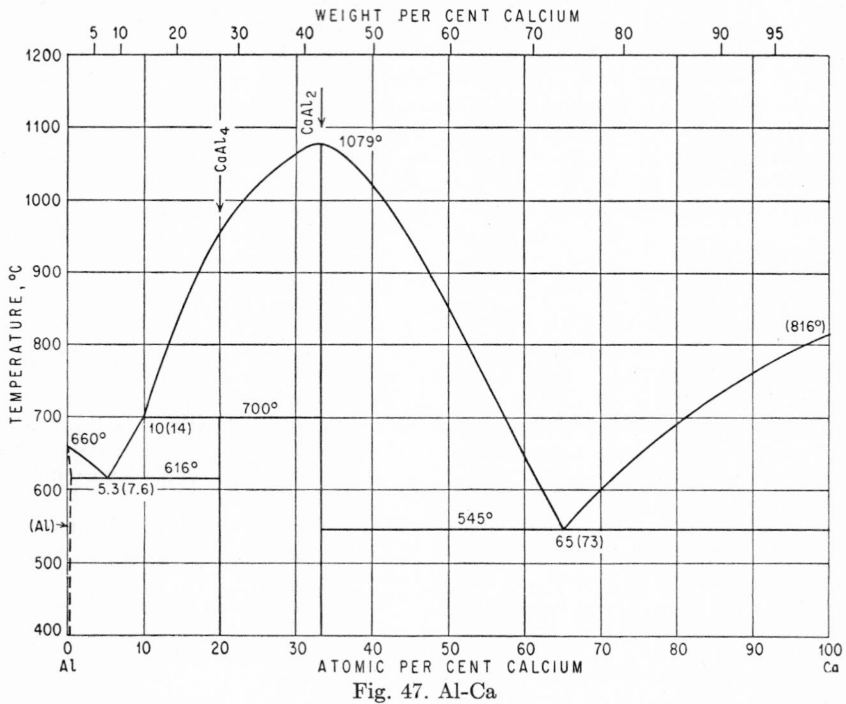

Fig. 47. Al-Ca

$a = 4.36$ A, $c = 11.09$ A, $c/a = 2.54$ [6]. $CaAl_2$ crystallizes in the $MgCu_2$ (C15) type, a = 8.038 A [8]. Both phases have a very narrow range of homogeneity.

Supplement. [9] observed no influence of Ca on the lattice parameter of Al and concluded that the solubility data given by [6] are incorrect.

1. L. Donski, *Z. anorg. Chem.*, **57**, 1908, 201–205.
2. The existence of $CaAl_3$ had been assumed earlier by H. Schlegel, dissertation, Leipzig, 1906.
3. Earlier, L. Stockem (*Metallurgie*, **3**, 1906, 149) had reported that both metals alloy in all proportions. See also K. Arndt, *Ber. deut. chem. Ges.*, **38**, 1905, 1972–1974.
4. K. Matsuyama, *Science Repts. Tôhoku Univ.*, **17**, 1928, 783–789. The components used had the following purity: Al, 0.3% Fe, 0.3% Si; Ca, 0.14% Fe, 0.11% Si, 0.15% Al, 1.17% Mg. According to G. Doan (*Z. Metallkunde*, **18**, 1926, 350–355)

and J. D. Grogan (*J. Inst. Metals*, **37**, 1927, 77–89), Ca in Al-Ca alloys containing Si is present as $CaSi_2$, which is practically insoluble in Al.
5. G. Bozza and C. Sonnino, *Giorn. chim. ind. ed appl.*, **10**, 1928, 443–449.
6. H. Nowotny, E. Wormnes, and A. Mohrnheim, *Z. Metallkunde*, **32**, 1940, 39–42.
7. J. D. Edwards and C. S. Taylor, *Trans. Am. Electrochem. Soc.*, **50**, 1926, 391–397. Al of 99.8% purity was used.
8. H. Nowotny and A. Mohrnheim, *Z. Krist.*, **A100**, 1939, 540–542.
9. G. Falkenhagen and W. Hofmann, *Z. Metallkunde*, **43**, 1952, 72.

Al-Cb Aluminum-Columbium

See aluminum-niobium (Al-Nb).

$\bar{1}.3802$
0.6198

Al-Cd Aluminum-Cadmium

The two metals are almost completely immiscible in the liquid state [1–3]. [3] stated that the melting point of Al is not noticeably affected by Cd. However, [4]

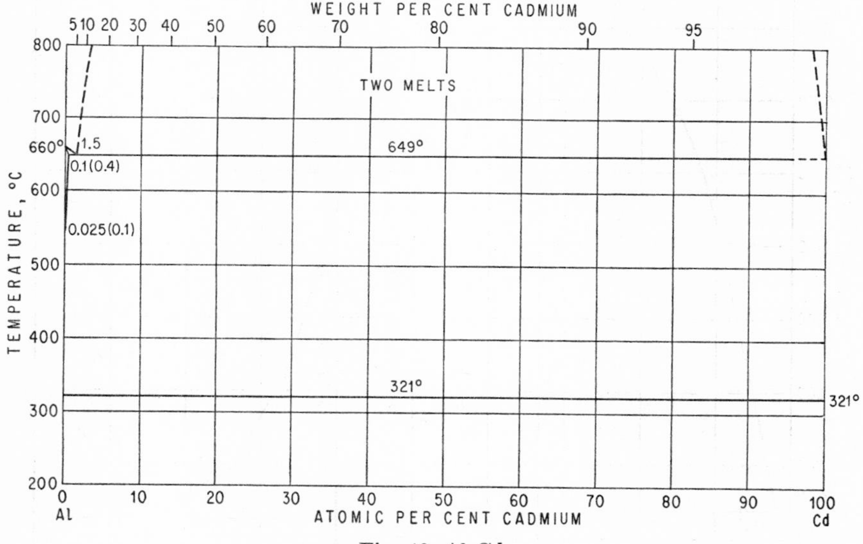

Fig. 48. Al-Cd

found a decrease from 660 to 649°C by 5–7 wt. (1.3–1.8 at.) % Cd, which is in agreement with the calculated value of the monotectic point, 6.7 wt. (1.7 at.) % Cd [5]. The melting point of Cd is not lowered by aluminum additions [6, 3, 4]. The solid solubility of Cd in Al, reported to be less than 1 wt. (<0.25 at.) % at 550°C and less than 0.2 wt. (0.05 at.) % at 200°C [4, 7], was more accurately determined by [5]. It is about 0.4, 0.25, and 0.1 wt. % (0.1, 0.06, and 0.025 at. %) at 640, 600, and 530°C, respectively (Fig. 48). Extrapolation of the calculated saturation curve indicates a solubility of 0.0002 wt. % at 165°C [5].

1. C. R. A. Wright, *J. Chem. Soc.*, **11**, 1892, 492–494; **13**, 1894, 1014–1019.
2. W. Campbell and J. A. Mathews, *J. Am. Chem. Soc.*, **24**, 1902, 255–256.

3. A. G. C. Gwyer, *Z. anorg. Chem.*, **57,** 1908, 149–151.
4. M. Hansen and B. Blumenthal, *Metallwirtschaft,* **10,** 1931, 925–927; Al of 99.9% purity was used.
5. H. K. Hardy, *J. Inst. Metals,* **80,** 1951-1952, 431–434; high-purity Al was used.
6. C. T. Heycock and F. H. Neville, *J. Chem. Soc.,* **61,** 1892, 911.
7. B. Blumenthal and M. Hansen, *Metallwirtschaft,* **11,** 1932, 671–674.

$\bar{1}.2845$
0.7155

Al-Ce Aluminum-Cerium

The first thermal and microscopic investigation [1] yielded a phase diagram [2] characterized by the existence of the five following intermediate phases: (*a*) $CeAl_4$

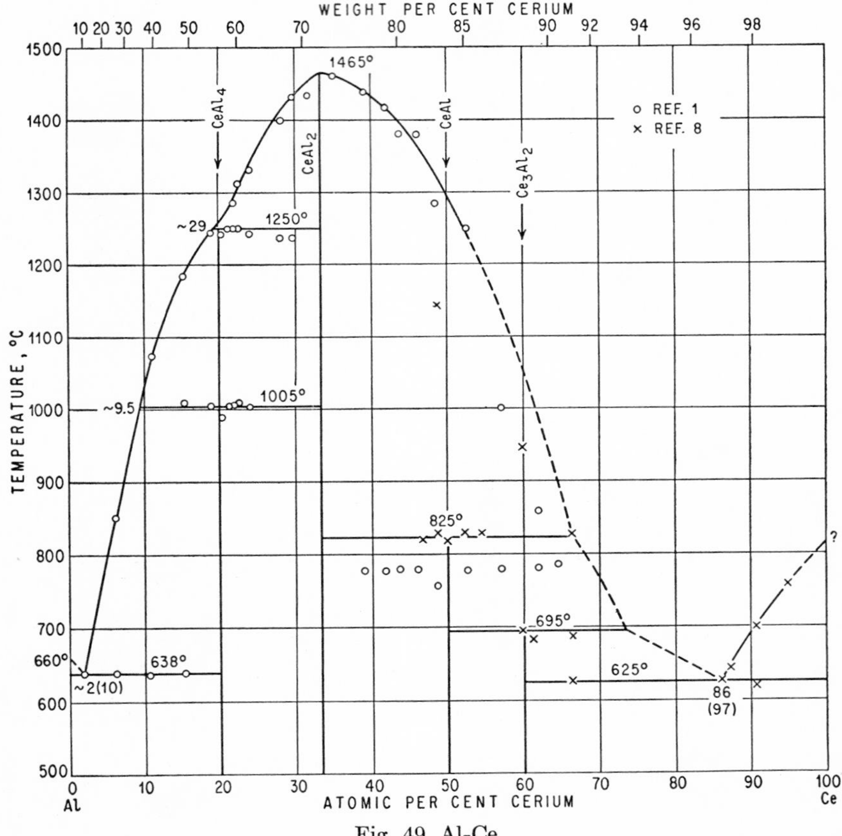

Fig. 49. Al-Ce

(56.49 wt. % Ce) [3], formed by the peritectic reaction of melt (29 at. % Ce) with $CeAl_2$ at 1250°C and undergoing a transformation at 1005°C; (*b*) $CeAl_2$ (72.20 wt. % Ce), having a maximum melting point of 1465°C; (*c*) CeAl (83.85 wt. % Ce) formed by the peritectic reaction of melt (about 65 at. % Ce) with $CeAl_2$ at 780°C; (*d*) Ce_2Al (91.22 wt. % Ce) formed peritectically, melt (69 at. % Ce) + CeAl, at 593°C; and (*e*) Ce_3Al (93.97 wt. % Ce) melting congruently at 614°C. Three eutectics were reported to exist: Al-$CeAl_4$ at approximately 2 at. % Ce, 637°C; Ce_2Al-Ce_3Al at 70.7 at. % Ce, 542°C; and Ce_3Al-Ce at 86 at. % Ce, 588°C.

Results of a thermal study of the constitution of the Al-rich alloys [4] were interpreted to indicate the existence of solid solutions up to at least 11 wt. % Ce, rather than a eutectic Al-$CeAl_4$. However, later investigations [5, 6] confirmed the findings of the first study [1]. The solid solubility of Ce in Al was given as less than 0.05 wt. % Ce [7] on the basis of conductivity measurements or as undetectable by lattice-parameter measurements [6].

Checking the constitution of the Ce-rich alloys up to 53 at. % Al, using a cerium of greater purity, [8] found that the compounds Ce_2Al and Ce_3Al reported earlier [1] do not exist. Instead, the compound Ce_3Al_2 (88.62 wt. % Ce) was claimed to be formed peritectically at 695°C (Fig. 49) and to form a eutectic with Ce at 86 at. (97 wt.) % Ce, 625°C. The temperature of the peritectic reaction by which CeAl is formed was determined to be 825°C, instead of 780°C.

$CeAl_4$ is b.c. tetragonal of the $D1_3$ type, a = 4.374 A, c = 10.12 A, c/a = 2.314 [6]. $CeAl_2$ has a structure of the $MgCu_2$ (C15) type, with a = 8.055 A [9] or a = 8.11 A [6].

1. R. Vogel, *Z. anorg. Chem.*, **75,** 1912, 41–57.
2. The cerium used was only of about 93.5% purity and contained mainly Nd, Pr, and La. The effect of the contaminants was considered by basing the calculation of the ratio Al:Ce on the values for pure Ce.
3. W. Muthmann and H. Beck, *Liebigs Ann.*, **331,** 1904, 47–50.
4. O. Barth, *Metallurgie*, **9,** 1912, 274–276.
5. K. L. Meissner, *Metall u. Erz*, **21,** 1924, 41–44.
6. H. Nowotny, *Z. Metallkunde*, **34,** 1942, 22–24.
7. J. Schulte, *Metall u. Erz*, **18,** 1921, 236–240.
8. R. Vogel and Th. Heumann, *Z. Metallkunde*, **35,** 1943, 29–42.
9. H. J. Wallbaum, *Z. Krist.*, **103,** 1941, 147–148.

$\bar{1}$.6606
0.3394

Al-Co Aluminum-Cobalt

Since the first study of the phase diagram [1] which established its principal features, parts of the system have been investigated more thoroughly. These investigations are concerned with (*a*) the alloys rich in aluminum [2], (*b*) the phase relations in the alloys with more than 50 at. % Co [3–6], (*c*) the number and composition of the intermediate phases [7, 2], (*d*) the effect of aluminum on the polymorphic [3, 8, 5, 6] and magnetic transformation [3, 4, 5, 9, 6] of Co, and (*e*) the crystal structure of the intermediate phases [10, 7, 11–15]. The diagram presented in Fig. 50 is essentially based on the studies by [1, 2, 6, 7].

In addition to the three intermediate phases suggested by [1], Co_3Al_{13} [16], later identified to be Co_2Al_9 [2, 12], Co_2Al_5, and CoAl = β', qualitative X-ray study of slowly cooled alloys [7] revealed the existence of two additional phases, Co_4Al_{13} and possibly a phase having a composition close to $CoAl_3$. The latter was omitted from Fig. 50 because of lack of definite identification. At present, there is no evidence as to how and at what temperature the phases Co_4Al_{13} and $CoAl_3$ are formed.

The solid solubility of Co in Al is less than 0.02 wt. % at 657°C [2]. Likewise, the solubility of Al in ϵ-Co is practically nil [7].

Crystal Structures. Co_2Al_9 [18.18 at. (32.68 wt.) % Co] is monoclinic with 4 Co and 18 Al atoms per unit cell and a = 8.5565 A, b = 6.290 A, c = 6.2130 A, β = 94.760° [13]; see also [12, 14, 15]. Co_4Al_{13} [23.52 at. (40.20 wt.) % Co] gave a complicated powder pattern different from those of Co_2Al_9 and Co_2Al_5 [7]. Co_2Al_5

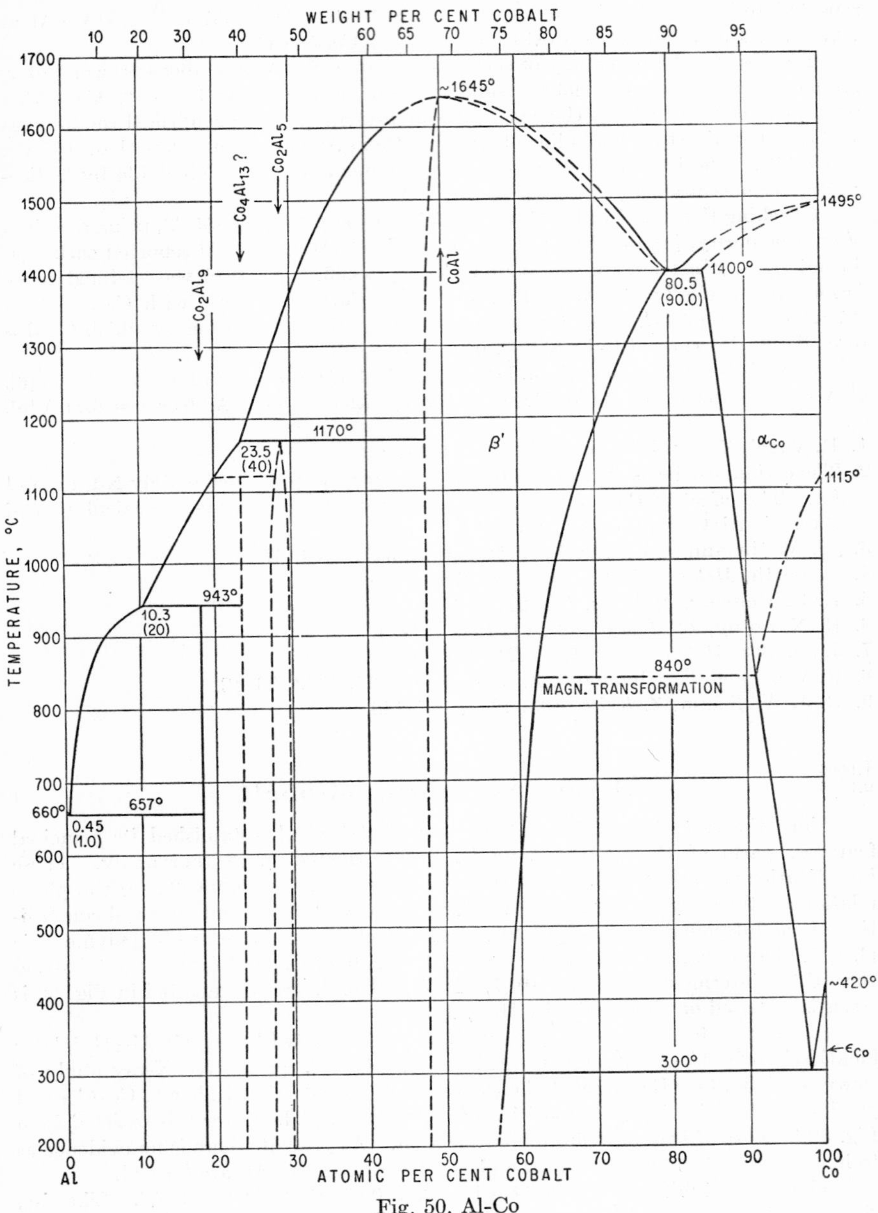
WEIGHT PER CENT COBALT
10 20 30 40 50 60 65 70 75 80 85 90 95
1700
1600
1500
1400
1300
1200
1100
1000
900
800
700
600
500
400
300
200
TEMPERATURE, °C
~1645°
Co_4Al_{13} ?
Co_2Al_5
Co_2Al_9
CoAl
1495°
1400°
80.5
(90.0)
1170°
23.5
(40)
β'
$α_{Co}$
1115°
943°
10.3
(20)
840°
MAGN. TRANSFORMATION
657°
660°
0.45
(1.0)
~420°
$ε_{Co}$
300°
0 10 20 30 40 50 60 70 80 90 100
Al
Co
ATOMIC PER CENT COBALT

Fig. 50. Al-Co

[28.57 at. (46.63 wt.) % Co] has the hexagonal $D8_{11}$ type of structure with 28 atoms per unit cell, a = 7.671 A, c = 7.608 A, c/a = 0.99 [11]. CoAl (β') has the ordered b.c.c. structure of the CsCl type [10, 7]; the lattice parameter reaches a maximum at 50 at. %, a = 2.862 A [7].

1. A. G. C. Gwyer, *Z. anorg. Chem.*, **57**, 1908, 140–147.
2. W. L. Fink and H. R. Freche, *Trans. AIME*, **99**, 1932, 141–148.
3. U. Haschimoto, *Kinzoku-no-Kenkyu*, **9**, 1932, 65–68.
4. W. Köster, *Arch. Eisenhüttenw.*, **7**, 1933-1934, 263.
5. U. Haschimoto, *Nippon Kinzoku Gakkai-Shi*, **1**, 1937, 177–190.
6. J. Schramm, *Z. Metallkunde*, **33**, 1941, 381-387.
7. A. J. Bradley and G. C. Seager, *J. Inst. Metals*, **64**, 1939, 81–88.
8. W. Köster and E. Wagner, *Z. Metallkunde*, **29**, 1937, 230–232.
9. W. Köster and E. Gebhardt, *Z. Metallkunde*, **30**, 1938, 281–286.
10. W. Ekman, *Z. physik. Chem.*, **B12**, 1930, 57–78.
11. A. J. Bradley and C. S. Cheng, *Z. Krist.*, **A99**, 1938, 480–487.
12. A. M. B. Parker, *Nature*, **156**, 1945, 783.
13. A. M. B. Douglas, *Nature*, **162**, 1948, 565–566; *Acta Cryst.*, **3**, 1950, 19–24.
14. G. V. Raynor and M. B. Waldron, *Nature*, **162**, 1948, 566; *Phil. Mag.*, **40**, 1949, 198–205.
15. G. V. Raynor and P. C. L. Pfeil, *J. Inst. Metals*, **73**, 1947, 609–624.
16. O. Brunck, *Ber. deut. chem. Ges.*, **34**, 1901, 2734.

$\bar{1}$.7150
0.2850

Al-Cr Aluminum-Chromium

Thermal and microscopic studies in the range up to 23 at. % Cr [1] proved that the phase diagram, outlined by [2], showing a wide miscibility gap in the liquid state was incorrect. Two intermediate phases of the approximate composition $CrAl_6$(?) [14.29 at. (24.32 wt.) % Cr] and $CrAl_4$ (32.52 wt. % Cr) were suggested to be formed by peritectic reactions at 725 and 1011°C, respectively [1]. A very careful investigation of the Al-rich alloys up to about 1.5 at. % Cr, prepared using high-purity Al, showed that the α solid solution does not enter into a eutectic [1], but rather a peritectic equilibrium at 661°C, with the peritectic melt containing 0.41 wt. (0.22 at.) % Cr [3]; see inset of Fig. 51. The phase which is in equilibrium with the α solid solution was determined by chemical analysis of isolated crystals to be $CrAl_7$ [12.5 at. (21.59 wt.) % Cr] [3]. This composition was verified by several later investigators [4–6] and, therefore, appears to be fully established. According to work by [7], covering the composition range up to 50 wt. % Cr, two compounds, $CrAl_4$ (32.52 wt. % Cr) and $CrAl_2$ (49.08 wt. % Cr), are formed peritectically at 803 and 1018°C, respectively. However, this is not consistent with the results reported by [1] and [4].

A comprehensive X-ray powder study [4] of more than 70 alloys with up to 86 at. (92 wt.) % Cr, annealed at and quenched from 600, 710, 800, 850, 890, 930, 1000, and 1100°C as well as "slowly cooled [from about 800°C] to room temperature [10°C per hr]," gave "approximate limits of temperature and composition" for each of the phases shown in the "tentative" diagram (Fig. 51), "the most important feature of which is the phase sequence below 600°C obtained from the slowly cooled alloys." The phase boundaries are considered to be "probably accurate to about 1.0 wt. % just below 850°C." "Above 850°C the results must be accepted with some caution," probably because of "partial decomposition on quenching"(!). The solidus and liquidus lines "are to some extent speculative."

The "tentative" phase diagram, outlined by [4] on a wt. % scale, is given in Fig. 51 in the form presented by the authors except that the phase boundaries are

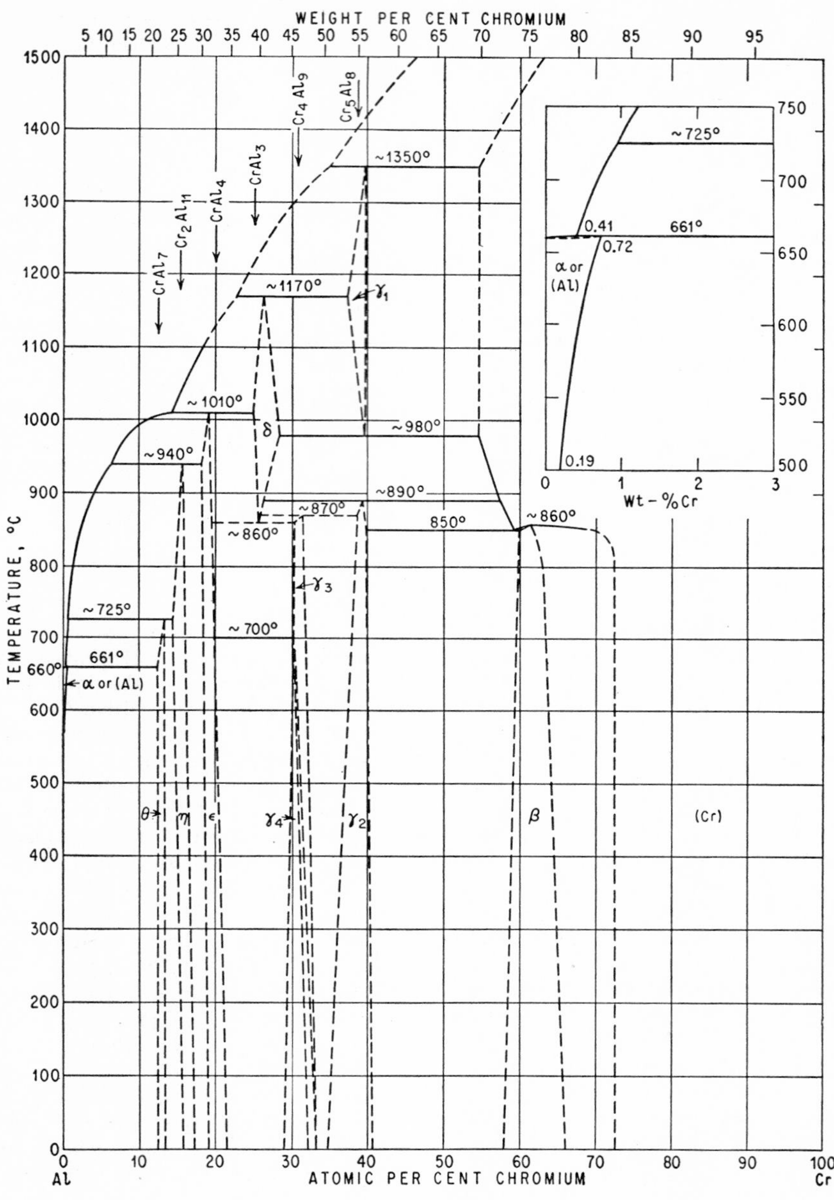
WEIGHT PER CENT CHROMIUM
5 10 15 20 25 30 35 40 45 50 55 60 65 70 75 80 85 90 95
TEMPERATURE, °C
1500
1400
1300
1200
1100
1000
900
800
700
600
500
400
300
200
100
0
660°
CrAl₇
Cr₂Al₁₁
CrAl₄
CrAl₃
Cr₄Al₉
Cr₅Al₈
~1350°
~1170°
γ₁
~1010°
δ
~980°
~940°
~890°
~870°
~860°
850°
~860°
γ₃
~725°
~700°
661°
α or (Al)
θ
η
ε
γ₄
γ₂
β
(Cr)
0
10
20
30
40
50
60
70
80
90
100
Al
ATOMIC PER CENT CHROMIUM
Cr
750
~725°
700
0.41
661°
0.72
650
α or (Al)
600
550
0.19
500
0
1
2
3
Wt-%Cr

Fig. 51. Al-Cr

drawn as dashed rather than solid curves. The temperatures of the three-phase equilibria can be considered only rough approximations. The phase designations and "approximate" chemical formulas were adopted from the original paper. For further details, see Crystal Structures.

The limit of the solid solution of Cr in Al has been determined by various methods. Some of the results differ widely [8, 7, 9] and are inconsistent with the results by others [3, 10, 11, 5, 6]. The findings by [3] (530–630°C) and [6] (375–620°C), both based on micrographic studies, agree very well. The solubility values due to [10] (275–660°C), based on X-ray and electrical-resistance work, lie at consistently higher (0.06–0.1 wt. %) chromium contents. The values by [11] (500°C, 0.3 wt. %; 600°C, 0.56 wt. %) are even a little higher, whereas the solubility at 650°C (0.75 wt. %) given by [5] is in substantial agreement with the data by [3, 10, 6]. The most probable solubility values in wt. % are as follows: 661°C, 0.72 (extrapolated); 630°C, 0.58; 600°C, 0.45; 550°C, 0.30; 500°C, 0.19; 400°C, 0.06; and 300°C, 0.015 (extrapolated). On very rapid solidification, aluminum tends to form supersaturated solid solutions with chromium up to about 1.6 wt. (0.83 at.) % Cr [5] and even up to 5.5 wt. (2.85 at.) % Cr [12]. This explains why [9] found a solubility of 1.63 at. (3.1 wt.) % Cr.

Crystal Structures. Lattice parameters of the α solid solution are reported by [11, 5, 12]; see also [13]. The θ phase ($CrAl_7$) was shown to have not an orthorhombic (pseudohexagonal) structure as asserted by [4], but a monoclinic structure, with $a = 20.47$ A, $b = 7.64$ A, $c = 25.36$ A, $\beta = 155°10'$ [5].

The η phase is based upon the composition Cr_2Al_{11} [15.38 at. (25.95 wt.) % Cr [4]; see also [14]. The structure of this phase, which extends at 680°C from 24 to 26.5 wt. (14.07 to 15.75 at.) % Cr, as well as that of the ϵ phase, is apparently very complicated, whereas the δ phase has a fairly simple powder pattern [4]. ϵ and δ correspond to the empirical formulas $CrAl_4$ (32.52 wt. % Cr) and $CrAl_3$ (39.12 wt. % Cr), respectively.

The phases γ_1, γ_2, γ_3, and γ_4 "have crystal structures which in their general form resemble that of a γ-brass" [4]. The information regarding the γ_1 phase "is so scanty that its position in the equilibrium diagram must be regarded as somewhat speculative" [4]. The crystal structure of the γ_2 phase, which is based on the composition Cr_5Al_8 [38.46 at. (54.65 wt.) % Cr] is rhombohedral, "but the positions of the atoms are almost identical with that in γ-brass" [4]. Its lattice dimensions are $a = 7.805$ A, $\alpha = 109°7.6'$ if the pattern is indexed as a cell with 26 atoms, or $a = 9.051$ A, $\alpha = 89°16.4'$ if indexed as a cell with 52 atoms [15]. [9] have reported that alloys with 30.9, 38.4, and 45.3 at. % Cr contain a phase (apparently γ_2) with a b.c.c. structure of the γ-brass type, $a = 9.12$ A (30.9 at. % Cr) and $a = 9.02$ A (45.4 at. % Cr). In agreement with Fig. 51, the alloy with 38.4 at. % Cr was found to be single-phase. The phases γ_3 and γ_4 "must be represented by the same formula," Cr_4Al_9 [30.76 at. (46.14 wt.) % Cr], which corresponds to the center of these phases [4]. The $\gamma_4 \rightarrow \gamma_3$ transformation is apparently of the order $\rightarrow$ disorder type.

Lattice spacings of the Cr-rich solid solution are given by [4]. Within a certain range of composition (Fig. 51) the b.c.c. primary solid solution transforms (at 850–860°C) into the tetragonal β phase. Its structure ($MoSi_2$ type) can be regarded as a deformed b.c.c. structure, due to an ordering process in which a superlattice is formed. The ideal composition of β is Cr_2Al (79.40 wt. % Cr) and lies somewhere outside the β field. At 78.5 wt. (about 65.5 at.) % Cr the dimensions of the unit cell containing 2 molecules Cr_2Al are $a = 3.004$ A, $c = 8.647$ A, $c/a = 2.878$ [15].

1. M. Goto and G. Dogane, *Nippon Kogyokwaishi*, **43**, 1927, 931; abstract in *J. Inst. Metals*, **43**, 1930, 446. K. Honda, *Proc. World Eng. Congr., Tokyo*, 1929, Report 658, pp. 24–25.

2. G. Hindrichs, *Z. anorg. Chem.*, **59**, 1908, 430–437.
3. W. L. Fink and H. R. Freche, *Trans. AIME*, **104**, 1933, 325–334.
4. A. J. Bradley and S. S. Lu, *J. Inst. Metals*, **60**, 1937, 319–337.
5. W. Hofmann and H. Wiehr, *Z. Metallkunde*, **33**, 1941, 369–372.
6. G. V. Raynor and K. Little, *J. Inst. Metals*, **71**, 1945, 481–489.
7. S. Hori, *Sumitomo Kinzoku Kogyo Kenkyu Hokoku*, **2**, 1935, 351–372; *Tetsu-to-Hagane*, **22**, 1936, 194 (quoted from [6]).
8. P. Röntgen and W. Koch, *Z. Metallkunde*, **25**, 1933, 182–185.
9. A. Knappwost and H. Nowotny, *Z. Metallkunde*, **33**, 1941, 153–157.
10. W. Koch and H. Winterhager, *Metallwirtschaft*, **17**, 1938, 1159–1163.
11. W. Hofmann and R. W. Herzer, *Metallwirtschaft*, **19**, 1940, 141–143.
12. G. Falkenhagen and W. Hofmann, *Z. Metallkunde*, **43**, 1952, 69–81.
13. G. V. Raynor, *Phil. Mag.*, **36**, 1945, 770–777.
14. G. V. Raynor and K. Little, *J. Inst. Metals*, **71**, 1945, 493–524.
15. A. J. Bradley and S. S. Lu, *Z. Krist.*, **96**, 1937, 20–37.

$\bar{1}.3075$
0.6925

Al-Cs Aluminum-Cesium

It has been reported [1, 2] that Al and Cs are almost immiscible in both the liquid and solid states, although a small zone of liquid miscibility exists at the aluminum end (0.05 wt. % Cs?) [2].

1. J. Czochralski and J. Kakzynski, *Wiadamosci Inst. Metallurgii i Metaloznawstawa*, **4**, 1934, 18 (quoted in [2]).
2. L. F. Mondolfo, "Metallography of Aluminum Alloys," pp. 15–16, John Wiley & Sons, Inc., New York, 1943.

$\bar{1}.6280$
0.3720

Al-Cu Aluminum-Copper

Figures 52 and 53 represent "the most probable composite equilibrium diagram" as proposed on the basis of a critical evaluation of data published up to 1942 [1]. The reader interested in a detailed information is referred to this publication, which contains a table showing the estimated accuracy within which the phase boundaries have been determined. "Between the limits 0–20 wt. (0–37 at.) % and 30–100 wt. (50.2–100 at.) % Al the equilibrium diagram is accurately established. In the range 20–30 wt. (37–50.2 at.) % Al, however, the equilibrium relations are complex and may not yet be fully understood" [1]. For a previous literature review and bibliography up to 1935, see [2].

After the fundamental, though still quite incomplete, thermal and microscopic investigations of the whole system by [3, 4] and particularly [5], the phase relationships have been clarified by the works of [6–11]. The diagram given by [12] is much less complete than those of previous investigators.

As the studies proceeded, the constitution in the range 14–30 wt. (27.7–50.2 at.) % Al was found to be more and more complex. Much additional and confirmatory information was contributed by investigations of the crystal structures, especially those of [13–20]. There was considerable disagreement as regards the interpretation of various three-phase equilibria and solid-state transformations in the region 30–50 at. % Al. Those shown in Figs. 52 and 53 are due to [1].

The Range 0–34 At. (0–18 Wt.) % Al. In plotting the phase boundaries in Fig. 52, [1] gave preference to the results of [6, 8, 9, 11]. The latter worker discovered the high-temperature phase X which forms peritectically at 1036°C and decomposes eutectoidally at 963°C. The $\alpha/(\alpha + \beta)$ and $\alpha/(\alpha + \gamma_2)$ boundaries were determined by micrographic analysis [6, 9] and lattice-spacing measurements [21]. Results are in good agreement, the estimated accuracy being ±0.1–0.2 wt. (about ±0.2–0.4 at.) % [1]. According to [22], a decrease in solubility between the eutectoid temperature and 350°C does not occur. [23] believed they found indications for the existence of an ordered structure in the α phase; see also [22, 24]. The boundaries of the phases β [6, 9, 11], X [11], γ_1 [6, 9], and γ_2 [6, 9, 20] have been established with an accuracy of ±0.2 wt. % or better [1].

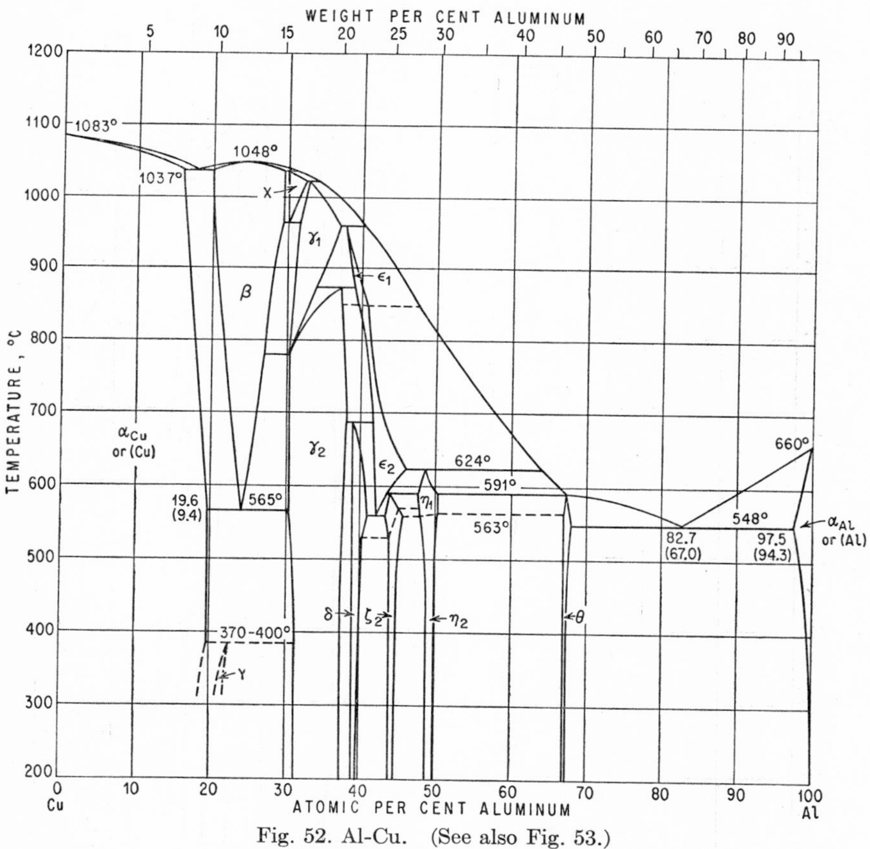

Fig. 52. Al-Cu. (See also Fig. 53.)

The temperature of the eutectoid $\beta \rightleftarrows \alpha + \gamma_2$, determined by numerous investigators [3, 5, 25, 26, 27, 6, 8, 28, 29, 9, 10, 30, 12, 31, 23, 32, 33, 34], was reported at varying temperatures, since the decomposition of β is prone to supercooling. Careful work gave the following temperatures: 568 ± 2°C [30], 560–565°C [31], 565.5 ± 1°C [33], and 565 ± 2°C [34]. The eutectoid composition of 11.8 wt. (23.97 at.) % Al, found by [6], was confirmed by [31]. According to [33] and [34] this point lies at 11.80 ± 0.15 wt. (23.97 ± 0.26 at.) % Al and 11.96 ± 0.05 wt. (24.23 ± 0.08 at.) % Al, respectively.

Recently, [35] gave microscopical and X-ray evidence for the existence of an additional stable phase, formed as the result of a peritectoid reaction between the

α and γ_2 phases at a temperature between 370 and 400°C. The composition is not yet accurately determined, but probably lies below 11.3 wt. (23.1 at.) % Al (Fig. 52).

The Range 34–67.2 At. (18–46.5 Wt.) % Al. The liquidus and solidus curves of the phases γ_1, ϵ_1, ϵ_2, and η_1 are substantially based on the work of [7–9]. As to the solid state, the composite phase relations proposed by [1], on the basis of the data of

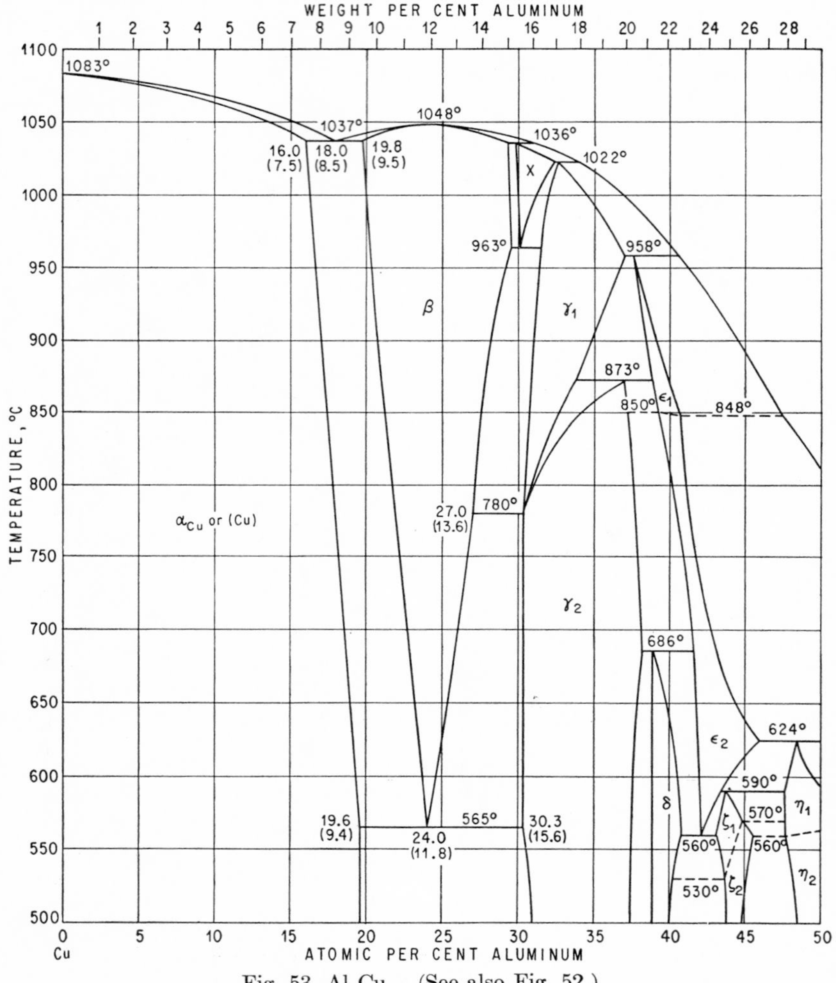

Fig. 53. Al-Cu. (See also Fig. 52.)

[7–10], have been accepted. Space does not permit analyzing the differing interpretations of the results of these and other investigators.

The transformation $\gamma_1 \rightleftarrows \gamma_2$ ranging between 780 and 873°C was reported by [25, 6, 7, 36, 8, 9, 10]. The phase relations shown are due to [8–10]. All the more recent studies agree in that the 958°C horizontal is a peritectic reaction resulting in the formation of ϵ_1 [7–10]. Also, it appears to be well established that the ϵ phase undergoes a transformation at about 850°C [7, 9] whose nature is not known. Contrary to this, [5, 7, 8, 10] had suggested these thermal effects to correspond to a peritectic reaction.

The existence of the η phase was fully established by microscopical and thermal studies [5, 7–10], as well as X-ray work [16, 17, 19, 20]. The horizontal at 624°C corresponds to the peritectic formation of η [5, 8–10], rather than the transformation $\epsilon \rightleftarrows \eta$ [7].

The region between γ_2 and η, at temperatures below 700°C, was the main point of controversy. It was soon recognized [16, 17, 19, 37] that the constitution was more complex than shown in the diagram of [7] which indicates no additional phase, stable at room temperature, between γ_2 and η. The existence of the ζ phase, first identified by [16], was verified by [9, 20]. Its homogeneity range was given by [9] as 43.7–44.6 at. (24.8–25.5 wt.) % Al, while [20], on the basis of X-ray work on slowly cooled alloys, concluded it to be located between 42.1 and 43.4 at. (23.6 and 24.6 wt.) % Al.

The δ phase was introduced by [8], who believed that it was formed peritectically at 847°C. Its existence was confirmed by [9]. X-ray investigations of alloys covering the range of the γ_2 and δ phases in Figs. 52 and 53 [13, 15, 17, 19, 20] have shown that there are only slight modifications of the γ-brass type of structure within this entire region. According to the most complete study [20], the following ranges of stability of three slightly different structures might be distinguished: (*a*) 31.3–35.3 at. (16.2–18.8 wt.) %, (*b*) 35.3–38.1 at. (18.8–20.7 wt.) %, and (*c*) 38.1–40.9 at. (20.7–22.7 wt.) % Al. The transition from one structure to the other was suggested to be continuous; at any rate, two-phase regions should be extremely narrow [20]; see also [38]. In contrast to this, [9] concluded the third of the above-mentioned regions (38.1–40.9 at. % Al) to be a totally distinct phase, δ.

The Range 67.2–100 At. (46.5–100 Wt.) % Al. The solidification was investigated by [39, 40, 5, 3, 4, 41, 42, 7, 8, 43, 44, 45, 9]. The liquidus in Fig. 52 has an accuracy of ±2–3°C [1]. The eutectic temperature is 548 ± 1°C, and the eutectic composition is generally accepted as 82.7 at. (67.0 wt.) % Al [7]. A possibly more accurate figure is that of [45*a*], 82.6 at. (66.85 wt.) % Al.

The θ phase, formerly regarded as a phase of the fixed composition $CuAl_2$, is of variable composition [17, 19, 45, 9]. It crystallizes directly from the melt [3, 5, 7, 9], rather than being formed peritectically [4, 8, 43]. The alloy with 67.2 at. (46.5 wt.) % Al has a constant freezing point of 591°C. The homogeneity range of θ was accurately determined micrographically [45]. At 548°C it extends from 67.1 (46.4) to 68.05 at. (47.5 wt.) % Al and at 400°C from 66.8 (46.1) to 67.35 at. (46.7 wt.) % Al [45]. The ideal composition $CuAl_2$ [66.67 at. (45.92 wt.) % Al] lies slightly outside the homogeneity region.

The solid solubility of Cu in Al was investigated by [46, 47, 41, 8, 48–51, 45, 52–60]. Between 400 and 548°C, the results of [48, 51, 45, 53, 56, 57, 60] agree within ±0.1 wt. (about ±0.045 at.) % Cu. At 300°C they vary between 0.35 (0.15) [51] and 0.75 wt. (0.32 at.) % Cu [48]. The following solubilities in at. (wt.) % Cu may be regarded as most reliable: 2.50 (5.70) at 548°C, 1.76 (4.05) at 500°C, 1.10 (2.50) at 450°C, 0.60 (1.40) at 400°C, 0.37 (0.85) at 350°C, 0.19 (0.45) at 300°C, and about 0.04–0.08 (0.1–0.2) at 250°C.

The solidus curve of the α_{Al} phase is a straight line over the composition in wt. % [8, 48, 45, 61].

Crystal Structures. As regards the α_{Cu} phase, see [62, 63, 13, 64, 19, 21, 65]; for lattice constants, see especially [21, 65].

The β phase [18, 19, 66, 67, 68] is b.c.c., A2 type [66–68], with a = 2.95 A at 600°C [66].

The structure of γ_2 [13–20, 69, 70, 38, 68] is substantially isotypic with γ-brass [75], a = 8.704 A [69]. Slight variations of this structure have been observed [13, 15,

17, 19, 20, 70, 38]. According to [20], "three types of structure are so closely related that no two-phase region can be found between them." (*a*) The cubic structure (Cu_9Al_4) exists from 31.3 to 35.3 at. % Al. (*b*) The first modification ($Cu_{32}Al_{19}$ = 37.25 at. % Al), stable in the range 35.3–38.1 at. % Al, is monoclinic with 102 atoms per unit cell [70]. (*c*) The second modification ($Cu_{30}Al_{20}$), stable between 38.1 and 40.9 at. % Al, has a structure yet unknown in detail. In Figs. 52 and 53, the latter is presented as a separate phase, δ [9].

ϵ_2 is cubic, similar to, but simpler than, the γ_2 phase [19].

ζ_1 was reported to be hexagonal, a = 8.09$_6$ A, c = 9.99$_9$ A, c/a = 1.23$_5$ [16] and ζ_2 to be monoclinic with 21 atoms per unit cell ($Cu_{12}Al_9$ = 42.86 at. % Al), a = 7.07 A, b = 4.08 A, c = 10.02 A, β = 90°38′ [20].

η_1 is orthorhombic with a = 4.095 A, b = 12.02 A, c = 8.652 A [16], and η_2 is centered-orthorhombic with 20 atoms per unit cell ($Cu_{10}Al_{10}$) and a = 6.89 A, b = 4.09 A, c = 9.89 A [20].

The structure of $CuAl_2$ = θ is the prototype of the C16-type structure [13, 71, 19, 72], a = 6.066 A, c = 4.874 A, c/a = 0.803$_5$ [19].

For lattice constants of the α_{Al} phase, see [51, 52, 71, 60, 72, 65, 73, 74], especially [60].

1. G. V. Raynor, Annotated Equilibrium Diagram Series, no. 4, Institute of Metals, London, 1944.
2. M. Hansen, "Der Aufbau der Zweistofflegierungen," pp. 98–108, Springer-Verlag OHG, Berlin, 1936.
3. H. C. H. Carpenter and C. A. Edwards, 8th Report to the Alloys Research Committee, *Inst. Mech. Engrs. (London)*, **72,** 1907, 57–269; see also *Rev. mét.*, **5,** 1908, 415–421; *Metallurgie*, **6,** 1909, 296–302; *J. Inst. Metals*, **1,** 1909, 114–116; **13,** 1915, 261–262.
4. A. G. C. Gwyer, *Z. anorg. Chem.*, **57,** 1908, 114–126.
5. B. E. Curry, *J. Phys. Chem.*, **11,** 1907, 425–436; see also *Metallurgie*, **6,** 1909, 296–302; *J. Inst. Metals*, **13,** 1915, 253.
6. D. Stockdale, *J. Inst. Metals*, **28,** 1922, 273–286; discussion, pp. 287–296.
7. D. Stockdale, *J. Inst. Metals*, **31,** 1924, 275–289; discussion, pp. 290–293.
8. M. Tazaki, *Kinzoku-no-Kenkyu*, **2,** 1925, 490–495; see [2].
9. C. Hisatsune, *Mem. Coll. Eng. Kyoto Univ.*, **8,** 1934, 74–91.
10. K. Matsuyama, *Kinzoku-no-Kenkyu*, **11,** 1934, 461–490. (Paper was not available.)
11. A. G. Dowson, *J. Inst. Metals*, **61,** 1937, 197–204.
12. W. Broniewski, S. Jelnicki, and M. Skwara, *Compt. rend.*, **207,** 1938, 233–235.
13. E. R. Jette, G. Phragmén, and A. Westgren, *J. Inst. Metals*, **31,** 1924, 193–206; discussion, pp. 206–215.
14. A. J. Bradley, *Phil. Mag.*, **6,** 1928, 878–888.
15. A. Westgren and G. Phragmén, *Metallwirtschaft*, **7,** 1928, 701.
16. G. D. Preston, *Phil. Mag.*, **12,** 1931, 980–993.
17. A. Westgren, *Trans. AIME*, **93,** 1931, 21–23.
18. I. Obinata, *Mem. Ryojun Coll. Eng.*, **3,** 1931, 285–294, 295–298; *Nature*, **126,** 1930, 809.
19. A. J. Bradley and P. Jones, *J. Inst. Metals*, **51,** 1933, 131–157; discussion, pp. 157–162.
20. A. J. Bradley, H. J. Goldschmidt, and H. Lipson, *J. Inst. Metals*, **63,** 1938, 149–161.
21. I. Obinata and G. Wassermann, *Naturwissenschaften*, **21,** 1933, 382–385.

22. V. P. Tarasova, *Vestnik. Moskov. Univ.*, **1947**, 105–107; *Chem. Abstr.*, **42**, 1948, 1172.
23. F. Bollenrath and W. Bungardt, *Z. Metallkunde*, **35**, 1943, 153–156.
24. H. Masumoto, H. Saito, and M. Takahashi, *Nippon Kinzoku Gakkai-Shi*, **18**, 1954, 98–100; *Met. Abstr.*, **22**, 1954, 177.
25. J. H. Andrew, *J. Inst. Metals*, **13**, 1915, 249–260.
26. P. Braesco, *Ann. phys.*, **14**, 1920, 5–75.
27. T. Matsuda, *Science Repts. Tôhoku Univ.*, **11**, 1922, 237–251.
28. I. Obinata, *Mem. Ryojun Coll. Eng.*, **2**, 1929, 205–225.
29. C. S. Smith and W. E. Lindlief, *Trans. AIME*, **104**, 1933, 69–105.
30. J. L. Bray, M. E. Carruthers, and R. H. Heyer, *Trans. AIME*, **122**, 1936, 337–346.
31. W. Hume-Rothery, G. V. Raynor, P. W. Reynolds, and H. K. Packer, *J. Inst. Metals*, **66**, 1940, 217–218.
32. D. J. Mack, *Trans. AIME*, **175**, 1948, 240–255.
33. D. L. Thomas and D. R. F. West, Research Correspondence, *Suppl. to Research*, **6**, no. 12, 1953, 61 S–62 S; see also [35].
34. R. Haynes, *J. Inst. Metals*, **83**, 1954-1955, 105–114.
35. D. R. F. West and D. L. Thomas, *J. Inst. Metals*, **83**, 1954-1955, 505–507.
36. D. Stockdale, *Trans. Faraday Soc.*, **19**, 1923, 135–139.
37. M. L. V. Gayler, *J. Inst. Metals*, **51**, 1933, 157.
38. H. Bittner and H. Nowotny, *Monatsh. Chem.*, **83**, 1952, 1308–1313.
39. H. Le Chatelier, *Bull. soc. encour. ind. natl.*, **10**, 1895, 569; *Z. anorg. Chem.*, **57**, 1908, 122.
40. W. Campbell and J. A. Mathews, *J. Am. Chem. Soc.*, **24**, 1902, 264–266; *J. Franklin Inst.*, **153**, 1902, 121; *Z. anorg. Chem.*, **57**, 1908, 123.
41. B. Otani and T. Hemmi, abstracted in *J. Inst. Metals*, **28**, 1922, 643.
42. K. E. Bingham and J. L. Haughton, *J. Inst. Metals*, **29**, 1923, 80, 95.
43. H. Nishimura, *Mem. Coll. Eng. Kyoto Imp. Univ.*, **5**, 1927, 64–66.
44. G. P. Kulbusch, abstracted in *J. Inst. Metals*, **44**, 1930, 485.
45. D. Stockdale, *J. Inst. Metals*, **52**, 1933, 111–116.
45*a*. D. Stockdale, *J. Inst. Metals*, **43**, 1930, 193–211.
46. P. D. Merica, R. G. Waltenberg, and J. R. Freeman, *Natl. Bur. Standards (U.S.) Sci. Papers* 337, 1919; *Trans. AIME*, **64**, 1920, 9–15; *Z. Metallkunde*, **13**, 1921, 575–576.
47. W. Rosenhain, S. L. Archbutt, and D. Hanson, 11th Report Alloys Research Committee, *Inst. Mech. Engrs. (London)*, **1921**, 200; see also *J. Inst. Metals*, **29**, 1923, 509; **40**, 1928, 299.
48. E. H. Dix and H. H. Richardson, *Trans. AIME*, **73**, 1926, 560–580; *Z. Metallkunde*, **18**, 1926, 196–197.
49. P. Saldau and N. N. Anisimov, *Izvest. Inst. Fiz.-Khim. Anal.*, **3**, 1926, 485; abstracted in *J. Inst. Metals*, **40**, 1928, 496.
50. A. v. Zeerleder and M. Bosshard, *Z. Metallkunde*, **19**, 1927, 462–464.
51. W. Stenzel and J. Weerts, *Metallwirtschaft*, **12**, 1933, 353–356, 369–374.
52. A. Phillips and R. M. Brick, *J. Franklin Inst.*, **215**, 1933, 557–577.
53. H. Auer, *Z. Metallkunde*, **28**, 1936, 164–175.
54. M. I. Zakharova, *Izvest. Sektora Fiz.-Khim. Anal.*, **10**, 1938, 113–118.
55. H. Borchers and H. Kremer, *Aluminium-Arch.*, no. 33, 1941.
56. R. H. Brown, W. L. Fink, and M. S. Hunter, *Trans. AIME*, **143**, 1941, 117–119.
57. H. Borchers and H. J. Otto, *Z. Metallkunde*, **34**, 1942, 90–93.
58. G. Borelius, J. Anderson, and K. Gullberg, *Ing. Vetenskaps Akad. Handl.* 169, 1943; see also *Trans. AIME*, **191**, 1951, 482.

59. H. W. L. Phillips, *J. Inst. Metals*, **74**, 1947-1948, 34.
60. E. C. Ellwood and J. M. Silcock, *J. Inst. Metals*, **74**, 1948, 457–467, 721–724.
61. J. A. Verö, *Acta Tech. Acad. Sci. Hungar.*, **2**, 1951, 97–113.
62. E. C. Bain, *Chem. Met. Eng.*, **28**, 1923, 21–24.
63. E. A. Owen and G. D. Preston, *Proc. Phys. Soc.*, **36**, 1923, 14–30.
64. S. Sekito, *Science Repts. Tôhoku Univ.*, **18**, 1929, 59–77.
65. A. P. Guljaev and E. F. Trusova, *Zhur. Tekh. Fiz.*, **20**, 1950, 66–78.
66. G. Wassermann, *Metallwirtschaft*, **12**, 1933, 358; **13**, 1934, 133–139.
67. A. H. Jay, *J. Inst. Metals*, **51**, 1933, 161.
68. P. Michel, *Compt. rend.*, **236**, 1953, 820–822.
69. J. B. Nelson and D. P. Riley, *Proc. Phys. Soc.*, **57**, 1945, 160–177.
70. A. J. Bradley, *Nature*, **168**, 1951, 661.
71. A. Phillips and R. M. Brick, *Trans. AIME*, **111**, 1934, 94–112.
72. H. J. Axon and W. Hume-Rothery, *Proc. Roy. Soc. (London)*, **A193**, 1948, 1–24.
73. E. C. Ellwood, *J. Inst. Metals*, **80**, 1951-1952, 605–608.
74. D. W. Smith, unpublished work; see W. L. Fink in "Physical Metallurgy of Aluminum Alloys," p. 49, The American Society for Metals, Cleveland, Ohio, 1949.
75. See W. Hume-Rothery, J. O. Betterton, and J. Reynolds, *J. Inst. Metals*, **80**, 1951-1952, 609–616.

$\bar{1}.6840$
0.3160

Al-Fe Aluminum-Iron

The Range 0–54 At. % Al. The liquidus and solidus curve of the α phase in Fig. 54 was plotted using the highest liquidus temperatures and smallest solidification intervals reported [1–4].

The vertex of the γ loop was found as about 1.2 wt. (2.4 at.) % Al [3] and about 1.0 wt. (2.0 at.) % Al [5]; see data points in the inset of Fig. 54. The $\gamma/(\alpha + \gamma)$ and $\alpha/(\alpha + \gamma)$ boundaries calculated by [6] extend to about 1.2 at. (0.6 wt.) % and 2.0 at. (1.0 wt.) % Al, respectively, at about 1150°C.

Because of the lack of reliable experimental data [1–4], the location of the high-Fe boundary of the $\alpha + \epsilon$ and $\alpha + \zeta$ fields is uncertain. In addition, information relative to the probable connection of the latter phase boundary with the disorder $\rightarrow$ order transformation $\alpha \rightarrow \beta_2$ = FeAl is still missing. The $\alpha/(\alpha + \epsilon)$ boundary is merely based on the thermal results of [4]; its extension to lower temperatures is tentative. Both [3] and [7] found that the solubility of Al in α-Fe decreased with fall in temperature. It is unknown whether the composition FeAl lies inside or outside the heterogeneous field.

The Order-Disorder and Magnetic Transformations. First indication of the occurrence of a transformation in the α solid solution was the detection of an ordered b.c.c. structure of the CsCl (B2) type [8]. The existence of two superlattices based on the ideal compositions Fe_3Al = β_1 and FeAl = β_2 was established by [9, 10]. Alloys with 0–51 at. % Al were (*a*) quenched from 700 and 600°C and (*b*) slowly cooled from 750°C. Up to about 18.5 at. % Al the atomic distribution in the b.c.c. lattice of α-Fe was random, independent upon heat-treatment. In the range 18.5–25 at. % Al the atomic arrangement was random after quenching from 600°C or higher, but ordered after cooling to room temperature, with the degree of ordering increasing to 92% at 25 at. % Al. The superlattice $FeAl_3$ gradually emerging from the random arrangement is of the BiF_3 (DO_3) type, and the lattice constant twice that of the α phase. Between 25 and 34 at. % Al the structure was of the $FeAl_3$ type (gradually changing over to the FeAl type) after slow cooling and of the FeAl (B2) type after

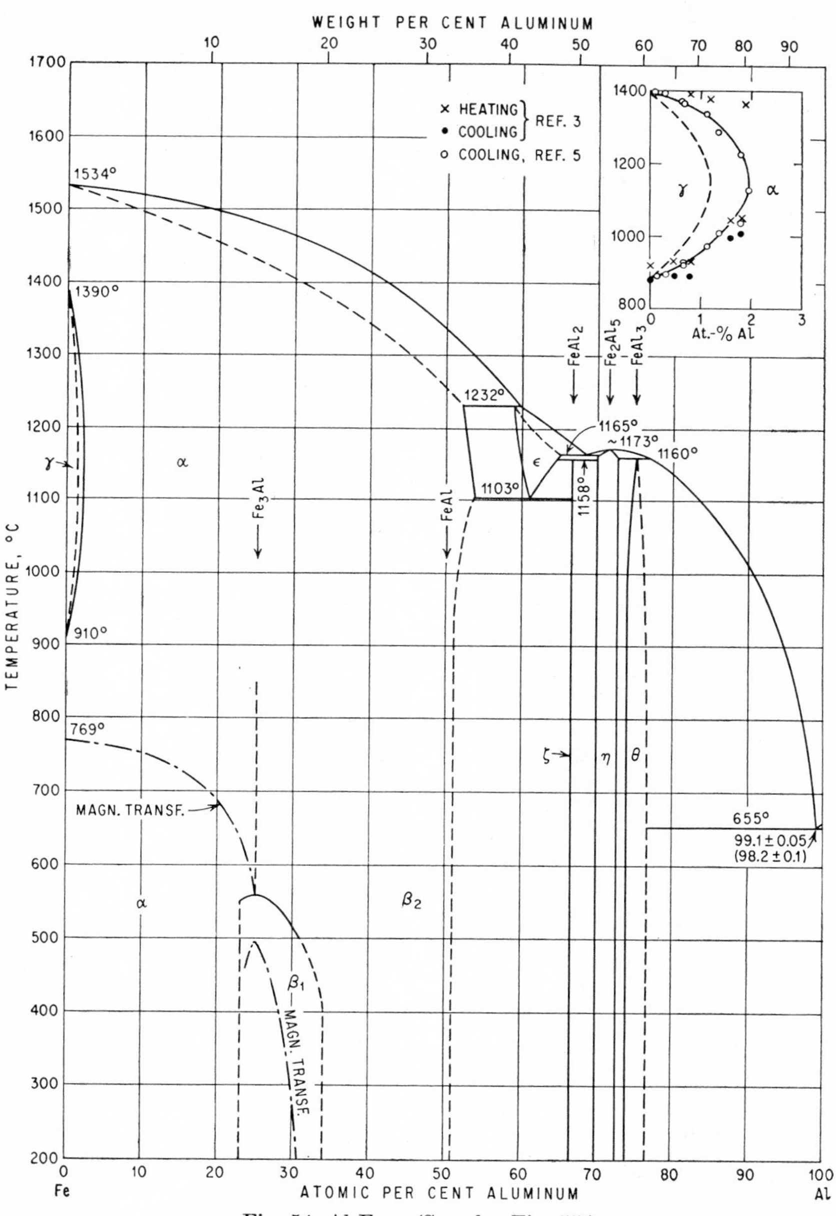

Fig. 54. Al-Fe. (See also Fig. 55.)

quenching from 600°C and higher temperatures. Above 34 at. % Al the structure was solely of the FeAl type, after both slow cooling and quenching from 600 and 700°C. These findings indicate that the Fe_3Al superlattice is formed below 600°C, whereas the FeAl superlattice is stable up to 700°C; however, the formation of the latter superlattice might not be suppressed by quenching. With the heat-treatments used in these studies, no regions containing both α and $FeAl_3$ structures and both $FeAl_3$ and FeAl structures could be found. [9, 10] claimed, therefore, that the transition of one type of structure to the other was continuous. Additional work, using clearly defined long-time anneals, is necessary to prove whether this is true.

[11–14] studied the order-disorder transformation in the $FeAl_3$ range by means of thermal [12], specific-heat [12, 14], dilatometric [12], resistivity [11, 13, 14], and magnetic methods [12, 14]. The transition temperatures reported by these workers

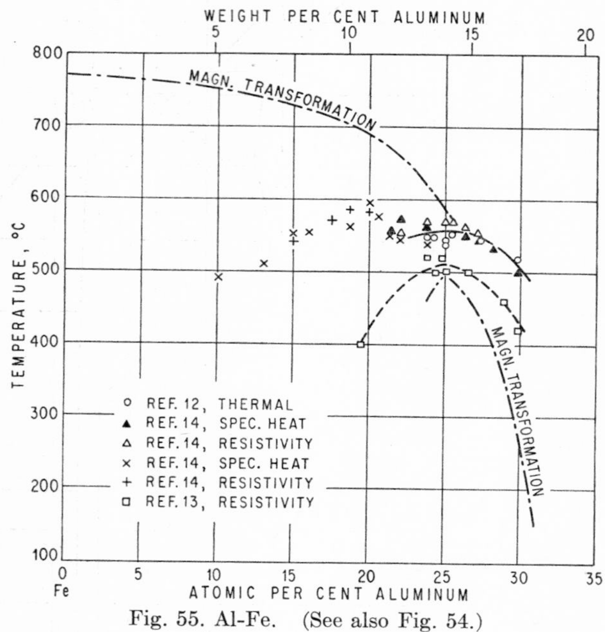

Fig. 55. Al-Fe. (See also Fig. 54.)

are shown in Fig. 55 by means of different data points. Figure 55 also contains the curves of the Curie points of the disordered α phase and the ordered $FeAl_3$ structure based on the fairly closely agreeing results of [12, 15, 13, 16]. Curie temperatures were also determined by [3, 17, 14].

As shown in Fig. 55, [14] found transition temperatures in the range 10–24 at. % Al which were claimed to belong to a "short-range ordered superstructure of $Fe_{13}Al_3$ (18.75 at. % Al)." This type of structure was found to coexist with the Fe_3Al superstructure in alloys containing more than 21.4 at. % Al. A new phase diagram was proposed for the range 10–30 at. % Al.

Although considerable information as to the order-disorder and magnetic transformations is available, more work is needed. As a consequence, the transformation curves and dashed structure boundaries in Fig. 54, proposed in principle by [18], are to be regarded as tentative. Transition temperatures of the order-disorder transformation $FeAl(\beta_2) \rightarrow \alpha$ have not been determined as yet. For more complete information of the transformations in the range 0–50 at. % Al, the papers by [9–24] should be consulted.

The Range 54–77 At. % Al. The solidification equilibria suggested by [1–4] are greatly at variance. For detailed information see the partial diagrams presented in [25]. In contrast to [2, 3], who found only one phase of variable composition between 68 and 76 at. % Al, [4, 26, 27] gave evidence of the existence of two phases in this region, $Fe_2Al_5 = \eta$ and $FeAl_3 = \theta$. In addition to the ϵ phase which forms peritectically at 1230 [2], 1207 [3], 1232 [4], or 1210°C [26] and decomposes eutectoidally at about 1100 [2], 1080 [3, 26], or 1103°C [4], another phase between 65 and 70 at. % Al was detected by [4]. This phase, ζ, was shown to be formed by the peritectoid reaction $\epsilon + \eta \rightarrow \zeta$ at 1158°C, i.e., only about 7°C below the $\epsilon + \eta$ eutectic [4]. Because of the incompleteness of this reaction on relatively rapid cooling and insufficient annealing, no single-phase alloy could be obtained in this range. [4, 28] assumed this phase to contain about 65 at. % Al; however, [25, 26] suggested it to be somewhat richer in Al and to correspond to the composition $FeAl_2$. Confirmatory X-ray evidence was given by [27].

The phase relationships shown in Fig. 54 are based mainly on the work of [4]. However, instead of a liquidus maximum at $FeAl_3$, 1161°C and a eutectic $FeAl_3$-Fe_2Al_5 at about 74 at. (58 wt.) % Al, 1159°C, [7] suggested the peritectic reaction: melt + $Fe_2Al_5 \rightleftarrows FeAl_3$. The thermal data of [2] and the microstructure of the alloys with 72.8 at. % Al [7] and 73.7 at. % Al [4], showing peritectic rather than eutectic structures, are in favor of this generally accepted revision.

The intermediate phase richest in Al has been widely accepted to correspond to the composition $FeAl_3$ [1, 3, 4, 26]. [27] claimed to have found strong X-ray evidence that $FeAl_3 = \theta$ decomposed at some temperature below 600°C to give a mixture of Fe_2Al_5 and Fe_2Al_7 (77.78 at. % Al), the latter being stable between about 77.5 and 78.6 at. (62.5 and 64 wt.) % Al. The powder pattern of Fe_2Al_7 was reported to differ only slightly from that of $FeAl_3$, the difference being clearly defined as second-order effects [27]. According to [18], no indication of a thermal effect was found on cooling curves, however. These results could not be reproduced by recent careful X-ray studies [29]. There were no differences in the powder patterns of specimens (including a single crystal) containing 76.4–76.8 at. % Al, quenched from 900°C or subjected to a long-time anneal at 470°C. The conclusion that $FeAl_3$ does not decompose was also drawn by [30]. On the other hand, there is some indication that the composition of the intermediate phase coexisting with the Al-rich solid solution is lower in Fe than previously assumed: [31], who analyzed primary crystals extracted from alloys, found only compositions near Fe_2Al_7. Additional work is necessary to clarify whether these and other findings [32, 33] indicate that the phase "$FeAl_3$" is actually richer in Al or has a wider homogeneity range. Until then the formula $FeAl_3$ should be retained for convenient reference.

The boundaries of the $Fe_2Al_5 = \eta$ phase given by various authors [4, 26, 27, 34] differ somewhat, ranging from 53–55.5 wt. % Al on the Fe side to 55.5–57 wt. % Al on the Al side. As a compromise, 53–56 wt. (70–72.5 at.) % Al was accepted in Fig. 54.

The Range 77–100 At. % Al. The liquidus curve is based on the closely agreeing thermal data of [2, 4, 35]. The eutectic temperature was given as 646–649 [1], 648 [36, 3], 652 [2], 650 [37], 655 [38], 653 [4], and 654°C [35, 39] and the eutectic composition in wt. % Fe as 2.0 [36], 2.5 [37], about 1.7 [38], about 2.2 [40], 1.9 [4], and 2.0 [35]. There appears to be little doubt that these differences are mainly caused by the use of Al of different purity. Also, the tendency of the melt to undercool might have affected the results. The most reliable data indicate the eutectic point to be located at 1.8 ± 0.1 wt. (0.9 ± 0.05 at.) % Fe and 654–655°C.

Results as to the solid solubility of Fe in Al were chiefly reported by [38, 41, 42]. Those of [42], based on lattice-spacing measurements, are most reliable and complete: 0.052 wt. % Fe at 655°C, 0.025% at 600°C, 0.006% at 500°C, and practically nil at 400–450°C (at. % = ½ wt. % Fe). With extremely high cooling rates on solidification, supersaturated solutions containing as much as 0.17 wt. % Fe can be obtained [43].

Crystal Structures. Lattice parameters of alloys in the range 0–50 at. % Al were reported by [44, 26, 10], those of [10] being most reliable. The structural changes connected with the disorder → order transformations $\alpha \rightarrow Fe_3Al(=\beta_1)$ and $\alpha \rightarrow FeAl(=\beta_2)$ were elucidated by [9, 10]. The Fe_3Al structure is of the BiF_3 ($D0_3$) type, with a lattice constant twice that of the α phase, and the FeAl structure is isotypic with the CsCl (B2) structure. Additional information is given in a preceding paragraph.

The ϵ phase was suggested to be b.c.c. with 16 atoms per unit cell and the ζ phase ($FeAl_2$) to be rhombohedral with 18 atoms per unit cell [26, 45]. The structure of the latter phase was claimed to be more complicated [27].

The η phase (Fe_2Al_5), first believed to be monoclinic with 56 atoms per unit cell [26, 45], was reported to be orthorhombic, with $a = 7.67_5$ A, $b = 6.40_3$ A, $c = 4.20_3$ A for 72.0 ± 5 at. % Al [46].

The θ phase ($FeAl_3$) was previously believed to be orthorhombic [26, 45, 47, 48]. A complete structure analysis, using specimens with about 76.5 at. % Al, revealed the structure to be monoclinic with 100 atoms per unit cell and $a = 15.489$ A, $b = 8.083$ A, $c = 12.476$ A, $\beta = 107°43' \pm 1'$ [49].

The existence of an unstable phase Fe_2Al, claimed to be present in a complex Fe-Ni-Cr alloy, was reported by [50]. It was stated to be isotypic with $MgZn_2$.

1. A. G. C. Gwyer, *Z. anorg. Chem.*, **57,** 1908, 126–133.
2. N. S. Kurnakow, G. Urasow, and A. Grigorjew, *Z. anorg. Chem.*, **125,** 1922, 207–227; *Izvest. Inst. Fiz.-Khim. Anal.*, **1,** 1919, 15.
3. M. Isawa and T. Murakami, *Kinzoku-no-Kenkyu*, **4,** 1927, 467–477.
4. A. G. C. Gwyer and H. W. L. Phillips, *J. Inst. Metals*, **38,** 1927, 35–44, 83.
5. F. Wever and A. Müller, *Mitt. Kaiser-Wilhelm-Inst. Eisenforsch. Düsseldorf*, **11,** 1928, 220–223; *Z. anorg. Chem.*, **192,** 1930, 340–345.
6. W. Oelsen, *Stahl u. Eisen*, **69,** 1949, 468–474.
7. N. W. Ageew and O. I. Vher, *J. Inst. Metals*, **44,** 1930, 84–85.
8. W. Ekman, *Z. physik. Chem.*, **B12,** 1931, 57–78; see also A. Westgren, *Metallwirtschaft*, **9,** 1930, 923; and *Z. Metallkunde*, **22,** 1930, 372.
9. A. J. Bradley and A. H. Jay, *Proc. Roy. Soc. (London)*, **A136,** 1932, 210–232.
10. A. J. Bradley and A. H. Jay, *J. Iron Steel Inst.*, **125,** 1932, 339–357.
11. C. Sykes and H. Evans, *Proc. Roy. Soc. (London)*, **A145,** 1934, 529–539.
12. C. Sykes and H. Evans, *J. Iron Steel Inst.*, **131,** 1935, 225–247.
13. A. T. Grigoriev and N. M. Gruzdeva, *Izvest. Sektora Fiz.-Khim. Anal., Inst. Obshchei Neorg. Akad. Nauk S.S.S.R.*, **14,** 1941, 245–253.
14. H. Saito, *Nippon Kinzoku Gakkai-Shi*, **14**(5), 1950, 1–6, 6–11.
15. W. Sucksmith, *Proc. Roy. Soc. (London)*, **A171,** 1939, 525–540.
16. W. D. Bennett, *J. Iron Steel Inst.*, **171,** 1952, 372–380.
17. M. Fallot, *Ann. phys.*, **6,** 1936, 356–365.
18. A. J. Bradley and A. Taylor, *Proc. Roy. Soc. (London)*, **A166,** 1938, 353–375.
19. W. L. Bragg, *Nature*, **131,** 1933, 751.
20. K. Schäfer, *Naturwissenschaften*, **21,** 1933, 207.
21. C. Sykes and J. W. Bampfylde, *J. Iron Steel Inst.*, **130,** 1934, 389–410.
22. W. Hume-Rothery and H. M. Powell, *Z. Krist.*, **91,** 1935, 23–47.

23. S. Matsuda, *J. Phys. Soc. Japan*, **6,** 1951, 131–135.
24. H. Sato, *Science Repts. Research Insts. Tôhoku Univ.*, **A3,** 1951, 13–23.
25. M. Hansen, "Der Aufbau der Zweistofflegierungen," pp. 108–115, Springer-Verlag OHG, Berlin, 1936.
26. A. Osawa, *Science Repts. Tôhoku Univ.*, **22,** 1933, 803–819.
27. A. J. Bradley and A. Taylor, *J. Inst. Metals*, **66,** 1940, 53–65.
28. A. G. C. Gwyer, H. W. L. Phillips, and L. Mann, *J. Inst. Metals*, **40,** 1928, 302, 358.
29. P. J. Black, *Acta Cryst.*, **8,** 1955, 175–182.
30. G. V. Raynor, C. R. Faulkner, J. D. Noden, and A. R. Harding, *Acta Met.*, **1,** 1953, 629–635.
31. M. Armand, *Compt. rend.*, **235,** 1952, 1506–1508.
32. G. V. Raynor and P. C. L. Pfeil, *J. Inst. Metals*, **73,** 1946-1947, 397–419.
33. H. Nowotny, K. Komarek, and J. Kromer, *Berg. u. Hüttenmänn. Monatsh.*, **96,** 1951, 161–169.
34. E. Gebhardt and W. Obrowski, *Z. Metallkunde*, **44,** 1953, 158–159.
35. R. S. Archer and W. L. Fink, *J. Inst. Metals*, **40,** 1928, 356–357.
36. W. Rosenhain, S. L. Archbutt, and D. Hanson, 11th Report Alloys Research Committee, *Inst. Mech. Engrs. (London)*, **1921,** 211–212; *Z. Metallkunde*, **18,** 1926, 65.
37. E. Wetzel, *Metallbörse*, **13,** 1923, 738.
38. E. H. Dix, *Proc. Am. Soc. Testing Materials*, **25**(2), 1925, 120–129.
39. E. Degischer, *Aluminium-Arch.*, no. 18, 1939, p. 12; H. Hanemann and A. Schrader, *Z. Metallkunde*, **31,** 1939, 183; *Aluminium*, **21,** 1939, 382.
40. G. Masing and O. Dahl, *Wiss. Veröffentl. Siemens-Konzern*, **5,** 1926, 152–159; *Z. anorg. Chem.*, **154,** 1926, 189–196.
41. A. Roth, *Z. Metallkunde*, **31,** 1939, 299–301.
42. J. K. Edgar, *Trans. AIME*, **180,** 1949, 225–229.
43. G. Falkenhagen and W. Hofmann, *Z. Metallkunde*, **43,** 1952, 69–81.
44. Z. Nishiyama, *Science Repts. Tôhoku Univ.*, **18,** 1929, 381–387.
45. A. Osawa, *Kinzoku-no-Kenkyu*, **10,** 1933, 432–445; *Metals & Alloys*, **5,** 1934, MA 154.
46. K. Schubert and coworkers, *Naturwissenschaften*, **40,** 1953, 437; see also K. Schubert and M. Kluge, *Z. Naturforsch.*, **8a,** 1953, 755–776.
47. E. Bachmetew, *Z. Krist.*, **89,** 1934, 575–586; **88,** 1934, 179–181.
48. G. Phragmén, *J. Inst. Metals*, **77,** 1950, 489.
49. P. J. Black, *Acta Cryst.*, **8,** 1955, 43–48, 175–182.
50. H. J. Beattie and F. L. VerSnyder, *Trans. ASM*, **45,** 1953, 397–423.

$\bar{1}.5877$
0.4123

Al-Ga Aluminum-Gallium

Two investigations of the system by means of thermal analysis covering the range up to 95 and 80 wt. (about 88 and 61 at.) % Ga, respectively [1, 2], yielded entirely different phase diagrams. Whereas [1] asserted that he found three intermediate phases (Al_2Ga, AlGa, $AlGa_2$), the work by [2], corroborated by X-ray analysis, showed that no compounds exist (Fig. 56). Later, [3] admitted that severe experimental errors had been made and that the diagram by [2] was correct.

The solidus temperature is somewhat uncertain; the temperature data vary between 26 [2] and 27 or 29°C [1, 3]. The composition of the eutectic is unknown but will lie very close to 100% Ga.

It has been concluded from thermal-analysis data and microscopic investigation that about 13 wt. (5.5 at.) % Ga is soluble in solid aluminum at the eutectic temperature [2]. However, [1] reported that the solid solubility, if any, is small. Also, electrical-conductivity measurements indicated the absence of a noticeable solubility [4]. Al-rich alloys proved not age-hardenable [5]. The lattice parameter of an alloy with 0.53 at. % Ga was found to be practically identical with that of pure Al [6].

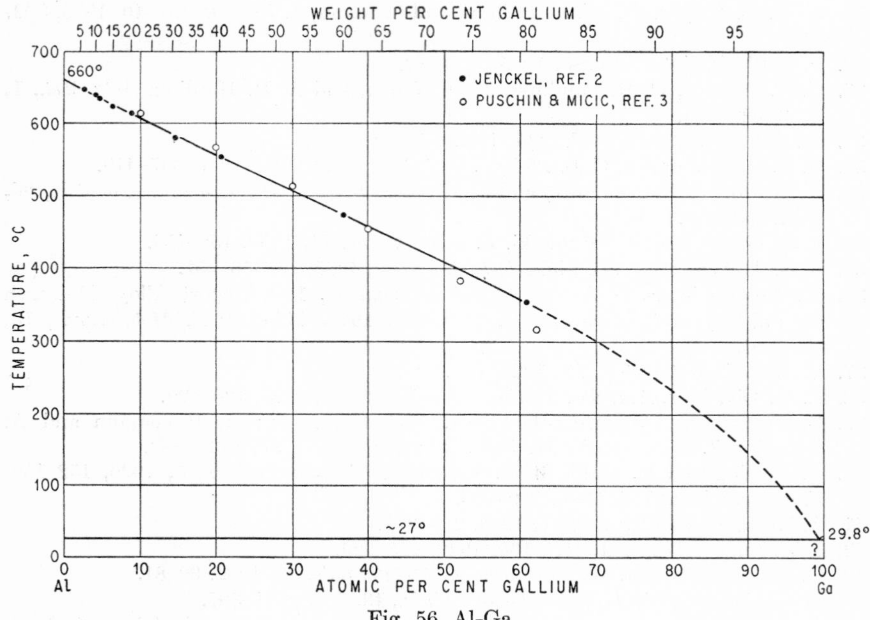

Fig. 56. Al-Ga

1. N. A. Puschin and V. Stajic, *Z. anorg. Chem.*, **216,** 1933, 26–28; see also N. A. Puschin, S. Stepanovic, and V. Stajic, *Z. anorg. Chem.*, **209,** 1932, 329–334.
2. E. Jenckel, *Z. Metallkunde*, **26,** 1934, 249–250.
3. N. A. Puschin and O. D. Micic, *Z. anorg. Chem.*, **234,** 1937, 233–234.
4. H. Röhrig, *Z. Metallkunde*, **26,** 1934, 250–251.
5. W. Kroll, *Metallwirtschaft*, **11,** 1932, 436.
6. E. A. Owen, Y. H. Liu, and D. P. Morris, *Phil. Mag.*, **39,** 1948, 831–845.

$\bar{1}.5701$
0.4299

Al-Ge Aluminum-Germanium

Thermal analysis was used to establish the phase relationships, covering the composition range up to 36 at. (60 wt.) % Ge [1] and 100% Ge [2] (Fig. 57). The eutectic was found to lie at 55 wt. (31.2 at.) % Ge, 423°C, and 29.5 at. (53.0 wt.) % Ge, 424°C, respectively. By means of qualitative X-ray investigations after annealing for several weeks and months, the solid solubility of Ge in Al was determined to be 1.7–2.05 at. (4.4–5.3 wt.) % at 395°C, 0.5–0.6 at. (1.4–1.65 wt.) % at 294°C, and approximately 0.2 at. (0.5 wt.) % Ge at 177°C [2]. At 424°C, the solubility was estimated at 2.8 ± 0.2 at. (7.2 ± 0.5 wt.) % Ge. Ge dissolves less than 3–4 at. % Al (estimated) [2].

For the lattice spacing of the solid solutions up to 1.98 at. (5.15 wt.) % Ge, see

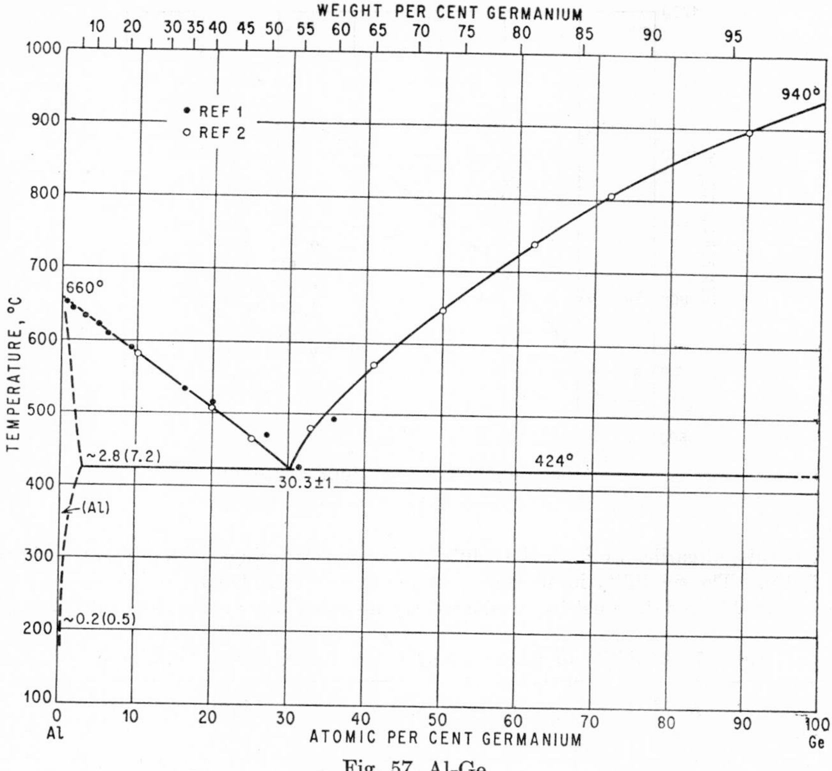

Fig. 57. Al-Ge

[3]. These data indicate that at least 2 at. % Ge is soluble in Al at approximately 500°C.

[4] has shown that the solubility of Ge in liquid Al can be approximated by the laws of strictly regular solutions.

1. W. Kroll, *Metall u. Erz*, **23,** 1926, 682–684; Al containing 0.25% Si, 0.45% Fe was used.
2. H. Stöhr and W. Klemm, *Z. anorg. Chem.*, **241,** 1939, 305–313; high-purity Al was used.
3. H. J. Axon and W. Hume-Rothery, *Proc. Roy. Soc.* (*London*), **A193,** 1948, 1–24.
4. C. D. Thurmond, *J. Phys. Chem.*, **57,** 1953, 827–830.

1.4276
$\bar{2}$.5724

Al-H Aluminum-Hydrogen

The literature on the solubility of hydrogen in liquid and/or solid aluminum has been reviewed in numerous publications, of which those by [1–7] are the most important ones. The most accurate data on the solubility in liquid Al are due to [8]; see Table 7. Results by [9, 10] are in good agreement with these findings, whereas work by [1, 11–14] resulted in considerably lower solubilities at temperatures close to the melting point of Al (Fig. 58). See also [16].

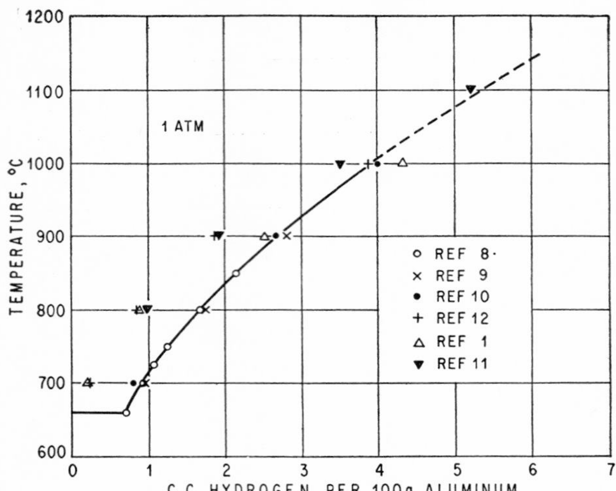

Fig. 58. Al-H. Solubility isobar (1 atm).

Results of earlier work on the solubility in solid Al were considerably at variance [4, 5, 15]. The solubility in dependence upon temperature has been determined only once [8]. These data must be considered the most reliable at present (Table 7).

Table 7. Solubility of Hydrogen at 1 Atm in 99.9985% Aluminum [8]

Solid state			Liquid state		
Temperature, °C	cc/100 g	At. %	Temperature, °C	cc/100 g	At. %
300	0.001	2.2×10^{-7}	660	0.69	1.55×10^{-4}
400	0.005	1.1×10^{-6}	700	0.92	2.07×10^{-4}
500	0.0125	2.8×10^{-6}	725	1.07	2.40×10^{-4}
600	0.026	5.85×10^{-6}	750	1.23	2.77×10^{-4}
660	0.036	8.1×10^{-6}	800	1.67	3.76×10^{-4}
			850	2.15	4.84×10^{-4}

1. P. Röntgen and H. Braun, *Metallwirtschaft*, **11**, 1932, 459–463.
2. P. E. Chrétien, H. A. Nipper, and E. Piwowarsky, *Aluminum-Arch.*, **23**, 1939.
3. L. W. Eastwood, "Gas in Light Alloys," pp. 19–23, John Wiley & Sons, Inc., New York, 1946.
4. D. P. Smith, "Hydrogen in Metals," pp. 283–285, University of Chicago Press, Chicago, 1948.
5. Y. Dardel, *Trans. AIME*, **180**, 1949, 273–286; see also *Trans. AIME*, **175**, 1948, 497–516.
6. R. Castro and M. Armand, *Rev. mét.*, **46**, 1949, 594–615.
7. H. Kostron, *Z. Metallkunde*, **43**, 1952, 269–284, 373–387.
8. C. E. Ransley and H. Neufeld, *J. Inst. Metals*, **74**, 1948, 599–620; see also pp. 781–786.

9. W. Baukloh and F. Oesterlen, *Z. Metallkunde*, **30**, 1938, 386–389.
10. W. R. Opie and N. J. Grant, *Trans. AIME*, **188**, 1950, 1237–1247.
11. P. Röntgen and F. Möller, *Metallwirtschaft*, **12**, 1934, 81–83, 97–100.
12. L. L. Bircumshaw, *Trans. Faraday Soc.*, **31**, 1935, 1439–1443; *Phil. Mag.*, **1**, 1926, 510–522.
13. P. Röntgen and H. Winterhager, *Aluminium*, **21**, 1939, 210–213; H. Winterhager, *Aluminium-Arch.*, **12**, 1938.
14. W. Baukloh and M. Redjali, *Metallwirtschaft*, **21**, 1942, 683–688.
15. R. Eborall and C. E. Ransley, *J. Inst. Metals*, **71**, 1945, 525–552.
16. W. Hofmann and J. Maatsch, *Z. Metallkunde*, **47**, 1956, 89–95.

$\bar{1}.1287$
0.8713

Al-Hg Aluminum-Mercury

The liquidus curve (Fig. 59) was determined by [1] [thermal analysis, 0–95.45 at. (0–99.4 wt.) % Hg] and [2] [chemical analysis, 17.3–92.5 at. (60.8–98.95 wt.) % Hg]. The findings are in excellent agreement. In addition, [3] determined the liquidus curve (solubility of Al in liquid Hg) between 312 and 76°C by chemical analysis and found the following values in wt. % Hg: 312°C, 0.18; 260°C, 0.11; 160°C, 0.035; 125°C, 0.024; 102°C, 0.016; and 76°C, 0.009. They agree substantially with those reported by [4]: 300°C, 0.17; 200°C, 0.069; 150°C, 0.034; 100°C, 0.012; and 60°C, 0.006. The fact that the eutectic temperature is practically identical with the melting point of Hg [1] is in accord with the extremely low solubility of Al in Hg at room temperature, given as about 0.002 [5] and 0.0023 wt. % (extrapolated) [3]; see also [6–8].

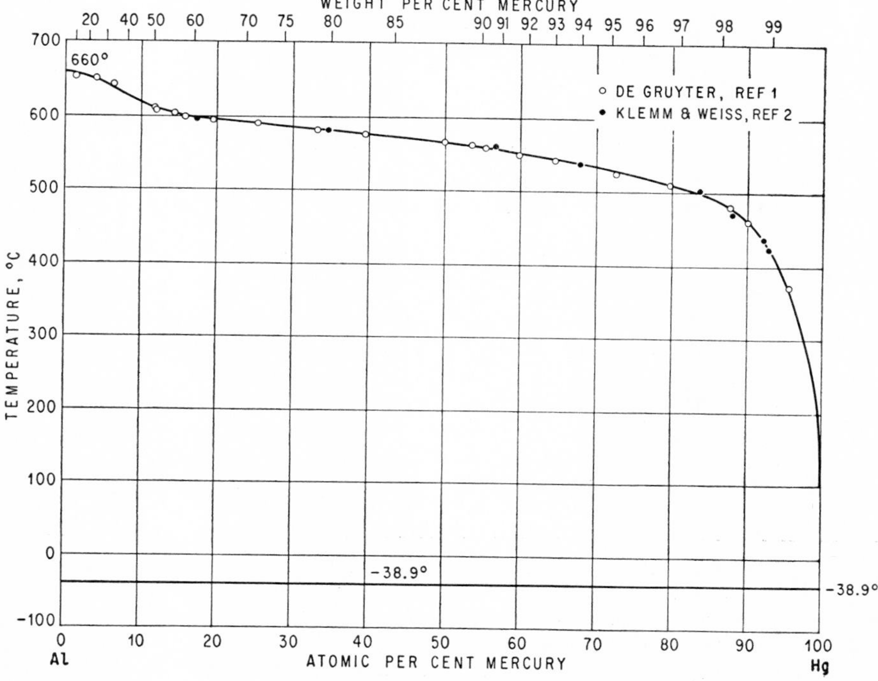

Fig. 59. Al-Hg

Data as to the solid solubility of Hg in Al at room temperature differ widely, since equilibrium is not easily attained and complete separation of the liquid and solid phases is difficult to achieve. [5, 9, 8] reported that the solubility is practically nil. On the other hand, [10] (emf measurements) and [11] (separation of phases) concluded that 0.78 at. (5.5 wt.) % Hg and even 8 at. (39.3 wt.) % Hg, respectively, are soluble in aluminum at room temperature.

Concerning measurements of the emf of aluminum amalgams, the reader must be referred to the literature [12, 13, 10, 9, 8].

1. C. J. de Gruyter, *Rec. trav. chim.*, **44,** 1925, 937–969; A. Smits and C. J. de Gruyter, *Proc. Koninkl. Akad. Wetenschap., Amsterdam*, **23,** 1921-1922, 966–968; *Afd. Natuurk. Akad. Wetenschap., Amsterdam*, **29,** 1920, 747–749. See also A. Smits, *Z. Elektrochem.*, **30,** 1924, 424.
2. W. Klemm and P. Weiss, *Z. anorg. Chem.*, **245,** 1940, 285–287.
3. H. A. Liebhafsky, *J. Am. Chem. Soc.*, **71,** 1949, 1468–1470.
4. W. Schmidt, *Metall*, **3,** 1949, 10–13.
5. I. Fogh, *Kgl. Danske Videnskab. Selskab. Mat.-fys. Medd.*, **3,** 1921, no. 15, p. 6.
6. W. J. Humphreys, *J. Chem. Soc.*, **69,** 1896, 1679–1691.
7. F. Mylius and F. Rose, *Z. Instrumentenk.*, **13,** 1893, 81. Solubility at room temperature is given as "at least 0.0011 wt. % Al."
8. R. Müller, *Z. Elektrochem.*, **35,** 1929, 240–249. Solubility at room temperature is given as <0.027 wt. % Al.
9. A. Dadieu, *Monatsh. Chem.*, **47,** 1926, 497–510.
10. A. Smits and H. Gerding, *Z. Elektrochem.*, **31,** 1925, 304–308; H. Gerding, *Z. physik. Chem.*, **A151,** 1930, 190–218.
11. C. J. de Gruyter, *Rec. trav. chim.*, **44,** 1925, 937–969.
12. R. Kremann and R. Müller, *Z. Metallkunde*, **12,** 1920, 289–303.
13. A. Smits, *Z. Elektrochem.*, **30,** 1924, 423–435.

$\bar{1}.3712$
0.6288

Al-In Aluminum-Indium

Within a period of 6 years, the solidification diagram (Fig. 60) has been determined six times [1–6]. Results, summarized in Table 8, are in substantial agreement.

Table 8. Characteristic Temperatures in °C and Approximate Compositions in At. % In (Wt. % in Parentheses)

	Ref. 1	Ref. 2	Ref. 3	Ref. 4	Ref. 5	Ref. 6
Monotectic temperature	634°C	635°C	640°C	638°C	....	639°C
Monotectic Al-rich melt..	4.7	∼3.5	∼6	∼4	∼3	
	(17.5)	(∼13)	(∼21)	(∼15)	(∼12)	
Monotectic In-rich melt..	>90	∼90	∼94	.....	∼88.5	89
	(>97.5)	(∼98)	(∼98.5)	.....	(∼97)	(97.2)
Solidus temperature.....	156°C	156°C	∼155°C	156°C	(∼155°C)	

The solubility of Al in liquid In was determined by chemical analysis of samples from melts, equilibrated at various temperatures [6] and found as follows, in wt. (at.) % Al: 200°C, <0.05 (<0.2); 300°C, <0.05 (<0.2); 400°C, 0.24 (1.01); 500°C, 0.64 (2.68); 550°C, 0.93 (3.85); 600°C, 1.68 (6.77); 625°C, 2.18 (8.65). Also, the diagram published by [4] indicates that the solubility of Al in liquid In was determined by these workers; however, no tabulated data are given.

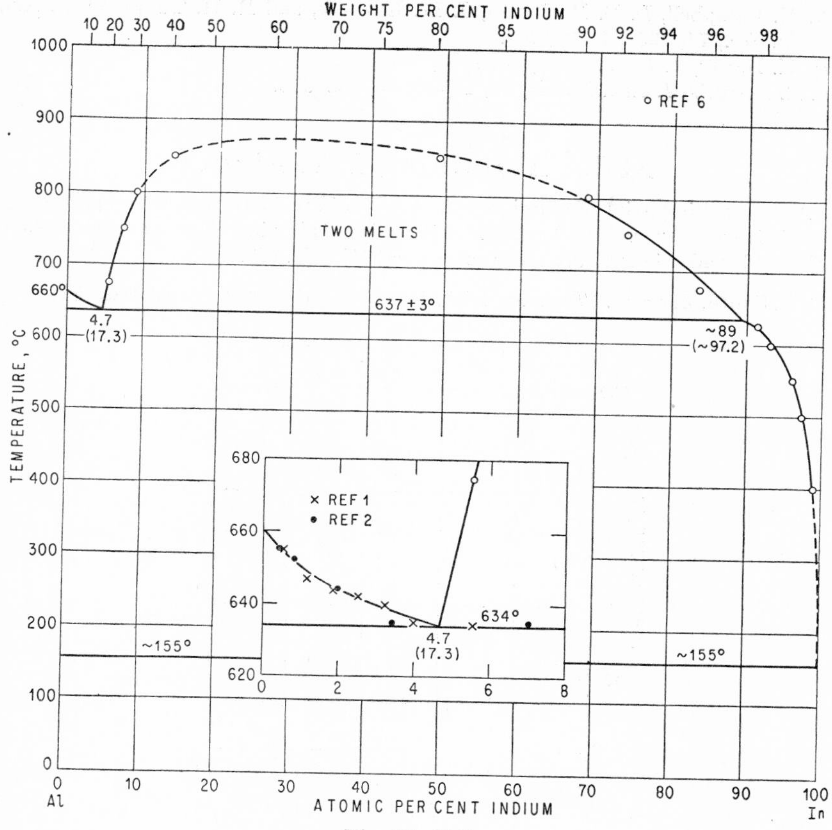

Fig. 60. Al-In

The boundary of the miscibility gap in the liquid state was established by the direct-sampling method [6]. Figures (see data points in Fig. 60) were reported to be reliable up to 800°C. The critical point was found by the rectilinear diameter relationship to be located at approximately 61 wt. (27 at.) % In, 875°C [6].

The solid solubility of In in Al, determined micrographically, was given as about 0.06, 0.07, 0.11 wt. (0.014–0.026 at.) % In at 560°C, 590°C, and the monotectic temperature [7].

Note Added in Proof. [8] found the solubility of In in Al to be 0.085 wt. (0.020 at.) % at 560°C and 0.17_5 wt. (0.04_1 at.) % In at the monotectic temperature; he gave reasons why the data of [7] are considered as incorrect.

1. E. Raub and M. Engel, *Z. Metallkunde*, **37,** 1946, 148–149.
2. S. Valentiner and I. Puzicha, *Z. Metallkunde*, **38,** 1947, 127–128.
3. W. Klemm and H. A. Klein, *Z. anorg. Chem.*, **256,** 1948, 240–241.
4. H. M. Davis and G. H. Rowe, unpublished work; see "Metals Handbook," 1948 ed., p. 1162, The American Society for Metals, Cleveland, Ohio.
5. S. A. Pogodin and I. S. Shumova, *Izvest. Sektora Fiz.-Khim. Anal.*, **17,** 1949, 200–203.

6. A. N. Campbell, L. B. Buchanan, J. M. Kuzmak, and R. H. Tuxworth, *J. Am. Chem. Soc.*, **74,** 1952, 1962–1966.
7. H. K. Hardy, *J. Inst. Metals*, **80,** 1952, 431–434.
8. L. E. Samuels, *J. Inst. Metals*, **84,** 1955-1956, 333–336.

$\bar{1}.8389$
0.1611

Al-K Aluminum-Potassium

At the melting point of aluminum the two liquid metals appear to be practically insoluble in each other. Neither the melting point of aluminum nor that of potassium is affected by additions of the other component [1] (Fig. 61).

1. D. P. Smith, *Z. anorg. Chem.*, **56,** 1908, 112–113. Al of 99.4% was used.

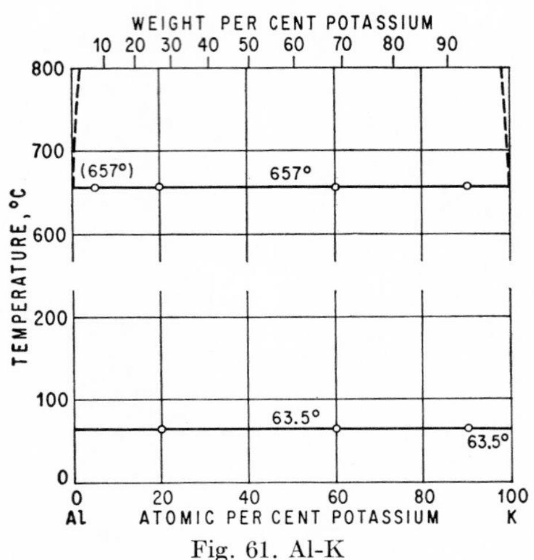

Fig. 61. Al-K

$\bar{1}.2883$
0.7117

Al-La Aluminum-Lanthanum

By thermal analysis and microscopic examination a phase diagram was obtained [1] characterized by the existence of the three intermediate phases $LaAl_4$ (56.29 wt. % La), formerly isolated from an Al-rich alloy [2]; $LaAl_2$ (73.03 wt. % La); and LaAl (83.74 wt. % La) (Fig. 62). The Al-rich part of the system, not considered by this work, was the subject of a study involving alloys up to 10 wt. (2.1 at.) % La, using thermal-analysis, thermoresistometric, and parametric investigations, and La of 99.86% purity [3]. Because of the spread in data points (see inset of Fig. 62), the eutectic point could not be located accurately but probably lies in the neighborhood of 2.5 at. (11.5 wt.) % La at 640–642°C. Verifying this, the eutectic temperature was found to be 643°C [4]. No solid solubility of La in Al could be detected [3, 4].

Checking the constitution of the La-rich alloys, using La containing 0.55% Fe, 1.02% Mg, and 0.05% Si, it was found [5] that an additional compound, La_3Al_2 (88.54 wt. % La), exists which is formed peritectically at 700°C (Fig. 62). Whereas the earlier work [1] showed a eutectic LaAl-La to exist at 80 at. (95.4 wt.) % La, 542°C (see data points), it is now claimed that the La_3Al_2-La eutectic is located at 627°C; the eutectic composition was not determined [5]. It is difficult to understand how one author found eutectic arrests at 542°C in taking cooling curves of 22 alloys, whereas another was unable to detect this thermal effect.

$LaAl_4$ is b.c. tetragonal of the $D1_3$ type and isomorphous with $CeAl_4$, $BaAl_4$, and others; a = 4.42 A, c = 10.2 A, c/a = 2.31 [4], or a = 4.4 A, c = 10.2 A [6]. $LaAl_2$ has a structure of the $MgCu_2$ (C15) type, a = 8.18 A [7], a = 8.17 A [4], or a = 8.13 A [8].

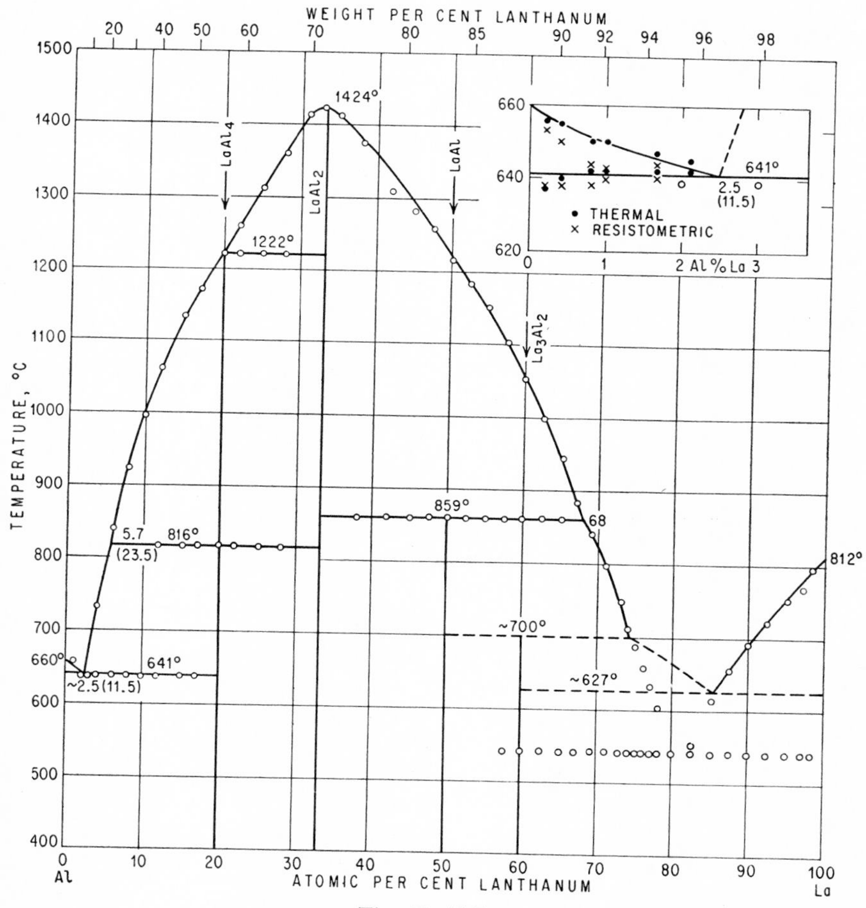

Fig. 62. Al-La

1. G. Canneri, *Metallurgia ital.*, **24**, 1932, 3–7. The lanthanum was 99.6% pure.
2. W. Muthmann and H. Beck, *Liebigs Ann.*, **331**, 1904, 46–57.
3. F. Weibke and W. Schmidt, *Z. Elektrochem.*, **46**, 1940, 357–364.
4. H. Nowotny, *Z. Metallkunde*, **34**, 1942, 22–24.
5. R. Vogel and T. Heumann, *Z. Metallkunde*, **35**, 1942, 29–42.
6. A. Rossi, *Atti reale accad. Lincei*, **17**, 1933, 182–185.
7. H. Nowotny, *Naturwissenschaften*, **29**, 1941, 654.
8. G. E. R. Schulze, *Z. Krist.*, **A104**, 1942, 257–260.

0.5897
$\bar{1}$.4103

Al-Li Aluminum-Lithium

The study of the constitution of the Al-rich alloys up to about 30 at. % Li [1], 35 at. % Li [2], and 46 at. % Li [3] by thermal and microscopic analysis showed that a eutectic exists between an Al-rich solid solution and a compound, Li_2Al_3 [2] or more likely LiAl (20.46 wt. % Li) [3]. The eutectic was reported to lie at about 22.5 at. (7 wt.) % Li, 590°C [1]; 24.75 at. (7.8 wt.) % Li, 598°C [2]; and 23.5 at. (7.3 wt.) % Li, 590°C [3].

The principal features of the first comprehensive phase diagram, determined by thermal analysis [4], are (*a*) a eutectic (α + LiAl) at 30 at. (9.9 wt.) % Li, 600°C; (*b*) the decomposition (on heating) of the phase LiAl at 698°C into two melts with about 44.5 at. (17 wt.) % and 60 at. (28 wt.) % Li; (*c*) the peritectic formation of the phase

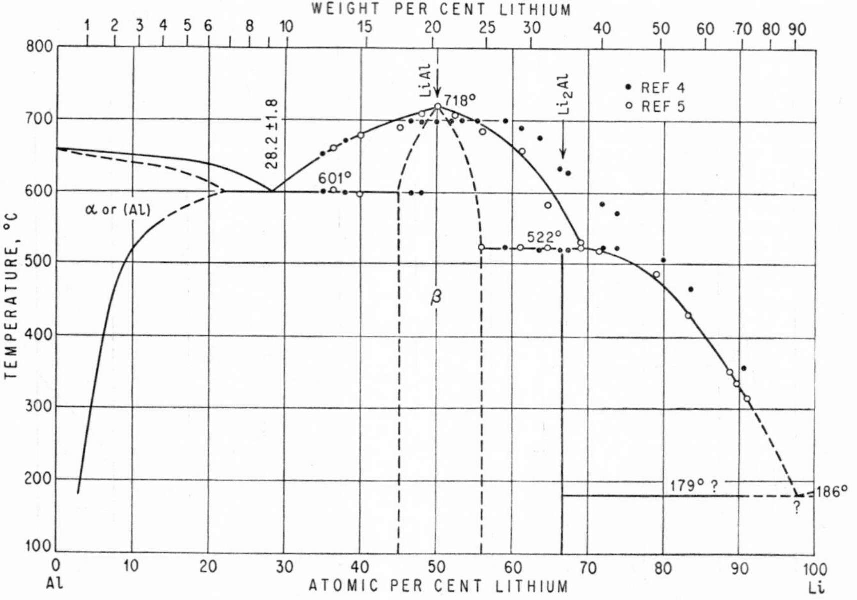

Fig. 63. Al-Li. (See Note Added in Proof.)

Li_2Al (33.97 wt. % Li) at 521°C, with the peritectic melt containing about 78 at. (47.5 wt.) % Li; and (*d*) the Li_2Al-Li eutectic at nearly 100% Li, 179°C.

The diagram suggested by [5] on the basis of thermal and microscopic analysis substantially agrees with that by [4], with the exceptions that (*a*) a miscibility gap in the liquid state could not be confirmed and (*b*) the Li_2Al-Li eutectic was found at 94 at. (80 wt.) % Li, 170.6°C. The α-LiAl eutectic was located at 26.3 at. (8.4 wt.) % Li, 602°C, and the composition of the peritectic melt at 69.5 at. (~37 wt.) % Li, 523°C. The difference in the liquidus curve of the two investigations is indicated in Fig. 63 by the data points between 35 and 90 at. % Li.

The solid solubility of Li in Al was determined by means of thermoresistometric analysis [4], electrical-resistance measurements of quenched alloys [6], and micrographic investigation [5]. The results differ by 4.5 at. % at 300°C and 1.8 at. % at 550°C; however, those by [6] can be regarded as most reliable. They are in at. % Li (wt. % in parentheses): 550°C, 12.7 (3.6); 500°C, 9.1 (2.5); 400°C, 6.3 (1.7); 300°C, 4.5 (1.2); and 200°C, 3.22 (0.85). The maximum solubility at the eutectic temperature was given as about 18 [4], 17.6 [5], and 22 at. % Li [6]. In Fig. 63, 22 at. % Li was adopted as a tentative value.

According to [5], the compound LiAl forms solid solutions (between about 45 and 56 at. % Li) with both Al and Li, whereas [4] suggested that the homogeneity range extends only between 50 and 56 at. % Li. However, in both studies no attempt was made to fix the boundaries accurately.

Crystal Structures. LiAl, first suggested to have a b.c.c. structure of the CsCl (B2) type [7], crystallizes in the b.c.c. NaTl (B32) type of structure with a = 6.37 A [8], a = 6.38 A [9]. For lattice spacing of the α phase see [10].

Note Added in Proof. Micrographic analysis gave the following solid solubilities of Li in Al in at. (wt.) %: 450°C, 5.94 (1.60); 400°C, 3.85 (1.02); 300°C, 1.23 (0.32); 200°C, 0.23 (0.06); 100°C, 0.019 (0.005) [11]. It appears that these values rather than those of [6] represent equilibrium conditions.

1. J. Czochralski and E. Rassow; see J. Czochralski, "Moderne Metallkunde," p. 36, Springer-Verlag OHG, Berlin, 1924.
2. P. Assmann, *Z. Metallkunde*, **18,** 1926, 51–52.
3. A. Müller, *Z. Metallkunde*, **18,** 1926, 231–232.
4. G. Grube, L. Mohr, and W. Breuning, *Z. Elektrochem.*, **41,** 1935, 880–883.
5. F. I. Shamray and P. Ya. Saldau, *Izvest. Akad. Nauk S.S.S.R., Otdel. Khim. Nauk*, **1937,** 631–640.
6. H. Vosskühler, *Metallwirtschaft*, **16,** 1937, 907–909.
7. S. Pastorello, *Gazz. chim. ital.*, **61,** 1931, 47–51.
8. E. Zintl and G. Woltersdorf, *Z. Elektrochem.*, **41,** 1935, 876–879.
9. G. Komovskiy and A. Maximov, *Z. Krist.*, **92,** 1935, 274–283.
10. H. J. Axon and W. Hume-Rothery, *Proc. Roy. Soc.* (*London*), **A193,** 1948, 1–24.
11. S. K. Nowak, *Trans. AIME*, **206,** 1956, 553–556.

0.0451
$\bar{1}$.9549

Al-Mg Aluminum-Magnesium

The literature has been critically reviewed by [1] and [2], who considered the results published up to 1944 and 1951, respectively. The composite diagrams based on these evaluations differ only in minor points. Apart from the phase relationships in the solid state between 40 and 50 at. % Mg, the constitution is well established (Fig. 64).

Most of the results of the early thermal studies [3–6] are obsolete [7]. The essential features of the phase diagram are due to [8]; however, much additional work was necessary to establish the phase boundaries in the present form.

In the following the most important literature is summarized. For a more detailed discussion the reader is referred to the publications of [1, 2]; that of [2] contains a complete bibliography.

The Range 0–37.4 At. (35.0 Wt.) % Mg. For the liquidus curve [8–13] preference has been given to the data of [13], since they were obtained with alloys prepared using high-purity metals. The $\alpha + \beta$ eutectic [4, 5, 8, 14, 9, 10, 13] was placed by [1] at 37.4 at. (35.0 wt.) % Mg, 450 ± 1°C, and by [2] at 37.5 at. (35.1 wt.) % Mg, 450°C, i.e., at a practically identical composition and temperature. [13] found 37.3 at. (34.9 wt.) % Mg, 450°C.

The solidus curve of the α phase was determined by [8, 9, 15, 11, 13]. The curve plotted in Fig. 64 is based on the micrographic work of [13]; these results closely agree with those determined by [15] using the thermoresistometric method.

The solid solubility of Mg in Al was investigated by [16, 8, 7, 14, 17, 18, 19, 20, 9, 21, 15, 11, 22]; however, only the results of [14, 17, 19, 20, 15] have to be considered here (Table 9). The solubility boundaries reported by these workers are based on micrographic [14, 17, 20], lattice-spacing [19, 15], and thermoresistometric studies [15], respectively. The data of [14, 17, 19] are in substantial agreement (within

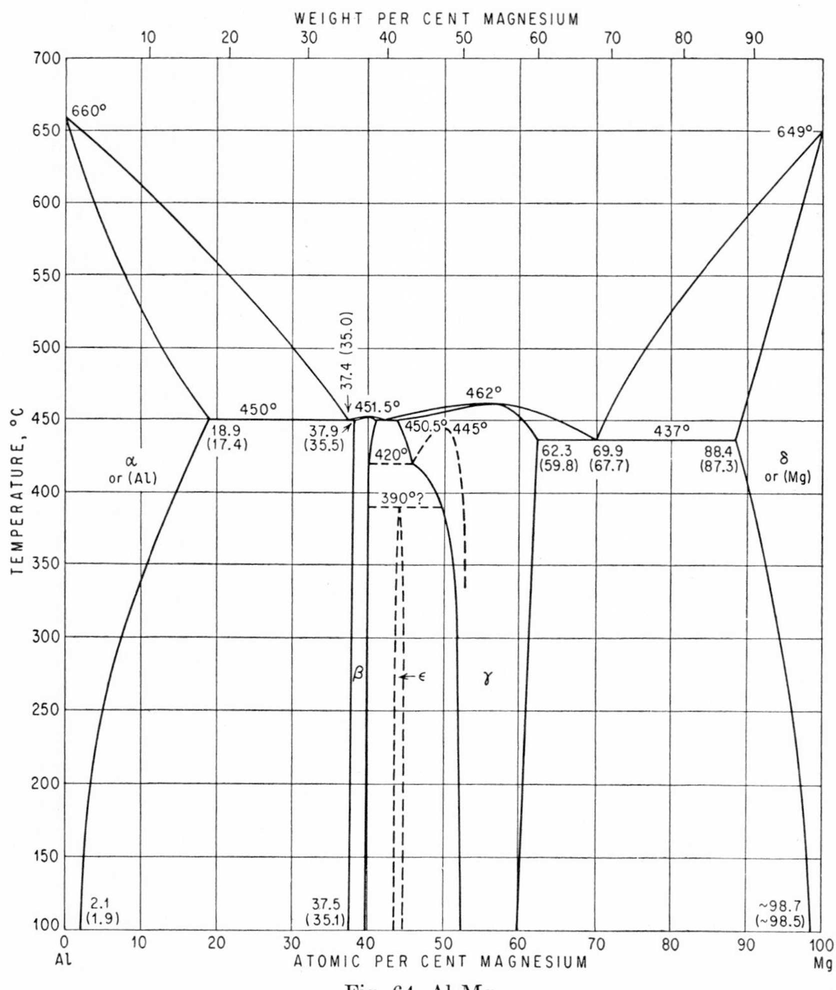

Fig. 64. Al-Mg

about ±0.3 at. %) at temperatures above 250°C; they also agree with those of [20] at 350°C and higher temperatures [23]. In contrast to this, [15] found considerably higher solubilities between 350 and 450°C and lower solubilities below 250°C. They claimed that their data were more reliable, pointing out that the X-ray results of [19] were affected by high quenching stresses. [2] gave weight to the solubility curve of [15] while [1] gave preference to the lower solubilities above 300°C (Table 9). The solubility curve of Fig. 64 was plotted using the data of [15], placing greater reliance on this X-ray investigation [24].

The Range 37.4–69.9 At. (35.0–67.7 Wt.) % Mg. *Solidification.* Whereas [9] assumed two peritectic reactions (which were not experimentally observed) at 449–450 and 454–455°C, [10], in substantial agreement with [8], concluded the existence of a flat maximum at 451.5–452°C and a eutectic at 450–450.5°C. Both [1] and [2] gave preference to the data of [10], since they were obtained with larger melts and at a low cooling rate of 0.5°C per min. Also, there is no evidence for the crystallization of three intermediate phases in this range, as suggested by [9].

The composition of the maximum was placed at 40.2 [1] and 40–41 at. % Mg [2] and that of the eutectic at 41.7 [1] and 42.5 at. % Mg [2]. In view of the small liquidus temperature differences and very narrow solidification intervals, the uncertainty in placing these two compositions is not surprising. The maximum of the γ phase was given as 56.5–57.1 at. %, 462°C [1] and 55–56 at. % Mg, 461°C [2]. The

Table 9. Solid Solubility of Mg in Al in At. % Mg (Wt. % in Parentheses)

°C	Ref. 14	Ref. 17	Ref. 19	Ref. 20	Refs. 15, 2	Ref. 1
450	16.25	~16.35	16.85	16.25	18.9	16.7
	(14.9)	(~15)	(15.35)	(14.9)	(17.4)	(15.35)
400	12.6	12.9	13.25	13.0	14.7	12.9
	(11.5)	(11.8)	(12.05)	(11.9)	(13.5)	(11.8)
350	9.55	10.0	10.05	9.9	10.9	9.8
	(8.7)	(9.1)	(9.05)	(9.0)	(9.9)	(8.9)
300	7.1	7.4	7.05	8.0	7.4	7.1
	(6.4)	(6.7)	(6.25)	(7.25)	(6.7)	(6.4)
250	5.4	~5.2	4.88	6.6	4.9	4.9
	(4.9)	(~4.7)	(4.38)	(6.0)	(4.4)	(4.4)
200	4.4	3.2–4.4	3.80	5.3	3.4	3.4
	(4.0)	(2.9–4.0)	(3.38)	(4.8)	(3.1)	(3.1)
150	.....	....	3.35	......	2.6	2.6
			(2.95)	......	(2.3)	(2.3)
100	.....	....	......	......	2.1	2.1
					(1.9)	(1.9)

$\gamma + \delta$ eutectic was concluded to be located at 69.8–69.9 at. % Mg and 436–437°C [1, 2]. The solidus in the range 37.4–69.9 at. % Mg is based on the analysis of cooling-curve data [10, 25]. They are more consistent than the results of micrographic work [8, 9].

Solid-state Equilibria. There is great disagreement in the literature regarding the phase relationships between 40 and 53 at. % Mg. Those shown in Fig. 64, substantially based on the investigations of [8, 25, 26], were proposed by [1] as the most probable form at the present. With only slight modifications, they were also accepted by [2]. Additional work, preferably by high-temperature X-ray analysis, is necessary to clarify (*a*) the nature of the transformation in the γ phase reaching a maximal temperature of 445°C at 50 at. % Mg and (*b*) the mode of formation of ϵ and its range of stability.

Since the work of [8] the existence of the phases β and γ has been established. Their ranges of homogeneity were studied by [27, 8, 7, 28, 6, 29, 9, 21, 10, 26] and [8, 9, 10, 25, 26, 30], respectively. The boundaries presented in Fig. 64 were selected by [1].

After [29, 31] found indications for the phase relations being more complicated than shown in the diagram of [8], the existence of an additional phase K, located between β and γ, was first reported by [9]. It was claimed to be formed peritectically at about 455°C. [32, 33] believed they confirmed this phase by means of microscopic [32] and X-ray investigations [33]. [32] claimed to have shown that the crystallization of K from the melt could be suppressed, resulting in the formation of a metastable equilibrium. Also, [34], by means of X-ray analysis, interpreted their results on the basis of the diagram of [9] but believed they found evidence that the K phase was either not stable below about 300°C or closely approaching the composition of γ with fall in temperature.

In addition to the K phase, [34] reported the formation below 400°C of another phase designated as ϵ in Fig. 64. The diffraction pattern of this phase was identical with that reported by [33], indicating that the latter was not the K phase. The absence of K was conclusively established by [10, 25, 26], on the basis of extensive thermal, micrographic, and X-ray studies. Neither did [35] find the K phase when studying the diffusion of Mg in Al at 420°C. The existence of ϵ was confirmed, but concluded to be a metastable phase [25]. In the region 41.3–43.8 at. % Mg, [36] found a phase which is probably identical with ϵ. After quenching from 400°C, alloys in this range were single-phase; annealing at a lower temperature (not stated) yielded a two-phase $\beta + \gamma$ structure. One might conclude from this that ϵ is stable only within a restricted temperature range. See also [67].

The nature of the transformation in the γ phase is unknown. [25] suggested it to be of the order-disorder type; however, there are indications that the constitution in this range is different [8, 34, 25]; see the discussion of [1, 2].

The Range 69.9–100 At. % Mg. For the liquidus curve [8, 9, 10, 37] preference has been given to the results of [37]. They are in very good agreement with temperatures extrapolated to 0% Cd from liquidus points of the ternary system Mg-Al-Cd [38]. The accurate micrographic data of [37] were used to plot the solidus curve of the δ phase, which was also determined by [8, 9, 11, 39].

The solid solubility of Al in Mg was the subject of micrographic [8, 40, 41, 42, 39] and X-ray investigations [43, 44, 45]. [37] determined the maximal solubility at 437°C as about 11.6 at. % Al. At the present time, the solubilities selected by [2] are to be regarded as the most probable data: 11.6 (12.7) at 437°C, 9.7 (10.7) at 400°C, 7.4 (8.2) at 350°C, 5.6 (6.2) at 300°C, 3.8 (4.2) at 250°C, 2.6 (2.9) at 200°C, 1.8 (2.0) at 150°C, and about 1.3 at. (1.5 wt.) % Al at 100°C.

The boiling points of alloys with 20, 40, 60, and 80 wt. % Mg were reported as 1300, 1200, 1150, and 1115°C [46]; see also [47].

Crystal Structures. Lattice parameters of the α phase have been reported by [48, 49, 50, 19, 51, 52, 53, 15, 54–62]; those of [19, 15, 55] cover the entire composition range. The data by [61, 62] are to be regarded as most accurate.

The β phase, first claimed to be hexagonal [33] or to be isomorphous with Cu_4Cd_3 [34], was found to have a complex f.c.c. structure with 1,173 atoms per unit cell and a = 28.13 A (?) [63]. The ϵ phase appears to have a complex structure of low symmetry [33, 34, 25]. The powder pattern is simpler than that of γ [36]. The γ phase is isotypic with α-Mn (A12 type) [64, 33, 34, 30]. The structure was suggested to be based on the ideal composition $Mg_{17}Al_{12}$ (58.62 at. % Mg). The lattice constant [33, 30] varies linearly from a = 10.469 A at 51.6 at. % Mg to a = 10.591 A at 61.5 at. % Mg [30]. Alloys with 43–46 at. % Mg, after quenching from the γ field, showed a distorted form of the α-Mn type of structure [34].

For lattice constants of the δ phase [48, 43, 44, 65, 66], see especially the data of [65].

1. G. V. Raynor, Annotated Equilibrium Diagram Series, no. 5, Institute of Metals, London, 1945.
2. K. Eickhoff and H. Vosskühler, *Z. Metallkunde*, **44**, 1953, 223–231.
3. O. Boudouard, *Compt. rend.*, **132**, 1901, 1325–1327.
4. G. Grube, *Z. anorg. Chem.*, **45**, 1905, 225–237.
5. G. Eger, *Intern. Z. Metallog.*, **4**, 1913, 42–46.
6. G. G. Urasow, *Izvest. Inst. Fiz.-Khim. Anal.*, **2**, 1924, 480–481.
7. B. Otani, *Kogyo Kwagaku Zasshi* (*J. Soc. Chem. Ind. Japan*), **25**, 1922, 36–52. The original paper was not available; see *J. Inst. Metals*, **28**, 1922, 643–644.
8. D. Hanson and M. L. V. Gayler, *J. Inst. Metals*, **24**, 1920, 201–227.
9. M. Kawakami, Anniversary Volume Dedicated to K. Honda, *Science Repts. Tôhoku Univ.*, **1936**, 727–747; abstracted in *J. Inst. Metals, Met. Abstr.*, **1**, 1934, 169. For the diagram see [32, 34].
10. N. S. Kurnakov and V. I. Mikheeva, *Izvest. Sektora Fiz.-Khim. Anal.*, **10**, 1938, 37–66.
11. N. S. Kurnakov and V. I. Mikheeva, *Izvest. Sektora Fiz.-Khim. Anal.*, **13**, 1940, 201–208.
12. E. Butchers, G. V. Raynor, and W. Hume-Rothery, *J. Inst. Metals*, **69**, 1943, 210–211.
13. E. Butchers and W. Hume-Rothery, *J. Inst. Metals*, **71**, 1945, 295–296.
14. E. H. Dix and F. Keller, *Trans. AIME, Inst. Metals Div.*, **1929**, 351–365; *Z. Metallkunde*, **21**, 1929, 205–206.
15. G. Siebel and H. Vosskühler, *Z. Metallkunde*, **31**, 1939, 359–362.
16. P. D. Merica, R. G. Waltenberg, and J. R. Freeman, *Natl. Bur. Standards (U.S.) Sci. Papers* 337, 1919, pp. 115–119; *Trans. AIME*, **64**, 1921, 15–21.
17. M. Hansen, unpublished work; see M. Hansen, "Der Aufbau der Zweistofflegierungen," p. 124, Springer-Verlag OHG, Berlin, 1936.
18. K. Miyazaki, *Kinzoku-no-Kenkyu*, **6**, 1929, 124–126; *J. Inst. Metals*, **41**, 1929, 438.
19. E. Schmid and G. Siebel, *Z. Metallkunde*, **23**, 1931, 202–204.
20. P. Saldau and L. N. Sergeev, *Metallurg*, **1934**, 67–70.
21. W. L. Fink and L. A. Willey, *Trans. AIME*, **124**, 1937, 85–86.
22. H. Borchers and H. J. Otto, *Aluminium*, **24**, 1942, 265–267.
23. See fig. 5 of [21].
24. The lattice parameters of the α phase closely agree with those reported by [51]; see fig. 1 of [15].
25. N. S. Kurnakov and V. I. Mikheeva, *Izvest. Sektora Fiz.-Khim. Anal.*, **13**, 1940, 209–224.
26. N. S. Kurnakov and V. I. Mikheeva, *Izvest. Sektora Fiz.-Khim. Anal.*, **10**, 1938, 5–35.
27. R. Vogel, *Z. anorg. Chem.*, **107**, 1919, 267–271.
28. W. Sander and K. L. Meissner, *Z. Metallkunde*, **16**, 1924, 13–14.
29. T. Halstead and D. P. Smith, *Trans. Am. Electrochem. Soc.*, **49**, 1926, 291–312.
30. E. S. Makarov, *Doklady Akad. Nauk S.S.S.R.*, **74**, 1950, 935–938.
31. R. F. Mehl, *Trans. Am. Electrochem. Soc.*, **46**, 1924, 164–176.
32. W. Köster and W. Dullenkopf, *Z. Metallkunde*, **28**, 1936, 309–312.
33. K. Riederer, *Z. Metallkunde*, **28**, 1936, 312–317.
34. F. Laves and K. Moeller, *Z. Metallkunde*, **30**, 1938, 232–235.
35. W. Bungardt and F. Bollenrath, *Z. Metallkunde*, **30**, 1938, 377–383.
36. E. M. Savitskii and M. A. Tylkina, *Doklady Akad. Nauk S.S.S.R.*, **67**, 1949, 81–83; *Chem. Abstr.*, **43**, 1949, 7398.

37. W. Hume-Rothery and G. V. Raynor, *J. Inst. Metals*, **63**, 1938, 201–209.
38. J. L. Haughton and R. J. M. Payne, *J. Inst. Metals*, **57**, 1935, 293–294.
39. H. Adenstedt and J. R. Burns, *Trans. ASM*, **43**, 1951, 873–886.
40. W. Schmidt (P. Spitaler), *Z. Metallkunde*, **19**, 1927, 452–455; see also *J. Inst. Metals*, **38**, 1927, 197.
41. P. Saldau and M. Zamotorin, *J. Inst. Metals*, **48**, 1932, 221–225; *Izvest. Inst. Fiz.-Khim. Anal.*, **7**, 1935, 21–30.
42. S. Ishida, *J. Mining Inst. Japan*, **46**, 1930, 245–268; *Met. Abstr.*, **1**, 1934, 417.
43. E. Schmid and H. Seliger, *Metallwirtschaft*, **11**, 1932, 409–411; *J. Inst. Metals*, **48**, 1932, 226.
44. E. Schmid and G. Siebel, *Z. Physik*, **85**, 1933, 37–41.
45. M. I. Zacharowa and W. K. Tschikin, *Z. Physik*, **95**, 1935, 769–774; M. I. Zacharowa, *Izvest. Sektora Fiz.-Khim. Anal.*, **10**, 1938, 113–118.
46. W. Leitgebel, *Z. anorg. Chem.*, **202**, 1931, 305–324.
47. A. Schneider and U. Esch, *Z. Elektrochem.*, **45**, 1939, 888–893.
48. E. A. Owen and G. D. Preston, *Proc. Phys. Soc.* (*London*), **36**, 1923, 25–28.
49. Z. Nishiyama, *Science Repts. Tôhoku Univ.*, **18**, 1929, 388.
50. G. Wassermann, *Z. Metallkunde*, **22**, 1930, 158.
51. R. M. Brick, A. Phillips, and A. J. Smith, *Trans. AIME*, **117**, 1935, 102–117.
52. P. Vachet, *Rev. mét.*, **32**, 1935, 614–626; *Rev. aluminium*, **12**, 1935, 3087–3099.
53. J. J. Trillat and M. Paič, *Rev. aluminium*, **15**, 1938, 1109–1116.
54. P. Lacombe and G. Chaudron, *Compt. rend.*, **207**, 1938, 860–862.
55. H. Küstner, *Z. Metallkunde*, **34**, 1942, 114–116.
56. P. Lacombe, *Rev. mét.*, **41**, 1944, 217–226.
57. H. J. Axon and W. Hume-Rothery, *Proc. Roy. Soc.* (*London*), **A193**, 1948, 1–24.
58. W. Hume-Rothery and T. H. Boultbee, *Phil. Mag.*, **40**, 1949, 71–80.
59. J. E. Dorn, P. Pietrokowsky, and T. E. Tietz, *Trans. AIME*, **188**, 1950, 933–943.
60. A. P. Guljaev and E. F. Trusova, *Zhur. Tekh. Fiz.*, **20**, 1950, 66–78.
61. D. M. Poole and H. J. Axon, *J. Inst. Metals*, **80**, 1951-1952, 599–604.
62. E. C. Ellwood, *J. Inst. Metals*, **80**, 1951-1952, 605–608.
63. H. Perlitz, *Nature*, **154**, 1944, 606.
64. F. Laves, K. Löhberg, and P. Rahlfs, *Nachr. Ges. Wiss. Göttingen, Jahresber. Geschäftsjahr Math.-physik. Kl., Fachgruppen* II, **1**, 1934, 67.
65. G. V. Raynor, *Proc. Roy. Soc.* (*London*), **A180**, 1942, 107–121.
66. R. S. Busk, *Trans. AIME*, **188**, 1950, 1460–1464; **194**, 1952, 207–209.
67. J. B. Clark and F. N. Rhines, *Trans. AIME*, **209**, 1957, 1–7.

$\bar{1}.6912$
0.3088

Al-Mn Aluminum-Manganese

Alloys Up to 25 At. (40.43 Wt.) % Mn. Although the constitution of this range has been studied, more or less comprehensively, by many investigators [1–13], there are still several points which have not been fully clarified. This holds especially for the region 20–25 at. % Mn. The findings of [2, 4–10] were reviewed by [11]; the reader is referred to this paper for a detailed discussion of the data available up to 1942. Discrepancies in the interpretation of microstructures and other information are due to a strong departure from equilibrium caused by the facts that (*a*) all alloys above 2.08 at. (4.1 wt.) % Mn undergo one or more peritectic reactions during solidification and (*b*) undercooling of the primary and secondary constituents occurs [14].

The phase diagram presented in Fig. 65 is essentially based on the results reported by [9, 11] and obtained by thermal and micrographic analysis. The solid solubility of Mn in Al will be dealt with separately (see below).

The composition of the intermediate phase richest in Al is no longer in doubt, since this has been conclusively established by various methods to be the compound $MnAl_6$ (25.34 wt. % Mn) [9–12], instead of $MnAl_7$ [6, 8], $MnAl_5$ [3, 7], or $MnAl_4$ [4]. $MnAl_6$ is formed peritectically at a temperature which is reported as 670–710°C.

Also, it is generally agreed, on the basis of the work of [9–11], that the intermediate phase formed peritectically at 810–830°C has a composition very close to

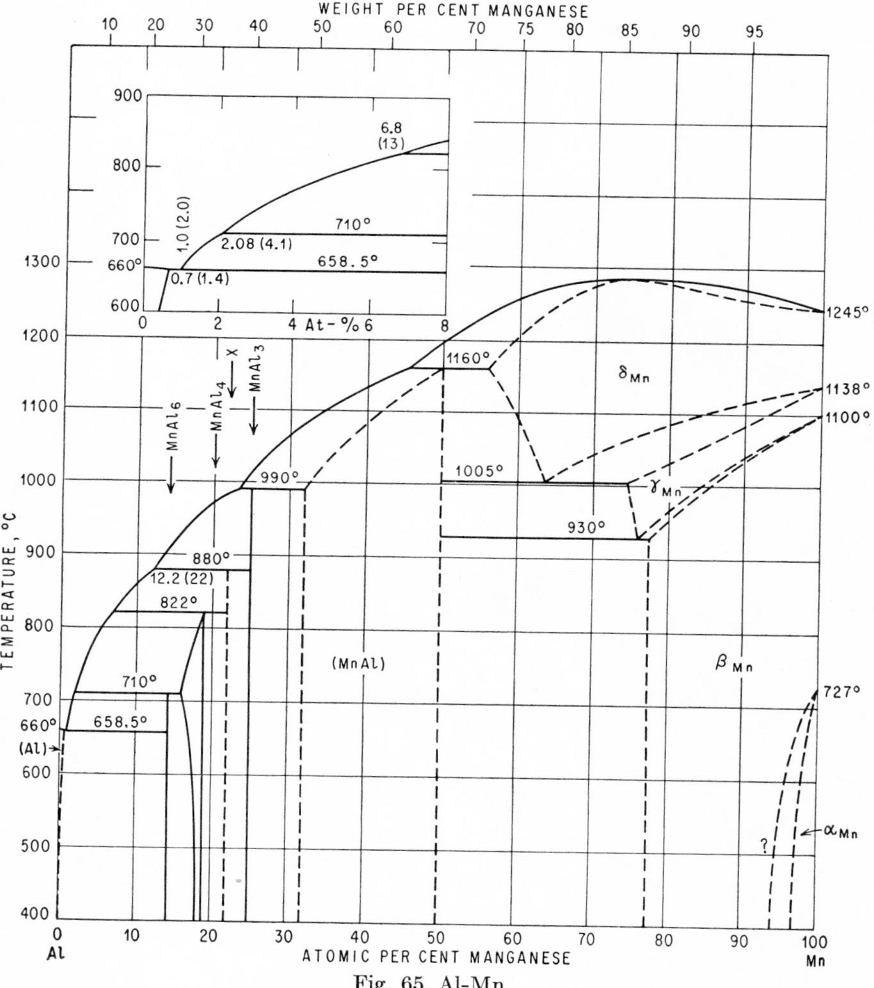

Fig. 65. Al-Mn

$MnAl_4$ (33.73 wt. % Mn). The boundary of the $MnAl_4$ phase field (Fig. 65) was outlined micrographically by [11].

The identity of the phase X, formed peritectically at 880°C, which was first conclusively established to exist by [11], is still unknown. It will lie somewhere between 20 and 25 at. (33.73 and 40.43 wt.) % Mn. There is no simple atomic proportion in this range (e.g., Mn_2Al_7 and Mn_3Al_{10}). It has been suggested by [11] that this phase is identical with crystals of monoclinic or triclinic habit reported to occur on the surface of ingots with about 20–40 wt. % Mn [10, 14*a*].

The existence of the phase generally designated as $MnAl_3$ (40.43 wt. % Mn) has long been assumed [15, 16, 1, 2, 17, 6, 7, 8, 18, 19]; however, the exact composition of this phase and its homogeneity range are still unknown. [10] has isolated crystals from a Mn-rich master alloy and found their composition as 41 wt. % Mn, corresponding to $MnAl_{2.93}$ (25.45 at. % Mn).

In addition to the peritectic reaction at 880°C recorded by [9] and especially [11], other thermal effects at 920°C were observed, but not investigated, by [9]. They could not be confirmed by [11], however. It would appear, therefore, that there is no additional phase between X and $MnAl_3$ [14*a*].

In an attempt to clarify the constitution of the region up to about 50 wt. (33 at.) % Mn, [13] has used thermal-analysis methods exclusively. Because of the incompleteness of the reactions during solidification, such an attempt will always be unsuccessful. The various versions of proposed phase relations [13] are of little significance and cannot discredit those suggested by [9, 11], based on conclusively established

Table 10. Solid Solubility of Mn in Al in Wt. %

Temperature, °C	Ref. 9	Ref. 21	Ref. 22	Ref. 23
650	1.69*	1.38	1.23	1.46†
600	1.04	0.96	0.85	0.76
550	0.64	0.60	0.54	0.51
500	0.36	0.34	0.34	0.36

* Extrapolated from 626°C.
† Extrapolated from 640°C.

data. The apparent reliquefaction that occurs on cooling of alloys with about 25–37 wt. (14–22.5 at.) % Mn through the temperature range 680–700°C is not constitutional in origin [13], but due to the difference in crystal habit between $MnAl_4$ present as the primary phase above 700°C and $MnAl_6$ formed peritectically and stable at temperatures below 680°C [11].

Thermal analysis has shown [2] that the miscibility gap in the liquid state at about 7.5–27 [1] or 5–29 at. % Mn [3] does not exist. This was confirmed by subsequent investigations [6, 7, 9, 11]. The liquidus curve between 0 and 25 at. % Mn shown in Fig. 65 is based on solubility determinations [9, 11] and thermal-analysis work [9, 11].

Not considering previous determinations, the eutectic point was placed at 2.2 wt. %, 657°C [5]; 1.95 wt. %, 658.5°C [9]; 1.99 wt. % Mn, 658.5°C [36]; and 1.9 wt. % Mn, 658°C [11]; though up to 2.2 wt. % may be found in the eutectic of slowly cooled alloys because of undercooling [11]. In this range the composition in at. % is very nearly half of that in wt. %.

Data of the solid solubility of Mn in Al were reported by [5, 20, 9, 21–23]. Those by [5, 20] are considerably too low because of the high Fe content of the alloys used. The other values, which are based on electrical-resistivity isotherms [9, 21], thermo-electric-force isotherms [23], and micrographic work [22], are scattered in the ranges given in Table 10. As [22] pointed out, it is difficult to account for the difference in their values and those of [9], since both used practically the same grade of high-purity Al and electrolytic Mn.

As shown by [10, 24, 25], solid Al can dissolve considerably larger amounts of Mn if melts are very rapidly quenched. The supersaturation, as indicated by lattice-parameter measurements, can be as high as 9.2 wt. (4.7 at.) % [26], i.e., 6.7 times higher than the maximal equilibrium solubility of about 1.4 wt. (0.7 at.) % Mn.

Mention should be made of the occurrence of a metastable phase G, probably having a composition close to $MnAl_{12}$ (7.69 at. % Mn). It was found to form in chill-cast alloys on annealing at 550°C and lower temperatures. The decomposition of G into stable (Al) and $MnAl_6$ takes place extremely slowly [26]. The G phase is perhaps a relatively "stable" transition phase which forms when supersaturated solid solutions of the type mentioned above are annealed.

Alloys with 25–100 At. % Mn. Phase relationships were first determined by [7], using thermal, thermoresistometric, dilatometric, and metallographic methods. The diagram was corrected and supplemented by [19], whose thermal and micrographic data were used to draw the partial diagram in Fig. 65. The phase boundaries are shown as dashed curves, since they are essentially based on thermal-analysis data. Solubility changes with temperature were not determined.

The diagram suggested by [7] is characterized by the existence of a phase ranging in composition between 25 and 50 at. % Mn. However, a two-phase field between 25 and 32 at. % Mn was found by X-ray work [8]. It was confirmed by [19], who established the two peritectic equilibria at 1160 and 990°C; only the latter had been detected by [7]. Additional corrections by [19] deal with the effect of Al on the $\gamma \rightleftarrows \delta$ and $\beta \rightleftarrows \gamma$ transformations of Mn. The (undetermined) solubility of Al in α-Mn is quite restricted [8, 19]. The γ_{Mn} phase cannot be retained by quenching [27].

Crystal Structures. Lattice parameters of the (Al) phase were reported by [10, 11, 24, 25; see also 35]; the last data include parameters of supersaturated solutions with up to 4.7 at. % Mn [25]. $MnAl_6$ is orthorhombic [10, 29, 30] with 28 atoms per unit cell [30] and $a = 6.498$ A, $b = 7.552$ A, $c = 8.870$ A [30]; see also [31]. $MnAl_4$ is hexagonal, $a = 28.41$ A, $c = 12.38$ A, $c/a = 0.436$ [10]; and $MnAl_3$ is orthorhombic, $a = 14.82$ A, $b = 12.63$ A, $c = 12.46$ A [10]. The phase (MnAl) is b.c.c. of the B2 type [32], with perhaps $a = 2.98$ A [33]. Lattice spacings of β_{Mn} alloys were given by [8].

The metastable G phase ($MnAl_{12}$) was reported to be primitive cubic with probably 12 formula weights per unit cell and $a = 13.28$ A [34].

1. G. Hindrichs, *Z. anorg. Chem.*, **59**, 1908, 441–448.
2. W. Rosenhain and F. C. A. H. Lantsberry, *Proc. Inst. Mech. Eng.*, **74**, 1910, 252–254.
3. M. Goto and T. Mishima, *Nippon Kogyo Kwaishi*, **41**, 1925, no. 477, 1–17; *Japan J. Eng.*, **5**, 1925, 48; *J. Inst. Metals*, **38**, 1927, 410.
4. W. Krings and W. Ostmann, *Z. anorg. Chem.*, **163**, 1927, 145–154.
5. E. H. Dix and W. D. Keith, *Proc. AIME, Inst. Metals Div.*, 1927, 315–333; *Z. Metallkunde*, **19**, 1927, 497–498.
6. E. Rassow, *Hauszeitschr. V.A.W. Erftwerk A.G. Aluminium*, **1**, 1929, 187–190.
7. T. Ishiwara, *Science Repts. Tôhoku Univ.*, **19**, 1930, 500–504; *Kinzoku-no-Kenkyu*, **3**, 1926, 13.
8. A. J. Bradley and P. Jones, *Phil. Mag.*, **12**, 1931, 1137–1152.
9. E. H. Dix, W. L. Fink, and L. A. Willey, *Trans. AIME*, **104**, 1933, 335–352.
10. W. Hofmann, *Aluminium*, **20**, 1938, 865–872.
11. H. W. L. Phillips, *J. Inst. Metals*, **69**, 1943, 275–291.
12. G. V. Raynor and W. Hume-Rothery, *J. Inst. Metals*, **69**, 1943, 415–421.
13. H. Fournier, *Publs. sci. et tech. ministère air (France)*, 1945, no. 195.
14. H. W. L. Phillips, *J. Inst. Metals*, **69**, 1943, 287–288.

14a. On the other hand, the monoclinic or triclinic phase was claimed to be richer, rather than lower, in Mn than $MnAl_3$ [28]. The microstructure of a cast alloy with 31 wt. (18.1 at.) % Mn was interpreted to show this phase as the primary constituent surrounded by successive peritectic envelopes of $MnAl_3$, $MnAl_4$, and $MnAl_6$. This would mean that $MnAl_3$, instead of X, is formed peritectically at 880°C and that the monoclinic or triclinic phase is formed at 990°C.
15. F. Wöhler and F. Michel, *Liebigs Ann.*, **115,** 1860, 104.
16. L. Guillet, *Compt. rend.*, **134,** 1902, 237–238; *Bull. soc. encour. ind. natl.*, **103,** 1902, 249–252.
17. W. Broniewski, *Ann. chim. et phys.*, **25,** 1912, 103–106.
18. G. V. Akimov and A. S. Oleshko, *Zhur. Fiz. Khim.*, **3,** 1932, 338; *Korrosion u. Metallschutz,* **10,** 1934, 133.
19. W. Köster and W. Bechthold, *Z. Metallkunde,* **30,** 1938, 294–296.
20. M. Bosshard, *Alluminio,* **1,** 1932, 363.
21. E. Fahrenhorst and W. Hofmann, *Metallwirtschaft,* **19,** 1940, 891–893.
22. E. Butchers and W. Hume-Rothery, *J. Inst. Metals,* **71,** 1945, 87–91.
23. I. L. Rogelberg and E. S. Shpichinetsky, *Zavodskaya Lab.*, **14,** 1948, 1216–1218.
24. W. Hofmann, *Abhandl. braunschweig. wiss. Ges.*, **1,** 1949, 83–88.
25. G. Falkenhagen and W. Hofmann, *Z. Metallkunde,* **43,** 1952, 69–81.
26. K. Little, G. V. Raynor, and W. Hume-Rothery, *J. Inst. Metals,* **73,** 1947, 83–90; discussion, pp. 747–749; see also [34].
27. U. Zwicker, *Z. Metallkunde,* **42,** 1951, 246–253.
28. H. Hanemann and A. Schrader, "Atlas Metallographicus," vol. III, part 1, pp. 58–65, Verlagsbuchhandlung Gebrüder Borntraeger, Berlin, 1941.
29. G. Phragmén, *J. Inst. Metals,* **77,** 1950, 498.
30. A. D. I. Nicol, *Acta Cryst.*, **6,** 1953, 285–293.
31. K. Schubert and M. Kluge, *Z. Naturforsch.*, **8a,** 1953, 755–756.
32. A. Westgren, *Z. Metallkunde,* **22,** 1930, 372; *Metallwirtschaft,* **9,** 1930, 923; W. Ekman, *Z. physik. Chem.*, **B12,** 1931, 57–78.
33. H. O. Dorum, see *Strukturbericht,* **2,** 1928-1932, 684–685.
34. K. Little and W. Hume-Rothery, *J. Inst. Metals,* **74,** 1948, 521–524.
35. G. V. Raynor, *Phil. Mag.*, **36,** 1945, 770–777.
36. E. Butchers, G. V. Raynor, and W. Hume-Rothery, *J. Inst. Metals,* **69,** 1943, 211–213, 226.

$\bar{1}.4490$
0.5510

Al-Mo Aluminum-Molybdenum

Aluminum-rich Alloys. A hypothetical diagram (0–35 wt. % Mo), based on thermal analysis (no individual data were given) and microscopic investigation of as-cast alloys, showed three horizontals at 660, 735, and possibly 1130°C. At 735°C the phase $MoAl_4$ (?) (47.06 wt. % Mo) was suggested to be formed peritectically. A second intermediate phase was found, but no attempt was made to identify it [1]. Earlier, [2] claimed to have isolated crystals of the composition $MoAl_4$ from an Al-rich alloy.

Reinvestigation of the system [3] yielded a diagram (0–60 wt. % Mo) characterized by two horizontals at 660 and 703°C, the latter being the temperature of the peritectic formation of $MoAl_5$ [16.67 at. (41.56 wt.) % Mo]. In addition, the phase $MoAl_3$ (54.24 wt. % Mo) is shown. The original paper was not available [4].

Another tentative diagram (0–70 wt. % Mo) was proposed by [5]. It shows the horizontals due to [1] and the two phases $MoAl_5$ and $MoAl_2$ (64.0 wt. % Mo). No detailed information of this work was available [4].

Alloys with 42, 50, 55, and 63 wt. % Mo prepared by hot pressing of powder mixtures at temperatures between 1300 and 1450°C were examined metallographically. Again, two intermediate phases, possibly $MoAl_5$ and $MoAl_2$, of which $MoAl_5$ is formed peritectically at 735°C, were shown to exist [6].

In summary, it appears to be well established that two intermediate phases occur in the range 0–40 at. % Mo. Their compositions lie in the neighborhood of

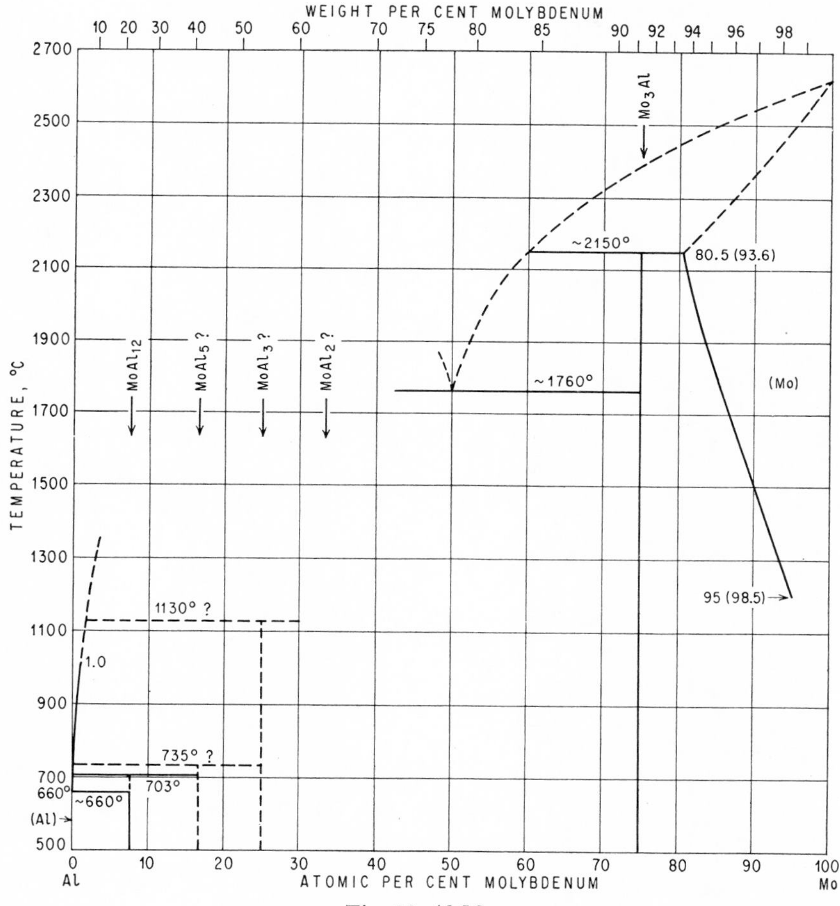

Fig. 66. Al-Mo

$MoAl_5$ and $MoAl_3$ (or $MoAl_2$) (but cf. Supplement). The liquidus curve between 660 and 1000°C shown in Fig. 66 is due to [3]. The solid solubility of Mo in Al was reported as about 0.01–0.02 wt. % Mo at 560°C [7] and about 0.2 wt. % Mo at 660°C [8]. The horizontal at 660°C must be regarded as a peritectic involving the reaction L (~0.1 wt. % Mo) + $MoAl_5 \rightleftarrows$ (Al) [8] (cf. Supplement).

Molybdenum-rich Alloys. A tentative diagram of the partial system between 46 and 100 at. % Mo (Fig. 66) was outlined by [9]. The solid-solubility curve is based on lattice-parametric work, and the intermediate phase coexisting with the

Mo-rich solid solution was identified by X-ray study as the peritectically formed compound Mo_3Al (91.43 wt. % Mo). Mo_3Al forms a eutectic (approximately 50 at. % Mo) with another compound of unknown composition.

It is not possible as yet to combine the tentative diagrams of the two partial systems. The question in particular is whether the phase in equilibrium with Mo_3Al is $MoAl_3$ or $MoAl_2$ or another phase richer in molybdenum than $MoAl_2$. The liquidus curve must reach a maximum in the range 25–40 at. % Mo.

Mo_3Al has the structure of the "β-W" (A15) type, $a = 4.95$ A [9]. For the lattice parameter of the Mo-rich solid solution, see [9].

Supplement. In analogy to WAl_{12}, there exists a compound of the composition $MoAl_{12}$ (7.69 at. % Mo) which was prepared by powder methods [10]. To insert this newly detected phase into the constitutional diagram, peritectic reactions at 1130 ($MoAl_3$?), 735 ($MoAl_5$?), and 703°C ($MoAl_{12}$) have been assumed in Fig. 66.

$MoAl_{12}$ is b.c.c. with $a = 7.573$ A, 26 atoms per unit cell, and is isotypic with WAl_{12}.

1. H. Reimann, *Z. Metallkunde,* **14,** 1922, 119–123.
2. F. Wöhler and F. Michel, *Liebigs Ann.,* **115,** 1860, 103.
3. K. Yamaguchi and K. Simizu, *Nippon Kinzoku Gakkai-Shi,* 4, 1940, 390–392.
4. Data were taken from [6].
5. W. D. Walther, unpublished work, Department of Metallurgy, Massachusetts Institute of Technology, Cambridge, Mass., 1950.
6. R. L. Wachtell, AF Technical Report 6601, parts 1 and 2, 1951-1952.
7. P. Röntgen and W. Koch, *Z. Metallkunde,* **25,** 1933, 182–185.
8. L. F. Mondolfo, "Metallography of Aluminum Alloys," pp. 30–31, John Wiley & Sons, Inc., New York, 1943.
9. J. L. Ham; J. L. Ham and A. J. Herzig, 1st and 2d Annual Report on Project NR 031-331, Climax Molybdenum Company, 1950 and 1951, respectively. J. L. Ham, *Trans. Am. Soc. Mech. Eng.,* **73,** 1951, 727–728.
10. J. Adam and J. B. Rich, *Acta Cryst.,* **7,** 1954, 813–816.

0.2847
$\bar{1}.7153$

Al-N Aluminum-Nitrogen

The literature on this system was reviewed by [1–3]. Nitrogen is practically insoluble in both liquid and solid Al [4, 5] and in molten Al is present only in the form AlN (34.18 wt. % N). At 1200–1500°C the solubility of AlN was reported to be less than 0.004 wt. % (?) [6]. The nitriding of Al was studied by [7–9]. AlN is hexagonal of the wurtzite (B4) type, $a = 3.12$ A, $c = 4.99$ A, $c/a = 1.60$ [10]; $a = 3.110$ A, $b = 4.975$ A, $c/a = 1.60$ [11]. For the preparation and properties of AlN, see [12]. The nitrogen content of high-purity Al was found to be 0.0001 wt. % [13].

1. L. W. Eastwood, "Gases in Light Alloys," p. 26, John Wiley & Sons, Inc., New York, 1946.
2. R. Castro and M. Armand, *Rev. mét.,* **46,** 1949, 594–615.
3. H. Kostron, *Z. Metallkunde,* **43,** 1952, 270.
4. W. Claus et al., *Z. Metallkunde,* **21,** 1929, 268–270.
5. P. Röntgen and H. Braun, *Metallwirtschaft,* **11,** 1932, 471–472.
6. J. Czochralski, *Z. Metallkunde,* **14,** 1922, 278–281.
7. P. Laffitte and P. Grandadam, *Compt. rend.,* **200,** 1935, 1039–1041.
8. P. Grandadam, *Ann. chim. (Paris),* **4,** 1935, 83–146.
9. P. Laffitte and E. Elchardus, *Rev. ind. minérale,* **1936,** 861–867.
10. H. Ott, *Z. Physik,* **22,** 1924, 201–214.

11. M. v. Stackelberg and K. F. Spiess, *Z. physik. Chem.*, **A175**, 1935, 127–139.
12. "Gmelins Handbuch der anorganischen Chemie," System No. 35B(1), pp. 132–149, Verlag Chemie, G.m.b.H., Berlin, 1934.
13. H. A. Sloman, *J. Inst. Metals*, **71**, 1945, 404.

0.0694
$\bar{1}$.9306

Al-Na Aluminum-Sodium

According to [1], neither the melting point of Al nor that of Na [2] is affected by addition of the other component; i.e., there is no noticeable mutual solubility in the liquid state. However, reinvestigation of the Al-rich alloys established that a true monotectic exists, found by thermal analysis to lie at 0.18 wt. % Na and a temperature 1.2°C below the melting point of Al (660.2°C) [3] (Fig. 67).

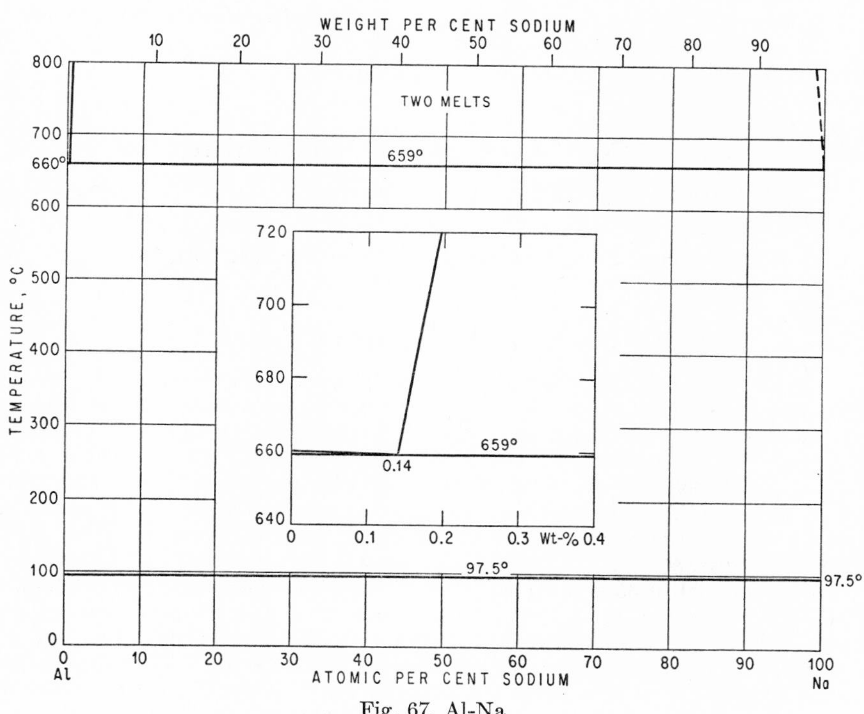

Fig. 67. Al-Na

The solubility of Na in liquid Al (in wt. % Na) was found to increase with temperature from about 0.10% at 700°C to 0.13% at 800°C [4] and from 0.14% (monotectic point) to 0.25% at 775°C [5]. On the other hand, [3] found a slight decrease with temperature, from 0.18% at 659°C to 0.155% at 800°C.

The solid solubility of Na in Al, reported by [4] to be practically nil, was given as <0.003 wt. % at 659°C [3] and approximately 0.002 wt. % at 550–650°C [5]. It probably decreases slightly with falling temperature [3, 5].

1. C. H. Mathewson, *Z. anorg. Chem.*, **48**, 1906, 192–193.
2. C. T. Heycock and F. H. Neville, *J. Chem. Soc.*, **55**, 1889, 668.

3. W. L. Fink, L. A. Willey, and H. C. Stumpf, *Trans. AIME*, **175,** 1948, 364–371.
4. E. Scheuer, *Z. Metallkunde*, **27,** 1935, 83–85. Al of 99.7% purity was used.
5. C. E. Ransley and H. Neufeld, *J. Inst. Metals*, **78,** 1950-1951, 25–46.

$\bar{1}.4630$
0.5370

Al-Nb Aluminum-Niobium

The crystal structure of the compound $NbAl_3$ (53.44 wt. % Nb) [1] is tetragonal of the $TiAl_3$ ($D0_{22}$) type, with a = 5.438, c = 8.601 A, c/a = 1.582, 4 formula weights per unit cell [2]. X-ray investigation of a Nb-rich alloy with approximately 10 at. (~3 wt.) % Al [3] did not reveal the composition of the intermediate phase coexisting with Nb or a Nb-rich solid solution [4].

1. L. Marignac, *Compt. rend.*, **66,** 1868, 180–183.
2. G. Brauer, *Z. anorg. Chem.*, **242,** 1939, 1–22; *Naturwissenschaften*, **26,** 1938, 710.
3. W. v. Bolton, *Z. Elektrochem.*, **13,** 1907, 146.
4. S. v. Olshausen, *Z. Krist.*, **61,** 1925, 475–478.

$\bar{1}.2719$
0.7281

Al-Nd Aluminum-Neodymium

The compound NdAl (84.25 wt. % Nd) has the structure of the CsCl (B2) type, a = 3.74 A [1].

1. C. W. Stillwell and E. E. Jukkola, *J. Am. Chem. Soc.*, **56,** 1934, 56–57.

$\bar{1}.6625$
0.3375

Al-Ni Aluminum-Nickel

Since the first investigation of the phase diagram [1], which established its principal features, phase relationships have undergone considerable changes in parts, especially above 50 at. % Ni. The diagram presented in Fig. 68 is largely based on the work by [2] [0–18 wt. (9.2 at.) % Ni], [3] (X-ray analysis of the whole system using slowly cooled powders), [4] (thermal and micrographic analysis of alloys with 5–95 wt. % Ni), [5] (thermal and micrographic analysis of alloys with more than 75 wt. % Ni), [6] (formation of the Ni_3Al phase), and [7] (phase boundaries in the range 70–90 at. % Ni).

Alloys with 0–50 At. % Ni. The partial phase diagram is well established. There is good agreement as to the liquidus curve [2, 4, 8]. The eutectic was placed at 6 wt. (2.85 at.) % Ni, 630°C [1]; 5.3 wt. (2.5 at.) % Ni, 633°C [9]; 5.7 wt. (2.7 at.) % Ni, 640°C [2]; and 6.3–6.4 wt. (3.0–3.06 at.) % Ni, 640°C [10]. The solid solubility of Ni in Al was found to be about 0.05 wt. (0.023 at.) % at 640°C, 0.028 wt. (0.013 at.) % at 600°C, and 0.006 wt. (0.003 at.) % Ni at 500°C [2]. In good agreement, [11] found 0.01–0.02 wt. (0.0045–0.009 at.) % Ni at 560°C.

$NiAl_3$ (42.03 wt. % Ni) [12, 2, 3, 13, 4] is of singular composition [3, 13]. Its temperature of peritectic formation was determined as 842°C [1] and 854°C [4], with the peritectic melt containing about 28 wt. (15.1 at.) % [1] or 28.4 wt. (15.3 at.) % Ni [4].

The intermediate phase formed peritectically at 1132–1133°C [1, 4] was first suggested to be $NiAl_2$ [1] but later identified as based on the ideal composition Ni_2Al_3 (59.19 wt. % Ni) [3, 13, 4]. Its range of homogeneity [55.3–60 wt. (36.3–40.8 at.) % Ni] was determined micrographically at temperatures between 1120 and 600°C [4]; see also [3]. The β' phase, based on the composition NiAl (68.51 wt. % Ni), has a

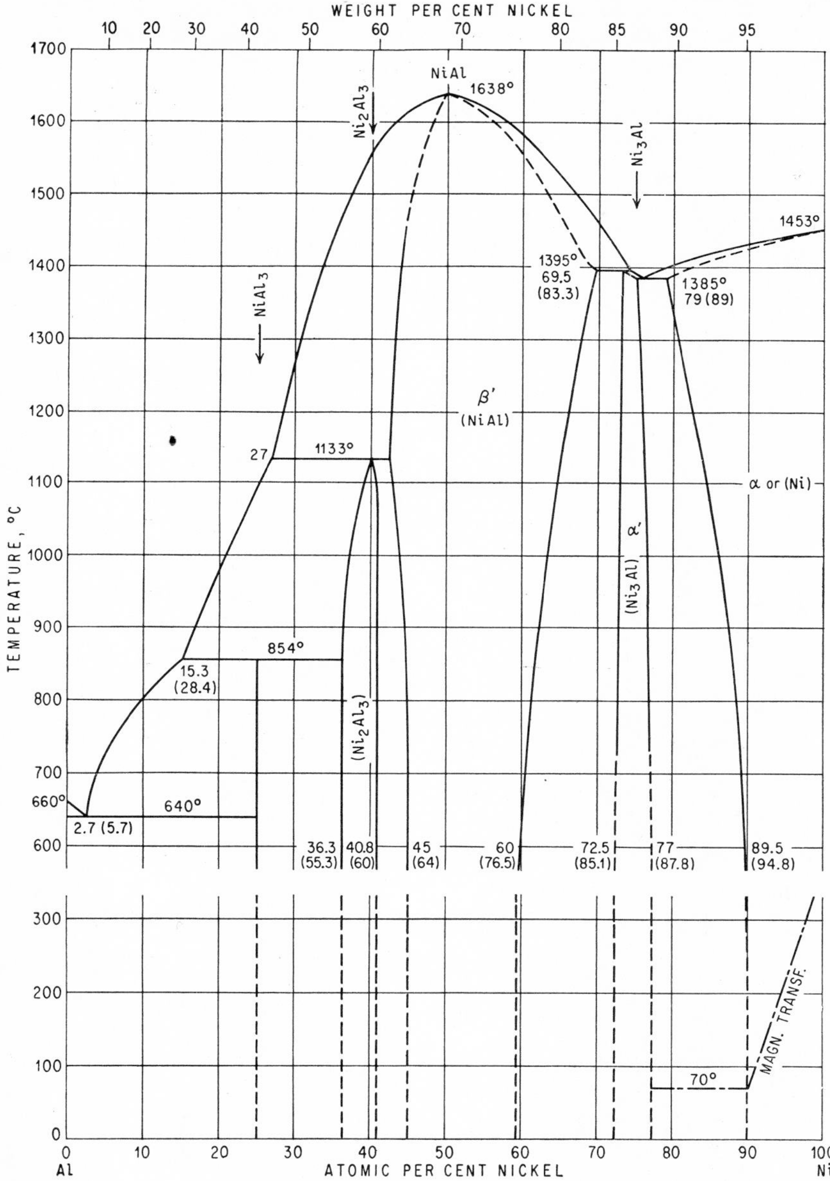

Fig. 68. Al-Ni. (See also Fig. 69.)

maximum melting point of 1638°C [4]. Its boundary on the Al-rich side was determined micrographically for temperatures between 1420 and 600°C [4]; see also [3].

Alloys with 50–100 At. % Ni. The constitution in this range of composition, which proved to be much more complicated than was assumed earlier [1], was the subject of numerous investigations [14, 15, 3, 4, 16, 5, 17, 6, 7]. The existence of the phase based on the ideal composition Ni_3Al (86.71 wt. % Ni), first detected by [14], was confirmed by all later investigators, with the exception of [16]. There is great discrepancy, however, as to the equilibria involving its formation, its range of homogeneity, as well as to the $\beta'/(\beta' + \alpha')$ and $\alpha/(\alpha + \alpha')$ boundaries.

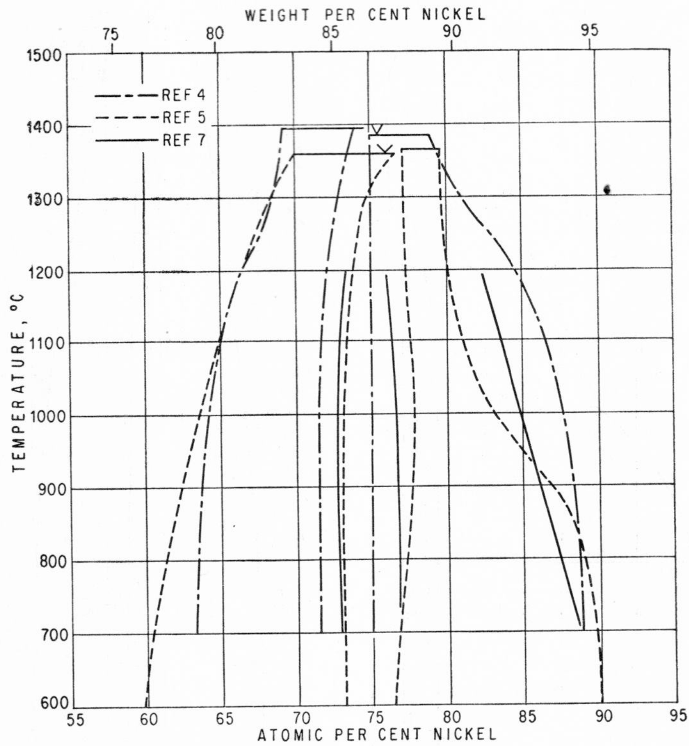

Fig. 69. Al-Ni. (See also Fig. 68.)

The diagrams given by [14] and [16] are highly improbable and incompatible with all later findings. [14] found phases of compositions Ni_3Al and Ni_5Al; and [16] concluded that the Ni_5Al phase was, in fact, Ni_4Al and that the Ni_3Al phase did not exist.

[3] suggested that the Ni_3Al (α') phase is continuous with the α phase at temperatures above 1100°C, i.e., that the $\alpha + \alpha'$ field takes the form of a closed loop. This was not confirmed, however, by [3–7, 17]. Whereas the diagram by [4] shows the two three-phase equilibria liq. $+ \beta' \rightleftarrows \alpha'$ at 1395°C and liq. $\rightleftarrows \alpha + \alpha'$ at 1385°C, [5] concluded that the reactions occurring at the solidus were liq. $+ \alpha \rightleftarrows \alpha'$ at 1362°C and liq. $\rightleftarrows \alpha' + \beta'$ at 1360°C. This discrepancy appears to have been cleared up [6] in favor of the results by [4].

The phase boundaries in this region are presented in Figs. 68 and 69. In the main diagram (Fig. 68) the three-phase equilibria by [4] and the phase boundaries by [5, 7] have been adopted, with the exception of the $\alpha/(\alpha + \alpha')$ boundary, which represents a compromise. The phase boundaries suggested by [17] were not considered, as they were obtained by extrapolation from an attempt to smooth out the boundaries within the ternary system Al-Fe-Ni.

Temperatures of the magnetic transformation of the α phase were determined by [1, 18, 19, 5]; those by [5] were adopted in Fig. 68.

Crystal Structures. $NiAl_3$ has an orthorhombic structure and is the prototype of the $D0_{20}$ type, a = 6.611 A, b = 7.366 A, c = 4.812 A [13, 3]; see also [20, 21]. Ni_2Al_3 is hexagonal, of the $D5_{13}$ type, a = 4.036 A, c = 4.900 A, c/a = 1.214, at the stoichiometric composition; for change of parameters with composition see [13, 3]. NiAl (β') is b.c.c. of the CsCl (B2) type, a = 2.887 A at 49.8 at. % Ni [3]; for change of parameter with composition see [3, 22]. See also earlier work by [23, 15, 24] and further studies by [25–27]. The α' phase (Ni_3Al) is f.c.c. of the Cu_3Au ($L1_2$) type [15, 3, 4, 21], a = 3.589 A at 75 at. % Ni [3]. Lattice parameters of the α phase have been reported by [15, 3].

Supplement. [28] measured the electrical resistivity and thermal dilatation of alloys containing up to 35 wt. % Al. The isotherms of these properties exhibit sharp minima at the compositions NiAl and Ni_3Al. From the dilatation data, [28] suggest that Ni_3Al undergoes a solid-state transformation at about 600°C.

1. A. G. C. Gwyer, *Z. anorg. Chem.*, **57**, 1908, 133–140.
2. W. L. Fink and L. A. Willey, *Trans. AIME*, **111**, 1934, 293–303.
3. A. J. Bradley and A. Taylor, *Proc. Roy. Soc. (London)*, **A159**, 1937, 56–72.
4. W. O. Alexander and N. B. Vaughan, *J. Inst. Metals*, **61**, 1937, 247–260.
5. J. Schramm, *Z. Metallkunde*, **33**, 1941, 347–355.
6. R. W. Floyd, *J. Inst. Metals*, **80**, 1951-1952, 551–553.
7. A. Taylor and R. W. Floyd, *J. Inst. Metals*, **81**, 1952-1953, 25–32.
8. H. W. L. Phillips, *J. Inst. Metals*, **68**, 1942, 28–30.
9. K. E. Bingham and J. L. Haughton, *J. Inst. Metals*, **29**, 1923, 80.
10. A. G. C. Gwyer, *J. Inst. Metals*, **61**, 1937, 260–261.
11. P. Röntgen and W. Koch, *Z. Metallkunde*, **25**, 1933, 184.
12. O. Brunck, *Ber. deut. chem. Ges.*, **34**, 1901, 2734.
13. A. J. Bradley and A. Taylor, *Phil. Mag.*, **23**, 1937, 1049–1067.
14. I. Iitaka, *Tetsu-to-Hagane*, **10**, 1924, 1–33.
15. A. Westgren and A. Almin, *Z. physik. Chem.*, **B5**, 1929, 14–28.
16. N. Nishimura and S. Watanabe, *Suiyokwai Shi*, **9**, 1937, 153–158; *Japan Nickel Rev.*, **5**, 1937, 552; *Mem. Coll. Eng. Kyoto Imp. Univ.*, **10**, 1938, 131–135.
17. A. J. Bradley, *J. Iron Steel Inst.*, **163**, 1949, 19–56.
18. C. Manders, *Ann. phys.*, **5**, 1936, 193–195.
19. V. Marian, *Ann. phys.*, **7**, 1937, 489.
20. G. V. Raynor and M. B. Waldron, *Phil. Mag.*, **40**, 1949, 198–205.
21. L. Vegard; see *Structure Repts.*, **11**, 1947-1948, 27.
22. L. N. Guseva, *Doklady Akad. Nauk S.S.S.R.*, **77**, 1951, 415–418.
23. K. Becker and F. Ebert, *Z. Physik*, **16**, 1923, 165–169.
24. W. Ekman, *Z. physik. Chem.*, **B12**, 1931, 57–58.
25. I. Isaichev and V. M. Iretsky, *Zhur. Tekh. Fiz.*, **10**, 1940, 316–322.
26. N. V. Ageev and L. N. Guseva, *Izvest. Akad. Nauk S.S.S.R., Otdel. Khim. Nauk*, **1949**, no. 3, 225–233.
27. L. N. Guseva, *Doklady Akad. Nauk S.S.S.R.*, **77**, 1951, 615–616.
28. I. I. Kornilov and R. S. Mints, *Doklady Akad. Nauk S.S.S.R.*, **88**, 1953, 829–832.

$\bar{1}.0563$
0.9437

Al-Np Aluminum-Neptunium

There are three Al-Np compounds which are isostructural with the corresponding Al-U compounds. $NpAl_4$ (68.7 wt. % Np) is b.c. orthorhombic with 4 formula weights per unit cell, and $a = 4.42$ A, $b = 6.26$ A, $c = 13.71$ A. $NpAl_3$ (74.5 wt. % Np) has a simple cubic structure of the Cu_3Au ($L1_2$) type, $a = 4.262$ A. $NpAl_2$ (81.5 wt. % Np) is f.c.c. of the $MgCu_2$ (C15) type, $a = 7.785$ A. Evidence was obtained that no other Al-Np compounds exist [1]. If heated in high vacuum at 1150°C, $NpAl_4$ decomposes to form $NpAl_3$. The latter decomposes to form $NpAl_2$ if heated to 1375°C.

1. O. J. C. Runnalls, *Trans. AIME,* **197,** 1953, 1460–1462.

0.2269
$\bar{1}.7731$

Al-O Aluminum-Oxygen

In molten Al, oxygen is present only in the form of Al_2O_3 (47.1 wt. % O). Its solubility is suggested to be <0.04 wt. % [1] and is probably close to 0.003 wt. % [2]; see also [3, 4]. Melting points in the range 45.4–60 at. (33–47.1 wt.) % O were determined by [5]. The oxygen content of high-purity Al was found to be 0.00015–0.0003 wt. % [6].

1. R. Sterner-Rainer and W. Ehrenberg, *Z. Metallkunde,* **23,** 1931, 276.
2. G. Jander and W. Brösse, *Z. angew. Chem.,* **41,** 1928, 702–704.
3. R. Castro and M. Armand, *Rev. mét.,* **46,** 1949, 594–615.
4. H. Kostron, *Z. Metallkunde,* **43,** 1952, 270.
5. E. Baur and R. Brunner, *Z. Elektrochem.,* **40,** 1934, 154–158.
6. H. A. Sloman, *J. Inst. Metals,* **71,** 1945, 404.

$\bar{1}.9400$
0.0600

Al-P Aluminum-Phosphorus

Aluminum forms only one phosphide, AlP (53.45 wt. % P) [1]. Products reported earlier [2] were mixtures of AlP with Al and Al_2O_3. If phosphorus is added to molten aluminum, the phosphide which starts to form at temperatures below 800°C rises to the surface and may be retained in the cast metal in a quantity of the order of 0.08 wt. % [3]. AlP does not melt or decompose at temperatures up to 1000°C [1]. It has the cubic structure of the ZnS (B3) type, $a = 5.46$ A [4] or 5.43 A [5].

1. W. E. White and A. H. Bushey, *J. Am. Chem. Soc.,* **66,** 1944, 1666–1672.
2. L. Franck, *Chem.-Ztg.,* **22,** 1898, 237–240.
3. J. Czochralski, *Z. Metallkunde,* **15,** 1923, 277–282.
4. V. M. Goldschmidt, *Skrifter Norske Videnskaps-Akad., Oslo, Mat. Naturv. Kl.,* no. 8, 1927.
5. L. Passerini, *Gazz. chim. ital.,* **58,** 1928, 655–664.

$\bar{1}.1146$
0.8854

Al-Pb Aluminum-Lead

The existence of a wide miscibility gap in the liquid state has long been known [1–5]; in fact, thermal-analysis data [5] indicated that neither the melting point of Al [5] nor that of Pb [5, 6] is lowered by additions of the other component. Later, [7, 8] showed that the melting point of Al is lowered by 1.5°C to a monotectic point (see Table 11). Likewise, the melting point of Pb (327.3°C) is lowered by 0.021 wt.

(0.16 at.) % Al to 326.8°C (eutectic point) [9]. [10] found the eutectic point at 0.025 wt. (0.19 at.) % Al, 326.2°C.

Results of determinations of the solubility of Pb in liquid Al (Table 11) by thermal analysis [8] and chemical analysis [9, 11–13] agree in that most of them show an appreciable increase with temperature. Data by [9] differ from these. The curve given in Fig. 70 is based on the work by [8].

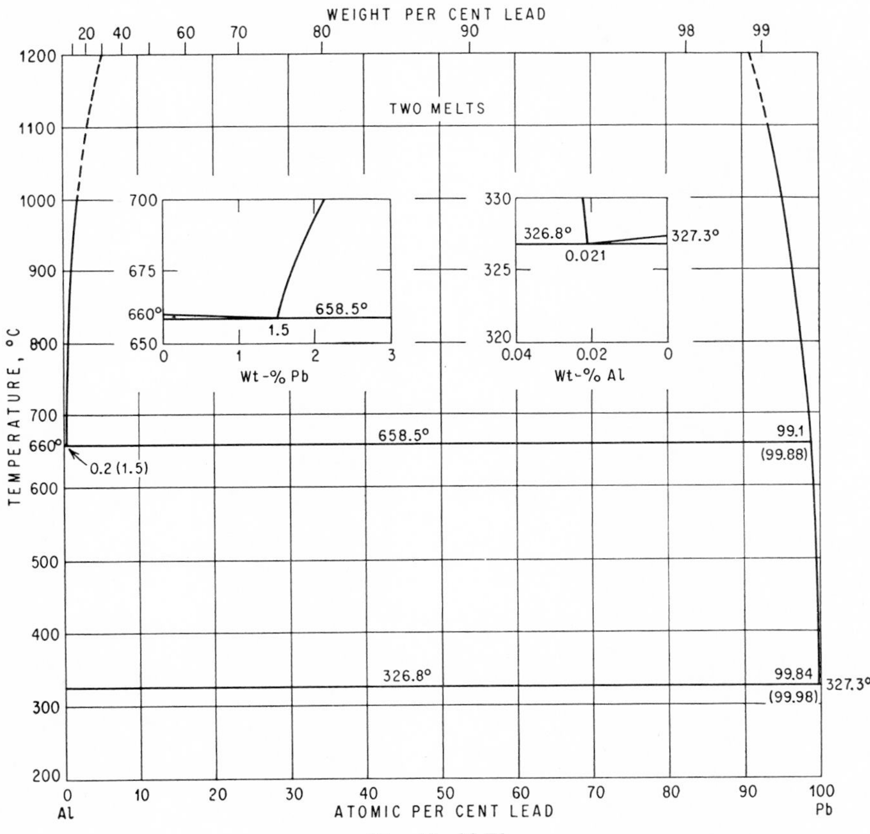

Fig. 70. Al-Pb

The solubility of Al in liquid Pb, as determined between 350 and 1100°C by [10], in wt. % Al (at. % in parentheses) is as follows: 350°C, 0.03 (0.25); 400°C, 0.04 (0.3); 500°C, 0.09 (0.7); 600°C, 0.10 (0.8); 700°C, 0.19 (1.45); 900°C, 0.50 (3.7); and 1100°C, 0.93 (6.7). Solubilities reported by [11] and [14] are higher: 650°C, 0.4 (3.0); 1000°C, 1.1 (1.8) [14]; 1020°C, 1.4 (9.8); and 1200°C, 2.6 (17.0) [11]. According to [9], there is no change in solubility between the eutectic temperature and 800°C, 0.021 wt. (0.16 at.) % Al.

The solubility of Pb in solid Al is not higher than 0.2 wt. (0.025 at.) % Pb at the monotectic temperature [8]; see also [15]. The solid solubility of Al in Pb is practically nil [9, 10].

1. C. R. A. Wright, *J. Chem. Soc.*, **11**, 1892, 492–494; **13**, 1894, 1014–1017.
2. W. Campbell and J. A. Mathews, *J. Am. Chem. Soc.*, **24**, 1902, 255–256.

3. L. Guillet, *Bull. soc. encour. ind. natl.*, **101,** 1902, 257–259.
4. H. Pécheux, *Compt. rend.*, **138,** 1904, 1042–1044; see also **143,** 1906, 397–398.
5. A. G. C. Gwyer, *Z. anorg. Chem.*, **57,** 1908, 147–149.
6. C. T. Heycock and F. H. Neville, *J. Chem. Soc.*, **61,** 1892, 888.
7. M. Hansen and B. Blumenthal, *Metallwirtschaft*, **10,** 1931, 925–927.
8. L. W. Kempf and K. R. Van Horn, *Trans. AIME*, **133,** 1939, 81–92.
9. A. N. Campbell and R. W. Ashley, *Can. J. Research*, **B18,** 1940, 281–287.
10. Y. Dardel, *Light Metals*, **9,** 1946, 220–222; see also *Z. Metallkunde*, **39,** 1948, 214.
11. W. Claus and I. Herrmann, *Metallwirtschaft*, **18,** 1939, 957–960.
12. H. Bauer, *Aluminium-Arch.*, **24,** 1939; see also [13].
13. M. Païc, *Rev. mét.*, **44,** 1947, 363–370.
14. K. Schneider, *Z. Metallkunde*, **39,** 1948, 349.
15. E. A. Owen, Y. H. Liu, and D. P. Morris, *Phil. Mag.*, **39,** 1948, 831–845.

Table 11. Solubility of Lead in Liquid Aluminum in Wt. % Pb (At. % in Parentheses)

°C	Ref. 8	Ref. 12*	Ref. 9	Ref. 13	Ref. 11
658.5 (monotectic point)	1.52 (0.2)	. . .	1.1 (0.15)	1.2 (0.16)	
700	2.1 (0.3)	1.8 (0.25)	1.25 (0.17)	1.6 (0.22)	
800	4.3 (0.6)	3.6 (0.5)	1.8 (0.24)	3.6 (0.5)	
900	8.0 (1.1)	7.0 (0.95)			
1000	12.6 (1.9)				
1100	. . . . .	. . . . .	. . . . .	. . . . .	18 (2.8)
1200	. . . . .	. . . . .	. . . . .	. . . . .	30 (5.3)

* Data were taken from a diagram in the paper by [9].

$\bar{1}.4029$
0.5971

Al-Pd Aluminum-Palladium

The phase relationships presented in Fig. 71 are based on thermal, thermoresistometric, and metallographic studies of alloys prepared under careful experimental conditions [1]. The boundaries of the three areas of solid solution were located by microscopic examination of samples quenched from 950, 800, and 650°C; however, no detailed data are given for this part of the investigation. Lattice-parameter measurements did not indicate a solid solubility of Pd in Al, but in this case X-ray analysis is too insensitive a method.

Pd_2Al_3 (72.50 wt. % Pd) is hexagonal of the Ni_2Al_3 ($D5_{13}$) type, a = 4.22 A, c = 5.15 A, c/a = 1.22 [2]. The high-temperature form of the (PdAl) phase is b.c.c. of the CsCl (B2) type, a = 3.04 A at 45 at. % Pd (after quenching from 700°C) [2]. The powder pattern of the low-temperature modification of (PdAl) is very complicated [1].

The formulas $PdAl_3$, PdAl, and Pd_2Al correspond to the compositions 56.86, 79.81, and 88.78 wt. % Pd, respectively.

Supplement. [3] prepared several Pd-Al alloys and observed the following compounds: $PdAl_3$, orthorhombic, a = 7.085 A, b = 7.531 A, c = 5.087 A, space group

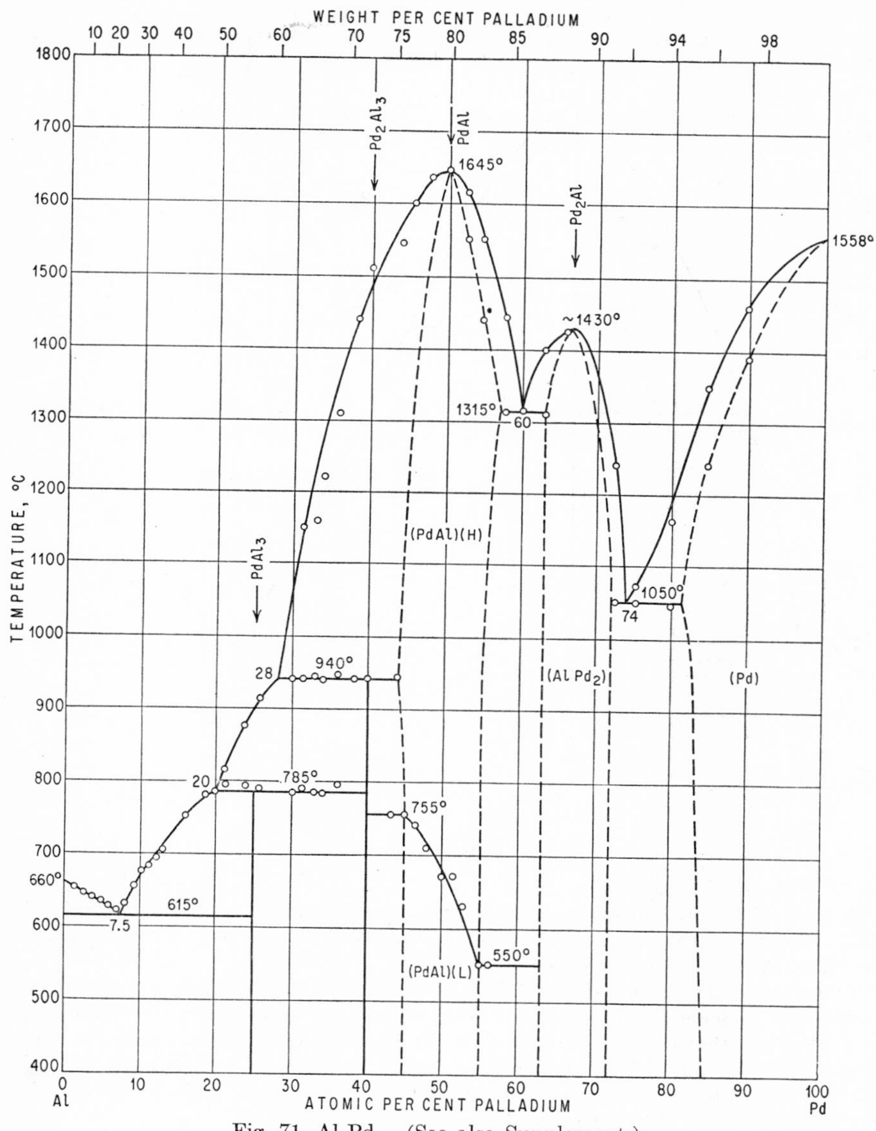

Fig. 71. Al-Pd. (See also Supplement.)

D_{2h}^{12} or C_{2v}^{10}; Pd_2Al_3, hexagonal, a = 4.216 A, c = 5.166 A, c/a = 1.225 (cf. above); PdAl, low-temperature form hexagonal, a = 3.964 A, c = 5.610 A, c/a = 1.415.

Unaware of the systematic study by [1], [4] investigated the entire system using thermal, microscopic, and X-ray analysis. The phase diagram arrived at differs in several respects from Fig. 71. It shows the three (stoichiometric) compounds $PdAl_3$, Pd_2Al_3, and PdAl, all formed peritectically, at 794, 910, and about 1300°C, respectively. PdAl is shown to undergo a polymorphic transformation at 1045°C; the Al-rich eutectic lies at 7 at. % Pd, 630°C; and Pd takes into solid solution about 20 at.

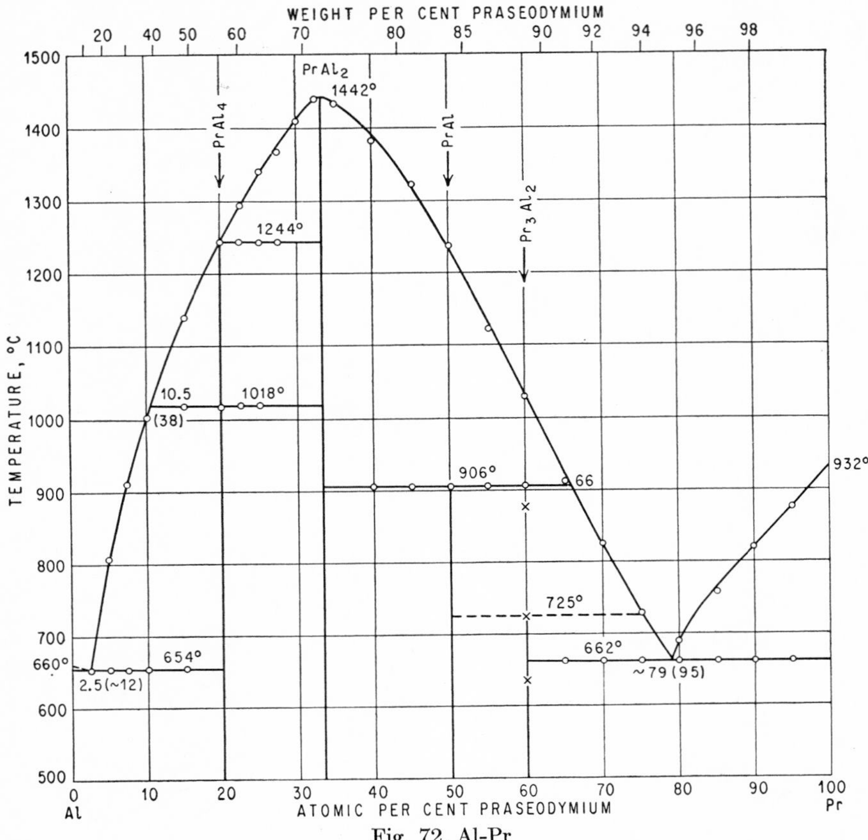

Fig. 72. Al-Pr

% Al. The lattice parameters reported for $PdAl_3$ and Pd_2Al_3 are nearly identical with those of [3]; the high-temperature modification of PdAl is said to be of the CsCl (B2) type, $a = 3.046$ A (cf. above), and the low-temperature form to be hexagonal, $a = 3.959$ A, $c = 5.614$ A, $c/a = 1.420$.

The phase diagram of [1] is certainly the more reliable one. Whereas [4] melted their alloys under a salt slag, [1] protected their melts by carefully purified argon. Also, [4] admit that the constitution of Pd-rich alloys may be more complicated than is shown in their phase diagram.

According to [5], the room-temperature modification of PdAl is monoclinic (26 atoms per unit cell) rather than hexagonal (cf. above).

1. G. Grube and R. Jauch, "Festschrift aus Anlass des 100-jährigen Jubiläums der Firma W. C. Heraeus G.m.b.H., Platinschmelze Hanau," pp. 52–68, Brönners Druckerei, Frankfurt am Main, 1951.
2. H. Pfisterer and K. Schubert, *Naturwissenschaften*, **37**, 1950, 112–113; see also data in [1].
3. Y. P. Simanov, *Zhur. Fiz. Khim.*, **27**, 1953, 1503–1509; *Chem. Abstr.*, **49**, 1955, 3606.
4. V. A. Nemilov, A. T. Grigorev, and T. A. Strunina, *Izvest. Sektora Platiny*, no. **28**, 1954, pp. 256–259.
5. K. Schubert and P. Esslinger, private communication.

Ī.2821
0.7179

Al-Pr Aluminum-Praseodymium

Thermal analysis and microscopic examination yielded the phase diagram given in Fig. 72. Solubilities in the solid state have not been investigated. Whereas the first study [1] showed only the three compounds $PrAl_4$ (56.64 wt. % Pr), $PrAl_2$ (72.32 wt. % Pr), and PrAl (83.94 wt. % Pr) to exist, later work of a cursory nature [2] indicated the existence of an additional compound, Pr_3Al_2 (88.68 wt. % Pr), formed peritectically at 725°C. The temperatures of two three-phase equilibria differ as follows: peritectic reaction $PrAl_2$ + melt ⇄ PrAl, 906 [1] vs. 878°C [2]; eutectic crystallization melt ⇄ Pr_3Al_2 + Pr, 662 [1] vs. 635°C [2].

1. G. Canneri, *Alluminio,* **2,** 1933, 87–89.
2. R. Vogel and T. Heumann, *Z. Metallkunde,* **35,** 1943, 29–42.

Ī.1405
0.8595

Al-Pt Aluminum-Platinum

According to the thermal-analysis data [1] covering the range up to 35.6 at. (80 wt.) % Pt (Fig. 73), three phases crystallize from hypereutectic melts. The phase

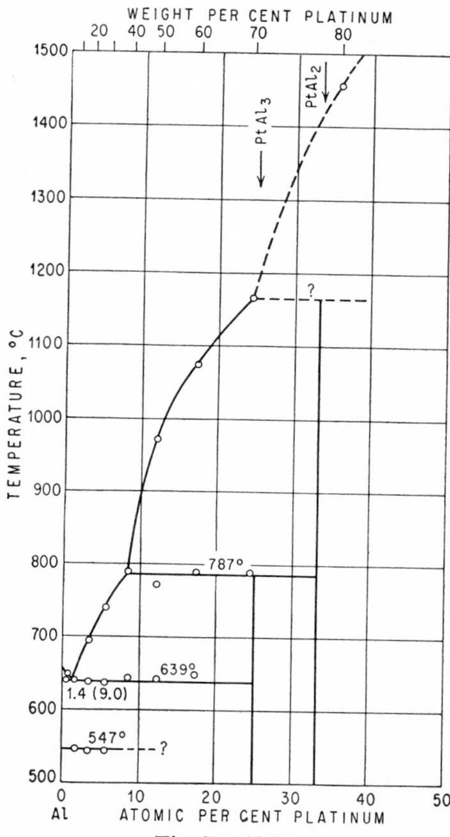

Fig. 73. Al-Pt

richest in Al is very likely $PtAl_3$ (70.69 wt. % Pt) [1, 2]. The phase which reacts at 787°C with melt to form $PtAl_3$ was identified by X-ray analysis to be $PtAl_2$ (78.35 wt. % Pt) [3]. The temperature of the peritectic reaction involving $PtAl_2$ is unknown and may be higher than that suggested in Fig. 73 (1164°C). The nature of the transformation observed at 547°C is not known.

In contradiction to Fig. 73, [4] reported, with no experimental evidence given, that the intermediate phase richest in Al is $PtAl_6$ [14.29 at. (54.68 wt.) % Pt], which is followed by $PtAl_3$.

$PtAl_2$ has the cubic lattice of the CaF_2 (C1) type, a = 5.922 A [3].

1. Chouriguine, *Rev. mét.*, **9,** 1912, 874–883; *Compt. rend.*, **155,** 1912, 156.
2. O. Brunck, *Ber. deut. chem. Ges.*, **34,** 1901, 2735. The author reported isolation of crystals of the composition Pt_3Al_{10}(68.46 wt. % Pt) from an alloy with about 86 wt. % Pt.
3. E. Zintl, A. Harder, and W. Haucke, *Z. physik. Chem.*, **B35,** 1937, 354–362.
4. L. F. Mondolfo, "Metallography of Aluminum Alloys," pp. 35–36, John Wiley & Sons, Inc., New York, 1943.

$\bar{1}$.0526
0.9474

Al-Pu Aluminum-Plutonium

The following structural results have been reported [1]: Pu_3Al (3.63 wt. % Al), partially ordered tetragonal $SrPb_3$ type (tetragonally distorted Cu_3Au type), with $a = 4.499 \pm 2$ A, $c = 4.538 \pm 2$ A ($c/a > 1$), or tetragonal with $a = 4.530$ A, $c = 4.475$ A ($c/a < 1$); ~PuAl (10.1 wt. % Al), probably one or more forms of a distorted CsCl structure; $PuAl_2$ (18.4 wt. % Al), cubic $MgCu_2$ (C15) type, with a varying between 7.838 ± 1 A (Pu-rich) and 7.848 ± 1 A (Al-rich); $PuAl_3$ (25.3 wt. % Al), hexagonal, $a = 6.08 \pm 1$ A, $c = 14.40 \pm 3$ A; $PuAl_4$ (31.1 wt. % Al), b.c. orthorhombic, $a = 4.41$ A, $b = 6.29$ A, $c = 13.79$ A, isotypic with UAl_4.

1. Review by A. S. Coffinberry and M. B. Waldron in "Metallurgy and Fuels," Progress in Nuclear Energy, ser. V, vol. 1, pp. 388–403, Pergamon Press Ltd., London, 1956. See also the somewhat earlier review (by A. S. Coffinberry and F. H. Ellinger) in "Proceedings of the International Conference on the Peaceful Uses of Atomic Energy," vol. 9, pp. 138–146, Geneva, 1955.

$\bar{1}$.4992
0.5008

Al-Rb Aluminum-Rubidium

According to [1, 2], there is a very limited, if any, miscibility of Rb in the liquid state and no solid solubility of Rb in Al.

1. J. Czochralski and J. Kakzynski, *Wiadamósci Inst. Metallurgii i Metaloznawstawa*, **4,** 1934, 18 (quoted from [2]).
2. L. F. Mondolfo, "Metallography of Aluminum Alloys," pp. 36–37, John Wiley & Sons, Inc., New York, 1943.

$\bar{1}$.9250
0.0750

Al-S Aluminum-Sulfur

Three entirely different tentative phase diagrams have been suggested for the system Al-Al_2S_3. According to an abstract of a paper by [1], the two compounds AlS (54.31 wt. % S) and Al_2S_3 (64.07 wt. % S) exist. It is assumed that (a) AlS has a maximum melting point of 2100°C, (*b*) Al_2S_3 is formed by peritectic reaction of AlS at 1100°C, and (*c*) the Al-AlS eutectic lies close to pure Al and about 650°C.

On the other hand, [2] reported that a miscibility gap in the liquid state exists between Al and Al_2S_3 (melting point about 1200°C) which extends from a composition very close to pure Al—0.08 wt. % S [3]—to approximately 61 wt. (about 57 at.) % S.

A slightly modified version of the hypothetical diagram suggested by [4] is given in Fig. 74. It is based on cooling-curve data and the examination of products obtained by sulfurization of Al at various temperatures.

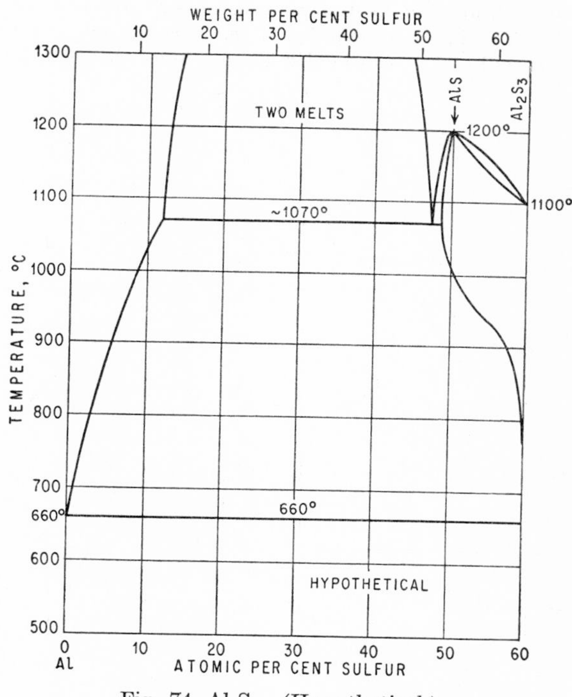

Fig. 74. Al-S. (Hypothetical.)

Supplement. Preparation and properties of Al_2S_3 were extensively studied by [5–7]. Al_2S_3 (melting point 1130°C [6], 1100°C [7]) undergoes a reversible polymorphic transformation at about 1000°C [7]. α-Al_2S_3, stable at room temperature, is hexagonal, $a = 6.423$ A, $c = 17.83$ A, $c/a = 2.775$, with $6Al_2S_3$ per unit cell. The high-temperature form, γ-Al_2S_3, can be retained by quenching and has the rhombohedral corundum structure with $a = 6.86$ A, $\alpha = 56°16'$. A third modification (called β-Al_2S_3), of the wurtzite type, appears to form only in the presence of carbon [7]. The transformation in Al_2S_3 has not been considered in Fig. 74.

A "subsulfide" Al_2S is stable in the gaseous state only [8].

1. T. Murakami and N. Shibata, *Nippon Kinzoku Gakkai-Shi*, **4**, 1940, 221–228.
2. R. Vogel and F. Hillen, *Arch. Eisenhüttenw.*, **15**, 1942, 551–555
3. J. Czochralski, *Z. Metallkunde*, **15**, 1923, 277–282.
4. E. J. Kohlmeyer and H. W. Retzlaff, *Z. anorg. Chem.*, **261**, 1950, 248–254.

5. J. Flahaut, *Compt. rend.*, **232**, 1951, 334–336.
6. J. Flahaut, *Compt. rend.*, **232**, 1951, 2100–2102.
7. J. Flahaut, *Ann. chim. (Paris)*, **7**, 1952, 632–696.
8. W. Klemm et al., *Z. anorg. Chem.*, **255**, 1948, 287–293.

$\bar{1}.3455$
0.6545

Al-Sb Aluminum-Antimony

The constitution of this system, the principal features of which were first recognized by [1], has been studied many times, using thermal analysis and microscopic examination [2–11]. The investigations by [4, 8, 9] cover only limited ranges of composition. The work by [2, 3] can be disregarded, since they found two maxima,

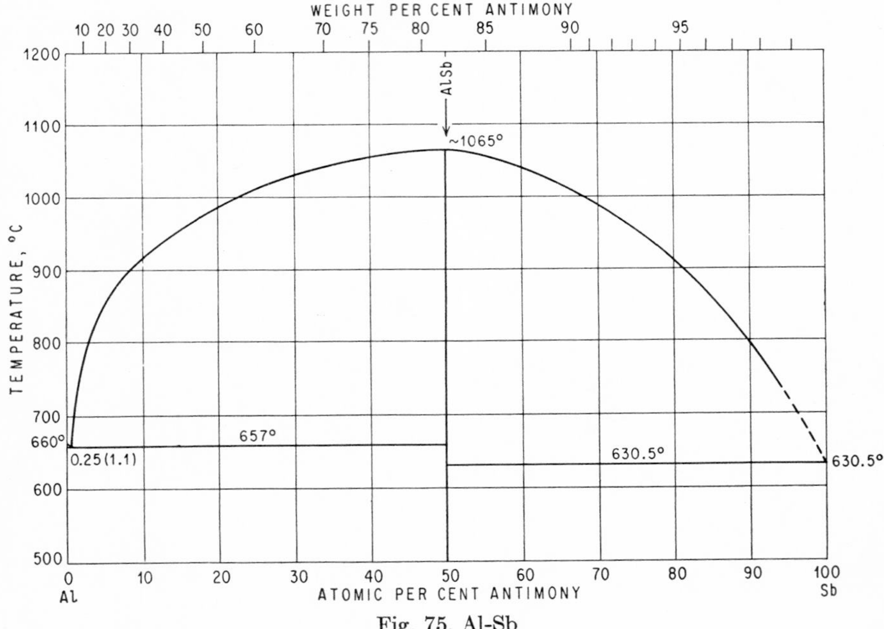

Fig. 75. Al-Sb

one at about 33–34 wt. (about 10 at.) % Sb, 950–980°C, and the other at the composition AlSb (81.86 wt. % Sb). [4] showed that the first maximum was due to the fact that homogeneous liquid solutions form very slowly when the two components are mixed.

The liquidus curves reported by [5, 6, 9, 10, 11] differ somewhat; however, the data by [5] may be regarded as the most reliable ones (Fig. 75). The melting point of AlSb was found to lie between 1050 and 1080°C [12, 3, 5, 10, 11]. The composition and temperature of the Al-AlSb eutectic was reliably determined to lie at 1.1 wt. (0.25 at.) % Sb, 657°C [8]. The solidus temperature between 50 and 100 at. % Sb was reported to be identical to the melting point of Sb [5, 6] or to lie at 555 [10] or 624°C [11]. Although in Fig. 75 the eutectic is shown to coincide with pure Sb, it may well be close to 99 wt. (95.6 at.) % Sb [11] and at a temperature somewhat below the melting point of Sb [3, 11].

The solid solubility of Sb in Al in the range 200–645°C is smaller than 0.10 wt. % Sb [8]. Whereas [5] asserted that the AlSb phase is of singular composition (with no evidence given), [11] reported indication of a certain range of homogeneity.

AlSb has the cubic structure of the ZnS (B3) type, a = 6.138 A [13], a = 6.10 A [14].

Supplement. In papers dealing with the semiconducting properties of AlSb, the following new data were reported: melting point 1060°C [15]; lattice parameter 6.1361 ± 3 A [16], 6.0959 A (at 26°C) [17].

1. C. R. A. Wright, *J. Soc. Chem. Ind.*, **11,** 1892, 493–494.
2. H. Gautier, *Compt. rend.*, **123,** 1896, 109–112; *Bull. soc. encour. ind. natl.*, **1,** 1896, 1313.
3. J. A. Mathews, *J. Franklin Inst.*, **153,** 1902, 121–123; W. Campbell and J. A. Mathews, *J. Am. Chem. Soc.*, **24,** 1902, 259–264.
4. G. Tammann, *Z. anorg. Chem.*, **48,** 1906, 53–60.
5. G. G. Urasow, *Zhur. Russ. Fiz.-Khim. Obshchestva*, **51,** 1919, 461–471; *Izvest. Inst. Fiz.-Khim. Anal.*, **1,** 1921, 461–471.
6. M. Goto, *Kinzoku-no-Kenkyu*, **4,** 1927, *Abstr. Sect.*, 34.
7. T. Matsukawa, *Suiyokwai Shi*, **5,** 1928, 596–603.
8. E. H. Dix, F. Keller, and L. A. Willey, *Trans. AIME*, **93,** 1931, 396–402.
9. E. Loofs-Rassow, *Hauszeitschr. V.A.W. Erftwerk A.G. Aluminium*, **3,** 1931, 20–23.
10. J. Veszelka, *Mitt. berg-u. hüttenmänn. Abt. kgl. ungar. Hochschule für Berg- u. Forstwesen Sopron*, **1931,** 193–201.
11. W. Guertler and A. Bergmann, *Z. Metallkunde*, **25,** 1933, 82–84.
12. E. van Aubel, *Compt. rend.*, **132,** 1901, 1266–1267.
13. E. A. Owen and G. D. Preston, *Proc. Phys. Soc.* (*London*), **36,** 1924, 341–348; *Nature*, **113,** 1924, 914.
14. V. M. Goldschmidt; see *Strukturbericht*, **1,** 1913–1928, 141.
15. H. Welker, *Z. Naturforsch.*, **8a,** 1953, 249.
16. R. K. Willardson, A. C. Beer, and A. E. Middleton, *J. Electrochem. Soc.*, **101,** 1954, 355.
17. R. F. Blunt, H. P. R. Frederikse, J. H. Becker, and W. R. Hosler, *Phys. Rev.*, **96,** 1954, 578.

$\bar{1}.5336$
0.4664

Al-Se Aluminum-Selenium

The diagram of Fig. 76 is based on thermal-analysis data [1]. The uncorrected data points indicate that the maximum of the liquidus curve lies at a temperature above 953°C between 50 and 60 at. % Se. Because of considerable volatilization of selenium during melting, the liquidus temperatures based on the intended compositions (the alloys were not analyzed) may not be real; i.e., the liquidus curve may run at higher temperatures. The exact composition of the intermediate phase, therefore, cannot be deduced from the data given. Since the analysis of segregated primary compound crystals gave a mean selenium content of 79.9 wt. % Se, the authors suggested the compound Al_3Se_4 (79.60 wt. % Se). However, the composition Al_2Se_3 (81.45 wt. % Se) seems more likely because a corresponding phase exists in the related systems Al-S and Al-Te. The compound Al_2Se_3 has also been identified in synthetic studies [2–4].

Supplement. Al_2Se_3 has also been prepared by [5–7]. According to [7], the compound exists in two polymorphic forms. The modification stable at room temperature possesses the hexagonal wurtzite (B4) type of structure with a = 3.89 A,

$c = 6.30$ A, $c/a = 1.62$ and 2 Se + 1⅓ Al atoms per unit cell. Superstructure lines reported by [6] could not be observed by [7]. A second modification, yielding a very complex powder pattern, was obtained as a sublimation product (monotropic form?).

A "subselenide" Al_2Se appears to be stable in the gaseous state only [5–7].

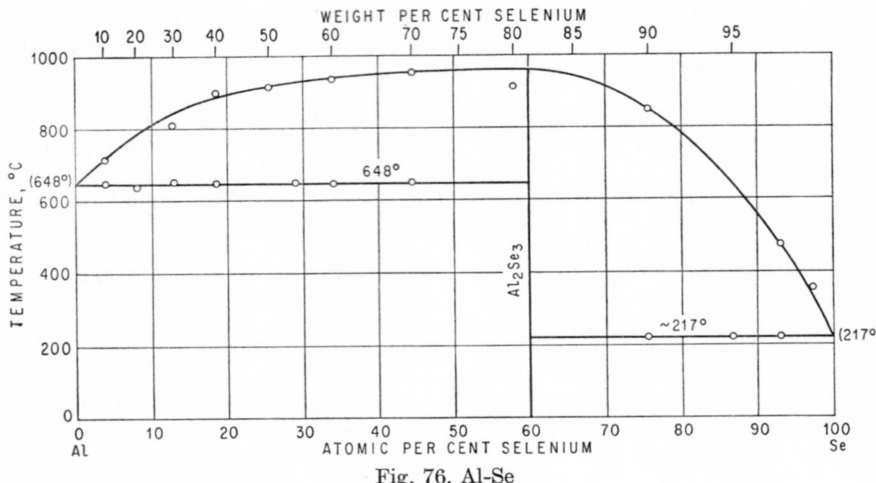

Fig. 76. Al-Se

1. M. Chikashige and T. Aoki, *Mem. Coll. Sci., Kyoto Imp. Univ.*, **2,** 1917, 249–254. The melting point of Al was found as 648°C. Alloys were melted in a CO_2 atmosphere; they were not analyzed.
2. H. Fonzes-Diacon, *Compt. rend.*, **130,** 1900, 1315.
3. C. Matignon, *Compt. rend.*, **130,** 1900, 1393.
4. L. Moser and E. Doctor, *Z. anorg. Chem.*, **118,** 1921, 285–286.
5. W. Klemm, K. Geiersberger, B. Schaeler, and H. Mindt, *Z. anorg. Chem.*, **255,** 1948, 287–293.
6. K. Geiersberger and H. Galster, *Angew. Chem.*, **64,** 1952, 81 (quoted by [7]).
7. A. Schneider and G. Gattow, *Z. anorg. Chem.*, **277,** 1954, 49–59.

$\bar{1}.9825$
0.0175

Al-Si Aluminum-Silicon

Solidification. The liquidus curve of the hypoeutectic alloys shown in Fig. 77 is based on the data of [1] and that of the hypereutectic alloys on the temperatures reported by [2, 3, 1, 4].

The eutectic temperature was determined by [2, 3, 5, 6, 7, 8, 1, 9, 10, 4, 11] and the eutectic composition by [2, 3, 5, 6, 7, 12, 8, 1, 9, 4, 11]. The most reliable values are 577°C [7, 1, 9] and 11.6 [7], 11.7 [1, 11], and 11.8 wt. % Si [9], which correspond to 11.2–11.4 at. % Si. (Cf. Supplement.)

Solidus temperatures of the (Al) phase were determined dilatometrically [13, 10] and by means of tests of tensile strength vs. temperature [14]. The latter values were used to draw the solidus curve in the upper inset of Fig. 77.

Solid State. The solubility of Si in Al was the subject of numerous investigations. Individual values reported in the older literature are not considered here, since they are obsolete. The temperature dependence of the solubility was determined by [8, 15, 16, 9, 17, 13, 10, 18–21] using various methods, with the exception of the lattice-spacing method. Data reported by [9, 10, 13, 15, 17–21] were summarized and tabulated by [22].

The solubility curve determined micrographically by [9] is generally considered

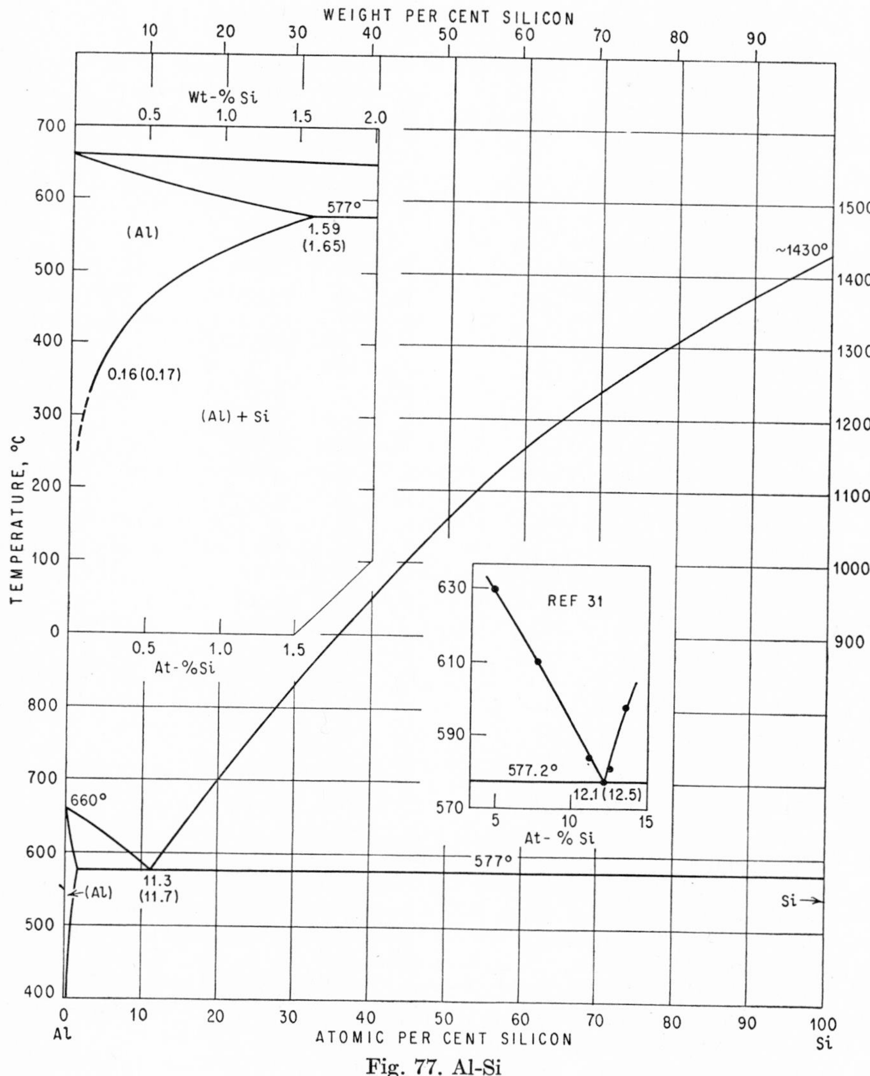

Fig. 77. Al-Si

to have the highest degree of accuracy, since these workers used high-purity Al and annealing times long enough to ensure equilibrium at temperatures down to at least 350°C. These solubilities in wt. % (at. %) are as follows: 577°C, 1.65 (1.59); 550°C, 1.30 (1.25); 500°C, 0.80 (0.77); 450°C, 0.48 (0.46); 400°C, 0.29 (0.28); and 350°C,

0.17 (0.16). For 300 and 250°C, solubilities of 0.10 and 0.05 wt. % Si were reported [9]; however, on the basis of thermodynamic calculation, solubilities of about 0.06 and 0.008 wt. % Si are more likely [23]; see also [24].

According to lattice-parameter measurements, Al is insoluble in Si [25, 26]; see also [27].

Lattice parameters of the (Al) phase with up to 0.93 at. (0.97 wt.) % Si were determined by [28]; see also [29].

Supplement. Two redeterminations [30, 31] of the eutectic data yielded the following results: 12.5 wt. (12.1 at.) % Si, 577.2°C [30]; slightly less than 12.5 wt. % Si, 577.2°C [31]. The thermal-analysis data of [31] are plotted in the lower inset of Fig. 77.

1. A. G. C. Gwyer and H. W. L. Phillips, *J. Inst. Metals*, **36**, 1926, 294–295.
2. W. Fraenkel, *Z. anorg. Chem.*, **58**, 1908, 154–158.
3. C. E. Roberts, *J. Chem. Soc.*, **105**(2), 1914, 1383–1386.
4. W. Broniewski and Smialowski, *Rev. mét.*, **29**, 1932, 542–552.
5. W. Rosenhain, S. L. Archbutt, and D. Hanson, 11th Report of the Alloys Research Committee, *Inst. Mech. Engrs. (London)*, **1921**, 221.
6. D. Hanson and M. L. V. Gayler, *J. Inst. Metals*, **26**, 1921, 323–324.
7. J. D. Edwards, *Chem. Met. Eng.*, **28**, 1923, 165–169.
8. B. Otani, *J. Inst. Metals*, **36**, 1926, 243–245.
9. E. H. Dix and A. C. Heath, *Trans. AIME*, **78**, 1928, 164–194.
10. L. Losana and R. Stratta, *Metallurgia ital.*, **23**, 1931, 193–197.
11. H. Koto, "Anniversary Volume M. Chikashige," pp. 303–304, Kyoto Imperial University, 1930.
12. E. Rassow, *Z. Metallkunde*, **15**, 1923, 106.
13. L. Anastasiadis, *Z. anorg. Chem.*, **179**, 1929, 145–154; see also discussion by W. Köster, *Z. anorg. Chem.*, **181**, 1929, 295–297.
14. A. R. E. Singer and S. A. Cottrell, *J. Inst. Metals*, **73**, 1947, 33–54.
15. W. Köster and F. Müller, *Z. Metallkunde*, **19**, 1927, 52–55.
16. A. G. C. Gwyer and H. W. L. Phillips, *J. Inst. Metals*, **38**, 1927, 31–35.
17. A. G. C. Gwyer, H. W. L. Phillips, and L. Mann, *J. Inst. Metals*, **40**, 1928, 300–302.
18. P. J. Saldau and M. W. Danilovich, *Izvest. Inst. Fiz.-Khim. Anal.*, **6**, 1933, 81; *Legkie Metal.*, **1932**, no. 9, 12–19.
19. A. Durer, *Z. Metallkunde*, **32**, 1940, 280–281.
20. H. Borchers and H. J. Otto, *Aluminium*, **24**, 1942, 219–221.
21. W. D. Treadwell and R. Walti, *Helv. Chim. Acta*, **25**, 1942, 1154–1162.
22. H. W. L. Phillips, *J. Inst. Metals*, **72**, 1946, 158.
23. G. Tammann and W. Oelsen, *Z. anorg. Chem.*, **186**, 1930, 285.
24. W. L. Fink and H. Freche, *Trans. AIME*, **111**, 1934, 304–317.
25. W. Hofmann, in H. Hanemann and A. Schrader, "Atlas Metallographicus," vol. III, part 1, 69, Verlagsbuchhandlung Gebrüder Borntraeger; see also [26].
26. G. Phragmén, *J. Inst. Metals*, **77**, 1950, 498.
27. W. L. Fink and K. R. Van Horn, *Trans. AIME*, **93**, 1931, 385, 394.
28. H. J. Axon and W. Hume-Rothery, *Proc. Roy. Soc. (London)*, **A193**, 1948, 1–24; W. Hume-Rothery and T. H. Boultbee, *Phil. Mag.*, **40**, 1949, 71–80.
29. A. P. Guljaev and E. F. Trusova, *Zhur. Tekh. Fiz.*, **20**, 1950, 66–78; *Structure Repts.*, **13**, 1950, 6.
30. Hanford Atomic Products Operations of the General Electric Co.; quoted by [31].
31. C. M. Craighead, E. W. Cawthorne, and R. I. Jaffee, *Trans. AIME*, **203**, 1955, 81–82.

$\bar{1}.3566$
0.6434

Al-Sn Aluminum-Tin

The liquidus curve, in whole or in part, was determined by [1–10]. The results by [6, 8, 10], especially, are in substantial agreement. The curve presented in Fig. 78 is based on the recent accurate work by [10]; since tabulated data were lacking, the curve was taken graphically from a diagram drawn in wt. %.

The eutectic temperature was found by various investigators to be 232 [6], 230–231 [8], 229 [1, 7, 9], 228–228.5 [4], and 228.3°C [10]. The eutectic composition (in wt. % Sn) was placed at 99.5 [1], 99.42 [9], 98.7 [11], and about 99.5 [10], the last value being given in Fig. 78.

Earlier estimates for the solubility of Sn in Al show very great discrepancies: values (in wt. % Sn) vary between nil [6, 8, 12], 2 [13], 2 at room temperature [14],

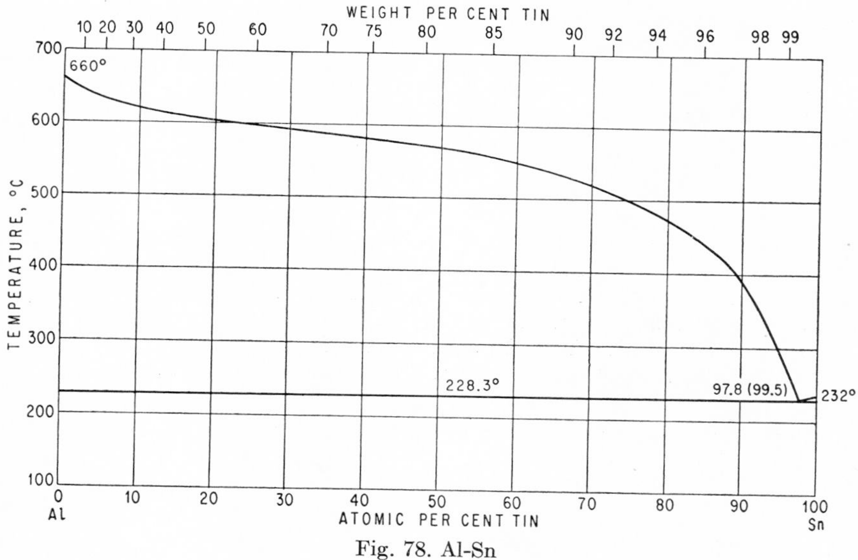

Fig. 78. Al-Sn

10 [3], and even 20 [5]. In more recent determinations by various methods, the solubility limit (in wt. % Sn) was placed as follows: (*a*) 0.05–0.06 at 500°C and smaller at both higher and lower temperatures (not more than a few thousandths of 1% at the eutectic temperature) [15, 16]; (*b*) 0.06 at 510°C, 0.045 at 400°C, and 0.025 at 300°C [17]; (*c*) of the order of 0.01 at 210–100°C [18]; (*d*) just below 0.05 at 530°C and well below 0.02 at 165°C [10]; (*e*) about 0.05 at 650°C, 0.09 at 620°C, 0.08 at 590°C, 0.06 at 560°C, 0.04 at 530°C, and negligibly small at 165°C [19]. The last-named values should be considered the most complete and reliable ones.

The solid solubility of Al in Sn has not been determined accurately as yet. For lattice parameters of Al-rich alloys, see [10, 18].

Supplement. [20] reported the following eutectic data: 228°C, 97.63 at. % Sn. The solid solubility of Sn in Al was found to be constant at 0.10 wt. (0.02_3 at.) % Sn between 530 and 640°C [21]. Reasons were given why the data of [19] are considered to be incorrect.

1. C. T. Heycock and F. H. Neville, *J. Chem. Soc.*, **57**, 1890, 385–386; **65**, 1894, 65–76.
2. Roland-Gosselin and H. Gautier, Société d'encouragement pour l'industrie nationale, Paris, Commission des alliages, "Contribution à l'étude des alliages," pp. 111–112, Typ. Chamerot et Renouard, Paris, 1901; *Bull. soc. encour. ind. natl.*, **1**, 1896, 1311.
3. W. Campbell and J. A. Mathews, *J. Am. Chem. Soc.*, **24**, 1902, 258–259; J. A. Mathews, *J. Franklin Inst.*, **153**, 1902, 123.
4. W. C. Anderson and G. Lean, *Proc. Roy. Soc. (London)*, **72**, 1903, 277–284.
5. E. S. Shepherd, *J. Phys. Chem.*, **8**, 1904, 233–274.
6. A. G. C. Gwyer, *Z. anorg. Chem.*, **49**, 1906, 311–316.
7. R. Lorenz and D. Plumbridge, *Z. anorg. Chem.*, **83**, 1913, 243–245.
8. E. Crepaz, *Giorn. chim. ind. ed appl.*, **5**, 1923, 115–122.
9. L. Losana and E. Carozzi, *Gazz. chim. ital.*, **53**, 1923, 546–547.
10. A. H. Sulley, H. K. Hardy, and T. H. Heal, *J. Inst. Metals*, **76**, 1949-1950, 269–281.
11. K. Kaneko and M. Kamiya, *Nippon Kogyo Kwaishi*, **40**, 1924, 509–516; taken from abstract in *J. Inst. Metals*, **36**, 1926, 436.
12. A. Müller, dissertation, Göttingen, 1926, p. 25.
13. M. Goto and T. Mishima, *Nippon Kogyo Kwaishi*, **39**, 1923, 714–721; taken from abstract in *J. Inst. Metals*, **36**, 1926, 433.
14. M. I. Zamotorin, *Trudy Leningrad Ind. Inst.*, **1936**, no. 4, 23; *Metallurg*, **1936**, no. 11, 103.
15 Unpublished work by W. L. Fink, reported by N. P. Nielsen and R. J. Nekervis, ASM "Metals Handbook," 1948 ed., p. 1166.
16. W. L. Fink in "Physical Metallurgy of Aluminum Alloys," p. 62, The American Society for Metals, Cleveland, Ohio, 1949.
17. Unpublished work by H. Y. Hunsicker, reported by [19].
18. T. A. Badaeva and R. I. Kuznetsova, *Doklady Akad. Nauk S.S.S.R.*, **72**, 1950, 507–509.
19. H. K. Hardy, *J. Inst. Metals*, **80**, 1951-1952, 431–434.
20. O. Hájiček, *Hutnické Listy*, **3**, 1948, 265–270; *Chem. Abstr.*, **43**, 1949, 4935.
21. L. E. Samuels, *J. Inst. Metals*, **84**, 1955-1956, 333–336.

$\bar{1}.4884$
0.5116

Al-Sr Aluminum-Strontium

The following conclusions [1] have been drawn from the results of X-ray studies of alloys with up to 50 at. (76.46 wt.) % Sr: (*a*) The solid solubility of Sr in Al is apparently negligibly small. (*b*) There are two intermediate phases, based on the compositions $SrAl_4$ (44.81 wt. % Sr) and SrAl. $SrAl_4$ possesses a homogeneity range which extends, at 700°C, from about 15 to 25.7 at. % Sr. SrAl dissolves some Sr and decomposes at temperatures up to at least 300°C into $SrAl_4$ and Sr. (*c*) The Al-$SrAl_4$ eutectic lies below 0.3 at. (about 1.0 wt.) % Sr. The eutectic temperature was not determined.

$SrAl_4$ is tetragonal of the $BaAl_4$ ($D1_3$) type, a = 4.46 A, c = 11.07 A, c/a = 2.48; and SrAl is b.c.c. of an unknown type with a = 15.8 A (a = 16.4 A if saturated with Sr) and 116 atoms per unit cell. The structure is probably a superstructure of the CsCl type.

1. H. Nowotny and H. Wesenberg, *Z. Metallkunde*, **31**, 1939, 363–364.

$\bar{1}.1736$
0.8264

Al-Ta Aluminum-Tantalum

The existence of the compound $TaAl_3$ (69.09 wt. % Ta), already suggested by [1], was established by [2]. It is isotypic ($D0_{22}$) with $TiAl_3$, a = 5.433 A, c = 8.553 A, c/a = 1.574 [2].

1. L. Marignac, *Compt. rend.*, **66,** 1868, 180–183.
2. G. Brauer, *Z. anorg. Chem.*, **242,** 1939, 1–22.

$\bar{1}.3252$
0.6748

Al-Te Aluminum-Tellurium

The phase diagram (Fig. 79) is based on thermal-analysis work [1]. Whether Al_2Te_3 (87.65 wt. % Te), identified earlier by [2], forms solid solutions also with Al has not been investigated. The composition of the solid solution at 414°C was determined only approximately, using the halting times of the eutectic crystallization at this temperature. Data on the transformation $\gamma \rightleftarrows \gamma'$ are very scanty; no thermal effects

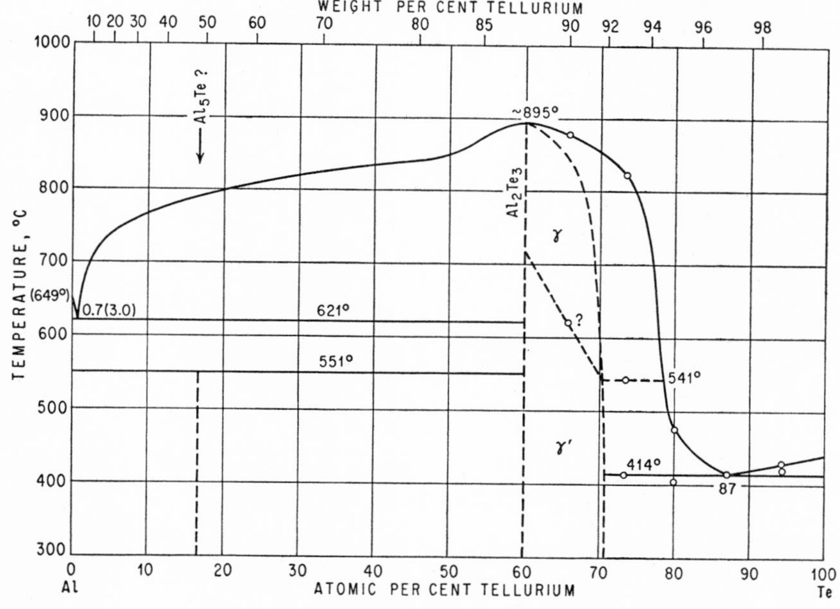

Fig. 79. Al-Te

were found in the phase field melt + Al_2Te_3. The transformation taking place at about 551°C in alloys with 0–85 wt. % Te was suggested to be due to the reaction $13Al + Al_2Te_3 \rightleftarrows 3\ Al_5Te$ [16.67 at. (48.61 wt.) % Te], since the largest thermal effect was found between 47 and 50 wt. % Te. Indeed, the alloy with 47 wt. % Te proved to be nearly single-phase. However, additional study appears to be necessary to clear up the nature of the reaction and the composition of the phase formed. Contrary to Fig. 79, [3] have asserted that Te is insoluble in liquid Al.

Supplement. The reinvestigation of the Al-Te system has been announced by [4], who studied the sublimation behavior of Al_2Te_3.

1. M. Chikashige and J. Nose, *Mem. Coll. Sci., Kyoto Imp. Univ.*, **2,** 1917, 227–232. The melting point of Al was found as 649°C. Alloys were melted in hydrogen atmosphere.
2. C. Whitehead, *J. Am. Chem. Soc.*, **17,** 1895, 849; see also L. Moser and K. Ertl, *Z. anorg. Chem.*, **118,** 1921, 271–273.
3. F. T. Sisco and M. R. Whitmore, *Ind. Eng. Chem.*, **16,** 1924, 838–841.
4. W. Klemm and H. Mindt, *Z. anorg. Chem.*, **255,** 1948, 287–293.

$\bar{1}.0653$
0.9347

Al-Th Aluminum-Thorium

Thermal [1, 2] and microscopic [1] investigations were used to establish the phase relationships up to 56 wt. (13 at.) % Th [1] and 50 wt. (10.4 at.) % Th [2] (Fig. 80). Results as to the liquidus curve are in substantial agreement; the eutectic point was given as 25.6 wt. (3.8 at.) % Th, 630°C [1], and 25.0 wt. (3.7 at.) % Th, 634°C [2].

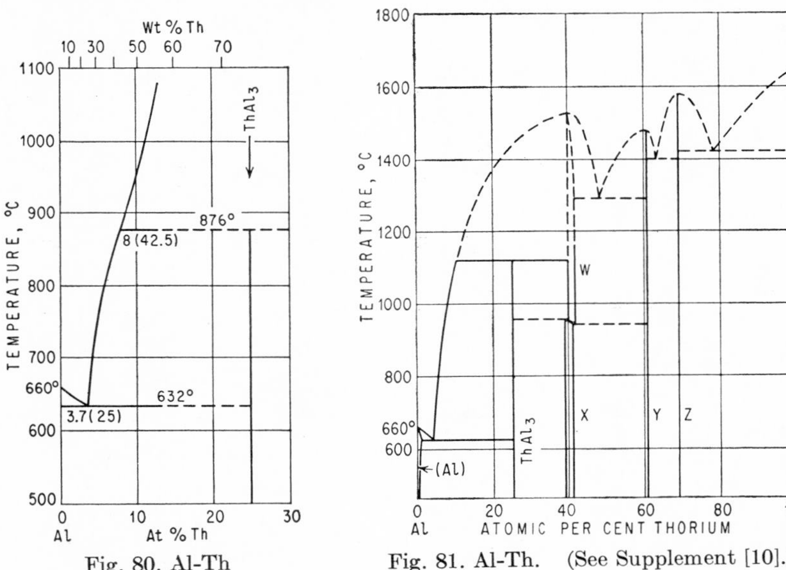

Fig. 80. Al-Th

Fig. 81. Al-Th. (See Supplement [10].)

There is difference of opinion as regards the composition of the phase formed peritectically at 880 [1] or 872°C [2]. Whereas [1] suggests that this is the compound $ThAl_3$ (74.15 wt. % Th), [2] believe it to have the composition $ThAl_7$ [12.5 at. (55.14 wt.) % Th]. The existence of $ThAl_7$ is not compatible with the fact that crystals of $ThAl_3$ have been isolated from alloys with a lower Al content [1, 3]. The phase crystallizing primarily above 876°C is not known. Perhaps its composition lies in the region of ThAl [4]. The solid solubility of Th in Al, first believed to be 0.24 at. (2.0 wt.) % Th at 610°C and smaller at lower temperatures [2], was later found to be negligibly small [5, 6], definitely below 0.01 wt. % at 620°C in Al of 99.99% purity [6]; see also [7].

$ThAl_3$ has a hexagonal structure with $a = 6.49$ A, $c = 4.61$ A, and 8 atoms per unit cell [8].

Supplement. In a review on thorium metallurgy [9], the phase diagram shown in Fig. 81 was published without any comments. The original investigation appears to be still classified. The diagram exhibits the following compounds: $ThAl_3$, Th_2Al_3 (high- and low-temperature form), $\sim Th_3Al_2$, $\sim Th_2Al$.

In the course of an X-ray investigation of the Th-Al system, [10] have discovered the existence of the following six intermetallic compounds: $ThAl_3$, hexagonal structure with $a = 6.499$ A, $c = 4.626$ A, and 8 atoms per unit cell (cf. above); $ThAl_2$, hexagonal AlB_2 (C32) type of structure, $a = 4.393$ A, $c = 4.164$ A; $\sim Th_4Al_7$, stable only in a narrow temperature range at about 1300°C, tetragonal unit cell with $a = 9.86$ A, $c = 7.81$ A; ThAl, orthorhombic with $a = 11.45$ A, $b = 4.42$ A, $c = 4.19$ A, and 8 atoms per unit cell; Th_3Al_2, probably stable only above 1100°C, tetragonal structure with $a = 8.13$ A, $c = 4.22$ A, and 10 atoms per unit cell; Th_2Al, tetragonal, $a = 7.62$ A, $c = 5.86$ A, 12 atoms per unit cell.

By means of neutron diffraction, [11] confirmed the C32 type of structure proposed by [10] for $ThAl_2$.

Note Added in Proof. The crystal structures of $ThAl_2$, Th_3Al_2, and Th_2Al reported by [12] are in agreement with those found by [11].

1. A. Leber, *Z. anorg. Chem.*, **166,** 1927, 16–26.
2. G. Grube and L. Botzenhardt, *Z. Elektrochem.*, **48,** 1942, 418–425.
3. O. Hönigschmid, *Compt. rend.*, **142,** 1906, 280–281.
4. B. W. Mott and H. R. Haines, *J. Inst. Metals*, **80,** 1951-1952, 629–636.
5. G. Grube, *Z. Elektrochem.*, **49,** 1943, 57.
6. H. Bückle, *Z. Metallkunde*, **37,** 1946, 43–47.
7. J. D. Grogan and T. H. Schofield, Aeronautical Research Committee, Reports and Memoranda 1253, 1929; abstract in *J. Inst. Metals*, **44,** 1930, 488.
8. G. Brauer, *Naturwissenschaften*, **26,** 1938, 710.
9. J. R. Keeler, *Chem. Engr. Progr., Symposium Ser.* no. 11, **50,** 1954, 57–61 (Nuclear Engineering, Part I, Thorium Metallurgy). The diagram bears the source reference Battelle Memorial Institute, July 1, 1952.
10. P. B. Braun and J. H. N. van Vucht, *Acta Cryst.*, **8,** 1955, 117, 246.
11. A. F. Andresen and J. A. Goedkoop, *Acta Cryst.*, **8,** 1955, 118.
12. J. R. Murray, *J. Inst. Metals*, **84,** 1955-1956, 91–96.

$\bar{1}.7507$
0.2493

Al-Ti Aluminum-Titanium

Alloys with 0–75 At. % Al ($TiAl_3$). The phase relationships in this range of composition were established by [1] following work by [2] and [3], who located the phase boundaries by X-ray diffraction at 750°C [2] and by micrographic analysis between 1000 and 1100°C [3]. The agreement is fairly good. Work by [4] was done with somewhat contaminated alloys and therefore need not be considered here.

The phase diagram of the partial system (Fig. 82) is based on a thorough micrographic analysis of heat-treated arc-melted alloys prepared from iodide Ti [1]. As compared with the tentative diagram by [3], which shows the peritectic reaction liq. $+ \beta \rightleftarrows \alpha$ at about 1630°C, data by [1] show conclusively that the $\alpha + \beta$ field gives rise to a peritectoid reaction ($\beta + \gamma \rightleftarrows \alpha$) at approximately 1240°C and that there is the peritectic reaction liq. $+ \gamma \rightleftarrows \beta$ at about 1460°C. The boundaries of the $\alpha + \beta$ field, determined by [3] up to 1050°C, differ from those by [1] (Fig. 82) by less than 1 wt. % Al. (cf. Supplement.) The solidus temperatures obtained by incipient melting studies [1] were accurate to the order of ± 15°C.

Alloys with 75–100 At. % Al. Early investigations [5–7] have shown that the phase coexisting with Al is the compound $TiAl_3$ (62.82 wt. % Al). The liquidus and solidus curves of the Al-rich alloys were determined by [6–8]. On the basis of these studies, which were confined to a maximum Ti content of 37.3 wt. % ($TiAl_3$), with a liquidus point of 1355°C, it was concluded that $TiAl_3$ melts without decomposition. However, there is no doubt that $TiAl_3$ participates in a peritectic reaction, at about 1340°C [1]. $TiAl_3$ has a certain range of homogeneity [9, 1], which becomes narrower with fall in temperature [1].

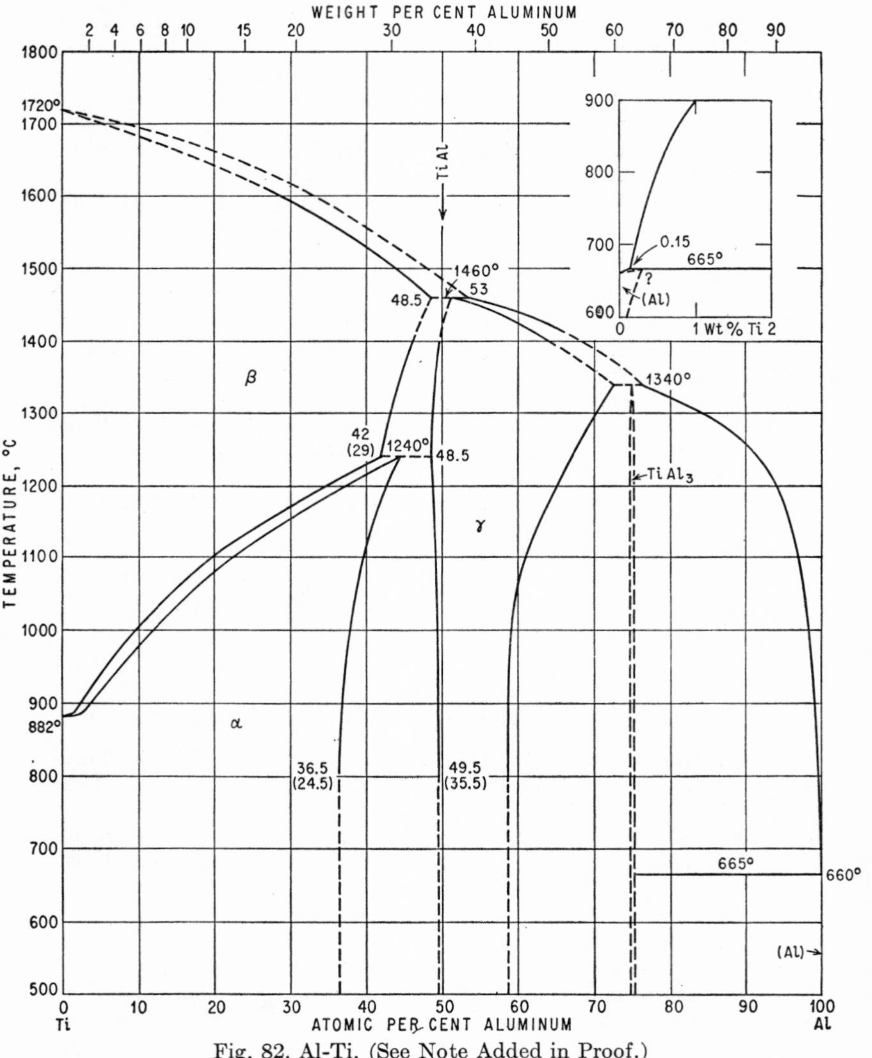

Fig. 82. Al-Ti. (See Note Added in Proof.)

The solidus temperature was formerly believed to coincide with the melting point of Al [6, 7] or to lie only very slightly below this temperature [8]. At present, it is generally accepted that a peritectic reaction liq. (0.15 wt. % Ti) + $TiAl_3 \rightleftarrows$ (Al) exists [10, 11] at 665°C [8, 12].

It has been reported [10, 11] that the limit of solid solubility is 0.24 wt. (about 0.14 at.) % Ti at 510°C and probably about 1.2 wt. (?) (0.7 at.) % Ti at 665°C. Data based on microhardness measurements [13] are not compatible with those by [10, 11] in that they indicate solubility limits of 0.11, 0.08, and 0.07 wt. % Ti at 600, 500, and 400°C, respectively. [14] published a diagram showing the maximum solubility to be 0.28 wt. % Ti. On very rapid solidification, supersaturated solid solutions of Ti in Al are formed [15–18, 13], up to 0.32 wt. (0.2 at.) % Ti [18].

The lattice parameters of the h.c.p. α solid solution were determined by [19, 20], those of the (Al) phase by [18]. The crystal structure of the γ phase is based on the composition TiAl (36.03 wt. % Al). It is ordered f.c. tetragonal of the CuAu type [3, 2, 1] with a = 3.99 A, c = 4.07 A, c/a = 1.02 [1]; for change of parameters with composition, see [3, 2, 1]. Either the ordered structure is stable up to high temperatures, or the disorder → order transformation, if any, cannot be suppressed by quenching.

$TiAl_3$ has a tetragonal lattice ($D0_{22}$ type), a = 5.435 A, c = 8.591 A, c/a = 1.581 [8]; a = 5.436 A, c = 8.596 A, c/a = 1.581 [21].

Supplement. [22] made a detailed study of the effect of small additions of Al (up to 7 at.%) on the transformation in Ti by means of the "hydrogen-pressure" method. It was found that the first additions of Al do not greatly increase the transformation temperature (Fig. 82). Above 4 at. % Al, the boundaries of the $\alpha + \beta$ field are in excellent agreement with those determined micrographically by [1].

[23] studied the mechanism by which the γ-phase lattice (CuAu type) adjusts itself to an increasing Al content. X-ray diffraction intensity measurements showed that there is substitution of Ti by Al atoms.

Note Added in Proof. More recent investigations [24–28] have shown that the primary solid solution of Al in α-Ti extends to only about 5.5 wt. (9.5 at.) % Al at lower temperatures and that one or two intermediate phases exist in the composition range previously assumed to be single-phase α. (Ti_3Al) is of the Mg_3Cd ($D0_{19}$) type.

1. E. S. Bumps, H. D. Kessler, and M. Hansen, *Trans. AIME*, **194**, 1952, 609–614.
2. P. Duwez and J. L. Taylor, *Trans. AIME*, **194**, 1952, 70–71.
3. H. R. Ogden, D. J. Maykuth, W. L. Finlay, and R. I. Jaffee, *Trans. AIME*, **191**, 1951, 1150–1155.
4. W. Gruhl, *Metall*, **6**, 1952, 134–135; see also *Trans. AIME*, **194**, 1952, 1213–1214.
5. W. Manchot and P. Richter, *Liebigs Ann.*, **357**, 1907, 140.
6. E. van Erkelenz, *Metall u. Erz*, **20**, 1923, 206–210.
7. W. Manchot and A. Leber, *Z. anorg. Chem.*, **150**, 1926, 26–34.
8. W. L. Fink, K. R. Van Horn, and P. M. Budge, *Trans. AIME*, **93**, 1931, 421–436.
9. K. Schubert, *Z. Metallkunde*, **41**, 1950, 420–421.
10. W. L. Fink and L. A. Willey, "Metals Handbook," 1948 ed., p. 1167, American Society for Metals, Cleveland, Ohio.
11. W. L. Fink, in "Physical Metallurgy of Aluminum Alloys," pp. 68–70, American Society for Metals, Cleveland, Ohio, 1949.
12. H. Nishimura and E. Matumoto (*Nippon Kinzoku Gakkai-Shi*, **4**, 1940, 339–343) gave the composition of the peritectic melt at 665°C as 0.05 wt. % Ti.
13. H. Bückle, *Z. Metallkunde*, **37**, 1946, 43–47.
14. L. F. Mondolfo, "Metallography of Aluminum Alloys," pp. 45–47, John Wiley & Sons, Inc., New York, 1943.
15. H. Bohner, *Z. Metallkunde*, **26**, 1934, 268–271.
16. H. Hanemann and A. Schrader, "Atlas Metallographicus," vol. III, part 1, Verlagsbuchhandlung Gebrüder Borntraeger, Berlin, 1943.
17. W. Hofmann, *Abhandl. braunschweig. wiss. Ges.*, **1**(1), 1949, 83–88.

18. G. Falkenhagen and W. Hofmann, *Z. Metallkunde,* **43,** 1952, 69–81.
19. W. Rostoker; see [1].
20. W. Rostoker, *Trans. AIME,* **194,** 1952, 212–213.
21. G. Brauer, *Z. anorg. Chem.,* **242,** 1939, 1–22.
22. A. D. McQuillan, *J. Inst. Metals,* **83,** 1955, 181–184.
23. R. P. Elliott and W. Rostoker, *Acta Met.,* **2,** 1954, 884–885.
24. E. Ence and H. Margolin, *Trans. AIME,* **209,** 1957, 484–485. (Results first reported in 1955.)
25. K. Sagel, E. Schulz, and U. Zwicker, *Z. Metallkunde,* **46,** 1956, 529–534. See the tentative diagram in this paper.
26. F. A. Crossley and W. F. Carew, *Trans. AIME,* **209,** 1957, 43–46.
27. K. Anderko, K. Sagel, and U. Zwicker, *Z. Metallkunde,* **48,** 1957, 57–58. (Results first reported in 1956.)
28. D. Clark and J. C. Terry, Letter to the Editor, *Bull. Inst. Metals,* **3,** 1956, 116.

$\bar{1}.1206$
0.8794

Al-Tl Aluminum-Thallium

Aluminum and thallium appear to be entirely insoluble in each other in the liquid state, as indicated by the fact that the melting points of Al and Tl are not affected by additions of the other component [1, 2]; see Fig. 83 with the data points reported by [2].

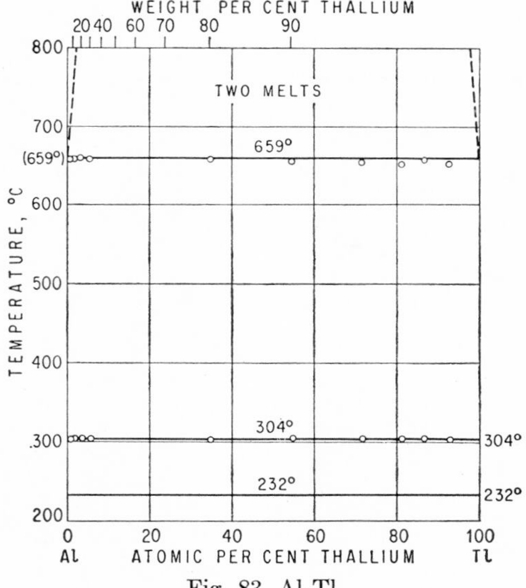

Fig. 83. Al-Tl

1. F. Doerinckel, *Z. anorg. Chem.,* **48,** 1906, 188–190. The metals used were of a low degree of purity. Their melting points were found to be 654 and 315°C.
2. E. Raub and M. Engel, *Z. Metallkunde,* **37,** 1946, 148–149. The melting points of the components were found to be 659 and 304°C.

$\bar{1}.0543$
0.9457

Al-U Aluminum-Uranium

The diagram of Fig. 84 is based on thermal (for the three-phase equilibria) [1], microscopic [1], and X-ray analysis data [1–3]. The liquidus curve between 43 and 100 at. % U is hypothetical, with the exception that the melting point of UAl_2 was

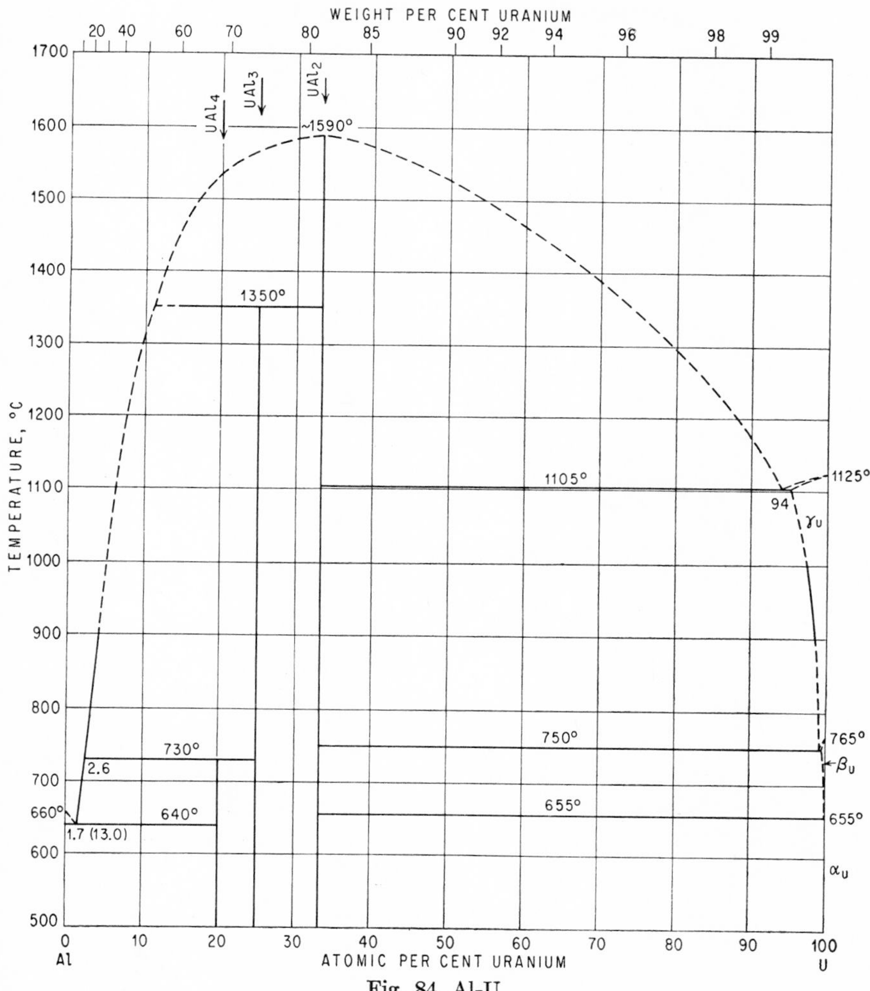

Fig. 84. Al-U

determined to be approximately 1590°C. Between 0 and 43 at. % U the liquidus curve is based on chemical-analysis work. Only the solid solubility of Al in γ-U was approximately determined metallographically.

The crystal structures of the three intermediate phases are as follows: UAl_4 (68.81 wt. % U), b.c. orthorhombic with 4 formula weights per unit cell, $a = 4.41$ A,

$b = 6.27$ A, $c = 13.71$ A [3]; UAl_3 (74.63 wt. % U), f.c.c. of the Cu_3Au ($L1_2$) type, $a = 4.287$ A [2]; UAl_2 (81.52 wt. % U), f.c.c. of the $MgCu_2$ (C15) type, $a = 7.811$ A [2].

Supplement. The solubility curve of Al in γ-U has been determined metallographically [4]. The solubility found is somewhat higher than that shown in Fig. 84: 0.7, 0.5, 0.35, and 0.25 wt. (5.9, 4.2, 3.0, and 2.2 at.) % Al at 1100, 1000, 900, and 800°C, respectively. The eutectoid point is located near 0.2 wt. (1.7 at.) % Al.

[5] reported that an addition of 0.4 wt. % Al to U made it possible to retain the β-U structure at room temperature.

1. P. Gordon and A. R. Kaufmann, *Trans. AIME*, **188,** 1950, 182–194.
2. R. E. Rundle and A. S. Wilson, *Acta Cryst.*, **2,** 1949, 148–150.
3. B. S. Borie, *Trans. AIME*, **191,** 1951, 800–802.
4. G. Cabane, "Proceedings of the International Conference on the Peaceful Uses of Atomic Energy," vol. 9, pp. 120–121, Geneva, 1955.
5. M. Englander and J. Lehmann, "Proceedings of the International Conference on the Peaceful Uses of Atomic Energy," vol. 9, pp. 126, 128, 136, Geneva, 1955.

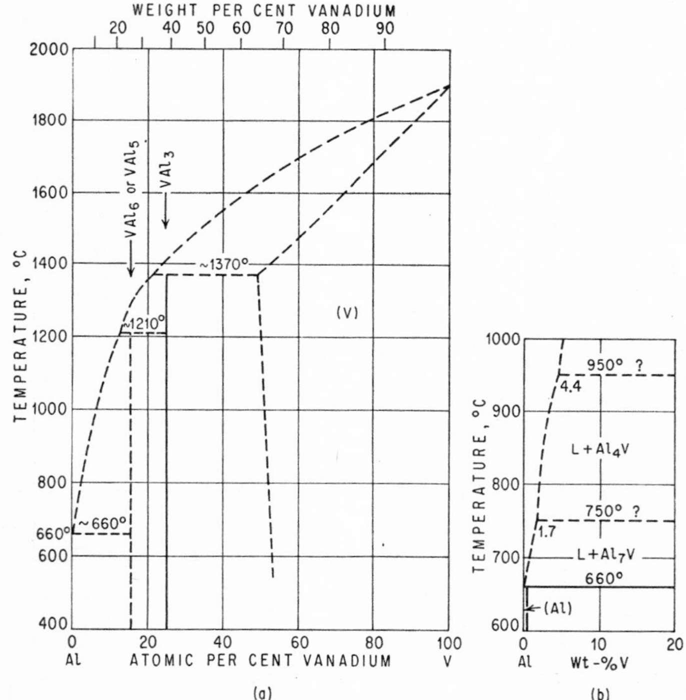

Fig. 85. Al-V. (See also Fig. 86.)

$\bar{1}.7239$
0.2761

Al-V Aluminum-Vanadium

By microscopic examination of alloys prepared by aluminothermic reduction of V_2O_5, the existence of the intermediate phases VAl_3, VAl_2 or V_2Al_3, VAl, and V_2Al was suggested [1]. These conclusions were based on inconclusive evidence, however.

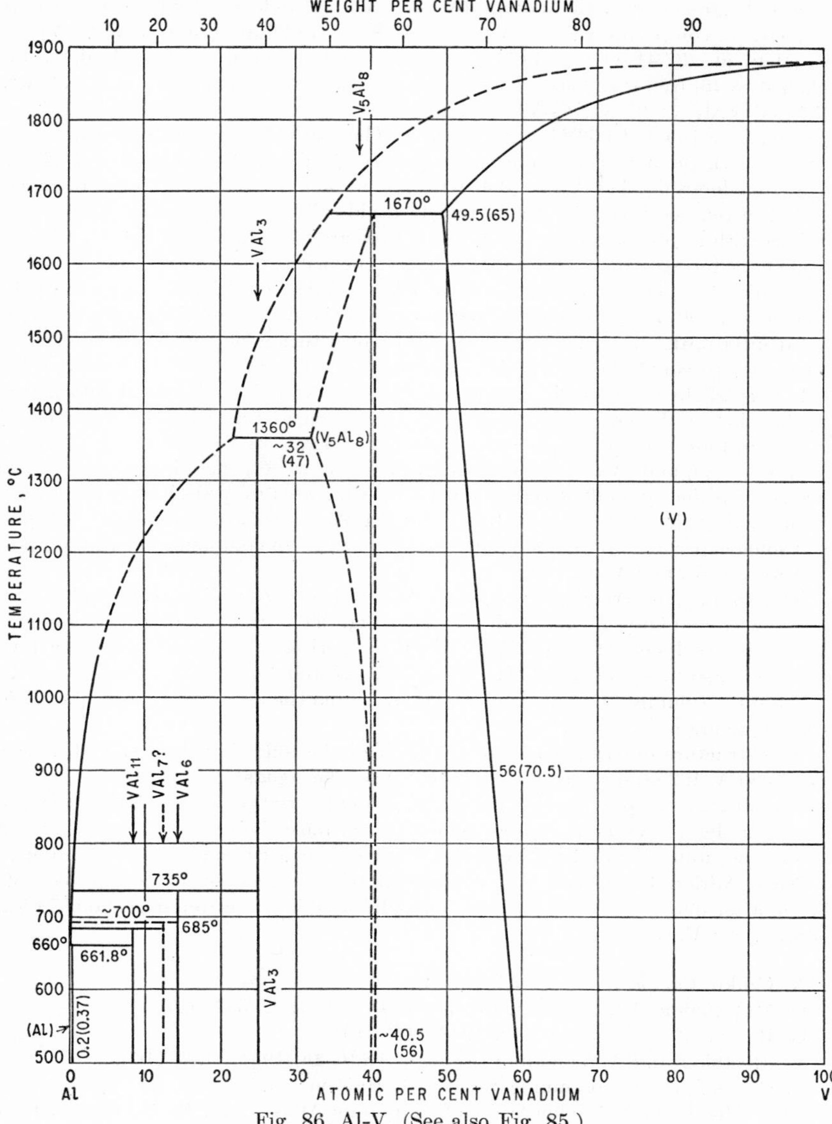

Fig. 86. Al-V. (See also Fig. 85.)

Crystals of the approximate composition of VAl_3 (38.63 wt. % V) were isolated from an alloy with 18.5 at. % V [1], and a residue having the approximate composition of VAl was isolated from an alloy with 64 at. % V [2]. Later, VAl_3 was definitely established by determining its crystal structure [3].

The tentative diagram shown in Fig. 85*a* is based on microscopic examination of 12 arc-cast alloys with 12.5–98 at. % V, after annealing at 900°C for 1 week and quenching [4]. The solid solubility of 50–55 at. (about 35–39 wt.) % Al in V is in substantial agreement with findings by [5]. The Al-rich compound could not be identified; its composition lies at or between VAl_6 [14.29 at. (23.94 wt.) % V] and VAl_5 [16.67 at. (27.41 wt.) % V] [4]. A tentative diagram of the Al-rich alloys was published by [6] (see Fig. 85*b*), with the two probable phases VAl_7 [12.5 at. (21.25 wt.) % V] and VAl_4 (32.07 wt. % V) being given as the phases formed peritectically at approximately 750 and 950°C, respectively. The 660°C horizontal is shown to be a peritectic with the melt containing about 0.1 wt. % V. The solid solubility of V in Al has been determined as 0.2 at. (0.37 wt.) % V at 630°C and does not seem to change with fall in temperature [7]. By extremely rapid solidification, supersaturated solid solutions with as much as 0.55 at. (1.0 wt.) % V are formed [8]; see also [9, 10].

VAl_3 is tetragonal of the $TiAl_3$ ($D0_{22}$) type with $a = 5.345$ A, $c = 8.322$ A, $c/a = 1.557$ [3]. The powder pattern of VAl_5 or VAl_6 could be indexed on a cubic lattice, $a = 14.50$ A [4]. For lattice parameters of the Al-rich solid solution, see [7, 8].

Supplement. [11] have examined the constitution of the entire system by means of microscopic, thermal, and X-ray analysis. The results of this investigation, which were slightly modified (as regards Al-rich alloys) by additional information published in the discussion on this paper, are shown in Fig. 86. [11] reported the existence of four intermediate phases: VAl_{11} (α) (8.33 at. % V), $\sim VAl_6$ (β) (14.29 at. % V), VAl_3 (γ), and (V_5Al_8) (δ) (38.46 at. % V); however, there is considerable evidence [12, 13] that actually a fifth compound exists, possibly VAl_7, having a peritectic horizontal very close to that of VAl_{11} [13] (in Fig. 86, ~700°C is assumed). [14] published some experimental data supporting the existence of compounds of the composition VAl_{11}, VAl_6, and VAl_3.

The solidus temperature for Al-rich alloys (661.8°C) as well as the liquidus up to 1050°C was carefully established by [12] (Fig. 86). Serious discrepancies exist between the results of [11] and [12] as regards the peritectic temperatures of the VAl_{11} and VAl_6 phases: 685 and 735°C [11] (Fig. 86), 727 and >850°C [12], respectively.

As to the solubility of V in solid Al, [11] accepted the results of [7] (see above) for his phase diagram.

The structure determinations of [11] yielded the following results: VAl_{11}, f.c.c., $a = 14.586$ A, 192 atoms per unit cell [15]; $\sim VAl_6$, hexagonal, $a = 7.718$ A, $c = 17.15$ A; VAl_3, structure as previously reported [3], a small variation in the c lattice constant indicates a slightly variable composition; (V_5Al_8), b.c.c., γ-brass type, $a = 9.207$ A, 52 atoms per unit cell; (V), lattice parameter increases with increasing Al content.

Note Added in Proof. [16] reported lattice parameters of the V-rich solid solution after quenching from 980°C. The solid solubility was found as about 53.5 at. (~38 wt.) % Al.

1. N. Czako, *Compt. rend.*, **156,** 1913, 140–142.
2. C. Matignon and E. Monnet, *Compt. rend.*, **134,** 1902, 542–545.
3. G. Brauer, *Z. Elektrochem.*, **49,** 1943, 208–210.
4. W. Rostoker and A. Yamamoto, *Trans. ASM*, **46,** 1954, 1136–1163.
5. P. Duwez, private communication to M. Hansen.
6. L. F. Mondolfo, "Metallography of Aluminum Alloys," pp. 48–49, John Wiley & Sons, Inc., New York, 1943.
7. A. Roth, *Z. Metallkunde*, **32,** 1940, 356–359.
8. G. Falkenhagen and W. Hofmann, *Z. Metallkunde*, **43,** 1952, 69–81.
9. Anon., *Aluminium Broadcast*, **3,** 1931, no. 5, 12–13; abstract in *J. Inst. Metals*, **50,** 1932, 472. A solid solubility of 0.65 wt. (0.35 at.) % V was reported.
10. C. Panseri and M. Monticelli, *Alluminio*, **17,** 1948, 335–338.
11. O. N. Carlson, D. J. Kenney, and H. A. Wilhelm, *Trans. ASM*, **47,** 1955, 520–536; discussion, pp. 537–542.
12. R. P. Elliott and L. F. Mondolfo, discussion on [11], pp. 538–540.

13. Carlson, Kenney, Wilhelm, and J. F. Smith, discussion on [11], pp. 541–542.
14. F. R. Morral, discussion on [11], p. 538.
15. In referring to the cubic VAl_{5-6} phase claimed by [4], [11] stated that "comparison of their data for the $Al_{5-6}V$ with the results obtained on the $Al_{11}V$ of this investigation indicates strongly that these are one and the same phase."
16. C. B. Jordan and P. Duwez, *Trans. ASM*, **48**, 1956, 789–790.

$\bar{1}.1664$
0.8336

Al-W Aluminum-Wolfram

The existence of the intermediate phases WAl_7 [1], WAl_5 [1], WAl_4 [2, 3], WAl_3 [3], and W_2Al [3] was suggested; however, the conclusions were based on inconclusive and partially contradictory evidence.

The diagram of Fig. 87 is due to [4]. It is based on thermal-analysis data

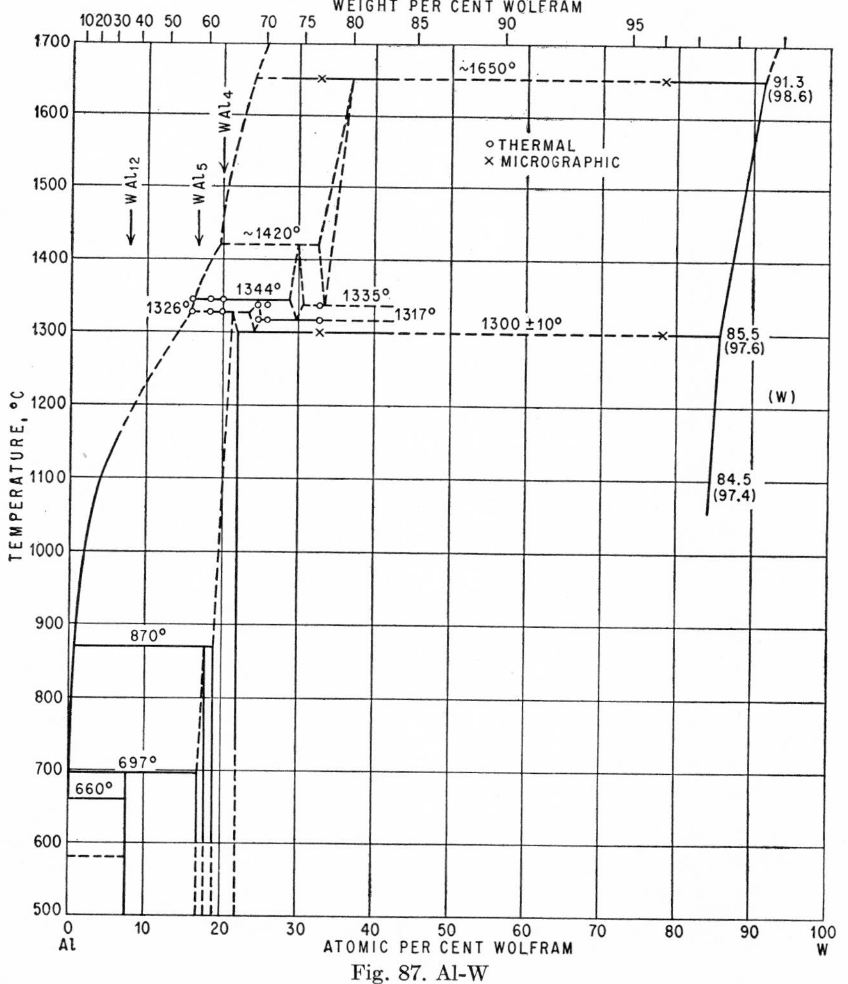

Fig. 87. Al-W

(liquidus curve determined up to 6 at. % W) and microscopic investigation of heat-treated alloys prepared by melting (0–30 wt. % W) and sintering (30–100 wt. % W) in vacuum at temperatures between 1200 and 1550°C, according to the composition [5]. The phase relations and compositions of the three intermediate phases in the range 0–20 at. % W appear to be well established. These phases were designated with chemical formulas WAl_{12} [7.69 at. (36.23 wt.) % W], WAl_5 [16.67 at. (57.69 wt.) % W], and WAl_4 (63.02 wt. % W); the author used the Greek letters γ, δ, and ϵ, respectively. The interpretation of the transformations observed in the alloys with about 56.5–77 wt. (about 16–33 at.) % W is largely hypothetical. There is some evidence for the formation of another phase by reaction of Al and WAl_{12} at temperatures below 580°C, but this reaction is not reversible. The solid solubility of W in Al at 650°C is reported to be 1.5 wt. (about 0.25 at.) % W.

WAl_4 was reported to have a monoclinic structure with a large unit cell [4].

Supplement. According to [6], WAl_{12} has a b.c.c. structure with a = 7.580 A and two WAl_{12} units per cell.

1. J. A. Mathews and W. Campbell, *J. Am. Chem. Soc.*, **24**, 1902, 253–266.
2. F. Wöhler and F. Michel, *Liebigs Ann.*, **115**, 1860, 103.
3. L. Guillet, *Compt. rend.*, **132**, 1901, 1112–1115.
4. W. D. Clark, *J. Inst. Metals*, **66**, 1940, 271–286.
5. Alloys were prepared in 1 wt. % intervals in the ranges 33–40, 55–75, and 94–100 wt. % W (see Fig. 87).
6. J. Adam and J. B. Rich, *Acta Cryst.*, **7**, 1954, 813–816.

$\bar{1}$.6156
0.3844

Al-Zn Aluminum-Zinc

In reviewing the literature, one recognizes two distinct periods of endeavor to establish the constitution of this system. The transition lies in the year 1932 when it was first proposed [1] that there was no peritectically formed intermediate phase of variable composition, but a closed miscibility gap between Al-base solid solutions of the same structural type. In 1944, a composite equilibrium diagram was suggested by [2]. Figure 88 differs only insignificantly from this diagram and that given by [3].

Prior to 1932, numerous phase diagrams were published [4–15], all of which are obsolete, with the exception of the liquidus curves and temperatures of the three-phase equilibria. The diagrams of [9, 12, 13] are very similar in that they show (a) an intermediate phase of variable composition around 60 at. % Zn, generally denoted β, which is formed by a peritectic reaction at about 443°C of melt with the Al-base solid solution containing about 48 at. (69 wt.) % Zn; (b) a eutectic of β and the Zn-base solid solution at 88.7 at. (95 wt.) % Zn, 380°C; and (c) a eutectoid decomposition of β, with the eutectoid point at about 60 at. (78.4 wt.) % Zn, 270–280°C.

Since 1932, most attention was devoted to the determination of (a) the solidus curve, (b) the solid solubility of Zn in Al, including the boundary of the closed miscibility gap above 275°C, and (c) the solid solubility of Al in Zn.

The Liquidus Curve [17, 18, 4–9, 11–16, 19–27]. The curve shown in Fig. 88 is that given by [2]. It is based on the results of [21], 0–4.4 at. (10 wt.) %; [20], 0–49 at. (70 wt.) %; [19], 21.6–90.8 at. (40–96 wt.) %; [14], 88.7–100 at. (95–100 wt.) % Zn. The more recent data of [22, 23] are in excellent agreement with this curve; those of [24, 25, 26] are several °C lower. Liquidus temperatures reported by [27] are themselves inconsistent and too high.

The Solidus Curve. The formerly accepted peritectic horizontal at about 443°C [6, 8, 10–14] does not exist. This was first conclusively established by [19], after the work of [28] already had shown that a three-phase equilibrium at this temperature was unlikely to occur [29]. The small heat changes observed on cooling were interpreted to be caused by the marked change in the composition of the phase crystallizing from the liquid taking place over a narrow range of temperature [19]. Similar conditions exist in the system Au-Pt, to which the reader is referred. According to [30], the thermal effect is caused by segregation during cooling.

The solidus curve of the α phase was determined, in full or in part, by [10, 13, 14, 19, 20, 31, 32, 21, 23, 25, 26, 27, 33, 34], using various classical methods as well as high-temperature X-ray analysis [32, 34]. Results covering the entire composition range [13, 19, 20, 26, 27, 34] agree in that they show the curve to have an inflection point between 45 and 50 at. % Zn. In the region 0–15 at. % Zn, temperatures by

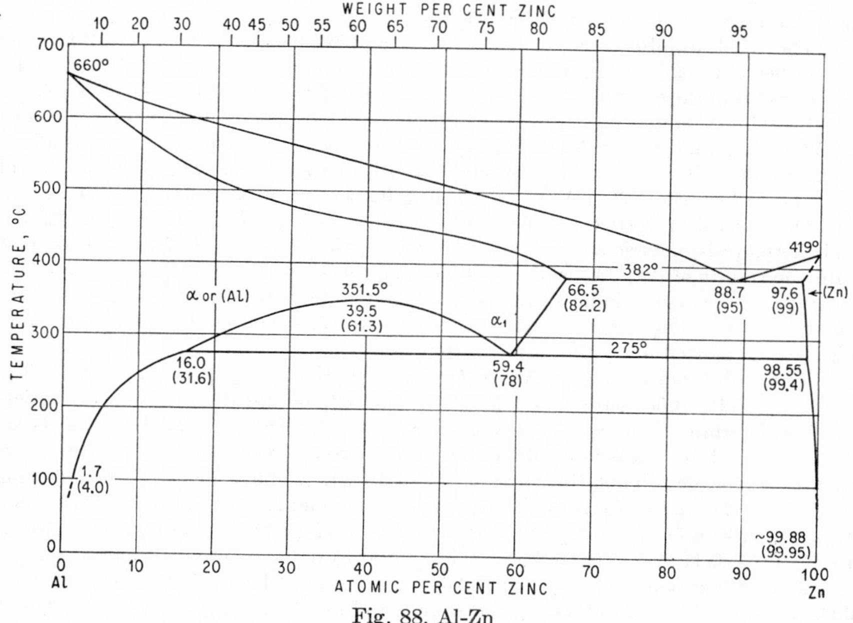

Fig. 88. Al-Zn

[23], based on micrographic work, are to be given preference. Between 15 and 66.5 at. % Zn, the micrographic data of [19, 26] and the X-ray data of [34] agree remarkably well; temperatures differ by about 10°C in the range 20–35 at. % Zn and to a lesser degree between 35 and 60 at. % Zn. Above 15 at. % Zn, the curve shown in Fig. 88 is a compromise of the results of [26, 34]. The dilatometrically determined solidus temperatures given by [33] lie at much lower temperatures above 40 at. % Zn and on a nearly straight line connecting the points 38 at. % Zn, 452°C, and 62 at. % Zn, 394°C.

There has been speculation as to the reason for the unusual shape of the solidus of the α phase. Whereas [19] suggested the formation of a metastable equilibrium, [35] believed the shape to be caused by the existence of an ordered structure, thermally stable up to the beginning of melting. The latter is very unlikely, since no indication of an ordered structure was found by other investigators [28, 34, 36]. Nevertheless, the question arises whether the solidus above 15 at. % Zn, shown in Fig. 88, represents the state of equilibrium, i.e., whether, under equilibrium conditions (absence of segregation), the curve might be located at higher temperatures and have no inflection point. It should be stated, however, that the specimens used [19, 26, 34] had been subjected to prolonged homogenization anneals at temperatures of 350–375°C. [37] gave microscopic evidence that the forged specimens did not show coring after annealing for several days at 365°C and quenching. Specimens used by [26] and [34] were given anneals of 14 days at 350°C (apparently in the as-cast state) and 3 weeks at 375°C, respectively. It appears, therefore, that the curve shown is an equilibrium curve.

The solidus curve of the Zn-rich phase has not been determined as yet.

The Solid Solubility of Zn in Al. Results of investigations prior to [28] differed considerably [8, 10, 11, 38, 13, 39, 14, 40, 1]; those of [1], covering the temperature range 160–345°C and based on lattice-spacing determinations of quenched alloys, appeared to be most reliable [41].

The whole solubility curve between 2.3 and 64.4 at. % Zn (125–382°C) was established by [28], using thermoresistometric and micrographic methods. The critical point of the closed region of immiscibility was found as 38.5 at. (60.3 wt.) % Zn and 353°C. A more comprehensive determination of the boundary in the range 27–52.5 at. % Zn by means of measurements of resistance vs. temperature of 34 alloys is due to [36]. This curve lies only about 2–6°C below that of [28], with the critical point being located at 39.5 ± 0.2 at. (61.3 ± 0.2 wt.) % Zn and 351.5 ± 0.4°C.

Data based on lattice-spacing determinations at high temperatures [42, 32, 34] and thermoresistometric studies [43], particularly those of [34], are in fair agreement with the results of [28, 36], while results of [44], based on measurements of susceptibility vs. temperature, indicate the critical point to be located at about 29–38 at. % Zn and 338°C. For a few individual data see [45–47]. The solid-solubility curve in Fig. 88 was plotted using the results of [28, 36].

The Solid Solubility of Al in Zn was investigated by [13, 14, 48–54]. The results of [52] (250–350°C) and [53] (100–350°C), obtained by means of lattice-spacing determinations, are in very good agreement. Data reported by [51] (100–350°C), based on measurements of magnetic susceptibility vs. temperature, are slightly higher, but do not differ by more than 0.1 wt. % above 150°C. The following solubility values in at. (wt.) % Al are the most reliable: 2.4 (1.0) at 380°C, 2.1 (0.88) at 350°C, 1.7 (0.7) at 300°C, 1.2 (0.5) at 250°C, 0.77 (0.32) at 200°C, 0.43 (0.18) at 150°C, 0.24 (0.1) at 100°C, and approximately 0.15 (0.05) at room temperature.

The boiling points of melts with 2.6 (6), 9.35 (20), 21.6 (40), 38.2 (60), 49 (70), 62.3 (80), and 78.8 at. (90 wt.) % Zn were found as 1400, 1175, 1050, 990, 970, 950, and 926°C, respectively [55].

Crystal Structures. The structural identity of the Al-rich terminal solid solution and the formerly assumed intermediate phase β was recognized by [56, 58, 1, 59, 42, 57, 60] and all later investigators. Lattice spacings of the α phase, not considering individual values, were reported by [61, 40, 1, 59, 42, 45, 32, 35, 62, 63, 64, 34], including parameters for elevated temperatures [40, 59, 42, 45, 32, 35, 63, 34]. The most comprehensive X-ray study of the α phase is that of [34], whose paper gives data for temperatures ranging between 25 and 620°C. Results indicate marked anomalies in the lattice parameter, also observed by [35].

For lattice constants of the Zn-rich solid solution, see [59, 60, 57, 49, 50, 52, 53].

1. E. Schmid and G. Wassermann, *Metallwirtschaft*, **11**, 1932, 386–387; *Z. Metallkunde*, **26**, 1934, 145–150.
2. G. V. Raynor, Annotated Equilibrium Diagram Series, no. 1, The Institute of Metals, London, 1944.

3. E. Gebhardt, *Z. Metallkunde*, **40,** 1949, 463–464.
4. E. S. Shepherd, *J. Phys. Chem.*, **9,** 1905, 504–512.
5. D. Ewen and T. Turner, *J. Inst. Metals*, **4,** 1910, 140–156.
6. W. Rosenhain and S. L. Archbutt, *Phil. Trans. Roy. Soc.* (*London*), **A211,** 1911, 315–343; *J. Inst. Metals*, **6,** 1911, 236–250.
7. G. Eger, *Int. Z. Metallog.*, **4,** 1913, 35–41.
8. O. Bauer and O. Vogel, *Mitt. deut. Materialprüfungs-Anst.*, **33,** 1915, 146–168; *Int. Z. Metallog.*, **8,** 1916, 101–132.
9. A. S. Fedorow, *Zhur. Russ. Fiz.-Khim. Obshchestva*, **49,** 1917, 394–407; abstract with diagram in *Chem. Zentr.*, **94**(4), 1923, 716.
10. D. Hanson and M. L. V. Gayler, *J. Inst. Metals*, **27,** 1922, 267–294.
11. T. Hemmi, *J. Soc. Chem. Ind. Japan*, **25,** 1922, 411–424; see *Chem. Abstr.*, **17,** 1923, 374 (with diagram).
12. E. Crepaz, *Giorn. chim. ind. ed appl.*, **5,** 1923, 285–286.
13. T. Tanabe, *J. Inst. Metals*, **32,** 1924, 415–427.
14. T. Isihara, *Science Repts. Tôhoku Univ.*, **13,** 1924, 427–442; *J. Inst. Metals*, **33,** 1925, 73–89; *Science Repts. Tôhoku Univ.*, **15,** 1926, 209–224.
15. O. Tiedemann, *Z. Metallkunde*, **18,** 1926, 18–21, 221–223; see also *Z. physik. Chem.*, **A191,** 1942, 133–144.
16. Although not published until 1937, the diagram determined by W. Broniewski, J. Kucharski, and W. Winawer (*Rev. mét.*, **34,** 1937, 449–461) has all the features of the diagrams of [9, 12, 13].
17. Roland-Gosselin; see H. Gautier, *Bull. soc. encour. ind. natl.*, **1,** 1896, 1308; Société d'encouragement pour l'industrie nationale, Paris, Commission des alliages, "Contribution à l'étude des alliages," p. 103, Typ. Chamerot et Renouard, Paris, 1901.
18. C. T. Heycock and F. H. Neville, *J. Chem. Soc.*, **71,** 1897, 389.
19. M. L. V. Gayler and E. G. Sutherland, *J. Inst. Metals*, **63,** 1938, 123–137.
20. T. Morinaga, *Nippon Kinzoku Gakkai-Shi*, **3,** 1939, 216–221.
21. E. Butchers, G. V. Raynor, and W. Hume-Rothery, *J. Inst. Metals*, **69,** 1943, 213.
22. E. Pelzel and H. Schneider, *Z. Metallkunde*, **35,** 1943, 124–127.
23. E. Butchers and W. Hume-Rothery, *J. Inst. Metals*, **71,** 1945, 296–297.
24. I. S. Solet and H. W. St. Clair, *U.S. Bur. Mines Rept. Invest.* 4553, 1949.
25. E. Pelzel, *Z. Metallkunde*, **40,** 1949, 134–136.
26. E. Gebhardt, *Z. Metallkunde*, **40,** 1949, 136–140.
27. V. S. Lyashenko, *Izvest. Akad. Nauk S.S.S.R., Otdel.* (*Khim.*) *Nauk*, **1951,** 242–254.
28. W. L. Fink and L. A. Willey, *Trans. AIME*, **122,** 1936, 244–260.
29. See discussions on the papers by [19, 28].
30. T. Morinaga, *Nippon Kinzoku Gakkai-Shi*, **4,** 1940, 216–220.
31. W. Guertler, H. Krause, and F. Voltz, *Metallwirtschaft*, **18,** 1939, 97–100.
32. E. C. Ellwood, *J. Inst. Metals*, **66,** 1940, 87–96.
33. J. A. Verö, *Acta Techn. Acad. Sci. Hung.*, **2,** 1951, 97–113.
34. E. C. Ellwood, *J. Inst. Metals*, **80,** 1951-1952, 217–224.
35. D. A. Petrov and T. A. Badaeva, *Zhur. Fiz. Khim.*, **21,** 1947, 785–797.
36. A. Münster and K. Sagel, *Z. physik. Chem.*, **7,** 1956, 296–316.
37. M. L. V. Gayler and E. G. Sutherland, *J. Inst. Metals*, **63,** 1938, 146–147.
38. W. Sander and K. L. Meissner, *Z. Metallkunde*, **14,** 1922, 385–387.
39. H. Nishimura, *Mem. Coll. Eng. Kyoto Imp. Univ.*, **3,** 1924, 133–163.
40. W. L. Fink and K. R. Van Horn, *Trans. AIME*, **99,** 1932, 132–140.
41. See fig. 80 of M. Hansen, "Der Aufbau der Zweistofflegierungen," Springer-Verlag OHG, Berlin. 1936.

42. E. A. Owen and L. Pickup, *Phil. Mag.*, **20,** 1935, 761–777.
43. G. Borelius and L. E. Larsson, *Arkiv. Mat. Astron. Fysik,* **35A,** 1948; see also G. Borelius, *Trans. AIME,* **191,** 1951, 482.
44. H. Auer and K. E. Mann, "Vorträge der Hauptversammlung der deutschen Gesellschaft für Metallkunde," 1938, pp. 48–52; see also "Gmelins Handbuch der anorganischen Chemie," System No. 35A (5), pp. 685–686, Verlag Chemie, G.m.b.H., Berlin, 1937.
45. I. Obinata, M. Hagiya, and S. Itimura, *Tetsu-to-Hagane,* **22,** 1936, 622–629.
46. H. Borchers and H. Egler, *Aluminium-Arch.*, no. 19, 1939.
47. H. Borchers and H. Kremer, *Aluminium-Arch.*, no. 33, 1941.
48. W. M. Peirce, *Trans. AIME,* **68,** 1932, 773–775.
49. A. Burkhardt, *Z. Metallkunde,* **28,** 1936, 299–308.
50. M. L. Fuller and R. L. Wilcox, *Trans. AIME,* **122,** 1936, 231–243.
51. H. Auer and K. E. Mann, *Z. Metallkunde,* **28,** 1936, 323–326.
52. K. Löhberg, *Z. Metallkunde,* **32,** 1940, 86–90.
53. W. Hofmann and G. Fahrenhorst, *Z. Metallkunde,* **42,** 1950, 460–463.
54. A. Pasternak, *Bull. intern. acad. polon. sci., Classe sci. math. nat., sér. A,* **1951,** 177–192; *Chem. Abstr.*, **48,** 1954, 1098.
55. W. Leitgebel, *Z. anorg. Chem.*, **202,** 1931, 305–324; see also *Z. Elektrochem.*, **43,** 1937, 509–518.
56. G. Edmunds, Master's thesis, Yale University, New Haven, Conn., May, 1929, quoted from [57].
57. M. L. Fuller and R. L. Wilcox, *Trans. AIME,* **117,** 1935, 338–354.
58. M. v. Schwarz and O. Summa, *Metallwirtschaft,* **11,** 1932, 369–371.
59. E. A. Owen and J. Iball, *Phil. Mag.*, **17,** 1934, 433–457.
60. G. F. Kossolapow and A. K. Trapesnikow, *Metallwirtschaft,* **14,** 1935, 45–46.
61. Z. Nishiyama, *Science Repts. Tôhoku Univ.*, **18,** 1929, 387.
62. H. J. Axon and W. Hume-Rothery, *Proc. Roy. Soc. (London),* **A193,** 1948, 1–24.
63. W. Hume-Rothery and T. H. Boultbee, *Phil. Mag.*, **40,** 1949, 71–80.
64. A. P. Guljaev and E. F. Trusova, *Zhur. Tekh. Fiz.*, **20,** 1950, 66–78.

$\bar{1}.4710$
0.5290

Al-Zr Aluminum-Zirconium

The phase diagram shown in Fig. 89 is based on metallographic and limited thermal studies [1] using arc-cast alloys prepared with iodide Zr. Alloys were analyzed. The range 0–10 wt. (27.3 at.) % Al was carefully investigated by means of micrographic analysis between 700 and 1300°C. The solid solubility of Al in α-Zr in wt. % Al (at. % in parentheses) is 0.35 (1.2), 1.0 (3.3), and approximately 3.5 (11) at 700, 800, and 940°C, respectively. The peritectoid formation of Zr_2Al (12.88 wt. % Al) between 1300 and 1200°C and of Zr_3Al (8.97 wt. % Al) between 1000 and 950°C was well established.

Because of the large number of intermediate phases in the range 30–100 at. % Al, complicated phase relationships, and sluggish diffusion rates, the examination of numerous as-cast alloy structures was more instrumental in establishing diagram features than the study of annealed and quenched specimens. The phase designated Zr_5Al_3 [37.5 at. (15.07 wt.) % Al] has a minimum temperature of stability. It is likely that it decomposes eutectoidally somewhere between 1100 and 600°C into Zr_2Al and Zr_3Al_2 (16.47 wt. % Al). Zr_4Al_3 [42.86 at. (18.16 wt.) % Al], ZrAl (22.83 wt. % Al), $ZrAl_2$ (37.17 wt. % Al), and $ZrAl_3$ (47.01 wt. % Al) were identified by one-phase structures. The composition of the phase designated Zr_2Al_3 (30.73 wt. % Al) is tentative; an alternative composition is Zr_3Al_4 [57.14 at. (28.28 wt.) % Al]. The

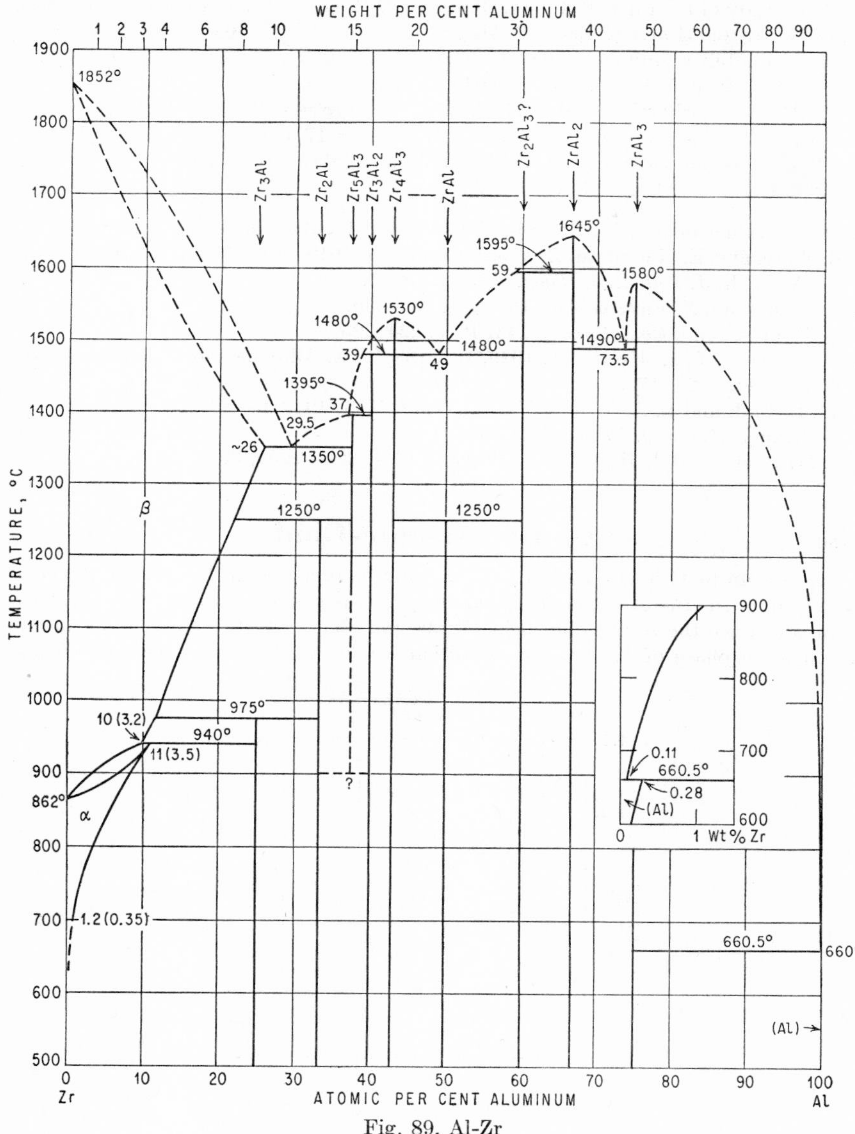

Fig. 89. Al-Zr

temperatures of the three-phase equilibria at 1350, 1395, 1480, 1595, and 1490°C and the maximum liquidus temperatures of 1530, 1645, and 1580°C were determined by thermal analysis (accuracy ±10°C).

The existence of the compounds Zr_3Al_4 [2], $ZrAl_2$ [3], and $ZrAl_3$ [4] had been reported earlier; however, the homogeneity of the isolated residues of these approximate compositions was not proved. Metallographic investigation of alloys with more than 60 at. % Al did not give conclusive results [5].

The study of the range 0–1.9 wt. (0.68 at.) % Zr showed that a peritectic reaction liq. [0.11 wt. (0.034 at.) % Zr] + $ZrAl_3 \rightleftarrows$ (Al) with 0.28 wt. (0.084 at.) % Zr exists. The solid-solubility limits are 0.050, 0.084, 0.15, and 0.23 wt. (0.015, 0.025, 0.045, and 0.068 at.) % Zr at 500, 550, 600, and 640°C, respectively [6].

$ZrAl_2$ is orthorhombic, a = 10.42 A, b = 7.22 A, c = 4.98 A [1]. $ZrAl_3$ is tetragonal [7, 6, 1] of the $D0_{23}$ type with a = 4.013 A, c = 17.320 A, c/a = 4.316 [7].

The compound richest in Zr, Zr_3Al, was shown [8] to have the cubic Cu_3Au ($L1_2$) type of structure with a = 4.372 ± 3 A.

1. D. J. McPherson and M. Hansen, *Trans. ASM*, **46,** 1954, 354–371.
2. L. Weiss and E. Neumann, *Z. anorg. Chem.*, **65,** 1910, 258–259; J. W. Marden and M. N. Rich, *J. Ind. Eng. Chem.*, **12,** 1920, 651.
3. E. Wedekind, *Z. Elektrochem.*, **10,** 1904, 331–335.
4. O. Hönigschmid, *Compt. rend.*, **143,** 1906, 224–226.
5. C. Sykes, *J. Inst. Metals*, **41,** 1929, 181–188; T. E. Allibone and C. Sykes, *J. Inst. Metals*, **39,** 1928, 173.
6. W. L. Fink and L. A. Willey, *Trans. AIME*, **133,** 1939, 69–80.
7. G. Brauer, *Z. anorg. Chem.*, **242,** 1934, 1–22.
8. J. H. Keeler and J. H. Mallery, *Trans. AIME*, **203,** 1955, 394.

$\bar{1}$.5796
0.4204

As-Au Arsenic-Gold

Alloys up to 11.5 wt. (25.5 at.) % As could be prepared [1] by melting the components in open crucibles [2]. Thermal analysis was possible up to 8.3 wt. (19.2 at.) % As and gave the results shown in Fig. 90 [1]. The eutectic between Au and an intermediate phase of unknown composition would supposedly lie between 40 and

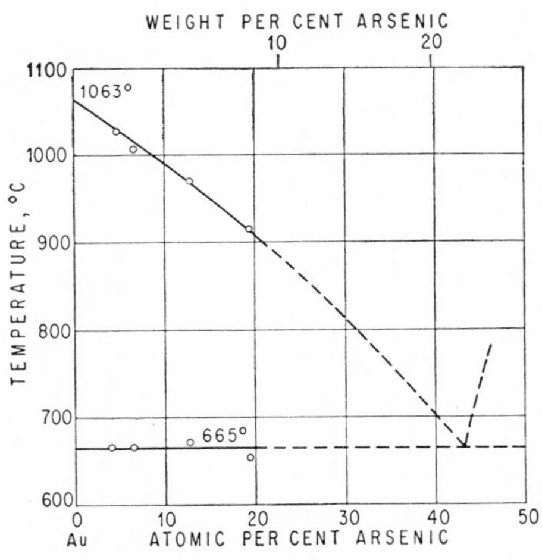

Fig. 90. As-Au

50 at. % As, provided that these alloys richer in As can be prepared under ordinary conditions. The alloys investigated consisted of primary Au crystals and eutectic.

Lattice-parameter measurements of heat-treated alloys with 0.59 and 1.60 at. % As indicate a solid solubility of about 0.2 at. (0.08 wt.) % As at 611°C and apparently no solubility at 400°C [3].

1. A. P. Schleicher, *Int. Z. Metallog.*, **6**, 1914, 18–22.
2. Earlier experiments were unsuccessful because of the volatility of arsenic: K. Friedrich, *Metallurgie*, **5**, 1908, 593, 603.
3. E. A. Owen and E. A. O. Roberts, *J. Inst. Metals*, **71**, 1945, 231.

$\bar{1}.7367$
0.2633

As-Ba Arsenic-Barium

Barium arsenide, Ba_3As_2 (26.66 wt. % As), has been prepared by reduction of $Ba_3(AsO_4)_2$ with carbon [1] and direct synthesis from the elements [2].

1. P. Lebeau, *Compt. rend.*, **129**, 1899, 48–49; *Ann. chim. et phys.*, **25**, 1902, 480–483.
2. F. Weibke, dissertation, Techn. Hochsch. Hannover, 1930, p. 8.

$\bar{1}.5544$
0.4456

As-Bi Arsenic-Bismuth

Small As additions continually lower the melting point of Bi to a eutectic at about 2.2 at. (0.8 wt.) % As, 265°C (based on a melting point of bismuth of 266.5°C) [1]. Cooling curves of melts with 13–92 at. % As, prepared in evacuated and sealed

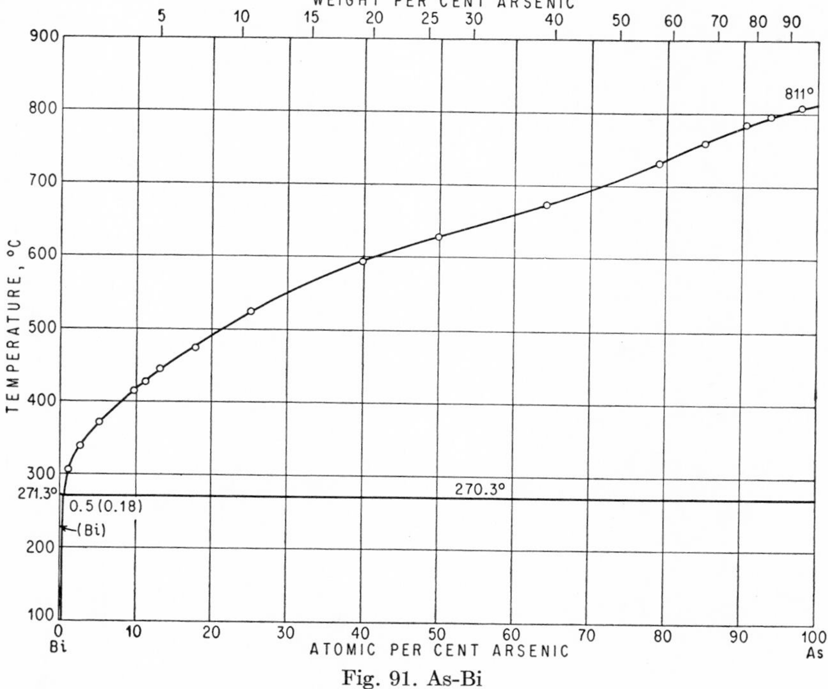

Fig. 91. As-Bi

porcelain tubes, showed that "the solidification diagram essentially consists of a liquidus curve ascending from the melting point of bismuth to that of arsenic and a (eutectic) horizontal lying close to the melting point of bismuth" [2]. No numerical data were given. An earlier investigation of alloys in the range of 0–33 at. % As seemed to indicate that the components are partially (above about 8 at. % As) insoluble in the liquid state, resulting in the formation of two layers [3]. However, microstructures of ingots of previously homogeneous melts proved that the heavy segregation observed in the Bi-rich alloys had taken place only during solidification [2]. Therefore, the formation of layers [3] was due to an insufficient mixing of the melts [2]. The solid solubility of As in Bi is only slight [4].

Supplement. The phase diagram in Fig. 91 was established by thermal, dilatometric, and metallographic work; the alloys were melted in sealed-off evacuated silica bulbs [5]. Hardness measurements showed the solid solubility of As in Bi to be 0.42 (0.15) at the eutectic temperature, 0.3 (0.11) at 200°C, 0.24 (0.09) at 100°C, and 0.2 at. (0.07 wt.) % As at room temperature.

1. C. T. Heycock and F. H. Neville, *J. Chem. Soc.*, **61**, 1892, 894.
2. W. Heike, *Intern. Z. Metallog.*, **6**, 1914, 209–211.
3. K. Friedrich and A. Leroux, *Metallurgie*, **5**, 1908, 148–149.
4. W. Trzebiatowski and E. Bryjak, *Z. anorg. Chem.*, **238**, 1938, 255–267.
5. G. A. Geach and R. A. Jeffrey, *J. Metals*, **5**, 1953, 1084 (preliminary publication); private communication to the author.

0.7950
$\bar{1}$.2050

As-C Arsenic-Carbon

In boiling arsenic (probably sublimizing is meant) only "traces or unweighably small amounts of carbon are soluble" [1].

1. O. Ruff and B. Bergdahl, *Z. anorg. Chem.*, **106**, 1919, 91.

0.2716
$\bar{1}$.7284

As-Ca Arsenic-Calcium

Calcium arsenide, Ca_3As_2 (55.48 wt. % As), has been prepared by reduction of $Ca_3(AsO_4)_2$ with carbon and direct synthesis [1].

1. P. Lebeau, *Compt. rend.*, **128**, 1899, 95–98; *Ann. chim. et phys.*, **25**, 1902, 477–479.

As-Cb Arsenic-Columbium

See As-Nb, Arsenic-Niobium.

$\bar{1}$.8237
0.1763

As-Cd Arsenic-Cadmium

The melting point of Cd is lowered by 0.3 wt. (0.45 at.) % As for about 1°C (eutectic point) [1]. Thermal investigation [2] of a wider range of composition (0–52.7 at. % As) confirmed the existence of the compound Cd_3As_2 (30.76 wt. % As), already reported earlier [3], with a maximum melting point of 712°C. The liquidus curve obtained agrees fairly well with that determined by a later investigation [4]. Between 0 and 34 at. % As the solidus temperature was found to be 320°C.

A thorough thermal analysis, supplemented by microscopic investigations [4], yielded the diagram given in Fig. 92. In alloys with As contents higher than that of Cd_3As_2, an unstable and a stable system exist, the former being due to the fact that the crystallization of the compound $CdAs_2$ (57.13 wt. % As) tends to supercool considerably.

The Unstable System. On "ordinary" cooling of the melts with more than 48 at. % As (without inoculation), the cooling curves show three effects: (*a*) the primary crystallization of Cd_3As_2, (*b*) its allotropic transformation at 578°C, and (*c*) a halting point, at which temperature a eutectic is formed which consists of Cd_3As_2 and probably pure As. As indicated by the data points (x), this eutectic crystallization takes place only after considerable supercooling. The pure eutectic solidifies at

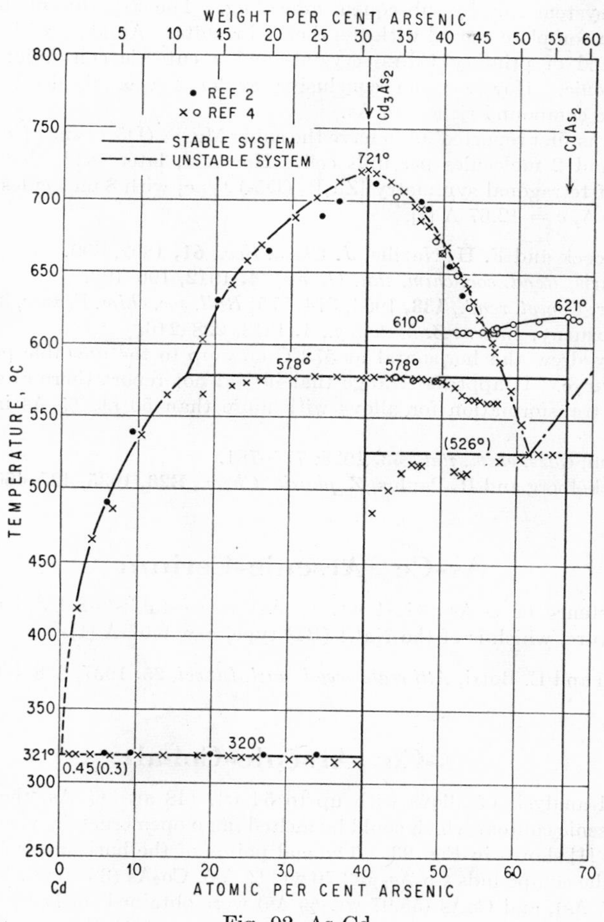

Fig. 92. As-Cd

526°C and contains about 61.5 at. % As. On cooling, either simultaneously with the eutectic crystallization or at somewhat higher temperatures, sudden heat evolutions take place which cause a temperature increase of 15–25°C and sometimes even as high as 80°C. If the temperature jump occurs before the eutectic crystallization, the thermal effect due to the latter is missing; e.g., the reaction in the melt takes place under spontaneous formation of the solid phases of the stable system Cd_3As_2-$CdAs_2$ However, if the temperature jump does not occur, the eutectic Cd_3As_2-As(?) is formed

The Stable System. If the melts are inoculated on cooling with small pieces of the alloy concerned, the stable system is formed immediately. The equilibrium curves of this system are the two liquidus curves corresponding to the primary crystallization of Cd_3As_2 and $CdAs_2$, the eutectic horizontal at 610°C with the eutectic at about 56 at. % As and the transformation horizontal at 578°C [5]. In alloys with less than approximately 50 at. % As, inoculation of the melts proved to be without effect; e.g., in these alloys the unstable system remains preserved and can be changed to the stable system only by subsequent annealing. The structure of the alloys was found to be in complete accord with the thermal results. An alloy with about 72 at. % As consisted of primary $CdAs_2$ crystals and a eutectic containing $CdAs_2$ and probably arsenic. However, no conclusive evidence was obtained that $CdAs_2$ actually is the compound richest in As.

Cd_3As_2 was first reported [6] to have the cubic Mg_3P_2 ($D5_5$) type of structure with $a = 6.30$ A and 2 molecules per unit cell. However, later work showed that the structure is of tetragonal symmetry [Zn_3P_2 ($D5_9$) type] with 8 molecules per unit cell and $a = 8.96$ A, $c = 12.67$ A [7].

1. C. T. Heycock and F. H. Neville, *J. Chem. Soc.*, **61,** 1892, 899.
2. P. de Cesaris, *Rend. soc. chim. ital.* (*Rome*), **4,** 1912, 196–199.
3. A. Granger, *Compt. rend.*, **138,** 1904, 574–575; *Bull. soc. chim. France*, **31,** 1904, 368.
4. S. F. Zemczuzny, *Intern. Z. Metallog.*, **4,** 1913, 228–246.
5. Zemczuzny drew the horizontal at 578°C only up to the unstable portion of the liquidus curve. It appears strange that he did not report thermal effects of the allotropic transformation for alloys with more than 50 at. % As inoculated on cooling.
6. L. Passerini, *Gazz. chim. ital.*, **58,** 1928, 775–781.
7. M. v. Stackelberg and R. Paulus, *Z. physik. Chem.*, **B28,** 1935, 427–460.

$\bar{1}.7280$
0.2720

As-Ce Arsenic-Cerium

The existence of CeAs (34.84 wt. % As) was established by determining its crystal structure, which is of the NaCl (B1) type, $a = 6.07$ A [1].

1. A. Iandelli and E. Botti, *Atti reale accad. nazl. Lincei*, **25,** 1937, 498–502.

0.1041
$\bar{1}.8959$

As-Co Arsenic-Cobalt

Thermal analysis of alloys with up to 54 wt. (48 at.) % As, the composition highest in arsenic content which could be melted in an open crucible, yielded the phase relationships [1] shown in Fig. 93. The end points of the horizontals and hence the formulas of the compounds Co_5As_2 (33.70 wt. % As), Co_2As (38.86 wt. % As), Co_3As_2 (45.87 wt. % As), and CoAs (55.97 wt. % As) were obtained solely by extrapolation of the halting times of the eutectic crystallization at 918°C, the peritectic reactions at 923, 958, and 1014°C, as well as the polymorphic transformations of Co_5As_2 and Co_3As_2 at about 828 and about 909°C, respectively. From determinations of the ignition temperature it was concluded [2] that the intermediate phases are of variable composition.

The diagram also shows the thermal-analysis data of another study [3] of alloys with As contents up to 25 at. %, which included the effect of As on the temperatures of the magnetic and polymorphic transformations of Co (see also Supplement). The indicated maximum solid solubility of As in Co of 7 wt. (5.6 at.) % As can be con-

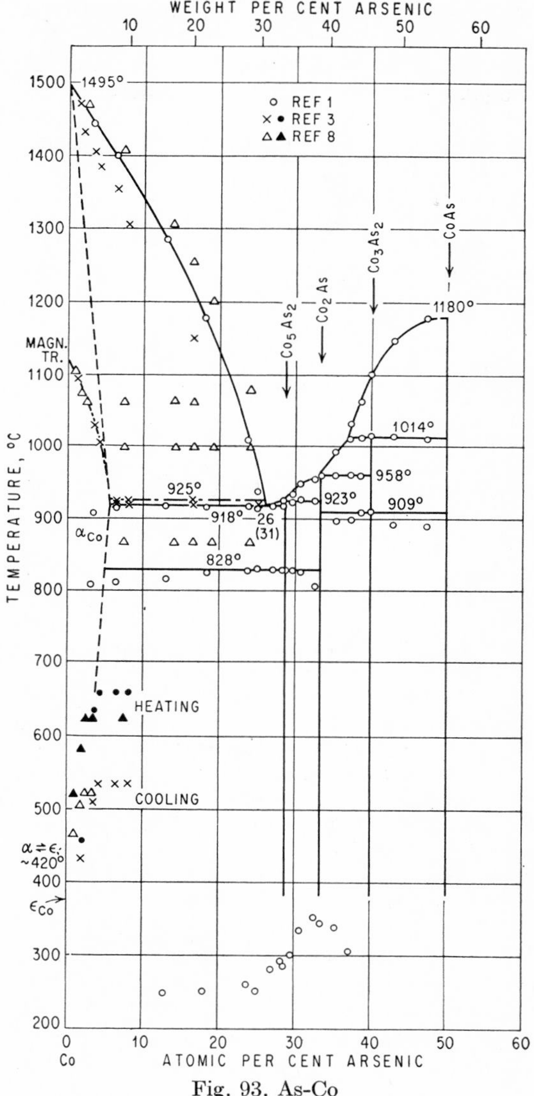

Fig. 93. As-Co

sidered only an approximation [3]. The nature of the thermal effects at temperatures from 248 to 352°C between 13 and 37 at. % As is unknown. The assumption that they are due to a polymorphic transformation of Co_2As [1] appears doubtful, because such a transformation would not occur in alloys with less than 28.6 at. % As. However, as it had been stated [1] that the microscopic examination confirmed the results of the thermal analysis, it does not appear justifiable at present to assume a transformation involving the formation of new phases.

By reactions of Co with As at temperatures of 800–1400, 600–800, 400–600, and below 400°C, products were formed having compositions of Co_3As_2 (melting point about 1000°C), CoAs, Co_2As_3 (65.59 wt. % As), and $CoAs_2$ (71.77 wt. % As), respectively. No analytical data were given, however [4]. For information on the As-rich cobalt arsenides, see [5].

CoAs is orthorhombic, of the MnP (B31) type, a = 5.97 A, b = 5.16 A, c = 3.52 A [6]. Concerning the composition and structure of the Co-As minerals (speiskobalt, skutterudite), see [5, 7].

Supplement. The effect of As additions up to 28.5 wt. (23.9 at.) % on the melting point and transformation points of Co was also studied by [8]. According to the data points in Fig. 93, the horizontals corresponding to the magnetic transformation (1063°C), eutectic crystallization (997°C), and transformation of Co_5As_2 (866°C) lie at considerably higher temperatures than those reported by the other investigators [1, 3].

1. K. Friedrich, *Metallurgie*, **5**, 1908, 150–157.
2. M. J. Kochner, *Doklady Akad. Nauk S.S.S.R.*, **73**, 1950, 1197–1199.
3. W. Köster and W. Mulfinger, *Z. Metallkunde*, **30**, 1938, 348–350.
4. F. Ducelliez, *Compt. rend.*, **147**, 1908, 424–426.
5. "Gmelins Handbuch der anorganischen Chemie," System No. 58, pp. 33–36, Verlag Chemie G.m.b.H., Berlin, 1932.
6. K. E. Fylking, *Ark. Kemi, Mineral. Geol.*, **B11**, 1935, no. 48.
7. I. Oftedal, *Z. Krist.*, **66**, 1928, 517–546.
8. U. Haschimoto, *Nippon Kinzoku Gakkai-Shi*, **1**, 1937, 177–190.

0.1584
$\bar{1}$.8416

As-Cr Arsenic-Chromium

X-ray investigations [1] have shown that the following chromium arsenides exist: Cr_2As (41.87 wt. % As), Cr_3As_2 (48.98 wt. % As), and CrAs (59.02 wt. % As). Magnetic measurements [2] of preparations with 25–50 at. % As at temperatures between −183 and 350°C confirmed this and indicated that the phase based on the composition Cr_3As_2 has a certain range of homogeneity, between about 41 and 42 or (at lower temperatures) 40 and 42 at. % As, whereas Cr_2As appears to have a fixed composition [1, 2]. The preparation richest in As obtained was $CrAs_{0.95}$—48.72 at. % As [2]. Therefore, it is doubtful whether the compound Cr_2As_3 (68.36 wt. % As), reported earlier [3], actually does exist. There is X-ray evidence that As is soluble in solid Cr to approximately 2 wt. (1.4 at.) % [1].

Cr_2As is of the tetragonal Cu_2Sb (C38) type, a = 3.620 A, c = 6.346 A, c/a = 1.753 [1]. CrAs has an orthorhombic structure of the MnP (B31) type, a = 3.486 A, b = 6.222 A, c = 5.741 A [1], and not of the NiAs type, as suggested earlier [4]. Cr_3As_2 probably has a transition type of structure between those of Cr_2As and CrAs [1].

1. H. Nowotny and O. Årstad, *Z. physik. Chem.*, **B38**, 1938, 461–465.
2. H. Haraldsen and E. Nygaard, *Z. Elektrochem.*, **45**, 1939, 686–688.
3. T. Dieckmann and O. Hanf, *Z. anorg. Chem.*, **86**, 1914, 291–295.
4. W. F. de Jong and H. W. V. Willems, *Physica*, **7**, 1927, 74–79.

0.0715
$\bar{1}$.9285

As-Cu Arsenic-Copper

The constitution of this system (Fig. 94) is well established up to the composition Cu_3As (28.21 wt. % As), with the exception of the solidus curve of the copper-rich solid solution. The liquidus is based on the work by [1–4]. The solid solubility of

As in Cu was the subject of various papers [5–9] and discussions [10–12]. Work by [5, 6] can be disregarded [11]. The remaining data, obtained by lattice-parameter measurements [7–9], agree within 0.1–0.2 at. % As between 689 [8.0 wt. (6.85 at.) % As] and 400°C [7.7 wt. (6.6 at.) % As]. According to [7], the solubility decreases linearly from 6.75 at. % As at 680°C to 6.4 at. % As at 300°C. The findings by [9] show a slight discontinuous decrease from 6.7 at. % As at about 380°C to 6.2 at. % As at 300°C and 5.9 at. % As at 215°C. This discontinuous change at 380°C indicates a transformation of the Cu_3As phase for which there is X-ray crystallographic evidence (see below).

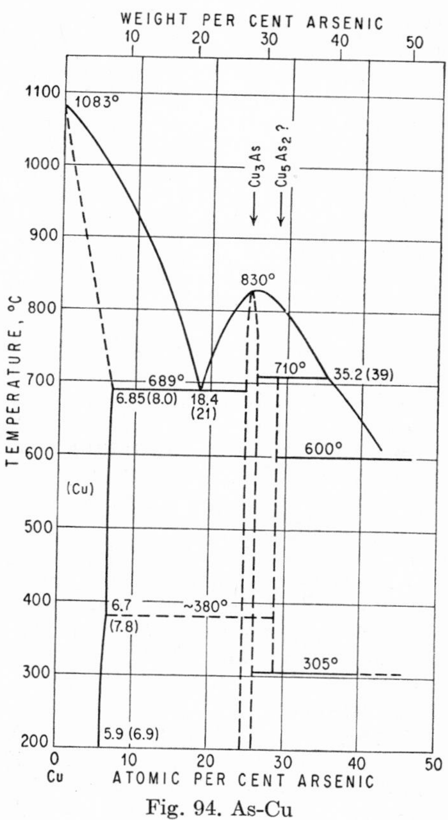

Fig. 94. As-Cu

The phase relationships in the composition range above 28 wt. % As are still uncertain. Systematic micrographic or X-ray analyses of well-annealed and quenched specimens are lacking. There is, however, agreement as to the liquidus curve and the existence and temperature of the three horizontals at 710, 600, and 305°C [2–4]. Also, there is no doubt that the first intermediate phase includes and is based on the composition Cu_3As [13–18]. The last study of this region dates as far back as 1910.

Whereas [2, 3] assumed that the thermal effect at 710°C (Fig. 94) is due to the peritectic reaction Cu_3As + melt [about 35 at. (39 wt.) % As] $\rightleftarrows$ Cu_5As_2 (32.05 wt. % As), [4] have suggested this reaction to be a polymorphic transformation of the solid solution based on the compound Cu_5As_2 [19]. Both interpretations, however, were offered without sufficient experimental evidence. This holds especially as to the composition of the phase designated as Cu_5As_2, which appears to be chosen arbitrarily. Also, the alloy of this composition was shown to be duplex in structure [15, 16]. Likewise, the authors [2–4] were not able to give an interpretation of the reaction at 305°C.

In the absence of conclusive evidence it has been assumed (Fig. 94) that the Cu_5As_2 (?) phase—or some other phase, e.g., Cu_2As (37.09 wt. % As)—is formed peritectically at 710°C and decomposes at 305°C into a eutectoid of Cu_3As and an As-rich phase of unknown composition. Although the formation of a eutectoid has not been observed as yet, there are some indications of considerable structural changes taking place in the solid state [2, 3, 14, 16]. It is likely that the observed microstructures [2, 3] and X-ray patterns [16] were far from representing equilibrium conditions, because of the incompleteness of the reactions at 710 and 305°C.

The mineral domeykite, Cu_3As, occurs in two modifications, (*a*) a cubic form (16 molecules per unit cell, $a = 9.611$ A), apparently stable at high temperatures, which transforms on heating at 225°C into (*b*) a hexagonal form of the Cu_3P ($D0_{21}$) type, $a = 7.102$ A, $c = 7.246$ A, $c/a = 1.020$ [17]. This structure was also found for synthetically prepared Cu_3As [16, 17]. Lattice parameters of Cu_3As were also given by [16]. The polymorphic transformation of Cu_3As probably occurs at about 380°C, as indicated by the change in direction of the solid-solubility curve of As in Cu at this temperature [9, 10].

1. A. H. Hiorns, *J. Soc. Chem. Ind. (London)*, **25,** 1906, 616–622; *Electrochemist & Metallurgist*, **3,** 1904, 648–655.
2. K. Friedrich, *Metallurgie*, **2,** 1905, 477–495.
3. K. Friedrich, *Metallurgie*, **5,** 1908, 529–535.
4. G. D. Bengough and B. P. Hill, *J. Inst. Metals*, **3,** 1910, 34–71.
5. D. Hanson and C. B. Marryat, *J. Inst. Metals*, **37,** 1927, 121–143.
6. W. Hume-Rothery, G. W. Mabbott, and K. M. Channel-Evans, *Phil. Trans. Roy. Soc. (London)*, **A233,** 1934, 1–97.
7. J. C. Mertz and C. H. Mathewson, *Trans. AIME*, **124,** 1937, 59–77.
8. E. A. Owen and E. W. Roberts, *Phil. Mag.*, **27,** 1939, 294–327.
9. E. A. Owen and V. W. Rowlands, *J. Inst. Metals*, **66,** 1940, 361–378.
10. E. A. Owen and D. P. Morris, *J. Inst. Metals*, **76,** 1949-1950, 158–159.
11. W. Hume-Rothery, *J. Inst. Metals*, **76,** 1949-1950, 682.
12. E. A. Owen and D. P. Morris, *J. Inst. Metals*, **76,** 1949-1950, 701–702.
13. G. A. Koenig, *Z. Krist.*, **38,** 1903, 529.
14. N. Puschin and E. Dischler, *Z. anorg. Chem.*, **80,** 1913, 65–70.
15. F. Machatschki, *Nachr. Jahresber. Mineral. Beil. Bd.*, **59,** 1929, 137.
16. N. Katoh, *Z. Krist.*, **76,** 1930, 228–234 (with additional references).
17. B. Steenberg, *Ark. Kemi, Mineral. Geol.*, **A12,** 1938, no. 26.
18. V. I. Mikheev; see *Structure Repts.*, **12,** 1949, 166 (with additional references).
19. The diagram given by Bengough and Hill is in contradiction to the phase rule.

0.1275
$\bar{1}.8725$

As-Fe Arsenic-Iron

Alloys Up to 33.3 At. % As (Fe_2As) (Fig. 95). Liquidus temperatures, as determined by [1] and [2], agree well. The eutectic temperature was found to be 833–835 [1], 827 [2], and 840°C [3]; and the eutectic composition was located at 30 [1], 31.3 [2], and 29.8 wt. % As [3] (24.2, 25.2, and 24.0 at. % As).

[2] suggested—without convincing evidence—that the temperature of the $\delta \rightleftarrows \gamma$ transformation of Fe is raised by As, giving rise to the peritectic reaction melt + $\delta \rightleftarrows \gamma$ at about 1440°C. However, [4] reported that a γ loop exists, in accordance with findings by [5] that the temperature of the $\alpha \rightleftarrows \gamma$ transformation is increased by As additions. From diffusion studies the saturated γ solid solution was estimated to contain about 3.75 wt. (2.8 at.) % As at 1150°C [6]. By means of dilatometric investigations the limits of the $\alpha + \gamma$ field were found on heating to lie at about 1.5 and 2.5 wt. (1.1 and 1.9 at.) % As at 1000°C and at about 2.4 and 3.3 wt. (1.8 and 2.5 at.) % As at 1100°C [7]. [3] extrapolated from micrographic studies, using

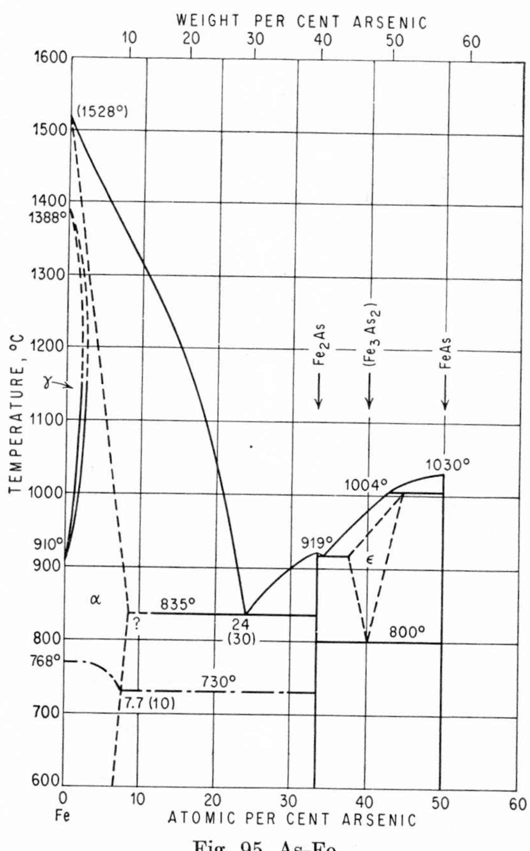

Fig. 95. As-Fe

alloys with 0.14 wt. % C, that the saturated solution of As in γ-Fe in the binary system is 2.2 wt. (1.65 at.) % As at 1150°C.

The boundary of the solid solution of As in α-Fe has not been determined accurately as yet. The solubility at the eutectic temperature was located at 8 wt. (~6.0 at.) % [1], 6.8 wt. (5.1 at.) % [2], and 11 wt. (8.4 at.) % As [3]. However, these values do not represent equilibrium conditions. Information of the effect of As on the Curie temperature of Fe is not conclusive [5, 8, 2]; see Supplement.

Alloys with 33.3 to 50.0 At. % As (FeAs). The phase relations in this composition range are not fully established as yet. On the basis of thermal analysis and metallographic work, [1] constructed a diagram which violates the phase theory. The peritectic formation of Fe_5As_4 at 1004°C and its decomposition at 800°C were indicated. In addition, the existence of the phase Fe_3As_2 (47.21 wt. % As), stable below 800°C, was shown; but [9], using X-ray analysis, was unable to find such a phase between Fe_2As and FeAs. He proved, however, that there is an additional intermediate phase, stable only at high temperatures, which cannot be retained on quenching from temperatures as high as 875°C since its (eutectoid) decomposition (at about 800°C) cannot be suppressed. There is some evidence [1] that this phase is of variable composition. Based on this information [1, 9] the partial diagram presented in Fig. 95 was proposed [10]. It is not clear whether Fe_2As has a maximal melting point (919°C) or is formed by a peritectic reaction at 917–919°C; see the Supplement.

Crystal Structures. The lattice parameter of α-Fe is extended by As additions [9, 7]. Fe_2As (40.14 wt. % As), having a very narrow range of homogeneity, is tetragonal of the Cu_2Sb (C38) type, $a = 3.634$ A, $c = 5.985$ A, $c/a = 1.647$ [9, 11]. FeAs is orthorhombic of the MnP (B31) type with $a = 6.028$ A, $b = 5.434$ A, $c = 3.373$ A [9, 12]. Its homogeneity range is unknown since no alloys with more than 50 at. % As were investigated. The structure of $FeAs_2$, prepared by [13], was studied in the form of the mineral löllingite [14, 15]. It is orthorhombic of the FeS_2 (C18) type, $a = 2.86$ A, $b = 5.26$ A, $c = 5.93$ A [15].

Supplement. The Curie point of the α phase is nearly constant at 768°C up to 4 wt. (3 at.) % As and then decreases to about 10 wt. (7.65 at.) % As at 730°C, indicating the approximate solubility at this temperature [16]. The temperatures of the three three-phase equilibria in the range 33.3–50 at. % As were given as 922, 1002, and 824°C. The eutectic, eutectoid, and peritectic melt were located at 41.7 (34.8), 47.2 (40.0), and 49.4 wt. (42.1 at.) % As, respectively. The maximal melting point of Fe_2As was given as 930°C [17].

1. K. Friedrich, *Metallurgie*, **4**, 1907, 129–137.
2. P. Oberhoffer and A. Gallaschik, *Stahl u. Eisen*, **43**, 1923, 398–400.
3. H. Sawamura and T. Mori, *Mem. Fac. Eng. Kyoto Univ.*, **14**, 1952, 129–144.
4. F. Wever, *Naturwissenschaften*, **17**, 1929, 304–309; *Arch. Eisenhüttenw.*, **2**, 1928-1929, 739–746.
5. F. Osmond, *Compt. rend.*, **110**, 1890, 346–348.
6. W. D. Jones, *J. Iron Steel Inst.*, **130**, 1934, 429–437.
7. V. N. Svechnikov and V. N. Gridnev, *Metallurg*, **1938**, no. 1, 13–19.
8. J. Liedgens, *Stahl u. Eisen*, **32**, 1912, 2099–2115.
9. G. Hägg, *Z. Krist.*, **68**, 1928, 470–471; **71**, 1929, 134–136; *Nova Acta Soc. Regiae Sci. Upsaliensis*, (4)**7**, 1929, no. 1, 44–70.
10. M. Hansen, "Der Aufbau der Zweistofflegierungen," pp. 180–184, Springer-Verlag OHG, Berlin, 1936.
11. M. Elander, G. Hägg, and A. Westgren, *Arkiv Kemi, Mineral. Geol.*, **12B**, 1936, no. 1.
12. K. E. Fylking, *Arkiv Kemi Mineral. Geol.*, **11B**, 1935, no. 48.
13. S. Hilpert and T. Dieckmann, *Ber. deut. chem. Ges.*, **44**, 1911, 2378–2385.
14. W. F. de Jong, *Physica*, **6**, 1926, 325–332.
15. M. J. Buerger, *Z. Krist.*, **82**, 1932, 165–187.
16. H. Sawamura and T. Mori, *Mem. Fac. Eng. Kyoto Univ.*, **16**, 1954, 182–189.
17. H. Sawamura et al., *Tetsu-to-Hagane*, **39**, 1953, 776–777.

0.0312
$\bar{1}.9688$

As-Ga Arsenic-Gallium

The phase diagram (Fig. 96) was established by thermal analysis of alloys prepared in evacuated sealed-off silica bulbs [1]. Results confirmed previous findings

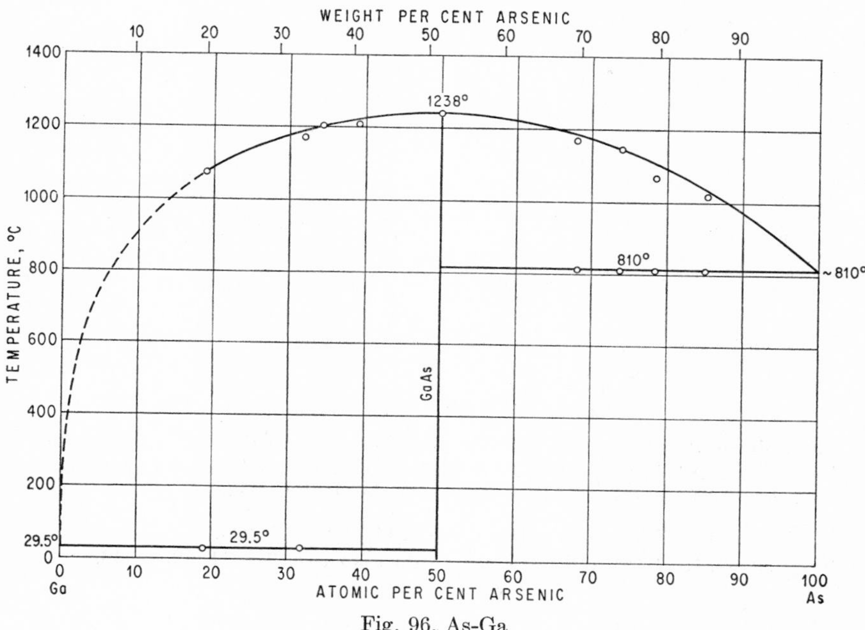

Fig. 96. As-Ga

that a 2 wt. (1.85 at.) % addition of As has no effect on the melting point of Ga [2]. GaAs (51.79 wt. % As) is isotypic with zincblende (B3 type), $a = 5.646$ A [3].

1. W. Köster and B. Thoma, *Z. Metallkunde*, **46,** 1955, 291–293.
2. R. M. Evans and R. I. Jaffee, *Trans. AIME*, **194,** 1952, 153.
3. V. M. Goldschmidt, *Skrifter Akad. Oslo*, **1926,** no. 8, 34, 110; *Ber. deut. chem. Ges.*, **60,** 1927, 1289; *Trans. Faraday Soc.*, **25,** 1929, 277.

0.0136
$\bar{1}.9864$

As-Ge Arsenic-Germanium

The thermal-analysis data (Fig. 97) were obtained by taking cooling curves of alloys prepared in sealed silica bulbs. The data points had to be taken graphically from a small diagram as tabulated data are missing. The approximate phase limits at 685°C (indicated in Fig. 97) were determined by qualitative X-ray analysis of specimens annealed for several months at this temperature and quenched. No two-phase region could be detected between the two intermediate phases of variable composition based on the compositions GeAs (50.78 wt. % As) and $GeAs_2$ (67.36 wt. % As), respectively [1].

The X-ray powder patterns of the phases GeAs and $GeAs_2$ are very complicated and quite similar; they resemble somewhat the arsenic pattern [1].

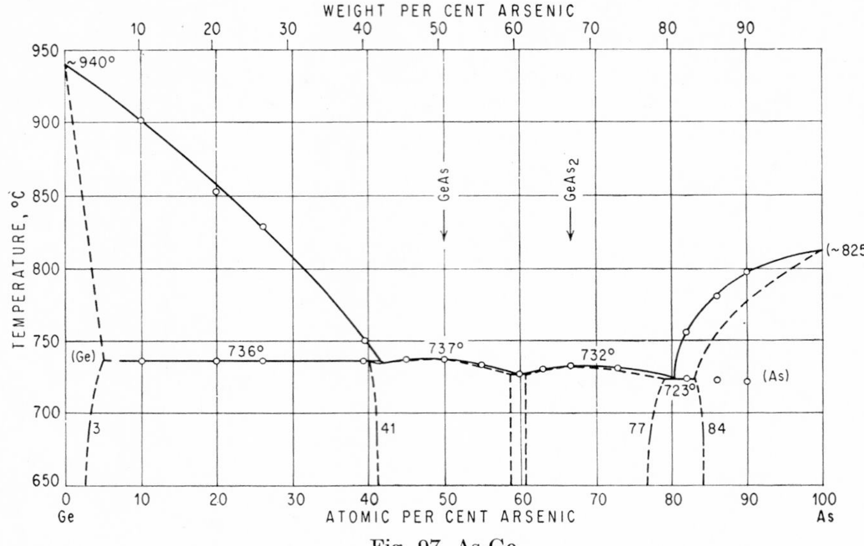

Fig. 97. As-Ge

Supplement. In a preliminary publication, [2] gave the lattice parameters of GeAs as a = 22.84 A, b = 3.78 A, c = 9.45, β = 43.97°, and those of $GeAs_2$ as a = 3.721 A, b = 10.12 A, c = 14.74 A. The structure will be reported later.

1. H. Stöhr and W. Klemm, *Z. anorg. Chem.*, **244,** 1940, 205–223.
2. K. Schubert, E. Dörre, and E. Günzel, *Naturwissenschaften,* **41,** 1954, 448.

$\bar{1}$.5722
0.4278

As-Hg Arsenic-Mercury

Various authors have reported that As is insoluble in Hg [1–3], even at its boiling point [1], or that the solubility is "extremely small" [4]. The compound Hg_3As_2 (19.93 wt. % As) has been prepared chemically [5, 6].

1. W. Ramsay, *J. Chem. Soc.*, **55,** 1889, 521–536.
2. B. Neumann, *Z. physik. Chem.*, **14,** 1894, 220.
3. W. J. Humphreys, *J. Chem. Soc.*, **69,** 1896, 1685.
4. G. Tammann and J. Hinnüber, *Z. anorg. Chem.*, **160,** 1927, 256.
5. A. Partheil and E. Amost, *Ber. deut. chem. Ges.*, **31,** 1898, 394–395.
6. F. Dumesnil, *Compt. rend.*, **152,** 1911, 868–869.

$\bar{1}$.8148
0.1852

As-In Arsenic-Indium

The phase diagram in Fig. 98 is based on thermal, microscopic, and X-ray investigations [1]. Alloys were prepared in evacuated and sealed capsules. The phase InAs (39.49 wt. % As) has a fixed composition. The composition of the In-InAs eutectic was claimed to be 0.02 wt. (0.03 at.) % As.

InAs is isotypic with zincblende (B3 type), a = 6.04_8 A [2], a = 6.058_4 A [1].

1. T. S. Liu and E. A. Peretti, *Trans. ASM*, **45,** 1953, 677–685.
2. A. Iandelli, *Gazz. chim. ital.*, **71,** 1941, 58–62.

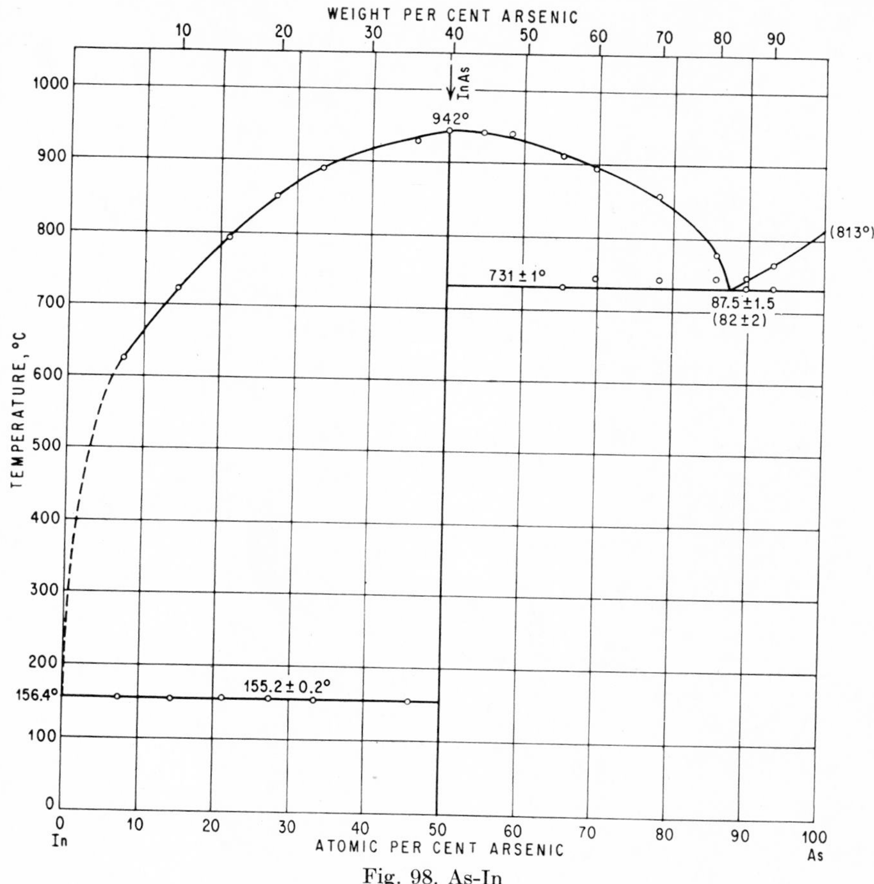

Fig. 98. As-In

$\bar{1}.5888$
0.4112

As-Ir Arsenic-Iridium

$IrAs_2$ (43.69 wt. % As) has been prepared by synthesis from the elements and heating of $IrCl_3$ with arsenic in a hydrogen stream at 500–600°C [1].

1. L. Wöhler and K. F. A. Ewald, *Z. anorg. Chem.*, **199**, 1931, 58–60.

0.2824
$\bar{1}.7176$

As-K Arsenic-Potassium

K_3As (38.98 wt. % As), prepared by melting of As with an excess of K in a closed container and various chemical reactions [1], has a hexagonal structure of the Na_3As ($D0_{18}$) type, $a = 5.794$ A, $c = 10.242$ A, $c/a = 1.768$ [2].

1. See chemical handbooks, e.g., "Gmelins Handbuch der anorganischen Chemie," System No. 22, p. 1015, Verlag Chemie G.m.b.H., Berlin, 1938.
2. G. Brauer and E. Zintl, *Z. physik. Chem.*, **B37**, 1937, 323–352.

$\bar{1}.7318$
0.2682

As-La Arsenic-Lanthanum

LaAs (35.03 wt. % As) is isotypic with NaCl (B1 type), $a = 6.13_7$ A [1].

1. A. Iandelli and E. Botti, *Atti reale accad. nazl. Lincei*, **25**, 1937, 498–502.

1.0332
$\bar{2}.9668$

As-Li Arsenic-Lithium

Li_3As (78.25 wt. % As), the preparation of which is described in various chemical handbooks [1], is isotypic with Na_3As ($D0_{18}$ type), $a = 4.396$ A, $c = 7.826$ A, $c/a = 1.780$ [2].

1. "Gmelins Handbuch der anorganischen Chemie," System No. 20, p. 248, Verlag Chemie G.m.b.H., Berlin, 1927.
2. G. Brauer and E. Zintl, *Z. physik. Chem.*, **B37**, 1937, 323–352.

0.4886
$\bar{1}.5114$

As-Mg Arsenic-Magnesium

The existence of the arsenide Mg_3As_2 (67.25 wt. % As), already claimed by [1], has been established by determination of its crystal structure, which is of the Mn_2O_3 ($D5_3$) type, b.c.c. with 16 molecules per unit cell, $a = 12.35_5$ A [2]. An earlier lattice-structure determination [3] was proved erroneous [2]. The compound melts at about 800°C [3]. An addition of As slightly increases the a-parameter of Mg; however, it was concluded that the solid solubility was exceedingly small [4].

1. J. Parkinson, *J. Chem. Soc.*, **20**, 1867, 127, 309.
2. E. Zintl and E. Husemann, *Z. physik. Chem.*, **B21**, 1933, 138–155; see also R. Paulus, *Z. physik. Chem.*, **B22**, 1933, 305–322.
3. G. Natta and L. Passerini, *Gazz. chim. ital.*, **58**, 1928, 541–550.
4. R. S. Busk, *Trans. AIME*, **188**, 1950, 1460–1464.

0.1347
$\bar{1}.8653$

As-Mn Arsenic-Manganese

Thermal-analysis data [1] obtained with alloys prepared using manganese of 97.4% purity [2] are given in the phase diagram of Fig. 99, which is characterized by the existence of Mn_2As (40.54 wt. % As) and MnAs (57.69 wt. % As) [1, 3]. The lengths of the halting times observed for the eutectic crystallization at 870°C indicated that Mn_2As forms solid solutions with As. The thermal effects found in the solid alloys with about 37–42 at. % As were suggested [1] to be due to the formation of another intermediate phase, presumably Mn_3As_2 (47.62 wt. % As): $Mn_2As + MnAs = Mn_3As_2$. However, there is no direct evidence for such a reaction as yet [4].

In addition, the existence of the phase Mn_3As (31.25 wt. % As), identified by X-ray analysis, has been reported [5]. If the phase diagram in Fig. 99 is assumed to be valid, Mn_3As could form only in the solid state, although crystals of Mn_3As were obtained by growth from the melt [5]. This discrepancy needs further study. There is probably a peritectic reaction in the range 20–33 at. % As.

By means of magnetic measurements and determinations of the Curie temperatures the phase boundaries between 33 and 50 at. (40.2 and 57.7 wt.) % As have been fixed as shown in the lower portion of Fig. 99 [6]. This necessitates the existence of an intermediate phase between Mn_2As and MnAs—either Mn_4As_3 (42.86 at. % As) [6] or $Mn_{29}As_{21}$ (42.0 at. % As) [7]—and of solid solutions of Mn_2As with As (33.3–40 at. % As) as well as MnAs with Mn (46.5–50 at. % As). Some of the results and interpretations of [6, 7] were disputed by [8], on the basis of Curie-point determinations in the range 47.4–53.7 at. % As. The existence of the phase Mn_8As_7 (46.67 at. % As) was suggested.

No information regarding the effect of As on the transformation temperatures of Mn is available. Arsenic additions do not stabilize the f.c. tetragonal γ-Mn phase at room temperature which, if alloyed with the proper elements, is obtained from the

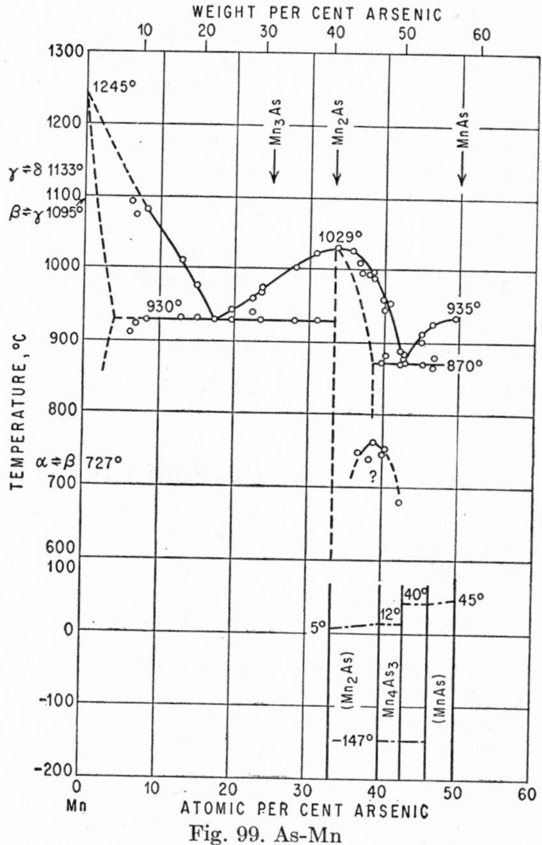

Fig. 99. As-Mn

f.c.c. high-temperature γ-Mn phase on quenching. Instead, the β-Mn phase was obtained [9].

Crystal Structures. Mn_3As is pseudo-tetragonal orthorhombic with $a = b =$ 3.788 A, $c = 16.2_9$ A, and 4 molecules per unit cell [5]. Its structure is closely related to those of Mn_2As and γ-Mn [5]. Mn_2As is tetragonal of the Cu_2Sb (C38) type [10, 5], $a = 3.76_9$ A, $c = 6.27_8$ A, $c/a = 1.666$ [10]. The structure of MnAs has been described as either the hexagonal NiAs (B8) or orthorhombic MnP (B31) type. NiAs type: $a = 3.72_3$ A, $c = 5.71_5$ A, $c/a = 1.535$ [11, 5]; $a = 3.725$ A, $c = 5.713$ A, $c/a = 1.534$ [6]. MnP type: $a = 6.39$ A, $b = 5.64$ A, $c = 3.63$ A [12]. For additional X-ray data on MnAs, see [13, 14].

1. P. Schoen, *Metallurgie*, **8,** 1911, 739–741.
2. The manganese contained 1.11% SiO_2, 0.34% Cu, 0.48% Fe, 0.44% Al.
3. S. Hilpert and T. Dieckmann, *Ber. deut. chem. Ges.*, **44,** 1911, 2378–2385.
4. E. Wedekind, *Z. Elektrochem.*, **11,** 1905, 850–851; *Physik. Z.*, **7,** 1906, 805–806.
5. H. Nowotny, R. Funk, and J. Pesl, *Monatsh. Chem.*, **82,** 1951, 513–519.
6. C. Guillaud and J. Wyart, *Compt. rend.*, **219,** 1944, 393–395; *Rev. mét.*, **45,** 1948, 271–276.
7. C. Guillaud, *J. recherches centre natl. recherche sci., Labs. Bellevue (Paris)*, **1947,** 15–21; *Ann. phys.*, **4,** 1949, 689–703.
8. K. H. Sweeny and A. B. Scott, *J. Chem. Phys.*, **22,** 1954, 917–921.
9. U. Zwicker, *Z. Metallkunde*, **42,** 1951, 246–252.
10. H. Nowotny and F. Halla, *Z. physik. Chem.*, **B36,** 1937, 322–324.
11. I. Oftedal, *Z. physik. Chem.*, **132,** 1928, 208–216.
12. K. E. Fylking, *Arkiv Kemi, Mineral. Geol.*, **B11,** 1935, no. 48.
13. C. Guillaud, *J. phys. radium*, **12,** 1951, 223–227.
14. B. T. M. Willis and H. P. Rooksby, *Proc. Phys. Soc. (London)*, **B67,** 1954, 290–296.

$\bar{1}.8925$
0.1075

As-Mo Arsenic-Molybdenum

$MoAs_2$ (60.96 wt. % As) was prepared by heating Mo powder with As in a sealed tube at 570°C and distilling off the excess of As [1].

1. E. Heinerth and W. Biltz, *Z. anorg. Chem.*, **198,** 1931, 171.

0.5129
$\bar{1}.4871$

As-Na Arsenic-Sodium

Na_3As (52.06 wt. % As) was prepared by reaction of Na with As in liquid NH_3 [1]. By potentiometric titration of a solution of Na in liquid NH_3 with a solution of As_2S_3 in liquid NH_3, indications have been obtained for the existence of Na_3As, Na_3As_3, Na_3As_7, and $NaAs_5$ [2]. The structure of Na_3As is the prototype of the $D0_{18}$ type of structure, a = 5.098 A, c = 9.00 A, c/a = 1.765 [3].

1. C. Hugot, *Compt. rend.*, **127,** 1898, 553; **129,** 1899, 604. See also P. Lebeau, *Compt. rend.*, **130,** 1900, 502.
2. E. Zintl, J. Goubeau, and W. Dullenkopf, *Z. physik. Chem.*, **A154,** 1931, 32–33.
3. G. Brauer and E. Zintl, *Z. physik. Chem.*, **B31,** 1937, 323–352.

$\bar{1}.9065$
0.0935

As-Nb Arsenic-Niobium

By heating of the elements in a sealed tube at 600°C a product of the composition $NbAs_{1.80}$ (64.29 at. % As) was obtained [1].

1. E. Heinerth and W. Biltz, *Z. anorg. Chem.*, **198,** 1931, 175.

$\bar{1}.7154$
0.2846

As-Nd Arsenic-Neodymium

NdAs (34.97 wt. % As) is isotypic with NaCl (B1 type), a = 3.96_6 A [1].

1. A. Iandelli and E. Botti, *Atti reale accad. nazl. Lincei*, **25,** 1937, 638–640.

0.1060
$\bar{1}$.8940

As-Ni Arsenic-Nickel

The existence of the following compounds obtained by chemical preparation was reported: Ni_3As (29.85 wt. % As) [1], Ni_2As (38.96 wt. % As) [2], Ni_3As_2 (45.97 wt. % As) [1, 3, 4], NiAs (56.07 wt. % As) [4, 5], and $NiAs_2$ (71.85 wt. % As) [4, 5]. Ni_3As and Ni_2As have been proved nonexistent.

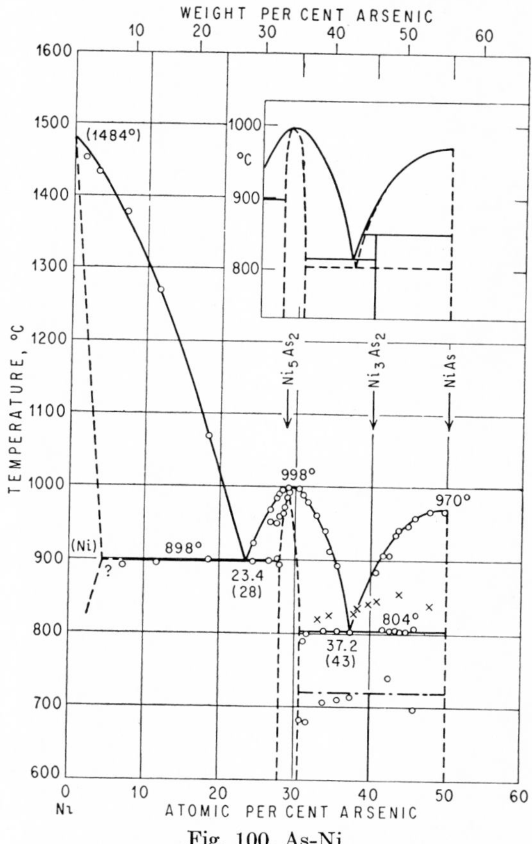

Fig. 100. As-Ni

The constitution of the alloys up to 49.6 at. % As (the alloy highest in As that could be melted in an open crucible), as established by thermal analysis and microscopic examination of alloys not heat-treated [6], is presented in Fig. 100. The solid solubility of As in Ni [5.5 wt. (4.4 at.) % As] and the range of homogeneity of the intermediate phase Ni_5As_2 (33.80 wt. % As) are based only on thermal-analysis data. As the melting point of Ni was assumed to be 1484°C rather than 1453°C, the thermal data would have to be corrected; however, this was not considered in Fig. 100.

Certain portions of the diagram are still obscure:

As indicated in Fig. 100, thermal effects between the liquidus and solidus have been observed in alloys with about 27–29 at. % As. They were interpreted as being due to an allotropic transformation of Ni_5As_2 [6]. However, this appears to be unlikely, as similar thermal effects would have been found also in alloys richer in As. Moreover, the presentation as given by the author [6] contradicts the laws of heterogeneous equilibria.

The cooling curves of melts with 30.5–48 at. % As, which were not inoculated on solidification, showed eutectic halting points at about 804°C and spontaneously occurring effects at temperatures between 680 and 740°C, preceded by great supercoolings (up to 53°C). In Fig. 100 both effects are indicated by circles. In case the melts were constantly inoculated on cooling, new thermal effects were found between the liquidus and eutectic temperatures, also preceded by considerable supercooling, at temperatures between 818 and 850°C (crosses in Fig. 100). Strange to say, the author [6] did not consider these latter effects in drawing the phase diagram. He concluded that the solid solution based on Ni_5As_2 and NiAs reacts peritectoidally to form the phase Ni_3As_2 (45.97 wt. % As).

However, the observed thermal effects can be interpreted much more simply if one assumes that the phase Ni_3As_2 is formed by a peritectic reaction of NiAs with melt at about 850°C. On rapid cooling of the melts, this reaction is prevented at first; i.e., the unstable system NiAs-Ni_5As_2 is formed. On further cooling, this system transforms by a spontaneous reaction, which, however, can proceed only partially into the stable systems NiAs-Ni_3As_2 and Ni_3As_2-Ni_5As_2. With inoculation of the melts, the peritectic reaction NiAs + melt → Ni_3As_2 is not suppressed; i.e., the stable system is formed at once. These phase relationships are schematically drawn in the inset of Fig. 100. The full curves represent the stable system and the dashed curves the unstable system.

In another paper the same author [7] thoroughly checked whether the compound Ni_2As (38.96 wt. % As)—analogous to Co_2As—exists. All the evidence obtained clearly indicated that this phase is missing.

Crystal Structures. The structure of NiAs is the prototype of the B8 type of structure [8–12], a = 3.617 A, c = 5.038 A, c/a = 1.393 [10], and a = 3.631 A, c = 5.049 A, c/a = 1.390 [12]. For the structure of natural $NiAs_2$ (rammelsbergite, pararammelsbergite), which is of the FeS_2 (C18) type, and natural Ni_3As_2 (maucherite), see [13–15] and [16], respectively. Ni_5As_2 has a "polyfluorite" type of structure, a = 6.80 A, c = 12.48 A [17].

1. A. Descamps, *Compt. rend.*, **1878,** 1065.
2. A. Granger, *Arch. sci. phys. et. nat.*, (4)**6,** 1898, 391.
3. A. Granger and G. Didier, *Compt. rend.*, **130,** 1900, 914–915.
4. E. Vigouroux, *Compt. rend.*, **147,** 1908, 426–428.
5. A. Beuttell, *Zentr. Mineral., Geol. u Paläontol.*, **1916,** 40–56.
6. K. Friedrich (F. Bennigson), *Metallurgie*, **4,** 1907, 202–216.
7. K. Friedrich, *Metallurgie*, **5,** 1908, 598, 601–603.
8. G. Aminoff, *Z. Krist.*, **58,** 1923, 203–219.
9. W. F. de Jong, *Physica*, **5,** 1925, 194–198.
10. N. Alsén, *Geol. Fören. i Stockholm Förh.*, **47,** 1925, 19–72.
11. W. F. de Jong and H. W. V. Willems, *Physica*, **7,** 1927, 74–79.
12. D. F. Hewitt, *Econ. Geol.*, **43,** 1948, 408–417.
13. M. A. Peacock and C. E. Michener, *Univ. Toronto Studies, Geol. Ser.*, no. 42, 1939, 95–112.
14. M. A. Peacock, *Am. Mineral.*, **24,** 1939, Dec. 2.
15. S. Kaiman, *Univ. Toronto Studies, Geol. Ser.*, no. 51, 1947, 49.
16. F. Laves, *Z. Krist.*, **90,** 1935, 279–282.
17. E. Hellner, *Fortschr. Mineral.*, **29–30,** 1951, 59–61.

0.3835
1̄.6165

As-P Arsenic-Phosphorus

Figure 101 shows the phase diagram proposed on the basis of thermal analysis, X-ray analysis, and visual observation of the beginning of solidification and melting [1]. The mixtures were melted in evacuated sealed silica bulbs. The constitution is similar to that of the system S-Se.

1. W. Klemm and I. v. Falkowski, *Z. anorg. Chem.*, **256,** 1948, 343–348.

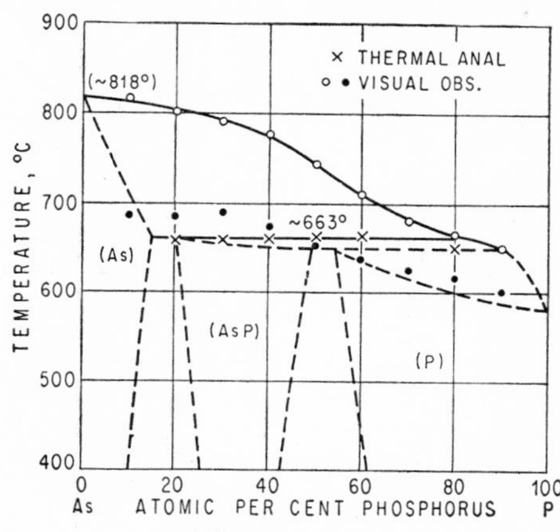

Fig. 101. As-P

1̄.5581
0.4419

As-Pb Arsenic-Lead

Arsenic additions lower the melting point of lead progressively to 292°C at 7.3 at. (2.8 wt.) % As [1]. Thermal analysis [2] established the phase diagram of Fig. 102 with the eutectic point at 2.8 ± 0.2 wt. (7.4 ± 0.5 at.) % As, 288–290°C [2, 3]. [4] located the eutectic point at 11.15 at. (4.3 wt.) % As, 292°C. Melting in evacuated porcelain bulbs made possible preparation of alloys with As contents as high as 82.5 at. % [2]. Under these conditions, the pressure at which the melts solidified increased with increasing As content. At atmospheric pressure, the As-rich portion of the diagram will be more complicated as As at this pressure does not melt but sublimes, at 616°C. At a pressure of 36 atm, As melts at 817°C.

Earlier as well as later studies [5, 6] were hampered by severe segregation, apparently due to insufficient mixing of the melts prior to cooling [2]. The absence of intermediate phases was noted also by others [7, 8]. The solid solubility of As in Pb is extremely small [9, 3]. By age-hardening tests the solubility was shown to be about 0.05 wt. (0.14 at.) % at 290°C and probably less than 0.01 wt. (0.03 at.) % As at room temperature [3].

1. C. T. Heycock and F. H. Neville, *J. Chem. Soc.*, **61,** 1892, 906.
2. W. Heike, *Intern. Z. Metallog.*, **6,** 1914, 49–57.
3. O. Bauer and W. Tonn, *Z. Metallkunde*, **27,** 1935, 183–187.
4. O. Hájiček, *Hutnické Listy*, **3,** 1948, 265–270.
5. K. Friedrich, *Metallurgie*, **3,** 1906, 41–52.
6. M. O. Faruq, *Proc. 15th Indian Sci. Congr.*, 1928, p. 176; abstract in *J. Inst. Metals*, **47,** 1931, 520.

7. S. F. Zemczuzny, *Zhur. Russ. Fiz.-Khim. Obshchestva*, **37**, 1905, 1283.
8. N. Puschin, *Zhur. Russ. Fiz.-Khim. Obshchestva*, **39**, 1907, 869–897.
9. K. S. Seljesater, *Trans. AIME, Inst. Metals Div.*, 1929, pp. 573–580.

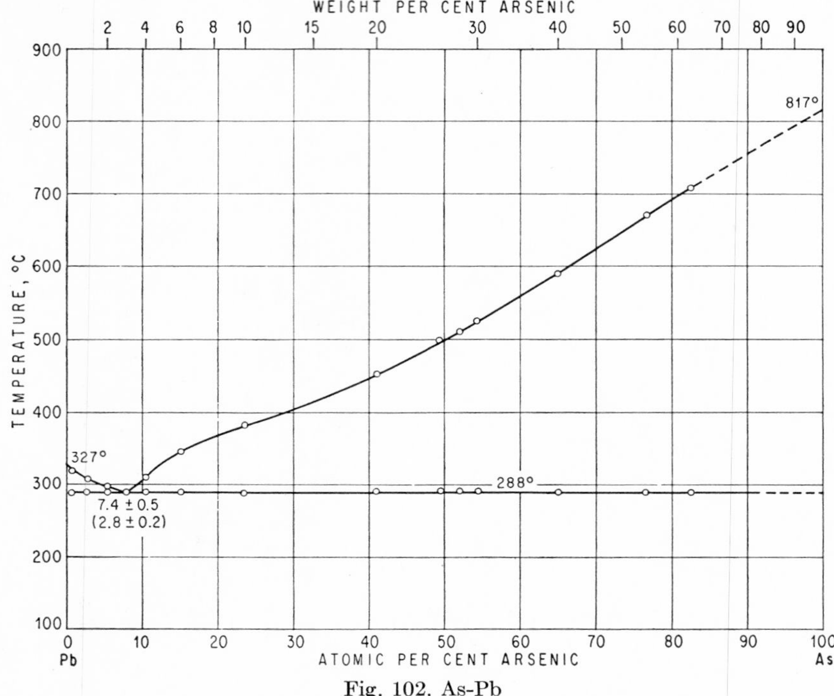

Fig. 102. As-Pb

$\bar{1}.8464$
0.1536

As-Pd Arsenic-Palladium

$PdAs_2$ (58.41 wt. % As), obtained by heating the elements in an evacuated silica tube, is isotypic with pyrite (C2 type), a = 5.982 A [1]. A reaction product with 50 at. % As proved to contain two phases; hence, PdAs (41.25 wt. % As) does not exist [1]. $PdAs_2$ melts, in a hydrogen atmosphere, at 680°C, giving off arsenic and possibly forming Pd_3As_2 (31.88 wt. % As) [2].

1. L. Thomassen, *Z. physik. Chem.*, **B4**, 1929, 279–281.
2. L. Wöhler and K. F. A. Ewald, *Z. anorg. Chem.*, **199**, 1931, 63–64.

$\bar{1}.7256$
0.2744

As-Pr Arsenic-Praseodymium

PrAs (34.71 wt. % As) is isotypic with NaCl (B1 type), a = 6.00_9 A [1].

1. A. Iandelli and E. Botti, *Atti reale accad. nazl. Lincei*, **25**, 1937, 498–502.

$\bar{1}.5840$
0.4160

As-Pt Arsenic-Platinum

The existence of the compound $PtAs_2$ (43.42 wt. % As), long since claimed [1, 2], has been established by showing that the crystal structures of the native platinum arsenide, sperrylite, which almost corresponds to the composition $PtAs_2$, and that of the substance of this composition prepared from the elements are identical [3–7]. The structure is of the FeS_2 (C2) type, with $a = 5.96_9$ A [6]. $PtAs_2$ crystals are formed by reaction of platinum black with an excess of arsenic in a bomb tube at 300–400°C [8, 9].

Thermal analysis yielded the phase diagram given in Fig. 103 [10]. The authors' conclusion that the intermediate phase coexisting with platinum is of the composition

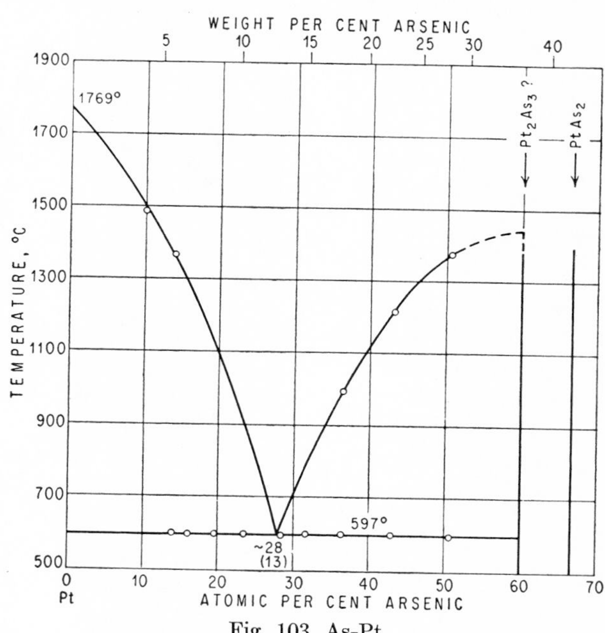

Fig. 103. As-Pt

Pt_2As_3 (36.53 wt. % As) is based solely on the insufficient criterion that the extrapolated halting time of the eutectic crystallization becomes zero at approximately 35 wt. % As. Whether Pt_2As_3 exists, besides $PtAs_2$, remains to be seen. The compound Pt_3As_2 (20.37 wt. % As), reported earlier [11], certainly does not exist [8]. Likewise, an alloy of the composition PtAs (27.73 wt. % As) was shown to have a duplex structure [6], in accordance with Fig. 103.

1. H. L. Wells, *Am. J. Sci.* (3)**37,** 1889, 69.
2. F. Roessler, *Z. anorg. Chem.*, **9,** 1895, 60–66.
3. L. S. Ramsdell, *Am. Mineralogist,* **10,** 1925, 281–304; **12,** 1927, 79.
4. W. F. de Jong, *Physica,* **5,** 1925, 292–301.
5. G. Aminoff and A. L. Parsons, *Am. Mineralogist,* **13,** 1928, 110.
6. L. Thomassen, *Z. physik. Chem.*, **B4,** 1929, 278–279.
7. F. A. Bannister and M. H. Hey, *Mining Mag.*, **23,** 1932, 188–206.
8. L. Wöhler, *Z. anorg. Chem.*, **186,** 1930, 324–336.
9. L. Wöhler and K. F. A. Ewald, *Z. anorg. Chem.*, **199,** 1931, 62–63.
10. K. Friedrich and A. Leroux, *Metallurgie,* **5,** 1908, 148–149.
11. D. Tivoli, *Gazz. chim. ital.*, **14,** 1884, 488.

$\bar{1}.4961$
0.5039

As-Pu Arsenic-Plutonium

The compound PuAs (23.9 wt. % As) has the cubic NaCl (B1) type of structure, a = 5.855 ± 4 A [1].

1. A. S. Coffinberry and F. H. Ellinger, "Proceedings of the International Conference on the Peaceful Uses of Atomic Energy," vol. 9, pp. 138–146, Geneva, 1955; also "Metallurgy and Fuels," p. 391, Progress in Nuclear Energy, ser. V, vol. 1, Pergamon Press Ltd., London, 1956.

$\bar{1}.6043$
0.3957

As-Re Arsenic-Rhenium

Tensiometric investigations of the system showed that, in equilibrium with arsenic vapor, there exists only one arsenide, which has the composition of about $ReAs_{2.3}$ (or Re_3As_7) [1].

1. F. Wiechmann, M. Heimburg, and W. Biltz, *Z. anorg. Chem.*, **240,** 1939, 129–138.

$\bar{1}.8621$
0.1379

As-Rh Arsenic-Rhodium

The synthesis of $RhAs_2$ (59.28 wt. % As) by heating finely divided Rh with As is difficult to achieve [1, 2]. The compound can be easily obtained, however, by heating $RhCl_3$ with As in a hydrogen stream at 700°C.

1. L. Wöhler and K. F. A. Ewald, *Z. anorg. Chem.*, **199,** 1931, 61–62.
2. After heating seven times, the reaction product, freed from excess of As, contained only 42.5 wt. % As. Although this corresponds to the formula RhAs (42.13 wt. % As), the composition was obtained only accidentally since a product of this As content could not be reproduced.

$\bar{1}.8672$
0.1328

As-Ru Arsenic-Ruthenium

The preparation of $RuAs_2$ (59.57 wt. % As) by heating finely divided Ru with As is difficult to achieve [1, 2]. The compound can be easily obtained, however, by heating $RuCl_3$ with As in a hydrogen stream.

1. L. Wöhler and K. F. A. Ewald, *Z. anorg. Chem.*, **199,** 1931, 61–62.
2. After the first treatment the reaction product contained only 32.7 wt. % As; after the twelfth and fortieth repetition the As content increased to 53.7 and 59.6 wt. %, respectively.

0.3685
$\bar{1}.6315$

As-S Arsenic-Sulfur

Although this system is not a metallic one, it is treated here as it is so far the only completely investigated system with arsenic, the phase diagram of which has been determined for a *constant* pressure (1 atm), considering the vapor phase. Figure 104 shows the diagram [1].

Of the three As-S compounds, As_2S_2 (29.98 wt. % S), As_2S_3 (39.10 wt. % S), and As_2S_5 (51.70 wt. % S), only the first two could be identified by thermal analysis. (Apparently As_2S_5 can be prepared only by precipitation from an aqueous solution of H_3AsO_4 with H_2S.) Above 55 at. % S the liquidus temperatures could not be determined as these melts have such a high viscosity that they do not crystallize; i.e., the equilibrium liquid ⇄ solid is approached extremely slowly. Therefore, the melting point of As_2S_3 could be measured only by using native As_2S_3 crystals. The "liquidus" points of the melts with 71 and 80 at. % S are actually hardening or softening points obtained by qualitative viscosity determinations. They did not indicate the formation of As_2S_5 either.

The curve of the boiling points (composition of melts at the beginning of boiling) shows that As_2S_2 is strongly dissociated in the molten state, as indicated by the large

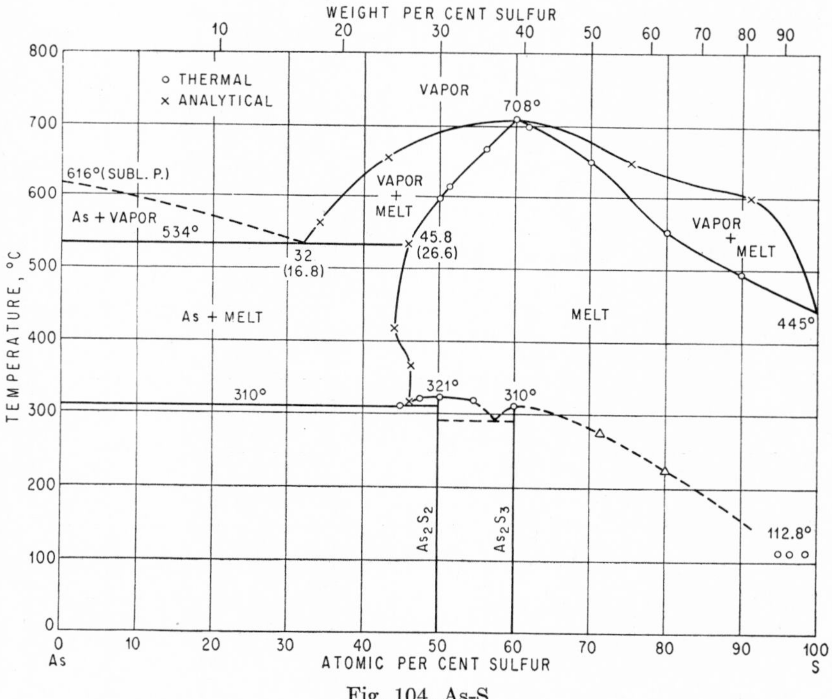

Fig. 104. As-S

difference between the temperatures of the beginning and end of boiling at this composition. On the other hand, As_2S_3 does not dissociate since the melt of this composition distills over without change in composition. The curve of the end of boiling (vapor curve) was obtained by determining analytically the composition of the vapor in equilibrium with a boiling liquid of known composition. In the range of 0–45.8 at. % S there exists a nonvariant equilibrium at 534°C between arsenic (solid), saturated melt (45.8 at. % S), and vapor of constant composition (32 at. % S). The curve descending from the sublimation point of arsenic (616°C) toward 534°C, which represents the composition of the vapor in equilibrium with solid arsenic, was not determined. For further details, see the original paper, which includes a discussion of earlier work on the system.

1. W. P. A. Jonker, *Z. anorg. Chem.*, **62**, 1909, 89–107.

Ī.7890
0.2110

As-Sb Arsenic-Antimony

The results of the first thermal investigation between 0 and 35 wt. % As [1], using melts prepared in open crucibles, are given in the inset of Fig. 105. Between about 22 and 27 at. % (about 15 and 18.5 wt. %) As the liquidus temperature was found to be practically constant at 612°C, with the minimum occurring at approximately 25.5 at. (17.4 wt.) % As. Another paper [2] dealt with thermal analysis of alloys

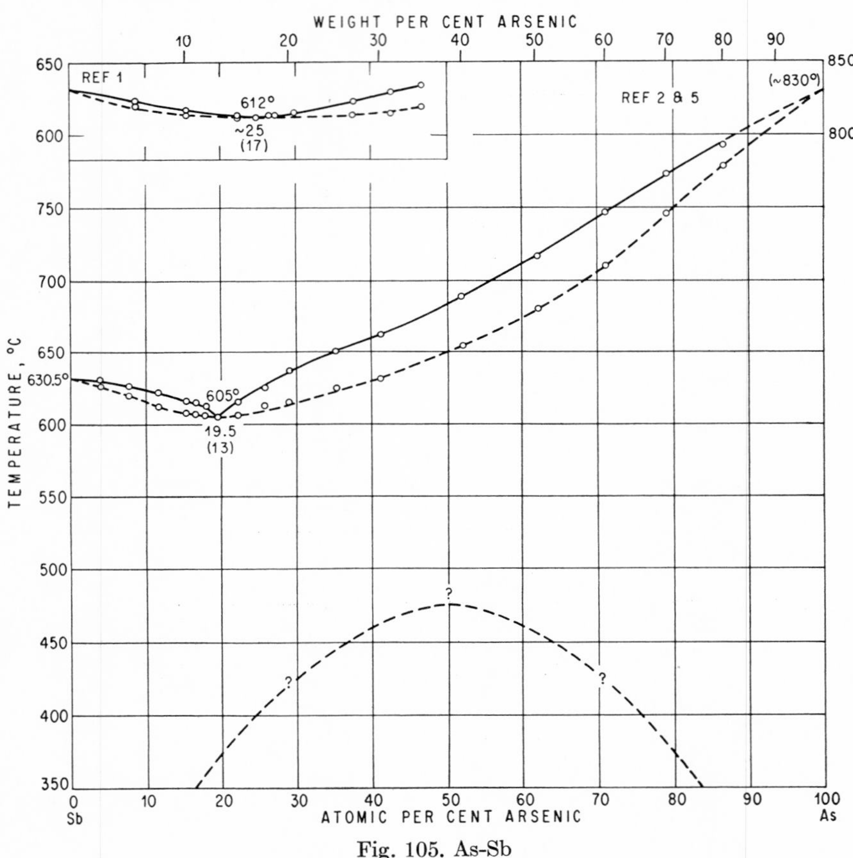

Fig. 105. As-Sb

up to 80 wt. (86.7 at.) % As, prepared by melting in evacuated glass tubes (Fig. 105); i.e., the melts richer in As solidified under a pressure increasing with the As content. At atmospheric pressure the As-rich portion of the diagram will not be so simple as shown in Fig. 105, as As, at ordinary pressure, does not melt but sublimes at 616°C. Therefore, the liquidus curve will cut the vapor (sublimation) curve before it reaches the arsenic axis (see As-S, page 176).

Both studies agree in that a series of solid solutions crystallizes from the melt. This was confirmed by the microstructure of the alloys [1, 2]. Quantitatively, however, results differ considerably: the minimum of the liquidus curve obtained by the second investigation is more pronounced. Also, in alloys beyond the composition of the minimum, the liquidus in the first paper nearly coincides with the solidus in the second.

The occurrence of a minimum melting point in a continuous series of solid solutions often is characteristic of the existence of a transformation in the solid state (see Au-Cu, Au-Ni). Some indications of a decomposition of the solid solutions have, indeed, been found by microscopic investigation [3] of the mineral allemontite, a natural As-Sb alloy varying in composition between 25 and 35 wt. % As. However, this was not verified by an X-ray study [4], from which it was concluded that the solid solution does not break up on annealing at 200–400°C for 1–3 weeks. Although one phase in allemontite has a composition corresponding to the formula AsSb, no evidence was found for the existence of this compound [4].

In a later paper [5], lattice-parameter measurements were reported of synthetic alloys annealed at 200–400°C for long periods of time (up to 100 days). Results were interpreted to indicate the occurrence of an extremely sluggish transformation between 550 and 400°C, involving decomposition of the high-temperature solid solution into two limiting solid solutions containing approximately between 20 and 25 at. % (13 and 17 wt. %) As and 70 at. % (59 wt. %) As. Also, a cubic (a = 11.08–11.14 A) intermediate phase of unknown but variable composition was claimed to have been formed occasionally in the range between 25 and 46 at. % (17 and 34 wt. %) As after annealing at 200–400°C. The appearance of this cubic phase was interpreted to be due to the formation of As_2O_3-Sb_2O_3 solid solutions [6]; however, this criticism was rejected by the investigators [7] as being entirely unproved. The dashed transformation curve inserted in Fig. 105 roughly outlines the composition and temperature range in which a transformation might occur.

Note Added in Proof. [8] redetermined the solidification diagram of the entire system and found the minimum to lie between 15 and 20 wt. (22 and 29 at.) % As at 612°C. X-ray patterns of apparently unannealed alloys confirmed the existence of a continuous solid solution.

1. N. Parravano and P. de Cesaris, *Intern. Z. Metallog.*, **2,** 1912, 70–75.
2. Q. A. Mansuri, *J. Chem. Soc.*, **1928**(2), 2107–2108.
3. G. Kalb, *Metall u. Erz*, **23,** 1926, 113–115.
4. P. Quensel, K. Ahlborg, and A. Westgren, *Geol. Fören. i Stockholm Förh.*, **59,** 1937, 135–144.
5. W. Trzebiatowski and E. Bryjak, *Z. anorg. Chem.*, **238,** 1938, 255–267.
6. P. E. Wretblad, *Z. anorg. Chem.*, **240,** 1939, 139–141.
7. W. Trzebiatowski, *Z. anorg. Chem.*, **240,** 1939, 142–144.
8. C. H. Shih and E. A. Peretti, *Trans. ASM*, **48,** 1956, 709–710.

0.4260
$\bar{1}$.5740

As-Si Arsenic-Silicon

Thermal-analysis data (Fig. 106) were obtained by taking heating and cooling curves of alloys prepared in sealed silica bulbs [1]. In the absence of tabulated data the data points had to be taken graphically from the diagram published. The existence of SiAs (72.73 wt. % As) was concluded from the maximum melting point of this alloy, that of $SiAs_2$ (84.21 wt. % As) from heating curves of alloys with 60 to 75 at. % As, annealed at a temperature 50–100°C below the peritectic temperature. Both intermediate phases have complicated X-ray powder patterns [1]. The lattice constants of SiAs were reported as a = 21.0 A, b = 3.7 A, c = 9.7 A, β = 45.0° [3]. As the lattice parameters of the elements do not change on alloying, the mutual solid solubility of the components appears to be very small. The compound $SiAs_6$ (94.12 wt. % As) reported earlier [2] does not exist.

1. W. Klemm and P. Pirscher, *Z. anorg. Chem.*, **247**, 1941, 211–220.
2. C. Winkler, *J. prakt. Chem.*, **91**, 1864, 193.
3. K. Schubert, E. Dörre, and E. Günzel, *Naturwissenschaften*, **41**, 1954, 448.

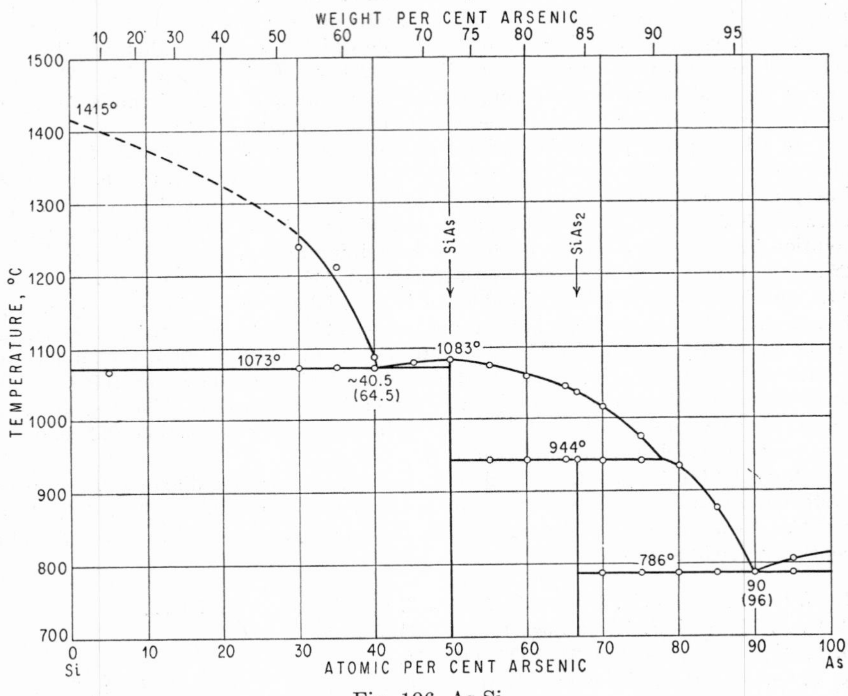

Fig. 106. As-Si

1̄.8001
0.1999

As-Sn Arsenic-Tin

The existence of the compound Sn_3As_2 (29.61 wt. % As), first established by [1], was corroborated by thermal and microscopic investigations [2, 3, 4] as well as X-ray analysis [5]. Sn_3As_2 is in equilibrium with Sn [2–5] and SnAs (38.69 wt. % As) [6, 2, 4, 5]. The solid solubility of As in Sn is vanishingly small [2–5]; that of Sn in As was found to be about 31 (22.1) [4], 30 (21.3) [5], and 32 wt. (22.9 at.) % [7], respectively. There is no evidence [1–5, 7] for the existence of the compounds Sn_6As (9.52 wt. % As) [8], Sn_4As_3 (32.14 wt. % As) [6], Sn_3As_4 (45.71 wt. % As) [9, 10], and Sn_2As_3 (48.63 wt. % As) [9, 10]. Potential measurements [11] are in agreement with the phase diagram.

The phase diagrams as determined by thermal and microscopic studies (Fig. 107) differ as follows: (*a*) The temperatures of the liquidus and horizontals obtained by [4] are higher than those found by [2]. (*b*) Sn_3As_2 was found to be formed peritectically [2] or to have a maximum melting point [4]. The microstructures published do not give conclusive evidence for the existence or absence of a Sn_3As_2-SnAs eutectic [2, 4];

however, the information available appears to be slightly in favor of a peritectic formation of Sn_3As_2 [12]. (*c*) Whereas [4] assumed the formation of solid solutions of As in SnAs between 38.7 and about 47 wt. % As (50–58.4 at. % As), the microstructures given by [2] and the X-ray data by [5] do not support this [12].

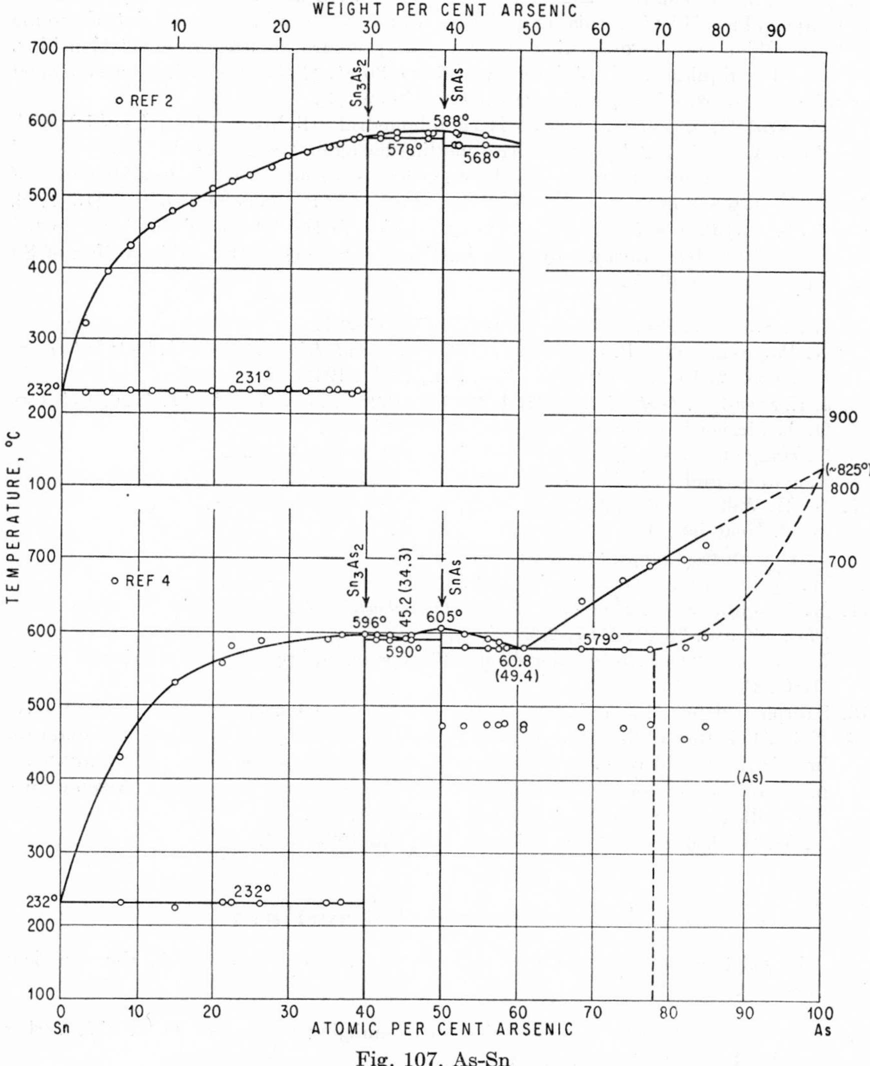

Fig. 107. As-Sn

Also, there is considerable discrepancy between the X-ray data of [5] and [7]. The latter denies the existence of Sn_3As_2 and assumes the formation of a solid solution of As in Sn up to 40 at. % As. However, there is no doubt that As is practically insoluble in Sn and that Sn_3As_2 is the phase coexisting with Sn [2–5]. Whereas [7] claimed to have found, by lattice-parameter measurements, a homogeneity range of the SnAs phase between 45.5 and 60.5 at. % As, [5] reported that there is no sign of variation of the parameter of this phase. The absence of a wide homogeneity range of SnAs is strongly supported by the microscopic evidence given by [2] and [4].

Figure 107 shows thermal effects at approximately 472°C between 50 and 84.6 at. % As [4]. It has been suggested [4] that at 472°C arsenic apparently is present as a vapor phase [13]; i.e., this temperature would be the sublimation point of arsenic in these alloys [14]. Concerning the effect of pressure on the phase relationships, see [4]. The liquidus and solidus curves above 86 at. % As (Fig. 107) were outlined by microscopic examination of quenched specimens [4].

Crystal Structures. Sn_3As_2 is rhombohedral (slightly distorted cubic) with $a = 4.090$ A, $c = 5.152$ A for the corresponding hexagonal cell. As the rhombohedral cell contains 7 atoms, the phase can be regarded as a solution of tin in a structure of the ideal composition Sn_4As_3 [5]. SnAs has the NaCl (B1) type of structure [15, 7, 5], $a = 5.71_9$ A [15], $a = 5.727$ A [5]. The variation in the lattice parameter reported by [7] could not be confirmed by [5]. Lattice parameters of the solid solution of Sn in As are given by [7, 5].

1. J. E. Stead, *J. Soc. Chem. Ind.*, **16,** 1897, 206–207.
2. N. Parravano and P. de Cesaris, *Atti reale accad. Lincei,* **20,** 1911, 593; *Intern. Z. Metallog.*, **2,** 1912, 1–12; *Gazz. chim. ital.*, **42**(1), 1912, 274.
3. J. E. Stead, *J. Inst. Metals,* **22,** 1919, 130–132; *Z. Metallkunde,* **12,** 1920, 134–135.
4. Q. A. Mansuri, *J. Chem. Soc.*, **123,** 1923, 214–223.
5. G. Hägg and A. G. Hybinette, *Phil. Mag.*, **20,** 1935, 913–929.
6. P. Jolibois and E. L. Dupuy, *Compt. rend.*, **152,** 1911, 1312–1314.
7. W. H. Willott and E. J. Evans, *Phil. Mag.*, **18,** 1934, 114–128.
8. W. P. Headden, *Am. J. Sci.*, (4)**5,** 1898, 95.
9. A. Descamps, *Compt. rend.*, **86,** 1878, 1066.
10. W. Spring, *Ber. deut. chem. Ges.*, **16,** 1883, 324.
11. N. Puschin, *Zhur. Russ. Fiz.-Khim. Obshchestva,* **39,** 1907, 528–566; abstract in *Chem. Zentr.*, **78**(2), 1907, 2027–2028.
12. M. Hansen, "Der Aufbau der Zweistofflegierungen," pp. 198–201, Springer-Verlag OHG, Berlin, 1936.
13. Samples quenched from below or above 472°C were denser or porous, respectively.
14. If 472°C is the sublimation point of arsenic in the alloys in question, the diagram in this range will be more complicated than shown in Fig. 107. At atmospheric pressure the portion above approximately 600°C is not realizable. Arsenic does not melt at normal pressure but sublimes at 616°C.
15. V. M. Goldschmidt, *Trans. Faraday Soc.*, **25,** 1929, 253.

$\bar{1}.9319$
0.0681

As-Sr Arsenic-Strontium

The existence of Sr_3As_2 (36.30 wt. % As) has been reported in the chemical literature [1].

1. "Gmelins Handbuch der anorganischen Chemie," System No. 29, p. 217, Verlag Chemie G.m.b.H., Berlin, 1931.

$\bar{1}.6172$
0.3828

As-Ta Arsenic-Tantalum

In the attempt to prepare $TaAs_2$ (45.30 wt. % As) by reaction of the elements in a sealed tube a product of the composition $TaAs_{1.4}$ (58.3 at. % As) containing unreacted tantalum was obtained. Under more satisfactory experimental conditions, a product richer in As might have been formed [1].

1. E. Heinerth and W. Biltz, *Z. anorg. Chem.*, **198,** 1931, 175.

$\bar{1}.7687$
0.2313

As-Te Arsenic-Tellurium

The liquidus curve given in Fig. 108 can be considered only a rough approximation; no individual data were reported [1]. The maximum at 362°C would suggest the existence of the compound As_2Te_3 (71.87 wt. % Te) [2], and the horizontal at 358°C indicates an extended miscibility gap in the liquid state. However, the data require verification.

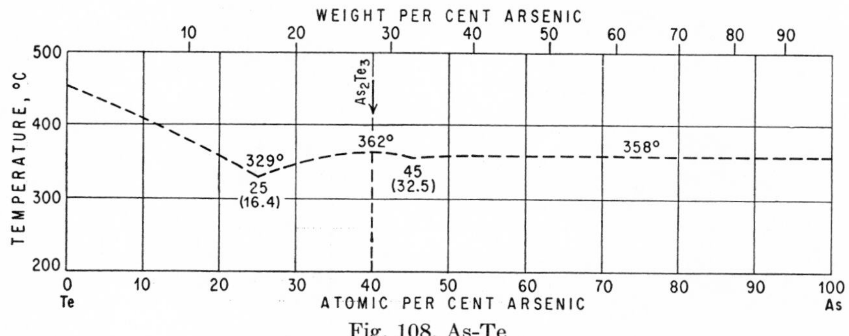

Fig. 108. As-Te

Supplement. As_2Te_3 is monoclinic, with a = 14.4 A, b = 4.05 A, c = 9.92 A, β = 97° [3].

1. H. Pélabon, *Compt. rend.*, **146,** 1908, 1397–1399; *Ann. chim et phys.* (8) **17,** 1909, 561–563.
2. Between 0 and 40 at. % As the diagram is analogous to that of the system Sb-Te. The formula of the compound As_2Te_3 also agrees with that of the Sb-Te compound.
3. J. Singer and C. W. Spencer, *Trans. AIME*, **203,** 1955, 144.

0.1942
$\bar{1}.8058$

As-Ti Arsenic-Titanium

The lowest per cent addition at which As appeared as insoluble compound in the microstructure of magnesium-reduced titanium-base alloys, hot-rolled and subsequently annealed at 790°C for ½ hr, was reported to be 0.26 wt. % As [1].

The existence of several Ti arsenides was reported [2, 3]; but apparently only one, TiAs (61.0 wt. % As), was identified. It is hexagonal, with 8 atoms per unit cell and a = 3.64_2 A, c = 12.06_4 A, c/a = 3.312 [3].

1. R. M. Goldhoff, H. L. Shaw, C. M. Craighead, and R. I. Jaffee, *Trans. ASM*, **45,** 1953, 941–971.
2. H. Nowotny, R. Funk, and J. Pesl, *Monatsh. Chem.*, **82,** 1951, 519.
3. W. Trzebiatowski and K. Lukaszewicz, *Roczniki Chem.*, **28,** 1954, 150–151; *Bull. acad. polon. sci., Classe III*, **2,** 1954, 277–279.

1.5641
0.4359

As-Tl Arsenic-Thallium

Thermal analysis [1] yielded the phase diagram given in Fig. 109. Alloys with 0–77 at. % As were melted in open crucibles under a salt layer; alloys richer in As, in evacuated (15 mm Hg) glass bulbs. Therefore, the diagram does not represent constant-pressure conditions. According to the investigator, the effect of pressure on the phase diagram is only slight, even above 600°C [2].

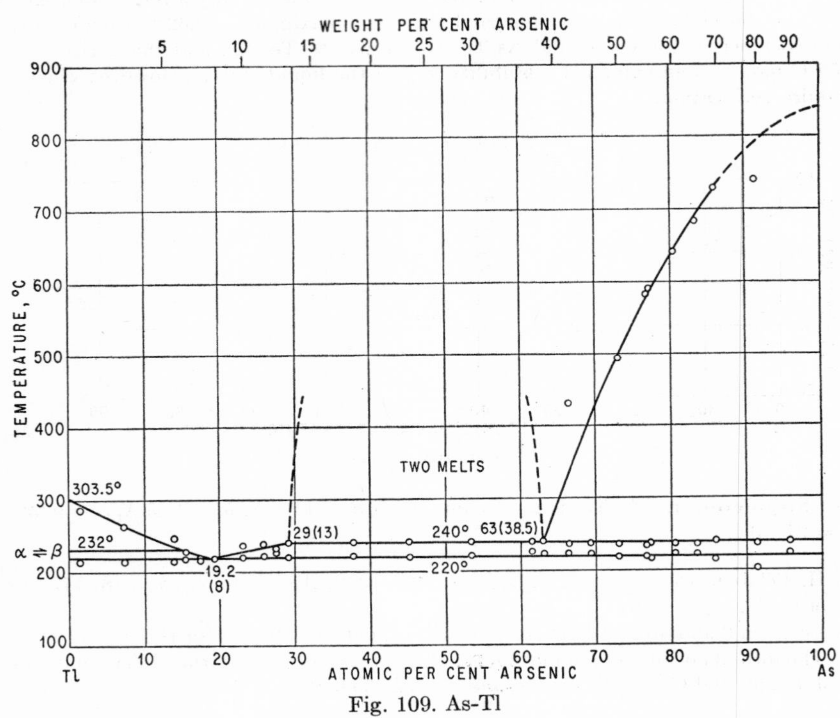

Fig. 109. As-Tl

1. Q. A. Mansuri, *J. Inst. Metals*, **28**, 1922, 453–468.
2. As As can be melted only under high pressure—at atmospheric pressure it sublimes at 616°C—the liquidus curve between 92 and 100 at. % As is realizable only under pressure much higher than that attainable in the investigation. Under a pressure of 35.8 atm As melts at 818°C.

1.4978
0.5022

As-U Arsenic-Uranium

The following phases have been identified by X-ray analysis: U_2As (13.59 wt. % As) [1], UAs (23.94 wt. % As) [1, 2], U_3As_4 (29.55 wt. % As) [2], and UAs_2 (38.62 wt. % As) [3]. UAs is isotypic with NaCl (B1 type), a = 5.767 A [1, 2]. U_3As_4, already reported by [4], is b.c.c. of the Th_3P_4 ($D7_3$) type, a = 8.507 A [2]. UAs_2 is isotypic with Cu_2Sb (C38 type), a = 3.954 A, c = 8.116 A, c/a = 2.053 [3].

1. See J. J. Katz and E. Rabinowitch, "Chemistry of Uranium," Part I, National Nuclear Energy Series, Div. VIII, vol. 5, pp. 165 and 241–242, McGraw-Hill Book Company, Inc., New York, 1951.
2. A. Iandelli, *Atti accad. nazl. Lincei, Rend., Classe sci. fis., mat. e nat.*, **13,** 1952, 138–143.
3. A. Iandelli, *Atti accad. nazl. Lincei, Rend., Classe sci. fis., mat. e nat.*, **13,** 1952, 144–151.
4. A. Colani, *Ann. chim. et phys.*, **12,** 1907, 93.

0.1674
$\bar{1}$.8326

As-V Arsenic-Vanadium

By various reactions, e.g., heating the elements in evacuated silica bulbs at 600–1000°C and distilling off the excess of arsenic, the compound VAs (59.92 wt. % As) has been prepared. It dissociates in vacuum above 1000°C to V_2As (42.37 wt. % As), which melts at 1345°C [1]. It has been stated that vanadium and arsenic "form several intermetallic phases" [2].

1. A. Morette, *Compt. rend.*, **212,** 1941, 639–641; *Bull. soc. chim. France*, **8,** 1942, 146–152.
2. H. Nowotny, R. Funk, and J. Pesl, *Monatsh. Chem.*, **82,** 1951, 519.

$\bar{1}$.6099
0.3901

As-W Arsenic-Wolfram

WAs_2 (44.89 wt. % As) has been prepared by reaction of WCl_6 with H_3As [1] and by heating the elements in a sealed tube at 620°C for 5 days [2]. The latter reaction product had the composition $WAs_{1.954}$ [66.15 at. (44.3 wt.) % As].

1. E. Defacqz, *Compt. rend.*, **132,** 1901, 138–139.
2. E. Heinerth and W. Biltz, *Z. anorg. Chem.*, **198,** 1931, 173.

0.0591
$\bar{1}$.9409

As-Zn Arsenic-Zinc

The phase diagram of Fig. 110 is based on the thermal data reported by [1] (0–12.4 at. % As) and [2] and microscopic examination of as-cast alloys [2]. Work by [3] was confined to alloys below 2 wt. % As.

The existence of Zn_3As_2 (43.31 wt. % As), which apparently has a very narrow homogeneity range [2], had been reported earlier [4, 5]. The region of the phase $ZnAs_2$ (69.92 wt. % As) given in Fig. 110 is only an approximation. The existence of the compound Zn_2As (36.42 wt. % As) [6] is in contradiction to Fig. 110.

The constitution of the systems As-Cd and As-Zn is very similar as regards compound formation and the fact that both Cd_3As_2 and Zn_3As_2 undergo a polymorphic transformation. However, there is no indication of an unstable system in the system As-Zn.

Crystal Structures. The structure of Zn_3As_2 was reported to be cubic with 2 molecules per unit cell, a = 5.82 A [7], and cubic of the Mn_2O_3 ($D5_3$) type, a = 11.76 A [8]. However, it was shown that the structure is actually tetragonal, of the Zn_3P_2 ($D5_9$) type, $a = 8.33_3$ A, c = 11.78 A, c/a = 1.414 [9]. $ZnAs_2$ is orthorhombic, 32 molecules per unit cell, $a = 7.73_5$ A, $b = 8.00_6$ A, c = 36.35 A [9].

1. K. Friedrich and A. Leroux, *Metallurgie*, **3,** 1906, 477–479. Alloys were prepared in open crucibles.

2. W. Heike, *Z. anorg. Chem.*, **118,** 1921, 264–268. Alloys were prepared in evacuated sealed porcelain crucibles.
3. T. Arnemann, *Metallurgie*, **7,** 1910, 201–211.
4. A. Descamps, *Compt. rend.*, **86,** 1878, 1066.
5. W. Spring, *Ber. deut. chem. Ges.*, **16,** 1883, 324.
6. G. A. König, *Z. Krist.*, **38,** 1904, 543.

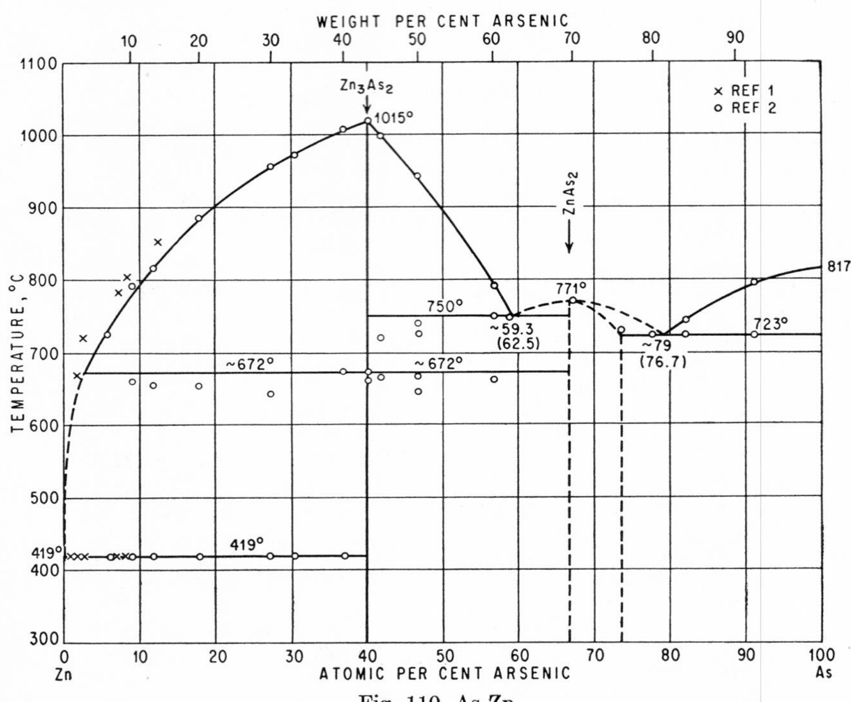

Fig. 110. As-Zn

7. G. Natta and L. Passerini, *Gazz. chim. ital.*, **58,** 1928, 541–550.
8. R. Paulus, *Z. physik. Chem.*, **B22,** 1933, 305–322.
9. M. v. Stackelberg and R. Paulus, *Z. physik. Chem.*, **B28,** 1935, 427–460.

1.2607
$\bar{2}$.7393

Au-B Gold-Boron

Experiments to cementate gold with boron were negative [1].

1. W. Loskiewicz, *Przeglad Górniczo-Hutniczy*, **21,** 1929, 583–611; abstract in *J. Inst. Metals*, **47,** 1931, 516–517.

0.1570
$\bar{1}$.8430

Au-Ba Gold-Barium

It was briefly reported that the compounds $BaAu_2$ (25.83 wt. % Ba) and Ba_2Au_3 (31.71 wt. % Ba) exist and that there are "one or more compounds richer in barium" [1]. On the other hand, the existence of a phase with the ideal composition $BaAu_5$

(12.23 wt. % Ba) was established by determining its crystal structure, which is of the $CaCu_5$ type, $a = 5.66$ A, $c = 4.57$ A (or kX?), $c/a = 0.81$. Apparently the actual composition of the phase is $BaAu_6$ (10.40 wt. % Ba) [2].

1. W. Biltz and F. Weibke, *Z. anorg. Chem.*, **236,** 1938, 12–23.
2. T. Heumann, *Nachr. Akad. Wiss. Göttingen, Math.-physik. Kl.*, **2,** 1950.

1.3400
$\bar{2}.6600$

Au-Be Gold-Beryllium

Thermal analysis of alloys, prepared using 98% pure Be and melting in Al_2O_3 crucibles in a hydrogen atmosphere, resulted in the diagram presented in Fig. 111 [1]. The composition of the alloys was estimated to be accurate within 1 at. %. In contrast to Fig. 111, [2] located the Au-$BeAu_3$ eutectic at 9.9 at. (0.5 wt.) % Be, 577°C.

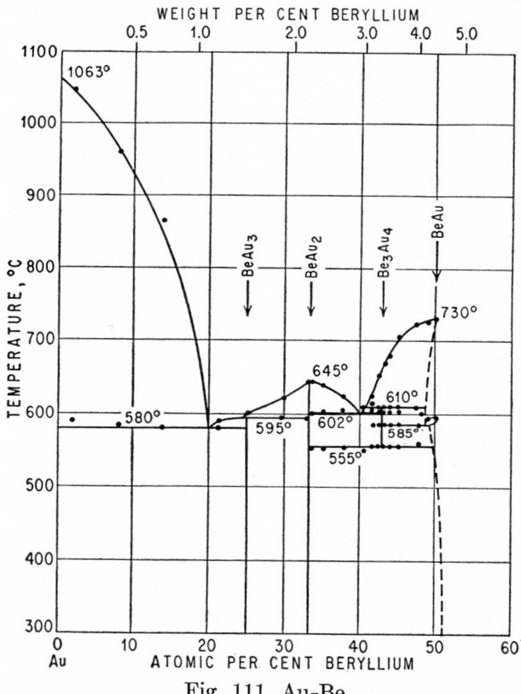

Fig. 111. Au-Be

The existence of the four intermediate phases $BeAu_3$ (1.50 wt. % Be), $BeAu_2$ (2.24 wt. % Be), Be_3Au_4 (3.32 wt. % Be), and BeAu (4.37 wt. % Be) and the eutectoid decomposition of Be_3Au_4 into $BeAu_2$ and BeAu were confirmed metallographically. BeAu apparently has a certain range of homogeneity and undergoes a transformation at 585–595°C [1].

An X-ray study of the whole system, using 42 alloys, disclosed the existence of the following compounds stable at room temperature: $BeAu_3$, $BeAu_2$, BeAu, Be_3Au (12.07 wt. % Be), and Be_5Au [83.34 at. (18.61 wt.) % Be] [3]. This confirms the work of [1] as to the composition range 0–50 at. % Be and that of [4] as to the existence of Be_5Au. No detailed data of the X-ray investigation [3] are available as yet, since this paper is only a preliminary publication.

An alloy with 0.9 at. (0.04 wt.) % Be, after cooling from the molten state, showed the presence of $BeAu_3$ crystals. This verifies earlier statements [5, 6] that the solid solubility of Be in Au is quite restricted. Also, lattice-spacing measurements showed that the solubility is probably only of the order of a few tenths of 1 at. % Be [7]. Theoretical considerations led [8] to conclude that up to about 3 at. % Au might be soluble in solid Be. This solid solution probably forms a relatively high melting eutectic with an intermediate phase (Be_5Au?).

Crystal Structures. BeAu is isotypic with FeSi (B20 type), a = 4.668 A [9]. Be_5Au has a f.c.c. structure with 24 atoms per unit cell, similar to that of the $MgCu_2$ (C15 type), a = 6.097 A (or 6.167 A?) [4]. As to the structure of a phase of the composition $Be_{3.51}Au$ (77.83 at. % Be) with a = 6.237 A, see [4].

1. O. Winkler, *Z. Metallkunde*, **30,** 1938, 171–173.
2. L. L. Stott, *Trans. AIME*, **122,** 1936, 57–73.
3. G. P. Chatterjee and S. S. Sidhu, *Phys. Rev.*, **76,** 1949, 175.
4. L. Misch, *Metallwirtschaft*, **14,** 1935, 897–899.
5. O. Loskiewicz, *Przeglad Górniczo-Hutniczy*, **21,** 1929, 583–611; abstract in *J. Inst. Metals*, **47,** 1931, 516–517.
6. L. Nowack, *Z. Metallkunde*, **22,** 1930, 99.
7. E. A. Owen and E. A. O. Roberts, *J. Inst. Metals*, **71,** 1945, 213–254.
8. G. V. Raynor, *J. Roy. Aeronaut. Soc.*, **50,** 1946, 410–413.
9. B. D. Cullity, *Trans. AIME*, **171,** 1947, 396–400.

$\bar{1}.9748$
0.0252

Au-Bi Gold-Bismuth

Thermal and microscopic studies [1] yielded a diagram showing a eutectic between Au and Bi at about 82 wt. (81.1 at.) % Bi, 241°C (based on a melting point of Bi of 266 instead of 271°C); i.e., no intermediate phase was detected by this investigation. [2] gave the eutectic point as 82.2 at. (81.3 wt.) % Bi, 240°C. The melting point of Bi, given as 267.5°C, was found to be lowered linearly by the addition of 4.5 wt. (4.75 at.) % Au [3, 4].

From the fact that Au-Bi alloys are superconductive, although the components are not superconductors, it was concluded [5] that an intermediate phase showing this phenomenon would exist. Indeed, superconductive crystals of the compound Au_2Bi (34.64 wt. % Bi) could be isolated from the eutectic by treatment with nitric acid [5]. Au_2Bi was found to be formed by a peritectic reaction at 373°C [6] (Fig. 112). Earlier, a residue isolated from an alloy with 97 wt. % Bi by leaching with nitric acid was found to correspond with the composition Au_3Bi (26.15 wt. % Bi). This was erroneously interpreted to indicate the existence of an intermediate phase of this composition [7].

The existence of solid solutions of Bi in Au seemed to be indicated by the presence of cored crystals in an alloy with 4 wt. % Bi [1]. However, lattice-parameter measurements of alloys annealed at 200, 380, and 500°C, respectively, showed the solid solubility to be practically nil [8], in accordance with other observations [6]. Based on micrographic investigations [9] in the temperature range 1040–500°C, the solid solubility of Bi in Au in at. % and wt. % was found to be <0.02 at 1040°C, 0.04 at 1000°C, 0.07 at 900°C, 0.03 at 800 and 700°C, and <0.03 at 600 and 500°C; i.e., the solubility boundary is retrograde.

Au_2Bi, formerly reported to be tetragonal [5], has the f.c.c. structure of the $MgCu_2$ (C15) type, a = 7.958 A [6, 10].

Supplement. According to [11], Au_2Bi decomposes at low temperature.

1. R. Vogel, *Z. anorg. Chem.*, **50,** 1906, 145–151.
2. O. Hajiček, *Hutnické Listy*, **3,** 1948, 265–270.

3. C. T. Heycock and F. H. Neville, *J. Chem. Soc.*, **1892,** 897.
4. C. T. Heycock and F. H. Neville, *J. Chem. Soc.*, **1894,** 65–76.
5. W. J. de Haas and F. Jurriaanse, *Naturwissenschaften*, **19,** 1931, 706.
6. F. Jurriaanse, *Z. Krist.*, **90,** 1935, 322–329.
7. F. Roessler, *Z. anorg. Chem.*, **9,** 1895, 70–72.
8. E. A. Owen and E. A. O'Donnell Roberts, *J. Inst. Metals*, **71,** 1945, 213–254.
9. E. Raub and A. Engel, *Z. Metallkunde*, **37,** 1946, 76–81.
10. W. J. de Haas and F. Jurriaanse, *Proc. Koninkl. Akad. Wetenschap. Amsterdam*, **35,** 1932, 748–750.
11. N. E. Alekseevskii, G. S. Zhdanov, and N. N. Zhuravlev, *Zhur. Eksptl. i Teoret. Fiz.*, **25,** 1953, 123–126; *Chem. Abstr.*, **49,** 1955, 5050.

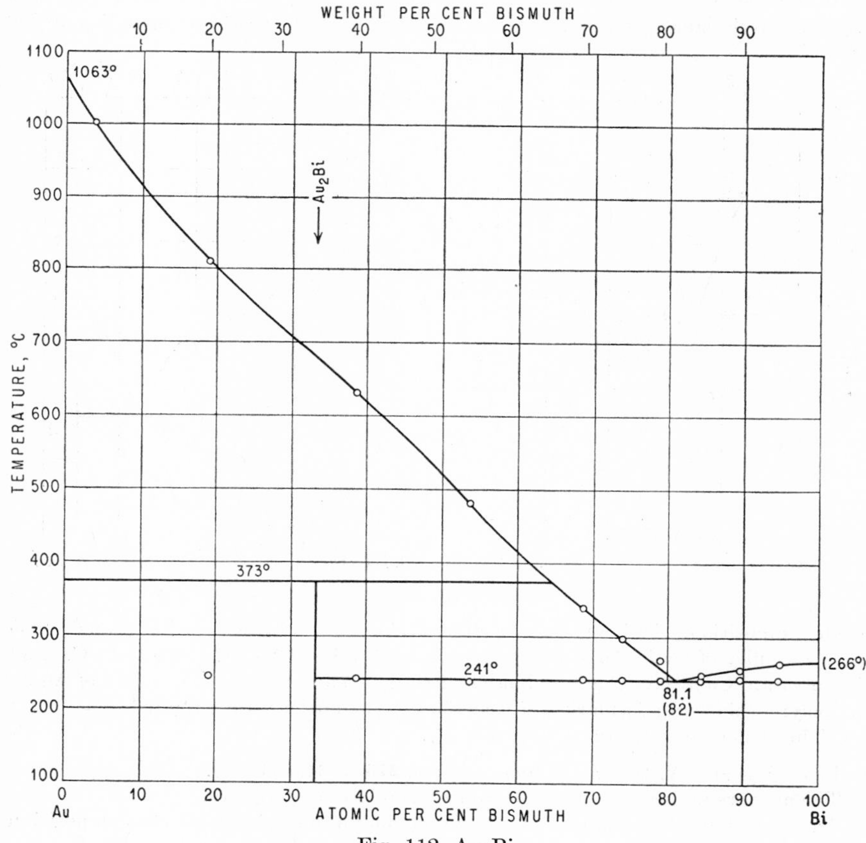

Fig. 112. Au-Bi

1.2154
$\bar{2}.7846$

Au-C Gold-Carbon

At its boiling point (~2970°C), gold dissolves carbon, which crystallizes on cooling of the melt in the form of graphite [1]. The solubility has been reported to be only very small ("traces or unweighably small amount") [2], or about 0.3 wt. % [3].

1. H. Moissan, *Compt. rend.*, **141,** 1905, 977–983.
2. O. Ruff and B. Bergdahl, *Z. anorg. Chem.*, **106,** 1919, 91.
3. W. Hempel, *Z. angew. Chem.*, **17,** 1904, 324.

0.6920
$\bar{1}$.3080

Au-Ca Gold-Calcium

Figure 113 shows the phase diagram which is exclusively based on thermal-analysis data [1]. Besides the phases of apparently fixed composition, $CaAu_4$ (4.83 wt. % Ca), $CaAu_3$ (6.35 wt. % Ca), $CaAu_2$ (9.23 wt. % Ca), Ca_4Au_3 (21.32 wt. % Ca) [2], and Ca_2Au (28.90 wt. % Ca), there is a phase of variable composition between 49 and 55.5 at. % Ca. The maximum melting point occurs at a composition corresponding to the formula $Ca_{10}Au_9$ [52.63 at. (18.43 wt.) % Ca] [3]; however, the investigators preferred to designate this phase as $Ca_{1.11}Au$ (18.4 wt. % Ca), corresponding to its

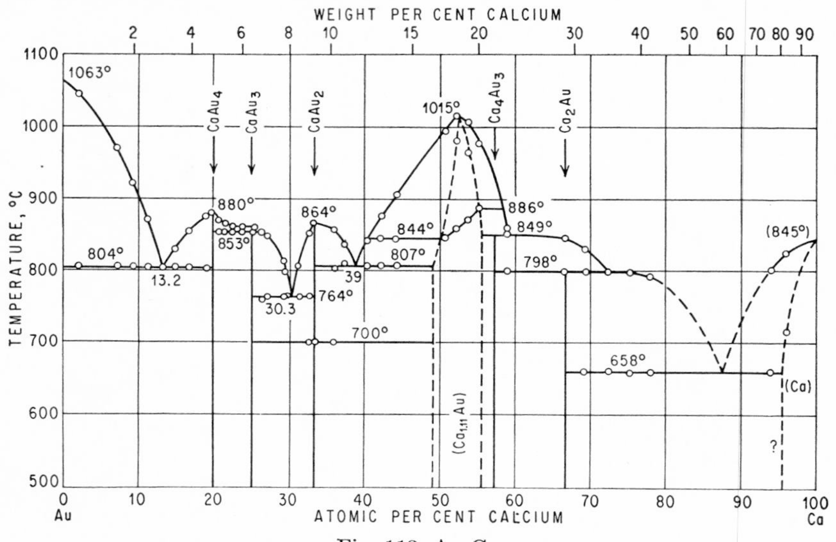

Fig. 113. Au-Ca

mean composition. This phase, as well as $CaAu_2$, exists in two modifications. The solid solubility of Ca in Au at 800°C is estimated to be <0.3 wt. (1.8 at.) % Ca, and that of Au in Ca at 658°C approximately 19 wt. (4.5 at.) % Au. The effect of Au on the polymorphic transformation of Ca (450°C) has not been investigated.

The structure of $CaAu_3$ is not cubic [4].

1. F. Weibke and W. Bartels, *Z. anorg. Chem.*, **218,** 1934, 241–248.
2. The authors prefer the "less obligatory formulation" $Ca_{1.33}Au$.
3. Analyses showed conclusively that the maximum does not lie at the composition CaAu (16.89 wt. % Ca).
4. A. Iandelli, *Rend. seminar. fac. sci. univ. Cagliari*, **19,** 1949, 133–139.

0.2441
$\bar{1}$.7559

Au-Cd Gold-Cadmium

Literature on Constitution Up to 1939. Early papers deal with the freezing-point depression of Cd by addition of Au [1] and with the precipitation of $AuCd_3$

(63.10 wt. % Cd) crystals from an aqueous $AuCl_3$ solution by immersing metallic Cd [2]. The findings of [2] were corroborated by [3].

Solidification equilibria were first worked out by [4]. His (obsolete) phase diagram showed a (Au) primary solid solution (up to 28 at. % Cd), a β phase between about 43 and 64.5 at. % Cd formed by a peritectic reaction at 623°C, and a stoichiometric $AuCd_3$ phase, forming a eutectic with Cd at 303°C, 87 wt. (92.1 at.) % Cd.

A reinvestigation of the phase diagram by [5], using thermal and resistometric analysis as principal tools, yielded essentially the same type of diagram. Among others, the following deviations were found: (*a*) The (Au) and the β phase are separated by a eutectic rather than a peritectic miscibility gap, with the eutectic point at 612°C, 43 at. % Cd; (*b*) the liquidus has a maximum at 50 at. %, 627°C, indicating that the β phase is based on the composition AuCd (36.31 wt. % Cd); (*c*) there occur solid-state reactions in the (Au) phase field near the composition Au_3Cd (15.97 wt. % Cd) at 450 and 135°C; (*d*) the $AuCd_3$ phase possesses a certain range of homogeneity; and (*e*) the Cd-rich eutectic lies at 308°C, 92.2 at. % Cd.

A very careful thermal and micrographic study of the range 52–100 at. % Cd by [6] led to the detection of the phases δ, δ', and γ' (designation according to [7]) which had been overlooked by [4, 5]. The exact phase boundaries in this composition range were established (Fig. 114), with the exception of the Cd-rich boundary of the β phase which was drawn by [6] according only to the thermal data of the nonvariant reactions. The Cd-rich eutectic was found at 309°C, 92.05 at. % Cd.

Emf measurements at temperatures between 250 and 500°C were carried out by [7]. The transformations $\gamma \rightleftarrows \gamma'$ at 340–333°C (suggested by [8]) and $\epsilon \rightleftarrows \epsilon'$ at 269°C were found (Fig. 114) besides reactions at the composition Au_3Cd at 420°C and in the β-phase field at 267°C (see below). The phase limits derived from these potential measurements agree very well with those reported by [6] for compositions above 50 at. % Cd.

In a review on X-ray work it was stated that [9] found 11 different phases in the Au-Cd system. This number agrees with the findings of [7], which are, however, still incomplete (see below).

Electrical and heat-conductivity measurements of (Au) phase alloys with up to 10.5 wt. % Cd were carried out by [10].

Data as to the solubility of Au in solid Cd: 3.5 at. % at 309°C, 2.1 at. % at 240°C [6], and about 2 at. % at "room temperature" (electrical resistivity and emf measurements) [5].

More Recent Investigations. Crystal Structures. The Au-rich part of the phase diagram has been extensively studied since 1940. In many respects conflicting results have been obtained.

0–40 *At. %* Cd. X-ray measurements by [11] on carefully annealed and quenched specimens showed that the extent of the Au-rich primary solid solution is not so great as was assumed by former authors and that phase regions, which had not previously been mapped, exist in this part of the diagram (Fig. 114). The α_1 phase, peritectoidally formed at 425°C, has a tetragonal structure (slightly distorted f.c.c.) with $a = 4.1157$ A, $c = 4.1309$ A, $c/a = 1.0037$ at 25.02 at. % Cd. No superstructure lines due to ordering were observed by [11]. The structure of the α_2 phase was found to be h.c.p. with $a = 2.9137$ A, $c = 4.7948$ A, $c/a = 1.6456$ at 30.35 at. % Cd. Parameter values were recorded in [11] for many alloys in the α, α_1, and α_2 regions.

Another X-ray study [12] substantiated the results of [11] as far as parameters of the α phase, phase boundaries [especially $\alpha/(\alpha + \alpha_2)$], and the homogeneity range of the α_2 phase are concerned. The following deviating results were obtained: (*a*) The structure of α_1 (called α' by [12]) is ordered, $a = 4.111$ A, $c = 4.132$ A, $c/a = 1.005$ at 24.6 at. % Cd. An ordered structure for α_1 is also suggested by measurements of the electrical resistivity [5], and the modulus of elasticity [13] as well as by X-ray results of [13]. (*b*) Superlattice reflections were also observed in the photographs of the cubic α phase near Au_3Cd up to annealing temperatures of 550°C [14]. The α''-phase field assumed by [12] is, however, thermodynamically impossible. The formation of a metastable superstructure in quenching an alloy from the disordered α field was detected by [15] in the Au-Zn system; likewise, curves of modulus of elasticity vs.

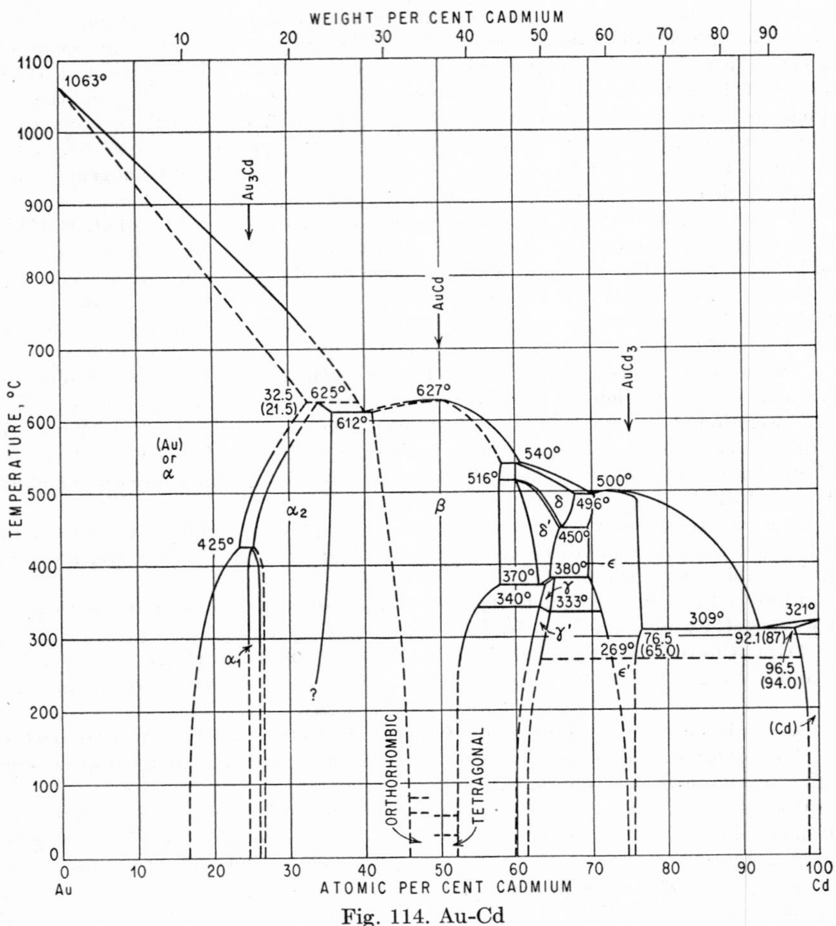

Fig. 114. Au-Cd

temperature measured by [13] on the Au_3Cd alloy quenched from 550°C suggest the existence of a (metastable) ordering reaction in quenched α alloys. Therefore, from the evidence available, there is no necessity to assume an α''-phase field. (*c*) The lattice dimensions of the α_2 phase differ considerably from those reported by [11]. Broadened and weakened reflections in the Au-rich part of the α_2 region suggest that there are occasional faults of the Co type [16] in the h.c.p. structure.

[12] further pointed out that the existence of a small ($\alpha + \alpha_2$) field was indicated by the emf measurements by [7] mentioned above. The reliability of these measure-

ments [7] is further shown by the fact that the "$\alpha/(\alpha + \beta)$" boundary of [7] agrees well with the $\alpha_2/(\alpha_2 + \beta)$ boundary determined roentgenographically by [11, 12].

The finding by [13] that a 25 at. % Cd alloy possesses a weakly tetragonally distorted and disordered f.c.c. lattice between 475 and 415°C is at variance with the results of other authors.

40–60 *At.* % Cd. The view by [17], based on X-ray measurements of quenched alloys, that there exists—between the α_2 and β phases—a rhombohedral α_3 phase extending from 46 to 48.5 at. % at 200°C and from 41 to 47.5 at. % Cd at 600°C was not corroborated by [12]. Although [12] could not account satisfactorily for some weak lines in their powder patterns, the α_3 phase cannot be looked upon as established; the more reliable emf measurements by [7] do not give any indication of its existence. Therefore, no α_3-phase field is shown in Fig. 114. The discrepancy in the lattice-parameter values measured on quenched alloys in the β-phase region [18] makes it advisable, as [12] also pointed out, to use the emf data by [7] in drawing the boundaries of the β-phase field [19] (Fig. 114).

The β phase, having a CsCl (B2) structure [20, 21, 13, 17, 12, 22] with a = 3.312 A at 50.1 at. % Cd [12], is stable only at elevated temperatures. In cooling, two different diffusionless transformations were observed: in alloys containing about 50 at. % Cd a tetragonal structure is formed at 30°C [13, 23–25], whereas in alloys with about 47.5 at. % Cd an orthorhombic structure forms at about 60°C [12, 22–24]. On heating, the transformations to the CsCl structure occur at somewhat higher temperatures (hysteresis). The structure of the tetragonal phase has not been established as yet. [13] reported a b.c. tetragonal cell with a = 5.084 A, c = 4.496 A, c/a = 0.88, and [25] the tetragonal axial ratio c/a = 0.985. As regards the orthorhombic structure, it had already been evaluated by [21] and is known as the B19 (AuCd) prototype of structure. However, [21], on the basis of emf measurements [7], thought the orthorhombic structure to be stable up to 267°C. The lattice parameters according to [22] are a = 3.1539 ± 5 A, b = 4.7644 ± 5 A, c = 4.8643 ± 5 A.

It is probable that the orthorhombic and the tetragonal phases are only metastable and tend to decompose via nucleation and growth into more stable products as all other known "martensites" do [26].

No corresponding structural change could be found by [12] for the effect observed by [7] at 267°C in the β-phase field, which was corroborated (at 290°C) by dilatometric measurements [12]. According to calorimetric measurements [27], the β phase becomes disordered below its melting point; owing to the rapid evaporation of Cd at elevated temperatures, this could not be examined by X-ray measurements [12].

According to [28], the phase ϵ' in Fig. 114 ($\sim AuCd_3$) has a structure similar to that of the γ phase (called γ'' by [28]) in the system Au-Zn (probably cubic, about 90 atoms per unit cell).

Measurements of the modulus of elasticity [13] suggest that the $\epsilon \rightleftarrows \epsilon'$ transformation at 269°C is of the order-disorder type.

[20] reported in 1928 that a h.c.p. structure was found in the Au-Cd system; in 1936, before the existence of h.c.p. α_2 was established, [29] suggested that this structure might belong to the γ' phase. However, it may well be that [20] already had the α_2 phase in hand.

1. C. T. Heycock and F. H. Neville, *J. Chem. Soc.*, **61**, 1892, 902, 914.
2. F. Mylius and O. Fromm, *Ber. deut. chem. Ges.*, **27**, 1894, 636–637.
3. P. A. Thiessen and J. Heumann, *Z. anorg. Chem.*, **209**, 1932, 325–327.
4. R. Vogel, *Z. anorg. Chem.*, **48**, 1906, 333–346.
5. P. Saldau, *Intern. Z. Metallog.*, **7**, 1915, 3–34; *Zhur. Russ. Fiz.-Khim. Obshchestva*, **49**, 1917-1918, 449–548; **55**, 1924, 275–286; see also *J. Inst. Metals*, **41, 1929,** 175–176 (discussion).

6. P. J. Durrant, *J. Inst. Metals*, **41**, 1929, 139–171.
7. A. Ölander, *J. Am. Chem. Soc.*, **54**, 1932, 3819–3833.
8. M. L. V. Gayler, *J. Inst. Metals*, **41**, 1929, 172–173 (discussion).
9. E. R. Jette, quoted by A. Westgren, *Z. Metallkunde*, **22**, 1930, 368–373.
10. E. Sedström, *Ann. Physik*, **59**, 1919, 134–144; see also J. O. Linde, *Ann. Physik*, **15**, 1932, 233–234.
11. E. A. Owen and E. A. O'Donnell Roberts, *J. Inst. Metals*, **66**, 1940, 389–400.
12. A. Byström and K. E. Almin, *Acta Chem. Scand.*, **1**, 1947, 76–89.
13. W. Köster and A. Schneider, *Z. Metallkunde*, **32**, 1940, 156–159.
14. *Structure Repts.*, **11**, 1947-1948, 55, stated: "Ordering near Au_3Cd (α'') was observed by means of a high-temperature camera." However, in studying the original publication it was concluded that the use of a high-temperature camera was limited to the investigation of β-phase alloys and that the superlattice reflections were observed by means of quenched alloys.
15. W. Köster, *Z. Metallkunde*, **32**, 1940, 151–156.
16. O. S. Edwards and H. Lipson, *Proc. Roy. Soc. (London)*, **A180**, 1942, 268; A. J. C. Wilson, *Proc. Roy. Soc. (London)*, **A180**, 1942, 277.
17. E. A. Owen and W. H. Rees, *J. Inst. Metals*, **67**, 1941, 141–151.
18. The results of [12] and [17] differ considerably, the lattice parameter of the cubic β phase decreasing in [17] with increasing Cd content, and in [12] increasing, "which seems a priori more likely as the usual atomic radius of Cd is greater than that of Au" (*Structure Repts.*, **11**, 1947-1948, 56).
19. This makes it necessary to shift the eutectic point, 43 at. % Cd, 612°C, according to [5], to a somewhat higher Au content. It is essential to redetermine the liquidus and solidus curves in this part of the diagram.
20. A. Westgren and G. Phragmén, *Metallwirtschaft*, **7**, 1928, 702.
21. A. Ölander, *Z. Krist.*, **83**, 1932, 145–148.
22. L. C. Chang, *Acta Cryst.*, **4**, 1951, 320–324.
23. L. C. Chang and T. A. Read, *Trans. AIME*, **191**, 1951, 47–52.
24. T. A. Read et al., *U.S. Atomic Energy Comm., Publ.* NYO-3961, 1953; *Met. Abstr.*, **21**, 1953, 250.
25. C. W. Chen, Ph.D. dissertation, Columbia University, New York, 1954. *Dissertation Abstr.*, **14**, 1954, 1193–1194; *Chem. Abstr.*, **48**, 1954, 13,392.
26. M. Cohen, The Martensite Transformation, in "Phase Transformations in Solids," pp. 588–659, especially pp. 592–593, 624–625, John Wiley & Sons, Inc., New York, 1951.
27. O. Kubaschewski, *Z. physik. Chem.*, **192**, 1943, 292–308.
28. A. Westgren and G. Phragmén, *Phil. Mag.*, **50**, 1925, 339 (footnote).
29. M. Hansen, "Der Aufbau der Zweistofflegierungen," Springer-Verlag OHG, Berlin, 1936.

0.1484
$\bar{1}$.8516

Au-Ce Gold-Cerium

Thermal-analysis data and microstructures of 12 alloys, prepared with cerium of 99.8+% purity, were used to outline roughly the phase diagram given in Fig. 115 [1]. Alloy compositions tested were selected under the assumption that gold forms the same compounds with cerium as with lanthanum and praseodymium: $CeAu_3$ (19.15 wt. % Ce), $CeAu_2$ (26.22 wt. % Ce), CeAu (41.54 wt. % Ce), and Ce_2Au (58.70 wt. % Ce). This was claimed to be true.

1. R. Vogel and T. Heumann, *Z. Metallkunde*, **35**, 1943, 29–42.

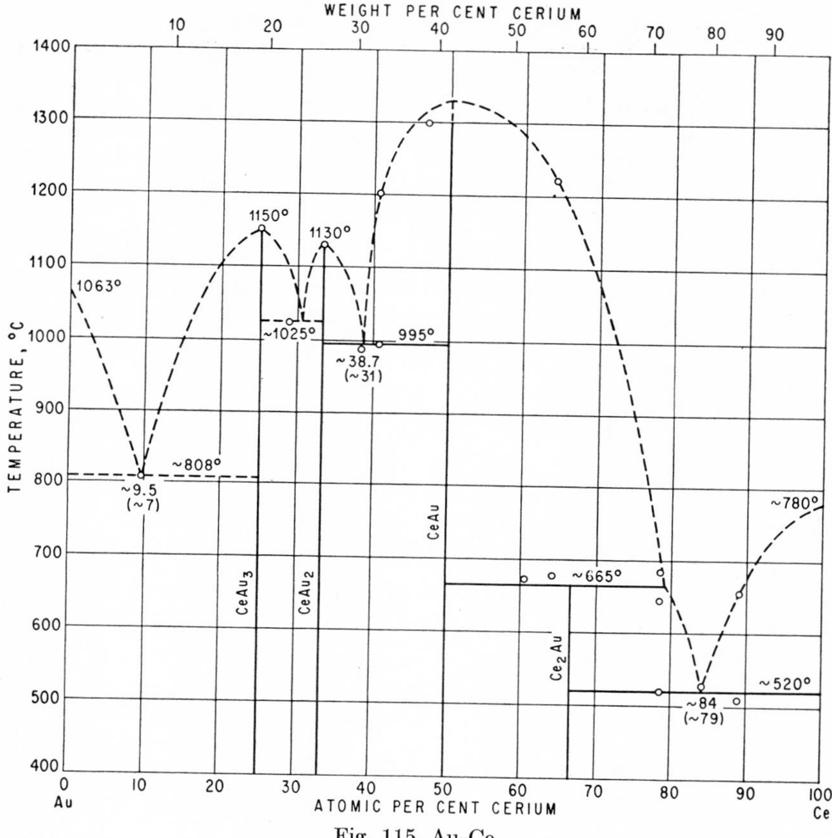

Fig. 115. Au-Ce

0.5245
$\bar{1}.4755$

Au-Co Gold-Cobalt

The phase diagram in Fig. 116 is largely due to the work by [1], results of which are in substantial agreement with those of an earlier investigation [2]. The eutectic temperature was reported to be 997 [2], 995 [1], and 1005°C [3]. The mutual solid solubility of the elements, including the solidus curve of the Co-rich phase, was determined by lattice-parametric analysis [1]. An investigation of the effect of Au on the polymorphic and magnetic transformations of Co (Fig. 116) led to the conclusion that the solid solubility of Au in Co is 6 wt. (1.9 at.) % at the eutectic temperature and approximately 3.5 wt. (1.1 at.) % at 400°C and below [3]. These values, however, as well as that reported by [4] (less than 0.37 wt. % Co), do not represent equilibrium conditions and are definitely less reliable than those by [1], given in Fig. 116. For an additional study of the effect of Au on the polymorphic transformation of Co, using supersaturated alloys quenched from 1200°C, see [5]. Lattice parameters of the Au-rich and Co-rich phases were reported by [1]; in both cases there is a strong positive deviation from additivity.

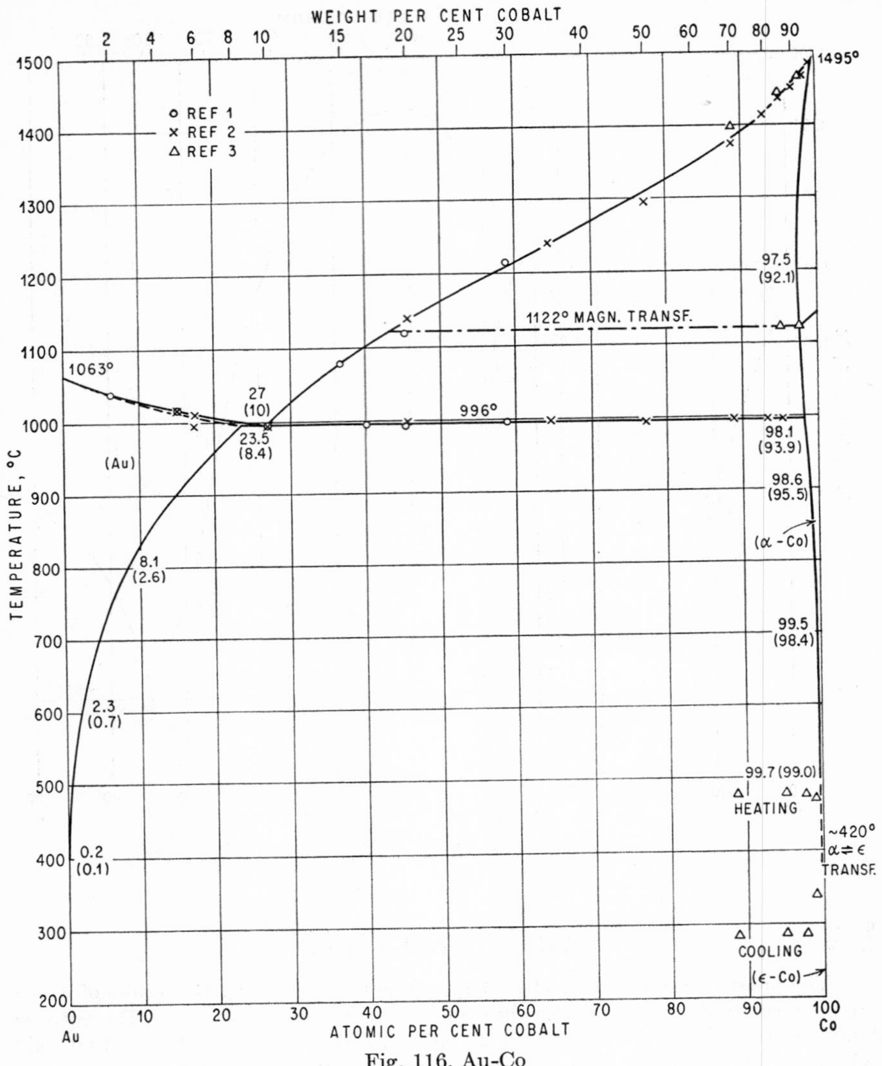

Fig. 116. Au-Co

1. E. Raub and P. Walter, *Z. Metallkunde*, **41**, 1950, 234–235.
2. W. Wahl, *Z. anorg. Chem.*, **66**, 1910, 60–72.
3. U. Haschimoto, *Nippon Kinzoku Gakkai-Shi*, **1**, 1937, 177–190.
4. E. Hildebrand, *Ann. Physik*, **30**, 1937, 593–608.
5. W. Köster and E. Horn, *Z. Metallkunde*, **43**, 1952, 333.

0.5788
$\bar{1}$.4212

Au-Cr Gold-Chromium

Figure 117 shows the phase diagram as published by [1]. Although the authors have tried to minimize contamination by melting in hydrogen atmosphere, the alloys nevertheless contained carbon and nitrogen as main contaminants [2]. The liquidus temperatures of the Cr-rich alloys, therefore, cannot be considered the liquidus temperatures of the binary system Au-Cr. This is also indicated by the far too low melting point of Cr, determined as 1573°C instead of about 1880°C.

The nearly equal liquidus temperatures found in the range 90–100 at. (70–100 wt.) % Cr led the investigators to believe at first that there is a miscibility gap in the liquid state, also indicated by the formation of two layers. It was proved, however, that this latter phenomenon was due to insufficient mixing because larger pieces of Cr did not dissolve readily.

The dashed phase boundaries in the region 14–44 at. % Cr have not been determined experimentally. The two facts (*a*) that Au-rich alloys decompose with fall in

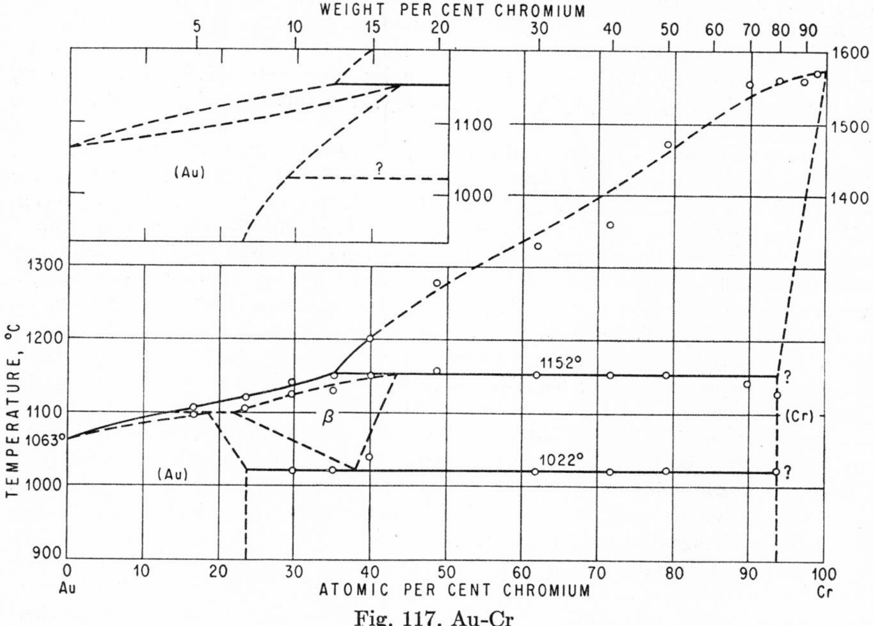

Fig. 117. Au-Cr

temperature by rejecting the Cr-rich phase and (*b*) that the thermal effect at the constant temperature of 1022°C [found to be largest at about 38 at. (14 wt.) % Cr] indicated a three-phase equilibrium permit no other interpretation than that presented in Fig. 117. It should be noted that the photomicrographs given by [1] are not characteristic of a eutectoid decomposition of the β phase. They would better account for the precipitation of (Cr) crystals from the (Au) solid solution, as indicated by the diagram in the inset of Fig. 117. However, if this interpretation is assumed, the occurrence of the transformation at 1022°C cannot be accounted for.

1. R. Vogel and E. Trilling, *Z. anorg. Chem.*, **129**, 1913, 276–292.
2. For experimental details see original paper.

0.1714
$\bar{1}$.8286

Au-Cs Gold-Cesium

The existence of a phase with the average composition $Cs_{0.77}Au$ (43.5 at. % Cs) [1] and of the compound CsAu (40.26 wt. % Cs) [2] has been reported.

1. W. Biltz and F. Weibke, *Z. anorg. Chem.*, **236**, 1938, 12–23.
2. A. Sommer, *Nature*, **152**, 1943, 215.

0.4919
$\bar{1}$.5081

Au-Cu Gold-Copper

Solidification. Thermal analyses of the whole composition range were carried out by [1] (14 alloys, liquidus only), [2] (10 alloys, liquidus and approximate solidus), and [3] (17 alloys, liquidus and solidus); liquidus data of some Cu-rich alloys (up to 3.3 at. % Au) were determined by [4]. The solidification curves in Fig. 118 are those of [3]; the Cu-rich liquidus lies up to 30 and 40°C, respectively, above the data given by [1, 2]. Data reported for the minimum are 82 wt. (59.5 at.) % Au, 905°C (thought to be a eutectic) [1]; 82 wt. % Au, 884°C [2]; 80 wt. (56.3 at.) % Au, 889°C [3].

The existence of a continuous series of solid solutions below the solidification range—concluded by [2]—was also indicated by the results of microscopic [5–8], conductometric [5, 9, 10, 11, 6, 7], X-ray [12–18, 7, 19], thermo-emf [20], and hardness [21] investigations.

Transformations in the Solid State. In 1916, [22] detected by means of thermal, conductometric, and hardness studies solid-state reactions based on the compositions Cu_3Au (50.85 wt. % Au) and CuAu (75.63 wt. % Au). [23, 14] recognized that they are caused by tendencies toward ordering in the atomic arrangement at lower temperatures. The Cu-Au system became the favorite object for experimental as well as theoretical order-disorder transformation studies; accordingly, a vast amount of literature on the constitution of this system and the thermodynamics, statistics, kinetics, and mechanisms of the ordering reactions has been published. In this review only those papers which have a direct bearing on the phase diagram could be considered; for the other fields of research, reference is made to the reviews by [24–26].

At variance with the preponderant views held in the 1940s that the order-disorder transformations in the Cu-Au system were homogeneous changes of state (e.g., [27, 28]) ("Curie point of order"), more recent experimental work has accumulated considerable evidence that those transitions are actually true "classical" or "Gibbsian" phase changes (especially [29–32]).

Besides Cu_3Au and CuAu, Cu_3Au_2 (67.42 wt. % Au) and $CuAu_3$ (90.30 wt. % Au) have been considered as formulas on which superlattices may be based. Whereas this appears to be well established for the latter, recent publications agree that there is no distinctive composition Cu_3Au_2.

The phase boundaries shown in Fig. 118 are mainly based on the data by [32], who made electrical-resistivity (and some X-ray) measurements on 24 carefully equilibrated alloys covering the range 19.5–70 at. % Au. The same type of constitution had already been established by [6] by means of thermoresistometric (heating curves) work. (However, neither [6] nor [32] found the CuAu II phase.)

(Cu_3Au). Transformation temperatures were measured by means of the following methods (underlined references deal only with a single alloy at or near the stoichiometric composition): thermal [22, 3], resistometric [10, 6, 11, 7, 8, 33, 34, 35, 36, 37, 38, 30, 32], emf [39, 40], thermo-emf [38], dilatometric [6, 11, 41, 37], modulus of elasticity [42, 43, 44, 45, 46], X-ray [47, 48, 49, 50], calorimetric [51, 37], and Hall effect [52, 36]. The best values for the temperature of the maximum at the composition Cu_3Au [53] lie within the range 388 ± 3°C; in Fig. 118, 390°C [32] has been chosen. For evidence of the "classical" character of the disorder-order transition, see [29, 30, 32, 40, 54, 50, 55].

(Cu_3Au_2) (?). The curves of solid-state transformation vs. composition measured thermoresistometrically by [8] and [7] showed maxima at 360 and 417°C, respectively, at about 40 at. % Au, and the existence of a (Cu_3Au_2) phase was suggested. [8] assumed this phase to be stable only in a very limited temperature range. In an X-ray investigation, [56] were unable to find any clear indication for a distinct phase at 40 at. % Au. Their resistivity isotherm (200°C) showed only a weak bend, whereas

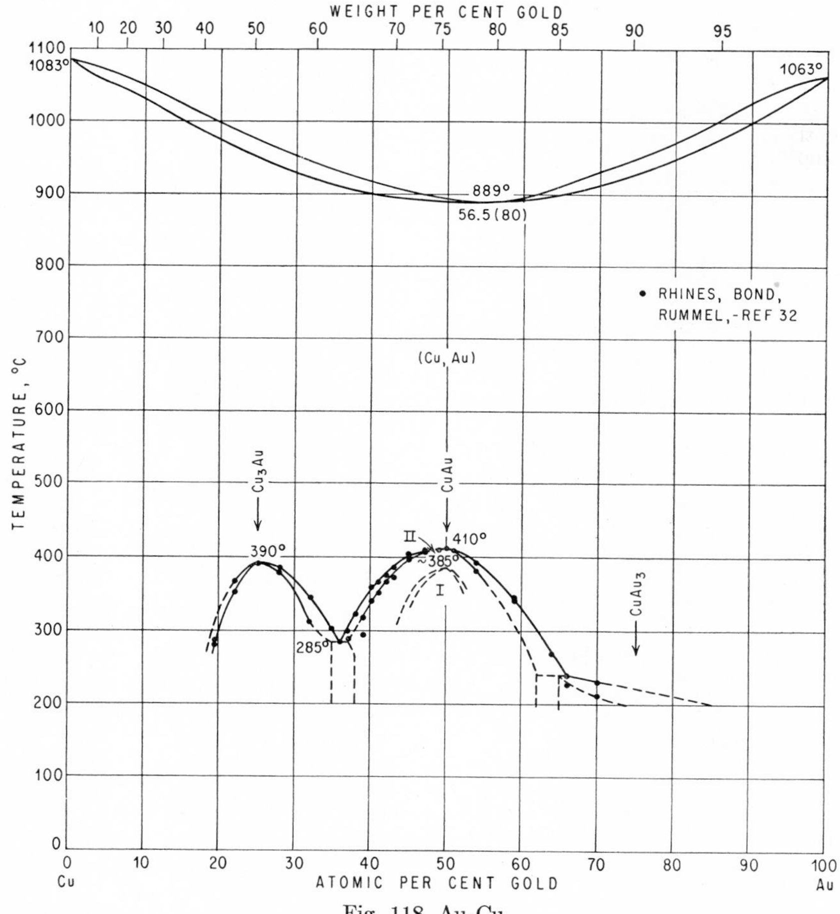

Fig. 118. Au-Cu

those of [7] exhibited pronounced minima at this composition. [39] found the breaks in their emf-vs.-temperature curves to be in general agreement with the resistometric results of [7] (but did not consider this a conclusive proof for the existence of a Cu_3Au_2 phase). On the other hand, the investigations by [57, 3], [58], and especially the recent ones by [31] (X-ray), [32] (resistivity, X-ray), [40] (emf), and [54], found no support for the existence of Cu_3Au_2 as a separate phase; therefore, such a phase has not been included in Fig. 118.

(CuAu). Ordered (CuAu) exists in two modifications, the tetragonal CuAu I and the orthorhombic CuAu II forms. The latter was detected only in 1936 [56]; former investigators considered the tetragonal phase to be stable up to the disorder temperatures. It is now well established [59, 60, 61, 36, 62, 63, 64, 40, 54, 65] that CuAu II is not just a metastable transition phase, but a stable one which has to be included in the phase diagram.

Order-disorder transformation temperatures were measured by the following methods (for the meaning of underlined reference numbers, see above): thermal [22, 3], resistometric [10, 6, 11, 8, 7, 34, 36, 37, 62, 32], emf [39, 40], dilatometric [6, 11, 37], modulus of elasticity [42], X-ray [61, 66, 31, 65], electron diffraction [64], calorimetric [37, 63], and Hall effect [36]. Most values for the temperature of the maximum at the composition CuAu [53] lie within the range 415 ± 10°C; in Fig. 118, 410°C has been chosen. For experimental evidence of the "classical" character of the order-disorder transition, see [31, 64, 32, 40, 54].

The constitution of the CuAu I ⇄ CuAu II equilibrium has been established only at higher temperatures as yet. [56], who obtained CuAu II by quenching alloys with 47–53 at. % from about 420°C or by prolonged tempering of alloys with 36–47 and 53–65 at. % Au at 200–400°C, stated that for alloys in a certain distance of the composition CuAu the orthorhombic form "appears to be the stable form even at the lowest temperatures." Values reported for the transition temperature at the stoichiometric composition vary between 370 and 400°C [42, 61, 36, 62, 63, 31, 40, 65]. The value 410°C reported by [37] is certainly too high. The curves shown in Fig. 118 are essentially those of [40], derived from emf measurements; however, they have been slightly lowered in order to conform with a maximum temperature of 385°C (instead of 400°C) at 50 at. %. Two-phase regions between the CuAu I and CuAu II phase fields have also been reported by [54].

(Cu_3Au + CuAu) *Region.* According to [6, 32], there is a eutectoid miscibility gap between the (Cu_3Au) and (CuAu) phase fields. As for the location of the eutectoid, the following data were found: ~37 at. % Au, 300°C [6]; 36 at. % Au, 285°C (Fig. 118) [32]. [6] assumed the miscibility gap to extend from 34 to 38 at. % Au; [32] estimated from X-ray work 35–40 at. % Au at 270°C.

($CuAu_3$). Between about 65 and 80 at. % Au another ordered phase exists which has a higher electrical resistivity than the conjugate disordered phase obtained by quenching [67, 56, 68, 69, 32]. Superlattice lines due to ordering were observed in X-ray [56, 68, 69] as well as electron-diffraction [70, 71] studies. According to resistometric work by [6, 7, 32], this phase is formed by a peritectoid reaction at about 250°C (260°C [6, 7]; 230–240°C [32]). There appears to be no maximum in the transformation curves at 75 at. % Au.

Crystal Structures. The lattice constant of the f.c.c. (Cu, Au) phase shows a weak positive deviation from additivity [12–16, 18, 7, 19]. The ordered cubic structure of the (Cu_3Au) phase forms the $L1_2$ prototype of structure [14, 72, 17, 73]. The CuAu I phase is tetragonal, $L1_0$ (CuAu) type of structure [14, 74, 17, 75–77].

According to [56], the structure of CuAu II is orthorhombic and closely related to the tetragonal CuAu I structure. It can be thought of as consisting of 10 tetragonal cells side by side, with Cu and Au atoms reversing positions after five cells (antiphase domains) [78]. These units, however, are no longer exactly tetragonal; the b/a ratio differs slightly from unity [78*a*]. X-ray and electron-diffraction studies by [59–61] and [79–81, 64], respectively, substantially corroborated the results of [56]. [81, 64] detected periodic defects at the boundaries of the antiphase domains.

As for ($CuAu_3$), superlattice lines have been observed by [56, 68, 69] (X-ray) and [70, 71] (electron diffraction). The structure has not been reliably analyzed as yet. [56] found that not all powder lines of a 69.2 at. % Au alloy could be indexed with a cubic cell containing 4 atoms.

Weak effects found by [6, 7] in thermoresistometric curves between about 150 and 280°C in the composition range 18–65 at. % Au were ascribed by [7] to transformations in the ordered phases. This interpretation is certainly incorrect; those effects are probably caused by the onset of a measurable decrease in long-range order on heating.

In electron-diffraction studies on thin CuAu films annealed at 450–500°C, [82] observed a zincblende (B3) type of pattern. This finding is still unconfirmed.

For information on physical properties of Cu-Au alloys, reference is made to [26].

1. W. C. Roberts-Austen and T. K. Rose, *Proc. Roy. Soc. (London)*, **A67**, 1900, 105–112.
2. N. S. Kurnakov and S. F. Zemczuzny, *Z. anorg. Chem.*, **54**, 1907, 158–165; *Zhur. Russ. Fiz.-Khim. Obshchestva*, **39**, 1907, 211.
3. W. Broniewski and K. Wesolowski, *Compt. rend.*, **198**, 1934, 370–372; *Ann. acad. sci. tech. Varsovie*, **1**, 1935, 44–69.
4. C. T. Heycock and F. H. Neville, *Phil. Trans. Roy. Soc. (London)*, **A189**, 1897, 69.
5. A. Matthiessen, *Pogg. Ann.*, **110**, 1860, 217–218; see also *Z. anorg. Chem.*, **54**, 1907, 164.
6. G. Grube, G. Schönmann, F. Vaupel, and W. Weber, *Z. anorg. Chem.*, **201**, 1931, 41–74.
7. M. LeBlanc and G. Wehner, *Ann. Physik*, **14**, 1932, 481–509.
8. J. L. Haughton and R. J. M. Payne, *J. Inst. Metals*, **46**, 1931, 457–480.
9. E. Sedström, *Ann. Physik*, **75**, 1924, 549–555.
10. G. Borelius, C. H. Johansson, and J. O. Linde, *Ann. Physik*, **86**, 1928, 291–318.
11. N. S. Kurnakov and N. V. Ageev, *J. Inst. Metals*, **46**, 1931, 481–501, 502–506; **48**, 1932, 312–313; *Izvest. Inst. Fiz.-Khim. Anal.*, **6**, 1933, 25–46.
12. F. Kirchner, *Ann. Physik*, **69**, 1922, 77–79.
13. H. Lange, *Ann. Physik*, **76**, 1925, 480–482.
14. C. H. Johansson and J. O. Linde, *Ann. Physik*, **78**, 1925, 439–460; **82**, 1927, 452–453.
15. L. Vegard and H. Dale, *Z. Krist.*, **67**, 1928, 157–161.
16. A. E. van Arkel and J. Basart, *Z. Krist.*, **68**, 1928, 475–476.
17. M. LeBlanc, K. Richter, and E. Schiebold, *Ann. Physik*, **86**, 1928, 929–1005.
18. C. S. Smith, *Mining and Met.*, **9**, 1928, 458–459.
19. L. Vegard and A. Kloster, *Z. Krist.*, **89**, 1934, 560–574.
20. E. Rudolfi, *Z. anorg. Chem.*, **67**, 1910, 88–90.
21. N. S. Kurnakov and S. F. Zemczuzny, *Z. anorg. Chem.*, **60**, 1908, 18–19; *Zhur. Russ. Fiz.-Khim. Obshchestva*, **39**, 1907, 1148.
22. N. S. Kurnakov, S. F. Zemczuzny, and M. Zasedatelev, *J. Inst. Metals*, **15**, 1916, 305–331; *Zhur. Russ. Fiz.-Khim. Obshchestva*, **47**, 1915, 871.
23. E. C. Bain, *Chem. Met. Eng.*, **28**, 1923, 67–68; *Trans. AIME*, **68**, 1923, 637–638.
24. F. C. Nix and W. Shockley, *Revs. Mod. Phys.*, **10**, 1938, 1–71.
25. C. S. Barrett, "Structure of Metals," 2d ed., McGraw-Hill Book Company, Inc., New York, 1952.
26. "Gmelins Handbuch der anorganischen Chemie," System No. 62 (Gold), pp. 882–930, Verlag Chemie, G.m.b.H., Weinheim/Bergstrasse, Germany, 1954.
27. D. Harker, *Trans. ASM*, **32**, 1944, 210–234.
28. "Metals Handbook," 1948 ed., p. 1171, American Society for Metals, Cleveland, Ohio.
29. F. N. Rhines, *J. Metals (AIME)*, **2**, 1950, 1216.

30. F. N. Rhines and J. B. Newkirk, *Trans. ASM*, **45,** 1953, 1029–1046; discussion, pp. 1046–1055.
31. J. B. Newkirk, *Trans. AIME,* **197,** 1953, 823–826; discussion, *Trans. AIME,* **200,** 1954, 673–675.
32. F. N. Rhines, W. E. Bond, and R. A. Rummel, *Trans. ASM*, **47,** 1955, 578–597.
33. C. Sykes and H. Evans, *J. Inst. Metals,* **58,** 1936, 255–281.
34. T. C. Wilson, *Phys. Rev.*, **56,** 1939, 598–611.
35. Y. Takagi and T. Sato, *Proc. Phys.-Math. Soc. Japan,* **21,** 1939, 251–258.
36. S. K. Sidorov, *Izvest. Akad. Nauk S.S.S.R., Ser. Fiz.*, **11,** 1947, 511–517; quoted from Gmelin [26], p. 890.
37. M. Hirabayashi, S. Nagasaki, and H. Maniwa, *Nippon Kinzoku Gakkai-Shi,* **14B,** 1950, 1–5.
38. T. Sato, *J. Phys. Soc. Japan,* **5,** 1950, 268–272; *Met. Abstr.*, **19,** 1952, 708.
39. F. Weibke and U. v. Quadt, *Z. Elektrochem.*, **45,** 1939, 715–727.
40. R. A. Oriani, *Acta Met.*, **2,** 1954, 608–615.
41. F. C. Nix and D. MacNair, *Phys. Rev.*, **60,** 1941, 320–329.
42. W. Köster, *Z. Metallkunde,* **32,** 1940, 145–150.
43. S. Siegel, *Phys. Rev.*, **57,** 1940, 537–545.
44. S. Siegel, *J. Chem. Phys.*, **8,** 1940, 860–866.
45. G. E. Bennett and R. M. Davies, *J. Inst. Metals,* **75,** 1949, 759–776, especially 770.
46. N. W. Lord, *J. Chem. Phys.*, **21,** 1953, 692–699.
47. Z. W. Wilchinsky, *J. Appl. Phys.*, **15,** 1944, 806–812.
48. E. A. Owen and Y. H. Liu, *Phil. Mag.*, **38,** 1947, 354–360.
49. D. T. Keating and B. E. Warren, *J. Appl. Phys.*, **22,** 1951, 286–290.
50. F. E. Jaumot and C. H. Sutcliffe, *Acta Met.*, **2,** 1954, 63–74.
51. C. Sykes and F. W. Jones, *Proc. Roy. Soc. (London)*, **A157,** 1936, 213–233.
52. A. Komar and S. Sidorov, *Zhur. Tekh. Fiz.*, **11,** 1941, 711–713; *Met. Abstr.*, **9,** 1942, 8.
53. Grube et al. [6] found the maxima of transformation temperatures for the ordered phases based on Cu_3Au and CuAu at 26 and 49 at. % Au, respectively.
54. J. H. Hollomon and D. Turnbull, *U.S. Atomic Energy Comm., Publ.* SO-2031, 1954; *Met. Abstr.*, **22,** 1954, 20.
55. J. B. Newkirk, *Acta Met.*, **2,** 1954, 644–645. (Alternate interpretation of the data of Jaumot and Sutcliffe [50].)
56. C. H. Johansson and J. O. Linde, *Ann. Physik,* **25,** 1936, 1–48.
57. V. Popisil, *Ann. Physik,* **18,** 1933, 497–514.
58. The heat-treatments used by [3] have been criticized by LeBlanc and Wehner; see the discussion in *Ann. Physik,* **23,** 1935, 570; **25,** 1936, 757–760.
59. R. Hultgren and L. Tarnopol, *Nature,* **141,** 1938, 473–474.
60. R. Hultgren and L. Tarnopol, *Trans. AIME,* **133,** 1939, 228–237.
61. O. Källbäck, J. Nyström, and G. Borelius, *Ing. Vetenskaps Akad. Handl.*, no. 157, 1941; *Met. Abstr.*, **9,** 1942, 8.
62. J. Nyström, *Arkiv Fysik,* **2,** 1950, 151–159 (in English).
63. M. Hirabayashi, *Nippon Kinzoku Gakkai-Shi,* **15,** 1951, 565–571; see also [31], discussion.
64. S. Ogawa and D. Watanabe, *J. Phys. Soc. Japan,* **9,** 1954, 475–488; see also [31], discussion.
65. B. W. Roberts, *Acta Met.*, **2,** 1954, 597–603.
66. N. N. Buinov, *Zhur. Eksptl. i Teoret. Fiz.*, **17,** 1947, 41–46; *Chem. Abstr.*, **42,** 1948, 1099.
67. W. Weber, thesis, Stuttgart, 1927; quoted by [56].
68. M. Hirabayashi, *J. Phys. Soc. Japan,* **6,** 1951, 129–130.
69. M. Hirabayashi, *Nippon Kinzoku Gakkai-Shi,* **16,** 1952, 67–72.

70. S. Ogawa and D. Watanabe, *J. Appl. Phys.*, **22,** 1951, 1502.
71. S. Ogawa and D. Watanabe, *J. Phys. Soc. Japan*, 7, 1952, 36–40.
72. G. Phragmén, *Tek. Tidskr.*, **56,** 1926, 81–85; *Fysisk Tids.*, **24,** 1926, 40–41.
73. G. Sachs and J. Weerts, *Z. Physik*, **67,** 1931, 507–515.
74. W. Gorsky, *Z. Physik*, **50,** 1928, 64–81.
75. K. Oshima and G. Sachs, *Z. Physik*, **63,** 1930, 210–223.
76. U. Dehlinger and L. Graf, *Z. Physik*, **64,** 1930, 359–377; *Z. Metallkunde*, **24,** 1932, 248–253.
77. G. D. Preston, *J. Inst. Metals*, **46,** 1931, 477–478.
78. Johansson and Linde [56] found indications that the identity period may vary with composition and temperature.
78*a*. "A most remarkable fact about the orthorhombic phase is its behavior at room temperature, where the tetragonal phase is presumably stable. After 6 months at 20°C, the small splitting of the tetragonal lines, which indicates that the structure is orthorhombic, becomes much greater. The change is thus not toward becoming tetragonal, which would be expected, but in the opposite direction." Quoted from [60].
79. L. H. Germer and F. E. Haworth, *Phys. Rev.*, **57,** 1940, 354.
80. S. Ogawa and D. Watanabe, *Acta Cryst.*, **5,** 1952, 848–849.
81. S. Ogawa and D. Watanabe, *Acta Cryst.*, **7,** 1954, 377–378.
82. O. Eisenhut and E. Kaupp, *Z. Elektrochem.*, **37,** 1931, 466–472.

0.5479
$\bar{1}$.4521

Au-Fe Gold-Iron

The essential features of the phase diagram were established by [1]; however, the ranges of solid solubility given were only a rough approximation. Neither the temperature of the $\alpha \rightleftarrows \gamma$ transformation of Fe nor that of its Curie point were found to be noticeably affected by Au. Later, a modified diagram was given by [2], based on unpublished work by [3]. It is characterized by (*a*) the existence of the intermediate phase Fe_3Au (45.93 wt. % Fe), presumably formed by a peritectoid reaction of the two primary solid solutions at about 850°C; (*b*) a minimum in the $\alpha \rightleftarrows \gamma$ transformation curve at about 95 wt. (98.5 at.) % Fe, 830°C; and (*c*) the increase of the $\gamma \rightleftarrows \delta$ transformation temperature of Fe. However, all subsequent investigations [4–7] did not verify the existence of Fe_3Au. Also, the minimum in the $\alpha \rightleftarrows \gamma$ transformation curve could not be confirmed [7]. [8] confirmed that the Curie point of Fe is not affected by Au.

The diagram presented in Fig. 119 is based on comprehensive thermal, micrographic, X-ray, and dilatometric studies [7]. The solid solubility of Fe in Au and that of Au in Fe were determined lattice-parametrically in the temperature range 400–1100°C and 500–1350°C, respectively. The results are in excellent agreement with the solubility curves, determined earlier [4] by lattice-parameter measurements, between 333 and 724°C (Fe in Au) and between 600 and 724°C (Au in Fe).

For lattice parameters of the Au-rich and Fe-rich solid solutions, see [4, 7].

Note Added in Proof. [9] suggested, on the basis of potential measurements, that (*a*) the solubility of Au in α-Fe at 850°C is less than 1.5 at. % (in good agreement with [7], who found 1.3 at. % Au), and that (*b*) the eutectoid temperature is 855–865°C, as compared with 903°C, according to [7]. The eutectoid composition was calculated to be 97.8 at. % Fe vs. 97.7 at. % Fe, according to [7].

1. E. Isaac and G. Tammann, *Z. anorg. Chem.*, **53,** 1907, 291–297.
2. L. Nowack, *Z. Metallkunde*, **22,** 1930, 97.
3. F. Wever, unpublished work.
4. E. R. Jette, W. L. Bruner, and F. Foote, *Trans. AIME*, **111,** 1934, 354–359.

5. S. T. Pan, A. R. Kaufmann, and F. Bitter, *J. Chem. Phys.*, **10,** 1942, 318–321.
6. A. R. Kaufmann, S. T. Pan, and J. R. Clark, *Revs. Modern Phys.*, **17,** 1945, 87–92.
7. E. Raub and P. Walter, *Z. Metallkunde*, **41,** 1950, 234–238.
8. M. Fallot, *Ann. phys.*, **6,** 1936, 376–381.
9. L. L. Seigle, *Trans. AIME*, **206,** 1956, 91–97.

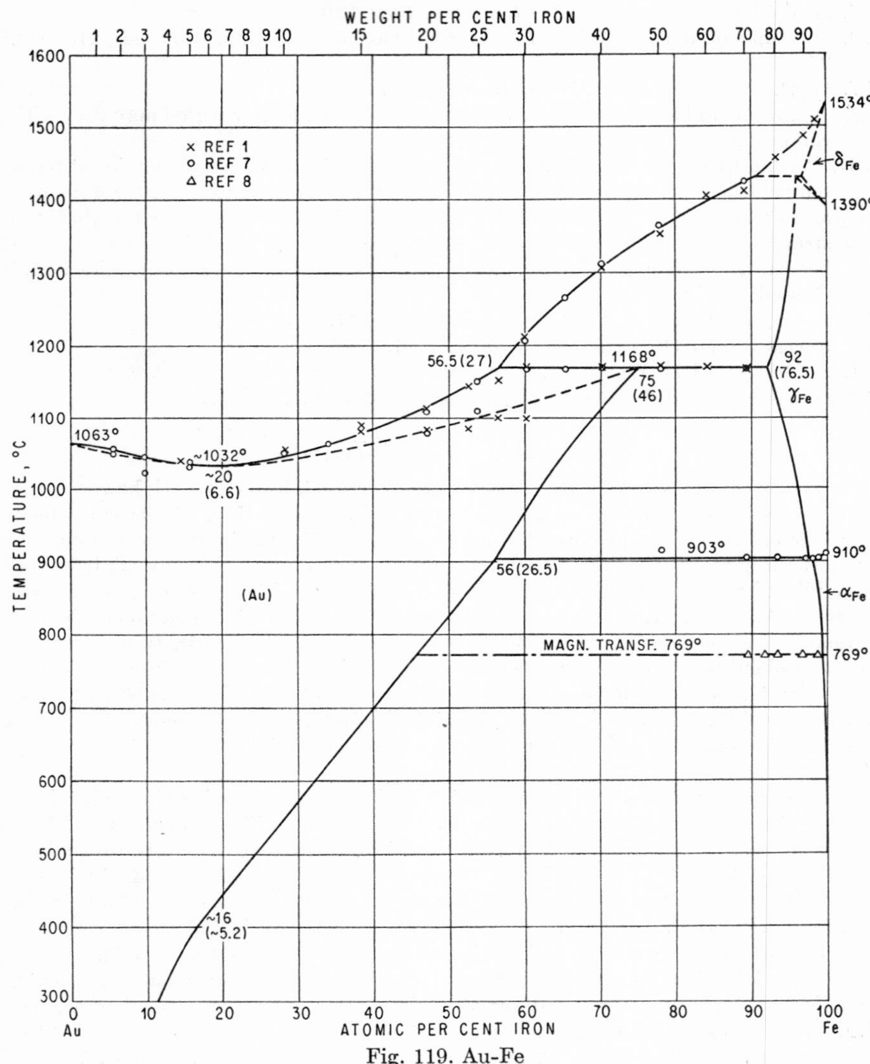

Fig. 119. Au-Fe

0.4516
$\bar{1}.5484$

Au-Ga Gold-Gallium

The phase relationships as presented in Fig. 120 are mainly based on thermal, besides microscopic and X-ray, investigations [1]. They are characterized by the existence of four intermediate phases: (*a*) a phase, β, stable above 275°C, ranging in composition between about 26.5 at. (11.3 wt.) and 29.2 at. (12.7 wt.) % Ga; it was interpreted to be based on the composition Au_3Ga (10.54 wt. % Ga) [2]; (*b*) a phase, γ, stable between about 29.8 at. (13.0 wt.) and 30.8 at. (13.6 wt.) % Ga, to which the formula Au_7Ga_3 (13.16 wt. % Ga) has been assigned [2]; (*c*) AuGa (26.12 wt. % Ga) and (*d*) $AuGa_2$ (41.42 wt. % Ga), both apparently of fixed composition.

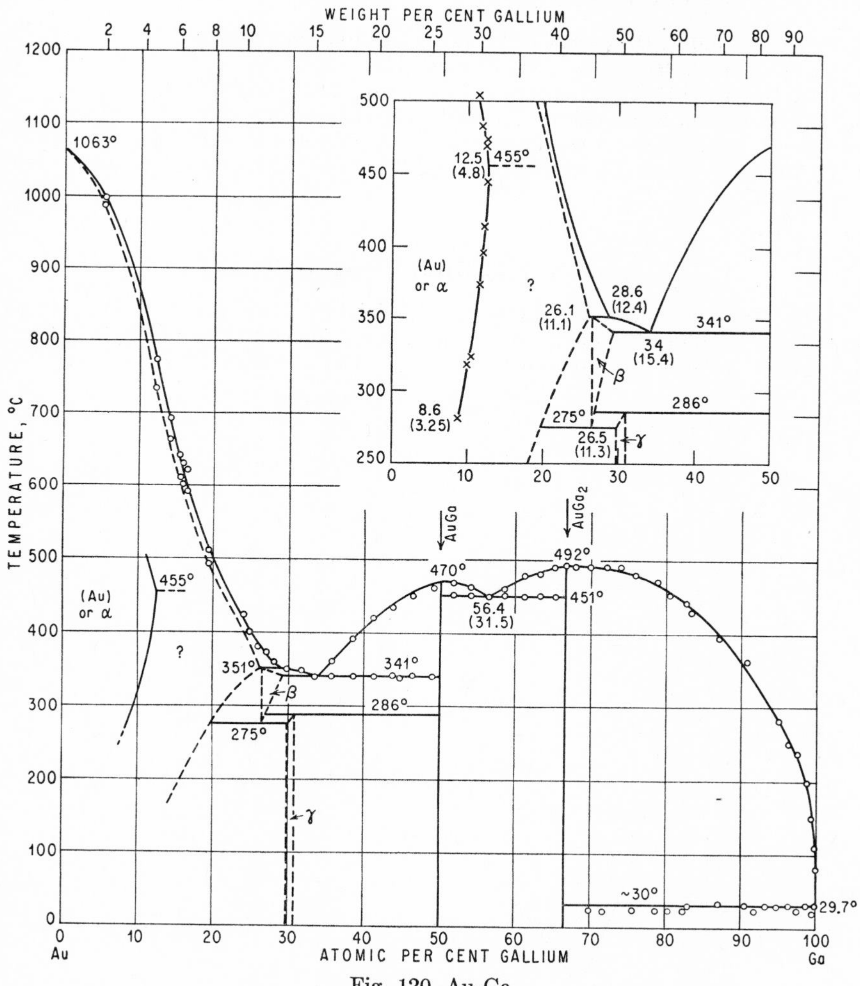

Fig. 120. Au-Ga

The boundary of the Au-rich solid solution was only approximately outlined parametrically for temperatures of 352°C [26.1 at. (11.1 wt.) % Ga] and 200°C [15.7 at. (6.2 wt.) % Ga] [1]. More recently, a careful determination, also by lattice-spacing measurements [3], indicated a much lower solubility, given in at. % Ga (wt. % in parentheses) as follows: 500°C, 11.15 (4.25); 455°C, 12.5 (4.8); 400°C, 11.8 (4.52); 350°C, 10.9 (4.15); 300°C, 9.4 (3.54); and 270°C, 8.1 (3.05) (see inset of Fig. 120).

As shown by additional X-ray work [3], there appears to be no doubt that this phase boundary represents equilibrium conditions and that in the earlier investigation [1] the existence of an intermediate phase between 13 and 26 at. % Ga, which probably is formed by a peritectic reaction at 455°C, was overlooked.

AuGa has a structure of the MnP (B31) type, a = 6.397 A, b = 6.267 A, c = 3.421 A [4]. $AuGa_2$ is isotypic with CaF_2 (C1 type), a = 6.075 A [5]. The γ phase is a **21:13** electron compound [6].

1. F. Weibke and E. Hesse, *Z. anorg. Chem.*, **240,** 1939, 289–299.
2. H. Pfisterer, *Z. Metallkunde*, **41,** 1950, 95–96.
3. E. A. Owen and E. A. O. Roberts, *J. Inst. Metals*, **71,** 1945, 213–254.
4. H. Pfisterer and K. Schubert, *Z. Metallkunde*, **41,** 1950, 358–367.
5. E. Zintl, A. Harder, and W. Haucke, *Z. physik. Chem.*, **B35,** 1937, 354–362.
6. W. Hume-Rothery, J. A. Betterton, and J. Reynolds, *J. Inst. Metals*, **80,** 1951-1952, 609–616.

0.4340
$\bar{1}$.5660

Au-Ge Gold-Germanium

The liquidus and solidus curves presented in Fig. 121 are based on thermal-analysis data [1]; for the liquidus curve they are considered to be accurate to ±15°C. The eutectic was formerly reported as lying at 24 at. (10.4 wt.) % Ge [2]. On the

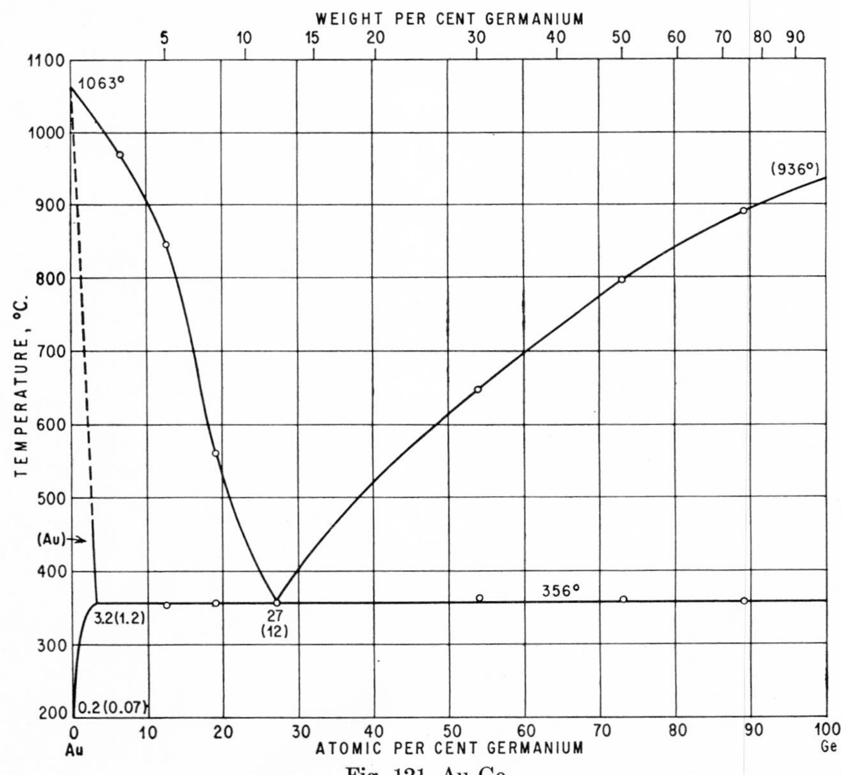

Fig. 121. Au-Ge

basis of limited microscopic data, the solid solubility of Ge in Au, earlier reported to be at least 0.6 wt. (1.6 at.) % Ge [3], was estimated as about 0.8 wt. (2.1 at.) % Ge at 350°C [1]. A more careful determination of the solubility curve by lattice-spacing measurements resulted in the following values in at. % Ge (wt. % in parentheses): 450°C, 2.8 (1.05); 400°C, 3.0 (1.13); 363°C (eutectic temperature), 3.12 (1.18); 350°C, 2.2 (0.83); 300°C, 0.9 (0.33); 250°C, 0.4 (0.15); and 200°C, 0.2 (0.07) [4]. They are estimated to be accurate to about ±0.3 at. (±0.1 wt.) %. No information is available as to the solid solubility of Au in Ge.

1. R. I. Jaffee, E. M. Smith, and B. W. Gonser, *Trans. AIME*, **161**, 1945, 366–372.
2. R. Schwarz, *Z. angew. Chem.*, **48**, 1935, 219–223; E. Einecke, *Chem. Ztg.*, **61**, 1937, 989.
3. J. O. Linde, *Ann. Physik*, **15**, 1932, 233–234.
4. E. A. Owen and E. A. O. Roberts, *J. Inst. Metals*, **71**, 1945, 213–254.

2.2914
$\bar{3}$.7086

Au-H Gold-Hydrogen

Hydrogen does not dissolve in either liquid or solid gold [1–3].

1. A. Sieverts and W. Krumbhaar, *Ber. deut. chem. Ges.*, **43**, 1910, 893–900.
2. W. Davies, *Phil. Mag.*, **17**, 1934, 233–251.
3. D. P. Smith, "Hydrogen in Metals," p. 282, The University of Chicago Press, Chicago, 1948.

$\bar{1}$.9926
0.0074

Au-Hg Gold-Mercury

Literature reviews have been published by [1–4]; special reference is made to [4].

Liquidus Curve. Several investigations were confined to chemoanalytical determinations of the solubility of Au in liquid Hg at normal temperatures [5–8]. The value of 0.1306 wt. (0.1329 at.) % Au at 20°C [9] must be considered the most accurate. The results of very exact and numerous solubility determinations—65 values between 7 and 80°C [9], 65 values between 80 and 200°C [10], 50 values between 200 and 300°C [11], 42 values between 280 and 400°C [2], and 70 values between 190 and 300°C [12]—yield a smooth curve which is characterized by a slight change in direction at 310°C and about 77.2 at. (77.5 wt.) % Hg [2] (Fig. 122). Solubility data by [8] covering the temperature range 18 to 410°C deviate considerably from this curve for temperatures above 250°C but have to be considered inferior. This also holds for some of the thermoanalytical data by [13] and especially for those by [1] which deviate entirely from all others [14]. Thermal-analysis data in the temperature range −40 to 515°C were reported by [15, 3].

In summarizing one may say that in the temperature range 0–250°C the chemoanalytical data are to be preferred. In the range 280–380°C the solubility data by [2] agree very well with those by [13] obtained by thermal analysis, whereas the thermal-analysis data by [15, 3] lie up to 30°C above the curve due to [2] (see Fig. 122). The temperature of the nonvariant equilibrium at 310°C found by [15, 3] agrees with the temperature of the break point mentioned above [2]. However, the end point of this horizontal representing the composition of the melt was found as 77.2 at. % Hg by [2] but as 86.5 at. % Hg by [15, 3].

The branch of the liquidus curve corresponding to primary crystallization of Hg is extremely small [16, 13, 1, 15, 3]. The depression of the freezing point of Hg is probably not greater than 0.1°C [15, 3].

Solubility of Hg in Solid Au. The saturation limit determined by conductivity measurements to be about 10 wt. (at.) % Hg at 200°C [17] and 15 or 17 wt. (at.) % Hg, respectively, at 390–400°C by thermal analysis [1, 15] does not represent equilibrium conditions. Lattice-parameter measurements of alloys after long-time annealings [18–22] resulted in solubility values which lie within a scattering band of

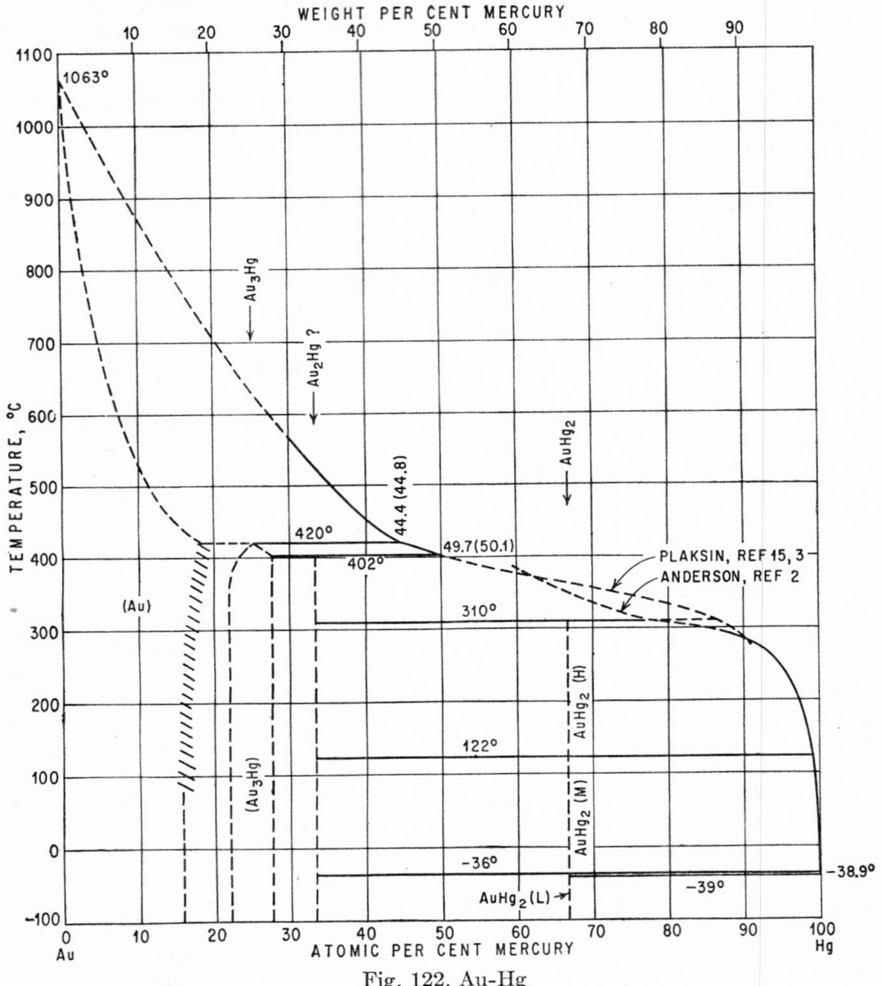

Fig. 122. Au-Hg

approximately ±1 at. (wt.) % at 100–400°C (Fig. 122); 17.3–19.3 at. (17.5–19.6 wt.) % at 400°C; 16.1–17.9 at. (16.3–18.2 wt.) % at 300°C; 15.6–16.8 at. (15.8–17.0 wt.) % at 200°C; and about 15–17 at. (wt.) % Hg at 100°C. Hg vapor-tension measurements by [23] indicated a solubility of about 18 wt. (at.) % Hg at 250–315°C, in fair agreement with the X-ray results.

[24] determined the lattice-parameter-vs.-composition curve of the (Au) phase on alloys prepared by the mutual reduction of mixed metal ion solutions; the curve is in very good agreement with that found by [18].

Intermediate Phases. There is strong evidence as regards the existence of a phase around the composition Au_3Hg (25.32 wt. % Hg) [13, 23, 18, 2, 25, 19, 3]. It has a h.c.p. structure [18, 19] with at least partial ordering [26] and is homogeneous between about 21.3 and 27.4 at. % [19] or 22 and 27.5 at. % Hg [3]. Lattice parameters at 25.3 at. % Hg: $a = 2.916$ A, $c = 4.801$ A, $c/a = 1.647$ [19].

On the other hand, there is great discrepancy as to the existence of one or more phases with higher Hg content. The following compositions have been suggested: Au_2Hg (33.72 wt. % Hg) [15, 3], AuHg (?) (50.43 wt. % Hg) [19], Au_2Hg_3 (60.41 wt. % Hg) [13, 18], $AuHg_2$ (67.05 wt. % Hg) [1, 15, 18, 19, 3], Au_2Hg_5 (71.77 wt. % Hg) [1, 19], and $AuHg_4$ (80.27 wt. % Hg) [1, 19].

Based on thermal and microscopic studies the first rather complete diagram was published in 1929 [15] and revised by the same author in 1938 [3]. The latter version (Fig. 122) is characterized by the existence of the three intermediate phases (Au_3Hg), Au_2Hg, and $AuHg_2$, all of which are formed peritectically, at 420, 402, and 310°C, respectively. $AuHg_2$ undergoes transformations at 122 and −36°C. Mention should also be made of the diagram suggested by [2], based on the results of [10, 15, 9, 11]. It deviates considerably from that given in Fig. 122 in that it shows only two intermediate phases: Au_3Hg, which forms peritectically at 421°C and transforms at 310°C, and Au_2Hg_3, which forms peritectically at 124°C.

Powder X-ray analyses of the entire system [18, 19] did not help to clarify the constitution. Whereas one author [18] reported the existence of Au_3Hg, Au_2Hg_3, and perhaps $AuHg_2$, the other [19] suggested that five intermediate phases exist: Au_3Hg, AuHg (?), $AuHg_2$, Au_2Hg_5, and $AuHg_4$.

Without being satisfactory in all points the diagram of Fig. 122 should be regarded as the most probable one at present. It still appears uncertain, however, whether Au_3Hg and $AuHg_2$ (or Au_2Hg_3) are the only intermediate phases and whether the composition of a third one is close to either Au_2Hg or AuHg.

Electron Diffraction. A large number of complex amalgam patterns were obtained in the investigation of thin gold amalgam films by the electron-diffraction method [27]. No agreement was found between these results and those of the X-ray work by [18, 19]. To the amalgam richest in Hg a simple cubic lattice with $a =$ 17.83 A (!) was ascribed. However, a recalculation by [28] yielded a hexagonal lattice with $a = 5.16$ A and a cubic lattice (γ-brass type) with $a = 9.9$ A.

1. S. A. Braley and R. F. Schneider, *J. Am. Chem. Soc.*, **43**, 1921, 740–746.
2. J. T. Anderson, *J. Phys. Chem.*, **36**, 1932, 2145–2165.
3. I. N. Plaksin, *Izvest. Sektora Fiz.-Khim. Anal.*, **10**, 1938, 129–159 (in Russian).
4. "Gmelins Handbuch der anorganischen Chemie," System No. 62, pp. 812–825, Verlag Chemie, G.m.b.H., Weinheim/Bergstrasse, Germany, 1954.
5. T. H. Henry, *Phil. Mag.*, **9**, 1855, 458.
6. Kasanzeff, abstracted in *Ber. deut. chem. Ges.*, **11**, 1878, 1255.
7. J. Gouy, *J. phys.*, **4**, 1895, 320–321.
8. G. T. Britton and J. W. McBain, *J. Am. Chem. Soc.*, **48**, 1926, 593–598.
9. A. A. Sunier and C. M. White, *J. Am. Chem. Soc.*, **52**, 1930, 1842–1850.
10. A. A. Sunier and E. B. Gramkee, *J. Am. Chem. Soc.*, **51**, 1929, 1703–1708.
11. A. A. Sunier and L. G. Weiner, *J. Am. Chem. Soc.*, **53**, 1931, 1714–1721.
12. G. Mees, *J. Am. Chem. Soc.*, **60**, 1938, 870–871; the results agree excellently with those by Sunier and Weiner [11].
13. N. Parravano, *Gazz. chim. ital.*, **48**(2), 1918, 123–138; *Z. Metallkunde*, **12**, 1920, 113–115.
14. The liquidus point of the alloy with 83.45 wt. % Hg was given as 318°C: E. D. Eastman and J. H. Hildebrand, *J. Am. Chem. Soc.*, **36**, 1914, 2020–2030.

15. I. N. Plaksin, *Zhur. Russ. Fiz.-Khim. Obshchestva,* **61,** 1929, 521–534 (in Russian); abstracts in *Z. Metallkunde,* **24,** 1932, 89; *J. Inst. Metals,* **39,** 1928, 498; **41,** 1929, 459. The thermal data were retabulated in [3].
16. G. Tammann, *Z. physik. Chem.,* **3,** 1889, 445.
17. N. Parravano and P. Jovanovich, *Gazz. chim. ital.,* **49**(1), 1919, 1–6; *Z. Metallkunde,* **12,** 1920, 115–116.
18. A. Pabst, *Z. physik. Chem.,* **B3,** 1929, 443–455.
19. S. Stenbeck, *Z. anorg. Chem.,* **214,** 1933, 16–26.
20. M. I. Zakharova, *Zhur. Tekh. Fiz.,* **7,** 1937, 171–174.
21. H. M. Day and C. H. Mathewson, *Trans. AIME,* **128,** 1938, 261–280.
22. E. A. Owen and E. A. O'Donnell Roberts, *J. Inst. Metals,* **71,** 1945, 213–254.
23. W. Biltz and F. Meyer, *Z. anorg. Chem.,* **176,** 1928, 27–32.
24. F. Hund and H. Mosthaf, *Naturwissenschaften,* **39,** 1952, 209.
25. R. Kremann, R. Baum, and L. Lämmermayr, *Monatsh. Chem.,* **61,** 1932, 315–329.
26. C. Wagner and W. Schottky, *Z. physik. Chem.,* **B11,** 1930, 208.
27. A. E. Aylmer, G. I. Finch, and S. Fordham, *Trans. Faraday Soc.,* **32,** 1936, 864–871.
28. N. A. Shishakov, *Bull. acad. sci. U.R.S.S., Classe sci. chim.,* **1941,** 683–689 (in Russian, summary in English); *Chem. Abstr.,* **37,** 1943, 2326.

0.2351
$\bar{1}$.7649

Au-In Gold-Indium

The phase relationships presented in Fig. 123 are based on results of thermal, microscopic, and X-ray analysis [1]. Five intermediate phases are shown to exist and assumed to have average compositions corresponding to the formulas Au_4In (12.70 wt. % In), Au_3In (16.25 wt. % In), Au_7In_3 (20.0 wt. % In), AuIn (36.79 wt. % In), and $AuIn_2$ (53.80 wt. % In) [1].

The solid solubility of In in Au, only outlined by [1, 2], was more reliably positioned by lattice-parameter measurements with an accuracy of about 0.15 at. % [3]. The solubility curve thus determined changes its direction at approximately 682°C, whereas according to [1] this change in direction should be expected to occur at the peritectic temperature, 647°C. The solubility in at. % In (wt. % in parentheses) was found to be 11.3 (6.9) at 767°C, 12.64 (7.8) at 700°C, 12.57 (7.7) at 650°C, 11.6 (7.1) at 560°C, 10.96 (6.7) at 482°C, and 10.36 (6.3) at 406°C.

The limits of the ζ phase were not determined accurately. Alloys with 15.3 and 20.0 at. % In proved to be single-phase after annealing at 340–450°C, whereas the alloy with 21.8 at. % In was heterogeneous. ζ is a 3:2 electron compound [4].

The conclusion that a phase designated as δ—γ according to [1]—exists within a very narrow temperature range and with a composition close to that of Au_3In is solely based on the thermal-analysis data shown in the inset of Fig. 123. It was suggested that these data could be interpreted only by assuming a phase to be formed peritectically at 488°C and to decompose eutectoidally at 482°C. No direct evidence was given.

The γ phase—δ according to [1]—undergoes a transformation at 373–364°C; it is probably of the order-disorder type. Its range of homogeneity, not positioned accurately, narrows down with fall in temperature. If one prefers to assign a formula to it, Au_9In_4 [30.77 at. (20.55 wt.) % In] appears to be preferable to Au_7In_3 [30.0 at. (20.0 wt.) % In] because its structure is of the γ-brass type [5, 6].

Whereas AuIn has a homogeneity range of about 49.1–50.3 at. (36.0-37.0 wt.) % In [1], $AuIn_2$ apparently is of fixed composition.

Crystal Structures. Lattice parameters of the (Au) phase were reported by [1, 3]. The ζ phase is h.c.p. with $a = 2.91_6$ A, $c = 4.76$ A, $c/a = 1.60$ on the Au side

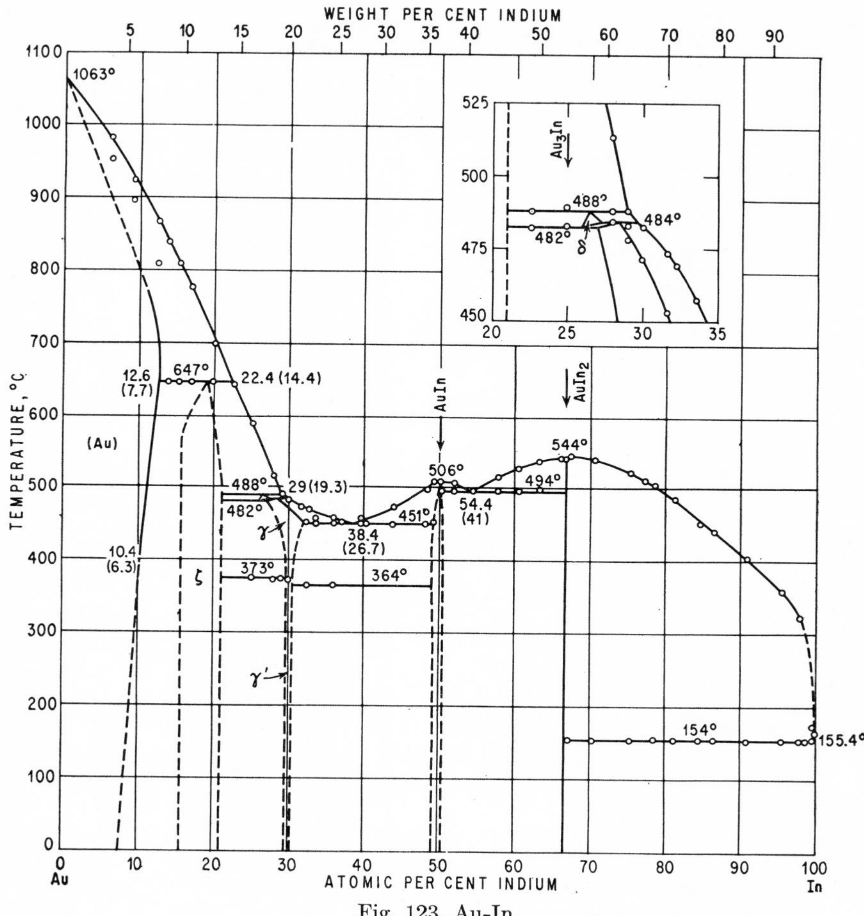

Fig. 123. Au-In

and $a = 2.91_4$ A, $c = 4.78_7$ A, $c/a = 1.643$ on the In side [1]. The γ' phase has the γ-brass type of structure [1, 5, 6], $a = 9.82$ A [5]. AuIn has a triclinic structure [7], and $AuIn_2$ is cubic of the CaF_2 (C1) type, $a = 6.515$ A [8], 6.519 A [1].

1. O. Kubaschewski and F. Weibke, *Z. Elektrochem.*, **44**, 1938, 870–877.
2. F. Weibke, *Z. Elektrochem.*, **46**, 1940, 346–348.
3. E. A. Owen and E. A. O. Roberts, *J. Inst. Metals*, **71**, 1945, 213–254.
4. W. Hume-Rothery, P. W. Reynolds, and G. V. Raynor, *J. Inst. Metals*, **66**, 1940, 191–207.
5. E. Hellner and F. Laves, *Z. Naturforsch.*, **2a**, 1947, 181.
6. W. Hume-Rothery, J. A. Betterton, and J. Reynolds, *J. Inst. Metals*, **80**, 1951-1952, 609–616.
7. K. Schubert et al., *Naturwissenschaften*, **40**, 1953, 437.
8. E. Zintl, A. Harder, and W. Haucke, *Z. physik. Chem.*, **B35**, 1937, 354–362.

0.0091
$\bar{1}.9909$

Au-Ir Gold-Iridium

According to [1, 2], Ir does not dissolve in liquid Au. However, [3] has prepared alloys with up to 2.76 at. (2.71 wt.) % Ir. From electrical-resistance measurements of specimens quenched from 900–950°C, it was concluded that the solid solubility of Ir in Au, even at this temperature, is exceedingly small [3].

1. E. Matthey, *Proc. Roy. Soc. (London)*, **47**, 1890, 180.
2. H. Rössler, *Chem. Ztg.*, **24**, 1900, 733–735.
3. J. O. Linde, *Ann. Physik*, **10**, 1931, 69.

0.7027
$\bar{1}.2973$

Au-K Gold-Potassium

Two compounds have been identified and their temperature stability determined: KAu_4 (4.72 wt. % K) and KAu_2 (9.02 wt. % K) [1]. The homogeneity range of KAu_2 is appreciably extended toward higher potassium contents; no quantitative data are given. KAu_4 also is of variable composition; however, its homogeneity range appears to be narrower.

There is some evidence that a eutectic exists at approximately 3.0 at. (13.5 wt.) % Au and 58–59°C [2]. This is consistent with the statement that the melting point of potassium is lowered by 4.7 wt. % Au for 1.8°C [3]. The microstructure of an alloy containing 0.1 wt. % K has been described by [4].

The structure of KAu_2 apparently is similar to that of $NaAu_2$ [1].

1. V. Quadt, F. Weibke, and W. Biltz, *Z. anorg. Chem.*, **232**, 1937, 297–306.
2. N. Brinckert and O. Hannebohn; see [1].
3. C. T. Heycock and F. H. Neville, *J. Chem. Soc.*, **55**, 1889, 676.
4. T. Andrews, *Engineering*, **66**, 1898, 541–542.

0.1522
$\bar{1}.8478$

Au-La Gold-Lanthanum

The phase diagram presented in Fig. 124 is based on thermal-analysis data [1]. The existence of the four intermediate phases $LaAu_3$ (19.02 wt. % La), $LaAu_2$ (26.05 wt. % La), LaAu (41.33 wt. % La), and La_2Au (58.47 wt. % La) was confirmed by microexamination of alloys cooled from the molten state. The solubilites in the solid state have not been investigated.

1. G. Canneri, *Metallurgia ital.*, **23**, 1931, 819–822. Purity of La: 99.6%.

1.4536
$\bar{2}.5464$

Au-Li Gold-Lithium

The b.c.c. structure of the CsCl (B2) type suspected to occur in the 50 at. % Li alloy could not be found. An alloy with 52 at. % Li gave an X-ray pattern which could not be indexed, whereas an alloy with 44 at. % Li showed the X-ray pattern of the f.c.c. lattice with the dimension approaching that of the gold lattice [1].

1. E. Zintl and G. Brauer, *Z. physik. Chem.*, **B20**, 1933, 245–271.

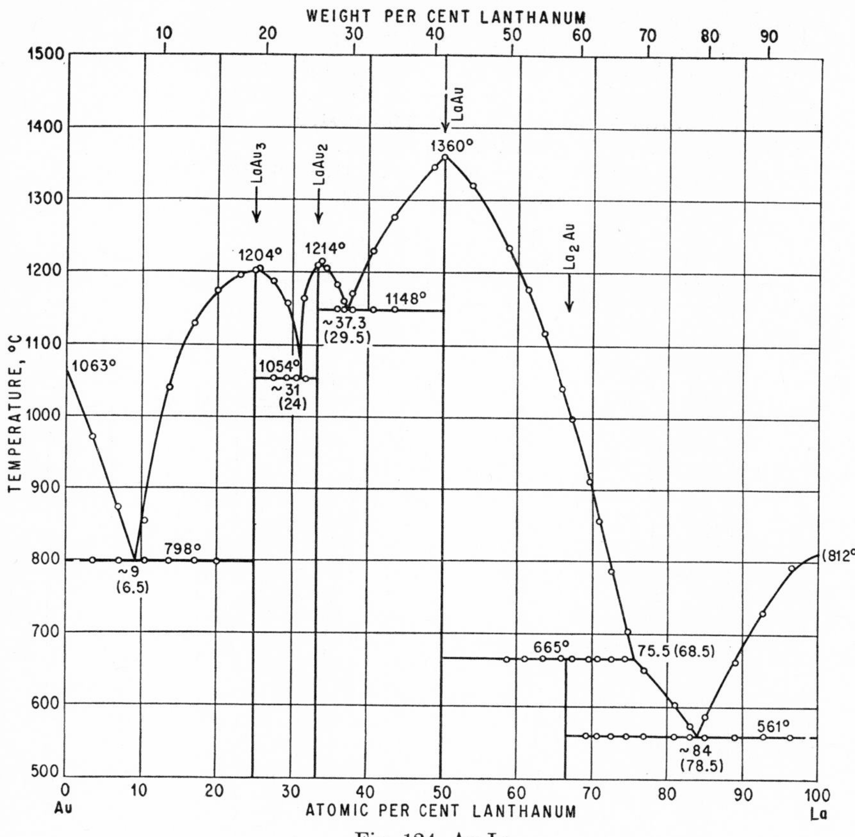

Fig. 124. Au-La

0.9090
$\bar{1}.0910$

Au-Mg Gold-Magnesium

The phase relationships were established by means of thermal analysis and metallographic examination [1, 2]. Figure 125 is almost entirely based on the more reliable data by [2], who used 107 alloys, all of which were analyzed, whereas [1] used only 25 alloys, only a few of which were analyzed. The largest discrepancy between the two papers is concerned with the constitution between Mg_2Au (19.79 wt. % Mg) and Mg_3Au (27.01 wt. % Mg), involving the peritectic formation of Mg_5Au_2 (23.57 wt. % Mg) at 796°C and its eutectoid decomposition at 721°C. In a common paper [3] the authors have agreed on the presentation shown in Fig. 125.

On the basis of micrographic [4] and lattice-parameter work [5], the solid solubility of Au in Mg at the eutectic temperature was estimated to be approximately 0.1 at. (0.8 wt.) % Au. The c/a ratio of Mg is slightly decreased by Au [5, 6]. The β' phase, based on the composition MgAu, has the CsCl (B2) type of structure, a = 3.266 A at 48.7 at. % Mg [7]. Mg_3Au is hexagonal of the Na_3As ($D0_{18}$) type, a = 4.64 A, c = 8.46 A, c/a = 1.82 [8].

1. R. Vogel, *Z. anorg. Chem.*, **63,** 1909, 169–183.
2. G. G. Urasow, *Z. anorg. Chem.*, **64,** 1909, 375–396.
3. G. G. Urasow and R. Vogel, *Z. anorg. Chem.*, **67,** 1910, 442–447.
4. W. Hume-Rothery and E. Butchers, *J. Inst. Metals*, **60,** 1937, 345–350.
5. F. Foote and E. R. Jette, *Trans. AIME*, **143,** 1941, 124–131.
6. R. S. Busk, *Trans. AIME*, **188,** 1950, 1460–1464.
7. G. Brauer and W. Haucke, *Z. physik. Chem.*, **B33,** 1936, 304–310.
8. K. Schubert and K. Anderko, *Z. Metallkunde*, **42,** 1951, 321–325.

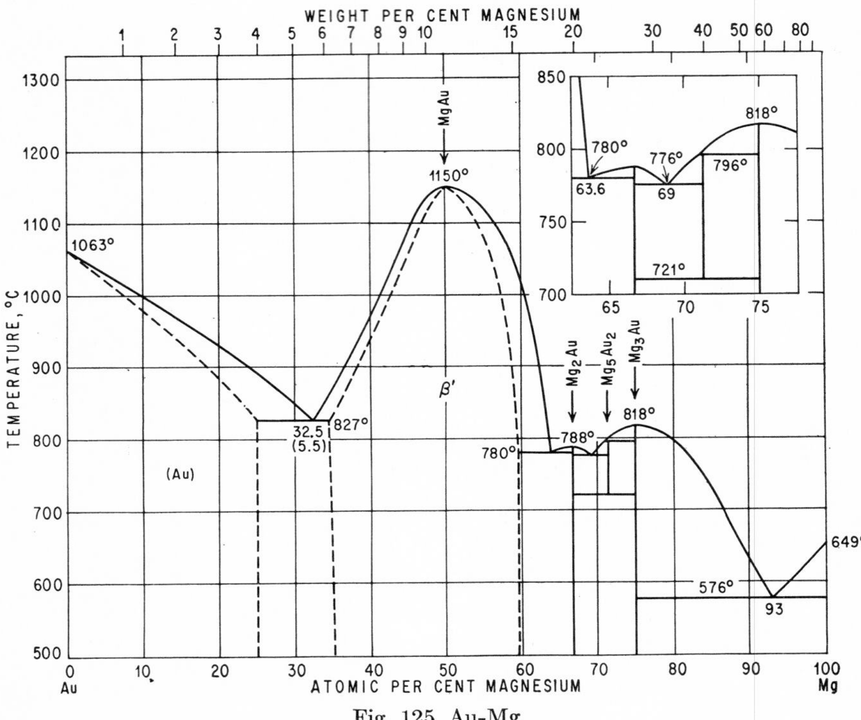

Fig. 125. Au-Mg

0.5551
$\bar{1}$.4449

Au-Mn Gold-Manganese

Until the publication of a recent paper [1], on which the diagram of Fig. 126 is based, information on the constitution of this system was very contradictory. Three general types of phase diagrams have been suggested, with characteristic points as given in Table 12.

According to type I, the gold-base solid solution extends to high manganese contents [at 1000°C, 95.1 at. (84.5 wt.) % Mn] [2], or Au and Mn are even assumed to form a continuous series of solid solutions [3, 4]. With this type the liquidus and solidus curves are shown to have, besides a maximum at 50 at. (78.21 wt.) % Au, two minima as given in Table 12. In the diagram of type II [5], the two minima of the liquidus curve are due to eutectics. The diagram of type III [6] is essentially a combination of the two other types in that the first minimum is assumed to be a eutectic whereas the second is interpreted as a minimum over a series of solid solutions (Table 12). In the earlier papers the transformations in the solid state were not found at all [3] or only partially detected [5, 6, 2, 4]. Also, the effect of Au on the transformation points of manganese was not considered [5, 3, 4] or only in a cursory way [2, 6]. A transformation between 600 and 700°C involving alloys around the composition $MnAu_3$ (91.50 wt. % Au) or possibly $MnAu_2$ (87.78 wt. % Au), first detected by [6], was confirmed by [2] and [4] and interpreted to be due to the formation of the phase $MnAu_3$ or $MnAu_2$, respectively. [4] first reported a transformation at about 640°C in the range of composition of Mn_3Au (54.48 wt. % Au).

Table 12. Characteristics of the System Au-Mn

Ref.	E = eutectic M = minimum of liquidus and solidus	Melting point of MnAu (78.21 wt. % Au), °C	Purity of manganese used, wt. %
3*	M: 29.5 at. (60 wt.) % Au, 1065°C M: 75.3 at. (91.5 wt.) % Au, 945°C	1195	91.7
4	M: ~20 at. (~47 wt.) % Au, 1050°C M: ~70 at. (~89.5 wt.) % Au, 960°C	1280	~99
5	E: 24.7 at. (54 wt.) % Au, 1085°C E: 70 at. (89.5 wt.) % Au, 990°C	1225	Unknown
6	E: 26.3 at. (56 wt.) % Au, 1073°C M: 67 at. (88 wt.) % Au, 977°C	1237	~97
1	E: 28.5 at. (59 wt.) % Au, 1095°C E: 68 at. (88.5 wt.) % Au, 960°C	1260	Electrolytic Mn

* The authors investigated only two alloys in the range 0–30 at. % Au.

The entire phase diagram was redetermined considering the possible sources of error (such as volatilization of Mn and reaction of the melts with the crucible material) and using electrolytic Mn. Thermal, micrographic, and X-ray analyses were used as principal methods [1]. The liquidus and solidus curves of this diagram (Fig. 126) are based on earlier investigations, including that by [7] for alloys up to 55 at. % Au. The existence of the intermediate phases Mn_3Au, $MnAu_2$, and $MnAu_3$ for which there was earlier evidence was confirmed, but phase relationships in the solid state were established for the first time.

Crystal Structures. Most of the lines of the Mn_3Au pattern could be indexed with a tetragonal cell [1]. According to high-temperature powder patterns [1], β is b.c.c (A2 type) [8] with $a = 3.25_5$ A [9] at 49.7 at. % Au, 710–820°C. At temperatures below 600°C, [2] found at 50 at. % Au a tetragonally distorted CsCl structure with $a = 3.29$ A, $c = 3.15$ A, $c/a = 0.96$. [1] observed above 50 at. % Au the β_1 phase with an axial ratio $c/a < 1$, and below 50 at. % Au the β_2 phase with $c/a > 1$. Apparently there is no continuous transition between β_1 and β_2, but a two-phase field could not be detected roentgenographically.

The structure of $MnAu_2$ could not be determined; however, there are indications that it has a tetragonal substructure cell [1]. The transformation in the range of $MnAu_3$ results in the formation of a complicated superstructure, the substructure of which is f.c. tetragonal, with $a = 4.108$ A, $c = 3.994$ A, $c/a = 0.972$ [2].

The f.c.c. γ_{Mn} solid solution may transform on quenching to a tetragonal lattice ($c/a = 0.95$–0.96) in alloys with less than 5 at. % Au [1, 10]. According to X-ray studies, the solid solubility of Au in β-Mn and α-Mn is "very small," probably below 1 at. % [1]. X-ray data of 25 alloys with 26.5–98.7 at. % Mn after various heat-treatments were tabulated by [1].

Note Added in Proof. [11] detected the ordered phase $MnAu_4$ (93.49 wt. % Au), formed on cooling at 420°C; it is f.c. tetragonal with $c/a < 1$ and has a complicated superstructure. Accordingly, the curves of the transformation Au ⇄ $MnAu_3$

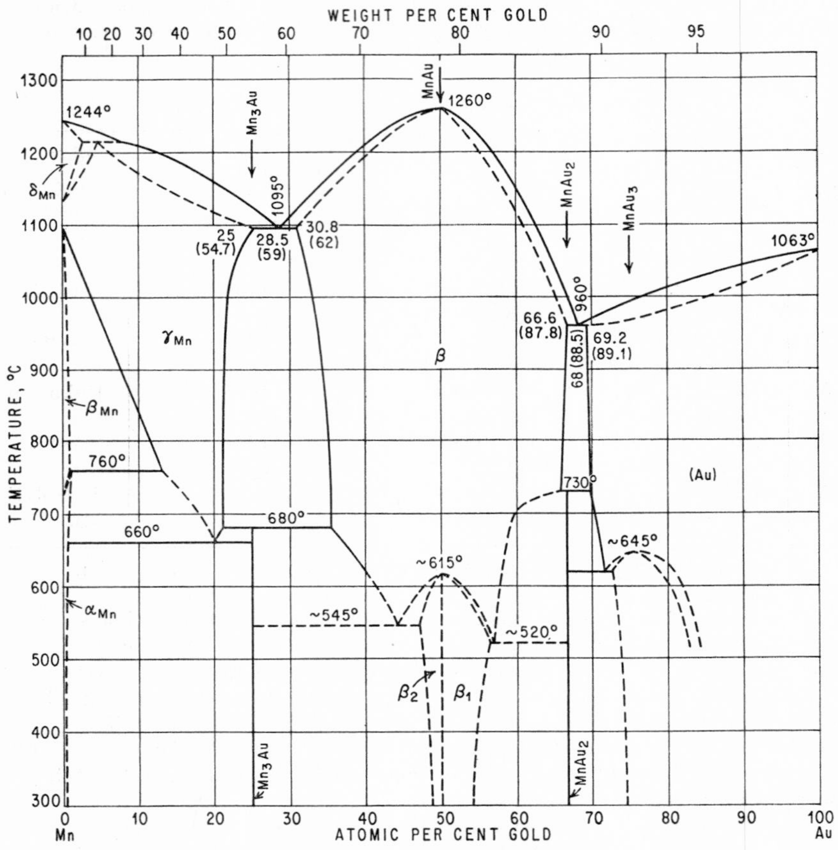

Fig. 126. Au-Mn. (See Note Added in Proof.)

(Fig. 126) will not cut 80 at. % Au. The temperatures of the transformations β ⇄ (MnAu) and β + (Au) ⇄ $MnAu_2$ were found as slightly above 640°C and about 740°C, respectively.

1. E. Raub, U. Zwicker, and H. Baur, *Z. Metallkunde*, **44**, 1953, 312–320.
2. H. Bumm and U. Dehlinger, *Metallwirtschaft*, **13**, 1934, 23–25.
3. L. Hahn and S. Kyropoulos, *Z. anorg. Chem.*, **95**, 1916, 105–114.
4. V. A. Nemilov and A. A. Rudnickij, *Compt. rend. acad. sci. U.R.S.S.*, **3**(8), 1935, 351–354 (in German); also *Izvest. Sektora Platiny*, **13**, 1936, 129–143.
5. N. Parravano and U. Perret, *Gazz. chim. ital.*, **45**(1), 1915, 293–303; *Z. Metallkunde*, **14**, 1922, 73–74.

6. H. Moser, E. Raub, and E. Vincke, *Z. anorg. Chem.*, **210,** 1933, 67–76; see also E. Raub, "Die Edelmetalle und ihre Legierungen," p. 179, footnote, Springer-Verlag OHG, Berlin, 1940.
7. H. Baur; see [1].
8. In [1] β is said to have the "b.c.c. cesium chloride lattice." However, since it is also stated that the formation of β_1 and β_2 "has to be ascribed essentially to an ordering of the atoms in the b.c.c. lattice of the β mixed crystals," it is quite obvious that β actually is disordered (A2 type).
9. Lattice parameters in [1] are given in "10^{-8} cm." According to a private communication by U. Zwicker, the values are actually in kX units.
10. U. Zwicker, *Z. Metallkunde*, **42,** 1951, 246, 327.
11. A. Kussmann and E. Raub, *Z. Metallkunde*, **47,** 1956, 9–15.

0.3129
$\bar{1}$.6871

Au-Mo Gold-Molybdenum

X-ray analysis [1] showed that Au and Mo do not form an intermediate phase [2]. The eutectic temperature was found as 1054°C. The solid solubility of Mo in Au was determined by means of lattice-parameter measurements. X-ray photographs were taken at temperatures of 200–800°C as well as at room temperature, after quenching from 600, 800, and 1000°C. Results of both series of experiments agree very well. The solubility was found in at. % Mo (wt. % in parentheses) as follows: 1054°C, 1.25 (0.61) (extrapolated); 1000°C, 1.2 (0.59); 800°C, 1.05–1.1 (0.52–0.54); 600°C, 0.9–0.95 (0.44–0.46); 400°C, 0.9 (0.44); 200°C, 0.7 (0.34). The solubility of Au in solid Mo appears to be exceedingly small. The solidus temperature of a 0.1 at. % Mo alloy was determined as 1058°C [1].

1. G. A. Geach and D. Summers-Smith, *J. Inst. Metals*, **82,** 1953-1954, 471–474.
2. This could be concluded already from work by Dreibholz, *Z. physik. Chem.*, **108,** 1924, 4–5.

1.1485
$\bar{2}$.8515

Au-N Gold-Nitrogen

Nitrogen is not dissolved by solid or liquid gold (investigated up to 1300°C) [1].

1. A. Sieverts and W. Krumbhaar, *Ber. deut. chem. Ges.*, **43,** 1910, 894; F. J. Toole and M. F. G. Johnson, *J. Phys. Chem.*, **37,** 1933, 331-346.

0.9332
$\bar{1}$.0668

Au-Na Gold-Sodium

By thermal analysis (see data points in Fig. 127) and analysis of isolated crystals, the existence of the phase $NaAu_2$ (5.51 wt. % Na) was established. No liquidus temperatures were reported for the range 65.8–95.5 at. (18.3–71.2 wt.) % Na. $NaAu_2$ forms a eutectic with Au at about 17 at. (2.3 wt.) % Na, 876°C, and was assumed to form a eutectic with Na also [1]. The temperature and composition of this Na-rich eutectic were determined by careful thermal work to lie at 82°C and 96.7 at. (77.3 wt.) % Na [2, 3]; see inset of Fig. 127 [4].

The existence of $NaAu_2$ was confirmed by X-ray studies [5–7] and claimed to have a homogeneity range from 32 to above 43 at. % Na [6]; [5] gave 38–40 at. % Na as the Na-rich solubility limit. Two additional phases, not found by thermal analysis [1], were detected by X-ray work [6], one of which is Na_2Au (18.91 wt. % Na). This phase was isolated from Na-rich alloys by extraction with liquid NH_3 and reported to dissolve Na up to 70–71 at. % Na. The composition of the third intermediate phase could not be established. It was shown to occur in rapidly cooled alloys with 50–70 at. % Na (10.4–20.6 wt. % Na), either together with $NaAu_2$ (between 50 and 55 at. % Na) or with Na_2Au (between 60 and 70 at. % Na). This phase is perhaps

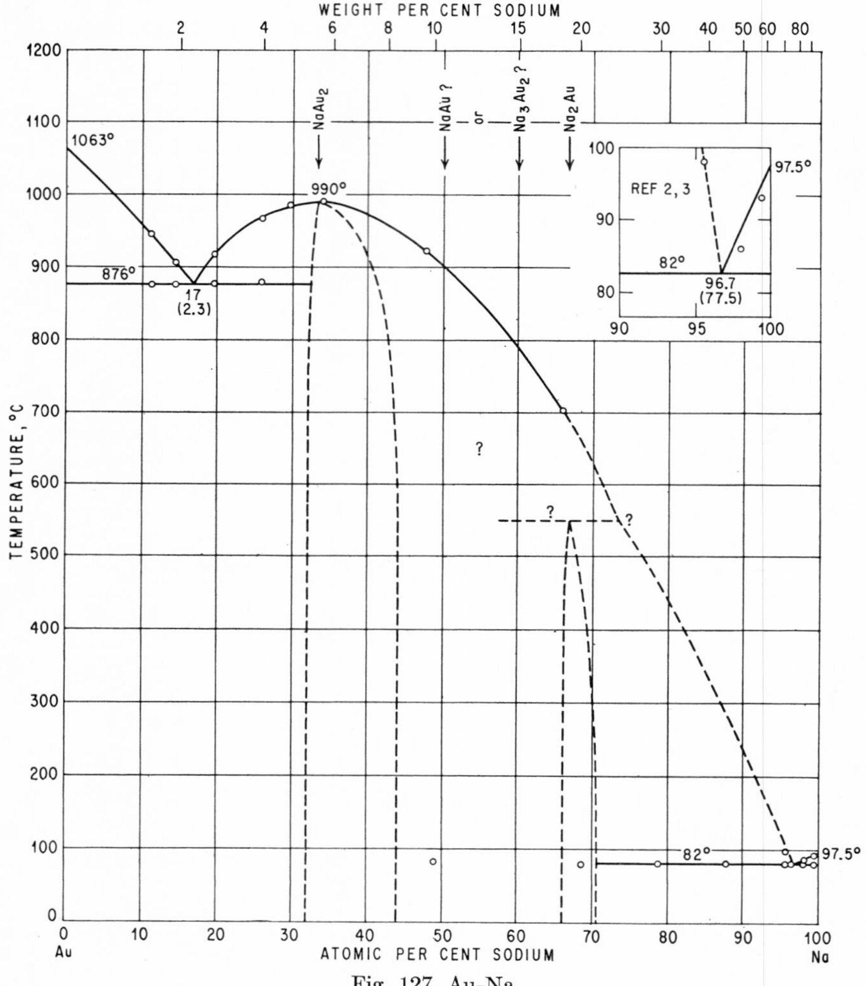

Fig. 127. Au-Na

stable only at elevated temperatures [6]. Simple atomic ratios within the composition range in question are NaAu (10.44 wt. % Na) and Na_3Au_2 (14.89 wt. % Na). If one assumes that the two liquidus data points at about 48 and 66 at. % Na (Fig. 127) are correct, it appears that both the phase of unknown composition and Na_2Au are formed by peritectic reactions. The solid solubility of Na in Au is considered to be extremely small [6].

[8] found indications for the existence of a compound of the approximate composi-

tion NaAu which is apparently stable only in a solution of NH_3. Also, vapor-deposition studies of Na on Au led to the conclusion that NaAu is formed [9].

Crystal Structures. $NaAu_2$ is isotypic with $MgCu_2$ (C15 type) [5–7], $a = 7.806$ A [5], $a = 7.817$ A at 38.1 at. % Na [6], and $a = 7.803$ A [7]. The structure of Na_2Au is tetragonal of the $CuAl_2$ (C16) type, $a = 7.417$ A, $c = 5.522$ A, $c/a = 0.745$ [6].

1. C. H. Mathewson, *Intern. Z. Metallog.*, **1**, 1911, 81–88.
2. C. T. Heycock and F. H. Neville, *J. Chem. Soc.*, **55**, 1889, 668–671.
3. C. T. Heycock and F. H. Neville, *J. Chem. Soc.*, **73**, 1898, 716–718.
4. G. Tammann (*Z. physik. Chem.*, **3**, 1889, 441–449) also determined the effect of Au (up to 0.5 at. %) on the melting point of Na.
5. U. Quadt, E. Weibke, and W. Biltz, *Z. anorg. Chem.*, **232**, 1937, 297–306.
6. W. Haucke, *Z. Elektrochem.*, **43**, 1937, 712–719; *Naturwissenschaften*, **25**, 1937, 61.
7. H. Perlitz and E. Aruja, *Z. Krist.*, **100**, 1938, 157–166; *Naturwissenschaften*, **25**, 1937, 461.
8. E. Zintl, J. Goubeau, and W. Dullenkopf, *Z. physik. Chem.*, **A154**, 1931, 1–46, especially pp. 15 and 44.
9. A. Sommer, *Nature*, **152**, 1943, 215.

0.5264
$\bar{1}.4736$

Au-Ni Gold-Nickel

The **liquidus** of Fig. 128 was drawn by graphical interpolation of the thermal results of several authors [1–4]. The accuracy may be higher than ±10°C. The minimum was located at about 25 wt. (53 at.) % Ni, 950°C [1]; 15 wt. (37 at.) % Ni, 955°C [2]; between 15 and 20 wt. (37 and 46 at.) % Ni [3, 4], 950°C [3] and 945–948°C [4], respectively.

Solidus. In older papers [1, 2] a eutectic between Au- and Ni-rich primary solid solutions was assumed. [5, 4] showed, however, that there is a continuous solid solution at higher temperatures. The solidus was determined from heating curves [5] of homogenized alloys and by measuring the solidification intervals on cooling curves [4]. The results deviate strongly between 20 and 100 wt. % Ni, as can be seen from Fig. 128. A decision in favor of one of the two curves is not possible since normally the solidification intervals found on cooling are greater than those found on heating; here we find the reverse order. A rough computation of the solidus near pure Ni by means of the van't Hoff relation is in favor of the results of [5].

Solid Alloys. The existence of a miscibility gap with a critical point was first recognized by microscopic observations of annealed and quenched alloys containing 10–80 wt. % Ni [5] (Fig. 128). The eutectoid-like appearance of the microstructures of two-phase alloys misled some authors [4, 6] to draw erroneous phase diagrams. The form of the miscibility gap was further determined by means of resistivity-temperature curves (heating rate 2°C per min) [7], roentgenographic [8, 9], electric [9], magnetic [9–11], and emf measurements [12] (Fig. 128); electrical-conductivity measurements by [13] on alloys up to 14 wt. % Ni did not yield unanimous results.

The kinetics of precipitation were studied microscopically, roentgenographically, and magnetically and by resistivity and hardness measurements by [9, 14]; magnetically by [15, 10]; and by means of resistivity measurements and thermodynamic computations (spinodal) by [16].

Ni enters the Au lattice as a weak paramagnetic component at room temperature up to the solubility limit [17].

The Curie point of Ni is depressed by addition of Au. Measurements on quenched homogeneous alloys showed that the Curie point reaches room temperature at about

34 wt. (63 at.) % Ni [9]. For alloys in the state of equilibrium, however, the loss of ferromagnetism is indicated by the dash-dot line in Fig. 128 [7, 9].

The miscibility gap in the solid state has stimulated thermodynamic and atomistic investigations on the reason why it occurs. Thermodynamic data, derived from emf measurements between 700 and 900°C, indicate that the lattice-distortion energy,

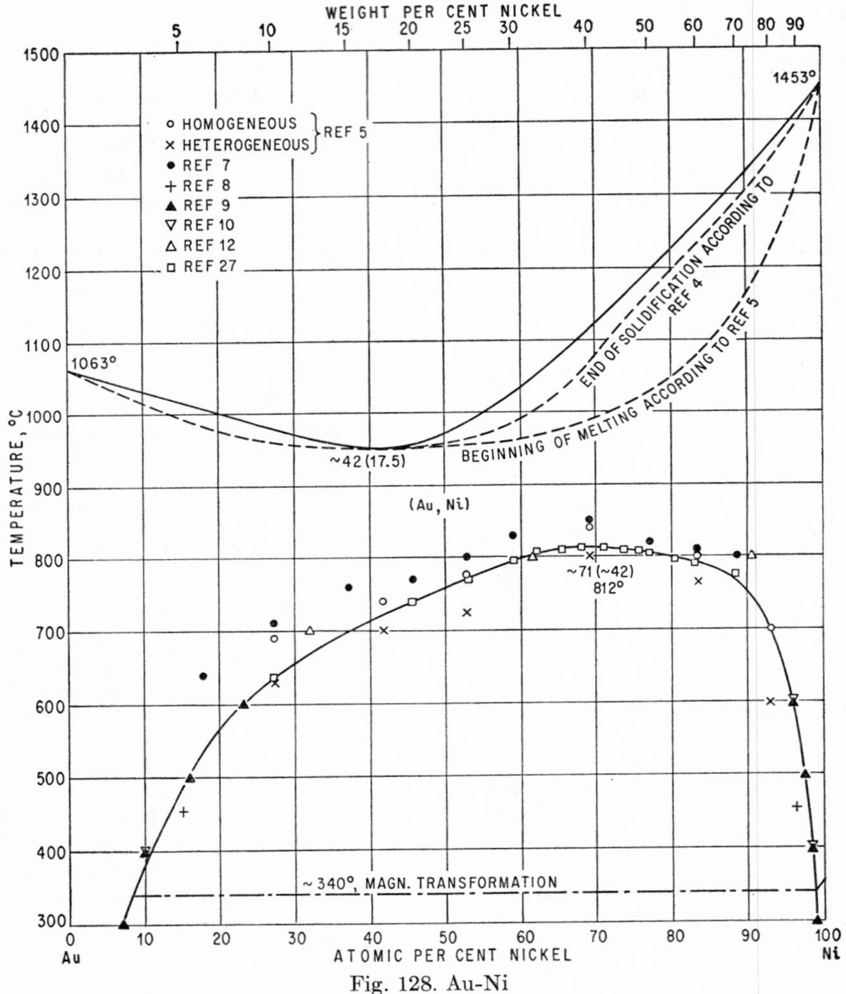

Fig. 128. Au-Ni

resulting from the size difference between Ni and Au atoms, rather than the energetic favoring of bonds between like atoms is responsible for the miscibility gap at lower temperatures [12]. In agreement with that, the measurement of diffuse X-ray scattering [18, 19, 19*a*] showed that short-range order rather than clustering of like atoms exists at 900°C.

Thermodynamic computations of the shape of the liquidus and solidus were made by [20].

The lattice parameters of quenched Au-Ni alloys show a positive deviation from Vegard's law [9, 21]. At about 60 at. % Ni, [9] observed—besides an anomalous high lattice parameter—some very weak additional lines and ascribed them to an unstable phase ($Ni_{2 \text{ or } 3}Au$) [22]. Several weaker anomalies in lattice spacing and density in the solid solution were ascribed [21] to Brillouin zone overlaps accompanied by the formation of vacant lattice sites.

Supplement. Magnetic measurements on quenched and on annealed (temperature?) Au-Ni alloys containing up to 40 at. % Au were made by [23]. The values reported for the solid solubility of Au in Ni at ordinary temperature (about 5 at. %) and for the temperature of the magnetic transformation within the miscibility gap (317°C) show that true equilibrium was not reached in these experiments (cf. Fig. 128). In the investigation of quenched alloys, it was found that at 71 at. % Ni the Curie-point curve reaches room temperature (taken from a graph; cf. above).

A detailed comparison of thermodynamic and X-ray data by [24] confirmed the conclusion previously mentioned that the strain energy plays a much more significant role than the chemical bonding energy in the thermodynamic behavior of solid Au-Ni alloys. For calorimetric investigations of a 48.3 at. % Ni alloy, see [25, 26].

Note Added in Proof. The boundary of the miscibility gap between 27 and 88.5 at. % Ni was redetermined by means of resistivity measurements on 14 carefully equilibrated alloys [27]. By virtue of the quality and number of these measurements it appears justifiable to base the miscibility curve of Fig. 128 on these data rather than draw an interpolating curve. X-ray work [27] failed to give any indications of the unstable phase suggested by [9]. See also [28].

1. M. Levin, *Z. anorg. Chem.*, **45,** 1905, 238–242.
2. P. De Cesaris, *Gazz. chim. ital.*, **43**(2), 1913, 609–611.
3. W. Fraenkel and A. Stern, *Z. anorg. Chem.*, **151,** 1926, 105–108.
4. H. Hafner (u. W. Heike), dissertation, Freiberg i. Sa., Germany, 1927.
5. W. Fraenkel and A. Stern, *Z. anorg. Chem.*, **166,** 1927, 161–164.
6. W. Heike and H. Kessner, *Z. anorg. Chem.*, **182,** 1929, 272–280.
7. G. Grube and F. Vaupel, *Z. physik. Chem.* (Bodenstein-Festband), **1931,** 187–197. The specimens for the resistivity measurements had been annealed (24 hr, 950°C) and slowly cooled (7 weeks) to room temperature. [7] admit that their data, since measured in heating, may be somewhat too high.
8. E. M. Wise, *Trans. AIME*, **83,** 1929, 384–403.
9. W. Köster and W. Dannöhl, *Z. Metallkunde*, **28,** 1936, 248–253.
10. W. Gerlach, *Z. Metallkunde*, **40,** 1949, 281–289.
11. A. I. Swartz, M.S. thesis, Massachusetts Institute of Technology, Cambridge, Mass., 1951.
12. L. L. Seigle, M. Cohen, and B. L. Averbach, *Trans. AIME*, **194,** 1952, 1320–1327.
13. B. Beckmann, *Arkiv Mat. Astron. Fysik*, **7,** 1912, 1–18.
14. W. Köster and A. Schneider, *Z. Metallkunde*, **29,** 1937, 103–104.
15. W. Gerlach, *Z. Metallkunde*, **29,** 1937, 102–103.
16. G. Borelius, *Trans. AIME*, **191,** 1951, 477–484.
17. E. Vogt and H. Krueger, *Ann. Physik*, **18,** 1933, 755–770.
18. P. A. Flinn and B. L. Averbach, *Phys. Rev.*, **83,** 1951, 1070.
19. B. L. Averbach et al., *U.S. Atomic Energy Comm., Publ.* NYO-581, 1951, and NYO-582, 1952; *Met. Abstr.*, **19,** 1952, 830; **20,** 1953, 757.

19*a*. P. A. Flinn, B. L. Averbach, and M. Cohen, *Acta Met.*, **1,** 1953, 664–673.

20. Y. E. Geguzin and B. Y. Pines, *Doklady Akad. Nauk S.S.S.R.*, **75,** 1950, 387–390 (in Russian); *Met. Abstr.*, **19,** 1952, 646.
21. C. E. C. Ellwood and K. Q. Bagley, *J. Inst. Metals*, **80,** 1952, 617–619.

22. Mention of the phase NiAu, which really does not exist, in M. Hansen, "Der Aufbau der Zweistofflegierungen," Springer-Verlag OHG, Berlin, 1936, is traceable to a misprint in the abstract *J. Inst. Metals*, **50**, 1932, 477, concerning the work of A. Westgren and W. Ekman, *Arkiv Kemi, Mineral. Geol.*, **B10**, 1930, 1–6, a fact which [9] have pointed out. The compound NiAl should replace NiAu.
23. V. Marian, *Ann. phys.*, **7**, 1937, 459–527.
24. B. L. Averbach, P. A. Flinn, and M. Cohen, *Acta Met.*, **2**, 1954, 92–100.
25. W. Desorbo, *Acta Met.*, **3**, 1955, 227–231. (Low-temperature heat capacity.)
26. R. A. Oriani, *Acta Met.*, **3**, 1955, 232–235. (Heat capacity at high temperatures and the entropy of formation.)
27. A. Münster and K. Sagel, to be published in *Z. physik. Chem.*, 1958.
28. C. Ang, J. Sivertsen, and C. Wert, *Acta Met.*, **3**, 1955, 558–565.

1.0908
$\bar{2}.9092$

Au-O Gold-Oxygen

Liquid gold does not dissolve measurable amounts of oxygen [1]. Measurements of the solubility of oxygen in solid gold between 300 and 900°C yielded poorly reproducible results which apparently were within the limit of error [2].

1. A. Sieverts and W. Krumbhaar, *Ber. deut. chem. Ges.*, **43**, 1910, 893–900.
2. F. J. Toole and F. M. G. Johnson, *J. Phys. Chem.*, **37**, 1933, 331–346.

0.0157
$\bar{1}.9843$

Au-Os Gold-Osmium

Electrical-resistance measurements of alloys with up to 2.96 at. (2.87 wt.) % Os, quenched from 900–950°C, showed that the solid solubility of Os in Au is negligibly small [1].

1. J. O. Linde, *Ann. Physik*, **10**, 1931, 69.

0.8039
$\bar{1}.1961$

Au-P Gold-Phosphorus

Phosphorus vapor is dissolved by molten gold and ejected on solidification [1]. Claims that preparations of the compositions Au_2P_3 (19.07 wt. % P) [2] and Au_3P_4 (17.32 wt. % P) [3] were compounds appeared questionable, as the homogeneity of these reaction products was not proved. Tensimetric analysis verified the existence of the phosphide Au_2P_3 and showed that no compound richer in phosphorus exists [4]. It was concluded that Au_2P_3 is the only gold phosphide which can be prepared by reaction of the elements at normal pressure. There is no solid solubility of Au in Au_2P_3 and of P in Au [4].

1. P. Hautefeuille and A. Perrey, *Compt. rend.*, **98**, 1884, 1378.
2. A. Schrötter, *Ber. Wien. Akad.*, **2**, 1849, 301–306.
3. A. Granger, *Compt. rend.*, **124**, 1897, 498–499.
4. H. Haraldsen and W. Biltz, *Z. Elektrochem.*, **37**, 1931, 502–508; H. Haraldsen, *Skrifter Norske Videnskaps-Akad. Oslo Mat. Naturv. Kl.*, **9**, 1932, 1–63.

$\bar{1}.9785$
0.0215

Au-Pb Gold-Lead

The phase relationships as presented in Fig. 129 are based on thermal analysis and metallographic examination [1, 2]. The existence of the two intermediate phases

Au_2Pb (34.44 wt. % Pb) and $AuPb_2$ (67.76 wt. % Pb) was confirmed by X-ray analysis [3–6]. On cold working, Au_2Pb decomposes into Au and $AuPb_2$; annealing at 400°C returns the mixture to Au_2Pb. This indicates that Au_2Pb is unstable below some unknown temperature; the very low rate of decomposition is increased by cold work [6].

The solid solubility of Pb in Au, reported to lie between 0.005 and 0.06 wt. % Pb at 650°C [7], was redetermined recently. By micrographic work, solubilities were found of about 0.1 wt. % between 950 and 700°C and <0.05 and <0.04 wt. % at

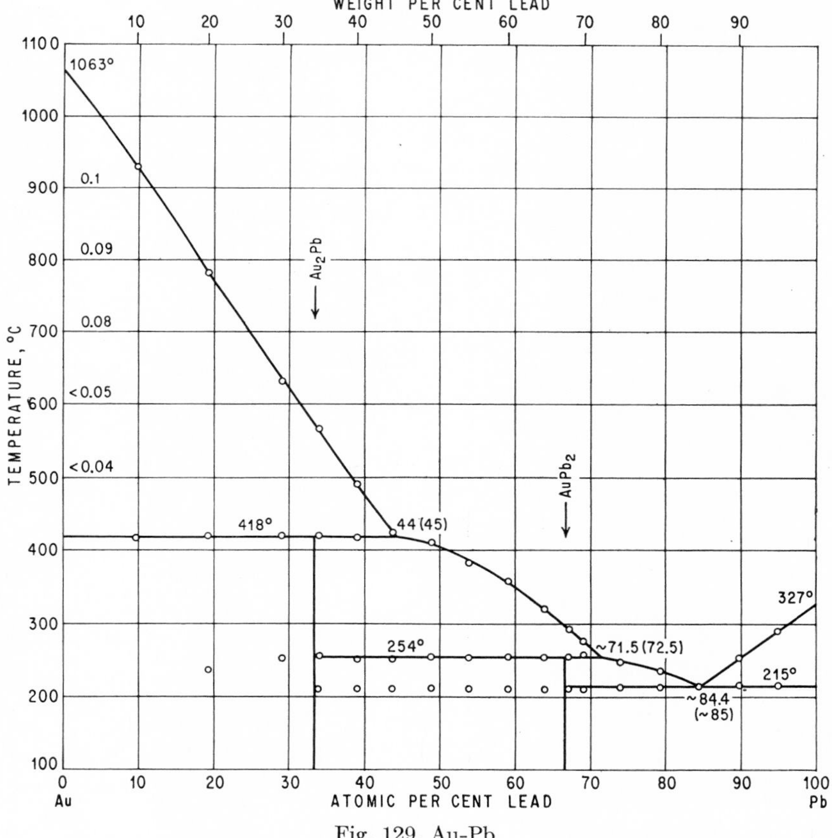

Fig. 129. Au-Pb

600 and 500°C, respectively [8]. According to lattice-parameter measurements there is no solid solubility between 200 and 500°C [9]. In accordance with this, alloys with 0.04 to 1.0 wt. (0.95 at.) % Pb proved to be two-phase [10]. As a result of diffusion tests the solid solubility of Au in Pb is 0.03 at. % at 170°C and 0.08 at. (0.076 wt.) % Au at 200°C [11].

Au_2Pb is isotypic with $MgCu_2$ (C15 type), $a = 7.92_6$ A [3], $a = 7.94_6$ A [6]. $AuPb_2$ has a b.c. tetragonal structure of the $CuAl_2$ (C16) type, $a = 7.32$ A, $c = 5.65$ A, $c/a = 0.77$ [4, 5]. Both components are appreciably soluble in $AuPb_2$ [4], which is still highly ordered near its melting point [12].

1. R. Vogel, *Z. anorg. Chem.*, **45,** 1905, 11–23.
2. Concerning the effect of Pb on the melting point of Au, see W. C. Roberts-Austen, *Proc. Roy. Soc. (London)*, **49,** 1891, 347–356. As to the effect of Au on the melting point of Pb, see C. T. Heycock and F. H. Neville, *J. Chem. Soc.*, **61,** 1892, 909–912; **65,** 1894, 65–76.
3. H. Perlitz, *Acta et Comment. Univ. Tartuensis,* **A27,** 1934; *Met. Abstr.*, **2,** 1935, 15, 222.
4. G. Mets and H. Torgren, *Tallinna Tehnikaülikooli Toimetused,* sec. A, no. 14, 1940; *Met. Abstr.*, **8,** 1941, 58.
5. H. J. Wallbaum, *Z. Metallkunde,* **35,** 1943, 218–221.
6. O. J. Kleppa and D. F. Clifton, *Acta Cryst.*, **4,** 1951, 74.
7. L. Nowack, *Z. anorg. Chem.*, **154,** 1926, 395–398; *Z. Metallkunde,* **19,** 1927, 241–244.
8. E. Raub and A. Engel, *Z. Metallkunde,* **37,** 1946, 76–81.
9. E. A. Owen and E. A. O. Roberts, *J. Inst. Metals,* **71,** 1945, 213–218, 246–254.
10. V. W. Gerlach and E. Schweitzer, *Z. anorg. Chem.*, **173,** 1928, 110.
11. W. Seith and A. Keil, *Z. physik. Chem.*, **B22,** 1933, 350–358; W. Seith and J. G. Laird, *Z. Metallkunde,* **24,** 1932, 195; W. Seith and H. Etzold, *Z. Elektrochem.*, **40,** 1934, 829–833.
12. O. Kubaschewski, *Z. physik. Chem.*, **192,** 1943, 292–308.

0.2668
$\bar{1}$.7332

Au-Pd Gold-Palladium

According to the phase diagram by [1] (Fig. 130), which was confirmed in a somewhat cursory manner (without giving individual data) by [2], the two metals form a

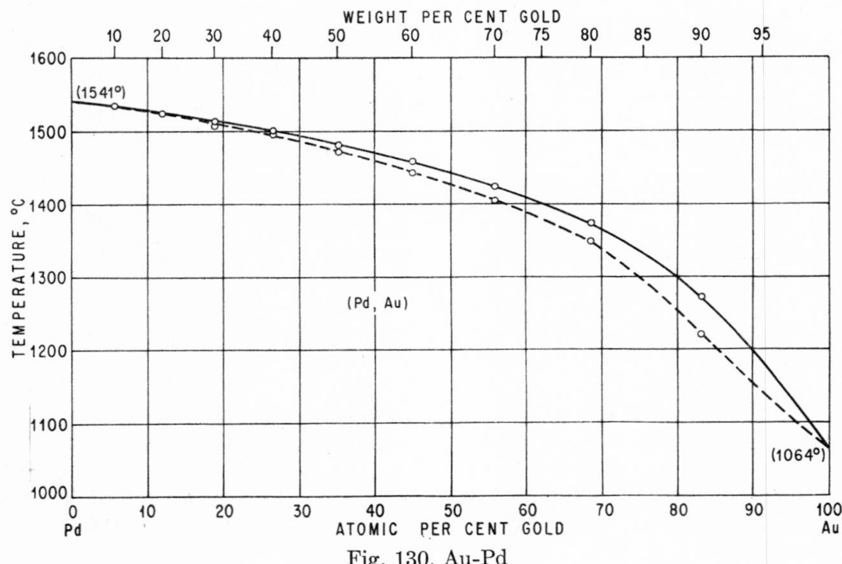

Fig. 130. Au-Pd

continuous series of solid solutions. Information as to the change of various properties as a function of composition tends to confirm that the system is devoid of transformations in the solid state. Properties investigated are electrical conductivity

[3–6], temperature coefficient of electrical resistance [3], thermal conductivity [7, 3], coefficient of linear expansion [8], lattice parameter [9, 10], elastic properties [6], tensile strength [11], and optical reflectivity [6]. The lattice parameter changes practically linearly with the composition in at. % [10].

However, from data of the thermal emf by [3] and measurements of the Peltier effect (using the alloys prepared by [4]) it was concluded [12] that "loose compounds" exist in the solid state. The curves showing the dependence of both properties were interpreted to consist of straight-line positions rather than being catenary curves. Also, [13], in referring to the electrical resistivity and temperature coefficient of resistance curves, pointed out that the "peculiar slope" in both curves in the neighborhood of 70 wt. % Au requires explanation. Likewise, the isotherms of the thermal emf [14] show a change of slope in the neighborhood of 70 wt. % Au [13]. [15] determined the magnetic susceptibility of the whole series of alloys at 20 and −183°C. The atomic susceptibility was shown to deviate strongly from the law of mixtures toward diamagnetism.

Thermodynamic computations of the liquidus and solidus and studies of Young's modulus are due to [16] and [17], respectively.

Supplement. The lattice parameters of 10 Pd-Au alloys containing 10–90 at. % Au have been measured by [18].

1. R. Ruer, *Z. anorg. Chem.*, **51,** 1906, 391–396.
2. W. Fraenkel and A. Stern, *Z. anorg. Chem.*, **166,** 1927, 164.
3. W. Geibel, *Z. anorg. Chem.*, **69,** 1911, 43–46.
4. E. Sedström, dissertation, Stockholm, 1924.
5. F. E. Carter, *Trans. AIME*, **78,** 1928, 775–776.
6. H. Röhl, *Ann. Physik*, **18,** 1933, 155–168.
7. F. A. Schulze, *Physik. Z.*, **12,** 1911, 1028–1031.
8. C. H. Johansson, *Ann. Physik*, **76,** 1925, 452–453.
9. S. Holgersson and E. Sedström, *Ann. Physik*, **75,** 1924, 149–150.
10. W. Stenzel and J. Weerts, "Festschrift zum 50-jährigen Bestehen der Platinschmelze G. Siebert G.m.b.H.," pp. 288–299, Hanau, 1931; *Z. Metallkunde*, **24,** 1932, 139–140.
11. E. M. Wise, W. G. Crowell, and J. T. Eash, *Trans. AIME, Inst. Metals Div.*, **99,** 1932, 363; see also [13].
12. G. Borelius, *Ann. Physik*, **53,** 1917, 615–628.
13. R. F. Vines, "The Platinum Metals and Their Alloys," pp. 110–112, The International Nickel Co., Inc., New York, 1941.
14. C. S. Sivil; see [13].
15. E. Vogt, *Ann. Physik*, **14,** 1932, 1–39.
16. Y. E. Geguzin and B. Ya. Pines, *Doklady Akad. Nauk. S.S.S.R.*, **75,** 1950, 387–390, 535–538; *Met. Abstr.*, **19,** 1952, 646.
17. Y. Shibuya, *Nippon Kinzoku Gakkai-Shi*, **16,** 1952, 235–238; *Met. Abstr.*, **20,** 1953, 472.
18. V. G. Kuznecov, *Izvest. Sektora Platiny, Akad. Nauk. S.S.S.R.*, **1946,** no. 20, 5–20. *Structure Repts.*, **10,** 1945-1946, 54.

0.1459
$\bar{1}$.8541

Au-Pr Gold-Praseodymium

Thermal analysis of the system [1] indicated the existence of the four intermediate phases $PrAu_4$ (15.16 wt. % Pr), $PrAu_2$ (26.32 wt. % Pr), PrAu (41.68 wt. % Pr), and Pr_2Au (58.83 wt. % Pr); see Fig. 131. A cursory check [2] using two alloys with about 18.5 and 25 at. % Pr showed, however, that the composition of the compound richest in gold is $PrAu_3$ (19.24 wt. % Pr) rather than $PrAu_4$ (Fig. 131), proving that Pr forms the same intermediate phases with Au as Ce and La. The solid solubility of Pr in Au was estimated to be approximately 0.4 wt. (0.55 at.) % Pr [1].

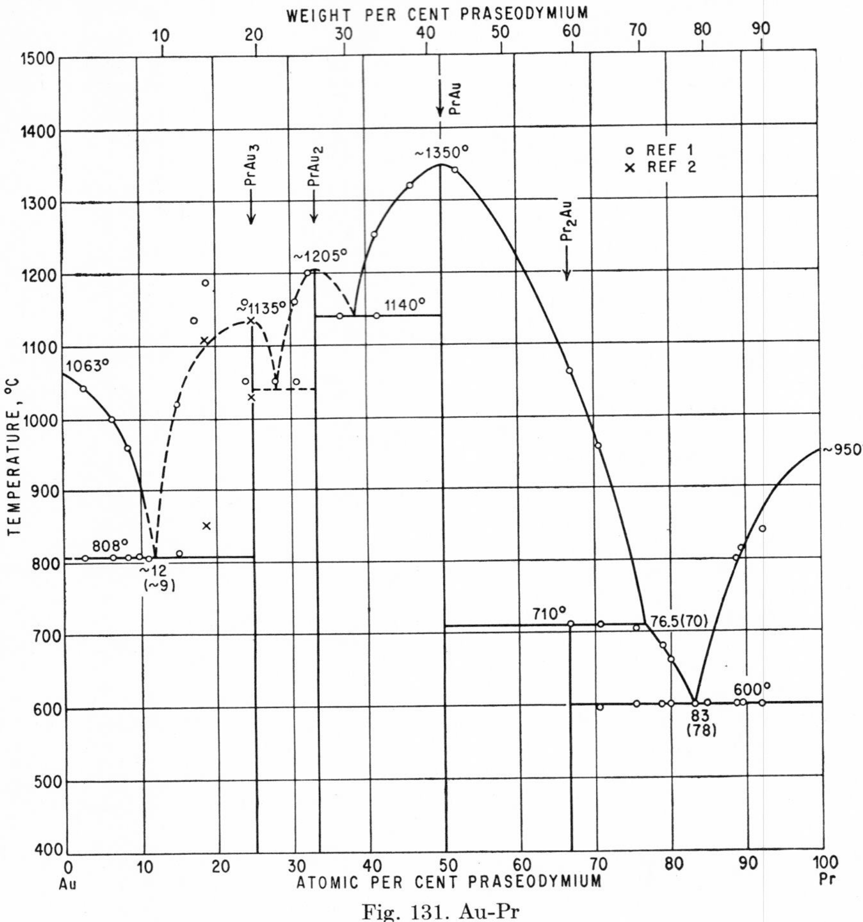

Fig. 131. Au-Pr

1. A. Rossi, *Gazz. chim. ital.*, **64**, 1934, 748–757.
2. R. Vogel and T. Heumann, *Z. Metallkunde*, **35**, 1943, 29–42.

0.0044
$\bar{1}.9956$

Au-Pt Gold-Platinum

The controversy which existed as regards the constitution of this system has been cleared by a recent investigation [1]; the phase diagram (Fig. 132) is now fully established.

Figure 132 shows the liquidus temperatures determined by [2, 3, 1]; those by [4] have been omitted as they scatter considerably in the range 20–50 at. % Pt although, on the whole, they exhibit the same trend.

[2] concluded from cooling-curve data that a continuous series of solid solutions crystallizes from Au-Pt melts. As to be expected, the temperatures of the end of solidification thus found are much too low. On the other hand, melting points of alloys with 5-95 at. % Pt determined by [5] in 1879, using the crude method of observing the beginning of deformation on annealing, are far too high. [4] reported that

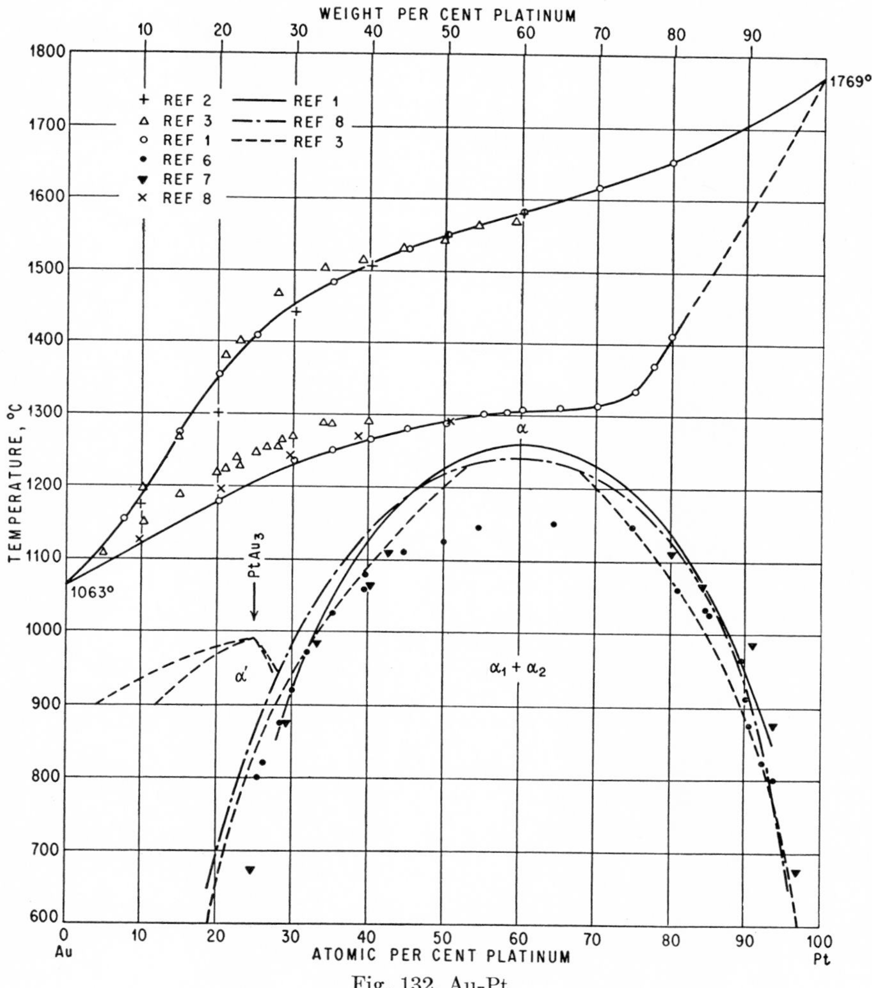

Fig. 132. Au-Pt

cooling curves of melts with about 20-61.5 at. % Pt showed arrests at approximately 1290°C. It was suggested that a peritectic equilibrium between melt, Au-rich, and Pt-rich solid solutions occurs, indicating that the two metals are not soluble in each other in all proportions but form a miscibility gap extending from about 25 to 80 at. % Pt.

However, [6], who first determined the solubility curve in the solid state (see below), gave evidence for a two-phase field bounded by a continuous curve with a critical point at approximately 65 at. % Pt, 1160°C, above which an uninterrupted series of solid solutions exists. Results of similar investigations by [7] and [8] are in agreement with this type of phase relations. The latter investigator determined solidus temperatures, in the range up to about 51 at. % Pt, by measurements of electrical resistance vs. temperature taken on heating; see data points in Fig. 132.

[3] confirmed the occurrence of weak thermal effects between 1295 and 1301°C, in the range 21–59 at. % Pt, and, therefore, were led to believe that the peritectic reaction suggested by [4] exists. The solidus temperatures between 0 and 40 at. % Pt (see data points in Fig. 132), determined by heating and electrical-resistance-vs.-temperature curves (on heating), in connection with the boundaries of the two-phase field in the solid state, seemed to point to Au-rich and Pt-rich solid solutions of about 59 and 61 at. % Pt, respectively, at the temperature of the supposed peritectic horizontal.

Confirming the findings by [6–8], [1] have given conclusive microscopic and X-ray evidence for the existence of only one phase at temperatures above 1260°C. The solidus points, determined by micrographic studies, are shown in Fig. 132, indicating a gently sloping curve having upper limits of 1315°C at 70% and 1305°C at 55 at. % Pt. The existence of a solidus so very nearly horizontal over such a wide range of composition will, of course, be reflected in the cooling and heating curves of all alloys with less than about 60 at. % Pt, and these, as has been recorded by [3, 4], show an arrest at just below 1300°C. Applying equations for the change in chemical potentials, [9] have calculated the phase boundaries. The liquidus and solidus curves thus obtained agree in shape with those based on experimental data. This holds especially for the flat portion of the solidus between 1300 and 1200°C; see also [10].

The boundary of the two-phase region ($\alpha_1 + \alpha_2$) has been determined, using lattice-parametric [6, 7, 3, 1], resistivity-vs.-composition [6, 8, 11], and thermoresistometric methods [8, 1]. Figure 132 shows the curves (without data points) according to [8, 3, 1] and the data points according to [6, 7]. Previously, [4] had shown, by microscopic examination and hardness tests, that a heterogeneous field exists and extends from about 25 to 80 at. % Pt. [12] had reported that alloys with 20 and 25 wt. % Pt showed age hardening at 550°C whereas the alloy with 15 wt. % Pt did not.

The boundary by [6] closes at too low a temperature, apparently because, in the range 50–70 at. % Pt, the samples (wires) had not been quenched rapidly enough to retain fully the high-temperature single-phase structure. The upper portion of the solubility curve was determined by electrical-resistance-vs.-temperature measurements [8, 1]. The temperature of the critical point thus found differs by only 20°C (Fig. 132). Also, the data by [3] indicate that the solubility curve closes at a temperature well above 1220°C.

As may be seen from Fig. 132, the various solubility curves are, on the whole, in substantial agreement, with the exception of the upper portion of the curve by [6]. In the Pt-rich range, the curves of [8] and [1] nearly coincide; however, at the Au-rich side the curve by [8] is at slightly lower Pt contents than that by [1] but, at temperatures below 900°C, close to the curve of [3]. [10] has shown by thermodynamic analysis that here the data by [8] may deserve some preference, although he emphasizes that the analysis cannot pretend to decide between the experimental results. The calculations show that the phase diagram "leads to self-consistent energy values from all the data it contains, i.e., the liquidus, solidus, and solubility curves are in accordance with a positive heat of solution in both the liquid and solid phases" [10].

[3] proved that no ordered phases PtAu and Pt_3Au, suspected by [6] to be formed at 400°C, exist. However, they found X-ray evidence for the formation of an apparently ordered phase α' ($PtAu_3$) below 1000°C, after severe cold working and annealing at 900°C, corresponding to $PtAg_3$ (Fig. 132).

The kinetics of two-phase formation of quenched single-phase alloys were studied by [13, 11]; see also [14, 15].

Crystal Structures. Lattice parameters of homogenized alloys covering the whole range of composition were determined by [6, 3, 1]. There is a slight deviation from the straight-line relation toward lower lattice constants [3, 1]. The parameter of the α' phase, after annealing at 900°C, increases from a = 3.873 A (or kX?) at 5 at. % Pt to 3.918 A (?) at 25 at. % Pt; the parameter of the α phase decreases from a = 4.062 A (?) at 5 at. % Pt to 4.027 A (?) at 25 at. % Pt [3].

1. A. S. Darling, R. A. Mintern, and J. C. Chaston, *J. Inst. Metals,* **81,** 1952-1953, 125–132.
2. F. Doerinckel, *Z. anorg. Chem.,* **34,** 1907, 345–349.
3. G. Grube, A. Schneider, and U. Esch, "Festschrift aus Anlass des 100-jährigen Jubiläums der Firma W.C. Heraeus G.m.b.H.," pp. 20–42, 1951.
4. A. T. Grigoriev, *Izvest. Inst. Izucheniyu Platiny,* **6,** 1928, 184–194; *Z. anorg. Chem.,* **178,** 1929, 97–107.
5. T. Erhard and A. Schertel, *Jahr. Berg-u. Hüttenwes. in Sachsen* (Saxony), **17,** 1879, 163; see also *Stahl u. Eisen,* **19,** 1899, 27.
6. C. H. Johansson and J. O. Linde, *Ann. Physik,* **5,** 1930, 762–792.
7. W. Stenzel and J. Weerts, "Festschrift zum 50-jährigen Bestehen der Platinschmelze G. Siebert G.m.b.H.," pp. 300–308, Hanau, 1931.
8. C. G. Wictorin, dissertation, University of Stockholm, 1947; *Arkiv Mat. Astron. Fysik,* **B36,** 1949, no. 9.
9. G. Scatchard and W. J. Hamer, *J. Am. Chem. Soc.,* **57,** 1935, 1809–1811.
10. H. K. Hardy, *J. Inst. Metals,* **81,** 1952-1953, 599–600.
11. C. G. Wictorin, *Ann. Physik,* **33,** 1938, 509–516.
12. L. Nowack, *Z. Metallkunde,* **22,** 1930, 97–98.
13. C. H. Johansson and O. Hagsten, *Ann. Physik,* **28,** 1937, 520–527.
14. G. Borelius, *Ann. Physik,* **28,** 1937, 507–509; **33,** 1938, 517–531.
15. G. Borelius, *Trans. AIME,* **191,** 1951, 477–484.

0.3630
$\bar{1}$.6370

Au-Rb Gold-Rubidium

The existence of the compound $RbAu_2$ (17.81 wt. % Rb), having only a narrow range of homogeneity (not determined), has been established. Its X-ray pattern is different from that of KAu_2. Apparently a compound $RbAu_4$, analogous with KAu_4, does not exist [1]. The formation of the compound RbAu (30.24 wt. % Rb) was reported by [2].

1. H. J. Ehrhorn and F. Weibke, *Z. anorg. Chem.,* **232,** 1937, 307–312.
2. A. Sommer, *Nature,* **152,** 1943, 215.

0.2824
$\bar{1}$.7176

Au-Rh Gold-Rhodium

Rhodium is reported to raise the melting point of gold [1, 2]. Microscopic and X-ray studies showed that this system consists of the terminal solid solutions of the elements, with no intermediate phase existing [2, 3]. From lattice-spacing measurement of alloys in an undefined condition, it was concluded that the solid solubility of Rh in Au is between 2.2 wt. (4.1 at.) and 4.9 wt. (9.0 at.) % Rh and that of Au in Rh between 2 wt. (1.1 at.) and 4.5 wt. (2.4 at.) % Au [3]. A more reliable value for the solid solubility of Rh in Au, about 0.3 wt. (0.56 at.) % Rh at 900°C, resulted from electrical-resistance measurements of alloys with up to 1.6 wt. (3.0 at.) % Rh, quenched from 900°C [4].

1. W. H. Wollaston, *Phil. Trans. Roy. Soc. (London)*, **1804,** 425.
2. H. Rössler, *Chem.-Ztg.*, **24,** 1900, 733–735.
3. R. W. Drier and H. L. Walker, *Phil. Mag.*, **16,** 1933, 294–298.
4. J. O. Linde, *Ann. Physik*, **10,** 1931, 69.

0.2876
$\bar{1}$.7124

Au-Ru Gold-Ruthenium

Electrical-resistance measurements of alloys with up to 2.76 at. (1.44 wt.) % Ru, quenched from 900–950°C, showed that the solid solubility of Ru in Au is negligibly small [1]. An X-ray photograph of an alloy with 5 wt. % Ru did not reveal the presence of an intermediate phase but possibly gave indication of some small solid solubility [2].

1. J. O. Linde, *Ann. Physik*, **10,** 1931, 69.
2. J. L. Byers, AIME Preprint no. 10, 1932.

0.7889
$\bar{1}$.2111

Au-S Gold-Sulfur

Experiments to prepare Au-S and Au-Se melts by adding sulfur or selenium, respectively, to molten gold were unsuccessful. The sulfur and selenium vaporized before noticeable amounts had been dissolved [1, 2].

1. K. Friedrich, *Metallurgie*, **5,** 1908, 593.
2. H. Pélabon, *Ann. chim. et phys.*, **17,** 1909, 566.

0.2094
$\bar{1}$.7906

Au-Sb Gold-Antimony

By means of thermal analysis and microscopic examination, the phase relationships shown in Fig. 133 were first established by [1] and, within the range 43–90 at. % Sb, verified by [2]. The existence of the only intermetallic phase, $AuSb_2$ (55.26 wt. % Sb), was confirmed by measurements of the thermoelectric force [3] and the electrical resistance [2], as well as by X-ray analysis [4–6]. $AuSb_2$ was claimed to undergo transformations at about 355 and 405°C [6, 7] and to be of practically singular composition [5]. The compound Au_3Sb (17.07 wt. % Sb) reported earlier [8] does not exist.

After only qualitative statements as to the solid solubility of Sb in Au had been available [9, 5], the solubility was determined parametrically (with an accuracy of ±0.05 at. %) to be 1.12 (0.7), 1.10 (0.68), 0.91 (0.56), 0.34 at. (0.21 wt.) %, and 0.0% at 600, 500, 400, 300, and 200°C, respectively [10]. The solubility limit was drawn by [10] to show a change in direction at about 430°C, indicating a three-phase equilibrium at this temperature, for which there is no evidence in the solidification diagram. The existence of a retrograde solubility curve appears more likely. The formation of Sb-rich solid solutions was indicated by electrical-conductivity measurements [2]; however, lattice-parameter measurements showed the solubility to be negligible at about 300°C [5].

Lattice parameters of the gold-rich solid solution were given by [10]. $AuSb_2$ is cubic of the FeS_2 (C2) type, with a lattice parameter reported to be 6.6_5 [4, 6], 6.66 [5], and 6.657 A [11].

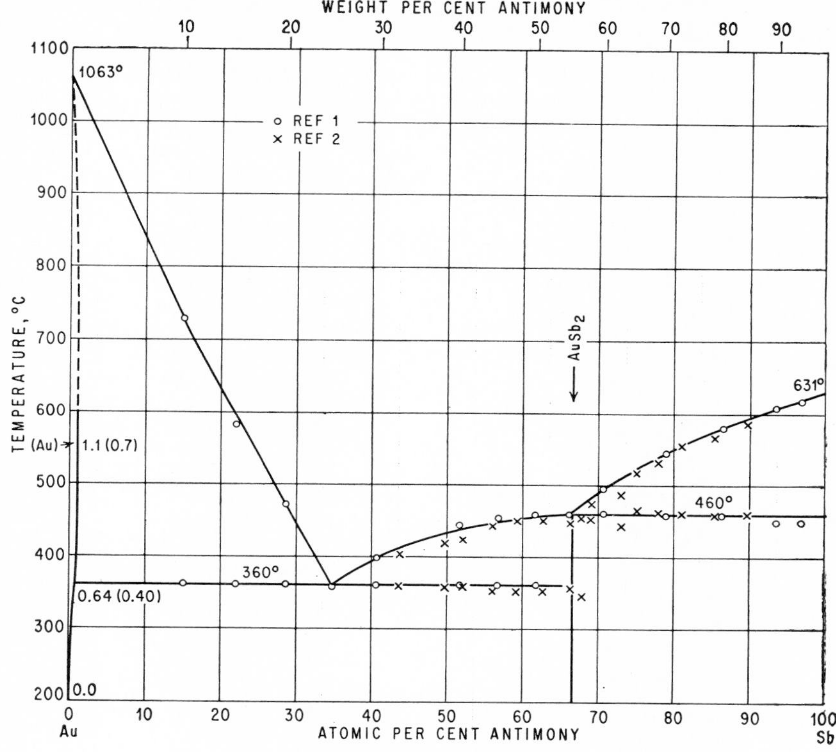

Fig. 133. Au-Sb

1. R. Vogel, *Z. anorg. Chem.*, **50,** 1906, 151–157.
2. A. T. Grigorjew, *Ann. inst. platine* (*U.S.S.R.*), **1929,** no. 7, 45–51; *Z. anorg. Chem.*, **209,** 1932, 289–294.
3. W. Haken, *Ann. Physik*, **32,** 1910, 328–329.
4. I. Oftedal, *Z. physik. Chem.*, **135,** 1928, 291–299.
5. O. Nial, A. Almin, and A. Westgren, *Z. physik. Chem.*, **B14,** 1931, 81–82.
6. J. Bottema and F. M. Jaeger, *Proc. Koninkl. Akad. Wetenschap., Amsterdam*, **35,** 1932, 916–928; abstracts in *J. Inst. Metals*, **53,** 1933, 13–14; *Chem. Zbl.*, **1933**(1), 389–390.
7. E. Rosenbohm and F. M. Jaeger, *Proc. Koninkl. Akad. Wetenschap., Amsterdam*, **39,** 1936, 366.
8. F. Roessler, *Z. anorg. Chem.*, **9,** 1895, 72–73.
9. L. Nowack, *Z. Metallkunde*, **19,** 1927, 241.
10. E. A. Owen and E. A. O. Roberts, *J. Inst. Metals*, **71,** 1945, 240–242.
11. A. R. Graham and S. Kaiman, *Am. Mineral.*, **37,** 1952, 416–469. See also J. Sobotka, *Rozpravyčesk. Akad. věd.*, **64**(7), 1954, 43–60; *Chem. Abstr.*, **49,** 1955, 2955 (a = 6.656 A).

0.3975
$\bar{1}$.6025

Au-Se Gold-Selenium

See Au-S, page 230.

0.8464
$\bar{1}.1536$

Au-Si Gold-Silicon

The phase diagram (Fig. 134) is due to [1]; the eutectic point was determined only by extrapolation of the two branches of the liquidus curve and, therefore, is rather uncertain. In agreement with Fig. 134 is an earlier report [2] that on solidification of Au-rich melts primary Au crystals precipitate. Lattice-parameter measurements

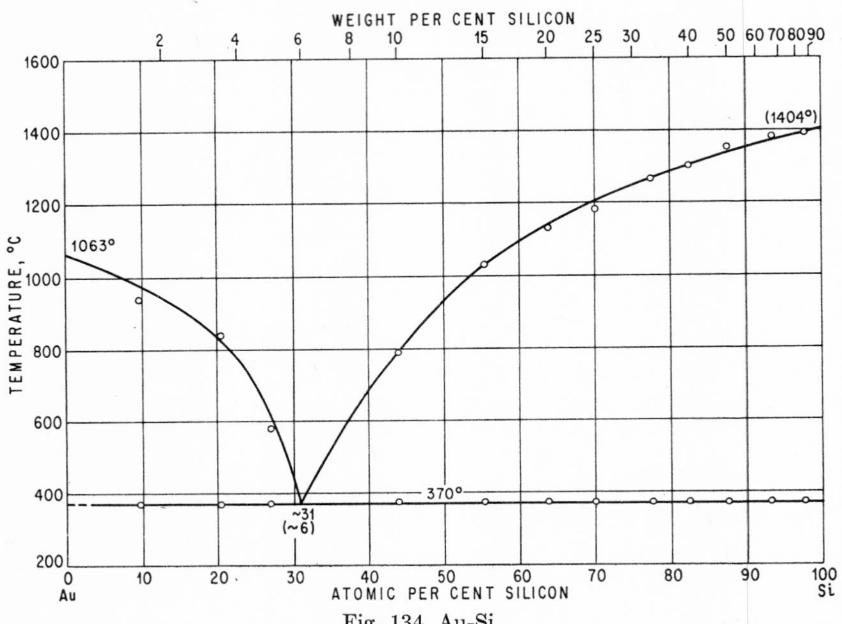

Fig. 134. Au-Si

indicated that there is no mutual solubility of the components [3]. However, [4] concluded from diffusion tests that Si dissolves to a limited extent in Au.

1. C. di Capua, *Rend. accad. nazl. Lincei*, **29**, 1920, 111–114.
2. E. Vigouroux, *Ann. chim. et phys.*, **12**, 1897, 170–171.
3. E. R. Jette and E. B. Gebert, *J. Chem. Phys.*, **1**, 1933, 753–755.
4. W. Loskiewicz, *Przeglad Górniczo-Hutniczy*, **21**, 1929, 583–611; abstract in *J. Inst. Metals*, **47**, 1931, 516–517.

0.2205
$\bar{1}.7795$

Au-Sn Gold-Tin

The entire diagram was investigated using thermal and microscopic methods [1], and the three intermediate phases AuSn (37.57 wt. % Sn), $AuSn_2$ (54.62 wt. % Sn), and $AuSn_4$ (70.65 wt. % Sn) were found. Strong peritectic coring was observed in the formation of $AuSn_4$. Indications for the existence of AuSn had been reported earlier [2–4]. Investigations by [5–8] gave no reliable information as to the composition of intermediate phases. The $AuSn_4$-Sn eutectic was carefully determined to lie at 90.5 wt. (94.1 at.) % Sn, 214°C [9], which agrees fairly well with 90.0 wt. (93.7 at.) % Sn, 217°C, as given by [1]. The existence of AuSn and $AuSn_2$ was corroborated by emf measurements [10]. $AuSn_4$, however, could not be detected by this method, apparently because of the disturbance of equilibrium in the crystallization of alloys with 60–80 wt. % Sn [11].

By powder-pattern X-ray analysis another intermediate phase (ζ) was found which is homogeneous between about 12 and 16 at. (7.5 and 10.3 wt.) % Sn [12].

The boundary of the gold-rich primary solid solution between 388 and 718°C was determined by very careful X-ray analysis [13]. The maximum solubility was

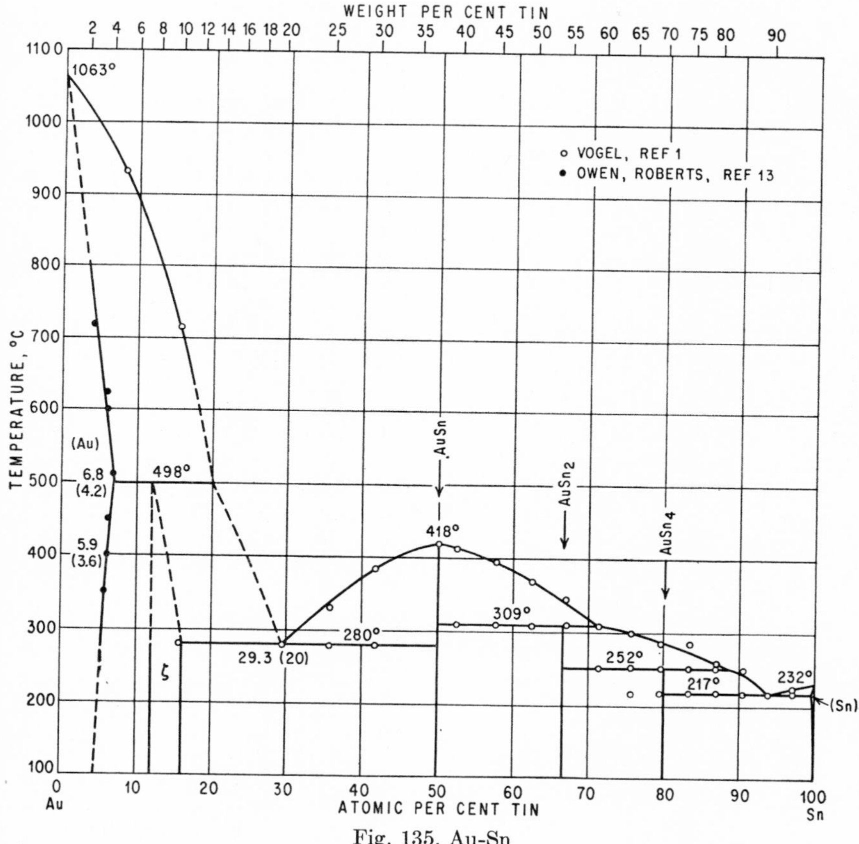

Fig. 135. Au-Sn

found to be 6.8_5 at. (4.2 wt.) % Sn at 498 ± 10°C. At 450, 400, and 350°C solubilities are 6.3_0, 5.9_4, and 5.6_6 at. (3.9, 3.6, and 3.5 wt.) % Sn, respectively. Some observations indicate the temperature of maximum solubility (about 500°C) to be that of the peritectic reaction liq. + (Au) ⇄ ζ. Solubilities above 500°C (Fig. 135) do not claim to be very accurate.

The solid solubility of Au in Sn at 200°C was found by various indirect methods to be about 0.2 at. (0.3 wt.) % Au [14].

Crystal Structures. Lattice parameters of the (Au) phase were determined by [13]. The ζ phase has a h.c.p. structure with probably statistical atom distribution (A3 type) and with $a = 2.90-2.94$ A, $c = 4.78-4.76$ A, $c/a = 1.65$–1.62 [12]. AuSn has the structure of the NiAs (B8) type [15, 12, 16], with a = 4.323 A, c = 5.523 A, c/a = 1.278 [12]. $AuSn_2$ has a very narrow range of homogeneity; its translation group is orthorhombic primitive with a = 6.85 A, b = 7.00 A, c = 11.78 A [17]. The structure of $AuSn_4$ is of the same type as $PtSn_4$ and $PdSn_4$ and has an orthorhombic unit cell (containing 20 atoms) with a = 6.44 A, b = 6.48 A, c = 11.60 A [18]. There is a relationship to the $CuAl_2$ (C16) type of structure [18]. The range of homogeneity of $AuSn_4$ is also very narrow [12, 19].

1. R. Vogel, *Z. anorg. Chem.*, **46,** 1905, 60–75.
2. A. Matthiessen, *Pogg. Ann.*, **110,** 1860, 214–215.
3. A. P. Laurie, *Phil. Mag.*, **33,** 1892, 94–99.
4. A. P. Laurie, *J. Chem. Soc.*, **65,** 1894, 1037–1038.
5. A. Matthiessen and M. v. Bose, *Proc. Roy. Soc. (London)*, **11,** 1861, 430.
6. A. Matthiessen and M. Holzmann, *Pogg. Ann.*, **110,** 1860, 31–32.
7. E. Maey, *Z. physik. Chem.*, **38,** 1901, 295.
8. J. F. Spencer and M. E. John, *Proc. Roy. Soc. (London)*, **116,** 1927, 61–72.
9. C. T. Heycock and F. H. Neville, *J. Chem. Soc.*, **55,** 1889, 667; **57,** 1890, 378; and especially **59,** 1891, 936–966.
10. N. A. Puschin, *Zhur. Russ. Fiz.-Khim. Obshchestva*, **39,** 1906, 353–399; abstract in *Chem. Zentr.*, **1907**(2), 1319; *Z. anorg. Chem.*, **56,** 1908, 1–45.
11. G. Tammann, *Z. anorg. Chem.*, **107,** 1919, 155–156.
12. S. Stenbeck and A. Westgren, *Z. physik. Chem.*, **B14,** 1931, 91–96.
13. E. A. Owen and E. A. O'Donnell Roberts, *J. Inst. Metals*, **71,** 1945, 213–254.
14. E. Jenckel and L. Roth, *Z. Metallkunde*, **30,** 1938, 135–144.
15. G. D. Preston and E. A. Owen, *Phil. Mag.*, **4,** 1927, 133–147.
16. J. A. Bottema and F. M. Jaeger, *Proc. Koninkl. Akad. Wetenschap., Amsterdam*, **35,** 1932, 916–928.
17. K. Schubert and U. Rösler, *Naturwissenschaften*, **40,** 1953, 437.
18. K. Schubert and U. Rösler, *Z. Metallkunde*, **41,** 1950, 298–300; *Z. Naturforsch.*, **5a,** 1950, 127.
19. G. Tammann and H. J. Rocha, *Z. anorg. Chem.*, **199,** 1931, 292–294.

0.1890
$\bar{1}$.8110

Au-Te Gold-Tellurium

The phase diagram was determined by [1–3]. Disregarding quantitative differences (Table 13), results agree substantially. Figure 136 is based on the thermal data

Table 13. Characteristic Temperatures in °C and Compositions in Wt. % Te (At. % in Parentheses) of the System Au-Te

	Ref. 1	Ref. 2	Ref. 3
Au-$AuTe_2$ eutectic	432°C 40 (50.7)	452°C 44 (54.7)	447°C ~42 (53)
Melting point of $AuTe_2$	452°C	472°C	464°C
$AuTe_2$-Te eutectic	80 (86)	84 (89)	82.5 (88)
Melting point of Te	440°C	452°C	451°C

by [3], which are to be regarded as the more reliable ones. The diagram by [1] agrees very well with that of Fig. 136 as to the liquidus curve between 0 and 50 at. % Te; however, liquidus temperatures above 50 at. % Te are too low, apparently because they are based on too low a melting point of tellurium (Table 13). [2] did not give tabulated data, but only described the course of the liquidus curve. However, as the Au-rich melts were not heated high enough, the liquidus temperatures in the range 0–50 at. % Te were not observed; and the eutectic horizontal was erroneously given as the curve of the beginning of crystallization.

Microscopic studies [4] and measurements of electrochemical potential [4] verified the results of thermal analysis and confirmed the existence of $AuTe_2$ (56.41 wt. % Te).

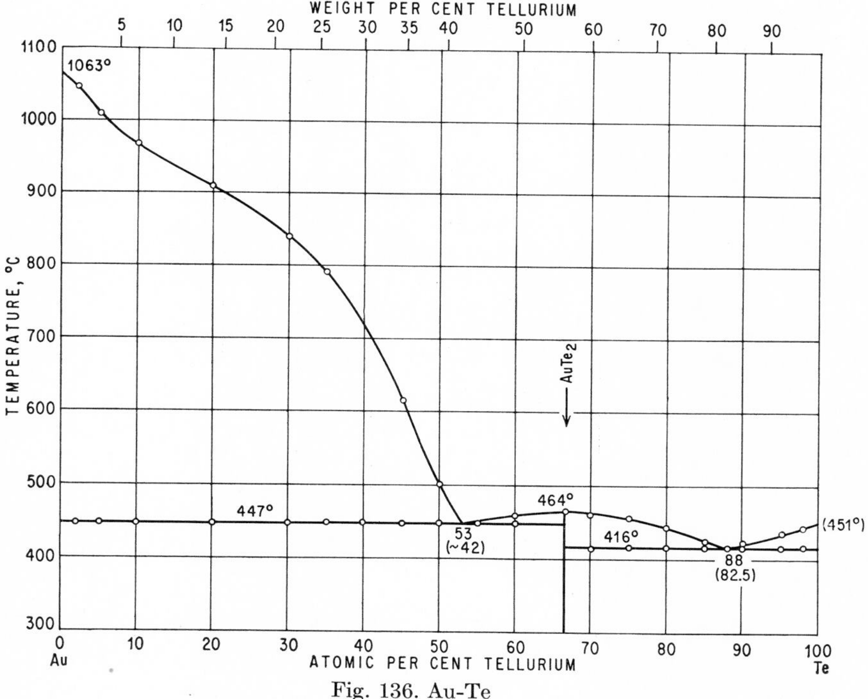

Fig. 136. Au-Te

The compound Au_2Te (24.45 wt. % Te) reported earlier [5, 6] does not exist. The solid solubility of Te in Au appears to be very small as an addition of only 0.1 wt. % Te is detrimental to the rollability of gold, because of the presence of the brittle phase $AuTe_2$ [7]. An alloy with 0.01 wt. % Te could be rolled satisfactorily [7].

$AuTe_2$ occurs in nature as the minerals calaverite and krennerite. The former crystallizes monoclinically, with a = 7.19 A, b = 4.41 A, c = 5.08 A, β approximately 90°C, and 2 molecules per unit cell [8], whereas the latter has an orthorhombic structure, a = 16.54 A, b = 8.82 A, c = 4.46 A, 8 molecules per unit cell [9].

1. T. K. Rose, *Trans. (Brit.) Inst. Mining Met.*, **17**, 1907-1908, 285; abstract (with diagram) in *J. Soc. Chem. Ind.*, **27**, 1908, 229.
2. H. Pélabon, *Compt. rend.*, **148**, 1909, 1176–1177; *Ann. chim. et phys.*, **17**, 1909, 564–566.

3. G. Pellini and E. Quercigh, *Atti accad. nazl. Lincei*, **19**(2), 1910, 445–449.
4. M. Coste, *Compt. rend.*, **152,** 1911, 859–862.
5. J. Margottet, *Ann. sci. école norm. sup.*, **8,** 1879, 247.
6. R. Brauner, *Monatsh. Chem.*, **10,** 1889, 411–457.
7. L. Nowack, *Z. Metallkunde*, **19,** 1927, 241.
8. G. Tunnell and C. J. Ksanda, *J. Wash. Acad. Sci.*, **25,** 1935, 32–33.
9. G. Tunnell and C. J. Ksanda, *J. Wash. Acad. Sci.*, **26,** 1936, 507–509.

$\bar{1}.9292$
0.0708

Au-Th Gold-Thorium

The partial system up to 40 at. % Th (Fig. 137) was studied by thermal analysis of alloys prepared from pressed pellets of powder mixtures melted under argon [1].

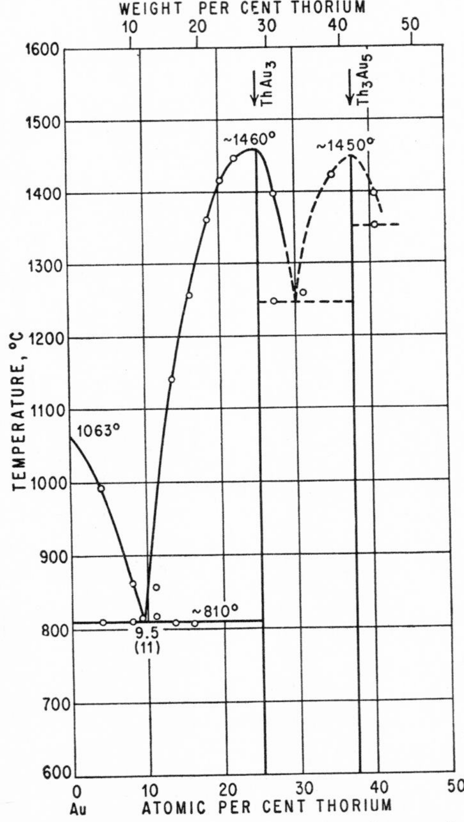

Fig. 137. Au-Th. (See also Supplement.)

The alloys richer in Th contained an unstated amount of ThO_2. All alloys were analyzed, and the oxide content was considered in computing the metal ratio. Alloys low in Th were melted in Pythagoras crucibles; those richer in Th, in corundum crucibles. Reaction of the melts with the crucible material at the high temperatures made the investigation of alloys richer in Th impossible. Solid solubility of Th in Au is practically nil; heat-treated samples of an alloy with 0.05 wt. % Th proved to be heterogeneous. An alloy with 40.5 at. (44.5 wt.) % Th contained an intermediate phase with more than 40 at. % Th. $ThAu_3$ (28.18 wt. % Th) and Th_3Au_5 (41.39 wt. % Th) are unstable in air.

Supplement. Th_2Au (70.19 wt. % Th) was identified by determining the crystal structure; it is tetragonal of the $CuAl_2$ (C16) type, $a = 7.42$ A, $c = 5.95$ A, $c/a = 0.802$ [2].

1. E. Raub and M. Engel, *Z. Elektrochem.*, **49**, 1943, 487–493.
2. J. R. Murray, *J. Inst. Metals*, **84**, 1955-1956, 91–96.

0.6146
$\bar{1}.3854$

Au-Ti Gold-Titanium

Alloys with 0–50 At. (19.5 Wt.) % Ti. Figure 138 shows the results of thermal analysis of alloys melted in Al_2O_3 crucibles under argon, using Ti of 98.6% purity [1]. The solid solubility of Ti in Au was determined micrographically, using alloys prepared with Mg-reduced Ti of higher purity, and found to be in at. % Ti (wt. % in parentheses) as follows: 8.5 (2.2), 7.4 (1.9), 6.7 (1.7), 4.8 (1.2), 3.6 (0.9), 2.4 (0.6), and 1.8 (0.45) % Ti at 1000, 950, 900, 800, 700, 600, and 500°C, respectively [1]. The compound $TiAu_6$ (3.89 wt. % Ti) was identified by a one-phase structure of an alloy of this composition, after annealing at 900°C. An alloy of the composition $TiAu_2$ did not show a one-phase structure "due to unavoidable contamination at the high melting temperature" [1]. However, the thermal data (Fig. 138) indicate that the liquidus has a maximum at the composition $TiAu_2$ (10.83 wt. % Ti). No evidence was found [1] for the compound $TiAu_3$ (7.49 wt. % Ti) reported by [2]. Work still in progress at the time of writing [3] confirmed the existence of $TiAu_2$ and that of a eutectic with the next intermediate phase richer in Ti, TiAu (19.54 wt. % Ti).

Alloys with 50–100 At. % Ti. The partial diagram covering the region 0–6 at. (21 wt.) % Au (Fig. 138) is due to [4]. It is based on micrographic analysis of alloys prepared using iodide titanium. According to [3], the eutectoid temperature was found as about 810°C and the solubility of Au in β-Ti at 1000°C as 5.0 at. (17.9 wt.) % Au, both values being lower than those of [4]. The solubility of Au in α-Ti was found to be less than 2.0 at. (7.75 wt.) % Au at 700°C, which is in agreement with [4].

Electrical-resistivity-vs.-temperature curves of six alloys with 0.75–9.4 at. % Au made it possible to sketch a tentative phase diagram [5]. The eutectoid temperature of 825°C and composition of approximately 15 wt. (4.1 at.) % Au as well as the $\beta/(\alpha + \beta)$ boundary are in good agreement with the findings of [4]. Otherwise there are deviations.

The intermediate phase Ti_3Au (42.15 wt. % Ti) was first reported by [6] and confirmed by [7]. It is accepted as being the compound richest in Ti [4]. According to [3], Ti_3Au forms eutectics with both TiAu and the β-Ti solid solution.

Crystal Structures. The lattice parameter of the (Au) phase was given by [1]. The structures of $TiAu_6$ and $TiAu_2$ were reported to be f.c. tetragonal, $a = 4.06$ A (?), $c/a = 0.97$, and h.c.p., $a = 2.78$ A (?), $c/a = 1.71$, respectively [1]. $TiAu_3$, assumed to exist by [2] but not confirmed by [1], was claimed to be isotypic with $TiCu_3$ (deformed h.c.p. with ordered structure) [2]. Ti_3Au, first reported to be isotypic with Cu_3Au [6], was shown to have the structure of the "β-W" (A15) type, $a = 5.094$ A [7].

Note Added in Proof. Phase relationships in the partial system $TiAu_2$-Ti were established over the temperature range 400–1500°C by means of micrographic analysis, using alloys based on iodide-Ti [8]. Main features of the diagram are: (*a*) $TiAu_2$, TiAu (has homogeneity range with marked temperature dependence), and Ti_3Au have open maximum melting temperatures of 1455, 1490, and 1395°C, respectively; (*b*) eutectic $TiAu_2$-(TiAu) at about 40 at. (13.9 wt.) % Ti, 1385°C; eutectic

(TiAu)-Ti_3Au at about **67** at. (33 wt.) % Ti, 1310°C; eutectic Ti_3Au-β [with 85 at. (57.9 wt.) % Ti] at about 79 at. (47.7 wt.) % Ti, 1367°C; (*c*) eutectoid $\beta_{Ti} \rightleftarrows Ti_3Au + \alpha_{Ti}$ at 95.6 at. (84.1 wt.) % Ti, 833°C; (*d*) solubility of Au in α-Ti decreases from 1.7 at. (6.6 wt.) % Au at 833°C to about 0.7 at. (2.8 wt.) % Au at 600°C.

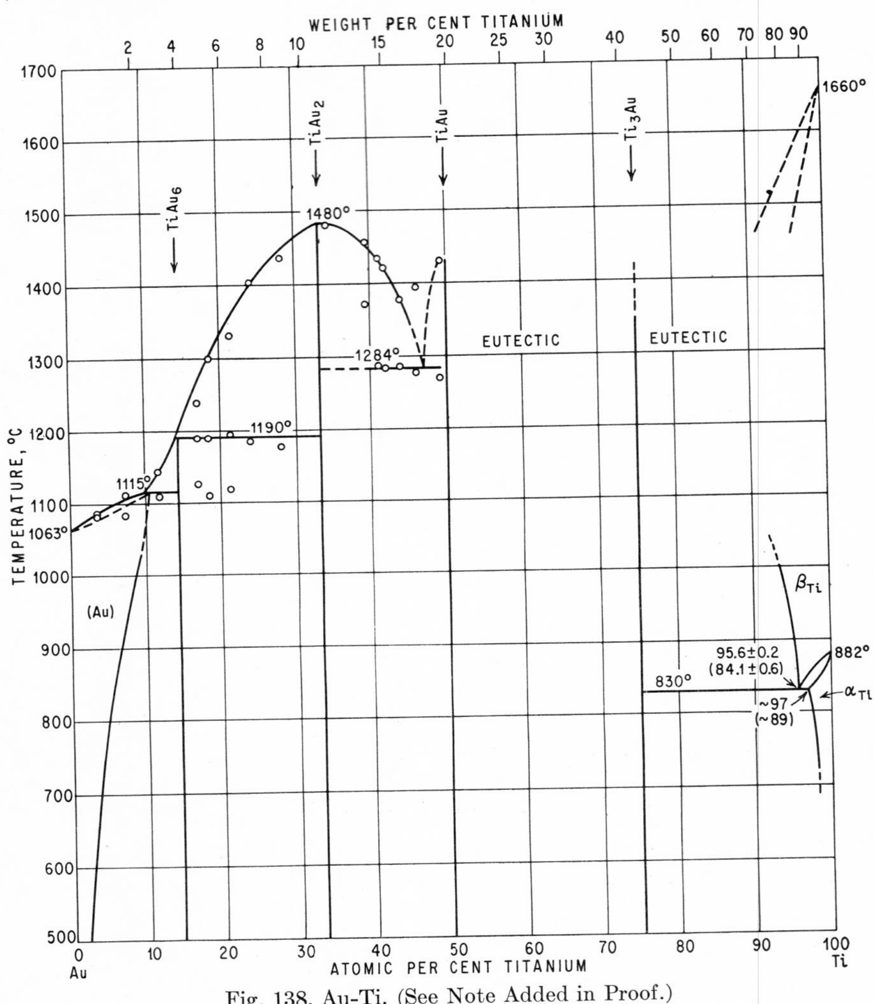

Fig. 138. Au-Ti. (See Note Added in Proof.)

1. E. Raub, P. Walter, and M. Engel, *Z. Metallkunde*, **43**, 1952, 112–118.
2. H. J. Wallbaum, *Naturwissenschaften*, **31**, 1943, 91–92.
3. Work by Materials Section, Jet Propulsion Laboratory, California Institute of Technology, 1953-1954; quoted from WADC Technical Report 54-502, pp. 8–10, September, 1954.
4. M. K. McQuillan, *J. Inst. Metals*, **82**, 1953-1954, 511–512.
5. P. Farrar, senior thesis, College of Engineering, New York University, New York, 1954; quoted from WADC Technical Report 54-502, pp. 8–10, September, 1954.

6. F. Laves and H. J. Wallbaum, *Naturwissenschaften*, **27**, 1939, 674–675.
7. P. Duwez and C. B. Jordan, *Acta Cryst.*, **5**, 1952, 213–214.
8. P. Pietrokowsky, E. P. Frink, and P. Duwez, *Trans. AIME*, **206**, 1956, 930–935.

$\bar{1}.9844$
0.0156

Au-Tl Gold-Thallium

The liquidus curve and eutectic shown in Fig. 139 are based on thermal analysis of only six alloys [1]. Earlier, [2] had reported that Au additions of up to 6.9 at. % Au continually lower the melting point of Tl (~98% pure), given as 301°C, to 261°C. [3] located the eutectic point at 71.55 wt. (70.8 at.) % Tl, 142°C.

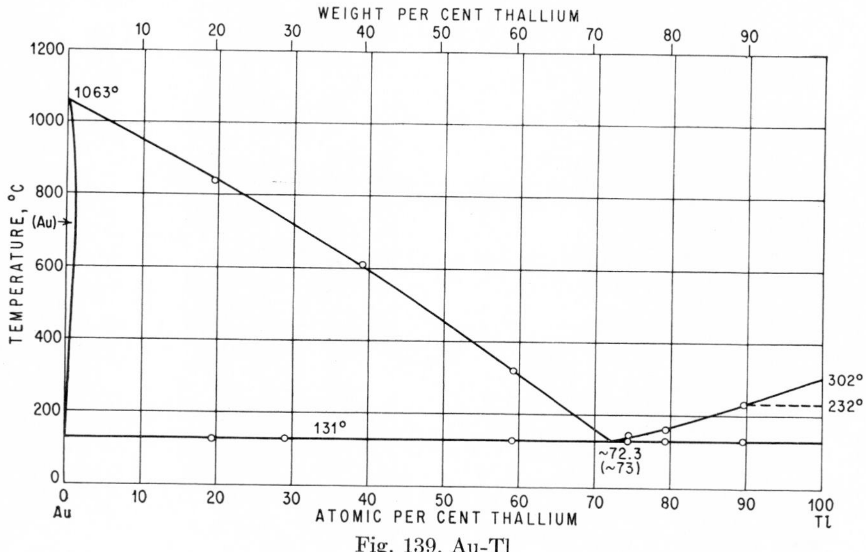

Fig. 139. Au-Tl

The solid solubility of Tl in Au was determined parametrically and found to be in at. % and wt. % Tl as follows: 0.4_8, 0.7_5, 0.7_9, 0.9_0, 0.8_2, 0.7_1, 0.5_1, 0.3_5, 0.2_1, and 0.1_7 at 1040, 1000, 900, 800, 600, 500, 450, 400, 300, and 200°C, respectively, indicating that the saturation curve is retrograde, with the maximum solubility occurring at about 800°C [4].

Lattice constants of the Au-rich terminal solid solution show a negative deviation from additivity [4].

1. M. Levin, *Z. anorg. Chem.*, **45**, 1905, 31–38.
2. C. T. Heycock and F. H. Neville, *J. Chem. Soc.*, **65**, 1894, 33.
3. O. Hajiček, *Hutnické Listy*, **3**, 1948, 265–270.
4. E. Raub and A. Engel, *Z. Metallkunde*, **37**, 1946, 76–81.

$\bar{1}.9182$
0.0818

Au-U Gold-Uranium

The phase diagram of Fig. 140 was established by [1] by correlation of thermal, X-ray, and microscopic data. The thermal data are plotted in the figure; the 76.1 at. % Au alloy was not melted at 1450°C, the limiting maximum temperature of the thermal-analysis furnace used. The interplanar spacings of the two existing compounds

U_2Au_3 (δ) (55.41 wt. % Au) and UAu_3 (ε) (71.30 wt. % Au) were listed, but the crystal structures were not evaluated. UAu_3 appears to be of low symmetry.

Previously it had been reported [2] that about 2 at. (1.7 wt.) % Au was soluble in α-U.

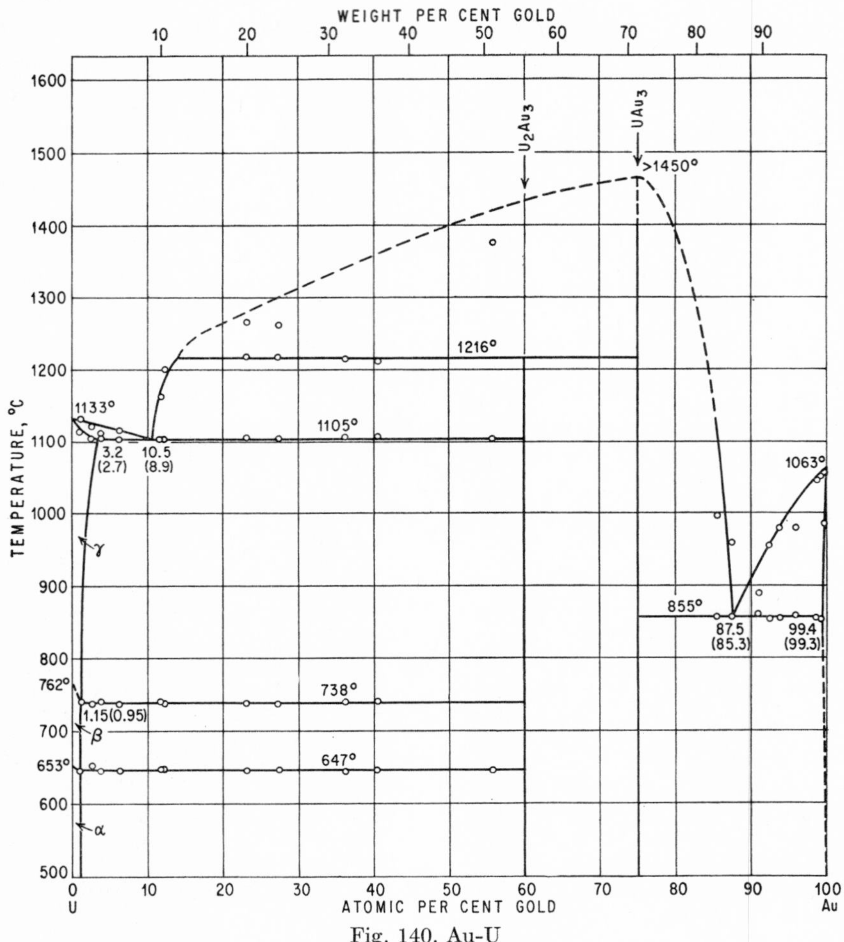

Fig. 140. Au-U

1. R. W. Buzzard and J. J. Park, *J. Research Natl. Bur. Standards*, **53**, 1954, 291–296.
2. J. J. Katz and E. Rabinowitch, "Chemistry of Uranium," Part I, Div. VIII, vol. 5, p. 175, National Nuclear Energy Series, McGraw-Hill Book Company, Inc., New York, 1951.

0.5878
.4122

Au-V Gold-Vanadium

The solid solubility of V in Au in the temperature range 500–1008°C was determined by lattice-parameter measurements [1]. The maximum solubility was interpolated to be about 17.5 at. (5.2 wt.) % V at about 970°C (Fig. 141).

1. D. Summers-Smith, *J. Inst. Metals*, **83**, 1954-1955, 189–190.

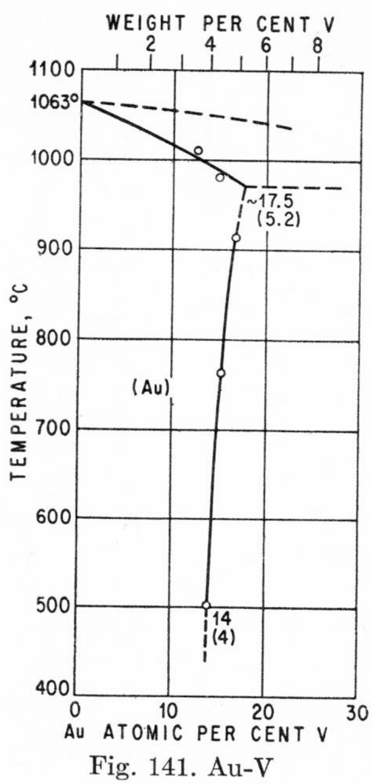

Fig. 141. Au-V

0.4795
$\bar{1}.5205$

Au-Zn Gold-Zinc

A comprehensive investigation of this system by thermal and micrographic analysis as well as measurements of electrical conductivity as a function of composition and temperature is due to [1]. According to an earlier thermal and microscopic study [2], the liquidus and solidus curves lie at considerably higher temperatures (15–30°C and more). This is best indicated by Table 14, which shows the temperatures in °C of the three-phase equilibria. In Fig. 142 the data by [1] were adopted to maintain, as far as possible, the unity of this much more detailed investigation, which, besides, included the solid-state transformations not found by [2].

Table 14. Temperatures in °C of the Three-phase Equilibria

	Ref. 1	Ref. 2	Ref. 3
Eutectic: melt $\rightleftarrows \alpha + \beta'$	642	672	
Eutectic: melt $\rightleftarrows \beta' + \gamma$	626	651	
Peritectic: melt $+ \gamma \rightleftarrows \epsilon$	475	490	
Peritectic: melt $+ \epsilon \rightleftarrows \eta$	423	438	432

The transformations in the α phase—425 and 270°C at the composition Au_3Zn (9.95 wt. % Zn) according to the measurements of conductivity of [1]—were confirmed by measurements of the modulus of elasticity and damping capacity (at 420 and 275°C) [4] and of the thermal dilatation (at 415 and 260°C) [5]. According to X-ray studies by [6], the solid solubility of Zn in Au at 500°C lies between 29.5 and 33.6 at. (12.0 and 14.3 wt.) % Zn, in good agreement with the findings of [1].

In the β' range, the isotherms of the electrical conductivity have a sharp maximum at the composition AuZn (24.90 wt. % Zn), even at temperatures close to the melting point, indicating that the ordered structure is stable up to this temperature [1]. The limits of the phase, given by [1] as 38 and 57 at. (17 and 30.5 wt.) % Zn "at room temperature," were found by microscopic examination to be approximately 48 and 52.5 at. (23.4 and 26.8 wt.) % Zn after "slow cooling" [7] (Fig. 142).

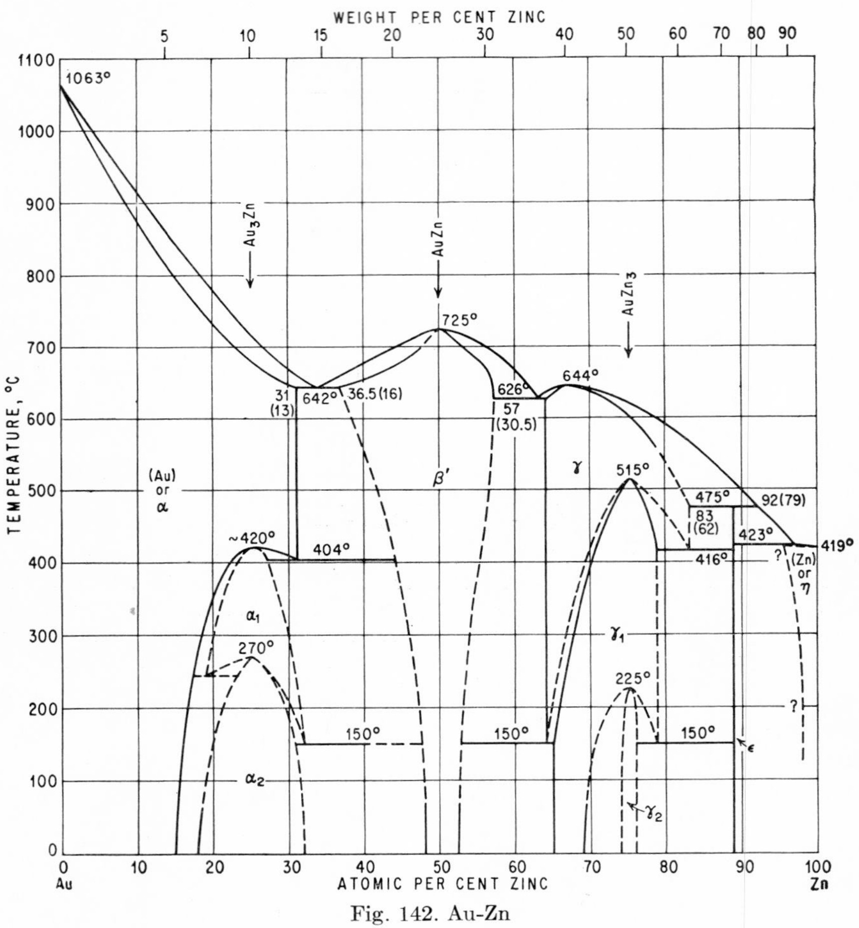

Fig. 142. Au-Zn

The temperatures of the transformations in the γ phase reach maxima at 515 and 225°C at the composition corresponding to the formula $AuZn_3$ (49.86 wt. % Zn) [1]. The upper transformation was also observed (at 520°C) by [4] (modulus of elasticity). The γ_1 field extends at room temperature from 65 to 69 at. (38 to 42.5 wt.) % Zn [1, 8]. For the extent of the γ_2 field, the values 73–78 at. % [1] and 75 ± 0.5 at. % Zn [8] were reported; the values chosen for Fig. 142 are 74–76 at. % Zn. For magnetic measurements in the composition range of the γ phases, see [9].

According to [1], the homogeneity range of the ϵ phase is very narrow (88.8–89.0 at. % Zn); [8], however, reported the boundaries 84 and 89 at. % Zn.

The solubility of Au in solid Zn has not yet been reliably determined. For "room temperature" the values 3 and 2 at. %, respectively, have been reported by [1] and [8].

Effects found on curves of electrical resistance vs. temperature of Zn-rich alloys have been attributed to a solid-state transformation of Zn [1]; such a transformation, however, does not exist [10].

Crystal Structures. Lattice parameters of the f.c.c. (Au) or α phase were determined by [6] on alloys quenched from 500°C. By quenching the alloy Au_3Zn from 450°C a metastable superstructure of the f.c.c. lattice is formed (α' phase) [4].

According to X-ray investigations at elevated temperatures [5], the α_1 phase has a tetragonal f.c. lattice with 4 atoms per unit cell and $a = 4.026$ "10^{-8} cm," $c = 4.107$ "10^{-8} cm" (probably kX), $c/a = 1.020$ at 300°C. [11] found, however, that this cell is only a substructure cell, the true structure being derived from the $L1_2$ (Cu_3Au) type by the regular disposition of atomic ordering faults (*Verschiebungsebenen*). See also [18].

The structure of the α_2 phase is not known with certainty as yet. Two different tetragonal f.c. unit cells were suggested with $a = 3.948$ "10^{-8} cm," $c = 8.306$ "10^{-8} cm" (probably kX), $c/2{:}a = 1.052$ [5], and $a = 4.023$ kX, $c = 4.144$ kX, $c/a = 1.05$ [4].

The β' phase has the structure of the CsCl (B2) type [8, 12, 13, 7], $a = 3.128$ A at 49.5 at. % Zn [7]. Lattice parameters over the whole range of homogeneity at room temperature were given by [7]. There is ample evidence that the ordered structure is at least partially retained up to temperatures close to the melting point [1, 13, 7, 14].

The phases γ, γ_1, and γ_2 probably all have cubic structures [8]. For the γ phase, which can be retained on quenching, a unit cell with 90 atoms and $a = 11.19$ A was suggested. γ_1 has the γ-brass structure ($D8_{1-3}$ type), with $a = 9.241$ A at 68.1 at. % Zn [8]; see also [15, 16]. The striking displacement of the homogeneity range to higher Zn contents—the composition corresponding to the electron: atom ratio of 21:13 lies in the ($\beta' + \gamma_1$) field—is attributed to the high stability of the β' phase and the formation of a defect structure [17]. The γ_2 phase probably has a cubic cell, with 32 atoms and $a = 7.90$ A at 75.4 at. % Zn [8].

The structure of the ϵ phase is h.c.p. (A3 type), with $a = 2.813$ A, $c = 4.378$ A, $c/a = 1.555$ at 88.8 at. % Zn [8]. The random atomic distribution, at temperatures close to the melting point, is confirmed by measurements of the melting entropy [14]. It is remarkable that the ϵ phase, presumably a 7:4 electron compound, is also displaced to much higher Zn contents.

According to [8], the parameter a of the hexagonal Zn lattice is raised and c is lowered by addition of Au.

1. P. Saldau, *J. Inst. Metals*, **30,** 1923, 351–400; *Z. anorg. Chem.*, **141,** 1925, 325–362.
2. R. Vogel, *Z. anorg. Chem.*, **48,** 1906, 319–332.
3. C. T. Heycock and F. H. Neville, *J. Chem. Soc.*, **71,** 1897, 419.
4. W. Köster, *Z. Metallkunde*, **32,** 1940, 151–156.
5. E. Raub, P. Walter, and A. Engel, *Z. Metallkunde*, **40,** 1949, 401–405.
6. E. A. Owen and E. A. O'Donnell Roberts, *J. Inst. Metals*, **71,** 1945, 213–254.
7. N. V. Ageev and D. N. Shoykhet, *Izvest. Sektora Fiz.-Khim. Anal.*, **13,** 1940, 165–170 (in Russian).
8. A. Westgren and G. Phragmén, *Phil. Mag.*, **50,** 1925, 311–341; *Strukturbericht*, **1,** 1913-1928, 559–560.
9. H. Nowotny and H. Bittner, *Monatsh. Chem.*, **81,** 1950, 887–906.
10. W. H. Pierce, E. A. Anderson, and P. van Dyk, *J. Franklin Inst.*, **200,** 1925, 349; J. R. Freeman, P. F. Brandt, and F. Sillers, *Sci. Papers Bur. Standards*, no. **522,** 1926; F. Simon and E. Vohsen, *Z. physik. Chem.*, **133,** 1928, 165–187.

11. K. Schubert, B. Kiefer, and M. Wilkens, *Z. Naturforsch.*, **9a**, 1954, 987–988.
12. E. A. Owen and G. D. Preston, *Phil. Mag.*, **2**, 1926, 1266–1270.
13. E. A. Owen and J. G. Edmunds, *Proc. Phys. Soc. (London)*, **50**, 1938, 389–397.
14. O. Kubaschewski, *Z. physik. Chem.*, **192**, 1943, 292–308.
15. A. J. Bradley and J. Thewlis, *Proc. Roy. Soc. (London)*, **A112**, 1926, 678–692.
16. A. J. Bradley and C. H. Gregory, *Phil. Mag.*, **12**, 1931, 143–162.
17. W. Hume-Rothery, J. O. Betterton, and J. Reynolds, *J. Inst. Metals*, **80**, 1952, 609–616.
18. K. Schubert, B. Kiefer, M. Wilkens, and R. Haufler, *Z. Metallkunde*, **46**, 1955, 692–715.

0.3348
$\bar{1}.6652$

Au-Zr Gold-Zirconium

The system was studied by thermal analysis up to 45.5 at. (27.8 wt.) % Zr; however, data are reliable only up to about 23 at. % Zr because of quite heavy reaction

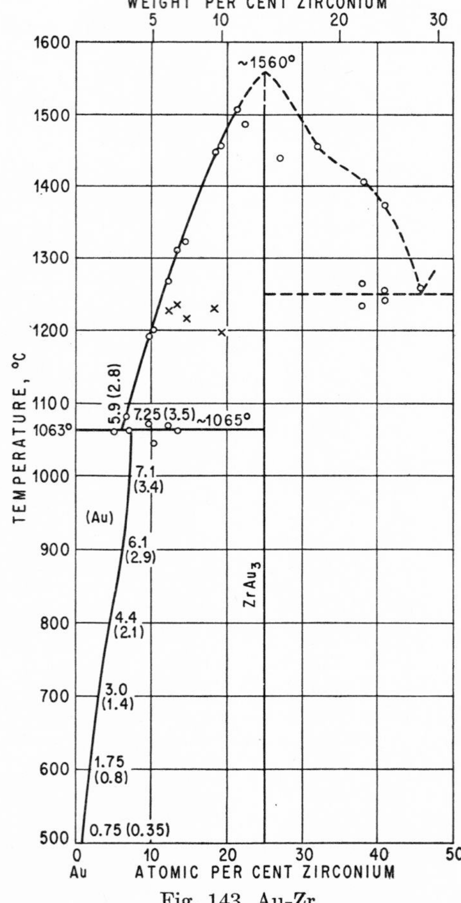

Fig. 143. Au-Zr

of the melts with the various crucible materials used. The presence of reaction products was also the reason why, between 12 and 19 at. % Zr, additional thermal effects were observed at a temperature between 1196 and 1235°C (crosses in Fig. 143).

The solid solubility of Zr in Au was determined roentgenographically and micrographically. Because of rapid precipitation of $ZrAu_3$ from the supersaturated solid solution on quenching and other reasons, the maximum solubility at 1065°C cannot be considered fully reliable. $ZrAu_3$ (13.36 wt. % Zr) is practically of singular composition [1].

Lattice parameters of the Au-rich solid solution show a negative deviation from additivity [1].

1. E. Raub and M. Engel, *Z. Metallkunde*, **39**, 1948, 172–177.

$\bar{2}$.8964
1.1036

B-Ba Boron-Barium

Borides of Ba, Ca, Ce, Er, Gd, La, Nd, Pr, Sr, Th, Y, and Yb having the composition XB_6 (85.71 at. % B) are all isostructural. Their structure is of the CaB_6 ($D2_1$) type, cubic with 1 formula weight per unit cell. The parameters are given in Table 15 [1, 2]; see also [3].

Table 15. Parameters of Isostructural Borides

Compound	Wt. % B	Parameter, A	Compound	Wt. % B	Parameter, A
CaB_6*	61.83	4.16 [1] 4.11 [2]	PrB_6	31.54	4.129 [1]
SrB_6	42.56	4.20 [1] 4.20 [2]	NdB_6	31.03	4.126 [1] 4.08 [2]
BaB_6	32.09	4.29 [1] 4.39 [2]	GdB_6	29.27	4.13 [2]
YB_6	42.20	4.08 [2]	ErB_6	27.97	4.110 [1] 4.06 [2]
LaB_6	31.85	4.153 [1] 4.17 [2]	YbB_6	27.28	4.14 [2]
CeB_6	31.66	4.129 [1] 4.13 [2]	ThB_6	21.86	4.16 [2] 4.33 [1]

* [4] found a = 4.153 A.

Supplement. [5] redetermined the lattice parameter of BaB_6: $a = 4.268_0$ A.

1. M. v. Stackelberg and F. Neumann, *Z. physik. Chem.*, **B12**, 1932, 314–320; see also F. Laves, *Z. physik. Chem.*, **B22**, 1933, 114–116.
2. G. A. Allard, *Bull. soc. chim. France*, **51**, 1932, 1213–1215.
3. R. Kiessling, *Acta Chem. Scand.*, **4**, 1950, 209–227.
4. L. Pauling and S. Weinbaum, *Z. Krist.*, **87**, 1934, 181–182.
5. P. Blum and F. Bertaut, *Acta Cryst.*, **7**, 1954, 81–86.

$\bar{2}$.7141
1.2859

B-Bi Boron-Bismuth

[1] added "amorphous" boron to molten Bi, Cu, Pb, Sn, and Tl and heated the melts in a magnesia tube to 1500–1600°C in a hydrogen stream. "In all cases the

melting point of the metals was not noticeably changed after heating and the boron was hardly wetted by the metals. Microscopically no second constituent could be detected. Therefore, amorphous boron apparently does not dissolve in the metals mentioned even at 1500–1600°C." Also, [2] found that B does not react with Bi and Cu at very high temperatures.

1. H. Giebelhausen, *Z. anorg. Chem.*, **91**, 1915, 261–262.
2. S. A. Tucker and H. R. Moody, *Proc. Chem. Soc., London*, **17**, 1901, 129–130; *J. Chem. Soc.*, **81**, 1902, 17.

$\bar{1}.4313$
0.5687

B-Ca Boron-Calcium

See B-Ba, page 245.

Supplement. The lattice parameter of CaB_6 has been redetermined by [1] (a = 4.15 A) and [2] (a = 4.145_0 A).

1. N. V. Belov and V. I. Mokeeva, *Trudy Inst. Krist., Akad. Nauk S.S.S.R.*, no. 5, 1949, 13–68; *Chem. Abstr.*, **47**, 1953, 3648.
2. P. Blum and F. Bertaut, *Acta Cryst.*, **7**, 1954, 81–86.

$\bar{2}.8877$
1.1123

B-Ce Boron-Cerium

Concerning the boride CeB_6, see B-Ba, page 245. The system was studied by X-ray diffraction methods, and the existence and structure of CeB_6 were confirmed, a = 4.139 A [1]. The only other boride found was the tetragonal phase CeB_4 (23.60 wt. % B), which is isostructural with ThB_4 and ZrB_4 (a = 7.205 A, c = 4.090 A, c/a = 0.568) [2, 1] and in equilibrium with Ce [1]. There are indications that (a) the Ce-CeB_4 eutectic lies close to pure Ce and (b) a peritectic liq. + $CeB_6 \rightleftarrows CeB_4$ lies at a temperature much higher than 2000°C [1].

Supplement. The lattice parameter of CeB_6 has been redetermined by [3]: a = 4.141_0 A.

1. L. Brewer, D. L. Sawyer, D. H. Templeton, and C. H. Dauben, *J. Am. Ceram. Soc.*, **34**, 1951, 173–179.
2. A. Zalkin and D. H. Templeton, *Acta Cryst.*, **6**, 1953, 269–272; *J. Chem. Phys.*, **18**, 1950, 391.
3. P. Blum and F. Bertaut, *Acta Cryst.*, **7**, 1954, 81–86.

$\bar{1}.2638$
0.7362

B-Co Boron-Cobalt

The borides Co_2B (8.41 wt. % B) and CoB (15.51 wt. % B) were identified by determining their structures [1], after their existence had already been claimed by [2] and [3], respectively. The phase diagram of Fig. 144 is due to [4]; results of another thermal study [5] were not available. The eutectic point was located at 5.5 wt. (about 24 at.) % B [4] and 3.9 wt. (about 18 at.) % B [5], respectively [6]. The solid solubility of B in Co was estimated to be approximately 1 wt. (5.2 at.) % B [4] at the eutectic temperature. The temperature of the polymorphic transformation of Co (about 420°C) is lowered to 360°C on heating and to 260°C on cooling [4]. Co_2B has a Curie point at 510°C [4].

Co_2B is b.c. tetragonal of the $CuAl_2$ (C16) type, a = 5.016 A, c = 4.220 A, c/a = 0.84; and CoB is orthorhombic of the FeB (B27) type, a = 5.253 A, b = 3.043 A, c = 3.956 A [1].

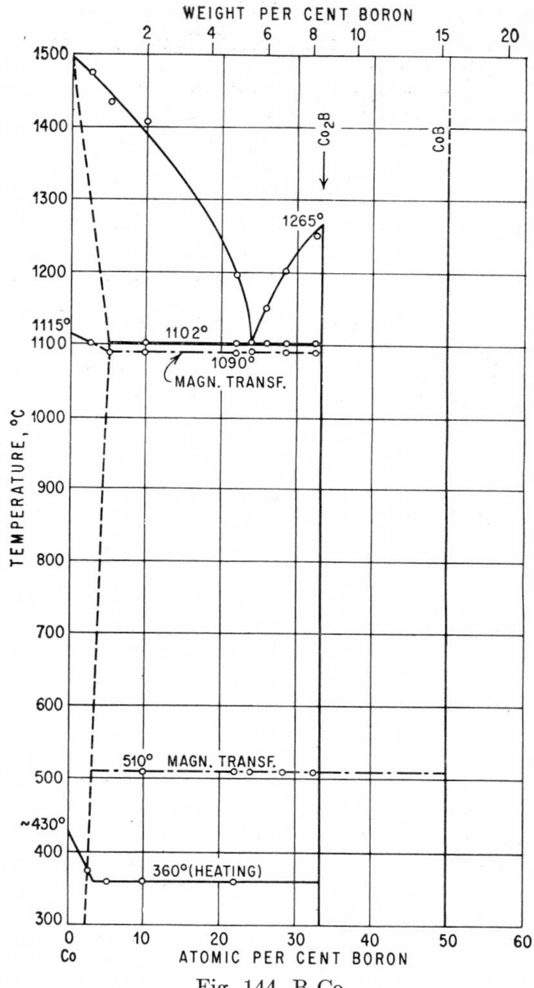

Fig. 144. B-Co

1. T. Bjurström, *Arkiv Kemi, Mineral. Geol.*, **A11**, 1933, no. 5, 1–12.
2. H. Moissan, *Compt. rend.*, **122,** 1896, 424.
3. A. Binet du Jassonneix, *Compt. rend.*, **145,** 1907, 240–241.
4. W. Köster and W. Mulfinger, *Z. Metallkunde*, **30,** 1938, 348–350.
5. N. P. Chizherskiy and B. A. Shmelev, *Trudy Moskov. Inst. Stali im. I. V. Stalina*, **17,** 1940, 3–39; *Met. Abstr.*, **10,** 1943, 277. The system was studied up to 9.9 wt. (about 37.5 at.)% B.
6. According to [3] the eutectic point would lie below about 5 wt. % B.

$\bar{1}.3181$
0.6819

B-Cr Boron-Chromium

By X-ray investigations [1, 2] the borides Cr_2B (9.42 wt. % B), Cr_3B_2 (12.18 wt. % B), CrB (17.22 wt. % B), Cr_3B_4 (21.72 wt. % B), and CrB_2 (29.38 wt. % B) were identified [3]. Results disagree with those by [4] in so far as the latter had concluded from X-ray phase analysis that only one phase, CrB, exists in the range 39.5–54.5 at. % B; i.e., Cr_3B_2 would be missing.

Cr_2B is orthorhombic (or very nearly orthorhombic), a = 14.7 A, b = 7.34 A, c = 4.29 A [2]. CrB is orthorhombic [1, 5] with a = 2.969 A, b = 7.858 A, c = 2.932 A [1] or a = 2.96 A, b = 7.81 A, c = 2.94 A [5]. Cr_3B_4 is isostructural with Mn_3B_4 and Ta_3B_4; the axes of the orthorhombic cell are a = 2.984 A, b = 13.02 A, c = 2.953 A [2]. CrB_2 is hexagonal (C32 type) with a = 2.969 A, c = 3.066 A, c/a = 1.03 [1]. The homogeneity range of CrB and CrB_2 is reported to be narrow [1]. The solid solubility of B in Cr appears to be very low [1].

The melting points of CrB and CrB_2 were reported to be 1550 ± 50 and 1850 ± 50°C, respectively [6].

Supplement. [7] obtained several Cr borides by electrolysis of salt melts containing borates and Cr oxide. The following compounds were isolated and identified by chemical as well as X-ray analysis: Cr_4B, Cr_2B, Cr_5B_3, CrB, and Cr_3B_4. Cr_4B and Cr_5B_3 had not previously been isolated. "Cr_3B_2" was found to be a mixture of Cr_2B and CrB.

[8] analyzed the structure of the borides Cr_4B, Cr_2B, and Cr_5B_3. Cr_4B is isomorphous with orthorhombic Mn_4B, a = 4.26_2 A, b = 7.38_2 A, c = 14.71 A; Cr_2B has the tetragonal $CuAl_2$ (C16) type of structure, a = 5.18_0 A, c = 4.31_6 A; Cr_5B_3 was found to be tetragonal, a = 5.46 A, c = 10.64 A.

[9] determined the melting point of CrB_2 as 1900°C (cf. above).

1. R. Kiessling, *Acta Chem. Scand.*, **3,** 1949, 595–602.
2. L.-H. Andersson and R. Kiessling, *Acta Chem. Scand.*, **4,** 1950, 160–164.
3. H. Moissan (*Compt. rend.*, **119,** 1894, 185), S. A. Tucker and H. R. Moody (*J. Chem. Soc.*, **81,** 1902, 14–17), A. Binet du Jassonneix (*Compt. rend.*, **143,** 1906, 897–899, 1149–1151), and E. Wedekind and K. Fetzer (*Ber. deut. chem. Ges.*, **40,** 1907, 297–301) had already concluded from chemical studies that the compound CrB exists. However, they did not prove the homogeneity of a product of this composition. Binet du Jassonneix also reported the existence of CrB_2.
4. S. J. Sindeband, *Trans. AIME*, **185,** 1949, 198–202.
5. A. J. Frueh, *Acta Cryst.*, **4,** 1951, 66–67; see also [4].
6. R. Kieffer, F. Benesovsky, and E. R. Honak, *Z. anorg. Chem.*, **268,** 1952, 191–200.
7. J. L. Andrieux and S. Marion, *Compt. rend.*, **236,** 1953, 805–807; *Met. Abstr.*, **21,** 1954, 1006.
8. F. Bertaut and P. Blum, *Compt. rend.*, **236,** 1953, 1055–1056.
9. B. Post, F. W. Glaser, and D. Moskowitz, *Acta Met.*, **2,** 1954, 20–25, especially 23.

$\bar{1}$.2312
0.7688

B-Cu Boron-Copper

See B-Bi, page 245. Diffusion tests of B in Cu had a negative result [1].

Supplement. Recent work [2, 3] has shown that—in contrast to earlier findings—Cu and B do alloy. [3], who also worked out various methods of preparation of Cu-B alloys, established the phase diagram up to 2 wt. (11 at.) % B by means of thermal, microscopic, conductometric, hardness, and X-ray studies (Fig. 145). A eutectic was found at 2 wt. (10.7 at.) % B, 1060 ± 2°C; to higher B contents, the liquidus rises probably very steeply [4]. There exists a B-rich compound for which chemical analyses suggest the formula CuB_{22}. The complex powder pattern of this phase could not be indexed. The solubility of B in solid Cu is small; it decreases from 0.09 wt. (0.53 at.) % at 1060°C to 0.06 wt. (0.35 at.) % B at "room temperature."

1. W. Loskiewicz, *Przeglad Góriczo-Hutniczy,* **21,** 1929, 583–611; abstract in *J. Inst. Metals,* **47,** 1931, 516–517.
2. H. Silliman, U.S. Patent 2,195,433, 1943 (see [3]).
3. F. Lihl and O. Feischl, *Metall,* **8,** 1954, 11–19.
4. The solubility of B in liquid Cu at 1200°C is estimated to be about 2.2 wt. (11.7 at.) %.

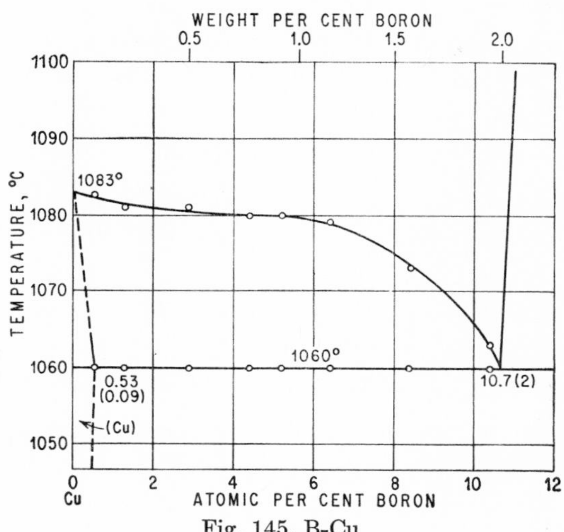

Fig. 145. B-Cu

$\bar{2}.8110$
1.1890

B-Er Boron-Erbium

See B-Ba, page 245.

$\bar{1}.2872$
0.7128

B-Fe Boron-Iron

The general features of the phase diagram have been established by thermal and microscopic investigations up to 8.5 wt. (about 32.5 at.) % B [1] and up to 11.5 wt. (about 41 at.) % B [2]. Results are in substantial agreement qualitatively but disagree as to temperatures and compositions of characteristic points. Both works agree especially in that (*a*) the solid solubility of B in γ-Fe increases with decrease in temperature and (*b*) the $\gamma \rightleftarrows \alpha$ transformation is lowered by B, giving rise to a eutectoid decomposition: $\gamma \rightleftarrows \alpha +$ boride (Fe_5B_2 [1], Fe_2B [2]) at 713 [1] and 760°C [2], respectively.

[3] have pointed out that the results by [1, 2] were in error and masked by the presence of appreciable amounts of C, Al, and Si as contaminants. The "idealized" diagram suggested on the basis of thermal, microscopic, and X-ray studies, using alloys much higher in purity [4] than those in the previous investigations, is represented in Fig. 146, except for the very low boron range and the eutectic temperature, which are based on more recent work (see below). The solid solubility of B in δ-, γ-, and α-iron was found to lie in the range 0.1–0.15 wt. % B, with the maximal solubility in γ-Fe being probably slightly smaller than in δ- and α-iron. The $\gamma \rightleftarrows \alpha$ transformation was shown to be raised to a peritectoid equilibrium at 915°C. The intermediate

phases were identified as Fe_2B (8.83 wt. % B) and FeB (16.23 wt. % B), thus confirming earlier results [5, 6].

The solid solubility of B in α-Fe and γ-Fe was the subject of several more recent investigations [7–11]. Whereas [7] was unable to detect any solubility of B in α-Fe with the aid of X-ray diffraction studies, [8–11] used more sensitive methods, i.e., essentially chemical and spectroscopic analysis of saturated solutions obtained by diffusion at different temperatures. Alloys were of high purity.

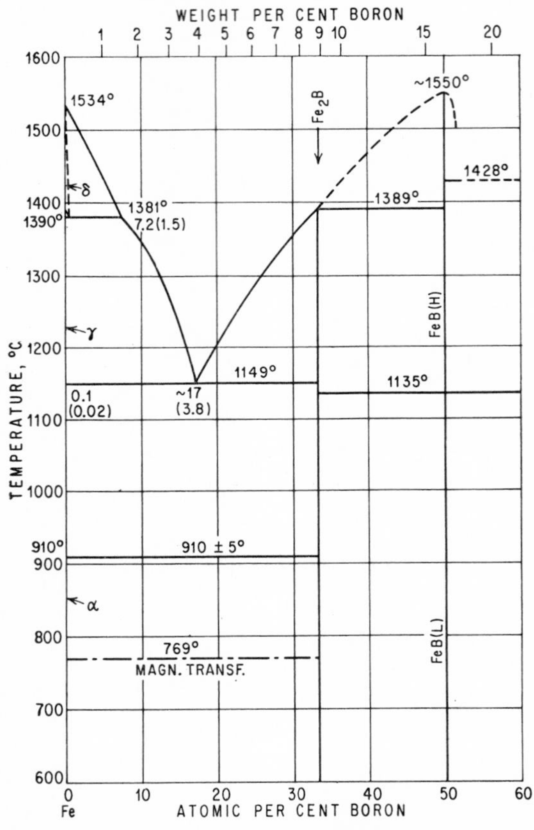

Fig. 146. B-Fe. (See also Fig. 147.)

The solubility boundaries as determined by [9–11] are summarized in Fig. 147 (in wt. %). Considering the experimental difficulties and very low absolute solubilities, results agree fairly well, although the solubility at 1100°C according to [9] is nearly twice as large as that found by [11]. A special feature of the diagram proposed by [10] is the minimum in the γ field at approximately 0.001 wt. % B and 835°C, deduced from microscopic and thermal data. It is stated, however, that this portion of the diagram "should be regarded as only approximately correct" [10].

Crystal Structures. Fe_2B is b.c. tetragonal of the $CuAl_2$ (C16) type [3, 12, 13], with a = 5.109 A, c = 4.249 A, c/a = 0.832 [13]; see also the interpretation by [14]. FeB exists in two modifications [3]. The low-temperature form, tentatively suggested [3] to be tetragonal, is orthorhombic of the B27 type, with 4FeB per unit cell and

$a = 5.506$ A, $b = 4.061$ A, $c = 2.952$ A [13]; see also the structural interpretation by [15] and [16]. Both borides are of singular composition.

Supplement. On the basis of diffusion data, [17] suggest that B forms a substitutional solid solution in α-Fe and an interstitial solid solution in γ-Fe. The same tentative conclusion had been advanced by [10].

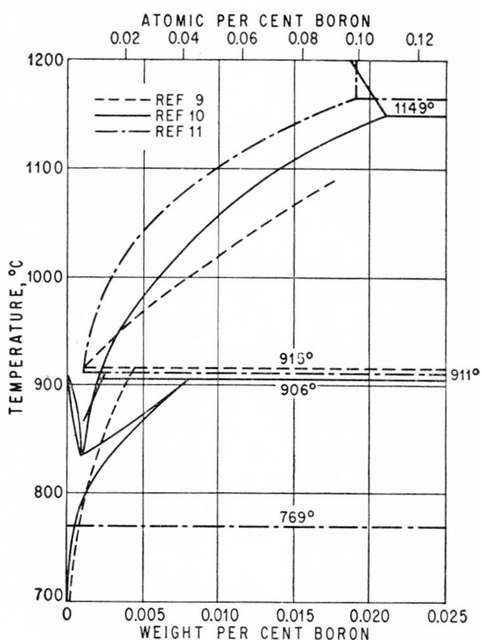

Fig. 147. B-Fe. (See also Fig. 146.)

1. G. Hannesen, *Z. anorg. Chem.*, **88**, 1914, 257–278.
2. N. Tschischewski and A. Herdt, *Zhur. Russ. Met. Obshchestva,* **1,** 1915, 533; abstracts in *Iron Age,* **98,** 1916, 396; *Rev. mét., Extraits,* **14,** 1917, 21–26; *J. Iron Steel Inst.,* **96,** 1917, 451.
3. F. Wever and A. Müller, *Mitt. Kaiser-Wilhelm-Inst. Eisenforsch. Düsseldorf,* **11,** 1930, 193–218.
4. The alloys were prepared, using electrolytic iron and ferroboron with 30.4% B, 0.06% C, 0.89% Si, 0.62% Mn, and 4.30% Al, in magnesia crucibles under hydrogen. The C content, on the average, was smaller than 0.01%; the Si content increased from 0.05% at 0.19% B to 0.54% at 17% B; the Al content increased from 0.06% to above 2% (all percentages in wt. %).
5. A. Binet du Jassonneix, *Compt. rend.,* **145,** 1907, 121–123. (Fe_2B, FeB.)
6. H. Moissan, *Compt. rend.,* **120,** 1895, 173–177. (FeB.)
7. A. K. Shevelev, *Zhur. Tekh. Fiz.,* **18,** 1948, 99–104.
8. M. E. Nicholson, *J. Metals,* **4,** 1952, 148 (note); see also preliminary diagram in [10].
9. P. E. Busby, M. E. Warga, and C. Wells, *Trans. AIME,* **197,** 1953, 1463–1468.
10. C. C. McBride, J. W. Spretnak, and R. Speiser, *Trans. ASM,* **46,** 1954, 499–520.
11. M. E. Nicholson, *Trans. AIME,* **200,** 1954, 185–190.

12. F. Wever, *Z. tech. Physik*, **10,** 1929, 137–138.
13. T. Bjurström and H. Arnfelt, *Z. physik. Chem.*, **B4,** 1929, 469–474.
14. G. Hägg, *Z. physik. Chem.*, **B11,** 1930, 152–162; **12,** 1931, 413–414.
15. S. B. Hendricks and P. R. Kosting, *Z. Krist.*, **74,** 1930, 517–522.
16. T. Bjurström, *Arkiv Kemi, Mineral. Geol.*, **A11,** 1933, no. 5, 1–22.
17. P. E. Busby and C. Wells, *Trans. AIME*, **200,** 1954, 972.

$\bar{2}.8386$
1.1614

B-Gd Boron-Gadolinium

See B-Ba, page 245.

Supplement. [1] have redetermined the lattice parameter of GdB_6: $a = 4.112_3$ A.

1. P. Blum and F. Bertaut, *Acta Cryst.*, **7,** 1954, 81–86.

$\bar{2}.7824$
1.2176

B-Hf Boron-Hafnium

HfB (5.71 wt. % B) and HfB_2 (10.81 wt. % B) were identified by X-ray diffraction methods, the former having a rather wide range of homogeneity. HfB is f.c.c. of the NaCl (B1) type, $a = 4.62$ A; HfB_2 is hexagonal and isotypic with ZrB_2, $a = 3.141$ A, $c = 3.470$ A, $c/a = 1.105$ [1]. The melting point of HfB_2 is 3250 ± 100°C [1], that of a boride of undefined compositon about 3100°C [2].

Supplement. For a structural discussion of the diboride see [3]. In this paper the melting point of HfB_2 is listed as 3240°C.

1. F. W. Glaser, D. Moskowitz, and B. Post, *Trans. AIME*, **197,** 1953, 1119–1120.
2. K. Moers (K. Becker), *Z. anorg. Chem.*, **198,** 1931, 243–275.
3. B. Post, F. W. Glaser, and D. Moskowitz, *Acta Met.*, **2,** 1954, 20–25.

$\bar{2}.7484$
1.2516

B-Ir Boron-Iridium

The borides Ir_3B_2 (3.60 wt. % B), IrB (5.31 wt. % B), and IrB_2 (10.08 wt. % B) were reported to have been identified by X-ray investigation [1].

1. J. H. Buddery and A. J. E. Welch, *Nature*, **167,** 1951, 362.

$\bar{2}.8915$
1.1085

B-La Boron-Lanthanum

See B-Ba, page 245.

Supplement. [1] have redetermined the lattice parameter of LaB_6: $a = 4.156_6$ A.

1. P. Blum and F. Bertaut, *Acta Cryst.*, **7,** 1954, 81–86.

$\bar{1}.2944$
0.7056

B-Mn Boron-Manganese

The existence of the borides MnB and MnB_2 was reported repeatedly [1]; however, the homogeneity of preparations of these compositions was not proved. By X-ray investigation of the system the following phases were established: Mn_4B (4.69 wt. % B), Mn_2B (8.97 wt. % B), MnB (16.46 wt. % B), and Mn_3B_4 (20.80 wt. % B) [2]. Crystal-structure determinations, using single crystals, had the following results: Mn_4B, probably having an extended homogeneity range, is orthorhombic with $8Mn_4B$ per unit cell and $a = 14.53$ A, $b = 7.293$ A, $c = 4.209$ A. Mn_2B has the tetragonal cell of the $CuAl_2$ (C16) type, $a = 5.148$ A, $c = 4.208$ A, $c/a = 0.817$. MnB is orthorhombic of the FeB (B27) type, $a = 5.560$ A, $b = 2.977$ A, $c = 4.145$ A; the orthorhombic cell reported by [3] was not confirmed. Mn_3B_4 is orthorhombic and isostructural with Ta_3B_4, $a = 3.032$ A, $b = 12.86$ A, $c = 2.960$ A. Mn_2B, MnB, and Mn_3B_4 appear to be of singular composition. With preparations with more than about 70 at. % B some weak interferences were observed which did not belong to the Mn_3B_4 phase. "It was not possible to decide whether these interferences belonged to a new phase (MnB_2?) or to impurities" [2].

1. A. Binet du Jassonneix, *Compt. rend.*, **139**, 1904, 1209–1211; **142**, 1906, 1336–1338. E. Wedekind and K. Fetzer, *Ber. deut. chem. Ges.*, **38**, 1905, 1228–1232; **40**, 1907, 1264–1266. See also J. Hoffmann, *Z. anorg. Chem.*, **66**, 1910, 361–399; F. Heusler and E. Take, *Trans. Faraday Soc.*, **8**, 1912, 180; R. Ochsenfeld, *Ann. Physik*, **12**, 1932, 354.
2. R. Kiessling, *Acta Chem. Scand.*, **4**, 1950, 146–159.
3. R. Hocart and M. Fallot, *Compt. rend.*, **203**, 1936, 1062–1064.

$\bar{1}.0522$
0.9478

B-Mo Boron-Molybdenum

The following borides have been identified: Mo_2B (5.34 wt. % B) [1–5], Mo_3B_2 (6.99 wt. % B) [4, 5], MoB (10.13 wt. % B) [1–5], MoB_2 (18.40 wt. % B) [6, 4, 5], and Mo_2B_5 [71.43 at. (22.01 wt.) % B] [2–4]. However, neither MoB_2 nor Mo_2B_5 exists in the stoichiometric composition. In addition, a high-temperature modification of the monoboride, called β-MoB, has been shown to exist [4].

The phase diagram (Fig. 148) was outlined on the basis of X-ray studies of heat-treated samples and some thermal data [4]. The first version [4] was corrected later [7] because of some conflict with the phase rule. The melting and transformation temperatures (Fig. 148) are assumed to be accurate within ± 50°C. However, [5] have reported that they found higher temperatures for some of the three-phase equilibria: for the peritectic reaction liq. + $Mo_3B_2 \rightleftarrows Mo_2B$, about 2120 [5] vs. about 2000°C [4]; for the peritectic reaction liq. + MoB $\rightleftarrows Mo_3B_2$, about 2250 [5] vs. about 2070°C [4]; for the eutectoid decomposition of Mo_3B_2, about 1950 [5] vs. about 1850°C [4]. The melting points of Mo_2B and MoB_2 were reported by [8] to be 1850 ± 50 and 2250 ± 60°C, respectively. The solid solubility of B in Mo is apparently very low [2].

Crystal Structures. Mo_2B is tetragonal of the $CuAl_2$ (C16) type, $a = 5.543$ A, $c = 4.735$ A, $c/a = 0.854$ [2, 4]; see also [3]. Mo_3B_2, stable only above 1850–1950°C [4, 5], is tetragonal and isotypic with Cr_3B_2. The unit cell of the low-temperature form of MoB is tetragonal with 8MoB and $a = 3.110$ A, $c = 16.95$ A, $c/a = 5.45$ [2, 7, 4]. Its range of homogeneity is about 48.8–51.5 at. % B [2]. The high-temperature modification of MoB (β-MoB in Fig. 148) is orthorhombic and isotypic with CoB, NbB, and TaB, $a = 3.16$ A, $b = 8.61$ A, $c = 3.08$ A [4]. The formula MoB_2 is used in Fig. 148 to designate the high-temperature phase shown at approximately 70 at. % B because its structure is based on this ideal composition. It is hexagonal of the AlB_2 (C32) type, $a = 3.05$ A, $c = 3.113$ A, $c/a = 1.02$ [6] and $a = 3.06$ A, $c = 3.10$ A, $c/a = 1.01$ [4]. Mo_2B_5 is also a defect structure and stable only with an excess of Mo atoms; however, the structure is definitely based on the ideal composition Mo_2B_5 [2]. The range of homogeneity was given as about 68.2–70 at. % B [4]. It is rhombohedral with the hexagonal axes $a = 3.011$ A, $c = 20.93$ A, $c/a = 6.95$ [2, 4].

Supplement. Metallographic, X-ray, and melting-point observations on a few Mo-rich Mo-B alloys were made by [9]. The $CuAl_2$ type of structure for the compound richest in Mo, Mo_2B, was corroborated. The eutectic composition was estimated to be 2.75 wt. (20.0 at.) % B; optical pyrometry placed the eutectic temperature at 2180°C (see inset in Fig. 148). A 0.69 wt. % B alloy quenched from 2180°C was found to have a slightly lower lattice parameter than that of the same sample slowly cooled. "This indicates a low solubility [of B in Mo] and, unexpectedly, a substitutional type of solid solution. It is not considered, however, that the nature of this solid solution has been established with certainty" [9].

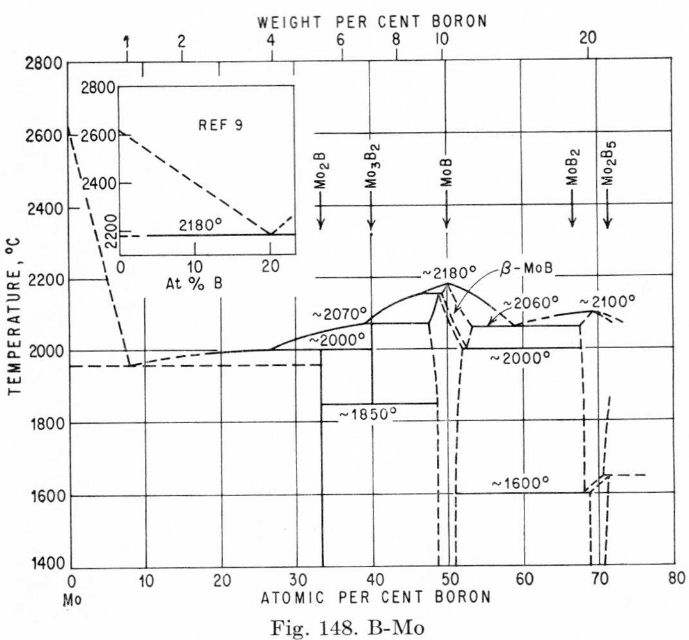

Fig. 148. B-Mo

According to [10], the electrical-resistivity-vs.-temperature curve of Mo_2B_5 showed an inflection at about 1600°C which coincides with the previously reported Mo_2B_5-MoB_2 transformation.

Vaporization studies of the Mo-B system were carried out by [11]. From the vapor-pressure–composition relationships, the stabilities and heats of formation of the various solid phases were established.

1. G. Weiss, *Ann. Chim.*, **1,** 1946, 446–525.
2. R. Kiessling, *Acta Chem. Scand.*, **1,** 1947, 893–916.
3. L. Brewer, D. L. Sawyer, D. H. Templeton, and C. H. Dauben, *J. Am. Ceram. Soc.*, **34,** 1951, 173–179.
4. R. Steinitz, I. Binder, and D. Moskovitz, *Trans. AIME*, **194,** 1952, 983–987.
5. P. W. Gilles and B. D. Pollock, *Trans. AIME*, **197,** 1953, 1537–1539.
6. F. Bertaut and P. Blum, *Acta Cryst.*, **4,** 1951, 72.
7. P. Rautala and R. Steinitz, *Trans. AIME*, **197,** 1953, 747.
8. R. Kieffer, F. Benesovsky, and E. R. Honak, *Z. anorg. Chem.*, **268,** 1952, 191–200.

9. Climax Molybdenum Company of Michigan, Arc-cast Molybdenum Base Alloys, pp. 94–99, Second Annual Report, 1951.
10. F. W. Glaser and D. Moskowitz, *Powder Met. Bull.*, **6**, 1953, 178–185.
11. P. W. Gilles and B. D. Pollock, *U.S. Atomic Energy Comm., Publ.* AECU-2894, 1954; *Met. Abstr.*, **22**, 1955, 651.

$\bar{1}.0662$
0.9338

B-Nb Boron-Niobium

By X-ray analysis, the existence of NbB (10.43 wt. % B) [1, 2], Nb_3B_4 (13.44 wt. % B) [1, 2], and NbB_2 (18.89 wt. % B) [3, 1, 2] was established [4]. No higher boride occurs [2]. In addition, there are two borides lower in boron [1, 2], NbB_m and NbB_n [2], tentatively designated as Nb_3B (3.74 wt. % B) and Nb_2B (5.50 wt. % B), respectively [5, 6]. The former is isotypic with TaB_m (Ta_3B?), but the latter has a structure different from that of TaB_n (Ta_2B?) [2]. NbB_n may have a primitive cubic lattice with a = 4.21 A [2]. It appears that NbB_m is not stable above 1930°C. Also, NbB_n and possibly Nb_3B_4 apparently have limited temperature ranges of stability, whereas NbB and NbB_2 are stable up to the melting point [2]. According to [6], NbB_2 decomposes at a temperature higher than 2900°C.

[1] found two phases, designated as β and β', in a preparation with about 10 at. % B, of which β' corresponds to a primitive cubic lattice with a = 4.210 A (TaB_n?). They concluded that β is stable at room temperature and β' at higher temperatures. [1] also reported a β'' phase to be present in quenched samples of 25–35 at. % B, which may be the same as NbB_m.

NbB, having a narrow homogeneity range, is orthorhombic and isotypic with CrB and TaB, with a = 3.298 A, b = 8.724 A, c = 3.137 A, according to [1], and a = 3.292 A, b = 8.713 A, c = 3.165 A, according to [2]. Nb_3B_4 is also orthorhombic and isotypic with Ta_3B_4 [1, 2], a = 3.305 A, b = 14.08 A, c = 3.137 A [1]. NbB_2 is hexagonal of the AlB_2 (C32) type [3, 2, 1]. It has a wide range of homogeneity, undetermined as yet [1, 2]; a = 3.110 A, c = 3.264 A, c/a = 1.05 when in equilibrium with Nb_3B_4, and a = 3.085 A, c = 3.311 A, c/a = 1.07 at the high boron side [2]. For the ideal composition, [1] gave a = 3.089 A, c = 3.303 A, c/a = 1.07.

Supplement. For a structural discussion of the diboride, see [7]; its melting point was found to be 3050°C.

1. L.-H. Andersson and R. Kiessling, *Acta Chem. Scand.*, **4**, 1950, 160–164.
2. L. Brewer, D. L. Sawyer, D. H. Templeton, and C. H. Dauben, *J. Am. Ceram. Soc.*, **34**, 1951, 173–179.
3. J. T. Norton, H. Blumenthal, and S. J. Sindeband, *Trans. AIME*, **185**, 1949, 749–751.
4. NbB_2 was first prepared by L. Andrieux, *Compt. rend.*, **189**, 1929, 1279.
5. P. Schwarzkopf and R. Kieffer, "Refractory Hard Metals," pp. 288–291, The Macmillan Company, New York, 1953.
6. F. W. Glaser, *Trans. AIME*, **194**, 1952, 391–396.
7. B. Post, F. W. Glaser, and D. Moskowitz, *Acta Met.*, **2**, 1954, 20–25.

$\bar{2}.8750$
1.1250

B-Nd Boron-Neodymium

See B-Ba, page 245.

Supplement. [1] have redetermined the lattice parameter of NdB_6: a = 4.128_4 A.

1. P. Blum and F. Bertaut, *Acta Cryst.*, **7**, 1954, 81–86.

$\bar{1}.2657$
0.7343

B-Ni Boron-Nickel

The phase diagram (Fig. 149), showing the existence of the borides Ni_2B (8.44 wt. % B) [1], Ni_3B_2 (10.95 wt. % B), NiB (15.57 wt. % B) [2], and Ni_2B_3? (21.66 wt. % B), is based on thermal-analysis work [3]. As tabulated data were missing, the data points had to be taken graphically from the original and converted into at. % [4]. Between the phases Ni_2B and Ni_3B_2, [3] found the solidification and melting processes to be irreversible, probably because of the formation of metastable conditions on cooling from the melt. Figure 149 represents the data found on heating after complete solidification.

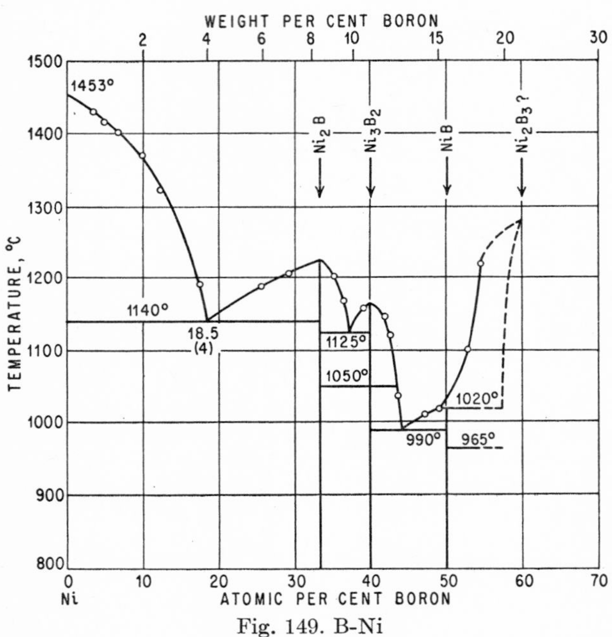

Fig. 149. B-Ni

X-ray analysis of alloys up to 20 wt. (57.6 at.) % B [5] indicated the existence of several phases; however, only Ni_2B could be identified by structure determination. Later X-ray work confirmed the presence of at least four intermediate phases [6, 7]: one with about 25–30 at. % B (Ni_3B?), Ni_2B, Ni_3B_2, and NiB [6]; the existence of NiB was confirmed by [7]. A phase with a lower B content than Ni_2B was not detected by [3] (Fig. 149).

Ni_2B is b.c. tetragonal of the $CuAl_2$ (C16) type, $a = 4.990$ A, $c = 4.245$ A, $c/a = 0.85$ [5]. NiB is orthorhombic of the CrB type, $a = 2.925$ A, $b = 7.396$ A, $c = 2.966$ A [7]. The powder photographs of Ni_3B (?) and Ni_3B_2 could not be interpreted [6].

1. Ni_2B had been isolated from an alloy with 5 wt. % B by A. Binet du Jassonneix, *Compt. rend.*, **145**, 1907, 240–241.
2. NiB was reported by H. Moissan (*Compt. rend.*, **122**, 1896, 424) to have been isolated from an alloy with about 10 wt. % B. According to Fig. 149 this is not possible.
3. H. Giebelhausen, *Z. anorg. Chem.*, **91**, 1915, 257–261.

4. Since melts with more than 27 wt. % B could not be prepared, [1] concluded that NiB_2 (26.94 wt. % B) might exist.
5. T. Bjurström, *Arkiv Kemi, Mineral. Geol.*, **11**, 1933, no. 5, 1–12.
6. L.-H. Andersson and R. Kiessling, *Acta Chem. Scand.*, **4**, 1950, 160–164.
7. P. Blum, *J. phys. radium*, **13**, 1952, 430–431.

$\bar{2}.7550$
1.2450

B-Os Boron-Osmium

The system was studied by X-ray analysis [1] over the full range of composition, with the result that the phases OsB (5.38 wt. % B), OsB_2 (10.22 wt. % B), and a phase of unknown composition richer in boron than OsB_2 were identified. OsB is cubic, $a = 7.04$ A [1].

1. J. H. Buddery and A. J. E. Welch, *Nature*, **167**, 1951, 362.

$\bar{2}.7178$
1.2822

B-Pb Boron-Lead

See B-Bi, page 245.

$\bar{1}.0061$
0.9939

B-Pd Boron-Palladium

According to [1], alloys with up to at least 7 at. (0.75 wt.) % B are single-phase and alloys with 13.8 and 16.6 at. (1.6 and 2.0 wt.) % B two-phase, after annealing at 700°C. By X-ray analysis, covering the full range of composition, a phase of the approximate composition Pd_3B_2 (?) (6.33 wt. % B) was detected [2]. It is hexagonal, $a = 6.49$ A, $c = 3.43$ A, $c/a = 0.529$ [2]. It has been reported that certain Pd-B alloys have low (eutectic) melting points [3].

1. A. Sieverts and K. Brüning, *Z. physik. Chem.*, **168**, 1934, 412.
2. J. H. Buddery and A. J. E. Welch, *Nature*, **167**, 1951, 362.
3. H. H. Kahlenberg, *Trans. Am. Electrochem. Soc.*, **47**, 1925, 23–63.

$\bar{2}.8853$
1.1147

B-Pr Boron-Praseodymium

See B-Ba, page 245.

$\bar{2}.7437$
1.2563

B-Pt Boron-Platinum

The formation of platinum borides was recognized by early workers [1, 2, 4], and the existence of PtB (?) (5.25 wt. % B) [4] and Pt_2B_3 (7.67 wt. % B) [3] was suggested without proof of their homogeneity. Also, it was observed that the melting point of Pt is considerably lowered by B additions [4].

By X-ray analysis, covering the full range of composition, an intermediate phase in the range 40–50 at. % B—Pt_3B_2 (3.56 wt. % B) or PtB—was found. The structure of PtB (?) was tentatively given as tetragonal, $a = 2.78$ A, $c = 2.96$ A, $c/a = 1.06$. Between Pt and this phase a "complex region" was observed [5].

1. H. Moissan, *Compt. rend.*, **114**, 1892, 320.
2. F. Wöhler and H. Sainte-Claire Deville, *Compt. rend.*, **43**, 1856, 1090.
3. C. A. Martius, *Liebigs Ann.*, **109**, 1859, 81.
4. A. Binet du Jassonneix, *Ann. chim. et phys.*, **17**, 1909, 212.
5. J. H. Buddery and A. J. E. Welch, *Nature*, **167**, 1951, 362.

$\bar{1}.0218$
0.9782

B-Rh Boron-Rhodium

By X-ray studies, covering the full range of composition, the phases Rh_2B (4.99 wt. % B), RhB (9.51 wt. % B), and a phase higher in boron content than RhB_2 were found to exist [1]. Between Rh_2B and RhB there appears to be an additional phase. Rh_2B is orthorhombic with 8 Rh atoms per unit cell and a = 5.42 A, b = 3.98 A, c = 7.44 A [2].

1. J. H. Buddery and A. J. E. Welch, *Nature,* **167,** 1951, 362.
2. R. W. Mooney and A. J. E. Welch, *Acta Cryst.,* **7,** 1954, 49–53.

$\bar{1}.0269$
0.9731

B-Ru Boron-Ruthenium

According to [1], the borides Ru_2B (5.05 wt. % B), RuB (9.62 wt. % B), Ru_2B_3 (13.76 wt. % B), and RuB_2 (17.55 wt. % B) were detected by X-ray studies covering the entire range of composition. RuB is cubic, a = 6.98 A.

Supplement. [2] prepared Ru_2B and stated that this compound yields a complex X-ray powder pattern.

1. J. H. Buddery and A. J. E. Welch, *Nature,* **167,** 1951, 362.
2. G. F. Hardy and J. K. Hulm, *Phys. Rev.,* **93,** 1954, 1004–1016, especially p. 1009.

$\bar{1}.5857$
0.4143

B-Si Boron-Silicon

The powder patterns of preparations with 50 at. (27.8 wt.) % B and 80 at. (60.6 wt.) % B, homogenized at 1725°C, showed "strong lines of pure Si with some weak lines probably due to B. No evidence of silicon boride was found" [1]. According to [2], the lattice of Si is contracted by boron additions, the solid solution being of the substitutional type. Alloys with more than 0.5 wt. (1.3 at.) % B were two-phase [2].

1. L. Brewer, D. L. Sawyer, D. H. Templeton, and C. H. Dauben, *J. Am. Ceram Soc.,* **34,** 1951, 173–179.
2. G. L. Pearson and J. Bardeen, *Phys. Rev.,* **75,** 1949, 867–868.

$\bar{2}.9598$
1.0402

B-Sn Boron-Tin

See B-Bi, page 245.

$\bar{1}.0916$
0.9084

B-Sr Boron-Strontium

See B-Ba, page 245.

Supplement. [1] have redetermined the lattice parameter of SrB_6: a = 4.198_4 A.

1. P. Blum and F. Bertaut, *Acta Cryst.,* **7,** 1954, 81–86.

$\bar{2}.7768$
1.2232

B-Ta Boron-Tantalum

After [1] had identified the boride TaB_2 (10.69 wt. % B) [2], the whole system was studied by X-ray analysis [3, 4]. [3] established the existence of Ta_2B (2.90 wt. % B),

TaB (5.64 wt. % B), Ta_3B_4 (7.39 wt. % B), and TaB_2. These results were confirmed by [4], who found an additional phase TaB_m, possibly Ta_3B (1.96 wt. % B) [5]. No phase higher in boron than TaB_2 appears to exist [4]. The melting point of TaB_2 was given as >3000°C [6].

There is an appreciable solid solubility of B in Ta at temperatures above 950°C, as indicated by the increase of the lattice parameter of Ta from a = 3.303 A to a = 3.321 A, after quenching from 1270°C. This was verified qualitatively by [4].

Ta_2B, supposed to exist only below a certain temperature, could not be obtained in a pure state; up to 14 at. % B, it occurs with (Ta) and above 14 at. % B with TaB [3]. [4] found Ta_2B and (Ta) in sintered samples with 20 and 25 at. % B, a mixture of Ta_2B, TaB, and (Ta) in samples with 33 and 40 at. % B, and Ta_2B and TaB in samples up to almost 50 at. % B. They were not able to show whether Ta_2B is stable only at high temperatures and decomposes on cooling or is formed by a peritectoid reaction, i.e., decomposes on heating. On the other hand, TaB_m (Ta_3B?) appears to be stable only at very high temperatures, as it was found in samples prepared by heating mixtures of Ta and Ta_2B to 1950°C and rapidly quenching [4].

TaB_m (Ta_3B) is isotypic with the corresponding phase in the system Nb-B [4]. Ta_2B is tetragonal of the $CuAl_2$ (C16) type, a = 5.778 A, c = 4.864 A, c/a = 0.842 [3]; a = 5.785 A, c = 4.867 A, c/a = 0.841 [4]. TaB, having a narrow range of homogeneity, is orthorhombic and isotypic with CrB (4TaB per unit cell), a = 3.276 A, b = 8.669 A, c = 3.157 A [3]. The parameters of Ta_3B_4, also orthorhombic (6 Ta atoms per unit cell), are a = 3.29 A, b = 14.0 A, c = 3.13 A [3]. The homogeneity range is narrow. TaB_2 has a wide range of homogeneity, extending from about 64 to 72 at. % B [3, 4]. It is hexagonal of the AlB_2 (C32) type [1, 3, 4]; a = 3.078 A, c = 3.265 A, c/a = 1.06 at the ideal composition [3]. The a axis decreases and the c axis increases from the low-boron limit to the high-boron limit of the range [3, 4].

Supplement. As for the homogeneity range and the lattice parameters of the diboride phase, [7] stated that their results were similar to those previously reported (cf. above). The melting point of TaB_2 was found to be 3200°C [7].

1. J. T. Norton, H. Blumenthal, and S. J. Sindeband, *Trans. AIME*, **185**, 1949, 749–751.
2. TaB_2 was already reported by L. Andrieux, *Compt. rend.*, **189**, 1929, 1279; and P. M. McKenna, *Ind. Eng. Chem.*, **28**, 1936, 767.
3. R. Kiessling, *Acta Chem. Scand.*, **3**, 1949, 603–615.
4. L. Brewer, D. L. Sawyer, D. H. Templeton, and C. H. Dauben, *J. Am. Ceram. Soc.*, **34**, 1951, 173–179.
5. J. W. Glaser, *Trans. AIME*, **194**, 1952, 391–396.
6. R. Kieffer, F. Benesovsky, and E. R. Honak, *Z. anorg. Chem.*, **268**, 1952, 191–200.
7. B. Post, F. W. Glaser, and D. Moskowitz, *Acta Met.*, **2**, 1954, 20–25.

$\bar{2}$.6685
1.3315

B-Th Boron-Thorium

As regards the boride ThB_6, already suggested by [1], see B-Ba, page 245. The only other boride existing is ThB_4 (15.72 wt. % B), identified by X-ray investigation [2, 3]. ThB_4 is probably the phase reported by [4] to be present in a preparation with 50 at. % B. Whereas [4] found indication of a certain solid solubility of B in Th, [2] stated that the pure metal coexists with ThB_4. There are indications that the Th-ThB_4 eutectic melts at about 1550°C and that the melting point of ThB_4 is considerably higher than 2500°C [2].

Powder patterns of preparations with 80 at. % B or less had, in addition to the

ThB_4 lines (observed between 40 and 80 at. % B), lines of a f.c.c. phase the parameter of which decreased from a = 5.655 A at 25 at. % B to a = 5.632 A at 50 at. % B. The strength of the lines decreased as the B content increased. It is believed [2] that this phase is not a thorium boride but a solid solution of ThO_2 (f.c.c., a = 5.598 A) with ThB_4. It is perhaps identical with the f.c.c. phase of unknown composition detected by [5] and supposed to be ThB_4.

ThB_4 is tetragonal with $4ThB_4$ per unit cell, a = 7.256 A, c = 4.113 A, c/a = 0.567 [3]. For the lattice parameter of ThB_6, see B-Ba, page 245. In addition, [6] found a = 4.113 A.

Supplement. [7] redetermined the lattice parameter of ThB_6: a = 4.113_2 A.

1. A. Binet du Jassonneix, *Compt. rend.*, **141,** 1905, 191–193; *Bull. soc. chim. France*, **35,** 1906, 278–280.
2. L. Brewer, D. L. Sawyer, D. H. Templeton, and C. H. Dauben, *J. Am. Ceram. Soc.*, **34,** 1951, 173–179.
3. A. Zalkin and D. H. Templeton, *Acta Cryst.*, **6,** 1953, 269–272; *J. Chem. Phys.*, **18,** 1950, 391.
4. L.-H. Andersson and R. Kiessling, *Acta Chem. Scand.*, **4,** 1950, 160–164.
5. G. Hägg, *Metallwirtschaft,* **10,** 1931, 387–390.
6. F. Bertaut and P. Blum, *Compt. rend.*, **234,** 1952, 2621.
7. P. Blum and F. Bertaut, *Acta Cryst.*, **7,** 1954, 81–86.

$\bar{1}.3539$
0.6461

B-Ti Boron-Titanium

The first systematic investigation, covering the range 2–40 wt. (8.3-74.5 at.) % B, was carried out by [1] by means of X-ray analysis of preparations obtained by sintering mixtures of the elements. The existence of two borides was claimed: TiB (18.43 wt. % B) of cubic ZnS (B3) type of structure and TiB_2 (31.12 wt. % B) with the hexagonal structure of the AlB_2 (C32) type and a certain range of homogeneity. Both phases had been reported earlier by [2] and [3, 4], respectively. Indications of another boride, richer in boron than TiB_2, were also found [1]. Recently, the occurrence of such a higher boride [5] was reported by [5, 6], and [7] identified the phase Ti_2B_5 [71.43 at. (36.09 wt.) % B].

In a study similar to that by [1], the existence of TiB could not be confirmed, and no phase with a B content higher than that of TiB_2 could be found [8]. Weak X-ray patterns of a f.c.c. phase (a = 4.25 A) were suggested to be due to the presence of TiN as a contaminant.

The existence of TiB, stable at room temperature, was also doubted by [9], who pointed out that the interstices of the zincblende structure do not provide sufficient space to accommodate B atoms. They suggested that the data for the TiB phase [1] are compatible with a f.c.c. NaCl-type structure. However, as even the octahedral interstices are too small for B atoms at room temperature, TiB would be stable only at higher temperatures and, under equilibrium conditions, decompose into the stable phases α-Ti and TiB_2. Later [7, 10] and especially [14] established the monoboride as a phase stable at room temperature.

The change of the lattice parameters of Ti-rich alloys would suggest that approximately 28.6 at. (8.3 wt.) % B is soluble in α-Ti [1]. However, [1] found that between 9.1 at. (2.2 wt.) % B and 44.4 at. (15.3 wt.) % B a superlattice of the h.c.p. structure of α-Ti is present. In this structure, the a axis was assumed to be twice as large as that of α-Ti, and the c axis to be the same. Within a certain scattering of the values, the a and c axes of α-Ti increase up to approximately 28.6 at. % B and remain constant up to 44.4 at. % B. It was claimed that there was a continuous transition between the α phase present up to 4.8 or 6.5 at. % B (1.1 or 1.6 wt. %) and the superlattice structure [1].

The presence of extra lines in the X-ray diffraction pattern of alloys with 1–5 wt. % B was confirmed [11]. However, as no evidence was found for a shift in the lattice constants of α-Ti (indicating that there is little, if any, solubility of B in α-Ti), it was concluded, adopting a suggestion by [12], that the extra lines are not superlattice lines but are more probably the result of a second phase of a hexagonal structure, almost identical in cell dimensions with those of α-Ti. This suggested the existence of Ti_2B (10.15 wt. % B) [12]. Indeed, [7] reported the existence of Ti_2B, having a tetragonal structure.

The high solid solubility of B in α-Ti according to [1] is in contradiction to the values reported by [12*a*], i.e., less than 0.1 wt. (0.43 at.) % B in both α- and β-Ti, based on results of metallographic examination. Also, [11] found a second phase to be present in iodide titanium–base alloys containing as little as 0.4 wt. (1.7 at.) % B, after annealing at 850°C. They were able to show by micrographic work that B additions do not measurably affect the transformation temperature of Ti. Recently, [13] reported that the lowest per cent addition at which boron appeared as an insoluble compound in the microstructure of magnesium-reduced titanium-base alloys, hot-rolled and annealed for ½ hr at 790°C, was 0.067 wt. % B.

[10] investigated the constitution in the range 0–33 wt. (68.5 at.) % B, using arc-melted iodide titanium–base alloys, and outlined the phase boundaries according to Fig. 150. Data between 750 and 1400°C are based on microscopic studies, complemented by X-ray analysis, of heat-treated alloys. Higher temperature data, including the phase fields containing melt as one phase present, are based on microstructures and X-ray diffraction patterns of specimens quenched after short time exposure to temperatures up to about 2400°C. It was claimed that equilibrium was obtained in alloys with up to 9 or 10 wt. (30.7 or 33 at.) % B. "For alloys of higher B contents, annealing at 1400°C (for ½ hr) and above modified the as-cast structure somewhat, but heat-treated specimens usually revealed the complete history of solidification. However, even though the as-cast structure could not be removed, as-cast microstructures, melting-point data, and differences in amounts of dominant phases and variations in X-ray line intensities, both observed after heat-treatment, suggested the diagram in Fig. [150]."

The diagram by [10] includes the borides Ti_2B (shown to be stable only at temperatures above about 1800°C), TiB, and TiB_2. The boride Ti_2B_5 lies outside the range of composition investigated. The solid solubility of B in α-Ti and β-Ti was found, in agreement with results by other investigators [12, 13], to be less than 0.05 wt. % at temperatures between 750 and 1300°C and slightly above 0.1 wt. % B at the eutectic temperature, 1670 ± 25°C. TiB_2 was found to be of variable composition; its melting point, 2790°C, is due to [6].

Crystal Structures. Ti_2B is tetragonal with $a = 6.11$ A, $c = 4.56$ A, $c/a = 0.746$ [7]. [10] stated that the pattern could be indexed in the two following ways: (1) $a = 6.10$ A, $c = 4.53$ A, $c/a = 0.743$ or (2) $a = 5.24$ A, $c = 7.60$ A, $c/a = 1.45$. Unit cell (1) agrees closely with the parameters reported by [7], and "appears to yield somewhat closer agreement between observed and calculated d-values. However, the fit of the unit cell (2) lines is closer in the back reflection region."

TiB, earlier reported to be cubic of the zincblende [1, 2] or NaCl [9, 7, 10] type, is orthorhombic of the FeB (B27) type, $a = 6.12$ A, $b = 3.06$ A, $c = 4.56$ A [14]. The crystal structure of TiB_2 is that of the AlB_2 (C32) type [1, 15, 16, 8, 10], $a = 3.030$ A, $c = 3.227$ A, $c/a = 1.065$ [1, 15, 16, 8]. Ti_2B_5 is hexagonal and isotypic with W_2B_5, $a = 2.98$ A, $c = 13.98$ A, $c/a = 4.691$ [7].

Supplement. [17] attributed a change of slope at higher temperatures in an electrical-resistivity-vs.-temperature curve for TiB_2 to a secondary conduction mechanism superimposed on metallic behavior at and near room temperature. [18] discussed some structural and physical properties of diborides; the melting point of TiB_2 was found to be 2920°C (cf. above).

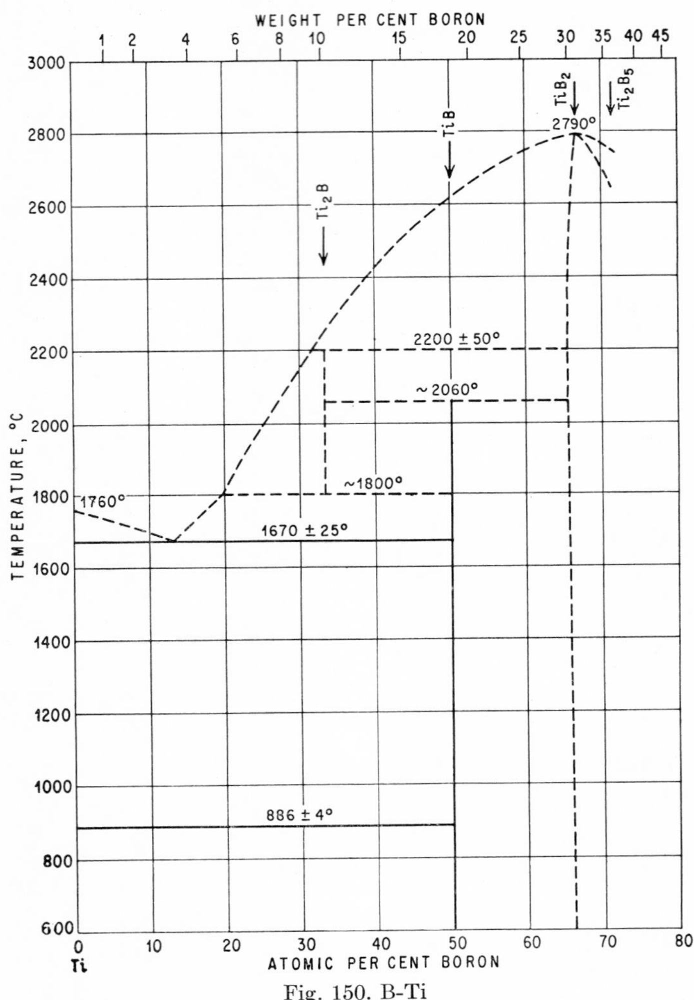

Fig. 150. B-Ti

1. P. Ehrlich, *Z. anorg. Chem.*, **259,** 1949, 1–41.
2. R. Juza and H. Hahn, *Z. anorg. Chem.*, **244,** 1940, 133.
3. L. Andrieux, *Ann. Chim.*, **12,** 1929, 423; *Rev. mét.*, **45,** 1948, 49.
4. P. M. McKenna, *Ind. Eng. Chem.*, **28,** 1936, 767.
5. H. M. Greenhouse, O. E. Accountius, and H. H. Sisler, *J. Am. Chem. Soc.*, **73,** 1951, 5086–5087.
6. F. W. Glaser, *Trans. AIME,* **194,** 1952, 391–396.
7. B. Post and F. W. Glaser, *J. Chem. Phys.*, **20,** 1952, 1050–1051.

8. L. Brewer, D. L. Sawyer, D. H. Templeton, and C. H. Dauben, *J. Am. Ceram. Soc.*, **34**, 1951, 173–179.
9. L. H. Andersson and R. Kiessling, *Acta Chem. Scand.*, **4**, 1950, 160–164.
10. A. E. Palty, H. Margolin, and J. P. Nielsen, *Trans. ASM*, **46**, 1954, 312–328.
11. H. R. Ogden and R. I. Jaffee, *Trans. AIME*, **191**, 1951, 335–336.
12. M. Hansen, private communication to R. I. Jaffee.
12*a*. C. M. Craighead, O. W. Simmons, and L. W. Eastwood, *Trans. AIME*, **188**, 1950, 485–513.
13. R. M. Goldhoff, H. L. Shaw, C. M. Craighead, and R. I. Jaffee, *Trans. ASM*, **45**, 1953, 941–965.
14. B. F. Decker and J. S. Kasper, *Acta Cryst.*, **7**, 1954, 77–80.
15. W. H. Zachariasen, *Acta Cryst.*, **2**, 1949, 94–99.
16. J. T. Norton, H. Blumenthal, and S. J. Sindeband, *Trans. AIME*, **185**, 1949, 749–751.
17. F. W. Glaser and D. Moskowitz, *Powder Met. Bull.*, **6**, 1953, 178–185.
18. B. Post, F. W. Glaser, and D. Moskowitz, *Acta Met.*, **2**, 1954, 20–25.

$\bar{2}.7238$
1.2762

B-Tl Boron-Thallium

See B-Bi, page 245.

$\bar{2}.6575$
1.3425

B-U Boron-Uranium

The three intermediate phases UB_2 (8.33 wt. % B), UB_4 (15.38 wt. % B), and UB_{12} [92.31 at. (35.29 wt.) % B] were identified by X-ray diffraction methods, after the existence of UB_2 and UB_4 had been suggested by [1] and [2], respectively. The temperature of the U-UB_2 eutectic is probably close to the melting point of U (1132°C) and that of the UB_2-UB_4 eutectic well above 1565°C. The melting point of UB_2 appears to be considerably lower than that of UB_4 [3].

UB_2 is hexagonal of the AlB_2 (C32) type [4, 3], with $a = 3.136$ A, $c = 3.988$ A, $c/a = 1.272$ [3]. UB_4 is tetragonal and isotypic with ThB_4, $a = 7.075$ A, $c = 3.979$ A, $c/a = 0.562$ [5]; $a = 7.066$ A, $c = 3.97$ A, $c/a = 0.56$ [6]. UB_{12} is f.c.c., with $4UB_{12}$ per unit cell and isotypic with ZrB_{12}, $a = 7.473$ A [6].

Supplement. [7] gave a detailed description of the structures of UB_4 and UB_{12}. The lattice parameters were redetermined: $a = 7.080$ A, $c = 3.978$ A for tetragonal UB_4; $a = 7.468$ A for cubic UB_{12}.

1. E. Wedekind and O. Jochem, *Ber. deut. chem. Ges.*, **46**, 1913, 1204–1205.
2. L. Andrieux, *Ann. chim.*, **12**, 1929, 423.
3. L. Brewer, D. L. Sawyer, D. H. Templeton, and C. H. Dauben, *J. Am. Ceram. Soc.*, **34**, 1951, 173–179.
4. A. H. Daane and N. C. Baenziger, *U.S. Atomic Energy Comm., Rept.* ISC-53, 1949.
5. A. Zalkin and D. H. Templeton, *Acta Cryst.*, **6**, 1953, 269–272; *J. Chem. Phys.*, **18**, 1950, 381.
6. F. Bertaut and P. Blum, *Compt. rend.*, **229**, 1949, 666–667.
7. P. Blum and F. Bertaut, *Acta Cryst.*, **7**, 1954, 81–86.

$\bar{1}.3271$
0.6729

B-V Boron-Vanadium

The existence of VB (17.52 wt. % B) [1, 2] and VB_2 (29.81 wt. % B) [3] has been reported. VB is orthorhombic of the CrB type, $a = 3.10$ A, $b = 8.17$ A, $c = 2.98$ A

[2]; VB_2 is hexagonal of the AlB_2 (C32) type, a = 2.998 A, c = 3.057 A, c/a = 1.02 [3].

An alloy with about 0.5 wt. (2.3 at.) % B was shown to have a small amount of eutectic, and an alloy with about 5 wt. (20.0 at.) % B to contain primary crystals surrounded by a peritectic residue and eutectic. This would indicate that the eutectic lies at considerably less than 5 wt. % B and that the VB phase is formed by the reaction liq. + $VB_2 \rightarrow$ VB at a temperature determined to be approximately 1780°C [4]. The melting point of VB_2 was given as 2100 ± 60°C [5]; see also [6].

Supplement. [7] confirmed the orthorhombic structure for VB found by [2] and reported the following lattice parameters: 3.058, 8.026, and 2.971 A. Furthermore, [7] detected two new phases "which were observed in several specimens of the over-all composition V_2B prepared by arc-melting, followed by prolonged annealing, in some cases, at temperatures of 1500 and 1700°C. These specimens were shown by X-ray diffraction analysis to contain two new compounds of unknown crystal structure. Only one of these phases was present in the unannealed arc melts; all of the annealed specimens contain both phases in roughly equal amounts."

[8] discussed some structural and physical properties of diborides; the melting point of VB_2 was found to be 2400°C (cf. above). The boride V_3B_4 (22.07 wt. % B) was reported by [9]. Its structure is orthorhombic, a = 3.030 A, b = 13.18 A, c = 2.986 A, and isotypic with Cr_3B_4, Mn_3B_4, Nb_3B_4, and Ta_3B_4.

1. E. Wedekind and C. Horst, *Ber. deut. chem. Ges.*, **46**, 1913, 1203–1204.
2. H. Blumenthal, *J. Am. Chem. Soc.*, **74**, 1952, 2942.
3. J. T. Norton, H. Blumenthal, and S. J. Sindeband, *Trans. AIME*, **185**, 1949, 749–751.
4. W. Rostoker and A. Yamamoto, *Trans. ASM*, **46**, 1954, 1136–1163.
5. R. Kieffer, F. Benesovsky, and E. R. Honak, *Z. anorg. Chem.*, **268**, 1952, 191–200.
6. F. W. Glaser, *Trans. AIME*, **194**, 1952, 391–396.
7. G. F. Hardy and J. K. Hulm, *Phys. Rev.*, **93**, 1954, 1004–1016, especially 1009.
8. B. Post, F. W. Glaser, and D. Moskowitz, *Acta Met.*, **2**, 1954, 20–25.
9. D. Moskowitz, *Trans. AIME*, **206**, 1956, 1325.

$\bar{2}$.7696
1.2304

B-W Boron-Wolfram

The following borides were identified by determining their crystal structures, after the existence of WB [1] and WB_2 [2, 3] had been claimed earlier: W_2B (2.86 wt. % B) [4, 5], having a very narrow range of homogeneity; WB (5.56 wt. % B) [4, 5] with a homogeneity range of about 48.0–50.5 at. % B; and W_2B_5 [71.43 at. (12.82 wt. %) B] [4, 5], also of variable composition. In addition, the existence of a high-temperature modification of WB, stable above about 1850 ± 50°C, was reported [6, 7]. WB_2 [2, 3] has not been verified. The solid solubility of B in W is apparently very low [4]. The temperatures of the eutectics in this system are considerably above 2000°C [5]. The melting points of W_2B and WB were reported to be 2770 ± 80 and 2860 ± 80°C [8], respectively. The melting point of a boride of unknown composition, probably WB, was given as 2920 ± 50°C [9]. W_2B_5 appears to melt undecomposed at 2980°C [7].

W_2B is tetragonal of the $CuAl_2$ (C16) type, a = 5.564 A, c = 4.740 A, c/a = 0.852 [4]; see also [5]. The low-temperature form of WB also has a tetragonal structure (MoB type), with a = 3.115 A, c = 16.93 A, c/a = 5.44 [4], whereas the high-temperature form, corresponding to β-MoB, is orthorhombic (CrB type), a = 3.19 A, b = 8.40 A, c = 3.07 A [6]. The phase W_2B_5 has a hexagonal defect structure,

$a = 2.982$ A, $c = 13.87$ A, $c/a = 4.65$. Its structure is closely related to that of Mo_2B_5, but the two phases are not isotypic [4].

It appears that the systems W-B and Mo-B are closely related. This may also hold for the phase diagrams (see Fig. 148).

Supplement. [10] found the melting point of W_2B_5 to be 2200°C (cf. above). They suggested, however, that W_2B_5—in analogy to Mo_2B_5—may transform to WB_2 near the melting point.

1. G. Weiss, *Ann. chim.*, **1**, 1946, 446–525.
2. S. A. Tucker and H. R. Moody, *Proc. Chem. Soc.*, **17**, 1901, 129–131.
3. E. Wedekind, *Ber. deut. chem. Ges.*, **46**, 1913, 1206.
4. R. Kiessling, *Acta Chem. Scand.*, **1**, 1947, 893–916.
5. L. Brewer, D. L. Sawyer, D. H. Templeton, and C. H. Dauben, *J. Am. Ceram. Soc.*, **34**, 1951, 173–179.
6. B. Post and F. W. Glaser, *J. Chem. Phys.*, **20**, 1952, 1050–1051.
7. F. W. Glaser, *Trans. AIME*, **194**, 1952, 391–396.
8. R. Kieffer, F. Benesovsky, and E. R. Honak, *Z. anorg. Chem.*, **268**, 1952, 191–200.
9. K. Moers (C. Agte), *Z. anorg. Chem.*, **198**, 1931, 243–275.
10. B. Post, F. W. Glaser, and D. Moskowitz, *Acta Met.*, **2**, 1954, 20–25.

$\bar{1}.2188$
0.7812

B-Zn Boron-Zinc

Amorphous boron is not wetted by boiling zinc [1].

1. H. Giebelhausen, *Z. anorg. Chem.*, **91**, 1915, 262.

$\bar{1}.0741$
0.9259

B-Zr Boron-Zirconium

The existence of the following borides has been established by determining their crystal structures: ZrB (10.60 wt. % B) [1–3], ZrB_2 (19.17 wt. % B) [4–7], and ZrB_{12} [92.31 at. (58.74 wt.) % B] [1, 8]. Earlier, [9, 10] had assumed the phase Zr_3B_4, however, without proving its homogeneity.

The phase diagram shown in Fig. 151 was outlined by [3]. Mixtures of zirconium hydride and B were prepared ranging in 2 wt. % steps from 2 to 75 wt. (14–96 at.) % B. Samples were heated to various temperatures differing by 50°C increments or less. Visual examination of cooled specimens indicated when specimens had melted. The melting point of ZrB_2, 3040 ± 50°C, is in good agreement with that reported by [11], about 3000°C, for a boride which was probably ZrB_2, although its structure was reported to be cubic rather than hexagonal.

ZrB, stable only between about 1250 and 800°C, is cubic of the NaCl (B1) type, $a = 4.65$ A [3]. The structure of ZrB_2, which has a narrow homogeneity range [5, 7], is hexagonal of the AlB_2 (C32) type. Lattice dimensions reported by various authors [5–7] agree very well and fall within the ranges $a = 3.168$–3.170 A, $c = 3.528$–3.533 A, $c/a = 0.114$. ZrB_{12} is f.c.c., with $4ZrB_{12}$ per unit cell, and isotypic with UB_{12}, $a = 7.408$ A [8].

The solid solubility of B in α-Zr was estimated to be about 1 at. % B, based on lattice parameters reported to increase from $a = 3.229$ A, $c = 5.139$ A for α-Zr to $a = 3.249$ A, $c = 5.203$ A for the saturated solid solution [5].

Supplement. Some structural and physical properties of the diboride were discussed by [12].

1. F. W. Glaser, *Trans. AIME*, **194**, 1952, 391–396.
2. B. Post and F. W. Glaser, *J. Chem. Phys.*, **20**, 1952, 1050–1051.

3. F. W. Glaser and B. Post, *Trans. AIME*, **197**, 1953, 1117–1118.
4. P. M. McKenna, *Ind. Eng. Chem.*, **28**, 1936, 767–772. The product contained 1.09% C.
5. R. Kiessling, *Acta Chem. Scand.*, **3**, 1949, 90–91.
6. J. T. Norton, H. Blumenthal, and S. J. Sindeband, *Trans. AIME*, **185**, 1949, 749–751.
7. L. Brewer, D. L. Sawyer, D. H. Templeton, and C. H. Dauben, *J. Am. Ceram. Soc.*, **34**, 1951, 173–179.
8. B. Post and F. W. Glaser, *Trans. AIME*, **194**, 1952, 631–632.
9. S. A. Tucker and H. R. Moody, *Proc. Chem. Soc.*, **17**, 1901, 129–130.
10. E. Wedekind, *Ber. deut. chem. Ges.*, **46**, 1913, 1201–1203.
11. K. Moers (K. Becker), *Z. anorg. Chem.*, **198**, 1931, 243–275.
12. B. Post, F. W. Glaser, and D. Moskowitz, *Acta Met.*, **2**, 1954, 20–25.

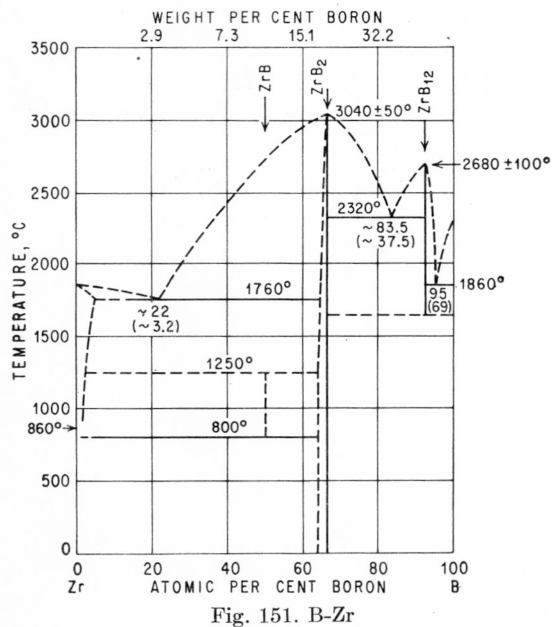

Fig. 151. B-Zr

$\bar{1}.8177$
0.1823

Ba-Bi Barium-Bismuth

Early melting experiments [1] resulted in crystalline products which oxidized quickly and, with Ba contents over 5 wt. %, were easily decomposed by water.

The phase diagram of Fig. 152 was established, by means of a thermal and microscopic investigation [2], only up to about 30 at. % Ba [3]; thermal arrests for alloys with higher Ba contents were too weak [4].

Calorimetric measurements [5] suggest the existence of the stoichiometric compound Ba_3Bi_2 (49.64 wt. % Ba); the heat of formation has a sharp maximum at this composition.

Some hardness data [2], measured on as-cast alloys, seem to indicate a solubility of Ba in Bi of about 0.3 wt. (0.5 at.) %.

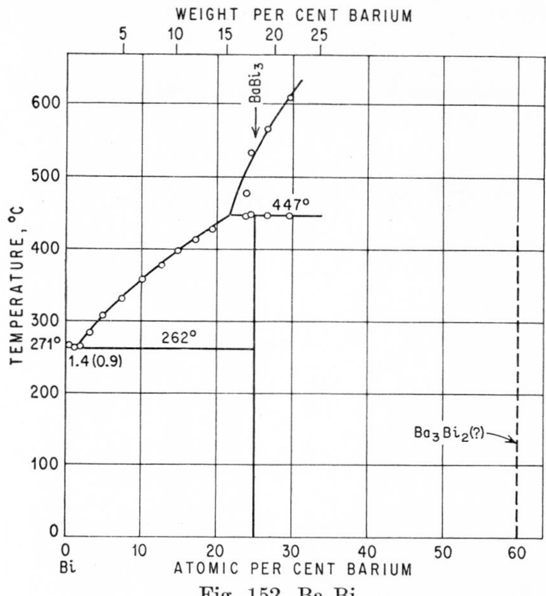

Fig. 152. Ba-Bi

Supplement. According to microscopic observations and cursory X-ray work, the compound $BaBi_3$ has cubic symmetry [6].

1. "Gmelins Handbuch der anorganischen Chemie," System No. 30, p. 359, Verlag Chemie, G.m.b.H., Berlin, 1932.
2. G. Grube and A. Dietrich, *Z. Elektrochem.*, **44**, 1938, 755–758. The alloys were melted in argon-filled sealed iron crucibles. The melting point of the 98.7 wt. % Ba used was 676°C.
3. The concentration values given in the text for the eutectic and peritectic points (1.9 and 13.61 wt. % Ba, respectively) do not agree with the plotted thermal data (cf. Fig. 152).
4. For the same reason the formula of the peritectically formed compound $BaBi_3$ (17.15 wt. % Ba) could not be determined from the length of the peritectic arrests. "Since a short arrest was observed at 17.01 wt. % Ba on the eutectic horizontal and only a weak break at 17.62 wt. % Ba, the formula of the compound should be $BaBi_3$ (17.15 wt. % Ba)."
5. O. Kubaschewski and H. Villa, *Z. Elektrochem.*, **53**, 1949, 32–40.
6. J. K. Hulm, Westinghouse Electric Corporation, Research Laboratories, East Pittsburgh, Pa.; private communication to M. Hansen, 1955.

1.0583
$\bar{2}.9417$

Ba-C Barium-Carbon

BaC_2 (14.88 wt. % C), whose melting point lies somewhere between 1770 and 2300°C [1], is strongly attacked by water. Above about 150°C, it has a cubic face-centered structure, with a = 6.55 A [2], and at room temperature the tetragonal CaC_2 (C11) type of structure with a = 6.23 A, c = 7.07 A, c/a = 1.134 [3] (pseudocubic face-centered cell). Molten BaC_2 dissolves carbon [4].

1. "Gmelins Handbuch der anorganischen Chemie," System No. 30, p. 300, Verlag Chemie, G.m.b.H., Berlin, 1932.
2. M. A. Bredig, *J. Phys. Chem.*, **46**, 1942, 801–819.
3. M. von Stackelberg, *Z. physik. Chem.*, **B9**, 1930, 437–475.
4. H. M. Kahn, *Compt. rend.*, **144**, 1907, 198; see [1], p. 301.

0.5349
$\bar{1}.4651$

Ba-Ca Barium-Calcium

A roentgenographic investigation (powder method) of slowly cooled alloys showed [1] that the room-temperature modifications of the two metals (Ca: f.c.c., Ba: b.c.c.) have a great mutual solubility. The solid solutions are separated by a narrow heterogeneous field between about 32 and 36 at. (62 and 66 wt.) % Ba [2]. The lattice parameter of pure Ba is practically unchanged after addition of 15 at. % Ca; at higher Ca contents it drops steeply to the value expected from additivity.

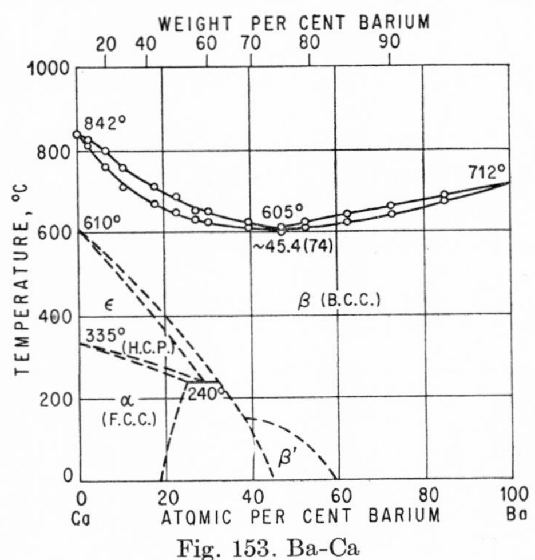

Fig. 153. Ba-Ca

Supplement. A thermal and roentgenographic investigation of the Ca-Ba system was carried out by [3] (Fig. 153). The liquidus and solidus curves were determined thermally, while the boundaries of the solid-phase fields were established by X-ray room- and high-temperature (up to 620°C) work.

Three allotropic modifications of Ca (f.c.c., h.c.p., and b.c.c.; cf. Fig. 153) but only one of Ba (b.c.c.) were observed. This permits the occurrence of a continuous series of solid solutions at higher temperatures. Solution of Ba in Ca increases the sluggishness of the transformations in Ca, and, therefore, the solid-phase boundaries are considered to be of no high accuracy. The two-phase region at room temperature ($\alpha + \beta'$) extends from about 45 to 73 wt. (19–44 at.) % Ba (cf. [1]).

Extra lines were observed in alloys with 26–58 at. % Ba, and the existence of a "diamond cubic" superlattice, i.e., NaTl (B32) type, has been suggested (β' in Fig. 153).

1. W. Klemm and G. Mika, *Z. anorg. Chem.*, **248**, 1941, 155–166. The alloys were melted from the elements (98–99 wt. % pure) in argon-filled iron crucibles.
2. The assumption of [1] that these phase limits correspond to the equilibrium state at

500–600°C obviously does not consider the existence of transformations in both elements (cf. Supplement).
3. E. A. Sheldon, dissertation, Syracuse University, Syracuse, N.Y., 1949. The alloys were melted in argon-filled iron crucibles.

0.0871
$\bar{1}$.9129

Ba-Cd Barium-Cadmium

Some work on preparation by [1] gives no information as to the constitution of Ba-Cd alloys. More recently a compound $BaCd_{11}$ [91.67 at. (90.00 wt.) % Cd] was evaluated by X-ray work [2]. It has a b.c. tetragonal structure with a = 12.02 A, c = 7.74 A, c/a = 0.644, and 48 atoms per unit cell.

Supplement. The compound BaCd (45.00 wt. % Cd) was identified by microscopy as well as determination of its crystal structure: CsCl (B2) type, a = 4.215 A [3].

1. H. Gautier, *Compt. rend.*, **134**, 1902, 1054–1056, 1109.
2. M. J. Sanderson and N. C. Baenziger, *Acta Cryst.*, **6**, 1953, 627–631.
3. R. Ferro, *Acta Cryst.*, **7**, 1954, 781.

0.3908
$\bar{1}$.6092

Ba-Fe Barium-Iron

There is no reliable evidence that Fe and Ba alloy. In the oldest literature the existence of Fe-Ba alloys is assumed [1]. According to [2], Ba is insoluble in solid iron. In preparing Ba in an iron crucible, [3] observed a contamination of the Ba by 0.4 wt. % Fe, evidently in metallic form.

[4] observed no diffusion of molten Ba into a soft steel (0.08–0.15 wt. % C) at 1200°C.

1. "Gmelins Handbuch der anorganischen Chemie," System No. 59 A(9), p. 1826, Verlag Chemie, G.m.b.H., Berlin, 1939.
2. F. Wever, *Naturwissenschaften*, **17**, 1929, 304–309; *Arch. Eisenhüttenw.*, **2**, 1928-1929, 739–746.
3. A. Guntz, *Ann. chim. et phys.*, **4**, 1905, 16.
4. N. V. Ageev and M. I. Zamotorin, *Izvest. Leningrad. Politech. Inst., Otdel. Mat. Fiz. Nauk*, **31**, 1928, 183–197.

2.1344
$\bar{3}$.8656

Ba-H Barium-Hydrogen

See Ca-H, page 398.

$\bar{1}$.8355
0.1645

Ba-Hg Barium-Mercury

The determinations of the solubility of Ba in liquid mercury between 0 and 99°C by means of chemical analysis yielded results which were in substantial agreement [1-3] (Fig. 154). At 0, 20, and 50°C the solubility is 0.16, 0.33, and 0.63 wt. (0.24, 0.48, and 0.92 at.) % Ba.

The composition of the solid phase in equilibrium with the saturated melts was found by various mechanical-separation methods to be close to $BaHg_{13}$ (5.00 wt. % Ba) if separated between 0 and 30°C [1, 2], $BaHg_{12}$ (5.40 wt. % Ba) [4, 1–3], and $BaHg_{11}$ (5.86 wt. % Ba) [5]. For the latter, an incongruent melting point of 165°C was determined [5]. Because it is difficult to obtain a complete separation of the liquid and solid phases, the highest barium content for the intermediate phase appears to be the most likely one. The formula $BaHg_{11}$ (8.33 at. % Ba) has been confirmed by a more recent investigation in which electrolytically prepared crystals were recrystallized by melting in an excess of saturated melt [6]. According to goniometric, densimetric, and X-ray examinations, $BaHg_{11}$ has a cubic structure with a = 9.62 A and 36 atoms per unit cell [6].

The existence of the compounds $BaHg_5$ (12.04 wt. % Ba) and BaHg (40.64 wt. % Ba) was claimed by [5]; however, the criteria given are unreliable.

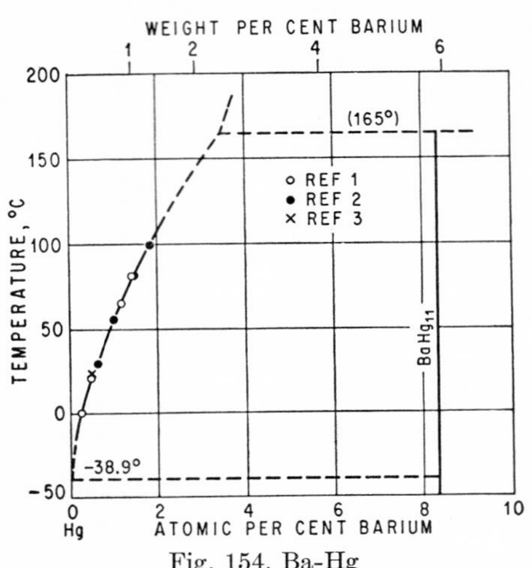

Fig. 154. Ba-Hg

Supplement. [7] determined the crystal structure of $BaHg_{11}$ (lattice parameter and unit-cell content as previously reported; see above). [7] admits the possibility that the structure may be slightly rhombohedrally deformed.

[8] identified the compound BaHg by microscopy as well as determining its crystal structure: CsCl (B2) type, a = 4.133 A.

1. W. Kerp, *Z. anorg. Chem.*, **17**, 1898, 303–305.
2. W. Kerp, W. Böttger, and H. Iggena, *Z. anorg. Chem.*, **25**, 1900, 44–53.
3. G. McPhail Smith and H. C. Bennett, *J. Am. Chem. Soc.*, **32**, 1910, 622–626.
4. A. Guntz and J. Férée, *Bull. soc. chim. France*, **15**, 1896, 834.
5. G. Langbein, dissertation, Königsberg, 1900. Data are given in Gmelin-Kraut, "Handbuch der anorganischen Chemie," vol. 5, pp. 1073–1078, Carl Winter's Universitätsbuchhandlung, Heidelberg, 1914.
6. G. Peyronel and E. Pacilli, *Gazz. chim. ital.*, **73**, 1943, 29–36.
7. G. Peyronel, *Gazz. chim. ital.*, **82**, 1952, 679–690; *Chem. Abstr.*, **47**, 1953, 9710.
8. R. Ferro, *Acta Cryst.*, **7**, 1954, 781.

0.7519
$\bar{1}$.2481

Ba-Mg Barium-Magnesium

The diagram of Fig. 155 is the result of a thermal, microscopic, and roentgenographic study by [1]. There are three intermediate phases: Mg_9Ba (38.56 wt. % Ba) and Mg_2Ba (73.85 wt. % Ba), with open maxima, and one peritectically formed compound, presumably Mg_4Ba (58.54 wt. % Ba).

Mg_9Ba and Mg_4Ba had already been detected in a thermal (up to 83 wt. % Ba) and microscopic (up to 40 wt. % Ba) investigation by [2]. However, the liquidus of their diagram above 30 wt. % Ba deviates appreciably from that shown in Fig. 155—best illustrated by the melting point of 854°C for the assumed compound Mg_5Ba_2 (69 wt. % Ba)—and there is no doubt that most of the thermal results of [2] are distorted [3]. It may also be mentioned that some of the microstructures published by [2] fit much better in the diagram of [1] than in their own [4]. The existence of Mg_2Ba—suggested by [2]—was first demonstrated by [5], who showed this compound to have the hexagonal $MgZn_2$ (C14) type with a = 6.649 A, c = 10.676 A, c/a = 1.606.

The liquidus points of nine alloys containing between 6 and 27 at. % Ba were measured by [9] using thermal analysis and were found to be in good agreement with the liquidus of [1].

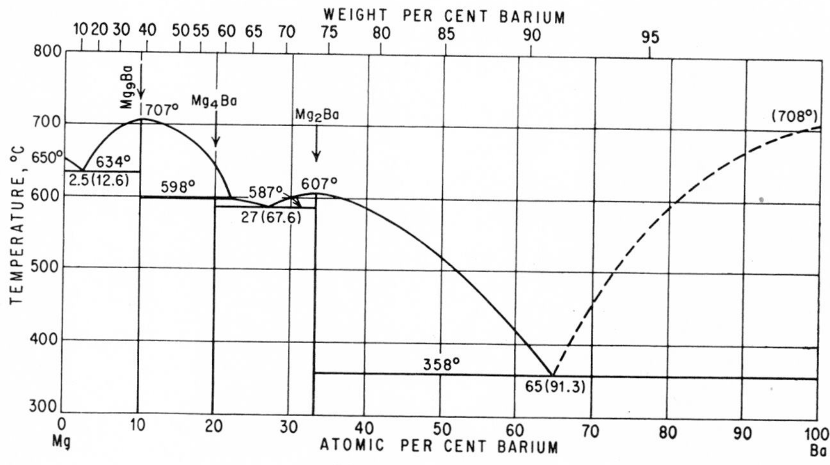

Fig. 155. Ba-Mg

Solubility of Ba in Solid Mg. [2, 6] found Ba to be practically insoluble in solid Mg. More precise microscopic investigations by [1] revealed a solubility of 1.8×10^{-3} at. (0.08 wt.) % Ba at 620°C; precision lattice-parameter measurements [7] are also in favor of a certain low solubility.

Supplement. According to [8], compounds of the formulas Mg_9Ba and Mg_9Ba_2 exist. Mg_9Ba_2 is probably identical with Mg_4Ba.

1. W. Klemm and F. Dinckelacker, *Z. anorg. Chem.*, **255,** 1947, 2–12. The barium used was 99.1 wt. % pure. The melting was carried out in argon-filled iron crucibles. With the exception of its characteristic points, which are tabulated, the liquidus had to be redrawn from a small-scale plot.
2. G. Grube and A. Dietrich, *Z. Elektrochem.*, **44,** 1938, 755–767.
3. It was shown [1] by stirring that all alloys were molten at 720°C.
4. The slowly cooled 37 and 40 wt. % Ba alloys show no peritectic coring, which is to be expected from the diagram of [2].
5. E. Hellner and F. Laves, *Z. Krist.*, **A105,** 1943, 134–143.
6. F. Weibke and W. Schmidt, *Z. Elektrochem.*, **46,** 1940, 357–364.
7. R. S. Busk, *Trans. AIME*, **188,** 1950, 1460–1464.
8. W. C. Zeek, unpublished work, quoted in E. M. Flanigen, thesis, Syracuse University, Syracuse, N.Y., 1952.
9. K. Anderko, *Trans. AIME*, **209,** 1957, 612.

0.9915
$\bar{1}$.0085

Ba-N Barium-Nitrogen

See the chemical literature. Recent papers: [1, 2].

1. V. A. Russell, Thermal Study of Barium–Barium Nitride System, M.S. thesis, Syracuse University, 1949.
2. S. M. Ariya and E. A. Prokofeva, *Sborník Statei Obshchei Khim., Akad. Nauk S.S.S.R.*, **1,** 1953, 9–18; *Chem. Abstr.*, **48,** 1954, 12522.

0.3693
$\bar{1}$.6307

Ba-Ni Barium-Nickel

Homogeneous alloys up to 0.20 wt. % Ba were made by [1]. According to a discussion on [1], molten Ni does not dissolve any Ba; molten Ba, however, dissolves a certain amount of Ni.

1. D. W. Randolph, *Trans. Electrochem. Soc.*, **66,** 1934, 85–90; cf. also discussion following by W. Kroll.

0.6468
$\bar{1}$.3532

Ba-P Barium-Phosphorus

See the chemical literature. Recent paper: [1].

1. S. A. Shchukarev, M. P. Morozova, and E. A. Prokofeva, *Zhur. Obshcheĭ Khim.*, **24,** 1954, 1277–1278; *Chem. Abstr.*, **49,** 1955, 2239. (BaP_3, BaP_2.)

$\bar{1}$.8214
0.1786

Ba-Pb Barium-Lead

The phase diagram of Fig. 156 was established by means of thermal and microscopic analysis as well as measurement of resistivity-temperature curves [1]. There exist the compounds Ba_2Pb (43.00 wt. % Pb), BaPb (60.04 wt. % Pb)—which undergoes a polymorphic transformation at 546°C—and $BaPb_3$ (81.90 wt. % Pb). The resistivity-temperature curves indicate that Ba_2Pb possesses a certain homogeneity range (see Fig. 156). The concentration of the Pb-rich eutectic, 95.5 wt. (93 at.) % Pb [2], was already found by [3] and [4], at temperatures of 282 and 291°C, respectively.

The solubility of Ba in solid Pb is ~0.5 wt. (0.8 at.) % at 293°C (resistivity, age-hardening) [1], 0.42 wt. (0.7 at.) % at 130°C (resistivity) [1], and 0.02 wt. (0.03 at.) % at room temperature (microscopic and age-hardening studies) [5]. On the other hand, Pb showed no solubility in solid Ba [1]. The heat of formation has a sharp maximum at the composition Ba_2Pb [6]. No literature on crystal structures was found.

1. G. Grube and A. Dietrich, *Z. Elektrochem.*, **44,** 1938, 755–767. The alloys were melted from the elements in argon-filled sealed iron crucibles. The melting point of the barium used (98.7 wt. % pure) was 676°C.
2. This concentration fits better in the liquidus determined by [1], who themselves assumed it to be 93.8 wt. % Pb.
3. J. Czochralski and E. Rossow, *Z. Metallkunde,* **12,** 1920, 337–340.
4. W. A. Cowan, L. D. Simpkins, and G. O. Hiers, *Trans. Am. Electrochem. Soc.*, **40,** 1921, 237–258; *Chem. Met. Eng.*, **25,** 1921, 1182–1184.
5. E. Schmid, *Z. Metallkunde,* **35,** 1943, 85–92.
6. O. Kubaschewski and H. Villa, *Z. Elektrochem.*, **53,** 1949, 32–40.

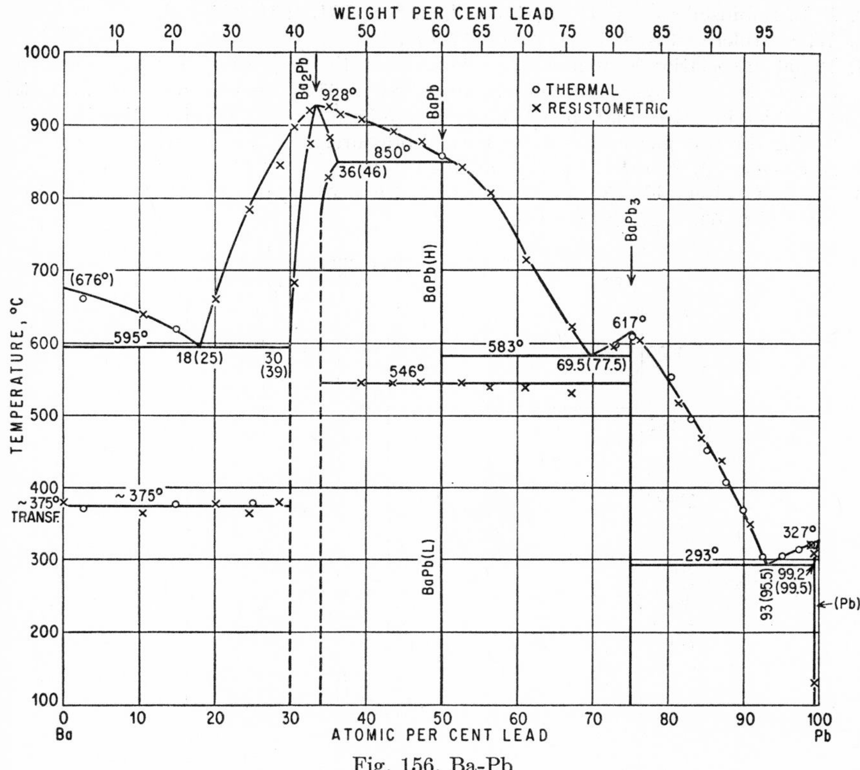

Fig. 156. Ba-Pb

Ī.8473 0.1527 Ba-Pt Barium-Platinum

The compound $BaPt_5$, stable only at higher temperatures, has the hexagonal $CaCu_5$-type structure (6 atoms per unit cell) with $a = 5.505$ A, $c = 4.342$ A (from kX), $c/a = 0.789$. The decomposition of this phase could be partially prevented by quenching. The temperature of decomposition was not determined [1].

1. T. Heumann, *Nachr. Akad. Wiss. Göttingen*, **2**(1), 1950, A, 1–6.

0.6318 Ī.3682 Ba-S Barium-Sulfur

Besides the well-known BaS (18.93 wt. % S; melting point >2000°C), the compounds Ba_2S (10.45 wt. % S) [1], BaS_2 (31.83 wt. % S) [2], and BaS_3 (41.19 wt. % S) [2, 3] have been prepared by melting processes. The cursory [4] partial phase diagram of Fig. 157 is based on thermal work by [2].

BaS has the NaCl (B1) type of structure [5, 6], with $a = 6.381 \pm 3$ A [6]. BaS_3 forms the orthorhombic $D0_{17}$-prototype structure [3], with $a = 8.34$ A, $b = 9.66$ A, $c = 4.83$ A, and 16 atoms per unit cell. BaS_2 probably undergoes a polymorphic transformation at 664°C [2] (Fig. 157).

1. A. Guntz and F. Benoit, *Bull. soc. chim. France*, (4)**35**, 1924, 719; abstract in "Gmelins Handbuch der anorganischen Chemie," System No. 30, p. 251, Verlag Chemie, G.m.b.H., Berlin, 1932.

2. P. L. Robinson and W. E. Scott, *J. Chem. Soc.*, **134,** 1931, 693–709.
3. W. S. Miller and A. J. King, *Z. Krist.*, **A94,** 1936, 439–446.
4. Rapid dissociation at temperatures lower than the respective melting points was observed with alloys of a S content below 32 wt. (67 at.) % or above 44 wt. (77 at.) %, which indicates that the gas phase has to be taken into consideration in establishing the true phase diagram at normal pressure.
5. S. Holgersson, *Z. anorg. Chem.*, **126,** 1923, 179.
6. V. M. Goldschmidt, *Skrifter Akad. Oslo*, **1926,** no. 8, pp. 43, 84, 146; *Ber. deut. chem. Ges.*, **60,** 1927, 1274, 1289.

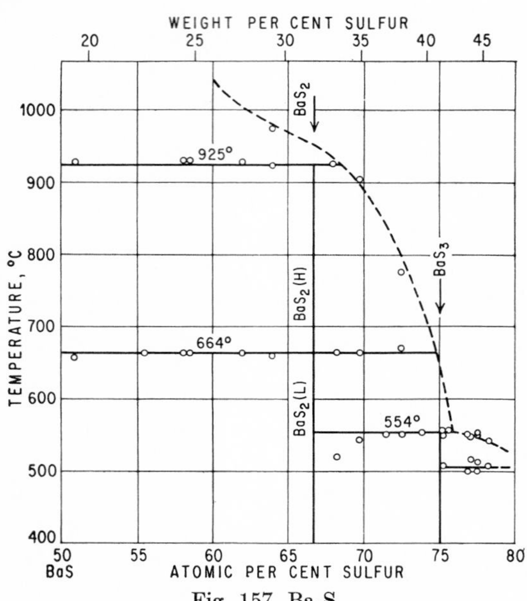

Fig. 157. Ba-S

0.0524
$\bar{1}$.9476

Ba-Sb Barium-Antimony

The constitution of this system is not yet known. Older work on preparation [1] showed that alloys containing more than 5 wt. % Ba decompose H_2O rapidly. The calorimetric measurement of the heat of formation indicated a maximum at the stoichiometric composition Ba_3Sb_2 [2].

1. For a short discussion, see "Gmelins Handbuch der anorganischen Chemie," System No. 30, p. 355, Verlag Chemie, G.m.b.H., Berlin, 1932.
2. O. Kubaschewski and H. Villa, *Z. Elektrochem.*, **53,** 1949, 32–40.

0.2404
$\bar{1}$.7596

Ba-Se Barium-Selenium

BaSe (36.51 wt. % Se) and BaTe (48.16 wt. % Te) have, like BaS, the NaCl (B1) type of structure [1]. The lattice parameters are as follows: BaSe, $a = 6.589 \pm 2$ A [2]; BaTe, $a = 6.999 \pm 3$ A [3].

1. *Strukturbericht,* **1,** 1913-1928, 135, 766; **2,** 1928-1932, 231.
2. V. M. Goldschmidt, *Skrifter Akad. Oslo,* **1926,** no. 8, pp. 43, 71, 146.
3. V. M. Goldschmidt, *Z. Krist.,* **69,** 1929, 411–414.

0.6893
$\bar{1}$.3107

Ba-Si Barium-Silicon

The silicides BaSi (16.98 wt. % Si) [1], $BaSi_2$ (29.02 wt. % Si) [2–4, 1], and $BaSi_3$ (38.02 wt. % Si) [1] have been prepared. Some evidence for the existence of additional compounds, still richer in Si—Ba_2Si_7 [77.78 at. (41.72 wt.) % Si] and $BaSi_4$ (44.99 wt. % Si)—is due to [1].

1. L. Wöhler and W. Schuff, *Z. anorg. Chem.,* **209,** 1932, 33–59.
2. C. S. Bradley, *Chem. News,* **82,** 1900, 149–150.
3. O. Hönigschmid, *Monatsh. Chem.,* **30,** 1909, 497–508.
4. R. Frilley, *Rev. mét.,* **8,** 1911, 531–532.

0.0634
$\bar{1}$.9366

Ba-Sn Barium-Tin

The phase diagram up to 30 wt. % Ba was established by means of thermal and microscopic work [1] (Fig. 158). The composition of the compounds $BaSn_5$ (18.80 wt. % Ba) and $BaSn_3$ (27.84 wt. % Ba) was determined by the lengths of the arrests at 422°C as well as by microscopic study of the annealed alloys. The existence of the

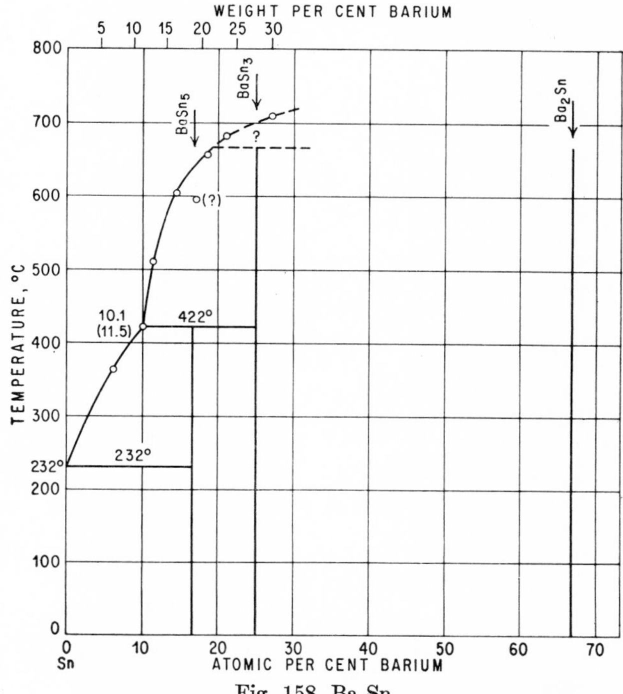

Fig. 158. Ba-Sn

stoichiometric compound Ba_2Sn (69.82 wt. % Ba) is indicated by a maximum in the heat of formation at this composition (cf. Ba_2Pb, Ca_2Sn) [2].

1. K. W. Ray and R. G. Thompson, *Metals & Alloys*, **1**, 1930, 314–316. The alloys were produced by electrolysis and analyzed for their Ba content.
2. O. Kubaschewski and H. Villa, *Z. Elektrochem.*, **53**, 1949, 32–40.

0.1952
$\bar{1}$.8048

Ba-Sr Barium-Strontium

A roentgenographic investigation (powder method) of 12 slowly cooled Ba-Sr alloys [1] showed that Sr (f.c.c.) and the room-temperature modification of Ba (b.c.c.) have a great mutual solubility. The solid solutions are separated by a narrow heterogeneous field between about 24 and 30 at. (33 and 40 wt.) % Ba (cf. Ba-Ca). A lattice-parameter-vs.-composition curve is given.

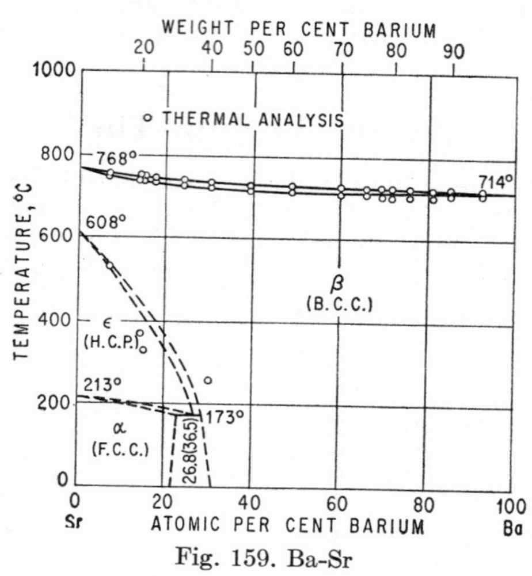

Fig. 159. Ba-Sr

Supplement. A thermal and roentgenographic investigation of the Sr-Ba system was carried out by [2] (Fig. 159). Liquidus and solidus were determined thermally, while the boundaries of the solid-phase fields were established mainly by X-ray room- and high-temperature work.

Three allotropic modifications of Sr (f.c.c., h.c.p., and b.c.c.; Fig. 159) but only one of Ba (b.c.c.) were observed (cf. Ba-Ca). This allows the existence of a continuous series of solid solutions at higher temperatures. The $(\alpha + \beta)$ two-phase region at room temperature extends from about 22 to 30.7 at. (30.7–41 wt.) % Ba (cf. [1]).

1. W. Klemm and G. Mika, *Z. anorg. Chem.*, **248**, 1941, 155–166. The alloys were melted from the elements (98–99 wt. % pure) in argon-filled iron crucibles.
2. G. Hirst, dissertation, Syracuse University, Syracuse, N.Y., 1953. The alloys were melted from the elements (99.5 wt. % pure) in steel crucibles under argon. Chemical analyses were made in intervals to check the compositions.

0.0320
$\bar{1}$.9680

Ba-Te Barium-Tellurium

See Ba-Se, page 274.

$\bar{1}.8274$
0.1726

Ba-Tl Barium-Thallium

A compound BaTl, if it does exist, has neither the β-brass (B2) nor the NaTl (B32) type of structure [1].

1. E. Zintl and G. Brauer, *Z. physik. Chem.*, **B20**, 1933, 245–271.

0.3224
$\bar{1}.6776$

Ba-Zn Barium-Zinc

Early papers [1] deal with preparation. The Zn-rich part of the phase diagram was established by thermal and microscopic analysis and hardness measurements [2] (Fig. 160). The compound adjacent to zinc, $BaZn_{13}$ [7.14 at. (13.91 wt.) % Ba] according to [3], was shown to have the cubic $NaZn_{13}$ ($D2_3$) type of structure with a = 12.35 A [3].

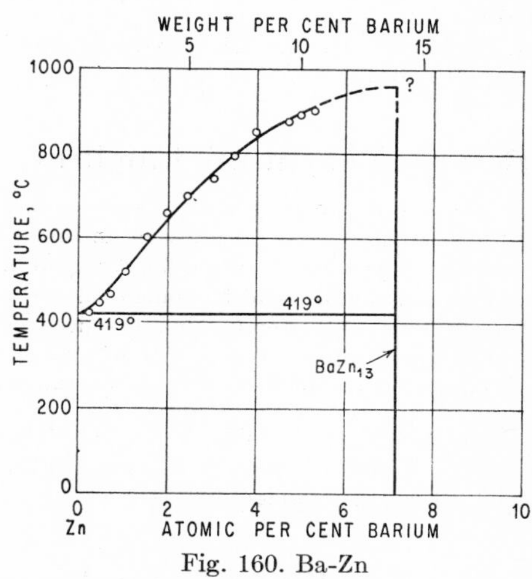

Fig. 160. Ba-Zn

Supplement. In exploratory work on the Ba-Zn system, [4] detected—besides $BaZn_{13}$—two additional intermediate phases, $BaZn_5$ and presumably BaZn. $BaZn_5$ was found to be orthorhombic, a = 5.37 A, b = 10.85 A, c = 8.50 A, with 24 atoms per unit cell.

1. See "Gmelins Handbuch der anorganischen Chemie," System No. 32, p. 328, Verlag Chemie, G.m.b.H., Leipzig-Berlin, 1924.
2. I. I. Kornilov, *Tsvet. Metally*, **10,** 1935, 73–84 (in Russian). The alloys were prepared electrolytically.
3. J. A. A. Ketelaar, *J. Chem. Phys.*, **5,** 1937, 668.
4. N. C. Baenziger, E. J. Duwell, and J. W. Conant, *U.S. Atomic Energy Comm., Publ.* C 00-127, 1954.

$\bar{2}.6347$
1.3653

Be-Bi Beryllium-Bismuth

A single experiment by [1] indicated that Be and Bi may be immiscible in the liquid state.

1. A. R. Kaufmann, P. Gordon, and D. W. Lillie, *Trans. ASM*, **42,** 1950, 801.

$\bar{1}.8753$
0.1247

Be-C Beryllium-Carbon

Microstructures of Be containing carbon were published by [1, 2]. According to these investigations, C appears to have a negligible solid solubility in Be. The carbide precipitate is Be_2C (39.99 wt. % C) [3, 1, 2], which has the CaF_2 (C1) type of structure, with $a = 4.34 \pm 1$ A [4], corroborated by [5]. The melting point of Be_2C is near 2400°C [6].

1. H. A. Sloman, *J. Inst. Metals*, **49,** 1932, 370.
2. A. R. Kaufmann, P. Gordon, and D. W. Lillie, *Trans. ASM*, **42,** 1950, 785–844.
3. P. Lebeau, *Compt. rend.*, **121,** 1895, 496–499.
4. M. v. Stackelberg, *Z. Elektrochem.*, **37,** 1931, 542–545; M. v. Stackelberg and F. Quatram, *Z. physik. Chem.*, **B27,** 1934, 50–52.
5. R. J. Teitel, *U.S. Atomic Energy Comm., Publ.* AECD-2251, 1948; abstract in *Structure Repts.*, **12,** 1949, 22.
6. M. W. Mallet, E. A. Durbin, M. C. Udy, D. A. Vaughan, and E. J. Center, *U.S. Atomic Energy Comm., Publ.* BMI/MWM/5, 1953.

$\bar{1}.3519$
0.6481

Be-Ca Beryllium-Calcium

According to some microscopic work [1], Ca appears to have a negligible solid solubility in Be. A 1.45 wt. (0.33 at.) % Ca alloy in an undefined state showed a second phase of cubic form, evidently an intermetallic compound.

The composition of this compound could be concluded perhaps from a previous alloying test [2]. On heating Be with molten Ca (in a magnesia crucible under argon) a regulus was obtained which was composed of a core of pure Be and a very brittle alloy skin of the composition 71.2% Be, 26.7 wt. % Ca [3] ($\sim Be_{10-13}Ca$).

1. A. R. Kaufmann, P. Gordon, and D. W. Lillie, *Trans. ASM*, **42,** 1950, 785–844.
2. W. Kroll and E. Jess, *Wiss. Veröffentl. Siemens-Konzern*, **10**(2), 1931, 30.
3. The Ca used contained 2.1 wt. % "foreign metals."

Be-Cb Beryllium-Columbium

See beryllium-niobium (Be-Nb).

$\bar{2}.8083$
1.1917

Be-Ce Beryllium-Cerium

An alloy in an undefined state containing 0.6 wt. % (0.04 at.) % Ce exhibited a second phase [1].

The intermetallic phase $Be_{13}Ce$ (54.45 wt. % Ce) has the $NaZn_{13}$ ($D2_3$) type of structure [2, 3], with $a = 10.375 \pm 1$ A [2].

1. A. R. Kaufmann, P. Gordon, and D. W. Lillie, *Trans. ASM*, **42,** 1950, 785–844.
2. N. C. Baenziger and R. E. Rundle, *Acta Cryst.*, **2,** 1949, 258.
3. W. C. Koehler, J. Singer, and A. S. Coffinberry, *Acta Cryst.*, **5,** 1952, 394.

$\bar{1}.1845$
0.8155

Be-Co Beryllium-Cobalt

After it was recognized [1] that the constitution of Co-rich Co-Be alloys is analogous to that of Ni-rich Ni-Be alloys, more detailed analyses were carried out by [2]

Table 16. Characteristics of the System Be-Co

	Ref. 2	Ref. 3
Eutectic (Co + BeCo)........	4.2 wt. (22.3 at.) % Be 1115°C	2.9 wt. (16.3 at.) % Be 1159°C
Saturation limit of the primary solid solution at the eutectic temperature...............	2.5 wt. (14.3 at.) % Be	1.59 wt. (9.5 at.) % Be
Solubility of Be at "room temperature".................	1 wt. (6 at.) % Be	1.1 wt. (6.8 at.) % Be
Magnetic-transformation temperature in the two-phase field (Co + BeCo).........	950°C	952°C
$\alpha \rightarrow \epsilon'$ transformation in the two-phase field (Co + BeCo):		
Heating...................	550°C	579°C
Cooling...................	265°C	305°C

(thermal, microscopic, and magnetic work, up to 10 wt. % Be) and [3] (with dilatometric studies in addition, investigated up to 4 wt. % Be). Characteristic data found by [2, 3] are listed in Table 16. It may be seen that [2, 3] agree on the broadening of the hysteresis range of the Co transformation by addition of Be [4].

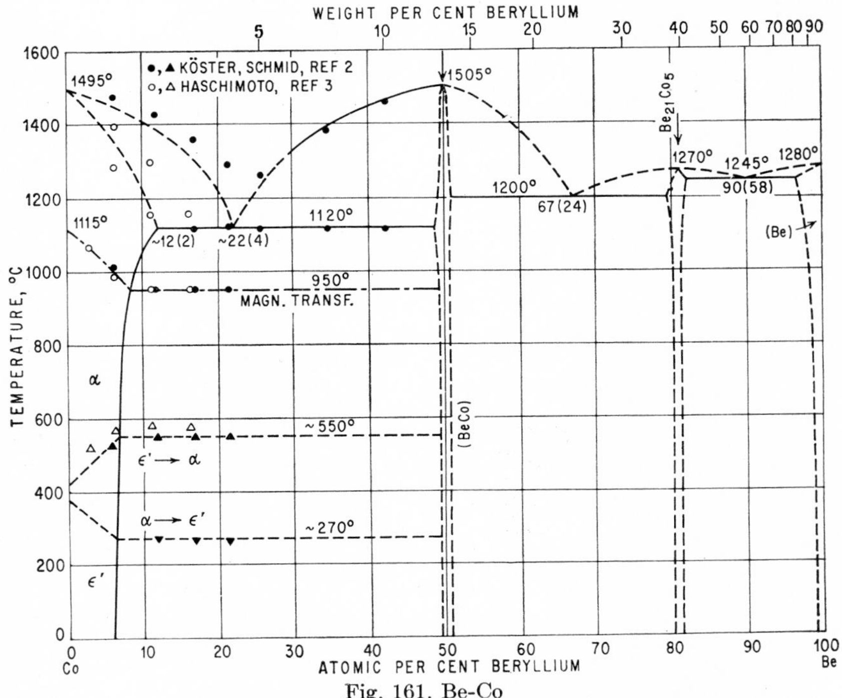

Fig. 161. Be-Co

Based on microscopic and roentgenographic studies as well as thermal determinations of characteristic points, the entire phase diagram was outlined by [5] (Fig. 161). Besides BeCo (13.27 wt. % Be), already reported by [6], the intermediate phase $Be_{21}Co_5$ [80.77 at. (39.1 wt.) % Be] was found (cf. [6]). For the Co-rich eutectic (cf. above) the data 26 at. (5.1 wt.) % Be, 1125°C were given by [5].

According to [7], solid Be dissolves 10–15 wt. (1.7–2.6 at.) % Co at 1050°C; separations at the grain boundaries of slowly cooled alloys indicate a decrease in solubility with falling temperature.

Crystal Structures. BeCo was shown to have the CsCl (B2) type of structure, with a = 2.61 A [6, 5]. Variations in the lattice parameter of slowly cooled alloys with the composition indicate the existence of a certain homogeneity range. $Be_{21}Co_5$ has the same structure as $Be_{21}Ni_5$ (deformed γ-brass type), with a = 7.68 A [5].

1. G. Masing, *Z. Metallkunde*, **20,** 1928, 21.
2. W. Köster and E. Schmid, *Z. Metallkunde*, **29,** 1937, 232–233. Transformation in Co: on heating, 430°C; on cooling, 395°C (taken from a graph).
3. U. Haschimoto, *Nippon Kinzoku Gakkai-Shi*, **2,** 1938, 70–71 (in Japanese, table and diagram in English); see also **1,** 1937, 177. The alloys were melted in graphite crucibles in a H_2 or N_2 atmosphere. Transformation in Co (99.9 wt. % pure): on heating, 467°C; on cooling, 426°C.
4. In Fig. 161 the hysteresis range for the transformation in pure Co was chosen between 420 and 380°C (cf. W. Köster, *Z. Metallkunde*, **43,** 1952, 299).
5. G. Venturello and A. Burdese, *Alluminio*, **20,** 1951, 558–564 (in Italian). Both Be and Co were 99.8 wt. % pure. The (25) alloys were melted in alumina crucibles in an inert atmosphere.
6. L. Misch, *Metallwirtschaft*, **15,** 1936, 163–166. Besides BeCo, an intermediate phase with a higher Be content was found.
7. A. R. Kaufmann, P. Gordon, and D. W. Lillie, *Trans. ASM*, **42,** 1950, 801–802.

$\bar{1}.2388$
0.7612

Be-Cr Beryllium-Chromium

The phase diagram up to 71 at. % Be (Fig. 162) has been established by [1], using thermal, microscopic, and X-ray methods. The solidus of the (Cr) phase below 1650°C was approximately determined by means of quenching experiments (microscopic "composition brackets"). The intermediate phase Be_2Cr (25.73 wt. % Be), detected by [2, 3], was shown [1] to have a small range of homogeneity only on its Be-rich side [4] with some change in solubility above 1400°C [5]. A 71 at. % Be alloy consisted of Be_2Cr with a considerable amount of a second brittle phase which has not been identified [1]. A 98.1 at. (90 wt.) % Be alloy was reported to be two-phase with 30–40% eutectic network [6]. It may be tentatively suggested, therefore, that—in analogy to the Be-Mo system—there exists a compound $Be_{13}Cr$ (92.86 at. % Be).

The solid solubility of Be in Cr was shown by means of X-ray back-reflection and micrographic studies on quenched alloys [1] to be 9.2, 8.6, 5.5, and 2.8 at. (1.7, 1.6, 1.0, and 0.5 wt.) % Be at 1500 (6 hr), 1435 (12 hr), 1300 (24 hr), and 1160°C (56 hr), respectively.

Crystal Structure. The lattice parameter of the (Cr) phase decreases with increasing Be content (positive deviation from additivity) [1]. Be_2Cr has the hexagonal $MgZn_2$ (C14) type of structure [2, 3, 1], with a = 4.27 A, c = 6.92 A, c/a = 1.62 (at 66.7 at. % Be) [1]. A 71 at. % Be alloy (beyond the homogeneity limit) "showed a distinctly smaller lattice parameter" [1].

1. A. R. Edwards and S. T. M. Johnstone, *Aeronaut. Research Lab., Dept. of Supply (Australia), Rept.* SM 161, 1950; also *J. Inst. Metals*, **84,** 1956, 313–317. Chromium: no metallic impurities, 0.06 oxygen, <0.001 nitrogen (wt. %); beryllium: 0.84 wt. % impurities (mainly Si, Al). The alloys were prepared and examined in an atmosphere of hydrogen or argon and analyzed for Be.
2. L. Misch, *Metallwirtschaft,* **15,** 1936, 163–166.
3. M. I. Zakharova and P. I. Dalnov, *Zhur. Tekh. Fiz.*, **8,** 1938, 252–255 (in Russian); also *Tech. Phys. U.S.S.R.*, **5**(3), 1938, 184–188 (in English).
4. [3] claimed to have prepared homogeneous alloys (monocrystals) containing only 56.5 at. (18.3 wt.) % Be. Another alloy with 68.8 at. % Be showed the same lattice parameters.
5. An alloy with 67.1 at. % Be was homogeneous at 1500°C but showed small traces of a second phase if annealed at 1435°C or lower [1].
6. A. R. Kaufmann, P. Gordon, and D. W. Lillie, *Trans. ASM*, **42,** 1950, 801.

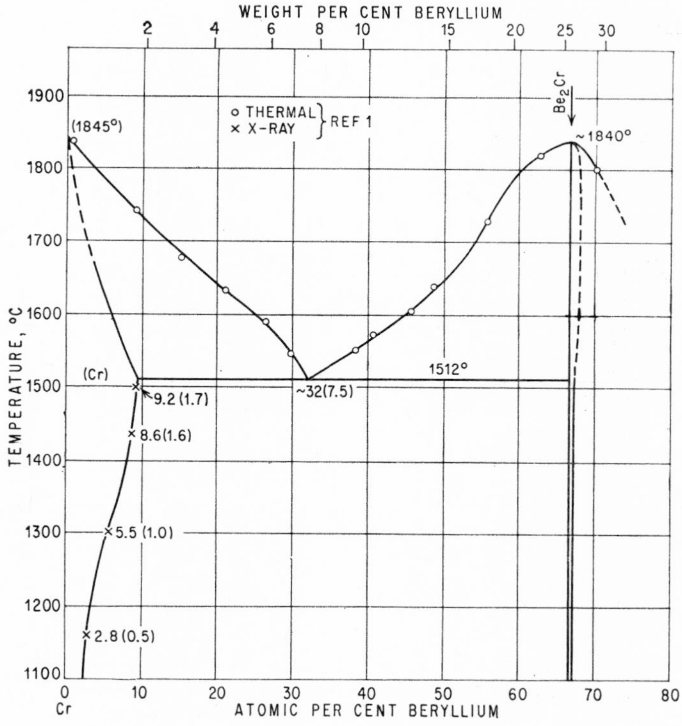

Fig. 162. Be-Cr

1̄.1518
0.8482

Be-Cu Beryllium-Copper

Data dealing with the constitution of this system, published prior to 1943, have been critically reviewed by [1]. The reader is expressly referred to this publication. On the basis of more recent work, certain corrections of the diagram proposed by [1] had to be made.

The α Phase. The liquidus was determined by [2–5]. Results of [2, 3, 4] are in good agreement, whereas the curve according to [5] lies at higher temperatures, with a maximal difference of about 40°C at 15 at. % Be. The curve shown in Fig. 163 is that given by [2, 3]. The solidus is known only qualitatively; the data reported [3, 5] were obtained by the analysis of cooling curves only. In Fig. 163 the results of

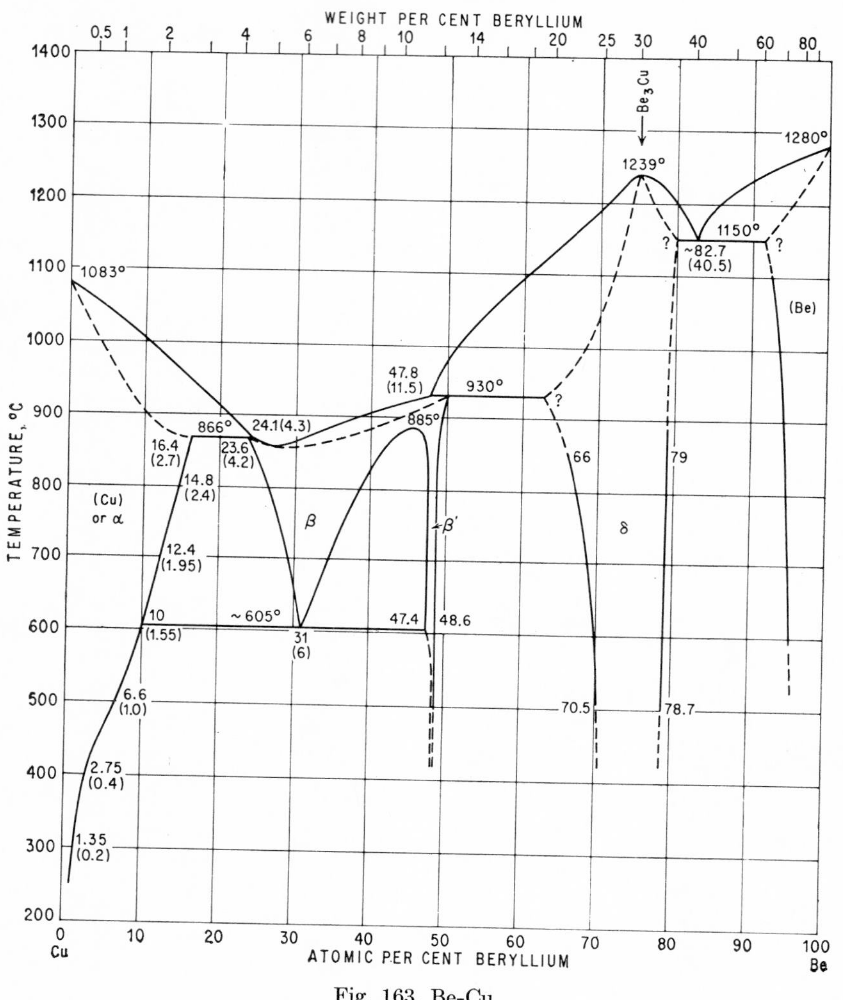

Fig. 163. Be-Cu

[3] have been preferred since they indicate higher temperatures; also, they are more consistent with the maximal solid solubility of Be in Cu at the peritectic temperature.

The boundary of the solid solubility was determined, in full or in part, by [6, 3, 7, 4, 5, 8, 9], using isothermal conductivity [6], micrographic [3, 4, 9], lattice-parametric [7], dilatometric [5, 8], and thermoresistometric methods [5]. The solubility at the eutectoid temperature, long accepted as about 575°C, is well established as 1.4 wt. (9.2 at.) % Be. At higher temperatures, the results of [6, 3, 5, 9] agree within approximately 0.3 wt. % and those of [6, 3, 9] within about 0.1–0.15 wt. %. There is excellent agreement between the data due to [6, 9], which have been used in Fig. 163. The solubility curve based on lattice-parameter determinations is steeper, reaching a maximal solubility of 2.1 wt. (13.1 at.) % Be at the peritectic temperature. Below the eutectoid temperature, results of the lattice-parameter study [7] are the most reliable ones; they closely agree with those of [8].

The β and β' Phases. The liquidus and (thermally determined) solidus curves in the range 4.3–10 wt. (24.1–44 at.) % Be agree on the whole [2, 3, 5, 9]. The minimum was found at 5.1–5.3 wt. (27.4–28.2 at.) % Be and about 847–860° C [3, 5, 9]. [4] reported that there was no minimum, i.e., that the horizontal at 863–870°C [2–5, 9] was a eutectic rather than a peritectic.

[2–5] claimed the existence of a peritectic horizontal at 920°C between about 11 and 11.3 wt. (46.5 and 47.3 at.) % Be [3], corresponding to the reaction: melt + β' $\rightleftarrows$ β [10]. No evidence for such a three-phase equilibrium was found by [9]. Instead, it was shown by micrographic analysis that there was a continuous transition $\beta \rightleftarrows \beta'$. Previously, [11] had already proposed that the boundary of the $\beta + \beta'$ field was a closed curve with a critical point.

The $\beta/(\alpha + \beta)$ boundary was determined by means of micrographic work [6, 3, 4, 9]. Results are in substantial agreement, especially those of [3, 6], shown in Fig. 163. For the boundary of the β field on the Be-rich side [6, 3, 4, 9], the most reliable micrographic data of [9] have been used. The critical point lies at 45–46 at. (10.4–10.8 wt.) % Be and about 885°C [9].

The composition of the eutectoid ($\alpha + \beta'$) is generally accepted as 6.0 wt. (31.0 at.) % Be [2, 6, 3]. 6.2 wt. (31.7 at.) % Be was proposed by [4]. The eutectoid temperature was reported as 573 [4], 576 [3], and 578°C [2, 5], on the basis of cooling-curve data. According to [12, 13], these values are too low. As more reliable temperatures, 601–618°C (on cooling and heating) [12], 614°C (heating) [13], 598°C (cooling) [13], and 608°C (isothermal-decomposition studies) [13] were given. The solid solubility of Be in Cu at 605°C is about 1.55 wt. (10.0 at.) % Be.

On the basis of thermal investigations [3], the homogeneity range of the β' phase [10] was assumed to extend from about 47.3 to 49.7 at. (11.3–12.3 wt.) % Be. [9] determined the upper portion of the β' field more accurately by means of micrographic work (Fig. 163). At 840°C it extends from 47.2 to 48.8 at. (11.25–11.9 wt.) % Be. For temperatures below the eutectoid decomposition of β, evidence was found of a narrowing of the β' field [4].

The δ Phase. The liquidus curve corresponding to the primary crystallization of δ was determined by [5]. Up to about 70 at. % Be, data by [2, 3] are in good agreement. The maximum at the composition Be_3Cu (29.85 wt. % Be) was found as about 1215 [3] and 1239°C [5]. The solidus curve [5] can be considered only a rough approximation. The temperature of the peritectic reaction, melt + $\delta \rightleftarrows \beta$, was found as 930 [2, 5, 9] and 933°C [3]. The solid-state boundaries were determined by means of microexamination of alloys annealed at and quenched from 850, 700, and 500°C [14].

The Range 82.5–100 At. % Be. Above 77.5 at. % Be only the thermal data reported by [5] are available. The boundary of the solid solution of Cu in Be was outlined by [15], using micrographic and X-ray analysis.

Crystal Structures. The lattice-constant-vs.-composition curve of the α phase was determined by [7]. For individual data see [16–18]. The β phase is b.c.c., disordered [19, 17], with $a = 2.79_5$ A at 35.4 at. % Be and 750°C (17). β', formerly designated as γ, is b.c.c., ordered (CsCl, B2 type) [16, 17, 20], $a = 2.69-2.70$ A (Be-rich) [16, 20]. The δ phase is isotypic with $MgCu_2$ (C15 type) [20, 5], $a = 5.952$ A (Cu-rich) [20, 5] and $a = 5.899$ A (Be-rich) [5]. As to evidence of short-range order in the β phase, see [21].

1. G. V. Raynor, The Equilibrium Diagram of the System Beryllium-Copper, Annotated Equilibrium Diagrams, no. 7, The Institute of Metals, London, 1949.
2. G. Oesterheld, *Z. anorg. Chem.*, **97**, 1916, 14–27.
3. H. Borchers, *Metallwirtschaft*, **11**, 1932, 317–321, 329–330.
4. K. Iwase and N. Okamoto, *Nippon Kinzoku Gakkai-Shi*, **5**, 1941, 82–84.
5. L. Losana and G. Venturello, *Alluminio*, **11**, 1942, 8–16.
6. G. Masing and O. Dahl, *Wiss. Veröffentl. Siemens-Konzern*, **8**, 1929, 94–100.
7. H. Tanimura and G. Wassermann, *Z. Metallkunde*, **25**, 1933, 179–181.
8. H. Borchers and H. J. Otto, *Metallwirtschaft*, **21**, 1942, 215–217.
9. N. K. Abrikosov, *Izvest. Sektora Fiz.-Khim. Anal.*, **21**, 1952, 101–115.
10. β' was formerly designated as γ.
11. W. Gruhl and G. Wassermann, *Metall*, **5**, 1951, 93–98, 141–145.
12. H. Thomas, *Z. Metallkunde*, **36**, 1944, 136–140.
13. R. H. Fillnow and D. J. Mack, *Trans. AIME*, **188**, 1950, 1229–1236.
14. N. K. Abrikosov, *Izvest. Sektora Fiz.-Khim. Anal.*, **21**, 1952, 116–120.
15. A. R. Kaufmann, P. Gordon, and D. W. Lillie, *Trans. ASM*, **42**, 1950, 785–844.
16. O. Dahl, E. Holm, and G. Masing, *Wiss. Veröffentl. Siemens-Konzern*, **8**, 1929, 154–186.
17. G. F. Kossolopow and A. K. Trapesnikow, *Metallwirtschaft*, **14**, 1935, 45–46.
18. J. Jitaka and S. Miyake, *Nature*, **137**, 1936, 457.
19. G. Wassermann, *Z. Metallkunde*, **26**, 1934, 257.
20. L. Misch, *Z. physik. Chem.*, **B29**, 1935, 42–58.
21. W. Hume-Rothery, P. W. Reynolds, and G. V. Raynor, *J. Inst. Metals*, **66**, 1940, 191–207.

$\bar{1}.2078$
0.7922

Be-Fe Beryllium-Iron

Fe-rich Alloys. The constitution of Fe-rich Fe-Be alloys was first investigated by thermal methods. [1] studied in particular the solidification equilibria in the range 0–62 at. % Be, and [2] the liquidus, the "γ loop," and the magnetic transformation in alloys with up to 13 at. % Be (see Fig. 164). The results of an investigation of alloys containing up to 54 at. % Be by means of several standard methods [3] and those of diffusion experiments [4] are in fair agreement with those of [1, 2]. The position of the Fe-rich eutectic in Fig. 164 has been taken from a more recent paper [5]; [1] found ~38 at. (9 wt.) % Be, 1155°C and [3] 41 at. (10 wt.) % Be, 1150°C as eutectic data.

The (maximum) solubility of Be in ferrite at the eutectic temperature was reported to be 6.5 wt. (30 at.) % Be by [1], and 7.5–8.0 wt. (33.4–35.0 at.) % Be by [3]. [3] claim to have determined the ferrite boundary down to room temperature by means of microscopic, roentgenographic, and dilatometric methods and assume a "room-temperature" solubility of 4.5–5.0 wt. (22.6–24.6 at.) % Be. They do not, however, mention any annealing treatment of their specimens (the published microstructures show cast alloys!); therefore their alloys appear to have been in a state of nonequilibrium. Age-hardening experiments by [6–9] strongly suggest that at room temperature the ferrite boundary lies at a considerably lower Be content [below 2 wt. (11 at.) % Be ?].

Be-rich Alloys. Be-rich alloys, especially, were the object of a recent thermal, microscopic, and X-ray diffraction study [5] (Fig. 164). The existence of three intermediate phases, reported previously by [1, 3, 10] (Be_2Fe) and [10] (Be_5Fe and another phase of higher Be content), was corroborated; and it was shown that two of them have wide ranges of homogeneity. The ternary character of the solidification equilibria of Be-rich alloys, caused by an unidentified impurity, has been eliminated in Fig. 164. There is fair agreement concerning a limited solid solubility of Fe in Be

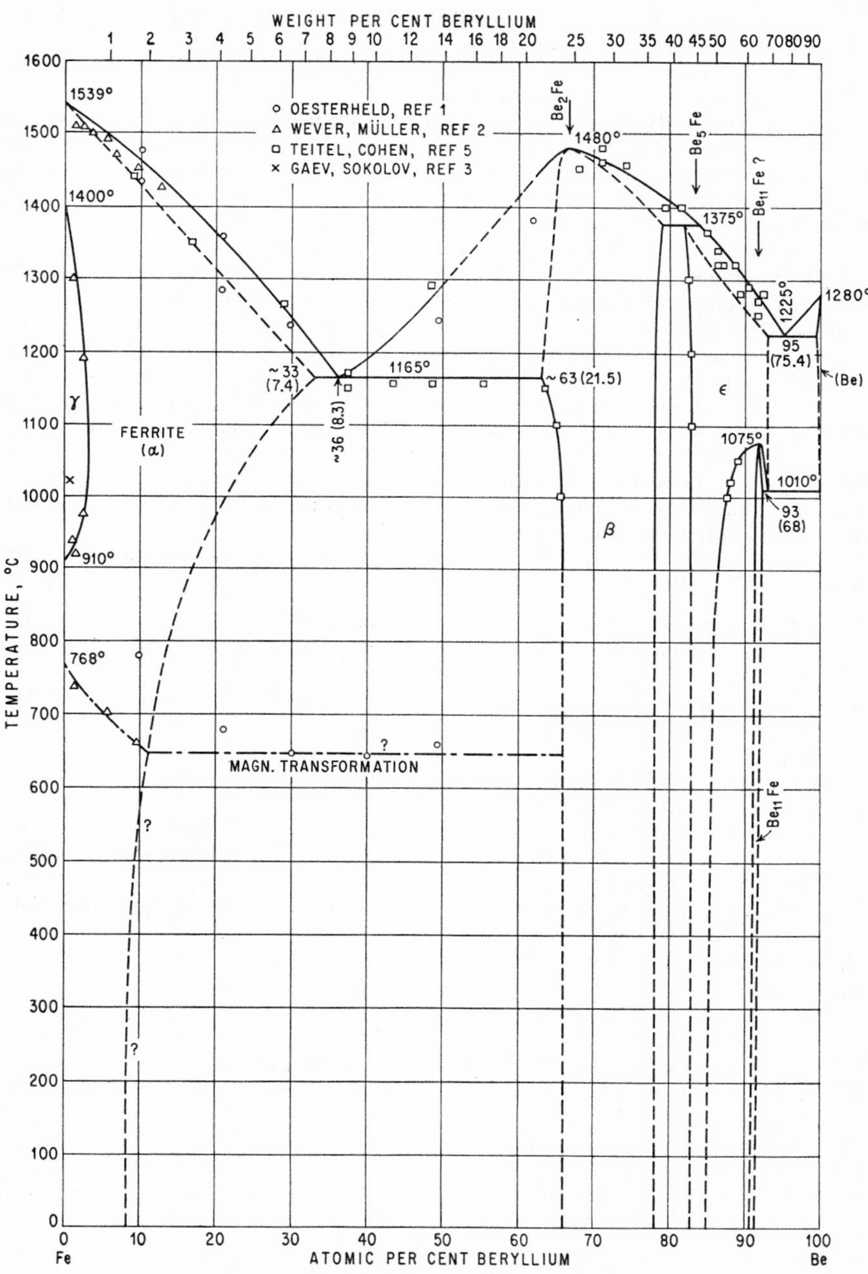
WEIGHT PER CENT BERYLLIUM
1 2 3 4 5 6 7 8 9 10 12 14 16 18 20 25 30 35 40 45 50 60 70 80 90
TEMPERATURE, °C
1600
1500
1400
1300
1200
1100
1000
900
800
700
600
500
400
300
200
100
0
1539°
1400°
910°
768°
○ OESTERHELD, REF 1
△ WEVER, MÜLLER, REF 2
□ TEITEL, COHEN, REF 5
× GAEV, SOKOLOV, REF 3
Be_2Fe
1480°
Be_5Fe
1375°
$Be_{11}Fe$?
1225°
1280°
95
(75.4)
(Be)
~33
(7.4)
~36 (8.3)
1165°
~63 (21.5)
γ
FERRITE
(α)
ε
1075°
1010°
93
(68)
β
MAGN. TRANSFORMATION
?
$Be_{11}Fe$
0 10 20 30 40 50 60 70 80 90 100
Fe
Be
ATOMIC PER CENT BERYLLIUM

Fig. 164. Be-Fe

11, 5, 12]; according to [5] the solubility at 1225°C is 0.4 at. (2.5 wt.) % Fe but decreases greatly at lower temperatures.

Crystal Structures. The lattice parameter of the ferrite phase decreases with increasing Be content [3, 5]; see also [2].

The homogeneity range of the "β" phase [5] includes the composition Be_2Fe (24.40 wt. % Be). The β phase was shown to have the hexagonal $MgZn_2$ (C14) type of structure [10, 3, 5], with a = 4.212 A, c = 6.853 A, c/a = 1.626 at 66.7 at. % Be [5]. The variation of the lattice parameters with composition was measured by [5]. Be_2Fe is ferromagnetic up to 521–524°C [10].

The "ϵ" phase [5] has a wide range of homogeneity at elevated temperatures [82–93 at. (42-68 wt.) % Be]; at lower temperatures the range narrows but still includes the composition Be_5Fe [83.33 at. (44.65 wt.) % Be]. This phase has the cubic $MgCu_2$ (C15) type of structure [10], with a = 5.884 A at the composition Be_5Fe [5]. For the variation of the lattice parameter with composition, obtained by solution quenching, see [5]. Ferromagnetism was found at liquid-air temperature only [10].

A further intermediate phase was located by [5] between about 91.8 and 92.2 at. (64.4–65.6 wt.) % Be. This corresponds to the formula $Be_{11}Fe$ [91.66 at. (64.0 wt.) % Be] or possibly $Be_{12}Fe$ [92.30 at. (66.0 wt.) % Be]. The powder pattern could be indexed with a hexagonal cell with a = 4.13 A, c = 10.71 A, c/a = 2.59 [5], containing 18 atoms (1½ formula weights: substructure or random arrangement of Fe and Be). No variation of parameter with composition was noted.

No change in lattice parameters could be detected in the very limited (Be) primary solid solution [5].

1. G. Oesterheld, *Z. anorg. Chem.*, **97**, 1916, 32–37. The Be was 99.5 wt. % pure the iron used melted at 1510°C and contained 0.07 C, 0.06 Si, 0.10 Mn, 0.03 (P, S, Cu) (wt. %).
2. F. Wever and A. Müller, *Mitt. Kaiser-Wilhelm-Inst. Eisenforsch. Düsseldorf*, **11**, 1929, 218–219; *Z. anorg. Chem.*, **192**, 1930, 337–340. Fe contained 0.01 C, 0.10 Si (wt. %).
3. J. S. Gaev and R. S. Sokolov, *Metallurg*, **4**, 1937, 42–48 (in Russian).
4. J. Laissus, *Rev. mét.*, **32**, 1935, 293, 351, 401.
5. R. J. Teitel and M. Cohen, *Trans. AIME*, **185**, 1949, 285–296 (the diagrams captioned as fig. 7 and fig. 9 must be interchanged!); discussion, *Trans. AIME*, **188**, 1950, 1028–1029. Also R. J. Teitel, *U.S. Atomic Energy Comm., Publ.* AECD-2251, 1948; abstract in *Structure Repts.*, **12**, 1949, 23–25. Raw materials used: Be 99.4, Fe 99.97 wt. % pure.
6. G. Masing, *Z. Metallkunde*, **20**, 1928, 21.
7. W. Kroll, *Wiss. Veröffentl. Siemens-Konzern*, **8**, 1929, 223–224.
8. K. Riedel, dissertation, Technische Hochschule, Aachen, 1929; not available for the author; abstract in "Gmelins Handbuch der anorganischen Chemie," System No. 59 A, p. 1815, Verlag Chemie, G.m.b.H., Berlin, 1939.
9. W. Aichholzer, *Berg-u. hüttenmänn. Monatsh. Montan. Hochschule Leoben*, **93**, 1948, 100–114; see also *Metal Progr.*, **57**, 1950, 254.
10. L. Misch, *Z. physik. Chem.*, **B29**, 1935, 42–58.
11. H. A. Sloman, *J. Inst. Metals*, **49**, 1932, 370–371.
12. A. R. Kaufmann, P. Gordon, and D. W. Lillie, *Trans. ASM*, **42**, 1950, 785–844.

$\bar{1}.1115$
0.8885

Be-Ga Beryllium-Gallium

"At 600°C Be dissolves in Ga to the extent of 4 ppm [4 × 10^{-4} wt. %], but Ga also diffuses very slowly into the Be to form a reaction zone [believed to consist of a protective compound]. At 800°C . . . the reaction layer . . . has an apparent solubility in the Ga of 0.9% by weight of Be" [1].

1. Quoted from L. R. Kelman, W. D. Wilkinson, and F. L. Yaggee, Resistance of Materials to Attack by Liquid Metals, ANL-4417, pp. 111–113, Argonne National Laboratory, July, 1950.

$\bar{1}.0939$
0.9061

Be-Ge Beryllium-Germanium

A 10 wt. (1.4 at.) % Ge alloy, as cooled from the melt, was two-phase with a light-colored compound at grain boundaries and within grains [1].

1. A. R. Kaufmann, P. Gordon, and D. W. Lillie, *Trans. ASM*, **42**, 1950, 801.

0.9514
$\bar{1}.0486$

Be-H Beryllium-Hydrogen

[1] "thought that he detected partial combination on heating the metal in hydrogen. The evidence is hardly conclusive; and, in the absence of other information, it can only be surmised that the behavior of beryllium is somewhat similar to that of magnesium" [2].

For the hydride BeH_2, a nonvolatile white powder, see the chemical literature.

1. C. Winkler, *Ber. deut. chem. Ges.*, **24**, 1891, 1966–1984, especially 1972–1973.
2. D. P. Smith, "Hydrogen in Metals," p. 282, University of Chicago Press, Chicago, 1948.

$\bar{2}.6525$
1.3475

Be-Hg Beryllium-Mercury

The solubility of Be in Hg has been estimated as increasing uniformly from 0.01×10^{-4} at 100°C to 0.4×10^{-4} wt. % Be at 800°C (0.22×10^{-4} and 8.9×10^{-4} at. % Be, respectively) [1]. Older information [2] said that Be forms no amalgam and seems to be insoluble in Hg.

1. L. R. Kelman, W. D. Wilkinson, and F. L. Yaggee, Resistance of Materials to Attack by Liquid Metals, ANL-4417, pp. 66–67, Argonne National Laboratory, July, 1950 (private communication by L. F. Epstein, Knolls Atomic Power Laboratory).
2. S. Bodforss, *Z. physik. Chem.*, **124**, 1926, 68.

$\bar{2}.6691$
1.3309

Be-Ir Beryllium-Iridium

According to an optical and X-ray diffraction investigation the compound Be_2Ir (91.46 wt. % Ir) exists. Its complex powder pattern is similar to that of Be_2Rh [1].

1. L. Misch, *Metallwirtschaft*, **15**, 1936, 163–166.

$\bar{1}.5689$
0.4311

Be-Mg Beryllium-Magnesium

According to microscopic [1–3] and thermal [1] investigations molten Mg—even if heated up to its boiling point (1110°C)—does not dissolve any Be (melting point

1280°C). Various other attempts [3] to alloy Be and Mg under normal pressure were also unsuccessful: pouring of fused Be into fused Mg followed by annealing, annealing of powder compacts, electrolytic deposition of Be on fused Mg, and reduction of Be fluoride by fused Mg.

The rate of oxidation of high-Mg alloys is decreased by small additions of Be [4, 5], although a spectral analysis [5] revealed only traces (some thousandths of a per cent) of dissolved (or only adsorbed?) Be.

Heating both metals up to the melting point of Be under hydrogen and a pressure of more than 100 atm in a ZrO_2 crucible [6] resulted in a microscopically homogeneous Mg phase containing (according to spectroscopic data) 0.5 wt. (1.34 at.) % Be, besides Be as a second constituent. The X-ray investigation, however, did not indicate any solubility of Be in Mg, since the lattice parameters of the 0.5 wt. % Be alloy were identical with those of pure Mg.

A cursory microscopic investigation of the solubility of Mg in solid Be [7] suggests a value of at least 0.05 wt. %. The short note by [8] that "Be can contain Mg in alloy form up to at least 1 wt. %" needs confirmation. The preparation of the compound $Be_{13}Mg$ by powder-metallurgical methods has been reported by [9]. It has the cubic $NaZn_{13}$ ($D2_3$) type of structure with $a = 10.166 \pm 5$ A.

1. G. Oesterheld, *Z. anorg. Chem.*, **97,** 1916, 14.
2. W. Kroll and E. Jess, *Wiss. Veröffentl. Siemens-Konzern,* **10,** 1931, 29–30.
3. R. J. M. Payne and J. L. Haughton, *J. Inst. Metals*, **49,** 1932, 363–364.
4. K. V. Peredelski, *Legkie Metal.*, **5,** 1936, 39–45 (in Russian); *Chem. Abstr.*, **31,** 1937, 1747.
5. F. Sauerwald, *Z. anorg. Chem.*, **258,** 1949, 296–306.
6. V. A. Pereslegin, *Trudy Tsentr. Gosudarst. Nauch.-Issledovatel Inst.*, **1937,** 162–173; abstract in *Met. Abstr.*, **7,** 1940, 109; "Gmelins Handbuch der anorganischen Chemie," System No. 27 A, pp. 456–459, Verlag Chemie, G.m.b.H., Weinheim/Bergstrasse, Germany, 1952.
7. A. R. Kaufmann, P. Gordon, and D. W. Lillie, *Trans. ASM*, **42,** 1950, 798–799.
8. C. B. Sawyer and B. Kjellgren, *Metals & Alloys*, **11,** 1940, 166.
9. T. W. Baker and J. Williams, *Acta Cryst.*, **8,** 1955, 519.

$\bar{1}.2151$
0.7849

Be-Mn Beryllium-Manganese

An alloy containing 10 wt. (1.8 at.) % Mn, as cooled from the melt, was two-phase (10 to 15% compound in the microstructure) [1].

The intermetallic phase Be_2Mn (75.29 wt. % Mn) has the hexagonal $MgZn_2$ (C14) type of structure with the lattice parameters $a = 4.240$ A, $c = 6.924$ A, $c/a = 1.632$ [2].

1. A. R. Kaufmann, P. Gordon, and D. W. Lillie, *Trans. ASM*, **42,** 1950, 801.
2. L. Misch, *Metallwirtschaft,* **15,** 1936, 163–166.

$\bar{2}.9728$
1.0272

Be-Mo Beryllium-Molybdenum

Microscopic and X-ray investigations of alloys produced by melting [1, 2] or diffusion [2] techniques agree on the existence of two intermediate phases. Be_2Mo (84.18 wt. % Mo) has the hexagonal $MgZn_2$ (C14) type of structure [1] with the lattice parameters $a = 4.433$ A, $c = 7.341$ A [2]; and the tetragonal [1] compound $Be_{13}Mo$ 7.14 at. (45.01 wt.) % Mo] [2] has 56 atoms in the unit cell and the lattice parameters

a = 10.27 A, c = 4.29 A [2]. The space group most probable for $Be_{13}Mo$ is $P42(D_4^1)$ [2].

An alloy containing 1 wt. % Mo was said to be two-phase with a eutectic [3].

The effect of small Be additions on some physical and technological properties of Mo was studied by [4]. Microscopic, hardness, and X-ray investigations indicate a certain solubility of Be in solid Mo. "It is evident that Mo will dissolve, at temperatures near the solidus, approximately 0.086 wt. [0.9 at.] % Be and that this quantity may be held in supersaturated solution near room temperature when cooled at rates prevalent in the cast ingot." Beyond the solid-solubility limit a eutectic at the grain boundaries was observed.

Supplement. In continuation of the work by [4], [5] outlined the Mo-rich portion of the phase diagram. There is a eutectic between the Mo phase and the compound Be_2Mo. The eutectic composition was estimated from microstructures to be approximately 6.5 wt. (42.5 at.) % Be. Initial melting tests placed the eutectic temperature slightly below 3400°F (1870°C). A tentative solvus line was deduced from hardness measurements: the solubility of Be in solid Mo (containing up to 0.05 wt. % carbon) decreases from 0.05 wt. (0.53 at.) % Be at the eutectic temperature to 0.04 wt. (0.42 at.) % at 1740°C and 0.02 wt. (0.2 at.) % Be at 1480°C. The lattice parameter of Mo is decreased by addition of Be, indicating the formation of a substitutional solid solution. The C14 structure for Be_2Mo was corroborated. A reinvestigation of the compound richest in Be showed it to have the composition $Be_{12}Mo$ rather than $Be_{13}Mo$. The correct structure is b.c. tetragonal, a = 7.271 ± 5 A, c = 4.234 ± 5 A, with 26 atoms per unit cell [6].

1. L. Misch, *Metallwirtschaft*, **15**, 1936, 163–166.
2. S. G. Gordon, J. A. McGurty, G. E. Klein, and W. J. Koshuba, *Trans. AIME*, **191**, 1951, 637–638.
3. A. R. Kaufmann, P. Gordon, and D. W. Lillie, *Trans. ASM*, **42**, 1950, 801.
4. Climax Molybdenum Company of Michigan, Arc-cast Molybdenum-base Alloys, Annual Report 031-331, April, 1950.
5. Climax Molybdenum Company of Michigan, Arc-cast Molybdenum-base Alloys, Second Annual Report 034-401, 1951.
6. R. F. Raeuchle and F. W. v. Batchelder, *Acta Cryst.*, **8**, 1955, 691–694.

$\bar{1}.8085$
0.1915

Be-N Beryllium-Nitrogen

Be to which 1 wt. (0.65 at.) % N was added showed characteristic needlelike inclusions (very probably Be_3N_2). Nitrogen thus appears to have a negligible solid solubility in Be [1].

Beryllium nitride, Be_3N_2 (50.89 wt. % N), melts at about 2200°C [2] but is appreciably volatile even below this temperature [2, 3]. Be_3N_2 has the cubic Mn_2O_3 ($D5_3$) type of structure [4, 3], with a = 8.15 ± 1 A [4].

For the rate of formation from the elements and considerations of Be_3N_2 stability, see [5].

1. A. R. Kaufmann, P. Gordon, and D. W. Lillie, *Trans. ASM*, **42**, 1950, 785–844, especially 796–800.
2. F. Fichter and E. Brunner, *Z. anorg. Chem.*, **93**, 1915, 84–94.
3. P. Chiotti, *J. Am. Ceram. Soc.*, **35**, 1952, 123–130.
4. M. v. Stackelberg and R. Paulus, *Z. physik. Chem.*, **B22**, 1933, 305–322.
5. E. A. Gulbransen and K. F. Andrew, *J. Electrochem. Soc.*, **97**, 1950, 383–395.

$\bar{2}.9868$
1.0132

Be-Nb Beryllium-Niobium

An alloy containing 10 wt. (1.1 at.) % Nb, as cooled from the melt, was two-phase (compound at grain boundaries) [1].

1. A. R. Kaufmann, P. Gordon, and D. W. Lillie, *Trans. ASM*, **42**, 1950, 801.

$\bar{1}.1863$
0.8137

Be-Ni Beryllium-Nickel

With detailed previous work on Ni-rich alloys [1–3] and an X-ray diffraction analysis [4] of the system as a basis, [5] established the phase diagram using mainly thermal, microscopic, and roentgenographic methods (Fig. 165). There exist two

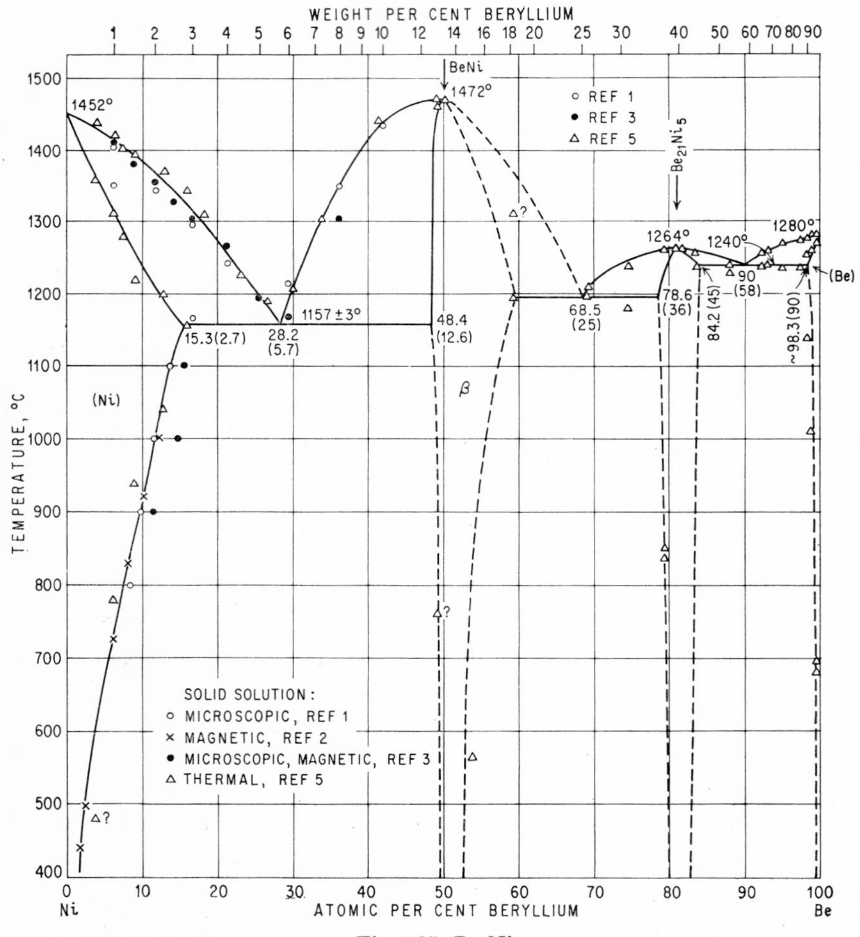

Fig. 165. Be-Ni

congruently melting compounds, BeNi (13.31 wt. % Be) and $Be_{21}Ni_5$ [80.77 at. (39.21 wt.) % Be], both of which exhibit limited ranges of homogeneity. The results of [1, 3, 5] agree fairly well on the position of the Ni-rich eutectic [28.2 at. (5.7 wt.) % Be; 1157 ± 3°C, in Fig. 165]. The solid solubility of Be in Ni was determined by various methods as indicated in Fig. 165. The values of [1] (between 1100 and 800°C) and [2] (between 440 and 1000°C) are the most reliable ones; the interpolated curve of Fig. 165 shows a solubility of 11.7, 7.3, 3.7, and 1.8 at. (2.0, 1.2, 0.6, and 0.3 wt.) % Be at 1000, 800, 600, and 440°C, respectively, which makes precipitation hardening [1–3] feasible. The solubility of Ni in solid Be is about 10 wt. (1.7 at.) % [5, 6].

Crystal Structures. The lattice parameter of the Ni primary solid solution decreases with increasing Be content [3, 5]. According to [2], the Curie point of Ni falls linearly with increasing Be (at. %) content and reaches 50°C at the (supersaturated) 13.5 at. (2.3 wt.) % Be alloy [7]. BeNi has the CsCl (B2) type of structure [4]. The lattice parameter decreases with increasing Be content: a = 2.62 A at 50.2 at. (13.4 wt.) % Be [4], a = 2.61 A at 52.5 at. (14.5 wt.) % Be [5]. $Be_{21}Ni_5$ has a deformed γ-brass type of structure [4, 5] with the pseudo-cubic lattice parameter a = 7.58 A at 41–42 wt. (81.9–82.5 at.) % Be [4]. The lattice parameters of the Be primary solid solution increase with increasing Ni content [5].

1. G. Masing and O. Dahl, *Wiss. Veröffentl. Siemens-Konzern*, **8,** 1929, 211–214.
2. W. Gerlach, *Z. Metallkunde*, **29,** 1937, 124–131; the values plotted in Fig. 165 were taken from a graph.
3. M. Okamoto, *Nippon Kinzoku Gakkai-Shi*, **3,** 1939, 444–448 (in Japanese; summary and diagrams in English). In the discussion of the paper by Misch [4], Okamoto erroneously refers to "Be_2Ni_5" instead of $Be_{21}Ni_5$.
4. L. Misch, *Z. physik. Chem.*, **B29,** 1935, 42–58.
5. L. Losana and C. Goria, *Alluminio*, **11,** 1942, 17–22 (in Italian). On p. 22, left column, second line, read $NiBe_4$ instead of NiBe. No third compound was found in this paper as was erroneously reported in *Met. Abstr.*, **11,** 1944, 216.
6. A. R. Kaufmann, P. Gordon, and D. W. Lillie, *Trans. ASM*, **42,** 1950, 801–802.
7. Okamoto [3] also measured Curie points of several alloys. The high value 90°C, found for an alloy with 2.7 wt. (15.3 at.) % Be, may be due to incomplete quenching, a source of error which Gerlach [2] noted at higher Be contents.

$\bar{2}.5801$
1.4199

Be-Np Beryllium-Neptunium

An exploratory X-ray analysis by [1] showed that only one intermediate phase exists at room temperature in the Be-Np system. This phase has a composition of approximately $Be_{13}Np$ [7.14 at. (66.9 wt.) % Np] and is not melted by heating to 1400°C. It is f.c.c., with X-ray reflections comparable in intensity with those from $Be_{13}U$ ($NaZn_{13}$ or $D2_3$ type). The lattice parameter decreases from 10.266 A on the Be-rich side to 10.256 A on the Np-rich side, indicating a small range of homogeneity.

1. O. J. C. Runnalls, *Acta Cryst.*, **7,** 1954, 222–223.

$\bar{1}.7508$
0.2492

Be-O Beryllium-Oxygen

The brittleness of even high-purity beryllium has often been attributed to an oxygen content (as solute or as BeO), but unequivocal evidence for this view is lacking [1]. X-ray measurements [2] indicate that the solid solubility of oxygen in Be is very restricted. A eutectic (Be + BeO) network, observed by [3] in a 0.2 wt. % O alloy, could not be found by [4]. The long-known [5] compound BeO has the wurtzite (B4) type of structure [6, 7], with $a = 2.70_0$ A, $c = 4.40_1$ A, $c/a = 1.63$ [8]. The data for the melting point of BeO reported by [9–12] lie within the range 2300–2525°C. For kinetic studies of the oxidation of beryllium, see [13, 14].

Supplement. [15] have redetermined the lattice parameters of BeO: $a = 2.698$ A, $c = 4.380$ A, $c/a = 1.623$ (26°C). [16] studied, by means of electron diffraction, the orientation and growth of oxide layers formed by thermal and anodic oxidation of beryllium surfaces.

1. See A. R. Kaufmann, P. Gordon, and D. W. Lillie, *Trans. ASM*, **42**, 1950, 785–844, especially 786–787, for a short discussion of the literature (up to 1948) involved.
2. See [1], pp. 803–805.
3. H. A. Sloman, *J. Inst. Metals*, **49**, 1932, 365–390.
4. See [1], p. 796.
5. See "Gmelins Handbuch der anorganischen Chemie," System No. 216, p. 82, Verlag Chemie, G.m.b.H., Berlin, 1930.
6. W. L. Bragg, *Phil. Mag.*, (6)**40**, 1920, 180.
7. For further references, see *Strukturbericht*, **1**, 1913-1928, 115–117.
8. H. Nitka, *Naturwissenschaften*, **29**, 1941, 336–337.
9. O Ruff, H Seiferheld, and J. Suda, *Z. anorg. Chem.*, **82**, 1913, 375.
10. E. Tiede and E. Birnbräuer, *Z. anorg. Chem.*, **87**, 1914, 155.
11. F. Fichter and E. Brunner, *Z. anorg. Chem.*, **93**, 1915, 86.
12. O. Ruff and G. Lauschke, *Z. anorg. Chem.*, **97**, 1916, 80.
13. E. A. Gulbransen and K. F. Andrew, *J. Electrochem. Soc.*, **97**, 1950, 383–395.
14. D. Cubicciotti, *J. Am. Chem. Soc.*, **72**, 1950, 2084–2086.
15. H. E. Swanson and E. Tatge, Standard X-ray Powder Diffraction Patterns, *Natl. Bur. Standards (U.S.) Circ.* 539 (I), p. 36, 1953.
16. I. S. Kerr and H. Wilman, *J. Inst. Metals*, **84**, 1956, 379–385.

$\bar{2}.6757$
1.3243

Be-Os Beryllium-Osmium

An alloy with 10 wt. (0.52 at.) % Os, as cooled from the melt, was two-phase with 20–30% eutectic [1]. According to an optical and X-ray investigation [2], the compound Be_2Os (91.34 wt. % Os) exists. Its powder pattern is similar to that of Be_2Ru.

1. A. R. Kaufmann, P. Gordon, and D. W. Lillie, *Trans. ASM*, **42**, 1950, 801.
2. L. Misch, *Metallwirtschaft*, **15**, 1936, 163–166.

$\bar{1}.4639$
0.5361

Be-P Beryllium-Phosphorus

The compound Be_3P_2 (69.61 wt. % P) was shown to have the cubic Mn_2O_3 ($D5_3$) type of structure, with $a = 10.17 \pm 3$ A [1]. It decomposes quickly in moist air.

1. M. v. Stackelberg and R. Paulus, *Z. physik. Chem.*, **B22**, 1933, 305–322.

$\bar{2}.9267$
1.0733

Be-Pd Beryllium-Palladium

Thermal analysis of alloys, prepared using Be of about 98 wt. % purity and melted in corundum crucibles in a reduced hydrogen atmosphere, gave the phase diagram presented in Fig. 166 [1]. The composition of the alloys was estimated to be accurate within 1 at. %. The existence of the phases $BePd_3$ (2.74 wt. % Be), $BePd_2$ (4.05 wt. % Be), and BePd (7.79 wt. % Be) was confirmed by metallographic studies; that of BePd had been reported before [2]. The compositions of the other three intermediate

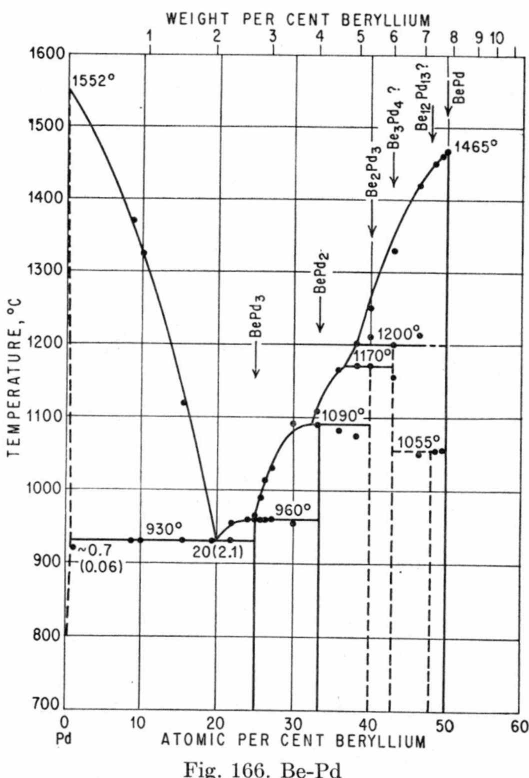

Fig. 166. Be-Pd

phases indicated in Fig. 166 were tentatively given as Be_2Pd_3 (5.33 wt. % Be), $BePd_{1.3}$ (Be_3Pd_4?) with 6.10 wt. % Be, and $BePd_{1.08}$ ($Be_{12}Pd_{13}$?) with 7.25 wt. % Be [3]. Some evidence of the peritectoid reaction at 1055°C was found; however, this was not conclusive as regards the composition of the phase formed. The solid solubility of Be in Pd was found to be between 0.6 and 1.0 at. (0.05 and 0.09 wt.) % at 950°C and less than 0.6 at. % at 400°C. Another compound, Be_5Pd (29.69 wt. % Be), was identified by determination of its crystal structure [4].

A "rough determination" of the solid solubility of Pd in Be, using microexamination and lattice-parameter measurements on quenched alloys, gave the following compositions in at. % Pd (wt. % in parentheses): 1200°C, 3.2 (28.0); 1000°C, 1.4 (14.4); 800°C, 0.9 (9.7); 600°C, 0.7 (7.7) [5].

The crystal structure of BePd is of the CsCl (B2) type, a = 2.819 A [2]; that of Be_5Pd, similar to the $MgCu_2$ (C15) type, a = 5.994 A [4].

1. O. Winkler, *Z. Metallkunde,* **30,** 1938, 170–171.
2. L. Misch, *Metallwirtschaft,* **15,** 1936, 163–166.
3. In the original paper, the compositions of $BePd_{1.3}$ and $BePd_{1.08}$ were erroneously given as $PdBe_{1.3}$ and $PdBe_{1.08}$, respectively.
4. L. Misch, *Metallwirtschaft,* **14,** 1935, 897–899.
5. A. R. Kaufmann, P. Gordon, and D. W. Lillie, *Trans. ASM,* **42,** 1950, 785–844.

$\bar{2}.6643$
1.3357

Be-Pt Beryllium-Platinum

The system Pt-Be has not been systematically investigated as yet. Alloys with a low Be content were studied with regard to microstructures and several mechanical and electrical properties by [1, 2]. According to microscopic observations [1], the solubility of Be in solid Pt is 0.06 wt. (1.28 at.) % Be and changes little with temperature [3].

An alloy with 80 wt. (98.9 at.) % Be was two-phase with 20% compound as a eutectic network [4]. In cursory X-ray work on the system [5], only one compound—composition about Pt_5Be_{21} [6]—could be detected, the structure of which seems to be a deformed γ-brass type.

1. K. W. Fröhlich, *Degussa Metallber.*, **1941,** 114–135 (not available to the author), abstract in "Gmelins Handbuch der anorganischen Chemie," System No. 68 A (6), pp. 748–749, Verlag Chemie, G.m.b.H., Weinheim/Bergstrasse, Germany, 1951.
2. V. A. Nemilov and A. A. Rudnitsky, *Izvest. Sektora Platiny,* **21,** 1948, 239–241 (in Russian). Five alloys with up to 0.58 wt. (11.21 at.) % Be were prepared and annealed for 6 days at 1200°C (cooling rate?).
3. Mainly on the basis of hardness measurements on annealed alloys (cooling rate?), [2] concluded a solubility of 0.25 wt. (5.15 at.) % Be. Their results, however, can also be interpreted by a much lower solubility limit (~0.15 wt. % Be).
4. A. R. Kaufmann, P. Gordon, and D. W. Lillie, *Trans. ASM,* **42,** 1950, 801.
5. L. Misch, *Metallwirtschaft,* **15,** 1936, 165.
6. Short note in L. Misch, *Z. physik. Chem.,* **B29,** 1935, 42.

$\bar{2}.5765$
1.4235

Be-Pu Beryllium-Plutonium

According to [1, 2], $Be_{13}Pu$ has the cubic $NaZn_{13}$ ($D2_3$) type of structure. The compound exhibits a measurable range of homogeneity, with a = 10.284 ± 1 A at the Be-rich limit and a = 10.278 ± 1 A at the Pu-rich limit [1]. The phase diagram shown in Fig. 166*a* is due to [2].

1. Review by A. S. Coffinberry and M. B. Waldron in "Metallurgy and Fuels," Progress in Nuclear Energy, ser. V, vol. 1, pp. 354–410, Pergamon Press Ltd., London, 1956. See also the somewhat earlier review (by A. S. Coffinberry and F. H. Ellinger) in "Proceedings of the International Conference on the Peaceful Uses of Atomic Energy," vol. 9, pp. 138–146, Geneva, 1955.
2. S. T. Konobeevsky, "Proceedings of the Academy of Science of the U.S.S.R. on the Peaceful Uses of Atomic Energy," Chemical Science Volume, p. 362, 1955; *Met. Abstr.,* **24,** 1956, 153. Phase diagram in wt. % reproduced in [1] ("Metallurgy and Fuels").

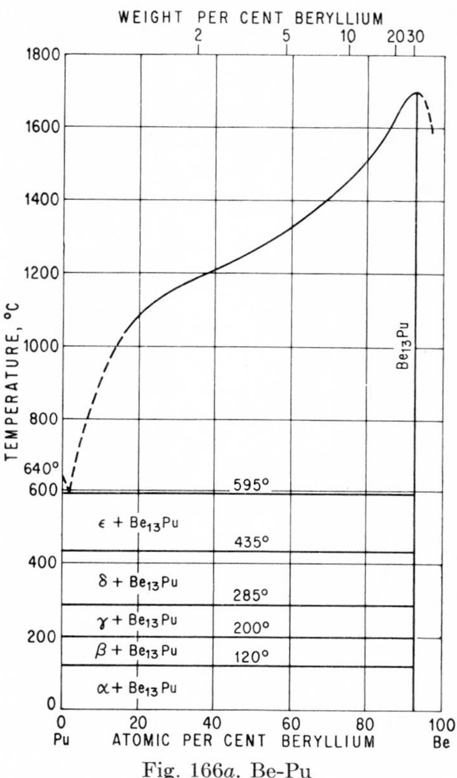

Fig. 166*a*. Be-Pu

$\bar{2}.6846$
1.3154

Be-Re Beryllium-Rhenium

The existence of the compound Be_2Re (91.18 wt. % Re) was indicated by X-ray investigation. Its structure belongs to the hexagonal $MgZn_2$ (C14) type, with a = 4.354 A, c = 7.101 A, c/a = 1.631 [1].

1. L. Misch, *Metallwirtschaft,* **15,** 1936, 163–166.

$\bar{2}.9424$
1.0576

Be-Rh Beryllium-Rhodium

An alloy with 10 wt. (1 at.) % Rh, as cooled from the melt, was two-phase with a eutectic network [1]. According to an optical and X-ray investigation [2], a compound Be_2Rh (85.10 wt. % Rh) exists. Its complex powder pattern, is similar to that of Be_2Ir.

1. A. R. Kaufmann, P. Gordon, and D. W. Lillie, *Trans. ASM,* **42,** 1950, 801.
2. L. Misch, *Metallwirtschaft,* **15,** 1936, 163–166.

$\bar{2}.9476$
1.0524

Be-Ru Beryllium-Ruthenium

An alloy with 10 wt. (1 at.) % Ru, as cooled from the melt, was two-phase and showed, besides about 20% eutectic, a fine precipitate in primary Be [1]. According to an optical and roentgenographic investigation [2], a compound Be_2Ru (84.94 wt. % Ru) exists. Its powder pattern is similar to that of Be_2Os.

1. A. R. Kaufmann, P. Gordon, and D. W. Lillie, *Trans. ASM*, **42,** 1950, 801.
2. L. Misch, *Metallwirtschaft*, **15,** 1936, 163–166.

$\bar{1}.4488$
0.5512

Be-S Beryllium-Sulfur

An alloy in an undefined state containing 0.05 wt. (0.015 at.) % S was microscopically homogeneous [1]. The compound BeS (78.06 wt. % S) has the zincblende (B3) type of structure, with $a = 4.86$ A [2].

1. A. R. Kaufmann, P. Gordon, and D. W. Lillie, *Trans. ASM*, **42,** 1950, 799.
2. W. Zachariasen, *Z. physik. Chem.*, **119,** 1926, 210.

$\bar{1}.0575$
0.9425

Be-Se Beryllium-Selenium

An alloy in an undefined state containing 0.05 wt. (0.006 at.) % Se was microscopically homogeneous [1]. The compound BeSe (89.75 wt. % Se) has the zincblende (B3) type of structure, with $a = 5.139 \pm 4$ A [2].

1. A. R. Kaufmann, P. Gordon, and D. W. Lillie, *Trans. ASM*, **42,** 1950, 799.
2. W. Zachariasen, *Z. physik. Chem.*, **124,** 1926, 440.

$\bar{1}.5063$
0.4937

Be-Si Beryllium-Silicon

The simple eutectic diagram by [1] (Fig. 167) differs from vague older reports [2] and a later theoretical paper [3] on the existence of a silicide. It has been repeatedly corroborated, however, that elementary Si is in equilibrium with Be [4, 5]. According to microscopic [4, 5] and roentgenographic [6] work, there seems to be only a negligible solubility of Si in solid Be. The solid solubility of Be in Si (utilized in semiconductor work) has apparently not been determined accurately as yet.

Supplement. [7] redetermined the eutectic data: 38.5 at. (66.1 wt.) % Si, 1090°C (cf. Fig. 167).

1. G. Masing and O. Dahl, *Wiss. Veröffentl. Siemens-Konzern*, **8,** 1929, 255–256; Be 99.8 wt. % pure, Si from Kahlbaum.
2. "Gmelins Handbuch der anorganischen Chemie," System No. 26, pp. 161–162, Verlag Chemie, G.m.b.H., Berlin, 1930.
3. G. V. Raynor, *J. Roy. Aeronaut. Soc.*, **50,** 1946, 410.
4. H. A. Sloman, *J. Inst. Metals*, **49,** 1932, 365–388.
5. A. R. Kaufmann, P. Gordon, and D. W. Lillie, *Trans. ASM*, **42,** 1950, 785–844, especially 795–801.
6. See [5], pp. 803–805.
7. O. Hájiček, *Hutnické Listy*, **3,** 1948, 265–270.

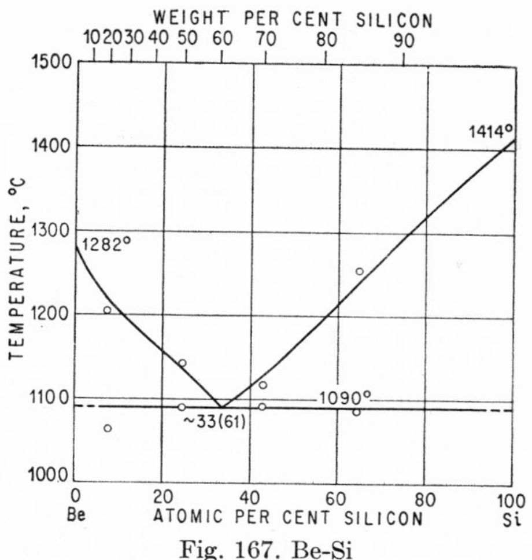

Fig. 167. Be-Si

$\bar{2}.8804$
1.1196

Be-Sn Beryllium-Tin

No alloying of Be and Sn can be attained below the melting point of Be (1280°C) [1]. From microscopic observations of alloys heated beyond the melting point of Be [1, 2], it can be concluded that a very extended miscibility gap occurs in the molten state. Solid Sn dissolves a few tenths of a (wt.) per cent of Be, as chemical and spectrographical analyses and changes in some mechanical properties indicate [1].

1. W. Guertler and M. Pirani, *Intern. Tin Research Develop. Council, Tech. Publ.* A, no. 50, 1937. (Contains several 50:50 wt. % microstructures etched in various ways.)
2. A. R. Kaufmann, P. Gordon, and D. W. Lillie, *Trans. ASM*, **42**, 1950, 801.

$\bar{2}.6975$
1.3025

Be-Ta Beryllium-Tantalum

An alloy containing 1 wt. (0.05 at.) % Ta, as cooled from the melt, was two-phase and showed a eutectic network [1]. A compound, probably tetragonal, was observed by [2] in this system; the composition (Be-rich?) is not reported.

1. A. R. Kaufmann, P. Gordon, and D. W. Lillie, *Trans. ASM*, **42**, 1950, 801.
2. L. Misch, *Metallwirtschaft*, **15**, 1936, 163–166.

$\bar{2}.8490$
1.1510

Be-Te Beryllium-Tellurium

BeTe (93.40 wt. % Te) has—in analogy to the sulfide and selenide—the zincblende (B3) type of structure with $a = 5.626 \pm 6$ A [1].

1. W. Zachariasen, *Z. physik. Chem.*, **124**, 1926, 277–284.

$\bar{2}.5892$
1.4108

Be-Th Beryllium-Thorium

An alloy containing 10 wt. (0.4 at.) % Th, as cooled from the melt, was two-phase with 5–10% angular compound in long stringers [1]. An intermetallic compound

$Be_{13}Th$ [7.14 at. (66.44 wt.) % Th] was found to have the cubic $NaZn_{13}$ ($D2_3$) type of structure [2, 3], with $a = 10.395 \pm 1$ A [2].

Supplement. [4] used (presumably unpublished) data by [5] to construct the phase diagram between pure Th and the compound $Be_{13}Th$. The liquidus drops from the melting point of Th (~1800°C) to a eutectic point at 1.75 wt. (31.5 at.) % Be, 1215°C, between Th and $Be_{13}Th$. The melting point of $Be_{13}Th$ is not given. No solid solubilities are shown.

1. A. R. Kaufmann, P. Gordon, and D. W. Lillie, *Trans. ASM*, **42,** 1950, 801.
2. N. C. Baenziger and R. E. Rundle, *Acta Cryst.*, **2,** 1949, 258.
3. W. C. Koehler, J. Singer, and A. S. Coffinberry, *Acta Cryst.*, **5,** 1952, 394.
4. J. R. Keeler, *Chem. Eng. Progr., Symposium Ser.*, **50,** no. 11, pp. 57–61 (Thorium Metallurgy), 1954.
5. H. A. Wilhelm et al., Iowa State College; F. Foote et al., Argonne National Laboratory.

$\bar{1}.2745$
0.7255

Be-Ti Beryllium-Titanium

By X-ray diffraction studies of sintered preparations with up to 65 wt. % Be, two intermediate phases were identified: BeTi (15.84 wt. % Be) or Be_3Ti_4 (12.37 wt. % Be), and Be_2Ti (27.34 wt. % Be), the latter being apparently of slightly variable composition [1]. X-ray patterns of preparations of the compositions Be_4Ti (42.94 wt. % Be) and $Be_{10}Ti$ (65.30 wt. % Be) revealed the existence of two additional beryllides with unknown composition and structure [1]. The compound $Be_{12}Ti$ (69.31 wt. % Be) has been identified by determination of its crystal structure [2].

It was reported that 2 wt. % Be lowers the melting point of Ti to approximately 1300°C [3]; liquid phase was found to be present in samples (prepared by powder metallurgy) with 3 and 10 wt. % Be on sintering at 975 and 950°C, respectively [4]. The solubility of Be in β-Ti at 950°C "apparently is between 1 and 2 wt. %"; that of Be in α-Ti, even at temperatures near the transformation temperature, "is considerably less than 1 wt. %" [5] but higher than 0.15 wt. % Be at 790°C [6]. The microstructure of an alloy with nominal 1 wt. % Ti showed considerable quantities of a second phase in a eutectic-like arrangement [7].

Crystal Structures. The structure of the intermediate phase coexisting with α-Ti, BeTi or Be_3Ti_4, could not be determined [1]; however, "a simple AB structure certainly does not exist." Be_2Ti is of the $MgCu_2$ (C15) type, $a = 6.448$ A [8] or $a = 6.44_0$ A [1]. $Be_{12}Ti$ is hexagonal with 48 "molecules" per unit cell, $a = 29.44 \pm 1$ A, $c = 7.33 \pm 1$ A, $c/a = 0.249$ [2].

1. P. Ehrlich, *Z. anorg. Chem.*, **259,** 1949, 1–41.
2. R. F. Raeuchle and R. E. Rundle, *Acta Cryst.*, **5,** 1952, 85–93.
3. W. Kroll, *Z. Metallkunde*, **29,** 1937, 189–192.
4. E. I. Larsen, E. F. Swazy, L. S. Busch, and R. H. Freyer, *Metal Progr.*, **55,** 1949, 359–361.
5. C. M. Craighead, O. W. Simmons, and L. W. Eastwood, *Trans. AIME*, **188,** 1950, 485–513.
6. R. M. Goldhoff, L. H. Shaw, C. M. Craighead, and R. I. Jaffee, *Trans. ASM*, **45,** 1953, 941–965.
7. A. R. Kaufmann, P. Gordon, and D. W. Lillie, *Trans. ASM*, **42,** 1950, 785–844.
8. L. Misch, *Metallwirtschaft*, **15,** 1936, 163–166.

$\bar{2}.5782$
1.4218

Be-U Beryllium-Uranium

The diagram of Fig. 168 is based mainly on thermal analysis and metallographic examination of as-cast samples [1]. The formula $Be_{13}U$ [7.14 at. (67.00 wt.) % U] of the sole compound, already established by X-ray [2, 3] and neutron-diffraction [3] work, was corroborated, whereas [4] gave the formula Be_9U. The constitution of the partial system Be-$Be_{13}U$ (see Fig. 168) should be reinvestigated as it appears somewhat doubtful whether thermal analysis, with optical pyrometer (as used for temperatures above 1300°C), can register effects which indicate formation of two liquids. In

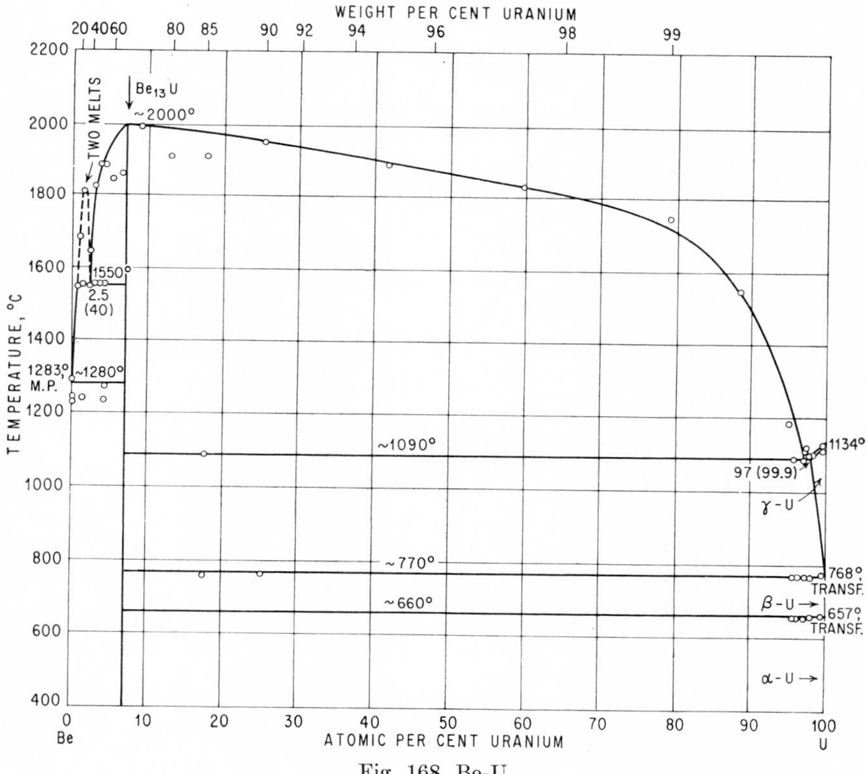

Fig. 168. Be-U

addition, it is difficult to ascertain a miscibility gap in the liquid state with a composition range as small as 1.5 at. %. Thermal arrests at about 1250°C with Be and Be-rich alloys (Fig. 168) are ascribed [1] either to impurities or to an allotropic transformation of Be. (Cf. Be-Fe.)

No solid solubility of U in Be [5, 1], or of Be in α- and β-U [1], could be observed. However, γ-U dissolves about 0.06 wt. ($\sim$1.6 at.) % Be at 1075°C, as X-ray investigations of quenched alloys indicate [1].

$Be_{13}U$ has the cubic $NaZn_{13}$ ($D2_3$) type of structure [2, 3]. The published lattice parameters—10.256 [2], 10.28 [3], and 10.370 A [1]—differ considerably, probably because of a homogeneity range of the phase [6].

1. R. W. Buzzard, *J. Research Natl. Bur. Standards*, **50**, 1953, 63–67; also R. W. Buzzard, J. T. Sterling, E. A. Buzzard, and J. H. Darr, *U.S. Atomic Energy Comm., Publ.* AECD-3417, 1952. The alloys were prepared in beryllia crucibles in an inert atmosphere. For an older tentative phase diagram, see R. W. Buzzard and H. E. Cleaves, *J. Metallurgy and Ceram.* (TID 65), no. 1, pp. 26–27 July, 1948 (not available to the author).
2. N. C. Baenziger and R. E. Rundle, *Acta Cryst.*, **2**, 1949, 258.
3. W. C. Koehler, J. Singer, and A. S. Coffinberry, *Acta Cryst.*, **5**, 1952, 394.
4. Battelle Memorial Institute, Manhattan Project Report CT-1009, 1943.
5. A. R. Kaufmann, P. Gordon, and D. W. Lillie, *Trans. ASM*, **42**, 1950, 801.
6. E. Gordon, Manhattan Project Report CT-3459, 1946; discussed in [2].

$\bar{1}.2477$
0.7523

Be-V Beryllium-Vanadium

According to microscopic and X-ray investigations [1] of 10 alloys containing up to 25 wt. (65 at.) % Be, a eutectic between V and Be_2V exists at about 15 wt. (50 at.) % Be. The solid solubility of Be in V was found to be less than 0.8 wt. (4.4 at.) % Be at 900°C. The eutectic temperature lies, as tentative measurements showed, at about 1650°C [1]. Be_2V has the hexagonal $MgZn_2$ (C14) type of structure [2, 1], with a = 4.394 A, c = 7.144 A, c/a = 1.629 [2].

1. W. Rostoker and A. Yamamoto, *Trans. ASM*, **46**, 1954, 1136–1163.
2. L. Misch, *Metallwirtschaft*, **15**, 1936, 163–166.

$\bar{2}.6902$
1.3098

Be-W Beryllium-Wolfram

An alloy with 1 wt. (0.05 at.) % W, as cooled from the melt, was two-phase with a eutectic [1]. Besides Be_2W (91.08 wt. % W)—with the hexagonal $MgZn_2$ (C14) type of structure [2] and a = 4.446 A, c = 7.289 A, c/a = 1.639—another intermediate phase exists [2] in the high-Be range with a tetragonal lattice of the dimensions a = 10.14 A, c = 4.23 A, c/a = 0.416. Since its X-ray pattern is like that of a similarly located phase in the Be-Mo system [2], which more recently was shown to have the formula $Be_{13}Mo$ (see Be-Mo), it may be suggested that the W compound likewise has the formula $Be_{13}W$ [7.14 at. (61.07 wt.) % W].

[3] found that a low-melting alloy was formed on depositing Be on a W filament at 1230°C.

1. A. R. Kaufmann, P. Gordon, and D. W. Lillie, *Trans. ASM*, **42**, 1950, 801.
2. L. Misch, *Metallwirtschaft*, **15**, 1936, 163–166.
3. L. Hackspill and J. Besson, *Bull. soc. chim. France*, **16**, 1949, 113.

$\bar{1}.1394$
0.8606

Be-Zn Beryllium-Zinc

On adding Zn to molten Be, an alloy with 0.6 wt. % Zn was prepared [1].

1. A. R. Kaufmann, P. Gordon, and D. W. Lillie, *Trans. ASM*, **42**, 1950, 801.

$\bar{2}.9948$
1.0052

Be-Zr Beryllium-Zirconium

The tentative phase diagram of Fig. 169 is based mainly on a preliminary microscopic and roentgenographic investigation of sintered alloys [1]. In agreement with other sources [2, 3] a eutectic in the Zr-rich region was found; the data given are ~95 wt. (65 at.) % Zr, 950°C [1]; 94–94.5 wt. (61–63 at.) % Zr, 980°C [3]. Another eutectic probably exists very close to pure Be [1, 4].

Of the four intermediate phases which [1] claim to have found, two—the lowest and the highest in Zr—certainly correspond to the known phases $Be_{13}Zr$ [7.14 at. (43.76 wt.) % Zr] [5] and Be_2Zr (83.50 wt. % Zr) [6]. At 50 at. % Zr a two-phase field was observed by [6, 1].

According to photomicrographs [4] and lattice-parameter measurements [7], Zr appears to have a negligible solid solubility in Be. Scanty microscopic evidence for solid Zr dissolving Be to some extent is given by [1]; the eutectoid decomposition of the (Zr) high-temperature phase as shown in Fig. 169 was suggested by [8].

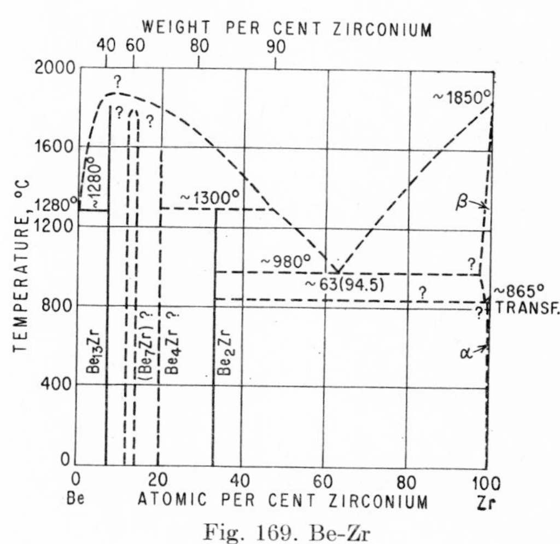

Fig. 169. Be-Zr

$Be_{13}Zr$ possesses the cubic $NaZn_{13}$ ($D2_3$) type of structure [5, 9], with a = 10.047 A [5]. More recently the compound Be_2Zr has been found [10] from powder diagrams to have the hexagonal AlB_2 (C32) type of structure, with a = 3.82 A, c = 3.24 A, c/a = 0.848.

1. H. H. Hausner and H. S. Kalish, *Trans. AIME*, **188**, 1950, 59–66; discussion, *Trans. AIME*, **188**, 1950, 1369–1371.
2. C. T. Anderson, E. T. Hayes, A. H. Roberson, and W. J. Kroll, *U.S. Bur. Mines, Rept. Invest.* 4658, 1950.
3. B. W. Mott, quoted by P. C. L. Pfeil, Atomic Energy Research Establishment (England), Report MT/N-11, 1952.
4. A. R. Kaufmann, P. Gordon, and D. W. Lillie, *Trans. ASM*, **42**, 1950, 785–844, especially 799–801.
5. N. C. Baenziger and R. E. Rundle, *Acta Cryst.*, **2**, 1949, 258.
6. L. Misch, *Metallwirtschaft*, **15**, 1936, 163–166.
7. See [4], pp. 803–805.
8. M. Hansen; see discussion on [1].
9. W. C. Koehler, J. Singer, and A. S. Coffinberry, *Acta Cryst.*, **5**, 1952, 394.
10. J. W. Nielsen and N. C. Baenziger, *Acta Cryst.*, **7**, 1954, 132–133.

1.2406
$\bar{2}.7594$

Bi-C Bismuth-Carbon

The solubility of carbon in liquid bismuth was found to be in wt. % (at. % in parentheses) 0.023, 0.0168, and 0.012 (0.40, 0.29, and 0.21) at 1490 (boiling point), 1408, and 1385°C, respectively [1], and 0.00030, 0.00024, 0.00020, and 0.00016 (0.0052, 0.0042, 0.0035, and 0.0028) at 750, 500, 400, and 300°C, respectively [2]. "The method of analysis for carbon used by [1] was rather insensitive and the experimental technique is questionable. Because of this and the wide intervening temperature range between the temperatures used by them and those used for the present study, no comparison of results was attempted" [2].

On solidification of the melt the dissolved carbon crystallizes as graphite [1].

1. O. Ruff and B. Bergdahl, *Z. anorg. Chem.*, **106**, 1919, 91.
2. C. B. Griffith and M. W. Mallett, *J. Am. Chem. Soc.*, **75**, 1953, 1832–1834.

0.7172
$\bar{1}.2828$

Bi-Ca Bismuth-Calcium

The constitution of the Ca-Bi system was determined by thermal and microscopic work [1] (Fig. 170). Scanty thermal results obtained earlier by [2] in the region near pure Bi (three-phase horizontals at 498 and 265°C) are in fair agreement.

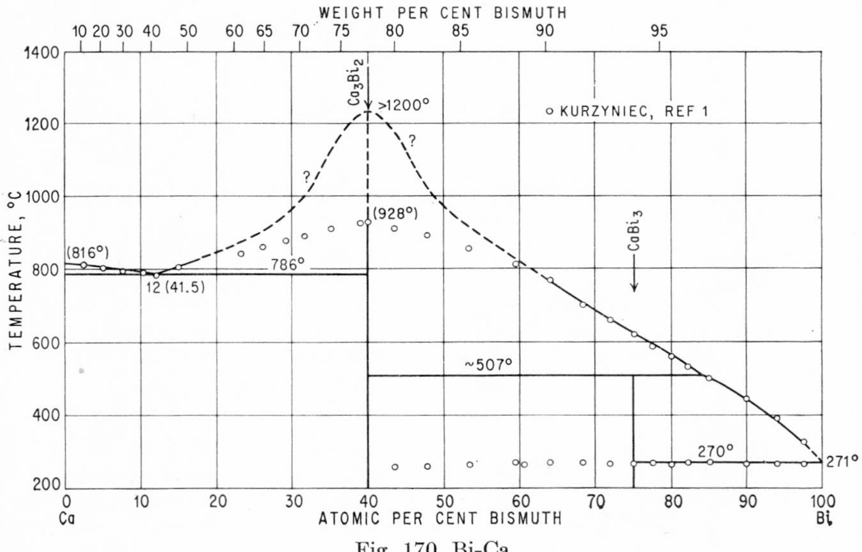

Fig. 170. Bi-Ca

There exist two compounds: the stoichiometric Ca_3Bi_2 (77.66 wt. % Bi) and $CaBi_3$ (93.99 wt. % Bi). The rate of the peritectic reaction at 507°C is very slow as the eutectic arrests in the (Ca_3Bi_2 + $CaBi_3$) phase field indicate (Fig. 170). Because of the high nitrogen content of the calcium used (see [1]), the temperatures of the liquidus and the eutectic found in the Ca-rich region were certainly too low (cf. Ca-N). In the light of this and especially of a recent finding [3] that an alloy of the composition Ca_3Bi_2 was not yet molten at 1200°C (see Fig. 170), a redetermination of the liquidus of the entire diagram is advisable. There is the possibility—paralleling the case of the Mg-Bi and Mg-Sb systems—that the "melting point" of Ca_3Bi_2, reported by [1] as 928°C, is only a transformation point of this phase. $CaBi_3$ was first detected by emf measurements [4]; the further supposed compounds Ca_2Bi_3 and $CaBi_2$, however, were not corroborated by [1].

According to [5], the structure of $CaBi_3$ is not cubic.

1. E. Kurzyniec, *Bull. intern. acad. polon. sci., sér. A*, **1931**, 31–58 (in German). Ca (containing 0.91 wt. % N) and Bi (99.8 wt. % pure) were alloyed under argon in sealed iron crucibles.
2. L. Donski, *Z. anorg. Chem.*, **57**, 1908, 214–216.
3. W. Köster and F. Sautter, *Z. Erzbergbau, Berg-u. Metallhüttenw.*, **5**, 1952, 303–307.
4. R. Kremann, H. Wostall, and H. Schöpfer, *Forschungsarb. Metallkunde u. Röntgenmetallog.*, **1922**, no. 5.
5. A. Iandelli, *Rend. seminar. fac. sci. univ. Cagliari*, **19**, 1949, 133–139.

0.2694
$\bar{1}.7306$

Bi-Cd Bismuth-Cadmium

The liquidus of Fig. 171 was obtained by graphical interpolation of numerous thermal data [1–9] and has an accuracy of better than ±5°C. A slight break at 161°C

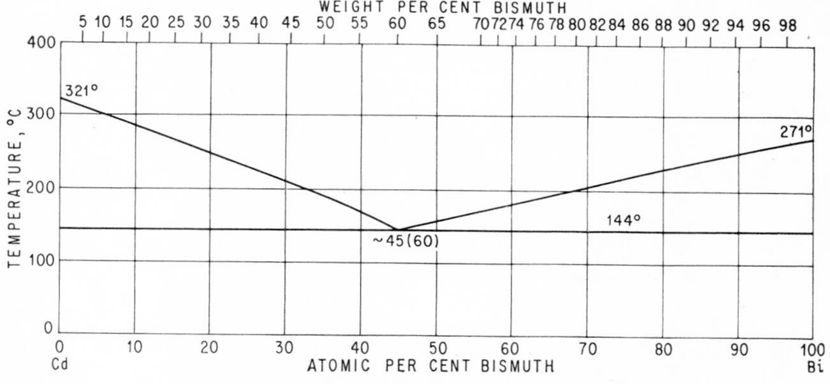

Fig. 171. Bi-Cd. (See also Supplement.)

near the composition Cd_3Bi_2 (due to nonideal activities in the melt; not shown in Fig. 171) was found by [10]. There have been many determinations for the position of the eutectic [2–4, 6–9, 11]; the most probable values are 144 ± 2°C, ~60 wt. (45 at.) % Bi. The crystallographic relations between the components of the eutectic were determined by X-ray work [12].

From the most reliable present data it appears certain that there are only very restricted solubilities in the solid state. According to [13], the solubility of Bi in Cd at the eutectic temperature is less than 0.05 wt. (0.03 at.) %. As a function of the composition, density [14, 15], specific heat [6,15], hardness [16], and emf (against Cd) [17, 18] all yield straight lines; electrical resistivity and its temperature coefficient [15], thermopower (against Cu) [15], Hall coefficient [15], and magnetic susceptibility [19] yield smooth curves with no measurable discontinuities [20]. A few other papers—older ones especially—report discontinuities in some property-composition curves, however [21, 5, 22]. [23] has measured the electrical-resistivity-vs.-temperature curve of an ~3 at. % Cd alloy and concluded a certain solubility of Cd in Bi. Thermal investigations and microscopic observations on unannealed alloys are not conclusive. A precision X-ray study of this problem should be made.

Thermodynamic computations of the phase diagram are due to [24–27]. [26, 27] based their work on measurements of the specific heat of several alloys from the solid to the liquid state.

Supplement. Two recent redeterminations of the eutectic temperature by [28] (146°C) and [29, 30] (146–147°C) suggest that 146°C is a better value than the average figure 144°C mentioned above (the data by [3, 4, 6–9, 11] were 146, 139, 146, 145, 142, 147, and 144°C, respectively). For the eutectic composition, [28] found 45.40 at. % Bi [31], and [29] 46.0 ± 1.5 at. (61.4 ± 1 wt.) % Bi, in good agreement with Fig. 171.

For the derivation of thermodynamic data and of liquidus points [32] of Bi-Cd alloys from calorimetric measurements ("thermodynamische Analyse"), see [29, 30].

1. C. T. Heycock and F. H. Neville, *J. Chem. Soc.*, **61,** 1892, 895–904.
2. A. W. Kapp, *Ann. Physik,* (4)**6,** 1901, 754–773.
3. A. Stoffel, *Z. anorg. Chem.*, **53,** 1907, 148–149.
4. A. Portevin, *Rev. mét.*, **4,** 1907, 389–394.
5. E. Rudolfi, *Z. anorg. Chem.*, **67,** 1910, 80–83.
6. M. Levin and H. Schottky, *Ferrum,* **10,** 1912-1913, 198–200.
7. C. H. Mathewson and W. M. Scott, *Intern. Z. Metallog.*, **5,** 1914, 15–16.
8. G. I. Petrenko and A. S. Fedorow, *Intern. Z. Metallog.*, **6,** 1914, 212–216.
9. F. Wüst and R. Durrer, "Temperatur-Wärmeinhaltskurven wichtiger Metalllegierungen," p. 22, Berlin, 1921. See also in [24].
10. N. W. Taylor, *J. Am. Chem. Soc.*, **45,** 1923, 2865.
11. F. Guthrie, *Phil. Mag.*, (5)**17,** 1884, 462.
12. M. Straumanis and N. Brakšs, *Z. physik. Chem.*, **B38,** 1937, 140–155.
13. G. Tammann and H. J. Rocha, *Z. Metallkunde,* **25,** 1933, 133–134. (Detection of small amounts of eutectics in metals by determination of tensile strength as a function of temperature.)
14. A. Matthiessen, *Pogg. Ann.*, **110,** 1860, 190.
15. S. Gabe and E. J. Evans, *Phil. Mag.*, **19,** 1935, 773–787.
16. C. di Capua and M. Arnone, *Rend. accad. nazl. Lincei,* **33,** 1924, 28–31.
17. M. Herschkowitsch, *Z. physik. Chem.*, **27,** 1898, 141.
18. P. Fuchs, *Z. anorg. Chem.*, **109,** 1920, 85–86.
19. T. Gnesotto and M. Binghinotto, *Atti reale ist. veneto sci., lettere ed arti,* **69,** 1910, 1382; quoted by [15]. The authors are of the opinion that the steepness of the curve in the region of high Bi content indicates the possibility of a slight solid solubility of Cd in Bi.
20. From the steep decline in the electrical resistivity, among other things, of Bi on small additions of Cd, [15] have suggested that a certain solid solubility may exist. However, if the reciprocal resistivity (conductivity) is plotted against volume per cent, there is no marked deviation from the curvature normal for mechanical mixtures.
21. A. Battelli, *Atti reale ist. veneto sci., lettere ed arti,* (6)**5,** 1886-1887, 1137; *Wied. Ann. Beibl.*, **12,** 1888, 269. See also W. Broniewski, *Rev. mét.*, **7,** 1910, 353–354.
22. W. Schischokin and W. Agejewa, *Z. anorg. Chem.*, **193,** 1930, 237.
23. N. Thompson, *Proc. Roy. Soc. (London)*, **A155,** 1936, 120.
24. V. Fischer, *Z. tech. Physik,* **6,** 1925, 146–148.
25. K. Honda and T. Ishigaki, *Science Repts. Tôhoku Imp. Univ.*, **14,** 1925, 221.
26. S. Nagasaki and E. Fujita, *Nippon Kinzoku Gakkai-Shi,* **16,** 1952, 313–321 (in Japanese; summary and diagrams in English).

27. Y. E. Geguzin and B. Y. Pines, *Zhur. Fiz. Khim.*, **25,** 1951, 1300–1305; *Chem. Abstr.*, **47,** 1953, 7985.
28. O. Hájiček, *Hutnické Listy*, **3,** 1948, 265–270 (not available to the author); *Chem. Abstr.*, **43,** 1949, 4935.
29. W. Oelsen, K. Bierett, and G. Schwabe, *Archiv Eisenhüttenw.*, **27,** 1956, 607–620.
30. W. Oelsen, W. Tebbe, and O. Oelsen, *Archiv Eisenhüttenw.*, **27,** 1956, 689–694.
31. Obviously by mistake, 45.40% Cd (instead of Bi) is given in the abstract.
32. Of different sets of data plotted in [29, 30], those in fig. 23 of [29] are considered the best.

0.1736
$\bar{1}$.8264

Bi-Ce Bismuth-Cerium

The constitution of Ce-Bi alloys was investigated by [1] using thermal and microscopic methods. Four compounds exist, three of which have relatively high melting points: Ce_3Bi (33.21 wt. % Bi), Ce_4Bi_3 [42.86 at. (52.80 wt.) % Bi], CeBi (59.86 wt. % Bi), and $CeBi_2$ (74.89 wt. % Bi). Although the peritectic reactions are sluggish and, in [1], gave rise to severe disturbances in equilibrium, the assumed constitution has a high degree of certainty, as can be verified by observing the course of solidification in microstructures. Besides the thermal points plotted in Fig. 172, [1] observed small heat effects between 830 and 860°C on cooling curves of alloys with 0–25 wt. % Bi and attributed them to impurities (La, Pr, Nd) in the Ce used [2]. There is no mention of the Ce transformations.

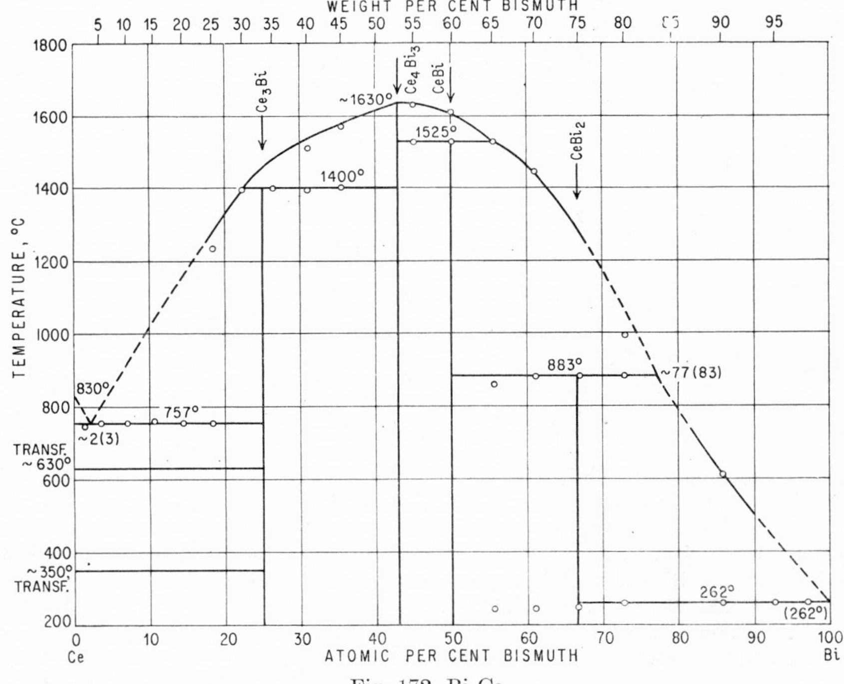

Fig. 172. Bi-Ce

According to [3], CeBi has the NaCl (B1) type of structure, with $a = 6.50$ A.

1. R. Vogel, *Z. anorg. Chem.*, **84**, 1914, 327–339. The heated metals react violently.
2. The purity of the Ce was not given; the metal used was probably the same as that of former investigations with a purity of only about 93.5 wt. % Ce.
3. A. Iandelli and E. Botti, *Atti reale accad. nazl. Lincei*, **25**, 1937, 233–238; abstract in *Strukturbericht*, **5**, 1937, 45.

0.5497
$\bar{1}.4503$

Bi-Co Bismuth-Cobalt

Figure 173 shows the phase diagram of [1] determined by thermal analysis. The following comments may be made: (*a*) Since the melting point found for Co was 55°C

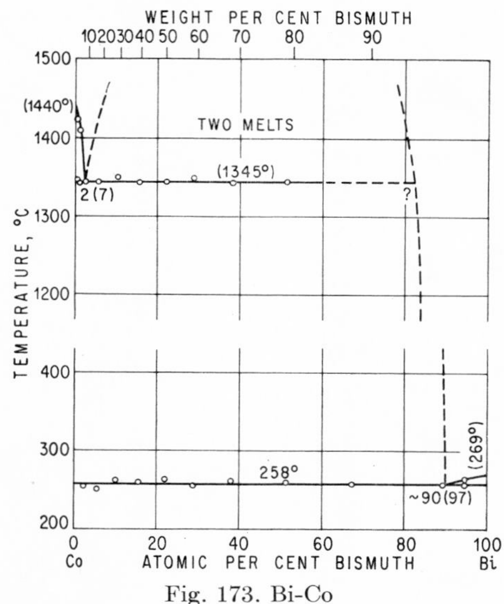

Fig. 173. Bi-Co

too low, the monotectic horizontal may similarly be about 50°C higher than 1345°C. (*b*) The extent of the miscibility gap in the liquid state at the temperature of the monotectic [~7–94 wt. (2–82 at.) % Bi] was determined by the thermal arrests of the monotectic reaction. Because of the wide extrapolation (Fig. 173), the assumed composition of the Bi-rich melt is very uncertain. (*c*) The eutectic temperature was observed at 251–262°C (average, 258°C).

The existence of a miscibility gap in the liquid state was corroborated by [2]; an alloy with 12 wt. (3.7 at.) % Bi showed two liquid layers at 1450°C.

Emf measurements [3] and extraction tests [2] (with HNO_3) gave no evidence for the existence of an intermediate compound.

[4] concluded from dilatometric and magnetic measurements—which indicated a rise in the transformation point "on heating" of Co (from 467 to 490°C) but no influence on the transformation point "on cooling" (425°C) by alloying with Bi—that there is a solid solubility of about 1 wt. (0.3 at.) % Bi at 490°C. This could not be corroborated by [5], who observed no influence of Bi on either the transformation or the Curie point of Co and assumed that there is no solid solubility. An alloy containing 0.25 wt. (0.07 at.) % Bi was reported to be two-phase. A sensitive magnetic method indicated

a solid solubility of Co in Bi of 9×10^{-3} to 10×10^{-3} wt. (0.032–0.035 at.) % Co [6].

1. A. Lewkonja, *Z. anorg. Chem.*, **59**, 1908, 315–318. Analysis of the Co used: 98.04 Co, 1.62 Ni, 0.17 Fe, residue 0.04 (wt. %); the alloys were melted in porcelain tubes under N_2 or H_2.
2. F. Ducelliez, *Bull. soc. chim. France*, (4)**5**, 1909, 61–62.
3. F. Ducelliez, *Bull. soc. chim. France*, (4)**7**, 1910, 199.
4. U. Haschimoto, *Nippon Kinzoku Gakkai-Shi*, **1**, 1937, 177–190 (in Japanese). The Co had a purity of 99.9 wt. %.
5. W. Köster and E. Horn, *Z. Metallkunde*, **43**, 1952, 333–334.
6. G. Tammann and W. Oelsen, *Z. anorg. Chem.*, **186**, 1930, 279–280.

0.6041
$\bar{1}$.3959

Bi-Cr Bismuth-Chromium

Some thermal and microscopic work on badly contaminated alloys (a "protective" nitrogen atmosphere was used in preparation) was carried out by [1]. The thermal data plotted in Fig. 174 indicate immiscibility in both the liquid and the solid

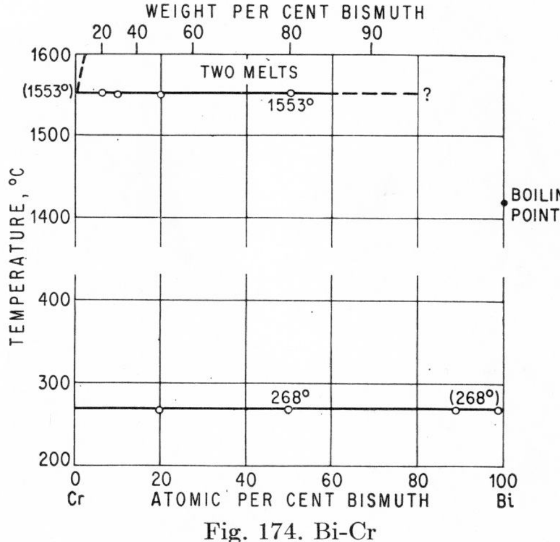

Fig. 174. Bi-Cr

state. The melting point of Cr observed lies more than 300°C below the true value (~1880°C), mainly because of the nitrogen content (cf. Cr-N).

1. R. S. Williams, *Z. anorg. Chem.*, **55**, 1907, 23–24; analysis of the Cr: 98.97 Cr, 0.67 Fe, 0.30 $Cr_2O_3 + SiO_2$ (wt. %). The alloys were prepared in porcelain or magnesia crucibles in a nitrogen atmosphere. "In the attempt to melt the alloys with 70 and 80 wt. % Cr considerable volatilization of Bi (boiling point 1420°C) occurred."

0.1966
$\bar{1}$.8034

Bi-Cs Bismuth-Cesium

The investigation of thin Cs-Bi layers (for use in photoelectric cells) showed that the metals alloy easily [1] and that there exists a compound Cs_3Bi (34.39 wt. % Bi) 2, 3] and probably one of the formula CsBi (61.13 wt. % Bi) also [3].

1. P. Görlich, *Z. Physik*, **101**, 1936, 337.
2. A. Sommer, *Nature*, **148**, 1941, 468.
3. A. Sommer, *Proc. Phys. Soc.* (*London*), **55**, 1943, 145–154.

0.5171
$\bar{1}$.4829

Bi-Cu Bismuth-Copper

Liquidus. Most of the thermal data measured by [1–4] for the liquidus are plotted in Fig. 175. The curve of [5] (no single values given) and the values of [6] were

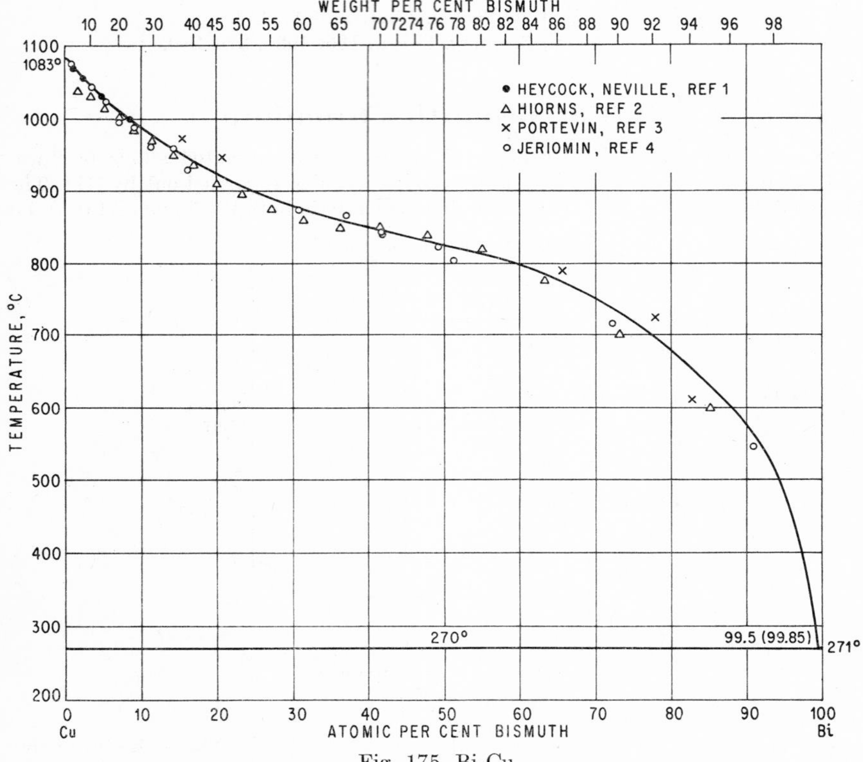

Fig. 175. Bi-Cu

not used in drawing the liquidus as these results, in parts, show considerable deviations from those of the other authors. Numerous liquidus points below 620°C due to a chemical-analytical method by [7] were used but not plotted in Fig. 175. A eutectic exists very near to pure Bi [0.6 at. (0.2 wt.) % [8]; 0.5 at. (0.15 wt.) % Cu [7]] at a temperature about 1° below the melting point of Bi [8]. Microscopic observations [3, 4] agree well with this composition. The absence of a Cu-Bi compound was corroborated roentgenographically [9, 10]. For a short discussion on a miscibility gap in the liquid state, which had been erroneously assumed earlier, and an intermediate phase, see [10*a*].

Solid Solubilities. According to roentgenographic work [9], the mutual solubility of Bi and Cu is less than 0.5 at. %. For alloys annealed 4 hr at 980°C and then quenched, [11] found microscopically a solubility of only about 0.002 wt. (6 ×

10^{-4} at.) % Bi in Cu; on the other hand, [12] asserted by reason of some indirect methods (for instance, embrittlement tests) that the solubility increases from less than 0.001 wt. (3×10^{-4} at.) % Bi at 600°C to about 0.01 wt. (3×10^{-3} at.) % Bi at 800°C. Whether these values are compatible—which would be so if the solubility curve has a maximum between 800 and 980°C (see [13, 14])—must be decided by further investigations. [13] found microscopically a maximum solubility of 0.06 at. (0.2 wt.) % Bi (above 600°C?), which is probably too high.

For the great influence of Bi on the workability and the mechanical properties of Cu, see [12, 15, 11] and the older papers discussed there. The grain-boundary diffusion of molten Bi in Cu was quantitatively investigated by [16].

The magnetic anisotropy of Bi crystals remains unchanged by alloying small quantities of Cu [17]; this indicates that there is no solid solubility.

1. C. T. Heycock and F. H. Neville, *Phil. Trans. Roy. Soc. (London)*, **A189,** 1897, 46.
2. A. H. Hiorns, *Trans. Faraday Soc.*, **1,** 1905, 179–186; *J. Soc. Chem. Ind.*, **25,** 1906, 618.
3. A. Portevin, *Rev. mét.*, **4,** 1907, 1077–1080.
4. K. Jeriomin, *Z. anorg. Chem.*, **55,** 1907, 412–414.
5. W. C. Roberts-Austen, *Engineering*, **55,** 1893, 660.
6. Roland-Gosselin; see in H. Gautier, *Bull. soc. encour. ind. natl.*, **1,** 1896, 1309; Société d'encouragement pour l'industrie nationale, Paris, Commission des alliages, "Contribution à l'étude des alliages," pp. 109–110, Typ. Chamerot et Renouard, Paris, 1901.
7. O. J. Kleppa, *J. Am. Chem. Soc.*, **74,** 1952, 6050.
8. C. T. Heycock and F. H. Neville, *J. Chem. Soc.*, **61,** 1892, 893.
9. W. F. Ehret and R. D. Fine, *Phil. Mag.*, (7)1930, 551–558.
10. R. Eborall, *J. Inst. Metals*, **70,** 1944, 435–446.

10*a*. M. Hansen, "Der Aufbau der Zweistofflegierungen," Springer-Verlag OHG, Berlin, 1936.

11. D. Hanson and G. W. Ford, *J. Inst. Metals*, **37,** 1927, 169–178.
12. E. Voce and A. P. C. Hallowes, *J. Inst. Metals*, **73,** 1947, 323–376.
13. E. Raub and A. Engel, *Z. Metallkunde*, **37,** 1946, 76–81.
14. W. Hume-Rothery, *J. Inst. Metals*, **69,** 1943, 230–231; discussion remark.
15. C. Blazey, *J. Inst. Metals*, **46,** 1931, 359–367.
16. E. Scheil and K. E. Schiessl, *Z. Naturforsch.*, **4a,** 1949, 524–526.
17. A. Goetz and A. B. Focke, *Phys. Rev.*, **45,** 1934, 179.

0.5731
$\bar{1}.4269$

Bi-Fe Bismuth-Iron

By means of cooling curves [1] (Fig. 176) and X-ray powder patterns [2], no liquid or solid [3] solubility could be found. Using a very sensitive magnetic method in studying the liquid solubility in Bi, [4] reported solubilities of 2×10^{-4} and 4×10^{-4} wt. (7.5×10^{-4} and 15×10^{-4} at.) % Fe at 400 and 600°C, respectively. Whereas [5] could not observe any diffusion of Bi into low-carbon steel at 1000°C, [6] considered the wettability of electrolytic Fe above 600°C as proof of Bi diffusion.

The fact that minute quantities of Bi can cause hot shortness in iron [7] is claimed to be due to a certain solubility of Bi in molten iron [8] or to an increased absorption of oxygen caused by the (insoluble) Bi [7].

1. E. Isaak and G. Tammann, *Z. anorg. Chem.*, **55,** 1907, 59–61.
2. G. Hägg, *Z. Krist.*, **68,** 1928, 472; *Nova Acta Regiae Soc. Sci. Upsaliensis*, (4)7, 1929, no. 1, p. 89 (in English).

3. F. Wever, *Arch. Eisenhüttenw.*, **2,** 1928-1929, 739–746.
4. G. Tammann and W. Oelsen, *Z. anorg. Chem.*, **186,** 1930, 277–279.
5. N. W. Ageew and M. Zamotorin, *Izvest. Leningrad. Politekh. Inst., Otdel. Mat. Fiz. Nauk*, **31,** 1928, 183–196 (in Russian); abstract in *J. Inst. Metals*, **44,** 1930, 556.
6. G. Tammann and A. Rühenbeck, *Z. anorg. Chem.*, **223,** 1935, 192–196.
7. O. Thallner, *Stahl u. Eisen*, **27,** 1907, 1684.
8. W. Guertler, "Handbuch der Metallographie," vol. 1, part 1, p. 580, Verlagsbuchhandlung Gebrüder Borntraeger, Berlin, 1912.

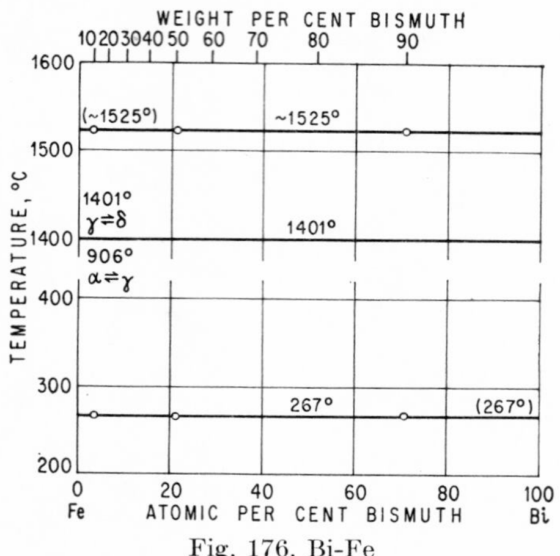

Fig. 176. Bi-Fe

0.4768
$\bar{1}.5232$

Bi-Ga Bismuth-Gallium

The phase diagram of Fig. 177 is due to a thermal investigation by [1]. A large miscibility gap exists in the liquid state; the composition of the Ga-rich melt at the

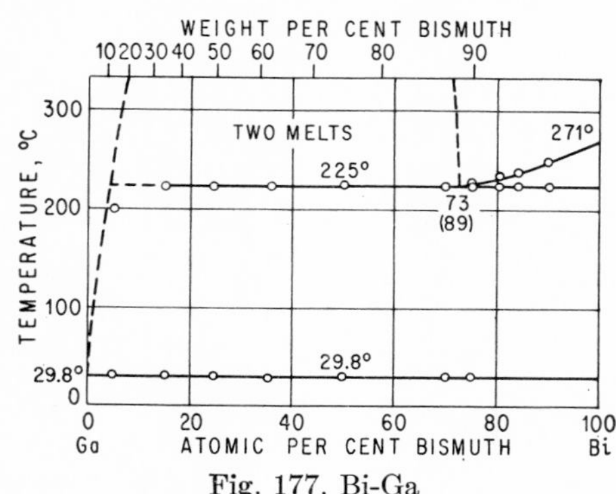

Fig. 177. Bi-Ga

monotectic temperature (225°C) cannot be fixed by the data presented. No depression of the melting point of Ga (29.8°C) by alloying Bi could be observed by [1], which means that Bi is insoluble in liquid Ga near its melting point. (It seems that the small depression observed by [2] is not real but is rather due to a supercooling effect.) From the effect of 0.5 at. % Ga on the electrical-resistivity-vs.-temperature curve of a

Bi single crystal, it appears possible that there is an extremely small solid solubility [3].

1. N. A. Puschin, S. Stepanovic, and V. Stajic, *Z. anorg. Chem.*, **209,** 1932, 329–334.
2. W. Kroll, *Metallwirtschaft,* **11,** 1932, 435–437.
3. N. Thompson, *Proc. Roy. Soc. (London)*, **A155,** 1936, 120.

0.4592
$\bar{1}$.5408

Bi-Ge Bismuth-Germanium

The phase diagram of Fig. 178 is the result of thermal-analysis work [1, 2] corroborated by metallographic [1] and qualitative X-ray studies [2]. As tabulated data

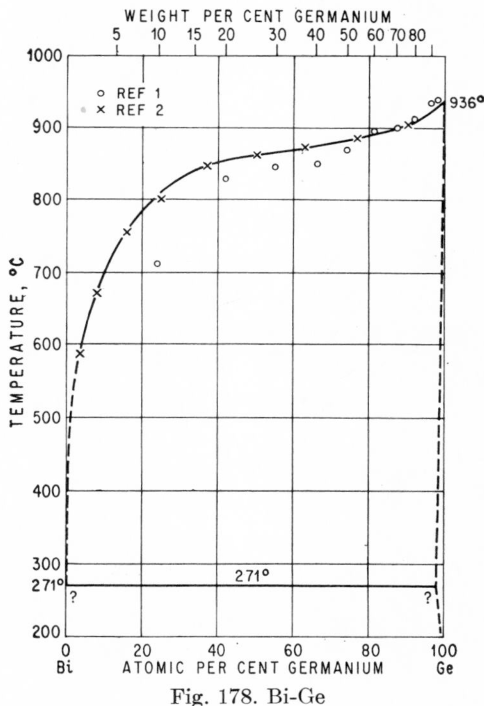

Fig. 178. Bi-Ge

were lacking, the data points of the liquidus curve had to be taken from small diagrams and cannot be considered very accurate. According to both papers, the eutectic consists of practically pure Bi [3]. [2] reported that qualitative X-ray investigations indicated solid solubilities of Ge in Bi and Bi in Ge of about 1.5 at. (0.54 wt.) % Ge and about 2 at. (5.4 wt.) % Bi, respectively, at 250°C, after annealing for several months at this temperature and quenching.

1. K. Ruttewit and G. Masing, *Z. Metallkunde,* **32,** 1940, 52–61.
2. H. Stöhr and W. Klemm, *Z. anorg. Chem.*, **244,** 1940, 205–233.
3. C. D. Thurmond, *J. Phys. Chem.*, **57,** 1953, 827–830.

Bi-H Bismuth-Hydrogen

According to [1], hydrogen is not dissolved by either solid or liquid Bi (investigated up to 600°C). A solid hydride apparently does not exist; the claim of [2] to have prepared solid Bi_2H_2 could not be corroborated by more recent work [3].

1. A. Sieverts and W. Krumbhaar, *Ber. deut. chem. Ges.*, **43,** 1910, 896.
2. E. J. Weeks and J. G. F. Druce, *Rec. trav. chim.*, **44,** 1925, 970; *J. Chem. Soc.*, **127,** 1925, 1799.
3. C. Brinc, G. Dallinga, and R. J. F. Nivard, *Rec. trav. chim.*, **68,** 1949, 234–236; abstract in *Structure Repts.*, **12,** 1949, 17.

0.0178
$\bar{1}.9822$

Bi-Hg Bismuth-Mercury

The liquidus, almost in its entirety, was determined by [1], corroborating former results on the Bi end of the diagram [2] and at room temperature [3] (Fig. 179).

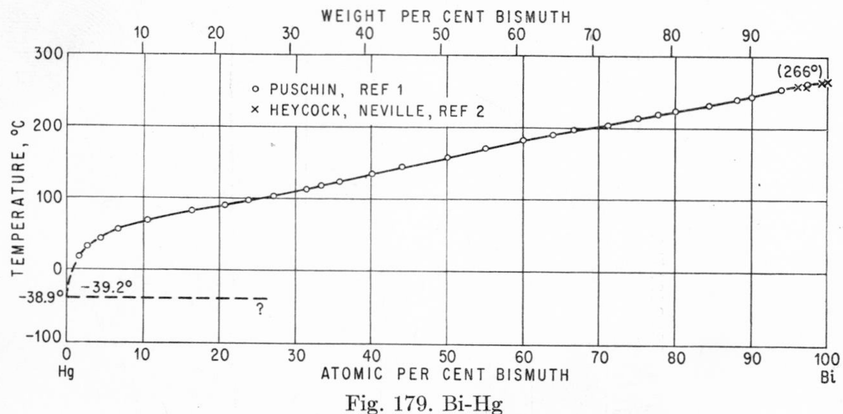

Fig. 179. Bi-Hg

According to [1], 1.4 at. (wt.) % Bi dissolves in Hg at 18°C. [4] investigated the influence of small additions of Bi on the solidification point of Hg and found that it is depressed for 0.15, 0.30, and 0.30°C by 0.05, 0.108, and 0.226 wt. (0.05, 0.103, and 0.216 at.) % Bi, respectively. The last two values obviously correspond to the end of solidification and indicate the existence of a eutectic at 0.1 wt. (at.) % Bi and −39.2°C. There is further indication of a solidus temperature near the solidification point of Hg for alloys up to 25 wt. % Bi [5, 6]. [1] could not determine the solidus temperatures of his alloys. Since the liquidus as well as emf measurements did not exhibit any evidence of compound formation and since crystal forms like those of pure Bi were observed in an alloy with 31 wt. % Bi, [1] concluded the existence of a simple eutectic system. This, however, is not yet proved, as the results of [1] do not exclude the possible existence of a compound formed below room temperature [7].

1. N. A. Puschin, *Z. anorg. Chem.*, **36,** 1903, 201–254.
2. C. T. Heycock and F. H. Neville, *J. Chem. Soc.*, **61,** 1892, 897.
3. Gouy, *J. Phys.*, (3)**4,** 1895, 320–321.
4. G. Tammann, *Z. physik. Chem.*, **3,** 1889, 444.
5. D. Mazzotto, *Atti ist. veneto sci., lettere ed arti,* (7)**4,** 1892–1893, 1311, 1527; abstract in *Z. physik. Chem.*, **13,** 1894, 572.
6. K. Bornemann, *Metallurgie*, **7,** 1910, 109–110.

7. The statement of Crookewitt (*J. prakt. Chem.*, **45**, 1848, 87) that a residue separated from the liquid amalgam had nearly the composition $HgBi_2$ is unimportant as the method used is not reliable.

0.2604
$\bar{1}$.7396

Bi-In Bismuth-Indium

The phase diagram (Fig. 180) was established by [1] (23 alloys) and [2] (33 alloys), using thermal, microscopic, and [2] X-ray powder methods. Some discrepancies concerning the type of three-phase equilibria occurring were clarified in discussions (see [1, 2]). There exist two compounds (first reported by [3]): In_2Bi (47.66 wt. % Bi) and InBi (64.56 wt. % Bi). According to thermal data [2], the maximum solubility of Bi in In at the eutectic temperature is 20.5 wt. (12.4 at.) % Bi. X-ray specimens, water-quenched (from 68°C) and held for 3 months at room temperature, indicated a

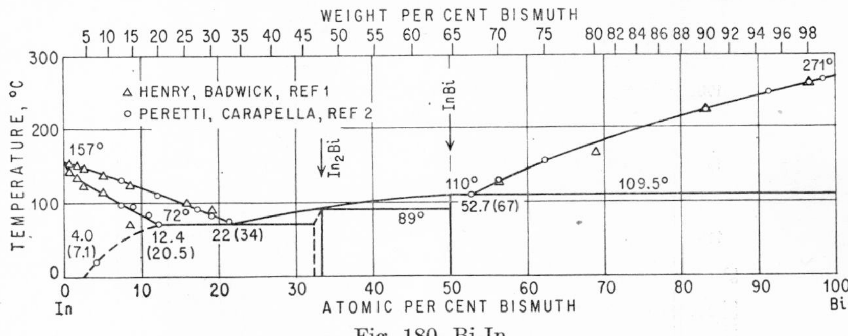

Fig. 180. Bi-In

solubility limit of 7.1 wt. (4.0 at.) % Bi [2]; the authors believe that the decomposition was not prevented by the above quenching treatment and the value obtained represents room-temperature equilibrium (Fig. 180).

There are no data concerning the solubility of In in solid Bi. From the effect of 0.5 at. % In on the electrical-resistivity-vs.-temperature curve of Bi [4] and from the qualitative statement of [5] that In hardens Bi, one may conclude that there is some solid solubility.

Crystal Structures. The axial ratio of In increases with increasing Bi content [2].

In_2Bi has, according to powder photographs [6], a hexagonal structure with $a = 5.498$ A, $c = 3.291$ A, $c/a = 0.598$, and 3 atoms per unit cell. [The point positions of the ordered AlB_2 (C32) type of structure are occupied at random.] The solubility range was reported to lie within about 46.5–47.5 wt. (32.3–33.2 at.) % Bi [2].

InBi has the tetragonal PbO (B10) type of structure [7] with $a = 5.015 \pm 3$ A, $c = 4.781 \pm 3$ A, $c/a = 0.953$. No solubility range was detected by X-rays [2] (see also discussion on [2]).

Note Added in Proof. The B10 type of structure of InBi has been corroborated by [8], who found the lattice parameters $a = 5.000$ A, $c = 4.773$ A, $c/a = 0.955$.

1. O. H. Henry and E. L. Badwick, *Trans. AIME*, **171**, 1947, 389–393; In 99.95 wt. %, Bi 99.99 wt. % pure. Discussion (especially F. N. Rhines), pp. 394–395.
2. E. A. Peretti and S. C. Carapella, *Trans. ASM*, **41**, 1949, 947–958. In 99.97 wt. %, Bi 99.95 wt. % pure. Discussion, pp. 958–960.
3. S. A. Dubinskij, dissertation, Institute of General and Inorganic Chemistry, Academy of Sciences, U.S.S.R., (1947); not available to the author, reference from [6, 7].

4. N. Thompson, *Proc. Roy. Soc.* (*London*), **A155,** 1936, 120.
5. A. W. Downes and L. Kahlenberg, *Trans. Am. Electrochem. Soc.*, **63,** 1933, 157.
6. E. S. Makarov, *Doklady Akad. Nauk S.S.S.R.*, **68,** 1949, 509–510 (in Russian); abstract in *Structure Repts.*, **11,** 1947-1948, 46–48.
7. E. S. Makarov, *Doklady Akad. Nauk S.S.S.R.*, **59,** 1948, 899 (in Russian); abstract, see [6].
8. W. P. Binnie, *Acta Cryst.*, **9,** 1956, 686–687.

0.0344
$\bar{1}$.9656

Bi-Ir Bismuth-Iridium

An attempt to alloy Bi and Ir at 800°C was unsuccessful [1]. [2], however, showed that alloying can be attained by heating the components to 1450–1500°C

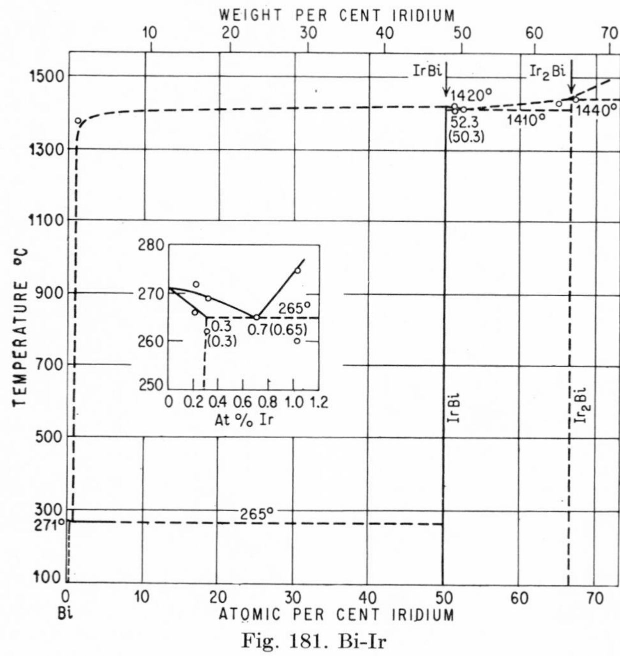

Fig. 181. Bi-Ir

for some hours (boiling point of Bi, 1420°C) and established a partial phase diagram in a very cursory manner by means of thermal, microscopic, and hardness methods (Fig. 181). Two compounds have been reported: IrBi (48.02 wt. % Ir) and Ir_2Bi (64.88 wt. % Ir); and two eutectics: at 0.7 at. (0.65 wt.) % Ir and 265°C and 52.3 at. (50.3 wt.) % Ir and 1410°C. The assumed course of the liquidus between the Bi-rich eutectic and IrBi (supported by only one thermal effect at 1380°C) is most improbable. The peritectic reaction at about 1440°C (Fig. 181) has been inserted; the author [2] merely reports 1440°C as the melting point of Ir_2Bi. The solubility of Ir in solid Bi is assumed to be 0.3 at. (wt.) % Ir at 265°C.

1. L. Wöhler and L. Metz, *Z. anorg. Chem.*, **149,** 1925, 310.
2. P. S. Belonogov, *Izvest. Sektora Fiz.-Khim. Anal.*, **11,** 1938, 36–47 (in Russian). The main impurities of the Ir sponge (Heraeus) were Rh, Ru, and Fe up to 0.15

wt. %. A layer of $BaCl_2$ 3–4 cm thick was used to prevent oxidation and evaporation of Bi.

0.7280
$\bar{1}.2720$

Bi-K Bismuth-Potassium

From thermal data, [1] concluded the existence of four compounds, namely: K_3Bi (64.05 wt. % Bi), K_3Bi_2 (78.09 wt. % Bi), K_9Bi_7 (?) [43.75 at. (80.61 wt.) % Bi], and KBi_2 (91.44 wt. % Bi) (Fig. 182). Microscopic work was restricted to a few

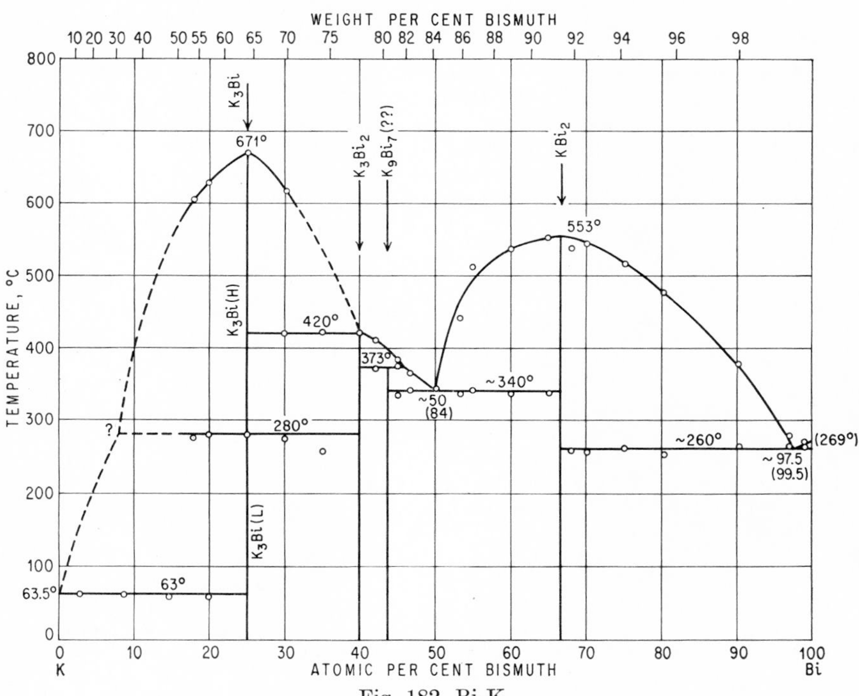

Fig. 182. Bi-K

alloys because of porosity and quick oxidation of the specimens; especially in the most interesting range, 40–50 at. % Bi, no clear microstructures could be obtained.

The existence of K_3Bi was corroborated by [2]. Emf measurements [3] gave indications for the existence of K_3Bi and KBi_2 only.

K_3Bi undergoes a polymorphic transformation at 280°C [1], accompanied by an increase in volume on cooling. The room-temperature modification was shown [4] to have the hexagonal Na_3As ($D0_{18}$) type of structure with $a = 6.191$ A, $c = 10.955$ A, $c/a = 1.770$. KBi_2 has the $MgCu_2$ (C15) type of structure with $a = 9.520 \pm 5$ A [5].

1. D. P. Smith, *Z. anorg. Chem.*, **56,** 1908, 125–129. The alloys were prepared in vessels of high-melting glass under hydrogen.
2. A. G. Vournasos, *Compt. rend.*, **152,** 1911, 714–715.
3. R. Kremann, J. Fritsch, and R. Riebl, *Z. Metallkunde,* **13,** 1921, 71–73. The emf of the element Bi|2/1,000 n KCl in pyridin|$K_{(1-x)}Bi_x$ was measured.
4. G. Brauer and E. Zintl, *Z. physik. Chem.*, **B37,** 1937, 323–352.
5. E. Zintl and A. Harder, *Z. physik. Chem.*, **B16,** 1932, 206–212. The range of

homogeneity is small; deviations of 3 at. % from the chemical formula already yield foreign lines.

0.1774
$\bar{1}.8226$

Bi-La Bismuth-Lanthanum

LaBi (60.07 wt. % Bi) has the NaCl (B1) type of structure, with $a = 6.57_8$ A [1].

1. A. Iandelli and E. Botti, *Atti reale accad. nazl. Lincei*, **25**, 1937, 233–238.

1.4788
$\bar{2}.5212$

Bi-Li Bismuth-Lithium

The phase diagram (Fig. 183) was established by means of measurements of thermal and electrical conductivity (the latter between 42.5 and 100 at. % Bi) [1].

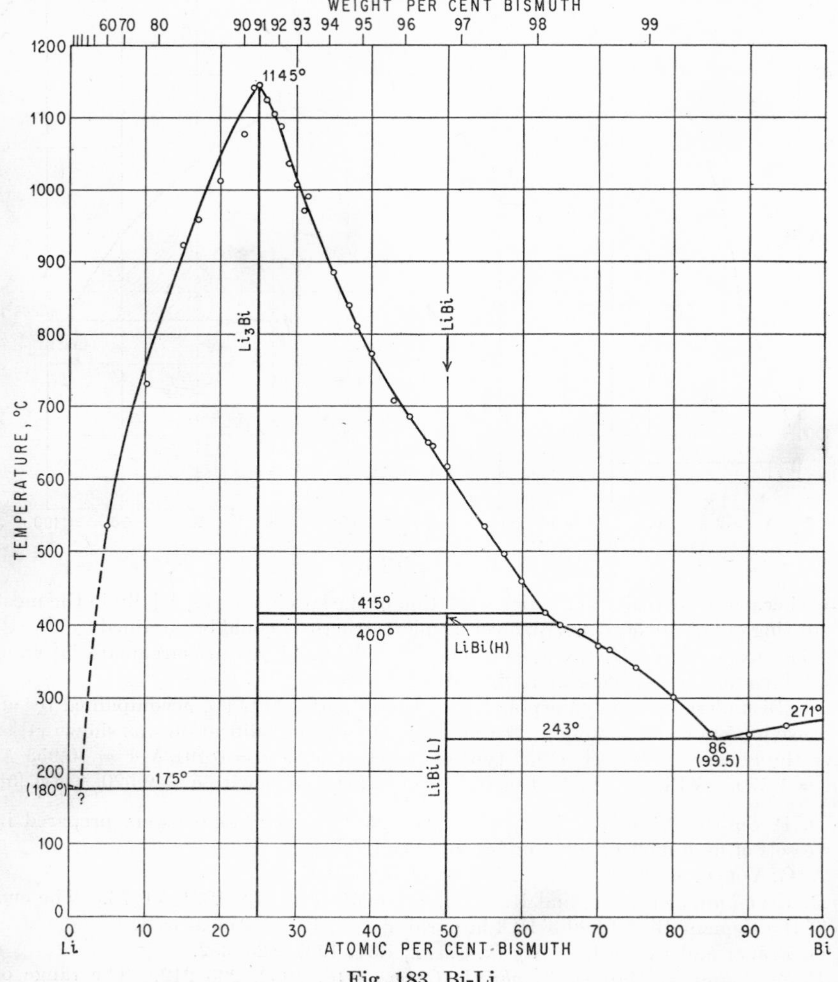

Fig. 183. Bi-Li

Two compounds exist, corroborated by a roentgenographic analysis [2]: Li_3Bi (90.94 wt. % Bi) and the peritectically formed LiBi (96.79 wt. % Bi), which shows a polymorphic transformation at 400°C (covered by numerous measurement points). Both compounds have a very narrow range of homogeneity. Primary solid solutions could not be observed.

Li_3Bi has the cubic BiF_3 (DO_3) type of structure, with a = 6.722 A and 16 atoms per unit cell [2]. The room-temperature modification of LiBi possesses the tetragonal CuAu ($L1_0$) type of structure, with a = 3.368 A, c = 4.256 A, c/a = 1.264 [2] (for the 2-atom unit cell).

1. G. Grube, H. Vosskühler, and H. Schlecht, *Z. Elektrochem.*, **40,** 1934, 270–274.
2. E. Zintl and G. Brauer, *Z. physik. Chem.*, **B20,** 1933, 245–271; *Z. Elektrochem.*, **41,** 1935, 297–303.

0.9342
$\bar{1}.0658$

Bi-Mg Bismuth-Magnesium

The constitution of Mg-Bi alloys was outlined by thermal, microscopic [1, 2], and conductivity [3] measurements. Since more recent and more detailed investigations are available, reference is made to [3*a*] and [3*b*] for a discussion of these older papers.

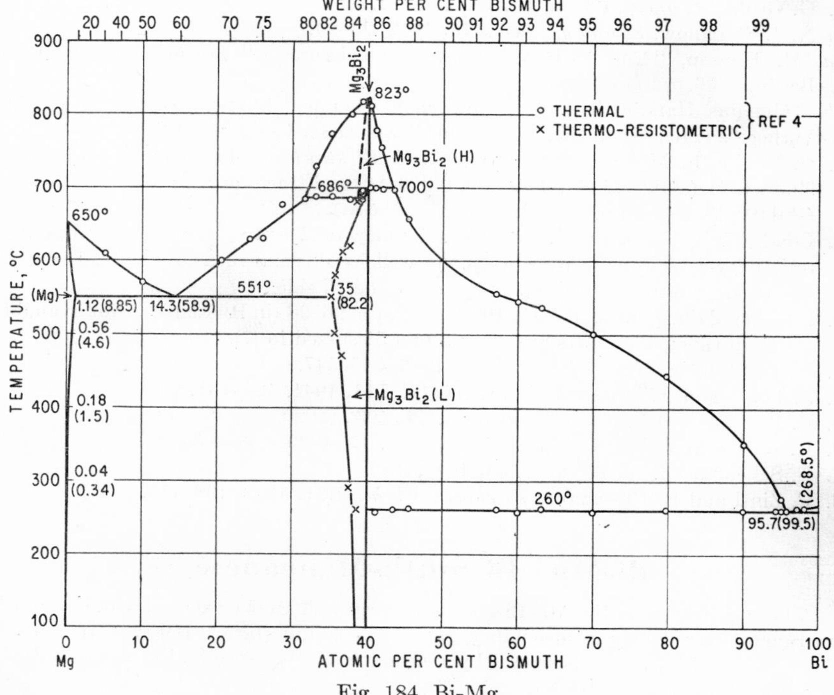

Fig. 184. Bi-Mg

The phase diagram of Fig. 184 is based mainly on a thermal and microscopic investigation and on resistivity measurements [4]. The only compound, Mg_3Bi_2 (85.14 wt. % Bi), was found to have a polymorphic transformation at about 700°C and to form a solid solution at least with Mg [5]. Liquidus [6] and solidus [6, 7] up to about 12 wt. (1.6 at.) % Bi were further investigated by thermal [6] and resistivity [7] measurements. Values for the solidus according to [7] are 0.24, 0.50, and 0.80 at. (2.00, 4.15, and 6.45 wt.) % Bi at 625, 600, and 575°C, respectively. The position of the Mg-rich eutectic as indicated in Fig. 184 [4] was corroborated by some cooling curves [7]. As to the Bi-rich eutectic, there are reported the values 97 at. (99.6 wt.) % Bi at 261°C [1] and 95.7 at. (99.5 wt.) % Bi at 260°C [4]. The solid solubility of Bi in Mg was investigated by resistivity [4, 7] and X-ray precision [8] measurements as well as thermodifferential, microscopic, and hardness methods [6]. The results of [7] and [8] are by far the most reliable ones and differ at most (at 525°C) by only 0.1 at. %. The most probable values are 1.12, 0.75, 0.56, 0.33, 0.18, 0.04, and 0.01 at. (8.85, 6.1, 4.6, 2.8, 1.5, 0.34, and 0.09 wt.) % Bi at 551, 525, 500, 450, 400, 300, and 200°C, respectively. Lattice-parameter measurements [9] indicate that there is probably only a negligible solubility of Mg in solid Bi. Emf measurements show deviations from the phase diagram of Fig. 184 only in the range 70–100 at. (95–100 wt.) % Bi, which are probably traceable to passivity phenomena [10].

Crystal Structures. Measurements of the lattice parameters of the Mg primary solid solution are due to [8, 11]. Mg_3Bi_2 (L) was shown to have the hexagonal La_2O_3 ($D5_2$) type of structure, with $a = 4.675$ A, $c = 7.416$ A, $c/a = 1.586$ [12].

1. N. J. Stepanow, *Zhur. Russ. Fiz. Khim. Obshchestva*, **37,** 1905, 1285.
2. G. Grube, *Z. anorg. Chem.*, **49,** 1906, 83–87.
3. N. J. Stepanow, *Z. anorg. Chem.*, **78,** 1912, 25–29.

3*a*. M. Hansen, "Der Aufbau der Zweistofflegierungen," Springer-Verlag OHG, Berlin, 1936.

3*b*. "Gmelins Handbuch der anorganischen Chemie," System No. 27 A, p. 420, Verlag Chemie, G.m.b.H., Weinheim/Bergstrasse, Germany, 1952.

4. G. Grube, L. Mohr, and R. Bornhak, *Z. Elektrochem.*, **40,** 1934, 143–150. Bi: 99.9 wt. % (+0.0024 Cu, 0.035 Pb, 0.026 Ag, and traces of Fe); Mg: 99.93 wt. % (+0.018 Si, 0.052 Pb, and traces of Cu and Al).
5. Conductivity isotherms indicate a solubility of 2 at. % Bi in the compound; resistivity-vs.-temperature curves, as well as thermal measurements, however, do not show any solid-solubility range on the Bi side.
6. I. G. Schulgin, *Tsvetnaya Met.*, **1940,** no. 9, pp. 96–98 (in Russian). Not available to the author; the results are plotted and discussed in [7].
7. H. Vosskühler, *Metallwirtschaft*, **22,** 1943, 545–547.
8. F. Foote and E. R. Jette, *Trans. AIME*, **143,** 1941, 124–131.
9. E. R. Jette and F. Foote, *Phys. Rev.*, **39,** 1932, 1020.
10. R. Kremann and H. Eitel, *Z. Metallkunde*, **12,** 1920, 363–365.
11. E. Scheufele, *Z. Metallkunde*, **33,** 1941, 219.
12. E. Zintl and E. Husemann, *Z. physik. Chem.*, **B21,** 1933, 138–155.

0.5803
$\bar{1}.4197$

Bi-Mn Bismuth-Manganese

The constitution of the Mn-Bi system is fairly well known except for the composition range covered by the miscibility gap in the liquid state. Below that gap the results obtained on alloys prepared by melting (and therefore badly segregated) do not reveal the true constitution.

The melting point of Bi is lowered about 7°C by small additions of Mn [1, 2]; a eutectic lies between 0.5 and 0.8 wt. (1.9–3.0 at.) % Mn [2]. Values for the maximal extent of the miscibility gap are 0.5–77 wt. (0.1–47 at.) % Bi [2], 7-70 wt. (1.9–38 at.) % Bi [3, 4]. From scanty experimental work, [1] concluded that only one compound exists, namely, MnBi (79.19 wt. % Bi), formed peritectically at about 450°C. That MnBi at least is the intermediate phase richest in Bi was confirmed by [2, 3, 5, 6]. The formulas richer in Bi named by [7, 8] are incorrect.

Mainly on the basis of cooling curves, [2] arrived at quite different views as to the constitution of the Mn-Bi system (Fig. 185). Five nonvariant equilibrium lines (at 1255, 1043, 597, 445, and 262°C) were found, all still observable even at low Bi contents because of considerable disturbances of equilibrium. Those at 1043, 597, and 445°C were ascribed to the formation of three intermediate phases—besides MnBi, the compounds X and Y lower in Bi. The composition of the X and Y phases could not be established. The impurity of the Mn used (and further contamination of the alloys by crucible materials) makes it probable that weak thermal effects (especially

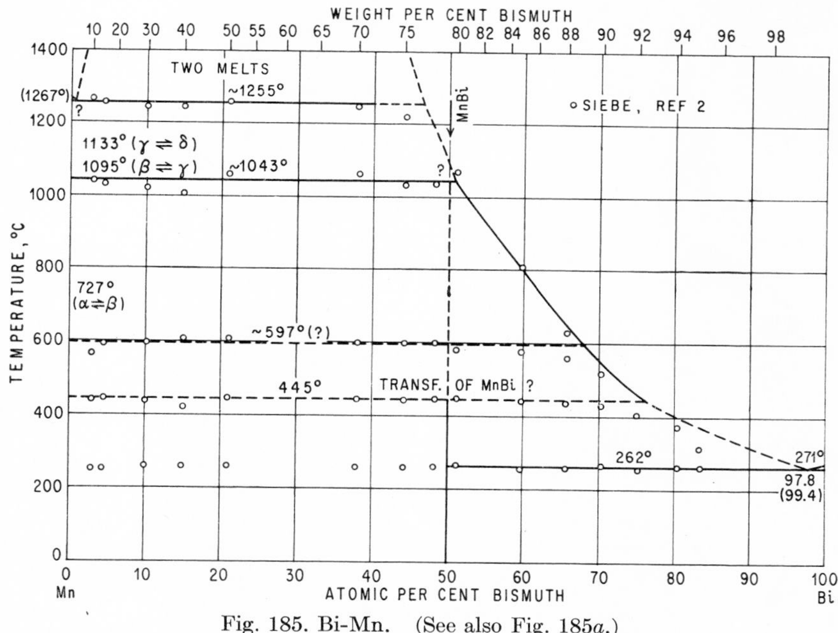

Fig. 185. Bi-Mn. (See also Fig. 185*a*.)

those near 600°C) as well as the different kinds of crystals observed microscopically are caused by reactions outside the binary system.

The view that MnBi is the only intermediate phase has support from the following observations of other authors: (*a*) [9] and [3] found well-shaped crystals of a composition near MnBi between the layers segregated in the liquid state. (*b*) [6] prepared alloys containing 25–75 at. % Bi by sintering of powder mixtures at temperatures up to 600°C and found microscopically and roentgenographically no other intermediate phases besides MnBi.

In drawing the tentative phase diagram of Fig. 185, these points were considered, along with assumptions [2, 5] and observations [10] that MnBi undergoes transformations and observations by [8] indicating that MnBi may decompose only at temperatures as high as 900–1200°C (but see Supplement). Further investigations are necessary in order to establish the true constitution. For preparation of Mn-rich alloys, sintering of master alloy-manganese mixtures is suggested.

MnBi was shown to have the NiAs (B8) type of structure [11, 6], with $a = 4.26_9$

A, $c = 6.08_2$ A, $c/a = 1.42_4$ [6]. According to [10], an alloy quenched from 400°C (nonferromagnetic) exhibits a strongly contracted c axis: $a = 4.33$ A, $c = 5.84$ A, $c/a = 1.35$.

The carrier of ferromagnetism of Mn-Bi alloys [12, 13] is unequivocally the compound MnBi [6]. The Curie point of this phase has been reported to be 343 ± 2°C [6], 350°C [14], and 340–360°C [10].

Mn and Bi are not expected to form extended primary solid solutions because of the unfavorable size factors.

Supplement. The results of a magnetic and X-ray analysis by [15] indicate that stoichiometric MnBi is the only intermediate phase in the Mn-Bi system.

[16] made a high-temperature X-ray study of MnBi samples (containing some excess Bi) up to 450°C. At 320°C, near the magnetic transition point, a discontinuous change in both the a and c axes was observed, a increasing by 1.5% and c decreasing by

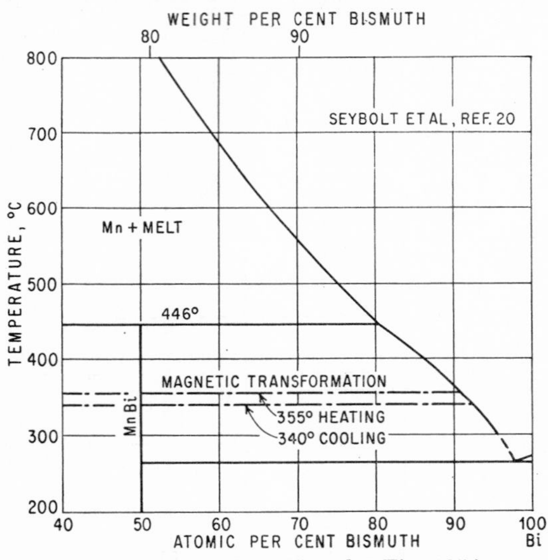

Fig. 185*a*. Bi-Mn. (See also Fig. 185.)

3% on heating through this temperature (cf. the results of [10]). At 435°C MnBi decomposed. This finding may be taken as a corroboration of the peritectic reaction assumed by early investigators [1, 2] at 445–450°C. Therefore, MnBi does not undergo merely a magnetic solid-state transition at this temperature, as was suggested by [7, 10, 17] and assumed in Fig. 185.

The magnetostructural transformation at ~350°C in MnBi was further studied by [18–21]. [20] also redetermined the solidification equilibria up to 800°C by means of differential thermal-analysis and liquid-saturation tests (Fig. 185*a*). "No attempt was made to check the eutectic composition. Siebe's [2] estimate of 0.5–0.8 wt. % Mn is probably not far off" [20].

1. E. Bekier, *Intern. Z. Metallog.*, **7**, 1914, 83–92. The Mn used contained 1.60 Fe, 1.52 Si, 0.10 Cu (wt. %).
2. P. Siebe, *Z. anorg. Chem.*, **108**, 1919, 161–171. The Mn used contained 3 Al, 0.5 Si, 0.6 Fe (wt. %). The alloys were melted in Pythagoras crucibles under H_2.
3. N. Parravano and U. Perret, *Gazz. chim. ital.*, **45**(1), 1923, 390–394; the Mn contained 1 wt. % Si.

4. [1] observed a high content of Si in the Mn-rich melt. The contaminants, which other authors also found in abundance, certainly affect the extension of the gap.
5. E. Montignie, *Bull. soc. chim. France*, **5,** 1938, 343–344.
6. K. Thielmann, *Ann. Physik*, **37,** 1940, 41–62.
7. U. Fürst and F. Halla, *Z. physik. Chem.*, **B40,** 1938, 285–307.
8. F. Halla and E. Montignie, *Z. physik. Chem.*, **B42,** 1939, 153–154.
9. E. Wedekind and T. Veit, *Ber. deut. chem. Ges.*, **44,** 1911, 2665–2666.
10. C. Guillaud, *J. phys. radium*, **12,** 1951, 143, 223; the author ascribes the disappearance of ferromagnetism (~350°C) to a transformation of the first order into the antiferromagnetic state and denotes the thermal effect at 445°C as the transition antiferromagnetism → normal paramagnetism.
11. R. Hocart and C. Guillaud, *Compt. rend.*, **209,** 1939, 443.
12. F. Heusler, *Ges. Bef. ges. Naturw. zu Marburg*, **13,** 1904.
13. S. Hilpert and T. Dieckmann, *Ber. deut. chem. Ges.*, **44,** 1911, 2831–2835.
14. F. Galperin, *Doklady Akad. Nauk S.S.S.R.*, **75,** 1950, 647–650 (in Russian).
15. C. Guillaud, *J. recherches centre natl. recherche sci., Labs. Bellevue (Paris)*, 1947, 15–21.
16. B. T. M. Willis and H. P. Rooksby, *Proc. Phys. Soc. (London)*, **67B,** 1954, 290–296.
17. J. S. Smart, *Phys. Rev.*, **90,** 1953, 55–58.
18. A. J. P. Meyer and P. Taglang, *J. phys. radium*, **12,** 1951, 63 pp. (quoted from [20], discussion).
19. R. R. Heikes, *Phys. Rev.*, **99,** 1955, 446.
20. A. U. Seybolt, H. Hansen, B. W. Roberts, and P. Yurcisin, *Trans. AIME*, **206,** 1956, 606–610; discussion, pp. 1406–1408.
21. B. W. Roberts, *Phys. Rev.*, **104,** 1956, 606–616. Neutron-diffraction study of MnBi from 4.2°K to 420°C (quoted from [20], discussion).

0.3381
$\bar{1}$.6619

Bi-Mo Bismuth-Molybdenum

Mo-Bi alloys were prepared by reduction of the metal oxides with carbon [1].

1. C. L. Sargent, *J. Am. Chem. Soc.*, **22,** 1900, 783–790.

1.1738
$\bar{2}$.8262

Bi-N Bismuth-Nitrogen

Neither solid nor liquid Bi (investigated up to 600°C) dissolves nitrogen [1].

1. A. Sieverts and W. Krumbhaar, *Ber. deut. chem. Ges.*, **43,** 1910, 894.

0.9585
$\bar{1}$.0415

Bi-Na Bismuth-Sodium

[1] investigated the influence of small additions of Na on the freezing point of Bi and found a depression of 8.3°C at 3.86 at. (0.44 wt.) % Na. The results of thermal analyses by [2, 3] are plotted in Fig. 186 and indicate the existence of two compounds, Na_3Bi (75.19 wt. % Bi) and NaBi (90.09 wt. % Bi). Another interesting method for preparing these compounds is the use of solutions in liquid NH_3 [4–8]. The two-phase field (NaBi + Bi) was corroborated roentgenographically [8]. Emf measurements of [9] are not suitable for checking the phase diagram.

Na_3Bi was shown [10] to have the hexagonal Na_3As ($D0_{18}$) type of structure with $a = 5.459$ A, $c = 9.674$ A, $c/a = 1.772$. NaBi possesses the CuAu ($L1_0$) type of structure, with $a = 3.47$ A, $c = 4.81$ A, $c/a = 1.39$ (for the 2-atom unit cell) [8].

1. C. T. Heycock and F. H. Neville, *J. Chem. Soc.*, **61,** 1892, 892.
2. N. S. Kurnakow and A. N. Kusnetzow, *Z. anorg. Chem.*, **23,** 1900, 455–462.
3. C. H. Mathewson, *Z. anorg. Chem.*, **50,** 1906, 187–192. The alloys were prepared in an iron crucible (for Na_3Bi) or in refractory glass tubes under hydrogen. The compositions of all alloys were corrected for oxidation losses (Na) according to analyses made on some of the alloys.
4. A. Joannis, *Compt. rend.*, **114,** 1892, 587.
5. P. Lebeau, *Compt. rend.*, **130,** 1900, 504.
6. C. A. Kraus and H. F. Kurtz, *J. Am. Chem. Soc.*, **47,** 1925, 43.
7. E. Zintl, J. Goubeau, and W. Dullenkopf, *Z. physik. Chem.*, **A154,** 1931, 1–46.
8. E. Zintl and W. Dullenkopf, *Z. physik. Chem.*, **B16,** 1932, 183–194.
9. R. Kremann, J. Fritsch, and R. Riebl, *Z. Metallkunde*, **13,** 1921, 66–71.
10. G. Brauer and E. Zintl, *Z. physik. Chem.*, **B37,** 1937, 323–352.

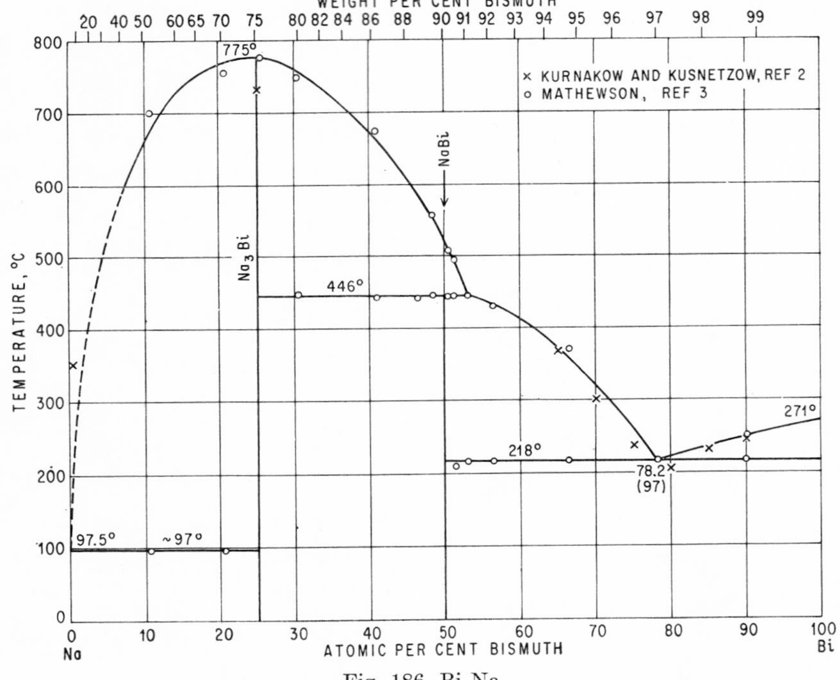

Fig. 186. Bi-Na

0.5516
$\bar{1}.4484$

Bi-Ni Bismuth-Nickel

Information on the phase diagram Ni-Bi is due to thermal and microscopic investigations [1, 2], as well as to a roentgenographic and microscopic study of [3]. The results agree on the main features of the system, especially as to the existence of two compounds, ~NiBi (78.08 wt. % Bi) and $NiBi_3$ (91.44 wt. % Bi). The thermal results of [1, 2] are plotted in Fig. 187, indicating that there are only some quantitative discrepancies as to the position of the liquidus and of the upper peritectic [4]. Strong peritectic coring makes determination of the formulas of the two compounds very difficult [5]. X-ray and microscopic work [3] on eight alloys of 8.6–94.5 at. % Bi (some annealed for 7 days at 400 or 600°C) indicated that the compound (NiBi) has a range of homogeneity—narrow at lower temperatures but broadening at elevated temperatures—which "as good as certain" lies below 50 at. % Bi. Also, the formula $NiBi_3$ for the Bi-rich compound was confirmed. The widely differing atomic diameters of Ni and Bi suggest that the primary solid solutions are very restricted. This conclusion is confirmed by lattice-parameter determinations which indicate no shifting of lines with composition [3]. The results of qualitative measurements of [2], showing a lowering of the Curie point of Ni by 20–25°C for 0.3 at. (1.06 wt.) % Bi, should therefore be checked.

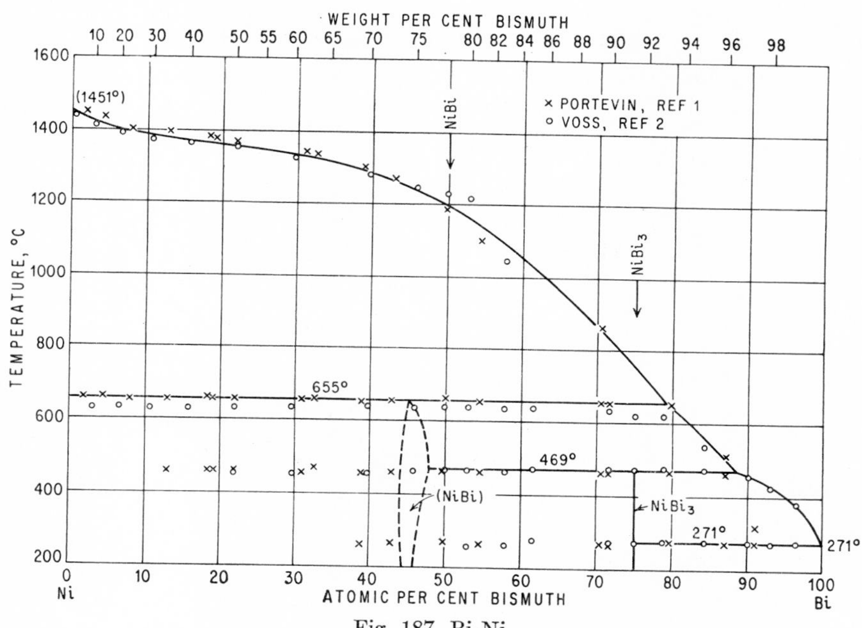

Fig. 187. Bi-Ni

Crystal Structures. (NiBi) was shown to have the NiAs (B8) type of structure [3]. The lattice parameters of alloys quenched from 500°C decrease with increasing Bi content: $a = 4.078$ A, $c = 5.36$ A, $c/a = 1.314$ on the Ni-rich side; $a = 4.069$ A, $c = 5.36$ A, $c/a = 1.317$ on the Bi-rich side. It is assumed [3] that the excess Ni atoms (see above) are located at random in interstitial positions.

$NiBi_3$ forms soft needles of hexagonal cross section and yields complex powder patterns, the lines of which show no shifting with composition [3].

Supplement. $NiBi_3$ was reported to be orthorhombic, with $a = 11.44$ A, $b = 9.01$ A, $c = 4.1$ A [6].

1. A. Portevin, *Compt. rend.*, **145,** 1907, 1168–1170; *Rev. mét.*, **5,** 1908, 110–120. The alloys were made with Ni of unknown composition in MgO-lined crucibles under an atmosphere of illuminating gas (*Leuchtgas*).
2. G. Voss, *Z. anorg. Chem.*, **57,** 1908, 52–58. The nickel used had 1.86 Co, 0.47 Fe (wt. %). Alloying was accomplished in porcelain crucibles under nitrogen.
3. G. Hägg and G. Funke, *Z. physik. Chem.*, **B6,** 1930, 272–283. The nickel contained 0.2 wt. % Co as the largest impurity. Alloying was accomplished under nitrogen.

4. For the upper peritectic, [1] reports 655 and [2] 638°C. The latter, however, remarks that his value may be too low because of supercooling effects. The values of [1] and [2] for the two other horizontal equilibrium lines (469 and 271°C in Fig. 187) agree within the limits of possible experimental error.
5. The disturbances of equilibrium are also evident from the thermal arrests plotted in Fig. 187.
6. G. S. Zhdanov, V. P. Glagoleva, N. N. Zhuravlev, and Y. N. Venevtsev, *Zhur. Eksptl. i Teoret. Fiz.*, **25**, 1953, 115–122. (Not available to the author.) *Chem. Abstr.*, **49**, 1955, 5050.

1.1160
$\bar{2}$.8840

Bi-O Bismuth-Oxygen

The solubility of oxygen in liquid Bi was determined by measurements of the Bi-Bi_2O_3 equilibrium [1]. The data given in Table 17 were obtained. These values fall on a straight line when the logarithm of the solubility is plotted vs. the reciprocal of absolute temperature.

Table 17. The Solubility of Oxygen in Liquid Bismuth

°C	Wt. % O	At. % O
400	0.00034	0.0044
500	0.0012	0.016
600	0.0029	0.038
675	0.0096	0.12
750	0.016	0.20

According to [2], the existence of an oxide BiO is improbable.

1. C. B. Griffith and M. W. Mallett, *J. Am. Chem. Soc.*, **75**, 1953, 1832.
2. O. Glemser and M. Filcek, *Z. anorg. Chem.*, **269**, 1952, 99–101.

0.8291
$\bar{1}$.1709

Bi-P Bismuth-Phosphorus

After saturation of molten Bi with P at 800°C, the as-cooled sample contained about 0.1 wt. (0.7 at.) % P in elementary form [1]. Neither by direct synthesis nor by indirect methods has a Bi phosphide been prepared with certainty [2].

1. A. Stock and F. Gomolka, *Ber. deut. chem. Ges.*, **42**, 1909, 4519–4521.
2. See chemical handbooks, for instance, "Gmelins Handbuch der anorganischen Chemie," System No. 19, pp. 193–194, Verlag Chemie, G.m.b.H., Berlin, 1927.

0.0037
$\bar{1}$.9963

Bi-Pb Bismuth-Lead

The phase diagram of Fig. 188 was drawn using data of the following papers: liquidus [1], eutectic [1, 2], solidus of (Pb) phase [3, 4], solid-solubility boundary of (Pb) phase [5], peritectic temperature [3, 4], boundaries of ϵ phase [4, 6], and solid-solubility boundary of (Bi) phase [3, 2]. Complete or partial determinations of the liquidus were carried out by [7–11, 1, 3, 12, 2, 4, 6]. The smooth curve of [1] (14 data points) still seems to be the most reliable. The reported data for the eutectic vary between 54.7 and 57.2 at. (54.9–57.4 wt.) % Bi and 124 and 125.5°C [13, 8, 14, 10, 11, 1, 3, 12, 2, 4, 15], those for the peritectic between 182 and 189°C [3, 12, 2, 4, 6]. Thermodynamic computations of some liquid-solid equilibria, based on emf measurements, are due to [3].

The discussion as to the existence of an intermediate phase ([16–18]; see also [12]) was positively decided by X-ray [19] and superconductivity [20] work. There is still no agreement, however, as to the homogeneity range of this phase (ϵ in Fig. 188). For room temperature the roentgenographic and resistometric data by [4] [23.5–33.2 at. (23.7–33.4 wt.) % Bi], which agree fairly well with older roentgenographic data [19], seem to be the most reliable ones. The extrapolated [4] value for the most Bi-rich composition at the eutectic temperature (>40% Bi) agrees with microscopic observations on quenched alloys [21] but contradicts the value 36.5 at. (36.7 wt.) % Bi based on emf measurements [3].

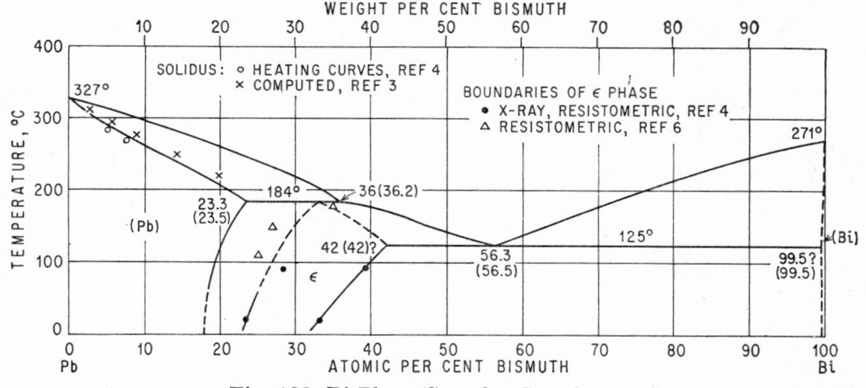

Fig. 188. Bi-Pb. (See also Supplement.)

Solid Solubility of Bi in Pb. Back-reflection X-ray work [5] on 13 carefully annealed alloys yielded the following solubility data: 23.25, 22.5, 22.0, 21.7, 20.6, 20.3, and 19.05 wt. (23.1, 22.35, 21.85, 21.6, 20.5, 20.2, and 18.9 at.) % Bi at 180, 170, 160, 150, 130, 120, and 100°C, respectively. Some X-ray and resistivity work by [4] agrees well with the foregoing data; however, there are other papers which report a much smaller solubility [2] or even an increasing solubility with decreasing temperature [12]. Numerous measurements of physical properties are obsolete or yield, if any, only qualitative information: density [22–27], specific heat [28], thermoelectric power [29, 30, 27], magnetic susceptibility [31, 32], electrochemical potential [33–36], hardness [21, 25, 37–39], electrical or thermal conductivity [40, 21, 41, 27], and Hall effect [27]. Tensile strength, elongation, and hardness of alloys up to 50 wt. % Bi were measured by [42].

Solid Solubility of Pb in Bi. In spite of the large number of publications, the extent of the Bi primary solid solution is not yet known precisely.

From roentgenographic [43] and emf [3] measurements it can be concluded that the solubility at the eutectic temperature is not less than 0.5 at. (wt.) % Pb and not much larger than 1.0 at. (wt.) % Pb, which margin agrees with some older data [21, 44] and also with the resistometric value, 0.5 wt. (at.) % Pb, by [2]. [45] microscopically found the solubility to be less than 0.1 wt. (at.) % Pb; their radiographic data are inconclusive (see also [46]).

Further measurements (cf. the general statement in the preceding section) are electrical conductivity [40, 27, 47], thermal conductivity [40], magnetic susceptibility

[31, 32, 48], thermoelectric power [29, 27], density [22–25, 27], specific heat [28], electrochemical potential [33-36], hardness [21, 25, 37–39], and Hall effect [27].

Crystal Structure. Lattice parameters of the (Pb) phase were measured by [19, 12, 2] and especially [5], those of the ϵ phase by [19, 2, 4]. The ϵ phase has a h.c.p. (A3) type of structure [19, 12, 4, et al.], with $a = 3.48$ A, $c = 5.78$ A, $c/a = 1.66$ [19, 4] at 25 wt. (24.8 at.) % Bi.

Supplement. For boiling points of some Pb-Bi alloys, see [49]. [50] measured lattice parameters of Pb-Bi alloys containing 0–40 at. % Bi. "The lattice spacings determined . . . are in qualitative agreement with the less detailed work of [4] and with the results published by [5]. . . . Six alloys with compositions in the range 21.88–26.84 at. % Bi were micrographically duplex after annealing for 19 days at 115–117°C; the diffraction patterns showed the presence of a h.c.p. structure, which was obtained pure at 28.83 and 30.78 at. % Bi. The lattice spacings of the former alloy were: $a = 3.5013$ kX, $c = 5.7054$ kX, $c/a = 1.6295$ at 20°C. Three alloys exceeding a Bi content of 34.81 at. % contained the h.c.p. phase and the Bi-rich solid solution. These results are consistent with the equilibrium diagram proposed by [4]" [51].

1. W. E. Barlow, *J. Am. Chem. Soc.*, **32,** 1910, 1394–1395; *Z. anorg. Chem.*, **70,** 1911, 183–184.
2. T. Takase, *Nippon Kinzoku Gakkai-Shi*, **1,** 1937, 143–150.
3. H. S. Strickler and H. Seltz, *J. Am. Chem. Soc.*, **58,** 1936, 2084–2093.
4. H. von Hofe and H. Hanemann, *Z. Metallkunde*, **32,** 1940, 112–117. The linear extrapolation of the Pb-rich boundary of the "β" phase, resulting in the value 32.3 wt. % Bi at the peritectic temperature (fig. 9 of the original paper), was not based on the correct value, 28.5 wt. % Bi at 92°C, from fig. 2! In "Metals Handbook" (1948 ed., The American Society for Metals, Cleveland, Ohio), the value 32.3 wt. % Bi is erroneously taken as the liquid composition point of the peritectic.
5. M. Hayasi, *Nippon Kinzoku Gakkai-Shi*, **3,** 1939, 123–125 (in Japanese; tables and diagrams in English).
6. Y. Mitani, *Repts. Osaka Ind. Research Inst.*, **3,** 1951, 36–38.
7. E. Wiedemann, *Wied. Ann.*, **20,** 1883, 236–243.
8. D. Mazzotto, *Mem. reale ist. Lombardo*, (3)**16,** 1886, 1; *Wied. Ann. Beibl.*, **11,** 1887, 231.
9. C. T. Heycock and F. H. Neville, *J. Chem. Soc.*, **61,** 1892, 910–911.
10. A. W. Kapp, dissertation, Königsberg, 1901; *Ann. Physik*, (4)**6,** 1901, 760, 769. See also [26].
11. G. Charpy, Société d'encouragement pour l'industrie nationale, Paris, Commission des alliages, "Contribution à l'étude des alliages," p. 220, Typ. Chamerot et Renouard, Paris, 1901.
12. N. S. Kurnakow and V. A. Ageeva, *Izvest. Akad. Nauk S.S.S.R.* (*Chim.*), (4) **1937,** 735–742 (in Russian; summary in German).
13. F. Guthrie, *Phil. Mag.*, (5)**17,** 1884, 464.
14. D. Mazzotto, *Nuovo cimento*, (5)**18**(2), 1909, 180–196.
15. Unpublished data of W. C. Smith; see "Metals Handbook," 1948 ed., The American Society for Metals, Cleveland, Ohio.
16. G. Tammann and H. Schimpff, *Z. Elektrochem.*, **18,** 1912, 595.
17. K. Bux, *Z. Physik*, **14,** 1923, 316–327.
18. C. Benedicks, *Z. Metallkunde*, **25,** 1933, 200–201.
19. D. Solomon and W. Morris-Jones, *Phil. Mag.*, **11,** 1931, 1090–1103.
20. W. Meissner, H. Franz, and H. Westerhoff, *Ann. Physik*, **13,** 1932, 979–984.
21. W. Herold, *Z. anorg. Chem.*, **112,** 1920, 131–154.

22. E. Maey, *Z. physik. Chem.*, **50,** 1905, 216–217.
23. E. S. Shepherd, *J. Phys. Chem.*, **6,** 1902, 522–523.
24. O. Richter, *Ann. Physik*, (4)**42,** 1913, 779–795.
25. J. Goebel, *Z. Metallkunde*, **14,** 1922, 390–392.
26. A. Stoffel, *Z. anorg. Chem.*, **53,** 1907, 149–151; see remark as to Shepherd's observation.
27. W. R. Thomas and E. J. Evans, *Phil. Mag.*, **16,** 1933, 329–353.
28. O. Richter, *Ann. Physik*, (4)**39,** 1912, 1590–1608; **42,** 1913, 779–795.
29. A. Battelli, *Atti reale ist. veneto*, (6)**5,** 1886-1887, 1137; *Wied. Ann. Beibl.*, **12,** 1888, 269. See also W. Broniewski, *Rev. mét.*, **7,** 1910, 352–353.
30. C. R. Darling and R. H. Rinaldi, *Proc. Phys. Soc. (London)*, **36,** 1924, 281–287.
31. H. Endo, *Science Repts. Tôhoku Univ.*, **14,** 1925, 498–499; K. Honda and H. Endo, *J. Inst. Metals*, **37,** 1927, 34–36.
32. Y. Shimizu, *Science Repts. Tôhoku Univ.*, **2,** 1932, 842–843.
33. A. P. Laurie, *J. Chem. Soc.*, **65,** 1894, 1034.
34. E. S. Shepherd, *J. Phys. Chem.*, **7,** 1903, 15–17.
35. N. Puschin, *Zhur. Russ. Fiz.-Khim. Obshchestva*, **39,** 1907, 869.
36. R. Kremann and A. Langbauer, *Z. anorg. Chem.*, **127,** 1923, 240.
37. C. di Capua and M. Arnone, *Rend. reale accad. nazl. Lincei*, **33,** 1924, 28–31.
38 A. Mallock, *Nature*, **121,** 1928, 827.
39. V. P. Schischokin and V. Ageeva, *Izvest. Metally*, **1932,** 119–136; abstract, *J. Inst. Metals*, **53,** 1933, 552.
40. F. A. Schulze, *Ann. Physik*, **9,** 1902, 580–581.
41. G. Tammann and H. Rüdiger, *Z. anorg. Chem.*, **192,** 1930, 9–13.
42. J. G. Thompson, *J. Research Natl. Bur. Standards*, **5,** 1930, 1085.
43. E. R. Jette and F. Foote, *Phys. Rev.*, **39,** 1932, 1018–1020.
44. W. Guertler, *Z. anorg. Chem.*, **51,** 1906, 411.
45. G. Tammann and G. Bandel, *Z. Metallkunde*, **25,** 1933, 153–156.
46. W. Seith and A. Keil, *Z. Metallkunde*, **26,** 1934, 68–69.
47. N. Thompson, *Proc. Roy. Soc. (London)*, **A155,** 1936, 111–123.
48. A. Goetz and A. B. Focke, *Phys. Rev.*, **45,** 1934, 170 (measurements on single crystals).
49. W. Leitgebel, *Z. anorg. Chem.*, **202,** 1931, 305–324.
50. C. Tyzack and G. V. Raynor, *Acta Cryst.*, **7,** 1954, 505–510.
51. According to the diagram proposed by von Hofe and Hanemann [4], alloys between 34.8 and 40 at. % Bi—if successfully quenched from about 115°C—should be single-phase (!); see Fig. 188.

0.2920
$\bar{1}$.7080

Bi-Pd Bismuth-Palladium

The influence of small additions of Pd on the melting point of Bi was investigated by [1], who found a continuous depression up to 5°C at 1.16 wt. (2.25 at.) % Pd without reaching a eutectic. By treating an alloy of about 5 wt. (9 at.) % Pd with dilute HNO_3, needles of the composition $PdBi_2$ were obtained [2].

A preliminary study of the Pd-Bi alloys, mainly by microscopic and roentgenographic methods, is due to [3]. The work is limited to a description of 26 alloys, most of them annealed for ¾–1½ hr, and does not attempt to determine the phase relationships. Based on the published microstructures and X-ray photographs and on more recent work [4], the tentative phase diagram of Fig. 189 is suggested. The following intermediate phases exist with certainty: Pd_5Bi_3 [37.5 at. (54.03 wt.) % Bi] with a polymorphic transformation [4] between 400 and 500°C, PdBi (66.20 wt. % Bi), and two phases—probably high- and low-temperature modifications—near the composition $PdBi_2$ (79.66 wt. % Bi). A further compound perhaps exists between 25 and 37.5 at. % Bi [3, 5] (cf. Supplement). With the exception of the Pd primary solid solution, which extends almost up to 25 at. % Bi, all phases have very narrow ranges of homogeneity [3].

Crystal Structures. The lattice parameter of the (Pd) phase increases with increasing Bi content [3]. Pd_5Bi_3 (H) has the NiAs (B8) type of structure, with $a = 4.5_1$ A, $c = 5.8_2$ A, $c/a = 1.29$ [4], and Pd_5Bi_3 (L) is a lower-symmetry variation of it [4]. The cleavage of PdBi and some unsatisfactory single-crystal photographs admit orthorhombic or monoclinic symmetry. The provisional cell dimensions do not show orthohexagonal relations [3]. The X-ray powder pattern of $PdBi_2$ (H) is relatively simple, suggesting a hexagonal or tetragonal structure [3]. The rotation photograph of $PdBi_2$ (L) can be interpreted by a monoclinic lattice, with $a = 12.74$ A,

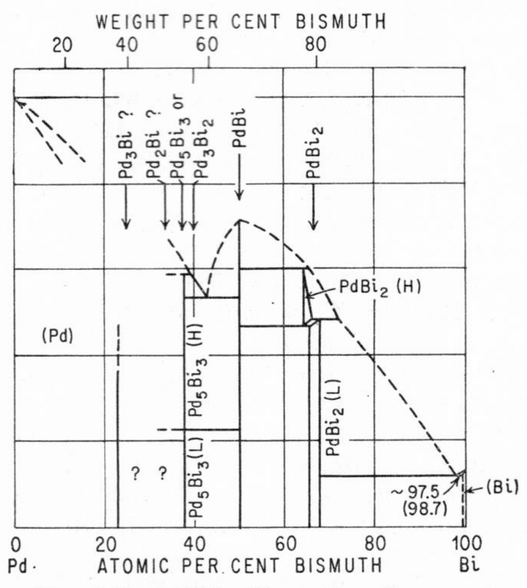

Fig. 189. Bi-Pd. (Tentative diagram.)

$b = 4.29$ A, $c = 5.67$ A, $\beta = 102°52'$, and 12 atoms per unit cell, or by an orthorhombic lattice, with $a' = 24.85$ A, $b' = 4.29$ A, $c' = 5.67$ A, and 24 atoms per unit cell [3]. The very slightly contracted Bi pattern of a two-phase alloy indicates a small solubility of Pd in solid Bi [3]. Lines and intensities of powder patterns of the compounds are reported by [3].

Supplement. [6] made an X-ray and metallographic analysis of the entire system and reported the existence of the following intermediate phases: Pd_3Bi, Pd_2Bi, Pd_3Bi_2, PdBi, α-$PdBi_2$ (low-temperature modification), and β-$PdBi_2$ (high-temperature modification). The following structural data were given: PdBi, orthorhombic, $a = 7.203$ A, $b = 8.707$ A, $c = 10.662$ A, 32 atoms per unit cell; α-$PdBi_2$, monoclinic, $a = 12.74$ A, $b = 4.25$ A, $c = 5.665$ A, $\beta = 102°35'$, 12 atoms per unit cell (cf. above); β-$PdBi_2$, tetragonal, $a = 3.362$ A, $c = 12.983$ A, 6 atoms per unit cell.

1. C. T. Heycock and F. H. Neville, *J. Chem. Soc.*, **61**, 1892, 894.
2. F. Roessler, *Z. anorg. Chem.*, **9**, 1895, 70.
3. S. V. Burr and M. A. Peacock, *Univ. Toronto Studies, Geol. Ser.*, no. 47, 1942, 19–31. The elements (Pd stated to be over 99 wt. % pure) were mixed in different proportions and heated in sealed, evacuated quartz tubes in an open gas

flame till alloying occurred. The melt was kept liquid for 10–15 min and shaken. Most of the products were held for various periods at temperatures somewhat below the melting point and then allowed to cool in air. Other heat-treatments were occasionally given. Abstract, "Gmelins Handbuch der anorganischen Chemie," System No. 68, pp. 575–578, Verlag Chemie, G.m.b.H., Weinheim/Bergstrasse, Germany, 1951.
4. K. Schubert and H. Beeskow, *Naturwissenschaften*, **40,** 1953, 269.
5. Prismatic crystals in a microstructure of a 33.3 at. % Bi alloy seem to indicate a further phase, but X-ray powder patterns do not support this conclusion.
6. N. N. Zhuravlev and G. S. Zhdanov, *Zhur. Eksptl. i Teoret. Fiz.*, **25,** 1953, 485–490; D. M. Kheiker, G. S. Zhdanov, and N. N. Zhuravlev, *Zhur. Eksptl. i Teoret. Fiz.*, **25,** 1953, 621–627; L. S. Zevin, G. S. Zhdanov, and N. N. Zhuravlev, *Zhur. Eksptl. i Teoret. Fiz.*, **25,** 1953, 751–754. (Not available to the author.) *Chem. Abstr.*, **49,** 1955, 4349–4350.

$\bar{1}.9979$
0.0021

Bi-Po Bismuth-Polonium

The solid solubility of Po in Bi is in the order of 5×10^{-10} at. (wt.) % [1].

1. G. Tammann and A. v. Löwis of Menar, *Z. anorg. Chem.*, **205,** 1932, 145–162.

0.1712
$\bar{1}.8288$

Bi-Pr Bismuth-Praseodymium

PrBi (59.73 wt. % Bi) was shown to have the NaCl (B1) type of structure, with $a = 6.46_1$ A [1].

1. A. Iandelli and E. Botti, *Atti reale accad. nazl. Lincei*, **25,** 1937, 233–238; abstract in *Strukturbericht*, **5,** 1937, 45.

0.0296
$\bar{1}.9704$

Bi-Pt Bismuth-Platinum

No systematic work has been done on this system as yet. Addition of 0.2–1.2 at. (0.19–1.12 wt.) % Pt depresses the solidification point of Bi by an average of 2°C [1]; thus the eutectic seems to lie below 0.2 at. % Pt. Alloying some Bi with Pt decreases only slightly the temperature coefficient of the electrical resistance [2], from which one may conclude that there is only a negligible solid solubility of Bi in Pt [3]. The existence of the compounds PtBi (51.71 wt. % Bi) [4], $PtBi_2$ (68.15 wt. % Bi) [5, 6], and $PtBi_3$ (76.25 wt. % Bi) [7] has been reported. PtBi has the NiAs (B8) type of structure [4] (no parameters given); and $PtBi_2$, the cubic pyrite (C2) type of structure, with $a = 6.69_7$ A [6].

1. C. T. Heycock and F. H. Neville, *J. Chem. Soc.*, **61,** 1892, 896.
2. C. Barus, *Am. J. Sci.*, (3)**36,** 1888, 427–442.
3. W. Guertler, "Handbuch der Metallographie," vol. 1, part 1, p. 622, Verlagsbuchhandlung Gebrüder Borntraeger, Berlin, 1913.
4. A. Harder; see E. Zintl and H. Kaiser, *Z. anorg. Chem.*, **211,** 1933, 128.
5. F. Roessler, *Z. anorg. Chem.*, **9,** 1895, 68–69.
6. H. J. Wallbaum, *Z. Metallkunde*, **35,** 1943, 200.
7. N. Alekseevsky, *Zhur. Eksptl. i Teoret. Fiz.*, **18,** 1948, 101–102. Not available to the author. *Met. Abstr.*, **19,** 1952, 579.

$\bar{1}.9418$
0.0582

Bi-Pu Bismuth-Plutonium

PuBi has the cubic NaCl (B1) type of structure, with $a = 6.350 \pm 1$ A. There is metallographic and thermal-analysis evidence of the existence of a compound $PuBi_2$ [1]. The solubility of Pu in liquid Bi is 0.7, 1.4, 3.6, 8.7, and 16 wt. (0.6, 1.2, 3.2, 7.7, and 14.3 at.) % Pu at 400, 500, 600, 700, and 800°C, respectively [2].

1. A. S. Coffinberry and F. H. Ellinger, "Proceedings of the International Conference on the Peaceful Uses of Atomic Energy," vol. 9, pp. 138–146, Geneva, 1955. "Metallurgy and Fuels," Progress in Nuclear Energy, ser. V, vol. 1, pp. 391, 398, Pergamon Press Ltd., London, 1956.
2. Taken from a graph in B. R. T. Frost, *Nuclear Eng.*, November, 1956, p. 336.

0.3883
$\bar{1}.6117$

Bi-Rb Bismuth-Rubidium

Photoelectric measurements indicate the existence of a compound Rb_3Bi (44.91 wt. % Bi) [1].

1. A. Sommer, *Nature*, **148**, 1941, 468.

0.3077
$\bar{1}.6923$

Bi-Rh Bismuth-Rhodium

From extraction tests the existence of the compounds $RhBi_4$ (10.96 wt. % Rh) [1, 2] and $RhBi_2$ (19.76 wt. % Rh) [2] was concluded.

The constitution of the Rh-Bi alloys up to about 60 at. % Rh was investigated by thermal and microscopic methods [3] (Fig. 190). A eutectic at 0.7 wt. (1.4 at.) % Rh is clearly indicated by the microstructures of an alloy of this composition; the thermal arrests found for the three-phase equilibrium [x in Fig. 190; spreading between 244 and 279°C (!)], however, mostly lie above the melting point of Bi, which would require a peritectic instead of a eutectic reaction. Herein lies a contradiction for which the temperature measurement may be responsible. Besides $RhBi_4$ and $RhBi_2$, the compound RhBi (33.00 wt. % Rh) was found. Very weak (~390°C) and clearer (~310°C) thermal effects near $RhBi_4$, as well as very weak ones (~498°C) near $RhBi_2$, are presumably due to polymorphic transformations of these compounds. The peritectic reactions, especially the one at 433°C, are very sluggish. It is not yet known with certainty whether RhBi is the compound richest in Rh.

Crystal Structure. RhBi was shown to have the NiAs (B8) type of structure, with $a = 4.08_3$ A, $c = 5.66_8$ A, $c/a = 1.388$ [4]. No indications for a polymorphic transformation of this compound were detected.

1. H. Rössler, *Chem. Ztg.*, **24**(2), 1900, 734–735.
2. L. Wöhler and L. Metz, *Z. anorg. Chem.*, **149**, 1925, 309–313.
3. E. J. Rode, *Izvest. Inst. Platiny*, **7**, 1929, 21–31 (in Russian); abstract in "Gmelins Handbuch der anorganischen Chemie," System No. 68 A, pp. 550–552, Verlag Chemie, G.m.b.H., Weinheim/Bergstrasse, Germany, 1951. The alloys were prepared in graphite crucibles under carbon powder or $BaCl_2$. Twelve microstructures are given.
4. H. Pfisterer and K. Schubert, *Z. Metallkunde*, **41**, 1950, 365.

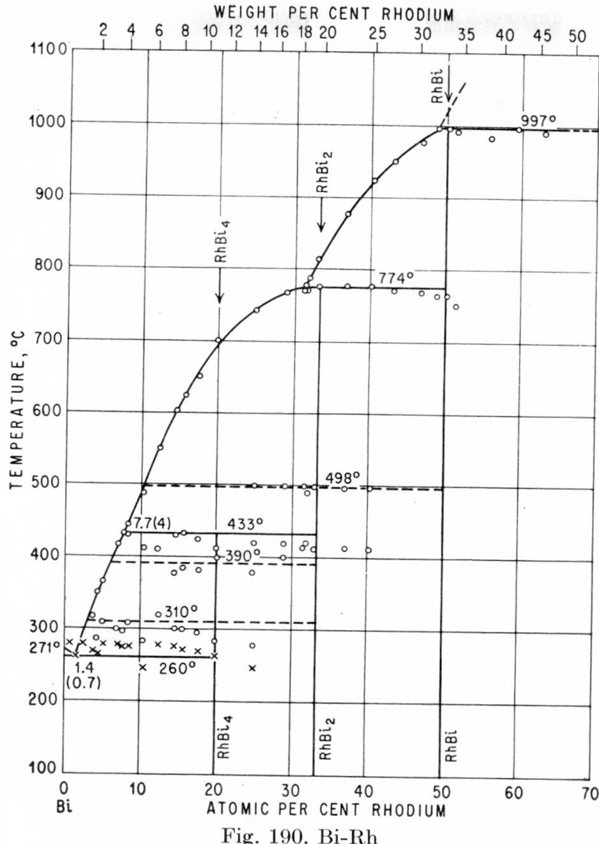

Fig. 190. Bi-Rh

0.3128
$\bar{1}.6872$

Bi-Ru Bismuth-Ruthenium

Ru is not attacked by molten Bi at 800°C [1]. An alloy of the composition RuBi showed no ferromagnetism at the temperature of liquid hydrogen [2].

1. L. Wöhler and L. Metz, *Z. anorg. Chem.*, **149,** 1925, 312.
2. L. Thomassen, *Z. physik. Chem.*, **B2,** 1929, 378.

0.8141
$\bar{1}.1859$

Bi-S Bismuth-Sulfur

The phase diagram up to 52.4 at. % S was established by means of cooling curves and microscopic observations [1]. The compound which crystallizes from Bi-rich melts was shown to be Bi_2S_3 (18.71 wt. % S) [2, 1]; the compound BiS [3–5] does not exist (see also [6, 7]). The equilibria involving the vapor phase were tentatively drawn in Fig. 191 to accommodate the observation that Bi_2S_3 decomposes upon heating [1]. Liquid S does not dissolve any Bi [1].

Crystal Structure. Bi_2S_3 (occurring naturally as bismuthite) has the orthorhombic Sb_2S_3 ($D5_8$) type of structure [8], with a = 11.15 A, b = 11.29 A, c = 3.98 A [8, 9].

1. A. H. W. Aten, *Z. anorg. Chem.*, **47**, 1905, 386–398. The alloys were enclosed in glass tubes.
2. F. Roessler, *Z. anorg. Chem.*, **9**, 1895, 44–46.
3. R. Schneider, *Pogg. Ann.*, **88**, 1853, 43; **97**, 1856, 480–482; *J. prakt. Chem.*, (2)**58**, 1898, 562; **60**, 1899, 524–543.
4. W. Herz and A. Guttmann, *Z. anorg. Chem.*, **53**, 1907, 71–73.
5. H. Pélabon, *Compt. rend.*, **137**, 1903, 648–650; *J. chim. phys.*, **2**, 1904, 324.
6. L. Vanino and F. Treubert, *Ber. deut. chem. Ges.*, **32**, 1899, 1078–1081.
7. H. Pélabon, *Ann. chim. et phys.*, (8)**17**, 1909, 546.
8. W. Hofmann, *Z. Krist.*, **86**, 1933, 225–245.
9. J. Garrido and R. Feo, *Bull. soc. franç. minéral.*, **61**, 1938, 196–204.

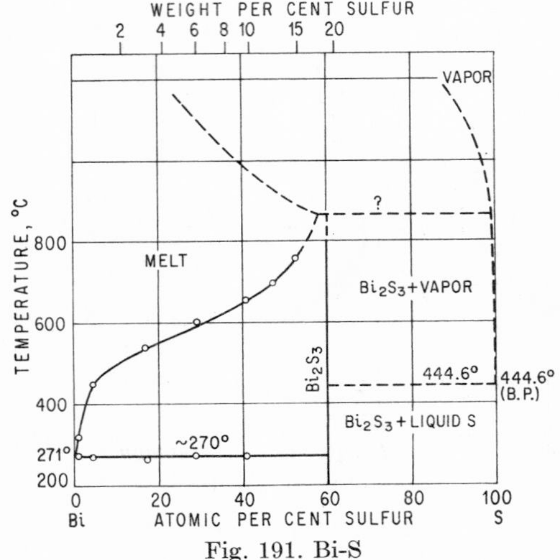

Fig. 191. Bi-S

0.2346
1.7654

Bi-Sb Bismuth-Antimony

Liquidus. The liquidus of Fig. 192 was drawn by a graphical interpolation of data from thermal [1–4] and thermoresistometric [5] determinations. With the exception of the range near pure Bi, the curve may have an accuracy better than ±5°C.

Solidus. Strong coring, even in slowly cooled alloys, leads to disturbances of equilibrium which falsify the solidus as an almost straight line up to about 75 at. % Sb [3, 4]; the possible existence of a three-phase horizontal and a miscibility gap in the solid state was therefore under discussion for some time. From microscopic observations on unannealed [6] and annealed [7, 8, 4, 5] alloys, however, a continuous solid solution could be concluded; and this view could be corroborated beyond any doubt by thermoresistometric determinations of the solidus [5, 9], as well as by roentgenographic work [10, 11].

Lattice Parameters. The parameters vary linearly with composition in at. % [10, 11]. A rough determination of d_{110} values was made by [12].

Other Physical Properties. Measurements of numerous physical properties as a function of composition have little interest from a metallurgical point of view because the alloys used certainly were not in a state of equilibrium. An exception, however, is the more recent determination of the magnetic susceptibility [13–15], the results of which are in agreement with a continuous solid solution. For the older literature, see [16]. Thermodynamic computations of [17] are obsolete.

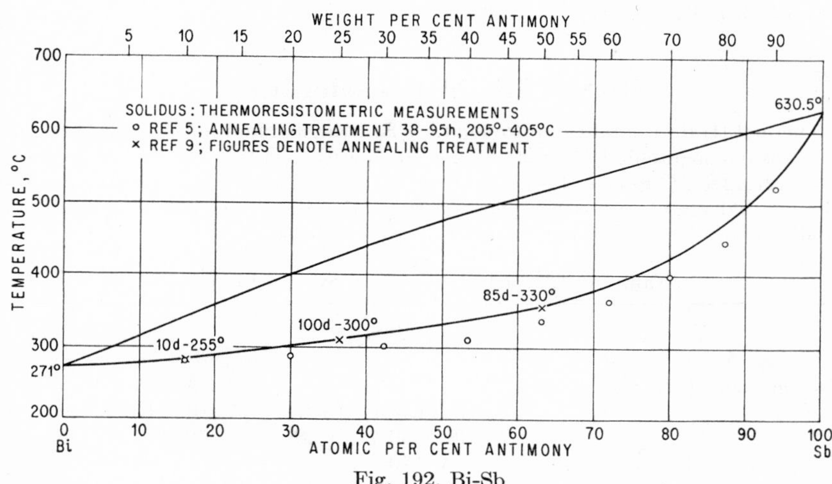

Fig. 192. Bi-Sb

Supplement. For boiling points of some Bi-Sb alloys, see [18]. Based on specific-heat measurements, [19] calculated the liquidus and the solidus curves.

1. C. T. Heycock and F. H. Neville, *J. Chem. Soc.*, **61,** 1892, 896.
2. H. Gautier and Roland-Gosselin, *Bull. soc. encour. ind. natl.*, **1,** 1896, 1314; Société d'encouragement pour l'industrie nationale, Paris, Commission des alliages, "Contribution à l'étude des alliages," p. 114, Typ. Chamerot et Renouard, Paris, 1901.
3. K. Hüttner and G. Tammann, *Z. anorg. Chem.*, **44,** 1905, 131–144.
4. M. Cook, *J. Inst. Metals,* **28,** 1922, 421–436; discussion, pp. 437–445.
5. B. Otani, *Science Repts. Tôhoku Univ.*, **13,** 1925, 293–297.
6. G. Charpy, *Bull. soc. encour. ind. natl.*, **2,** 1897, 384; Société d'encouragement pour l'industrie nationale, Paris, Commission des alliages, "Contribution à l'étude des alliages," pp. 138–139, Typ. Chamerot et Renouard, Paris, 1901.
7. Ssaposhnikow, *Zhur. Russ. Fiz.-Khim. Obshchestva,* **40,** 1908, 665 (in Russian); quoted from [4].
8. N. Parravano and E. Viviani, *Atti reale accad. nazl. Lincei,* (5)**19**(1), 1910, 835–840; *Gazz. chim. ital.*, **40**(2), 1910, 446.
9. G. Masing, P. Rahlfs, and W. Schaarwächter, *Z. Metallkunde,* **40,** 1949, 333–334.
10. E. G. Bowen and W. Morris-Jones, *Phil. Mag.*, (7)**13,** 1932, 1029–1032.
11. W. F. Ehret and M. B. Abramson, *J. Am. Chem. Soc.*, **56,** 1934, 385–388.
12. H. von Hofe and H. Hanemann, *Z. Metallkunde,* **32,** 1940, 115–116.
13. H. Endo, *Science Repts. Tôhoku Univ.*, **16,** 1927, 225–227; see also K. Honda and H. Endo, *J. Inst. Metals,* **37,** 1927, 38.
14. Y. Shimizu, *Science Repts. Tôhoku Univ.*, **21,** 1932, 836–838.
15. A. Goetz and A. B. Focke, *Phys. Rev.*, **45,** 1934, 180.

16. "Gmelins Handbuch der anorganischen Chemie," System No. 19, pp. 198-202, Verlag Chemie, G.m.b.H., Berlin, 1927; M. Hansen, "Der Aufbau der Zweistofflegierungen," Springer-Verlag OHG, Berlin, 1936.
17. C. Yap, *Trans. AIME*, **93**, 1931, 185–206.
18. W. Leitgebel, *Z. anorg. Chem.*, **202**, 1931, 305–324.
19. Y. E. Geguzin and B. Y. Pines, *Zhur. Fiz. Khim.*, **26**, 1952, 27–30; *Chem. Abstr.*, **47**, 1953, 7985.

0.4227
$\bar{1}$.5773

Bi-Se Bismuth-Selenium

Thermal [1] (in the range up to 60 at. % Se), thermal and microscopic [2, 3], as well as roentgenographic [4] investigations agree on the existence of two compounds [5] BiSe (27.42 wt. % Se) and Bi_2Se_3 (36.17 wt. % Se) and other characteristic features of the phase diagram (Fig. 193). Some quantitative discrepancies in the results of [2] and [3] can be seen from Fig. 193 as well as from Table 18.

Table 18. Characteristics of the System Bi-Se

	Ref. 2	Ref. 3
Peritectic (melt + $Bi_2Se_3 \rightleftarrows$ BiSe)..	605°C ~26.5–36.2 wt. (48.8–60 at.) % Se	599–605 (602)°C ~27.4–36.2 wt. (50–60 at.) % Se
Melting point of Bi_2Se_3...........	706°C	688°C
Monotectic......................	615–622 (618)°C	602–609 (605)°C
Miscibility gap at monotectic temperature	50–>95 wt. (72.6–>98 at.) % Se	51–~91 wt. (73.4–~96.4 at.) % Se
End of solidification of Se-rich alloys........................	216–218 (217)°C	150–170°C

According to [1–3] the Bi-rich eutectic practically coincides with pure Bi. Since measurements of susceptibility [6] and resistivity [7] on single crystals indicate a certain solid solubility of Se in Bi [8] [about 1 at. (0.4 wt.) % Se according to [7]], the liquidus near pure Bi should be determined. Thermal effects at about 420°C on cooling curves of some of the alloys which contained BiSe were observed by [3] and interpreted as a polymorphic transformation of this compound. As to the extension of the monotectic toward high Se content, the larger value of [2] seems to be more reliable. The Se-rich eutectic, which according to [2] coincides with pure Se, was drawn by [3] near pure Se but at about 160°C (supercooling?).

Crystal Structures. The structure of Bi_2Se_3 is isotypic with that of Bi_2Te_3 and Bi_2Te_2S [9, 10; 11]; the rhombohedral lattice constants are 9.84 ± 2 A, α = 24.4° [10].

For indexing attempts of BiSe (L), see [4].

Note Added in Proof. In an electron-diffraction investigation of Bi-Se alloy films, [12, 13] observed three intermediate phases: rhombohedral Bi_2Se_3 (C33 type) and Bi_3Se_4 and cubic (NaCl type) BiSe. Variations of the lattice constants with composition indicate that these phases are of variable composition.

1. H. Pélabon, *J. chim. phys.*, **2**, 1904, 328–330; see also *Ann. chim. et phys.*, (7)**25**, 1902, 432, and (8)**17**, 1909, 549.
2. N. Parravano, *Gazz. chim. ital.*, **43**(1), 1913, 201–209.

3. N. Tomoshige, *Mem. Coll. Sci. Kyoto Imp. Univ.*, **4**, 1919, 55–60.
4. N. Parravano and V. Cagliotti, *Gazz. chim. ital.*, **60**, 1930, 923–933; also *Strukturbericht*, **2**, 1928-1932, 203–204.
5. Both compounds had already been prepared much earlier.
6. A. Goetz and A. B. Focke, *Phys. Rev.*, **45**, 1934, 170.
7. N. Thompson, *Proc. Roy. Soc. (London)*, **A155**, 1936, 111–123.
8. The (cursory) X-ray work of [4] does not indicate any solid solutions.
9. E. Dönges, *Z. anorg. Chem.*, **265**, 1951, 56–61.
10. K. Schubert, K. Anderko, and P. Esslinger, *Z. Metallkunde*, **44**, 1953, 461; see remark concerning the Bi_2Te_2S structure.
11. This type of crystal structure was erroneously named C33 (a C type should have the composition AB_2); see *Strukturbericht*, **7**, 1939, 110 (footnote).
12. S. A. Semiletov, *Trudy Inst. Krist., Akad. Nauk S.S.S.R.*, **10**, 1954, 76–83; *Chem. Abstr.*, **50**, 1956, 1403.
13. S. A. Semiletov and Z. G. Pinsker, *Doklady Akad. Nauk S.S.S.R.*, **100**, 1955, 1079–1082; *Chem. Abstr.*, **50**, 1956, 5359.

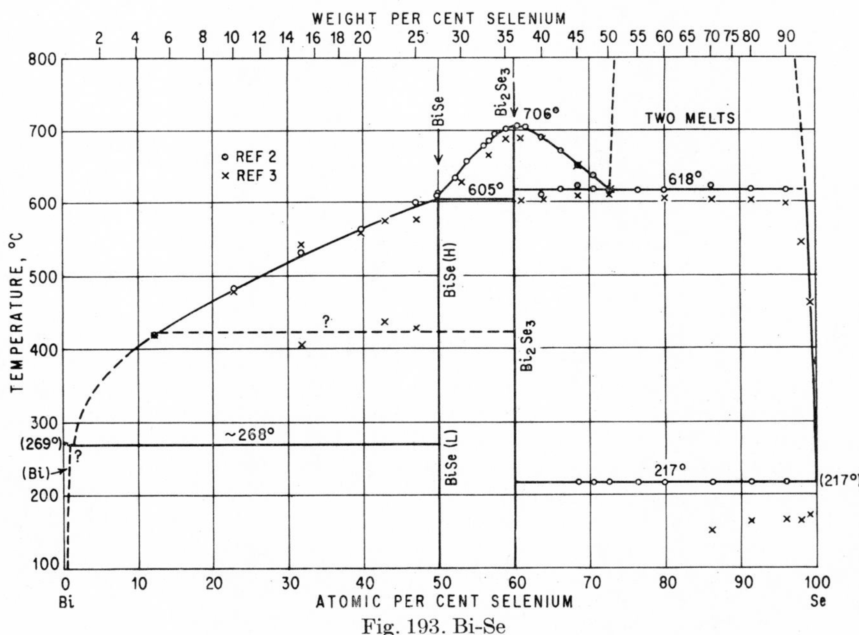

Fig. 193. Bi-Se

0.8716
$\bar{1}.1284$

Bi-Si Bismuth-Silicon

The phase diagram of Fig. 194 is due to a thermal investigation by [1]. The miscibility gap in the liquid state was found to extend from 2 wt. (13 at.) % to practically 100% Si at the monotectic temperature. The position of the eutectic is not known; the small depression (3–4°C) of the melting point of Bi may be caused by impurities in the Si used [1]. There is agreement that no intermediate phases exist [2, 1, 3]. As to solid solutions, 0.2–0.8 wt. (1.5–5.7 at.) % Si could not be detected microscopically [1]; according to X-ray results [3], however, the mutual solid solubilities, if any exist, are very small: <0.2 at. (0.03 wt.) % Si in Bi and <0.1 at. (0.7 wt.) % Bi in Si.

1. R. S. Williams, *Z. anorg. Chem.*, **55,** 1907, 21–23. Si, 98.07 wt. % pure. The alloys were prepared in porcelain crucibles under N_2 at 1500°C; Bi sublimated on the cold crucible walls to a small extent.
2. E. Vigouroux, *Compt. rend.*, **123,** 1896, 115.
3. E. R. Jette and E. B. Gebert, *J. Chem. Phys.*, **1,** 1933, 753–755.

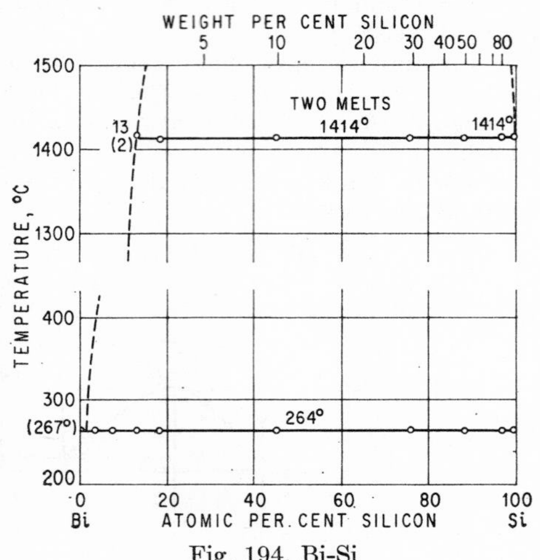

Fig. 194. Bi-Si

0.2457
$\bar{1}$.7543

Bi-Sn Bismuth-Tin

Liquidus. Determinations of the entire liquidus are due to [1–7]; liquidus points of single alloys were measured by [8–12], among others. For the Sn branch of the liquidus there is good agreement among the several investigators, but discrepancies up to 40°C occur on the Bi side which are mainly due to supercooling effects. For Fig. 195 the curve of [7] was chosen as the most reliable one; the position of its Bi branch above that of [6] has support by thermodynamic computations based on emf measurements [13]. More recent values for the eutectic are 60 wt. (46 at.) % Bi at 140°C [5], 54 wt. (40 at.) % Bi at 140°C [12], 58 wt. (44 at.) % Bi at 139°C [6], and 57 wt. (43 at.) % Bi at 138.5°C [7] (Fig. 195).

Solid Alloys. To account for effects in various physical-properties-vs.-temperature curves around 100°C, particularly with Sn-rich alloys, nonvariant reactions (eutectoid, peritectoid) were assumed [14–17]. However, later investigations, [18, 19, 7] especially, have shown that no intermediate phases occur and that those effects were due to a strong decrease in solubility of Bi in Sn, this interpretation having been suggested already by [20].

(Sn). An excellent determination of the solidus (heating curves combined with microscopic work) was made by [7] (Fig. 195). The solidus was also determined by thermal [21] and thermoresistometric [6] methods, as well as computed from emf data [13]; single points were reported by [22, 23].

The solvus curve was investigated by the microscopic composition-bracket method [7] at 18 different temperatures, which work outweighs single values reported by [20, 24, 19, 23]. The solubilities are 9.6, 6.9, 4.8, 3.0, 1.5, and 0.5 at. (15.8, 11.6, 8.2, 5.25, 2.7, and 1.0 wt.) % Bi at 120, 100, 80, 60, 40, and 20°C, respectively (taken from a graph).

(Bi). The existence of a limited primary solid solution on the Bi side can be concluded from, among other evidence, many of the numerous measurements of physical properties listed below. From these papers, as well as microscopic [25, 4], thermal [21], thermoresistometric [6], and X-ray [26] measurements, however, no reliable quantitative conclusions can be drawn. In [26*a*], a solubility of about 1 wt. (1.7 at.) % Sn at the eutectic temperature was assumed as the most probable value. In the meantime, widely differing values for that solubility were reported: about 0.1 wt. (0.18 at.) % Sn from microscopic work on long-annealed (1 month, 138°C) and quenched alloys [7] and 2.4 wt. (4.1 at.) % Sn from emf measurements [13];

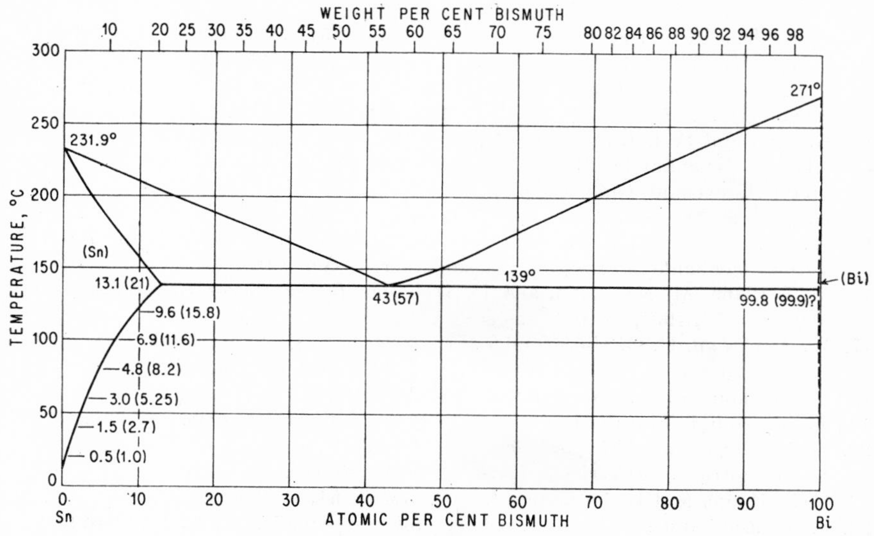

Fig. 195. Bi-Sn

further, [27] could produce single crystals with a Sn content as high as 5 at. (2.9 wt.) % Sn (supersaturated?). For Fig. 195 the value of [7] was preferred as that due to the most direct method.

Physical Properties. Because of the indefinite states of the alloys measured (especially in older works [28]), the insensitivity of the particular property, or the inadequate number of specimens, the following measurements do not give quantitative contributions to the phase diagram: electrical conductivity [29–31, 8, 32–34, 16, 35, 36], heat conductivity [32], thermoelectric power [37, 33, 16, 35], magnetic susceptibility [38–41, 27], Hall effect [35], density [42, 39, 35], emf [43–45], and hardness [46, 47]. Specific heats of several alloys with various compositions were measured by [23] continuously from the solid to the liquid state.

Supplement. [48] reported the eutectic data 139°C, 43.4 at. % Bi, which are in excellent agreement with those in Fig. 195. The eutectic alloy was studied by [49] by means of X-ray diffraction. [50] found that under sufficiently high pressure a new phase appears, believed to be the intermetallic compound SnBi. The lattice spacings of Sn-Bi alloys containing up to about 5 at. % Bi were measured by [51]. Both *a* and *c* of tetragonal Sn are considerably expanded by addition of Bi; the axial ratio

c/a is decreased. On the basis of measurements of electrical properties, [52] estimated the solubility of Sn in solid Bi to be 1.5 at. %.

1. F. Rudberg, *Pogg. Ann.*, **18,** 1830, 240.
2. D. Mazzotto, *Mem. reale ist. Lombardo,* **16,** 1886, 1; see also K. Bornemann, *Metallurgie,* **8,** 1911, 277, and later papers of Mazzotto.
3. A. W. Kapp, Dissertation, Königsberg, 1901; also *Ann. Physik,* **6,** 1901, 759, 769.
4. W. v. Lepkowski, *Z. anorg. Chem.*, **59,** 1908, 286–289.
5. J. Würschmidt, *Z. Physik,* **5,** 1921, 39–47.
6. H. Endo, *Science Repts. Tôhoku Univ.*, **14,** 1925, 489–495; *Kinzoku-no-Kenkyu,* **2,** 1925, 682–691 (in Japanese).
7. A. C. Davidson, *Tech. Pub. Int. Tin Research Develop. Council,* **A77,** 1938.
8. C. L. Weber, *Wied. Ann. Physik,* **34,** 1888, 580.
9. C. T. Heycock and F. H. Neville, *J. Chem. Soc.*, **57,** 1890, 384.
10. C. T. Heycock and F. H. Neville, *J. Chem. Soc.*, **61,** 1892, 896.
11. A. Stoffel, *Z. anorg. Chem.*, **53,** 1907, 147–148.
12. K. Gilbert, *Z. Metallkunde,* **14,** 1922, 249–251.
13. H. Seltz and F. J. Dunkerley, *J. Am. Chem. Soc.*, **64,** 1942, 1392–1395; see also "Thermodynamics in Physical Metallurgy," pp. 56–62, American Society for Metals, Cleveland, Ohio, 1950.
14. K. Bornemann, *Metallurgie,* **8,** 1911, 279.
15. W. Guertler, "Handbuch der Metallographie," vol. 1, part 1, pp. 736–742, Verlagsbuchhandlung Gebrüder Borntraeger, Berlin, 1912.
16. M. LeBlanc, M. Naumann, and D. Tschesno, *Ber. K. Sächs. Ges. Wiss., Math.-phys. Kl.*, **79,** 1927, 71–106.
17. W. A. Cowan, G. O. Hiers, and F. H. Edwards, "National Metals Handbook," p. 1379, American Society for Steel Treating, Cleveland, Ohio, 1933.
18. D. Solomon and W. Morris-Jones, *Phil. Mag.*, (7)**11,** 1931, 1090–1103.
19. T. Sato and T. Matuhasi, *Nippon Kinzoku Gakkai-Shi,* **2,** 1938, 592–597.
20. D. Mazzotto, *Intern. Z. Metallog.*, **4,** 1913, 273–294.
21. D. Mazzotto, *Nuovo cimento,* (5)**18**(2), 1909, 180–196.
22. C. E. Homer and H. Plummer, *J. Inst. Metals,* **64,** 1939, 169.
23. S. Nagasaki and E. Fujita, *Nippon Kinzoku Gakkai-Shi,* **16,** 1952, 313–321.
24. E. Jenckel and L. Roth, *Z. Metallkunde,* **30,** 1938, 135–144.
25. W. Guertler, *Z. anorg. Chem.*, **51,** 1906, 411.
26. E. R. Jette and F. Foote, *Phys. Rev.*, **39,** 1932, 1018–1020.

26*a*. M. Hansen, "Der Aufbau der Zweistofflegierungen," Springer-Verlag OHG, Berlin, 1936.

27. A. Goetz and A. B. Focke, *Phys. Rev.*, **45,** 1934, 170–199, especially 181 (crystal diamagnetism of Bi crystals).
28. For the measurements of electrical properties before 1907, see also the compilation by W. Guertler, *Z. anorg. Chem.*, **51,** 1906, 408–411; **54,** 1907, 68–69.
29. A. Matthiessen, *Pogg. Ann.*, **110,** 1860, 212–213.
30. A. Righi, *J. phys.*, (2)**3,** 1884, 355.
31. A. v. Ettinghausen and W. Nernst, *Wied. Ann. Physik,* (2)**33,** 1888, 474.
32. F. A. Schulze, *Ann. Physik,* **9,** 1902, 555–589.
33. A. Bucher, *Z. anorg. Chem.*, **98,** 1916, 117–126.
34. Künzel-Mehner, Dissertation, Leipzig, 1920.
35. W. R. Thomas and E. J. Evans, *Phil. Mag.*, **17,** 1934, 65–83.
36. N. Thompson, *Proc. Roy. Soc. (London),* **A155,** 1936, 111–123.
37. C. Hutchins, *Am. J. Sci.*, (3)**48,** 1894, 226; see W. Broniewski, *Rev. mét.*, **7,** 1910, 353.

38. H. Endo, *Science Repts. Tôhoku Univ.*, **14**, 1925, 489–495; **16**, 1927, 229; K. Honda and H. Endo, *J. Inst. Metals*, **37**, 1927, 36.
39. J. F. Spencer and M. E. John, *Proc. Roy. Soc.* (*London*), **A116**, 1927, 69–70; J. F. Spencer, *J. Soc. Chem. Ind.*, **50**, 1931, 37–39.
40. A. Goetz and A. B. Focke, *Phys. Rev.*, **38**, 1931, 1569–1572.
41. Y. Shimizu, *Science Repts. Tôhoku Univ.*, **21**, 1932, 840–842.
42. E. S. Shepherd, *J. Phys. Chem.*, **6**, 1902, 523–526.
43. A. P. Laurie, *J. Chem. Soc.*, **65**, 1894, 1031.
44. E. S. Shepherd, *J. Phys. Chem.*, **7**, 1903, 15–16.
45. N. A. Puschin, *Z. anorg. Chem.*, **56**, 1908, 24–26.
46. C. di Capua, *Rend. accad. nazl. Lincei*, **33**(1), 1924, 141–144.
47. W. Schischokin and W. Agejewa, *Z. anorg. Chem.*, **193**, 1930, 237–238.
48. O. Hájiček, *Hutnické Listy*, **3**, 1948, 265–270.
49. S. V. Avakyan, E. N. Kislyakova, and N. F. Lashko, *Zhur. Fiz. Khim.*, **24**, 1950, 1057–1060; *Chem. Abstr.*, **45**, 1951, 1000 (Nature of eutectic alloys. Monocrystalline phase in binary eutectics).
50. P. W. Bridgman, *Bull. soc. chim. Belges*, **62**, 1953, 26–33 (in English); *Met. Abstr.*, **21**, 1953, 243.
51. J. A. Lee and G. V. Raynor, *Proc. Phys. Soc.* (*London*), **B67**, 1954, 737–747.
52. G. A. Ivanov and A. R. Regel, *Zhur. Tekh. Fiz.*, **25**, 1955, 49–65; *Met. Abstr.*, **23**, 1955, 282–283.

0.3775
$\bar{1}$.6225

Bi-Sr Bismuth-Strontium

According to microscopic observations and cursory X-ray work, the compound $SrBi_3$ (87.74 wt. % Bi) has cubic symmetry [1].

1. J. K. Hulm, Westinghouse Electric Corporation Research Laboratory, East Pittsburgh, Pa.; private communication to M. Hansen, 1955.

0.2143
$\bar{1}$.7857

Bi-Te Bismuth-Tellurium

As to the type of the phase diagram, the thermal results of [1–5] are in full agreement with each other. The small discrepancies in the results of [2, 3, 5], who used a pure Te (melting point, 447–452°C), can be seen from the following values: Bi-Bi_2Te_3 eutectic, 1–1.5 wt. (1.6–2.4 at.) % Te, 263–267°C; melting point of Bi_2Te_3, 583–586°C; Bi_2Te_3-Te eutectic, 85 wt. (90.3 at.) % Te, 410–413°C. The liquidus of Fig. 196 was taken mainly from [3, 5].

Microscopic [1, 4, 5] and qualitative X-ray [5] investigations corroborated the existence of only one compound, Bi_2Te_3 (47.80 wt. % Te). From thermal results [3–5] as well as measurements of physical properties (electrical resistivity [6, 4], thermoelectric power [6, 3, 5], magnetic susceptibility [7–10], Hall effect [11], and hardness [12]), a broad range of homogeneity of this phase can be concluded. The reported saturation limits [36–40 and 53–55 wt. (48–52 and 65–67 at.) % Te, respectively] have no quantitative importance; only [12] examined annealed alloys. The unbroken-line portion of the solidus of (Bi_2Te_3) in Fig. 196 was determined by means of resistivity-vs.-temperature curves [4]; it is nevertheless uncertain whether he found equilibrium data.

From curves of thermoelectric power [6, 3, 5], electrical resistivity [13] and magnetic susceptibility [7, 8, 14, 15], as well as thermal and microscopic observations [4, 5], a certain solid solubility of Te in Bi can be concluded, which according to [13] lies between 0.1–0.2 at. (0.06–0.12 wt.) % Te. The solid solubility of Bi in Te seems to be negligible.

Magnetic susceptibilities of liquid alloys were measured by [16].

Crystal Structure. Bi_2Te_3 was shown to have a rhombohedral structure [17–19], with a = 10.47 A, α = 24°8′ [17]; see also [20]. It is isotypic with that of Bi_2Te_2S and Bi_2Se_3 (C33 type); see discussion of system Bi-Se.

The wide range of homogeneity of this phase forms the basis of several binary and more complex minerals [21].

Note Added in Proof. On the basis of measurements of electrical properties, [22] estimated the solubility of Te in solid Bi to be 0.25 at. % (cf. above). The electrical conductivity and thermo-emf were measured by [23] over the whole composition range on cast as well as annealed specimens. A more detailed study of the Bi_2Te_3

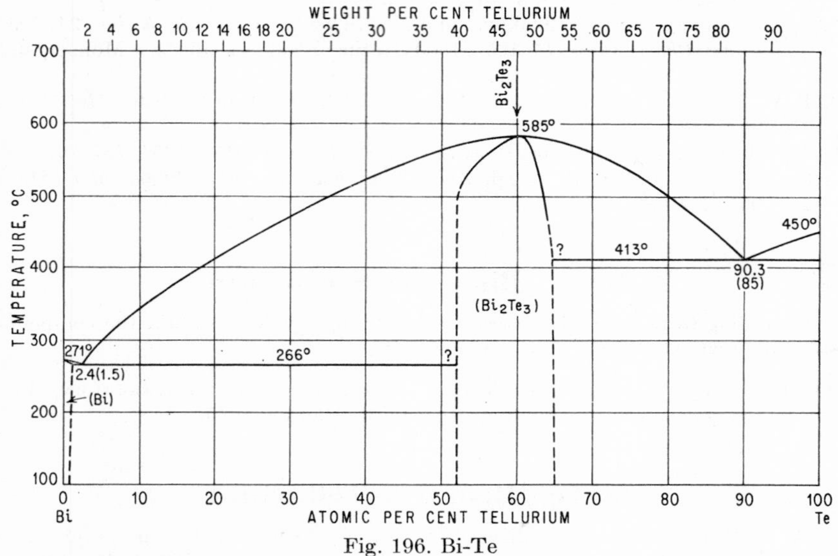

Fig. 196. Bi-Te

region (temperature dependence of thermo-emf, electrical conductivity, Hall effect) was made by [24]. The abstracts available give no constitutional details.

In an electron-diffraction investigation of Bi-Te alloy films, [25] observed, besides solid solutions of Bi in Bi_2Te_3, a cubic NaCl (B1) type BiTe phase with a = 6.47 A.

Based on an X-ray ionization spectrometer investigation of 12 Bi-Te alloys containing 47–49 wt. % Te, [26] suggest the existence of two polymorphic modifications of Bi_2Te_3. Possible structures are discussed in relation to published accounts of the Bi_2Te_3 structure.

1. K. Mönkemeyer, *Z. anorg. Chem.*, **46,** 1905, 415–422.
2. H. Pélabon, *Compt. rend.*, **146,** 1908, 1397–1400; *Ann. chim. et phys.*, (8)**17,** 1909, 526–566.
3. For the results of the work of F. Körber, some together with Saemann, in 1914 and 1918, see [5].
4. H. Endo, *Science Repts. Tôhoku Univ.*, **14,** 1925, 507–510.
5. F. Körber and U. Haschimoto, *Z. anorg. Chem.*, **188,** 1930, 114–126. For Te, a 98.68 wt. % stock material was distilled in vacuo; all other authors made no note concerning the purity of their Te.

6. W. Haken, *Ann. Physik*, **32,** 1910, 319–323.
7. K. Honda and T. Soné, *Science Repts. Tôhoku Univ.*, **2,** 1913, 12–13.
8. H. Endo, *Science Repts. Tôhoku Univ.*, **14,** 1925, 479–512; see also K. Honda and H. Endo, *J. Inst. Metals*, **37,** 1927, 38–45.
9. C. E. Mendenhall and W. F. Lent, *Phys. Rev.*, **32,** 1911, 406. These authors found no indication for the compound in their susceptibility-composition curve, owing certainly to the fact that they investigated only a few alloys in the critical region.
10. Y. Shimizu, *Science Repts. Tôhoku Univ.*, **21,** 1932, 846–847.
11. G. C. Trabacchi, *Atti accad. nazl. Lincei*, **24**(1), 1915, 809–812; *Nuovo cimento*, **9,** 1915, 95–98.
12. V. P. Schischokin and V. Ageeva, *Tsvetnye Metally*, **1932,** 119–136 (in Russian). Not available to the author; abstract in *J. Inst. Metals*, **53,** 1933, 552.
13. N. Thompson, *Proc. Roy. Soc. (London)*, **A155,** 1936, 111–123 (electrical resistance of single crystals).
14. Shimizu [10] could not corroborate the break in the curve of susceptibility vs. composition at 0.5–1 wt. % Te previously found by Endo [8]. See also some remarks on this in [15], pp. 173–174.
15. A. Goetz and A. B. Focke, *Phys. Rev.*, **45,** 1934, 170–199 (susceptibility of single crystals).
16. H. Endo, *Science Repts. Tôhoku Univ.*, **16,** 1927, 206–209; see also K. Honda and H. Endo, *J. Inst. Metals*, **37,** 1927, 38–45.
17. P. W. Lange, *Naturwissenschaften*, **27,** 1939, 133–134.
18. M. A. Peacock and L. G. Berry, *Univ. Toronto Studies, Geol. Ser. No.* 44, 1940, pp. 47–69.
19. C. Frondel, *Am. J. Sci.*, **238,** 1940, 880–888.
20. E. Dönges, *Z. anorg. Chem.*, **265,** 1951, 56–61.
21. See, for instance, H. V. Warren and M. A. Peacock, *Univ. Toronto Studies, Geol. Ser. No.* 49, 1945, pp. 55–69.
22. G. A. Ivanov and A. R. Regel, *Zhur. Tekh. Fiz.*, **25,** 1955, 49–65; *Met. Abstr.*, **23,** 1955, 282–283.
23. F. I. Vasenin, *Zhur. Tekh. Fiz.*, **25,** 1955, 397–401; *Met. Abstr.*, **23,** 1956, 416.
24. R. M. Vlasova and L. S. Stil'bans, *Zhur. Tekh. Fiz.*, **25,** 1955, 569–576; *Met. Abstr.*, **23,** 1956, 416.
25. S. A. Semiletov, *Trudy Inst. Krist., Akad. Nauk S.S.S.R.*, **10,** 1954, 76–83; *Chem. Abstr.*, **50,** 1956, 1403.
26. F. I. Vasenin and P. F. Konovalov, *Zhur. Tekh. Fiz.*, **26,** 1956, 1406–1414; *Met. Abstr.*, **24,** 1957, 350.

$\bar{1}.9544$
0.0456

Bi-Th Bismuth-Thorium

"The diagram for this system (Fig. 196*a*) is based on an investigation by [1] and some later work by [2]. [1] assigned the formulae Th_2Bi and $ThBi_3$ to the two compounds but Bryner's [2] later work indicates that the correct formula of the Bi-rich compound more probably is Th_3Bi_5. [2] also redetermined the liquidus line on the Bi-side of the diagram. The central part of the diagram is extremely difficult to work out as both of the two compounds are unstable in air" [3].

1. H. B. Johnson, unpublished work, Iowa State College, Ames Laboratory, 1950.
2. J. S. Bryner, unpublished work, Brookhaven National Laboratory, 1955.
3. Quoted from Nuclear Metallurgy, *AIME, IMD Special Report Series*, no. 1, pp. 54–55, 1955.

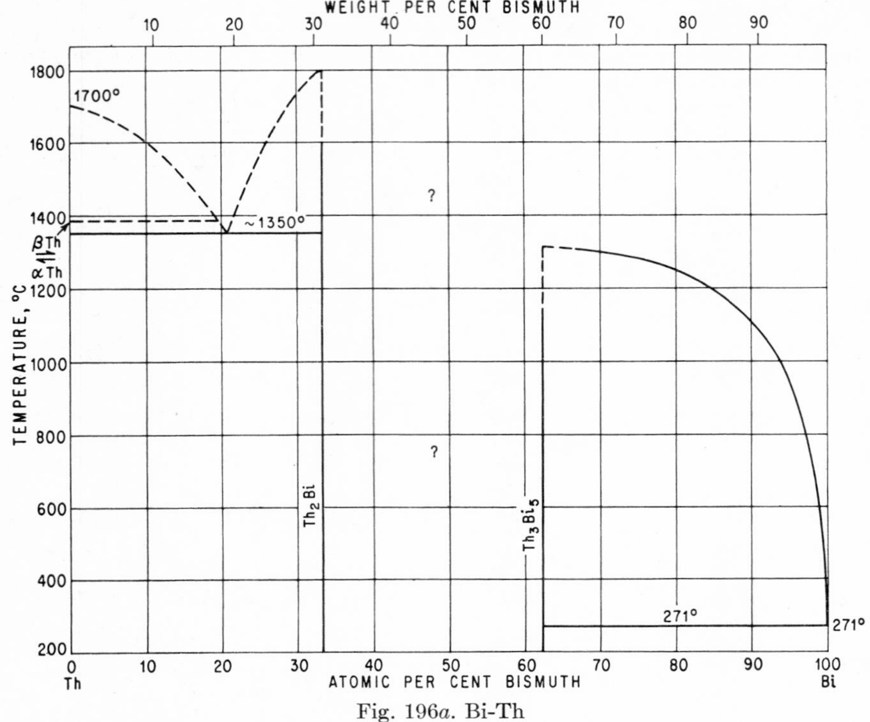

Fig. 196*a*. Bi-Th

0.6398
$\bar{1}$.3602

Bi-Ti Bismuth-Titanium

Bismuth and titanium form only one intermediate phase, Ti_4Bi (47.83 wt. % Ti). Its X-ray diffraction patterns are very similar to those of Ti_4Pb; however, Ti_4Bi apparently has a symmetry lower than hexagonal [1, 2]. Bismuth contents up to 0.3 wt. % do not appear to alter the microstructure of commercially pure titanium after quenching from temperatures between 790 and 900°C [3].

1. H. Nowotny and J. Pesl, *Monatsh. Chem.*, **82,** 1951, 336–343.
2. H. Nowotny and J. Pesl, *Monatsh. Chem.*, **82,** 1951, 344–347.
3. Battelle Memorial Institute, Summary Report—Part III, Preparation and Evaluation of Titanium Alloys, July 30, 1949.

0.0097
$\bar{1}$.9903

Bi-Tl Bismuth-Thallium

Liquidus. Thermal determinations of the liquidus (partial [1, 2]; complete [3, 4]) as well as thermoresistometric measurements [5] are in good agreement [6]. In Fig. 197 an interpolated curve was drawn. A very flat maximum near pure Tl (0.5–1°C above the melting point of Tl) reported by [2–4] is improbable and may be due to a slight supercooling of pure Tl.

Solid State. (The reader is referred to the legend of Fig. 197.) Physical properties interesting from the standpoint of constitution were measured: electrical conductivity [7, 4, 5, 8], hardness [4, 9], magnetic susceptibility [10, 11], thermo-emf

[12], and emf [13, 14]. The view of [15, 5], based chiefly on conductivity measurements, that the α phase includes the X-phase region (Fig. 197) and extends below room temperature was not confirmed by emf measurements [14] and other data [16] which indicate a heterogeneous (eutectic) range. This is supported by the fact that α-Tl is b.c.c., whereas α_1 (and probably X) are f.c.c. On the other hand, results of thermal [3, 4] and thermoresistometric studies [5] of the $\alpha \rightleftarrows \epsilon$ transformation (Fig. 197) are not compatible with a temperature of the eutectoid reaction $\alpha \rightleftarrows \epsilon + X$ as high as 182°C, as indicated by emf measurements [14]. This temperature was certainly obtained by too wide an extrapolation.

The phase boundaries in Fig. 197 represent a compromise of the foregoing findings and also consider X-ray data [17] indicating that ϵ and α_1 coexist at room temperature. The experimental data suggesting a transformation $X \rightleftarrows \alpha_1$ are quite scanty; no explanation as to the nature of this transformation can be offered at this time (see Crystal Structures).

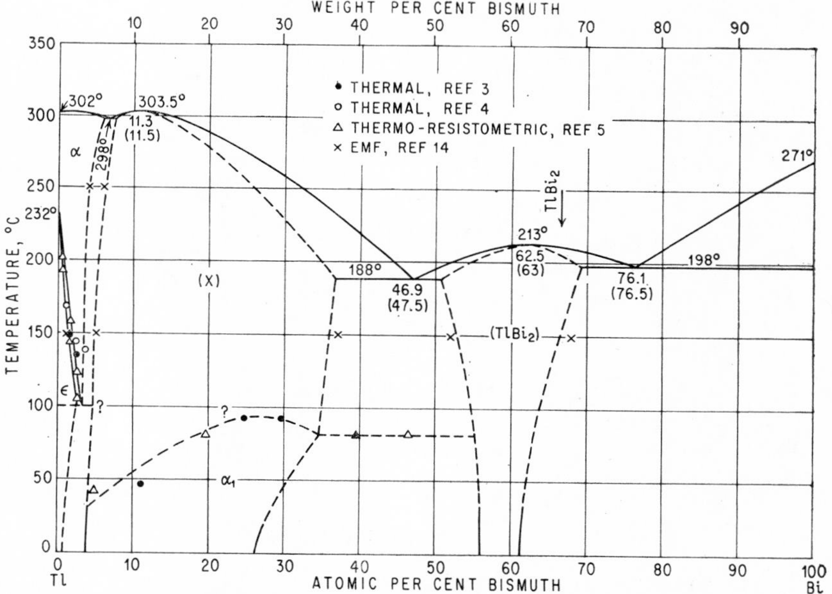

Fig. 197. Bi-Tl. Note Added in Proof: The high-temperature form of Tl should be designated β rather than α, since it is now well established that this modification is b.c.c. However, to maintain uniformity of text and diagram, the designation was not changed in the text.

The Bi-rich boundary of the X and α_1 phases was drawn in consideration of the emf value at 150°C (37 at. % Bi) [14] and an X-ray value at room temperature (see Crystal Structures) [17]. The microstructure of a 41 wt. (40.5 at.) % Bi alloy [4] also supports a solubility decreasing with falling temperature.

The homogeneity region of the ($TlBi_2$) phase is not yet known accurately. From the location of breaks in hardness and electrical-conductivity isotherms, [4] conclude that the homogeneity range [55–64 at. (55.5–64.5 wt.) % Bi], determined by duration of eutectic arrests, remains unchanged down to room temperature. On the other hand, parameter measurements on alloys annealed at 140°C (and subsequently slowly cooled) [18] showed a phase width of 56.0–61.4 at. (56.5–61.9 wt.) % Bi only and emf measurements at 150°C [14], one of 52–68 at. (52.2–68.5 wt.) % Bi. In Fig. 197 the logically assumed tendency toward broadening of the phase field with increasing temperature is indicated. The conclusion of conductivity measurements [4], that the homogeneity range is not represented by any distinct stoichiometric composition, was verified by the structure determination.

Conductivity isotherms of [4] recognize no solubility of Tl in Bi, those of [5] a slight one of about 0.5 (at., wt.) % Tl; according to X-ray work [14] the solubility is "very small."

Crystal Structures. In quenched (temperature?) 2 and 4 wt. % Bi alloys, [19] observed a (metastable) f.c.c. phase. The α_1 phase has a f.c.c. lattice [17, 8] and a lattice parameter varying between 4.852 and 4.938 A in its homogeneity range of 4–27 at. % Bi [17]. There is no evidence for an ordered arrangement of the atoms according to conductivity [4, 5] and emf [14] measurements. (X-ray methods give no information because of the very similar scattering power of Tl and Bi.)

As to ($TlBi_2$), the hexagonal structure proposed by [17] could be corroborated by [18]. The lattice parameters vary as follows: a = 5.653–5.675 A, c = 3.382–3.377 A, c/a = 0.598–0.595 in the homogeneity range mentioned above [18]. The unit cell contains 3 atoms. According to [18], the point positions of the AlB_2 (C32) type of structure are occupied probably at random; from emf measurements, [17] concludes a certain order in the atomic arrangement.

The observation of [20], that an alloy with about 10 wt. % Bi possesses a f.c.c. lattice, agrees with the phase diagram of Fig. 197. There is no confirmation, however, of a CsCl (B2) structure observed by [20] at Bi contents higher than 10 wt. %; it is very unlikely that the X phase possesses this structure.

1. C. T. Heycock and F. H. Neville, *J. Chem. Soc.*, **61**, 1892, 895.
2. C. T. Heycock and F. H. Neville, *J. Chem. Soc.*, **65**, 1894, 34.
3. M. Chikashige, *Z. anorg. Chem.*, **51**, 1906, 328–335.
4. N. S. Kurnakow, S. F. Zemczuzny, and V. Tararin, *Z. anorg. Chem.*, **83**, 1913, 200–227.
5. W. Guertler and A. Schulze, *Z. physik. Chem.*, **106**, 1923, 1–17.
6. M. Chikashige [3] erroneously stated a discrepancy in the reported positions for the maximum at 304°C in a comparison of his results with those of [2]; he did not realize that [2] used the number of Bi atoms per 100 atoms Tl rather than at. % as concentration unit.
7. A. E. Whitford, *Phys. Rev.*, **35**, 1912, 144.
8. N. S. Kurnakow, W. A. Agejewa, and N. W. Agejew, *Ízvest. Inst. Fiz.-Khim. Anal.*, **7**, 1935, 49–58 (in Russian).
9. V. P. Schischokin and W. A. Agejewa, *Tsvetnye Metally*, no. 2, 1932, pp. 119–136 (in Russian).
10. C. E. Mendenhall and W. F. Lent, *Phys. Rev.*, **32**, 1911, 412.
11. A. W. David and J. F. Spencer, *Trans. Faraday Soc.*, **32**, 1936, 1512–1516.
12. E. van Aubel, *Bull. acad. Belg.*, (5)**12**, 1926, 563 (thermo-emf of cast alloys).
13. R. Kremann and A. Lobinger, *Z. Metallkunde*, **12**, 1920, 249–251.
14. A. Ölander, *Z. physik. Chem.*, **A169**, 1934, 260–268.
15. W. Guertler, "Handbuch der Metallographie," vol. 1, part 1, pp. 544–548, Verlagsbuchhandlung Gebrüder Borntraeger, Berlin, 1912.
16. From measurements of the electrical conductivity [4] and its temperature coefficient as well as from microscopic observations [3], a heterogeneous field (eutectic) was concluded.

17. A. Ölander, *Z. Krist.*, **89**, 1934, 89–92.
18. E. S. Makarov, *Doklady Akad. Nauk S.S.S.R.*, **74**, 1950, 935–938 (in Russian).
19. S. Sekito, *Z. Krist.*, **74**, 1930, 193–195, 200.
20. V. M. Goldschmidt, *Z. physik. Chem.*, **133**, 1928, 409–411 (measurements of T. Barth).

$\bar{1}.9434$
0.0566

Bi-U Bismuth-Uranium

The existence of the following U-Bi compounds has been reported in the open literature: UBi (46.75 wt. % Bi) [1–3], probably U_4Bi_5 [55.55 at. (52.31 wt.) % Bi] [2, 3], U_3Bi_4 [57.14 at. (53.92 wt.) % Bi] [3], and UBi_2 (63.71 wt. % Bi) [1–4].

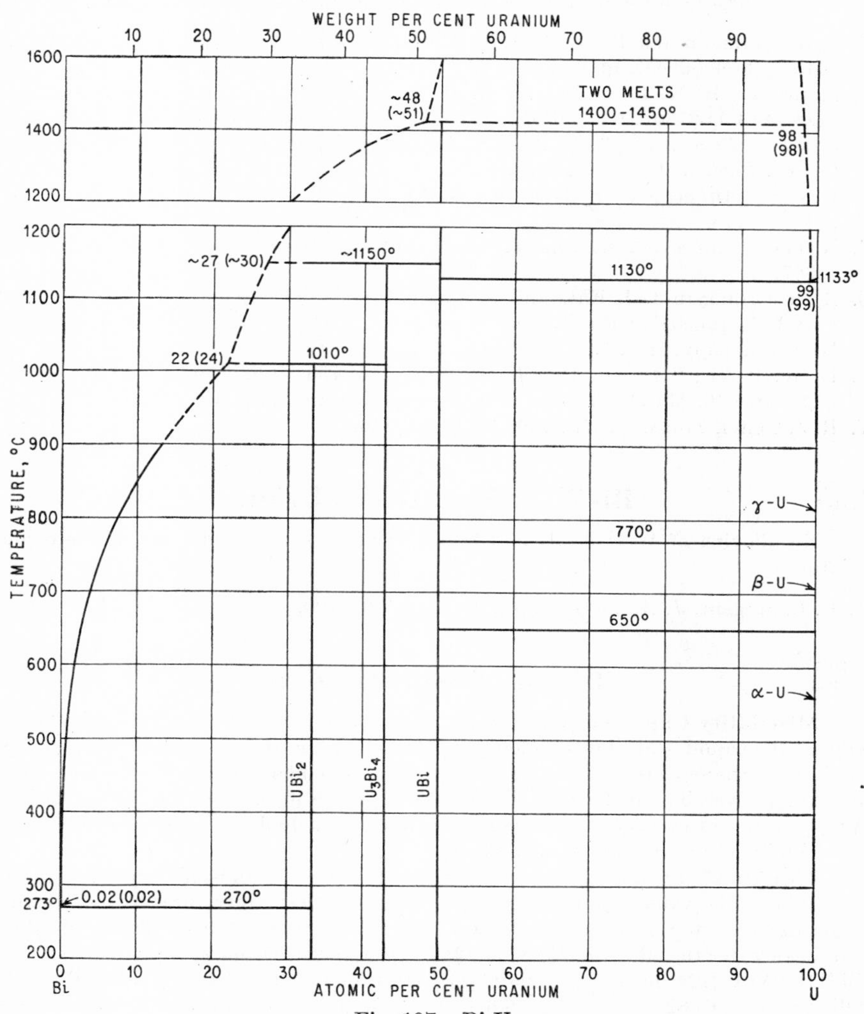

Fig. 197*a*. Bi-U

The paper of [1], containing a partial phase diagram worked out by cooling curves and microscopic examination, is still classified. Solid solubilities in the elements [5, 3] as well as in the intermediate phases [3] were found to be nil at lower temperatures.

Crystal Structures. UBi is cubic of the NaCl (B1) type [2, 3], with a = 6.364 ± 4 A [2]. U_3Bi_4 has the cubic Th_3P_4 ($D7_3$) type of structure, with a = 9.350 A [3], and UBi_2 the tetragonal Cu_2Sb (C38) type of structure [4], with a = 4.445 A, c = 8.908 A, c/a = 2.004.

Note Added in Proof. The phase diagram of Fig. 197*a*, incorporating findings by [1–4, 6, 7], was published by [7]. There is disagreement as to the crystal structure of UBi. Whereas [2, 3] claimed it to be of the NaCl type, [7], on the basis of neutron diffraction, arrived at a b.c. tetragonal cell with a = 11.12 A, c = 10.55 A, containing 48 atoms. The high-temperature region of the diagram needs further study.

1. D. H. Ahmann and R. R. Baldwin, *U.S. Atomic Energy Comm., Publ.* CT-2961, 1945 (still classified); quoted from [2].
2. L. Brewer, R. K. Edwards, and D. H. Templeton, *U.S. Atomic Energy Comm., Publ.* AECD-2730, 1949, also known as AECU-653 and UCRL-433. Abstract, *Structure Repts.*, **12,** 1949, 29–30. Also R. K. Edwards, *U.S. Atomic Energy Comm., Publ.* AECD-3394, 1949.
3. R. Ferro, *Atti accad. nazl. Lincei, Rend.*, **13,** 1952, 401–405; *Chem. Abstr.*, **47,** 1953, 11114–11115.
4. R. Ferro, *Atti accad. nazl. Lincei, Rend.*, **14,** 1953, 89–94; *Chem. Abstr.*, **47,** 1953, 10307.
5. [1], according to J. J. Katz and E. Rabinowitch, "The Chemistry of Uranium," Part I, National Nuclear Energy Series, Div. VIII, vol. 5, p. 175, McGraw-Hill Book Company, Inc., New York, 1951.
6. D. W. Bareis, "Reactor Handbook," vol. 2, p. 753, *U.S. Atomic Energy Comm.*, AECD-3646, March, 1955 (solubility of uranium in liquid Bi). Quoted from [7].
7. R. J. Teitel, *Trans. AIME*, **209,** 1957, 131–136.

0.0555
$\bar{1}$.9445

Bi-W Bismuth-Wolfram

No alloying of W and Bi could be achieved by carbon reduction of the oxide mixture [1].

1. C. L. Sargent, *J. Am. Chem. Soc.*, **22,** 1900, 783–790.

0.5047
$\bar{1}$.4953

Bi-Zn Bismuth-Zinc

Miscibility Gap. Early papers [1, 2] dealing with the extent of the miscibility gap in the liquid state are obsolete. In Fig. 198, the data obtained by chemical analyses of the separated layers [3, 4], by emf measurements [5], and by a differential thermal analysis [6] are plotted. The results of [5] and [6] are—except for the critical temperature—in excellent agreement, and they are undoubtedly more reliable than those obtained by the analytical method.

Liquidus. By means of cooling curves [7, 8], all liquid-solid equilibria were determined; the values for the monotectic and eutectic temperatures as well as the respective concentrations in Fig. 198 were taken from these papers. Values of other authors are, for the eutectic, 251°C [9], 248°C [10], ~240°C and 97.2 wt. (91.6 at.) % Bi [11], 254°C [12]; monotectic, 395°C [11], 416°C and 1 wt. (0.3 at.) % Bi [12]; miscibility gap, 1.95–82.5 wt. (0.62–59.6 at.) % Bi near the monotectic temperature [13, 14].

Solid Solubilities. These values have not yet been determined accurately. For the solubility of Zn in Bi the following data were reported: magnetic-susceptibility curves (heat-treatment?), ~2 wt. (6.1 at.) % Zn [15]; microscopic observations on annealed (temperature?) alloys, ~4 wt. (11.8 at.) % Zn (?) [16], 2.7 wt. (8.1 at.) % Zn [17]; emf-composition curve (annealed alloys), 2.7 wt. % Zn [17]; microscopic observation of a slowly cooled alloy, <0.2 wt. (0.6 at.) % Zn [13].

The solubility of Bi in Zn is very small [15]; a slowly cooled alloy with 0.1 wt. (0.03 at.) % Bi was heterogeneous [12].

The following measurements agree with the phase diagram of Fig. 198: emf [18, 17, 19], magnetic susceptibility [15, 20], and specific volume [21].

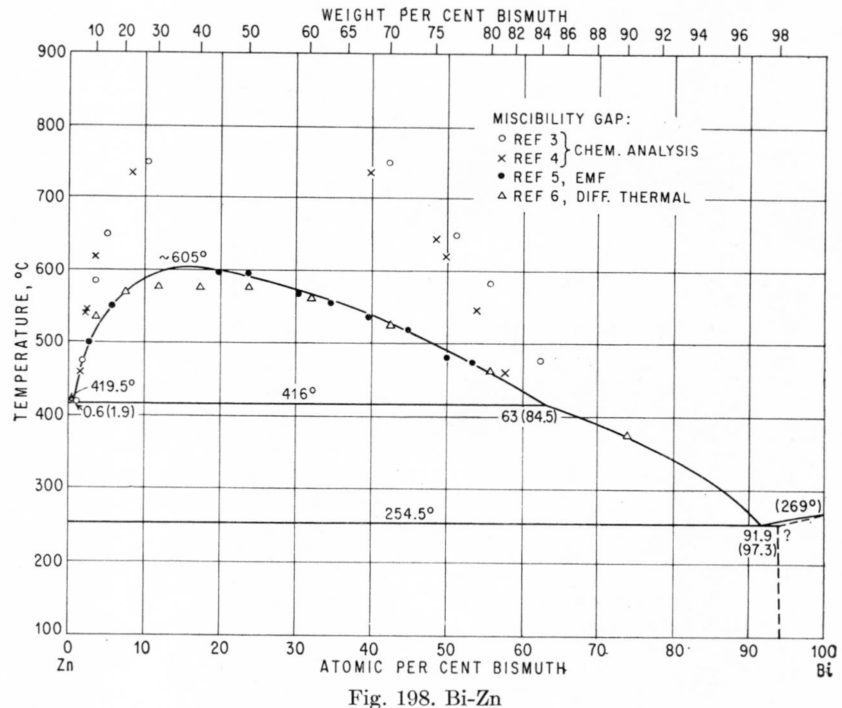

Fig. 198. Bi-Zn

Supplement. In resistivity measurements on alloys quenched from 255, 240, and 232°C, no solubility of Bi in solid Zn was observed [22].

1. A. Matthiessen and M. v. Bose, *Proc. Roy. Soc. (London)*, **11**, 1861, 430.
2. C. R. A. Wright and C. Thompson, *Proc. Roy. Soc. (London)*, **49**, 1890–1891, 156–158.
3. W. Spring and L. Romanoff, *Z. anorg. Chem.*, **13**, 1897, 29–35.
4. K. Hass and K. Jellinek, *Z. anorg. Chem.*, **212**, 1933, 356–361.
5. O. J. Kleppa, *J. Am. Chem. Soc.*, **74**, 1952, 6052–6056.
6. H. Johnen, Dissertation, Münster (Westfalen), 1952; W. Seith, H. Johnen, and J. Wagner, *Z. Metallkunde*, **46**, 1955, 773–779.
7. C. T. Heycock and F. H. Neville, *J. Chem. Soc.*, **61**, 1892, 893.
8. C. T. Heycock and F. H. Neville, *J. Chem. Soc.*, **71**, 1897, 390–392, 399–400.

9. F. Rudberg, *Pogg. Ann.*, **18,** 1830, 247.
10. F. Guthrie, *Phil. Mag.*, (5)**17,** 1894, 462.
11. H. Gautier and Roland-Gosselin, *Bull. soc. encour. ind. natl.*, **1,** 1896, 1308; Société d'encouragement pour l'industrie nationale, Paris, Commission des alliages, "Contribution a l'étude des alliages," pp. 108–109, Typ. Chamerot et Renouard, Paris, 1901 (beginning of solidification of alloys with 40–100 wt. % Bi).
12. P. T. Arnemann, *Metallurgie*, **7,** 1910, 206, 209.
13. C. H. Mathewson and W. M. Scott, *Intern. Z. Metallog.*, **5,** 1914, 1–15.
14. The Bi-rich layer contained small drops of Zn-rich melt; this explains the deviation from the composition value of Heycock and Neville.
15. H. Endo, *Science Repts. Tôhoku Univ.*, **14,** 1925, 501–502.
16. B. E. Curry, *J. Phys. Chem.*, **13,** 1909, 601–605.
17. P. Fuchs, *Z. anorg. Chem.*, **109,** 1920, 86–88.
18. M. Herschkowitsch, *Z. physik. Chem.*, **27,** 1898, 145.
19. R. Kremann, A. Langbauer, and H. Rauch, *Z. anorg. Chem.*, **127,** 1923, 231–232.
20. N. Thompson, *Proc. Roy. Soc. (London)*, **A155,** 1936, 111–123.
21. E. Maey, *Z. physik. Chem.*, **50,** 1905, 215.
22. A. Pasternak, *Bull. intern. acad. polon. sci., Classe sci. math. nat.*, sér. A, **1951,** 177–192; *Chem. Abstr.*, **48,** 1954, 1098.

$\bar{1}.4766$
0.5234

C-Ca Carbon-Calcium

CaC_2 (37.47 wt. % C) can exist in four types of structure [1]: CaC_2 IV, f.c.c. with a = 5.93 A, stable above 450°C; CaC_2 I, f.c. tetragonal with $4CaC_2$ per unit cell, a = 5.49 A, c = 6.38 A, c/a = 1.162 [1, 2], stable between 25 and 450°C; CaC_2 II, having an unknown structure of lower symmetry, metastable between 25 and 200°C and stable below 25°C; CaC_2 III, of unknown structure, metastable below 450°C and unstable above 450°C; see also [3, 4]. The tetragonal form is the one usually obtained technically.

1. M. A. Bredig, *J. Phys. Chem.*, **46,** 1942, 801–819.
2. M. v. Stackelberg, *Z. physik. Chem.*, **B9,** 1930, 437–475.
3. H. H. Franck, M. A. Bredig, G. Hoffmann, and H. Füldner, *Z. anorg. Chem.*, **232,** 1937, 61–74.
4. H. H. Franck, M. A. Bredig, and Kin-Hsing Kou, *Z. anorg. Chem.*, **232,** 1937, 75–111.

C-Cb Carbon-Columbium

See carbon-niobium (C-Nb).

$\bar{1}.0287$
0.9713

C-Cd Carbon-Cadmium

Boiling cadmium, mercury, tin, and zinc dissolve only "traces or unweighable amounts" of carbon which precipitate on cooling in the form of graphite [1, 2].

1. O. Ruff and B. Bergdahl, *Z. anorg. Chem.*, **106,** 1919, 91–92.
2. This was already shown to be the case for tin by H. Moissan, *Compt. rend.*, **125,** 1898, 840.

$\bar{2}.9330$
1.0670

C-Ce Carbon-Cerium

The crystal structure of CeC_2 (14.63 wt. % C) is tetragonal of the CaC_2 (C11) type, a = 5.49 A, c = 6.49 A, c/a = 1.18 [1].

[2] reported that "an investigation of the Ce-C system in the range between 50 and 67 at. % C showed the phases CeC, Ce_2C_3, and CeC_2 plus several unidentified phases. The Ce_2C_3 (b.c.c. Pu_2C_3 type) lattice constant was a = 8.455 ± 0.008 A [3]. The CeC (NaCl type) lattice constant was a = 5.130 ± 0.002 A [3]. The Ce-C samples were golden brown in color."

1. M. v. Stackelberg, *Z. physik. Chem.*, **B9**, 1930, 437–475; *Z. Elektrochem.*, **37,** 1931, 542–545.
2. O. H. Krikorian, *U.S. Atomic Energy Comm., Publ.* UCRL-2888, 1955, pp. 30–31.
3. C. H. Dauben, University of California Radiation Laboratory, private communication (to O. H. Krikorian).

$\bar{1}.3091$
0.6909

C-Co Carbon-Cobalt

According to [1], a solid solution of C in Co forms a eutectic with graphite (Fig. 199*a*). This was verified by reinvestigations [2, 3]. The composition and temperature of the eutectic were found (or assumed) as follows: between 2.6 and 3.1 wt. (11.6 and 13.5 at.) % C and 1274–1317°C (based on a melting point of Co of 1448°C) [1]; at

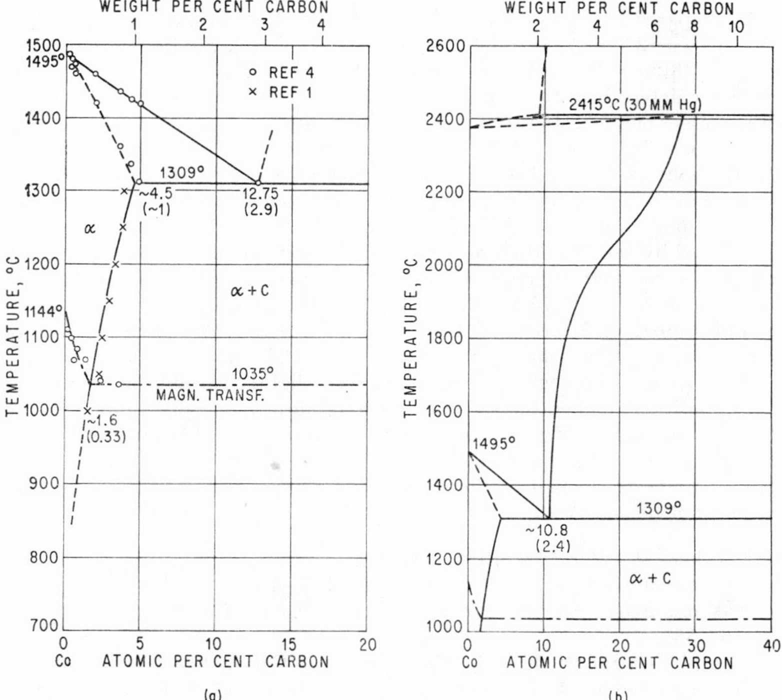

Fig. 199. C-Co

2.9 wt. (12.75 at.) % C, 1309°C [2, 4]; and at 2.6 wt. (11.6 at.) % C, 1315°C [3]. According to [5], the eutectic temperature is 1321°C.

The solid solubility of C in Co (Fig. 199*a*) was determined by chemical determination of carbon in solution in samples annealed at and quenched from various temperatures. After slow cooling, only about 0.1 wt. (0.48 at.) % C was in solution [1]. Values reported by [2] are slightly higher; however, they do not represent equilibrium conditions. Lattice-parameter measurements of Co powder, carburized at about 220°C, gave no evidence of any solid solubility of C in Co [6].

The Curie point of Co was found to be lowered to 1035 [2, 4], 1075 [3], or 1045°C [7]. Data as to the effect of C on the temperature of the $\epsilon \rightleftarrows \alpha$ transformation of Co [2–4, 7] differ considerably, probably because of different degrees of supersaturation. Temperatures for the two-phase field were reported as 465 [3], 502 [4], and 390°C [7] on heating and 380 [3], 341 [4], and 90°C [7] on cooling.

The existence of a cobalt carbide, persisting only at moderate temperatures, was reported by several investigators. On carburizing finely divided Co with CO at 225–230°C, [8] prepared Co_2C (9.25 wt. % C). This was confirmed by [6], whereas [9], using illuminating gas as the carburizing medium at temperatures of 500–800°C, claimed the existence of Co_3C (6.36 wt. % C), reported to be isotypic with Fe_3C (a = 4.53 A, b = 5.09 A, c = 6.74 A). Results of studies of the methane-hydrogen equilibrium over Co at 350–900°C [10, 11] were interpreted as showing that a carbide Co_nC exists which is formed by the eutectoid decomposition of the (α-Co) phase with about 0.1 wt. (0.48 at.) % C into ϵ-Co and carbide at 685°C. The occurrence of a carbide, slightly more stable than Ni_3C, was also assumed on the basis of studies of graphite formation in Co-C alloys [5].

Co_2C, prepared by carburizing with CO at 220°C, is orthorhombic and isotypic with Co_2N, a = 2.896 A, b = 4.446 A, c = 4.371 A [12]. Similar parameters were reported by [13]: a = 2.891 A, b = 4.463 A, c = 4.369 A.

[14] have determined the solubility of C in molten Co at a pressure of 30 mm Hg up to the boiling point (2415 ± 10°C) of the melt saturated with C, containing 7.4 wt. (28.1 at.) % C (Fig. 199*b*). Melts were quenched, and the carbon content was determined. At about 2100°C and 6 wt. % C the solubility curve has an inversion point, similar to that in the system C-Ni but not as pronounced. The authors suggested that (*a*) the melt contains Co_3C molecules and (*b*) the further increase in solubility may be caused by the formation of a carbide richer in carbon. The solubility curve of graphite in the melt intersects the eutectic horizontal at about 10.8 at. (2.4 wt.) % C as compared with 12.75 at. (2.9 wt.) % C in Fig. 199*a*.

Supplement. [15] prepared Co_2C and studied its thermal decomposition by means of X-ray and thermomagnetic analyses. Co_2C, isotypic with Co_2N (cf. above), has a certain range of homogeneity as indicated by slightly variable lattice parameters and decomposition temperatures (~300°C in vacuo) among different preparations. Equilibrium solubilities of C in Co were not determined; it is interesting to note, however, that obviously highly supersaturated solid solutions of C in hexagonal Co occur in the course of the preparation as well as decomposition of Co_2C. A compound Co_3C could not be observed.

1. G. Boecker, *Metallurgie*, **9,** 1912, 296–303.
2. U. Haschimoto, *Kinzoku-no-Kenkyu*, **9,** 1932, 57–73.
3. S. Takeda, *Science Repts. Tôhoku Imp. Univ., K. Honda Anniv. Vol.*, **1936,** pp. 864–881.
4. U. Haschimoto and N. Kawai, *Nippon Kinzoku Gakkai-Shi*, **2,** 1938, 26–28.
5. H. Morrogh and W. J. Williams, *J. Iron Steel Inst.*, **155,** 1947, 321–371.
6. L. J. E. Hofer and W. C. Peebles, *J. Am. Chem. Soc.*, **69,** 1947, 893–899.

7. W. Köster and E. Schmid, *Z. Metallkunde*, **29**, 1937, 232–233.
8. H. A. Bahr and V. Jessen, *Ber. deut. chem. Ges.*, **63**, 1930, 2226.
9. W. F. Meyer, *Z. Krist.*, **97**, 1937, 145–169; *Metallwirtschaft*, **17**, 1938, 413–416.
10. R. Schenck, F. Krägeloh, and F. Eisenstecken, *Z. anorg. Chem.*, **164**, 1927, 313.
11. R. Schenck and H. Klas, *Z. anorg. Chem.*, **178**, 1929, 146–156.
12. J. Clarke and K. H. Jack, *Chem. Ind.*, **1951**, 1004–1005.
13. R. Juza and H. Puff, *Naturwissenschaften*, **38**, 1951, 331.
14. O. Ruff and F. Keilig, *Z. anorg. Chem.*, **88**, 1914, 410–423; *Ber. deut. chem. Ges.*, **45**, 1913, 3142; *Ferrum*, **13**, 1915-1916, 108–109.
15. J. Drain, *Ann. chim. (Paris)*, **8**, 1953, 900–953.

$\bar{1}.3634$
0.6366

C-Cr Carbon-Chromium

The phase diagram in Fig. 200 is based on the findings of numerous investigations which established the existence of the three carbides, $Cr_{23}C_6$ [20.69 at. (5.68 wt.) % C; formerly designated as Cr_4C, with 5.46 wt. % C] [1–7], Cr_7C_3 (9.00 wt. % C) [3, 5–8], Cr_3C_2 (13.34 wt. % C) [9, 10, 3, 4, 11–13, 7], as well as the phase relationships [4–7]. The data points indicated are due to [7]; they were obtained by thermal analysis under conditions which prevented contamination from the atmosphere.

The existence of a monocarbide CrC (18.76 wt. % C), suggested by [6], was denied by [7], who pointed out that the constituent observed by [6] was probably CrN. The fact that the primary constituent in alloys with more than about 40 at. % C was found to be graphite [10] is also in contradiction to the existence of CrC. The curve of the melts saturated with graphite [10] intersects the liquidus curve of Cr_3C_2 at a point which is in accord with the Cr-rich portion of the diagram (Fig. 200).

The three carbides were first positively identified by the X-ray work of [3]. Several previous investigators, using microexamination and isolation of crystals as methods of investigation, concluded the existence of Cr_5C_2 [28.57 at. (8.46 wt.) % C] instead of Cr_7C_3 [10, 11] and also Cr_2C [10, 14]; see also [12].

On rapid solidification, the peritectic formation of Cr_4C is suppressed, resulting in the formation of a metastable equilibrium. This is the reason why [11] believed the carbide richest in Cr was $Cr_5C_2(Cr_7C_3)$ instead of Cr_4C. Further evidence for the metastable system Cr-Cr_7C_3 was presented by [5, 15], on the basis of thermal [5], microscopic [5], and X-ray data [15] (Fig. 200). Results of [10] may also have been affected by metastable conditions in the alloys examined.

The composition of the eutectic Cr-$Cr_{23}C_6$ (in wt. % C) was reported as about 1.7 [2], 4.3 [10], slightly above 3.3 [4, 16], 4.25 [5], 3.7 [6], and 3.2% [7]; and the eutectic temperature as about 1543 [17], 1510 [5], 1485 [6], and 1498°C [7].

The temperature of the peritectic formation of $Cr_{23}C_6$ was reported as 1530 [6] and 1518°C [7] and that of the peritectic reaction, melt + $Cr_3C_2 \rightleftarrows Cr_7C_3$, as 1665 [5] and 1780°C [7]. The solubility of C in Cr, if any, is certainly exceedingly small [3].

Thermal effects at approximately 1505°C in the range 9–14.5 wt. % C were believed to be caused by a polymorphic transformation of Cr_3C_2 [6]. Corresponding effects at 1525 ± 10°C were also observed by [7] and interpreted as being due to the peritectic reaction at about 1518°C (Fig. 200).

Crystal Structure. $Cr_{23}C_6$, first identified as Cr_4C [3], is complex f.c.c. with 116 atoms per unit cell ($D8_4$ type), a = 10.66 A [18]; see also [19, 20]. Cr_7C_3 is hexagonal, with 80 atoms per unit cell, a = 14.01 A, c = 4.53 A, c/a = 0.322 [3, 21]. Cr_3C_2 is orthorhombic with 20 atoms per unit cell ($D5_{10}$ type) and a = 2.83 A, b = 5.53 A, c = 11.48 A [13]. [22] reported the monocarbide CrC, perhaps stable at temperatures above 2000°C, to be cubic (NaCl type?), a = 3.62 A.

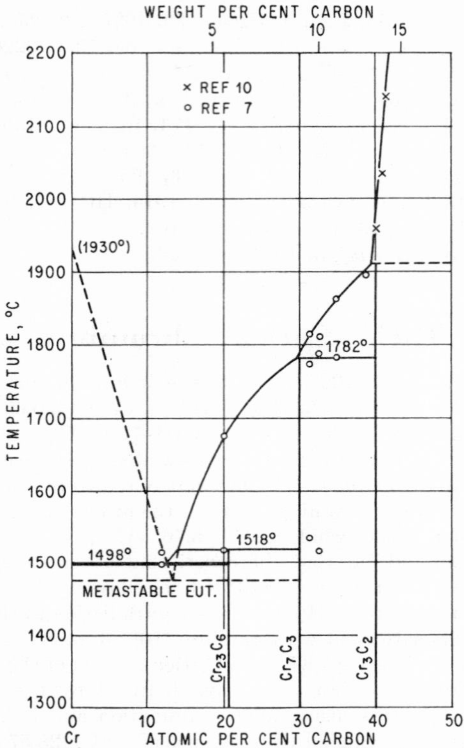

Fig. 200. C-Cr. (See also Note Added in Proof.)

Note Added in Proof. According to microscopic examination of equilibrated and quenched alloys, the solubility of C in Cr decreases from 0.32 wt. (1.4 at.) % at 1498°C to 0.006 wt. (0.026 at.) % C at 900°C [23].

1. H. Moissan, *Compt. rend.*, **119**, 1894, 185; **125**, 1897, 839.
2. T. Murakami, *Science Repts. Tôhoku Univ.*, **7**, 1918, 263.
3. A. Westgren and G. Phragmén, *Svensk Vetenskaps Akad. Handl.*, **2**, 1925, no. **5**.
4. A. Westgren, G. Phragmén, and T. Negresco, *J. Iron Steel Inst.*, **117**, 1928, 386–387.
5. E. Friemann and F. Sauerwald, *Z. anorg. Chem.*, **203**, 1931, 64–74.
6. K. Hatsuta, *Kinzoku-no-Kenkyu*, **8**, 1931, 81–88; *Technol. Repts. Tôhoku Univ.*, **10**, 1932, 680–688.
7. D. S. Bloom and N. J. Grant, *Trans. AIME*, **188**, 1950, 41–46.
8. W. Crafts and J. L. Lamont, *Trans. AIME*, **185**, 1949, 957–967.
9. H. Moissan, *Compt. rend.*, **116**, 1893, 349; *Ann. chim. et phys.*, **8**, 1896, 559.
10. O. Ruff and T. Foehr, *Z. anorg. Chem.*, **104**, 1918, 27–46.
11. R. Kraiczek and F. Sauerwald, *Z. anorg. Chem.*, **185**, 1930, 193–216; see also [16].
12. R. Schenck, F. Kurzen, and H. Wesselkock, *Z. anorg. Chem.*, **203**, 1931, 169–176.
13. K. Hellbom and A. Westgren, *Svensk Kem. Tidskr.*, **45**, 1933, 141–150.
14. K. Nischk, *Z. Elektrochem.*, **29**, 1923, 384–387; see discussion by O. Ruff, *Z. Elektrochem.*, **29**, 1923, 469–470.

15. F. Sauerwald, W. Teske, and G. Lempert, *Z. anorg. Chem.*, **210**, 1933, 21–23.
16. A. Westgren and G. Phragmén, *Z. anorg. Chem.*, **187**, 1930, 401–403.
17. A. v. Vegesack, *Z. anorg. Chem.*, **154**, 1926, 41–42.
18. A. Westgren, *Jernkontorets Ann.*, **117**, 1933, 501–512.
19. V. I. Arkharov, I. S. Kvater, and S. T. Kiselev, *Izvest. Akad. Nauk S.S.S.R.* (*Tekh.*), **1947**, 749–756.
20. A. G. Allten, J. G. Y. Chow, and A. Simon, *Trans. ASM*, **46**, 1954, 948–972.
21. A. Westgren, *Jernkontorets Ann.*, **119**, 1935, 231–240.
22. W. Epprecht, *Chimia*, **5**, 1951, 49–60.
23. W. H. Smith, *Trans. AIME*, **209**, 1957, 47–49.

$\bar{1}.2765$
0.7235

C-Cu Carbon-Copper

Earlier data on the solubility of carbon in molten copper, now obsolete, were reviewed and tabulated by [1]. By careful investigation the solubility, in wt. % C, was determined to be about 0.0001 at 1100°C, 0.00015 at 1300°C, 0.0005 at 1500°C, and 0.003 at 1700°C [1]. As carbon does not diffuse through solid copper, the solubility must be exceedingly small [2].

1. M. B. Bever and C. F. Floe, *Trans. AIME*, **166**, 1946, 128–141.
2. W. Baukloh and F. Springorum, *Z. anorg. Chem.*, **230**, 1937, 315–320.

$\bar{1}.3325$
0.6675

C-Fe Carbon-Iron*

The literature published prior to 1936 has been reviewed by [1, 2].

Figures 201 and 202 show the phase diagrams on the wt. % and at. % scale, respectively. Phase relations are presented in the form of the generally accepted double diagram. As usual, the curves of the metastable system Fe-Fe_3C are drawn in solid lines and those of the stable system Fe-graphite in dashed lines, except for the stable equilibria between melt and the phases δ and γ.

There has been some difference of opinion as to whether the double diagram is justified [1]. This point is not being treated here; however, it should be stressed that there is no reasonable argument against not using the double diagram, since it offers a convenient means of representing the observed behavior.

THE METASTABLE SYSTEM Fe-Fe_3C

Solidification. The liquidus curve AB of the δ phase (Fig. 203) is practically a straight line connecting the melting point of Fe with the end point of the peritectic horizontal $\delta + \text{melt} \rightleftarrows \gamma$, whose temperature was found as 1487 [3, 4], 1493 [5], and 1495°C [6], based on a melting point of Fe of 1528 [3, 4], 1533 [5], and 1537°C [6]. The point H was placed at 0.07 wt. (0.33 at.) % C [3, 4] and 0.10 wt. (0.46 at.) % C [5]; the point J at about 0.13 wt. (0.60 at.) % C [4], 0.16 wt. (0.74 at.) % C [5], and 0.18 wt. (0.83 at.) % C [3]; and the point B at about 0.36 wt. (1.65 at.) % C [3], 0.38 wt. (1.75 at.) % C [4], 0.4 wt. (1.83 at.) % C [7], 0.51 wt. (2.33 at.) % C [5], and 0.71 wt. (3.22 at.) % C [6]. In plotting Figs. 201 and 202, weight was given to the results of [5] which were obtained using high-purity alloys. The melting point of Fe was accepted as 1534°C [8].

* The author is indebted to Prof. E. Scheil and Dr. E. Saftig for having reviewed the more recent literature. Portions of this text are based on their manuscript.

The liquidus curve *BC* of the γ phase (Fig. 204) was determined, in part or in full, by [9, 10, 11, 3, 7, 12, 6, 4, 5], data points from the papers of [7, 4, 5] being indicated in Fig. 204. The liquidus in Figs. 201 and 202 is based on the results of [7, 5] which appear most reliable.

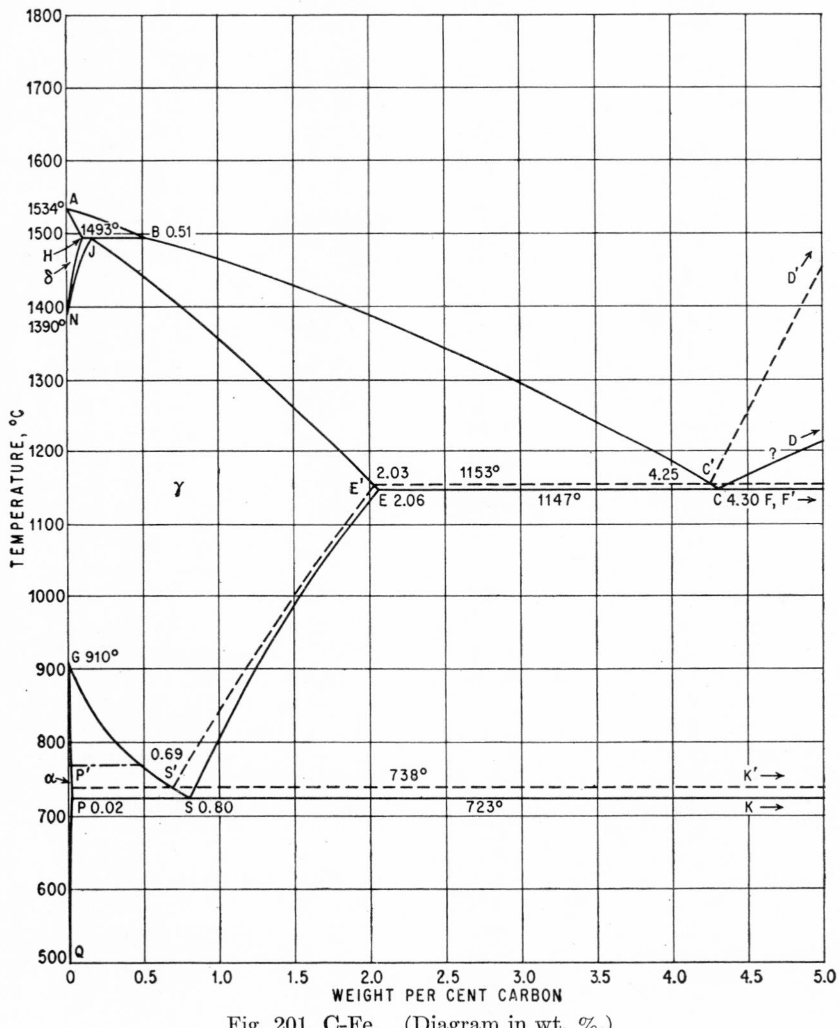

Fig. 201. C-Fe. (Diagram in wt. %.)

The liquidus curve *CD* of Fe_3C is unknown. A few points up to 5 wt. (19.7 at.) % C were reported by [9, 13, 4], but it is questionable whether they represent points on the liquidus of Fe_3C [14]. The curve shown in Figs. 201 and 202 is schematic. According to [15], it was plotted to intersect the solubility curve of graphite in Fe about 10°C below the temperature of the $\gamma + Fe_3C$ eutectic. The melting point of Fe_3C is indeterminable, since the carbide decomposes at high temperatures [14].

The solidus of the δ phase has not yet been accurately determined, but only outlined by [5], based on some thermal data (Fig. 203).

As regards the solidus of the γ phase, there are two groups of results which differ widely. Within each group agreement is remarkably good. According to [10, 16, 6, 17, 5], the curve *JE* is a fairly straight line on the wt. % scale [18]. On the other hand, [11, 19, 20, 13, 4] found a curve strongly convex to the composition axis. The findings of the second group appear unlikely, especially in view of the location of the point *E'* (see below). Therefore, weight has been given to the curve of [5], which was extrapolated to the point *E'*, based on the data of [21, 22]. Also, [23, 24, 1] gave preference to the nearly straight-line solidus.

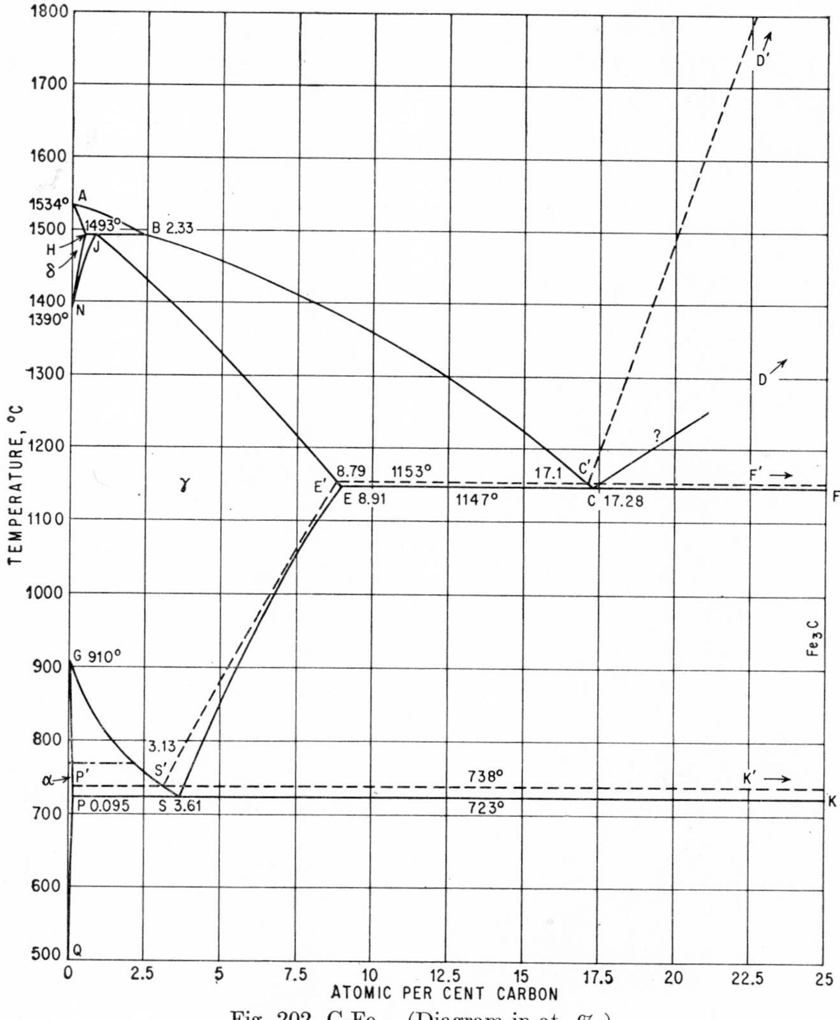

Fig. 202. C-Fe. (Diagram in at. %.)

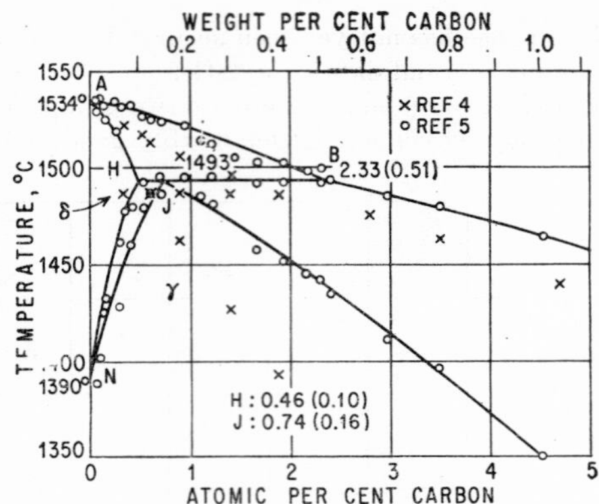

Fig. 203. C-Fe. δ-phase region.

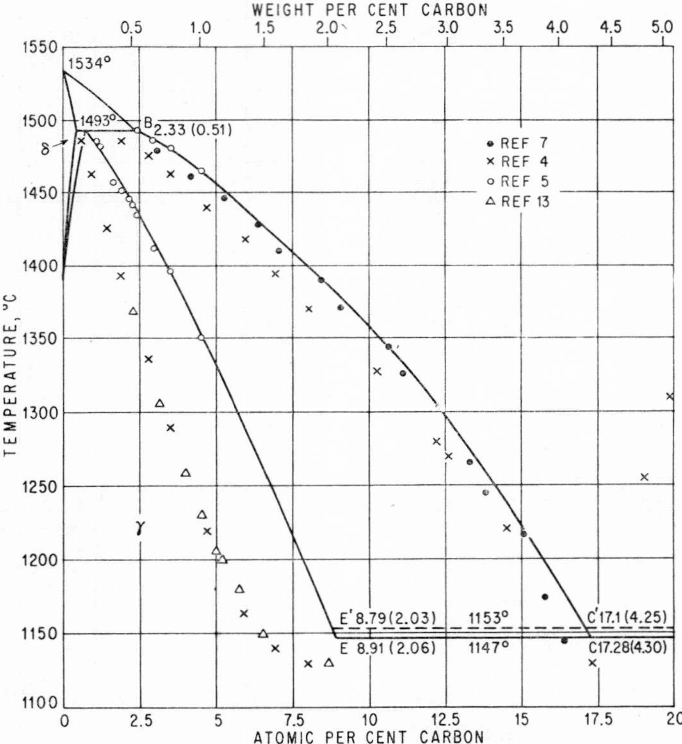

Fig. 204. C-Fe. Liquidus and solidus of γ phase.

The temperature of the eutectic horizontal *ECF* (melt $\rightleftarrows \gamma + Fe_3C$) was reported as 1120–1140 [9], 1110–1146 [10], 1093–1134 [11], 1145 [7, 25], <1140 [20], 1130 [13, 4], and 1147°C [26]. Although [1] selected 1130°C and this temperature has been widely accepted since then, especially in the American literature, the present writer believes that 1145°C, based on the accurate work of [7, 25], is more reliable. [27] found

1130°C on cooling and 1151°C on heating. 1145°C agrees closely with the recently found value of 1147°C [26].

The composition of the $\gamma + Fe_3C$ eutectic (ledeburite) has been generally accepted as 4.3 wt. (17.28 at.) % C, on the basis of the results of [28]. Previous work had indicated about 4.2 wt. (16.93 at.) % C [29, 11, 7].

Transformations in the Solid State. Temperatures of the $\delta \rightleftarrows \gamma$ transformation were determined by [3, 6, 5]. The curves *NH* and *NJ* in Figs. 201 and 202 were taken from the work of [5]; see Fig. 203.

The effect of carbon on the $\gamma \rightleftarrows \alpha$ transformation point (A_3 transformation, curve *GS*) was the subject of numerous investigations, using thermal [9, 10, 30–35], micro-

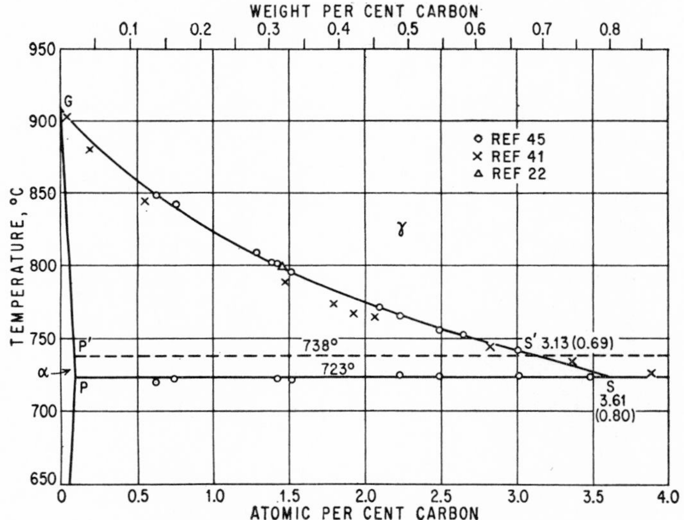

Fig. 205. C-Fe. $\gamma \rightleftarrows \alpha$ transformation (metastable system).

graphic [36, 37], thermoresistometric [38, 39], thermomagnetometric [40, 12, 13, 41], dilatometric [42, 43, 44, 41, 45, 46], and thermoelectric methods [47]. Figure 205 shows the data points of [41, 45]. They were obtained by determining dilatometrically the transformation temperatures on different heating and cooling rates and extrapolating to the rate zero. The data of [45], obtained with alloys prepared by carburization of high-purity Fe, are doubtless the most accurate ones [48]. A point at 800°C, due to [22], is in excellent agreement (Fig. 205).

Using data of the transformation heat of Fe, [24] calculated the saturation curve *GS* and found it to be in good agreement with experimental results. The same holds for the calculations of [49]—using the specific-heat measurements of pure Fe by [50]—based on his theory of the constitution of austenite [51].

The temperature of the eutectoid horizontal *PSK* (A_1 transformation) was reported as 718 [32], 710–721.5 [33], 721 ± 3 [7, 42], 722 [44], 726 [41], and 723 ± 2°C [45], the latter value being accepted in Figs. 201 and 202. The composition of the pearlite eutectoid in wt. % C was given as 0.95 [36], 0.93 [30], 0.91 [42], 0.90 [52], 0.89 [10, 37, 38], 0.86 [40, 43, 44, 41], 0.85 [9], 0.82 [35], 0.80 ± 0.1 [45], and 0.75 wt. % C [32, 34]. 0.86 wt. % C was selected by [24] and 0.83% by [1]. Accepting the *GS* line by [45] as most reliable, the eutectoid point is located at 0.80 ± 0.01 wt. (3.61 ± 0.04 at.) % C.

The saturation curve *ES* of the γ phase (A_{cm} transformation) was determined by means of thermal [10, 35], microscopic [9, 53, 37, 54, 45], chemical [11], thermoresistometric [38, 20], thermomagnetometric [13, 41], and dilatometric methods [42, 43, 41, 45]. None of these widely differing curves, including the most recent one [45], can be regarded as representing the metastable boundary of austenite. As shown in Fig. 206, the boundary *ES* according to [45] intersects the well-established boundary *E'S'* of the stable system at about 940°C. This would signify a change in stability between graphite and Fe_3C in equilibrium with the γ phase and require either (*a*) another stability change below the point *E* or (*b*) the cementite eutectic *ECF* to lie at a higher temperature than the graphite eutectic *E'C'F'*. The first

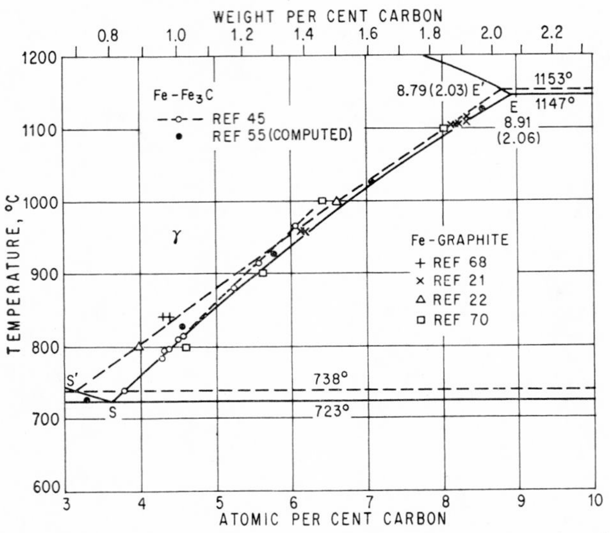

Fig. 206. C-Fe. High-carbon boundary of γ phase (metastable and stable systems).

case of a twofold stability change appears quite unlikely, and the second case is incompatible with the measurements of [28, 27], according to which the melting point of the γ + graphite eutectic lies distinctly above that of the eutectic $\gamma + Fe_3C$.

In view of this serious discrepancy, [55] computed the curve *ES* by thermodynamic methods. The boundary thus obtained (Fig. 206) indicates that the solubility of Fe_3C is greater than that of graphite in the entire temperature range. The curve *ES* in the diagrams of Figs. 201, 202, and 206 was plotted using the data points of [45] in the lower temperature range. The upper portion is hypothetical and was drawn so as not to intersect the curve *E'S'*. Both curves are much closer than assumed previously. As to the location of the point *E*, see the paragraph below dealing with the curve *E'S'*.

Previous data relative to the saturation curve *PQ* of α for Fe_3C, summarized by [1, 2, 56], are obsolete. They were found investigating alloys of commercial purity and using inadequate methods. Results of more recent studies are shown in Fig. 207. Considering the very low solubility, the data of [57, 58, 56] are in good agreement. They are based on damping [57, 58] and diffusion measurement [56]. Results of calorimetric studies of the precipitation of Fe_3C from α [59] disagree considerably. According to [59], the discrepancy is outside the limits of the accidental errors of the calorimetric measurements and "there must be some methodical error in the way of measuring or computing the solubility" either in the work of [59] or in that of [56–58]. (See Note Added in Proof.)

The saturation curve *GP* was determined by [56], using diffusion studies. Results agree with those of [22], based on measurements of the gas equilibria CO-CO_2 and CH_4-H_2, and are consistent with the data of [56, 57] on the solubility curve *PQ* (Fig. 207). Results reported by [60] indicate a solubility of about 0.03 wt. (0.14 at.) % C at 723°C.

The Curie point of Fe (A_2 transformation) is not affected by carbon additions.

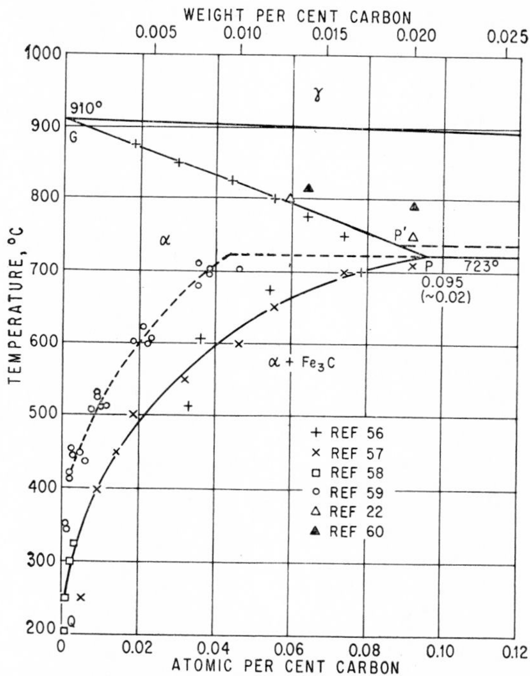

Fig. 207. C-Fe. α-phase region (metastable system).

THE STABLE SYSTEM Fe-GRAPHITE

Solidification. The temperature of the eutectic horizontal *E'C'F'* (melt ⇄ γ + graphite) was found as 1153 [7], 1154–1155 [25], 1155 [27], and 1153°C [26]. This is appreciably higher than has been accepted by [1], viz., 1135°C, or 5°C above the temperature of *ECF* selected by him. The difference between the temperature of *ECF* and *E'C'F'* is 7 ± 2°C [7, 25, 26]. The composition of the graphite eutectic *C'* was placed at 4.25 wt. (17.1 at.) % C [28], as compared with 4.15 wt. (16.75 at.) % C according to [7]. The former value is based on the extrapolation of the solubility curve *C'D'* of graphite in Fe and agrees closely with 4.27 wt. (17.18 at.) % C found in the same way by [61].

The liquidus curve *C'D'* of graphite was determined by [62–65, 28, 66, 67, 61]. Results of [28, 66] are in excellent agreement; differences lie within the experimental error. Also, the data of [61, 67] corroborate those of [28, 66]. In Fig. 208 the curve—

practically a straight line—is shown up to 2000°C, with data points of the most reliable investigations, those of [66] being omitted since they were not tabulated. As to determinations at higher temperatures, see [63, 64, 65].

Transformations in the Solid State. Prior to the work of [68, 21, 22], great differences existed as regards the solubility of graphite in austenite (curve $E'S'$). As a consequence, the point E', assuming a eutectic temperature of about 1153°C, had been placed at considerably differing compositions, between approximately 1.35 and 1.9 wt. (5.98 and 8.26 at.) % C [11, 69, 70, 68]. [21] has shown that the solubility was appreciably higher at temperatures above 1100°C than had been believed previously, although [70] already had found 1.84 wt. % C to be soluble at 1100°C. His data points, together with those of [70, 68, 22], are indicated in Fig. 206. Accepting the point S' as 0.69 wt. (3.13 at.) % C and 738°C [68], a nearly straight line for the boundary $E'S'$ can be plotted. E' then would be located at about 2.03 wt. (8.79 at.) % C and 1153°C. The point E must lie at a somewhat higher carbon concentration; it was assumed at 2.06 wt. (8.91 at.) % C and 1147°C. On the basis of lattice-parameter measurements of austenite in rapidly solidified alloys, [71] suggested that 1.7–1.8 wt. % C for the point E was more likely than a composition close to 2.0 wt. % C. There is no doubt, however, that the consistent results of [70, 21] and especially those of [22] indicate the higher solubility at the eutectic temperature.

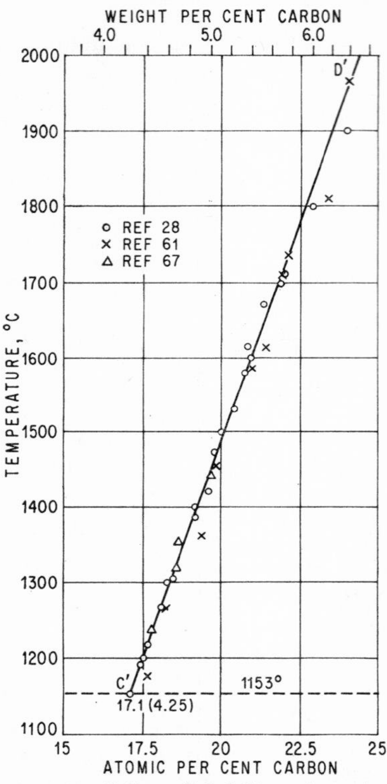

Fig. 208. C-Fe. Solubility of graphite in liquid iron.

The temperature of the eutectoid horizontal $P'S'K'$ was reported as 746°C on heating and 720°C on cooling (mean value 733°C) [14] and 738 ± 3°C [68]. The point S' was concluded to lie at 0.7 wt. (3.17 at.) % C [14], 0.85 wt. (3.83 at.) % C [70], and 0.69 wt. (3.13 at.) % C [68]. The latter value, deduced from the curve GS [45] and the eutectoid temperature of 738°C [68], was accepted in Figs. 201 and 202.

The solubility of graphite in α-Fe (curve $P'Q'$) was computed by thermodynamic methods [72]. Results indicate that also in this temperature range Fe_3C is metastable with respect to graphite.

Crystal Structures. *Austenite.* The crystal structure of austenite (f.c.c.) and the increase of its lattice constant with increasing carbon content were first studied on alloy steels [73]. For lattice-constant data of fairly pure austenitic Fe-C alloys, see especially [74–78, 71, 79]. The conclusion of [80], based on data collected from the literature, that the lattice parameter of austenite is lower in the presence of martensite (which exerts a compressive stress on the austenite) has been questioned [81] and could not be corroborated by measurements on an Fe-Ni alloy [82].

By means of careful measurement of X-ray diffraction intensities, [83] gave direct evidence for the location of the carbon atoms in the octahedral (largest) interstices of the f.c.c. (A1 type) Fe lattice; because of the restricted solubility, one position in twelve is occupied, at the most.

Ferrite. Since the solubility of C in b.c.c. (A2 type) α-Fe is very restricted, the ferrite phase has practically the lattice constant [84–86] and the magnetic transition temperature (768°C) of pure α-Fe. From measurements of the broadening of the X-ray diffraction lines (due to the local strains around the occupied interstices), [87] concluded that at least 85% of the carbon atoms occupy octahedral interstices in the α-Fe lattice (cf. Martensite).

Martensite. Rapid quenching from the austenitic region of the phase diagram results in the formation of metastable martensite. This phase is tetragonal with the Fe atoms on a b.c. tetragonal lattice and can be considered as supersaturated and distorted ferrite. The variation of the lattice dimensions with the carbon content was measured, among others, by [86, 88, 89, 74–76, 90, 78, 91]. The a axis decreases slightly with increasing carbon content whereas the c axis and the axial ratio c/a increase.

There is as yet no direct evidence as to which interstices the carbon atoms occupy in the Fe lattice. Proposals have been made by [86, 74, 76, 92–94], and it is most probable that the carbon atoms are in selected octahedral interstices—see especially [93, 94]—which positions are expected if the carbon atoms do not move from the interstices they occupy in the parent phase, f.c.c. austenite.

As an explanation for certain changes in physical properties of quenched carbon steels during low-temperature annealing, [95, 96, 75, 97] suggested the existence of a distinct cubic β-martensite phase. Evidence against this view has been put forth by [76], who suggested that "the structural element described as cubic martensite probably consists of tetragonal martensite in which the tetragonal symmetry is not observable owing to low carbon content." Referring to this controversy, [98] stated: "Hägg's view is supported by more recent workers [99–101] who conclude that the first stage of tempering involves the formation of ferrite and a precipitated phase which is not cementite but which is too finely dispersed to yield an X-ray diffraction pattern. . . . New X-ray observations by [98] now give direct evidence that the loss of tetragonality of martensite during the first tempering stage is due to the precipitation of a close-packed hexagonal iron carbide which it is proposed should be named ϵ-iron carbide, or ϵ-Fe_3C, because of its structural similarity with ϵ-Fe_3N. ϵ-Iron carbide is formed as a coherent transitional phase. . . ." See also the remarks by [102] on cubic martensite.

Cementite, Fe_3C. Contributions to the evaluation of the crystal structure of cementite have been made by [103–109]. The structure arrived at by [108] has been confirmed and refined by Fourier methods [109]: orthorhombic $D0_{11}$ prototype. Unit cell dimensions have also been measured by [110–114]; those of [112]—a = 4.5244 ± 5 A, b = 5.0885 ± 5 A, c = 6.7431 ± 5 A—are, within experimental error, in agreement with those found by [109, 111]. At elevated temperatures, cementite appears to be of slightly variable composition [111, 115, 116].

Cementite is ferromagnetic, its Curie point (A_0 transition) being at 210–215°C [117–124, 116].

The above-mentioned h.c.p. transitional phase ϵ-Fe_3C has the lattice parameters a = 2.73 A, c = 4.33 A, c/a = 1.58 [98]. Reflections observed on X-ray photographs of tempered martensite by [125, 100, 115] and ascribed by them to a new iron carbide were shown [98] to correspond to a certain stage in the structural transition ϵ-iron carbide—cementite.

Higher Carbides, Fe_2C *and* FeC. The existence of a (metastable) compound Fe_2C has been well established [126–129, 109, 130, 112, 131, 132]. [112] observed a narrow range of homogeneity near 31 at. % C and ascribed, therefore, the formula $Fe_{20}C_9$; the powder pattern was tentatively indexed with a large hexagonal or orthorhombic unit cell. According to [131, 132], Fe_2C exists in two modifications, one of them being h.c.p. with $a = 2.755$ A, $c = 4.349$ A, $c/a = 1.579$ [132].

The composition of the "FeC" carbide reported by [133] needs corroboration.

Note Added in Proof. The solubility of carbon in α-Fe has been redetermined between 440 and 710°C by [134] by means of measurements of elastic relaxation. The following solubility data, in wt. % (at. % in parentheses), were obtained: 0.0197 (0.092), 0.0130 (0.060), 0.0092 (0.043), 0.0068 (0.032), 0.0050 (0.023), 0.0036 (0.017), 0.0027 (0.013), and 0.0022 (0.010) at 713, 660, 606, 568, 534, 499, 468, and 444°C, respectively. Above 550°C, they are in good agreement with the solid-line curve of Fig. 207; at lower temperatures they indicate a somewhat lower solubility.

[134] stated that "below 230°C the phase that appears in equilibrium with carbon is, as recently found, not cementite but Fe_2C." He computed, on the basis of measurements by [135] of the evolution of heat during precipitation of Fe_2C, the solubility of C in α-Fe between 230 and about 120°C. The fact that two carbides, Fe_3C and Fe_2C, enter into equilibrium with α-Fe explains, according to [134], the much too low solubilities derived by [59] from their calorimetric measurements.

1. S. Epstein, "The Alloys of Iron and Carbon," vol. I, "Constitution," McGraw-Hill Book Company, Inc., New York, 1936.
2. M. Hansen, "Der Aufbau der Zweistofflegierungen," pp. 360–371, Springer-Verlag OHG, Berlin, 1936; see also [24].
3. R. Ruer and R. Klesper, *Ferrum*, **11**, 1913-1914, 257–261.
4. S. Umino, *Science Repts. Tôhoku Univ.*, **23**, 1935, 720–725.
5. F. Adcock, *J. Iron Steel Inst.*, **135**, 1937, 281–292.
6. J. H. Andrew and D. Binnie, *J. Iron Steel Inst.*, **119**, 1929, 309–358.
7. R. Ruer and F. Goerens, *Ferrum*, **14**, 1916-1917, 161–177.
8. G. B. Harris and W. Hume-Rothery, *J. Iron Steel Inst.*, **174**, 1953, 212–218.
9. W. C. Roberts-Austen, *Proc. Inst. Mech. Eng.*, **1897**, 31–100; **1899**, 35–102.
10. H. C. H. Carpenter and B. F. E. Keeling, *J. Iron Steel Inst.*, **65**, 1904, 224–242.
11. N. Gutowsky, *Metallurgie*, **6**, 1909, 731–743; *Stahl u. Eisen*, **29**, 1909, 2066–2068.
12. K. Honda and H. Endo, *Science Repts. Tôhoku Univ.*, **16**, 1927, 627–637.
13. K. Honda and H. Endo, *Science Repts. Tôhoku Univ.*, **16**, 1927, 235–244; *J. Inst. Metals*, **37**, 1927, 45–49.
14. R. Ruer, *Z. anorg. Chem.*, **117**, 1920, 249–261.
15. M. Hillert, *Acta Met.*, **3**, 1955, 37–38.
16. O. W. Ellis, *Iron Steel Inst.* (*London*), *Carnegie Schol. Mem.*, **15**, 1926, 195–215.
17. W. E. Jominy, *Trans. ASST*, **16**, 1929, 372–392.
18. Results of [16] and [17] are based on rather crude methods of investigation, viz., forgeability tests and investigations of the burning and overheating of steel, respectively. The curve shown in the figures was obtained by means of heating curves [5].
19. G. Asahara, *Sci. Papers Inst. Phys. Chem. Research* (*Tokyo*), **2**, 1924, 420–425.
20. S. Kaya, *Science Repts. Tôhoku Univ.*, **14**, 1925, 529–536.
21. R. W. Gurry, *Trans. AIME*, **150**, 1942, 147–153.
22. R. P. Smith, *J. Am. Chem. Soc.*, **68**, 1946, 1163–1175.
23. O. W. Ellis, *Metals & Alloys*, **1**, 1930, 462–464.
24. F. Körber and W. Oelsen, *Arch. Eisenhüttenw.*, **5**, 1931-1932, 569–578.
25. E. Piwowarsky, *Stahl u. Eisen*, **54**, 1934, 82–84.
26. Unpublished work by L. S. Darken, quoted from L. S. Darken and R. W. Gurry, *Trans. AIME*, **191**, 1951, 1015–1018.

27. T. Kasé, *Science Repts. Tôhoku Univ.*, **19**, 1930, 17–35.
28. R. Ruer and J. Biren, *Z. anorg. Chem.*, **113**, 1920, 98–112.
29. C. Benedicks, *Metallurgie*, **3**, 1906, 393–395, 425–441, 466–476.
30. E. Heyn, *Verh. Ver. Befördg. Gewerbfl.*, **83**, 1904, 355–397; see also E. Heyn, "Die Theorie der Eisen-Kohlenstoff-Legierungen," pp. 15–18, Springer-Verlag OHG, Berlin, 1924.
31. K. Honda, *Science Repts. Tôhoku Univ.*, **2**, 1913, 203.
32. G. Rümelin and R. Maire, *Ferrum*, **12**, 1914-1915, 141–154.
33. P. Bardenheuer, *Ferrum*, **14**, 1916-1917, 129–133, 145–151.
34. E. Maurer (M. Hetzler), *Stahl u. Eisen*, **41**, 1921, 1696–1706.
35. R. H. Harrington and W. P. Wood, *Trans. ASST*, **18**, 1930, 632–654.
36. P. Goerens and H. Meyer, *Metallurgie*, **7**, 1910, 307–312.
37. P. Saldau and P. Goerens, *Zhur. Russ. Met. Obshchestva*, **1914**(1), 789–824; *Rev. mét., Extraits*, **14**, 1917, 65–74; *Stahl u. Eisen*, **38**, 1918, 15–17.
38. P. Saldau, *Zhur. Russ. Met. Obshchestva*, **1915**(1), 655–690; *Stahl u. Eisen*, **38**, 1918, 39–40.
39. S. Iitaka, *Science Repts. Tôhoku Univ.*, **6**, 1917, 172.
40. K. Honda, *Science Repts. Tôhoku Univ.*, **5**, 1916, 285.
41. T. Sato, *Technol. Repts. Tôhoku Univ.*, **8**, 1929, 27–52.
42. S. Konno, *Science Repts. Tôhoku Univ.*, **12**, 1923, 127–136.
43. F. Stäblein, *Stahl u. Eisen*, **46**, 1926, 101–104.
44. H. Esser, *Stahl u. Eisen*, **47**, 1927, 337–344.
45. R. F. Mehl and C. Wells, *Trans. AIME*, **125**, 1937, 429–469.
46. J. B. Austin; see discussion on the paper of [45].
47. J. F. T. Berliner, *Natl. Bur. Standards (U.S.), Sci. Papers* 484, **19**, 1923-1924, 347–356.
48. Even at a heating and cooling rate of only ⅛°C per min, the hysteresis was still 8, 12, 18, and 19°C at 0.16, 0.28, 0.49, and 0.66 wt. % C, respectively.
49. E. Scheil, unpublished work.
50. P. R. Pallister, *J. Iron Steel Inst.*, **161**, 1949, 87–90.
51. E. Scheil, *Arch. Eisenhüttenw.*, **22**, 1951, 37–52.
52. A. Meuthen, *Ferrum*, **10**, 1912-1913, 1–21.
53. N. J. Wark, *Metallurgie*, **8**, 1911, 704–713; *Stahl u. Eisen*, **31**, 1911, 2108–2109.
54. N. Tschischewsky and N. Schulgin, *J. Iron Steel Inst.*, **95**, 1917, 189–198.
55. L. S. Darken and R. W. Gurry, *Trans. AIME*, **191**, 1951, 1015–1018.
56. J. K. Stanley, *Trans. AIME*, **185**, 1949, 752–761.
57. L. J. Dijkstra, *Trans. AIME*, **185**, 1949, 252–260.
58. C. A. Wert, *Trans. AIME*, **188**, 1950, 1243–1244.
59. G. Borelius and S. Berglund, *Arkiv Fysik*, **4**, 1951, 173–182; see also G. Borelius, *Trans. AIME*, **191**, 1951, 484.
60. W. A. Pennington, *Trans. ASM*, **41**, 1949, 213–256; see also *Trans. ASM*, **37**, 1946, 48–91.
61. J. Chipman et al., *Trans. ASM*, **44**, 1952, 1215–1230.
62. H. Hanemann, *Stahl u. Eisen*, **31**, 1911, 333–336.
63. O. Ruff and O. Goecke, *Metallurgie*, **8**, 1911, 417–421; see also O. Ruff, *Metallurgie*, **8**, 1911, 456–464, 497–508.
64. N. F. Wittorf, *Z. anorg. Chem.*, **79**, 1911, 1–70.
65. O. Ruff and W. Bormann, *Z. anorg. Chem.*, **88**, 1914, 397–409.
66. K. Schichtel and E. Piwowarsky, *Arch. Eisenhüttenw.*, **3**, 1929-1930, 139–147.
67. J. A. Kitchener, J. O'M. Bockris, and D. A. Spratt, *Trans. Faraday Soc.*, **48**, 1952, 608–617.
68. C. Wells, *Trans. ASM*, **26**, 1938, 289–344.

69. R. Ruer and N. Iljin, *Metallurgie*, **8**, 1911, 97–101.
70. E. Söhnchen and E. Piwowarsky, *Arch. Eisenhüttenw.*, **5**, 1931-1932, 111–120.
71. G. Falkenhagen and W. Hofmann, *Arch. Eisenhüttenw.*, **23**, 1952, 73–74.
72. M. Hillert, *Acta Met.*, **2**, 1954, 11–14.
73. For literature see *Strukturbericht*, **1**, 1913-1928, 576.
74. E. Öhman, *J. Iron Steel Inst.*, **123**, 1931, 445–463.
75. K. Honda and Z. Nishiyama, *Science Repts. Tôhoku Univ.*, **21**, 1932, 299–331.
76. G. Hägg, *J. Iron Steel Inst.*, **130**, 1934, 439–451.
77. W. J. Wrazej, *Nature*, **163**, 1949, 212–213; *Structure Repts.*, **12**, 1949, 45–46.
78. J. Mazur, *Nature*, **166**, 1950, 828; *Structure Repts.*, **13**, 1950, 62.
79. The data by [74–76] are listed in [1], p. 213.
80. E. Houdremont and O. Krisement, *Arch. Eisenhüttenw.*, **24**, 1953, 60.
81. E. Scheil, private communication to M. Hansen.
82. Measurements by E. Saftig; private communication by E. Scheil to M. Hansen.
83. N. J. Petch, *J. Iron Steel Inst.*, **145**, 1942, 111–123.
84. F. Wever, *Z. Elektrochem.*, **30**, 1924, 376–382; F. Wever and P. Rütten, *Mitt. Kaiser-Wilhelm-Inst. Eisenforsch. Düsseldorf*, **6**, 1924, 1–6.
85. W. L. Fink and E. D. Campbell, *Trans. ASST*, **9**, 1926, 717–748.
86. N. Seljakov, G. Kurdjumov, and N. Goodtzov, *Z. Physik*, **45**, 1927, 384–408.
87. G. K. Williamson and R. E. Smallman, *Acta Cryst.*, **6**, 1953, 361–362.
88. G. Kurdjumov and E. Kaminski, *Nature*, **122**, 1928, 475–476.
89. S. Sekito, *Science Repts. Tôhoku Univ.*, **18**, 1929, 69–77.
90. G. Kurdjumov and L. Lyssak, *Zhur. Tekh. Fiz.*, **16**, 1946, 1307–1318; *Structure Repts.*, **10**, 1945-1946, 35–37.
91. M. P. Arbuzov, *Doklady Akad. Nauk S.S.S.R.*, **74**, 1950, 1085–1087; *Structure Repts.*, **13**, 1950, 62.
92. C. H. Johansson, *Arch. Eisenhüttenw.*, **11**, 1937-1938, 241.
93. N. J. Petch, *J. Iron Steel Inst*, **147**, 1943, 221–227.
94. H. Lipson and A. M. B. Parker, *J. Iron Steel Inst.*, **149**, 1944, 123–141.
95. K. Honda and S. Sekito, *Science Repts. Tôhoku Univ.*, **17**, 1928, 743–760; *Nature*, **121**, 1928, 744.
96. K. Honda, *Science Repts. Tôhoku Univ.*, **18**, 1929, 503–516.
97. K. Honda and Z. Nishiyama, *Trans. ASST*, **20**, 1932, 464–470.
98. K. H. Jack, *Acta Cryst.*, **3**, 1950, 392–394.
99. D. P. Antia, S. G. Fletcher, and M. Cohen, *Trans. ASM*, **32**, 1944, 290 (periodical incorrectly given by [98]).
100. M. Arbuzov and G. Kurdjumov, *J. Phys. U.S.S.R.*, **5**, 1941, 101.
101. G. Kurdjumov and L. Lyssak, *J. Iron Steel Inst.*, **156**, 1947, 29–36.
102. C. Zener, *Trans. AIME*, **167**, 1946, 570–572.
103. A. Westgren and G. Phragmén, *Z. physik. Chem.*, **102**, 1922, 1–25; *J. Iron Steel Inst.*, **105**, 1922, 241–273; **109**, 1924, 159–174.
104. A. Westgren, G. Phragmén, and T. Negresco, *J. Iron Steel Inst.*, **117**, 1928, 383–400.
105. F. Wever, *Mitt. Kaiser-Wilhelm-Inst. Eisenforsch. Düsseldorf*, **4**, 1923, 67–80.
106. S. Shimura, *Proc. World Eng. Congr., Tokyo*, **34**, 1929, 223–225; *Proc. Imp. Acad. (Tokyo)*, **6**, 1930, 269–271; *J. Fac. Eng. Univ. Tokyo*, **20**, 1931, 1–53; abstract in *Strukturbericht*, **2**, 1928-1932, 303–304.
107. S. B. Hendricks, *Z. Krist.*, **74**, 1930, 534–545.
108. A. Westgren, *Jernkontorets Ann.*, **87**, 1932, 457–468; abstract in *Strukturbericht*, **2**, 1928-1932, 304.
109. H. Lipson and N. J. Petch, *J. Iron Steel Inst.*, **142**, 1940, 95–103.

110. W. Hume-Rothery, G. V. Raynor, and A. T. Little, *J. Iron Steel Inst.*, **145**, 1942, 143–149.
111. N. J. Petch, *J. Iron Steel Inst.*, **149**, 1944, 143–150.
112. K. H. Jack, *Proc. Roy. Soc. (London)*, **A195**, 1948, 56–61.
113. H. J. Goldschmidt, *Metallurgia*, **40**, 1949, 103–104.
114. E. G. Azincev and M. P. Arbuzov, *Zhur. Tekh. Fiz.*, **20**, 1950, 32–37; *Structure Repts.*, **13**, 1950, 63–64.
115. I. V. Isajcev, *Zhur. Tekh. Fiz.*, **17**, 1947, 839–854; *Structure Repts.*, **11**, 1947-1948, 68–69.
116. R. Bernier, *Ann. chim.*, **6**, 1951, 104–161.
117. S. W. J. Smith, *Proc. Phys. Soc. (London)*, **25**, 1912, 77–81 (210°C).
118. K. Honda and H. Takagi, *Science Repts. Tôhoku Univ.*, **4**, 1915, 161–168 (210°C).
119. K. Honda and T. Murakami, *Science Repts. Tôhoku Univ.*, **6**, 1917, 23 (215°C).
120. T. Ishiwara, *Science Repts. Tôhoku Univ.*, **6**, 1917, 285–294 (210°C).
121. G. Tammann and K. Ewig, *Stahl u. Eisen*, **42**, 1922, 772; *Z. anorg. Chem.*, **167**, 1927, 390 (210°C).
122. E. Lehrer, *Z. tech. Physik*, **9**, 1928, 142 (215°C).
123. A. Mittasch and E. Kuss, *Z. Elektrochem.*, **34**, 1928, 167 (215°C).
124. A. Travers and R. Diebold, *Compt. rend.*, **205**, 1937, 797–799 (210°C).
125. M. Arbuzov and G. Kurdjumov, *Zhur. Tekh. Fiz.*, **10**, 1940, 1093.
126. W. Gluud, K. V. Otto, and H. Ritter, *Ber. deut. chem. Ges.*, **62**, 1929, 2483; *Ber. Ges. Kohlentech.*, **3**, 1931, 40.
127. U. Hofmann and E. Groll, *Z. anorg. Chem.*, **191**, 1930, 414–428.
128. H. A. Bahr and V. Jessen, *Ber. deut. chem. Ges.*, **66**, 1933, 1238.
129. G. Hägg, *Z. Krist.*, **A89**, 1934, 92–94.
130. K. H. Jack, *Nature*, **158**, 1946, 60–61.
131. H. Pichler and H. Merkel, Thesis, Kaiser-Wilhelm-Institut Kohlenforschung, quoted by [132].
132. L. J. E. Hofer, E. M. Cohn, and W. C. Peebles, *J. Am. Chem. Soc.*, **71**, 1949, 189–195.
133. H. C. Eckstrom and W. A. Adcock, *J. Am. Chem. Soc.*, **72**, 1950, 1042–1043.
134. E. Lindstrand, *Acta Met.*, **3**, 1955, 431–435.
135. O. Krisement, *Arkiv Fysik*, **7**, 1953, 353. Quoted in [134].

$\bar{2}.8277$
1.1723

C-Hf Carbon-Hafnium

The existence of HfC (6.30 wt. % C), first prepared by [1, 2], was established by determining its crystal structure [3]. Like TiC and ZrC, the phase HfC will have a certain homogeneity range. Also, it may form a eutectic with carbon (see C-Ti). Its melting point was found to be 3890 ± 150°C [4]. HfC is cubic of the NaCl (B1) type with a parameter reported to be a = 4.70 [5], a = 4.467 [6], a = 4.64 [7], a = 4.635 [8], and a = 4.6365 A [9].

Note Added in Proof. [10] found evidence of a possible HfC-C eutectic at 2800°C and of a variable composition of HfC. The extrapolated parameter for HfC of theoretical composition was 4.641 ± 1 A.

1. C. Agte and K. Moers, *Z. anorg. Chem.*, **198**, 1931, 236–238.
2. K. Moers, *Z. anorg. Chem.*, **198**, 1931, 243–252, 262–275.
3. K. Becker, see [4].
4. C. Agte and H. Alterthum, *Z. tech. Physik*, **11**, 1930, 185.
5. K. Becker, *Physik. Z.*, **34**, 1933, 185–198.

6. P. M. McKenna, *Ind. Eng. Chem.*, **28,** 1936, 767.
7. F. W. Glaser, D. Moskowitz, and B. Post, *Trans. AIME*, **197,** 1953, 1119–1120.
8. L. M. Doney, *Ceramic Age*, **63,** 1954 (March), 21.
9. C. E. Curtis, L. M. Doney, and J. R. Johnson, *J. Am. Ceram. Soc.*, **37,** 1954, 464–465.
10. P. G. Cotter and J. A. Kohn, *J. Am. Ceram. Soc.*, **37,** 1954, 415–420.

$\bar{2}.7772$
1.2228

C-Hg Carbon-Mercury

See C-Cd, page 348.

$\bar{2}.7938$
1.2062

C-Ir Carbon-Iridium

Molten iridium, osmium, palladium, platinum, rhodium, and ruthenium dissolve appreciable amounts of carbon, which precipitate on cooling in the form of graphite. The percentage of carbon, obtained by determining the graphite content in the solidified samples, generally increases with the temperature of the melt; however, no temperature measurements were made [1, 2].

C-Ir: Carbon contents up to 1.19 wt. % were found [1]; at the boiling point of Ir the carbon absorption was 2.8 wt. % [2].

C-Os: At the boiling point 3.9–4 wt. % C was dissolved [2]. By treating osmium powder with methane at 1200 and 2000°C, no roentgenographically detectable change of the lattice parameters was observed [3].

C-Pd: At least 2.45 wt. % C was taken in solution [1].

C-Pt: See page 377.

C-Rh: The solubility of carbon in molten Rh was found to vary between 1.42 and 7.38 wt. % [1]. At the boiling point, however, only 2.19 wt. % C was dissolved [2] (?).

C-Ru: At the boiling point, 4.8 wt. % C was dissolved [2].

1. H. Moissan, *Compt. rend.*, **123,** 1896, 16–18.
2. H. Moissan, *Compt. rend.*, **142,** 1906, 189–195.
3. W. Trzebiatowski, *Z. anorg. Chem.*, **233,** 1937, 376–384.

$\bar{2}.9368$
1.0632

C-La Carbon-Lanthanum

The crystal structure of LaC_2 (14.74 wt. % C) is tetragonal of the CaC_2 (C11) type, $a = 5.55$ A, $c = 6.56$ A, $c/a = 1.18$ (pseudocubic cell) [1].

1. M. v. Stackelberg, *Z. physik. Chem.*, **B9,** 1930, 437–475; *Z. Elektrochem.*, **37,** 1931, 542–545.

$\bar{1}.6936$
0.3064

C-Mg Carbon-Magnesium

As regards the preparation and properties of the magnesium carbides Mg_2C_3 (42.56 wt. % C) and MgC_2 (49.69 wt. % C), see [1–5, 7]. MgC_2 is stable up to approximately 600°C, at which temperature it decomposes: $2MgC_2 \rightarrow Mg_2C_3 + C$. Mg_2C_3 decomposes at about 660°C into Mg vapor and C [5]; see also [7].

MgC_2 is tetragonal, with 12 atoms per unit cell, $a = 5.55$ A, $c = 5.03$ A, $c/a = 0.906$; and Mg_2C_3 is hexagonal, with 40 atoms per unit cell, $a = 7.45$ A, $c = 10.61$ A, $c/a = 1.424$ [6].

Mg_2C_3 and MgC_2 decompose water, forming allylene and acetylene, respectively.

1. "Gmelins Handbuch der anorganischen Chemie," System No. 27B (2), pp. 299–301, Verlag Chemie, G.m.b.H., Berlin, 1939.
2. J. Novák, *Z. physik. Chem.*, **73**, 1910, 513–546.
3. H. H. Franck, M. A. Bredig, and Kin-Hsing Kou, *Z. anorg. Chem.*, **232**, 1937, 110–111.
4. W. H. C. Rueggeberg, *J. Am. Chem. Soc.*, **65**, 1943, 602–607; see also M. A. Bredig, *J. Am. Chem. Soc.*, **65**, 1943, 1482–1483.
5. P. Ehrlich, unpublished research, *FIAT Rev. Ger. Sci.*, 1939-1946, Inorganic Chemistry, Part II, p. 191.
6. F. Irmann, *Helv. Chim. Acta*, **31**, 1948, 1584–1602, 2263.
7. A. Schneider and J. F. Cordes, *Z. anorg. Chem.*, **279**, 1955, 94–103.

$\bar{1}.3397$
0.6603

C-Mn Carbon-Manganese

The carbide Mn_3C (6.79 wt. % C) has long been regarded as the manganese carbide lowest in carbon content and, in fact, the sole intermediate phase occurring in this system. It was claimed to be obtained as a product of chemical reactions [1, 2]. Also, an alloy of this composition was reported to be single-phase [3, 4], and carburization studies [5] indicated its existence. However, since the work of [6] it is generally accepted that Mn_3C does not exist (but see Supplement).

The phase diagrams proposed by [3] and [4], both showing the phase Mn_3C, are obsolete. At the time these investigations were carried out, it was not known that Mn is polymorphous.

Considerable progress was made when Mn-C alloys were studied roentgenographically. It was found [7] that C is interstitially soluble in Mn. [8] briefly reported that, according to work of [9], published much later [6], the carbides Mn_4C (5.18 wt. % C) and Mn_7C_3 (8.57 wt. % C) occur. For some time the existence of Mn_3C was still maintained, and even its crystal structure was given [8]. [10] reported that the structure of the carbide lowest in C content was based on the composition $Mn_{23}C_6$ [20.69 at. (5.40 wt.) % C] rather than Mn_4C. Both formula weights differ by only 0.69 at. (0.22 wt.) %.

The phase diagram published by [11] resembles that presented in Fig. 209 in many respects; however, it does not indicate the existence of the low carbide $Mn_{23}C_6$ (or Mn_4C), which was already established prior to this work. The phase relations involving the phases based on α-, β-, and γ-Mn were found to be those shown in Fig. 209.

The diagram in Fig. 209 was established by means of thermal, dilatometric, magnetic, and micrographic analysis of high-purity alloys [12]. The composition of the phases of the three-phase equilibria in wt. % (at. % in parentheses) is as follows:

1235°C: δ_{Mn}, 0.12 (0.55); γ_{Mn}, 0.25 (1.15); melt, 0.8 (3.55)
1260°C: melt, 2.7 (11.2); γ_{Mn}, 2.95 (12.1); ϵ, 3.05 (12.5)
1340°C: melt, 8.1 (28.7); ϵ, 8.4 (29.5); (Mn_7C_3), 8.75 (30.5)
1010°C: ϵ, 3.4 (13.8); ($Mn_{23}C_6$), 5.18 (20.0); (Mn_7C_3), 6.9 (25.3)
950°C: γ_{Mn}, 2.3 (9.7); ϵ, 3.0 (12.3); ($Mn_{23}C_6$), 4.8 (18.8)
857°C: β_{Mn}, 0.05 (0.23); α_{Mn}, 0.90 (4.0); γ_{Mn}, 1.1 (4.8)
820°C: α_{Mn}, 1.0 (4.4); γ_{Mn}, 1.3 (5.7); ($Mn_{23}C_6$), 4.2 (16.7)

The first carbide was designated as Mn_4C, and the peritectoid point at 1010°C was placed at this composition. The composition $Mn_{23}C_6$ lies just outside the homogeneity range at high temperature The characteristic composition of the second carbide was suggested to be Mn_2C (9.85 wt. % C); however, the crystal structure points to Mn_7C_3.

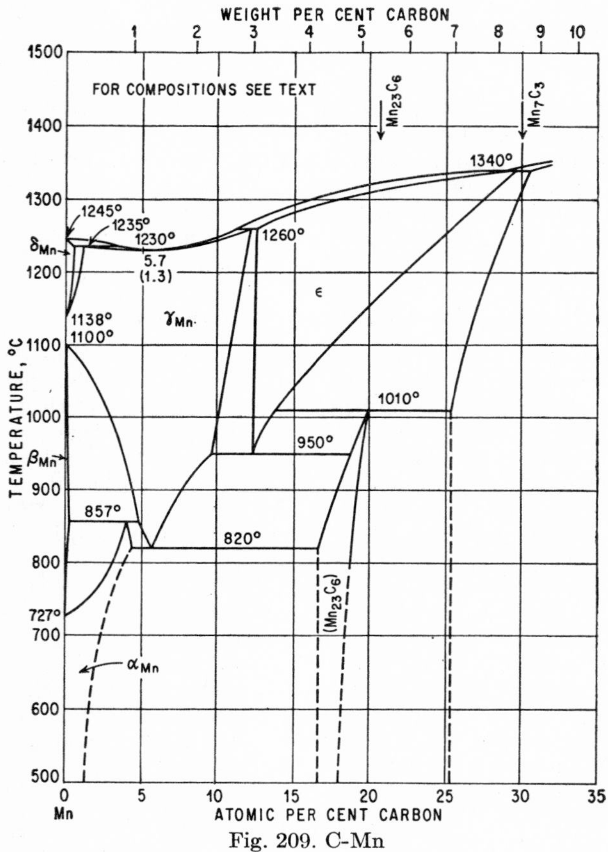

Fig. 209. C-Mn

The solubility of graphite in liquid Mn up to 1525°C was determined by [13]. Their data are incompatible with the fact that liquidus temperatures could be determined for alloys with as high as 8.6 wt. % C [12].

Crystal Structures. $Mn_{23}C_6$ is cubic of the $D8_4$ type, 116 atoms per unit cell, a = 10.585 A [10, 6]. Mn_7C_3 is hexagonal, with 80 atoms per unit cell, a = 13.90 A, c = 4.54 A, c/a = 0.327 [6].

Supplement. In a preliminary account [14] of an X-ray study of the Mn-C system by [15], it is said that [15] found no less than five intermediate carbide phases, the formulas and crystallographic data of which have been summarized in Table 19. It is to be expected that the phase relations will be discussed in the forthcoming publication.

As regards the ferromagnetic "Mn_4C" phase reported by [16], [14] suggest that this carrier of ferromagnetism is actually an interstitial carboxide $Mn_4(C,O)$—not stable below 800°C—rather than a pure carbide phase. However, [17], who tried to avoid contamination of his alloys, found an alloy of the composition Mn_4C to be weakly ferromagnetic after quenching from 1050°C and to contain, besides some $Mn_{23}C_6$ phase, 70% of a f.c.c phase (a = 3.869 kX) [18]. It was found that small quantities of elements such as In or Sn can readily stabilize ferromagnetic "Mn_4C."

Table 19. Characteristics of the Manganese Carbides [14]

Carbide	Symmetry	Lattice constants	Number of formula units per unit cell	Stability range	Remarks
$Mn_{23}C_6$..	Cubic	a = 10.61 A	4	Up to 1025°C	Isomorphous with $Cr_{23}C_6$
Mn_7C_2? (22.22 at. % C)	Unknown	Unknown	...	850–1000°C	As the structure of this carbide has not yet been determined, the formula Mn_7C_2 cannot be considered certain
Mn_3C...	Orthorhombic	a = 4.530 A b = 5.080 A c = 6.772 A	4	950–1050°C	Isomorphous with Fe_3C
Mn_5C_2.. (28.57 at. % C)	Monoclinic	a = 5.086 A b = 4.573 A c = 11.66 A β = 97.75°	4	Up to 1050°C	
Mn_7C_3..	Trigonal	a = 13.90 A c = 4.54 A	8	Up to 1100°C	Isomorphous with Cr_7C_3

1. L. Troost and Hautefeuille, *Compt. rend.*, **80,** 1875, 964; H. Moissan, *Compt. rend.*, **122,** 1896, 421–423; **125,** 1897, 839–841; Gin and Leleux, *Compt. rend.*, **126,** 1898, 749–750.
2. O. Ruff and E. Gersten, *Ber. deut. chem. Ges.*, **46,** 1913, 400–406.
3. A. Stadeler, *Metallurgie*, **5,** 1908, 260–267, 281–288.
4. K. Kido, *Science Repts. Tôhoku Univ.*, **9,** 1920, 305–310.
5. R. Schenck, N. G. Schmahl, and O. Ruetz, *Z. Elektrochem.*, **42,** 1936, 569.
6. E. Öhman, *Jernkontorets Ann.*, **128,** 1944, 13–16.
7. A. Westgren and G. Phragmén, *Z. Physik*, **33,** 1925, 785–786.
8. B. Jacobson and A. Westgren, *Z. physik. Chem.*, **B20,** 1933, 362.
9. E. Öhman, see [6].
10. A. Westgren, *Jernkontorets Ann.*, **117,** 1933, 501–512.
11. R. Vogel and W. Döring, *Arch. Eisenhüttenw.*, **9,** 1935-1936, 247–252; *Metallurgist*, **10,** 1936, 102–104.

12. M. Isobe, *Science Repts. Research Inst. Tôhoku Univ.*, **A3,** 1951, 468–490.
13. O. Ruff and W. Bormann, *Z. anorg. Chem.*, **88,** 1914, 365–385; see also *Ber. deut. chem. Ges.*, **45,** 1912, 3142; *Ferrum*, **13,** 1915-1916, 106.
14. K. Kuo and L. E. Persson, *J. Iron Steel Inst.*, **178,** 1954, 39–44 (system Fe-Mn-C).
15. K. Kuo, to be published.
16. E. Wedekind (1911), quoted in R. M. Bozorth, "Ferromagnetism," p. 337, D. Van Nostrand Company, Inc., Princeton, N.J., 1951.
17. E. R. Morgan, *Trans. AIME*, **200,** 1954, 983–988.
18. The claim that this alloy decomposes on slow cooling to $Mn_{23}C_6$ and β-Mn is at variance with Fig. 209.

$\bar{1}.0975$
0.9025

C-Mo Carbon-Molybdenum

The existence of the carbide Mo_2C (5.89 wt. % C), already claimed by [1] and also reported by [2–4], was definitely established by determining the crystal structure, which was found to be a h.c. packing of Mo atoms with interstitial C atoms [5]. Since then, Mo_2C has been confirmed and identified by many investigators [6–18]. The homogeneity range of this phase was given as about 5.5–6 wt. (31.75–33.75 at.) % C [6] and about 5.3–6 wt. (31–33.75 at.) % C [15]. In good agreement with this, [5] reported the lower limit to lie at about 31 at. (5.3 wt.) % C.

The question whether the monocarbide (11.12 wt. % C) does exist has been a controversial subject for some time. Its occurrence was claimed by [19, 2, 4, 8, 9]; however, [5], using X-ray analysis, were unable to find a second intermediate phase in alloys up to 39 at. % C. Also, [6] reported that, according to microscopic studies, graphite appears to be the primary phase in alloys above 6 wt. % C. Whereas results by [11, 12, 20] had to be considered inconclusive, X-ray data given by [13] and [15] and other evidence [15] showed that a second intermediate phase exists. Results by [13] and [15] disagree, however, in that the former believed this phase to be hexagonal whereas the latter asserted that the structure was different. [16] also reported the phase MoC to exist, but their preparations were not proved homogeneous. [21] showed that, although MoC does not exist as a stable separate phase, it can be stabilized as a solid solution with WC. Recently, [22] found evidence for a phase, probably f.c.c. in structure, having a composition close to that of MoC, and [18] gave conclusive evidence for the existence of MoC, identified by roentgenographic methods.

The *tentative* phase diagram of Fig. 210 was constructed by [15] on the basis of micrographic and X-ray data. Earlier, the melting points of the intermediate phases had been given as follows: Mo_2C, 2230–2330°C [4], 2690 ± 50°C [9]; MoC, 2570°C [4], 2695 ± 50°C [9]. The eutectic Mo-Mo_2C was located at about 4 wt. (~25 at.) % [6] and about 1.8 wt. (~12.5 at.) % C [15], the latter value being better founded.

The solid solubility of C in Mo, assumed by [6] and [15] to be approximately 0.3 and 0.1 wt. % C, respectively, was determined more accurately to lie between 0.005 and 0.009 wt. % at 1650°C, 0.012 and 0.013 wt. % at 1925°C, and 0.018 and 0.022 wt. % C at 2200°C [23].

Crystal Structures. The lattice parameter of Mo (a = 3.14664 A) is increased by 0.018 wt. % C to a = 3.14768 A [24]. This is proof of the interstitial type of solid solution, as already asserted by [25] in discussing the contradictory data by [15].

Lattice parameters of the h.c.p. structure (2 Mo atoms per unit cell) of Mo_2C, being isotypic with W_2C, agree quite well [5, 13, 26, 14, 15, 27, 18]. According to [18], they are a = 3.002 A, c = 4.724 A, c/a = 1.574. [17] found evidence of a f.c.c. modification of Mo_2C with a = 4.15 A.

The hexagonal phase MoC is isotypic with WC, 1 Mo atom per unit cell. Its parameters are a = 2.898 A, c = 2.809 A, c/a = 0.969 [18], in fairly good agreement with values reported earlier [13]. There are indications of a f.c.c. phase, probably also of the composition MoC, with a = 4.28 A [22]. Besides the stable phase MoC, [18] found another phase, likewise hexagonal, with a = 2.932 A, c = 10.97 A, c/a =

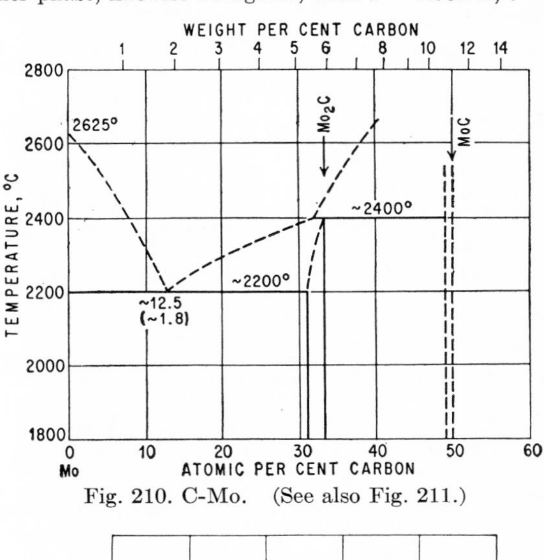

Fig. 210. C-Mo. (See also Fig. 211.)

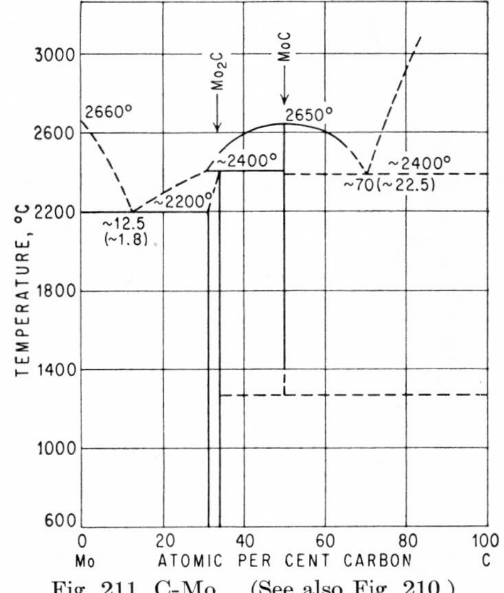

Fig. 211. C-Mo. (See also Fig. 210.)

3.742, and 4 Mo atoms per unit cell. It is also presumably of the composition MoC, but apparently not a stable phase.

Supplement. [28] measured the melting points of six alloys containing between 40 and 75 at. % C and detected a eutectic at about 70 at. (22.5 wt.) % C and ~2400°C. They found indications that MoC decomposes on cooling somewhere between 1900 and 700°C (Fig. 211). As regards the crystal structure of MoC, none of the previously reported unit cells or structures could be confirmed. [28] found a hexagonal unit cell, a = 3.01 A, c = 14.61 A, c/a = 4.86, containing—if no defects occur—12 atoms. A tentative structure is given. [28] suggest that the structural investigations of [18] were made on oxygen-contaminated alloys. [29] reported 2405°C as the melting point for Mo_2C.

1. H. Moissan, *Compt. rend.*, **116**, 1893, 1225; **120**, 1895, 1320–1326; **125**, 1897, 839–844.
2. S. Hilpert and M. Ornstein, *Ber. deut. chem. Ges.*, **46**, 1913, 1669–1675.
3. K. Nischk, *Z. Elektrochem.*, **29**, 1923, 387–388.
4. E. Friederich and L. Sittig, *Z. anorg. Chem.*, **144**, 1925, 183–184, 189.
5. A. Westgren and G. Phragmén, *Z. anorg. Chem.*, **156**, 1926, 27–36.
6. T. Takei, *Science Repts. Tôhoku Univ.*, **17**, 1928, 939–944.
7. S. Sekito, see [6], footnote 10.
8. A. A. Ravdel, *Zhur. Russ. Fiz.-Khim. Obshchestva*, **62**, 1930, 515–522.
9. C. Agte and H. Alterthum, *Z. tech. Physik*, **11**, 1930, 185.
10. S. L. Hoyt, *Trans. AIME*, **89**, 1930, 9–58.
11. R. Schenck, F. Kurzen, and H. Wesselkock, *Z. anorg. Chem.*, **203**, 1932, 183–185.
12. W. Meissner, H. Franz, and H. Westerhoff, *Ann. Physik*, **13**, 1932, 543–548; **17**, 1933, 599–601; W. Meissner and H. Franz, *Z. Physik*, **65**, 1930, 45–47.
13. H. Tutiya, *Sci. Papers Inst. Phys. Chem. Research, Tokyo*, **19**, 1932, 384–392.
14. V. Adelsköld, A. Sundelin, and A. Westgren, *Z. anorg. Chem.*, **212**, 1933, 401.
15. W. P. Sykes, K. R. Van Horn, and C. M. Tucker, *Trans. AIME*, **117**, 1935, 173–186.
16. G. Weiss, *Ann. chim. (Paris)*, **1**, 1946, 446; J. L. Andrieux and G. Weiss, *Bull. soc. chim. France*, **15**, 1948, 598.
17. J. J. Lander and L. H. Germer, *Trans. AIME*, **175**, 1948, 660–663.
18. K. Kuo and G. Hägg, *Nature*, **170**, 1952, 245–246.
19. H. Moissan and M. K. Hoffmann, *Ber. deut. chem. Ges.*, **37**, 1904, 3324–3327; *Compt. rend.*, **138**, 1904, 1558.
20. K. Becker and F. Ebert, *Z. Physik*, **31**, 1925, 268–272.
21. W. Dawihl, *Z. anorg. Chem.*, **262**, 1950, 212–217.
22. H. Nowotny and R. Kieffer, *Z. anorg. Chem.*, **267**, 1952, 261–264.
23. W. E. Few and G. K. Manning, *Trans. AIME*, **194**, 1952, 271–274.
24. R. Speiser, J. W. Spretnak, W. E. Few, and R. M. Parke, *Trans. AIME*, **194**, 1952, 275–277.
25. J. T. Norton, *Trans. AIME*, **117**, 1935, 187–188.
26. H. H. Lester, quoted from J. L. Gregg, "The Alloys of Iron and Molybdenum," p. 59, McGraw-Hill Book Company, Inc., New York, 1932.
27. H. Nowotny and R. Kieffer, *Z. Metallkunde*, **38**, 1947, 257–265.
28. H. Nowotny, E. Parthé, R. Kieffer, and F. Benesovsky, *Monatsh. Chem.*, **85**, 1954, 255–272.
29. G. A. Geach and F. O. Sones, *Plansee Proc.*, **1955**, 80–91; *Met. Abstr.*, **24**, 1957, 366.

$\bar{1}.1115$
0.8885

C-Nb Carbon-Niobium

The monocarbide NbC (11.45 wt. % C), long since known [1–3], has the cubic NaCl (B1) type of structure. Its lattice parameter was determined by [4–10] and is very likely a = 4.470 A [6, 9, 10]. [11] claimed that the formula Nb_4C_3 [42.86 at.

(8.84 wt.) % C], in analogy with V_4C_3, was more nearly correct; however, the lattice parameter (a = 4.47 A) was found to coincide with that of NbC. Possibly a homogeneity range (subtraction-type structure) including the composition Nb_4C_3 exists. The melting point of NbC was found to be 3700–3800°C [12] or 3500 ± 125°C [13]. [14] briefly reported that he was able to identify Nb_2C (6.07 wt. % C), which is isostructural with Nb_2N. Also, [6] believed that he found indication of the existence of a lower carbide (Nb_4C or Nb_3C).

Supplement. [15] reported a = 4.469 A as the lattice parameter of NbC. The existence of a hexagonal Nb_2C phase was confirmed by [16, 17].

[18] studied the entire composition range by means of X-ray, chemical, and microscopic analyses. The alloys were prepared chiefly by the sintering of powder compacts at 1600–1700°C; some samples prepared at 1200°C by carburation of Nb

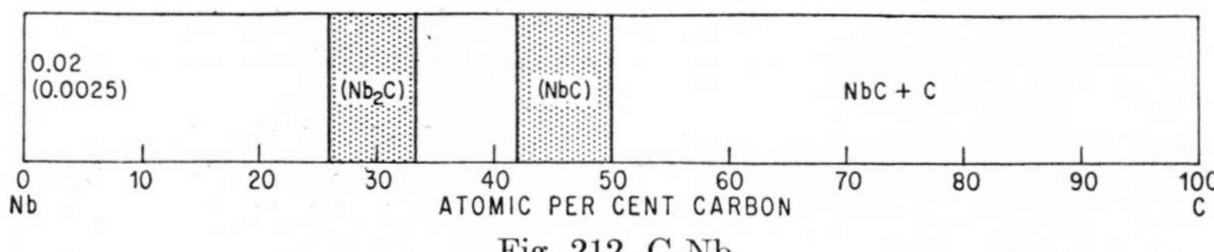

Fig. 212. C-Nb

by means of CH_4 yielded identical results. The solubility of C in solid Nb was microscopically determined to be 0.02 at. (0.0025 wt.) %. The "subcarbide" Nb_2C has a range of homogeneity between $NbC_{0.35}$ and $NbC_{0.50}$ (25.9 and 33.3 at. % C). The Nb atoms form a h.c.p. lattice (as with Nb_2N); the carbon atoms are presumably distributed at random over the largest holes of this lattice. The lattice parameters decrease with decreasing carbon content from a = 3.119 to 3.111 kX and c = 4.953 to 4.945 kX. (NbC) is homogeneous between $NbC_{0.72}$ and $NbC_{1.00}$ (41.9 and 50.0 at. % C). The lattice parameter of the NaCl-type structure decreases with decreasing carbon content from 4.457 to 4.424 kX. Above 50 at. % C, free carbon coexists with the NbC phase (see the schematic diagram of Fig. 212).

1. A. Joly, *Ann. sci. école norm.*, **6**, 1877, 145.
2. E. Friederich and L. Sittig, *Z. anorg. Chem.*, **144**, 1925, 182–183, 189.
3. C. Agte and K. Moers, *Z. anorg. Chem.*, **198**, 1931, 236–238.
4. K. Becker and F. Ebert, *Z. Physik*, **31**, 1925, 268–272.
5. P. M. McKenna, *Ind. Eng. Chem.*, **28**, 1936, 767–772.
6. J. S. Umanski, *Zhur. Fiz. Khim.*, **14**, 1940, 332–339.
7. H. Nowotny and R. Kieffer, *Z. Metallkunde*, **38**, 1947, 257–265.
8. H. Krainer and K. Konopicky, *Berg. hüttenmänn. Monatsh.*, **92**, 1947, 166–178.
9. J. T. Norton and A. L. Mowry, *Trans. AIME*, **185**, 1949, 133–136.
10. P. Duwez and F. Odell, *J. Electrochem. Soc.*, **97**, 1950, 299–304.
11. H. Eggers and W. Peter, *Mitt. Kaiser-Wilhelm-Inst. Eisenforsch. Düsseldorf*, **20**, 1938, 205–211.
12. E. Friederich and L. Sittig, *Z. anorg. Chem.*, **145**, 1925, 245; *Z. Physik*, **31**, 1925, 814; see also [2].
13. C. Agte and H. Alterthum, *Z. tech. Physik*, **11**, 1930, 185.
14. G. Brauer, *Z. Elektrochem.*, **46**, 1940, 401 (footnote).
15. A. E. Kovalski and J. S. Umanski, *Zhur. Fiz. Khim.*, **20**, 1946, 769–772; *Structure Repts.*, **10**, 1945-1946, 39.
16. G. F. Hardy and J. K. Hulm, *Phys. Rev.*, **93**, 1954, 1010.
17. O. H. Krikorian, *U.S. Atomic Energy Comm., Publ.* UCRL-2888, April, 1955, p. 30.
18. G. Brauer, H. Renner, and J. Wernet, *Z. anorg. Chem.*, **277**, 1954, 249–257.

$\bar{1}.3110$
0.6890

C-Ni Carbon-Nickel

It has long been known [1] that molten Ni can dissolve substantial amounts of C, which precipitate as graphite on cooling [2–4]. The solidification of Ni-rich melts was investigated by thermal analysis [5–7]; results deviate considerably because of the tendency to supercooling. The liquidus branch of the Ni-rich phase given in Fig. 213 was obtained by graphical interpolation of the data reported [5–7]. The Ni-C eutectic was found as follows: 2–2.5 wt. (9.1–11.2 at.) % C, 1307–1318°C (average 1314°C) [5, 8]; 2.2 wt. (9.9 at.) % C, 1304–1325°C (average 1312°C) [6]; 2.22 wt. (10.0 at.) % C, 1305–1318°C (average 1313°C) [7]. See also the interesting studies on graphite formation in Ni-C alloys by [9], who gave the eutectic temperature as about 1326°C.

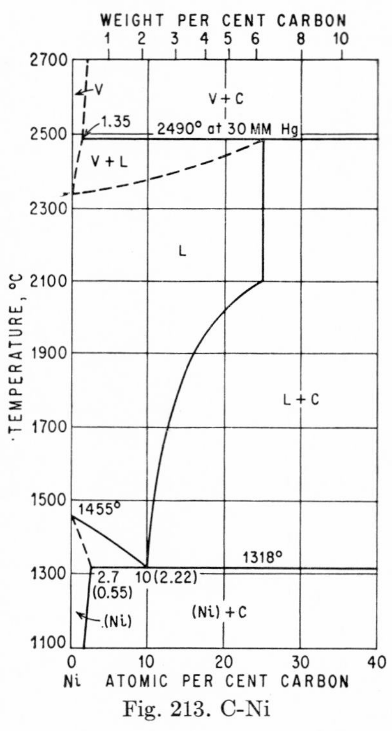

Fig. 213. C-Ni

On the basis of the duration of the halting time of the eutectic crystallization the solid solubility of C in Ni at 1318°C was assumed to be 0.3–0.4 wt. % [6] and 0.55 wt. (2.7 at.) % C [7]. [10] gave 0.52 wt. % C. The value of 0.55 wt. % is in good agreement with determinations by [11], which indicate the solubility, in wt. % (at. % in parentheses), as 0.5 (2.4) at 1265°C, 0.4 (1.95) at 1170°C, 0.3 (1.45) at 1055°C, 0.2 (0.95) at 925°C, 0.1 (0.48) at 745°C, and 0.08 (0.4) at 700°C. Between 1000 and 1200°C these data are in excellent agreement with those reported by [12], in wt. % (at. % in parentheses): 0.65 (3.1) at 1318°C, 0.39 (1.9) at 1200°C, 0.29 (1.4) at 1100°C, and 0.23 (1.1) at 1000°C.

The liquidus temperatures representing the beginning of crystallization of graphite (Fig. 213) were obtained by solubility determinations [13, 14, 6, 7], especially [13, 7]. Results are in good accord. The composition of the vapor in equilibrium with the saturated melt [6.4 wt. (25.0 at.) % C], boiling at 2490°C at 30 mm Hg pressure, was given as 0.28 wt. (1.4 at.) % C [6].

The composition of the point where the solubility of C in molten Ni changes in slope at 2100°C coincides nearly with that of Ni_3C (6.39 wt. % C). [13] concluded from this fact that Ni_3C exists but is stable only at these high temperatures and decomposes already on quenching from the molten state. In the structure of some quenched samples a constituent, besides (Ni) and C, was observed which could have been the carbide. This would mean that by very rapid cooling the decomposition of Ni_3C can, at least partially, be suppressed. Recently, [9] have definitely shown the presence of a new phase, probably Ni_3C, in rapidly quenched melts; addition of CaSi can stabilize the carbide. It cannot be decided as yet whether there is a temperature range within which Ni_3C is a stable phase or the carbide is a metastable phase at any temperature.

According to [15], Ni_3C is rather persistent above 1600°C and below 300°C. To obtain the carbide it is necessary to cool rapidly from 2000 to 1000°C. [16] prepared a carbide Ni_xC by treating Ni with CO at about 250°C; the decomposition temperature was given as about 700°C. By the same method [17] prepared a carbide, identified as Ni_3C, which decomposed at temperatures as low as 380–420°C; see also [18].

According to [19], Ni_3C has a h.c.p. structure of the Ni atoms, $a = 2.651$ A, $c = 4.338$ A, $c/a = 1.636$. On the other hand, [20] reported Ni_3C to have the orthorhombic structure of the Fe_3C ($D0_{11}$) type. The latter structure "is considered as much more probably the correct one" [21] and appears to have been widely accepted [22, 23].

Supplement. The interstitial A3-type structure for Ni_3C reported by [19] has support from the X-ray work by [24, 25]; these authors gave the following lattice parameters: $a = 2.6449$ A, $c = 4.3296$ A, $c/a = 1.637$ [24]; $a = 2.64_6$ A, $c = 4.32_0$ A, $c/a = 1.63_3$ [25]. According to [24], Ni_3C is metastable at any temperature; "its observed 'stability' is purely a result of the small rate of decomposition at low temperatures. This 'stability' is a kinetic effect and is, therefore, subject to the mode of preparation of the carbide, addition of inhibitors or accelerators of the decomposition, thermal treatment, etc."

In a thermomagnetic study, [25] observed the beginning of decomposition of Ni_3C on heating between 210 and 235°C, dependent on the previous heat-treatment. From this he concluded a certain homogeneity range, with Ni_3C considered to be the upper limit of carbon content. In light of the view by [24] mentioned above, and the fact that no measurable shift in the positions of the powder lines could be observed, this conclusion is not convincing. [25] observed a rather extended, but obviously highly supersaturated, terminal solid solution of C in f.c.c. Ni.

[26] reported that Ni_3C decomposed completely when heated at 370°C for 24 hr. [27] suggested that the carbide phase reported by [16] to be stable below 419°C (decomposition observed at 700°C; see above) was not identical with that investigated by [26]. In this connection it is interesting to note that the existence of a carbide Ni_6C has been claimed [28, 29, 24].

The assumption of [9] that the formation of graphite in eutectic Ni-C alloys involves a metastable (Ni + Ni-carbide) eutectic as an intermediate stage was disproved by [30], who gave strong evidence for a direct graphite crystallization. The carbide phase observed by [9] (see above) is considered [30] to be the product of a solid-state reaction.

[31] prepared Ni_3C by cementation, in a current of CO, of thin films of Ni at 250–400°C. Above 400°C Ni_3C did not form. [31] assume that at this temperature another carbide forms, possibly Ni_3C_2, as suggested by [17], which is unstable below 400°C. The (hexagonal) lattice constants of Ni_3C were found, by electron diffraction, to be $a = 2.631$ A, $c = 4.314$ A, $c/a = 1.640$.

1. The older literature has been reviewed by [2, 6, 7].
2. W. Hempel, *Z. angew. Chem.*, **17,** 1904, 300–301.
3. E. Heyn, *Stahl u. Eisen*, **26,** 1906, 1390.
4. N. S. Kurnakow and S. F. Zemczuzny, *Z. anorg. Chem.*, **54,** 1907, 151.
5. K. Friederich and A. Leroux, *Metallurgie*, **7,** 1910, 10–13.
6. O. Ruff and W. Bormann, *Z. anorg. Chem.*, **88,** 1914, 386–396; *Ber. deut. chem. Ges.*, **45,** 1912, 3142; *Ferrum*, **13,** 1915-1916, 108.
7. T. Kasé, *Science Repts. Tôhoku Univ.*, **14,** 1925, 187–193.
8. Alloys contained 0.2–0.6 wt. % Fe.
9. H. Morrogh and W. J. Williams, *J. Iron Steel Inst.*, **155,** 1947, 341–357; see also **156,** 1947, 491–496; **157,** 1947, 193–197.

10. T. Mishima, *World Eng. Congr. Tokyo, Paper* 716, 1929; abstract in *J. Inst. Metals,* **47,** 1931, 76.
11. J. J. Lander, H. E. Kern, and A. L. Beach, *J. Appl. Phys.,* **23,** 1952, 1305–1309.
12. T. E. Kihlgren and J. T. Eash, unpublished work by R. H. Schaefer, "Metals Handbook," 1948 ed., p. 1183, The American Society for Metals, Cleveland, Ohio.
13. O. Ruff and W. Martin, *Metallurgie,* **9,** 1912, 143–148.
14. O. Ruff and E. Gersten, see *Metallurgie,* **9,** 1912, 145 (footnote); *Z. anorg. Chem.,* **88,** 1914, 393.
15. E. Briner and R. Senglet, *J. chim. phys.,* **13,** 1915, 351–375.
16. G. Meyer and F. E. C. Scheffer, *Rec. trav. chim.,* **46,** 1927, 1–7; see also F. E. C. Scheffer, T. Dokkum, and J. Al, *Rec. trav. chim.,* **45,** 1926, 803.
17. H. A. Bahr and T. Bahr, *Ber. deut. chem. Ges.,* **61,** 1928, 2177–2183; **63,** 1930, 99–102.
18. J. Schmidt (E. Osswald), *Z. anorg. Chem.,* **216,** 1933, 85–98.
19. B. Jacobson and A. Westgren, *Z. physik. Chem.,* **B20,** 1933, 361–367.
20. R. Kohlhaas and W. F. Meyer, *Metallwirtschaft,* **17,** 1938, 786–790.
21. H. J. Goldschmidt, *J. Iron Steel Inst.,* **160,** 1948, 347.
22. "Metals Reference Book," p. 216, Butterworth & Co. (Publishers) Ltd., London, 1949.
23. W. Epprecht, *Chimia,* **5,** 1951, 49–60.
24. L. J. E. Hofer, E. M. Cohn, and W. C. Peebles, *J. Phys. & Colloid Chem.,* **54,** 1950, 1161–1169.
25. R. Bernier, *Ann. chim. (Paris),* **6,** 1951, 104–161.
26. L. C. Browning and P. H. Emmett, *J. Am. Chem. Soc.,* **74,** 1952, 1680–1682.
27. G. Meyer and F. E. C. Scheffer, *J. Am. Chem. Soc.,* **75,** 1953, 486.
28. H. Schirrmacher, Dissertation, 1936, Wilhelms Universität Münster, Westfalen; quoted in [29].
29. J. A. Tebboth, *J. Soc. Chem. Ind.,* **67,** 1948, 65.
30. E. Scheil, *Arch. Eisenhüttenw.,* **25,** 1954, 71–76.
31. S. Oketani et al., *Nippon Kinzoku Gakkai-Shi,* **18,** 1954, 325–328, 329–332 (in Japanese); *Met. Abstr.,* **23,** 1956, 444.

$\bar{2}.7048$
1.2952

C-Np Carbon-Neptunium

The crystal structure of NpC (about 4.8 wt. % C) was found to be of the NaCl (B1) type, $a = 5.004$ A [1].

[2] reported the preparation and identification of NpC_2 (isomorphous with the corresponding uranium compound), Np_2C_3, and NpC.

1. D. H. Templeton and C. H. Dauben, *U.S. Atomic Energy Comm., Publ.* AECD-3443 (also UCRL-1886), 1952.
2. I. Sheft and S. Fried, *J. Am. Chem. Soc.,* **75,** 1953, 1236–1237. "In each case the identity of the compound was established from an analysis of the X-ray diffraction pattern by Prof. W. H. Zachariasen of the Physics Department of the University of Chicago."

$\bar{2}.8003$
1.1997

C-Os Carbon-Osmium

See C-Ir, page 366.

$\bar{2}.7631$
1.2369

C-Pb Carbon-Lead

According to [1], molten lead dissolves 0.024, 0.046, and 0.094 wt. % C at 1170, 1415, and 1555°C (boiling point), respectively. On cooling, the carbon precipitates in the form of graphite.

1. O. Ruff and B. Bergdahl, *Z. anorg. Chem.*, **106**, 1919, 91.

$\bar{1}.0514$
0.9486

C-Pd Carbon-Palladium

See C-Ir, page 366.

$\bar{2}.7890$
1.2110

C-Pt Carbon-Platinum

Like all other metals of the platinum group (see C-Ir), molten Pt dissolves carbon which precipitates on cooling in the form of graphite. Carbon contents of 1.2 wt. % [1] and 1.45 wt. % [2] were found to be dissolved. According to [3], the freezing point of Pt was found to be reduced to 1734°C when melted in a graphite crucible. The metal contained 1.2 wt. % C, and the structure showed primary graphite in a eutectic-like matrix. As the lattice parameter was slightly larger, the solid solubility of C was estimated to be about 0.25 wt. (about 4 at.) % C [3]. At 800–900°C no diffusion of C in Pt was observed when the metal was treated with carbon-producing vapors [4].

Supplement. [5] made an electron-diffraction study of a Pt carbide produced by alternate deposition of C and Pt on a thin collodion film and heating this preparation in vacuo for several hours at 1100°C. The final film was 100 A thick. The diffraction patterns showed—in addition to the lines of a f.c.c. Pt lattice shrunk by 1%—the primitive cubic superlattice lines. It appears that C replaces a number of Pt atoms in the lattice.

1. W. Hempel, *Z. angew. Chem.*, **17**, 1904, 321–323.
2. H. Moissan, *Compt. rend.*, **116**, 1893, 608–611; **142**, 1906, 189–195.
3. L. J. Collier, T. H. Harrison, and W. G. A. Taylor, *Trans. Faraday Soc.*, **30**, 1934, 581–587.
4. G. Tammann and K. Schönert, *Z. anorg. Chem.*, **122**, 1922, 28–29.
5. H. König, *Naturwissenschaften*, **38**, 1951, 154–155; *Chem. Abstr.*, **46**, 1952, 1837.

$\bar{2}.6957$
1.3043

C-Pu Carbon-Plutonium

The existence of PuC (about 4.7 wt. % C) and Pu_2C_3 (about 6.9 wt. % C) has been established by determining the crystal structures. PuC is f.c.c. of the NaCl (B1) type, a = 4.920 A [1]; Pu_2C_3 is cubic with 8 molecules per unit cell, a = 8.145 A [2], and isotypic with U_2C_3.

1. W. H. Zachariasen, in "The Transuranium Elements," Part II, National Nuclear Energy Series, Div. IV, vol. 14B, pp. 1448–1450, McGraw-Hill Book Company, Inc., New York, 1949; also W. H. Zachariasen, *Acta Cryst.*, **2**, 1949, 388–390.
2. W. H. Zachariasen, *Acta Cryst.*, **5**, 1952, 17–19.

$\bar{2}.8093$
1.1907

C-Re Carbon-Rhenium

Rhenium powder, heated in methane at 800–2200°C, absorbs about 0.9 wt. % C, as indicated by an increase in the lattice spacing. In carbon monoxide, coarse-grained Re powder also absorbs about 1 wt. % C at 450–1100°C, with the increase in lattice spacing being greater; on heating at higher temperatures, however, this effect disappears. Activated Re forms a carbide at 470–600°C in carbon monoxide; this decomposes into C and the saturated solid solution at above 1600°C [1]. See also [2].

1. W. Trzebiatowski, *Z. anorg. Chem.*, **233,** 1937, 376–384.
2. R. Schenck, F. Kurzen, and H. Wesselkock, *Z. anorg. Chem.*, **203,** 1931, 159–187.

$\bar{1}.0671$
0.9329

C-Rh Carbon-Rhodium

See C-Ir, page 366.

$\bar{1}.0722$
0.9278

C-Ru Carbon-Ruthenium

See C-Ir, page 366.

The preparation of a Ru carbide of unknown composition by reaction of Ru oxide with graphite at about 2500°C has been claimed by [1].

1. J. C. McLennan, J. F. Allen, and J. O. Wilhelm, *Trans. Roy. Soc. Can.*, **25**(3), 1931, 15; abstract in "Gmelins Handbuch der anorganischen Chemie," System No. 63, p. 79, Verlag Chemie, G.m.b.H., Berlin, 1938.

$\bar{2}.9940$
1.0060

C-Sb Carbon-Antimony

According to [1], molten antimony dissolves 0.033, 0.068, and 0.094 wt. % C at 1055, 1265, and 1327°C (boiling point), respectively. On cooling, the carbon precipitates in the form of graphite.

1. O. Ruff and B. Bergdahl, *Z. anorg. Chem.*, **106,** 1919, 91.

$\bar{1}.4254$
0.5746

C-Sc Carbon-Scandium

A scandium carbide, the more probable formula of which was proposed to be Sc_4C_3 (16.69 wt. % C), was reported by [1]. According to [2], there is indication of the existence of ScC (21.08 wt. % C), having a structure of the NaCl (B1) type, and probably also of Sc_2C (11.78 wt. % C) with a h.c.p. structure, in analogy with V_2C.

1. E. Friederich and L. Sittig, *Z. anorg. Chem.*, **144,** 1925, 186–187.
2. B. Jacobson and A. Westgren, *Z. physik. Chem.*, **B20,** 1933, 361–363.

$\bar{1}.6310$
0.3690

C-Si Carbon-Silicon

Two possible phase diagrams for the system C-Si have been suggested by [1] (Fig. 214). Indications for a nonvariant reaction involving vapor, CSi, and Si-rich melt were found in analyses of residues of vaporizing experiments [1]. The decomposition temperature of CSi lies at about 2700°C [2, 1]. X-ray and optical investigations by [1] demonstrated that only one compound, CSi (70.05 wt. % Si), exists and that neither this compound nor Si forms solid solutions of measurable extent.

Carborundum, Silicon Carbide, CSi. A digest of the chemical, physical, structural, and mechanical characteristics of silicon carbide, based on a study of the widely dispersed literature, has recently been published by [3].

CSi, in addition to its strength at high temperatures, refractoriness, and high thermal conductivity, has a low coefficient of linear thermal expansion and a high degree of chemical stability. However, it is attacked by many basic oxides and slags as well as some gases and molten metals (Fe, Co, Ni, Cr, Pt).

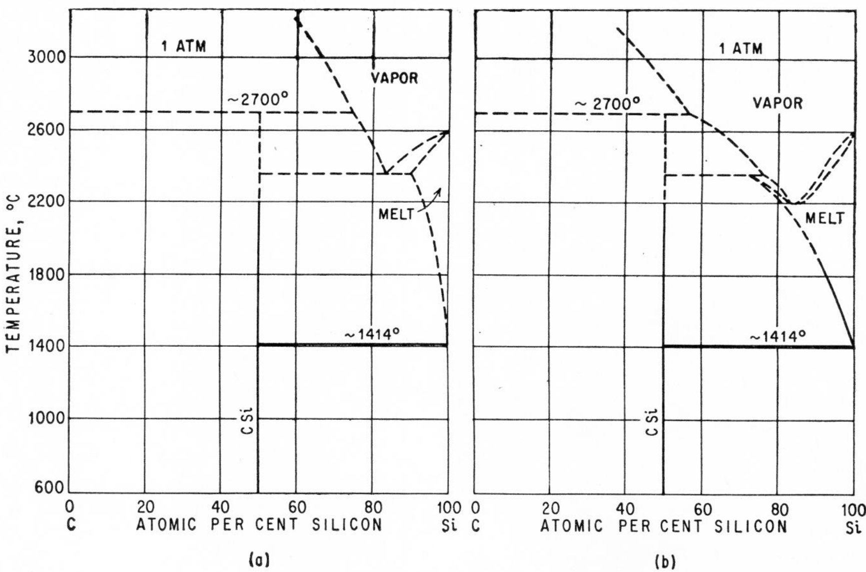

Fig. 214. C-Si

CSi is a unique compound as far as its great number of polymorphic forms (not considered in Fig. 214) is concerned. [4] stated, "With the discovery of each new polymorph of silicon carbide it becomes more evident that there is no limit to the possible modifications of this substance. Up to the present time, the structures of 15 types have been established." Literature on crystal structure: [5, 6–13, 4].

1. H. Nowotny, E. Parthé, R. Kieffer, and F. Benesovsky, *Monatsh. Chem.*, **85**, 1954, 255–272.
2. O. Ruff, *Trans. Electrochem. Soc.*, **68**, 1935, 87–109.
3. C. G. Harman and W. G. Mixer, *U.S. Atomic Energy Comm., Publ.* BMI-748, 1952.
4. R. S. Mitchell, *J. Chem. Phys.*, **22**, 1954, 1977–1983.
5. *Strukturbericht*, **1**, 1913-1928, 144–146; **2**, 1928-1932, 238; **3**, 1933-1935, 261. *Structure Repts.*, **10**, 1945-1946, 87–88; **11**, 1947-1948, 226–235; **12**, 1949, 137–138; **13**, 1950, 166–167.
6. L. S. Ramsdell and J. A. Kohn, *Acta Cryst.*, **4**, 1951, 75.
7. L. S. Ramsdell and J. A. Kohn, *Acta Cryst.*, **4**, 1951, 111–113.
8. E. B. Gasilova, M. S. Beletskii, and M. I. Sokhor, *Doklady Akad. Nauk S.S.S.R.*, **82**, 1952, 57–60.
9. E. B. Gasilova and M. I. Sokhor, *Doklady Akad. Nauk S.S.S.R.*, **82**, 1952, 249–251.

10. L. S. Ramsdell and J. A. Kohn, *Acta Cryst.*, **5**, 1952, 215–224.
11. H. N. Baumann, *J. Electrochem. Soc.*, **99**, 1952, 109–114.
12. L. S. Ramsdell and R. S. Mitchell, *Am. Mineral.*, **38**, 1953, 56–59.
13. H. Jagodzinski, *Acta Cryst.*, **7**, 1954, 300.

$\bar{1}.0051$
0.9949

C-Sn Carbon-Tin

See C-Cd, page 348.

$\bar{1}.1369$
0.8631

C-Sr Carbon-Strontium

SrC_2 (21.51 wt. % C) exists in three different modifications: (*a*) a high-temperature form, stable above 370°C, cubic with $a = 6.25$ A [1]; (*b*) a f.c. tetragonal structure (C11 type), stable between −30 and 370°C, $a = 5.82$ A, $c = 6.69$ A, $c/a = 1.150$ [1, 2]; (*c*) a modification of lower symmetry stable below −30°C [1].

1. M. A. Bredig, *J. Phys. Chem.*, **46**, 1942, 801–819.
2. M. v. Stackelberg, *Z. physik. Chem.*, **B9**, 1930, 437–475; *Z. Elektrochem.*, **37**, 1931, 542–545.

$\bar{2}.8222$
1.1778

C-Ta Carbon-Tantalum

The existence of TaC (6.23 wt. % C) is well established [1–9], especially since the crystal structure was determined [10, 11]. The less stable carbide Ta_2C (3.21 wt. % C) was first reported by [8] and confirmed by [9]. The "tentative" phase diagram presented in Fig. 215 was constructed from metallographic, X-ray, and melting-point data [9]. Earlier, the melting point of TaC was reported to be 3730–3830°C [2] and 3880 ± 150°C [12]. Ta_2C and TaC are said to be able to dissolve 0.2 wt. (about 1.6 at.) % Ta and 1.2 wt. (about 6 at.) % Ta, respectively [9].

The crystal structure of TaC is cubic of the NaCl (B1) type [10, 11, 13, 8, 14–19]; the most probable lattice parameter appears to be $a = 4.45_{4-7}$ A [8, 15, 19]. Ta_2C is hexagonal of the L′3 (interstitial A3) type [8, 9], with $a = 3.097$ A, $c = 4.94$ A, $c/a = 1.59$ [8]. According to [8], Ta_2C, like W_2C, occurs in two modifications.

Supplement. [20] determined by means of parametric work the extent of the (TaC) phase as 40–49.7 at. % C. As regards the (Ta_2C) phase, it was found that it contains slightly less carbon than the stoichiometric content and has a very small composition range (no shift in the high-angle diffraction lines could be observed).

[21] examined 16 Ta-C alloys by means of chemical and X-ray analyses and reported the following results: (*a*) Solid Ta dissolves about 3 at. % C; the lattice parameter of b.c.c. Ta decreases from $a = 3.306_5$ to 3.303_5 A by addition of C. (*b*) The h.c.p. (Ta_2C) phase extends from $TaC_{0.38}$ to $TaC_{0.50}$ (27.5–33.3 at. % C), $a = 3.101_2$–3.104_2 A, $c = 4.937_2$–4.941_0 A. (*c*) (TaC) with NaCl-type structure is homogeneous between $TaC_{0.58}$ and $TaC_{0.91}$ (36.7–47.6 at. % C), $a = 4.420_6$–4.456_4 A.

No higher carbide was found. [22] found the melting point of TaC to be 3540°C.

1. A. Joly, *Ann. sci. école norm.*, **6**, 1877, 148.
2. E. Friederich and L. Sittig, *Z. anorg. Chem.*, **144**, 1925, 174–181.
3. A. E. van Arkel and J. H. de Boer, *Z. anorg. Chem.*, **148**, 1925, 347–348.
4. K. Becker and H. Ewest, *Z. tech. Physik*, **11**, 1930, 148–150, 216–220.
5. C. Agte and K. Moers, *Z. anorg. Chem.*, **198**, 1931, 236–238.
6. K. Moers, *Z. anorg. Chem.*, **198**, 1931, 252.

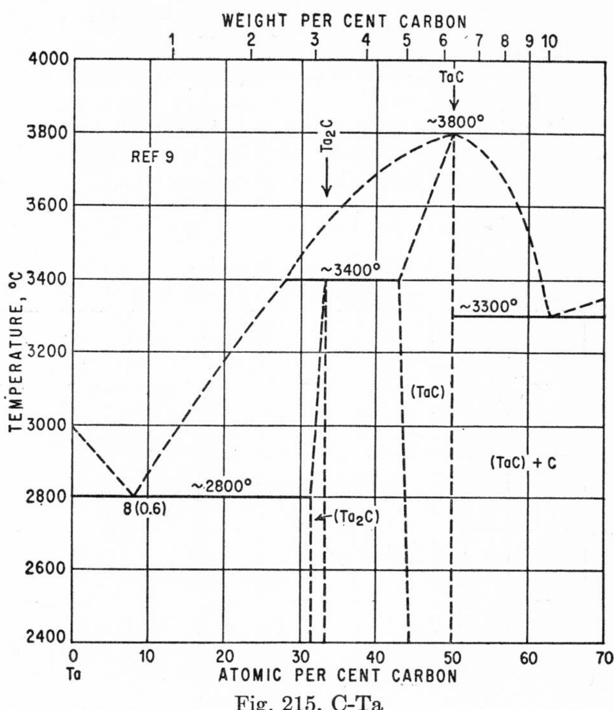

Fig. 215. C-Ta

7. F. C. Kelley, *Trans. ASST*, **19**, 1932, 233–246.
8. W. G. Burgers and J. C. M. Basart, *Z. anorg. Chem.*, **216**, 1934, 209–222.
9. F. H. Ellinger, *Trans. ASM*, **31**, 1943, 89–102.
10. A. E. van Arkel, *Physica*, **4**, 1924, 286–301.
11. K. Becker and F. Ebert, *Z. Physik*, **31**, 1925, 268–272.
12. C. Agte and H. Alterthum, *Z. tech. Physik*, **11**, 1930, 185.
13. M. v. Schwarz and O. Summa, *Metallwirtschaft*, **12**, 1933, 298.
14. L. P. Molkov and A. V. Chochlova, *Redkie Metal.*, 1935, no. 1, 24–30.
15. P. M. McKenna, *Ind. Eng. Chem.*, **28**, 1936, 767–772.
16. A. E. Kovalski and J. S. Umanski, *Zhur. Fiz. Khim.*, **20**, 1946, 769–772.
17. H. Krainer and K. Konopicky, *Berg. hüttenmänn. Monatsh.*, **92**, 1947, 166–178.
18. H. Nowotny and R. Kieffer, *Z. Metallkunde*, **38**, 1947, 257–265.
19. J. T. Norton and A. L. Mowry, *Trans. AIME*, **185**, 1949, 133–136.
20. J. G. McMullin and J. T. Norton, *Trans. AIME*, **197**, 1953, 1205–1208.
21. V. I. Smirnova and B. F. Ormont, *Doklady Akad. Nauk S.S.S.R.*, **96**, 1954, 557–560; *Met. Abstr.*, **22**, 1955, 533.
22. G. A. Geach and F. O. Jones, *Plansee Proc.*, **1955**, 80–91; *Met. Abstr.*, **24**, 1957, 366.

$\bar{2}.7138$
1.2862

C-Th Carbon-Thorium

The existence of ThC_2 (9.38 wt. % C), first prepared by [1, 2], was established by determining the crystal structure [3]. ThC (4.92 wt. % C) was found by [4] and confirmed by [5].

[6] have presented a tentative phase diagram that is largely based on metallographic and melting-point data; see also [5]. According to this diagram it is assumed that a continuous series of solid solutions crystallizes from melts between Th and ThC_2. Further, it is assumed that at some lower temperature these solid solutions decompose into a Th-rich phase and a phase based on ThC, on the one hand, and into two phases nearly corresponding to the compositions ThC and ThC_2, respectively, on the other. Accordingly, the tentative diagram shows two miscibility gaps with critical points at approximately 2 wt. (about 29 at.) % C, about 1975°C, and between about 6 and 7 wt. (about 55–60 at.) % C.

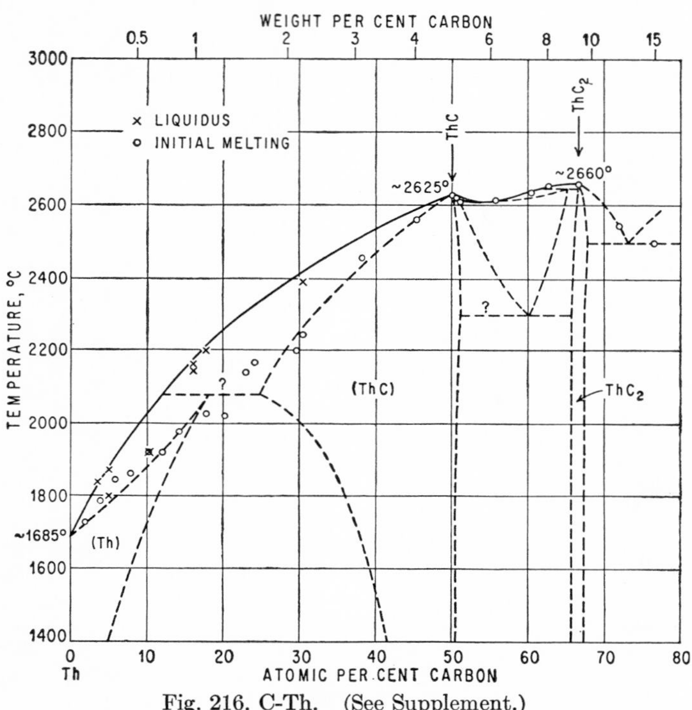

Fig. 216. C-Th. (See Supplement.)

It is not conceivable that such a constitution exists since a continuous transition from the structure of Th via that of ThC to that of ThC_2 cannot occur. With the limited information presented in two papers [5, 6], the diagram shown in Fig. 216 has been constructed. This tentative diagram, of course, should be regarded as only a rough approximation and is presented here to give a conception of what the phase relationships could be. Although there is no evidence as yet as to the existence of the carbide Th_2C_3 (7.20 wt. % C), this phase may exist, in analogy with U_2C_3.

Crystal Structures. The (ThC) phase has the structure of the NaCl (B1) type, with a = 5.34 A [6]. As to earlier parameter data and the variation of the parameter with composition, see [6]. The structure of ThC_2 was first reported by [3] to be f.c. tetragonal of the CaC_2 type; however, [4] showed this to be incorrect, suggesting an orthorhombic structure. [7] established, by X-ray and neutron diffraction, a monoclinic structure, with a = 6.53 A, b = 4.24 A, c = 6.56 A, β = 104°, and 4 molecules per unit cell.

Supplement. More recent studies [8] have shown that C lowers the melting point and raises the newly detected allotropic transition (1380 ± 20°C) of Th. This results in a peritectic reaction β-Th (~2 at. % C) + melt (~4.5 at. % C) ⇄ α-Th (~3.5 at. % C) about 50°C below the melting point of Th.

New values of the solubility of C in solid Th have been reported [9]: 0.91, 0.57, 0.43, and 0.35 wt. (15.1, 10.0, 7.7, and 6.4 at.) % C at 1215, 1018, 800, and "20°C."

1. L. Troost, *Compt. rend.*, **116,** 1893, 1227–1230.
2. H. Moissan and A. Étard, *Compt. rend.*, **122,** 1896, 573–577; *Ann. chim. et phys.*, **12,** 1897, 427–432.
3. M. v. Stackelberg, *Z. physik. Chem.*, **B9,** 1930, 437–475; *Z. Elektrochem.*, **37,** 1931, 542–545.
4. N. C. Baenziger and D. Trieck, unpublished (1945); quoted by [6].
5. H. A. Wilhelm, P. Chiotti, A. I. Snow, and A. H. Daane, *J. Chem. Soc., Suppl. Issue* 2, 1949, pp. S318–321.
6. H. A. Wilhelm and P. Chiotti, *Trans. ASM,* **42,** 1950, 1295–1310.
7. E. B. Hunt and R. E. Rundle, *J. Am. Chem. Soc.*, **73,** 1951, 4777–4781.
8. P. Chiotti (unpublished), quoted in Nuclear Metallurgy, *AIME, IMD Special Report Series*, no. 1, p. 55, 1955.
9. D. Peterson and R. Mickelson (unpublished information, February, 1954), quoted by H. A. Saller and F. A. Rough, *U.S. Atomic Energy Comm., Publ.* BMI-1000, June 1, 1955.

$\bar{1}.3992$
0.6008

C-Ti Carbon-Titanium

The phase diagram in Fig. 217 is essentially based on micrographic and melting-point data of arc-melted alloys prepared by using iodide Ti [1]. The monocarbide (20.05 wt. % C) [2], which is of variable composition, is the only intermediate phase occurring. Ti_2C (11.14 wt. % C), suspected by [3], could not be found [4, 1]. Apparently there is no solid solubility of C in TiC [4]. The melting point of TiC was reported to be 3160 [5], 3140 [6], 3030 [41], and 3250°C [7]. TiC and carbon form a eutectic [8, 9] close to 30 wt. (63 at.) % C and above 2400°C [8].

Prior to the work by [1] a systematic X-ray analysis was carried out using alloys prepared by sintering mixtures of Ti and TiC [4]. Lattice-parameter measurements indicated a rather high solubility [about 7.5 at. (2.0 wt.) % C] of C in α- Ti. This was not confirmed by others [10, 1, 11]. The phase (TiC) was located between 22 and 50 at. (6.5 and 20 wt.) % C (temperature?), whereas [1] found the TiC/(TiC + β_{Ti}) boundary as given in Fig. 217. [12, 13] reported the lower boundary of (TiC) to lie at about 6.8 wt. (22.5 at.) % C, according to parametric work. Although this is in good agreement with the value by [4], preference should be given to the data by [1] as these were found by bracketing the boundary at various temperatures.

The solid solubility of C in α-Ti, reported by [10] to lie between 0.3 and 0.5 wt. % C at the peritectoid temperature (900–920°C), was determined parametrically and micrographically. It is about 0.48, 0.27, and 0.12 wt. (1.8, 1.1, and 0.45 at.) % C at 920, 800, and 600°C, respectively [1]. The solubility of C in β-Ti, determined micrographically, was found as follows (in wt. %): 920°C, about 0.15; 1400°C, about 0.27; and 1750°C, about 0.8 (0.6, 1.1, and 3.1 at. %).

Crystal Structures. The structure of TiC is of the NaCl (B1) type [14]. The lattice parameter has been determined many times [14–27, 4, 28–30, 12, 31, 32]; that of the stoichiometric composition can be assumed to be a = 4.329 ± 0.001 A [16, 18, 22, 26, 27, 28, 30]. For the change of the parameter with composition, see [4, 22, 33].

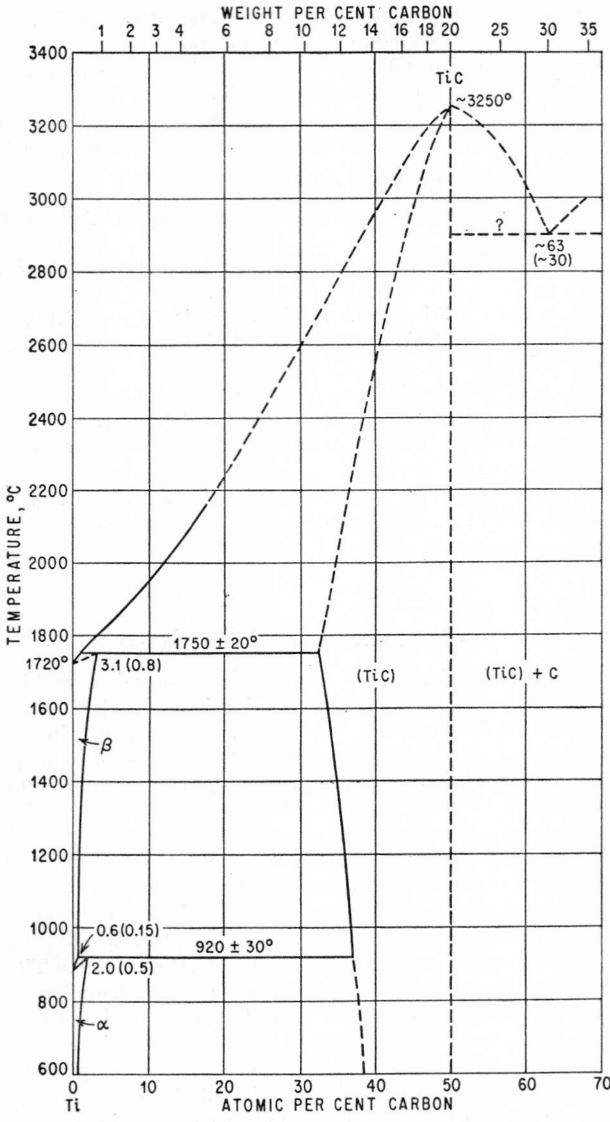

Fig. 217. C-Ti. (See also Supplement.)

The lattice spacing of the α-Ti solid solution was determined by [4, 34, 1]; data by [34] and [1] agree very well [35].

Supplement. According to [36], "there seems to be no reason to doubt that the true upper limit of the (TiC) phase is at 50 at. %"; see also [37]. The eutectic between TiC and C was observed further by [37–39], with the data >2850°C [37], ~30 wt. % C [39] given. The results of microscopic work by [39] on low-carbon alloys—especially up to 1 wt. % C—are in keeping with Fig. 217; chill-cast

13 and 15 wt. (37.3 and 41.3 at.) % C alloys were found to be two-phase and one-phase, respectively.

For a study of how the microstructure and the mechanical properties of Ti-C alloys (with up to 0.47 wt. % C) are affected by various heat-treatments, see [40]. [42] studied the rate of diffusion of carbon in α- and β-Ti and determined, by chemical analysis of slices of the diffusion samples having the composition of the solubility limit, the solubility between 1150 and 736°C. Below 1050°C the results are in complete agreement with those given by [1]. Deviations from the curve of [1] between 1050 and 1150°C toward higher solubilities are assumed to be caused by oxygen contamination of the diffusion samples.

1. I. Cadoff and J. P. Nielsen, *Trans. AIME,* **197,** 1953, 248–252.
2. As to the extensive literature on the preparation and properties of TiC, see P. Schwarzkopf and R. Kieffer, "Refractory Hard Metals," pp. 67–88, The Macmillan Company, New York, 1953.
3. B. Jacobson and A. Westgren, *Z. physik. Chem.,* **B20,** 1933, 362.
4. P. Ehrlich, *Z. anorg. Chem.,* **259,** 1949, 1–41.
5. E. Friederich and L. Sittig, *Z. anorg. Chem.,* **144,** 1925, 171.
6. C. Agte and K. Moers, *Z. anorg. Chem.,* **198,** 1931, 233.
7. Unpublished work, quoted from P. Schwarzkopf and R. Kieffer, see [2], p. 87.
8. Unpublished work, Armour Research Foundation, 1952.
9. R. Kieffer, unpublished work, quoted from P. Schwarzkopf and R. Kieffer, see [2], p. 84.
10. R. I. Jaffee, H. R. Ogden, and D. J. Maykuth, *Trans. AIME,* **188,** 1950, 1261–1266.
11. R. M. Goldhoff, H. L. Shaw, C. M. Craighead, and R. I. Jaffee, *Trans. ASM,* **45,** 1953, 953.
12. D. V. Ragone, B.S. thesis, Department of Metallurgy, Massachusetts Institute of Technology, Cambridge, Mass., 1951.
13. G. P. W. Rengstorff, M.S. thesis, Department of Metallurgy, Massachusetts Institute of Technology, Cambridge, Mass., 1947.
14. A. E. van Arkel, *Physica,* **4,** 1924, 286–301.
15. K. Becker and F. Ebert, *Z. Physik,* **31,** 1925, 268–272.
16. L. R. Brantley, *Z. Krist.,* **77,** 1931, 505–506.
17. M. v. Schwarz and O. Summa, *Z. Elektrochem.,* **38,** 1932, 743–744.
18. W. Burgers and J. C. M. Basart, *Z. anorg. Chem.,* **216,** 1934, 209–222.
19. G. E. Meerson, *Redkie Metal.,* **4,** 1935, 6.
20. W. Hofmann and A. Schrader, *Arch. Eisenhüttenw.,* **10,** 1936-1937, 65–66.
21. W. Dawihl and W. Rix, *Z. anorg. Chem.,* **244,** 1940, 191–197.
22. J. S. Umanski and S. S. Khidekel, *Zhur. Fiz. Khim.,* **15,** 1941, 983–996.
23. W. Hume-Rothery, G. V. Raynor, and A. T. Little, *J. Iron Steel Inst.,* **145,** 1942, 129–139.
24. A. E. Kovalski and J. S. Umanski, *Zhur. Fiz. Khim.,* **20,** 1946, 769.
25. H. Nowotny and R. Kieffer, *Z. Metallkunde,* **38,** 1947, 257–265.
26. H. Krainer and K. Konopicky, *Berg. hüttenmänn. Monatsh.,* **92,** 1947, 166–178.
27. A. G. Metcalfe, *J. Inst. Metals,* **73,** 1947, 591–607.
28. J. T. Norton and A. L. Mowry, *Trans. AIME,* **185,** 1949, 133–136.
29. H. J. Goldschmidt, *J. Iron Steel Inst.,* **163,** 1949, 384; *Metallurgia,* **40,** 1949, 103–104.
30. P. Duwez and F. Odell, *J. Electrochem. Soc.,* **97,** 1950, 299–304.
31. H. J. Beattie and F. L. VerSnyder, *Trans. ASM,* **45,** 1953, 406.
32. A. Münster and K. Sagel, *Z. Elektrochem.,* **57,** 1953, 571–579.

33. I. Cadoff and J. P. Nielsen, see WADC Technical Report 53–41, 1953, p. 29.
34. W. L. Finlay and J. A. Snyder (H. T. Clark), *Trans. AIME*, **188**, 1950, 277–286.
35. Data by [34] have to be replotted from hardness-vs.-parameter and hardness-vs.-composition curves.
36. J. G. McMullin and J. T. Norton, *Trans. AIME*, **197**, 1953, 1205–1208.
37. H. Nowotny, E. Parthé, R. Kieffer, and F. Benesovsky, *Z. Metallkunde*, **45**, 1954, 97–101.
38. E. Stover, quoted in [36].
39. R. J. Van Thyne and H. D. Kessler, *Trans. AIME*, **200**, 1954, 197–198.
40. H. R. Ogden, R. I. Jaffee, and F. C. Holden, *Trans. AIME*, **203**, 1955, 73–80.
41. G. A. Geach and F. O. Jones, *Plansee Proc.*, **1955**, 80–91; *Met. Abstr.*, **24**, 1957, 366.
42. F. C. Wagner, E. J. Bucur, and M. A. Steinberg, *Trans. ASM*, **48**, 1956, 742–761.

$\bar{2}.7028$
1.2972

C-U Carbon-Uranium

The first carbide reported to exist was U_2C_3 (7.04 wt. % C) [1]. However, later investigators [2–4] showed this phase to be more likely UC_2 (9.17 wt. % C) until [5] definitely established the existence of UC_2 by X-ray study; see also [6–9]. Later, UC (4.80 wt. % C) [6, 7, 9] and U_2C_3 [10] were also shown to exist. U_2C_3 was first suspected to be stable only at high temperatures, above 2000°C [6, 9]; but [10] have given evidence that it is formed by the peritectoid reaction $UC + UC_2 \rightarrow U_2C_3$ below 1800°C. UC was confirmed by [11, 12] and U_2C_3 by [12]. The results by [8] which indicated a wide range of homogeneity of UC_2 down to at least 26 at. % C, eliminating the existence of UC, could not be confirmed [13]. There is ample evidence that no carbides lower than UC and higher than UC_2 exist and some evidence that C is soluble in UC_2 at high temperatures [6, 9].

By means of careful micrographic analysis and melting-point determinations (chemical analysis of melts saturated with C), a phase diagram was constructed by [12]. Figure 218 is based on these data; however, the range between UC and UC_2 at temperatures above 1800°C has been somewhat modified. On the basis of the rather convincing evidence offered [12], the authors proposed that UC and UC_2 are completely soluble in each other at temperatures above about 2000°C. Their original diagram shows, therefore, a two-phase field (UC + UC_2) with a critical point at approximately 56 at. % C and 2225°C. Since this seems unlikely (although the structures of UC and UC_2 have certain similarities), their diagram was modified as presented in Fig. 218. As already suggested earlier [14], it shows the existence of an intermediate phase between UC and UC_2, having a wide range of homogeneity, which decomposes eutectoidally. At some lower temperature, 1775°C, UC and UC_2 react to form U_2C_3. A similar constitution occurs in other systems, e.g., the system Ag-Al.

[12] stated that a continuous transition between the phases UC and UC_2 is "a rather startling phenomenon, in view of the difference between the structures of the two phases." They referred to the phase diagram of the system Mn-Ni established by [15] as an example where complete miscibility exists between f.c.c. nickel and f.c. tetragonal γ-Mn. Since it is known that γ-Mn is also a f.c.c. phase which becomes tetragonally distorted only on quenching [16], the constitution of the system Mn-Ni cannot be offered as a support for the constitution of the partial system UC-UC_2 proposed by [12].

It may be that the partial diagram U-UC (Fig. 218) would also need some modification if evidence of solid solubility of U in UC is found later. In that case, phase relationships may be similar to those in the partial system Th-ThC (Fig. 216).

Crystal Structures. UC has the cubic NaCl (B1) type of structure, with $a = 4.961$ A [6], $a = 4.965$ A [7]. U_2C_3 is b.c.c., 8 molecules per unit cell, $a = 8.088$ A [10]. The structure of UC_2 is of the CaC_2 (C11a) type, b.c. tetragonal, with $a = 3.524$ A, $c = 5.999$ A, and $2UC_2$ per unit cell [6]. The structure may also be described as a f.c. tetragonal cell with $4UC_2$ and $a = 4.96$ A, $c = 5.95$ A [12].

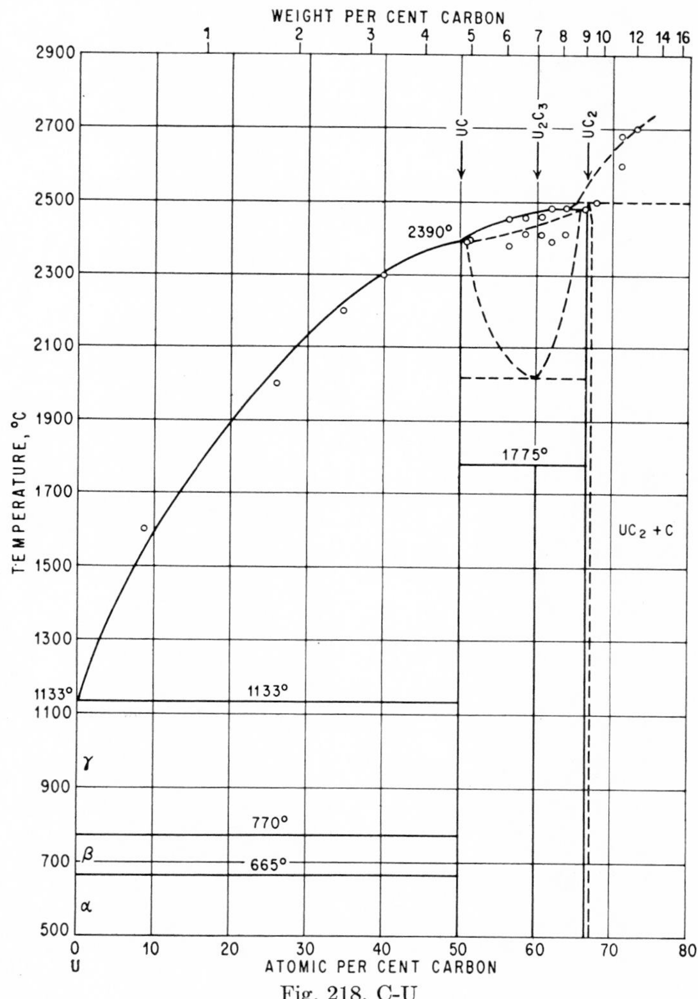

Fig. 218. C-U

Supplement. The following data were reported recently: melting point of UC, 2590 ± 50°C (initial melting, optical pyrometer) [17]; lattice parameter of UC, $a = 4.951$ A [18], 4.952 A [19]. The room-temperature stability of U_2C_3 was verified by [20].

1. H. Moissan, *Compt. rend.*, **116,** 1893, 347, 1433; **122,** 1896, 274–280.
2. P. Lebeau, *Compt. rend.*, **152,** 1911, 955–958.

3. O. Ruff and A. Heinzelmann, *Z. anorg. Chem.*, **72,** 1911, 72–73.
4. O. Heusler, *Z. anorg. Chem.*, **154,** 1926, 353.
5. G. Hägg (H. Arnfelt), *Z. physik. Chem.*, **B12,** 1931, 33.
6. R. E. Rundle, N. C. Baenziger, A. S. Wilson, and R. A. McDonald, *J. Am. Chem. Soc.*, **70,** 1948, 99–105.
7. L. M. Litz, A. B. Garrett, and F. C. Croxton, *J. Am. Chem. Soc.*, **70,** 1948, 1718–1722.
8. U. Esch and A. Schneider, *Z. anorg. Chem.*, **257,** 1948, 254–266.
9. H. A. Wilhelm, P. Chiotti, A. I. Snow, and A. H. Daane, *J. Chem. Soc. Suppl. Issue* 2, 1949, pp. S318–321.
10. W. Mallet, A. F. Gerds, and D. A. Vaughan, *J. Electrochem. Soc.*, **98,** 1951, 505-509.
11. P. Chiotti, *J. Am. Ceram. Soc.*, **35,** 1952, 123.
12. W. Mallet, A. F. Gerds, and H. R. Nelson, *J. Electrochem. Soc.*, **99,** 1952, 197–204.
13. Their preparations contained up to about 9 at. % O and 9 at. % N.
14. See J. J. Katz and E. Rabinowitch, "The Chemistry of Uranium," Part I, National Nuclear Energy Series, Div. VIII, vol. 5, pp. 215–226, McGraw-Hill Book Company, Inc., New York, 1951.
15. W. Köster and W. Rauscher, *Z. Metallkunde*, **39,** 1948, 178–184.
16. U. Zwicker, *Z. Metallkunde*, **42,** 1951, 246–252, 327–330.
17. P. Chiotti, *J. Am. Ceram. Soc.*, **35,** 1952, 123–130.
18. J. H. Carter and A. H. Daane, U.S. Patent 2,569,225; *Chem. Abstr.*, **46,** 1952, 405.
19. G. F. Hardy and J. K. Hulm, *Phys. Rev.*, **93,** 1954, 1010.
20. M. D. Burdick et al., *J. Research Natl. Bur. Standards*, **54,** 1955, 217–229; *Met. Abstr.*, **23,** 1956, 728.

$\bar{1}.3724$
0.6276

C-V Carbon-Vanadium

The existence of VC (19.08 wt. % C), first reported by [1] and confirmed by [2, 3], was definitely established by determining the crystal structure, which is of the NaCl (B1) type [4]. Many attempts have been made to determine the composition of vanadium carbide by isolating from vanadium-alloyed steels [5–9, 11, 12] and cast iron [10]. Besides the questionable formula V_2C_3 [5], the composition V_4C_3 [42.86 at. (15.02 wt.) % C] has been found repeatedly [6–12]. However, as the structures of the compositions V_4C_3 and VC are identical [11] and of the NaCl type, it appears justified to assume that the phase VC has a wide range of homogeneity, the lower limit of which may correspond approximately to the composition V_4C_3 (subtraction-type solid solution) [11], and that in vanadium steels only the unsaturated carbide of the composition V_4C_3 occurs; see also [13–15].

The first systematic investigation of the system [9], by X-ray and microscopic methods, showed that two intermediate phases exist in the range of composition studied (6–44.5 at. % C). They were assumed to have the compositions V_5C [16.67 at. (4.50 wt.) % C] and V_4C_3. V_5C (?) was reported to have a h.c.p. structure and V_4C_3 a f.c.c. structure similar to that found for VC [4]. The solid solubility of C in V was reported to be very small, which is in agreement with findings by [16] that C does not diffuse into V at 800–980°C.

The phase V_5C was disputed by [17, 18], who suggested the h.c.p. phase to have the composition V_2C (10.54 wt. % C), in analogy with Mo_2C. They also regard the f.c.c. phase to be based on the composition VC rather than V_4C_3. In addition, [18] has pointed out that, by contamination with oxygen and nitrogen, the composition of the two phases may be shifted to lower C concentrations. VC, VN, and VO are isostructural.

Recently, the existence of V_2C was confirmed by [19]. Arc-melted alloys with 0.4–19 wt. (1.6–50 at.) % C were annealed at 900°C for 1 week and quenched. The *tentative* phase diagram, based on metallographic work, is given in Fig. 219, which indicates a wide range of homogeneity of VC and a eutectic V-V_2C at 3.5 wt. (13 at.) % C, about 1650°C. Melting-point data reported by [2] are also shown in Fig. 219. The melting point of VC was found to be 2750 [2] and 2830°C [3].

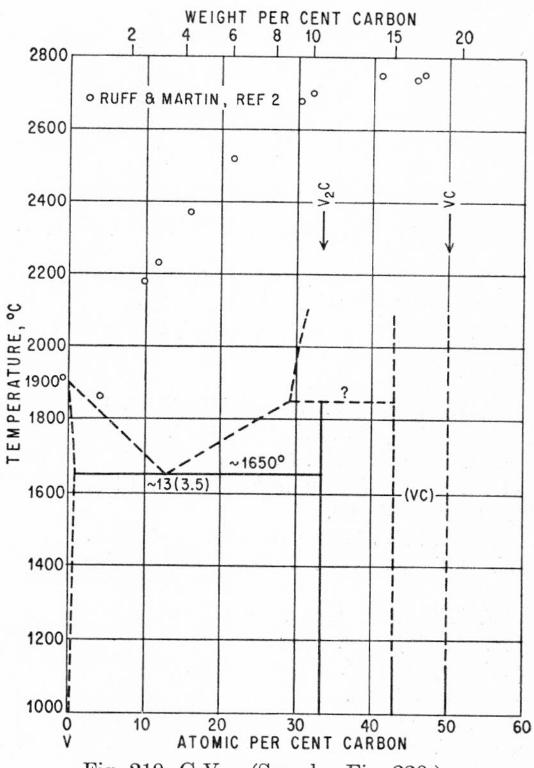

Fig. 219. C-V. (See also Fig. 220.)

Crystal Structures. The f.c.c. NaCl-type structure of the VC phase has been confirmed repeatedly (4, 7, 9–11, 20–25). The most probable lattice parameter of the stoichiometric composition appears to be a = 4.168–4.169 A [24, 23]. The phase V_2C is h.c.p. (L'3 type), with parameters of about $a = 2.87$ A, $c = 4.55$ A, $c/a = 1.59$ [9] and $a = 2.90$ A, $c = 4.51$ A, $c/a = 1.56$ [19].

Supplement. The existence of a hexagonal V_2C phase was confirmed by [26]. [27] studied V-C alloys (prepared in vacuo and slowly cooled down to room temperature) by means of X-ray analysis and reported the following results (see Fig. 220): (*a*) The solid solubility of C in V at about 1000°C is 1 at. (0.2 wt.) % C. (*b*) The (V_2C) phase (L'3 type) is homogeneous between 27 and 33.3 at. % C, a = 2.881–2.906 A, c = 4.547–4.597 A, c/a = 1.578–1.576. (*c*) The phase field of the (VC) phase of NaCl-type structure extends between about 43 and 49 at. % C, a = 4.136–4.182 A; however, "the fact that the formula VC could not be attained may be due to low reaction velocities."

The occurrence of three intermediate phases has been reported by [28]: $VC_{0.42-0.50}$ (29.6–33.3 at. % C), h.c.p.; $VC_{0.5-0.7}$ (33.3–41 at. % C), cubic, a = 4.115–4.130 kX; $VC_{0.7-0.96}$ (41–49 at. % C), cubic, a = 4.150–4.160 kX. Since the investigated alloys contained appreciable amounts of oxygen, these results are of minor importance. It is well established by previous work that only two intermediate phases occur in the binary V-C system.

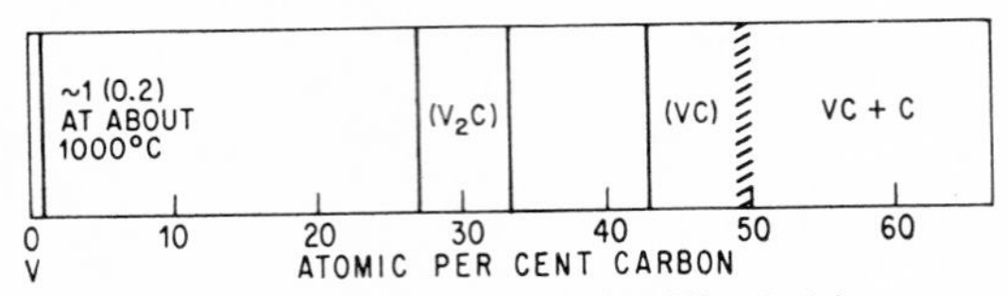

Fig. 220. C-V. (See also Fig. 219.)

1. H. Moissan, *Compt. rend.*, **116**, 1893, 1225; **122**, 1896, 1297; *Z. anorg. Chem.*, **14**, 1897, 174.
2. O. Ruff and W. Martin, *Z. angew. Chem.*, **25**, 1912, 53–56.
3. E. Friederich and L. Sittig, *Z. anorg. Chem.*, **144**, 1925, 173–174.
4. K. Becker and F. Ebert, *Z. Physik*, **31**, **1925**, **268–272**; K. Becker, *Physik. Z.*, **34**, 1933, 185.
5. P. Pütz, *Metallurgie*, **3**, 1906, 651.
6. J. O. Arnold and A. A. Read, *J. Iron Steel Inst.*, **85**, 1912, 219–222.
7. E. Maurer, *Stahl u. Eisen*, **45**, 1925, 1629–1632.
8. H. Krainer and R. Mitsche, *Arch. Eisenhüttenw.*, **20**, 1925, 197–198.
9. A. Osawa and M. Oya, *Kinzoku-no-Kenkyu*, **5**, 1928, 434–442; *Science Repts. Tôhoku Univ.*, **19**, 1930, 95–108.
10. A. Morette, *Bull. soc. chim. France*, **5**, 1938, 1063–1069.
11. E. Maurer, W. Döring, and H. Pulewka, *Arch. Eisenhüttenw.*, **13**, 1939-1940, 337–344.
12. W. Crafts and J. L. Lamont, *Trans. AIME*, **188**, 1950, 561–574.
13. R. Vogel and E. Martin, *Arch. Eisenhüttenw.*, **4**, 1930-1931, 487–495.
14. F. Wever, A. Rose, and H. Eggers, *Mitt. Kaiser-Wilhelm-Inst. Eisenforsch. Düsseldorf*, **18**, 1936, 239–246.
15. W. Bischof, *Arch. Eisenhüttenw.*, **8**, 1934-1935, 255–258.
16. G. Tammann and K. Schönert, *Z. anorg. Chem.*, **122**, 1922, 28–30.
17. A. Westgren, *Metallwirtschaft*, **9**, 1930, 921–924.
18. G. Hägg, *Z. physik. Chem.*, **B12**, 1933, 51 (footnote).
19. W. Rostoker and A. Yamamoto, *Trans. ASM*, **46**, 1954, 1136–1163.
20. W. Dawihl and W. Rix, *Z. anorg. Chem.*, **244**, 1940, 191–197.
21. H. Nowotny and R. Kieffer, *Z. Metallkunde*, **38**, 1947, 257–265.
22. H. Krainer and K. Konopicky, *Berg. hüttenmänn. Monatsh.*, **92**, 1947, 166–178.
23. V. I. Arkharov, I. S. Kvater, and S. T. Kiselev, *Izvest. Akad. Nauk S.S.S.R. (Tekh.)*, **1947**(6), 749–756; see *Structure Repts.*, **11**, 1947-1948, 63.
24. J. T. Norton and A. L. Mowry, *Trans. AIME*, **185**, 1949, 133–136.
25. P. Duwez and F. Odell, *J. Electrochem. Soc.*, **97**, 1950, 299–304.
26. G. F. Hardy and J. K. Hulm, *Phys. Rev.*, **93**, 1954, 1004–1016.
27. N. Schönberg, *Acta Chem. Scand.*, **8**, 1954, 624–626.
28. M. A. Gurevich and B. F. Ormont, *Doklady Akad. Nauk S.S.S.R.*, **96**, 1954, 1165–1168; *Met. Abstr.*, **22**, 1955, 663.

$\bar{2}.8149$
1.1851

C-W Carbon-Wolfram

Information as to the constitution of this system was reviewed by [1] and, more recently, by [2].

The fact that wolfram forms two carbides, W_2C (3.16 wt. % C) and WC (6.13 wt. % C), was first recognized by [3] and firmly established through the work of [4–13] and subsequent investigators, especially [14]. Earlier, the existence of WC had been reported by [15–19] and that of W_2C by [20].

In the first systematic study of the system, [18] believed they found evidence for the existence of W_3C, besides WC and another unidentified carbide between W_3C and WC. Also, [8] discussed the possibility that there could exist a carbide with a carbon content between W_2C and WC. The formation of one or perhaps two unstable carbides (W_5C_2 or W_3C_2?) at about 700°C was suggested by [21], based on carburizing and decarburizing tests.

[6–9] reported a polymorphic transformation of W_2C to occur at about 2400°C. The crystal structure of the high-temperature form β-W_2C was not elucidated. Recently, [22] gave evidence for a f.c.c. W_2C modification being present in wolfram layers prepared by decomposition of wolfram carbonyl. Whether this modification is identical with the β-W_2C phase according to [6–9] remains to be seen. Mechanical treatment changes the unstable β-W_2C into α-W_2C.

In an attempt to establish phase relations, [12] determined the melting points of alloys with carbon contents up to 9 wt. % C. The data points given by this author are shown in Fig. 221. It was concluded that W_2C has a maximal melting point and that WC melts under decomposition. The diagram in Fig. 221 is that suggested by [12] with the exception of the homogeneity range of W_2C which is based on X-ray diffraction studies by [23].

The melting point of W_2C was reported as 2880°C [3], 2860 ± 50°C [24], 2730 ± 15°C [10], 2750 ± 50°C [12], and 2730°C [25]; and that of WC as 2650 ± 50°C [18], 2780°C [3], 2880°C [26], 2870 ± 50°C [24], 2600°C [12], and 2630°C [25]. The eutectic W-W_2C was placed at about 1.4 wt. (17.9 at.) % C, 2690°C [18], and about 1.5 wt. (18.9 at.) % C, 2475° [12]. The eutectic W_2C-WC was reported to be located at about 4.5 wt. (41.9 at.) % C, 2525°C [12], and 4.1 wt. (39.6 at.) % C [27].

Crystal Structures. The low-temperature modification of W_2C is hexagonal with two W and one C atom per unit cell. Lattice constants were determined by [4, 8, 28–30]. The most probable values are a = 2.994 ± 0.002 A, c = 4.724 ± 0.002 A, c/a = 1.578. The high-temperature modification β-W_2C was said to give a similar X-ray diagram as α-W_2C except that many lines are missing [8]. A f.c.c. form of W_2C with a = 4.16 A was reported by [22]; see above. There is also a f.c.c. phase Mo_2N (see Mo-N).

WC has a simple hexagonal structure [4, 8]. Lattice constants were reported by [4, 8, 11, 31, 32, 28, 30, 33–35]. The most probable values are a = 2.906 A, c = 2.837 A, c/a = 0.976. The structure consists of two interpenetrating simple hexagonal lattices of the W and C atoms [4, 36], rather than one similar to the NiAs-type structure [37].

Supplement. [38] found that the c axis of the W_2C lattice decreases with decreasing carbon content whereas a remains practically unchanged. Although the somewhat scattered data do not allow fixing phase boundaries, they show clearly that the stoichiometric composition lies within the (W_2C + WC) field as was claimed by [23] (Fig. 221).

1. M. Hansen, "Der Aufbau der Zweistofflegierungen," pp. 386–392, Springer-Verlag OHG, Berlin, 1936.

2. P. Schwarzkopf and R. Kieffer, "Refractory Hard Metals," pp. 153–161, The Macmillan Company, New York, 1953.
3. M. R. Andrews, *J. Phys. Chem.*, **27**, 1923, 270–283; M. R. Andrews and S. Dushman, *J. Franklin Inst.*, **192**, 1921, 545–546.
4. A. Westgren and G. Phragmén, *Z. anorg. Chem.*, **156**, 1926, 27–36.
5. K. Becker and R. Hölbling, *Z. angew. Chem.*, **40**, 1927, 512–513.
6. F. Skaupy (K. Becker), *Z. Elektrochem.*, **33**, 1927, 512–513.
7. K. Becker, *Z. Elektrochem.*, **34**, 1928, 640–642.

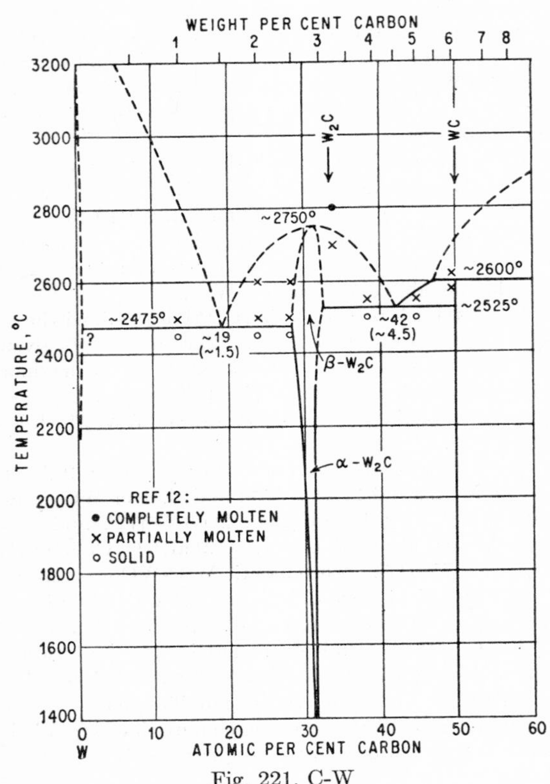

Fig. 221. C-W

8. K. Becker, *Z. Physik*, **51**, 1928, 481–489.
9. K. Becker, *Z. Metallkunde*, **20**, 1928, 437–441.
10. B. T. Barnes, *J. Phys. Chem.*, **33**, 1929, 688–691.
11. J. L. Gregg and C. W. Küttner, *Trans. AIME*, **88**, 1929, 581–590.
12. W. P. Sykes, *Trans. ASST*, **18**, 1930, 968–991.
13. A. A. Ravdel, *Zhur. Russ. Fiz.-Khim. Obshchestva* (*Chast. Fiz.*), **62**, 1930, 515–522.
14. C. W. Horsting, *J. Appl. Phys.*, **18**, 1947, 95–102.
15. P. Williams, *Compt. rend.*, **126**, 1898, 410–412.
16. S. Hilpert and M. Ornstein, *Ber. deut. chem. Ges.*, **46**, 1913, 1669–1675.
17. J. O. Arnold and A. A. Read, *Engineering*, **117**, 1914, 434–435.
18. O. Ruff and R. Wunsch, *Z. anorg. Chem.*, **85**, 1914, 292–328.

19. A. Hultgren, "Metallographic Study of Tungsten Steels," p. 50, John Wiley & Sons, Inc., New York, 1920.
20. H. Moissan, *Compt. rend.*, **116,** 1893, 1225–1227; **123,** 1896, 13–16; **125,** 1897, 839–844.
21. R. Schenck, F. Kurzen, and H. Wesselkock, *Z. anorg. Chem.*, **203,** 1932, 177–183.
22. J. J. Lander and L. H. Germer, *Trans. AIME,* **175,** 1948, 648–689.
23. J. T. Norton, unpublished research, quoted from [2].
24. C. Agte and H. Alterthum, *Z. tech. Physik,* **11,** 1930, 185.
25. L. Brewer, L. A. Bromley, P. W. Gilles, and N. L. Lofgren, in "The Chemistry and Metallurgy of Miscellaneous Materials: Thermodynamics," National Nuclear Energy Series, Div. IV, vol. 19B, pp. 40ff, McGraw-Hill Book Company, Inc., New York, 1950; quoted from [2].
26. E. Friederich and L. Sittig, *Z. anorg. Chem.*, **144,** 1925, 184–185.
27. W. Dawihl, see *FIAT Rev. Ger. Sci.*, 1939-1946, General Metallurgy, **1948,** p. 94.
28. H. Krainer and K. Konopicky, *Berg. hüttenmänn. Monatsh.*, **92,** 1947, 166–178.
29. V. I. Arkharov, I. S. Kvater, and S. T. Kiselev, *Izvest. Akad. Nauk S.S.S.R.* (*Tekh.*), **1947**(6), 749–756.
30. A. G. Metcalfe, *J. Inst. Metals,* **73,** 1947, 591–603.
31. V. Adelsköld, A. Sundelin, and A. Westgren, *Z. anorg. Chem.*, **212,** 1933, 401–409.
32. A. E. Kovalski and J. S. Umanski, *Zhur. Fiz. Khim.*, **20,** 1946, 773–778.
33. H. Nowotny and R. Kieffer, *Z. Metallkunde,* **38,** 1947, 257–265.
34. H. Krainer, *Arch. Eisenhüttenw.*, **21,** 1950, 119–127.
35. L. Brewer, D. L. Sawyer, D. H. Templeton, and C. H. Dauben, *J. Am. Ceram. Soc.*, **34,** 1951, 174.
36. H. Pfau and W. Rix, *Z. Metallkunde,* **45,** 1954, 116–118.
37. G. Hägg, *Z. physik. Chem.*, **B12,** 1931, 33–56.
38. H. Nowotny, E. Parthé, R. Kieffer, and F. Benesovsky, *Z. Metallkunde,* **45,** 1954, 97, 99.

$\bar{1}$.2641
0.7359

C-Zn Carbon-Zinc

See C-Cd, page 348.

$\bar{1}$.1194
0.8806

C-Zr Carbon-Zirconium

ZrC (11.63 wt. % C) is well established [1–10] and appears to be the only zirconium carbide existing. No evidence was found for the ZrC_2 assumed by [11, 3]. The melting point of ZrC was found to be 3100–3200 [4], 3530 ± 125 [12], and more recently 3175 ± 50°C [13]. Carbon additions lower the melting point to about 2430°C [7], indicating that, as in the system C-Ti, a eutectic ZrC-C will exist. In general, the phase diagram may be similar to that of the system C-Ti.

ZrC is cubic of the NaCl (B1) type structure. Lattice constants were determined by [14, 15, 6, 9, 16–19]. The newer data [9, 16–19] vary between 4.678 A [17] and 4.695 ± 0.001 A [9, 19], the latter value being more likely as ZrC forms solid solutions with Zr [9].

1. H. Moissan and F. Lengfeld, *Compt. rend.*, **122,** 1896, 651–654.
2. E. Wedekind, *Ber. deut. chem. Ges.*, **43,** 1910, 290; *Chem. Ztg.*, **30,** 1906, 938; **31,** 1907, 654.
3. O. Ruff and R. Wallstein, *Z. anorg. Chem.*, **128,** 1923, 96.
4. E. Friederich and L. Sittig, *Z. anorg. Chem.*, **144,** 1925, 171–173.

5. A. E. van Arkel and J. H. de Boer, *Z. anorg. Chem.*, **148**, 1925, 347–348.
6. C. H. Prescott, *J. Am. Chem. Soc.*, **48**, 1926, 2534–2550.
7. C. Agte and K. Moers, *Z. anorg. Chem.*, **198**, 1931, 236–238.
8. K. Moers, *Z. anorg. Chem.*, **198**, 1931, 248–251.
9. W. G. Burgers and J. C. M. Basart, *Z. anorg. Chem.*, **216**, 1934, 209–222.
10. For additional literature on the preparation and properties of ZrC, see P. Schwarzkopf and R. Kieffer, "Refractory Hard Metals," pp. 89–97, The Macmillan Company, New York, 1953.
11. L. Troost, *Compt. rend.*, **116**, 1893, 1228.
12. C. Agte and H. Alterthum, *Z. tech. Physik*, **11**, 1930, 185.
13. Unpublished work by Laboratories of American Electro Metal Corp., quoted from P. Schwarzkopf and R. Kieffer [10].
14. A. E. van Arkel, *Physica*, **4**, 1924, 286–301.
15. K. Becker and F. Ebert, *Z. Physik*, **31**, 1925, 268–272.
16. A. E. Kovalski and J. S. Umanski, *Zhur. Fiz. Khim.*, **20**, 1946, 769–772.
17. H. Nowotny and R. Kieffer, *Z. Metallkunde*, **38**, 1947, 257–265.
18. J. T. Norton and A. L. Mowry, *Trans. AIME*, **185**, 1949, 133–136.
19. P. Duwez and F. Odell, *J. Electrochem. Soc.*, **97**, 1950, 299–304.

$\bar{1}.5521$
0.4479

Ca-Cd Calcium-Cadmium

The phase diagram of Fig. 222 is due to a cursory (thermal) investigation [1], the only exception being that according to more recent roentgenographic work [2] the formula of the intermediate phase richest in Cd is assumed to be $CaCd_2$ (84.87 wt. % Cd) instead of $CaCd_3$ (89.38 wt. % Cd), suggested by [1] and [5].

Evidence for a compound Ca_3Cd_2 (65.15 wt. % Cd) [3] consists only of weak

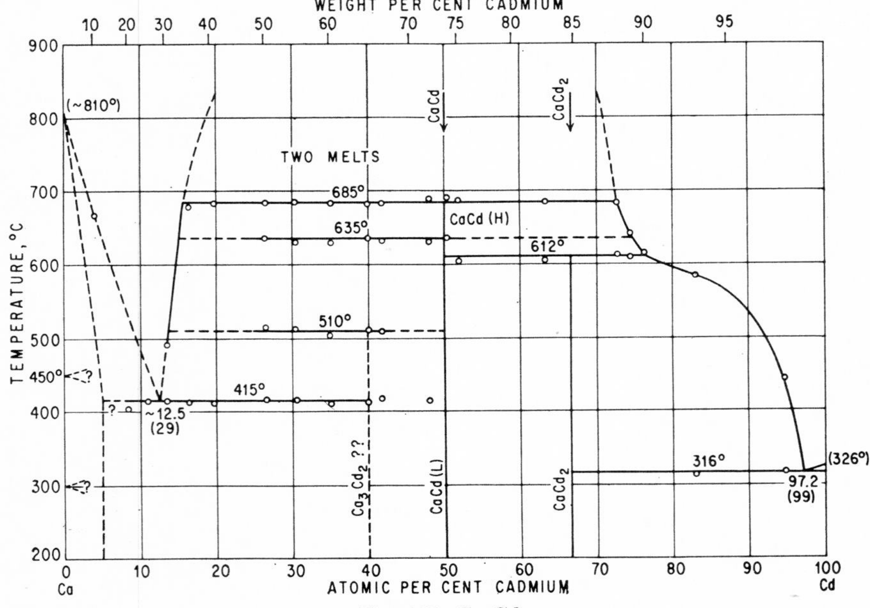

Fig. 222. Ca-Cd

thermal effects at 510°C; microscopic work could not be carried out because of the reactivity of the alloys. The existence of the compound CaCd (73.72 wt. % Cd) is well established by the thermal results; its transformation (635°C), strange to say, could not be observed in alloys with excess Cd. The compound richest in Cd (see above) which forms peritectically at 612°C presumably has a transformation point at about 590°C [1]. The Ca primary solid solution was only concluded from the extrapolated end of the eutectic at 415°C. Emf measurements up to 15 wt. (33 at.) % Ca are due to [4].

CaCd was shown to have the CsCl (B2) type of structure, with $a = 3.83_8$ A [5]. According to [2], $CaCd_2$ has the hexagonal $MgZn_2$ (C14) type of structure, with $a = 5.99_3$ A, $c = 9.65_4$ A, $c/a = 1.61$.

[5] claimed to have prepared the compound $CaCd_3$, the structure of which is "non-cubic."

1. L. Donski, *Z. anorg. Chem.*, **57**, 1908, 193–199. The alloys were prepared in open glass tubes without any protection, using calcium with 0.55 wt. % Al + Fe, 0.28 wt. % SiO_2, and an unknown nitrogen content. The alloys certainly contained much nitrogen (see Ca-N).
2. H. Nowotny, *Z. Metallkunde*, **37**, 1946, 31–34.
3. Erroneously named Ca_2Cd_3 by Donski.
4. R. Kremann, H. Wostall, and H. Schöpfer, *Forschungsarb. Metallkunde*, **1922**, no. 5. These measurements give no information as to the formula of the phase richest in Cd.
5. A. Iandelli, *Rend. seminar. fac. sci. univ. Cagliari*, **19**, 1949, 133–139.

$\bar{1}.4564$
0.5436

Ca-Ce Calcium-Cerium

According to an abstract, [1] established the Ca-Ce phase diagram by microscopic analysis. There are two eutectics, containing ~0.6 at. % Ce (843°C) and ~0.3 at. % Ca (801.8°C), respectively.

1. G. D. Zverev, *Doklady Akad. Nauk S.S.S.R.*, **104**, 1955, 242–245; *Met. Abstr.*, **23**, 1956, 811.

$\bar{1}.8325$
0.1675

Ca-Co Calcium-Cobalt

According to dilatometric and magnetic measurements [1], Ca raises the transformation temperature of Co (Fig. 223). The solid solubility of Ca in Co decreases from 7.9 at. (5.5 wt.) % Ca at 626°C to about 6.3 at. (4.4 wt.) % Ca at "room temperature."

1. U. Haschimoto, *Nippon Kinzoku Gakkai-Shi*, **1**, 1937, 177–190.

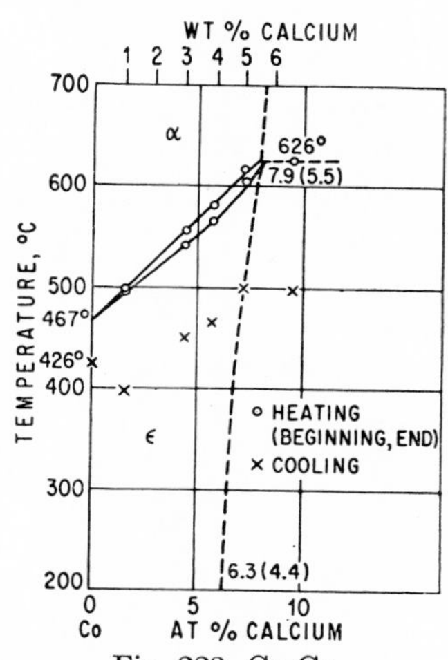

Fig. 223. Ca-Co

$\bar{1}.7999$
0.2001

Ca-Cu Calcium-Copper

The phase diagram of Fig. 224 is based mainly on a thermal and microscopic investigation by [1] and on roentgenographic work (as to the formula $CaCu_5$ for the Cu-rich compound [2]) by [3, 4] and is in agreement with some older observations [5, 6], [6] giving 920°C as the temperature of the Cu-rich eutectic. However, the Ca-rich part of the diagram was redrawn tentatively, since the constitution suggested by [1] is thermodynamically impossible (cf. inset of Fig. 224 and [7]). A one-phase field obviously exists around 30 wt. % Cu [1] (Ca_4Cu = 28.38 wt. % Cu) [8]. A certain range of homogeneity for $CaCu_5$ [83.33 at. (88.80 wt.) % Cu] can be concluded from coring, observed by [1]. A reinvestigation of this system is desirable.

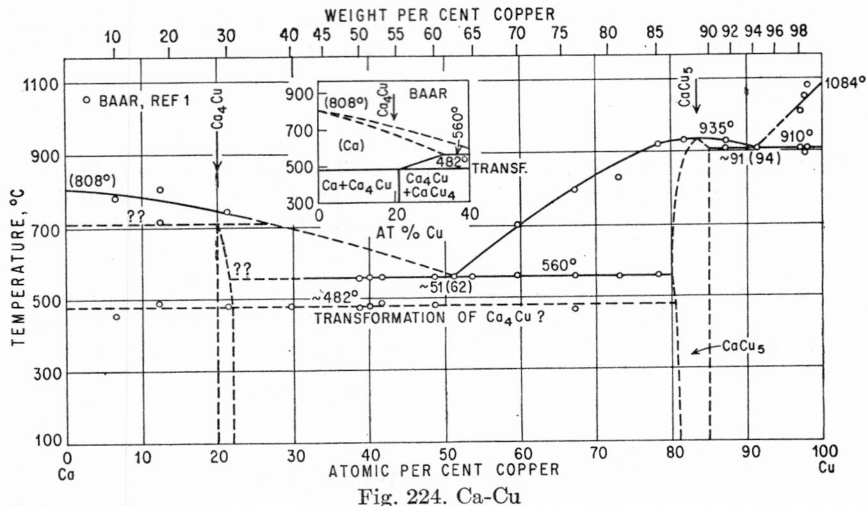

Fig. 224. Ca-Cu

Solid Solubility of Ca in Cu. Whereas [9] concluded from microscopic examinations that a wrought alloy with 0.06 wt. (0.1 at.) % Ca annealed for 1 hr at 800°C is two-phase, [10] claimed an alloy with 0.47 wt. (0.74 at.) % Ca to be one-phase. A similar discrepancy exists as to the effect of Ca on the electrical conductivity of Cu; [9] observed a weak, [10] a strong depression. The fact that the lattice parameter of Cu is not changed by addition of Ca [4] is in favor of the findings of [9].

Crystal Structure. $CaCu_5$ has a hexagonal structure [3, 4] with a = 5.092 A, c = 4.086 A, c/a = 0.803 at 83.1 at. % Cu [3] and 6 atoms per unit cell ("$CaCu_5$ type").

1. N. Baar, *Z. anorg. Chem.*, **70,** 1911, 377–383. The alloys were prepared in glass tubes under H_2 using a calcium with 0.55 Al + Fe, 0.28 Si (wt. %), and an unknown nitrogen content (see Ca-N).
2. Baar [1] assumed the formula $CaCu_4$ and Ssyromjatnikow [10] mentioned $CaCu_3$.
3. W. Haucke, *Z. anorg. Chem.*, **244,** 1940, 17–22.
4. H. Nowotny, *Z. Metallkunde*, **34,** 1942, 247–253.
5. L. Stockem, *Metallurgie*, **3,** 1906, 148–149.
6. L. Donski, *Z. anorg. Chem.*, **57,** 1908, 218.
7. An extended primary solid solution as assumed by [1] above 482°C is very improbable considering the unfavorable size factor.
8. A compound near 50 at. % Cu has been assumed by [4].
9. E. E. Schuhmacher, W. C. Ellis, and J. F. Eckel, *Trans. AIME*, **89,** 1930, 151–161. The Ca content of the alloys was determined. Spectrographic analysis showed only faint traces of impurities.

10. R. R. Ssyromjatnikow, *Metallurg*, **6**, 1931, 466–485 (in Russian); abstract in *Chem. Zbl.*, **1932**(2), 3615–3616; *J. Inst. Metals*, **53**, 1933, 182. The 0.47 wt. % Ca alloy contained 0.34 wt. % Fe. This may account for its relatively high electrical resistance (annealing treatment: 700°C in H_2, 15 min, slowly cooled).

$\bar{1}.8559$
0.1441

Ca-Fe Calcium-Iron

Varied series of attempts to dissolve Ca in liquid iron had negative results; only ingots prepared under increased pressure to avoid volatilization of Ca (boiling point, 1440°C) contained some calcium, which probably was only mechanically enclosed [1–3]. Experiments by [4, 5] are inconclusive as to the question whether Fe and Ca do or do not alloy.

Diffusion tests at 750–1000°C [1, 2, 6] were also negative [7].

In the light of the foregoing results the formation of a compound at temperatures as low as 380–540°C, which was concluded [8] from an exothermic reaction in a pressed and heated mixture of iron powder and Ca filings, is very unlikely.

1. C. Quasebart, *Metallurgie*, **3**, 1906, 28–29; see only experiments 4–7 of this paper in which low-carbon iron was used.
2. O. P. Watts, *J. Am. Chem. Soc.*, **28**, 1906, 1152–1155. Fe: 0.03–0.04 C, 0.01–0.09 Si (wt. %).
3. A. Hirsch and J. Aston, *Trans. Am. Electrochem. Soc.*, **13**, 1908, 143–150. Electrolytic iron was used.
4. A. Ledebur, *Stahl u. Eisen*, **22**, 1902, 710–713.
5. L. Stockem, *Metallurgie*, **3**, 1906, 147–148.
6. N. W. Ageew and M. Zamotorin, *Izvest. Leningrad. Politech. Inst., Otdel. Mat. Fiz. Nauk*, **31**, 1928, 183–196 (in Russian, abstract following in English). Fe: 0.08 wt. % C.
7. See also F. Wever, *Naturwissenschaften*, **17**, 1929, 304–309; *Arch. Eisenhüttenw.*, **2**, 1928-1929, 739–746.
8. G. Tammann and K. Schaarwächter, *Z. anorg. Chem.*, **167**, 1927, 405.

$\bar{1}.7596$
0.2404

Ca-Ga Calcium-Gallium

The solidus temperature of an alloy with 2 wt. (3.4 at.) % Ca was the melting point of pure Ga [1]. The compound $CaGa_2$ (77.67 wt. % Ga) has the hexagonal AlB_2 (C32) type of structure, with $a = c = 4.323$ A, $c/a = 1.00$ [2].

1. R. M. Evans and R. I. Jaffee, *Trans. AIME*, **194**, 1952, 153–156. The components were heated at 370°C for 5 hr.
2. F. Laves, *Naturwissenschaften*, **31**, 1943, 145.

$\bar{1}.7420$
0.2580

Ca-Ge Calcium-Germanium

The existence of the compounds CaGe (64.43 wt. % Ge) [1] and $CaGe_2$ (78.36 wt. % Ge) [2] has been reported. $CaGe_2$ has the rhombohedral $CaSi_2$ (C12) type of structure, with $a = 10.51$ A, $\alpha = 21°42'$ [2].

More recently, the crystal structures of the compounds Ca_2Ge and CaGe have been determined by X-ray single-crystal work: Ca_2Ge, orthorhombic $PbCl_2$ (C23) type, $a = 9.069 \pm 9$ A, $b = 7.734 \pm 7$ A, $c = 4.834 \pm 4$ A [3]; CaGe, orthorhombic, isotypic with CaSi, $a = 4.001 \pm 1$ A, $b = 4.575 \pm 2$ A, $c = 10.845 \pm 1$ A [4].

1. P. Royen and R. Schwarz, *Z. anorg. Chem.*, **211**, 1933, 412–422.
2. H. J. Wallbaum, *Naturwissenschaften*, **32**, 1944, 76.
3. P. Eckerlin and E. Wölfel, *Z. anorg. Chem.*, **280**, 1955, 321–331.
4. P. Eckerlin, H. J. Meyer, and E. Wölfel, *Z. anorg. Chem.*, **281**, 1955, 322–328.

1.5995
$\bar{2}$.4005

Ca-H Calcium-Hydrogen

CaH_2 (4.79 wt. % H), SrH_2 (2.25 wt. % H), and BaH_2 (1.45 wt. % H) were shown to have an orthorhombic structure—C29 type—with the lattice parameters (± 0.005 for a, b; ± 0.003 for c) $a = 5.948$ A, $b = 6.852$ A, $c = 3.607$ A, for CaH_2; $a = 6.377$ A, $b = 7.358$ A, $c = 3.883$ A, for SrH_2; $a = 6.802$ A, $b = 7.845$ A, $c = 4.175$ A, for BaH_2. Heated in a hydrogen atmosphere, the compounds dissociate only above 1000°C [1].

1. E. Zintl and A. Harder, *Z. Elektrochem.*, **41**, 1935, 34–52.

$\bar{1}$.3006
0.6994

Ca-Hg Calcium-Mercury

From a review of the present literature the existence of three compounds in the Hg-rich part of the phase diagram can be concluded. There is still some doubt, however, as to the formulas of these compounds.

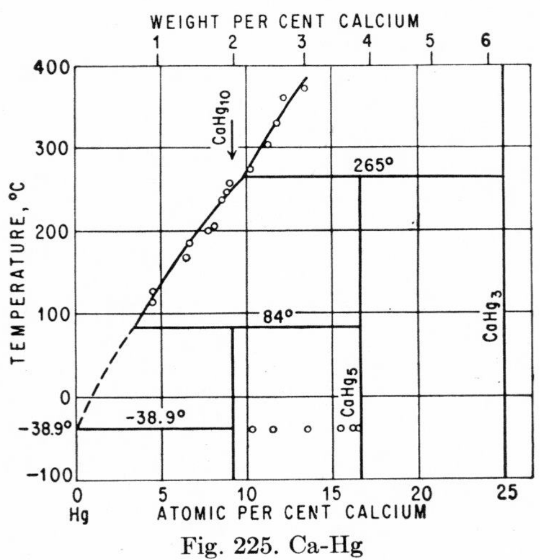

Fig. 225. Ca-Hg

The diagram of Fig. 225 is based on a careful thermal and microscopic analysis [1]. There are shown the compounds $CaHg_3$ (6.24 wt. % Ca), $CaHg_5$ [16.66 at. (3.84 wt.) % Ca], and $CaHg_{10}$ [9.09 at. (1.96 wt.) % Ca] [2]. Of these formulas $CaHg_5$ could be corroborated by chemical analysis of single crystals. Another thermal analysis (0–6.5 wt. % Ca) is due to [3]. These authors found nonvariant reactions at 266 and −39.8°C and assumed the existence of $CaHg_2$ (9.08 wt. % Ca) and $CaHg_4$ (4.76 wt. % Ca). At higher Ca contents their liquidus is shifted—in comparison with that of [1]—to somewhat lower temperatures which, according to [1], may be due to a certain inaccuracy in their chemical analyses and may account for the finding of $CaHg_4$ instead of $CaHg_5$.

Compound formulas were also concluded from work on preparation—Ca_3Hg_4 [4], doubted by [5]; $CaHg_5$ [6]; $CaHg_8$ [7]; see also [8]; from studying the raising of the boiling point of Hg by Ca—$CaHg_{10}$ [9]; and from emf measurements—$CaHg_4$ and $CaHg_9$ (2.17 wt. % Ca) [10]; $CaHg_4$ [11].

According to [3], liquid Hg dissolves 0.3 wt. (1.48 at.) % Ca at 25°C. From their emf measurements, [11] concluded a certain solubility of Hg in solid Ca.

More recently the compounds CaHg (16.65 wt. % Ca) and $CaHg_3$ have been prepared for X-ray investigations [12]. CaHg was shown to have the CsCl (B2) type of structure, with $a = 3.75_8$ A, and $CaHg_3$ was assumed to have a non cubic structure.

1. A. Eilert, *Z. anorg. Chem.*, **151,** 1926, 96–104. Ca: 99.2 wt. % pure. Preparation and cooling of the alloys took place under dried and purified CO_2 in a closed glass tube.
2. As to the peritectic reaction at 84°C, the author remarks only that there were weak thermal arrests in the composition range 0.9–3.8 wt. (4.4–16.5 at.) % Ca, with a maximum at 1.9 wt. (8.8 at.) % Ca.
3. L. Cambi and G. Speroni, *Atti reale accad. Lincei,* (5)**23**(2), 1914, 599–605.
4. J. Ferrée, *Compt. rend.*, **127,** 1898, 618–620.
5. W. Kerp, W. Böttger, and H. Iggena, *Z. anorg. Chem.*, **25,** 1900, 32–33.
6. J. Schürger, *Z. anorg. Chem.*, **25,** 1900, 425–429.
7. H. Moissan and Charanne, *Compt. rend.*, **140,** 1905, 125.
8. By reduction of mercuric salts by Ca in liquid ammonia as solvent, a precipitate of the composition Ca:Hg = 3:2 was obtained. C. A. Kraus and H. F. Kurtz, *J. Am. Chem. Soc.*, **47,** 1925, 43–60.
9. E. Beckmann and O. Liesche, *Z. anorg. Chem.*, **89,** 1914, 171–190.
10. L. Cambi, *Atti reale accad. Lincei,* (5)**23**(2), 1914, 606–611.
11. R. Kremann, H. Wostall, and H. Schöpfer, *Forschungsarb. Metallkunde,* **1922,** no. 5.
12. A. Iandelli, *Rend. seminar. fac. sci. univ. Cagliari,* **19,** 1949, 133–139.

$\bar{1}.5431$
0.4569

Ca-In Calcium-Indium

The compound CaIn (74.12 wt. % In), having a noncubic structure, was prepared by [1]. An alloy of the composition $CaIn_3$ was heterogeneous, and its powder pattern showed In lines.

1. A. Iandelli, *Rend. seminar. fac. sci. univ. Cagliari,* **19,** 1949, 133–139.

$\bar{1}.4602$
0.5398

Ca-La Calcium-Lanthanum

Cursory thermal and microscopic work on the system Ca-La is due to [1]. Besides the melting points of both metals, only weak thermal effects in cooling curves could be observed; in the range 15–82 wt. (5–57 at.) % La an effect at 680°C occurs. A eutectic, however, could not be observed microscopically. A miscibility gap in the liquid state certainly exists.

1. W. von Mässenhausen, *Z. Metallkunde,* **43,** 1952, 53–54. Besides commercial Ca (melting point 809°C), only a very impure La was available.

$\overset{0.7616}{\bar{1}.2384}$

Ca-Li Calcium-Lithium

The thermal data plotted in Fig. 226 are due to [1]. The author assumes a compound $LiCa_2$ (92.03 wt. % Ca), which conclusion is based essentially on a hardness maximum near that composition. X-ray work [2], however, revealed without any reasonable doubt the existence of a compound Li_2Ca (74.28 wt. % Ca). In Fig. 226 this phase is shown to be formed at the peritectic temperature. The position of the hardness maximum cannot be accounted for; perhaps it may be due to strong segregation.

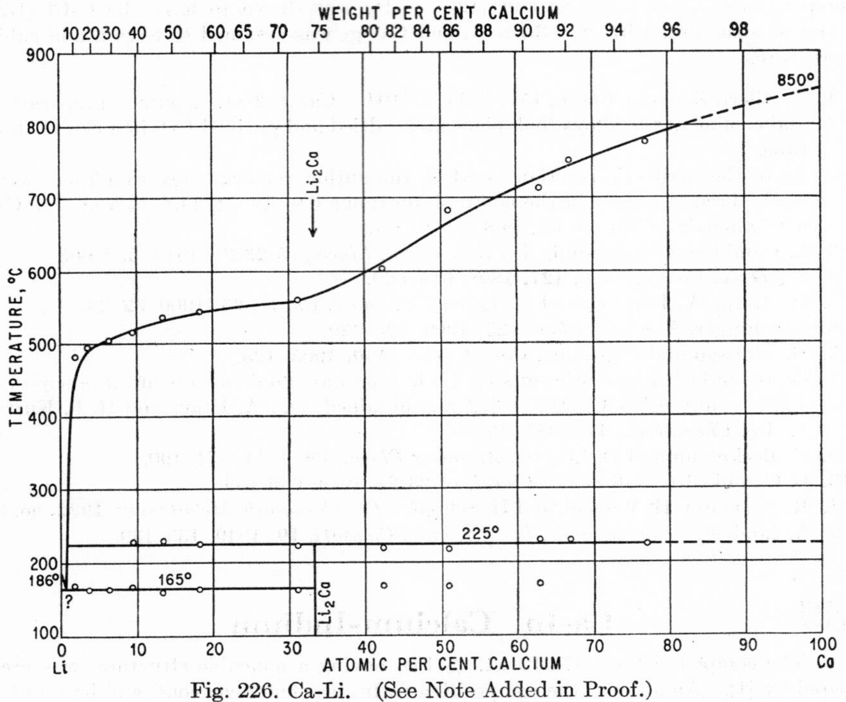

Fig. 226. Ca-Li. (See Note Added in Proof.)

Li_2Ca has, analogous to Mg_2Ca, the hexagonal $MgZn_2$(C14) type of structure, with $a = 6.260 \pm 8$ A, $c = 10.25 \pm 2$ A, $c/a = 1.637$ [2].

Note Added in Proof. Unfortunately unaware of the paper by [2], [3] reinvestigated the phase diagram using thermal-differential analysis and hardness measurements of as-cast samples. Peritectic and eutectic arrests were found at 230.9 and 141.8°C, respectively, the eutectic point being at 7.7 at. (32.5 wt.) % Ca (cf. Fig. 226). From the magnitude of the isothermal reactions at various compositions, the solid solubilities of Ca in Li and of Li in Ca are estimated to be 2.1 and 3 at. %, respectively. Thermal and hardness data point to a composition near LiCa for the intermediate phase.

1. M. I. Zamotorin, *Metallurg,* no. 1, 1938, 96–99 (in Russian). Li: 99.6 wt. %, Ca: 99.6 wt. % (nitrogen content probably very small). The alloys were prepared and measured under an argon atmosphere.

2. E. Hellner and F. Laves, *Z. Krist.*, **105**, 1943, 134–143.
3. M. R. Wolfson, *Trans. ASM*, **49**, 1957, 794–804. Li and Ca both 99% pure. Thermal analyses were run under an atmosphere of Navy grade A helium.

0.2170
1̄.7830

Ca-Mg Calcium-Magnesium

Liquidus. Determinations of the liquidus, in whole or in part, were carried out by [1], 14 alloys; [2], 12 alloys; [3], 0–55 wt. % Ca, 20 alloys; [4], 0–26 wt. % Ca, 27 alloys; [5], 10–73 at. (16–82 wt.) % Ca, 11 alloys.

In the range Mg-Mg_2Ca the results of [3] are in very good agreement with those of [4] and [5], and a graphical interpolation of these data was used in Fig. 227. In

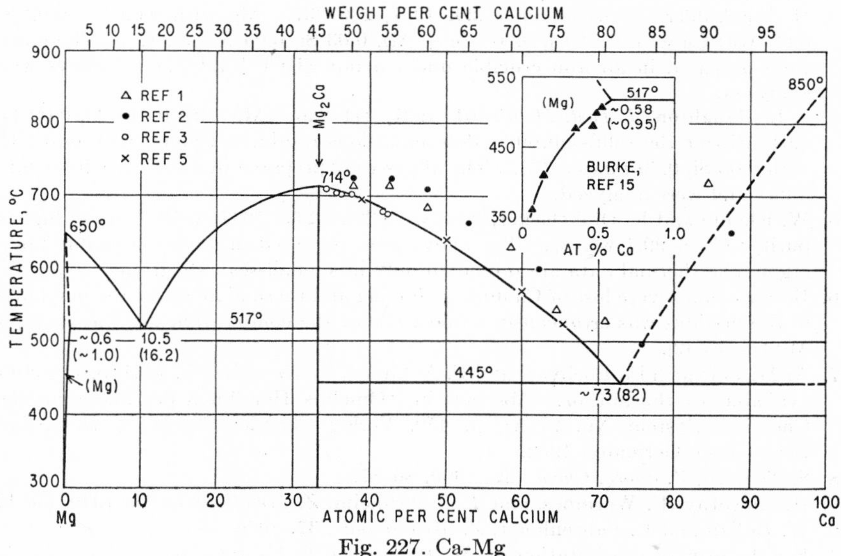

Fig. 227. Ca-Mg

the range Mg_2Ca-Ca all available data were plotted, and the most probable course of the liquidus was drawn [6]. Data for the Mg-rich eutectic were given by [1, 7, 2–5]; the most reliable data are 16.2 wt. (10.5 at.) % Ca, 517°C [4] (Fig. 227), and 16.3 wt. (10.6 at.) % Ca, 516°C [3]. The following data were reported for the Ca-rich eutectic: 450–457°C [8]; 78.7 wt. (69.2 at.) % Ca, 446°C [1]; 77 wt. (67 at.) % Ca, 445°C [7]; 82 wt. (73.7 at.) % Ca, 460°C [2], 445°C [3]; 73 at. (81.7 wt.) % Ca [5]. See Fig. 227 for the most reliable data.

The solid solubility of Ca in Mg was investigated by various methods: microscopically [3, 4, 9, 10], roentgenographically [9], and resistometrically [10]. Mainly because of rapid oxidation and the difficult chemical analysis of the alloys, the results differ considerably, the extreme values for the maximum solubility at the eutectic temperature being 0.78 wt. (0.47 at.) % Ca [3] and 1.8 wt. (1.1 at.) % Ca [4]. According to the more recent data by [9, 10], the solubility decreases from 1.2–1.4 wt. (0.7–0.85 at.) % Ca at 517°C to about 0.8 wt. (0.5 at.) % Ca at 300°C [11]. For a graphical comparison of all results, see [10]. (See Supplement.)

No information as to the solubility of Mg in solid Ca is available. For emf measurements (indicating erroneously the formula Mg_4Ca_3 for the only compound), see [12].

Crystal Structure. [13] concluded that the change in the parameters of Mg caused by addition of Ca is too small to be found (see, however, [9]). Mg_2Ca (45.18 wt. % Ca) was shown to have the $MgZn_2$ (C14) type of structure, with a = 6.23 A, c = 10.12 A, c/a = 1.62 [14], and has a very narrow range of homogeneity [5].

Supplement. The solubility of Ca in solid Mg between 508 and 365°C has been redetermined by [15] by means of quantitative metallography. The results, plotted in the inset of Fig. 227, are in relatively close agreement with those of [3] and suggest that the data of [4, 9, 10] are too high.

1. N. Baar, *Z. anorg. Chem.*, **70,** 1911, 362–366.
2. R. Pâris, *Publ. sci. et tech. ministère air (France)*, no. 45, 1934, 39–41.
3. H. Vosskühler, *Z. Metallkunde,* **29,** 1937, 236–237. Mg with 0.02 Fe, 0.02 Si, 0.03 Mn; Ca with 0.44 Cl, 0.06 (Fe + Al), 0.03 Si, 0.05 N (wt. %). The alloys were prepared in an iron crucible under a flux (KCl + LiCl). All alloys were analyzed.
4. J. L. Haughton, *J. Inst. Metals,* **61,** 1937, 241–246. Mg with 0.018 Al, 0.02 Fe, 0.013 Si (for the solid-solubility determinations resublimed metal was used); Ca with 0.05 Si, 0.29 Fe (wt. %). The alloys were prepared in a steel crucible under a flux and were analyzed.
5. W. Klemm and F. Dinkelacker, *Z. anorg. Chem.*, **255,** 1947, 2–12. Mg of highest purity; Ca ~ 99.1 wt. %. The alloys were prepared in an iron crucible under argon; the thermal data are presented only in a small-scale diagram.
6. Because of a severe loss of Ca during alloying and thermal analysis, the maximum in the liquidus was erroneously found to lie at the composition Mg_4Ca_3 by [1] or Mg_5Ca_3 by [2].
7. A. M. Bočvar (also Bochvar) and F. A. Lunev, *Tsvetnye Metal.*, 1931, p. 1138; not available to the author. Abstract in "Gmelins Handbuch der anorganischen Chemie," System No. 27(A), p. 459, Verlag Chemie, G.m.b.H., Weinheim/Bergstrasse, Germany, 1952.
8. S. Tamaru, *Z. anorg. Chem.*, **62,** 1909, 86–87.
9. H. Nowotny, E. Wormnes, and A. Mohrnheim, *Z. Metallkunde,* **32,** 1940, 39–42.
10. W. Bulian and E. Fahrenhorst, *Z. Metallkunde,* **37,** 1946, 70.
11. For the form of precipitation see W. Bulian and E. Fahrenhorst, *Z. Naturforsch.*, **1,** 1946, 263–267.
12. R. Kremann, H. Wostall, and H. Schöpfer, *Forschungsarb. Metallkunde,* **1922,** no. 5.
13. R. S. Busk, *Trans. AIME,* **188,** 1950, 1460–1464.
14. H. Witte, *Naturwissenschaften,* **25,** 1937, 795.
15. E. C. Burke, *Trans. AIME,* **203,** 1955, 285–286.

0.4566
$\bar{1}$.5434

Ca-N Calcium-Nitrogen

Calcium reacts with nitrogen at elevated temperatures to form the nitride Ca_3N_2 (18.90 wt. % N). Traces of Na—always present in commercial calcium—increase the rate of this reaction [1, 2]. Pure commercial calcium generally contains 0.3–0.6 wt. (0.85–1.7 at.) % N (2–3 wt. % nitride) and has a melting point of about 810°C [3]. Similar temperatures were often quoted in the older literature [4] as "melting point" of Ca. Sublimed calcium, however, with a nitrogen content of only 0.05–0.08 wt.

(0.14–0.23 at.) % N (0.3–0.4 wt. % nitride), was shown to have a melting point of 848–849°C [3, 5]; an extrapolation yields 851 ± 1°C as the melting point of chemically pure Ca [3].

The partial diagram of Fig. 228 is based on some thermal work by [3], [6]. The eutectic is assumed to lie between 3 and 4 wt. % nitride [1.61–2.14 at. (0.57–0.76 wt.) % N] [7]. Ca_3N_2 was found to exist in two [2, 8] or even three [9] polymorphic forms; according to [2], the compound is pseudohexagonal, with a = 3.560 A, c = 4.12 A, c/a = 1.157, when prepared at 300°C, but cubic—Mn_2O_3 ($D5_3$) type, with a = 11.42 ± 1 A [8]—when prepared at 800°C (irreversible transition pseudohexagonal → cubic).

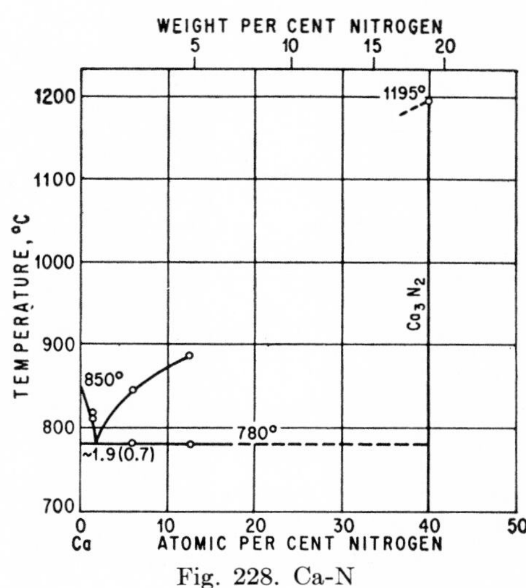

Fig. 228. Ca-N

From the work by [3] it follows that especially the liquidus temperatures of the binary systems of Ca must be corrected in all cases where melts were made using commercial calcium and without protective atmospheres.

1. A. von Antropoff and E. Germann, *Z. physik. Chem.*, **137,** 1928, 209–237. See also H. Moissan, *Compt. rend.*, **127,** 1898, 497.
2. H. Hartmann and H. J. Fröhlich, *Z. anorg. Chem.*, **218,** 1934, 190–192.
3. A. von Antropoff and F. Falk, *Z. anorg. Chem.*, **187,** 1930, 405–416.
4. For literature, see [3] and W. Hume-Rothery, *J. Inst. Metals,* **35,** 1926, 330–331.
5. F. Hoffmann and A. Schulze, *Z. Metallkunde,* **27,** 1935, 155–158.
6. The lower arrest in the 1.4 at. (0.49 wt.) % N alloy (Fig. 228) may be due to contaminations (Si, Al, Fe, Mg, $MgCl_2$) present in the commercial calcium used for this alloy. For the preparation of the other alloys, a sublimed Ca was used. H. Moissan (*Compt. rend.*, **127,** 1896, 495, 584) found the melting point of Ca_3N_2 at about 1200°C.
7. For a microstructure of an alloy with 25 wt. % nitride, see [1].
8. R. Paulus, *Z. physik. Chem.*, **B22,** 1933, 305–322.
9. H. H. Franck, M. A. Bredig, and G. Hofmann, *Naturwissenschaften,* **21,** 1933, 330–331.

0.2413
$\bar{1}$.7587

Ca-Na Calcium-Sodium

For the investigation of the system Na-Ca, two principal methods were used: (*a*) thermal analysis for the liquidus and the nonvariant reactions [1, 2] and (*b*) determination of the equilibrium of the reaction $Ca + 2NaCl \rightleftarrows CaCl_2 + 2Na$ at different temperatures for the extension of the miscibility gap [3] in the liquid state [4, 2; 5].

The results of [2] are the more reliable ones and were used for Fig. 229. According to [2], the melting point of Na is depressed only 0.025°C by Ca.

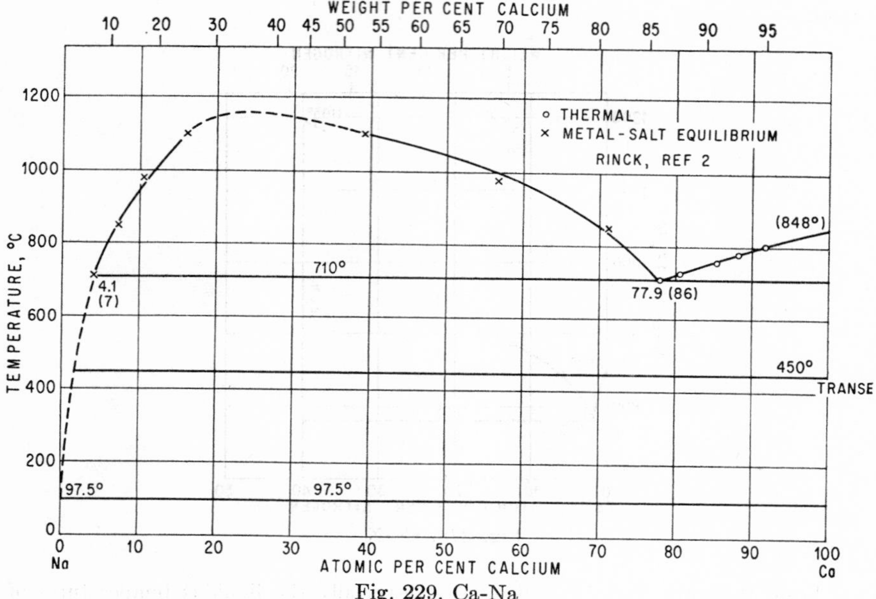

Fig. 229. Ca-Na

1. R. Lorenz and R. Winzer, *Z. anorg. Chem.*, **179**, 1929, 281–286. The alloys were prepared in steel bombs, using a 98.76 wt. % pure Ca.
2. E. Rinck, *Compt. rend.*, **192**, 1931, 1378–1381. Distilled Ca (melting point 848°C; see Ca-N) was used; the alloys were prepared in an argon atmosphere. The data had to be taken from a graph.
3. J. Metzger (*Liebigs Ann.*, **355**, 1907, 141) observed two layers in a solidified 50:50 wt. % alloy because of a miscibility gap in the liquid state. The interpretation given by Metzger is wrong.
4. R. Lorenz and R. Winzer, *Z. anorg. Chem.*, **181**, 1929, 193–202.
5. The indirect way via the ternary system Na-Ca-Cl had to be used because the segregation of the two metallic liquids is a very sluggish process.

$\bar{1}$.8344
0.1656

Ca-Ni Calcium-Nickel

The compound $CaNi_5$ [83.33 at. (87.98 wt.) % Ni] has a hexagonal structure ($CaCu_5$ type), with a = 4.960 A, c = 3.948 A, c/a = 0.796, and 6 atoms per unit cell [1].

1. H. Nowotny, *Z. Metallkunde*, **34**, 1942, 247–253.

1.2865
0.7135

Ca-Pb Calcium-Lead

Thermal analyses were carried out by [1] (59–100 at. % Pb), [2] (0–57 at. % Pb), [3] (95–100 at. % Pb, precision measurements), and [4] (63–100 at. % Pb). Nearly all data [5] are plotted in Fig. 230. The thermal results and some microscopic work [1, 2, 4] indicate the existence of three compounds, Ca_2Pb (72.11 wt. % Pb), CaPb (83.79 wt. % Pb), and $CaPb_3$ (93.94 wt. % Pb), which is confirmed by emf [7] and

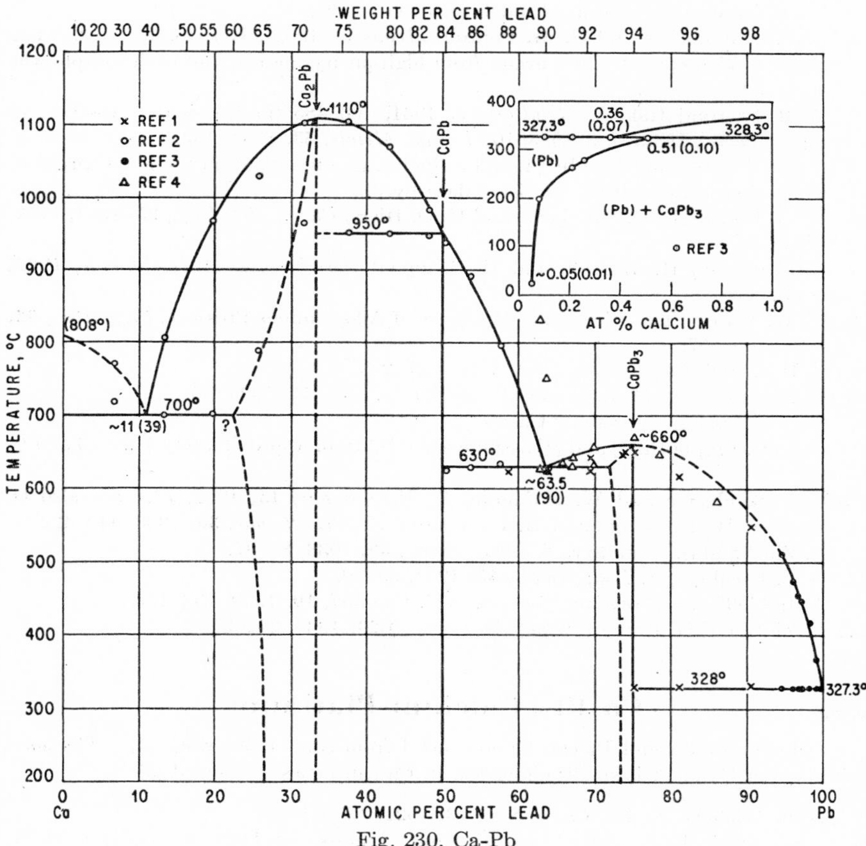

Fig. 230. Ca-Pb

resistivity [8] measurements. A compound Ca_2Pb_3, assumed by [9, 10], does not exist. There is some evidence that Ca_2Pb (thermal [2], see Fig. 230) and $CaPb_3$ (thermal and microscopic [1, 4]) may dissolve some Ca [11, 12].

Solid Solubility of Ca in Pb (cf. inset of Fig. 230). The maximum solubility near the peritectic temperature was shown to be 0.10 wt. (0.51 at.) % Ca by microscopic examination. The value for 25°C was found to be approximately 0.01 wt. (0.05 at.) % Ca by both age-hardening and conductivity experiments; three points, 0.016, 0.04, and 0.05 wt. (0.08, 0.21, and 0.26 at.) % Ca at 200, 265, and 280°C, respectively, were found by age-hardening experiments [3]. Supersaturated solid solutions with nearly 1 at. (0.2 wt.)% Ca could be produced [13] by extremely high quenching rates.

Crystal Structures. $CaPb_3$ was shown to have the Cu_3Au ($L1_2$) type of structure [14, see also 15, 16], with $a = 4.901 \pm 3$ A (at 75.2 at. % Pb) [14]. [16] stated the structure of CaPb to be not cubic, and [17], to be not of the β-brass or NaTl type.

1. L. Donski, *Z. anorg. Chem.*, **57**, 1908, 208–211.
2. N. Baar, *Z. anorg. Chem.*, **70**, 1911, 372–377. Both authors used a calcium with 0.55 (Al + Fe), 0.28 Si (wt. %). Baar prepared his alloys in glass tubes under H_2; certainly they contained much nitrogen (see Ca-N).
3. E. E. Schumacher and G. M. Bouton, *Metals & Alloys*, **1**, 1930, 405–409. The alloys were prepared under argon from high-purity metals, and each sample was analyzed.
4. R. R. Ssyromjatnikow, *Metallurg*, **6**, 1931, 466–485 (in Russian). Abstract in *Chem. Zentr.*, **1932**(2), 3615–3616; *J. Inst. Metals*, **53**, 1933, 182.
5. Omitted were some data by [4] and a single value by [6], all in the low-Ca region, which do not fit with the precision data by [3].
6. W. A. Cowan, L. D. Simpkens, and G. O. Hiers, *Chem. Met. Eng.*, **25**, 1921, 1182, 1184.
7. R. Kremann, H. Wostall, and H. Schöpfer, *Forschungsarb. Metallkunde*, **1922**, no. 5.
8. C. W. Ufford, The Electrical Resistance of Alloys under Pressure, *Phys. Rev.*, **32**, 1928, 505–507.
9. L. Hackspill, *Compt. rend.*, **143**, 1906, 227–229.
10. C. A. Kraus and H. F. Kurtz, *J. Am. Chem. Soc.*, **47**, 1925, 56.
11. Cf. the analogous system Ca-Sn.
12. As to $CaPb_3$, both [1] and [4] assume a solubility of approximately 1 wt. (3 at.) % Ca.
13. G. Falkenhagen and W. Hofmann, *Z. Metallkunde*, **43**, 1952, 73. See also O. Heckler, W. Hofmann, and H. Hanemann, *Z. Metallkunde*, **30**, 1938, 419–422.
14. E. Zintl and S. Neumayr, *Z. Elektrochem.*, **39**, 1933, 86–97.
15. G. S. Farnham, *J. Inst. Metals*, **55**, 1934, 69–70.
16. A. Iandelli, *Rend. seminar. fac. sci. univ. Cagliari*, **19**, 1949, 133–139.
17. E. Zintl and G. Brauer, *Z. physik. Chem.*, **B20**, 1933, 245–271.

$\bar{1}.3124$
0.6876

Ca-Pt Calcium-Platinum

Alloying of Ca and Pt can be concluded from some older work [1]. The solid solubility of Ca in Pt is small; alloys low in Ca can be age-hardened [2].

1. See E. Wichers, *J. Am. Chem. Soc.*, **43**, 1921, 1268–1273.
2. C. S. Sivil, in R. F. Vines, "The Platinum Metals and Their Alloys," pp. 56–57, International Nickel Co., New York, 1941.

0.0969
$\bar{1}.9031$

Ca-S Calcium-Sulfur

CaS (44.44 wt. % S) has the NaCl (B1) type of structure [1–5], with $a = 5.6836$ A [5]. Polysulfides could not be prepared thermally from the elements [6].

1. H. Küstner, *Physik. Z.*, **23**, 1922, 257–262.
2. S. Holgersson, *Z. anorg. Chem.*, **126**, 1923, 179.
3. W. P. Davey, *Phys. Rev.*, **21**, 1923, 213.

4. I. Oftedal, *Z. physik. Chem.*, **128,** 1927, 154–158.
5. W. Primak, H. Kaufman, and R. Ward, *J. Am. Chem. Soc.*, **70,** 1948, 2043–2046.
6. P. L. Robinson and W. E. Scott, *J. Chem. Soc.*, **134,** 1931, 693–709.

$\bar{1}$.5174
0.4826

Ca-Sb Calcium-Antimony

On investigating the solidification of the alloys with 0–9 wt. (0–23 at.) % Ca, [1] detected a eutectic (Fig. 231). The existence of a compound Ca_3Sb_2 (33.06 wt. % Ca)

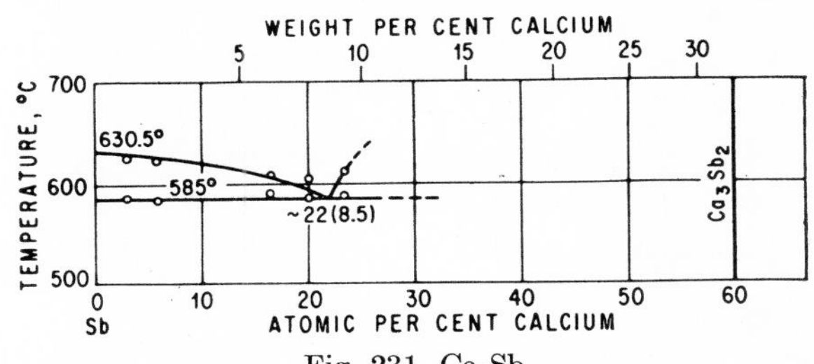

Fig. 231. Ca-Sb

is indicated by a maximum in the heat-of-formation curve [2]. Note the analogy with the Mg-Sb system.

1. L. Donski, *Z. anorg. Chem.*, **57,** 1908, 216–217. For the preparation of the alloys, see Ca-Cd.
2. O. Kubaschewski and A. Walter, *Z. Elektrochem.*, **45,** 1939, 732–740.

$\bar{1}$.7055
0.2945

Ca-Se Calcium-Selenium

CaSe (66.33 wt. % Se) has the NaCl (B1) type of structure, with $a = 5.924 \pm 3$ A [1, 2].

1. W. P. Davey, *Phys. Rev.*, **21,** 1923, 213.
2. I. Oftedal, *Z. physik. Chem.*, **128,** 1927, 154–158.

0.1544
$\bar{1}$.8456

Ca-Si Calcium-Silicon

Older Literature. Besides $CaSi_2$ [1–14] a further compound was assumed to exist in some [9–13] of the older papers on preparation, for which the formula Ca_3Si_2 [10, 12, 13] or $Ca_{11}Si_{10}$ [11] was named. From a thermal and microscopic analysis, greatly hindered by experimental difficulties, [15] concluded the existence of the compound $CaSi_2$. Only a few of the thermal data were plotted in Fig. 232.

Phase Diagram. The phase diagram of Fig. 232 is based on work by [16–21]. A thermal analysis in the range 20–90 at. % Si carried out by [17] showed clearly the existence of three compounds, Ca_2Si (25.95 wt. % Si), CaSi (41.20 wt. % Si), and $CaSi_2$ (58.36 wt. % Si) [22]. This has support from emf measurements by [17] (Ca_2Si, CaSi), work on preparation by [16] (CaSi, $CaSi_2$) and [19] (Ca_2Si, CaSi), and X-ray work by [18] ($CaSi_2$), [19] (Ca_2Si, CaSi), and [21] (CaSi). The heat-of-formation-vs.-composition curve shows breaks at Ca_2Si and CaSi [20].

Crystal Structures. The powder pattern of Ca_2Si could be indexed with a f.c.c. unit cell, with $a = 4.74_3$ A [19]. CaSi was shown to have an orthorhombic structure [21], with $a = 3.91 \pm 4$ A, $b = 4.59 \pm 5$ A, $c = 10.795 \pm 8$ A (probably kX), and 8 atoms per unit cell. A polymorphic transformation—which could explain slight differences in the powder patterns of annealed and unannealed samples [19, 21]—was not detected in the single-crystal work of [21].

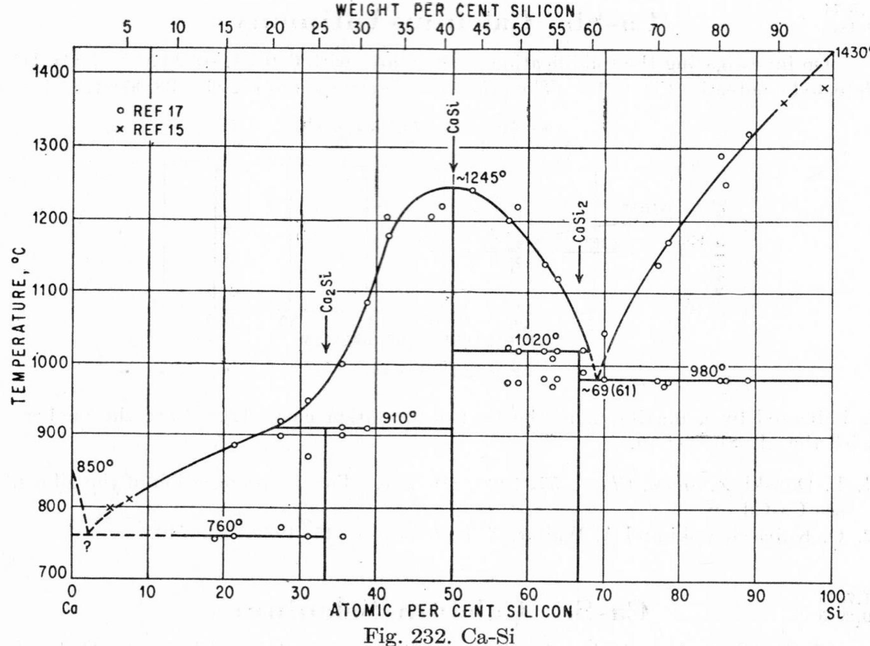

Fig. 232. Ca-Si

$CaSi_2$ forms the rhombohedral C12 prototype of structure, with $a = 10.4$ A, $\alpha = 21°30'$ [18]. Standard powder methods gave no indications of the existence of any solid solutions [19].

Supplement. [23] indexed powder patterns of Ca_2Si with a tetragonal unit cell, $a = 9.43$ A, $c = 10.19$ A, $c/a = 1.08$ (cf. Crystal Structures). According to single-crystal work by [24], Ca_2Si is orthorhombic of the $PbCl_2$ (C23) type, with $a = 9.002 \pm 16$ A, $b = 7.667 \pm 8$ A, $c = 4.799 \pm 6$ A. Indications were found of the existence of a cubic modification, somewhat richer in Ca, with an a spacing twice that reported by [19].

1. F. Wöhler, *Liebigs Ann.*, **127,** 1863, 257.
2. G. de Chalmot, *Am. Chem. J.*, **18,** 1896, 319.
3. E. Jüngst and R. Mewes, *Chem. Zentr.*, **1905**(1), 195.
4. C. B. Jacobs and C. S. Bradley, *Chem. News*, **82,** 1900, 149.
5. H. Moissan and W. Dilthey, *Ann. chim. et phys.*, **26,** 1902, 289; *Compt. rend.*, **134,** 1902, 503.
6. T. Goldschmidt, *Z. Elektrochem.*, **14,** 1908, 561.
7. Eichel, Dissertation, Dresden, 1909.
8. R. Frilley, *Rev. mét.*, **8,** 1911, 526–530.
9. H. Le Chatelier, *Bull. soc. chim. France*, (3)**17,** 1897, 793.
10. L. Hackspill, *Bull. soc. chim. France*, (4)**3,** 1908, 619.

11. R. Formhals, Dissertation, Giessen, 1909; A. Kolb, *Z. anorg. Chem.*, **64,** 1909, 342–367; **68,** 1910, 297–300.
12. A. Burger, Dissertation, Basel, 1907.
13. O. Hönigschmid, *Monatsh. Chem.*, **30,** 1909, 497; *Z. anorg. Chem.*, **66,** 1910, 414–417.
14. See the compilation by L. Baraduc-Muller, *Rev. mét.*, **7,** 1910, 692–695.
15. S. Tamaru, *Z. anorg. Chem.*, **62,** 1909, 81–88.
16. L. Wöhler and F. Müller, *Z. anorg. Chem.*, **120,** 1921, 49–70; see also L. Wöhler and W. Schuff, *Z. anorg. Chem.*, **209,** 1932, 33–59.
17. L. Wöhler and O. Schliephake, *Z. anorg. Chem.*, **151,** 1926, 1–11. Ca: 98.45 wt. %, Si: 99.48 wt. % pure; crucibles: clay + alumina.
18. J. Böhm and O. Hassel, *Z. anorg. Chem.*, **160,** 1927, 152–164.
19. V. Louis and H. H. Franck, *Z. anorg. Chem.*, **242,** 1939, 117–127. The authors have claimed that $CaSi_2$ is the only stable compound at room temperature (see also abstract in *Met. Abstr.*, **6,** 1939, 451). This assumption is not in accordance with the phase diagram.
20. O. Kubaschewski and H. Villa, *Z. Elektrochem.*, **53,** 1949, 32–40.
21. E. Hellner, *Z. anorg. Chem.*, **261,** 1950, 226–236.
22. The thermal data indicate the incompleteness of the peritectic reactions.
23. G. Busch, P. Junod, U. Katz, and U. Winkler, *Helv. Phys. Acta*, **27,** 1954, 193–195 (in French).
24. P. Eckerlin and E. Wölfel, *Z. anorg. Chem.*, **280,** 1955, 321–331.

$\bar{1}.5285$
0.4715

Ca-Sn Calcium-Tin

Based on some thermal [1] and thermal and microscopic [2] work on Sn-rich alloys (up to about 0.2 and 40 at. % Ca, respectively), the phase diagram was further established by [3] by means of a careful thermal and microscopic investigation (Fig. 233). Three compounds, Ca_2Sn (59.69 wt. % Sn), CaSn (74.76 wt. % Sn), and $CaSn_3$ (89.88 wt. % Sn), were found in agreement with the results of [2] ($CaSn_3$) and emf measurements by [4] (CaSn, $CaSn_3$, and probably Ca_2Sn). No evidence of the existence of solid solutions could be found by [3]; from hardness and resistivity measurements the solubility of Ca in Sn at 200°C was concluded to lie below 0.1 at. (0.03 wt.) % Ca [5]. Whether a break at 20 at. % Sn in the curve of heat of formation vs. composition is real was not decided by the author [6]. It could indicate the existence of an additional phase [7].

$CaSn_3$ was shown to have the Cu_3Au ($L1_2$) type of structure, with a = 4.742 A [8]. According to [9], the structure of CaSn is not cubic.

[10] indexed powder patterns of Ca_2Sn with a tetragonal unit cell, a = 12.15 A, c = 11.93 A, c/a = 0.98.

Note Added in Proof. According to X-ray single-crystal work, CaSn is orthorhombic, isotypic with CaSi, with $a = 4.349 \pm 4$ A, $b = 4.821 \pm 4$ A, $c = 11.52 \pm 2$ A [11].

1. C. T. Heycock and F. H. Neville, *J. Chem. Soc.*, **57,** 1890, 384.
2. L. Donski, *Z. anorg. Chem.*, **57,** 1908, 212–214.
3. W. Hume-Rothery, *J. Inst. Metals*, **35,** 1926, 319–335. The calcium used contained about 0.6 wt. % Fe and traces of Na, but the Fe did not enter into the alloys unless they contained free Ca. The alloys were prepared and the cooling curves taken in a closed silica tube in an atmosphere of commercial argon (with 10% nitrogen) (see Ca-N). In all alloys, the Ca and Sn content was determined.

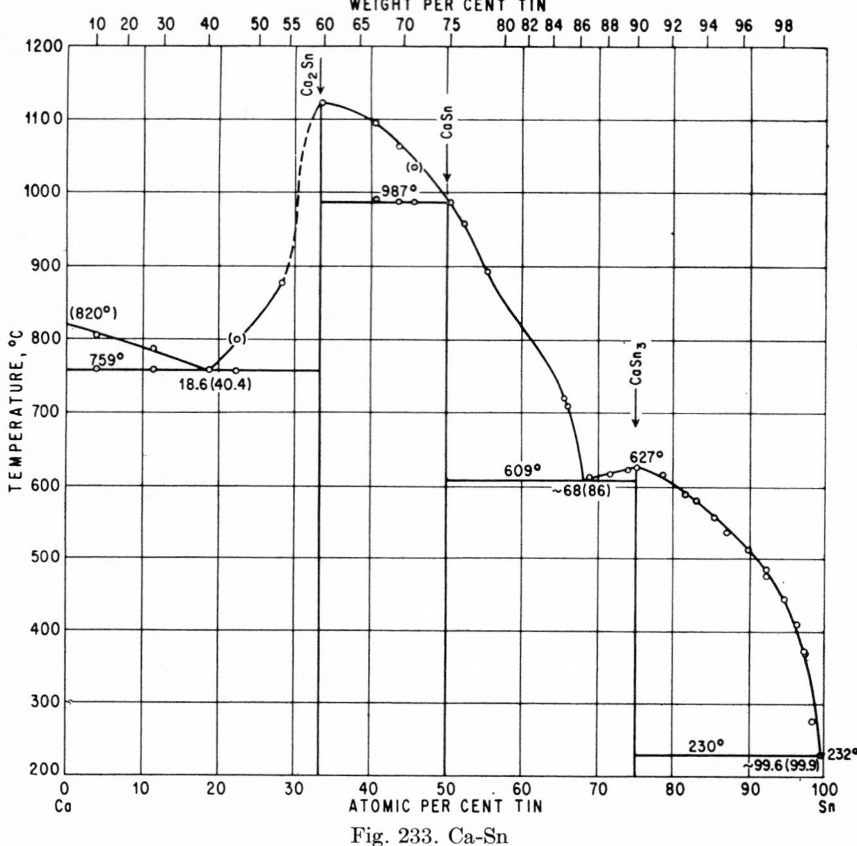

Fig. 233. Ca-Sn

4. R. Kremann, H. Wostall, and H. Schöpfer, *Forschungsarb. Metallkunde,* **1922,** no. 5.
5. E. Jenckel and L. Roth, *Z. Metallkunde,* **30,** 1938, 135–144.
6. O. Kubaschewski and H. Villa, *Z. Elektrochem.,* **53,** 1949, 32–40.
7. C. A. Kraus and H. F. Kurtz (*J. Am. Chem. Soc.,* **47,** 1925, 43–60) produced a precipitate of the composition Ca_5Sn by reduction of stannous iodide by calcium in liquid ammonia solution.
8. E. Zintl and S. Neumayr, *Z. Elektrochem.,* **39,** 1933, 86–97.
9. A. Iandelli, *Rend. seminar. fac. sci. univ. Cagliari,* **19,** 1949, 133–139.
10. G. Busch, P. Junod, U. Katz, and U. Winkler, *Helv. Phys. Acta,* **27,** 1954, 193–195 (in French).
11. P. Eckerlin, H. J. Meyer, and E. Wölfel, *Z. anorg. Chem.,* **281,** 1955, 322–328.

$\bar{1}.6603$
0.3397

Ca-Sr Calcium-Strontium

X-ray powder analyses of slowly cooled Ca-Sr alloys showed [1, 2] that these elements form a continuous series of solid solutions at lower temperatures. The lattice-parameter-vs.-composition (in at. %) curve of [1] is slightly convex (contraction); that of [2] is a straight line.

1. W. Klemm and G. Mika, *Z. anorg. Chem.*, **248,** 1941, 155–166.
2. A. J. King, *J. Am. Chem. Soc.*, **64,** 1942, 1226–1227.

$\bar{1}.4970$
0.5030

Ca-Te Calcium-Tellurium

CaTe (76.10 wt. % Te) has the NaCl (B1) type of structure, with $a = 6.358 \pm 8$ A [1, 2].

1. V. M. Goldschmidt, Geochem. Verteilungsgesetze VII, VIII. *Skrifter Norske Videnskaps-Akad. Oslo, I. Mat. Naturv. Kl.*, **1926,** no. 2, and **1927,** no. 8.
2. I. Oftedal, *Z. physik. Chem.*, **128,** 1927, 154–158.

$\bar{1}.2925$
0.7075

Ca-Tl Calcium-Thallium

The diagram of Fig. 234 is the result of thermal and microscopic investigations by [1] (Tl-rich alloys only) and [2]. According to [2], three compounds exist, CaTl (83.60 wt. % Tl), Ca_3Tl_4 [57.14 at. (87.18 wt.) % Tl], and $CaTl_3$ (93.86 wt. % Tl), the existence of CaTl [3] and $CaTl_3$ [1, 4] being corroborated by other authors.

The extent of the solid solutions formed by Ca and CaTl was concluded [2] only from the end points of the eutectic horizontal at 692°C (9 and 41 at. = 33 and 78 wt.

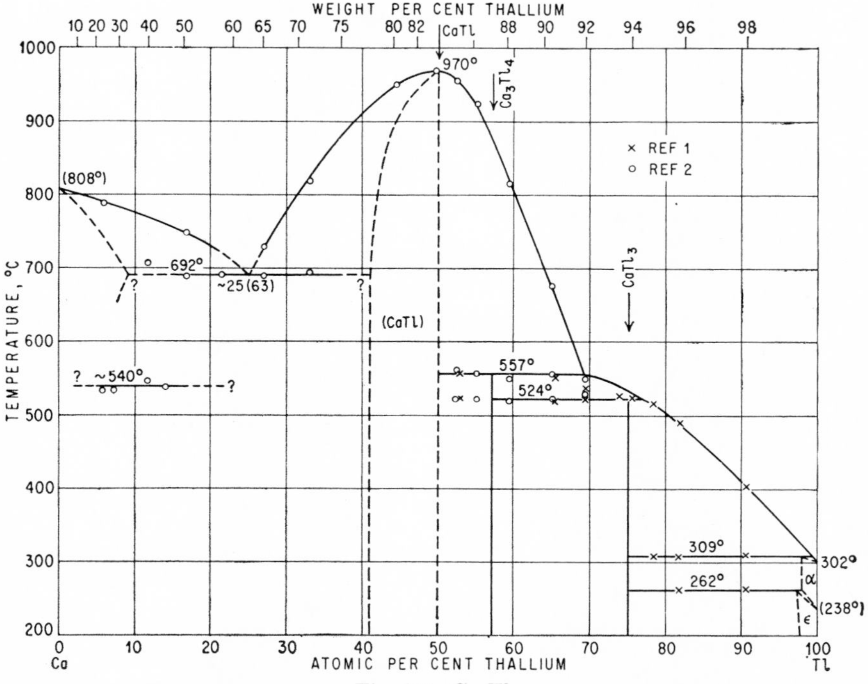

Fig. 234. Ca-Tl

% Tl), which were extrapolated from the lengths of the thermal arrests of a few alloys. An additional problem in the constitution of the Ca-rich alloys is the unknown nature of some thermal arrests near 540°C (Fig. 234).

The increase in the melting point and in the transformation temperature of Tl by addition of Ca [1] indicates a certain—not precisely determined—solid solubility of Ca in Tl.

CaTl has the CsCl (B2) type of structure, with a = 3.855 ± 4 A (at 48.3 at. % Tl) [3], and $CaTl_3$ the Cu_3Au ($L1_2$) type of structure, with a = 4.804 A [4].

1. L. Donski, *Z. anorg. Chem.*, **57**, 1908, 206–208.
2. N. Baar, *Z. anorg. Chem.*, **70**, 1911, 366–372.
   [1] and [2] used calcium with 0.55 Al + Fe, 0.28 Si (wt. %); from the melting point (808°C) of this material a heavy contamination by nitrogen can be concluded (see Ca-N). Baar prepared his melts in glass tubes under H_2. Alloys with 45–80 wt. % Tl could not be examined microscopically because of quick oxidation.
3. E. Zintl and G. Brauer, *Z. physik. Chem.*, **B20**, 1933, 245–271.
4. E. Zintl and S. Neumayr, *Z. Elektrochem.*, **39**, 1933, 86–97.

$\bar{1}$.2262
0.7738

Ca-U Calcium-Uranium

In attempts to prepare Ca-U alloys, [1] found no reaction between the two metals after as much as 24 hr at 800°C.

1. D. H. Ahmann, unpublished information (1945). Quoted in *U.S. Atomic Energy Comm., Publ.* BMI-1000, 1955.

$\bar{1}$.3383
0.6617

Ca-W Calcium-Wolfram

According to experiments by [1], Ca and W do not form alloys.

1. D. Kremer, *Abhandl. Inst. Metallhütt. u. Elektromet. Tech. Hochsch. Aachen*, **1**, 1916, no. 2, 7–8.

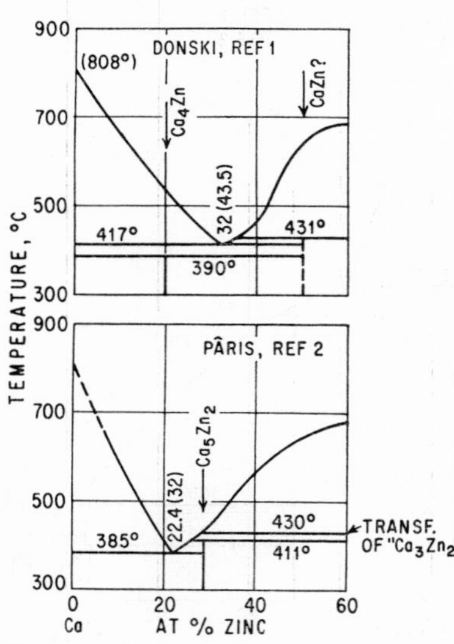

Fig. 235. Ca-Zn. (See also Fig. 236.)

$\bar{1}$.7875
0.2125

Ca-Zn Calcium-Zinc

Thermal and microscopic investigations by [1, 2], which agreed on the constitution of the Zn-rich alloys [compounds Ca_2Zn_3, $CaZn_4$, and $CaZn_{10}$ and a eutectic at ~83.5 wt. (75.6 at.) % Zn, 633°C] but disagreed as to that of the Ca-rich alloys (see Fig. 235), gave rather inaccurate results since the alloys were not analyzed chemically. More recent roentgenographic and microscopic work in the Zn-rich range has shown that the true formulas of the three compounds named are $CaZn_2$ (76.53 wt. % Zn) [3], $CaZn_5$ [83.33 at. (89.08 wt.) % Zn] [3, 4], and $CaZn_{13}$ [92.86 at. (95.50 wt.) % Zn] [5, 3, 4; cf. 6] (Fig. 236), thus being lower in Ca content than assumed by [1, 2]. The existence of a compound CaZn (61.99 wt. % Zn), tentatively concluded by [1], was corroborated by [4]. According to microscopic work [4], solid Zn does not dissolve Ca.

The Ca-rich part of the phase diagram needs careful reinvestigation. In Fig. 236 it has been drawn favoring the results of [2]. The formula Ca_5Zn_2 (28.57 at. % Zn) corresponds to 39.48 wt. % Zn.

Further Literature (of Minor Interest). Some alloys with a low Ca content were prepared by [7]. Emf measurements [8] indicated the compounds Ca_4Zn, Ca_2Zn_3, $CaZn_4$, and $CaZn_{10}$. On reduction of Zn cyanide by Ca in liquid ammonia, a metallic precipitate of the composition Ca_7Zn is formed [9].

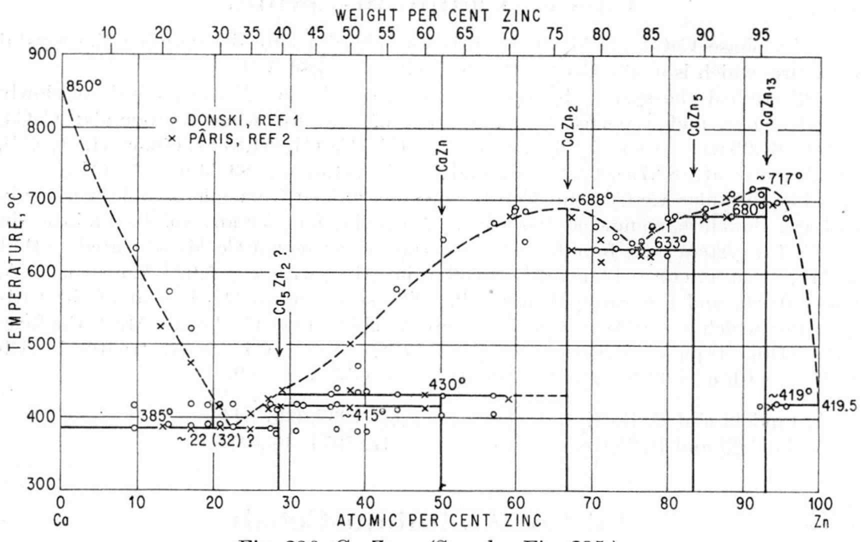

Fig. 236. Ca-Zn. (See also Fig. 235.)

Crystal Structures. $CaZn_5$ has a hexagonal structure [3, 4], with $a = 5.416$ A, $c = 4.191$ A, $c/a = 0.774$ [3], and 6 atoms per unit cell ("$CaCu_5$ type"). $CaZn_{13}$ has the cubic $NaZn_{13}$ ($D2_3$) type of structure, with $a = 12.154 \pm 5$ A [5].

1. L. Donski, *Z. anorg. Chem.*, **57,** 1908, 185–193. For the experimental technique used see Ca-Cd.
2. R. Pâris, *Publ. sci. et tech. ministère air (France)*, no. 45, 1934, pp. 41–44. The alloys were prepared in sealed iron crucibles in 1 atm of their own vapors.
3. W. Haucke, *Z. anorg. Chem.*, **244,** 1940, 17–22. As to $CaZn_2$ and $CaZn_{13}$ this paper gives only preliminary information about a systematic roentgenographic investigation, the full report of which probably could not be published because of the war.
4. H. Nowotny, *Z. Metallkunde*, **34,** 1942, 247–253. Besides a corroboration of the structure of $CaZn_5$ as given by [3], this paper reports the results of a cursory microscopic and roentgenographic investigation of alloys up to 50 wt. (62 at.) % Ca. The compounds $CaZn_{13}$, $CaZn_5$, and CaZn were found.
5. J. A. A. Ketelaar, *J. Chem. Phys.*, **5,** 1937, 668.
6. As early as 1869, G. vom Rath, *Pogg. Ann.*, **136,** 1869, 434, named $CaZn_{12}$ as formula for the compound richest in Zn.

7. T. H. Norton and E. Twitchell, *Am. Chem. J.*, **10**, 1888, 70. An alloy with 2.28 wt. % Ca showed the melting point of pure Zn.
8. R. Kremann, H. Wostall, and H. Schöpfer, *Forschungsarb. Metallkunde*, **1922**, no. 5.
9. C. A. Kraus and H. F. Kurtz, *J. Am. Chem. Soc.*, **47**, 1925, 43–60.

Cb- Columbium-

For binary alloys of columbium, see the respective system of niobium (Nb).

$\bar{1}.9043$
0.0957

Cd-Ce Cadmium-Cerium

The phase CeCd (44.51 wt. % Cd) was identified by determining the crystal structure, which is of the CsCl (B2) type, with a = 3.86 A [1].

[2] studied the system by means of microscopic and X-ray analysis of slowly cooled alloys and reported the existence of the following compounds: CeCd, $CeCd_2$ (61.60 wt. % Cd), $CeCd_3$ (70.64 wt. % Cd), Ce_2Cd_9 [81.81 at. (78.30 wt.) % Cd], $CeCd_6$ [85.71 at. (82.79 wt.) % Cd], and $CeCd_{11}$ [91.67 at. (89.83 wt.) % Cd].

The formulas Ce_2Cd_9 and $CeCd_6$ were ascertained from microscopic work only and are, therefore, somewhat tentative. No indications for any solid solubilities in the Ce-Cd system were found. The structure of hexagonal $CeCd_2$ is related to that of AlB_2 (z parameter 0.42 instead of 0.50 as in AlB_2), with a = 5.073 A, c = 3.450 A, c/a = 0.680, and 3 atoms per unit cell. The space group D_{3d}^3 is that of the CdI_2 type (for which $z \sim 0.25$). $CeCd_3$ has presumably—like (Ce, La, Pr)Mg_3—the f.c.c. BiF_3 (DO_3) type of structure, with a = 7.228 A. $CeCd_{11}$ is isostructural with $BaHg_{11}$, with a = 9.319 A and 36 atoms in the cubic unit cell.

1. A. Iandelli and E. Botti, *Gazz. chim. ital.*, **67**, 1937, 638–644.
2. A. Iandelli and R. Ferro, *Gazz. chim. ital.*, **84**, 1954, 463–478.

0.2804
$\bar{1}.7196$

Cd-Co Cadmium-Cobalt

Cooling curves of cadmium melts to which up to 10 wt. % Co had been added showed only eutectic arrests at 316°C, as compared with the melting point of Cd given as 322°C. Structures indicated the presence of an intermediate phase in a eutectic matrix [1].

In an attempt to prepare Co-rich alloys, [2] heated the components in evacuated sealed porcelain bulbs up to 1200°C. No alloying was noted. On the other hand, [3] reported that the temperature of the polymorphic transformation of Co (given as 467°C) is increased by 0.3 and 1.5 wt. (0.16 and 0.8 at.) % Cd to 472 and 482°C, respectively, if determined on heating. It was concluded that about 0.5 wt. (0.27 at.) % Cd is soluble in this temperature range.

An alloy of the composition $Cd_{21}Co_5$ (11.1 wt. % Co) was claimed [4] to have the structure of γ-brass.

[5] tried to prepare Cd-Co alloys of various compositions by the mixing of Cd and Co amalgams and distilling off mercury at 360°C. No alloying was observed. Also, an attempt failed to prepare the intermediate phase claimed by [4] by immersion of Co in a Cd melt of 700°C.

1. K. Lewkonja, *Z. anorg. Chem.*, **59**, 1908, 322–323.
2. W. Köster and E. Horn, *Z. Metallkunde*, **43**, 1952, 333–334.
3. U. Hashimoto, *Nippon Kinzoku Gakkai-Shi*, **2**, 1938, 67–77.

4. A. Westgren and W. Ekman, *Arkiv Kemi, Mineral. Geol.*, **B10**, 1930, no. 11, pp. 1–6.
5. F. Lihl and E. Buhl, *Z. Metallkunde*, **46**, 1955, 787–791.

0.3347
$\bar{1}$.6653

Cd-Cr Cadmium-Chromium

Chromium does not dissolve in liquid cadmium heated to 650°C [1].

1. G. Hindrichs, *Z. anorg. Chem.*, **59**, 1908, 427–428. The chromium used contained 1.2% Fe and 0.3% Si as chief contaminants.

$\bar{1}$.9272
0.0728

Cd-Cs Cadmium-Cesium

The compound $CsCd_{13}$ [92.86 at. (91.66 wt.) % Cd] was identified by determination of the crystal structure: cubic of the $NaZn_{13}$ ($D2_3$) type, 8 molecules per unit cell, a = 13.92 A [1]. See also Cd-K.

1. E. Zintl and W. Haucke, *Z. Elektrochem.*, **44**, 1938, 104–111.

0.2478
$\bar{1}$.7522

Cd-Cu Cadmium-Copper

The constitution of this system is well established; Fig. 237 is based on the careful thermal and microscopic studies by [1], with the exception of the boundary of the Cu-rich solid solution (see below). Prior to this work, the phase diagram was determined by [2]. The liquidus curves of both investigations are in excellent agreement. However, [2] had overlooked the existence of the phases Cu_4Cd_3 [42.86 at. (57.02 wt.) % Cd] and $CuCd_3$ (84.15 wt. % Cd) [3].

As far as Cu_4Cd_3 is concerned, this is not surprising since this phase does not form on solidification but appears only after prolonged annealing at 450–500°C. At a moderate rate of cooling, as is used for taking cooling curves, the temperature of the peritectic reaction (Cu) + liq. $\rightleftarrows$ Cu_2Cd is depressed to ~ 544°C, which temperature is below the equilibrium temperature (547°C) for the peritectic reaction Cu_2Cd + liq. $\rightleftarrows$ Cu_4Cd_3. Under these conditions the formation of Cu_4Cd_3 does not occur, and Cu_2Cd continues to separate from the melt until, at ~ 540°C, a metastable eutectic of Cu_2Cd and γ (formerly designated as δ) is formed. This is indicated by dashed lines in Fig. 237.

The maximum of the liquidus curve is very flat; both [1] and [2] assumed it to be at the composition Cu_2Cd_3 (72.63 wt. % Cd). However, the structure of the γ phase is based on the composition Cu_5Cd_8 [61.54 at. (73.89 wt.) % Cd] as it is of the γ-brass ($D8_2$) type [4, 5].

The existence of all four intermediate phases was confirmed by X-ray analysis [5]. Cu_2Cd and $CuCd_3$ seem to have extremely narrow ranges of homogeneity [1, 5].

The solid solubility of Cd in Cu was determined by lattice-parameter measurements [5, 6] and micrographic work [7, 6], covering the ranges 300–500 [5], 250–550 [7], and 300–1000°C [6]. At 500°C the solubility values agree within 2.2–2.7 wt. (1.28–1.55 at.) % Cd [1, 5–7]. The solubility at 300°C was found as 0.5 [5], 0.9 [6], and 1.3 wt. % [7] or 0.3, 0.52, and 0.75 at. % Cd; the lowest value may be the most probable one. [6] showed the solubility curve to be of the retrograde type, with the maximal solubility of about 4.5 wt. (2.6 at.) % at 650°C and 3.7, 3.9, and 1.6 wt. (2.1, 2.3, and 0.9 at.) % Cd at 550, 800, and 1000°C, respectively.

The $CuCd_3$-Cd eutectic was accurately determined as being located at 1.2 wt. (2.1 at.) % Cu, 314°C [8]. The solid solubility of Cu in Cd was reported as about

0.07 wt. (0.12 at.) % at 300°C [1], in good agreement with 0.05–0.1 wt. (0.09–0.17 at.) % at 270°C found by [9].

The boiling point of Cd (768°C) is raised by 19 wt. (~29 at.) % Cu to 801°C and by 31.5 wt. (~45 at.) % Cu to 820°C [10].

Crystal Structures. The lattice parameter of Cu (a = 3.6147 A) is increased to a = 3.6285 A by 1.76 at. % Cd [11]; see also [6]. The γ phase (formerly designated

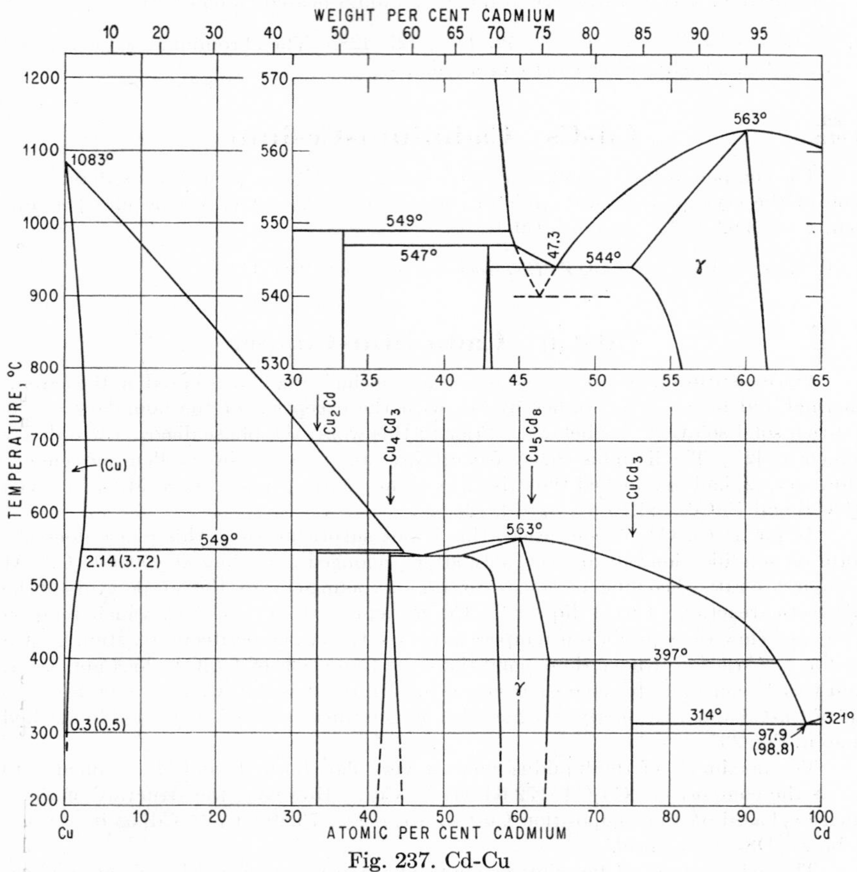

Fig. 237. Cd-Cu

as δ) is isotypic with γ-brass and, therefore, based on the composition Cu_5Cd_8 [4, 5], with a lattice constant of a = 9.654 A at 60.5–63 at. % Cd [4] and a = 9.615 A at 60.4 at. % Cd [5]. The atomic distribution differs from that in the structure of Au_5Zn_8 [4]. Powder patterns of the other intermediate phases are very complex [5]. The structure of Cu_2Cd was reported as that of the $MgZn_2$ (C14) type, a = 4.96 A, c = 7.99 A, c/a = 1.61 [12].

Supplement. In the phase diagram of Fig. 237 the homogeneity range of the γ phase extends up to 64 at. % Cd. According to [13], the limiting composition lies at a somewhat higher Cd content, about 66 at. (77.5 wt.) % (taken from a graph in wt. %). These authors made microscopic observations and measurements of magnetic susceptibility on some Cu-Cd alloys annealed at 400°C for 48 hr.

1. C. H. M. Jenkins and D. Hanson, *J. Inst. Metals*, **31**, 1924, 257–270.
2. R. Sahmen, *Z. anorg. Chem.*, **49**, 1906, 301–310.
3. The existence of a peritectically formed Cu-rich intermediate phase could already be concluded from microscopic studies by H. Le Chatelier, *Compt. rend.*, **130**, 1900, 87.
4. A. J. Bradley and C. H. Gregory, *Phil. Mag.*, **12**, 1931, 143–162.
5. E. A. Owen and L. Pickup, *Proc. Roy. Soc.* (*London*), **A139**, 1933, 526–541.
6. E. Raub, *Z. Metallkunde*, **38**, 1947, 119–120.
7. S. A. Pogodin, V. I. Mikheeva, and G. A. Kagan, *Izvest. Inst. Fiz.-Khim. Anal.*, **7**, 1935, 39–47.
8. C. T. Heycock and F. H. Neville, *J. Chem. Soc.*, **61**, 1892, 898.
9. G. Tammann and A. Heinzel, *Z. anorg. Chem.*, **176**, 1928, 148–149.
10. K. Bornemann and K. Wagenmann, *Ferrum*, **11**, 1913-1914, 289–314, 330–343.
11. E. A. Owen, *J. Inst. Metals*, **73**, 1947, 471–489.
12. P. I. Kripiakevich, E. J. Gladyshevskii, and E. E. Cherkashin, *Doklady Akad. Nauk S.S.S.R.*, **82**, 1952, 253–256; for errata, cf. **85**, 1952, 324.
13. H. Nowotny and H. Bittner, *Monatsh. Chem.*, **81**, 1950, 887–906, especially 893.

0.3038
$\bar{1}$.6962

Cd-Fe Cadmium-Iron

Cd is regarded as being insoluble in solid Fe [1]. No diffusion of Cd into Fe was observed at 750°C [2]. Inactivity of steel in molten Cd was also reported by [3]. [4] introduced small amounts of Fe into molten Cd at 400 and 700°C and found, by means of magnetic measurements, a solubility of 2 to 3 $\times$ 10^{-4} wt. % Fe.

The existence of an intermediate phase $FeCd_2$ has been suggested [5]; see also [6]. However, X-ray investigation showed the supposed compound to be a mixture of Cd and Fe [7]. No alloying of Cd and Fe was observed after Cd and Fe amalgams had been mixed and mercury distilled off at 360°C [8].

1. F. Wever, *Arch. Eisenhüttenw.*, **2**, 1928-1929, 739–746; *Naturwissenschaften*, **17**, 1929, 304–309.
2. N. W. Ageew and M. I. Zamotorin, *Ann. inst. polytech. Leningrad, sec. math. phys. sci.*, **31**, 1928, 15–28; abstract in *J. Inst. Metals*, **44**, 1930, 556.
3. E. J. Daniels, *J. Inst. Metals*, **46**, 1931, 87.
4. G. Tammann and W. Oelsen, *Z. anorg. Chem.*, **186**, 1930, 277–279.
5. E. J. Daniels, *J. Inst. Metals*, **49**, 1932, 178–179.
6. E. Isaac and G. Tammann, *Z. anorg. Chem.*, **55**, 1907, 61–62.
7. E. Scheil, *Z. Metallkunde*, **38**, 1947, 320.
8. F. Lihl and E. Buhl, *Z. Metallkunde*, **46**, 1955, 787–791.

0.2074
$\bar{1}$.7926

Cd-Ga Cadmium-Gallium

The phase diagram of Fig. 238 is based on thermal-analysis data [1]. In contrast to these findings, [2] reported that the melting point of Ga is continuously lowered by Cd additions until a eutectic is reached at 9 wt. (about 6 at.) % Cd and 18°C and that the liquidus temperature of the alloy with 14 wt. (about 9 at.) % Cd is 29°C. As to a similar discrepancy, see the systems Ga-Sn and Ga-Zn.

1. N. A. Puschin, S. Stepanović, and V. Stajić, *Z. anorg. Chem.*, **209**, 1932, 329–334.
2. W. Kroll, *Metallwirtschaft.*, **11**, 1932, 435–437.

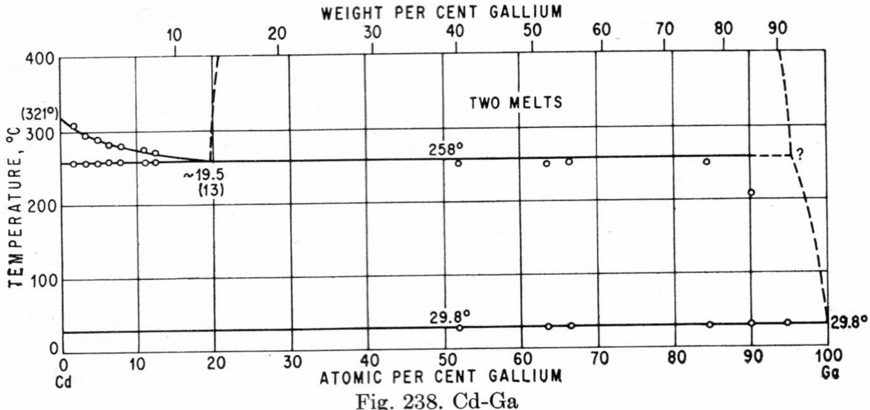

Fig. 238. Cd-Ga

0.1899
$\bar{1}.8101$

Cd-Ge Cadmium-Germanium

The phase diagram of Fig. 239 was established by [1] by means of thermal and microscopic analysis. The alloys were melted in open porcelain crucibles under a protective atmosphere. The liquidus above 35 wt. % Ge could not be determined because of volatilization of Cd.

1. H. Spengler, *Metall*, **8**, 1954, 937.

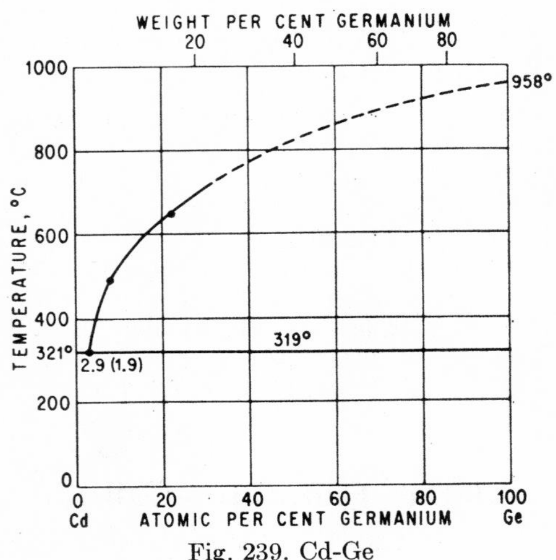

Fig. 239. Cd-Ge

2.0473
$\bar{3}.9527$

Cd-H Cadmium-Hydrogen

Neither solid nor molten Cd (investigated up to 400°C) dissolves hydrogen [1].

1. A. Sieverts and W. Krumbhaar, *Ber. deut. chem. Ges.*, **43**, 1910, 896.

$\bar{1}.7484$
0.2516

Cd-Hg Cadmium-Mercury

In the following presentation the phase diagram shown in Fig. 240, essentially due to the work by [1] and [2], will serve as a basis to discuss briefly the results of the many investigations dealing with the constitution of this system.

Liquidus Curve. Data covering the ranges 15–99.5 at. % Hg [1] and 1.7–79.5 at. % Hg [3] agree very well and indicate a slight change in direction at 36.5 at. % Hg, representing the composition of the melt of the peritectic equilibrium at 188°C. Various portions of the liquidus curve were determined by means of (*a*) thermal analysis between 99.691 and 99.927 wt. % Hg (indicating that 0.073 wt. % Cd raises the melting point of Hg 0.4°C) [4], 0–5.5 at. % Hg [5], 20–80.5 at. % Hg [6], 76.5–86.3 at. % Hg [7], and 73.4–100 at. % Hg [2]; (*b*) emf measurements between 0 and 65°C [8], 0 and 41°C [9]; and (*c*) chemical analysis between 0 and 99°C [10]. In addition, individual data, especially close to room temperature, were obtained by various methods [11–13, 1, 14, 7]. The peritectic reaction at −34°C was detected by [2]. [15] reported, without giving data, that he confirmed the liquidus curve of [1] in the range 34–74 at. % Hg.

The question whether Cd-Hg compounds exist was the subject of numerous papers in the older literature. The results of all these early studies are obsolete since the work by [1, 3] has proved that none of the compounds proposed can exist. More recently, however, it has been claimed, on the basis of roentgenographic investigation of alloys with 25 and 30 at. % Hg annealed at 147°C, that the phase Cd_3Hg (37.30 wt. % Hg) exists [16]. The very complex powder pattern excludes the possibility that these alloys consist of a mixture of the phases (Cd) and ω. The pattern was indexed as a b.c. tetragonal lattice, with a = 16.56 A, c = 12.11 A, c/a = 0.731, and 152 atoms (38 molecules Cd_3Hg) per unit cell. The fact that [17] found the X-ray pattern of the alloy with 27 at. % Hg, after annealing at 170°C, showed only the (Cd) phase was suggested by [16] to be due to the difference in annealing temperatures used (?). If the data reported by [16] are correct, either the phase relations would be quite different from those shown in Fig. 240 or one would have to assume that the compound Cd_3Hg is formed by reaction of (Cd) and ω at a temperature between 170 and 147°C.

Solidus. The solidus curves of the (Cd) and ω phases were determined dilatometrically [1]; see data points in Fig. 240. It appears doubtful that they represent equilibrium conditions. Also, a number of solidus points can be derived from emf isotherms for temperatures between 0 and 75°C [13, 3, 8, 7, 15]. In reviewing these latter data, [15] suggested the solidus temperatures in this temperature range to be best represented by a straight line connecting the points 71.3 at. % Hg, 60° C, and 82 at. % Hg, 0°C. [18] gave the solidus point of the alloy with 76.5 at. % Hg as 25°C and claimed this was an equilibrium point [19].

The Boundaries of the Miscibility Gaps. The compositions of the (Cd) and ω phases at the peritectic temperature of 188°C were obtained by extrapolation of the solidus curves as approximately 23.5 and 27 at. % Hg. The same uncertainty holds for the saturation limits of the (Cd) and ω phases, as shown by the rather widely scattered data points in Fig. 240 which are based on emf isotherms for 20°C [3], 25, 50, and 75°C [1], i.e., temperatures where the rate of diffusion is very slow; see also [20].

The extent of the peritectic horizontal at −34°C [2] is doubtful as the low rate of diffusion will not permit equilibrium to be attained. The course of the solidus curve of the ω phase [1, 15] would indicate a saturated solution of about 88 at. (93 wt.) % Hg at −34°C. The maximal thermal effect at −34°C was found at 96.5 at. % Hg [2].

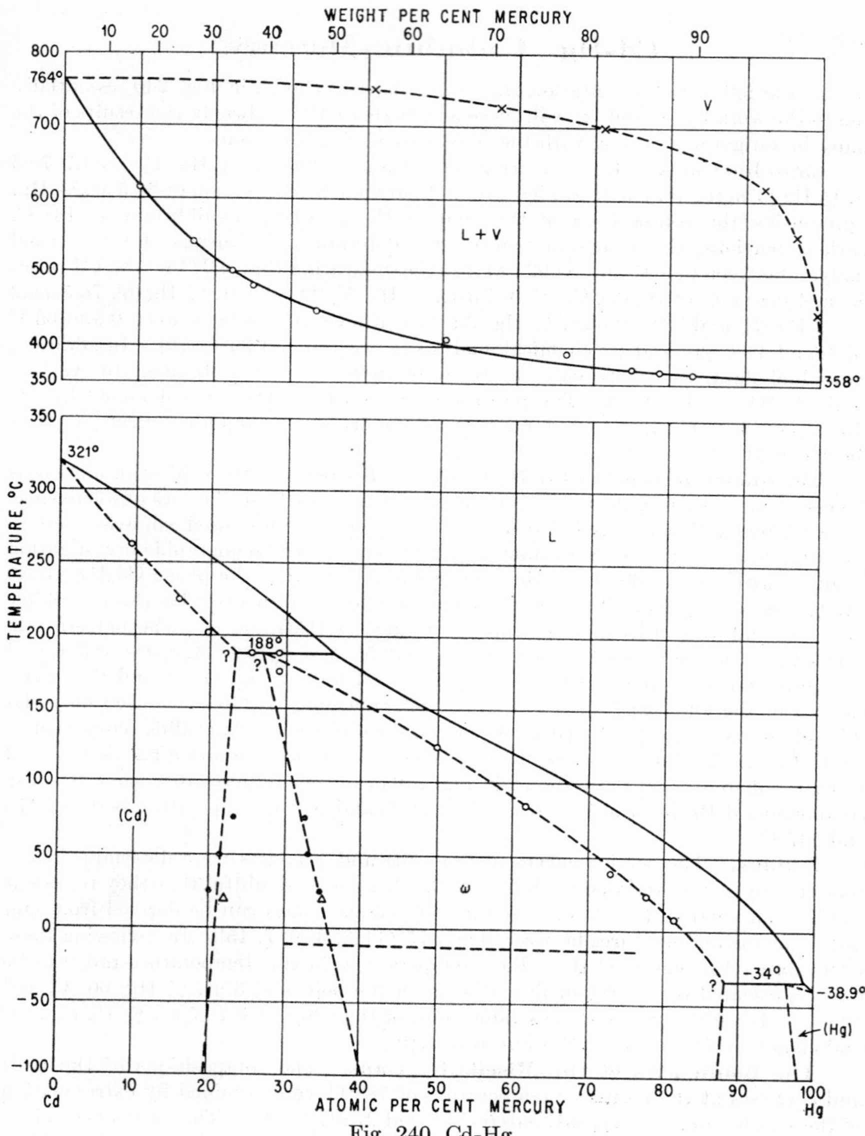

Fig. 240. Cd-Hg

Microscopic Investigations. Alloys with up to 37 at. % Hg were regarded to be solid solutions of Hg in Cd, those higher in Hg content solid solutions of Cd in Hg [3]. Microscopic studies and hardness measurements of cast alloys (!) indicated the two-phase field to be between approximately 18.5 and 36 at. % Hg [21].

A transformation in the ω phase was found by thermal analysis to occur at −10.5 to −14°C between 30 and 76 at. % Hg. No structural change was found by X-ray diffraction [2].

The temperatures of the beginning of boiling of Cd-Hg melts were determined and the temperatures of the end of boiling (vapor curve) were calculated by [22]; see Fig. 240.

Crystal Structures. The existence of the intermediate phase ω, first indicated by qualitative X-ray work [23, 24], was established by [17] and confirmed by [25]. According to [17], the dimensions of the tetragonal unit cell containing 2 atoms (a = 3.94 A, c = 2.91 A, c/a = 0.739) do not change with composition. However, [25] reported data showing the lattice spacings to change from a = 3.964 A, c = 2.849 A, c/a = 0.719 at 37 at. % Hg, to a = 3.926 A, c = 2.906 A, c/a = 0.740, at 73.7 at. % Hg. Lattice spacings of the h.c.p. (Cd) phase were given by [25]. As to the structure of the allegedly existing compound Cd_3Hg, see above.

Supplement. [26] redetermined the change in axial ratio with composition of the ω phase and found that—at variance with the results of previous investigators, especially those of [25]—the ratio decreases with increasing Hg content. This is in keeping with theoretical expectations [26].

1. H. C. Bijl, *Z. physik. Chem.*, **41**, 1902, 641–671.
2. R. F. Mehl and C. S. Barrett, *Trans. AIME*, **89**, 1930, 575–588.
3. N. A. Puschin, *Z. anorg. Chem.*, **36**, 1903, 201–254.
4. G. Tammann, *Z. physik. Chem.*, **3**, 1889, 445.
5. C. T. Heycock and F. H. Neville, *J. Chem. Soc.*, **61**, 1892, 888.
6. E. Jänecke, *Z. physik. Chem.*, **60**, 1907, 409.
7. A. Schulze, *Z. physik. Chem.*, **105**, 1923, 177–203.
8. F. E. Smith, *Natl. Phys. Lab. Collected Researches*, **6**, 1910, 137–163; *Phil. Mag.*, **19**, 1910, 250; *Z. physik. Chem.*, **95**, 1920, 293.
9. A. L. T. Moesveld and W. A. T. de Meester, *Z. physik. Chem.*, **130**, 1927, 146–153.
10. W. Kerp, W. Böttger, and H. Iggena, *Z. anorg. Chem.*, **25**, 1900, 59–67.
11. Gouy, *J. phys.*, **4**, 1895, 320–321.
12. W. J. Humphreys, *J. Chem. Soc.*, **69**, 1896, 1679–1691.
13. W. Jaeger, *Wied. Ann.*, **65**, 1898, 106–110.
14. G. A. Hullet and R. E. de Lury, *J. Am. Chem. Soc.*, **30**, 1908, 1811.
15. C. E. Teeter, *J. Am. Chem. Soc.*, **53**, 1931, 3927–3940.
16. N. W. Taylor, *J. Am. Chem. Soc.*, **54**, 1932, 2713–2720.
17. R. F. Mehl, *J. Am. Chem. Soc.*, **50**, 1928, 381–390.
18. T. W. Richards, H. L. Frevert, and C. E. Teeter, *J. Am. Chem. Soc.*, **50**, 1928, 1293–1302.
19. According to [18], attainment of equilibrium at 25°C would require several years.
20. G. Tammann and C. F. Marais, *Z. anorg. Chem.*, **138**, 1924, 162–166.
21. G. Tammann and Q. A. Mansuri, *Z. anorg. Chem.*, **132**, 1923, 69–70.
22. E. Kordes and F. Raaz, *Z. anorg. Chem.*, **181**, 1929, 225–236.
23. C. v. Simson, *Z. physik. Chem.*, **109**, 1924, 195–197.
24. C. E. Teeter [15] has evaluated data by [23].
25. T. Johnson and L. Vegard, *Skrifter Norske Videnskaps-Akad.*, **1947**, no. 2, 27–33; see *Structure Repts.*, **11**, 1947-1948, 57.
26. K. Schubert, U. Rösler, W. Mahler, E. Dörre, and W. Schütt, *Z. Metallkunde*, **45**, 1954, 643–647.

$\bar{1}.9910$
0.0090

Cd-In Cadmium-Indium

The liquidus curve in Fig. 241 is based on tabulated data by [1]. Earlier, [2] published a liquidus curve the hypoeutectic branch of which lies at much lower temperatures, perhaps due to considerable supercooling. The eutectic point was placed at 74.6 at. % In, 122.5°C [2], and 74 at. % In, 123.1°C [1].

The solid solubility of In in Cd, not determined accurately as yet, appears to be slight, i.e., certainly less than 1% at the eutectic temperature [2]. This is also indicated by measurements of electrical resistance [1]. The solubility of Cd in In was given as 18 at. % Cd at 120°C and about 15 at. % at "room temperature" [2]. [1] reported that a solubility of 18 at. % Cd was too high; however, the conclusion was

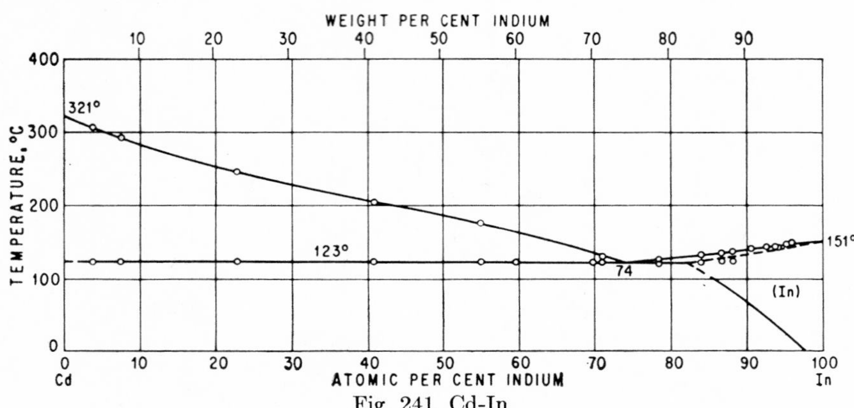

Fig. 241. Cd-In

based on thermal-analysis data only. More accurate (roentgenographic) determinations are due to [3], indicating that the solubility is 14.5 ± 0.5 at. % Cd at 100°C (X-ray work at this temperature) and 4.5 ± 0.5 at. % Cd at 20°C. The latter value was corroborated by [4].

1. S. Valentiner (A. Grönefeld), *Z. Metallkunde*, **35**, 1943, 250–253.
2. C. L. Wilson and O. J. Wick, *Ind. Eng. Chem.*, **29**, 1937, 1164–1166.
3. W. Betteridge, *Proc. Phys. Soc. (London)*, **50**, 1938, 519–524; the solubility data were ascertained by plotting the volume of the unit cell against composition. In the measurements at 100°C a continuous transition tetragonal → cubic was observed (at about 4.5 at. % Cd).
4. S. C. Carapella and E. A. Peretti, *Trans. ASM*, **43**, 1951, 854–856.

0.4586
$\bar{1}$.5414

Cd-K Cadmium-Potassium

The phase diagram (Fig. 242) is based on thermal-analysis data [1], with the exception that the intermediate phase, formerly believed to be either KCd_{11} or KCd_{12} [1], was identified as KCd_{13} [92.86 at. (97.34 wt.) % Cd] [2, 3]. [1] assumed also the compound KCd_7 [87.5 at. (95.26 wt.) % Cd], supposedly formed by the peritectic reaction L + KCd_{13} → KCd_7 at about 473°C. However, this conclusion was based solely on the dependence of the halting times of the secondary crystallization upon composition. Data points shown in the inset of Fig. 242 do not indicate two horizontals at 468 and 473°C. It should be mentioned also that the temperatures of the beginning and end of solidification varied considerably because of strong supercooling [4].

KCd_{13} is cubic of the $NaZn_{13}$ ($D2_3$) type, 8 molecules per unit cell, a = 13.80 A [2] or a = 13.81 A [3].

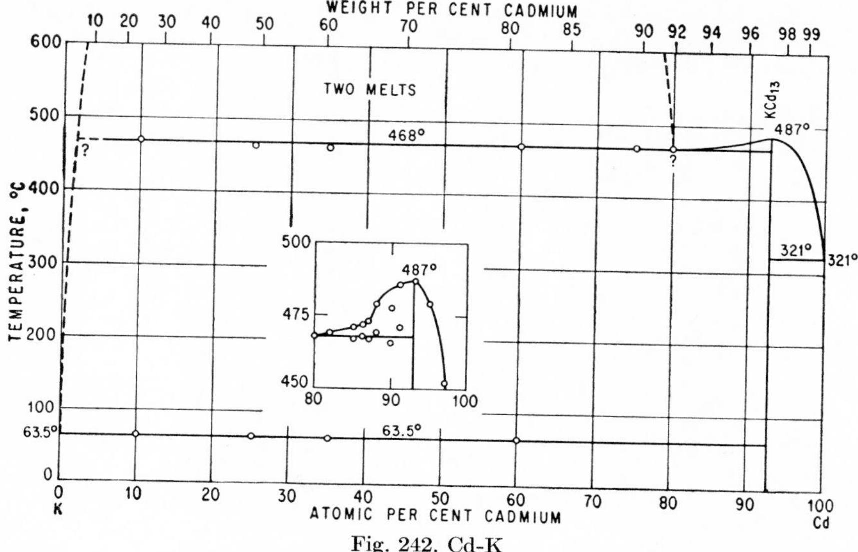

Fig. 242. Cd-K

1. D. P. Smith, *Z. anorg. Chem.*, **56**, 1908, 119–125.
2. E. Zintl and W. Haucke, *Z. Elektrochem.*, **44**, 1938, 104–111.
3. J. A. A. Ketelaar, *J. Chem. Phys.*, **5**, 1937, 668.
4. The solidification of the melts with 3.7, 4.5, and 4.9 wt. % K started after supercoolings of 35, 18, and 11°C, respectively. The secondary crystallization supercooled for 4–6°C, i.e., was of a magnitude as large as the difference between 468 and 473°C.

$\bar{1}.9080$
0.0920

Cd-La Cadmium-Lanthanum

The compound LaCd (44.73 wt. % Cd) having a melting point of 920°C [1] has been identified by determining its crystal structure, which is of the CsCl (B2) type, a = **3.89$_7$** A [2]. The Laves phase $LaCd_2$ (61.81 wt. % Cd) probably exists also [3].

[4] reported that (*a*) $LaCd_2$ is hexagonal, a = 5.075 A, c = 3.458 A, c/a = 0.681, and isostructural with $CeCd_2$ (see Cd-Ce); (*b*) $LaCd_{11}$ [91.67 at. (89.90 wt.) % Cd] has the same crystal structure as $BaHg_{11}$, with a = 9.339 A and 36 atoms in the cubic unit cell.

1. L. Rolla and A. Iandelli, *Ricerca sci.*, **20**, 1941, 1223.
2. A. Iandelli and E. Botti, *Gazz. chim. ital.*, **67**, 1937, 638–644.
3. H. Nowotny, *Z. Metallkunde*, **37**, 1946, 34.
4. A. Iandelli and R. Ferro, *Gazz. chim. ital.*, **84**, 1954, 463–478.

1.2094
$\bar{2}.7906$

Cd-Li Cadmium-Lithium

The phase relationships presented in Fig. 243 are based on comprehensive investigations [1] by means of thermal [2], resistometric [3], and dilatometric [3] analysis [4].

The existence of the various intermediate phases was corroborated by roentgenographic studies [5–9], especially [9]. Earlier work [10] agrees fairly well with Fig. 243 as far as the liquidus curve is concerned but is obsolete otherwise.

The powder patterns of the elevated temperature phase X, studied at 250°C using an alloy with 12.4 at. % Li, was tentatively interpreted as "a superimposition of the pattern of the (Cd) phase and that of a cubic phase with a = 4.683 A" [9].

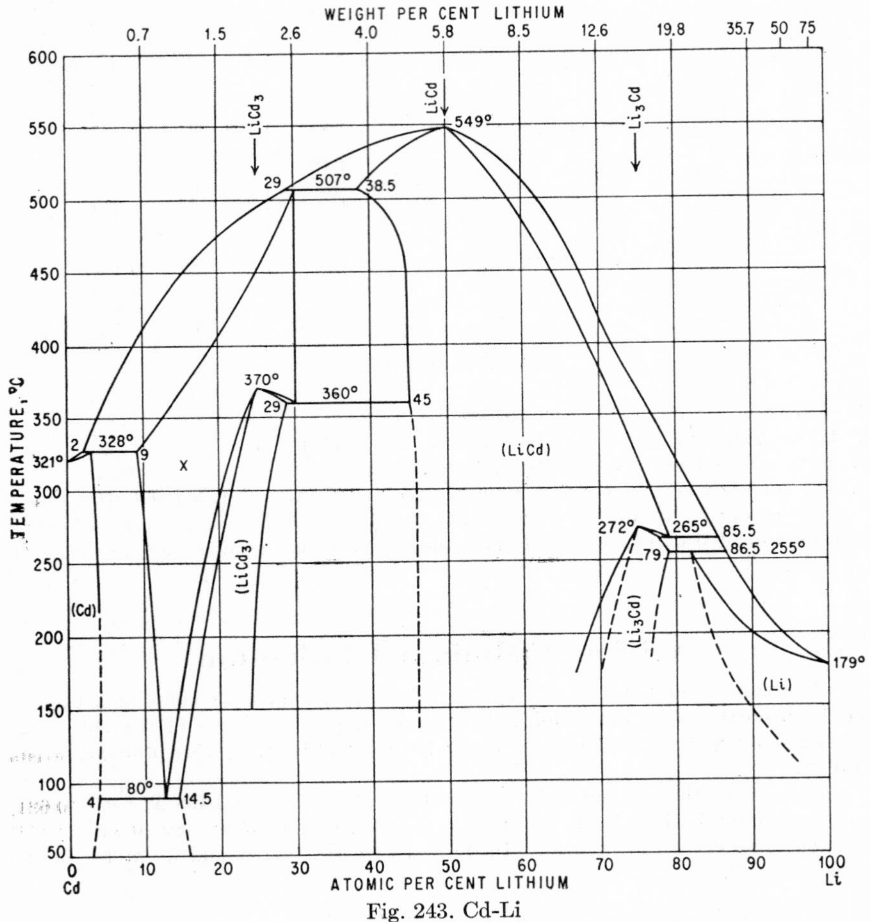

Fig. 243. Cd-Li

($LiCd_3$) is h.c.p. (A3 type) with statistical atomic distribution, 2 atoms per unit cell, a = 3.086 A, c = 4.899 A, c/a = 1.587 [9]. The structure of (LiCd), suggested by [5, 6] to be b.c.c. of the CsCl (B2) type with a = 3.33 A, was established to be of the NaTl (B32) type, a = 6.700 A [7–9]. (Li_3Cd) is f.c.c. (A1 type) with statistical atomic distribution, a = 4.259 A [9].

1. G. Grube, H. Vosskühler, and H. Vogt, *Z. Elektrochem.*, **38**, 1932, 869–880.
2. Alloys were prepared, using Li of 99%, in Fe crucibles under argon.

3. After slow cooling from high temperatures, resistance-vs.-temperature curves and dilatation curves were taken at rates of heating of 10°C per 6 min and 10°C per 8 min, respectively, in argon atmosphere.
4. All data points are shown in Fig. 187 of M. Hansen, "Der Aufbau der Zweistofflegierungen," p. 428, Springer-Verlag OHG, Berlin, 1936.
5. A. Baroni, *Atti reale accad. Lincei*, **18,** 1933, 41–44.
6. A. Baroni, *Z. Elektrochem.*, **40,** 1934, 565.
7. E. Zintl and G. Brauer, *Z. physik. Chem.*, **B20,** 1933, 245–271.
8. E. Zintl and A. Schneider, *Z. Elektrochem.*, **40,** 1934, 107; **39,** 1933, 95.
9. E. Zintl and A. Schneider, *Z. Elektrochem.*, **41,** 1935, 294–297.
10. G. Masing and G. Tammann, *Z. anorg. Chem.*, **67,** 1910, 194–197.

0.6648
$\bar{1}.3352$

Cd-Mg Cadmium-Magnesium

Solidification. Liquidus and solidus temperatures of alloys covering the entire range of composition were reported by [1–7, 31]. Disregarding a number of points by [1], the agreement as to the liquidus curve is, with the exception of the range 65–85 at. % Cd, quite good. Results by [4, 6, 7] are concordant within several °C over wide ranges of composition.

The solidus curves given by [2, 3, 5, 7] were determined by thermal analysis [2, 3, 5] or measurements of specific heat vs. temperature [7], whereas those due to [4, 6] are based on more exact micrographic work. The latter two agree within 5–15°C, except for the region of approximately 65–80 at. % Cd.

The assumption that the liquidus and solidus curves touch at 50 at. % [2, 3, 5] is obsolete. The same holds for the existence of (*a*) a compound $MgCd_2$ of fixed composition, assumed to be formed peritectically at 379°C and giving rise to another peritectic reaction at 358°C [4]; (*b*) a narrow two-phase field between 67 and 72 at. % Cd, originating from a peritectic reaction involving the terminal solid solutions [8]; and (*c*) two or even four peritectic reactions corresponding to the existence of one or three intermediate solid solutions, suggested on the basis of an interpretation of earlier work [9].

It was demonstrated [6] that the supposed existence of $MgCd_2$ had been erroneously concluded from microscopic examination of specimens containing a contaminant formed by atmospheric attack. "$MgCd_2$" had not been detected previously [2, 3, 10] and was denied later by [11, 8, 12, 5, 13].

Confirming earlier suggestions [2, 10, 3, 5], it has been conclusively established [6, 7, 14] that Mg and Cd form a continuous series of solid solutions at temperatures between the solidus and about 250°C. The liquidus and solidus curves presented in Fig. 244 are based on the excellent work by [6]; they are to be regarded as final. For the deduction of the solidification diagram from thermodynamic data, see [7]. Thermodynamic properties of the disordered solid solutions were determined by [14].

Transformations. The transformations in the solid state have been investigated extensively by (*a*) determination of transformation temperatures [2, 10, 3, 15, 4, 8, 16, 5, 17–20], (*b*) study of the crystal structure [11, 12, 21–26], and (*c*) determination of the specific heat and coefficient of thermal expansion vs. temperature [18–20, 27] and thermodynamic studies [20].

Whereas earlier investigators [2, 3, 15, 5] believed the transformation to be a polymorphic change of the compound MgCd, [4, 11] recognized the transformation in the vicinity of 50 at. % to be of the order-disorder type. This was verified by [12] and all later researchers using X-ray and other methods. [8] and especially [12] were the first who established that order-disorder transitions also occur in the region of the

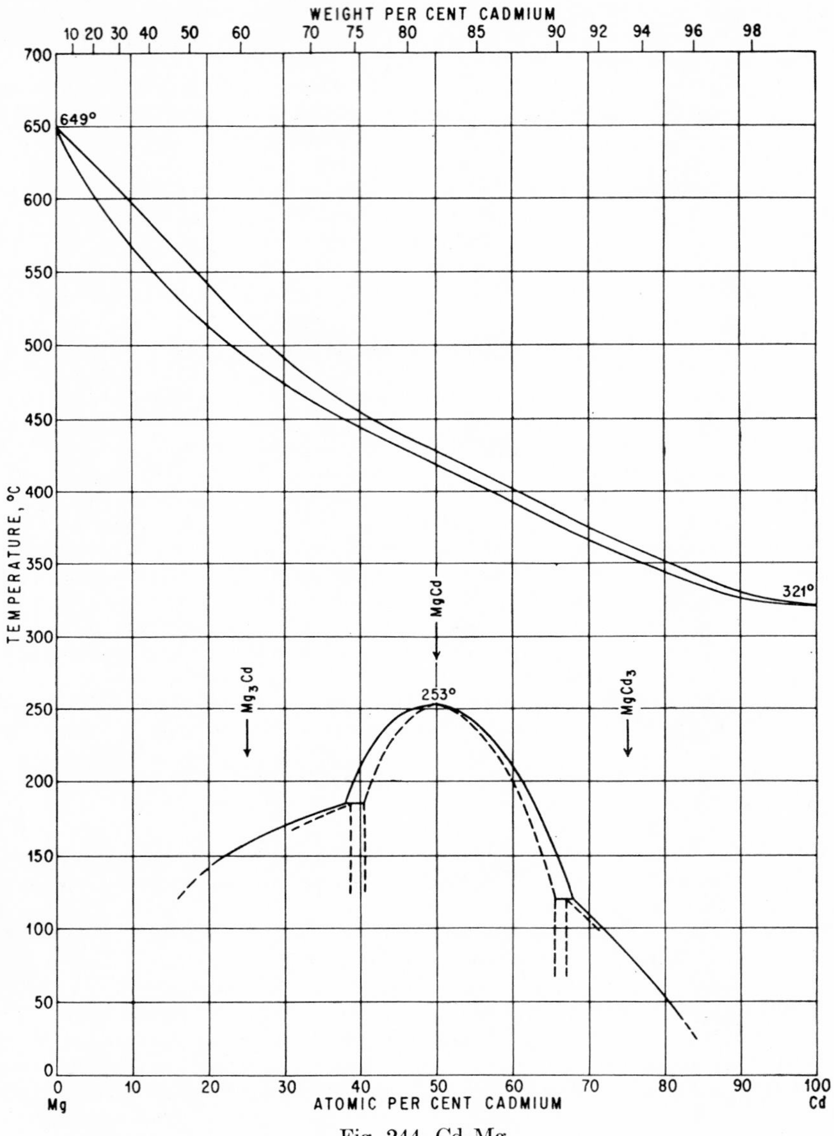

Fig. 244. Cd-Mg

compositions Mg_3Cd and $MgCd_3$. Ordering in quenched alloys takes place at a comparatively high rate [17, 28–30], even on storage at room temperature [21, 31, 6, 24].

Disorder-order temperatures of alloys having the compositions Mg_3Cd, MgCd, and $MgCd_3$ were reported as follows: Mg_3Cd, 150–160°C [8, 18, 24, 38]; MgCd, 246–258°C [2, 3, 10, 15, 4, 16, 8, 29, 18]; and $MgCd_3$, 80–94°C [8, 17, 18, 26]. Transformation temperatures of alloys covering a wider composition range were especially determined by [8, 19, 20]. They indicate that there is only a maximum at 50 at. %. The whole transformation curve apparently consists of three branches with points of inflection at approximately 38 and 68 at. % Cd [8, 19, 20]. However, [5] reported a maximal transition point at $MgCd_3$, 80°C.

The upper transformation curve of the end of disordering presented in Fig. 244 is a compromise of most of the data available, especially those of [8, 19, 20], which are in substantial agreement. Temperatures reported were obtained by means of thermal analysis [2, 3, 4, 29] and dilatometric [15, 8], resistometric [8, 18], and specific-heat-vs.-temperature measurements [19, 20]. Some of the data are not tabulated and had to be calculated from small-scale diagrams [19, 20].

[8, 20] also reported temperatures of the beginning of disordering, the latter authors only for the region MgCd; see also [32]. For the most part, these temperatures lie 5–20°C below the upper transition curve, according to the composition. On the other hand, it was concluded from measurements of heat capacity that the disordering process in the 50 at. % alloy starts at about −40°C [33] and in the 25 at. % Cd and 75 at. % Cd alloys at about 0°C and between −90 and +60°C, respectively [34]. This seems unlikely because it would mean that the order-disorder transitions in the system Cd-Mg would differ profoundly from those in the system Au-Cu.

Since the ordered states Mg_3Cd and $MgCd_3$ differ crystallographically from the ordered state MgCd, it seems reasonable that heterogeneous fields must exist. Indeed, there is evidence in diffraction patterns of the coexistence of $MgCd_3$ with the Cd-base terminal solid solution [8, 26]. Likewise, there will be two-phase regions separating the ordered and disordered phase fields. Inflections in the upper transition curve might indicate that the Mg_3Cd and $MgCd_3$ phases form peritectoidally [8]. Experimental evidence suggested that both curves touch at the compositions Mg_3Cd and $MgCd_3$ [8].

The boiling points of melts with 20, 40, 60, 80, 90, and 95 wt. % Cd were determined as about 1080, about 1050, 980, 872, 827, and 793°C, respectively [35].

Crystal Structures. The most accurate lattice parameters of the disordered solid solution stable above about 250°C are those of [6], obtained by exposure to 310°C (whole composition range), and [36] (2–16 at. % Cd); see also [21]. Papers dealing with X-ray investigations of the ordered structures are listed above. Apparently some of these data [12, 21] are affected by the susceptibility of the alloys to atmospheric attack [6, 13].

The ordered structure MgCd was found to be orthorhombic [21, 25, 37] of the AuCd (B19) type, $a = 5.005$ A, $b = 3.222$ A, $c = 5.270$ A [25], and $a = 4.99_9$ A, $b = 3.22_0$ A, $c = 5.26_8$ A [37].

Mg_3Cd and $MgCd_3$ are isotypic with Ni_3Sn ($D0_{19}$ type) [12, 21, 23, 26]. Mg_3Cd: $a = 6.31$ A, $c = 5.08$ A, $c/a = 0.805$ [23, 26]; $MgCd_3$: $a = 6.23$ A, $c = 5.04$–5.05 A, $c/a = 0.809$–0.811 [21, 26].

Note Added in Proof. The formation of order in the alloy Mg_3Cd was studied by [38] using X-ray and optical-diffraction experiments. The results support the view that ordering progresses by the interchange of nearest-neighbor atoms (in such a way that Cd atoms avoid each other) throughout the crystal and not by the growing of ordered nuclei.

1. O. Boudouard, *Compt. rend.*, **134**, 1902, 1431; *Bull. soc. chim. France*, **27**, 1902, 854–858.
2. G. Grube, *Z. anorg. Chem.*, **49**, 1906, 72–77.
3. G. Bruni and C. Sandonnini, *Z. anorg. Chem.*, **78**, 1912, 277–281.
4. W. Hume-Rothery and S. W. Rowell, *J. Inst. Metals*, **38**, 1927, 137–154.
5. N. I. Stepanov and I. I. Kornilov, *Izvest. Sektora Fiz.-Khim. Anal.*, **10**, 1938, 67–77.

6. W. Hume-Rothery and G. V. Raynor, *Proc. Roy. Soc. (London)*, **A174**, 1940, 471–486.
7. M. Hirabayashi, *Nippon Kinzoku Gakkai-Shi* (*J. Japan. Inst. Metals*), **16**, 1952, 295–299.
8. G. Grube and E. Schiedt, *Z. anorg. Chem.*, **194**, 1930, 190–222.
9. E. Jänecke, "Kurzgefasstes Handbuch aller Legierungen," pp. 46, 50, Otto Spamer, Leipzig, 1937; "Handbuch aller Legierungen," pp. 333–336, Carl Winter, Heidelberg, 1949; E. Jänecke, L. Neundeubel, and K. Rumpf, *Z. Metallkunde*, **30**, 1938, 424–429.
10. G. G. Urasow, *Z. anorg. Chem.*, **73**, 1912, 31–47.
11. G. Natta, *Ann. chim. appl.*, **18**, 1928, 135–188.
12. U. Dehlinger, *Z. anorg. Chem.*, **194**, 1930, 223–238.
13. G. E. R. Schulze, *Z. Metallkunde*, **32**, 1940, 252.
14. F. A. Trumbore, W. E. Wallace, and R. S. Craig, *J. Am. Chem. Soc.*, **74**, 1952, 132–136.
15. J. Valentin, *Rev. mét.*, **23**, 1926, 216–218.
16. M. Smialowski, *Prace Zakladu Metalurgicznego Politechniki Warsawskie*, **4**, 1934, 100–139; *Met. Abstr.*, **4**, 1937, 623.
17. N. I. Stepanov and I. I. Kornilov, *Izvest. Sektora Fiz.-Khim. Anal.*, **10**, 1938, 79–95, 97–112.
18. S. Nagasaki, M. Hirabayashi, and H. Nagasu, *Nippon Kinzoku Gakkai-Shi*, **13**, 1949, 1–6.
19. M. Hirabayashi, S. Nagasaki, H. Maniwa, and H. Nagasu, *Nippon Kinzoku Gakkai-Shi*, **13**, 1949, 6–10.
20. M. Hirabayashi, H. Maniwa, and S. Nagasaki, *Nippon Kinzoku Gakkai-Shi*, **14**, 1950, 6–8.
21. K. Riederer, *Z. Metallkunde*, **29**, 1937, 423–426.
22. I. Igarashi and M. Kosaki, *Sumimoto Kinzoku Kogyo Kenkyu Hokoku*, **3**, 1939, 410–418; *Met. Abstr.*, **6**, 1939, 238.
23. N. V. Ageev and D. L. Ageeva, *Izvest. Akad. Nauk S.S.S.R.* (*Khim.*), **1946**, 143–152; *Structure Repts.*, **10**, 1945-1946, 32–33.
24. H. Steeple and L. Lipson, *Nature*, **167**, 1951, 110–111.
25. H. Steeple, *Acta Cryst.*, **5**, 1952, 247–249; *Nature*, **167**, 1951, 481.
26. D. A. Edwards, W. E. Wallace, and R. S. Craig, *J. Am. Chem. Soc.*, **74**, 1952, 5256–5261.
27. M. Hirabayashi, *Nippon Kinzoku Gakkai-Shi*, **15**, 1951, 237–241.
28. N. I. Stepanov and S. A. Burlach, *Compt. rend. acad. sci. U.R.S.S.*, **4**, 1935, 147–151.
29. I. I. Kornilov, *Izvest. Akad. Nauk S.S.S.R.* (*Khim.*), **1937**, 313–331.
30. I. I. Kornilov, *Izvest. Akad. Nauk S.S.S.R.* (*Khim.*), **1938**, 437–449.
31. M. Goto, M. Nito, and H. Asada, *Rep. Aeronaut. Research Inst. Tokyo Imp. Univ.*, **12**, 1937, 163–180.
32. M. Hirabayashi, *Trans. AIME*, **200**, 1954, 673–674.
33. C. B. Satterthwaite, R. S. Craig, and W. E. Wallace, *J. Am. Chem. Soc.*, **76**, 1954, 232–238.
34. L. W. Coffer, R. S. Craig, C. A. Krier, and W. E. Wallace, *J. Am. Chem. Soc.*, **76**, 1954, 241–244.
35. W. Leitgebel, *Z. anorg. Chem.*, **202**, 1931, 305–324.
36. G. V. Raynor, *Proc. Roy. Soc.* (*London*), **A174**, 1940, 457–471.
37. K. Schubert, U. Rösler, W. Mahler, E. Dörre, and W. Schütt, *Z. Metallkunde*, **45**, 1954, 643–647.
38. H. Steeple and I. G. Edmunds, *Acta Cryst.*, **9**, 1956, 934–941.

0.3110
$\bar{1}.6890$

Cd-Mn Cadmium-Manganese

Powdered electrolytic Mn, when heated in molten Cd up to about 800°C in an open crucible and to 1200°C in an evacuated sealed silica bulb, does not form alloys with cadmium. The lattice parameter of α-Mn was unchanged whereas that of Cd decreased slightly because of a slight solubility of Mn [1].

1. U. Zwicker, *Z. Metallkunde,* **41,** 1950, 399–400.

0.9044
$\bar{1}.0956$

Cd-N Cadmium-Nitrogen

Nitrogen does not dissolve in solid and liquid Cd (investigated up to 400°C) [1].

1. A. Sieverts and W. Krumbhaar, *Ber. deut. chem. Ges.,* **43,** 1910, 894.

0.6891
$\bar{1}.3109$

Cd-Na Cadmium-Sodium

The constitution was studied by thermal analysis several times [1–4] (Fig. 245). Also, the effect of small Cd additions on the melting point of Na [5, 6] and of small Na additions on the melting point of Cd [7] was carefully investigated.

After a cursory determination of the liquidus curve by [1] the same authors [2] carried out a comprehensive thermal analysis (55 melts). The liquidus curve was found to have two maxima, at $NaCd_6$ [14.29 at. (3.30 wt.) % Na], 363.5°C, and $NaCd_2$ (9.28 wt. % Na), 384°C, and three eutectics at 5.5 at. (1.2 wt.) % Na, 291°C; 19.2 at. (4.7 wt.) % Na, 351°C; and 99.29 at. (96.65 wt.) % Na, 95.4°C. The last eutectic point is in good agreement with that found earlier [5, 6]: 99.42 at. (97.35 wt.) % Na, 95.4–95.7°C.

The phase diagram published by [3] is based on data of the cooling curves of 21 melts with 5 to 85 at. % Na. The existence of $NaCd_2$ was confirmed (melting point 385°C), but instead of $NaCd_6$ the phase $NaCd_5$ [16.67 at. (3.93 wt.) % Na] was assumed (melting point 360°C). According to the more numerous thermal data in this range given by [2] (alloys were analyzed after the test), the maximum melting point at $NaCd_6$ appears to be more likely. However, it must be mentioned that the maximum is very flat: between 12.4 and 17.3 at. % Na the liquidus temperature was found to lie in the interval 358.5–363.5°C [2].

[8] have titrated a solution of Na in NH_3 with a solution of CdI_2 in NH_3. "The results appear to indicate a phase between $NaCd_7$ and $NaCd_5$." Likewise, emf measurements [9] do not permit a decision between $NaCd_6$ and $NaCd_5$ since a discontinuous voltage change—besides one at 34 at. % Na ($NaCd_2$)—was found between 15 and 20 at. % Na. [10] even assumed, without giving a reason, the existence of $NaCd_4$.

In a cursory reinvestigation of the system between 1 and 32.6 at. % Na, using thermal analysis, [4] confirmed the phase $NaCd_2$ (melting point about 390°C) and found the liquidus points of the melts $NaCd_6$ and $NaCd_5$ as 358 and 350°C, respectively. This is in favor of the findings by [2]. He also confirmed the existence of a range of homogeneity of the phase $NaCd_6$, already reported by [2] on the basis of thermal data and metallographic examination.

The diagram by [3] further deviates from that by [2] in that it shows a miscibility gap in the liquid state which extends at 332°C (monotectic temperature) from about 60 to about 70 at. % Na. It is somewhat doubtful whether this miscibility gap exists as [2]—in their otherwise very careful work—did not observe thermal effects due to a

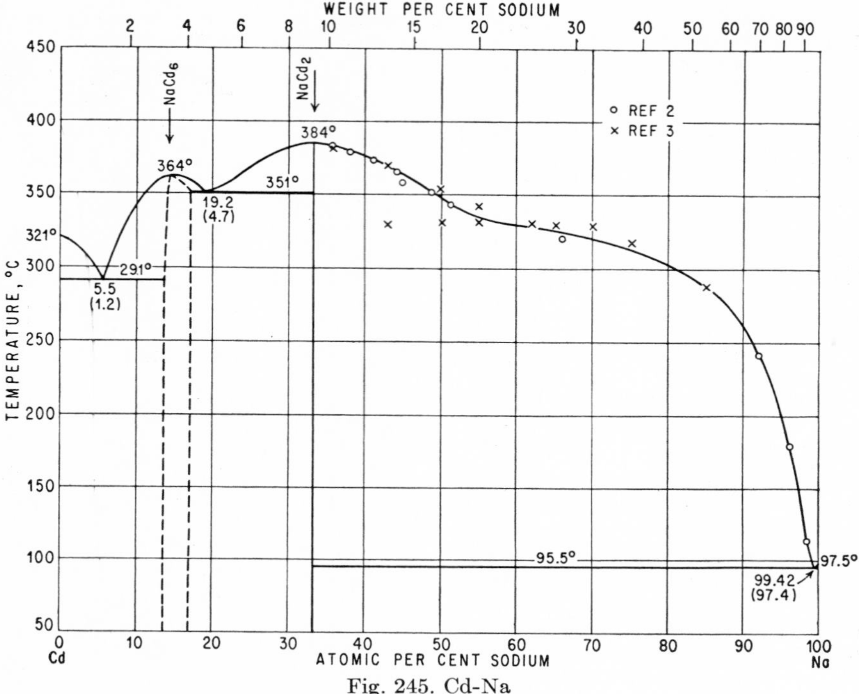

Fig. 245. Cd-Na

monotectic reaction. According to [2], the liquidus has a point of inflection at approximately 50 at. % Na.

[11] has tried to determine the crystal structure of $NaCd_2$.

1. N. S. Kurnakow and A. N. Kusnetzow, *Z. anorg. Chem.*, **23,** 1900, 455–462.
2. N. S. Kurnakow and A. N. Kusnetzow, *Z. anorg. Chem.*, **52,** 1907, 173–185.
3. C. H. Mathewson, *Z. anorg. Chem.*, **50,** 1906, 180–187.
4. S. Allaria, *Atti reale accad. sci. Torino,* **78,** 1942-1943, 145–158.
5. G. Tammann, *Z. physik. Chem.*, **3,** 1889, 447.
6. C. T. Heycock and F. H. Neville, *J. Chem. Soc.*, **55,** 1889, 673.
7. C. T. Heycock and F. H. Neville, *J. Chem. Soc.*, **61,** 1892, 897.
8. E. Zintl, J. Goubeau, and W. Dullenkopf, *Z. physik. Chem.*, **A154,** 1931, 43.
9. R. Kremann and P. v. Reininghaus, *Z. Metallkunde,* **12,** 1920, 285–287.
10. E. Jänecke, *Z. Metallkunde,* **20,** 1928, 117.
11. L. Pauling, *J. Am. Chem. Soc.*, **45,** 1923, 2779–2780.

0.2822
$\bar{1}$.7178

Cd-Ni Cadmium-Nickel

Figure 246 shows the thermal-analysis data points obtained by [1] and [2]. The partial phase diagrams constructed from these data and the results of microscopic examination agree in principal but deviate in detail. Whereas [1] assumed the existence of $NiCd_4$ (11.55 wt. % Ni), [2] considered the formula $NiCd_7$ [12.5 at. (6.94 wt.) % Ni] more likely. As alloys with 12 [1] and 7 wt. % Ni [2] (20.7 and 12.6 at. % Ni) were found to consist nearly of one phase, the phase coexisting with

Ni is of variable composition. According to [3], it is based on the ideal composition Ni_5Cd_{21} [19.23 at. (11.06 wt.) % Ni] because roentgenographic investigation of an alloy with 17.5 at. % Ni showed the crystal structure to be of the γ-brass ($D8_{1-3}$) type, a = 9.781 A; see also [4, 5].

[2] established a eutectic at 0.25 wt. (0.47 at.) % Ni and 318°C; [1] had falsely reported the solidus temperature to coincide with the melting point of Cd, 321°C. The peritectic temperature was found as 502 [1] and 490°C [2], respectively. The nature of the transformation at 405°C [1], which is characterized by an increase in volume with fall in temperature, is unknown; perhaps it is the polymorphic transformation of another intermediate phase X since no mention was made of the formation or decomposition of a phase other than X.

Supplement. The existence of a γ-brass phase in the Ni-Cd system has been confirmed by [6]. On the basis of measurements of magnetic susceptibility on a few alloys annealed at 400°C for 48 hr, [6] suggested the high Cd boundary of the γ phase to be located at 5 wt. (9.2 at.) % Ni (cf. Fig. 246). [7] prepared alloys containing between 1.7 and 99 wt. % Ni by gradual vaporization of Cd from a Cd-rich alloy as well as by thermal decomposition of Cd-Ni amalgams (residual Hg in solid solution). X-ray examination showed three intermediate phases to exist: γ, at about 18.5 at. (10.5 wt.) % Ni, with a structure similar to the γ-brass type, $a = 2 \times 9.753$ kX; γ_1, homogeneous between about 29 and 29.5 at. (17.5 and 18 wt.) % Ni, with γ-brass structure, a = 9.655 kX; and β (~NiCd), homogeneous between about 47 and 48.5 at. (31.5 and 33 wt.) % Ni. No indications of mutual solid solubilities of the elements were found. The state of the alloys investigated by [7] corresponds to an equilibrium temperature of 300–400°C.

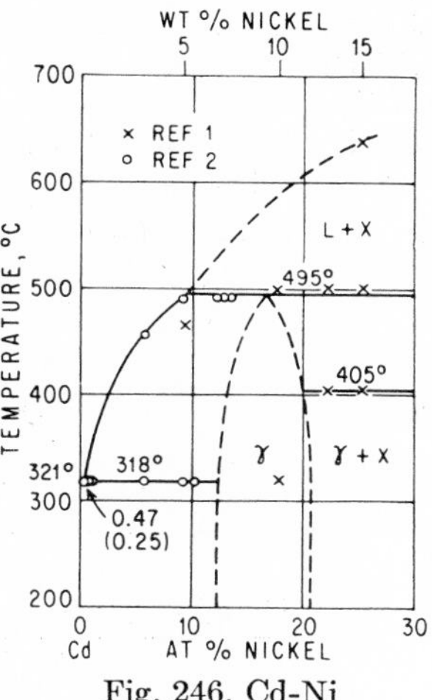

Fig. 246. Cd-Ni

1. G. Voss, *Z. anorg. Chem.*, **57**, 1908, 69–70.
2. C. E. Swartz and A. J. Phillips, *Trans. AIME*, **111**, 1934, 333–336.
3. W. Ekman, *Z. physik. Chem.*, **B12**, 1931, 69–77; see also A. Westgren, *Z. Metallkunde*, **22**, 1930, 372.
4. A. Roux and J. Cournot, *Rev. mét.*, **26**, 1929, 655–661.
5. U. Dehlinger and H. Nowotny, *Z. Metallkunde*, **35**, 1943, 151–152.
6. H. Nowotny and H. Bittner, *Monatsh. Chem.*, **81**, 1950, 887–906, especially 902.
7. F. Lihl and E. Buhl, *Z. Metallkunde*, **46**, 1955, 787–791.

0.8467
$\bar{1}$.1533

Cd-O Cadmium-Oxygen

CdO (12.46 wt. % O) is isotypic with NaCl (B1) type. The parameter was reported as 4.69–4.73 A [1]. In a study of the oxidation of molten Cd at 400–550°C an oxide was detected which was not identical with CdO [2].

Supplement. Precision measurements of the lattice parameter of CdO: a = 4.6943 A [3]; a = 4.6953 A (at 27°C) [4].

1. *Strukturbericht*, **1**, 1913-1928, 120; **2**, 1928-1932, 218.
2. W. Gruhl and G. Wassermann, *Z. Metallkunde*, **41**, 1950, 178–184.

3. J. C. Felipe, *Rev. real acad. cienc. exact., fís. y nat., Madrid,* **34,** 1940, 180–195. Quoted in [4]; *Chem. Abstr.,* **45,** 1951, 3215.
4. H. E. Swanson and R. K. Fuyat, *Natl. Bur. Standards (U.S.), Circ.* 539(2), 1953, 27–28.

0.5598
$\bar{1}$.4402

Cd-P Cadmium-Phosphorus

The existence of the phosphides Cd_3P_2 (15.52 wt. % P) and CdP_2 (35.53 wt. % P) was established by structure determinations. The structure of Cd_3P_2 is not cubic of the Mn_2O_3 ($D5_3$) type, as first reported by [1], but tetragonal of the Zn_3P_2 ($D5_9$) type, with 8 molecules per unit cell and a = 8.76 A, c = 12.30 A, c/a = 1.404 [2]. The small cubic cell (2 molecules per unit cell and a = 6.07 A) suggested by [3] was refuted by [2]. CdP_2 is tetragonal, with 8 molecules per unit cell and a = 5.29 A, c = 19.74 A, c/a = 3.73 [2].

1. R. Paulus, *Z. physik. Chem.,* **B22,** 1933, 305–322.
2. M. v. Stackelberg and R. Paulus, *Z. physik. Chem.,* **B28,** 1935, 427–460.
3. L. Passerini, *Gazz. chim. ital.,* **58,** 1928, 655–664, 775.

$\bar{1}$.7344
0.2656

Cd-Pb Cadmium-Lead

The liquidus curve (Fig. 247) is known with great accuracy; the results of the various determinations which cover either the whole range of composition [1–5] or

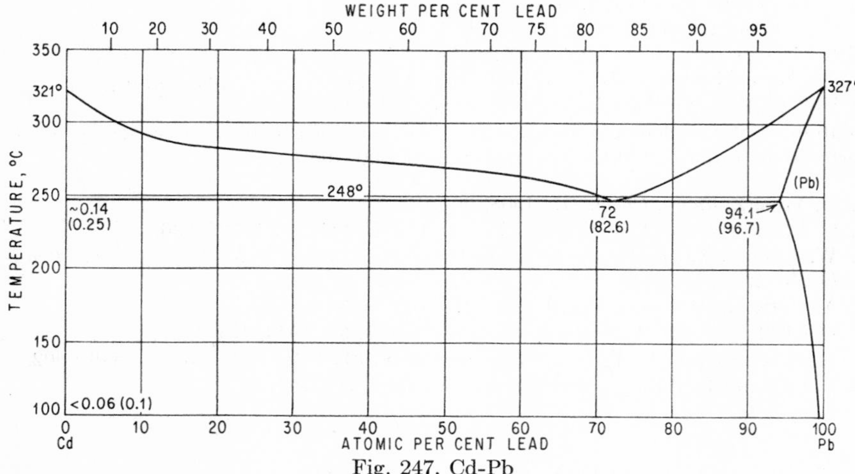

Fig. 247. Cd-Pb

only portions of it [6–8, 25] agree within about 5°C [9]. The eutectic temperature was found to be 249 [1], 249–252, average 250 [2], 247.3 [3], 248 [8, 25], 245 [4], and 247°C [5] and the eutectic composition about 69.8 [1], 67.2 [2], 72.0 [3], 71.9 [8, 10, 25], and 72.15 at. % Pb [5].

From measurements of the electrical conductivity [11, 12], density [13, 8], hardness [8, 14, 15], and electromotive force [16, 17], it can be concluded that the solid solubility of Pb in Cd is only very slight. By means of a special microscopic method [18, 19] the solubility was found to be >0.1 wt. % Pb at 270°C [18], 0.2–0.3 wt.

(0.11–0.16 at.) % Pb at 248°C [19], and less than 0.1 wt. (0.06 at.) % Pb after "slow cooling" [18, 19].

Measurements of the electrical conductivity (or resistivity) [14, 19, 20], electromotive force [16, 21, 17, 20], hardness [22, 8, 14, 19, 15], and thermoelectric force [23] indicate the existence of lead-rich solid solutions. However, the findings do not permit quantitative conclusions to be drawn. [20] reported solubilities of <2 wt. (<3.6 at.) % at 50°C and 3–6 wt. (5.4–10.5 at.) % Cd at 200°C. By means of measurements of hardness and electrical resistance, [24] found 3.1 wt. (5.6 at.) % Cd at 248°C, 2.3 wt. (4.2 at.) % at 200°C, 1.7 wt. (3.1 at.) % at 150°C, and 1.2 wt. (2.2 at.) % Cd at 100°C. The more accurate values are those by [25], based on measurements of electrical resistance and microscopic studies of long-annealed and quenched samples: 3.3 wt. (5.9 at.) % at 248°C, 1.6 wt. (2.9 at.) % at 200°C, 0.8 wt. (1.45 at.) % at 150°C, and 0.4 wt. (0.7 at.) % Cd at 100°C. [25] also determined the solidus curve of the Pb-rich solid solutions, using micrographic analysis.

Supplement. For boiling points of some Cd-Pb alloys, see [26]. By means of measurements of electrical resistivity on quenched alloys, [27] determined the solubility of Cd in solid Pb at 232°C to 4.5 at. % Cd. This value is in excellent agreement with the solvus curve of [25] (Fig. 247). The lattice parameter of Pb is only slightly decreased by addition of Cd [28]; see also [24]. The work by [27, 28] showed that quenched supersaturated (Pb) phase alloys can be kept undecomposed only for restricted periods of time. The publication of liquidus and solidus data of high accuracy, measured by "quantitative thermal analysis," has been announced [29].

1. A. W. Kapp, *Ann. Physik,* **6,** 1901, 764–770.
2. E. Jänecke, *Z. physik. Chem.,* **60,** 1907, 399, 409.
3. W. E. Barlow, *J. Am. Chem. Soc.,* **32,** 1910, 1392–1394; *Z. anorg Chem.,* **70,** 1911, 181–183.
4. C. diCapua, *Rend. accad. nazl. Lincei,* **31**(1), 1922, 162–164.
5. E. Abel, O. Redlich, and J. Adler, *Z. anorg. Chem.,* **174,** 1928, 265–268.
6. C. T. Heycock and F. H. Neville, *J. Chem. Soc.,* **61,** 1892, 903, 907; **65,** 1894, 65–76.
7. A. Stoffel, *Z. anorg. Chem.,* **53,** 1907, 151–152.
8. J. Goebel, *Z. Metallkunde,* **14,** 1922, 388–390.
9. For data points see M. Hansen, "Der Aufbau der Zweistofflegierungen," p. 440, Springer-Verlag OHG, Berlin, 1936.
10. M. Cook, *J. Inst. Metals,* **31,** 1924, 297.
11. A. Matthiessen, *Pogg. Ann.,* **110,** 1860, 208.
12. B. Beckman, *Arkiv Mat., Astron. Fysik,* **7,** 1912, 1.
13. A. Matthiessen and M. Holzmann, *Pogg. Ann.,* **110,** 1860, 33.
14. C. diCapua and M. Arnone, *Rend. accad. nazl. Lincei,* **33**(1), 1924, 293–297.
15. W. Schischokin and W. Agejewa, *Z. anorg. Chem.,* **193,** 1930, 240.
16. M. Herschkowitsch, *Z. physik. Chem.,* **27,** 1898, 140–141.
17. R. Kremann and H. Langbauer, *Z. anorg. Chem.,* **127,** 1923, 240.
18. G. Tammann and A. Heinzel, *Z. anorg. Chem.,* **176,** 1928, 148.
19. G. Tammann and H. Rüdiger, *Z. anorg. Chem.,* **192,** 1930, 3–9.
20. B. G. Petrenko and E. E. Cherkashin, *Ukrain. Khem. Zhur.,* **12,** 1937, 385–396; *Met. Abstr.,* **5,** 1938, 148.
21. P. Fuchs, *Z. anorg. Chem.,* **109,** 1920, 84–85.
22. P. Ludwik, *Z. anorg. Chem.,* **94,** 1916, 168, 174–175.
23. A. Battelli, *Atti ist. Veneto,* **5,** 1887; *Wied. Ann. Beibl.,* **12,** 1888, 269; see also W. Broniewski, *Rev. mét.,* **1910,** 354–357.
24. E. Jenckel and H. Mäder, *Metallwirtschaft,* **16,** 1937, 499–502.

25. E. C. Rollason and V. B. Hysel, *J. Inst. Metals,* **63,** 1938, 191–200.
26. W. Leitgebel, *Z. anorg. Chem.,* **202,** 1931, 305–324.
27. A. Pasternak, *Bull. intern. acad. polon. sci., Classe sci. math. nat.,* sér. A, **1951,** 177–192 (in English).
28. C. Tyzack and G. V. Raynor, *Acta Cryst.,* **7,** 1954, 505–510, especially 508.
29. W. Oelsen et al., *Arch. Eisenhüttenw.,* **27,** 1956, 487–511, especially 511.

0.0226
$\bar{1}$.9774

Cd-Pd Cadmium-Palladium

The phase diagram as shown in Fig. 248 was outlined by means of thermal, powder X-ray, micrographic, and magnetic investigations [1]. Only between 70 and

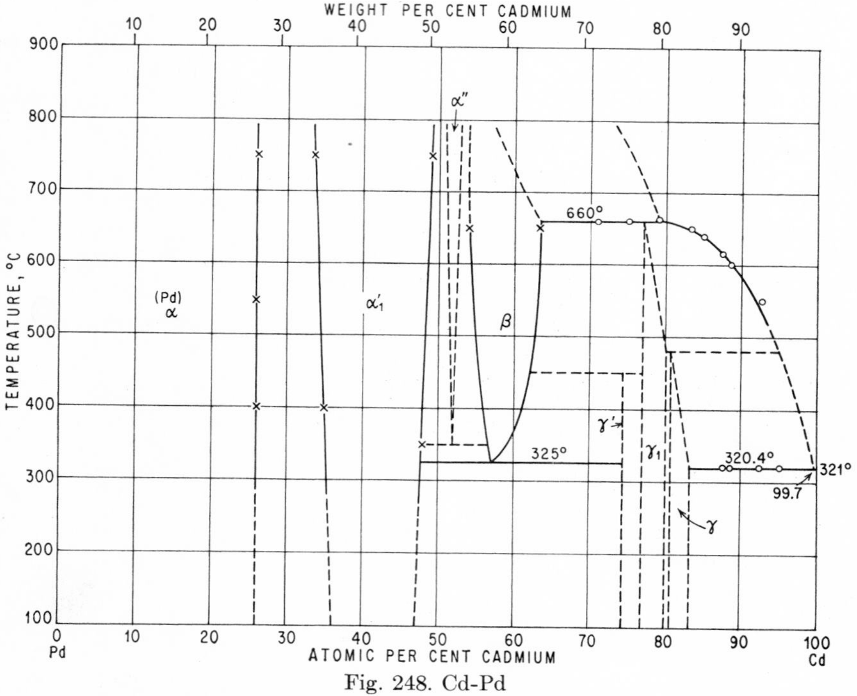

Fig. 248. Cd-Pd

100 at. % Cd could alloys be melted in open crucibles; those higher in Pd were prepared under nitrogen pressure of 150–190 atm. Alloys were annealed at and quenched from various temperatures between 300 and 750°C, according to the composition; annealing times are not given.

The α_1' phase is f.c. tetragonal of the CuAu ($L1_0$) type [2, 1], with $a = 4.28_6$ A, $c = 3.62_2$ A, $c/a = 0.845$ [1] at 49.0 at. % Cd. Its homogeneity range, as well as that of the (Pd) phase, was determined by lattice-parameter measurements (see data points in Fig. 248).

The boundaries of the b.c.c. β phase were placed at 750, 650, and 400°C by lattice-parameter and magnetic-susceptibility measurements. Its lattice spacing is $a = 3.25_3$ A at 58.8 at. % Cd [1]. The powder photograph of the phase designated as α'' is somewhat related to that of α_1'. In case it is a metastable phase, as considered possible by the authors, a phase field should not have been inserted in the diagram.

Three different kinds of b.c.c. Hume-Rothery γ phases ($D8_{1-3}$ type) [3] were found to exist: (*a*) the γ phase proper, stable between about 80.7 and 83.3 at. % Cd, with a = 9.96 A at 82.5 at. % Cd (Pt_5Cd_{21} corresponds to 80.77 at. % Cd); (*b*) the γ_1 phase between about 77 and 80 at. % Cd, probably of lower symmetry, with a = 9.93 A at 78.0 at. % Cd; and (*c*) the γ' phase representing still another modification. Measurements of the magnetic susceptibility [4, 1] proved helpful in locating the boundaries between these slightly different phases.

According to [5], a eutectic point lies at 0.3 wt. % Pd and a temperature 0.6°C below the melting point of Cd.

1. H. Nowotny, A. Stempfl, and H. Bittner, *Monatsh. Chem.*, **82,** 1951, 949–958.
2. H. Nowotny, E. Bauer, and A. Stempfl, *Monatsh. Chem.*, **81,** 1950, 1164.
3. W. Hume-Rothery, M. O. Betterton, and J. Reynolds, *J. Inst. Metals,* **80,** 1951-1952, 609–616.
4. H. Nowotny and H. Bittner, *Monatsh. Chem.*, **81,** 1950, 887–906.
5. C. T. Heycock and F. H. Neville, *J. Chem. Soc.*, **61,** 1892, 900.

$\bar{1}.7286$
0.2714

Cd-Po Cadmium-Polonium

The solid solubility of Po in Cd was reported as 2.3×10^{-11} wt. % [1].

1. G. Tammann and A. v. Löwis of Menar, *Z. anorg. Chem.*, **205,** 1932, 145–162.

$\bar{1}.9018$
0.0982

Cd-Pr Cadmium-Praseodymium

The phase PrCd (44.37 wt. % Cd) has the b.c.c. structure of the CsCl (B2) type, a = 3.82 A [1].

The results of a structural investigation by [2] are as follows: $PrCd_2$ (61.47 wt. % Cd) is hexagonal, a = 5.035 A, c = 3.466 A, c/a = 0.688, and isostructural with $CeCd_2$ and $LaCd_2$ (see Cd-Ce); $PrCd_3$ (70.52 wt. % Cd) has presumably the f.c.c. BiF_3 ($D0_3$) type of structure, with a = 7.200 A; $PrCd_{11}$ [91.67 at. (89.77 wt.) % Cd] is isostructural with $BaHg_{11}$, with a = 9.306 A and 36 atoms in the cubic unit cell.

1. A. Iandelli and E. Botti, *Gazz. chim. ital.*, **67,** 1937, 638–644.
2. A. Iandelli and R. Ferro, *Gazz. chim. ital.*, **84,** 1954, 463–478.

$\bar{1}.7602$
0.2398

Cd-Pt Cadmium-Platinum

From a study of the phase relations in the range 50–100 wt. % Cd, [1] concluded the existence of two intermediate phases, $PtCd_2$ (53.52 wt. % Cd) and Pt_2Cd_9 [81.82 at. (72.15 wt.) % Cd], with melting points of 725 and 615°C, respectively. No detailed data were reported. A eutectic point was found at 2 wt. (1.2 at.) % Pt, 315°C [1]. This does not agree with findings by [2]: 0.24 wt. (0.14 at.) % Pt and a temperature of 0.6°C below the melting point of Cd. [3] gave 314°C as the eutectic temperature.

The phase diagram of Fig. 249 was outlined by [3]. For experimental details, see Cd-Pd. The homogeneity ranges of the (Pt), or α, and α_1' phases were determined parametrically (see data points in Fig. 249) and corroborated by measurements of the magnetic susceptibility and by micrographic work. There are indications of a transformation $\alpha \rightleftarrows \alpha'$ in the range of Pt_3Cd between 600 and 550°C. The structure of the α_1' phase is of the CuAu ($L1_0$) type, $a = 4.17_4$ A, $c/a = 0.914$ at 48.5 at. % Cd. The phase designated as ($PtCd_2$) has a structure related to that of the AlB_2 (C32) type. Three different Hume-Rothery γ phases [4] were detected and their approximate homogeneity ranges determined: (*a*) the γ phase proper, stable between 82.5 and 85.5 at. % Cd, with $a = 9.89_7$ A at 83.9 at. % Cd; (*b*) the γ_2 (72–74 at. % Cd) and (*c*) the γ_1 (75.5–77 at. % Cd) phases which are modifications of the γ structure.

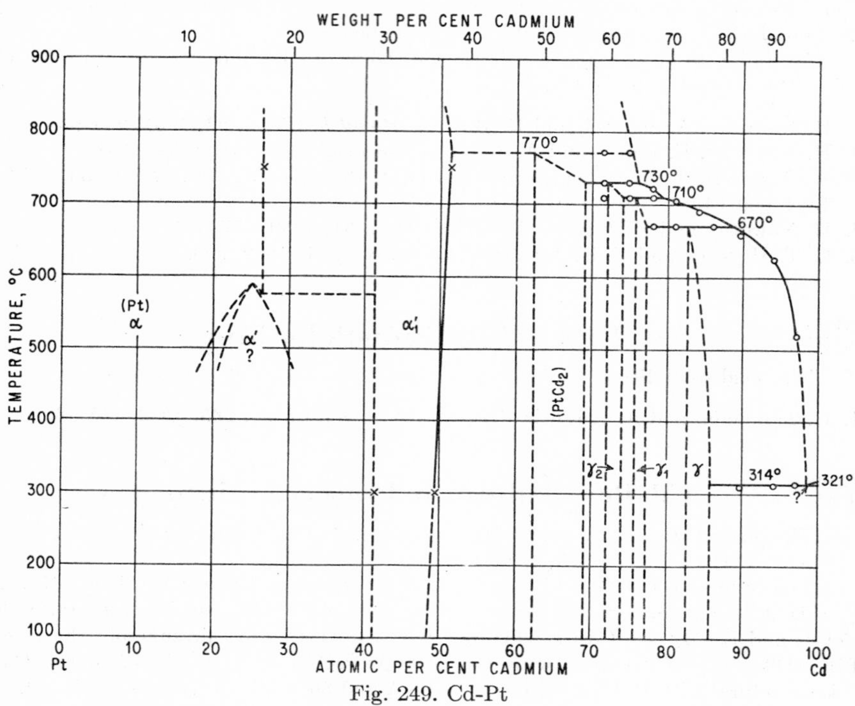

Fig. 249. Cd-Pt

1. K. W. Ray, *Proc. Iowa Acad. Sci.*, **38**, 1931, 166.
2. C. T. Heycock and F. H. Neville, *J. Chem. Soc.*, **61**, 1892, 901.
3. H. Nowotny, E. Bauer, A. Stempfl, and H. Bittner, *Monatsh. Chem.*, **83**, 1952, 221–236. Preliminary magnetic and X-ray results were published in H. Nowotny and H. Bittner, *Monatsh. Chem.*, **81**, 1950, 904, and H. Nowotny, E. Bauer, and A. Stempfl, *Monatsh. Chem.*, **81**, 1950, 1164, respectively.
4. W. Hume-Rothery, J. O. Betterton, and J. Reynolds, *J. Inst. Metals*, **80**, 1951-1952, 609–616.

0.1189
$\bar{1}$.8811

Cd-Rb Cadmium-Rubidium

The phase $RbCd_{13}$ [92.86 at. (94.47 wt.) % Cd] was identified by determining the crystal structure: cubic of the $NaZn_{13}$ ($D2_3$) type, 8 molecules per unit cell, $a = 13.91$ A [1]. See also Cd-K.

1. E. Zintl and W. Haucke, *Z. Elektrochem.*, **44**, 1938, 104–111.

0.0383
$\bar{1}$.9617

Cd-Rh Cadmium-Rhodium

It is probable that, in analogy with Cd-Pd, Cd-Pt, Pd-Zn, Pt-Zn, and Rh-Zn, a phase exists which is based on the composition Rh_5Cd_{21} [80.77 at. (82.10 wt.) % Cd] and has the structure of γ-brass ($D8_{1-3}$ type) [1] [2].

1. A. Westgren and W. Ekman, *Arkiv Kemi, Mineral Geol.*, **10B**, 1930, no. 11, 1–6.
2. See also the systems mentioned.

0.5448
$\bar{1}$.4552

Cd-S Cadmium-Sulfur

CdS (22.02 wt. % S) exists in two modifications. If prepared by chemical reaction (e.g., $CdSO_4 + H_2S$) the cubic modification having the structure of the zincblende (B3) type is formed [1, 2]. On annealing at 700–800°C in an atmosphere of sulfur vapor it transforms into the hexagonal form of the wurtzite (B4) type [1]. [3] believed they obtained only the hexagonal form, but since they did not use X-ray methods for identification they were unable to recognize the seemingly amorphous CdS as the cubic form. Heating and cooling curves indicated no transformation temperature up to 1000°C [3]. CdS starts to sublime at 980°C [4] and melts at approximately 1750°C under a pressure of 100 atm [5].

The phase diagram can be expected to be similar to that of the system Cd-Se.

The unit cells have the following dimensions: cubic modification, a = 5.83 A [1, 6, 7]; hexagonal modification, a = 4.150 A, c = 6.737 A [1], c/a = 1.623 [1, 8].

1. F. Ulrich and W. Zachariasen, *Z. Krist.*, **62,** 1925, 260–273, 614.
2. J. Böhm and H. Niclassen, *Z. anorg. Chem.*, **132,** 1923, 7.
3. E. T. Allen and J. L. Crenshaw, *Z. anorg. Chem.*, **79,** 1913, 147–155, 183–185.
4. W. Biltz, *Z. anorg. Chem.*, **59,** 1908, 278–279.
5. E. Tiede and A. Schleede, *Ber. deut. chem. Ges.*, **53,** 1920, 1720.
6. V. M. Goldschmidt, see *Strukturbericht*, **1,** 1913-1928, 129.
7. W. J. Müller and G. Löffler, *Z. angew. Chem.*, **46,** 1933, 538–539.
8. W. L. Bragg, *Phil. Mag.*, **39,** 1920, 647.

$\bar{1}$.9653
0.0347

Cd-Sb Cadmium-Antimony

As Fig. 250 shows, this system is characterized by the existence of a metastable system.

The first two thermal (and microscopic) investigations [1, 2] yielded almost identical thermal data; however, the authors had different conceptions as to the conditions of stability of the intermediate phases occurring, Cd_3Sb_2 (41.93 wt. % Sb) and CdSb (52.00 wt. % Sb). [1] regarded the phase Cd_3Sb_2 to be metastable within the whole temperature and composition range and—under certain conditions of cooling—found it to crystallize as the primary phase only from melts with about 30–52 at. % Sb. Stabilization takes place by a subsequent transformation by which Cd_3Sb_2 completely disappears: $Cd_3Sb_2 = 2CdSb + Cd$ (between 0 and 40 at. % Sb) and $Cd_3Sb_2 + Sb = 3CdSb$ (between 40 and 100 at. % Sb). On the other hand, [2] assumed Cd_3Sb_2 to have a stable homogeneity region, to precipitate as the primary stable phase from melts with 7 to 34 at. % Sb, and to be formed by the peritectic reaction: melt (34 at. % Sb) + CdSb = Cd_3Sb_2 at about 410°C. According to later studies [3–6], there is no doubt that Cd_3Sb_2 is a metastable phase [7].

The stable and metastable diagram presented in Fig. 250 is based on work by [4]. The former was obtained by means of electrical-resistance-vs.-temperature curves and the latter by means of cooling curves of melts which were not inoculated on solidification. The diagram deviates from that by [1] only in so far as it shows that Cd_3Sb_2 (*a*) can also crystallize from melts with less than about 30 at. % Sb, and (*b*) forms a metastable eutectic at 285°C. On cooling, at temperatures between about 250 and 350°C, the metastable system transforms into the stable one (Fig. 250).

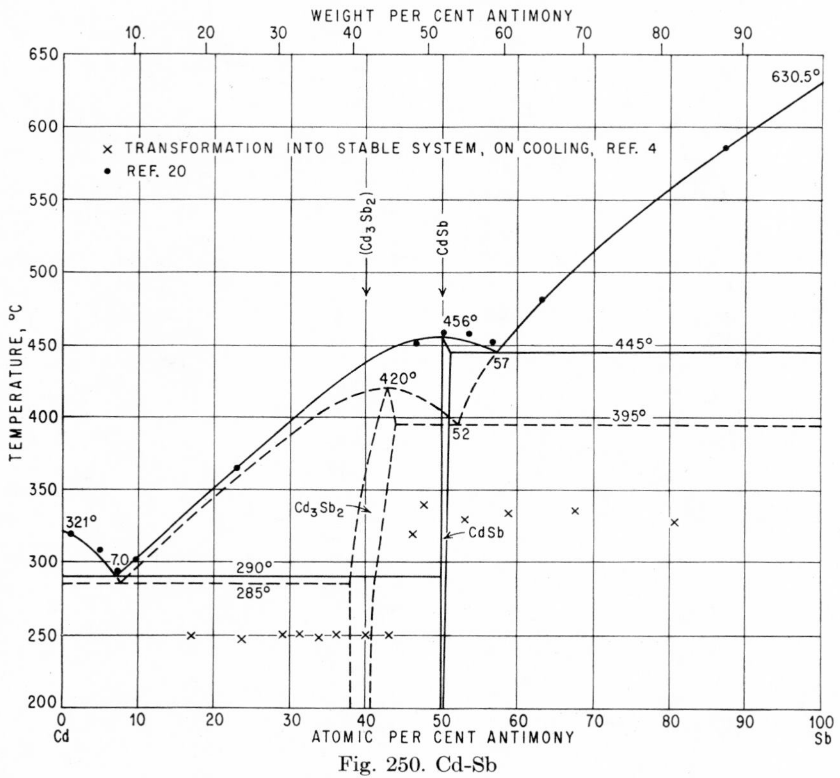

Fig. 250. Cd-Sb

The characteristic points of the stable and metastable systems are summarized in Table 20, the data of [2] being given on the basis of the interpretation of [1].

The solid solubility of Sb in Cd and Cd in Sb has not been determined as yet. According to measurements of the freezing-point lowering of Cd by Sb [8], thermoelectric force [9, 10], electrical and thermal conductivity [10], and magnetic susceptibility [11], all of which indicate no perceptible solubility, it is certainly very small. This also holds apparently for the solubility of Cd in Sb [10]; only measurements of magnetic susceptibility indicate that solid solutions are formed [11]. [4] reported a solubility of 0.15 wt. % Cd at 445°C and less at lower temperatures.

Isotherms of the thermoelectric force [9, 10], electrical and thermal conductivity [10], Hall effect [10], magnetic susceptibility [11, 12], and electromotive force [13] clearly show that CdSb, having a limited range of solubility of Sb, is the only stable phase. From emf measurements [6] the homogeneity range of CdSb was concluded to be 50–50.5 at. % Sb, as compared with 47–50.5 at. % Sb (at 290°C) shown in the diagram of [4]. In Fig. 250 the Sb-rich boundary of the CdSb phase is that of [4], whereas the Cd-rich boundary has been assumed at 50 at. % (see Supplement). Measurements of magnetic susceptibility [11] and a thermodynamic study [14] of liquid alloys proved the presence of undissociated CdSb molecules.

Table 20. Characteristic Compositions in At. % Sb and Temperatures in °C

	Ref. 1	Ref. 2	Ref. 3	Ref. 4
Stable system:				
Cd-CdSb eutectic	295°C	290°C	292°C	290°C
	7.4	7.0	7.0	7.0
Melting point of CdSb	~465°C	455°C	458°C	456°C
CdSb-Sb eutectic	~455°C	445°C	446°C	445°C
	58	58	58.9	57
Metastable system:				
Cd-Cd_3Sb_2 eutectic	....	.....	.....	285°C
				7.7
Melting point of Cd_3Sb_2*	~423°C	423°C	424°C	420°C
	~43	.....	.....	~42.5
Cd_3Sb_2-Sb eutectic	408°C	402°C	402°C	395°C
	52	50.5	51.3	52

* The maximum of the metastable liquidus curve does not coincide with the composition Cd_3Sb_2 (40 at. % Sb) but lies at a somewhat higher Sb content [1, 4].

Crystal Structures. The structure of CdSb (first erroneously reported to be hexagonal [15] or orthorhombic, with a = 6.53 A, b = 8.62 A, c = 4.17 A, and 8 atoms per unit cell [5]) is orthorhombic with 16 atoms per unit cell [16, 17]. Lattice dimensions were found to be a = 6.403 A, b = 8.337 A, c = 8.509 A [16], or a = 6.471 A, b = 8.253 A, c = 8.526 A [17]. The structure of the metastable phase Cd_3Sb_2 was reported to be monoclinic, 20 atoms per unit cell, a = 7.21 A, b = 13.54 A, c = 6.17 A, β = 100°14′ [18].

Supplement. The investigation of the electrical properties of semiconducting CdSb by [19] has provided further evidence that this intermediate phase does not form any solid solution with excess Cd. Thermal data for the stable liquidus measured by [20] have been plotted in Fig. 250. As can be seen, they are in good agreement with the liquidus of [4].

1. W. Treitschke, *Z. anorg. Chem.*, **50**, 1906, 217–225.
2. N. S. Kurnakow and N. S. Konstantinow, *Z. anorg. Chem.*, **58**, 1908, 12–22; *Zhur. Russ. Fiz.-Khim. Obshchestva*, **37**, 1905, 580.
3. E. Abel, O. Redlich, and J. Adler, *Z. anorg. Chem.*, **174**, 1928, 257–264.
4. T. Murakami and T. Shinagawa, *Kinzoku-no-Kenkyu*, **5**, 1928, 283–300.
5. F. Halla and J. Adler, *Z. anorg. Chem.*, **185**, 1929, 184–192.
6. A. Ölander, *Z. physik. Chem.*, **A173**, 1935, 284–294.
7. Data points of the phase diagrams by [1, 2, 4] are shown in Figs. 193 and 194 of M. Hansen, "Der Aufbau der Zweistofflegierungen," pp. 444–450, Springer-Verlag OHG, Berlin, 1936.
8. C. T. Heycock and F. H. Neville, *J. Chem. Soc.*, **61**, 1892, 901.
9. A. Battelli, see W. Broniewski, *Rev. mét.*, **7**, 1910, 358–360.

10. A. Eucken and G. Gehlhoff, *Verh. deut. physik. Ges.*, **14**, 1912, 169–182; *Z. Metallkunde*, **12**, 1920, 194–196; *Z. anorg. Chem.*, **159**, 1927, 336–338.
11. H. Endo, *Science Repts. Tôhoku Univ.*, **16**, 1927, 220–222; K. Honda and H. Endo, *J. Inst. Metals*, **37**, 1927, 39.
12. F. L. Meara, *Phys. Rev.*, **37**, 1931, 467; *Physica*, **2**, 1932, 33–41.
13. R. Kremann and J. Gmachl-Pammer, *Z. Metallkunde*, **12**, 1925, 241–245. See also W. Jenge, *Z. anorg. Chem.*, **118**, 1921, 111–114.
14. H. Seltz and B. J. DeWitt, *J. Am. Chem. Soc.*, **60**, 1938, 1305–1308.
15. M. Chikashige and T. Yamamoto, "Anniversary Volume Dedicated to Masumi Chikashige," pp. 195–200, Kyoto Imperial University, 1930; see *Strukturbericht*, **2**, 1928-1932, 750.
16. A. Ölander, *Z. Krist.*, **91**, 1935, 243–247.
17. K. E. Almin, *Acta Chem. Scand.*, **2**, 1948, 400–407.
18. F. Halla, H. Nowotny, and H. Tompa, *Z. anorg. Chem.*, **214**, 1933, 196–197.
19. E. Justi and G. Lautz, *Z. Naturforsch.*, **7a**, 1952, 191–200, 602–613, especially 200, 609.
20. H. J. Fisher and A. Phillips, *Trans. AIME*, **200**, 1954, 1062. (Viscosity and density of liquid Pb-Sn and Sb-Cd alloys, pp. 1060–1070.)

0.1534
$\bar{1}$.8466

Cd-Se Cadmium-Selenium

According to [1], molten Cd and Se do not mutually dissolve. Solidified samples consist of two layers at the interface of which CdSe (41.26 wt. % Se) is present. Cooling curves of mixtures in various proportions, therefore, show thermal arrests at the freezing points of the elements; and the existence of CdSe, having a melting

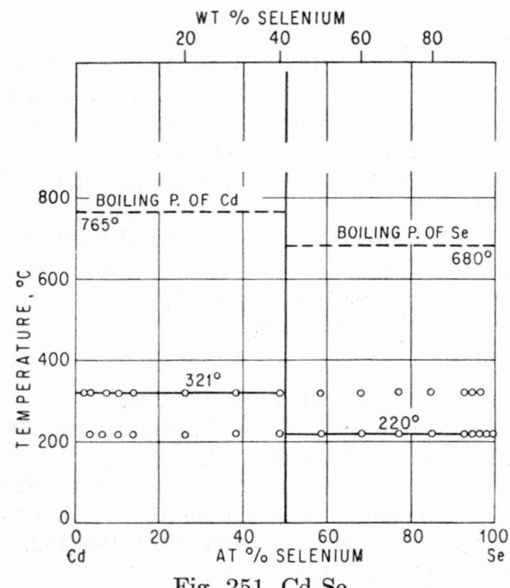

Fig. 251. Cd-Se

point of above 1350°C, is not indicated. The amount of CdSe formed increases as the reaction temperature increases; however, this temperature is limited by the boiling point of Se, 680°C (Fig. 251).

CdSe exists in two modifications. If formed by direct combination of the elements, it is hexagonal of the wurtzite (B4) type [2, 3], with $a = 4.31$ A, $c = 7.02$ A, $c/a = 1.630$ [3]. CdSe prepared by chemical reaction (e.g., $CdSO_4 + H_2Se$) is cubic of the zincblende (B3) type, $a = 6.05$ A [2]. Some information as to the transformation of the modifications was given by [4].

Supplement. The existence of only one compound in the system Cd-Se, CdSe, was confirmed by X-ray analysis [5].

1. M. Chikashige and R. Hitosaka, *Mem. Coll. Sci. Kyoto Univ.*, **2,** 1917, 239–244.
2. V. M. Goldschmidt, see *Strukturbericht,* **1,** 1913-1928, 136.
3. W. Zachariasen, *Z. physik. Chem.*, **124,** 1926, 436–448.
4. S. Nagata and K. Agata, *J. Phys. Soc. Japan*, **6,** 1951, 523–524.
5. L. S. Palatnik and V. V. Levitin, *Doklady Akad. Nauk S.S.S.R.*, **96,** 1954, 975–978; *Met. Abstr.*, **22,** 1955, 651.

0.6022
$\bar{1}.3978$

Cd-Si Cadmium-Silicon

From lattice-parameter measurements it was concluded that the components do not form solid solutions [1].

1. E. R. Jette and E. B. Gebert, *J. Chem. Phys.*, **1,** 1933, 753–755.

$\bar{1}.9764$
0.0236

Cd-Sn Cadmium-Tin

Liquidus Curve. Thermal data covering the whole range of composition were reported by [1–5] and of portions of the liquidus by [6–9]. The liquidus curve shown in Fig. 252 is largely based on the data by [5]. The eutectic temperature was found to lie at 177–178 [1], 175–177 [8], 177 [2–4], 182–183 [10], and 176°C [5, 9, 11]; and the eutectic composition at approximately 71.5 wt. (70.4 at.) % [1], 66–68 wt. (64.7–66.8 at.) % [2], 70 wt. (68.8 at.) % [4], 67.75 wt. (66.54 at.) % [11], and 67.0 wt. (65.75 at.) % Sn [5]. The value 67.75 wt. (66.54 at.) % Sn has to be regarded as the most accurate. For thermal-data points, see [12].

The Transformation at 133°C. [8] first observed the occurrence of a transformation in the solid state, at 122°C on cooling and at 135°C on heating, in alloys between at least 10.5 and 97.5 wt. (10.0–97.4 at.) % Sn. Since then, this transformation has been the subject of numerous and extensive investigations [2, 13–17, 4, 18, 10, 5, 9]. Various interpretations have been offered as to the nature of this transformation. It has been assumed to be due to (*a*) the peritectoid formation of the intermediate phase $CdSn_4$ [8, 15, 18], (*b*) a (nonexistent) polymorphic transformation of tin giving rise to a eutectoid decomposition of the high-temperature solid solution of Cd in Sn [2, 13, 14, 4], (*c*) the termination of metastable conditions [such as formed by supercooling of the precipitation of (Cd) from the Sn-rich phase] [17], and (*d*) the existence of a eutectoid equilibrium: Sn-rich solid solution with 95–97.5 wt. % Sn $\rightleftarrows$ (Cd) + Sn-rich solid solution with 98.5–98.75 wt. % Sn [10, 5]. A detailed discussion [12] of these investigations is not warranted as it has been established by [9] and confirmed by [19] that the thermal effect is due to the eutectoid decomposition of a Sn-rich intermediate phase β, formed peritectically at 223°C (Fig. 252). The range of homogeneity of the β phase was determined by thermal, thermoresistometric, and microscopic analysis to be 98 wt. (97.95 at.) % at 223°C to 94.4 wt. (94.15 at.) % Sn at 177°C [9]. [19] found slightly different values by means of determinations of the temperature of embrittlement of samples on heating: 98.6 wt. (98.55 at.) % at 223°C and 94.8 wt. (94.55 at.) % Sn at 177°C. The eutectoid point was located at 95.0 wt.

(94.77 at.) % Sn [9]. "The earlier diagrams by [5] and [10] can be explained partly as representing metastable conditions which are very easily produced, both on heating and on cooling; thus, the peritectic reaction at 223°C is only observed on cooling curves taken at very slow rates of cooling" [9].

Solid Solubility of Sn in Cd. The existence of a narrow range of solid solubility of Sn in Cd was already indicated by measurements of the thermoelectric force [20, 16, 18], electrical conductivity [16, 18], and hardness [21, 22]. According to [5], the solubility is less than 0.5 wt. (at.) % Sn at 160°C and more than 0.2 wt. % Sn at 173°C. [23], who also determined the solidus curve of the (Cd) phase, found the solubility to be approximately 0.25 wt. (at.) % Sn at 176°C.

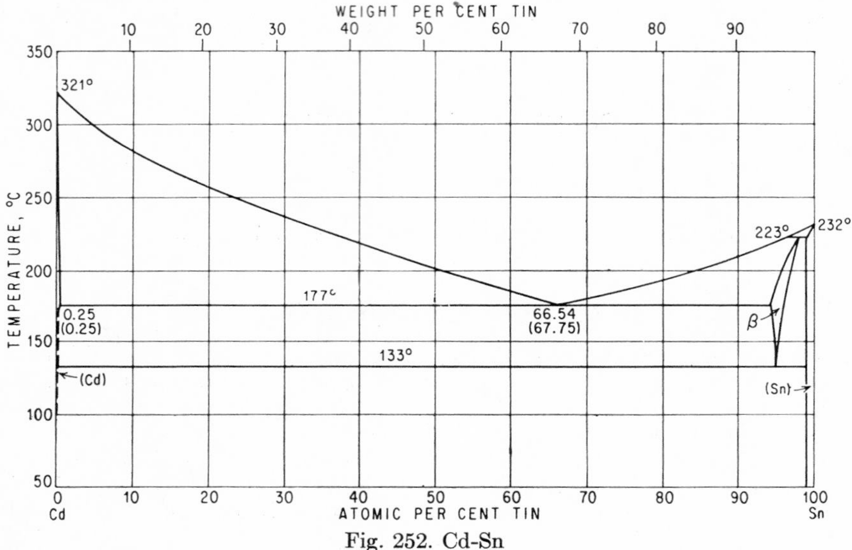

Fig. 252. Cd-Sn

Solid Solubility of Cd in Sn. Solid solubilities have been reported by [2, 4, 14, 18]; however, in view of the metastable conditions capable of forming in this region, the data are meaningless. According to micrographic and resistometric work by [9], the solubility limit is about 0.9 wt. (at.) % Cd at 223°C and 1.1 wt. (at.) % Cd at 133 and 100°C. [10] gave 1.5 wt. % Cd at 130°C.

Crystal Structure. High-temperature X-ray studies [24, 25] revealed the β phase to have a simple hexagonal structure, with a = 3.2328 A, c = 3.0023 A, c/a = 0.92870, at 4.89 at. % Cd, 176°C, according to [24].

Supplement. [26] reported 177°C, 68.9 at. (?) % Sn as eutectic data. A rather unusual (and questionable) solubility determination of Cd in solid Sn was carried out by [27]. The surface of Sn-rich Cd-Sn alloy samples annealed at and quenched from 105°C was treated with concentrated nitric acid, which operation is said to attack the (Cd) but not the (Sn) phase (?). The Cd content of the (Sn) phase—which remained in relief—was then spectroscopically found to be about 0.3 at. (wt.) %. [28] found anomalies in the lattice-spacing-vs.-composition curves of the tetragonal (Sn) phase which are presumably associated with the development of vacant lattice sites.

1. A. W. Kapp, *Ann. Physik*, **6**, 1901, 762, 770–771.
2. A. P. Schleicher, *Intern. Z. Metallog.*, **2**, 1912, 76–89.

3. R. Lorenz and D. Plumbridge, *Z. anorg. Chem.*, **83**, 1913, 234–236.
4. A. Fedorow, *J. chim. Ukraine*, **2**, 1926, 69–74; *J. Inst. Metals*, **39**, 1928, 502.
5. D. Hanson and W. T. Pell-Walpole, *J. Inst. Metals*, **56**, 1935, 162–182.
6. C. T. Heycock and F. H. Neville, *J. Chem. Soc.*, **61**, 1892, 901.
7. C. T. Heycock and F. H. Neville, *J. Chem. Soc.*, **57**, 1890, 383.
8. A. Stoffel, *Z. anorg. Chem.*, **53**, 1907, 140–147, 167.
9. D. Hanson and W. T. Pell-Walpole, *J. Inst. Metals*, **59**, 1936, 281–300.
10. Y. Matuyama, *Science Repts. Tôhoku Univ.*, **20**, 1931, 649–680.
11. D. Stockdale, *J. Inst. Metals*, **43**, 1930, 198–211.
12. M. Hansen, "Der Aufbau der Zweistofflegierungen," pp. 451–458, Springer-Verlag OHG, Berlin, 1936.
13. W. Guertler, "Handbuch der Metallographie," vol. 1, pp. 710–711, Verlagsbuchhandlung Gebrüder Borntraeger, Berlin, 1912; *Intern. Z. Metallog.*, **2**, 1912, 90–102, 172–177.
14. D. Mazzotto, *Intern. Z. Metallog.*, **4**, 1913, 13–27, 273–294.
15. M. Padoa and F. Bovini, *Gazz. chim. ital.*, **44**, 1914, 528–534.
16. A. Bucher, *Z. anorg. Chem.*, **98**, 1916, 106–117.
17. Künzel-Mehner, Dissertation, Leipzig, 1920, according to [18].
18. M. LeBlanc, M. Naumann, and D. Tschesno, *Ber. Verhandl. K. sächs. Ges. Wiss., Math.-phys. Kl.*, **79**, 1927, 72–106, especially 99–106.
19. C. E. Homer and H. Plummer, *J. Inst. Metals*, **64**, 1939, 169–200.
20. A. Battelli, see W. Broniewski, *Rev. mét.*, **7**, 1910, 356.
21. C. diCapua, *Rend. reale accad. nazl. Lincei*, **33**(1), 1923, 141–144.
22. W. Schischokin and W. Agejewa, *Z. anorg. Chem.*, **193**, 1930, 240.
23. D. Stockdale, *J. Inst. Metals*, **56**, 1935, 184–185.
24. G. V. Raynor and J. A. Lee, *Acta Met.*, **2**, 1954, 616–620.
25. K. Schubert, U. Rösler, W. Mahler, E. Dörre, and W. Schütt, *Z. Metallkunde*, **45**, 1954, 643–647.
26. O. Hájiček, *Hutnické Listy*, **3**, 1948, 265–270; *Chem. Abstr.*, **43**, 1949, 4935.
27. H. Triché, *Compt. rend.*, **227**, 1948, 52–54.
28. J. A. Lee and G. V. Raynor, *Proc. Phys. Soc. (London)*, **B67**, 1954, 737–747.

0.1082
$\bar{1}.8918$

Cd-Sr Cadmium-Strontium

The thermal-analysis data tabulated by [1] are shown in Fig. 253 on a wt. % scale. The authors pointed out that "due to the lack of precision of the apparatus used and the complexity of the phase relations in the solidified alloys, the thermal diagram cannot be satisfactorily presented. At least one intermetallic compound, probably $Cd_{12}Sr$ (6.10 wt. % Sr), is present, and there seem to be indications of another, possibly CdSr (43.81 wt. % Sr). Two eutectics are noted, one at 0.9 wt. % and one at 13.8 wt. % Sr. The transitions in the solid phase are not clearly defined but may involve several forms of CdSr. In all slowly cooled specimens with less than 6 wt. % Sr a large percentage of metallic cadmium is present."

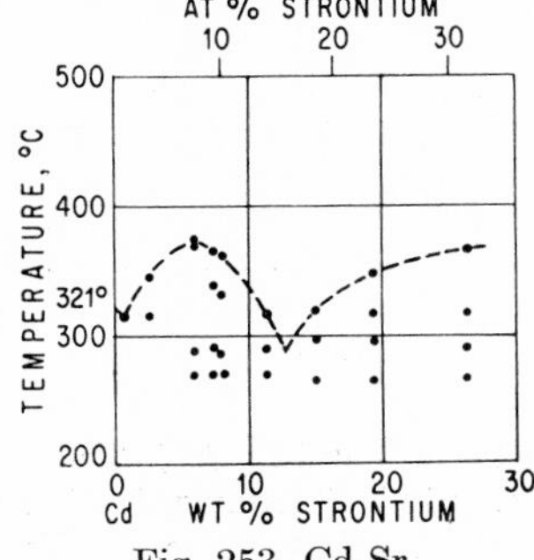

Fig. 253. Cd-Sr

There are a number of contradictory statements in the paper. For instance, in their discussion of the photomicrographs (covering alloys up to 26.4 wt. % Sr) the authors do not mention the assumed compound $SrCd_{12}$ but state that "annealing the alloys containing more than 8 wt. % Sr gave massive cadmium crystals" (!).

The photomicrographs presented allow some tentative conclusions to be drawn as to the possible constitution. At 0.7 wt. % Sr (unannealed) a few crystals of a compound (characterized by the shape) are embedded in a matrix of Cd. (However, [1] regard the matrix to be the eutectic with 0.9 wt. % Sr and the primary crystals to be Cd.) The same characteristic primary crystals are present in alloys with 7.9, 8.1, 10.9, and 11.4 wt. % Sr; their amount increases with the Sr content. Therefore, the existence of $SrCd_{12}$ is impossible, and the composition of the intermediate phase coexisting with Cd must be well above 11 wt. % Sr. This phase is probably formed by a peritectic reaction as the structure of the alloy with 14.6 wt. % Sr shows peritectic encasements [2].

Supplement. [3] identified the compound $SrCd_{11}$ by determination of its crystal structure: b.c. tetragonal, a = 12.02 A, c = 7.69 A, 48 atoms per unit cell, isotypic with $BaCd_{11}$. [4] prepared the compound SrCd and found it to have the CsCl (B2) type of structure, with a = 4.011 A.

1. H. C. Hodge and six coworkers, *Metals & Alloys*, **2,** 1931, 355–357.
2. Photomicrographs of alloys with 19.3 and 26.4 wt. % Sr are not clearly reproduced.
3. M. J. Sanderson and N. C. Baenziger, *Acta Cryst.*, **6,** 1953, 627–631.
4. R. Ferro, *Acta Cryst.*, **7,** 1954, 781.

$\bar{1}$.9449
0.0551

Cd-Te Cadmium-Tellurium

Thermal and microscopic studies [1] yielded the phase diagram shown in Fig. 254 and confirmed the existence of CdTe (53.17 wt. % Te) [2]. The liquidus temperatures

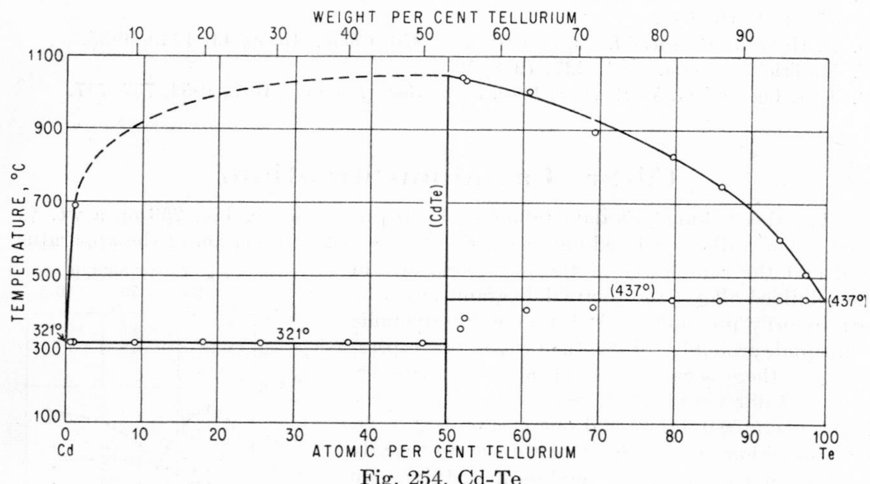

Fig. 254. Cd-Te

between 1 and 52 at. % Te could not be determined because of uncontrollable evaporation of Cd (boiling point 765°C). CdTe is cubic of the zincblende (B3) type, with a = 6.47_7 A [3]. See also Te-Zn.

Note Added in Proof. [4] studied thin Cd-Te alloy films by means of electron diffraction. No indications of the formation of solid solutions of CdTe with Te or

Cd were observed. Besides the well-known cubic, a hexagonal form of CdTe was observed: $a = 4.57$ A, $c = 7.47$ or 11.27 A.

1. M. Kobayashi, *Z. anorg. Chem.*, **69,** 1911, 1–6. Alloys with 52–97.3 at. % Te were analyzed.
2. J. Margottet, *Compt. rend.*, **84,** 1877, 1294–1295; C. Fabre, *Compt. rend.*, **105,** 1887, 279; C. A. Tibbals, *J. Am. Chem. Soc.*, **31,** 1909, 908; L. M. Dennis and R. P. Anderson, *J. Am. Chem. Soc.*, **36,** 1914, 887.
3. W. Zachariasen, *Norsk Geol. Tidsskr.*, **8,** 1926, 302–306; *Z. physik. Chem.*, **124,** 1926, 277–284.
4. S. A. Semiletov, *Trudy Inst. Krist., Akad. Nauk S.S.S.R.*, no. 11, 1955, 121–123; *Chem. Abstr.*, **50,** 1956, 16250.

0.3705
$\bar{1}.6295$

Cd-Ti Cadmium-Titanium

The existence of the compound Ti_2Cd (53.99 wt. % Cd) was established, apparently on the basis of X-ray diffraction studies of alloys prepared by powder metallurgy [1].

As the boiling point of Cd (765°C) is considerably lower than the melting point of Ti (1670°C), Ti-rich alloys cannot be prepared by melting both metals together under normal conditions of pressure.

1. California Institute of Technology, Letter Report No. 10, on Contract No. DA-04-495-ORD 18, to Watertown Arsenal, December, 1952.

$\bar{1}.7403$
0.2597

Cd-Tl Cadmium-Thallium

The liquidus curve was carefully determined using 42 melts [1] (Fig. 255). [2] have studied the effect of up to 3.85 at. % Tl on the melting point of Cd, and [3]

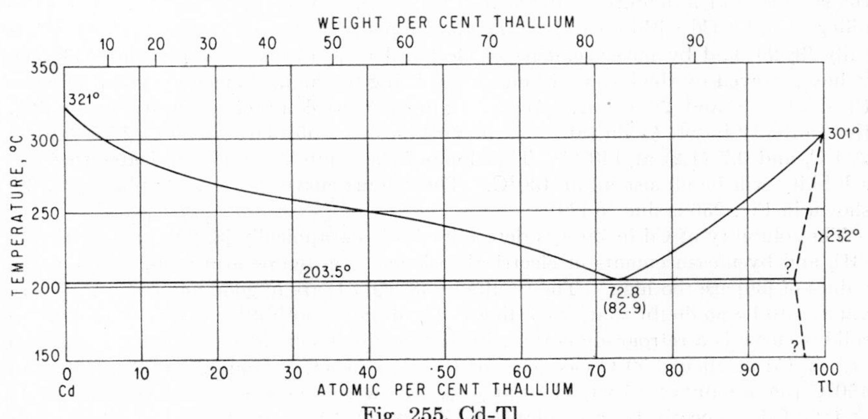

Fig. 255. Cd-Tl

and [4] have made cursory checks. The eutectic was located as follows: **72.8 at. (82.9 wt.) % Tl, 203.5°C [1]**, and 71.4 at. (81.9 wt.) % Tl, 203.5°C [4].

The mutual solid solubility of the elements has not been investigated as yet. On the basis of the eutectic halting times the solubility of Cd in Tl at 203°C was estimated to be about **2.5** wt. (4.3 at.) % Cd [3]. There are some further undefined indications of a slight solid solubility of Cd in Tl [5–7]. The effect of Cd additions on the temperature of the polymorphic transformation of Tl (232°C) and the nature of a thermal effect at 150°C, found over practically the whole range of composition [4], are unknown.

1. N. S. Kurnakow and N. A. Puschin, *Z. anorg. Chem.*, **30**, 1902, 101–108.
2. C. T. Heycock and F. H. Neville, *J. Chem. Soc.*, **61**, 1892, 903.
3. C. diCapua, *Rend. reale accad. nazl. Lincei*, **32**, 1923, 282–285.
4. N. S. Kurnakow and N. I. Korenew, *Izvest. Inst. Fiz.-Chim. Anal.*, **6**, 1933, 50.
5. C. diCapua, *Rend. reale accad. nazl. Lincei*, **32**, 1923, 343–346.
6. F. L. Meara, *Physics*, **2**, 1932, 33–41.
7. A. W. David and J. F. Spencer, *Trans. Faraday Soc.*, **32**, 1936, 1512–1516.

0.2354
$\bar{1}$.7646

Cd-Zn Cadmium-Zinc

After [1] had investigated the effect of small additions of Zn on the melting point of Cd, it was first established by [2] that the components form a eutectic type of phase diagram. The liquidus curve published by [3], based on cooling-curve data of 43 melts, has not been surpassed in accuracy by the later more or less comprehensive reinvestigations [4–9]. All data points [10], with the exception of the considerably deviating ones by [2], lie within a scattering band having a width of about 10–15°C.

The eutectic temperature varies essentially between 260 and 270°C [3–9, 11, 12] and the eutectic composition between 17 and 17.6 wt. (26–26.8 at.) % Zn [3–5, 7, 8, 13]. The most accurate value [13] must be considered to be 17.4 wt. (26.5 at.) % Zn.

Numerous investigations, using various methods, have indicated the existence of Cd-rich [14, 4, 15–19, 5, 20, 6, 7, 21–23, 12] and Zn-rich solid solutions [14, 4, 15–19, 6, 7, 21, 24–27, 22, 23, 28, 12]. However, no reliable quantitative conclusions as to the solubility at a definite temperature can be drawn from these results. The solubility of Zn in Cd within a wider range of temperature was determined micrographically [8, 29] and by measurements of electrical resistance vs. temperature [30], the values reported by the latter without doubt being too high. The agreement between the data of [8] and [29] is fair. According to the more comprehensive studies by [29], the solubility in wt. % Zn (at. % in parentheses) is 2.95 (5.0) at 266°C, 1.75 (2.9) at 200°C, and 0.7 (1.2) at 140°C. The slope of the solubility curve indicates that the solubility will be almost nil at 100°C. The solidus curve of the (Cd) phase [29, 30] shown in Fig. 256 is due to [29].

The solubility of Cd in Zn was determined micrographically [8, 29], parametrically [31], and by measurements of electrical resistance vs. temperature [30]. Again, the values of [30] are too high. The results by [8, 29, 31] are in good agreement. There appears to be no doubt that, according to the data by both [8] and [29], the solvus + solidus curve is a retrograde curve with the maximal solubility of about 2.5 wt. (1.5 at.) % Cd at about 350°C (see inset of Fig. 256 on wt. % scale). At 266, 200, and 150°C the amounts 2.15 wt. (1.30 at.) %, 1.75 wt. (1.05 at.) %, and 0.69 wt. (0.41 at.) % Cd, respectively, are soluble in Zn [8, 29, 31]. At 100°C the solubility is practically nil [31]; see also [32, 33].

Supplement. For boiling points of some Cd-Zn alloys, see [34]. [35] reported 263°C, 26.59 at. % Zn as eutectic data.

The solubility of Cd in solid Zn has been reinvestigated by resistometric [36] and X-ray parametric [37] measurements on quenched alloys. [36] reported the value

1.25 at. (2.13 wt.) % Cd at 255°C, and [37] the following solubility data in at. % (wt. % in parentheses): 1.08, 1.07, 0.97, 0.92, 0.69, 0.49, 0.38, 0.28, 0.20, and 0.05 (1.84, 1.83, 1.66, 1.57, 1.18, 0.84, 0.65, 0.48, 0.34, and 0.09) at 271, 266, 259, 253, 235, 210, 190, 172, 152, and 94°C, respectively. These data were also plotted in the inset of Fig. 256. It appears from the results of [37] that the solubility at the eutectic temperature is somewhat lower than previous authors assumed. [38] demonstrated the efficiency of their "quantitative thermal analysis" (combined calorimetric and thermal analysis) by measurements on Cd-Zn alloys. The liquidus and eutectic (17.5 wt. % Zn, 264°C) data obtained are in good agreement with Fig. 256. It is

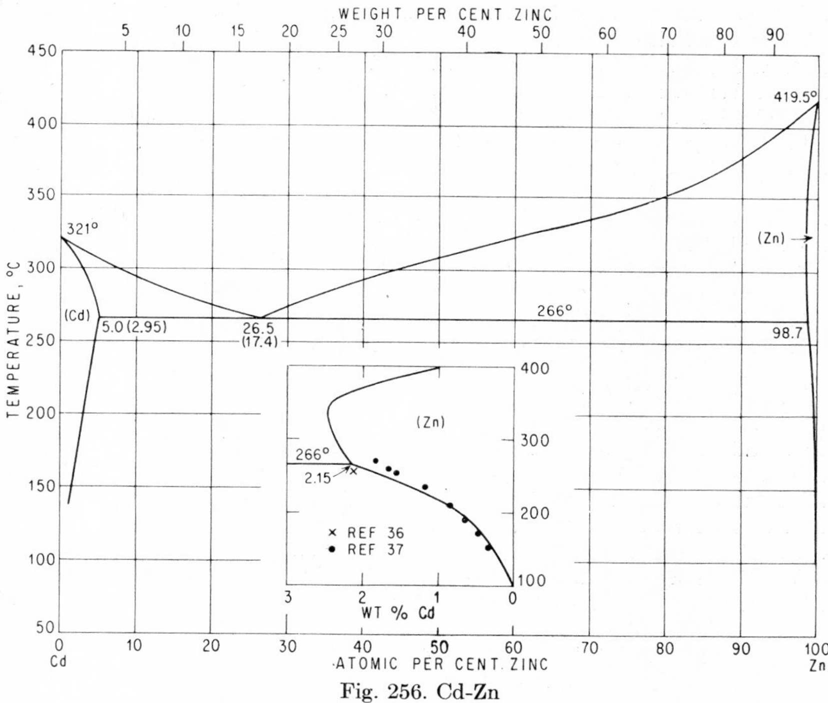

Fig. 256. Cd-Zn

remarkable that this also holds for the solidus and solvus data, although [38] do not claim these to be equilibrium values.

1. C. T. Heycock and F. H. Neville, *J. Chem. Soc.*, **61**, 1892, 899.
2. Roland-Gosselin and H. Gautier, *Bull. soc. encour. ind. natl.*, (5) **1**, 1896, 1307; Société d'encouragement pour l'industrie nationale, Paris, Commission des alliages, "Contribution à l'étude des alliages," p. 107, Typ. Chamerot et Renouard, Paris, 1901.
3. C. T. Heycock and F. H. Neville, *J. Chem. Soc.*, **71**, 1897, 387–388.
4. G. Hindrichs, *Z. anorg. Chem.*, **55**, 1907, 415–418.
5. G. Bruni, C. Sandonnini, and E. Quercigh, *Z. anorg. Chem.*, **68**, 1910, 75–78.
6. R. Lorenz and D. Plumbridge, *Z. anorg. Chem.*, **83**, 1913, 231–233; **85**, 1914, 435–436.

7. C. H. Mathewson and W. M. Scott, *Intern. Z. Metallog.*, **5**, 1914, 16–17.
8. C. H. M. Jenkins, *J. Inst. Metals*, **36**, 1926, 63–97.
9. P. T. Arnemann (*Metallurgie*, **7**, 1910, 204–205) determined only three liquidus points.
10. M. Hansen, "Der Aufbau der Zweistofflegierungen," pp. 462–467, Springer-Verlag OHG, Berlin, 1936.
11. M. Cook, *J. Inst. Metals*, **31**, 1924, 299.
12. M. Le Blanc and H. Schöpel, *Z. Elektrochem.*, **39**, 1933, 695–701.
13. D. Stockdale, *J. Inst. Metals*, **43**, 1930, 193–211.
14. A. Matthiessen, *Pogg. Ann.*, **110**, 1860, 207.
15. A. Sapozhnikov and M. Sakharov, *Zhur. Russ. Fiz.-Khim. Obshchestva*, **39**, 1907, 907–914; see also [21].
16. N. A. Puschin, *Z. anorg. Chem.*, **56**, 1908, 26–27.
17. N. S. Kurnakow and S. F. Zemczuzny, *Z. anorg. Chem.*, **60**, 1908, 32, footnote; see also N. S. Kurnakow and A. N. Achnarasow, *Z. anorg. Chem.*, **125**, 1922, 191.
18. B. E. Curry, *J. Phys. Chem.*, **13**, 1909, 589–605.
19. E. Rudolfi, *Z. anorg. Chem.*, **67**, 1910, 75–78.
20. G. Bruni and C. Sandonnini, *Z. anorg. Chem.*, **78**, 1912, 273–275.
21. A. Glasunow and M. Matweew, *Intern. Z. Metallog.*, **5**, 1914, 113–121.
22. W. Schischokin and W. Agejewa, *Z. anorg. Chem.*, **193**, 1930, 242.
23. F. L. Meara, *Phys. Rev.*, **37**, 1931, 467; *Physics*, **2**, 1932, 33–41.
24. C. Benedicks and R. Arpi, *Z. anorg. Chem.*, **88**, 1914, 237–254.
25. P. Ludwik, *Z. anorg. Chem.*, **94**, 1916, 177–178.
26. K. E. Bingham, *J. Inst. Metals*, **24**, 1920, 337–338, 340.
27. W. M. Peirce, *Trans. AIME*, **68**, 1923, 769–771.
28. M. Straumanis, *Z. physik. Chem.*, **148**, 1930, 124.
29. D. Stockdale, *J. Inst. Metals*, **44**, 1930, 75–80.
30. G. Grube and A. Burkhardt, *Z. Metallkunde*, **21**, 1929, 231–232.
31. W. Boas, *Metallwirtschaft*, **11**, 1932, 603–604.
32. M. Straumanis, *Metallwirtschaft*, **13**, 1933, 175–176.
33. R. Chadwick, *J. Inst. Metals*, **51**, 1933, 114.
34. W. Leitgebel, *Z. anorg. Chem.*, **202**, 1931, 305–324.
35. O. Hájiček, *Hutnické Listy*, **3**, 1948, 265–270; *Chem. Abstr.*, **43**, 1949, 4935.
36. A. Pasternak, *Bull. intern. acad. polon. sci., Classe sci. math. nat.*, sér. A, **1951**, 177–192 (in English).
37. J. R. Brown, *J. Inst. Metals*, **83**, 1954, 49–52.
38. W. Oelsen, O. Oelsen, and G. Heynert, *Arch. Eisenhüttenw.*, **27**, 1956, 549–556.

0.0907
$\bar{1}.9093$

Cd-Zr Cadmium-Zirconium

A cursory X-ray analysis of alloys prepared by solid-state diffusion had the following results [1]: (*a*) In Zr-rich alloys, annealed at 700°C, a f.c.c. intermediate phase (Cu type, a = 4.3768 A) occurs, which may contain about 60 at. % Cd, certainly not more than 67 at. % Cd. (*b*) Data from alloys with 67, 75, and 80 at. % Cd indicated that two phases were present, the f.c.c. phase and a phase whose lines appeared to be split from the cubic phase, possibly of b.c. tetragonal symmetry, a = 4.4184 A, c = 4.3008 A, c/a = 0.973. The unit cells of both structures contain 4 atoms randomly distributed.

In the systems Ti-Zn and Zr-Zn, the phases $TiZn_2$, $TiZn_3$, and $ZrZn_2$ have been identified.

1. P. Pietrokowski, *Trans. AIME*, **200**, 1954, 219–226.

0.3761
$\bar{1}.6239$

Ce-Co Cerium-Cobalt

Only 17 alloys, prepared with Ce of 98–99% purity, were used to outline the phase relationships by thermal analysis. Data points in Fig. 257, given to show the limited extent of the work, were taken graphically from a smaller diagram and should

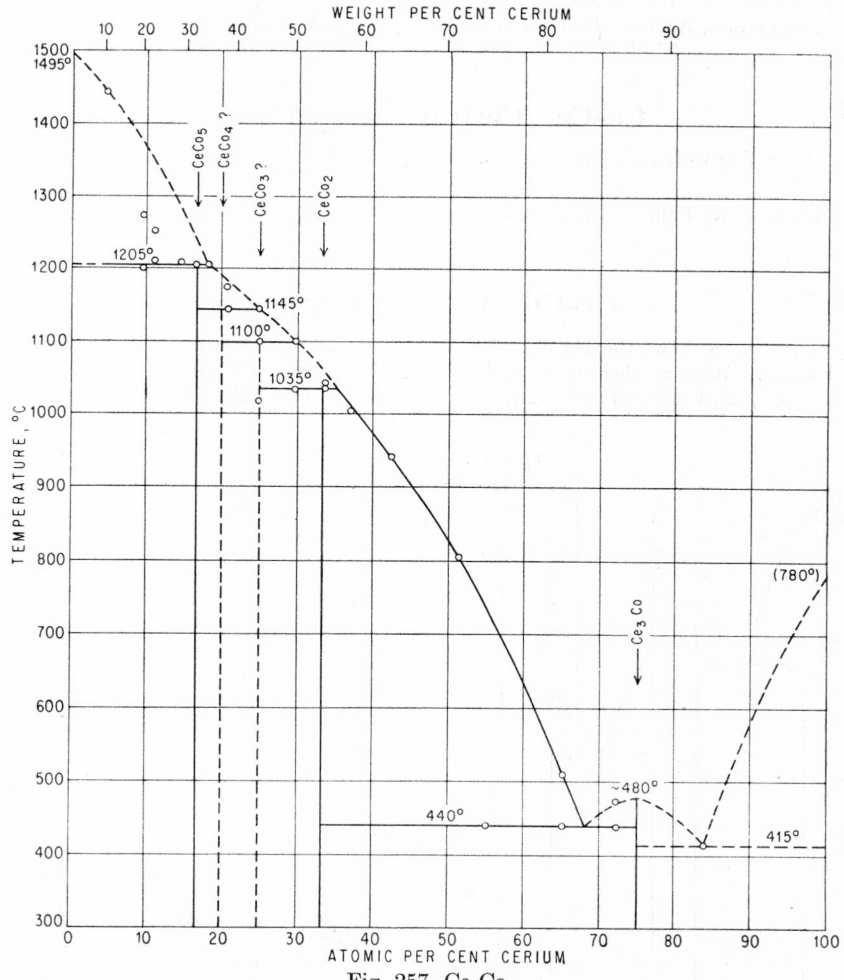

Fig. 257. Ce-Co

not be considered quantitative. Metallographic examination was impossible because most of the alloys decomposed to powder. The existence of five intermediate phases was claimed: $CeCo_5$ (32.23 wt. % Ce), $CeCo_4$? (37.28 wt. % Ce), $CeCo_3$? (44.21 wt. % Ce), $CeCo_2$ (54.31 wt. % Ce), and Ce_3Co (87.70 wt. % Ce) [1]. Of these, $CeCo_5$ and $CeCo_2$ have been identified by crystal-structure studies [2]. Whereas the existence of Ce_3Co is very likely (Fig. 257), that of phases having compositions $CeCo_4$ and $CeCo_3$ is somewhat uncertain.

The structure of $CeCo_5$ is of the $CaCu_5$ type [2, 3], $a = 4.955$ A, $c = 4.055$ A [3]. $CeCo_2$ is isotypic with $MgCu_2$ (C15 type), $a = 7.159$ A [2].

1. R. Vogel (W. Fülling), *Z. Metallkunde*, **38,** 1947, 97–103.
2. W. Fülling, K. Möller, and R. Vogel, *Z. Metallkunde*, **34,** 1942, 253–254.
3. T. Heumann, *Nachr. Akad. Wiss. Göttingen, Math.-phys. Kl.*, **1948,** 21–26.

0.4304
$\bar{1}.5696$

Ce-Cr Cerium-Chromium

According to [1], Ce forms no compounds with Cr.

1. R. Vogel (W. Fülling), *Z. Metallkunde*, **38,** 1947, 102.

0.3435
$\bar{1}.6565$

Ce-Cu Cerium-Copper

Figure 258 shows the phase diagram based on thermal and microscopic investigations [1]. Whereas there is no doubt about the existence of $CeCu_6$ [14.29 at. (26.88 wt.) % Ce] [2], $CeCu_4$ (35.54 wt. % Ce) [2, 3], and $CeCu_2$ (52.44 wt. % Ce), that of

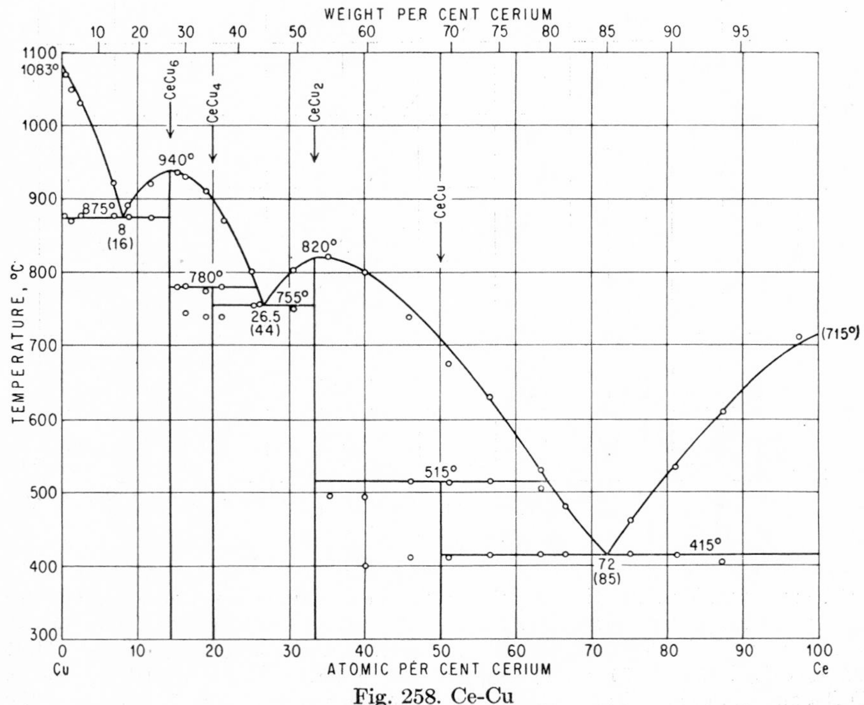

Fig. 258. Ce-Cu

CeCu (68.80 wt. % Ce) is at least very likely, as shown by microexamination of annealed specimens. An alloy with 0.4 wt. (~ 0.2 at.) % Ce proved to be two-phase after cooling from the melt.

$CeCu_6$ is orthorhombic, with 4 formula weights per unit cell and a = 8.08 A, b = 5.09 A, c = 10.17 A [2]. The structure of $CeCu_4$ (more precisely written $Ce_{1.2}Cu_{4.8}$) is of the $CaCu_5$ type, a = 5.141 A, c = 4.132 A [3]; a = 5.150 A, c = 4.102 A [2].

1. F. Hanaman, *Intern. Z. Metallog.*, **7**, 1915, 174–212. The cerium metal used was 96.7% pure and had a freezing point of 715°C. Alloys were analyzed.
2. A. Byström, P. Kierkegaard, and O. Knop, *Acta Chem. Scand.*, **6**, 1952, 709–719.
3. T. Heumann, *Nachr. Akad. Wiss. Göttingen, Math.-phys. Kl.*, **1948**, 21–26.

0.3995
$\bar{1}$.6005

Ce-Fe Cerium-Iron

Figure 259 shows the thermal-data points reported by [1]. The author assumed that the two intermediate phases, formed peritectically at about 1090 and 773°C,

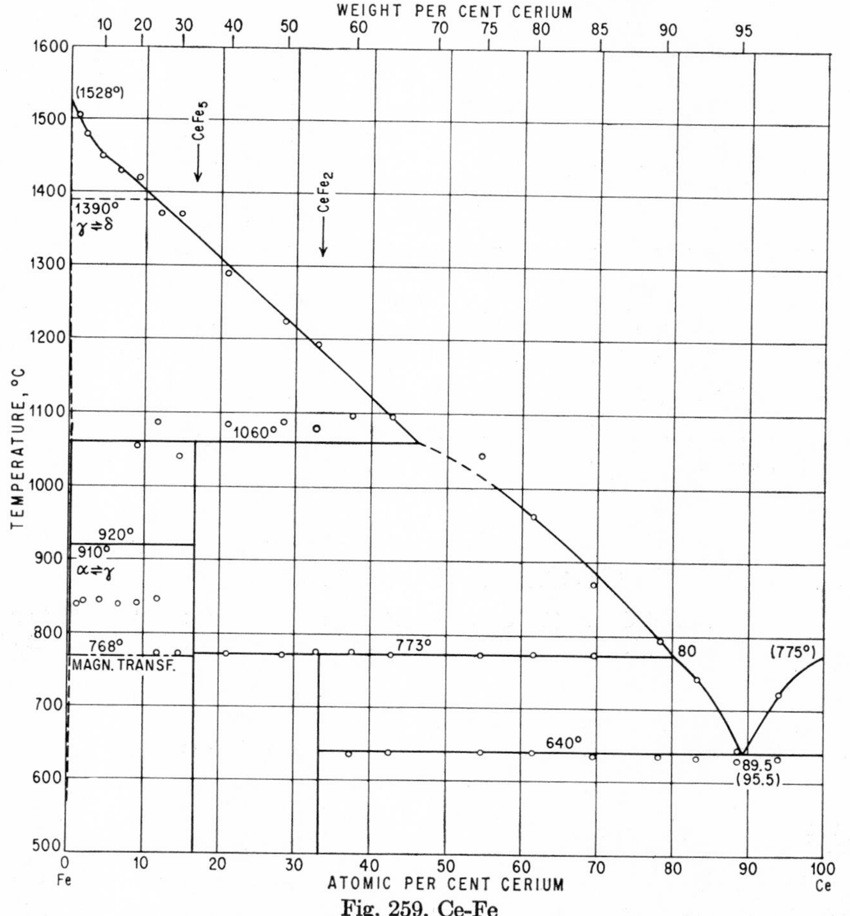

Fig. 259. Ce-Fe

have the compositions Ce_2Fe_5 [28.57 at. (50.09 wt.) % Ce] and $CeFe_2$ (55.64 wt. % Ce). However, this conclusion, based only on the duration of the thermal arrests at these temperatures, was not convincing since the incompleteness of the peritectic reactions resulted in structures containing three phases (between about 42 and 78 at. % Ce) or even four phases (between about 7 and 29 at. % Ce). Indeed, [2] have identified the Fe-rich intermediate phase as $CeFe_5$ [16.67 at. (33.41 wt.) % Ce] [3]. The existence of $CeFe_2$ was confirmed, and it was found to melt incongruently at 1060 ± 5°C [2].

From microexaminations of alloys annealed for a short time at 850–1100°C, [1] concluded that approximately 12–15 wt. (5.2–6.6 at.) % Ce was soluble in α- and γ-Fe in this temperature range. [2] have established conclusively that the solubility limit between 815 and 1015°C lies close to 0.4 wt. (0.16 at.) % Ce. The temperature of the $\alpha \rightleftarrows \gamma$ transformation of Fe (910°C) is raised to about 920°C, giving rise to the peritectoid equilibrium $\gamma + CeFe_5 \rightleftarrows \alpha$ [2].

All Ce-Fe alloys are ferromagnetic at room temperature [1]. The phase $CeFe_2$ was reported to have a Curie point at 116°C [1]. On the other hand, [4] found that the Curie point of alloys with more than 70 at. % Ce lies at 2°C and that of an alloy with about 62 at. % Ce at about 200°C.

The crystal structure of $CeFe_5$ was tentatively given as simple hexagonal, with a = 4.900 A, c = 4.136 A, c/a = 0.844, based on the indexing of 32 out of a total of 39 reflections. The structure is believed to be somewhat related to, but not identical with, that of the $CaCu_5$ type although $CeCo_5$ and $CeNi_5$, which could logically be isomorphous with $CeFe_5$, were reported to be isotypic with $CaCu_5$. Structure determinations of $CeCo_5$ [5] and $CeNi_5$ [6] are regarded as unreliable [2].

$CeFe_2$ is isotypic with $MgCu_2$ (C15 type) [7, 2], a = 7.302 A [2].

1. R. Vogel, *Z. anorg. Chem.*, **99**, 1917, 25–49. Alloys were prepared using partly Ce of unknown purity and partly Ce of 95.6%; they were not analyzed.
2. J. O. Jepson and P. Duwez, Progress Report 20-217, Jet Propulsion Laboratory, California Institute of Technology, Mar. 30, 1954; also *Trans. ASM*, **47**, 1955, 543–553. High-purity Ce was used; alloys were arc-melted and analyzed.
3. As flints usually contain 30 wt. % Fe, they consist of $CeFe_2$ and Ce and perhaps some $CeFe_5$.
4. J. R. Clark, S. T. Pan, and A. R. Kaufmann, *Phys. Rev.*, **63**, 1943, 139.
5. T. Heumann, *Nachr. Akad. Wiss. Göttingen, Math.-phys. Kl.*, **1948**, 21–26.
6. H. Nowotny, *Z. Metallkunde*, **34**, 1942, 247–253.
7. W. Fülling, K. Möller, and R. Vogel, *Z. Metallkunde*, **34**, 1942, 253–254.

0.3032
$\bar{1}$.6968

Ce-Ga Cerium-Gallium

The structure of $CeGa_2$ (49.88 wt. % Ga) is hexagonal of the AlB_2 (C32) type, a = 4.312 A, c = 4.316 A, c/a = 1.00_1 [1]. Ce does not lower the melting point of Ga [2].

1. F. Laves, *Naturwissenschaften*, **31**, 1943, 145.
2. R. M. Evans and R. I. Jaffee, *Trans. AIME*, **194**, 1952, 153–154.

2.1431
$\bar{3}$.8569

Ce-H Cerium-Hydrogen

The absorption of hydrogen by Ce at 1 atm and temperatures up to 1200°C and various lower pressures was studied by [1]. As Ce is an exothermic occluder [2], the solubility decreases with rise in temperature. At room temperature it absorbs approximately 2 wt. % H or, very roughly, 1,500 volumes [2]. For additional information and references, see the review by [2]. See also [3–5].

1. A. Sieverts and G. Müller-Goldegg, *Z. anorg. Chem.*, **131,** 1923, 65–95; A. Sieverts, *Z. Metallkunde,* **21,** 1929, 37–42; A. Sieverts and A. Gotta, *Z. Elektrochem.*, **32,** 1926, 105–109; *Z. anorg. Chem.*, **172,** 1928, 1-31.
2. D. P. Smith, "Hydrogen in Metals," pp. 15–17, 28, 178–180, University of Chicago Press, Chicago, 1948.
3. R. Viallard, *Ann. chim. (Paris)*, **20,** 1945, 5–72.
4. K. Dialer, *Monatsh. Chem.*, **79,** 1948, 296–310.
5. K. Dialer and W. Rothe, *Z. Elektrochem.*, **59,** 1955, 970–976.

$\bar{1}.8442$
0.1558

Ce-Hg Cerium-Mercury

Measurements of the vapor pressure of cerium amalgams with 12–57 wt. (16.3–65.5 at.) % Ce at 340°C [1] have shown that the vapor pressure between 0 and 12 wt. (16.3 at.) % Ce is practically that of Hg and decreases discontinuously to a value which is constant between 15.2 wt. (20.4 at.) % and 57 wt. (65.5 at.) % Ce, thus indicating the existence of $CeHg_4$ (14.87 wt. % Ce). This phase is only slightly soluble in liquid Hg (at 340°C) since the lowering of the vapor pressure of Hg in the range up to 12 wt. % Hg is only small. Also, [1] have shown that no intermediate phase richer in Ce exists. X-ray diffraction studies [2] were inconclusive; however, the authors believe that there are indications of an additional intermediate phase richer in Ce.

The work by [1] does not exclude the existence of one or more (peritectically formed) compounds richer in Hg which are stable below 340°C. According to [3], $CeHg_4$ decomposes at 470°C.

Supplement. According to an abstract, [4] prepared and analyzed chemically the compound CeHg. "The Debye-Scherrer powder X-ray diagram shows . . . a centered cubic lattice with one rare earth and Hg atom per unit cell." Presumably the structure is of the CsCl type. The lattice parameter is $a = 3.808$ A.

1. W. Biltz and F. Meyer, *Z. anorg. Chem.*, **176,** 1928, 32–38.
2. A. Iandelli and E. Botti, *Gazz. chim. ital.*, **67,** 1937, 638–644.
3. P. T. Daniltchenko, see abstract in *J. Inst. Metals*, **50,** 1932, 540.
4. A. Iandelli and R. Ferro, *Atti accad. nazl. Lincei, Rend., Classe sci. fis., mat. e nat.*, **10,** 1951, 48–52; *Chem. Abstr.*, **45,** 1951, 7842.

0.0867
$\bar{1}.9133$

Ce-In Cerium-Indium

The phase diagram (Fig. 260) is almost exclusively based on the thermal-data points indicated [1]. Because of rapid decomposition of the samples, microscopic examination was reliable only in the ranges 0–25 and 70–100 at. % In. The thermal effect at 1105°C in the alloy with 45 at. % In is believed to be caused by a polymorphic transformation of CeIn (45.02 wt. % In). The composition in wt. % In of the other compounds is Ce_3In, 21.44%; Ce_2In, 29.05%; Ce_2In_3, 55.13%; $CeIn_3$, 71.07%. The structure of $CeIn_3$ is f.c.c. of the Cu_3Au ($L1_2$) type, $a = 4.68$ A [1].

1. R. Vogel and H. Klose, *Z. Metallkunde*, **45,** 1954, 633–638.

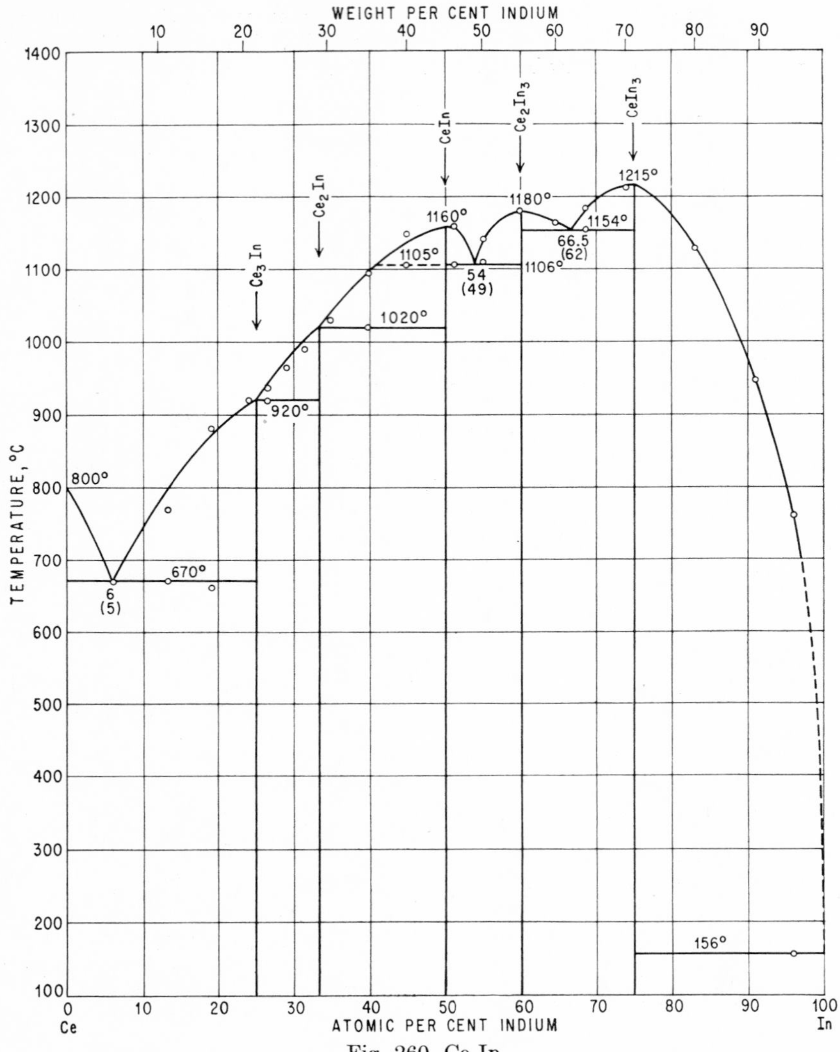

Fig. 260. Ce-In

0.0038
$\bar{1}.9962$

Ce-La Cerium-Lanthanum

According to thermal analysis and microexamination [1], the α, β, and γ modifications of Ce [2] and La form continuous series of solid solutions [3]. The diagram in Fig. 261 does not show the $\alpha \rightleftarrows \beta$ transformation curves of the solid solutions since these transformation points of the two metals are not yet well established. They are believed to lie at about 300°C(Ce) and about 350°C (La) [1].

1. R. Vogel and H. Klose, *Z. Metallkunde*, **45**, 1954, 633.
2. There is disagreement in the literature as to the crystal structure of Ce at normal temperatures. The h.c.p. "α-Ce" claimed by [1], among others, is considered to be a metastable form according to A. W. Lawson and T. Y. Tang (*Phys. Rev.*, **76**, 1949, 301).
3. Metals of the following purity were used: Ce, 99.5%; La, 97%, with 1.4% Fe, 0.3% Si, 0.3% Mg, 1% ?.

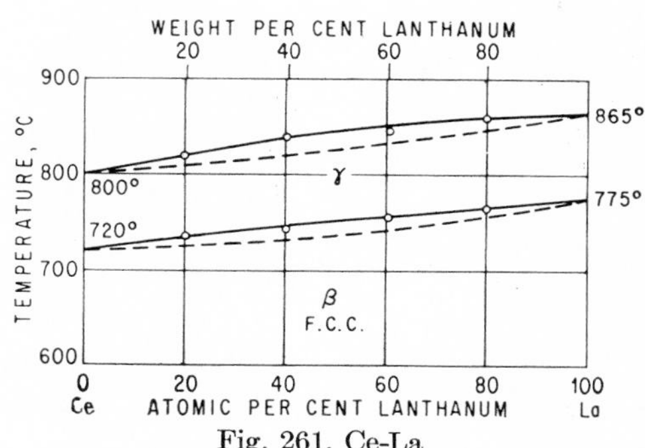

Fig. 261. Ce-La

0.7606
$\bar{1}$.2394

Ce-Mg Cerium-Magnesium

The first investigation [1] of the whole system by thermal and microscopic studies, using Ce of only 93.5% purity, established the existence of three intermediate phases: MgCe (85.21 wt. % Ce) with a maximum melting point of 740°C, Mg_3Ce (65.76 wt. % Ce) having a maximum melting point of 780°C and taking about 12 at. % Mg into solid solution, and Mg_9Ce (39.03 wt. % Ce) formed by a peritectic reaction at 622°C of Mg_3Ce with melt containing 34 wt. (8.2 at.) % Ce. In addition, $MgCe_4$ (95.84 wt. % Ce) was claimed to exist and to decompose into MgCe and Ce at 497°C; however, the evidence was inconclusive [2].

A thorough reinvestigation of the range 55 wt. (17.5 at.) to 100% Ce by thermal, micrographic, and X-ray methods [3] revealed the phase relationships presented in Fig. 262. The cerium metal used was reported as containing 0.14% Mg, 0.14% Fe, and 0.02% Si. The thermal effects at 490°C were interpreted as corresponding to the eutectoid decomposition of a solid solution based on a high-temperature modification of Ce. $MgCe_4$ was shown to be nonexistent. The compound Mg_2Ce (74.23 wt. % Ce), reported earlier [4], was confirmed; it decomposes into Mg_3Ce and MgCe at about 615°C. Mg_3Ce forms solid solutions with Mg, the limit of which was found as 21.7 at. (61.5 wt.) % Ce at 625°C and more than 24.4 at. (65 wt.) % Ce at 370°C. The existence of the three intermediate phases stable at room temperature was verified by magnetic investigations [5]. There is some indication that Mg_9Ce undergoes a polymorphic change [1].

The Mg-rich alloys with up to 10.4 at. (40 wt.) % Ce were the subject of a thermal and microscopic study [6]. The results used to plot this portion of the diagram (Fig. 262) differ from those of [1] but deserve preference.

The solid solubility of Ce in Mg was determined micrographically [6] and thermoresistometrically [7]. The solubility in at. (wt.) % was given as follows: 0.28 (1.6) at

590°C, 0.18 (1.0) at 550°C, 0.09 (0.5) at 500°C, and <0.03 (0.15) at 340°C [6]; and 0.39 (2.2) at 585°C, 0.34 (1.9) at 570°C, 0.26 (1.5) at 530°C, 0.16 (0.9) at 500°C, and 0.09 (0.5) at 440°C [7]. According to [8], the solubility is even lower: 0.15 (0.85) % at 575°C.

Crystal Structures. Mg_3Ce: BiF_3 (DO_3) type, a = 7.388 A [9] and a = 7.436 A [3]. Mg_2Ce: $MgCu_2$ (C15) type, a = 8.73 A [4] and a = 8.70 A [3]. MgCe: CsCl (B2) type, a = 3.906 A [10] and a = 3.892 A [11]. For the effect of Ce on the lattice parameters of Mg, see [12].

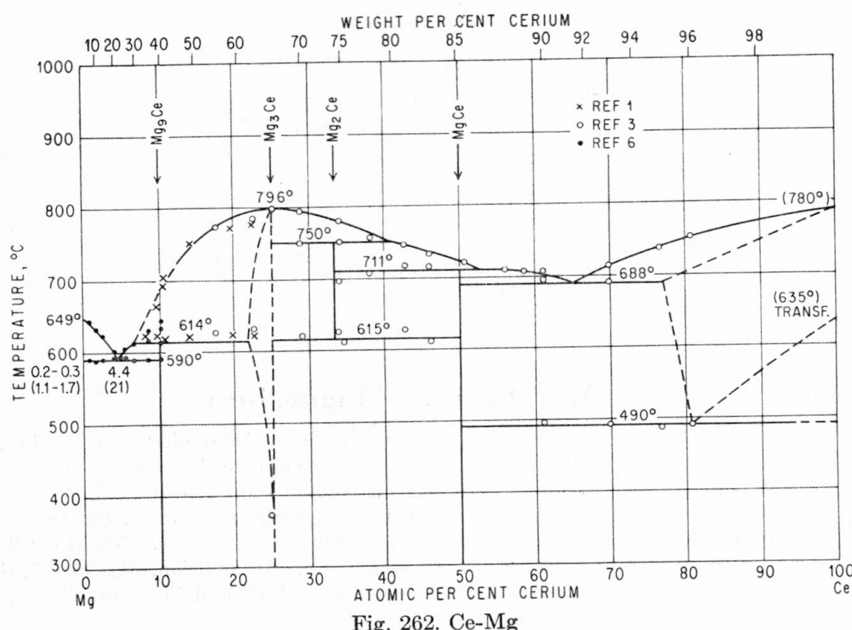

Fig. 262. Ce-Mg

1. R. Vogel, *Z. anorg. Chem.*, **91**, 1915, 277–298.
2. M. Hansen, "Der Aufbau der Zweistofflegierungen," pp. 472–474, Springer-Verlag OHG, Berlin, 1936.
3. R. Vogel and T. Heumann, *Z. Metallkunde*, **38**, 1947, 1–8.
4. F. Laves, *Naturwissenschaften*, **31**, 1943, 96.
5. F. Mahn, *Ann. phys.*, **3**, 1948, 393–458; *Rev. mét.*, **46**, 1949, 365–369.
6. J. L. Haughton and T. H. Schofield, *J. Inst. Metals*, **60**, 1937, 339–344.
7. F. Weibke and W. Schmidt, *Z. Elektrochem.*, **46**, 1940, 357–364.
8. Unpublished work of Dow Chemical Company, Midland, Mich., quoted from Summary Report, Part I, June 30, 1955; Contract DA-11-022-ORD-1645, Project TB4-15.
9. A. Rossi and A. Iandelli, *Atti accad. nazl. Lincei, Rend.*, **19**, 1934, 415–420.
10. H. Nowotny, *Z. Metallkunde*, **34**, 1942, 247–253.
11. A. Rossi, *Gazz. chim. ital.*, **64**, 1934, 774–778.
12. R. S. Busk, *Trans. AIME*, **188**, 1950, 1460–1464.

0.4067
$\bar{1}.5933$

Ce-Mn Cerium-Manganese

As was stated previously [1], this system has a miscibility gap in the liquid state smaller than that in the system La-Mn (page 886). The phase diagram in Fig. 263 is based on thermal-analysis data [2]. The monotectic and eutectic equilibria were found at temperatures varying between 987 and 1023°C and 595 and 620°C, respectively.

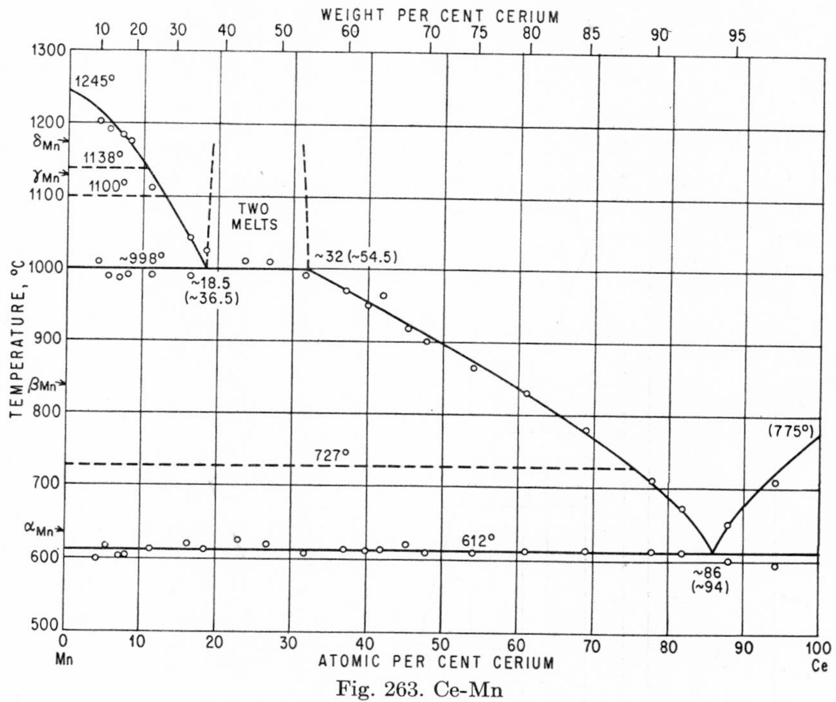

Fig. 263. Ce-Mn

1. L. Rolla and A. Iandelli, *Ber. deut. chem. Ges.*, **75**, 1942, 2094.
2. A. Iandelli, *Atti accad. nazl. Lincei, Rend.*, **13**, 1952, 265–268.

1.0002
$\bar{2}.9998$

Ce-N Cerium-Nitrogen

CeN (9.09 wt. % N) is isotypic with NaCl (B1 type), $a = 5.02$ A [1].

1. A. Iandelli and E. Botti, *Atti reale accad. nazl. Lincei*, **25**, 1937, 129–132.

0.3780
$\bar{1}.6220$

Ce-Ni Cerium-Nickel

The phase diagram (Fig. 264) was determined by thermal analysis and corroborated by microscopic examination of about 26 alloys prepared using Ce of 98–99% purity. In the absence of tabulated data, the data points shown were taken from a smaller diagram and, therefore, should not be considered quantitative. The existence of the intermediate phases $CeNi_5$ (32.32 wt. % Ce), $CeNi_2$ (54.42 wt. % Ce), CeNi (70.48 wt. % Ni), and Ce_3Ni (87.75 wt. % Ce) was well established. However, the compositions of the two additional phases, given as $CeNi_4$ (37.38 wt. % Ce) and $CeNi_3$ (44.32 wt. % Ce), are uncertain because of the incompleteness of the peritectic reactions at 1065, 930, and 830°C [1].

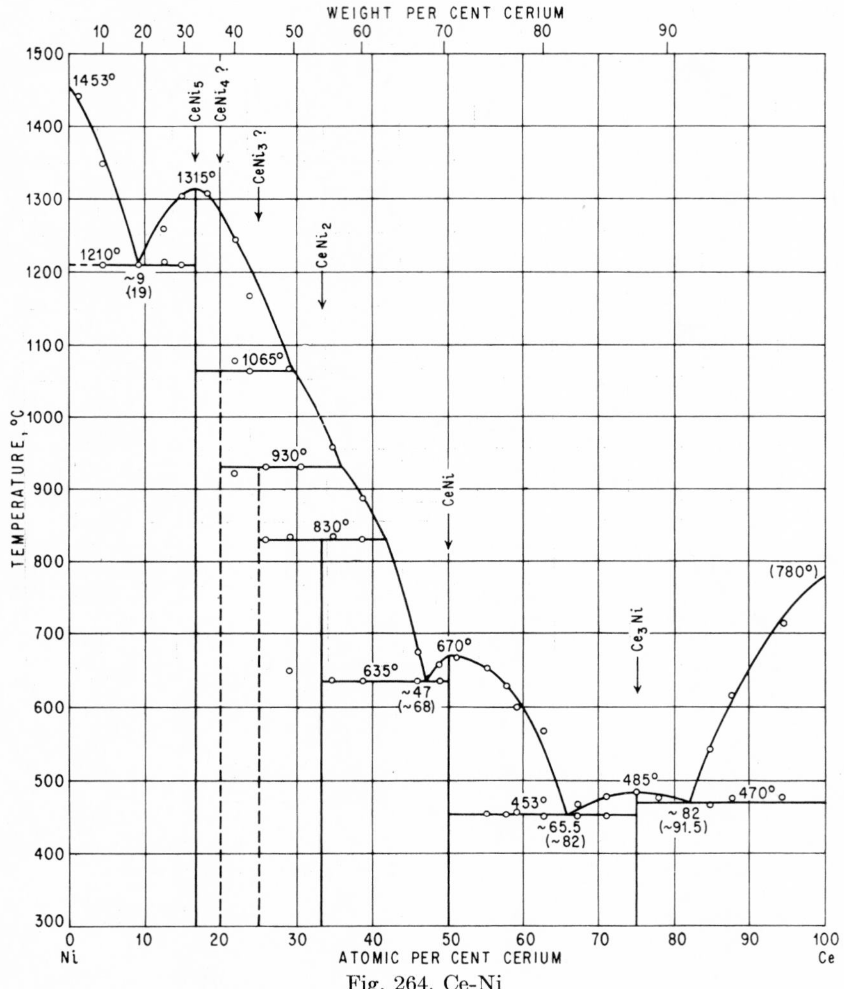

Fig. 264. Ce-Ni

$CeNi_5$ is hexagonal of the $CaCu_5$ type, a = 4.874 A, c = 4.004 A, $2c/a$ = 1.644 [2]; however, see Ce-Fe. $CeNi_2$ is isotypic with $MgCu_2$ (C15 type), a = 7.192 A [2] or a = **7.204** A [3].

1. R. Vogel (W. Fülling), *Z. Metallkunde*, **38**, 1947, 97–103.
2. H. Nowotny, *Z. Metallkunde*, **34**, 1942, 247–253.
3. W. Fülling, K. Möller, and R. Vogel, *Z. Metallkunde*, **34**, 1942, 253–254.

0.6555
$\bar{1}.3445$

Ce-P Cerium-Phosphorus

CeP (18.1 wt. % P) is isotypic with NaCl (B1 type), $a = 5.91$ A [1].

1. A. Iandelli and E. Botti, *Atti reale accad. nazl. Lincei*, **24**, 1936, 459–464.

$\bar{1}.8301$
0.1699

Ce-Pb Cerium-Lead

An early study of the system was unsuccessful [1]; it was merely reported that the phase diagram appears to be similar to that of the system Ce-Sn. Later, the existence of the compound $CePb_3$ (81.60 wt. % Pb) was established by determination of its

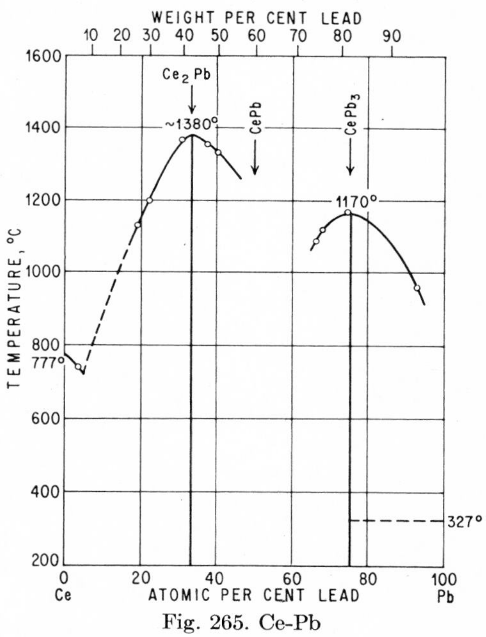

Fig. 265. Ce-Pb

crystal structure [2]. It was also reported that a compound CePb (59.66 wt. % Pb) having a b.c.c. lattice could not be found [3]. A more recent thermal and microscopic investigation yielded the incomplete diagram given in Fig. 265 [4]. The existence of the two compounds indicated, Ce_2Pb (42.51 wt. % Pb) and $CePb_3$, appears to be fully established. The Ce-Ce_2Pb eutectic lies close to 7.5 wt. (5.2 at.) % Pb, as found by microscopic examination.

Since there is in the systems La-Pb (page 888) and Pb-Pr (page 1096), besides the corresponding intermediate phases X_2Pb and XPb_3, a third phase XPb and since the constitution of the systems of La, Ce, and Pr with many other metals is quite similar, with regard to the number and formulas of the intermediate phases [4], it appears very likely that a compound CePb does exist.

$CePb_3$ has a structure of the Cu_3Au ($L1_2$) type, $a = 4.874$ A [2].

1. R. Vogel, *Z. anorg. Chem.*, **72**, 1911, 320.
2. E. Zintl and S. Neumayr, *Z. Elektrochem.*, **39**, 1933, 86–97.

3. E. Zintl and G. Brauer, *Z. physik. Chem.*, **B20,** 1933, 245–271.
4. R. Vogel and T. Heumann, *Z. Metallkunde*, **35,** 1943, 29–42.

$\bar{1}.8560$
0.1440

Ce-Pt Cerium-Platinum

The phase $CePt_2$ (73.59 wt. % Pt) has the cubic structure of the $MgCu_2$ (C15) type, a = 7.730 A [1].

1. W. Zachariasen, *Acta Cryst.*, **2,** 1949, 388–390.

0.6405
$\bar{1}.3595$

Ce-S Cerium-Sulfur

The crystal structure of Ce_2S_3 (25.56 wt. % S) is cubic of the defective Th_3P_4 ($D7_3$) type, a = 8.6347 A, with 16 S and 10⅔ Ce atoms per unit cell. The homogeneity range extends to the composition Ce_3S_4 (23.38 wt. % S), where all metal sites are filled (28 atoms per unit cell), a = 8.6250 A [1, 2]. At S contents below Ce_3S_4 the phase is in equilibrium with CeS (18.62 wt. % S), which is of the NaCl type, a = 5.778 A [3].

1. W. H. Zachariasen, *Acta Cryst.*, **2,** 1949, 57–60.
2. In *Structure Repts.*, **12,** 179, the parameter is erroneously given as a = 8.6093 A, computed from a = 8.6076 kX.
3. W. H. Zachariasen, *Acta Cryst.*, **2,** 1949, 291–296.

0.0610
$\bar{1}.9390$

Ce-Sb Cerium-Antimony

CeSb (46.49 wt. % Sb) has the cubic structure of the NaCl (B1) type, a = 6.11 A [1].

1. A. Iandelli and E. Botti, *Atti reale accad. nazl. Lincei*, **25,** 1937, 498–502.

0.6980
$\bar{1}.3020$

Ce-Si Cerium-Silicon

The partial phase diagram of Fig. 266 (Si-rich alloys) is due to [1], who concluded from the extrapolation of the eutectic halting times that the intermediate phase, observed in the microstructure, is probably of the composition CeSi (16.70 wt. % Si). However, by X-ray studies the existence of $CeSi_2$ (28.62 wt. % Si), already reported by [2], was established [3, 4].

$CeSi_2$ is b.c. tetragonal of the $ThSi_2$ type, with 4 $CeSi_2$ per unit cell. Lattice dimensions were found to be a = 4.16 A, c = 13.90 A, c/a = 3.34 [3], and a = 4.156 A, c = 13.84 A, c/a = 3.33 [4].

Supplement. [5] prepared Ce-Si samples by reacting small pieces of Ce rod with Si powder at about 1700°K (~1430°C). "They all had the appearance of being fused or partially fused. The X-ray patterns were rather poor, and the samples were inhomogeneous mixtures; however, it was established that several lower silicides exist. These are provisionally assigned the formulas $CeSi_{0.35}$ [26 at. % Si], $CeSi_{0.5}$ [Ce_2Si], and $CeSi_{0.75}$ [43 at. % Si]. The $CeSi_{0.35}$ phase appeared along with CeO_2 in all of its samples. The CeO_2 probably came about from oxidation of Ce metal during preparation of the specimens for X-ray study in spite of the precautions. Therefore, it is believed that $CeSi_{0.35}$ is the lowest silicide. The $CeSi_{0.5}$ appeared in samples con-

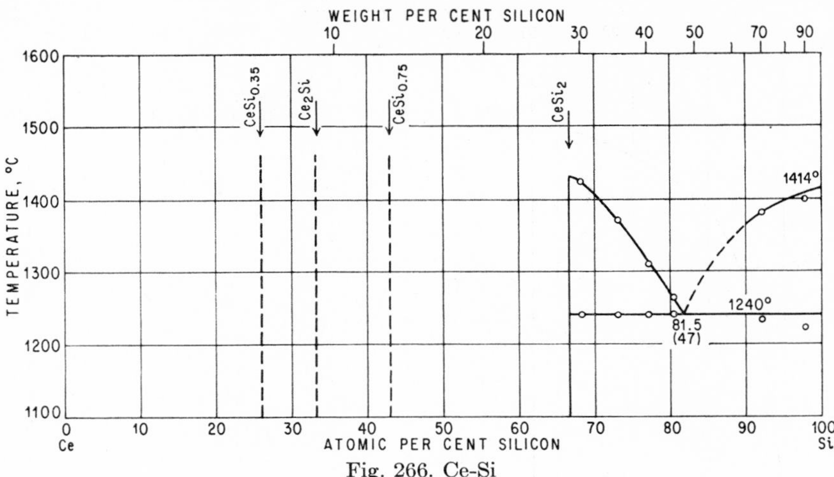

Fig. 266. Ce-Si

taining 33 to 37 at. % Si and $CeSi_{0.75}$ appeared in samples containing 40 to 50 at. % Si." As for the $CeSi_2$ phase, [5] reported the lattice parameters $a = 4.175 \pm 2$ A, $c = 13.848 \pm 6$ A.

1. R. Vogel, *Z. anorg. Chem.*, **84,** 1913, 323–327. Apparently Ce of low purity (93.5%?) was used.
2. J. Sterba, *Compt. rend.*, **135,** 1902, 170–172.
3. W. Zachariasen, *Acta Cryst.*, **2,** 1949, 94–99.
4. G. Brauer and H. Haag, *Z. anorg. Chem.*, **267,** 1952, 198–212.
5. O. H. Krikorian, *U.S. Atomic Energy Comm., Publ.* UCRL-2888, pp. 16–17, 1955. In a 1954 publication (L. Brewer and O. Krikorian, UCRL-2544) the formulas Ce_3Si, Ce_2Si, and CeSi had been assigned to the lower silicides.

0.0721
$\bar{1}.9279$

Ce-Sn Cerium-Tin

The constitution was studied by thermal analysis and metallographic examination [1], with Ce of only 93.5% purity [2] and a melting point of 830°C. No thermal effects were found for the secondary crystallization of the alloys between 0 and 33 at. % Sn. Also, the composition of the Sn-rich eutectic was not determined. The existence of the intermediate phases Ce_2Sn (29.75 wt. % Sn), Ce_2Sn_3 (55.96 wt. % Sn), and $CeSn_2$ (62.88 wt. % Sn) resulted from the maximum melting points at these compositions (Fig. 267). Reinvestigation of the composition range between 68.5 and 80.5 at. % Sn, however, revealed the existence of $CeSn_3$ (71.76 wt. % Sn) rather than $CeSn_2$ [3]. Thus, earlier findings [4] which, on the basis of crystal-structure determinations, proved the compound $CeSn_3$ to exist were confirmed. The solid solubility of Ce in Sn at 200°C was reported to be definitely smaller than 0.1 at. (wt.) % Ce [5].

$CeSn_3$ is isotypic with Cu_3Au ($L1_2$ type), $a = 4.720$ A [4].

1. R. Vogel, *Z. anorg. Chem.*, **72,** 1911, 319–328.
2. Main contaminants were neodymium and praseodymium.
3. R. Vogel and T. Heumann, *Z. Metallkunde*, **35,** 1943, 29–42.
4. E. Zintl and E. Neumayr, *Z. Elektrochem.*, **39,** 1933, 86–97.
5. E. Jenckel and L. Roth, *Z. Metallkunde*, **30,** 1938, 135–144.

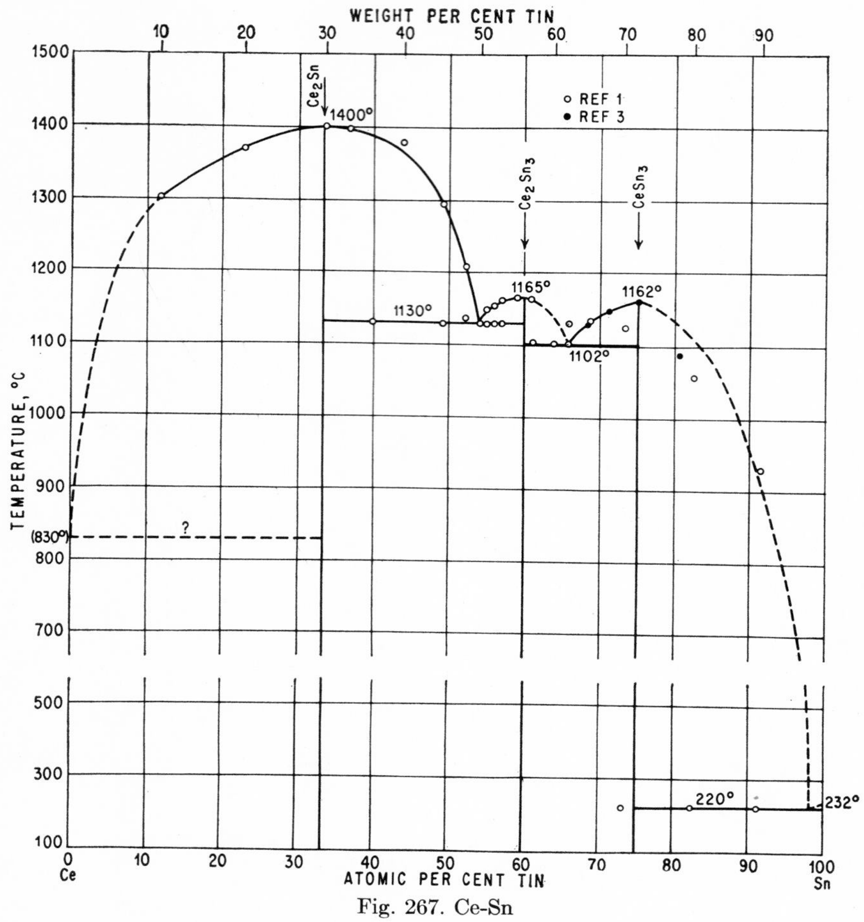

Fig. 267. Ce-Sn

0.0406
$\bar{1}.9594$

Ce-Te Cerium-Tellurium

CeTe (47.66 wt. % Te) was prepared [1] by heating to redness an equal mole mixture of powdered Ce and Te. If excess Te is used, it is volatilized, but excess Ce will result in a solution of Ce in CeTe.

1. E. Montignie, *Bull. soc. chim. France,* **1947,** 748–749; *Chem. Abstr.,* **42,** 1948, 2535.

$\bar{1}.7808$
0.2192

Ce-Th Cerium-Thorium

"[1] investigated a series of Th-Ce alloys and found that Ce and α-Th are completely soluble. He prepared the alloys by co-reduction at temperatures of 1000 to 1400°C. Earlier experiments by [2] also indicated complete solubility in both the liquid and solid states in this system. There are no compounds in the Th-Ce system" [3].

1. D. Peterson and R. Mickelson, unpublished information, Mar. 31, 1952.
2. F. Foote, unpublished information, April, 1945.
3. Quoted from H. A. Saller and F. A. Rough, *U.S. Atomic Energy Comm., Publ.* BMI-1000, June, 1955.

0.4662
$\bar{1}.5338$

Ce-Ti Cerium-Titanium

It has been stated that the tendency to form intermediate phases decreases greatly in the systems of Ce (and other rare-earth metals such as La and Pr) with Ni,

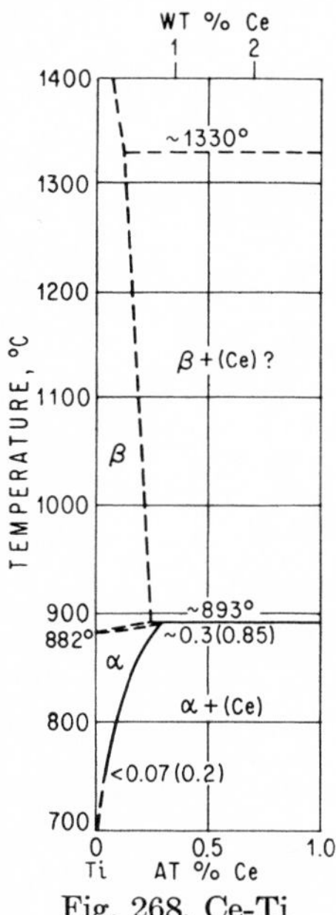

Fig. 268. Ce-Ti

Co, Fe, Mn, Cr, and Ti in the order given, so that no such phase will exist in the systems Ce-Ti, La-Ti (see page 893), and Pr-Ti [1].

This was verified by a preliminary study [2] of the system by means of micrographic analysis [for the range up to 9 wt. (3.3 at.) % Ce] and X-ray work [for iodide titanium-base arc-melted alloys up to 50 wt. (about 25 at.) % Ce]. The phase coexisting with α-Ti was identified as Ce (or a Ce-rich solid solution). According to Fig. 268, about 0.85 wt. (0.3 at.) % Ce is soluble in α-Ti at the peritectoid temperature and less than 0.2 wt. (0.07 at.) % Ce at 750°C.

Some indication was found of the existence of a miscibility gap in the liquid state. If so, the monotectic temperature would be located above 1330°C since at this temperature incipient melting occurred; i.e., the Ce-rich phase is stable up to about 1330°C.

1. R. Vogel, *Z. Metallkunde,* **38,** 1947, 102–103.
2. J. L. Taylor, Progress Report 20-207, Feb. 11, 1954, ORDCIT Project, Contract DA-04-495-Ord 18. *Trans. AIME,* **209,** 1957, 94–96.

Ī.8361
0.1639

Ce-Tl Cerium-Thallium

The phase diagram given in Fig. 269 is essentially based on the thermal-analysis data indicated [1]. The compounds Ce_2Tl (42.17 wt. % Tl), CeTl (59.33 wt. % Tl), and $CeTl_3$ (81.40 wt. % Tl) correspond to the compounds of lanthanum and praseodymium with thallium.

1. R. Vogel and T. Heumann, *Z. Metallkunde*, **35**, 1943, 29–42.

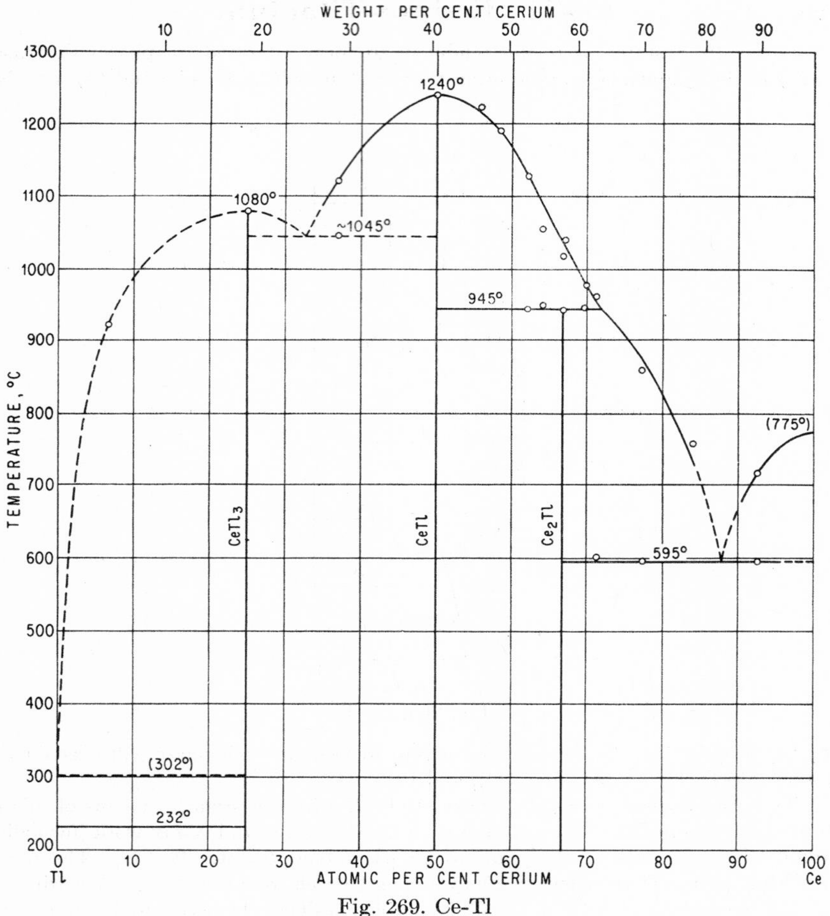

Fig. 269. Ce-Tl

Ī.7698
0.2302

Ce-U Cerium-Uranium

The solubility of Ce in α-U and γ-U has been reported to be nil [1].

According to a review by [2], "U and Ce are only partially miscible in the liquid state [3–5]. There is evidence of a eutectic at the U end of the system, perhaps at about 1000°C [4, 5]. The solubility of Ce in U at the eutectic temperature is less than

0.5 wt. %, and of U in Ce about 0.1 wt. % [3]. There are no compounds in the U-Ce system."

1. J. J. Katz and E. Rabinowitch, "The Chemistry of Uranium," Part I, National Nuclear Energy Series, Div. VIII, vol. 5, p. 175, McGraw-Hill Book Company, Inc., New York, 1951.
2. H. A. Saller and F. A. Rough, *U.S. Atomic Energy Comm., Publ.* BMI-1000, June, 1955.
3. National Physical Laboratory, United Kingdom, unpublished information, May, 1949.
4. M. Neher, B. D. Cullity, and A. R. Kaufman, unpublished information, January, 1945.
5. A. B. Greninger and F. Foote, unpublished information, May, 1945.

0.3311
$\bar{1}.6689$

Ce-Zn Cerium-Zinc

The diagram of Fig. 270 [1] appears to be well established up to about 10 at. % Ce. It shows the existence of the phases $CeZn_{11}$ [8.33 at. (16.31 wt.) % Ce] and $CeZn_9$ (19.23 wt. % Ce), both of which are stable within a certain range of homogeneity, indicated in Fig. 270 but not accurately determined. See also the system La-Zn (page 895).

At Ce concentrations above 11 at. %, five three-phase equilibria were found: at about 942, 870, 840, 817, and 790°C. The thermal effects at 942, 870, and 790°C

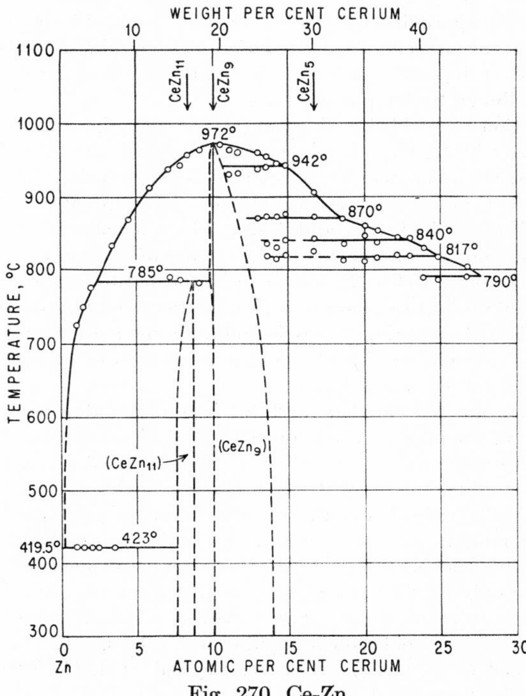

Fig. 270. Ce-Zn

were reported to be quite large and are probably due to the peritectic formation of intermediate phases, one of which very likely is $CeZn_5$ (30.0 wt. % Ce), possibly formed at 870°C.

$CeZn_{11}$ is isotypic with $BaCd_{11}$, b.c. tetragonal with 48 atoms per unit cell and a = 10.66 A, c = 6.86 A, c/a = 0.644 [2]. $CeZn_9$ and $LaZn_9$ appear to have the same structure [1]. The existence of the phase CeZn (68.19 wt. % Ce) was established by determining its crystal structure, which is of the CsCl (B2) type, a = 3.71 A [3].

From measurements of the electrochemical potential [4], it was concluded that the phases Ce_2Zn (81.08 wt. % Ce) and Ce_4Zn (89.55 wt. % Ce) exist [5]. However, equivalent compounds of La and Zn have not been detected (see La-Zn).

1. J. Schramm, *Z. Metallkunde*, **33**, 1941, 358–360.
2. M. J. Sanderson and N. C. Baenziger, *Acta Cryst.*, **6**, 1953, 627–631.
3. A. Iandelli and E. Botti, *Gazz. chim. ital.*, **67**, 1937, 638–644.
4. F. Clotofski, *Z. anorg. Chem.*, **114**, 1920, 16–23.
5. W. Muthmann and H. Beck, *Liebigs Ann.*, **331**, 1904, 46–57.

0.0543
$\bar{1}$.9457

Co-Cr Cobalt-Chromium

Phase Relationships. After the first cursory study [1] of the constitution, [2, 3] published a phase diagram which, although based on comprehensive investigations, could be considered only a qualitative outline of phase fields. The diagram [4] is characterized by the existence of two intermediate phases, a phase of singular composition CrCo (46.88 wt. % Cr) and a phase of variable composition [later designated as σ (Fig. 271)] with approximately 55–62 wt. (58–65 at.) % Cr. No attempt was made to determine accurately the boundaries of the primary solid solutions and intermediate phase σ; the rate of diffusion, even at higher temperatures, proved to be extremely low, as shown by later studies [5].

A reinvestigation [6] yielded a diagram [4], the essential features of which agree with those found by [2, 3], with the exception that only one intermediate phase, σ, with about 53.5–62 wt. (56.6–65 at.) % Cr, was shown to exist.

[21] published a phase diagram which, according to a brief statement, was drawn on the basis of original work and data published earlier [2, 3, 6]. Essentially, the diagram seems to be a combination of the results of [2, 6]. It is characterized by the existence of the phase CrCo [2], shown to be formed by a peritectoid reaction of α and σ at about 1260°C and to decompose again eutectoidally into α and σ at approximately 940°C. The designation CrCo is followed by a question mark, probably indicating that no conclusive evidence for the occurrence of this phase, reported by [2] and not detected by [6], was found.

The phase diagram presented in Fig. 271 is basically due to [7]. The liquidus and solidus curves of the alloys up to about 65 wt. % Cr were taken from the work of [2, 8]. The eutectic temperature, found to be 1408 [2], 1393 [6], and 1401°C [8], was placed at the mean value, 1400°C [7]. [2, 6, 7] have located the eutectic composition at 42 wt. (45 at.) % Cr; [8] found the possibly more accurate value 42.5 wt. (45.5 at.) % Cr. The boundaries of the phases ϵ, α, δ, σ, and β were outlined by micrographic analysis of alloys [9], quenched from various temperatures between 600 and 1300°C after annealing for only 50–65 hr, regardless of temperature. Admittedly, this time is by far insufficient to reach equilibrium; however, a special graphical method used was claimed to permit extrapolation to equilibrium values. The homogeneity range of σ was given as 56.6–61 at. % Cr. However, beyond 39 at. % Cr the boundaries are said to be in error by 2 wt. (2 at.) % Cr.

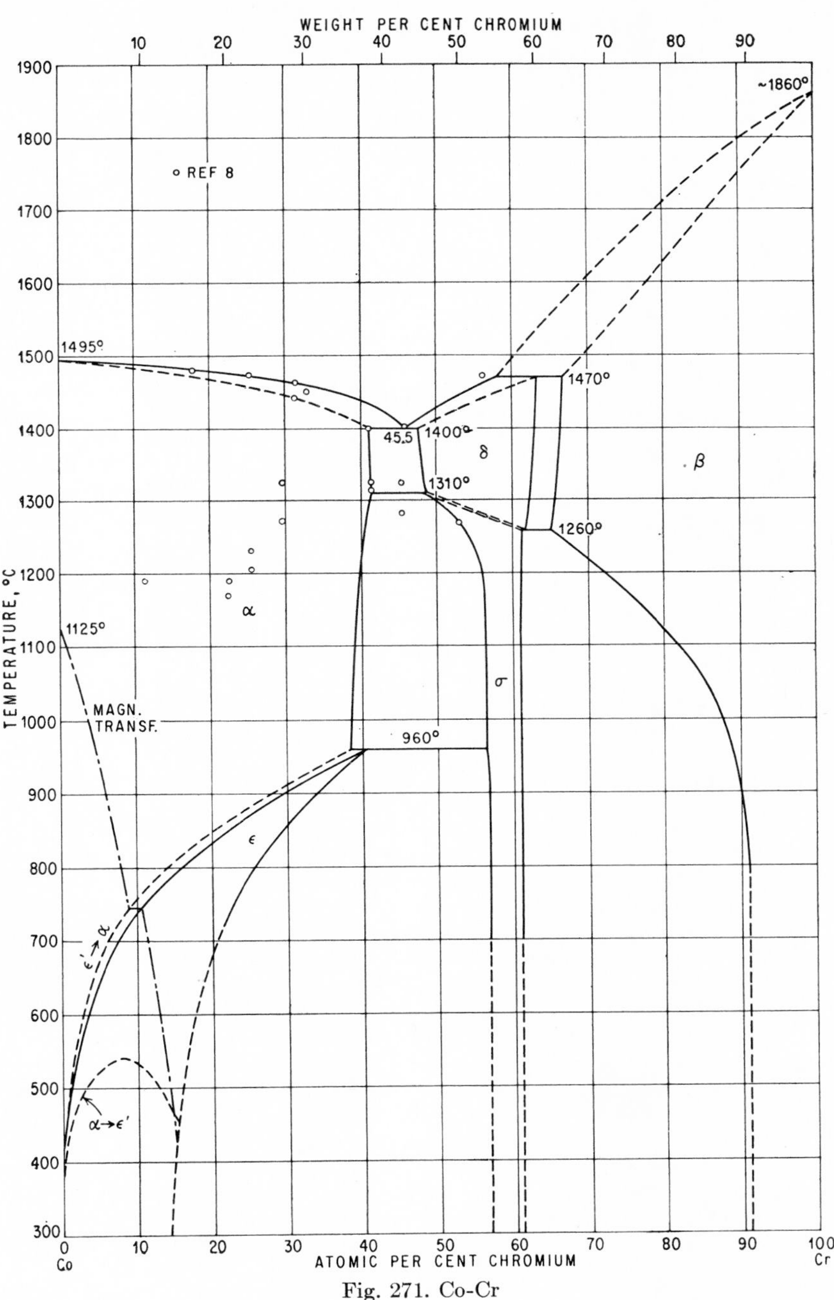
WEIGHT PER CENT CHROMIUM
10
20
30
40
50
60
70
80
90
1900
1800
1700
1600
1500
1400
1300
1200
1100
1000
900
800
700
600
500
400
300
TEMPERATURE, °C
~1860°
o REF 8
1495°
1470°
45.5
1400°
δ
β
1310°
1260°
α
1125°
MAGN.
TRANSF.
σ
960°
ε
ε′→α
α→ε′
0
Co
10
20
30
40
50
60
70
80
90
100
Cr
ATOMIC PER CENT CHROMIUM

Fig. 271. Co-Cr

The partial diagram proposed by [8] is in substantial agreement with that of [7]. The main difference lies in the assumption of a polymorphic transformation of Co at 1120–1145°C and minor changes connected with this transformation [10]; see data points in Fig. 271.

The magnetic transformation of the Co-rich solid solutions was studied by [1, 11, 2, 3, 6]. There is substantial agreement of the results; in fact, data by [2, 3] and [6] agree very well. The temperatures of the transformation $\epsilon' \rightarrow \alpha$ [2, 3, 6] do not agree for alloys beyond 5 wt. % Cr. The transformation curves shown in Fig. 271 were averaged from the data by [2, 3, 6] and differ from those proposed by [7]; the latter apparently are not based on experimental data.

The σ Phase. It was first suggested by [12] that the phase in the range 56.6–61 at. % Cr [7] is a σ phase [13]. This was confirmed by [14–16] and later workers. [17] determined the homogeneity range at 1200°C to be 58.6–63 at. % Cr. [15] proposed a tetragonal cell, with $a = 6.205$ A, $c = 9.030$ A, $c/a = 1.455$; and [16] tried to index the powder pattern on an orthorhombic cell. [18] proposed a tetragonal cell with 30 atoms and referred to the great similarity to the structure of β-U. Later, they gave the parameter as $a = 8.80$ A, $c = 4.56$ A, $c/a = 0.518$ for a single crystal with 56.4 at. % Cr [19]. [20] reported the parameters of a single crystal with 52.8 at. % Cr to be $a = 8.75$ A, $c = 4.54$ A, $c/a = 0.519$ (16 Cr and 14 Co atoms per unit cell, corresponding to 53.34 at. % Cr). More recently the authors of [18, 19] have published a refinement of the σ-phase structure [22]. There is some evidence for an ordered distribution of Co and Cr atoms in the atomic sites.

1. K. Lewkonja, *Z. anorg. Chem.*, **59**, 1908, 323–327.
2. F. Wever and U. Haschimoto, *Mitt. Kaiser-Wilhelm-Inst. Eisenforsch. Düsseldorf*, **11**, 1929, 293–308.
3. F. Wever and H. Lange, *Mitt. Kaiser-Wilhelm-Inst. Eisenforsch. Düsseldorf*, **12**, 1930, 353–363.
4. The diagram is reproduced in the paper by [7].
5. J. W. Weeton, *Trans. ASM*, **44**, 1952, 436–449.
6. Y. Matsunaga, *Kinzoku-no-Kenkyu*, **8**, 1931, 549–561.
7. A. R. Elsea, A. B. Westerman, and G. K. Manning, *Trans. AIME*, **180**, 1949, 579–602; see discussion, *Trans. AIME*, **185**, 1949, 298–300.
8. A. G. Metcalfe, *Trans. AIME*, **197**, 1953, 357–364.
9. Alloys, prepared using electrolytic Co and Cr, were melted in vacuum or under purified argon. Heat-treatment was carried out under protective conditions.
10. For a discussion of earlier work on the proposed upper polymorphic transformation point of Co, see A. G. Metcalfe, *Proc. World Met. Congr.*, 1952, pp. 717–731 (numerous references). See also J. B. Newkirk and A. H. Geisler, *Acta Met.*, **1**, 1953, 456; A. G. Metcalfe, *Acta Met.*, **1**, 1953, 609.
11. K. Honda, *Ann. Physik*, **32**, 1910, 1009–1010.
12. A. H. Sully and T. J. Heal, *Research*, **1**, 1948, 288. See also A. H. Sully, *Nature*, **167**, 1951, 365-366; *J. Inst. Metals*, **80**, 1951-1952, 173–179.
13. Beyond 39 at. % Cr the boundaries are in error by 2 at. (2 wt.) % Cr [7].
14. P. A. Beck and W. D. Manly, *Trans. AIME*, **185**, 1949, 354.
15. P. Duwez and S. R. Baen, *ASTM Spec. Tech. Publ.* 110, 1950, pp. 48–54.
16. P. Pietrokowsky and P. Duwez, *Trans. AIME*, **188**, 1950, 1283–1284.
17. S. Rideout, W. D. Manly, E. L. Kamen, B. S. Lement, and P. A. Beck, *Trans. AIME*, **191**, 1951, 872–876.
18. G. J. Dickins, A. M. B. Douglas, and W. H. Taylor, *Nature*, **167**, 1951, 192.
19. G. J. Dickins, A. M. B. Douglas, and W. H. Taylor, *J. Iron Steel Inst.*, **167**, 1951, 27.

20. J. S. Kasper, B. F. Decker, and J. R. Belanger, *J. Appl. Phys.*, **22,** 1951, 361–362.
21. U. Haschimoto, *Nippon Kinzoku Gakkai-Shi,* **1,** 1937, 177–190.
22. G. J. Dickins, A. M. B. Douglas, and W. H. Taylor, *Acta Cryst.*, **9,** 1956, 297–303.

$\bar{1}.9674$
0.0326

Co-Cu Cobalt-Copper

Solidification. Figure 272 shows the liquidus temperatures reported by [1]. Between 0 and about 70 at. % Cu they are in excellent agreement with those found previously by [2]. Both investigations have conclusively established that the liquidus curve has a slight slope and that there is no miscibility gap in the liquid state, as was assumed by [3, 4] and believed to extend from approximately 32 to 70 at. % Cu at the supposed monotectic temperature, 1370–1380°C [3]. Neither [3] nor [2, 1] found any indication of a separation of the melt into two layers. In addition, no monotectic reaction was observed [2, 1] in alloys with less than about 30 at. % Cu.

The temperature of the peritectic reaction (α-Co) + melt $\rightleftarrows$ (Cu) was reported as 1105–1108 [3], 1107 ± 4 [2], and 1110–1115°C [1]. The solidus curve of the α-Co primary solid solution was outlined by cooling-curve data and, between 1250 and 1110°C, by micrographic work [1].

Transformations and Solubilities in the Solid State. The effect of Cu on the temperatures of the magnetic and polymorphic transformation of Co was determined using magnetic [5, 1], resistometric [1], and thermal analyses [1]. Results are in substantial agreement. The Curie point of Co is lowered to 1040 [5] or 1065°C [1]. The hysteresis of the $\alpha \rightleftarrows \epsilon$ transformation increases with increasing Cu content, reaching a maximum in the two-phase field. The presentation in Fig. 272 is a compromise of the findings by [5, 1], those by [1] being more comprehensive [6].

The solubility limit of the α-Co primary solution was outlined micrographically between 1070 and 410°C [1]. [7] reported, on the basis of differential thermal analysis, that the solubility of Cu in Co at 305°C is not over 2 wt. (1.7 at.) % Cu. However, these transformation data do not represent equilibrium since the $\alpha \rightarrow \epsilon'$ transformation curve is a so-called realization rather than an equilibrium curve.

The solid solubility of Co in Cu has been established by micrographic [8] and magnetic studies [9] for the temperature ranges 600–1000°C and 440–1070°C, respectively. At 800°C, results agree very well. The solubility curve in Fig. 272 is based on the data of [9], which are more likely since they are more consistent with the fact that the solubility at 500°C is less than 0.12 wt. (0.13 at.) % Co [10]; [9] gave 0.26 wt. (0.28 at.) % at 500°C. In figures slightly rounded off, the solubility in wt. % Co (at. % in parentheses) is as follows: 1110°C, 5.1–5.2 (5.5–5.6) extrapolated; 1070°C, 4.5 (4.9); 1000°C, 3.65 (3.9); 900°C, 2.6 (2.8); 800°C, 1.7 (1.8); 700°C, 1.0 (1.05); 500°C, <0.1 (<0.1).

Results of electrical conductivity [11] and emf measurements [12], as well as X-ray studies [13], are in accord with Fig. 272.

1. U. Haschimoto, *Nippon Kinzoku Gakkai-Shi,* **1,** 1937, 19–26.
2. R. Sahmen, *Z. anorg. Chem.*, **57,** 1908, 1–9.
3. N. Konstantinov, *Rev. mét.*, **4,** 1907, 983–988; *Zhur. Russ. Fiz.-Khim. Obshchestva,* **39,** 1907, 771–777.
4. D. Iitsuka, *Mem. Coll. Sci. Kyoto Univ.*, **12,** 1929, 179–181.
5. W. Köster and E. Wagner, *Z. Metallkunde,* **29,** 1937, 230–232.
6. U. Haschimoto, *Nippon Kinzoku Gakkai-Shi,* **1,** 1937, 177–190.
7. A. A. Rudnitsky, L. A. Panteleimonov, V. V. Pimenova, and M. E. Berezkina, *Věstnik Moskov. Univ.*, **7,** no. 3, *Ser. Fiz.-Mat. i Estestven. Nauk,* no. 2, 1952, pp. 51–54.

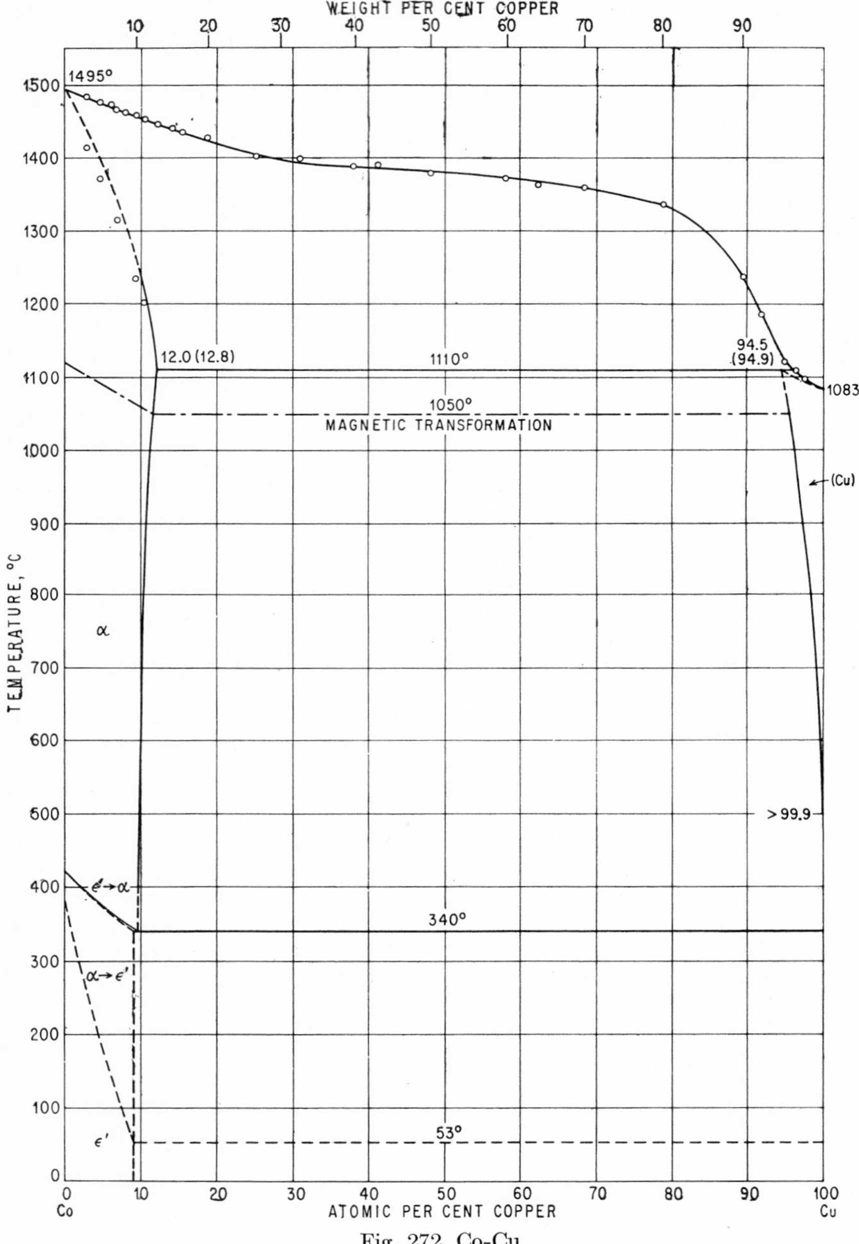
WEIGHT PER CENT COPPER
10
20
30
40
50
60
70
80
90
1495°
1500
1400
1300
1200
1100
1000
900
800
700
600
500
400
300
200
100
0
TEMPERATURE, °C
12.0 (12.8)
1110°
94.5
(94.9)
1083°
1050°
MAGNETIC TRANSFORMATION
(Cu)
α
>99.9
ε'→α
340°
α→ε'
53°
ε'
0
Co
10
20
30
40
50
60
70
80
90
100
Cu
ATOMIC PER CENT COPPER

Fig. 272. Co-Cu

8. M. G. Corson, *Proc. Inst. Metals, Div. AIME,* **1927,** 435–450; *Rev. mét.,* **27,** 1930, 95–101.
9. G. Tammann and W. Oelsen, *Z. anorg. Chem.,* **186,** 1930, 260–264.
10. E. Hildebrand, *Ann. Physik,* **30,** 1937, 593–608.
11. G. Reichardt, *Ann. Physik,* **6,** 1901, 832; see also *Z. anorg. Chem.,* **51,** 1906, 405-406; *Ann. Physik,* **32,** 1910, 332–333.
12. F. Ducelliez, *Compt. rend.,* **150,** 1910, 98–101; *Bull. soc. chim. France,* **7,** 1910, 196–199.
13. L. Vegard and H. Dale, *Z. Krist.,* **67,** 1928, 154–157.

0.0234
$\bar{1}$.9766

Co-Fe Cobalt-Iron

Solidification. Thermal-analysis data have been reported by [1–8]. Those by [3, 8] and [5, 7] deal only with Fe-rich or Co-rich alloys, respectively. The solidification interval is very narrow; between 0 and 30 at. % Co it is only 2–3°C [8]. There is a minimum in the liquidus curve between 60 and 70 at. % Co at 1477°C [2, 4] or probably at a slightly higher temperature, judging from the data by [8] and [7].

The temperature of the peritectic reaction, L + $\delta \rightleftarrows$ austenite, was found as 1493 [2], 1492 [3], 1496 [4], 1505 [6], and 1499°C [8]. Results by [8], which must be regarded as the most accurate for the composition range up to 31 at. % Co, show that the range of the peritectic reaction is smaller, i.e., 16.5–19.5 at. % Co, than assumed earlier: about 15–21 [2], about 16–23 [3], and about 14.5–26 at. % Co [6].

Solid-state Transformations. Temperatures of the austenite $\rightleftarrows \delta$ transformation were determined by [2–4, 6, 8]; those given by [8] were adopted in Fig. 273.

The austenite $\rightleftarrows$ ferrite and magnetic transformations were studied in great detail by means of thermal [2, 6, 9], magnetic [1, 2, 10–12], and dilatometric [10] analysis. In the range above about 70 at. % Co, where the polymorphic transformation temperature drops sharply and, therefore, dynamic methods proved to be unsuccessful, the lattice-parametric method [9] was used.

In general, transformation temperatures agree quite well. Temperature differences for the polymorphic transformation, being of the order of 20–25°C at most, are due to supercooling or overheating, the degree of which depends upon the rate of temperature change. The results by some investigations agree very well, e.g., those by [9] and [12]. The transformation curves presented in Fig. 273 are based on the work by [9]; mean values for heating and cooling were taken.

The width of the austenite + ferrite field below 900°C was determined by means of lattice-parameter measurements of alloys annealed at and quenched from 800, 680, and 575–580°C [9]. According to these data, the two-phase region extends from 75 to 88.5 at. % Co at 600°C, as compared with 73–81 at. % Co, based on microscopical studies [6]. The two-phase structure of alloy preparations with 80.2–88.7 at. % Co, obtained by reduction of Co-Fe formiate solid solutions, is consistent with the austenite + ferrite field shown in Fig. 273 [13]; see also [14].

Data as to the diffusionless transformation austenite—ϵ' in Co-rich solid solutions differ considerably. The transformation curve obtained on heating reaches room temperature at about 91 at. % [7] or about 93 at. % Co [10]. The curve shown in Fig. 273 was averaged from these data. The same was done for the transformation curve found on cooling [5, 7, 10].

Investigation of various physical properties of alloys covering the whole range of composition, including saturation magnetization [15–18, 11], electrical conductivity (or resistivity) [17, 19, 20, 11], thermal conductivity [17], thermoelectric force [21, 19], and thermal expansion [22], have shown that there are anomalies in the property-vs.-

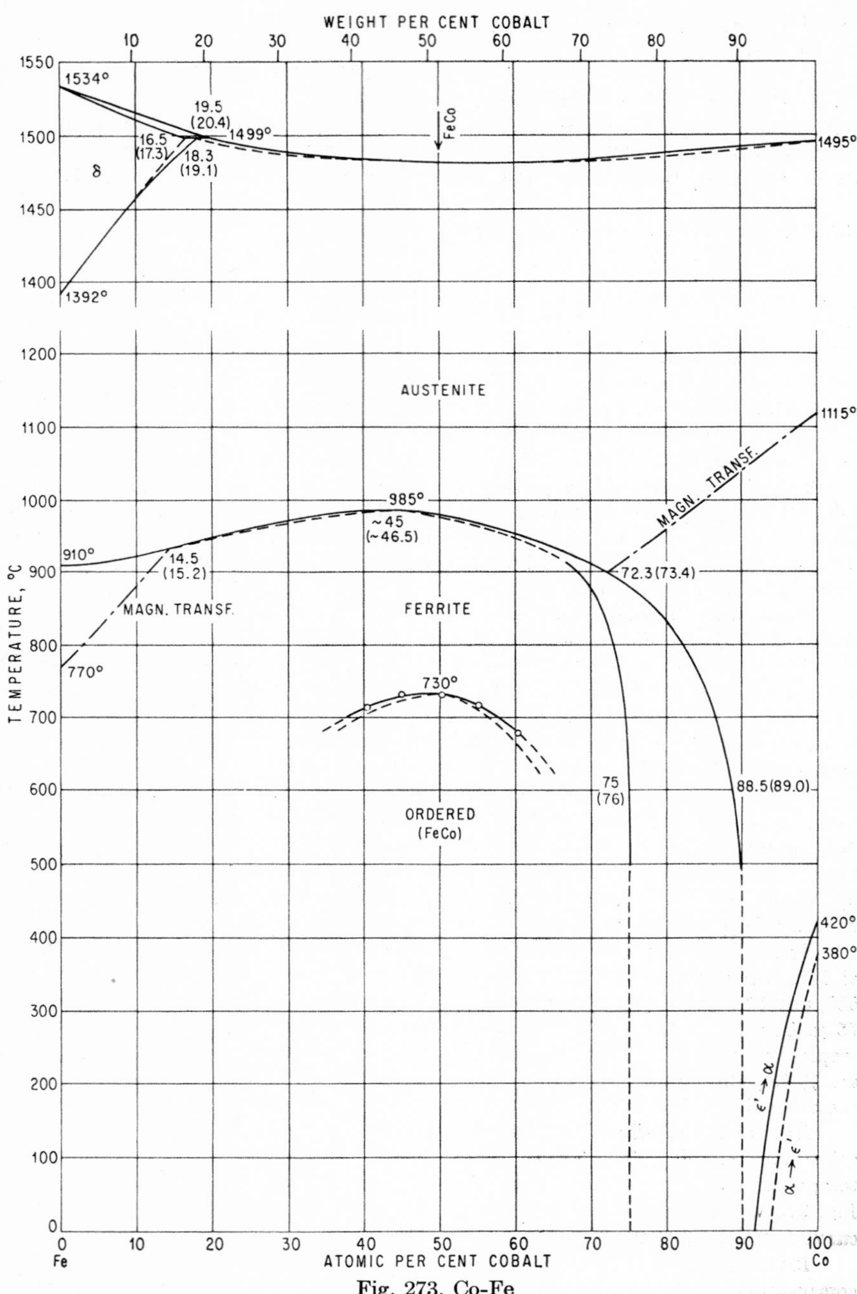
WEIGHT PER CENT COBALT
10
20
30
40
50
60
70
80
90
1550
1534°
19.5
(20.4)
FeCo
1499°
1500
16.5
(17.3)
18.3
(19.1)
1495°
δ
1450
1400
1392°
1200
AUSTENITE
1100
1115°
MAGN. TRANSF.
1000
985°
~45
(~46.5)
910°
14.5
(15.2)
72.3 (73.4)
900
MAGN. TRANSF.
FERRITE
TEMPERATURE, °C
800
770°
730°
700
600
75
(76)
88.5 (89.0)
ORDERED
(FeCo)
500
400
420°
380°
300
200
ε' → α
100
α → ε'
0
0
Fe
10
20
30
40
50
60
70
80
90
100
Co
ATOMIC PER CENT COBALT

Fig. 273. Co-Fe

composition curves at compositions of about 35% Co [15, 16], about 40% Co [17], and about 50% Co [19, 20, 11]. These are inconsistent with the existence of a series of ferritic solid solutions with statistical atom distribution; see also [23]. [11] have first proposed that the anomalous behavior is caused by the formation of an ordered structure based on the composition FeCo (51.35 wt. % Co).

Direct evidence for an order-disorder transformation was found by means of X-ray [24, 9] and neutron-diffraction studies [25] (see Crystal Structure). According to thermal-analysis data [9], ordering starts at about 730°C; thermal effects were found in the composition range 40.5–60.3 at. % Co (Fig. 273). Earlier, [24] had observed thermal effects in the 50:50 alloy at a much higher temperature, 952 and 919°C; however, the data of [9] are better substantiated.

Representative microstructures of alloys covering a wide composition range were published by [2, 6].

Crystal Structure. Lattice parameters of the ferrite and ϵ' phases were reported by [19, 26, 27, 24, 9, 41], those of b.c.c. solid solutions by [24, 9] being the most complete and accurate ones. Previously, [28] had used X-ray diffraction for phase identification. In the range 40–60% Co, the parameter-vs.-composition curve [24, 9] shows a deviation toward greater values from the continuous slope which would be expected if ordering were absent and was found to reach a small subsidiary maximum at 48 wt. (47.6 at.) % Co (alloys were slowly cooled from 600–650°C) [24]. Extra lines were observed in X-ray [24, 9] and neutron-diffraction photograms [25], which indicate the formation of an ordered b.c.c. superlattice structure of the CsCl (B2) type.

Physical Properties. The following studies of physical properties have been made: density [15, 19, 27], electrical conductivity (resistivity) [17, 21, 29, 20, 19, 11], thermal conductivity [17], thermal expansion [30, 31, 18, 32, 11, 22], thermoelectric force [21, 9], and magnetic properties [15–17, 33, 34, 19, 35, 11, 36, 37, 12, 38, 39, 23].

Supplement. Lattice spacings over the whole composition range were measured by [40] on solidified as well as electrodeposited alloys. Precision lattice-parameter measurements of ferrite containing up to 9.2 at. % Co were made by [42]. On the basis of measurements of various physical properties, Japanese investigators [43, 44] suggested the existence of superlattices based on the compositions Fe_3Co and $FeCo_3$.

Further results in support of the wide ferrite + austenite two-phase field shown in Fig. 273 at lower temperatures were reported by [41] (reduction of Co-Fe formiate solid solutions), [45] (alloys prepared by thermal decomposition of Co-Fe amalgams), and [46] (heat-treatment of heavily worked specimens). According to [41], the two-phase field extends even from 74.5 to 93 wt. (73.5–92.5 at.) % Co at 300°C (the alloys were strongly contaminated by oxygen, however).

1. W. Guertler and G. Tammann, *Z. anorg. Chem.*, **45,** 1905, 203–224.
2. R. Ruer and K. Kaneko, *Ferrum,* **11,** 1913-1914, 33–39.
3. R. Ruer and R. Klesper, *Ferrum,* **11,** 1913-1914, 257–261.
4. T. Kasé, *Science Repts. Tôhoku Univ.*, **16,** 1927, 494–495.
5. U. Haschimoto, *Kinzoku-no-Kenkyu,* **9,** 1932, 63-64.
6. J. H. Andrew and C. G. Nicholson, *Iron Steel Inst. Spec. Rept.*, 1936, no. 14, pp. 93–96.
7. U. Haschimoto, *Nippon Kinzoku Gakkai-Shi,* **1,** 1937, 177–190.
8. G. B. Harris and W. Hume-Rothery, *J. Iron Steel Inst.*, **174,** 1953, 212–218.
9. W. C. Ellis and E. S. Greiner, *Trans. ASM,* **29,** 1941, 415–432.
10. H. Masumoto, *Science Repts. Tôhoku Univ.*, **15,** 1926, 469–476; *Trans. ASST,* **10,** 1926, 491–492.

11. A. Kussmann, B. Scharnow, and A. Schulze, *Z. tech. Physik,* **10,** 1932, 449–460; see also *Z. Metallkunde,* **25,** 1933, 145–146.
12. M. Fallot, *Métaux, corrosion, usure,* **18,** 1943, 214–219.
13. F. Lihl, *Metall,* **5,** 1951, 183–187.
14. O. S. Edwards, *J. Inst. Metals,* **67,** 1941, 70.
15. A. Preuss, *Trans. Faraday Soc.,* **8,** 1912, 57.
16. P. Weiss, *Trans. Faraday Soc.,* **8,** 1912, 149.
17. K. Honda, *Science Repts. Tôhoku Univ.,* **8,** 1919, 51–58.
18. H. Masumoto and S. Nara, *Science Repts. Tôhoku Univ.,* **16,** 1927, 335–336.
19. W. C. Ellis, *Rensselaer Polytech. Inst. Bull., Eng. Sci. Ser.,* no. 16, 1927.
20. A. Schulze, *Z. tech. Physik,* **8,** 1927, 425–427.
21. Mallet, Thesis, Rensselaer Polytechnic Institute, 1924, as quoted from [19].
22. M. F. Fine and W. C. Ellis, *Trans. AIME,* **175,** 1948, 242–254.
23. J. F. Libsch, E. Both, G. W. Beckman, D. Warren, and R. J. Franklin, *Trans. AIME,* **188,** 1950, 287–296.
24. J. W. Rodgers and W. R. Maddocks, *Iron Steel Inst. Spec. Rept.,* 1939, no. **24,** pp. 167–177.
25. C. G. Shull and S. Siegel, *Phys. Rev.,* **75,** 1949, 1008–1010; **74,** 1948, 1255.
26. A. Osawa, *Science Repts. Tôhoku Univ.,* **19,** 1930, 115–121.
27. Z. Nishiyama, *Science Repts. Tôhoku Univ.,* **18,** 1929, 359–400.
28. M. R. Andrews, *Phys. Rev.,* **18,** 1921, 245–254.
29. Holmes, Thesis, Rensselaer Polytechnic Institute, 1925, as quoted from [19].
30. K. Honda and Y. Okubo, *Science Repts. Tôhoku Univ.,* **13,** 1924, 106–107.
31. A. Schulze, *Physik. Z.,* **28,** 1927, 669–673.
32. H. Masumoto, *Science Repts. Tôhoku Univ.,* **20,** 1931, 101–123.
33. K. Honda and K. Kido, *Science Repts. Tôhoku Univ.,* **9,** 1920, 226–231.
34. A. Schulze, *Z. tech. Physik,* **8,** 1927, 500–501.
35. R. Forrer, *Compt. rend.,* **190,** 1930, 1284–1287.
36. Y. Masiyama, *Science Repts. Tôhoku Univ.,* **21,** 1932, 394–410.
37. S. Kaya and H. Sato, *Proc. Phys. Math. Soc. Japan,* **25,** 1943, 261–273.
38. A. Z. Zaimovski, *Izvest. Sektora Fiz.-Khim. Anal.,* **16,** 1946, 126–147.
39. F. Gal'perin, *Doklady Akad. Nauk S.S.S.R.,* **75,** 1950, 647–650.
40. K. Aotani, *Nippon Kinzoku Gakkai-Shi,* **14B,** 1950, no. 5; abstract in *Structure Repts.,* **13,** 1950, 86–87.
41. F. Lihl, H. Wagner, and P. Zemsch, *Z. Elektrochem.,* **56,** 1952, 619–624.
42. A. L. Sutton and W. Hume-Rothery, *Phil. Mag.,* **46,** 1955, 1295–1309.
43. T. Yokoyama, *Nippon Kinzoku Gakkai-Shi,* **17,** 1953, 259–263, 263–266 (in Japanese); *Met. Abstr.,* **23,** 1956, 421.
44. H. Masumoto et al., *Science Repts. Research Insts. Tôhoku Univ.,* **6,** 1954, 523–528 (in English); *Met. Abstr.,* **22,** 1955, 947.
45. F. Lihl, *Z. Metallkunde,* **46,** 1955, 434–441.
46. J. Papier, *Compt. rend.,* **242,** 1956, 2455–2457; *Met. Abstr.,* **24,** 1957, 471.

$\bar{1}.9270$
0.0730

Co-Ga Cobalt-Gallium

Alloys with 4, 9, and 10 wt. (3.4, 7.8, and 8.6 at.) % Ga were used to study the effect of Ga on the temperatures of the magnetic and polymorphic transformations of Co [1]. Results are presented in Fig. 274 [2]. According to [3], the phase CoGa has the CsCl (B2) type of structure with a = 2.87 A.

1. Heat-treatment is not given.
2. W. Köster and E. Horn, *Z. Metallkunde*, **43**, 1952, 333–334.
3. P. Esslinger and K. Schubert, *Z. Metallkunde*, **48**, 1957, 126–134.

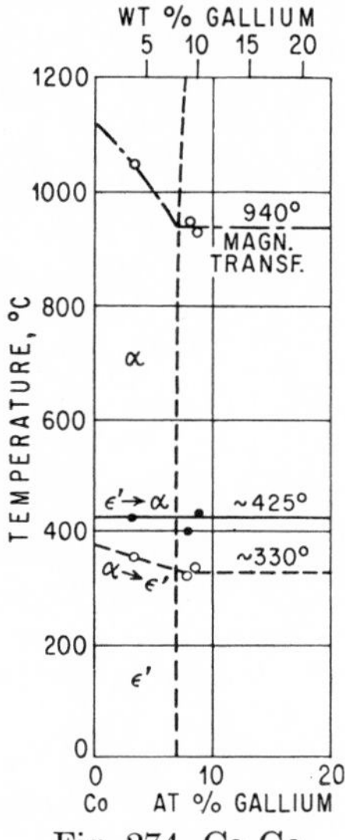

Fig. 274. Co-Ga

$\bar{1}.9095$
0.0905

Co-Ge Cobalt-Germanium

The phase diagram in Fig. 275 is based on thermal, microscopic, and roentgenographic investigations by [1], with the alloys used indicated by data points. Data as to the magnetic and polymorphic transformations of the Co-rich primary solid solution are due to [2].

Besides the phases Co_2Ge (38.11 wt. % Ge) and $CoGe_2$ (71.13 wt. % Ge) reported earlier [3, 4], the existence of CoGe (55.19 wt. % Ge) and Co_2Ge_3 (64.88 wt. % Ge) was established by this work. The boundaries of the intermediate phases have been outlined only and, therefore, are shown as dashed curves. The boundary of the Co-rich solid solution was not determined. The curves of the magnetic and $\epsilon' \rightarrow \alpha$ transformation indicate a solubility of about 10 at. (12 wt.) % Ge at 700–800°C. On the basis of lattice-parameter measurements, the solid solubility of Ge in α-Co is estimated to be approximately twice as large as that in ϵ-Co [1]. The solid solubility of Co in Ge is negligibly small [1].

The structure of (Co_2Ge) is of the "filled-up" NiAs (B8) type [3, 1, 5]. According to [1], however, this holds only for temperatures above 625–400°C (varying with composition). Below these temperatures a superstructure develops (ordering of the interstitial atoms), shown by extra lines in the powder patterns and thin markings in the microstructures of slowly cooled samples. The following B8-type lattice parameters were reported: a = 3.92 A, c = 5.03 A, c/a = 1.285 (27 wt. % Ge) [1]; a = 3.933 A, c = 5.013 A, c/a = 1.275 (36 at. % Ge) [3]; a = 3.918 A, c = 4.979 A, c/a = 1.271 (37 at. % Ge) [5].

$CoGe_2$ was the first representative of a new type of crystal structure. It has a

B-face centered orthorhombic, pseudotetragonal unit cell with 16 Ge and 7 Co atoms and $a = b = 5.68$ A, $c = 10.82$ A [6].

1. H. Pfisterer and K. Schubert, *Z. Metallkunde,* **40,** 1949, 378–383.
2. W. Köster and E. Horn, *Z. Metallkunde,* **43,** 1952, 333–334.
3. F. Laves and H. J. Wallbaum, *Z. angew. Mineral.,* **4,** 1941-1942, 17–46.
4. H. J. Wallbaum, *Z. Metallkunde,* **35,** 1943, 218.
5. L. Castelliz, *Monatsh. Chem.,* **84,** 1953, 767.
6. K. Schubert and H. Pfisterer, *Z. Metallkunde,* **41,** 1950, 433–441.

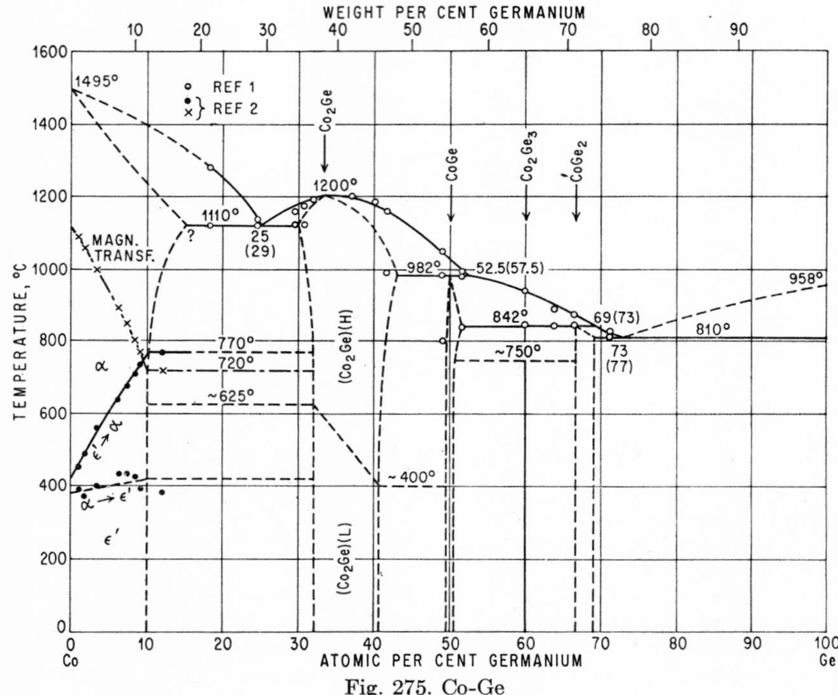

Fig. 275. Co-Ge

1.7670 2.2330 Co-H Cobalt-Hydrogen

The solubility of hydrogen in cobalt at 1 atm in the temperature range 600–1200°C was determined by [1, 2]; see also [3].

[4, 5] reported that heat-treating Co in an atmosphere of H_2 lowers the transformation temperatures T_c and T_r from 498 to 408°C and 447 to 240°C, respectively, and increases the lattice parameters of both Co modifications slightly.

1. A. Sieverts (P. Beckmann), *Z. physik. Chem.,* **60,** 1907, 169–201.
2. A. Sieverts and H. Hagen, *Z. physik. Chem.,* **A169,** 1934, 237–240.
3. D. P. Smith, "Hydrogen in Metals," pp. 34, 60, 262, University of Chicago Press, Chicago, 1948.
4. J. Drain, R. Bridelle, and A. Michel, *Bull. soc. chim. France,* **1954,** 828–830.

5. A. Michel, J. Drain, and R. Bridelle, *Compt. rend.*, **238,** 1954, 107–108; *Met. Abstr.*, **21,** 1954, 783.

$\bar{1}.5185$
0.4815

Co-Hf Cobalt-Hafnium

"An alloy of $HfCo_2$ [39.76 wt. % Co] has been identified as isomorphous with $MgCu_2$ [C15 type] with a lattice parameter of 6.908 kX [6.922 A]. The incipient melting temperature was determined as 1570°C. No allotropy was observed at 600, 800, 1000, 1200, 1400 or 1500°C" [1].

1. R. P. Elliott, Armour Research Foundation, Chicago, Ill., Technical Report 1, OSR Technical Note OSR-TN-247, August, 1954.

$\bar{1}.4681$
0.5319

Co-Hg Cobalt-Mercury

Cobalt "amalgam," prepared by electrolysis of an aqueous solution of a Co salt on a mercury cathode, was shown by X-ray analysis to consist of a suspension of finely divided Co in mercury [1–3, 8].

By a chemical-analytical method with a sensitivity of 8×10^{-5} wt. (27×10^{-5} at.) % Co, no cobalt could be detected in saturated Hg at room temperature [4]. Older indirect methods indicated a much higher solubility [5, 6], but the results either had no very reliable experimental basis [5] or were shown [7] to be inconclusive [6].

Supplement. A recent spectrographic analysis [8] indicated that the solubility of Co in Hg at 30°C is even less than 1×10^{-6} wt. (3.4×10^{-6} at.) %.

1. N. Katoh, *Nippon Kwagaku Kwai-Shi*, **64,** 1943, 1211–1212.
2. F. Pawlek, *Z. Metallkunde*, **41,** 1950, 451–453.
3. F. Lihl, *Z. Metallkunde*, **44,** 1953, 160–166.
4. N. M. Irvin and A. S. Russell, *J. Chem. Soc.*, **135,** 1932, 891–898.
5. G. Tammann and W. Oelsen (*Z. anorg. Chem.*, **186,** 1930, 280–281) computed a solubility of 6.2×10^{-2} wt. % Co from the value of the specific magnetization of a single alloy measured by H. Nagaoka (*Wied. Ann. Physik*, **59,** 1896, 66).
6. G. Tammann and K. Kollmann, *Z. anorg. Chem.*, **160,** 1927, 244–246.
7. E. Palmaer, *Z. Elektrochem.*, **38,** 1932, 70–76.
8. J. F. de Wet and R. A. W. Haul, *Z. anorg. Chem.*, **277,** 1954, 96–112.

$\bar{1}.7106$
0.2894

Co-In Cobalt-Indium

Alloys with 1–10 wt. (0.5–5.4 at.) % In "proved to be microscopically heterogeneous. Accordingly, neither the Curie nor the [polymorphic] transformation temperatures [of cobalt] were changed. It may be assumed, therefore, that indium is insoluble in cobalt" [1].

1. W. Köster and E. Horn, *Z. Metallkunde*, **43,** 1952, 333–334.

$\bar{1}.4846$
0.5154

Co-Ir Cobalt-Iridium

As shown by lattice-parametric and microscopic investigations [1], α-Co and Ir form a continuous series of solid solutions. The parameter of the f.c.c. phase changes on a straight line connecting the parameters of the components. The transformation curves (Fig. 276), based on magnetic and dilatometric analysis, are "realization curves" (not equilibrium curves), indicating the temperatures of the start of the diffusionless $\epsilon' \rightarrow \alpha$ and $\alpha \rightarrow \epsilon'$ transformations on heating and cooling, respectively. Above about 25 at. % Ir the curves are hypothetical. The limit of the ϵ' field, under the conditions tested, lies between 77 and 91 wt. (about 50 and 75 at.) % Ir. The curves designated as θ_α and $\theta_{\epsilon'}$ represent the Curie temperatures of the α and ϵ' phases, respectively.

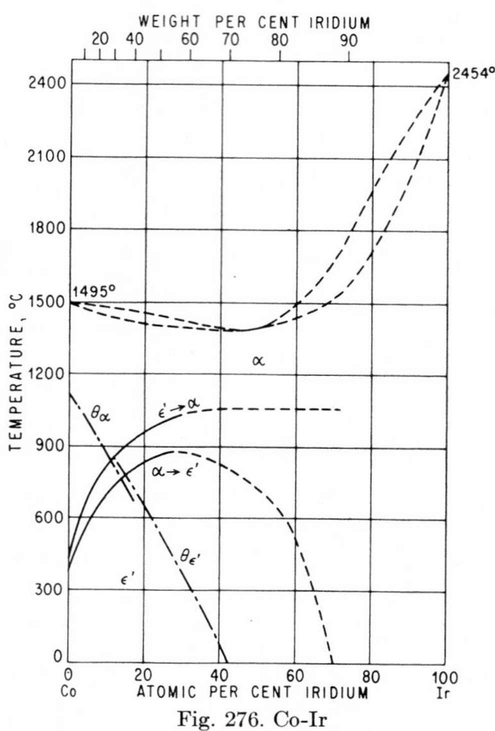

Fig. 276. Co-Ir

The hypothetical liquidus and solidus curves are shown to have a minimum as alloys with 70 and 77 wt. (41.6 and 50.6 at.) % Ir were partially molten on annealing at 1400°C.

1. W. Köster and E. Horn, *Z. Metallkunde*, **43**, 1952, 444–449.

0.1782
$\bar{1}$.8218

Co-K Cobalt-Potassium

The temperature of the $\epsilon \rightleftarrows \alpha$ transformation of Co (found to be 467°C on heating and 426°C on cooling) is raised (on heating) to 477°C and lowered (on cooling) to 342°C by the addition of approximately 0.25 wt. (0.38 at.) % K. Higher K contents do not affect the transformation point any further [1].

1. U. Haschimoto, *Nippon Kinzoku Gakkai-Shi*, **2**, 1938, 67–77.

0.9290
$\bar{1}.0710$

Co-Li Cobalt-Lithium

According to dilatation and magnetic measurements [1], the temperature of the $\epsilon \rightleftarrows \alpha$ transformation of Co is affected as shown in Fig. 277.

1. U. Haschimoto, *Nippon Kinzoku Gakkai-Shi,* **1,** 1937, 177–190.

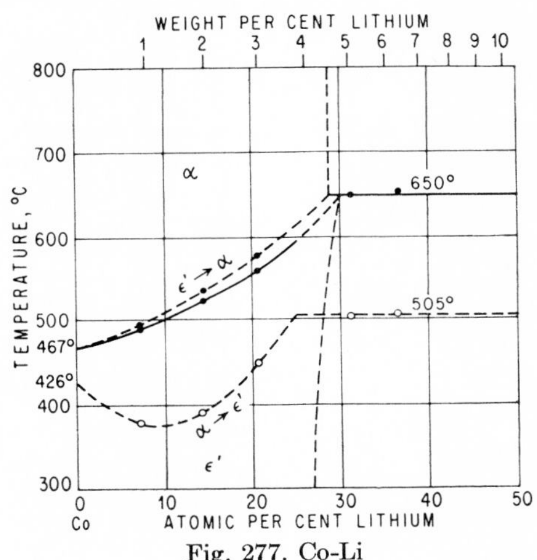

Fig. 277. Co-Li

0.3844
$\bar{1}.6156$

Co-Mg Cobalt-Magnesium

The melting point of Mg is lowered almost linearly by Co additions until a eutectic point is reached that was found to lie at about 5 wt. (2.2 at.) % Co, 635°C [1] and about 4.5 wt. (2.0 at.) % Co, 632°C [2]. On the basis of the microstructure, however, the eutectic composition would be located at a higher Co content [1, 2], since no entirely eutectic structure was observed in alloys up to 6 wt. (2.6 at.) % Co [2]. The composition of the phase coexisting with Mg was estimated to be Mg_2Co (54.79 wt. % Co) [2]. This phase is probably formed by a peritectic reaction at about 689°C [2] (Fig. 278). No change of the lattice spacing of Mg could be detected in Mg-rich alloys [2]. Also, measurements of electrical conductivity indicate that the solid solubility is negligibly small [3].

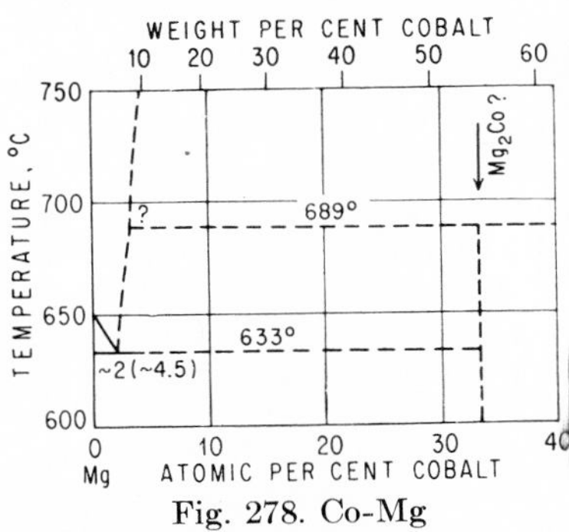

Fig. 278. Co-Mg

The temperature of the $\epsilon \rightleftarrows \alpha$ transformation of Co (found as 467°C on heating and 426°C on cooling) was reported to be lowered by Mg. On heating, the transformation of alloys with 2 and 3 wt. (0.9 and 1.4 at.) % Co was found to start at 318 and 287°C, respectively. On cooling, the corresponding temperatures were 223 and 181°C [4]. These findings are not compatible with those by [1–3] and, therefore, have not been incorporated in Fig. 278.

1. J. P. Wetherill, *Metals & Alloys*, **6**, 1935, 153–155.
2. E. M. Cramer, H. P. Nielsen, and F. W. Schonfeld, *Light Metal Age*, **5**, September, 1947, 6–9.
3. H. Vosskühler, see A. Beck, "Magnesium und seine Legierungen," p. 124, Springer-Verlag OHG, Berlin, 1939; A. Beck, "The Technology of Magnesium and Its Alloys," p. 122, F. A. Hughes and Co., Ltd., London, 1940.
4. U. Haschimoto, *Nippon Kinzoku Gakkai-Shi*, **1**, 1937, 177–190.

0.0306
$\bar{1}.9694$

Co-Mn Cobalt-Manganese

The first thermoanalytical investigation [1] seemed to indicate that a continuous series of solid solutions crystallizes from the melt, with a minimum in the liquidus curve at about 70 wt. % Mn, 1061°C. In accordance with this, the microstructure of alloys with 10–90 wt. % Mn, after annealing at 1000°C, proved to be "practically" homogeneous. The fact that Mn is polymorphic was not known at that time.

Liquidus points were also determined between 5 and 51 wt. % Mn [2], 5 and 30 wt. % Mn [3], and about 40–100 wt. % Mn [4]. The last authors detected that the $\gamma \rightleftarrows \delta$ transformation temperature of Mn is raised and leads to a peritectic reaction at about 1180°C, between about 88 and 93 wt. (88.7–93.4 at.) % Mn.

An X-ray analysis, covering the entire range of composition, was carried out by [5]. Sintered samples were annealed at and rapidly quenched from temperatures close to the solidus (1140–1200°C). The lattice-spacing-vs.-composition curve shows a gradual increase of the parameter of the f.c.c. α-Co solid solutions up to 90.6 at. (90.0 wt.) % Mn. Between 92 at. (91.5 wt.) % and 96 at. (95.7 wt.) % Mn, the f.c. tetragonal lattice of the quenched γ-Mn solid solutions was observed, with the axial ratio decreasing from 0.994 at 92 at. % to 0.935_5 at 100% Mn. It was concluded that there is a continuous transition from the f.c.c. α-Co into the f.c. tetragonal γ-Mn lattice. Earlier, [6] had assumed that a small two-phase region exists which gives rise to another peritectic equilibrium at about 1170°C. Results obtained in a study of the ternary system Co-Mn-Al supported this assumption.

Since it is known that γ-Mn is f.c.c. rather than f.c. tetragonal, the f.c. tetragonal lattice being formed on quenching [7], a continuous transition α-Co $\rightleftarrows$ γ-Mn is quite possible. The two-phase field $\alpha + \beta_{Mn}$, roughly outlined in Fig. 279, is based on data by [4, 8, 9]. According to [8], an alloy with 49.6 wt. (51.4 at.) % Mn was two-phase. Roentgenographic work [9] showed that alloys with 56–87 wt. (57.8–87.8 at.) % Mn contained the β-Mn phase, after cooling to room temperature. It appears possible that the $\alpha + \beta_{Mn}$ field is actually wider than shown in Fig. 279 and that the $\alpha/(\alpha + \beta_{Mn})$ boundary, especially, has to be shifted to lower Mn compositions.

The effect of Mn on the temperatures of the magnetic and polymorphic transformations of Co was determined by [2, 10, 3]. Results by [10] and [3] agree quite well; earlier determinations [2] are obsolete. The curves of the polymorphic transformation (Fig. 279) were found on heating [10, 3]. On cooling, the $\alpha \rightarrow \epsilon'$ transformation is undercooled to 85°C at 15 wt. (16 at.) % Mn [3].

Lattice parameters of the ϵ' and α solid solutions were determined by [8] and [8, 5], respectively. Both works show that the parameter-vs.-composition curve of the α phase slightly changes in direction at about 30 at. % Mn [8] or 40 at. % Mn [5], respectively, because of the transition from the ferromagnetic to the paramagnetic state. For X-ray studies of Mn-rich alloys, see [11, 7].

Supplement. From a study of the dependence of the paramagnetic susceptibility of the alloys MnCo, Mn_3Co (and Mn_2Co) on the annealing time and temperature, [12] suggested the existence of order-disorder transformations in these alloys (cf. Mn-Ni system). X-ray measurements on annealed samples showed a f.c.c. lattice, with a = 3.59 A, for MnCo and—at variance with the constitution shown in Fig. 279—a hexagonal lattice, with a = 2.41 A, c = 4.45 A (c/a = 1.846), for "Mn_3Co." No superstructure lines were observed [13].

[14] found an as-cast alloy of the composition $MnCo_2$ to be f.c.c., with a = 3.581 kX, in agreement with Fig. 279.

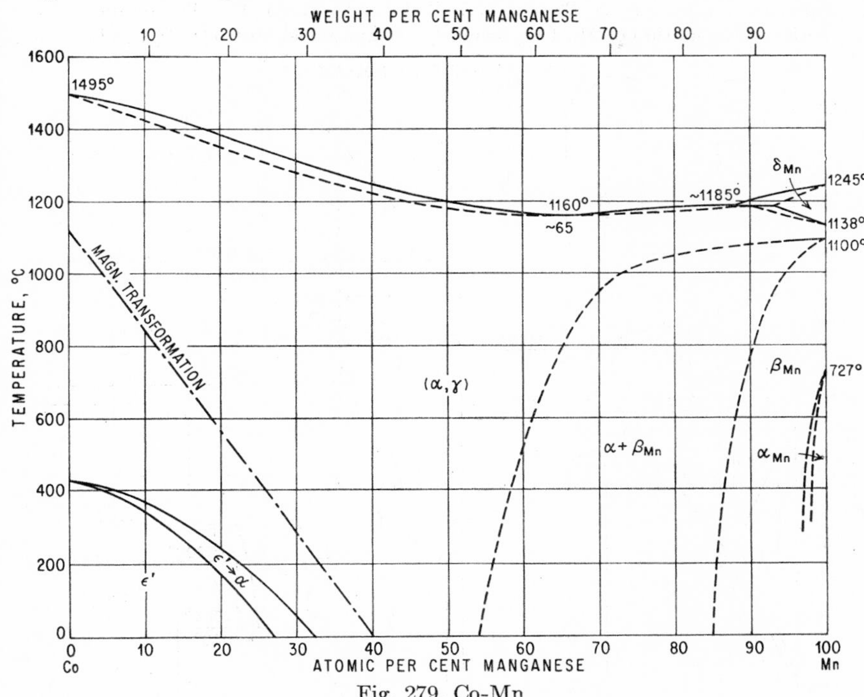

Fig. 279. Co-Mn

1. K. Hiege, *Z. anorg. Chem.*, **83**, 1913, 253–256. Liquidus temperatures between 0 and 50% Mn are too high since they were based on a melting point of Co of 1525°C.
2. U. Haschimoto, *Kinzoku-no-Kenkyu*, **9**, 1932, 64–65.
3. U. Haschimoto, *Nippon Kinzoku Gakkai-Shi*, **1**, 1937, 177–190.
4. Unpublished work by G. Grube and W. Fischer, 1946. For information see [5] and [6].
5. A. Schneider and W. Wunderlich, *Z. Metallkunde*, **40**, 1949, 260–263.
6. W. Köster and E. Gebhardt, *Z. Metallkunde*, **30**, 1938, 281–282.
7. U. Zwicker, *Z. Metallkunde*, **42**, 1951, 246–252.
8. W. Köster and W. Schmidt, *Arch. Eisenhüttenw.*, **8**, 1934-1935, 25–27.
9. B. Blumenthal, A. Kussmann, and B. Scharnow, *Z. Metallkunde*, **21**, 1929, 416. The microstructures are interchanged.
10. W. Köster and W. Schmidt, *Arch. Eisenhüttenw.*, **7**, 1933-1934, 121–126.
11. E. Persson and E. Oehmann, *Nature*, **24**, 1929, 333.
12. F. Galperin, *Doklady Akad. Nauk S.S.S.R.*, **77**, 1951, 1011–1014 (in Russian).
13. Note the similar scattering powers of Mn and Co.
14. R. P. Elliott, Armour Research Foundation, Chicago, Ill., Technical Report 1, OSR Technical Note OSR-TN-247, August, 1954, p. 15.

$\bar{1}.7884$
0.2116

Co-Mo Cobalt-Molybdenum

Solidification (Fig. 280). The liquidus and solidus curves in the range 0–50 at. % Mo have been determined with good agreement by [1–4]. The eutectic was found at 37 wt. (26.5 at.) % Mo [1–4], 35 wt. (25 at.) % Mo [5], and about 1335 [1], 1340 [2, 4], 1345 [5], and 1360 ± 5°C [3]. The solubility of Mo in Co at the eutectic temperature was reported as 26–28 wt. (17.7–19.3 at.) % Mo [1–4]. The temperature of the peritectic formation of Mo_6Co_7, formerly designated as MoCo [1–4], was located at

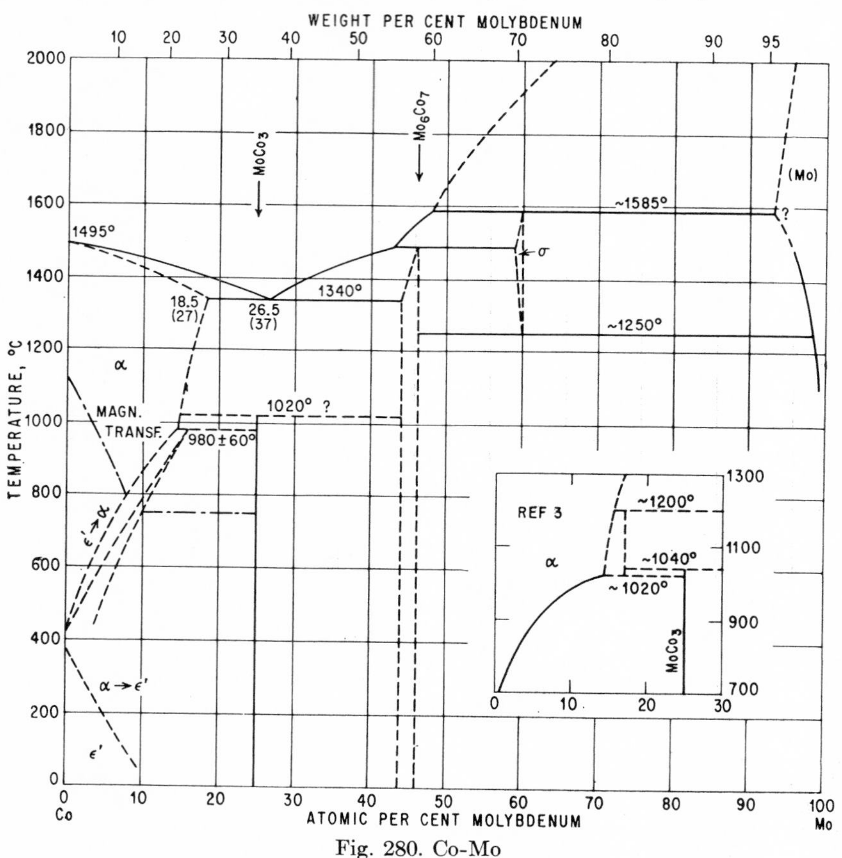

Fig. 280. Co-Mo

1485 [1, 4], 1500 [2], and 1550°C [3], and the peritectic melt at 58 [1, 2], 52 [3], and 56 wt. % [4] (46, 40, and 44 at. %).

The existence of a phase with approximately 60 at. % Mo and stable only at high temperatures was detected by [2] and confirmed by [3, 6]. This phase, having a σ-phase type structure [7, 8], is formed by a peritectic reaction at 1550 [2] or 1620°C [3] and decomposes eutectoidally into Mo_6Co_7 and (Mo) at 1340 [2] or 1250°C [3].

Solid State. An additional phase with about 32 wt. (22.5 at.) % Mo was found by [3] to be formed by a peritectoid reaction of (Co) with Mo_6Co_7 at approximately 1020°C. This phase was later identified by X-ray work as the compound $MoCo_3$ (35.18 wt. % Mo) [9]. [3] also believed they found evidence for the existence of a phase with about 25 wt. (17.0 at.) % Mo which is stable within a rather small temperature range (see inset in Fig. 280).

The effect of Mo on the temperature of the polymorphic and magnetic transformations of Co was studied by means of magnetometric and dilatometric analysis [2, 10, 4]. Results agree in that they show that the $\epsilon \rightarrow \alpha$ transformation is raised, giving rise to a peritectoid equilibrium $\alpha + Mo_6Co_7 \rightleftarrows \epsilon$ at 1040 [2], 915 [10], and 1014°C [4]. However, the transformation temperatures of the unsaturated solid solutions differ very considerably; e.g., the beginning of the $\epsilon' \rightarrow \alpha$ transformation at 10 wt. (6.4 at.) % Mo was found at about 800 [2], 620 [10], and 725°C [4].

The boundary of the solid solution of Mo in α-Co as reported by [3] (see inset in Fig. 280) is incompatible with the $\epsilon' \rightarrow \alpha$ transformation data [2, 10, 4] since it lies above the $\epsilon' \rightarrow \alpha$ transformation curves found by these investigators. This discrepancy can be clarified only by further studies.

The solid solubility of Co in Mo, as determined by the lattice-spacing method, was found to be 4.4, 2.8, 2.2, 1.5, and 0.95 at. (2.75, 1.75, 1.4, 0.95, and 0.6 wt.) % Co at 1480, 1375, 1300, 1200, and 1100°C, respectively [6, 11].

Crystal Structure. Since the powder pattern of $MoCo_3$ is identical with that of WCo_3 [9], this phase is isotypic with Ni_3Sn ($D0_{19}$ type). Mo_6Co_7 [46.15 at. (58.25 wt.) % Mo] is rhombohedral-hexagonal and isotypic with W_6Fe_7 ($D8_5$ type) [12, 13]. Lattice parameters are a = 8.980 A, α = 30°48′ (or a = 4.767 A, c = 25.65 A) at the ideal composition and a = 8.873 A, α = 30°53′ (or a = 4.725 A, c = 25.42 A) if saturated with Co [13].

1. U. Raydt and G. Tammann, *Z. anorg. Chem.*, **83**, 1913, 246–252.
2. T. Takei, *Kinzoku-no-Kenkyu*, **5**, 1928, 364–379.
3. W. P. Sykes and H. F. Graff, *Trans. ASM*, **23**, 1935, 249–283.
4. U. Haschimoto, *Nippon Kinzoku Gakkai-Shi*, **1**, 1937, 177–190.
5. A. G. Metcalfe, *Trans. AIME*, **197**, 1953, 357–364.
6. J. L. Ham, Climax Molybdenum Co., First Annual Report, Project NR 031-331, Apr. 1, 1950.
7. D. Summers-Smith, *Nature*, **168**, 1951, 786.
8. H. J. Goldschmidt, *Research*, **4**, 1951, 343.
9. M. M. Babich, E. N. Kislyakova, and J. S. Umanskiy, *Zhur. Tekh. Fiz.*, **8**, 1938, 122.
10. W. Köster and W. Tonn, *Z. Metallkunde*, **24**, 1932, 296–299.
11. J. L. Ham, *Trans. ASME*, **73**, 1951, 723–731.
12. M. M. Babich, E. N. Kislyakova, and J. S. Umanskiy, *Zhur. Tekh. Fiz.*, **9**, 1939, 533–536.
13. E. Henglein and H. Kohsok, *Rev. mét.*, **46**, 1949, 569–571.

0.6240
$\bar{1}$.3760

Co-N Cobalt-Nitrogen

According to [1], N_2 is practically insoluble in Co up to 1200°C. [2] have carried out roentgenographic investigations of Co-N preparations, obtained by nitriding Co powder with NH_3 at 380°C [3]. The nitrides Co_3N (7.34 wt. % N) and Co_2N (10.71 wt. % N) were identified and shown to have the following structures: Co_3N, stable between 7.7 and 8.0 wt. % N, hexagonal lattice of metal atoms, a = 2.663 A, c = 4.360 A, c/a = 1.637, at 7.7 wt. % N; Co_2N, rhombically deformed hexagonal packing of metal atoms, a = 2.848 A, b = 4.636 A, c = 4.339 A. The solid solubility of N in α-Co at 600°C was estimated to be 0.6 wt. (2.50 at.) % N. [4] showed that Co_2N is isostructural with Co_2C, a = 2.853_5 A, b = 4.605_6 A, c = 4.344_3 A. The arrangement of the interstitial N atoms is different from that in Fe_2N. Co_3N was prepared by thermal dissociation of Co_2N [2, 4]. See also [5].

Supplement. [6] reported that heat-treating Co in an atmosphere of N_2 lowers the transformation temperatures T_c and T_r from 498 to 454°C and 447 to 380°C, respectively, and increases the lattice parameters of both Co modifications slightly.

1. A. Sieverts and H. Hagen, *Z. physik. Chem.*, **A169,** 1934, 237–240.
2. R. Juza and W. Sachsze, *Z. anorg. Chem.*, **253,** 1945, 95–108.
3. G. Hägg (*Z. physik. Chem.*, **B6,** 1929, 221–232) was unable to nitride Co powder with NH_3 at 300–1000°C.
4. J. Clarke and K. H. Jack, *Chemistry & Industry,* **1951,** 1004–1005.
5. O. Schmitz-Dumont, H. Broja, and H. F. Piepenbrink, *Z. anorg. Chem.*, **253,** 1947, 118–135.
6. J. Drain, R. Bridelle, and A. Michel, *Bull. soc. chim. France,* **1954,** 828–830; see also A. Michel, J. Drain, and R. Bridelle, *Compt. rend.*, **238,** 1954, 107–108.

0.4087
$\bar{1}.5913$

Co-Na Cobalt-Sodium

According to [1], the temperature of the $\epsilon \rightleftarrows \alpha$ transformation of Co (found as 467°C on heating and 426°C on cooling) is raised on heating to 485°C and lowered on cooling to 420°C by 0.2 wt. (0.51 at.) % Na. The same transformation points were found for an alloy with 2 wt. (4.96 at.) % Co.

1. U. Haschimoto, *Nippon Kinzoku Gakkai-Shi,* **2,** 1938, 67–77.

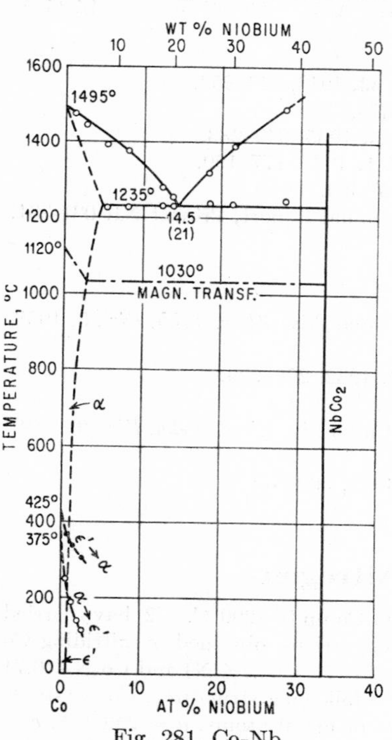

Fig. 281. Co-Nb

$\bar{1}.8024$
0.1976

Co-Nb Cobalt-Niobium

Thermal analysis (Fig. 281) showed that a Co-rich primary solid solution forms a eutectic with a phase the composition of which was assumed to be Nb_2Co_5 [28.57 at. (38.67 wt.) % Nb] with a melting point of about 1550°C [1]. X-ray analysis [2, 3] proved, however, that the intermediate phase in question is an AB_2 (Laves) phase and, therefore, of the ideal composition $NbCo_2$ (44.08 wt. % Nb). $NbCo_2$ has been claimed to exist in two "modifications," one at 33.3 at. % Nb being of the cubic $MgCu_2$ (C15) type and the other at about 27 at. % Nb having the hexagonal $MgNi_2$ (C36) type of structure [2, 3]. However, thermal analysis failed to prove the assumed peritectic formation of the latter phase [2]. Until the findings of [2, 3] have been checked, it appears justified [4] to present only the $NbCo_2$ phase at the ideal composition in the phase diagram (Fig. 281).

The curve of the Curie temperatures [1] indicates that approximately 4 wt. (2.6 at.) % Nb is soluble at 1030°C. With fall in temperature the solubility decreases to less than 3 wt. (2.0 at.) % Nb as an alloy of this composition shows age hardening [1]. The temperatures of the $\epsilon' \rightarrow \alpha$ and $\alpha \rightarrow \epsilon'$ transformations indicated in Fig. 281 do not represent equilibrium conditions. See also the effect of Ta on the $\epsilon \rightleftarrows \alpha$ transformation of Co (page 508).

The lattice spacings of the two "modifications" of $NbCo_2$ were reported [3] to be as follows: $MgCu_2$ type (33.3 at. % Nb), a = 6.758 A; $MgNi_2$ type (27.0 at. % Nb), a = 4.738 A, c = 15.46 A, $c/2a$ = 1.631 [4]. A brief examination did not indicate the existence of a σ phase [5].

Supplement. [6] stated, "Two alloys were prepared for investigation, $NbCo_2$ and $Nb_{0.81}Co_{2.19}$ [27 at. % Nb]. The incipient melting temperature of $NbCo_2$ was determined as 1570°C. The structure of $NbCo_2$ was found to be isomorphous with $MgCu_2$ with a lattice parameter of a = 6.755 kX. No allotropy was observed in the temperature range 600 to 1500°C. The $Nb_{0.81}Co_{2.19}$ alloy was found to be isomorphous with $NbCo_2$ at all temperatures from 800 to 1400°C. These experiments indicate no justification for the existence of the $MgNi_2$ structure in this system."

1. W. Köster and W. Mulfinger, *Z. Metallkunde*, **30**, 1938, 348–350.
2. H. J. Wallbaum, *Z. Krist.*, **103**, 1941, 391–402.
3. H. J. Wallbaum, *Arch. Eisenhüttenw.*, **14**, 1940-1941, 521–526.
4. It has been claimed by [2, 3] that also in the system Fe-Zr there are two phases, with 33.3 and 27 at. % Zr, having structures of the $MgCu_2$ and $MgNi_2$ types, respectively. However, C. B. Jordan and P. Duwez (Progress Report 20-196, June 16, 1953, ORDCIT Project, Contract DA-04-495-Ord-18) were unable to verify this and found only the $MgCu_2$ type of structure at both compositions; see Fe-Zr.
5. P. Greenfield and P. A. Beck, *Trans. AIME*, **200**, 1954, 253–257.
6. R. P. Elliott, Armour Research Foundation, Chicago, Ill., Technical Report 1, OSR Technical Note OSR-TN-247, August, 1954.

0.0018
$\bar{1}$.9982

Co-Ni Cobalt-Nickel

Co-Ni melts crystallize to form a continuous series of solid solutions (Fig. 282). Results of thermal analyses [1–4] are in satisfactory agreement and show that solidification takes place in a temperature interval of only a few degrees.

Curie temperatures were determined by [1, 2, 5, 4, 6], covering the whole range of composition, and by [7], between 0 and 30% Ni [8]. Agreement among those found by [2, 5, 7] is very good.

The effect of Ni on the temperature of the $\epsilon \rightleftarrows \alpha$ transformation of Co was determined by measuring property changes on heating and cooling [5, 7, 4]. The hysteresis of the diffusionless transformation increases with the Ni content. The curve found on heating reaches 30% Ni at a temperature between 100 and 150°C [5, 4]. Using a special technique, [9] determined a "strain-transformation curve" showing the temperatures at which deformation begins to transform ϵ' to α (both having the same composition), and vice versa. The two curves agree within experimental error and, between 15 and 30% Ni, nearly coincide with the $\epsilon' \rightarrow \alpha$ transition found by dynamic methods [5].

From measurements of the density [10] and various magnetic properties [11] of alloys cooled from 1000°C, it was concluded that, at lower temperatures, alloys with 22–32% Ni consist of a mixture of α and ϵ. Since the $\alpha \rightarrow \epsilon'$ transformation normally is not complete, a two-phase structure in these alloys does not necessarily represent equilibrium conditions. In preparations obtained by reduction of Co-Ni formiate solid solutions at about 350°C, [12] found the two-phase region to extend from 17 ± 1.5 to 32 ± 4% Ni. (See Supplement.)

Crystal Structure. Lattice dimensions of α and ϵ' alloys were determined by [13] and, more accurately, by [14]. Parameters of α alloys with 91–98% Ni were given by [15]. The parameter-vs.-composition curve of [14] shows a marked inflection in the region of 70% Ni and a second inflection in the region of 26% Ni, the latter being associated with the coexistence of hexagonal and cubic structures. There is additional evidence of superlattice formation in the vicinity of $CoNi_3$ [14] although superlattice lines are not observed owing to the very great similarity in the X-ray scattering factors of Co and Ni atoms.

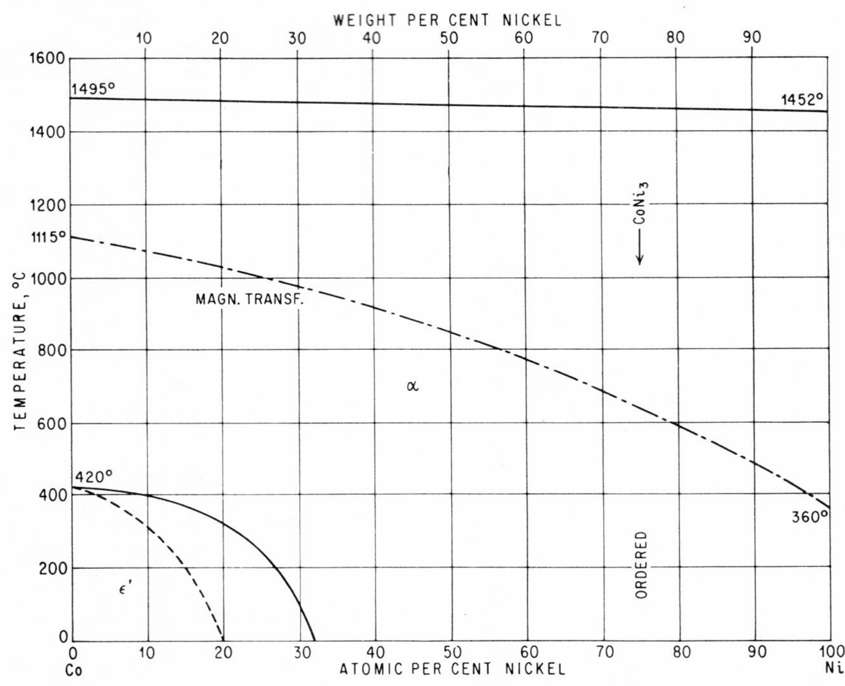

Fig. 282. Co-Ni. (See also Supplement.)

Physical Properties. A number of references for papers dealing with various physical properties may be given: electrical conductivity [16, 17, 6], temperature coefficient of electrical resistance [17, 6], thermal conductivity [16], thermoelectric force [6], thermal expansion [17–19, 6], density [10], and magnetic properties [20–22, 16, 17, 23–25, 6, 26, 11].

Supplement. Lattice parameters over the whole composition range were measured by [27] on solidified as well as electrodeposited alloys. [28] claims to have established the equilibrium boundary $\alpha/(\alpha + \epsilon)$ down to 250°C by means of the interesting "amalgam method" (thermal decomposition of Co-Ni amalgams and heat-treatment of the loose powders obtained): 0, 13, 24.5, 34, and 50 at. % Ni at 475, 450, 400, 350, and 250°C, respectively. The maximal solubility of Ni in ϵ-Co is estimated to be 5 at. %.

1. W. Guertler and G. Tammann, *Z. anorg. Chem.*, **42,** 1904, 353–362.
2. R. Ruer and K. Kaneko, *Metallurgie,* **9,** 1912, 419–422.

3. T. Kasé, *Science Repts. Tôhoku Univ.*, **16,** 1927, 491–513.
4. U. Haschimoto, *Nippon Kinzoku Gakkai-Shi*, **1,** 1937, 177–190.
5. H. Masumoto, *Science Repts. Tôhoku Univ.*, **15,** 1926, 463–468; *Trans. ASST*, **10,** 1926, 491–492.
6. W. Broniewski and W. Pietrik, *Compt. rend.*, **201,** 1935, 206–208.
7. U. Haschimoto, *Kinzoku-no-Kenkyu*, **9,** 1932, 63; see M. Hansen, "Der Aufbau der Zweistofflegierungen," p. 499, Springer-Verlag OHG, Berlin, 1936.
8. At 50 at. %, atomic and weight percentages differ by only 0.1 %.
9. J. B. Hess and C. S. Barrett, *Trans. AIME*, **194,** 1952, 645–647.
10. M. Yamamoto, *Science Repts. Research Insts. Tôhoku Univ.*, **A4,** 1952, 871–877.
11. M. Yamamoto, *Science Repts. Research Insts. Tôhoku Univ.*, **A4,** 1952, 14–27; numerous references are given.
12. F. Lihl, *Metall*, **5,** 1951, 183–187.
13. A. Osawa, *Science Repts. Tôhoku Univ.*, **19,** 1930, 110–115.
14. A. Taylor, *J. Inst. Metals*, **77,** 1950, 585–594.
15. T. H. Hazlett and E. R. Parker, *Trans. ASM*, **46,** 1954, 701–715.
16. H. Masumoto, *Science Repts. Tôhoku Univ.*, **16,** 1927, 321–332.
17. A. Schulze, *Z. tech. Phys.*, **8,** 1927, 423–425, 502; *Physik. Z.*, **28,** 1927, 669–673.
18. H. Masumoto and S. Nara, *Science Repts. Tôhoku Univ.*, **16,** 1927, 333–335.
19. H. Masumoto, *Science Repts. Tôhoku Univ.*, **20,** 1931, 101–123.
20. O. Bloch, *Ann. chim. et phys.*, **26,** 1912, 5–22.
21. P. Weiss and O. Bloch, *Compt. rend.*, **155,** 1912, 941–943.
22. P. Weiss, *Trans. Faraday Soc.*, **8,** 1912, 149.
23. H. Masumoto, *Science Repts. Tôhoku Univ.*, **18,** 1929, 195.
24. G. W. Elmen, *J. Franklin Inst.*, **207,** 1929, 538.
25. Y. Masiyama, *Science Repts. Tôhoku Univ.*, **22,** 1933, 340; **26,** 1937, 65.
26. J. J. Went, *Physica*, **17,** 1951, 98.
27. K. Aotani, *Nippon Kinzoku Gakkai-Shi*, **14B,** 1950, no. 5; abstract in *Structure Repts.*, **13,** 1950, 86–87.
28. F. Lihl, *Z. Metallkunde*, **46,** 1955, 434–441.

0.5663
$\bar{1}.4337$

Co-O Cobalt-Oxygen

There are three cobalt oxides, CoO (21.35 wt. % O), Co_3O_4 [57.14 at. (26.58 wt.) % O], and Co_2O_3 (28.94 wt. % O); the latter appears to be stable only in the form of the hydrate, $Co_2O_3.xH_2O$ [1]. Co_3O_4 dissociates at about 900°C, and CoO is stable up to its melting point, given as 1810°C [2] and 1935°C [3].

Liquid Co, if treated with O_2 at 1550–1700°C, dissolves up to 15 wt. % CoO (3.2 wt. % O) [2]. The melting point of Co is lowered to a eutectic point [4, 2] at approximately 1–1.2 wt. % CoO (0.21–0.25 wt. % O), 1451°C [2]; i.e., the constitution is similar to that of Ni-O (page 1024) (Fig. 283).

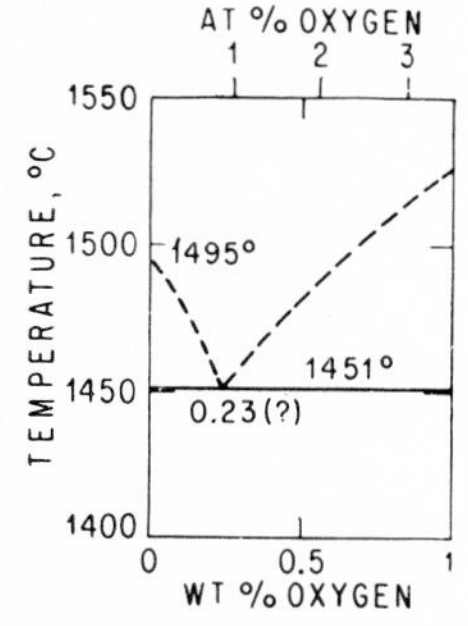

Fig. 283. Co-O

[4] determined the solid solubility of O in Co at temperatures between 600 and 1500°C. It was found (in wt. % O) to be 0.006 at 600°C, 0.009 at 700°C, 0.016 at 810°C, 0.010 at 875°C, 0.007 at 945°C, 0.008 at 1000°C, and 0.013 at 1200°C. From the discontinuous decrease above 875°C, it was concluded that Co undergoes a phase change at about 850°C [4].

CoO is isotypic with the NaCl (B1) structure, a = 4.26 A [5]. Co_3O_4 has the structure of the spinel ($H1_1$) type, $a = 8.0_9$ A [6]. [7] reported the lattice parameter $a = 4.2581 \pm 5$ A (20°C) for CoO.

Supplement. CoO is an antiferromagnetic compound. By means of measurements of magnetic susceptibility, its Curie-point temperature was found to be 271°K by [8] and 292°K by [9]. The magnetic transformation is accompanied by a tetragonal distortion of the NaCl-type structure [7, 10–12] which increases with decreasing temperature.

For the kinetics of the oxidation of Co see [13], which paper also gives a short review of previous studies.

1. See "Gmelins Handbuch der Anorganischen Chemie," 8th ed., System No. **58,** Part A(2), pp. 222–236, Verlag Chemie, G.m.b.H., Berlin, 1932.
2. P. Asanti and E. J. Kohlmeyer, *Z. anorg. Chem.*, **265,** 1951, 90–98.
3. H. v. Wartenberg and W. Gurr, *Z. anorg. Chem.*, **196,** 1931, 377.
4. A. U. Seybolt and C. H. Mathewson, *Trans. AIME*, **117,** 1935, 156–172.
5. *Strukturbericht*, **1,** 1913-1928, 123; **2,** 1928-1932, 222.
6. *Strukturbericht*, **1,** 1913-1928, 420.
7. N. C. Tombs and H. P. Rooksby, *Nature*, **165,** 1950, 442–443.
8. H. Bizette, *Ann. phys.*, **1,** 1946, 295–305.
9. F. Trombe (and H. La Blanchetais), *J. phys., radium*, **12,** 1951, 170–171.
10. S. Greenwald and J. S. Smart, *Nature*, **166,** 1950, 523–524.
11. J. S. Smart and S. Greenwald, *Phys. Rev.*, **82,** 1951, 113–114.
12. S. Greenwald, *Acta Cryst.*, **6,** 1953, 396–398.
13. E. A. Gulbransen and K. F. Andrew, *J. Elektrochem. Soc.*, **98,** 1951, 241–251.

$\bar{1}.4912$
0.5088

Co-Os Cobalt-Osmium

Lattice-parametric and microscopic studies [1] have shown that ϵ-Co and Os form an uninterrupted series of solid solutions. The lattice parameters of the h.c.p. ϵ phase lie on a straight line connecting those of the pure metals. The transformation curves up to 1200°C (Fig. 284) were determined by magnetic analysis of alloys homogenized at 1200°C for 4 days. Below about 1000°C they are "realization curves"; they represent the temperatures at which the diffusionless $\epsilon' \rightarrow \alpha$ and $\alpha \rightarrow \epsilon'$ transformations start on heating and cooling, respectively. Above 1000°C the boundaries are equilibrium curves. Phase relations above 1200°C are *hypothetical.* The fact that alloys with 30–73 at. (about 58–90 wt.) % Os could be homogenized only by annealing at temperatures about 100°C below the solidus indicates that the solidification interval of the ϵ solid solutions is broad. The curves designated as θ_α and θ_ϵ represent the Curie temperatures of the α and ϵ phases, respectively.

1. W. Köster and E. Horn, *Z. Metallkunde*, **43,** 1952, 444–449; "Festschrift aus Anlass des 100-jährigen Jubiläums der Firma W. C. Heraeus G.m.b.H.," pp. 114–120, Hanau, 1952.

0.2794
$\bar{1}.7206$

Co-P Cobalt-Phosphorus

The phase diagram (Fig. 285) is based on thermoanalytical and microscopic studies [1]. Alloys with more than 33.7 at. (21.2 wt.) % P could not be prepared by melting in an open crucible. The liquidus curve was confirmed "in general" by [2], who found the eutectic temperature as 1041°C; otherwise no data are given. The transformation of the phase Co_2P (20.81 wt. % P) at 920°C [1] could not be verified by [2] and [3] but was confirmed by [4], who located the eutectic temperature at 1024°C. CoP (34.45 wt. % P) was shown to exist by determination of its crystal structure [5]. The work by [1] and tensimetric analysis [6] established that none of the compounds reported earlier (Co_5P, Co_4P, Co_3P, Co_3P_2, Co_4P_3, and Co_2P_3) occurs and that the only phosphides existing are Co_2P, CoP, and CoP_3 (61.19 wt. % P) [6]. P is slightly soluble in Co_2P [3]. According to lattice-parameter measurements there is no solid solubility of P in ϵ-Co and α-Co [3]; an alloy with 0.25 wt. (0.45 at.) % P was found to be heterogeneous [2]. The temperature of the magnetic transformation of Co (1142°C) was found to be lowered to 1091°C [4]. Data were reported

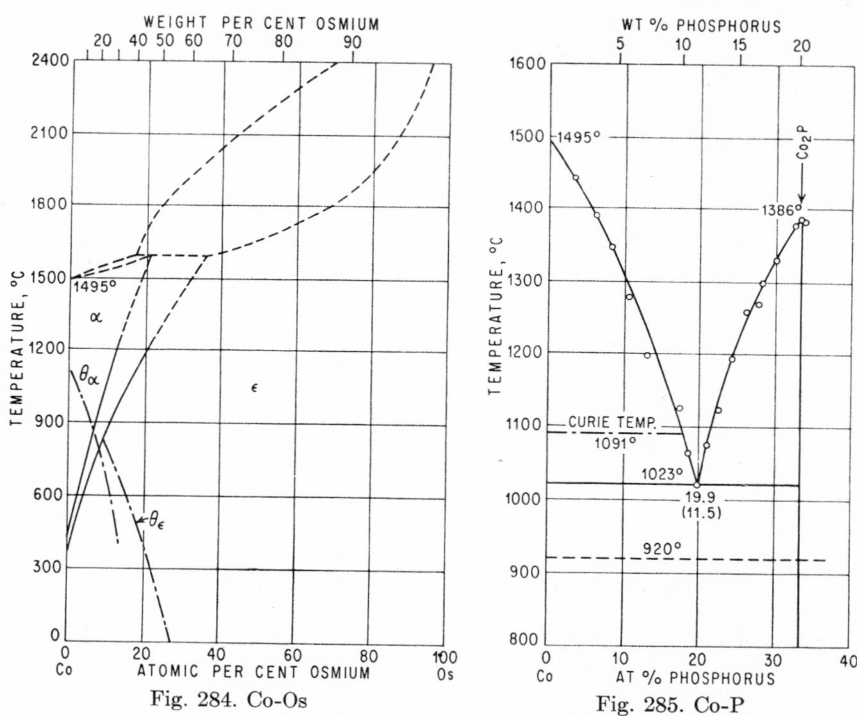

Fig. 284. Co-Os

Fig. 285. Co-P

which show that the $\epsilon \rightarrow \alpha$ transformation (found as 467°C) is raised to 525°C and the $\alpha \rightarrow \epsilon$ transformation (426°C) is lowered to 384°C [4]. The conclusion [4] that 0.35 wt. (0.63 at.) % P is soluble at about 500°C contradicts the findings by [2, 3].

Co_2P was reported to have a structure different from that of Fe_2P (C22) type [6]. According to [3], it is orthorhombic of the $PbCl_2$ (C23) type, with $a = 6.63$ A, $b = 5.67$ A, $c = 3.52$ A(?) at 20 wt. (32.2 at.) % P, and $a = 6.68$ A, $b = 5.75$ A, $c = 3.53$ A(?) at 22 wt. (34.9 at.) % P. CoP is isotypic with MnP (B31 type), $a = 5.599$ A, $b = 5.076$ A, $c = 3.281$ A [5]. CoP_3 may be isotypic with NiP_3 [6].

1. S. Zemczuzny and J. Schepelew, *Z. anorg. Chem.*, **64,** 1909, 245–257.
2. J. Berak, *Arch. Eisenhüttenw.*, **22,** 1951, 131–132.
3. H. Nowotny, *Z. anorg. Chem.*, **254,** 1947, 31–36.
4. U. Haschimoto, *Nippon Kinzoku Gakkai-Shi*, **2,** 1938, 67–77.
5. K. E. Fylking, *Arkiv Kemi, Mineral. Geol.*, **11B,** 1934, no. 48.
6. W. Biltz and M. Heimbrecht, *Z. anorg. Chem.*, **241,** 1939, 349–360.

$\bar{1}.4540$
0.5460

Co-Pb Cobalt-Lead

The phase diagram (Fig. 286) was obtained by thermal analysis of only five alloys [1]. The melting point of the 98% pure cobalt [2] was found as 1440°C (instead of 1495°C for the pure metal) and the temperature of the monotectic as 1438–1439°C. This indicates that the solubility of Pb in liquid Co can be only very slight. The solidified Co-rich layer contained 0.89 wt. % Pb + 1.92 wt. % contaminants (!). By extrapolation (!) of the monotectic halting times the composition of the Pb-rich layer at 1438–1439°C was located at approximately 97 wt. (90 at.) % Pb. Both compositions are very uncertain. There was some indication of a small solubility of Co in liquid Pb at its melting point; however, the investigation was not accurate enough to draw a conclusion as to the position of the eutectic.

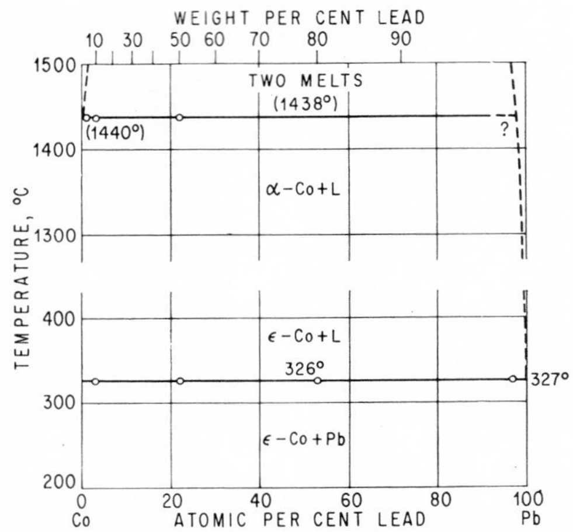

Fig. 286. Co-Pb. (See also Note Added in Proof.)

The solid solubility of Pb in Co is of the order of 11–12 $\times$ 10^{-4} wt. % Pb [3]. In agreement with this is the fact that neither the temperature of the magnetic transformation [1] nor that of the $\epsilon \rightleftarrows \alpha$ transformation [4] is noticeably affected by Pb additions. See also [5, 6].

Note Added in Proof. [7] investigated the solubility of Co in liquid Pb. Cooling curves showed the eutectic temperature to be less than 0.5°C below the melting point of Pb. According to chemical analysis of saturated melts, the solubility increases from 0.037 wt. (0.13 at.) % at 327°C to 0.053 wt. (0.19 at.) % at 500°C and 0.137 wt. (0.48 at.) % Co at 727°C.

1. K. Lewkonja, *Z. anorg. Chem.*, **59,** 1908, 312–315.
2. Analysis: 98.04% Co, 1.62% Ni, 0.17% Fe, 0.04% residue.
3. G. Tammann and W. Oelsen, *Z. anorg. Chem.*, **186,** 1930, 279–280.
4. U. Haschimoto, *Nippon Kinzoku Gakkai-Shi*, **2,** 1938, 67–77.
5. F. Ducelliez, *Bull. soc. chim. France*, **3,** 1908, 621–622.
6. F. Ducelliez, *Bull. soc. chim. France*, **7,** 1910, 201–202; *Compt. rend.*, **150, 1910,** 98–101.
7. E. Pelzel, *Metall*, **9,** 1955, 692–694.

$\bar{1}.7422$
0.2578

Co-Pd Cobalt-Palladium

The phase diagram of Fig. 287 is based on thermal analysis [1], thermoresistometric [1], and magnetometric [2] data. The minimum of the liquidus curve lies at or close to 50 at. % Pd, 1217°C. The Curie temperatures determined by [2] are slightly lower than those obtained by [1]. Earlier, [3] had determined the Curie

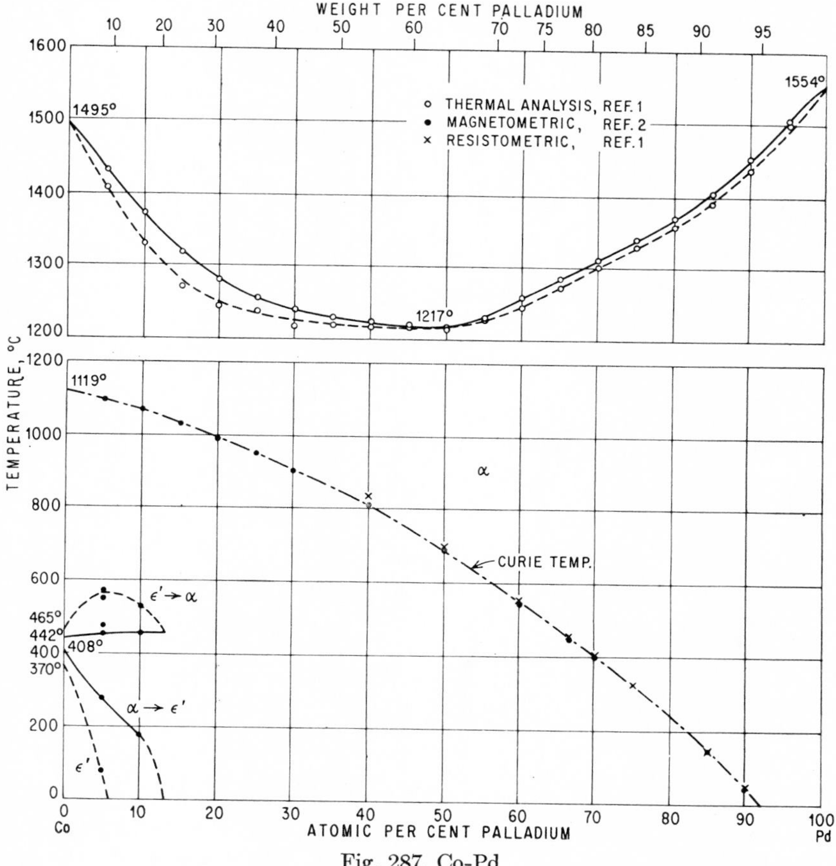

Fig. 287. Co-Pd

temperatures of two alloys with 90 and 95 wt. % (83.2 and 91.3 at.) % Pd to be 189 and 28°C, respectively, when annealed samples were used [4]. The polymorphic $\epsilon \rightleftarrows \alpha$ transformation is characterized by a strong hysteresis. The solid curves in Fig. 287 indicate the temperatures of the beginning of the $\epsilon' \rightarrow \alpha$ and $\alpha \rightarrow \epsilon'$ transformations, respectively. The dashed curves indicate the approximate temperatures of the end of the transformation. The irreversible transition probably is associated with the ferromagnetism of the h.c.p. ϵ-Co modification [5]. No evidence of a transformation in the α solid solutions (other than the $\alpha \rightleftarrows \epsilon'$ transformation) was found. However, the electrical-conductivity isotherms are somewhat distorted because of the magnetic transformation [1].

Supplement. [6] studied the hardness, electrical resistance, and microstructure of numerous Co-Pd alloys. According to an abstract (in which hardness and resistance data are listed), the temperature coefficient of electrical resistance showed a sharp minimum near 12 wt. % Pd and a sharp maximum near 25 wt. % Pd. The microstructures showed solid solution formation. For a discussion of the variation of the Curie temperature with composition, see [7].

1. G. Grube and H. Kästner, *Z. Elektrochem.*, **42,** 1936, 156–160.
2. G. Grube and O. Winkler, *Z. Elektrochem.*, **41,** 1935, 52–60.
3. F. W. Constant, *Phys. Rev.*, **36,** 1930, 1654–1660.
4. The Curie points of the cold-worked samples were found to be 238 and 86°C.
5. M. Fallot, *Compt. rend.*, **199,** 1934, 128.
6. V. A. Nemilov and L. A. Panteleimonov, *Izvest. Sektora Platiny*, **27,** 1952, 202–205 (not available to the author); abstract in *Chem. Abstr.*, **49,** 1955, 2143.
7. E. P. Wohlfarth, *Phil. Mag.*, **45,** 1954, 647–649.

$\bar{1}.4799$
0.5201

Co-Pt Cobalt-Platinum

Thermal analysis [1, 2] has shown that Co-Pt melts crystallize to form a continuous series of solid solutions [3, 4]. The liquidus curve has a flat minimum in the vicinity of 15 at. (37 wt.) % Pt, 1430° [2], or about 25 at. (52.5 wt.) % Pt, 1465°C [1].

A disorder-order transformation in the region of 50 at. % was first detected by means of X-ray work [5] and confirmed by [2, 6] and others [7, 8]. The phase diagram of Fig. 288 is based almost entirely on work by [2], who used thermal analysis and electrical-resistance, dilatation, and magnetic measurements as well as roentgenographic studies. The maximum of the order $\rightleftarrows$ disorder transformation curve lies at 50 at. % and 825°C [2]. The homogeneity range of the (CoPt) phase extends from ~42 to 74 at. (70.6–90.4 wt.) % Pt [2]. Additional transformation points were determined by means of X-ray and resistometric methods [7]. Later, [9] found that another order-disorder transformation occurs in the region of 75 at. (90.86 wt.) % Pt, with the maximum of the transformation curve lying between 700 and 800°C. No detailed information of this transition is available as yet; the transformation curves in Fig. 288 are hypothetical.

The mechanism and kinetics of the ordering reaction $\alpha \rightarrow$ (CoPt) were thoroughly studied by [2, 8]; see also [10]. Ordering goes through a two-phase stage. A thermodynamic treatment of the system is due to [11].

Crystal Structure. Lattice parameters of the f.c.c. α phase, covering the whole range of composition, were reported by [2], and parameters of α alloys with about 46, 52, and 58 at. % Pt were given by [7]. The phase (CoPt) has the f.c. tetragonal structure of the CuAu ($L1_0$) type [2, 6–8]. At 50 at. % Pt, the lattice dimensions of f.c.c. α and ordered (CoPt) are 3.751 A and a = 3.793 A, c = 3.675 A, c/a = 0.969, respectively [2]. For an alloy with 52.4 at. % Pt, parameters of a = 3.782 A and a = 3.812 A, c = 3.708 A, c/a = 0.973, respectively, were reported [7]. No indication of a b.c.c. superstructure [5, 6] was found [7]. The assumption [2] that there is a disordered f.c. tetragonal structure of the (CoPt) phase between 520 and 825°C could not be verified by X-ray investigations at temperatures between 500 and 720°C [7].

The 70 at. % Pt alloy, f.c.c. (a = 3.829 A) after quenching from temperatures above 800°C, forms a f.c.c. superlattice of the Cu_3Au ($L1_2$) type. After ordering at 700°C, the parameter of $CoPt_3$ was found to be a = 3.831 A [9]. The magnetic-transformation curve shown in Fig. 288 was measured [2] on alloys quenched from 1000°C.

Supplement. Curie-point temperatures measured by [12] on alloys with 75, 79, 83, and 87 at. % Pt quenched from 1200°C are in good agreement with the curve of [2] (Fig. 288). On slowly cooled alloys of the same compositions, the values +13, +33, +53, and −9°C were measured.

Theoretical discussions of the variation of the Curie temperature with concentration are due to [12, 13]. [14] determined the degree of long-range order as a function of temperature in the CoPt alloy by means of X-ray intensity and lattice tetragonality measurements. The critical temperature was found to be 833 ± 2°C.

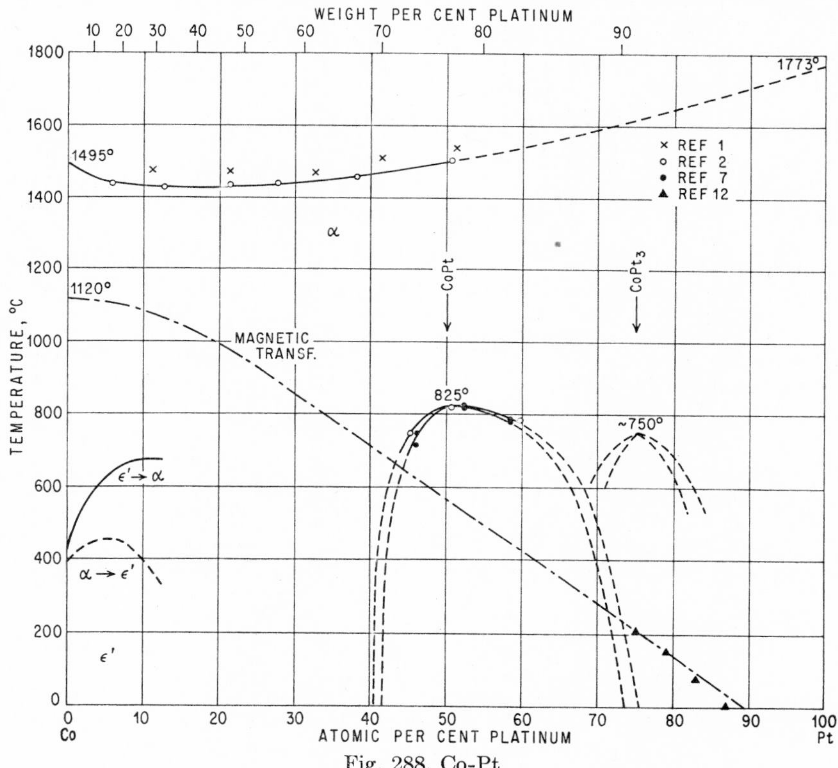

Fig. 288. Co-Pt

Diffuse X-ray scattering measurements on quenched disordered alloys containing 20–80 at. % Pt revealed a preference for unlike nearest-neighbors.

1. W. A. Nemilow, *Z. anorg. Chem.*, **213,** 1933, 283–291; *Izvest. Inst. Platiny,* **9,** 1932, 23–30.
2. E. Gebhardt and W. Köster, *Z. Metallkunde,* **32,** 1940, 253–261.
3. On the basis of microscopical examination, F. C. Carter (*Proc. AIME, Inst. Metals Div.,* **1928,** pp. 759–782) had already supposed an uninterrupted series of solid solutions.
4. The existence of complete miscibility could be concluded from magnetic studies of Pt-rich alloys by F. W. Constant, *Phys. Rev.,* **36,** 1930, 1654–1660.
5. W. Jellinghaus, *Z. tech. Physik,* **17,** 1936, 33–36.
6. R. Hultgren and R. I. Jaffee, *J. Appl. Phys.,* **12,** 1941, 501–502.

7. J. B. Newkirk, R. Smoluchowski, A. H. Geisler, and D. L. Martin, *J. Appl. Phys.*, **22,** 1951, 290–298.
8. J. B. Newkirk, A. H. Geisler, D. L. Martin, and R. Smoluchowski, *Trans. AIME*, **188,** 1950, 1249–1260.
9. A. H. Geisler and D. L. Martin, *J. Appl. Phys.*, **23,** 1952, 375.
10. L. Weil, *Compt. rend.*, **224,** 1947, 923–925.
11. R. A. Oriani, *Acta Met.*, **1,** 1953, 144–151.
12. A. W. Simpson and R. H. Tredgold, *Proc. Phys. Soc. (London)*, **B67,** 1954, 38–41. The temperature data had to be taken from a small graph.
13. E. P. Wohlfarth, *Phil. Mag.*, **45,** 1954, 647–649.
14. P. S. Rudman and B. L. Averbach, *Acta Met.*, **5,** 1957, 65–73.

$\bar{1}.3920$
0.6080

Co-Pu Cobalt-Plutonium

The existence of the following intermediate phases has been reported [1]: Pu_6Co, b.c. tetragonal, $a = 10.46 \pm 2$ A, $c = 5.33 \pm 1$ A, isotypic with U_6Mn; Pu_3Co (?), unsolved powder pattern; Pu_2Co, probably hexagonal Fe_2P (C22) type, $a = 7.902 \pm 4$ A, $c = 3.549 \pm 2$ A (Pu-rich), and $a \doteq 7.762 \pm 3$ A, $c = 3.649 \pm 2$ A (Co-rich); $PuCo_2$, cubic $MgCu_2$ (C15) type, $a = 7.081 \pm 1$ A (Pu-rich); $PuCo_3$, unsolved powder pattern (isostructural with $PuNi_3$); Pu_2Co_{17}, hexagonal Th_2Ni_{17} type, $a = 8.325 \pm 2$ A, $c = 8.104 \pm 3$ A.

1. Review by A. S. Coffinberry and M. B. Waldron in "Metallurgy and Fuels," Progress in Nuclear Energy, ser. V, vol. 1, Pergamon Press Ltd., London, 1956.

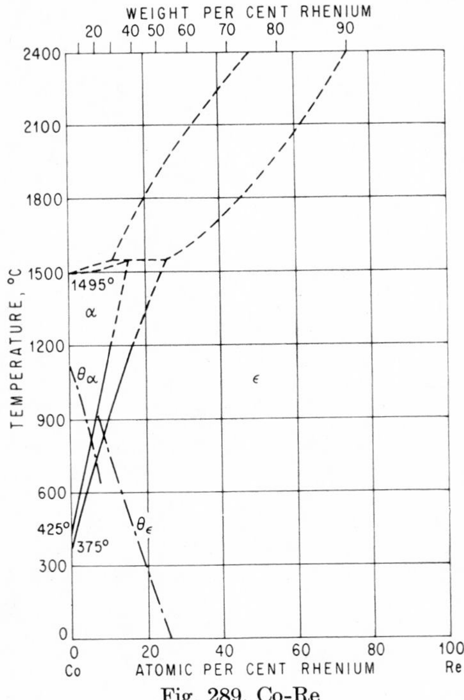

Fig. 289. Co-Re

$\bar{1}.5002$
0.4998

Co-Re Cobalt-Rhenium

According to lattice-parameter measurements and microscopic examination [1], ε-Co and Re form a continuous series of solid solutions. The transformation curves shown in Fig. 289 were determined by magnetic analysis of alloys annealed at 1200°C for 3 weeks and quenched. Below approximately 1000°C the boundaries between the ε and α fields are "realization curves" representing the temperatures at which the diffusionless $\epsilon' \rightarrow \alpha$ transformation and $\alpha \rightarrow \epsilon'$ transformation start on heating and cooling, respectively. Above about 1000°C the boundaries are equilibrium curves. The phase relations above 1200°C are *hypothetical*. The Curie temperatures, θ, of the ε and α solid solutions are also shown. The lattice constants of the ε phase lie on a straight line connecting the constants of the components.

1. W. Köster and E. Horn, *Z. Metallkunde*, **43**, 1952, 444–449.

$\bar{1}.7580$
0.2420

Co-Rh Cobalt-Rhodium

Lattice-parametric and microscopic investigations [1] have shown that α-Co and Rh form a continuous series of solid solutions. The parameters of the f.c.c. α phase

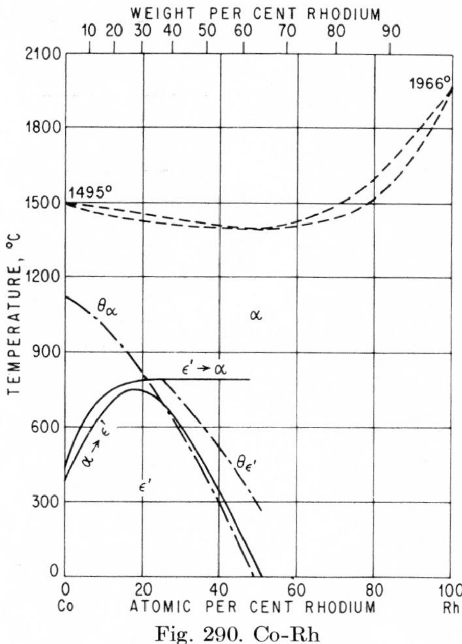

Fig. 290. Co-Rh

lie on a straight line connecting the parameters of the components. Up to 64 wt. (50.5 at.) % Rh the microstructure indicated that the alloys had undergone the polymorphic transformation $\alpha \rightarrow \epsilon'$.

The transformation curves in Fig. 290, obtained by magnetic and dilatometric analysis, are "realization curves" (not equilibrium curves), indicating the temperatures of the start of the diffusionless $\epsilon' \rightarrow \alpha$ and $\alpha \rightarrow \epsilon'$ transformations on heating and cooling, respectively. They show that the transformation hysteresis increases considerably as the Rh content increases so that the $\alpha \rightarrow \epsilon'$ transformation is suppressed to lower than room temperature. The curves designated as θ_α and $\theta_{\epsilon'}$ represent the Curie temperatures of the α and ϵ' phases, respectively.

The liquidus and solidus curves are *hypothetical.* As alloys with 70 wt. (57.2 at.) % and 84 wt. (75.2 at.) % Rh were partially molten on annealing at 1400°C, a minimum is shown.

1. W. Köster and E. Horn, *Z. Metallkunde,* **43,** 1952, 444–449.

$\bar{1}.7631$
0.2369

Co-Ru Cobalt-Ruthenium

Lattice-parameter measurements and microscopic examination have shown [1] that ϵ-Co and Ru form a continuous series of solid solutions. The transformation curves presented in Fig. 291 were determined by magnetic analysis of alloys annealed

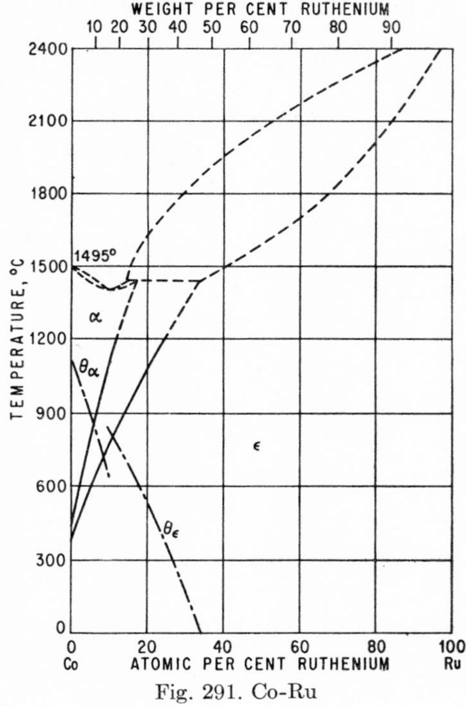

Fig. 291. Co-Ru

at and quenched from 1200°C. Below about 1000°C the boundaries between the ϵ and α fields are "realization curves," representing the temperatures at which the diffusionless $\epsilon' \rightarrow \alpha$ and $\alpha \rightarrow \epsilon'$ transformations start on heating and cooling, respectively. Above 1000°C the boundaries are equilibrium curves. The phase relations above 1200°C are *hypothetical.* As alloys with 9 and 12 at. % Ru were partially molten on annealing at 1400°C, there appears to be a minimum in the liquidus curve. The curves θ_ϵ and θ_α show the Curie temperatures of the ϵ and α phases. The lattice parameters of the ε phase lie on a straight line connecting the constants of the components.

1. W. Köster and E. Horn, *Z. Metallkunde,* **43,** 1952, 444–449.

0.2644
$\bar{1}$.7356

Co-S Cobalt-Sulfur

Partial System Co-CoS. Phase relationships (Fig. 292) were established, by means of thermal analysis, by [1] and confirmed by [2]. The temperatures of the

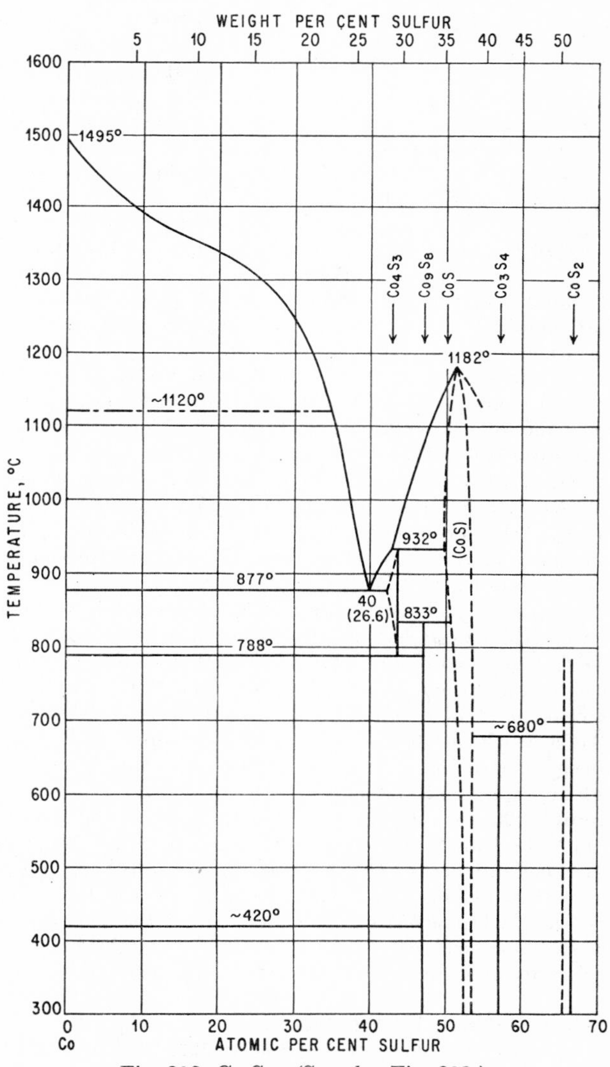

Fig. 292. Co-S. (See also Fig. 293.)

three-phase equilibria were given as follows: Co-(Co_4S_3) eutectic, 879°C [1], 874°C [2]; peritectic formation of (Co_4S_3), 935°C [1], 928°C [2]; eutectoid decomposition of (Co_4S_3), 788°C [1], 787°C [2]; peritectoid formation of Co_9S_8, 834°C [1], 832°C [2].

The high-temperature phase of variable composition, which is formed peritectically at about 932°C, has been designated as (Co_4S_3) [1, 2] although the (still unknown) crystal structure of this constituent may not be based on this composition. (Co_4S_3) cannot be retained on quenching [2]. The phase formed by the peritectoid reaction at 833°C was originally suggested to be Co_6S_5 [45.45 at. (31.20 wt.) % S] [1, 2]; however, [3] have shown that the crystal structure calls for the composition Co_9S_8 [47.04 at. (32.60 wt.) % S] [4]. On the other hand, [5] found indication of this phase having a slightly lower sulfur content, corresponding to the formula $Co_{15}S_{13}$ [46.41 at. (32.04 wt.) % S].

[1, 2] have determined the liquidus curve up to about 48.2 and 47.7 at. % S, respectively. According to their data, the liquidus point of the 50 at. % S melt would lie close to 1160°C (extrapolated), i.e., well above the value given by [6] as the melting point of CoS (1100°C). However, [5] have shown, by thermal analysis of melts contained in silica bombs, that the maximum of the liquidus curve lies at a composition beyond CoS, at 51.2 at. % S and 1182°C.

The alloy corresponding to the formula CoS is heterogeneous at lower temperatures and consists of Co_9S_8 (or $Co_{15}S_{13}$) crystals precipitated from a matrix of (CoS) [2, 3, 5, 7].

The solid solubility of S in Co must be very small [3] and was estimated [8] to be 0.04 wt. % S at about 450°C (?). Accordingly, the effect of S on the temperature of the $\epsilon \rightleftarrows \alpha$ transformation in Co is negligible [8]. Recently, [23] has shown that sulfur contents exceeding 0.005 wt. % cause hot brittleness of Co because of the formation of a low-melting eutectic grain-boundary network. Since the alloys tested were in the as-cast condition, the solid solubility of S in Co may be higher than that figure.

Partial System CoS-CoS₂. The constitution of this system (Fig. 292) is based on data by [9–11, 3, 12, 7, 5]. Between the phase of variable composition (CoS) and CoS_2 (52.11 wt. % S) exists the singular phase Co_3S_4 [57.14 at. (42.05 wt.) % S], first established by [11] and confirmed by [3, 12]. It is stable only up to about 680°C [12]. The boundaries of the phase (CoS) are only a rough approximation and are drawn according to data by [9, 3, 5, 7].

Crystal Structures. The structure of (Co_4S_3) is possibly the same as that of Ni_4S_3 (maucherite) [13]. Co_9S_8 has the $D8_9$ prototype of structure, with $a = 9.927$ A [11, 3], and was first assumed to be that of Co_4S_3 [14] and Co_6S_5 [2]. The phase is of singular composition, and the unit cell contains 68 atoms [11]. CoS is isotypic with NiAs (B8 type) [15, 16, 14, 3], $a = 3.374$ A, $c = 5.187$ A, $c/a = 1.537$ if saturated with Co and $a = 3.368$ A, $c = 5.170$ A, $c/a = 1.535$ if saturated with S [3]. Co_3S_4 has the $D7_2$ prototype of structure [17, 18, 3], $a = 9.401$ A [3]. CoS_2 is isotypic with FeS_2 (C2 type) [19, 3], $a = 5.535$ A [3].

Supplement. [20] took cooling curves of Co-S alloys and reported 1100°C as the melting point of CoS and 879°C, 40 at. % S as eutectic data. As a result of decomposition-pressure measurements [21] and some X-ray work, [22] revised the Co-S phase diagram as shown in Fig. 293. [22] found that (*a*) the lower boundary of the (CoS) phase is located at about 51 at. (36 wt.) % S, i.e., the homogeneity range does not include the stoichiometric composition; (*b*) (CoS) is unstable below approximately 460°C; (*c*) Co_3S_4 is decomposed at about 625 rather than 680°C (as shown in Fig. 292).

1. K. Friedrich, *Metallurgie*, **5**, 1908, 212–215.
2. O. Hülsmann and F. Weibke, *Z. anorg. Chem.*, **227**, 1936, 113–123.
3. D. Lundquist and A. Westgren, *Z. anorg. Chem.*, **239**, 1938, 85–88.

4. See also R. Schenck and P. v. d. Forst, *Z. anorg. Chem.*, **249,** 1942, 76–87.
5. W. Curlook and L. M. Pidgeon, *Can. Mining Met. Bull.*, **46,** 1953, 297–301; *Trans. Can. Inst. Mining Met.*, **56,** 1953, 133–137.
6. W. Biltz, *Z. anorg. Chem.*, **59,** 1908, 280–281.
7. R. Vogel and G. F. Hillner, *Arch. Eisenhüttenu.*, **24,** 1953, 133–135.
8. U. Haschimoto, *Nippon Kinzoku Gakkai-Shi*, **2,** 1938, 67–77.
9. O. Hülsmann, W. Biltz, and K. Meisel, *Z. anorg. Chem.*, **224,** 1935, 73–83 (tensimetric analysis).
10. H. Haraldsen, *Z. anorg. Chem.*, **224,** 1935, 85–92 (magnetic investigation).
11. M. Lindquest, D. Lundquist, and A. Westgren, *Svensk Kem. Tidskr.*, **48,** 1936, 156–160.
12. M. Heimbrecht, W. Biltz, and K. Meisel, *Z. anorg. Chem.*, **242,** 1939, 229–232 (tensimetric analysis).

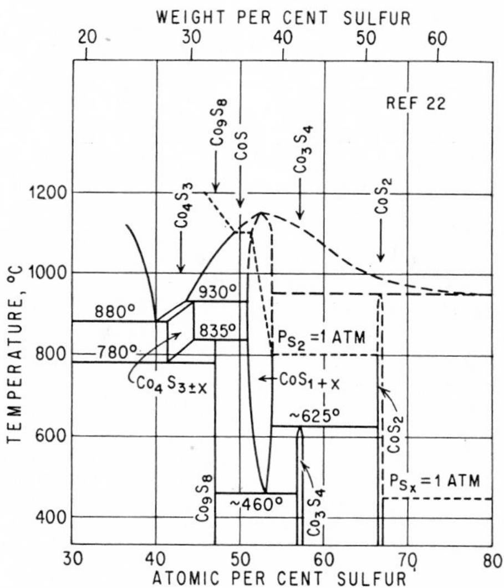

Fig. 293. Co-S. (See also Fig. 292.)

13. H. Strunz, "Mineralogische Tabellen," 2d ed., p. 71, Geest and Portig, Leipzig, 1949 (as quoted by [7]).
14. V. Caglioti and G. Roberti, *Gazz. chim. ital.*, **62,** 1932, 19–29.
15. N. Alsén, *Geol. Fören. i Stockholm Förh.*, **47,** 1925, 19–72.
16. W. F. deJong and H. W. V. Willems, *Physica*, **7,** 1927, 16.
17. G. Menzer, *Z. Krist.*, **64,** 1926, 506–507.
18. W. F. deJong and H. W. V. Willems, *Z. anorg. Chem.*, **161,** 1927, 311–315.
19. W. F. deJong and H. W. V. Willems, *Z. anorg. Chem.*, **160,** 1927, 185–189.
20. P. Asanti and E. J. Kohlmeyer, *Z. anorg. Chem.*, **265,** 1951, 90.
21. The decomposition pressures (at 700, 800, 850, 1000, and 1250°C) were measured by means of the reaction with hydrogen to form H_2S. From the variation of the decomposition pressures with composition, the presence and extent of the various sulfide phases were determined.
22. T. Rosenqvist, *J. Iron Steel Inst.*, **176,** 1954, 37–57.
23. D. L. Martin, *Trans. AIME*, **206,** 1956, 578–579.

$\bar{1}.6849$
0.3151

Co-Sb Cobalt-Antimony

Thermoanalytical studies (some of them supplemented by microscopic investigations) were carried out by [1–3], covering the whole range of composition, and [4, 5] in the ranges 7.9–42.1 and 0–7.9 at. % Sb, respectively. The characteristic points of the phase diagram, summarized in Table 21, and data points in Fig. 294 show that

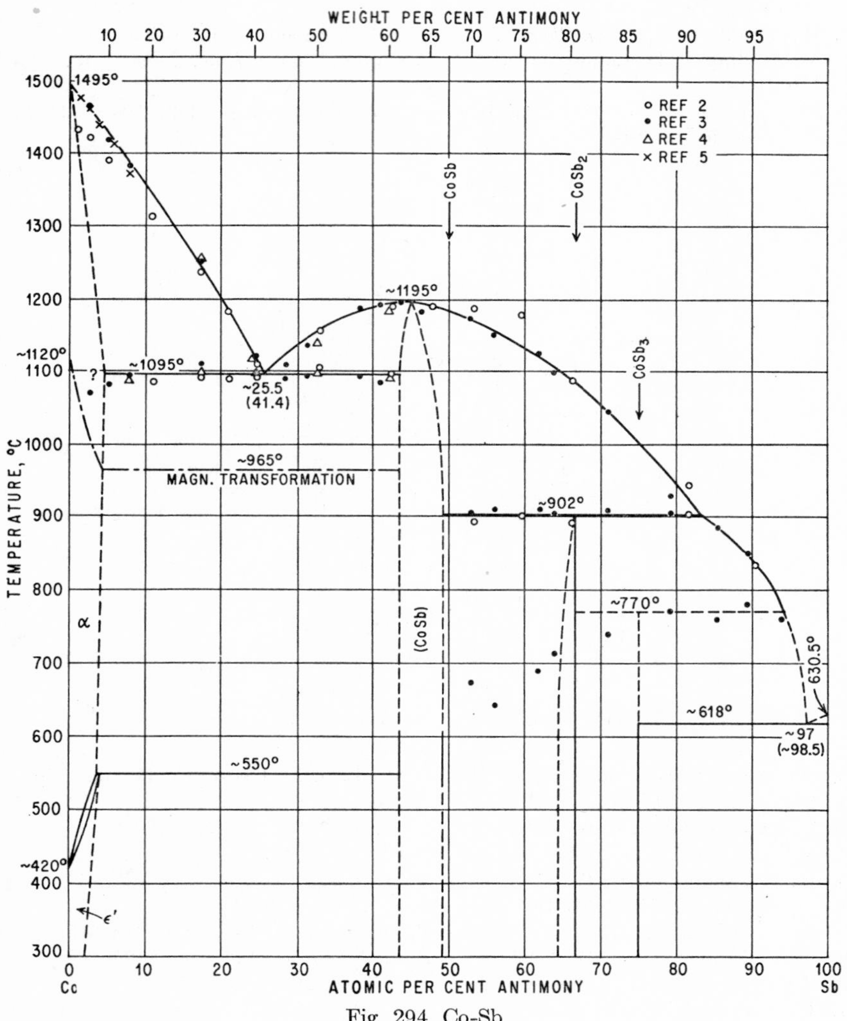

Fig. 294. Co-Sb

the temperature differences within one work and among the various investigations are considerable.

As to Fig. 294, the following comments may be made: (*a*) According to the data by [3] the maximum of the liquidus curve was found at 41–43.5 at. % rather than 50 at. % Sb. Data by [2] indicate a very flat maximum, apparently due to incorrect temperature measurements; temperatures between 60 and 75 wt. (42–59 at.) % Sb vary only within 11°C. The assumption [1–3] that the maximum would lie at the composition CoSb (67.38 wt. % Sb) appears to be no longer justified since [6] have shown a homogeneity range to exist between 43.4 and 49.2 at. (61.2 and 66.6 wt.) % Sb. See also the phase diagram of the system Co-Sn (Fig. 298). (*b*) The intermediate phase in equilibrium with Sb is the compound $CoSb_2$ (80.51 wt. % Sb), as identified by roentgenographic analysis [7] and a residue isolated from alloys with more than 81 wt. (67.4 at.) % Sb [8] (see Supplement). (*c*) In the range of about 70–95 wt. (53–90 at.) % Sb, [1] observed thermal effects at widely scattered temperatures between the peritectic at about 902°C and the eutectic at about 618°C. They may be due to a polymorphic transformation of $CoSb_2$ or the peritectic formation of an additional intermediate phase (see Supplement!). Since the other investigators [1, 2] did not find these effects, a horizontal at about 770°C is tentatively shown in Fig. 294. (*d*) The composition of the Co-rich eutectic, as given by [2, 3] in Table 21, was obtained by extrapolation of the halting times of the eutectic crystallization. However, the liquidus points [2, 3] are in favor of a composition higher in Sb, above 40 wt. (24.4 at.) % Sb. (*e*) The boundary of the solid solution of Sb in α- and ϵ-Co has not been determined as yet. According to [9], the solubility is approximately 8.5 wt. (4.3 at.) % Sb, after annealing for 6 hr at 1000°C and subsequent furnace cooling. On the basis of determinations of the effect of Sb on the transformation points of Co, the solubility was estimated to be about 8.7 wt. (4.4 at.) % Sb at 940°C and 7.2–7.8 wt. (3.6–4.0 at.) % Sb at 575°C [5]. (*f*) Results of studies of the effect of Sb on the Curie point and $\epsilon \rightleftarrows \alpha$ transformation of Co differ considerably. [9] found that about 4.3 at. % Sb lowers the Curie point from about 1120 to 990°C, raises the $\epsilon \rightarrow \alpha$ transformation from about 420 to 530°C, and lowers the $\alpha \rightarrow \epsilon$ transformation from about 380 to 190°C. According to [5], the Curie point of Co (1142°C) is lowered by 4.4 at. % Sb to 943°C, the $\epsilon \rightarrow \alpha$ transformation raised from 467 to 574°C by 3.6 at. % Sb, and the $\alpha \rightarrow \epsilon$ transformation raised from 426 to 461°C by 4.0 at. % Sb. [2] found the Curie point (1143°C) to be lowered to 940°C by about 6.5 at. % Sb.

Crystal Structures. (CoSb) has the structure of the NiAs (B8) type [10, 11, 7], a = 3.874 A, c = 5.193 A, c/a = 1.341 [10, 7]. $CoSb_2$ is isotypic with FeS_2 (C18 type), a = 3.21 A, b = 5.79 A, c = 6.43 A [7].

Table 21. Characteristic Temperatures in °C and Compositions in At. % Sb (Wt. % in Parentheses)

	Ref. 1	Ref. 2	Ref. 3	Ref. 4	Ref. 5
Melting point of Co..........	?	1440°	1505°	......	1492°
α-Co-(CoSb) eutectic........	1082°	$1093^{+12°}_{-7°}$	1089 ± 20°	......	1093°
	25.2	23.7	23.3	~25.2	
	(41)	(39)	(38.5)	(~41)	
Melting point of "CoSb".....	1238°	~1193°	~1190°		
Melt + (CoSb) ⇄ $CoSb_2$:					
Temperature..............	888°	898 ± 6°	906 ± 3°		
Melt.....................	84.9	83.1	81.3		
	(92)	(91)	(90)		
$CoSb_2$-Sb eutectic (see Supplement)	613°	616 ± 3°	625 ± 5°		
	97.1	~97.1	~100		
	(98.5)	(~98.5)	(~100)		

Supplement. In the range 50–100 at. % Sb, [12] found the following intermediate phases which were studied by magnetic, X-ray, and optical goniometer methods: (*a*) $CoSb_{2-x}$ ($x = 0.1$–0.2) with marcasite (C18) structure, $a = 5.596$ A, $b = 6.373$ A, $c = 3.370$ A (cf. above; crystallographic axes have been interchanged). From density measurements it seems likely that $CoSb_{2-x}$ is a substitutional solid solution where Co replaces Sb in the crystal structure. (*b*) $CoSb_3$ (not previously reported) with skutterudite ($CoAs_3$ or $D0_2$ type) structure, $a = 9.036$ A. For CoSb, [12] reported the lattice parameters $a = 3.880$ A, $c = 5.185$ A. In Fig. 294 it has been tentatively assumed that $CoSb_3$ is peritectically formed at ~770°C.

1. N. S. Podkopajew, *Zhur. Russ. Fiz.-Khim. Obshchestva,* **38,** 1906, 463. No detailed information is given.
2. K. Lewkonja, *Z. anorg. Chem.,* **59,** 1908, 305–312.
3. K. Lossew, *Zhur. Russ. Fiz.-Khim. Obshchestva,* **43,** 1911, 375–392.
4. W. Geller, *Arch. Eisenhüttenw.,* **13,** 1939, 263–266.
5. U. Haschimoto, *Nippon Kinzoku Gakkai-Shi,* **1,** 1937, 177–190.
6. N. V. Ageew and E. S. Makarow, *Izvest. Akad. Nauk S.S.S.R. (Khim.),* **1943,** 87–98.
7. U. Fürst and F. Halla, *Z. physik. Chem.,* **B40,** 1938, 285–307.
8. F. Ducelliez, *Compt. rend.,* **147,** 1908, 1048–1050. Melting point of CoSb was found as about 1200°C.
9. W. Köster and E. Wagner, *Z. Metallkunde,* **29,** 1937, 230–232.
10. I. Oftedal, *Z. physik. Chem.,* **128,** 1927, 135–153.
11. W. F. deJong and H. W. V. Willems, *Physica,* **7,** 1927, 74–79.
12. T. Rosenqvist, *Acta Met.,* **1,** 1953, 761–763.

$\bar{1}.8730$
0.1270

Co-Se Cobalt-Selenium

Figure 295 represents the phase relations due to [1]; data points are shown only for the solidification diagram. According to [1], the $\epsilon \rightarrow \alpha$ transformation of Co is raised from 467 to 520°C by 2.8 wt. (2.1 at.) % Se and the $\alpha \rightarrow \epsilon$ transformation lowered from 426 to 400°C by 1.6 wt. (1.2 at.) % Se, indicating that approximately these amounts are soluble at about 520 and 400°C, respectively. On the other hand, [2] found an alloy with 0.25 wt. (0.19 at.) % Se to be heterogeneous and, in agreement herewith, no effect of Se on the transformation points of Co. CoSe (57.26 wt. % Se) [3–5] undergoes a polymorphic transformation at 892°C [1]. The compound Co_2Se (40.11 wt. % Se) [5] does not exist.

CoSe is isotypic with NiAs [6, 7], $a = 3.621$ A, $c = 5.289$ A, $c/a = 1.461$ [6]. $CoSe_2$ has the structure of the FeS_2 (C2) type [8, 9], $a = 5.857$ A [9]. The homogeneity of preparations of the compositions Co_3Se_4 and Co_2Se_3 has not been proved [5].

1. U. Haschimoto, *Nippon Kinzoku Gakkai-Shi,* **2,** 1938, 67–77.
2. W. Köster and E. Horn, *Z. Metallkunde,* **43,** 1952, 333–334.
3. G. Little, *Liebigs Ann.,* **112,** 1859, 211.
4. C. Fabre, *Ann. chim. et phys.,* **10,** 1887, 505.
5. Fonces-Diacon, *Compt. rend.,* **131,** 1900, 704–705.
6. I. Oftedal, *Z. physik. Chem.,* **128,** 1927, 137–138.
7. W. F. deJong and H. W. V. Willems, *Physica,* **7,** 1927, 74–79.
8. W. F. deJong and H. W. V. Willems, *Z. anorg. Chem.,* **170,** 1928, 241–245.
9. S. Tengnér, *Z. anorg. Chem.,* **239,** 1938, 126–132.

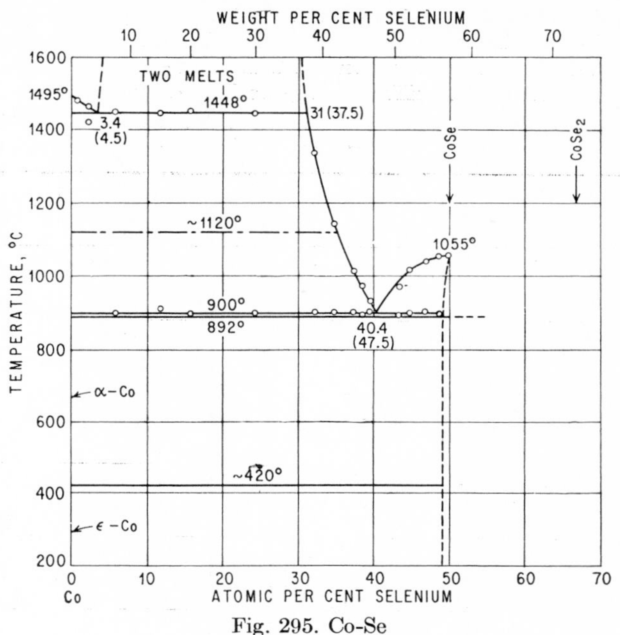

Fig. 295. Co-Se

0.3219
$\bar{1}$.6781

Co-Si Cobalt-Silicon

Except for the partial system Co-Co_2Si (Figs. 296 and 297), the constitution appears to be established.

Solidification. The first phase diagram proposed [2] was characterized by the existence of five intermediate phases: Co_2Si (19.24 wt. % Si), Co_3Si_2 (24.11 wt. % Si), CoSi (32.28 wt. % Si), $CoSi_2$ (48.80 wt. % Si), and $CoSi_3$ (58.84 wt. % Si). Co_2Si, CoSi, and $CoSi_3$ were reported to have maximal melting points of 1327, 1395, and 1306°C, respectively, and $CoSi_2$ to be formed peritectically at 1277°C. Co_3Si_2 was claimed to be formed by the peritectoid reaction $Co_2Si + CoSi \rightarrow Co_3Si_2$ at about 1215°C [2, 3]. The occurrence of Co_2Si [4], CoSi [5], and $CoSi_2$ [6] had already been suggested earlier, however, without conclusive proof of the homogeneity of alloys of these compositions.

The partial system Co-CoSi was reinvestigated by [7], using thermal and microscopic methods (Fig. 297). Two new reactions were detected at about 1210 and 1160°C which were interpreted to correspond to the peritectic formation and eutectoid decomposition of a new phase, Co_3Si (13.71 wt. % Si). The existence of Co_3Si_2 was not confirmed. Instead, the solid-state reaction between Co_2Si and CoSi—at 1215°C according to [2] and at 1208°C according to [7]—was concluded to correspond to the eutectoid decomposition of a solid solution based on a high-temperature modification of Co_2Si. Figure 297 indicates that the transformation point of Co_2Si, assumed at 1320°C (i.e., 12°C below its melting point), is lowered to the eutectoid temperature, 1208°C.

Both of these findings were verified by [1]. A comparison of Figs. 296 and 297 indicates the temperature differences found in the two investigations.

Still another version of the phase relationships in the partial system Co-Co_2Si was proposed by [8]. These authors reported that they did not find evidence for the existence of Co_3Si. Instead, they interpreted the 1160–1170°C horizontal as corresponding to the peritectoid equilibrium $\alpha + Co_2Si \rightleftarrows \epsilon$; see lower part of Fig. 297.

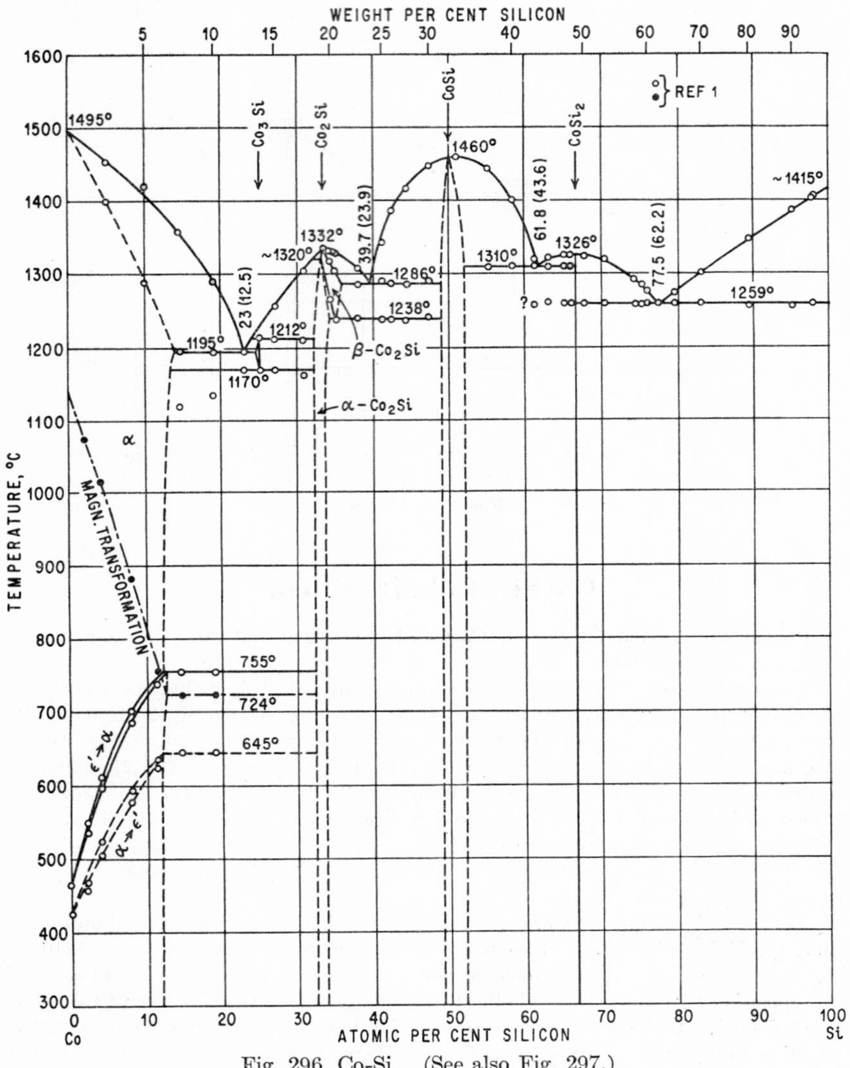

Fig. 296. Co-Si. (See also Fig. 297.)

Although their data appear to indicate that the polymorphic transformation of Co is raised to 1160°C, no explanation can be offered for the phase equilibrium at 1210–1212°C. Additional study is needed to clear up the discrepancies discussed.

The phase diagram of the partial system CoSi-Si (Fig. 296) is based on thermal and microscopic work [1]. It differs from the diagram of [2] in that it does not show the phase $CoSi_3$. The existence of this phase has long appeared doubtful as no other transition metal forms a trisilicide. [9] have verified by X-ray analysis and [10] by microscopic examination that $CoSi_3$ does not occur. Also, $CoSi_2$ was found to melt congruently at 1326°C [1], rather than being formed peritectically at 1277°C [2].

Transformations and Solubilities in the Solid State. The transformations of the intermediate phases have already been dealt with above. The effect of Si on the magnetic and polymorphic changes of Co was studied by [1] and [8]. Results

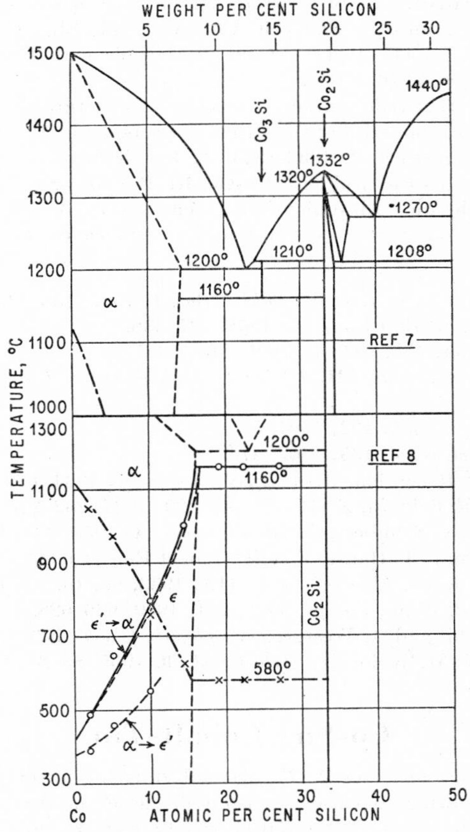

Fig. 297. Co-Si. (See also Fig. 296.)

differ greatly. According to [8], the $\epsilon \rightarrow \alpha$ transformation is raised to 1160°C, giving rise to the peritectoid reaction $\alpha + Co_2Si \rightleftarrows \epsilon$ (Fig. 297). Up to about 10 at. % Si the hysteresis of the transformations $\epsilon' \rightarrow \alpha$ and $\alpha \rightarrow \epsilon'$ increases, but at about 14.5 at. % Si and higher Si contents both transformations were found nearly to coincide; i.e., in this range the transformation apparently takes place by diffusion rather than a shear mechanism.

On the other hand, work by [1] indicated that the temperature of the $\epsilon \rightarrow \alpha$ transformation is increased only up to about 755°C (Fig. 296). Both studies [1, 8] agree in that the Curie point of Co is lowered by Si, although the Curie-point values of the saturated solution differ widely: 580 [8] vs. 724°C [1]. [2] reported the Curie point to be lowered only to approximately 1037°C.

The solid solubility of Si in α-Co and the ranges of homogeneity of Co_2Si and CoSi (Fig. 296) are to be regarded as tentative, although [1] reported some micrographic evidence for the extent of these phase fields. [1, 2] claimed, on the basis of thermal and microscopic data, that there is some solid solubility of Co in Si. X-ray evidence is conflicting [11, 12]; however, it appears quite unlikely that the solubility of Co in Si, if any, is appreciable.

Crystal Structures. Co_2Si is orthorhombic, with a = 7.109 A, b = 4.918 A, c = 3.738 A, and 12 atoms per unit cell (C37 type) [11, 13]. CoSi is isotypic with FeSi (B20 type), a = 4.447 A [11]. $CoSi_2$ is isotypic with CaF_2 (C1 type), with a = 5.367 A [9] or a = 5.376 A [12].

Supplement. [14] redetermined the crystal structure of Co_2Si from single-crystal X-ray data and showed that the structure proposed by [11, 13], which is known as C37 type (*Strukturbericht*, 3, 1933-1935, 32–33), is incorrect. According to the new results, Co_2Si is isostructural with Rh_2Ge, Rh_2B, and δ-Ni_2Si and has a distorted Ni_2In structure. The unit-cell dimensions given by [11, 13] remain valid; however, because of the desire to adhere to convention, the a and b axes have been interchanged by [14].

1. U. Haschimoto, *Nippon Kinzoku Gakkai-Shi,* **1,** 1937, 135–143.
2. K. Lewkonja, *Z. anorg. Chem.*, **59,** 1908, 327–338.
3. L. Baraduc-Muller, *Rev. mét.*, **7,** 1910, 707–711.
4. H. Moissan, *Compt. rend.*, **121,** 1895, 621; E. Vigouroux, *Compt. rend.*, **121,** 1895, 686; **142,** 1906, 635.
5. P. Lebeau, *Compt. rend.*, **132,** 1901, 556.
6. P. Lebeau, *Compt. rend.*, **135,** 1902, 475.
7. R. Vogel and K. Rosenthal, *Arch. Eisenhüttenw.*, **7,** 1934, 689–691.
8. W. Köster and E. Schmid, *Z. Metallkunde,* **29,** 1937, 232–233.
9. H. Pfisterer and K. Schubert, *Z. Metallkunde,* **41,** 1950, 433–441.
10. W. R. Johnson and M. Hansen, AF Technical Report 6383, June, 1951.
11. B. Borén, *Arkiv Kemi, Mineral. Geol.*, **11A,** 1933, no. 10, pp. 17–22.
12. F. Bertaut and P. Blum, *Compt. rend.*, **231,** 1950, 626–628.
13. B. Borén, S. Stahl, and A. Westgren, *Z. physik. Chem.*, **B29,** 1935, 231–235.
14. S. Geller (and V. M. Wolontis), *Acta Cryst.*, **8,** 1955, 83–87.

$\bar{1}$.6960
0.3040

Co-Sn Cobalt-Tin

Results of thermal analyses [1, 2], although differing in temperatures (see data points in Fig. 298) owing to various reasons [3], agree in that three intermediate phases were found to exist: Co_2Sn (50.17 wt. % Sn) [4], melting congruently at 1161 ± 10°C, and CoSn (66.82 wt. % Sn), formed peritectically at 936 ± 9°C and undergoing a polymorphic transformation at 525 ± 10°C. Measurements of the electromotive force [5, 6] indicated the existence of CoSn; see, however, [7].

A reinvestigation of the system by means of roentgenographic analysis [8] showed that there are four instead of three intermediate phases: (*a*) a phase, designated as γ by [8], with a homogeneity range of about 41–42 at. (58.4–59.3 wt.) % Sn [9]; (*b*) a polymorphic modification of this phase, γ′, stable below about 550°C; (*c*) the phase CoSn, being of singular composition; and (*d*) the phase $CoSn_2$ (80.11 wt. % Sn) [10], also of invariable composition. $CoSn_2$ was assumed to be formed peritectically at 525 ± 10°C, considered formerly [1, 2] to be the transformation temperature of CoSn. The temperature of peritectic formation of CoSn—945–950°C according to

[1] and 927°C according to [2]—was reported [8] to be in the vicinity of 850°C (?). The composition of the γ (γ') phase is not compatible with the fact that the maximum in the liquidus curve lies close to the composition Co_2Sn (or perhaps Co_3Sn_2). For this reason the upper portion of the γ-phase field may have the form shown in Fig. 298.

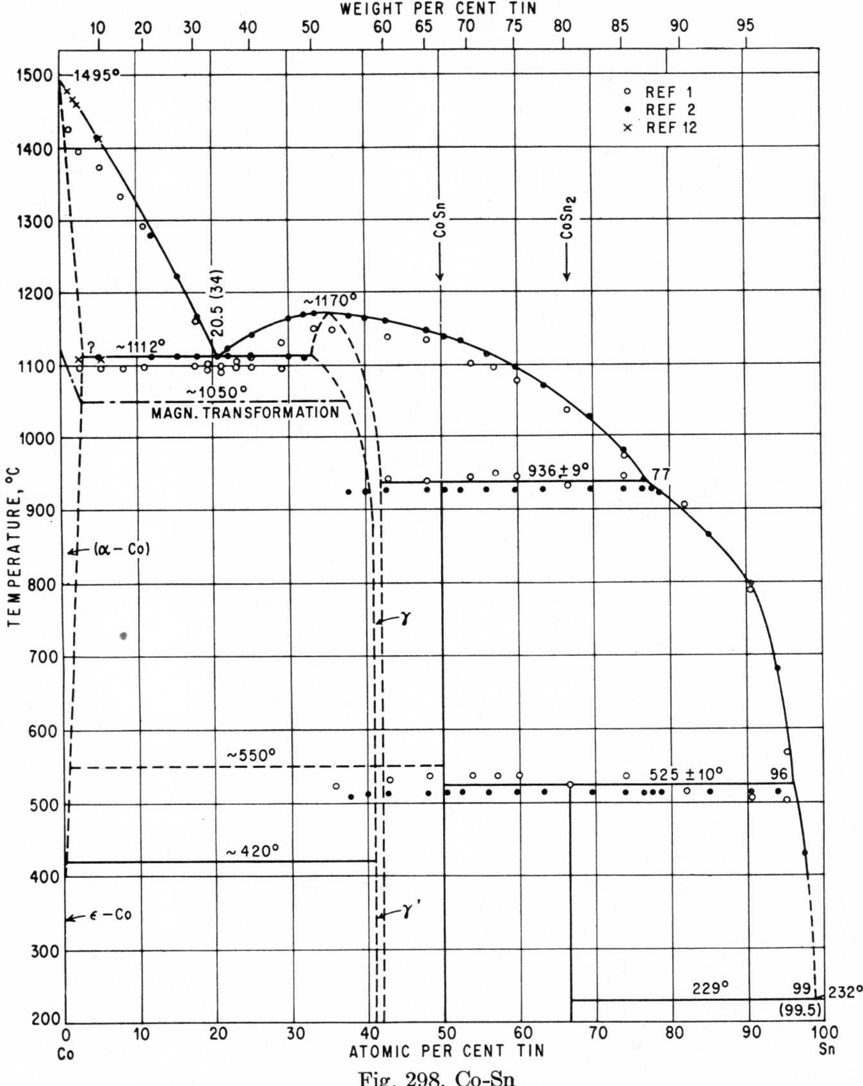

Fig. 298. Co-Sn

The solid solubility of Sn in ϵ- and α-Co has not been determined as yet. [8] estimated it to be about 2.5 at. (4.9 wt.) % at 1000°C and considerably smaller at lower temperatures. On the basis of data concerning the effect of Sn on the transformation points of Co, the solubility was estimated to be about 9 wt. (4.7 at.) %

at 1070°C [11], about 4 wt. (2 at.) % at 1033°C [12], about 5 wt. (2.5 at.) % at 230 ± 110°C [11], and about 3.3–3.5 wt. (1.7–1.8 at.) % Sn at 420 ± 115°C [12]. None of these values represents equilibrium conditions.

Data as to the effect of Sn on the magnetic and polymorphic transformations of Co differ widely. According to [11] and [12] the Curie point is lowered for 50 and 110°C by about 4.7 and 2 at. % Sn, respectively. The $\epsilon \rightleftarrows \alpha$ transformation was found to be lowered, on both heating and cooling, to 340 and 120°C, respectively [11]. However, [12] found it to be raised on heating to 536°C and lowered on cooling to 308°C.

Crystal Structures. The high-temperature γ phase is isotypic with NiAs (B8 type), a = 4.111 A, c = 5.183 A, c/a = 1.261 if saturated with Co at 850°C and a = 4.104 A, c = 5.171 A, c/a = 1.260 if saturated with Sn at 950°C. γ' appears to have an ordered structure based on the NiAs structure, possibly with a = 16.39 A, c = 5.208 A. The structure can also be ascribed to the related orthorhombic cell, with a = 8.20 A, b = 7.09 A, c = 5.21 A. Hexagonal CoSn is isotypic with PtTl (B35 type), a = 5.279 A, c = 4.259 A, c/a = 0.807; and the structure of $CoSn_2$ is tetragonal of the $CuAl_2$ (C16) type, a = 6.361 A, c = 5.452 A, c/a = 0.857 [8].

1. K. Lewkonja, *Z. anorg. Chem.*, **59,** 1908, 294–304.
2. S. F. Zemczuzny and S. W. Belynski, *Z. anorg. Chem.*, **59,** 1908, 364–370.
3. The melting point of Ni was assumed as 1451°C by [1] and as 1484°C by [2]. As a result, the melting point of Co was found as 1440 [1] and 1502°C [2], respectively. The purity of Co was given by [1] as 98.04% Co, 1.62% Ni, 0.17% Fe; [2] used Co of unknown purity. Since above 1000°C the temperatures reported by [2] lie higher and below 1000°C lower than those found by [1], other factors must have played a role.
4. Earlier, a phase of this composition had been isolated from alloys with 81–92 wt. % Sn by F. Ducelliez (*Compt. rend.*, **144,** 1907, 1432–1434; **145,** 1907, 431–433, 502–504). This was an accidental result since the compound $CoSn_2$ is present in the alloys used.
5. N. A. Puschin, *Zhur. Russ. Fiz.-Khim. Obshchestva*, **39,** 1907, 884.
6. F. Ducelliez, *Compt. rend.*, **150,** 1910, 98–101.
7. G. Tammann and A. Koch, *Z. anorg. Chem.*, **133,** 1924, 179–186.
8. O. Nial, *Z. anorg. Chem.*, **238,** 1938, 287–296; Dissertation, Stockholm, 1945, pp. 26–37; *Svensk Kem. Tidskr.*, **59,** 1947, 168.
9. Crystals of approximately this composition had been isolated from alloys with 9–57 wt. % Sn by F. Ducelliez; see [4].
10. The existence of $CoSn_2$ was already reported by A. S. Russell, T. R. Kennedy, and R. P. Lawrence, *J. Chem. Soc.*, **1934,** 1750–1754.
11. W. Köster and E. Wagner, *Z. Metallkunde*, **29,** 1937, 230–232.
12. U. Haschimoto, *Nippon Kinzoku Gakkai-Shi*, **2,** 1938, 67–77.

$\bar{1}$.5130
0.4870

Co-Ta Cobalt-Tantalum

The constitution was investigated by thermal analysis up to about 17 at. (38.5 wt.) % Ta [1] and up to about 50 at. (75.4 wt.) % Ta [2] (Fig. 299). Both studies showed that a eutectic exists which was located at 31 wt. (12.8 at.) % Ta, 1275°C [1], or 33 wt. (13.9 at.) % Ta, 1278°C [2]. Whereas [1] assumed the existence of Ta_2Co_5 [28.57 at. (55.11 wt.) % Ta] as the phase coexisting with the Co-rich solid solution, [2] concluded this phase to correspond to the composition TaCo (75.42 wt. % Ta). Neither composition is compatible with the fact that an AB_2 (Laves) phase, $TaCo_2$ (60.54 wt. % Ta), was found to occur [3, 4]. It has been reported [3, 4] that the phase

$TaCo_2$ can exist in two modifications, one with the ideal composition (33.3 at. % Ta), having the cubic structure of the $MgCu_2$ (C15) type, and the other, containing about 6 at. % Ta less, with the hexagonal structure of the $MgNi_2$ (C36) type. No indication of the latter phase was found by thermal analysis [3]; however, it was emphasized [3, 4] that X-ray evidence is beyond doubt. Until the findings by [3, 4] have been checked, it appears justified [5] to present a phase diagram showing only the $TaCo_2$ phase at the ideal composition (Fig. 299). (See Supplement.)

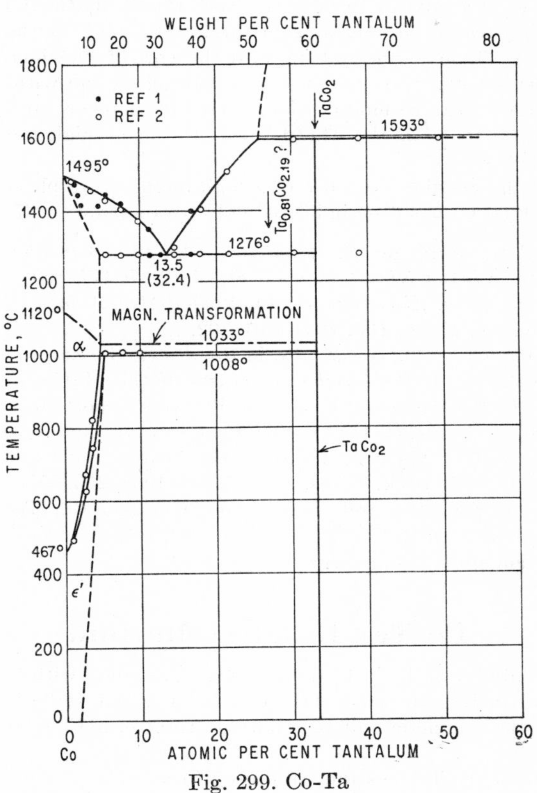

Fig. 299. Co-Ta

Both [1] and [2] have reported that the Curie point of Co is lowered by Ta to 1035 or 1031°C, respectively. However, there is great discrepancy as to the effect of Ta on the temperature of the $\epsilon \rightleftarrows \alpha$ transformation of Co. Whereas [1] found this transformation to be lowered, on both heating and cooling, the data by [2] show that the temperature is raised, on heating as well as cooling. In Fig. 299 the findings by [2] have been considered.

The solid solubility of Ta in ϵ-Co and α-Co has not been determined as yet. On the basis of the curve of the Curie temperatures the solubility limit at about 1000°C would lie between about 8.5 and 13 wt. (3–4.6 at.) % Ta [1, 2]. With fall in temperature the solubility decreases to at least 6 wt. (2 at.) % Ta [1], probably considerably lower.

The lattice spacings of the two "modifications" of $TaCo_2$ at 33.3 and 27.0 at. % Ta were reported [4] to be as follows: $MgCu_2$ type (33.3 at. % Ta), a = 6.732 A;

$MgNi_2$ type (27.0 at. % Ta), a = 4.731 A, c = 15.42 A, $c/2a$ = 1.63 [5]. A brief examination did not indicate the existence of a σ phase [6].

Supplement. [7] prepared two alloys for investigation, $TaCo_2$ and $Ta_{0.81}Co_{2.19}$ (27 at. % Ta). "The incipient melting temperature of $TaCo_2$ was determined as 1610°C. $TaCo_2$ was found to be isomorphous with $MgCu_2$, with a lattice parameter of 6.745 kX. There is no allotropy in the temperature range 600 to 1600°C [8]. The $Ta_{0.81}Co_{2.19}$ alloy was predominantly the $MgCu_2$ modification at 800°C. At 1000°C there was evidence of a mixture of both the $MgCu_2$ and $MgZn_2$ (C14) modifications. At 1200°C the alloy was a pure $MgZn_2$ structure. At 1400°C [8] the alloy was a pure $MgCu_2$ structure. These findings substantiate the work of [4], who reported allotropy in the off-composition alloy; however, they are in disagreement with [4] as to the crystal structure of the allotropic modification. Such phenomena can best be explained in assuming the existence of the $MgZn_2$ modification at high temperature on the Co-rich side of $TaCo_2$."

In Fig. 299 the possible existence of a high-temperature phase (stable between about 1000 and 1300°C) has been indicated only by an arrow at 27 at. % Ta.

1. W. Köster and W. Mulfinger, *Z. Metallkunde*, **30,** 1938, 348–350.
2. U. Haschimoto, *Nippon Kinzoku Gakkai-Shi*, **1,** 1937, 177–190.
3. H. J. Wallbaum, *Arch. Eisenhüttenw.*, **14,** 1940-1941, 521–526.
4. H. J. Wallbaum, *Z. Krist.*, **103,** 1941, 391–402.
5. It has been claimed by [3, 4] that also in the system Fe-Zr there are two phases, with 33.3 and 27 at. % Zr, having structures of the $MgCu_2$ and $MgNi_2$ types, respectively. C. B. Jordan and P. Duwez (Progress Report 20-196, June 16, 1953, ORDCIT Project Contract DA-04-495-Ord 18) were unable to verify this and found only the $MgCu_2$ type of structure at both compositions; see Fe-Zr.
6. P. Greenfield and P. A. Beck, *Trans. AIME*, **200,** 1954, 253–257.
7. R. P. Elliott, Armour Research Foundation, Chicago, Ill., Technical Report 1, OSR Technical Note OSR-TN-247, August, 1954, p. 22.
8. X-ray diffraction study of quenched samples.

$\bar{1}.6645$
0.3355

Co-Te Cobalt-Tellurium

The solid solubility of Te in Co is only slight; an alloy with 0.25 wt. (0.12 at.) % Te was found to be two-phase (heat-treatment is not stated). Therefore, the temperatures of the magnetic and polymorphic transformations are not noticeably affected [1].

CoTe (68.40 wt. % Te) has the structure of the NiAs (B8) type, a = 3.894 A, c = 5.371 A, c/a = 1.380 [2]. The structure of $CoTe_2$ (81.24 wt. % Te), which is of the CdI_2 (C6) type, forms continuously from the CoTe structure by partial loss of metal atoms. From CoTe to $CoTe_2$ lattice dimensions change from a = 3.890 A, c = 5.378 A, c/a = 1.383 to a = 3.792 A, c = 5.414 A, c/a = 1.428 [3]. On annealing at 250°C the CdI_2 structure of $CoTe_2$ transforms into the FeS_2 (C18) type of structure, a = 3.890 A, b = 5.312, c = 6.311 A [3].

Note Added in Proof. Ferromagnetism observed by [4, 5] in preparations of the composition CoTe could be traced by [6, 7] to the presence of metallic Co. A careful roentgenographic and magnetic reinvestigation of the system between 33.3 and 68.8 at. % Te has been carried out by [7]. Their results largely corroborate those of [3] but deviate in important details: (a) The phase field of the B8-C6 phase (β) does not include the CoTe and $CoTe_2$ compositions but extends at 600°C from 54.5 to 64.3 at. % Te and at 335°C from 56.5 to 63 at. % Te only. The lattice parameters change from a = 3.8937 A, c = 5.3763 A, at 54.55 at. % to a = 3.8017 A, c = 5.4094

A, at 64.29 at. % Te. (*b*) The C18-type phase (γ) possesses a narrow range of homogeneity (66.7–67.4 at. % Te) and is the only phase stable at the $CoTe_2$ composition. At the stoichiometric composition the lattice parameters are a = 6.3185 A, b = 5.3189 A, c = 3.8970 A. The solidus point of the CoTe alloy is 968°C; the melting point of $CoTe_2$ > 770°C [7].

1. W. Köster and E. Horn, *Z. Metallkunde*, **43**, 1952, 333–334.
2. I. Oftedal, *Z. physik. Chem.*, **128**, 1927, 135–153.
3. S. Tengnér, *Z. anorg. Chem.*, **239**, 1938, 126–132; *Naturwissenschaften*, **26**, 1938, 429.
4. F. M. Galperin and T. M. Perekalina, *Doklady Akad. Nauk S.S.S.R.*, **69**, 1949, 19 (quoted from [7]).
5. E. Uchida, *J. Phys. Soc. Japan*, **10**, 1955, 517–522; *Met. Abstr.*, **23**, 1956, 533.
6. E. Uchida, *J. Phys. Soc. Japan*, **11**, 1956, 465–466; *Met. Abstr.*, **24**, 1956, 20.
7. J. Haraldsen, F. Grønvold, and T. Hurlen, *Z. anorg. Chem.*, **283**, 1956, 143–164.

$\bar{1}.4047$
0.5953

Co-Th Cobalt-Thorium

The following intermediate phases have been identified by means of X-ray powder and single-crystal work [1]: Th_2Co_{17} [10.53 at. (31.66 wt.) % Th], monoclinic, a = 9.62 A, b = 8.46 A, c = 6.32 A, β = 99°06′, structure closely related to that of $ThCo_5$; $ThCo_5$ [16.67 at. (44.06 wt.) % Th], hexagonal, a = 5.01 A, c = 3.97 A, isotypic with $CaZn_5$; $ThCo_{2-3}$, hexagonal, a = 5.03 A, c = 8 or 9 × 24.54 A; ThCo (80.03 wt. % Th), orthorhombic, a = 3.74 A, b = 10.88 A, c = 4.16 A, 8 atoms per unit cell; Th_7Co_3 (90.81 wt. % Th), hexagonal, a = 9.83 A, c = 6.17 A, c/a = 0.628, 20 atoms per unit cell. [2] reported the lattice parameters a = 4.950 A, c = 4.039 A, c/a = 0.816 for $ThCo_5$.

1. J. V. Florio, N. C. Baenziger, and R. E. Rundle, *Acta Cryst.*, **9**, 1956, 367–372. See also the earlier publications: N. C. Baenziger, *U.S. Atomic Energy Commission, Publ.* AECD-3237, 1948; J. V. Florio, R. E. Rundle, and A. I. Snow, *Acta Cryst.*, **5**, 1952, 449–457.
2. T. Heumann, *Nachr. Akad. Wiss. Göttingen, Math.-physik. Kl.*, **1948**, 21–26.

0.0901
$\bar{1}.9099$

Co-Ti Cobalt-Titanium

Cobalt-rich Alloys. The constitution of the Co-rich alloys, based on thermal analysis and microscopic investigation [1] of alloys prepared using Ti of 95% purity, is shown in Fig. 300. The existence of $TiCo_2$ (28.89 wt. % Ti) [2, 3] and TiCo (44.83 wt. % Ti) [4, 3] was fully established by roentgenographic work. Earlier, the phase in equilibrium with the Co-rich solid solution [5] was suggested to be $TiCo_3$ [6].

According to the thermal effects reported [1], TiCo is formed by a peritectic reaction: liq. [41 wt. (46 at.) % Ti] + $Ti_2Co \rightarrow$ TiCo, at 1450°C. A similar reaction was claimed to exist [1] in the system Ni-Ti. However, since a later investigation has shown that TiNi has a maximum melting point, there may be considerable doubt about the peritectic formation of TiCo [7]. As to the occurrence of two modifications of $TiCo_2$, see below.

There is no reliable information concerning the solid solubility of Ti in Co. After annealing at 1000–1100°C and furnace cooling, approximately 7.2 wt. (8.7 at.) % Ti was found to be soluble [8]. Information on the effect of Ti on the polymorphic transformation is conflicting. Whereas [8] found that both the $\epsilon \rightarrow \alpha$ and $\alpha \rightarrow \epsilon$

transformations are lowered by Ti additions (Fig. 300), [9] reported the data presented in the inset of Fig. 300. The Curie point of Co is lowered to 890°C by 7.2 wt. (8.7 at.) % Ti [8] or to 908°C by 6.2 wt. (7.5 at.) % Ti [9].

Titanium-rich Alloys. As a systematic study is lacking, information on this portion of the diagram is very limited. [10] reported that about 5 wt. (4.1 at.) % Co lowers the melting point of Ti to approximately 1500°C. [1] claimed the existence of a eutectic, Ti-Ti_2Co, of unknown temperature and composition. According to

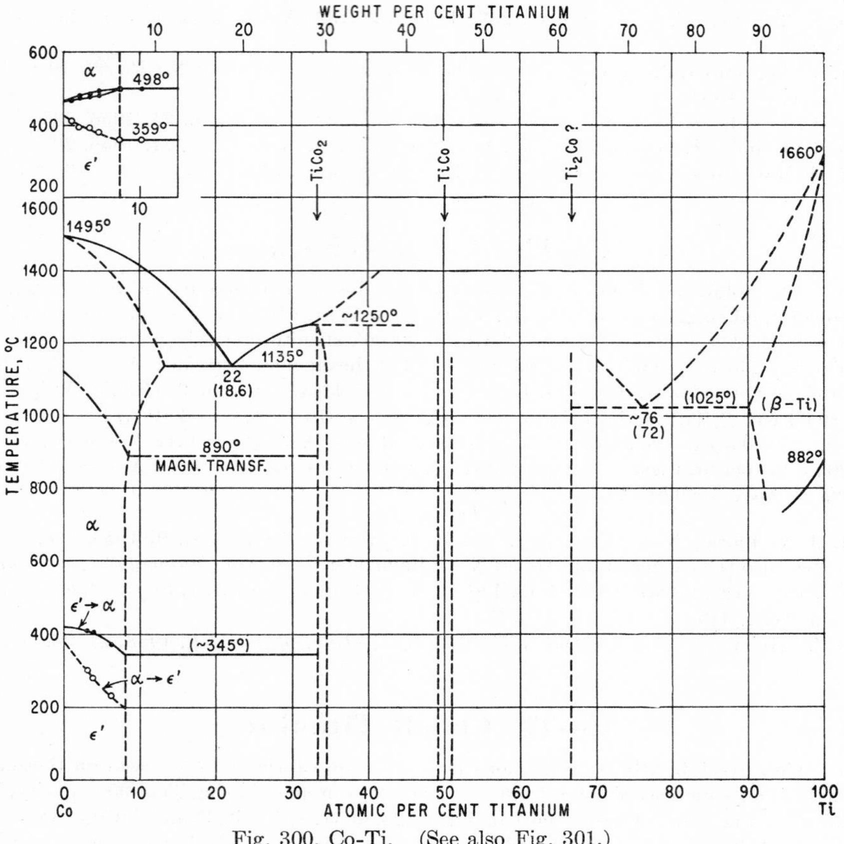

Fig. 300. Co-Ti. (See also Fig. 301.)

[11], this eutectic lies at about 72 wt. (76 at.) % Ti, 1025°C. The existence of a eutectic temperature between 1050 and 1200°C can also be concluded from sintering tests [12]. Whether a compound Ti_2Co actually exists appears doubtful and can be decided only by additional work (see Crystal Structures).

The boundaries of the (α-Ti + β-Ti) field were outlined by metallographic examination of magnesium-reduced Ti-base alloys up to 3 wt. (2.5 at.) % Co, quenched from temperatures between 790 and 900°C after a ½-hr anneal [13]. The alloys contained 0.3 wt. % W, on the average. [14] determined the effect of Co on the transformation point of Ti down to 735°C by measuring the hydrogen pressure in equilibrium with an extremely dilute solution of hydrogen in iodide titanium-base alloys as a function of temperature. This curve is shown in Fig. 300.

The solubility of Co in β-Ti at the eutectic temperature of about 1025°C was tentatively given as "in the vicinity of 6 wt. (5 at.) % Co." Considering the transformation curve by [14], however, the solubility limit would lie at a higher Co content. The temperature of the eutectoid α-Ti + Ti_2Co (?) must lie below 735°C [14].

Crystal Structures. The existence of two structural types of the phase $TiCo_2$, both stable at room temperature, was claimed by [2]: the $MgCu_2$ (C15) type stable in an excess of Ti (a = 6.704 A) and the $MgNi_2$ (C36) type stable in an excess of Co (a = 4.724 A, c = 15.40 A, c/a = 3.26). However, [3] found only the $MgNi_2$ type of structure, a = 4.729 A, c = 15.423 A, c/a = 3.261. TiCo is b.c.c. [4, 3] with a = 2.994 A [3]; no superlattice lines were found [3]. "This may be a result of too small a difference between the atomic scattering power of the components, and it does not prove the absence of ordering" (a = 2.991 A) [19].

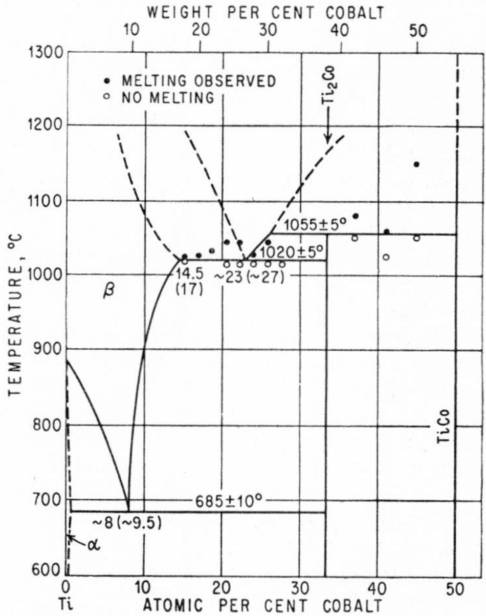

Fig. 301. Co-Ti. (See also Fig. 300.)

The phase Ti_2Co was reported to have a f.c.c. structure with 96 atoms per unit cell [4, 3], a = 11.306 A [3]. On the other hand, [15] found an alloy of the composition Ti_4Co_2O to have a structure of the Fe_3W_3C type (a = 11.318 A), the powder pattern of which is similar to that of the Ti_2Co phase reported by [3]. He was unable to find evidence for the existence of Ti_2Co. An alloy of this composition showed neither the Fe_3W_3C type of structure nor the b.c.c. structure of the TiCo phase. The predominant structure for the composition Ti_2Co was found to agree with that reported by [2] for the $TiCo_2$ phase having an excess of Ti ($MgCu_2$ type). Additional work is necessary to clear up the problem as to the existence of Ti_2Co (see Supplement).

Supplement. [16] stated: "In contradiction to previous work . . . it was found that $TiCo_2$ occurs only in the $MgCu_2$ modification with a lattice parameter of a = 6.692 kX. Extra lines are in several patterns, but these do not fit the $MgZn_2$ or $MgNi_2$ lattices. These lines may be accounted for by either the b.c.c. TiCo phase or h.c.p. Co. There is no allotropy in the temperature range 600 to 1200°C." (X-ray examination of quenched samples.)

[17] have thoroughly delineated the phase relationships in the 1–55 wt. % Co range. The constitution is shown in Fig. 301. Alloys were prepared from iodide Ti and 99.5 wt. % purity Co by arc melting. Metallographic and X-ray diffraction analysis of cast and heat-treated samples served as the principal tools of investigation.

The solubility of Co in α-Ti is less than 1 wt. (0.8 at.) % at 685°C; no alloys more dilute than 1 wt. % Co were prepared. The melting point of TiCo was not determined. Figure 301 is incompatible with the view of [1] that TiCo is formed by a peritectic reaction. Accordingly, it is still impossible to draw the entire diagram because of the uncertainty of this central region. It is believed that the relationships proposed by [17] are correct and that TiCo melts with an open maximum.

The work of [17] strongly supports the reported existence of Ti_2Co. In the discussion to [17], [18] cited additional work which confirms the existence of this phase and reverses his earlier statements [15].

1. H. J. Wallbaum, *Arch. Eisenhüttenw.*, **14,** 1940-1941, 521–526.
2. H. Witte and H. J. Wallbaum, *Z. Metallkunde,* **30,** 1938, 102; H. J. Wallbaum and H. Witte, *Z. Metallkunde,* **31,** 1939, 185–187.
3. P. Duwez and J. L. Taylor, *Trans. AIME,* **188,** 1950, 1173–1176.
4. F. Laves and H. J. Wallbaum, *Naturwissenschaften,* **27,** 1939, 674–675.
5. B. Egeberg, *Abhandl. Inst. Metallhütt. Elektrochem. T. H. Aachen,* **1,** 1915, 37–54.
6. W. Köster and W. Geller, *Arch. Eisenhüttenw.*, **8,** 1934-1935, 471–472; W. Köster and E. Wagner, *Z. Metallkunde,* **29,** 1937, 231.
7. Apparently Wallbaum's thermal data were influenced to a considerable degree by the presence of contaminants resulting from the use of impure Ti and crucibles which reacted with the melt.
8. W. Köster and E. Wagner, *Z. Metallkunde,* **29,** 1937, 230–232.
9. U. Haschimoto, *Nippon Kinzoku Gakkai-Shi,* **2,** 1938, 67–77.
10. W. Kroll, *Z. Metallkunde,* **29,** 1937, 189–192.
11. F. L. Orrell, Ohio State University, private communication, May, 1952.
12. E. Larsen, E. Swazy, L. Busch, and R. Freyer, *Metal Progr.*, **55,** 1949, 359–361.
13. C. M. Craighead, O. W. Simmons, and L. W. Eastwood, *Trans. AIME,* **188,** 1950, 485–513.
14. A. D. McQuillan, *J. Inst. Metals,* **80,** 1952, 363–368.
15. W. Rostoker, *Trans. AIME,* **194,** 1952, 209–210.
16. R. P. Elliott, Armour Research Foundation, Chicago, Ill., Technical Report 1, OSR Technical Note OSR-TN-247, August, 1954, pp. 26–27.
17. F. L. Orrell and M. G. Fontana, *Trans. ASM,* **47,** 1955, 554–564; discussion, p. 564.
18. W. Rostoker, see discussion on [17].
19. T. V. Philip and P. A. Beck, *Trans. AIME,* **209,** 1957, 1269–1271.

$\bar{1}.4600$
0.5400

Co-Tl Cobalt-Thallium

According to work by [1], it is likely that the phase diagram is similar to that of Co-Pb, with the exception that the portion above the boiling point of Tl (1460°C) cannot be realized under atmospheric pressure. It was found that pulverized Co dissolved in molten Tl at 900°C up to 2.9 wt. (0.86 at.) %. This percentage caused lowering of the melting and transformation points of Tl by 6 and 8°C, respectively. An alloy with 2.5 wt. (0.75 at.) % Co consisted of primary Tl crystals in a eutectic matrix; with a slightly higher Co content, two layers were formed.

1. K. Lewkonja, *Z. anorg. Chem.*, **59,** 1908, 318–319.

$\bar{1}.3937$
0.6063

Co-U Cobalt-Uranium

The phase diagram shown in Fig. 302 is due to [1]. No information on the experimental details was available.

The existence of the three intermediate phases UCo_2 (66.88 wt. % U), UCo (80.16 wt. % U), and U_6Co [85.71 at. (96.03 wt.) % U] has been corroborated by determination of their crystal structures [2]. UCo_2 is f.c.c. of the $MgCu_2$ (C15) type, $a = 6.9924$ A. At 1000°C the limit of its solid solution with Co lies at 29.2 at. % U. UCo is b.c.c. with a unique structure, eight groups UCo per unit cell, $a = 6.3557$ A. U_6Co is b.c. tetragonal of a new structural type with $4U_6Co$ per unit cell, $a = 10.36$ A, $c = 5.21$ A, $c/a = 0.503$.

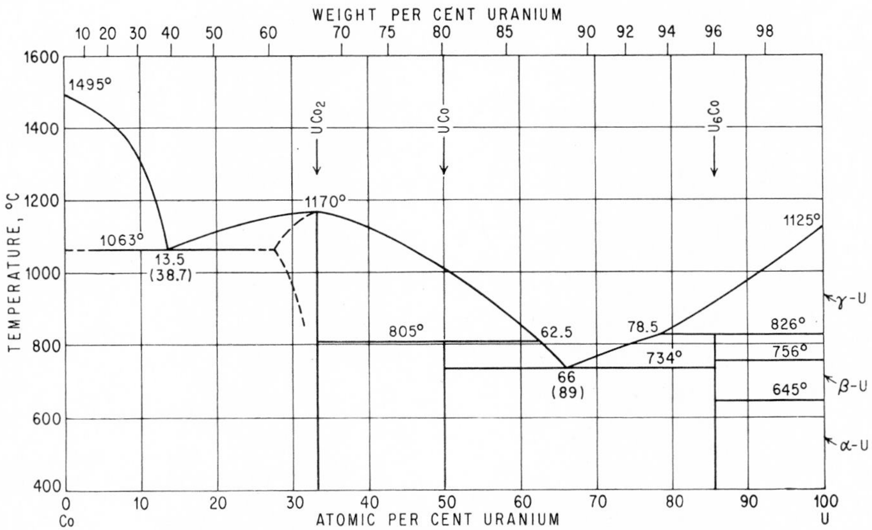

Fig. 302. Co-U. (See also Note Added in Proof.)

Note Added in Proof. A nearly complete Co-U phase diagram has also been established by [3]. The results of the two groups of investigators are in agreement on most of the system, but differences occur at the high-Co end. Thus, [3] assume the existence of an additional compound at about 26 at. % U stable up to just above 800°C. For further details reference is made to the compromise diagram recommended by [4].

1. W. K. Noyce and A. H. Daane, *U.S. Atomic Energy Commission, Publ.* AECD-2826, 1946. Diagram was taken from "Annual Review of Nuclear Science," vol. 1, p. 448, Stanford, 1952.
2. N. C. Baenziger, R. E. Rundle, A. I. Snow, and A. S. Wilson, *Acta Cryst.*, **3**, 1950, 34–40.
3. Atomic Energy Research Establishment, United Kingdom, unpublished information, July, 1952; quoted in [4].
4. H. A. Saller and F. A. Rough, Compilation of U.S. and U.K. U and Th Constitutional Diagrams, *U.S. Atomic Energy Commission, Publ.* BMI-1000, June, 1955.

0.0633
$\bar{1}$.9367

Co-V Cobalt-Vanadium

The effect of V on the Curie point and $\epsilon \rightleftarrows \alpha$ transformation of Co was determined by [1, 2]. Results of [1] are shown in Fig. 303. According to [2], 15 at. % V lowers the Curie point to 921°C, the $\alpha \rightarrow \epsilon$ transformation from 426 to 220°C, and raises the $\epsilon \rightarrow \alpha$ transformation from 467 to 707°C. The limit of the solid solubility of V in α-Co at 900°C lies close to 23 wt. (25.7 at.) % V [3].

The existence of a σ phase, predicted by [4], has been established [5–9]. Its homogeneity range was given as 40–45 to > 54 at. % V at 600–700°C [8] and 44.4–

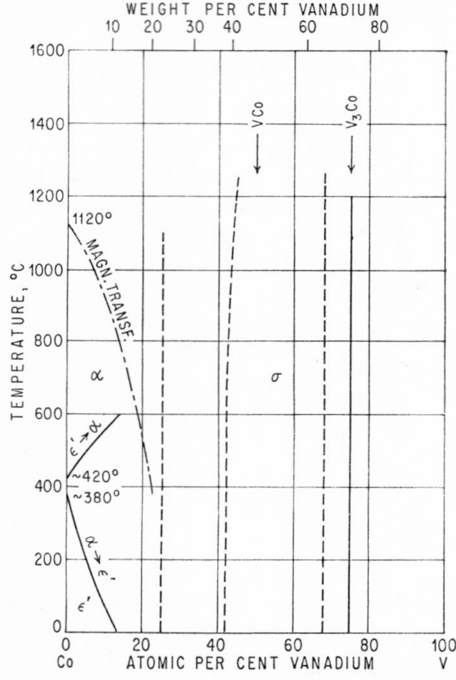

Fig. 303. Co-V. (See also Fig. 304.)

53.1 at. % V at 1200°C [9] (see Supplement). A tentative tetragonal cell was suggested by [5], $a = 6.246$ A, $c = 9.104$ A, $c/a = 1.458$. [6] have tried an orthorhombic indexing of the σ phase.

The phase V_3Co (72.17 wt. % V) was detected by [10]. The structure is of the "β-W" (A15) type, $a = 4.675$ A.

Supplement. In a correction to [9], it was stated [11] that the V-rich limit of the σ phase lies actually at 68 at. % V at 1200°C (Fig. 303). [12] studied an as-cast alloy of the composition VCo_2: "Metallographic examination showed this alloy to be two-phase. X-ray examination proved this to be body-centered cubic with a small amount of a second phase, dissimilar to the Laves-type."

The phase diagram of Fig. 304 was established by [13] using thermal, micrographic, X-ray, and magnetic analyses as principal tools. In the neighborhood of 25 at. % V the f.c.c. α solid solution becomes ordered below 1070°C: (VCo_3) phase. Very weak superstructure lines could be observed.

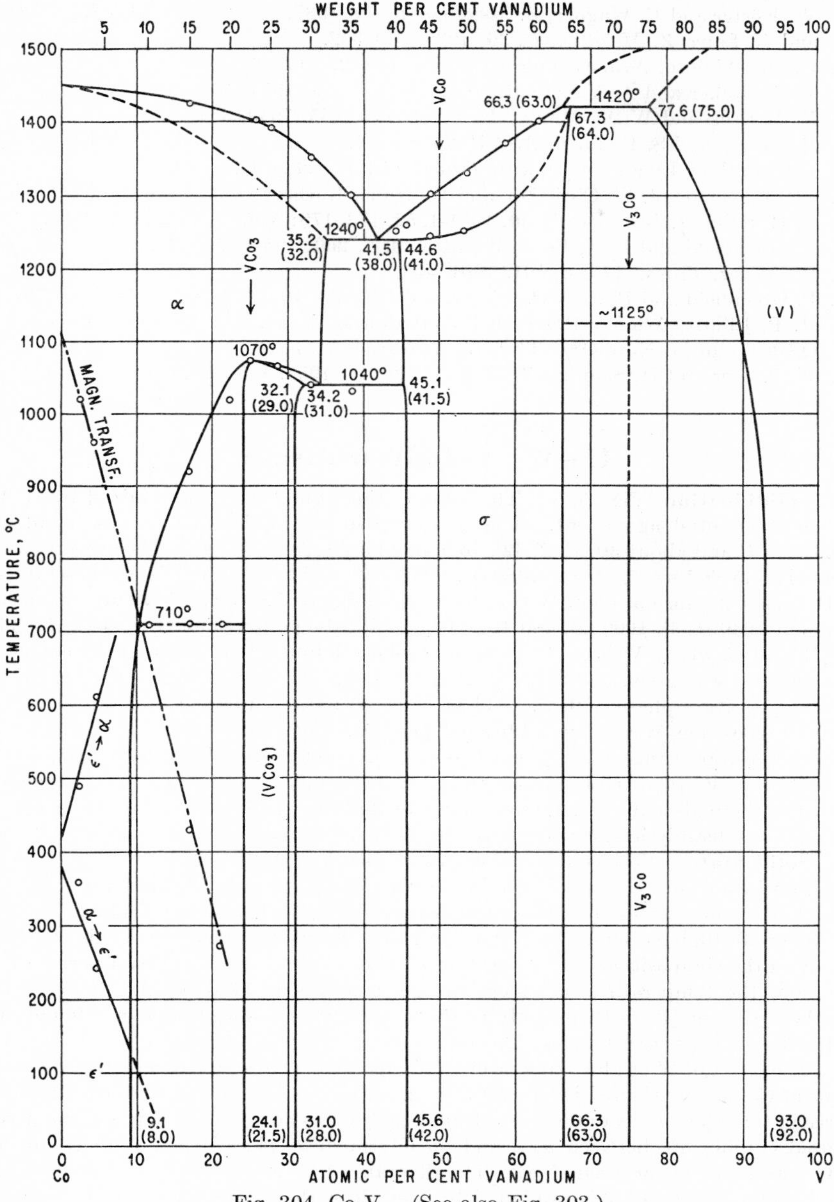

Fig. 304. Co-V. (See also Fig. 303.)

1. W. Köster and E. Wagner, *Z. Metallkunde,* **29,** 1937, 230–232; see also W. Köster and K. Lang, *Z. Metallkunde,* **30,** 1938, 350–352.
2. U. Haschimoto, *Nippon Kinzoku Gakkai-Shi,* **2,** 1938, 67–77.
3. W. Rostoker and A. Yamamoto, *Trans. ASM,* **46,** 1954, 1136–1163.
4. P. A. Beck and W. D. Manly, *Trans. AIME,* **185,** 1949, 354.
5. P. Duwez and S. R. Baen, *ASTM Spec. Tech. Publ.* 110, 1950, pp. 48–54.
6. P. Pietrokowsky and P. Duwez, *Trans. AIME,* **188,** 1950, 1283–1284.
7. W. B. Pearson, J. W. Christian, and W. Hume-Rothery, *Nature,* **167,** 1951, 110.
8. A. H. Sully, *J. Inst. Metals,* **80,** 1951-1952, 173–179; *Nature,* **167,** 1951, 365–366.
9. P. Greenfield and P. A. Beck, *Trans. AIME,* **200,** 1954, 253–257.
10. P. Duwez, *Trans. AIME,* **191,** 1951, 564.
11. P. Greenfield and P. A. Beck, *Trans. AIME,* **200,** 1954, 758 (work by J. Darby).
12. R. P. Elliott, Armour Research Foundation, Chicago, Ill., Technical Report 1, OSR Technical Note OSR-TN-247, August, 1954, p. 30.
13. W. Köster and H. Schmid, *Z. Metallkunde,* **46,** 1955, 195–197.

$\bar{1}.5058$
0.4942

Co-W Cobalt-Wolfram

Solidification (Fig. 305). The solidification equilibria as determined by [1–4] are in substantial agreement. The maximum in the liquidus curve was found at 1503 ± 5°C and about 30 [1], 27 [2], 16.5 [3], and 25 wt. % W [4]. The eutectic was located at 45 ± 1 wt. % W and 1480 [1], 1465 [2, 3], and 1477°C [4]. The temperature of the peritectic formation of W_6Co_7, formerly designated as WCo [1–4], was reported as about 1630 [1, 3], 1690 [2], and 1618°C [4], with the peritectic melt at about 51 [2], 61 [3], and 65 wt. % W [4]. The maximal solid solubility of W in Co at the eutectic temperature was given as 38 ± 2 wt. % W [1–4]. As [5] reported that there is a high-temperature phase with the σ-phase type structure, apparently of the approximate composition W_3Co_2, an additional peritectic horizontal would be required in this composition range (see system Co-Mo). Approximate solidus temperatures of sintered samples reported by [6, 7] can be regarded only as qualitative; in agreement with the thermal-analysis data they indicate that the solidus temperature does not change considerably in the range 0–70 wt. % W.

Solid State. Besides the phases W_6Co_7 [46.15 at. (72.77 wt.) % W] [8, 9] and σ [5], an intermediate phase richer in Co was found to exist. According to [2, 4], this phase approximates in composition the formula W_2Co_7 [22.22 at. (47.13 wt.) % W]. However, [8, 10] have shown by X-ray work that the crystal structure of this phase is based on the composition WCo_3 (50.98 wt. % W). WCo_3 is formed by a peritectoid reaction, the temperature of which was given as about 1100 [2] and 1065°C [3]. Neither [11] nor [4], in their study of the solid-state transformations, has found an indication of the peritectoid equilibrium involving the phase WCo_3.

The effect of W on the temperature of the polymorphic and magnetic transformations of Co, not considered in the work by [2], was investigated by [11, 3, 4]. Results agree in that they indicate the $\epsilon \rightarrow \alpha$ transformation to be raised, giving rise to a peritectoid equilibrium between α, ϵ, and an intermediate phase (then assumed to be WCo) at 1040 [11] or 1060°C [4]. The diagram of [3] also exhibits the peritectoid equilibrium; however, it is presented to coincide with the peritectoid formation of WCo_3 in a manner that is in conflict with the phase rule.

The solid solubility of W in α-Co was outlined by [2], on the basis of micrographic (1100–850°C) and change-of-electrical-resistance data (800–550°C). It was found to decrease from 32 wt. (13 at.) % W at 1100°C to about 21.5, 16.5, 10.5, 6.5, and 4.5 wt. (8, 6, 3.6, 2.2, and 1.5 at.) % W at 1050, 1000, 900, 800, and 700°C, respectively (see inset in Fig. 305). This solubility curve is incompatible with the $\epsilon \rightarrow \alpha$ transformation temperatures reported by [11, 3, 4] as it lies above the $\epsilon \rightarrow \alpha$ transformation curves shown in Fig. 305. Further work is necessary to resolve this discrepancy. If the solubility curve according to [2] is correct, the $\epsilon \rightarrow \alpha$ transformation would represent metastable conditions. This is difficult to conceive of since one could expect that the $\epsilon \rightarrow \alpha$ transformation approaches equilibrium, at least at temperatures above about 900°C.

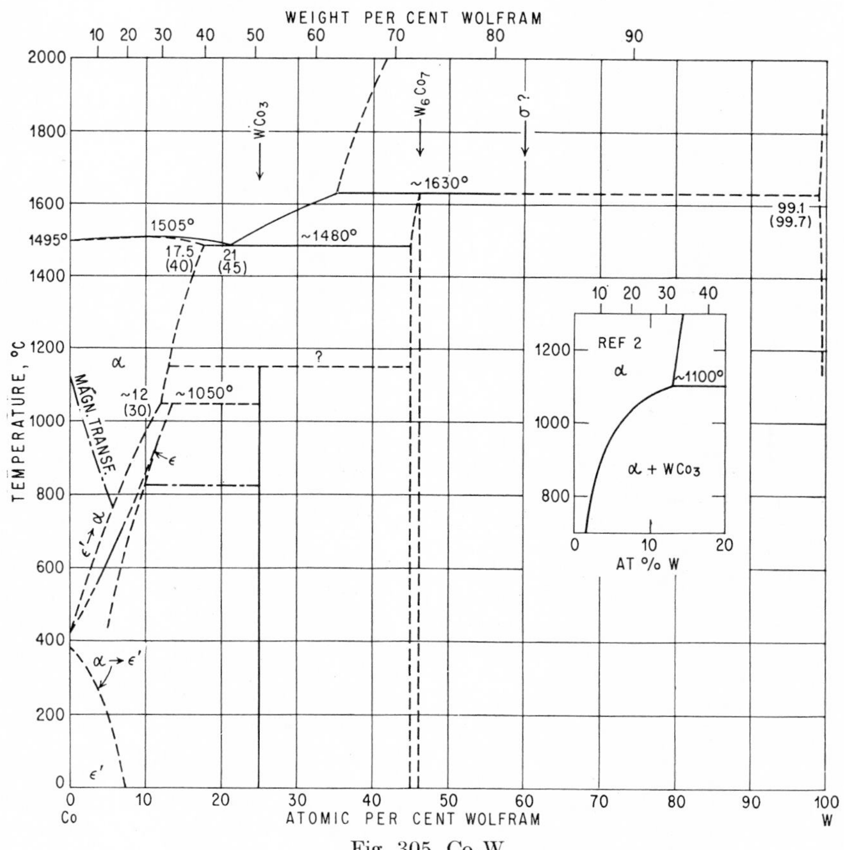

Fig. 305. Co-W

The martensitic-type transformation $\alpha \rightarrow \epsilon'$ was determined by [3, 4]. The curve in Fig. 305 is a compromise of these data.

Crystal Structures. WCo_3 is hexagonal of the Ni_3Sn ($D0_{19}$) type, $a = 5.13$ A, $c = 4.13$ A, $c/a = 0.805$ [8, 10]. W_6Co_7, having a certain range of homogeneity, is rhombohedral-hexagonal and isotypic with W_6Fe_7 ($D8_5$ type) [8, 9]. Parameters are $a = 8.94$ A, $\alpha = 30°42'$ (or $a = 4.732$ A, $c = 25.53$ A) at the Co-rich limit and $a = 9.01$ A, $\alpha = 30°40'$ (or $a = 4.761$ A, $c = 25.72$ A) at the W-rich limit [8]. A high-temperature phase, apparently of the approximate composition W_3Co_2, has the σ-type structure [5]. (See Supplement.)

Supplement. Exploratory work by [12] with a Co-W alloy of approximately 60 at. % W at 1300°C failed to confirm the existence of a σ phase claimed by [5].

1. K. Kreitz, *Metall u. Erz*, **19**, 1922, 137–140.
2. W. P. Sykes, *Trans. ASST*, **21**, 1933, 385–421.
3. S. T eda, *Science Repts. Tôhoku Univ., Honda Anniv. Vol.*, 1936, pp. 864–881.
4. U. Haschimoto, *Nippon Kinzoku Gakkai-Shi*, **1**, 1937, 177–190.
5. H. J. Goldschmidt, *Research*, **4**, 1951, 343.
6. W. Geiss and J. A. M. van Liempt, *Z. Metallkunde*, **19**, 1927, 113–114.
7. C. Agte, K. Becker, and v. Göler, *Metallwirtschaft*, **11**, 1932, 447–450.
8. A. Magneli and A. Westgren, *Z. anorg. Chem.*, **238**, 1938, 268–272.
9. M. M. Babich, E. N. Kislyakova, and J. S. Umanskiy, *Zhur. Tekh. Fiz.*, **9**, 1939, 533–536; *J. Physics (U.S.S.R.)*, **1**, 1939, 309–313.
10. M. M. Babich, E. Kislyakova, and J. S. Umanskiy, *Zhur. Tekh. Fiz.*, **8**, 1938, 119–121.
11. W. Köster and W. Tonn, *Z. Metallkunde*, **24**, 1932, 296–299.
12. P. Greenfield and P. A. Beck, *Trans. AIME*, **206**, 1956, 265–276.

$\bar{1}.9550$
0.0450

Co-Zn Cobalt-Zinc

The phase diagram shown in Fig. 306 is due to one investigator who reported his results in a series of five thorough studies by means of thermal [1, 2], micrographic [1, 3], and X-ray analysis [4, 3], as well as measurements of magnetic susceptibility vs. temperature [5]. (See Supplement.) Earlier, [6] had investigated the composition range 79.4–99 at. % Zn; however, his findings are obsolete. Also, [7] had detected a γ-brass type phase and reported its homogeneity range to be 78–85 at. % Zn at lower temperature and somewhat wider at 700°C and higher. The same phase was found by [8] in alloys with 76.7 and 81.3 at. % Zn. In alloys with less than 78 at. % Zn, a β-Mn type phase (β_1) was found to be present [7]. In general, these data were confirmed by [4].

The boundary of the solid solution of Zn in α-Co is based on lattice-parameter measurements (400–925°C) and micrographic work (600–950°C) [3]. The curve of the Curie points was determined in the ranges 0–47.4 at. % Zn [3], 0–18.4 at. % Zn [9], and 0–6 at. % Zn [10]. Data by [3] and [9] are practically identical. Results of investigations of the effect of Zn on the polymorphic transformation of Co differ widely. According to [9], the temperature of the $\epsilon \rightarrow \alpha$ transformation is raised and that of the $\alpha \rightarrow \epsilon$ transformation is lowered (Fig. 306), whereas [10] found both temperatures to be lowered, to 260°C by 8.2 at. % Zn and to 160°C by 6.0 at. % Zn, respectively.

The boundaries of the intermediate phases β_1, γ, γ_1, and γ_2 were determined by micrographic and/or parametric work [1, 3, 4]. Their homogeneity ranges are as follows: β_1, 47.9–56.5; γ, 75.2–85.4; γ_1, 87.4–88.6; γ_2, 91–92.8 at. % Zn (see Supplement). The boundaries of the high-temperature phases β and δ are based on thermal-analysis data. The solid solubility of Co in Zn appears to be extremely small [1, 11], but slightly more at higher than at lower temperatures [11]. γ_2 and Zn form a eutectic at a temperature only a fraction of **a** °C below the melting point of Zn [11, 1] and a composition below 0.03% Co [11].

The β_1 phase is ferromagnetic at room temperature; its Curie points vary between about 125 and 195°C according to [3] and between 129 and 198°C according to [12]. The latter workers concluded from their data that β_1 is homogeneous between 48 and 55 at. % Zn as compared with 48 and 56.5 at. % Zn according to [3]. (See Supplement.)

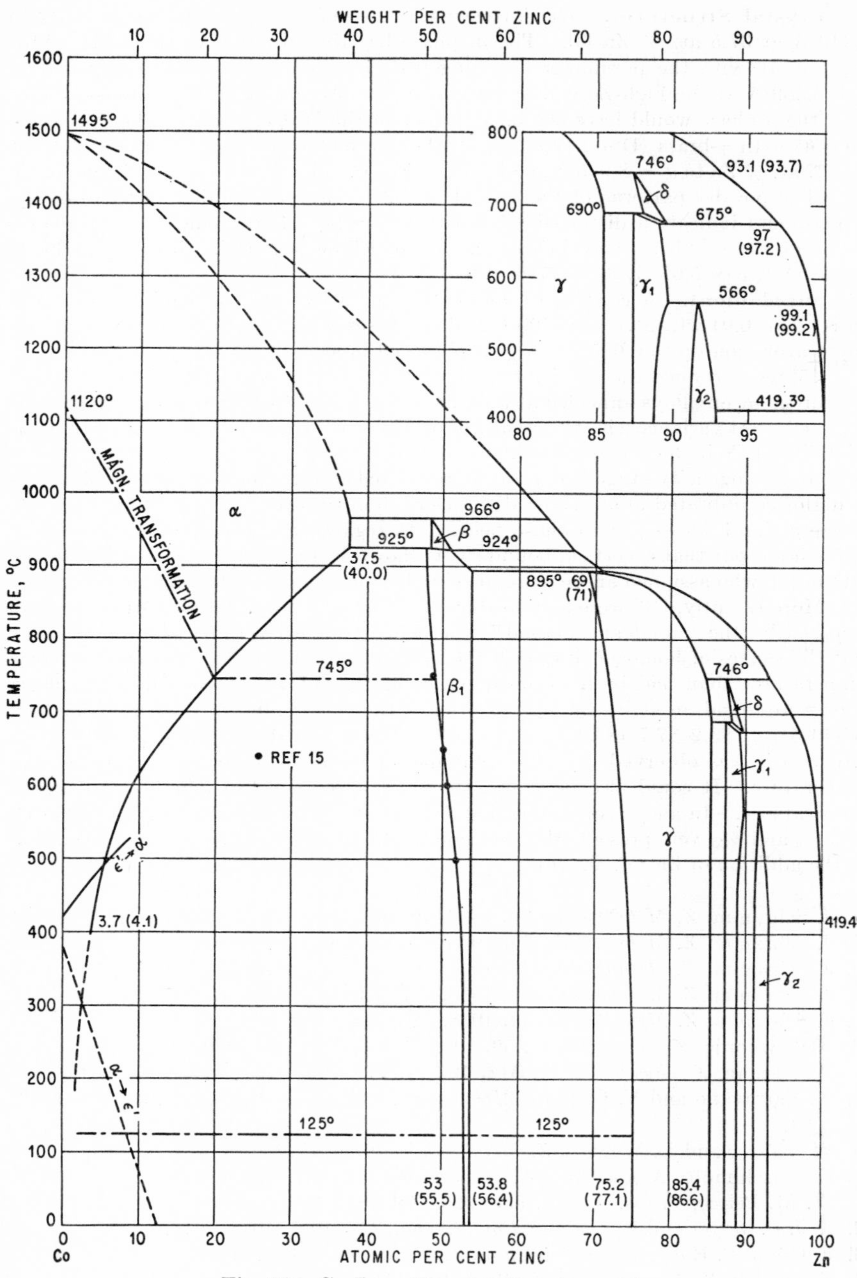

Fig. 306. Co-Zn. (See also Supplement.)

Crystal Structures. The lattice parameter of α-Co increases from 3.542 to 3.642 A at 37.5 at. % Zn [3]. The β_1 phase has the structure of the β-Mn (A13) type [7, 4], with the parameter changing from a = 6.314 A at the low-Zn end to a = 6.369 A at the high-Zn end of the range [4]. According to the Hume-Rothery rule, the β phase would have the b.c.c. lattice of the W (A2) type. The γ phase is isotypic with γ-brass ($D8_{1-3}$ type) [7, 4]; the parameter increases from a = 8.908 A at 76.2 at. % Zn to a = 8.985 A at 84.7 at. % Zn [4]. Similar values were reported by [7]. The powder patterns of the γ_1 and γ_2 phases are very similar to those of the γ phase and indicate a distorted γ-brass structure [4]. Later, single-crystal photograms of the γ_2 phase were indexed as a monoclinic unit cell, with a = 13.49 A, b = 7.50 A, c = 5.07 *A*, β = 127°5′ [13].

Supplement. According to microscopic work by [14], the solubility of Co in Zn is about 0.01 wt. (at.) % at 400°C and decreases slightly with fall in temperature. Supersaturation up to 0.035 wt. % was observed in as-cast alloys.

[15] redetermined the Co-rich boundary of the β_1 phase by means of Curie-point measurement on alloys annealed at and quenched from 750, 650, 600, and 500°C (see Fig. 306). At variance with the results of previous investigators it was found that the solubility of Co in β_1 decreases markedly with decreasing temperature and that, therefore, the homogeneity range of β_1 is rather restricted at lower temperatures. Measurements of equilibrated alloys showed the Curie temperature to be 125°C in both the (Co + β_1) and (β_1 + γ) two-phase fields. As regards the Zn-rich boundary of β_1, [15] pointed out that a vertical boundary as shown in Fig. 306 is in keeping with the data of [1], who assumed an increase in solubility of Zn in β_1 with falling temperature.

More recently, [16] were able to show, by means of X-ray investigation of alloys prepared by the "amalgam method" (thermal decomposition of Co-Zn amalgams), that the ranges of homogeneity of all the intermediate phases are narrower at lower temperatures than had been assumed by [1, 3, 4]. At 300°C, the following phase fields were found, in at. % Zn (wt. % in parentheses): ϵ_{Co}, 0–2.5 (0–3); β_1, 55.5–59.0 (58–61.5); γ, 83.2–87.7 (84.6–88.8); γ_1, 89.0 (90.0); γ_2, 91.0 (91.8). In agreement with [15] it was observed that the solubility of Co in β_1 increases with increasing temperature. It could also be shown that the phase field broadens toward higher temperatures. In alloys containing 5 and more wt. % Zn, after annealing at 380°C, both ϵ_{Co} and α_{Co} were present; this fact points to a lowering of the Co transformation by the addition of Zn (cf. above).

1. J. Schramm, *Z. Metallkunde,* **30,** 1938, 10–14.
2. J. Schramm, *Z. Metallkunde,* **30,** 1938, 131–135.
3. J. Schramm, *Z. Metallkunde,* **33,** 1941, 46–48.
4. J. Schramm, *Z. Metallkunde,* **30,** 1938, 122–130.
5. J. Schramm, *Z. Metallkunde,* **30,** 1938, 327–334.
6. K. Lewkonja, *Z. anorg. Chem.,* **59,** 1908, 319–322.
7. W. Ekman, *Z. physik. Chem.,* **B12,** 1931, 57–78.
8. N. Parravano and V. Caglioti, *Mem. accad. Italia, Cl. sci. fis. mat. nat.,* **3,** 1932, 1–21.
9. W. Köster and E. Wagner, *Z. Metallkunde,* **29,** 1937, 230–232.
10. U. Haschimoto, *Nippon Kinzoku Gakkai-Shi,* **1,** 1937, 177–190.
11. W. M. Peirce, *Trans. AIME,* **68,** 1923, 779–781.
12. A. J. P. Meyer and P. Taglang, *Compt. rend.,* **232,** 1951, 1914–1916.
13. F. Götzl, F. Halla, and J. Schramm, *Z. Metallkunde,* **33,** 1941, 375.
14. F. Pawlek, *Z. Metallkunde,* **36,** 1944, 105–111.
15. W. Köster and H. Schmid, *Z. Metallkunde,* **46,** 1955, 468–469.
16. F. Lihl and E. Weisbier, *Z. Metallkunde,* **46,** 1955, 579–581.

$\bar{1}.8103$
0.1897

Co-Zr Cobalt-Zirconium

Crystallization and transformation temperatures were determined in the range 0–16 wt. (11.0 at.) % Zr [1] and 0–0.4 wt. (0.26 at.) % Zr [2]. According to [1], there is a eutectic point at about 12 wt. (8.1 at.) % Zr, 1460°C; beyond 12 wt. % Zr the liquidus temperature increases steeply (Fig. 307). [2] found the eutectic temperature at 1473°C.

The intermediate phase coexisting with the primary solid solution was assumed to be $ZrCo_4$ (27.90 wt. % Zr), apparently based on the microstructure of the alloy with 16 wt. % Zr [1]. The solid solubility of Zr in Co was estimated to be about 2 wt. (1.3 at.) % Zr, independent of temperature [1]. The Curie temperature of Co was found to be not noticeably affected [1] and to be lowered to 1029°C (Co: 1142°C) [2]. The temperatures of the $\epsilon \rightarrow \alpha$ and $\alpha \rightarrow \epsilon$ transformations are lowered to 305 and 220°C by about 2 wt. % Zr according to [1]. [2] found the $\epsilon \rightarrow \alpha$ transformation to be raised from 467 to 502°C and the $\alpha \rightarrow \epsilon$ transformation to be lowered from 426 to about 350°C by about 0.18 wt. (0.12 at.) % Zr (solubility limit).

[3] have briefly described the microstructure of cast Zr-rich alloys with up to 46 wt. (57 at.) % Co; however, the information does not allow conclusions as to the intermediate phases present in this range of composition.

The existence of $ZrCo_2$ (43.63 wt. % Zr) was established by [4, 5]. The structure is of the $MgCu_2$ (C15) type, a = 6.901 A [5] or a = 6.94 A [6]. Whether $ZrCo_4$ [1] or $ZrCo_2$ is the phase in equilibrium with the primary Co-rich solid solution is unknown.

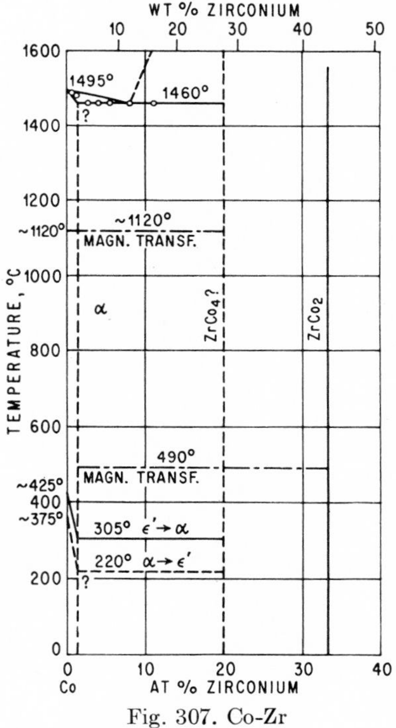

Fig. 307. Co-Zr

Supplement. A brief examination [7] of the constitution of Zr-rich alloys gave the following results: Up to 46 wt. (57 at.) % Co the alloys are two-phase, the second phase being an intermediate compound of low solubility in α-Zr. The intermediate phase lies between 19 and 45 wt. (26.6 and 56 at.) % Co, and a eutectic occurs on the Co-rich side of this phase.

[8] corroborated the $MgCu_2$ structure for $ZrCo_2$ (a = 6.954 A) and found its melting temperature to be 1560°C. No polymorphy was observed in the range 600–1500°C (X-ray examination of quenched specimens).

1. W. Köster and W. Mulfinger, *Z. Metallkunde,* **30,** 1938, 348–350.
2. U. Haschimoto, *Nippon Kinzoku Gakkai-Shi,* **2,** 1938, 67–77.
3. C. T. Anderson, E. T. Hayes, A. H. Roberson, and W. J. Kroll, *U.S. Bur. Mines Rept. Invest.* 4658, 1950.
4. H. J. Wallbaum, *Arch. Eisenhüttenw.,* **14,** 1941, 521–526.
5. H. J. Wallbaum, *Z. Krist.,* **103,** 1941, 391–402.

6. C. B. Jordan and P. Duwez, California Institute of Technology, Progress Report 20-196, June 16, 1953.
7. S. M. Shelton, *U.S. Atomic Energy Comm., Publ.* AF-TR-5932, 1949; *Met. Abstr.*, **21**, 1954, 869.
8. R. P. Elliott, Armour Research Foundation, Chicago, Ill., Technical Report 1, OSR Technical Note OSR-TN-247, August, 1954, p. 35.

$\bar{1}.9130$
0.0870

Cr-Cu Chromium-Copper

The phase diagram of Fig. 308 is mainly due to the work by [1, 2], most of the later investigations being restricted to Cu-rich alloys. The miscibility gap in the liquid state was suggested by [3] and confirmed by thermal [1, 2] and microscopic

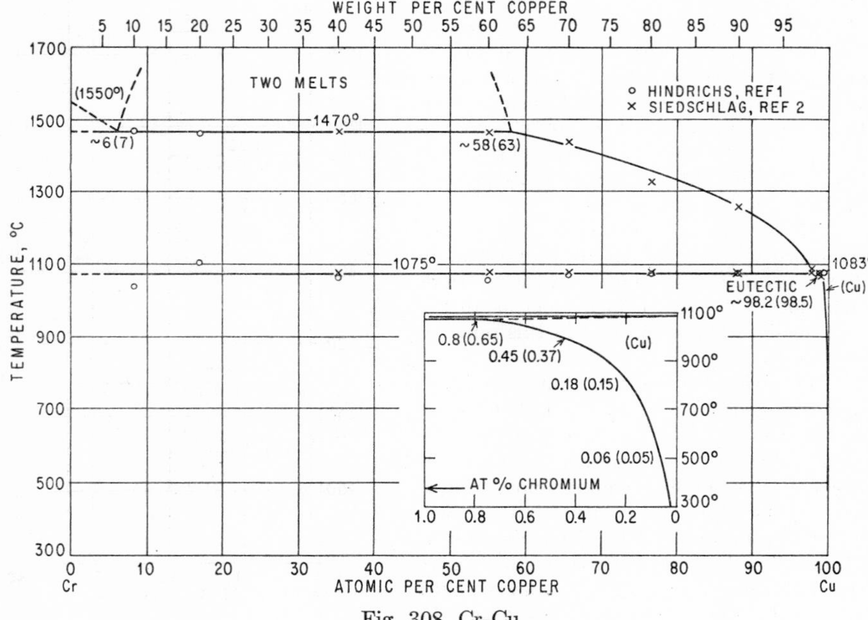

Fig. 308. Cr-Cu

[1, 4, 2] work; the only reliable data referring to its extent [about 7–63 wt. (6–58 at.) % Cu at the monotectic temperature] were given by [2] on the basis of chemical analyses of the separated layers of two alloys.

The work of [1] was done with a chromium containing 1.2 Fe and 0.3 Si (wt. %); alloys with up to 50 wt. % Cr were melted in porcelain, those richer in chromium in magnesia, crucibles in a nitrogen (!) atmosphere (see Cr-N). They were severely contaminated by chromium nitride, carbide, and reaction products of the melt with the crucible materials. This is clearly indicated by the erroneously low melting point of "chromium" observed (1550°C). [2] used somewhat better working conditions (hydrogen atmosphere, alumina-lined carbon crucibles) but chromium of only 97.8 wt. % purity.

The eutectic temperature was reported to be about 1075 [1, 2] and 1070°C [5], respectively. There is no such agreement, however, as to the data for the eutectic composition found by microscopic work, which lie between <0.5 and 1–2 wt. % Cr [1, 2, 6, 5]; 1.5 wt. (1.8 at.) % Cr seems to be the most reliable value.

The solid solubility of Cu in Cr has not yet been adequately investigated but seems to be very small [1]. The solubility of Cr in solid Cu has attracted more attention; and resistometric [7, 5], microscopic [6, 5, 8], chemical-analytical [6], and cursory roentgenographic [9] work has been done [10]. The experimental data of [6] and [5] are in good agreement [11] and indicate a rapid decrease in the solubility with the temperature: 0.65 ± 0.1, 0.37, 0.15, 0.05, and <0.03 wt. (0.8, 0.45, 0.18, 0.06, and <0.04 at.) % Cr at 1075, 1000, 800, 500, and 400°C, respectively (see inset in Fig. 308). As [9] have shown, supersaturation up to 1.45 wt. (1.8 at.) % Cr can be reached by extremely high quenching rates.

For more recent studies on age hardening of Cu-rich Cr-Cu alloys, see especially [5, 12, 13].

1. G. Hindrichs, *Z. anorg. Chem.*, **59,** 1908, 420–423.
2. E. Siedschlag, *Z. anorg. Chem.*, **131,** 1923, 173–178.
3. L. Guillet, *Rev. mét.*, **3,** 1906, 176. Microscopic work; it is not certain, however, that the chromium was molten at all.
4. D. F. McFarland and O. E. Harder, *Univ. Illinois Bull.*, **14,** 1916, no. 93. "Preliminary Study of the Alloys of Cr, Cu, and Ni"; microscopic, physical, and mechanical examinations of 9 binary Cu-Cr alloys.
5. W. R. Hibbard, F. D. Rosi, H. T. Clark, and R. I. O'Herron, *Trans. AIME,* **175,** 1948, 283–294. All heat-treatments were carried out in a nitrogen (!) atmosphere.
6. W. O. Alexander, *J. Inst. Metals,* **64,** 1939, 93–109.
7. M. A. Hunter and F. M. Sebast, *J. Am. Inst. Metals,* **11,** 1917-1918, 115.
8. W. R. Hibbard and N. K. Chen, *Trans. AIME,* **175,** 1948, 294–295.
9. G. Falkenhagen and W. Hofmann, *Z. Metallkunde,* **43,** 1952, 69–81.
10. M. G. Corson (*Trans. AIME,* **77,** 1927, 435; *Rev. mét.*, **27,** 1930, 86–95) reported—without giving experimental details—some rather high solubility data. They are plotted in [5].
11. Alexander [6] made a questionable extrapolation to the eutectic temperature. (See also C. S. Smith in "Metals Handbook," 1948 ed., p. 1193, The American Society for Metals, Cleveland, Ohio.)
12. G. Bunge, E. R. Honak, and W. Nielsch, *Z. Metallkunde,* **44,** 1953, 71–76.
13. W. Köster and W. Knorr, *Z. Metallkunde,* **45,** 1954, 350–356.

$\bar{1}.9691$
0.0309

Cr-Fe Chromium-Iron

Early Papers. The practical difficulties in preparing alloys of only moderate purity proved so formidable to the earlier workers [1–5] with restricted resources that their achievements are mainly of historical interest. The pronounced ability of Cr and its alloys to absorb C, O, and especially N (see Cr-N) and to react with refractory materials used then accounts for the very discordant results obtained. For a discussion of these papers, see [6, 7].

Solidification. Indications for the solidification of a continuous series of solid solutions from the melt had been found by [2]; and this was corroborated mainly by thermal [8–10], microscopic [11, 8–10], roentgenographic [12–16], and magnetic [11] investigations. For the liquidus and solidus of Fig. 309 the data given by [10] were used as the most reliable ones; those for the solidus—obtained by means of heating curves on annealed alloys—were looked upon as preliminary by [10]. The relatively wide temperature interval between liquidus and solidus with Cr-rich alloys, however, is supported by strong coring of slowly cooled alloys [10]. Also plotted in Fig. 309 are some recent liquidus data by [17]. For the temperature minimum [22 wt. (23 at.) % Cr, 1507°C according to [10]] the data given by [11, 8, 9] vary between 10 and 30 wt. (11–31.5 at.) % Cr, 1400–1490°C.

γ **Loop.** The existence of the γ loop was discovered by [18], mainly by means of microscopic examination of quenched low-carbon alloys. In spite of a large number of later papers [19, 13, 20, 21, 15, 22–24, 10, 25–32] which deal at least partially with the boundaries of the γ loop, its shape and extent are not yet established exactly. This is due chiefly to the sluggish character of the transformation—especially at higher Cr contents—and to the fact that the extent of the loop is greatly affected by impurities such as C and N.

In the following, a short description of the more important investigations is given (see also [6, 7]). [18]: Hardness and X-ray measurements in addition to the work mentioned above. [13]: Differential thermal and roentgenographic analysis on alloys containing up to 0.01 wt. % C (Fig. 310). [20]: Special dilatometric method on wires containing less than 0.01 wt. % C; the lower boundary of the loop (see Fig. 310) indicates an error of unknown origin. [21]: Differential thermal analysis on alloys with 0.03 wt. % C. [15]: High-temperature X-ray analysis. [22]: Dilatometric analysis, same alloys as [21]. [23]: Dilatometric studies, same alloys as [13]. [10]: Thermal (under vacuum) and dilatometric (under argon) measurements and some microscopic work on alloys with 0.03–0.04 wt. % C (possibly some carbon introduced by the cutting tool before chemical analysis). The data plotted in Fig. 310 are average values of thermal and dilatometric heating data. [25]: Dilatometric studies of alloys with ~0.03 wt. % C and up to 0.23 (!) wt. % Si. [26]: Diffusion experiments at 1200 and 1350°C. [27]: Mainly dilatometric studies on relatively pure alloys (~0.01 wt. % C) (Figs. 309 and 310). [28]: Dilatometric and magnetic measurements; same alloys as [25]. [30]: Solubility of hydrogen in some Fe-Cr alloys as a function of temperature. [32]: Dilatometric measurements. Only the Cr content of the alloys was analyzed. The experimental conditions are to be regarded as unsatisfactory. See Fig. 310 [33].

Only [13, 25, 28, 32, 33] found clear evidence for the existence of $(\alpha + \gamma)$ two-phase fields.

One factor—besides the quality of the experimental conditions—in selecting the data plotted in Fig. 310 was that, of dynamic data, only those points found on heating curves are reliable [25, 28]. As to the vertex of the γ loop, most authors who investigated low-carbon [34] alloys found a concentration near 12 wt. (12.8 at.) % Cr (<14, ~13, 12.2, between 11.3 and 12.6, 12.4, and 11.6 wt. % Cr according to [18, 13, 20, 23, 32, and 10] respectively). [27], however, claimed to have observed volume changes in carefully homogenized alloys containing up to about 18 wt. (19 at.) % Cr with the alloys held at about 1000°C for some hours (Figs. 309 and 310). Unfortunately, this paper seems to have been overlooked by later investigators so that these interesting results have not been checked. Values up to 17 wt. % Cr have been reported by [21, 15, 22, 26] but on alloys with somewhat higher carbon content [34].

The temperature minimum in the γ loop [35] was detected by [13] and located at 7–8 wt. (7.5–8.5 at.) % Cr, 840°C. Data found by other authors are ~10 wt. (10.7 at.) % Cr, 830°C [21]; 6.5–7.5 wt. (7–8 at.) % Cr, 850°C [10]; ~9 wt. (9.6 at.) % Cr, 830°C [28]; and 5.2 wt. (5.6 at.) % Cr, 796°C [32]. Since there is no investigation that can be regarded as definitive nor any general agreement, the plot of the γ loop in Fig. 309 is only tentative, with a slight preference for data by [10].

The σ Phase. Indications for the formation of an intermediate phase in the solid state in the equiatomic region were observed by [12, 36, 37], and this was definitely corroborated by [38–42] by means of several standard methods. The extremely

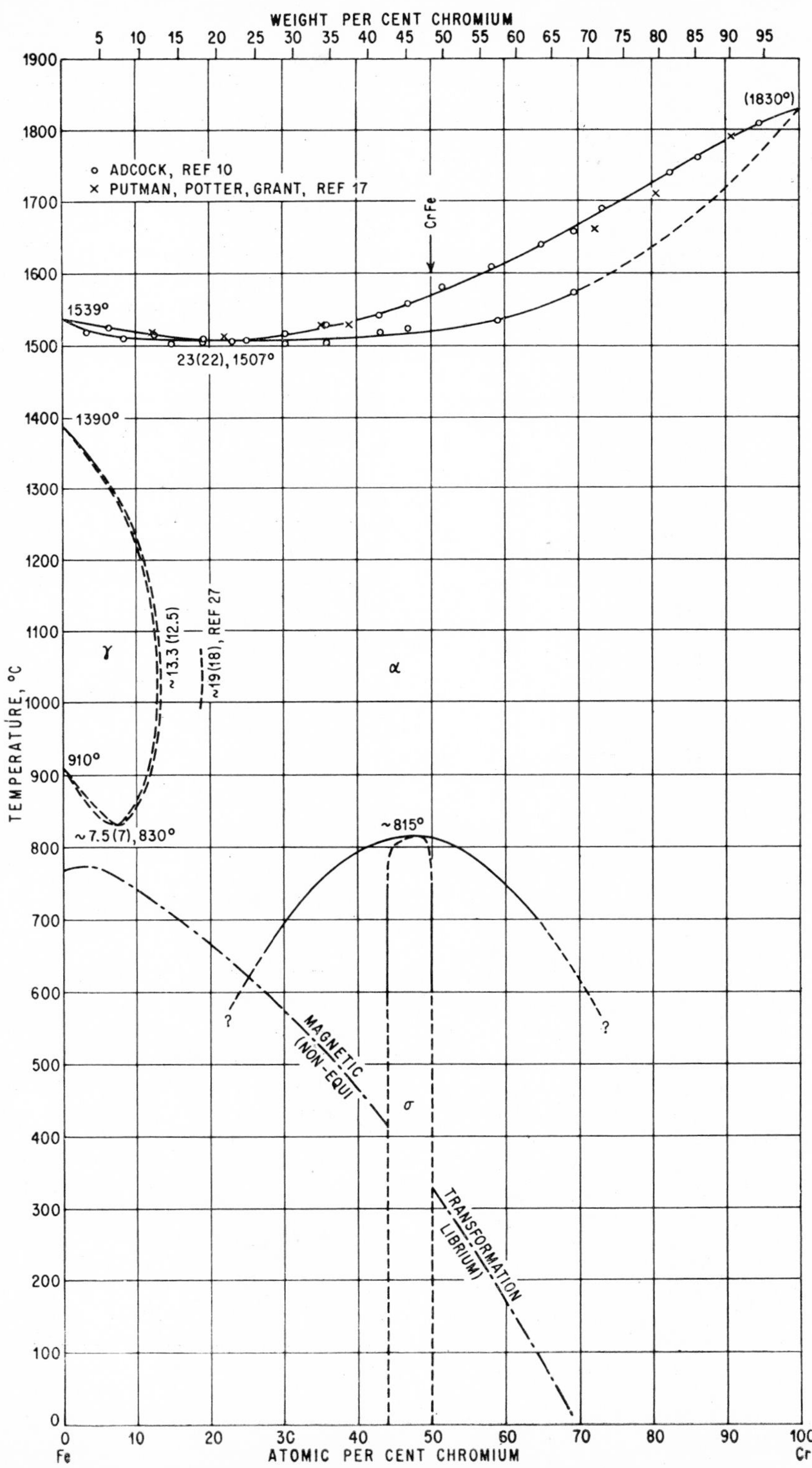

Fig. 309. Cr-Fe. (See also Fig. 310.)

low rate of formation of this phase in very pure and undeformed alloys accounts for the fact that [43] was unable to find it [42].

Contributions to the determination of phase boundaries in the σ-phase region were made by [39, 41, 44–48]. The most reliable data, extensively used in Fig. 309, are based on the roentgenographic and microscopic investigation by [45]. The upper limit of the σ phase was found to be below 820°C (815°C in Fig. 309); the σ phase at 600°C, likely to be present in alloys containing 26–71 ($\pm$1) at. (24.7–69.5 wt.) % Cr; and the σ-phase field itself, to extend from 44 to 50 at. (42.3–48.2 wt.) % Cr [49]. The upper part of the σ-phase field could not be outlined by [45]; its shape in Fig. 309 follows a suggestion by [50]. Some later studies by [51, 46, 47] at temperatures above

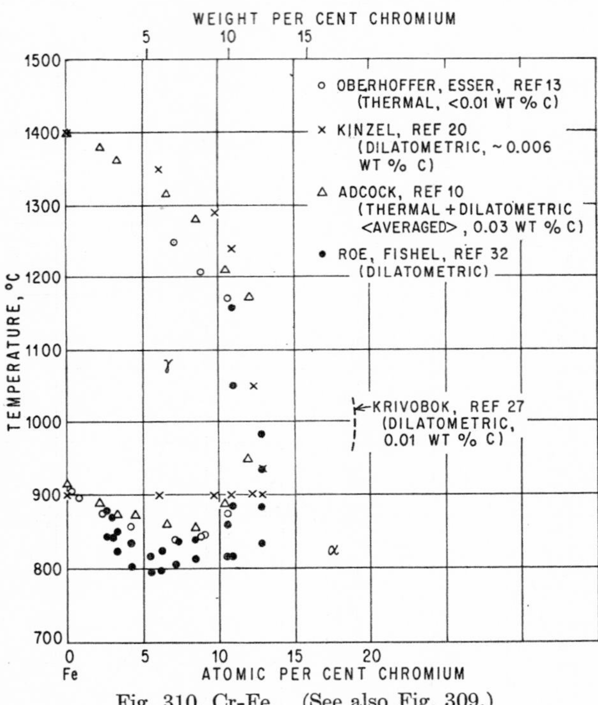

Fig. 310. Cr-Fe. (See also Fig. 309.)

600°C agree fairly well with the results of [45], but there is some evidence [47] that extremely long annealing (>100 days) at 600°C would shift the $\alpha/(\alpha + \sigma)$ boundary even below 26 at. % Cr (Fig. 309: ~24 at. (22.7 wt.) % Cr [52]).

Crystal Structure. Before single-crystal data were available, tentative unit cells with tetragonal [54], cubic [55], or orthorhombic [56] symmetry were proposed. A similarity between the σ-phase pattern and that of α-Mn was noted by [44]. Tables with interplanar spacings of the σ phase were published by [57, 54].

When single crystals became available, the unit cell was shown to have tetragonal symmetry [58–60]; the lattice parameters according to [58, 60] are (averaged values) $a = 8.794$ A, $c = 4.552$ A, $c/a = 0.518$. The unit cell contains 30 atoms. The structure of the σ phase has not yet been completely elucidated, but it appears that it is essentially the same as that of β-uranium [60–62].

There are some indications that the $\alpha \rightarrow \sigma$ transformation may involve a transition stage [44, 50]; see also Supplement.

The σ phase is brittle and—at least at ordinary temperatures [65, 66]—not ferromagnetic.

As to theoretical considerations about σ phases in general, reference is made to [67], who also give further literature.

Some Peculiarities in Fe-Cr Alloys. There are some peculiarities in Fe-rich Fe-Cr alloys which, at least in part, cannot be accounted for by the phase diagram of Fig. 309: age hardening [27], embrittlement [68; see also 27], breaks in magnetic-properties-vs.-concentration curves [69]. [69] suggested the formation of superstructure phases (see also [70]) $Fe_{15}Cr$ and Fe_7Cr.

[71] measured the rate of the isothermal transformation of quenched alloys containing 40.5–51.5 at. % Cr at 650°C by a magnetic method and found three maxima in the transformation-rate-vs.-composition curve. The conclusion that there exist, besides σ, two additional intermediate phases in the concentration range investigated, however, is not convincing. (See Supplement.)

The Magnetic Inversion. Curie-temperature-vs.-composition curves were determined by [11, 19, 13, 36, 21, 10, 69, 29, 71], those by [11, 36, 10] extending to below room temperature. The Curie points appear to reach a small maximum [11, 13, 36, 10, 29] as the Cr content is increased from zero before being lowered (Fig. 309). Because of the sluggish formation of the nonmagnetic σ phase, the data in the middle of the diagram depend heavily on the heat-treatment applied [10, 72, 69]—true equilibrium has not yet been reached at temperatures below about 650°C—and therefore data [10] on quenched (metastable ferritic) alloys were plotted in Fig. 309. For Fe-rich alloys, the data of [10] (thermal, dilatometric, and magnetic) are in fair agreement with those of [13, 21, 29].

Further Investigations. For the sake of completeness, reference is made below to some investigations which deal with physical properties of Cr-Fe alloys without giving information on phase boundaries: lattice parameters of ferrite [14, 16, 73, 27, 42, 74, especially 16], electrical resistivity [75, 76, 10, 77], magnetic properties [77, 78, 69, 74], thermoelectric force [76], density [76, 10], thermal-expansion coefficients [76, 28], and hardness [10, 27].

Supplement. The crystal structure of σ has been established by [79]. As had been previously suggested (see Crystal Structure), it is nearly identical with that of β-U. The tetragonal lattice parameters reported by [79] are $a = 8.800$ A, $c = 4.544$ A. Lattice parameters of ferrite were measured up to 4.5 at. % Cr by [80] and up to 5.37 at. % Cr (precision data) by [81].

[82] concluded from magnetic measurements on alloys annealed at and furnace-cooled from 800°C that the σ phase is homogeneous between 46 and 52 at. % Cr. [83] claimed, on the basis of microscopic observations on quenched specimens, that the σ-phase field extended from about 45 to 57 (!) at. % Cr, and up to 900°C at the equiatomic composition. The $\alpha \rightarrow \sigma$ transformation has been studied by [84–87, 83].

Long-time annealing of Fe-Cr alloys containing between about 10 and 80 wt. % Cr at 400–550°C, i.e., well below the temperature range of measurable rates of σ formation, hardens and embrittles these alloys. This phenomenon is known as "475°C (885°F) embrittlement" and has received much attention in recent years. A full review of the pertinent literature is beyond the scope of this book, and, in the following, only some general information will be given. It has been established that the reactions involved lead to metastable equilibria which, however, seem to be of a very complex nature. Whereas the results of some investigators (see [88, 89]) are compatible with a miscibility gap (extending from 10 to 80 wt. % Cr at 450°C, upper critical point at about 580°C [89]) in the metastable ferrite phase field, others found evidence of the existence of ordered phases. The formulas $CrFe_3$ [90, 83], $CrFe_2$ [83], CrFe [63, 64, 91, 92, 83], Cr_2Fe [83], and Cr_3Fe [83] have been named. (See also Some Pecularities in Fe-Cr Alloys.)

X-ray examination of Fe-Cr alloys prepared by mutual deposition of the components on Hg and distilling off Hg showed the presence of a heterogeneous mixture of Fe and Cr [93]. If, as is claimed by [93], this finding corresponds to the state of equilibrium at room temperature, it would imply that the σ phase is not stable at lower temperatures.

1. Investigations on high-carbon alloys: R. A. Hadfield and F. Osmond, *J. Iron Steel Inst.*, **1892**(2), 49–114, 115–131. For further references see M. Sack, *Z. anorg. Chem.*, **35,** 1903, 325.
2. The first investigation on relatively low-carbon alloys was carried out by W. Treitschke and G. Tammann, *Z. anorg. Chem.*, **55,** 1907, 402–411.
3. P. Monnartz, *Metallurgie,* **8,** 1911, 163–168.
4. E. Jänecke, *Z. Elektrochem.*, **23,** 1917, 49–55.
5. K. Fischbeck, *Stahl u. Eisen,* **44,** 1924, 716–717. (Attempt to correlate former results.)
6. M. Hansen, "Der Aufbau der Zweistofflegierungen," Springer-Verlag OHG, Berlin, 1936.
7. A. B. Kinzel and W. Crafts, "The Alloys of Iron and Chromium," vol. I, McGraw-Hill Book Company, Inc., New York, 1937.
8. E. Pakulla and P. Oberhoffer, *Ber. Werkstoffausschuss Verein. deut. Eisenhüttenleute,* no. 68, 1925, pp. 1–6.
9. A. v. Vegesack, *Z. anorg. Chem.*, **154,** 1926, 37–41.
10. F. Adcock, *J. Iron Steel Inst.*, **124,** 1931, 99–139; see also National Physical Laboratory, Report 1930, pp. 263–264.
11. T. Murakami, *Science Repts. Tôhoku Imp. Univ.*, **7,** 1918, 224–225, 264–266.
12. E. C. Bain, *Chem. & Met. Eng.*, **28,** 1923, 23.
13. P. Oberhoffer and H. Esser, *Stahl u. Eisen,* **47,** 1927, 2021–2031.
14. A. Westgren, G. Phragmén, and T. Negresco, *J. Iron Steel Inst.*, **117,** 1928, 385–386.
15. C. Kreutzer, *Z. Physik,* **48,** 1928, 560–564; P. Oberhoffer and C. Kreutzer, *Arch. Eisenhüttenw.*, **2,** 1928-1929, 451–453. See also [13].
16. G. D. Preston, *J. Iron Steel Inst.*, **124,** 1931, 139–141; *Phil. Mag.*, (7)**13,** 1932, 419–425.
17. J. W. Putman, R. D. Potter, and N. J. Grant, *Trans. ASM,* **43,** 1951, 824–847.
18. E. C. Bain, *Trans. ASST,* **9,** 1926, 9–32.
19. H. Esser and P. Oberhoffer, see [8].
20. A. B. Kinzel, *Trans. AIME,* **80,** 1928, 301–307.
21. E. Maurer and H. Nienhaus, *Stahl u. Eisen,* **48,** 1928, 999–1000.
22. F. Stäblein, *Arch. Eisenhüttenw.*, **3,** 1929-1930, 301–305.
23. K. Schroeter, see [22], p. 303.
24. A. Merz, *Arch. Eisenhüttenw.*, **3,** 1929-1930, 591–592.
25. E. C. Bain and R. H. Aborn, Constitution of Iron-Chromium Alloys in "Metals Handbook," 1933 and 1936, The American Society for Metals, Cleveland, Ohio.
26. L. C. Hicks, *Trans. AIME,* **113,** 1934, 163–172; discussion, pp. 172–178.
27. V. N. Krivobok, *Trans. ASM,* **23,** 1935, 1–60.
28. J. B. Austin and R. H. H. Pierce, *Trans. AIME,* **116,** 1935, 289–306.
29. J. Martelly, *Ann. phys.*, **9,** 1938, 318–333, especially 333.
30. A. Sieverts and H. Moritz, *Ber. deut. chem. Ges.*, **75,** 1942, 1726–1729.

31. J. W. Pugh and J. D. Nisbet, *Trans. AIME,* **188,** 1950, 268–276, especially 269–270; discussion, pp. 1356–1358.
32. W. P. Roe and W. P. Fishel, *Trans. ASM,* **44,** 1952, 1030–1040; discussion, pp. 1040–1046.
33. " . . . we realize that the inner loop [see Fig. 310] is not the equilibrium condition. By heating . . . until the transformation starts, and then holding at this temperature for considerable time . . . we found that the inner loop approaches more nearly the outer one." (Quoted from [32], p. 1046.)
34. Both C and N expand the γ loop to higher Cr contents. (See especially [18, 27].)
35. For an interesting diffusion test (microstructure showing two diffusion lines) see [26] with discussion (W. D. Jones).
36. P. Chevenard, *Trav. mém. bur. int. poids et mesures,* **12,** 1927. Not available to the author; abstract in *J. Inst. Metals,* **37,** 1927, 471–472.
37. E. C. Bain and W. E. Griffiths, *Trans. AIME,* **75,** 1927, 166–211.
38. F. Wever and W. Jellinghaus, *Mitt. Kaiser-Wilhelm-Inst. Eisenforsch. Düsseldorf,* **13,** 1931, 107.
39. F. Wever and W. Jellinghaus, *Mitt. Kaiser-Wilhelm-Inst. Eisenforsch. Düsseldorf,* **13,** 1931, 143–147.
40. S. Eriksson, *Jernkontorets Ann.,* **118,** 1934, 530–543.
41. N. N. Kurnakov and N. I. Korenev, *Izvest. Sektora Fiz.-Khim. Anal.,* **9,** 1936, 85–98.
42. E. R. Jette and F. Foote, *Metals & Alloys,* **7,** 1936, 207–210.
43. F. Adcock, *J. Iron Steel Inst.,* **124,** 1931, 147–149.
44. A. J. Bradley and H. J. Goldschmidt, *J. Iron Steel Inst.,* **144,** 1941, 273–283; discussion, pp. 284–288.
45. A. J. Cook and F. W. Jones, *J. Iron Steel Inst.,* **148,** 1943, 217–223; discussion, pp. 223–226. To reach equilibrium, filings for X-ray work were annealed up to 60 days at 600°C.
46. S. R. Baen and P. Duwez, *Trans. AIME,* **191,** 1951, 331–335.
47. F. J. Shortsleeve and M. E. Nicholson, *Trans. ASM,* **43,** 1951, 142–156, especially 152; discussion, pp. 156–160.
48. A. T. Grigor'ev, N. M. Gruzdeva, and I. A. Bondar, *Izvest. Sektora Fiz.-Khim. Anal.,* **21,** 1952, 132–143.
49. Russian authors [41, 48] assume that the range of homogeneity of the σ phase extends somewhat beyond 50 at. % Cr.
50. H. J. Goldschmidt, *J. Iron Steel Inst.,* **148,** 1943, 223–224 (correspondence).
51. W. P. Rees, B. D. Burns, and A. J. Cook, *J. Iron Steel Inst.,* **162,** 1949, 325–336.
52. For the influence of commercial impurities on the Fe-rich $\alpha/(\alpha + \sigma)$ boundary, see especially [53, 47].
53. J. J. Heger, *ASTM Symposium, Spec. Tech. Publ.* 110, 1950, pp. 75–78; discussion, pp. 79–81.
54. P. Duwez and S. Baen, *ASTM Symposium, Spec. Tech. Publ.* 110, 1950, pp. 48–54; discussion, pp. 55–60.
55. K. G. Carroll, see [54], pp. 59–60.
56. P. Pietrokowsky and P. Duwez, *Trans. AIME,* **188,** 1950, 1283–1284.
57. H. J. Goldschmidt, *Metallurgia,* **40,** 1949, 103.
58. D. P. Shoemaker and B. G. Bergman, *J. Am. Chem. Soc.,* **72,** 1950, 5793.
59. L. Menezes, J. K. Roros, and T. A. Read, *ASTM Symposium, Spec. Tech. Publ.* 110, 1950, pp. 71–74.
60. G. J. Dickins, A. M. B. Douglas, and W. H. Taylor, *J. Iron Steel Inst.,* **167,** 1951, 27–28; *Nature,* **167,** 1951, 192.

61. B. G. Bergman and D. P. Shoemaker, *J. Chem. Phys.*, **19,** 1951, 515.
62. J. S. Kasper, B. F. Decker, and J. R. Belanger, *J. Appl. Phys.*, **22,** 1951, 361–362 (σ phase in Co-Cr).
63. S. Takeda and N. Nagai, *Nippon Kinzoku Gakkai-Shi,* **13**(2), 1949, 26 (in Japanese). Quoted in [64].
64. M. Tagaya, S. Nenno, and Z. Nishiyama, *Mem. Inst. Sci. Ind. Research, Osaka Univ.,* **9,** 1952, 76–79 (in English); *Nippon Kinzoku Gakkai-Shi,* **B15,** 1951, 235–236.
65. K. W. J. Bowen and T. P. Hoar, *Research,* **3,** 1950, 484.
66. P. A. Beck, *Trans. AIME,* **194,** 1952, 420.
67. P. Greenfield and P. A. Beck, *Trans. AIME,* **200,** 1954, 253–257.
68. See [47]; discussion, p. 160.
69. M. Fallot, *Ann. phys.*, **6,** 1936, 305–387, especially 365–369.
70. V. N. Svechnikov, *Metallurg,* no. 4, 1938, pp. 45–49 (in Russian); *Met. Abstr.,* **6,** 1939, 508.
71. I. I. Kornilov and V. S. Mikheev, *Doklady Akad. Nauk S.S.S.R.*, **68,** 1949, 527–530.
72. E. Scheil, *Stahl u. Eisen,* **51,** 1931, 1577–1578.
73. Z. Nishiyama, *Science Repts. Tôhoku Imp. Univ.*, **18,** 1929, 377–387.
74. B. A. Johnson, in L. Vegard, *Skrifter Norske Videnskaps-Akad. Oslo, Mat.-Naturv. Kl.*, no. 2, 1947, pp. 47–56 (see *Structure Repts.*, **11,** 1947-1948, 89–90).
75. A. Hunter and F. H. Sebast, *J. Am. Inst. Metals,* **11,** 1917-1918, 115.
76. K. Ruf, *Z. Elektrochem.*, **34,** 1928, 813–818.
77. Fischer and F. Kapp, *Rensselaer Polytech. Inst. Bull. Eng. Sci. Ser.*, no. **28,** 1930, pp. 1–32.
78. C. E. Webb, *J. Iron Steel Inst.*, **124,** 1931, 141–145.
79. G. Bergman and D. P. Shoemaker, *Acta Cryst.*, **7,** 1954, 857–865.
80. A. P. Guljaev and E. F. Trusova, *Zhur. Tekh. Fiz.*, **20,** 1950, 66–78; abstract in *Structure Repts.*, **13,** 1950, 77.
81. A. L. Sutton and W. Hume-Rothery, *Phil. Mag.*, **46,** 1955, 1295–1309.
82. N. I. Korenev and E. I. Koreneva, *Izvest. Sektora Fiz.-Khim. Anal., Akad. Nauk S.S.S.R.*, **20,** 1950, 54–65; *Chem. Abstr.*, **48,** 1954, 8154.
83. S. Takeda and N. Nagai, *Mem. Fac. Eng., Nagoya Univ.*, **8**(1), 1956, 1–28 (in English).
84. G. Mima and S. Imoto, *Nippon Kinzoku Gakkai-Shi,* **17,** 1953, 549–552 (in Japanese); *Met. Abstr.*, **23,** 1956, 894.
85. P. Bastien and G. Pomey, *Compt. rend.*, **239,** 1954, 1797–1799.
86. E. Baerlecken and H. Fabritius, *Arch. Eisenhüttenw.*, **26,** 1955, 679–686.
87. P. Bastien and G. Pomey, *Compt. rend.*, **240,** 1955, 866–868.
88. R. M. Fisher, E. J. Dulis, and K. G. Carroll, *Trans. AIME,* **197,** 1953, 690–695.
89. W. Köster and A. v. Kienlin, *Arch. Eisenhüttenw.*, **27,** 1956, 793–799.
90. Y. Imai and K. Kumada, *Science Repts. Research Insts. Tôhoku Univ.*, **A5,** 1953, 218–226, 520–532.
91. P. Bastien and G. Pomey, *Compt. rend.*, **239,** 1954, 1636–1638.
92. G. Pomey and J. Philibert, *Compt. rend.*, **241,** 1955, 877–879.
93. F. Lihl, *Z. Metallkunde,* **46,** 1955, 434–441.

$\bar{1}.8552$
0.1448

Cr-Ge Chromium-Germanium

Three compounds were found by a roentgenographic investigation [1] of Cr-Ge alloys: Cr_3Ge (31.76 wt. % Ge) has the "β-W" (A15) type of structure, with a =

4.623 A; Cr_3Ge_2 (48.20 wt. % Ge) is isomorphous with Cr_3Si_2 (structure not yet known); and CrGe (58.26 wt. % Ge) has the FeSi (B20) type of structure, with a = 4.789 A. All three compounds are isotypic with the corresponding Si compounds.

1. H. J. Wallbaum, *Naturwissenschaften*, **32**, 1944, 76–77.

1.7126
$\bar{2}$.2874

Cr-H Chromium-Hydrogen

The literature up to 1946 has been competently reviewed by [1]. The solubility isobars at 1 atm pressure measured by [2, 3] between 450 and 1125° and 300 and 1200°C, respectively, are in almost perfect agreement and show the usual appearance of an endothermic solubility curve. For 400, 800, and 1200°C the solubilities are 76,920, 22,730, and 3,817 atoms Cr per atom H (0.13×10^{-2}, 0.44×10^{-2}, and 2.6×10^{-2} at. % H) [4]. [5] suggested, with the calorimetric measurements of [6] as a basis, "that the solubility passes through a minimum somewhere between 300°C and room temperature, and that below this minimum the process of absorption is therefore exothermic."

Electrolytic Chromium and Hydrogen. Electrolytic chromium may contain amounts of hydrogen much in excess of the solubility limit. For literature and an extensive discussion of the probable types of occlusion, reference is made to [1].

Electroplating factors can be adjusted so that chromium is deposited either in its normal (but slightly expanded) b.c.c. form [7–9], in a h.c.p. (A3) form [10, 11, 8, 12], or in the form of the α-Mn (A12) type of structure (observed only by [11]; cf. [13]), with a = 8.734 A. The hexagonal modification has been looked upon as stabilized by the presence of hydrogen [14, 15] or as an allotropic form of Cr [16–18].

In many respects, results of recent extensive work on chromium plating [19, 20], which indicate that chromium may be plated in an unstable hydride form, seem to replace older views. The existence of two hydrides has been reported: a hexagonal one (~CrH), corresponding to the "hexagonal Cr" mentioned above, which has the wurtzite (B4) type of structure, and a cubic one (~CrH_2 [21]), having the fluorite type of structure, with a = 3.860 A [20, 22].

1. D. P. Smith, "Hydrogen in Metals," University of Chicago Press, Chicago, 1948.
2. E. Martin, *Arch. Eisenhüttenw.*, **3**, 1929, 407–416; abridged in *Metals & Alloys*, **1**, 1929, 831–835.
3. L. Luckemeyer-Hasse and H. Schenck, *Arch. Eisenhüttenw.*, **6**, 1932, 209–214.
4. See [1], pp. 28, 73. Solubilities estimated from the curves of [2, 3].
5. D. P. Smith [1], p. 188.
6. A. Sieverts and A. Gotta, *Z. anorg. Chem.*, **172**, 1928, 1–31.
7. G. F. Hüttig and F. Brodkorb, *Z. anorg. Chem.*, **144**, 1925, 341–348 (with observations by G. Wilke on the expansion of the Cr lattice by hydrogen).
8. W. A. Wood, *Phil. Mag.*, **12**, 1931, 853–864.
9. W. A. Wood, *Phil. Mag.*, **23**, 1937, 984–988.
10. E. A. Ollard and A. J. Bradley, *Nature*, **117**, 1926, 122.
11. K. Sasaki and S. Sekito, *J. Soc. Chem. Ind. Japan* (*Suppl.*), **33**, 1930; *Trans. Electrochem. Soc.*, **59**, 1931, 437–443.
12. L. Wright, H. Hirst, and J. Riley, *Trans. Faraday Soc.*, **31**, 1935, 1253–1259.
13. S. A. Nemnonow, *Zhur. Tekh. Fiz.*, **18**, 1948, 239–245; abstract, *Structure Repts.*, **12**, 1949, 53. Review, no original work. Nemnonow considers Cr of the α-Mn type to be a transition phase between the hexagonal and cubic phases.
14. E. A. Ollard, *Met. Ind.* (*London*), **28**, 1926, 153.
15. C. G. Fink, *Trans. Electrochem. Soc.*, **59**, 1931, 437–443.
16. W. A. Wood, *Phil. Mag.*, **24**, 1937, 511–518.

17. W. Hume-Rothery, "Structure of Metals and Alloys," Institute of Metals, London, 1936.
18. L. Pauling, *J. Am. Chem. Soc.*, **69**, 1947, 542.
19. C. A. Snavely, *Trans. Electrochem. Soc.*, **92**, 1947, 537–576.
20. C. A. Snavely and D. A. Vaughan, *J. Am. Chem. Soc.*, **71**, 1949, 313–314.
21. A Cr:H ratio of 1:1.7 was found by an extraction test. No similar tests were undertaken for the hexagonal hydride as it was doubted that a sufficiently pure sample could be obtained [19].
22. The structure types named are postulated since the positions of the hydrogen atoms could not be established from X-ray data. The hexagonal hydride is assumed to exist between the compositions Cr_2H and CrH, the cubic hydride—cf. [21]—between CrH and CrH_2.

$\bar{1}.4642$
0.5358

Cr-Hf Chromium-Hafnium

"An alloy of [the composition] $HfCr_2$ (36.81 wt. % Cr) has been identified as isomorphous with $MgNi_2$ (C36 type) with lattice parameters of c = 16.325 kX and a = 5.047 kX, c/a = 3.234. The incipient melting temperature of this alloy was determined as 1480°C. There is a possibility of allotropy with a $MgZn_2$ modification at lower temperatures. The powder method, however, is too insensitive to ascertain this" [1].

1. R. P. Elliott, Armour Research Foundation, Chicago, Ill., Technical Report **1**, OSR Technical Note OSR-TN-247, August, 1954, p. 10.

$\bar{1}.4137$
0.5863

Cr-Hg Chromium-Mercury

The preparation of a "Cr-amalgam" has been repeatedly claimed [1, 2]. By mechanical separation from the liquid phase a residue of the composition CrHg was obtained by [2]; this finding, however, is not sufficient evidence for the existence of a compound of this composition.

On the basis of emf measurements, [3] assume the solubility of Cr in Hg at 18°C to be 3.1×10^{-11} wt. (11.9×10^{-11} at.) % Cr. According to a chemical analysis [4], the solubility is less than 5×10^{-5} wt. %.

Supplement. X-ray examination of the centrifugalized solid component of electrolytically prepared "Cr-amalgam" showed it to be b.c.c. Cr. Spectroscopic analysis indicated the solubility of Cr in liquid Hg at 30°C to be less than 4×10^{-7} wt. (16×10^{-7} at.) % [5].

1. H. Moissan, *Compt. rend.*, **88**, 1879, 180–183; *Ann. chim. et phys.*, (5)**21**, 1880, 250.
2. J. Férée, *Compt. rend.*, **121**, 1895, 822–824.
3. G. Tammann and J. Hinnüber, *Z. anorg. Chem.*, **160**, 1927, 257–259.
4. N. M. Irvin and A. S. Russell, *J. Chem. Soc.*, **135**, 1932, 891–898.
5. J. F. de Wet and R. A. W. Haul, *Z. anorg. Chem.*, **277**, 1954, 96–112.

$\bar{1}.4303$
0.5697

Cr-Ir Chromium-Iridium

A Cr-Ir alloy with about 10 wt. (29 at.) % Cr, prepared by sintering, proved to be ferromagnetic [1].

According to recent X-ray diffraction and microscopic work by [2], the system Cr-Ir has the following constitution: Ir dissolves about 27.5 at. (9.4 wt.) % Cr:

around 25 at. % Cr a ferromagnetic (cf. above) ordered phase of the Cu_3Au ($L1_2$) type forms at lower temperatures. The homogeneity range of a h.c.p. ϵ phase extends from 31.5 to 58 at. (10.7–27.1 wt.) % Cr; between 72.5 and 78.5 at. (41.5–49.6 wt.) % Cr the cubic (Cr_3Ir) phase with "β-W" (A15) type of structure is stable. Cr takes about 2.5 at. (8.7 wt.) % Ir into solid solution (at 900°C).

1. E. Friederich, *Z. tech. Physik,* **13,** 1932, 59.
2. E. Raub and W. Mahler, *Z. Metallkunde,* **46,** 1955, 210–215.

$\bar{1}.5733$
0.4267

Cr-La Chromium-Lanthanum

Some cursory alloying tests indicate that probably no intermediate phases occur in this system [1].

1. L. Rolla and A. Iandelli, *Ber. deut. chem. Ges.,* **75,** 1942, 2094–2095.

0.3301
$\bar{1}.6699$

Cr-Mg Chromium-Magnesium

The preparation of Cr-Mg alloys by reduction of Cr_2O_3 with an excess of Mg has been claimed by [1, 2]. [2] suggests the existence of several compounds (CrMg, Cr_2Mg_3, $CrMg_3$, and $CrMg_4$); the method used in deriving these formulas, however, is questionable [3].

1. J. Parkinson, *J. Chem. Soc.,* **20,** 1867, 117–131, especially 128.
2. E. Montignie, *Bull. soc. chim. France,* (5)**5,** 1938, 567–568.
3. There are maxima and minima in the "free chromium" (residue after treatment of the reaction product with hot dilute HCl) vs. composition curve.

$\bar{1}.9763$
0.0237

Cr-Mn Chromium-Manganese

The phase diagram of Fig. 311 is mainly based on the very careful thermal, micrographic, and roentgenographic investigations by [1, 2].

The constitution in the range 40–100 wt. % Mn was first outlined by thermal analysis and metallography by [3]; in the light of Fig. 311 the results—as available through an abstract—have to be considered incomplete, especially since the phase existing around 75 at. % Mn was not found. This phase, ($CrMn_3$) (76.01 wt. % Mn), was detected during a cursory thermal, microscopic, and roentgenographic study [5] which also showed that the extent of the Cr-primary solid solution decreases greatly with decreasing temperature.

The liquidus above ~1500°C, determined by using an optical pyrometer [1], has been checked by [6] by means of a Mo-W thermocouple (see Fig. 311) [7]. According to microscopic and roentgenographic work [8], the phase field of ($CrMn_3$) extends—in agreement with former results [2]—at 1000°C from 76 to 84 at. (77–85 wt.) % Mn. This compound undergoes a transformation ($\sigma \rightleftarrows \sigma'$) near 1000°C as [9, 10, 2] have shown by means of several standard methods; for Fig. 311 the data given by [2] were used.

As is indicated by the dashed lines below 800°C in the central part of Fig. 311, equilibrium could not be reached under conditions of annealing of the order of 1–2 months [2]. "In the range 800–600°C when the solubility limit of the (Cr) phase is exceeded, three-phase alloys of the (Cr) + σ' + α'-Mn [see Crystal Structures] type are produced on annealing previously homogenized specimens, and from these the α'-Mn

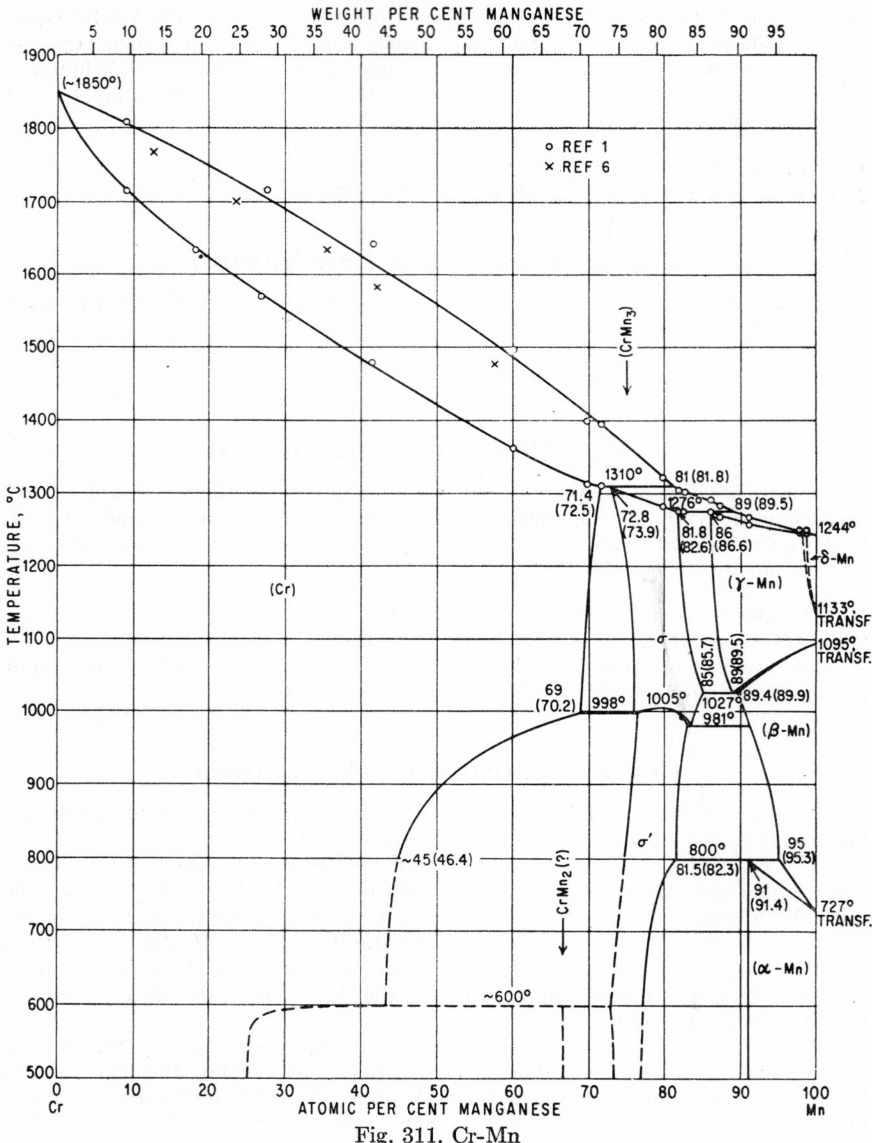

Fig. 311. Cr-Mn

phase gradually disappears on prolonged annealing so that there is little doubt that above 600°C conditions of true equilibrium involve two-phase (Cr) + σ' alloys. A horizontal almost certainly exists at 600°C, and below this temperature, three-phase alloys are again produced on annealing previously homogenized specimens and it seems almost certain that, under conditions of true equilibrium, the (Cr) phase is in equilibrium with the α'-Mn phase, but the transformations are so sluggish that the composition of the latter cannot yet be determined; $CrMn_2$ [Fig. 311] is a possible composition for the α'-Mn phase" [2, 11].

Crystal Structures. The increase in the lattice parameter of (Cr) with increasing Mn content was found to be very small [5, 1, 2]. The phase ($CrMn_3$) was shown to be a σ phase [12]; its transformation near 1000°C may be of the order-disorder type [10, 2].

The compound $CrMn_2$ (?) gave rise to a set of characteristic lines which were identified [2] as due to a structure similar to that of the α-Mn type; the symbol α'-Mn is used to emphasize this resemblance [2].

Addition of Cr decreases the lattice parameter of α-Mn as well as that of β-Mn [1].

Supplement. [13] studied a specimen of $CrMn_3$ by means of neutron diffraction after quenching from 1100°C as well as after reannealing at 850°C. A comparison of the observed intensities gives support to the view that ordering occurs below about 1000°C.

1. S. J. Carlile, J. W. Christian, and W. Hume-Rothery, *J. Inst. Metals,* **76,** 1949-1950, 169–194. For a correction, see W. Hume-Rothery and J. W. Christian, *Monthly J. Inst. Metals* (*News Section*), May, 1950, p. 152.
2. W. B. Pearson and W. Hume-Rothery, *J. Inst. Metals,* **81,** 1952-1953, 311–314. In both investigations [1, 2] the Cr and Mn used were high-purity electrolytic hydrogen-reduced metals and were melted in alumina or thoria-lined alumina crucibles in an atmosphere of hydrogen. In many cases the alloys were analyzed after the final annealing treatment.
3. I. I. Kornilov and A. I. Tat'yanchikova, *Doklady Akad. Nauk S.S.S.R.,* **50,** 1945, 223–225; abstracts in *Chem. Abstr.,* **44,** 1950, 6372; Associated Electrical Industries, Ltd., Research Report A 295, September, 1953. The alloys were melted in air and therefore certainly were strongly contaminated by nitrogen; cf. [4].
4. According to [1], Cr-Mn alloys absorb nitrogen readily at high temperatures. See also S. J. Carlile and W. Hume-Rothery, *J. Inst. Metals,* **76,** 1949-1950, 195 and 718 (discussion).
5. U. Zwicker, *Z. Metallkunde,* **40,** 1949, 377–378.
6. H. T. Greenaway, S. T. M. Johnstone, and M. K. McQuillan, *J. Inst. Metals,* **80,** 1951-1952, 109–114.
7. No correlation was found between the (low) gas contents (O, N) of the alloys and the differences between the two sets of results. The explanation for these differences is probably to be found in the different methods of temperature measurement used.
8. P. Greenfield and P. A. Beck, *Trans. AIME,* **200,** 1954, 254.
9. U. Zwicker, *Z. Metallkunde,* **42,** 1951, 277–278.
10. W. B. Pearson, *Nature,* **169,** 1952, 934.
11. For a discussion of the possibility that σ' decomposes eutectoidally into (Cr) and α'-Mn, see the original paper [2]. In this connection it is interesting to note that [3] assumed the formation of a phase CrMn by solid-state reaction.
12. W. B. Pearson, J. W. Christian, and W. Hume-Rothery, *Nature,* **167,** 1951, 110.
13. J. S. Kasper and R. M. Waterstrat, *Acta Cryst.,* **9,** 1956, 289–295.

$\bar{1}.7340$
0.2660

Cr-Mo Chromium-Molybdenum

By means of X-ray investigations of sintered [1, 2] and cast [3] alloys, covering the whole range of composition, it was established that there is a continuous series of solid solutions [9]. The existence of only one phase was confirmed for the regions

0–50 [4] and 50–100 at. % Mo [5]. No indication of a decomposition of the solid solutions at lower temperatures was found by X-ray studies after long annealings at 1200, 1000, 600 [1, 3], and 650°C [2].

Melting-point determinations by optical observation indicated the existence of a minimum at about 15 at. (25 wt.) % Mo, 1700°C [1, 3], based on a melting point of only 1770°C [1] for pure Cr. Recently, liquidus temperatures for the region 0–45 at. % Mo, determined by thermal analysis, have been reported [4]. They indicate the minimum to occur at about 12 at. (20 wt.) % Mo and 1860°C, in very good agreement with a similar study by [6] (Fig. 312).

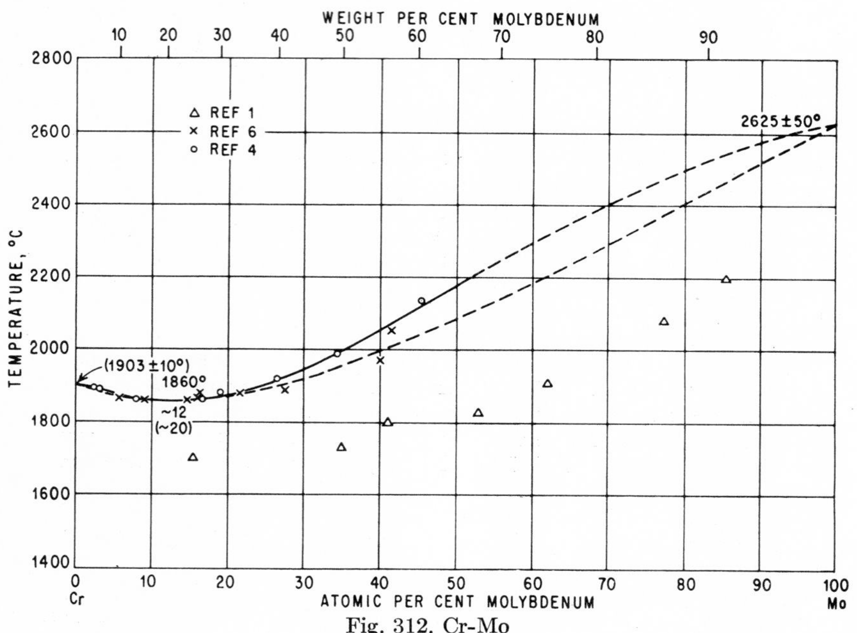

Fig. 312. Cr-Mo

Cooling and heating curves of alloys with up to 45 at. % Mo showed thermal effects at temperatures approximately 100 ± 30°C below the liquidus [4]; data points are not shown in Fig. 312. They were proposed "to be derived from the transformation detected in the pure chromium," at 1840 ± 15°C [4, 7]. "A phase diagram for the Cr-Mo system which conforms to the thermal analyses and includes the chromium phase change is not easily constructed; nevertheless, it is clear that the Cr-Mo phase diagram as previously drawn cannot be considered to be complete or correct" [4].

Lattice parameters were reported by [1, 2, 3, 8].

1. W. Trzebiatowski and H. Ploszek, *Naturwissenschaften,* **26,** 1938, 462; W. Trzebiatowski, H. Ploszek, and J. Lobzowski, *Anal. Chem.,* **19,** 1947, 93–95.
2. S. R. Baen and P. Duwez, *Trans. AIME,* **191,** 1951, 331–335.
3. O. Kubaschewski and A. Schneider, *Z. Elektrochem.,* **48,** 1942, 671–674.
4. D. S. Bloom and N. J. Grant, *Trans. AIME,* **200,** 1954, 261–268.
5. H. D. Kessler and M. Hansen, *Trans. ASM,* **42,** 1950, 1008–1030.
6. J. W. Putman, R. D. Potter, and N. J. Grant, *Trans. ASM,* **43,** 1951, 824–847; for data see fig. 5 of [4].

7. D. S. Bloom, J. W. Putman, and N. J. Grant, *Trans. AIME*, **194**, 1952, 626.
8. Alloys with up to 8 at. % Cr: Climax Molybdenum Co., Arc-cast Molybdenum-base Alloys, 1st Annual Report, Apr. 1, 1950.
9. Earlier work by E. Siedschlag (*Z. anorg. Chem.*, **131**, 1923, 191–196) is obsolete.

0.5697
$\bar{1}.4303$

Cr-N Chromium-Nitrogen

Preparative Work. The preparation of chromium nitrides by nitration of $CrCl_2$ or CrO_2Cl_2 [1–6] yielded very impure products. It proved more advantageous to let metallic chromium react with NH_3 or N_2 [7–12]; the nitrogen content of the products obtained varied owing to the different conditions of temperature and time applied, but CrN (21.22 wt. % N) could be isolated by [7, 8]. A compound Cr_3N_2 which [4, 9] claimed to have found was not corroborated by later work.

Solubility of N in Liquid Cr. Nitrogen greatly lowers the melting point of pure Cr. Chromium which had been melted under a "protective" $H_2 + N_2$ atmosphere dissolved 3.65 wt. % N and showed a melting point of about 1650°C [13]. [14] found about 1630°C as the melting point for chromium melted in air (2.1 wt. % N besides oxygen) and observed an additional thermal effect at 1580°C which has been ascribed to a eutectic. The solubility under atmospheric pressure is, however, limited. At 1600°C the melt in equilibrium with nitrogen gas of atmospheric pressure contains about 4.0 wt. (13–14 at.) % N [11, 15, 16, 17]. There is also agreement [11, 16] that the solubility decreases to about 3.5 wt. (12 at.) % N at 1750°C. For the influence of pressure (2 atm) on the solubility, see [11].

X-ray and Microscopic Work. [18] investigated eight alloys with up to 50 at. (21 wt.) % N—which were annealed 1 hr at 1000°C or 3 hr at 1250–1300°C in evacuated silica tubes—by the X-ray powder method with the following results: (*a*) No shifting of chromium lines occurs so that the solubility of N in solid Cr is very limited. This observation could be corroborated by [11]. (*b*) There exist two intermediate phases, Cr_2N (11.87 wt. % N) and CrN. As to Cr_2N, a h.c.p. structure and a homogeneity range between 11.3 and 11.9 wt. (32 and 33.3 at.) % N were found. CrN was shown to have the NaCl (B1) type of structure, with a = 4.15 A.

[16] reported the same unit cell for Cr_2N as [18], but [19] has shown that, as indicated by weak superstructure lines, the true unit cell is three times larger than that of the simple h.c.p. structure (with interstitial-at-random distribution of the N atoms) suggested before. The true unit cell has the lattice parameters a = 4.759–4.805 A, c = 4.438–4.479 A, c/a = 0.933–0.932 in its homogeneity range from 9.3 to 11.9 wt. (27.6–33.3 at.) % N (cf. above). The larger cell was also observed by [20]; see also [21].

The microstructures of Cr-N alloys prepared by melting methods and containing up to 4 wt. % N show a nitride phase [14, 13, 11, 22, 12] (evidently Cr_2N; see [22]) in a eutectic-like arrangement and as a needlelike precipitate from the (Cr) phase [11, 12]. It seems, therefore, that solid Cr dissolves a limited amount of N at higher temperatures. (See Note Added in Proof.)

Physicochemical and Physical Measurements. [3] observed that CrN starts to decompose at about 1450°C. From measurements of the dissociation tension of CrN at 800°C, [23] concluded that Cr and CrN form a continuous series of solid solutions. [24] took the same view on the basis of measurements of equilibrium pressures at different temperatures. The existence of solid solutions of N in Cr was also claimed by [25], based on electrical-conductivity measurements; this interpretation of the data, however, is not convincing. [26] measured pressure-concentration isotherms between 810 and 1000°C and [27] interpreted these data by assuming that Cr_2N and CrN form at temperatures below about 1000°C by decomposition of a solid solution (0 to at least 50 at. % N!) stable at higher temperatures. However, these views, which assume extended solid solutions, are in contradiction to the more direct (although limited) X-ray and microscopic evidence and are obsolete.

In the course of thermal decomposition of $Cr(NH_2)_3$, X-ray amorphous CrN and Cr_2N were observed [28]. Microhardness measurements on two-phase ($Cr + Cr_2N$) alloys were carried out by [12]. Variations in the hardness of both the Cr and the Cr_2N phase as a function of the annealing temperature were observed.

Melting Point of Chromium. Data for the melting point of chromium published in the voluminous literature up to 1935 vary from 1500 to 1920°C, the variation undoubtedly being due to contaminants (mainly N, O, C) and to faulty temperature measurement. For a critical discussion of these data, reference is made to [28*a*].

Several more recent determinations gave values as follows: 1890 ± 10 [29], 1860 [30], 1845 ± 10 [31], 1892 [32], and 1900 ± 10°C [33]. Taking into account that the value of [31] may be somewhat too low (see [34]) the value 1880 ± 20°C covers fairly well the data mentioned.

It is obvious today that most of the older work on Cr alloys was carried out under quite unsatisfactory experimental conditions (melting in air or under a "protective" nitrogen atmosphere, use of unsuitable crucible materials or of impure Cr; for details see the particular systems). The Cr-rich regions, at least, of many binary Cr systems should therefore be reinvestigated.

Note Added in Proof. [35] determined the solubility of nitrogen in solid Cr by means of measurement of pressure-composition isotherms and microscopic observations as 0.26, 0.16, 0.097, 0.056, and 0.028 wt. (0.96, 0.59, 0.36, 0.21, and 0.10 at.) % N at 1400, 1300, 1200, 1100, and 1000°C, respectively.

1. J. Liebig, *Pogg. Ann.*, **21,** 1831, 359; see also [2].
2. A. Schrötter, *Liebigs Ann.*, **37,** 1841, 148.
3. C. R. Ufer, *Liebigs Ann.*, **112,** 1859, 281.
4. Uhrlaub, Dissertation, University Göttingen, 1859.
5. A. Smits, *Rec. trav. chim.*, **15,** 1897, 136.
6. Guntz, *Compt. rend.*, **135,** 1902, 739.
7. F. Briegleb and A. Geuther, *Liebigs Ann.*, **123,** 1862, 239.
8. J. Férée, *Bull. soc. chim. France,* (3)**25,** 1901, 618.
9. G. G. Henderson and J. C. Galletly, *J. Soc. Chem. Ind.*, **27,** 1908, 388.
10. L. Duparc, P. Wenger, and W. Schusselé, *Helv. Chim. Acta*, **13,** 1930, 917–929.
11. R. M. Brick and J. A. Creevy, *AIME Metals Technol.*, **7,** 1940, TP 1165.
12. M. L. Korolev, *Izvest. Akad. Nauk S.S.S.R.* (*Tekh.*), **1953**(10), 1465–1470. Solid Cr was nitrided by NH_3 (800–900°C) and then melted. The sample contained 2.5 wt. % N and showed needlelike inclusions of Cr_2N.
13. F. Sauerwald and A. Wintrich, *Z. anorg. Chem.*, **203,** 1931, 73–74.
14. F. Adcock, *J. Iron Steel Inst.*, **114,** 1926, 117–126.
15. T. Saito, *Science Repts. Research Insts. Tôhoku Univ.*, **A1,** 1949, 419–424.
16. V. S. Mozgovoi and A. M. Samarin, *Doklady Akad. Nauk S.S.S.R.*, **74,** 1950, 729–732; *Izvest. Akad. Nauk S.S.S.R.* (*Tekh.*), **1950**(10), 1529–1536.
17. [16] give the most detailed information about the solubility of N in liquid Cr as a function of temperature (P_{N_2} = 1 atm; two measurements at each temperature): 1600°C, 4.036 and 4.133; 1650°C, 3.879 and 3.933; 1700°C, 3.854 and 3.827; 1725°C, 3.72 and 3.80; 1750°C, 3.55 and 3.533 (wt. % N).
18. R. Blix, *Z. physik. Chem.*, **B3,** 1929, 229–239.
19. S. Eriksson, *Jernkontorets Ann.*, **118,** 1934, 530–543; see *Strukturbericht*, **3,** 1933-1935, 586–587.

20. W. Hume-Rothery and W. B. Pearson, *J. Inst. Metals*, **76**, 1949-1950, 722–725 (discussion).
21. F. C. Blake and J. O. Lord (*Phys. Rev.*, **35**, 1930, 660) claimed to have found a h.c.p. structure for CrN. As their lattice parameters agree fairly well with those reported by [18] for Cr_2N, it is obvious that the authors assigned the wrong formula to their alloy.
22. A. H. Sully and T. J. Heal, *J. Inst. Metals*, **76**, 1949-1950, 719–722 (discussion).
23. E. Baur and G. L. Voerman, *Z. physik. Chem.*, **52**, 1905, 473.
24. I. Shukow, *Zhur. Russ. Fiz.-Khim. Obshchestva*, **40**, 1908, 457–459.
25. I. Shukow, *Zhur. Russ. Fiz.-Khim. Obshchestva*, **42**, 1910, 40–41.
26. G. Valensi, *J. chim. phys.*, **26**, 1929, 152–177, 202–218.
27. G. Tammann, *Z. anorg. Chem.*, **188**, 1930, 396–401.
28. O. Schmitz-Dumont, G. Broja, and H. F. Piepenbrink, *Z. anorg. Chem.*, **254**, 1947, 329–342.
28*a*. M. Hansen, "Der Aufbau der Zweistofflegierungen," Springer-Verlag OHG, Berlin, 1936.
29. G. Grube and R. Knabe, *Z. Elektrochem.*, **42**, 1936, 794–795.
30. S. J. Carlile, J. W. Christian, and W. Hume-Rothery, *J. Inst. Metals*, **76**, 1949-1950, 181–182.
31. H. T. Greenaway, S. T. M. Johnstone, and M. K. McQuillan, *J. Inst. Metals*, **80**, 1951-1952, 109–114.
32. J. W. Putman, R. D. Potter, and N. J. Grant, *Trans. ASM*, **43**, 1951, 834.
33. D. S. Bloom, J. W. Putman, and N. J. Grant, *Trans. AIME*, **194**, 1952, 626.
34. N. J. Grant and D. S. Bloom, *Trans. AIME*, **194**, 1952, 524 (discussion).
35. A. U. Seybolt and R. A. Oriani, *Trans. AIME*, **206**, 1956, 556–562.

$\bar{1}.7480$
0.2520

Cr-Nb Chromium-Niobium

According to a cursory, mainly roentgenographic, investigation [1] in which a niobium of only 86–90 wt. % purity [2] was used, the constitution of the Nb-Cr system appears to be much the same as that of the Ta-Cr system. The melting points are said to be about 50–100°C lower than in the Ta-Cr system, extending from 1750 to 1950°C for alloys containing up to 50 wt. % Nb. The only intermediate phase is located around the stoichiometric composition $NbCr_2$ (52.81 wt. % Cr) [3, 4]. The solubility of Nb in solid Cr is estimated to amount to several wt. % [1].

Crystal Structure. Addition of Nb causes a slight expansion of the Cr lattice [1]. $NbCr_2$ possesses the cubic $MgCu_2$ (C15) type of structure, with a = 6.990 A [3].

Supplement. [5] corroborated the $MgCu_2$-type structure for $NbCr_2$ (a = 6.985 A) and determined the incipient melting temperature of this alloy as 1710°C.

1. O. Kubaschewski and A. Schneider, *J. Inst. Metals*, **75**, 1948-1949, 410–411.
2. Remainder mainly Ta; traces of Ti.
3. P. Duwez and H. Martens, *Trans. AIME*, **194**, 1952, 72–74.
4. Kubaschewski and Schneider [1] had suggested the formula Nb_2Cr_3. In contrast to $TaCr_2$, the Nb compound undergoes no polymorphic transformation [3].
5. R. P. Elliott, Armour Research Foundation, Chicago, Ill., Technical Report 1, OSR Technical Note OSR-TN-247, August, 1954, p. 18.

$\bar{1}.9475$
0.0525

Cr-Ni Chromium-Nickel

Liquidus. Determinations of the liquidus or parts thereof were carried out by [1–7]; but, except for the work of [7] and, to some extent, that of [6], the experimental conditions (especially as to purity of chromium and precautions to keep nitrogen away) were poor. Unfortunately, [7] give no detailed data so that information from this source had to be taken from a small diagram. The liquidus in Fig. 313 is that of [7] with the exception that 1880°C (see Cr-N) was chosen as the melting point of Cr instead of 1930° plotted in [7] (which they later corrected to about 1900°C [8, 9]). Some of the liquidus data of [6] are also plotted in Fig. 313 (small circles). The values for the eutectic temperature reported by [2, 6, 7] lie in the range 1340–1346°C.

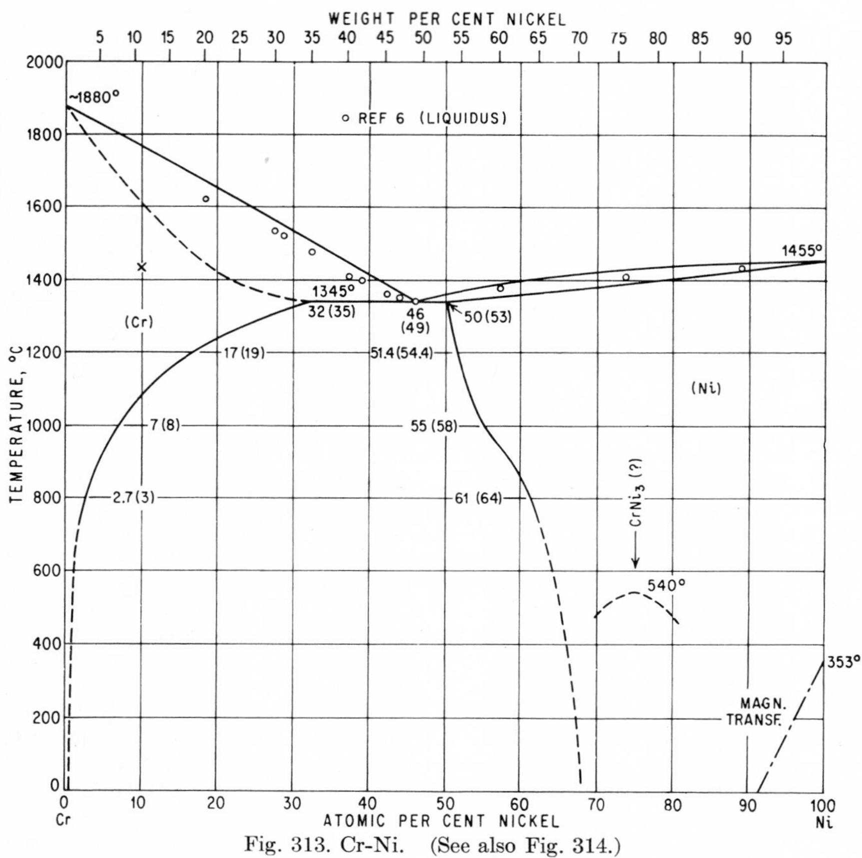

Fig. 313. Cr-Ni. (See also Fig. 314.)

Solidus. Solidus temperatures were determined by [2–4, 6], but only [6] investigated alloys below 20 wt. % Ni. For the Ni-rich branch in Fig. 313 the thermal and microscopic results of [6] were used. As to the Cr-rich branch, however, the curve found by [6] mainly by micrographic work, the general course of which is indicated by the single cross in Fig. 313, is not accepted since it presumably suffered from insufficient homogenization of the alloys. The curve shown in Fig. 313 rises somewhat more steeply from the end point of the eutectic horizontal than those found by [2, 3] but not nearly so steeply [10] as would be concluded from scanty thermal work in this region by [4].

Solid Alloys. The boundaries of both primary solid solutions in Fig. 313 down to 750°C were taken graphically from the paper of [11]. The boundary of the (Cr) phase below 1100°C, as measured roentgenographically by these authors, is believed to be correct to within ±1.0 at. % and is in satisfactory agreement with former X-ray work (between 524 and 1113°C) by [12]. On the other hand, this curve fits smoothly onto the boundary established micrographically by [3, 6] for temperatures above 1200°C. Data derived for 1100 and 1200°C from oxidation-equilibrium measurements [13] also fit fairly well on this curve.

The boundary of the (Ni) phase, as roentgenographically determined by [11] between 750 and 1200°C (believed to be accurate to within 0.5 at. % [14]), shows a change in slope at about 1000°C and approximates very closely, above 950°C, the micrographic curve of [6]; single microscopic data by [2, 15, 16] also agree with this boundary. The X-ray results of [12], which indicate a much higher solubility of Cr in Ni, were shown [6, 11] to be incorrect.

Numerous other papers dealing with the solid boundaries will only be mentioned, being now obsolete [17]: [18–23, 4, 24]. Lattice parameters of both primary solid solutions were measured by [12, 11].

An order-disorder transformation at about 540°C based on the composition $CrNi_3$ (77.20 wt. % Ni) is indicated by the results of electrical-resistivity [25, 26], X-ray [25, 11, 26], specific-heat [26], and dilatometric [25] measurements. The ordered phase has the cubic Cu_3Au ($L1_2$) type of structure [26]. (See Supplement.)

From theoretical considerations [27, 28] there is reason to suspect that a σ phase could form in the Cr-Ni system, and this is substantiated by the position and shape of σ-phase fields in ternary systems (see for instance [29, 30]). Extensive tests [11, 30], however, have failed to produce such a phase in Cr-Ni alloys as yet [31, 32].

More recently, the simple form of the diagram as shown in Fig. 313 has been questioned. [7, 33] concluded, mainly from roentgenographic and microscopic evidence, that a eutectoid at about 1180°C, 35 wt. (32.3 at.) % Ni exists; the allotropic transformation in Cr implied by the eutectoid reaction was reported [7, 9] to lie near 1840°C. The Cr high-temperature form was tentatively identified as a f.c.c. (A1) structure, with $a = 3.77$ A. Furthermore, it has been observed that alloys of "near-eutectoid composition" can form a metastable body-centered tetragonal phase ($c/a = 1.09$), and this may be the phase reported by [34, 35] as Cr_2Ni.

As to the high-temperature form of Cr and its solid-solution field, the evidence published is not wholly convincing, and corroboration by other authors is still lacking [11, 36–38]. It seems premature, therefore, to abandon the generally accepted phase diagram of Fig. 313.

The influence of Cr on the Curie temperature of Ni was investigated by [1, 39, 40, 2, 41, 42]. A straight line down to 15 at. % Cr, 0°K agrees very well, at least above 0°C, with the most recent data of both [41] and [42] (Fig. 313).

Ferromagnetism in Cr-rich alloys, which is connected with the metastable phase mentioned above, has been reported by [7].

Supplement. The property anomalies in Ni-rich alloys, which [25, 11, 26] suggested to be due to the existence of an ordered $CrNi_3$ phase, were also studied by [43] (X-ray), [44] (resistivity), [45] (resistivity, X-ray), [46] (metallography, X-ray, dilatometry, and resistivity), and [47] (specific heat). Based on their results and the fact that superstructure lines have never been positively detected unless the alloys contained some per cent of Al, [45] concluded that the existence of long-range order is very improbable and that only short-range order (below 800–900°C) exists in the neighborhood of $CrNi_3$ (see also [43]). [44] concluded "deviation from the at random atomic distribution" and [46, 47], formation of a "compound" $CrNi_3$. Since the resistivity of alloys with more than 30 at. % Cr decreased markedly during long-time annealing ("which cannot be explained by precipitation of either the Cr-rich phase or impurities"), [45] suggested that long-range ordering of the CuAu type might exist in these alloys.

[48] published the results of further work aimed to substantiate earlier claims (see above) for the existence of a high-temperature form of Cr and its solid-solution field (Fig. 314). However, confirmation by other authors is still missing [49].

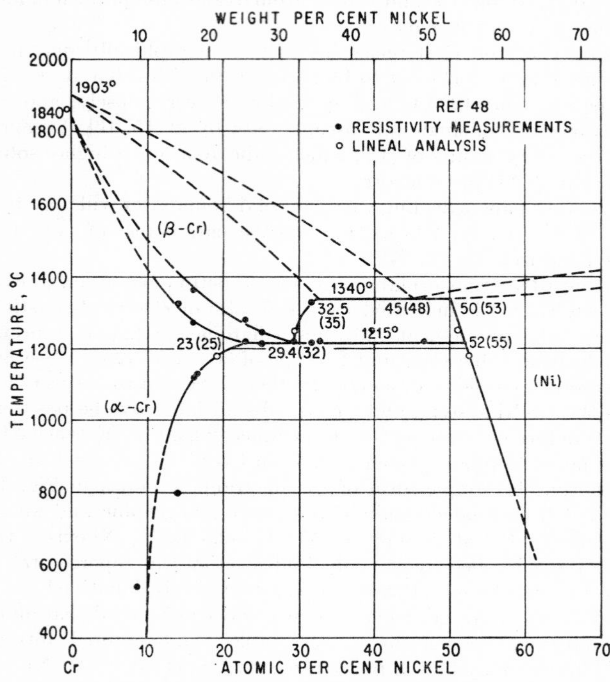

Fig. 314. Cr-Ni. (See also Fig. 313.)

Recently [50] reported that he found, by using a special X-ray technique, superstructure lines which are believed to belong to the ordered phase $CrNi_2$. The alloy investigated (29.2 wt. % Cr, 0.9 wt. % Mn, 0.65 wt. % Si, balance Ni) had been annealed for a long time at 500°C in hydrogen.

1. G. Voss, *Z. anorg. Chem.*, **57,** 1908, 58–61.
2. Y. Matsunaga, *Kinzoku-no-Kenkyu,* **6,** 1929, 207–218; *Japan Nickel Rev.*, **1,** 1933, 347.
3. S. Nishigori and M. Hamasumi, *Science Repts. Tôhoku Univ.*, **18,** 1929, 491–502; *Kinzoku-no-Kenkyu,* **6,** 1929, 219–228.
4. F. Wever and W. Jellinghaus, *Mitt. Kaiser-Wilhelm-Inst. Eisenforsch. Düsseldorf,* **13,** 1931, 93–108.
5. M. Tasaki, *Proc. World Eng. Cong., Tokyo, 1929,* **36,** 1931, 231–245 (in German).
6. C. H. M. Jenkins, E. H. Bucknall, C. R. Austin, and G. A. Mellor, *J. Iron Steel Inst.*, **136,** 1937, 187–220. Contains an almost complete review of investigations up to 1937.
7. D. S. Bloom and N. J. Grant, *Trans. AIME,* **191,** 1951, 1009–1014.

8. N. J. Grant and D. S. Bloom, *Trans. AIME,* **194,** 1952, 524 (discussion).
9. D. S. Bloom, J. W. Putman, and N. J. Grant, *Trans. AIME,* **194,** 1952, 626.
10. See [12] for a graphical comparison of the results of [2–4].
11. A. Taylor and R. W. Floyd, *J. Inst. Metals,* **80,** 1951-1952, 577–587.
12. E. R. Jette, V. H. Nordstrom, B. Queneau, and F. Foote, *Trans. AIME,* **111,** 1934, 361–371.
13. G. Grube and M. Flad, *Z. Elektrochem.,* **48,** 1942, 377–389; see J. L. Meijering (*Philips Research Repts.,* **3,** 1948, 281–302) for a critical discussion of this paper.
14. A. Taylor and R. W. Floyd, *J. Inst. Metals,* **80,** 1951-1952, 699 (discussion).
15. W. Rosenhain and C. H. M. Jenkins, *J. Iron Steel Inst.,* **121,** 1930, 231.
16. C. H. M. Jenkins, H. J. Tapsell, C. R. Austin, and W. P. Rees, *J. Iron Steel Inst.,* **121,** 1930, 246.
17. Much of the older work was done on alloys which certainly were not in equilibrium. The rate of diffusion in this system is slow even at high temperatures.
18. E. C. Bain, *Trans. AIME,* **68,** 1923, 631–633.
19. W. C. Phebus and F. C. Blake, *Phys. Rev.,* **25,** 1925, 107.
20. F. C. Blake and A. E. Focke, *Phys. Rev.,* **27,** 1926, 798–799.
21. F. C. Blake, J. O. Lord, and A. E. Focke, *Phys. Rev.,* **29,** 1927, 206–207.
22. E. J. Smithells, S. V. Williams, and J. W. Avery, *J. Inst. Metals,* **40,** 1928, 275–276.
23. S. Sekito and Y. Matsunaga, *Kinzoku-no-Kenkyu,* **6,** 1929, 229–233.
24. J. W. Pugh and J. D. Nisbet, *Trans. AIME,* **188,** 1950, 270, 274.
25. Z. Yano, *Rikwagaku-Kenkyu-jo Iho,* **19,** 1940, 110–117; *Japan Nickel Rev.,* **9,** 1941, 17.
26. A. Taylor and K. G. Hinton, *J. Inst. Metals,* **81,** 1952-1953, 169–180.
27. A. H. Sully, *Nature,* **167,** 1951, 365–366.
28. A. H. Sully, *J. Inst. Metals,* **80,** 1951-1952, 173–179.
29. P. A. Beck and W. D. Manly, *Trans. AIME,* **185,** 1949, 354.
30. D. S. Bloom and N. J. Grant, *Trans. AIME,* **200,** 1954, 266.
31. P. Chevenard (*Compt. rend.,* **174,** 1922, 109–112) suggested the existence of a compound Cr_3Ni_2 from dilatometric work on Ni-rich alloys.
32. See also H. J. Goldschmidt, *J. Inst. Metals,* **80,** 1951-1952, 696 (discussion).
33. See [30], pp. 262–263, 266.
34. F. C. Blake, J. O. Lord, W. C. Phebus, and A. E. Focke, *Phys. Rev.,* **31,** 1928, 305.
35. F. C. Blake and J. O. Lord, *Phys. Rev.,* **35,** 1930, 660.
36. See discussion on [11] in *J. Inst. Metals,* **80,** 1951-1952, 695, 699–700.
37. Discussion on [7] in *Trans. AIME,* **194,** 1952, 523–524 (especially P. Duwez and H. Martens).
38. J. O. McCaldin and P. Duwez, *Trans. AIME,* **200,** 1954, 619–620.
39. K. Honda, *Ann. Physik,* **32,** 1910, 1007–1009.
40. P. Chevenard, *J. Inst. Metals,* **36,** 1926, 46–53; *Rev. mét.,* **25,** 1928, 14–22; *Stahl u. Eisen,* **48,** 1928, 1045–1047; *Trav. mem. bur. int. poids et mesures,* **17,** 1927.
41. C. Sadron, *Compt. rend.,* **190,** 1930, 1339–1340.
42. V. Marian, *Ann. phys.,* **7,** 1937, 459–527, especially 502.
43. M. Silverstrone, Rensselaer Polytechnic Institute, Troy, N.Y. Abstract in Central Air Documents Office, Wright-Patterson Air Force Base, Dayton, Ohio, ATI-78124:

    "The properties of Nichrome V (approx. 80 wt. Ni and 20 wt. % Cr, with small amounts of other elements) were determined in the temperature range 400 to 800°C. A definite change in lattice parameter was found, closely paralleling the change in resistivity with temperature. This evidence, together with the results of other studies on Nichrome V, appears to indicate that a short range ordering phenomenon is taking place."

44. H. Thomas, *Z. Physik*, **129**, 1951, 219–232.
45. R. Nordheim and N. J. Grant, *J. Inst. Metals*, **82**, 1954, 440–444.
46. I. I. Kornilov and R. S. Mints, *Doklady Akad. Nauk S.S.S.R.*, **95**, 1954, 543–545; *Met. Abstr.*, **22**, 1955, 433.
47. H. Masumoto, M. Sugihara, and M. Takahashi, *Nippon Kinzoku Gakkai-Shi*, **18** (2), 1954, 85–87; *Met. Abstr.*, **22**, 1954, 183.
48. C. Stein and N. J. Grant, *Trans. AIME*, **203**, 1955, 127–134.
49. Unpublished microscopic and X-ray powder work by P. A. Beck on Cr-rich Cr-Ni alloys quenched from temperatures up to 1300°C failed to confirm the existence of a (Cr) high-temperature phase. (Private communication.)
50. G. Baer, *Naturwissenschaften*, **43**, 1956, 298.

0.5120
$\bar{1}$.4880

Cr-O Chromium-Oxygen

The Range Cr-Cr_2O_3. [1, 2] have shown by measurements of the equilibrium $2Cr + 1\frac{1}{2}O_2 \rightleftarrows Cr_2O_3$ [3], which were supported by roentgenographic and chemical analyses, that at least between 780 and 1300°C Cr and Cr_2O_3 (31.58 wt. % O) coexist without measurable mutual solubilities. The existence of a (presumably unstable) suboxide CrO (23.53 wt. % O) is indicated by work on preparation [4–6], but corroboration of its binary character seems necessary (see also [7]).

On the basis of scanty thermal [8], resistometric [8], and microscopic [9, 8] information it may be suggested that a miscibility gap in the liquid state occurs which extends down to about 1 wt. % Cr_2O_3 [0.3 wt. (1 at.) % oxygen] at the monotectic temperature (about 1780°C). An effect in the resistance-vs.-temperature curve of a 1 wt. % Cr_2O_3 alloy at about 1600°C has been interpreted [8] to be caused by a eutectic (certainly involving Cr and Cr_2O_3) [10].

Even below its melting point (about 2275°C [11]), Cr_2O_3 starts to decompose [12, 11, 13]; this partial dissociation [14] accounts for the fact that former authors [15, 16] reported much too low values for the melting point.

Crystal Structure. Cr_2O_3 has the rhombohedral Al_2O_3 ($D5_1$) type of structure [17, 18, 7], with the (hexagonal) lattice parameters a = 4.950 A, c = 13.665 A, c/a = 2.761 at 20°C [19] (taken from a graph). Cr_2O_3 is an antiferromagnetic substance having its Curie point at approximately 40°C [27, 20–22]. Relative intensity variations of X-ray diffraction lines, depending on the method of preparing the oxide, were observed by [23]. The powder patterns of CrO (see above) are similar to those of BeO and ZnO (wurtzite type) [6].

Cr_2O_3-CrO_3. The products which are obtained as intermediates in the thermal decomposition of CrO_3 to Cr_2O_3 have been studied by many investigators. For some of the more important papers, see [24–26] (in which additional references may be found).

Supplement. [28] identified by X-ray analysis an oxide of the probable formula Cr_3O with "β-W" (A15) type structure, a = 4.544 A. This oxide, however, could not be obtained in a pure state. The existence of an oxide CrO claimed by [6] could not be confirmed.

In a report on the exploratory investigation of the system Cr-Cr_2O_3-CaO, the following data for the binary Cr-Cr_2O_3 system were given [29]: (*a*) A miscibility gap in the liquid state extends from about 2 to about 70% Cr_2O_3 (0.6, 22 wt. % O, and 2, 48 at. % O, respectively) at the monotectic temperature of 1810°C; (*b*) a eutectic between Cr and Cr_2O_3 occurs at 80% Cr_2O_3 [25 wt. (52 at.) % O], 1660°C; (*c*) the melting point of Cr_2O_3 is 2400°C.

[30], who studied the phase relations in the Cr-Fe-O system by means of micro-

scopic, X-ray, and chemical analysis, concluded somewhat different solidification equilibria, characterized by the occurrence of a Cr_3O_4 high-temperature phase. Figure 315 was constructed from the qualitative information given by [30] and the monotectic and eutectic data of [29] (see above). Cr_3O_4 could be retained at room temperature and be indexed with a tetragonal cell, $a = 8.72$ A, $c/a = 0.86$. Figure 315 "indicates that in contradiction to the conventional assumptions Cr_2O_3 does not occur as a phase in equilibrium with the liquid metal" [30]. The solubility of oxygen

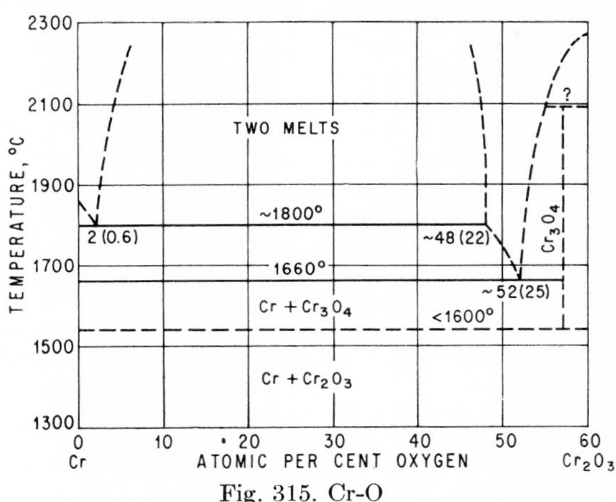

Fig. 315. Cr-O

in solid Cr as determined from diffusivity experiments [35] is approximately 0.03 wt. (0.1 at.) % at 1350°C and less at lower temperatures.

Recent papers on higher Cr oxides: [31–34].

1. G. Grube and M. Flad, *Z. Elektrochem.*, **45,** 1939, 835–837.
2. G. Grube and M. Flad, *Z. Elektrochem.*, **48,** 1942, 377–381.
3. More limited work of the same type was carried out by H. v. Wartenberg and S. Aoyama (*Z. Elektrochem.*, **33,** 1927, 144) and J. Granat (*Metallurg,* **11,** 1936, 35). The conclusion by Granat that above 1100°C CrO is produced as a first reduction product of Cr_2O_3 was not corroborated by [2].
4. J. Férée, *Bull. soc. chim. France,* (3)**25,** 1901, 620.
5. T. Dieckmann and O. Hanf, *Z. anorg. Chem.*, **86,** 1914, 301–304.
6. H. Lux and E. Proeschel, *Z. anorg. Chem.*, **257,** 1948, 73–78. "CrO" prepared by melting NaF with Cr_2O_3 at 1050°C in a nitrogen(!)-hydrogen atmosphere.
7. C. W. Stillwell, *J. Phys. Chem.*, **30,** 1926, 1441–1466.
8. G. Grube and R. Knabe, *Z. Elektrochem.*, **42,** 1936, 795–796, 800–802.
9. F. Adcock, *J. Iron Steel Inst.*, **115,** 1927, 369–392.
10. Other effects (between 1350 and 1580°C) have been observed by [8] in differently treated electrolytic chromium samples which are attributed to the oxygen content.
11. H. v. Wartenberg and H. J. Reusch, *Z. anorg. Chem.*, **207,** 1932, 12.
12. H. v. Wartenberg and H. Werth, *Z. anorg. Chem.*, **190,** 1930, 183–184.
13. H. v. Wartenberg and E. Prophet, *Z. anorg. Chem.*, **208,** 1932, 374–375.
14. H. v. Wartenberg et al. assume the formation of Cr_3O_4 but give no conclusive evidence.
15. O. Ruff, *Z. anorg. Chem.*, **82,** 1913, 373.
16. C. W. Kanolt, *Z. anorg. Chem.*, **85,** 1914, 1.
17. W. P. Davey, *Phys. Rev.*, **21,** 1923, 716.

18. W. H. Zachariasen, in V. M. Goldschmidt, T. Barth, and G. Lunde, *Skrifter Norske Videnskaps-Akad. Oslo, I. Mat.-Naturv. Kl.*, **1925**, no. 7; see *Strukturbericht*, **1**, 1913-1928, 266.
19. S. Greenwald, *Nature*, **168**, 1951, 379.
20. H. Bizette, *Ann. phys.*, **1**, 1946, 300.
21. M. Foëx and C. H. LaBlanchetais, *Compt. rend.*, **228**, 1949, 1579.
22. T. R. McGuire, in S. Greenwald and J. S. Smart, *Nature*, **166**, 1950, 523.
23. S. S. Sidhu and M. Darrin, *Phys. Rev.*, **58**, 1940, 206.
24. A. Simon and T. Schmidt, *Z. anorg. Chem.*, **153**, 1926, 191–218.
25. R. S. Schwartz, I. Fankuchen, and R. Ward, *J. Am. Chem. Soc.*, **74**, 1952, 1676–1677.
26. O. Glemser, U. Hauschild, and F. Trüpel, *Naturwissenschaften*, **40**, 1953, 317.
27. G. Foëx and M. Graff, *Compt. rend.*, **209**, 1939, 160–162.
28. N. Schoenberg, *Acta Chem. Scand.*, **8**, 1954, 221–225.
29. Y. I. Olshanskii, A. I. Tsvetkov, and V. K. Shlepov, *Doklady Akad. Nauk S.S.S.R.*, **96**, 1954, 1007–1009; *Chem. Abstr.*, **49**, 1955, 2169.
30. D. C. Hilty, W. D. Forgeng, and R. L. Folkman, *Trans. AIME*, **203**, 1955, 253–268.
31. M. Dominé-Bergès, *Compt. rend.*, **228**, 1949, 1435–1437.
32. A. Byström and K. A. Wilhelmi, *Acta Chem. Scand.*, **4**, 1950, 1131–1141.
33. S. M. Ariya, S. A. Shchukarev, and V. B. Glushkova, *Zhur. Obshchei Khim.*, **23**, 1953, 1241–1245.
34. O. Glemser, U. Hauschild, and F. Trüpel, *Z. anorg. Chem.*, **277**, 1954, 113–126.
35. D. Caplan and A. A. Burr, *Trans. AIME*, **203**, 1955, 1052.

$\bar{1}.4369$
0.5631

Cr–Os Chromium–Osmium

An exploratory investigation (nine alloys) of the Cr-Os system by means of X-ray and microscopic work as well as hardness measurements has been carried out by [1]. Os takes up to 51.7 at. (22.6 wt.) % Cr into solid solution; between 1350 and 600°C no appreciable change in solubility could be observed. The phase Cr_2Os, stable only at elevated temperatures, has the tetragonal σ-type structure, $a = 9.105$ A, $c/a = 0.516$. The phase Cr_3Os possesses a narrow range of homogeneity and is of the cubic "β-W" (A15) type of structure, $a = 4.684$ A. Cr takes about 9.5 at. (27.7 wt.) % Os into solid solution.

1. E. Raub, *Z. Metallkunde*, **48**, 1957, 53–56.

0.2251
$\bar{1}.7749$

Cr-P Chromium-Phosphorus

Older work on preparation indicated the existence of the compounds CrP (37.33 wt. % P) [1–4] and Cr_2P_3 [3], the latter of which could not be corroborated by later work [5].

Some thermal and microscopic work up to 15 wt. (23 at.) % P is due to [6] (see Fig. 316). The compound Cr_3P (16.56 wt. % P) [7, 6, 5] was shown to have a tetragonal structure [7], with $a = 9.145$ A, $c = 4.569$ A, $c/a = 0.499$, and 32 atoms per unit cell [isotypic with $(Mn, Fe, Ni)_3P$]. There is conflicting evidence as to the existence of Cr_2P (22.95 wt. % P); with roentgenographic analysis it was found by [5] but not by [7], and the work by [6] gives indirect evidence for its existence [8]. CrP, also found by [7, 5], has the orthorhombic MnP (B31) type of structure, with $a = 5.94$ A, $b = 5.37$ A, $c = 3.13$ A [7]. The compound CrP_2 (54.35 wt. % P) was prepared by

[5]. Vapor-pressure measurements of the system CrP-CrP_2 gave no evidence for a compound Cr_2P_3 (see above) and indicated that CrP_2 decomposes at about 700°C.

No solid solution of P in Cr could be observed by [6]. Representations of powder photographs of alloys up to CrP_2 are given by [5].

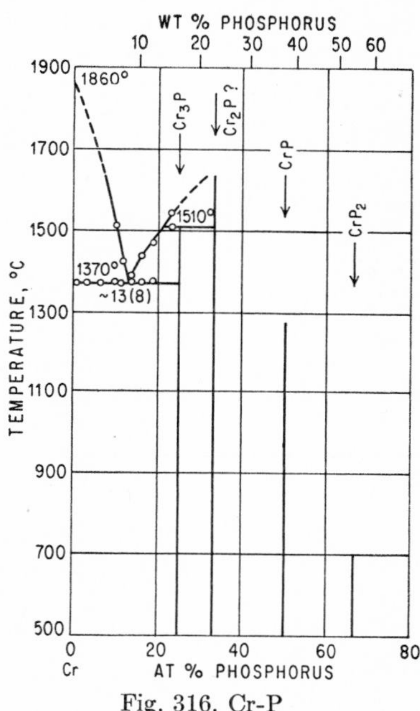

Fig. 316. Cr-P

Supplement. [9] redetermined the atomic-position parameters in the CrP structure. The existence of CrP_2 could be confirmed.

1. A. Granger, *Compt. rend.*, **124,** 1897, 190–191; *Ann. chim. et phys.*, (7) **14,** 1898, 38.
2. G. Maronneau, *Compt. rend.*, **130,** 1900, 658.
3. T. Dieckmann and O. Hanf, *Z. anorg. Chem.*, **86,** 1914, 291–295.
4. E. Heinerth and W. Biltz, *Z. anorg. Chem.*, **198,** 1931, 175.
5. F. E. Faller and W. Biltz, *Z. anorg. Chem.*, **248,** 1941, 209–228.
6. R. Vogel and G. W. Kasten, *Arch. Eisenhüttenw.*, **12,** 1939, 387–391.
7. H. Nowotny and E. Henglein, *Z. anorg. Chem.*, **239,** 1938, 14–16.
8. Extrapolated from results in the ternary system Fe-Cr-P (continuous range of solid solutions Fe_2P-Cr_2P).
9. N. Schönberg, *Acta Chem. Scand.*, **8,** 1954, 226–239.

$\bar{1}.3997$
0.6003

Cr-Pb Chromium-Lead

[1] concluded from scanty thermal work that Cr and Pb are not completely miscible in the liquid state (Fig. 317). This assumption, however, is almost wholly based on the fact that the liquidus points of the alloys with 50 and 75 wt. % Pb were found at the same temperature [2]. A microscopic examination was unsuccessful because the alloys proved to be very brittle; only weak indications of the formation of

layers and drops could be observed. The monotectic point at 9 at. (28 wt.) % Pb was obtained by extrapolation of the Cr-rich liquidus. The composition of the Pb-rich melt at the monotectic temperature is not known; [1] arbitrarily assumed that it lay between 75 and 100 wt. % Pb.

Because of unfavorable experimental conditions [1] (see Cr-Cu), the alloys were severely contaminated, which is clearly indicated by the erroneously low melting point of "chromium" observed (1550°C; see Cr-N).

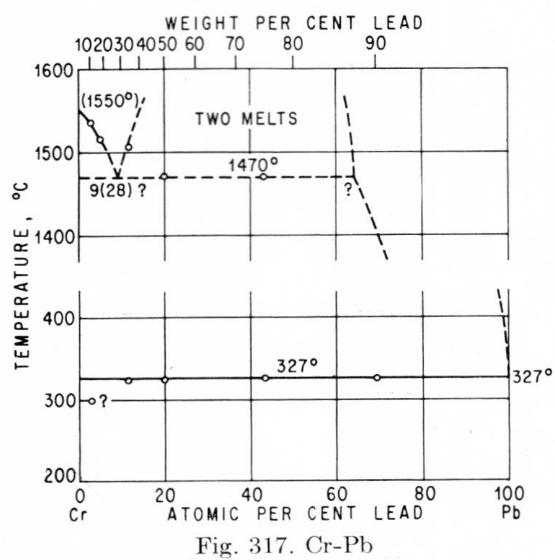

Fig. 317. Cr-Pb

From all the facts mentioned above it is obvious that new work is necessary to establish the true constitution of this system.

1. G. Hindrichs, *Z. anorg. Chem.*, **59,** 1908, 428–430.
2. W. Guertler ("Handbuch der Metallographie," vol. 1, pp. 569–571, Verlagsbuchhandlung Gebrüder Borntraeger, Berlin, 1912) pointed out that the thermal data of [1] may even be interpreted as belonging to a liquidus dropping steadily from the melting point of Cr to that of Pb.

$\bar{1}.6879$
0.3121

Cr-Pd Chromium-Palladium

The constitution of Cr-Pd alloys was investigated by [1] by means of thermal, X-ray, conductometric, microscopic, and hardness measurements (Fig. 318). For determination of the liquidus up to 40 at. % Pd a thermal-differential method had to be used owing to the small heat effects involved.

The electrical-conductivity isotherms between 100 and 1300°C form catenary curves between 40 and 100 at. % Pd, and the lattice parameter of Pd decreases continuously with additions up to 60 at. % Cr (Cr_3Pd_2). [1], therefore, assume a continuous series of solid solutions between the intermediate phase Cr_3Pd_2 (57.77 wt. % Pd)—marked by a pronounced melting-point maximum—and Pd, with a minimum in the liquidus at approximately 44 at. (62 wt.) % Pd and 1295°C [2]. This, however, would be a unique example of constitution. It would seem more probable that a very narrow heterogeneous field exists in the solid state around 44 at. % Pd and that the minimum of the liquidus is thus a eutectic point. A reinvestigation of the critical region should be made, as small discontinuities in the physical-properties-vs.-composition curves may have been overlooked by [1].

The conductivity isotherms indicate a solid solubility of Pd in Cr of about 5 at. (9.7 wt.) %, independent of temperature [3]. The hardness of slowly cooled alloys has a maximum at 40 at. % Pd.

Ordering reactions have not been detected by [1]; for some heat-treatments used see [3, 4].

According to [1], Cr_3Pd_2 has the Cu (A1) type of structure, with a = 3.846 A at 40 at. % Pd.

Supplement. The phase diagram of [1] has not been corroborated by differential thermal, microscopic, and X-ray diffraction analyses carried out by [5]. The

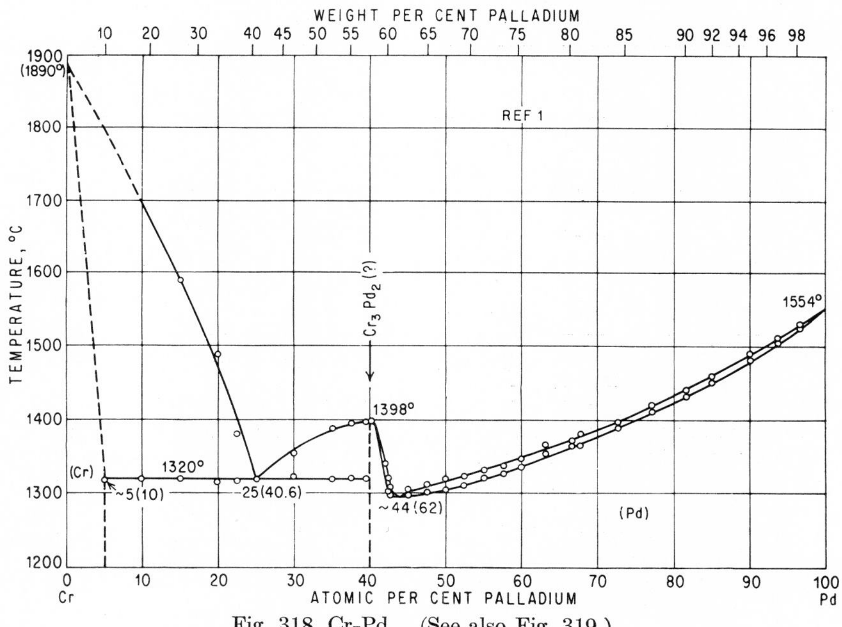

Fig. 318. Cr-Pd. (See also Fig. 319.)

diagram established by [5] (Fig. 319) shows no melting-point maximum, but a peritectoidally formed intermediate phase, CrPd, with a narrow range of homogeneity. The powder pattern of this phase could be indexed with a f.c. tetragonal cell of the dimensions a = 3.881 A, c/a = 0.975 (at 50.5 at. % Pd); no CuAu ($L1_0$) superstructure lines could be observed.

The results of a microscopic and roentgenographic investigation [6] of two Cr-Pd alloys with 40 and 50 at. % Pd, quenched from 1200°C, are in keeping with the new phase diagram.

1. G. Grube and R. Knabe, *Z. Elektrochem.*, **42**, 1936, 739–804. The alloys were prepared in alumina crucibles under purified argon. The Pd content of all alloys was determined.
2. These data have been interpolated in Fig. 318. In the original diagram, liquidus and solidus do not touch.
3. The conductivity isotherms have been derived from resistivity-vs.-temperature

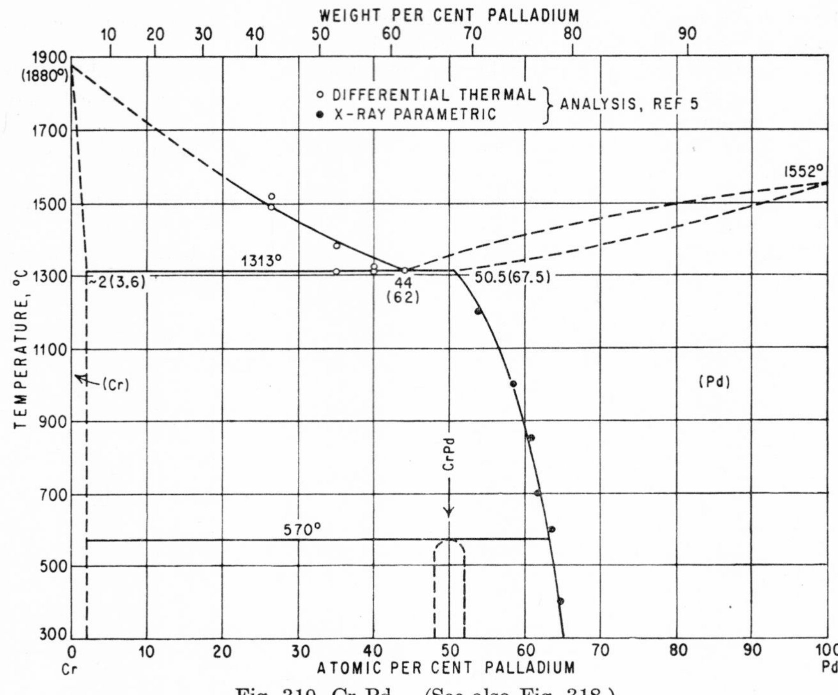

Fig. 319. Cr-Pd. (See also Fig. 318.)

curves (heating rate 15°C per 10 min) measured on alloys which had been annealed for 3 days at 800°C and cooled to room temperature over a period of 10 days. It is not certain whether equilibrium was reached at lower temperatures by this treatment.

4. X-ray powder patterns of briefly heat-treated alloys (4 hr, 200 and 400°C, slowly cooled) containing 30–67 at. % Pd showed no superstructure lines.
5. E. Raub and W. Mahler, *Z. Metallkunde,* **45,** 1954, 648–650.
6. P. Greenfield and P. A. Beck, *Trans. AIME,* **206,** 1956, 265–276.

$\bar{1}.4255$
0.5745

Cr-Pt Chromium-Platinum

Liquid-Solid Equilibria. The liquidus [1, 2] and solidus [2] data available are plotted in Fig. 320; the scattered points of [1], determined by a photocell method [3], are certainly too high. Between 80 and 85 at. % Cr a eutectic or peritectic reaction is assumed to occur [2].

The Solid State. The data obtained by microscopic [1, 4, 5, 2], resistometric [6, 1, 4, 5], magnetic [7, 5, 2], hardness [4, 2], roentgenographic [5, 2], and densimetric [5] work are still insufficient to establish the exact constitution of the system.

As shown in Fig. 320, an extended solid-solution region, originating from pure Pt, exists, the boundary of which was determined by microscopic and roentgenographic work [2] as lying at about 78 at. (48.6 wt.) % Cr at 1400°C and at about 71 at. (39.5 wt.) % Cr at temperatures below 1000°C (cf. Supplement). In the two-phase field beyond this boundary a second phase was observed roentgenographically in 84 [5] and 78.8 [2] at. % Cr alloys [8]. Its powder pattern could be indexed, partially [5] or entirely [2], as a b.c.c. structure with a lattice parameter a = 4.66 A [5] or 4.683 A [2], respectively. In alloys with 94 and 97 at. (80 and 90 wt.) % Cr [8], however, the (Pt) phase was found to be coexistent with Cr [5]. Later work proved this latter finding to be incorrect (see Supplement).

Superlattice and Ferromagnetism in the (Pt) Phase. In slowly cooled or annealed alloys with at least about 30 at. (10 wt.) % Cr (according to [4], other authors giving higher Cr contents), the microstructures indicate a transformation in

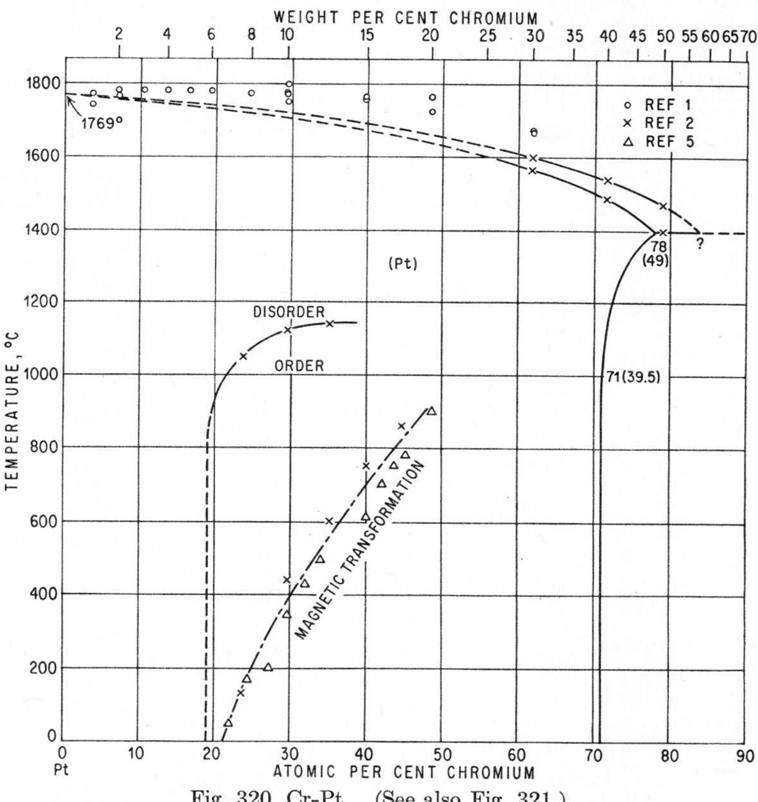

Fig. 320. Cr-Pt. (See also Fig. 321.)

the solid state. As X-ray work showed [5, 2], this transformation is of the order-disorder type, the cubic symmetry of the lattice remaining intact. Superstructure lines have been observed in annealed alloys containing 43–84 at. (17–60 wt.) % Cr by [5] and in alloys with 44.5 to only 61.6 at. (17.5–30 wt.) % Cr, slowly cooled from 1000°C, by [2]. A broadening and splitting up of powder lines in alloys around 75 at. % Cr has been reported by [5], but this could not be corroborated by [2].

Ferromagnetism in Pt-Cr alloys was first observed by [7]. The range of ferromagnetic alloys at room temperature was determined to be 22–48.5 at. (7–20 wt.) % Cr by [5], in fair agreement with data given by [2]. The same authors measured the Curie temperature as a function of composition (Fig. 320) and found that it disappears at about 900°C above 49 at. % Cr, the magnetic saturation showing a maximum at about 30 at (10 wt.) % Cr [5].

By rough quenching, [2] could metastabilize the disordered state at room temperature and show that ferromagnetism is limited to a certain region of the *ordered* phase. The order-disorder transformation temperature could so be determined for some compositions (Fig. 320).

It is not yet known on what stoichiometric composition the superlattice is based. The formulas $CrPt_3$ [2], CrPt [5, 2], and Cr_3Pt [5, 2] have been discussed [9].

The lattice parameter of the (Pt) solid solution decreases with increasing Cr content [5, 2]. The solid solubility of Pt in Cr at "room temperature" [10], which must be less than 3 at. % Pt (see above), cannot be estimated from the parameter data given by [5] as these data are not precise enough.

Supplement. The results of a new roentgenographic and microscopic investigation [11] are shown in the semiquantitative phase diagram of Fig. 321. No attempts were made to determine the temperature dependence of the phase boundaries. The

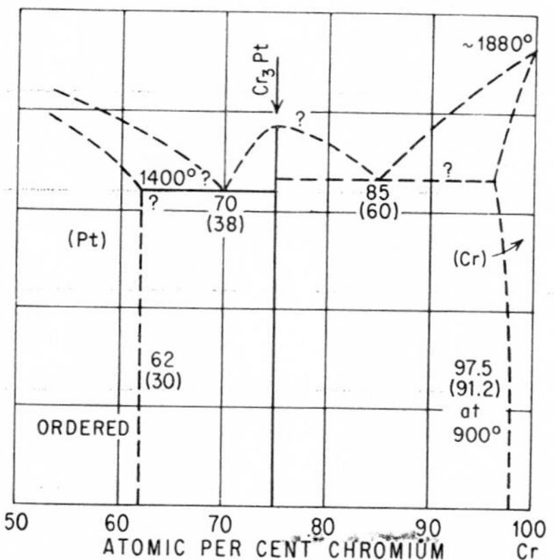

Fig. 321. Cr-Pt. (See also Fig. 320.)

solubility limit of the (Pt) phase was found at an appreciably lower Cr content than that reported by [2]. On both sides of the Cr_3Pt phase—with "β-W" (A15) type of structure—eutectics are formed.

A microscopic and roentgenographic study of three Cr-Pt alloys with 50, 63, and 75 at. % Cr quenched from 1000°C was made by [12]. These authors also found the Cr_3Pt phase having "β-W" structure; the 63 at. % Cr alloy was tentatively identified as Cu_3Au ($L1_2$) type ordered structure.

1. L. Müller, *Ann. Physik,* (5)**7,** 1930, 9–47.
2. E. Gebhardt and W. Köster, *Z. Metallkunde,* **32,** 1940, 262–264.
3. For a discussion of the possible sources of error (e.g., superficial oxidation of the melt) see the original paper.
4. W. A. Nemilow, *Z. anorg. Chem.,* **218,** 1934, 33–44.
5. A. Kussmann and E. Friederich, *Physik. Z.,* **36,** 1935, 185–192.
6. C. Barus, *Am. J. Sci.,* (3)**36,** 1888, 434; *Ann. Physik Beibl.,* **13,** 1889, 709. Measurements of the electrical resistance and its temperature coefficient of some Pt-rich Pt-Cr alloys defined only by their densities. W. Guertler ("Handbuch der Metallographie," vol. 1, p. 368, Verlagsbuchhandlung Gebrüder Borntraeger, Berlin, 1912) assumed that the above alloys contained up to 8 at. (2.2 wt.) % Cr.
7. E. Friederich, *Z. tech. Physik,* **13,** 1932, 59.

8. Heat-treatments of X-ray specimens: [5] probably annealed at 500°C and higher; [2] 24 hr, 1000°C, slowly cooled in air.
9. On the basis of a hardness study, [4] concluded that the compounds CrPt and Cr_2Pt are formed. His hardness-vs.-composition curve, however, could not be corroborated by [2].
10. [4] had erroneously assumed a continuous solid solution between Pt and Cr at high temperatures and a primary solid solution of Cr extending up to 25 at. % Pt at lower temperatures.
11. E. Raub and W. Mahler, *Z. Metallkunde*, **46,** 1955, 210–215.
12. P. Greenfield and P. A. Beck, *Trans. AIME,* **206,** 1956, 265–276.

$\bar{1}.3377$
0.6623

Cr-Pu Chromium-Plutonium

The solidification equilibria of the Pu-Cr system are of simple eutectic type with the eutectic composition close to pure Pu [<0.5 wt. (2.3 at.) % Cr]. The eutectic temperature is 615°C (melting point of Pu: 640°C). Addition of Cr does not appreciably reduce the phase-transition temperatures of Pu. No intermediate phases are formed [1].

1. S. T. Konobeevsky, Conference of the Academy of Sciences of the U.S.S.R. on the Peaceful Uses of Atomic Energy, July 1–5, 1955, Division of Chemical Science, pp. 209–210; English translation by U.S. Atomic Energy Commission, Washington, 1956.

$\bar{1}.4459$
0.5541

Cr-Re Chromium-Rhenium

Solidus points of some Cr-Re alloys, prepared by sintering, have been determined by [1] (Fig. 322). According to [2], who studied seven arc-melted Cr-Re alloys

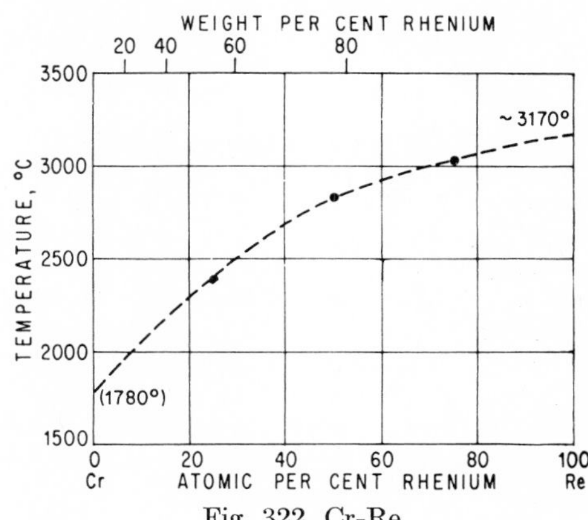

Fig. 322. Cr-Re

quenched from 1200°C, by microscopic and X-ray powder methods, there exists only one intermediate phase, of the σ structure, in this system. An alloy chemically analyzed as containing 63.2 at. % Re consisted entirely of the σ phase. Its homogeneity range appears to be narrow.

1. C. Agte, Dissertation, Technische Hochschule, Berlin, 1931, pp. 21–23; *Metallwirtschaft*, **10,** 1931, 789. Electrolytic Cr of 92 wt. % purity (+ Cr_2O_3) was used.

For the true melting point of Cr, see Cr-N. The melting point of Re was reported by Agte to be 3440 ± 60°K (3167 ± 60°C.)

2. P. Greenfield and P. A. Beck, *Trans. AIME*, **206**, 1956, 265–276.

$\bar{1}.7036$
0.2964

Cr-Rh Chromium-Rhodium

Exploratory work on the Cr-Rh system is due to [1, 2]. [1] studied 10 alloys with 5.8–89.7 at. % Cr quenched from 800–1300°C, and [2], 6 alloys with 25–75 at. % Cr quenched from 1200°C, by means of X-ray and microscopic work. [1] also measured microhardnesses. The following phases and ranges of homogeneity, in at. % Cr, were found: (Rh): 0–8 (900°C) [1], 0–>25 (1200°C) [2]; ϵ (h.c.p.): 24–60 (900°C) [1], 50–65 (1200°C) [2]; Cr_3Rh ("β-W" type) [2]; (Cr): 85.5–100 (1300°C), 94.5–100 (800°C) [1].

1. E. Raub and W. Mahler, *Z. Metallkunde*, **46**, 1955, 210–215.
2. P. Greenfield and P. A. Beck, *Trans. AIME*, **206**, 1956, 265–276.

$\bar{1}.7088$
0.2912

Cr-Ru Chromium-Ruthenium

Exploratory work on the Cr-Ru system is due to [1, 2]. [1] studied 10 alloys with 30–90 at. % Cr quenched from 650–1300°C, and [2], 5 alloys with 50–74 at. % Cr quenched from 1200°C, by means of X-ray and microscopic work. [1] also measured microhardnesses. The following phases and ranges of homogeneity, in at. % Cr, were found: (Ru): 0–40 (1300°C), 0–34 (900°C) [1], 0–>50 (1200°C) [2]; $\sim Cr_2Ru$ [1, 2], according to [2] of σ-type structure; Cr_3Ru ("β-W" type), formed below 950°C [1]; $\sim Cr_4Ru$, formed below 950°C [1]; (Cr): 82–100 (1300°C), 84–100 (900°C) [1], >74–100 (1200°C) [2].

1. E. Raub and W. Mahler, *Z. Metallkunde*, **46**, 1955, 210–215.
2. P. Greenfield and P. A. Beck, *Trans. AIME*, **206**, 1956, 265–276.

0.2100
$\bar{1}.7900$

Cr-S Chromium-Sulfur

In the older chemical literature the compounds CrS [1], Cr_3S_4 [2], and Cr_2S_3 [3] are mentioned.

The partial phase diagram of Fig. 323 is based on thermal, microscopic, and chemical work up to 50 at. % S carried out by [4] and on magnetic [5] and X-ray powder [6] investigations on 14 alloys covering the range 33.3–59.7 at. % S. [4] located the miscibility gap in the liquid state between 2.2 and 27.5 wt. (3.5 and 38.1 at.) % S at the monotectic temperature by chemical analysis of the separated layers. To account for weak thermal effects around 1150°C (see Fig. 323) and certain microscopic observations, [4] suggested that a Cr-rich sulfide may be formed in the solid state. X-ray results by [6], however, do not confirm this view, since below about 50 at. % S chromium lines were observed. They coincide with those of pure chromium so that the solid solubility of S in Cr must be small or even nil.

At 50 at. (38.14 wt.) % S the structure found by [6] was not the simple NiAs (B8) type as reported by [7] for CrS, but rather a hexagonal superstructure [8] of it, with a = 12.02 A, c = 11.54 A, c/a = 0.96. At about 52.4 at. % S, however, the superstructure lines have disappeared, and up to 54 at. % S a NiAs (B8) structure is found, with a = 3.45 A, c = 5.76 A, c/a = 1.67 at its S-rich boundary. No two-phase field could be detected roentgenographically between B8 and its superstructure so that the concentration range between about 50 and 54 at. % S [9] may be assumed homogeneous [(CrS) or α [6]].

Beyond a narrow two-phase field [about 54–55 at. (42–43 wt.) % S] a presumably monoclinic deformed B8 structure was observed, with $a = 5.98$ A, $b = 3.44$ A, $c = 5.68$ A, $\beta = 91°20'$ at its Cr-rich boundary. This structure persisted at 58.3 at. % S, but a 59.7 at. % S alloy again had the hexagonal B8 structure, with $a = 3.43$ A, $c = 5.56$ A, and $c/a = 1.62$. The experimental evidence does not allow a definite decision as to whether there is a two-phase field (Fig. 323) between the composition ranges of the monoclinic (β) and the B8 (γ) structures or a continuous structural transition; by analogy with the system Cr-Se the former alternative is preferred [10].

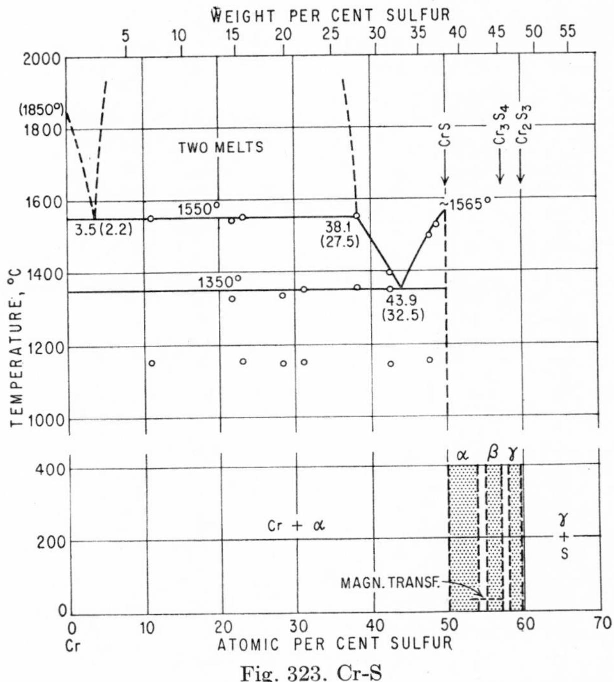

Fig. 323. Cr-S

Below about 50 and above 59.7 at. % S, two-phase fields (Cr + α) and (γ + sulfur), respectively, have been observed roentgenographically.

The volume of the unit cells of the intermediate phases always decreases with increasing sulfur content. The (CrS), or α, phase has both vacant Cr and S sites in its lattice at 50 at. % S, and both subtraction of Cr and addition of S atoms occur when the sulfur content increases. In the β and γ phases, however, only the subtraction mechanism was found [6].

Weak ferromagnetism was observed in alloys from 53.5 to 57.1 (especially 54) at. % S between room temperature and about −120°C [11] and in the 59.7 at. % S alloy below −150°C. For details, see the original paper ([5], also [10]).

Supplement. [12] measured the magnetization, electrical resistivity, and thermo-emf of two Cr-S alloys of the compositions $CrS_{1.17}$ and $CrS_{1.33}$ (53.9 and 57 at. % S, respectively) between about 100 and 300°K. In agreement with the results of [5], the magnetization was found to drop suddenly in cooling below about −100°C.

[13] observed weak ferromagnetism between $CrS_{1.17}$ and $CrS_{1.21}$ (53.9 and 54.8 at. % S), with a maximum at the composition $CrS_{1.181}$ (54.15 at. % S).

1. H. Moissan, *Compt. rend.*, **90,** 1880, 818; A. Mourlot, *Compt. rend.*, **121,** 1895, 943.
2. E. Wedekind and C. Horst, *Ber. deut. chem. Ges.*, **48,** 1915, 105–112.
3. Cr_2S_3 has long been known in the chemical literature; see "Gmelin-Kraut Handbuch," Carl Winter's Universitätsbuchhandlung, Heidelberg, 1912, and H. Moissan, *Compt. rend.*, **90,** 1880, 817; **119,** 1894, 189.
4. R. Vogel and R. Reinbach, *Arch. Eisenhüttenw.*, **11,** 1938, 457–462. Only Cr-rich alloys were melted under argon. The crucibles used (alumina, Pythagoras) were attacked by the melt.
5. H. Haraldsen and A. Neuber, *Naturwissenschaften*, **24,** 1936, 280; *Z. anorg. Chem.*, **234,** 1937, 337–352. The alloys were prepared in evacuated quartz tubes using electrolytic chromium. In the *Zeitschrift für anorganische Chemie* paper, susceptibility measurements by [2] are replotted.
6. H. Haraldsen, *Z. anorg. Chem.*, **234,** 1937, 372–390. For preparation of the alloys, see [5]. The powder samples were annealed at 1000 or 700°C; "it made no difference whether the samples were slowly cooled or quenched."
7. W. F. de Jong and H. W. V. Willems, *Physica*, **7,** 1927, 74–79.
8. Its unit cell is 24 times as large as that of the substructure.
9. According to the magnetic measurements [5], the (CrS), or α, phase extends down to 46–47 at. % S.
10. H. Haraldsen and F. Mehmed, *Z. anorg. Chem.*, **239,** 1938, 388–393 (comparison of Cr-S, Cr-Se, and Cr-Te).
11. The ferromagnetism disappears almost immediately on cooling to this temperature.
12. H. Watanabe and N. Tsuya, *Science Repts. Research Inst., Tôhoku Univ.*, **A2,** 1950, 503–506.
13. F. S. Smirnov, *Zhur. Tekh. Fiz.*, **23,** 1953, 50–55; *Chem. Abstr.*, **49,** 1955, 2799.

$\bar{1}.6306$
0.3694

Cr-Sb Chromium-Antimony

The phase diagram of Fig. 324 was established by thermal and microscopic work [1]. Strong peritectic coring was observed in the formation of $CrSb_2$ (82.40 wt. % Sb), but annealing at 660°C for 60 hr led to the equilibrium state. CrSb (70.07 wt. % Sb) was shown to have a range of homogeneity of about 2 wt. (at.) % on the Cr side; the Sb side was not investigated.

Cr seems to dissolve about 12 wt. (5.5 at.) % Sb as alloys containing about 11 and 13 wt. % Sb proved to be one- and two-phase, respectively, after being annealed 10 hr at 1050°C (cooling rate?) [1].

The low value for the melting point of Cr observed by [1] (1553°C, which is more than 300°C below the true melting point) indicates a severe contamination occurring during the melting process, probably mainly by nitrogen. The constitution of the Cr-rich alloys should therefore be reexamined [2].

Crystal Structure. CrSb was shown to have the NiAs (B8) type of structure [3–7]. The lattice parameters reported differ somewhat, certainly due to concentration differences. For an alloy containing 66.3 wt. (45.7 at.) % Sb, [7] found $a = 4.108$ A, $c = 5.440$ A, $c/a = 1.324$. CrSb is an antiferromagnetic substance with its transition temperature in the neighborhood of 400°C [8, 5–7].

$CrSb_2$ has the orthorhombic marcasite (C18) type of structure, with $a = 3.278$ A, $b = 6.031$ A, $c = 6.875$ A [5], and is also antiferromagnetic [8].

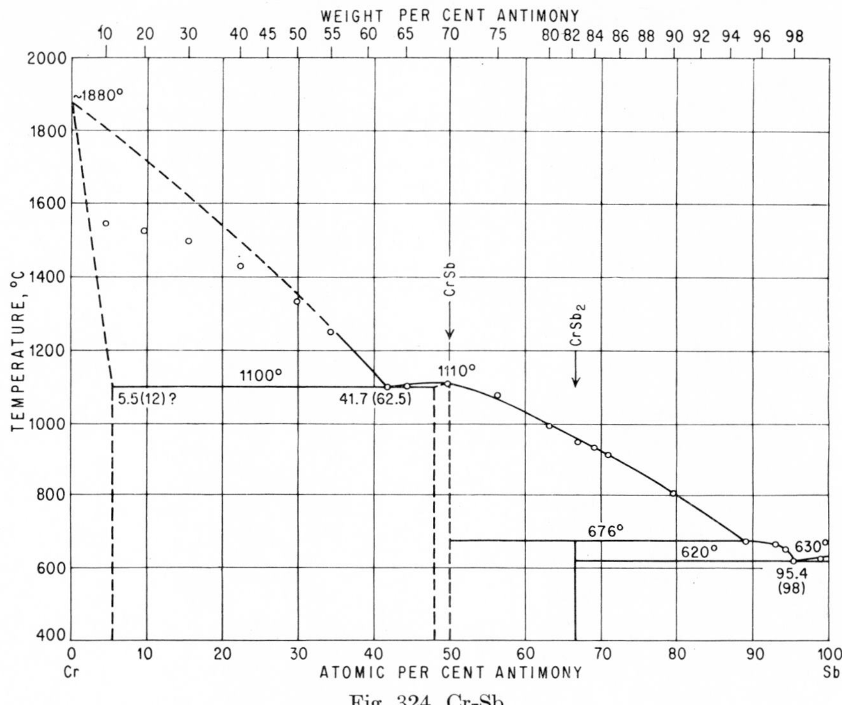

Fig. 324. Cr-Sb

1. R. S. Williams, *Z. anorg. Chem.*, **55**, 1907, 7–11. Analysis of chromium: 98.97 Cr, 0.67 Fe, 0.30 (Cr_2O_3 + SiO_2) (wt. %). The alloys, up to 50 wt. % Sb, were prepared in magnesia crucibles and, those between 50 and 100 wt. % Sb, in porcelain crucibles under nitrogen (!) (see Cr-N) in a carbon resistance furnace.
2. Curiously enough, [1] reports nothing on nonbinary heterogeneities.
3. I. Oftedal, *Z. physik. Chem.*, **128**, 1927, 135–153.
4. W. F. de Jong and H. W. V. Willems, *Physica*, **7**, 1927, 74–79.
5. H. Haraldsen and T. Rosenqvist, *Tidsskr. Kjemi, Bergvesen Met.*, **3**, 1943, 81–82; H. Haraldsen, T. Rosenqvist, and Grønvold, *Arch. Math. Naturvidenskab*, B. L. Nr. 4 (not available to the author); abstract in *Chem. Abstr.*, **39**, 1945, 4273; see also [6].
6. A. I. Snow, *Phys. Rev.*, **85**, 1952, 365; *Revs. Mod. Phys.*, **25**, 1953, 127.
7. B. T. M. Willis, *Acta Cryst.*, **6**, 1953, 425–426.
8. G. Foëx and S. Graff, *Compt. rend.*, **209**, 1939, 160.

$\bar{1}.8187$
0.1813

Cr-Se Chromium-Selenium

The selenides CrSe and Cr_2Se_3 were described by [1] in 1880.

In a roentgenographic (powder method) and magnetic investigation [2] of 13 alloys, three intermediate phases were found in the concentration range 50–60 at. % Se. Between 50 and about 53.5 at. (60.3–63.6 wt.) % Se the homogeneity region of a NiAs (B8) structure (detected at CrSe by [3]), with a = 3.69–3.68 A, c = 6.03–6.00 A, c/a = 1.63_3–1.62_9, has been observed (α phase [2]).

Separated from α by a narrow two-phase field, the β phase—a presumably monoclinic distorted B8 structure—extends approximately from 54.5 to 57 at. (64.5–66.8 wt.) % Se; the parameters a = 6.31 A, b = 3.61 A, c = 5.86 A, β = 91°30′ were measured at 57.1 at. % Se.

The γ phase has its homogeneity range between about 59 and 59.7 at. (68.6 and 69.2 wt.) % Se and has again the B8 type of structure, probably with vacant Cr sites in the lattice, with a = 3.61 A, c = 5.79–5.78 A, c/a = 1.60_4–1.60_1. Above 59.7 at. % Se, free selenium is present.

The magnetic measurements, carried out between −183 and 320°C, corroborate the roentgenographic phase limits very well; a distinct influence of temperature on the phase boundaries could be observed only with the β phase where its Se-rich boundary was found to change from about 58 at. % Se at room temperature to 57 at. % Se at 150 and 320°C.

In the concentration and temperature ranges investigated, no ferromagnetism could be detected (cf. Cr-S, Cr-Te).

1. H. Moissan, *Compt. rend.*, **90**, 1880, 819.
2. H. Haraldsen and F. Mehmed, *Z. anorg. Chem.*, **239**, 1938, 369–394. The alloys were prepared by heating weighed quantities of electrolytic chromium and pure selenium for 24 hr at 900°C in evacuated quartz tubes. The X-ray powder samples were annealed 3 days at 800–900°C and slowly cooled.
3. W. F. de Jong and H. W. V. Willems, *Physica*, 7, 1927, 74–79. For CrS, a = 3.60 A, c = 5.81 A, c/a = 1.62.

0.2675
$\bar{1}$.7325

Cr-Si Chromium-Silicon

In early work on preparation the following Cr silicides were claimed: Cr_3Si [1–3], Cr_2Si [4, 5, 2, 6, 7, 3], Cr_3Si_2 [2, 8, 9], CrSi [3], Cr_2Si_3 [3], $CrSi_2$ [10, 2, 3], and Cr_2Si_7 [3]. Not all these have been corroborated by more recent work, however.

Systematic investigations were made by [11] (roentgenographic), [12–14] (thermal [15], microscopic, hardness, and temperature coefficient of electrical resistance), and [16] (thermal [15], microscopic, and roentgenographic mainly). These authors agree upon the existence of Cr_3Si (15.26 wt. % Si), CrSi (35.07 wt. % Si), and $CrSi_2$ (51.93 wt. % Si); but there is conflicting evidence whether an additional compound may have the formula Cr_2Si (21.27 wt. % Si) or Cr_3Si_2 (26.47 wt. % Si) or whether even both may exist.

[11] found, at an unknown composition between Cr_3Si and CrSi, a compound (θ) which was said to be stable only below 1000°C; [17] ascribed to θ the formula Cr_3Si_2. The liquidus of [14] (only a few of the listed data were plotted in Fig. 325) shows a maximum at the composition Cr_2Si; but his resistivity measurements gave a strong indication for the existence of Cr_3Si_2 also. [16]—unfortunately unaware of [13, 14]—were also unable to establish the constitution between 30 and 40 at. % Si. They suggest the formula Cr_3Si_2 for the unknown compound, which is said to have "a rather extended homogeneity range which includes Cr_3Si_2 as well as Cr_2Si" and to melt, under decomposition, at 1560 ± 50°C, but also to be probably "unstable at temperatures above 850°C" (cf. [11]).

In Fig. 325 no assumptions on the constitution of solid alloys with compositions between Cr_3Si and CrSi are made; only the liquidus lines of [14] and [16] are shown in this region [18]. It may be suggested that a high- and a low-temperature phase exist.

Primary Solid Solutions. [11] stated that Cr dissolves a little Si [approximately 1 wt. (1.8 at.) % [19]] with lattice contraction. On the other hand, [14] found microscopically a solubility of 7.5 wt. (13 at.) % Si [20] between 750 and 1250°C; however, the investigation obviously suffered from an oxygen contamination of the alloys [21]. Cr appears to be insoluble in solid Si [11, 16].

Phase Diagram. Except for the region between Cr_3Si and CrSi discussed above, the diagrams given by [14] and [16] are at least in qualitative agreement (cf. the data plotted in Fig. 325). The Cr-rich eutectic in Fig. 325 was taken from [14]; [16] assumes this eutectic at about 9 at. % Si.

Crystal Structures. Cr_3Si has the cubic "β-W" (A15) type of structure [11, 16], with $a = 4.564 \pm 3$ A [11], CrSi the cubic FeSi (B20) type [11, 16], with $a = 4.629 \pm 2$ A [11], and $CrSi_2$ the hexagonal C40 prototype structure [11, 16], with $a = 4.431 \pm 5$ A, $c = 6.364 \pm 5$ A, $c/a = 1.44$ [11]. The θ phase (Cr_3Si_2? see above) may be orthorhombic [11]. Cr_3Si_2 was listed by [22] as isomorphous with Cr_3Ge_2 (unknown structure); see also [23].

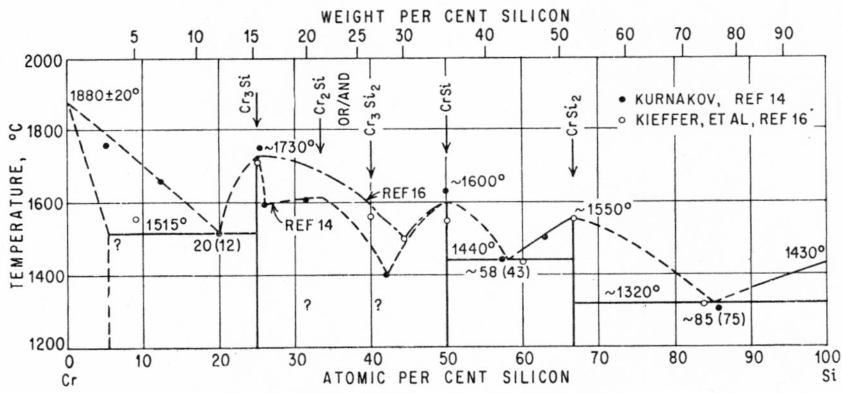

Fig. 325. Cr-Si. (See also Supplement.)

Supplement. [24] examined the microstructures of four alloys containing 52, 76, 83, and 98 wt. % Si. In agreement with the eutectic composition reported by [13, 14], the 76 wt. % Si alloy was found to be entirely eutectic.

According to single-crystal and powder X-ray work by [25], Cr_3Si_2 is tetragonal, with $a = 9.18$ A, $c = 4.65$ A, $c/a = 0.507$, and 6 formula weights per unit cell. However, [25] admit that Cr_5Si_3 (37.50 at. % Si) is also a possible formula for this compound ($4Cr_5Si_3$ per unit cell).

1. C. Zettel, *Compt. rend.*, **126,** 1898, 833–835.
2. P. Lebeau and J. Figueras, *Compt. rend.*, **136,** 1903, 1329–1331.
3. R. Frilley, *Rev. mét.*, **8,** 1911, 476–483.
4. H. Moissan, *Compt. rend.*, **121,** 1895, 624; *Ann. chim. et phys.*, (7)**9,** 1896, 292.
5. H. N. Warren, *Chem. News*, **78,** 1898, 318.
6. C. Matignon and R. Trannoy, *Compt. rend.*, **141,** 1905, 190.
7. L. Baraduc-Muller, *Rev. mét.*, **7,** 1910, 700–703.
8. E. Vigouroux, *Compt. rend.*, **144,** 1907, 83–85.
9. [7], pp. 698–700.
10. G. de Chalmont, *Am. Chem. J.*, **19,** 1897, 69–70.
11. B. Borén, *Arkiv Kemi, Mineral. Geol.*, **11A**(10), 1933, 2–10; *Strukturbericht*, **3,** 1933-1935, 628.
12. N. N. Kurnakov, *Compt. rend. acad. sci. U.R.S.S.*, **26,** 1940, 362–364 (in English); *Met. Abstr.*, **8,** 1941, 262.

13. N. N. Kurnakov, *Compt. rend. acad. sci. U.R.S.S.*, **34**, 1942, 110–113; *Met. Abstr.*, **10**, 1943, 243.
14. N. N. Kurnakov, *Izvest. Sektora Fiz.-Khim. Anal.*, **16**, 1948, 77–84. The alloys were melted from the elements (Si, 98.1 wt. % pure) in corundum crucibles under a flux. The temperature scale of the phase diagram (fig. 4 in the paper) is numbered 100° too high.
15. The melting points of high-melting (Cr-rich) alloys were approximately determined by incipient melting measurements.
16. R. Kieffer, F. Benesovsky, and H. Schroth, *Z. Metallkunde*, **44**, 1953, 437–442.
17. A. Andersen and E. Jette, *Trans. ASM*, **24**, 1936, 375.
18. In order not to complicate the diagram, both lines begin and end at averaged melting points for Cr_3Si and CrSi.
19. As quoted by]16].
20. [14] stated agreement with the results of [11], obviously equating—incorrectly—the first observation of Cr_3Si lines (at 8 wt. % Si by [11]) in powder patterns with the (Cr) phase limits.
21. Microstructures 1 and 3 in [14] appear to be exchanged.
22. F. Laves in C. J. Smithells, "Metals Reference Book," Butterworth & Co. (Publishers) Ltd., London, 1949.
23. H. J. Wallbaum, *Naturwissenschaften*, **32**, 1944, 76–77.
24. W. R. Johnson and M. Hansen, AF Technical Report 6383, June, 1951, Armour Research Foundation, Chicago, Ill.
25. E. Parthé, H. Schachner, and H. Nowotny, *Monatsh. Chem.*, **86**, 1955, 182–185.

$\bar{1}.6416$
0.3584

Cr-Sn Chromium-Tin

The phase diagram in Fig. 326, due to [1], cannot be considered that of the binary system Cr-Sn [2]. Earlier, [3] had reported that the electromotive force of alloys with 4–100 at. % Sn is practically that of Sn, indicating that no intermediate phase occurs. This was confirmed by roentgenographic studies [4]: powder photographs contained only the reflections of Cr and Sn in the same positions as those of the pure metals.

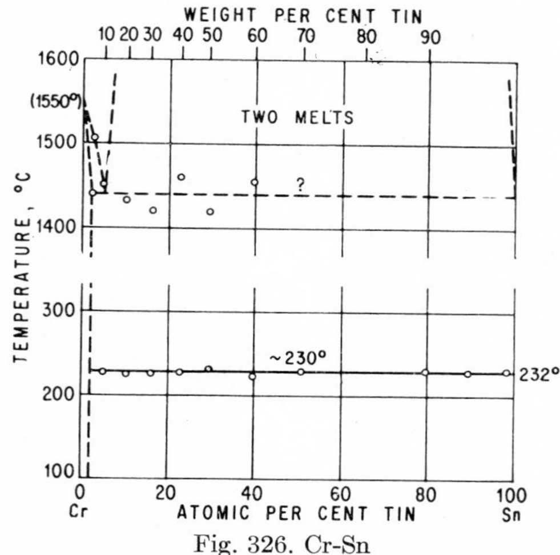

Fig. 326. Cr-Sn

1. G. Hindrichs, *Z. anorg. Chem.*, **59,** 1908, 416–420.
2. Alloys were prepared using commercial Cr with 98.7% Cr, (+ Fe, Si), without protective conditions. The melting point of this Cr was found as 1550°C, instead of about 1880°C.
3. N. Puschin, *Zhur. Russ. Fiz.-Khim. Obshchestva*, **39,** 1907, 869–897.
4. O. Nial, Dissertation, University Stockholm, 1945; *Svensk Kem. Tidskr.*, **59,** 1947, 165–183.

$\bar{1}.4587$
0.5413

Cr-Ta Chromium-Tantalum

A cursory thermal, microscopic, and roentgenographic analysis of the Ta-Cr system was carried out by [1]. The thermal data (see Fig. 327), measured by optical

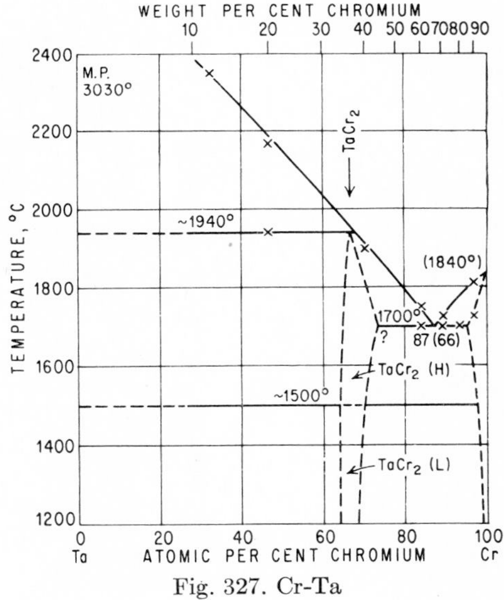

Fig. 327. Cr-Ta

pyrometer, are not claimed to be very accurate. There exists only one intermediate phase [1, 2] which is, according to [2], located around the stoichiometric composition $TaCr_2$ (36.51 wt. % Cr) [3].

The solid solubility of Ta in Cr was estimated from microscopic examination to be less than 5 wt. (1.5 at.) % Ta [1] at "room temperature."

Crystal Structure. Ta causes a very slight expansion of the Cr lattice [1]. $TaCr_2$ has a polymorphic transformation at a temperature between 1375 and 1590°C [2]. The high-temperature form, which can be retained at room temperature by quenching, has the hexagonal $MgZn_2$ (C14) type of structure, with a = 4.925 A, c = 8.062 A, c/a = 1.637, and the low-temperature form has the cubic $MgCu_2$ (C15) type of structure, with a = 6.961 A [2].

Supplement. [4] determined the incipient melting temperature of $TaCr_2$ as 2015°C. X-ray examination of $TaCr_2$ samples quenched from various temperatures between 600 and 2000°C indicated only the presence of the $MgZn_2$ form, a = 4.932 A, c = 8.082 A, c/a = 1.639.

1. O. Kubaschewski and A. Schneider, *J. Inst. Metals,* **75,** 1948-1949, 410–412; O. Kubaschewski and H. Speidel, *J. Inst. Metals,* **75,** 1948-1949, 418–419.
2. P. Duwez and H. Martens, *Trans. AIME,* **194,** 1952, 72–74.
3. Kubaschewski et al. [1] suggested a broad homogeneous field ("extending from about 62 wt. [32 at.] % Ta upwards") and assigned the formula Ta_2Cr_3.
4. R. P. Elliott, Armour Research Foundation, Chicago, Ill., Technical Report 1, OSR Technical Note OSR-TN-247, August, 1954, p. 23.

$\bar{1}.6102$
0.3898

Cr-Te Chromium-Tellurium

The most detailed investigation of the constitution of solid Cr-Te alloys (mainly from 50 to 60 at. % Te) was carried out by [1] using magnetic and X-ray powder methods. The results are analogous to those found in the systems Cr-S and Cr-Se and may be summarized as follows:

(*a*) The homogeneity range of the intermediate phase richest in Cr (α) extends from somewhat below 50 at. (71.04 wt.) % Te to about 54 at. (74.2 wt.) % Te. The hexagonal NiAs (B8) structure which was first reported by [2] could be corroborated, but there is perhaps a weak distortion of the lattice toward the Cr-rich boundary of the phase. The lattice parameters do not change much with the composition and are $a = 3.96$ A, $c = 6.17$ A, $c/a = 1.56$ at 50 at. % Te. [2] reported somewhat larger parameters.

(*b*) The powder patterns of 54.55, 57.1, and 58.33 at. % Te alloys could be indexed by a—presumably monoclinic—distorted B8 structure (β), the lattice parameters of which ($a = 3.93$ A, $b = 6.09$ A, $c = 6.85$ A, $\beta = 91°5'$ at 58.33 at. % Te) decrease with the Cr content.

(*c*) Around 60 at. % Te again a hexagonal B8 structure with the smaller parameters (indicating vacant Cr sites) $a = 3.92$ A, $c = 6.04$ A, $c/a = 1.54$ (at 60 at. % Te) is stable (γ phase). Its Te-rich boundary is not yet precisely known but certainly lies below 66.7 at. (83.1 wt.) % Te, since at this composition free Te was observed [2, 1]. Heterogeneous fields between the structures mentioned could not be observed by [1] but are assumed to exist [3].

(*d*) The magnetic investigations showed that—at least at low temperatures—all intermediate phases are ferromagnetic [4]. The Curie temperature is 70–80°C up to 57.1 at. % Te but then falls to 40 and −80°C at 58.33 and 60 at. % Te, respectively. The highest magnetization was observed at the upper boundary of the α phase (54 at. % Te) (cf. below).

The older magnetic results of [5], which in some points deviate from those of [1], need not be discussed in detail, since the alloys investigated (Cr_3Te-$CrTe_2$) apparently were not in complete equilibrium (see [1]).

More recently two further papers were published [6, 7], but unfortunately their authors had no knowledge of the work by [1]. [6, 7] corroborated the B8 structure for CrTe and found the maximum magnetization at 50 at. % Te. [6] stated that CrTe does not dissolve any Cr and gave 66°C as the Curie temperature for CrTe. The work of [7] on alloys from 1 to 70 at. % Te included microscopic (no microstructures published) as well as magnetic and X-ray studies. The recorded changes in lattice parameters and Curie temperatures with composition, though not as detailed, are in general much like those found by [1]. In view of the published evidence, the conclusion that a one-phase field extends from 15 (!) to 70 at. % Te is not at all convincing.

Supplement. The authors of ref. [7] obviously revised their results in an additional publication [8]. According to an abstract, they found alloys containing less than 50 at. % Te to have a Curie point at 91°C and to consist of mixtures of Cr and CrTe. These findings are in good agreement with those of [1, 6].

1. H. Haraldsen and A. Neuber, *Z. anorg. Chem.*, **234,** 1937, 353–371; for the 50 at. % alloy, see also H. Haraldsen and E. Kowalski, *Z. anorg. Chem.*, **224,** 1935, 333. The alloys were prepared by heating the elements (electrolytic Cr) for 24 hr or longer at 1000°C in evacuated quartz tubes. The alloys were not melted at this temperature.
2. I. Oftedal, *Z. physik. Chem.*, **128,** 1927, 135–153, especially 139.
3. H. Haraldsen and F. Mehmed, *Z. anorg. Chem.*, **239,** 1938, 388–393 (comparison of Cr-S, Cr-Se, and Cr-Te).
4. Ferromagnetism in Cr-Te alloys was detected by V. M. Goldschmidt (*Ber. deut. chem. Ges.*, **60**(1), 1927, 1287).
5. R. Ochsenfeld, *Ann. Physik*, **12,** 1932, 353–356.
6. C. Guillaud and S. Barbezat, *Compt. rend.*, **222,** 1946, 386–388; also C. Guillaud, *Compt. rend.*, **222,** 1946, 1224-1226.
7. F. M. Gal'perin and T. M. Perekalina, *Zhur. Eksptl. Teoret. Fiz.*, **19,** 1949, 470–472.
8. F. M. Gal'perin and T. M. Perekalina, *Doklady Akad. Nauk S.S.S.R.*, **69,** 1949, 19–22. *Met. Abstr.*, **20,** 1953, 1001 (in line 15 of this abstract, read °K instead of °C).

$\bar{1}.3504$
0.6496

Cr-Th Chromium-Thorium

Cr and Th form a simple eutectic system with the eutectic point located at about 60 at. (25 wt.) % Cr, 1235°C. There are no intermediate phases. The mutual solid solubility of the elements is quite limited [1].

1. H. A. Wilhelm and O. N. Carlson, unpublished information (February, 1946). Quoted from H. A. Saller and F. A. Rough, Compilation of U.S. and U.K. Uranium and Thorium Constitutional Diagrams, *U.S. Atomic Energy Comm., Publ.* BMI-1000, June, 1955.

0.0358
$\bar{1}.9642$

Cr-Ti Chromium-Titanium

First Investigations. [1] studied the Cr-rich part of the system and reported that a eutectic between a compound Ti_3Cr_2 and the Cr-rich solid solution exists at 57 wt. % Cr and 1400°C. This was not corroborated by later work. [2] concluded from sintering experiments that 10 wt. % Cr depresses the melting point of Ti to about 1200°C. A tentative phase diagram for Ti-rich alloys with up to 15 wt. % Cr, outlined by [3], showed that the transformation point of Ti is lowered by Cr additions. [4] presented photomicrographs of alloys with up to 18 wt. % Cr after various heat-treatments and concluded that about 10 wt. % or more Cr apparently stabilizes the β solid solution down to lower temperatures.

Phase Diagram. The Ti-Cr system has been studied in more detail over the whole range of composition by [5, 6] and in part by [7–10].

[5] published a "provisional" phase diagram which was based on micrographic and X-ray analysis of arc-melted alloys. Magnesium-reduced Ti was used.

The constitution between 0 and 30 at. % Cr and within the temperature range from 650 to 1100°C was studied by [7] by means of micrographic and X-ray analysis of arc-melted alloys based on iodide Ti. Furthermore, the intermediate phase was shown to have the ideal composition $TiCr_2$ (68.47 wt. % Cr).

[8] studied the range 0–76 wt. % Cr by means of micrographic analysis, supplemented by X-ray diffraction, detection of incipient melting, and thermal analysis.

The phase diagram published by [6] is based on micrographic and roentgenographic analysis (supplemented by dilatometric work) of arc-melted alloys for which iodide Ti (up to 30 wt. % Ti) and sponge Ti were used.

[9] determined the $\beta/(\alpha + \beta)$ boundary in the range of composition up to 5 at. % Cr by measuring the hydrogen pressure in equilibrium with an extremely dilute solution of hydrogen in the Ti-rich alloys (iodide Ti base) as a function of temperature.

The solid-state phase boundaries up to 14 at. % Cr were reinvestigated by [10] by means of micrographic analysis.

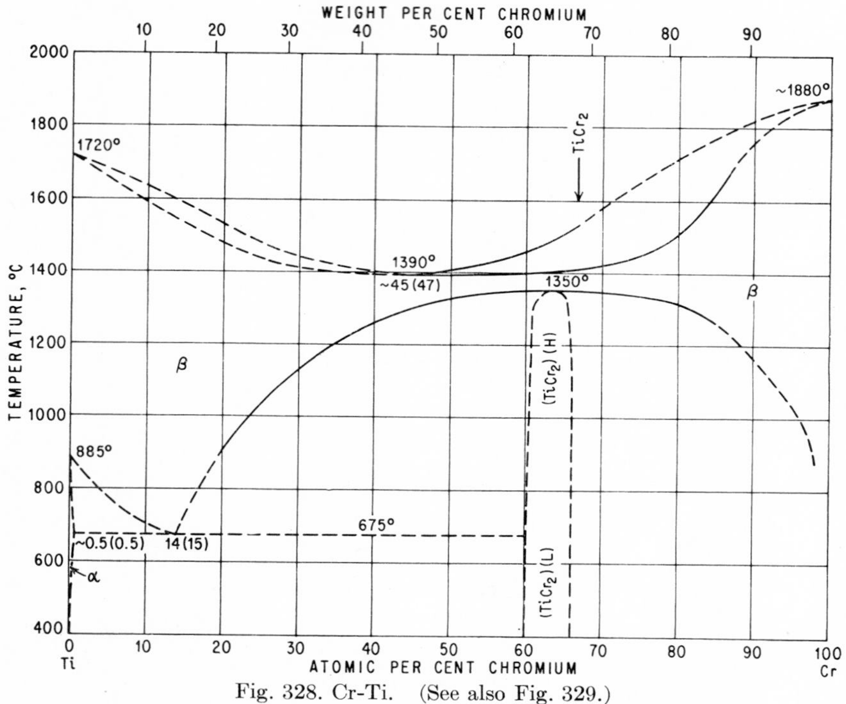

Fig. 328. Cr-Ti. (See also Fig. 329.)

Solidification. The existence of a continuous series of solid solutions at higher temperatures was recognized by [5] and corroborated by later workers. Liquidus data were reported only by [8], who took cooling curves of alloys containing between 40 and 68 wt. % Cr. Incipient-melting (solidus) data were microscopically determined by [5, 6, 8], whereas [5, 8] found the solidus to be almost horizontal between 40 and 70 wt. % Cr (Fig. 328), that of [6] rises smoothly on the Cr-rich side of the minimum. The data reported for the minimum in the liquidus-solidus curves are as follows: about 50 at. (52 wt.) % Cr, 1400°C [5]; 45 wt. (43 at.) % Cr, 1380°C [8]; and 50 wt. (48 at.) % Cr, 1400°C [6]. For Fig. 328, the values 45 at. (47 wt.) % Cr, 1390°C, were chosen.

Ti-rich Alloys. As has been stated above, the transformation point of Ti is lowered by Cr additions. The solubility of Cr in α-Ti is small, the following values having been reported (at. or wt. %): <1 [7], <0.5 [8], and about 0.5 [6]. The boundaries of the β phase as determined by various investigators can be seen from Fig. 329. Whereas the results of [6–8] are in substantial agreement, at least as far as the eutectoid data are concerned (see also Table 22), those of [10] place the eutectoid at a much lower temperature. The work by [10] was done with the objective of checking the prior hydrogen-pressure $\beta/(\alpha + \beta)$ boundary of [9] by micrography of instantaneously quenched specimens. [10] claims a very precise check between the instantaneously quenched micrographic specimens and the hydrogen-pressure measurements, including the unusual bulge in the $\beta/(\alpha + \beta)$ boundary presented. Since the projected boundary in this new determination would not be compatible with the accepted eutectoid point for this system, [10] employed two additional alloys richer in Cr (but hypoeutectoid) to investigate this point. The conclusion is advanced that the eutectoid temperature must be between room temperature and 548°C. This depends

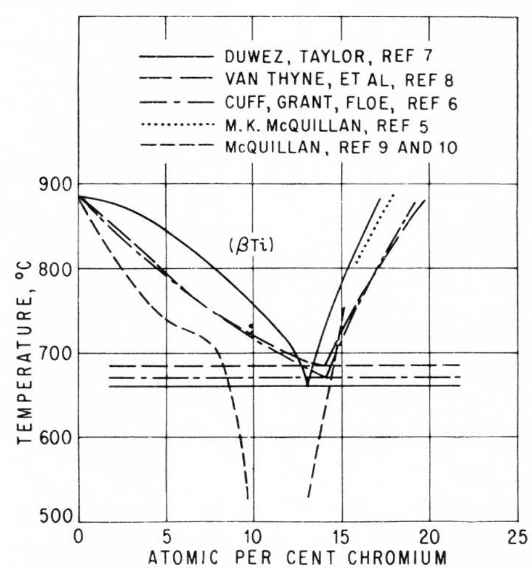

Fig. 329. Cr-Ti. (See also Fig. 328.)

upon an interpretation of the structures obtained upon annealing specimens below 650°C. The β phase is not retained as one should expect at these compositions, if the diagram of [10] is correct. [10] has proposed a theory of β-Ti electronic-state differences (which might act to give distinctly different bonding characteristics and lead to atom "clustering" on long-time annealing) in order to account for her microstructural interpretations.

More evidence is expected to accumulate now that this interesting theory has been promulgated; in the meantime, a eutectoid temperature level of about 675°C is preferred.

Table 22. Eutectoid Data

	Ref. 6	Ref. 7	Ref. 8	Ref. 10
Eutectoid composition	14 at. (15 wt.) % Cr	13 at. (14 wt.) % Cr	14 at. (15 wt.) % Cr	?
Eutectoid temperature	670°C	660°C	675–700°C (685°C)	Between 550°C and room temperature

[11] determined the temperature at which the martensitic type of transformation $\beta \rightarrow \alpha'$ takes place in alloys up to 4.2 wt. % Cr by thermal analysis at various rates of cooling. The minimum concentration for retaining the β structure by quenching lies at about 7 wt. (6.5 at.) % Cr [7, 8]. The effect of deformation on the transformation of a retained 10 wt. % Cr β-phase alloy was studied by [12].

Remainder of the Diagram. In the composition range between 50 and 75 at. % Cr the β phase is stable only in a narrow temperature range. At lower temperatures it decomposes, precipitating an intermediate phase based on the composition $TiCr_2$.

The $\beta/(\beta + TiCr_2)$ boundary between 14 and 60 at. % Cr in Fig. 328 is that given by [6], which, above 900°C, runs somewhat higher than that of [8] and somewhat lower than that of [5]. The boundary in the high-Cr range was taken from the work by [5]; it should be mentioned, however, that [6] reported a considerably higher solubility of Ti in Cr, approximately 15 wt. (16 at.) % Ti at 600°C.

The ($TiCr_2$) phase field appears to lie somewhere in the composition range 60–68 wt. (58–66 at.) % Cr. However, its width has not been determined with sufficient certainty as yet to draw even approximate phase boundaries.

To avoid a speculative presentation, the existence of a transformation in the ($TiCr_2$) phase (see below) was omitted from Fig. 328.

Crystal Structure. [7] demonstrated that the crystal structure of the intermediate phase is based on the ideal composition $TiCr_2$. It is f.c.c. ($MgCu_2$ or C15 type), with $a = 6.943$ A. Apparently the structure is stable only with an excess of Ti atoms. Their results were confirmed by [6], who reported a parameter of 6.92 A. [13] found that $TiCr_2$ occurs in two modifications, the transformation to the high-temperature form—hexagonal $MgZn_2$ (C14) type with $a = 4.932$ A, $c = 7.961$ A, $c/a = 1.614$—taking place somewhere between 1000 and 1300°C. The transformation has been corroborated by [14]. Lattice parameters of β solid-solution alloys were measured by [6, 7]. See also [15].

1. R. Vogel and B. Wenderott, *Arch. Eisenhüttenw.*, **14**, 1940, 279.
2. E. I. Larsen et al., *Metal Progr.*, **55**, 1949, 359–361.
3. C. M. Craighead, O. W. Simmons, and L. W. Eastwood, *Trans. AIME*, **188**, 1950, 485–513.
4. D. J. McPherson and M. G. Fontana, *Trans. ASM*, **43**, 1951, 1098–1125.
5. M. K. McQuillan, *J. Inst. Metals*, **79**, 1951, 379–390.
6. F. B. Cuff, N. J. Grant, and C. F. Floe, *Trans. AIME*, **194**, 1952, 848–853; discussion in *Trans. AIME*, **197**, 1953, 743–744.
7. P. Duwez and J. L. Taylor, *Trans. ASM*, **44**, 1952, 495–513; discussion, pp. 513–517.
8. R. J. Van Thyne, H. D. Kessler, and M. Hansen, *Trans. ASM*, **44**, 1952, 974–989.
9. A. D. McQuillan, *J. Inst. Metals*, **80**, 1952, 363–368.
10. M. K. McQuillan, *J. Inst. Metals*, **82**, 1954, 433–439; discussion, **82**, 1954, 644–652.
11. P. Duwez, *Trans. ASM*, **45**, 1953, 934–940.
12. H. M. Otte, *Nature*, **174**, 1954, 506.
13. B. W. Levinger, *Trans. AIME*, **197**, 1953, 196.
14. R. P. Elliott, Armour Research Foundation, Chicago, Ill., Technical Report 1, OSR Technical Note OSR-TN-247, August, 1954.
15. A. E. Austin and J. R. Doig, *Trans. AIME*, **209**, 1957, 27–30 (Structure of the transition phase omega in Ti-Cr alloys).

$\bar{1}.3394$
0.6606

Cr-U Chromium-Uranium

The solubilities of Cr in solid U are reported to be <2.5 at. (<0.6 wt.) % and <4 at. (<0.9 wt.) % Cr in β- and γ-U, respectively [1]. α-U dissolves no Cr, and solid Cr dissolves no U [1].

Cr depresses the U transitions considerably so that the high-temperature forms of U may be retained at room temperature by quenching [2–4]. The crystal structure of β-U has thus been deduced from room-temperature X-ray work on 1.4 at. % Cr single crystals [5, 6]. There has been some discussion, however, regarding the identity of the β-phase structure of the pure metal with that of the Cr alloy [7, 8].

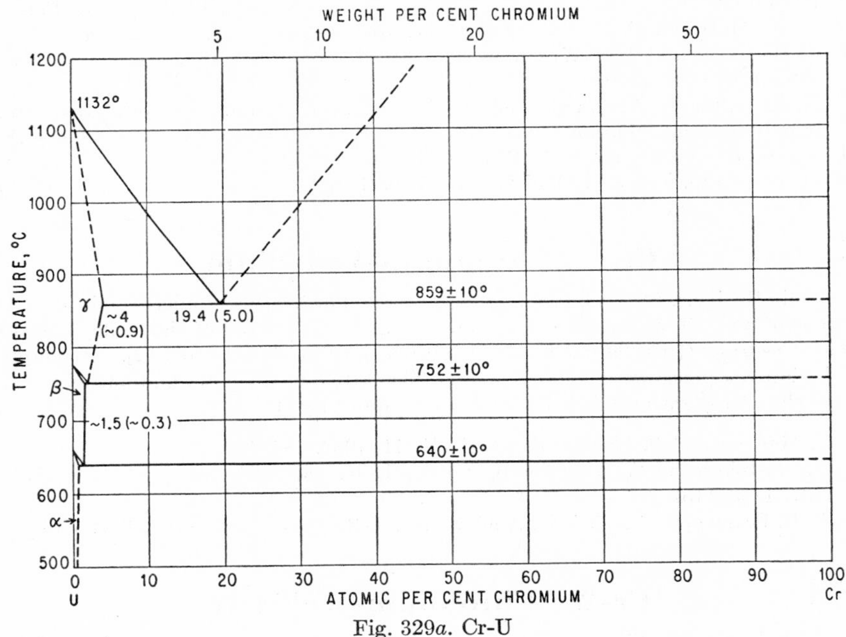

Fig. 329*a*. Cr-U

Supplement. A microscopic and X-ray examination [9] of eight Cr-U alloys varying in composition from 2 to 80 at. % Cr indicated that the two metals do not form any intermediate phase. The 20 at. % Cr alloy exhibited a completely eutectic structure. [10] studied metallographically the β → α transformation of U in low-Cr alloys.

The phase diagram of Fig. 329*a* is based largely on the work of [11], who studied the system using thermal, metallographic, and X-ray techniques. The eutectoid isotherms and the eutectic composition—737°, 612°C, and about 20 at. % Cr, respectively, according to [11]—have been revised by [12]. The solubility of Cr in β-U was estimated to be 1 at. % at 660°C [11] and 1.5 at. % at 720°C [13]. The solubility in α-U is less than that in β-U [11]. The isothermal transformation kinetics of β_U alloys have been studied by [13]. The solubility of U in solid Cr appears to be negligible [11].

1. J. J. Katz and E. Rabinowitch, "The Chemistry of Uranium," National Nuclear Energy Series, Div. VIII, vol. 5, p. 175, McGraw-Hill Book Company, Inc.,

New York, 1951. Information is based on Manhattan Project Report CT-3335 (not available to the author).

2. A. S. Wilson and R. E. Rundle, *U.S. Atomic Energy Comm., Publ.* AECD-2046, 1948.
3. A. S. Wilson and R. E. Rundle, *Acta Cryst.*, **2,** 1949, 126–127.
4. C. W. Tucker, *Trans. ASM,* **42,** 1950, 762–770.
5. C. W. Tucker, *Science,* **112,** 1950, 448; and *U.S. Atomic Energy Comm., Publ.* AECD-2957, 1950.
6. C. W. Tucker, *Acta Cryst.*, **4,** 1951, 425–431.
7. J. Thewlis, *Nature,* **168,** 1951, 198–199.
8. C. W. Tucker, *Acta Cryst.*, **5,** 1952, 395–396.
9. P. Gordon, *U.S. Atomic Energy Comm., Publ.* AECU-1833, 1952, pp. 151–155.
10. B. W. Mott and H. R. Haines, *Rev. mét.*, **51,** 1954, 614–616.
11. A. H. Daane and A. S. Wilson, work at Ames Laboratory (1947); *Trans. AIME,* **203,** 1955, 1219–1220.
12. H. A. Saller, F. A. Rough, and R. F. Dickerson, Battelle Memorial Institute, unpublished information (1953); quoted in H. A. Saller and F. A. Rough, *U.S. Atomic Energy Comm., Publ.* BMI-1000, June, 1955.
13. D. W. White, *Trans. AIME,* **203,** 1955, 1221–1228.

0.0090
$\bar{1}$.9910

Cr-V Chromium-Vanadium

Only the b.c.c. (A2) solid solution was detected in X-ray and microscopic investigations of nine alloys (10–90 at. %) annealed at 700°C [1] and of eight alloys [up to 40 wt. (~40 at.) % Cr] both in as-cast condition and annealed at 900°C [2]. Attempts to disclose a σ phase were fruitless [3, 1]. The lattice-parameter-vs.-atomic-per-cent curve shows a slight negative deviation from additivity [1].

1. H. Martens and P. Duwez, *Trans. ASM,* **44,** 1952, 484–493.
2. W. Rostoker and A. Yamamoto, *Trans. ASM,* **46,** 1954, 1136–1163, especially 1137, 1150–1151.
3. W. B. Pearson, J. W. Christian, and W. Hume-Rothery, *Nature,* **167,** 1951, 110.

$\bar{1}$.4515
0.5485

Cr-W Chromium-Wolfram

The diagram in Fig. 330 is based on the work by [1] (X-ray analysis) and [2] (thermal, micrographic, and X-ray analysis). The upper part of the miscibility gap above 1400°C was determined micrographically [2].

The existence of a complete series of solid solutions at high temperatures was also found by [3] (X-ray studies). On the other hand, [4], using alloys containing about 5 wt. % Al prepared by the aluminothermic method, assumed a eutectic at approximately 30 at. (60 wt.) % W between primary solid solutions, the W-rich of which was claimed to contain maximal 20 wt. (about 47 at.) % Cr. Reasons for the contradictory findings of [4] were given by [5]. The minimum in the liquidus curve at about 30 wt. (10.8 at.) % W, 1720 ± 50°C, reported by [3], on the basis of optical observations of incipient melting of sintered powder samples, was not confirmed by thermal analysis; see Fig. 330.

[6] computed a theoretical boundary of the two-phase field and found "almost certainly fortuitous" good agreement with the boundary due to [2], especially on the W side; see also [7, 8].

Lattice-spacing data were reported by [1–3].

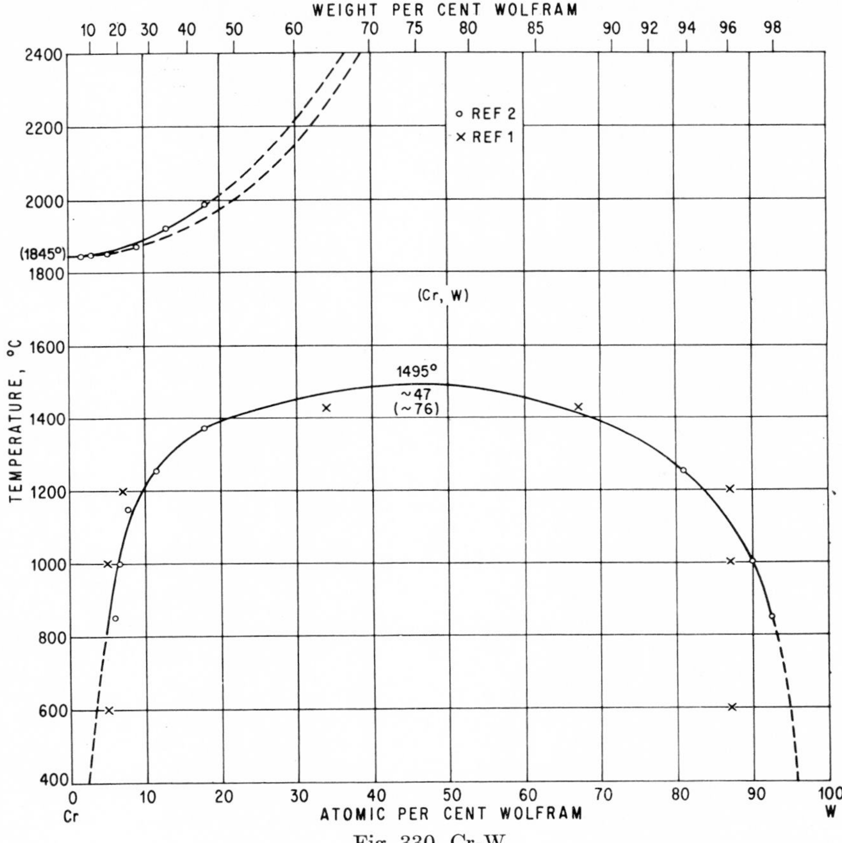

Fig. 330. Cr-W

1. W. Trzebiatowski, H. Ploszek, and J. Lobzowski, *Anal. Chem.*, **19**, 1947, 93–95.
2. H. T. Greenaway, *J. Inst. Metals*, **80**, 1951-1952, 589–592.
3. O. Kubaschewski and A. Schneider, *Z. Elektrochem.*, **48**, 1942, 671–674.
4. S. Isida, H. Asada, and S. Higasimura, *Rept. Aeronaut. Research Inst., Tokyo Imp. Univ.*, **13**, 1938, 195–210 (in Japanese).
5. F. Weibke and U. v. Quadt, *Z. Elektrochem.*, **46**, 1940, 635–641.
6. A. D. McQuillan, *J. Inst. Metals*, **80**, 1951-1952, 697–698.
7. H. T. Greenaway, *J. Inst. Metals*, **80**, 1951-1952, 698.
8. I. J. Polmear, *Bull. Inst. Metals*, **3**, 1956, 71–72.

$\bar{1}.9006$
0.0994

Cr-Zn Chromium-Zinc

According to [1–3], solid Cr dissolves very slowly in molten Zn, a fact which makes the preparation of even low-Cr alloys [3, 4], with no undissolved Cr, a difficult task. Another method for preparing low-Cr alloys, which was first used by [5], consists in reducing chromium chloride by molten Zn; if an operating temperature of

750–800°C [2–4] is chosen, the speed of reduction of the chloride equals the rate of solution of the Cr produced [2].

The available information on constitution is summarized in Fig. 331. There is strong microscopic [2–4] and roentgenographic [2, 4] evidence that the Zn-rich compound (θ), which was first observed by [5, 1], is the only intermediate phase occurring in this system. It has a hexagonal unit cell, with $a = 12.89$, $c = 30.5$ "10^{-8} cm" (presumably kX), $c/a = 2.37$ [2], containing about 287 atoms [3]. However, its exact composition has not been established yet. The values given [5, 3, 4] vary between 3.8 and 7 wt. % Cr, with 4–5 wt. (5–6.2 at.) % Cr [3, 4] as the most reliable ones; an extended homogeneity range does not exist [6]. The temperature of the decomposition of θ was determined by a microscopic method (464°C) [3]. For the eutectic, the data 0.25 wt. (0.31 at.) % Cr at 418.5°C [7] and 0.22 wt. (0.28 at.) % Cr at 415°C [3] were reported.

The solid solubility of Cr in Zn decreases from 0.03 wt. (0.04 at.) % Cr at the eutectic temperature (microscopically determined) [8] to 0.02 and 0.01 wt. (0.025 and 0.013 at.) % Cr at 375 and 350°C (microhardness method), respectively [3].

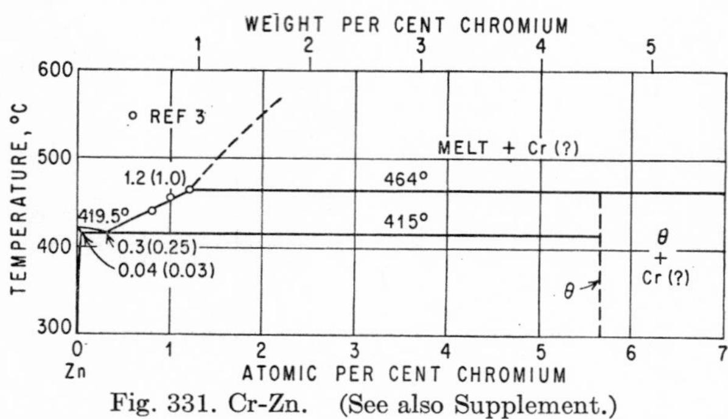

Fig. 331. Cr-Zn. (See also Supplement.)

Supplement. [9] prepared Cr-Zn alloys by electrolytic deposition of Cr at a liquid zinc amalgam electrode and subsequent distillation of the Hg at 360–380°C. X-ray diffraction analysis of 18 alloys varying in composition from 17 to 99 wt. % Zn yielded the following results: (*a*) At temperatures around 250°C, Cr takes into solid solution up to 30 wt. (25.5 at.) % Zn. (*b*) In agreement with previous findings, there exists only one intermediate phase. It is located at 4.5 wt. (5.6 at.) % Cr, and the formula $CrZn_{17}$ (288 atoms per unit cell) may be assigned to it.

The findings of [10], who dissolved small amounts of Cr in molten Zn at temperatures up to 850°C and studied the alloys by microscopic, X-ray, and chemical analysis, are in part at variance with those of previous investigators. Two intermediate phases could be isolated. The hexagonal θ phase with 4–5 wt. % Cr, $a = 12.95$ kX, is assumed—on the basis of its etching characteristics—to have an appreciable range of homogeneity (cf. above). Needles of another intermediate phase, Y, of presumably orthorhombic symmetry, analyzed 7–8 wt. % Cr. Air-cooled alloys showed Y crystals surrounded by peritectic rims of θ in a matrix of Zn. The binary character of these findings needs corroboration, since Cr of only 98 wt. % purity (main contaminant: 1.5% Fe) was used.

1. G. Hindrichs, *Z. anorg. Chem.*, **59,** 1908, 427.
2. H. Hanemann, *Z. Metallkunde*, **32,** 1940, 91–92.
3. T. Heumann, *Z. Metallkunde*, **39,** 1948, 45–52.

4. A. R. Harding and G. V. Raynor, *J. Inst. Metals*, **80**, 1951-1952, 436–439, 446.
5. H. Le Chatelier, *Compt. rend.*, **120**, 1895, 835.
6. [4], p. 446.
7. E. Weisse, A. Blumenthal, and H. Hanemann, *Z. Metallkunde*, **34**, 1942, 221. Redetermination of earlier [2] data.
8. [2] suggested about 0.1 wt. % Cr.
9. F. Lihl and P. Jenitschek, *Z. Metallkunde*, **45**, 1954, 686–689.
10. H. Hartmann , W. Hofmann, and D. Müller, *Abhandl. braunschweig. wiss. Ges.*, **7**, 1955, 100–106.

$\bar{1}.7560$
0.2440

Cr-Zr Chromium-Zirconium

Phase diagrams of the Zr-Cr system were published by [1], [2] (up to 66.7 at. % Cr), and [3]. The work of [1, 2] is based on alloys prepared [4] with magnesium-reduced zirconium, whereas [3] used a higher-purity iodide-zirconium crystal bar. The results of [2, 3] are in agreement on the general constitution of the Zr-rich region [5]; but [1] assumed a vastly different set of phase relationships [6] obviously due, in the main, to a misinterpretation of microstructures.

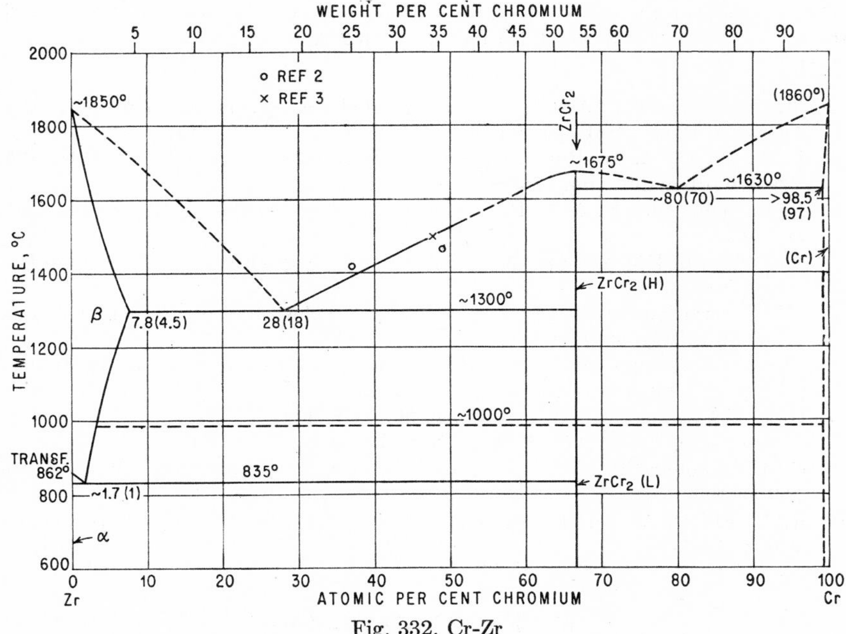

Fig. 332. Cr-Zr

Only a few data fixing the liquidus are available, as may be seen from Fig. 332. [2, 3] agree on the composition of the Zr-rich eutectic, but there are discrepancies as to the eutectic temperature—1380°C [2] (optical pyrometry), 1280°C [3] (mainly thermocouple)—and the melting point of the only compound, $ZrCr_2$ (53.27 wt. % Cr) (1650 ± 10°C, discussion on [2]; 1700 ± 25°C [3]). For these temperatures in Fig. 332 the data ~1300 and 1675°C, respectively, were chosen. The eutectic between $ZrCr_2$

and (Cr) has been located close to 80 at. (70 wt.) % Cr [1, 3]. Thermal analysis [3] yielded 1635 ± 15°C (Fig. 332), and microscopic examination of quenched alloys [1], about 1550°C as eutectic temperature.

The β-Zr solid-solution phase field in Fig. 332 is that of [3], which was carefully established by micrographic work up to 1600°C. A somewhat larger field was assumed by [2], based on micrographic work up to 1300°C [extrapolated maximum solubility 6.2 wt. (10.4 at.) % Cr at 1380°C; eutectoid at 1.8 wt. (3.1 at.) % Cr and 805°C]. A Cr content of more than 1.5 wt. % seems necessary for retaining β_{Zr} by normal quenching methods (see [3]).

X-ray [2] and microscopic [2, 3] work indicated that the solid solubility of Cr in α-Zr must be very small, and cursory microscopic work [1, 3] showed that of Zr in Cr to be less than 1.5 at. (~3 wt.) % Zr.

Crystal Structures. [7, 3] reported the hexagonal $MgZn_2$ (C14) type, and [2], the cubic $MgCu_2$ (C15) type of structure for the compound $ZrCr_2$. It was then shown by [8] that $ZrCr_2$ undergoes a polymorphic transformation, C14 being the low- and C15 the high-temperature (stable above about 950°C) modification. [9] confirmed the existence of a polymorphic transition but concluded the reverse order of these structure types and a transformation temperature probably near the melting point of $ZrCr_2$. The results of a more recent investigation [10]—like those of the preceding authors, based on room-temperature X-ray work—favor those of [8] (polymorphic temperature near 1000°C). The lattice parameters of both modifications according to [10] are a = 7.207 A (for C15) and a = 5.102 A, c = 8.238 A, c/a = 1.615 (for C14).

1. M. K. McQuillan, Aeronaut. Research Laboratory, Department of Supply, Australia, Report SM-165, January, 1951.
2. E. T. Hayes, A. H. Roberson, and M. H. Davies, *Trans. AIME,* **194,** 1952, 304–306; discussion, pp. 1211–1213.
3. R. F. Domagala, D. J. McPherson, and M. Hansen, *Trans. AIME,* **197,** 1953, 279–283.
4. The alloys of [1-3] were prepared by arc melting in an argon [1] or helium [2, 3] atmosphere.
5. Minor variations in the results as mentioned below may be attributed to some extent to the use of different zirconium stock.
6. Characterized by a solid solution of Cr in β-Zr extending up to about 60 at. % Cr at high temperatures.
7. H. J. Wallbaum, *Naturwissenschaften,* **30,** 1942, 149.
8. W. Rostoker, *Trans. AIME,* **197,** 1953, 304.
9. C. B. Jordan and P. Duwez, *Jet Propulsion Lab., Calif. Inst. Tech., Progress Rept.* 20-196, 1953.
10. R. P. Elliott, Armour Research Foundation of Illinois Institute of Technology, to be published.

0.3765
$\bar{1}$.6235

Cs-Fe Cesium-Iron

Cs at its boiling point (690°C) does not attack iron [1]. Cs is insoluble in solid iron [2].

1. O. Ruff and O. Johannsen, *Ber. deut. chem. Ges.*, **38,** 1905, 3602; abstract in "Gmelins Handbuch der anorganischen Chemie," System No. 59 (A), p. 1813, Verlag Chemie, G.m.b.H., Berlin, 1934-1939.
2. F. Wever, *Arch. Eisenhüttenw.*, **2,** 1928-1929, 739–746.

0.2802
$\bar{1}.7198$

Cs-Ga Cesium-Gallium

After condensation of gaseous Cs on Ga followed by mild heating to reach alloying, extraction by liquid NH_3 dissolved only Cs [1].

1. E. Zintl and H. Kaiser, *Z. anorg. Chem.*, **211,** 1933, 121–122.

0.2626
$\bar{1}.7374$

Cs-Ge Cesium-Germanium

See Cs-Si.

$\bar{1}.8212$
0.1788

Cs-Hg Cesium-Mercury

Thermal work on this system was carried out by [1] (Fig. 333). The authors did not take into consideration the halting times of the three-phase equilibria and did not determine the end of solidification of alloys with 40–70 wt. % Hg. Additional information on this system was obtained by a careful magnetic analysis [2]. The results of both studies as to existing compounds are summarized in Table 23.

Table 23. Reported Cs-Hg Compounds

Thermally and magnetically indicated	Only thermally indicated	Only magnetically indicated
		~CsHg (60.15 wt. % Hg)
Cs_3Hg_4 [57.14 at. (66.81 wt.) % Hg] [3] $CsHg_2$ (75.11 wt. % Hg) $CsHg_4$ (85.79 wt. % Hg) $CsHg_6$ [85.71 at. (90.05 wt.) % Hg]		
	$CsHg_x$ $(6 < x < 14)$	
		~$CsHg_{20}$ [95.24 at. (96.79 wt.) % Hg]

No reliable explanation can be given for the weak thermal effects observed at 188°C on cooling curves of alloys with 62.4–66.4 at. % Hg (Fig. 333). [1] suggested a polymorphic transformation of $CsHg_2$; however, the fact that no arrests occurred in alloys with an excess of Hg gives no support to that assumption.

Volume changes in alloys with 50–65 at. % Hg, observed on storing them at room temperature or at 100°C [2], may either indicate a solid-state transformation [2] or simply be due to the completion of undercooled peritectic reactions. There are some indications [1, 2] that $CsHg_6$ has a certain range of homogeneity.

The composition of the compound which forms peritectically at 12°C ($CsHg_x$, see Table 23) is not yet known. By mechanical separation of crystals from their Hg-rich mother liquor, the formula $CsHg_{12}$ [92.31 at. (94.77 wt.) % Hg] was concluded [4]. The existence of a compound still richer in Hg (~$CsHg_{20}$, see Table 23) is improbable.

Supplement. [5] reported briefly the finding of a cubic phase with the possible composition $CsHg_{12}$.

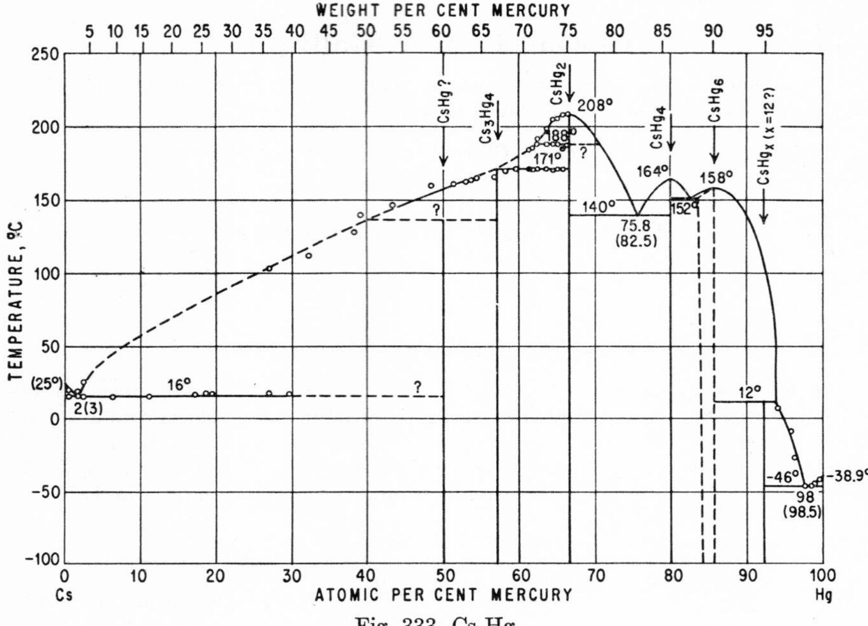

Fig. 333. Cs-Hg

1. N. S. Kurnakov and G. J. Zukovsky, *Z. anorg. Chem.*, **52,** 1907, 416–427. The numerous data points above 66.6 at. % Hg, fixing the liquidus and the three-phase equilibria at 140, 152, 12, and −46°C, were omitted from Fig. 333.
2. W. Klemm and B. Hauschulz, *Z. Elektrochem.*, **45,** 1939, 346–353. The alloys were prepared in evacuated glass tubes and annealed for 12 hr 30°C below their respective melting points.
3. The thermal results do not fix the exact composition and admit also the formula Cs_2Hg_3 (69.37 wt. % Hg) which was concluded in M. Hansen, "Der Aufbau der Zweistofflegierungen," Springer-Verlag OHG, Berlin, 1936.
4. G. McPhail Smith and H. C. Bennett, *J. Am. Chem. Soc.*, **32,** 1910, 622–626. This method tends to give formulas too high in Hg.
5. N. C. Baenziger, J. W. Nielsen, and E. J. Duwell, AEC Contract AT(11-1)-72 Project 4, University of Iowa, Ames, Iowa, Dec. 1, 1952.

0.0638
$\bar{1}.9362$

Cs-In Cesium-Indium

After condensation of gaseous Cs on In followed by mild heating to reach alloying, extraction by liquid NH_3 dissolved only Cs [1].

1. E. Zintl and H. Kaiser, *Z. anorg. Chem.*, **211,** 1933, 121–122.

0.5314
$\bar{1}.4686$

Cs-K Cesium-Potassium

Thermal analyses of the whole system were carried out by [1, 2]. There is agreement only as to the general constitution (Fig. 334); the data of [2]—which were also confirmed by measurements of electrical conductivity—are the more reliable ones and were used for the liquidus and solidus of Fig. 334. The conductivity-vs.-composition (catenary) curve at $-39°C$ has a minimum at 50 at. % [2].

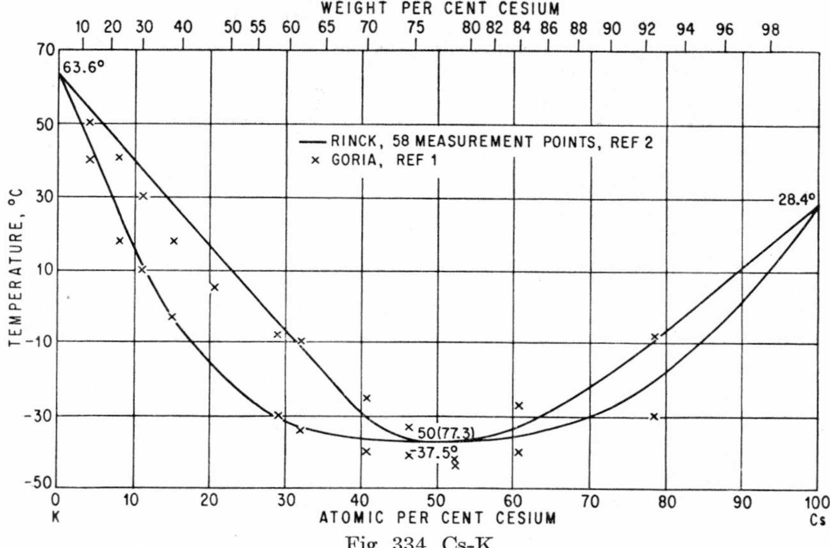

Fig. 334. Cs-K

Magnetic susceptibilities and lattice parameters of the solid solutions were measured by [3]; the latter change nearly linearly with the composition in at. %.

1. C. Goria, *Gazz. chim. ital.*, **65,** 1935, 1226–1230. Alloys were prepared and studied under an inert-gas atmosphere.
2. E. Rinck, *Compt. rend.*, **203,** 1936, 255–257. Alloys were prepared and studied in vacuum. Thermal data had to be taken from a graph.
3. B. Böhm and W. Klemm, *Z. anorg. Chem.*, **243,** 1939, 69–85.

1.2822
$\bar{2}$.7178

Cs-Li Cesium-Lithium

According to a thermal investigation [1], Cs and Li are immiscible in both the liquid and the solid state.

1. B. Böhm and W. Klemm, *Z. anorg. Chem.*, **243,** 1939, 69–85.

0.7619
$\bar{1}$.2381

Cs-Na Cesium-Sodium

Thermal analyses were carried out by [1, 2] (Fig. 335). Both authors agree upon the existence of one compound, Na_2Cs (74.29 wt. % Cs), which forms peritectically at $-8°C$ by a very sluggish reaction [3]. Data for the eutectic are 75 at. (94.5 wt.) % Cs, $-30°C$ [1], and ~70 at. (93 wt.) % Cs, $-28°C$ [2].

The compound Na_2Cs is also indicated by magnetic measurements [4]. It has not been possible yet to observe X-ray patterns of this compound, however, mainly because of the small extent to which it forms [4].

1. E. Rinck, *Compt. rend.*, **199**, 1934, 1217–1219. Alloys were prepared and studied in vacuum. Thermal data had to be taken from a graph.
2. C. Goria, *Gazz. chim. ital.*, **65**, 1935, 1226–1230. Alloys were prepared and studied under an inert-gas atmosphere.

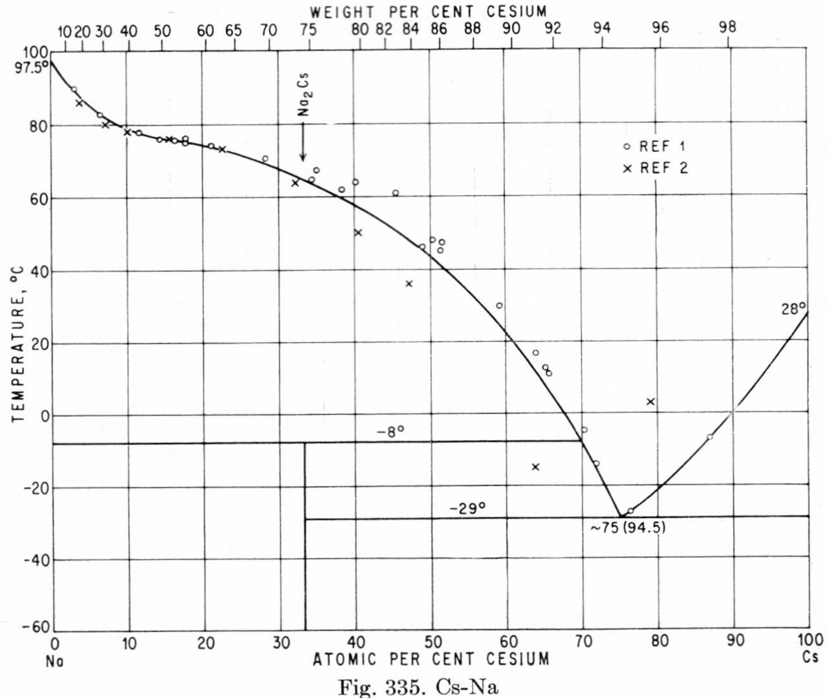

Fig. 335. Cs-Na

3. The eutectic arrest could still be observed in alloys with less than 10 at. % Cs.
4. B. Böhm and W. Klemm, *Z. anorg. Chem.*, **243**, 1939, 69–85.

0.1917
$\bar{1}$.8083

Cs-Rb Cesium-Rubidium

Rb and Cs form a continuous solid solution, as thermal [1], conductometric [1], and roentgenographic [2] work indicates beyond any doubt [3] (Fig. 336). The electrical-conductivity-vs.-composition (catenary) curves for the solid state have—like the liquidus—minima at 50 at. %. The lattice parameter of a 69 at. % Cs alloy lies nearly on the straight line connecting the parameters of the elements.

1. E. Rinck, *Compt. rend.*, **205**, 1937, 135–137. The alloys were prepared and studied in vacuum. The thermal data for Fig. 336 had to be taken from a graph.
2. B. Böhm and W. Klemm, *Z. anorg. Chem.*, **243**, 1939, 69–85.
3. C. Goria (*Gazz. chim. ital.*, **65**, 1935, 1226–1230) concluded from thermal measurements a simple eutectic system without solid solubilities (!). See the discussion of these results by Rinck [1].

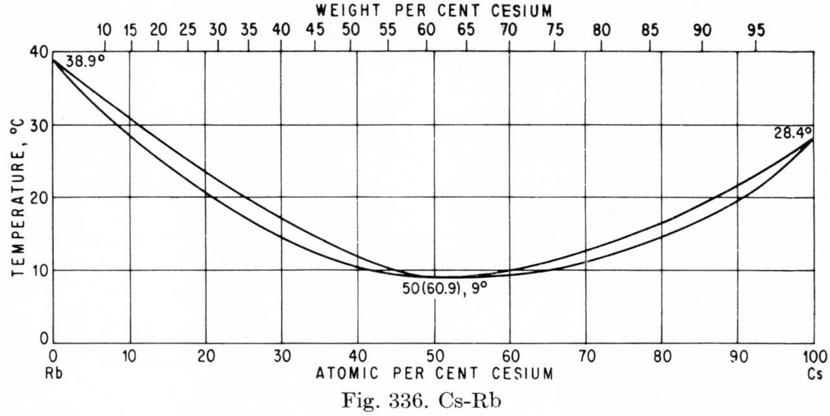

Fig. 336. Cs-Rb

0.0380
$\bar{1}$.9620

Cs-Sb Cesium-Antimony

Measurements of several properties of thin Cs-Sb layers (for use in photoelectric cells) indicate that the metals alloy easily [1–6] and that at least the compounds Cs_3Sb (23.40 wt. % Sb) [2, 3, 5, 6] and CsSb (47.81 wt. % Sb) [5, 6] exist.

1. P. Görlich, *Z. Physik,* **101,** 1936, 337.
2. A. Sommer, *Nature,* **148,** 1941, 468.
3. A. Sommer, *Proc. Phys. Soc.,* **55,** 1943, 145–154.
4. A. I. Frimer, *Doklady Akad. Nauk S.S.S.R.,* **63,** 1948, 255–257.
5. H. Miyazawa and S. Fukuhara, *J. Phys. Soc. Japan,* **7,** 1952, 645–647 (in English).
6. H. Miyazawa, K. Noga, S. Chikazumi, and A. Kobayashi, *J. Phys. Soc. Japan,* **7,** 1952, 647–648 (in English).

0.2261
$\bar{1}$.7739

Cs-Se Cesium-Selenium

Cs_2Se (22.90 wt. % Se) and Cs_2Te (32.44 wt. % Te) melt under decomposition at 660 and 680°C, respectively, when heated in vacuo [1]. A small excess of Cs atoms causes semiconductivity in both compounds [1], which yield complex powder patterns [1, 2].

The constitution of Cs-Se alloys higher in Se, which were magnetically investigated (up to $CsSe_2$) by [2], is not yet known.

1. A. Bergmann, *Z. anorg. Chem.,* **231,** 1937, 271.
2. W. Klemm, H. Sodomann, and P. Langmesser, *Z. anorg. Chem.,* **241,** 1939, 281–304.

0.6750
$\bar{1}$.3250

Cs-Si Cesium-Silicon

CsSi (17.45 wt. % Si) and CsGe (35.33 wt. % Ge) were prepared by direct synthesis. Both compounds are persistent only in an inert and dry atmosphere. On thermal decomposition, substances of approximate composition $CsSi_8$ [88.89 at. (62.84 wt.) % Si] and $CsGe_4$ (68.60 wt. % Ge) are obtained. Representations of powder patterns are given [1].

1. E. Hohmann, *Z. anorg. Chem.,* **257,** 1948, 113–126.

0.0177
$\bar{1}.9823$

Cs-Te Cesium-Tellurium

See Cs-Se.

0.0560
$\bar{1}.9440$

Cu-Fe Copper-Iron

The literature up to 1900 has been summarized by [1], and results of work prior to 1934 have been thoroughly reviewed by [2, 3].

Because of a great deal of conflicting evidence, the question of whether liquid Cu and Fe alloy in all proportions or form a wide miscibility gap has been a controversial subject for a long time. Already [4], and later [1], reported that with iron low in carbon the liquid metals are completely miscible but that higher undefined carbon contents cause separation into two layers. The latter result was in agreement with observations of numerous workers prior to 1906; see [2, 3]. The effect of carbon in promoting liquation was also recognized by later investigators [5, 6]; however, they claimed to have found evidence that liquid segregation was a characteristic feature of the binary system, too, although neither a horizontal portion of the liquidus curve nor a monotectic equilibrium had been detected [7, 5, 6].

To overcome this inconsistency with the phase rule, it was proposed [8] that the miscibility gap in the liquid state does not intersect the liquidus curve but closes at a temperature above it, i.e., has a lower critical point. Experimental evidence supporting this hypothesis was reported [9–11] and disputed [12, 13]; see also [14, 15].

On the basis of the work by [7] and especially [16] and others [17–19], it is now generally accepted that in moderately pure melts a miscibility gap does not exist. [18] reported that layer formation occurs if the carbon content is higher than 0.02–0.03%; silicon also causes liquid segregation [17].

The liquidus curve was determined by [7, 5, 6, 16, 18]. Results agree in that there is no horizontal portion; however, temperatures differ somewhat. The temperatures between 30 and 70 wt. % Fe according to [6] are the highest and those by [18] the lowest (maximal difference 23°C at 30 wt. % Fe and 8°C at 70 wt. % Fe). Between 90 and 100 wt. % Fe, temperatures by [16] lie 4–12°C above those by [6], with those by [7] in between. The slope of the liquidus is very small, the temperature differences per 10 wt. % between 30 and 70 wt. % Fe being 3–4 [6, 16], 4–6 [7], and 5–9°C [18]. The curve shown in Fig. 337 is based on the data by [16]. The temperature of the peritectic reaction involving the δ phase was found as 1477 [20] and 1484°C [16]. There is very good agreement as to the temperature of the peritectic reaction: melt + γ $\rightleftarrows$ (Cu) [21, 6, 16].

Temperatures given for the eutectoid equilibrium $\gamma \rightleftarrows \alpha$ + (Cu) differ widely because of a great hysteresis between heating and cooling. On cooling, the following temperatures were found: 791 [7], about 820 [6], 824 [22], and 755°C [16]; and on heating, about 850 [6], 856 [22], and 855–860°C [23], and the interval 745–860°C [16]. As accurate determinations were lacking, 835 ± 15°C was tentatively assumed in Fig. 337. This also holds for the eutectoid composition, given as about 2.5 wt. % [6] and 3.5 wt. % Cu [16].

The boundary of the (Cu) phase was determined by [24], [25], and [26], using micrographic, magnetic, and X-ray methods, respectively; see also [27]. Results agree fairly well. The curve given in Fig. 337 is based on the data by [25]. According to [24], the solubility falls nearly linearly from about 3.8 wt. % at 1094°C to about 0.3 wt. % Cu at 800°C. [26] reported solubilities of 2.5, 1.5, 0.9, and 0.5 wt. % Fe at 1025, 900, 800, and 700°C, respectively. The room-temperature solubility of 0.26 wt. % Fe assumed by [28] is definitely too high (and also in disagreement with work by [29]), since equilibrium will never be attained at this low temperature. At 200°C, a solubility of the order of 1.3×10^{-5} wt. % Fe could be expected [25]. As to the precipitation of α-Fe phase from (Cu) phase, reference is made to the papers by [30–33].

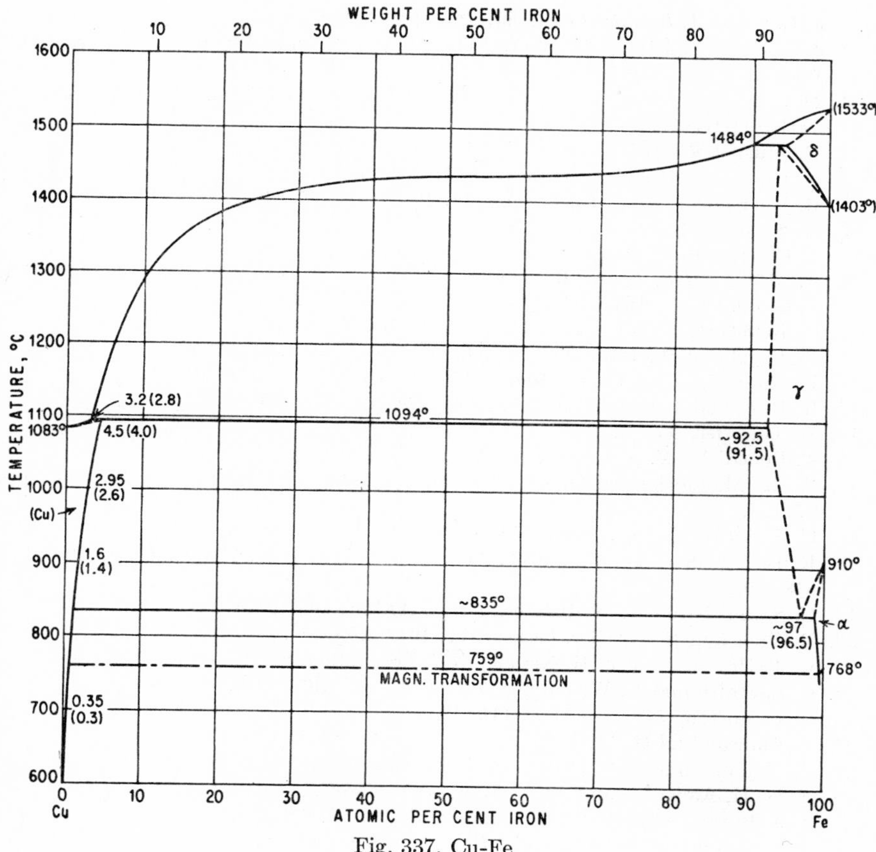

Fig. 337. Cu-Fe

The solubility of Cu in γ-Fe, not accurately determined as yet, has been reported as about 7.5 wt. (6.7 at.) % [16] or about 8.0 wt. (7.1 at.) % [11] at 1484 and 1430°C, respectively, and as about 7.5 wt. [16] or 8.5 wt. (7.6 at.) % Cu [11] at 1094°C. [34] stated: "Experiments on sintered compacts of Fe and Cu showed that the solubility of Cu in γ-Fe between the peritectic and eutectic temperatures may be considered lower than 8 wt. % at 1094°C and 4 wt. % at 850°, and that the copper solubility lines in γ-Fe and α-Fe may form a nearly continuous curve."

The solubility of Cu in α-Fe, found by measurements of electrical resistivity vs. composition to be about 3.0 wt. % at 800°C, 2.0 at 750°C, and 1.25 at 700°C, was considered to be too high [22] and incompatible with the composition of the eutectoid. Indeed, [23] reported much lower solubilities determined by the lattice-parametric method: about 1.4 wt. (1.2 at.) % at 850°C, 0.9 wt. (0.8 at.) % at 800°C, 0.5 wt. (0.4 at.) % at 750°C, and 0.3 wt. (0.25 at.) % Cu at 700°C.

Lattice parameters of the (Cu) phase were reported by [27, 28] and especially [26], and those of the α-Fe solid solution by [23]; see [34].

For a thermodynamic treatment of the α_{Fe} and γ_{Fe} phase boundaries, see [35].

1. J. E. Stead, *J. Iron Steel Inst.*, **60,** 1901, 104–121; *Engineering,* **72,** 1901, 851–853.
2. J. L. Gregg and B. N. Daniloff, "The Alloys of Iron and Copper," McGraw-Hill Book Company, Inc., New York, 1934.
3. M. Hansen, "Der Aufbau der Zweistofflegierungen," pp. 557–563, Springer-Verlag OHG, Berlin, 1936.
4. D. Mushet, *Phil. Mag.*, **6,** 1835, 81–85.
5. R. Ruer and K. Fick, *Ferrum,* **11,** 1914, 39–51.
6. R. Ruer and F. Goerens, *Ferrum,* **14,** 1916-1917, 49–61.
7. R. Sahmen, *Z. anorg. Chem.*, **57,** 1908, 9–20.
8. F. Ostermann, *Z. Metallkunde,* **17,** 1925, 278.
9. A. Müller, *Mitt. Kaiser-Wilhelm-Inst. Eisenforsch. Düsseldorf,* **9,** 1927, 173–175; *Z. anorg. Chem.*, **162,** 1927, 231–236; **169,** 1928, 272.
10. O. Reuleaux, *Metall u. Erz,* **24,** 1927, 99–100.
11. R. Vogel and W. Dannöhl, *Arch. Eisenhüttenw.*, **8,** 1934, 39–40.
12. R. Ruer, *Z. anorg. Chem.*, **164,** 1927, 336–376.
13. R. Ruer and J. Kuschmann, *Z. anorg. Chem.*, **153,** 1926, 260–262.
14. W. Guertler, see [8].
15. C. Benedicks, *Z. physik. Chem.*, **131,** 1928, 289–293.
16. W. R. Maddocks and G. E. Claussen, *Iron Steel Inst. Spec. Rept.* 14, 1936, pp. 97–124.
17. C. S. Smith, *J. Inst. Metals,* **54,** 1934, 251.
18. K. Iwase, M. Okamoto, and T. Amemiya, *Science Rept. Tôhoku Univ.*, **26,** 1938, 618–640.
19. C. S. Smith and E. W. Palmer, *Trans. AIME,* **188,** 1950, 1486–1499.
20. R. Ruer and R. Klesper, *Ferrum,* **11,** 1914, 259–260.
21. C. T. Heycock and F. H. Neville, *Phil. Trans. Roy. Soc. (London),* **A189,** 1897, 69.
22. C. S. Smith and E. W. Palmer, *Trans. AIME,* **105,** 1933, 133–164.
23. J. T. Norton, *Trans. AIME,* **116,** 1935, 386–394.
24. D. Hanson and G. W. Ford, *J. Inst. Metals,* **32,** 1924, 335–361.
25. G. Tammann and W. Oelsen, *Z. anorg. Chem.*, **186,** 1930, 267–277.
26. A. G. H. Anderson and A. W. Kingsbury, *Trans. AIME,* **152,** 1943, 38–40.
27. A. J. Bradley and H. J. Goldschmidt, *J. Inst. Metals,* **65,** 1939, 388–401.
28. T. S. Hutchison and J. Reekie, *Phys. Rev*, **83,** 1951, 854–855.
29. F. H. Hetherington and J. Reekie, *J. Appl. Phys.*, **22,** 1951, 1293–1294.
30. R. Gordon and M. Cohen, "Age Hardening of Metals," pp. 161–183, The American Society for Metals, Cleveland, Ohio, 1939-1940.
31. F. Bitter and A. R. Kaufmann, *Phys. Rev.*, **56,** 1939, 1044–1051.
32. C. S. Smith, *Phys. Rev.*, **57,** 1940, 337.
33. A. Knappwost, *Z. Metallkunde,* **45,** 1954, 137–142.
34. B. N. Daniloff, "Metals Handbook," 1948 ed., p. 1196, The American Society for Metals, Cleveland, Ohio.
35. W. Oelsen, *Stahl u. Eisen,* **69,** 1949, 468–475.

$\bar{1}.9597$
0.0403

Cu-Ga Copper-Gallium

By means of thermal analysis, supplemented by microscopic and roentgenographic investigations, the constitution of the entire system was first determined by

[1]. It is characterized by the existence of four peritectically formed intermediate phases β, γ, γ_2, and $CuGa_2$ (designation according to Fig. 338), of which β was shown to decompose eutectoidally at 620°C into α and γ, and γ to transform into γ_1. The phase relationships in the composition range 35–50 at. (37.1–52.3 wt.) % Ga were not fully established and were, as published by [1], in contradiction to the phase rule.

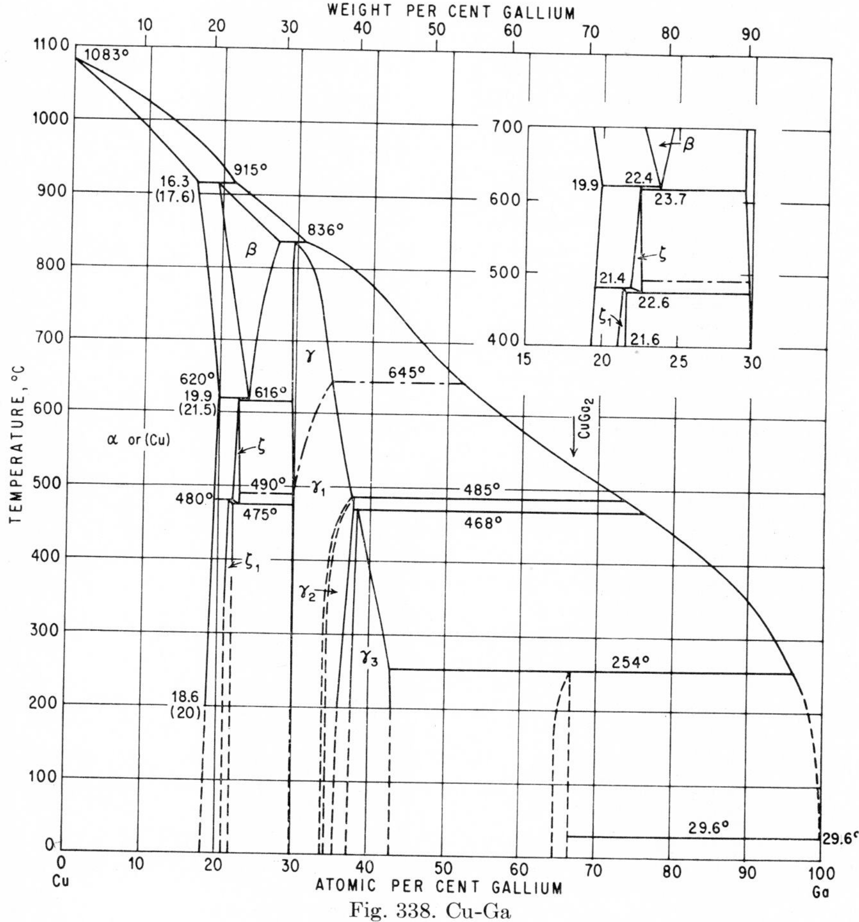

Fig. 338. Cu-Ga

Also, evidence for the composition of the intermediate phase richest in Ga, formed peritectically at 249°C and claimed to contain approximately 56 at. % Ga, was not conclusive.

Considerable modifications of the diagram by [1] were made by [2] and [3], covering the ranges 18–32 at. (19–34 wt.) % and 30–100 at. (32–100 wt.) % Ga, respectively. They are based on detailed determinations using thermal, micrographic, and lattice-spacing methods. The diagram in Fig. 338 was drawn on the basis of these findings.

The solidus and solubility curve of the α phase were determined micrographically by [4]. The solubility curve due to [5], based on parameter determinations, shows slightly lower solubilities, the differences being 0.2, 0.2, 0.55, and 0.2 at. % at 915, 800, 620, and 300°C, respectively.

The intermediate phase richest in copper is a 3:2 electron compound which exists in three distinct modifications β, ζ, and ζ_1 (see inset of Fig. 338) [2]. The β phase is b.c.c. [1, 12], and ζ and ζ_1 have h.c.p. structures [6]. The existence of a h.c.p. phase in this range was first reported by [7]. This indicated that the assumption of a eutectoid decomposition of β into $\alpha + \gamma$ [1] appeared doubtful. The eutectoid decomposition of β has recently been thoroughly studied [13].

The γ phase, based on the composition Cu_9Ga_4 [30.77 at. (32.78 wt.) % Ga], is a 21:13 electron compound [8]. γ_1, γ_2, and γ_3 are modifications of the γ-brass structure, the latter existing at temperatures above 490–645°C, according to the composition [3]. The $\gamma \rightleftarrows \gamma_1$ transformation is an order-disorder change [3]. The intermediate phase, formed peritectically at 254°C, does not contain 56 at. % Ga [1], but corresponds almost exactly to the composition $CuGa_2$ (68.69 wt. % Ga) [3]. A solid solubility of Cu in Ga was not detectable by X-ray diffraction [1].

Crystal Structures. Lattice-spacing data of the α solid solutions were reported by [1] and especially [9, 10]. β is b.c.c. [1]. Parameters of a h.c.p. phase, presumably ζ, in an alloy with 21 at. % Ga were given as a = 2.599 A, c = 4.237 A, c/a = 1.630 [7]. The structure of the alloys between about 30 and 43 at. % Ga was recognized to be of the γ-brass ($D8_{1-3}$) type [1, 7]. This range was thoroughly investigated by [3], who reported numerous lattice-spacing data of alloys of various compositions. For alloys with 29.68, 33.90, 35.89, and 42.64 at. % Ga, parameters were found to be a = 8.758, 8.7387 (maximal value), 8.7209, and 8.635 A, respectively [3]. In brief, the γ-brass structures of γ_1, γ_2, and γ_3 are characterized by slight differences in the intensities of some of the weaker X-ray diffraction lines [11]. The γ_1 phase contains the full number of 52 atoms per unit cell at its copper-rich boundary. Increasing the Ga content results in a slight decrease in the number of atoms per unit cell. On passing to the γ_2 and γ_3 fields, there is a marked diminution in the number of atoms per unit cell, and this dropping out of atoms occurs in such a way as to maintain a constant electron concentration [3].

$CuGa_2$ is tetragonal, with 3 atoms per unit cell and a = 2.836 A, c = 5.843 A, c/a = 2.06 [7].

1. F. Weibke, *Z. anorg. Chem.*, **220**, 1934, 293–311.
2. W. Hume-Rothery and G. V. Raynor, *J. Inst. Metals*, **61**, 1937, 205–222.
3. J. O. Betterton and W. Hume-Rothery, *J. Inst. Metals*, **80**, 1951-1952, 459–468.
4. W. Hume-Rothery, G. W. Mabbott, and K. M. Channel-Evans, *Phil. Trans. Roy. Soc. (London)*, **A233**, 1934, 1–97.
5. E. A. Owen and V. W. Rowlands, *J. Inst. Metals*, **66**, 1940, 361–378.
6. W. Hume-Rothery, P. W. Reynolds, and G. V. Raynor, *J. Inst. Metals*, **66**, 1940, 191–207.
7. E. Zintl and O. Treusch, *Z. physik. Chem.*, **B34**, 1936, 225–237.
8. W. Hume-Rothery, J. O. Betterton, and J. Reynolds, *J. Inst. Metals*, **80**, 1951-1952, 609–616.
9. W. Hume-Rothery, G. F. Lewin, and P. W. Reynolds, *Proc. Roy. Soc. (London)*, **A157**, 1936, 167–183.
10. E. A. Owen and E. W. Roberts, *Phil. Mag.*, **27**, 1939, 294–327.
11. The existence of two phases with closely related structures between 30 and 43 at. % Ga was already suspected by [1].
12. W. Hume-Rothery, G. V. Raynor, P. W. Reynolds, and H. K. Packer, *J. Inst. Metals*, **66**, 1940, 238.
13. C. W. Spencer and D. J. Mack, *J. Inst. Metals*, **84**, 1956, 461–466.

$\bar{1}.9421$
0.0579

Cu-Ge Copper-Germanium

In spite of considerable work, covering the entire range of composition [1, 2] and portions of the system [3–6], it was not until recently [7, 8] that the phase relationships and boundaries were fully established (Fig. 339). [1, 2], arriving at substantially similar results, have determined the complete liquidus curve and nearly all the temperatures of the three-phase equilibria. However, their interpretations of the phase

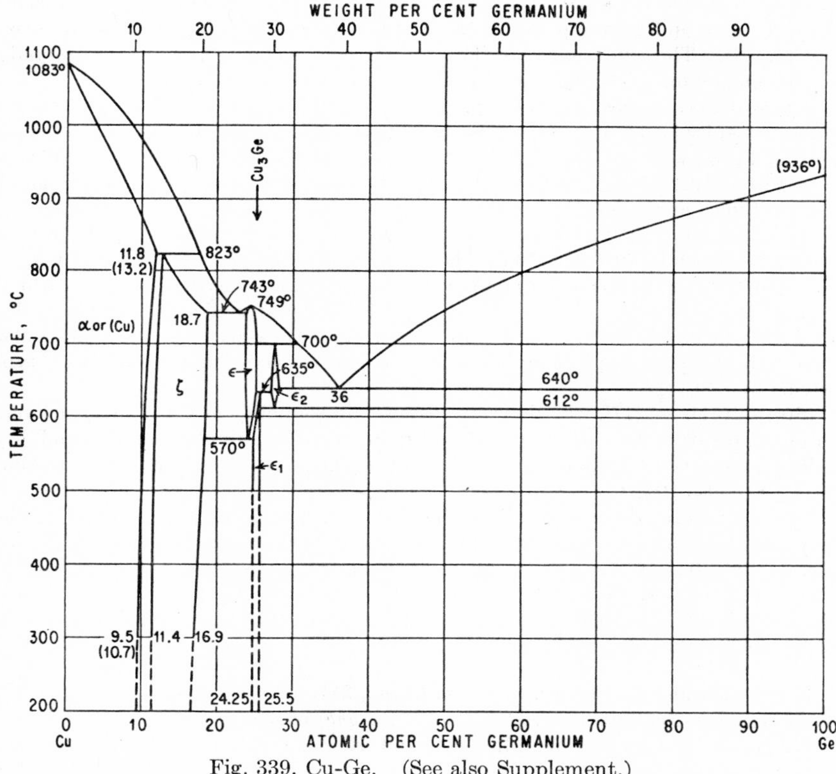

Fig. 339. Cu-Ge. (See also Supplement.)

transformations in the range 18–30 at. (20–33 wt.) % Ge, besides other minor points, were proved to be incorrect because of incompletely defined equilibrium relationships, based on thermal-analysis data only. [4], covering the range 14.4–31.7 at. % Ge, showed the ϵ phase (Cu_3Ge) to decompose eutectoidally into ζ and ϵ_1, at 20.8 at. % Ge and 570°C, and reported some additional information as to the transformation in the vicinity of 25 at. % Ge, later [7, 8] shown to be erroneous.

The primary solid-solubility curve was accurately determined [3, 5, 6] and the form of the ζ-phase area more accurately defined [5]. Corrections to the region between 18 and 30 at. % Ge according to [4] were made by [7, 8]. These workers established the existence of the peritectically formed (700°C) high-temperature phase ϵ_2, at a slightly higher Ge content than suggested by [1, 3]. The solidus curve of the (Ge) phase was calculated by [9]; see below.

Figure 339 is based on the data reported by [1–4]: liquidus temperatures; [3]: solidus curve of α phase; [3, 5, 6]: α-phase boundary; [5]: boundaries of the ζ phase; [7, 8]: range 18–30 at. % Ge; and [1, 2, 4]: range 30–100 at. % Ge. The α-phase boundaries due to [3, 5, 6] differ somewhat. That by [6], based on lattice-parameter measurements, indicates slightly lower solubilities than that by [5], which was determined micrographically. At 800, 600, and 400°C, differences are about 0.45, 0.25, and 0.55 at. % Ge, respectively; see also [10]. The phase boundaries in the range 18–30 at. % Ge due to [8] are in very good agreement with those according to [7]. At the time of writing, detailed data as to the work by [7] had not yet been reported. Results by [8] are based on high-temperature X-ray studies, those by [7] apparently on micrographic work. The solidus curve of the (Ge) phase, as calculated by [9], is retrograde, with the maximum solubility of about 0.00007 at. % Ge at about 880°C.

Crystal Structures. Lattice spacings covering the entire range of the α phase were reported by [4, 11, 12]. The ζ phase with 11.4–18.7 at. % Ge is h.c.p. (A3 type) [2, 13, 4, 14, 8], with a = 2.515 A, c/a = 1.638 at 11.9 at. % Ge; and a = 2.658 A, c/a = 1.61 at 19.3–19.4 at. % Ge [8]. It is to be considered a 3:2 electron compound based on the composition Cu_5Ge (16.67 at. % Ge) [15].

The high-temperature phase ϵ (Cu_3Ge) is h.c.p., with a = 4.20 A, c/a = 1.20 at 24.1 at. % Ge and 600–680°C [8]. The phase ϵ_1 (Cu_3Ge), formerly reported to be h. c. p. [16, 2, 13, 4], has an orthorhombically distorted A3-type structure, a = 2.645 A, b = 4.553 A, c = 4.202 A, at 25 at. % Ge [8]. [14] had suggested a monoclinically distorted A3 type of structure. The high-temperature phase ϵ_2 with about 27.5 at. % Ge is cubic; the structure is probably related to the γ-brass ($D8_2$) and $NiAl_3$ ($D5_{13}$) type of structures [8].

Supplement. [17], who determined the solid solubility of Cu in Ge as a function of temperature by both electrical and radioactivity methods, found a solubility maximum (less than 3 atoms per million of Ge, i.e., $<3.10^{-4}$ at. % Cu) at about 875°C. See also [18].

[19] have thoroughly redetermined the constitution in the region 22–100 at. % Ge. Their work "confirms the general form of the diagram proposed by [8] (Fig. 339) for the region 20–40 at. % Ge, but the phase boundaries of the ϵ, ϵ_1, and ϵ_2 phases (phase designation according to Fig. 339) lie at Ge contents higher by 1–2 at. %. It is possible that this difference is the result of loss of Ge during the annealing of the filings or powder used in the X-ray work of [8]" [19].

1. R. Schwarz and G. Elstner, *Z. anorg. Chem.*, **217**, 1934, 289–297.
2. H. Maucher, *Forschungsarb. Metallkunde u. Röntgenmetallog.*, no. 20, 1936.
3. W. Hume-Rothery, G. W. Mabbott, and K. M. Channel-Evans, *Phil. Trans. Roy. Soc.* (*London*), **A233,** 1934, 1–97.
4. F. Weibke, *Metallwirtschaft*, **15,** 1936, 301–303.
5. W. Hume-Rothery, G. V. Raynor, P. W. Reynolds, and H. K. Packer, *J. Inst. Metals*, **66,** 1940, 209–239.
6. E. A. Owen and V. W. Rowlands, *J. Inst. Metals*, **66,** 1940, 361–378.
7. W. Hume-Rothery and J. O. Betterton, unpublished work quoted by P. Greenfield and G. V. Raynor, *J. Inst. Metals*, **80,** 1951-1952, 377–378.
8. K. Schubert and G. Brandauer, *Z. Metallkunde*, **43,** 1952, 262–268.
9. C. D. Thurmond and J. D. Struthers, *J. Phys. Chem.*, **57,** 1953, 831–834.
10. E. A. Owen and D. P. Morris, *J. Inst. Metals*, **76,** 1949-1950, 707.
11. W. Hume-Rothery, G. F. Lewin, and P. W. Reynolds, *Proc. Roy. Soc.* (*London*), **A157,** 1936, 167–183.
12. E. A. Owen and E. W. Roberts, *Phil. Mag.*, **27,** 1939, 294–327.
13. F. Laves, private communication to F. Weibke, see [4].

14. H. Nowotny and K. Bachmayer, *Monatsh. Chem.*, **81,** 1950, 669–678.
15. W. Hume-Rothery, P. W. Reynolds, and G. V. Raynor, *J. Inst. Metals,* **66,** 1940, 191–207.
16. V. M. Goldschmidt, *Z. physik. Chem.*, **133,** 1928, 397–419.
17. C. S. Fuller, J. D. Struthers, J. A. Ditzenberger, and K. B. Wolfstirn, *Phys. Rev.*, **93,** 1954, 1182–1189.
18. R. J. Hodgkinson, *Phil. Mag.*, **46,** 1955, 410 (Maximum solubility 1 atom of Cu in 10^7 atoms of Ge).
19. J. Reynolds and W. Hume-Rothery, *J. Inst. Metals,* **85,** 1956, 119–127.

1.7996
$\bar{2}$.2004

Cu-H Copper-Hydrogen*

The solubility of gaseous hydrogen in Cu has been measured by [1–6]. The results of [1, 3, 5], especially those of [3, 5], are in good agreement; those of [2] are certainly incorrect (Fig. 340). At the melting point of Cu the solubility increases on heating from about 2.0 to 5.4 cc (NTP) per 100 g Cu [3]. It was shown by [1] that the solubility of hydrogen is proportional to the square root of the pressure (see also [8]).

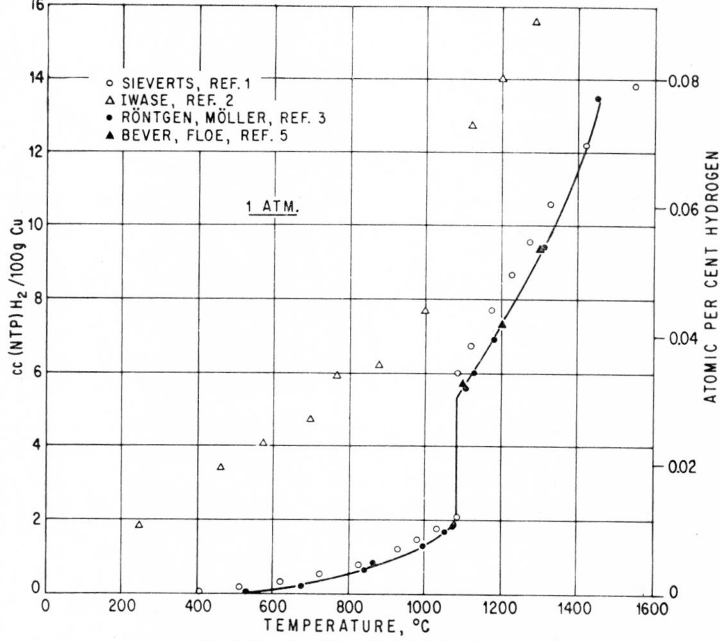

Fig. 340. Cu-H. Solubility isobar.

The only well-authenticated hydride of Cu is the cuprous hydride (CuH) of Wurtz, prepared by aqueous reaction [7, 9, 10]. The Cu atoms have a h.c.p. arrangement [9, 10], and the lattice parameters are $a = 2.89 \pm 3$ A, $c = 4.63 \pm 5$ A, $c/a = 1.60$. The deuteride has slightly larger lattice dimensions [10].

* The assistance of Dr. F. Wolstein in preparing this system is gratefully acknowledged.

Supplement. [11] determined the solubility of H_2 in Cu at 450, 550, and 680°C and 1 atm hydrogen pressure as 0.0149×10^{-2}, 0.0526×10^{-2}, and 0.1530×10^{-2} at. % H, respectively, in good agreement with the results of [3]. According to neutron and X-ray diffraction work by [12], both CuH and CuD possess the wurtzite (B4) structure, with a = 2.920 A, c = 4.614 A for CuH.

1. A. Sieverts, *Z. physik. Chem.*, **60,** 1907, 139–153. A. Sieverts and W. Krumbhaar, *Z. physik. Chem.*, **74,** 1910, 288–294; *Ber. deut. chem. Ges.*, **43,** 1910, 896–898. A. Sieverts, *Z. physik. Chem.*, **77,** 1911, 594–598; *Z. Metallkunde,* **21,** 1929, 40, 44.
2. K. Iwase, *Science Repts. Tôhoku Univ.*, **15,** 1926, 531–566 (in English).
3. P. Röntgen and F. Möller, *Metallwirtschaft,* **13,** 1934, 81–83, 97–100.
4. K. Iwase and M. Fukusima, *Nippon Kinzoku Gakkai-Shi,* **1,** 1937, 202–213 (in Japanese). Not available to the author. *Met. Abstr.*, **5,** 1938, 469.
5. M. Bever and C. Floe, *Trans. AIME,* **156,** 1944, 149–159.
6. Some (obsolete) papers of the nineteenth century are listed in the extensive review of [7], pp. 264–265.
7. D. P. Smith, "Hydrogen in Metals," University of Chicago Press, Chicago, 1948.
8. A. J. Phillips, *Trans. AIME,* **171,** 1947, 17–46.
9. H. Müller and A. J. Bradley, *J. Chem. Soc.*, **129,** 1926, 1669–1673.
10. J. C. Warf and W. Feitknecht, *Helv. Chim. Acta,* **33,** 1950, 613–639.
11. K. H. Lieser and H. Witte, *Z. physik. Chem.*, **202,** 1954, 321–351.
12. J. A. Goedkoop and A. F. Andresen, *Acta Cryst.*, **7,** 1954, 672; **8,** 1955, 118–119.

$\bar{1}.5512$
0.4488

Cu-Hf Copper-Hafnium

The diffraction pattern of an alloy of the composition $HfCu_2$ was complex and bore no similarity to Laves-type patterns [1].

1. R. P. Elliott, Armour Research Foundation, Chicago, Ill., Technical Report 1, OSR Technical Note OSR-TN-247, August, 1954, p. 11.

$\bar{1}.5007$
0.4993

Cu-Hg Copper-Mercury

The solubility of Cu in Hg at normal temperatures (in wt. %) was given as 0.001 [1], 0.0031 [2], 0.16 (?) [3], about 0.0024 [4], 0.0032 [5], and 0.002 [6]. The value 0.002–0.003 wt. % may be considered the most reliable. A very low solubility is also indicated by measurements of the electromotive force [7]. Solubilities at temperatures between 100 and 400°C were reported [8]; however, they were taken from a hypothetical diagram [14] and, therefore, are not based on experimental work.

The composition of the intermediate phase in equilibrium with the saturated solution of Cu in liquid Hg is not yet known with certainty. Products of the approximate composition Cu_3Hg_2 (67.84 wt. % Hg) [9] and CuHg (75.95 wt. % Hg) [10] have been isolated from Hg-rich amalgams, using high pressure. Measurements of the electromotive force [11-13] proved unsuccessful in establishing the presence of an intermediate phase.

On the basis of thermal and microscopic studies, [14] have outlined a substantially hypothetical phase diagram. The complex and somewhat conflicting phenomena observed could be interpreted only by assuming three intermediate phases which decompose on melting at 150, 115, and 96°C. The phase CuHg (melting point 96°C) was claimed to have been well established; however, the other two phases richer in Cu could not be identified. After annealing at 100°C for 40 hr, two intermediate phases were detected microscopically, CuHg and a yellowish phase richer in copper.

Results of roentgenographic investigations agree in that they have indicated the existence of only one intermediate phase (Fig. 341) having a structure very similar to that of γ-brass [15–19]. This structure was observed in preparations containing 2–65.9 at. % Hg, and the copper structure in those with 2–42.5 at. % Hg [16]. This would indicate the composition Cu_4Hg_3 [42.86 at. (70.31 wt.) % Hg], assuming that the preparations were in equilibrium. By separating the liquid phase from the solid in an amalgam with 90 wt. (74 at.) % Hg, under a pressure of 10,000 atm, a product containing 74.5 wt. (48 at.) % Hg was obtained, i.e., a composition close to CuHg [16]. Also, [20] have found only one intermediate phase with a composition close to Cu_4Hg_3 [21]. The X-ray pattern was indexed as a tetragonal structure; however, [16] has pointed out that such a structure is impossible. [18], in a study similar to that by [16], isolated a product with γ-brass structure and assumed the composition CuHg. Recently, [19] reported that by electrolysis of $CuSO_4$ against a

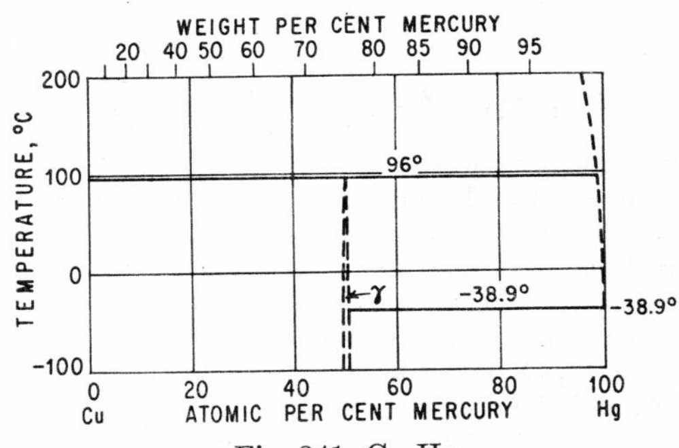

Fig. 341. Cu-Hg

mercury cathode a final product, having γ-brass structure, was obtained which contained 70.53 wt. % Hg, i.e., a composition corresponding to Cu_4Hg_3.

From microscopic examination, [14] concluded that 24 wt. (9 at.) % Hg is soluble in copper. However, the lattice parameter of the Cu phase in Cu amalgams was found to be that of Cu, indicating insolubility of Hg in Cu [16]. This is in agreement with the rather large size factor of 17%.

Crystal Structure. As mentioned above, the only intermediate phase is isotypic with γ-brass ($D8_2$ type). Since the lattice parameter—9.418–9.420 A according to [16, 17] and 9.425 A according to [18]—was found constant, the homogeneity range of the γ phase is narrow [16].

The ideal composition of the γ-brass type phase would be Cu_5Hg_8 (61.54 at. % Hg), with an electron concentration of 21:13. Assuming the composition Cu_4Hg_3 or CuHg to be correct, the γ phase would be stable with a much lower Hg content [22].

1. Gouy, *J. phys.*, **4,** 1895, 320–321.
2. W. J. Humphreys, *J. Chem. Soc.*, **69,** 1896, 247.
3. H. Iggena, Dissertation, Göttingen, 1899.
4. T. W. Richards and R. N. Garrod-Thomas, *Z. physik. Chem.*, **72,** 1910, 177–181.
5. G. Tammann and K. Kollmann, *Z. anorg. Chem.*, **160,** 1927, 246–248.
6. N. M. Irvin and A. S. Russell, *J. Chem. Soc.*, **1932,** 891–898.
7. J. F. Spencer, *Z. Elektrochem.*, **11,** 1905, 681–684.
8. W. Schmidt, *Metall,* **3,** 1949, 10–13.

9. J. P. Joule, *J. Chem. Soc.*, **16,** 1863, 378.
10. A. Guntz and de Greift, *Compt. rend.*, **154,** 1912, 357–358.
11. B. Neumann, *Z. physik. Chem.*, **14,** 1894, 211.
12. N. A. Puschin, *Z. anorg. Chem.*, **36,** 1903, 240–241.
13. E. Cohen, F. D. Chattaway, and W. Tombrock, *Z. physik. Chem.*, **60,** 1907, 715–718 (62.5–97 at. % Hg).
14. G. Tammann and T. Stassfurth, *Z. anorg. Chem.*, **143,** 1925, 357–369.
15. V. M. Goldschmidt, *Z. physik. Chem.*, **133,** 1928, 397–419.
16. N. Katoh, *Z. physik. Chem.*, **B6,** 1929, 27–39.
17. N. Katoh, *Bull. Chem. Soc. Japan*, **5,** 1930, 13–16.
18. F. Schossberger, *Z. physik. Chem.*, **B29,** 1935, 65–78.
19. F. Lihl, *Z. Metallkunde*, **44,** 1953, 160–166.
20. H. Terrey and C. M. Wright, *Phil. Mag.*, **6,** 1928, 1055–1069.
21. The authors erroneously gave the formula Cu_3Hg_4.
22. W. Hume-Rothery, J. O. Betterton, and J. Reynolds, *J. Inst. Metals*, **80,** 1951-1952, 609–616.

$\bar{1}.7433$
0.2567

Cu-In Copper-Indium

The first complete diagram was determined by [1], using thermal, microscopic, and roentgenographic methods. Except for a small alteration in the region 30–32 at. % In above 600°C, it has all the essential features of the diagram given in Fig. 342. However, some corrections have been made later by investigators who placed more emphasis on quantitative equilibrium relationships.

Figure 342 represents a combination of the results by [2, 3] on the α-phase boundary (see Supplement), [2] on the β-phase field, [4] on the region 25–35 at. % In (see inset of Fig. 342), and the original work by [1], as far as the composition range above 35 at. % In is concerned. The liquidus curve is due to [1], with the exception of the range 26.3–34.8 at. % In, which is based on data by [4]. The solidus curve of the α phase has not been determined as yet. Also, the region above 35 at. % In may undergo some revision if reinvestigated. This holds especially for the composition of the phase formed peritectically at 310°C which does not appear to be conclusively established.

Determinations of the α solid-solubility curve, using micrographic [2] and lattice-parametric methods [3], agree, within the limit of experimental error, between 710 and 574°C. Below this temperature, data by [3], indicating higher solubilities as shown in Fig. 342, appear to be more accurate [5]. (See Supplement.)

The range of the β phase, 18.1–24.5 at. % In at the solidus temperatures, was carefully determined micrographically by [2] and that between 25 and 35 at. % In above 150°C (see inset of Fig. 342) by [4], using the same method. The latter authors proved that a high-temperature phase in the range 30–32 at. % In, denoted ϵ by [1], is nonexistent. It was already found by [6] that γ and ϵ are identical in structure. A transformation of the η phase at 389°C [1] could not be confirmed either [4]. The kinetics of the eutectoid decomposition of β was studied by [7].

Crystal Structures. Lattice parameters of the α phase were reported by [1, 8]; the parameter of Cu (a = 3.6147 A) increases to a = 3.7158 A at 10.9 at. % In [8]. The β phase, a 3:2 electron compound of the ideal composition Cu_3In [9], is b.c.c., with $a = 3.01_4$ A after quenching [1] and a = 3.046 A at 672°C [10].

As shown by high-temperature X-ray work [6, 4], the γ phase is isotypic with γ-brass, a = 9.250 A for the alloy with 29.6 at. % In at 650°C [4]. The ideal composition Cu_9In_4 (30.77 at. % In) with an electron concentration of 21:13 [11] lies just outside the high-In boundary of the γ-phase field. The structure of the δ phase, earlier reported [1] to be quite similar to that of γ-brass [12], was proposed to be a superlattice of the NiAs-type structure [6]. Recently, [4] suggested a tetragonally deformed γ-brass structure, with a = 8.99 A, c = 9.16 A, c/a = 1.02.

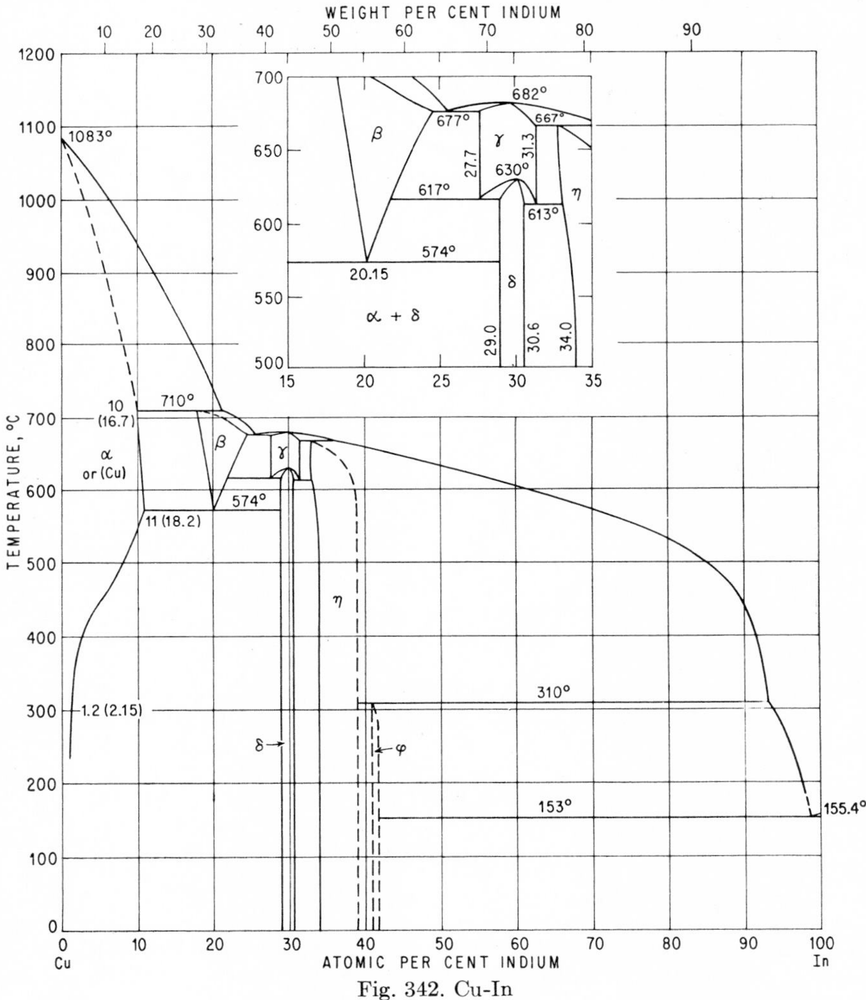

Fig. 342. Cu-In

The η phase is isotypic with NiAs (B8 type) [13, 14]. Lattice parameters reported are a = 4.277 A, c = 5.249 A, c/a = 1.227 at 35.6 at. % In [13]; a = 4.289 A, c = 5.263 A, c/a = 1.227 at 36 at. % In [14]; and a = 4.266 A, c = 5.282 A, c/a = 1.229 at 41.3 at. % In [14].

Supplement. [15] redetermined the α solvus curve by X-ray parametric work from the peritectic temperature (710°C) down to 250°C (previous determinations by [2, 3] were down to only 516 and 470°C, respectively). In the range from 710 to 575°C, the results of [15] are in very close agreement with those of [2, 3], and from

575 to 470°C, with those of [3]. The maximum solubility (at 575°C) was found to be 10.8_5 at. % In, and the solubility at the peritectic temperature to be 10.0_5 at. % In. At about 470°C the solvus shows a point of inflection (Fig. 342) "which would lead one to expect a transformation also at 470°C, but we have not investigated the point with sufficient thoroughness to arrive at a definite conclusion" [15].

1. F. Weibke and H. Eggers, *Z. anorg. Chem.*, **220,** 1934, 273–292; see also F. Weibke, *Z. Metallkunde*, **31,** 1939, 228–230.
2. W. Hume-Rothery, G. V. Raynor, P. W. Reynolds, and H. K. Packer, *J. Inst. Metals*, **66,** 1940, 209–239.
3. E. A. Owen and D. P. Morris, *J. Inst. Metals*, **76,** 1949-1950, 145–168; E. A. Owen and E. A. O'Donnell Roberts, *J. Inst. Metals*, **81,** 1952-1953, 479–480.
4. J. Reynolds, W. A. Wiesman, and W. Hume-Rothery, *J. Inst. Metals*, **80,** 1951-1952, 637–640.
5. See discussion of paper by Owen and Morris [3] in *J. Inst. Metals*, **76,** 1949-1950, 682, 707.
6. E. Hellner and F. Laves, *Z. Naturforsch.*, **2a,** 1947, 177–183.
7. C. W. Spencer and D. J. Mack, *J. Inst. Metals*, **82,** 1953-1954, 81–85.
8. E. A. Owen, *J. Inst. Metals*, **73,** 1947, 471–489.
9. W. Hume-Rothery, P. W. Reynolds, and G. V. Raynor, *J. Inst. Metals*, **66,** 1940, 191–207.
10. K. W. Andrews and W. Hume-Rothery, *Proc. Roy. Soc. (London)*, **A178,** 1941, 464–473.
11. W. Hume-Rothery, J. O. Betterton, and J. Reynolds, *J. Inst. Metals*, **80,** 1951-1952, 609–616.
12. H. Bittner and H. Nowotny, *Monatsh. Chem.*, **83,** 1952, 1308–1313.
13. F. Laves and H. J. Wallbaum, *Z. angew. Mineral.*, **4,** 1942, 17–46.
14. E. S. Makarov, *Bull. acad. sci. U.R.S.S., Classe sci. chim.*, **1943**(4), 264–270.
15. R. O. Jones and E. A. Owen, *J. Inst. Metals*, **82,** 1954, 445–448.

$\bar{1}.5173$
0.4827

Cu-Ir Copper-Iridium

Measurements of electrical conductivity of copper-rich alloys indicate that at least 0.48 at. (1.5 wt.) % Ir is soluble in Cu (after annealing at 800–950°C) [1].

1. J. O. Linde, *Ann. Physik*, **15,** 1932, 226.

$\bar{1}.6603$
0.3397

Cu-La Copper-Lanthanum

Thermal and microscopic investigations [1] indicated the existence of the four intermediate phases $LaCu_4$ (35.34 wt. % La), $LaCu_3$ (42.16 wt. % La), $LaCu_2$ (52.23 wt. % La), and LaCu (68.62 wt. % La). $LaCu_4$ and $LaCu_2$ were found to have maximum melting points of 902 and 834°C, respectively, whereas $LaCu_3$ and LaCu were found to melt under decomposition at 793 and 551°C, respectively. The three resulting eutectics were located at about 15 wt. (7.4 at.) % La, 851°C; about 45 wt. (27.3 at.) % La, 742°C; and about 86 wt. (73.8 at.) % La, 468°C.

On the assumption that Cu forms the same compounds with La as with Ce (page 450) and Pr (page 615), i.e., XCu_6, XCu_4, XCu_2, and XCu, the system was reinvestigated by thermal analysis, using only nine alloys between 16 and 50 wt. (8 and 31.4 at.) % La [2]. The data presented in Fig. 343, corroborated by metallographic studies, were claimed to prove that $LaCu_6$ (26.71 wt. % La) instead of $LaCu_4$ forms a eutectic with Cu at about 18 wt. (9 at.) % La, 840°C, and that $LaCu_4$ replaces

$LaCu_3$ as the phase formed peritectically at 793 [1] or 785°C [2]. No indication of the presence of $LaCu_5$ (30.42 wt. % La), identified by microscopic examination and crystal-structure determination [3], could be obtained by thermal analysis [2]; however, there was some microstructural evidence for the existence of an additional intermediate phase between $LaCu_6$ and $LaCu_4$ [2]. Also, [3] reported the existence of an additional phase ($LaCu_6$?) in alloys containing less than 28 wt. (15 at.) % La between Cu and $LaCu_5$.

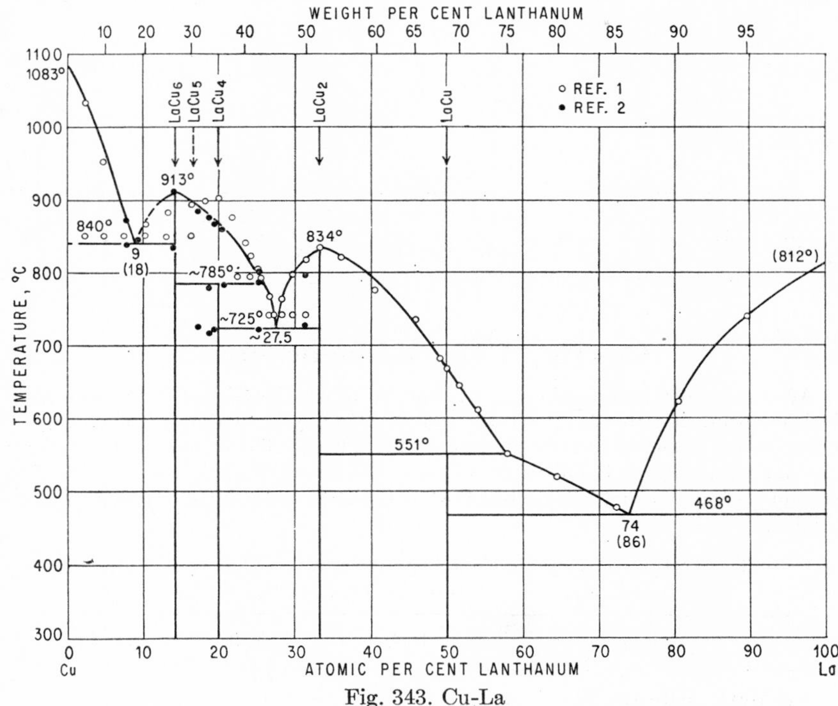

Fig. 343. Cu-La

$LaCu_5$ is hexagonal and isotypic with $CaCu_5$, with a = 5.092 A, c = 4.086 A, c/a = 0.802 according to [4]; and a = 5.169 A, c = 4.116 A, c/a = 0.796 according to [3]. $LaCu_4$ has a tetragonal structure [2].

1. G. Canneri, *Metallurgia ital.*, **23**, 1931, 813–815.
2. R. Vogel and T. Heumann, *Z. Metallkunde*, **35**, 1943, 29–42.
3. H. Nowotny, *Z. Metallkunde*, **34**, 1942, 247–253.
4. W. Haucke, *Z. anorg. Chem.*, **244**, 1940, 17–22.

0.9617
$\bar{1}$.0383

Cu-Li Copper-Lithium

The phase diagram shown in Fig. 344 is based on thermal-analysis data of six alloys. The absence of an intermediate phase was corroborated by X-ray analysis [1]. The system needs reexamination; see Ag-Li.

1. S. Pastorello, *Gazz. chim. ital.*, **60**, 1930, 988–992.

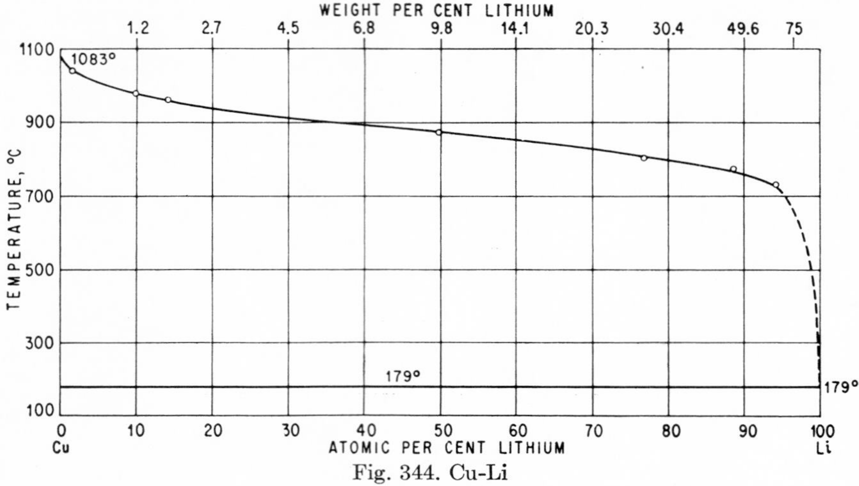

Fig. 344. Cu-Li

0.4171
$\bar{1}.5829$

Cu-Mg Copper-Magnesium

After a cursory investigation by [1], the entire solidification diagram was established independently by [2] (76 alloys) and [3] (24 alloys). Later, [4] reinvestigated the system by means of thermal analysis of 117 alloys. The characteristic compositions and temperatures found in these works are given in Table 24. The liquidus and solidus shown in Fig. 345 are based on the data by [4]. A paper by [25] was not available.

Table 24. Characteristic Compositions in Wt. % Mg (At. % in Parentheses) and Temperatures in °C of the System Cu-Mg

	Ref. 2	Ref. 3	Ref. 4
α-$MgCu_2$ eutectic	9.7	~9.5	9.7
	(21.9)	(21.5)	(21.9)
	725°C	728–730°C	722°C
$MgCu_2$	799°C	797°C	819°C
$MgCu_2$-Mg_2Cu eutectic	35.0	32.5–33	34.6
	(58.4)	(55.6–56.1)	(57.9)
	555°C	555°C	552°C
Mg_2Cu	570°C	570°C	568°C
Mg_2Cu-Mg eutectic	66.6	~69	69.3
	(83.9)	(85.3)	(85.5)
	480°C	485°C	485°C

The solid solubility of Mg in Cu is not accurately known as yet. Solubilities reported by [5–9] are based on inconclusive evidence since the alloys investigated were not in a defined structural state. [10], using measurements of electrical resistivity vs. composition of samples annealed at and quenched from various temperatures, found the solubility to decrease almost linearly from about 3 wt. (7.4 at.) % at 700°C to about 1.2 wt. (3.1 at.) % Mg at 400°C. According to micrographic work by [4], the solubility is 2.5–2.7 wt. (6.3–6.8 at.) % at 700°C, 2.4–2.5 wt. (6.0–6.3 at.) % Mg at 680°C, and slightly less at 500°C. In both investigations, heat-treatment was not such as to attain equilibrium [11]. The solidus curve of the α phase between 722 and 900°C was determined micrographically [4].

All investigators, whether they used micrographic [12, 13, 4], roentgenographic [7, 8], or electrical-conductivity methods [14, 15], agree that the solid solubility of Cu in Mg is very small. [12] established that there is a decrease in solubility with fall in temperature; however, determination is hindered by the fact that partial precipitation of Mg_2Cu already takes place on quenching. According to [12] and [15], the

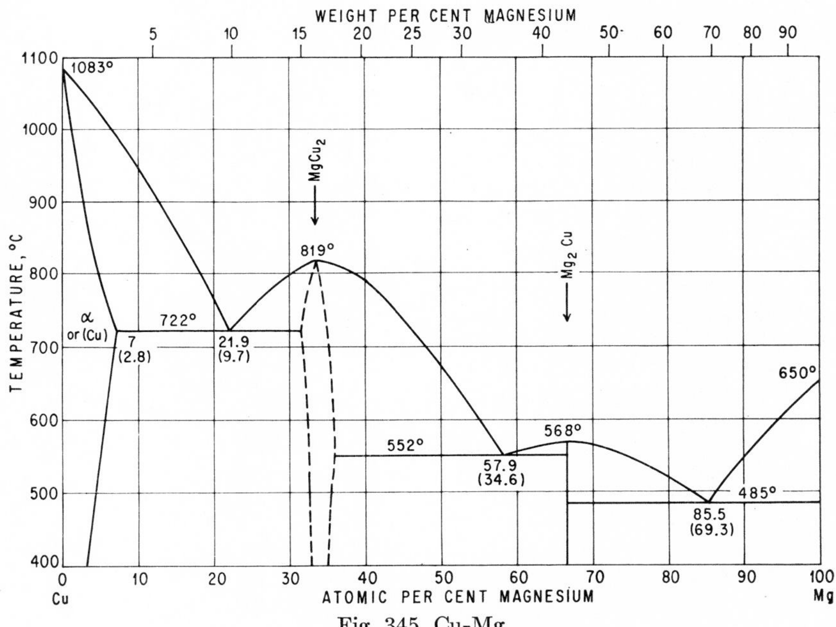

Fig. 345. Cu-Mg

solubility at the eutectic temperature is not higher than 0.4–0.5 or 0.55 wt. % Cu, respectively, and below 0.2 wt. % [12, 15] or above 0.1 wt. % [14] at some lower temperature [16]. [13] showed that the solubility at lower temperatures is more likely less than 0.02 wt. % Cu. Also, [4] concluded from microscopic work that it is about 0.03 wt. % Cu at 470–480°C. Solubility is not detectable by lattice-parameter determination [17].

The boiling point of Mg (1103 ± 5°C) is raised to 1174°C by 2.3 at. % Cu [18].

Crystal Structures. Lattice parameters of the α phase were reported by [7, 8, 19, 26]. $MgCu_2$ is the prototype of the C15 type of structure (f.c.c. with 8 $MgCu_2$ per unit cell) [20, 7, 8, 19]. Its homogeneity range extends from about 32.8 to 35.5 at. % Mg at 500°C and is considerably narrower at lower temperatures [19]. Therefore, the lattice parameters reported [20, 7, 8, 19, 21, 22] differ according to the composition and heat-treatment, which were not always defined. After quenching from 500°C, the parameter varies from a = **7.02$_7$** A (Cu side) to a = **7.05$_1$** A (Mg_2Cu side) [19].

Mg_2Cu, reported by [8] to be hexagonal, is f.c. orthorhombic (new type), with 48 atoms per unit cell and $a = 5.284$ A, $b = 9.07$ A, $c = 18.25$ A [7, 23]; see also [24].

1. O. Boudouard, *Bull. soc. encour. ind. natl.*, **102,** 1903, 200; *Compt. rend.*, **135,** 1902, 794–796; **136,** 1903, 1327–1329.
2. G. G. Urasow, *Zhur. Russ. Fiz.-Khim. Obshchestva*, **39,** 1907, 1566–1581.
3. R. Sahmen, *Z. anorg. Chem.*, **57,** 1908, 26–33.
4. W. R. D. Jones, *J. Inst. Metals*, **46,** 1931, 395–419.
5. L. Guillet, *Rev. mét.*, **4,** 1907, 622.
6. N. I. Stepanow, *Z. anorg. Chem.*, **78,** 1912, 17–22.
7. A. Runqvist, H. Arnfelt, and A. Westgren, *Z. anorg. Chem.*, **175,** 1928, 43–48.
8. G. Grime and W. Morris-Jones, *Phil. Mag.*, **7,** 1929, 1113–1134.
9. N. Ageew, *J. Inst. Metals*, **46,** 1931, 419–420.
10. O. Dahl, *Wiss. Veröffentl. Siemens-Konzern*, **6,** 1927, 222–225.
11. W. Hume-Rothery, *J. Inst. Metals*, **46,** 1931, 420–421; discussion of paper by [4].
12. M. Hansen, *J. Inst. Metals*, **37,** 1927, 93–100.
13. J. W. Jenkin, *J. Inst. Metals*, **37,** 1927, 100–101.
14. F. de Carli, *Metallurgia ital.*, **23,** 1931, 18–24.
15. N. I. Stepanov and I. I. Kornilov, *Izvest. Inst. Fiz.-Khim. Anal.*, **7,** 1935, 89–98.
16. J. A. Gann, *Trans. AIME*, **83,** 1929, 309–332.
17. R. S. Busk, *Trans. AIME*, **188,** 1950, 1460–1464.
18. A. Schneider and U. Esch, *Z. Elektrochem.*, **45,** 1939, 888–893.
19. V. G. Sederman, *Phil. Mag.*, **18,** 1934, 343–352.
20. J. B. Friauf, *J. Am. Chem. Soc.*, **49,** 1927, 3107–3110.
21. C. Goria and G. Venturello, *Metallurgia ital.*, **32,** 1940, 49.
22. F. Laves and H. J. Wallbaum, *Z. anorg. Chem.*, **250,** 1942, 113.
23. G. Ekwall and A. Westgren, *Arkiv Kemi, Mineral. Geol.*, **14B,** 1940, no. 7.
24. K. Schubert and K. Anderko, *Z. Metallkunde*, **42,** 1951, 321–325.
25. S. Ishida, *Nippon Kogyo Kwai Shi*, **45,** 1929, 256–268, 611–621, 786–790; *Met. Abstr.*, **1,** 1934, 417.
26. A. P. Guljaev and E. F. Trusova, *Zhur. Tekh. Fiz.*, **20,** 1950, 66. *Structure Repts.*, **13,** 1950, 103.

0.0632
$\bar{1}$.9368

Cu-Mn Copper-Manganese

This system has received the attention of numerous investigators, from the early part of this century up to the present time. In fact, it was not until quite recently that the problem whether Cu and γ-Mn form an uninterrupted series of solid solutions was conclusively settled. In the light of the results of a number of comprehensive and careful investigations, the bulk of the earlier papers is obsolete now and, therefore, need not be considered in detail. Work prior to 1935 has been reviewed by [1, 2].

Liquidus Curve. Temperatures of the beginning of solidification have been determined by [3–11]. In the range up to about 30 wt. % Mn, temperatures reported agree fairly well, the deviation being of the order of ±10°C or less. However, at higher Mn contents, liquidus points scatter within wide limits, of the order of 50–70°C, disregarding the greater deviation of data by [3, 4]. The liquidus curve of Fig. 346 is based on the data by [5, 6, 10, 11]. The minimum is generally assumed to lie at or close to 35 wt. % (38.4 at.) % Mn and 870 ± 5°C. In the range 70–100 wt. % Mn, the liquidus curve determined by [10] lies distinctly lower than those by [5, 11] and others. [10] assumed a change in direction at 75.5 wt. (78.1 at.) %Mn and 1115°C because melts higher in Mn solidify to form δ-Mn solid solutions. In this range, an averaging curve was drawn.

Solidus Curve. There is an even larger discrepancy for the solidus curve in the range above 35 wt. % Mn. The reason is that the majority of solidus points reported were determined by thermal analysis [4, 6, 8, 9] or measurements of electrical resistance vs. temperature [10], i.e., under nonequilibrium conditions. These findings indicate

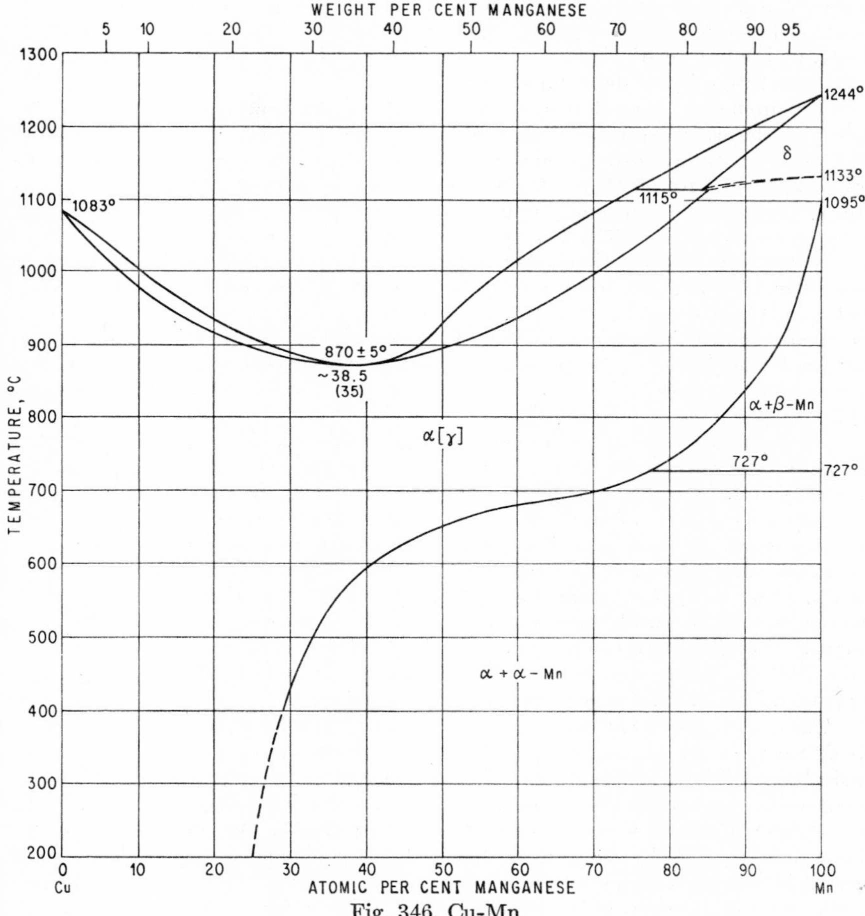

Fig. 346. Cu-Mn

that diffusion during solidification is extremely sluggish, resulting in large deviations from equilibrium. As a consequence, alloys show strong dendritic segregation, which in turn is responsible for the fact that some of the solid-solubility limits reported are in error, since equilibrium of such a structure requires deformation and long homogenization anneals. At present, the most reliable solidus curve is that by [12], which is based on incipient melting studies by means of micrographic investigation. Surprisingly, the solidus curve by [5] lies 10–15°C higher than that by [12] although the former was determined by thermal analysis. The fact that, under extreme deviation from equilibrium, a nearly horizontal portion of the solidus curve was found [6, 8, 9] led [8, 9] to assume that a eutectic [8] or peritectic horizontal [9] at 850 or 895°C, respectively, exists.

Solid-state Phase Boundaries. [5, 6] assumed that a continuous series of solid solutions exists and that no transformations take place in the solid state. This was also concluded by [13], on the basis of measurements of electrical resistivity. However, it was not known at that time that Mn exists in four polymorphic modifications. [8] and [9] published phase diagrams showing a eutectic or peritectic, respectively, and only a relatively slight increase in solid solubility of Mn in Cu with rise in temperature. These diagrams are obsolete.

Measurements of the electrical conductivity [14] and magnetic susceptibility [15] indicated the existence of limited solubility at lower temperatures. [16] was the first, however, to give direct evidence for a wide miscibility gap since X-ray photograms of alloys with 40–90 wt. % Mn showed interferences of Mn. This was confirmed by [17], who concluded from X-ray data that the solubility limit at lower temperatures was about 35 wt. % Mn. The same value was given by [18].

In 1929, [8] reported that γ-Mn-base solid solutions crystallize from melts with more than 75 wt. % Mn and can be retained by quenching and even by rather slow cooling. [19] showed that the axial ratio of the tetragonal γ-Mn solid solutions, obtained by quenching alloys with 85–97 wt. (86.8–97.4 at.) % Mn from temperatures between 900 and 1050°C, increases with increasing Cu content, approaching $c/a = 1$ at about 83 wt. (85 at.) % Mn. He suggested that Cu and γ-Mn might form an uninterrupted series of solid solutions.

The solubility limit in its presently accepted slope was first determined by [20], using the lattice-spacing method. Lattice parameters of alloys quenched from temperatures above the solubility curve indicated that a gradual transition from the f.c.c. Cu-base to the f.c. tetragonal γ-Mn-base solid solution appeared likely to occur at about 83 at. % Mn. However, [20] preferred not to assume a continuous transition of the two lattice types and proposed a narrow two-phase field at about 80–82 at. % Mn. The X-ray photogram of an alloy with 84 at. % Mn, quenched from 1000°C, showed lines of the two lattices, but [20] believed—erroneously—that this alloy was partially molten at the quenching temperature. Later, [10, 12] confirmed the general form of the lattice-parameter–composition curve of quenched solid solutions found by [20] and concluded that a two-phase field was missing.

The controversy was settled when [21] and later [22] showed, by high-temperature X-ray work, that γ-Mn is f.c.c. rather than f.c. tetragonal and that the tetragonal γ-Mn-base solid solution is a metastable phase formed on quenching by a diffusionless (martensitic) type of transformation [22]. This had been previously suggested by [23], on the basis of transformation structures observed by [24]. In contrast to [10, 12], X-ray data reported by [25] proved that both the cubic and tetragonal lattices are present in quenched alloys with about 84–86.5 at. % Mn. "The two-phase photographs are not, however, to be interpreted as a region in which cubic and tetragonal phases are in equilibrium; they are rather to be regarded as indicating a temperature interval over which transformation takes place" [22].

The general form of the boundary of the f.c.c. α phase (also designated as γ), already established by [20], was confirmed by [12] by means of comprehensive metallographic and roentgenographic studies. The micrographically determined curve is to be regarded as the most accurate one (Fig. 346). The solubility curve by [10] does not represent equilibrium relationships. The same holds for the curve between 400 and 700°C reported by [26] and even more so for the solubility at 800°C given by [27].

[20, 10, 12] agree in that there is only a very slight, if any, solid solubility of Cu in α-Mn and β-Mn. The effect of Cu on the $\gamma \rightleftarrows \delta$ transformation of Mn was investigated by [10]. It was found that the transformation temperature of Mn (given as 1162°C) is lowered by about 11 at. % Cu to 1115°C; at this temperature the reaction δ (88.5 at. %) $\rightarrow$ melt (78 at. %) + γ (89.5 at. % Mn) takes place. Since, in Fig. 346, the $\gamma \rightleftarrows \delta$ transformation was assumed at 1133°C, the transformation curve had to be slightly changed and, in this form, has to be regarded as tentative.

Crystal Structure. The most significant results of X-ray studies have been given in previous sections. Lattice parameters of quenched $\alpha(\gamma)$ solid solutions were reported by [19, 20, 28, 10, 29, 12, 25, 22]. High-temperature X-ray data were given by [21] and especially [22]. The martensitic-type transformation of the $\alpha(\gamma)$ phase on quenching (cubic $\rightarrow$ tetragonal) was studied in detail by [22].

Supplement. [30] measured the lattice parameters of alloys containing up to 4.1 at. % Mn. [31] studied solid Cu-Mn alloys, especially those containing 70–95 wt. % Mn, by observing changes of thermal expansion, electrical resistance, specific heat, lattice parameter, and age hardening caused by quenching and tempering. Anomalous changes of hardness were found in alloys containing ~10 wt. % Mn even after low-temperature annealing (cf. below).

From measurements of magnetic susceptibility and specific heat of Cu-rich alloys, [32] concluded the existence of short-range order, based on the composition Cu_4Mn or Cu_3Mn, at temperatures below about 300°C. The existence of an ordered phase Cu_3Mn had been suggested by [28] but denied by [33].

1. M. Hansen, "Der Aufbau der Zweistofflegierungen," pp. 576–584, Springer-Verlag OHG, Berlin, 1936.
2. W. Broniewski and S. Jaślan, *Ann. acad. sci. tech. Varsovie,* **3,** 1936, 141–143.
3. A. E. Lewis, *J. Soc. Chem. Ind.,* **21,** 1902, 842–844.
4. S. Wologdine, *Rev. mét.,* **4,** 1907, 25–38.
5. S. F. Zemczuzny, G. Urasow, and A. Rykowskow, *Zhur. Russ. Fiz.-Khim. Obshchestva,* **39,** 1907, 787–802; *Z. anorg. Chem.,* **57,** 1908, 253–266.
6. R. Sahmen, *Z. anorg. Chem.,* **57,** 1908, 20–26.
7. Only a few liquidus points were determined by N. Parravano [*Gazz. chim. ital.,* **42**(2), 1912, 385–394] and L. Rolla [*Gazz. chim. ital.,* **44**(1), 1914, 646–662].
8. T. Ishiwara and M. Isobe, *Kinzoku-no-Kenkyu,* **6,** 1929, 383–397; T. Ishiwara, *Proc. World Eng. Congr., Tokyo,* 1929, Paper 223; *Science Repts. Tôhoku Univ.,* **19,** 1930, 504–509; see also K. Honda, *Proc. World Eng. Congr., Tokyo,* 1929, Paper 658, p. 32.
9. W. Broniewski and S. Jaślan, *Ann. acad. sci. tech. Varsovie,* **3,** 1936, 141–154.
10. G. Grube, E. Oestreicher, and O. Winkler, *Z. Elektrochem.,* **45,** 1939, 776–784.
11. R. S. Dean and C. T. Anderson, 1941, see [12].
12. R. S. Dean, J. R. Long, T. R. Graham, E. V. Potter, and E. T. Hayes, *Trans. ASM,* **34,** 1945, 443–463; see also R. S. Dean, C. T. Anderson, and J. H. Jacobs, *Trans. ASM,* **29,** 1941, 881–898.
13. S. F. Zemczuzny, S. A. Pogodin, and W. A. Finkeisen, *Izvest. Inst. Fiz.-Khim. Anal.,* **2,** 1924, 405–449.
14. M. A. Hunter and F. M. Sebast, *J. Am. Inst. Metals,* **11,** 1917-1918, 115.
15. H. Endo, *Science Repts. Tôhoku Univ.,* **14,** 1925, 510–511.
16. E. C. Bain, *Trans. AIME,* **68,** 1923, 633; *Chem. & Met. Eng.,* **28,** 1923, 21–24.
17. R. A. Patterson, *Phys. Rev.,* **23,** 1924, 552; *Ind. Eng. Chem.,* **16,** 1924, 689–691.
18. O. Heusler, *Z. anorg. Chem.,* **159,** 1926, 38–39.
19. S. Sekito, *Z. Krist.,* **72,** 1929, 406–415.
20. E. Persson, *Nature,* **124,** 1929, 333–334; *Z. physik. Chem.,* **B9,** 1930, 25–42.

21. U. Zwicker, *Z. Metallkunde,* **42,** 1951, 327–330.
22. Z. S. Basinski and J. W. Christian, *J. Inst. Metals,* **80,** 1951-1952, 659–666.
23. C. Zener, "Elasticity and Anelasticity of Metals," pp. 160–162, University of Chicago Press, Chicago, 1948.
24. F. T. Worrell, *J. Appl. Phys.,* **19,** 1948, 929–933, 1139.
25. U. Zwicker, *Z. Metallkunde,* **42,** 1951, 246–253.
26. M. Kawasaki, K. Yamaji, and O. Izumi, *Science Repts. Research Insts. Tôhoku Univ.,* **A3,** 1951, 66–77.
27. N. I. Korenev, *Izvest. Sektora Fiz.-Khim. Anal.,* **11,** 1938, 47–63.
28. S. Valentiner and G. Becker, *Z. Physik,* **80,** 1933, 735–754.
29. L. D. Ellsworth and F. C. Blake, *J. Appl. Phys.,* **15,** 1944, 507–512.
30. A. P. Guljaev and E. F. Trusova, *Zhur. Tekh. Fiz.,* **20,** 1950, 66–78. *Structure Repts.,* **13,** 1950, 103.
31. M. Kawasaki, K. Yamaji, and O. Izumi, *Science Repts. Research Insts., Tôhoku Univ.,* **7A,** 1955, 443–454; *Met. Abstr.,* **24,** 1956, 21.
32. A. Kussmann and H. J. Wollenberger, *Naturwissenschaften,* **43,** 1956, 395.
33. T. Sato, *J. Phys. Soc. Japan,* **5,** 1950, 287 (quoted from [32]).

$\bar{1}.8210$
0.1790

Cu-Mo Copper-Molybdenum

Copper and molybdenum form a miscibility gap in the liquid state ranging from 0 to 100% Mo [1, 2]; see also [3]. By means of measurements of electrical conductivity, the solid solubility of Mo in Cu at 900°C was found to be "vanishingly small" [4].

1. E. Siedschlag, *Z. anorg. Chem.,* **131,** 1923, 196–202.
2. L. Dreibholz, *Z. physik. Chem.,* **108,** 1924, 214.
3. This is in accord with earlier experiments by C. Sargent (*J. Am. Chem. Soc.,* **22,** 1900, 783–790) and C. Lehmer (*Metallurgie,* **3,** 1906, 596–597), who reported that they were unsuccessful in alloying the two metals.
4. J. O. Linde, *Ann. Physik,* **15,** 1932, 231.

0.6567
$\bar{1}.3433$

Cu-N Copper-Nitrogen

According to [1], nitrogen is insoluble in solid and liquid copper (investigated up to 1400°C); see also [2]. There is no reaction of Cu with N_2 at temperatures up to 900°C [3].

Copper nitride, Cu_3N (6.85 wt. % N), has been prepared by passing NH_3 over finely divided CuO, Cu_2O, or CuF_2 at 250–280°C [4–7]. The formation of a nitride by treating Cu with NH_3 at about 900–1000°C was reported by [8–10]; see, however, [4].

The crystal structure of Cu_3N is cubic of the ReO_3 ($D0_9$) type, with 1 Cu_3N per unit cell and a = 3.814 A [6, 11].

1. A. Sieverts and W. Krumbhaar, *Z. physik. Chem.,* **74,** 1910, 280; *Ber. deut. chem. Ges.,* **43,** 1910, 894.
2. P. Röntgen and F. Möller, *Metallwirtschaft,* **13,** 1934, 31–83, 97–100.
3. P. Laffitte and G. Grandadam, *Compt. rend.,* **200,** 1935, 1039–1041.
4. A. Schrötter, *Liebigs Ann.,* **37,** 1841, 131.
5. A. Guntz and H. Bassett, *Bull. soc. chim.,* **35,** 1906, 201–207.
6. R. Juza and H. Hahn, *Z. anorg. Chem.,* **239,** 1938, 282–287.
7. R. Juza and H. Hahn, *Z. anorg. Chem.,* **241,** 1939, 172–178.
8. H. N. Warren, *Chem. News,* **55,** 1887, 156.

9. G. T. Beilby and G. G. Henderson, *J. Chem. Soc.*, **79,** 1901, 1245–1256.
10. C. Matignon and R. Trannoy, *Compt. rend.*, **142,** 1906, 1210–1211.
11. R. Juza, *Z. anorg. Chem.*, **248,** 1941, 118–120.

0.4414
$\bar{1}$.5586

Cu-Na Copper-Sodium

No indication of the existence of a Cu-Na compound was found on potentiometric titration of a solution of sodium in liquid ammonia with a solution of cuprous iodide in ammonia [1].

1. E. Zintl, J. Goubeau, and W. Dullenkopf, *Z. physik. Chem.*, **154,** 1931, 44.

$\bar{1}$.8350
0.1650

Cu-Nb Copper-Niobium

"The as-cast ingot of the $NbCu_2$ composition had limited ductility. Filings, annealed for 10 min at 800°C, gave diffraction patterns of f.c.c. copper and b.c.c. niobium. Microexamination showed a segregated structure, probably of a eutectic. From these data, it is concluded that the Nb-Cu system is probably simple eutectic" [1].

1. R. P. Elliott, Armour Research Foundation, Chicago, Ill., Technical Report 1, OSR Technical Note OSR-TN-247, August, 1954, p. 19.

0.0345
$\bar{1}$.9655

Cu-Ni Copper-Nickel

The liquidus curve in Fig. 347 was established by [1-3]; that by [4] lies at slightly lower temperatures. A previously determined curve [5] deviates to considerably higher temperatures [6]. The solidus curve is based only on cooling-curve data, which agree quite well.

The existence of a continuous series of solid solutions was confirmed by microscopic investigations [1–3, 7] and results of X-ray analyses [8–18], according to which all alloys investigated have the f.c.c. lattice of the components. The lattice parameters reported by [17] are to be regarded as the most accurate ones; they indicate a maximal contraction of the lattice of 0.11% at about 34 at. % Ni.

Curie temperatures were determined by (1, 19–22, 30) and, most complete and accurate [23], by [24], by means of magnetic (between 368 and −141°C) and resistance measurements (between 368 and −258°C). The Curie points lie on a straight line intersecting room temperature at about 70.2 at. (68.5 wt.) % Ni and reaching the absolute zero point at about 43.5 at. (41.5 wt.) % Ni (Fig. 347).

Since the physical properties will be somewhat affected by the magnetic transformation, it is to be expected that anomalies in property-vs.-composition curves will occur. There is only a very slight, if any, irregularity in the isotherms of the electrical conductivity [25, 26, 24] and thermal conductivity [26] which are catenary-type curves with a very flat minimum. On the other hand, the curves of the temperature coefficient of resistance [25, 26, 24], the thermoelectric force [25, 21, 27, 24], and especially that of its temperature coefficient [24] show strong deviations for temperature ranges in which the magnetic transformation takes place. For temperatures above the Curie point of Ni, however, thermoelectric-force–composition curves [21, 28] are also catenary-type curves. Also, the curve of the coefficient of thermal expansion shows small but distinctive irregularities [24, 29]. As to the change of physical properties with composition and temperature, the reader is especially referred to the work by [24]. There is no indication that the anomalies are caused by phase transformation.

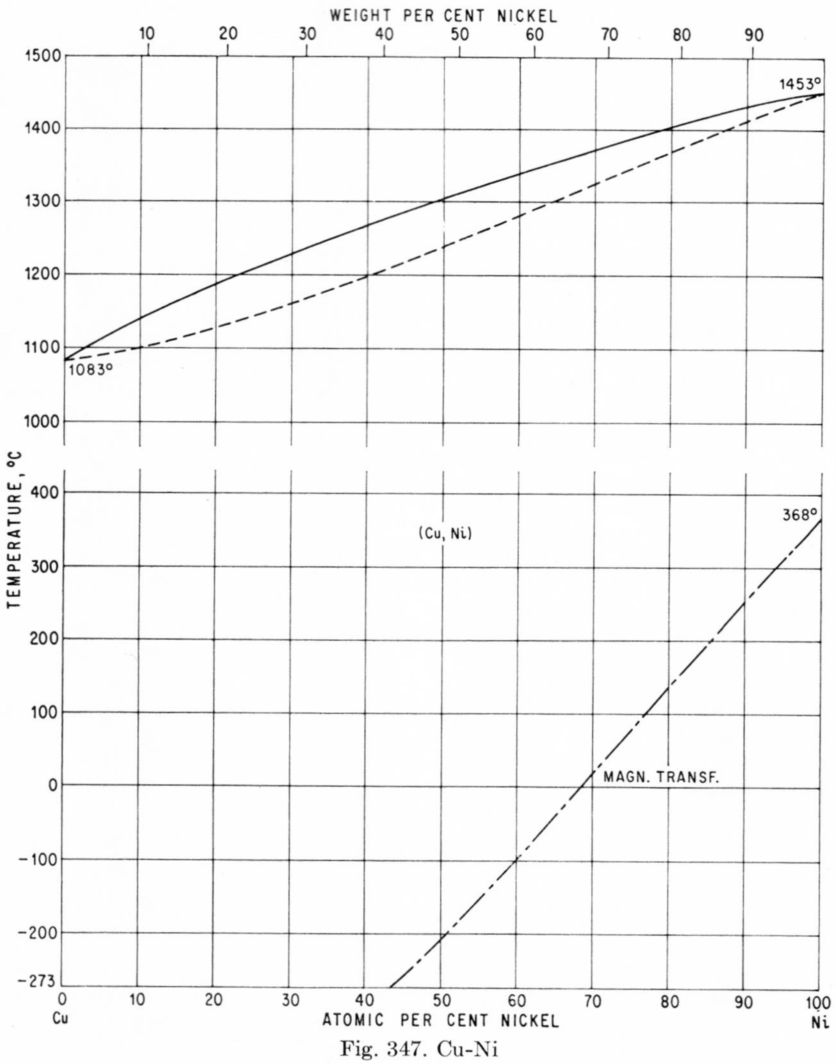

Fig. 347. Cu-Ni

In contrast to this, [20] concluded from magnetic measurements that a miscibility gap exists between about 47 and 52 at. % Ni. Also, the curve of the Curie temperatures showed a horizontal section between 47 and 51 at. % Ni, at −90 to −105°C. There is no doubt that these deviations were caused by inhomogeneity of the alloys, since they had not been heat-treated. This is also proved by the fact that Curie points were found in alloys with Ni contents as low as 37 and 42 at. % Ni, which, if in equilibrium, are paramagnetic down to extremely low temperatures (Fig. 347). See also X-ray evidence against a miscibility gap [17].

[21] suggested that the compound CuNi might exist. Thermo-emf vs. temperature (0–1000°C) and curves of the true temperature coefficient of electrical resistance vs. temperature (−200 to 1000°C) indicate, besides the effect of the magnetic transformation, a reversible anomalous change (inflection points) at 400–500°C, the maximum of which is reached at about 52 at. % Ni. This "X-anomaly" is interpreted to be due to a "quasi-reversible physicochemical reaction" beginning at low temperatures and ending at about 450°C. Within this temperature interval, the reaction would cause a decrease in electrical resistance which retards the normal increase with rise in temperature. [21] emphasizes also that the maximum of the electrical-resistance isotherms (−200 to 1000°C) becomes sharper and approaches the composition CuNi as the temperature decreases. However, this is no indication of the formation of a "compound" or an ordered structure since it would result in a relative minimum in electrical resistance.

No direct indication of an order-disorder change has been found as yet, possibly because the difference in scattering power of the Cu and Ni atoms is too small.

Supplement. [31] measured the lattice parameters of alloys with up to 5.0 at. % Ni. [32] determined the following Curie-point temperatures on annealed and slowly cooled specimens: 297, 237, 177, and 117°C at 6, 12, 18, and 24 wt. (5.6, 11.2, 16.9, and 22.6 at.) % Cu, respectively. These data are in good agreement with the curve shown in Fig. 347. [33] measured the lattice parameter at room temperature over the whole composition range. In agreement with [17] (cf. above), a slight contraction was found. The most marked change of slope occurs at the composition, where the Curie temperature of the alloys reaches room temperature.

1. W. Guertler and G. Tammann, *Z. anorg. Chem.*, **52,** 1907, 25–29.
2. N. S. Kurnakow and S. F. Zemczuzny, *Z. anorg. Chem.*, **54,** 1907, 151–155; *Zhur. Russ. Fiz.-Khim. Obshchestva,* **39,** 1907, 211–219.
3. V. E. Tafel, *Metallurgie,* **5,** 1908, 348–349.
4. P. de Cesaris, *Gazz. chim. ital.*, **43**(2), 1913, 365–379.
5. H. Gautier, *Bull. soc. encour. ind. natl.*, **1,** 1896, 1309–1310; *Compt. rend.*, **123,** 1896, 173–174.
6. C. T. Heycock and F. H. Neville [*Phil. Trans. Roy. Soc.* (*London*), **A189,** 1897, 69] found that the melting point of Cu is raised to 1100°C by 4.3 at. % Ni.
7. A. Krupkowski, *Rev. mét.*, **26,** 1929, 203–206.
8. E. C. Bain, *Trans. AIME,* **68,** 1923, 635–636; *Chem. & Met. Eng.*, **26,** 1922, 655; no lattice parameters are reported.
9. E. A. Owen and G. D. Preston, *Proc. Phys. Soc.* (*London*), **36,** 1923, 28–29.
10. H. Lange, *Ann. Physik,* **76,** 1925, 482–484.
11. A. Sacklowski, *Ann. Physik,* **77,** 1925, 260–264.
12. S. Holgersson, *Ann. Physik,* **79,** 1926, 46–49.
13. L. Vegard and H. Dale, *Z. Krist.*, **67,** 1928, 154–157.
14. S. Pienkowski, see A. Krupkowski, *Rev. mét.*, **26,** 1929, 206–207.
15. S. Sekito, *Science Repts. Tôhoku Univ.*, **18,** 1929, 59–68.
16. W. G. Burgers and J. C. M. Basart, *Z. Krist.*, **75,** 1930, 155–157.
17. E. A. Owen and L. Pickup, *Z. Krist.*, **88,** 1934, 116–121.
18. B. A. Johnsen and L. Vegard, *Skrifter Norske Videnskaps-Akad. Oslo, Mat. Naturv. Kl.*, **1947**(2), 47–56.
19. B. Hill, *Verh. deut. physik. Ges.*, **4,** 1902, 194.
20. R. Gans and A. Fonseca, *Ann. Physik,* **61,** 1920, 742–752.
21. P. Chevenard, *Chaleur & ind.*, July 4, 1923; see *J. Inst. Metals,* **36,** 1926, 53–62.

22. V. Marian, *Ann. phys.*, **7,** 1937, 459–527.
23. By long annealings, alloys were brought into equilibrium.
24. A. Krupkowski, *Rev. mét.*, **26,** 1929, 131–153, 193–208.
25. K. Feussner, *Verh. deut. physik. Ges.*, **10,** 1921, 109; K. Feussner and S. Lindeck, *Wiss. Abhandl. physik.-tech. Reichsanstalt,* **2,** 1895, 503–516; see also *Ann. Physik,* **32,** 1910, table XI.
26. E. Sedström, Dissertation, Stockholm, 1924; see also *Ann. Physik,* **59,** 1919, 134–144.
27. E. Sedström, Dissertation, Stockholm, 1924.
28. F. E. Bash, *Trans. AIME,* **64,** 1921, 239–260.
29. C. H. Johansson, *Ann. Physik,* **76,** 1925, 448–449.
30. M. A. Wheeler, *Phys. Rev.*, **56,** 1939, 1145.
31. A. P. Guljaev and E. F. Trusova, *Zhur. Tekh. Fiz.*, **20,** 1950, 66–78. *Structure Repts.*, **13,** 1950, 103.
32. K. Torkar and H. Götz, *Z. Metallkunde,* **46,** 1955, 374.
33. B. R. Coles, *J. Inst. Metals,* **84,** 1956, 346–348.

0.5989
$\bar{1}$.4011

Cu-O Copper-Oxygen*

The Partial System Cu-Cu_2O. *Alloys with up to* 1 *Wt.* (3.85 *At.*) % O. Liquidus temperatures in this range were determined by [1–4]. The eutectic temperature was found by all investigators to be 1065°C, and the eutectic composition was given as 0.39 wt. (1.54 at.) % O (3.49 wt. % Cu_2O) [1, 3, 4] and 0.5–0.56 wt. (1.96–2.2 at.) % O [2]. A tentative value of 0.47 wt. (1.85 at.) % O is based on microscopic and analytical work [5].

Various investigations indicated that the solid solubility of O in Cu was extremely small [1, 6–9], certainly lower than 0.009 wt. (0.036 at.) % O at 1000°C [6] and lower than 0.005 wt. (0.018 at.) % O at 500°C [9]. According to [10], the solubility increases from 0.007 wt. (0.028 at.) % at 600°C to about 0.009 (0.036) % at 800°C, 0.01 (0.04) % at 900°C, and 0.015 (0.06) % O at 1050°C. However, the data reported by [11], which are one order of magnitude smaller, appear to be most reliable: 0.007 (0.028) % at 1040°C, 0.0045 (0.018) % at 1000°C, 0.0027 (0.011) % at 900°C, 0.002 (0.008) % at 800°C, and approximately 0.0017 (0.007) % at 550°C. Using the same specimens but a different analytical method, [12] found still lower values: 0.005 (0.02) % at 1040°C, 0.001 (0.004) % at 700°C, and 0.0008 (0.003) % at 550°C. Data reported by [13] lie between those of [10] and [11]: 0.009 (0.036) % at 900°C and 0.004 (0.016) % at 600°C.

Alloys with up to 33.3 *At.* (11.18 *Wt.*) % O. The existence of a miscibility gap in the liquid state between Cu and Cu_2O, first recognized by [3], was verified by [4]. The monotectic temperature was found as 1195–1203°C [3] and 1195–1205°C [4]. The composition of the two melts being in equilibrium at this temperature was determined by analysis of the solidified layers as 2.26 wt. (8.4 at.) and 10.57 wt. (31.9 at.) % O [3]. On the basis of thermal data, [4] gave 1.5 (5.7) and 10.2 (31.1) % O. The composition of the two melts at 1254, 1340, and 1400°C was determined analytically after quenching from these temperatures [3]; see data points in Fig. 348. Since this method of determining the boundaries of a range of immiscibility in the liquid state is not reliable, these values have to be accepted with reservation. Indeed, the compositions of the Cu-rich melts thus determined are inconsistent with that one at 1200°C more reliably placed by thermal analysis at 1.5 wt. (5.7 at.) % O [4].

* The assistance of Dr. F. Wolstein in preparing this system is gratefully acknowledged.

The melting point of Cu_2O was reported as 1222 [14] and 1235°C [15].

The Partial System Cu_2O-CuO. To obtain information as to the constitution in this composition range, [16–19] and especially [20, 15] have carried out measurements of equilibrium (dissociation) pressures. On the basis of pressure-vs.-temperature curves, [15] have deduced the phase diagram. Thermal and microscopic work was done by [4]. Results of the afore-mentioned studies and that of [14] were used

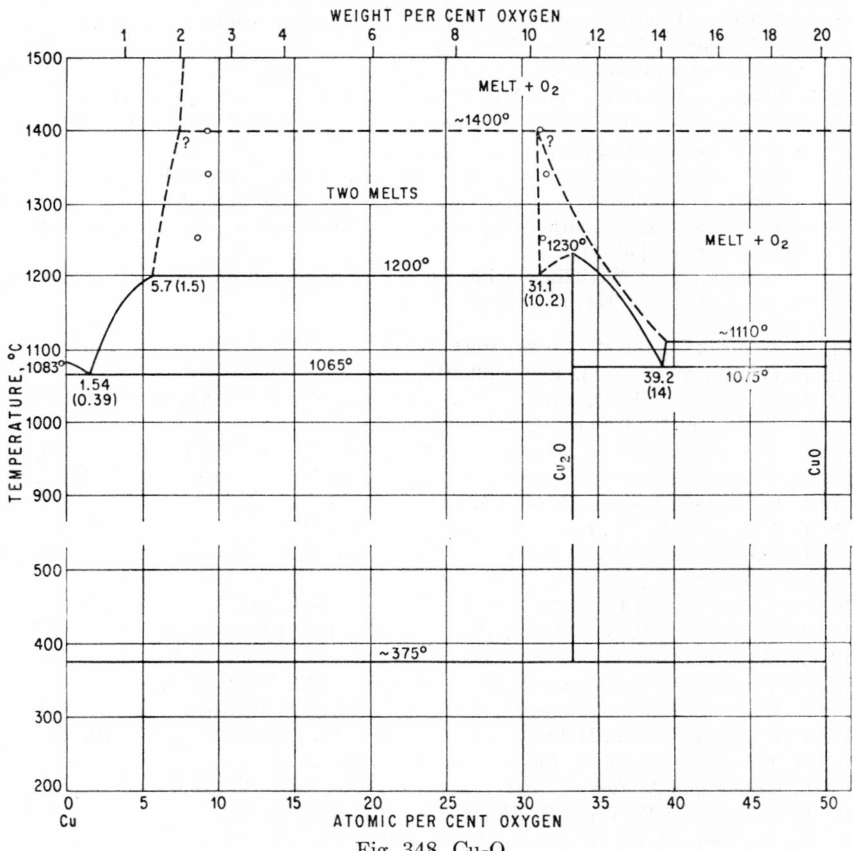

Fig. 348. Cu-O

by [4] to design the phase diagram for a pressure of 1 atm illustrated in Fig. 348 [21]. Cu_2O and CuO (20.12 wt. % O) are indicated as phases of fixed composition. This was corroborated by measurements of the magnetic susceptibility and X-ray studies [22]. The two compounds form a eutectic with 13.8–14.0 wt. (38.9–39.3 at.) % O [15, 4], corresponding to 68 wt. % Cu_2O, 32 wt. % CuO. The eutectic temperature was found as 1070 [19], 1080 [20], and 1075°C [4]; see also [23].

Since the decomposition of CuO into Cu_2O and O_2 is considerable even in the solid state—according to [15] its dissociation pressure at 1026°C equals the partial pressure of O_2 in the atmosphere (153 mm Hg), and at the eutectic temperature it is 402 mm Hg—the thermal equilibrium of mixtures of Cu_2O and CuO and, consequently, their behavior on heating must be highly dependent on the O_2 pressure existing over the mixture. At 1100°C the dissociation pressure is 1 atm. At this temperature, therefore, the still undecomposed CuO will melt, forming Cu_2O-rich liquid and O_2. At some undefined temperature the Cu_2O-rich melt will decompose, forming Cu-rich melt and O_2. In Fig. 348 this temperature was assumed at about 1400°C. As to the effect of pressure on the composition-temperature diagram, see [15, 4].

[4] concluded from microexamination of Cu_2O and Cu + Cu_2O alloys, slowly cooled as well as annealed at 300–350°C, that Cu_2O is unstable at temperatures below about 375°C and decomposes according to the equation $Cu_2O \rightarrow Cu + CuO$. The rate of this reaction is very low, however.

Crystal Structures. The structure of Cu_2O (cuprite) is the prototype of the cubic C3-type structure. The lattice parameter has been determined by [24–36]. The most accurate value is that reported by [36]: $a = 4.2696$ A at 26°C. The space group is still uncertain [24, 37].

CuO (tenorite), formerly believed to be triclinic [24], is monoclinic [38, 39], with $a = 4.684$ A, $b = 3.425$ A, $c = 5.129$ A, $\beta = 99°28'$ at 26°C, and 4 formula weights per unit cell [39]. As to another crystal form of CuO or an intermediate oxide between Cu_2O and CuO, see [40, 41].

The existence of a triclinic oxide Cu_4O was suggested on the basis of electron-diffraction studies [42]; see, however, [43].

1. E. Heyn, *Z. anorg. Chem.*, **39,** 1904, 1–23.
2. P. Dejean, *Rev. mét.*, **3,** 1906, 233–240.
3. R. E. Slade and F. D. Farrow, *Proc. Roy. Soc. (London)*, **A87,** 1912, 524–534.
4. R. Vogel and W. Pocher, *Z. Metallkunde*, **21,** 1929, 333–337, 368–371.
5. F. Johnson, *Metal Ind. (London)*, **27,** 1925, 208; *J. Inst. Metals*, **4,** 1910, 230.
6. D. Hanson, C. Marryat, and G. W. Ford, *J. Inst. Metals*, **30,** 1923, 197–227.
7. F. L. Antisell, *Trans. AIME*, **64,** 1921, 435.
8. R. P. Heuer, *J. Am. Chem. Soc.*, **49,** 1927, 2711–2720.
9. Unpublished work by T. Hewitt, see N. P. Allen and A. C. Street, *J. Inst. Metals*, **51,** 1933, 235.
10. F. N. Rhines and C. H. Mathewson, *Trans. AIME*, **111,** 1934, 337–353.
11. A. Phillips and E. N. Skinner, *Trans. AIME*, **143,** 1941, 301–308.
12. W. A. Baker, see [11].
13. M. Clasing and F. Sauerwald, *Z. anorg. Chem.*, **271,** 1952, 81–87.
14. R. Ruer and M. Nakamoto, *Rec. trav. chim.*, **42,** 1923, 675–685.
15. H. S. Roberts and F. H. Smyth, *J. Am. Chem. Soc.*, **43,** 1921, 1061–1079.
16. H. Debray and Joannis, *Compt. rend.*, **99,** 1884, 583.
17. L. Wöhler and A. Foss, *Z. Elektrochem.*, **12,** 1906, 781–786; see also [23].
18. L. Wöhler and W. Frey, *Z. Elektrochem.*, **15,** 1909, 34–38.
19. H. W. Foote and E. K. Smith, *J. Am. Chem. Soc.*, **30,** 1908, 1345–1346.
20. F. H. Smyth and H. S. Roberts, *J. Am. Chem. Soc.*, **42,** 1920, 2582–2607.
21. See also the compilation of data in M. Randall, R. F. Nielsen, and G. H West, *Ind. Eng. Chem.*, **23,** 1931, 391-393.
22. B. Reuter and X. Schröder, *Z. anorg. Chem.*, **277,** 1954, 146–155.
23. R. E. Slade and F. D. Farrow, *Z. Elektrochem.*, **18,** 1912, 817–818.
24. P. Niggli, *Z. Krist.*, **57,** 1922, 253–299.
25. W. P. Davey, *Phys. Rev.*, **19,** 1922, 248–251.
26. G. Greenwood, *Phil. Mag.*, **48,** 1924, 654–663.
27. J. Böhm, *Z. Krist.*, **64,** 1926, 550.
28. G. P. Thompson, *Proc. Roy. Soc. (London)*, **A128,** 1930, 649–661.
29. J. A. Darbishire, *Trans. Faraday Soc.*, **27,** 1931, 675–678.
30. M. C. Neuburger, *Z. Physik*, **67,** 1931, 845–850; *Z. Krist.*, **77,** 1931, 169–170.

31. W. Wrigge and K. Meisel, *Z. anorg. Chem.*, **203,** 1932, 312–320.
32. H. Germer, *Phys. Rev.*, **52,** 1937, 959–967.
33. N. Smith, *J. Am. Chem. Soc.*, **58,** 1937, 173–179.
34. T. Yamaguti, *Proc. Phys. Soc. Japan*, **20,** 1938, 230–241; *Strukturbericht,* **6,** 1938, 61.
35. H. Lal, *J. Sci. Ind. Research (India)*, **12B,** 1953, 424–430.
36. H. E. Swanson and R. K. Fuyat, *Natl. Bur. Standards (U.S.) Circ.* 539, II, 1953, p. 23.
37. T. Okada, *J. Phys. Soc. Japan*, **4,** 1949, 140–141; *Chem. Abstr.*, **44,** 1950, 10434.
38. G. Tunell, E. Posnjak, and C. J. Ksanda, *Z. Krist.*, **90,** 1935, 120–142.
39. H. E. Swanson and E. Tatge, *Natl. Bur. Standards (U.S.) Circ.* 539, I, 1953, p. 49.
40. C. A. Murison, *Phil. Mag.*, **17,** 1934, 96–98.
41. G. Honjo, *J. Phys. Soc. Japan*, **4,** 1949, 330; *Structure Repts.*, **12,** 1949, 301.
42. S. Yoshioka, *Nippon Kinzoku Gakkai-Shi*, **14B**(7), 1950, 24–33.
43. *Structure Repts.*, **13,** 1950, 104–105.

$\bar{1}.5238$
0.4762

Cu-Os Copper-Osmium

Measurements of electrical resistance of copper-rich alloys indicate that the solid solubility of Os in Cu is "vanishingly small" (after annealing at 900°C) [1].

1. J. O. Linde, *Ann. Physik,* **15,** 1932, 231.

0.3120
$\bar{1}.6880$

Cu-P Copper-Phosphorus

Partial System Cu-Cu_3P. Liquidus temperatures were determined by [1–4]; those reported by [4] should be regarded as the most reliable ones (Fig. 349). The temperature of the α-Cu_3P eutectic was found as 615 [1], 620 [2], 707 [3], about 710 [5], and 714°C [4], and the eutectic concentration as 9–10 wt. (16.9–18.6 at.) [1], 8.2 wt. (15.4 at.) [2, 3, 6], and 8.38 wt. (15.7 at.) % P [4]. The melting point of Cu_3P was reported to be 1005 [2], 1018 [3], and 1023°C [7].

The existence of Cu-rich solid solutions follows from the strong effect of P on the electrical conductivity [8–13] and thermal conductivity [9, 14, 12, 15] of Cu. However, no quantitative conclusions can be drawn from these and other data [3, 16, 17]. By means of micrographic investigations [18], the solubility was found to be slightly less than 0.5 wt. (1.0 at.) % at 280°C, and about 0.6 wt. (1.2 at.), 0.8 wt. (1.6 at.), and 1.15 wt. (2.3 at.) % P at 400°C, 600°C, and the eutectic temperature, respectively. [19] gave 1.3 wt. (2.6 at.) % as the solubility at 700°C. Much lower values were reported by [7]. The solubility curve in Fig. 349 is based on lattice-parameter determinations [20]. Solubilities are 0.6, 0.85, 1.1, 1.4, and 1.7 wt. (1.2, 1.7, 2.2, 2.8, and 3.4 at.) % at 300, 400, 500, 600, and 700°C, respectively. The solidus curve of the α phase was determined micrographically (21).

Partial System Cu_3P-P. [3] have shown that the liquidus temperature of alloys with 14.24–14.96 wt. (about 25.4–26.5 at.) % P lies at 1022–1024°C. This temperature was also assumed to be the boiling point of these melts which crystallize to form solid solutions of P in Cu_3P. By treating copper turnings with phosphorus vapor at 300–400°C, products with up to 27.4 wt. (43.6 at.) % P were obtained. If heated under charcoal, these products lose P until a limiting composition is reached, depending upon the annealing temperature: 15 wt. (26.6 at.) % at 800°C, 14.5 wt. (25.9 at.) % at 900–1000°C, and 14 wt. (25 at.) % P at 1100°C (melting).

Products richer in copper were prepared by [5], treating copper turnings with phosphorus vapor in a CO_2 atmosphere. They contained about 30.9 wt. (47.8 at.) % at 290°C, 25.9 wt. (41.7 at.) % at 400°C, 21.8 wt. (36.4 at.) % at 500°C, 17.1 wt. (29.8 at.) % at 600°C, and 14.8 wt. (26.3 at.) % P at 700°C. These temperatures represent the temperatures up to which the compositions mentioned are stable in an inert atmosphere without vaporization (Fig. 349).

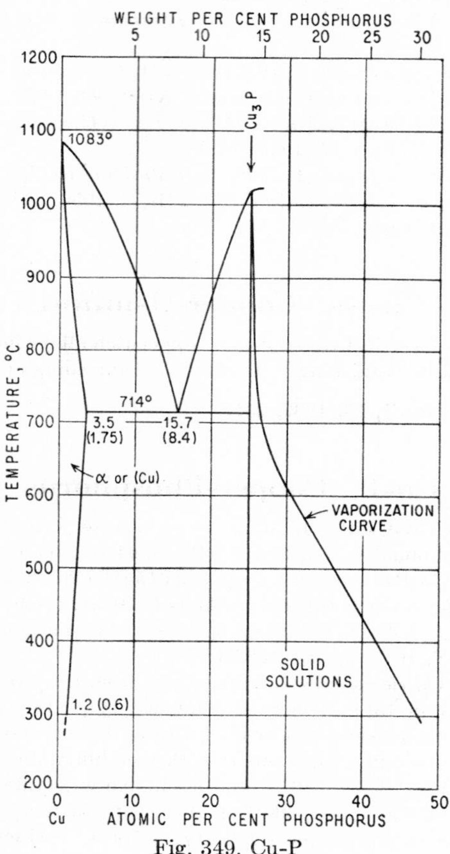

Fig. 349. Cu-P

By means of tensimetric analysis, corroborated by X-ray work, [22] has proved that besides Cu_3P there is only one other phosphide, CuP_2 (49.37 wt. % P). According to [7], preparations with up to 26 wt. (41.8 at.) % P were found to be single-phase, i.e., solid solutions of P in Cu_3P. There was no indication [7, 22] of the existence of Cu_5P_2 [3, 5], CuP [5], or any of the other phosphides, as had been reported in the older literature on the basis of inconclusive evidence.

Crystal Structures. Lattice parameters of the α phase were reported by [20]. Cu_3P is hexagonal, with 6 Cu_3P per unit cell and $a = 6.95$ A, $c = 7.12 \pm 0.02$ A, $c/a = 1.02$ (DO_{21} type) [23, 22, 24]. CuP_2 has a very complex powder pattern [22].

1. L. Guillet, *Génie civil*, **47**, 1905, 187; *Rev. mét.*, **2**, 1906, 171–173.
2. A. H. Hiorns, *J. Soc. Chem. Ind.*, **25**, 1906, 618, 620.

3. E. Heyn and O. Bauer, *Z. anorg. Chem.*, **52**, 1907, 129–151; *Metallurgie*, **4**, 1907, 242–247, 257–266.
4. W. E. Lindlief, *Metals & Alloys*, **4**, 1933, 85–87.
5. C. A. Edwards and A. J. Murphy, *J. Inst. Metals*, **27**, 1922, 183–213.
6. A. K. Huntington and C. H. Desch, *Trans. Faraday Soc.*, **4**, 1908, 51–58.
7. H. Moser, K. W. Fröhlich, and E. Raub, *Z. anorg. Chem.*, **208**, 1932, 226–227.
8. A. Matthiessen and M. Holzmann, *Pogg. Ann.*, **110**, 1860, 228; A. Matthiessen and C. Vogt, *Pogg. Ann.*, **122**, 1864, 19.
9. A. Rietzsch, *Ann. Physik*, **3**, 1900, 403–427.
10. E. Münker, *Metallurgie*, **9**, 1912, 195–197.
11. D. Hanson, S. L. Archbutt, and G. W. Ford, *J. Inst. Metals*, **43**, 1930, 50–51.
12. C. S. Smith, *Trans. AIME*, **93**, 1931, 176–184.
13. J. S. Smart and A. A. Smith, *Trans. AIME*, **166**, 1946, 144–155.
14. G. Pfleiderer, *Ges. Abhandl. z. Kenntnis Kohle*, **4**, 1919, 409–426.
15. D. Hanson and C. E. Rodgers, *J. Inst. Metals*, **48**, 1932, 37–42.
16. E. A. Lewis, *Engineering*, **76**, 1903, 753.
17. O. F. Hudson and E. F. Law, *J. Inst. Metals*, **3**, 1910, 163.
18. D. Hanson, S. L. Archbutt, and G. W. Ford, *J. Inst. Metals*, **43**, 1930, 54–55.
19. J. Verö, *Z. anorg. Chem.*, **213**, 1933, 258.
20. J. C. Mertz and C. H. Mathewson, *Trans. AIME*, **124**, 1937, 59–77.
21. D. K. Crampton, H. L. Burghoff, and J. T. Stacy, *Trans. AIME*, **137**, 1940, 357.
22. H. Haraldsen, *Z. anorg. Chem.*, **240**, 1939, 337–354.
23. B. Steenberg, *Arkiv Kemi, Mineral. Geol.*, **12A**, 1938, no. 26.
24. H. Nowotny and E. Henglein, *Monatsh. Chem.*, **69**, 1948, 385–393.

$\bar{1}.4866$
0.5134

Cu-Pb Copper-Lead

Liquidus temperatures covering the entire range of composition were reported by [1–7]; [2, 3] first recognized that molten Cu and Pb are not miscible in all proportions. The temperature of the monotectic was found as 952 [2], 954 [3], 953 [5], 949 [6], 957 [7], and 955–956°C [8]; and the monotectic point as about 38 wt. (15.8 at.) [2, 6], 40 wt. (17 at.) [3], 36 wt. (14.7 at.) [5], and 52 wt. (24.9 at.) % Pb [7]. The composition of the Pb-rich melt at the monotectic temperature was given as about 77 wt. (50.7 at.) [2], 85 wt. (63.6 at.) [3, 5], 86.5 wt. (66.4 at.) [6], 87 wt. (67.2 at.) [7], and 92.7 wt. (79.5 at.) % Pb [8].

Since the portion of the liquidus between the monotectic temperature and the melting point of Pb cannot be determined accurately by thermal analysis, [9] have carried out careful solubility determinations in the range 950–326°C. This curve (Fig. 350) intersects the monotectic horizontal at about 67 at. (87 wt.) % Pb, in good agreement with several formerly reported data. See also Supplement.

Data as to the extent of the miscibility gap in the liquid state above 954°C are highly contradictory. By means of electrical-resistance-vs.-temperature measurements, [8] have found solubility curves indicating a critical temperature well above 1500°C. On the other hand, [10], using chemical analysis to determine the composition of the two melts in equilibrium at 954, 975, 1000, and 1025°C, concluded that the critical point of the closed solubility curve was at about 64.5 wt. (35.8 at.) % Pb and 1025°C, i.e., only about 70°C above the monotectic temperature.

The copper-rich branch of the solubility curve determined by [11], using the same method, is in good agreement with that by [10]; the critical point would be located at about 65 wt. (36.3 at.) % Pb and 1000°C. By measurement of the local electrical resistance at a sequence of levels in a molten bath kept constant at various temperatures, [12] determined the temperature at which a discontinuous change, caused by the interface between the two liquids, is observed on a voltage-vs.-depth plot. He concluded that his findings substantiate the work by [11].

[13] has tried to outline the area of immiscibility by quenching droplets from various temperatures. It was concluded that the critical point would lie at about

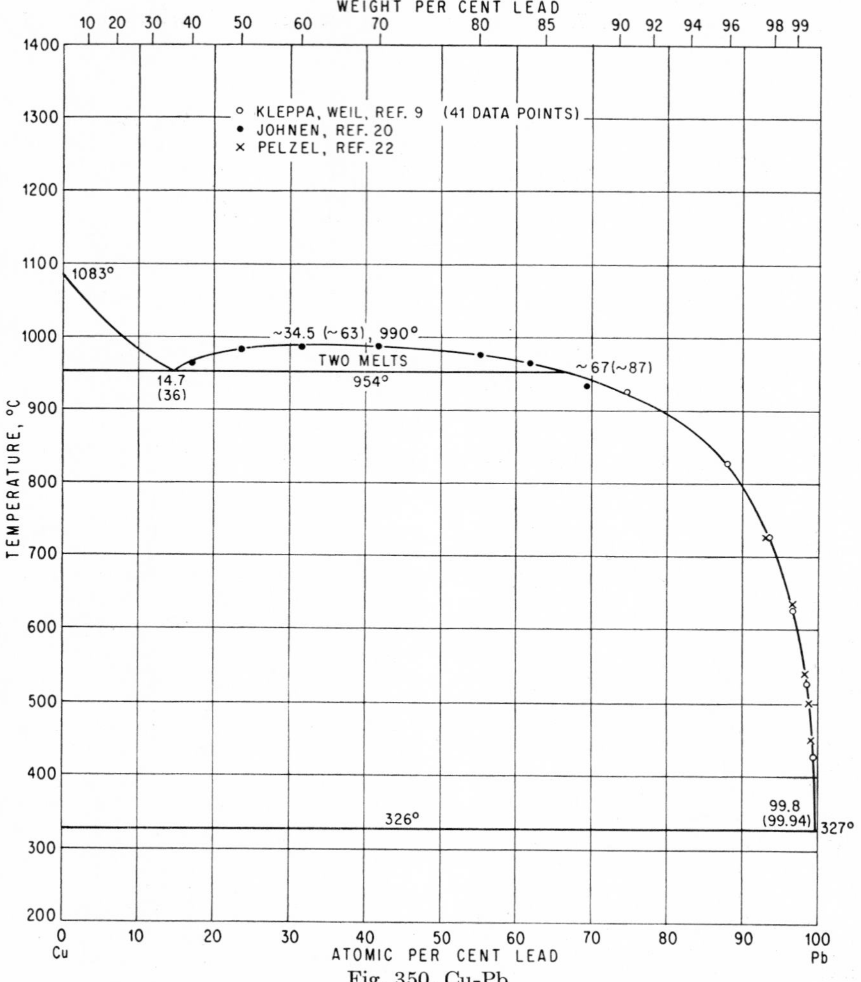

Fig. 350. Cu-Pb

75 wt. (47.9 at.) % Pb and 970°C, i.e., only 16°C above the monotectic temperature. However, these findings appear to be highly doubtful.

At this state of knowledge, it was difficult to decide which of the very conflicting results was right. Since it was believed that such a wide miscibility gap would not have a critical point lying only 50–70°C above the monotectic temperature, the results given by [8] were accepted as more likely than those reported by [10, 11]. Furthermore, [10, 11] did not examine whether the melt at temperatures above the supposed boundary was one-phase or consisted of an emulsion of two liquid phases. In an effort to explain the discrepancy, it was assumed [14, 15] that layer formation does not readily take place at temperatures above about 1000°C, but that the mixture of the two liquid phases, formed by separation of the homogeneous melt at much higher temperatures, is in the form of a finely divided emulsion. This would mean that the supposed boundary [10, 11, 12] would actually represent the temperatures at which layer formation becomes effective [15], rather than be an equilibrium curve.

Two recent investigations [20, 21], however, yielded further evidence for a relatively low critical temperature, so that the steep boundaries of the miscibility gap claimed by [8] have to be discarded. The data obtained by means of differential thermal analysis [20] are plotted in Fig. 350. They indicate the critical data 990°C, 63 wt. (34.3 at.) % Pb, and an extent of the miscibility gap from about 37 to about 86 wt. (15.3–65.3 at.) % Pb at the monotectic temperature (954–955°C). According to [21], the critical point lies below 1100°C. See also Supplement.

According to [16], the eutectic point lies at about 0.06 wt. (0.18 at.) % Cu and 326°C. This composition is in good agreement with the solubility determined by [9], 0.2 at. % Cu at 327°C. The solid solubility of Pb in Cu was estimated to be not higher than 0.09 at. (0.29 wt.) % Pb above 600°C, based on microscopic and roentgenographic work [17]. The solid solubility of Cu in Pb is less than 0.007 wt. % Pb [18]. In electrolytically deposited Cu-Pb alloys, [19] observed a solid solubility of about 10–12 wt. (3.3–4 at.) % Pb in Cu.

Supplement. [22] determined the solubility of Cu in liquid Pb at six temperatures between 370 and 727°C by chemical analysis of equilibrated and quenched melts. The data obtained are in good agreement with those of [9] (see Fig. 350). The claim that the eutectic composition is 0.11 wt. % Cu (instead of 0.06 wt. % according to [16, 9]) is not well founded.

Typical microstructures of Cu-Pb alloys were published and interpreted by [23]. This author determined the monotectic temperature as 955 ± 0.5°C and the extent of the miscibility gap at this temperature, by microscopic analysis, as about 37.8–85.2 wt. (15.7–63.8 at.) % Pb, in good agreement with [20]. He also gave some evidence for a low critical temperature, about 990°C.

1. H. Gautier, *Bull. soc. encour. ind. natl.*, **1**, 1896, 1310; Société d'encouragement pour l'industrie national, Paris, Commission des alliages, "Contribution à l'étude des alliages," pp. 93–118, Typ. Chamerot et Renouard, Paris, 1901.
2. W. C. Roberts-Austen, *Engineering*, **63**, 1897, 253–255; *Proc. Inst. Mech. Engrs. (London)*, **1897**, 51–52.
3. C. T. Heycock and F. H. Neville, *Phil. Trans. Roy. Soc. (London)*, **A189**, 1897, 42–45, 60–62.
4. A. H. Hiorns, *J. Soc. Chem. Ind.*, **25**, 1906, 618–619.
5. K. Friedrich and A. Leroux, *Metallurgie*, **4**, 1907, 299–302.
6. F. Giolitti and M. Marantonio, *Gazz. chim. ital.*, **40**(1), 1910, 51–59.
7. M. Nishikawa, *Suiyokwai Shi*, **8**, 1933, 239–243.
8. K. Bornemann and K. Wagenmann, *Ferrum*, **11**, 1913-1914, 291–293.
9. O. J. Kleppa and J. A. Weil, *J. Am. Chem. Soc.*, **73**, 1951, 4848–4850.
10. K. Friedrich and M. Waehlert, *Metall u. Erz*, **10**, 1913, 578–586.
11. S. Briesemeister, *Z. Metallkunde*, **23**, 1931, 226–228.
12. R. E. Bish, *Trans. AIME*, **194**, 1952, 81–82.
13. B. Bogitch, *Compt. rend.*, **161**, 1915, 416–417.

14. W. Claus, *Z. Metallkunde,* **23,** 1934, 264–266; *Kolloid Z.*, **57,** 1931, 14–16.
15. W. Claus, *Metallwirtschaft,* **13,** 1934, 226–227.
16. C. T. Heycock and F. H. Neville, *J. Chem. Soc.*, **61,** 1892, 905.
17. E. Raub and A. Engel, *Z. Metallkunde,* **37,** 1946, 76–81.
18. J. N. Greenwood and C. W. Orr, *Proc. Australasian Inst. Mining & Met.*, **109,** 1938, 1.
19. E. Raub and A. Engel, *Z. Metallkunde,* **41,** 1950, 485–491.
20. H. Johnen, Dissertation, Universität Münster, Westfalen, Germany, 1952. See also W. Seith, H. Johnen, and J. Wagner, *Z. Metallkunde,* **46,** 1955, 773–779.
21. E. Scheil and E. Ruoff (Dissertation, E. Ruoff), Technische Hochschule, Stuttgart, 1952. Quoted in E. Gebhardt and W. Obrowski, *Z. Metallkunde,* **45,** 1954, 333.
22. E. Pelzel, *Metall,* **9,** 1955, 692–694.
23. E. Pelzel, *Metall,* **10,** 1956, 1023–1028.

$\bar{1}.7749$
0.2251

Cu-Pd Copper-Palladium

According to the solidification diagram (Fig. 351), based on data by [1, 2], Cu-Pd melts crystallize, forming a continuous series of solid solutions. This was confirmed by measurements of the electrical resistivity [3–5] and magnetic susceptibility [5] as well as X-ray studies [6, 7] of alloys, covering the entire range of composition, after quenching from temperatures above the disorder-order range.

At lower temperatures, ordered structures based on the theoretical compositions $PdCu_3$ (35.89 wt. % Pd) and PdCu (62.68 wt. % Pd) are formed. These transformations have been studied by numerous investigators, using thermal [8, 10, 2], thermoresistometric [4, 8, 9], electrical-resistivity-vs.-composition [11, 3, 4, 5, 8, 10], and X-ray methods [6, 12, 3, 4, 7, 9].

Transformation temperatures were determined by thermal analysis [8, 10, 2] and measurements of electrical resistivity vs. temperature [4, 8]. The most reliable temperatures are those given by [8] since they were found by thermoresistometric curves taken at low heating and cooling rates (low hysteresis). The data points shown in Fig. 351 were selected by [9] from the more complex results of [8].

For the transformation in the low Pd range, the temperature maximum was found to lie at about 17 [2, 13], 22 [8], or 25 at. % Pt [4, 10]. According to the interpretation by [9], the data by [8] justify placing the very flat maximum between 15 and 20 at. % Pt. The composition range of this transformation extends from slightly below 10 to somewhat above 25 at. % Pd but definitely not above 30 at. % Pd [3, 4, 5, 8, 9, 14).

The maximum temperature of the transformation involving the ordered structure PdCu was reported to lie at about 37 [2, 15], 40–41 [8], 47 [10], or 50 at. % Pd [4]. It appears to be well established that the composition does not coincide with the ideal composition PdCu but lies well below that concentration, close to 40 at. % Pd. As to the composition range, the majority of investigators agree that it does not extend beyond 50 at. % Pd [3, 4, 5, 16, 17, 8]. However, [9] reported that the alloy with 55 at. % Pd, after cooling from 500 to 250°C over a period of 4 months, contained approximately 20% ordered PdCu and 80% disordered α solid solution.

The maximum degree of order, as indicated by the intensity of the superlattice lines or the minimum value of electrical resistivity, was reported to lie at about 17 [4, 5] or 15 at. % Pd [8] for the $PdCu_3$ range and about 45 [4], 46.5 [17], or 47 at. % Pd [5, 8] for the PdCu region. The compositions assumed by [10], 25 and 50 at. % Pd, respectively, are certainly too high.

As for the rate of the disorder-order transformation and dependence upon composition and temperature, see the papers by [3, 18, 9, 19] for $PdCu_3$ and those by [3, 5, 17, 20, 9, 19] for PdCu. In general, the rate of ordering is greater, the lower the Pd content.

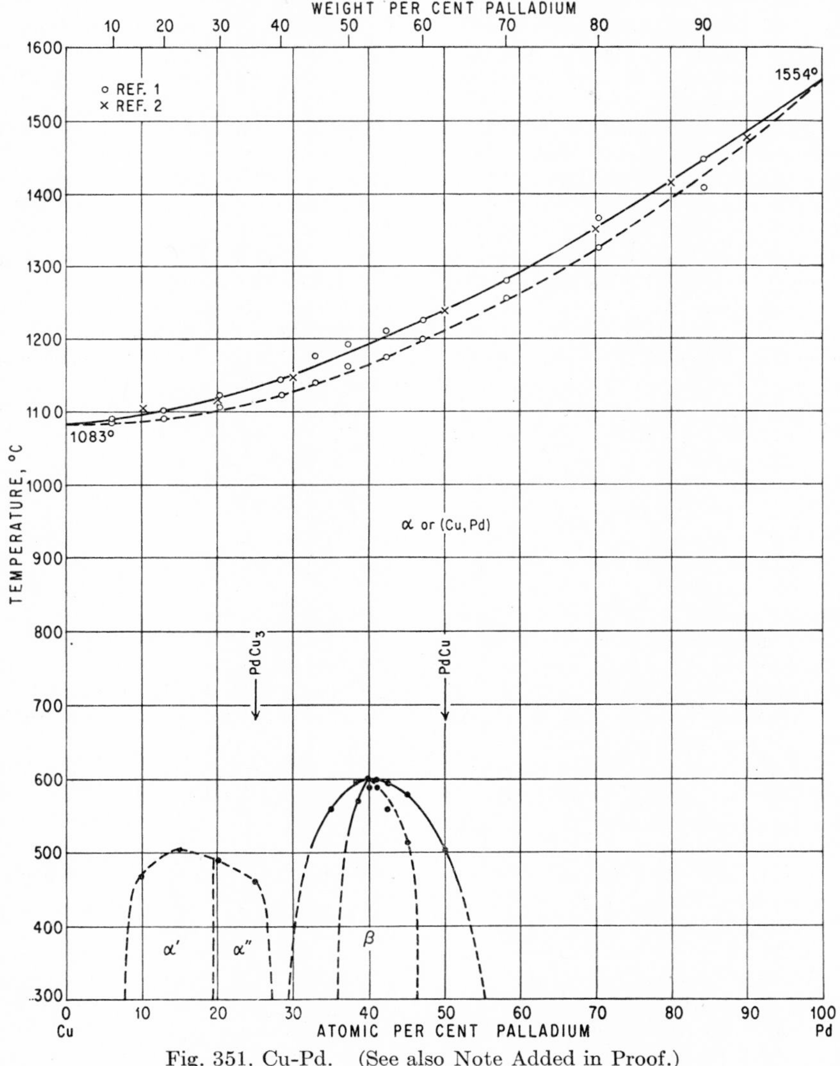

Fig. 351. Cu-Pd. (See also Note Added in Proof.)

Crystal Structure. Lattice parameters of a series of f.c.c. disordered α-phase alloys were reported by [6, 7]. For parameters of individual alloys, see [16, 20, 18].

Previously, the structure based on $PdCu_3$ was thought to be that of the Cu_3Au ($L1_2$) type within the entire range of about 10–25 at. % Pd [3, 4]. However, [9] reported that, on cooling below 460°C, alloys with 22, 24.4, and 25 at. % Pd showed the presence of a tetragonal structure, the amount of which increased progressively as the temperature was lowered. If the ordered structure is considered in terms of the original f.c.c. cell, the axial ratio would be $c/a = 0.986$ ($a = 3.714$ A), after slow cooling over 4 months to room temperature. A unit cell with 36 atoms and $a = 9.76$ A, $c = 7.33$ A was tentatively suggested. Alloys with 10–19 at. % Pd were confirmed to have the Cu_3Au type of lattice [7, 9] with the same parameter as that of the disordered state [7]. No indication of a two-phase transition within the entire $PdCu_3$ range was found [9, 19].

The disorder-order transformation in the region 30–50 at. % Pd is connected with a change in crystal structure, that of the ordered state being of the CsCl (B2) type [6, 12, 7, 9]; see also [20, 18]. As an example, the lattice parameter of the alloy with 45.5 at. % Pd was found as $a = 3.752$ A in the disordered and $a = 2.973$ A in the ordered state. Alloys with 38–45 at. % Pd were reported to consist of the b.c.c. structure only, after proper heat-treatment [3, 9]. According to [9], there are composition ranges in which the b.c.c. and f.c.c. structures coexist (Fig. 351).

Supplement. [21] took thermo-emf-vs.-temperature curves over the entire range of composition. According to an abstract, these curves consisted "in the regions of the chemical compounds $PdCu_5$ and Pd_3Cu_5" of three portions, corresponding to the ordered, disordering, and disordered structures. Temperatures of the beginning and ending of the ordering process (cooling rate?) are listed in the abstract; for the 25 at. % Pd alloy they are 490 and 365°C, respectively.

[22] studied the ordering reaction in the 20 at. % Pd ($PdCu_4$) alloy by means of X-ray and hardness measurements and microscopic observations. The transformation temperature was determined as 478°C; no evidence of a two-phase temperature interval was found. For the fully ordered phase, [22] proposed a tetragonal structure, $a = 5.826$ A, $c = 7.328$ A, with 20 atoms per unit cell.

[23], who studied several compositions within the α'' field, confirmed neither of the structures suggested by [9, 22]. Their results indicate a tetragonal unit cell which is based on a substructure cell of the Cu_3Au type. Whereas the a axis is that of the substructure, the c axis varies with composition, being eighteen times that of the substructure at 18.5 at. % and eight times at 25 at. % Pd (cf. the CuAu II structure).

In X-ray measurements at temperature and on quenched samples, [24] studied the variation of the degree of order with temperature for the $PdCu_3$ alloy. The transition cubic → tetragonal was observed at 460°C (in excellent agreement with Fig. 351); below 270°C the alloy was fully ordered. The axial ratio was taken as a measure of the degree of order; however, only the changes that took place in the main cubic lattice lines were considered.

Note Added in Proof. The antiphase domain structures of variable domain lengths reported by [23, 26] and [25, 27] for the α'' phase are in essential agreement. For structural details, reference has to be made to the original papers.

[28] measured various physical properties (lattice parameters, magnetic susceptibility, electrical resistivity, thermo-emf, and hardness) of alloys with 6–30 at. % Pd for a variety of heat-treatments and determined the effect of cold-work distortion on disordered and partially ordered samples. They found indication that ordering in this composition range is a classical phase change involving nucleation and growth and proposed new phase relations for the region of ordering. These are characterized by a eutectoid equilibrium between the ($PdCu_4$) and ($PdCu$) phases. The single phase field of ($PdCu_4$) is assumed to extend from about 10 to 29 at. % Pd [29] and up to 475°C. The work by [26], too, had shown that the Pd-rich boundary of the α'' phase in Fig. 351 is at too low a Pd content.

1. R. Ruer, *Z. anorg. Chem.*, **51,** 1906, 223–230.
2. V. A. Nemilov, A. A. Rudnitsky, and R. S. Polyakova, *Izvest. Sektora Platiny*, no. 24, 1949, pp. 26–34.
3. C. H. Johansson and J. O. Linde, *Ann. Physik*, **82,** 1927, 449–458.
4. G. Borelius, C. H. Johansson, and J. O. Linde, *Ann. Physik*, **86,** 1928, 299–318.
5. B. Swensson, *Ann. Physik*, **14,** 1932, 699–711.
6. S. Holgersson and E. Sedström, *Ann. Physik*, **75,** 1924, 150–162.
7. J. O. Linde, *Ann. Physik*, **15,** 1932, 249–251.
8. R. Taylor, *J. Inst. Metals*, **54,** 1934, 255–272.
9. F. W. Jones and C. Sykes, *J. Inst. Metals*, **65,** 1939, 419–433; see also A. J. Bradley, W. L. Bragg, and C. Sykes, *J. Iron Steel Inst.*, **13,** 1940, 121–142.
10. P. S. Belonogov, *Metallurg*, **11**(6), 1936, 92–95.
11. E. Sedström, Dissertation, Stockholm, 1924; see also *Ann. Physik*, **75,** 1924, 161; *Z. anorg. Chem.*, **159,** 1927, 339–341.
12. C. H. Johansson and J. O. Linde, *Ann. Physik*, **78,** 1925, 454–457.
13. Nemilov et al. assumed the phase $PdCu_5$ (16.67 at. % Pd).
14. H. J. Seemann, *Z. Physik*, **84,** 1933, 557–564.
15. Nemilov et al. assumed the phase Pd_3Cu_5 (37.5 at. % Pd).
16. H. Röhl, *Ann. Physik*, **18,** 1933, 155–168.
17. H. J. Seemann, *Z. Physik*, **88,** 1934, 14–24.
18. G. Rienäcker, G. Wessing, and G. Trautmann, *Z. anorg. Chem.*, **236,** 1938, 252–262.
19. A. Schneider, *Z. Elektrochem.*, **46,** 1940, 312–325.
20. L. Graf, *Physik. Z.*, **36,** 1935, 489–498.
21. A. A. Rudnitskii, *Izvest. Sektora Platiny*, **27,** 1952, 227–238. Not available to the author. *Chem. Abstr.*, **49,** 1955, 2143.
22. A. H. Geisler and J. B. Newkirk, *Trans. AIME*, **200,** 1954, 1076–1082. See also discussion, *Trans. AIME*, **203,** 1955, 710.
23. K. Schubert, B. Kiefer, and M. Wilkens, *Z. Naturforsch.*, **9a,** 1954, 987–988. See also discussion on [22].
24. D. M. Jones and E. A. Owen, *Proc. Phys. Soc. (London)*, **67B,** 1954, 297–303.
25. D. Watanabe, M. Hirabayashi, and S. Ogawa, *Acta Cryst.*, **8,** 1955, 510–512.
26. K. Schubert, B. Kiefer, M. Wilkens, and R. Haufler, *Z. Metallkunde*, **46,** 1955, 692–715.
27. D. Watanabe and S. Ogawa, *J. Phys. Soc. Japan*, **11,** 1956, 226–239; *Met. Abstr.*, **24,** 1956, 29.
28. F. E. Jaumot and A. Sawatzky, *Acta Met.*, **4,** 1956, 118–144.
29. In the light of the afore-mentioned structural findings this is certainly an over-simplification.

$\bar{1}.6541$
0.3459

Cu-Pr Copper-Praseodymium

Figure 352 shows the phase diagram which is based on thermal analysis and microscopic examination of alloys cooled from the molten state. Solubilities in the solid state have not been investigated [1]. The compositions of the four intermediate phases in wt. % Pr are as follows: $PrCu_6$, 26.99; $PrCu_4$, 35.67; $PrCu_2$, 52.58; and PrCu, 68.92.

1. C. Canneri, *Metallurgia ital.*, **26,** 1934, 869–871.

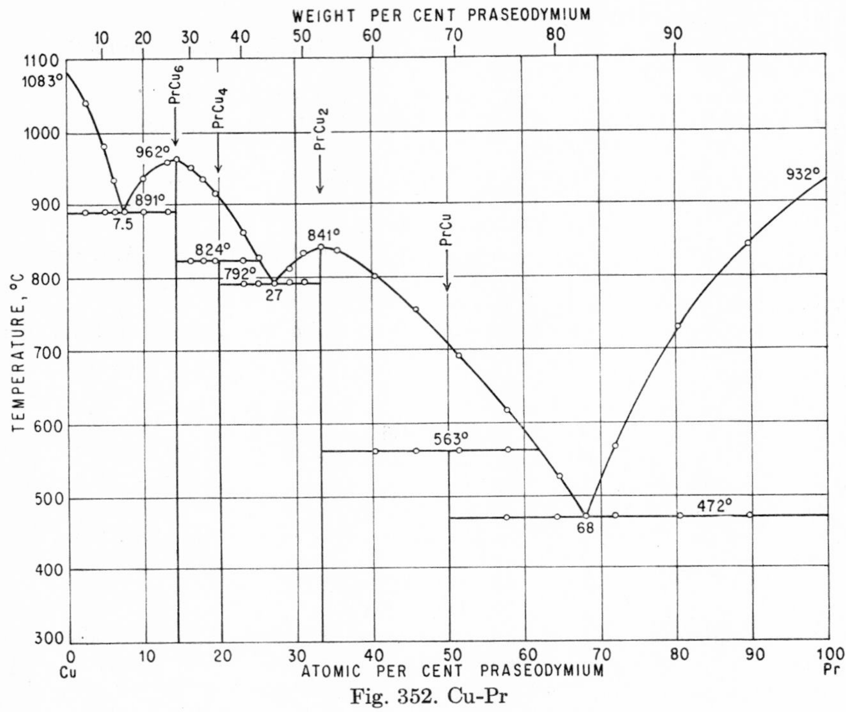

Fig. 352. Cu-Pr

Ī.5125
0.4875

Cu-Pt Copper-Platinum

The solidification diagram (Fig. 353) is established only up to 43 at. (70 wt.) % Pt [1]. However, measurements of the electrical conductivity (or resistivity) [2–4], temperature coefficient of resistance [4], and lattice parameters [2, 3, 5] of alloys, covering the entire range of composition [6], which were annealed at and quenched from temperatures above 800°C prove that Cu and Pt form a continuous series of solid solutions at high temperatures. Electrical-conductivity isotherms for temperatures of 800 and 900°C, derived from resistivity-vs.-temperature curves, verify this [5].

As Fig. 353 shows, order-disorder transformations take place in a very wide range of composition [7]. These changes were detected by [2], who, on the basis of measurements of electrical resistivity and X-ray investigations of alloys annealed at 400°C, distinguished between the ranges of approximately 10–26, 35–55, and 60–80 at. % Pt, which include the compositions $PtCu_3$, PtCu, and Pt_3Cu, respectively. [3] corroborated and supplemented these findings by similar studies, with emphasis on the crystal structure (see below). He proved that the composition range within which transformations occur extends from about 8.5 to above 90 at. % Pt with a gap around about 30 at. % Pt, as indicated by measurements of electrical resistivity after annealing at 300°C. [4] confirmed the transformations by measurements of electrical resistivity after various heat-treatments and determined, by thermal analysis, the transformation temperatures in the range 40–58 at. % Pt shown in Fig. 353.

By means of measurements of emf (vs. Cu) of alloys with 14–86 at. % Pt at tem-

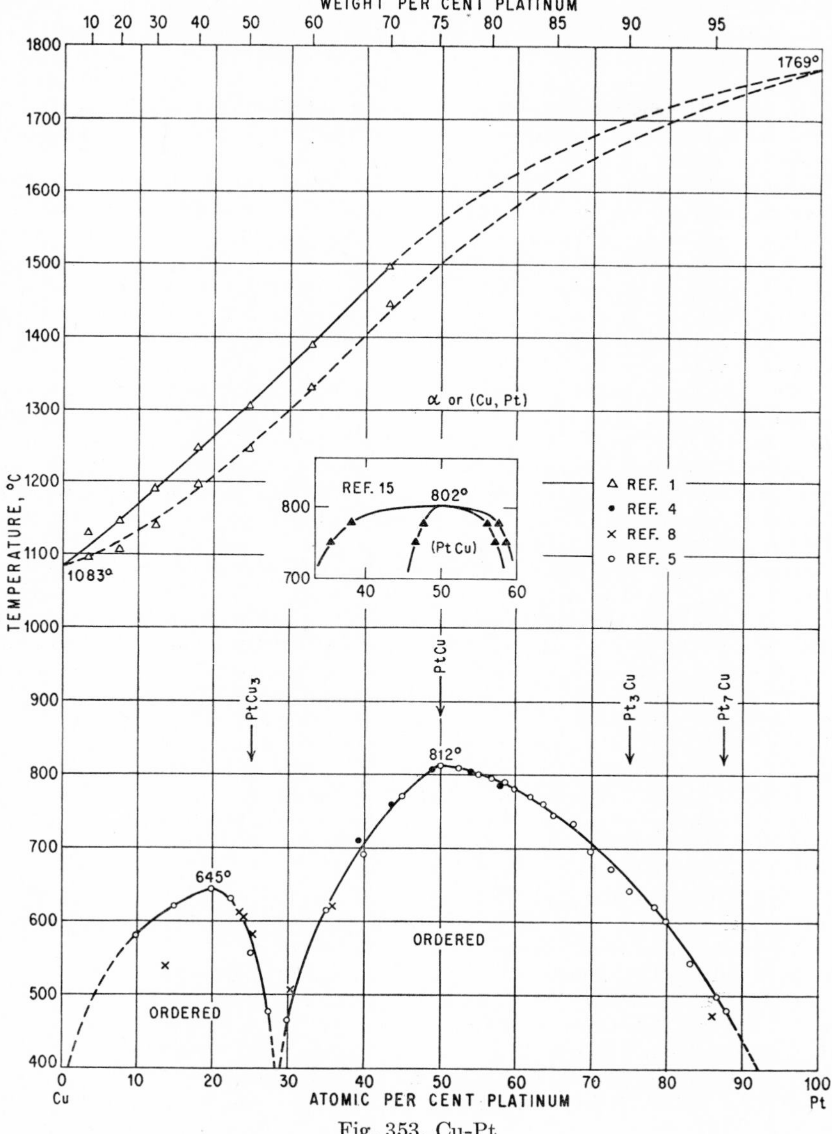

Fig. 353. Cu-Pt

peratures between about 400 and 650°C, [8] determined the transformation points indicated in Fig. 353. A maximum was found at about 24 at. % Pt, 615°C.

The most comprehensive investigation of the order-disorder change, covering the range 10–95.5 at. % Pt, is that by [5]. The transformation temperatures shown in Fig. 353 were obtained from electrical-resistivity-vs.-temperature curves taken on both heating and cooling. Electrical-conductivity isotherms indicated that the highest degree of order is attained at 22.5 (instead of 25 according to [3]), 50, 72.5, and 86 at. % Pt, corresponding to the ordered structures $PtCu_3$, PtCu, Pt_3Cu, and Pt_7Cu (87.5 at. % Pt).

Crystal Structures. X-ray investigations of a series of alloys [6] were carried out by [2, 3, 5], who reported lattice parameters of disordered and ordered alloys. $PtCu_3$ has a f.c.c. superstructure of the Cu_3Au ($L1_2$) type [2, 5] (see Supplement). The ordered phases PtCu, Pt_3Cu, and Pt_7Cu between 30 and 90 at. % Pt with maximal ordering at 50, 72.5, and 86 at. % Pt are similar in structure with a unit cell containing 32 atoms and a parameter twice as large as that of the disordered structure. The transition between the three structures is continuous. At the composition PtCu the unit cell is, for purely geometrical reasons, deformed in the direction of the space diagonal, resulting in a trigonal structure ($L1_1$ type) [5, 2]; see also [9]. This deformation exists only in the region 35–62 at. % Pt [5]. The ordered structure with the maximal ordering, Pt_5Cu_3 [2], could not be confirmed [5] by either X-ray or electrical-resistivity investigations.

The maximal ordering in the vicinity of the composition Pt_3Cu originates from uniform distribution of excess Pt atoms in the PtCu structure, resulting in a pseudocubic (orthorhombic) unit cell [5]. A new type of f.c.c. ordered structure (32 atoms per unit cell) for Pt_3Cu, reported to be stable in the range 63–88 at. % Pt, was suggested by [10]. The existence of the structure Pt_7Cu could not be definitely proved by X-ray investigation; a f.c.c. structure was suggested [5].

For additional information see [11–14].

Supplement. According to [15], the formation of ordered (PtCu) is a first-order reaction. The boundaries of the two-phase fields as obtained from the measurement of equilibrium pressures of oxidation are shown in the inset of Fig. 353. In alloys with 24–26 at. % Pt which had been annealed at 570 or 430°C, [16] observed a distortion of the $L1_2$-type structure stable at lower Pt contents. The distorted form exhibits a regular disposition of atomic ordering faults (antiphase domains).

1. F. Doerinckel, *Z. anorg. Chem.*, **54,** 1907, 335–338.
2. C. H. Johansson and J. O. Linde, *Ann. Physik*, **82,** 1927, 459–477.
3. J. O. Linde, *Ann. Physik,* **30,** 1937, 151–164.
4. N. S. Kurnakow and V. A. Nemilow, *Z. anorg. Chem.*, **210,** 1933, 1–12.
5. A. Schneider and U. Esch, *Z. Elektrochem.*, **50,** 1944, 290–301.
6. Studies of individual alloys are not considered.
7. E. Sedström (Dissertation, Stockholm, 1924) determined the electrical and thermal conductivity and thermo-emf of alloys with up to 40 at. % Pt and concluded that a miscibility gap exists between 10 and 20 at. % Pt.
8. F. Weibke and H. Matthes, *Z. Elektrochem.*, **47,** 1941, 421–432.
9. C. B. Walker, *J. Appl. Phys.*, **23,** 1952, 118–123.
10. Y. C. Tang, *Acta Cryst.*, **4,** 1951, 377–378.
11. G. Rienäcker and H. Hildebrandt, *Z. anorg. Chem.*, **248,** 1941, 52–64.
12. V. G. Kuznetsov, *Izvest. Sektora Fiz.-Khim. Anal.*, **16,** 1948, 150–167.
13. H. Sato, *Science Repts. Research Inst. Tôhoku Univ.*, **A4,** 1952, 160–163.
14. G. Fournet, *Compt. rend.*, **235,** 1952, 1377–1379.
15. P. Assayag and M. Dodé, *Compt. rend.*, **239,** 1954, 762–764.
16. K. Schubert, B. Kiefer, M. Wilkens, and R. Haufler, *Z. Metallkunde*, **46,** 1955, 692–715, especially 699.

Ī.4246
0.5754

Cu-Pu Copper-Plutonium

According to metallographic and X-ray diffraction work, three compounds exist to which the tentative formulas PuCu, $PuCu_3$, and $PuCu_7$ have been assigned [1].

1. Work at Los Alamos, quoted in "Metallurgy and Fuels," Progress in Nuclear Energy, ser. V, vol. 1, pp. 392, 399, Pergamon Press Ltd., London, 1956.

Ī.5328
0.4672

Cu-Re Copper-Rhenium

Several attempts to alloy Re and Cu at temperatures up to 2175°C were unsuccessful [1].

1. U. Holland-Nell and F. Sauerwald, *Z. anorg. Chem.*, **276**, 1954, 155–158.

Ī.7906
0.2094

Cu-Rh Copper-Rhodium

On the basis of the limited thermal, microscopic, X-ray, and hardness data reported by [1], it is not possible to arrive at a clear conception of the constitution of

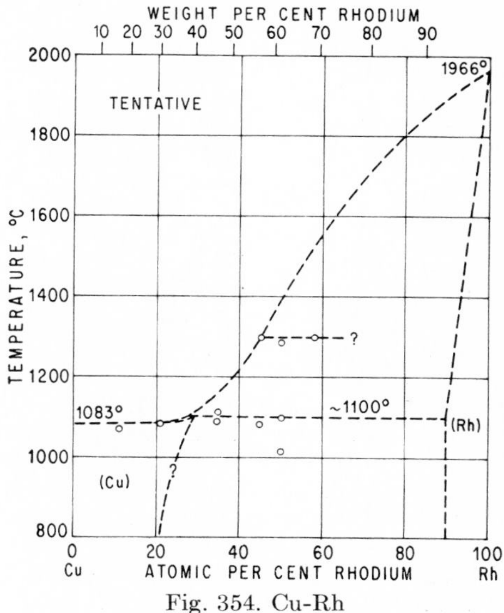

Fig. 354. Cu-Rh

this system. Very likely, the alloys investigated were not near the state of equilibrium since they were annealed for only relatively short times. However, it appears to be established that the components form primary solid solutions with approximately 0–20 and 90–100 at. % Rh. From measurements of hardness, X-ray, and other data of specimens annealed at 750–800°C for 72 hr, the author concluded that ordered phases occur at 50 and 75 at. % and possibly 25 at. % Rh. The structure of all alloys is reported to be f.c.c., and those corresponding to RhCu and Rh_3Cu are assumed to have practically the same lattice constant as the Rh-rich solid solution.

Perhaps the phase diagram is somewhat similar to that of the system Ag-Pt. Figure 354 shows the thermal-data points reported.

1. O. E. Svyagintsev and B. K. Brunovskiy, *Izvest. Sektora Platiny*, **12**, 1935, 37–66.

$\bar{1}.7957$
0.2043

Cu-Ru Copper-Ruthenium

Measurements of electrical resistance of Cu-rich alloys indicate that the solid solubility of Ru in Cu is "vanishingly small" (after annealing at 900°C) [1].

1. J. O. Linde, *Ann. Physik*, **15**, 1932, 231.

0.2970
$\bar{1}.7030$

Cu-S Copper-Sulfur*

The Partial System Cu-Cu_2S (20.15 Wt. % S) [1]. The existence of a monotectic and eutectic equilibrium between Cu and Cu_2S was first established by [2]. The eutectic point was placed at 0.77 wt. (1.51 at.) % S [2] and 0.88 wt. (1.73 at.) % S [3]. The eutectic temperature was reported as 1067 [2, 4, 5] and 1070°C [3] and the monotectic temperature as 1102 [2], 1105 [5], 1117 [6], and 1121°C [3].

The Cu-rich boundary of the region of liquid immiscibility was determined, using chemical analysis of liquid samples [7] and thermoresistometric studies [4]. Results indicate the composition of the melt at the monotectic temperature to be approximately 1.4 wt. (2.7 at.) % S [7] or 1.8 wt. (3.5 at.) % S [4]. The latter value agrees with that of [2], based on thermal-analysis data. It should be stated, however, that the S-rich boundary of the liquid miscibility gap according to [4] is greatly at variance with all other data (Fig. 355). In addition, [8] reported the solubility of S in liquid Cu at about 1200°C to be only 1.55 wt. (3.0 at.) % S, in good agreement with the results of [7].

The monotectic point was determined as 17.1 [2], 17.9 [4], about 19.6 [7], 19.8 [9, 5], and 19.8–19.9 wt. (32.85–32.99 at.) % S [10]. The excellent agreement of the thermal data of [5] and the S-rich boundary of the liquid miscibility gap, reliably determined by [10], is remarkable.

The melting point of Cu_2S was reported as 1130–1131°C [11, 12, 13]; for data of other workers, see [5, 11]. According to the accurate work of [5], the liquidus temperature of the melt of the composition Cu_2S is 1127°C and of that containing an excess of 0.10 wt. % S is 1129°C. Results indicate that the composition Cu_2S melts in an interval of 20°C.

The solid solubility of S in Cu was already noted by [2] to be very small. According to measurements of electrical resistance it is about 0.0005 wt. % S at 600°C and about 0.002 wt. (0.004 at.) % S at 800°C [14].

It has long been known that Cu_2S undergoes a transformation at 100 ± 3°C [15]. This was confirmed by numerous investigators whose data were critically reviewed by [16]. The crystal structures below and above the transformation temperature (105°C) were determined by [17]. The existence of an additional transformation at a higher temperature, suggested by [18], could not be confirmed [11, 19, 16].

The Partial System Cu_2S-CuS (33.54 Wt. % S). This system was the subject of numerous investigations by mineralogists and crystallographers. Phase relationships are still obscure, although the work of [16, 17] has much contributed to

*The assistance of Dr. F. Wolstein in preparing this system is gratefully acknowledged.

clear up the discrepancies as to the phases existing [11, 20–22] and their crystal structures [23–27, 19].

By means of high-temperature X-ray work, [16] has conclusively established that besides two polymorphic forms of Cu_2S (chalcocite), both of which are of variable composition, there is another intermediate phase of the approximate composition Cu_9S_5 [21.90 wt. (35.71 at.) % S], called digenite (Fig. 355). This phase probably also undergoes a transformation at about 78°C [16] or decomposes at this temperature into Cu_2S and CuS.

The solidification of melts containing up to 23 wt. (37.2 at.) % S was studied by [28, 11, 13, 5]. The concordant results of [13, 5] show that these melts solidify in a

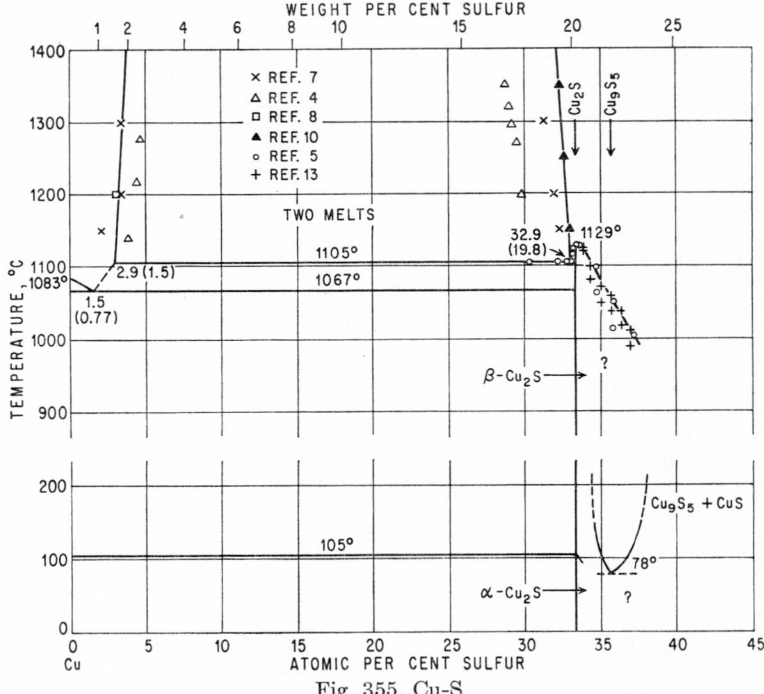

Fig. 355. Cu-S

temperature interval corresponding to the formation of solid solutions, presumably based on β-Cu_2S and Cu_9S_5.

Crystal Structures. β-Cu_2S, the high-temperature form of Cu_2S [29], is hexagonal, with a = 3.89 A, c = 6.68 A, c/a = 1.717, and 6 atoms per unit cell [17]; see also [30]. Previously, the cubic phase Cu_9S_5 was believed to be β-Cu_2S [23, 24, 27]; see also [31]. α-Cu_2S is orthorhombic [26, 27, 19, 17], with a = 11.90 A, b = 27.28 A, c = 13.41 A, and 96 formula weights per unit cell [17]. For additional data, see [32, 33].

The phase Cu_9S_5 (digenite) has a f.c.c. structure somewhat similar to that of the CaF_2 (C1) type [23, 24, 25, 19, 17], with a = 5.575 A at 170°C [19]; see also [34, 35].

The structure of CuS (covellite) is the prototype of the hexagonal C18 type of structure [26, 36], with a = 3.75 A, c = 16.2_3 A, c/a = 4.32, and 12 atoms per unit cell [36]; see also [37].

1. The papers of F. H. Edwards, *Trans. Inst. Min. Met. (London)*, **33**, 1924, 492–510, and G. Peÿronel and E. Pacilli, *Atti reale accad. Italia, Mem. classe sci. fis. mat. e nat.*, **14**, 1943, 203–223, were not available.
2. E. Heyn and O. Bauer, *Metallurgie*, **3**, 1906, 73–82.
3. G. G. Urasow, *Ann. inst. polytechn. Petrograd*, **23**, 1915, 593–627 (not available).
4. K. Bornemann and K. Wagenmann, *Ferrum*, **11**, 1913-1914, 276–282, 293–294, 303, 306, 310–313, 331–332.
5. E. Jensen, *Avhandl. Norske Videnskaps-Akad. Oslo, I. Mat.-Naturv. Kl.*, no. 6, 1947.
6. H. Schlegel and A. Schüller, *Z. Metallkunde*, **43**, 1952, 412–428.
7. K. Friedrich and M. Waehlert, *Metall u. Erz*, **10**, 1913, 976–979.
8. C. S. Smith, *Trans. AIME*, **128**, 1938, 325–334.
9. G. G. Urasow [3], quoted from [5].
10. R. Schuhmann and O. W. Moles, *Trans. AIME*, **191**, 1951, 235–241.
11. E. Posnjak, E. T. Allen, and H. E. Merwin, *Econ. Geol.*, **10**, 1915, 491–535; *Z. anorg. Chem.*, **94**, 1916, 95–138.
12. F. H. Edwards, *Trans. Inst. Min. Met. (London)*, **33**, 1924, 492–510.
13. J. G. Juokoff, *Metall u. Erz*, **26**, 1929, 137–141.
14. J. S. Smart and A. A. Smith, *Trans. AIME*, **166**, 1946, 144–155.
15. W. Hittorf, *Pogg. Ann.*, **84**, 1851, 1–28.
16. N. W. Buerger, *Econ. Geol.*, **36**, 1941, 19–44.
17. M. J. Buerger and N. W. Buerger, *Am. Mineralogist*, **27**, 1942, 216–217; **29**, 1944, 55–65; *Structure Repts.*, **12**, 1949, 156.
18. W. Mönch, *Neues Jahrb. Mineral. Beilage Bd.*, **20**, 1905, 365–435.
19. P. Rahlfs, *Z. physik. Chem.*, **B31**, 1936, 184–193.
20. A. M. Bateman, *Econ. Geol.*, **24**, 1929, 424–439.
21. W. Biltz and R. Juza, *Z. anorg. Chem.*, **190**, 1930, 173–176.
22. A. M. Bateman and S. G. Lasky, *Econ. Geol.*, **27**, 1932, 52–86.
23. T. Barth, *Zentr. Mineral. Geol. Paläontol.*, **A1926**, 284–286.
24. L. C. Ramsdell, *Am. Mineralogist*, **13**, 1928, 115.
25. G. P. Thomson, *Proc. Roy. Soc. (London)*, **A128**, 1930, 649–661.
26. N. Alsén, *Geol. Fören. i. Stockholm Förh.*, **53**, 1931, 111–120.
27. W. Kurz, *Z. Krist.*, **92**, 1935, 408–434.
28. K. Friedrich, *Metallurgie*, **5**, 1908, 52–53.
29. Rahlfs [19] called this modification α-Cu_2S.
30. R. Molé, *Ann. chim. (Paris)*, **9**, 1954, 145–180.
31. E. Hirahara, *J. Phys. Soc. Japan*, **2**, 1947, 211–213; **4**, 1949, 10–12.
32. R. Ueda, *J. Phys. Soc. Japan*, **4**, 1949, 287–292.
33. H. Wilman and A. P. B. Sinha, *Acta Cryst.*, **7**, 1954, 682–683.
34. P. Ramdohr, *Z. prakt. Geol.*, **51**, 1943, 1–9.
35. R. Hocart and R. Molé, *Compt. rend.*, **228**, 1949, 1138–1139.
36. I. Oftedal, *Z. Krist.*, **83**, 1932, 9–25.
37. H. S. Roberts and C. J. Ksanda, *Am. J. Sci.*, **17**, 1929, 489–503.

$\bar{1}.7176$
0.2824

Cu-Sb Copper-Antimony

Solidification. The liquidus and solidus curves of the whole system or of parts of it were determined by [1–11]. In Fig. 356 the data by [10, 11], obtained mainly by thermal and resistometric analyses, were used. The liquidus exhibits one maximum, one peritectic, and two eutectic points. The temperature and composition ranges, which cover the data reported by various authors for these characteristic points, may be seen from Table 25, which also considers the results of microscopic determinations of the eutectic compositions by [12, 13].

Table 25

(Cu) + β eutectic	29.5–32 wt. (18–19.7 at.) % Sb, 620–646°C
Maximum in the liquidus curve	40–45 wt. (25.8–30 at.) % Sb, 650–695°C
Peritectic (liquidus point, temperature)	60–62 wt. (44–46 at.) % Sb, 580–586°C
Cu_2Sb + Sb eutectic	75–76.5 wt. (61–63 at.) % Sb, 470–530°C

[3, 6] suggested the maximum in the liquidus to lie at the composition Cu_3Sb (38.98 wt. % Sb); however, the best experimental data point to a value of about 43 wt. (28.5 at.) % Sb, which is close to the composition Cu_5Sb_2 [28.57 at. (43.38 wt.) % Sb].

Solubility of Sb in Solid Cu. Figure 357 shows the results of more recent solubility determinations: [14], micrographic; [15], X-ray; [16], dilatometric, micrographic; [17–19], X-ray. The four crosses (×) in Fig. 357 indicate the general trend of the solubility curve established by [16, 17]; all other markers plotted constitute actual measurement data. The solubility curve chosen in the main diagram (Fig. 356) is also shown in Fig. 357; it follows closely the data by [15, 18, 19], [20]. Insufficient annealing times and an oxygen content of their alloys appear to be responsible for the apparent high solubilities found by [14, 16, 17] below 300°C. The following solubility determinations have mainly historical interest, and their results need not be considered here: conductometric [22, 23, 24, 25, 26, 27], microscopic [28, 13, 4, 6, 26, 10], X-ray parametric [29, 30].

Intermediate Phases. (*a*) 15–31 *At.* (25–46 *Wt.*) % Sb. For a detailed discussion of the early literature (up to 1930) on the constitution in this composition range, reference is made to the review by [31] which covers the papers by [22, 13, 2–6, 32, 7, 8, 33–35, 29, 9, 25, 30, 25*a*]. The phase diagram recommended by [31] showed a β high-temperature phase with a maximal extent of 32–48 wt. (20–32.5 at.) % Sb which transformed between 450 (Cu-rich) and 420°C (Sb-rich) into the ε room-temperature phase, homogeneous between about 31 and 39 wt. (19 and 25 at.) % Sb. The transformation equilibria shown in that phase diagram were based on the results of [6, 8, 9]. With the exception of a part of the Sb-rich β-phase boundary determined by [6, 8, 9], the limits of the β and ε phases were not considered by the reviewer [31] to have been reliably established.

At variance with the results of many previous investigators, [36] found by means of microscopic work in the assumed ε-phase region that long-annealed alloys containing 33–39 wt. (20.5–25 at.) % Sb are actually two-phase. Japanese investigators reexamined the entire phase diagram by means of thermal, microscopic, resistometric, dilatometric, and X-ray analysis. Their results were published in a series of papers [10, 16, 11, 17, 37], with the final diagram in [11]. As regards the composition range under discussion, four solid phases were claimed to exist: two high-temperature phases, η ("$Cu_{11}Sb_2$") and β (Fig. 356), and two phases stable at room temperature, ε (called δ by the Japanese authors), with a homogeneity range between 30 and 32 wt. (18.3 and 19.7 at.) % Sb, and θ ("$Cu_{11}Sb_4$"), homogeneous between about 39 and 40 wt. (25 and 26 at.) % Sb at normal temperature. Of these phases, only θ was not corroborated by later work.

[38] studied alloys with 30–40 wt. % Sb by means of microscopic and X-ray methods. According to their findings, the Cu-rich limit of the β phase is located at 35 wt. (22 at.) % Sb at 550°C, and the ϵ phase is homogeneous between 32.1 and 33.9 wt. (19.8 and 21.1 at.) % Sb below 300°C.

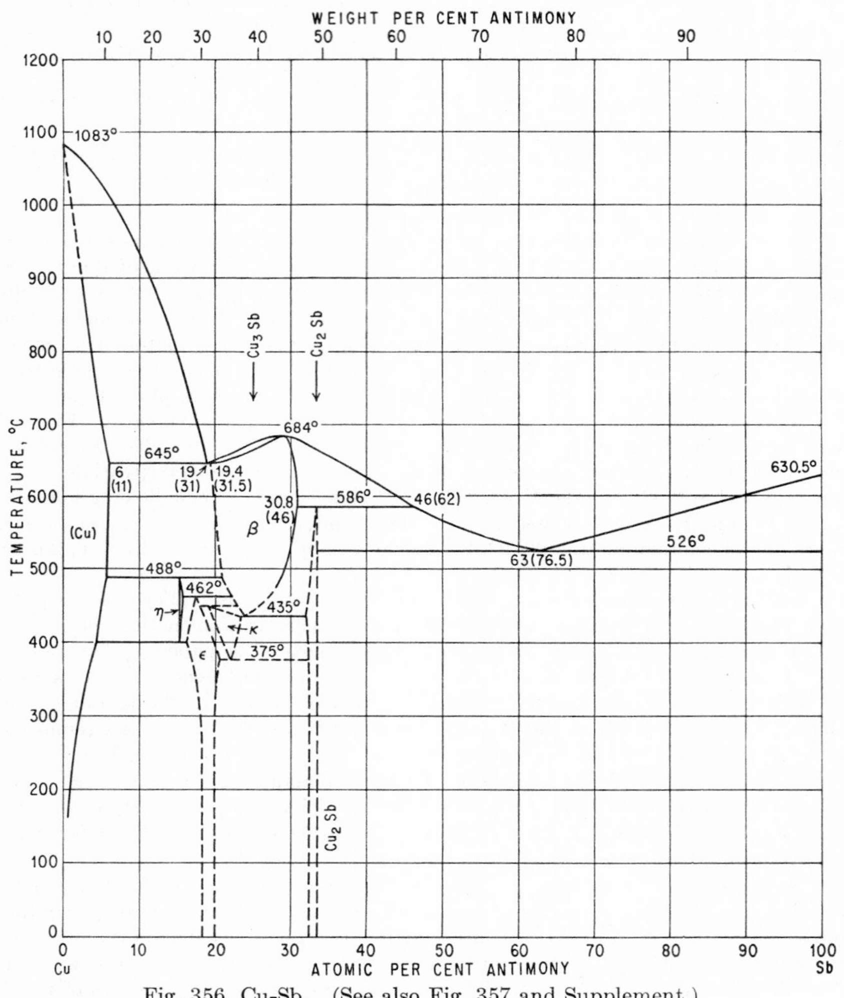

Fig. 356. Cu-Sb. (See also Fig. 357 and Supplement.)

[21] stated that their microscopic observations were at variance with the diagram worked out by the Japanese authors and published a preliminary diagram which shows no η or θ phase, but—besides β—a broad ϵ-phase field. From supporting X-ray work, [39] also concluded that there is no necessity for replacing the extended ϵ-phase field with a group of separated phases. [40], who carried out an X-ray, dilatometric, and microscopic investigation, also were unable to confirm the θ phase claimed by the Japanese workers. However, they corroborated the existence of the

η phase and found—by means of high-temperature X-ray as well as dilatometric work—besides η, β, and ϵ an additional intermediate phase, called κ (kappa). The limits of this phase in Fig. 356 are only tentative since no detailed study of the extent of the κ-phase field was made by [40].

The β-phase field shown in Fig. 356 is essentially that of [11]. They found the value 36.5 wt. % Sb for the eutectoid composition; other values reported are 41 wt. % Sb [31], 38 wt. (24 at.) % Sb [38] (Fig. 356), and ~39 wt. % Sb [21].

(*b*) Cu_2Sb. The formula Cu_2Sb (48.83 wt. % Sb) for the intermediate phase richest in Sb has been established by numerous investigators [22, 12, 13, 3, 4, 6, 32, 7,

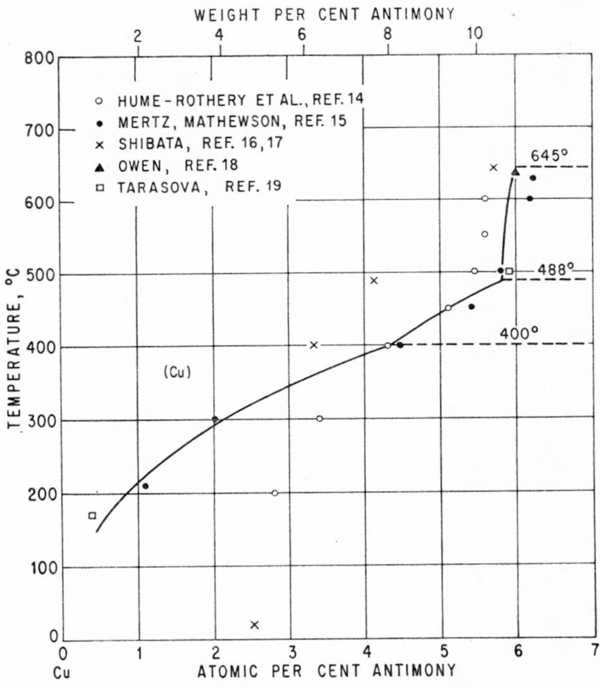

Fig. 357. Cu-Sb. (See also Fig. 356.)

35, 41, 29, 25, 9, 30, 42, 10, 11, 17, 21]. According to [10, 11, 17], this phase has a narrow homogeneity range between 48 and 49 wt. (32.5 and 33.4 at.) % Sb.

Solubility of Cu in Solid Sb. [30, 17, 43] have found that addition of Cu does not change the lattice dimensions of Sb. Therefore, the solubility of Cu in Sb appears to be negligible. Other investigating methods yielded the following results (solubility in wt. % Cu): microscopic <0.5 [13], 0.5 at 400°C [6], 0.2 at 526°C, and 0.1 at room temperature [10]; electrical properties 0 [22, 25]; magnetic 1.4 [44].

Crystal Structures (see also Supplement). The lattice parameter of Cu is increased by addition of Sb [29, 30, 15, 17, 18, 43]. The η phase, being homogeneous between 25.9 and 26.5 wt. (15.4 and 15.8 at.) % Sb at 462°C [11], has a h.c.p. (A3) structure, according to X-ray work at high temperature [40] and on quenched alloys [39, 40]. The lattice parameters for the 15.4 at. % Sb alloy at 480°C are $a = 2.71_0$ A, $c = 4.36_6$ A, $c/a = 1.61_1$ [40]. The orthorhombic (distorted hexagonal) structure reported by [17] was not corroborated.

Whereas several authors [35, 29, 30, 38, 21] ascribed the simple h.c.p. (A3) structure to the phase named ϵ in Fig. 356, the results of others suggest a more complex structure, probably a superstructure of the h.c.p. type. Thus, [17] reported the lattice parameters a = 10.836 kX, c = 8.611 kX, c/a = 0.7947, and a unit cell containing 54 atoms; [39] found a h.c.p. structure with a multiplied a axis when he investigated primary ϵ crystals separated out of a Pb-rich ternary melt; [43, 40] noted the occurrence of some lines in addition to those of the h.c.p. structure.

The κ (kappa) phase, stable between about 450 and 375°C, has, according to high-temperature X-ray patterns, an orthorhombic (distorted h.c.p.) structure with the lattice parameters $a = 2.77_4$ A, $b = 4.78_0$ A, $c = 4.35_6$ A [40].

The β high-temperature phase possesses a superstructure of the b.c.c. lattice, the f.c.c. BiF_3 ($D0_3$) type [38, 45]. Its lattice parameter—a = 6.01 A at the composition Cu_3Sb and 475°C [45]—decreases with increasing Sb content [38]. Previously, a b.c.c. [46, 39] or b.c. tetragonal [17] structure had been reported. Whereas [45] used high-temperature X-ray methods in evaluating the structure of β, [38] reached the same result in working with quenched alloys. This is worth noting, since several investigators [29, 30, 17, 37, 21, 39] have reported that the β phase cannot be retained by quenching. [37] observed metastable cubic and tetragonal phases when β alloys were quenched and tempered.

The structure of Cu_2Sb is tetragonal, with 6 atoms per unit cell [29, 30, 42], and forms the C38 prototype of structure [42], a = 4.000 A, c = 6.103 A, c/a = 1.525 [42]. More recently, [17] claimed to have found additional lines in the powder pattern which would make it necessary to enlarge the a axis by the factor $\sqrt{2}$.

Further Investigations. Thermal dilatation [47, 48], density [49, 50, 25, 17], emf [3]. The constitution of galvanic Cu-Sb alloy deposits was studied by [43].

Supplement. [51] studied, on thin films obtained by vacuum deposition, the formation and structures of Cu-Sb intermediate phases, as well as their transformations on heating and cooling, by means of a continuously recording electron-diffraction camera. They corroborated the simple h.c.p. (A3) structure for η, the f.c.c. $D0_3$ (or $L2_1$) type of structure for β [52], the superstructure cell for ϵ (called δ by [51]) of [17] (doubling of the c axis, quadrupling of the a axis of a simple h.c.p. cell), and the lattice parameters a = 4.00 A, c/a = 1.525 for Cu_2Sb. As regards the κ phase, [51] consider it not to be separated from ϵ by a two-phase field but to be the fully ordered version of the partially ordered hexagonal ϵ phase.

The eutectoid reaction $\kappa \rightleftarrows \epsilon + Cu_2Sb$ at 375°C (Fig. 356), denied by [51], has further support from microscopic and dilatometric work by [53]. These authors showed that the eutectoid decomposition is a sluggish reaction, which fact may explain why [51] failed to observe it. [53] found clear microscopic evidence of an ($\epsilon + \kappa$) two-phase field and determined the following phase boundaries at 390°C, in wt. % Sb (at. % in parentheses): ϵ, Sb-rich 31.6 (19.4); κ, 37.9–39.6 (24.2–25.5); Cu_2Sb, Cu-rich 49.5 (33.8).

Whereas [40] assumed tentatively the reaction $\beta \rightleftarrows \kappa + Cu_2Sb$ (Fig. 356), [53] found microscopic evidence that β decomposes eutectoidally at 436°C according to β (35.5 wt. % Sb) $\rightleftarrows \epsilon + \kappa$.

[54] stated briefly that a new phase with an A3 substructure is homogeneous at 21.5 at. % Sb and 350°C.

1. H. Le Chatelier, *Bull. soc. encour. ind. natl.*, **10,** 1895, 569; Société d'encouragement pour l'industrie nationale, Paris, Commission des alliages, "Contribution à l'étude des alliages," p. 394, Typ. Chamerot et Renouard, Paris, 1901. See also H. Gautier, *Bull. soc. encour. ind. natl.*, **1,** 1896, 1300; "Contribution à l'étude des alliages," pp. 99–100.

2. A. Stansfield, quoted by W. Campbell, *J. Franklin Inst.*, **154,** 1902, 209.
3. A. Baikov, *Zhur. Russ. Fiz.-Khim. Obshchestva*, **36,** 1904, 111–165; see also *Bull. soc. encour. ind. natl.*, **102,** 1903, 626.
4. A. H. Hiorns, *J. Soc. Chem. Ind.*, **25,** 1906, 617.
5. N. Parravano and E. Viviani, *Atti reale accad. Lincei*, **19,** 1910, 838–840.
6. H. C. H. Carpenter, *Intern. Z. Metallog.*, **4,** 1913, 300–321.
7. N. S. Kurnakov and K. F. Beloglazov, *Zhur. Russ. Fiz.-Khim. Obshchestva*, **48,** 1916, 700–701 (not available to the author). Abstract in *J. Inst. Metals*, **16,** 1916, 237–238.
8. H. Reimann, *Z. Metallkunde*, **12,** 1920, 321–331.
9. M. Tasaki, *Mem. Coll. Eng. Kyoto Imp. Univ.*, **12,** 1929, 230, 249.
10. T. Murakami and N. Shibata, *Science Repts. Tôhoku Imp. Univ.*, **25,** 1936, 527–568 (in English) (I. Thermal analysis and microscopic examination).
11. T. Murakami and N. Shibata, *Science Repts. Tôhoku Imp. Univ.*, **27,** 1938, 459–484 (in English) (III. Transformations in the solid state).
12. G. Charpy, *Compt. rend.*, **124,** 1897, 957–958. *Bull. soc. encour. ind. natl.*, **2,** 1897, 397–401; Société d'encouragement pour l'industrie nationale, Paris, Commission des alliages, "Contribution à l'étude des alliages," pp. 134–137, Typ. Chamerot et Renouard, Paris, 1901.
13. J. E. Stead, *J. Soc. Chem. Ind.*, **17,** 1898, 1111–1116; see also W. Campbell, *J. Franklin Inst.*, **154,** 1902, 209–211.
14. W. Hume-Rothery, G. W. Mabbott, and K. M. Ch. Evans, *Phil. Trans. Roy. Soc. (London)*, **A233,** 1934, 1–97.
15. J. C. Mertz and C. H. Mathewson, *Trans. AIME*, **124,** 1937, 59–77.
16. N. Shibata, *Science Repts. Tôhoku Imp. Univ.*, **27,** 1938, 189–209 (in English) (II. The constitution of Cu-rich Sb-Cu alloys).
17. A. Osawa and N. Shibata, *Science Repts. Tôhoku Imp. Univ.*, **28,** 1939, 1–19 (in English) (IV. An X-ray investigation of Cu-Sb alloys).
18. E. A. Owen, *J. Inst. Metals*, **73,** 1947, 471–489.
19. V. P. Tarasova, *Vestnik. Moskov. Univ.*, no. 4, 1947, 105–107.
20. A. Schrader and H. Hanemann [21] stated briefly that their own X-ray results agreed well with those of [15].
21. A. Schrader and H. Hanemann, *Z. Metallkunde*, **33,** 1941, 49–60.
22. G. Kamensky, *Phil. Mag.*, **17,** 1884, 270. See also W. Guertler, *Z. anorg. Chem.*, **51,** 1906, 418–420.
23. L. Addicks, *Trans. AIME*, **36,** 1906, 18–27.
24. A. H. Hiorns and S. Lamb, *J. Soc. Chem. Ind.*, **28,** 1909, 453.
25. E. Stephens and E. J. Evans, *Phil. Mag.*, **7,** 1929, 161–176.
25a. J. O. Linde, *Ann. Physik*, **8,** 1931, 124–128.
26. S. L. Archbutt and W. E. Prytherch, *J. Inst. Metals*, **45,** 1931, 278–281.
27. J. O. Linde, *Ann. Physik*, **15,** 1932, 219–233.
28. J. Arnold and J. Jefferson, *Engineering*, **61,** 1896, 177.
29. A. Westgren, G. Hägg, and S. Eriksson, *Z. physik. Chem.*, **B4,** 1929, 453-468.
30. E. V. Howells and W. Morris-Jones, *Phil. Mag.*, **9,** 1930, 993–1014.
31. M. Hansen, "Der Aufbau der Zweistofflegierungen," Springer-Verlag OHG, Berlin, 1936.
32. N. S. Kurnakov, P. Nabereznov, and V. Ivanov, *Zhur. Russ. Fiz.-Khim. Obshchestva*, **48,** 1916, 701; abstract in *J. Inst. Metals*, **16,** 1916, 237.
33. N. S. Kurnakov and K. F. Beloglazov, see *J. Inst. Metals*, **29,** 1923, 637, and *Rev. mét.*, **19,** 1922, 588–589.
34. N. S. Kurnakov and K. F. Beloglazov, *Izvest. Inst. Fiz.-Khim. Anal.*, **2,** 1924, 490–492; abstract in *J. Inst. Metals*, **36,** 1926, 440.

35. W. Morris-Jones and E. J. Evans, *Phil. Mag.*, **4,** 1927, 1302–1311.
36. W. Guertler and W. Rosenthal, *Z. Metallkunde,* **24,** 1932, 32.
37. A. Osawa and N. Shibata, *Science Repts. Tôhoku Imp. Univ.*, **28,** 1939, 197–216 (in English) (Structural changes due to quenching and tempering of the β phase of the Cu-Sb system).
38. N. V. Ageev and E. S. Makarov, *Izvest. Sektora Fiz.-Khim. Anal.*, **13,** 1940, 171–176.
39. W. Hofmann, *Z. Metallkunde,* **33,** 1941, 61–62.
40. K. Schubert and M. Ilschner, *Z. Metallkunde,* **45,** 1954, 366–370.
41. W. G. Davies and E. S. Keeping, *Phil. Mag.*, **7,** 1929, 150–153.
42. M. Elander, G. Hägg, and A. Westgren, *Arkiv Kemi, Mineral. Geol.*, **B12**(1), 1936.
43. E. Raub, *Z. Erzbergb. u. Metallhüttenw.*, **5,** 1952, 153–160.
44. H. Endo, *Science Repts. Tôhoku Imp. Univ.*, **14,** 1925, 501.
45. W. Hofmann, *Z. Metallkunde,* **33,** 1941, 373.
46. A. Westgren and G. Phragmén, *Metallwirtschaft,* **7,** 1928, 701.
47. H. Le Chatelier, *Compt. rend.*, **128,** 1899, 1444.
48. P. Braesco, *Compt. rend.*, **170,** 1920, 103–105.
49. G. Kamensky, *Proc. Phys. Soc. (London)*, **6,** 1883, 53.
50. E. Maey, *Z. physik. Chem.*, **50,** 1905, 204–206.
51. A. Boettcher and R. Thun, *Z. anorg. Chem.*, **283,** 1956, 26–48.
52. Occasionally noncubic superstructure lines were observed.
53. T. Heumann and F. Heinemann, *Z. Elektrochem.*, **60,** 1956, 1160–1169.
54. K. Schubert et al., *Naturwissenschaften,* **44,** 1957, 229–230.

$\bar{1}.9056$
0.0944

Cu-Se Copper-Selenium

The data points shown in Fig. 358 are due to [1], who studied the system by thermal analysis and microscopic examination in the range up to 43.6 wt. (38.4 at.) % Se. Melts higher in Se could not be prepared in an open crucible because of vaporization. According to [1], there "appears to be a certain tendency for layer formation in the liquid state"; however, no miscibility gap was shown in the original diagram, probably because the primary thermal effect was not found at a constant temperature. [Between 8.7 and 37 wt. (7.1 and 32.1 at.) % Se, liquidus temperatures of 1104–1109°C were reported.] However, since the microstructure of an alloy with 30.5 wt. % Se shows unmistakable evidence of two layers, with spherical inclusions of the other phase, a miscibility gap was assumed in Fig. 358.

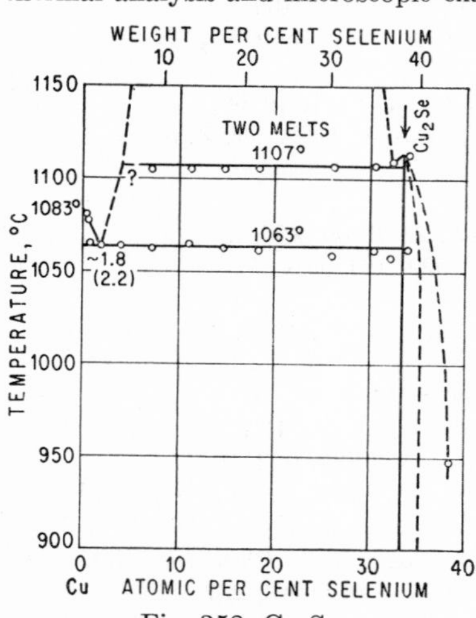

Fig. 358. Cu-Se

The solid solubility of Se in Cu is very small, considerably below 0.02 wt. % at temperatures up to 800°C [2]. [3] estimated the solubility (in wt. %) as approximately 0.015 at 800°C, 0.003 at 700°C, and 0.001 at 600°C, on the basis of measurements of electrical conductivity.

Intermediate Phases. (*a*) Cu_2Se. The structure of Cu_2Se (38.22 wt. % Se) was reported to be isotypic with CaF_2 (C1 type) [4, 5]. [6, 7] showed, however, that at room temperature Cu_2Se has a more complex—deformed cubic—structure. The form stable at elevated temperatures is cubic [6, 8, 7]—according to [6], f.c.c., with $a = 5.85$ A (at 170°C) and 12 atoms per unit cell [9]. As for the transformation temperature, 110°C had been reported in 1890 [10]. [7] observed the cubic form already at 55°C. According to [8], cubic $Cu_{1.96}Se$ (33.8 at. % Se) becomes tetragonal ($a = 11.51$ A, $c = 11.74$ A) below 103°C.

It appears that addition of Se lowers the transformation temperature so that the cubic structure becomes stable at room temperature [6]. Decrease in lattice parameter with increasing Se content [6, 8, 7] indicates subtraction of Cu atoms: $Cu_{2-x}Se$, $a = 5.740$ A at $x = 0.2$ (35.7 at. % Se) [8]. According to [7], the Se-rich boundary of this phase lies below 36.4 at. % Se (Cu:Se = 28:16). The conclusion of [11], based on emf measurements, that Se is soluble in Cu_2Se up to 63 at. % Se (!) is incorrect.

(*b*) Cu_3Se_2 [6, 12, 7]. Cu_3Se_2 (mineral umangite) is orthorhombic, with the lattice parameters $a = 4.28$ A, $b = 6.40$ A, $c = 12.46$ A, and 20 atoms per unit cell [7]. "There is also the possibility that umangite, like rickardite [$Cu_{4-x}Te_2$], is a grossly defective compound $Cu_{4-x}Se_2$ only approximating Cu_3Se_2 in composition" [7]. According to high-temperature X-ray work by [6], Cu_3Se_2 decomposes on heating into $Cu_{2-x}Se$ and CuSe below 170°C.

(*c*) CuSe [13–15, 6, 12]. According to [14], CuSe (55.41 wt. % Se) melts incongruently at approximately 700°C. This compound is hexagonal and most likely isotypic with CuS (B18 type), $a = 3.94$ A, $c = 17.25$ A, $c/a = 4.378$. There are indications of a superstructure with twelvefold multiplicity of the *a* axis [15].

The preparation of $Cu_{2-x}Se$ and $Cu_{4-x}Se_2$ by compression of Cu and Se powders and by reaction of CuCl with H_2Se has been described by [16–18].

1. K. Friedrich and A. Leroux, *Metallurgie*, **5**, 1908, 355–357.
2. R. Eborall, *J. Inst. Metals*, **70**, 1944, 435–446.
3. J. S. Smart and A. A. Smith, *Trans. AIME*, **166**, 1946, 144–155.
4. W. P. Davey, *Phys. Rev.*, **21**, 1923, 380.
5. W. Hartwig, *Z. Krist.*, **64**, 1926, 503–504.
6. P. Rahlfs, *Z. physik. Chem.*, **B31**, 1936, 157–194.
7. J. W. Earley, *Am. Mineralogist*, **35**, 1950, 338–364.
8. W. Borchert, *Z. Krist.*, **106**, 1945, 5–24. (Not available to the author.) *Chem. Abstr.*, **42**, 1948, 1099.
9. See *Structure Repts.*, **10**, 1945-1946, 49, for a criticism of this structure.
10. M. Bellati and S. Lussana, *Z. physik. Chem.*, **5**, 1890, 282.
11. H. Pélabon, *Compt. rend.*, **154**, 1912, 1415.
12. C. Goria, *Gazz. chim. ital.*, **70**, 1940, 461–471. *Met. Abstr.*, **10**, 1943, 347.
13. G. Little, *Liebigs Ann.*, **112**, 1859, 211.
14. H. Fonzes-Diacon, *Compt. rend.*, **131**, 1900, 1207.
15. J. W. Earley, *Am. Mineralogist*, **33**, 1948, 194; **34**, 1949, 435–440.
16. R. Hocart, R. Molé, and L. Schué-Muller, *Compt. rend.*, **233**, 1951, 661–662.
17. R. Molé and R. Hocart, *Bull. soc. chim. France*, **1954**, 977–980.
18. R. Molé, *Ann. chim.*, **9**, 1954, 145–180.

0.3545
$\bar{1}.6455$

Cu-Si Copper-Silicon

Papers published up to 1934 have been thoroughly reviewed by [1]; diagrams by [2–8] are presented.

Solidification. After a cursory study of the system up to 55 at. % Si by [9], the first thermally determined diagram, covering the entire composition range, was reported by [10]. The liquidus consists of four branches corresponding to the crystallization of α, β, η, and Si. The latter, which has not been determined again, is shown in Fig. 359. [2] detected the liquidus curve of the δ phase which was overlooked again by [3]. The liquidus curve and peritectic and eutectic equilibria, which are based on the careful work of [4, 5], covering the range 0–20 wt. (36 at.) % Si, were used in drawing the diagram in Fig. 359. Results of later investigations [7, 8] differ from those by [4, 5] only in minor details as regards temperature and composition of characteristic points; see [1].

The α Phase. The solidus curve (Fig. 359), based on micrographic work, is due to [4]. The curve of the solid solubility of Si in Cu was determined by [11, 4, 7, 12, 8, 13, 14, 15, 16]. Results by [11, 4, 7, 12], obtained by micrographic analysis, are obsolete; since these investigators did not recognize the existence of the κ phase, their solubility curves above 550–600°C lie close to the high-Si boundary of this phase. The micrographically determined curve by [15] (Fig. 359) is generally accepted at present. The curve by [16], based on lattice-spacing data, coincides with that by [15] at temperatures below 500°C, but deviates somewhat at higher temperatures. The curve reported by [8] lies rather close to those by [15, 16], at temperatures between 500 and 700°C. Solubilities determined parametrically by [13, 14] are lower under 700°C than those reported by [15]—largest deviation at 500°C is about 0.9 wt. (1.9 at.) %—but must be regarded as less accurate.

The κ Phase. The existence of an additional phase between 10 and 15 at. % Si, overlooked by all investigators prior to 1933, was first suspected by [17, 18]. [19] found that the solubility of Si at 725°C, based on lattice-parameter measurements, was only about 11.6 at. %, i.e., 2.4 at. % less than the solubility reported by [4], then accepted as most accurate data. By X-ray investigation, [8] proved beyond doubt that an additional phase was present around 12.5 at. % Si. Its existence was confirmed by [13, 20] and later investigators [14, 15, 16, 21, 22].

Although [8] had suggested the κ phase to be a high-temperature phase, [13] believed it to be stable down to room temperature. [20] first proved that κ decomposes eutectoidally into $\alpha + \gamma$. Its decomposition temperature was determined as 552 [15], 557 [16], 590 [14], and 555°C [21] and the eutectoid point as 11.05 [15], 10.8 [16], 12.2 [14], and 10.9 at. % Si [21]. The κ-phase field was established by [15, 16, 14, 21]; that shown in Fig. 359 is due to [15]. The $\kappa \rightarrow \alpha + \gamma$ transformation is extremely sluggish [13, 22]; it was studied by [23, 24].

Alloys with 15–30 At. % Si. Work by [10, 2, 3] need not be considered. By means of thermal and microscopic studies, [5] established the presently accepted phase relationships in this range, with the exception that the ϵ phase was not included. This phase was first detected by a comprehensive X-ray analysis [6]. It was suggested that ϵ is formed by a peritectoid reaction of γ and η at 620°C, since no X-ray evidence was found for the transformation of the η phase, first reported by [2] and confirmed by [15]. The partial phase diagram (0–35 at. % Si) by [7], based on thermal, resistometric, dilatometric, and microscopic data, is very similar to that by [15], with the exception that the ϵ phase [6] was introduced and shown to be formed by a peritectoid reaction of δ and η at 800°C (Fig. 359). In addition, the η phase was reported to undergo two transformations. Contributions to this composition range by [8, 25] are of minor importance.

The Si Phase. The solidus curve, as calculated by [26], is retrograde and indicates solubilities of about 0.003, 0.0008, and 0.00009 at. % Cu at about 1330, 1030, and 800°C, respectively.

Crystal Structures. Lattice parameters of the α phase were reported by [6, 19, 13, 16, 27]; the parameter increases from $a = 3.615$ A at 0% Si to $a = 3.622$ A at 11.7 at. % Si [6, 16].

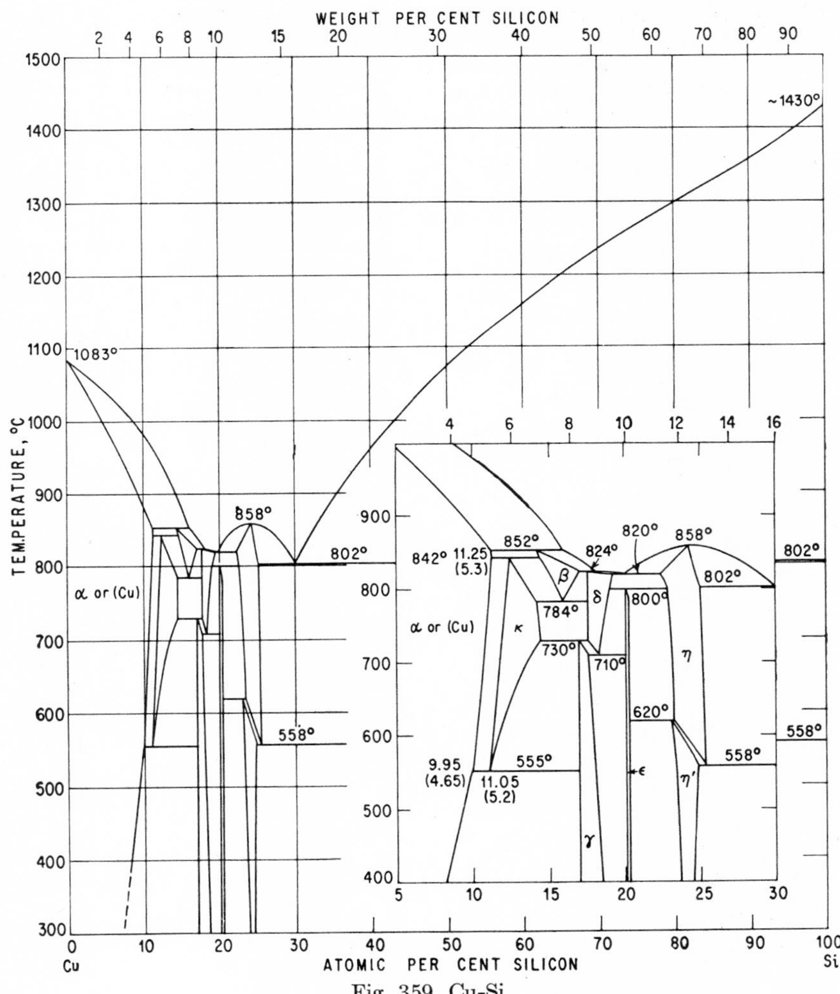

Fig. 359. Cu-Si

The κ phase is h.c.p. (A3 type) [6, 8, 13, 16], with $a = 2.559$ A, $c = 4.185$ A, $c/a = 1.635$ at 11.8 at. % Si and $a = 2.562$ A, $c = 4.182$ A, $c/a = 1.632$ at 14.0 at. % Si [16].

The β phase, a 3:2 electron compound [28], is b.c.c. of the A2 type, $a = 2.854$ A at 14.9 at. % Si [13].

The γ phase is cubic of the β-Mn (A13) type [6, 8, 29, 14], $a = 6.222$ A [5, 29], $a = 6.198$ A [14].

The high-temperature δ phase is considered to be a 21:13 electron compound ($Cu_{31}Si_8$) [30]. [6] suggested δ to have a deformed γ-brass structure; however, this conclusion was based on quenched alloys. According to [8], δ readily decomposes into γ + η on quenching. [14] gave a cubic structure ($D8_{1-3}$ type), with a = 8.506 A.

The ε phase ($Cu_{15}Si_4$) has a b.c.c. structure with 76 atoms per unit cell ($D8_6$ type), with a = 9.714 A [6, 31].

The η′ phase (Cu_3Si) has a slightly distorted γ-brass (trigonally distorted A2 type) structure [6, 32] and is isostructural with the high-temperature phase Cu_3Ge [32].

Supplement. Magnetic measurements, supplemented by metallographic and X-ray examination, were made by [33] on 10 alloys with 5.6–12.5 wt. % Si, both in the as-cast and the annealed (550°C) state. [33] found indications neither for the existence of the ε phase nor for a transformation in the η phase. As regards the δ phase, a simple superstructure of the B2 (CsCl) type was suggested.

The microstructures of as-cast alloys containing between 13 and 96 wt. % Si were found [34] to be in keeping with the phase diagram of Fig. 359.

1. M. Hansen, "Der Aufbau der Zweistofflegierungen," pp. 621–630, Springer-Verlag OHG, Berlin, 1936.
2. A. Sanfourche, *Rev. mét.*, **16,** 1919, 246–256.
3. K. Matuyama, *Science Repts. Tôhoku Univ.*, **17,** 1928, 665–673.
4. C. S. Smith, *J. Inst. Metals*, **40,** 1928, 359–371.
5. C. S. Smith, *Trans. AIME*, **83,** 1929, 414–439.
6. S. Arrhenius and A. Westgren, *Z. physik. Chem.*, **B14,** 1931, 66–79.
7. K. Iokibe, *Kinzoku-no-Kenkyu*, **8,** 1931, 433–456.
8. K. Sautner, *Forschungsarb. Metallkunde u. Röntgenmetallog.*, no. 9, 1933.
9. L. Guillet, *Rev. mét.*, **3,** 1906, 173–174.
10. E. Rudolfi, *Z. anorg. Chem.*, **53,** 1907, 216–227.
11. M. G. Corson, *Trans. AIME, Proc. Inst. Metals Div.*, **1927,** 435–440.
12. E. Crepaz, *Metallurgia ital.*, **23,** 1931, 711–716.
13. T. Isawa, *Nippon Kinzoku Gakkai-Shi*, **2,** 1938, 400–409; see [15].
14. T. Isawa, *Nippon Kinzoku Gakkai-Shi*, **4,** 1940, 398–404.
15. C. S. Smith, *Trans. AIME*, **137,** 1940, 313–329.
16. A. G. H. Andersen, *Trans. AIME*, **137,** 1940, 334–350.
17. E. Voce, *J. Inst. Metals*, **44,** 1930, 331–361; see also 387–388.
18. M. v. Schwarz, *Z. Metallkunde*, **24,** 1932, 124–126.
19. H. F. Kaiser and C. S. Barrett, *Phys. Rev.*, **37,** 1931, 1697.
20. M. Okamoto, *Science Repts. Tôhoku Univ.*, **27,** 1938, 155–161.
21. N. Takamoto, *Nippon Kinzoku Gakkai-Shi*, **4,** 1940, 198–200.
22. W. Hofmann, J. Ziegler, and H. Hanemann, *Z. Metallkunde*, **42,** 1951, 55–57; work done in 1941-1942.
23. W. R. Hibbard, G. H. Eichelman, and W. P. Saunders, *Trans. AIME*, **180,** 1949, 92–100.
24. A. D. Hopkins, *J. Inst. Metals*, **82,** 1953-1954, 163–165.
25. K. Matsuyama, *Kinzoku-no-Kenkyu*, **11,** 1934, 466–469.
26. C. D. Thurmond and J. D. Struthers, *J. Phys. Chem.*, **57,** 1953, 831–834.
27. A. P. Guljaev and E. F. Trusova, *Zhur. Tekh. Fiz.*, **20,** 1950, 66–78; *Structure Repts.*, **13,** 1950, 102.
28. W. Hume-Rothery, P. W. Reynolds, and G. V. Raynor, *J. Inst. Metals*, **66,** 1940, 191–207.
29. S. Fagerberg and A. Westgren, *Metallwirtschaft*, **14,** 1935, 265–267.
30. W. Hume-Rothery, J. O. Betterton, and J. Reynolds, *J. Inst. Metals*, **80,** 1951-1952, 609–616.

31. F. R. Morral and A. Westgren, *Arkiv Kemi, Mineral. Geol.*, **11B,** 1934, no. 37.
32. K. Schubert and G. Brandauer, *Z. Metallkunde,* **43,** 1952, 267–268.
33. H. Nowotny and H. Bittner, *Monatsh. Chem.*, **81,** 1950, 887–906.
34. W. R. Johnson and M. Hansen, Armour Research Foundation, Chicago, Ill., AF Technical Report 6383, June, 1951.

$\bar{1}.7286$
0.2714

Cu-Sn Copper-Tin

A detailed critical study of the literature up to 1935 was made by [1]. A few years later, a number of papers were published which made necessary major corrections of (*a*) the α-phase boundary below 520°C and (*b*) the phase relations in the range 12–25 at. (20.3–38.3 wt.) % Sn. They were considered by [2] in his critical review, including work published up to 1940. No corrections have been made since then. The composite equilibrium diagram in Fig. 360 is that proposed by [2].

The liquidus curve is well established. The whole curve was determined by [3-11] and portions thereof by [12–23]. According to [2], the curve shown in Fig. 360 is based on the closely agreeing results of [6, 8, 17, 22] up to approximately 50 wt. (35 at.) % Sn, [8, 16] for the range 50–90 wt. (35–83 at.) % Sn, and [15, 16, 23, 24] for the alloys richer in Sn. The eutectic point was reliably found as 98.6–98.7 at. (99.25–99.3 wt.) % Sn [23, 24], in very good agreement with data reported earlier [25, 15]. The liquidus curve proposed by [1] differs from that of [2] by less than 5°C.

The Range 0–9.1 At. (15.8 Wt.) % Sn. The solidus curve is based on reliable micrographic work [26]. The solid solubility of Sn in Cu at temperatures between 798 and 520°C was determined, in full or in part, using micrographic [26–30] and lattice-spacing investigations [31, 32, 34]. Results lie within a scattering band about 0.4 at. (0.65 wt.) % wide. Solubility data reported for temperatures below 520°C are based on micrographic [26, 27, 29, 30] and X-ray work [31–35]. Results of [26, 27, 29, 30] do not represent equilibrium conditions. The strong decrease in solubility becomes effective only on long annealing treatment after severe cold work. The solubilities according to [33, 35] are in close agreement; they differ by maximal 0.5 at. % at 300°C. At 170°C the solubility is not higher than 0.4 at. (0.74 wt.) % Sn and probably practically zero at 100°C [33].

The Range 9.1–25.9 At. (15.8–39.5 Wt.) % Sn. The constitution in this range at temperatures above 500°C was the subject of considerable controversy. Although there was agreement with regard to the existence of the phases β, γ, δ, and ϵ since the classical work of [36], the equilibrium relations suggested differed widely. Partial diagrams based on the studies of [5, 36, 37, 7, 13, 38, 8, 9, 17, 29, 10, 20, 22] are presented in [1].

The most controversial points were (*a*) whether β [36, 37, 38, 9, 17, 29, 10, 20] or γ [39, 8, 40, 22] decomposes into $\alpha + \delta$ and (*b*) the nature of the heat effect at 586°C. The latter has been attributed to (*a*) an $\alpha \rightarrow \alpha'$ transformation [9], (*b*) a $\beta \rightarrow \beta'$ transformation [26, 17, 41, 20, 42], or (*c*) a $\beta \rightarrow \alpha + \gamma$ eutectoid decomposition [39, 8, 40, 22]. The discrepancies were settled by the investigations of [43, 44, 45] which showed that the transformation at 586°C was a eutectoid decomposition $\beta \rightarrow \alpha + \gamma$ [46]. "The location of the ($\beta + \gamma$) field is not exactly established experimentally, but cannot differ greatly from that shown [in Fig. 360]" [2].

Based on the X-ray investigations of [47, 29, 42, 48, 49], it is believed that the β and γ phases are crystallographically similar; see [47, 48] and especially [50]. The fact that both phases decompose readily on quenching, forming metastable transition structures [49, 51–54], is probably responsible for the various forms of phase relations suggested [1].

The ζ phase was first shown to exist by [20] and confirmed by [55, 22, 44, 50]. Phase relations in this region suggested previously differed widely [1].

The δ phase was regarded by many earlier investigators as a phase of the fixed composition Cu_4Sn (31.83 wt. % Sn). This composition, however, is without any

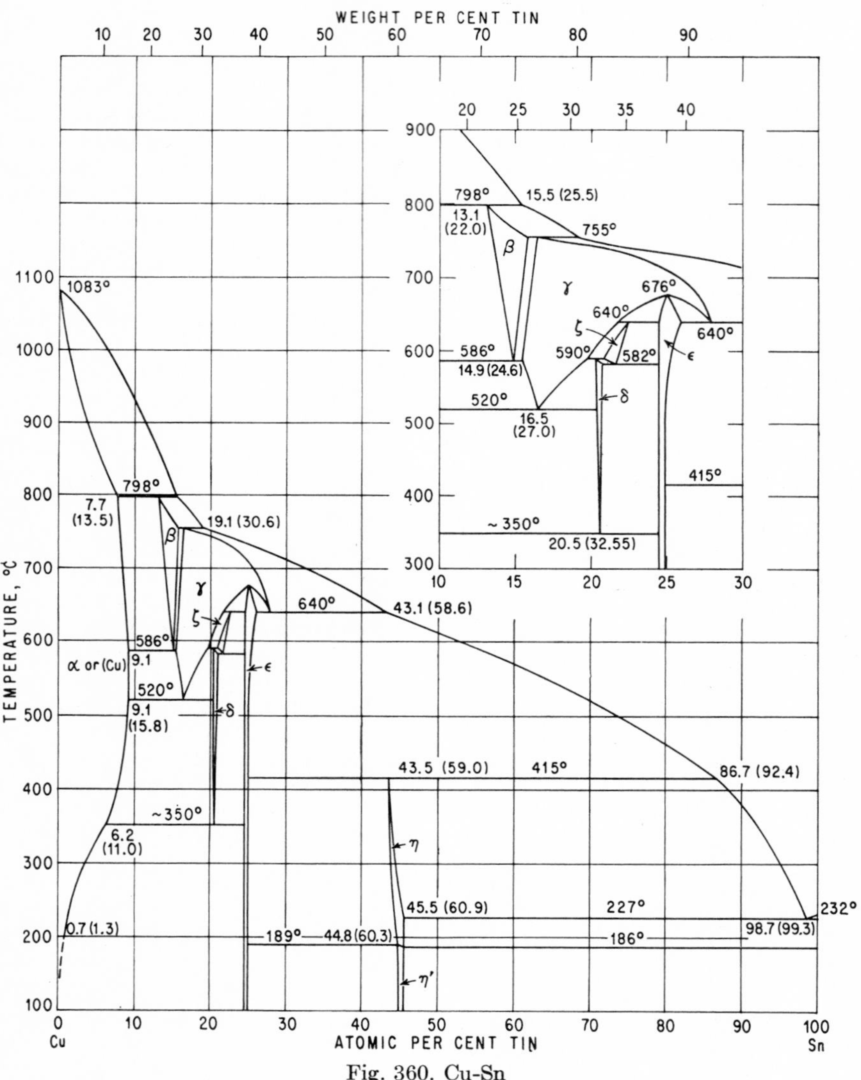

Fig. 360. Cu-Sn

doubt two-phase [56, 29, 20, 57]. The same holds for the composition $Cu_{41}Sn_{11}$ [21.15 at. (33.39 wt.) % Sn], suggested on the basis of X-ray studies [58]. Careful microscopic examination of alloys annealed for 3 weeks at an unstated temperature indicated that an alloy with 20.6 at. % Sn was homogeneous, whereas alloys containing 20.4 and 20.8 at. % Sn were heterogeneous [56]. The composition 20.6 at. % Sn corresponds to the formula $Cu_{31}Sn_8$ [20.51 at. (32.53 wt.) % Sn] suggested by [47]. The δ phase is of slightly variable composition [29, 20, 57]; the boundaries shown in Fig. 360 are based on lattice-spacing data for the temperature range 330–550°C [57]. The composition at 330°C corresponds to the formula $Cu_{31}Sn_8$, in concordance with the result of [56].

The δ phase decomposes into the eutectoid $\alpha + \epsilon$ at a temperature reported as about 380 [32, 33], 330 [57], 350–370 [35], and 325°C [48]. This transformation occurs extremely slowly, especially in massive specimens [59]. For most purposes, δ can be considered as stable below 350°C.

The ϵ phase, formed by transformation of γ, is based on the composition Cu_3Sn (38.37 wt. % Sn). There is close agreement as regards the range of homogeneity determined micrographically [60, 20] and by the lattice-spacing method [57]. For the temperature range 250–530°C, the boundaries shown in Fig. 360 are based on the most accurate data of [57]. At the temperature of the 640°C horizontal ($\gamma \rightleftarrows \epsilon$ + melt), the ϵ phase was found to extend to a composition between about 25.7 and 27.1 at. (39.2–41 wt.) % Sn [7–10, 20].

The Range 25.9–100 At. (39.5–100 Wt.) % Sn. The solidification equilibria were already established by [5, 6]; for detailed numerical data reported by these and later investigators, see [1]. The boundaries of the η phase in Fig. 360 are based on careful thermal, thermoresistometric, and microscopic studies [61]. In agreement with these findings, [8] reported that the composition of η crystals, isolated from alloys with 93–95 wt. % Sn, nearly corresponded to the formula Cu_6Sn_5 [45.45 at. (60.89 wt.) % Sn]. In addition, [10] found the alloy with 44.5 at. (60 wt.) % Sn to be single-phase at 250°C; see also [21, 45]. X-ray work also indicated the composition Cu_6Sn_5 to be characteristic for the structure of η [58]. η undergoes a transformation [61, 9, 16, 10, 21, 45] whose nature is still uncertain; it is probably concerned with a change in atomic arrangement [2].

The solid solubility of Cu in Sn at the eutectic temperature was found as 0.006 wt. (0.01 at.) % Cu [62]. The solidus of the (Sn) phase is a straight line [62].

Crystal Structures (see also Supplement). Lattice parameters of the α phase have been reported by [63, 64, 47, 29, 65, 66, 31, 32, 34, 35, 33, 67, 68]. The parameter at the saturation limit is a = 3.7053 A [67].

Both the β and γ phases are b.c.c. (A2 type), as shown by investigations of quenched alloys [47, 29, 69, 49] as well as high-temperature X-ray work [29, 48]. The lattice constant at room temperature increases from a = 2.981 A at 15.2 at. % Sn to a = 2.991 A at 17.2 at. % Sn [49]. An atomic rearrangement similar to a superstructure was suspected by [48] to be the cause of the transition $\beta \rightarrow \gamma$. According to [50], both phases might be distinguished by the absence of a Brillouin zone overlap in β and the presence of one in γ.

The crystal structure of δ is analogous to that of the γ-brass ($D8_2$) type [47, 58, 34, 57], with a = 17.951 A on the Cu-rich side and a = 17.960 A on the Sn-rich side [57]. The unit cell contains 416 = 8 × 52 atoms [47, 58]. As a 21:13 electron compound, its composition corresponds to the composition $Cu_{31}Sn_8$ (see above). According to [58], however, the unit cell probably contains 328 Cu atoms and 88 Sn atoms, indicating the ideal formula $Cu_{41}Sn_{11}$ [70]. See also [50].

The ζ phase is hexagonal, with a = 7.331 A, c = 7.870 A, c/a = 1.073, and 26 atoms per unit cell, suggesting the ideal composition $Cu_{20}Sn_6$ [23.08 at. (35.92 wt.) % Sn] [55]; see also [50]. The structure is related to the γ-brass type of structure.

The ϵ phase (Cu_3Sn) is orthorhombic, with 64 atoms per unit cell and a = 4.34 A, b = 5.56 A, c = 38.1_8 A [58], or a = 4.328 A, b = 5.521 A, c = 33.25 A [55]; see also [71, 50]. The structure can be regarded as a superlattice on a h.c.p. basis [72, 47, 73,

29, 34, 57], with lattice constants ranging from a = 2.753 A, c = 4.329 A, c/a = 1.572 (Cu-side), to a = 2.760 A, c = 4.329 A, c/a = 1.569 (Sn-side) [47, 34, 57].

The η' phase [47, 58, 55] has a structure that can be regarded as a superlattice on the NiAs (B8) type structure, with a = 20.89 A, c = 25.1_5 A, c/a = 1.204 [58], or a = 20.99 A, c = 25.48 A, c/a = 1.214 [55]. The dimensions of the hexagonal pseudo cell were reported as a = 4.198 A, c = 5.096 A, c/a = 1.214 [47]. The orthorhombic unit cell contains 230–250 Sn atoms and 300–280 Cu atoms, indicating the composition Cu_6Sn_5 [58, 55]. The high-temperature modification η is possibly isotypic with NiAs.

Supplement. [74] studied the superstructure of the ϵ (Cu_3Sn) phase by X-ray single-crystal work and reported the orthorhombic lattice parameters a = 5.53 A, b = 47.8 A, c = 4.34 A. For structural details see the original paper.

According to [75], who took X-ray powder patterns at 700°C of 15 alloys varying in composition from 11.8 to 25 at. % Sn, the γ high-temperature phase is ordered analogous to Fe_3Al ($D0_3$ type), with a = 6.116_6 A at 25 at. % Sn. See also [76].

1. M. Hansen, "Der Aufbau der Zweistofflegierungen," pp. 630–647, Springer-Verlag OHG, Berlin, 1936.
2. G. V. Raynor, Annotated Equilibrium Diagram Series, no. 2, The Institute of Metals, London, 1944.
3. H. Le Chatelier, *Compt. rend.*, April, 1894; *Bull. soc. encour. ind. natl.*, **1895,** 573.
4. A. Stansfield, *Proc. Inst. Mech. Engrs.* (*London*), **1895,** 269–279.
5. W. C. Roberts-Austen and A. Stansfield, *Proc. Inst. Mech. Engrs.* (*London*), **1897,** 67–69.
6. C. T. Heycock and F. H. Neville, *Phil. Trans. Roy. Soc.* (*London*), **A189,** 1897, 47–51, 62–66.
7. F. Giolitti and G. Tavanti, *Gazz. chim. ital.*, **38**(2), 1908, 209–239.
8. O. Bauer and O. Vollenbruck, *Z. Metallkunde,* **15,** 1923, 119–125, 191–195.
9. T. Isihara, *Science Repts. Tôhoku Univ.*, **13,** 1924, 75–100; *J. Inst. Metals,* **31,** 1924, 315–345; see also [18].
10. M. Tasaki, *Mem. Coll. Eng. Kyoto Imp. Univ.*, **A12,** 1929, 228–229.
11. W. Broniewski, J. T. Jablonski, and S. Maj, *Compt. rend.*, **202,** 1936, 305–307.
12. C. T. Heycock and F. H. Neville, *J. Chem. Soc.*, **57,** 1890, 379.
13. M. P. Slavinski, *Zhur. Russ. Met. Obshchestva,* **1,** 1913, 543–563; *Rev. mét.*, **12,** 1915, 405–409.
14. L. J. Gurevich and J. S. Hromatko, *Trans. AIME,* **64,** 1921, 233–235; *J. Inst. Metals,* **37,** 1927, 235.
15. H. J. Miller, *J. Inst. Metals,* **37,** 1927, 188–190.
16. F. H. Jeffery, *Trans. Faraday Soc.*, **23,** 1927, 563–570.
17. A. R. Raper, *J. Inst. Metals,* **38,** 1927, 217–231.
18. T. Isihara, *Science Repts. Tôhoku Univ.*, **17,** 1928, 927–937.
19. G. O. Hiers and G. P. de Forest, *Trans. AIME,* **89,** 1930, 207–218.
20. M. Hamasumi and S. Nishigori, *Tech. Repts. Tôhoku Univ.*, **10,** 1931, 131–187.
21. M. Hamasumi, *Kinzoku-no-Kenkyu,* **10,** 1933, 137–147; *J. Inst. Metals,* **53,** 1933, 550–551.
22. J. Verö, *Z. anorg. Chem.*, **218,** 1934, 402–424.
23. D. Hanson, E. J. Sandford, and H. Stevens, *J. Inst. Metals,* **55,** 1934, 119–121.
24. J. V. Harding and W. T. Pell-Walpole, *J. Inst. Metals,* **75,** 1948, 115–130.
25. T. E. Rooney, *J. Inst. Metals,* **25,** 1921, 333.
26. D. Stockdale, *J. Inst. Metals,* **34,** 1925, 111–120.
27. J. L. Haughton, *J. Inst. Metals,* **34,** 1925, 121–123.
28. M. Hansen, *Z. Metallkunde,* **19,** 1927, 407–409.

29. O. A. Carson, *Can. Mining Met. Bull.* 201, 1929, pp. 129–270.
30. J. T. Eash and C. Upthegrove, *Trans. AIME,* **104,** 1933, 228–231.
31. T. Isawa and I. Obinata, *Metallwirtschaft,* **14,** 1935, 185–188; *Mem. Ryojun Coll. Eng., Commemoration Vol. Dedicated to K. Inouye,* **1934,** 235–242.
32. S. T. Konobeevsky and V. P. Tarasova, *Zhur. Eksptl. i Teoret. Fiz.,* **4,** 1934, 272–291; *Physik. Z. Sowjetunion,* **4,** 1933, 571–575; **5,** 1934, 848–876.
33. S. T. Konobeevsky and V. P. Tarasova, *Zhur. Fiz. Khim.,* **9,** 1937, 681–692; *Acta Physicochim. U.R.S.S.,* **6,** 1937, 781–798; see also V. P. Tarasova, *Vestnik Moskov. Univ.,* no. 4, 1947, pp. 105–107.
34. E. A. Owen and J. Iball, *J. Inst. Metals,* **57,** 1935, 267–286.
35. C. Haase and F. Pawlek, *Z. Metallkunde,* **28,** 1936, 73–80.
36. C. T. Heycock and F. H. Neville, *Phil. Trans. Roy. Soc. (London),* **A202,** 1904, 1–69; *Proc. Roy. Soc. (London),* **A68,** 1901, 171–178; **69,** 1901, 320–329; **71,** 1903, 409–412.
37. E. S. Shepherd and E. Blough, *J. Phys. Chem.,* **10,** 1906, 630–653.
38. C. R. Corey, M.A. thesis, Columbia University, New York, 1915; see *Trans. AIME,* **73,** 1926, 1159, 1162.
39. S. L. Hoyt, *J. Inst. Metals,* **10,** 1913, 259–265; *Trans. AIME,* **60,** 1919, 198.
40. T. Matsuda, *Science Repts. Tôhoku Univ.,* **11,** 1922, 224–237; **17,** 1928, 141–161.
41. G. Shinoda, *Suiyokwai-Shi,* **5,** 1928, 687–694.
42. T. Isawa, *Mem. Ryojun Coll. Eng.,* **10,** 1937, 53–61.
43. M. Hamasumi and Y. Odamura, *Nippon Kinzoku Gakkai-Shi,* **1,** 1937, 165–167.
44. M. Hamasumi and N. Takamoto, *Nippon Kinzoku Gakkai-Shi,* **1,** 1937, 251–261.
45. M. Hamasumi, *Nippon Kinzoku Gakkai-Shi,* **2,** 1938, 147–161.
46. The existence of two peritectic reactions at 756 and 742°C in the range 15.5–20.8 at. % Sn [22] was disproved by [44].
47. A. Westgren and G. Phragmén, *Z. anorg. Chem.,* **175,** 1928, 80–89; see also *Z. Metallkunde,* **18,** 1926, 279–284.
48. M. Hamasumi and K. Morikawa, *Nippon Kinzoku Gakkai-Shi,* **2,** 1938, 39–44.
49. I. Isaichev, *Zhur. Tekh. Fiz.,* **9,** 1939, 1286–1292, 1867–1872.
50. K. Schubert and G. Brandauer, *Z. Metallkunde,* **43,** 1952, 262–268.
51. I. Isaichev and G. Kurdyumov, *Physik. Z. Sowjetunion,* **5,** 1934, 6–21; W. Bugakov, I. Isaichev, and G. Kurdyumov, *Physik. Z. Sowjetunion,* **5,** 1934, 22–30.
52. H. Imai and I. Obinata, *Mem. Ryojun Coll. Eng.,* **3,** 1930, 117–135; H. Imai and M. Hagiya, *Mem. Ryojun Coll. Eng.,* **5,** 1932, 77–89.
53. I. Isaichev and I. Salli, *Zhur. Tekh. Fiz.,* **10,** 1940, 752–756.
54. I. Isaichev, *Zhur. Tekh. Fiz.,* **17,** 1947, 829–834.
55. O. Carlsson and G. Hägg, *Z. Krist.,* **83,** 1932, 308–317.
56. D. Stockdale, see J. D. Bernal, *Nature,* **122,** 1928, 54.
57. E. A. Owen and E. C. Williams, *J. Inst. Metals,* **58,** 1936, 283–297.
58. J. D. Bernal, *Nature,* **122,** 1928, 54.
59. C. C. Wang and M. Hansen, *Trans. AIME,* **191,** 1951, 1212.
60. W. Hume-Rothery, *Phil. Mag.,* **8,** 1929, 114–121.
61. J. L. Haughton, *J. Inst. Metals,* **13,** 1915, 222–242; **25,** 1921, 309–330, 335–336; see also *Trans. Faraday Soc.,* **24,** 1928, 212–213.
62. C. E. Homer and H. Plummer, *J. Inst. Metals,* **64,** 1939, 169–200.
63. E. C. Bain, *Chem. Met. Eng.,* **28,** 1923, 21–24.
64. H. Weiss, *Proc. Roy. Soc. (London),* **A108,** 1925, 643–654.
65. S. Sekito, *Science Repts. Tôhoku Univ.,* **18,** 1929, 59–68.
66. R. F. Mehl and C. S. Barrett, *Trans. AIME,* **89,** 1930, 203–206.
67. E. A. Owen, *J. Inst. Metals,* **73,** 1947, 476–479.
68. A. P. Guljaev and E. F. Trusova, *Zhur. Tekh. Fiz.,* **20,** 1950, 66–78.

69. G. Shinoda, *Suiyokwai-Shi*, **7**, 1932, 367–372.
70. 416 is not a multiple of 39. The formula $Cu_{41}Sn_{11}$ does not exactly correspond to a valency electron concentration of 21:13.
71. J. O. Linde, *Ann. Physik*, **8**, 1931, 124–128.
72. E. C. Bain, *Chem. & Met. Eng.*, **28**, 1923, 69; *Ind. Eng. Chem.*, **16**, 1924, 692–698.
73. W. Morris-Jones and E. J. Evans, *Phil. Mag.*, **4**, 1927, 1302–1311.
74. K. Schubert, B. Kiefer, M. Wilkens, and R. Haufler, *Z. Metallkunde*, **46**, 1955, 692–715, especially 693, 709, 711–713.
75. H. Hendus and H. Knödler, *Acta Cryst.*, **9**, 1956, 1036.
76. H. Knödler, *Acta Cryst.*, **10**, 1957, 86–87.

$\bar{1}.5457$
0.4543

Cu-Ta Copper-Tantalum

According to [1], the solid solubility of Ta in Cu is very small [2].

Supplement. [3] stated: "The arc-melting technique was unsuccessful for this binary system, probably because of complete insolubility of the components and the wide difference in the melting temperatures. An X-ray diffraction pattern of the $TaCu_2$ composition of the inhomogeneous alloy gave parameters of unalloyed Cu and Ta indicating immiscibility. Another alloy was made by sintering a compact of Cu and Ta powder for 3 days at 1000°C. The powder pattern of this alloy indicated a b.c.c. pattern of Ta and a f.c.c. pattern of Cu. No intermetallic compounds were detected."

1. A. G. Dowson, Abstr. Diss. University Cambridge, 1936-1937, p. 116.
2. No detailed information was available.
3. R. P. Elliott, Armour Research Foundation, Chicago, Ill., Technical Report 1, OSR Technical Note OSR-TN-247, August, 1954, p. 23.

$\bar{1}.6972$
0.3028

Cu-Te Copper-Tellurium

Until recently, there was only one thermal and microscopic investigation which dealt with the constitution of this system [1]. The phase diagram based on this work has the following features: (*a*) a miscibility gap in the liquid state extending from about 1 to 32 at. % Te at the monotectic temperature, 1033°C; (*b*) the phase Cu_2Te (50.10 wt. % Te), either having a congruent melting point of approximately 875°C and forming a eutectic with Cu at 857°C or formed peritectically at this temperature; (*c*) the phase Cu_4Te_3 (60.10 wt. % Te) formed by the peritectic reaction Cu_2Te + melt (50 at. % Te) $\rightarrow$ Cu_4Te_3 at 623°C; and (*d*) a eutectic Cu_4Te_3-Te at 71 at. (83 wt.) % Te, 344°C. Thermal effects in the solid state were interpreted to be transformations of Cu_2Te at 387 and 360–345°C and of Cu_4Te_3 at 365°C. Temperatures given are based on melting points of Cu and Te of 1055 and 438°C, instead of 1083 and 453°C.

A reinvestigation by thermal analysis (and supplementary microscopic examination) gave the data points shown in Fig. 361 [2]. According to unpublished work, the monotectic lies at 4.3 at. (8.2 wt.) % Te, 1048°C, and the Cu-Cu_2Te eutectic at about 30.1 at. (46.4 wt.) % Te and probably only a few °C below the monotectic. This agrees very well with findings by [3], according to which the monotectic and eutectic temperatures are 1054 and 1053°C, respectively. Cu_2Te (having a certain range of homogeneity) was found to melt congruently at a temperature much higher than that found by [1]. In addition to Cu_2Te and Cu_4Te_3, a phase *X*, formed by a peritectic reaction at 727°C, was assumed, whose composition lies at about 36–37 at.

% Te [4]. Except for the thermal effects at 727°C, no conclusive evidence for the existence of the phase *X* was given. Therefore, this portion of the diagram has to be regarded as tentative. In contrast to [1], transformations of Cu_2Te were found at 555–400, 360, and 305°C (Fig. 361). Thermal effects at about 853°C in the range up to about 20 at. % Te could not be accounted for; at this temperature level, [1] assumed the temperature of the eutectic Cu-Cu_2Te (see above). [3] did not observe thermal effects in this temperature range.

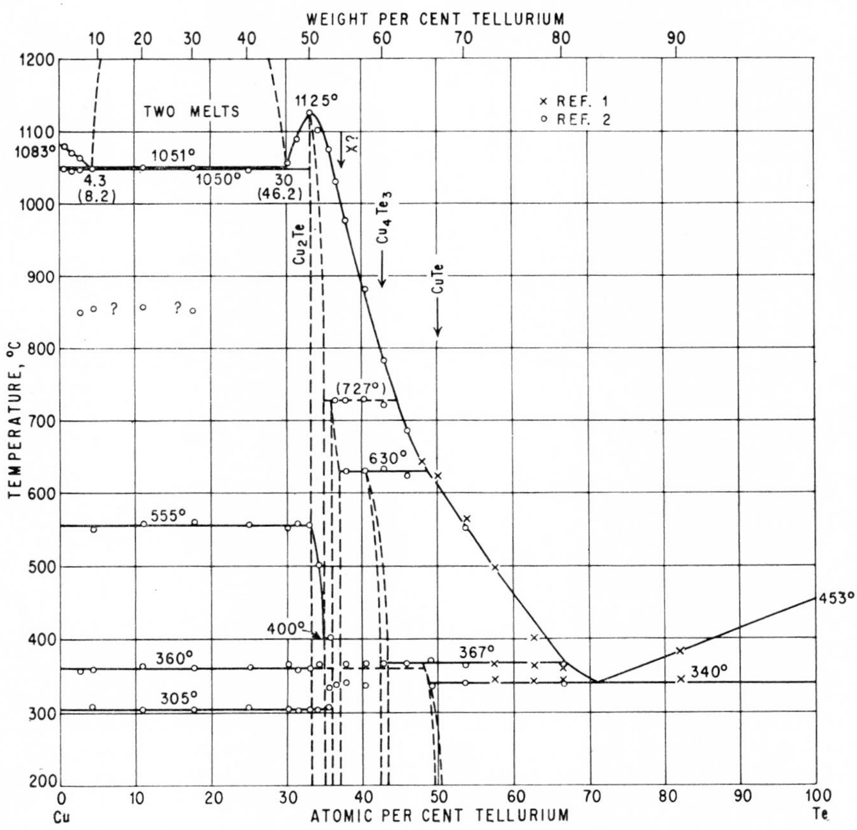

Fig. 361. Cu-Te. (See also Supplement.)

Metallographic and roentgenographic studies [5] established the existence of an additional phase, CuTe (66.76 wt. % Te), not detected by [1, 2] but already suggested by [6], on the basis of emf measurements and microscopic examination. CuTe is formed peritectically at 365–367°C, i.e., the temperature formerly [1, 2] assumed to be a transformation point of Cu_4Te_3. Both Cu_4Te_3 [2, 5] and CuTe [5] change their compositions with change in temperature; however, the homogeneity ranges have not been determined as yet. Dilatometric investigation [5] of an alloy with 34.5 at. % Te confirmed that Cu_2Te undergoes transformations at about 180, 305, 345, and 465°C, the last three temperatures being in good agreement with the data given by [2]. High-temperature X-ray work [5] proved that Cu_4Te_3 also has a transformation point, probably close to the peritectic temperature, 367°C. However, CuTe does not undergo a transformation [5].

The solid solubility of Te in Cu was reported by [3] to be considerably less than 0.02 wt. % at temperatures up to 800°C and estimated by [7] as about 0.0075 wt. % at 800°C, 0.0015% at 700°C, and 0.0003% at 600°C, on the basis of measurements of electrical conductivity.

Crystal Structures. X-ray photograms of Cu_2Te (35 at. % Te) taken at 640, 440, 330, and 270°C showed that the structure at 640°C (i.e., above the upper transformation at 555–400°C) is f.c.c., with 12 atoms per unit cell and a = 6.11 A [5]. The structure is isotypic with that of the high-temperature forms of Ag_2Te and Cu_2Se [8] and is not of the CaF_2 type. Diffraction patterns of the forms stable below the upper transformation point have the characteristic arrangement of the strong lines of the room-temperature form and differ only in the intensities of the weaker lines. For the room-temperature form, [9] gave a hexagonal structure, 6 atoms per unit cell, a = 4.245 A, c = 7.289 A, c/a = 1.717; however, in view of the number of thermal effects, the structure might be more complex (see Supplement). [8] noted that he obtained a number of different diagrams, depending upon composition, within the Cu_2Te phase. The existence of the X phase would make the problem even more complicated; see also [10].

Cu_4Te_3 (mineral rickardite) has a defect structure of the Cu_2Sb (C38) type, a = 3.98 A, c = 6.12 A, c/a = 1.54 [11]. CuTe is orthorhombic, with 4 atoms per unit cell and a = 3.16 A, b = 4.08 A, c = 6.93 A (new type) [5].

Supplement. [12, 13] prepared weissite ($Cu_{2-x}Te$) and rickardite ($Cu_{4-x}Te_2$) by compression and heating of a mixture of the components. [14] measured the dissociation pressures of Cu_2Te in the range 500–800°C and observed anomalous behavior in the neighborhood of 600 and 800°C. This is thought [14] to be due to hitherto unknown phase transformations. (The lower effect may correspond to the 555°C transformation shown in Fig. 361.) [15] prepared Cu-Te alloys by sintering powder compacts at 200°C and examined them by X-ray diffraction and ore microscopy. The following intermediate phases were found: Cu_2Te, with a range of homogeneity of 33.3–37.5 at. % Te; rickardite at 41.7 at. % Te (therefore designated as Cu_7Te_5); and stoichiometric CuTe. In agreement with [5], [15] considers the structure of the room-temperature form of Cu_2Te to be more complex than that given by [9] (which presumably belongs to a form stable at intermediate temperatures). For the room-temperature form, [15] suggests a superstructure with tripled a and c axes. The lattice parameters were found to decrease from a = 12.54 A, c = 21.71 A at 33.3 at. % to a = 12.45 A, c = 21.56 A at 37.5 at. % Te.

By means of thermal analysis, [16] determined the congruent melting point of Cu_2Te as 1111°C. A 50 wt. (33.2 at.) % Te alloy showed, on cooling, transformation effects at 560, 362, 316, and 260°C, in good agreement with the results of [2, 5]. [16] suggest that an increasing Te content depresses their 260°C effect to lower temperatures, reaching 180°C at 34.5 at. % Te [5].

1. M. Chikashige, *Z. anorg. Chem.*, **54,** 1907, 50–57.
2. O. Keymling, Thesis, Clausthal, 1952, unpublished.
3. R. Eborall, *J. Inst. Metals*, **70,** 1944, 435–436.
4. This corresponds to about the composition Cu_5Te_3.
5. K. Anderko and K. Schubert, *Z. Metallkunde*, **45,** 1954, 371–378; *Naturwissenschaften*, **40,** 1953, 269.
6. N. A. Puschin, *Z. anorg. Chem.*, **56,** 1908, 9–12.
7. J. S. Smart and A. A. Smith, *Trans. AIME*, **152,** 1943, 103–117.
8. P. Rahlfs, *Z. physik. Chem.*, **B31,** 1936, 157–194.

9. H. Nowotny, *Z. Metallkunde*, **37**, 1946, 40–42.
10. R. Hocart and R. Molé, *Compt. rend.*, **234**, 1952, 111–113.
11. S. A. Forman and M. A. Peacock, *Am. Mineralogist*, **34**, 1949, 441–451.
12. R. Molé and R. Hocart, *Bull. soc. chim. France*, **1954**, 977–980.
13. R. Molé, *Ann. chim.*, **9**, 1954, 145–180.
14. M. I. Kochnev and T. N. Zaidman, *Doklady Akad. Nauk S.S.S.R.*, **94**, 1954, 65–67. *Met. Abstr.*, **22**, 1954, 297.
15. I. Patzak, *Z. Metallkunde*, **47**, 1956, 418–420.
16. H. Gravemann and H. J. Wallbaum, *Z. Metallkunde*, **47**, 1956, 433–441.

$\bar{1}.4373$
0.5627

Cu-Th Copper-Thorium

Data as to the constitution of this system are contradictory. [1] reported that a Cu-rich solid solution with about 2–3 wt. (0.6–0.9 at.) % Th forms a eutectic with the phase $ThCu_3$ (54.91 wt. % Th). [2] studied the composition range up to 21.5 at. (50 wt.) % Th. The phase richest in Cu was assumed to be $ThCu_4$ (47.74 wt. % Th), which forms a eutectic at 7.5 at. (23 wt.) % Th, 940°C (see inset of Fig. 362).

The partial phase diagram proposed by [3] is characterized by the existence of the intermediate phases $ThCu_6$ (37.85 wt. % Th), $ThCu_3$, and probably Th_3Cu_5 (68.67 wt. % Th), having congruent melting points of about 1062, 1060, and 1030°C, respectively (Fig. 362). Eutectics were placed at approximately 8 at. (24.2 wt.) % Th, 946°C; 19.5 at. (46.9 wt.) % Th, 1020°C; 31 at. (62.1 wt.) % Th, 970°C; and 49 at. (77.8 wt.) % Th, 881°C. These results are in conflict with the existence of the phases

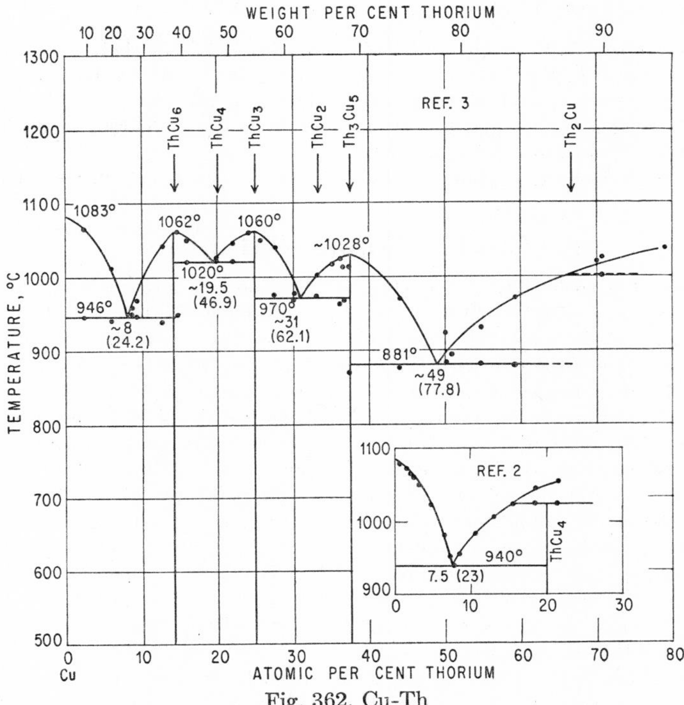

Fig. 362. Cu-Th

$ThCu_2$ (64.62 wt. % Th) and Th_2Cu (87.96 wt. % Th), which were identified by X-ray structural analysis [4].

No attempt is being made here to account for the widely differing results. It may be correct, however, to assume that the alloys studied [1–3]—especially those rich in Th [3]—were contaminated, although the investigators have endeavored to avoid reaction of the melts with the atmosphere and crucible materials.

The assumption of the existence of $ThCu_6$ is based on thermal data [3] (Fig. 362) and the microstructure of alloys with 37.2 and 40.5 wt. % Th [3]. Also, Ce, La, and Pr each form the corresponding compound with Cu. There is no indication, however, for the existence of $ThCu_6$ in the partial diagram by [2] (inset of Fig. 362). The next intermediate phase is very likely $ThCu_3$ [3] rather than $ThCu_4$ [2]. The phase Th_3Cu_5 perhaps has to be replaced by $ThCu_2$; they differ in composition by 4.2 at. % or 4.05 wt. %. On the other hand, however, phases corresponding to Th_3Cu_5 have also been found in the systems Ag-Th and Au-Th [3].

There is great discrepancy also as regards the solid solubility of Th in Cu. [2] reported lattice-spacing data which indicate a solubility of 4 wt. % at 900°C and 2.2 wt. % Th at 300°C. This is in agreement with results by [1], who observed pronounced age hardening of Cu-rich alloys. However, [3] claimed the solubility to be less than 0.1 wt. % Th at the eutectic temperature, 946°C. [5] noted there was reason to believe that work by [2] was not entirely accurate.

It was briefly stated that $ThCu_2$ and Th_2Cu are isotypic with AlB_2 and $CuAl_2$, respectively [4].

Supplement. Crystallographic data of $ThCu_2$ and Th_2Cu have recently been published by [6]: $ThCu_2$, hexagonal AlB_2 (C32) type, $a = 4.35$ A, $c = 3.47$ A; Th_2Cu, b.c. tetragonal $CuAl_2$ (C16) type, $a = 7.28$ A, $c = 5.74$ A.

1. T. Liepus, see W. Guertler, *Metallwirtschaft*, **19**, 1940, 435–444.
2. G. Grube and L. Botzenhardt, *Z. Elektrochem.*, **48**, 1942, 418–425.
3. E. Raub and M. Engel, *Z. Elektrochem.*, **49**, 1943, 487–493.
4. J. V. Florio, R. E. Rundle, and A. J. Snow, *Acta Cryst.*, **5**, 1952, 449–457.
5. G. Grube, *Z. Elektrochem.*, **49**, 1943, 57.
6. N. C. Baenziger, R. E. Rundle, and A. I. Snow, *Acta Cryst.*, **9**, 1956, 93–94.

0.1227
$\bar{1}.8773$

Cu-Ti Copper-Titanium

Alloys Up to 50 At. % Cu (Fig. 363). The two intermediate phases in this composition range, Ti_2Cu (39.88 wt. % Cu) and TiCu (57.02 wt. % Cu), were first identified by X-ray work [1]. [2] confirmed the existence of two phases by X-ray analysis but found the one with the lowest copper content to be Ti_3Cu rather than Ti_2Cu. Their homogeneity ranges were given as about 25–29 and 47–50 at. % Cu, respectively. Alloys were prepared by melting powdered electrolytic copper and titanium (from titanium hydride) in zirconia crucibles under vacuum. Samples were heat-treated by sealing in evacuated silica tubes maintained at temperatures up to 1200°C and quenching; however, at temperatures above 800°C reactions between the specimen and tube often occurred. This resulted in a great uncertainty in the determination of the homogeneity ranges of the intermediate phases.

[2] stated that the supposed phase Ti_2Cu is identical with the phase Ti_3Cu_3O (which may be of variable composition so as to include the composition Ti_4Cu_2O) detected by him in Ti-Cu alloys contaminated with oxygen. However, [3] was able to show that both Ti_2Cu and Ti_4Cu_2O exist and give almost identical diffraction patterns. These data appear to be in favor of the existence of Ti_2Cu rather than Ti_3Cu. On the other hand, however, [2] found his Ti_3Cu phase to have a tetragonal

structure which may be considered to be of a deformed Cu_3Au type. This structure is not compatible with the formula Ti_2Cu.

[4] investigated the region 30–55 at. % Cu, using alloys based on Ti of 98.6% purity and melted in corundum crucibles under argon. It was stated that Ti_3Cu,

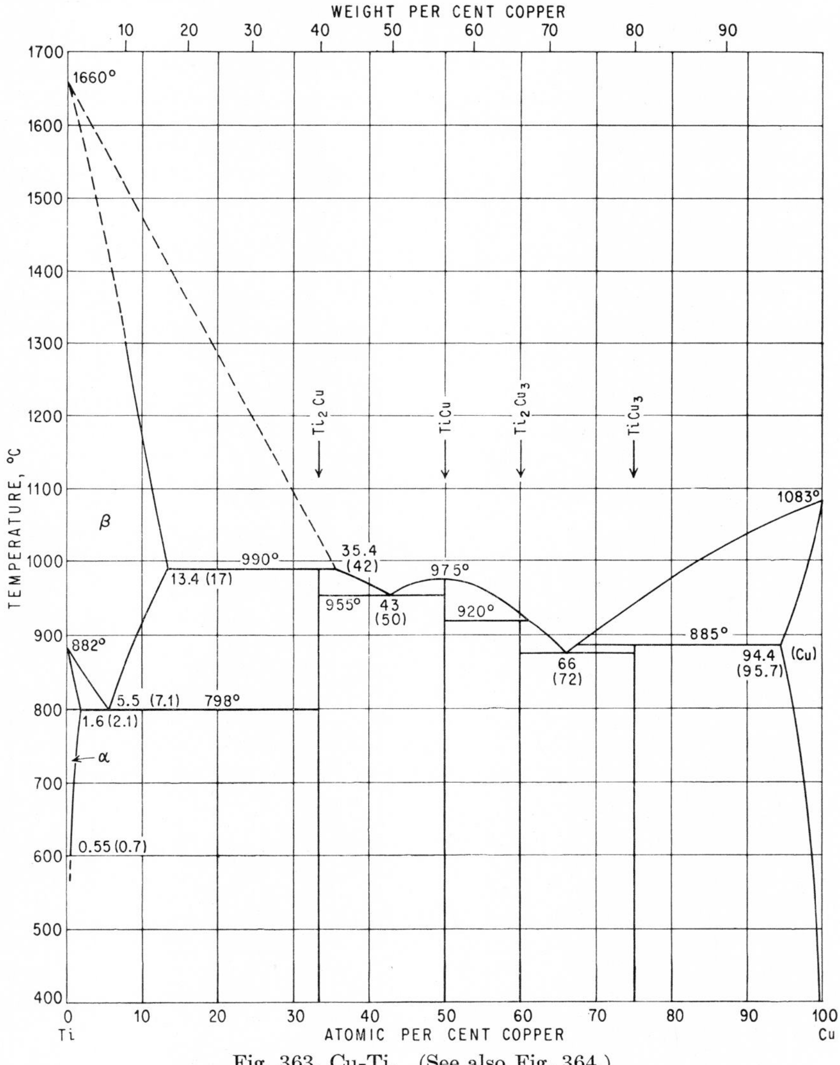

Fig. 363. Cu-Ti. (See also Fig. 364.)

besides TiCu, was confirmed although the composition Ti_3Cu lay outside the range of concentration investigated. Alloys in this range were contaminated because of reaction of the melts with the crucible material.

The partial diagram (0–50 at. % Cu) shown in Fig. 363 is based on micrographic, X-ray, and thermal analysis of arc-melted and homogenized alloys [5]. With the exception of the Ti-rich alloys up to 8 at. % Cu, which were based on iodide Ti, the alloys were prepared with magnesium-reduced Ti. No indication was found of the existence of the compound Ti_3Cu claimed by [2, 4]; the alloy of the composition Ti_2Cu was found to be single-phase.

The boundaries of the $(\alpha + \beta)$ field were first outlined by metallographic examination of magnesium-reduced Ti-base alloys up to about 3.7 at. % Cu, quenched from temperatures between 790 and 900°C after a 12-hr anneal [6]. [7] determined the constitution up to 5 at. % Cu (iodide Ti–base alloys) in the temperature range 900–750°C by means of measurement of hydrogen pressures in equilibrium with a very dilute solution of hydrogen in the alloys, as a function of temperature. He found that the β phase with about 7 at. (9 wt.) % Cu (extrapolated) decomposes eutectoidally at 776°C. The solubility of Cu in α-Ti at 776°C was found to be 1.25 at. (1.65 wt.) %. According to [5], the eutectoid point lies at 5.5 at. (7.1 wt.) % Cu, 798°C, and the maximal solubility of Cu in α-Ti at 1.6 at. (2.1 wt.) % Cu. For alloys based on magnesium-reduced Ti, [5] found the eutectoid at approximately 6.5 at. (8.5 wt.) % Cu, 785°C.

Alloys with 50–100 At. % Cu (Fig. 363). According to [2], there is, besides the phase with 47–50 at. % Cu (see above), another phase, also based on the ideal composition TiCu, having a homogeneity range of about 50–55 at. % Cu. [4] reported that the existence of these two phases was verified by X-ray analysis, agreeing with [2] in the assumption that the existence of two tetragonal structures in the neighborhood of 50 at. % appears to be most compatible with the diffraction patterns observed. Neither phase could be distinguished metallographically, however. In opposition to this, [5] found only one phase to occur around the composition 50 at. %; no homogeneity range was given.

The existence of the phase Ti_2Cu_3 (66.55 wt. % Cu) (Fig. 363), claimed only by [5], is based on thermal data (peritectic reaction at 920°C) as well as on the fact that an alloy of this composition gave a distinct X-ray pattern, although the microstructure appeared to be duplex. It seems possible that the Ti_2Cu_3 phase is identical with the Ti-rich "modification" of the TiCu phase according to [2, 4].

There is agreement as regards the composition of the intermediate phase, $TiCu_3$ (79.92 wt. % Cu), coexisting with the Cu-rich solid solution [1, 2, 4, 5]. Only [2] reported this phase to undergo an order-disorder transformation somewhere between 500 and 700°C. According to [4], its composition is shifted to higher Cu contents so that it could also be regarded as being based on the composition $TiCu_4$ (84.14 wt. % Cu).

Earlier, [8, 9] had studied the constitution of the Cu-rich alloys up to about 23 and 33 at. % Ti, respectively, by thermal analysis and microscopic investigations. The peritectic temperature was found as 900 [8] and 878°C [9], as compared with about 890 [4] and 885°C [5], but the phase formed was not identified. Earlier data as to the solid solubility of Ti in Cu [8, 9, 10] are obsolete since [4] accurately determined the solidus and solubility curve by means of micrographic and X-ray analysis. The solubility of Ti in Cu in wt. % (at. % in parentheses) is as follows: 3.8 (5.0) at 850°C, 3.0 (3.9) at 800°C, 1.8 (2.35) at 700°C, 1.1 (1.45) at 600°C, 0.6 (0.8) at 500°C, and 0.4 (0.5) at 400°C [4].

Crystal Structures. Ti_2Cu: f.c.c., 96 atoms per unit cell [1]. X-ray pattern was given by [5]. According to [2], this phase does not exist, but [3] reported Ti_2Cu as well as Ti_4Cu_2O to have a f.c.c. structure (Fe_3W_3C type), with a = 11.26 A and a = 11.49 A, respectively.

Ti_3Cu: f.c. tetragonal structure which may be considered to be of a deformed Cu_3Au ($L1_2$) type [2]. According to [3, 5], this phase does not exist.

TiCu: [2, 4] found two phases having this ideal composition. With an excess of

Ti, the phase was reported to be b.c. tetragonal of the B11 type, with $a = 3.108$ A, $c = 5.887$ A, $c/a = 1.894$, and $a = 3.118$ A, $c = 5.921$ A, $c/a = 1.899$, for the smallest and largest dimensions, respectively [2]. [4] found for an alloy with 50 at. %, $a = 3.115$ A, $c/a = 1.894$. With an excess of Cu, the phase was found to be isotypic with CuAu ($L1_0$ type), $a = 4.436$ A, $c = 2.813$ A, $c/a = 0.634$, and $a =$ 4.441 A, $c = 2.856$ A, $c/a = 0.643$, for the smallest and largest dimensions, respectively [2]. [4] reported $a = 4.422$ A, $c/a = 0.643$. X-ray pattern was given by [5].

Ti_2Cu_3: X-ray pattern was given by [5].

$TiCu_3$: deformed h.c.p. structure [1]. [2] reported two phases of this composition: a phase, stable at lower temperatures, with an ordered orthorhombic structure (slightly deformed h.c.p.), and a high-temperature phase with the corresponding

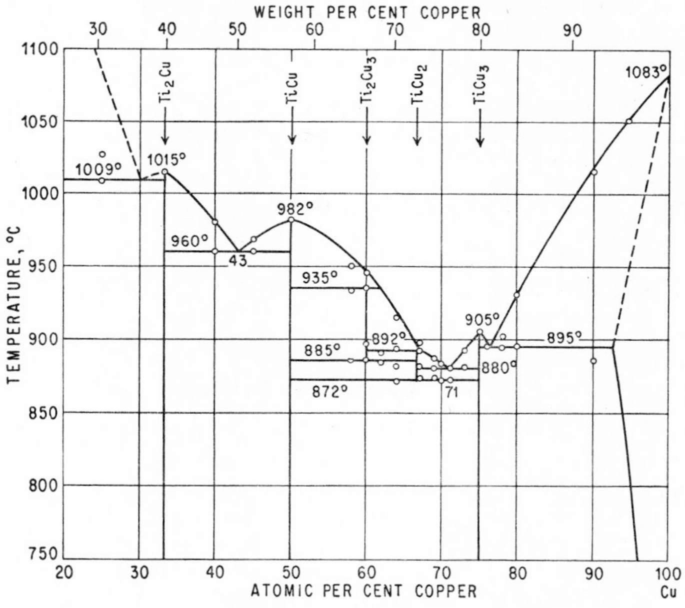

Fig. 364. Cu-Ti. (See also Fig. 363.)

disordered structure and the smallest and largest parameters $a = 2.572$ A, $b = 4.503$ A, $c = 4.313$ A, and $a = 2.585$ A, $b = 4.528$ A, $c = 4.351$ A, respectively. For an alloy with 75 at. % Cu, [4] gave $a = 5.152$ A, $b = 4.522$ A, $c = 4.338$ A.

For lattice parameters of the Cu-rich solid solution, see [4].

Supplement. The effect of heat-treatment upon the microstructures and mechanical properties of Ti-rich Ti-Cu alloys was studied by [11]. In agreement with the observations of [5], it was found that the β_{Ti} phase cannot be retained on buenching.

The diagram shown in Fig. 364 is based on thermal, X-ray, and metallographic analysis [12]. Alloys were prepared, using sponge Ti, by vacuum melting in refractory crucibles (Al_2O_3, ThO_2, and BeO). The proposed phase relationships disagree in several respects with those of [5]. Ti_2Cu_3 is reported to be an equilibrium phase over a limited temperature range (885–935°C), but capable of stabilization to lower temperatures. Ti_2Cu and $TiCu_3$ were shown to melt congruently rather than incongruently as claimed by [5]. A new phase, $TiCu_2$ (72.63 wt. % Cu), was found to exist over the temperature range 892–872°C but could not be stabilized to lower temperatures. [12] suggest that the discrepancies between their work and others lie in misinterpretation of data and uncorrelated results of previous investigators. The phase relationships in the vicinity of Ti_2Cu_3 and $TiCu_2$ appear to be worthy of more detailed scrutiny since it is impossible at this time to be certain which diagram is correct.

1. F. Laves and H. J. Wallbaum, *Naturwissenschaften,* **27,** 1939, 674–675; H. J. Wallbaum, *Naturwissenschaften,* **31,** 1943, 91–92. Evidence was found of "at least one compound" near 50 at. %.
2. N. Karlsson, *J. Inst. Metals,* **79,** 1951, 391–405.
3. W. Rostoker, *Trans. AIME,* **194,** 1952, 209–210.
4. E. Raub, P. Walter, and M. Engel, *Z. Metallkunde,* **43,** 1952, 112–118.
5. A. Joukainen, N. J. Grant, and C. F. Floe, *Trans. AIME,* **194,** 1952, 766–770.
6. C. M. Craighead, O. W. Simmons, and L. W. Eastwood, *Trans. AIME,* **188,** 1950, 485–513.
7. A. D. McQuillan, *J. Inst. Metals,* **79,** 1951, 73–88; see also A. D. McQuillan, *J. Inst. Metals,* **80,** 1951-1952, 363–368; **82,** 1953-1954, 47–48.
8. W. Kroll, *Z. Metallkunde,* **23,** 1931, 33–34.
9. F. R. Hensel and E. I. Larsen, *Trans. AIME,* **99,** 1932, 55–62.
10. E. E. Schumacher and W. C. Ellis, *Metals & Alloys,* **2,** 1931, 111.
11. F. C. Holden, A. A. Watts, H. R. Ogden, and R. I. Jaffee, *Trans. AIME,* **203,** 1955, 117–125.
12. W. Trzebiatowski, J. Berak, and T. Romotowski, *Roczniki Chem.,* **27,** 1953, 426–437.

$\bar{1}.4926$
0.5074

Cu-Tl Copper-Thallium

Thermal analysis, corroborated by metallographic examination [1], showed the existence of a miscibility gap in the liquid state (Fig. 365). [1] found the monotectic

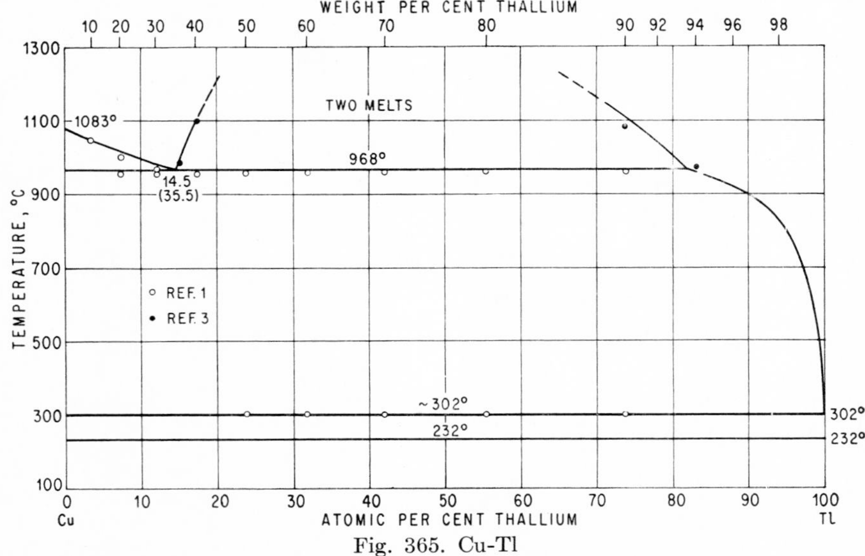

Fig. 365. Cu-Tl

point at 962°C, 34–35 wt. (13.8–14.3 at.) % Tl. The liquidus between the monotectic temperature and the melting point of Tl was obtained by solubility determinations [2]. According to a differential thermal analysis [3], the miscibility gap extends from 35.5 to 94 wt. (14.5–83 at.) % Tl at the monotectic temperature (968°C) (Fig. 365).

Average values of the solubility of Tl in solid Cu, determined [4] micrographically and by the lattice-spacing method, are as follows in wt. % (at. % in parentheses): 1050°C, 0.45 (0.14_5); 1000°C, 0.87 (0.27_5); 950°C, 0.80 (0.25); 900°C, 0.65 (0.21); 800°C, 0.29 (0.09_5); 700°C, 0.15 (0.05); 600°C, <0.1 (<0.03).

1. F. Doerinckel, *Z. anorg. Chem.*, **48,** 1906, 185–188.
2. O. J. Kleppa, *J. Am. Chem. Soc.*, **74,** 1952, 6047–6051. Data had to be converted from a small diagram.
3. H. Johnen, Dissertation, Universität Münster, Westfalen, Germany, 1952; W. Seith, H. Johnen, and J. Wagner, *Z. Metallkunde*, **46,** 1955, 773–779.
4. E. Raub and A. Engel, *Z. Metallkunde*, **37,** 1946, 76–81.

$\bar{1}.4263$
0.5737

Cu-U Copper-Uranium

The phase diagram of Fig. 366 is due to [1] and based on thermal and microscopic investigations of alloys prepared by melting under vacuum in zirconia or

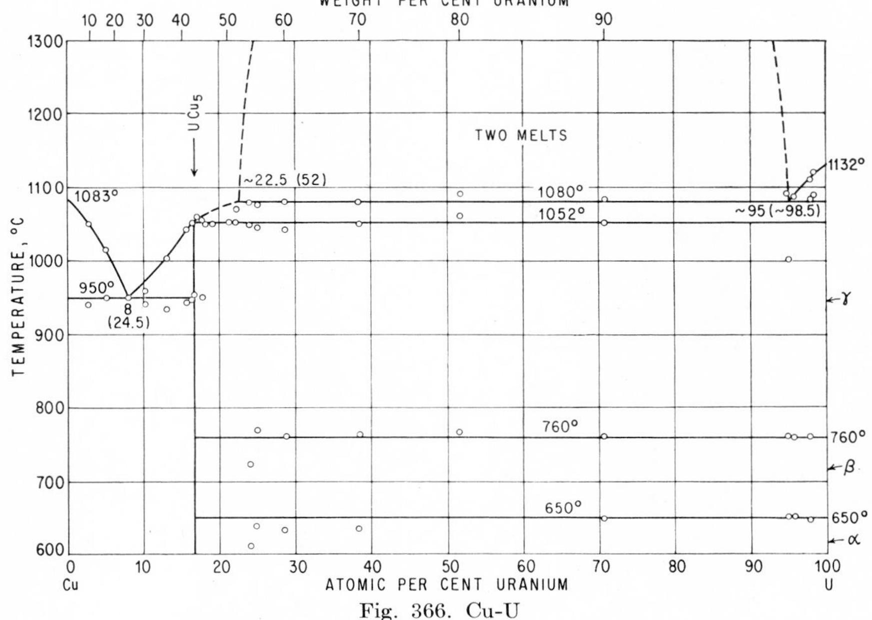

Fig. 366. Cu-U

beryllia crucibles. The existence of the phase UCu_5 (42.84 wt. % U), apparently formed peritectically at 1052°C, was confirmed by X-ray analysis [2]. Weak thermal effects observed at about 860°C in the composition range 29–98 at. % U could not be accounted for; it is probable that they were caused by an impurity introduced in preparing the alloys. Solid solubilities have not been thoroughly investigated.

Since the transformation temperatures of U are practically unaffected by Cu additions, it appears likely that the solubility of Cu in α-U, β-U, and γ-U is very slight.

UCu_5 has a f.c.c. structure, with 24 atoms per unit cell and a = 7.033–7.038 A, indicating a slight solubility of Cu. The structure is isotypic with Be_5Au and Be_5Pd and of the $MgCu_2$ (C15) type. Some Cu atoms occupy sites which would normally be occupied by U atoms. According to [3], there are indications that an additional compound, or compounds, may occur in this system.

1. H. A. Wilhelm and O. N. Carlson, *Trans. ASM*, **42,** 1950, 1311–1325.
2. N. C. Baenziger, R. E. Rundle, A. I. Snow, and A. S. Wilson, *Acta Cryst.*, **3,** 1950, 34–40.
3. Fulmer Research Institute, unpublished information, 1952-1953; Swansea University College, unpublished information, May, 1951; quoted in H. A. Saller and F. A. Rough, Compilation of U.S. and U.K. Uranium and Thorium Constitutional Diagrams, *U.S. Atomic Energy Comm., Publ.* BMI-1000, June, 1955.

0.0959
$\bar{1}$.9041

Cu-V Copper-Vanadium

There is confusion regarding the constitution of the copper-rich alloys. [1] studied alloys prepared by reduction of V_2O_5 with Al in the presence of Cu and concluded that (*a*) 6–7 wt. (7.4–8.6 at.) % V is soluble in solid copper; (*b*) alloys with 10–25 wt. (12.2–29.4 at.) % V consist of two layers, one being the solid solution and the other a bluish constituent of unknown composition; and (*c*) alloys with more than 25 wt. (29.4 at.) % V consist of two "compounds." Photomicrographs published, however, appear to indicate a different set of phase relationships. The 7 wt. % alloy, in particular, appears to consist of primary crystals in a eutectic matrix. This would be in agreement with a statement by [2] that a eutectic point exists at approximately 7 wt. % V. [3] heated aluminothermically prepared V of only 94.2% purity with molten copper to temperatures above the melting point of V. Cooling curves indicated that the liquid metals are practically immiscible. In all these cases, contaminants very likely have affected the constitution considerably. [4] found that the solid solubility of V in Cu was very small (no details were available).

[5] examined metallographically arc-melted V-rich alloys (prepared using 99.8+% vanadium) with 1–25 wt. (0.8–21.1 at.) % Cu (nominal composition) in the as-cast condition and after annealing at 900°C for 1 week. The 7.5 wt. (6.1 at.) % Cu alloy showed a cored solid solution, while the 10 wt. (8.2 at.) % and 25 wt. (21.1 at.) % Cu alloys gave evidence of Cu distributed interdendritically. The latter fact precludes the presence of intermediate phases. The solid-solubility limit of Cu in V at 900°C lies between 7.5 and 10 wt. % Cu.

Supplement. Miscroscopic and X-ray examination of an as-cast alloy of the composition VCu_2 showed the major constituents to be V and Cu solid solutions [6].

1. L. Guillet, *Rev. mét.*, **3,** 1906, 174–175.
2. G. L. Norris, *J. Franklin Inst.*, **171,** 1911, 580–581.
3. H. Giebelhausen, *Z. anorg. Chem.*, **91,** 1915, 256.
4. A. G. Dowson, Abstr. Diss. University Cambridge, 1936-1937, p. 116.
5. W. Rostoker and A. Yamamoto, *Trans. ASM*, **46,** 1954, 1136–1163.
6. R. P. Elliott, Armour Research Foundation, Chicago, Ill., Technical Report 1, OSR Technical Note OSR-TN-247, August, 1954, p. 31.

$\bar{1}.5384$
0.4616

Cu-W Copper-Wolfram

It is established that wolfram is insoluble in liquid copper [1]. So-called Cu-W "alloys" used technically are mixtures of the two metals prepared either by pressing the mixed metal powders and sintering in hydrogen above the melting point of copper or by absorption of molten copper into the pores of sintered wolfram bodies by capillary attraction.

1. L. Guillet, *Rev. mét.*, **3**, 1906, 176; O. Rumschöttel, *Metall u. Erz*, **12**, 1915, 45–50; D. Kremer, *Abhandl. Inst. Metallhütt. u. Elektromet. T. H. Aachen*, **1**(2), 1916, 10–11; K. Schröter, *Z. Metallkunde*, **23**, 1931, 197.

$\bar{1}.9876$
0.0124

Cu-Zn Copper-Zinc

The equilibrium diagram presented in Fig. 367 can be regarded as very well established. It differs only in minor points from those proposed by [1] and [2] on the basis of a critical review of the numerous data in the literature. For a detailed discussion of the results published prior to 1927, reference is made to the monograph of [3]. Data available up to 1934 and 1940 were considered by [1] and [2], respectively.

Liquidus Curve. Liquidus temperatures of the whole range of composition were reported by [4–14]. The liquidus curve between 54 and 100 at. % Zn was determined by [3], and the Zn-rich portion was also covered by the work of [15–18]. In drawing the liquidus in Fig. 367, preference was given to the data of [10, 3, 12, 13]; it is accurate within ±3°C or better [2].

The Peritectic Horizontals. $\alpha + Melt \rightleftarrows \beta$. The temperature was found as 890 [5], 893 [4, 10], 900 [13], 902 [12], 903 [11, 14], 905 [8, 19, 9], and 906°C [7, 20]. The points *B* [32.5 wt. (31.9 at.) % Zn] and *C* [36.8 wt. (36.1 at.) % Zn] were most reliably determined by [20] and [3, 13], respectively. The point *D* is located at 37.5 wt. (36.8 at.) % Zn [13].

$\beta + Melt \rightleftarrows \gamma$. The temperatures recorded vary between 833 and 840°C [5, 7–11, 3, 12–14]; 834 ± 1°C is to be regarded as most reliable [3, 12, 13]. Nearly all investigators agree in that the composition of the melt and γ is identical, within the limits of experimental error. The point *H* was placed at 62 [10], 61 [3], and 59.8 wt. % Zn [12, 13]. The latter figure (59.1 at. % Zn) was accepted in Fig. 367. The point *G* was most accurately determined as 56.5 wt. (55.8 at.) % Zn [3, 13].

$\gamma + Melt \rightleftarrows \delta$. The temperature of this three-phase equilibrium was given as 685 [4], 690 [5], 695 [10, 3], 698 [14], 699 [7], and 700°C [8, 9, 11–13]. The compositions of the points *L* and *M* were most reliably determined by [3], using micrographic analysis. They are located at 69.8 wt. (69.2 at.) % Zn and 73.0 wt. (72.45 at.) % Zn. The point *N* was found at 80.0 wt. (79.55 at.) % Zn [8, 9, 10, 12] and 80.5 wt. (80.05 at.) % Zn [11, 3, 13].

$\delta + Melt \rightleftarrows \epsilon$. The temperature was reported as 585 [5, 9], 590 [4], 592 [10], 594 [7, 3], 595 [8, 11], 597 [12], 598 [14], and 600°C [13]. The points *O* and *P* accepted in Fig. 367 are based on the micrographic work of [3]. For the point *Q*, found between 87.7 and 89.5 wt. % Zn, a composition of 88.0–88.5 wt. (87.7–88.2 at.) % Zn appears to be most reliable [8, 10, 3, 12, 13].

$\epsilon + Melt \rightleftarrows \eta$. The temperature was found as 421 [6], 423 [10, 3, 12], 423.5 [13], 424 [15, 21], 425 [7, 16, 8, 9, 11, 14, 18a], and 426°C [17]. The point *U* very likely lies at 87.5 wt. (87.2 at.) % Zn [3]. The composition of the point *V*, representing the maximal solubility of Cu in Zn, was found as 97.34 [22], 97.3 [23, 24], and 97.15 wt.

% Zn [25]. The point *W* was found as lying between 97.5 and 98.5 wt. % Zn [4, 15, 5, 7, 21, 10, 17, 11, 3, 12, 13, 14], with 98.3 wt. % Zn being the most reliable figure [13].

The Solidus Curves of the Terminal and Intermediate Phases. The solidus of the α phase shown in Fig. 367 is that determined micrographically by [26]. The solidus of the β phase was claimed practically to coincide with the liquidus up to about 47 wt. % Zn [12]. According to evidence by [13] and former workers there is no doubt, however, that a narrow freezing range exists throughout the β region.

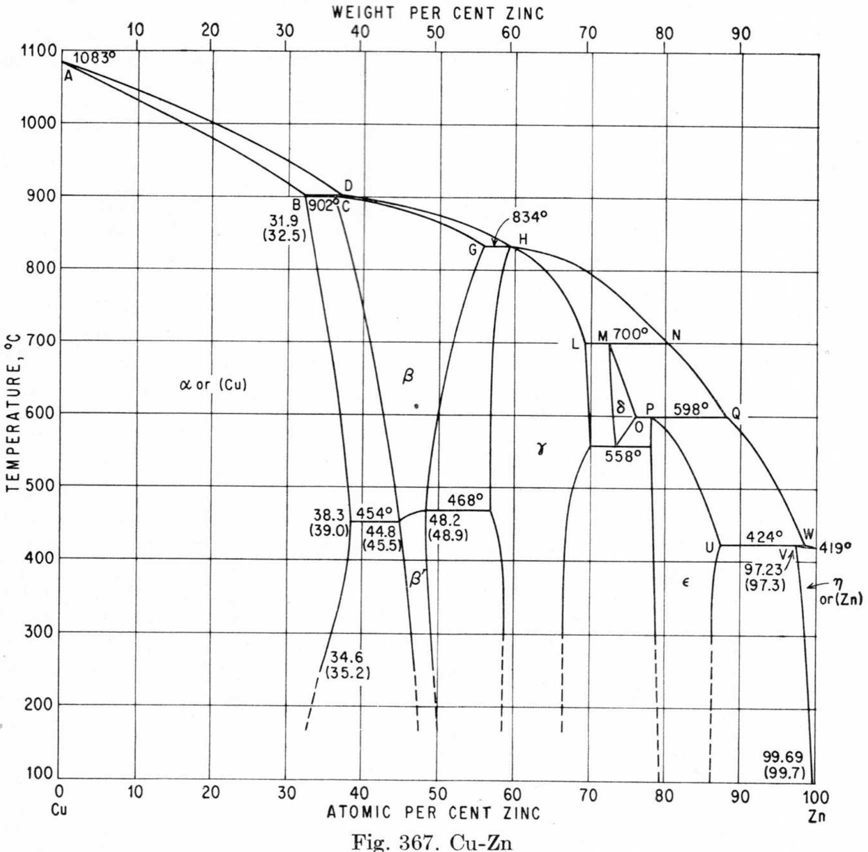

Fig. 367. Cu-Zn

The solidus curves of the γ and ϵ phases are based on thermal data [10, 12] and that of δ on micrographic results [3]. The freezing range of η is only a few °C wide.

The $\alpha/(\alpha + \beta)$ Boundary (Solubility of Zn in Cu). In the temperature range between the solidus and about 400°C, the boundary was determined micrographically [5, 19, 20, 10] and by means of X-ray work [27, 28]. Results differ considerably; see [1, 3]. The highest degree of accuracy has the curve due to [20] (Fig. 367). It was substantially confirmed by the results of [27, 29, 30] and to a somewhat lesser degree by those of [28]. [29] determined the curve in the range 590–400°C by means of micrographic analysis, and [30] fixed it below 700°C, using quantitative metallography. In addition, data reported by [31, 32, 33, 3, 34] agree with the findings of [20, 29, 30] in that the alloy with 39 wt. (38.3 at.) % Zn is single-phase at 400–450°C.

It was first conclusively shown by [35], on the basis of X-ray studies, that the solubility of Zn in Cu decreases at temperatures below 450°C from 38.75 wt. (38.1 at.) % at 400°C to 33.35 wt. (32.75 at.) % Zn at 167°C. This was verified by [36] and [30], using measurements of electrical conductivity [36], X-ray diffraction [36, 30], and quantitative metallography [30]. The data of [35, 36, 30] differ only slightly and allow one to draw the $\alpha/(\alpha + \beta)$ boundary below 400°C with fair accuracy (Fig. 367). Recently, [37] reported results which indicate that the solubility of Zn in Cu below 400°C is considerably lower than shown in Fig. 367, namely, about 38.5, 37.3, 35.4, 33.4, and 30.8 at. % Zn at 450, 400, 350, 300, and 250°C, respectively. See Supplement.

The $\beta/(\alpha + \beta)$ Boundary. Above 400°C this boundary was determined by micrographic [5, 19, 10, 3] and X-ray work [27, 28]. The portions 590–400 [29], 550–440 [38], 700–250 [30], and 450–250°C [37] were determined, using micrographic [29, 38], X-ray [30, 37], and quantitative metallographic methods [30]. For individual data, see [32, 33]. The curve shown in Fig. 367 is a compromise of the fairly consistent results of [19, 3, 29, 30, 37]. It should be emphasized here that the careful reexamination of the $\beta/(\alpha + \beta)$ boundary by [30] did not give evidence for an offset or clear change of slope at 454°C (see the paragraph on the $\beta \rightleftarrows \beta'$ transformation).

The $\beta/(\beta + \gamma)$ Boundary. Micrographic [5, 19, 10, 29, 38, 3], X-ray [27, 28, 30, 37], and quantitative metallographic studies [30] were used to determine this curve or parts thereof. For the differences between the results, see [1, 3]. In the range 500–400°C, the boundaries given by all investigators, with the exception of [38], lie between 49 and 50 wt. (48.3 and 49.3 at.) % Zn. In drawing the curve in Fig. 367, preference has been given to the findings of [3], for temperatures above 468°C, and [3, 29, 30], for the range below this temperature. There is strong evidence that below about 400°C the boundary shifts to higher Zn contents [27, 3, 30], resulting in a change of slope at about the transformation temperature of 468°C [30]. Results reported by [37, 38] are not compatible with those accepted in Fig. 367.

The $\gamma/(\beta + \gamma)$ boundary was determined by means of micrographic [39, 10, 3], X-ray [27], and quantitative metallographic studies [30], as well as emf measurements [40]. Results of [3, 27, 30] are in very good agreement (they differ by less than 0.5%), whereas those of [10, 40] are greatly at variance with the curve shown in Fig. 367.

The $\gamma/(\gamma + \delta)$ and $\gamma/(\gamma + \epsilon)$ boundaries were determined only by [10, 3], using micrographic analysis. Results agree on the whole. [27] reported the composition of the γ phase saturated at 500 and 380°C. The curves in Fig. 367 are based on the data of [3].

The δ Field. Prior to the work of [3], there were great discrepancies as to the boundaries of the δ phase. Those shown in Fig. 367 are based on the most comprehensive data of [3]. The eutectoid temperature lies probably between 555 [10, 3] and 560°C [12, 13].

The ϵ Field. The $\epsilon/(\epsilon + \gamma)$ boundary in Fig. 367 is a compromise of the data of [10, 3, 27]; they agree within ±0.5% at 400°C. The $\epsilon/(\epsilon + \eta)$ boundary is based on the closely agreeing results of [3, 27].

The $\eta/(\epsilon + \eta)$ Boundary (Solubility of Cu in Zn). The solubility curve was established by [22, 23, 24, 41, 25]. Results agree within ±0.1%, except for those of [41] at temperatures above 250°C. The most likely values in wt. % Cu (at. % in parentheses) are 0.3 (0.31), 0.9 (0.92), 1.65 (1.69), 2.5 (2.56), and 2.7 (2.77) at 100, 200, 300, 400, and 424°C, respectively.

The $\beta \rightleftarrows \beta'$ Transformation. Since the work of [42] it is conclusively established that this transformation, as suspected earlier [43, 44, 45, 1], is of the order-disorder type. It is also known that ordering cannot be suppressed by quenching from the β field [39, 43, 46].

The view of the nature of the transformation has changed over the years. The interpretation that it was a eutectoid decomposition $\beta \rightarrow \alpha + \gamma$ [47, 48, 49, 50] proved to be in error after it could be shown (*a*) that the β phase was formed at temperatures below 450°C [51, 52, 53] and (*b*) that the transformation occurred in the $\alpha + \beta$ alloys at a lower temperature than in the $\beta + \gamma$ alloys. Temperatures reported [54, 55, 39, 10, 29, 32, 3, 12, 56, 57] vary within certain limits; however, 454 ± 1 and 468 ± 1°C, respectively, may be accepted as most probable values [2]. These facts suggested the transformation to be a polymorphic change, although no difference in the crystal structure of the two phases could be found [58, 59, 45] by high-temperature X-ray diffraction [60]. As a result, the existence of a $\beta + \beta'$ region and discontinuous solubility changes of the β phase at the transformation temperatures were indicated in the phase diagram [19, 29, 32, 38, 3].

When study of the transformation characteristics by means of specific heat vs. temperature, thermoresistometric, and dilatometric work showed the transformation to be a progressive change, the effects found on cooling merely marking the beginning of the transformation [61, 62, 39, 9, 43, 58, 63, 59, 57], it was preferred not to present the transformation as a phase change. Instead, the transformation lines were drawn as dotted lines, and the β-phase boundaries were drawn without offsets at 454 and 468°C [39, 9, 10, 11, 46, 64, 1]. This was maintained [2, 30] also after evidence for the existence of an order-disorder transformation had been given [42, 65].

Presentation of order-disorder transformations in equilibrium diagrams is not being handled uniformly. While some investigators regard this type of transformation as a classical phase change, others do not. The careful investigation of [30], undertaken to prove or disprove the existence of an offset in the β-phase boundaries and, consequently, a two-phase region $\beta + \beta'$, showed that there was, within the limits of experimental error, no discontinuous composition displacement of the phase boundaries at the transition temperatures. Whether this means that there is none has to remain open. At any rate, the offset can only be of the order of 0.1 wt. (at.) %. See also [98].

Recently, [66], who postulated order-disorder transformations to be classical phase changes, published electrical-resistance-vs.-temperature curves of β alloys taken at a slow rate of heating. They were claimed to demonstrate a lower as well as an upper transition point. The lower and upper transition points of alloys with 47.25 and 47.79 wt. % Zn were claimed to be 447, 459°C, and 451, 465°C, respectively. In case the lower points were real, they may perhaps be taken as the lower limits of the $\beta + \beta'$ field, although they were found to lie 8 and 4°C below the transition temperature in the $\alpha + \beta$ field, 455°C. The lower points of alloys with 45.35 and 46.55 wt. % Zn, however, were found to occur at 429 and 431°C, i.e., at temperatures more than 20°C below the transition temperature in the $\alpha + \beta$ alloys. It appears, then, that these data cannot be accepted as conclusive proof for the existence of a classical phase change in the β phase.

A Supposed Transformation in the α Phase. On the basis of earlier data and thermoresistometric investigations [3], it had been concluded [3, 1] that no transformation occurred in the α phase. Recently, a slight anomaly of the specific heat in α-phase alloys at 200–260°C was believed to indicate a superstructure Cu_3Zn [67]. See Supplement.

The Transformation in the γ Phase. Thermoresistometric studies with alloys containing 58.6–64.4 wt. (57.9–63.7 at.) % Zn indicated a transformation at 280 [39] or 250–260°C [9]. By determining the change of the modulus of elasticity with temperature, [94] found the transformation to occur at 320°C. The nature of this transformation is unknown. Perhaps it corresponds to the change in symmetry reported by [28].

Boiling points of nine alloys with 6.5–88.5 wt. % Zn were measured by [68].

In alloys with about 38.5–41.6 wt. (37.9–40.9 at.) % Zn, a reversible martensitic-type transformation of the metastable β phase takes place, resulting in the formation of a phase (α') with a f.c. tetragonal structure [69–76]. According to [76], the M_s temperatures lie on a straight line extending from -20°C at 38.5 wt. % Zn to -131°C at 40.0 wt. % Zn.

Crystal Structures. Lattice parameters of α-phase alloys were reported by [77–81, 27, 28, 82–85, 95, 96]. The β phase is b.c.c. of the W (A2) type [86, 58, 59, 45], and the β' phase is ordered b.c.c. of the CsCl (B2) type [42, 65]. For lattice parameters of β', see [78, 79, 27, 45, 82]. The structure of the γ phase is the prototype of the so-called γ-brass structure ($D8_2$ type) [79]. A complete structure analysis was carried out by [87] which points to the ideal composition Cu_5Zn_8 (21:13 electron compound) [88]. Lattice parameters were reported by [79, 87, 28]. The γ phase saturated with Cu was claimed to have a lower, probably rhombohedral, symmetry [89, 28].

The δ phase was found to be b.c.c., with $a = 3.00$ A at 600°C for 74.5 at. % Zn [90]. The ϵ phase is h.c.p. of the Mg (A3) type, with parameters given by [78, 79, 91–93]. Lattice constants of the η phase were reported by [79, 22, 91, 23, 24, 41, 25].

Supplement. A redetermination of the $\alpha/(\alpha + \beta)$ boundary between 300 and 210°C by lattice-parameter measurements [96] indicated solubilities higher than those shown in Fig. 367. As to studies of possible ordering in α-brass, see [97].

1. M. Hansen, "Der Aufbau der Zweistofflegierungen," pp. 652–672, Springer-Verlag OHG, Berlin, 1936.
2. G. V. Raynor, Annotated Equilibrium Diagram Series, no. 3, The Institute of Metals, London, 1944.
3. O. Bauer and M. Hansen, "Der Aufbau der Kupfer-Zinklegierungen, eine Monographie," Springer-Verlag OHG, Berlin, 1927; excerpt in *Z. Metallkunde*, **19,** 1927, 423–434.
4. W. C. Roberts-Austen, *Proc. Inst. Mech. Engrs. (London)*, **1897,** 36–47; *Engineering*, **63,** 1897, 222–224, 253.
5. E. S. Shepherd, *J. Phys. Chem.*, **8,** 1904, 421.
6. O. Sackur, *Ber. deut. chem. Ges.*, **38,** 1905, 2186–2196.
7. V. E. Tafel, *Metallurgie*, **5,** 1908, 349–352, 375–383.
8. N. Parravano, *Gazz. chim. ital.*, **44**(2), 1914, 476–484.
9. H. Imai, *Science Repts. Tôhoku Univ.*, **11,** 1922, 313–332.
10. D. Iitsuka, *Mem. Coll. Sci. Kyoto Imp. Univ.*, **A8,** 1925, 179–212; *Z. Metallkunde*, **19,** 1927, 396–403.
11. E. Crepaz, *Ann. regia scuola ing., Padova*, **2,** 1926, 49–54.
12. R. Ruer and K. Kremers, *Z. anorg. Chem.*, **184,** 1929, 193–231.
13. J. Schramm, *Metallwirtschaft*, **14,** 1935, 995–1001, 1047–1050.
14. W. Broniewski, J. T. Jablonski, and S. Maj, *Compt. rend.*, **202,** 1936, 411–414.
15. C. T. Heycock and F. H. Neville, *J. Chem. Soc.*, **71,** 1897, 383, 419.
16. P. T. Arnemann, *Metallurgie*, **7,** 1910, 206–208.
17. J. L. Haughton and K. E. Bingham, *Proc. Roy. Soc. (London)*, **A99,** 1921, 47–68; W. Rosenhain, J. L. Haughton, and K. E. Bingham, *J. Inst. Metals*, **23,** 1920, 261–317.
18. G. Edmunds, *Trans. AIME*, **156,** 1944, 263–276.

18a. E. Weisse, A. Blumenthal, and H. Hanemann, *Z. Metallkunde*, **34,** 1942, 221.

19. C. H. Mathewson and P. Davidson, *J. Am. Inst. Metals*, **11,** 1917, 12–36.
20. R. Genders and G. L. Bailey, *J. Inst. Metals*, **33,** 1925, 213–221.
21. V. Jareš, *Z. Metallkunde*, **10,** 1919, 2–3.
22. M. Hansen and W. Stenzel, *Metallwirtschaft*, **12,** 1933, 539–542.

23. E. A. Anderson, M. L. Fuller, R. L. Wilcox, and J. L. Rodda, *Trans. AIME*, **111**, 1934, 264–292.
24. A. Burkhardt, *Z. Metallkunde*, **28**, 1936, 300.
25. W. Hofmann and G. Fahrenhorst, *Z. Metallkunde*, **41**, 1950, 460–462.
26. W. Hume-Rothery, G. W. Mabbott, and K. M. Channel-Evans, *Phil. Trans. Roy. Soc.* (*London*), **A233**, 1934, 74.
27. E. A. Owen and L. Pickup, *Proc. Roy. Soc.* (*London*), **A137**, 1932, 397–417; **140**, 1933, 179–191, 191–204; see also *J. Inst. Metals*, **55**, 1934, 215–228.
28. A. Johansson and A. Westgren, *Metallwirtschaft*, **12**, 1933, 385–387.
29. M. L. V. Gayler, *J. Inst. Metals*, **34**, 1925, 235–244.
30. L. H. Beck and C. S. Smith, *Trans. AIME*, **194**, 1952, 1079–1083.
31. O. W. Ellis, *Trans. AIME*, **70**, 1924, 389–390.
32. J. L. Haughton and W. T. Griffiths, *J. Inst. Metals*, **34**, 1925, 245–253.
33. O. W. Ellis and M. A. Haughton, *J. Inst. Metals*, **33**, 1925, 223–225.
34. F. Ostermann, *Z. Metallkunde*, **19**, 1928, 186.
35. S. T. Konobeevsky and V. Tarasova, *Physik. Z. Sowjetunion*, **10**, 1936, 427–428.
36. C. Haase and F. Pawlek, *Vorträge der Hauptversammlung* 1938 *der deutschen Gesellschaft für Metallkunde*, **1938**, 57–60.
37. F. Lihl, *Z. Metallkunde*, **46**, 1955, 441.
38. P. Saldau and I. Schmidt, *J. Inst. Metals*, **34**, 1925, 258–260; *Z. anorg. Chem.*, **173**, 1928, 273–286; *Z. Metallkunde*, **21**, 1929, 97–98.
39. T. Matsuda, *Science Repts. Tôhoku Univ.*, **11**, 1922, 251–268.
40. A. Ölander, *Z. physik. Chem.*, **A164**, 1933, 428–438.
41. K. Löhberg, *Z. Metallkunde*, **32**, 1940, 86–90.
42. F. W. Jones and C. Sykes, *Proc. Roy. Soc.* (*London*), **A161**, 1937, 440–446.
43. G. Tammann and O. Heusler, *Z. anorg. Chem.*, **158**, 1926, 349–358.
44. H. v. Steinwehr and A. Schulze, *Physik. Z.*, **35**, 1934, 385–397; *Z. Metallkunde*, **26**, 1934, 130–135.
45. E. A. Owen and L. Pickup, *Proc. Roy. Soc.* (*London*), **A145**, 1934, 258–267.
46. M. Hansen, *Z. Physik*, **59**, 1930, 466–496.
47. H. C. H. Carpenter and C. A. Edwards, *J. Inst. Metals*, **5**, 1911, 127–149, 158–193; *Intern. Z. Metallog.*, **1**, 1911, 156–172.
48. H. C. H. Carpenter, *J. Inst. Metals*, **7**, 1912, 70–88; *Intern. Z. Metallog.*, **2**, 1912, 129–149.
49. H. C. H. Carpenter, *J. Inst. Metals*, **8**, 1912, 51–58, 59–73.
50. J. H. Andrew and R. Hay, *J. Inst. Metals*, **34**, 1925, 185–187.
51. O. F. Hudson, *J. Inst. Metals*, **12**, 1914, 89–99; see also O. F. Hudson and R. M. Jones, *J. Inst. Metals*, **14**, 1915, 98–108.
52. H. Weiss, *Compt. rend.*, **171**, 1920, 108–111.
53. G. Masing, *Z. Metallkunde*, **16**, 1924, 96–98; *Z. anorg. Chem.*, **62**, 1909, 301–303.
54. M. P. Slavinsky, *Zhur. Russ. Met. Obshchestva*, **1**, 1914, 778.
55. Hatch, see [19].
56. R. Ruer, *Z. anorg. Chem.*, **209**, 1932, 364–368.
57. C. Sykes and H. Wilkinson, *J. Inst. Metals*, **61**, 1937, 223–239.
58. A. Phillips and L. W. Thelin, *J. Franklin Inst.*, **204**, 1927, 359–368.
59. v. Göler and G. Sachs, *Naturwissenschaften*, **16**, 1928, 412–416.
60. Under normal conditions of working, superstructure lines cannot be detected, since the Cu and Zn atoms have a nearly equal diffraction power; see [78] and [79].
61. P. D. Merica and L. W. Schad, *Natl. Bur. Standards* (*U.S.*), *Bull.*, **14**, 1919, 571–590.
62. P. Braesco, *Ann. phys.*, **14**, 1920, 5–75.
63. C. H. Johansson, *Ann. Physik*, **84**, 1927, 976–1008; *Z. anorg. Chem.*, **187**, 1930, 334–336.

64. O. Bauer and M. Hansen, *Z. Metallkunde*, **24**, 1932, 1–2.
65. H. Nowotny and A. Winkels, *Z. Physik*, **114**, 1939, 455–458.
66. F. N. Rhines and J. B. Newkirk, *Trans. ASM*, **45**, 1953, 1029–1046.
67. H. Masumoto, H. Saito, and M. Sugihara, *Nippon Kinzoku Gakkai-Shi*, **16**, 1952, 359–361; *Met. Abstr.*, **20**, 1953, 563.
68. W. Leitgebel, *Z. anorg. Chem.*, **202**, 1931, 305–324.
69. E. Kaminsky and G. Kurdjumow, *Metallwirtschaft*, **15**, 1936, 905–907; *Zhur. Tekh. Fiz.*, **6**, 1936, 984–988.
70. E. Kaminsky, *Zhur. Tekh. Fiz.*, **8**, 1938, 1781–1792; *Met. Abstr.*, **7**, 1940, 156.
71. I. Isaitchew and V. Miretsky, *Zhur. Tekh. Fiz.*, **8**, 1938, 1333–1339; *Met. Abstr.*, **7**, 1940, 156.
72. A. B. Greninger and V. G. Mooradian, *Trans. AIME*, **128**, 1938, 337–355.
73. I. Isaitchew, E. Kaminsky, and G. Kurdjumow, *Trans. AIME*, **128**, 1938, 361–367.
74. V. Ganenko and T. Sempur, *Zhur. Tekh. Fiz.*, **10**, 1940, 571–573; *Met. Abstr.*, **10**, 1943, 387.
75. G. Kurdjumow, *Zhur. Tekh. Fiz.*, **18**, 1948, 999; see [76].
76. A. L. Titchener and M. B. Bever, *Trans. AIME*, **200**, 1954, 303–304.
77. E. C. Bain, *Chem. Met. Eng.*, **28**, 1923, 21–22.
78. E. A. Owen and G. D. Preston, *Proc. Phys. Soc. (London)*, **36**, 1923, 49–66.
79. A. Westgren and G. Phragmén, *Phil. Mag.*, **50**, 1925, 311–341.
80. S. Sekito, *Science Repts. Tôhoku Univ.*, **18**, 1929, 59–68.
81. v. Göler and G. Sachs, *Z. Physik*, **55**, 1929, 581–620.
82. E. A. Owen and L. Pickup, *J. Inst. Metals*, **55**, 1934, 215–228.
83. E. A. Owen and L. Pickup, *Proc. Roy. Soc. (London)*, **A149**, 1935, 282–298.
84. W. Hume-Rothery, G. F. Lewin, and P. W. Reynolds, *Proc. Roy. Soc. (London)*, **A157**, 1936, 167–183.
85. E. A. Owen and E. W. Roberts, *Phil. Mag.*, **27**, 1939, 294–327.
86. M. R. Andrews, *Phys. Rev.*, **18**, 1921, 245–254.
87. A. J. Bradley and J. Thewlis, *Proc. Roy. Soc. (London)*, **A112**, 1926, 678–692; see also A. J. Bradley and C. H. Gregory, *Phil. Mag.*, **12**, 1931, 143–162.
88. W. Hume-Rothery, J. O. Betterton, and J. Reynolds, *J. Inst. Metals*, **80**, 1951–1952, 609–616.
89. A. Westgren, *Trans. AIME*, **93**, 1931, 20–21.
90. K. Schubert and E. Wall, *Z. Metallkunde*, **40**, 1949, 383–385.
91. E. A. Owen and L. Pickup, *Proc. Roy. Soc. (London)*, **A140**, 1933, 179–191.
92. K. Moeller, *Z. Metallkunde*, **35**, 1943, 27–28.
93. V. Montoro, *Metallurgia ital.*, **35**, 1943, 57–58.
94. W. Köster, *Z. Metallkunde*, **32**, 1940, 155.
95. A. P. Guljaev and E. F. Trusova, *Zhur. Tekh. Fiz.*, **20**, 1950, 66–78; *Structure Repts.*, **13**, 1950, 103.
96. S. Yamada, see *Structure Repts.*, **13**, 1950, 109.
97. D. T. Keating, *Acta Met.*, **2**, 1954, 885–887.
98. D. Balesdent, *Compt. rend.*, **242**, 1956, 116–119 (Experimental determination of the order of $\beta \rightarrow \beta'$ transformation by a chemical equilibrium method); *Met. Abstr.*, **23**, 1956, 1093–1094.

$\bar{1}.8430$
0.1570

Cu-Zr Copper-Zirconium

Figure 368 shows the phase diagram due to [1]. It is based on micrographic analysis, supplemented by thermal investigation, of arc-cast alloys prepared using iodide Zr crystal bar (99.8% pure). Temperatures indicated are reported to be accurate within ±10°C. The composition of the various eutectics as well as the intermediate phases Zr_2Cu (74.17 wt. % Zr), ZrCu (58.94 wt. % Zr), and Zr_2Cu_3 (48.90 wt. % Zr) are well established. The compositions of the two other intermediate phases richer in Cu, denoted as Zr_2Cu_5 (36.48 wt. % Zr) and $ZrCu_3$ (32.37 wt. % Zr), are open to some doubt since the phase formed peritectically at 1070°C might

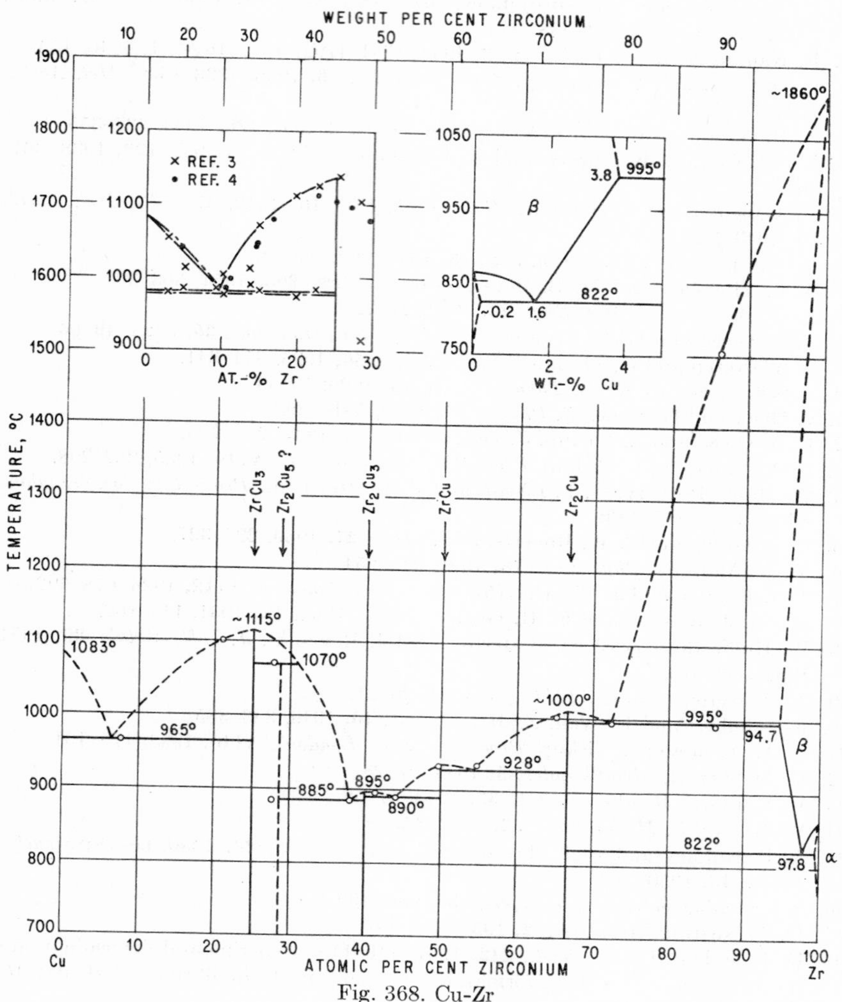

Fig. 368. Cu-Zr

alternatively be $ZrCu_3$ and that melting at about 1100°C might possibly be $ZrCu_4$ (26.41 wt. % Zr).

Prior work on this system was confined largely to Cu-rich [2–4] and Zr-rich alloys [5, 6], using techniques less suitable for these alloys than those used by [1]. According to [2], a eutectic Cu-$ZrCu_3$ exists at 12.5 wt. (9.1 at.) % Zr, 964°C. [3]

studied alloys up to 36 wt. (28.2 at.) % Zr by thermal analysis. They found the Cu-$ZrCu_3$ eutectic at 12.9 wt. (9.4 at.) % Zr, 980°C, and the melting point of $ZrCu_3$ as 1138°C. [4] confirmed the compound $ZrCu_3$ (melting point about 1115°C) and located the Cu-$ZrCu_3$ eutectic at 13.7 wt. (10.0 at.) % Zr, 977°C. Between $ZrCu_3$ and 60 at. % Zr only one eutectic was found at 51 wt. (42 at.) % Zr, 877°C; i.e., the phases Zr_2Cu_5, Zr_2Cu_3, and ZrCu were not detected (see inset of Fig. 368).

[5] examined microscopically as-cast alloys with 90–99.5 wt. % Zr, prepared by melting sponge Zr-base alloys in graphite crucibles. [6] outlined phase relationships in the range between Zr_2Cu (melting point about 1075°C) and Zr, using alloys based on commercially pure Zr and melted in graphite crucibles. The Zr_2Cu-β eutectic was placed at 80.3 wt. (74 at.) % Zr, 998°C, and the maximal solubility of Cu in β-Zr at about 7.5 wt. (10.4 at.) % Cu. [7] briefly reported identifying the compound Zr_3Cu; however, this phase definitely does not exist [1].

There is agreement as to the existence of the phases $ZrCu_3$ [2–4, 1] and Zr_2Cu [6, 1]. On the other hand, the composition of the Cu-$ZrCu_3$ eutectic differs widely, between 9.1–10 at. % Zr [2–4] and 6.5 at. % Zr [1].

The solid solubility of Zr in Cu was reported to be 0.9, 0.7, and 0.28 wt. (0.64, 0.5, and 0.2 at.) % Zr at 925, 825, and 600°C, respectively, based on micrographic work [3]. According to [4], the solubility is indeterminable parametrically. An alloy with 0.13 wt. % Zr, quenched from 940°C, proved to be two-phase [4]. This value is much lower than those reported by [3]. A decrease in solubility is indicated by age-hardening tests [8, 9].

According to [7], the (nonexistent) phase Zr_3Cu is tetragonal (distorted Cu_3Au type), a = 4.550 A, c = 3.726 A, c/a = 0.819. [1] reported nearly the same parameters for the f.c. tetragonal unit cell of Zr_2Cu, a = 4.545 A, c = 3.723 A, c/a = 0.819.

Supplement. According to [10], the X-ray diffraction pattern of the $ZrCu_2$ alloy was complex and bore no similarity to Laves-type patterns.

1. C. E. Lundin, D. J. McPherson, and M. Hansen, *Trans. AIME*, **197**, 1953, 273–278.
2. T. E. Allibone and C. Sykes, *J. Inst. Metals*, **39**, 1928, 176–179.
3. S. A. Pogodin, I. S. Shumova, and F. A. Kugucheva, *Compt. rend. acad. sci. U.S.S.R.*, **27**, 1940, 670–672; S. A. Pogodin and I. S. Shumova, *Izvest. Sektora Fiz.-Khim. Anal.*, **13**, 1940, 225–232; see also "Metals Handbook," 1948 ed., p. 1207, The American Society for Metals, Cleveland, Ohio.
4. E. Raub and M. Engel, *Z. Metallkunde*, **39**, 1948, 172–177.
5. C. T. Anderson, E. T. Hayes, A. H. Roberson, and W. J. Kroll, *U.S. Bur. Mines, Rept. Invest.* 4658, 1950.
6. R. N. Augustson, *U.S. Atomic Energy Comm., Publ.* AECD-3456 (and ISC-138), 1950.
7. N. Karlsson, *J. Inst. Metals*, **79**, 1951, 404 (footnote).
8. G. F. Comstock and R. E. Bannon, *Metals & Alloys*, **8**, 1937, 106.
9. F. R. Hensel, E. I. Larsen, and A. S. Doty, *Metals & Alloys*, **10**, 1939, 372–373.
10. R. P. Elliott, Armour Research Foundation, Chicago, Ill., Technical Report 1, OSR Technical Note OSR-TN-247, August, 1954, p. 36.

2.1784
3.8216

Eu-H Europium-Hydrogen

Europium deuteride, $EuD_{1.95}$, was shown to have the orthorhombic SrH_2 (C29) type of structure, with a = 6.21 A, b = 3.77 A, c = 7.16 A (each ±0.02 A) [1].

1. W. L. Korst and J. C. Warf, *Acta Cryst.*, **9**, 1956, 452–454.

$\bar{1}.9037$
0.0963

Fe-Ga Iron-Gallium

According to [1], Ga alloys easily with Fe. An alloy with 1.2 wt. % Ga could be rolled easily. No age-hardenability could be observed.

1. W. Kroll, *Metallwirtschaft*, **11**, 1932, 435–437.

$\bar{1}.5514$
0.4486

Fe-Gd Iron-Gadolinium

There are at least three intermediate phases, all of which are ferromagnetic [1]. The second of these is $GdFe_2$ (58.41 wt. % Gd), which is isotypic with $MgCu_2$ (C15 type), $a = 7.44$ A [2].

1. W. Klemm, *FIAT Rev. Ger. Sci.*, 1939-1946, Inorganic Chemistry, Part III, p. 255.
2. F. Endter and W. Klemm, *Z. anorg. Chem.*, **252**, 1944, 377–379.

$\bar{1}.8861$
0.1139

Fe-Ge Iron-Germanium

The phase diagram in Fig. 369 is due to [1]; data points were taken from a small figure. The solid-state phase boundaries have not been determined; however, micro-

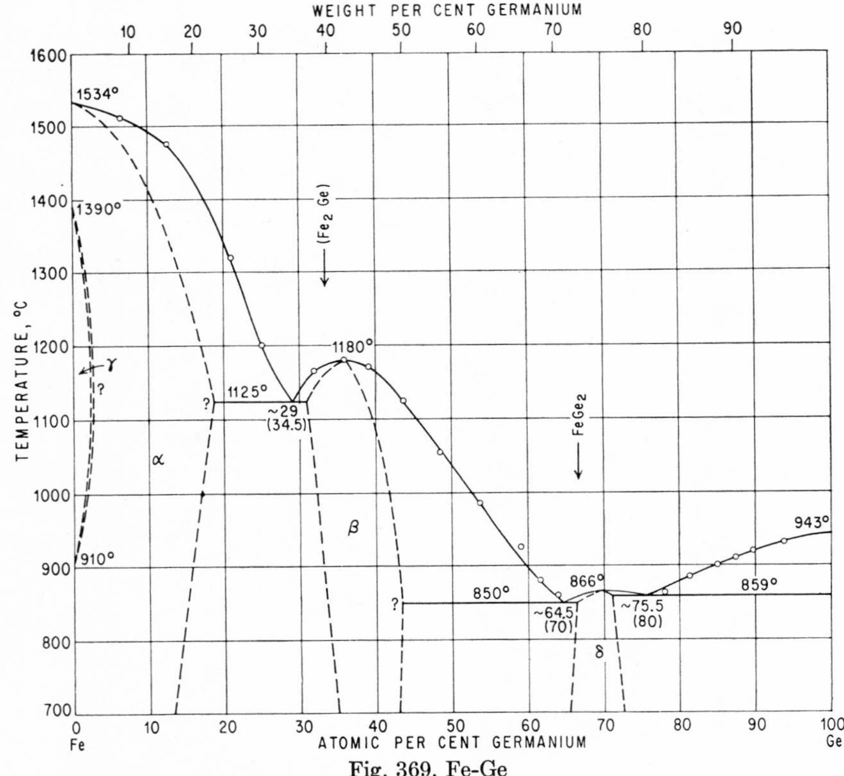

Fig. 369. Fe-Ge

scopic examination showed that the alloy of the composition Fe_2Ge, homogeneous after solidification, becomes heterogeneous on cooling by precipitating α crystals.

On rapid solidification of melts with more than 60 at. % Ge, the crystallization of the δ phase is suppressed and a metastable eutectic (β + Ge) formed at approximately 70 at. % Ge and 830°C.

Indication was found of a closed γ loop, which had been predicted by [2].

The β phase has the structure of the (partially filled) NiAs (B8) type, with a = 4.039 A, c = 5.032 A, c/a = 1.246 at 36.2 at. % Ge [3, 4]. δ ($FeGe_2$) is isotypic with $CuAl_2$ (C16 type), a = 5.911 A, c = 4.951 A, c/a = 0.838 [5].

1. K. Ruttewit and G. Masing, *Z. Metallkunde*, **32**, 1940, 52–61.
2. F. Wever, *Arch. Eisenhüttenw.*, **2**, 1928-1929, 739–746; *Naturwissenschaften*, **17**, 1929, 304–309.
3. F. Laves and H. J. Wallbaum, *Z. angew. Mineral.*, **4**, 1941-1942, 17–46.
4. For the alloy with 37 at. % Ge, L. Castelliz (*Monatsh. Chem.*, **84**, 1953, 765–776) found a = 4.017 A, c = 5.005 A, c/a = 1.246.
5. H. J. Wallbaum, *Z. Metallkunde*, **35**, 1943, 218–221.

1.7436
$\bar{2}$.2564

Fe-H Iron-Hydrogen

As shown in Fig. 370, the solubility isobars for 1 atm pressure determined by [1] and [2] are in very good agreement [3]. Also shown is the isobar according to [4], which agrees with those by [1, 2] up to about 1150°C. It indicates a solubility hysteresis at the $\alpha \rightleftarrows \gamma$ transformation and also a discontinuous change in solubility at the $\gamma \rightleftarrows \delta$ transformation characterized by a hysteresis. Neither had been found by [1]. The solubility curve of the δ phase forms a continuation of that of the α phase. Work by [5] confirmed qualitatively the presence of the upper discontinuity as well as the hystereses at both the $\alpha \rightleftarrows \gamma$ and $\gamma \rightleftarrows \delta$ transformations.

According to [6], the data by [1, 2] were "confirmed with satisfactory approximation, for temperatures between about 500 and 1200°C, by more recent studies of [7, 8, 5, 9, 10], and to 600°C by that of [11]. They may therefore be supposed to represent, with an accuracy which cannot at present readily be exceeded, the equilibrium values for occlusion of hydrogen by iron at atmospheric pressure over the range of temperatures indicated, except for the immediate vicinity of the A_3 transition near 910°C."

For additional information the reader is referred to the comprehensive treatment of the subject by [6] which contains numerous references to the literature.

Supplement. Data reported by [12] for the solubility of hydrogen in liquid iron under 760 mm Hg pressure at 1685 and 1560°C are plotted in Fig. 370. The location of hydrogen in the α-Fe lattice has been studied by [13], using X-ray diffraction.

1. A. Sieverts (E. Jurich), *Z. physik. Chem.*, **77**, 1911, 591–613.
2. E. Martin, *Arch. Eisenhüttenw.*, **3**, 1929-1930, 407–416; *Metals & Alloys*, **1**, 1929, 831–835.
3. Tabulated data in [6].
4. L. Luckemeyer-Hasse and H. Schenck, *Arch. Eisenhüttenw.*, **6**, 1932-1933, 209–214.
5. A. Sieverts, G. Zapf, and H. Moritz, *Z. physik. Chem.*, **A183**, 1938, 19–37.
6. D. P. Smith, "Hydrogen in Metals," pp. 39–58, 257–261, University of Chicago Press, Chicago, 1948.
7. F. Pihlstrand, *Jernkontorets Ann.*, **121**, 1937, 219–231.
8. W. Baukloh and R. Müller, *Arch. Eisenhüttenw.*, **11**, 1937, 509–514.
9. K. Iwase and M. Fukusima, *Science Repts. Tôhoku Univ.*, **27**, 1938, 162.

10. J. H. Andrew, H. Lee, and A. C. Quarrell, *J. Iron Steel Inst.*, **146,** 1942, 181–192.
11. M. H. Armbruster, *J. Am. Chem. Soc.*, **65,** 1943, 1043–1054; in this paper, solubilities according to [2, 4, 7, 8, 9, 5, 10, 11] at 300–900°C, 1 atm, are tabulated.
12. M. Karnaukhov and A. N. Morozov, *Izvest. Akad. Nauk S.S.S.R., Otdel. Tekh. Nauk,* **1948,** 1845-1855; *Chem. Abstr.*, **43,** 1949, 2490.
13. J. Plusquellec, P. Azou, and P. Bastien, *Compt. rend.*, **244,** 1957, 1195–1197.

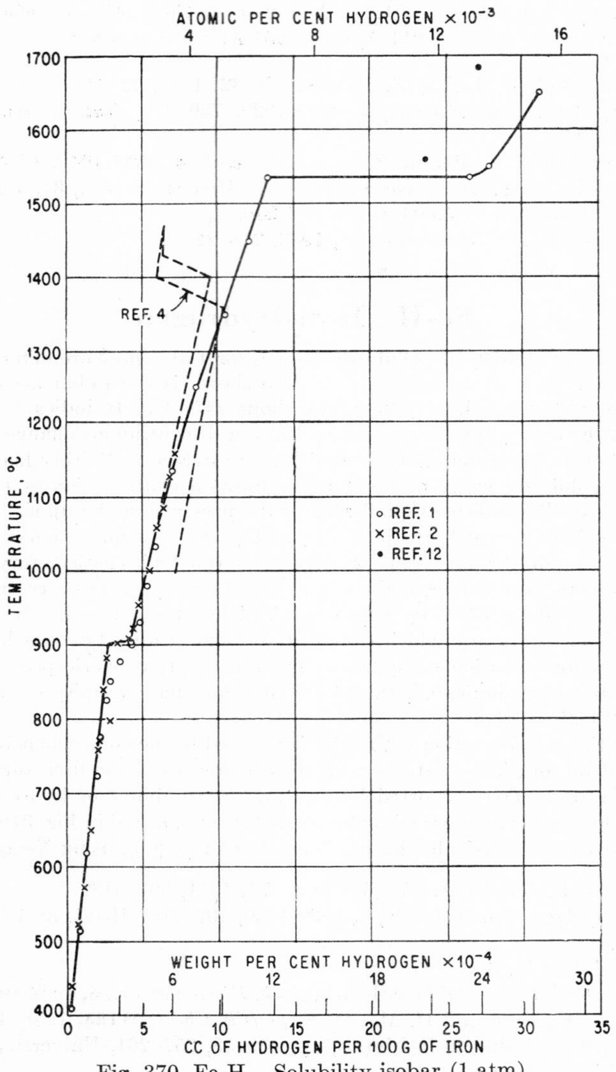

Fig. 370. Fe-H. Solubility isobar (1 atm).

$\bar{1}.4951$
0.5049

Fe-Hf Iron-Hafnium

"An alloy of [the composition] $HfFe_2$ exhibits a strong pattern of the $MgCu_2$ [C15] modification with a lattice parameter of a = 7.025 A, but there is also a faint but positive pattern of a phase isomorphous with $MgZn_2$ [C14] type. This duplex pattern occurs at all temperatures from 600 to 1400°C. The pattern of the 1600°C sample is entirely of the $MgZn_2$ modification, with parameters c = 8.103 A, a = 4.962 A, c/a = 1.633. The incipient melting temperature of $HfFe_2$ has been determined to be in the range of 1650°C" [1].

1. R. P. Elliott, Armour Research Foundation, Chicago, Ill., Technical Report 1, OSR Technical Note OSR-TN-247, August, 1954, p. 11 (X-ray examination of quenched samples).

$\bar{1}.4447$
0.5553

Fe-Hg Iron-Mercury

By means of roentgenographic investigation, analytical determination, and other methods, it has been conclusively established that iron amalgam is a suspension of finely divided Fe in Hg [1–5]. Reported values (in wt. %) of the solubility of Fe in Hg at room temperature differ very widely: <0.01 [6], 1.34×10^{-3} [7], 1.15×10^{-17} [8], 6.2×10^{-2} [9], 7×10^{-5} [2], $<1 \times 10^{-5}$ [10], and 0.15×10^{-5} [11]. The findings by [2, 10, 11] agree at least in that they are of about the same order of magnitude.

[2] reported that the solubility does not change noticeably between 20 and 211°C. [11] determined the solubility at various temperatures up to 700°C (at this temperature the Hg pressure is about 50 atm) and gave the following smoothed values, in 10^{-5} wt. %, derived from an equation based on the experimental results: 100°C, 0.19; 200°C, 0.30; 300°C, 0.54; 400°C, 1.1; 500°C, 2.1; 600°C, 4.5; and 700°C, 9.6.

According to [12], Hg is insoluble in solid Fe.

Supplement. According to colorimetric and spectrometric analyses [13], the solubility of Fe in Hg at room temperature is less than 5×10^{-7} wt. ($<18 \times 10^{-7}$ at.) %.

1. M. Rabinowitsch and P. B. Zywotinski, *Kolloid Z.*, **52,** 1930, 31–37.
2. E. Palmaer, *Z. Elektrochem.*, **38,** 1932, 70–76.
3. R. Brill and W. Haag, *Z. Elektrochem.*, **38,** 1932, 211–212.
4. F. Pawlek, *Z. Metallkunde*, **41,** 1950, 451–453.
5. F. Lihl, *Z. Metallkunde*, **44,** 1953, 160–166.
6. Gouy, *J. phys.*, **4,** 1895, 320–321.
7. T. W. Richards and R. N. Garrod-Thomas, *Z. physik. Chem.*, **72,** 1910, 181–182.
8. G. Tammann and K. Kollmann, *Z. anorg. Chem.*, **160,** 1927, 243–246.
9. G. Tammann and W. Oelsen, *Z. anorg. Chem.*, **186,** 1930, 280–281.
10. N. M. Irvin and A. S. Russell, *J. Chem. Soc.*, **1932,** 891–898.
11. A. L. Marshall, L. F. Epstein, and F. J. Norton, *J. Am. Chem. Soc.*, **72,** 1950, 3514–3516.
12. F. Wever, *Arch. Eisenhüttenw.*, **2,** 1928-1929, 739–746; *Naturwissenschaften*, **17,** 1929, 304–309.
13. J. F. de Wet and R. A. W. Haul, *Z. anorg. Chem.*, **277** 1954, 96–112.

$\bar{1}.4612$
0.5388

Fe-Ir Iron-Iridium

As suggested by [1], the system Fe-Ir has an open γ field; i.e., γ-Fe and Ir form a continuous series of solid solutions [2, 3]. Figure 371 shows the temperatures of the $\alpha' \rightarrow \gamma$ and $\gamma \rightarrow \alpha'$ transformations on heating and cooling, respectively, and the Curie points, determined by magnetic analysis [2]. Hardness and microstructure of alloys slowly cooled from 1300°C indicate the formation of an ordered structure based on the composition FeIr (77.57 wt. % Ir) [3].

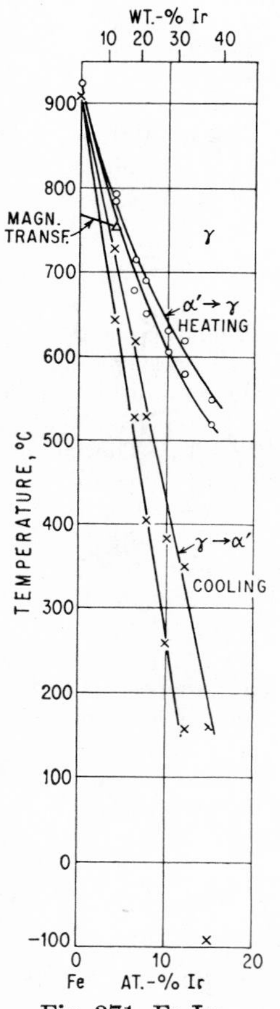

Fig. 371. Fe-Ir

1. F. Wever, *Arch. Eisenhüttenw.*, **2,** 1928-1929, 739–746; *Naturwissenschaften*, **17,** 1929, 304–309.
2. M. Fallot, *Ann. phys.*, **10,** 1938, 291–332; *Compt. rend.*, **205,** 1937, 517–518.
3. V. A. Nemilov and T. A. Vidusova, *Izvest. Sektora Platiny*, **20,** 1947, 240–244.

0.1548
$\bar{1}.8452$

Fe-K Iron-Potassium

K is regarded as being insoluble in solid Fe [1]; no diffusion of Fe in Armco iron at 1000–1300°C could be detected [2]. The boiling point of K (770°C) lies far below the melting point of Fe.

1. F. Wever, *Arch. Eisenhüttenw.*, **2,** 1928-1929, 739–746; *Naturwissenschaften*, **17,** 1929, 304–309.
2. W. D. Jones, *J. Iron Steel Inst.*, **130,** 1934, 436.

0.9057
$\bar{1}.0943$

Fe-Li Iron-Lithium

Li is regarded as being insoluble in solid Fe [1]. Boiling Li (boiling point 1370°C) reportedly attacks Fe [2]; however, no diffusion of Li into Fe at 1200°C was observed by [3].

1. F. Wever, *Arch. Eisenhüttenw.*, **2,** 1928-1929, 739–746; *Naturwissenschaften*, **17,** 1929, 304–309.
2. O. Ruff and O. Johannsen, *Ber. deut. chem. Ges.*, **38,** 1905, 3602.
3. N. W. Ageew and M. I. Zamotorin, *Ann. inst. polytech., Leningrad, Sec. math. phys. sci.*, **31,** 1928, 15–28; abstract in *J. Inst. Metals*, **44,** 1930, 556.

0.3611
$\bar{1}.6389$

Fe-Mg Iron-Magnesium

According to [1], Mg is insoluble in solid Fe. [2] concluded from magnetic measurements that there is at least a slight solubility. Also, emf measurements and other data appeared to indicate a certain solid solubility [3]. Primary crystals isolated from Mg-rich alloys might have contained some Mg [4]; however, this could not be definitely proved by X-ray investigation [4, 5].

The solubility of Fe in liquid Mg was determined by [6–9, 5]. Results (in wt. % Fe) are presented in Fig. 372. Up to 750°C, solubilities reported by [7, 9] agree well, and at higher temperatures, those by [7, 5] are in fair accord. However, considering the low absolute solubility, agreement of all data is quite satisfactory. Solubilities at 1000, 1100, and 1200°C were found to be 0.32, 0.56, and 0.84 wt. % Fe, respectively [7].

All the solubility (or liquidus) curves in Fig. 372 indicate that the solubility at the melting point of Mg has a finite value. If one assumes that Mg and Fe form a

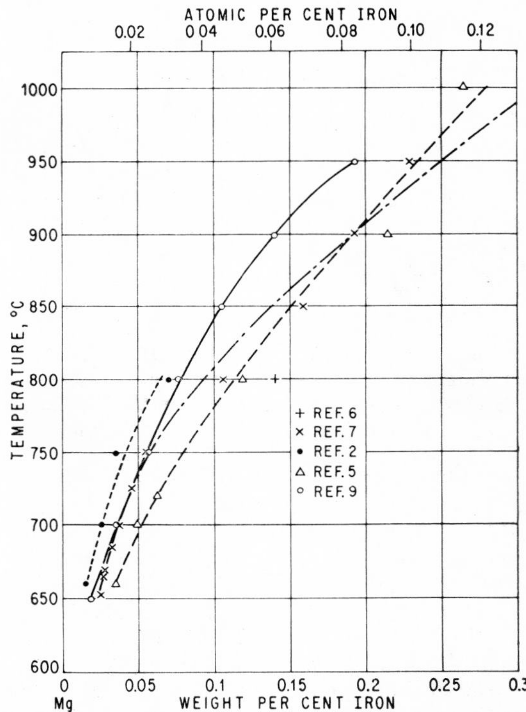

Fig. 372. Fe-Mg. Solubility of Fe in liquid Mg.

eutectic at a temperature close to the melting point of Mg (650°C), the eutectic composition would lie between 0.015 and 0.03 wt. (0.0065 and 0.013 at.) % Fe; see also [10].

Microstructures of Mg-rich alloys were published by [11, 7, 4, 12, 5].

1. F. Wever, *Arch. Eisenhüttenw.*, **2,** 1928-1929, 739–746; *Naturwissenschaften*, **17,** 1929, 304–309.
2. T. D. Yensen and N. A. Ziegler, *Trans. AIME*, **95,** 1931, 313–317.
3. R. Kremann and J. Lorber, *Monatsh. Chem.*, **35,** 1914, 603–614.
4. W. Bulian and E. Fahrenhorst, *Z. Metallkunde*, **34,** 1942, 166–170.
5. D. W. Mitchell, *Trans. AIME*, **175,** 1948, 570–578.
6. A Beerwald, *Z. Metallkunde*, **33,** 1941, 28–31.
7. E. Fahrenhorst and W. Bulian, *Z. Metallkunde*, **33,** 1941, 31–34; additional references are given.
8. A. Beerwald, *Metallwirtschaft*, **23,** 1944, 404–407.
9. G. Siebel, *Z. Metallkunde*, **39,** 1948, 22–27.

10. W. R. D. Jones, *Metallurgist*, **11**, 1937-1938, 157–158.
11. H. Vosskühler, in A. Beck, "Magnesium und seine Legierungen," p. 43, Springer-Verlag OHG, Berlin, 1939; "The Technology of Magnesium and Its Alloys," F. A. Hughes & Co., Ltd., London, 1940.
12. W. Bulian and E. Fahrenhorst, "Metallographie des Magnesiums und seiner technischen Legierungen," pp. 27–31, Springer-Verlag OHG, Berlin, 1942; 2d ed., 1949.

0.0072
$\bar{1}$.9928

Fe-Mn Iron-Manganese

Solidification (Fig. 373). Solidification temperatures covering the whole range of composition were determined by [1–3] and in the range 64–100 at. % Mn by [4]. Liquidus temperatures determined by [3, 4] agree within the limit of accuracy. However, results by [4] differ from those by [3] in that they show a flat minimum at about 91% Mn, 1243°C, and a peritectic horizontal (δ_{Mn} + melt $\rightleftarrows \gamma_{Mn}$) at about 1245°C, which extends over the approximate range 94–98% Mn. (These temperatures are based on a melting point of Mn of 1252 instead of 1245°C.)

Between 65.4 and 74.3 at. % Mn, [3] assumed a peritectic horizontal at 1270°C, based on (*a*) the determination of the liquidus and solidus and (*b*) the microscopic examination of specimens quenched from 1200 and 1100°C. [5] have shown, however, that the liquidus and solidus data of [3] also fit smooth curves and do not necessitate the assumption of the peritectic reaction γ_{Fe} + melt $\rightleftarrows \gamma_{Mn}$. Since γ-Fe and γ-Mn are isomorphous [6, 7], it is indeed possible that they form a continuous series of solid solutions (see below).

The solidification and transformation equilibria in the range 0–10% Mn above 1390°C are due to [3]. According to the thermal-analysis data, the peritectic horizontal lies at 1504°C; it extends from 1 to 8% Mn. Micrographic results would indicate the peritectic temperature as 1520°C; however, [3] apparently preferred 1504°C as the more likely temperature. Earlier, [2] had found the peritectic horizontal to extend over the range 2–12.5% Mn at a considerably lower temperature, 1455°C (melting point of Fe given as 1528°C). The (micrographically determined) solidus curve between 10 and 90% Mn is also due to [3].

Solid-state Transformations. Fe-*rich Alloys.* The transformations in these alloys were the subject of numerous investigations [2, 8–36]. Transformation temperatures were determined by [2, 8–11, 16, 17, 19, 23–27, 29, 32–34], especially [17, 24, 27, 32–34]. The crystal structure was studied by [13–15, 20, 31, 32, 35, 36]. Special reference is made to the papers by [13, 15, 27, 32–35].

Under normal conditions of cooling and heating within the region of the transformation temperature, the alloys remain in a metastable state since both the transformations $\gamma \rightleftarrows \alpha'$ and $\gamma \rightleftarrows \epsilon$ (Fig. 374) have the characteristics of martensitic transformations [32, 34, 35]. The transformation of γ on cooling, in alloys with more than 3.2% Mn [32], results in the formation of two martensitic products, according to the composition, (*a*) a supersaturated b.c.c. solution α' and (*b*) a supersaturated h.c.p. solid solution ϵ, both having the same composition as the γ solid solution from which they form. The temperature at which these structures start to form athermally (M_s temperature) is independent of the rate of cooling, at least within a wide range of cooling rates [32, 34]. On heating, the change is reversible; however, the temperature at which this occurs lies considerably higher than the M_s temperature, with the temperature hysteresis of the $\gamma \rightleftarrows \alpha'$ change increasing as the Mn content increases. The hysteresis of the $\gamma \rightleftarrows \epsilon$ transformation is much smaller (Fig. 374).

Temperatures of the $\gamma \rightleftarrows \alpha'$ transformation have been found in alloys up to approximately 13% Mn and those of the $\gamma \rightleftarrows \epsilon$ transformation down to about 7.5% Mn; i.e., alloys with 7.5–13% Mn may consist of ($\alpha' + \epsilon$) on cooling to room temperature. It has been reported, however, that α' could be detected in alloys up to 20% Mn [13]. The presence of ϵ was observed in alloys up to approximately 32% Mn

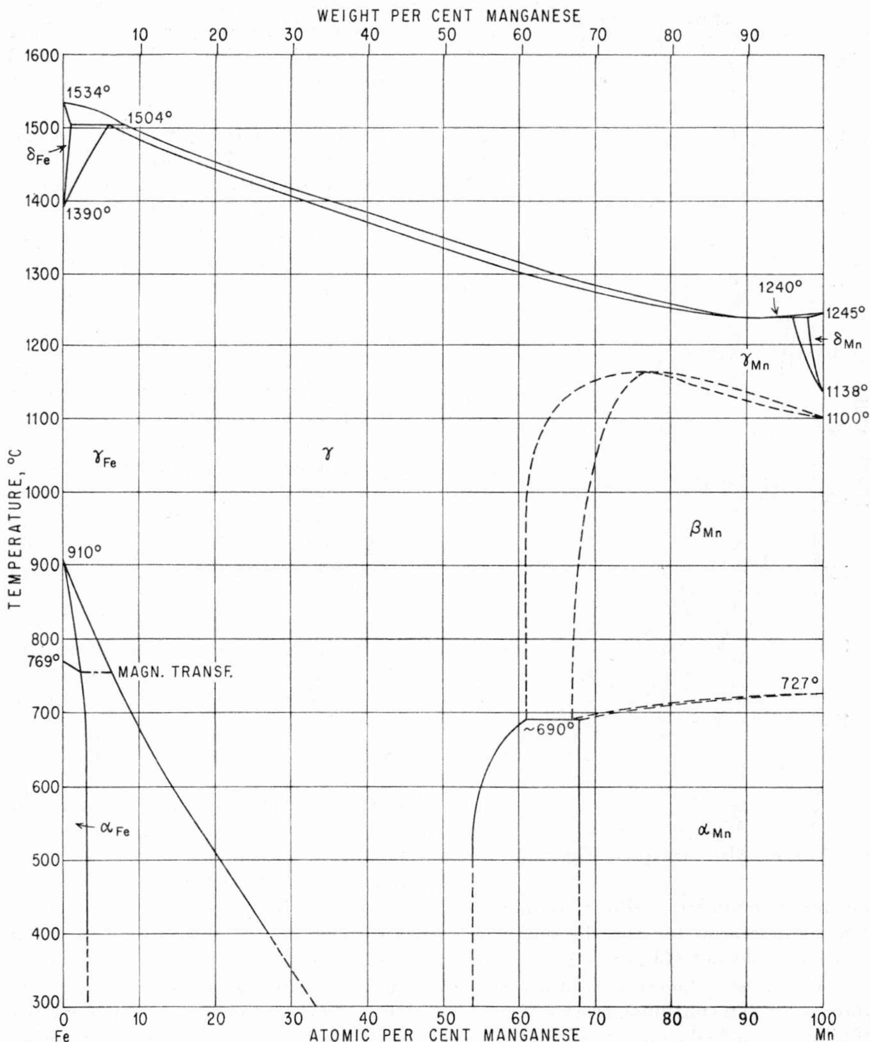

Fig. 373. Fe-Mn. (See also Fig. 374 and A. Hellawell and W. Hume-Rothery, *Phil. Trans. Roy. Soc. London*, **249A**, 1957, 417–459.)

[13, 20]. Some γ may be retained in alloys with more than 10% Mn. Results reported are not consistent since cold work increases the formation of ϵ and causes its rapid destruction on annealing. As to the conditions of formation of ϵ and its decomposition into α' (or α), see the papers by [13, 15, 20, 27, 31, 32, 35, 37].

The temperatures of the beginning of the $\gamma \rightarrow \alpha'$ transformation (M_s temperature) reported by various investigators differ somewhat, probably because of differences in purity of the alloys, inhomogeneity, and uncertainty in detecting the start of the reaction. However, the curve shown in Fig. 374, based on the data of [27, 32, 34] appears to have a high degree of accuracy. This does not hold for the curve of the end of the $\alpha' \rightarrow \gamma$ transformation. Also, the dashed curves are to be regarded as somewhat uncertain.

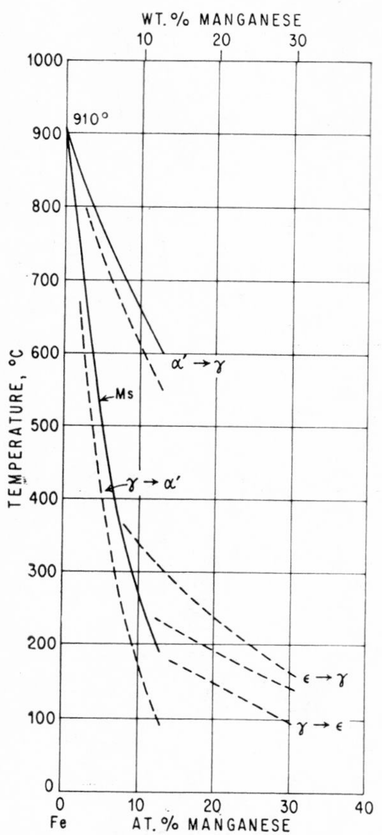

Fig. 374. Fe-Mn. (See also Fig. 373.)

Temperatures of the $\gamma \rightleftarrows \epsilon$ transformation scatter widely. The curves shown in Fig. 374 are a compromise of the data reported by [17, 27, 29, 32, 33].

The first attempt to determine the equilibrium boundaries of the $(\alpha + \gamma)$ field was made by [15] by means of the lattice-spacing method. However, the annealing times applied were not sufficient to attain equilibrium. The boundaries shown in Fig. 373 are based on the outstanding work of [32], who also used the lattice-parametric method. Specimens were annealed at various temperatures between 800 and 300°C for times varying between 1 day and 720 days. Even so, 2 years of annealing at 300°C was not sufficient to reach equilibrium. The boundaries calculated by [34, 38] are in very good agreement with those given by [32].

Mn-*rich Alloys.* The constitution in the range 50–100% Mn was studied by [15, 16, 3, 27, 39, 40], using mainly micrographic, X-ray, and thermal analysis. In addition, X-ray work was done by [41, 13, 14]. Also, reference is made to the discussion of the papers by [3, 27, 39].

The phase diagrams arrived at agree in many respects, especially in that they indicate wide regions of primary solid solutions of β-Mn and α-Mn. However, the locations of the phase boundaries differ considerably, especially at temperatures above 800°C. The main reason for the discrepancies appears to be the fact that the γ_{Mn} solid solution could not always be retained by quenching [42].

The most conflicting data are concerned with the question whether or not γ_{Fe} and γ_{Mn} form a continuous series of solid solutions. Lattice-parameter measurements of alloys quenched from the γ field have suggested that there is a continuous transition between the two γ modifications [14, 15, 27]. This appears quite possible since it has been established recently that both γ_{Fe} and γ_{Mn} are f.c.c. [6, 7]. On the other hand, two-phase structures were claimed to exist in the range 65–70% Mn at temperatures between 1100 and 1200°C [3, 40], in addition to a three-phase equilibrium at about 1030 [3] or 1060°C [40]. The location of the phase boundaries has been further affected by the assumption that β_{Mn} is stable up to temperatures of about 1180–1190°C [3, 27] rather than about 1100°C [40].

It seems to be impossible to draw final conclusions from the very conflicting data. However, considering most of the evidence presented, the writer has tried to arrive at a compromise shown in Fig. 373.

Crystal Structures. Lattice constants of the following phases have been reported: α_{Fe} and α' [13, 43, 14, 15, 20, 27, 31, 32, 44, 46]; ϵ [13, 14, 15, 20, 31, 35, 36]; γ [13, 14, 15, 20, 27, 31, 33]; β_{Mn} [13, 14, 15, 20, 3, 27, 42]; and α_{Mn} [15, 20, 3, 27, 42].

Supplement. [45] published a phase diagram which is essentially that of [3] modified by their own thermal work in the range 60–100 wt. % Mn and 1000–1300°C. It shows a miscibility gap between γ_{Fe} and γ_{Mn}, originating from the peritectic reaction γ_{Fe} + melt $\rightleftarrows$ γ_{Mn} at 1270°C. The γ_{Mn} phase is assumed to decompose eutectoidally at about 71 wt. (at.) % Mn, 1028°C. The $\gamma_{Mn} \rightleftarrows \beta_{Mn}$ transformation curves have a maximum at 89 wt. (at.) % Mn, 1150°C.

1. M. Levin and G. Tammann, *Z. anorg. Chem.*, **47,** 1905, 136–144.
2. G. Rümelin and K. Fick, *Ferrum*, **12,** 1915, 41–44.
3. M. L. V. Gayler (C. Wainwright), *J. Iron Steel Inst.*, **128,** 1933, 293–340.
4. M. Isobe, *Science Repts. Research Insts. Tôhoku Univ.*, **A3,** 1951, 151–154.
5. F. M. Walters and C. Wells, *Trans. ASM*, **23,** 1935, 747–750.
6. U. Zwicker, *Z. Metallkunde*, **42,** 1951, 327–330.
7. Z. S. Basinski and J. W. Christian, *J. Inst. Metals*, **80,** 1951-1952, 659–666.
8. E. Gumlich, *Wiss. Abhandl. physik.-tech. Reichsanstalt*, **4,** 1918, 377–384; see also A. Schulze, *Z. tech. Physik*, **9,** 1928, 340–343.
9. P. Dejean, *Compt. rend.*, **171,** 1920, 791–794.
10. H. Esser and P. Oberhoffer, *Ber. Werkstoffausschuss V. d. Eisenhüttenleute*, no. 69, 1925, pp. 6–7.
11. R. Hadfield, *J. Iron Steel Inst.*, **115,** 1927, 345–352.
12. C. R. Wohrmann, *Trans. AIME*, **80,** 1928, 197–228.
13. W. Schmidt, *Arch. Eisenhüttenw.*, **3,** 1929, 293–300.
14. A. Osawa, *Science Repts. Tôhoku Univ.*, **19,** 1930, 247–264.
15. E. Öhman, *Z. physik. Chem.*, **B8,** 1930, 81–110.
16. T. Ishiwara, *Science Repts. Tôhoku Univ.*, **19,** 1930, 509–519; *Kinzoku-no-Kenkyu*, **7,** 1930, 115–136; *Proc. World Eng. Congress, 1929*, **36,** 1931, 143–155.
17. H. Scott, *Trans. AIME*, **95,** 1931, 284–300.
18. F. M. Walters, *Trans. ASST*, **19,** 1931-1932, 577–589.
19. F. M. Walters and C. Wells, *Trans. ASST*, **19,** 1931-1932, 590–598.
20. M. Gensamer, J. F. Eckel, and F. M. Walters, *Trans. ASST*, **19,** 1931-1932, 599–607.
21. F. M. Walters and M. Gensamer, *Trans. ASST*, **19,** 1931-1932, 608–621.
22. V. N. Krivobok and C. Wells, *Trans. ASST*, **21,** 1933, 807–820.
23. F. M. Walters, *Trans. ASST*, **21,** 1933, 821–829.
24. F. M. Walters, *Trans. ASST*, **21,** 1933, 1002–1013.
25. F. M. Walters and J. F. Eckel, *Trans. ASST*, **21,** 1933, 1016–1020.
26. F. M. Walters and C. Wells, *Trans. ASST*, **21,** 1933, 1021–1027.
27. F. M. Walters and C. Wells, *Trans. ASM*, **23,** 1935, 727–746.
28. C. Wells and F. M. Walters, *J. Iron Steel Inst.*, **128,** 1933, 345–349.
29. E. Scheil, *Arch. Eisenhüttenw.*, **9,** 1935-1936, 115–116.
30. V. A. Nemilov and M. N. Putsykina, *Zhur. Priklad. Khim.*, **12,** 1939, 398–405; *Met. Abstr.*, **9,** 1942, 40 (paper was not available).
31. V. G. Kusnetzov and N. N. Evseeva, *Zhur. Priklad. Khim.*, **12,** 1939, 406–414.
32. A. R. Troiano and F. T. McGuire, *Trans. ASM*, **31,** 1943, 340–359.
33. A. T. Grigorev and D. L. Kudryavtsev, *Izvest. Sektora Fiz.-Khim. Anal.*, **16,** 1946, 70–81.

34. F. W. Jones and W. I. Pumphrey, *J. Iron Steel Inst.*, **163**, 1949, 121–131.
35. J. G. Parr, *J. Iron Steel Inst.*, **171**, 1952, 137–141.
36. J. G. Parr, *Acta Cryst.*, **5**, 1952, 842–843.
37. E. C. Bain, E. S. Davenport, and W. S. N. Waring, *Trans. AIME*, **100**, 1932, 228–249.
38. W. Rostoker, *Trans. AIME*, **191**, 1951, 1203–1205.
39. M. L. V. Gayler and C. Wainwright, *J. Iron Steel Inst.*, **135**, 1937, 269–273.
40. H. Yoshisaki, *Science Repts. Research Insts. Tôhoku Univ.*, **A3**, 1951, 137–150.
41. E. C. Bain, *Chem. & Met. Eng.*, **23**, 1923, 23.
42. See also U. Zwicker, *Z. Metallkunde*, **42**, 1951, 246–252.
43. Z. Nishiyama, *Science Repts. Tôhoku Univ.*, **18**, 1929, 359–400.
44. A. P. Guljaev and E. F. Trusova, *Zhur. Tekh. Fiz.*, **20**, 1950, 66–78. *Structure Repts.*, **13**, 1950, 77.
45. R. Vogel and J. Berak, *Arch. Eisenhüttenw.*, **23**, 1952, 217–223 (The System Fe-P-Mn).
46. A. L. Sutton and W. Hume-Rothery, *Phil. Mag.*, **46**, 1955, 1295–1309.

$\bar{1}.7650$
0.2350

Fe-Mo Iron-Molybdenum

The phase diagram in Fig. 375 is essentially based on the work by [1-3] and [4].

Alloys Up to 30 At. (42.4 Wt.) % Mo. Previously, the α and ϵ phases were assumed to form a eutectic at about 36 wt. (25 at.) % [1] or 40 wt. (28 at.) % Mo [4] and 1440°C. However, [3] showed the liquidus and solidus curve of α to have a minimum at approximately 35.5 wt. (24.3 at.) % Mo, 1440°C, and the three-phase equilibrium to be a peritectic at 1450°C, with the peritectic point (maximal solubility of Mo in α-Fe) at 37.5 wt. (25.9 at.) % Mo. The α solubility curve by [1] did not agree with that by [4] (both based on micrographic studies); however, the portion above 1200°C was later corrected [3] and found to be substantially in accordance with that by [4]. [5] confirmed that about 4 at. (6.7 wt.) % Mo is soluble at 650°C. The α solidus curve is due to [3].

The vertex of the γ loop was found to lie at about 4 (2.4) [1], 2.7 (1.6) [6], or 3 wt. (1.8 at.) % Mo [4]. According to [1, 6], the Curie point of Fe remains unaffected by Mo additions, whereas [4] observed it to be slightly but distinctly lowered (Fig. 375).

Alloys with 30–100 At. % Mo. The existence of the intermediate phases ϵ [1, 7, 4, 8] and σ (formerly designated as η) [7, 4, 2, 9] was established by microscopic [1, 4, 2] and X-ray investigations [7–9]. The composition of the Mo-rich boundary of ϵ is assumed to correspond approximately to the empirical formula Mo_2Fe_3 (53.39 wt. % Mo) [1, 4, 5]; however, the crystal structure is characterized by the ideal formula Mo_6Fe_7 [46.15 at. (59.56 wt.) % Mo] [8]. The homogeneity range of ϵ was reported to have a width of about 0.5 [1], 3 [4], or less than 2 wt. (or at.) % [5], extending toward Fe. The high-Mo boundary of σ was assumed to coincide with the composition MoFe (63.21 wt. % Mo) [4, 2]. The other two boundaries of the σ field, as shown in Fig. 375, are based on micrographic data [2].

[10] claimed to have isolated the compound $MoFe_2$ (46.21 wt. % Mo) from 0.1C-16Cr-25Ni-6Mo steel by electrolytic separation and found it to be isostructural with WFe_2. None of the previous investigators had found this phase; especially [8] reported that Fe-rich alloys, annealed for 1 month at 500–700°C, contain no intermediate phase other than ϵ.

The solid solubility of Fe in Mo was determined by lattice-parameter measurements to be 16.7, 11.0, 7.9, 6.0, and 4.5 at. (10.5, 6.7, 4.8, 3.6, and 2.7 wt.) % Fe at 1480, 1400, 1300, 1200, and 1100°C [11]. Other data [1, 5] are less accurate.

Crystal Structures. Lattice parameters of α alloys with up to 6 at. % Mo were reported by [12]. The ϵ phase, formerly believed to be hexagonal [7], is rhombohedral of the W_6Fe_7 ($D8_5$) type, a = 8.99 A, α = 30°38.6′ [8]. This structure was confirmed by [5, 13]. σ (MoFe) was shown to be isotypic with the σ phase of the

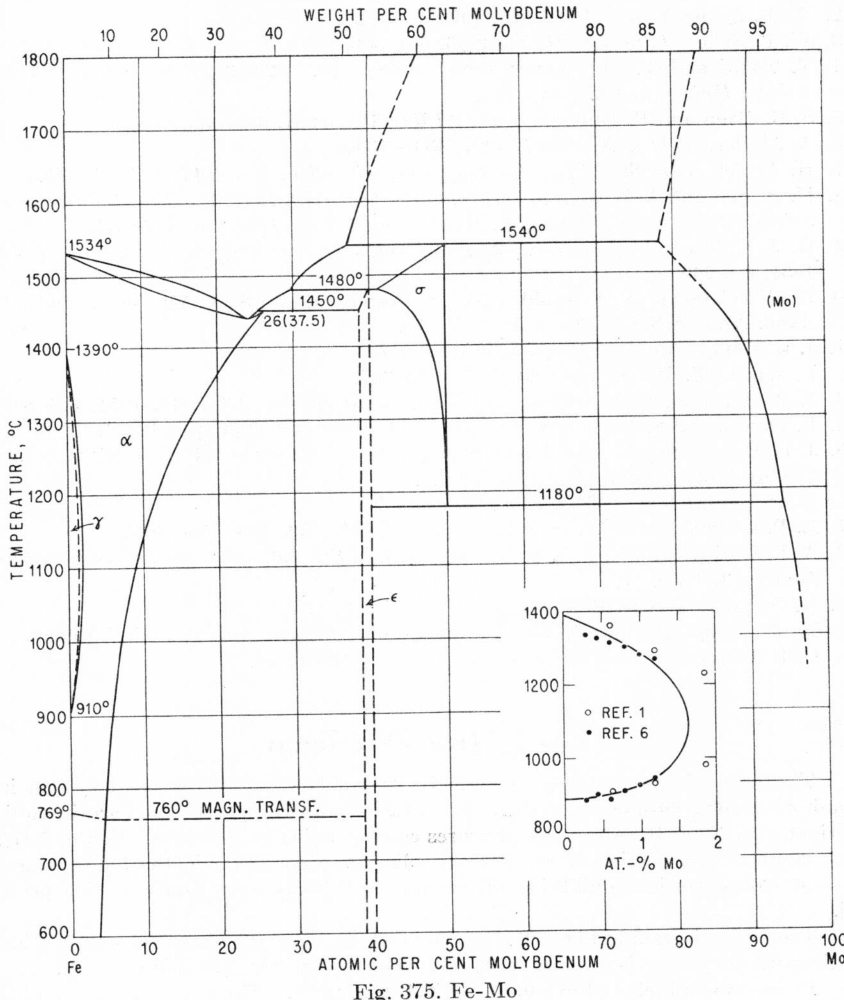

Fig. 375. Fe-Mo

Fe-Cr system [9, 13]; see also [14, 5]. For lattice parameters of the Mo-rich primary solid solution, see [15]. Qualitative X-ray investigations were carried out by [16, 17].

Supplement. [18] measured lattice parameters of α alloys with up to 3.8 at. % Mo.

The structure of the σ phase is tetragonal, with a = 9.188 A, c = 4.812 A, c/a = 0.5237 (50 at. % Mo), and 30 atoms per unit cell [19]. It is nearly identical with that of β-uranium.

[20] reported that the (diffuse and faint) diffraction patterns of the $MoFe_2$ alloy annealed at and quenched from 600 and 800°C were identical to the diffraction patterns of WFe_2 (see the findings of this author in the Fe-W system).

1. W. P. Sykes, *Trans. ASST*, **10,** 1926, 839–871, 1035.
2. W. P. Sykes, *Trans. ASST*, **16,** 1929, 358–369.
3. W. P. Sykes, *Trans. ASM*, **24,** 1936, 541–550.
4. T. Takai and T. Murakami, *Trans. ASST*, **16,** 1929, 339–358; *Science Repts. Tôhoku Univ.*, **18,** 1929, 135–153.
5. S. R. Baen and P. Duwez, *Trans. AIME*, **191,** 1951, 331–335.
6. A. Müller, *Stahl u. Eisen*, **47,** 1927, 1341–1342.
7. H. Arnfelt, *Iron Steel Inst. (London), Carnegie Schol. Mem.*, **17,** 1928, 13–31.
8. H. Arnfelt and A. Westgren, *Jernkontorets Ann.*, **119,** 1935, 185–196; A. Westgren, *Science Repts. Tôhoku Univ., K. Honda Anniv. Vol.*, 1936, pp. 852–863.
9. H. J. Goldschmidt, *Research*, **2,** 1949, 343–344; *Iron Steel Inst. Spec. Rept.* 43, 1951, pp. 249–257.
10. R. P. Zaletaeva, N. F. Lashko, M. D. Nesterova, and S. A. Yuganova, *Doklady Akad. Nauk S.S.S.R.*, **81,** 1951, 415–416.
11. J. L. Ham, *Trans. ASME*, **73,** 1951, 723–731.
12. F. Wever, *Z. Metallkunde*, **20,** 1928, 366–367.
13. J. W. Putman, R. D. Potter, and N. J. Grant, *Trans. ASM*, **43,** 1951, 824–847.
14. P. Duwez and S. Baen, *ASTM Spec. Tech. Publ.* 110, 1950, pp. 48–54.
15. J. L. Ham, Arc-cast Molybdenum-base Alloys, 1st Annual Report, NR 031–331, Climax Molybdenum Co., Apr. 1, 1950.
16. E. C. Bain, *Chem. & Met. Eng.*, **28,** 1923, 23.
17. E. P. Chartkoff and W. P. Sykes, *Trans. AIME*, **89,** 1930, 566–574.
18. A. P. Guljaev and E. F. Trusova, *Zhur. Tekh. Fiz.*, **20,** 1950, 66–78; see *Structure Repts.*, **13,** 1950, 77.
19. G. Bergman and D. P. Shoemaker, *Acta Cryst.*, **7,** 1954, 857.
20. R. P. Elliott, Armour Research Foundation, Chicago, Ill., Technical Report 1, OSR Technical Note OSR-TN-247, August, 1954, p. 16.

0.6006
$\bar{1}.3994$

Fe-N Iron-Nitrogen

The phase diagram shown in Fig. 376 does not represent the Fe-N system in equilibrium at atmospheric pressure but rather shows the phases that are in equilibrium with N_2 at the very high pressures corresponding to the dissociation of NH_3. The figure is a projection of the various solid-phase equilibria in the temperature-pressure-concentration equilibrium diagram onto a temperature-concentration plane [1].

The main diagram in Fig. 376 is drawn in wt. %, since the nitrogen concentration in the entire literature is given in wt. % N. The inset of Fig. 376 is in at. % N.

Phase Boundaries above 0.1 Wt. (0.4 At.) % N. The literature up to 1930 contains a variety of phase diagrams which differ significantly from one another [2–6]. Nevertheless, the existence of the intermediate phases $\gamma' = Fe_4N$ [2, 7, 4, 5, 6], ϵ [7, 4, 5], and $\zeta = Fe_2N$ [5] had been established.

The effort to determine the phase relationships came to a certain conclusion when, in 1930, [8] and [9] published phase diagrams which showed excellent agreement, although different methods—X-ray analysis [8] and magnetic analysis [9]—had been used. The only significant discrepancy between these diagrams consisted in the type of the three-phase equilibrium between γ, γ', and ϵ: [8] assumed that

γ' was formed by a peritectoid reaction ($\gamma + \epsilon \rightarrow \gamma'$), while [9] concluded the existence of a eutectoid equilibrium ($\epsilon \rightarrow \gamma + \gamma'$). Work by [10–14] decided in favor of the latter version.

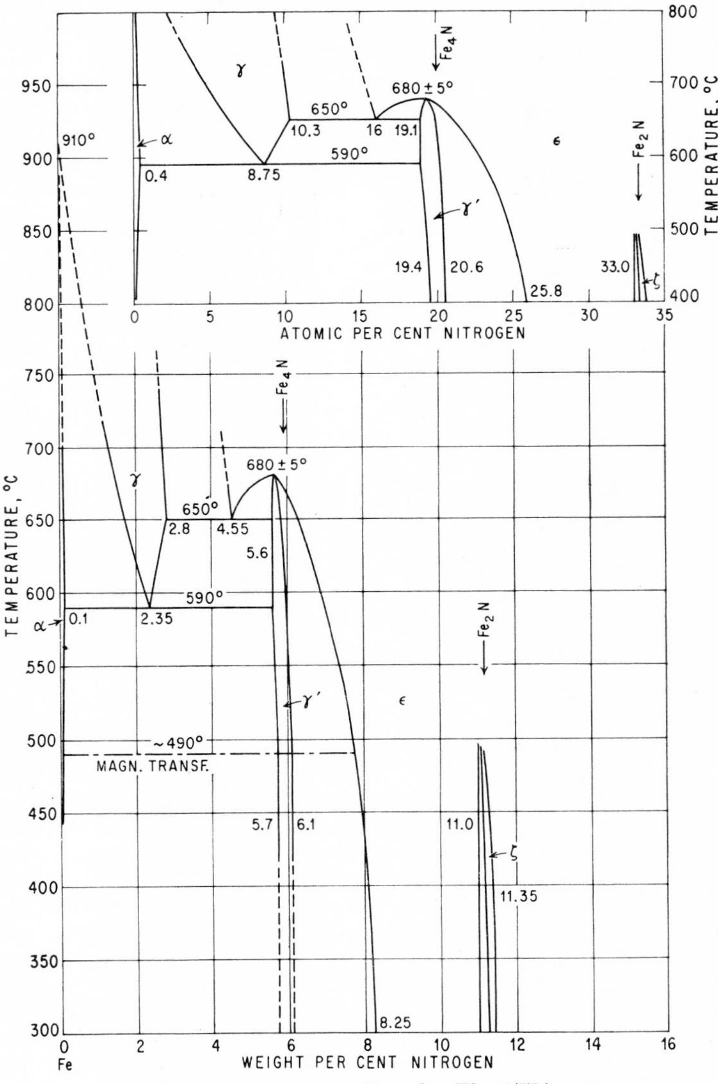

Fig. 376. Fe-N. (See also Fig. 377.)

With the exception of the α-phase boundary (see below), the phase boundaries determined by [8]—apart from the above-mentioned three-phase equilibrium—and [9] have been excellently confirmed by the work of [13] and [14]. The boundaries given by [13] are based on X-ray studies and a controlled nitrogenizing method, those by [14] on X-ray work, the latter being less complete.

As far as the homogeneity range of the γ' phase is concerned, there are slight differences between the findings of [8, 9, 14] on one side and those of [13] on the other; see Table 26. In the diagram of [13] both boundaries are shifted for about 0.2–0.25 wt. % N to lower N concentrations. According to [8, 9, 14], the composition Fe_4N (5.90 wt. % N) lies within the homogeneity range, whereas according to [13] it lies outside this range, at a distance of 0.10–0.15 wt. % N. In Fig. 376 preference was given to the data of [8, 9, 14] although it should be mentioned that there are points in favor of the data of [13]. The γ' phase has a magnetic transformation point at about 480 [2], 475–500 [9], or 488°C [15].

Table 26. The Boundaries of the γ' Field in Wt. % N

Boundary	Temp, °C	Ref. 8	Ref. 9	Ref. 13	Ref. 14
$\gamma'/(\gamma + \gamma')$	650	(5.75)	5.56	5.30	
$\gamma'/(\alpha + \gamma')$	590	5.60	5.56	5.30	
	550	....	5.64	5.45, 5.48	
	500	....	....	5.45	
	450	5.70	....	5.47	5.70
$\gamma'/(\gamma' + \epsilon)$	600	5.95	5.96	5.6	
	500	....	5.96	5.70, 5.80	
	450	6.02	5.96	....	6.10

The composition Fe_2N has been considered by many investigators as the high-N boundary of the ϵ phase rather than a separate phase. According to the findings of [5, 12] and especially [16], there is no doubt, however, that there is a separate phase, $\zeta = Fe_2N$. The latter author has shown by lattice-parameter determinations that the narrow two-phase field $(\epsilon + \zeta)$ extends at 400°C from 11.0 (33.02) to 11.1 wt. (33.23 at.) % N. The stoichiometric composition Fe_2N corresponds to 11.14 wt. % N.

The Solubility of N in α-Fe. Attempts to determine the solubility by means of lattice-parameter measurements failed, since no variation of the parameter on the N concentration of the α phase could be found [5, 6, 7, 17, 18, 13]. In view of these findings the results of [8], who, on the basis of parameter measurements, gave solubilities of 0.32, 0.42, 0.39, and 0.34 wt. % at 450, 590, 620, and 700°C, respectively, are obviously incorrect. [19], using dilatometric and thermomagnetic methods, found the solubility at the eutectoid temperature, 590°C, as 0.13 and at 550°C as 0.05 wt. % N.

In recent years, the solubility was determined by more reliable methods. The investigations covered the temperature ranges 200–580 [20], 450–700 [13], 172–575 [21], and 295–445°C [22] and were carried out using the following methods: mechanical relaxation [20, 22], isothermal calorimetry [21, 22], and controlled nitrogenizing [13]. Figure 377 shows the results. In the lower temperature range (up to 445°C), the data reported by [22] appear to be the most reliable ones, while at higher temperatures the extension of this phase boundary, lying between those of [20, 13] and [21], probably is to be preferred. The maximum solubility at 590°C is about 0.10 wt. % [13]. The solubilities at 100–400°C reported by [23], who used technical Fe as base material, are of the same order of magnitude as those of [20–22]: about 0.001, 0.005, 0.01, and 0.02 wt. % N at 100, 200, 300, and 400°C, respectively. See also Supplement.

Crystal Structures. *α Phase.* The lattice parameter of this phase is practically identical with that of α-Fe [5, 6, 7, 17, 18, 13]. Contrary results [8], indicating a solubility of 0.42 wt. % N at 590°C, are erroneous, since the maximum solubility is only 0.1 wt. (0.4 at.) % (see above).

γ Phase. This phase is isomorphous with that of the γ phase of the Fe-C system (austenite). The N atoms are randomly distributed in the interstices of the f.c.c. Fe lattice. Parameters in dependence upon the composition were reported by [8, 13, 24]. The parameter changes from a = 3.594 A at 0.81 wt. (3.14 at.) % N to a = 3.646 A at 2.33 wt. (8.67 at.) % N [24]. For individual values, see [5, 6, 25].

γ′ Phase. γ′ (= Fe_4N) has a f.c.c. structure with an ordered arrangement of the N atoms in the interstices [26, 13]; for a complete structure analysis, see [26]. The change of the parameter with composition was determined by [5, 7, 8, 26, 13]. According to [13], it varies from a = 3.791 A to a = 3.801 A. Numerous individual values have been reported [27, 6, 17, 28, 18, 29, 30, 15, 25].

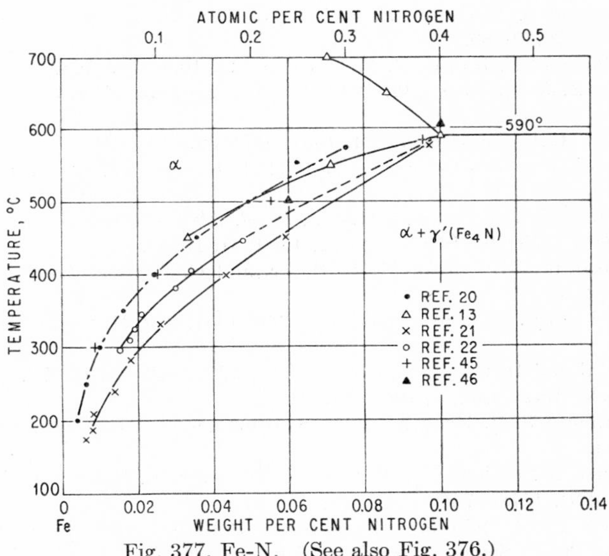

Fig. 377. Fe-N. (See also Fig. 376.)

ε Phase. This phase is h.c.p., with an ordered arrangement of the N atoms as determined by [16]. Lattice parameters in dependence upon the composition have been reported by [8, 18, 13] and especially [16]. They change from a = 2.660 A, c = 4.344 A, c/a = 1.633 at 5.7 wt. (19.42 at.) % to a = 2.764 A, c = 4.420 A, c/a = 1.599 at 11.0 wt. (33.02 at.) % N [16]. For additional data, see [5, 7, 6, 31, 28, 29, 30, 25].

ζ Phase. ζ (= Fe_2N), whose structure was first reported by [5], is orthorhombic, with an ordered arrangement of the N atoms as determined by [16, 26], a = 2.764 A, b = 4.829 A, c = 4.425 A [16]; see also [12]. As to the transition ε → ζ, see [16].

Additional Literature. As to the equilibria considering the gas phase and the effect of pressure, see the papers by [9, 17, 18, 32–34] and especially [35].

The absorption of nitrogen gas (at atmospheric pressure) in solid and liquid iron was investigated by [36–41] and [38, 42–44], respectively; see also Supplement.

Supplement. The solubility of N in α-Fe in equilibrium with Fe_4N was redetermined by [45, 46]; see also [47]. Results are indicated as data points in Fig. 377. The solubility in equilibrium with N-martensite, "Fe_8N" [48], was found as 0.00052, 0.0088, 0.055, and (0.20) wt. % N at 100, 200, 300, and 400°C, respectively [45].

The solubility (absorption) of nitrogen in equilibrium with N_2 was determined for the temperature ranges 500–900°C [45] and 650–1400°C [46].

1. See discussion in *Trans. AIME,* **188,** 1950, 1354.
2. A. Fry, *Stahl u. Eisen,* **43,** 1923, 1271–1279.
3. C. B. Sawyer, *Trans. AIME,* **69,** 1923, 798–828.
4. T. Murakami and S. Iwaizumi, *Kinzoku-no-Kenkyu,* **5,** 1928, 159.
5. G. Hägg, *Nature,* **121,** 1927, 826–827; **122,** 1928, 314, 962; *Nova Acta Regiae Soc. Sci. Upsaliensis,* (4)**7,** 1929, 6–22; *Z. physik. Chem.,* **B8,** 1930, 455–474.
6. S. Epstein, H. C. Cross, E. C. Groesbeck, and I. J. Wymore, *J. Research Natl. Bur. Standards,* **3,** 1929, 1005–1027; S. Epstein, *Trans. ASST,* **16,** 1929, 19–65.
7. A. Osawa and S. Iwaizumi, *Z. Krist.,* **69,** 1928, 26–34; *Science Repts. Tôhoku Univ.,* **18,** 1929, 79–89.
8. O. Eisenhut and E. Kaupp, *Z. Elektrochem.,* **36,** 1930, 392–404.
9. E. Lehrer, *Z. Elektrochem.,* **36,** 1930, 383–392, 460–473; *Z. tech. Phys.,* **10,** 1929, 183–185.
10. W. Köster, *Arch. Eisenhüttenw.,* **4,** 1931, 537–539.
11. S. Nishigori, *Kinzoku-no-Kenkyu,* **9,** 1932, 490–505; *Tech. Repts. Tôhoku Univ.,* **11,** 1933, 68–92.
12. I. E. Kontorovich and A. A. Sovalova, *Izvest. Akad. Nauk S.S.S.R., Otdel. Tekn. Nauk,* **1949,** 1675–1684; *Chem. Abstr.,* **45,** 1951, 10026.
13. V. G. Paranjpe, M. Cohen, M. B. Bever, and C. F. Floe, *Trans. AIME,* **188,** 1950, 261–267.
14. K. H. Jack, *Proc. Roy. Soc. (London),* **A208,** 1951, 200–215; *Acta Cryst.,* **5,** 1952, 404–411.
15. C. Guillaud and H. Creveaux, *Compt. rend.,* **222,** 1946, 1170–1172.
16. K. H. Jack, *Acta Cryst.,* **5,** 1952, 404–411.
17. P. H. Emmett, S. B. Hendricks, and S. Brunauer, *J. Am. Chem. Soc.,* **52,** 1930, 1456–1464.
18. S. Brunauer, M. E. Jefferson, P. H. Emmett, and S. B. Hendricks, *J. Am. Chem. Soc.,* **53,** 1931, 1778–1786.
19. A. Portevin and D. Séférian, *Compt. rend.,* **199,** 1934, 1613–1615; *Rev. mét.,* **33,** 1936, 705–720; D. Séférian, *Rev. ind. minérale,* **1936,** 901–914; *Chimie & industrie,* **37,** 1937, 426–439.
20. L. J. Dijkstra, *Trans. AIME,* **185,** 1949, 252–260.
21. G. Borelius, S. Berglund, and O. Avsan, *Arkiv Fysik,* **2,** 1950, 551–557; see also G. Borelius, *Trans. AIME,* **191,** 1951, 477–484.
22. H. U. Aström, *Arkiv Fysik,* **8,** 1954, 495–503; H. U. Aström and G. Borelius, *Acta Met.,* **2,** 1954, 547–549.
23. W. Köster, *Arch. Eisenhüttenw.,* **3,** 1930, 553–558, 637–658.
24. K. H. Jack, *Proc. Roy. Soc. (London),* **A208,** 1951, 200–215.
25. I. P. Kricevskii and N. E. Khazanova, *Zhur. Fiz. Khim.,* **21,** 1947, 719–733.
26. K. H. Jack, *Proc. Roy. Soc. (London),* **A195,** 1948, 34–40.
27. R. Brill, *Z. Krist.,* **68,** 1928, 379–384; *Naturwissenschaften,* **16,** 1928, 593–594.
28. A. Sieverts and F. Krüll, *Ber. deut. chem. Ges.,* **63,** 1930, 1071–1072; *Z. Elektrochem.,* **39,** 1933, 735–736.
29. S. Satoh, *Bull. Chem. Soc. Japan,* **7,** 1932, 315–333.
30. B. Jones, *J. Iron Steel Inst.,* **136,** 1937, 169–185.
31. S. B. Hendricks and P. B. Kosting, *Z. Krist.,* **74,** 1930, 511–533.
32. I. P. Kricevskii and N. E. Khazanova, *Zhur. Fiz. Khim.,* **19,** 1945, 676; **21,** 1947,

719–733; **24,** 1950, 1188–1196; *Chem. Abstr.*, **40,** 1946, 3379; **42,** 1948, 2166; **49,** 1955, 807.
33. I. P. Kricevskii and N. E. Khazanova, *Doklady Akad. Nauk S.S.S.R.*, **71,** 1950, 677–680; *Chem. Abstr.*, **45,** 1951, 1410.
34. A. V. Smirnov, *Zhur. Tekh. Fiz.*, **23,** 1953, 1400–1410; *Chem. Abstr.*, **49,** 1955, 2971.
35. V. G. Paranjpe and M. Cohen, *Trans. Indian Inst. Metals*, **5,** 1951, 173–185.
36. E. Martin, *Arch. Eisenhüttenw.*, **3,** 1929-1930, 407–416.
37. A. Sieverts, *Z. physik. Chem.*, **155,** 1931, 299–313.
38. A. Sieverts, G. Zapf, and H. Moritz, *Z. physik. Chem.*, **183,** 1938, 19–37.
39. K. Iwase and M. Fukusima, *Science Repts. Tôhoku Univ.*, **27,** 1939, 162–188.
40. I. Hayasi, *Tetsu-to-Hagane*, **26,** 1940, 101–122; *Chem. Abstr.*, **34,** 1940, 4713.
41. L. S. Darken, R. P. Smith, and E. W. Filer, *Trans. AIME*, **191,** 1951, 1174–1179.
42. J. Chipman and D. W. Murphy, *Trans. AIME*, **116,** 1935, 179–190.
43. T. Kootz, *Arch. Eisenhüttenw.*, **15,** 1941, 77–82.
44. T. Saito, *Science Repts. Research Insts. Tôhoku Univ.*, **A1,** 1949, 411–417.
45. J. D. Fast and M. B. Verrijp, *J. Iron Steel Inst.*, **180,** 1955, 337–343; *Acta Met.*, **3,** 1955, 203–204.
46. N. S. Corney and E. T. Turkdogan, *J. Iron Steel Inst.*, **180,** 1955, 344–348.
47. R. Rawlings and D. Tambini, *Acta Met.*, **3,** 1955, 212–213; the authors make erroneous statements as to the solubility reported by [20], who found 0.025 wt. % N at 400°C, rather than 0.0316 wt. % N.
48. K. H. Jack, *Acta Cryst.*, **3,** 1950, 392–394; *Proc. Roy. Soc.* (*London*), **A208,** 1951, 216–224; see also [24].

0.3854
$\bar{1}$.6146

Fe-Na Iron-Sodium

Na is considered to be insoluble in solid Fe [1]. As the boiling point of Na (892°C) lies far below the melting point of Fe, it cannot be determined under normal pressure whether Na is soluble in liquid Fe. Fe does not dissolve in detectable amounts in liquid Na up to 250°C [2]. According to [3], the solubility of Fe in liquid Na increases from 0.0005 wt. % at 225°C to 0.0013 wt. % at 500°C (wet chemical analyses). "Radiochemical determinations have given results lower by a factor of 400. No satisfactory explanation of the discrepancy is known" [4].

1. F. Wever, *Arch. Eisenhüttenw.*, **2,** 1928-1929, 739–746; *Naturwissenschaften*, **17,** 1929, 304–309.
2. C. T. Heycock and F. H. Neville, *J. Chem. Soc.*, **55,** 1889, 668.
3. L. F. Epstein, quoted in [4].
4. "Liquid Metals Handbook," Sodium (NaK) Supplement, 3d ed., pp. 15, 16, 22, U.S. Atomic Energy Commission and Department of the Navy, 1955.

$\bar{1}$.7790
0.2210

Fe-Nb Iron-Niobium

Solidification. The liquidus curve shown in Fig. 378 is based on the thermal-analysis data by [1]. It is in substantial agreement with that of [2] and the earlier results of [3] which cover only the range 0–16.6 at. % Nb. The maximum of the liquidus was found to lie at about 37.5 at. % Nb, 1650°C [2], and about 40 at. % Nb, 1650–1660°C [1]. The ($\delta + \epsilon$) eutectic was reported as 9.8 at. (15.3 wt.) % Nb, 1365°C [3]; 17.5 at. (26 wt.) % Nb, 1360°C [2]; and 11.3–11.7 at. (17.5–18 wt.) % Nb, 1356°C [1]. On the basis of microexamination, [4] and [5] located the eutectic point at 9.6–13 at. (15–20 wt.) % Nb and 10.6–15.1 at. (16.4–22.7 wt.) % Nb, respectively.

The ϵ phase was assumed to be the compound $NbFe_2$ [4], Nb_3Fe_5 [2], or Nb_2Fe_3 [3, 1]. Its crystal structure indicates that it is of the ideal composition $NbFe_2$ [6]. The homogeneity range shown in Fig. 378 does not represent equilibrium condition; it is based on thermal-analysis data only [1]. Whether the composition $NbFe_2$ is single-phase is unknown; [4] reported that the compound $NbFe_2$ dissolves Fe to a

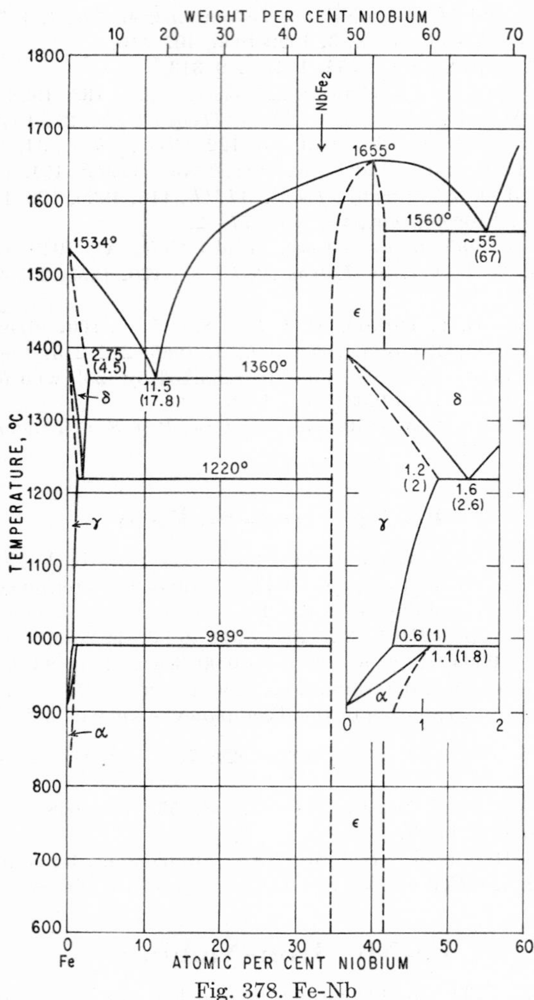

Fig. 378. Fe-Nb

limited extent; however, the alloys investigated were prepared by aluminothermic reduction and were not pure; see also [9].

A second eutectic was found at approximately 55–55.5 at. (67–67.5 wt.) % Nb, 1560°C [1], and 62 at. (73 wt.) % Nb, 1515°C [2]. Whereas [1] concluded the existence of another intermediate phase of unknown composition, [2] reported that the eutectic consists of ϵ and Nb.

Transformations. [2] assumed that a γ loop exists in this system, in accordance with [7]. However, the work by [1, 5] has established conclusively that there are two three-phase equilibria in which the γ phase participates: $\delta \rightleftarrows \gamma + \epsilon$ at 1220°C [1, 5] and $\gamma + \epsilon \rightleftarrows \alpha$ at about 965°C (on cooling) [1] or about 989°C (on heating) [5].

The composition of the $(\gamma + \epsilon)$ eutectoid was first reported as 10 wt. (6.3 at.) % Nb [1]. However, [5] have given evidence for the eutectoid to be located at about 2.6 wt. (1.6 at.) % Nb. Likewise, the composition of the peritectoidally formed α phase was first assumed to be about 5 wt. (~3 at.) % Nb [1] and later corrected to be about 1.8 wt. (1.1 at.) % Nb [5]. The partial diagram in the inset of Fig. 378 is based on thermal and micrographic findings [5]. It is in good agreement with the diagram suggested later by [8], on the basis of thermodynamic calculations, and the revised diagram by [9]. The latter shows only 4.5 wt. (2.75 at.) % Nb to be soluble in δ-Fe at the eutectic temperature, instead of 12 wt. (7.6 at.) % Nb [1].

$NbFe_2$ is isotypic with $MgZn_2$, a = 4.830 A, c = 7.882 A, c/a = 1.632 [6].

Supplement. [10] corroborated the $MgZn_2$-type structure for $NbFe_2$ and gave the lattice parameters a = 4.834 A, c = 7.880 A, c/a = 1.630. No polymorphy was observed in the temperature range 600–1600°C.

1. H. Eggers and W. Peter, *Mitt. Kaiser-Wilhelm-Inst. Eisenforsch.*, **20,** 1938, 199–203.
2. R. Vogel and R. Ergang, *Arch. Eisenhüttenw.*, **12,** 1938, 155–156.
3. N. M. Voronov, *Izvest. Akad. Nauk S.S.S.R.* (*Khim.*), **1937,** 1369–1379.
4. S. A. Pogodin, N. F. Blagov, and M. B. Reifman, *Metallurg*, **1937,** 3–8.
5. R. Genders and R. Harrison, *J. Iron Steel Inst.*, **140,** 1939, 29–37.
6. H. J. Wallbaum, *Z. Krist.*, **103,** 1941, 391–402; *Arch. Eisenhüttenw.*, **14,** 1941, 521–526.
7. F. Wever, *Arch. Eisenhüttenw.*, **2,** 1928-1929, 739–746; *Naturwissenschaften*, **17,** 1929, 304–309.
8. W. Oelsen, *Stahl u. Eisen*, **69,** 1949, 468–474.
9. W. Peter and W. A. Fischer, *Arch. Eisenhüttenw.*, **19,** 1948, 161–168; see also W. Peter, unpublished work abstracted in *FIAT Rev. Ger. Sci.*, 1939-1946, Inorganic Chemistry, Part III, pp. 265–266.
10. R. P. Elliott, Armour Research Foundation, Chicago, Ill., Technical Report 1, OSR Technical Note OSR-TN-247, August, 1954, p. 19.

$\bar{1}.9785$
0.0215

Fe-Ni Iron-Nickel

This system has received consideration from a large number of investigators, as indicated in reviews of published information covering the work up to 1936–1942 [1–4]. Recently [5] has given a composite equilibrium diagram based on the most reliable data published up to 1954 (Fig. 379).

Solidification. Liquidus temperatures of the whole range of composition or portions thereof were reported by [6–10] and [11–13], respectively. The curve shown in Fig. 379 is based on the findings of [13], 0–10% Ni; [8, 12], 10–30% Ni; and [8], 30–100% Ni [5].

Data as to the peritectic equilibrium δ + melt $\rightleftarrows \gamma$ differ considerably. The peritectic temperature was found as 1502 [8], 1455 [11], 1509 [10], 1494 [12], and 1512°C [13]. The peritectic horizontal was reported to extend over the region 3–6 wt. % Ni [8], 6 to about 35 wt. % Ni [11], 3.5–8 wt. % Ni [10], 3–12 wt. % Ni [12], and 3.4–6.2 wt. (3.24–5.9 at.) % Ni [13]. In Fig. 379 preference was given to the results of [13], which appear to be the most reliable [5].

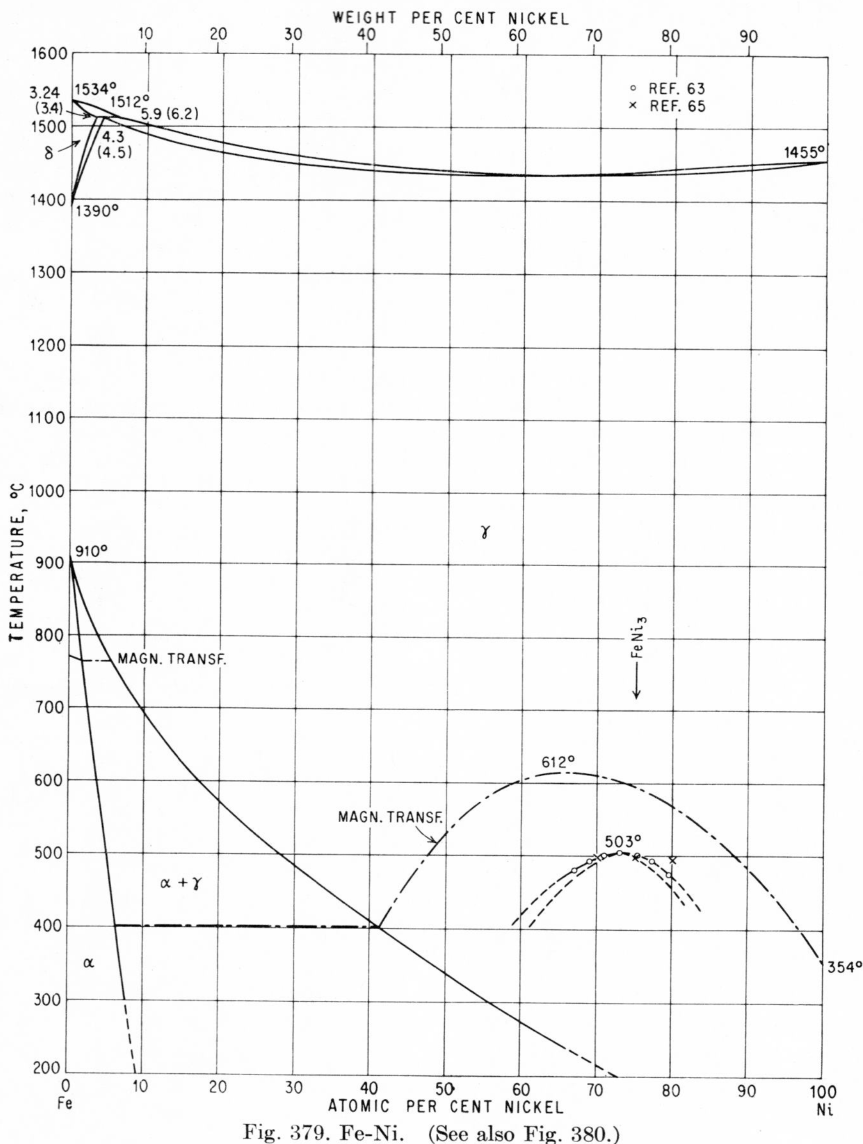

Fig. 379. Fe-Ni. (See also Fig. 380.)

The solidification interval of the γ phase above about 20% Ni was generally reported to have a width of only 5–10°C, on the basis of cooling-curve data. [14] have attempted to determine the solidus by means of micrographic studies. Although the form of this curve is consistent with that of the liquidus, it lies up to more than 10°C below the solidus shown in Fig. 379. The minimum in the liquidus curve, located at 1436°C and about 68 wt. (66.9 at.) % Ni [8], would require the liquidus and solidus to touch at this point. According to [14], however, the minimum of the solidus curve would lie at a temperature of 1420–1425°C.

Equilibrium in the Solid State. Owing to the formation of metastable structural states which vary widely according to the composition and heat-treatment—see, for instance, the papers of [15] and especially [16]— and the extreme sluggishness of diffusion at temperatures below 500°C [17, 18], the exact placing of the equilibrium phase boundaries has met with great experimental difficulties. All $\alpha \rightleftarrows \gamma$ transformation diagrams published prior to the work of [17] and a number of diagrams suggested thereafter [19–22] do not represent the equilibrium state. In particular, the complication of the tentative diagram suggested by [15] results from an attempt to fit metastable states into the equilibrium diagram.

The boundaries of the α and γ phases shown in Fig. 379 are based on the careful X-ray studies (largely by means of the vanishing-phase method) of alloys subjected to extremely long periods of annealing [23], especially at temperatures below 400°C [17, 18]. The boundaries established by [18] closely agree with those determined by [17], except for (*a*) the α boundary below 400°C and (*b*) the γ boundary in the temperature range 500–700°C (Table 27). The boundaries shown in Fig. 379 are those reported by [18].

For comparison, the compositions of the α and γ phases at various temperatures, claimed by [19, 21] to represent equilibrium states, are also included in Table 27. These compositions, determined by means of magnetic methods, indicate that the γ boundary of [19] agrees quite well with that of [18] in the range above 450°C, while the portion of the γ boundary between 350 and 550°C, reported by [21], differs considerably.

This also holds for the α boundary suggested by [19, 21, 22], as compared with those due to [17, 18]. There is no doubt that the latter deserve preference, since they are based on the investigation of alloys subjected to very long periods of annealing [23]. A few data by [24] and [16] also lend support to the α boundary of [17, 18].

Table 27. α- and γ-phase Boundary Compositions in At. % Ni at Various Temperatures in °C

Temperature, °C	α boundary				γ boundary			
	[17]	[18]	[19]	[21]	[17]	[18]	[19]	[21]
800	1.0	1.2	(2.2)		4.5	3.8	3.2	
700	2.5	2.5	(4)		9	9.4	9.2	
600	3.5	3.7	(5.6)	(7.5)	14	17.3	18.5	
550	(4.4)	4.4	(6.3)	(8.8)	18.5	22.2	24	
500	5.0	5.0	(7.1)	10.1	26.5	27.5	(29.5)	28.2
450	(5.7)	5.7	(7.8)	11.5	34	34.3	33–35	30.7
400	6.5	6.3	(8.5)	13.2	41.5	41.5	.....	32.7
350	5.8	6.9	...	(15.5)	49	48.5	.....	(34.4)
300	4.8	7.5*			56.5	56		

* Tentative figure.

The compositions of the α and γ phases, saturated at a temperature close to 250°C, were determined by means of the lattice-spacing method [25]. Alloys were prepared using a special process which results in the equilibrium state at this low temperature. The compositions found, 9.5 and 66 at. % Ni, are fairly consistent with the results of [17, 18].

[16] calculated the boundaries of the ($\alpha + \gamma$) field from the free-energy relationships of the α and γ phases. They are in excellent agreement with those presented in Fig. 379. According to these data, the γ boundary would reach 66.5 at. % Ni at 200°C and about 80 at. % Ni at 100°C; the α boundary is shifted to lower Ni contents at temperatures below 400°C. Also, the boundaries derived from free-energy curves calculated by [26] agree, on the whole, with those established by [17, 18]. The γ boundary was calculated also by [27].

Metastable States. From the technical point of view, the metastable states occurring in the Fe-rich alloys are of much greater importance than the equilibrium state. The existence of metastable conditions is evidenced by the fact that the $\alpha \rightleftarrows \gamma$ transformation is affected with a temperature hysteresis that increases with increasing Ni content. Figure 380 shows the transformation diagram (the so-called realization diagram) found on continuous heating and cooling. This diagram, which was determined by means of dilatometric analysis, is accepted as the most precise transformation diagram available at present [16]. In this case, the solid curves indicated, however, are not the boundaries of the two ($\alpha + \gamma$) fields existing on heating and cooling, but represent the temperatures corresponding to 10 and 90% transformation. The range of transformation is slightly wider than shown. In the range between the curves *a* and *c*, the alloys may be in either the γ or α state, according to the previous heat-treatment. With relatively fast rates of change of temperature, the transformation does not proceed here in either direction.

It appears now to be fully established that the $\alpha \rightleftarrows \gamma$ transformation on continuous heating and cooling has the features of a diffusionless, martensitic-type transformation. The temperatures of the transformations are independent of the rate of change of temperature between 2 and 150°C per min [16]. The α phase formed on continuous cooling is a supersaturated solution having the same composition as the γ phase from which it is formed. This phase, which gives broad lines in the X-ray spectrogram, is usually called α_2. Alloys containing up to approximately 27 at. % Ni, if quenched from temperatures above 500°C, are entirely (or predominantly) converted to α_2. In alloys richer in Ni, treated in the same way, the γ phase is entirely (or predominantly) retained. If, however, such alloys are cooled to liquid-air temperature (−180°C), it is apparent from Fig. 380 that some transformation of the γ phase to α_2 will occur.

Transformation temperatures found on continuous heating and cooling were determined by many workers, using thermal, dilatometric, thermoresistometric, thermomagnetic, and X-ray methods [28–31, 6, 7, 32–37, 9, 38–46, 16, 47]. On the whole, the results agree with those in Fig. 380; however, there are appreciable differences as regards the location of the temperatures of the beginning and end of the transformations, as well as the width of the transformation ranges. These discrep, ancies may be due to actual differences in the behavior of the alloys (purity-homogeneity, etc.) or to temperature gradients or to other experimental imperfections.

The mode and rate of the transformations occurring after various pretreatments were the subject of numerous studies. Since these subjects go beyond the scope of this work, the reader is referred to the papers of [40, 48–51, 42, 19, 21, 52, 53, 47]. It should be mentioned that the phenomena observed are not yet fully understood.

Order-Disorder Transformation. The occurrence of a superlattice structure in the region $FeNi_3$, first suggested by [54], has been established by X-ray, resistivity, magnetic, specific heat, dilatometric, etc., methods [54–65]. Results of [65] indicate that the order-disorder transformation covers a wide range of composition, about

50–80 at. % Ni. The transition temperatures indicated by data points in Fig. 379 are in very good agreement [63, 65]. [61] found 506°C at 75 at. % Ni.

Magnetic Transformations. The magnetic-inversion curve of the α phase is due to [38]. The Curie points of the γ phase [32, 37, 9, 38, 66, 40, 21] lie on a curve which reaches a maximum at 64–68 at. % Ni. The curve shown in Fig. 379 is based

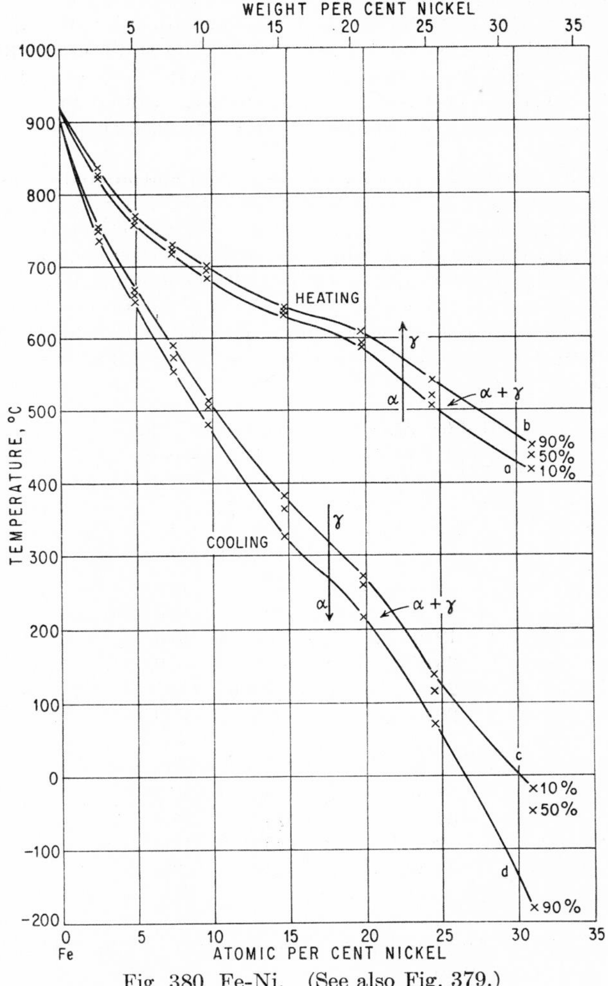

Fig. 380. Fe-Ni. (See also Fig. 379.)

on the most reliable data of [32, 38]. Under nonequilibrium conditions, in which the γ phase is retained in alloys with less than about 42 at. % Ni, this curve extends to lower temperatures, reaching approximately 28 at. % Ni at 100°C.

Crystal Structures. Numerous papers deal with the crystal structure of the Fe-Ni alloys [67–75, 49, 24, 76–80, 59, 58, 81–83, 65]. With the exception of the ordered $FeNi_3$ structure [58, 59, 65], which is of the Cu_3Au ($L1_2$) type, only the b.c.c.

α structure (A2 type) and the f.c.c. γ structure (A1 type) have been found. This disproves the existence of compounds such as Fe_2Ni, Fe_3Ni_2, $FeNi_2$, and $FeNi_4$, which had been previously suggested on the basis of physical properties–composition curves.

Lattice-parameter measurements have shown that additions of Ni to Fe [69, 24, 76, 78, 79, 83, 25, 84] and Fe to Ni [69, 71, 73, 75, 24, 76, 77, 79, 82, 65, 25] expand the lattices. The parameter of the γ phase was found to reach a maximal value at 37 [75] or 39 at. % Ni [24, 76, 79]. No satisfactory explanation for the existence of this maximum has been offered as yet. In the equilibrium state one would expect the parameters of the α and γ phases to be constant within the $(\alpha + \gamma)$ field and to change in the one-phase fields. This condition was, indeed, realized in the work of [25] which covers the entire range of composition, with the alloys being in equilibrium at about 250°C.

[65] reported lattice parameters of the disordered and ordered states in the region 50–80 at. % Ni. The 75 at. % Ni alloy was found to have a parameter of $a = 3.5544$ A and $a = 3.5522$ A in the disordered and ordered state, respectively.

Note Added in Proof. According to neutron-diffraction work [85], ordered $FeNi_3$ also possesses a magnetic superstructure. [86] redetermined $\alpha \rightleftarrows \gamma$ transformation temperatures of alloys with 9.5–33.2 at. % Ni by means of resistance-temperature curves (cooling and heating rate 5°C per min). The data obtained are listed in Table 27*a* (cf. Fig. 380). [86] also developed a general thermodynamic treatment of the Fe-Ni system.

Table 27*a*. Martensite-start (M_s) and Austenite-start (A_s) Temperatures in °C According to [86]

At. % Ni	M_s	A_s
9.5	525	680
14.5	350	625
19.0	210	570
23.75	120	510
28.0	7	425
29.3	−30	390
29.7	−42	365
30.7	−72	335
31.7	−115	315
32.4	−146	300
32.7	−180	300
33.0	−223	300

1. M. Hansen, "Der Aufbau der Zweistofflegierungen," pp. 696–703, Springer-Verlag OHG, Berlin, 1936.
2. C. H. Desch, *Iron Steel Inst. Spec. Rept.* 14, 1936, pp. 63–83. This paper contains a bibliography by W. T. Griffiths with more than 200 references.
3. J. S. Marsh, "The Alloys of Iron and Nickel," vol. I, pp. 24–55, Alloys of Iron Research Monograph Series, McGraw-Hill Book Company, Inc., New York, 1938.
4. C. Benedicks, *Arkiv. Mat. Astron. Fysik*, **A28,** no. 14, 1942.
5. R. W. Floyd, Institute of Metals Annotated Equilibrium Diagrams Series, no. 11, The Institute of Metals, London, 1955.
6. W. Guertler and G. Tammann, *Z. anorg. Chem.*, **45,** 1905, 205–216.

7. R. Ruer and E. Schüz, *Metallurgie*, **7**, 1910, 415–420.
8. D. Hanson and J. R. Freeman, *J. Iron Steel Inst.*, **107**, 1923, 301–314.
9. T. Kasé, *Science Repts. Tôhoku Univ.*, **14**, 1925, 173–187.
10. T. Kasé, *Science Repts. Tôhoku Univ.*, **16**, 1927, 492–494.
11. R. Vogel, *Z. anorg. Chem.*, **142**, 1925, 193–228; *Arch. Eisenhüttenw.*, **1**, 1927-1928, 605–611.
12. H. Bennek and P. Schafmeister, *Arch. Eisenhüttenw.*, **5**, 1931-1932, 123–125.
13. C. A. Bristow, *Iron Steel Inst. Spec. Rept.* 24, 1939, pp. 1–8.
14. C. H. M. Jenkins, E. H. Bucknall, C. R. Austin, and G. A. Mellor, *J. Iron Steel Inst.*, **136**, 1937, 188–193.
15. A. J. Bradley and H. J. Goldschmidt, *J. Iron Steel Inst.*, **140**, 1939, 11–27.
16. F. W. Jones and W. I. Pumphrey, *J. Iron Steel Inst.*, **163**, 1949, 121–131.
17. E. A. Owen and A. H. Sully, *Phil. Mag.*, **27**, 1939, 614–636; see also **31**, 1941, 314–318.
18. E. A. Owen and Y. H. Liu, *J. Iron Steel Inst.*, **163**, 1949, 132–137.
19. A. T. Pickles and W. Sucksmith, *Proc. Roy. Soc. (London)*, **A175**, 1940, 331–344.
20. G. Sachs and J. W. Spretnak, *Trans. AIME*, **145**, 1941, 340–354.
21. K. Hoselitz and W. Sucksmith, *Proc. Roy. Soc. (London)*, **A181**, 1943, 303–313.
22. K. Hoselitz, *J. Iron Steel Inst.*, **149**, 1944, 193–205; see also the discussion by A. H. Sully, pp. 205–208.
23. Periods of annealing: 6 days at 400°C, 186 days at 350°C, 265 days at 300°C [17]; 124 days at 400°C, 400–616 days at 350°C, 430 days at 300°C [18].
24. E. R. Jette and F. Foote, *Trans. AIME*, **120**, 1936, 259–272.
25. F. Lihl, *Arch. Eisenhüttenw.*, **25**, 1954, 475–478.
26. O. Kubaschewski and O. v. Goldbeck, *Trans. Faraday Soc.*, **45**, 1949, 948–960.
27. W. Rostoker, *Trans. AIME*, **191**, 1951, 1203–1205.
28. F. Osmond, *Compt. rend.*, **110**, 1890, 242–244; **118**, 1894, 532–534; **128**, 1899, 304–307, 1395–1398.
29. C. E. Guillaume, *Compt. rend.*, **124**, 1897, 176; **125**, 1897, 235; **126**, 1898, 738; **136**, 1903, 303–306.
30. L. Dumas, *Compt. rend.*, **130**, 1900, 1311–1314.
31. F. Osmond and G. Cartaud, *Rev. mét.*, **1**, 1904, 69–79.
32. F. Hegg, *Arch. sci. phys. nat. Genève*, **30**, 1910, 15–45; see P. D. Merica, *Chem. & Met. Eng.*, **24**, 1921, 377.
33. P. Chevenard, *Compt. rend.*, **159**, 1914, 175–178; *Rev. mét.*, **11**, 1914, 841.
34. P. Chevenard, *Compt. rend.*, **182**, 1926, 1388–1391.
35. P. Chevenard, *Trav. mém. bur. internat. poids et mesures*, **12**, 1927; abstract, *J. Inst. Metals*, **37**, 1927, 471.
36. K. Honda and H. Takagi, *Science Repts. Tôhoku Univ.*, **6**, 1918, 321–340.
37. D. Hanson and H. E. Hanson, *J. Iron Steel Inst.*, **102**, 1920, 39–60.
38. M. Peschard, *Rev. mét.*, **22**, 1925, 490–514, 581–609, 663–684.
39. K. Honda and S. Miura, *Science Repts. Tôhoku Univ.*, **16**, 1927, 745–753; *Trans. ASST*, **13**, 1928, 270–279.
40. G. Gossels, *Z. anorg. Chem.*, **182**, 1929, 19–27.
41. A. Merz, *Arch Eisenhüttenw.*, **3**, 1929-1930, 587–596.
42. F. Wever and H. Lange, *Mitt. Kaiser-Wilhelm-Inst. Eisenforsch.*, **18**, 1936, 217–225.
43. S. D. Smith, *Trans. ASM*, **26**, 1938, 255–262.
44. A. T. Grigoriev and D. L. Kudryavtsev, *Zhur. Priklad. Khim.*, **15**, 1942, 204–213.
45. S. Takeuchi, *Science Repts. Research Insts. Tôhoku Univ.*, **A1**, 1948, 43–49.
46. Y. Tino, *J. Phys. Soc. Japan*, **4**, 1949, 24–29; see also Y. Tino, *J. Sci. Research Inst. Tokyo*, **46**, 1952, 47–52.

47. G. Masing and O. Nickel, *Arch. Eisenhüttenw.*, **24**, 1953, 143–151.
48. L. Anastasiadis and W. Guertler, *Z. Metallkunde*, **23**, 1931, 189–190.
49. G. Wassermann, *Arch. Eisenhüttenw.*, **6**, 1932-1933, 347–351.
50. U. Dehlinger (H. Bumm), *Z. Metallkunde*, **26**, 1934, 112–116.
51. E. Scheil, *Arch. Eisenhüttenw.*, **9**, 1935-1936, 163–166.
52. N. P. Allen and C. C. Earley, *J. Iron Steel Inst.*, **166**, 1950, 281–288.
53. E. Scheil, *Arch. Eisenhüttenw.*, **24**, 1953, 153–160.
54. O. Dahl, *Z. Metallkunde*, **24**, 1932, 107–111; see also O. Dahl and J. Pfaffenberger, *Z. Metallkunde*, **25**, 1933, 241–245.
55. A. Kussmann, B. Scharnow, and W. Steinhaus, "Festschrift der Heraeus Vakuumschmelze," pp. 310–338, Albertis Hofbuchhandlung, Hanau (Main), 1933.
56. S. Kaya, *J. Fac. Sci. Hokkaido Imp. Univ.*, (2)**2**(2), 1938, 29–53; see *Z. Metallkunde*, **31**, 1939, 212–214.
57. S. Kaya and M. Nakayama, *Z. Physik*, **112**, 1939, 420–429.
58. P. Leech and C. Sykes, *Phil. Mag.*, **27**, 1939, 742–753.
59. F. E. Haworth, *Phys. Rev.*, **56**, 1939, 289; see also **54**, 1938, 693–698.
60. F. C. Nix, H. G. Beyer, and J. R. Dunning, *Phys. Rev.*, **58**, 1940, 1031–1034.
61. O. Kallbach, *Arkiv. Mat. Astron. Fysik*, **B34**, 1947, no. 17.
62. E. Josso, *Compt. rend.*, **229**, 1949, 594–596; *Rev. mét.*, **47**, 1950, 769–777.
63. E. Josso, *Compt. rend.*, **230**, 1950, 1467–1469.
64. F. Galperin, *Doklady Akad. Nauk S.S.S.R.*, **75**, 1950, 647–650.
65. R. J. Wakelin and E. L. Yates, *Proc. Phys. Soc. (London)*, **B66**, 1953, 221–240.
66. H. Scott, *Trans. ASST*, **13**, 1928, 829–846; the alloys contained 0.6–1.5% Mn.
67. M. R. Andrews, *Phys. Rev.*, **18**, 1921, 245–254.
68. F. Kirchner, *Ann. Physik*, **69**, 1922, 75–77.
69. L. W. McKeehan, *Phys. Rev.*, **21**, 1923, 402–407.
70. E. C. Bain, *Chem. & Met. Eng.*, **28**, 1923, 23–24; *Trans. AIME*, **68**, 1923, 633–634.
71. A. Osawa, *Science Repts. Tôhoku Univ.*, **15**, 1926, 387–398; *J. Iron Steel Inst.*, **113**, 1926, 447–456.
72. F. C. Blake and A. E. Focke, *Phys. Rev.*, **29**, 1927, 206–207.
73. A. O. Jung, *Z. Krist.*, **65**, 1927, 309–334.
74. O. L. Roberts and W. P. Davey, *Metals & Alloys*, **1**, 1930, 648–654.
75. G. Phragmén, *J. Iron Steel Inst.*, **123**, 1931, 465–477.
76. E. A. Owen, E. L. Yates, and A. H. Sully, *Proc. Phys. Soc. (London)*, **49**, 1937, 315–322.
77. E. A. Owen and E. L. Yates, *Proc. Phys. Soc. (London)*, **49**, 1937, 17–28.
78. E. A. Owen and E. L. Yates, *Proc. Phys. Soc. (London)*, **49**, 1937, 307–314.
79. A. J. Bradley, A. H. Jay, and A. Taylor, *Phil. Mag.*, **23**, 1937, 545–547.
80. E. A. Owen, *Phil. Mag.*, **29**, 1940, 553–567.
81. L. Vegard (S. Holstein-Hansen), *Skrifter Norske Videnskaps-Akad. Oslo, Mat.-Naturv. Kl.*, no. 2, 1947, pp. 56–65.
82. H. Hahn and H. Mühlberg, *Z. anorg. Chem.*, **259**, 1949, 121–134.
83. A. P. Guljaev and E. F. Trusova, *Zhur. Tekh. Fiz.*, **20**, 1950, 66–78.
84. A. L. Sutton and W. Hume-Rothery, *Phil. Mag.*, **46**, 1955, 1295–1309.
85. C. G. Shull and M. K. Wilkinson, *Phys. Rev.*, **97**, 1955, 304–310.
86. L. Kaufman and M. Cohen, *Trans. AIME*, **206**, 1956, 1393–1401.

0.5429
$\bar{1}.4571$

Fe-O Iron-Oxygen

The Range 0–50 At. % O. *Solubility of* O *in Liquid* Fe. According to early determinations, the solubility at or close to the melting point of Fe was found as

0.24 [1], 0.29 [2], 0.28 (at an unknown temperature) [3], 0.21 [4, 5], and 0.22 wt. % O [6]. [7] reported 0.24 wt. % at 1600°C. The solubility increases with rise in temperature. Work of [8] indicated an increase from about 0.22 wt. (0.76 at.) % at 1540°C to about 0.31 wt. (1.08 at.) % at 1600°C and 0.45 wt. (1.55 at.) % at 1700°C. Quite similar values were obtained by [9]; see [10]. Considerably lower solubilities were found by [10, 11]: about 0.18 wt. (0.63 at.) % at 1540°C [10, 11], about 0.22 wt. (0.76 at.) % [10] or 0.23 wt. (0.80 at.) % [11] at 1600°C, and about 0.30 wt. (1.04 at.) % [10] or 0.33 wt. (1.14 at.) % [11] at 1700°C. Although no obvious reason can be given for the disagreement, the data of [10, 11] are now generally accepted. Moreover, they are supported by findings of [12].

The melting point of Fe was found to be lowered by about 15 [5] or 11°C [13, 14]. At the monotectic temperature, 1523°C, the solubility is 0.16–0.17 wt. (0.56–0.59 at.) % O [10, 11] (Fig. 381).

The solubility of O *in solid* Fe was the subject of numerous studies [15–19, 5, 20–38] yielding highly discordant results [39]. For a review of the data as available at various times, see [40–46].

At the monotectic temperature the solubility was reported as 0.05–0.06 wt. (0.17–0.21 at.) % [33], whereas [38] believed it to be only 0.003–0.007 wt. (0.01_1–0.02_5 at.) %, i.e., about ten times smaller. According to [35], who emphasized the effect of impurities, the apparent maximum solubility in γ-Fe at 1345 ± 20°C was found as low as 0.003 ± 0.003 wt. (0.01 ± 0.01 at.) %. This value is regarded as an upper limit, and the true solubility in ideally pure γ-Fe is estimated to be even less than 0.001 wt. (0.003_5 at.) % [35]. Experiments by [34] show that at 950°C the solubility must be less than 0.007 wt. (0.02_5 at.) %.

[36] has claimed that the solubility in α-Fe increases from about 0.008 wt. (0.02_8 at.) % at 700°C to 0.018 wt. (0.06 at.) % at 800°C and 0.029 wt. (0.10 at.) % at 900°C. Contrary to this, [25] found <0.01 wt. (0.03_5 at.) % at 800–1000°C. The lowest value of [36]—0.008 wt. % at 700°C—is fairly consistent with the solubility "after slow cooling to room temperature" reported by [32], 0.003–0.006 wt. (0.01–0.02 at.) %. According to metallographic evidence, iron oxide dissolves and precipitates, depending upon heat-treatment below 900°C [37]; see also [47, 48].

[46] has tried to analyze the data of [35, 36]. He came to the conclusion that either the results of [36] are too high or those of [35] are too low. At any rate, it seems reasonable to accept that the solubility is much lower in γ-Fe than in $\alpha(\delta)$-Fe. See also Supplement.

The Range 50–57.14 At. % O (Fe_3O_4). The iron oxide lowest in oxygen was formerly considered to correspond to the stoichiometric composition FeO (22.27 wt. % O). It was believed to have a certain solution capacity for O and Fe [17, 19, 41]. However, [22] have shown that the compound FeO does not exist, but is stable only with an excess of oxygen. This phase of variable composition was denoted as wüstite [22]. All later investigators, with the exception of [49], have confirmed this result; see following section.

Solidification. The solidification equilibria were studied by [43, 26, 50, 51, 14]. Results of [43, 26] are greatly in error [52], whereas those of [50, 51, 14] indicate the existence of a eutectic, melt $\rightleftarrows$ γ-Fe + wüstite, and a peritectic, melt + Fe_3O_4 $\rightleftarrows$ wüstite. The eutectic temperature was found as 1370 [5], 1366–1369 [50], 1380 ± 5 [53], 1380 [51], 1369 [13], and 1371°C [14], and the peritectic temperature as 1430–1435 [50], 1432 [51], and 1424°C [14]. The liquidus and solidus curves for the region 50–57.14 at. % O shown in Figs. 381 and 382 are due to [14]. They are based on the determination of the equilibrium relations between partial pressure of O_2 and temperature and composition by equilibration of liquid iron oxide with several controlled gas mixtures.

The Wüstite Area. The homogeneity limits of the wüstite field were determined by [22, 26, 50, 53, 54, 49]. Results, shown in Fig. 382, have been critically reviewed, especially by [54, 49]. There is agreement among the findings of [22, 26, 50, 53, 54, 55] that the compound FeO does not occur. In contrast to this, [49] claimed to have

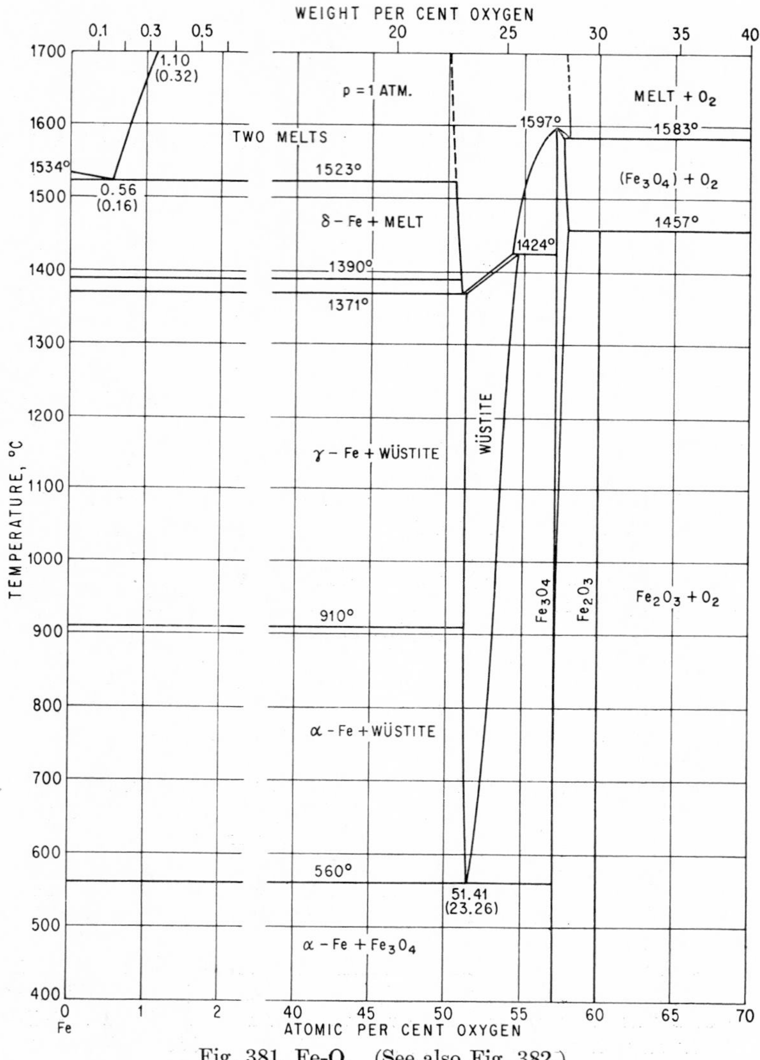

Fig. 381. Fe-O. (See also Fig. 382.)

proved that the low-oxygen boundary of the wüstite area practically coincided with the composition FeO [56]. Although it is difficult to dispute the evidence reported by [49], preference was given to the data of [54], since a low-oxygen boundary at 50 at. % O would be incompatible with the solidification equilibria due to [50, 51, 54]

(Fig. 382). The boundaries accepted in Fig. 381 are those based on the work of [54]. See Supplement.

The wüstite phase decomposes eutectoidally into α-Fe and Fe_3O_4 [57]. The reaction was studied by [57–63]. The eutectoid temperature was reported as 560–580°C [22, 49, 50, 53, 54, 57, 62] and the eutectoid composition as 23.1(51.18) [49, 50], 23.26 (51.41) [54], 23.6 (51.88) [53], and 24.0 wt. (52.43 at.) % O [22].

The Range 57.14–60.0 At. % O (Fe_2O_3). The constitution of the partial system Fe_3O_4-Fe_2O_3 has been investigated by [64–68, 50, 69–71, 14]. Special reference is made to the outstanding work of [69] and that of [67, 70, 71, 14]. For a review of the literature, including additional references, see [69, 71].

Melting Range. The melting point of Fe_3O_4 (27.64 wt. % O) was found as 1600 [64, 51], 1597 ± 2 [14], 1591 ± 5 [69], and 1590°C [66, 50]. The compound melts to a liquid of identical composition [69, 14], rather than incongruently [50, 51]. Oxygen additions lower the melting temperature [69, 70, 14]. Under an oxygen pressure of 1 atm, Fe_2O_3 (30.06 wt. % O) decomposes into (Fe_3O_4) and O_2 at 1455 [67], 1452 ± 5 [69], 1459 [71], and 1457 ± 2°C [14]. Under an oxygen pressure of 159 mm Hg (partial pressure of O_2 in dry air), this decomposition occurs at 1388 ± 2 [69] or 1392 ± 2°C [14]. At a pressure of more than 1 atm, (Fe_3O_4) and Fe_2O_3 are believed to form a eutectic [69, 70, 71].

Solid State. The homogeneity region of the (Fe_3O_4) phase (magnetite) was determined by [67, 68, 50, 69, 70, 71, 14]. In the temperature range 1200–1450°C, the results of [69, 71, 14] agree within about 0.1 wt. (0.12 at.) % O. The solubility curve shown in Figs. 381 and 382 is due to [14].

Results as regards the homogeneity region of Fe_2O_3 or hematite [67, 68, 50, 69, 70, 71] differ considerably. According to [70, 71], the solubility is about 0.1–0.3 wt. (0.12–0.35 at.) % Fe at 1300°C and about 0.45 wt. (0.53 at.) % Fe at 1450°C. Contrary to this, [69] found the solubility of Fe to be less than 0.02 wt. (0.024 at.) % in the entire temperature range investigated, 1200–1450°C. Also, [67] reported the solubility of Fe in Fe_2O_3 as practically zero at 1150–1200°C. The data of [69] were accepted by [14] (Fig. 381).

Crystal Structures. Wüstite has the NaCl (B1) type structure [72]. Lattice parameters were reported by [72, 53, 73, 74, 75, 59, 76–82], especially [53, 73, 78, 80, 82]. The lattice spacing decreases nearly linearly with increasing O content, from $a = 4.311$ A at 23.21 wt. (51.34 at.) % O to $a = 4.281$ A at 24.28 wt. (52.81 at.) % O [53, 73]. As regards the interpretation of the structure (subtraction-type solid solution in FeO), see [73, 83, 84]. At temperatures below −70°C the cubic structure transforms to a rhombohedral structure [80, 85, 86]. At high O contents a transition into a tetragonal structure occurs [87].

Fe_3O_4 (magnetite) is isotypic with spinel ($H1_1$ type), with lattice parameters reported by [88, 89, 90, 74, 91, 75, 92, 93, 81, 94], $a = 8.397$ A [92, 81], $a = 8.3940$ A [93, 94]. As to the interrelation of the crystal structures of FeO and Fe_3O_4, see [84]. At temperatures below −154°C the cubic structure transforms into either a rhombohedral [93, 95] or an orthorhombic arrangement [94]; see also [96]. See Supplement.

Fe_2O_3 (hematite) exists in a stable form α-Fe_2O_3 and a metastable ferromagnetic form γ-Fe_2O_3. α-Fe_2O_3 is rhombohedral of the α-Al_2O_3, corundum ($D5_1$) type, with parameters reported by [97–105, 81], $a = 5.4271$ A, $\alpha = 55°15.8'$ [105]. The parameters of the hexagonal unit cell with 30 atoms were reported by [102, 106, 107, 105], $a = 5.0345$ A, $c = 13.749$ A, $c/a = 2.731$ [105].

γ-Fe_2O_3, formed by oxidation of Fe_3O_4 at temperatures between 200 and 400°C [108, 92, 109], has a spinellike structure with 21⅓ Fe atoms, 32 O atoms, and 2⅔ vacant positions per unit cell [110, 92, 111], $a = 8.34$ A [92, 111]. There is indication of a more complex structure [111]. Another hexagonal ferromagnetic modification, δ-Fe_2O_3, was claimed by [112]: $a = 5.10$ A, $c = 4.42$ A, $c/a = 0.866$, and 10 atoms per unit cell.

Supplement. The range of stability of the wüstite phase was redetermined by [113, 114] (between 600 and 1300°C) and [115] (between 650 and 1000°C). The results of these investigations are in very good agreement and indicate that the range

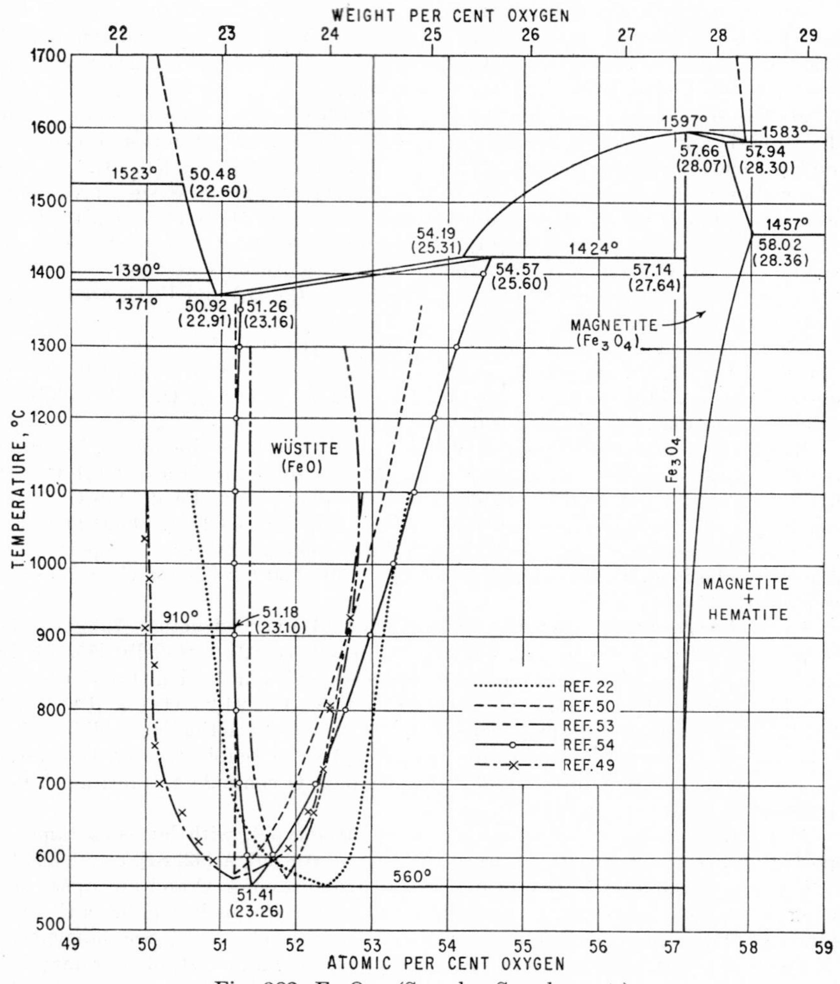

Fig. 382. Fe-O. (See also Supplement.)

of homogeneity lies at slightly higher oxygen contents than was found by [54]: 51.4–53.3 and 51.6–52.6 at. % O at 1000 and 700°C, respectively [116].

According to [117, 118], the solubility of O in well-recrystallized Fe of high purity is practically nil at 700 and 850°C. [119] have given further evidence that the low-temperature modification of magnetite is of orthorhombic (or lower) symmetry.

1. A. Ledebur, *Stahl u. Eisen*, **3**, 1883, 502.
2. L. Romanoff, *Stahl u. Eisen*, **19**, 1899, 267.
3. W. Austin, *J. Iron Steel Inst.*, **92**, 1915, 157–161.
4. Bureau of Standards, *Chem. & Met. Eng.*, **26**, 1922, 778.
5. F. S. Tritton and D. Hanson, *J. Iron Steel Inst.*, **110**, 1924, 90–128.
6. C. H. Herty, *Trans. AIME*, **73**, 1926, 1107–1131.
7. H. Le Chatelier, *Rev. mét.*, **9**, 1912, 514.
8. C. H. Herty and J. M. Gaines, *Trans. AIME*, **80**, 1928, 142–154.
9. F. Körber and W. Oelsen, *Mitt. Kaiser-Wilhelm-Inst. Eisenforsch.*, **14**, 1932, 181–204; *Stahl u. Eisen*, **52**, 1932, 133–142.
10. J. Chipman and K. L. Fetters, *Trans. ASM*, **29**, 1941, 953–967.
11. C. R. Taylor and J. Chipman, *Trans. AIME*, **154**, 1943, 228–245.
12. M. N. Dastur and J. Chipman, *Trans. AIME*, **185**, 1949, 441–445.
13. J. Chipman and S. Marshall, *J. Am. Chem. Soc.*, **62**, 1940, 299–305.
14. L. S. Darken and R. W. Gurry, *J. Am. Chem. Soc.*, **68**, 1946, 798–816.
15. H. Le Chatelier and B. Bogitch, *Compt. rend.*, **167**, 1918, 472–477; *Rev. mét.*, **16**, 1919, 129.
16. J. E. Stead, *J. Iron Steel Inst.*, **103**, 1921, 271–275.
17. A. Matsubara, *Trans. AIME*, **67**, 1922, 3–55; *Z. anorg. Chem.*, **124**, 1922, 39–55.
18. H. Monden, *Stahl u. Eisen*, **43**, 1923, 784–785.
19. E. D. Eastman and R. M. Evans, *J. Am. Chem. Soc.*, **46**, 1924, 888–903.
20. A. Wimmer, *Stahl u. Eisen*, **45**, 1925, 74.
21. P. Oberhoffer, H. J. Schiffler, and W. Hessenbruch, *Arch. Eisenhüttenw.*, **1**, 1927-1928, 57–68.
22. R. Schenck and T. Dingmann, *Z. anorg. Chem.*, **166**, 1927, 113–154; see also **171**, 1928, 239–257.
23. R. Schenck, T. Dingmann, P. H. Kirscht, and H. Wesselkock, *Z. anorg. Chem.*, **182**, 1929, 97–117; *Stahl u. Eisen*, **50**, 1930, 1530–1531.
24. W. Krings and J. Kempkens, *Z. anorg. Chem.*, **183**, 1929, 225–250; **190**, 1930, 313–320.
25. H. Dünwald and C. Wagner, *Z. anorg. Chem.*, **199**, 1931, 342–346; see also **201**, 1931, 188–192.
26. H. Schenck and F. Hengler, *Arch. Eisenhüttenw.*, **5**, 1931-1932, 209–214.
27. H. Esser, *Z. anorg. Chem.*, **202**, 1931, 73–76.
28. N. A. Ziegler, *Trans. ASST*, **20**, 1932, 73–84; see also T. D. Yensen and N. A. Ziegler, *Trans. AIME*, **116**, 1935, 397–404.
29. J. Reschka, *Mitt. Forsch.-Inst. Ver. Stahlwerke*, **3**, **1932**, **1–18**.
30. R. Schenck, T. Dingmann, P. H. Kirscht, and A. Kortengräber, *Z. anorg. Chem.*, **206**, 1932, 73–96, 208, 255–256.
31. H. Esser and H. Cornelius, *Stahl u. Eisen*, **53**, 1933, 534–535; *Metals & Alloys*, **4**, 1933, 121–122.
32. H. A. Sloman, *J. Iron Steel Inst.*, **143**, 1941, 311.
33. W. A. Fischer and H. vom Ende, *Arch. Eisenhüttenw.*, **21**, 1950, 297–304.
34. W. P. Rees and B. E. Hopkins, *J. Iron Steel Inst.*, **172**, 1952, 403–409; quoted from [46].
35. J. A. Kitchener, J. O'M. Bockris, M. Gleiser, and J. W. Evans, *Acta Met.*, **1**, 1953, 93–101; *Trans. Faraday Soc.*, **48**, 1952, 995–997.
36. A. U. Seybolt, *Trans. AIME*, **200**, 1954, 641–644; see also discussion, **203**, 1955, 697.
37. A. U. Seybolt, *Trans. AIME*, **200**, 1954, 979–982.
38. F. Wever, W. A. Fischer, and H. Engelbrecht, *Stahl u. Eisen*, **74**, 1954, 1521–1526.

39. See tabulated data in [45].
40. K. Schönert, *Z. anorg. Chem.*, **154,** 1926, 220–225.
41. C. Benedicks and H. Löfquist, "Non-metallic Inclusions in Iron and Steel," pp. 47–63, John Wiley & Sons, Inc., New York, 1931; see also *Z. Ver. deut. Ing.*, **71,** 1927, 1576–1577; *Z. anorg. Chem.*, **171,** 1928, 231–238.
42. O. C. Ralston, *U.S. Bur. Mines Bull.* 296, 1929.
43. C. H. Mathewson, E. Spire, and W. E. Milligan, *Trans. ASST*, **19,** 1931, 66–88.
44. J. Klärding, *Stahl u. Eisen*, **52,** 1932, 785.
45. M. Hansen, "Der Aufbau der Zweistofflegierungen," pp. 703–710, Springer-Verlag OHG, Berlin, 1936.
46. J. L. Meijering, *Acta Met.*, **3,** 1955, 157–162.
47. W. Hessenbruch, *Trans. ASST*, **20,** 1932, 88–89.
48. R. Castro and A. Portevin, *Rev. mét.*, **39,** 1942, 225–232.
49. J. Bénard, *Bull. soc. chim. France*, **16,** 1949, D 109–116; see also J. Bénard, *Compt. rend.*, **205,** 1937, 912–914; *Ann. chim. (Paris)*, **12,** 1939, 5–92.
50. L. B. Pfeil, *J. Iron Steel Inst.*, **123,** 1931, 237–255.
51. R. Vogel and E. Martin, *Arch. Eisenhüttenw.*, **6,** 1932-1933, 108–111.
52. See fig. 284 in [45].
53. E. R. Jette and F. Foote, *Trans. AIME*, **105,** 1933, 276–284.
54. L. S. Darken and R. W. Gurry, *J. Am. Chem. Soc.*, **67,** 1945, 1398–1412.
55. E. E. Wood and J. B. Ferguson, *J. Wash. Acad. Sci.*, **26,** 1936, 289–293.
56. Also, P. Günther and H. Rehaag (*Z. anorg. Chem.*, **243,** 1939, 60–68) reported the preparation of stoichiometric FeO by decomposition of ferrous oxalate under vacuum at about 850°C.
57. G. Chaudron, *Compt. rend.*, **172,** 1921, 152–155.
58. G. Chaudron and H. Forestier, *Compt. rend.*, **178,** 1924, 2173–2176.
59. J. Bénard and G. Chaudron, *Compt. rend.*, **202,** 1936, 1336–1338.
60. H. Nowotny and F. Halla, *Z. anorg. Chem.*, **230,** 1936, 95–96.
61. J. Bénard, *Ann. chim. (Paris)*, **12,** 1939, 5–92.
62. G. Chaudron and J. Bénard, *Bull. soc. chim. France*, **16,** 1949, D 109–116.
63. L. Castelliz, W. de Sutter, and F. Halla, *Monatsh. Chem.*, **85,**1954, 487–490.
64. E. J. Kohlmeyer, *Metall u. Erz*, **10,** 1913, 453–455.
65. R. B. Sosman and J. C. Hostetter, *J. Am. Chem. Soc.*, **38,** 1916, 807–833.
66. J. C. Hostetter and H. S. Roberts, *J. Am. Ceram. Soc.*, **4,** 1921, 927–938.
67. R. Ruer and M. Nakamoto, *Rec. trav. chim.*, **42,** 1923, 675–682.
68. L. B. Pfeil, *J. Iron Steel Inst.*, **119,** 1929, 501–547; discussion, pp. 548–560.
69. J. W. Greig, E. Posnjak, H. E. Merwin, and R. B. Sosman, *Am. J. Sci.*, **30,** 1935, 239–316.
70. J. White, R. Graham, and R. Hay, *J. Iron Steel Inst.*, **131,** 1935, 91–113; J. White, *Iron Steel Inst. (London), Carnegie Schol. Mem.*, **27,** 1938, 1–75.
71. N. G. Schmahl, *Z. Elektrochem.*, **47,** 1941, 821–835.
72. R. W. G. Wyckhoff and E. D. Crittenden, *J. Am. Chem. Soc.*, **47,** 1925, 2876–2882; *Z. Krist.*, **63,** 1926, 144–147.
73. E. R. Jette and F. Foote, *J. Chem. Phys.*, **1,** 1933, 29–36.
74. B. S. Ellefson and N. W. Taylor, *J. Chem. Phys.*, **2,** 1934, 58–64.
75. W. P. Kasanzev, *Z. physik. Chem.*, **174,** 1935, 370–383.
76. J. Bénard, *Compt. rend.*, **205,** 1937, 912–914.
77. A. H. Jay and K. W. Andrews, *Nature*, **154,** 1944, 116.
78. J. Bénard, *Bull. soc. chim. France*, **16,** 1949, D 109–116.
79. R. Collongues and G. Chaudron, *Compt. rend.*, **231,** 1950, 143–145.
80. B. T. M. Willis and H. P. Rooksby, *Acta Cryst.*, **6,** 1953, 827–831.
81. G. Bitsianes and T. L. Joseph, *Trans. AIME*, **197,** 1953, 1641–1647.

82. J. Bénard, *Acta Cryst.*, **7**, 1954, 214.
83. G. Hägg, *Trans. AIME*, **105**, 1933, 287.
84. H. J. Goldschmidt, *J. Iron Steel Inst.*, **146**, 1942, 157–174.
85. N. C. Tombs and H. P. Rooksby, *Nature*, **165**, 1950, 442–443.
86. N. P. Rooksby and N. C. Tombs, *Nature*, **167**, 1951, 364.
87. R. C. Collongues, *Acta Cryst.*, **7**, 1954, 213.
88. W. H. Bragg, *Phil. Mag.*, **30**, 1915, 305–315; *Nature*, **95**, 1915, 561.
89. R. W. G. Wyckhoff and E. D. Crittenden, *J. Am. Chem. Soc.*, **47**, 1925, 2868–2876.
90. S. Holgersson, *Lunds Univ. Årsskr.*, **23**, 1927, no. 9; *Strukturbericht*, **1**, 1913-1928, 416–417.
91. O. Krause, *Ber. deut. keram. Ges.*, **15**, 1934, 101–110.
92. G. Hägg, *Z. physik. Chem.*, **B29**, 1935, 95–103.
93. N. C. Tombs and H. P. Rooksby, *Acta Cryst.*, **4**, 1951, 474–475.
94. S. C. Abrahams and B. A. Calhoun, *Acta Cryst.*, **6**, 1953, 105–106.
95. H. P. Rooksby and B. T. M. Willis, *Acta Cryst.*, **6**, 1953, 565–566.
96. E. J. W. Verwey, P. W. Haayman, and F. C. Romeijn, *J. Chem. Phys.*, **15**, 1947, 181–187.
97. W. P. Davey, *Phys. Rev.*, **21**, 1923, 716.
98. C. Mauguin, *Compt. rend.*, **178**, 1924, 785–787.
99. L. Pauling and S. B. Hendricks, *J. Am. Chem. Soc.*, **47**, 1925, 781–790.
100. V. M. Goldschmidt, T. Barth, and G. Lunde, see *Strukturbericht*, **1**, 1913-1928, 266.
101. E. A. Harrington, *Am. J. Sci.*, **13**, 1927, 467–479.
102. W. H. Zachariasen, *Skrifter Norske Videnskaps-Akad. Oslo*, **1928**, no. 4; see *Strukturbericht*, **2**, 1928-1932, 310.
103. R. Brill, *Z. Krist.*, **83**, 1932, 323–325.
104. S. Katzoff and E. Ott, *Z. Krist.*, **86**, 1933, 311–312.
105. B. T. M. Willis and H. P. Rooksby, *Proc. Phys. Soc. (London)*, **B65**, 1952, 950–954.
106. N. Parravano and G. Malquori, *Anal. españ. fís. y quím.*, **27**, 1929, 454–459; *Strukturbericht*, **2**, 1928-1932, 310.
107. N. V. Belov and V. I. Mokeeva, *Trudy Inst. Krist. Akad. Nauk S.S.S.R.*, **1949**, no. 5, 13–68; *Chem. Abstr.*, **47**, 1953, 3648.
108. L. A. Welo and O. Baudisch, *Phil. Mag.*, **50**, 1925, 399–408.
109. R. Fricke and W. Zerrweck, *Z. Elektrochem.*, **43**, 1937, 52–65.
110. E. J. W. Verwey, *Z. Krist.*, **91**, 1935, 65–69; E. J. W. Verwey and J. H. de Boer, *Rec. trav. chim.*, **55**, 1936, 531–540.
111. R. Haul and T. Schoon, *Z. physik. Chem.*, **B44**, 1939, 216–226.
112. O. Glemser and E. Gwinner, *Z. anorg. Chem.*, **240**, 1939, 161–166.
113. J. Aubry and F. Marion, *Compt. rend.*, **240**, 1955, 1770–1771; see also discussion by G. Chaudron, pp. 1771–1772.
114. F. Marion, *Doc. métallurg.*, **24**, 1955, 87–136; quoted from [115].
115. H. Engell, *Arch. Eisenhüttenw.*, **28**, 1957, 109–115.
116. Taken from a graph in wt. % in [115].
117. R. Sifferlen, *Compt. rend.*, **240**, 1955, 2526–2528.
118. R. Sifferlen, *Compt. rend.*, **244**, 1957, 1192–1193.
119. S. C. Abrahams and B. A. Calhoun, *Acta Cryst.*, **8**, 1955, 257–260.

$\bar{1}.4678$
0.5322

Fe-Os Iron-Osmium

Since Os is h.c.p., neither γ-Fe nor α-Fe can form a continuous series of solid solutions with Os. According to [1], an extended γ field, similar to that in the system C-Fe, can be expected. Figure 383 shows the temperatures of the $\alpha' \rightarrow \gamma$ and $\gamma \rightarrow \alpha'$ transformations on heating and cooling, respectively, and the Curie points, determined by magnetic analysis [2]. As there is no indication of a eutectoid decomposition of the γ solid solution, it appears that γ is stable down to room temperature.

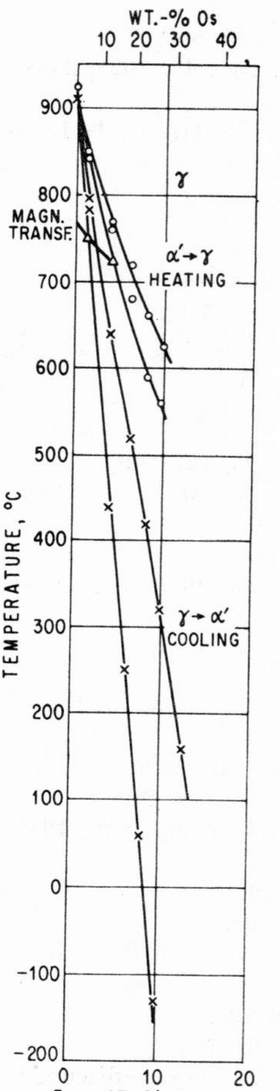

Fig. 383. Fe-Os

1. F. Wever, *Arch. Eisenhüttenw.*, **2,** 1928-1929, 739–746; *Naturwissenschaften*, **17,** 1929, 304–309.
2. M. Fallot, *Ann. phys.*, **10,** 1938, 291–332; *Compt. rend.*, **205,** 1937, 227–230.

0.2560
$\bar{1}.7440$

Fe-P Iron-Phosphorus

The constitution appears to be definitely established. Phase equilibria could be determined only for the composition range up to 30 wt. (43.6 at.) % P (Fig. 384). However, it has been conclusively shown [1, 2] that, besides the compounds Fe_3P (15.60 wt. % P) and Fe_2P (21.71 wt. % P), there are three higher phosphides, namely, FeP (35.68 wt. % P), FeP_2 (52.59 wt. % P), and a phase homogeneous between 72 and 79.5 at. % P. As to the older literature on iron phosphides, see the reviews by [3, 4].

Solidification. The main features of the constitution were already recognized by [5], who examined the microstructure of alloys up to about 36 at. % P. Solidification temperatures were determined by [6, 7, 4, 8, 9]. The liquidus and solidus temperatures reported by [6] were shown to be incorrect [7, 4] because the solidification of Fe-P melts is characterized by an exceptionally great susceptibility to undercooling. In addition, melts with more than 10 wt. (16.7 at.) % P and especially those containing over 15 wt. (24.1 at.) % P tend to form unstable states if solidification takes place at too high a rate or without inoculation [4, 8, 9]. Under such conditions, the primary crystallization and peritectic formation of Fe_3P are suppressed and, instead, an unstable eutectic ($\alpha + Fe_2P$) tends to form at approximately 945°C. On cooling to this or some higher temperature, equilibrium is spontaneously restored, as indicated by a sudden rise in temperature of more than 100°C.

Except for slight temperature differences, the solidification diagrams of [4, 8, 9] are in substantial agreement. The liquidus and solidus curves shown in Fig. 384 are based on the heating- and cooling-curve data by [8]. The solidus curve of the α phase was outlined micrographically [8].

Transformations in the Solid State. According to [10, 11], alloys with 1.36 and 0.83 wt. % P do not undergo the $\alpha \rightleftarrows \gamma$ transformation. [12] showed by means of dilatometric tests that the $\gamma \rightleftarrows \delta$ transformation is lowered to 1355°C by 0.26 wt. (0.47 at.) % P and the $\alpha \rightleftarrows \gamma$ transformation raised to about 1000°C by 0.3 wt. (0.54 at.) % P; no transformation was observed at 0.42 wt. (0.75 at.) % P.

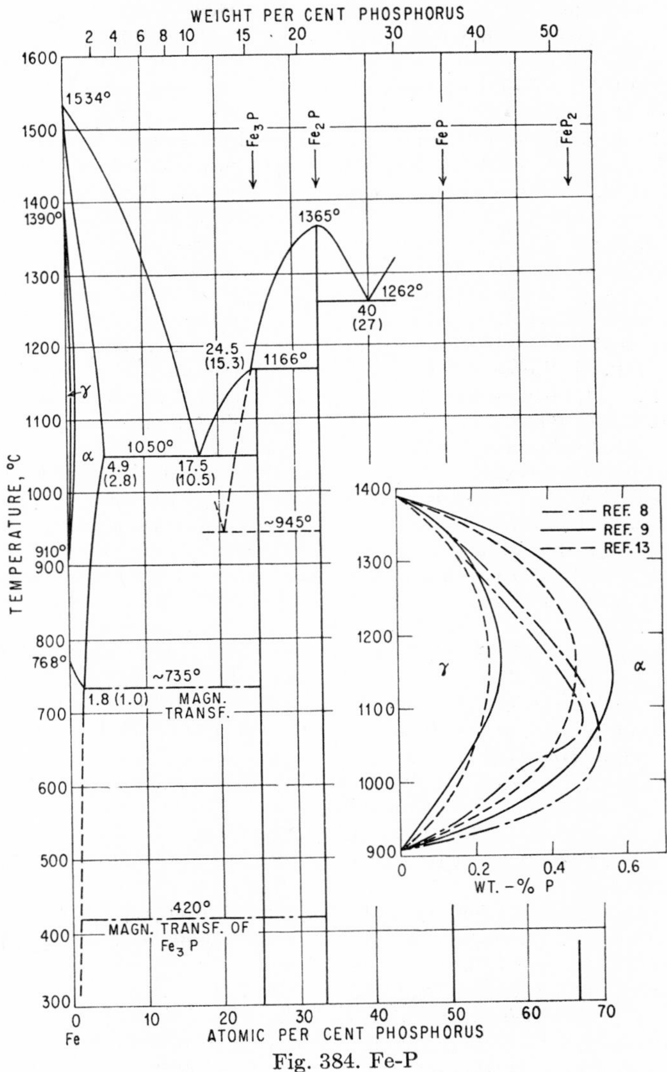

Fig. 384. Fe-P

The existence of a closed $\alpha(\delta) \rightleftarrows \gamma$ transformation curve was confirmed by [8, 9, 13]. They determined the boundaries of the $(\alpha + \gamma)$ field by micrographic [8, 9, 13] and thermal analysis [8]; see inset in Fig. 384, with composition in wt. % P. Results by [9, 13] agree substantially and appear to deserve preference over those by [8]. The vertex of the (inner) $\gamma/(\alpha + \gamma)$ boundary can be assumed as close to 0.25 wt.

(0.45 at.) % P, 1150°C [9, 13]. [14] stated that thermodynamic considerations would indicate a lower composition, perhaps about 0.1 wt. (0.18 at.) % P. However, [14a] found that, by exercising special care in choosing the experimental point which is to be used as a basis for subsequent calculation, much better agreement with the experimental results of [9] is obtainable. The vertex of the (outer) $\alpha/(\alpha + \gamma)$ boundary lies at approximately 0.5 wt. (0.9 at.) % P, 1150°C [9, 13].

The Curie point of Fe was found to be lowered to about 710–720 [10], 740 [15], 745 [8], or 720°C [9]. Fe_3P undergoes a magnetic transformation at 440 [3] or 420°C [8], and Fe_2P at 80°C [3].

The solid solubility of P in α-Fe was found to decrease with temperature [16, 7, 17, 15]. More accurate solubility determinations by means of micrographic work are due to [18, 8, 9]. For temperatures above 850°C, results are in good agreement; below this temperature (down to about 700°C) those by [8] are to be regarded as most reliable (Fig. 384). For lower temperatures, only three single observations are known. [18a] reported that, after annealing for 5 days at 400°C, specimens with 0.6 wt. (1.1 at.) % P showed precipitation of Fe_3P under the microscope. [18b] interpreted some observations on the effects of P on the magnetic properties of Fe as indicating that the solubility at "room temperature" does not exceed 0.015 wt. (0.026 at.) % P. [19] observed no age hardening with an alloy containing 1.1 wt. (2.0 at.) % P. [14] calculated the following solubilities: 1.26, 0.83, and 0.25 wt. (2.25, 1.5, and 0.45 at.) % P at 800, 700, and 500°C, respectively. However, the reliability of these computed values has been questioned by [14a].

Crystal Structures [20]. The lattice parameter of α-Fe is not affected by taking P into solution [21]. Fe_3P is b.c. tetragonal, with 24 atoms per unit cell and a = 9.108 A, c = 4.455 A, c/a = 0.4891 [21]. Fe_2P is hexagonal, with 9 atoms per unit cell (C22 type) [21–24] and a = 5.864 A, c = 3.460 A, c/a = 0.5901 [21]. FeP is isotypic with MnP (B31 type), a = 5.794 A, b = 5.187 A, c = 3.095 A [25]. FeP_2 is isotypic with FeS_2 (C18 type), a = 2.730 A, b = 4.985 A, c = 5.668 A [26].

1. W. Franke, K. Meisel, and R. Juza, *Z. anorg. Chem.*, **218,** 1934, 346–359.
2. M. Heimbrecht and W. Biltz, *Z. anorg. Chem.*, **242,** 1939, 233–236.
3. H. Le Chatelier and S. Wologdine, *Compt. rend.*, **149,** 1909, 709–714.
4. N. Konstantinow, *Z. anorg. Chem.*, **66,** 1910, 209–227.
5. J. E. Stead, *J. Iron Steel Inst.*, **58,** 1900, 60–84.
6. B. Saklatwalla, *J. Iron Steel Inst.*, **77,** 1908, 92–103; *Metallurgie,* **5,** 1908, 331–336.
7. E. Gercke, *Metallurgie,* **5,** 1908, 604–609.
8. J. L. Haughton, *J. Iron Steel Inst.*, **115,** 1927, 417–433.
9. R. Vogel (H. Gontermann), *Arch. Eisenhüttenw.*, **3,** 1929-1930, 369–371.
10. J. O. Arnold, *J. Iron Steel Inst.*, **45,** 1894, 144.
11. Von Schwarze, Dissertation, Aachen, 1924, quoted from [12].
12. H. Esser and P. Oberhoffer, *Ber. Verein deut. Eisenhüttenl. Werkstoffausschuss Ber.*, no. 69, 1925, pp. 5–6.
13. P. Roquet and G. Jegaden, *Rev. mét.*, **48,** 1951, 712–721.
14. W. Oelsen, *Stahl u. Eisen,* **69,** 1949, 468–474.
14a. H. Schrader and B. N. Bose, *Trans. Indian Inst. Metals,* **6,** 1952, 104–136.
15. J. L. Haughton and D. Hanson, *J. Iron Steel Inst.*, **97,** 1918, 413–414.
16. J. E. Stead, *J. Iron Steel Inst.*, **58,** 1900, 60–84; **91,** 1915, 140–198.
17. J. E. Stead, *J. Iron Steel Inst.*, **97,** 1918, 398–405.
18. H. Hanemann and H. Voss, *Zentr. Hütten- u. Walzwerke,* **31,** 1927, 245–248.
18a. R. Vogel and H. Bauer, *Arch. Eisenhüttenw.*, **5,** 1931, 276 (quoted from [14a]).
18b. T. D. Yensen, *Trans. Am. Inst. Elec. Eng.*, **43,** 1924, 145 (quoted from [14a]).
19. W. Köster, *Arch. Eisenhüttenw.*, **4,** 1930-1931, 609–611.

20. X-ray work by C. Kreutzer (*Z. Physik*, **48,** 1928, 564–565) and P. Oberhoffer and C. Kreutzer (*Arch. Eisenhüttenw.*, **2,** 1928–1929, 454–455) is obsolete.
21. G. Hägg, *Z. Krist.*, **68,** 1928, 470; *Nova Acta Regiae Soc. Sci. Upsaliensis*, (4)**7,** 1929, 26–43.
22. J. B. Friauf, *Trans. ASST*, **17,** 1930, 499–508.
23. H. Nowotny and E. Henglein, *Monatsh. Chem.*, **69,** 1948, 385–393.
24. S. B. Hendricks and P. R. Kosting, *Z. Krist.*, **74,** 1930, 522–533.
25. K. E. Fylking, *Arkiv Kemi, Mineral. Geol.*, **11B,** no. 48, 1935.
26. K. Meisel, *Z. anorg. Chem.*, **218,** 1934, 360–364.

$\bar{1}.4306$
0.5694

Fe-Pb Iron-Lead

Thermal analysis [1] has shown that the liquid metals are mutually practically insoluble (Fig. 385) [2]. Inability to alloy was also observed by [3]. Therefore, it is

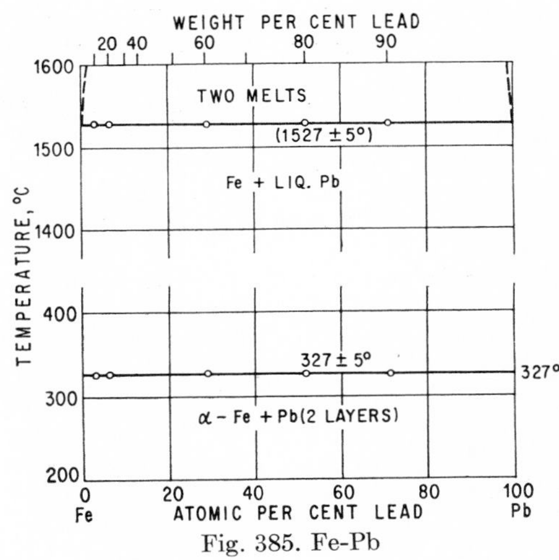

Fig. 385. Fe-Pb

also to be expected that Pb is insoluble in solid Fe. Indeed, it was reported that Pb did not diffuse into Fe at 1000°C [4] and 1000–1300°C [5]. The solid solubility of Fe in Pb, on the basis of magnetic measurements, was given as 2–3 × 10^{-4} wt. % [6]. A compound $FePb_2$ (88.12 wt. % Pb), suggested by [7], does not exist [8, 9].

1. E. Isaac and G. Tammann, *Z. anorg. Chem.*, **55,** 1907, 58–59.
2. Arrests on cooling curves were found within ±5°C of the freezing points of Fe (given as 1527°C) and Pb.
3. A. Stavenhagen and E. Schuchard, *Ber. deut. chem. Ges.*, **35,** 1902, 910.
4. N. W. Ageew and M. I. Zamotorin, *Ann. inst. polytech. Leningrad, sect. math. phys. sci.*, **31,** 1928, 15–28; abstract in *J. Inst. Metals*, **44,** 1930, 556.
5. W. D. Jones, *J. Iron Steel Inst.*, **130,** 1934, 436.
6. G. Tammann and W. Oelsen, *Z. anorg. Chem.*, **186,** 1930, 277–279.
7. E. J. Daniels, *J. Inst. Metals*, **49,** 1932, 179–180.
8. H. Nowotny and K. Schubert, *Z. Metallkunde*, **37,** 1946, 22.
9. E. Scheil, *Z. Metallkunde* **38,** 1947, 320.

$\bar{1}.7189$
0.2811

Fe-Pd Iron-Palladium

Solidification. Figure 386 shows the data points for the beginning and end of solidification, obtained by means of cooling curves [1]. Since the liquidus temperature between about 37 and 61 at. % Pd differs by only 20°C, the minimum is very flat; it appears to lie in the vicinity of 50 at. %, 1310°C. The effect of Pd on the $\gamma \rightleftarrows \delta$ transformation of Fe is unknown.

$\alpha \rightleftarrows \gamma$ **Transformation.** The temperatures of the $\gamma \rightarrow \alpha$ transformation, determined thermally [1], are indicated in Fig. 386. They differ considerably from those reported by [2], in that the latter, determined by magnetic analysis, are constant at about 669–695°C up to at least 25 at. % Pd. Also, the $\alpha \rightarrow \gamma$ transformation temperatures were found at a constant temperature, 840–850°C, between 5.5 and 15 at. % Pd. The same holds for the Curie points at 756°C up to 33.3 at. % Pd [2].

[3] have tried to outline the $\gamma/(\alpha + \gamma)$ boundary by X-ray investigation of powder samples quenched from various temperatures. Their data are not too convincing since the γ phase with up to at least 20 at. % Pd is not retained on quenching from 800°C but transforms to a martensitic b.c.c. structure. However, at 750°C alloys with 5.5, 10, and 20 at. % Pd were two-phase. The $(\alpha + \gamma)$ field is believed to extend beyond 40 at. % Pd [3]. The dashed $\gamma/(\alpha + \gamma)$ boundary in Fig. 386 is a compromise between the data by [2, 3] and can be regarded only as an approximation.

The solubility of Pd in α-Fe was estimated from lattice-parameter data to be 2.4, 1.8, and 0.9 at. (4.5, 3.4, and 1.7 wt.) % Pd at 740, 690, and 500°C, respectively [3]. According to [2], the solubility at about 850°C appears to be higher, about 5 at. % Pd.

Order-Disorder Transformations. [1] first reported the existence of a transformation in the range 62.8–74.4 at. % Pd and gave transformation temperatures (by cooling curves) with a maximum at 71.5–73.8 at. % Pd and 810°C (see data points in Fig. 386), indicating an order-disorder transformation based on $FePd_3$. [4] discovered another order-disorder change in the range of 50 at. % Pd by X-ray investigation and found the ordered structure to be isotypic with CuAu.

The existence of both changes was confirmed by [5–7]. Also, the curve of the Curie points of annealed alloys in the temperature range 485 to −35°C is in accord with these transformations (Fig. 386). [3] roughly outlined the range of existence of the two ordered structures by X-ray studies, showing maximal temperatures of about 775°C at both 60 and 70 at. % Pd. [7] reported the composition range of the ordered structure FePd to extend from about 40 to about 62.6 at. % Pd and gave the transformation points, based on X-ray work, shown in Fig. 386, with a maximum at 60–61.5 at. % Pd and 880°C.

Crystal Structures. The lattice constant of the α phase increases from $a = 2.866$ A at 0% Pd to $a = 2.879$ A at 2.4 at. % Pd [3]. The lattice parameters of the f.c.c. γ phase (after quenching) reported by [5–7] agree on the whole and show a strong positive deviation from the straight-line relationship.

The ordered structure of FePd is f.c. tetragonal of the CuAu ($L1_0$) type [4, 6, 7], with $a = 3.860$ A, $c = 3.730$ A, $c/a = 0.966$ at 51.9 at. % Pd, the axial ratio decreasing with higher Pd content [6]. On the other hand, [4] gave $c/a = 1.03$ at 50 at. %; and [7], $c/a = 1.027$ at 62.6 at. % Pd. [5] believed, erroneously, that FePd was cubic.

$FePd_3$ is isotypic with Cu_3Au ($L1_2$ type) [6, 7], with $a = 3.848$–3.851 A at 75 at. % Pd [6, 7].

For theoretical reasons, [8] suggested the existence of an ordered structure based on Fe_3Pd.

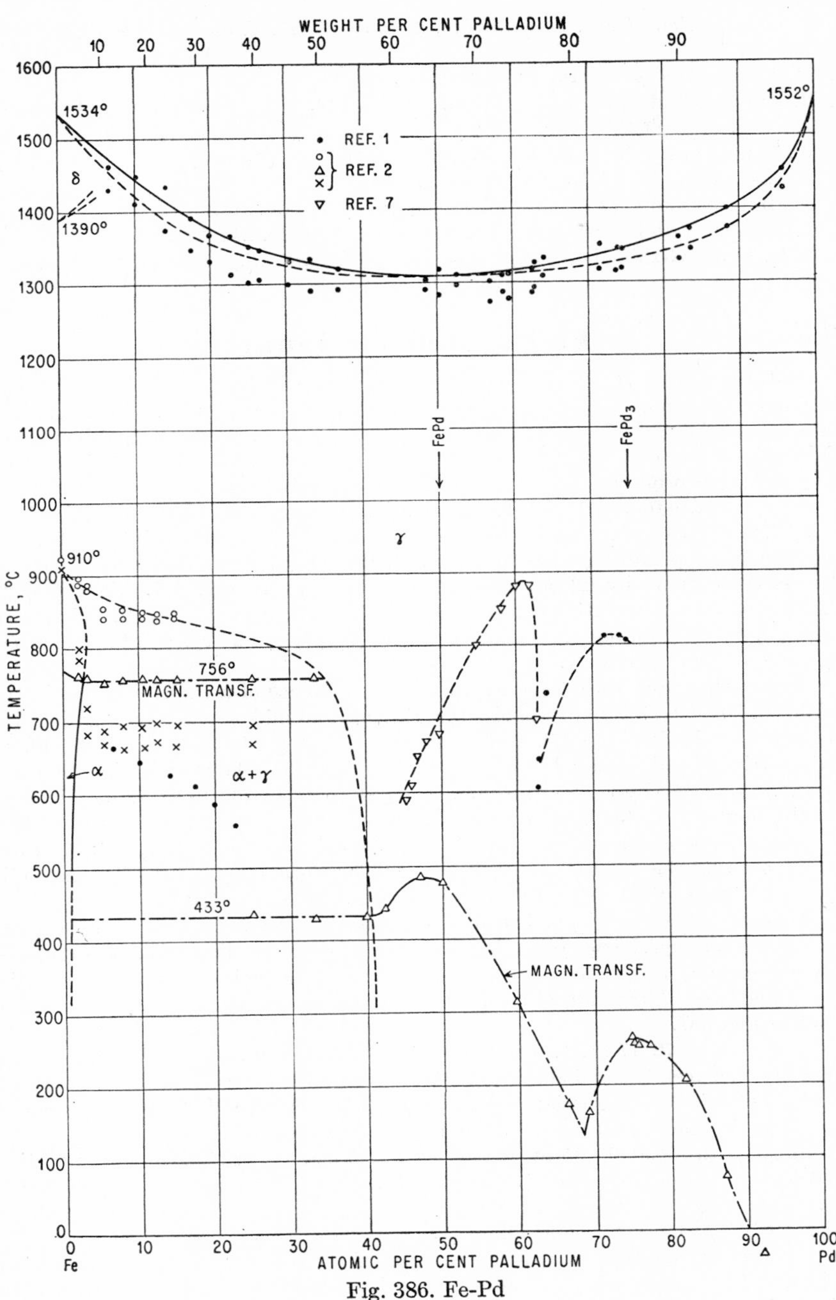
WEIGHT PER CENT PALLADIUM
10
20
30
40
50
60
70
80
90
1600
1500
1400
1300
1200
1100
1000
900
800
700
600
500
400
300
200
100
0
1534°
1552°
1390°
δ
REF. 1
REF. 2
REF. 7
FePd
FePd3
γ
910°
756°
MAGN. TRANSF.
α
α+γ
433°
MAGN. TRANSF.
TEMPERATURE, °C
0
Fe
10
20
30
40
50
60
70
80
90
100
Pd
ATOMIC PER CENT PALLADIUM

Fig. 386. Fe-Pd

1. A. T. Grigorjew, *Izvest. Inst. Platiny*, **8**, 1931, 25–37; *Z. anorg. Chem.*, **209**, 1932, 295–307.
2. M. Fallot, *Ann. phys.*, **10**, 1938, 291–332.
3. R. Hultgren and C. A. Zapffe, *Trans. AIME*, **133**, 1939, 58–68; *Nature*, **142**, 1938, 395–396.
4. W. Jellinghaus, *Z. tech. Physik*, **17**, 1936, 33–36.
5. R. Hocart and M. Fallot, *Compt. rend.*, **204**, 1937, 1465–1467.
6. R. Hultgren and C. A. Zapffe, *Z. Krist.*, **99**, 1938, 509–512.
7. G. Alaverdov and S. Shavlo, *Zhur. Tekh. Fiz.*, **9**, 1939, 211–214.
8. N. V. Grum-Grzhimailo, *Compt. rend. acad. sci. U.R.S.S.*, **33**, 1941, 237–240.

$\bar{1}.4565$
0.5435

Fe-Pt Iron-Platinum

Solidification. Cooling-curve data in Fig. 387 are due to [1]. The solidification interval was found as practically nil between 0 and about 10 at. % Pt. A minimum, assumed to occur at about 6 at. % Pt [1], is not shown in Fig. 387, although one might exist because of the peritectic reaction δ_{Fe} + melt $\rightleftarrows \gamma$. The effect of Pt on the $\gamma \rightleftarrows \delta$ transformation of Fe has not been investigated yet.

$\alpha \rightleftarrows \gamma$ **Transformation.** Transformation temperatures were determined by [1–5]. The most comprehensive data are those by [3, 4]; see data points in Fig. 387. Transformation points reported by [1, 5], on the whole, follow the same trend [6], whereas those found by [2] on heating scatter widely; see Fig. 387. According to [4], the $\gamma \rightarrow \alpha'$ transformation reaches room temperature at 21.2 at. % Pt, and according to [5], the transformation of the alloys with 22, 25, and 26 at. % Pt starts at 50, 20, and −60°C, respectively.

Order-Disorder Transformations. The existence of a transformation in the vicinity of 50 at. % Pt, first recognized by thermal and microscopic investigations [1], was confirmed by hardness [7], as well as X-ray [8, 2, 9–11, 5], magnetic [2, 10, 5], dilatometric [5], and electrical-resistance studies [7, 12, 5]. Transformation temperatures were reported by [2, 5] (Fig. 387). According to [10, 5], the maximum lies considerably higher than found previously [2]. The transformation $\gamma \rightarrow$ FePt cannot be fully suppressed on quenching from the γ field according to [1, 7, 2, 10]; however, [5] reported that the γ phase can be retained. The Curie points of the ferromagnetic superlattice phase FePt reach a maximum at about 50 at. %, 480°C.

Magnetic, dilatometric, and roentgenographic investigations indicated the existence of another superlattice phase based on the composition Fe_3Pt [5]. Its range of existence extends from about 19–33 at. % Pt; i.e., there is overlapping with the $\alpha' \rightarrow \gamma$ and $\gamma \rightarrow \alpha'$ transformations. Fe_3Pt is ferromagnetic; its Curie points are shown in Fig. 387.

In the range 72–77 at. % Pt, a third order-disorder transformation takes place somewhere between 700 and 800°C [5]. The superlattice is based on the composition $FePt_3$. $FePt_3$ is paramagnetic; however, alloys between 80 and 93 at. % Pt become ferromagnetic when cooled to approximately −100°C [5].

Alloys with about 25–35 at. (54–65 wt.) % Pt are characterized by a negative coefficient of thermal expansion [12, 13] which is largest after quenching from 800°C [12]. The Invar effect is caused by the extremely high magnetostriction of the alloys and exceeds that found in Fe-Ni alloys [12].

Crystal Structures. Lattice parameters of the α and γ phases were reported by [2] and [2, 5], respectively. FePt, previously assumed to be disordered b.c.c. [2], is isotypic with CuAu [10, 11, 5], a = 2.719 A, c = 3.722 A, c/a = 1.369 [10]. [5]

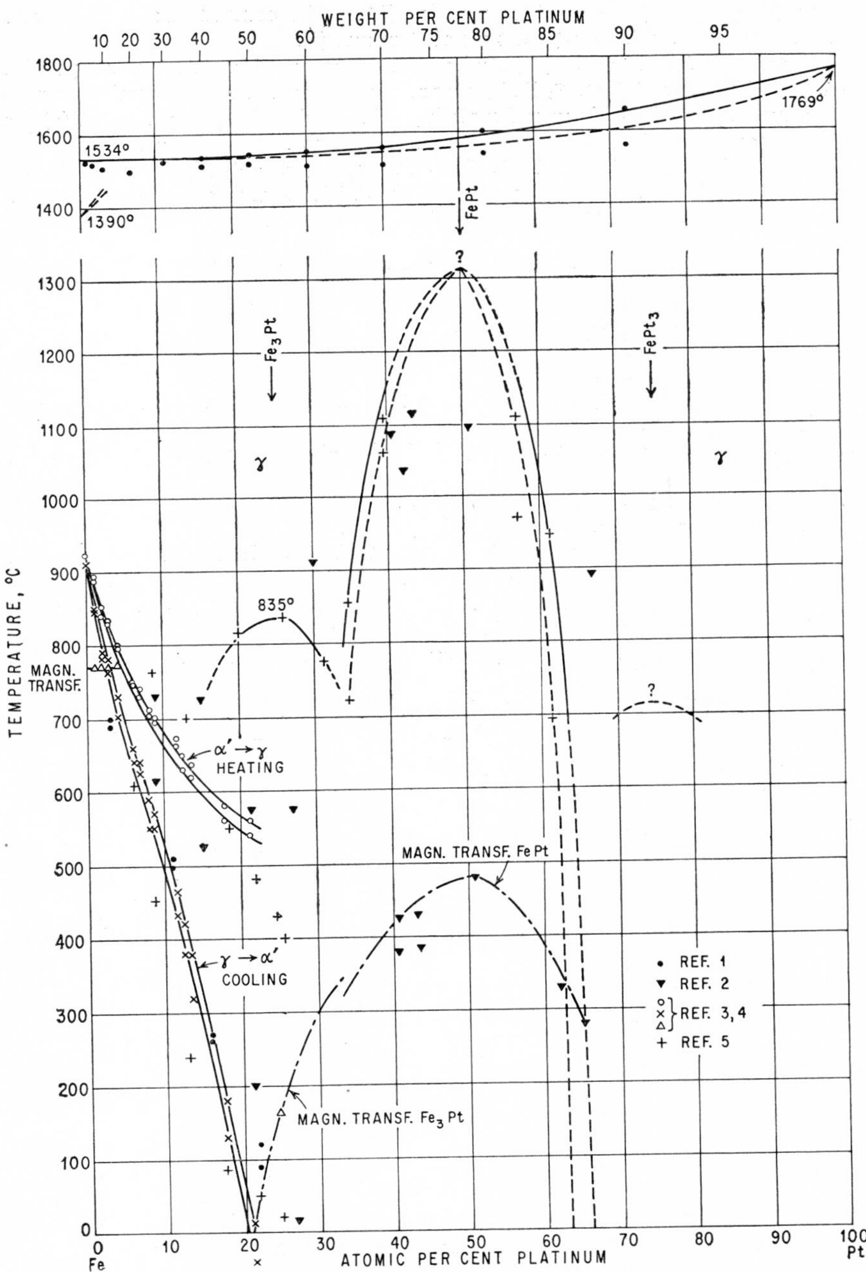

Fig. 387. Fe-Pt. (See also Note Added in Proof.)

reported $a = 3.87$ A, $c/a = 0.973$; see also [14]. The ordered phases Fe_3Pt and $FePt_3$ are both of the Cu_3Au ($L1_2$) type of structure [5].

Note Added in Proof. [15] studied the phase transformations in alloys with 14.4, 24.2, 27.1, and 30.0 at. % Pt by means of resistometric and magnetic measurements, supplemented by X-ray and metallographic work. They suggest that (*a*) the critical temperature for long-range-order formation at Fe_3Pt is about 100°C lower than reported by [5] (cf. Fig. 387); (*b*) a eutectoid $\gamma \rightleftarrows \alpha + (Fe_3Pt)$ exists in the neighborhood of 15 at. % Pt and somewhat above 550°C; and (*c*) the $\gamma \rightarrow \alpha'$ realization curve passes through 25 at. % Pt at 0°C.

1. E. Isaac and G. Tammann, *Z. anorg. Chem.*, **55,** 1907, 63–71.
2. L. Graf and A. Kussmann, *Physik. Z.* **36,** 1935, 544–551.
3. J. Martelly, *Ann. phys.*, **9,** 1938, 318–323.
4. M. Fallot, *Ann. phys.*, **10,** 1938, 291–332; *Compt. rend.*, **199,** 1934, 128–129.
5. A. Kussmann and G. v. Rittberg, *Z. Metallkunde*, **42,** 1950, 470–477.
6. The data points by [5] are not exact as they had to be taken from a small and distorted diagram.
7. V. A. Nemilov, *Izvest. Inst. Platiny*, **7,** 1912, 1–12; *Z. anorg. Chem.*, **204,** 1932, 49–59.
8. U. Dehlinger and L. Graf, *Z. Physik*, **64,** 1930, 359.
9. W. Jellinghaus, *Z. tech. Physik*, **17,** 1936, 33–36.
10. H. Lipson, D. Schoenberg, and G. V. Stupart, *J. Inst. Metals*, **67,** 1941, 333–340.
11. L. Weil, *Compt. rend.*, **224,** 1947, 923–925.
12. A. Kussmann, M. Auwärter, and G. v. Rittberg, *Ann. Physik*, **4,** 1948, 174–182; see also A. Kussmann, *Physik. Z.*, **38,** 1937, 41–42.
13. H. Masumoto and T. Kobayashi, *Science Repts. Research Insts. Tôhoku Univ.*, **A2,** 1950, 856–860.
14. F. A. Bannister and M. H. Hey, *Mineralog. Mag.*, **23,** 1932, 188–206.
15. A. E. Berkowitz, F. J. Donahoe, A. D. Franklin, and R. P. Steijn, *Acta Met.*, **5,** 1957, 1–12.

$\bar{1}.3686$
0.6314

Fe-Pu Iron-Plutonium

The phase diagram of Fig. 387*a* [1] was established by means of metallographic, thermal, X-ray, dilatometric, and microhardness work on alloy samples weighing at the most some hundreds of milligrams [2]. There are two compounds in this system [2–4]. Pu_6Fe [14.29 at. (3.75 wt.) % Fe], which has good ductility [2], is isotypic with b.c. tetragonal U_6Mn. The lattice parameters reported are $a = 10.403$ A, $c = 5.348$ A [2]; $a = 10.41 \pm 1$ A, $c = 5.359 \pm 4$ A [3]; and $a = 10.40 \pm 2$ A, $c = 5.345 \pm 5$ A [4].

$PuFe_2$ (31.8 wt. % Fe) is of the cubic $MgCu_2$ (C15) type of structure [2–5]. The lattice-parameter values reported vary between 7.150 and 7.191 A, which may indicate a certain range of homogeneity.

1. The diagram had to be reproduced from a small-scale plot in wt. % [2, 6].
2. S. T. Konobeevsky, Conference of the Academy of Sciences of the U.S.S.R. on the Peaceful Uses of Atomic Energy, Division of Chemical Science, July 1–5, 1955; English translation by U.S. Atomic Energy Commission, Washington, 1956.
3. Work at Los Alamos, quoted in [6].
4. Work at Harwell, quoted in [6].
5. Work at Chalk River, quoted in [6].

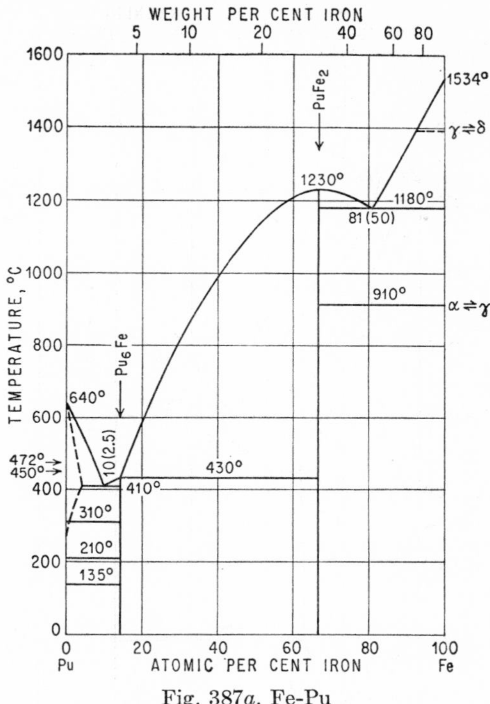

Fig. 387*a*. Fe-Pu

6. "Metallurgy and Fuels," Progress in Nuclear Energy, ser. V, vol. 1, pp. 353–410, Pergamon Press Ltd., London, 1956.

$\bar{1}.8152$
0.1848

Fe-Rb Iron-Rubidium

According to [1], Rb is regarded as insoluble in solid Fe. As the boiling point of Rb (680°C) lies far below the melting point of Fe, it cannot be determined under normal pressure whether Rb is soluble in liquid Fe.

1. F. Wever, *Arch. Eisenhüttenw.*, **2,** 1928-1929, 739–746; *Naturwissenschaften*, **17,** 1929, 304–309.

$\bar{1}.4768$
0.5232

Fe-Re Iron-Rhenium

The phase diagram of the partial system Fe-Re_2Fe_3 (68.98 wt. % Re) presented in Fig. 388 is based on thermal analysis and microscopic data [1]. X-ray powder patterns, obtained by exposure samples at high temperatures, corroborate the existence of the phases shown. Although the diagram cannot be regarded as representing equilibrium conditions, at least it outlines the phase relationships shown. According to [2], Re_2Fe_3 is of the σ-type structure.

1. H. Eggers, *Mitt. Kaiser-Wilhelm-Inst. Eisenforsch. Düsseldorf*, **20,** 1938, 147–152. Alloys melted under H_2 were analyzed.

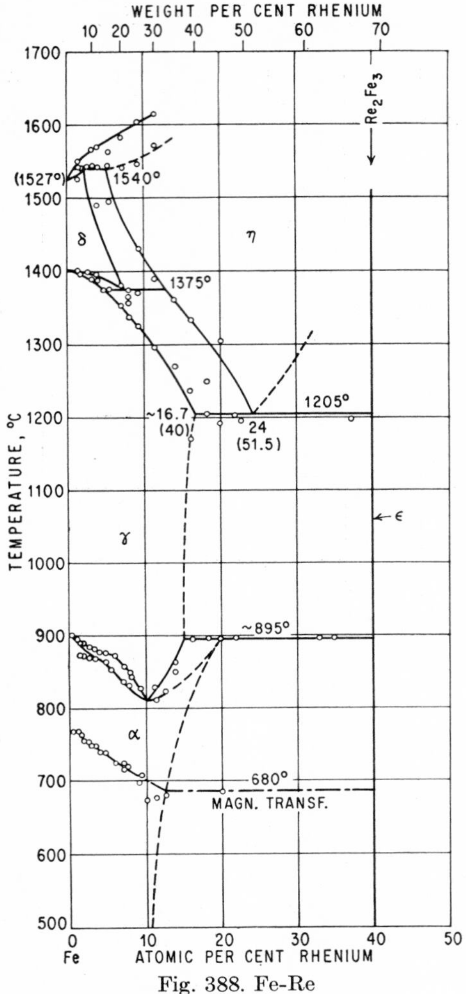

Fig. 388. Fe-Re

2. J. Niemiec and W. Trzebiatowski, *Bull. acad. polon. sci.*, **4**, 1956, 601–603; *Met. Abstr.*, **24**, 1957, 819.

$\bar{1}.7346$
0.2654

Fe-Rh Iron-Rhodium

As the low-temperature modification of Rh, stable up to approximately 1150°C, is f.c.c., it is possible that it forms a continuous series of solid solutions with γ-Fe.

Figure 389 shows the temperatures of the $\alpha' \rightarrow \gamma$ transformation on heating and the $\gamma \rightarrow \alpha'$ transformation on cooling, as well as the Curie points, determined by magnetic analysis [1]. Since both pairs of curves have a minimum at 15–20 at. % Rh, it seems probable that, under equilibrium conditions, the $\alpha \rightleftarrows \gamma$ transformation reaches a maximum at higher Rh contents, similar to the relations in the system Co-Fe. However, in contrast to Co-Fe, work by [2] has shown that (*a*) above 50 at. % Rh the Curie temperature becomes constant, indicating a two-phase field, and (*b*) another change occurs in the vicinity of 50 at. % Rh, with transformation temperatures rising from about −100°C at 50 at. % Rh to about 135°C at 55–56 at. % Rh

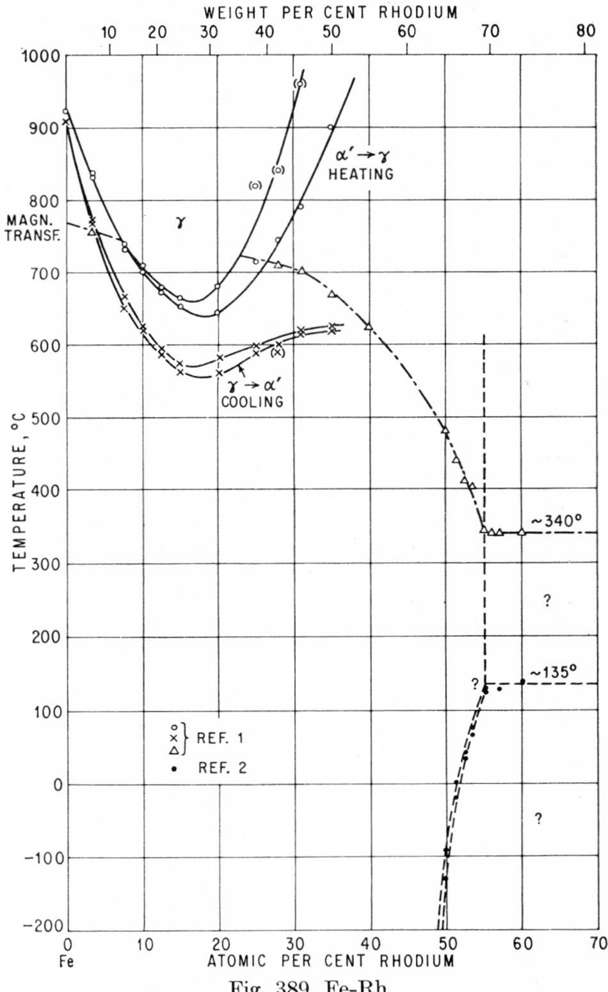

Fig. 389. Fe-Rh

(Fig. 389). The latter change is characterized by a strong increase in magnetization as the temperature rises (!). X-ray investigation indicated some change in the lattice structure of alloys with 40–64 at. % Rh, according to the heat-treatment. However, the information available is too scanty to permit a plausible interpretation.

1. M. Fallot, *Ann. phys.*, **10,** 1938, 291–332; *Compt. rend.*, **205,** 1937, 227–230.
2. M. Fallot and R. Hocart, *Rev. sci.*, **77,** 1939, 498–499.

1.7397
0.2603

Fe-Ru Iron-Ruthenium

Since Ru is h.c.p., it cannot form a continuous series of solid solutions with either γ-Fe or α-Fe. According to [1], an extended γ field, similar to that in the system C-Fe, can be expected. However, as there is no indication of a eutectoid decomposition of the γ solid solution (Fig. 390), it appears that γ is stable down to room temperature.

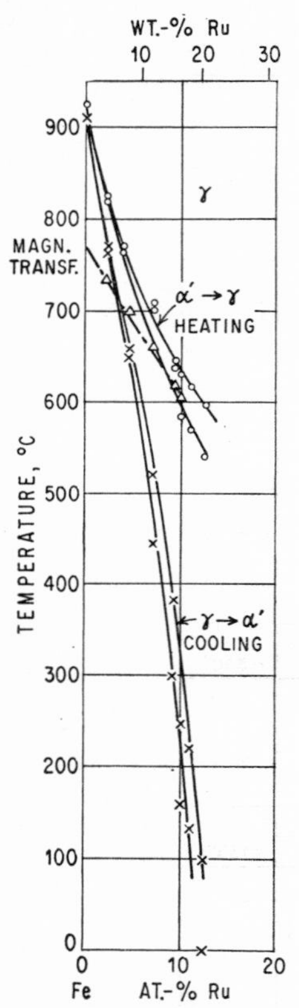

Fig. 390. Fe-Ru

Figure 390 shows the temperatures of the $\alpha' \rightarrow \gamma$ and $\gamma \rightarrow \alpha'$ transformations on heating and cooling, respectively, and the Curie points, determined by magnetic analysis [2, 3].

1. F. Wever, *Arch. Eisenhüttenw.*, **2**, 1928-1929, 739–746; *Naturwissenschaften*, **17**, 1929, 304–309.
2. J. Martelly, *Ann. phys.*, **9**, 1938, 318–333 (Alloys with 2.24 and 4.51 at. % Ru).
3. M. Fallot, *Ann. phys.*, **10**, 1938, 291–332; *Compt. rend.*, **205**, 1937, 227–230.

0.2410
1.7590

Fe-S Iron-Sulfur

Partial System Fe-FeS (36.48 Wt. % S). *Solidification.* The liquidus curve of Fe shown in Fig. 391 is a compromise of the data of [1–4]. Previously, [5] had found that Fe and (FeS) alloy in all proportions and form a eutectic. In contrast to this, [6] had concluded from thermal data that there was a miscibility gap in the liquid state extending at 1400°C from about 3 to 29 wt. (5–41.5 at.) % S, although formation of two layers could not be detected [7].

The eutectic temperature was found at 983 [1], 985 [2, 3], and 988°C [8] and the eutectic composition at >30 wt. (42.7 at.) % S [9], 30.5 wt. (43.3 at.) % S [3], 30.8 wt. (43.7 at.) % S [2], and 31.2 wt. (44.1 at.) % S [8]; see also [10].

As indicated by the data points in Fig. 391, the liquidus temperatures of the (FeS) phase, reported by [2, 3, 11], are about 20–25°C higher throughout than those determined by [1, 8, 12]. There is agreement, however, that the liquidus still rises beyond the composition FeS [1, 2, 11, 12, 13]. [8] extended his study to a composition of 40 wt. (53.7 at.) % S, using melts in sealed silica tubes. He found the liquidus to have a maximum at 1190°C and 38.0–38.5 wt. (51.6–52.2 at.) % S, corresponding to a formula $FeS_{1.09}$ or $Fe_{12}S_{13}$. The composition FeS melts over a temperature range of approximately 100°C. The solidus point of the composition 36.75 wt. (50.3 at.) % S lies at 1100°C [14]. At the eutectic temperature the phase (FeS) contains about 36.4 wt. (49.9 at.) % S, indicating that the solubility of Fe in FeS, if any, is only very slight [14].

The solid solubility of S *in* Fe in the temperature range 900–1500°C was determined using measurements of the chemical activity of S in the solid solution by means of equilibration with a mixture of H_2 and H_2S [15]. The solubility in γ-Fe at 1000, 1200, and 1335°C was determined by means of a similar method [16]. Results of both investigations are in good agreement; see inset of Fig. 391. They show that the $\gamma \rightleftarrows \delta$ (A4) transformation of Fe is lowered to about 1365°C [15]. This is in

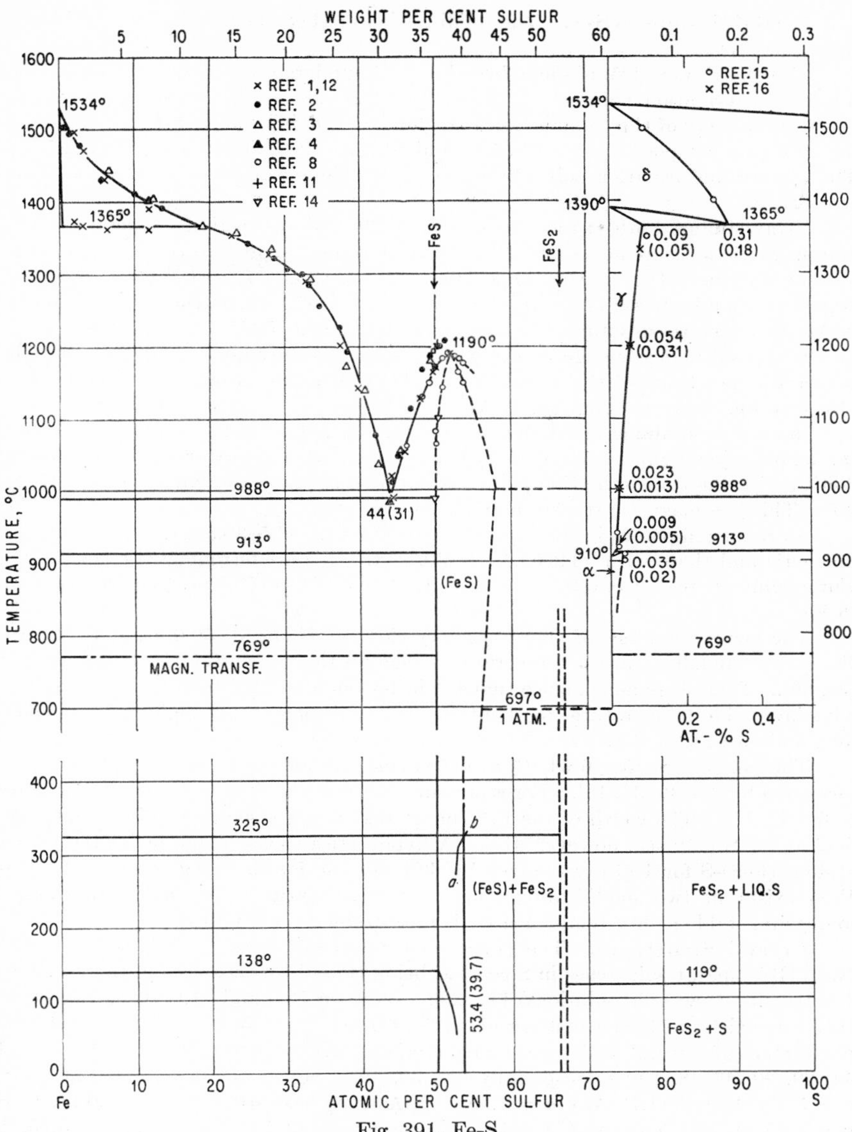

Fig. 391. Fe-S

accord with (*a*) thermal data of [1], which can be interpreted as indicating the reaction $\delta \rightleftarrows \gamma$ + melt to occur at 1366^{+8}_{-5}°C [17], and (*b*) qualitative observations that S depresses the A4 transformation [18, 19]. The $\alpha \rightleftarrows \gamma$ (A3) transformation point is slightly raised [15, 20] rather than lowered [21, 19]. Previous work on the effect of S on the transformation points of Fe [6, 1, 2, 3] and the solid solubility of S in Fe [22–24] is obsolete. In view of the very low solubility at temperatures below 900°C, it is to be expected that the Curie point of Fe is not affected by S.

Partial System FeS-S. The solidification of (FeS) melts was discussed in the preceding section.

The phase (FeS) [25] has long been known to undergo two transformations. The lower transformation point [5, 6, 26, 1, 2, 13, 27, 28] is now accepted as lying at 138°C in an alloy of the composition FeS [28]. It is depressed by S additions, reaching 50°C at a composition between 52.4 and 52.7 at. (38.7 and 39.0 wt.) % S [27, 28]. The upper transformation point [2] lies at 325°C [28] and is unaffected by the S content [27, 28].

X-ray studies, both at room temperature [29] and at elevated temperatures [30], and magnetic measurements [28, 29] have given some insight into the nature of these transformations [31]. In the range 50.0–51.0 at. (36.5–37.4 wt.) % S, the lower transformation is characterized by the disappearance of a superstructure based on the NiAs (B8) type of structure, whereas in the range 51.0–52.7 at. (37.4–39.0 wt.) % S only a discontinuous change of the c axis and a discontinuous change of the temperature dependence of the a axis of the NiAs structure occur [30]. The latter is also true for the transformation at 325°C [30].

In alloys containing more than about 51.2 at. % S an additional transformation, taking place in a temperature interval, was observed at temperatures between 110 and 300°C, with the transformation point being raised as the S content increases [27, 28, 30]. This is a magnetic transformation [28–30].

At room temperature, alloys with 50.0–52.2 at. (36.5–38.5 wt.) % S are paramagnetic and those with 52.2–53.4 at. (39.7 wt.) % S are ferromagnetic [28, 29, 25]. Curie points increase from 270°C at 52.4 at. % to 325°C at 53.5 at. % S [28] (curve *ab* in Fig. 391).

The low-S boundary of (FeS) lies very close to, if not at, 50.0 at. % S [32], and the high-S boundary, in the temperature range up to at least 325°C, at 53.4 at. % S [29, 30]. There is some evidence that the latter boundary shifts to higher S contents with increasing temperature: to about 54.5 at. % at 565°C [13], 55.5 at. % at 650°C [33], and 56.8 at. % S at 700–800°C [14]; see also [34].

The phase FeS_2 (53.46 wt. % S) decomposes on heating. Vapor pressures were measured by [35–38, 34, 14]. For a pressure of 1 atm the decomposition temperature is 697°C [14]. The width of the FeS_2-phase region was suggested to be about 66.0–66.7 at. % S [34]; see, however, [14]. A hypothetical phase diagram for the partial system (FeS)-S for 1 atm was given by [39] and one for changing pressures by [14]. FeS_2 exists in two modifications, marcasite and pyrite. The transformation of marcasite, stable at low temperatures, is measurable at 450°C [13].

Crystal Structures. The (FeS) phase above 51.0 at. % S is isotypic with NiAs (B8 type), the increase in S content being caused by vacancies in the Fe lattice (subtraction-type solid solutions) [33, 29]. Between 50.0 and 51.0 at. % S a superstructure exists, with 24 atoms per unit cell at 50 at. % S [33, 29]. Numerous lattice parameters of natural and synthetic materials have been reported [40–42, 33, 43, 44, 29, 30, 45–48, 19]; see especially [29, 30, 49]. Superstructure at 50.0 at. % S: $a = 5.97_6$ A, $c = 11.76$ A, $c/a = 1.96_8$. NiAs-type base structure: at 50.0 at. % S, $a = 3.45_0$ A, $c = 5.88_2$ A, $c/a = 1.70_5$ (pseudo cell); at 50.98 at. % S, $a = 3.45_6$ A, $c = 5.79_8$ A, $c/a = 1.67_8$; at 53.50 at. % S, $a = 3.43_3$ A, $c = 5.69_8$ A, $c/a = 1.66_0$ [29].

FeS_2 exists in two modifications, cubic pyrite (C2 type) and orthorhombic marcasite (C18 type). Lattice constant of pyrite [50–56], $a = 5.405$ A [56]. Lattice constants of marcasite [57, 58], $a = 4.445$ A, $b = 5.425$ A, $c = 3.388$ A [58].

1. K. Friedrich, *Metallurgie*, **7**, 1910, 257–261.
2. R. Loebe and E. Becker, *Z. anorg. Chem.*, **77**, 1912, 301–319; E. Becker, *Stahl u Eisen*, **32**, 1912, 1017–1021.
3. K. Miyazaki, *Science Repts. Tôhoku Univ.*, **17**, 1928, 877–881.

4. Y. I. Ol'shanskii, *Doklady Akad. Nauk S.S.S.R.*, **80,** 1951, 893–896.
5. H. Le Chatelier and M. Ziegler, *Bull. soc. encour. ind. natl.*, **1902**(2), 368.
6. W. Treitschke and G. Tammann, *Z. anorg. Chem.*, **49,** 1906, 320–335.
7. Since [6] used technical-grade FeS containing oxide, it was believed that immiscibility came from the oxide content [9, 1, 2]. See, however, the contrary findings of B. Bogitch (*Compt. rend.*, **182,** 1926, 217–219) and C. Benedicks (*Z. physik. Chem.*, **131,** 1928, 288–289), who reported that immiscibility can be caused by either carbon or silicon.
8. E. Jensen, *Am. J. Sci.*, **240,** 1942, 695–709.
9. M. Ziegler, *Rev. mét.*, **6,** 1909, 459–475.
10. The paper by F. H. Edwards (*Trans. Inst. Mining Met.*, **33,** 1924, 492–510) was not available.
11. K. Bornemann, *Metallurgie*, **5,** 1908, 63–64.
12. K. Friedrich, *Metallurgie*, **5,** 1908, 55–56.
13. E. T. Allen, J. L. Crenshaw, J. Johnston, and E. S. Larsen, *Am. J. Sci.*, **33,** 1912, 169–236; *Z. anorg. Chem.*, **76,** 1912, 201–273.
14. T. Rosenquist, *J. Iron Steel Inst.*, **176,** 1954, 37–57.
15. T. Rosenquist and B. L. Dunicz, *Trans. AIME*, **194,** 1952, 604–608.
16. E. T. Turkdogan, S. Ignatowicz, and J. Pearson, *J. Iron Steel Inst.*, **180,** 1955, 349–354.
17. These thermal data are one of the earliest indications of the existence of the $\gamma \rightleftarrows \delta$ transformation in Fe.
18. A. Heinzel, *Arch. Eisenhüttenw.*, **2,** 1928-1929, 747.
19. R. Vogel and G. F. Hillner, *Arch. Eisenhüttenw.*, **24,** 1953, 133–141.
20. It was already predicted by F. Wever (*Arch. Eisenhüttenw.*, **2,** 1928-1929, 739–746) that S forms a contracted γ region as shown in Fig. 391.
21. J. H. Andrew and D. Binnie, *J. Iron Steel Inst.*, **119,** 1929, 346–352.
22. J. O. Arnold and A. McWilliam, *J. Iron Steel Inst.*, **55,** 1899, 85–106.
23. A. Fry, *Stahl u. Eisen*, **43,** 1923, 1039–1044.
24. C. R. Wohrmann, *Trans. ASST*, **14,** 1928, 255–276.
25. Designation of the minerals in this range is not uniform. Some authors prefer to call all the solid solutions of S in FeS pyrrhotite and the stoichiometric composition FeS troilite. According to others, troilite comprises the (FeS) compositions which are paramagnetic and pyrrhotite those which are ferromagnetic.
26. F. Rinne and H. E. Boeke, *Z. anorg. Chem.*, **53,** 1907, 338–343.
27. H. S. Roberts, *J. Am. Chem. Soc.*, **57,** 1935, 1034–1038.
28. H. Haraldsen, *Z. anorg. Chem.*, **231,** 1937, 78–96.
29. H. Haraldsen, *Z. anorg. Chem.*, **246,** 1941, 169–194.
30. H. Haraldsen, *Z. anorg. Chem.*, **246,** 1941, 195–226.
31. Phenomena are more complex than presented here.
32. The lower transformation point of an alloy with 49.7 at. % S was found as 145°C, i.e., 7°C higher than for the 50.0 at. % alloy [28]. This might indicate a slight solubility of Fe in FeS; see also [27].
33. G. Hägg and I. Sucksdorff, *Z. physik. Chem.*, **B22,** 1933, 444–452.
34. R. Juza and W. Biltz (K. Meisel), *Z. anorg. Chem.*, **205,** 1932, 273–286.
35. E. T. Allen and R. H. Lombard, *Am. J. Sci.*, **43,** 1917, 175–195.
36. M. G. Raeder, *Norske Videnskaps Selskab, Forh.*, **2,** 1929, 151–154.
37. L. D'Or, *J. chim. phys.*, **27,** 1930, 239–249; **28,** 1931, 377–408; *Compt. rend.*, **190,** 1930, 1296–1298.
38. F. de Rudder, *Bull. soc. chim. France*, **47,** 1930, 1225–1254.
39. M. Hansen, "Der Aufbau der Zweistofflegierungen," p. 726, Springer-Verlag OHG, Berlin, 1936.
40. W. F. de Jong and H. W. V. Willems, *Physica*, **7,** 1925, 74–79.

41. N. Alsén, *Geol. Fören. i. Stockholm Förh.*, **47**, 1925, 19–72.
42. F. Heide, E. Herschkowitsch, and E. Preuss, *Chemie der Erde*, **7**, 1932, 483–502.
43. A. Michel and G. Chaudron, *Compt. rend.*, **198**, 1934, 1913-1915; **203**, 1936, 1004–1006.
44. S. S. Sidhu and V. Hicks, *Phys. Rev.*, **52**, 1937, 667; **53**, 1938, 207.
45. G. T. Lindroth, *Jernkontorets Ann.*, **130**, 1946, 27–75; *Tek. Tidskr.*, **76**, 1946, 383–387.
46. M. J. Buerger, *Am. Mineralogist*, **32**, 1947, 411; *Geol. Soc. Am., Bull.*, **56**, 1945, 1150.
47. A. R. Graham, *Am. Mineralogist*, **34**, 1949, 462–464.
48. F. Bertaut, *Compt. rend.*, **234**, 1952, 1295–1297.
49. *Structure Repts.*, **11**, 1947-1948, 246–252.
50. W. L. Bragg, *Phil. Mag.*, **40**, 1920, 168–169.
51. L. S. Ramsdell, *Am. Mineralogist*, **10**, 1925, 281–304.
52. W. F. de Jong, *Physica*, **7**, 1927, 23–28.
53. I. Oftedal, *Z. physik. Chem.*, **135**, 1928, 291–299.
54. H. M. Parker and W. J. Whitehouse, *Phil. Mag.*, **14**, 1932, 939–961.
55. P. F. Kerr, R. J. Holmes, and M. S. Knox, *Am. Mineralogist*, **30**, 1945, 498–504.
56. D. Lundquist, *Arkiv Kemi, Mineral Geol.*, **24A**, 1947, no. 22, 1–12; no. 23, 1–7.
57. M. J. Buerger, *Am. Mineralogist*, **16**, 1931, 361–395.
58. M. J. Buerger, *Z. Krist.*, **97**, 1937, 504–513.

$\bar{1}.6615$
0.3385

Fe-Sb Iron-Antimony

Solidification. Thermal-analysis data, covering the whole composition range [1] or parts thereof [2, 3], are in substantial agreement, as shown by the data points in Fig. 392. The maximum of the liquidus curve was found at about 40 at. % Sb, 1014°C [1], and about 44 at. % Sb, 1020°C [2]. The temperature of the (α-Fe)-ϵ eutectic was reported as 1002°C [1, 2] and about 988°C [3]. The Sb-rich part of the solidification diagram is based on the work of [1].

Solid-state Equilibria. *γ Loop and Magnetic Transformation.* The existence of a closed γ field, first recognized by [4], was confirmed by later investigators [2, 5–8]. The vertex of the $\alpha/(\alpha + \gamma)$ boundary was reported to lie at about 4.5 wt. (2.1 at.) % Sb, 1150°C [5], or between 1.5 and 3 wt. (0.7–1.4 at.) % Sb [6]. The maximum solubility of Sb in γ-Fe was estimated as approximately 2 wt. (0.93 at.) % Sb [2]. [7] attempted to outline the $(\alpha + \gamma)$ field up to 1150°C; the limit of the γ-phase field at 1150°C lies at about 2 wt. (0.93 at.) % Sb, while at the same temperature that of the two-phase field is reached at about 4.5 wt. (2.1 at.) % Sb.

The Curie point of Fe does not appear to be affected by Sb [9, 6].

The solid solubility of Sb *in* α-Fe has not yet been determined accurately. Data reported differ very widely, owing to the undefined state of the alloys examined. According to X-ray work of [10], the solubility limit at some lower temperature lies at 3 at. (6.3 wt.) % Sb. On the basis of microscopic observations, [3] reported a solubility of about 3.8 at. (8 wt.) % Sb at the eutectic temperature and about 2.4 at. (5 wt.) % Sb at "room temperature." [6] outlined the solubility boundary in the range 950–1150°C by means of microexamination. He gave solubilities of approximately 25 wt. (13.3 at.) % at 1130°C (solidus point), 35 wt. (19.8 at.) % at the eutectic temperature, and 12 wt. (5.9 at.) % Sb at 950°C. The latter values are tentatively indicated in Fig. 392.

The ϵ phase was first assumed to be a solid solution of Sb in the compound Fe_3Sb_2 [1]. As shown by work of [11, 10, 12, 2, 3, 13], however, the homogeneity range is shifted with decreasing temperature to higher Sb contents. At the eutectic temperature, the composition of ε can be assumed to lie at 36–37 at. (55–56 wt.) % Sb [1, 2]. [10] found, by lattice-parameter measurements, that the homogeneity range extended from about 43 to 46 at. (62.2–65 wt.) % Sb at 600°C. These results were reported to have been verified [2]. From lattice-spacing data of alloys quenched from 800°C,

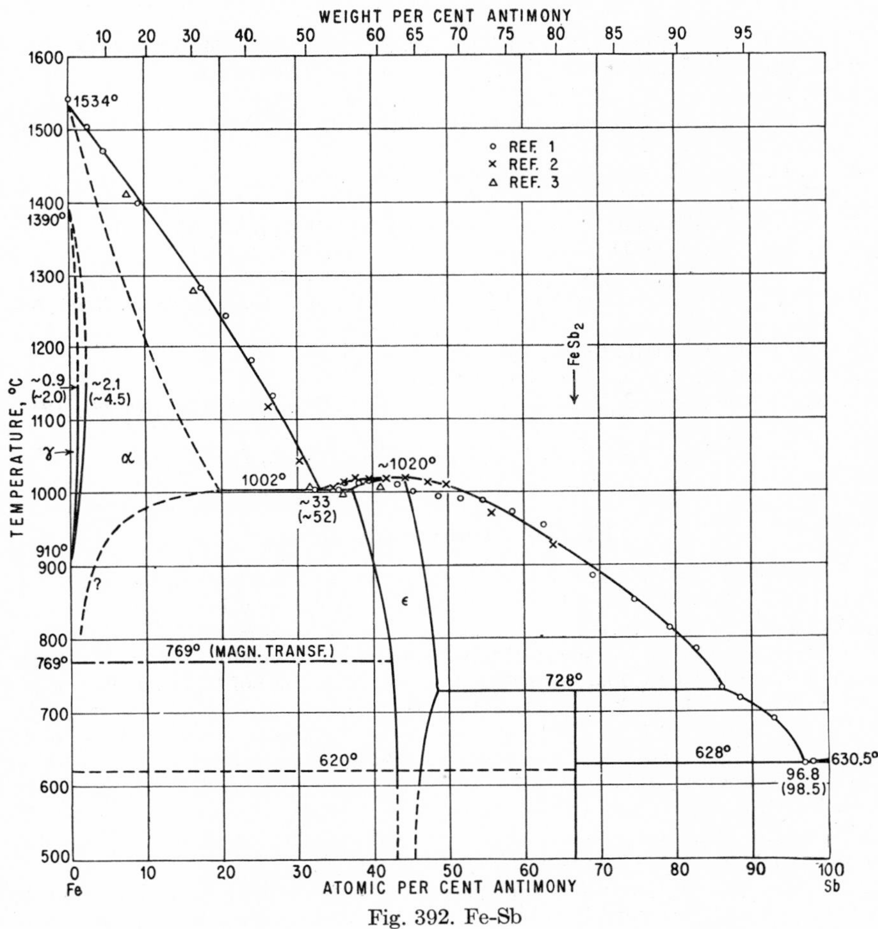

Fig. 392. Fe-Sb

[13] concluded the ε phase to extend from 42–48 at. (61.2–66.8 wt.) % Sb. The existence of a transformation at 620 ± 5°C was detected by [6]; the nature of this change is unknown.

The Range 50–100 *At.* % Sb. The phase $FeSb_2$ (81.35 wt. % Sb) was identified metallographically by [1] and confirmed by [10, 14], who determined the structure. The solid solubility of Fe in Sb is practically nil [10].

Crystal Structures. Lattice parameters of the (α-Fe) phase were reported by [10]. The ε phase is isotypic with NiAs (B8 type) [11, 10, 12, 13]; the following lattice constants were found: $a = 4.13$ A, $c = 5.17_5$ A, $c/a = 1.253$ at 40 at. % Sb

[11, 12]; $a = 4.11_3$ A, $c = 5.15_5$ A, $c/a = 1.253$ at 43 at. % Sb [10]; $a = 4.07_2$ A, $c = 5.14_0$ A, $c/a = 1.262$ at 50 at. % Sb [11]. For additional data (not tabulated), see [13]. The values indicate that the lattice dimensions increase as the Fe content increases. This suggests that the excess Fe atoms are located in the interstices of the NiAs lattice. $FeSb_2$ is isotypic with marcasite FeS_2 (C18 type), with $a = 3.195$ A, $b = 5.831$ A, $c = 6.53$ A [10] or $a = 3.204$ A, $b = 5.83$ A, $c = 6.55$ A [14].

1. N. S. Kurnakow and N. S. Konstantinow, *Z. anorg. Chem.*, **58**, 1908, 1–12.
2. R. Vogel and W. Dannöhl, *Arch. Eisenhüttenw.*, **8**, 1934-1935, 39–40.
3. W. Geller, *Arch. Eisenhüttenw.*, **13**, 1939, 263–266.
4. F. Wever, *Arch. Eisenhüttenw.*, **2**, 1928-1929, 739–746; *Naturwissenschaften*, **17**, 1929, 304–309.
5. W. D. Jones, *J. Iron Steel Inst.*, **130**, 1934, 429–437.
6. P. Fournier, *Rev. chim. ind.*, **44**, 1935, 195–199.
7. V. N. Svechnikov and V. N. Gridnev, *Metallurg*, **1938**, no. 1, 13–19.
8. B. Jones and J. D. D. Morgan, *J. Iron Steel Inst.*, **140**, 1939, 115–136.
9. A. Portevin, *Rev. mét.*, **8**, 1911, 312–314.
10. G. Hägg, *Nova Acta Regial Soc. Sci. Upsaliensis*, (4)**7**, 1929, 71–88; *Z. Krist.*, **68**, 1928, 471–472.
11. I. Oftedal, *Z. physik. Chem.*, **128**, 1927, 135–153.
12. I. Oftedal, *Z. physik. Chem.*, **B4**, 1929, 67–70.
13. N. V. Ageev and E. S. Makarov, *Izvest. Akad. Nauk S.S.S.R. (Khim.)*, **1943**, 87–98.
14. T. Rosenquist, *Acta Met.*, **1**, 1953, 761–763; *Acta Cryst.*, **7**, 1954, 636.

$\bar{1}.8496$
0.1504

Fe-Se Iron-Selenium

Fe forms two selenides, FeSe (58.57 wt. % Se) [1–4] and $FeSe_2$ (73.87 wt. % Se) [5]. Other selenides, reported in the literature but not identified as homogeneous phases [6, 1], do not exist [4].

X-ray analysis [4] has shown that FeSe exists in two modifications: (*a*) α-FeSe, stable below 300–600°C and having the tetragonal structure of the PbO (B10) type, $a = 3.773$ A, $c = 5.529$ A, $c/a = 1.466$; and (*b*) β-FeSe, isotypic with NiAs (B8 type) [3, 4] and stable at higher temperatures as well as, with an excess of Se, at room temperature (Fig. 393). The β phase forms subtraction-type solid solutions with Se up to 57.5 at. % Se. Between 53.7 at. % Se and the saturation limit, the hexagonal structure is deformed into a monoclinic structure. Lattice parameters are, for the NiAs type, $a = 3.644$ A, $c = 5.970$ A, $c/a = 1.638$ [4, 7]; for the monoclinic structure, $a = 6.260$ A, $b = 3.588$ A, $c = 5.821$ A, $\beta = 90.98°$ [4].

Neither [8] nor [4] were able to prepare $FeSe_2$; however, [5] succeeded in obtaining a preparation of this composition by annealing the components for 4 months at 250°C. The diselenide is isotypic with FeS_2 (C18 type), $a = 3.582$ A, $b = 4.801$ A, $c = 5.727$ A.

Note Added in Proof. Recently, Fe-Se alloys have been studied by Japanese investigators [9, 10]. In the following, quotations from abstracts of these papers are given.

[9] made thermal, X-ray, and magnetic analyses of alloys with 48.8–57.4 at. % Se over the temperature range 20–500°C. "The results show that at room temperature, within the Se concentration range 48.8–53.1 at. %, the alloys comprise two phases, α and β, of which α has the PbO-type structure and β the NiAs-type. These two phases transform through a eutectoid reaction into a γ phase with a NiAs-structure of slightly different lattice parameters from those of the β-phase; the eutectoid

temperature is 350°C. The solubility limit on the Fe side of this phase occurs at the composition corresponding to 53.1 at. % Se. The ferrimagnetism of the system originates at the β(NiAs-type) phase; the α phase is feebly paramagnetic or antiferromagnetic."

[10] made a structural study of iron selenides $FeSe_x$. "X-ray analyses, for values of $x = 8/7, 7/6, 6/5, 5/4$, and $4/3$ [53.3, 53.8, 54.5, 55.5, and 57.1 at. % Se], the range within which defects of Fe atoms are found to be ordered, revealed new phases in what was previously considered a single-phase region. For $x = 8/7$, all the distances between neighbouring defects are $\sim 2a$, a denoting a dimension of the unit cell. In the case of the ortho-hexagonal and pseudo-orthohexagonal unit cells, for $x = 8/7$ and $4/3$,

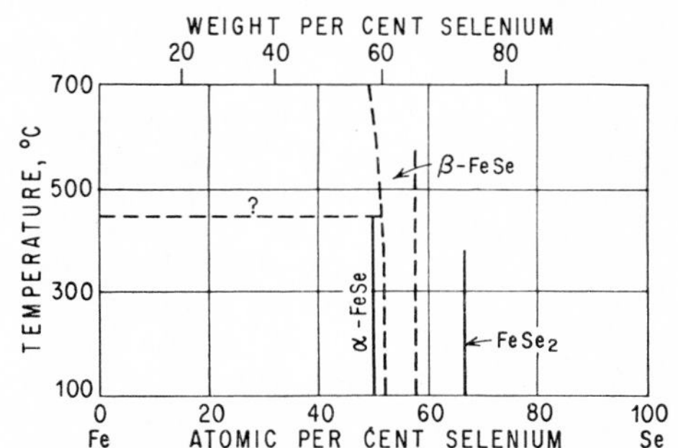

Fig. 393. Fe-Se. (See also Note Added in Proof.)

resp., the unit-cell dimensions of the superstructure resulting from ordering are twice as large as those of the fundamental structure along the a and b axes, three times as large along the c axis for $x = 8/7$, and only twice as large along the c axis for $x = 4/3$."

1. H. Fonces-Diacon, *Compt. rend.*, **130,** 1900, 1710–1712.
2. L. Moser and E. Doctor, *Z. anorg. Chem.*, **118,** 1921, 285–287.
3. N. Alsén, *Geol. Fören. i Stockholm Förh.*, **47,** 1925, 19.
4. G. Hägg and A.-L. Kindström, *Z. physik. Chem.*, **B22,** 1933, 453–464.
5. S. Tengnér, *Z. anorg. Chem.*, **239,** 1938, 126–132.
6. G. Little, *Liebigs Ann.*, **112,** 1859, 211.
7. Alsén gave $a = 3.62$ A, $c = 5.88$ A, $c/a = 1.62$.
8. W. F. de Jong and H. W. V. Willems, *Z. anorg. Chem.*, **170,** 1928, 241–245.
9. T. Hirone and S. Chiba, *J. Phys. Soc. Japan*, **11,** 1956, 666–670; *Met. Abstr.*, **24,** 1957, 356.
10. A. Okazaki and K. Hirakawa, *J. Phys. Soc. Japan*, **11,** 1956, 930–936; *Met Abstr.*, **24,** 1957, 714–715.

0.2985
$\bar{1}.7015$

Fe-Si Iron-Silicon

Early work on preparation and residue analytical investigations aimed at determining the composition of Fe silicides have essentially only historical interest nowadays. The compounds Fe_3Si, Fe_5Si_2, Fe_2Si, Fe_3Si_2, FeSi, $FeSi_2$, and $FeSi_3$ were claimed to be found; for the original literature, see the reviews by [1, 2, 10]. Mention should also be made here of the investigations by [3], according to which there exists $FeSi_2$ but not $FeSi_3$; [4], who tried to establish the composition of silicides by density measurements; and [5], who concluded the existence of Fe_2Si and FeSi from magnetic measurements.

Thanks to a series of valuable investigations of the constitution by means of thermal, microscopic, roentgenographic, magnetic, and dilatometric analyses—which will be briefly discussed below—the phase diagram can be looked upon as essentially established. However, there are still regions which need careful reinvestigation.

Phase diagrams derived from literature studies were published by [2, 6–10]. Those of [2, 7] are obsolete; this holds only for parts of the others which differ in details from each other as well as from the recommended phase diagram of Fig. 394. Remarks by [11–15] helped to clarify some conflicting points.

The Partial System Fe-FeSi. *The solidification curves* were investigated by [16], [17] (only up to 1.2 wt. % Si), [18–21]. The results differ appreciably, certainly mainly because of the high impurity content of most of the alloys. Despite this there is, however, strong indication that the liquidus and solidus curves either meet or approach closely somewhere between 12 and 15 wt. (21.5–26 at.) % Si. Following interpretations by [8] and [21], a peritectic reaction has been assumed in Fig. 394. Such a reaction can account for the peculiarity in the melting range as well as serve as end point for the equilibrium lines of the ordering reaction in the solid state (see below). See, however, Supplement.

Except for the peritectic reaction, the liquidus and solidus in Fig. 394 are largely based on the data by [20], whose alloys were of relatively high purity. According to [20], the eutectic point is located at 20 wt. (33. at.) % Si, 1195°C, and the melting point of FeSi (33.46 wt. % Si) at 1410°C. Data reported by other authors are—eutectic: 22 wt. % Si, 1240°C [16]; 23 wt. % Si, 1205°C [18]; 22.3 wt. % Si, 1230°C [19]; 21.2 wt. % Si [22]; 22 wt. % Si, 1200°C [21]; melting point of FeSi: 1443 [16], 1430 [18], 1463 [19], and 1400°C [21].

γ Loop. The influence of Si on the transformations in solid Fe has been studied [23–25, 16, 26, 17, 27, 28, 18, 29–33, 22, 34, 20, 35, 36] and discussed [37, 2, 11, 12, 14, 15, 7] by many authors. The existence of a closed "γ loop" had been suggested by [14, 15] and was corroborated by later work: thermal [31, 20], roentgenographic [34], dilatometric [32, 36], and magnetic [35]. Critical data of the more important papers are collected in Table 28.

Table 28

Ref.	Method	Vertex of the loop
31	Thermal	1.8 wt. % Si
34	High-temperature X-ray (disappearing-phase method)	2.5 wt. % Si, 1160°C
35	Magnetic	2.15 wt. (4.2 at.) % Si, ~1170°C
36	Dilatometric	1.80 wt. % Si, 1115°C

The loop in Fig. 394 is that given by [35]. He stated, "It may be significant that the results of [32, 17, 31], obtained using measurements on bulk properties of the materials, all tend to agree fairly well with the present work, while those of [34], obtained from the use of a surface property, do not. It is to be expected that any error arising from the loss of active Si by slight oxidation during prolonged heating in even a high vacuum will be the most marked in surface measurements. It would have the effect of displacing the apparent phase boundaries towards higher Si contents."

α, α', and α'' Phases. In approaching the composition Fe_3Si (14.36 wt. % Si), ordering takes place as shown by the appearance of superlattice lines above about 6.5 wt. (12 at.) % Si [33, 22, 38, 39]. The ordered region has been considered by [21] to be a separate phase [α'] (see also [8, 38, 41]). At variance with this, in many (even

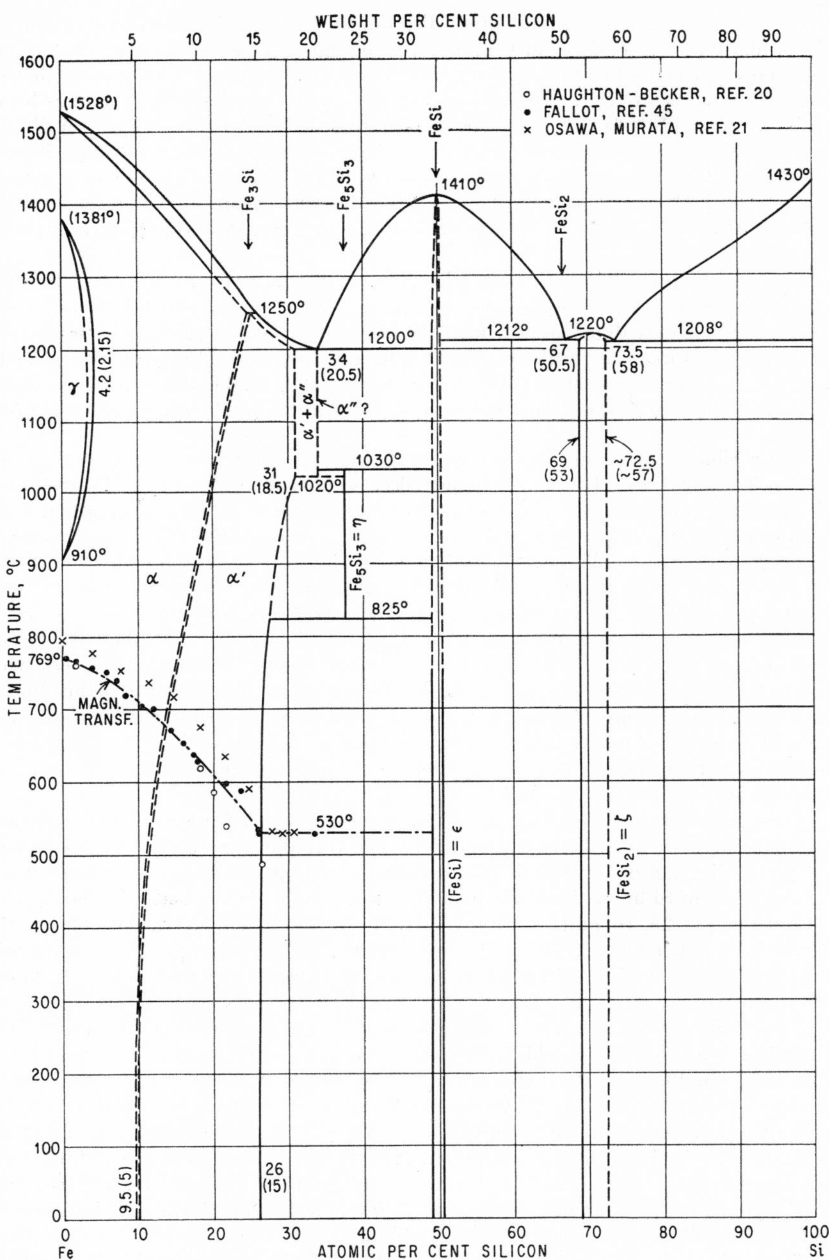

Fig. 394. Fe-Si. (See also Supplement.)

more recent) publications the transition disorder-order is assumed to occur in homogeneous phase and, therefore, in the diagrams of those investigators [18, 42, 20, 43, 39, 44] no α'-phase field appears (i.e., their Si-rich α boundary corresponds to the Si-rich α' boundary of [21]). In drawing Fig. 394 the interpretation by [21] has been preferred. [21] observed the $\alpha \rightleftarrows \alpha'$ transformation up to 1100°C by means of differential dilatometric measurements and found also thermal evidence for a peritectic reaction at 1250°C (see Supplement). The $(\alpha + \alpha')$ field must be very narrow. According to [21], the solubility of Si in α-Fe decreases from 14.4 wt. (25 at.) % Si at 1250°C to 7.5 wt. (14 at.) % Si at 700°C, and 5 wt. (9.5 at.) % Si at "room temperature."

The Si-rich boundary of the ordered region (α' in Fig. 394) has been investigated by many authors: [18], [42], [20] (micrographic), [41] (X-ray), [45] (magnetic), [43] (X-ray), [21] (X-ray, dilatometric, microscopic), [39] (X-ray), and [46] (magnetic). For low temperatures, the results of [45, 43, 39] agree upon a solubility of about 15 wt. (26 at.) % Si. Between 600 and 1000°C the data by [20, 43, 39] are in fair agreement (in Fig. 394 those of [39] were used); those of other authors [18, 21, 46] are certainly too high.

The region between 30 and 40 at. (18–25 wt.) % Si and 1000 and 1200°C needs careful reinvestigation. [18, 20, 43] reported the α phase to extend up to 23, 18.5, and 20.4 wt. % Si, respectively. [21] assumed a broad α''-phase field between about 18.6 and 23 wt. % Si which has been criticized by [39]. The α'' phase shown in Fig. 394 is that suggested by [39]. These authors found, in alloys of about 19.5 and 20 wt. % Si quenched from temperatures above 1030°C, two phases of identical structure but different composition ($\alpha' + \alpha''$ according to the nomenclature used in Fig. 394). The composition of the α'' phase was determined to be about 20.4 wt. (33.8 at.) % Si. These findings were corroborated by the X-ray and magnetic work of [46, 44], who reported 20.8 wt. % Si as the composition of the α'' phase. [39] stated that the almost exact equality of the eutectic and reaction temperatures and also of the compositions of the eutectic and α'' makes it highly probable that α'' is formed either by a peritectic reaction at a temperature slightly greater than the eutectic temperature or by a polymorphic reaction [46a].

Since the existence of the α'' phase has been concluded from work on quenched alloys only, it appears a little doubtful whether this phase actually belongs to the equilibrium diagram or whether it is only a metastable phase formed on quenching.

The effect of Si upon the Curie-point temperature of Fe has been studied by the early authors [23, 25, 16] and later by [18, 31, 20, 45, 21, 46, 68]. The data of [20, 45, 21] have been plotted in Fig. 394. [45] interpreted breaks in his curve at the compositions $Fe_{15}Si$, Fe_7Si, and Fe_3Si as indicating the existence of superstructures based on these compositions.

The Fe_5Si_3 (η) *Phase.* The reaction at 1030°C was first observed by [47]. [18] interpreted it as the peritectoid formation of the compound Fe_3Si_2. [43] and later authors [48, 39, 46, 44] showed this phase to be stable only in the temperature range 1030–825°C. It can, however, be retained metastable at room temperature and is ferromagnetic below about 90°C (data by [18, 20, 45, 46] are 90, 82, 93, and 120°C, respectively). Already [43] have observed that "Fe_3Si_2" contains somewhat more Fe than is required by this formula. It was then shown by [21, 48, 49, 39] that the correct formula of this compound is Fe_5Si_3 [37.50 at. (23.18 wt.) % Si]. No indications for any extended homogeneity range were observed.

The FeSi (ϵ) *Phase.* The composition of the FeSi phase was found to be variable within narrow limits [18, 50, 22, 21]. According to microscopic and X-ray work by [21], the homogeneity range extends from 49 to 50.5 at. % Si (taken from a diagram).

Crystal Structures. Lattice parameters in the α- and α′-phase regions were measured by [22, 51, 38, 41, 21, 39, 52]. The most reliable curves are those by [38, 39]; they show a break at about 5 wt. (9.5 at.) % Si which, according to Fig. 394, is caused by the $\alpha \rightarrow \alpha'$ transition. Ordering, which leads to a doubling of the lattice parameter of the simple b.c.c. α cell, is complete at the composition Fe_3Si: BiF_3 (DO_3) type [33, 38, 21, 39]. "If more silicon is added, the alloy being maintained as a single phase by quenching from a high temperature, the ordering continues exactly as in the Fe-Al system" [39].

Fe_5Si_3 (η) has a hexagonal [21, 48, 49] structure of the Mn_5Si_3 ($D8_8$) type [49], with $a = 6.755$ A, $c = 4.717$ A, $c/a = 0.6983$ [49].

The structure of FeSi constitutes the cubic B20 prototype [50, 22, 53, 54, 39, 55]; $a = 4.489 \pm 5$ A [55].

The kinetics of the solid-state reactions in Fe-rich Fe-Si alloys have been studied by [48, 44].

The Partial System FeSi-Si. There was early agreement that only one intermediate phase, zeta (ζ), exists in this partial system. However, the views of the investigators differ widely as far as the formation equilibria and composition of this phase are concerned.

The assumption that the composition of the very brittle ζ phase corresponds to, or at least includes, the formula $FeSi_2$ (50.14 wt. % Si) [56, 57, 50, 58, 59, 6, 60, 8, 61, 62] was not corroborated; it has been definitely established that the composition range lies above this composition [19, 22, 20, 21, 63, 64]. Data reported for its homogeneity range are 53.5–56.5 wt. % Si, micrographic [20]; 53.0–57.5 wt. % Si, X-ray, microscopic [21]; and 53–59 wt. % Si, X-ray [63]. In Fig. 394 the boundaries of the ζ phase have been placed at 53 and 57 wt. (69 and 72.5 at.) % Si.

It appears now well established that the ζ phase melts with an open maximum and enters into eutectic equilibria with FeSi and Si [59, 22, 20, 21, 64]. Data reported for the (FeSi + ζ) eutectic are 45.7 wt. % Si, 1205°C [59]; 50.5 wt. % Si [22]; 51 wt. % Si, 1212°C [20]; and ~50 wt. % Si [64]; those for the (ζ + Si) eutectic are 55.4 wt. % Si, 1215°C [59]; 61 wt. % Si [22]; 58 wt. % Si, 1208°C [20]; and 58 wt. % Si [64]; and those for the melting point of the ζ phase are 1275 [59] and 1220°C [20]. In Fig. 394 the data 50.5 wt. (67 at.) % Si, 1212°C; 58 wt. (73.5 at.) % Si, 1208°C; and 1220°C have been chosen. In favor of a peritectic formation of the ζ phase were [50, 6, 60, 8, 63]. [63] claimed to have found that on tempering the ζ phase (prepared with Si of only 96.4 wt. % purity) it is decomposed.

According to [50, 22], the ζ phase has a tetragonal structure, with $a = 2.692$ A, $c = 5.137$ A, $c/a = 1.908$ [22], and 3 atoms per unit cell.

Solubility of Fe *in Solid* Si. According to X-ray work by [22], the solubility is negligible (which is to be expected from the nature of Si). On the other hand, [66] claimed to have found roentgenographically a certain solubility. [58] concluded from density measurements (!) a solubility of 5 wt. % Fe, [60] from microscopic work one of 4 wt. % Fe. All these conflicting data are of little value since they were obtained with impure silicon.

Supplement. Calorimetric [67] and dilatometric [65] measurements indicate that the ζ phase undergoes transformations at about 908 and 650°C.

By means of electrical and thermal-conductivity measurements as well as X-ray work on quenched samples, [68] determined the following $\alpha \rightleftarrows \alpha'$ disorder-order transformation temperatures: 900, 990, 1040, 1060, 1100, 1120, 1120, and 1120°C at 10.9, 13.8, 16.1, 18.2, 22.0, 25.5, 26.9, and 27.9 at. % Si, respectively. Below 1100°C they are higher than those of [21] (Fig. 394) which were measured at temperature. This is to be expected considering the possibility of ordering during the quench. On the other hand, the results of [68] clearly show that the ordered phase is not stable up to the melting range as is assumed in Fig. 394.

In agreement with [21], onset of order at ordinary temperatures was observed by [68] at 9.5 at. % Si.

1. L. Baraduc-Muller, *Rev. mét.*, **7**, 1910, 718–735.
2. W. Guertler, *Metallographie,* **1**, 1917, 658–673.
3. M. v. Schwarz, *Ferrum,* **11**, 1913-1914, 80–90, 112–117.
4. R. Frilley, *Rev. mét.*, **8**, 1911, 492–501.
5. A. Jouve, *Compt. rend.*, **134**, 1902, 1577–1579.
6. F. Körber, *Z. Elektrochem.*, **32**, 1926, 371–376.
7. M. G. Corson, *Trans. AIME,* **80**, 1928, 249–300.
8. B. Stoughton and E. S. Greiner, *Trans. AIME*, **90**, 1930, 155–191.
9. E. S. Greiner, J. S. Marsh, and B. Stoughton, "The Alloys of Iron and Silicon," Alloys of Iron Research, Monograph Series, McGraw-Hill Book Company, Inc., New York, 1933.
10. M. Hansen, "Der Aufbau der Zweistofflegierungen," Springer-Verlag OHG, Berlin, 1936.
11. W. Guertler, *Stahl u. Eisen*, **42**, 1922, 667.
12. K. Honda and T. Murakami, *J. Iron Steel Inst.*, **107**, 1923, 546–547. *Science Repts. Tôhoku Imp. Univ.*, **12**, 1924, 258–260.
13. G. Tammann, "Lehrbuch der Metallographie," p. 289, Leopold Voss, Leipzig, 1923.
14. P. Oberhoffer, *Stahl u. Eisen*, **44**, 1924, 979.
15. P. Oberhoffer, "Das technische Eisen," pp. 103–105, Springer-Verlag OHG, Berlin, 1925.
16. W. Guertler and G. Tammann, *Z. anorg. Chem.*, **47**, 1905, 163–179.
17. R. Ruer and R. Klesper, *Ferrum,* **11**, 1913-1914, 259.
18. T. Murakami, *Science Repts. Tôhoku Imp. Univ.*, **10**, 1921, 79–92; **16**, 1927, 481–482.
19. N. S. Kurnakov and G. Urasov, *Z. anorg. Chem.*, **123**, 1922, 92–107.
20. J. L. Haughton and M. L. Becker, *J. Iron Steel Inst.*, **121**, 1930, 315–335.
21. A. Osawa and T. Murata, *Nippon Kinzoku Gakkai-Shi*, **4**, 1940, 228–242. In Japanese; tables and summary in English.
22. G. Phragmén, *J. Iron Steel Inst.*, **114**, 1926, 397–403. See also *Stahl u. Eisen*, **47**, 1927, 193–195.
23. F. Osmond, *J. Iron Steel Inst.*, **37**, 1890, 62.
24. J. O. Arnold, *J. Iron Steel Inst.*, **45**, 1894, 143.
25. T. Baker, *J. Iron Steel Inst.*, **64**, 1903, 322–325.
26. G. Charpy and A. Cornu, *Compt. rend.*, **156**, 1913, 1240–1243.
27. W. E. Ruder, *Trans. AIME*, **47**, 1914, 569–583.
28. A. Sanfourche, *Compt. rend.*, **167**, 1919, 683–685.
29. P. Oberhoffer and A. Heger, *Stahl u. Eisen*, **43**, 1923, 1474–1476.
30. Büscher, Thesis, Technische Hochschule, Aachen, 1921; quoted by [29].
31. F. Wever and P. Giani, *Mitt. Kaiser-Wilhelm-Inst. Eisenforsch. Düsseldorf*, **7**, 1925, 59–68. F. Wever, *Z. anorg. Chem.*, **154**, 1926, 297–304; *Stahl u. Eisen*, **45**, 1925, 1208–1210.
32. H. Esser and P. Oberhoffer, *Ber. Verein deut. Eisenhüttenleute, Werkstoffausschuss Ber.*, no. 69, 1925.
33. G. Phragmén, *Stahl u. Eisen*, **45**, 1925, 299–300.
34. C. Kreutzer, *Z. Physik*, **48**, 1928, 558–560; P. Oberhoffer and C. Kreutzer, *Arch. Eisenhüttenw.*, **2**, 1929, 450–451; *Stahl u. Eisen*, **49**, 1929, 189–190.

35. J. Crangle, *Brit. J. Appl. Phys.*, **5**, 1954, 151–154.
36. G. G. Bentle, Dissertation, Vanderbilt University, Nashville, Tenn., 1954 (not available to the author); *Dissertation Abstr.*, **14**, 1954, 1914; G. G. Bentle and W. P. Fishel, *Trans. AIME*, **206**, 1956, 1345–1348.
37. G. Gontermann, *Z. anorg. Chem.*, **59**, 1908, 384–387.
38. E. R. Jette and E. S. Greiner, *Trans. AIME*, **105**, 1933, 259–274.
39. M. C. M. Farquhar, H. Lipson, and A. R. Weill, *J. Iron Steel Inst.*, **152**, 1945, 457--472. See also [40].
40. A. R. Weill, *Rev. mét.*, **42**, 1945, 266–270. (Shortened version of [39].)
41. C. P. Yap, *J. Phys. Chem.*, **37**, 1933, 951–967.
42. H. Hanemann and H. Voss, *Zentr. Hütten-u. Walzwerke*, **31**, 1927, 259–262.
43. E. S. Greiner and E. R. Jette, *Trans. AIME*, **125**, 1937, 473–481.
44. K. M. Guggenheimer and H. Heitler, *Trans. Faraday Soc.*, **45**, 1949, 137–145.
45. M. Fallot, *Ann. phys.*, **6**, 1936, 305–387.
46. K. M. Guggenheimer, H. Heitler, and K. Hoselitz, *J. Iron Steel Inst.*, **158**, 1948, 192–199.
46a. Ordinary metallographic methods have failed to confirm the existence of α'': J. E. Hurst and R. V. Riley, *J. Iron Steel Inst.*, **155**, 1947, 172–178.
47. A. Sanfourche, *Rev. mét.*, **16**, 1919, 217–224.
48. H. Lipson and A. R. Weill, *Trans. Faraday Soc.*, **39**, 1943, 13–18.
49. A. R. Weill, *Nature*, **152**, 1943, 413.
50. G. Phragmén, *Jernkontorets Ann.*, **1923**, 121–131; see *Stahl u. Eisen*, **45**, 1925, 51–52.
51. Z. Nishiyama, *Science Repts. Tôhoku Imp. Univ.*, **18**, 1929, 385–386.
52. Y. P. Selisskij, *Zhur. Fiz. Khim.*, **20**, 1946, 597–604; *Structure Repts.*, **10**, 1945-1946, 17.
53. F. Wever and H. Möller, *Z. Krist.*, **75**, 1930, 362–365; *Naturwissenschaften*, **18**, 1930, 734–735.
54. B. Borén, *Arkiv. Kemi, Mineral. Geol.*, **11A**, 1933, 1.
55. L. Pauling and A. M. Soldate, *Acta Cryst.*, **1**, 1948, 212–216.
56. L. Baraduc-Muller, *Rev. mét.*, **7**, 1910, 764–786.
57. A. T. Lowzow, *Chem. & Met. Eng.*, **24**, 1921, 481–484.
58. O. Hengstenberg, *Stahl u. Eisen*, **44**, 1924, 914–915.
59. M. Bamberger, O. Einerl, and J. Nussbaum, *Stahl u. Eisen*, **45**, 1925, 141–144.
60. T. Murakami, *Science Repts. Tôhoku Imp. Univ.*, **16**, 1927, 475–489.
61. C. Bedel, *Compt. rend.*, **195**, 1932, 329–330.
62. C. Bedel, *Compt. rend.*, **196**, 1933, 262–264.
63. N. V. Ageev, N. S. Kurnakov, L. N. Guseva, and O. K. Konenko-Gracheva, *Metallurg*, no. 1, 1940, 5–12; *Chem. Abstr.*, **36**, 1942, 6122.
64. W. R. Johnson and M. Hansen, Air Force Technical Report 6383, Armour Research Foundation, Chicago, Ill., June, 1951.
65. P. V. Gel'd and N. N. Serebrennikov, *Doklady Akad. Nauk S.S.S.R.*, **97**, 1954, 827–830; *Met. Abstr.*, **23**, 1956, 537.
66. A. Osawa, see Murakami [60].
67. N. N. Serebrennikov and P. V. Gel'd, *Doklady Akad. Nauk S.S.S.R.*, **97**, 1954, 695–698; *Met. Abstr.*, **23**, 1956, 422.
68. F. W. Glaser and W. Ivanick, *Trans. AIME*, **206**, 1956, 1290–1295; discussion (H. Sato), **209**, 1957, 529–530.

$\bar{1}.6726$
0.3274

Fe-Sn Iron-Tin

Since the first thermal and microscopic investigation by [1], this system has been the subject of several studies by means of thermal [2, 3] and microscopic [2–4] analysis and extensive X-ray work covering the whole range of composition [5–7]. In addition, contributions have been made by [8–16]. The literature has been thoroughly reviewed repeatedly [17, 7, 18], lastly in 1942 by [18], to which the reader is referred. It is still somewhat doubtful, however, whether the diagram presented in Fig. 395, which has been combined from the results by [3, 5, 6, 7, 16], can be regarded as final. This holds especially for the phase relations in the composition range 20–50 at. % Sn, as well as some of the temperatures of the three-phase equilibria.

The essential points under discussion for a period of time were concerned with (*a*) the existence of a miscibility gap in the liquid state and (*b*) the composition of the intermediate phases. Work by [1, 8, 3, 9, 5, 6, 16] has established beyond doubt that there is a range of immiscibility in the liquid state. For this reason, thermal analysis and micrographic studies of alloys solidified from the melt have to be considered inferior methods of investigation for phase identification, since none of the numerous reactions in the partially and fully solid states will be completed under normal conditions of cooling and heat-treatment. Therefore, investigations which have made use of alloys prepared by annealing powdered compacts deserve preference. Also, X-ray diffraction as a means of phase identification is superior to other methods.

The existence of the following phases has been reported in the modern literature [19]: Fe_3Sn (41.47 wt. % Sn) [1, 2, 13, 7]; Fe_2Sn (51.52 wt. % Sn) [3, 5, 4, 6, 7]; Fe_3Sn_2 (58.62 wt. % Sn) [6, 7]; a "NiAs" phase (γ) with about 43.5–44 at. (62–62.5 wt.) % Sn [5, 6, 7]; FeSn (68.0 wt. % Sn) [3, 4, 12, 6, 7, 22]; $FeSn_2$ (80.95 wt. % Sn) [2, 3, 5, 4, 13, 6, 14, 7, 15, 22]. There is a discrepancy as to whether the intermediate phase richest in iron is Fe_3Sn or Fe_2Sn. It appears that this problem has been solved in favor of Fe_3Sn [13, 7]; see below.

The effect of Sn on the polymorphic transformations of Fe was studied by [2, 11]. According to [2], 1 wt. (0.5 at.) % Sn raises the temperature of the $\alpha \rightleftarrows \gamma$ transformation about 40°C and lowers that of the $\gamma \rightleftarrows \delta$ transformation about 140°C. The γ loop closes at approximately 1.9 wt. (0.9 at.) % Sn (estimated) [20]. According to [11], the γ loop lies between 2 and 2.5 wt. (1–1.3 at.) % Sn at 1150°C.

The Curie point of Fe is but slightly depressed by addition of Sn [21]; see the data points plotted in Fig. 395.

The boundary of the primary solid solution shown in Fig. 395 was determined micrographically [3]. On the basis of lattice parameters [5], the solubilities at 900 and 680°C were estimated to be about 9.8 and 4.9 at. % Sn, respectively. The liquidus curve between the monotectic temperature and the melting point of Sn is due to [16]; it was obtained by solubility determinations. The melting point of Sn is raised to 650 and 800°C by additions of 0.3 wt. (0.6 at.) and 2.0 wt. (4.2 at.) % Fe, respectively.

Crystal Structures. The parameter of α-Fe (a = 2.867 A) [5] is increased to a = 2.931 A by 9.8 at. % Sn [5]. The intermediate phase richest in Fe is Fe_3Sn according to [13, 7] and Fe_2Sn according to [5, 6]. [7] made clear that the hexagonal structure of Fe_2Sn reported by [5, 6] actually must be that of Fe_3Sn. It is isotypic with Ni_3Sn and Mg_3Cd ($D0_{19}$ type), a = 5.458 A, c = 4.361 A, c/a = 0.799 [13, 7]. Fe_3Sn_2, reported by [6] to be hexagonal, with a = 21.36 A, c = 4.390 A, c/a = 0.205, was shown to have a monoclinic structure with 40 atoms per unit cell and a = 13.53 A, b = 5.34 A, c = 9.20 A, β = 103° [7]. The lattice dimensions of the γ, or "NiAs," phase, with a composition close to 43.5–44 at. % Sn [6, 7], increase from a = 4.230 A, c = 5.208 A, c/a = 1.231 at the Fe-rich side to a = 4.233 A, c = 5.213 A, c/a = 1.232

at the Sn-rich side of the homogeneity range, which is estimated to be 0.3–1.0 at. % wide [13, 7].

FeSn is isotypic with CoSn (B35 type) and of fixed composition. Lattice parameters reported by [5–7] are a = 5.300 ± 0.002 A, c = 4.449 ± 0.002 A, c/a = 0.839.

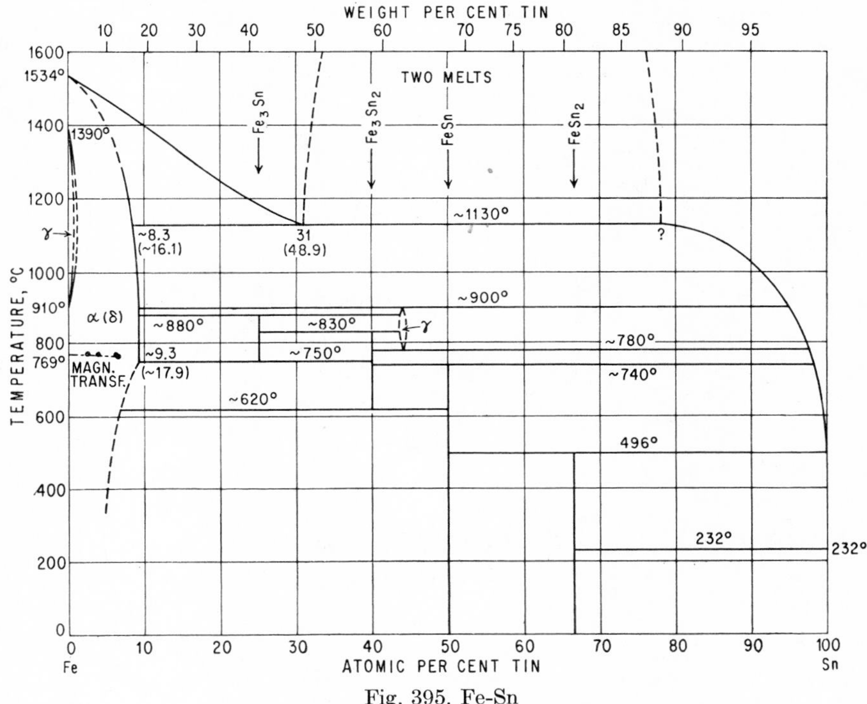

Fig. 395. Fe-Sn

$FeSn_2$, reported by [5] to be hexagonal, is tetragonal of the $CuAl_2$ (C16) type, a = 6.533 ± 0.01 A, c = 5.321 ± 0.015 A, c/a = 0.815 [13–15, 7].

1. E. Isaac and G. Tammann, *Z. anorg. Chem.*, **53,** 1907, 281–291.
2. F. Wever and W. Reinecken, *Mitt. Kaiser-Wilhelm-Inst. Eisenforsch. Düsseldorf*, **7,** 1925, 69–79; *Z. anorg. Chem.*, **151,** 1926, 349–372; F. Wever, *Z. anorg. Chem.*, **154,** 1926, 294.
3. C. A. Edwards and A. Preece, *J. Iron Steel Inst.*, **124,** 1931, 41–66.
4. W. D. Jones and W. E. Hoare, *J. Iron Steel Inst.*, **129,** 1934, 273–280.
5. W. F. Ehret and A. F. Westgren, *J. Am. Chem. Soc.*, **55,** 1933, 1339–1351.
6. W. F. Ehret and D. H. Gurinsky, *J. Am. Chem. Soc.*, **65,** 1943, 1226–1230.
7. O. Nial, Dissertation, University Stockholm, 1945; *Svensk Kem. Tidskr.*, **59,** 1947, 165–170.
8. R. Ruer and J. Kuschmann, *Z. anorg. Chem.*, **153,** 1926, 260–262.
9. C. O. Bannister, *J. Iron Steel Inst.*, **124,** 1931, 68.
10. W. E. Hoare, *J. Iron Steel Inst.*, **129,** 1934, 253–264.
11. W. D. Jones, *J. Iron Steel Inst.*, **130,** 1934, 435.
12. O. Nial, *Z. anorg. Chem.*, **238,** 1938, 287–296.
13. O. Nial, *Arkiv Kemi, Mineral. Geol.*, **17B,** 1943, no. 11.

14. H. J. Wallbaum, *Z. Metallkunde*, **35,** 1943, 218–221.
15. H. Nowotny and K. Schubert, *Z. Metallkunde*, **37,** 1946, 17–23.
16. A. N. Campbell, J. H. Wood, and G. B. Skinner, *J. Am. Chem. Soc.*, **71,** 1949, 1729–1733.
17. M. Hansen, "Der Aufbau der Zweistofflegierungen," pp. 739–747, Springer-Verlag OHG, Berlin, 1936.
18. O. E. Romig, *Metal Progr.*, **42,** 1942, 899–904.
19. For the composition of the intermediate phases reported earlier, see [17].
20. According to [2], the transformation points are largely affected by carbon. An alloy with 1 wt. % Sn (prepared using Fe with 0.02 wt. % C) showed an increase in the $\gamma \rightarrow \delta$ point of about 120°C if 0.04 wt. % C was added.
21. L. Néel, *Ann. phys.*, **18,** 1932, 84–87; see also M. Fallot, *Ann. phys.*, **6,** 1936, 381–384.
22. F. Lihl, *Z. Metallkunde*, **46,** 1955, 439.

$\bar{1}.8044$
0.1956

Fe-Sr Iron-Strontium

Sr is regarded as being insoluble in solid Fe [1]. The boiling point of Sr (1380°C) lies below the melting point of Fe; therefore, it cannot be determined under normal pressure whether Sr is soluble in liquid Fe.

1. F. Wever, *Arch. Eisenhüttenw.*, **2,** 1928-1929, 739–746; *Naturwissenschaften*, **17,** 1929, 304–309.

$\bar{1}.4896$
0.5104

Fe-Ta Iron-Tantalum

The existence of a eutectic between an Fe-rich solid solution and an intermediate phase, both of undetermined composition, was first reported by [1]. From a cursory microscopic study, using only six alloys having Ta contents ranging between 2 and 48 at. % Ta, [2] concluded that the intermediate phase was the compound TaFe, having a melting point of about 1700°C. The eutectic was placed between 9 and 24 at. % Ta, and its melting point was given as about 1400°C. Previously, [3] had reported that there was a closed γ loop extending up to not more than 1.35 at. (4.24 wt.) % Ta.

Thermal and microscopic work complemented by measurements of electrical resistivity [4], covering the range up to 13.8 at. (34.2 wt.) % Ta, showed the eutectic to be located at 10 at. (26.5 wt.) % Ta and about 1410°C (Fig. 396). The intermediate phase was assumed to have the composition TaFe and the solubility of Ta in α-Fe to be of the order of 2 at. (6.2 wt.) % at the eutectic temperature and 0.4 at. (1.3 wt.) % Ta at "room temperature."

The solid-state transformations in the range up to 2.5 at. (7.7 wt.) % Ta were established by [5], using thermal and microscopic analysis. Data points are indicated in the inset of Fig. 396. The Fe-rich eutectic was found to lie between 5.8 (16.6) and 7.4 at. (20.6 wt.) % Ta (closer to 7.4 at. %) at approximately 1450°C and the solubility at this temperature to be about 2.3 at. (7.1 wt.) % Ta. In contrast to earlier suggestions, the intermediate phase, probably of slightly variable composition, was identified as $TaFe_2$ (61.82 wt. % Ta). This was confirmed by determination of the crystal structure, which is of the $MgZn_2$ (C14) type, with $a = 4.81$ A, $c = 7.85_5$ A, $c/a = 1.63_3$ according to [6] and $a = 4.827$ A, $c = 7.838$ A, $c/a = 1.624$ according to [7]. The melting point of $TaFe_2$ was found as 1775°C [7] and the Ta-rich eutectic to contain about 55 at. (80 wt.) % Ta [5].

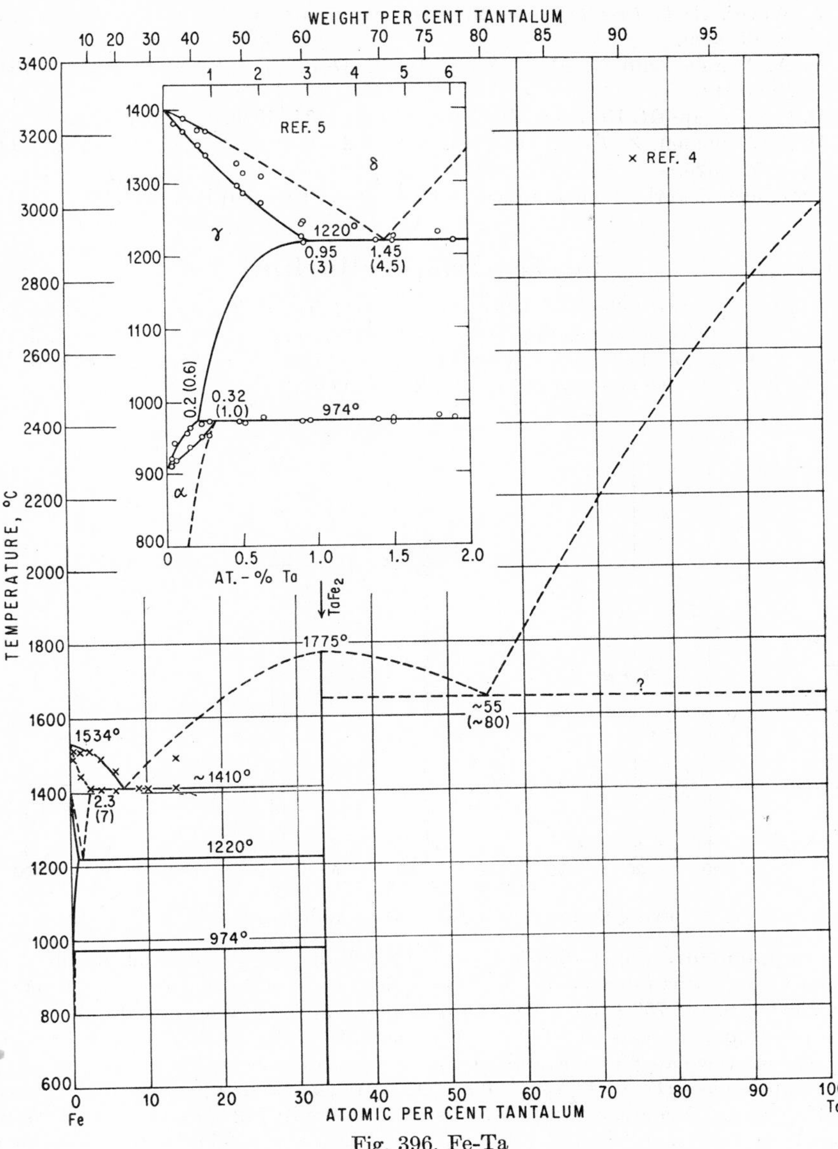

Fig. 396. Fe-Ta

There is pretty good agreement as to the composition of the Fe-rich eutectic: about 10 at. (26.5 wt.) % [4] and about 7 at. (19.6 wt.) % Ta [5]; the eutectic temperature: about 1400 [2], 1410 [4], and about 1450°C [5]; and the solubility at the eutectic temperature: about 2 at. (6.2 wt.) % [4] and 2.4 at. (7.4 wt.) % Ta [5].

1. J. Laissus, *Rev. mét.*, **24**, 1927, 387–395.
2. W. Jellinghaus, *Z. anorg. Chem.*, **223**, 1935, 362–364.

3. F. Wever, *Arch. Eisenhüttenw.*, **2**, 1928-1929, 739–746; *Naturwissenschaften*, **17**, 1929, 304–309.
4. V. A. Nemilov and N. M. Voronov, *Izvest. Akad. Nauk S.S.S.R. (Khim.)*, **1938** (4), 905–912.
5. R. Genders and R. Harrison, *J. Iron Steel Inst.*, **134**, 1936, 173–209.
6. H. J. Wallbaum, *Z. Krist.*, **103**, 1941, 391–402; see also *Arch. Eisenhüttenw.*, **14**, 1941, 521–526.
7. Unpublished work, Armour Research Foundation, Chicago, Ill., 1954.

$\bar{1}.6411$
0.3589

Fe-Te Iron-Tellurium

Iron forms two tellurides, FeTe (69.55 wt. % Te) [1–3] and $FeTe_2$ (82.04 wt. % Te) [4]. FeTe is hexagonal of the NiAs (B8) type, a = 3.81 A, c = 5.67 A, c/a = 1.488 [3]; and $FeTe_2$ is isotypic with FeS_2 (C18 type), a = 3.857 A, b = 5.351 A, c = 6.273 A [4]. Te does not diffuse into Fe at 1000°C [5].

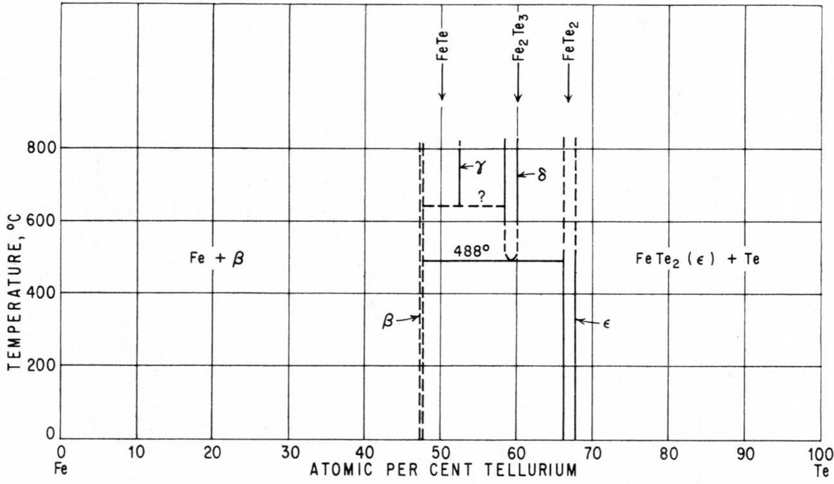

Fig. 397. Fe-Te. (See also Note Added in Proof.)

Supplement. [6] studied by X-ray examination Fe-Te alloys of 29 different compositions in the range 33–75 at. % Te which had been slowly cooled or quenched from 300, 600, and 750°C [7]. The following results were obtained (see also the tentative phase diagram of Fig. 397): (*a*) The solubility of Te in solid Fe is very small. (*b*) The intermediate phase richest in Fe, β, has a narrow range of homogeneity close to the composition $FeTe_{0.90}$ (47.4 at. % Te). Its tetragonal structure is closely related to the PbO (B10) type and can be considered as an AB-interstitial structure with the formula $Fe_{1.11}Te$ or Fe_9Te_8. The lattice constants a = 3.823_0 A, c = 6.276_7 A were measured at the Fe-rich limit, and a = 3.819_8 A, c = 6.280_5 A at the Te-rich limit. (*c*) The γ phase, $FeTe_{1.1}$ (52.5 at. % Te), is stable in the temperature range around 750°C. Its structure is not yet known; powder lines are listed in [6]. (*d*) The δ phase, $FeTe_{1.4}$–$FeTe_{1.5}$ (58.3–60.0 at. % Te), is stable above 488°C. The structure is of the subtractive NiAs (B8) type, with a = 3.816_2 A, c = 5.654_8 A at 60 at. % Te; however, a monoclinic deformation is observed at the Fe-rich side of the homogeneity range. (*e*) The ε phase, $FeTe_2$, with orthorhombic marcasite (C18) type of structure has a range of homogeneity between $FeTe_{1.95}$ and $FeTe_{2.10}$ (66.1 and 67.7 at. % Te). The lattice constants are $a = 6.264_1$ A, $b = 5.260_6$ A, $c = 3.872_4$ A for $FeTe_{1.90}$, and $a = 6.274_7$ A, $b = 5.278_8$ A, $c = 3.863_6$ A for $FeTe_{2.10}$. (*f*) The solubility of Fe in solid Te is small.

Note Added in Proof. [8] studied the system by X-ray, thermal, and thermomagnetic analysis at temperatures up to 1000°C. Three peritectically formed compounds (which correspond to the β, δ, and ϵ phases of [6]) were detected: antiferromagnetic (θ = 130°K) $FeTe_{0.95}$ (48.7 at. % Te); antiferromagnetic (θ = 527°C) $FeTe_{1.50}$ (60 at. % Te) of the NiAs structure which decomposes eutectoidally into $FeTe_{0.95}$ and $FeTe_{2.00}$ at 520°C (cf. [6]); and antiferromagnetic (θ = 67°K) $FeTe_{2.00}$ of the marcasite type.

1. C. Fabre, *Compt. rend.*, **105,** 1887, 277.
2. L. Moser and K. Ertl, *Z. anorg. Chem.*, **118,** 1921, 271–273.
3. I. Oftedal, *Z. physik. Chem.*, **132,** 1928, 208–216.
4. S. Tengnér, *Z. anorg. Chem.*, **239,** 1938, 126–132.
5. N. W. Ageev and M. I. Zamotorin, *Ann. inst. polytech., Leningrad, sec. math. phys. sci.*, **31,** 1928, 15–28; abstract in *J. Inst. Metals*, **44,** 1930, 556.
6. F. Gronvold, H. Haraldsen, and J. Vihovde, *Acta Chem. Scand.*, **8,** 1954, 1927–1942 (in English).
7. Most of the alloys were prepared from the elements at 900°C. At this temperature all preparations with more than 44.5 at. % Te were molten.
8. S. Chiba, *J. Phys. Soc. Japan*, **10,** 1955, 837–842; *Met. Abstr.*, **23,** 1956, 537–538.

$\bar{1}.3813$
0.6187

Fe-Th Iron-Thorium

The following intermediate phases have been identified [1] by X-ray analysis: Th_7Fe_3 (90.65 wt. % Th), hexagonal with a = 9.85 A, c = 6.15 A, c/a = 0.624, 20 atoms per unit cell, isotypic with Th_7Co_3 and Th_7Ni_3; $ThFe_3$ (58.07 wt. % Th), probably hexagonal with a = 5.22 A, c = 24.96 A; $ThFe_5$ [16.67 at. (45.39 wt.) % Th], hexagonal with a = 5.13 A, c = 4.02 A, isotypic with $CaZn_5$; and Th_2Fe_{17} [10.53 at. (32.84 wt.) % Th], monoclinic with a = 9.68 A, b = 8.56 A, c = 6.46 A, β = 99°20′, structure related to that of $ThFe_5$.

The compound Th_7Fe_3 appears to be formed by a peritectic reaction between 850 and 1000°C [2]. A eutectic has been reported at 54 at. % Th and 860°C [3]. There is evidence of some solubility of Fe in Th [2] and of Th in Fe [4], but the regions of both solid solutions are quite limited.

1. J. V. Florio, N. C. Baenziger, and R. E. Rundle, *Acta Cryst.*, **9,** 1956, 367–372. Earlier publications: N. C. Baenziger, *U.S. Atomic Energy Comm., Publ.* AECD-3237, 1948; J. V. Florio, R. E. Rundle, and A. I. Snow, *Acta Cryst.*, **5,** 1952, 449–457.
2. F. Foote, unpublished information, 1945, quoted in [5].
3. H. A. Wilhelm, unpublished information, 1946, quoted in [5].
4. H. W. Russell, unpublished information, 1946, quoted in [5].
5. H. A. Saller and F. A. Rough, Compilation of U.S. and U.K. Uranium and Thorium Constitutional Diagrams, *U.S. Atomic Energy Comm., Publ.* BMI-1000, 1955.

0.0667
$\bar{1}.9333$

Fe-Ti Iron-Titanium

Partial System Fe-TiFe. From thermal and microscopic studies of alloys with up to 21.5 wt. (24.2 at.) % Ti, [1] concluded that the phase $TiFe_3$ (22.23 wt. % Ti) forms a eutectic with the primary solid solution of Fe [about 6 wt. (6.9 at.) % Ti] at about 13.2 wt. (15 at.) % Ti, 1298°C. Since the alloys were contaminated with Al and Si (besides TiN) the solidification temperatures reported were considerably too low. [2] claimed that the structure of $TiFe_3$ was similar to that of $TiAl_3$.

However, [3], who investigated the range of 18–38 wt. (20.4–41.7 at.) % Ti by means of thermal, microscopic, and X-ray methods, definitely established that the compound $TiFe_3$ does not exist. The phase coexisting with the Fe-rich solid solution was found to be $TiFe_2$ (30.01 wt. % Ti), having a maximum melting point of 1530°C [3]. The temperature of the eutectic was located at 1310 [3], 1350 [4], and 1340 ± 10°C [5].

[6] presented a phase diagram in which he adopted the phase relationships between Fe and $TiFe_2$ as obtained by [1, 3]. In addition to $TiFe_2$, the phases TiFe (46.17 wt. % Ti) and Ti_2Fe (63.17 wt. % Ti), reported to have been identified by X-ray analysis [7], were shown to exist. In contrast to the findings of [3], it was claimed that $TiFe_2$ does not have a maximum melting point, but is formed by the peritectic reaction: melt + TiFe → $TiFe_2$ at 1520°C. However, as has been shown by [8], there is no doubt that $TiFe_2$ melts without decomposition (Fig. 398).

The occurrence of a closed γ field, first assumed by [9, 10], was confirmed by [11, 4, 12], who reported the vertex of the γ loop to be located at about 0.6 wt. (0.7 at.) % [11], slightly below 0.8 wt. (0.93 at.) % [4], and about 0.75 wt. (0.87 at.) % Ti [12]. The latter workers determined the boundary of the γ loop by means of dilatation measurements [12]; see Fig. 398. The Curie point of Fe is lowered to 690°C between 0.4 and 2.6 wt. % Ti in alloys with 0.1 wt. % C [13]. The solubility of Ti in α-Fe decreases from 6.3 wt. (7.2 at.) % Ti at 1350°C to about 2.5 wt. (2.9 at.) % Ti at 600°C [4].

Partial System TiFe-Ti. [14] outlined the boundaries of the $(\alpha + \beta)$ field by metallographic examination of Mg-reduced Ti-base alloys up to 2.2 wt. % Fe in the temperature range 900–790°C. [15] determined the $\beta/(\alpha + \beta)$ boundary down to 800°C by means of measurements of hydrogen pressures in equilibrium with a very dilute solution of hydrogen in the alloys as a function of temperature (Fig. 398).

The phase relations in the composition range up to the phase TiFe, as shown in Fig. 398, were first established by [16], by means of micrographic and X-ray analysis. His findings were verified by the work of [17, 8], who used micrographic and thermal analysis as the chief methods of investigation. Data by [17, 8] are based on alloys prepared using iodide Ti. As indicated by the inset of Fig. 398 (composition in wt. %), there are only minor quantitative discrepancies between the results of [16, 17, 8]. They also agree in that no evidence was found for the existence of the phase Ti_2Fe [7, 6, 18], either by micrographic analysis or by thermal analysis and X-ray studies (see below).

On the basis of X-ray investigations, [18] had claimed that TiFe decomposes on cooling between 800 and 650°C into $TiFe_2$ and Ti_2Fe. However, [8], using TiFe samples annealed for 10 days at 800, 650, and 500°C and quenched, did not find X-ray evidence for the instability of TiFe at lower temperatures. Also, [16] reported that only TiFe could be identified by X-ray analysis of alloys with 46–75 wt. (49.8–77.8 at) % Ti after prolonged annealing at temperatures between 560 and 650°C.

[19] determined the temperature at which the martensitic type of transformation, $\beta \rightarrow \alpha'$, takes place in alloys up to 4.6 wt. (4.0 at.) % Fe by thermal analysis. The minimum concentration for retaining the β structure by quenching lies between 3.1 and 4.1 wt. (2.6–3.5 at.) % Fe [20].

Crystal Structures. $TiFe_2$ is isotypic with $MgZn_2$ (C14 type), with parameters reported as a = 4.78 A, c = 7.80_6 A, c/a = 1.633 [3] and a = 4.779 A, c = 7.761 A, c/a = 1.624 [18]. TiFe was reported to be b.c.c. [7, 18, 16]; whereas [7] claimed the

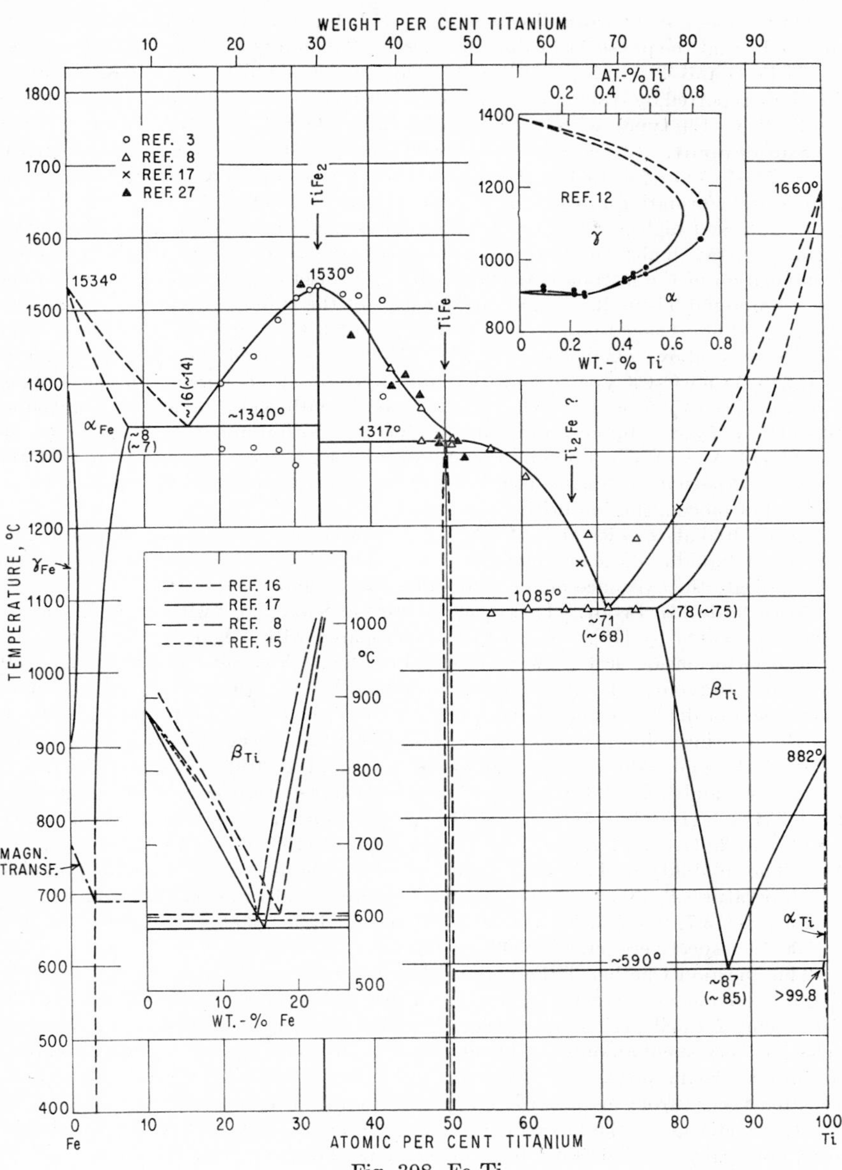

Fig. 398. Fe-Ti

CsCl (B2) type of structure, [18] found no superlattice lines due to ordering [21]. The lattice constant was given as a = 2.975 A [18], a = 2.97 A [16] (see Note Added in Proof). Ti_2Fe was claimed to be f.c.c., with 96 atoms per unit cell [7, 18], a = 11.328 A [18]. [16, 22] gave the parameter-vs.-composition relationship of the β phase.

On the basis of the work discussed above [16, 17, 8], as well as that of [23], there is no doubt that the phase Ti_2Fe does not exist. [23] found that alloys of the composition Ti_3Fe_3O and Ti_4Fe_2O yielded single-phase patterns which were almost identical with those reported by [18] for Ti_2Fe and could be identified as a structure isomorphous with Fe_3W_3C ($E9_3$ type), with a = 11.17 A for Ti_3Fe_3O and a = 11.298 A for Ti_4Fe_2O.

Supplement. [24] corroborated the $MgZn_2$-type structure for $TiFe_2$ (a = 4.760 A, c = 7.831 A, c/a = 1.645) and stated that this compound undergoes no polymorphic transformation in the temperature range 600–1400°C.

[25] studied high-purity Ti-rich Ti-Fe alloys made by a technique of levitation melting. X-ray high-temperature work placed the temperature of the eutectoid decomposition of the β_{Ti} phase at 625 ± 10°C (cf. inset of Fig. 398). As to the disputed compound Ti_2Fe, [25] found powder lines of the cubic structure (a = 11.34 A) after annealing crushed-powder specimens at 1000°C. They suggest that Ti_2Fe is stable only above about 600°C. For TiFe, [25] reported the lattice parameter a = 2.978 A; no CsCl-type superlattice lines were observed. [See also the findings on martensite (α') formation which do not agree with those of previous authors.] [26] isolated, from aluminothermically produced ferrotitanium, f.c.c. single crystals (a = 11.184 A) of the approximate composition $(Ti,Al)_2Fe$ and suggested that the peritectic formation of binary Ti_2Fe is easily supercooled but will occur when a certain level of inoculating impurities is present.

Arc-melted alloys with about 25–60 wt. % Ti were studied by [27]. The optically determined liquidus data are plotted in Fig. 398. The microstructures of as-cast alloys indicate the existence of the compounds $TiFe_2$ and TiFe.

Note Added in Proof. Lattice parameters of 8 α_{Fe} alloys with up to 2.26 at. % Ti were measured by [28]. A contraction compared with Vegard's rule was found. The heat of formation of TiFe was measured by [29]. According to an abstract, [30] "undertook to remove the uncertainties with regard to the composition and the mode of formation of definite compounds in the system Ti-Fe; they also proposed to determine the limiting solubility of Ti in Fe. They confirmed the existence of two compounds, TiFe and $TiFe_2$, melting at about 1500 and 1480°C, respectively [cf. Fig. 398]. Ti_2Fe and $TiFe_3$, described by certain authors, were not found. Eutectics (β_{Ti} + TiFe), (TiFe + $TiFe_2$), and ($TiFe_2$ + α_{Fe}) contain respectively 68, 37.5, and 17.5 wt. (71.2, 41.2, and 19.8 at.) % Ti and melt at 1100, 1280 and 1298°C [cf. Fig. 398]. The solubility of Ti in α_{Fe} was measured by the method of quenching at various temperatures." The limiting solubilities were found to be 12, 8.5, 7.5, 5.0, 4.0, and 2.5 wt. (13.7, 9.8, 8.6, 5.8, 4.6, and 2.9 at.) % Ti at 1200, 1100, 1000, 900, 800, and 500°C, respectively [cf. Fig. 398].

The question whether Ti_2Fe does or does not exist has been reexamined by [31] using microscopic and X-ray techniques. These workers tend to the view that Ti_2Fe exists, being formed peritectoidally in the vicinity of 1000°C. However, the possibility that very small amounts of oxygen may be sufficient to shift the alloy composition into a ternary phase field involving the Ti_4Fe_2O phase (cf. Crystal Structures) prevented them "from making the final conclusion that Ti_2Fe does exist."

[32] redetermined the lattice parameter of TiFe as 2.976 ± 1 A. "Superlattice lines were again not definitely detectable. This, however, may be a result of too small a difference between the atomic scattering power of the components, and it does not prove the absence of ordering."

1. J. Lamort, *Ferrum*, **11,** 1913-1914, 225–234.
2. W. Jellinghaus, *Z. anorg. Chem.*, **227,** 1936, 62–64.
3. H. Witte and H. J. Wallbaum, *Z. Metallkunde*, **30,** 1938, 100–102.
4. W. Tofaute and A. Büttinghaus, *Arch. Eisenhüttenw.*, **12,** 1938-1939, 33–37.

5. I. S. Gaev, *Metallurg,* **1934,** 19–33.
6. H. J. Wallbaum, *Arch. Eisenhüttenw.,* **14,** 1940-1941, 521–526.
7. F. Laves and H. J. Wallbaum, *Naturwissenschaften,* **27,** 1939, 674–675.
8. W. J. Fretague, C. S. Barker, and A. E. Peretti, Air Force Technical Report 6597, Parts 1 and 2, November, 1951; March, 1952.
9. F. Wever, *Arch. Eisenhüttenw.,* **2,** 1928-1929, 739–746; *Naturwissenschaften,* **17,** 1929, 304–309.
10. A. Michel and P. Bénazet, *Rev. mét.,* **27,** 1930, 326–333.
11. V. N. Svechnikov and V. N. Gridnev, *Domez,* no. 2. 1935; *Chem. Abstr.,* **29,** 1935, 6547.
12. W. P. Roe and W. P. Fishel, *Trans. ASM,* **44,** 1952, 1030–1040.
13. A. Portevin, *Rev. mét.,* **6,** 1909, 1355–1358.
14. C. M. Craighead, O. W. Simmons, and L. W. Eastwood, *Trans. AIME,* **188,** 1950, 485–513.
15. A. D. McQuillan, *J. Inst. Metals,* **79,** 1951, 73–88; see also A. D. McQuillan, *J. Inst. Metals,* **80,** 1952, 363–368.
16. H. W. Worner, *J. Inst. Metals,* **79,** 1951, 173–188.
17. R. J. VanThyne, H. D. Kessler, and M. Hansen, *Trans. ASM,* **44,** 1952, 495–513.
18. P. Duwez and J. L. Taylor, *Trans. AIME,* **188,** 1950, 1173–1176
19. P. Duwez, *Trans. ASM,* **45,** 1953, 934–940.
20. H. W. Worner, *J. Inst. Metals,* **80,** 1952, 213–216.
21. Incorrectly, [18] named the disordered A2 structure "CsCl type."
22. B. W. Levinger, *Trans. AIME,* **197,** 1953, 195.
23. W. Rostoker, *Trans. AIME,* **194,** 1952, 209–210.
24. R. P. Elliott, Armour Research Foundation, Chicago, Ill., Technical Report 1, OSR Technical Note OSR-TN-247, August, 1954, p. 28.
25. D. H. Polonis and J. G. Parr, *Trans. AIME,* **200,** 1954, 1148–1154. See also discussion, *Trans. AIME,* **203,** 1955, 718.
26. W. Gruhl and D. Ammann, *Arch. Eisenhüttenw.,* **25,** 1954, 599–600.
27. H. Nishimura and K. Kamei, *Bull. Eng. Research Inst., Kyoto Univ.,* **6,** 1954, 38–42 (in Japanese; tables and inscriptions in English).
28. A. L. Sutton and W. Hume-Rothery, *Phil. Mag.,* **46,** 1955, 1295–1309.
29. O. Kubaschewski and W. A. Dench, *Acta Met.,* **3,** 1955, 339–346.
30. I. I. Kornilov and N. G. Boriskina, *Doklady Akad. Nauk S.S.S.R.,* **108,** 1956, 1083–1085; *Titanium Abstract Bull.* (issued by Imperial Chemical Industries Ltd., Birmingham, England), **2,** 1956, 2009.
31. E. Ence and H. Margolin, *Trans. AIME,* **206,** 1956, 572–577. Discussion (W. Rostoker and R. J. VanThyne; D. H. Polonis and J. G. Parr; author's reply), *Trans. AIME,* **206,** 1956, 1417–1419.
32. T. V. Philip and P. A. Beck, *Trans. AIME,* **209,** 1957, 1269–1271.

$\bar{1}.4366$
0.5634

Fe-Tl Iron-Thallium

According to [1], Fe and Tl do not react with each other, not even at the boiling point of Tl (1460°C) (see also [2]). Whether the two liquid metals are soluble in each other cannot be determined at ordinary pressure, since the melting point of Fe is higher than the boiling point of Tl.

1. E. Isaac and G. Tammann, *Z. anorg. Chem.,* **55,** 1907, 61.
2. F. Wever, *Arch. Eisenhüttenw.,* **2,** 1928-1929, 739–746.

$\bar{1}.3703$
0.6297

Fe-U Iron-Uranium

Two thermal and microscopic investigations [1, 2] yielded concordant results, except for minor temperature differences, (Fig. 399). The intermediate phases UFe_2 (68.07 wt. % U) and U_6Fe (96.23 wt. % U) were also identified by X-ray analysis

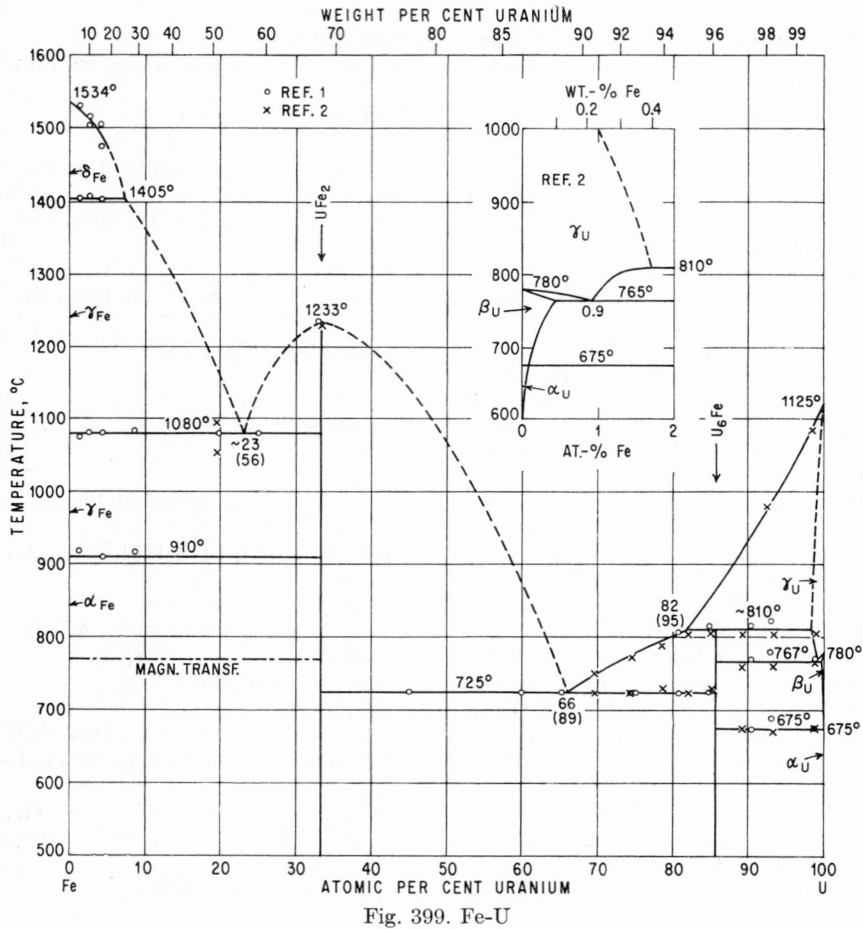

Fig. 399. Fe-U

[1, 3, 4]. The partial diagram in the inset of Fig. 399 is due to [2] and based on micrographic studies. Also, [1] reported the maximal solubility of Fe in γ-U to be 1.5–2.0 at. (0.35–0.48 wt.) % and the eutectoid to lie slightly below 1.0 at. (0.24 wt.) % Fe. The solubility of Fe in β-U is about 0.42 at. (0.1 wt.) % Fe at 765°C and less than 0.06 at. (0.014 wt.) % Fe at 650°C. The solubility in α-Fe is negligibly small [1, 2].

UFe_2 is isotypic with $MgCu_2$ (C15 type), with a = 7.04–7.05 A [1, 4]. U_6Fe is tetragonal (new type), with 28 atoms per unit cell and a = 10.31 A, c = 5.24 A, c/a = 0.508 [3].

1. P. Gordon and A. R. Kaufmann, *Trans. AIME*, **188**, 1950, 182–194.
2. J. D. Grogan, *J. Inst. Metals*, **77**, 1950, 571–580.
3. N. C. Baenziger, R. E. Rundle, A. I. Snow, and A. S. Wilson, *Acta Cryst.*, **3**, 1950, 34–40.
4. C. J. Clews, *J. Inst. Metals*, **77**, 1950, 577–580.

0.0399
$\bar{1}$.9601

Fe-V Iron-Vanadium

In 1950 a review of previous work was made by [1].

Solidification. Solidification temperatures of alloys with up to about 70 wt. % V were determined by [2, 3]. These authors agree upon the crystallization of a continuous series of solid solutions and upon the existence of a minimum in the liquidus —32 wt. (34 at.) % V, 1435°C [2]; 31 wt. (33 at.) % V, 1468°C [3] (Fig. 400)—with strong coring observed by [3]. Both liquidus and solidus should be redetermined and completed using high-purity alloys. As to the melting point of high-purity vanadium, the value 1900 ± 25°C was found by [4] and corroborated by [5].

The γ Loop. The effect of V on the allotropic transformations of Fe has been studied by [6–9, 3, 10–13], [8] having been the first to recognize the existence of a closed γ loop. The main features of the more important investigations may be seen from Table 29. The results are in rather poor agreement, mainly because of the impurity content of the alloys and the sluggish character [13] of the transformation. Carbon displaces—by neutralizing V by carbide formation—the limits of the loop to higher V contents [11]. The most recent investigation [13], a careful [14] dilatometric analysis of 17 alloys, indicates a temperature minimum in the loop at about 0.2 wt. (at.) % V, 896°C (Fig. 400), that had not been found by previous authors. [13] made no measurements above 1200°C; in the inset of Fig. 400 their data as well as the A_4 data of [3] are plotted. Besides [13], [9] also found indications for an $(\alpha + \gamma)$ two-phase field of measurable extent.

Table 29

Author	Method	Impurities	Vertex of the loop
Maurer [8].............	Thermal	0.04–0.09 wt. % C	2.1 wt. (2.3 at.) % V; 1.2 wt. (1.3 at.) % V according to the interpretation by [3]
Oya [9].................	Thermal	Aluminothermic	2.5 wt. (2.7 at.) % V
Wever-Jellinghaus [3]...	Thermal	0.01 wt. % C	1.1 wt. (1.2 at.) % V
Vogel-Martin [10].......	Thermal	Unknown	1.8 wt. (2 at.) % V
Hougardy [11]..........	Dilatometric, hardness	(Influence of carbon on the extent of the loop)	
Abram [12].............	Thermal	0.06 wt. % C in alloys with up to 0.56 wt. % V	~2 wt. (at.) % V
Lucas-Fishel [13]........	Dilatometric	Armco iron, high-purity vanadium as stock materials	~1.5 wt. (1.6 at.) % V

The σ (VFe) Phase. The existence of an intermediate phase in the solid state in the equiatomic region was discovered by [3] by means of thermal, microscopic,

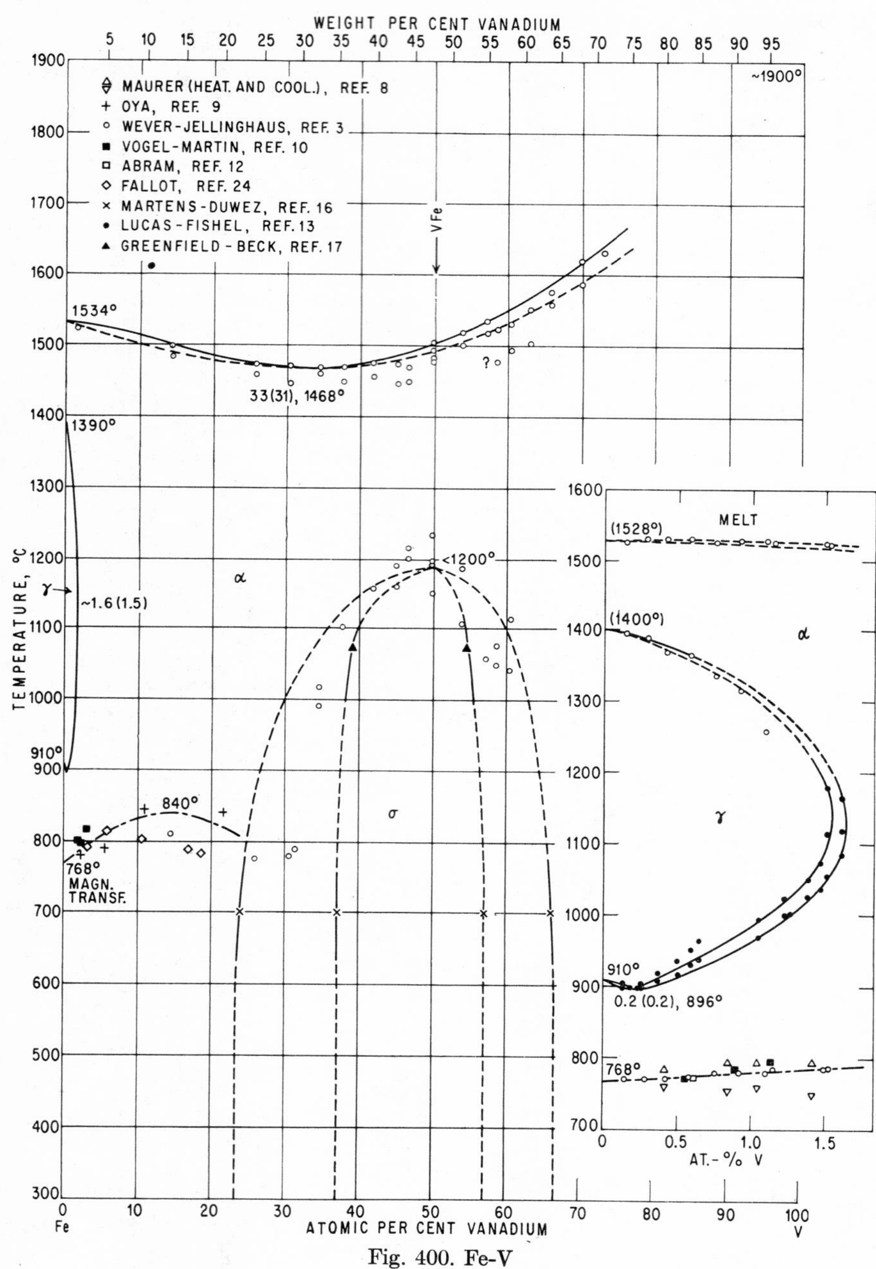
WEIGHT PER CENT VANADIUM
MAURER (HEAT. AND COOL.), REF. 8
OYA, REF. 9
WEVER-JELLINGHAUS, REF. 3
VOGEL-MARTIN, REF. 10
ABRAM, REF. 12
FALLOT, REF. 24
MARTENS-DUWEZ, REF. 16
LUCAS-FISHEL, REF. 13
GREENFIELD-BECK, REF. 17
~1900°
VFe
1534°
33(31), 1468°
1390°
~1.6 (1.5)
<1200°
840°
910°
768°
MAGN.
TRANSF.
TEMPERATURE, °C
ATOMIC PER CENT VANADIUM
Fe
V
MELT
(1528°)
(1400°)
910°
0.2 (0.2), 896°
768°
AT.-% V

Fig. 400. Fe-V

and roentgenographic work. Cooling curves showed definite arrests between 29 and 60 wt. % V (Fig. 400), but these have only qualitative significance because of the sluggish character of the transformation (see also [15]). According to [15a], σ forms only below 1200°C. X-ray [16, 17] and microscopic [17] investigations of annealed alloys showed that the homogeneity range of σ extends from 37 to 57 at. (35–55 wt.) % V at 700°C [16] and from 39 to 54.5 (±1) at. (37–52 wt.) % V at 1075°C [17]. The boundaries between the α and ($\alpha + \sigma$) regions at 700°C were located at 24 and 66 at. (22.5 and 64 wt.) % V [16] (Fig. 400).

It was pointed out by [18, 19] that VFe is isomorphous with the σ-CrFe phase, which latter possesses a tetragonal structure with 30 atoms per unit cell (see Cr-Fe). For VFe the lattice parameters a = 8.95 A, c = 4.62 A, c/a = 0.516 were found [16] (taken from a graph). Earlier attempts to index a VFe pattern [20, 21] are obsolete. See Note Added in Proof.

The magnetic transition in α solid-solution alloys was measured by [7, 22, 8, 9, 3, 10, 23, 12, 24]. There is no doubt that, contrary to the results of [22] which show a continuous decrease of the A_2 point with increasing V content, the magnetic transition goes through a maximum (Fig. 400) which, according to [9, 23], lies at approximately 15 at. (14 wt.) % V [25]. The agreement of the data by different authors is unusually poor above 5 at. % V. Above about 20 at. % V this is attributable to the fact that the alloys measured were not in equilibrium; therefore, these data were omitted from Fig. 400.

According to [3], the magnetic transition in quenched α solid-solution alloys reaches room temperature at approximately 58 wt. (60 at.) % V. The magnetic behavior of the σ phase is not yet clarified; indications for Curie points at 20–50°C or −41°C were reported by [3] and [24], respectively. See Note Added in Proof.

Further Investigations. The variation of the lattice parameter of the α phase with composition indicates a strong negative deviation from Vegard's rule [26, 3, 16, 32]. [27] measured the specific resistance, its temperature coefficient, the thermal expansion, and the density of alloys with 6, 12, and 18 wt. % V; [28] determined lattice constants, Young's modulus, and density of alloys up to 8 wt. % V.

A microscopic and roentgenographic examination [29] of V-rich alloys (90–100 wt. % V) verified the expected homogeneity.

Note Added in Proof. In an equiatomic alloy quenched from the α range, a metastable CsCl-type ordered structure forms at around 600°C preceding the formation of the stable σ phase [30, 31].

Recent measurements by [33] showed that the σ alloy with 47.6 at. % V becomes ferromagnetic below 203°K (−70°C). A neutron-diffraction study [34] yielded definite evidence for atomic ordering in the σ-phase structure. The 60 at. % V alloy investigated (containing 1.5 wt. % Si impurity) remained non(ferro)magnetic down to 4.2°K.

1. W. B. Pearson, *J. Iron Steel Inst.*, **164**, 1950, 149–159.
2. R. Vogel and G. Tammann, *Z. anorg. Chem.*, **58**, 1908, 79–82. The alloys contained about 1 wt. % Si.
3. F. Wever and W. Jellinghaus, *Mitt. Kaiser-Wilhelm-Inst. Eisenforsch. Düsseldorf*, **12**, 1930, 317–322. The alloys were melted in magnesia or Pythagoras crucibles and contained between 0.03 and 0.13 wt. % C and between 0.42 and 2.01 wt. % Si.
4. H. K. Adenstedt, J. R. Pequignot, and J. M. Raymer, *Trans. ASM*, **44**, 1952, 990–1003, especially 1002.
5. D. J. McPherson, discussion on [4], *Trans. ASM*, **44**, 1952, 1029.
6. P. Pütz, *Metallurgie*, **3**, 1906, 635–638, 649–656. The author concluded a rise

of A_3 with increasing V content, but the cooling curves given (perhaps confused) indicate rather a depression of A_3.
7. A. Portevin, *Rev. mét.*, **6,** 1909, 1352–1355.
8. E. Maurer, *Stahl u. Eisen*, **45,** 1925, 1629–1632.
9. M. Oya, *Science Repts. Tôhoku Imp. Univ.*, **19,** 1930, 235–245; *Kinzoku-no-Kenkyu*, **5,** 1928, 349–356.
10. R. Vogel and E. Martin, *Arch. Eisenhüttenw.*, **4,** 1930-1931, 487–495.
11. H. Hougardy, *Arch. Eisenhüttenw.*, **4,** 1931, 497–503.
12. H. H. Abram, *J. Iron Steel Inst.*, **130,** 1934, 351–375.
13. W. R. Lucas and W. P. Fishel, *Trans. ASM*, **46,** 1954, 277–289.
14. The alloys, however, were melted in air and analyzed only for their V content.
15. A. H. Sully, *J. Inst. Metals,* **80,** 1951, 173–179.
15a. P. A. Beck, private communication. See also [30].
16. H. Martens and P. Duwez, *Trans. ASM*, **44,** 1952, 484–493.
17. P. Greenfield and P. A. Beck, *Trans. AIME,* **200,** 1954, 253–257.
18. F. Wever and W. Jellinghaus, *Mitt. Kaiser-Wilhelm-Inst. Eisenforsch. Düsseldorf,* **13,** 1931, 93–109, 143–147.
19. K. W. Andrews, *Research,* **1,** 1948, 478–479.
20. P. Duwez and S. R. Baen, *ASTM Symposium*, *Spec. Tech. Publ.* 110, 1950, pp. 48–54.
21. P. Pietrokowsky and P. Duwez, *Trans. AIME,* **188,** 1950, 1283–1284.
22. K. Honda, *Ann. Physik,* **32,** 1910, 1010–1011.
23. D. L. Edlund, unpublished thesis, Massachusetts Institute of Technology, Cambridge, Mass., 1934. (Not available to the author.) The Curie-temperature-vs.-composition curve reproduced in the 1948 edition of "Metals Handbook" (p. 1219, The American Society for Metals, Cleveland, Ohio) apparently is based on Edlund's data; it shows a maximum at about 13 wt. (14 at.) % V, 839°C.
24. M. Fallot, *Ann. phys.*, **6,** 1936, 305–387.
25. Fallot [24] found the maximum near 6.25 at. % V and suggested the formation of a superstructure, VFe_{15}.
26. A. Osawa and M. Oya, *Science Repts. Tôhoku Imp. Univ.*, **18,** 1929, 727–731. *Kinzoku-no-Kenkyu*, **6,** 1929, 234–236.
27. K. Ruf, *Z. Elektrochem.*, **34,** 1928, 813–818.
28. Z. Nishiyama, *Science Repts. Tôhoku Imp. Univ.*, **18,** 1930, 359–400.
29. W. Rostoker and A. Yamamoto, *Trans. ASM*, **46,** 1954, 1136–1163.
30. P. A. Beck et al., *Trans. AIME*, **206,** 1956, 148–149.
31. T. V. Philip and P. A. Beck, *Trans. AIME*, **209,** 1957, 1269–1271.
32. A. L. Sutton and W. Hume-Rothery, *Phil. Mag.*, **46,** 1955, 1295–1309.
33. M. V. Nevitt and P. A. Beck, *Trans. AIME*, **203,** 1955, 669–674.
34. J. S. Kasper and R. M. Waterstrat, *Acta Cryst.*, **9,** 1956, 289–295.

$\bar{1}.4824$
0.5176

Fe-W Iron-Wolfram

Early studies of the Fe-W system consisted chiefly of attempts to identify intermediate phases by residue analysis [1–5]; and the compounds WFe_3 [3, 5], WFe_2 [2], W_2Fe_3 [4], and W_2Fe [1] were claimed to be present. For details of these papers, see [6]. As is well known, this chemical method gives no reliable results. (This is even more true when impure alloys are used.) Later systematic work has shown that compounds WFe_3 and W_2Fe certainly do not exist.

Detailed reviews of the literature up to 1933 and 1935, respectively, have been published by [6] and [7].

Solidification. Thermal-analysis and/or initial-melting determinations have been carried out by [8–15] in more or less extended composition ranges. Whereas most of the investigators [9–11, 16, 12, 13] who studied the binary system reported a eutectic in the Fe-rich portion of the phase diagram, those [17, 14] who worked on ternary systems containing Fe and W agreed upon a peritectic reaction involving the α phase. This could be corroborated by [15], who observed the melting temperatures of Fe-W wire specimens containing up to 35 wt. (14 at.) % W; he found a peritectic horizontal at 1540°C and a minimum in the solidus curve at 15 wt. (5 at.) % W, 1525°C (Fig. 401). [14] had reported 1530°C as the peritectic temperature.

The following temperatures have been reported for the upper peritectic horizontal in Fig. 401: 1500 [9], 1660 [10], 1640 [11, 12] (Fig. 401), and 1675°C [13]. The Fe-rich end of this horizontal is not exactly known; whereas the thermal data by [10] point to a composition of 43 wt. (18.6 at.) % W, [18] recommended 50 wt. (23.3 at.) % W in a recently published phase diagram which is mainly based on his own numerous publications. In Fig. 401 the value 45 wt. (20 at.) % W has been chosen. (See Note Added in Proof.)

A tentative diagram for the high-temperature region involving the gas phase (boiling point of Fe: 2740°C!) may be found in [13].

γ Loop. It was known early that the $\gamma \rightleftarrows \delta$ and $\alpha \rightleftarrows \gamma$ transformations in Fe are lowered [8] and raised [19, 8, 10], respectively, by the addition of W. The existence of a closed γ loop was recognized by [11]; his thermal (cooling) data as well as most of his later micrographic data [20] are plotted in the inset of Fig. 401, which also gives the results of dilatometric [16] and H_2-solubility [21] measurements. The γ phase cannot be retained at room temperature even by a drastic quench [20].

For thermodynamic computations of the γ-loop boundaries, see [22, 23].

Solubility of W in Solid Fe. There is unusually good agreement in the literature as far as the maximum solubility of W in Fe is concerned. A value of 33 wt. (13 at.) % has been reported by [9, 11, 16, 15]. The solubility decreases with decreasing temperature [24, 9, 11, 16, 13, 15]; the more recent data are plotted in Fig. 401. From the values reported in the temperature range 600–750°C (lying between <0.3 [13] and 2.6 at. % W [11]), that of [25]—1.4 at. (4.5 wt.) % W at 700°C, determined by X-ray parametric work—appears to be the most reliable one.

According to [8, 10, 16], the Curie temperature of the α phase is essentially independent of the W content.

Intermediate Phases. There has been much discussion concerning how many and which compounds exist in the Fe-W system. It appears from recent experimental results that the correct answer has not been found as yet.

The results of early residue analyses have already been dealt with. In a cursory microscopic study, [24] claimed to have found two compounds, the compositions of which could not be reliably established. The phase diagrams worked out by [9] and [10] showed only one intermediate phase, WFe_2 (62.22 wt. % W). [26] concluded from exploratory X-ray work the existence of a hexagonal compound WFe which was, however, not corroborated by later investigations. [11] established by careful thermal and microscopic work the existence of the compound W_2Fe_3 (68.70 wt. % W). From an X-ray examination, [27] concluded that the system contains two intermediate phases, WFe_2 (ϵ) and W_2Fe_3 (ζ), both having narrow ranges of homogeneity. These phases could be chemically isolated. The temperature of formation of WFe_2 was found to lie between 1000 and 1450°C.

[16, 13] confirmed W_2Fe_3 but claimed to have disproved the existence of WFe_2. [16] suggested—from admittedly meager microscopic evidence—the existence of a high-temperature phase stable above 1550°C and containing between 80 and 90 wt.

% W. This X phase, however, definitely does not exist, as subsequent work has shown [28, 12, 13].

[12] made a special study of the intermediate phases in the Fe-W system, using X-ray and microscopic methods. Their results, used in drawing Fig. 401, confirmed

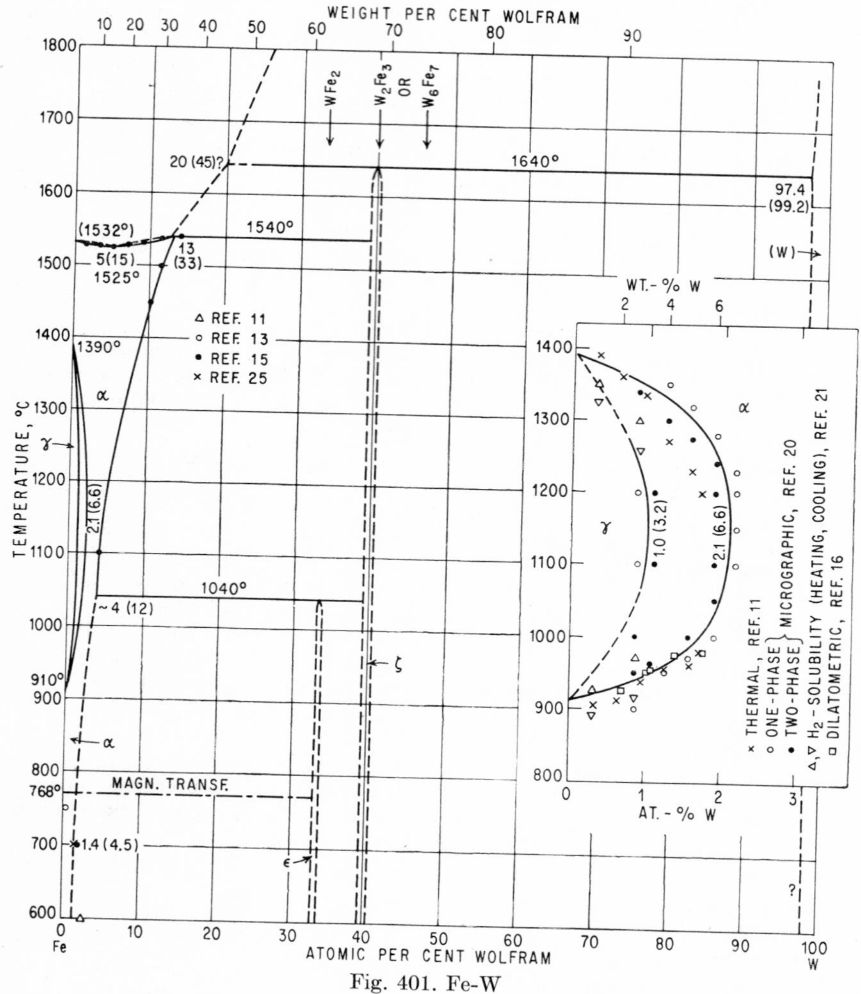

Fig. 401. Fe-W

the existence of WFe_2 (formed peritectoidally at 1040°C) and W_2Fe_3. The homogeneity ranges of these phases (indicated by line shifting) could not be exactly determined; however, there is some microscopic evidence that the composition range of W_2Fe_3 lies between 68 and 69 wt. (39.2 and 40.3 at.) % W at 1600°C. Some disconcerting findings (see the original paper) may be due to the sluggishness of solid-state reactions in Fe-W alloys [29].

Based on the X-ray data of [27], [30] determined the crystal structure of the two intermediate phases (see Crystal Structures). Their work points to an "ideal" composition W_6Fe_7 [46.15 at. (73.84 wt.) % W] for the compound ζ, called W_2Fe_3 heretofore. Apparently the compound exists only in a composition range outside the stoichiometric formula. [30] also stated, "Technical ferro-tungsten with about 85 wt. % W has been found to consist of WFe_2 and W. W_6Fe_7 is accordingly not stable at low temperatures. Its decomposition rate is, however, evidently very low" [31].

[32] claimed in a short note that a high-temperature σ phase exists in the Fe-W system (no composition or temperatures given). Exploratory work by [33] on a 60 at. % W alloy failed to detect a σ phase.

The results of some X-ray work by [34]—if confirmed by announced further studies—would appreciably complicate the rather simple constitution shown in Fig. 401. [34] carried out several experiments to produce the ε phase (WFe_2) by annealing sintered Fe-W mixtures. He stated:

"At least three distinct structures were observed . . . all form modifications of [ζ] [35] by the omission or reordering of W atoms in the lattice, so that a sequence of closely related but distinctive phases is formed. The identification of any of the present types of [ζ] with Sykes's compound Fe_2W does not appear justified. Slight discrepancies in the interplanar spacings also exist with Arnfelt's [27] compounds. These results, however, are compatible if they are regarded as representing different stages in successive structural transformations, possibly with Fe_3W_2 and Fe_2W as end compositions. The further possibility that several allotropic forms of [ζ] exist cannot be excluded."

[34] also suggested that the compound richest in Fe may have the composition W_3Fe_7 (30 at. % W).

Solubility of Fe in Solid W. [11] first estimated that the solubility was 1.2 wt. % at 1600°C, but in subsequent microscopic work [36] he found that it was less than 1.1 wt. (3.5 at.) % at this temperature. In a recent review [18], he suggested a value of 0.8 wt. (2.6 at.) % Fe at 1640°C. There appears to be little change in the solubility with temperature [11].

Crystal Structures. Solute W atoms increase the lattice parameter of the α-iron phase [27, 37, 36, 38, 12, 13, 25, 39]. According to [36, 12], the W lattice is expanded by the addition of Fe atoms, indicating that the solution of Fe in W is of the interstitial type. At variance with this, [13] reported a decrease in the W parameter by dissolving Fe.

[27] found the following hexagonal unit cells for the compounds WFe_2 and W_2Fe_3, respectively: $a = 4.727$ kX, $c = 7.704$ kX, $c/a = 1.630$ (12 atoms per unit cell), and $a = 4.731$ kX, $c = 25.76$ kX, $c/a = 5.440$ (trigonal Laue symmetry, 39–40 atoms per unit cell; corroborated by [38]).

[12] reported that the interplanar spacings computed from the patterns of alloys containing the ε and ζ phases agreed well with those found by [27].

[13] claimed for W_2Fe_3 a h.c.p. unit cell containing 10 atoms with parameters nearly identical with those found by [27] for WFe_2.

[30] determined the crystal structure of the two intermediate phases ε and ζ from the X-ray data of [27]. The crystallographic data reported for WFe_2 (ε) are those of the hexagonal $MgZn_2$ (C14) type of structure. As to the structure of the ζ phase, [30] found the true unit cell to be rhombohedral (having a volume one-third as great as the corresponding hexagonal one assumed by [27, 38]), with $a = 9.04$ A, $\alpha = 30°30.5'$. There are 13 atoms per unit cell, which fact points to the ideal formula W_6Fe_7 ($D8_5$ type).

As mentioned above, [34] observed "at least three distinct structures" between 30 and 40 at. % W: besides ζ two structures closely related to ζ, showing different intensities and certain line displacements.

[40] prepared WFe_2 by powder-metallurgical techniques. "After sintering at 1200°C, specimens were annealed at 600, 800, and 1000°C. Low-temperature annealing developed a predominant phase which could be indexed as isomorphous with $MgZn_2$, with a = 4.735 kX, c = 7.706 kX, c/a = 1.627. Many weaker lines could not be accounted for, indicating the possibility that this structure is more complex than the $MgZn_2$ type."

In summary, it may be said that the constitution in the composition range 30–50 at. % W may be more complex than is shown in Fig. 401. (Decomposition of ζ below 1000°C? Existence of another intermediate phase?) A careful systematic reinvestigation using long-time annealings should be made.

Note Added in Proof. [41] reexamined some Fe-rich alloys by means of thermal and microscopic analysis and located the Fe-rich end of the 1640°C peritectic horizontal at 39 wt. (16.3 at.) % W.

1. T. Poleck and B. Grützner, *Ber. deut. chem. Ges.*, **26**, 1893, 35–38.
2. H. Behrens and A. R. van Linge, *Rec. trav. chim.*, **13**, 1894, 155.
3. A. Carnot and E. Goutal, *Compt. rend.*, **125**, 1897, 213–216.
4. E. Vigouroux, *Compt. rend.*, **142**, 1906, 1197–1199.
5. T. Swinden, *J. Iron Steel Inst.*, **73**, 1907, 292; *Metallurgie*, **6**, 1909, 720.
6. J. L. Gregg, "The Alloys of Iron and Tungsten," Alloys of Iron Research, Monograph Series, McGraw-Hill Book Company, Inc., New York, 1934.
7. M. Hansen, "Der Aufbau der Zweistofflegierungen," Springer-Verlag OHG, Berlin, 1936.
8. H. Harkort, *Metallurgie*, **4**, 1907, 617–631, 639–647, 673–682.
9. K. Honda and T. Murakami, *Science Repts. Tôhoku Imp. Univ.*, **6**, 1918, 264–271.
10. S. Ozawa, *Science Repts. Tôhoku Imp. Univ.*, **11**, 1922, 333–340; see also *Stahl u. Eisen*, **46**, 1926, 1834–1835.
11. W. P. Sykes, *Trans. AIME*, **73**, 1926, 968–1008.
12. W. P. Sykes and K. R. van Horn, *Trans. AIME*, **105**, 1933, 198–212; discussion, pp. 212–214.
13. O. Landgraf, *Forschungsarb. Metallkunde u. Röntgenmetallog.*, **12**, 1934, 1–46.
14. K. Winkler and R. Vogel, *Arch. Eisenhüttenw.*, **6**, 1932-1933, 165–172.
15. W. P. Sykes, *Trans. ASM*, **24**, 1936, 541–550.
16. S. Takeda, *Kinzoku-no-Kenkyu*, **6**, 1929, 298–308 (in Japanese); *Tech. Repts. Tôhoku Imp. Univ.*, **9**, 1930, 447–461 (in English).
17. W. Köster and W. Tonn, *Arch. Eisenhüttenw.*, **5**, 1931-1932, 431–440.
18. W. P. Sykes, "Metals Handbook," 1948 ed., p. 1220, The American Society for Metals, Cleveland, Ohio.
19. F. Osmond, *Compt. rend.*, **104**, 1887, 985.
20. W. P. Sykes, *Trans. AIME*, **95**, 1931, 307–311.
21. A. Sieverts and H. Moritz, *Ber. deut. chem. Ges.*, **75B**, 1942, 1726–1729.
22. W. Oelsen, *Stahl u. Eisen*, **69**, 1949, 468–474.
23. W. Rostoker, *Trans. AIME*, **191**, 1951, 1203–1205.
24. D. Kremer, *Abhandl. Inst. Metallhütt. u. Elektromet. Tech. Hochsch. Aachen*, **1**, 1916, 14–18. (Microscopic investigation of nine aluminothermic alloys containing 8.8–80.5 wt. % W.)
25. C. S. Smith, *J. Appl. Phys.*, **12**, 1941, 817–822 (Precipitation Hardening in the Fe-W System).
26. E. C. Bain, *Chem. & Met. Eng.*, **28**, 1923, 23.
27. H. Arnfelt, *Iron Steel Inst. (London), Carnegie Schol. Mem.*, **17**, 1928, 1–13.
28. W. P. Sykes, *Trans. ASST*, **16**, 1929, 368.

29. This sluggishness may also explain the failure of some previous authors to detect the WFe_2 phase.
30. H. Arnfelt and A. Westgren, *Jernkontorets Ann.*, **119,** 1935, 185–196; A. Westgren, *Science Repts. Tôhoku Imp. Univ., K. Honda Anniv. Vol.*, 1936, pp. 852–863 (in English).
31. See also discussion on [12], p. 213.
32. H. J. Goldschmidt, *Research*, **4,** 1951, 343.
33. P. Greenfield and P. A. Beck, *Trans. AIME*, **206,** 1956, 265.
34. H. J. Goldschmidt, *J. Iron Steel Inst.*, **170,** 1952, 189–204, especially 197–199.
35. Called ξ (xi) by [34].
36. E. P. Chartkoff and W. P. Sykes, *Trans. AIME*, **89,** 1930, 566–573.
37. Z. Nishiyama, *Science Repts. Tôhoku Imp. Univ.*, **18,** 1929, 377–386.
38. A. Osawa and S. Takeda, *Kinzoku-no-Kenkyu*, **8,** 1931, 181–196; abstract in *J. Inst. Metals*, **47,** 1931, 534.
39. A. P. Guljaev and E. F. Trusova, *Zhur. Tekh. Fiz.*, **20,** 1950, 66–78. *Structure Rept.*, **13,** 1950, 77.
40. R. P. Elliott, Armour Research Foundation, Chicago, Ill., Technical Report 1, OSR Technical Note OSR-TN-247, 1954.
41. R. Schneider and R. Vogel, *Arch. Eisenhüttenw.*, **26,** 1955, 483–484.

$\bar{1}.9316$
0.0684

Fe-Zn Iron-Zinc

The phase diagram presented in Fig. 402 is that proposed by [1]. It is given as "the most probable composite diagram, based on reliable and recent work." Data reported up to 1948 have been considered; newer results are given below. The literature published up to 1935 has been reviewed by [2].

Solidification. The liquidus temperatures reported by [2] were obtained by solubility determinations. They have been used to draw the liquidus curve between 419 and 900°C (Figs. 402 and 403). Solubility determinations in the range up to 600°C [3] closely agree with these findings. Liquidus temperatures based on cooling curves [4–8] do not represent equilibrium conditions; they lie to the Fe-rich side of the curve shown in Figs. 402 and 403. The peritectic melts at 672 and 782°C were found to contain 3.0 wt. (3.5 at.) % and 7.4 wt. (8.55 at.) % Fe [2] vs. 3.7 wt. (4.3 at.) % and 8.8 wt. (10.15 at.) % Fe [8]. The discrepancy of the data reported by [4–7] is still greater. For the range 900–1100°C the temperatures obtained by thermal analysis of melts in closed containers [8] are fairly consistent with an extrapolation of the results of [2].

The temperatures of the various peritectic reactions were found as follows: (*a*) 766 [7], 777 [5], 780 [6–10], and 782°C [2]; (*b*) 647 [7], 662 [5, 6], 668 [8, 11, 9, 10], and 672°C [2]; (*c*) 620°C [9, 10]; and (*d*) 530°C [11, 10]. The solidus of the γ phase was outlined micrographically [8], while that of the phases Γ and δ_1 are based on data obtained by means of heating curves [8, 11], microexamination [8, 11], and measurements of magnetic susceptibility vs. temperature [10].

The three-phase equilibrium involving nearly pure Zn was claimed to be a eutectic. The temperature was found to lie about 0.05°C below the melting point of Zn and the eutectic composition at about 0.018 wt. (0.021 at.) % Fe [2], in good agreement with 0.012 wt. (0.014 at.) % Fe according to [3, 12]. 0.09 wt. (0.10_5 at.) % Fe was reported by [8].

Solid-state Phase Boundaries. *α and γ Phases.* The solubility curves for Zn in α-Fe and γ-Fe were determined micrographically [8] and by lattice-spacing measurements [13]. Results of the latter are more reliable. Little evidence was given by [8] for the boundaries of the $(\alpha + \gamma)$ field as plotted in Fig. 402. They can be regarded only as an approximation. The general form of the phase equilibria in this region of composition, however, resembles that suggested by earlier workers [14, 7]. There is agreement with regard to the temperature of the eutectoid horizontal, about 623°C [14, 7, 8].

In contrast to these results, [15, 16] believed they found evidence for the existence of a closed γ field extending to about 17.5 wt. (15.4 at.) % Zn at 1150°C. Diffusion tests [15] and dilatometric study [16], respectively, were used in this work.

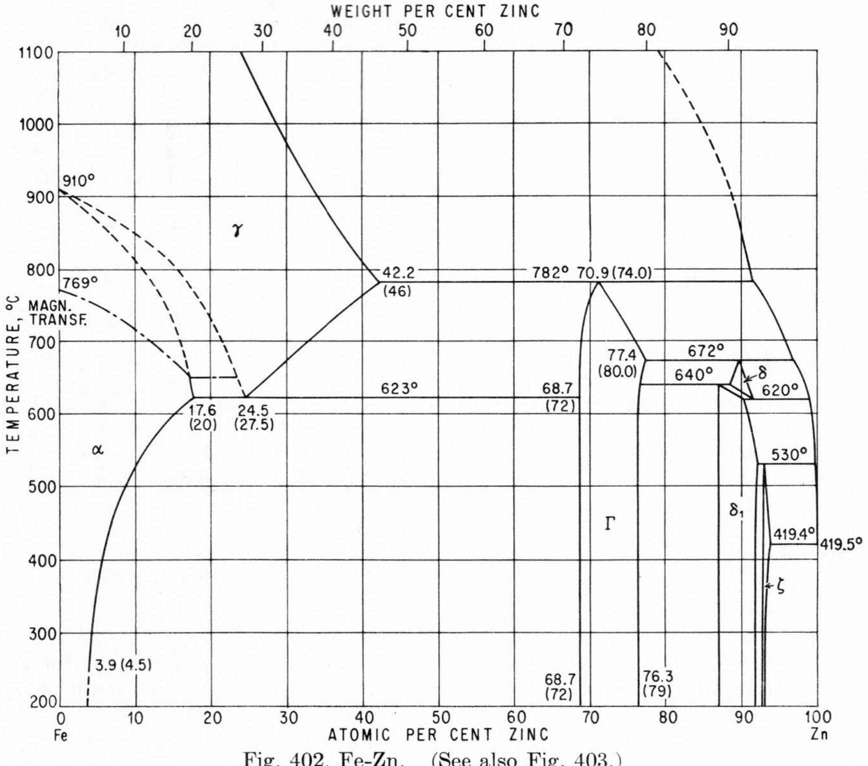

Fig. 402. Fe-Zn. (See also Fig. 403.)

The whole composition range up to about 25 at. % Zn requires more investigation, although the solubility boundaries of the γ and α phases appear to be well established [13]. According to [17], the solubility at 250°C is 4.5 wt. (3.9 at.) % Zn.

Γ *Phase.* The homogeneity range of this phase, previously denoted $FeZn_3$ [5, 6, 18, 19, 14, 20], Fe_3Zn_{10} (76.92 at. % Zn) [21], and Fe_5Zn_{21} (80.77 at. % Zn) [22], was determined micrographically as 73–80 wt. (69.8–77.4 at.) % Zn [8] and by the lattice-spacing method as 72–79 wt. (68.7–76.3 at.) % Zn [13]. This proved that the approximate range of 77–81 at. % Zn, based on X-ray work [22], was in error. The boundaries at about 250°C were found as 70–77 wt. (66.6–74.1 at.) % Zn [17]. At higher temperatures the homogeneity range expands somewhat toward higher Zn concentrations [9].

The δ_1 *phase,* recognized as a phase of variable composition by previous investigators and designated as $FeZn_7$ (87.5 at. % Zn) [5, 6, 18, 14, 21, 7, 20] and $FeZn_{10}$ (90.91 at. % Zn) [4, 23, 24, 25], was first reported by [8] to be stable between 88.5 and 93.7 wt. (86.9–92.7 at.) % Zn. After the detection of the ζ phase with 94 wt. (93.1 at.) % Zn [11], the boundary on the Zn-rich side was placed at 93.0 wt. (91.9 at.) % Zn.

δ_1 undergoes a transformation $\delta_1 \rightleftarrows \delta$ at temperatures between about 640 and 620°C [9]. The precise location of the phase boundaries in this composition and temperature range is not conclusively established. The evaluation of all data available, however, is in favor of the boundaries shown in Figs. 402 and 403 [9].

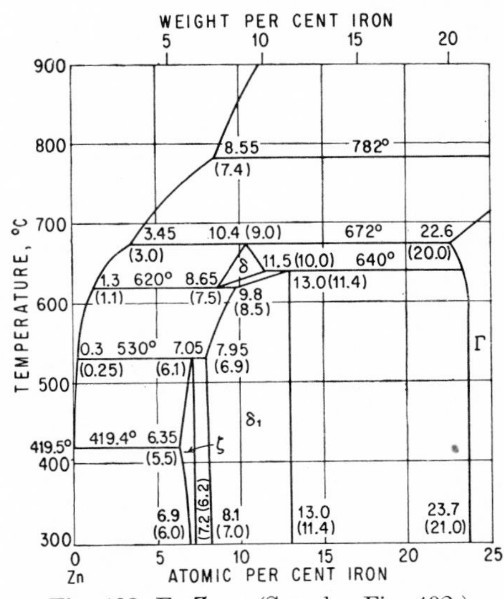

Fig. 403. Fe-Zn. (See also Fig. 402.)

Using a special method of preparation which results in the formation of alloys being in equilibrium at 250–300°C, [17] were unable to detect the δ_1 phase. This would indicate that δ_1 is unstable at temperatures below about 300°C. [17] proposed that δ_1 probably decomposes eutectoidally into Γ and ζ [26] (see next paragraphs).

The existence of *the* ζ *phase* was established by X-ray, microscopic, and magnetic investigations [11, 10, 13]. It was shown that the peritectic formation at approximately 530°C, under normal conditions of cooling from the liquid state, is almost completely suppressed. This was the reason why the ζ phase had been overlooked by previous investigators, with the probable exception of [18], who had reported, on the basis of metallographic studies of galvanized coatings, that three rather than two intermediate phases occurred. [27] confirmed that ζ is not present at temperatures above 530°C.

Microexamination of heat-treated alloys indicated the homogeneity range of ζ to be 93.8–94.0 wt. (92.8–93.1 at.) % Zn [11]. In great contrast to this, [17] reported the ζ phase at 250°C to be stable at 89 wt. (87.4 at.) % Zn; i.e., between 530 and 250°C ζ would change its composition from about 93 to about 87.4 at. % Zn. [87.4 at. % Zn differsf rom the $\delta_1/(\Gamma + \delta_1)$ boundary according to [8] by only 0.5 at. %.] This is incompatible with all the findings of [8–11]. No explanation for the discrepancy can be offered at present.

(Zn) *Phase.* The solid solubility of Fe in Zn should, of course, be lower in Fe content than the composition of the Zn-rich eutectic, 0.012–0.018 wt. % Fe [2, 3, 12]. A maximum solubility of 0.008 wt. % Fe was reported by [28]. By metallographic methods, [2] showed that the solubility was between 0.0009 and 0.0028 wt. % Zn at 150–400°C; see also [29]. Previous results are obsolete [30, 31].

Magnetic Transformation. The effect of Zn on the Curie point of Fe was studied by [14, 7, 8, 32]. Agreement between these findings is poor. In Fig. 402 those of [8] have been used to draw the magnetic-transformation curve.

Crystal Structures. Lattice parameters of the α phase were reported by [21] and especially [13]. The latter author gave parameters of highly supersaturated α (up to 41 at. % Zn), obtained by quenching from the γ field at 775°C.

The Γ phase is isotypic with γ-brass ($D8_{1-3}$ type) [21, 22, 13]. The parameter varies linearly from $a = 8.974$ A at 68.7 at. % Zn to $a = 9.018$ A at 76.3 at. % Zn [13].

The δ_1 phase was first claimed to be h.c.p. of the Mg type [21]. However, the powder pattern was found to be much more complicated and somewhat similar to that of Γ [13]. According to [33], the unit cell containing 555 ± 8 atoms has the dimensions $a = 12.83$ A, $c = 57.7$ A; see also [34].

The ζ phase is monoclinic, with 28 atoms per unit cell ($2 \times FeZn_{13}$) and $a = 13.68$ A, $b = 7.63$ A, $c = 5.07$ A, $\beta = 128°44'$ [35, 34]. [36] also found that the c dimension of the unit cell was 5.07 A.

1. G. V. Raynor, Institute of Metals Annotated Equilibrium Diagram Series, no. 8, The Institute of Metals, London, 1951.
2. E. C. Truesdale, R. L. Wilcox, and J. L. Rodda, *Trans. AIME,* **122,** 1936, 192–228.
3. G. Edmunds, *Trans. AIME,* **156,** 1944, 263–276.
4. S. Wologdine, *Rev. mét.,* **3,** 1906, 701–708.
5. A. v. Vegesack, *Z. anorg. Chem.,* **52,** 1907, 34–40.
6. P. T. Arnemann, *Metallurgie,* **7,** 1910, 203–204, 208–209.
7. Y. Ogawa and T. Murakami, *Tech. Repts. Tôhoku Univ.,* **8,** 1929, 53–69.
8. J. Schramm, *Z. Metallkunde,* **28,** 1936, 203–207.
9. J. Schramm, *Z. Metallkunde,* **30,** 1938, 131–135.
10. J. Schramm, *Z. Metallkunde,* **30,** 1938, 327–334.
11. J. Schramm, *Z. Metallkunde,* **29,** 1937, 222–224.
12. M. W. Vinaver, *J. Inst. Metals,* **81,** 1953, 709–710; see also C. W. Roberts, *J. Inst. Metals,* **81,** 1953, 710.
13. J. Schramm, *Z. Metallkunde,* **30,** 1938, 122–130.
14. U. Raydt and G. Tammann, *Z. anorg. Chem.,* **83,** 1913, 257–266.
15. W. D. Jones, *J. Iron Steel Inst.,* **130,** 1934, 429–437.
16. V. N. Svechnikov and V. N. Gridnev, *Metallurg,* **12**(1), 1937, 35–39.
17. F. Lihl and A. Demel, *Z. Metallkunde,* **43,** 1952, 307–310.
18. W. Guertler, *Intern. Z. Metallog.,* **1,** 1911, 353–375.
19. E. Vigouroux, F. Ducelliez, and A. Bourbon, *Bull. soc. chim. France,* **11,** 1912, 480–485.
20. H. Grubitsch, *Monatsh. Chem.,* **60,** 1932, 165–180.
21. A. Osawa and Y. Ogawa, *Z. Krist.,* **68,** 1928, 177–188; *Science Repts. Tôhoku Univ.,* **18,** 1929, 165–176.
22. W. Ekman, *Z. physik. Chem.,* **B12,** 1931, 57–78.
23. F. Taboury, *Compt. rend.,* **159,** 1914, 241–243.
24. E. Lehmann, *Physikal. Z.,* **22,** 1921, 601–603.
25. C. W. Stillwell and G. L. Clark, *Ind. Eng. Chem., Anal. Ed.,* **2,** 1930, 266–272.
26. See the polemic discussion of H. Bablik, F. Götzl, R. Kubaczka, and F. Lihl, *Z. Metallkunde,* **44,** 1953, 391–392.
27. E. Scheil and H. Wurst, *Z. Metallkunde,* **29,** 1937, 224–229.

28. W. M. Peirce and L. H. Marshall, "National Metals Handbook," p. 1425, American Society for Steel Treating, Cleveland, Ohio, 1933.
29. F. Pawlek, *Z. Metallkunde*, **36,** 1944, 105–112.
30. W. M. Peirce, *Trans. AIME*, **68,** 1923, 771–772.
31. R. Chadwick, *J. Inst. Metals*, **51,** 1933, 114.
32. M. Fallot, *Ann. phys.*, **7,** 1937, 420–428.
33. H. Bablik, F. Götzl, and F. Halla, *Z. Metallkunde*, **30,** 1938, 249–252.
34. F. Götzl, F. Halla, and J. Schramm, *Z. Metallkunde*, **33,** 1941, 375.
35. F. Halla, R. Weil, and F. Götzl, *Z. Metallkunde*, **31,** 1939, 112.
36. V. Montoro, *Ricerca sci.*, **2,** 1937, 449–450.

$\bar{1}.7869$
0.2131

Fe-Zr Iron-Zirconium

Thermal analysis [1] yielded the data points shown in Fig. 404. The investigators concluded that the only intermediate phase existing corresponds to the composition Zr_2Fe_3, with an estimated melting point of 1620–1640°C. However, [2] has identified the phase by X-ray analysis as $ZrFe_2$ (44.95 wt. % Zr). It was claimed [2, 3] that $ZrFe_2$ can exist in two forms, both being stable at room temperature. One form, having the cubic $MgCu_2$ (C15) type of structure, was said to exist at the stoichiometric composition and the other, stable with an excess of approximately 6–8 at. % Fe, to be isostructural with the hexagonal phase $MgNi_2$ (C36 type). [4] could not verify this; reexamination of the structure of two alloys with 27 and 33.3 at. % Zr proved that both compositions exhibit only the $MgCu_2$ structure.

Prior to [1], the microstructure of as-cast alloys with up to 30 wt. (20.8 at.) % Zr was examined by [5]. By chemical analysis of an alloy with purely eutectic structure the eutectic point was located at 12.1 wt. (7.8 at.) % Zr [6]. Since [1] did not analyze their alloys and a certain loss of Zr by oxidation had been observed [5], it may very well be that the liquidus curve of the δ phase has to be shifted to somewhat lower Zr contents.

As shown in Fig. 404, the $\delta \rightleftarrows \gamma$ transformation is lowered. The temperature of the resulting reaction $\delta \rightleftarrows \gamma +$ melt could not be separated from that of the eutectic crystallization and was therefore assumed to be about 1335°C. The $\gamma \rightleftarrows \alpha$ transformation was also found to be lowered; however, results scatter considerably [1, 5]. The Curie point of Fe is not affected by Zr additions [1, 5]. The solubility of Zr in γ-Fe and α-Fe was not determined; that in α-Fe probably is less than 0.3 wt. (0.18 at.) % Zr at 835°C [1].

Between $ZrFe_2$ and Zr, [1] had tentatively assumed (*a*) a eutectic at approximately 88 wt. (81.7 at.) % Zr, 1350°C, and (*b*) an increase of the transformation temperature $\alpha \rightleftarrows \beta$ to a peritectoid equilibrium at 1000°C. Since the alloys were prepared by melting in alumina crucibles, they must have been heavily contaminated.

The portion of the diagram for Zr-rich alloys (Fig. 404) is mainly based on micrographic and thermal-analysis data [7]. Alloys prepared using magnesium-reduced Zr were melted in graphite crucibles. The solubility of Fe in α-Zr was not determined accurately but can be estimated to be of the order of 0.02 wt. % at 800 ± 5°C, the eutectoid temperature determined by thermal analysis. The eutectoid temperature of high-purity alloys can be expected to be lower than 800°C. Microexamination and X-ray work appeared to confirm the existence of only one intermediate phase, i.e., that identified as $ZrFe_2$ [2, 4].

$ZrFe_2$ is isotypic with $MgCu_2$, a = 7.05 A [2, 4, 7].

Supplement. Like [4], [8] was unable to find any evidence for the $MgNi_2$ modification claimed by [2, 3]. An incipient melting temperature of 1645°C was found for $ZrFe_2$ [8].

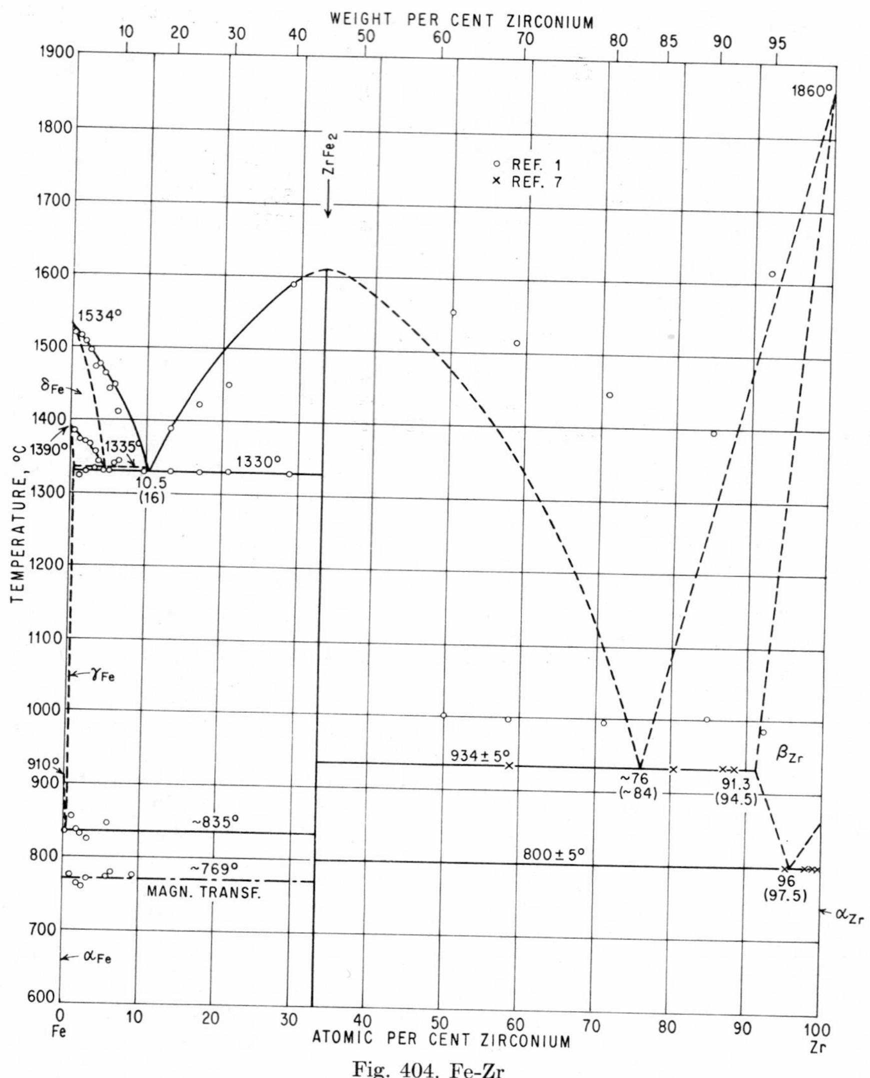

Fig. 404. Fe-Zr

1. R. Vogel and W. Tonn, *Arch. Eisenhüttenw.*, **5**, 1931-1932, 387–389.
2. H. J. Wallbaum, *Z. Krist.*, **103**, 1941, 391–402.
3. H. J. Wallbaum, *Arch. Eisenhüttenw.*, **14**, 1940-1941, 521–526.
4. C. B. Jordan and P. Duwez, California Institute of Technology, Progress Rept. 20-196, June 16, 1953.
5. T. E. Allibone and C. Sykes, *J. Inst. Metals*, **39**, 1928, 182–185; see also C. Sykes, *J. Inst. Metals*, **41**, 1929, 179–181.
6. E. S. Davenport and W. P. Kiernan (*J. Inst. Metals*, **39**, 1928, 189) reported an alloy with 5.6 wt. % Zr to consist of Fe and eutectic.

7. E. T. Hayes, A. H. Roberson, and W. L. O'Brien, *Trans. ASM*, **43,** 1951, 888–904.
8. R. P. Elliott, Armour Research Foundation, Chicago, Ill., Technical Report 1, OSR Technical Note OSR-TN-247, August, 1954, p. 37.

$\bar{1}.9824$
0.0176

Ga-Ge Gallium-Germanium

The phase diagram was established by [1] by means of thermal analysis. Figure 405 also shows the results of supplementary work on Ga-rich [2] and Ge-rich [3] alloys. The eutectic practically coincides with the melting point of pure Ga [4].

According to roentgenographic work [3] on quenched alloys, Ge dissolves 2 wt. (2.1 at.) % Ga at 600°C and 2.5 wt. (2.6 at.) % at 780°C. This indicates a retrograde solidus curve.

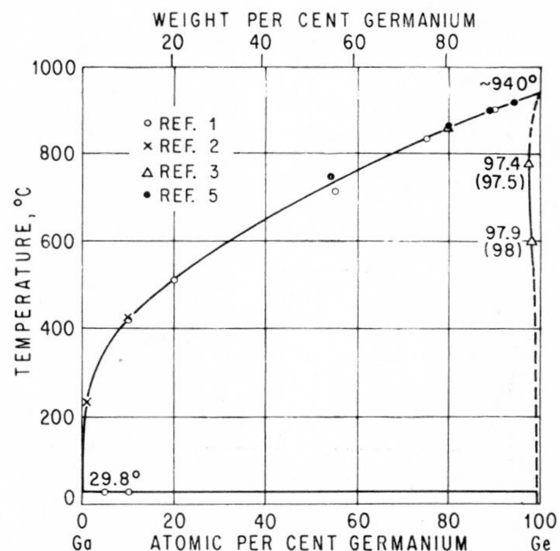

Fig. 405. Ga-Ge. (See also Supplement.)

Supplement. [5] measured (*a*) the melting point of Ge (937.2 ± 0.5°C), (*b*) the liquidus points of four Ga-Ge alloys (Fig. 405), and (*c*) the lattice constants of a series of Ga-Ge alloys containing up to 1.3 at. % Ga quenched from temperatures between 600 and 900°C. The solubility of Ga in solid Ge at these temperatures is suggested to be approximately 1 at. % (cf. above) [6].

In extremely dilute solutions of Ga in Ge, [7] determined the distribution coefficient C_s/C_l, C_s and C_l, respectively, being the atomic concentrations of Ga in the solid and the liquid phase (see also [5]). [8] determined the liquidus and eutectic temperatures of 32 Ga-Ge alloys. [8] also found, in agreement with [5], that Ga slightly expands the lattice of Ge.

1. W. Klemm and E. Hohmann, *Z. anorg. Chem.*, **256,** 1948, 244. The alloys were prepared and examined in an evacuated quartz vessel. The melting point of the carefully purified germanium was found to be 940–941°C.
2. P. H. Keck and J. Broder (*Phys. Rev.*, **90,** 1953, 521–522) examined five melts previously saturated at temperatures up to 400°C. The liquidus temperatures for the 1 and 10 at. % Ge alloys plotted in Fig. 405 were taken from a graphical representation of the measurement data.
3. E. S. Greiner, *J. Metals*, **4,** 1952, 1044 (short note).

4. See C. D. Thurmond (*J. Phys. Chem.*, **57,** 1953, 827–830) for an estimation of the eutectic point based on a regular solution equation for the liquidus.
5. E. S. Greiner and P. Breidt, *Trans. AIME*, **203,** 1955, 187–188.
6. It appears to the reviewer that the lattice parameters plotted can also be interpreted as showing the solubility to be > 1.3 at. %, as was previously [3] suggested.
7. J. A. Burton et al., *J. Chem. Phys.*, **21,** 1953, 1991.
8. N. de Roche, *Z. Metallkunde*, **48,** 1957, 59–60.

$\bar{1}.5410$
0.4590

Ga-Hg Gallium-Mercury

[1] dissolved Ga to the extent of about 0.8 wt. (2.3 at.) % in boiling Hg, and [2], in a magnetic study, prepared an amalgam containing 0.24 wt. (0.7 at.) % Ga at

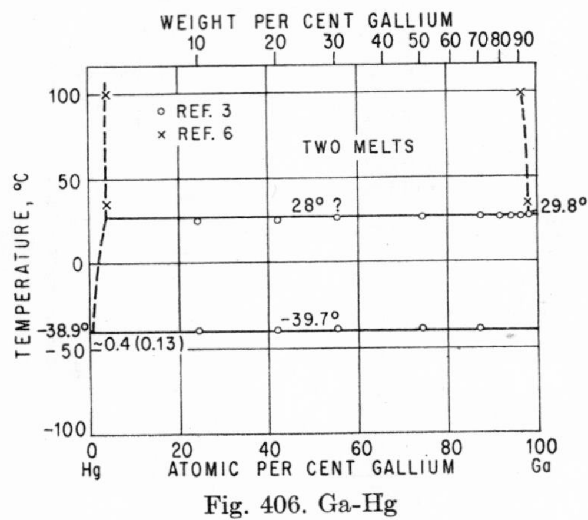

Fig. 406. Ga-Hg

19°C. Neither author, however, attempted to determine the solubility. [3] made a thermal analysis (Fig. 406) and concluded that the metals are at their freezing points either insoluble or very slightly soluble in one another [4]. [5] observed a depression of 0.8 ± 0.2°C of the freezing point of Hg and concluded a solubility of about 0.13 wt. (0.4 at.) % Ga.

More recently, solubility values for 35 and 100°C were reported by [6], based on a direct method [7]–Hg-rich layer: 1.3 and 1.4 wt. (3.6 and 3.9 at.) % Ga; Ga-rich layer: 93.5 and 91.4 wt. (98.0 and 97.8 at.) % Ga (at 35 and 100°C, respectively).

1. W. Ramsay, *J. Chem. Soc.*, **55,** 1889, 521.
2. W. G. Davies and E. S. Keeping, *Phil. Mag.*, **7,** 1929, 145–153.
3. N. A. Puschin, S. Stepanovíc, and V. Stajíc, *Z. anorg. Chem.*, **209,** 1932, 329–334.
4. The slight depression of the freezing point of Ga (the upper thermal arrests were observed at 25–28°C) was considered due more likely to supercooling (which readily occurs) than to Hg in solution. The lower thermal arrests were observed at −39°C.
5. E. S. Gilfillan and H. E. Bent, *J. Am. Chem. Soc.*, **56,** 1934, 1661–1663.
6. W. M. Spicer and H. W. Bortholomay, *J. Am. Chem. Soc.*, **73,** 1951, 868–869.
7. The liquids were mixed and allowed to equilibrate (up to 2 months). Weighed samples were then removed from each layer, and the Ga was dissolved out. The remaining Hg was weighed.

$\bar{1}.7836$
0.2164

Ga-In Gallium-Indium

The general character of the constitution of Ga-In alloys was concluded [1] from the somewhat qualitative results of an investigation by [2] (temperatures of incipient melting and of completion of melting) on four alloy compositions. More recently this system was studied by means of thermal [3, 4], microscopic [4], and electrical-resistance–temperature [5] methods. The concave-upward In-rich branch of the liquidus reported by [3] (dashed line in Fig. 407 [6]) could not be corroborated by

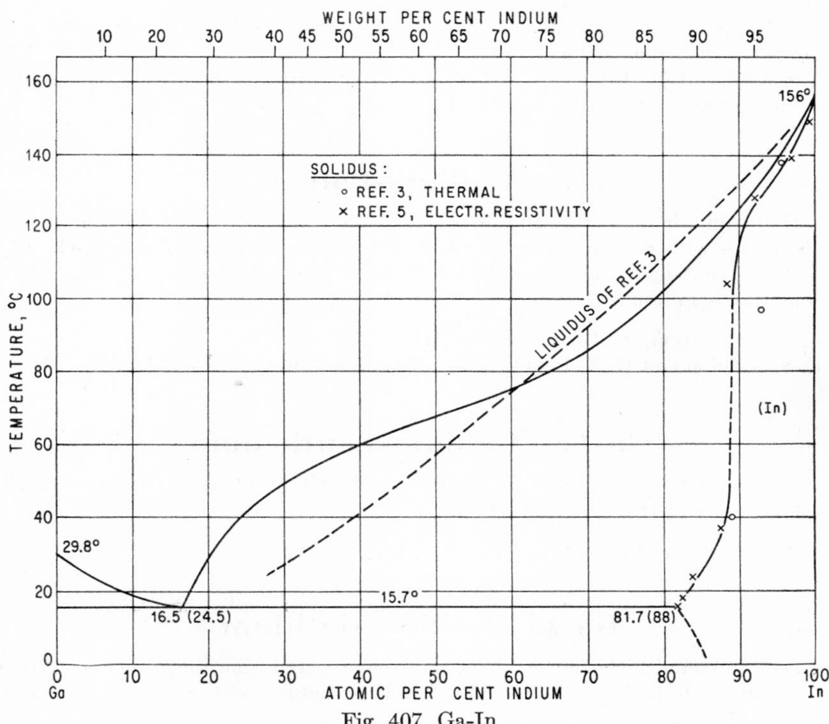

Fig. 407. Ga-In

[4, 5]; these authors found an inflection in the liquidus at about 70 wt. % In. Their liquidus curves agree fairly well [7] with each other with exception of the range between the eutectic and the inflection composition where the curve of [4] runs up to about 10°C below that of [5]. The data for the unbroken-line liquidus in Fig. 407 were taken mainly from the paper of [5]; only in the range mentioned above was an interpolation made between the results of [4, 5].

As to the eutectic, the data 24 wt. (16.1 at.) % In [3, 4], 24.8 wt. (16.7) at. % In [5], and 16°C [3], 15.7°C [4, 5] were reported. For Fig. 407, 24.5 wt. (16.5 at.) % In, 15.7°C, was chosen. The solid solubility of In in gallium is less than 0.5 wt. (0.3 at.) % [5].

The solidus of the (In) phase was investigated by [3, 5] (Fig. 407). According to [5], who annealed their alloys for at least 15 hr, the maximum solubility at the eutectic temperature is about 12 wt. (18.3 at.) % Ga.

1. M. Hansen, "Der Aufbau der Zweistofflegierungen," Springer-Verlag OHG, Berlin, 1936.
2. Lecoq de Boisbaudran, *Compt. rend.*, **100,** 1885, 701–703.
3. S. J. French, D. J. Saunders, and G. W. Ingle, *J. Phys. Chem.*, **42,** 1938, 265–274. Thermal-differential measurements. In general (with the exception of Ga-rich alloys), both heating and cooling curves were taken.
4. J. P. Denny, J. H. Hamilton, and J. R. Lewis, *Trans. AIME*, **194,** 1952, 39–42. (See also a discussion remark by J. P. Denny, *Trans. AIME*, **194,** 1952, 1184.)
5. W. J. Svirbely and S. M. Selis, *J. Phys. Chem.*, **58,** 1954, 33–35. The results of [3] are inaccurately replotted in fig. 3 on p. 35.
6. There is no convincing evidence given in [3] for their placing of the liquidus in the range between about 30 and 75 wt. % In.
7. The results of [3, 4] were published only in the form of small-scale plots, which makes quantitative comparisons rather difficult.

0.2512
$\bar{1}$.7488

Ga-K Gallium-Potassium

"Potassium in ammonia solution was found to react slowly with gallium at room temperatures with the formation of potassium amide and a gray-black K-Ga alloy of high Ga content" [1]. On the basis of emf-measurement data, [2] assumed the solubility of K in liquid Ga to be about 4×10^{-6} wt. (7×10^{-6} at.) % at 32°C.

1. F. W. Bergstrom, *J. Am. Chem. Soc.*, **47,** 1925, 1839.
2. E. S. Gilfillan and H. E. Bent, *J. Am. Chem. Soc.*, **56,** 1934, 1661–1663.

$\bar{1}$.7006
0.2994

Ga-La Gallium-Lanthanum

$LaGa_2$ (50.09 wt. % Ga) has the hexagonal AlB_2 (C32) type of structure, with $a = 4.33$ A, $c = 4.41$ A, $c/a = 1.02$ [1].

1. F. Laves, *Naturwissenschaften*, **31,** 1943, 145.

1.0020
$\bar{2}$.9980

Ga-Li Gallium-Lithium

The compound LiGa (90.95 wt. % Ga) was prepared from the elements at 700°C and was shown to have the NaTl (B32) type of structure, with $a = 6.207 \pm 7$ A [1].

1. E. Zintl and G. Brauer, *Z. physik. Chem.*, **B20,** 1933, 245–271.

0.4574
$\bar{1}$.5426

Ga-Mg Gallium-Magnesium

Liquidus. The results of thermal analyses by [1, 2] can be seen from Fig. 408. The work by [2] from which the data for the Mg-rich eutectic were taken (the less exact data of [1] are 20 at. % Ga, 424°C) is restricted to Mg-rich alloys. In both papers, supercooling effects are mentioned. Whereas the highest solidus temperature for Ga-rich alloys reported by [1] is 28°C (supercooling?), [3] found by thermal analysis of a 2 wt. % Mg alloy that Mg does not lower the melting point of Ga.

Solid Alloys. Age hardening of a 4.6 wt. % Ga alloy was observed by [4]; this indicated the existence of a (Mg) primary solid solution. Its extension between 640 and 192°C was determined by micrographic work [2]. The solidus falls linearly (composition in at. %) to the eutectic temperature where the maximum solubility amounts to 3.14 at. (8.50 wt.) % Ga; the solubilities at 400, 300, and 200°C are 2.6, 1.0, and 0.3 at. (7.1, 2.8, and 0.8 wt.) % Ga, respectively (taken from a graph).

The existence of four intermediate phases was first recognized roentgenographically by [5] and later corroborated by thermal [1] and X-ray [6] work. These phases are Mg_5Ga_2 [28.57 at. (53.41 wt.) % Ga], Mg_2Ga (58.91 wt. % Ga), MgGa (74.14 wt. % Ga), and $MgGa_2$ (85.15 wt. % Ga). Only [1] identified the compound richest in Ga. For Mg_2Ga a range of homogeneity extending from about 31 to 37 at. (56–63 wt.) % Ga was reported [5].

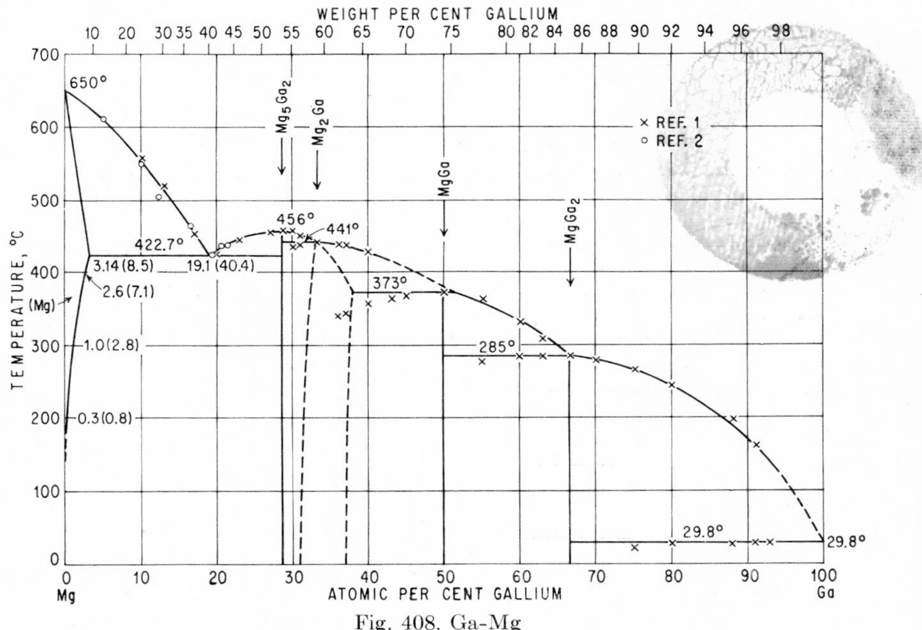

Fig. 408. Ga-Mg

Crystal Structures. Lattice parameters of the (Mg) phase (up to 2.50 at. % Ga) were measured by [7]. Mg_5Ga_2 has a body-centered orthorhombic translation group [6, 8], with a = 13.72 A, b = 7.02 A, c = 6.02 A [8], and 28 atoms per unit cell. [8] published an approximate description of the structure but gave no experimental details.

Mg_2Ga has a hexagonal translation group [5, 6], with a = 7.861 A, c = 6.958 A, c/a = 0.885 [5], and 18 atoms per unit cell. Its structure is only partially known [5], but seems to be closely related to the Fe_2P (C22) type [9].

MgGa and $MgGa_2$ yielded complex powder patterns [5, 6] which could not be analyzed.

1. N. A. Puschin and O. D. Micić, *Z. anorg. Chem.*, **234,** 1937, 229–232. The alloys were prepared in supremax-glass tubes under paraffin or RbCl + LiCl.
2. W. Hume-Rothery and G. V. Raynor, *J. Inst. Metals*, **63,** 1938, 201–226. Materials used were Ga, 99.88; Mg, 99.95 (wt. %). The alloys were melted under a standard flux and analyzed for both metals.
3. R. M. Evans and R. I. Jaffee, *Trans. AIME*, **194,** 1952, 153–156.
4. W. Kroll, *Metallwirtschaft*, **11,** 1932, 435–437.

5. K. Weckerle, Dissertation, Freiburg i. Br., Germany, 1935. This paper was not available. Information was taken from abstracts in "Gmelins Handbuch der anorganischen Chemie," System No. 36, pp. 59–60, Verlag Chemie, G.m.b.H, Berlin, 1936; System No. 27 (A), pp. 570–573, 1952; and *Z. Metallkunde*, **41,** 1950, 191–192.
6. W. Haucke, *Naturwissenschaften*, **26,** 1938, 577–578.
7. G. V. Raynor, *Proc. Roy. Soc. (London)*, **A180,** 1942, 107–121.
8. E. Hellner, *Fortschr. Mineral.*, **29–30,** 1950-1951, 58.
9. E. Hellner, *Fortschr. Mineral.*, **27,** 1948, 32–33.

0.1036
$\bar{1}$.8964

Ga-Mn Gallium-Manganese*

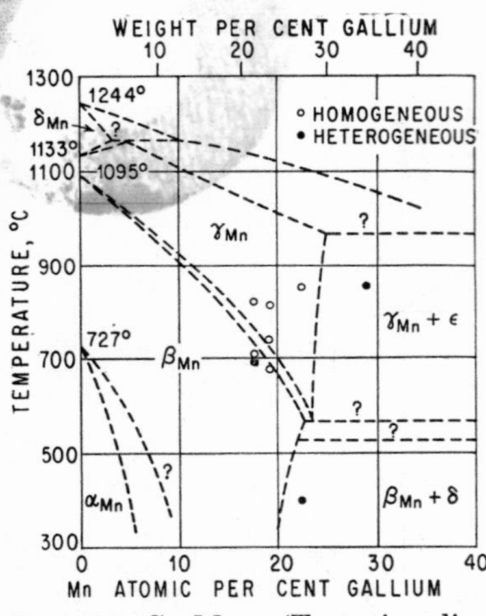

Fig. 409. Ga-Mn. (Tentative diagram.)

The Mn-rich region of the system Ga-Mn was studied by [1] in a cursory manner. On the basis of X-ray work at high temperatures (up to 820°C) and on alloys quenched from 400–1000°C, [1] concluded that the $\beta \rightleftarrows \gamma$ transition in Mn is greatly depressed by the addition of Ga. The occurrence of a tetragonal δ phase (composition not given, $a = 3.88$ A, $c/a = 0.906$), besides β-Mn in a 22.5 at. % Ga alloy quenched from 400°C, indicates a decomposition of the γ phase somewhere between 750 and 400°C. The γ phase is in equilibrium with a hexagonal ϵ phase (composition not given, $a = 2.65$ A, $c/a = 0.662$). In Fig. 409 a tentative diagram was drawn.

In quenched alloys the unstable tetragonal form of γ-Mn (see Cu-Mn) was observed up to 21.7 at. % Ga.

1. U. Zwicker, *Z. Metallkunde*, **42,** 1951, 248, 329.

$\bar{1}$.8613
0.1387

Ga-Mo Gallium-Molybdenum

"Mo reacts with Ga at . . . 600 and 800°C . . . to form more than one reaction product, one of which is a solid solution" [1].

1. Quoted from L. R. Kelman, W. D. Wilkinson, and F. L. Yaggee, "Resistance of Materials to Attack by Liquid Metals," ANL-4417, Argonne National Laboratory, July, 1950.

0.6970
$\bar{1}$.3030

Ga-N Gallium-Nitrogen

Ga does not react with molecular nitrogen at temperatures from 500 to 1000°C [1]. By the action of NH_3 gas on Ga at temperatures around 1000°C, the nitride GaN

* The assistance of W. Thiele and Dr. U. Zwicker in preparing this system is gratefully acknowledged.

(16.73 wt. % N) is formed [1–3], which sublimes without decomposition above 800°C [1].

GaN has the wurtzite (B4) type of structure [2, 3], with a = 3.19 A, c = 5.18 A, c/a = 1.625 [3].

1. W. C. Johnson, J. B. Parsons, and M. C. Crew, *J. Phys. Chem.*, **36,** 1932, 2651–2654.
2. J. V. Lirman and H. S. Zhdanov (also Shdanov), *Acta Physicochim. U.R.S.S.* **6,** 1937, 306; abstract in *Chem. Abstr.*, **31,** 1937, 8296; *Chem. Zentr.*, **1937**(1), 4739.
3. R. Juza and H. Hahn, *Z. anorg. Chem.*, **239,** 1938, 282–287.

0.4817
$\bar{1}$.5183

Ga-Na Gallium-Sodium

Until recently, no systematic work on the constitution had been published. There are observations that the metals alloy easily on warming to 100 or 200°C [1, 2]

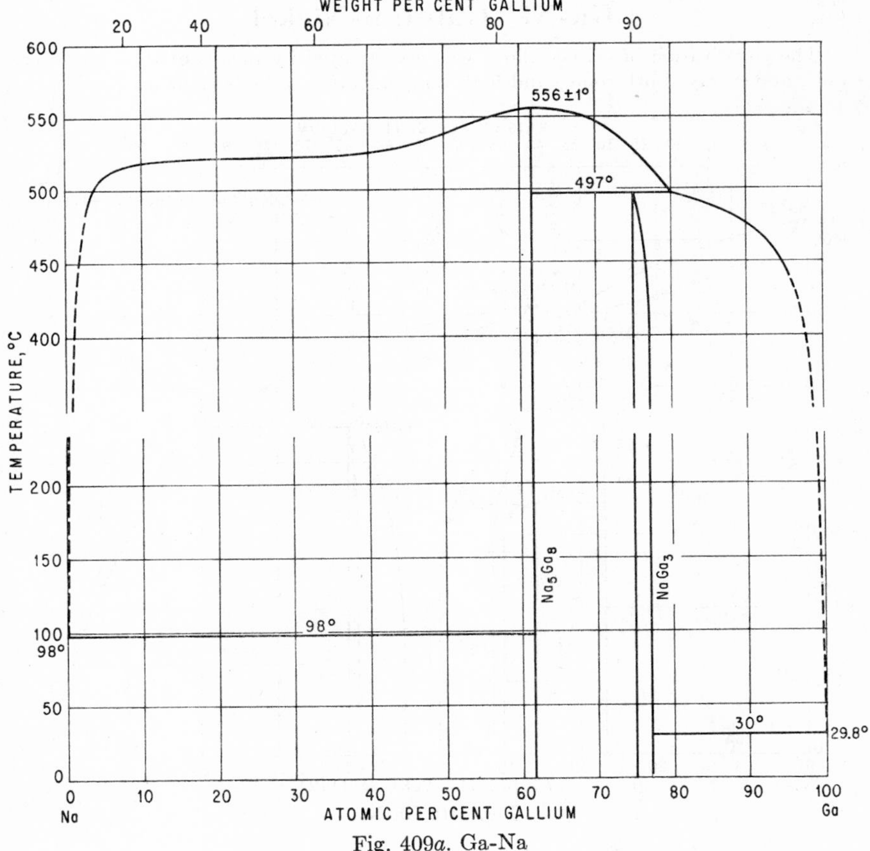

Fig. 409*a*. Ga-Na

and that a high-melting compound [3, 2] forms under evolution of an appreciable heat of reaction [2]. On the basis of emf-measurement data, [2] assumed the solubility of Na in liquid Ga to be about 0.001 wt. (0.003 at.) % at 30°C and suggested that perhaps a second compound may exist.

Note Added in Proof. The phase diagram of Fig. 409*a* has been established by [4] using thermal, microscopic, and chemical analysis. There are two intermediate phases: Na_5Ga_8 (61.54 at. % Ga), melting at 556 ± 1°C, and $NaGa_3$, homogeneous between 75 and 77 at. % Ga, formed peritectically at 497°C. The mutual solid solubilities of the elements are very restricted. The solidus temperatures of Na- and Ga-rich alloys practically coincide with the melting points of Na (98°C) and Ga (29.80°C), respectively.

1. E. Zintl and H. Kaiser, *Z. anorg. Chem.*, **211,** 1933, 121.
2. E. S. Gilfillan and H. E. Bent, *J. Am. Chem. Soc.*, **56,** 1934, 1661–1663.
3. N. A. Puschin, S. Stepanović, and V. Stajić, *Z. anorg. Chem.*, **209,** 1932, 333.
4. P. Feschotte and E. Rinck, *Compt. rend.*, **243,** 1956, 1525–1528.

0.0748
$\bar{1}$.9252

Ga-Ni Gallium-Nickel

The constitution of Ni-Ga alloys was investigated by means of thermal, microscopic, and X-ray (both room- and high-temperature) methods [1] (Fig. 410).

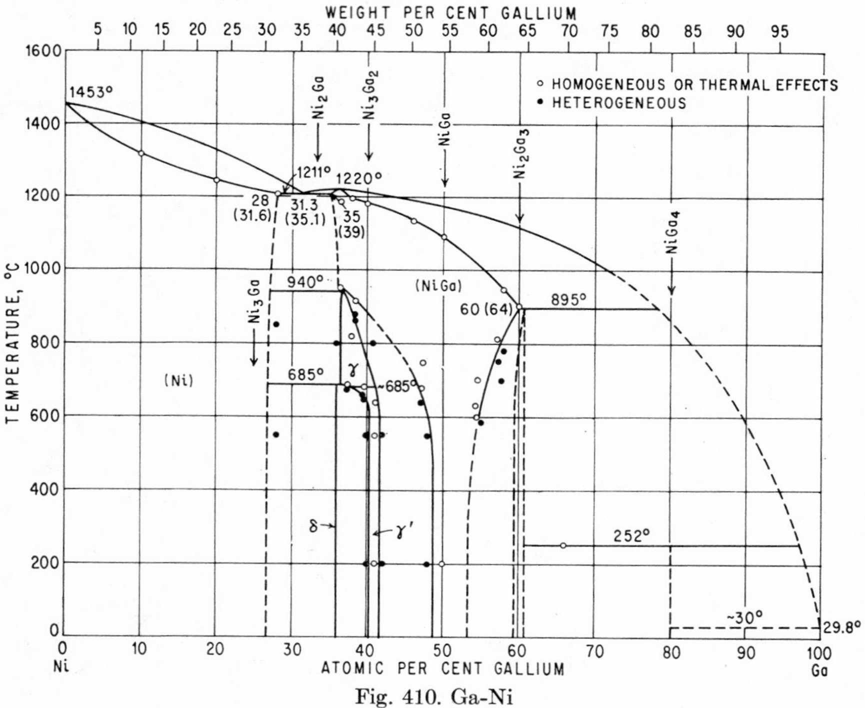

Fig. 410. Ga-Ni

The (Ni) primary solid solution extends up to 28 at. (31.6 wt.) % Ga; microscopic work (annealing treatment?) indicates that probably only a small decrease in solubility occurs with falling temperature.

Six intermediate phases were identified. (NiGa) (54.3 wt. % Ga) extends its range of homogeneity greatly at higher temperatures. The γ and δ ($\sim Ni_2Ga$) phases form at 940 and 685°C, respectively, in the solid state; the range of the former includes the composition Ni_3Ga_2 (44.2 wt. % Ga), but the latter is homogeneous only at 36 at. (40.1 wt.) % Ga. At about 685°C, γ transforms into γ' (see Crystal Structures), which phase is homogeneous somewhere between 40 and 42 at. (44.2 and 46.2 wt.) % Ga below about 650°C. Ni_2Ga_3 (64.05 wt. % Ga), which has a narrow range of homogeneity, and another phase with a composition of about $NiGa_4$ (82.6 wt. % Ga) are formed peritectically.

Crystal Structures. (NiGa) has the CsCl (B2) type of structure, with a = 2.88 A [2] at 50 at. %. The γ high-temperature phase has a "filled" NiAs (B8) type of structure, with a = 4.00 A, c = 4.98 A, c/a = 1.25 (at 36 at. % Ga). The γ' (Ni_3Ga_2) phase shows additional lines besides those of the B8 type, and the structure is probably based on a superlattice. The δ ($\sim Ni_2Ga$) phase has presumably the tetragonal In (A6) type, with a = 3.76 A, c = 3.39 A, c/a = 0.903 (for the face-centered cell). Ni_2Ga_3 crystallizes in the hexagonal Ni_2Al_3 ($D5_{13}$) type, with a = 4.06 A, c = 4.90 A, c/a = 1.21. $\sim NiGa_4$ shows a pattern similar to that of γ-brass, with a = 8.42 A. A density determination indicates a "γ-brass type with voids" (only 36–38 atoms per unit cell). The lattice parameter of Ni increases by alloying with Ga.

Supplement. [3] reported the finding of a Ni_3Ga phase with the Cu_3Au ($L1_2$) superlattice structure. According to [3], "the ordered phase has a relatively small homogeneous range, while on the Ni-rich side a wide heterogeneous range is found at low temperatures, producing a considerable difference of lattice parameter of the two coexisting cubic phases."

1. The phase diagram was first published in E. Hellner and F. Laves, *Z. Naturforsch.*, **2a,** 1947, 177–183 ("Kristallchemie des In und Ga in Legierungen mit einigen Übergangselementen"). Experimental details were given in E. Hellner, *Z. Metallkunde*, **41,** 1950, 480–484. Most of the data for Fig. 410 had to be taken from a graph.
2. The spacings given in [1] as "Å" are certainly kX and were therefore converted into "true A."
3. W. B. Pearson, *Nature*, **173,** 1954, 364.

0.3524
$\bar{1}$.6476

Ga-P Gallium-Phosphorus

GaP (30.76 wt. % P) has the cubic ZnS (B3) type of structure, with a = 5.447 ± 6 A [1]. According to [2], the solubility of Ga in yellow phosphorus at 45°C is between 0.01 and 0.1 μg Ga per ml P_4.

1. V. M. Goldschmidt, *Skrifter Akad. Oslo*, no. 8, 1926, pp. 33, 34, 110.
2. R. G. Armstrong and G. J. Rotariu, *J. Am. Chem. Soc.*, **76,** 1954, 5350–5351.

$\bar{1}$.5270
0.4730

Ga-Pb Gallium-Lead

The phase diagram of Fig. 411 was established by thermal analysis [1]. There is an extended miscibility gap in the liquid state with a monotectic point at about 95 wt. (86.5 at.) % Pb, 317°C, but the extension of the gap on the Ga side is not yet accurately known. Pb has no effect on the melting point of Ga [1] (see Supplement). [2] determined the solidification temperatures of alloys up to 24 wt. % Pb and found values between 24 and 30°C; this is certainly due to supercooling effects which readily occur with Ga.

Ga is practically insoluble in solid lead since it was found [2] that even minute Ga contents make lead brittle (due to free Ga).

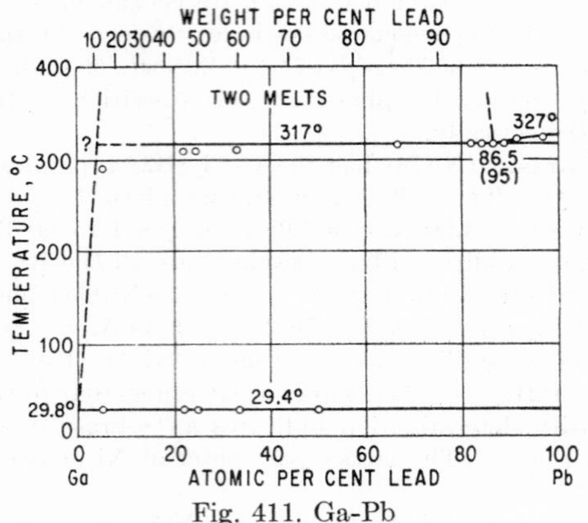

Fig. 411. Ga-Pb

Supplement. Based on a strong inflection below the liquidus point (29.77°C) in the heating curve of commercial Ga containing 0.05 wt. % Pb, [3] concluded the existence of a (Ga + Pb) eutectic at 29.4°C [4].

1. N. A. Puschin, S. Stepanović, and V. Stajić, *Z. anorg. Chem.*, **209,** 1932, 329–334.
2. W. Kroll, *Metallwirtschaft,* **11,** 1932, 435–437.
3. J. H. Hamilton, J. R. Lewis, and J. P. Denny, *U.S. Atomic Energy Comm., Publ.* NP-1092, 1950.
4. [1], in the thermal analysis of the system, observed cooling-curve arrests at 29.2 and 29.5°C which were explained on the basis of supercooling.

$\bar{1}.8152$
0.1848

Ga-Pd Gallium-Palladium

In a cursory report by [1], four intermediate phases are listed: ~Pd_3Ga (17.89 wt. % Ga), ~Pd_2Ga (24.63 wt. % Ga), PdGa (39.52 wt. % Ga), and ~Pd_3Ga_7 (60.39 wt. % Ga). PdGa was shown to crystallize in the cubic FeSi (B20) type with a = 4.89 A [1], and there are indications that Pd_3Ga_7 is isotypic with the cubic [CaF_2 (C1) related] structure of Ir_3Sn_7 [1, 2]. According to [3], Pd_2Ga is of the orthorhombic Co_2Si (C37) type, with $a = 7.81_4$ A, $b = 5.49_3$ A, $c = 4.06_4$ A.

1. E. Hellner and F. Laves, *Z. Naturforsch.*, **2a,** 1947, 177–183, especially 179, 182.
2. K. Schubert and H. Pfisterer, *Z. Metallkunde,* **41,** 1950, 438–439.
3. K. Schubert et al., *Naturwissenschaften,* **44,** 1957, 229–230.

$\bar{1}.6944$
0.3056

Ga-Pr Gallium-Praseodymium

The constitution of this system was established by [1] by means of thermal, microscopic, and X-ray powder work (Fig. 412). There exist four intermediate phases, to which the formulas Pr_3Ga (14.16 wt. % Ga), Pr_3Ga_2 (24.80 wt. % Ga), PrGa (33.10 wt. % Ga), and $PrGa_2$ (49.73 wt. % Ga) were ascribed.

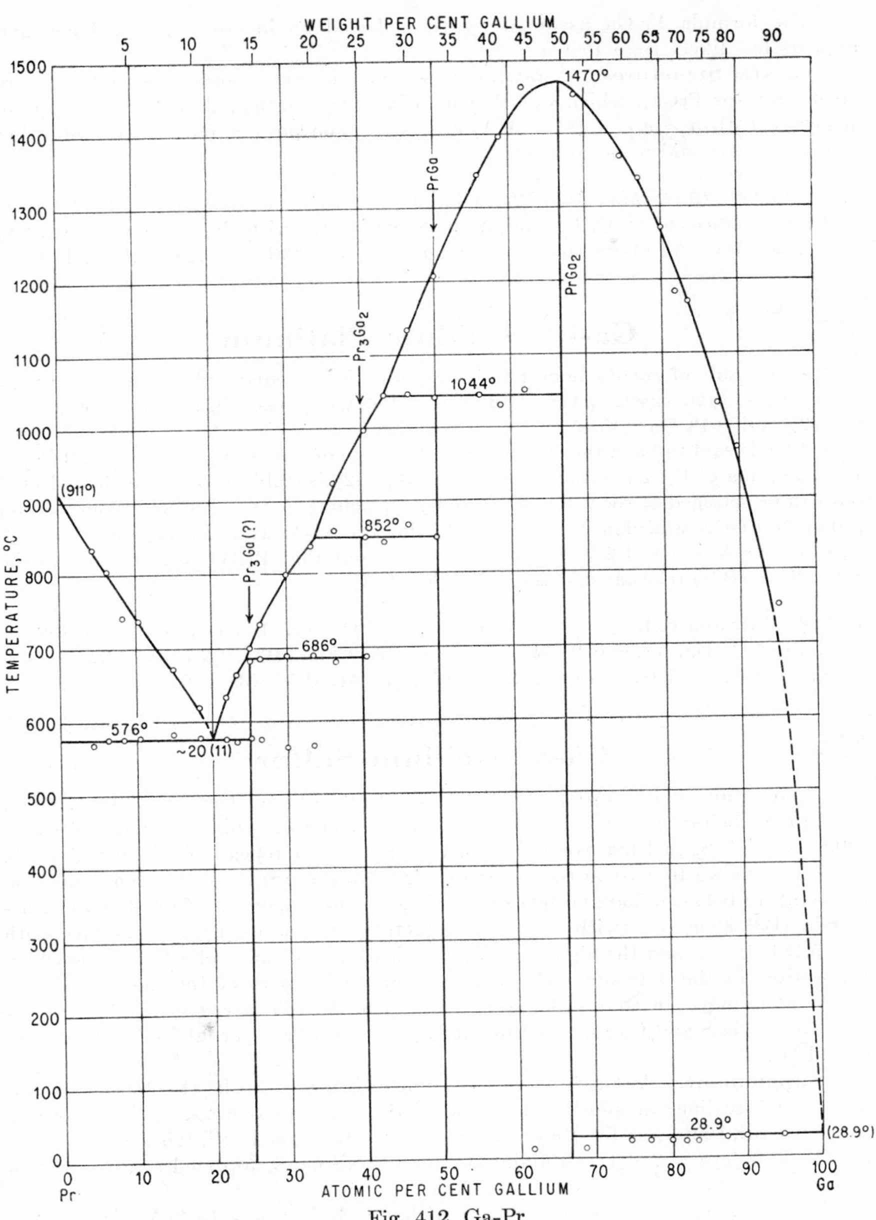
WEIGHT PER CENT GALLIUM
5
10
15
20
25
30
35
40
45
50
55
60
65
70
75
80
90
1470°
PrGa
PrGa2
Pr3Ga2
1044°
(911°)
852°
Pr3Ga (?)
686°
576°
~20 (11)
28.9°
(28.9°)
TEMPERATURE, °C
1500
1400
1300
1200
1100
1000
900
800
700
600
500
400
300
200
100
0
0
Pr
10
20
30
40
50
60
70
80
90
100
Ga
ATOMIC PER CENT GALLIUM

Fig. 412. Ga-Pr

The formula Pr_3Ga assumed by [1] is based on inconclusive evidence and appears less likely than Pr_2Ga.

Crystal Structures. X-ray analyses were run on all specimens; but data are given only for $PrGa_2$, which was shown to have the hexagonal AlB_2 (C32) type of structure with $a = c = 4.284 \pm 4$ A, $c/a = 1$. The powder patterns of the other compounds are said to be complex.

1. A. Iandelli, *Gazz. chim. ital.*, **79,** 1949, 70–79. Materials: electrolytic Pr, 99.5 wt. %, with traces of Si, C, Fe, and W; 99.85 wt. % Ga with chief impurities Zn, Cu, Pb, and Fe. All alloys (prepared under a NaCl + $BaCl_2$ flux) were analyzed.

$\bar{1}.5528$
0.4472

Ga-Pt Gallium-Platinum

As a result of roentgenographic work, four intermediate phases are known at present: PtGa (26.31 wt. % Ga) [1], Pt_2Ga_3 (34.88 wt. % Ga) [1], $PtGa_2$ (41.66 wt. % Ga) [2], and $\sim Pt_3Ga_7$ (45.45 wt. % Ga) [1].

PtGa is said to have the cubic FeSi (B20) type of structure, with $a = 4.91$ A [1]. $PtGa_2$, with a CaF_2 (C1) structure ($a = 5.923$ A) [2], is stable only above about 150°C but can be retained at room temperature by quenching. As one of its decomposition products, Pt_2Ga_3 was identified by [1] as having the Ni_2Al_3 ($D5_{13}$) type, with $a = 4.23$ A, $c = 5.18$ A, $c/a = 1.223$. There are indications that Pt_3Ga_7 is isotypic with the cubic [CaF_2 (C1) related] structure of Ir_3Sn_7 [1, 3].

1. E. Hellner and F. Laves, *Z. Naturforsch.*, **2a,** 1947, 177–183, especially 180, 182.
2. E. Zintl, A. Harder, and W. Haucke, *Z. physik. Chem.*, **B35,** 1937, 354–362.
3. K. Schubert and H. Pfisterer, *Z. Metallkunde*, **41,** 1950, 438–439.

0.3373
$\bar{1}.6627$

Ga-S Gallium-Sulfur

Preparation and properties of the sulfides of gallium—Ga_2S (18.70 wt. % S, decomposes before melting), GaS (31.50 wt. % S, melting point 965°C), and Ga_2S_3 (40.83 wt. % S, melting point ~1250°C)—have been repeatedly described [1–4]. Ga_2S_3 was shown to exist in two polymorphic modifications, with the transformation temperature between 550 and 600°C [5]. The room-temperature form has the zincblende (B3) structure (with part of the metal positions empty at random), with $a = 5.181 \pm 4$ A; and the high-temperature form, which can be effectively quenched, crystallizes in the wurtzite (B4) structure (also with part of the metal positions empty at random), with $a = 3.685 \pm 5$ A, $c = 6.028 \pm 6$ A, $c/a = 1.636$. Powder patterns of Ga_2S and GaS may be found in [2]; for GaS a layer lattice has been suggested [4].

Supplement. A Ga_2S_3 specimen annealed for 1 week at 1150°C showed superstructure lines in addition to those of the wurtzite structure [6]. According to a structural analysis [7], they correspond to a hexagonal cell, with $a = 6.37_0$ A, $c = 18.05$ A, $c/a = 2.833$, containing 6 molecules of Ga_2S_3 (ordered arrangement of the vacant metal lattice sites).

GaS has a hexagonal layer structure, with $a = 3.57_8$ A, $c = 15.4_7$ A, $c/a = 4.323$, and 8 atoms per unit cell [7].

1. A. Brukl and G. Ortner, *Naturwissenschaften*, **18,** 1930, 393.
2. A. Brukl and G. Ortner, *Monatsh. Chem.*, **56,** 1930, 358–364.
3. W. C. Johnson and B. Warren, *Naturwissenschaften*, **18,** 1930, 666.

4. W. Klemm and H. U. v. Vogel, *Z. anorg. Chem.*, **219,** 1934, 45–64.
5. H. Hahn and W. Klingler, *Z. anorg. Chem.*, **259,** 1949, 135–142.
6. H. Hahn, *Angew. Chem.*, **64,** 1952, 203.
7. H. Hahn, *Angew. Chem.*, **65,** 1953, 538.

$\bar{1}.7579$
0.2421

Ga-Sb Gallium-Antimony

Sb does not lower the melting point of gallium [1]. GaSb (63.58 wt. % Sb) was prepared from the elements and was shown to have the cubic ZnS (B3) type of structure, with $a = 6.105 \pm 6$ A [2].

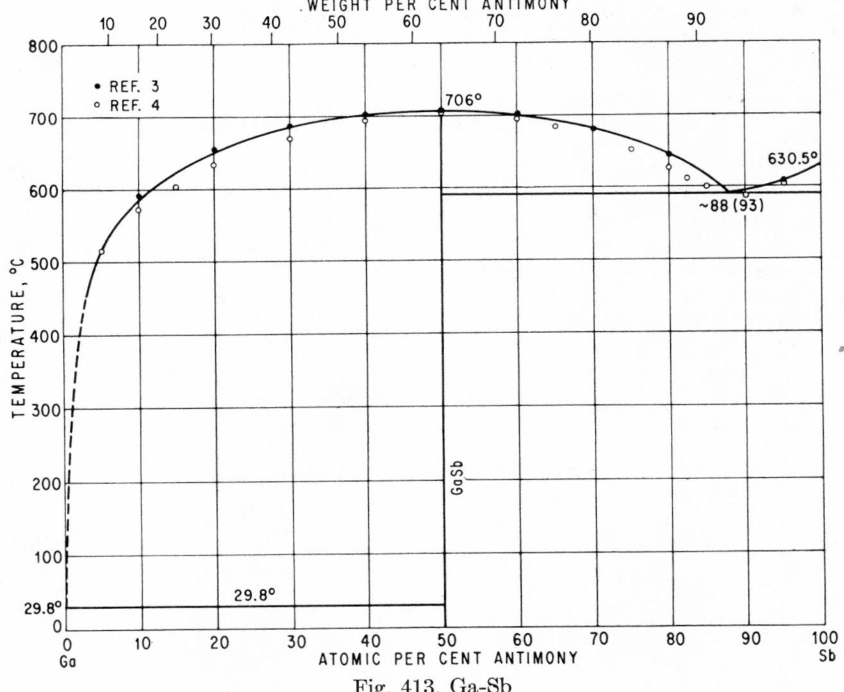

Fig. 413. Ga-Sb

Supplement. The constitution of the Ga-Sb system has been established by [3, 4] by means of thermal, microscopic, and X-ray analysis (Fig. 413). The results of the two investigations are in good agreement, as may be seen from the liquidus data [5], plotted in Fig. 413, and from Table 30.

Table 30

	Ref. 3	Ref. 4
Eutectic (Ga + GaSb)....	Close to pure Ga, 29.8°C	Close to pure Ga, 29.5°C
Melting point of GaSb....	705.9°C	703°C
Eutectic (GaSb + Sb)....	88.2 at. % Sb, 589.8 ± 0.5°C	87 at. % Sb, 583°C

The thermal data of [3] are considered to be the more accurate ones since [3] used specimens of greater weight and, as to the freezing of the Ga-rich eutectic, seeded the melt to prevent supercooling. Addition of Ga did not alter the lattice parameters of Sb [3].

1. R. M. Evans and R. I. Jaffee, *Trans. AIME*, **194**, 1952, 153–156. The solidus temperature of a 2 wt. % Sb alloy was the melting point of gallium.
2. V. M. Goldschmidt, *Skrifter Akad. Oslo*, **1926**, no. 8, 35, 110.
3. I. G. Greenfield and R. L. Smith, *Trans. AIME*, **203**, 1955, 351–353.
4. W. Köster and B. Thoma, *Z. Metallkunde*, **46**, 1955, 291–293.
5. In both papers the thermal data are given only in small graphs. The data of [3] plotted in Fig. 413 are interpolated values taken from the graph; those of [4] are actual measurement data available to the author in tabulated form.

$\bar{1}.9460$
0.0540

Ga-Se Gallium-Selenium

The preparation and some properties of Ga_2Se (36.15 wt. % Se, probably unstable), GaSe (53.11 wt. % Se, melting point 960 ± 10°C), and Ga_2Se_3 (62.94 wt. % Se, melting point >1020°C) were described by [1]. Ga_2Se_3 was shown to have the zincblende (B3) structure, with part of the metal positions empty at random and $a = 5.429 \pm 5$ A [2].

Supplement. [3] stated briefly that long-time annealing of Ga_2Se_3 resulted in the appearance of superstructure lines in addition to those of the zincblende structure (cf. Ga-S).

[4] reported GaSe to have a hexagonal layer structure, with $a = 3.75_2$ A, $c = 15.95$ A, $c/a = 4.251$, and 8 atoms per unit cell (space group D_{6h}^4), and to be isotypic with GaS. However, according to [5], two different phases of closely related crystal structures exist in the equiatomic region: a hexagonal one, with $a = 3.74$ A, $c = 15.92$ A, and 8 atoms per unit cell (space group C_{3h}^1; cf. above) between 48 and <51 at. % Se; and—separated by a narrow two-phase field—a rhombohedral one, $a = 3.75$ A, $c = 23.91$ A, with 12 atoms per unit cell, somewhere between 52 and 53.5 at. % Se.

There are indications [5] for a eutectic between GaSe and Ga_2Se_3 and for a homogeneity range of Ga_2Se_3 toward compositions richer in Se. No Ga_2Se phase was observed at alloying temperatures down to 400°C.

1. W. Klemm and H. U. v. Vogel, *Z. anorg. Chem.*, **219**, 1934, 45–64.
2. H. Hahn and W. Klingler, *Z. anorg. Chem.*, **259**, 1949, 135–142.
3. H. Hahn, *Angew. Chem.*, **64**, 1952, 203.
4. H. Hahn, *Angew. Chem.*, **65**, 1953, 538.
5. K. Schubert, E. Dörre, and M. Kluge, *Z. Metallkunde*, **46**, 1955, 216–224.

0.3948
$\bar{1}.6052$

Ga-Si Gallium-Silicon

The phase diagram was established by [1] by means of thermal analysis. Figure 414 shows also the results of supplementary work on Ga-rich alloys by [2]. The eutectic coincides practically with the melting point of pure Ga [3, 4]. Solid solubilities were not determined by [1]; [5] found, by measurements of electrical conductivity, that silicon single crystals grown from the melt contained approximately 10^{-3} at. % Ga.

1. W. Klemm and E. Hohmann, *Z. anorg. Chem.*, **256,** 1948, 239–252, especially 243–244.
2. P. H. Keck and J. Broder (*Phys. Rev.*, **90,** 1953, 521–522) determined the Si content (by weight difference or spectroscopically) of three melts saturated at different temperatures. The data for Fig. 414 were taken from a graph.
3. R. M. Evans and R. I. Jaffee (*Trans. AIME*, **194,** 1952, 153–156) observed—in agreement with [1]—that the melting point of Ga is not lowered by addition of Si.

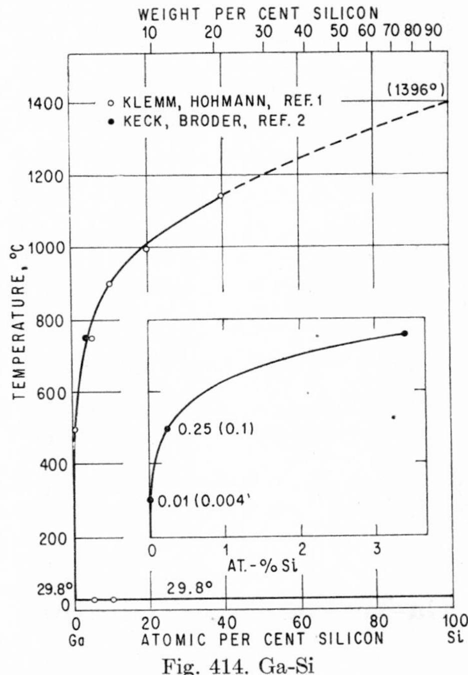

Fig. 414. Ga-Si

4. See C. D. Thurmond, *J. Phys. Chem.*, **57,** 1953, 827–830, for an estimation of the eutectic point based on a regular solution equation for the liquidus.
5. [2], p. 522.

$\bar{1}.7689$
0.2311

Ga-Sn Gallium-Tin

The phase diagram of Fig. 415 is based on thermal-analysis data by [1]. According to them the eutectic lies at about 8 wt. (5 at.) % Sn, 20°C. Some liquidus points reported by [2] which do not fit well (supercooling?) in the accepted diagram are also plotted in Fig. 415. Solid solubilities were not investigated by [1]. The results of magnetic measurements [3] on tin single crystals containing up to 0.89 wt. % Ga might indicate a certain solubility of Ga in solid Sn.

1. N. A. Puschin, S. Stepanović, and V. Stajić, *Z. anorg. Chem.*, **209,** 1932, 329–334.
2. W. Kroll, *Metallwirtschaft*, **11,** 1932, 435–437.
3. H. J. Hoge, *Phys. Rev.*, **48,** 1935, 615–619.

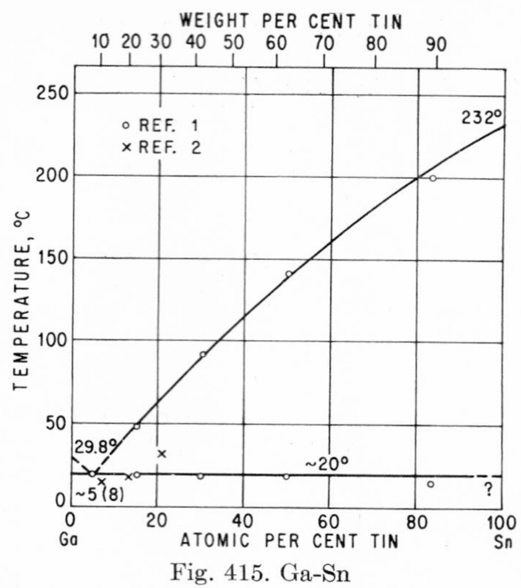

Fig. 415. Ga-Sn

$\bar{1}.5860$
0.4140

Ga-Ta Gallium-Tantalum

"Static tests indicate that the apparent solubility of Ta in Ga is 0.01 wt. (0.004 at.) % at 450°C, 0.58 wt. (0.22 at.) % at 600°C, and 1.3 wt. (0.5 at.) % at 800°C" [1].

1. Quoted from L. R. Kelman, W. D. Wilkinson, and F. L. Yaggee, Resistance of Materials to Attack by Liquid Metals, Argonne National Laboratory, ANL-4417, July, 1950.

$\bar{1}.7375$
0.2625

Ga-Te Gallium-Tellurium

The constitution of Ga-Te alloys was established by [1] by means of thermal and qualitative X-ray analysis (Fig. 416) [2]. The compounds GaTe (64.67 wt. % Te) and Ga_2Te_3 (73.30 wt. % Te) were identified. "Not quite clarified is the constitution of the Te-rich region. It seems possible that a polytelluride exists which has, however, a relatively low rate of formation" [1]. The horizontal at 470°C, therefore, was indicated by a dashed line by [1]. It can, however, be concluded with a high degree of certainty—because the longest thermal arrest was observed at 75 at. % Te—that at 470°C the compound $GaTe_3$ (84.59 wt. % Te) is formed peritectically.

Ga_2Te_3 was shown to have the zincblende (B3) structure (with part of the metal positions empty at random), with $a = 5.886 \pm 5$ A [3].

Supplement. [4] stated briefly that long-time annealing of Ga_2Te_3 resulted in the appearance of superstructure lines in addition to those of the zincblende structure (cf. Ga-S, Ga-Se). According to [5, 6], the structure of GaTe is monoclinic. However, the unit cells reported are not identical: $a = 12.7_6$ A, $b = 4.0_3$ A, $c = 14.99$ A, $\beta = 103.9°$ (24 atoms per unit cell) [5]; $a = 23.81$ A, $b = 4.076$ A, $c = 10.48$ A, $\beta = 45.4°$ [6].

1. W. Klemm and H. U. v. Vogel, *Z. anorg. Chem.*, **219**, 1934, 45–64.
2. The data for Fig. 416 had to be taken from a small-scale graph.

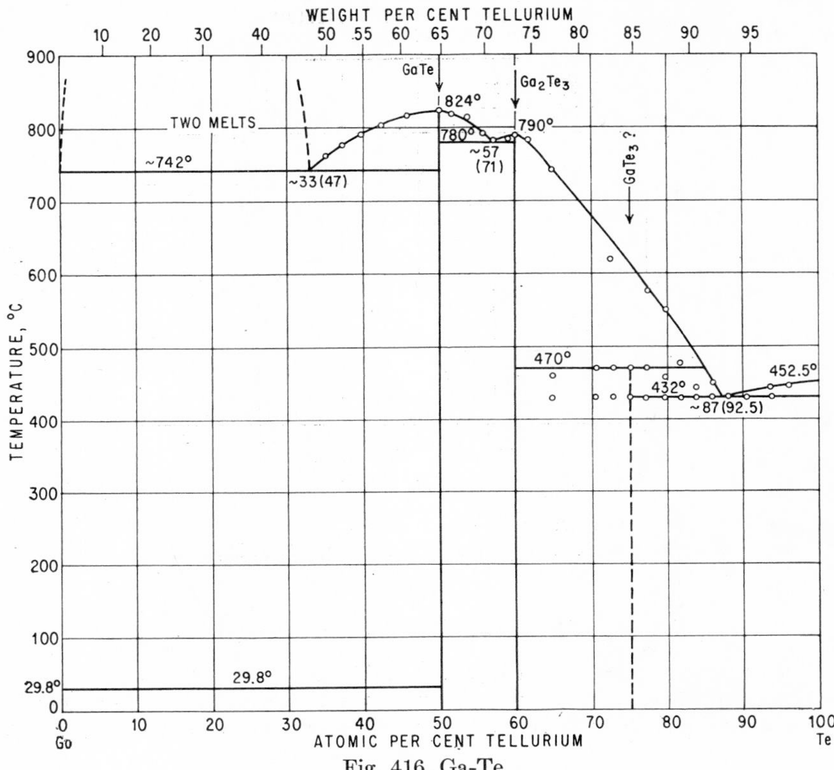

Fig. 416. Ga-Te

3. H. Hahn and W. Klingler, *Z. anorg. Chem.*, **259**, 1949, 135–142.
4. H. Hahn, *Angew. Chem.*, **64**, 1952, 203.
5. H. Hahn, *Angew. Chem.*, **65**, 1953, 538.
6. K. Schubert, E. Dörre, and E. Günzel, *Naturwissenschaften*, **41**, 1954, 448.

0.1630
$\bar{1}$.8370

Ga-Ti Gallium-Titanium

According to microscopic and roentgenographic investigations of the alloy $TiGa_3$ (81.36 wt. % Ga), there exists an intermediate phase of this composition which crystallizes in the tetragonal $TiAl_3$ ($D0_{22}$) type [1] of structure, with $a = 5.559 \pm 3$ A, $c = 8.109 \pm 5$ A, $c/a = 1.459$ [2]. A compound of the approximate composition Ti_2Ga (42.12 wt. % Ga) was shown to have the hexagonal Ni_2In (filled-up NiAs) type of structure, with $a = 4.51$ A, $c = 5.50$ A, $c/a = 1.22$ [3].

The partial phase diagram of Fig. 416*a* is the result of a recent micrographic and X-ray powder investigation of Ti-rich alloys [4]. As in the systems Ti-Al and Ti-In, an ordered phase (α_2) of the hexagonal Mg_3Cd ($D0_{19}$) type was found to occur around 75 at. % Ti, with $a = 5.76$ A, $c = 4.64$ A, $c/a = 0.805$ at 24 at. % Ga.

1. *Strukturbericht*, **7**, 1939, 13, 100–101.
2. H. J. Wallbaum, *Z. Metallkunde*, **34**, 1942, 118–119.

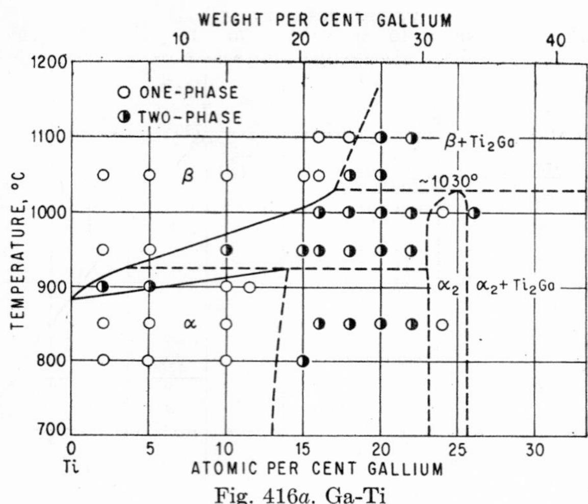

Fig. 416*a*. Ga-Ti

3. K. Anderko, *Naturwissenschaften*, **44**, 1957, 88.
4. K. Anderko and U. Zwicker, *Naturwissenschaften*, **44**, 1957, 510.

$\bar{1}$.5329
0.4671

Ga-Tl Gallium-Thallium

The thermal data plotted in Fig. 417 are due to a thermal investigation by [1], in which both cooling and heating curves were taken. There is an extensive miscibility gap in the liquid state; the monotectic point is said to lie at 94 at. (97.9 wt.) % Tl

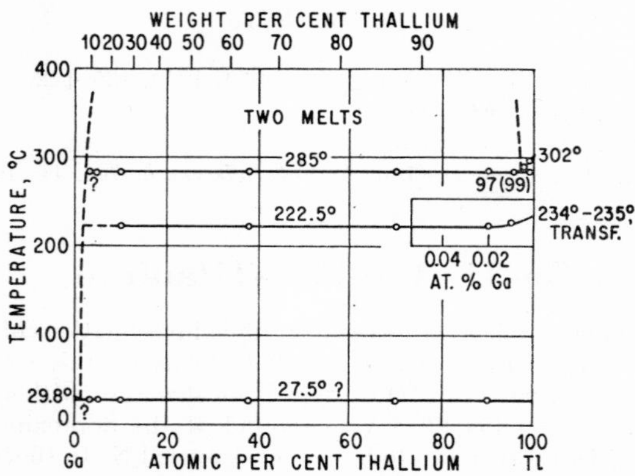

Fig. 417. Ga-Tl. (See also Note Added in Proof.)

[2], 285°C, but the (certainly small) solubility of Tl in liquid Ga at this temperature is not accurately known. A depression in the melting point of Ga by Tl was observed by [1] (eutectic <2.7 at. % Tl, 27.5°C; see Fig. 417) but not by [3], who made a thermal analysis of a 2 wt. (0.7 at.) % Tl alloy.

From the depression of the transformation temperature of Tl by Ga [4] (see Fig. 417, inset), a solid solubility of about 0.015 at. (0.005 wt.) % Ga in Tl was concluded by [1].

Note Added in Proof. [5] reexamined the system by means of thermal analysis. The following characteristic data were found (cf. Fig. 417): extension of the miscibility gap at the monotectic temperature (280°C), 5.5–99 wt. (2–97 at.) % Tl; eutectic point 27.3 ±0.5°C (on heating), 0.5 wt. (0.2 at.) % Tl; transformation point of Tl (in equilibrium with Ga-rich melt), 224°C.

1. W. Klemm and E. Orlamünder, *Z. anorg. Chem.*, **256,** 1948, 239–252, especially 241–242.
2. In the diagram given by [1], however, the monotectic point is shown at about 97 at. % Tl; this position is in better agreement with the plotted liquidus data.
3. R. M. Evans and R. I. Jaffee, *Trans. AIME*, **194,** 1952, 153–156.
4. Previous heat-treatment: 24 hr, 250°C, quenched.
5. H. Spengler, *Z. Metallkunde*, **46,** 1955, 464.

$\bar{1}.4666$
0.5334

Ga-U Gallium-Uranium

The (incomplete) phase diagram of Fig. 417*a* has been reproduced from a review by [1], to whom much hitherto unpublished information was available.

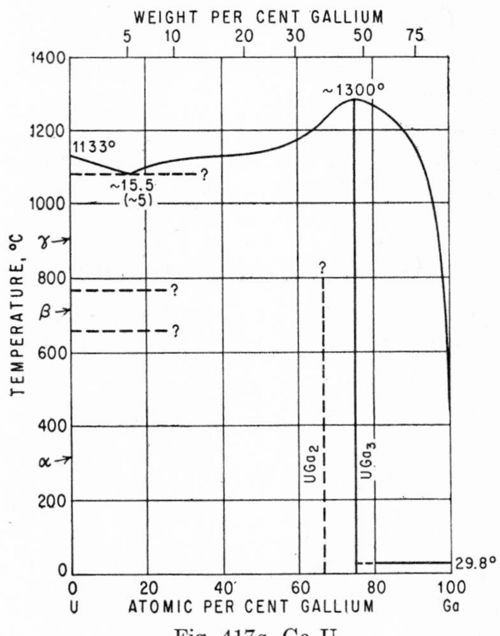

Fig. 417*a*. Ga-U

The liquidus has been determined by [2] using a rather crude experimental method; nevertheless, "the resulting data connect nicely with filtration data obtained by [3] and [4] for temperatures of 700 and 500°C at the Ga end of the system" [1]. [2] found the melting point of Ga virtually unaffected by small additions of U. On the other hand, it was reported by [5] that a eutectic occurs at 86 at. % Ga and about 28°C. This is in serious disagreement with the liquidus as shown.

Evidence of the existence of two compounds has been reported: UGa_3 [6, 7] and UGa_2 [8, 5]. "Judging from the liquidus and a reported melting point of about 1300°C for UGa_3 [5], it seems likely that UGa_3 melts congruently. The compound UGa_2, if present, may be stable to some lower temperature'' [1].

The solubility of Ga in solid U [4] and of U in solid Ga [2] is very restricted.

UGa_3 (46.76 wt. % Ga) is of the Cu_3Au ($L1_2$) type of structure [6, 7], a = 4.249 A [6], a = 4.2475 A [7].

1. H. A. Saller and F. A. Rough, Compilation of U.S. and U.K. Uranium and Thorium Constitutional Diagrams, *U.S. Atomic Energy Comm., Publ.* BMI-1000, June, 1955.
2. R. I. Jaffee, R. M. Evans, E. O. Fromm, and B. W. Gonser, unpublished information (January, 1950); quoted in [1].
3. E. E. Hayes and P. Gordon, unpublished information (July, 1948); quoted in [1].
4. W. D. Wilkinson, unpublished information (March, 1948); quoted in [1].
5. Atomic Energy Research Establishment, United Kingdom, unpublished information (May, 1951); quoted in [1].
6. A. Iandelli and R. Ferro, *Ann. chim. (Rome)*, **42,** 1952, 598–606.
7. B. R. Frost and J. T. Maskrey, *J. Inst. Metals*, **82,** 1953, 171–180.
8. A. J. Dempster, unpublished information (March, 1948); quoted in [1].

$\bar{1}.5787$
0.4213

Ga-W Gallium-Wolfram

"Spectrographic analysis of Ga that had been heated with W at 815°C indicates a solubility in Ga of 0.001 to 0.008 wt. %" [1].

1. R. I. Jaffee et al. (BMI-T-17, 1949), quoted from L. R. Kelman, W. D. Wilkinson, and F. L. Yaggee, Resistance of Materials to Attack by Liquid Metals, Argonne National Laboratory, ANL-4417, July, 1950.

0.0279
$\bar{1}.9721$

Ga-Zn Gallium-Zinc

The phase diagram of Fig. 418 is based on the thermal results of [1]. The eutectic lies at about 95 wt. (at.) % Ga, 25°C. Solid solubilities were not investigated, but

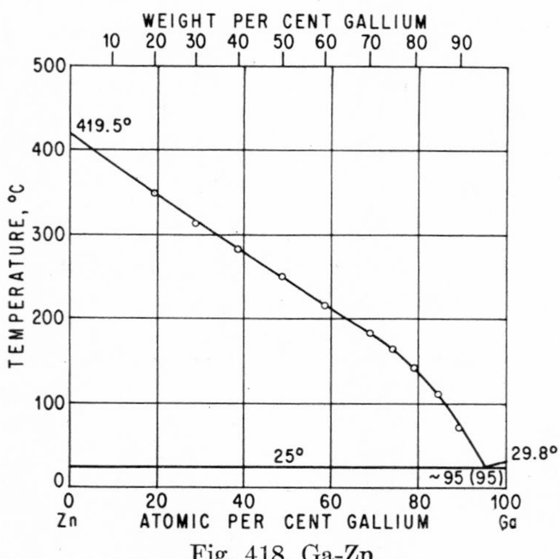

Fig. 418. Ga-Zn

that of Ga in Zn must be small since a 0.5 wt. % Ga alloy cannot be hot-rolled "because liquid Ga or a Ga-rich eutectic separates along the grain boundaries" [2]. [2] made also some thermal measurements mainly on alloys up to 28 wt. % Zn [3].

1. N. A. Puschin, S. Stepanović, and V. Stajić, *Z. anorg. Chem.*, **209,** 1932, 329–334.
2. W. Kroll, *Metallwirtschaft,* **11,** 1932, 435–437.
3. According to these measurements, the melting point of Ga is lowered to 21°C by 9 wt. % Zn, and with still increasing Zn content the "liquidus" temperature rises to 25°C at 28 wt. % Zn. A comparison of these results with Fig. 418 shows that [2] actually measured solidus temperatures of his higher-Zn alloys and also that some supercooling probably occurred.

$\bar{1}.8833$
0.1167

Ga-Zr Gallium-Zirconium

The compound $ZrGa_3$ (69.63 wt. % Ga) has the tetragonal $TiAl_3$ ($D0_{22}$) type [1] of structure, with $a = 5.616 \pm 4$ A, $c = 8.729 \pm 4$ A, $c/a = 1.554$ [2].

1. *Strukturbericht,* **7,** 1939, 13, 100–101.
2. H. J. Wallbaum, *Z. Metallkunde,* **34,** 1942, 118–119.

2.1922
$\bar{3}.8078$

Gd-H Gadolinium-Hydrogen

Both hydrogen and deuterium react with Gd at temperatures above 150°C. The products obtained are Gd_2H_3 and Gd_2D_3, respectively. If these compounds are cooled in the presence of hydrogen and deuterium, respectively, GdH_2 and GdD_2 are obtained. Although stable at room temperature, these compounds dissociate at temperatures above 250°C, yielding the original products Gd_2H_3 and Gd_2D_3 [1]. According to [2], GdH_3 is isostructural with PuH_3.

1. R. Viallard, *Ann. chim. (Paris),* **20,** 1945, 5–72; *Chem. Abstr.,* **40,** 1946, 4283.
2. Short statement by A. S. Coffinberry and M. B. Waldron in "Metallurgy and Fuels," p. 401, Progress in Nuclear Energy, ser. V, vol. 1, Pergamon Press Ltd., London, 1956.

0.8097
$\bar{1}.1903$

Gd-Mg Gadolinium-Magnesium

From investigations of the magnetic properties of Gd-Mg alloys, [1] concluded that the intermediate phases Mg_9Gd, Mg_3Gd, MgGd, and $MgGd_9$ exist.

1. F. Gaume-Mahn, *Compt. rend.,* **232,** 1951, 1815–1816; **235,** 1952, 352–354; **237,** 1953, 702–704; *Bull soc. chim. France,* **21,** 1954, 569–575.

0.4558
$\bar{1}.5442$

Gd-Mn Gadolinium-Manganese

There are at least four intermediate phases, three of which are ferromagnetic [1]. One phase was identified as $GdMn_2$ (58.8 wt. % Gd), which is isotypic with $MgCu_2$ (C15 type), $a = 7.75$ A [2].

1. W. Klemm, *FIAT Rev. Ger. Sci.,* 1939-1946, Inorganic Chemistry, Part III, pp. 255–256.
2. F. Endter and W. Klemm, *Z. anorg. Chem.,* **252,** 1944, 377–379.

1.0492
$\bar{2}$.9508

Gd-N Gadolinium-Nitrogen

GdN (8.2 wt. % N) is cubic of the NaCl (B1) type, a = 4.99 A [1].

1. F. Endter, *Z. anorg. Chem.*, **257,** 1948, 127–130.

0.4271
$\bar{1}$.5729

Gd-Ni Gadolinium-Nickel

$GdNi_5$ (34.8 wt. % Gd) is isotypic with $CaCu_5$, a = 4.91 A, c = 3.99 A, c/a = 0.813 [1].

1. F. Endter and W. Klemm, *Z. anorg. Chem.*, **252,** 1943, 64–66.

$\bar{1}$.6091
0.3909

Ge-Hf Germanium-Hafnium

$HfGe_2$ (44.84 wt. % Ge) is of the orthorhombic $ZrSi_2$ (C49) type of structure, with a = 3.8154 A, b = 15.004 A, c = 3.7798 A [1].

1. J. F. Smith and D. M. Bailey, *Acta Cryst.*, **10,** 1957, 341–342.

$\bar{1}$.5586
0.4414

Ge-Hg Germanium-Mercury

The solubility of Ge in liquid Hg is very small and becomes appreciable only above 250°C. Measurements of electrical resistance indicated that at least 0.027 wt. % Ge is soluble at 300°C [1].

1. T. I. Edwards, *Phil. Mag.*, **2,** 1926, 15–17.

$\bar{1}$.8012
0.1988

Ge-In Germanium-Indium

The solidification diagram in Fig. 419 is due to [1]. The liquidus curve was complemented by determinations of the solubility of Ge in liquid In in the temperature range of about 300–500°C [2]. The liquidus curve and eutectic point were calculated by [3]; the latter coincides, for all practical purposes, with 100% In.

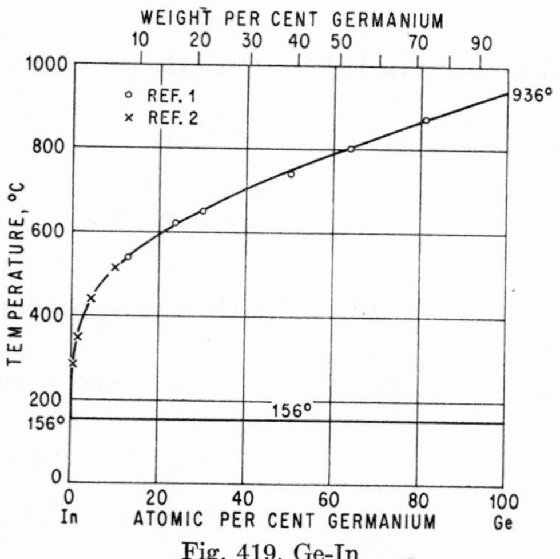

Fig. 419. Ge-In

1. W. Klemm and L. Klemm, *Z. anorg. Chem.*, **256**, 1948, 244–245.
2. P. H. Keck and J. Broder, *Phys. Rev.*, **90**, 1953, 521–522.
3. C. D. Thurmond, *J. Phys. Chem.*, **57**, 1953, 827–830.

$\bar{1}.5752$
0.4248

Ge-Ir Germanium-Iridium

IrGe (27.32 wt. % Ge) has the MnP (B31) type structure, $a = 6.281$ A, $b = 5.611$ A, $c = 3.490$ A [1]. Ir_3Ge_7 (46.73 wt. % Ge), the intermediate phase richest in Ge, is b.c.c. with 40 atoms per unit cell and isotypic with Ir_3Sn_7, $a = 8.753$ A [2].

1. H. Pfisterer and K. Schubert, *Z. Metallkunde*, **41**, 1950, 358–367.
2. K. Schubert and H. Pfisterer, *Z. Metallkunde*, **41**, 1950, 433–441.

0.2688
$\bar{1}.7312$

Ge-K Germanium-Potassium

KGe (65.00 wt. % Ge) and RbGe (45.93 wt. % Ge) were prepared by direct synthesis. Both compounds react with air and water. On thermal decomposition in a high vacuum, substances of the approximate composition KGe_4 (88.13 wt. % Ge) and $RbGe_4$ (77.26 wt. % Ge) are formed. Representations of the rather complex powder photographs of all four phases are given; it appears that KGe is isotypic with RbGe but not with NaGe [1].

1. E. Hohmann, *Z. anorg. Chem.*, **257**, 1948, 113–126.

0.4750
$\bar{1}.5250$

Ge-Mg Germanium-Magnesium

The phase diagram in Fig. 420 is based on thermal and microscopic investigations [1]. The Mg-Mg_2Ge eutectic was accurately determined by thermal analysis as

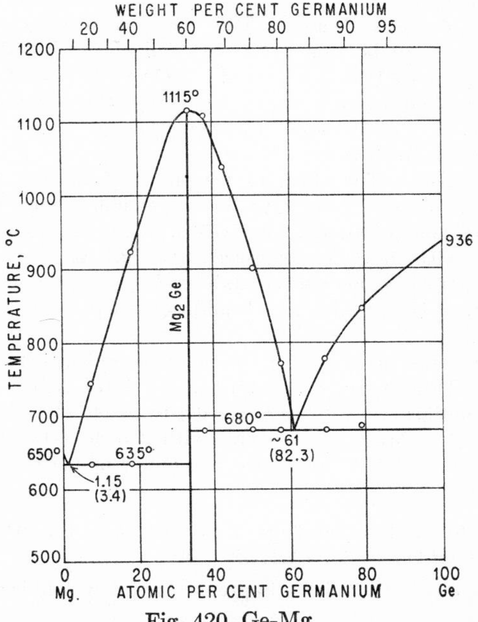

Fig. 420. Ge-Mg

being located at 1.15 at. (3.4 wt.) % Ge and 634.7°C [2]. This is in good agreement with 1.2–1.5 at. % Ge, 635°C, reported by [1]. Lattice-parameter measurements indicated that the solid solubility of Ge in Mg can be only very slight at 600°C and negligible at lower temperatures [2]. According to micrographic work, it is definitely smaller than 0.1 at. % at 400–600°C [1].

Mg_2Ge (59.88 wt. % Ge) is isotypic with CaF_2 (C1 type) [3, 1, 4, 5], $a = 6.390 \pm 0.003$ A [3, 1, 4].

1. W. Klemm and H. Westlinning, *Z. anorg. Chem.*, **245**, 1940, 365–380.
2. G. V. Raynor, *J. Inst. Metals*, **66**, 1940, 403–426.
3. E. Zintl and H. Kaiser, *Z. anorg. Chem.*, **211**, 1933, 125–131.
4. G. Brauer and J. Tiesler, *Z. anorg. Chem.*, **262**, 1950, 319–327.
5. G. Busch and U. Winkler, *Helv. Phys. Acta*, **26**, 1953, 578–583.

0.1211
$\bar{1}.8789$

Ge-Mn Germanium-Manganese*

The constitution of the whole system (Fig. 421) was established by thermal, microscopic, and roentgenographic investigations [1]. A more detailed study of the range 0–15.5 at. % Ge was carried out by [2], using thermal, micrographic, and high-temperature X-ray analysis. Thermal-analysis data for the range 68–100 at. % Ge, not covered by [1], were reported by [3]. The eutectic temperature was found as 720°C as compared with 697°C [1]; consequently, the eutectic composition would lie close to 56 at. % Ge (extrapolated), instead of 52.5 at. % Ge according to [1].

Microexamination of heat-treated samples showed that the phase Mn_5Ge_3 [37.5 at. (44.23 wt.) % Ge] is of variable composition, with the homogeneity range extending with increase in temperature. The other three intermediate phases, $Mn_{3.25}Ge$ [23.53 at. (28.91 wt.) % Ge], Mn_5Ge_2 [28.57 at. (34.58 wt.) % Ge], and Mn_3Ge_2 (46.84 wt. % Ge), are almost of singular composition. Mn_5Ge_2 undergoes a transformation at about 630°C.

The boundaries of the α, β, and γ fields were outlined by micrographic work. In addition, the $\gamma/(\beta + \gamma)$ boundary was determined by high-temperature X-ray investigation of alloys with 11.0 and 12.5 at. % Ge at 320–830°C. Results showed that the γ phase is stable down to room temperature within a narrow range of composition. No evidence for a eutectoid decomposition $\gamma \rightarrow \beta + Mn_{3.25}Ge$ was found after annealing at 500°C for 3 weeks [2].

Crystal Structures. The effect of Ge on the f.c.c. structure of γ-Mn was studied by [4]. Alloys up to 12.7 at. % Ge, after quenching from the γ field, showed a metastable f.c. tetragonal structure, alone (7.75–9.5 at. % Ge) or together ($c/a = 0.97$ remaining constant) with the f.c.c. structure (10.5–12.7 at. % Ge). In alloys with 13–16.6 at. % Ge, the f.c.c. structure was stabilized by quenching. At the saturation limit the parameter of γ is $a = 3.72$ A. Ge additions increase the lattice parameter of α-Mn and β-Mn [1].

The crystal structure of $Mn_{3.25}Ge$, based on the ideal composition Mn_3Ge, is h.c.p. of the Ni_3Sn ($D0_{19}$) type, $a = 5.347$ A, $c = 4.374$ A, $c/a = 0.818$ [1]. The structures of both modifications of Mn_5Ge_2 seem to be closely related with the NiAs (B8) type structure [1]. Mn_5Ge_3 is isotypic with Mn_5Si_3 ($D8_8$ type), $a = 7.184$ A, $c = 5.053$ A, $c/a = 0.703$ [5]. The lattice parameter of Ge is not affected by Mn additions [1].

1. U. Zwicker, E. Jahn, and K. Schubert, *Z. Metallkunde*, **40**, 1949, 433–436.
2. U. Zwicker, *Z. Metallkunde*, **42**, 1951, 327–330.

* The assistance of Dr. U. Zwicker in preparing this system is gratefully acknowledged.

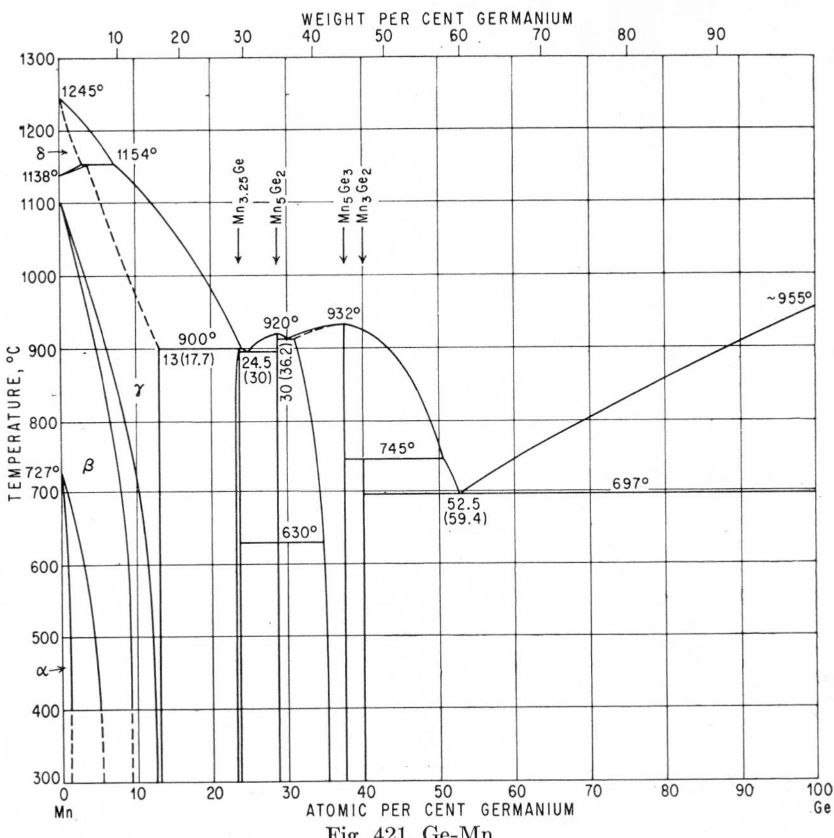

Fig. 421. Ge-Mn

3. J. H. Downing and D. Cubicciotti, *J. Am. Chem. Soc.*, **73**, 1951, 4025.
4. U. Zwicker, *Z. Metallkunde*, **42**, 1951, 246–252.
5. L. Castelliz, *Monatsh. Chem.*, **84**, 1953, 765–776.

$\bar{1}.8789$
0.1211

Ge-Mo Germanium-Molybdenum

[1, 2] have sintered mixtures of Mo and Ge powder in various proportions and identified, by X-ray studies, the phases Mo_3Ge (20.14 wt. % Ge), Mo_3Ge_2 (33.53 wt. % Ge), Mo_2Ge_3 (53.16 wt. % Ge), and two polymorphic modifications of $MoGe_2$ (60.21 wt. % Ge). Mo_3Ge is isotypic with Mo_3Si ("β-W" type), $a = 4.933$ A (or kX?). The structures of Mo_3Ge_2, Mo_2Ge_3, and that of the low-temperature form of $MoGe_2$ are neither cubic, tetragonal, nor hexagonal. Mo_3Ge_2 appears to be isostructural with Mo_3Si_2 [3]. The high-temperature modification of $MoGe_2$ is isotypic with $MoSi_2$ (C11b type), tetragonal with 6 atoms per unit cell and $a = 3.313$ A, $c = 8.195$ A (or kX?), $c/a = 2.474$.

No indication of melting was found in alloys with up to 67 at. % Ge when sintered at temperatures as high as 1350°C. The melting point of Mo_3Ge lies above 1750°C.

1. A. W. Searcy, R. J. Peavler, and H. J. Yearian, *J. Am. Chem. Soc.*, **74**, 1952, 566–567.

2. A. W. Searcy and R. J. Peavler, *J. Am. Chem. Soc.*, **75,** 1953, 5657–5659.
3. H. J. Wallbaum, *Naturwissenschaften*, **32,** 1944, 76–77.

0.7146
$\bar{1}$.2854

Ge-N Germanium-Nitrogen

According to [1], Ge_3N_4 (20.46 wt. % N) has a rhombohedral structure of the Be_2SiO_4 ($S1_3$) type, with a = 8.59 A, α = 107°48′, and 6 molecules per unit cell. This was disputed by [2], who proposed an orthorhombic structure, with a = 13.84 A, b = 4.06 A, c = 8.18 A, and 12 Ge_3N_4 per unit cell. [3] reported, however, that his findings were in accordance with the structure suggested by [1]. For earlier work on Ge_3N_4, see [4]. [5] reported the formation of Ge_3N_2 (11.4 wt. % N) on heating Ge in N_2 at 800–950°C.

1. R. Juza and H. Hahn, *Naturwissenschaften*, **27,** 1939, 32; *Z. anorg. Chem.*, **244,** 1940, 129–132.
2. W. C. Leslie, K. G. Carroll, and R. M. Fish, *Trans. AIME*, **194,** 1952, 204–206.
3. P. Duwez, unpublished work; discussion of paper by [2], *Trans. AIME*, **194,** 1952, 1201.
4. W. C. Johnson, *J. Am. Chem. Soc.*, **52,** 1930, 5160–5165; R. Schwarz and P. W. Schenk, *Ber. deut. chem. Ges.*, **63**(1), 1930, 300.
5. J. R. Hart; see W. C. Johnson, *J. Am. Chem. Soc.*, **52,** 1930, 5164, footnote.

0.4993
$\bar{1}$.5007

Ge-Na Germanium-Sodium

NaGe (75.94 wt. % Ge) was prepared by direct synthesis [1]. The compound persists only in an inert dry atmosphere and dissociates completely if heated in a high vacuum to 480°C. A representation of the powder pattern is given. For earlier investigations of NaGe, see [2].

1. E. Hohmann, *Z. anorg. Chem.*, **257,** 1948, 113–126.
2. L. M. Dennis and N. A. Skow, *J. Am. Chem. Soc.*, **52,** 1930, 2369–2372; E. Zintl and H. Kaiser, *Z. anorg. Chem.*, **211,** 1933, 120–121; W. C. Johnson and A. C. Wheatley, *Z. anorg. Chem.*, **216,** 1934, 282–287; C. A. Kraus and E. S. Carney, *J. Am. Chem. Soc.*, **56,** 1934, 765–768.

$\bar{1}$.8929
0.1071

Ge-Nb Germanium-Niobium

$NbGe_2$ (60.98 wt. % Ge) is isotypic with $CrSi_2$ (C40 type), a = 4.967 A, c = 6.784 A, c/a = 1.366 [1]. In both arc-melted and sintered specimens, [2] found—besides the digermanide—evidence for a compound of approximate composition Nb_2Ge.

Note Added in Proof. [3] examined by X-ray diffraction 26 alloys prepared by heating powder mixtures in vacuo. The following intermediate phases were found: Nb_3Ge, stable at least up to 1910°C; a phase within the composition range $NbGe_{0.54\pm0.06}$ (32.4–37.5 at. % Ge), stable at least up to 1910°C; $\sim Nb_3Ge_2$, stable at least up to 1650°C; and $NbGe_2$. Nb_3Ge was shown to be of the cubic "β-W" (A15) type of structure, with a = 5.168 ± 2 A. For $NbGe_2$ the lattice parameters a = 4.966 ± 3 A, c = 6.781 ± 3 A were found, in excellent agreement with [1]. For the crystal structures of $NbGe_{0.54}$ and $\sim Nb_3Ge_2$, see [4].

1. H. J. Wallbaum, *Naturwissenschaften*, **32,** 1944, 76–77.
2. G. F. Hardy and J. K. Hulm, *Phys. Rev.*, **93,** 1954, 1014–1015.
3. J. H. Carpenter and A. W. Searcy, *J. Am. Chem. Soc.*, **78,** 1956, 2079–2081.

4. H. Nowotny, A. W. Searcy, and J. E. Orr, *J. Phys. Chem.*, **60**, 1956, 677; quoted in [3].

0.0924
$\bar{1}$.9076

Ge-Ni Germanium-Nickel

The phase diagram in Fig. 422 was established by means of thermal and microscopic investigations [1]. The boundaries in the solid state do not represent equi-

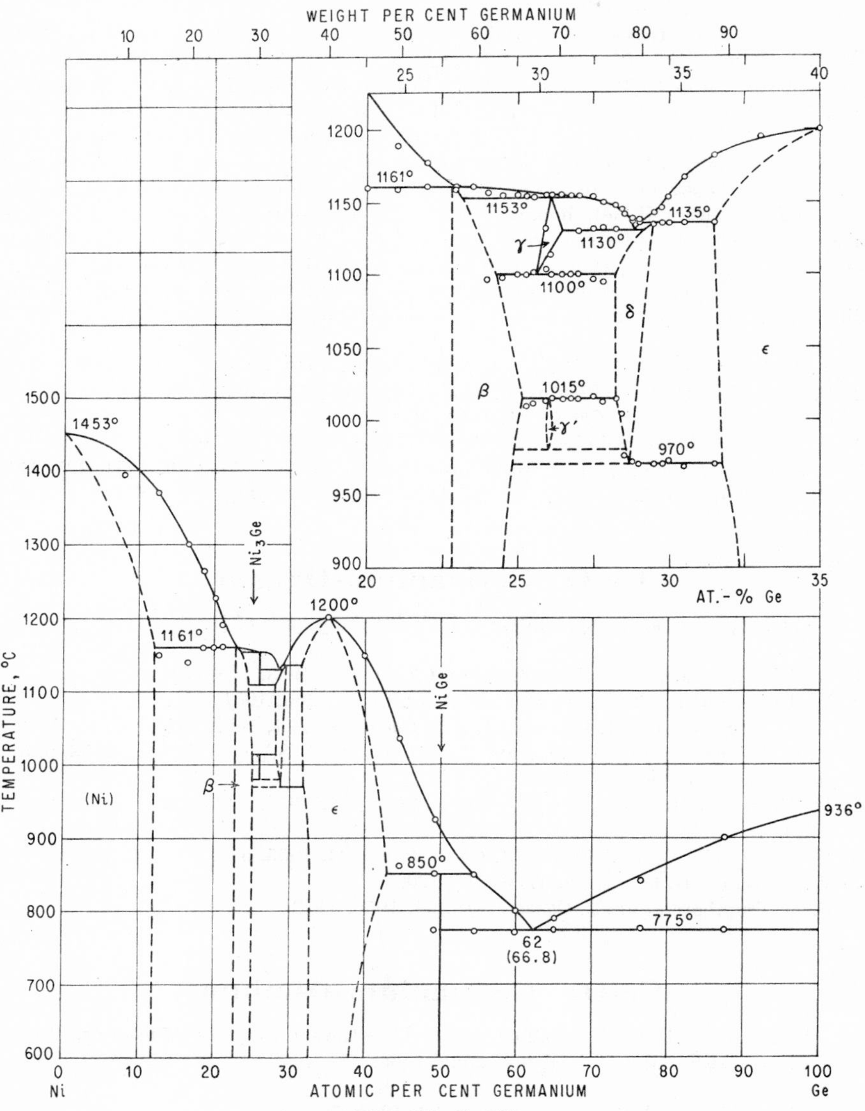

Fig. 422. Ge-Ni

librium conditions but were drawn principally on the basis of the thermal data only. The existence of the phases γ and γ', both being stable only within a limited temperature range, was verified by micrographic work: An alloy with 25.95 at. (30.25 wt.) % Ge was shown to be single-phase after quenching from 1145°C (γ) and 1010°C (γ'), but two-phase after quenching from 1050°C ($\beta + \delta$) and 950°C ($\beta + \epsilon$). No thermal arrests were found which correspond to (*a*) the eutectoid decomposition of γ' (assumed at about 980°C) or (*b*) the eutectoid decomposition of δ at 970°C between about 25 and 28.7 at. % Ge. The phase relationships in this composition range, therefore, are to be considered tentative.

Crystal Structures. [2] found the structure of the ϵ phase to be of the NiAs (B8) type, with $a = 3.955$ A, $c = 5.047$ A, $c/a = 1.276$ at 35 at. % Ge and $a = 3.849$ A, $c = 4.998$ A, $c/a = 1.299$ at 37 at. % Ge. For the alloy with 37 at. % Ge, [3] reported $a = 3.920$ A, $c = 5.046$ A, $c/a = 1.287$. The alloy with 35 at. % Ge, after annealing at 450°C, showed a superlattice. The powder pattern of the δ phase (about 29 at. % Ge) was reported to be very similar to that of the ϵ phase with 35 at. % Ge, suggesting that the structure of δ is an ordered and/or distorted version of that of ϵ [2].

The β phase (Ni_3Ge), homogeneous between about 22.9 and 24.8 at. (26.8–29.0 wt.) % Ge, has the structure of the Cu_3Au ($L1_2$) type, $a = 3.57$ A [4]. NiGe is isotypic with MnP (B31 type), $a = 5.811$ A, $b = 5.381$ A, $c = 3.428$ A [4].

Supplement. According to [5], who measured the solid solubility and the diffusion of Ni in Ge, the solidus has a retrograde character.

1. K. Ruttewit and G. Masing, *Z. Metallkunde*, **32**, 1940, 52–61.
2. F. Laves and H. J. Wallbaum, *Z. angew. Mineral.*, **4**, 1942, 17–46.
3. L. Castelliz, *Monatsh. Chem.*, **84**, 1953, 765–776.
4. H. Pfisterer and K. Schubert, *Z. Metallkunde*, **41**, 1950, 358–367.
5. F. van der Maesen and J. A. Brenkman, *Philips Research Rept.*, **9**, 1954, 225–230. Not available to the author. *Chem. Abstr.*, **49**, 1955, 2144.

0.6568
$\bar{1}.3432$

Ge-O Germanium-Oxygen

The literature on the system Ge-O includes the papers of [1–8].

1. R. Schwarz and E. Haschke, *Z. anorg. Chem.*, **252**, 1943, 170–172.
2. W. Buess and H. v. Wartenberg, *Z. anorg. Chem.*, **266**, 1951, 281.
3. W. L. Jolly and W. M. Latimer, *J. Am. Chem. Soc.*, **74**, 1952, 5757.
4. H. E. Swanson and E. Tatge, *Natl. Bur. Standards (U.S.), Circ.* 539 (I), 1953, p. 51. (Lattice parameters of GeO_2.)
5. M. Hoch and H. L. Johnston, *J. Chem. Phys.*, **22**, 1954, 1376–1377. (The Ge-O system.)
6. E. S. Candidus and D. Tuomi, *J. Chem. Phys.*, **23**, 1955, 588.
7. W. H. Baur, *Acta Cryst.*, **9**, 1956, 515–520.
8. F. A. Trumbore et al., *J. Chem. Phys.*, **24**, 1956, 1112.

$\bar{1}.5817$
0.4183

Ge-Os Germanium-Osmium

According to [1], $OsGe_2$ (43.29 wt. % Ge) has the same crystal structure as $RuSi_2$, which is of a still undetermined tetragonal type.

1. H. J. Wallbaum, *Naturwissenschaften*, **32**, 1944, 76–77.

0.3699
$\bar{1}.6301$

Ge-P Germanium-Phosphorus

Tensimetric and X-ray investigations established the existence of only one phosphide, GeP (29.90 wt. % P) [1].

1. M. Zumbusch, M. Heimbrecht, and W. Biltz, *Z. anorg. Chem.*, **242**, 1939, 237–238.

$\bar{1}.5445$
0.4555

Ge-Pb Germanium-Lead

The solidification diagram by [1] (Fig. 423) was complemented by the liquidus points of three Pb-rich alloys [2]. The liquidus curve and the eutectic point were calculated by [3]; the latter coincides, for all practical purposes, with 100% Pb.

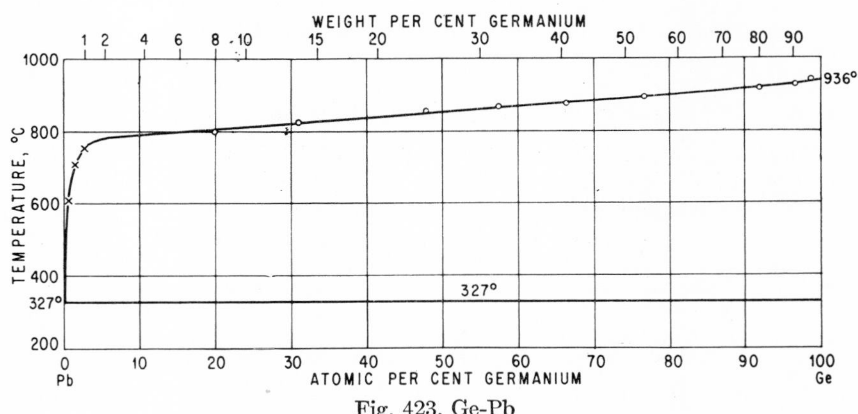

Fig. 423. Ge-Pb

1. T. R. Briggs and W. S. Benedict, *J. Phys. Chem.*, **34**, 1930, 173–177.
2. K. Ruttewit and G. Masing, *Z. Metallkunde*, **32**, 1940, 52–61.
3. C. D. Thurmond, *J. Phys. Chem.*, **57**, 1953, 827–830.

$\bar{1}.8328$
0.1672

Ge-Pd Germanium-Palladium

Limited microscopic examination [1] has revealed, besides primary Pd solid solutions, the existence of the intermediate phases Pd_5Ge_2 [28.57 at. (21.39 wt.) % Ge], Pd_2Ge (25.39 wt. % Ge), PdGe (40.49 wt. % Ge), and possibly Pd_4Ge, the latter being stable only at high temperatures. Pd_5Ge_2 is in equilibrium with the Pd-rich solid solution. Pd_2Ge apparently melts congruently and reacts with the melt to give PdGe, which forms a eutectic with Ge at approximately 70 wt. (77.5 at.) % Ge.

Pd_2Ge is isotypic with Fe_2P (C22 type), $a = 6.67$ A, $c = 3.39$ A, $c/a = 0.507$ [2]. The structure of PdGe is of the MnP (B31) type, $a = 6.259$ A, $b = 5.782$ A, $c = 3.481$ A [1].

1. H. Pfisterer and K. Schubert, *Z. Metallkunde*, **41**, 1950, 358–367.
2. K. Anderko and K. Schubert, *Z. Metallkunde*, **44**, 1953, 307–312.

$\bar{1}.7120$
0.2880

Ge-Pr Germanium-Praseodymium

$PrGe_2$ (50.75 wt. % Ge) is tetragonal, with 12 atoms per unit cell and a = 4.253 A, c = 13.940 A, c/a = 3.277 [1].

1. A. Iandelli, *Atti accad. nazl. Lincei, Rend. Classe sci. fis. mat. e nat.*, **6,** 1949, 727–734.

$\bar{1}.5704$
0.4296

Ge-Pt Germanium-Platinum

According to microscopic and X-ray work, the phases Pt_2Ge (15.68 wt. % Ge), PtGe (27.11 wt. % Ge), and Pt_2Ge_3 (35.81 wt. % Ge) exist [1]. PtGe is isotypic with MnP (B31 type), a = 6.088 A, b = 5.733 A, c = 3.701 A [1]. Pt_2Ge is isotypic with Fe_2P (C22 type), a = 6.68 A, c = 3.53 A, c/a = 0.528 [2].

1. H. Pfisterer and K. Schubert, *Z. Metallkunde*, **41,** 1950, 358–367.
2. K. Anderko and K. Schubert, *Z. Metallkunde*, **44,** 1953, 307–312.

$\bar{1}.4825$
0.5175

Ge-Pu Germanium-Plutonium

The existence of the following intermediate phases has been reported [1]: Pu_2Ge_3 of a distorted AlB_2 (C32) type of structure, $PuGe_2$ of the tetragonal $ThSi_2$ type of structure, and $PuGe_3$ of the cubic Cu_3Au ($L1_2$) type of structure, with a = 4.223 ± 1 A.

1. "Metallurgy and Fuels," pp. 393, 399, 400, 402, Progress in Nuclear Energy, ser. V, vol. 1, Pergamon Press Ltd., London, 1956.

$\bar{1}.9291$
0.0709

Ge-Rb Germanium-Rubidium

See Ge-K, page 765.

$\bar{1}.5907$
0.4093

Ge-Re Germanium-Rhenium

According to an X-ray diffraction and vapor-pressure investigation [1] of the Re-Ge system, only one intermediate phase, $ReGe_2$ (43.80 wt. % Ge), is formed. This compound decomposes peritectically at 1132 ± 15°C to solid Re and nearly pure liquid Ge. Its powder pattern is said to be complex.

1. A. W. Searcy, R. A. McNess, and J. M. Criscione, *J. Am. Chem. Soc.*, **76,** 1954, 5287–5289.

$\bar{1}.8485$
0.1515

Ge-Rh Germanium-Rhodium

An X-ray diffraction analysis [1] of the region 25.0–66.7 at. % Ge revealed the existence of four distinct intermediate phases. The compositions and structures of three of these phases are discussed in [1].

The structure of Rh_2Ge (26.08 wt. % Ge) is orthorhombic, with the lattice parameters a = 5.44 A, b = 7.57 A, c = 4.00 A, and 12 atoms per unit cell. It is a distorted Ni_2In-type structure and isotypic with that of Rh_2B.

The structure of Rh_5Ge_3 [37.50 at. (29.74 wt.) % Ge] is orthorhombic, with

a = 5.42 A, b = 10.32 A, c = 3.96 A, and 16 atoms per unit cell. "There does not appear to be any other reported compound which is isostructural with Rh_5Ge_3" [1].

RhGe (41.37 wt. % Ge) has the orthorhombic MnP (B31) type of structure, with a = 5.70 A, b = 6.48 A, c = 3.25 A.

1. S. Geller, *Acta Cryst.*, **8,** 1955, 15–21.

$\bar{1}.8536$
0.1464

Ge-Ru Germanium-Ruthenium

$RuGe_2$ (58.81 wt. % Ge) was reported to have the same crystal structure as $RuSi_2$, which is of a still undetermined tetragonal type [1].

1. H. J. Wallbaum, *Naturwissenschaften*, **32,** 1944, 76–77.

0.3549
$\bar{1}.6451$

Ge-S Germanium-Sulfur

The structures of GeS (30.64 wt. % S) and GeS_2 (46.91 wt. % S) form the B16 and C44 prototypes, respectively. Both compounds are orthorhombic: GeS, 4 molecules per unit cell, a = 4.30 A, b = 10.44 A, c = 3.65 A [1]; GeS_2, 24 molecules per unit cell, a = 11.68 A, b = 22.39 A, c = 6.87 A [2].

1. W. H. Zachariasen, *Phys. Rev.*, **40,** 1932, 917–922.
2. W. H. Zachariasen, *J. Chem. Phys.*, **4,** 1936, 618–619; *Phys. Rev.*, **49,** 1936, 884; see also W. C. Johnson and A. C. Wheatley, *Z. anorg. Chem.*, **216,** 1934, 274.

$\bar{1}.7754$
0.2246

Ge-Sb Germanium-Antimony

The phase diagram of Fig. 424 is the result of thermal-analysis work [1, 2], corroborated by metallographic [1] and X-ray studies [2]. The data points given

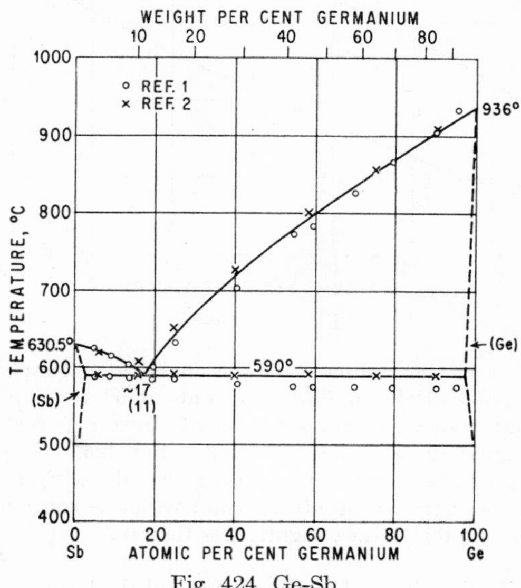

Fig. 424. Ge-Sb

had to be taken from small diagrams and therefore cannot be considered very accurate. The two branches of the liquidus curve intersect at approximately 17 at. % Ge [1] or 13 at. % Ge [2]. The difference is probably due to segregation on solidification. The eutectic temperature was given as 588 [1] and 592°C [2].

Qualitative X-ray investigations and measurements of the magnetic susceptibility indicated a solid solubility of approximately 2.5 at. (1.5 wt.) % Ge and 2.4 at. (4 wt.) % Sb at 540°C, after annealing for 4 months and quenching. The latter value is in considerable disagreement with the solidus curve of the (Ge) phase calculated by [3]. According to these data, the solidus is retrograde, indicating solubilities of about 0.02, 0.003, and 0.00009 at. % Sb at 880, 730, and 580°C, respectively.

1. K. Ruttewit and G. Masing, *Z. Metallkunde*, **32,** 1940, 52–61.
2. H. Stöhr and W. Klemm, *Z. anorg. Chem.*, **244,** 1940, 205–223.
3. C. D. Thurmond and J. D. Struthers, *J. Phys. Chem.*, **57,** 1953, 831–834.

0.4124
$\bar{1}.5876$

Ge-Si Germanium-Silicon

The diagram in Fig. 425 was established by thermal and X-ray analysis [1]. Because of the absence of diffusion during solidification, the solidus temperature

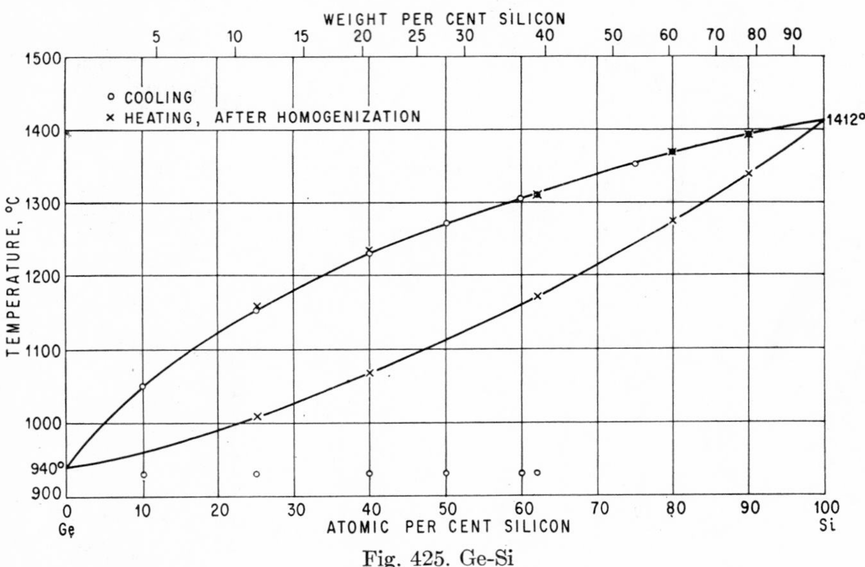

Fig. 425. Ge-Si

on normal cooling from the molten state was found at a temperature about 10°C below the melting point of Ge, in alloys up to about 60 at. % Si. Heating curves taken after elaborate homogenization treatments over a period of many months yielded the solidus and liquidus points indicated in Fig. 425. No phase changes were detected roentgenographically after annealing for several months at 925, 715, 295, and 177°C. The curve of lattice constants vs. composition is continuous and shows an average contraction of 0.009 A (i.e., slightly less than 0.2%) over the whole composition range.

[2] calculated the liquidus and solidus and found the latter to coincide with that

presented in Fig. 425. The liquidus lies somewhat below the experimentally determined curve.

Supplement. [3] measured the lattice constant and density of Ge-Si alloys containing up to 86 at. % Si. The variation of lattice constant with composition was found to agree within experimental error with that obtained by [1]. Electrical properties of Ge-Si alloys have been studied by [4].

1. H. Stöhr and W. Klemm, *Z. anorg. Chem.*, **241,** 1939, 305–323.
2. C. D. Thurmond, *J. Phys. Chem.*, **57,** 1953, 827–830.
3. E. R. Johnson and S. M. Christian, *Phys. Rev.*, **95,** 1954, 560–561.
4. A. Levitas, *Phys. Rev.*, **99,** 1955, 1810–1814.

$\bar{1}.7865$
0.2135

Ge-Sn Germanium-Tin

The phase diagram in Fig. 426, based on the thermal data indicated [1, 2], was corroborated by X-ray analysis [1]. The solidus temperature was found to coincide

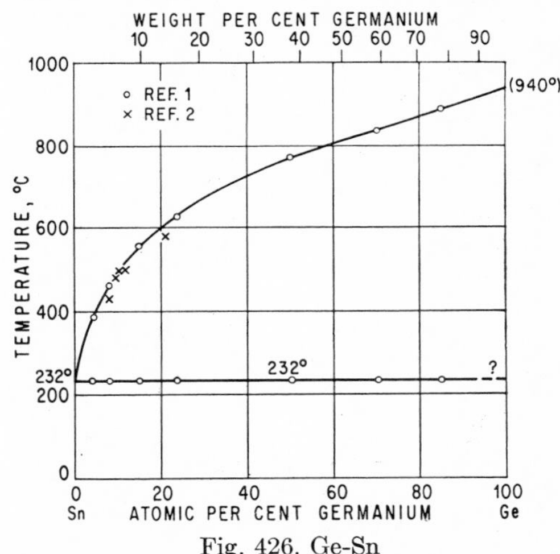

Fig. 426. Ge-Sn

with the melting point of Sn [1, 2] within the limit of accuracy of ±1°C [1]; see also [3].

On the basis of qualitative X-ray studies, using the disappearing-phase method, the mutual solid solubility, after annealing at 195°C for 4½ months and quenching, was believed to be definitely smaller than 0.6 at. % Ge and smaller than 1 at. % Sn. Alloys with 1.0 at. % Ge and 1.6 at. % Sn were found to be two-phase [1]. Spectrographic analysis showed the Sn-rich matrix to contain 0.01–0.1 wt. (0.16 at.) % Ge [2]. According to a brief abstract of unpublished work by [4], the existence of solid solutions of Sn in Ge was established by the lattice-spacing method; no quantitative data are given, however.

As to the effect of Ge on the transformation of white to gray Sn, see [5].

1. H. Stöhr and W. Klemm, *Z. anorg. Chem.*, **241,** 1939, 305–323.
2. W. Guertler and M. Pirani, *Tech. Publ. Internat. Tin Research Develop. Council*, ser. A, no. 50, 1937.
3. C. D. Thurmond, *J. Phys. Chem.*, **57,** 1953, 827–830.

4. L. Dowell and K. Lark-Horowitz, *Proc. Indiana Acad. Sci.*, **56,** 1946, 237.
5. R. R. Rogers and J. F. Fydell, *J. Electrochem. Soc.*, **100,** 1953, 161–164.

$\bar{1}.6036$
0.3964

Ge-Ta Germanium-Tantalum

$TaGe_2$ (44.53 wt. % Ge) is isotypic with $CrSi_2$ (C40 type), a = 4.958 A, c = 6.751 A, c/a = 1.362 [1].

Supplement. [2] observed, in addition to the digermanide, an intermediate phase of the approximate composition Ta_2Ge.

[3] studied the Ta-Ge system by means of X-ray diffraction and dissociation-pressure analysis. The compounds Ta_5Ge, Ta_2Ge, and $TaGe_2$ were observed. As to Ta_2Ge, four different forms are claimed to exist. $TaGe_2$ was found to decompose at 1280 ± 20°C to Ta_2Ge and Ge-rich liquid.

1. H. J. Wallbaum, *Naturwissenschaften*, **32,** 1944, 76–77.
2. G. F. Hardy and J. K. Hulm, *Phys. Rev.*, **93,** 1954, 1014–1015.
3. J. M. Criscione, Dissertation, Purdue University, Lafayette, Ind., 1954. Not available to the author. See *Dissertation Abstr.*, **14,** 1954, 1671.

$\bar{1}.7551$
0.2449

Ge-Te Germanium-Tellurium

The phase diagram in Fig. 427 is due to [1]. The composition of GeTe (63.74 wt. % Te) was found to coincide practically with the peritectic melt at 725 ± 3°C. However, the microstructure of an alloy with 43.3 at. % Te indicated the existence of a Ge-GeTe eutectic estimated to be located at about 45 at. % Te [2]. Since this is not compatible with the thermal data of [1], reinvestigation is necessary to clear up the discrepancy.

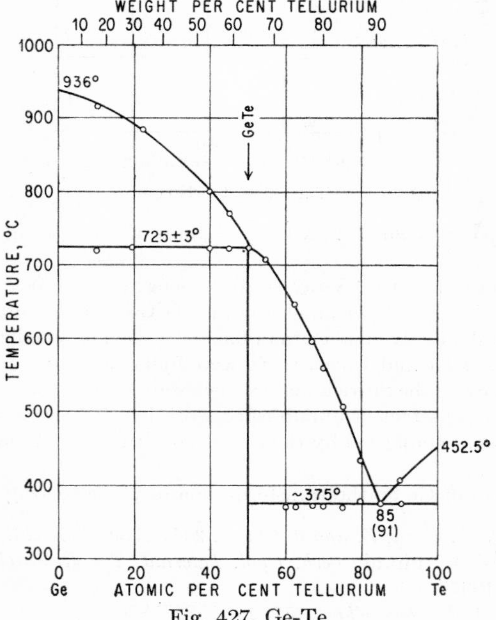

Fig. 427. Ge-Te

GeTe exists in two modifications. The low-temperature form has a trigonally distorted NaCl (B1) structure, a = 5.986 A, α = 88.35°C. High-temperature X-ray work (temperature not stated) revealed the high-temperature modification to be isotypic with NaCl [2].

1. W. Klemm and G. Frischmuth, *Z. anorg. Chem.*, **218,** 1934, 249–251.
2. K. Schubert and H. Fricke, *Z. Metallkunde*, **44,** 1953, 457–461.

0.1806
$\bar{1}$.8194

Ge-Ti Germanium-Titanium

The existence of the compounds Ti_5Ge_3 (47.63 wt. % Ge) [1] and $TiGe_2$ (75.20 wt. % Ge) [2] was established by determination of their crystal structures. Ti_5Ge_3 is

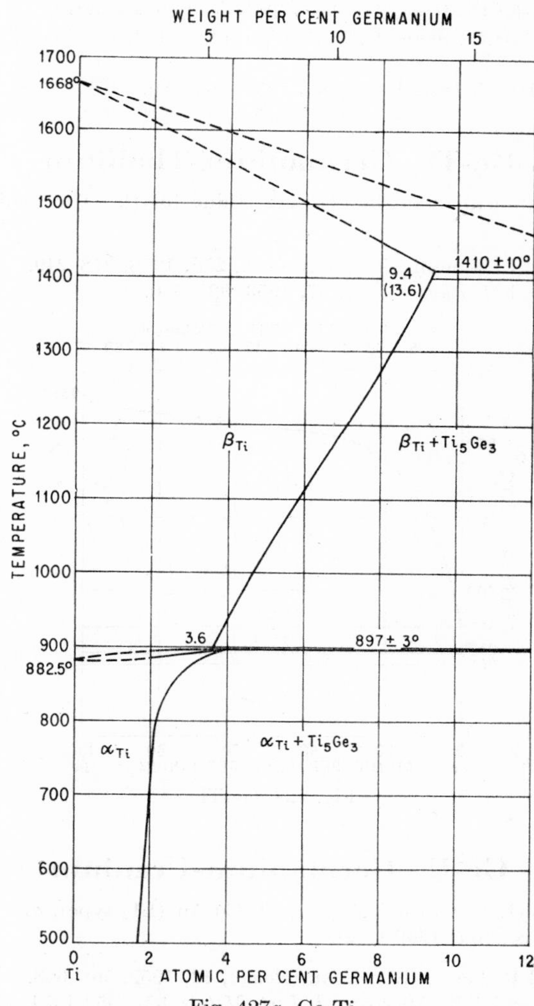

Fig. 427*a*. Ge-Ti

isotypic with Mn_5Si_3 ($D8_8$ type), a = 7.552 A, c = 5.234 A, c/a = 0.693 [3]. $TiGe_2$ has the $TiSi_2$ (C54) type of structure, a = 8.594 A, b = 5.030 A, c = 8.864 A.

Microexamination of alloys based on magnesium-reduced Ti has shown that at least 1 wt. (0.65 at.) % Ge is soluble, both in α-Ti at 790°C and in β-Ti at 925°C [4]. In agreement with this, an alloy with 0.24 wt. (0.16 at.) % Ge, after annealing at 790°C, was found to be one-phase [5].

Note Added in Proof. The partial phase diagram of Fig. 427*a* has been established by [6] by means of micrographic analysis of 14 alloy compositions. The location of the eutectic (β_{Ti} + Ti_5Ge_3) is estimated at about 16 at. % Ge.

1. P. Pietrokowsky and P. Duwez, *Trans. AIME*, **191,** 1951, 772–773.
2. H. J. Wallbaum, *Naturwissenschaften*, **32,** 1944, 76–77.
3. Computed from values probably based on kX units.
4. Battelle Memorial Institute, Air Force Technical Report 6218, Part 2, June, 1950.
5. R. M. Goldhoff, H. L. Shaw, C. M. Craighead, and R. I. Jaffee, *Trans. ASM*, **45,** 1953, 941–965.
6. M. K. McQuillan, *J. Inst. Metals*, **83,** 1955, 485–489.

$\bar{1}.5505$
0.4495

Ge-Tl Germanium-Thallium

The solidification diagram in Fig. 428 is due to [1]. The liquidus curve was calculated by [2].

1. W. Klemm and L. Klemm, *Z. anorg. Chem.*, **256,** 1948, 248–249.
2. C. D. Thurmond, *J. Phys. Chem.*, **57,** 1953, 827–830.

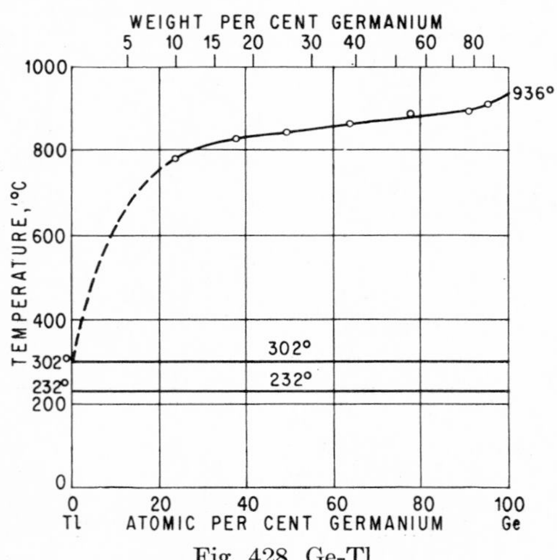

Fig. 428. Ge-Tl

$\bar{1}.4842$
0.5158

Ge-U Germanium-Uranium

UGe_3 (47.78 wt. % Ge) is isotypic with Cu_3Au ($L1_2$ type), a = 4.206 A [1, 2]. Its melting point is about 1200°C [2].

1. A. Iandelli and R. Ferro, *Ann. chim. (Rome)*, **42,** 1952, 598–608.
2. B. R. T. Frost and J. T. Maskrey, *J. Inst. Metals*, **82,** 1953-1954, 171–180.

0.1538
$\bar{1}.8462$

Ge-V Germanium-Vanadium

V_3Ge (32.20 wt. % Ge) has the "β-W" (A15) type of structure a = 4.769 A [1].

An X-ray study by [2] indicated the presence of several new phases in addition to V_3Ge; "but since none of these were superconducting above 1.20°K, they were not investigated further."

1. H. J. Wallbaum, *Naturwissenschaften,* **32,** 1944, 76–77.
2. G. F. Hardy and J. K. Hulm, *Phys. Rev.,* **93,** 1954, 1014.

$\bar{1}.5963$
0.4037

Ge-W Germanium-Wolfram

It has been briefly stated that W-Ge compounds differ in composition and structure from W-Si compounds [1].

[2] were unable to find any W-Ge compound in the X-ray examination of sintered specimens of different Ge contents.

1. H. J. Wallbaum, *Naturwissenschaften,* **32,** 1944, 76–77.
2. G. F. Hardy and J. K. Hulm, *Phys. Rev.,* **93,** 1954, 1014–1015.

0.0455
$\bar{1}.9545$

Ge-Zn Germanium-Zinc

The phase diagram in Fig. 429 is based on thermal and microscopic investigations [1, 2]. The mutual solid solubilities of Ge and Zn are extremely small.

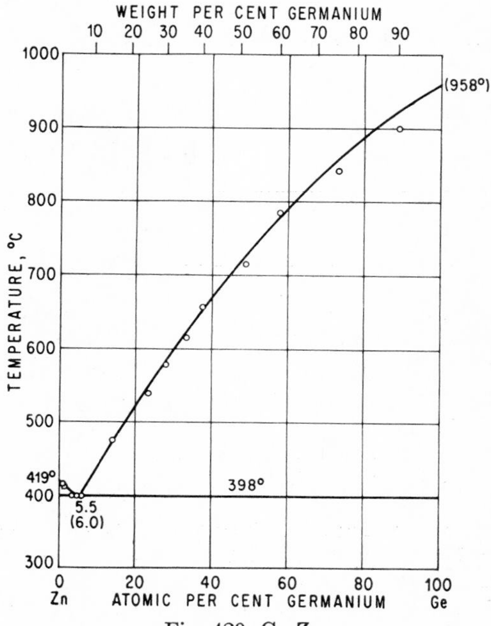

Fig. 429. Ge-Zn

1. E. Gebhardt, *Z. Metallkunde*, **34**, 1942, 255–257.
2. Data points were taken from a small diagram. A melting point of Ge of 936°C would fit the upper portion of the liquidus better than one of 958°C.

$\bar{1}.9008$
0.0992

Ge-Zr Germanium-Zirconium

The compound $ZrGe_2$ (61.42 wt. % Ge) was identified by determination of its crystal structure [1]: orthorhombic $ZrSi_2$ (C49) type, $a = 3.81$ A, $b = 15.04$ A, $c = 3.77$ A.

The phase diagram of Fig. 430 is largely based on thermal, microscopic, and X-ray work by [2]. Whereas both the Zr- and Ge-rich ends of the diagram appear

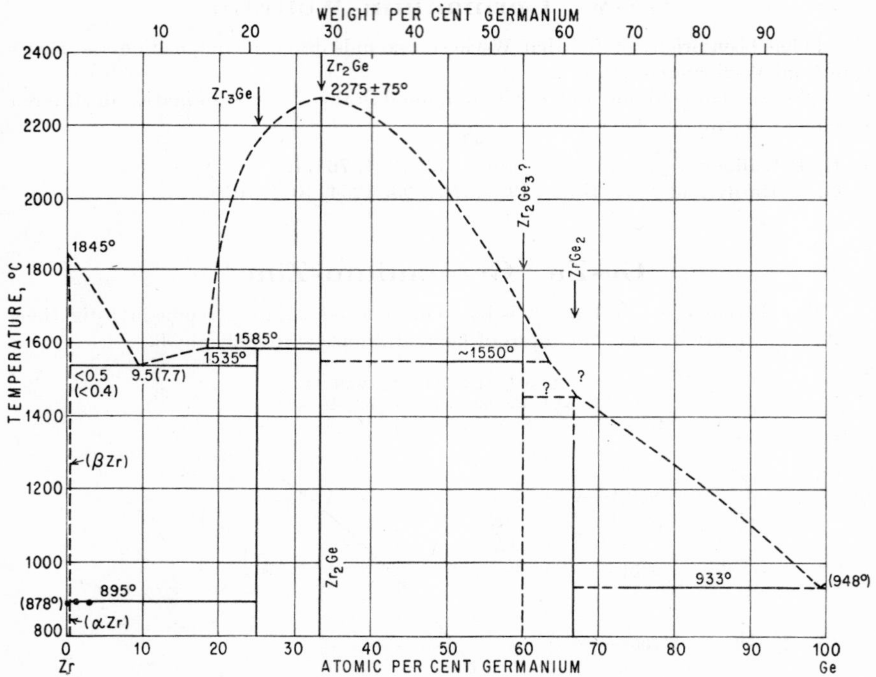

Fig. 430. Ge-Zr. (For a revised diagram, see Note Added in Proof.)

to be fairly well established, this does not hold for the region between 33.3 and about 70 at. % Ge. In this region, the phase diagram of [2] shows only one compound, with the tentative composition Zr_2Ge_3. However, [2] admits that an additional intermediate phase may exist. Since there is no reason to doubt the existence of $ZrGe_2$ claimed by [1], this phase is shown in Fig. 430. Its peritectic formation is purely hypothetical.

The following unit cells have been suggested by [2]—Zr_3Ge: hexagonal, $a = 8.14$ A, $c = 7.71$ A; Zr_2Ge: tetragonal, $a = 6.52$ A, $c = 5.38$ A; Zr_2Ge_3 ("or Zr_3Ge_5, $ZrGe_2$"): tetragonal, $a = 6.06$ A, $c = 5.29$ A.

Note Added in Proof. The phase diagram published recently by [3] deviates in the following ways from Fig. 430, which, as stated above, is based mainly on the cursory work of [2]: (*a*) The maximum solid solubility of Ge in Zr is between 0.5 and 1.0 wt. % Ge; (*b*) the compound Zr_5Ge_3, melting at 2330°C, replaces Zr_2Ge; (*c*) there is a compound of the tentative formula ZrGe, formed peritectically at 2240°C; and (*d*) the fourth intermediate phase (besides Zr_3Ge, Zr_5Ge_3, and ~ZrGe) is $ZrGe_2$, formed peritectically at 1520°C.

Zr_5Ge_3 appears to be isotypic with Ti_5Sn_3, Ti_5Ge_3, and Ti_5Si_3 (hexagonal $D8_8$ type), with a = 7.99 A, c = 5.54 A [3]. ZrGe is said to yield a complex powder pattern [3]. The indexing attempts of "Zr_2Ge" and "Zr_2Ge_3" by [2] are obsolete.

Single-crystal work by [4] confirmed the structure of $ZrGe_2$ derived by [1] on the basis of powder diffraction patterns. [4] determined the precision lattice parameters a = 3.7893 A, b = 14.975 A, c = 3.7606 A.

1. H. J. Wallbaum, *Naturwissenschaften*, **32,** 1944, 76–77.
2. P. E. Armstrong, M.S. thesis, Iowa State College, Ames, Iowa, 1952. The alloys were prepared in a helium arc furnace from sponge or crystal bar Zr and Ge powder of 99.9 wt. % purity.
3. O. N. Carlson, P. E. Armstrong, and H. A. Wilhelm, *Trans. ASM*, **48,** 1956, 843–854.
4. J. F. Smith and D. M. Bailey, *Acta Cryst.*, **10,** 1957, 341–342.

$\bar{3}.7516$
$\bar{2}.2484$

H-Hf Hydrogen-Hafnium

The phases found at various compositions in a roentgenographic and microscopic study [1] of slowly cooled and annealed (24 hr., 350°C) alloys may be seen from Table 31 and Fig. 431 (see also [2]). The three intermediate phases have the following structures: The pseudocubic phase, homogeneous somewhere between 60.5 and 63 at. % H, has a tetragonal unit cell, with a = 4.718 A, c = 4.683 A, c/a = 0.993 [2]. The cubic phase, homogeneous between about 63 and 65 at. % H, has a face-centered structure, with a = 4.708 A (at 63.0 at. % H). The hydrogen-rich tetragonal phase is stable around the ideal composition HfH_2 (1.12 wt. % H) [3] and is isotypic with ZrH_2 [1–3]; the body-centered unit cell has the dimensions a = 3.461 A, c = 4.395 A, c/a = 1.270 at 66.4 at. % H [3]. The solid solubility of H in Hf at room temperature seems to be well below 2 at. % H [1].

Table 31

Composition	At. % H	Phases
$HfH_{0.023}$	2.25	Hf + pseudocubic
$HfH_{0.27}$	21.4	Hf + pseudocubic
$HfH_{0.55}$	35.0	Hf + pseudocubic
$HfH_{0.85}$	46.0	Hf + pseudocubic
$HfH_{0.995}$	49.9	Hf + pseudocubic
$HfH_{1.36}$	57.6	Pseudocubic + Hf
$HfH_{1.53}$	60.5	Pseudocubic + Hf
$HfH_{1.70}$	63.0	Cubic
$HfH_{1.80}$	64.8	Cubic
$HfH_{1.78-1.86}$	64–65	Cubic + tetragonal
$HfH_{1.87}$	65.3	Tetragonal
$HfH_{1.98}$	66.4	Tetragonal

Note Added in Proof. [4] determined the location of deuterium atoms in the cubic and in the tetragonal phase by means of X-ray and neutron-diffraction work on alloy samples containing 61.9 at. % D ($HfD_{1.628}$) and 66.4 at. % D ($HfD_{1.983}$), respectively. The lattice constants found are $a = 4.680 \pm 3$ A and $a = 4.887 \pm 3$ A, $c = 4.345 \pm 3$ A, $c/a = 0.889$ (face-centered cell), respectively (cf. above). The cubic phase is of the CaF_2 (C1) type with a defect in D or H atoms. The tetragonal structure, already reported by [3], was refined. The transition pseudocubic-cubic (at about 62 at. % H) is considered [4] to be a continuous one (second-order transition).

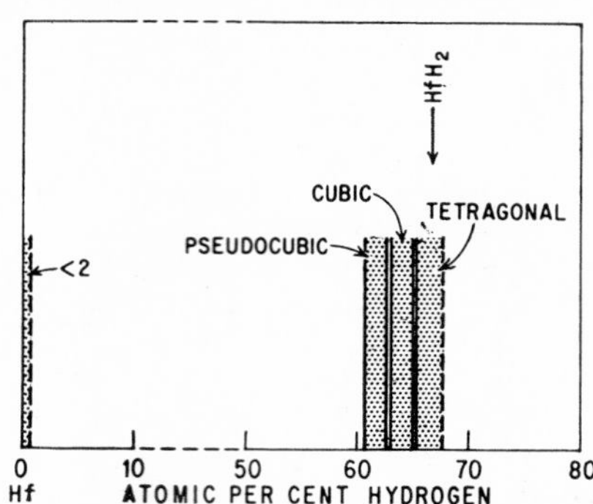

Fig. 431. H-Hf. (See also Note Added in Proof.)

The two-phase field shown in Fig. 431 is, therefore, incorrect. Under certain conditions of preparation, a transition phase is formed in this composition range [4].

1. S. S. Sidhu and J. C. McGuire, *J. Appl. Phys.*, **23,** 1952, 1257–1261. Major impurity of hafnium used was about 5 at. % Zr.
2. S. S. Sidhu, *Trans. ASM*, **46,** 1954, 652–654 (Discussion remark).
3. S. S. Sidhu, *Acta Cryst.*, **7,** 1954, 447–449.
4. S. S. Sidhu, L. Heaton, and D. D. Zauberis, *Acta Cryst.*, **9,** 1956, 607–614.

$\bar{3}.7177$
2.2823

H-Ir Hydrogen-Iridium

For a review of the literature on H-Ir [1-4], reference is made to [5].

1. F. Rother, *Ber. K. sächs. Akad. Wiss.*, **64,** 1912, 5–12.
2. A. Gutbier and W. Schieferdecker, *Z. anorg. Chem.*, **184,** 1929, 305–332.
3. I. I. Zhukov, *Izvest. Inst. Fiz.-Khim. Anal.*, **3,** 1926-1927, 461–462, 600–640. *Chem. Abstr.*, **21,** 1927, 3801; **22,** 1928, 4399.
4. M. Volmer and H. Wick, *Z. physik. Chem.*, **A172,** 1933, 429–447.
5. D. P. Smith, "Hydrogen in Metals," University of Chicago Press, Chicago, 1948.

$\bar{3}.8607$
2.1393

H-La Hydrogen-Lanthanum

The ability of La to absorb hydrogen with formation of dark friable products was known early [1–3]. Formulas ascribed to these certainly very impure reaction products were LaH_2 [1] or LaH_3 [3]. A systematic study (isotherms, isobar, heat of formation, densities, and kinetics) of the absorption of hydrogen by La-rich *Mischmetall*—84 wt. % La (+Pr, Nd, Y earth), 10 wt. % Ce, traces Fe, Al, Si—was made by [4–7]. The solubility at 1 atm hydrogen pressure amounts to about 220 cc

(NTP) H_2 per g metal [2.0 wt. (73 at.) % H] at room temperature [8] and decreases with increasing temperature. These authors tended to the view that solid solutions, rather than phases of fixed composition, are formed (see also [9]).

X-ray evidence for the existence of an intermediate hydride phase was first given by [10, 11]; a specimen containing 200 cc H_2 per g metal ($LaH_{2.5}$) yielded a pattern which could be indexed by a f.c.c. cell, with a = 5.63–5.64 A [11] (cf. below).

The preparation of a deuteride LaD_3 was reported by [12]; it is supposed to be analogous to the (unconfirmed) hydride LaH_3 (2.13 wt. % H; 240 cc H_2 per g La) of [3].

The results of the most recent study [13], which yielded pressure-temperature-composition and X-ray data as well as valuable microscopic evidence, may be summarized as follows: Below about the composition LaH_2 (1.43 wt. % H; 160 cc H_2 per g La) [14] the alloys consist of two phases, metal + hydride. The hydride (LaH_2) has a f.c.c. structure, with $a \cong 5.66$ A, and may dissolve further hydrogen; at the composition $LaH_{2.39}$ (70.6 at. % H) the lattice constant has fallen to 5.63 A. "It seems that pressures of hydrogen greater than 1 atm would be necessary to reach a solid composition of LaH_3" [13].

In spite of some unsatisfactory features of this work [13], the existence of the (La + LaH_2) two-phase field and of a certain homogeneity range of the hydride phase is well established. There is no information as to the influence of hydrogen on the allotropic transformation(s) in La. (See Note Added in Proof.)

Supplement. [15] reacted La with H (and D) and reported lattice parameters between a = 5.650 A and a = 5.695 A for the f.c.c. hydrides formed. Also, lines of a phase with a = 5.87 A (or with lower symmetry than f.c.c.) were observed.

Note Added in Proof. New X-ray analyses of the H-La system are due to [16, 17]. [16] observed a f.c.c. hydride phase with the composition range $LaH_{2.0}$ to $LaH_{3.0}$ and with the lattice constant varying from a = 5.66 A to a = 5.60 A. [17] gives parameters of La and its hydride under various conditions of heat-treatment and hydrogen pressure. The h.c.p. room-temperature structure of La was found to disappear at temperatures above 240°C in the presence of even a trace of H and not to reappear even at the temperature of liquid N; it is replaced by a f.c.c. lattice of a = 5.29 A (high-temperature form of La).

According to X-ray and neutron-diffraction work [18], LaH_2 has the f.c.c. CaF_2 (C1) type of structure with $a = 5.667 \pm 1$ A. Additional hydrogen in the $LaH_{>2}$ compositions is statistically distributed in the octahedral interstices of this structure. In agreement with the results of other authors [11, 13, 16], it was found that the lattice parameter decreases with increasing H content.

1. C. Winkler, *Ber. deut. chem. Ges.*, **24,** 1891, 873–899.
2. C. Matignon, *Compt. rend.*, **131,** 1900, 891–892.
3. W. Muthmann et al., *Liebigs Ann.*, **325,** 1902, 263–291.
4. A. Sieverts and G. Müller-Goldegg, *Z. anorg. Chem.*, **131,** 1923, 65–95.
5. A. Sieverts and E. Roell, *Z. anorg. Chem.*, **146,** 1925, 149–165.
6. A. Sieverts and A. Gotta, *Z. Elektrochem.*, **32,** 1926, 105–109.
7. A. Sieverts and A. Gotta, *Z. anorg. Chem.*, **172,** 1928, 1–31.
8. In good agreement with [3].
9. D. P. Smith, "Hydrogen in Metals," University of Chicago Press, Chicago, 1948.
10. A. Rossi, *Atti congr. nazl. chim. pura ed appl. 4th Congr. Rome, 1932,* **1933,** 593–594; *Chem. Abstr.*, **29,** 1935, 2873.
11. A. Rossi, *Nature,* **133,** 1934, 174.
12. R. Viallard and P. Jaszczyn, *Compt. rend.*, **228,** 1949, 485–487.
13. R. N. Mulford et al., *U.S. Atomic Energy Comm., Publ.* LAMS-1374, 1952. "Pre-

liminary investigations of the La-H and La-N systems." The lanthanum used contained large amounts of nonmetallic inclusions (presumably nitrides, oxides, and carbides). See *J. Phys. Chem.*, **59,** 1955, 1222–1226.
14. Because of the impure metal used, concentration data given by [13] are not claimed to have a high accuracy.
15. B. Dreyfus-Alain, *Compt. rend.*, **235,** 1952, 540–542, 1295; **236,** 1953, 1265–1266. B. Dreyfus-Alain and R. Viallard, *Compt. rend.*, **237,** 1953, 806–808.
16. B. Stalinski, *Bull. acad. polon. sci.*, **3,** 1955, 613–614; *Met. Abstr.*, **23,** 1956, 894.
17. B. Dreyfus-Alain, *Ann. phys.*, **10,** 1955, 305–362; *Met. Abstr.*, **24,** 1956, 84.
18. C. E. Holley et al., *J. Phys. Chem.*, **59,** 1955, 1226–1228.

$\bar{2}$.6175
1.3825

H-Mg Hydrogen-Magnesium

Liquid Mg. The solubility of hydrogen in liquid (commercial) magnesium at 760°C (saturated by bubbling H_2 through the melt) was determined by [1], using a chlorination-oxidation method [2], to be 41 cc per 100 g metal ($\sim 9 \times 10^{-2}$ at. % H). This value is reliable also in the light of the results of some earlier attempts [3, 4] to approximate such data.

Solid Mg. No systematic work has been carried out to determine the solubility of hydrogen as a function of temperature and pressure.

Molecular hydrogen dissolves very slowly in the compact metal [5, 4, 6]. If, however, ionic or atomic hydrogen is supplied, for instance by an ionizing discharge [5, 7, 8] or simply by reaction of magnesium with the normal, humid atmosphere [3, 4], hydrogen is absorbed easily and in considerable quantities [9]. [11] concluded that only nondiffusive occlusion or submicroscopic entrapment occurs, but the results of more recent—mainly electrical and roentgenographic—work on Mg-rich alloys [4] indicate that a true (interstitial) solution—to the extent of 15–20 cc per 100 g of metal, in fair agreement with data of [3, 12]—is formed.

For information on MgH_2, see the chemical literature.

1. F. Sauerwald, *Z. anorg. Chem.*, **258,** 1949, 27–32.
2. F. Sauerwald, *Z. anorg. Chem.*, **256,** 1948, 217–225.
3. H. Winterhager, *Aluminium-Arch.*, no. 12, 1938.
4. R. S. Busk and E. G. Bobalek, *Trans. AIME*, **171,** 1947, 261–275.
5. A. L. Reimann, *Phil. Mag.*, **16,** 1933, 673–686.
6. It was found by [4] that storage of degassed metal in a H_2-filled desiccator for as long as 6 months did not affect its hydrogen content.
7. O. Masaki and Y. Morimoto, *J. Sci. Hiroshima Univ.*, **A8,** 1938, 113–120; *Chem. Abstr.*, **32,** 5696.
8. L. F. Ehrke and C. M. Slack, *J. Appl. Phys.*, **11,** 1940, 129–137.
9. The presence of such sources of ionic or atomic hydrogen is assumed to be equivalent to high pressure of molecular hydrogen (cf. [4, 10]).
10. D. P. Smith, "Hydrogen in Metals," pp. 145–147, University of Chicago Press, Chicago, 1948.
11. D. P. Smith; see [10], pp. 4, 191–192, 282–283.
12. E. G. Bobalek and S. A. Shrader, *Ind. Eng. Chem., Anal. Ed.*, **17,** 1945, 544–553.

$\bar{2}$.2636
1.7364

H-Mn Hydrogen-Manganese

A detailed review of the literature was made by [1].

Gas-Metal Equilibria. The capacity of Mn for solution of hydrogen is far higher than that of its neighboring elements Cr, Fe, Co, Ni. The occurrence of a solubility minimum at about 500°C (see Fig. 432) is a characteristic feature of this element (cf. Cr-H), extraordinary in other respects also. Solubility measurements were made by [2–8], the outstanding papers being those of [5, 8]. Figure 432 shows the isobar for 760 mm Hg hydrogen pressure measured by [5] on a vacuum-distilled Mn [9]; the three allotropic transformations of the metal are clearly indicated [10]. The results of [8] show the same general type of curve and—except for temperatures below 500°C, where [8] gives solubilities of 21.6 and 11.4 cc per 100 g Mn at 24 and 500°C, respectively—the solubility data also are in fair agreement. The solubilities at lower temperatures should be redetermined. Extended solubility tables and comparisons of the results of several investigators with regard to the temperatures of transition of Mn were given by both [5, 8]. "Occlusion of hydrogen in manganese

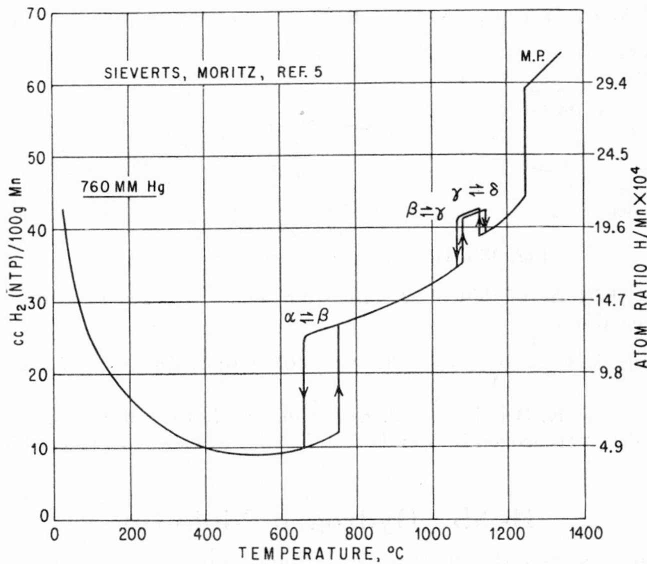

Fig. 432. H-Mn. Solubility isobar (1 atm).

has several times been held to render this metal ferromagnetic, but more recent evidence seems to contradict this" [1]; see [11, 12].

Surcharging. In electrolytic manganese, quantities of hydrogen much beyond the true solubility were observed, up to 1,200, 1,260, and 615 cc H_2 per 100 g Mn by [13, 14, 15], respectively; see also [7]. Extension of the lattice of α-Mn [15] indicates that, partially at least, the occluded hydrogen is in supersaturated solid solution.

1. D. P. Smith, "Hydrogen in Metals," University of Chicago Press, Chicago, 1948.
2. L. Troost and P. Hautefeuille, *Ann. chim. et phys.*, **7,** 1876, 155–177.
3. E. Wedekind and T. Veit, *Ber. deut. chem. Ges.*, **41,** 1908, 3771.
4. L. Luckemeyer-Hasse and H. Schenck, *Arch. Eisenhüttenw.*, **6,** 1932, 209–214.
5. A. Sieverts and H. Moritz, *Z. physik. Chem.*, **A180,** 1937, 249–263.
6. K. Iwasé and M. Fukusima, *Nippon Kinzoku Gakkai-Shi*, **1,** 1937, 202–213 (in Japanese). Not available to the author. *Met. Abstr.*, **5,** 1938, 469.
7. E. V. Potter, E. T. Hayes, and H. C. Lukens, *Trans. AIME*, **161,** 1945, 373–381.
8. E. V. Potter and H. C. Lukens, *Trans. AIME*, **171,** 1947, 401–412; discussion, pp. 412–415.

9. Supplied by Heraeus Vakuumschmelze, Hanau/Main.
10. The hysteresis loops are due to sluggishness in the transformation (cf. [8]).
11. M. A. Wheeler, *Phys. Rev.*, **49**, 1936, 642–643.
12. [1], pp. 185, 281–282.
13. H. Bockshammer, Dissertation, Stuttgart, 1921; quoted by [5].
14. H. H. Oakes and W. E. Bradt, *Trans. Electrochem. Soc.*, **69**, 1936, 567–584.
15. E. V. Potter and R. W. Huber, *Phys. Rev.*, **68**, 1945, 24–29.

$\bar{2}.0214$
1.9786

H-Mo Hydrogen-Molybdenum

Figure 433 shows the results of two determinations [1, 2] of the solubility of hydrogen in solid Mo (at 1 atm). As for a decision between these two curves, it can only

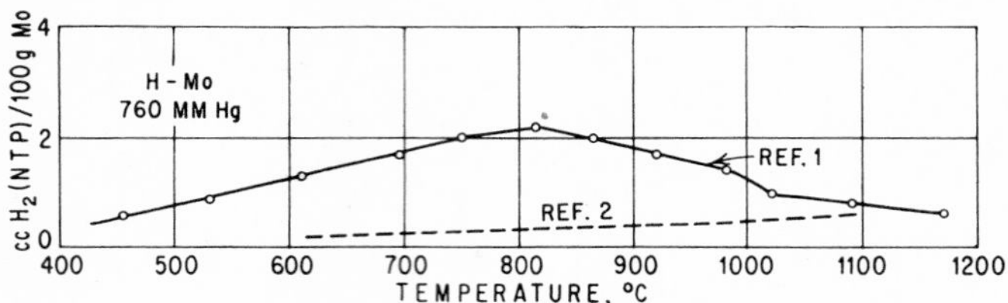

Fig. 433. H-Mo. Solubility isobar (1 atm).

be said that the form of that of [2] agrees far better with general considerations on occlusion (see also [3]).

1. E. Martin, *Arch. Eisenhüttenw.*, **3**, 1929, 407–416; *Metals & Alloys*, **1**, 1929, 831–835.
2. A. Sieverts and K. Brüning, *Arch. Eisenhüttenw.*, **7**, 1934, 641–645.
3. D. P. Smith, "Hydrogen in Metals," University of Chicago Press, Chicago, 1948.

$\bar{2}.0354$
1.9646

H-Nb Hydrogen-Niobium

Niobium was early shown [1] to have a large capacity for absorbing hydrogen. Solubility isotherms and isobars were measured by [3, 4]; [4] determined an isobar for deuterium also. Figure 434 shows the isobar for 1 atm hydrogen pressure, measured on 98.5 wt. % pure niobium [5] (for room-temperature saturation, cf. below).

A systematic X-ray study of both H-Nb and D-Nb was recently made by [6]. As can be seen from Fig. 435, the regions of homogeneity of the Nb (α) phase extend up to 10 at. % H ($NbH_{0.11}$)—in agreement with the results of previous X-ray work by [7]—and 8.25 at. % D ($NbD_{0.09}$), respectively. At approximately the compositions $Nb(H,D)_{0.7}$ (41 at. % H,D), the homogeneity ranges of a second phase [β or Nb(H,D)] begin, which extend up to the highest hydrogen (deuterium) contents obtainable. The maximum was determined only for hydrogen by slow cooling of the alloy and found to be $NbH_{0.94}$ [48.5 at. % H, 112 cc (NTP) H_2 per g Nb].

The existence of a NbH phase had been previously suggested by [8, 9]; the lattices assumed by these authors, however—b.c.c. [8], f.c.c. [9]—were not corroborated by [6], who indexed their (NbH and NbD) patterns with an orthorhombic (pseudo-b.c.c.) unit cell (cf. H-Ta), which for $NbH_{0.89}$ has the parameters a = 4.84 A, b = 4.90 A, c = 3.45 A. The kinetics of the reaction of niobium with hydrogen and the stability of the reaction products to high vacuum were studied by [10, 11]. Reviews on H-Nb were made by [12, 2].

Supplement. [13] studied nine Nb-H alloys with compositions up to $NbH_{0.86}$ (46 at. % H) by X-ray and magnetic analysis. In substantial agreement with the results of [6], the ($\alpha + \beta$) two-phase field was found to extend from $NbH_{0.10}$ to $NbH_{0.57}$ (9.1–36.4 at. % H). However, [13] assume the β phase to be b.c.c. [14] showed by

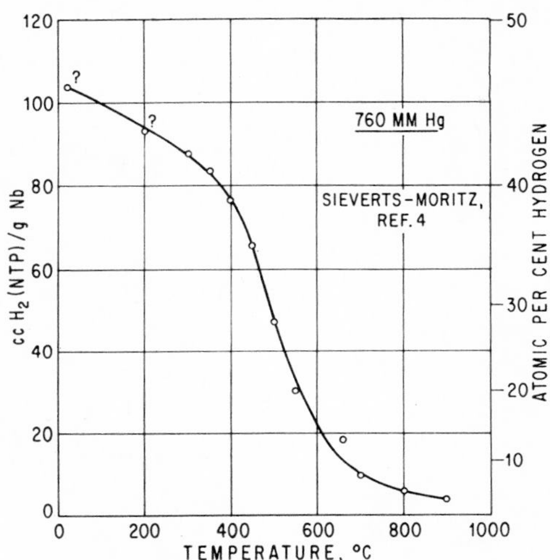

Fig. 434. H-Nb. Solubility isobar (1 atm). (See also Fig. 435.)

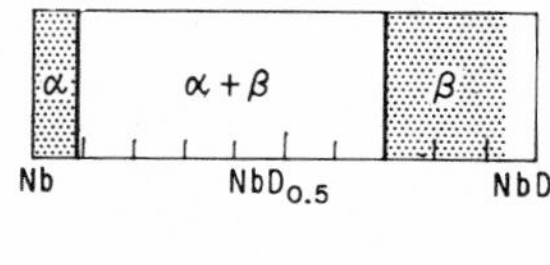

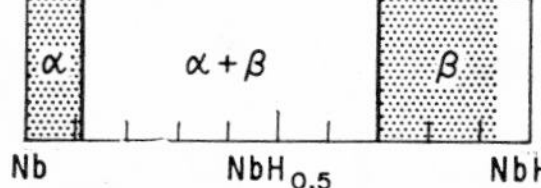

Fig. 435. H-Nb. (See also Fig. 434.)

means of X-ray work that the solubility of atomic H in Nb is greater than when the gas is in the molecular state.

1. For six references (1864–1907), see [2], p. 169.
2. D. P. Smith, "Hydrogen in Metals," University of Chicago Press, Chicago, 1948.
3. H. Hagen and A. Sieverts, *Z. anorg. Chem.*, **185,** 1930, 225–238.
4. A. Sieverts and H. Moritz, *Z. anorg. Chem.*, **247,** 1941, 124–130.
5. Remainder chiefly Ta.
6. G. Brauer and R. Hermann, *Z. anorg. Chem.*, **274,** 1953, 11–23.
7. F. H. Horn and W. T. Ziegler, *J. Am. Chem. Soc.*, **69,** 1947, 2762–2769.
8. I. Umanskii, *Zhur. Fiz. Khim.*, **14,** 1940, 332–339. *Chem. Abstr.*, **36,** 1942, 4001.
9. G. Aschermann, E. Friederich, E. Justi, and J. Kramer, *Physik. Z.*, **42,** 1941, 349–360.
10. E. A. Gulbransen and K. F. Andrew, *J. Electrochem. Soc.*, **96,** 1949, 364–376.
11. E. A. Gulbransen and K. F. Andrew, *Trans. AIME*, **188,** 1950, 586–599.

12. R. H. Myers, *Proc. Australian Inst. Mining & Met.*, **129,** 1943, 55–79.
13. W. Trzebiatowski and B. Stalinski, *Bull. acad. polon. sci.*, *Classe III*, **1,** 1953, 317–318 (in English).
14. V. M. Vukanović et al., *Compt. rend.*, **241,** 1955, 1298–1299; *Met. Abstr.*, **23,** 1956, 871–872.

$\bar{3}.8443$
2.1557

H-Nd Hydrogen-Neodymium

For a review of the literature on H-Nd [1–3], see [4]. Recent publications: [5, 6].

1. C. Matignon, *Compt. rend.*, **131,** 1900, 891–892.
2. W. Muthmann and H. Beck, *Liebigs Ann.*, **331,** 1904, 58–59.
3. A. Sieverts and E. Roell, *Z. anorg. Chem.*, **150,** 1926, 261–276.
4. D. P. Smith, "Hydrogen in Metals," University of Chicago Press, Chicago, 1948.
5. R. N. R. Mulford and C. E. Holley, *J. Phys. Chem.*, **59,** 1955, 1222–1226.
6. C. E. Holley et al., *J. Phys. Chem.*, **59,** 1955, 1226–1228.

$\bar{2}.2349$
1.7651

H-Ni Hydrogen-Nickel

For a comprehensive discussion of the literature on H-Ni, including the subjects of diffusion and surcharging, reference is made to [1]. The following includes only a selection of this literature.

The solubility of hydrogen in Ni was first adequately determined by [2–6] in the range 200–1600°C; later data by [7–12] are in fair agreement with their results [13]. The isobar for 1 atm pressure [14] shows the following increase in solubility with increasing temperature: 0.16, 0.51, and 1.7 relative volumes (9.3×10^{-3}, 3.0×10^{-2}, and 0.10 at. %) hydrogen at 210, 620, and 1453°C (melting point of Ni), respectively; at the melting point of Ni the solubility approximately doubles and then continues to increase at a somewhat accelerated rate.

Like many other metals, Ni has the ability to take up hydrogen much in excess of the equilibrium quantity, when in a cold-worked condition or while being subjected to cathodic or chemically liberated hydrogen. For details of these surcharging effects, see [1].

"Hexagonal Ni" [15–18] appears to be a hydrogen-metastabilized modification of Ni rather than a hydride (cf. [19, 20]).

The solubilities of deuterium [21, 11] and tritium [11] in solid Ni were also measured.

1. D. P. Smith, "Hydrogen in Metals," University of Chicago Press, Chicago, 1948.
2. A. Sieverts (with experiments by P. Beckmann), *Z. physik. Chem.*, **60,** 1907, 169–201.
3. A. Sieverts and J. Hagenacker, *Ber. deut. chem. Ges.*, **42,** 1909, 338–347.
4. A. Sieverts and W. Krumbhaar, *Ber. deut. chem. Ges.*, **43,** 1910, 893–900.
5. A. Sieverts (with W. Krumbhaar and E. Jurisch), *Z. physik. Chem.*, **77,** 1911, 591–613.
6. A. Sieverts, *Z. Metallkunde*, **21,** 1929, 37–44 (Review).
7. L. Luckemeyer-Hasse and H. Schenck, *Arch. Eisenhüttenw.*, **6,** 1932, 209–214.
8. J. Smittenberg, *Rec. trav. chim.*, **52,** 1933, 112–122, 339–351; **53,** 1934, 1065–1083; *Nature*, **133,** 1934, 872.
9. K. Iwase and M. Fukusima, *Nippon Kinzoku Gakkai-Shi*, **1,** 1937, 202–213.
10. M. H. Armbruster, *J. Am. Chem. Soc.*, **65,** 1943, 1043–1054.

11. N. J. Hawkins, *U.S. Atomic Energy Comm., Publ.* KAPL-868, 1953. *Met. Abstr.*, **21,** 1953, 310.
12. K. H. Lieser and H. Witte, *Z. physik. Chem.*, **202,** 1954, 321–351.
13. A convenient tabular comparison of all but the latest findings, calculated 1 atm pressure for temperatures of 300–900°C, is found in [10].
14. Plotted in [1], p. 63.
15. G. Bredig and R. Allolio, *Z. physik. Chem.*, **126,** 1927, 41–71.
16. S. Valentiner and G. Becker, *Naturwissenschaften*, **17,** 1929, 639–640 (Questioning the hexagonal Ni lattice of [15]).
17. W. Büssem and F. Gross, *Z. Physik*, **87,** 1934, 778–799.
18. J. Terminasov and M. S. Beleckij, *Doklady Akad. Nauk S.S.S.R.*, **63,** 1948, 411–413 (in Russian). *Structure Repts.*, **11,** 1947-1948, 127.
19. G. Masing, "Lehrbuch der allgemeinen Metallkunde," pp. 476–477, Springer-Verlag, Berlin, 1950.
20. Smith ([1], especially pp. 70–71) came to the conclusion that the interpretations of the (rather diffuse) X-ray patterns in terms of a hexagonal lattice are erroneous. This view, however, is not convincing.
21. A. Sieverts and W. Danz, *Z. anorg. Chem.*, **247,** 1941, 141–144.

$\bar{3}.7242$
2.2758

H-Os Hydrogen-Osmium

[1] observed no cathodic absorption by osmium. The sorption of hydrogen of atmospheric pressure by fine osmium black was studied by [2] up to 208°C; annealed (sintered) material showed appreciably lower sorption values.

1. R. Böttger, *J. prakt. Chem.*, (2)**9,** 1874, 193–199.
2. S. Gutbier and W. Schieferdecker, *Z. anorg. Chem.*, **184,** 1929, 305–332.

$\bar{3}.6398$
2.3602

H-Pa Hydrogen-Protactinium

PaH_3 (1.29 wt. % H) is structurally analogous to UH_3 [1].

1. P. Sellers, S. Fried, R. Elson, and W. Zachariasen, *U.S. Atomic Energy Comm., Publ.* AECD-3167, 1952. *Met. Abstr.*, **20,** 1952, 229.

$\bar{3}.6870$
2.3130

H-Pb Hydrogen-Lead

The extensive and controversial literature [1] on the occlusion of hydrogen by solid lead has been competently reviewed by [2], who reached the following conclusion: "We may perhaps sum up the rather unsatisfactory evidence with regard to lead-hydrogen as indicating only an extremely small diffusive occlusion, if any, and that probably confined to temperatures not far below melting. In view of the well-known self-annealing character of the metal, doubt may be entertained of its capacity for rift occlusion at the temperatures indicated; and further experiments are to be desired."

An investigation of the solubility of hydrogen in molten lead [3] was made more recently by [4], using a direct method, in the range of 500–900°C and 100–800 mm Hg. At 1 atm the solubility increases from $\sim$0.1 cc of hydrogen per 100 g of lead (2×10^{-3} at. % H) at 500°C to $\sim$1.25 cc (2.3×10^{-2} at. % H) at 900°C (cc data taken from a graph).

Note Added in Proof. In a careful redetermination of the solubility of hydrogen in liquid Pb at 600°C, the value reported by [4], 0.25 cc H_2 per 100 g Pb, could not be corroborated by [5]. [5] found the solubility at this temperature to be less than 0.01 cc H_2 per 100 g Pb. Likewise, [6] observed that Pb does not absorb hydrogen, even on melting in a hydrogen atmosphere.

1. For 15 references, see [2].
2. D. P. Smith, "Hydrogen in Metals," University of Chicago Press, Chicago, 1948, especially pp. 191–192, 286.
3. No measurable solubility was found in either the solid or the liquid state (investigated up to 600°C) by A. Sieverts and W. Krumbhaar, *Ber. deut. chem. Ges.*, **43**, 1910, 896.
4. W. R. Opie and N. J. Grant, *Trans. AIME*, **191**, 1951, 244–245; correction on p. 528.
5. W. Hofmann and J. Maatsch, *Z. Metallkunde*, **47**, 1956, 89–95.
6. W. Mannchen and M. Baumann, *Metall*, **9**, 1955, 686–688.

$\bar{3}.9753$
2.0247

H-Pd Hydrogen-Palladium

H-Pd is by far the most thoroughly studied of all hydrogen systems. The extensive literature involved has recently been reviewed [1–3], and for detailed discussions reference is made to these sources. In the following only a brief description of the phase constitution is given, and only a few original papers are quoted.

Solubility isotherms (Fig. 436) and X-ray results show clearly the existence of two solid solutions, named α and β, under normal conditions. The saturated Pd (α) phase ($\alpha_{sat.}$) has a lattice parameter of a = 3.902 A [5, 6], as compared with a = 3.891 A for pure Pd. The β phase also possesses the f.c.c. (A1) arrangement of the Pd atoms with hydrogen in interstitial positions. An enlarged parameter value, a = 4.026 A, which increases further with the hydrogen content, corresponds to its inferior limit, β_i. At higher temperatures the miscibility gap between α and β narrows, and they coalesce in a critical point for which the data 295°C, 19.87 atm, H/Pd = 0.27 (21 at. % H) were found by [7].

The concentrations which correspond to the above lattice constants are not precisely known; $\alpha_{sat.}$ ~ 0.025, β_i = 0.6–0.8, atom ratio H/Pd (~2.4 and 37.5–44.5 at. % H, respectively) [8–10]. The β phase is assumed [8] to dissolve interstitially up to about 1,300 relative volumes (H/Pd = 1.0, 50 at. % H); additional hydrogen—up to a total of 2,800 volumes (H/Pd = 2.2, 69 at. % H)—may further be rift-occluded in a highly ionized state if conditions favorable to large rift occlusion are met. (See Note Added in Proof.)

Isobars for ordinary pressure are plotted in Figs. 437 and 438 (volumes measured at NTP). The upper branch of Fig. 438 relates to the β phase and the lower branch (as well as the whole curve in Fig. 437) to the α phase, while the region with two generally vertical branches (corresponding to absorption and evolution of hydrogen) is that of the transition $\alpha \rightleftarrows \beta$, with which hysteresis is associated.

The system deuterium-palladium shows great similarity to H-Pd, forming analogous α and β phases, with closely corresponding intervals of immiscibility (critical data: 276°C, 35 atm, D/Pd = 0.25 or 20 at. % D [14]). Isobars are plotted in Figs. 437 and 438.

Recent theoretical calculations on H,D-Pd are due to [15–17].

Note Added in Proof. [18] determined, on Pd wire specimens, the effect of pressure on the solubility of molecular hydrogen in the β phase. With the hydrogen pressure increasing from 1 to 700 atm, the solubility increased, in atom ratio H/Pd (at. % H in parentheses), from 0.700 to 0.876 (41.2 to 46.7) at 15°C, and from 0.560 to 0.774 (35.9 to 43.6) at 88°C.

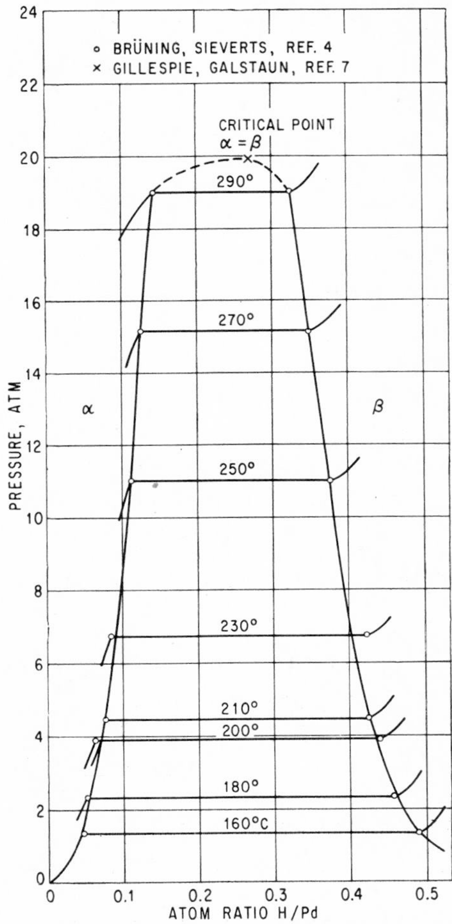

Fig. 436. H-Pd. (See also Figs. 437 and 438.)

1. E. Raub, "Die Edelmetalle und ihre Legierungen," pp. 279–286, Springer-Verlag OHG, Berlin, 1940.
2. "Gmelins Handbuch der anorganischen Chemie," System No. 65, pp. 116–256, Verlag Chemie, G.m.b.H., Weinheim/Bergstrasse, 1942.
3. D. P. Smith, "Hydrogen in Metals," University of Chicago Press, Chicago, 1948.
4. H. Brüning and A. Sieverts, *Z. physik. Chem.*, **163,** 1933, 409–441.
5. E. A. Owen and J. I. Jones, *Proc. Phys. Soc.* (*London*), **49,** 1937, 587–602, 603–610. (Lattice parameters in the original paper presumably in kX units.)
6. [3], pp. 86, 110.
7. L. J. Gillespie and L. S. Galstaun, *J. Am. Chem. Soc.*, **58,** 1936, 2565–2573.

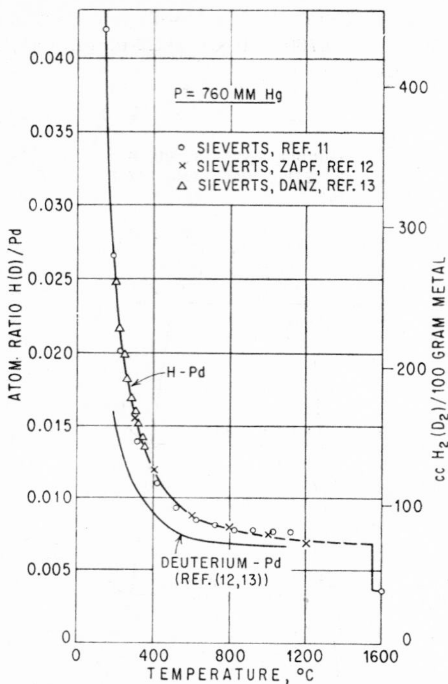

Fig. 437. H-Pd. Solubility isobar (1 atm). (See also Figs. 436 and 438.)

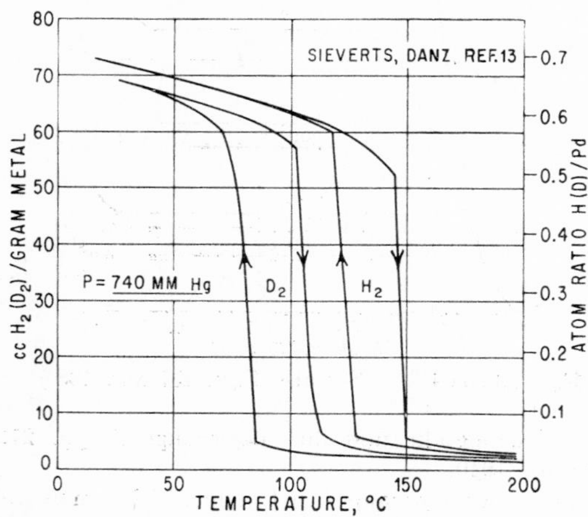

Fig. 438. H-Pd. Solubility isobar (740 mm Hg). (See also Figs. 436 and 437.)

8. G. A. Moore, *Trans. Electrochem. Soc.*, **75**, 1939, 237–267.
9. [2], pp. 193–195.
10. [3], pp. 86, 111.
11. A. Sieverts, *Z. Metallkunde*, **21**, 1929, 44.
12. A. Sieverts and G. Zapf, *Z. physik. Chem.*, **A174**, 1935, 359–364.

13. A. Sieverts and W. Danz, *Z. physik. Chem.*, **B34,** 1936, 158–159; **38,** 1938, 46–60.
14. L. J. Gillespie and W. R. Downs, *J. Am. Chem. Soc.*, **61,** 1939, 2496–2502.
15. J. R. Lacher, *Proc. Roy. Soc.* (*London*), **161A,** 1937, 525–545.
16. A. Harashima, T. Tanaka, and K. Sakaoku, *J. Phys. Soc. Japan*, **3,** 1948, 208–213 (in English) (H-Pd).
17. T. Tanaka, K. Sakaoku, and A. Harashima, *J. Phys. Soc. Japan*, **3,** 1948, 213–218 (D-Pd).
18. P. S. Perminov, A. A. Orlov, and A. N. Frumkin, *Doklady Akad. Nauk S.S.S.R.*, **84,** 1952, 749–752; *Met. Abstr.*, **23,** 1956, 426.

$\bar{3}.8545$
2.1455

H-Pr Hydrogen-Praseodymium

Literature on H-Pr may be found in [1–4], which have been reviewed by [5]; in addition, X-ray work was done by [6]. See [7, 8].

1. C. Matignon, *Compt. rend.*, **131,** 1900, 891–892.
2. W. Muthmann and H. Beck, *Liebigs Ann.*, **331,** 1904, 58–59.
3. A. Sieverts and E. Roell, *Z. anorg. Chem.*, **150,** 1926, 261–276.
4. A. Sieverts and A. Gotta, *Z. anorg. Chem.*, **172,** 1928, 1–31.
5. D. P. Smith, "Hydrogen in Metals," University of Chicago Press, Chicago, 1948.
6. A. Rossi, *Nature,* **133,** 1934, 174; abstract in *Strukturbericht,* **3,** 1933-1935, 263.
7. R. N. R. Mulford and C. E. Holley, *J. Phys. Chem.*, **59,** 1955, 1222–1226.
8. C. E. Holley et al., *J. Phys. Chem.*, **59,** 1955, 1226–1228.

$\bar{3}.7129$
2.2871

H-Pt Hydrogen-Platinum

According to the most reliable data available [1], the solubility of hydrogen in compact Pt [2] is relatively small (see Table 32). "Whether any of the small occlusion is due to solution in the lattice is not wholly certain, particularly because of the lack of adequate X-ray investigation [3], but it seems probable that, in platinum, only rift occlusion occurs" [4].

For a more comprehensive discussion of the extensive literature on H-Pt, see [5, 6].

Table 32. Solubility of Hydrogen in Pt at 1 Atm

Temperature, °C	cc/g $\times 10^3$	Atomic ratio, H/Pt	At. % hydrogen
409	0.67	0.12×10^{-4}	0.12×10^{-2}
827	1.00	0.18×10^{-4}	0.18×10^{-2}
1033	2.36	0.41×10^{-4}	0.41×10^{-2}
1136	4.00	0.70×10^{-4}	0.70×10^{-2}
1239	6.11	1.07×10^{-4}	1.07×10^{-2}
1342	9.32	1.64×10^{-4}	1.64×10^{-2}

1. A. Sieverts and E. Jurisch, *Ber. deut. chem. Ges.*, **45,** 1912, 221–229.
2. An estimation—based on thermodynamics—of the solubility of hydrogen in liquid Pt at the melting point was made by Y. L. Yao (*J. Chem. Phys.*, **21,** 1953, 1308–1309).

3. "X-ray observations by A. Osawa, *Science Repts. Tôhoku Imp. Univ.*, **14**, 1925, 43–45, on Pt black, and by G. Bredig and R. Allolio, *Z. physik. Chem.*, **126**, 1927, 41–71, on sputtered films, were both interpreted by their authors as showing some distension of the lattice and hence formation of solid solution of hydrogen in Pt. The last-named observers noted marked line broadening; and, in view of this and of the early dates for such work, some doubt seems warranted as to the indications." ([6], p. 267).
4. [6], p. 78.
5. "Gmelins Handbuch der Anorganischen Chemie," System No. 68 C, pp. 5–25, Verlag Chemie, G.m.b.H., Weinheim/Bergstrasse, 1940.
6. D. P. Smith, "Hydrogen in Metals," University of Chicago Press, Chicago, 1948.

$\bar{3}.6196$
2.3804

H-Pu Hydrogen-Plutonium

Pu forms a hydride of variable composition (approximately PuH_{2-3}) ([1], see also [2]). The substitution of deuterium in place of hydrogen yielded nearly identical results [1].

Note Added in Proof. [3] measured pressure-composition isotherms of the Pu-H and Pu-D systems in the temperature ranges 400–800°C and 600–800°C, respectively. The extent of the horizontal portions of the isotherms indicates a certain solubility of H in solid (and liquid) Pu and the following lower limits of composition of (PuH_2), in atom ratio H/Pu (at. % H in parentheses): 1.88, 1.86, 1.85, 1.80, and 1.75 (65.3, 65.0, 64.9, 64.3, and 63.7) at 500, 600, 700, 750, and 800°C, respectively. The decomposition pressure of PuH_2 reaches 251 mm Hg at 800°C.

PuH_2 is of the f.c.c. CaF_2 (C1) type of structure, with $a = 5.359$ A [4, 5]. It takes additional hydrogen into solid solution up to the composition $PuH_{2.7}$ (73 at. % H), the lattice parameter being lowered to $a = 5.34$ A at $PuH_{2,5}$ [5]. At 75 at. % H, hexagonal PuH_3 occurs [3, 5], with $a = 3.78$ A, $c = 6.76$ A [5], and 8 atoms per unit cell.

The chemical properties of plutonium hydride (of the approximate composition $PuH_{2.7}$) were studied by [6].

1. I. B. Johns, *U.S. Atomic Energy Comm., Publ.* MDDC-717, 1944 (Pressure-temperature-composition relationships).
2. J. E. Burke, *U.S. Atomic Energy Comm., Publ.* AECD-2124, 1944.
3. R. N. R. Mulford and G. E. Sturdy, *J. Am. Chem. Soc.*, **77**, 1955, 3449–3452.
4. F. H. Ellinger, unpublished information, quoted in [3].
5. Review by A. S. Coffinberry and M. B. Waldron in "Metallurgy and Fuels," pp. 393, 400, 401, Progress in Nuclear Energy, ser. V, vol. 1, Pergamon Press Ltd., London, 1956.
6. F. Brown, H. M. Ockenden, and G. A. Welch, *J. Chem. Soc.* (*London*), **1955**, 3932–3936.

$\bar{3}.7332$
2.2668

H-Re Hydrogen-Rhenium

Re annealed at 1200°C in hydrogen shows the same lattice parameters as when annealed at 1300°C in vacuo [1].

1. W. Trzebiatowski, *Z. anorg. Chem.*, **233**, 1937, 377.

$\bar{3}.9910$
2.0090

H-Rh Hydrogen-Rhodium

"[1] observed no direct occlusion; [2] found very little absorption by rhodium sponge, when heated in hydrogen, whence nonocclusion by this metal in the dense state might be inferred. [3] reported considerable absorptions, but this was not confirmed by [4]. The conflicting data suggest a highly variable occlusive capacity of Rh" [5].

1. A. Sieverts and E. Jurisch, *Ber. deut. chem. Ges.*, **45**, 1912, 221–229.
2. A. Gutbier and O. Maisch, *Ber. deut. chem. Ges.*, **52**, 1919, 1368–1374. Cf. A. Gutbier and W. Schieferdecker, *Z. anorg. Chem.*, **184**, 1929, 305–332.
3. I. I. Zhukov, *Izvest. Inst. Fiz.-Khim. Anal.*, **3**, 1926, 114–141, 461–462; *Chem. Abstr.*, **21**, 1927, 3800; **22**, 1928, 4399.
4. I. F. Adadurov and N. I. Pevny, *Russ. J. Appl. Chem.*, **10**, 1937, 1216–1219; *Met. Abstr.*, **5**, 1938, 759.
5. D. P. Smith, "Hydrogen in Metals," University of Chicago Press, Chicago, 1948.

$\bar{3}.9961$
2.0039

H-Ru Hydrogen-Ruthenium

"[1], in experiments directed only to the polarization capacity of ruthenium cathodes after charging under high hydrogen pressures, seem to have obtained indication of a considerable cathodic occlusion" [2].

The sorption (presumably for the most part adsorption) of hydrogen of atmospheric pressure by fine ruthenium black was studied by [3] up to 186°C.

1. L. Cailletet and E. Collardeau, *Compt. rend.*, **119**, 1894, 830–834.
2. D. P. Smith, "Hydrogen in Metals," pp 190–191, University of Chicago Press, Chicago, 1948.
3. A. Gutbier and W. Schieferdecker, *Z. anorg. Chem.*, **184**, 1929, 305–332.

$\bar{3}.9180$
2.0820

H-Sb Hydrogen-Antimony

"[1] found no occlusion; but [2] reported solubilities of 0.62 at 200°C and 1.4 at 400°C, in cubic centimeters per 100 g [6.7×10^{-3} and 1.5×10^{-2} at. %, respectively], the hydrogen being measured at NTP. In view of the well-known formation of stibine, any true occlusion seems unlikely" [3].

[4] claimed to have prepared a solid hydride "Sb_2H_2," but the existence of this compound was denied by [5].

1. A. Sieverts and W. Krumbhaar, *Ber. deut. chem. Ges.*, **43**, 1910, 893–900.
2. K. Iwasé, *Science Repts. Tôhoku Univ.*, **15**, 1926, 531–566 (in English).
3. D. P. Smith, "Hydrogen in Metals," University of Chicago Press, Chicago, 1948, especially pp. 191–192, 286.
4. E. J. Weeks and J. G. F. Druce, *Rec. trav. chim.*, **44**, 1925, 970; *J. Chem. Soc.*, **127**, 1925, 1069–1072.
5. C. Brink, G. Dallinga, and R. J. F. Nivard, *Rec. trav. chim.*, **68**, 1949, 234–236; abstract in *Structure Repts.*, **12**, 1949, 17.

$\bar{3}.8261$
2.1739
H-Sm Hydrogen-Samarium

Sm combines with hydrogen at elevated temperatures, yielding a product which is easily dissociated by higher heating [1]. SmH_2 has been shown to be, like other rare-earth hydrides, of the f.c.c. CaF_2 (C1) type of structure, $a = 5.376 \pm 3$ A [2]. According to [3], SmH_3 is isostructural with PuH_3.

1. C. Matignon, *Compt. rend.*, **131,** 1900, 891–892.
2. C. E. Holley et al., *J. Phys. Chem.*, **59,** 1955, 1226–1228.
3. Short statement by A. S. Coffinberry and M. B. Waldron in "Metallurgy and Fuels," p. 401, Progress in Nuclear Energy, ser. V, vol. 1, Pergamon Press Ltd., London, 1956.

$\bar{3}.9290$
2.0710
H-Sn Hydrogen-Tin

A certain solubility of hydrogen in liquid Sn was found by [2, 3], contrary to the negative result of [1], who made measurements up to 800°C. The following data were reported by [2]: 0.44, 0.57, 0.70, 0.93, 1.40, and 1.46 cc hydrogen per 100 g Sn (NTP) at 400, 553, 704, 798, 900, and 1005°C, respectively; and [3] found no solubility at 306°C, but 0.42 cc per 100 g Sn (extrapolated, NTP) at 800°C.

As to solid Sn, no occlusion of hydrogen was found by [1, 2]. On the other hand, [4] reported diffusion of hydrogen through a tin cathode, and [5] found a measurable hydrogen content in pure commercial tin.

It was concluded by [6] that "the occurrence of any diffusive occlusion by tin is doubtful, so far as present evidence indicates."

1. A. Sieverts and W. Krumbhaar, *Ber. deut. chem. Ges.*, **43,** 1910, 896.
2. K. Iwasé, *Science Repts. Tôhoku Univ.*, **15,** 1926, 531–566. Tin of "extra purity" was used.
3. L. L. Bircumshaw, *Phil. Mag.*, **1,** 1926, 510–522. The tin used contained 0.13 wt. % impurities (As, Pb, Cu, Fe, Ni, and Zn).
4. D. Alexejew and L. Savinina, *Zhur. Russ. Fiz.-Khim. Obshchestva*, **56,** 1924, 560.
5. W. Hessenbruch, *Z. Metallkunde*, **21,** 1929, 46–55.
6. D. P. Smith, "Hydrogen in Metals," University of Chicago Press, Chicago, 1948, especially pp. 191–192, 286.

$\bar{2}.0608$
1.9392
H-Sr Hydrogen-Strontium

See Ca-H.

$\bar{3}.7461$
2.2539
H-Ta Hydrogen-Tantalum

For a detailed discussion of the rather extensive literature on H-Ta up to 1946, reference is made to [1]. This author concluded that "with regard to the constitution of Ta-H, there still exists some uncertainty. Sievert's isobar [see Fig. 439] indicates a close analogy to Pd-H, with α and β phases only and an interval of immiscibility marked by the drop of the isobar; and most of our incidental information appears to be in keeping with this simple constitution" [2].

The existence of only two solid phases at room temperature, (Ta) and one intermediate phase (hydride), was corroborated by a recent X-ray study [3]. With regard to the (Ta) phase there is conflicting evidence as far as the maximum amount of hydrogen taken in solid solution is concerned (lattice parameter of pure Ta, $a = 3.297$ kX): ~12 at. %, $a = 3.354$ kX, 20°C [4]; 16.7 at. %, 400–500°C [5]; >34 at. %, $a = 3.399$ kX, 20°C [6]; ~17 at. % H, $a = 3.33$ kX, 20°C [3]. For Fig. 440, a room-temperature solubility of 20 at. % H was chosen. (See Note Added in Proof.)

The hydride (β) phase has, according to [4, 3], an orthorhombic (deformed b.c.c.) structure, with $a = 4.67$–4.79 A, $b = 4.67$–4.78 A, $c = 3.39$–3.44 A [3]. Its homogeneity range extends from about $TaH_{0.6}$ (37.5 at. % H) [3] to the highest

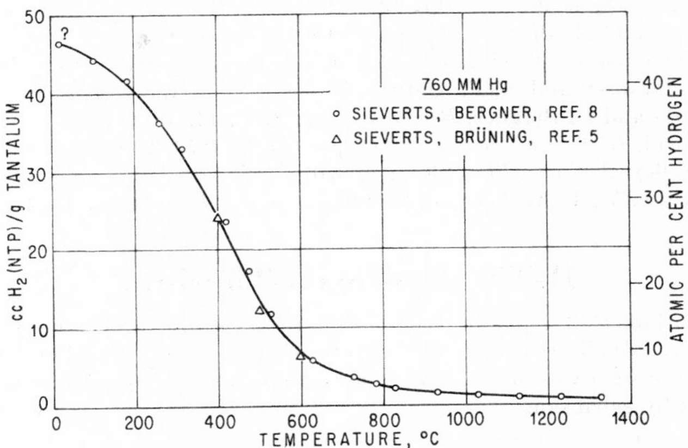

Fig. 439. H-Ta. Solubility isobar (1 atm). (See also Fig. 440.)

hydrogen content obtainable, 47.3 at. % ($TaH_{0.9}$) according to [4]. A second hydride phase, hexagonal Ta_2H, suggested by [4], was not corroborated by [3], who ascribed the weak diffraction lines observed to a nitride (Ta_2N).

The kinetics of the reaction of Ta with hydrogen and the stability of the reaction products to high vacuum were studied by [7].

According to [3], the deuterium-tantalum system is very similar to H-Ta (Fig. 440).

Supplement. Lattice constants and magnetic susceptibilities of 15 Ta-H alloys with up to 43 at. % H ($TaH_{0.75}$) prepared at 300°C and below were measured by

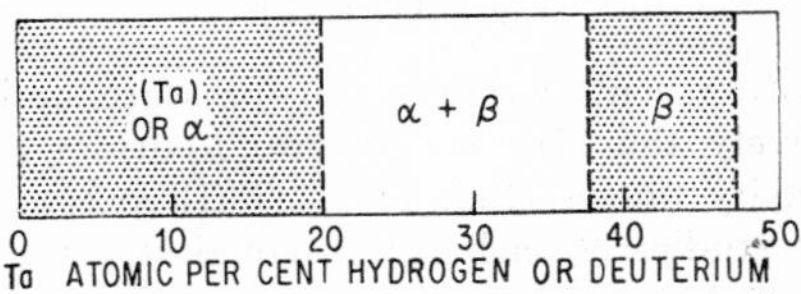

Fig. 440. H-Ta. (See also Fig. 439, Supplement, and Note Added in Proof.)

[9]. The heterogeneous field between the (Ta) phase and the orthorhombic β phase was found to extend from $TaH_{0.37}$ to $TaH_{0.47}$ (27–32 at. % H) (cf. above).

Note Added in Proof. On the basis of X-ray and electrical-resistance measurements as well as heat-capacity data by [10], [11] advanced the following conception of the phase relationships in the temperature range −145 to +70°C: the solubility of H in Ta, being at least 40 at. % above 50°C, decreases to 10 and 0 at. % at 0 and −145°C, respectively, by precipitating a hydride (β phase) of the approximate composition Ta_2H. Ta_2H is b.c. tetragonal, $a = 3.38$ A, $c = 3.41$ A; however, "as the hydrogen concentration in the β phase is increased beyond that of Ta_2H, the lattice appears to gradually distort into a f.c. orthorhombic structure."

The strongly temperature-dependent solubility of H in Ta claimed by these authors could account for the afore-mentioned conflicting solubility data.

1. D. P. Smith, "Hydrogen in Metals," University of Chicago Press, Chicago, 1948; contains 23 references (pp. 177, 366).
2. [1], p. 172.
3. G. Brauer and R. Hermann, *Z. anorg. Chem.*, **274,** 1953, 11–23.
4. G. Hägg, *Z. physik. Chem.*, **B11,** 1930-1931, 433–454.
5. A. Sieverts and H. Brüning, *Z. physik. Chem.*, **A174,** 1935, 365–369.
6. F. H. Horn and W. T. Ziegler, *J. Am. Chem. Soc.*, **69,** 1947, 2762–2769.
7. E. A. Gulbransen and K. F. Andrew, *Trans. AIME*, **188,** 1950, 586–599.
8. A. Sieverts and E. Bergner, *Ber. deut. chem. Ges.*, **44,** 1911, 2394–2402.
9. B. Stalinski, *Bull. acad. polon. sci., Classe III*, **2,** 1954, 245–247 (in English).
10. K. K. Kelley, *J. Chem. Phys.*, **8,** 1940, 316.
11. T. R. Waite, W. E. Wallace, and R. S. Craig, *J. Chem. Phys.*, **23,** 1956, 634.

$\bar{3}.6378$
2.3622

H-Th Hydrogen-Thorium

Early authors [1–3] reported that thorium reacts with hydrogen up to the composition ThH_2 or even ThH_4. On the basis of solubility studies, [4, 5] favored the concept that the system was of the solid-solution type; and [6] found that absorption is accompanied by expansion.

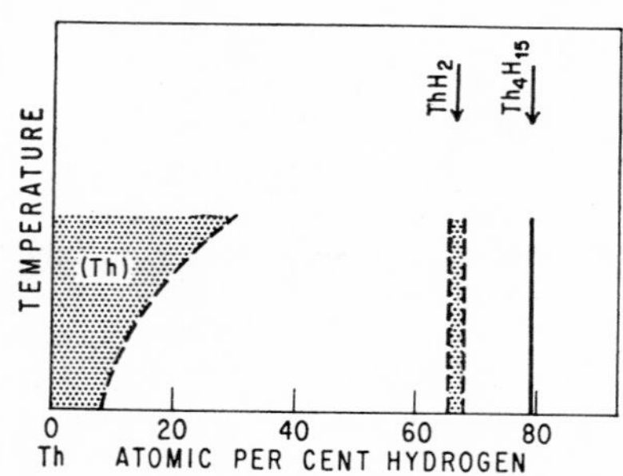

Fig. 441. H-Th

The results of more recent careful vapor-pressure and X-ray measurements [7–10] showed conclusively the existence of two hydrides [11] having presumably rather narrow homogeneity ranges (Fig. 441). ThH_2 possesses, according to X-ray and neutron-diffraction studies [9], a b.c. tetragonal structure (isotypic with HfH_2, ZrH_2), with $a = 4.10 \pm 3$ A, $c = 5.03 \pm 3$ A, $c/a = 1.227$, and 6 atoms per unit cell.

The higher hydride, Th_4H_{15} [$ThH_{3.75}$, 78.95 at. % H, 180 cc (NTP) H_2 per g Th] according to [10], was shown to have a cubic body-centered structure, with $a = 9.11 \pm 2$ A and 76 atoms (4 Th_4H_{15}) per unit cell [10].

The measurements of dissociation pressure of [8] indicate that the solubility of hydrogen in solid thorium increases from about 13 at. % at 650°C to about 23 at. % at 900°C.

1. C. Winkler, *Ber. deut. chem. Ges.*, **24,** 1891, 873–899.
2. C. Matignon, *Compt. rend.*, **131,** 1900, 891–892.
3. C. Matignon and M. Delepine, *Compt. rend.*, **132,** 1901, 36–38, 232.

4. A. Sieverts and E. Roell, *Z. anorg. Chem.*, **153,** 1926, 289–308.
5. A. Sieverts and A. Gotta, *Z. anorg. Chem.*, **172,** 1928, 1–31.
6. A. Sieverts, A. Gotta, and S. Halberstadt, *Z. anorg. Chem.*, **187,** 1930, 156–164.
7. R. W. Nottorf, A. S. Wilson, R. E. Rundle, A. S. Newton, and J. E. Powell, Manhattan Project Report CC-2722.
8. M. W. Mallett and I. E. Campbell, *J. Am. Chem. Soc.*, **73,** 1951, 4850–4852.
9. R. E. Rundle, C. G. Shull, and E. O. Wollan, *Acta Cryst.*, **5,** 1952, 22–26.
10. W. H. Zachariasen, *Acta Cryst.*, **6,** 1953, 393–395.
11. For preparation of powdered Th via the hydrides by suitable temperature and pressure changes, see P. Chiotti and B. A. Rogers, *U.S. Atomic Energy Comm., Publ.* AECD 2974, 1950.

$\bar{2}.3231$
1.6769

H-Ti Hydrogen-Titanium*

A comprehensive account of the literature on the Ti-H system up to 1946 has been given by [1].

The phase diagram of Fig. 442 is that given by [2]; it is taken chiefly from [3] but with modifications indicated by the work of [2].

The technique employed by [3] was the measurement of equilibrium pressures of a known volume of hydrogen in a system with Ti, at a series of temperatures. Iodide Ti, containing less than 0.07 wt. % combined Fe, Si, Mg, Sb, and Cu, was used. According to this diagram, the Ti-H system is of the eutectoid type. The β phase decomposes at about 38 at. % H and 320°C into ($\alpha + \gamma$) (cf. Fig. 442). The γ phase has a variable solubility extending from 48 to 63 at. % H at the eutectoid temperature, and from about 60 to 63 at. % H at 600°C. The maximum solubility of H in α-Ti is 7.9 at. % at the eutectoid temperature. The maximum solubility of H in β-Ti is 49 at. % at 640°C, the boundary being cut off by the 1-atm isobar at this temperature. [3] checked the diagram derived from measurements of equilibrium pressure by microexamination of alloys annealed at 750°C, slowly cooled to 300°C, and held for 24 hr. The results were in close agreement [4].

The modifications made by [2] are all in the low-temperature portion of the diagram and are as follows: (*a*) addition of the curve showing the solubility limit of hydrogen in α-Ti, (*b*) inclusion of two eutectoid horizontals to represent the hysteresis observed in thermal-analysis and dilatometric studies, (*c*) a shift of the eutectoid composition to higher hydrogen concentrations, and (*d*) an increase in the hydrogen concentration of the lower limit of the hydride phase. Alterations (*a*), (*c*), and (*d*) are based on microscopic work; as to (*d*), the structures "indicated that the lower limit of the γ-hydride phase at both 300°C and room temperature was above 50 at. % H."

High-purity iodide Ti was employed by [2]. The solubility of H in α-Ti below 125°C was found to be between 0.05 and 0.14 at. (0.001 and 0.0029 wt.) %. This points to an unusual discontinuity in the solubility curve at about 125°C (see Note Added in Proof). It might be thought to portend an invariant reaction. [2] recognized this and made efforts to reveal the existence of a lower hydride but were unsuccessful.

The decrease in solubility of H in α-Ti with decreasing temperature gives rise to "line markings" in microstructures [5, 2]. The hydride phase (γ) cannot be retained

* The text of this system was taken chiefly from Constitution of Titanium Alloy Systems, WADC Technical Reports 53-41, 1953, and 54-502, 1954. The permission of the author of this section, Dr. D. J. McPherson, to use his text is gratefully acknowledged.

in solid solution by quenching in cold water. It precipitates as a finely dispersed phase, either during the quench or on subsequent aging at room temperature.

[6] have examined the pressure-concentration relationships of hydrogen, deuterium, and tritium with Ti in the same manner as [3, 4]. Hydrogen was reacted with a relatively impure Ti powder at 300, 400, and 500°C. The experiments at 500°C were repeated with Mg-reduced Ti foil. The results with foil agree quite well with the phase boundaries deduced by [4] for Kroll-process Ti. Because of the

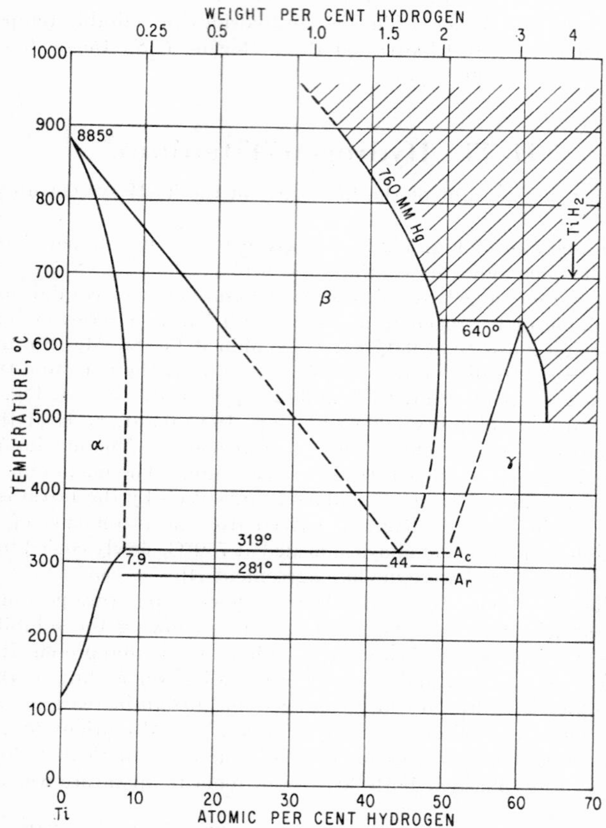

Fig. 442. H-Ti. (See also Noted Added in Proof.)

low purity of Ti used by [6], the quantitative values reported are not included in this review, but it is important that the general type of phase relationships in this system has been confirmed independently once again.

Comparison with Other Literature. [7] have compiled a survey of the literature on H in Ti. Their report discusses the other works which bear on the Ti-H system, such as those by [8–12]. Compositional limits of the $(\beta + \gamma)$ field for various isotherms may be deduced from the pressure studies of [8, 10, 11], and these are in generally good agreement with the boundaries given by [3]. Detailed comparisons are presented by [7].

The early X-ray work by [9], indicating a solubility of H in Ti of 33 at. %, was for many years the only information available on this system. This value is certainly erroneous as far as solubility of H in α-Ti is concerned. Insufficient information on the purity of materials and absorption temperatures used by [9] is given, to be certain; but if this system behaves like many other Ti eutectoid systems, he probably absorbed hydrogen in the β phase and cooled rapidly enough to preclude decomposition of the hypoeutectoid alloys into ($\alpha + \gamma$), obtaining instead metastable α', which, on X-ray examination, seemed to indicate a wide homogeneous field of α solid solutions. The same explanation certainly holds for the solubility value 31 at. % reported more recently by [13].

The γ Phase and Crystal Structures. Considerable confusion and misinformation exist in the literature concerning solid-state intermediate phases in the Ti-H system. The formulas Ti_2H, TiH, and TiH_2 have been used in connection with phases presumed to exist. Ti_2H, corresponding to 33 at. % H, was the limiting boundary of the terminal Ti solid solution determined by [9]. He characterized the limiting composition by use of this formula but did not imply that the phase existed per se. At any rate, his boundary has since been shown to be far too high, so that Ti_2H, as used by [14] and others, can be discounted as nonexistent. [9] also found a single-phase region extending between compositions corresponding to TiH and (nearly) TiH_2. This is in good agreement with later results (see above). Assigning one or the other of these formulas to the single broad phase which exists is rather meaningless. Positioning the H atoms in the lattice by X-ray studies has been impossible thus far. [9] postulated that the zincblende structure would represent the lower end of this homogeneous region, while the fluorite (CaF_2) structure would represent the upper boundary.

There is conflicting experimental evidence whether the homogeneity range of the γ phase extends nearly [9, 3] or wholly [13] up to the composition TiH_2, or whether even a distinct phase TiH_2 exists. [11] made the claim of preparing stoichiometric TiH_2 [66.67 at. (4.04 wt.) % H], which they report to be unstable below about 400°C. The compound, according to [11], is obtained only as a metastable, or transitory, phase on cooling and decomposes over a period of days when stored at room temperature. They state that there is unequivocal evidence that the composition and gross properties are definite and reproducible and that the X-ray pattern contains moderately strong lines in addition to those of the γ phase. [11] used only commercial-purity Ti in this study and pressures no greater than 800 mm Hg. [12] tried to fortify the evidence for a TiH_2 phase by conducting a high-pressure study. They obtained a pressure plateau in the isotherms above about 64.5 at. % H and at pressures of the order of 10,000 cm Hg, which presumably would indicate the presence of a two-phase field ($\gamma + TiH_2$).

According to [3], the presence of H has no measurable effect on the lattice parameter of the α-Ti phase. The structure of the γ phase is f.c.c., and increasing hydrogen content has been shown to increase the parameter of this lattice. Lattice parameters of the γ phase were measured by [9, 15, 16, 3, 11, 13]. (See Note Added in Proof.)

Kinetics and Diffusion. The kinetics of the reaction of Ti with H_2 as a function of time, temperature, and pressure were studied by [17, 18]; the diffusion of hydrogen in α- and β-Ti was studied by [19].

Note Added in Proof. According to an X-ray and neutron-diffraction study [20] of specimens finally annealed at 350°C, the composition range of the f.c.c. phase extends from approximately 60 to 66 at. % H or D, the structure being of the CaF_2 (C1) type, with $a = 4.440 \pm 3$ A (at $TiD_{1.971}$). In approaching the composition TiD_2, broadening of diffraction lines occurs, indicating deformation to a f.c. tetragonal cell.

On the basis of X-ray and neutron-diffraction work as well as metallographic observations, [21] suggested a tetragonal phase (substructure f.c. tetragonal, with $a = 4.42$ A, $c = 4.18$ A, $c/a = 0.946$ for the deuteride) containing more than 62 at. % H or D to be in equilibrium with α_{Ti} at 100–255°C.

A magnetic investigation of the Ti-H system has been made by [22]. Results are discussed in [23]. Electrical-resistivity–concentration isotherms of β-phase alloys have been measured by [24].

[25] obtained data on the solubility of H in α_{Ti} between 100 and 300°C in measurements of mechanical damping (heating rate 3°C per min). The solubility curve obtained shows no discontinuity around 125°C (cf. above) and, at higher temperatures, runs at somewhat lower hydrogen contents than that of [2] plotted in Fig. 442.

1. D. P. Smith, "Hydrogen in Metals," University of Chicago Press, Chicago, 1948.
2. G. A. Lenning, C. M. Craighead, and R. I. Jaffee, *Trans. AIME*, **200,** 1954, 367–376.
3. A. D. McQuillan, *Proc. Roy. Soc.* (*London*), **A204,** 1950, 309–323.
4. In a later investigation (*J. Inst. Metals*, **79,** 1951, 371–378), A. D. McQuillan repeated these studies, using Mg-reduced Ti of commercial purity. The region of the system which could be determined was much reduced by two factors: longer times to achieve equilibrium at low temperatures and greater hydrogen equilibrium pressures, so that the 1-atm isobar occurred at lower temperatures and concentrations of hydrogen. It was determined that the $\alpha \rightarrow \beta$ transformation occurred over a temperature interval of 850–955°C and, as in other systems, the $(\alpha + \beta)$ field was considerably broadened.
5. C. M. Craighead, G. A. Lenning, and R. I. Jaffee, *Trans. AIME*, **194,** 1952, 1317–1319.
6. R. M. Haag and F. J. Shipko, *U.S. Atomic Energy Comm., Publ.* KAPL-955, 1953; *J. Am. Chem. Soc.*, **78,** 1956, 5155–5159.
7. G. A. Lenning, C. M. Craighead, and R. I. Jaffee, Battelle Memorial Institute, Interim Report 1, Contract DA 33-019-ORD-220 with Watertown Arsenal, April, 1952.
8. L. Kirschfeld and A. Sieverts, *Z. physik. Chem.*, **A145,** 1929, 227–240.
9. G. Hägg, *Z. physik. Chem.*, **B11,** 1930-1931, 433–454; see also **B12,** 1931, 33–56.
10. C. F. P. Bevington, S. L. Martin, and D. H. Mathews, *Proc. Intern. Congr. Pure and Appl. Chem. 11th Congr. London*, 1947, pp. 3–16.
11. T. R. P. Gibb and H. W. Kruschwitz, *J. Am. Chem. Soc.*, **72,** 1950, 5365–5369.
12. T. R. P. Gibb, J. J. McSharry, and R. W. Bragdon, *J. Am. Chem. Soc.*, **73,** 1951, 1751–1755.
13. A. Chretien, W. Freundlich, and M. Bichara, *Compt. rend.*, **238,** 1954, 1423–1424.
14. C. S. Barrett, "Structure of Metals," 2d ed., p. 244, McGraw-Hill Book Company, Inc., New York, 1952.
15. J. Fitzwilliam, A. Kaufmann, and C. Squire, *J. Chem. Phys.*, **9,** 1941, 678–682.
16. P. Ehrlich, *Angew. Chem.*, **59A,** 1947, 163.
17. E. A. Gulbransen and K. F. Andrew, *J. Electrochem. Soc.*, **96,** 1949, 364–376.
18. E. A. Gulbransen and K. F. Andrew, *Trans. AIME*, **185,** 1949, 741–748.
19. R. J. Wasilewski and G. L. Kehl, *Metallurgia*, **50,** 1954, 225–230.
20. S. S. Sidhu, L. Heaton, and D. D. Zauberis, *Acta Cryst.*, **9,** 1956, 607–614.
21. L. D. Jaffe, *Trans. AIME*, **206,** 1956, 861.
22. W. Trzebiatowski and B. Stalinski, *Bull. acad. polon. sci.*, (III)**1,** 1953, 131; quoted in [23].
23. E. P. Wohlfarth, *Acta Met.*, **4,** 1956, 225–227.
24. S. L. Ames and A. D. McQuillan, *Acta Met.*, **4,** 1956, 602–610.
25. W. Köster, L. Bangert, and M. Evers, *Z. Metallkunde*, **47,** 1956, 564–570.

$\bar{3}.6930$
2.3070

H-Tl Hydrogen-Thallium

According to [1], neither solid nor liquid Tl (investigated up to 600°C) dissolves any hydrogen. See also [2].

For preparation and properties of the unstable volatile hydride TlH, see the chemical literature.

1. A. Sieverts and W. Krumbhaar, *Ber. deut. chem. Ges.*, **43**, 1910, 896.
2. D. P. Smith, "Hydrogen in Metals," University of Chicago Press, Chicago, 1948, especially pp. 191–192, 285–286.

$\bar{3}.6268$
2.3732

H-U Hydrogen-Uranium

The solubility of hydrogen in uranium was detected by [1], the formation of a hydride by [2]. A systematic study on both the H-U and D-U systems was initiated

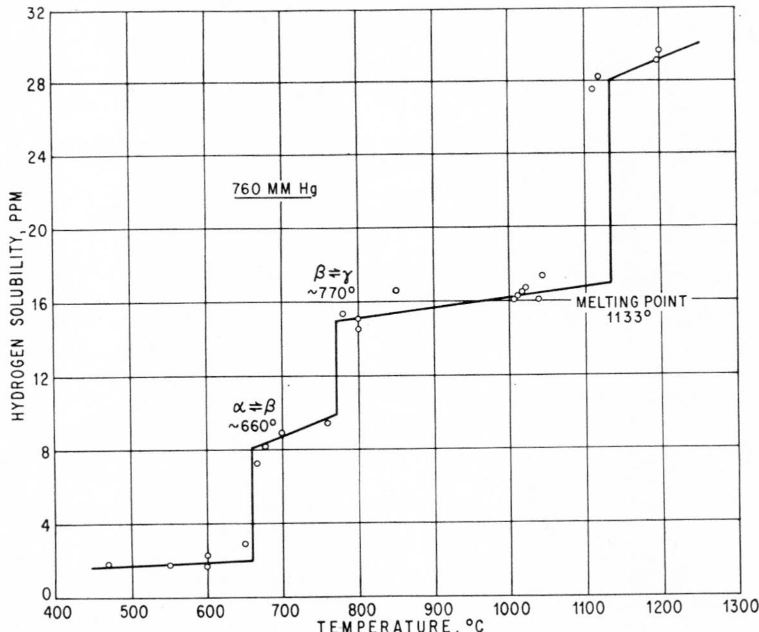

Fig. 443. H-U. Solubility isobar (1 atm). (See also Fig. 443*a*.)

by the wartime Manhattan Project; a review of this work has been given by [3], together with a list of all project reports, the substance of which is now available in the open literature (especially [4–6]). In the following, reference is made to those compilations rather than to the original project literature.

Solubility of Hydrogen in Solid and Liquid Uranium. The solubility isobar for hydrogen pressure of 1 atm is shown in Fig. 443 [7]. It shows that the equilibrium content of hydrogen in the α-uranium phase is about 2 ppm (0.05 at. %) and changes slightly with temperature. The $\alpha \rightarrow \beta$ transformation, at 660°C, causes the solubility to increase from 2 to about 8 ppm (0.05 and 0.2 at. %, respectively);

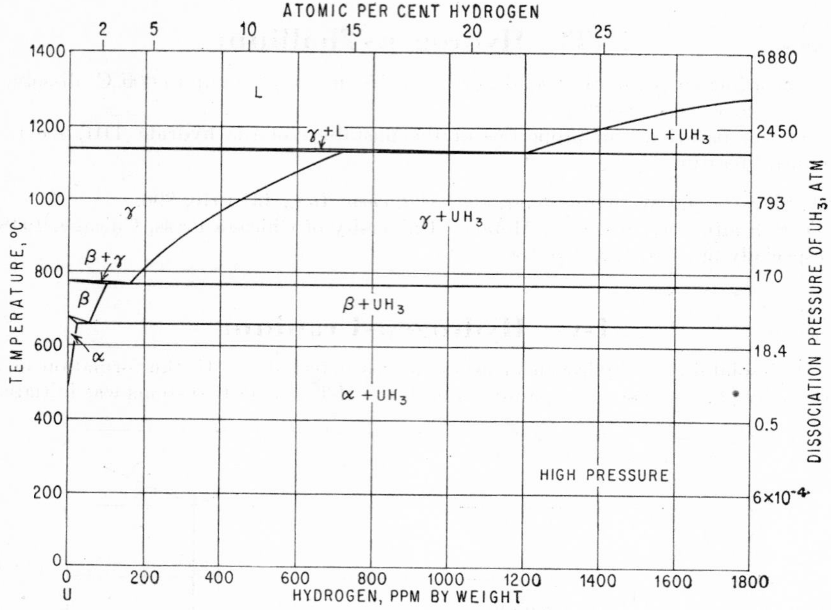

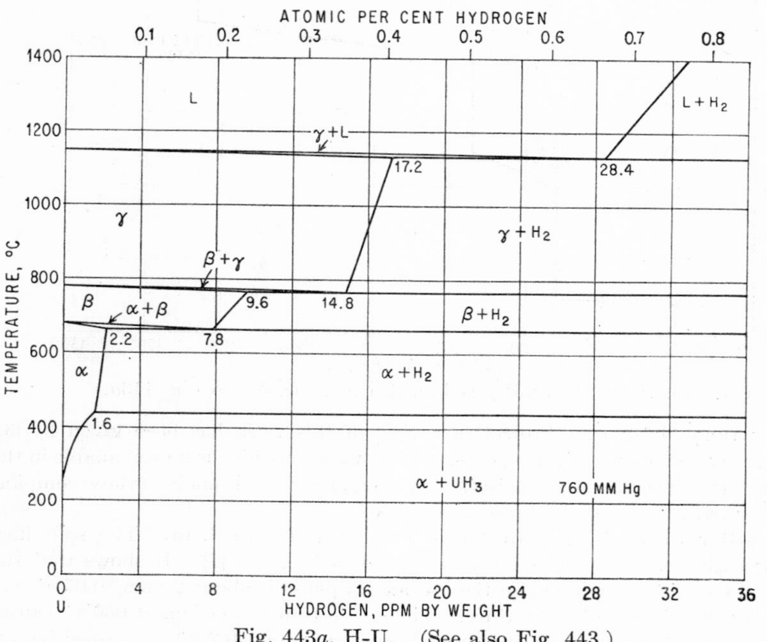

Fig. 443*a*. H-U. (See also Fig. 443.)

and the β-γ transformation, at about 770°C, leads to another increase to 15 ppm (0.35 at. % H). Melting, which occurs at 1133°C, increases the solubility from 17 to 28 ppm (0.40 to 0.66 at. % H), and the hydrogen content of the liquid metal continues to increase rapidly with temperature, reaching 30 ppm (0.71 at. % H) at 1250°C.

Uranium Hydride and Deuteride. Detailed studies of the preparation, composition, and physical and chemical properties, especially of the hydride, have been carried out since 1943 [9–11, 4–6, 3, 12, 13]. The hydride was shown to have the formula UH_3 (1.25 wt. % H) [4, 3]; neither hydrides prepared at high hydrogen pressures nor those prepared with excess of metal showed any alteration in the lattice spacings [11]. However, [13] concluded from pressure measurements that a dilute solid solution of U in UH_3 is formed [14]. The decomposition temperature for the hydride at 1 atm hydrogen pressure given as 430–436°C by [3, 12] obviously refers to the U-saturated hydride.

The structures of UH_3 and UD_3 were studied by X-ray [11, 3] as well as neutron [15] diffraction. The compounds have primitive cubic structures, with $a = 6.6445 \pm 8$ A [11] for UH_3 and $a = 6.633 \pm 2$ A [11] for UD_3 (32 atoms per unit cell).

Considerations by [16] in support of the hydrogen positions originally proposed by [11] are obsolete.

The occurrence of a second crystal form of UH_3 was discovered by [17, 18]. It appears under certain conditions of preparation at lower temperatures, has a simple cubic lattice, with $a = 4.16$ A and 8 atoms per unit cell, and is believed [18] to be a metastable form. It is interesting to note in this connection that UH_3 is said [19] to have ferromagnetic properties below 173°K.

Supplement. [20] studied microscopically the UH_3 inclusions in metallic uranium and measured the formation and decomposition temperatures of the hydride under various (elevated) pressures. A tentative temperature-composition-pressure diagram [21] is given.

[22] measured the solubility of H in U at 295°C under various low pressures.

Note Added in Proof. The phase diagrams of Fig. 443*a* are due to [23]. They are given on a wt. % scale (1 ppm $= 10^{-4}$ wt. %) to facilitate comparison with Fig. 443. The lower diagram shows phase relations at 1 atm hydrogen pressure, the upper one those at pressures above the dissociation pressures of UH_3 [24].

1. A. Sieverts and E. Bergner, *Ber. deut. chem. Ges.*, **45,** 1912, 2576–2583.
2. F. H. Driggs, U.S. Patents 1,816,830 and 1,835,024 (1929).
3. J. J. Katz and E. Rabinowitch, "The Chemistry of Uranium," Part I, National Nuclear Energy Series, Div. VIII, vol. 5, pp. 183–213, McGraw-Hill Book Company, Inc., New York, 1951.
4. F. H. Spedding et al., *Nucleonics*, **4**(1), 1949, 4–15 (Preparation, Composition, Physical Properties of Uranium Hydride).
5. A. S. Newton et al., *Nucleonics*, **4**(2), 1949, 17–25 (Radiochemical and Chemical Properties of Uranium Hydride).
6. J. C. Warf et al., *Nucleonics*, **4**(3), 1949, 43–47 (Uranium Hydride: Dispersions in Mercury).
7. This figure, based on project work by the Battelle group, was taken from [3]; however, the melting point of U is shown at 1133°C [8] rather than below 1100°C as in the original diagram.
8. [3], pp. 150–152.
9. J. Gueron and L. Yaffe, *Nature*, **160,** 1947, 575.
10. J. E. Burke and C. S. Smith, *J. Am. Chem. Soc.*, **69,** 1947, 2500–2502.
11. R. E. Rundle, *J. Am. Chem. Soc.*, **69,** 1947, 1719–1723.
12. H. E. Flotow and B. M. Abraham, *U.S. Atomic Energy Comm., Publ.* AECD-3074,

1951. *Nuclear Sci. Abstr.*, **5**, 1951, 445. "Dissociation pressures of uranium hydride and uranium tritide."
13. T. R. P. Gibb, J. J. McSharry, and H. W. Kruschwitz, *J. Am. Chem. Soc.*, **74**, 1952, 6203–6207.
14. The first appearance of the uranium phase "depends on the temperature; for low temperatures it appears as soon as small amounts of hydrogen have been withdrawn [from $UH_{3.00}$], whereas at 650°C the uranium phase does not appear until approximately 8% of the original hydrogen content has been removed" [13].
15. R. E. Rundle, *J. Am. Chem. Soc.*, **73**, 1951, 4172–4174.
16. L. Pauling and F. J. Ewing, *J. Am. Chem. Soc.*, **70**, 1948, 1660–1661.
17. R. Caillat, H. Coriou, and P. Perio, *Compt. rend.*, **237**, 1953, 812–813.
18. R. N. R. Mulford, F. H. Ellinger, and W. H. Zachariasen, *J. Am. Chem. Soc.*, **76**, 1954, 297–298.
19. W. Trzebiatowski, A. Sliwa, and B. Stalinski, *Roczniki Chem.*, **26**, 1952, 110–112 (English summary).
20. H. Mogard and G. Cabane, *Rev. mét.*, **51**, 1954, 617–622.
21. The diagram is based on the assumption that Sievert's square-root relation is valid over a wide pressure range.
22. H. C. Mattraw, *J. Phys. Chem.*, **59**, 1955, 93–94.
23. M. Mallett, unpublished information (1954). Diagrams reproduced in H. A. Saller and F. A. Rough, Compilation of U.S. and U.K. Uranium and Thorium Constitutional Diagrams, *U.S. Atomic Energy Comm., Publ.* BMI-1000, June, 1955.
24. Attention should be paid to the fact that the upper diagram does not represent phase relations at *constant* pressure; rather the phase boundaries shown are projections of phase boundaries that actually lie on a three-dimensional pressure-concentration-temperature surface.

$\bar{2}.2963$
1.7037

H-V Hydrogen-Vanadium

The literature on the H-V system has been reviewed recently [1, 2]. The information available is insufficient to elucidate the constitution of the system.

Early observations on the interaction of vanadium and hydrogen were made by [3–5]. The absorption of hydrogen is accompanied by a decrease in density [6, 7] and was found to be exothermic [6], in accordance with the decline in solubility with rising temperature. Solubility isotherms and isobars up to 1100°C were measured by [8, 7]; below 400°C, equilibrium could not be reached so that the smooth slope of all these curves does not exclude the possible existence of definite hydrides (cf. [7, 9, 10]). [11] studied absorption products by X-ray diffraction; new lines were observed but they were too weak and diffuse to allow interpretation. No shift in the lines of the V phase could be observed. The highest absorption at room temperature and 1 atm hydrogen pressure was observed by [11], namely, 50 at. % or 218 cc (NTP) H_2 per g vanadium. (Cf. H-Nb, H-Ta systems.)

1. D. P. Smith, "Hydrogen in Metals," University of Chicago Press, Chicago, 1948.
2. W. B. Pearson, *J. Iron Steel Inst.*, **164**, 1950, 149–159.
3. H. Roscoe, *Phil. Trans. Roy. Soc. (London)*, **159**, 1869, 691; *Liebigs Ann. Ergänzungsbd.*, **7**, 1870, 70.
4. W. Muthmann, L. Weiss, and R. Riedelbauch, *Liebigs Ann.*, **355**, 1907, 85–92.
5. W. Prandl and H. Manz, *Z. anorg. Chem.*, **79**, 1913, 209–222.
6. A. Sieverts and A. Gotta, *Z. anorg. Chem.*, **172**, 1928, 1–31.
7. L. Kirschfeld and A. Sieverts, *Z. Elektrochem.*, **36**, 1930, 123–129.

8. H. Huber, L. Kirschfeld, and A. Sieverts, *Ber. deut. chem. Ges.*, **59**, 1926, 2891–2896.
9. K. Iwasé and M. Fukusima, *Nippon Kinzoku Gakkai-Shi*, **1**, 1937, 202–213. Not available to the author. *Chem. Abstr.*, **32**, 1938, 5750.
10. A. Sieverts and H. Moritz, *Z. anorg. Chem.*, **247**, 1941, 124–130. (V, Nb, Ta)-H systems compared.
11. G. Hägg, *Z. physik. Chem.*, **B11**, 1930-1931, 433–454. Vanadium 96.8 wt. % pure.

$\bar{3}.7388$
2.2612

H-W Hydrogen-Wolfram

W does not occlude hydrogen from the gas in a quantity measurable by ordinary methods [1–5]. Reports on the existence of a hydride [6, 7] are not convincing (see also [8]).

1. A. Sieverts and E. Bergner, *Ber. deut. chem. Ges.*, **44**, 1911, 2394–2402 (measurements up to 1500°C).
2. C. Agte, H. Becker-Rose, and G. Heyne, *Z. angew. Chem.*, **38**, 1925, 1128; "Gmelins Handbuch der anorganischen Chemie," System No. 54, p. 106, Verlag Chemie, G.m.b.H., Berlin, 1933.
3. E. Martin, *Arch. Eisenhüttenw.*, **3**, 1929, 407–416; *Metals & Alloys*, **1**, 1929-1930, 831–835.
4. K. Iwasé and M. Fukusima, *Nippon Kinzoku Gakkai-Shi*, **1**, 1937, 202–213.
5. D. P. Smith, "Hydrogen in Metals," University of Chicago Press, Chicago, 1948.
6. T. J. Dillon, *Proc. Phys. Soc. (London)*, **41**, 1929, 546.
7. G. Stead and B. Trevelyan, *Phil. Mag.*, **48**, 1924, 978.
8. C. J. Smithells, *Trans. Faraday Soc.*, **17**, 1921, 491.

$\bar{2}.0545$
1.9455

H-Y Hydrogen-Yttrium

Absorption of hydrogen by yttrium (formation of Y_2H_3?) was observed by [1].

1. C. Winkler, *Ber. deut. chem. Ges.*, **24**, 1891, 1966–1984, especially 1983–1984.

$\bar{3}.7653$
2.2347

H-Yb Hydrogen-Ytterbium

Ytterbium deuteride (maximum D content attained, $YbD_{1.98}$) was shown to have the orthorhombic SrH_2 (C29) type of structure with $a = 5.871$ A, $b = 3.561$ A, $c = 6.763$ A (each $\pm$ 0.005 A) [1].

1. W. L. Korst and J. C. Warf, *Acta Cryst.*, **9**, 1956, 452–454.

$\bar{2}.1880$
1.8120

H-Zn Hydrogen-Zinc

Early authors [1] claimed that zinc was incapable of occlusion of hydrogen. [3] found solubilities of 0.55, 1.16, and 1.73 cc (NTP) H_2 per 100 g Zn (3.2×10^{-3}, 6.8×10^{-3}, 10×10^{-3} at. % H) at 206, 310, and 400°C, respectively, under 1 atm hydrogen pressure; and [4, 5] showed Zn to be permeable to hydrogen between 300°C and its melting point. The absorption of hydrogen was shown [6] to be most intensive at the melting point of the metal.

"Retention of hydrogen by electrodeposited zinc has been noted by several observers and the commercial metal appears commonly to contain this gas. The nature of the retention as a nondiffusive occlusion is indicated by the results of [7], who found that quantities as great as 1 cc per g, or 7 volumes, were only partially removed by heating in vacuo but were gradually set free on repeated severe working and heating. The general occlusive behavior of zinc is seemingly similar to that of the better-investigated aluminum" [8].

As for ZnH_2, a nonvolatile white powder, see [9].

Note Added in Proof. [10] found the solubility of hydrogen in molten Zn at 516°C to be as low as 1.8 ($\pm$1) $\times$ 10^{-3} cc per 100 g of metal.

1. For four references (1874–1910), see [2].
2. D. P. Smith, "Hydrogen in Metals," University of Chicago Press, Chicago, 1948.
3. K. Iwasé, *Science Repts. Tôhoku Univ.*, **15,** 1926, 531–566.
4. H. G. Deming and B. C. Hendricks, *J. Am. Chem. Soc.*, **45,** 1923, 2857–2864.
5. B. C. Hendricks and R. R. Ralston, *J. Am. Chem. Soc.*, **51,** 1929, 3278–3285.
6. M. S. Kashkhozhev, *Metallurg,* no. 12, 1939, pp. 19–34 (in Russian).
7. H. Winterhager, *Aluminium-Arch.*, no. 12, 1938.
8. Quoted from [2], p. 283.
9. G. D. Barbaras, C. Dillard, A. E. Finholt, T. Wartik, K. E. Wilzbach, and H. I. Schlesinger, *J. Am. Chem. Soc.*, **73,** 1951, 4585–4590.
10. W. Hofmann and J. Maatsch, *Z. Metallkunde,* **47,** 1956, 89–95.

$\bar{2}$.0434
1.9566

H-Zr Hydrogen-Zirconium

The large capacity of Zr for dissolving hydrogen and the decrease in density which accompanies absorption were observed early [1]. Solubility isotherms and isobars were determined by [3, 4], those of [4] being the more reliable ones. The isobar for 760 mm Hg hydrogen pressure reproduced in Fig. 444 shows a saturation value of 240 cc per g Zr ($ZrH_{1.95}$, 66.1 at. % H) at room temperature. The pressure-composition isotherms suggest a terminal solid solution up to about 50 at. % H at temperatures above 700°C; the pressures at lower temperatures were below the lower limit of measurement.

[5] showed, by studies at low pressures, that hydrogen is more soluble in the β (high-temperature) form than in the α form of Zr and that the transition point is lowered by the presence of hydrogen.

An X-ray investigation of the system on slowly cooled preparations (which were not in equilibrium) was made by [6]. He concluded the existence of five solid phases but admitted that their boundaries as shown in Fig. 445 are very uncertain. For the intermediate phases, the following lattices were found:

β or "Zr_4H": f.c.c., with a = 4.664 kX
γ or "Zr_2H": h.c.p., with a = 3.335–3.339 kX, c = 5.453–5.455 kX
δ or "ZrH": f.c.c., with a = 4.765–4.768 kX
ϵ or "ZrH_2": f.c. tetragonal, a = 4.964 kX, c = 4.440 kX, c/a = 0.894 (see also [7])

In an extensive review, [8] considered the hydride phases of [6] to be unstable distortion products of α- and β-Zr and proposed a constitution similar to that of H-Pd. This view, however, can certainly not be maintained in the light of more recent work.

In corroboration [9] of the cursory results of [6], ZrH_2 was shown [10], by means of X-ray and neutron diffraction, to have a b.c. tetragonal structure, with $a = 3.520 \pm 3$ A, $c = 4.449 \pm 3$ A, $c/a = 1.264$, and 6 atoms per unit cell (isotypic with HfH_2 and ThH_2).

X-ray studies by [11, 12]—of Zr specimens reacted with hydrogen to produce a gradient zone of hydrogen in the surface—indicated a constitutional analogy to the H-Hf system, since tetragonal ZrH_2, a cubic phase (assumed to be Hägg's ZrH), and

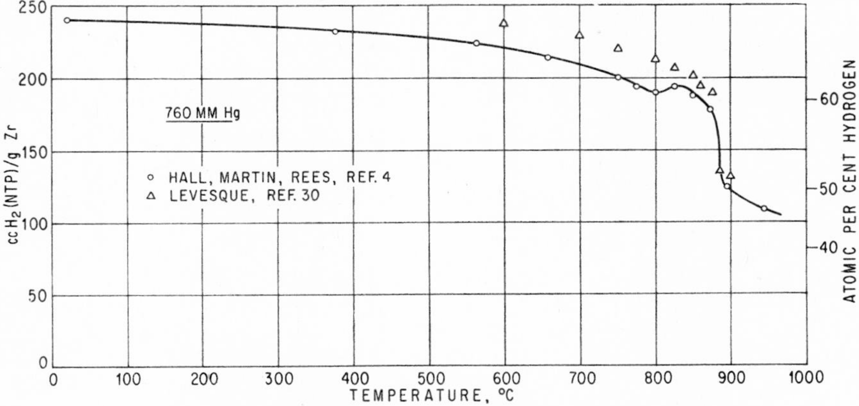

Fig. 444. H-Zr. Solubility isobar (1 atm).

a tetragonal (pseudocubic) phase [13] were observed: the last—according to electron-diffraction evidence [11]—is presumably in equilibrium with Zr. No hexagonal compound, corresponding to Hägg's [6] Zr_2H, could be found.

The solubility values of hydrogen in α-Zr given below (see also Fig. 446) were obtained by [11] in diffusion studies by extrapolation [14] of concentration-penetration curves back to the boundary between the α solid solution and a hydrogen-rich layer at the surface of the specimen: 5.2, 3.7, 2.6, and 0.1 at. % at 500, 450, 400°C, and room

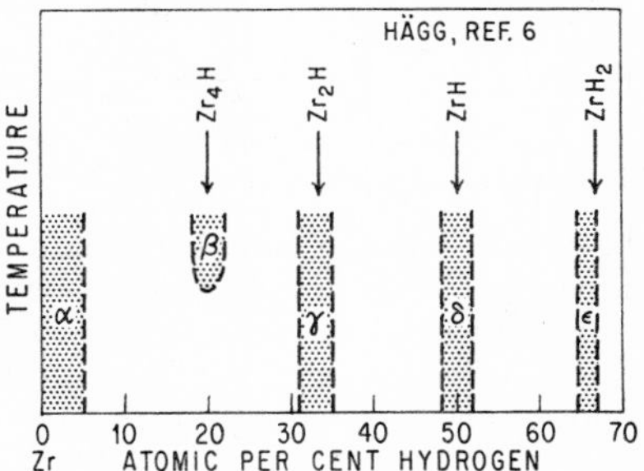

Fig. 445. H-Zr. (See main diagram in Fig. 447.)

temperature, respectively. The results of quenching experiments [11] agree roughly with these data.

The kinetics of the reaction of hydrogen with Zr were studied by [15–19, 12], at low pressures (getter!) by [15–17]; see also [20]. Calorimetric and densimetric measurements were made by [21, 22], magnetic studies by [7]. Another review, besides that of [8], on the literature up to about 1947 was made by [23]. A statistical-mechanical interpretation of the solubility of H in Zr was given by [24, 25].

Supplement. According to recent work, the constitution of Zr-H is very similar to that of the Ti-H system.

[26, 27] measured decomposition pressures in the temperature range 325–550°C and made X-ray studies on specimens prepared at temperatures of 300°C and lower (see lower diagram of Fig. 446). Besides the f.c.c. δ and the f.c. tetragonal ϵ phase, the X-ray work [26] indicated the existence of a f.c. tetragonal γ' phase of the probable composition Zr_2H [29]. γ' was observed, however, only in small or trace amounts together with the α and δ phases, and no evidence was found for this phase in the decomposition-pressure studies at higher temperatures.

The partial phase diagram for the high-temperature region proposed by [28] (see upper diagram of Fig. 446) is mainly based on pressure-composition isotherms for

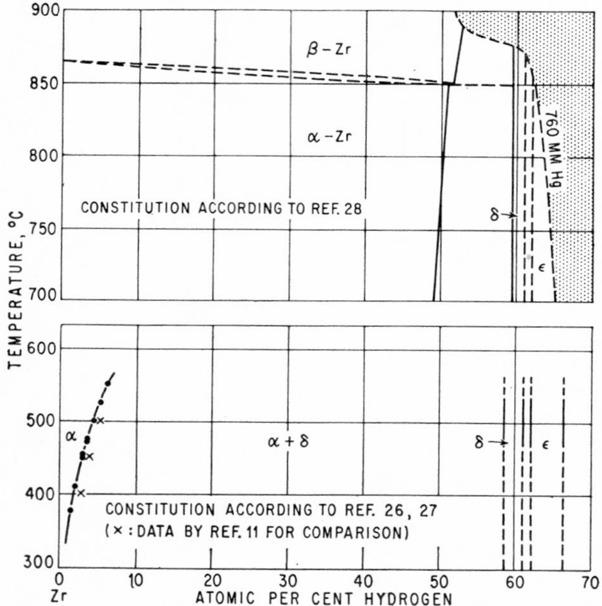

Fig. 446. H-Zr. (See main diagram in Fig. 447.)

the temperature range 600–900°C. Since the authors admitted that their method of measurement was unsatisfactory for the detection of narrow two-phase fields at low pressures, it came as no surprise when later investigators could not corroborate their positioning of the $\alpha + \beta$ field. The values obtained by [28] for the total uptake of hydrogen by Zr at 1 atm pressure were about 10% higher than those given by [4] (see Fig. 444 [30]).

The close similarity of the Zr-H and Ti-H phase diagrams was made evident by a high-temperature X-ray study up to 850°C of nine alloy compositions between 20 and 65 at. % H [32]. The eutectoid $\beta_{Zr} \rightleftarrows \alpha_{Zr}$ + hydride was located at 42 ± 3 at. % H, 560 ± 10°C. The solid solubility of hydrogen in β-Zr was found to exceed 50 at. % at 850°C. [32] suggest that only one hydride phase (called γ) exists, the composition range being approximately 60–67 at. % H at room temperature and 55–62 at. % H at 600°C (see data points in Fig. 447). "Changes in composition shift the axial ratio of the tetragonal structure through unity, so that two tetragonal hydrides and one cubic hydride appear to exist as different phases" [33].

Measurements of hydrogen equilibrium pressure for a series of Zr-H alloys in the composition range 1–44 at. % H at temperatures between 500 and 950°C were made by [34]. The results of this investigation, together with selected data of other authors, are plotted in Fig. 447, which gives a fairly complete picture of the constitution of the Zr-H system.

The role of surface films in the reaction of Zr with hydrogen was studied by [31].

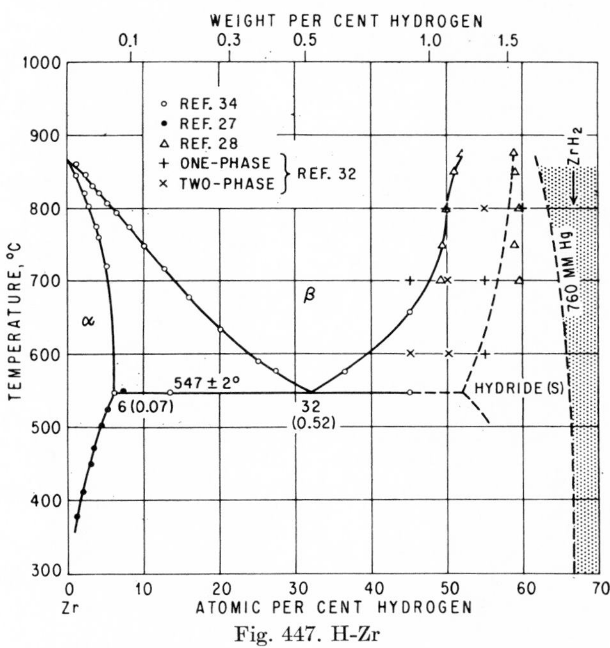

Fig. 447. H-Zr

1. See [2] for six references (1890–1921).
2. D. P. Smith, "Hydrogen in Metals," pp. 162–163, University of Chicago Press, Chicago, 1948.
3. A. Sieverts and E. Roell, *Z. anorg. Chem.*, **153,** 1926, 289–308.
4. M. N. A. Hall, S. L. H. Martin, and A. L. G. Rees, *Trans. Faraday Soc.*, **41,** 1945, 306–316.
5. J. H. De Boer and J. D. Fast, *Rec. trav. chim.*, **55,** 1936, 350–366.
6. G. Hägg, *Z. physik. Chem.*, **B11,** 1930-1931, 433–454.
7. J. Fitzwilliam, A. Kaufmann, and C. Squire, *J. Chem. Phys.*, **9,** 1941, 678–682.
8. [2], especially pp. 156–167.
9. The axes of [6] have to be transformed from f.c.t. to b.c.t.
10. R. E. Rundle, C. G. Shull, and E. O. Wollan, *Acta Cryst.*, **5,** 1952, 22–26.
11. C. M. Schwartz and M. W. Mallett, *Trans. ASM*, **46,** 1954, 640–651; discussion, pp. 652–654.
12. J. Belle, B. B. Cleland, and M. W. Mallett, *J. Electrochem. Soc.*, **101,** 1954, 211–214.
13. No attempt was made to verify the composition of this phase or that of the cubic phase labeled ZrH by Hägg. S. S. Sidhu, in a discussion on [11], suggested that the pseudocubic phase is identical with Hägg's "Zr_4H."

14. This extrapolation is based on the assumption that the concentration at the surface-layer–α-solid-solution interface represents saturation at the temperature of the experiment.
15. L. F. Ehrke and C. M. Slack, *J. Appl. Phys.*, **11**, 1940, 129–137.
16. S. Hukagawa and J. Nambo, *Electrotech. J. Japan*, **5**, 1941, 27–30.
17. W. G. Guldner and L. A. Wooten, *J. Electrochem. Soc.*, **93**, 1948, 223–235.
18. E. A. Gulbransen and K. F. Andrew, *J. Electrochem. Soc.*, **96**, 1949, 364–376.
19. E. A. Gulbransen and K. F. Andrew, *Trans. AIME*, **185**, 1949, 515–525.
20. [4], pp. 313–314.
21. A. Sieverts and A. Gotta, *Z. anorg. Chem.*, **172**, 1928, 1–31.
22. A. Sieverts, A. Gotta, and S. Halberstadt, *Z. anorg. Chem.*, **187**, 1930, 156–164.
23. C. F. P. Bevington, S. L. Martin, and D. H. Mathews, *Proc. Intern. Congr. Pure and Appl. Chem., 11th Congr., London*, 1947, pp. 3–16.
24. S. L. H. Martin and A. L. G. Rees, *Trans. Faraday Soc.*, **50**, 1954, 343–352.
25. A. L. G. Rees, *Trans. Faraday Soc.*, **50**, 1954, 335–342.
26. E. A. Gulbransen and K. F. Andrew, *J. Electrochem. Soc.*, **101**, 1954, 474–480; discussion, **102**, 1955, 357.
27. E. A. Gulbransen and K. F. Andrew, *Trans. AIME*, **203**, 1955, 136–144.
28. R. K. Edwards, P. Levesque, and D. Cubicciotti, *J. Am. Chem. Soc.*, **77**, 1955, 1307–1311.
29. According to [26], the γ' phase was first recognized by K. H. Jack.
30. Data taken from P. Levesque, Dissertation, Illinois Institute of Technology, Chicago, Ill., June, 1953.
31. E. A. Gulbransen and K. F. Andrew, *Rev. mét.*, **51**, 1954, 101–107.
32. D. A. Vaughan and J. R. Bridge, *Trans. AIME*, **206**, 1956, 528–531.
33. The tetragonal hydride with $c/a > 1$, the first to form when Zr reacts with H, is said to be metastable below the eutectoid temperature.
34. C. E. Ells and A. D. McQuillan, *J. Inst. Metals*, **85**, 1956, 89–96.

0.8659
$\bar{1}.1341$

Hf-Mg Hafnium-Magnesium

According to microscopic and chemical analyses [1], Hf and Mg do not alloy, even if kept for long periods at high temperatures (~1000°C).

1. F. Sauerwald, *Z. anorg. Chem.*, **258**, 1949, 296–300.

0.5121
$\bar{1}.4879$

Hf-Mn Hafnium-Manganese

"An alloy of $HfMn_2$ has been identified as isomorphous with $MgNi_2$ [C36 type], with lattice parameters $c = 16.334$ kX, $a = 5.006$ kX, $c/a = 3.263$. At lower temperatures there seems to be a trend to the $MgZn_2$ structure [C14 type], but this could not be firmly established because of the diffuseness of the patterns at the lower temperatures. The incipient melting temperature was determined as 1585°C"[1].

1. R. P. Elliott, Armour Research Foundation, Chicago, Ill., Technical Report 1, OSR Technical Note OSR-TN-247, August, 1954. (X-ray examination of quenched alloys.)

0.2698
$\bar{1}.7302$

Hf-Mo Hafnium-Molybdenum

"The melting temperature of the $HfMo_2$ compound has been determined as being in excess of 2300°C, the limit of the incipient melting-temperature apparatus.

At all temperatures between 600 and 2200°C this compound is isomorphous with the $MgNi_2$ modification [C36 type]. At the intermediate temperatures 1200, 1400, and 1600°C there is evidence from intensification of certain lines that the structure consists of a mixture of both the $MgNi_2$ and $MgCu_2$ [C15 type] modifications. The lattice constant for the $MgCu_2$ modification is a = 7.562 kX; the lattice constants for the $MgNi_2$ modification are c = 17.312 kX, a = 5.330 kX, c/a = 3.248" [1].

1. R. P. Elliott, Armour Research Foundation, Chicago, Ill., Technical Report 1, OSR Technical Note OSR-TN-247, August, 1954. (X-ray examination of quenched alloys.)

1.1055
2.8945

Hf-N Hafnium-Nitrogen

HfN [1, 2] (7.27 wt. % N) has a metallic character [1], melts at 3310°C [3], and has a f.c.c. structure, with a = 4.52 ± 2 A [4] [apparently, like ZrN, NaCl (B1) type].

1. A. E. van Arkel and J. H. de Boer, *Z. anorg. Chem.*, **148,** 1925, 347.
2. K. Moers, *Z. anorg. Chem.*, **198,** 1931, 255.
3. O. Ruff, *Congr. chim. ind.*, *15th Congr.*, *Bruxelles*, 1935, p. 68; *Chim. Ind.*, **35,** 1936, 46. From "Gmelins Handbuch der anorganischen Chemie," System No. 43, p. 50, Verlag Chemie, G.m.b.H., Berlin, 1941.
4. F. W. Glaser, D. Moskowitz, and B. Post, *Trans. AIME*, **197,** 1953, 1119–1120.

0.2838
$\bar{1}$.7162

Hf-Nb Hafnium-Niobium

A b.c.c. solid solution exists at 1000°C [1] between Hf and Nb from 30 at. (18 wt.) to 100% Nb, and a two-phase region (hexagonal Hf plus cubic solid solution) extends from an undetermined concentration up to about 30 at. % Nb. This result is interpreted [1] as an indirect proof that the high-temperature form of Hf [2] is b.c.c. and suggests that at temperatures above the transformation of Hf the two metals are completely soluble in each other [3].

Supplement. [4] examined an alloy of the composition $HfNb_2$ by X-ray diffraction. Patterns of filings annealed 10 min at 800°C and quenched indicated a mixture of a b.c.c. and a h.c.p. phase.

1. P. Duwez, *J. Appl. Phys.*, **22,** 1951, 1174–1175. The Zr content of the van Arkel process hafnium used is not given but was estimated by other authors on the basis of lattice-parameter data (see Hf-Zr). Nine compositions between 10 and 77 at. % Nb were investigated.
2. For a discussion of the allotropic transformation of Hf, see Hf-Zr.
3. The high-temperature b.c.c. phase in the alloys rich in Hf cannot be retained by quenching [1].
4. R. P. Elliott, Armour Research Foundation, Chicago, Ill., Technical Report 1, OSR Technical Note OSR-TN-247, August, 1954.

0.4833
$\bar{1}$.5167

Hf-Ni Hafnium-Nickel

"An alloy of the [$HfNi_2$] composition was difficult to fracture and powder. The diffraction pattern showed a multitude of lines but was poorly defined. There was no visible similarity to typical Laves-type patterns. No allotropy was exhibited in the temperature range 600°C to the determined incipient melting temperature of 1790°C. Comparing this alloy with similar alloys in the Ti-Ni and Zr-Ni systems, a hexagonal pattern with similar parameters to $TiNi_3$, with $c = 8.3$ kX, $c/a \sim 1.6$, could be identified" [1].

1. R. P. Elliott, Armour Research Foundation, Chicago, Ill., Technical Report 1, OSR Technical Note OSR-TN-247, August, 1954.

1.0478
$\bar{2}.9522$

Hf-O Hafnium-Oxygen

Hf forms a stable dioxide (melting point ~2780°C). For further information on HfO_2 (15.19 wt. % O), see the chemical literature. [1] studied the oxidation rate of Hf (major impurity 5% Zr) over the temperature range 350–1200°C in oxygen at 10–760 mm Hg pressure.

1. W. W. Smeltzer and M. T. Simnad, *Acta Met.*, **5**, 1957, 328–334.

0.8033
$\bar{1}.1967$

Hf-Si Hafnium-Silicon

By means of powder metallurgy, [1] prepared two silicides which were identified as HfSi (13.59 wt. % Si) and $HfSi_2$ (23.93 wt. % Si) by analogy with the corresponding Zr silicides. HfSi is said to be hexagonal, with $a = 6.86$ A, $c = 12.60$ A, $c/a = 1.84$, and $HfSi_2$ to be orthorhombic, with $a = 3.67$ A, $b = 14.56$ A, $c = 3.64$ A. However, since recent work has shown ZrSi to be actually of the orthorhombic FeB (B27) type (see Si-Zr), the hexagonal unit cell claimed for HfSi needs corroboration.

Note Added in Proof. $HfSi_2$ is of the orthorhombic $ZrSi_2$ (C49) type of structure [2, 3], with $a = 3.677$ A, $b = 14.550$ A, $c = 3.649$ A [3]. According to [4], a compound Hf_5Si_3 exists.

1. B. Post, F. W. Glaser, and D. Moskowitz, *J. Chem. Phys.*, **22**, 1954, 1264.
2. P. G. Cotter, J. A. Kohn, and R. A. Potter, *J. Am. Ceram. Soc.*, **39**, 1956, 11–12; *Met. Abstr.*, **24**, 1957, 569.
3. J. F. Smith and D. M. Bailey, *Acta Cryst.*, **10**, 1957, 341–342.
4. B. Post and F. W. Glaser, private communication (1954) to R. Kieffer and F. Benesovsky [see *Symposium on Powder Metallurgy*, 1954, *Iron Steel Inst.* (*London*), *Spec. Rept.* 58].

$\bar{1}.9945$
0.0055

Hf-Ta Hafnium-Tantalum

"An alloy prepared at the $HfTa_2$ composition could not be fractured and crushed. The powder diffraction pattern of filings annealed for 10 min at 800°C proved to be a mixture of the b.c.c. and h.c.p. structures. Hf and Ta are probably miscible in all proportions in the b.c.c. modification, with Ta stabilizing the low-temperature h.c.p. structure of Hf" [1].

1. R. P. Elliott, Armour Research Foundation, Chicago, Ill., Technical Report 1, OSR Technical Note OSR-TN-247, August, 1954.

0.5715
$\bar{1}.4285$

Hf-Ti Hafnium-Titanium

It was stated by [1] that Hf "with the same type of structure, transition [2], and favorable atom size has complete solubility in Ti under all conditions, that is, in both the α and β phases." No experimental evidence of this type of phase relationship has been published as yet, however.

Supplement. [3] reported that filings of an alloy of the composition $HfTi_2$ annealed for 10 min at 800°C had a h.c.p. structure. This finding is in keeping with the hypothesis of [1].

1. B. W. Gonser, *Ind. Eng. Chem.*, **42**, 1950, 222–226.
2. For some experimental evidence that the high-temperature phase of Hf is, like that of Ti, b.c.c. (A2 type), see Hf-Nb.
3. R. P. Elliott, Armour Research Foundation, Chicago, Ill., Technical Report 1, OSR Technical Note OSR-TN-247, August, 1954.

0.5447
$\bar{1}.4553$

Hf-V Hafnium-Vanadium

According to [1], HfV_2 (36.33 wt. % V) is isotypic with $MgCu_2$ (C15 type), a = 7.386 A. The incipient-melting temperature was determined as 1500°C. No polymorphic transformation could be observed between 600 and 1400°C.

1. R. P. Elliott, Armour Research Foundation, Chicago, Ill., Technical Report 1, OSR Technical Note OSR-TN-247, August, 1954.

$\bar{1}.9872$
0.0128

Hf-W Hafnium-Wolfram

In sintering rods of HfO_2 (containing W) in a hydrogen atmosphere, the oxide is partially reduced and a solid solution of Hf in W is formed [1].

Supplement. "An alloy of the HfW_2 composition was prepared by pressing the metal powders and sintering. The crystal structure was indexed as isomorphous with $MgCu_2$ [C15 type], with a = 7.556 kX. To establish possible allotropy, a specimen was annealed at 2000°C and found to be of the same structure although the $MgCu_2$-type pattern was more prominent. Faint extra lines were observed but did not indicate the existence of any other allotropic modification" [2].

1. J. A. M. van Liempt, *Nature,* **115,** 1925, 194.
2. R. P. Elliott, Armour Research Foundation, Chicago, Ill., Technical Report 1, OSR Technical Note OSR-TN-247, August, 1954.

0.2918
$\bar{1}.7082$

Hf-Zr Hafnium-Zirconium

The curve of the temperature coefficient of the electrical resistance [1] shows that at ordinary temperatures Zr and Hf form a continuous series of solid solutions [2]. At elevated temperatures both Zr and Hf undergo allotropic transformations. "It was always assumed, and experiments [see Hf-Nb] give some evidence, that the crystal structure of the high-temperature modification of Hf is the same as that of Zr and Ti, i.e., body-centered cubic. Considering the small difference in atomic volume between Zr and Hf, this would almost certainly imply that these metals form an uninterrupted series of solid solutions also in their cubic forms" [3]. The tentative diagram of [3], therefore, shows a narrow (b.c.c. + h.c.p.) two-phase field, connecting both element transformations. There is, however, great uncertainty as to the temperature of the allotropic transformation of Hf. Independent determinations by [4–6], carried out on hafnium with a more or less high Zr content [7], had the following results: [4], between 1330 and 1630°C; [5], 1310 ± 10°C; [6], 1950 ± 100°C (extrapolated [9] to pure Hf using his own data and those of [4, 5]). In the schematic diagram of Fig. 448, 1700°C was assumed.

Supplement. [10] found the alloy $ZrHf_2$ to be h.c.p. after annealing at 800 and 600°C, which is in keeping with Fig. 448. For lattice parameters and thermal-expansion coefficients of low Hf zirconium in the range 0–600°C, see [11]. In the course of work on the Zr-H system, [12] recently observed that small additions of Hf (approximately 1.5 at. %) cause a depression of the transformation temperature of

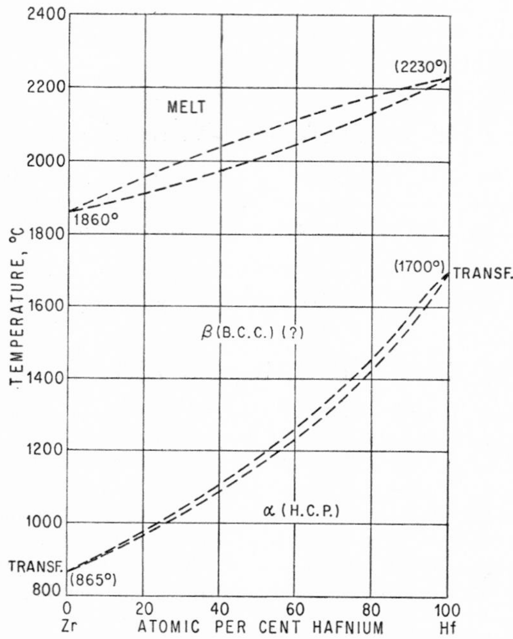

Fig. 448. Hf-Zr. Schematic diagram (see also Supplement).

Zr and concluded that the $\alpha + \beta$ region in the Zr-Hf system passes through a minimum point. (See Ti-Zr.)

1. A. E. van Arkel, *Metallwirtschaft*, **13**, 1934, 514.
2. The properties of some Zr-Hf alloys were described by J. H. de Boer and J. D. Fast, *Z. anorg. Chem.*, **187**, 1930, 203–204.
3. J. D. Fast, *J. Appl. Phys.*, **23**, 1952, 350–351.
4. C. Zwikker, *Physica*, **6**, 1926, 361–365 (in Dutch). Resistivity, thermal dilatation, thermionic emission.
5. P. Duwez, *J. Appl. Phys.*, **22**, 1951, 1174–1175. Special thermal technique.
6. See [3], measurements of electrical resistance.
7. The hafnium used by Fast [3] contained 3 wt. (5.7 at.)% Zr; that used by Zwikker [4] had an even "much greater" [3] Zr content. The Zr content of the van Arkel process hafnium used by Duwez [5] has been assumed to be as high as 19 at. % [3] or only 6 at. % [8]; these values, however, were derived from lattice-parameter data, the units for which (A or kX) may have been confused [8].
8. R. B. Russell, *J. Appl. Phys.*, **24**, 1953, 232–233.

9. Judging from the uncertainties mentioned [7], the extrapolation may prove to be of no great value.
10. R. P. Elliott, Armour Research Foundation, Chicago, Ill., Technical Report 1, OSR Technical Note OSR-TN-247, August, 1954.
11. R. B. Russell, *Trans. AIME*, **200**, 1954, 1045–1052 (Coefficients of Thermal Expansion for Zr).
12. C. E. Ells and A. D. McQuillan, *J. Inst. Metals*, **85**, 1956, 95.

0.2426
$\bar{1}.7574$

Hg-In Mercury-Indium

The solubility of In in Hg, determined by chemical analysis, was reported to increase from 1.23 wt. (2.14 at.) % at 0°C to 1.31 wt. (2.27 at.) % In at 50°C [1]. In very great contrast to this, [2] found the solubility by thermal analysis to be

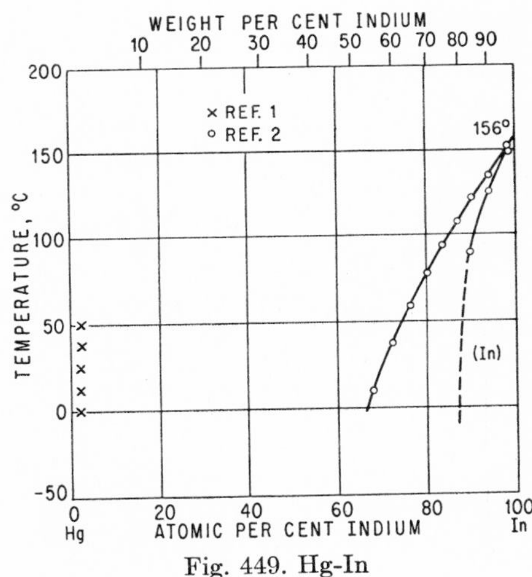

Fig. 449. Hg-In

54.9 wt. (68.04 at.) % In at 10°C. Their liquidus curve is presented in Fig. 449. In agreement with these results, alloys with 15.6 wt. (24.4 at.) % and 36.8 wt. (50.4 at.) % In were found to consist of only liquid phase at 25°C.

Results by [1] and [2] could be compatible if the liquidus of this system were similar to that of Hg-Tl, with a maximum somewhere between 20 and 60 at. % In. However, [2] reported that no thermal effects for the beginning of solidification were found in the range 5–50 wt. (8.4–63.6 at.) % In. It is not known whether the cooling curves in this composition range were followed down to room temperature. The experiments with the alloys containing 15.6 and 36.8 wt. % In, however, would not suggest the existence of a maximum above 25°C.

Supplement. Lattice spacings of 11 alloys with 0.44–9.98 at. % Hg were determined by [3], after annealing at 94 and 60°C, according to composition. Hg increases the a axis and decreases the c axis of In, so that the axial ratio of In (c/a = 1.07586) decreases to a minimum observed value of 1.0429 at 6.08 at. % Hg. At higher Hg contents the axial ratio is unity, and the system is analogous to the systems Cd-In and In-Tl. The alloy with 6.45 at. % Hg, which, after quenching from 94°C, was f.c.c., transformed into the tetragonal form after 1 month at room temperature.

1. W. G. Parks and W. G. Moran, *J. Phys. Chem.*, **41**, 1937, 343–349.
2. W. M. Spicer and C. J. Banick, *J. Am. Chem. Soc.*, **75**, 1953, 2268–2269.
3. C. Tyzack and G. V. Raynor, *Trans. Faraday Soc.*, **50**, 1954, 675–684.

0.7102
$\bar{1}$.2898

Hg-K Mercury-Potassium

Chemical Investigations. The solubility of K in Hg at room temperature was reported as 0.46 wt. (2.3 at.) % [1], 0.395 wt. (2.0 at.) % [2], and 0.38 wt. (1.9 at.) % K (extrapolated value) [3], and at 0°C as 0.265 wt. (1.34 at.) % K [1]. The solubility

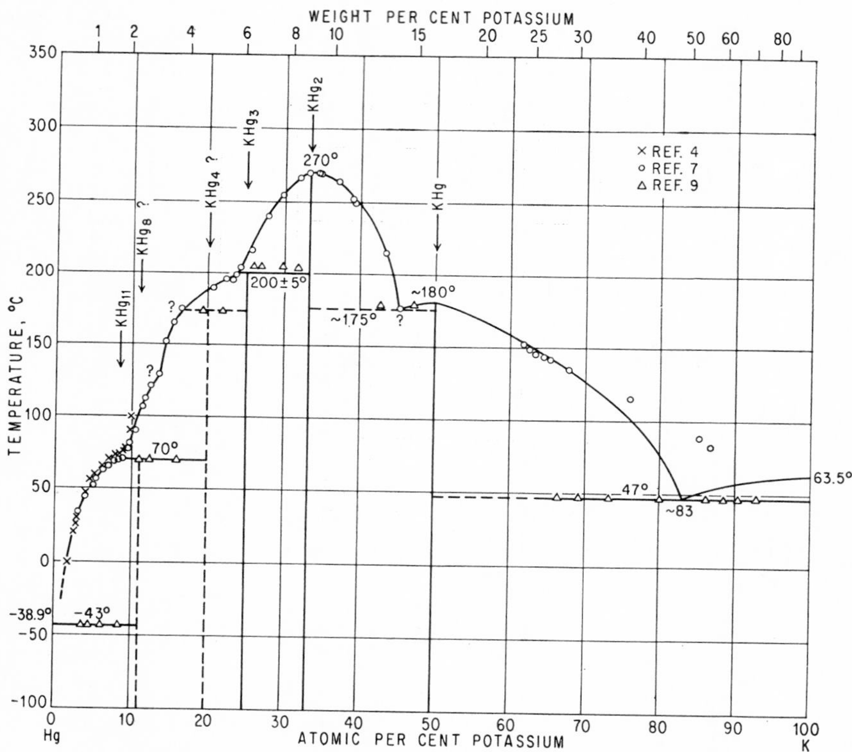

Fig. 450. Hg-K. (See also Supplement and Note Added in Proof.)

between 0 and 100°C was determined by [4]; see crosses in Fig. 450. The curve has a kink at 1.9 wt. (9 at.) % K, 75°C, indicating the composition of a peritectic melt.

Determinations of the solid phase in equilibrium with the saturated liquid at room temperature yielded the compositions KHg_{12} [1, 4], KHg_{10} [2, 5], and a value between KHg_{12} and KHg_{11} [6]. [4] assumed, on the basis of determinations in the range 0 and 100°C, that below 0°C KHg_{14}, between 0 and 71°C KHg_{10}, and above 75°C perhaps KHg_9 or KHg_8 are in equilibrium with the melt. All compositions higher in Hg content than KHg_{10} (1.91 wt. % K) are not compatible with the solubility curve of [4] and the thermal data of [7]. Determinations of the specific volume of alloys with 0–55 at. % K were claimed to indicate the compounds KHg_{11} and KHg_5 (3.75 wt. % K), besides KHg_3 (6.10 wt. % K), KHg_2 (8.88 wt. % K), and KHg (16.31 wt. % K) [3].

Phase Diagram. *Alloys with* 0–33.3 *At.* % K. The liquidus temperatures shown as circles in Fig. 450 are due to [7]. The curve between Hg and KHg_2 is characterized by three points of inflection at 8.7 at. % K, 70°C; about 13.6 at. % K, 129°C; and 23.4 at. % K, 195°C. They were interpreted to correspond to three peritectic melts participating in reactions to form the phases KHg_{10} (or KHg_9), KHg_5 (or KHg_6), and KHg_3. However, since [7] determined only liquidus temperatures, final conclusions as to the composition of the phases formed at 70 and 129°C cannot be drawn. The composition of the peritectic melts at these temperatures would permit other formulas as well: e.g., for the reaction at 70°C, not only KHg_{10} or KHg_9 but also KHg_8 and KHg_7; and for the reaction at 129°C, not only KHg_6 or KHg_5 but also KHg_4.

In interpreting these data, [8] assumed that besides KHg_2 only the phases KHg_9 and KHg_3 occur. Indeed, the inflection point at 13.6 at. % K, 129°C, is much less pronounced than the other two.

Redetermination of the liquidus by [9] confirmed the existence of three intermediate phases between Hg and KHg_2. Unfortunately, no tabulated data were given; only a few data points (triangles in Fig. 450) were inserted in the diagram. The peritectic temperatures were found as 70, 173, and 204°C; and the peritectic horizontals were drawn to intersect the liquidus at the compositions of the three phases assumed: KHg_9, K_2Hg_9 [18.18 at. (4.15 wt.) % K], and KHg_3. Later, the same author [10] offered another interpretation of the experimental data. It was proposed that KHg_8 should be substituted for KHg_9 and KHg_4 for K_2Hg_9. Also, all three phases existing in this composition range were assumed to melt congruently: KHg_8 at 70°C, KHg_4 at 182°C, and KHg_3 at 201°C.

The Hg-rich eutectic was located at 0.4 wt. (2.0 at.) % K, −43°C [9]. This composition appears to be too high since the solubility curves based on the data of [4, 7] point to a lower concentration of K [11].

In view of the conflicting and incomplete information, no final conclusions can be drawn as to the composition of the phases between 0 and 25 at. % K and the type of phase relationships in this range. The presentation in Fig. 450 is still to be regarded as tentative.

Alloys with 33.3–100 *At.* % K. As indicated by the data points in Fig. 450, the investigation of [7] was incomplete. He assumed the phase KHg to exist and to be formed peritectically. However, the peritectic melt was shown to lie at about 45.5 at. % K rather than 50 at. % or higher.

[9] first suggested the composition KHg to coincide with that of the (peritectic) melt at 178°C, but later [10] preferred to indicate a flat maximum at this temperature and a KHg_2-KHg eutectic at approximately 48 at. % K, 175°C. The eutectic KHg-K was first shown to lie at about 82 at. % K, 47°C [9], and later given as 84.5 at. % K, 50°C [10]. The thermal data of [7] would suggest a eutectic point above 90 at. % K. (See Note Added in Proof.)

Measurements of the magnetic susceptibility [12] of the whole system gave inconclusive results as regards the composition of the various phases, except KHg_3, KHg_2, and KHg. Results were claimed to indicate the existence of an additional phase in the vicinity of 57 at. % K (K_4Hg_3?).

The X-ray structure of the molten alloy KHg_2 was investigated by [13]. [14] reported that KHg_{11} is cubic, with 36 atoms per unit cell.

Supplement. The findings of [14] were corroborated by [15, 16], who showed KHg_{11} to be isostructural with $BaHg_{11}$ (cubic, 36 atoms per unit cell). Further crystallographic data: KHg_2, orthorhombic, $a = 8.10$ A, $b = 5.16$ A, $c = 8.77$ A, 4 formula weights per unit cell [16, 17]; KHg, triclinic, $a = 6.62$ A, $b = 6.91$ A, $c = 6.94$ A, $\alpha = 106°$, $\beta = 102°19'$, $\gamma = 92°48'$, 4 formula weights per unit cell [15]; $\sim K_5Hg_7$ (41.7 at. % K), orthorhombic, $a = 9.99$ A, $b = 19.23$ A, $c = 8.25$ A, 4 formula weights per unit cell [16, 17].

The finding of the phases KHg_{11} and $\sim K_5Hg_7$ makes a redetermination of the solidification equilibria very desirable.

Note Added in Proof. [17] determined, on single crystals, the crystal structures of KHg_2 and KHg. For KHg, the revised triclinic lattice constants $a = 6.59$ A, $b = 6.76$ A, $c = 7.06$ A, $\alpha = 106°5'$, $\beta = 101°52'$, $\gamma = 92°47'$ are given. The lattice parameter of cubic KHg_{11} was found to be $a = 9.6455 \pm 15$ A. [18] redetermined the liquidus of K-rich alloys. The eutectic (KHg-K) was found to lie at 94.1 at. % K, 47.55°C.

1. W. Kerp, *Z. anorg. Chem.*, **17**, 1898, 300–303.
2. A. Guntz and J. Férée, *Compt. rend.*, **131**, 1900, 183–184.
3. E. Maey, *Z. physik. Chem.*, **29**, 1899, 119–138.
4. W. Kerp, W. Böttger, and H. Winter, *Z. anorg. Chem.*, **25**, 1900, 19–29.
5. G. P. Grimaldi, *Atti accad. nazl. Lincei*, **4**, 1887, 71.
6. G. McPhail Smith and H. C. Bennett, *J. Am. Chem. Soc.*, **32**, 1910, 622–626.
7. N. S. Kurnakow, *Z. anorg. Chem.*, **23**, 1900, 441–455.
8. G. Tammann, *Z. anorg. Chem.*, **37**, 1903, 303–313.
9. E. Jänecke, *Z. physik. Chem.*, **58**, 1907, 245–249.
10. E. Jänecke, *Z. Metallkunde*, **20**, 1928, 113–114.
11. G. Tammann (*Z. physik. Chem.*, **3**, 1889, 443) reported that the melting point of Hg is lowered for 1.24°C by 0.136 wt. (0.69 at.) % K. The eutectic temperature was not reached.
12. W. Klemm and B. Hauschulz, *Z. Elektrochem.*, **45**, 1939, 346–354.
13. F. Sauerwald and W. Teske, *Z. anorg. Chem.*, **210**, 1933, 247–256.
14. J. A. A. Ketelaar, *J. Chem. Phys.*, **5**, 1937, 668 (footnote).
15. N. C. Baenziger and E. J. Duwell, *Acta Cryst.*, **7**, 1954, 635.
16. N. C. Baenziger, E. J. Duwell, and J. W. Conant, *U.S. Atomic Energy Comm., Publ.* COO-127, 1954.
17. E. J. Duwell and N. C. Baenziger, *Acta Cryst.*, **8**, 1955, 705–710.
18. A. Roeder and W. Morawietz, *Z. Metallkunde*, **47**, 1956, 734.

0.1596
$\bar{1}.8404$

Hg-La Mercury-Lanthanum

X-ray analysis of 30 alloys with more than 10 at. % La revealed the existence of the phases $LaHg_4$ (14.76 wt. % La), $LaHg_3$ (18.75 wt. % La), $LaHg_2$ (25.72 wt. % La), and LaHg (40.91 wt. % La). In addition, there are several unidentified phases richer in Hg and probably La_3Hg (67.51 wt. % La) [1]. The existence of $LaHg_4$ was already established by [2]. The solubility of La in Hg was found to increase from 0.0064 at. (0.0045 wt.) % at 0°C to 0.0133 at. (0.0092 wt.) % at 25°C and 0.0246 at. (0.0171 wt.) % La at 50°C [3].

$LaHg_4$ is cubic, $a = 3.656$ A [4]; $LaHg_3$ is hexagonal, $a = 3.404$ A, $c = 4.951$ A, $c/a = 1.454$; $LaHg_2$ is hexagonal of the AlB_2 (C32) type (?), $a = 4.948$ A, $c = 3.633$ A, $c/a = 0.734$; LaHg is b.c.c. (B2 type), $a = 3.837$ A [1].

Supplement. One of the unidentified Hg-rich phases mentioned above was shown [5] to have the composition $LaHg_2$ and to decompose above 250°C. It has an orthorhombic crystal structure.

1. A. Iandelli and R. Ferro, *Atti accad. nazl. Lincei, Rend. Classe sci. fis. mat. e nat.*, **10,** 1951, 48–52; **11,** 1951, 85–93; see also A. Iandelli and E. Botti, *Gazz. chim. ital.*, **67,** 1937, 638–644.
2. P. T. Daniltchenko, *Zhur. Obshchěi Khim.*, **1,** 1931, 467–474; abstract in *J. Inst. Metals*, **50,** 1932, 540.
3. W. G. Parks and J. L. Campanella, *J. Phys. Chem.*, **40,** 1936, 333–341.
4. [1] discuss also a unit cell, perhaps more likely, of $a = 3 \times 3.656 = 10.968$ A with 54 atoms (γ-brass type).
5. A. Iandelli and R. Ferro, *Gazz. chim. ital.*, **84,** 1954, 478.

$\overline{1}.4610$
$\overline{2}.5390$

Hg-Li Mercury-Lithium

Chemical Investigations. The solubility of Li in Hg at room temperature was determined as 0.04 [1], about 0.032 [2], 0.036 [3], and 0.047 wt. % Li [4] (about 1.15, 0.92, 1.04, and 1.34 at. %); mean value 0.039 wt. (1.12 at.) % Li. At 65, 81, and 100°C, [1] found solubilities of 0.10, 0.11, and 0.13 wt. (2.80, 3.08, and 3.64 at.) % Li, respectively. Solubilities in this temperature range, based on thermal-analysis data, are considerably higher, i.e., roughly 2.7 at. (0.096 wt.) % [5] or 3.5 at. (0.125 wt.) % Li [6] at 20°C, and 5.4 at. % [5] or 8.1 at. % [6] at 100°C. With the present type of system, solubilities based on chemical analysis, if carried out carefully, should be more accurate.

The phase coexisting with the saturated Hg-rich liquid solution at room temperature was determined by separating the liquid from the solid phase. Thus, compositions corresponding to $LiHg_5$ (0.69 wt. % Li) [7, 1, 2] and $LiHg_3$ (1.14 wt. % Li) [4] were found. The latter phase was established by thermal analysis [5, 6].

Phase Diagram. The solidification of the entire range of composition was investigated by [5, 6], using 90 and 67 alloys, respectively. Figure 451 shows only a selection of representative liquidus data points.

Between 0 and 50 at. % Li results are in good agreement, except for relatively slight temperature differences. The composition of the Hg-rich eutectic at −42°C [5, 6] is not known exactly but certainly lies closer to the extrapolated value of 0.6 at. (0.02 wt.) % Li [6] than to 0.97 at. (0.03 wt.) % Li [5]. The temperatures of the peritectic formation of $LiHg_3$ and $LiHg_2$ (1.70 wt. % Li) were found as 240 [5] or 235°C [6] and 338 [5] or 340°C [6], respectively, and the melting point of LiHg (3.34 wt. % Li) as 600 [5] or about 590°C [6].

Between 50 and 100 at. % Li the findings of [5] were not conclusive although indications were found of the existence of Li_2Hg (6.46 wt. % Li) and Li_3Hg (9.40 wt. % Li). [6] established that Li_3Hg melts without decomposition at 375°C, only 4°C above the temperature of the Li_2Hg-Li_3Hg eutectic. Li_2Hg is formed peritectically at 379 [5] or 375°C [6]. The thermal effect at 166 [5] or 164°C [6] was suggested by [5] to represent a polymorphic transformation of Li_3Hg. However, [6] have proved it to be the temperature of the peritectic formation of Li_6Hg (17.19 wt. % Li). Liquidus temperatures between 85 and 100 at. % Li deviate greatly. According to [5], the Li-rich eutectic would lie at about 97.5 % Li, 162°C, whereas [6] located it at about 92.5 at. % Li, 160°C.

Thermal-analysis data indicate that LiHg forms solid solutions with both Hg and Li [5, 6]. The homogeneity range extends from approximately 38 at. % Li at 339°C to approximately 62 at. % Li at 375°C. [6] also assumed, on the basis of thermal data, that about 2.5 at. % Hg was soluble in Li at 160°C.

Crystal Structures. $LiHg_3$ is hexagonal of the Ni_3Sn ($D0_{19}$) type, $a = 6.253$ A, $c = 4.804$ A, $c/a = 0.768$ [8]. LiHg is isotypic with CsCl (B2 type), $a = 3.294$ A at 50.9 at. % Li [9], and Li_3Hg is isotypic with BiF_3 ($D0_3$ type), $a = 6.597$ A [8].

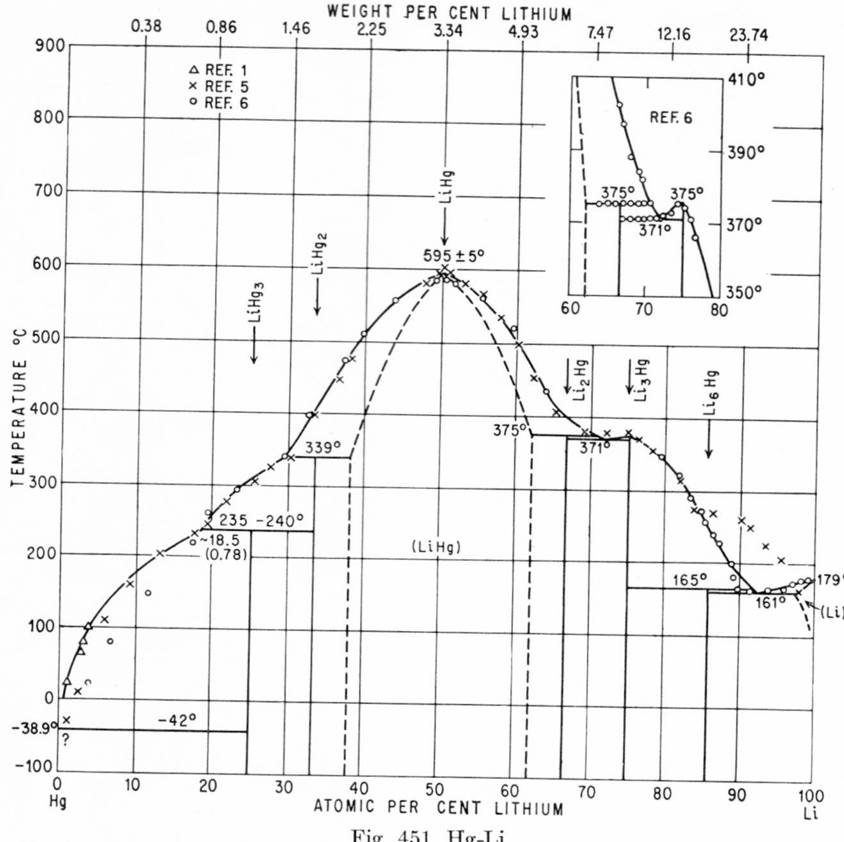

Fig. 451. Hg-Li

1. W. Kerp, W. Böttger, and H. Winter, *Z. anorg. Chem.*, **25,** 1900, 16–19.
2. E. Maey, *Z. physik. Chem.*, **29,** 1899, 119–138.
3. T. W. Richards and R. N. Garrod-Thomas, *Z. physik. Chem.*, **72,** 1910, 182–185.
4. G. McPhail Smith and H. C. Bennett, *J. Am. Chem. Soc.*, **31,** 1909, 804; **32,** 1910, 622–626; *Z. anorg. Chem.*, **74,** 1912, 172–173.
5. G. J. Zukowsky, *Z. anorg. Chem.*, **71,** 1911, 403–418.
6. G. Grube and W. Wolf, *Z. Elektrochem.*, **41,** 1935, 675–679.
7. A. Guntz and J. Férée, *Bull. soc. chim. France*, **15,** 1896, 834.
8. E. Zintl and A. Schneider, *Z. Elektrochem.*, **41,** 1935, 771–774.
9. E. Zintl and G. Brauer, *Z. physik. Chem.*, **B20,** 1933, 245–271.

0.9164
$\bar{1}$.0836

Hg-Mg Mercury-Magnesium

The solubility of Mg in Hg at temperatures close to room temperature was determined by chemical analysis [1, 2], emf measurements [3], and thermal analysis [4, 5] to be 0.313 wt. (2.52 at.) % [1], 0.323 (2.60) % [2], <0.3 (2.4) % [3], and 0.31 (2.5) % at 17°C [4], and 0.34 (2.74) % at 15°C [5]. By measurements of electrical conductivity, [6] found 0.265 wt. (2.15 at.) % at 4°C. The solid phase isolated from the liquid solution at room temperature was found to correspond closely to the compositions $MgHg_6$ [1] and $MgHg_4$ [6], indicating that the separation was incomplete (see below).

Alloys with 0–50 At. % Mg (Fig. 452). The constitution of this partial system is based on the thermal-analysis data of [4, 7, 5], which are in very good agreement. The phase formed peritectically at 169–171°C [4, 7, 5] was assumed to be $MgHg_2$ (5.72 wt. % Mg) [4, 7] or Mg_2Hg_5 [5]. X-ray investigation showed conclusively

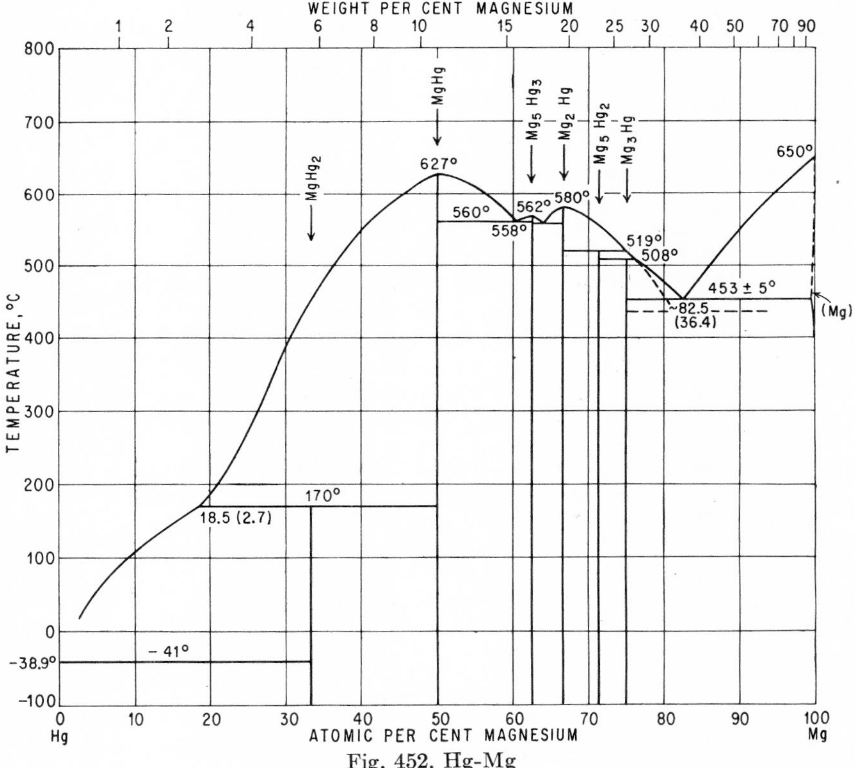

Fig. 452. Hg-Mg

that it is $MgHg_2$ [8]. The eutectic temperature of −41°C was determined only by [4]; the eutectic composition is unknown. The existence of MgHg (10.81 wt. % Mg), first reported by [9] to melt congruently, was confirmed by [7, 5, 8]; melting point 624°C [7], about 630°C [5].

In the range 30–45 at. % Mg, [4] found thermal effects at about 415°C, suggesting the existence of a peritectic horizontal. However, [7, 5] were unable to verify this effect.

Alloys with 50–100 At. % Mg (Fig. 452). From thermal-analysis data, [7] concluded the existence of the phases Mg_3Hg_2 (15.39 wt. % Mg) and Mg_2Hg (19.51 wt. % Mg) with congruent melting points of about 562 and 573°C and Mg_7Hg_3 (22.05 wt. % Mg) and Mg_3Hg (26.67 wt. % Mg) which are formed peritectically at about 519 and 508°C, respectively. [5] confirmed the phase Mg_2Hg (melting point 570°C) but was unable to verify a phase in the vicinity of 60 at. % Mg. However, X-ray work revealed the existence of Mg_5Hg_3 [62.5 at. (16.81 wt.) % Mg] [8]. Also, the thermal data of [7] fit better with the composition Mg_5Hg_3 than with Mg_3Hg_2.

Between 67 and 75 at. % Mg, [5] found only one peritectic horizontal at 509°C, corresponding to the formation of Mg_3Hg. X-ray analysis [8] showed that there is another phase between Mg_2Hg and Mg_3Hg which has the composition Mg_5Hg_2 [71.43 at. (23.26 wt.) % Mg] instead of Mg_7Hg_3 [7].

The Mg_3Hg-Mg eutectic lies at about 83 at. (37 wt.) % Mg, 458°C [7], or about 82 at. (35.5 wt.) % Mg, 448°C [5]. According to [7], the peritectic reaction at 508°C may be suppressed, resulting in a metastable Mg_5Hg_2-Mg eutectic at 435°C.

The solid solubility of Hg in Mg could not be determined parametrically [8, 10] since both elements have nearly equal atomic radii. Micrographic work showed that about 3 wt. (0.4 at.) % Hg is soluble at the eutectic temperature and about 0.9 wt. (0.12 at.) % at 400°C [10].

Crystal Structures. $MgHg_2$ is isotypic with $MoSi_2$ (C11 type), a = 3.838 A, c = 8.799 A, c/a = 2.293 [8]. MgHg is b.c.c. of the CsCl (B2) type, a = 3.449 A [12, 8]. Mg_5Hg_3 is isotypic with Mn_5Si_3 ($D8_8$ type) [11], a = 8.260 A, c = 5.931 A, c/a = 0.718 [8, 11]. Mg_3Hg is isotypic with Na_3As ($D0_{18}$ type) [11], a = 4.868 A, c = 8.656 A, c/a = 1.778 [8, 11].

1. W. Kerp, W. Böttger, and H. Iggena, *Z. anorg. Chem.*, **25,** 1900, 33–35.
2. A. G. Loomis, *J. Am. Chem. Soc.*, **44,** 1922, 10.
3. R. Kremann and R. Müller, *Z. Metallkunde*, **12,** 1920, 307–312.
4. L. Cambi and G. Speroni, *Atti accad. nazl. Lincei*, **24**(1), 1915, 734–738.
5. P. T. Daniltschenko, *Zhur.Russ. Fiz.-Khim. Obshchestva*, **62**(1), 1930, 975–988.
6. P. Bachmetjew and J. Wzarow, *Zhur. Russ. Fiz.-Khim. Obshchestva*, **25**(1), 1893, 115, 219.
7. R. P. Beck, *Rec. trav. chim.*, **41,** 1922, 353–399; A. Smits and R. P. Beck, *Proc. Akad. Wetenschap. Amsterdam*, **23,** 1921-1922, 975–976.
8. G. Brauer and R. Rudolph, *Z. anorg. Chem.*, **248,** 1941, 405–424.
9. G. J. Zukowsky, *Z. anorg. Chem.*, **71,** 1911, 418.
10. H. Nowotny, *Z. Metallkunde*, **37,** 1946, 130–136.
11. G. Brauer, H. Nowotny, and R. Rudolph, *Z. Metallkunde*, **38,** 1947, 81–84.
12. G. Brauer and W. Haucke, *Z. physik. Chem.*, **B33,** 1936, 304–310.

0.5625
$\bar{1}$.4375

Hg-Mn Mercury-Manganese

The solubility of Mn in Hg at room temperature was reported to be (in wt. %) 3.8×10^{-3} [1], 2.5×10^{-4} [2], 1×10^{-3} [3], and 3×10^{-3} [4], most likely $1\text{-}3 \times 10^{-3}$.

Electrolytically prepared amalgam with 22 wt. (50.8 at.) % Mn was claimed [5] to contain, besides α-Mn, a phase with an undetermined f.c.c. structure, a = 4.685 A. On heating at 450°C in vacuum the intermediate phase decomposed, leaving behind a residue of α-Mn. The intermediate phase could not be observed by [6]. Exposure of amalgam with 4–6 wt. (13–19 at.) % Mn to 350–400°C in a hydrogen atmosphere yielded a residue of α-Mn.

[7, 4] claimed to have isolated the phase Mn_2Hg_5 [28.57 at. (9.87 wt.) % Mn]. This phase was found to be stable only up to 90°C [4]. Between 86 and 100°C the phase MnHg (21.50 wt. % Mn) was assumed to be stable [4]. These findings require verification [8].

Supplement. In good agreement with previous data, [9] found colorimetrically a solubility of 1.7×10^{-3} wt. % Mn in Hg at room temperature. [9] also briefly mentioned finding the compounds MnHg and $MnHg_4$.

Note Added in Proof. The existence of the compounds MnHg and Mn_2Hg_5 has been independently corroborated by [10, 11]. MnHg exists at temperatures <265°C [10], and Mn_2Hg_5 is formed peritectically ($MnHg + Hg \rightleftarrows Mn_2Hg_5$) at 75°C [10], 72–73°C [11]. MnHg has the B2 (CsCl) type of structure, with $a = 3.315$ A [10], $a = 3.318 \pm 2$ A [11], and Mn_2Hg_5 has a tetragonal structure, with $a = 9.74 \pm 1$ A, $c = 3.00 \pm 1$ A, $c/a = 0.308$ [11]. A tentative equilibrium diagram is given in [10].

1. A. Campbell, *J. Chem. Soc.*, **125,** 1924, 1713–1716.
2. G. Tammann and J. Hinnüber, *Z. anorg. Chem.*, **160,** 1927, 251–254.
3. N. M. Irvin and A. S. Russell, *J. Chem. Soc.*, **1932,** 891–898.
4. H. D. Royce and L. Kahlenberg, *Trans. Electrochem. Soc.*, **59,** 1931, 126–132.
5. F. Pawlek, *Z. Metallkunde*, **41,** 1950, 451–453.
6. F. Lihl, *Z. Metallkunde*, **44,** 1953, 160–166.
7. O. Prelinger, *Monatsh. Chem.*, **14,** 1893, 353.
8. H. Nowotny, *Z. Metallkunde*, **37,** 1946, 130–136.
9. J. F. deWet and R. A. W. Haul, *Z. anorg. Chem.*, **277,** 1954, 96–112.
10. F. Lihl, *Monatsh. Chem.*, **86,** 1955, 186–190; *Met. Abstr.*, **23,** 1955, 292.
11. J. F. deWet, *Angew. Chem.*, **67,** 1955, 208.

0.3203
$\bar{1}.6797$

Hg-Mo Mercury-Molybdenum

The solubility of Mo in Hg at room temperature was reported as 2×10^{-5} wt. % [1].

By squeezing off the liquid phase from an amalgam with 2 wt. % Mo with increasing pressures, residues of the approximate compositions $MoHg_9$, Mo_2Hg_3, and $MoHg_2$ were obtained [2]. Results do not permit the conclusion that phases of these compositions exist; they only illustrate the progressive removal of the liquid phase from the liquid-solid mixture.

1. N. M. Irvin and A. S. Russell, *J. Chem. Soc.*, **1932,** 891–898.
2. J. Férée, *Compt. rend.*, **122,** 1896, 733.

0.9407
$\bar{1}.0593$

Hg-Na Mercury-Sodium

Chemical Investigations. The solubility of Na in Hg at temperatures between 0 and 100°C was determined by [1, 2]. Results, in wt. % Na, are shown in the inset of Fig. 453. At room temperature the solubility was found as 0.64 wt. (5.3 at.) % (25°C) [1], 0.65 wt. (5.4 at.) % (25°C) [2], 0.57 wt. (4.8 at.) % [3], 0.62 wt. (5.15 at.) % (extrapolated) [4], and 0.6 wt. (4.97 at.) % (16.5°C, thermal) [5].

Experiments to determine the composition of the phase in equilibrium with the saturated liquid solution at room temperature, by separating it from the melt, are only of historical interest. Only one investigator [3] succeeded, by using high pressure, in obtaining a residue of the composition $NaHg_4$ (2.79 wt. % Na). Otherwise, compositions nearly corresponding to $NaHg_6$ [6, 2] and $NaHg_5$ [7, 1, 8] were found.

Phase Diagram (Fig. 453). The phase relations were established by the thermal studies of [5, 9, 10], using 74, about 100, and 90 alloys, respectively. Only [9] determined the solidus temperatures of all alloys investigated. The liquidus temperatures differ somewhat in the range 2.5–15 wt. (18–60 at.) % Na but are in excellent agreement otherwise. This also holds for the temperatures of the three-phase equilibria and the compositions of the peritectic melts (see Table 33). [11] reported that the

Table 33. Characteristic Temperatures in °C and Compositions in At. % Na (Wt. % in Parentheses) of the System Hg-Na

	Ref. 5	Ref. 9	Ref. 10
Eutectic Hg-$NaHg_4$:			
Temperature*	....	−48.2	−46.8
Composition	....	2.8 (0.33)	2.7 (0.32)
Peritectic, $NaHg_4$:			
Temperature	155	159	156
Composition	17.95 (2.45)	18.1 (2.5)	18.0 (2.45)
$NaHg_2$	346	360	354
Peritectic, Na_7Hg_8:			
Temperature	218	227	222
Composition	47.6 (9.45)	48.1 (9.65)	47.6 (9.45)
Peritectic, NaHg:			
Temperature	210	219	212
Composition	50.6 (10.55)	50.9 (10.65)	51–51.5 (10.7–10.9)
Peritectic, Na_3Hg_2:			
Temperature	(119)	123	119
Composition	....	61.9 (15.7)	63.0 (16.3)
Peritectic, Na_5Hg_2:			
Temperature	67	66	66
Composition	71.9 (22.8)	71.8 (22.7)	?
Peritectic, Na_3Hg:			
Temperature	....	34	34
Composition	....	84.1 (38.0)	83.4–83.7 (36.7–37.3)
Eutectic, Na_3Hg-Na:			
Temperature	21.25	21.4	21.4
Composition	85.05 (39.6)	85.2 (39.8)	85.2 (39.8)

* [17] found −48°C.

melting point of Hg is lowered 2.2°C by 0.11 wt. (1.0 at.) % Na. The liquidus points of a number of Na-rich alloys [11, 12] agree very well with the results of [9, 10].

[5, 9, 10] agree in that the phases $NaHg_2$ (5.42 wt. % Na) and NaHg (10.28 wt. % Na) exist. Their differences in interpretation may be summarized as follows: (*a*) The phase formed peritectically at 157 ± 2°C was assumed by [5] to be either $NaHg_4$ or $NaHg_3$. [9, 10] agree in that it is $NaHg_4$. (*b*) No composition for the phase formed at 223 ± 5°C was given by [5]. [9] assumed $Na_{12}Hg_{13}$ [48.0 at. (9.57 wt.) % Na] and [10] proposed Na_7Hg_8 [46.67 at. (9.12 wt.) % Na]. This phase undergoes a transformation at 180°C [9]. (*c*) Since [5] did not determine the temperatures of the end of solidification, he overlooked the peritectic reaction at 121 ± 2°C. [9, 10] agree that the composition of the phase formed at this temperature is Na_3Hg_2 (14.67 wt. % Na). Recently, this was verified by X-ray analysis [13]. (*d*) The phase formed peritectically at 66–67°C was assumed by [5] to be either Na_2Hg (18.65 wt. % Na) or Na_5Hg_2 (22.28 wt. % Na). [9] suggested the composition Na_5Hg_2 whereas [10] claimed the composition Na_3Hg (25.59 wt. % Na). The thermal data of [10] are uncertain in this range. Two sets of liquidus temperatures were reported, one showing an open maximum at about Na_3Hg and the other indicating a peritectic melt. Since both [5] and [9] found a point of inflection at 71.8–71.9 at. % Na, a phase of the composition Na_3Hg cannot be formed at 66°C. Indeed, the compound Na_5Hg_2 was identified by X-ray investigation [13]. (*e*) [5] overlooked the three-phase equilibrium at 34°C. [9] interpreted this transformation as a peritectic reaction yielding the phase Na_3Hg. [10] believed the transformation to be a polymorphic change of Na_3Hg; however, see above.

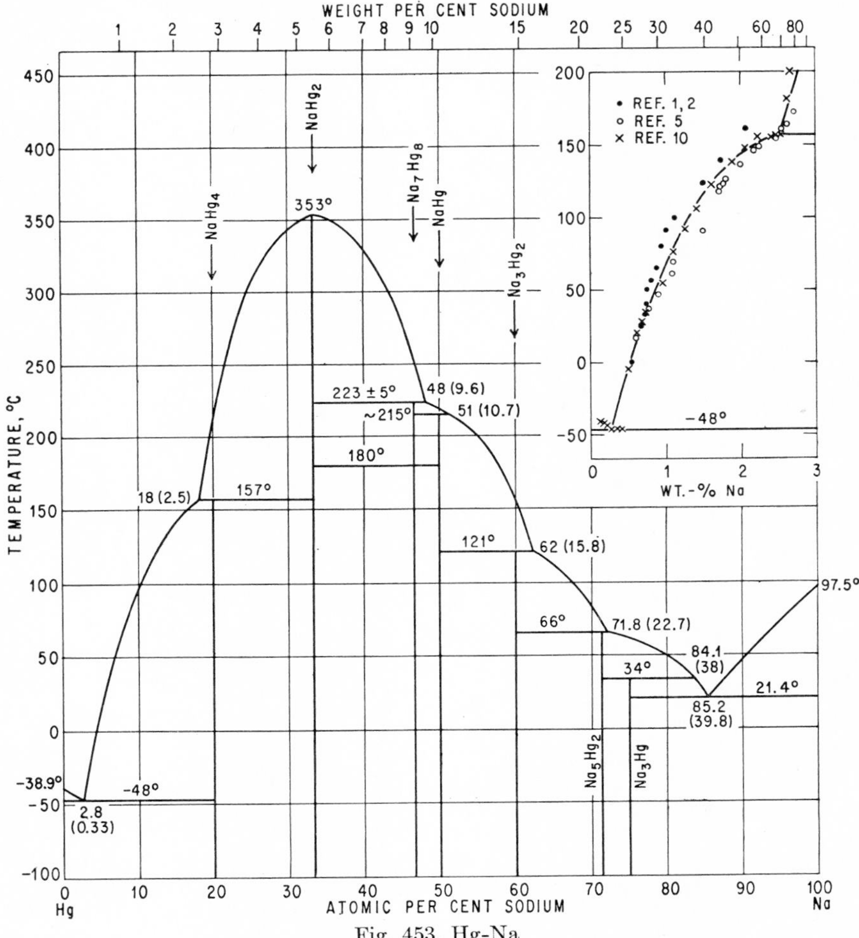

Fig. 453. Hg-Na

[9] reported that Na_5Hg_2 undergoes transformations at 60 and 49°C. [10] also found thermal effects at nearly the same temperatures. Because of the unknown nature of these changes, they have been omitted in Fig. 453.

In 1928, [14] published a diagram which was based on the thermal analysis of about 100 alloys (no tabulated data are given). It was reported that only the phases $NaHg_4$, $NaHg_2$, NaHg, and Na_3Hg were found and that they probably melt congruently [15]. Later, the same author [16] adopted largely the diagram by [10].

Measurements of magnetic susceptibility vs. composition [18] were found to be in accord with the existence of the seven compounds shown in Fig. 453 and to indicate the occurrence of $NaHg_3$. No evidence for this phase was detected by thermal analysis [5, 9, 10, 14].

Crystal Structures. $NaHg_4$ is hexagonal, with $a = 61.5$ A, $c = 9.7$ A; the unit cell could contain about 230 units of $NaHg_4$. $NaHg_2$ is isotypic with AlB_2 (C32 type), $a = 5.029$ A, $c = 3.230$ A, $c/a = 0.642$. NaHg is end-centered orthorhombic, 16 atoms per unit cell, $a = 7.19$ A, $b = 10.79$ A, $c = 5.21$ A. Na_3Hg_2 has a primitive tetragonal unit cell with 20 atoms, $a = 8.52$ A, $c = 7.80$ A, $c/a = 0.916$. Na_5Hg_2 is rhombohedral, with $a = 18.52$ A, $\alpha = 29°23'$ (hexagonal cell dimensions are $a = 9.39$ A, $c = 53.1$ A) [19].

1. W. Kerp, *Z. anorg. Chem.*, **17**, 1898, 288–300.
2. W. Kerp, W. Böttger, and H. Winter, *Z. anorg. Chem.*, **25**, 1900, 7–16.
3. A. Guntz and J. Férée, *Compt. rend.*, **131**, 1900, 182–184.
4. E. Maey, *Z. physik. Chem.*, **29**, 1899, 129.
5. N. S. Kurnakow, *Z. anorg. Chem.*, **23**, 1900, 441–455.
6. Kraut and Popp, *Liebigs Ann.*, **159**, 1871, 188.
7. G. P. Grimaldi, *Atti accad. nazl. Lincei*, **4**, 1887, 32.
8. G. McPhail Smith and H. C. Bennett, *J. Am. Chem. Soc.*, **32**, 1910, 622–626.
9. A. Schüller, *Z. anorg. Chem.*, **40**, 1904, 385–399.
10. E. Vanstone, *Trans. Faraday Soc.*, **7**, 1911, 42–63; *Chem. News*, **103**, 1911, 181–185, 198–200, 207–209.
11. G. Tammann, *Z. physik. Chem.*, **3**, 1889, 443, 447.
12. C. T. Heycock and F. H. Neville, *J. Chem. Soc.*, **55**, 1889, 672.
13. N. C. Baenziger, J. W. Nielsen, and E. J. Duwell, *U.S. Atomic Energy Comm., Publ.* COO-126, 1953; see also [19].
14. E. Jänecke, *Z. Metallkunde*, **20**, 1928, 113–115.
15. E. Jänecke (*Z. physik. Chem.*, **57**, 1907, 510) determined the melting points of $NaHg_2$ and NaHg as 350 and 217°C.
16. E. Jänecke, "Kurzgefasstes Handbuch aller Legierungen," pp. 263–265, Carl Winter, Heidelberg, 1949.
17. R. C. Rodgers, *Phys. Rev.*, **8**, 1916, 259.
18. W. Klemm and B. Hauschulz, *Z. Elektrochem.*, **45**, 1939, 346–353.
19. J. W. Nielsen and N. C. Baenziger, *Acta Cryst.*, **7**, 1954, 277–282.

0.1432
$\bar{1}.8568$

Hg-Nd Mercury-Neodymium

The existence of $NdHg_4$ (15.24 wt. % Nd) [1] and NdHg (41.83 wt. % Nd) [2] has been reported. NdHg is isotypic with CsCl (B2 type), $a = 3.772$ A.

1. P. T. Daniltchenko, *Zhur. Obshchĕi Khim.*, **1**, 1931, 467–474; abstract in *J. Inst. Metals*, **50**, 1932, 540.
2. A. Iandelli and R. Ferro, *Atti accad. nazl. Lincei, Rend. Classe sci. fis. mat. e nat.*, **10**, 1951, 48–52.

0.5338
$\bar{1}.4662$

Hg-Ni Mercury-Nickel

The solubility of Ni in Hg at room temperature was reported as (in wt. %) 5.9×10^{-4} [1], 1.4×10^{-4} [2], and 2×10^{-5} [3].

It appears that only one intermediate phase exists [4], which was identified as $NiHg_4$ (6.82 wt. % Ni) [4, 5]. It is apparently stable up to 200–300°C [4, 6] and has a homogeneity range of about 20–25 at. % Ni, which would include the composition

$NiHg_3$ (8.89 wt. % Ni) suggested earlier [6]. Initially, the powder pattern was indexed as a primitive cubic structure, with $a = 3.00_6–3.01_0$ A [6, 4]. However, a more detailed study revealed the existence of a b.c.c. superstructure isotypic with that of $PtHg_4$ (10 atoms per unit cell), $a = 6.016$ A [5]. Work by [7] did not give conclusive results as to the composition and structure of the intermediate phase.

Supplement. According to a recent careful analysis [8], the solubility of Ni in Hg at room temperature is less than 2×10^{-6} wt. ($<7 \times 10^{-6}$ at.) %.

1. G. Tammann and K. Kollmann, *Z. anorg. Chem.*, **160**, 1927, 244–246.
2. E. Palmaer, *Z. Elektrochem.*, **38**, 1932, 70–76.
3. N. M. Irvin and A. S. Russell, *J. Chem. Soc.*, **1932**, 891–898.
4. F. Lihl, *Z. Metallkunde*, **44**, 1953, 160–166.
5. F. Lihl and H. Nowotny, *Z. Metallkunde*, **44**, 1953, 359; see also E. Bauer, H. Nowotny, and A. Stempfl, *Monatsh. Chem.*, **84**, 1953, 692–700.
6. R. Brill and W. Haag, *Z. Elektrochem.*, **38**, 1932, 212.
7. F. Pawlek, *Z. Metallkunde*, **42**, 1950, 451–453.
8. J. F. deWet and R. A. W. Haul, *Z. anorg. Chem.*, **277**, 1954, 96–112.

$\bar{1}.9859$
0.0141

Hg-Pb Mercury-Lead

Liquidus. Liquidus temperatures were determined for the composition ranges (in at. %) 0–0.35 [1], 93.7–99.3 [2], 64.3–94.9 [3], 3.3–97.4 [4], 20–95 [5], and 1.5–4.3 [6]. The data points, shown in Fig. 454, indicate excellent agreement, except for those by [3] which are about 10–30°C too low. Liquidus temperatures determined by [7] were not available.

In the temperature range 20–70°C, the chemically determined solubilities by [6] are to be regarded as the most accurate data (see inset). Liquidus points reported by [1] show that the freezing point of Hg (−38.9°C) is lowered by 0.015 wt. (at.) % for only 0.02°C and then raised 0.027, 0.37, 0.89, 1.24, and 1.30°C by 0.07, 0.165, 0.24, 0.32, and 0.35 at. % Pb, respectively. This would indicate the existence of a peritectic equilibrium at about −37.6°C. The liquidus temperatures by [6] appear to be in accord with these findings (see inset).

Other solubility data are less certain. The following values (in at. %), based on chemical determination [8] and emf measurements [4, 9, 10], have been recorded: 1.26 at 15–18°C [8], <1.75 at 20°C [4], close to 1.16 at 18°C [9], below 1.94 at 29 and 15.5°C [10], and slightly above 0.97 at 0°C [10].

Solidus. Solidus temperatures in the low-temperature range were determined by means of thermoresistometric analysis on heating [11] and cooling curves [3]. The temperatures by [11] would suggest the existence of a peritectic horizontal between −37 and −37.7°C, whereas those by [3], for the region 1.9–64.3 at. % Pb, scatter between −37.4 and −40.9°C, with a mean value of −39.1°C. The latter data do not appear to be reliable since the liquidus temperatures reported by the same authors are not correct either (Fig. 454).

The thermal effect in the vicinity of −39°C [3] was observed only up to 64.3 at. % Pb; at 69.3 at. % and higher Pb contents it was missing. Indeed, alloys with more than 69 at. % Pb proved to be single-phase [3]. A wide range of solid solubility of Hg in Pb is also indicated by the approximate temperatures of the end of solidification between 75 and 95 at. % Pb [5] as well as by emf measurements at temperatures of 30, 15.5, 0 [10], and 20°C [4]; see data points in Fig. 454.

X-ray investigations proved that at least 20 at. % Hg is soluble in Pb [12], resulting in a decrease in the lattice constant of Pb by 1.6%. The curve of the temperatures of superconductivity vs. composition suggests a solubility limit of

60–65 wt. % Pb [13, 14]. The value of about 75 at. %, assumed by [15] on the basis of microexamination, is too high since the alloys were not in equilibrium, as shown by [16]. (See Supplement.)

As discussed earlier, there is good reason to assume that a peritectic horizontal exists at about −37.6°C. No information is available, however, as to whether this peritectic reaction is characterized by the formation of an intermediate phase [liq. + (Pb) → X], as in the system Hg-Sn, or an Hg-rich solid solution [liq. + (Pb) → (Hg)].

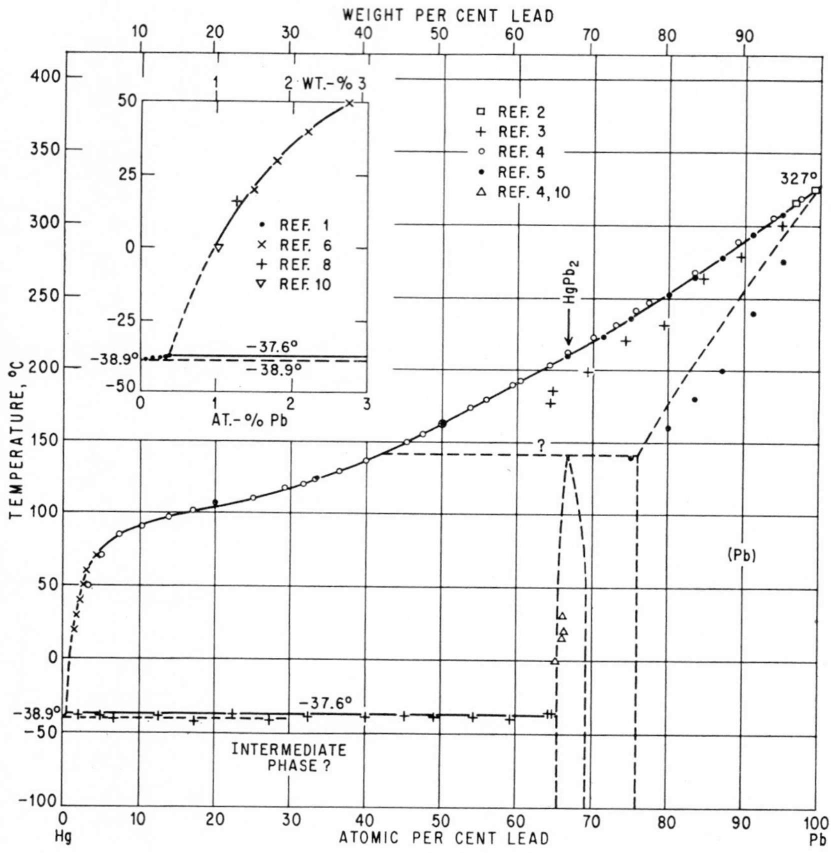

Fig. 454. Hg-Pb. (See also Supplement.)

If the very slight lowering of the freezing point of Hg, reported by [1], is real, the reaction would be definitely of the former type. But even if this is not the case, the existence of a peritectically formed intermediate phase is not ruled out but, in fact, appears to be more likely than the existence of an appreciable solid solubility of Pb in Hg. Earlier assumptions of a compound—Hg_3Pb_2 [17], HgPb [18], and $HgPb_2$ [3]—are based on inconclusive evidence.

Supplement. [19] examined, by X-ray and microscopic analysis, alloys with 66–100 at. % Pb quenched from 93–95°C. The lattice parameter of Pb was found to decrease linearly by addition of Hg. The 33.87 at. % Hg (66.13 at. % Pb) alloy was homogeneous and gave a diffraction pattern corresponding to a f.c. tetragonal structure, with $a = 4.9715$ kX, $c = 4.5027$ kX, $c/a = 0.9057$. The 25.85 at. % Hg (74.15 at. % Pb) alloy was metallographically duplex, giving a double diffraction pattern f.c. tetragonal + (Pb). "An intermediate phase thus exists at 33 at. % Hg, and may be analogous to the f.c.c. $NaPb_3$, which is stable over the range 27–33 at. % Na in the system Pb-Na. Since the new phase is stable at 93–95°C, it must be formed peritectically from the liquid and the Pb-rich solid solution at some higher temperature [see Fig. 454]. If the linearity of the lattice spacings of the Pb-rich solution with composition continues up to the solubility limit of Hg in Pb, the approximate solid solubility is 24 at. % Hg, at 93–95°C" [19].

1. G. Tammann, *Z. physik. Chem.*, **3,** 1889, 444–445.
2. C. T. Heycock and F. H. Neville, *J. Chem. Soc.*, **61,** 1892, 910.
3. H. Fay and E. North, *Am. Chem. J.*, **25,** 1901, 216–231.
4. N. A. Puschin, *Z. anorg. Chem.*, **36,** 1903, 201–254.
5. E. Jänecke, *Z. physik. Chem.*, **60,** 1907, 400.
6. H. E. Thompson, *J. Phys. Chem.*, **39,** 1935, 655–664.
7. D. Mazzotto, *Atti ist. veneto*, **4,** 1892-1893, 1311, 1527; abstract in *Z. physik. Chem.*, **13,** 1894, 571–572.
8. Gouy, *J. phys.*, **4,** 1895, 320–321.
9. J. F. Spencer, *Z. Elektrochem.*, **11,** 1905, 683.
10. J. J. Babinski, Dissertation, University of Leipzig, 1906; see also G. Timofejew, *Z. physik. Chem.*, **78,** 1912, 304.
11. G. Gressmann, *Phys. Rev.*, **9,** 1899, 20; *Physik. Z.*, **1,** 1900, 345.
12. C. v. Simson, *Z. physik. Chem.*, **109,** 1924, 198.
13. W. Meissner, H. Franz, and H. Westerhoff, *Ann. Physik*, **13,** 1932, 521–524.
14. C. Benedicks, *Z. Metallkunde*, **25,** 1933, 199–200.
15. G. Tammann and Q. A. Mansuri, *Z. anorg. Chem.*, **132,** 1923, 67–68.
16. G. Tammann and H. Rüdiger, *Z. anorg. Chem.*, **192,** 1930, 29–33.
17. Joule, *Chem. Gazz.*, **1850,** 399.
18. Crookewitt, *J. prakt. Chem.*, **45,** 1848, 87.
19. C. Tyzack and G. V. Raynor, *Acta Cryst.*, **7,** 1954, 505–510.

0.2742
$\bar{1}.7258$

Hg-Pd Mercury-Palladium

An X-ray investigation [1] of Pd amalgams with 11.2–34.8 wt. (19.2–50 at.) % Pd, prepared by amalgamation of Pd powder and annealing at 400°C for 2 hr and subsequently at 100°C for 200 hr, had the following results: The patterns of the products with 17 and 20 wt. (27.7 and 32.0 at.) % Pd showed the presence of a structure related to that of γ-brass. At 23 and 26 wt. (35.9 and 39.8 at.) % Pd this phase coexists with one whose structure appears to be somewhat similar to that of $PtZn_2$. At 29 wt. (43.4 at.) % Pd the latter structure together with that of the CuAu ($L1_0$) type was observed. The 34.8 wt. (50 at.) % Pd alloy showed nearly the pure pattern of the CuAu-type structure, with $a = 4.284$ A, $c = 3.692$ A, $c/a = 0.862$. These data were concluded to indicate the existence of the phases Pd_2Hg_5 (28.57 at. % Pd), Pd_2Hg_3, and PdHg. Also, indications were found of a fourth phase with approximately 25 at. % Pd.

Measurements of the magnetic susceptibility [2] were reported to be in accord with the phase limits presented in Fig. 455.

Partially amalgamated Pd leaf gave a complex electron-diffraction pattern [3].

Supplement. [4] studied the mineral potarite of the approximate composition PdHg (plus trace of Au) and thought it to be probably isotypic with cubic FeSi (B20 type). Attempts to synthesize PdHg were unsuccessful.

1. H. Bittner and H. Nowotny, *Monatsh. Chem.*, **83,** 1952, 287–289.
2. H. Bittner and H. Nowotny, *Monatsh. Chem.*, **83,** 1952, 1308–1313.
3. A. E. Aylmer, G. I. Finch, and S. Fordham, *Trans. Faraday Soc.*, **32,** 1936, 864–871.
4. M. A. Peacock, *Univ. Toronto Studies, Geol. Ser.*, **49,** 1945, 71–73; *Structure Repts.*, **10,** 1945-1946, 73–74.

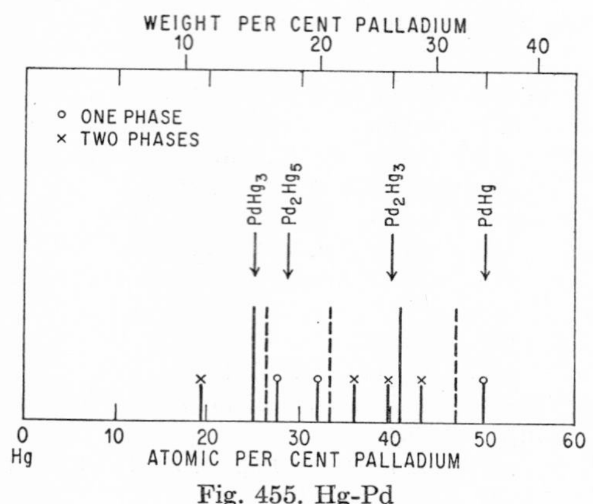

Fig. 455. Hg-Pd

0.1534
$\bar{1}.8466$

Hg-Pr Mercury-Praseodymium

The existence of $PrHg_4$ (14.94 wt. % Pr) has been reported [1]; see also system Hg-La. PrHg (41.26 wt. % Pr) is b.c.c. (B2 type), a = 3.791 A [2].

1. P. T. Daniltchenko, *Zhur. Obshchĕi Khim.*, **1,** 1931, 467–474; abstract in *J. Inst. Metals*, **50,** 1932, 540.
2. A. Iandelli and R. Ferro, *Atti accad. nazl. Lincei, Rend. Classe sci. fis. mat. e nat.*, **10,** 1951, 48–52.

0.0118
$\bar{1}.9882$

Hg-Pt Mercury-Platinum

By X-ray analysis of alloys covering the whole composition range in increments of 5 at. % [1], three intermediate phases have been identified: $PtHg_4$ (19.57 wt. % Pt), $PtHg_2$ (32.73 wt. % Pt), and PtHg (49.32 wt. % Pt), each being practically of singular composition. The solid solubility of Hg in Pt was determined parametrically as about 18.5 at. % Hg (after annealing at about 450°C).

Earlier, [2] had studied the system by thermal and qualitative X-ray methods. It was claimed that about 23 at. % Hg was soluble in Pt and that three phases of the approximate compositions PtHg, Pt_2Hg, and Pt_3Hg existed. The proposed tentative diagram [2] shows these phases to be formed by peritectic reactions at approximately 160, 245, and 480°C, respectively. The solubility of Pt in Hg (in at. % Pt) was reported as 0.10, 0.20, 0.91, 0.98, 1.08, 1.20, and 1.77 at 24, 54, 71, 101, 144, 171, and 200°C, respectively. These values indicate an inexplainable abrupt solubility increase between 54 and 71°C.

On the assumption that the thermal effects reported by [2] represent true phase transformations, a tentative phase diagram is given in Fig. 456 which combines the phase equilibria suggested by [2] with the phases identified by [1]. This diagram is to be considered with great reservation.

Crystal Structures. Lattice constants of Pt-rich solid solutions were given by [1]. $PtHg_4$ is b.c.c., with 10 atoms per unit cell (new type), $a = 6.18_6$ A. $PtHg_2$ is tetragonal, with 3 atoms per unit cell and $a = 4.68_7$ A, $c = 2.91_3$ A, $c/a = 0.621$.

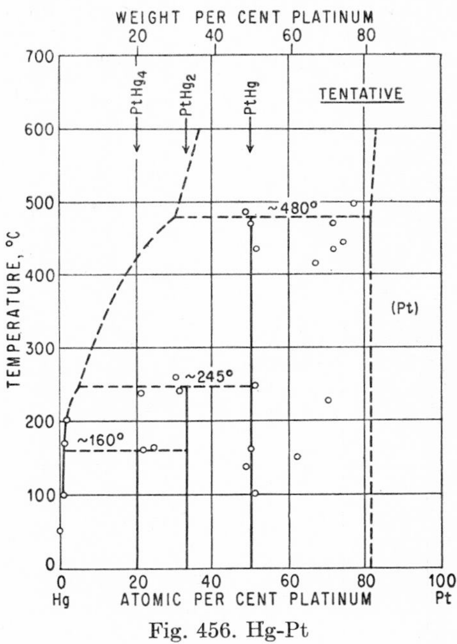

Fig. 456. Hg-Pt

The structure represents a transition between the C1- and C16-type structures. PtHg is isotypic with CuAu ($L1_0$ type), $a = 4.20$ A, $c = 3.82_5$ A, $c/a = 0.909$ [1].

1. E. Bauer, H. Nowotny, and A. Stempfl, *Monatsh. Chem.*, **84**, 1953, 692–700; see also **84**, 1953, 211–212.
2. I. N. Plaksin and N. A. Suvorovskaya, *Izvest. Sektora Platiny*, **18**, 1945, 67–76; *Acta Physicochim. U.R.S.S.*, **13**, 1940, 83–96; *Compt. rend. acad. sci. U.R.S.S.*, **27**, 1940, 460–463; *Zhur. Fiz. Khim.*, **15**, 1941, 978–980.

$\bar{1}.9240$
0.0760

Hg-Pu Mercury-Plutonium

According to [1], $PuHg_3$ and $PuHg_4$ are isostructural with the corresponding U compounds.

1. A. S. Coffinberry and M. B. Waldron, Review of the Physical Metallurgy of Plutonium, in "Metallurgy and Fuels," Progress in Nuclear Energy, vol. 1, Pergamon Press Ltd., London, 1956.

0.3705
$\bar{1}.6295$

Hg-Rb Mercury-Rubidium

The solubility of Rb in Hg at 0 [1], 19.5 [2], and 25°C [1] was found as 0.92 wt. (2.15 at.) %, 1.21 wt. (2.80 at.) %, and 1.37 wt. (3.15 at.) % Rb, in good agreement with 1.43 wt. (3.31 at.) % at 26.4°C, based on cooling-curve data [3].

Liquidus temperatures in the range 3.3–14.6 at. % Rb were reported by [3] (see crosses in Fig. 457), and the whole solidification diagram was determined by [4]. In the composition range covered by both investigations the results of [3] appear

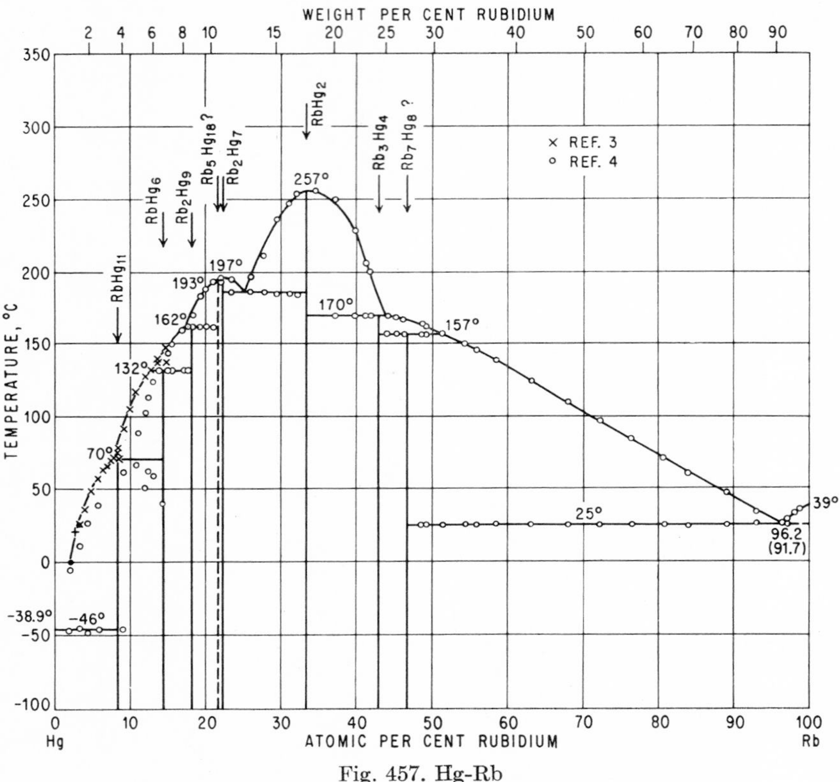

Fig. 457. Hg-Rb

to be more accurate since (*a*) higher liquidus temperatures were found for a given composition, probably indicating a greater sensitivity of the temperature recording, and (*b*) they agree very well with the chemical determinations of the solubility [1, 2], which generally give reliable results.

According to the data of [3], the composition of the intermediate phase richest in Hg, formed peritectically at 70 [3] or 67°C [4], could lie between $RbHg_{11}$ (8.33 at. % Rb) and $RbHg_7$ (12.5 at. % Rb), whereas those by [4] would be compatible only with compositions between $RbHg_9$ (10.0 at. % Rb) and $RbHg_7$. [4] assumed $RbHg_9$, i.e., a composition practically coinciding with that of the peritectic melt at 67°C, according to their data. This assumption is based only on the extrapolation of the eutectic halting times at −46°C. Measurements of magnetic susceptibility, covering the whole range of composition, would suggest $RbHg_{10}$ [5]. The composition $RbHg_{12}$ [1, 2], found by isolating the solid phase at room temperature, is inconsistent with the thermal data, indicating that this method is successful only if high pressure is used to separate liquid from solid. (See Supplement.)

For the phase formed peritectically at 137 [3] or 132°C [4], both [3] and [4] assumed the composition $RbHg_6$ [14.29 at. (6.63 wt.) % Rb], the latter on the basis of the duration of the peritectic reaction. The existence of Rb_2Hg_9 [18.18 at. (8.65 wt.) % Rb] appears to be certain [4, 5], although the composition $RbHg_4$ would have the preference of a simpler atomic ratio.

The first maximum of the liquidus was found at 22.0 at. % Rb, 196°C, suggesting the composition Rb_2Hg_7 [22.22 at. (10.85 wt.) % Rb]. The assumption of another phase only slightly different in composition, Rb_5Hg_{18} [21.74 at. (10.58 wt.) % Rb], is based on only one secondary thermal effect at 193°C of the 22.0 at. % Rb alloy.

Besides the compound $RbHg_2$ (17.56 wt. % Rb), characterized by a maximum melting point, two intermediate phases richer in Rb are formed at 170 and 157°C. The former was assumed to be Rb_3Hg_4 [42.86 at. (24.21 wt.) % Rb]. This composition is in accord with the measurements of magnetic susceptibility of [5]. The existence of Rb_7Hg_8 [46.67 at. (27.15 wt.) % Rb] is based on inconclusive evidence; the composition RbHg (29.88 wt. % Rb) would also be consistent with the thermal data. The magnetic-susceptibility-vs.-composition curve has two breaks in this composition range, at about Rb_7Hg_8 and RbHg [5], in disagreement with the thermal-analysis data.

Supplement. According to [6], $RbHg_{11}$ has a structure similar to that of $BaHg_{11}$ (the latter being cubic, with 36 atoms per unit cell). The lattice parameter of $RbHg_{11}$ is $a = 9.734 \pm 2$ A [7].

1. W. Kerp, W. Böttger, and H. Winter, *Z. anorg. Chem.*, **25**, 1900, 29–31.
2. G. McPhail Smith and H. C. Bennett, *J. Am. Chem. Soc.*, **32**, 1910, 622–626.
3. N. S. Kurnakow and G. J. Zukowsky, *Z. anorg. Chem.*, **52**, 1907, 427–428.
4. W. Biltz, F. Weibke, and H. Eggers, *Z anorg. Chem.*, **219**, 1934, 119–128.
5. W. Klemm and B. Hauschulz, *Z. Elektrochem.*, **45**, 1939, 346–353.
6. N. C. Baenziger and E. J. Duwell, *Acta Cryst.*, **7**, 1954, 635.
7. E. J. Duwell and N. C. Baenziger, *Acta Cryst.*, **8**, 1955, 709.

0.0321
$\bar{1}.9679$

Hg-Re Mercury-Rhenium

Electrolysis of an aqueous solution of $KReO_4$ with Hg as cathode yielded Re amalgam [1]. No attack of Re powder by Hg, heated to 300°C in reducing atmosphere, was observed by [2].

1. H. Hölemann, *Z. anorg. Chem.*, **202**, 1931, 277–291.
2. G. Heyne and K. Moers, *Z. anorg. Chem.*, **196**, 1931, 157.

0.7963
$\bar{1}.2037$

Hg-S Mercury-Sulfur

HgS (13.78 wt. % S) exists in two, perhaps three [1], modifications: metacinnabarite is isotypic with zincblende (B3 type), $a = 5.86$ A [2–6]; cinnabar is the prototype of the B9 type, hexagonal with 3 HgS per unit cell, $a = 4.16$ A, $c = 9.53$ A, $c/a = 2.29$ [7, 4, 5, 8, 9]. The latter is the stable form of HgS at all temperatures up to the sublimation point reported as 580°C [1] or 446 ± 10°C [10]. According to [1], the third form also has a hexagonal structure.

Supplement. [11] has redetermined the lattice parameters ($a = 4.14_6$ A, $c = 9.49_7$ A) and the atom positions of the cinnabar structure.

1. E. T. Allen and J. L. Crenshaw, *Z. anorg. Chem.*, **79,** 1912, 155–171, 185–189.
2. N. H. Kolkmeijer, J. M. Bijvoet, and A. Karssen, *Rec. trav. chim.*, **48,** 1924, 678–679, 894–896.
3. W. Lehmann, *Z. Krist.*, **60,** 1924, 379–413.
4. S. v. Olshausen, *Z. Krist.*, **61,** 1925, 463–514.
5. H. E. Buckley and W. S. Vernon, *Mineralog. Mag.*, **20,** 1925, 382–392.
6. W. Hartwig, *Sitzgsber. preuss. Akad. Wiss.*, **10,** 1926, 79–80.
7. C. Mauguin, *Compt. rend.*, **176,** 1923, 1483–1486.
8. W. F. de Jong and H. W. V. Willems, *Physica,* **6,** 1926, 129–136.
9. B. Gossner and F. Mussgnug, *Zentr. Mineral.*, **1927A,** 410–413.
10. W. Biltz, *Z. anorg. Chem.*, **59,** 1908, 279.
11. K. L. Aurivillius, *Acta Chem. Scand.*, **4,** 1950, 1413–1436; abstract in *Structure Repts.*, **13,** 1950, 179.

0.2168
$\bar{1}.7832$

Hg-Sb Mercury-Antimony

The solubility of Sb in Hg at 18°C was reported as 2.9×10^{-5} wt. % [1]. Earlier observations had indicated that Sb is practically insoluble in Hg at room temperature [2].

1. G. Tammann and J. Hinnüber, *Z. anorg. Chem.*, **160,** 1927, 254–256.
2. W. J. Humphreys, *J. Chem. Soc.*, **69,** 1896, 1686; B. Neumann, *Z. physik. Chem.*, **14,** 1894, 219.

0.4049
$\bar{1}.5951$

Hg-Se Mercury-Selenium

The crystal structure of synthetically prepared HgSe (28.24 wt. % Se) was determined by [1] and that of natural HgSe (tiemannite) by [2, 3]. The compound is isotypic with zincblende (B3 type) [1–3], $a = 6.08$ A [1, 3].

As to the behavior of melts containing an excess of Se over the composition HgSe, see [4]. The HgSe-Se eutectic appears to lie very close to 100% Se.

Supplement. [5] redetermined the lattice parameter of HgSe by electron diffraction: $a = 6.074 \pm 6$ A.

1. W. H. Zachariasen, *Z. physik. Chem.*, **124,** 1926, 436–448.
2. W. F. de Jong, *Z. Krist.*, **63,** 1926, 466–471.
3. W. Hartwig, *Sitzgsber. preuss. Akad. Wiss.*, *Phys.-math. Kl.*, **10,** 1926, 79–80.
4. G. Pellini and R. Sacerdoti, *Gazz. chim. ital.*, **40**(2), 1910, 42–46.
5. U. Zorll, *Z. Physik,* **138,** 1954, 167–169; *Chem. Abstr.*, **48,** 1954, 13330.

0.8538
$\bar{1}.1462$

Hg-Si Mercury-Silicon

Silicon is not attacked by mercury at room temperature [1] and elevated temperature [2]. Confirming these observations, [3] found no attack at temperatures up to the boiling point of Hg.

1. W. Guertler, "Handbuch der Metallographie," vol. 1, part 2, p. 702, Verlagsbuchhandlung Gebrüder Borntraeger, Berlin, 1917.
2. Winkler, *J. prakt. Chem.*, **91,** 1864, 193.

3. Unpublished work by L. F. Epstein, quoted from L. R. Kelman, W. D. Wilkinson, and F. L. Yaggee, "Resistance of Materials to Attack by Liquid Metals," p. 67, Argonne National Laboratory, July, 1950. The observations by [1, 2] were erroneously reported here as indicating attack.

0.2279
$\bar{1}.7721$

Hg-Sn Mercury-Tin

Liquidus. The liquidus curve shown in Fig. 458 is based on the thermal data by [1, 2] and is known with great accuracy, as indicated by the data points. Additional data which cover limited composition ranges [3, 4] or alloys of various selected compositions [5–8] are in very good agreement with this curve. This also holds for the liquidus determined by [9], for which no tabulated data were published. Results of work by [10] were not available.

The solubility of Sn in Hg at normal temperatures was determined as 1.00 at. (0.6 wt.) % at 15–18°C [11], 0.97 at. % at 15°C, 1.21 at. % at 25°C [2], 1.05 at. % at 14°C, and 1.24 at. % at 25.4°C [7].

Liquidus points in the temperature range down to the melting point of Hg were determined by chemical and thermal analysis [2]. They indicate that there is a peritectic equilibrium at −34.6°C and a eutectic practically coinciding with 100% Hg (see inset).

Intermediate Phases. Prior to 1924, it had been assumed, on the basis of the work by [12, 1, 2, 13], that (*a*) the only intermediate phase (of unknown composition) is formed peritectically at −34.6°C and (*b*) at room temperature Hg-rich melt is in equilibrium with Sn-rich primary solid solution. The saturation limit of the latter phase was given as about 94 or 99 at. % [2] or 88 at. % Sn [13]. However, there was already some evidence which pointed to more complicated phase relations, e.g., the fact that, on the basis of emf measurements, the solubility limit of Hg in Sn was assumed to be 99 at. % Sn [2], whereas the peritectic reaction at −34.6°C could be detected thermally only up to 75.6 at. % Sn [2] and dilatometrically up to about 85 at. % Sn [2].

Direct evidence for the existence of a Sn-rich intermediate phase was first given by [14] and confirmed by [15], by means of X-ray analysis. This phase was found to be present in alloys with about 91–99 at. % [14] and to be homogeneous at about 91–93 at. % [14], 90–94 at. % Sn [15].

Unpublished work by [9] showed that three intermediate phases exist: (*a*) a high-temperature β phase of variable composition around 96.5 at. (94 wt.) % Sn; (*b*) a γ phase homogeneous in the range 89–94.5 at. (82.75–91 wt.) % Sn [16]; and (*c*) the phase $HgSn_3$, stable below −34.6°C (Fig. 458). The γ phase is identical with that detected by [14, 15]. It was recently confirmed by [17, 23].

[18] established the presence of a fourth peritectically formed δ phase, as indicated by the occurrence of thermal effects at about 90°C in alloys with 20, 30, and 50 wt. % Sn. The constitution of the ternary system Ag-Hg-Sn also indicates that this phase exists [19].

As no details of the work by [9] are available, it is unknown whether this investigator has established the existence of $HgSn_3$ or assumed this composition as a likely one, on the basis of previously published data. If one assumes that $HgSn_3$ exists, the δ phase would lie between $HgSn_3$ and γ. In an alloy with 78 at. % Sn, [17] found the γ phase coexisting with Hg-rich melt; however, the amount of δ phase formed at 90°C could have been too small to be detected by X-ray diffraction, because of incompleteness of the peritectic reaction.

If, as appears likely, the phase $HgSn_3$ has not been established, the phases formed at 90 and −34.6°C might well have compositions richer in Hg. Since the rate of transformation will be very low—even on prolonged aging the peritectic reactions will not be completed—the determination of their compositions, especially of the phase stable below −34.6°C, is a difficult experimental task. The thermal and dilatometric studies of [2] are of little value here, since the effects observed do not represent equilibrium conditions. Although [2] reported that the maximal thermal

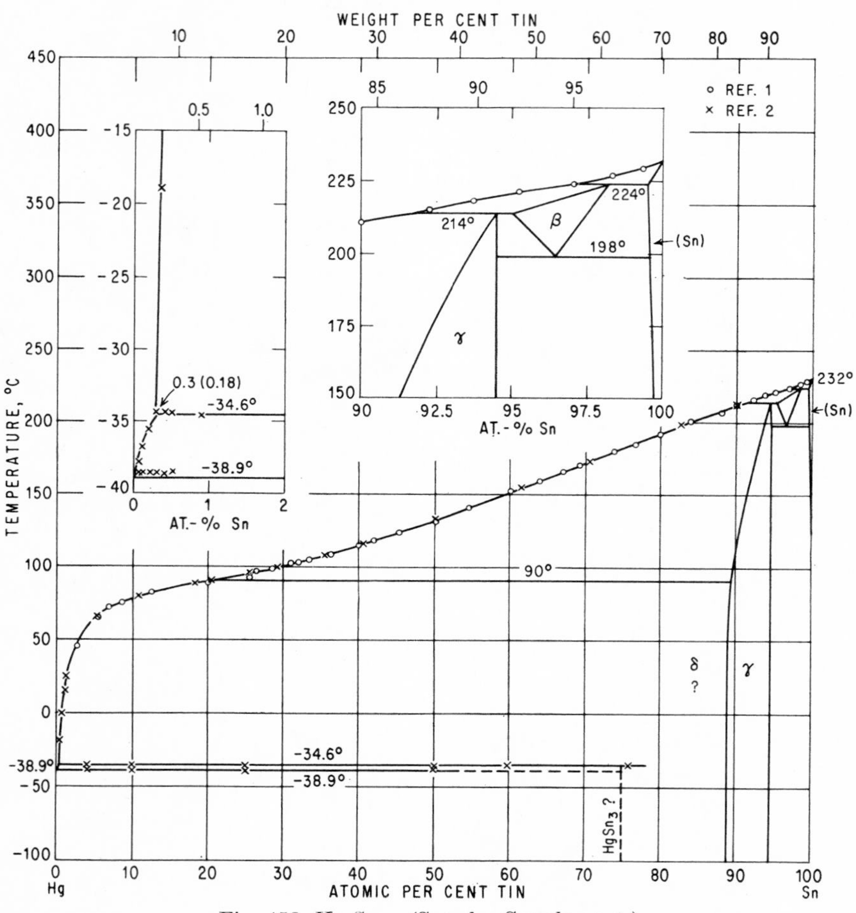

Fig. 458. Hg-Sn. (See also Supplement.)

effect was found at 25 at. % Sn (dilatometric investigations gave inconclusive results), he assumed 60 at. % Sn as the most likely composition. Others have preferred the composition Hg_3Sn [20, 21] or HgSn [22] as more likely.

Crystal Structure. The γ phase has a simple hexagonal structure [14, 15, 17, 23], a = 3.2127 A, c = 2.9916 A, c/a = 0.9312 at 7.19 at. % Hg [23]; for parameters of additional alloys, see [14, 15, 17]. At 8.8 at. % Hg, [15] found the structure to be orthorhombically distorted, a = 5.559 A, b = 3.202 A, c = 2.987 A. Perhaps the latter structure is that of the δ phase. β has a structure very similar to that of γ; it is also simple hexagonal, $a = 3.2415$ A, $c = 3.0065$ A, $c/a = 0.9275$ at 188°C [23]. Lattice parameters of the (Sn) phase were reported by [17] and [24].

Supplement. On the basis of high-temperature X-ray work in the β-phase region, [25] suggested that β and γ are one and the same phase, the γ phase shifting its homogeneity range to higher Sn contents at elevated temperatures.

1. N. A. Puschin, *Z. anorg. Chem.*, **36,** 1903, 201–254.
2. W. J. van Heteren, *Z. anorg. Chem.*, **42,** 1904, 129–173.
3. G. Tammann, *Z. physik. Chem.*, **3,** 1889, 445.
4. C. T. Heycock and F. H. Neville, *J. Chem. Soc.*, **57,** 1890, 383.
5. E. Wiedemann, *Wied. Ann.*, **3,** 1878, 249.
6. C. Cattaneo, *Wied. Ann. Beibl.*, **14,** 1890, 1188.
7. R. A. Joyner, *J. Chem. Soc.*, **99,** 1911, 195–208.
8. M. L. V. Gayler, *J. Inst. Metals*, **60,** 1937, 403.
9. W. E. Prytherch, unpublished work; the diagram proposed is shown in *J. Inst. Metals*, **60,** 1937, 381, 403.
10. D. Mazzotto, *Atti inst. veneto*, **4,** 1892-1893, 1311, 1527; abstract, *Z. physik. Chem.*, **13,** 1894, 571–572.
11. Gouy, *J. phys.*, **4,** 1895, 320–321.
12. B. Neumann, *Z. physik. Chem.*, **14,** 1894, 217.
13. G. Tammann and Q. A. Mansuri, *Z. anorg. Chem.*, **132,** 1923, 66–67.
14. C. v. Simson, *Z. physik. Chem.*, **109,** 1924, 187–192.
15. S. Stenbeck, *Z. anorg. Chem.*, **214,** 1933, 23–26.
16. Compositions and temperatures shown in Fig. 458 were taken from the published diagram.
17. L. Løvold-Olsen; see L. Vegard, *Skrifter Norske Videnskaps-Akad. Oslo, Mat. Naturv. Kl.*, no. 2, 1947, 32–37; quoted from *Structure Repts.*, **11,** 1947-1948, 163.
18. M. L. V. Gayler, *J. Inst. Metals*, **60,** 1937, 403.
19. M. L. V. Gayler, *J. Inst. Metals*, **60,** 1937, 379–400.
20. K. Bornemann, *Metallurgie*, **7,** 1909, 108–109.
21. M. Hansen, "Der Aufbau der Zweistofflegierungen," pp. 806–814, Springer-Verlag OHG, Berlin, 1936.
22. W. Guertler, "Handbuch der Metallographie," vol. 1, part 1, p. 719, Verlagsbuchhandlung Gebrüder Borntraeger, Berlin, 1912.
23. G. V. Raynor and J. A. Lee, *Acta Met.*, **2,** 1954, 616–620.
24. J. A. Lee and G. V. Raynor, *Proc. Phys. Soc.* (*London*), **B67,** 1954, 737–747 (Lattice constants of a 0.72 at. % Hg alloy).
25. K. Schubert, U. Rösler, W. Mahler, E. Dörre, and W. Schütt, *Z. Metallkunde*, **45,** 1954, 643–647.

0.3597
$\bar{1}.6403$

Hg-Sr Mercury-Strontium

The solubility of Sr in Hg increases from 0.73 wt. (1.65 at.) % at 0°C to 1.79 wt. (4.00 at.) % at 64.5°C [1]. At room temperature the solubility was found as 1.04 wt. (2.34 at.) % (20°C) [1] and 1.12 wt. (2.50 at.) % (23°C) [2]. The solubility curve in Fig. 459 suggests that the solubility just above the freezing point of Hg is still appreciable; i.e., there is a eutectic of Hg and an intermediate phase.

Several investigators have attempted to determine the phase in equilibrium with the saturated solution at room temperature [1–6] and temperatures of 46–81°C [4, 1] by separating the liquid and solid phases. Thus, compositions nearly corresponding to $SrHg_{12}$ (3.51 wt. % Sr) [1, 2] and $SrHg_{11}$ (3.82 wt. % Sr) [3–6] have been found. At 46–81°C, [1] obtained residues ranging in composition between about

$SrHg_8$ (5.18 wt. %) and $SrHg_7$ (5.87 wt. %). However, none of these compositions is based on conclusive evidence since it was not certain whether the separation was complete. Vacuum distillation of Hg-rich amalgams yielded residues of the compositions Sr_2Hg_5 (14.87 wt. % Sr) [1, 6] and $SrHg_6$ (6.79 wt. % Sr) [6].

Supplement. In a short note, [7] stated that $SrHg_{11}$ has a structure similar to that of $BaHg_{11}$ reported by [8]. According to [10], $SrHg_{11}$ is isostructural with $BaHg_{11}$, KHg_{11}, and $RbHg_{11}$. The lattice parameter of the cubic unit cell, containing 36 atoms, is 9.5099 ± 8 A.

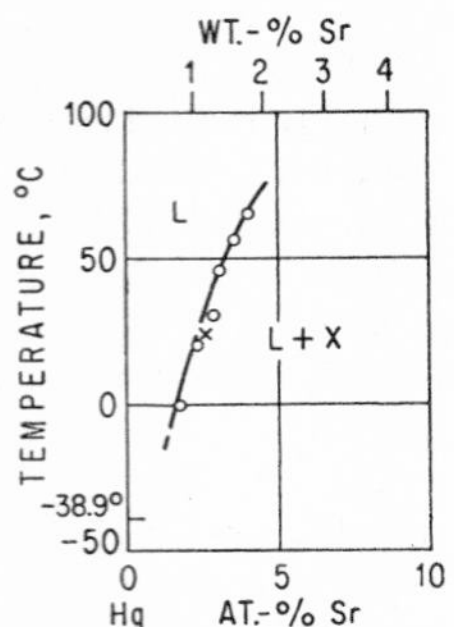

Fig. 459. Hg-Sr. (See also Supplement.)

[9] prepared the compound SrHg by alloying the elements in pure iron vessels and showed it to have the CsCl (B2) type of structure, with a = 3.930 A.

1. W. Kerp, W. Böttger, and H. Iggena, *Z. anorg. Chem.*, **25,** 1900, 35–44.
2. G. McPhail Smith and H. C. Bennett, *J. Am. Chem. Soc.*, **32,** 1910, 622–626.
3. A. Guntz and J. Férée, *Bull. soc. chim. France*, **17,** 1897, 390.
4. W. Kerp, *Z. anorg. Chem.*, **17,** 1898, 305–308.
5. G. Langbein, Dissertation, University of Königsberg, 1900.
6. A. Guntz and G. Roederer, *Bull. soc. chim. France*, **35,** 1906, 494–503.
7. N. C. Baenziger and E. J. Duwell, *Acta Cryst.*, **7,** 1954, 635.
8. G. Peyronel, *Gazz. chim. ital.*, **82,** 1952, 679 (see Ba-Hg).
9. R. Ferro, *Acta Cryst.*, **7,** 1954, 781.
10. E. J. Duwell and N. C. Baenziger, *Acta Cryst.*, **8,** 1955, 709.

0.0450
$\bar{1}$.9550

Hg-Ta Mercury-Tantalum

Attempts to amalgamate Ta have been unsuccessful. No attack could be detected, not even at higher temperatures [1].

1. W. v. Bolton, *Z. Elektrochem.*, **11,** 1905, 51.

0.1965
$\bar{1}$.8035

Hg-Te Mercury-Tellurium

The solidification temperatures of melts with 60–100 at. (48.8–100 wt.) % Te were determined by [1] (Fig. 460). The eutectic halting times at 411 ± 3°C indicate that the phase richest in Te is HgTe (38.88 wt. % Te), which has long been known. Its melting point cannot be determined under atmospheric pressure since decomposition starts at about 550°C (in vacuum at about 370°C).

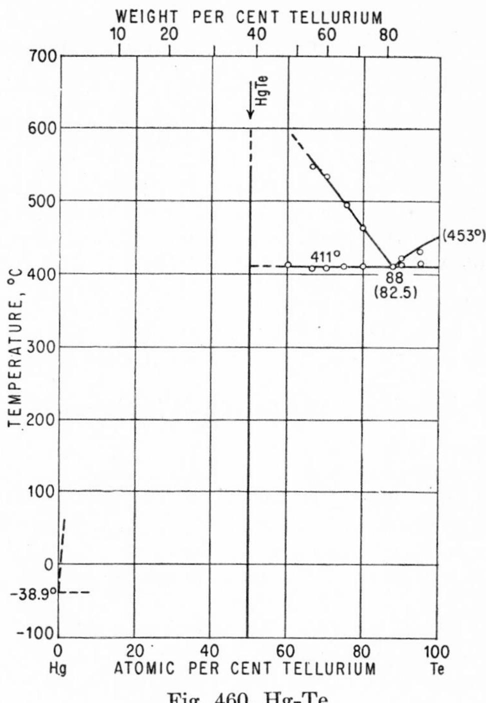

Fig. 460. Hg-Te

HgTe is isotypic with zincblende (B3 type), a = 6.44–6.45 A [2].

Supplement. [3] redetermined the lattice parameter of HgTe by electron diffraction: a = 6.429 ± 6 A.

1. G. Pellini and C. Aureggi, *Gazz. chim. ital.*, **40**(2), 1910, 42–49; *Atti accad. nazl. Lincei*, **18**(2), 1909, 211–217.
2. W. F. de Jong, *Z. Krist.*, **63**, 1926, 466–472; W. H. Zachariasen, *Z. physik. Chem.*, **124**, 1926, 277–284; W. Hartwig, *Sitzgsber. preuss. Akad. Wiss.*, **10**, 1926, 79–80.
3. U. Zorll, *Z. Physik*, **138**, 1954, 167–169; *Chem. Abstr.*, **48**, 1954, 13330.

$\bar{1}$.9366
0.0634

Hg-Th Mercury-Thorium

The solubility of Th in Hg at 25°C was reported to be not greater than 0.014 at. (0.016 wt. %) [1]. [2] reported that "at approximately the composition $ThHg_3$ a hexagonal phase, a = 3.38 A, c = 4.72 A [c/a = 1.40], occurs with z = ½ [i.e., two atoms per cell]. Intensity data are compatible with a disordered h.c.p. structure, but probably do not exclude ordering. The lattice constants vary somewhat from sample to sample and presumably the composition is variable over a range, but the solubility limits have not been established. The compound resembles UHg_3 [3]."

1. W. G. Parks and G. E. Prime, *J. Am. Chem. Soc.*, **58**, 1936, 1413–1414.
2. N. C. Baenziger, R. E. Rundle, and A. I. Snow, *Acta Cryst.*, **9**, 1956, 93–94.
3. R. E. Rundle and A. S. Wilson, *Acta Cryst.*, **2**, 1949, 148–150.

0.6220
$\bar{1}.3780$

Hg-Ti Mercury-Titanium

The solubility of Ti in Hg at room temperature was reported as 1×10^{-5} wt. % [1].

Three intermediate phases have been identified: TiHg (19.27 wt. % Ti) and two modifications of Ti_3Hg (41.74 wt. % Ti), designated as γ-Ti_3Hg and δ-Ti_3Hg. TiHg has the f.c. tetragonal structure of the CuAu ($L1_0$) type, a = 4.256 A, c = 4.041 A, c/a = 0.949. γ-Ti_3Hg, stable between 540 and 760°C, is isotypic with "β-W" (A15 type), a = 5.189 A. δ-Ti_3Hg, found to be stable at 816°C, is isotypic with Cu_3Au ($L1_2$ type), a = 4.165 A [2].

1. N. M. Irvin and A. S. Russell, *J. Chem. Soc.*, **1932,** 891–898.
2. P. Pietrokowsky, *Trans. AIME,* **200,** 1954, 219–226.

$\bar{1}.9919$
0.0081

Hg-Tl Mercury-Thallium

The liquidus curve (Fig. 461) of the entire system was determined by [1–3]. Also, liquidus temperatures for the following composition ranges in at. % Tl [4] were reported: 0–0.48 [5], 16.7–42.3 [6], 82–100 [7], 19.1–43.3 [8]. Results are in excellent agreement, with the exception of the data for the range 10–40 at. % Tl given by [1].

Special attention was paid to the question of the composition at which the maximum is located. In contrast to [1], who found it at 33.3 at. % Tl (Hg_2Tl), all other investigators agree in that it lies within the following ranges in at. % Tl: 25–31.7 [2], 27.1–30.1 [3], 26–31 [6], and 26.4–30.2 [8]. In these regions, the liquidus temperature was found to differ by about 1 [2], less than 0.5 [3], 0.6 [6], and 0.5°C [8], indicating that the maximum is very flat. Since the simplest atomic proportion in the ranges mentioned is Hg_5Tl_2 (28.57 at. % Tl), it was believed [3, 6, 8] that this composition was characteristic for the intermediate phase of variable composition (see below). The absolute maximum was found by all four investigators to lie in the small range 28.6–29.1 at. % Tl at 14.4–15°C.

There is also very good agreement as to the location of the two eutectics; compositions are given in at. % Tl. (Hg + α) eutectic: 8.34, −60°C [1]; 8.0, −60°C [2]; and 8.56, −59°C [3]. (α + β_{Tl}) eutectic: 40.0, 3.5°C [1]; about 41.5, 2°C [2]; 40.0, 0.6°C [3]; 40.5, 0.9°C [6]; and 40.2, 1.1°C [8].

The f.c.c. α phase was located between about 20 and 30 at. % Tl, on the basis of thermal data [2, 3] and emf measurements [9]. The crystal structure does not indicate the composition Hg_5Tl_2 to be characteristic for this phase (see below).

[2, 3] concluded from thermal data that 14 or 18 at. % Hg, respectively, was soluble in Tl at the eutectic temperature. [7] found, by means of emf measurements at 20°C, that (*a*) the terminal solid solution α_{Tl} extends up to about 4 wt. (at.) % Hg and (*b*) the phase β_{Tl} is stable between about 85 wt. (84.75 at.) % and 90.5 wt. (90.35 at.) % Tl (Fig. 461). In agreement with this, [9] found alloys with 86 and 90 at. % Tl to be b.c.c. (β_{Tl}).

Emf measurements in the ranges in wt. % Tl 0–56 at 18°C [10]; 0–75 at 0, 17, and 30°C [11]; 0–50 at −80, 0, and 37°C [12]; and 0–50 at 20, 30, and 40°C [6] are in general agreement with the phase relations presented in Fig. 461.

Crystal Structures. The α phase is f.c.c., probably with random atomic distribution (A1 type), a = 4.673 A at 29 at. % Tl [9], a = 4.68 A [13]; see also [14]. For the b.c.c. phase β_{Tl}, the following parameters were reported: a = 3.819 A at 86 at. % and a = 3.827 A at 90 at. % Tl [9]. X-ray diffraction of an alloy with

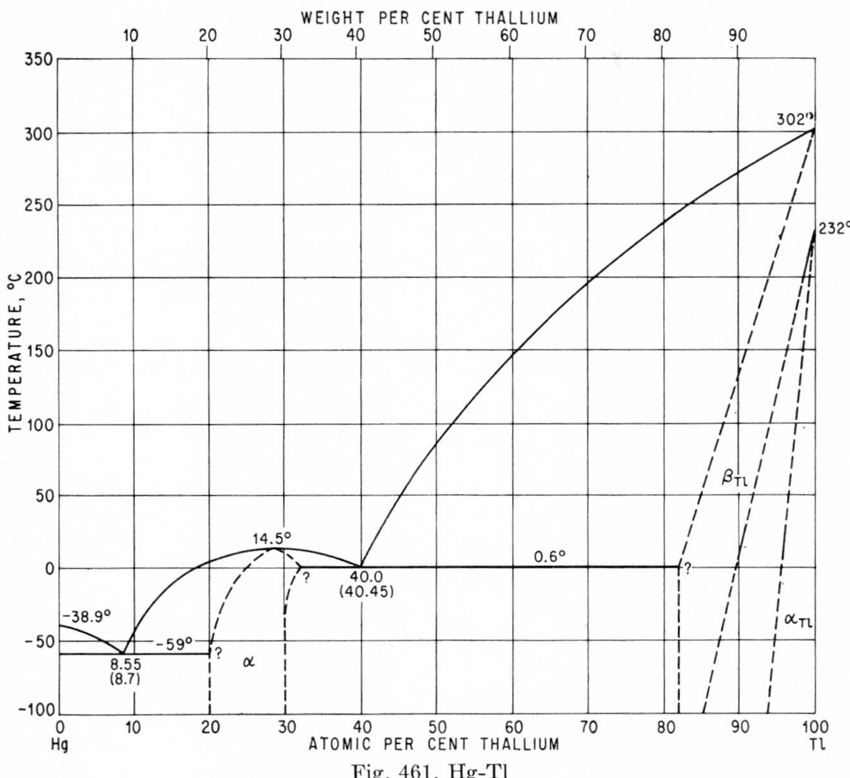

Fig. 461. Hg-Tl

29 at. % Tl at a temperature 10°C below its melting point gave no indication of a transformation in that temperature range as supposed by [3]. As regards X-ray studies of liquid alloys, see [15, 16].

1. N. S. Kurnakow and N. A. Puschin, *Z. anorg. Chem.*, **30,** 1902, 101–108.
2. P. Pawlowitch, *Zhur. Russ. Fiz. Khim. Obshchestva*, **47,** 1915, 29–46.
3. G. D. Roos, *Z. anorg. Chem.*, **94,** 1916, 358–370.
4. At. % and wt. % differ by maximal 0.5%.
5. G. Tammann, *Z. physik. Chem.*, **3,** 1889, 443.
6. T. W. Richards and F. Daniels, *J. Am. Chem. Soc.*, **41,** 1919, 1732–1767.
7. T. W. Richards and C. P. Smyth, *J. Am. Chem. Soc.*, **44,** 1922, 524–545.
8. N. A. Puschin, *Bull. soc. chim. Belgrade*, **14,** 1949, 101–103.
9. A. Ölander, *Z. physik. Chem.*, **171,** 1934, 425–435.
10. J. F. Spencer, *Z. Elektrochem.*, **11,** 1905, 681–684.
11. J. J. Babinski, Dissertation, Leipzig, 1906.
12. A. Sucheni, *Z. Elektrochem.*, **12,** 1906, 729–731.
13. E. Osswald and F. Sauerwald, *Science Repts. Tôhoku Univ., K. Honda Anniv. Vol.*, 1936, pp. 931–932.
14. K. Schubert, *Z. Metallkunde*, **39,** 1948, 88–96.
15. F. Sauerwald and W. Teske, *Z. anorg. Chem.*, **210,** 1933, 247–256.
16. F. Sauerwald and E. Osswald, *Z. anorg. Chem.*, **257,** 1948, 195–198.

$\bar{1}.9256$
0.0744

Hg-U Mercury-Uranium

The phase diagram in Fig. 462 is due to [1]. It was established by means of thermal, X-ray, and low-temperature micrographic analysis. Special experimental

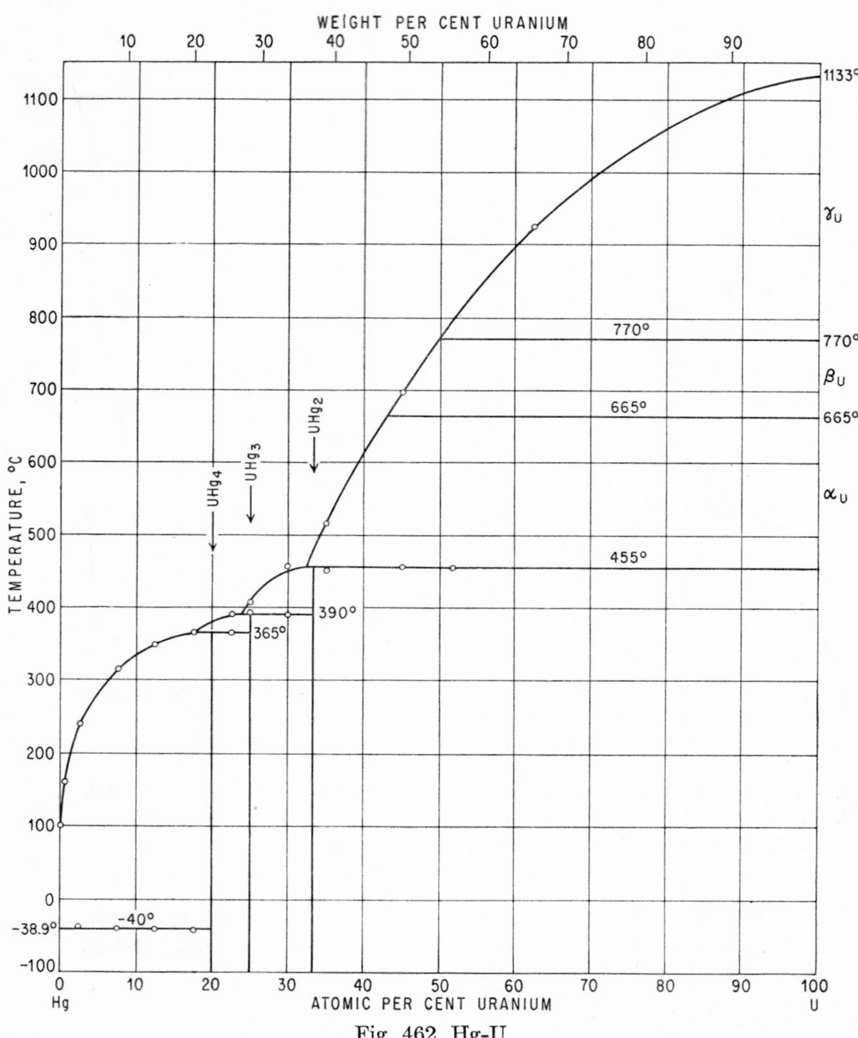

Fig. 462. Hg-U

techniques in preparation and examination of the alloys had to be used. Since at all, except very dilute, concentrations the alloys were pyrophoric, all operations had to be carried out in vacuum.

The existence of the three intermediate phases UHg_4 (22.88 wt. % U), UHg_3 (28.35 wt. % U), and UHg_2 (37.24 wt. % U) [5] had been established earlier by X-ray work [2]. The mutual solid solubility of the components is very small; the transformation temperatures of U are not affected by Hg additions [1].

The solubility of U in liquid Hg at room temperature was reported (in wt. %) as 1.4×10^{-4} [3] and 1×10^{-5} [4].

Crystal Structures. The structure of UHg_4 could not be established because of experimental difficulties [1, 2]. [2] suggested a b.c. pseudocubic cell, with $a = 3.63$ A [6]. UHg_3 is hexagonal, with 2 atoms per unit cell and $a = 3.32$–3.33 A, $c = 4.88$–4.89 A, $c/a = 1.47$ [2, 1]. UHg_2 is isotypic with AlB_2 (C32 type), $a = 4.98$–4.99 A, $c = 3.22$–3.23 A, $c/a = 0.647$ [2, 1].

1. B. R. T. Frost, *J. Inst. Metals*, **82**, 1953-1954, 456–462.
2. R. E. Rundle and A. S. Wilson, *Acta Cryst.*, **3**, 1949, 148–150.
3. G. Tammann and J. Hinnüber, *Z. anorg. Chem.*, **160**, 1927, 260–261.
4. N. M. Irvin and A. S. Russell, *J. Chem. Soc.*, **1932**, 891–898.
5. Whereas [1] studied the system by sealing samples and allowing the pressure to rise, [7], in another study of this system, worked at a pressure of 1 atm. The temperatures of decomposition into α-U and Hg vapor of the compounds UHg_4, UHg_3, and UHg_2 were found to be 360–365, 390, and 455°C, respectively.
6. According to E. Bauer, H. Nowotny, and A. Stempfl (*Monatsh. Chem.*, **84**, 1953, 692–700) UHg_4 is probably isotypic with $PtHg_4$.
7. D. H. Ahmann, R. R. Baldwin, and A. S. Wilson, *U.S. Atomic Energy Comm., Publ.* CT-2960, Oct. 27, 1945.

0.5952
$\bar{1}.4048$

Hg-V Mercury-Vanadium

The solubility of V in Hg at room temperature was reported as 5×10^{-5} wt. % [1]. [2], using an electrochemical method, did not succeed in determining the solubility.

1. N. M. Irvin and A. S. Russell, *J. Chem. Soc.*, **1932**, 891–898.
2. G. Tammann and J. Hinnüber, *Z. anorg. Chem.*, **160**, 1927, 256–257.

0.0377
$\bar{1}.9623$

Hg-W Mercury-Wolfram

According to [1], the solubility of W in Hg at room temperature is 1×10^{-5} wt. %. Using an electrochemical method, [2] were unable to arrive at a reliable solubility value.

1. N. M. Irvin and A. S. Russell, *J. Chem. Soc.*, **1932**, 891–898.
2. G. Tammann and J. Hinnüber, *Z. anorg. Chem.*, **160**, 1927, 260.

0.4869
$\bar{1}.5131$

Hg-Zn Mercury-Zinc

Liquidus temperatures were determined for the composition ranges (in at. % Zn) 0.31–0.81 [1], 4.7–13.5 [2], 2.6–94.9 [3], 4.05–18.9 [4], and 0.31–3.10 [5], by means of chemical [2, 4] and thermal analysis [1, 3, 5]; see Fig. 463. The solubility of Zn in Hg at room temperature was found to be 1.8 wt. (5.3 at.) % at 15–18°C [5a], 2–2.1 wt. (5.85–6.15 at.) % at 20°C [2], about 2.2 wt. (6.4 at.) % at 20°C [3], and 1.99 wt. (5.83 at.) % at 20°C [4]. The last value appears to be the most accurate one. The eutectic point was determined as 0.56 wt. (1.70 at.) % Zn, −41.6°C [5].

[3, 6] claimed, on the basis of microexamination and other data, that there is no intermediate phase. The assumption of the existence of Hg_2Zn_3 [7] and $HgZn_2$ [8, 9],

reported in the early literature, is based on inconclusive evidence, although later work showed that intermediate phases of approximately these compositions actually appear to exist.

First proof of the existence of a three-phase equilibrium was already presented by [10]. Measurements of electrical resistance on heating and cooling between 15 and

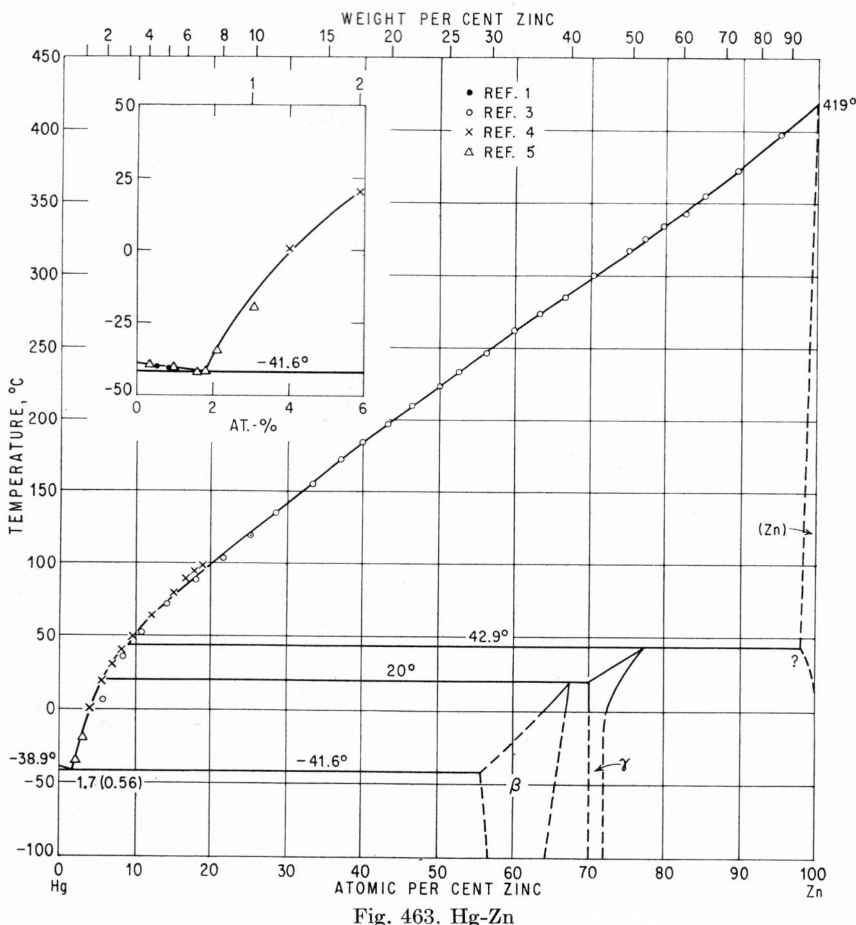

Fig. 463. Hg-Zn

100°C of alloys with about 24, 49, and 60.5 at. % Zn showed that the reaction is characterized by a considerable temperature hysteresis, because of a very low rate of diffusion. Breaks found on heating indicated two transformation points at about 35 and 60–75°C. Also, [11], studying the change of the electrical resistance on heating of alloys with about 33.3, 50, and 66.7 at. % Zn, found a pronounced drop in resistance at 60–70°C. On cooling to 30–40°C, the reverse transformation was not observable because of supercooling. [12] found that the coefficient of thermal expansion reaches a maximum at about 70°C.

Dilatometric work, with an alloy of 10 wt. % Zn, indicated that the lower transformation point reported by [10] lies, under conditions approaching equilibrium, at 19–20°C and the upper point at 42.9°C [13]. The same investigators have carried out comprehensive and accurate emf measurements at 0, 12, 25, 35, and 50°C, using alloys covering the range 7.25–99.5 at. % Zn. The limits of the two intermediate phases β and γ shown in Fig. 463 are based on breaks in the emf isotherms. It is doubtful whether these phase boundaries represent equilibrium conditions, since it would require a very long time to complete the peritectic reaction at 43°C and, more so, that at 20°C.

X-ray analysis [14, 15] confirmed the existence of one of the two intermediate phases; the other is stable only below room temperature. Lines of this phase were detected in photograms of alloys with 65–90 at. % Zn [14] and in the range 51–88 at. % Zn [15]. The composition of this phase was tentatively assumed to correspond to $HgZn_3$ [14] or Hg_3Zn_8 (72.73 at. % Zn) [15], i.e., compositions near that found by [13].

The solid solubility of Hg in Zn was reported as about 13 at. % Hg [6] and 0.5–1.5 at. % Hg [13], on the basis of microexamination and emf measurements, respectively. The lattice-parameter data given by [15] would indicate a solubility of 2–3 at. % Hg. [17] concluded that the solubility limit at "room temperature" lies at 0.25 wt. (0.08 at.) % Hg.

Crystal Structure. Lattice parameters of the Zn-rich solid solution were reported by [15]. Unit cell dimensions of the intermediate phase were given as $a = 2.7$ A, $c = 5.4$ A, $c/a = 2.0$ [14] and $a = 2.713$ A, $c = 5.480$ A, $c/a = 2.020$ [15].

1. G. Tammann, *Z. physik. Chem.*, **3**, 1889, 443.
2. W. Kerp and W. Böttger (H. Iggena), *Z. anorg. Chem.*, **25**, 1900, 54–59.
3. N. A. Puschin, *Z. anorg. Chem.*, **36**, 1903, 201–254.
4. E. Cohen and K. Inouye, *Z. physik. Chem.*, **71**, 1910, 625–635.
5. V. Peshkov, *Acta Physicochim. U.R.S.S.*, **21**, 1946, 109–134; *Zhur. Fiz. Khim.*, **20**, 1946, 835; see also [16].

5a. Gouy, *J. phys.* **4**, 1895, 320–321.

6. G. Tammann and Q. A. Mansuri, *Z. anorg. Chem.*, **132**, 1923, 68–69.
7. Crookewitt, *J. prakt. Chem.*, **45**, 1848, 87.
8. Joule, *Chem. Gaz.*, **1850**, 399.
9. L. Schüz, *Wied. Ann.*, **46**, 1892, 177.
10. R. S. Willows, *Phil. Mag.*, **48**, 1899, 433; see also [13].
11. Lohr, Dissertation, University Erlangen, 1914, quoted from A. Schulze, Die elektrische u. thermische Leitfähigkeit, in W. Guertler, "Handbuch der Metallographie," pp. 609–610, Verlagsbuchhandlung Gebrüder Borntraeger, Berlin, 1925.
12. J. Würschmidt, *Ber. deut. physik. Ges.*, **14**, 1912, 1065–1087.
13. E. Cohen and P. J. H. van Ginneken, *Z. physik. Chem.*, **75**, 1911, 437–493.
14. C. v. Simson, *Z. physik. Chem.*, **109**, 1924, 192–195.
15. L. Løvold-Olsen in L. Vegard, *Skrifter Norske Videnskaps-Akad. Oslo, Math. Naturv. Kl.*, no. 2, 1947, pp. 32–37; see *Structure Repts.*, **11**, 1947-1948, 165.
16. O. Hájiček (*Hutnické Listy*, **3**, 1948, 265–270) reported the eutectic as 3.25 at. % Zn, −43°C.
17. R. Chadwick, *J. Inst. Metals*, **51**, 1933, 114.

0.3423
$\bar{1}.6577$

Hg-Zr Mercury-Zirconium

The intermediate phases $ZrHg_3$ (13.16 wt. % Zr), ZrHg (31.26 wt. % Zr), and Zr_3Hg (57.70 wt. % Zr) have been identified by determining their crystal structures. $ZrHg_3$ is isotypic with Cu_3Au ($L1_2$ type), $a = 4.365$ A. ZrHg is f.c. tetragonal of the

CuAu ($L1_0$) type, a = 3.15 A, c = 4.17 A, c/a = 1.32. Zr_3Hg is isotypic with "β-W" (A15 type), a = 5.558 A [1]. See also Hg-Ti.

The solubility of Zr in Hg was reported as 5×10^{-4} wt. % at 350°C and 16×10^{-4} wt. % at 550°C [2].

1. P. Pietrokowsky, *Trans. AIME*, **200,** 1954, 219–226.
2. Quoted from L. R. Kelman, W. D. Wilkinson, and F. L. Yaggee, "Resistance of Materials to Attack by Liquid Metals," p. 68, Argonne National Laboratory, July, 1950.

1.2184
$\bar{2}$.7816

In-Li Indium-Lithium

The phase diagram was outlined by thermal analysis [1]; alloys were not analyzed. Data points in Fig. 464 indicating solidus temperatures were found on heating. The

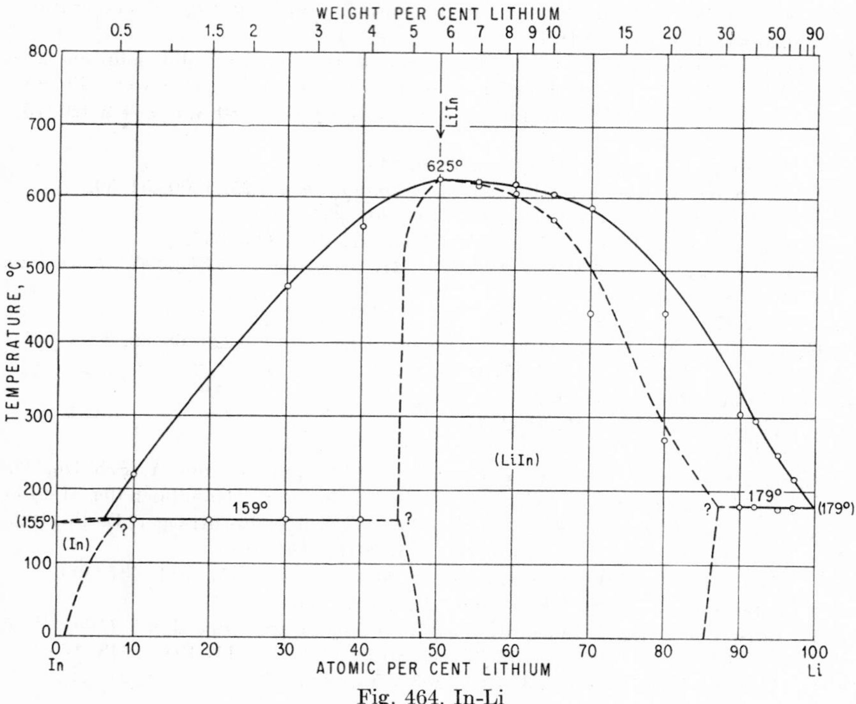

Fig. 464. In-Li

authors assumed that the phase LiIn (5.70 wt. % Li) forms a continuous series of solid solutions with Li. Since this is impossible, the diagram in Fig. 464 was drawn to show a two-phase field at the Li-rich end.

LiIn is isotypic with NaTl (B32 type), a = 6.80_0 A [2].

1. G. Grube and W. Wolf, *Z. Elektrochem.*, **41,** 1935, 675–681.
2. E. Zintl and G. Brauer, *Z. physik. Chem.*, **B20,** 1933, 245–271.

0.6738
$\bar{1}.3262$

In-Mg Indium-Magnesium

Phase relations in the range 0–50 at. % In (Fig. 465) are based on two prominent papers. Liquidus, solidus, and solubility limits in the range 0–25 at. % In were established by means of thermal and micrographic investigations [1]. The constitution between 25 and 50 at. % In was investigated using micrographic analysis and

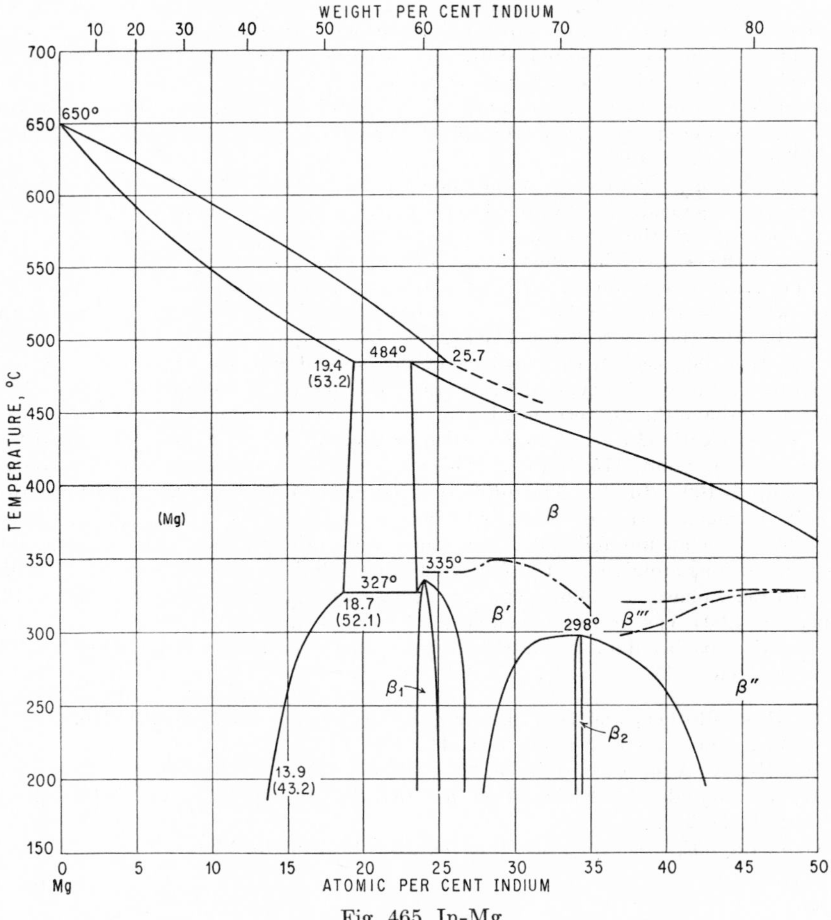

Fig. 465. In-Mg

X-ray analysis at ordinary and high temperatures (300–350°C) [2]. For additional X-ray studies, see below.

The f.c.c. (disordered) β phase is believed to extend to 100% In, merging into the tetragonal structure of In at higher In concentrations [2]. β with 23.55 at. % In decomposes eutectoidally at 327°C into (Mg) and β_1. The latter phase, characterized by a maximum transformation temperature of 335°C at 23.9 at. % In, is stable between 23.5 and 25.0 at. % In at 200°C. The β_2-phase field, having a maximum at 34.3 at. % In, 298°C, lies between about 34.0 and 34.5 at. % In at 200°C.

High-temperature X-ray work verified the existence of β_1 and β_2 and showed that order-disorder transformations occur in the composition range investigated, 24–49 at. % In. The boundaries between the regions of disordered and ordered structures were determined with an accuracy of $\pm 2°C$ or less; no hysteresis was observed. Results showed that four different structures exist, denoted as β, β', β'', and β'''. β is f.c.c. disordered; β' is f.c.c. ordered of the Cu_3Au ($L1_2$) type; β'' is f.c. tetragonal ordered of the CuAu ($L1_0$) type with an axial ratio of about 0.96; and β''' is orthorhombic, believed to be a stable structure in the temperature range shown [2]. The c/a ratio of β''' is close to 1; faint superlattice lines were observed.

The β_1 phase has a complex diffraction pattern not yet elucidated. [3] reported the existence of a phase designated as Mg_5In_2 (28.57 at. % In) with a b.c. orthorhombic structure, 28 atoms per unit cell. According to Fig. 465, this composition has a f.c.c. ordered structure of the Cu_3Au type at temperatures above 250°C and is two-phase below 250°C.

The β_2 phase also has a complex powder pattern [2]. [3] claimed the existence of Mg_2In, hexagonal with 18 atoms per unit cell and isomorphous with Mg_2Ga. It is not known whether this phase corresponds to β_2; the latter phase does not lie exactly at the composition Mg_2In.

For an alloy with 50 at. % In, [3, 2] reported a tetragonal structure of the CuAu type, with $a = 3.24_7$, $c = 4.39$ A, $c/a = 1.352$ for the unit cell with 2 atoms [3] and $c/a = 0.96$ for the f. c. tetragonal cell [2].

Further, [3] assumed the existence of the phase $MgIn_2$ (or $MgIn_3$) with a cubic structure of the Cu_3Au type and $a = 4.61$ A. According to [2], this alloy can be expected to have the f.c. tetragonal structure of the CuAu type.

An alloy with 92.7 at. % In, annealed for 1 month at 100°C, proved to be single-phase (melting point, 175–180°C). This suggests that the β phase stretches continuously up to 100% In [2]. The diffraction pattern of this alloy was found to be that of a f.c. tetragonal structure, similar to that of In [2].

Results of an unpublished X-ray study were briefly abstracted; unfortunately no compositions are given. The following intermediate phases were reported: a disordered f.c.c. phase, $a = 4.65_7$ A; a f.c.c. phase with a superstructure of the $NaPb_3$ type, $a = 4.604$ A; a f.c. tetragonal structure, $c/a = 0.96$ [4].

Lattice-spacing data of Mg-rich primary solid solutions with 0.63–11.65 at. % In were given by [5].

1. W. Hume-Rothery and G. V. Raynor, *J. Inst. Metals,* **63,** 1938, 201–216.
2. G. V. Raynor, *Trans. Faraday Soc.,* **44,** 1948, 15–28.
3. W. Haucke, *Naturwissenschaften,* **26,** 1938, 577–578.
4. W. Klemm and H. A. Klein, *FIAT Rev. Ger. Sci.,* 1939-1946, Inorganic Chemistry, Part III, p. 249.
5. G. V. Raynor, *Proc. Roy. Soc. (London),* **A174,** 1940, 457–471.

0.3200
$\bar{1}.6800$

In-Mn Indium-Manganese*

Thermal analysis, micrographic, and X-ray data were used to draw the phase diagram in Fig. 466 [1]. Phase relations involving the solid solutions of δ-Mn and γ-Mn were outlined only by thermal investigations. The microstructure and X-ray patterns of alloys with 3.8–40 at. % In, annealed at 800°C and quenched, showed the solubility of In in β-Mn to be about 14 at. (25.4 wt.) % In and the only intermediate

* The assistance of Dipl.-Phys. W. Thiele and Dr. U. Zwicker in preparing this system is gratefully acknowledged.

phase occurring to correspond to the composition Mn_3In (41.05 wt. % In). The β solid solution with about 12 at. (22.2 wt.) % In decomposes eutectoidally at 590°C into Mn_3In and an α-Mn solid solution of unknown composition.

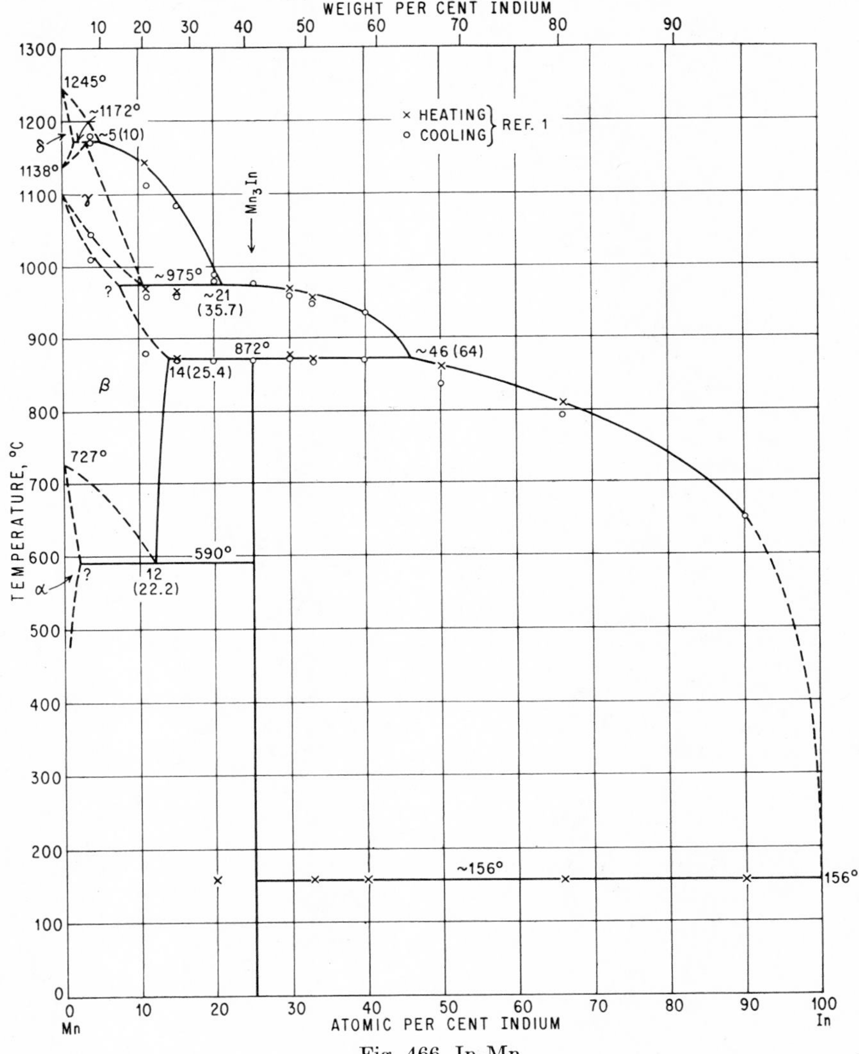

Fig. 466. In-Mn

The phase Mn_3In, which is of practically singular composition, has the structure of the γ-brass ($D8_{1-3}$) type, a = 9.43_5 A [1]. The lattice dimensions of both β-Mn and α-Mn increase with increasing In content. The solid solubility of In in α-Mn was found parametrically to decrease with temperature.

From magnetic measurements, [2] and [3] concluded the existence of the phases Mn_2In or Mn_4In, respectively, which disagrees with Fig. 466. Since [1] reported that the alloy with 25 at. % In, after homogenization at 800°C, proved to be nearly single-phase and contained only traces of β, the discrepancy may be due to the difficulty of attaining equilibrium in alloys with about 20–40 at. % In, because of the peritectic reaction at 872°C and the very broad solidification interval.

1. U. Zwicker, *Z. Metallkunde*, **41**, 1950, 399–401.
2. W. V. Goeddel and D. M. Yost, *Phys. Rev.*, **82**, 1951, 555.
3. S. Valentiner, *Z. Metallkunde*, **44**, 1953, 259–260.

0.9134
$\bar{1}$.0866

In-N Indium-Nitrogen

InN (10.88 wt. % N) has the structure of the wurtzite (B4) type, $a = 3.53_7$ A, $c = 5.70$ A, $c/a = 1.611$ [1].

1. R. Juza and H. Hahn, *Z. anorg. Chem.*, **239**, 1938, 282–287.

0.6981
$\bar{1}$.3019

In-Na Indium-Sodium

NaIn (16.69 wt. % Na) is isotypic with NaTl (B32 type), $a = 7.312$ A [1]. According to [2], who determined the effect of additions up to 2.1 wt. (0.48 at.) % In on the freezing point of Na, a eutectic lies at 96.0–96.2°C and between 1.2 and 2 wt. (0.28 and 0.45 at.) % In.

[3] reported solidification temperatures of alloys with up to 17.1 wt. (50.8 at.) % Na. However, these apparently not very accurate data do not permit drawing of conclusions as to the intermediate phases and phase relationships in this range. Between 33 and 50 at. % Na a constant "liquidus" temperature of 420 ± 1°C was recorded.

1. E. Zintl and S. Neumayr, *Z. physik. Chem.*, **B20**, 1933, 272–275.
2. C. T. Heycock and F. H. Neville, *J. Chem. Soc.*, **55**, 1889, 676.
3. M. F. W. Heberlein, *Trans. ASM*, **44**, 1952, 545–548.

0.2912
$\bar{1}$.7088

In-Ni Indium-Nickel

The phase diagram presented in Fig. 467 is due to [1], with the exception of the (Ni) phase boundary which was determined parametrically by [2]. It is based on thermal analysis and X-ray investigations, complemented by micrographic work in the temperature range 740-855°C, used to outline the ranges of the two intermediate phases of variable composition, ϵ and β [3], and to corroborate the results obtained by the other methods.

The temperature of the [(Ni) + ϵ] eutectic was given by [2] as 883°C, as compared with 908°C by [1]. The intermediate phases whose compositions correspond to atomic proportions, Ni_3In (39.46 wt. % In), NiIn (66.16 wt. % In), Ni_2In_3 (74.59 wt. % In), and Ni_3In_7 (82.02 wt. % In), are shown as singular phases although they may have (undetermined) narrow ranges of homogeneity. In the original diagram, the phase designated as Ni_3In_7 was shown to contain about 72 at. % rather than 70 at. % In (see below).

Crystal Structures. Lattice constants of the (Ni) phase were reported by [2]; see also [1, 4]. Ni_3In is isotypic with Ni_3Sn ($D0_{19}$ type), hexagonal with 8 atoms per unit cell, $a = 5.32$ A, $c = 4.24$ A, $c/a = 0.797$ [4, 1]. The structure of the ϵ phase [3], also sometimes designated as Ni_2In, is of the NiAs (B8) type [5, 6, 7], $a = 4.179$ A,

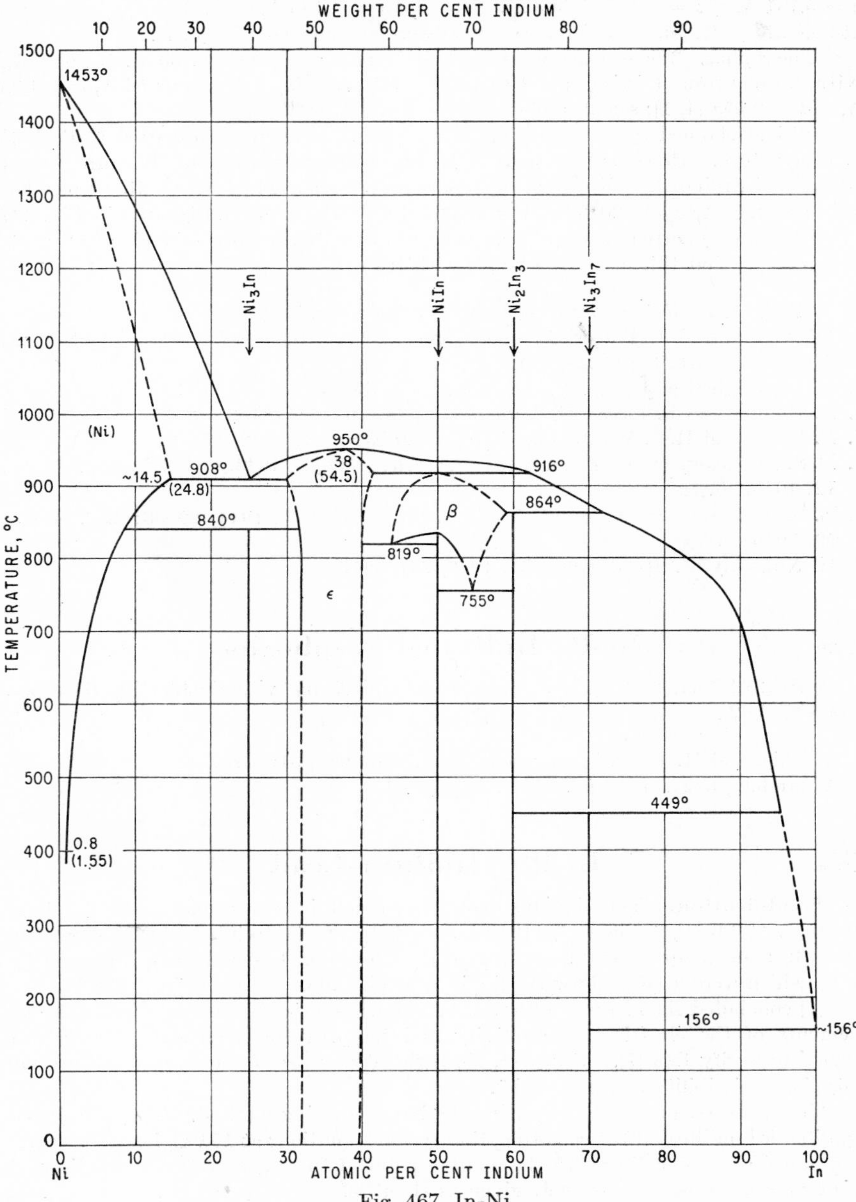
WEIGHT PER CENT INDIUM
10
20
30
40
50
60
70
80
90
1500
1400
1300
1200
1100
1000
900
800
700
600
500
400
300
200
100
0
TEMPERATURE, °C
1453°
(Ni)
Ni_3In
NiIn
Ni_2In_3
Ni_3In_7
950°
~14.5
908°
(24.8)
38
(54.5)
916°
840°
864°
β
819°
ε
755°
449°
0.8
(1.55)
156°
~156°
0
Ni
10
20
30
40
50
60
70
80
90
100
In
ATOMIC PER CENT INDIUM
Fig. 467. In-Ni

$c = 5.131$ A, $c/a = 1.228$ at 33.3 at. % In and $a = 4.186$ A, $c = 5.163$ A, $c/a = 1.233$ at 35.6 at. % In [5]. Lattice parameters were also reported by [6, 7].

The β phase [3] is isotypic with CsCl (B2 type), $a = 3.09$ A at 50 at. % In [4, 1]. NiIn, formed from β, is isotypic with CoSn (B35 type) [6, 4, 1], $a = 4.54$ A, $c = 4.34$ A, $c/a = 0.958$ [4, 1]; see also [6].

[8] had claimed the existence of $NiIn_2$ with a tetragonally distorted CaF_2 (C1) type structure. However, [4, 1] have proved that this phase and structure do not occur. Ni_2In_3 has the Ni_2Al_3 ($D5_{13}$) type structure, hexagonal with 5 atoms per unit cell, $a = 4.39$ A, $c = 5.20$ A, $c/a = 1.206$ [4, 1]. The phase designated in Fig. 467 as Ni_3In_7 was reported to have the structure of "γ brass with voids," $a = 9.18$ A. It is possible that this phase is isotypic with Ir_3Sn_7.

1. E. Hellner, *Z. Metallkunde*, **41**, 1950, 401–406; see also [4].
2. F. Weibke and I. Pleger, *Z. anorg. Chem.*, **231**, 1937, 197–216; F. Weibke, *Z. Metallkunde*, **31**, 1939, 228–230.
3. In the original papers [1, 4], the ϵ phase was designated as β and the β phase as δ.
4. E. Hellner and F. Laves, *Z. Naturforsch.*, **2a**, 1947, 177–183.
5. F. Laves and H. J. Wallbaum, *Z. angew. Mineral.*, **4**, 1942, 17–46.
6. E. S. Makarov, *Izvest. Akad. Nauk. S.S.S.R. (Khim.)*, **1943**, 264–270; *Met. Abstr.*, **12**, 1945, 10.
7. E. S. Makarov, *Izvest. Akad. Nauk. S.S.S.R. (Khim.)*, **1944**, 29–33; *Met. Abstr.*, **12**, 1945, 148.
8. H. Nowotny, *Z. Metallkunde*, **34**, 1942, 237–241.

0.5688
$\bar{1}$.4312

In-P Indium-Phosphorus

InP (21.25 wt. % P) [1] was reported to be isotypic with zincblende (B3 type), $a = 5.87$ A [2].

1. A. Thiel and H. Koelsch, *Z. anorg. Chem.*, **66**, 1910, 319–320.
2. A. Iandelli, *Gazz. chim. ital.*, **71**, 1941, 58–62.

$\bar{1}$.7434
0.2566

In-Pb Indium-Lead

Solidification. The liquidus was determined in the ranges 0–100% [1–3], 30–40 at. % Pb [4], 15–40 at. % Pb [5], and 11–28 at. % Pb [6]. Results agree substantially; the temperatures according to [1, 4] lie 5–10°C below those reported by [2, 3], which were used to draw the liquidus in Fig. 468.

[1] concluded that In and Pb form a continuous series of solid solutions. Measurements of the electrical conductivity and temperature coefficient of resistance seemed to verify this [7]. However, since the structures of the two metals are not isotypic, a miscibility gap must exist.

This was first proved by [4], using the lattice-spacing method. They suggested a peritectic equilibrium between the primary solid solutions at 154°C, i.e., a temperature practically coinciding with the melting point of In. The saturated solutions at this temperature were tentatively given as about 32 and 38 at. % Pb.

Reinvestigation by thermal analysis indicated the existence of two peritectic horizontals, at 171.9 and 159.2°C [2]. Similar results were reported by [3], 173.6 and 159.4°C, and by [6], 170 and 160°C. [5], using heating curves after homogenization, found the upper horizontal at 180°C; the lower horizontal lay outside the composition range investigated thermally.

As for the extent of the horizontals, it should be noted that cooling-curve data will indicate that the peritectic reactions occur within composition ranges considerably wider than under equilibrium conditions. Indeed, [2] observed the lower reaction in melts with up to 29.4 at. % Pb [8] and the upper reaction in melts with as high as 41 at. % Pb. Consequently, the solidus temperatures reported [2, 3, 6], with the possible exception of those by [5] shown in Fig. 468, do not represent equilibrium temperatures.

Solid State. (See also Supplement and Note Added in Proof.) The existence of an intermediate phase of variable composition, evidenced by the occurrence of two peritectic equilibria and measurements of electrical resistivity [9], was confirmed by X-ray analysis [2, 5, 6]. This phase is f.c. tetragonal, with $c/a = 0.93$ as compared with $c/a = 1.075$–1.078 reported for In. The homogeneity range has not yet been determined accurately. The following approximate regions in at. % Pb were

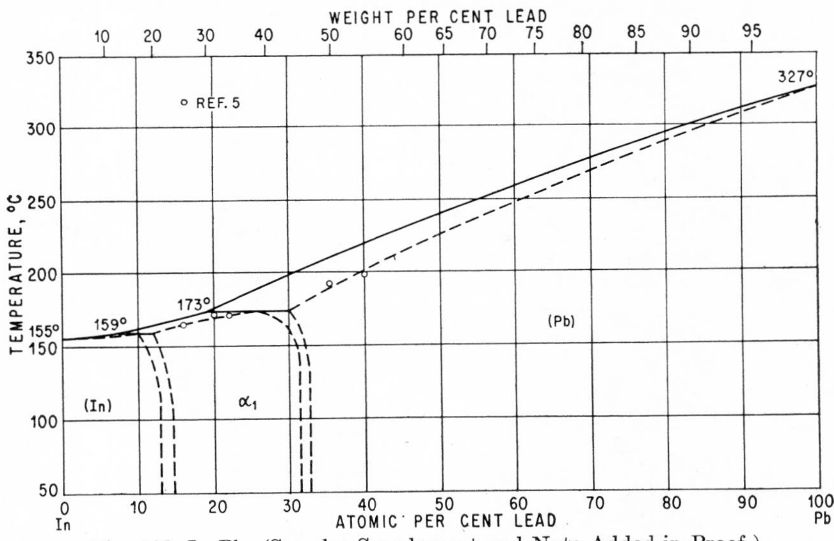

Fig. 468. In-Pb. (See also Supplement and Note Added in Proof.)

given: 17–25 [2], 18.5–29 (based on thermal data) [3], 12–31 [5], and 14–30 [6]. In Fig. 468 the tentative boundaries suggested by [5] were accepted. The partial phase diagram proposed by [6] is similar; it also shows two-phase fields having a width of only 1–2 at. %.

Crystal Structures. The lattice constants of In, given as $a = 4.60$ A, $c/a = 1.075$, are increased to about $a = 4.63$ A, $c/a = 1.08$ by 15.5 at. % Pb. The unit cell dimensions of the intermediate phase α_1 vary from $a = 4.87$ A, $c/a = 0.93$ at 15.5 at. % Pb to $a = 4.89$ A, $c/a = 0.93$ at 31 at. % Pb. The lattice parameter of the (Pb) phase decreases from $a = 4.94$ A at 0% In to about 4.81 A at 73 at. % In [2]. The lattice-spacing-vs.-composition curve for the range 0–14.2 at. % In is linear [10].

Supplement. Lattice spacings of alloys with up to 27 at. % Pb were determined by [11], after annealing at and quenching from 150°C. The axial ratio of In (1.07586) is progressively increased by Pb to a maximal value of 1.0866 at 12.3 at. % Pb (but see Note Added in Proof), whereas the c/a value of the intermediate phase first decreases from 0.9327 at 14.3 at. % Pb to 0.9274 at 20.6 at. % and then increases to 0.9284 at 27 at. % Pb. At 13.07 at. % Pb both f.c. tetragonal structures were observed. "Assuming equilibrium conditions, therefore, any two-phase region between these two phases must be narrow (approximately 1.5 at. % in width)." The 31.06 at. % Pb alloy contained small amounts of the (Pb) phase. For results as to the transformation f.c.t. ⇄ f.c.t., the reader is referred to the original paper.

[12] measured the lattice parameters of cast In-Pb alloys containing 90–100 wt. % Pb and studied the diffusion of In into Pb. According to [13], there is a minimum in the liquidus at 2.5 wt. (1.4 at.) % Pb and about 0.15°C below the freezing point of In.

Note Added in Proof. At variance with previous work [11] which had indicated that the lattice parameters *a*, *c*, and *c*/*a* of In increase continuously with the addition of Pb, [14] found that *c* and *c*/*a* go through maxima at 10 and 7 at. % Pb, respectively. The (In) and α_1 phases were observed to coexist in the composition ranges (in at. % Pb) 12.0–12.7 at 145°C and 12.7–13.7 at 20°C; however, conclusive evidence as regards the order of the transformation could not be obtained.

1. N. S. Kurnakow and N. A. Puschin, *Z. anorg. Chem.*, **52,** 1907, 442–445.
2. S. Valentiner and A. Haberstroh, *Z. Physik,* **110,** 1938, 727–741; **111,** 1939, 212–214.
3. H. M. Davis and G. H. Rowe, unpublished work; see "Metals Handbook," 1948 ed., p. 1223, The American Society for Metals, Cleveland, Ohio.
4. N. Ageew and V. Ageewa, *J. Inst. Metals,* **59,** 1936, 311–316.
5. W. Klemm and H. Volk, *Z. anorg. Chem.*, **256,** 1947, 246–248.
6. V. S. Kogan and B. Y. Pines, *Doklady Akad. Nauk S.S.S.R.*, **87,** 1952, 771–773.
7. N. S. Kurnakow and S. F. Zemczuzny, *Z. anorg. Chem.*, **64,** 1909, 149–183.
8. If the slope of the solidus is small—as is here the case in the range 0–30 at. % Pb—cooling curves may indicate a peritectic reaction, although no such reaction exists (see Au-Pt).
9. S. Valentiner, *Z. Physik,* **115,** 1940, 11–16.
10. C. Tyzack and G. V. Raynor, *Acta Cryst.*, **7,** 1954, 505–510.
11. C. Tyzack and G. V. Raynor, *Trans. Faraday Soc.*, **50,** 1954, 675–684.
12. T. Voyda, *Ann. Proc. Tech. Sessions Am. Electroplaters' Soc.*, **1946,** 33–48. See *Structure Repts.*, **10,** 1945-1946, 58.
13. A. N. Campbell, R. M. Screaton, T. P. Schaefer, and C. M. Hovey, *Can. J. Chem.*, **33,** 1955, 511–526.
14. A. Moore, J. Graham, G. K. Williamson, and G. V. Raynor, *Acta Met.*, **3,** 1955, 579–589.

0.0316
$\bar{1}$.9684

In-Pd Indium-Palladium

[1] reported the existence of the following phases, identified by X-ray analysis: (*a*) a saturated primary Pd-rich solid solution with about 20 at. (21.2 wt.) % In, at which composition the lattice constant of Pd has expanded by 1.45%; (*b*) a phase of the composition Pd_3In (26.39 wt. % In), f.c. tetragonal of the In (A6) type (?), $a = 4.07$ A, $c = 3.80$ A, $c/a = 0.934$; (*c*) a phase of the approximate composition Pd_2In (34.97 wt. % In) which apparently forms in the solid state; (*d*) PdIn (51.82 wt. % In), isotypic with CsCl (B2 type), $a = 3.25_7$ A; (*e*) Pd_2In_3 (61.73 wt. % In), isotypic with Ni_2Al_3 ($D5_{13}$ type), hexagonal with 1 formula weight per unit cell, $a = 4.53$ A, $c = 5.50$ A, $c/a = 1.214$; (*f*) a phase of the approximate composition of 75 at. % In (76.34 wt. %) having a "γ-brass structure with voids," $a = 9.44$ A. This latter phase is probably isotypic with Ir_3Sn_7 (see In-Pt). According to [2], Pd_2In is

of the orthorhombic Co_2Si (C37) type of structure, with a = 8.24 A, b = 5.61 A, c = 4.22 A.

1. E. Hellner and F. Laves, *Z. Naturforsch.*, **2a,** 1947, 177–183.
2. K. Schubert et al., *Naturwissenschaften*, **44,** 1957, 229–230.

$\bar{1}.7692$
0.2308

In-Pt Indium-Platinum

The following intermediate phases have been identified by X-ray analysis: (*a*) $PtIn_2$ (54.05 wt. % In), isotypic with CaF_2 (C1) type, a = 6.366 A [1]; this phase is stable only at high temperatures [1] and decomposes to form (*b*) Pt_2In_3 (46.86 wt. % In) [2] which is isotypic with Ni_2Al_3 ($D5_{13}$ type), hexagonal with 1 formula weight per unit cell, a = 4.53 A, c = 5.51 A, c/a = 1.125 [2]; (*c*) Pt_3In_7 (57.84 wt. % In), isotypic with Ir_3Sn_7, b.c.c. with 40 atoms per unit cell, a = 9.435 A [3].

1. E. Zintl, A. Harder, and W. Haucke, *Z. physik. Chem.*, **B35,** 1937, 354–362.
2. E. Hellner and F. Laves, *Z. Naturforsch.*, **2a,** 1947, 177–183.
3. K. Schubert and H. Pfisterer, *Z. Metallkunde*, **41,** 1950, 433–441.

$\bar{1}.6814$
0.3186

In-Pu Indium-Plutonium

According to [1], Pu_3In is of the partially ordered Cu_3Au ($L1_2$) type of structure, with a = 4.703 ± 2 A.

1. A. S. Coffinberry and M. B. Waldron, Review of the Physical Metallurgy of Plutonium, in "Metallurgy and Fuels," Progress in Nuclear Energy, ser. V, vol. 1, Pergamon Press Ltd., London, 1956.

0.0473
$\bar{1}.9527$

In-Rh Indium-Rhodium

The compound RhIn has been reported to be of the CsCl (B2) type of structure, with a = 3.20 A [1].

1. K. Schubert et al., *Naturwissenschaften*, **44,** 1957, 229–230.

0.5538
$\bar{1}.4462$

In-S Indium-Sulfur

The phase relations in the partial system In-In_2S_3 (29.53 wt. % S) (Fig. 469) were determined by thermal analysis and microscopic and qualitative roentgenographic studies [1]. No evidence was found for the existence of the sulfide In_2S (12.26 wt. % S) assumed by [2, 3]. This is in accord with findings by [4].

The composition of the phases formed peritectically at 770 and 840°C was tentatively given as In_5S_6 [54.55 at. (25.11 wt.) % S] and In_3S_4 [57.14 at. (27.15 wt.) % S], respectively [1]; see Supplement. The latter phase was found to be unstable at room temperature; thermal effects observed at 370°C were interpreted to indicate the temperature of its decomposition into In_5S_6 and In_2S_3. The melting points of InS (21.84 wt. % S) and In_2S_3 (29.53 wt. % S) were previously reported as 692 and 1050°C, respectively [5].

In_2S_3 exists in two modifications [6, 1]. The low-temperature form is f.c.c., a = 5.37 A, and appears to be isotypic with γ'-Al_2O_3. The S atoms are in cubic close packing, with about 70% of the In atoms randomly distributed in the octahedral holes and the rest in the tetrahedral holes. Above 300°C it transforms irreversibly into the high-temperature form, which has a spinel-type structure as in γ-Al_2O_3, a = 10.74 A [6].

Supplement. The compound In_4S_5 [55.56 at. (25.89 wt.) % S] was identified by determining its crystal structure, which is monoclinic, with a = 9.10 A, b = 3.88 A, c = 17.69 A, β = 108.15° [7]. Therefore, in Fig. 469 the composition of the phase formed at 770°C has been assumed to be In_4S_5 rather than In_5S_6 as suggested by [1]. InS is orthorhombic, with 8 atoms per unit cell, a = 3.94 A, b = 4.44 A, c = 10.64 A [7].

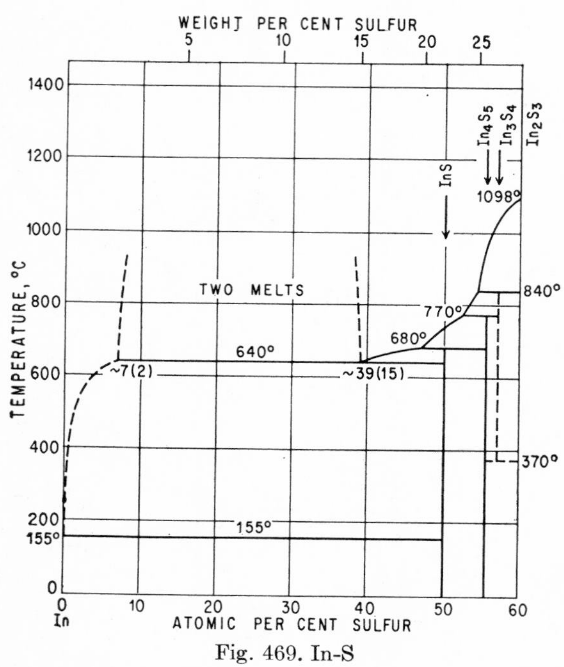

Fig. 469. In-S

1. M. F. Stubbs, J. A. Schufle, A. J. Thompson, and J. M. Duncan, *J. Am. Chem. Soc.*, **74**, 1952, 1441–1443; see also A. J. Thompson, M. F. Stubbs, and J. A. Schufle, *J. Am. Chem. Soc.*, **76**, 1954, 341–343.
2. A. Thiel, *Z. anorg. Chem.*, **40**, 1904, 324–327; A. Thiel and H. Koelsch, *Z. anorg. Chem.*, **66**, 1910, 313–315; A. Thiel and H. Luckmann, *Z. anorg. Chem.*, **172**, 1928, 353–371.
3. W. Klemm and H. U. v. Vogel, *Z. anorg. Chem.*, **219**, 1934, 45–64.
4. H. Hahn, personal communication to [1].
5. A. Thiel and H. Luckmann, *Z. anorg. Chem.*, **172**, 1928, 353–371.
6. H. Hahn and W. Klingler, *Z. anorg. Chem.*, **260**, 1949, 97–109.
7. K. Schubert, E. Dörre, and E. Günzel, *Naturwissenschaften*, **41**, 1954, 448.

$\bar{1}$.9743
0.0257

In-Sb Indium-Antimony

The phase diagram was established by [1, 2]; data points for the liquidus are shown in Fig. 470. The temperature differences are 10°C at most. The eutectic temperatures were found as about 155 [1], 154.8°C [2], and 505–507 [1], 494°C [2]. The eutectic compositions were placed at 0.7 wt. (0.66 at.) % Sb [2], as determined by metallographic examination, and 71.6 wt. (70.4 at.) % Sb [1], 69.5 wt. (68.3 at.) % Sb [2], the latter value being based on microexamination [3]. The melting point of InSb (51.48 wt. % Sb) was given as 536 [1] and 525°C [2]. The mutual solid solubility of the elements appears to be vanishingly small [2].

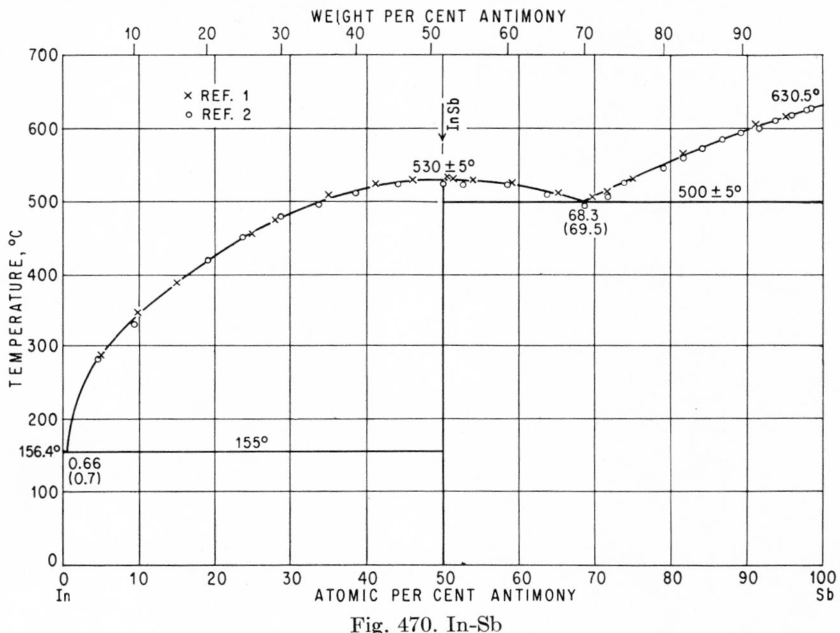

Fig. 470. In-Sb

InSb, which is of fixed composition [2], is isotypic with zincblence (B3 type), with $a = 6.46_5$ A [4], $a = 6.47_4$ A [5], or $a = 6.476$ A [6, 2].

1. S. A. Pogodin and S. A. Dubinsky, *Izvest. Sektora Fiz.-Khim. Anal.*, **17**, 1949, 204–208.
2. T. S. Liu and E. A. Peretti, *Trans. ASM*, **44**, 1952, 539–545.
3. If the liquidus temperature of the 69.7 at. % Sb alloy given by [1] is attributed to the liquidus curve of Sb rather than that of InSb, the eutectic composition would be shifted to about 69.3 at. (70.5 wt.) % Sb, i.e., closer to the value according to [2].
4. V. M. Goldschmidt, *Skrifter Norske Videnskaps-Akad. Oslo, Math. Naturv. Kl.*, no. 2, 1926; *Trans. Faraday Soc.*, **25**, 1929, 253–283.
5. A. Iandelli, *Gazz. chim. ital.*, **71**(1), 1941, 58–62.
6. T. S. Liu and E. A. Peretti, *Trans. AIME*, **191**, 1951, 791.

0.1624
$\bar{1}.8376$

In-Se Indium-Selenium

The indium selenides In_2Se (25.60 wt. % Se) [1], InSe (40.76 wt. % Se) [1], and In_2Se_3 (50.79 wt. % Se) [2, 1] have been identified. Melting points are InSe, 660 ± 10°C; In_2Se_3, 890 ± 10°C [1].

In_2Se is orthorhombic, with $a = 4.073$ A, $b = 12.26$ A, $c = 15.26$ A. InSe has a rhombohedral structure, with $a = 4.02$ A, $c = 25.05$ A [3].

According to [4], In_2Se_3 exists in two modifications, the high-temperature form being probably hexagonal, with $a = 3.99$ A, $c = 19.0$ A, $c/a = 4.76$, and $4In_2Se_3$ per unit cell.

1. W. Klemm and H. U. v. Vogel, *Z. anorg. Chem.*, **219,** 1934, 45–64.
2. A. Thiel and H. Koelsch, *Z. anorg. Chem.*, **66,** 1910, 315–317.
3. K. Schubert, E. Dörre, and E. Günzel, *Naturwissenschaften*, **41,** 1954, 448.
4. H. Hahn, *Angew. Chem.*, **65,** 1953, 538.

0.6112
$\bar{1}.3888$

In-Si Indium-Silicon

The solidification diagram in Fig. 471 is due to [1]. The liquidus curve was calculated by [2] and complemented by determination of the solubility of Si in liquid In in the temperature range of about 550–1000°C [3].

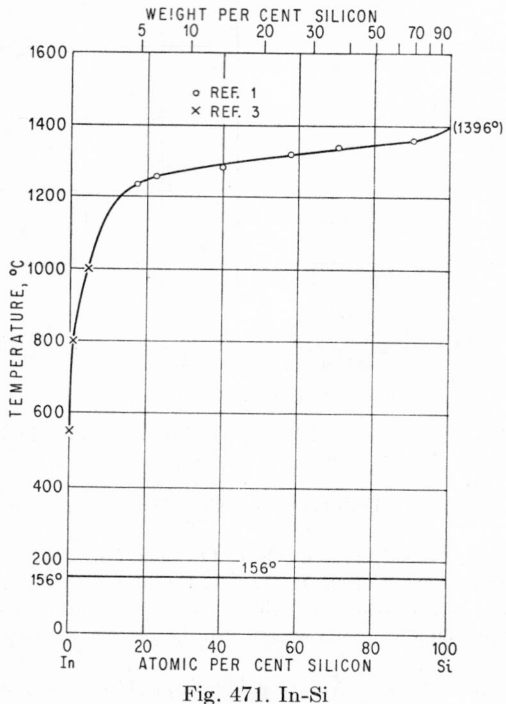

Fig. 471. In-Si

1. W. Klemm and L. Klemm, *Z. anorg. Chem.*, **256,** 1948, 244–245.
2. C. D. Thurmond, *J. Phys. Chem.*, **57,** 1953, 827–830.
3. P. H. Keck and J. Broder, *Phys. Rev.*, **90,** 1953, 521–522.

$\bar{1}.9853$
0.0147

In-Sn Indium-Tin

Although the constitution has been extensively studied by means of thermal [1–3], micrographic [3], and roentgenographic methods [1, 2], there are still some doubts as to certain phase relations (Fig. 472).

The solidification was investigated by [1, 2, 3], using 32, 18, and 61 alloys, respectively. Liquidus temperatures agree within a few °C between 0 and 48 at. % Sn; temperature differences are slightly larger for the Sn-rich alloys. The eutectic was found at 48 at. %, 117°C [1]; 48.5 at. %, 116°C [2]; 47.2 at. % Sn, 117°C [3].

[1] recognized by X-ray work that two intermediate phases of variable composition occur. They were found to be present in alloys with approximately 15–70 and

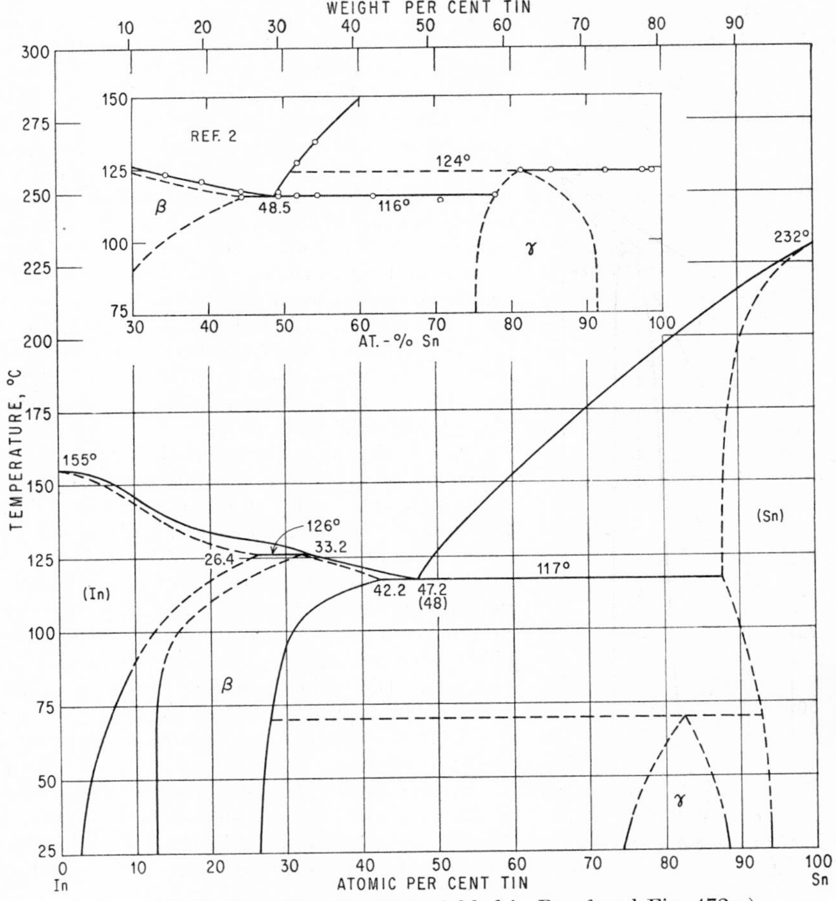

Fig. 472. In-Sn. (See also Note Added in Proof and Fig. 472*a*.)

70–93 at. % Sn, respectively. The homogeneity ranges of the primary solid solutions and β and γ phases suggested by [2] are based on lattice-spacing data of alloys aged for 2 years at room temperature. The following ranges were given: (In), 0–12 at. %; β, 14–28 at. %; γ, 75–88 at. %; (Sn), about 96.7–100 at. % Sn. No attempt was made to determine the phase boundaries at various temperatures. It was assumed that β was formed by a peritectic reaction at 126.5°C, with the peritectic horizontal extending from about 25.5 to 29 at. % Sn. Cooling-curve data indicated a reaction at 124°C in alloys with 81.3–98.7 at. % Sn but not in alloys lower in Sn. This thermal effect was interpreted as representing the peritectic reaction liq. + (Sn) $\rightleftarrows \gamma$ (see inset in Fig. 472), but this was not substantiated.

The main diagram in Fig. 472 is based on the findings of [3]. The solid-phase boundaries were established by micrographic analysis; the dashed curves are stated as being somewhat doubtful. The peritectic reaction liq. + (In) $\rightleftarrows \beta$ was located at 126°C between 27 and 34 wt. (26.4 and 33.3 at.) % Sn, based on liquation data. However, there was some indication of this reaction occurring at lower Sn compositions, for instance, at ~ 132°C, 18–24 wt. % Sn. As to the formation of the γ phase, the investigators, disregarding the effects at 124°C [2], preferred to present this phase as being formed by a peritectoid reaction at some temperature below 80°C; however, [2] reported that γ is formed on annealing at 100°C. The solid solubility of In in Sn, after aging for 3 years at room temperature, was given as about 6.5 wt. (6.7 at.) % In, which is in substantial agreement with the boundary reported by [1], about 7 at. % In, but not with that of [2], 3.3 at. % In.

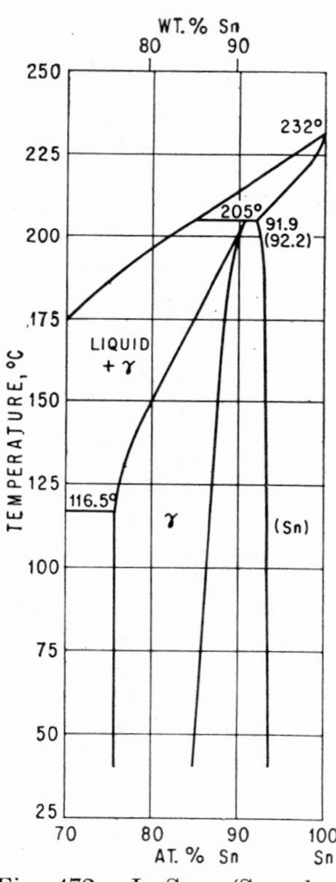

Fig. 472*a*. In-Sn. (See also Fig. 472.)

Crystal Structures. The dimensions of the f.c. tetragonal lattice of In change from $a = 4.599$ A, $c = 4.947$ A, $c/a = 1.076$ to $a = 4.575$ A, $c = 5.020$ A, $c/a = 1.097$ at 9.6 at. % Sn, those and of the tetragonal lattice of Sn from $a = 5.831$ A, $c = 3.181$ A, $c/a = 0.546$ to $a = 5.828$ A, $c = 3.182$ A, $c/a = 0.546$ at 5.2 at. % In [2].

The lattice of β was reported to be simple f.c. tetragonal [1, 4], which is equivalent to the b.c. tetragonal lattice with 2 atoms per unit cell [2]. This simple structure, however, was disputed by [5] on the basis of powder-line intensities [5a].

The γ phase was reported to be b.c. tetragonal and similar to that of Sn [1]. [2] showed this to be incorrect and suggested a simple hexagonal lattice with 1 atom per unit cell. [5, 6, 7] verified the latter structure. Dimensions change from $a = 3.216$ A, $c = 2.998$ A, $c/a = 0.932$ at 77.66 at. % Sn to $a = 3.221$ A, $c = 3.00$ A, $c/a = 0.931$ at 89.65 at. % Sn [2]; see also [6].

Supplement. High-temperature X-ray work by [7] suggests that the β-phase field extends, at elevated temperatures, to higher compositions of both In and Sn than is shown—according to [3]—in Fig. 472.

[8] observed anomalies in the lattice-parameter–composition curves of (Sn)-phase alloys between 1.8 and 2.4 at. % In which are, as evidenced by measurements of density, associated with the development of vacant lattice sites.

Note Added in Proof. A reinvestigation of Sn-rich alloys by means of thermal, microscopic, and X-ray analysis as well as incipient-melting tests yielded the partial phase diagram of Fig. 472*a* [9]. The γ phase was found to be formed by a peritectic reaction at 205°C. "The evidence for this reaction is to some extent indirect, since no thermal arrest was observed in spite of many attempts to find one. The reasons for interpreting the experimental results as indicating a peritectic reaction are that there is a slight inflection in the liquidus curve and a marked inflection in the solidus at 205°C. In addition, fracture tests at elevated temperatures indicate a small horizontal in the solidus curve between 8 and 9 wt. % In" [9].

1. S. Valentiner, *Z. Metallkunde*, **32,** 1940, 31–35; see also *Z. Physik*, **115,** 1940, 11–46 (Electrical resistivity).
2. C. G. Fink, E. R. Jette, S. Katz, and F. J. Schnettler, *Trans. Electrochem. Soc.*, **88,** 1945, 229–241; preliminary results in *Trans. Electrochem. Soc.*, **75,** 1939, 463–467.
3. F. N. Rhines, W. M. Urquhart, and H. R. Hoge, *Trans. ASM*, **39,** 1947, 694–711.
4. W. Klemm and E. Orlamünder, *Z. anorg. Chem.*, **256,** 1948, 245–246.
5. R. M. Screaton and R. B. Ferguson, *Acta Cryst.*, **7,** 1954, 364–365.
5a. The criticism by [5] may be questioned since it appears possible that the observed intensity ratios were influenced by the exterior shape of the filings (disturbance of the at-random orientation).
6. G. V. Raynor and J. A. Lee, *Acta Met.*, **2,** 1954, 616–620.
7. K. Schubert, U. Rösler, W. Mahler, E. Dörre, and W. Schütt, *Z. Metallkunde*, **45,** 1954, 643–647.
8. J. A. Lee and G. V. Raynor, *Proc. Phys. Soc. (London)*, **B67,** 1954, 737–747.
9. J. C. Blade and E. C. Ellwood, *J. Inst. Metals*, **85,** 1956, 30–32.

$\bar{1}.9539$
0.0461

In-Te Indium-Tellurium

The diagram in Fig. 473, based on thermal-analysis data, is due to [1]; see Supplement, however. Besides the phases In_2Te (35.73 wt. % Te), InTe (52.65 wt. % Te) [2, 3], and In_2Te_3 (62.52 wt. % Te), there is a higher telluride which was not identified. The holding times of the peritectic reaction at 455°C and eutectic crystallization at 427°C point to a composition close to 75 at. % Te (see Supplement).

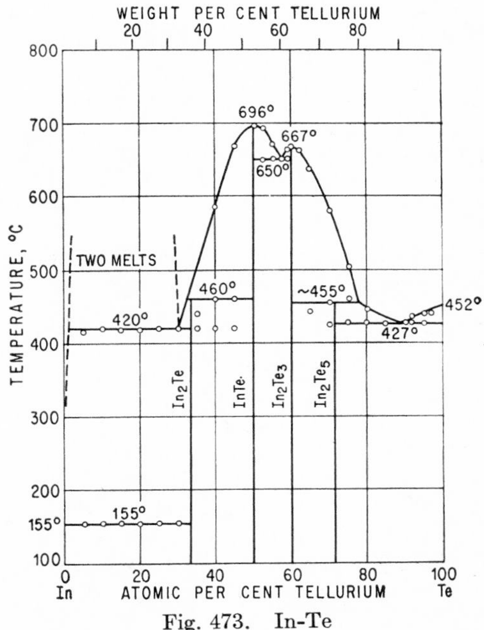

Fig. 473. In-Te

InTe is isotypic with TlSe (B37 type), $a = 8.43_7$ A, $c = 7.13_9$ A, $c/a = 0.845$ [4, 7]. In_2Te_3, isotypic with the low-temperature form of Ga_2S_3, has the structure of zincblende (B3 type), with some metal positions empty at random, $a = 6.146$ A (or kX?) [5].

Supplement. The phase formed peritectically at about 455°C was identified as In_2Te_5 [71.43 at. (73.55 wt.) % Te]. It is monoclinic, $a = 4.38$ A, $b = 16.11$ A, $c = 13.37$ A, $\beta = 92.05°$ [6]. In_2Te is orthorhombic, with $a = 4.46$ A, $b = 12.61$ A, $c = 15.35$ A [6].

[8] reported that long-time annealing of In_2Te_3 results in the appearance of superstructure lines, probably due to ordering of the vacancies in the zincblende structure. According to single-crystal work by [9], the structure of In_2Te_3 is f.c.c., with $a = 18.40 \pm 4$ A and 35.6 ± 0.5 molecules per unit cell (space group F43m).

1. W. Klemm and H. U. v. Vogel, *Z. anorg. Chem.*, **219,** 1934, 45–64.
2. A. Thiel and H. Koelsch, *Z. anorg. Chem.*, **66,** 1910, 317–319.
3. M. Renz, *Ber. deut. chem. Ges.*, **37,** 1904, 2112.
4. K. Schubert et al., *Naturwissenschaften*, **40,** 1953, 269.
5. H. Hahn and W. Klingler, *Z. anorg. Chem.*, **260,** 1949, 97–109.
6. K. Schubert, E. Dörre, and E. Günzel, *Naturwissenschaften*, **41,** 1954, 448.
7. K. Schubert, E. Dörre, and M. Kluge, *Z. Metallkunde*, **46,** 1955, 216–224.
8. H. Hahn, *Angew. Chem.*, **64,** 1952, 203.
9. H. Inuzuka and S. Sugaike, *Proc. Japan Acad.*, **30,** 1954, 383–386; *Met. Abstr.*, **22,** 1954, 193.

0.3794
$\bar{1}.6206$

In-Ti Indium-Titanium

[1] reported that the intermediate phase richest in In is $TiIn_3$; a literature reference is not given. No publication could be found in which reference is made to the existence of this phase. According to [2], who investigated four alloys in the composition range 0–30 wt. % In by means of metallographic examination of heat-treated samples, In lowers the α-β transformation temperature of Ti until a eutectoid β [~25 wt. (12 at.) % In] $\rightleftarrows \alpha$ + compound is reached somewhere between 800 and 850°C. The solubility of In in α-Ti at 800°C was bracketed between 4 and 9 wt. % In.

The latter result could not be corroborated by [3], who reported the maximal solubility of In in α-Ti to be about 20 wt. (9.5 at.) %. The intermediate phase coexisting with Ti (up to at least 950°C [4]) was found [3] to be (Ti_3In) of the hexagonal Mg_3Cd ($D0_{19}$) type (with $a = 5.89$ A, $c = 4.76$ A, $c:a/2 = 1.62$ at 21 at. % In [4]). In alloys with higher (not analyzed) In contents an intermediate phase was observed yielding the X-ray pattern of a primitive cubic structure with $a = 4.22$ A [5].

1. E. Jänecke, "Kurzgefasstes Handbuch aller Legierungen," p. 318, Carl Winter, Heidelberg, 1949.
2. C. M. Jackson, thesis, College of Engineering, New York University, May, 1954; quoted in WADC Technical Report 54-502, Armour Research Foundation, September, 1954.
3. K. Anderko, K. Sagel, and U. Zwicker, *Z. Metallkunde*, **48,** 1957, 57–58.
4. H. Böhm and K. Anderko, unpublished information (1956).
5. K. Anderko, *Naturwissenschaften*, **44,** 1957, 88.

$\bar{1}.7493$
0.2507

In-Tl Indium-Thallium

Solidification. The liquidus curve shown in Fig. 474 is based on 66 liquidus temperatures reported by [1]. It is characterized by a very flat minimum at about

12 at. (19.6 wt.) % Tl and a temperature only 2.2°C below the melting point of In (given as 156.4°C). A peritectic reaction at 171°C was observed in the composition range of about 38.5–58 at. (52.7–71.1 wt.) % Tl. It is to be expected that, under equilibrium conditions, the peritectic horizontal will extend to less than 58 at. % Tl.

An earlier cursory determination of the liquidus, using only six alloys, has already indicated a peritectic equilibrium between the two primary solid solutions at 180°C, with the peritectic horizontal assumed to extend from about 45.5 to 56 at. % Tl [2].

Solid State. *Alloys with* 0–30 *At.* % Tl. Lattice-parameter measurements of alloys with 5.9, 12.3, 19.4, and 27.3 at. % Tl revealed that Tl additions cause the axial ratio of the f.c. tetragonal lattice of In (given as 1.075) to approach unity. The alloy with 27.3 at. % Tl proved to be f.c.c., $a = 4.76$ A [1]. These observations were verified by [3], who reported that the tetragonal structure, as evidenced by determination of the lattice constants, was stable up to about 25 at. % Tl. Since no two-phase region between both structures could be found, it was concluded that the transition f.c.t. → f.c.c. was continuous [3].

Comprehensive X-ray studies at 24°C by [4] confirmed the gradual decrease of the axial ratio of the f.c.t. (In) phase to $c/a = 1.023$ at 22.24 at. % Tl; at 22.73 at. % Tl the structure was found to be f.c.c. The transformation temperatures in the range 18–23 at. % Tl, indicated as data points in Fig. 474, were determined by microexamination at elevated temperature and lattice-parameter measurements (on heating and cooling) [5]. In the phase diagram proposed by [4], no two-phase region was shown separating the two solutions, "since all diffraction patterns at room temperature contain[ed] only lines from one structure or the other, never from both." It was concluded that the transformation f.c.t. ⇄ f.c.c. was of the second order; i.e., the phases in equilibrium do not differ in composition. However, it was also stated that the results at room temperature indicated "no two-phase region wider than about 0.5 at. % Tl."

The conclusions of [4] were criticized by [6], who pointed out that the experimental evidence presented by [4] was not sufficient to decide whether the transformation was of the first or second order; however, preference was given to the former. Also, [7] claimed, on the basis of dilatometric studies, that the phase change was of the first order. "Although it seems questionable whether the heterogeneous equilibrium of a martensitic reaction may be described" [4], two phase boundaries are indicated in Fig. 474. (See Note Added in Proof.)

The crystallography of the cubic ⇄ tetragonal transformation was the subject of investigations and discussions [8–15]. The reader is referred to these publications.

Alloys with 30–100 *At.* % Tl. Room-temperature X-ray analysis [1] had the following results: (*a*) The f.c.c. α phase was present in alloys with up to at least 69 at. % Tl; (*b*) alloys with 62–85 at. % Tl were found to contain a b.c.c. phase which coexisted with α between 62 and 69 at. % and around 85 at. % Tl with the primary solid solution of the low-temperature modification of Tl. The interpretation of these findings was based on the erroneous assumption that the high-temperature form of Tl was f.c.c. instead of b.c.c., as was established later [16]; therefore, [1] was unable to arrive at plausible phase relations in the range 60–90 at. % Tl.

[17] reevaluated the findings of [1] and suggested a tentative phase diagram, similar to that in Fig. 474, showing the two-phase field ($\alpha + \beta_{Tl}$) to extend from the region 38.5–58 at. % Tl at the peritectic temperature of 171°C to approximately 62–69 at. % Tl at "room temperature." The latter compositions constitute the approximate boundaries of the range within which [1] had observed both the f.c.c. structure of α and the b.c.c. structure of β_{Tl}.

The partial diagram of the region 30–100 at. % Tl (Fig. 474) was proposed by [4]. It was derived from the solidification data of [1] and the data points shown; the latter were obtained by X-ray work [4] and thermal analysis [1]. The suggested eutectoid decomposition of the β_{Tl} phase at a temperature close to room temperature is hypothetical since the location of the $\beta_{Tl}/(\beta_{Tl} + \alpha_{Tl})$ boundary is quite vague. As indicated in Fig. 474, it was assumed [4] that the thermally determined transformation temperatures in the range 91.2–100 at. % Tl [1], marked by crosses, represent temperatures found on heating. These data points, however, are average values of temperatures found on heating and cooling, the hysteresis being unknown. Also, the data points at 75 and 77 at. % Tl are rather uncertain since [4] found the 75% alloy to contain a small amount of α_{Tl}, whereas [1] reported the 77% alloy to be single-phase, both based on diffraction results. Therefore, it may be possible that the $(\beta_{Tl} + \alpha_{Tl})$ field is actually shifted to higher Tl contents, resulting in a more restricted

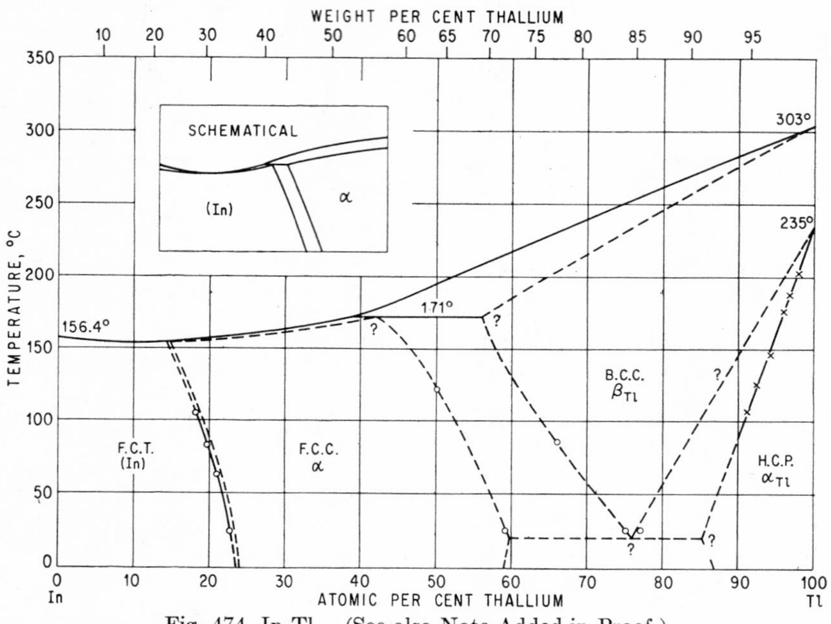

Fig. 474. In-Tl. (See also Note Added in Proof.)

α_{Tl} field and a sharper slope of the $\beta_{Tl}/(\alpha_{Tl} + \beta_{Tl})$ boundary. This would mean that the β_{Tl} field would be stable down to much lower than ordinary temperatures.

Crystal Structures. The structure of the various phases has been discussed above and is also indicated in Fig. 474. Lattice constants were reported by [1, 3, 4].

Note Added in Proof. [18] measured the variation of lattice parameters with temperature for alloys with 19.0, 20.4, and 21.7 at. % Tl. A finite range of coexistence of the two phases in the transformation f.c.t. → f.c.c. could be observed. As regards the type of transformation, the following conclusion was reached: "As it is unlikely that the reaction can be of the second order the structures observed in the two-phase region must result from an approximation to conditions demanded in a first-order transformation. Since no evidence for diffusion has been obtained, the phases observed are necessarily metastable," and, therefore, representation of a two-phase field in an equilibrium diagram, on the basis of the findings, is impossible. "Nevertheless, the presence of an equilibrium two-phase field is demanded by thermodynamic considerations" [18].

The question as to the stable equilibrium conditions was left open by this study. Apparently equilibrium cannot be easily reached because of the very small difference in the free energies of the two phases.

1. S. Valentiner, *Z. Metallkunde*, **32,** 1940, 244–248.
2. N. S. Kurnakow and N. A. Puschin, *Z. anorg. Chem.*, **52,** 1907, 445–446.
3. E. S. Makarov, *Izvest. Akad. Nauk S.S.S.R.* (*Khim.*), **1950,** 485–491.
4. L. Guttman, *Trans. AIME*, **188,** 1950, 1472–1477.
5. J. W. Stout and L. Guttman (*Phys. Rev.*, **88,** 1952, 713–714) found indication of the transformation of the 30 at. % Tl alloy at about 60°K.
6. A. H. Geisler and D. L. Martin, *Trans. AIME*, **191,** 1951, 1056–1057.
7. M. W. Burkart and T. A. Read, *Trans. AIME*, **197,** 1953, 1516–1524.
8. J. S. Bowles, C. S. Barrett, and L. Guttman, *Trans. AIME*, **188,** 1950, 1478–1485.
9. A. H. Geisler and D. L. Martin, *Trans. AIME*, **191,** 1951, 1057–1060; L. Guttman, *Trans. AIME*, **191,** 1951, 1060.
10. Z. S. Basinski and J. W. Christian, *J. Inst. Metals*, **80,** 1951-1952, 659–666.
11. A. H. Geisler, *Acta Met.*, **1,** 1953, 260–281.
12. Z. S. Basinski and J. W. Christian, *Acta Met.*, **1,** 1953, 759–761.
13. M. W. Burkart and T. A. Read, *Trans. AIME*, **197,** 1953, 1516–1524; see also M. S. Wechsler, D. S. Lieberman, and T. A. Read, *Trans. AIME*, **197,** 1953, 1503–1515.
14. Z. S. Basinski and J. W. Christian, *Acta Met.*, **2,** 1954, 148–166.
15. A. H. Geisler, *Acta Met.*, **2,** 1954, 639–642.
16. H. Lipson and A. R. Stokes, *Nature*, **148,** 1941, 437.
17. W. Klemm and E. Orlamünder, *Z. anorg. Chem.*, **256,** 1948, 242–243.
18. A. Moore, J. Graham, G. K. Williamson, and G. V. Raynor, *Acta Met.*, **3,** 1955, 579–589.

$\bar{1}.6831$
0.3169

In-U Indium-Uranium

UIn_3 (59.12 wt. % In), the intermediate phase richest in In [1], is isotypic with Cu_3Au ($L1_2$ type), a = 4.588 A [1]; a = 4.6013 A [2].

1. A. Iandelli and R. Ferro, *Ann. chim.* (*Rome*), **42,** 1952, 598–608.
2. B. R. I. Frost and J. T. Maskrey, *J. Inst. Metals*, **82,** 1953-1954, 171–180.

0.2444
$\bar{1}.7556$

In-Zn Indium-Zinc

The solidification diagram was determined by [1], [2] (11 alloys) and [3] (48 alloys) (Fig. 475). The liquidus curve reported by [1] deviates considerably from those by [2] and [3], the latter being in very good agreement, except for the range 0–30 at. % Zn. [4] calculated the liquidus on the basis of activities derived from emf measurements in the temperature range 360–515°C. Results agree substantially with those of [2, 3].

The eutectic temperature was found as 143.5 [1, 3] and 141.5°C [2], and the eutectic composition as about 4 wt. (6.7 at.) [1], about 8 at. (4.75 wt.) [2], 2.8 wt. (4.8 at.) [3], and 2 ± 0.1 wt. (3.4 at.) % Zn [5]. The microscopically determined composition of 2 wt. (3.4 at.) % [5] is not consistent with the numerous liquidus temperatures of hypo- and hypereutectic alloys reported by [3]. At least in part, this may be due to supercooling in the as-cast alloys examined by [2], resulting in shifting the eutectic to a lower Zn concentration.

The mutual solid solubility is not accurately known. [3] concluded from metallographic work that approximately 1.2 wt. (2.06 at.) % Zn is soluble at the eutectic temperature, 0.9 wt. (1.55 at.) % at 140°C, and 0.4 wt. (0.7 at.) % at 120°C. Similarly, the solubility of In in Zn was given as about 0.3 wt. (0.17 at.) at 143°C, 0.2–0.1 wt. (0.12–0.06 at.) at 140°C, and <0.1 wt. % In at 120°C [3].

Note Added in Proof. [6] redetermined the solidification equilibria by means of calorimetric measurements ("quantitative thermal analysis") and derived thermodynamic data, such as activities and entropies of mixing. The Zn branch of the liquidus curve of [2] could be corroborated within ±5°C. The eutectic was located at 1.8 ± 0.1 wt. (3.1 at.) % Zn, 144 ± 1°C. The solid solubilities of Zn in In and of In in Zn at the eutectic temperature were estimated to be about 0.45 at. (0.2 wt.) % and about 0.65 at. (1.1 wt.) %, respectively.

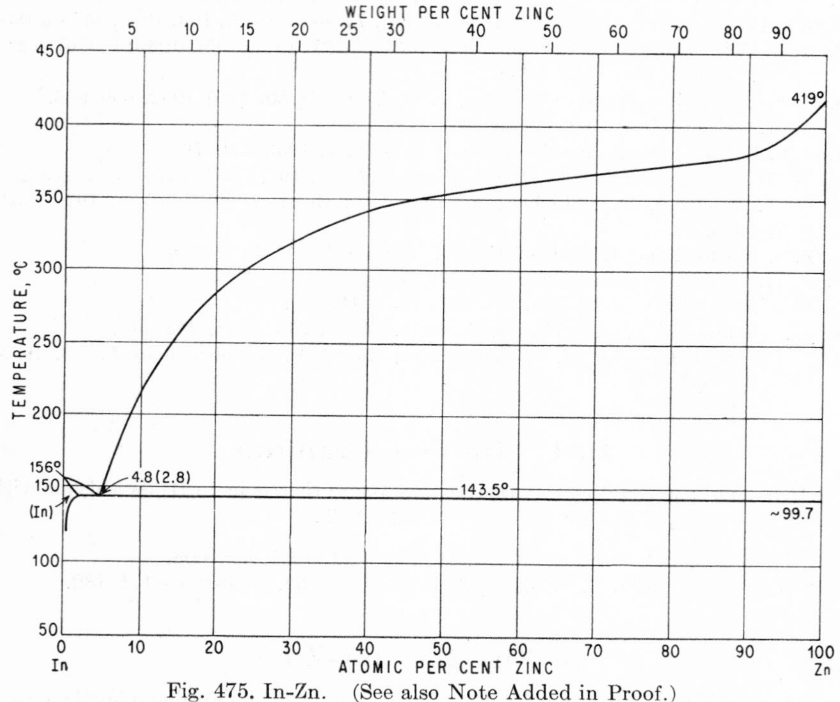

Fig. 475. In-Zn. (See also Note Added in Proof.)

1. C. L. Wilson and E. A. Peretti, *Ind. Eng. Chem.*, **28,** 1936, 204–205.
2. S. Valentiner (A. Grönefeld), *Z. Metallkunde,* **35,** 1943, 250–253.
3. F. N. Rhines and A. H. Grobe, *Trans. AIME,* **156,** 1944, 253–262.
4. W. T. Svirbely and S. M. Selis, *J. Am. Chem. Soc.*, **75,** 1953, 1532–1535.
5. S. C. Carapella and E. A. Peretti, *Trans. AIME,* **188,** 1950, 890–891.
6. W. Oelsen and P. Zühlke, *Archiv Eisenhüttenw.*, **27,** 1956, 743–752.

0.8998
$\bar{1}$.1002

Ir-Mg Iridium-Magnesium

Ir additions to Mg were found to increase slightly the axial ratio of Mg. Whether this really indicates a slight solid solubility is not certain, however [1].

1. R. S. Busk, *Trans. AIME,* **188,** 1950, 1460–1464.

0.5460
$\bar{1}$.4540

Ir-Mn Iridium-Manganese

An exploratory X-ray and microscopic study [1] of the Mn-Ir system yielded the phase relations shown in the partial phase diagram of Fig. 476. The γ_{Mn} phase becomes ordered (Cu_3Au type) below 900°C in the neighborhood of the composition Mn_3Ir. The structure of the β_1 phase is primitive tetragonal, with a = 2.73 A, c/a = 1.355 at 50 at. % Ir. There is microscopic evidence that above 940°C a high-temperature modification, presumably of the cubic CsCl (B2) type, exists.

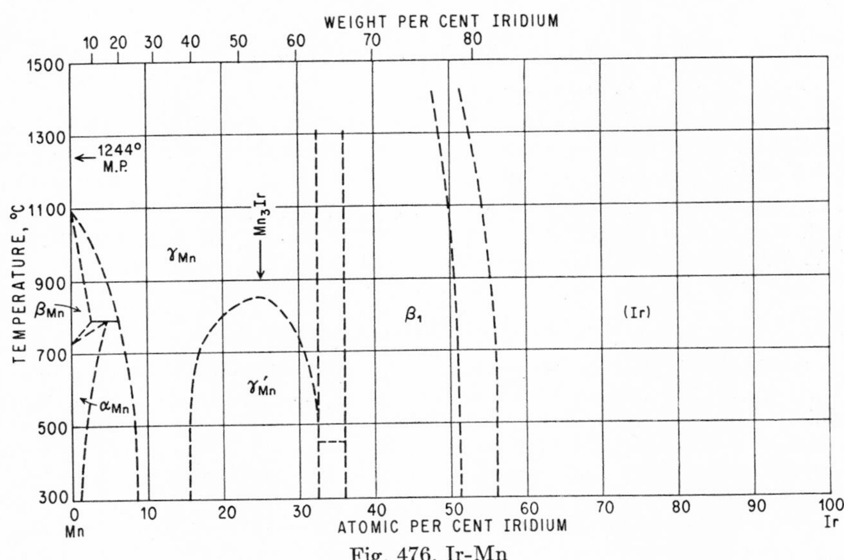

Fig. 476. Ir-Mn

The saturation limit of the extended solid solution of Mn in Ir was located at 52 and 56 at. % Ir, respectively, at 1350 and 650°C. No indications for ordering could be found. X-ray data of 14 alloys with 6–91.5 at. % Ir after various heat-treatments were tabulated by [1].

1. E. Raub and W. Mahler, *Z. Metallkunde*, **46**, 1955, 282–290.

0.3037
$\bar{1}$.6963

Ir-Mo Iridium-Molybdenum

Lattice-parameter measurements and microexamination of alloys with 5.2–82.4 at. (2.6–70 wt.) % Mo, annealed at various temperatures between 800 and 1600°C, revealed the existence of the following phases: (*a*) Ir-rich solid solutions extending up to 16.5 at. (8.9 wt.) % Mo, (*b*) a hexagonal phase designated as ϵ, ranging in composition from 21.5 to 59 at. (12.0–41.7 wt.) % Mo at 1200°C and from 21.5 to 48 at. (12.0–31.6 wt.) % Mo at 800°C; (*c*) the singular phase Mo_3Ir (59.85 wt. % Mo), which coexists with nearly pure Mo, since a solubility of Ir in Mo was not detectable parametrically.

The lattice constant of Ir increases to a = 3.845 A at the saturation limit. The ϵ phase is h.c.p., with a = 2.736 A, c = 4.378 A, c/a = 1.600 at 21.5 at. % Mo and

$a = 2.771$ A, $c = 4.436$ A, $c/a = 1.601$ at 57.3 at. % Mo. Mo_3Ir is isotypic with "β-W" (A15 type), $a = 4.959$ A (mean) [1].

1. E. Raub, *Z. Metallkunde*, **45**, 1954, 23–30.

0.3177
$\bar{1}.6823$

Ir-Nb Iridium-Niobium

Nb_3Ir is of the cubic "β-W" (A15) type of structure, with $a = 5.131 \pm 1$ A [1].

1. S. Geller, B. T. Matthias, and R. Goldstein, *J. Am. Chem. Soc.*, **77**, 1955, 1502–1504.

1.0817
$\bar{2}.9183$

Ir-O Iridium-Oxygen

On annealing Ir in air, IrO_2 is formed. At temperatures above 900°C [1], the oxide decomposes and the surface becomes metallic again. Volatilization of the oxide with Pt-Ir alloys appears to become effective above 1200°C [2]. IrO_2 is isotypic with TiO_2 (C4 type), $a = 4.50$ A, $c/a = 0.700$ [3].

1. S. Pastorello, *Rend. accad. nazl. Lincei*, **7**, 1928, 754–757.
2. G. Masing, K. Eckhardt, and K. Kloiber, *Z. Metallkunde*, **32**, 1940, 122–124.
3. V. M. Goldschmidt; see *Strukturbericht*, **1**, 1913-1928, 213.

0.0066
$\bar{1}.9934$

Ir-Os Iridium-Osmium

Osmiridium is a natural alloy, with Os contents varying roughly between 30 and 65 wt. % [1] and smaller or larger amounts of Pt, Rh, and Ru [2]. From X-ray studies, [3–5] concluded that the f.c.c. Ir-rich solid solution in these natural alloys extends up to 31–35 wt. % Os [5]; above this composition, apparently only the h.c.p. structure of the Os-rich phase was observed [2, 3], with an axial ratio varying irregularly between 1.58 and 1.62 [3, 6].

X-ray investigation and microexamination of a number of synthetic Ir-Os alloys in the as-cast condition showed that alloys with 38.7, 59.4, and 79 wt. % Os were two-phase [7]. It can be concluded that the constitution is characterized by the existence of a miscibility gap between the primary solid solutions, whose boundaries lie somewhere between 24 and 39 wt. % and 79 and 100 wt. % Os [1].

The melting point of Ir was reported to be raised several hundred °C by a few per cent Os [8]. For the microstructure of a natural alloy, see [9].

1. At. % and wt. % differ by maximal 0.4%.
2. O. E. Zvyagintsev, *Z. Krist.*, **83**, 1932, 172–186.
3. O. E. Zvyagintsev and B. K. Brunovskiy, *Z. Krist.*, **83**, 1932, 187–192.
4. O. E. Zvyagintsev and B. K. Brunovskiy, *Z. Krist.*, **93**, 1936, 229–237.
5. O. E. Zvyagintsev, *Compt. rend. acad. sci. U.R.S.S.*, **18**, 1938, 295–297; *Izvest. Sektora Fiz.-Khim. Anal.*, **16**, 1943, 220–228.
6. G. Aminoff and G. Phragmén, *Z. Krist.*, **56**, 1921, 510–514.
7. H. C. Vacher, C. J. Bechtoldt, and E. Maxwell, *Trans. AIME*, **200**, 1954, 80.
8. H. v. Wartenberg, H. Werth, and H. J. Reusch, *Z. Elektrochem.*, **38**, 1932, 50.
9. E. Raub and G. Buss, *Z. Elektrochem.*, **46**, 1940, 195–202.

0.7948
$\bar{1}.2052$

Ir-P Iridium-Phosphorus

Measurements of vapor pressure and X-ray analysis have shown that there are only two Ir phosphides, Ir_2P (7.41 wt. % P) and IrP_2 (24.29 wt. % P) [1], the latter being stable up to about 1230°C. Ir_2P possibly melts congruently at about 1350°C and appears to form a eutectic with Ir at about 1300°C, containing between 20 and 28 at. (3.86–5.87 wt.) % P. An alloy with 16.7 at. (3.11 wt.) % P melts at about 1400–1480°C [1].

Ir_2P is isotypic with CaF_2 (C1 type), a = 5.54_6 A [2]. IrP_2 has a complex diffraction pattern [1], similar to that of RhP_2 [3].

1. K. H. Söffge, M. Heimbrecht, and W. Biltz, *Z. anorg. Chem.*, **243,** 1940, 297–306.
2. M. Zumbusch, *Z. anorg. Chem.*, **243,** 1940, 322–329.
3. F. E. Faller, E. F. Strotzer, and W. Biltz, *Z. anorg. Chem.*, **244,** 1940, 317–328.

$\bar{1}.9694$
0.0306

Ir-Pb Iridium-Lead

According to [1], there is a peritectically formed phase with the NiAs (B8) type of structure, a = 3.993 A, c = 5.566 A, c/a = 1.394. It very likely has the composition IrPb (51.76 wt. % Pb).

1. H. Pfisterer and K. Schubert, *Z. Metallkunde*, **41,** 1950, 358–367.

$\bar{1}.9952$
0.0048

Ir-Pt Iridium-Platinum

The liquidus shown in Fig. 477 is due to [1]. The temperatures reported, based on a melting point of Ir of 2340°C, were adjusted to conform with the more recently determined melting point of about 2454°C. The solidus is hypothetical; its distance from the liquidus may be larger than shown, since cast alloys show very pronounced coring [2].

The existence of an unbroken series of solid solutions at high temperatures is corroborated by the microstructure [1, 3], measurements of the temperature coefficient of electrical resistance [3], and the lattice parameter [4, 8], which changes linearly with composition.

The observation that the tensile strength of hard-drawn wires with 15–25% Ir increases on annealing at 700–800°C [5] was interpreted as indicating age-hardening phenomena [6]. Confirmation of a transformation was obtained by a study of the effect of quenching temperature, varied between ~ 600 and 1100°C, on the electrical resistivity and tensile strength of alloys with 10–40% Ir [7]. Results suggest the transformation temperatures shown in Fig. 477. The nature of this change has not been established, but X-ray investigations suggest that it more likely involves ordering than precipitation of a second phase. On the other hand, the microstructure of a natural Pt-Ir alloy (composition?) [7] was found to be two-phase, with the minor phase as a precipitate in the matrix.

Note Added in Proof. [8] have established, mainly by means of resistometric, microscopic, dilatometric, and X-ray work, that the solid-state effects mentioned above are due to the formation of a miscibility gap below about 975°C rather than to ordering. Data of [8], together with the transformation data of [7], are plotted in Fig. 477*a*. The exact positioning of the miscibility curve is made difficult by the very sluggish rates of precipitation.

1. L. Müller, *Ann. Physik*, **7,** 1930, 9–47; O. Feussner and L. Müller, "Heraeus Festschrift," pp. 14–15, Hanau (Main), 1930.
2. E. Raub and G. Buss, *Z. Elektrochem.*, **46,** 1940, 195–202.
3. V. A. Nemilov, *Z. anorg. Chem.*, **204,** 1932, 41–48; *Izvest. Inst. Platiny*, **7,** 1929, 13–20.

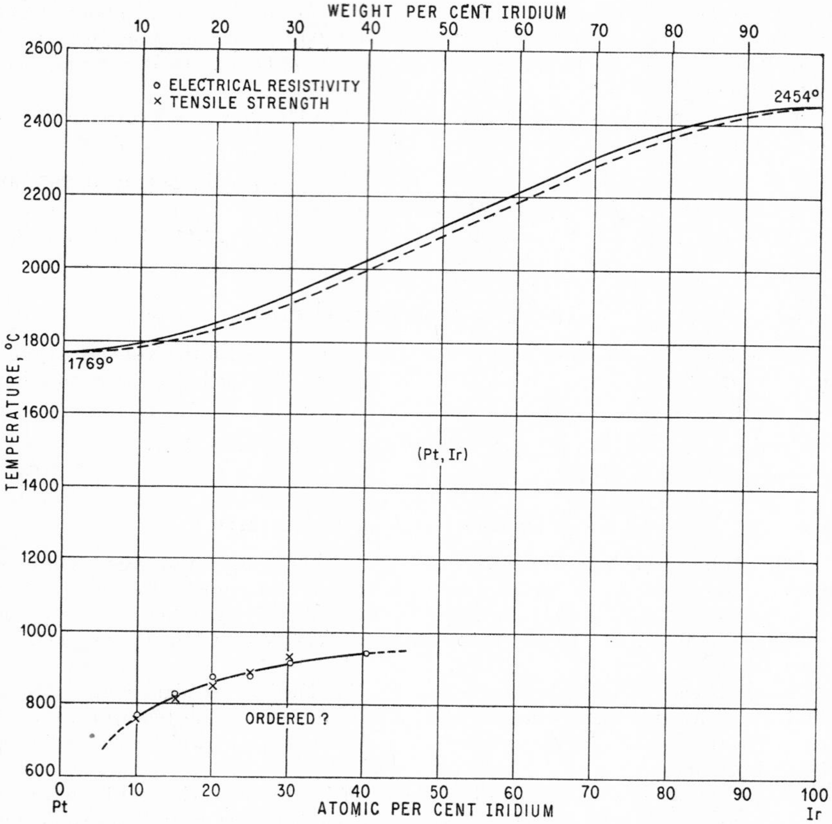

Fig. 477. Ir-Pt. (See also Fig. 477*a*.)

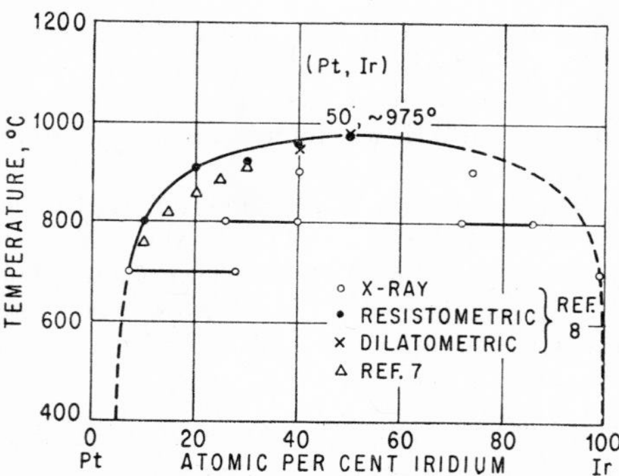

Fig. 477*a*. Ir-Pt. (See also Fig. 477.)

4. J. Weerts (F. Beck), *Z. Metallkunde*, **24**, 1932, 139–140.
5. W. Geibel, *Z. anorg. Chem.*, **70**, 1911, 246–251.
6. L. Nowack, *Z. Metallkunde*, **22**, 1930, 140.
7. G. Masing, K. Eckhardt, and K. Kloiber, *Z. Metallkunde*, **32**, 1940, 122–124.
8. E. Raub and W. Plate, Z. *Metallkunde*, **47**, 1956, 688–693.

0.0156
$\bar{1}$.9844

Ir-Re Iridium-Rhenium

The thermo-emf of various Ir-Re alloys was measured (against W and Ir) by [1] up to 2450°C. In supplementary X-ray and microscopic work it was found that (*a*) Re dissolves about 40 and 50 wt. (39 and 49 at.) % Ir at 2000°C and "room temperature," respectively, and (*b*) Ir takes about 5 wt. (at.) % Re into solid solution at "room temperature."

1. G. Haase and G. Schneider, Z. *Physik*, **144**, 1956, 256–262.

0.2733
$\bar{1}$.7267

Ir-Rh Iridium-Rhodium

The constitution of this system has not been studied as yet. Since both metals are f.c.c. and their atomic diameters differ by only about 0.9%, it is likely that they form an unbroken series of solid solutions.

0.2785
$\bar{1}$.7215

Ir-Ru Iridium-Ruthenium

It has been reported that a few per cent Ru raises the melting point of Ir by several hundred °C [1].

1. H. v. Wartenberg, H. Werth, and H. J. Reusch, *Z. Elektrochem.*, **38**, 1932, 50.

0.7797
$\bar{1}$.2203

Ir-S Iridium-Sulfur

X-ray investigations and measurements of vapor pressure at various temperatures between 880 and 1073°C showed that Ir forms the sulfides Ir_2S_3 (19.94 wt. % S), IrS_2 (24.93 wt. % S), and Ir_3S_8 [72.73 at. (30.69 wt.) % S] [1]. IrS (14.24 wt. % S) has also been reported; it decomposes in N_2 atmosphere above 750°C into Ir and S [2]. Preparations with 72.1–72.6 at. % S were found to have a structure similar to that of FeS_2 (pseudopyrite) [1].

1. W. Biltz, J. Laar, P. Ehrlich, and K. Meisel, *Z. anorg. Chem.*, **233**, 1937, 257–281.
2. L. Wöhler, K. Ewald, and H. G. Krall, *Ber. deut. chem. Ges.*, **66**, 1933, 1638.

0.3884
$\bar{1}$.6116

Ir-Se Iridium-Selenium

The existence of the selenides Ir_2Se_3 (38.02 wt. % Se) [1], $IrSe_2$ (44.99 wt. % Se) [2], and $IrSe_3$ (55.09 wt. % Se) [3], prepared by reaction of Se with $IrCl_3$, has been reported. A preparation of the composition $IrSe_{2.89}$ (74.29 at. % Se) was suggested to have a structure similar to that of pyrite, FeS_2 [3].

1. C. Chabrié and A. Bouchonnet, *Compt. rend.*, **137**, 1903, 1059–1061.
2. L. Wöhler, K. Ewald, and H. G. Krall, *Ber. deut. chem. Ges.*, **66**, 1933, 1638–1652.

3. W. Biltz, F. W. Wrigge, P. Ehrlich, and K. Meisel, *Z. anorg. Chem.*, **233**, 1937, 282–285.

0.8372
$\bar{1}.1628$

Ir-Si Iridium-Silicon

In a preliminary note, the existence of the phases Ir_3Si_2 (8.84 wt. % Si), IrSi (12.70 wt. % Si), Ir_2Si_3 (17.91 wt. % Si), and $IrSi_3$ (30.38 wt. % Si), identified by X-ray analysis, has been reported [1]. In addition, there is another unidentified phase richer in Ir than Ir_3Si_2. The latter silicide is hexagonal, $a = 3.96_8$ A, $c = 5.13$ A, $c/a = 1.293$ [1].

The existence of $IrSi_3$ may be somewhat doubtful, since no other transition element forms a higher silicide than XSi_2.

1. J. H. Buddery and A. J. E. Welch, *Nature*, **167**, 1951, 362.

0.2113
$\bar{1}.7887$

Ir-Sn Iridium-Tin

[1] melted a quantity of the mineral iridosmium (see Ir-Os, page 870) with six times as much Sn and heated the alloy with dilute HCl. Cubes with 56.5 wt. % Sn were isolated, nearly corresponding to the composition $IrSn_2$ (55.15 wt. % Sn) [2]. In a similar way, [3] isolated crystals of the approximate composition $IrSn_3$ (64.84 wt. % Sn).

By X-ray analysis, [4] identified the phases $IrSn_2$ and Ir_3Sn_7 (58.92 wt. % Sn), the latter being the intermediate phase richest in Sn occurring in this system. $IrSn_2$ is isotypic with CaF_2 (C1 type), $a = 6.338$ A. Ir_3Sn_7 is cubic, with 40 atoms per unit cell and $a = 9.360$ A.

X-ray study [5] of two alloys with about 20 and 26 wt. % Sn (28.9 and 36.4 at. % Sn), said to be "not in equilibrium," revealed the presence of a phase which probably corresponds to the composition IrSn (38.07 wt. % Sn). It is isotypic with NiAs (B8 type), $a = 3.988$ A, $c = 5.567$ A, $c/a = 1.396$.

Sn is soluble in solid Ir to an unknown amount [5], but Ir is practically insoluble in solid Sn [4].

1. H. St. Claire Deville and H. Debray, *Ann. chim. et phys.*, **56**, 1859, 385.
2. Os and Ir do not form a compound.
3. H. Debray, *Compt. rend.*, **104**, 1887, 1470.
4. O. Nial, Dissertation, University of Stockholm, 1945; *Svensk Kem. Tidskr.*, **59**, 1947, 172–183.
5. H. Nowotny, K. Schubert, and U. Dettinger, *Z. Metallkunde*, **37**, 1946, 137–145.

0.1799
$\bar{1}.8201$

Ir-Te Iridium-Tellurium

The existence of $IrTe_2$ (56.9 wt. % Te) [1] and $IrTe_3$ (66.5 wt. % Te) [1, 2] has been reported in the chemical literature. A systematic study of the system, using preparations obtained synthetically, has not been carried out as yet. A product of the composition $IrTe_{2.89}$ (74.29 at. % Te) was suggested to have a structure similar to that of pyrite, FeS_2 [2].

1. L. Wöhler, K. Ewald, and H. G. Krall, *Ber. deut. chem. Ges.*, **66**, 1933, 1638–1652.
2. W. Biltz, F. W. Wrigge, P. Ehrlich, and K. Meisel, *Z. anorg. Chem.*, **233**, 1937, 282–285.

0.6054
$\bar{1}$.3946

Ir-Ti Iridium-Titanium

[1] investigated the possibility of the occurrence of AB_2 and AB_3 types of phases in this system and reported that $TiIr_2$ and $TiIr_3$ could not be detected.

1. H. J. Wallbaum, *Naturwissenschaften*, **31**, 1943, 91–92.

$\bar{1}$.9091
0.0909

Ir-U Iridium-Uranium

There is evidence that β-U may be stabilized by quenching an alloy containing 2 at. % Ir [1]. The intermediate phase UIr_2 (61.86 wt. % Ir) is of the f.c.c. $MgCu_2$ (C15) type of structure, with $a = 7.4939 \pm 5$ kX (at 20°C) [2].

1. A. U. Seybolt, unpublished information (1944); quoted in H. A. Saller and F. A. Rough, *U.S. Atomic Energy Comm., Publ.* BMI-1000, June, 1955.
2. T. J. Heal and G. I. Williams, *Acta Cryst.*, **8**, 1955, 494–498.

0.0212
$\bar{1}$.9788

Ir-W Iridium-Wolfram

Measurements of lattice parameters and microexamination of homogenized alloys with 5.25–95.2 at. (5–95 wt.) % W, prepared by melting, showed the existence of one intermediate phase, designated as β. It ranges in composition from 23.3 at. (22.5 wt.) to 56 at. (55.2 wt.) % W, after annealing at 1400–1800°C. The solid solubility of W in Ir was found to decrease from 19.8 at. (19 wt.) % at 1800°C to 13.7 (13.3), 12.6 (12.4), and 11.5 (11.3) % W at 1300, 1100, and 900°C, respectively. The solubility of Ir in W was estimated as about 2.5 at. (2.6 wt.) % after annealing at 1600°C [1].

19.8 at. % W increases the lattice constant of Ir to $a = 3.850$ A. The β phase is h.c.p., with $a = 2.736$ A, $c/a = 1.602$ at the Ir side and $a = 2.764$ A, $c/a = 1.611$ at the W side of the homogeneity range [1].

1. E. Raub and P. Walter, "Festschrift aus Anlass des 100-jährigen Jubiläums der Firma W. C. Heraeus G.m.b.H.," pp. 124–146, Hanau, 1951.

0.3257
$\bar{1}$.6743

Ir-Zr Iridium-Zirconium

$ZrIr_2$ (19.11 wt. % Zr) is isotypic with $MgZn_2$ (C14 type) [1].

1. H. J. Wallbaum, *Naturwissenschaften*, **30**, 1942, 149.

0.7508
$\bar{1}$.2492

K-Li Potassium-Lithium

According to the thermal results (six alloys) of [1], Li and K do not alloy. The depression of the melting point of Li by addition of K [2] as reported by [3] was not corroborated.

1. B. Böhm and W. Klemm, *Z. anorg. Chem.*, **243**, 1939, 69–85.
2. The phase diagram of [3] shows a monotectic horizontal at 166°C.
3. G. Masing and G. Tammann, *Z. anorg. Chem.*, **67**, 1910, 187–190.

0.2062
$\bar{1}.7938$

K-Mg Potassium-Magnesium

[1] determined the temperatures of the beginning and the end of solidification of three alloys with 6, 53, and 85 wt. % Mg and found arrests only at the melting points of K and Mg. It follows that liquid K and liquid Mg are practically insoluble in each other.

1. D. P. Smith, *Z. anorg. Chem.*, **56,** 1908, 113–114.

0.2305
$\bar{1}.7695$

K-Na Potassium-Sodium

Thermal analyses of the K-Na system were carried out by [1–4]; individual melting and solidification points were determined by [5–11] among others. The results are in substantial agreement [12]. The liquidus in Fig. 478 was taken from the review in [13] and is mainly based on the (vacuum) work by [4]. The formula of the only compound, Na_2K (45.95 wt. % K), was first established by [2]. The existence

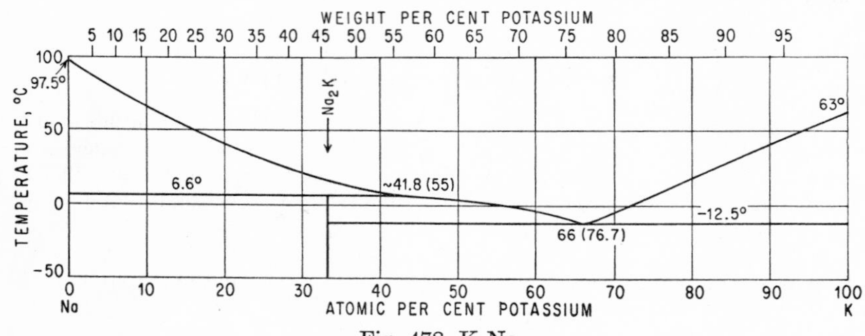

Fig. 478. K-Na

of solid solutions has been suggested by [1, 14]; however, representations of powder patterns published by [15] show no clear line shifting. It is of interest to note that the "size factors" [16] for the system K-Na are unfavorable and therefore any extended primary solid solutions are improbable.

Na_2K was shown [17] to have the hexagonal $MgZn_2$ (C14) type of structure, with $a = 7.50$ A, $c = 12.30$ A, $c/a = 1.64$.

The influence of pressure (up to 10,000 kg per cm²) on the K-Na phase diagram was investigated by [18] using measurements of resistivity [19].

Several authors [20–22] concluded the existence of molecules of an intermetallic compound in the melt; however, it was shown—in agreement with the modern viewpoint—by an X-ray diffraction analysis [23] that no permanent compact molecules exist in liquid K-Na alloys.

1. N. S. Kurnakow and N. A. Puschin, *Z. anorg. Chem.*, **30,** 1902, 109–112.
2. G. L. C. M. van Rossen Hoogendijk van Bleiswijk, *Z. anorg. Chem.*, **74,** 1912, 152–156.
3. E. Jänecke, *Z. Metallkunde*, **20,** 1928, 115.
4. E. Rinck, *Compt. rend.*, **197,** 1933, 49–51.
5. E. Hagen, *Wied. Ann.*, **19,** 1883, 472.
6. G. Tammann, *Z. physik. Chem.*, **3,** 1889, 446.
7. C. T. Heycock and F. H. Neville, *J. Chem. Soc.*, **55,** 1889, 674.

8. M. Rosenfeld, *Ber. deut. chem. Ges.*, **24**, 1891, 1658.
9. K. Siebel, *Ann. Physik*, **60**, 1919, 260–278.
10. S. Walters and R. R. Miller, *Ind. Eng. Chem.* (*Anal. Ed.*), **18**, 1946, 468–469.
11. C. T. Ewing, R. S. Hartmann, H. B. Atkinson, and R. R. Miller, Naval Research Laboratory Report C-3105; quoted from [13].
12. For the peritectic and eutectic points the following concentrations were given: 40.05 [1], 42 [2], and 40–42 [4] at. % K and 66.6 [1], 66.6 [2], and 66 [4] at. % K, respectively.
13. "Liquid Metals Handbook," U.S. Atomic Energy Commission and Department of the Navy, June, 1950.
14. K. Bornemann, *Metallurgie*, **6**, 1909, 239.
15. B. Böhm and W. Klemm, *Z. anorg. Chem.*, **243**, 1939, 69–85. This paper also contains a magnetic analysis of the system.
16. See, e.g., W. Hume-Rothery and G. V. Raynor, "The Structure of Metals and Alloys," 3d ed., The Institute of Metals, London, 1954.
17. F. Laves and H. J. Wallbaum, *Z. anorg. Chem.*, **250**, 1942, 110–120.
18. C. H. Kean, *Phys. Rev.*, **55**, 1939, 750–754.
19. Seven liquidus points for atmospheric pressure agree with the thermal results of [4].
20. A. Joannis, *Ann. chim. et phys.*, (6)**12**, 1887, 358 (Heat of formation).
21. K. Banerjee, *Indian J. Phys.*, **3**, 1929, 399 (X-ray diffraction); cf. C. W. Heaps, *Phys. Rev.*, (2)**48**, 1935, 491.
22. R. Kremann, M. Pestemer, and H. Schreiner, *Rec. trav. chim.*, **51**, 1932, 557 (Viscosity).
23. N. S. Gingrich and R. E. Henderson, *J. Chem. Phys.*, **20**, 1952, 1117–1120.

$\bar{1}.2758$
0.7242

K-Pb Potassium-Lead

The diagram of Fig. 479 was worked out by [1]. The determination of the diagram by thermal analysis was hindered by experimental difficulties (see the original paper) as well as those resulting from the constitution. The compositions of the various compounds could not be given with certainty in all cases because of great disturbances of equilibrium, caused by the presence of the miscibility gap in the liquid state, by the formation—certainly incomplete—of a compound from two melts of widely different specific weights, and by three peritectic horizontals in close succession [2].

The existence of the miscibility gap in the liquid state [3] and of the compound KPb_2 (91.37 wt. % Pb) appears to be well established, that of the compounds K_2Pb (72.60 wt. % Pb) and KPb_4 (95.50 wt. % Pb) rather probable. The thermal effects at 376°C indicate the peritectic formation of an additional compound, X, with a composition between K_2Pb and KPb_2 (KPb = 84.12 wt. % Pb?). It is suggested by [1] that at this or a somewhat higher temperature K_2Pb undergoes a polymorphic transformation; however, this assumption is not well founded.

Emf measurements [4] on annealed alloys gave indications for the existence of K_2Pb and KPb_4, but not for X and KPb_2.

[5] did not observe age-hardening with Pb-rich alloys and concluded that the solubility of K in solid Pb is negligible [6].

A compound K_4Pb_9, stable obviously only in liquid NH_3 solution, was reported by [8].

Note Added in Proof. KPb_2 is of the hexagonal $MgZn_2$ (C14) type of structure, with a = 6.66 A, c = 10.76 A, c/a = 1.61_4 [9]. This compound appears to be unstable at lower temperatures since slow cooling from the homogenizing temperature (300°C) resulted in an X-ray pattern exhibiting chiefly the lines of KPb_4. The pattern of KPb_4 could tentatively be indexed with a b.c.c. cell of the lattice parameter $a = 12.31$ A [9].

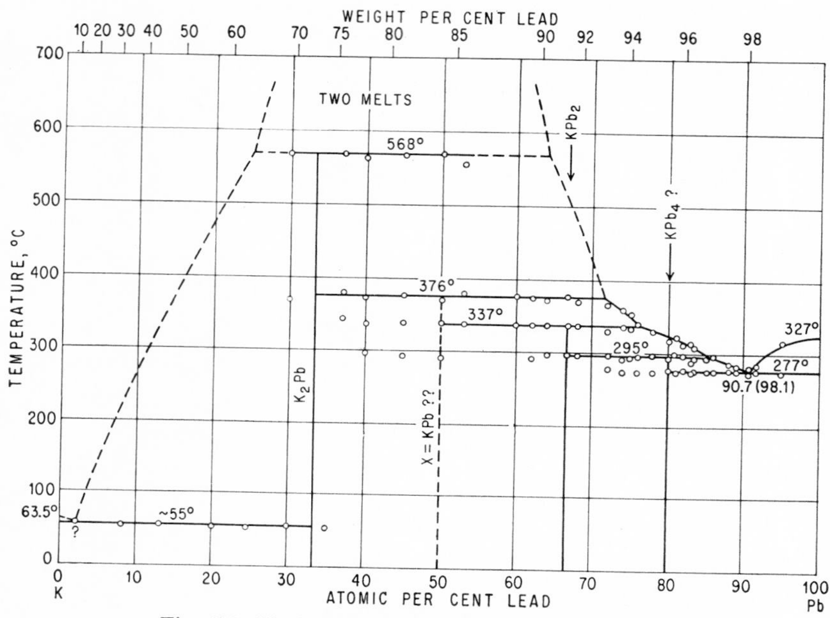

Fig. 479. K-Pb. (See also Note Added in Proof.)

1. D. P. Smith, *Z. anorg. Chem.*, **56,** 1908, 133–139. No microstructures are given.
2. Note the thermal effects plotted in Fig. 479.
3. The extension of the horizontal line at 568°C in Fig. 479 is based only on a rough estimation of the lengths of thermal arrests.
4. R. Kremann and E. Pressfreund, *Z. Metallkunde,* **13,** 1921, 19–21.
5. G. Tammann and H. Rüdiger, *Z. anorg. Chem.*, **192,** 1930, 26–29.
6. The "size factors" [7] for the K-Pb system are unfavorable.
7. See, e.g., W. Hume-Rothery and G. V. Raynor, "The Structure of Metals and Alloys," 3d ed., The Institute of Metals, London, 1954.
8. E. Zintl, J. Goubeau, and W. Dullenkopf, *Z. physik. Chem.*, **A154,** 1931, 39–40.
9. D. Gilde, *Z. anorg. Chem.*, **284,** 1956, 142–143.

$\bar{1}.3016$
0.6984

K-Pt Potassium-Platinum

No systematic investigation of the K-Pt system has been carried out as yet. Pt is attacked by K vapor as well as molten K [1–3]. The attack by liquid K begins at 400°C; above 500°C the solubility of Pt increases greatly; on cooling, the dissolved Pt separates again [4].

1. "Collected Works of Sir Humphry Davy," vol. 5, pp. 231, 232, 245, Smith, Elder and Co., London, 1840.
2. J. Dewar and A. Scott, *Chem. News,* **40,** 1879, 294.

3. V. Meyer, *Ber. deut. chem. Ges.*, **13,** 1880, 391.
4. L. Hackspill, *Proc. Intern. Congr. Appl. Chem., 7th Congr. London,* 1909, II, p. 267; abstract in "Gmelins Handbuch der anorganischen Chemie," System No. 68 A(6), p. 747, Verlag Chemie, G.m.b.H., Weinheim/Bergstrasse, 1951.

$\bar{1}.6603$
0.3397

K-Rb Potassium-Rubidium

Measurements of conductivity and flow pressure on K-Rb alloys were carried out by [1], and a continuous series of solid solutions was concluded. This was corroborated by [2] in a thermal and microscopic investigation; a minimum in the liquidus was found at 66.7 at. (81.4 wt.) % Rb, 32.8°C. On the other hand, [3] suggested from his thermal results a eutectic system, with the eutectic point at 85 wt. (72.2 at.) % Rb, 34°C, and a miscibility gap in the solid state between 78 and 90 wt. (62–80.5 at.) % Rb, but felt that his conclusion should be checked by X-ray analysis [4].

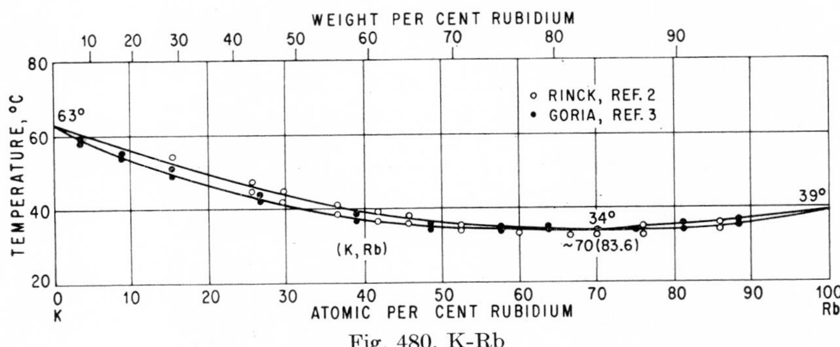

Fig. 480. K-Rb

Considering the small difference in the metallic radii of K and Rb (about 7%), there is little doubt that K and Rb are completely miscible in the solid state [5]. In Fig. 480 the minimum in the liquidus has been placed at 70 at. (83.6 wt.) % Rb, 34°C.

1. N. S. Kurnakow and A. J. Nikitinsky, *Z. anorg. Chem.*, **88,** 1914, 151–160.
2. E. Rinck, *Compt. rend.*, **200,** 1935, 1205–1206.
3. C. Goria, *Gazz. chim. ital.*, **65,** 1935, 865–870.
4. Because of the very flat minimum in the liquidus, thermal analysis alone cannot give any decisive evidence.
5. B. Böhm and W. Klemm, *Z. anorg. Chem.*, **243,** 1939, 69–85.

$\bar{1}.5067$
0.4933

K-Sb Potassium-Antimony

The phase diagram for K-Sb (Fig. 481) was established by thermal analysis [1] and shows the existence of two compounds, K_3Sb (50.93 wt. % Sb) and KSb (75.69 wt. % Sb). K_3Sb was shown [2] to have the hexagonal Na_3As ($D0_{18}$) type of structure, with a = 6.037 A, c = 10.717 A, c/a = 1.775. Whether the compounds have slightly variable compositions is not known. Because of the very unfavorable "size factor" [3], more than a negligible solubility of K in solid Sb is not to be expected.

1. N. Parravano, *Gazz. chim. ital.*, **45,** 1915, 485–489. The alloys were prepared under H_2.

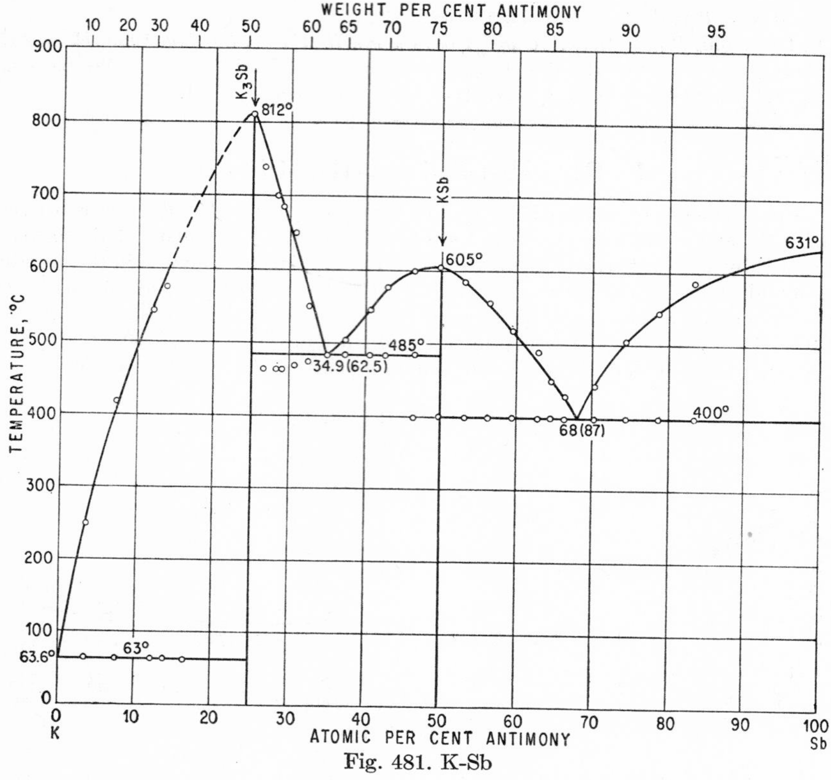

Fig. 481. K-Sb

2. G. Brauer and E. Zintl, *Z. physik. Chem.*, **B37**, 1937, 323–352.
3. See, e.g., W. Hume-Rothery and G. V. Raynor, "The Structure of Metals and Alloys," 3d ed., The Institute of Metals, London, 1954.

$\bar{1}.6948$
0.3052

K-Se Potassium-Selenium

In the chemical literature the preparation of the compounds K_2Se (50.24 wt. % Se), K_2Se_2 (66.88 wt. % Se), K_2Se_3 (75.18 wt. % Se), K_2Se_4 (80.15 wt. % Se), and K_2Se_5 [71.43 at. (83.47 wt.) % Se] has been described. K_2Se was shown [1] to have the CaF_2 (C1) type of structure, with a = 7.691 A.

The compounds named were confirmed by thermal and roentgenographic analyses [2] (Fig. 482); however, the constitution in the composition range between K_2Se and K_2Se_2 could not be clarified. It is assumed [2] that K_2Se dissolves Se—under distortion of the cubic lattice symmetry—up to about $K_2Se_{1.2}$. Thermal effects between 40 and 50 at. % Se (see Fig. 482) could not be interpreted by [2] but may be taken as an indication for the existence of an additional phase.

K_2Se, K_2Se_2, K_2Se_3, and K_2Se_4 are diamagnetic [2].

1. E. Zintl, A. Harder, and B. Dauth, *Z. Elektrochem.*, **40**, 1934, 588–593.
2. W. Klemm, H. Sodomann, and P. Langmesser, *Z. anorg. Chem.*, **241**, 1939, 281–304.

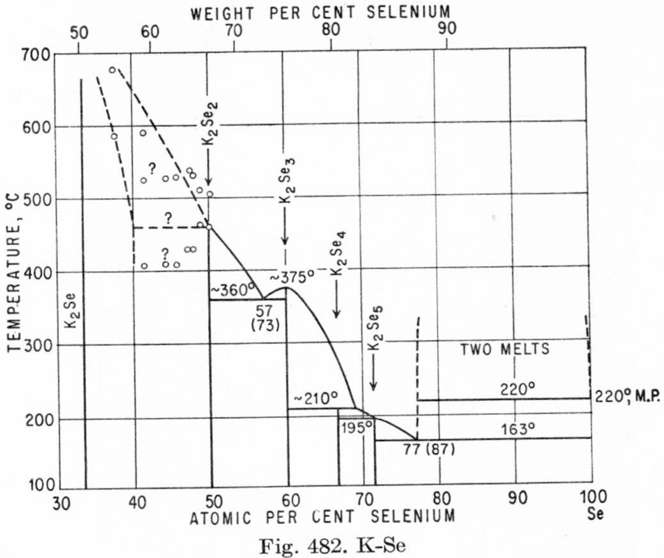

Fig. 482. K-Se

0.1436
$\bar{1}$.8564

K-Si Potassium-Silicon

KSi (41.81 wt. % Si) was prepared by direct synthesis. On thermal decomposition a substance of approximate composition KSi_8 (85.18 wt. % Si) is obtained. Representations of powder patterns are given [1].

1. E. Hohmann, *Z. anorg. Chem.*, **257,** 1948, 113–126.

$\bar{1}$.5177
0.4823

K-Sn Potassium-Tin

The phase diagram of Fig. 483 was drawn according to thermal data given by [1]. There are only a few liquidus data, and the constitution between 50 and 67 at. % Sn could not be clarified because of great experimental difficulties at higher temperatures (boiling point of K: 770°C).

The existence of the four compounds K_2Sn (60.29 wt. % Sn), KSn (75.22 wt. % Sn), KSn_2 (85.86 wt. % Sn), and KSn_4 (92.39 wt. % Sn) was concluded from the lengths of thermal arrests at 535, 600, and 414°C [2] and has some corroboration by emf measurements [3]. Whether KSn forms peritectically at 670°C or undergoes there only a polymorphic transformation is unknown, since it is uncertain to which compound (KSn, KSn_2, or one of intermediate composition) the maximum in the liquidus belongs.

The thermal data do not exclude the existence of restricted primary solid solutions; however, it may be noted that the "size factors" [4] for K-Sn are very unfavorable.

[5] observed the formation of a K-Sn compound soluble in liquid NH_3 and ascribed the formula K_4Sn_8. (= KSn_2).

1. D. P. Smith, *Z. anorg. Chem.*, **56,** 1908, 129–133. The alloys were melted in glass crucibles in a hydrogen atmosphere. The microstructure was not investigated.

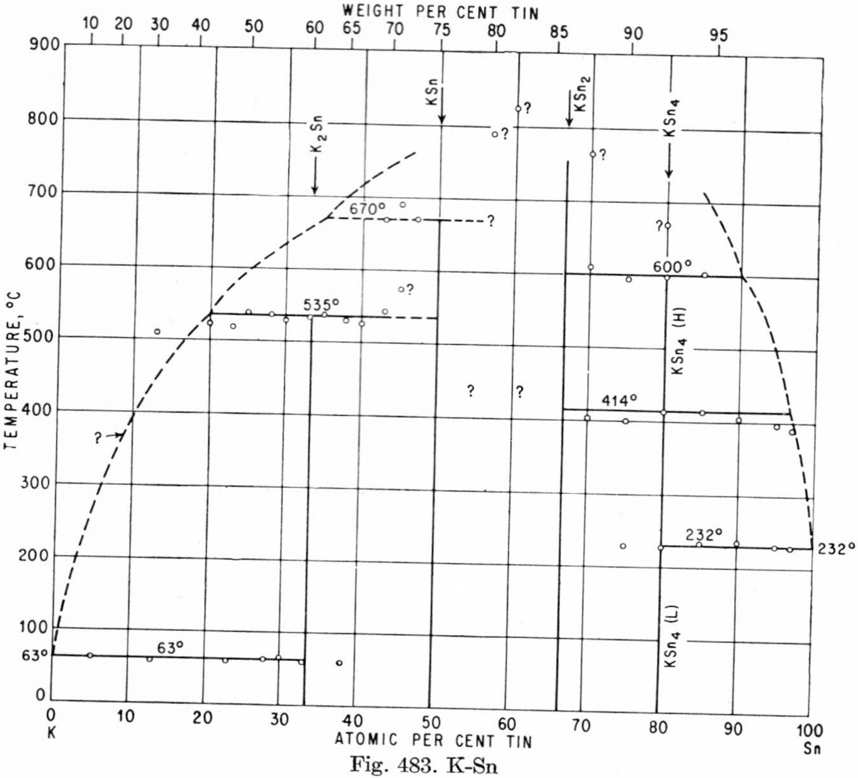

Fig. 483. K-Sn

2. KSn_4 is assumed to undergo a polymorphic transformation at 414°C.
3. R. Kremann and E. Pressfreund, *Z. Metallkunde*, **13**, 1921, 21–24.
4. See, e.g., W. Hume-Rothery and G. V. Raynor, "The Structure of Metals and Alloys," 3d ed., The Institute of Metals, London, 1954.
5. F. W. Bergstrom, *J. Phys. Chem.*, **30**, 1926, 12.

$\bar{1}.4863$
0.5137

K-Te Potassium-Tellurium

The preparation of the compounds K_2Te (62.01 wt. % Te) and K_2Te_3 (83.04 wt. % Te) has been described in the chemical literature. K_2Te has the CaF_2 (C1) type of structure [1, 2], with 8.168 A [1]; it is assumed by [2] that K_2Te dissolves Te—under distortion of the cubic lattice symmetry—up to about the composition $K_2Te_{1.3}$ (cf. K-Se).

There is X-ray and magnetic evidence [2] for the existence of a further compound, K_2Te_2 (76.55 wt. % Te).

1. E. Zintl, A. Harder, and B. Dauth, *Z. Elektrochem.*, **40**, 1934, 588–593.
2. W. Klemm, H. Sodomann, and P. Langmesser, *Z. anorg. Chem.*, **241**, 1939, 281–304.

$\bar{1}.2817$
0.7183

K-Tl Potassium-Thallium

The phase diagram of Fig. 484 was worked out by [1] using thermal analysis exclusively. As the investigation is particularly limited to the determination of liquidus temperatures, a conclusive decision on the constitution is not possible.

A compound KTl (83.94 wt. % Tl) is clearly indicated by the maximum in the liquidus. [1] suggested the existence of a second compound, K_2Tl (72.33 wt. % Tl), based on the break in the liquidus at 33 at. % Tl. This break probably indicates a peritectic reaction near 242°C; however, instead of K_2Tl another formula lying between 33 and 50 at. % Tl—e.g., K_3Tl_2—could be ascribed with equal justification to this second intermediate phase.

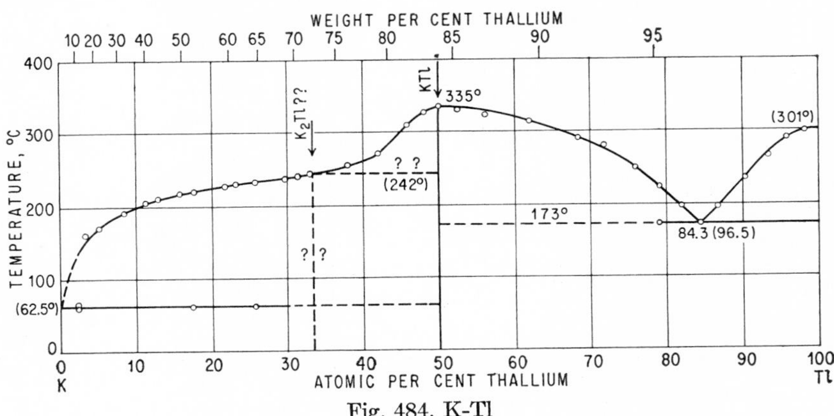

Fig. 484. K-Tl

Emf measurements by [2] failed to give any constitutional information, probably owing to the formation of nobler protective films.

KTl has neither the β-brass nor the NaTl (B32) type of structure [3].

1. N. S. Kurnakow and N. A. Puschin, *Z. anorg. Chem.*, **30,** 1902, 87–101. The alloys were melted under petroleum jelly or paraffin in iron crucibles. Oxidation losses were extremely small.
2. R. Kremann and E. Pressfreund, *Z. Metallkunde,* **21,** 1921, 24–27.
3. E. Zintl and G. Brauer, *Z. physik. Chem.*, **B20,** 1933, 245–271.

$\bar{1}.7767$
0.2233

K-Zn Potassium-Zinc

According to the thermal results of [1] (Fig. 485) a Zn-rich compound is formed by reaction of two immiscible melts consisting practically of the pure metals. [1] considered some peculiarities in the cooling curves [2] of alloys with 60–91 at. % Zn as evidence for transformations of the compound; however, a simpler explanation would be to assume that these peculiarities are caused by a disturbance of equilibrium, which is to be expected with this type of phase relationship.

The compound (named KZn_{12} by [1]) was shown more recently [3–5] to have the composition KZn_{13} [92.86 at. (95.60 wt.) % Zn] and the cubic $NaZn_{13}$ ($D2_3$) type of structure, with $a = 12.360 \pm 5$A [3], $a = 12.38$ A [4, 5]. Emf measurements by [6] failed to yield constitutional information.

1. D. P. Smith, *Z. anorg. Chem.*, **56,** 1908, 114–119. The alloys were melted in a hydrogen atmosphere and stirred by a glass rod during the solidification.
2. Repeated supercooling and recalescence; cf. the thermal data plotted in Fig. 485.
3. J. A. A. Ketelaar, *J. Chem. Phys.*, **5,** 1937, 668.
4. E. Zintl and W. Haucke, *Naturwissenschaften*, **25,** 1937, 717.
5. E. Zintl and W. Haucke, *Z. Elektrochem.*, **44,** 1938, 104–111.
6. R. Kremann and A. Mehr, *Z. Metallkunde*, **12,** 1920, 453–455.

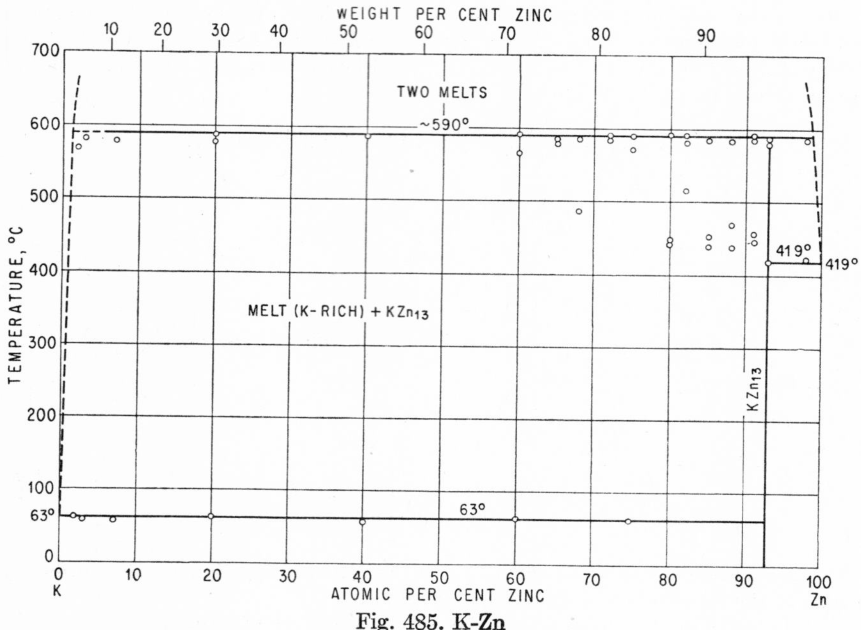

Fig. 485. K-Zn

0.7568
$\bar{1}$.2432

La-Mg Lanthanum-Magnesium

From thermal-analysis data and microscopic examination of alloys cooled from the molten state, essentially the following conclusions had been drawn [1]: (*a*) Four intermediate phases exist: Mg_9La (38.83 wt. % La), Mg_3La (65.57 wt. % La), MgLa (85.10 wt. % La), and $MgLa_4$ (95.81 wt. % La). (*b*) There is a range of liquid immiscibility which extends from about 15 to 37 at. % La at 766°C. At this temperature, the two melts react to form the phase Mg_3La. (*c*) MgLa forms solid solutions with Mg between 40 and 50 at. % La. (*d*) $MgLa_4$ decomposes at 503°C into MgLa and La.

Checking the existence of a miscibility gap in the liquid state, [2] reported that no indication of the formation of two layers could be detected. Instead, a very flat maximum in the liquidus curve at the composition Mg_3La was suggested.

Reinvestigation of the composition range 20–100 at. % La by thermal and micrographic analysis [3] revealed the phase relations presented in Fig. 486. The existence of the phase Mg_2La (74.07 wt. % La), reported earlier [4], was confirmed; it decomposes into MgLa and Mg_3La at about 626°C. Mg_3La forms solid solutions with Mg which yield Mg_9La crystals with fall in temperature. $MgLa_4$ was shown not to exist.

The thermal effects at about 530°C were interpreted to be due to the eutectoid decomposition of a solid solution based on a high-temperature modification of La.

The constitution of the range 0–12.4 wt. (2.45 at.) % La was investigated by [5]; see inset of Fig. 486 in wt. % La. The solid solubility of La in Mg, based on thermo-

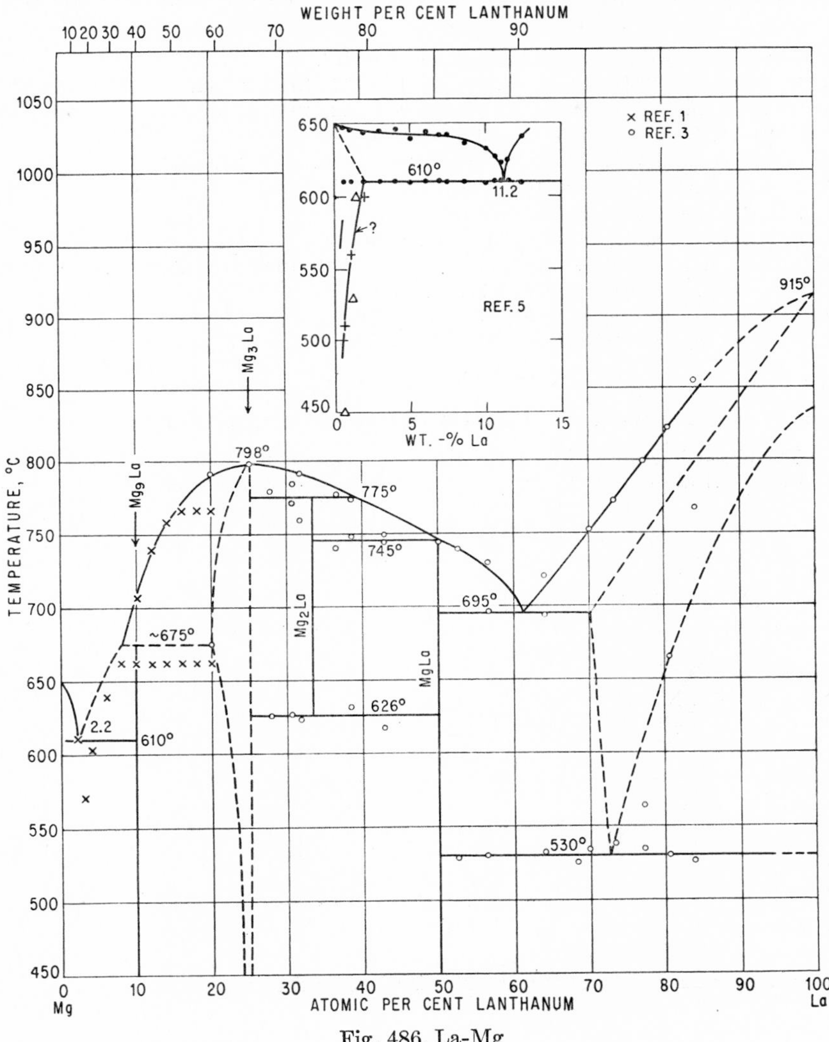

Fig. 486. La-Mg

resistivity work, was found to be 0.5, 1.0, and 1.9 wt. (0.09, 0.17, and 0.34 at.) % La at 510, 560, and 600°C, respectively. Lattice-parameter measurements showed solubilities of about 0.45, 1.1, and 1.4 wt. (0.08, 0.19, and 0.24 at.) % La at 450, 530, and 600°C, respectively. According to [6], the solubility is considerably smaller, as indicated by measurements of electrical conductivity. Recently, [7] gave a solubility of 0.4 wt. (0.07 at.) % La at 575°C (see inset).

Crystal Structures. The effect of La on the lattice dimensions of Mg was determined by [5, 8]. Mg_3La is cubic of the BiF_3 ($D0_3$) type, with 16 atoms per unit cell, a = 7.493 A [9]. The lattice constants of alloys with 20.1 and 24.0 at. % La were given as a = 7.465 A and a = 7.480 A, respectively [3]. Mg_2La is isotypic with $MgCu_2$ (C15 type), a = 8.79 A [4]. MgLa is isotypic with CsCl (B2 type), a = 3.96 A [10]; a = 3.973 A [11].

1. G. Canneri, *Metallurgia ital.*, **23**, 1931, 810–813.
2. R. Vogel and T. Heumann, *Z. Metallkunde*, **35**, 1943, 29–42.
3. R. Vogel and T. Heumann, *Z. Metallkunde*, **38**, 1947, 1–8.
4. F. Laves, *Naturwissenschaften*, **31**, 1943, 96.
5. F. Weibke and W. Schmidt, *Z. Elektrochem.*, **46**, 1940, 359–362.
6. T. E. Leontis, *Trans. AIME*, **185**, 1949, 968–983.
7. Dow Chemical Co., unpublished work, 1953.
8. R. S. Busk, *Trans. AIME*, **188**, 1950, 1460–1464.
9. A. Rossi and A. Iandelli, *Atti reale accad. Lincei*, **19**, 1934, 415–420; see also A. Rossi, *Gazz. chim. ital.*, **64**, 1934, 774–778.
10. A. Rossi, *Gazz. chim. ital.*, **64**, 1934, 774–778.
11. H. Nowotny, *Z. Metallkunde*, **34**, 1942, 247–253.

0.4030
$\bar{1}$.5970

La-Mn Lanthanum-Manganese

The phase diagram in Fig. 487 is due to [1]. La of 99.5% purity and electrolytic Mn were used to prepare the alloys. Figure 487 deviates from the original diagram

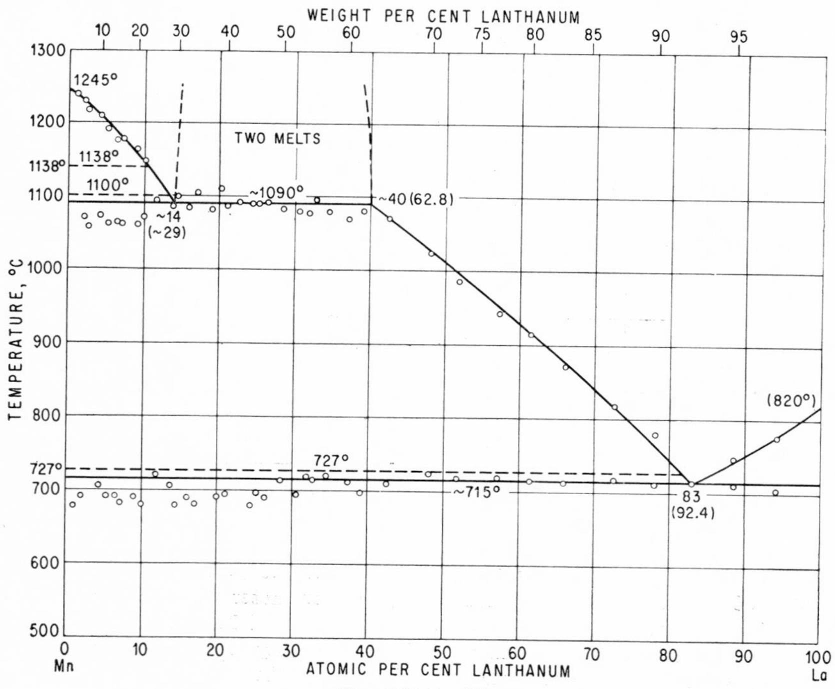

Fig. 487. La-Mn

in that the monotectic and eutectic temperatures were assumed as 1090 and 715°C instead of 1081 and 701°C, respectively. X-ray analysis confirmed that no intermediate phase exists. Measurements of paramagnetic susceptibility are also in accord with this diagram [2].

1. L. Rolla and A. Iandelli, *Ber. deut. chem. Ges.*, **75**, 1942, 2091–2095.
2. A. Serres, *J. phys. radium*, **13**, 1952, 46–47.

0.9964
$\bar{1}$.0036

La-N Lanthanum-Nitrogen

LaN (9.16 wt. % N) is isotypic with NaCl (B1 type), a = 5.286 A [1]; a = 5.29 A [2]; a = 5.295 A [3]; a = 5.305 A [4].

1. A. Iandelli and E. Botti, *Atti reale accad. Lincei, Rend.*, **25**, 1937, 129–132.
2. P. Chiotti, *J. Am. Ceram. Soc.*, **35**, 1952, 123–130.
3. R. A. Young and W. T. Ziegler, *J. Am. Chem. Soc.*, **74**, 1952, 5251–5253.
4. R. N. Mulford, *U.S. Atomic Energy Comm., Publ.* LAMS-1374, 1952; *Met. Abstr.*, **20**, 1952, 242.

0.7811
$\bar{1}$.2189

La-Na Lanthanum-Sodium

Attempts to elucidate the system by thermal analysis and microexamination were unsuccessful because of experimental difficulties, etc. In addition, the lanthanum used was heavily contaminated [1].

1. W. v. Mässenhausen, *Z. Metallkunde*, **43**, 1952, 53–54.

0.3742
$\bar{1}$.6258

La-Ni Lanthanum-Nickel

The phase relationships presented in Fig. 488 were outlined by thermal analysis of only nine alloys prepared using La of 98–99% purity. The composition of the alloys was selected on the assumption that the compositions of the intermediate phases are the same as those of the system Ce-Ni studied in more detail (see Ce-Ni). In the absence of tabulated data, the data points shown had to be taken from a smaller figure and, therefore, cannot be considered quantitative. It was concluded that six intermediate phases exist: La_3Ni (12.34 wt. % Ni), LaNi (29.70 wt. % Ni), $LaNi_2$ (45.80 wt. % Ni), $LaNi_3$? (55.90 wt. % Ni), $LaNi_4$? (62.82 wt. % Ni), and $LaNi_5$ (67.87 wt. % Ni) [1].

$LaNi_2$ and $LaNi_5$ were identified by crystal-structure investigations. $LaNi_2$ is isotypic with $MgCu_2$ (C15 type), a = 7.26_2 A [1]. The hexagonal structure of $LaNi_5$ is of the $CaCu_5$ type, a = 4.962 A, c = 4.008 A, $2c/a$ = 1.616 [2].

X-ray work indicated that solid solubility of La in Ni is not detectable [2]. [3] concluded from magnetic measurements that 9 at. % La is soluble in Ni; however, this appears to be doubtful. Since the atomic diameters of Ni and La differ greatly, [2] would have noticed an appreciable change in the lattice constant of Ni on formation of solid solutions.

1. R. Vogel (W. Fülling), *Z. Metallkunde*, **38**, 1947, 97–103.
2. H. Nowotny, *Z. Metallkunde*, **34**, 1942, 247–253.
3. J. Wucher, *J. phys. radium*, **13**, 1952, 278–282.

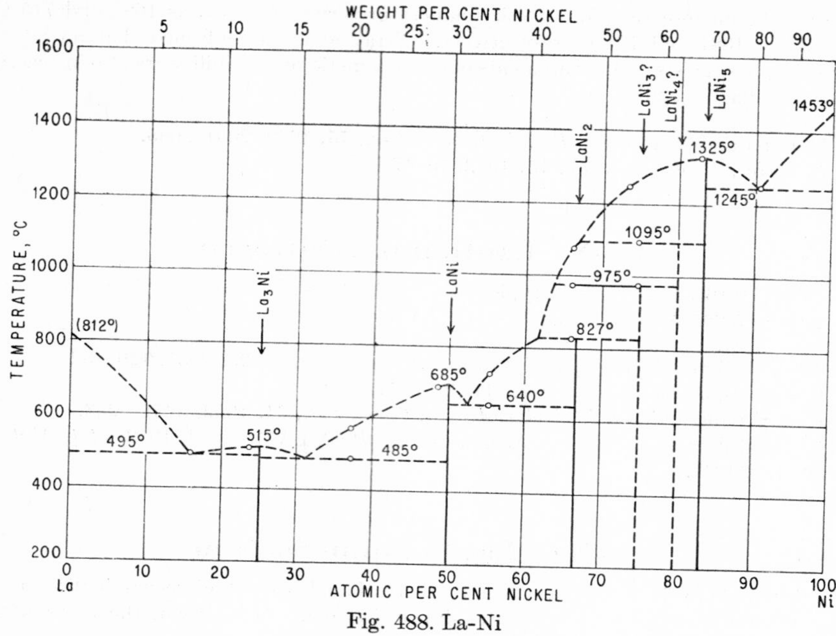

Fig. 488. La-Ni

0.9386
$\bar{1}.0614$

La-O Lanthanum-Oxygen

The structure of La_2O_3 is the prototype of the $D5_2$-type structure, hexagonal with 1 La_2O_3 per unit cell, $a = 3.9_4$ A, $c = 6.1_3$ A, $c/a = 1.55_7$ [1]; see also [2]. Another modification is cubic of the Mn_2O_3 ($D5_3$) type, $a = 11.4_2$ A [3].

Supplement. [4] redetermined the lattice parameters of hexagonal La_2O_3: $a = 3.9373$ A, $c = 6.1299$ A (at 26°C). According to [5], La_2O_3 melts at 2210 ± 20°C.

1. W. Zachariasen, *Z. physik. Chem.*, **123,** 1926, 134–150; *Z. Krist.*, **70,** 1929, 187–189.
2. W. C. Koehler and E. O. Wollan, *Acta Cryst.*, **6,** 1953, 741–742.
3. K. Löhberg, *Z. physik. Chem.*, **B28,** 1935, 402–407.
4. H. E. Swanson, R. K. Fuyat, and G. M. Ugrinic, *Natl. Bur. Standards (U.S.), Circ.* 539, III, 1954, p. 33.
5. W. A. Lambertson and F. H. Gunzel, *U.S. Atomic Energy Comm., Publ.* AECD-3465, 1952.

0.6518
$\bar{1}.3482$

La-P Lanthanum-Phosphorus

LaP (18.23 wt. % P) is isotypic with NaCl (B1 type), $a = 6.02_5$ A [1].

1. A. Iandelli and E. Botti, *Atti reale accad. nazl. Lincei, Rend.*, **24,** 1936, 459–464.

$\bar{1}.8264$
0.1736

La-Pb Lanthanum-Lead

Thermal analysis of the whole system [1] showed the existence of the three phases La_2Pb (42.72 wt. % Pb) [2], LaPb (59.86 wt. % Pb), and $LaPb_2$ (74.89 wt. % Pb).

The phase $LaPb_3$ (81.73 wt. % Pb), rather than $LaPb_2$, was identified by determination of its crystal structure [3], however. By quantitative metallographic investigation [4], the existence of $LaPb_3$ could be verified (Fig. 489). Solubilities in the solid state have not been investigated.

$LaPb_3$ is isotypic with Cu_3Au ($L1_2$ type), a = 4.90_3 A [3].

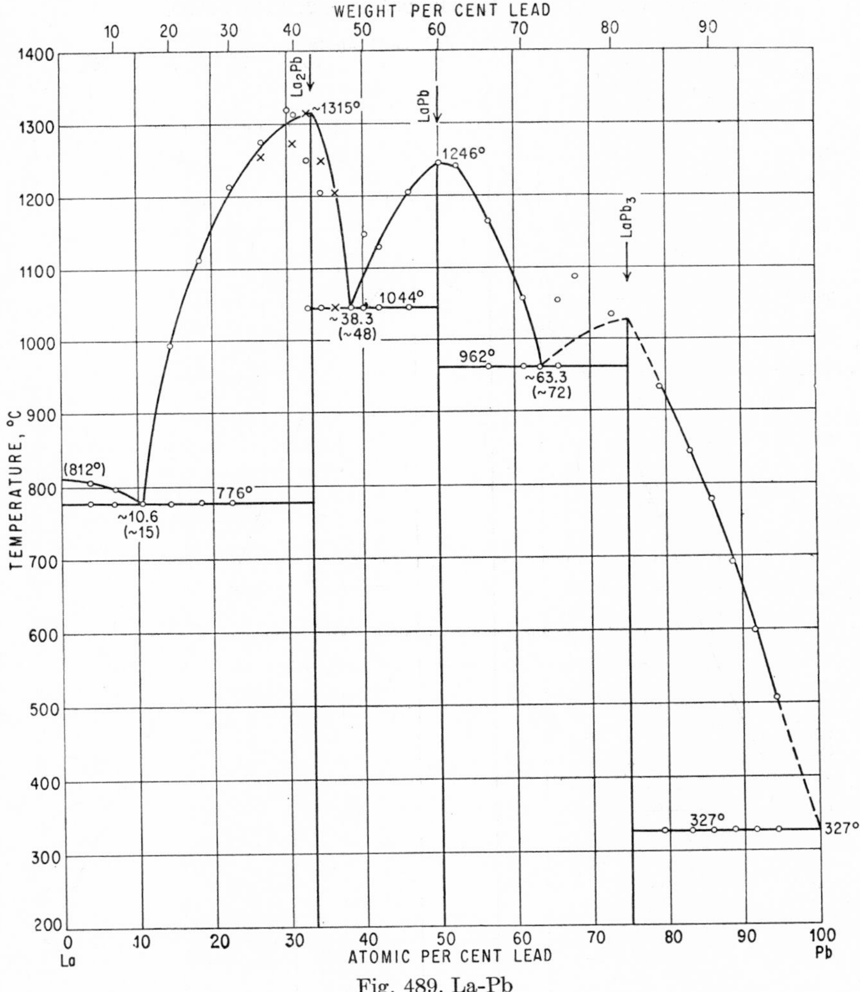

Fig. 489. La-Pb

1. G. Canneri, *Metallurgia ital.*, **23**, 1931, 805–806.
2. In the original paper by Canneri, the maximum melting point of La_2Pb was shown to lie at 39–40 wt. % Pb (circles in Fig. 489) rather than at 42.7 wt. % Pb. Revised liquidus temperatures (crosses in Fig. 489) were obtained from the author.
3. A. Rossi, *Atti reale accad. Lincei, Rend.*, **17**, 1933, 839–846; *Gazz. chim. ital.*, **64**, 1934, 832.
4. R. Vogel and T. Heumann, *Z. Metallkunde*, **35**, 1943, 29–42.

0.6367
$\bar{1}.3633$

La-S Lanthanum-Sulfur

La_2S_3 (25.72 wt. % S) is isotypic with Ce_2S_3, cubic with 16 S atoms and $10\frac{2}{3}$ La atoms per unit cell, a = 8.723 A [1].

1. W. H. Zachariasen, *Acta Cryst.*, **1**, 1948, 265–268; **2**, 1949, 57–60.

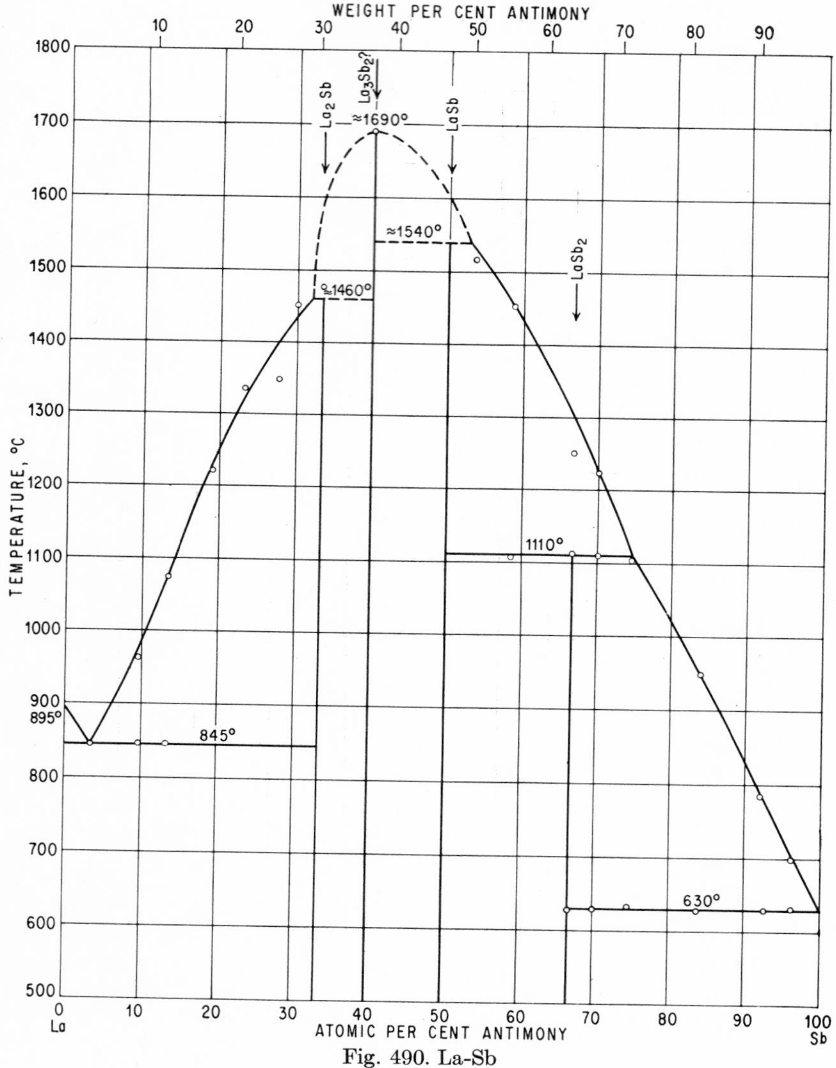

Fig. 490. La-Sb

0.0573
$\bar{1}.9427$

La-Sb Lanthanum-Antimony

The phase diagram in Fig. 490 was obtained by thermal and microscopic investigations [1]. In spite of a number of experimental difficulties, such as reaction of the melts with the crucible material (especially in the range 30–60 at. % Sb), the composition of the four intermediate phases indicated, La_2Sb (30.47 wt. % Sb), La_3Sb_2 (36.88 wt. % Sb), LaSb (46.71 wt. % Sb), and $LaSb_2$ (63.68 wt. % Sb), appears to be fairly certain, perhaps with the exception of La_3Sb_2. LaSb had been identified earlier [2]. It is isotypic with NaCl (B1 type), a = 6.488 A.

1. R. Vogel and H. Klose, *Z. Metallkunde*, **45**, 1954, 633–638.
2. A. Iandelli and E. Botti, *Atti reale accad. nazl. Lincei, Rend.*, **25**, 1937, 498–502.

0.6942
$\bar{1}.3058$

La-Si Lanthanum-Silicon

$LaSi_2$ (28.80 wt. % Si) is isotypic with $ThSi_2$, b.c. tetragonal, with 12 atoms per unit cell, a = 4.28 A, c = 13.75 A, c/a = 3.21 [1]; a = 4.38 A, c = 13.59 A, c/a = 3.10 [2, 3].

1. G. Brauer and H. Haag, *Naturwissenschaften*, **37**, 1950, 210–211; *Z. anorg. Chem.*, **267**, 1952, 198–212.
2. F. Bertaut and P. Blum, *Acta Cryst.*, **3**, 1950, 319.
3. According to G. Brauer and H. Haag, *Z. anorg. Chem.*, **267**, 1952, 198–212, footnote 21, the original values were given in kX units.

0.0683
$\bar{1}.9317$

La-Sn Lanthanum-Tin

Thermal analysis [1] indicated the existence of the three intermediate phases La_2Sn (29.23 wt. % Sn), La_2Sn_3 (56.17 wt. % Sn), and $LaSn_2$ (63.09 wt. % Sn). The identification of the phase $LaSn_3$ (71.94 wt. % Sn) by determination of its crystal structure [2], however, was in contradiction to that of $LaSn_2$. Reinvestigation of the system between 52 and 90 wt. (about 56–91.5 at.) % Sn by thermal analysis [3] revealed the existence of $LaSn_3$ rather than $LaSn_2$ (Fig. 491). No information is available as to the solubilities in the solid state.

$LaSn_3$ is isotypic with Cu_3Au ($L1_2$ type), a = 4.78_2 A [2].

1. G. Canneri, *Metallurgia ital.*, **23**, 1931, 806–809.
2. A. Rossi, *Atti reale accad. Lincei, Rend.*, **17**, 1933, 839–846; *Gazz. chim. ital.*, **64**, 1934, 832.
3. R. Vogel and T. Heumann, *Z. Metallkunde*, **35**, 1943, 29–42.

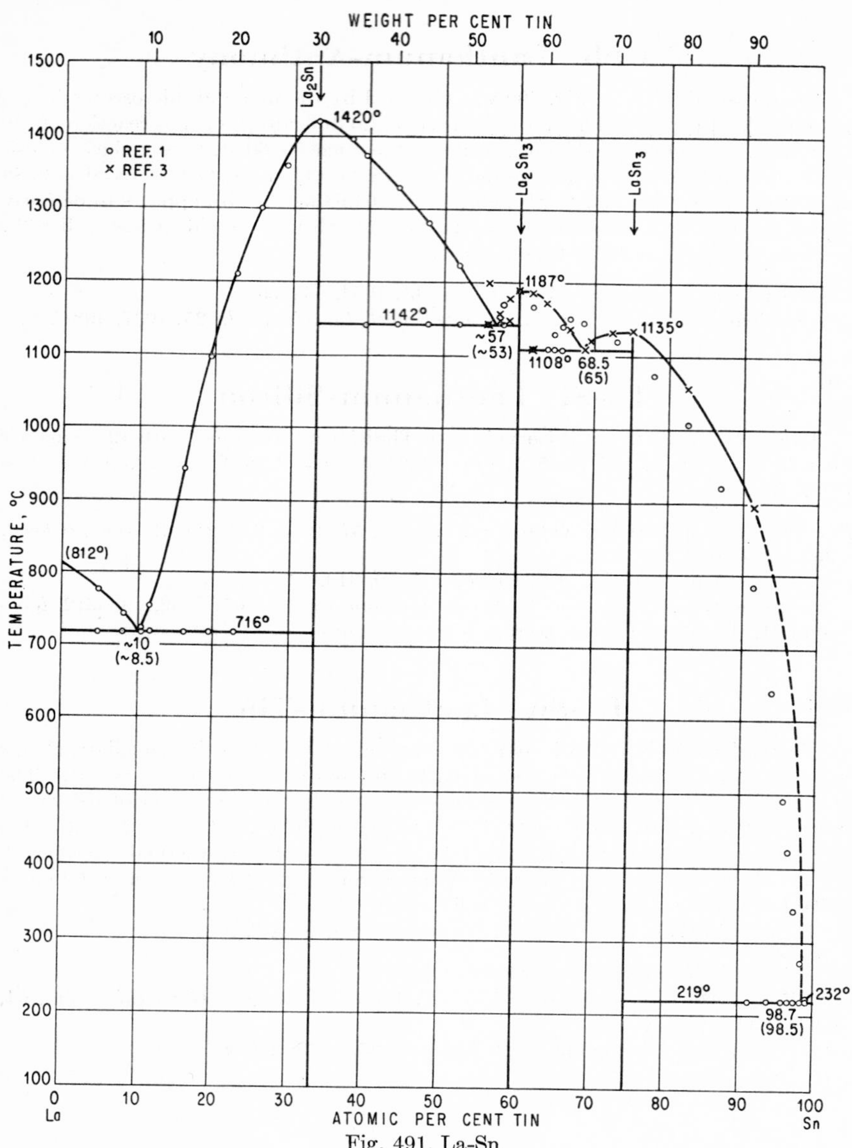
WEIGHT PER CENT TIN
10
20
30
40
50
60
70
80
90
1500
1400
1300
1200
1100
1000
900
800
700
600
500
400
300
200
100
TEMPERATURE, °C
La_2Sn
La_2Sn_3
$LaSn_3$
o REF. 1
× REF. 3
1420°
1187°
1142°
~57
(~53)
1108°
68.5
(65)
1135°
(812°)
716°
~10
(~8.5)
219°
232°
98.7
(98.5)
0
La
10
20
30
40
50
60
70
80
90
100
Sn
ATOMIC PER CENT TIN

Fig. 491. La-Sn

$\bar{1}.7770$
0.2230

La-Th Lanthanum-Thorium

"[1] studied a series of Th-La alloys metallographically and found that La and α-Th are completely soluble. The alloys were prepared by coreduction at 1000 to 1400°C. There are no compounds in this system" [2].

1. D. Peterson and R. Mickelson, unpublished information (March, 1952).
2. Quoted from H. A. Saller and F. A. Rough, Compilation of U.S. and U.K. Uranium and Thorium Constitutional Diagrams, *U.S. Atomic Energy Comm., Publ.* BMI-1000, June, 1955.

0.4624
$\bar{1}.5376$

La-Ti Lanthanum-Titanium

[1] have heated mixtures of Ti and La (melting point, 826°C) in a vacuum to 1500–1600°C. The samples contained unreacted Ti in a possibly eutectic matrix. It was stated that Ti and La probably do not form a compound and that there might be a miscibility gap in the liquid state.

1. L. Rolla and A. Iandelli, *Ber. deut. chem. Ges.*, **75**, 1942, 2094–2095.

$\bar{1}.8323$
0.1677

La-Tl Lanthanum-Thallium

Study of the system by thermal analysis [1] established the existence of the three intermediate phases La_2Tl (42.38 wt. % Tl) [2], LaTl (59.54 wt. % Tl), and $LaTl_3$ (81.53 wt. % Tl), all of which were found to melt without decomposition (see data points in Fig. 492). The fact that the phases Ce_2Tl (page 464) and Pr_2Tl (page 1134) are formed by the peritectic reaction liq. + XTl → X_2Tl, rather than having maximum melting points, suggested that the phase La_2Tl might also be formed peritectically. A cursory check [3] by taking cooling curves of two melts with 33.3 and 40.5 at. % Tl indicated that this appears to be the case (see full data points in Fig. 492). It is, of course, difficult to understand how the numerous thermal effects at higher temperatures in the range 20–45 wt. % Tl, reported by [1], could have been observed. In Fig. 492, both data are included to show the degree of discrepancy between the two findings.

$LaTl_3$ is h.c.p. of the Mg (A3) type, $a = 3.45_7$ A, $c = 5.53$ A, $c/a = 1.60$ [4]. LaTl is isotypic with CsCl (B2 type), $a = 3.92$ A [5].

1. G. Canneri, *Metallurgia ital.*, **23**, 1931, 809–810.
2. In the original paper by Canneri, the maximum melting point of La_2Tl was shown to occur at 47.5–48 wt. % Tl (circles in Fig. 492), rather than at 42.4 wt. % Tl. Revised liquidus temperatures (crosses in Fig. 492) were obtained from the author.
3. R. Vogel and T. Heumann, *Z. Metallkunde*, **35**, 1943, 29–42.
4. A. Rossi, *Gazz. chim. ital.*, **64**, 1934, 955–957.
5. Quoted from "Metals Reference Book," p. 187, Butterworths Scientific Publications, London, 1949. No publication dealing with the structure of LaTl could be found.

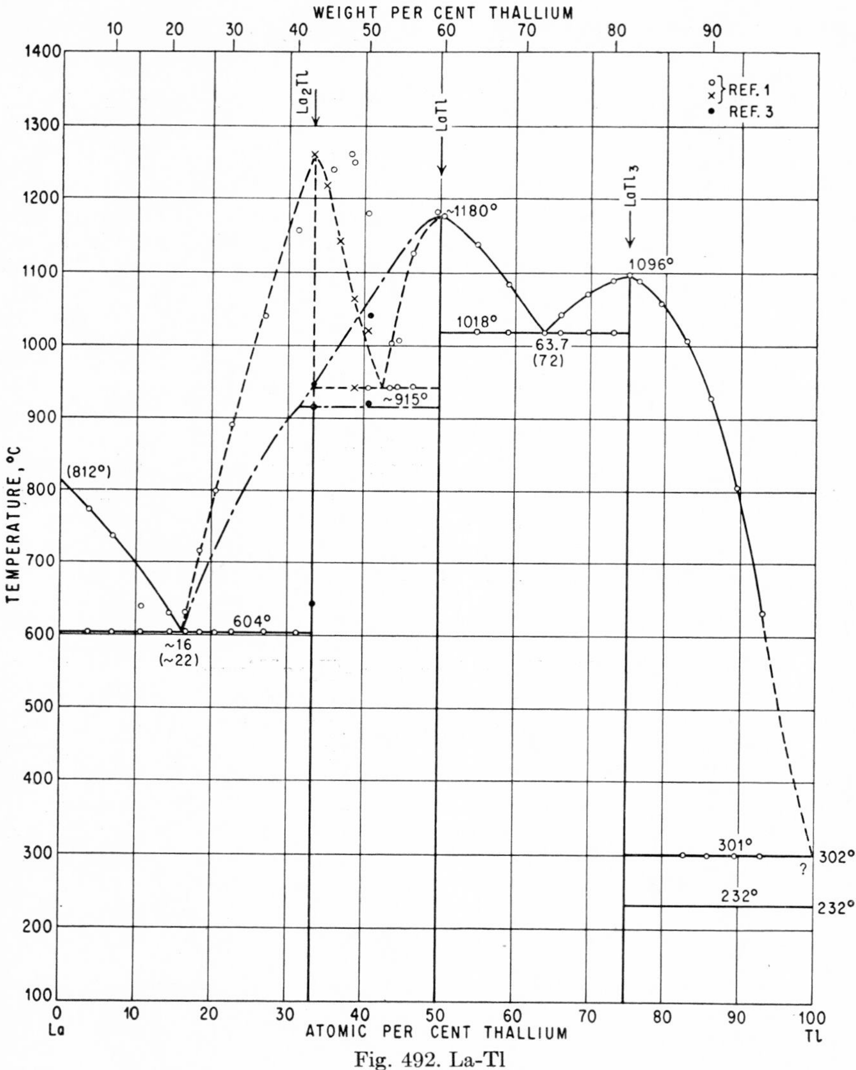

Fig. 492. La-Tl

$\bar{1}.7660$
0.2340

La-U Lanthanum-Uranium

"La, like Pr and Nd, is reported to be much less soluble in U than is Ce. The solubility of Ce is reported to be 0.5 wt. % near the melting point of U, and Ce is only partially miscible with U in the liquid state. Probably there are no compounds in the U-La system [1]" [2].

1. National Physical Laboratory, United Kingdom, unpublished information (1949-1950).

2. Quoted from H. A. Saller and F. A. Rough, Compilation of U.S. and U.K. Uranium and Thorium Constitutional Diagrams, *U.S. Atomic Energy Comm., Publ.* BMI-1000, June, 1955.

0.3273
$\bar{1}$.6727

La-Zn Lanthanum-Zinc

The phase relationships have been investigated twice, between 0 and 100% Zn [1] and in the region 76–100 at. % Zn [2] (Figs. 493 and 494), using thermal analysis and microexamination. The existence of the phases LaZn (32.00 wt. % Zn) [3],

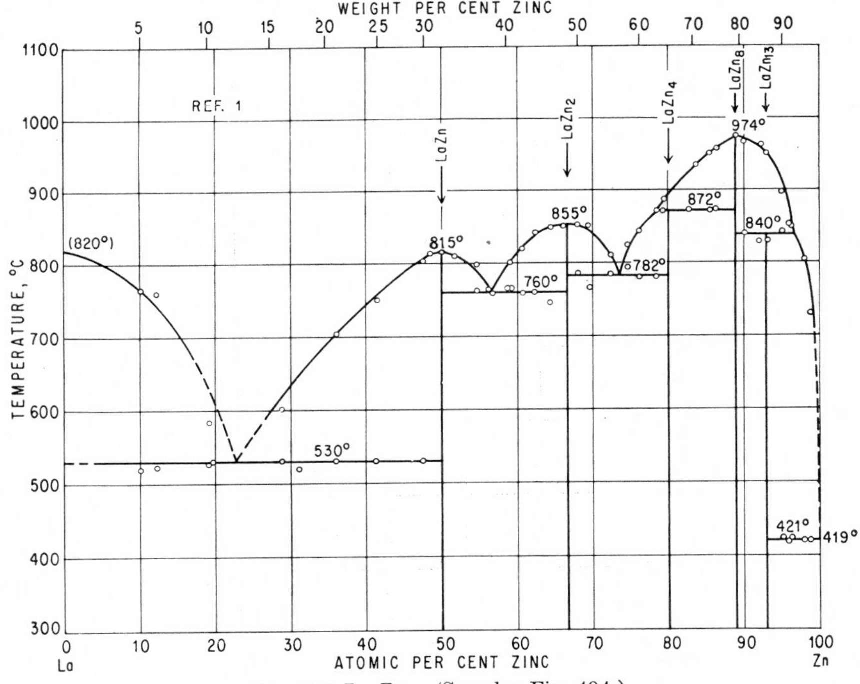

Fig. 493. La-Zn. (See also Fig. 494.)

$LaZn_2$ (48.49 wt. % Zn) [4], and $LaZn_{11}$ [91.67 at. (83.81 wt.) % Zn] [5] has been verified by determining their crystal structures. In addition to the phases shown in Figs. 493 and 494, the compound $LaZn_5$ [83.33 at. (70.18 wt.) % Zn] has been identified by X-ray work [6].

The concepts of the Zn-rich portion of the system (see Figs. 493 and 494) differ in that the two intermediate phases richest in Zn have been assumed to be $LaZn_{13}$ [92.86 at. (85.95 wt.) % Zn] and $LaZn_8$ [88.88 at. (79.01 wt.) % Zn] [1], as opposed to $LaZn_{11}$ and $LaZn_9$ [90.0 at. (80.90 wt.) % Zn] [2]. According to [2], both phases form homogeneity ranges, indicated in Fig. 494 but not accurately determined.

[2] found thermal effects at 930°C, not observed by [1], in the region 84.5–88 at. % Zn (Fig. 494), which were interpreted as indicating the formation of another phase besides the one formed at 872 [1] or 852°C [2]. This, however, cannot be the phase $LaZn_5$ [6]. On the other hand, the existence of $LaZn_4$ (Fig. 493) is incompatible with

the diagram of Fig. 494 because the Zn content of the peritectic melt (about 80.7 at. % Zn) is higher than that of $LaZn_4$ (80.0 at. % Zn), although the difference in composition is small. It must be concluded, therefore, that (*a*) $LaZn_4$ probably does not exist and will have to be replaced by $LaZn_5$ and (*b*) an additional phase occurs between $LaZn_5$ and $LaZn_8$ [1] or $LaZn_9$ [2]. Also, $LaZn_{13}$ has to be replaced by $LaZn_{11}$, since the latter was identified by X-ray analysis [5]. $LaZn_{11}$ undergoes a polymorphic transformation at about 710°C, involving the formation of a structure stable below 710°C and different from that of $CeZn_{11}$ (page 465).

The temperature of the peritectic reaction $LaZn_9$ + melt $\rightleftarrows$ $LaZn_{11}$ was found to be 853°C on heating and 825–830°C on cooling. The corresponding peritectic

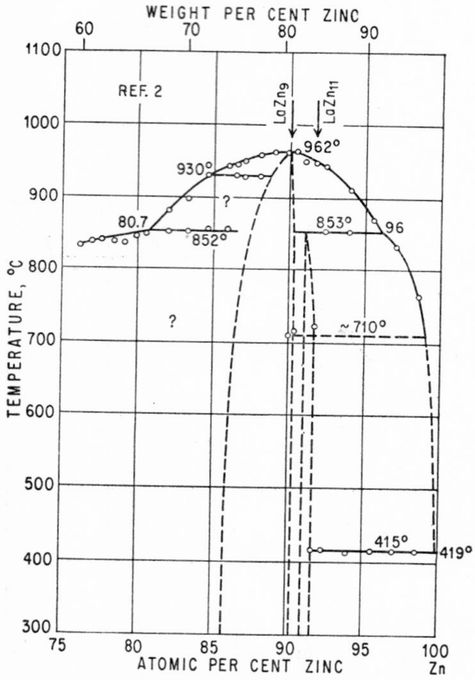

Fig. 494. La-Zn. (See also Fig. 493.)

temperature, according to [1], was determined to lie between these limits, i.e., about 840°C. A further divergence exists as regards the solidus temperature of the Zn-rich alloys. Whereas 421°C (Fig. 493) would indicate a peritectic equilibrium, 415°C (Fig. 494) would point to a eutectic.

Crystal Structures. LaZn is isotypic with CsCl (B2 type), a = 3.76 A [3], and $LaZn_2$ appears to be a Laves phase [4]. $LaZn_5$ has the hexagonal structure of the $CaCu_5$ type, a = 5.427 A, c = 4.225 A [6]. $LaZn_9$ is isomorphous with $CeZn_9$ [2]. $LaZn_{11}$ is isotypic with $BaCd_{11}$, b.c. tetragonal, with 48 atoms per unit cell, a = 10.68 A, c = 6.87 A, c/a = 0.643 [5].

1. L. Rolla and A. Iandelli, *Ricerca sci.*, **20,** 1941, 1216–1226.
2. J. Schramm, *Z. Metallkunde*, **33,** 1941, 358–360.
3. A. Iandelli and E. Botti, *Gazz. chim. ital.*, **67,** 1937, 638–644.

4. F. Laves, *Naturwissenschaften*, **27**, 1939, 65.
5. M. J. Sanderson and N. C. Baenziger, *Acta Cryst.*, **6**, 1953, 627–631.
6. H. Nowotny, *Z. Metallkunde*, **34**, 1942, 247–253.

$\bar{1}$.4554
0.5446

Li-Mg Lithium-Magnesium

Cursory microscopic and thermal work on Li-Mg alloys was carried out by [1].

Liquidus. Thermal-analysis determinations of the liquidus were made by [2–5]. The general form of the liquidus as shown in Fig. 495 (maximum, eutectic) is well established by the work of [2, 5]; it may be mentioned that [3] suggested a peritectic rather than a eutectic reaction, and [4] a second three-phase horizontal (peritectic) at

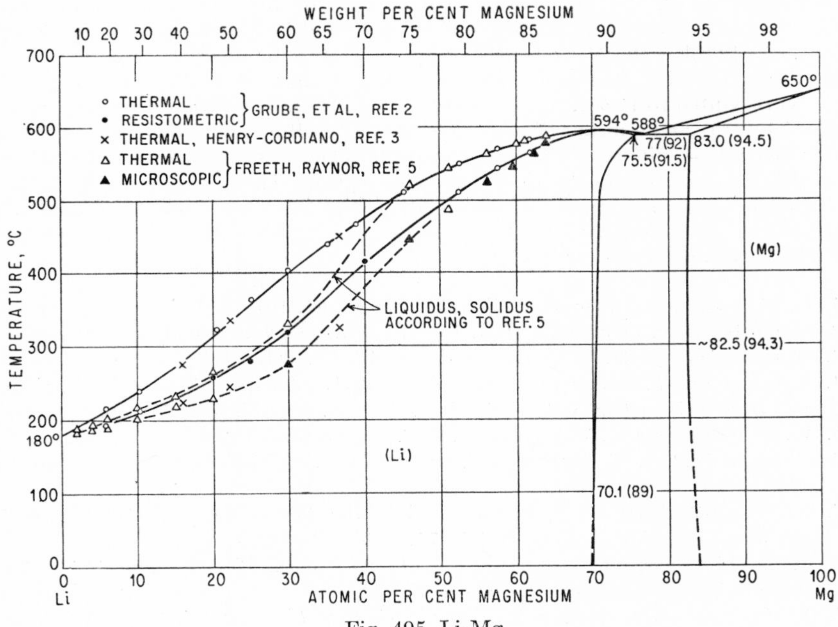

Fig. 495. Li-Mg

182°C near pure Li. In Fig. 495 some data of [2, 3, 5] are plotted to show the disagreement of results in the Li-rich region; since the experimental conditions of [2] and [5] were comparable, no explanation for this inconsistency can be offered. The most reliable data for the maximum are ~70.8 at. % Mg, 594°C [5]; those for the eutectic, 77 at. (92 wt.) % Mg, 588°C [5] [6].

Solidus. Data for the solidus have been obtained by thermal [2–5], microscopic [7, 5], and resistometric [2] analyses. As in the case of the liquidus, there is no agreement in the literature as to the exact position of the solidus in Li-rich alloys; since no claims for precision in this region were made, an average position between the solid- and broken-line curves in Fig. 495 may be suggested as a compromise. As to the Mg-rich solidus, the results of [2, 7, 5] are in substantial agreement.

Solid Solubilities. The view of [4] that the solubility of Mg in Li is small and that the wide solid solution is an intermediate phase (cf. Liquidus) based on the compound $LiMg_2$ has not been corroborated by more recent work. Alloys up to 70 at. %

Mg have the b.c.c. structure [8–11], and no two-phase field could be observed at high-Li contents [5].

Detailed micrographic studies of the (Li)-phase boundary were made by [12, 5]; that shown in Fig. 495 was taken from [5]. The marked curvature above 500°C had not been found by earlier workers [2–4]. As to the boundary of the (Mg) phase, the metallographic results of [7] are in good agreement with those of X-ray back-reflection work by [13]. The solubility diminishes slightly at higher temperatures (0.4–0.5 at. % over the whole range of temperature concerned [7]).

The existence of a superstructure in the b.c.c. (Li) phase had been suggested by [2, 7]; however, [14] gave some evidence that extra lines in powder patterns arise from contamination and not from superlattice formation [17].

The lattice-parameter-vs.-composition curve for the (Li) phase shows a marked minimum at 50 at. % [11] (see also [2]) which may be due to a Brillouin-zone effect [11]. For low-temperature transformations in Li and Li-Mg alloys, see [10, 15, 18]; 12 at. % Mg raises the M_s transformation temperature to about 130°K [10].

For preparation, fabrication, and properties of Mg-Li base alloys, see [16].

Note Added in Proof. The structures of Li-Mg solid solutions have been investigated by [19, 20] by means of measurements on the Bragg reflections [19] as well as measurements of diffuse X-ray scattering [20]. There is short-range order but no superlattice formation in the wide (Li)-phase region. The minimum in the lattice-parameter-vs.-composition curve was found at 35 at. % Mg at both 20 and −183°C (cf. above). Addition of Li to Mg decreases the *a* and *c* dimensions of the unit cell as well as the axial ratio *c/a* [13, 21, 19].

1. G. Masing and G. Tammann, *Z. anorg. Chem.*, **67,** 1910, 197–198.
2. G. Grube, H. v. Zeppelin, and H. Bumm, *Z. Elektrochem.*, **40,** 1934, 160–164.
3. O. H. Henry and H. V. Cordiano, *Trans. AIME*, **111,** 1934, 319–332.
4. P. Saldau and F. Shamrey, *Z. anorg. Chem.*, **224,** 1935, 388–398; see also F. Shamrey, *Bull. Science U.S.S.R., Sect. Chem. Sci.*, no. 6, 1947, pp. 605–616.
5. W. E. Freeth and G. V. Raynor, *J. Inst. Metals*, **82,** 1953-1954, 575–580.
6. [2] found 78.2 at. % Mg, 587.5°C, and [4] 73 at. % Mg, 588°C, for the eutectic point.
7. W. Hume-Rothery, G. V. Raynor, and E. Butchers, *J. Inst. Metals*, **71,** 1945, 589–601; discussion, *J. Inst. Metals*, **72,** 1946, 538–542.
8. A. Baroni, *Congr. intern. quim. pura y apl. 9th Congr., Madrid*, **2,** 1934, 464–470; abstract in *Strukturbericht*, **3,** 1933-1935, 633.
9. A. Maximov and G. Komovskij, *Zhur. Tech. Fiz.* (*U.S.S.R.*), **6,** 1936, 1612. Not available to the author. Abstract, "Gmelins Handbuch der anorganischen Chemie," System No. 27 (A), pp. 439–444, Verlag Chemie, G.m.b.H., Weinheim/Bergstrasse, 1952.
10. C. S. Barrett and O. R. Trautz, *Trans. AIME*, **175,** 1948, 579–601.
11. D. W. Levinson, *Acta Met.*, **3,** 1955, 294–295.
12. J. A. Catterall, *Nature*, **169,** 1952, 336.
13. W. Hofmann, *Z. Metallkunde*, **28,** 1936, 160–163.
14. R. L. P. Berry and G. V. Raynor, *Nature*, **171,** 1953, 1078–1079.
15. C. S. Barrett and D. F. Clifton, *Trans. AIME*, **188,** 1950, 1329–1332.
16. J. H. Jackson, P. D. Frost, A. C. Loonam, L. W. Eastwood, and C. H. Lorig, *Trans. AIME*, **185,** 1949, 149–168.
17. B. L. Averbach et al., *U.S. Atomic Energy Comm., Publ.* NYO-7036, 1953; *Met. Abstr.*, **22,** 1954, 52. F. H. Herbstein and B. L. Averbach, *Acta Cryst.*, **9,** 1956, 91–92.
18. C. S. Barrett, *Acta Met.*, **4,** 1956, 528–531.

19. F. H. Herbstein and B. L. Averbach, *Acta Met.*, **4,** 1956, 407–413.
20. F. H. Herbstein and B. L. Averbach, *Acta Met.*, **4,** 1956, 414–420.
21. R. S. Busk, *Trans. AIME,* **188,** 1950, 1460.

$\bar{1}.4797$
0.5203

Li-Na Lithium-Sodium

[1] reported that 0.5 wt. (1.6 at.) % Li lowers the freezing point of Na by 2°C. Thermal analyses of the whole system [2, 3] agree on an extended solubility gap in the

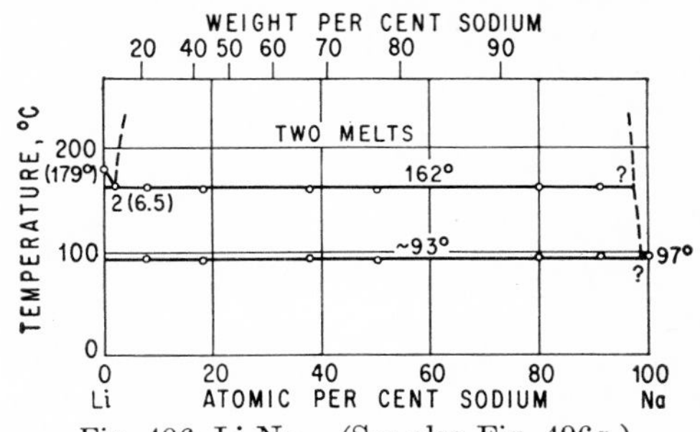

Fig. 496. Li-Na. (See also Fig. 496*a*.)

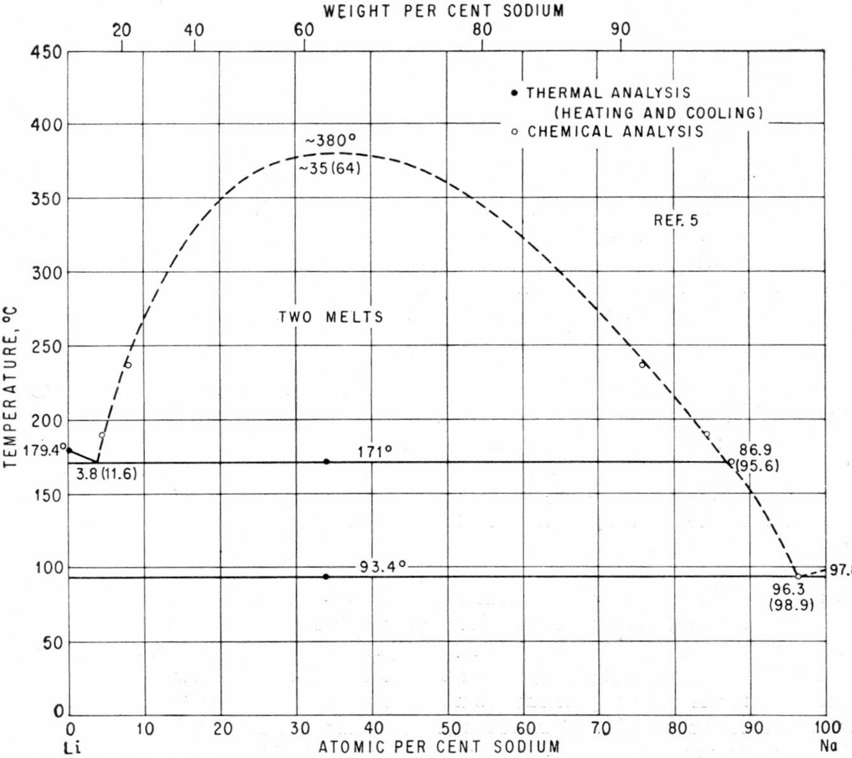

Fig. 496*a*. Li-Na. (See also Fig. 496.)

liquid state and a mutual depression of the freezing points of the elements. The data of [2] are plotted in Fig. 496; those of [3] locate the two horizontal lines at about 166 and 91°C, respectively [4].

Supplement. The Li-Na system was reinvestigated by [5] (Fig. 496*a*) by means of thermal analysis and chemical analysis of the immiscible liquids sampled at temperature. Thermodynamic principles were used to extrapolate the immiscibility loop beyond the points determined experimentally.

1. C. T. Heycock and F. H. Neville, *J. Chem. Soc.*, **55**, 1889, 675.
2. G. Masing and G. Tammann, *Z. anorg. Chem.*, **67**, 1910, 187–190.
3. B. Böhm and W. Klemm, *Z. anorg. Chem.*, **243**, 1939, 69–85.
4. In the paper of [3] the thermal data—of six alloys between 20 and 80 at. % Na—are plotted in a very small diagram.
5. O. N. Salmon and D. H. Ahmann, *J. Phys. Chem.*, **60**, 1956, 13–14.

$\bar{2}.5250$
1.4750

Li-Pb Lithium-Lead

A preliminary thermal and microscopic investigation of Pb-rich alloys (97.85–100 wt. % Pb) was carried out by [1]. The phase diagram shown in Fig. 497 is mainly based on careful thermal and thermoresistometric work (the latter only on alloys containing 40–100 at. % Pb) by [2]. Data for the Pb-rich eutectic, 83.7 at. % Pb, 231°C [1]; 83.0 at. % Pb, 235°C [2]; and 83.6 at. % Pb, 235.5°C [3], are in substantial agreement.

[2] concluded the existence of five intermediate phases (the first four on the basis of thermal evidence only): Li_4Pb (88.18 wt. % Pb), Li_7Pb_2 [22.22 at. (89.51 wt.) % Pb], Li_3Pb (90.87 wt. % Pb), Li_5Pb_2 [28.57 at. (92.27 wt.) % Pb], and LiPb (96.76 wt. % Pb). A systematic X-ray investigation of the Li-rich region would, therefore, be desirable, all the more since [4] claimed that only one compound—$Li_{10}Pb_3$ [23.08 at. (89.96 wt.) % Pb]—exists in the region occupied by Li_4Pb, Li_7Pb_2, and Li_3Pb in the phase diagram of [2]. As insufficient evidence is given by [4] to justify such a major change, Fig. 497 shows only the compound $Li_{10}Pb_3$ rather than Li_7Pb_2. (See Note Added in Proof.)

Solubility of Li in Solid Pb. The thermoresistometric and thermal measurements of [2] indicate the following solubilities: 3.2, 2.0, and 1.0 at. (0.11, 0.07 and 0.03 wt.) % Li at 235, 212, and 120°C, respectively. Hardness measurements on long-annealed (up to 25 days) and quenched alloys [3] yielded the data 2.1, 1.8, and 1.2 at. (0.07, 0.06, and 0.04 wt.) % Li at 200, 170, and 120°C, respectively (Fig. 497). [3] extrapolated their results (by means of the Schroeder–Le Chatelier equation) to the eutectic and to room temperature, finding the values 2.6 and 0.35 at. (0.09 and 0.01 wt.) % Li, respectively. In Fig. 497 the solubility at the eutectic temperature is assumed to be 3 at. (0.1 wt.) % Li.

The influence of Li on the hardness of Pb was also investigated by [5], that on the recrystallization of Pb by [6].

Crystal Structures. According to [4], the "Li-richest compound" $Li_{10}Pb_3$ has the cubic Cu_9Al_4 ($D8_3$) type of structure, with a = 10.102 A (see Note Added in Proof). The transformation in the (LiPb) phase, observed resistometrically (but not thermally) (Fig. 497) by [2], was further investigated by [7, 8], using mainly roentgenographic [7, 8], resistometric [7], and dilatometric [7] measurements. Both [7, 8] reported that the powder patterns are not changed in passing through the transformation temperature, which means that the arrangement of the Pb atoms is unaffected [9]. [8] assumes a primitive cubic unit cell, with a = 3.529 A [CsCl (B2) structure], but [7] a b.c.c. arrangement of the Pb atoms, with a = 5.025 A (45 at. % Pb) [10].

The conclusion of [8], however, is much better founded and, therefore, the transformation mechanism suggested by [7]—essentially a gradual rearrangement of the Li atoms—must be reconsidered.

For crystallochemical considerations on $Li_{10}Pb_3$ and/or (LiPb), see also [11–14].

Note Added in Proof. [15] studied alloys of compositions between Li_4Pb and Li_5Pb_2 by means of thermal and X-ray diffraction analysis. In agreement with [2] and in contrast to [4], four intermediate phases were found to exist in this composition range. Li_4Pb was tentatively indexed with a f.c.c. cell of the lattice parameter

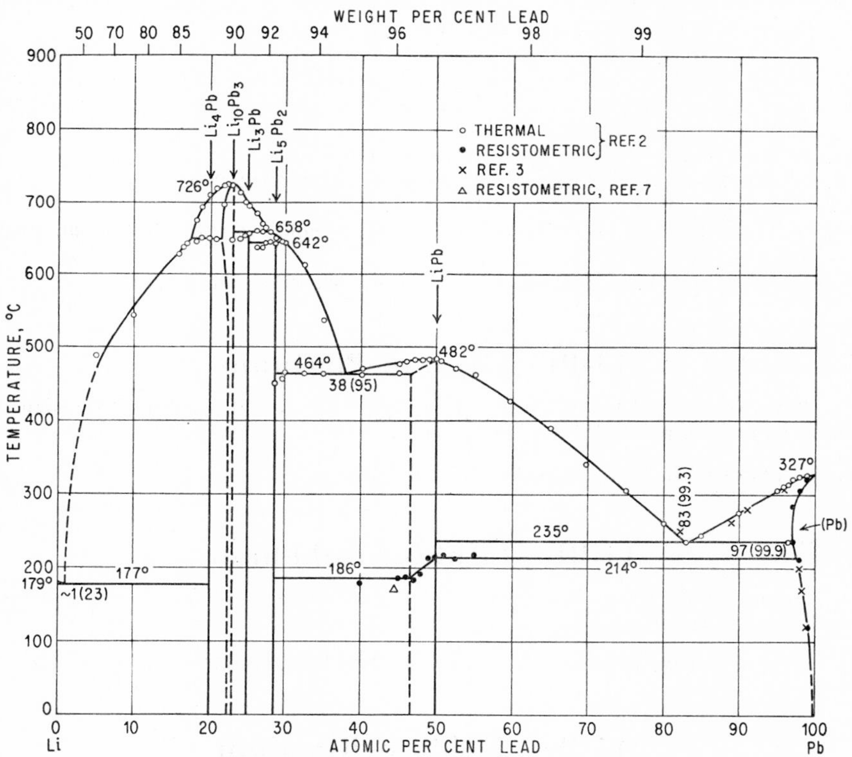

Fig. 497. Li-Pb. (See also Note Added in Proof.)

$a = 20.0$ A. The structure of Li_7Pb_2 was found to be similar to the hexagonal Na_3As ($D0_{18}$) type, with $a = 4.751 \pm 2$ A, $c = 8.589 \pm 4$ A, $c/a = 1.808$; and that of Li_3Pb was found to be of the f.c.c. Li_3Bi ($D0_3$) type, with $a = 6.687 \pm 3$ A. Powder patterns of alloys of about the composition Li_5Pb_2 "are moderately complex and were not indexed. Subsequent single-crystal analysis has shown this material to have a monoclinic cell with a probable formula Li_8Pb_3 [27.2 at. % Pb]. $Li_{10}Pb_3$ as described by [4] was not observed in any of the diffraction patterns obtained."

1. J. Czochralski and E. Rassow, *Z. Metallkunde*, **19**, 1927, 111–112.
2. G. Grube and H. Klaiber, *Z. Elektrochem.*, **40**, 1934, 745–754; the alloys were melted from the elements (Li: 99 wt. % pure) in iron crucibles in an argon atmosphere.

3. S. A. Pogodin and E. S. Shpichinetsky, *Izvest. Sektora Fiz.-Khim. Anal.*, **15**, 1947, 88–95 (in Russian).
4. M. A. Rollier and E. Arreghini, *Z. Krist.*, **101**, 1939, 470–482 (in Italian, summary in German).
5. K. v. Hanffstengel and H. Hanemann, *Z. Metallkunde*, **30**, 1938, 50–51.
6. W. Hofmann and H. Hanemann, *Z. Metallkunde*, **30**, 1938, 47–49.
7. T. C. Wilson, *J. Chem. Phys.*, **8**, 1940, 13–19.
8. H. Nowotny, *Z. Metallkunde*, **33**, 1941, 388.
9. With the relatively very small diffracting power of the Li atoms, their location cannot be detected by standard methods.
10. Both unit cells mentioned yield nearly the same set of powder lines. However, for the (321) b.c.c. reflection, for example, there is no equivalent in the primitive cubic pattern, and [8] actually did not observe that reflection. Unfortunately, [7] did not list his diffraction lines.
11. J. Gundermann, *Z. Metallkunde*, **34**, 1942, 120.
12. U. Dehlinger and H. Nowotny, *Z. Metallkunde*, **34**, 1942, 200.
13. K. Schubert, *Z. Metallkunde*, **39**, 1948, 91.
14. W. Hume-Rothery, J. O. Betterton, and J. Reynolds, *J. Inst. Metals*, **80**, 1952, 609–616.
15. A. Zalkin and W. J. Ramsey, *J. Phys. Chem.*, **60**, 1956, 234–236.

$\bar{2}.5508$
1.4492

Li-Pt Lithium-Platinum

"It was observed that Pt reacts with molten Li . . . at a temperature slightly above its melting point (186°C)" [1].

1. A. v. Grosse, *Z. Naturforsch.*, **8b**, 1953, 535 (footnote).

$\bar{2}.9095$
1.0905

Li-Rb Lithium-Rubidium

According to a thermal analysis by [1], Li and Rb do not alloy. Thermal arrests were found at the freezing-point temperatures of the pure metals.

1. B. Böhm and W. Klemm, *Z. anorg. Chem.*, **243**, 1939, 69–85, especially 71–72.

$\bar{2}.5711$
1.4289

Li-Re Lithium-Rhenium

The preparation of lithium rhenide (LiRe), probably as a hydrate, from aqueous solution in the form of a white crystalline solid is described by [1].

1. A. v. Grosse, *Z. Naturforsch.*, **8b**, 1953, 533–536 (in English).

$\bar{2}.7559$
1.2441

Li-Sb Lithium-Antimony

Li_3Sb (85.40 wt. % Sb), first prepared by [1], was shown [2] to occur in two modifications: up to at least 650°C the cubic BiF_3 ($D0_3$) type, with a = 6.572 A, while at higher temperatures the hexagonal Na_3As ($D0_{18}$) type of structure, with a = 4.710 A, c = 8.326 A, c/a = 1.768, is stable. According to [2], the melting point of this compound lies between 1150 and 1300°C; [1] had reported it to be somewhat higher than 950°C.

1. P. Lebeau, *Compt. rend.*, **134,** 1902, 231–233, 284–286.
2. G. Brauer and E. Zintl, *Z. physik. Chem.*, **B37,** 1937, 323–352.

$\bar{2}.9440$
1.0560

Li-Se Lithium-Selenium

Li_2Se (85.05 wt. % Se) and Li_2Te (90.19 wt. % Te) were shown to have the fluorite (C1) type of structure, with $a = 6.017$ A and $a = 6.517$ A, respectively [1].

1. E. Zintl, A. Harder, and B. Dauth, *Z. Elektrochem.*, **40,** 1934, 588–593.

$\bar{1}.3928$
0.6072

Li-Si Lithium-Silicon

The compound Li_3Si (57.43 wt. % Si) reported by [1, 2] could not be corroborated by [3], who found, by means of chemical and X-ray analysis, that the compounds Li_4Si (or $Li_{15}Si_4$ [4]) and Li_2Si (66.93 wt. % Si) exist. $Li_{15}Si_4$ may be isotypic with $Na_{15}Sn_4$ [4]. The diffusion of Li into silicon single crystals was measured by [5].

1. H. Moissan, *Compt. rend.*, **134,** 1902, 1083.
2. E. A. Boom, *Doklady Akad. Nauk S.S.S.R.*, **67,** 1949, 871–874 (in Russian). *Met. Abstr.*, **19,** 1951, 275.
3. W. Klemm and M. Struck, *Z. anorg. Chem.*, **278,** 1955, 117–121.
4. A. Weiss and A. Weiss, quoted in [3].
5. C. S. Fuller and J. A. Ditzenberger, *Phys. Rev.*, **91,** 1953, 193.

$\bar{2}.7669$
1.2331

Li-Sn Lithium-Tin

The constitution of the Li-Sn system was first investigated by [1] using thermal and microscopic analyses, and the existence of the compounds Li_4Sn, Li_3Sn_2, and Li_2Sn_5 was concluded. Thermal and roentgenographic analyses were carried out by [2], who found only four different powder patterns which were ascribed to Li, Li_4Sn, Li_3Sn_2, and Sn. It was assumed that the compound richest in Sn—"$LiSn_4$"—decomposes at lower temperatures. [3] reported, on the basis of X-ray powder work, that a 50 at. % Sn alloy was "obviously not homogeneous."

A careful thermal and thermoresistometric [4] analysis of 58 alloys [5] revealed, however, a rather more complicated constitution. According to Fig. 498, there exist six intermediate phases: Li_4Sn (81.04 wt. % Sn), Li_7Sn_2 [22.22 at. (83.01 wt.) % Sn], Li_5Sn_2 [28.57 at. (87.25 wt.) % Sn], Li_2Sn (89.53 wt. % Sn), LiSn (94.48 wt. % Sn), and $LiSn_2$ (97.16 wt. % Sn). For Li_7Sn_2 there is thermal, for LiSn resistometric, evidence for a restricted homogeneity range.

No primary solid solutions could be found by [2, 5]; [6] reported one of Sn to be <0.1 at. % Li at 200°C. The average pressure coefficients of electrical resistance to 12,000 kg per cm^2 for six Li-Sn alloys were determined by [7].

It would be valuable to corroborate the existence and the formulas of the six intermediate phases in Fig. 498 by X-ray analysis.

1. G. Masing and G. Tammann, *Z. anorg. Chem.*, **67,** 1910, 190–194.
2. A. Baroni, *Rend. accad. Lincei, Roma,* (6)**16,** 1932, 153–158.
3. E. Zintl and G. Brauer, *Z. physik. Chem.*, **B20,** 1933, 245–271.
4. Thermoresistometric analysis only from 34 to 100 at. % Sn.
5. G. Grube and E. Meyer, *Z. Elektrochem.*, **40,** 1934, 771–777. The alloys were melted in iron crucibles in an argon atmosphere. Most of the alloys were analyzed after the thermal analysis.

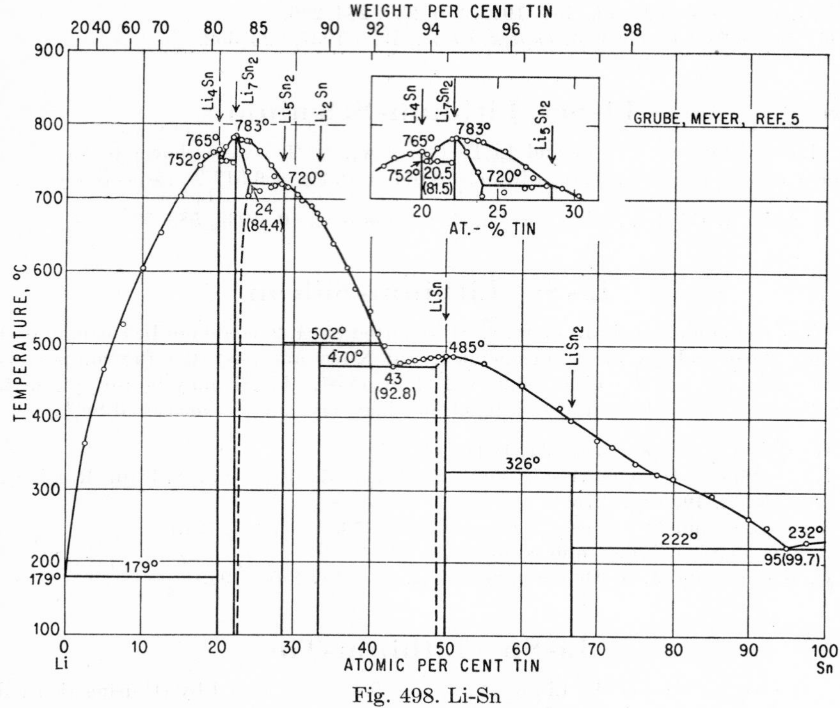

Fig. 498. Li-Sn

6. E. Jenckel and L. Roth, *Z. Metallkunde*, **30**, 1938, 135–144.
7. C. W. Ufford, *Phys. Rev.*, **32**, 1928, 505–507.

$\bar{2}.7355$
1.2645

Li-Te Lithium-Tellurium

See Li-Se.

$\bar{2}.5309$
1.4691

Li-Tl Lithium-Thallium

The phase diagram of Fig. 499 was established by means of careful thermal and thermoresistometric analyses [1]. There exist the compounds Li_4Tl (88.04 wt. % Tl), Li_3Tl (90.76 wt. % Tl), Li_5Tl_2 [28.57 at. (92.18 wt.) % Tl], Li_2Tl (93.64 wt.% Tl), and LiTl (96.72 wt. % Tl). LiTl, which has an appreciable homogeneity range on its Li-rich side, was shown [2] to have the CsCl (B2) type of structure, with a = 3.431 ± 3 A (50.3 at. % Tl).

1. G. Grube and G. Schaufler, *Z. Elektrochem.*, **40**, 1934, 593–600. The alloys were melted in iron crucibles in an argon atmosphere and analyzed for both Li and Tl contents.
2. E. Zintl and G. Brauer, *Z. physik. Chem.*, **B20**, 1933, 245–271.

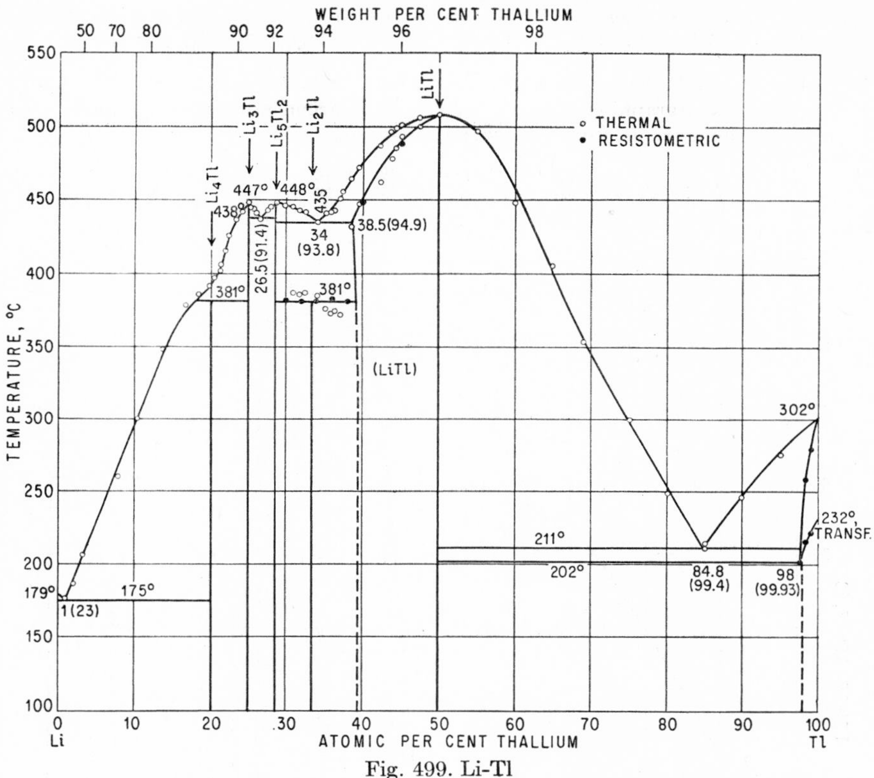

Fig. 499. Li-Tl

$\bar{1}.0259$
0.9741

Li-Zn Lithium-Zinc

The data plotted in Fig. 500 are due to a thermal and thermoresistometric analysis by [1]. According to these authors (and their nomenclature) there exist the eight intermediate phases δ, δ'', δ', $LiZn_2$, γ, γ', β, and β'. The phase-field limits in Fig. 500 are mainly those of [1]; at lower temperatures a few changes [2] and supplements [3] have been made to accommodate the roentgenographic results of [4] which confirm, however, the existence of five intermediate room-temperature phases [δ'' (LiZn), δ' (Li_2Zn_3), $LiZn_2$, γ' (Li_2Zn_5), and β'].

The investigations of [5] (thermal and microscopic, from 50 to 100 at. % Zn) and [6] (thermal and roentgenographic) are, for the most part, obsolete [7]. For the Zn-rich eutectic the data 94.5 at. % Zn, 403°C [5]; 95 at. % Zn, 403°C [1]; and 96.5 at. % Zn (0.38 wt. % Li), 403.25°C [8] were reported; for Fig. 500, the values 96 at. % Zn and 403°C were chosen. At room temperature the primary solid solutions are of negligible extent [1, 6, 4].

Crystal Structures. δ'' (LiZn) has the NaTl (B32) type of structure [9, 4], with $a = 6.221 \pm 5$ A (51.3 at. % Zn) [9]. The structures of the δ' and $LiZn_2$ (94.96 wt. % Zn) phases could not be elucidated from powder patterns [4]; however, it was claimed by [6] that Li_2Zn_3 ($=\delta'$?) has very probably a cubic structure ($a = 4.27$ A).

For γ' a hexagonal pseudo cell with $a = 4.371$ A, $c = 2.515$ A, $c/a = 0.575$

(71.9 at. % Zn) was suggested by [4]. β′ has a h.c.p. structure with statistical atom distribution and a = 2.788 A, c = 4.394 A, c/a = 1.576 (89.5 at. % Zn) [4].

1. G. Grube and H. Vosskühler, *Z. anorg. Chem.*, **215**, 1933, 211–224. The alloys were melted in graphite or iron crucibles in an argon atmosphere. Alloys used in the resistometric measurements were analyzed.

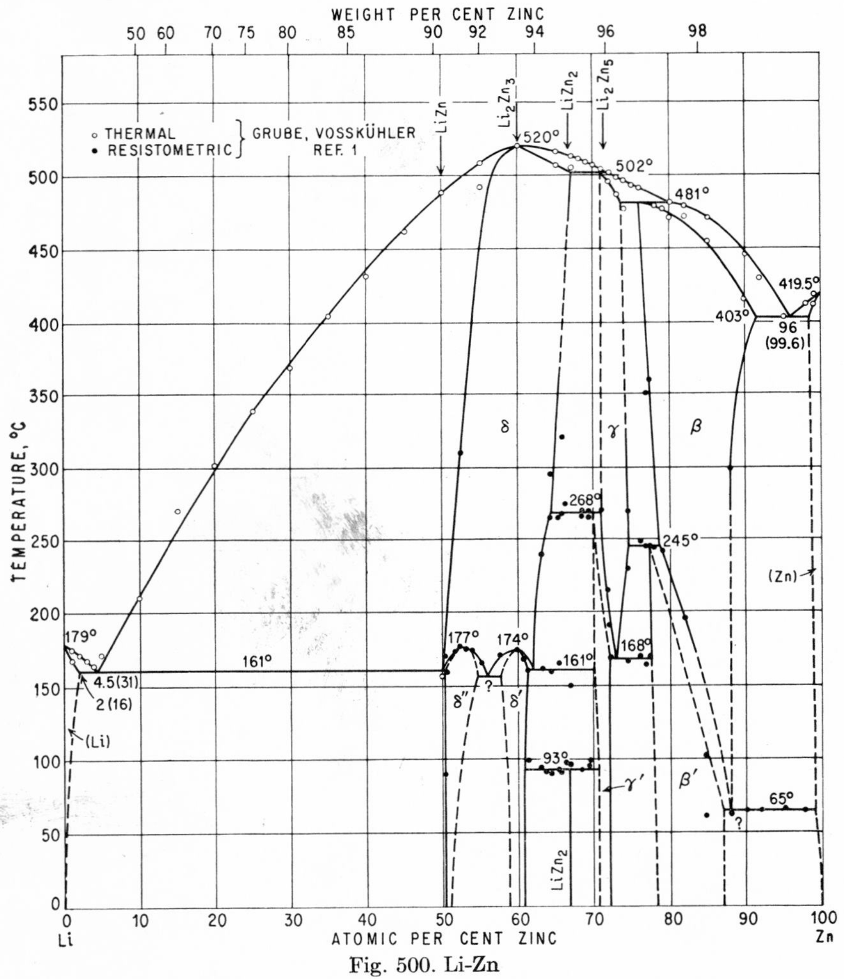

Fig. 500. Li-Zn

2. The β′-phase field of [1] extends between about 78 and 84 at. % Zn at room temperature; however, [4] observed at 89.5 at. % Zn an X-ray pattern containing only β′ lines. In Fig. 500 the Zn-rich β′ limit was drawn at 87 at. % Zn.
3. In the diagram of [1] the boundaries of δ″ and δ′ phases are not extended to lower temperatures. Alloys with 51, 54, and 58 at. % Zn were found by [4] to be two-phase at room temperature.
4. A. Zintl and A. Schneider, *Z. Elektrochem.*, **41**, 1935, 764–767.

5. W. Fraenkel and R. Hahn, *Metallwirtschaft,* **10,** 1931, 641–642.
6. A. Baroni, *Congr. intern. quim. pura y apl., 9th Congr. Madrid,* **2,** 1934; abstract, *Strukturbericht,* **3,** 1933-1935, 633–634.
7. [5] reported one intermediate phase, (Li_2Zn_3), with a wide range of homogeneity, and [6] two compounds, Li_2Zn_3 and $LiZn_4$, both of strictly stoichiometric composition.
8. E. Weisse, A. Blumenthal, and H. Hanemann, *Z. Metallkunde,* **34,** 1942, 221; see also *Techn. Berichte Zinkberatungsstelle,* **1948,** no. 3, 1–13.
9. E. Zintl and G. Brauer, *Z. physik. Chem.,* **B20,** 1933, 251.

$\bar{1}.6462$
0.3538

Mg-Mn Magnesium-Manganese

Solidification. It has been conclusively established that there is a peritectic equilibrium at the Mg-rich end of the system [1–5], rather than a eutectic as was assumed earlier [6–9].

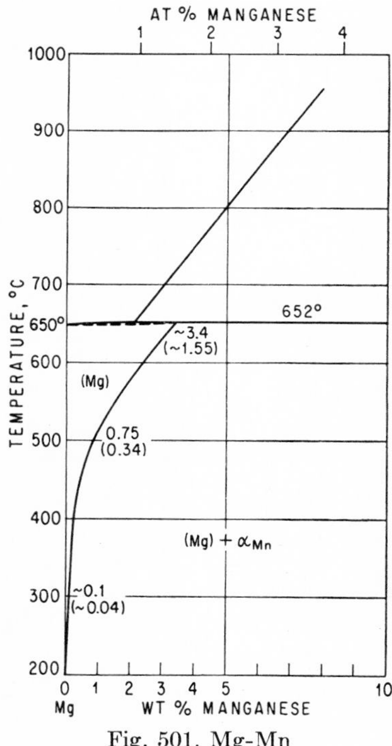

Fig. 501. Mg-Mn

The solubility of Mn in liquid Mg was determined by chemical analysis of samples taken from melts equilibrated at various temperatures [1–5]. The liquidus curve shown in Fig. 501 (in wt. %) is that reported by [4], whose data are very consistent and, in addition, average the slightly scattered data of the other investigators. The composition of the peritectic melt is 2.05 ± 0.05 wt. (about 0.92 ± 0.02 at.) % Mn [1–4]. The peritectic temperature was determined as 651°C (based on a melting point of Mg of 649.5°C) [1] and 653 ± 0.5°C [5].

The primary phase crystallizing from melts with more than 2.05 wt. % Mn was recognized roentgenographically as α-Mn [1, 3, 10, 11] or β-Mn [12, 5], depending on the composition and the rate of cooling from the molten state [13]. The existence of an intermediate phase (Mg_9Mn), assumed to be formed peritectically at 726°C [7], could not be confirmed [1–5, 13]. The occurrence of a miscibility gap in the liquid state at temperatures around 1300°C [8, 9] was due to heavy contamination with iron introduced by reaction with the iron container.

Solid Solubility. The existence of Mg-rich solid solutions, recognized by [14–17], was established by determining the solubility boundary, using the lattice-parametric method [6]. It was located as follows: 3.1 wt. (1.4 at.) % at 635°C, 2.35 (1.05) at 600°C, 1.5 (0.67) at 550°C, 0.75 (0.34) at 500°C, 0.25 (0.11) at 400°C, and about 0.1 (0.04) at 300°C [6]. According to [1], the solubility is smaller: about 2.45 wt. (1.1 at.) % at 651°C, 2.06 (0.92) at 620°C, 1.0 (0.45) at 540°C, and 0.25 (0.11) at 445°C. However, the data of [6] deserve preference since in this case the X-ray method appears to be more reliable than the metallographic method [1]. For additional (less important) data, see [8, 9].

The *a* as well as the *c* axis of Mg decreases by dissolving Mn [6, 18].

1. J. D. Grogan and J. L. Haughton, *J. Inst. Metals,* **69,** 1943, 241–248.
2. A. Beerwald, *Metallwirtschaft,* **23,** 1944, 404–407.
3. N. Tiner, *Trans. AIME,* **161,** 1945, 351–359.
4. G. Siebel, *Z. Metallkunde,* **39,** 1948, 22–27.
5. A. Schneider and H. Stobbe-Scholder, *Metall,* **4,** 1950, 178–183; for data, see also [4].
6. E. Schmid and G. Siebel, *Metallwirtschaft,* **10,** 1931, 923–925; *Z. Elektrochem.,* **37,** 1931, 455–459.
7. H. Sawamoto, *Suiyokwai Shi,* **8,** 1935, 763–768.
8. M. Goto, M. Nito, and H. Asada, *Rept. Aeronaut. Research Inst. Tokyo Imp. Univ.,* **12,** 1937, 163–318.
9. S. Ishida, *Rept. Aeronaut. Research Inst. Tokyo Imp. Univ.,* no. 280, 1944, pp. 397–405.
10. R. D. Heidenreich, L. Sturkey, and H. L. Woods, *J. Appl. Phys.,* **17,** 1946, 127–136.
11. W. Bulian and E. Fahrenhorst, "Metallographie des Magnesiums und seiner technischen Legierungen," 2d ed., p. 20, Springer-Verlag OHG, Berlin, 1949.
12. E. F. Bachmetev and J. M. Golovchinev, *Acta Physicochim. U.R.S.S.,* **2,** 1935, 571–574; *Zhur. Fiz. Khim.,* **6,** 1935, 597–600.
13. H. Nowotny, *Z. Metallkunde,* **37,** 1946, 130–136.
14. J. A. Gann, *Trans. AIME,* **83,** 1929, 309–332.
15. H. E. Bakken and R. T. Wood, "American Society for Steel Treating Handbook," p. 560, The American Society for Steel Treating, Cleveland, Ohio, 1929.
16. G. W. Pearson, *Ind. Eng. Chem.,* **22,** 1930, 367–370.
17. W. Mannchen, *Z. Metallkunde,* **23,** 1931, 193–196.
18. R. S. Busk, *Trans. AIME,* **188,** 1950, 1460–1464.

$\bar{1}.4039$
0.5961

Mg-Mo Magnesium-Molybdenum

Mo does not alloy with Mg [1]. On arc melting of Mo with Mg additions, only traces of Mg were recoverable [2]. Mg is ineffective as a deoxidizer of Mo [2].

1. F. Sauerwald, *Z. anorg. Chem.,* **258,** 1949, 296–300.
2. Climax Molybdenum Co., Arc-cast Molybdenum-base Alloys, First and Second Annual Reports (NR 031-331), 1950 and 1951.

0.2396
$\bar{1}.7604$

Mg-N Magnesium-Nitrogen

Mg_3N_2 (27.44 wt. % N) is cubic (16 Mg_3N_2 per unit cell) of the Mn_2O_3 ($D5_3$) type, a = 9.95 A [1]; a = 9.97 A [2]. As to the reaction of N_2 and NH_3 with Mg, see the chemical handbooks.

1. G. Hägg, *Z. Krist.*, **74**, 1930, 95–99; **82**, 1932, 470–472.
2. E. Zintl, reported by M. v. Stackelberg and R. Paulus, *Z. physik. Chem.*, **B22**, 1933, 305–322.

0.0243
$\bar{1}.9757$

Mg-Na Magnesium-Sodium

The phase diagram in Fig. 502 is due to [1]. The composition of the Na-rich melt, which coexists at 638°C with the Mg-rich melt containing about 2 wt. (2.1 at.)

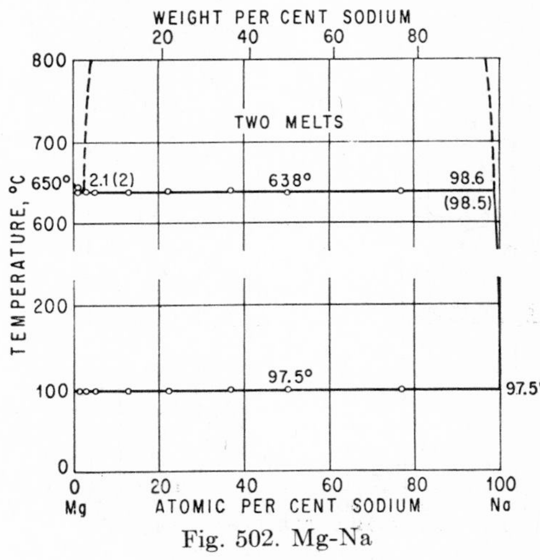

Fig. 502. Mg-Na

% Na, was determined by chemical analysis of the solid Na-rich layer to be approximately 98.5 wt. (98.6 at.) % Na.

1. C. H. Mathewson, *Z. anorg. Chem.*, **48**, 1906, 193–195. As alloys were prepared in glass tubes, there was some reaction of the melts with the container material.

$\bar{1}.6174$
0.3826

Mg-Ni Magnesium-Nickel

Figure 503 shows the thermal-analysis data points due to [1, 2]. In the composition range 0–34 at. (55.4 wt.) % Ni covered by both investigators, the results of [2] are to be regarded as more accurate. The phase boundaries in the Ni-rich portion of the system are only outlined. Since the liquidus temperature in the region 70–85 wt. (49.2–70.1 at.) % Ni was found to be nearly constant at 1142 ± 3°C, [1] concluded that a miscibility gap existed and that the two melts reacted to form the phase $MgNi_2$ (82.84 wt. % Ni). The formation of layers, however, was not observed by [1], nor by later workers who prepared alloys of the composition $MgNi_2$. The melting point of $MgNi_2$ was found by [3] as about 1143°C, in accord with that reported by [1], 1145°C. The temperature of the $MgNi_2$-Ni eutectic, found as 1082°C by [1], was determined more accurately as 1095°C [4].

The solid solubility of Ni in Mg was reported to be below 0.25 wt. % Ni [5] and more accurately as less than 0.1 wt. (0.04 at.) % Ni at 500°C [2]. The *a* and *c* parameters of Mg are slightly affected by Ni additions [6]. The solid solubility of Mg in Ni is also less than 0.1 wt. (0.24 at.) % Mg, after annealing at 1100°C [4].

Crystal Structures. Mg_2Ni (54.68 wt. % Ni) has a hexagonal structure of a new type which is related to that of $CuAl_2$ (C16), 18 atoms per unit cell, a = 5.19 A, c = 13.22 A, c/a = 2.547 [7]. [8] had erroneously reported the existence of Mg_3Ni,

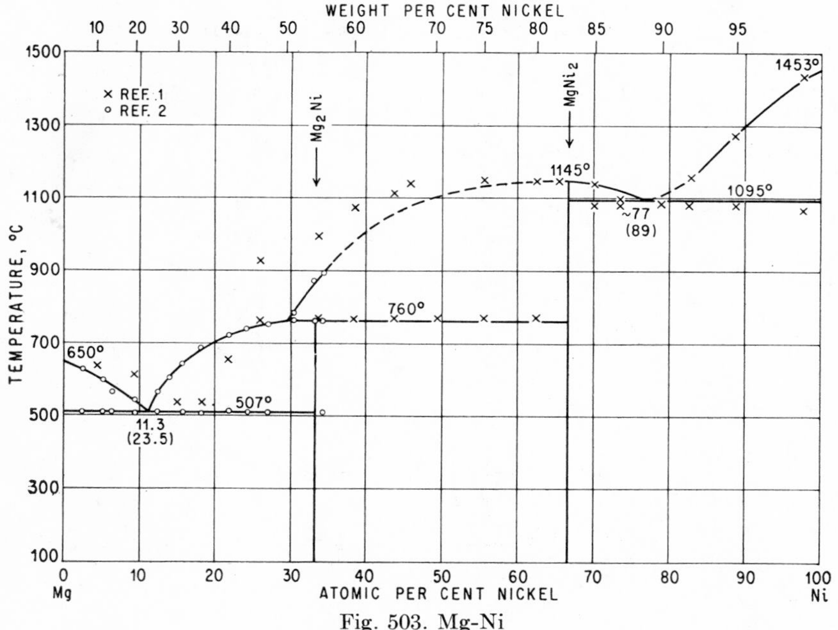

Fig. 503. Mg-Ni

with a = 5.27 A, c = 13.3₃ A, c/a = 2.529. Actually he had determined the lattice constants of Mg_2Ni.

The structure of $MgNi_2$ is the prototype of the C36 type of structure, a = 4.815 A, c = 15.80 A, c/a = 3.281 [9]; a = 4.812 A, c = 15.85 A, c/a = 3.294 [10].

1. G. Voss, *Z. anorg. Chem.*, **57**, 1908, 61–67. The nickel used contained 1.86% Co, 0.47% Fe.
2. J. L. Haughton and R. J. M. Payne, *J. Inst. Metals*, **54**, 1934, 275–283.
3. K. H. Lieser and H. Witte, *Z. Metallkunde*, **43**, 1952, 396–401.
4. P. D. Merica and R. G. Waltenberg, *Natl. Bur. Standards* (*U.S.*), *Technol. Paper* 281, 1925.
5. J. A. Gann, *Trans. AIME*, **83**, 1929, 309–332.
6. R. S. Busk, *Trans. AIME*, **188**, 1950, 1460–1464.
7. K. Schubert and K. Anderko, *Z. Metallkunde*, **42**, 1951, 321–325.
8. E. F. Bachmetev, *Acta Physicochim. U.R.S.S.*, **2**, 1935, 567–570; *Zhur. Fiz. Khim.*, **6**, 1935, 590–592.

9. F. Laves and H. Witte, *Metallwirtschaft,* **14,** 1935, 645–649.
10. E. F. Bachmetev, *Metallwirtschaft,* **14,** 1935, 1001–1002; *Acta Physicochim. U.R.S.S.,* **3,** 1935, 531; see also [8].

0.1818
$\bar{1}$.8182

Mg-O Magnesium-Oxygen

MgO (39.68 wt. % O) is cubic of the NaCl (B1) type. The lattice constant, determined by numerous investigators [1], is a = 4.208 A [2]. As to an electron-diffraction study of the thermal oxidation of Mg at room temperature, see [3].

Supplement. [4] redetermined the lattice constant of MgO: 4.213 A (25°C).

1. *Strukturbericht,* **1,** 1913-1928, 117–118; "Gmelins Handbuch der anorganischen Chemie," 8th ed., System No. 27 (B), pp. 17–19, Verlag Chemie, G.m.b.H., Berlin, 1939.
2. A. H. Jay and K. W. Andrews, *J. Iron Steel Inst.,* **152,** 1946, 15–18; *Nature,* **154,** 1944, 116.
3. L. de Brouckère, *J. Inst. Metals,* **71,** 1945, 131–147, 603–611.
4. H. E. Swanson and E. Tatge, *Natl. Bur. Standards (U.S.), Circ.* 539, I, 1953, 37.

$\bar{1}$.1068
0.8932

Mg-Os Magnesium-Osmium

It was found that Os readily alloys with Mg and that a Mg-rich eutectic exists [1].

1. F. Sauerwald, *Z. anorg. Chem.,* **258,** 1949, 296–300.

$\bar{1}$.8950
0.1050

Mg-P Magnesium-Phosphorus

Mg_3P_2 (45.92 wt. % P) is cubic, with 16 formula weights per unit cell, and isotypic with Mn_2O_3 ($D5_3$ type) [1, 2], a = 12.03 A [2]. Earlier work by [3] is obsolete.

1. E. Zintl and E. Husemann, *Z. physik. Chem.,* **B21,** 1933, 138–155.
2. M. v. Stackelberg and R. Paulus, *Z. physik. Chem.,* **B22,** 1933, 305–322.
3. L. Passerini, *Gazz. chim. ital.,* **58,** 1928, 655–664.

$\bar{1}$.0696
0.9304

Mg-Pb Magnesium-Lead

Liquidus temperatures covering the whole composition range were determined by [1, 2], and for the range 0–21.7 at. (0–70.2 wt.) % Pb by [3, 4]. [5] investigated the effect of small additions of Mg on the melting point of Pb. Data points by [1–4] are given in Fig. 504.

The temperature of the (Mg)-Mg_2Pb eutectic was found as 459 [1], 475 [2], 470 [6], 466 [3], and 465°C [4] and its composition as 19.2 at. (67 wt.) % Pb [1], 20 at. (68.1 wt.) % Pb [2], and 19.1 at. (66.8 wt.) % Pb [3, 4]. The temperature of the Mg_2Pb-(Pb) eutectic was reported as about 247 [1], 253 [2], and 249°C [6], and its composition as about 79 at. (97 wt.) % Pb [1] and, more accurately, as 84.3 at. (97.8 wt.) % Pb [2].

The solid solubility of Pb in Mg, already indicated by the work of [7, 8], was thoroughly determined for the temperature ranges above 100 [3], 150 [9], and 200°C [4], using electrical conductivity vs. composition [3], micrographic [4], and lattice-parametric methods [9]. At temperatures between 250 and 400°C, the data by [3, 4] are in excellent agreement; they differ by only 0.1 at. %. Those of [9] are higher by less than about 0.5 at. %. Above 400°C, the values differ more, as indicated by the solubilities at the eutectic temperature: 6.67 at. (37.8 wt.) [9], 7.75 at. (41.7 wt.) [4], and 9.1 at. (46.2 wt.) % Pb [3]. The value of [4] appears to be the most accurate one. Results by [3] and [9] are in excellent agreement between 150 and 250°C. The solubility at 200°C was given as 0.5 at. (4.1 wt.) [4], 0.9 at. (7.2 wt.) [3], and 0.95 at. (7.55 wt.) % Pb [9].

The solidus curve of the (Mg) phase was determined by means of micrographic analysis [4] and thermoresistometric studies [3]. The former (Fig. 504) represents equilibrium conditions.

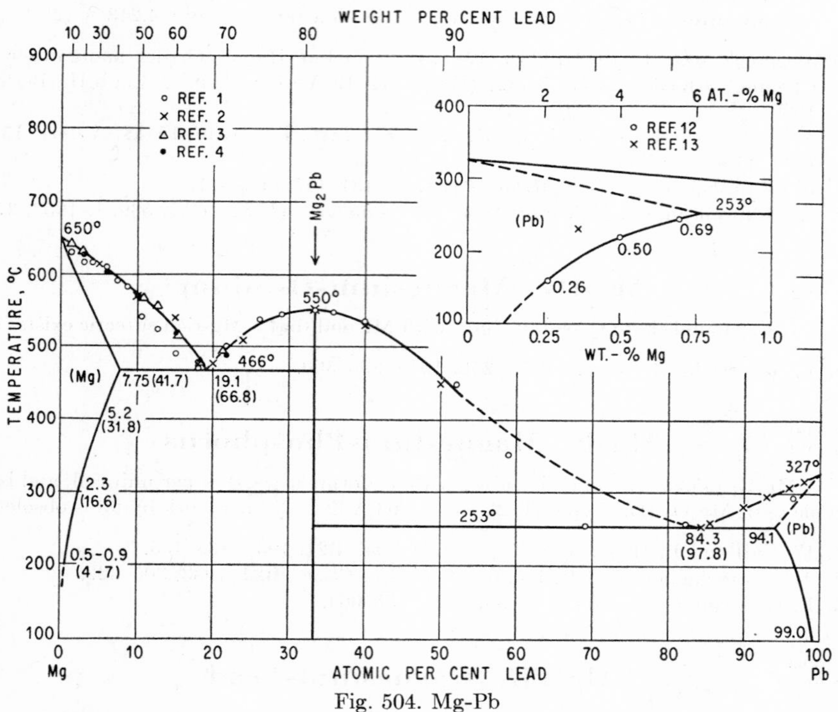

Fig. 504. Mg-Pb

The existence of Pb-rich solid solutions was recognized by [10, 11]. [12] reported solubility values based on measurements of hardness. Their findings are shown in the inset of Fig. 504 in wt. % Mg, together with a value [0.36 wt. (3.0 at.) % Mg] reported by [13], based on measurements of resistance after annealing at 232°C for 125 days.

Mg_2Pb is to be regarded as a phase of practically fixed composition. Its homogeneity range, if any, is at most a few tenths of 1 at. % [14]. On the basis of measurements of hardness, [23] claimed a wider homogeneity range.

As to the boiling points of melts in the range 7–32 at. % Pb, see [15, 16].

Crystal Structure. Mg_2Pb is isotypic with CaF_2 (C1 type) [17–19], a = 6.81–6.86 A [17–19, 14, 20, 21]. Lattice spacings of the Mg-rich solid solution were reported by [9, 22].

1. G. Grube, *Z. anorg. Chem.*, **44,** 1905, 117–130.
2. N. S. Kurnakow and N. J. Stepanow, *Z. anorg. Chem.*, **46,** 1905, 177–192.

3. H. Vosskühler, *Z. Metallkunde*, **31,** 1939, 109–111.
4. G. V. Raynor, *J. Inst. Metals*, **66,** 1940, 403–426.
5. C. T. Heycock and F. H. Neville, *J. Chem. Soc.*, **61,** 1892, 904–905.
6. E. Abel, O. Redlich, and F. Spausta, *Z. anorg. Chem.*, **190,** 1930, 82.
7. N. J. Stepanow, *Z. anorg. Chem.*, **60,** 1908, 209–229; **78,** 1912, 11–13.
8. M. Hansen; see W. Schmidt, *Z. Metallkunde*, **19,** 1927, 455.
9. F. Foote and E. R. Jette, *Trans. AIME*, **143,** 1941, 124–131.
10. D. Stenquist, *Z. Metallkunde*, **13,** 1921, 245.
11. J. Goebel, *Z. Metallkunde*, **14,** 1922, 360–361.
12. N. S. Kurnakow, S. A. Pogodin, and T. A. Vidusova, *Izvest. Inst. Fiz.-Khim. Anal.*, **6,** 1933, 266–267; *Izvest. Sektora Fiz.-Khim. Anal.*, **15,** 1947, 74–79.
13. A. Pasternak, *Bull. intern. acad. polon. sci., Classe sci. math. nat.*, **A1951,** 177–192.
14. W. Klemm and H. Westlinning, *Z. anorg. Chem.*, **245,** 1940, 365–380.
15. W. Leitgebel, *Z. anorg. Chem.*, **202,** 1931, 305–324.
16. A. Schneider and U. Esch, *Z. Elektrochem.*, **45,** 1939, 888–893.
17. A. Sacklowski, *Ann. Physik*, **77,** 1925, 264–271.
18. J. B. Friauf, *J. Am. Chem. Soc.*, **48,** 1926, 1906–1909.
19. E. Zintl and H. Kaiser, *Z. anorg. Chem.*, **211,** 1933, 113–131.
20. A. H. Geisler, C. S. Barrett, and R. F. Mehl, *Trans. AIME*, **152,** 1943, 201–223.
21. G. Brauer and J. Tiesler, *Z. anorg. Chem.*, **262,** 1950, 319–327.
22. G. V. Raynor, *Proc. Roy. Soc. (London)*, **A180,** 1942, 107–121.
23. S. A. Pogodin, L. M. Kefeli, and E. S. Berkavich, *Izvest. Sektora Fiz.-Khim. Anal.*, **17,** 1949, 193–199.

$\bar{1}.3578$
0.6422

Mg-Pd Magnesium-Palladium

Pd was reported to increase very slightly the *a* parameter value of Mg, although the atomic radius of Pd is less than that of Mg. The *c/a* ratio of Mg is slightly decreased by Pd [1].

1. R. S. Busk, *Trans. AIME*, **188,** 1950, 1460–1464.

$\bar{1}.2370$
0.7630

Mg-Pr Magnesium-Praseodymium

From thermal-analysis data and microscopic examination of alloys cooled from the molten state, essentially the following conclusions had been drawn [1]: (*a*) There are three intermediate phases, Mg_3Pr (65.89 wt. % Pr), which forms a eutectic with Mg at about 5 at. (23.5 wt.) % Pr, 593°C; MgPr (85.28 wt. % Pr); and $MgPr_4$ (95.86 wt. % Pr). (*b*) Mg_3Pr and MgPr form a series of solid solutions. A continuous transition between these two phases, however, is impossible; besides, alloys in this range of composition showed two-phase structures. (*c*) $MgPr_4$ (or $MgPr_3$) decomposes into MgPr and Pr at 528°C.

Later, it was reported [2] that—in analogy with the systems Ce-Mg and La-Mg—the compound Mg_9Pr (39.17 wt. % Pr) exists. It was found to be formed by the peritectic reaction liq. (approx. 30 wt. % Pr) + $Mg_3Pr \rightarrow Mg_9Pr$ at 635°C.

A more detailed thermal and microscopic investigation of the systems Ce-Mg and La-Mg [3] established the phase relationships presented in Figs. 262 and 486 with the phase Mg_2X existing. As there is very close resemblance in the constitution of the alloy systems of Ce, La, and Pr with other metals [2], the phase diagram of the

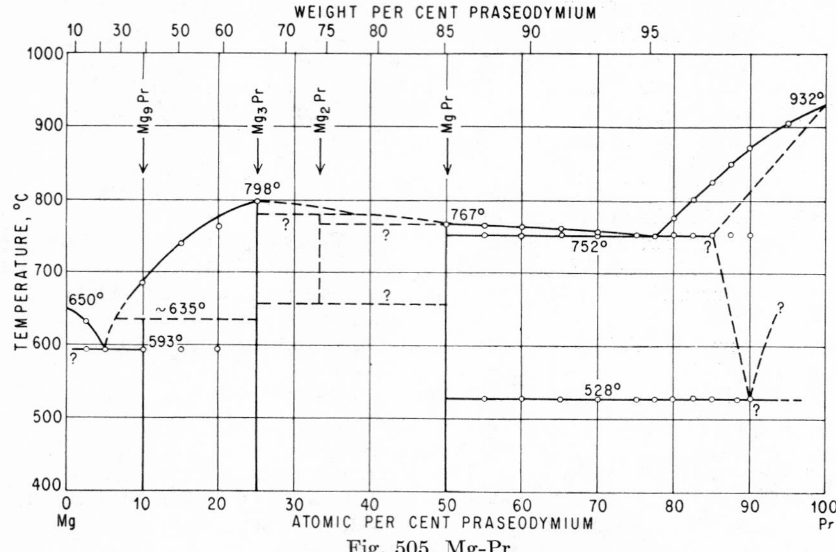

Fig. 505. Mg-Pr

system Mg-Pr will be very similar to those shown in Figs. 262 and 486. The phase relations between Mg_3Pr and Pr given in Fig. 505 are drawn schematically.

Mg_3Pr has a cubic structure of the BiF_3 (DO_3) type, a = 7.39 A [4]. The structure of MgPr is b.c.c. of the CsCl (B2) type, a = 3.89 A [5].

1. G. Canneri, *Metallurgia ital.*, **25,** 1933, 250–252.
2. R. Vogel and T. Heumann, *Z. Metallkunde,* **35,** 1943, 29–42.
3. R. Vogel and T. Heumann, *Z. Metallkunde,* **38,** 1947, 1–8.
4. A. Rossi and A. Iandelli, *Atti reale accad. Lincei,* **19,** 1934, 415–420.
5. A. Rossi and A. Iandelli, *Atti reale accad. Lincei,* **18,** 1933, 156–161.

$\bar{1}.0954$
0.9046

Mg-Pt Magnesium-Platinum

When Mg vapor in a stream of hydrogen reacted with Pt till a constant weight was attained, a product of the approximate composition Mg_2Pt (80.06 wt. % Pt) was obtained, the homogeneity of which, however, was not proved [1]. Pt was claimed to increase the c/a ratio of Mg very slightly [2].

1. W. R. Hodgkinson, R. Waring, and A. P. H. Desborough, *Chem. News,* **80,** 1899, 185.
2. R. S. Busk, *Trans. AIME,* **188,** 1950, 1460–1464.

$\bar{1}.0076$
0.9924

Mg-Pu Magnesium-Plutonium

Intermediate phases of the approximate compositions Pu_2Mg and $PuMg_2$ "were formed under conditions such that knowledge of their compositions could be deduced only from their X-ray diffraction patterns. The apparent fluorite [C1 type] structure of Pu_2Mg seems to establish its composition rather firmly, but the composition of $PuMg_2$ must be regarded as questionable" [1]. The lattice parameter of Pu_2Mg

is $a = 7.34 \pm 1$ A. The pattern of $PuMg_2$ (?) was tentatively indexed with a hexagonal cell, $a = 13.8$ A, $c = 9.7$ A [1].

1. A. S. Coffinberry and M. B. Waldron, Review of Physical Metallurgy of Plutonium, in "Metallurgy and Fuels," Progress in Nuclear Energy, ser. V, vol. 1, Pergamon Press Ltd., London, 1956.

$\bar{1}.3735$
0.6265

Mg-Rh Magnesium-Rhodium

Rh was found to decrease very slightly the *a* parameter and to increase the axial ratio of Mg [1].

1. R. S. Busk, *Trans. AIME*, **188**, 1950, 1460–1464.

$\bar{1}.8799$
0.1201

Mg-S Magnesium-Sulfur

MgS (56.87 wt. % S) is isotypic with NaCl (B1 type), $a = 5.09$ A [1]; $a = 5.15$ A [2]; $a = 5.200$ A [3]; $a = 5.1913$ A [4].

1. S. Holgersson, *Z. anorg. Chem.*, **126**, 1923, 179–182.
2. S. Holgersson, *Lunds Univ. Årsskr.*, **23**, 1927, no. 9, 17.
3. E. Broch, *Z. physik. Chem.*, **127**, 1927, 446–454.
4. W. Primak, H. Kaufmann, and R. Ward, *J. Am. Chem. Soc.*, **70**, 1948, 2043–2046.

$\bar{1}.3005$
0.6995

Mg-Sb Magnesium-Antimony

The first phase diagram, outlined by thermal analysis, showed the phase Mg_3Sb_2 (76.95 wt. % Sb) with a melting point of 961°C [1]. The existence of a high-temperature modification, having a melting point of 1228°C, was overlooked. Figure 506 represents the revised diagram, based on thermal-analysis data [2]. The two eutectic temperatures had been reported as about 627 [1], 622 [3], and 631°C [4], and about 594 [1] and 591°C [3].

The solid solubility of Sb in Mg was given as less than 0.1 wt. (0.02 at.) % up to 500°C and less than 0.2 wt. (0.04 at.) % Sb at 550°C, based on microexamination of annealed alloys [4]. [5] concluded from lattice-parameter measurements that the solubility is "vanishingly small."

The boiling-point curve tends to approach a maximum at about 1800°C at the composition Mg_3Sb_2, indicating that this phase does not dissociate in the melt [6].

α-Mg_3Sb_2 is isotypic with La_2O_3, hexagonal with 5 atoms per unit cell ($D5_2$ type), $a = 4.582$ A, $c = 7.244$ A, $c/a = 1.581$ [7]; see also [8]. β-Mg_3Sb_2 was suggested to have the cubic structure of Mn_2O_3 ($D5_3$ type) [9].

1. G. Grube, *Z. anorg. Chem.*, **49**, 1906, 87–91.
2. G. Grube and R. Bornhak, *Z. Elektrochem.*, **40**, 1934, 140–142.
3. E. Abel, O. Redlich, and F. Spausta, *Z. anorg. Chem.*, **190**, 1930, 81.
4. W. R. D. Jones and L. Powell, *J. Inst. Metals*, **67**, 1941, 177–188.
5. R. S. Busk, *Trans. AIME*, **188**, 1950, 1460–1464.
6. W. Leitgebel, *Z. Elektrochem.*, **43**, 1937, 509–518; *Z. anorg. Chem.*, **202**, 1931, 305–324.
7. E. Zintl and E. Husemann, *Z. physik. Chem.*, **B21**, 1933, 138–155.
8. T. A. Kontorova, *Zhur. Tekh. Fiz.*, **18**, 1948, 1478–1484.
9. E. Zintl, *Z. Elektrochem.*, **40**, 1934, 142.

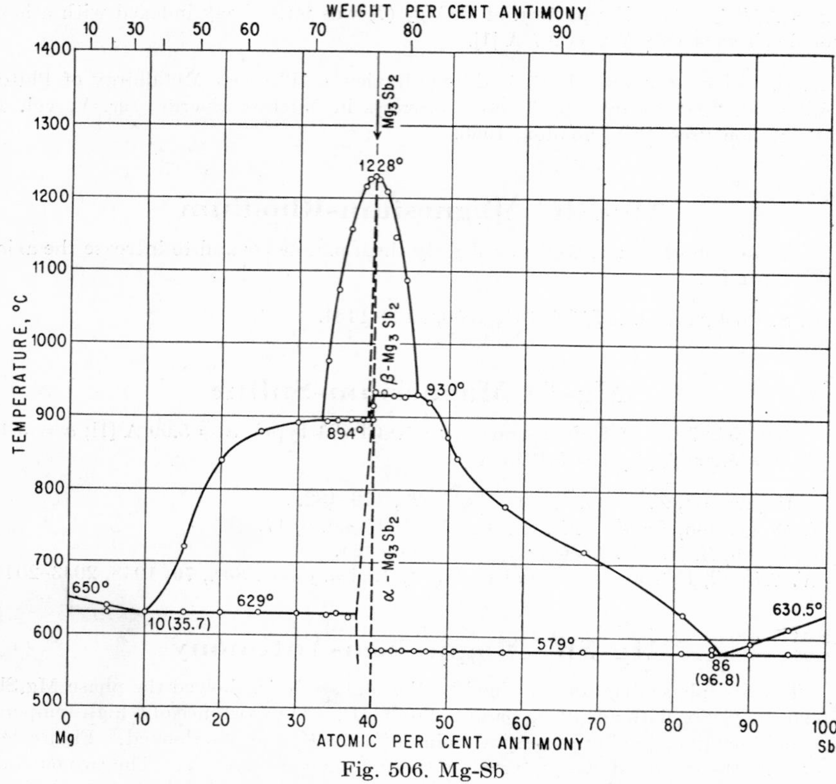

Fig. 506. Mg-Sb

$\bar{1}.4886$
0.5114

Mg-Se Magnesium-Selenium

MgSe (76.45 wt. % Se) is isotypic with NaCl (B1 type), a = 5.46 A [1].

1. E. Broch, *Z. physik. Chem.*, **127**, 1927, 446–454.

$\bar{1}.9374$
0.0626

Mg-Si Magnesium-Silicon

The fact that Mg and Si form only the compound Mg_2Si (36.61 wt. % Si) was first conclusively established by microexamination of alloys with 0.38–77.2 wt. % Si [1]. The solidification of alloys, covering a wide composition range, was studied by [2, 3]. The liquidus temperatures differ by about 30°C (Fig. 507). This discrepancy is difficult to understand. [2] used two grades of Si, one which contained as much as 6 wt. % Fe and 1.7 wt. % Al and another which was 99.2 wt. % pure. Liquidus temperatures of alloys prepared using these two grades differ only slightly (Fig. 507). The silicon used by [3] contained 99.48 wt. % Si, 0.52 wt. % SiO_2.

The temperature of the Mg-Mg_2Si eutectic was reported as about 645 [2], 625 [3], 640 [4], 632 [5], 642 [6], and 637.6°C [7]. The last value is to be regarded as the most accurate. The composition of this eutectic in wt. % Si (at. % in parentheses) was given as 1.37–2 (1.18–1.7) [1], about 4 (3.48) [2], about 1.4 (1.22) [8], 3.2 (2.8) [4], 2.4 (2.08) [5], 1.21 (1.05) [6], and 1.34 (1.16) [7]. The last value is to be considered the most reliable (see inset of Fig. 507).

The Mg_2Si-Si eutectic was reported as being located at about 950°C, about 58 wt. (54.5 at.) % Si (extrapolated) [2], and 920°C, 57 wt. (53.5 at.) % Si [3].

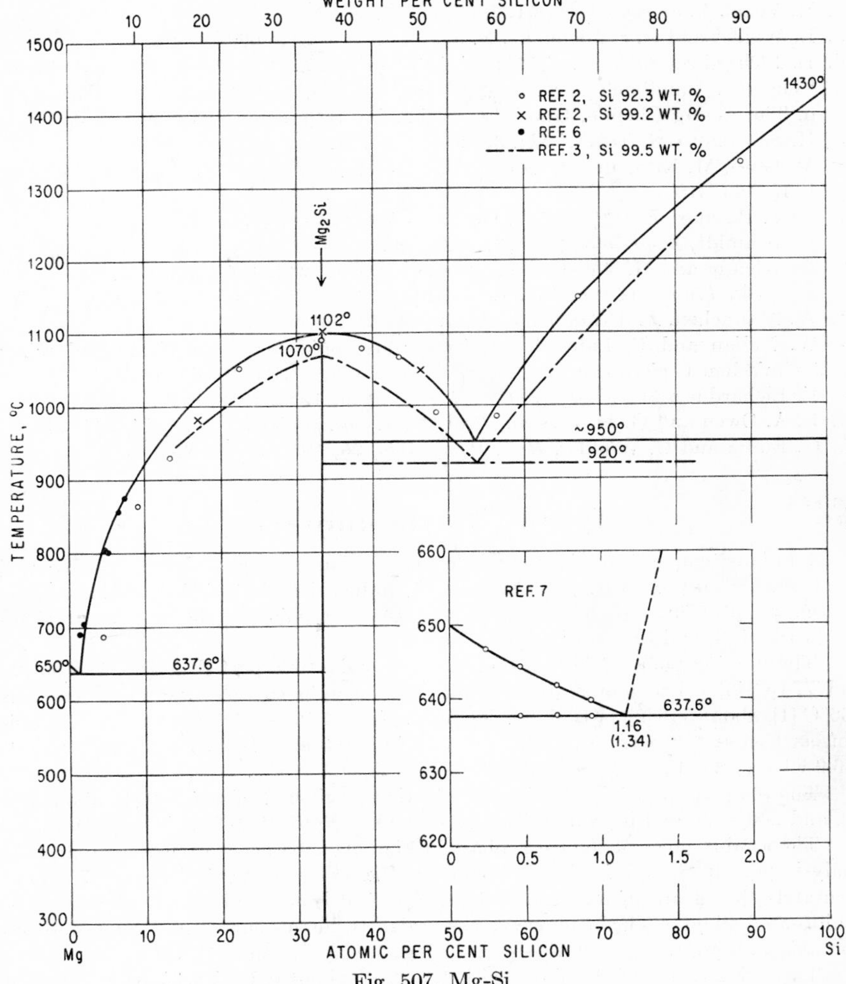

Fig. 507. Mg-Si

The solid solubility of Si in Mg was given as less than 0.1 at. (0.12 wt.) %, since an alloy of this composition proved roentgenographically and microscopically to be two-phase [9]. [7] estimated the solubility to be of the order of 0.003 at. (wt.) % at 600°C, based on lattice-parameter measurements. It was established that there is a slight increase in solubility between 450 and 600°C [7]. [10] concluded from lattice-parameter measurements that the solubility is "vanishingly small"; see also [11, 12].

As to the existence of MgSi, assumed to be formed by thermal decomposition of

Mg_2Si, see [3]. The existence of an unstable system involving the phase Mg_4Si, claimed to form a eutectic with Mg, was reported by [4, 13].

Mg_2Si is isotypic with CaF_2 (C1 type), a = 6.404 A [14]; a = 6.351 A [9, 15].

1. P. Lebeau and P. Bossuet, *Rev. mét.*, **6,** 1909, 273–278; *Compt. rend.*, **146,** 1908, 282–284.
2. R. Vogel, *Z. anorg. Chem.*, **61,** 1909, 46–53.
3. L. Wöhler and O. Schliephake, *Z. anorg. Chem.*, **151,** 1926, 11–20.
4. E. Elchardus, *Publ. sci. et tech. ministère air (France)*, 1935, no. 70; quoted from "Gmelins Handbuch der anorganischen Chemie," System No. 27, Part A(3), p. 376, Verlag Chemie, G.m.b.H., Weinheim/Bergstrasse, Germany, 1952.
5. H. Sawamoto, *Suiyokwai Shi*, **8,** 1935, 713–727.
6. M. Goto, M. Nito, and H. Asada, *Rept. Aeronaut. Research Inst. Tokyo Univ.*, **12,** 1937, 163–318.
7. G. V. Raynor, *J. Inst. Metals*, **66,** 1940, 403–426.
8. W. Schmidt, *Z. Metallkunde*, **19,** 1927, 452.
9. W. Klemm and H. Westlinning, *Z. anorg. Chem.*, **245,** 1940, 365–380.
10. R. Busk, *Trans. AIME*, **188,** 1950, 1460–1464.
11. W. Mannchen, *Z. Metallkunde*, **23,** 1931, 193–196.
12. W. Bulian and E. Fahrenhorst, "Metallographie des Magnesiums und seiner technischen Legierungen," p. 27, Springer-Verlag OHG, Berlin, 1942.
13. E. Elchardus and P. Laffitte, *Compt. rend.*, **200,** 1935, 1938–1940.
14. E. A. Owen and G. D. Preston, *Proc. Phys. Soc. (London)*, **36,** 1924, 341–348.
15. G. Busch and U. Winkler, *Helv. Phys. Acta*, **26,** 1953, 578–583.

$\bar{1}.3115$
0.6885

Mg-Sn Magnesium-Tin

Liquidus temperatures for the whole composition range were determined by [1–3] and for the range 0–11.8 at. % Sn by [4]. [5] studied the effect of Mg additions on the melting point of Sn. Figure 508 shows only the data points of the more comprehensive work of [3] and those of [5].

The melting point of Mg_2Sn (70.93 wt. % Sn) was reported as 783 [1], 795 [2], and 778°C [3]. The temperature of the (Mg)-Mg_2Sn eutectic was determined as 566°C [1], about 581 [2], 561 [3], 560 [6], 560.6 [4], and 561–562°C [7], and its composition as 11.6 at. (39 wt.) [1], 12 at. (40 wt.) [2], 10.5 at. (36.4 wt.) [3], and 10.7 at. (36.9 wt.) % Sn [4].

The temperature of the Mg_2Sn-Sn eutectic was given as about 209 [1], about 203 [2], and 200°C [3] and its composition as very close to 91 at. (98 wt.) % Sn [1–3].

The solidus curve of the (Mg) phase in Fig. 508 is based on careful micrographic analysis [4]. [6] determined this boundary thermoresistometrically and [7], more accurately, by means of measurements of electrical conductivity vs. composition of equilibrated alloys. The latter curve has a slightly convex curvature to the composition axis and deviates from that shown in Fig. 508 by maximal 10°C.

The solidification of the alloys with 1–3.2 at. (about 0.2–0.7 wt.) % Mg showed an irregularity [3]; see inset in Fig. 508. In contrast to [5], who found a straight liquidus curve in this range, 0–7.1 at. % Mg, [3] observed an additional thermal effect (+ in Fig. 508) slightly below the liquidus. Since small solidified droplets of a second melt were present in the microstructure, this phenomenon was interpreted as indicating the existence of a very narrow miscibility gap in the liquid state. For the main diagram in Fig. 508, the data by [5] were adopted.

The solid solubility of Sn in Mg, recognized earlier by [8, 9], was determined by [6, 4, 7, 10]; a study by Japanese investigators [11] was not available. Whereas the data by [6] do not represent equilibrium conditions, those by [4, 7, 10] are in excellent agreement. Between 400 and 500°C, they differ only by about 0.1–0.2 at. (approximately 0.4–0.9 wt.) %, and even less at lower temperatures. The solubility in at. % Sn (wt. % in parentheses) at the eutectic temperature was given as 3.44 (14.8) [6], 3.35 (14.45) [4], 3.54 (15.2) [7], and 3.41 (14.7) [10], and at 300°C as 0.35 (1.65) [4], 0.29 (1.35) [7], and 0.30 (1.40) [10]. The findings of [4] are based on micrographic analysis, those of [7] on measurements of electrical conductivity vs. composition, and those of [10] presumably also on micrographic work. On the whole, the solubility curve given by [7] lies between the other two and was therefore used to draw this phase boundary in Fig. 508.

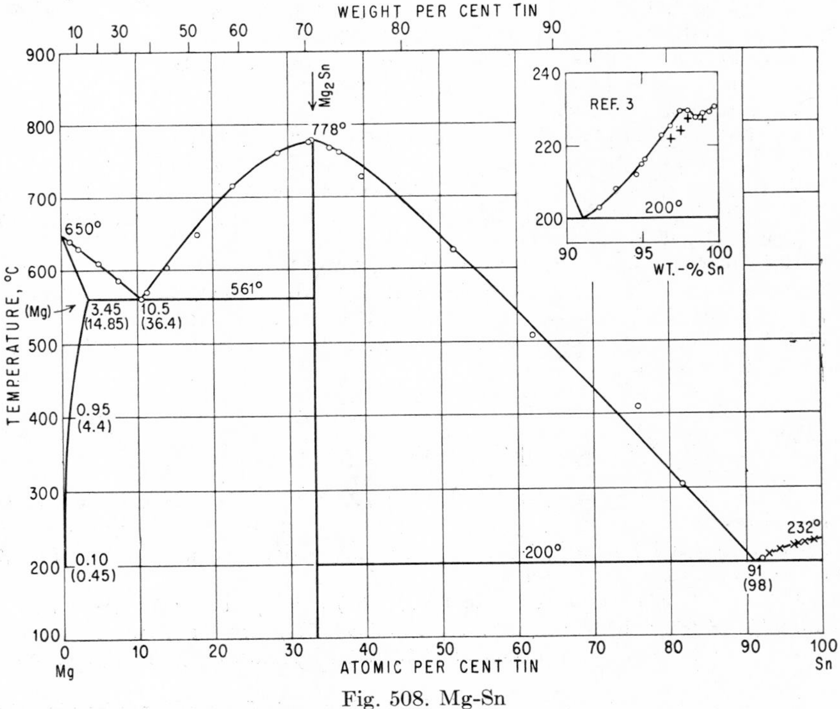

Fig. 508. Mg-Sn

The solid solubility of Mg in Sn has not been determined accurately, but appears to be very restricted. [12] suggested a solubility of 0.1–0.2 at. (about 0.02–0.04 wt.) % Mg at 200°C and about 0.1 at. % at 160°C, based on hardness and recrystallization tests.

Mg_2Sn can be considered a phase of truly fixed composition; no evidence was found for the existence of solid solutions [3, 13, 6, 24].

The boiling temperature of melts with 0, 6.1, 19.3, and 33.2 at. % Sn were reported as 1103, 1107, 1136, and 1171°C (± 5°C), respectively [14].

Crystal Structure. Mg_2Sn is cubic of the CaF_2 (C1) type [15–17], with a lattice constant of a = 6.76–6.77 A [15–22]. Lattice spacings of the (Mg) phase with up to 2.47 at. % Sn were reported by [23].

1. G. Grube, *Z. anorg. Chem.*, **46**, 1905, 76–84.
2. N. S. Kurnakow and N. J. Stepanow, *Z. anorg. Chem.*, **46**, 1905, 181–184.

3. W. Hume-Rothery, *J. Inst. Metals,* **35,** 1926, 336–347.
4. G. V. Raynor, *J. Inst. Metals,* **66,** 1940, 403–426.
5. C. T. Heycock and F. H. Neville, *J. Chem. Soc.,* **57,** 1890, 381.
6. G. Grube and H. Vosskühler, *Z. Elektrochem.,* **40,** 1934, 566–570.
7. H. Vosskühler, *Metallwirtschaft,* **20,** 1941, 805–808.
8. N. J. Stepanow, *Z. anorg. Chem.,* **78,** 1912, 13–17.
9. J. A. Gann, *Trans. AIME,* **83,** 1929, 309–332.
10. L. A. Willey, unpublished work; see "Metals Handbook," 1948 ed., p. 1227, The American Society for Metals, Cleveland, Ohio.
11. H. Nishimura and K. Tanaka, *Suiyokwai Shi,* **10,** 1940, 343–350.
12. E. Jenckel and L. Roth, *Z. Metallkunde,* **30,** 1938, 135–144.
13. W. Hume-Rothery, *J. Inst. Metals,* **38,** 1927, 127–131.
14. A. Schneider and U. Esch, *Z. Elektrochem.,* **45,** 1939, 888–893.
15. L. Pauling, *J. Am. Chem. Soc.,* **45,** 1923, 2777–2780.
16. A. Sacklowski, *Ann. Physik,* **77,** 1925, 264–272.
17. E. Zintl and H. Kaiser, *Z. anorg. Chem.,* **211,** 1933, 113–131.
18. W. Klemm and H. Westlinning, *Z. anorg. Chem.,* **245,** 1941, 365–380.
19. A. H. Geisler, C. S. Barrett, and R. F. Mehl, *Trans. AIME,* **152,** 1943, 201–223.
20. J. T. Norton, quoted by W. D. Robertson and H. H. Uhlig, *Trans. AIME,* **180,** 1949, 345–355.
21. G. Brauer and J. Tiesler, *Z. anorg. Chem.,* **262,** 1950, 319–327.
22. G. Busch and U. Winkler, *Helv. Phys. Acta,* **26,** 1953, 578–583.
23. G. V. Raynor, *Proc. Roy. Soc.* (*London*), **A180,** 1942, 107–121.
24. S. A. Pogodin, L. M. Kefeli, and E. S. Berkavich, *Izvest. Sektora Fiz.-Khim. Anal.,* **17,** 1949, 193–199.

$\bar{1}.4433$
0.5567

Mg-Sr Magnesium-Strontium

The phase diagram in Fig. 509 is due to [1, 2] (but see Supplement). Within the composition range covered by both studies, 0–40 wt. (0–15.6 at.) % Sr, results agree very well. Since [1] investigated 18 alloys in this region, as compared with 7 by [2], the phase boundaries shown are based on the former data.

The melting point of Mg_9Sr (28.59 wt. % Sr) was reported as 606 [1] and 603°C [2], the Mg-Mg_9Sr eutectic as 582°C, 5.9 at. (18.4 wt.) % Sr [1], and 585°C, 6 at. % Sr [2], and the Mg_9Sr-Mg_4Sr eutectic as 592°C, 15.5 at. (39.8 wt.) % Sr [1], and 587°C, 15.5 at. % Sr [2].

The phase between Mg_9Sr and Mg_2Sr (64.31 wt. % Sr) could not be identified conclusively because of the incompleteness of the peritectic reaction at 598°C [2]. The authors believe, however, that the composition Mg_4Sr (47.39 wt. % Sr) is very likely. In addition, thermal effects at 604°C were observed in alloys with about 21–30 at. % Sr. Perhaps they represent another peritectic reaction resulting in the formation of the phase Mg_3Sr [2]. This appears quite probable, since [2] were unable to homogenize the alloy corresponding to Mg_4Sr by annealing for several days at 550–600°C, although its composition lies close to that of the peritectic melt, about 18 at. % Sr. The structure of this alloy may have remained heterogeneous because of the presence of Mg_3Sr crystals formed at 604°C (see Supplement).

The solid solubility of Sr in Mg at 450–570°C was reported as 0.11–0.15 wt. (0.03–0.04 at.) % Sr [1].

Mg_2Sr is isotypic with $MgZn_2$ (C14 type), a = 6.439 A, c = 10.494 A, c/a = 1.630 [3]. The phase MgSr, reported to be b.c.c. of the CsCl (B2) type, a = 3.908 A [4], does not exist according to the thermal analysis by [2]; see Fig. 509.

Supplement. [5] studied the system Mg-Sr by means of thermal, microscopic, and X-ray analysis over the whole range of composition. The thermal data, except those in the range 0–15 at. % Sr, are plotted in Fig. 509. For the Mg-rich eutectics the values 586°C, 18.6 wt. (6.0 at.) % Sr, and 592°C, 39.0 wt. (15.1 at.) % Sr are given, and 609°C as the melting point of Mg_9Sr, in good agreement with the results of [1, 2]. [5] prefers the formula Mg_9Sr_2 (18.18 at. % Sr) to Mg_4Sr for the compound formed peritectically at 599°C and ascribes a second peritectic reaction at 608°C (cf. above)

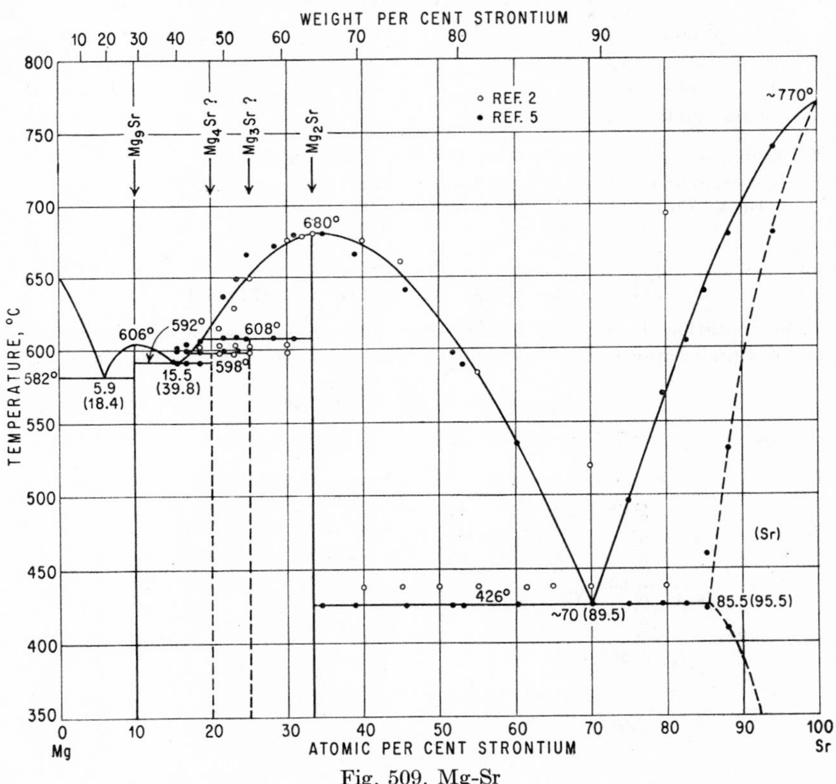

Fig. 509. Mg-Sr

to the formation of a Mg_3Sr phase. Confirming the results of [1], [5] metallographically found the limit of solid solubility of Sr in Mg to be 0.11 wt. (0.03 at.) %.

For the constitution of the Sr-rich region in Fig. 509, the data of [5] have been accepted because of the larger number of alloys studied. [2] had reported a eutectic between Mg_2Sr and pure Sr at 438°C, ~65 at. % Sr.

[5] made tentative descriptions of the crystallography of the intermediate phases. He calls Mg_2Sr hexagonal, with a = 5.41 "A", c = 8.76 "A" (probably kX); but unquestionably the determination by [3] is correct. Both "Mg_9Sr_2" and Mg_3Sr are claimed to be hexagonal with parameters a = 6.63 "A," c = 10.52 "A," and a = 6.76 "A," c = 10.78 "A"; and Mg_9Sr is claimed to be f.c.c., with a = 5.26 "A." In view of the number of atoms per "molecule" and the size of the atoms, the unit cell for Mg_9Sr is suspect.

1. H. Vosskühler, *Metallwirtschaft*, **18**, 1939, 377–378.
2. W. Klemm and F. Dinkelacker, *Z. anorg. Chem.*, **255**, 1947, 2–12.
3. E. Hellner and F. Laves, *Z. Krist.*, **105**, 1943, 134–143.
4. H. Nowotny, *Z. Metallkunde*, **34**, 1942, 247–253.
5. J. P. Ray, M.S. thesis, Syracuse University, Syracuse, N.Y., 1947.

$\bar{1}.2801$
0.7199

Mg-Te Magnesium-Tellurium

MgTe (83.99 wt. % Te) has the hexagonal structure of ZnS, wurtzite (B4 type), $a = 4.5_3$ A, $c = 7.3_4$ A, $c/a = 1.62$ [1]; $a = 4.53$ A, $c = 7.38$ A (or kX?), $c/a = 1.62_9$ [2]. Lattice-parameter measurements showed that the solid solubility of Te in Mg is "vanishingly small" [3].

1. W. Zachariasen, *Z. physik. Chem.*, **128**, 1927, 417–420.
2. W. Klemm and K. Wahl, *Z. anorg. Chem.*, **266**, 1951, 289–292.
3. R. S. Busk, *Trans. AIME*, **188**, 1950, 1460–1464.

$\bar{1}.0202$
0.9798

Mg-Th Magnesium-Thorium

The tentative diagram in Fig. 510 is based on an unpublished investigation carried out under appropriate experimental conditions [1]. The solidus temperatures

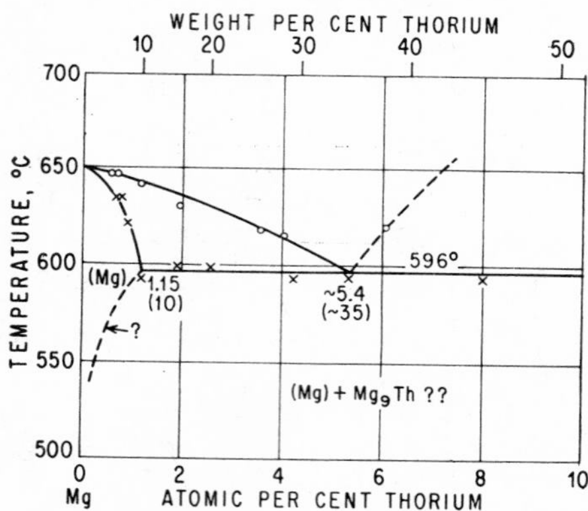

Fig. 510. Mg-Th. (See also Fig. 510*a* and Note Added in Proof.)

indicated were obtained by micrographic work. The solid solubility of Th in Mg below the eutectic temperature was not determined as yet; equilibrium can be attained only by very long annealings. The composition of the intermediate phase coexisting with (Mg) was estimated to be about 55 wt. (11.4 at.) % Th. A possible composition of Mg_9Th corresponds to a Th content of 51.5 wt. %.

Earlier, [2] had determined the solidus temperatures of alloys with 1–50 wt. % Th using as-cast specimens. It was concluded that the solubility at the eutectic temperature, about 587°C, was between 5.8 and 10.2 wt. (about 0.65–1.2 at.) % Th.

Papers by [3, 4] are of some interest, although they do not furnish additional data as to the constitution.

Note Added in Proof. The phase relations shown in Fig. 510*a* are chiefly those established in an unpublished investigation [5]. No experimental details were available to the reviewer. Whereas the phase diagram of [5] shows the two intermediate phases Mg_2Th (17.32 wt. % Mg) and Mg_3Th, later work [6] suggested that there exists only the compound Mg_2Th, which undergoes a polymorphic transformation somewhere between 700 and 800°C. The Mg-rich portion of Fig. 510*a* was taken from Fig. 510.

According to [5], the solubility of Mg in Th appears to be very low, since no change in the lattice parameter of Th could be observed at room temperature.

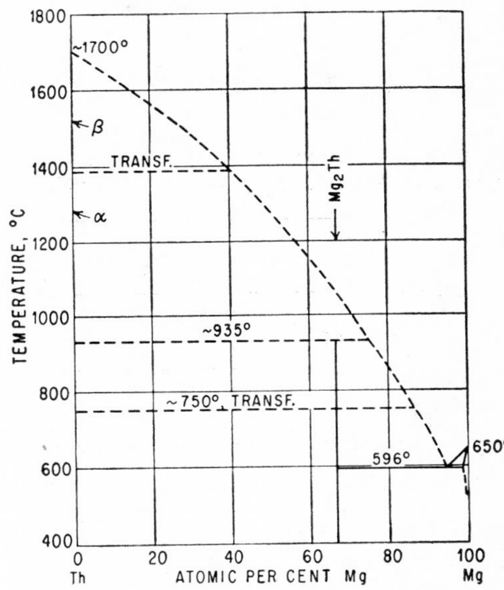

Fig. 510*a*. Mg-Th. (See also Fig. 510.)

The high-temperature form of Mg_2Th is isotypic with f.c.c. $MgCu_2$ (C15 type), a = 8.570 A, and the low-temperature form is of the hexagonal $MgNi_2$ (C36) type, with a = 6.086 A, c = 19.64 A, c/a = 3.23 [6].

1. A. Jones and R. R. Nash, WADC Report 53-113, October, 1953.
2. T. E. Leontis, *Trans. AIME*, **194**, 1952, 287–294.
3. F. A. Fox, *J. Inst. Metals*, **73**, 1946, 223–228.
4. F. Sauerwald, *Z. anorg. Chem.*, **258**, 1949, 296–300.
5. D. Peterson, unpublished information (March, 1954); quoted in H. A. Saller and F. A. Rough, Compilation of U.S. and U.K. Uranium and Thorium Constitutional Diagrams, *U.S. Atomic Energy Comm., Publ.* BMI-1000, June, 1955.
6. D. T. Peterson, P. F. Diljak, and C. L. Vold, *Acta Cryst.*, **9**, 1956, 1036–1037.

$\bar{1}$.7056
0.2944

Mg-Ti Magnesium-Titanium

The solubility of Ti in liquid Mg was determined by [1, 2]. As shown in Fig. 511, results differ considerably; at 700°C the solubility according to [2] is about twice and at 800°C about ten times that reported by [1]. The latter investigators saturated Mg melts with Ti at various temperatures and took liquid samples for chemical analysis. This method appears to be more reliable than that used by [2], which is based on the assumption that Ti, which was dissolved in liquid Mg, is soluble in sulfuric acid, whereas the excess Ti is not [2a]. It seems more probable that the sulfuric acid–

soluble portion of Ti is actually that which formed a solid solution with Mg during cooling. Table 34 gives the solubilities found by [1].

Table 34. Solubility of Ti in Liquid Mg According to [1]

Temp., °C	Wt. % Ti	At. % Ti
651	0.0025	0.0013
700	0.0066	0.0033
775	0.011	0.0056
850	0.015	0.0076

On the basis of lattice-parameter measurements and the results of [1], [4] suggested that there is, as in the system Mg-Zr, a peritectic equilibrium at the Mg-rich

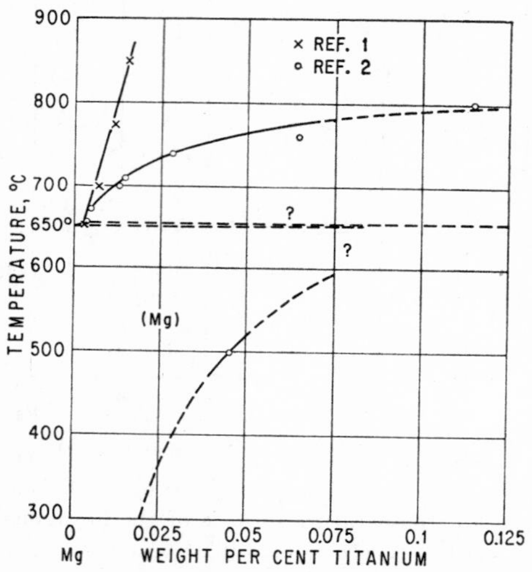

Fig. 511. Mg-Ti. (See also Supplement.)

end of the system. This was verified by [2], who determined the solid solubility as approximately 0.045 wt. % at 500°C. There was indication of a decrease in solubility with temperature. For some unknown lower temperature, the solubility was given as about 0.0135 wt. %, after annealing for 30 days at 100°C [2].

Supplement. The effect of Mg on the α-β transformation of Ti was investigated by [5]. Alloys containing up to 1.5 wt. (2.9 at.) % Mg were made from Ti sponge and high-purity Mg by sheet-rolling compressed compacts and annealing them for periods of 24 to 120 hr in a helium atmosphere. It was found that Mg is soluble in solid Ti to the extent of at least 1.5 wt. % in both the α and β phases. The $(\alpha + \beta)$ gap was found to widen with increasing Mg content in that its upper and lower boundaries are raised and lowered, respectively, in regard to the transformation point in pure Ti. Since oxygen—which raises the transformation—was known to be present in the alloys, the above results indicate that in binary alloys Mg lowers the $\alpha \rightleftarrows \beta$ transformation of Ti.

1. K. T. Aust and L. M. Pidgeon, *Trans. AIME,* **185,** 1949, 585–587.
2. H. Eisenreich, *Metall,* **7,** 1953, 1003–1006.

2a. Ti was introduced by [2] by reduction of $TiCl_4$ (see also [3]), the melts were rapidly solidified from various temperatures, and the Ti soluble in sulfuric acid was determined analytically. [2] claims that dissolving of metallic Ti does not give equilibrium solubilities.
3. J. D. Grogan and T. H. Schofield, *J. Inst. Metals,* **51,** 1933, 123–128.
4. R. S. Busk, *Trans. AIME,* **188,** 1950, 1460–1464.
5. J. W. Fredrickson, *Trans. AIME,* **203,** 1955, 368.

$\bar{1}.0755$
0.9245

Mg-Tl Magnesium-Thallium

After an outline of the phase relationships by thermal analysis [1], the diagram was corrected and complemented by [2], using thermal and thermoresistometric

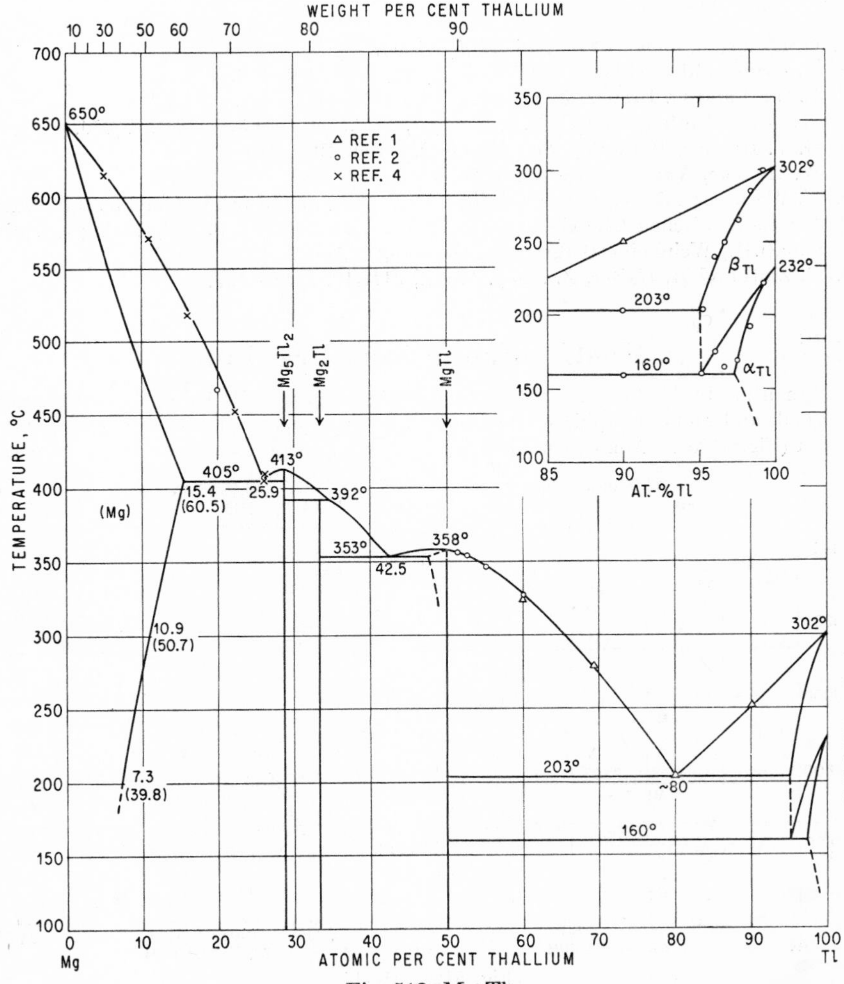

Fig. 512. Mg-Tl

methods. This work showed that the intermediate phase richest in Mg was not Mg_8Tl_3 but Mg_5Tl_2 [28.57 at. (77.07 wt.) % Tl]. Also, the intermediate phase richest in Tl, first assumed to be Mg_3Tl_2, was found to be MgTl (89.37 wt. % Tl), which had been identified earlier [3]. [4] redetermined the liquidus in the region 0–26 at. % Tl and established the solidus and solubility curves of the (Mg) phase by micrographic analysis. Figure 512 shows the diagram based on the findings by [2, 4]. For the sake of clarity, the numerous data points in the range 26–50 at. % Tl [2] have been omitted.

Crystal Structure. The lattice spacings of the (Mg) phase up to 9.1 at. % Tl were determined by [5]. Mg_5Tl_2 is isotypic with Mg_5Ga_2, b.c. orthorhombic, with 28 atoms per unit cell; lattice constants were not determined [6]. Mg_2Tl (80.78 wt. % Tl) is isotypic with Mg_2Ga, hexagonal, with 18 atoms per unit cell [6], a = 8.121 A, c = 7.352 A, c/a = 0.905 [7]; see also [8]. MgTl is b.c.c. of the CsCl (B2) type, a = 3.635 A [4].

1. G. Grube, *Z. anorg. Chem.*, **46,** 1905, 84–93.
2. G. Grube and J. Hille, *Z. Elektrochem.*, **40,** 1934, 101–106.
3. E. Zintl and G. Brauer, *Z. physik. Chem.*, **B20,** 1933, 258.
4. W. Hume-Rothery and G. V. Raynor, *J. Inst. Metals*, **63,** 1938, 201–226.
5. G. V. Raynor, *Proc. Roy. Soc.* (*London*), **A180,** 1942, 107–121.
6. W. Haucke, *Naturwissenschaften*, **26,** 1938, 577–578.
7. K. Weckerle, Dissertation, Freiburg i. Br., 1935; quoted from "Gmelins Handbuch der anorganischen Chemie," System No. 27, Part A (4), p. 580, Verlag Chemie, G.m.b.H., Weinheim/Bergstrasse, Germany, 1952.
8. E. Zintl and H. Kaiser, *Z. anorg. Chem.*, **211,** 1933, 126–127.

$\bar{1}.0092$
0.9908

Mg-U Magnesium-Uranium

In a study of the constitution of the Mg-U system [1] by analytical, X-ray, thermal, and metallographic methods, the following phase relationships were established (Fig. 512*a*). Almost complete liquid immiscibility was found up to a tempera-

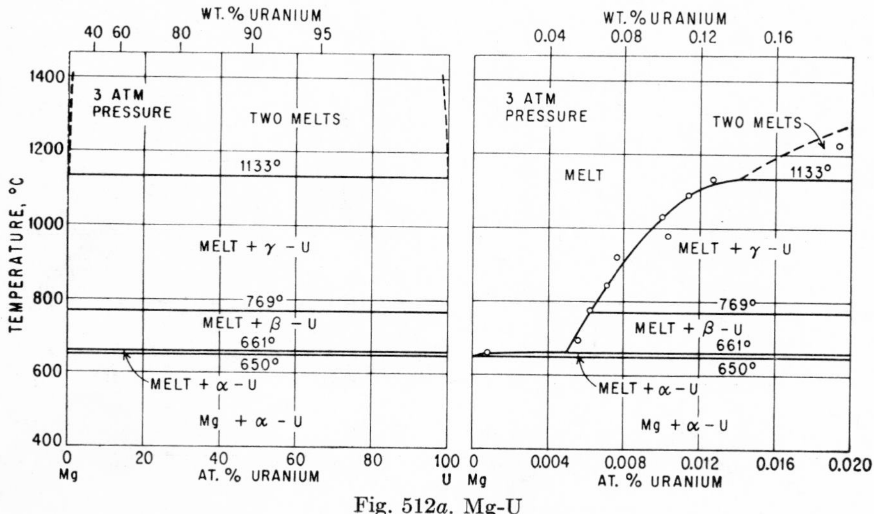

Fig. 512*a*. Mg-U

ture of 1255°C, and the compositions of the liquids which coexist at 1135°C and 3 atm pressure are approximately 0.14 wt. (0.014 at.) % U in Mg and 0.004 wt. (0.04 at.) % Mg in U. The solubility of U in Mg decreases to nearly 0.05 wt. (0.005 at.) % at 675°C and to ~0.0005 wt. (0.00005 at.) % at 650°C. U has little or no effect on the melting point of Mg, and Mg does not influence the U transformation temperatures to any detectable degree.

Previously, [2, 3] had reported that U does not alloy with Mg, and [4], that "it is extremely difficult to prepare U containing more than a few parts per million of Mg."

1. P. Chiotti, G. A. Tracy, and H. A. Wilhelm, *Trans. AIME*, **206**, 1956, 562–567.
2. F. Sauerwald, *Z. anorg. Chem.*, **258**, 1949, 296–300.
3. A. Iandelli and R. Ferro, *Ann. chim.* (*Rome*), **42**, 1952, 598–606.
4. J. J. Katz and E. Rabinowitch, "The Chemistry of Uranium," Part I, National Nuclear Series, Div. VIII, vol. 5, p. 177, McGraw-Hill Book Company, Inc., New York, 1951.

$\bar{1}.1213$
0.8787

Mg-W Magnesium-Wolfram

W does not affect the lattice-parameter values of Mg [1]. According to [2, 3], it does not alloy with Mg.

1. R. S. Busk, *Trans. AIME*, **188**, 1950, 1460–1464.
2. D. Kremer, *Abhandl. Inst. Metallhütt. u. Elektromet. Tech. Hochsch., Aachen*, **1**, 1916, 8–9.
3. F. Sauerwald, *Z. anorg. Chem.*, **258**, 1949, 296–300.

$\bar{1}.4370$
0.5630

Mg-Y Magnesium-Yttrium

It was found that Y "alloys relatively easily" with Mg [1].

1. F. Sauerwald, *Z. anorg. Chem.*, **258**, 1949, 296–300.

$\bar{1}.5705$
0.4295

Mg-Zn Magnesium-Zinc

Partial System Mg-$MgZn_2$. The liquidus shown in Fig. 513 is based substantially on the data of [1, 2], which cover the ranges 0–66.7 and 40–66.7 at. % Zn, respectively. The eutectic point was found at 30.2 at. % Zn, 342°C [1]. Previously, solidification temperatures were reported by [3–6]. All these early investigators concluded that there was only one intermediate phase, $MgZn_2$ (84.32 wt. % Zn), which formed a eutectic with Mg at 29.6 at. %, 332°C [3]; 28.5 at. %, 344°C [4]; 27.9 at. %, 340°C [5]; and 27.9 at. %, 355°C [6]. Later work, however, showed that the phase relations were more complicated and indicated the existence of at least two, possibly three, intermediate phases between Mg and $MgZn_2$.

The Phase MgZn (72.89 *Wt.* % Zn). [1] first reported the occurrence of a reaction at 356°C in the range of about 33–61 at. % Zn, but was unable to find structural evidence for a peritectic reaction. [2] confirmed these thermal effects and showed them to be due to the peritectic formation of the phase MgZn. The existence of this phase was confirmed by [13, 7–10] but overlooked or denied by [11, 12]. There is no agreement as to the temperature range of stability. Whereas [7] claimed MgZn to be stable only between approximately 350 and 310°C, [2] proved by micrographic work, after long-time annealing, that MgZn was stable at least down to 200°C. Also, the phase diagram proposed by [8] shows MgZn, formed peritectically at 366°C, as a phase stable at ordinary temperature.

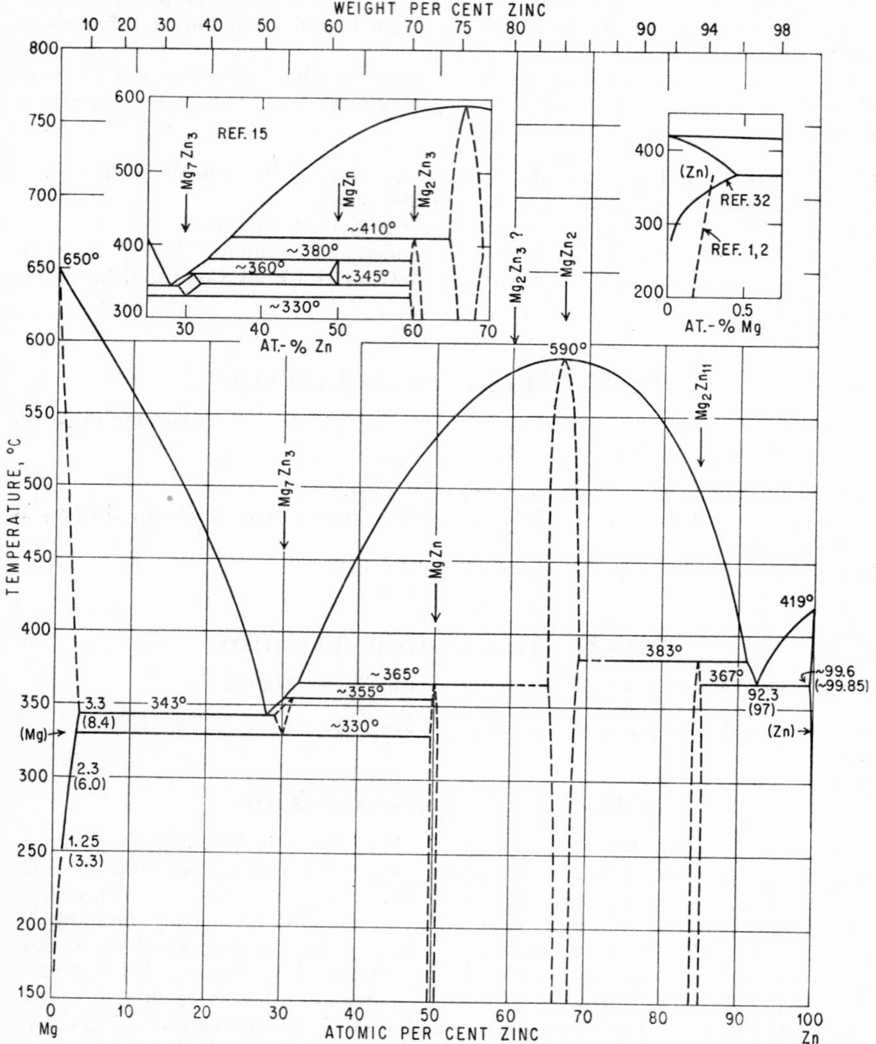

Fig. 513. Mg-Zn. (See also Fig. 514, Supplement, and Note Added in Proof.)

The Phase Mg_2Zn_3 (80.13 *Wt.* % Zn). This phase was first reported to be formed by peritectic reaction of $MgZn_2$ with melt at 410°C [11]. It was claimed to be identified by micrographic and X-ray studies [7]; however, no evidence for its existence was found by [1, 2, 8, 10]. Both [11] and [7] assumed Mg_2Zn_3 to be in equilibrium with the Mg-rich primary solid solution and to be stable down to room temperature.

The Phase "Mg_7Zn_3" (53.53 *Wt.* % Zn). The existence of this phase was first reported by [11, 14] and conclusively established by [7, 8, 15, 16]. Additional evidence was found by [17, 18]. Under equilibrium conditions, the phase is stable only within a narrow temperature range; it decomposes eutectoidally at approximately 330 [11] or 335°C [8]. It can be easily retained at room temperature, since its decomposition is a very sluggish reaction [15, 16]. The transformation found by [12] at about 260°C was probably due to the breakdown of the retained phase "Mg_7Zn_3" when heated to this temperature, at which the rate of decomposition was commensurate with the rate of heating.

The exact location and homogeneity range of this phase have not been determined as yet. Nor is it known whether it forms a eutectic with (Mg) [8, 15] or an intermediate phase [11] and whether its decomposition products are (Mg) + Mg_2Zn_3 [11, 7] or (Mg) + MgZn [8].

The presentation of the phase relations in the range 30–66.7 at. % Zn (Fig. 513) should be considered tentative (see Supplement and Note Added in Proof). Only one investigator claimed the existence of both MgZn and Mg_2Zn_3 [7]; others have reported the existence of either MgZn [2, 8] or Mg_2Zn_3 [11]. [17] accepted the diagram proposed by [7] but did not give additional information on the problem. The phase relations suggested by [15] are shown in the upper left inset of Fig. 513.

Solubility of Zn *in* Mg. Solubility data were reported by [19, 20, 1, 11, 12, 21, 14, 22, 23]; those by [22], based on lattice-spacing determinations, are to be regarded as most accurate, although it is doubtful that the solubility at temperatures below 250°C represents equilibrium conditions. The solubility in at. % (wt. % in parentheses) is as follows: 340°C, 3.3 (8.4); 300°C, 2.3 (6.0); 250°C, 1.25 (3.3); 200°C, 0.75 (2.0); and 150°C, 0.65 (1.7) [22].

The solidus curve of the (Mg) phase has not been accurately determined as yet; data reported disagree considerably and do not represent equilibrium conditions [19, 1, 11, 12, 24].

There is also a discrepancy on the question of whether the phase $MgZn_2$ is of fixed composition [2] or variable composition [1, 12, 25].

Partial System $MgZn_2$-Zn. Formerly, this partial system was believed to consist of the phases $MgZn_2$ and Zn [3, 4, 26, 6]. The existence of $MgZn_5$ [83.33 at. (93.07 wt.) % Zn], detected by [1], was confirmed by [2, 12, 13, 27, 9, 10, 28] but could not be verified by [8]. X-ray work [27, 28] suggested the more appropriate composition Mg_2Zn_{11} [84.62 at. (93.67 wt.) % Zn].

The reported temperatures and compositions of the eutectic vary between 363 and 372°C and 89.9 and 92.5 at. % Zn [4, 26, 6, 1, 2, 11, 13, 8, 29], respectively. The peritectic temperature was given as 381–386 [1], 380.5 [2], and 379°C [13].

The solid solubility of Mg in Zn was investigated by [30, 1, 2, 12, 31, 32]. Data reported by [33, 34] appear to be based on the work of [2, 31]. According to [1, 2], the solubility at the eutectic temperature is about 0.3 at. (0.11 wt.) % Mg and about 0.18 at. (0.065 wt.) % Mg at 200°C [2]. [32] gave the following solubility values in wt. % Mg (at. % in parentheses): 0.16–0.17 (0.43–0.45), 0.12 (0.32), 0.05 (0.13), 0.02 (0.05), and 0.008 (0.02) at 364, 350, 325, 300, and 200°C, respectively. (See upper right inset in Fig. 513.) [31] reported less than 0.002 wt. % Mg at "room temperature"; see also [35]. The solidus point of the alloy with 0.18 at. % Mg was found as 405°C [2].

Boiling temperatures of melts with 25–96 wt. % Zn were determined by [36].

Crystal Structures. Lattice constants of the Mg-rich solid solution were determined by [22, 37]; see also [38]. The crystal structure of MgZn was investigated by [39]. Results were criticized by [40] and were also disputed by [41]. See Supplement.

$MgZn_2$ is hexagonal, with 12 atoms per unit cell (C14 type) [42–45]; a = 5.16 A,

$c = 8.49_7$ A, $c/a = 1.646$ [42]; $a = 5.18$ A, $c = 8.51_7$ A, $c/a = 1.644$ [43]; $a = 5.22$ A, $c = 8.55_7$ A, $c/a = 1.639$ [44].

The structure of Mg_2Zn_{11} (formerly designated as $MgZn_5$) was claimed to be hexagonal [39]; see, however, [40, 27]. [27, 28] showed it to be cubic, with 39 atoms per unit cell, $a = 8.55$ A; it is isotypic with $Mg_2Cu_6Al_5$ [28].

Supplement. The constitution of the Mg-Zn system in the controversial range between 30 and 66.7 at. % Zn has been carefully reinvestigated [46] by microscopic, thermal, and X-ray analysis as well as incipient-melting tests (Fig. 514). The existence of both the compounds MgZn and Mg_2Zn_3, formed peritectically at 349 ± 2 and ~410°C, respectively, has been clearly established. MgZn is stable down to at least

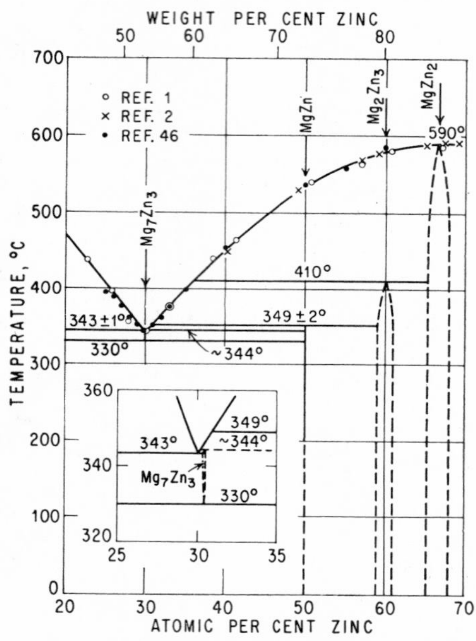

Fig. 514. Mg-Zn. (See also Fig. 513.)

200°C; and, therefore, the Mg_7Zn_3 phase (the composition of which practically coincides with that of the eutectic) decomposes eutectoidally into (Mg) + MgZn. The powder pattern obtained from a MgZn specimen which was microscopically proved to be one-phase showed little resemblance with the pattern on which [39] based his structure determination of "MgZn."

Note Added in Proof. Constitutional work in the central region of the Mg-Zn phase diagram has been carried out recently also by [47, 48]. Their results are in essential agreement with those of [46]. The following findings are of special interest: (*a*) Both the MgZn and Mg_2Zn_3 phases have some, probably small, homogeneity range at elevated temperatures [47]; (*b*) the MgZn-phase field appears to lie nearer to 74.5 wt. (52.1 at.) % Zn than to 72.9 wt. (50 at.) % Zn [47]; (*c*) the X-ray patterns of Mg_7Zn_3, MgZn, and Mg_2Zn_3 are even more complex than reported by [46]; see [47, 48]; (*d*) the temperature of the eutectoid decomposition of Mg_7Zn_3 (which was not redetermined by [46]) is below 330°C [47, 48], probably close to 325°C [47]; (*e*) there exists a metastable modification of MgZn [47]; and (*f*) MgZn is stable at least down to 93°C [47].

1. R. Chadwick, *J. Inst. Metals,* **39,** 1928, 285–298.
2. W. Hume-Rothery and E. O. Rounsefell, *J. Inst. Metals,* **41,** 1929, 119–138.
3. O. Boudouard, *Compt. rend.,* **139,** 1904, 424–426.
4. G. Grube, *Z. anorg. Chem.,* **49,** 1906, 77–83.
5. G. Bruni and C. Sandonnini, *Z. anorg. Chem.,* **78,** 1912, 276–277.
6. G. Eger, *Intern. Z. Metallog.,* **4,** 1913, 46–50.
7. F. Laves, *Naturwissenschaften,* **26,** 1939, 454–455.
8. G. G. Urasov, N. A. Filin, and A. V. Shashin, *Metallurg,* **15,** 1940, 3–11.
9. E. Pelzel and F. Sauerwald, *Z. Metallkunde,* **33,** 1941, 229–232.
10. E. M. Savitskii and V. V. Baron, *Doklady Akad. Nauk S.S.S.R.,* **64,** 1949, 693–696.
11. T. Takei, *Kinzoku-no-Kenkyu,* **6,** 1929, 177–185.
12. G. Grube and A. Burkhardt, *Z. Elektrochem.,* **35,** 1929, 315–332.
13. M. I. Zaharova and A. B. Mladzeevskiy, *Izvest. Sektora Fiz.-Khim. Anal.,* **9,** 1936, 193–202.
14. S. Ishida, *Nippon Kogyo Kwai-Shi,* **45,** 1929, 256–268, 611–621, 786–790.
15. W. Köster and F. Müller, *Z. Metallkunde,* **39,** 1948, 352–353.
16. W. Köster, *Z. Metallkunde,* **41,** 1950, 37–39.
17. G. Edmunds and M. L. Fuller, unpublished work; see "Metals Handbook," 1939 ed., pp. 1750–1751, The American Society for Metals, Cleveland, Ohio.
18. Unpublished X-ray work, The Dow Chemical Company; see "Metals Handbook," 1948 ed., p. 1227, The American Society for Metals, Cleveland, Ohio.
19. M. Hansen, *J. Inst. Metals,* **39,** 1928, 298–300; W. Schmidt (M. Hansen), *Z. Metallkunde,* **19,** 1927, 454–455.
20. B. Stoughton and M. Miyake, *Trans. AIME,* **73,** 1926, 556–557.
21. J. A. Gann, *Trans. AIME,* **83,** 1929, 309–332.
22. E. Schmid and H. Seliger, *Metallwirtschaft,* **11,** 1932, 409–411; *Z. Elektrochem.,* **37,** 1931, 455–459.
23. M. Goto, M. Nito, and H. Asada, *Rept. Aeronaut. Research Inst. Tokyo Univ.,* **12,** 1937, 163–318.
24. H. Adenstedt and J. R. Burns, *Trans. ASM,* **43,** 1951, 873–886.
25. A. A. Botschwar and I. P. Welitschko, *Z. anorg. Chem.,* **210,** 1933, 164–165.
26. G. Bruni, C. Sandonnini, and E. Quercigh, *Z. anorg. Chem.,* **68,** 1910, 78–79.
27. F. Laves and S. Werner, *Z. Krist.,* **95,** 1936, 114–128.
28. S. Samson, *Acta Chem. Scand.,* **3,** 1949, 835–843.
29. E. Weisse, A. Blumenthal, and H. Hanemann, *Z. Metallkunde,* **34,** 1942, 221.
30. W. M. Peirce, *Trans. AIME,* **68,** 1923, 781–782.
31. E. A. Anderson and J. L. Rodda, "Metals Handbook," 1939 ed., pp. 1750–1751, The American Society for Metals, Cleveland, Ohio.
32. K. Löhberg, *J. Inst. Metals,* **81,** 1952-1953, 680.
33. A. S. Kenneford, *J. Inst. Metals,* **73,** 1947, 450.
34. C. W. Roberts, *J. Inst. Metals,* **81,** 1952-1953, 308.
35. R. Chadwick, *J. Inst. Metals,* **51,** 1933, 114.
36. W. Leitgebel, *Z. anorg. Chem.,* **202,** 1931, 305–324.
37. E. Schmid and G. Siebel, *Z. Physik,* **85,** 1933, 37–41.
38. K. Löhberg, *Z. Metallkunde,* **40,** 1949, 68–72.
39. L. Tarschisch, *Z. Krist.,* **86,** 1933, 423–438.
40. L. W. McKeehan, *Z. Krist.,* **91,** 1935, 501–503.
41. F. Laves, private communication to M. Hansen.
42. J. B. Friauf, *Phys. Rev.,* **29,** 1927, 34–40.
43. L. Tarschisch, A. T. Titow, and F. K. Garjanow, *Phys. Z. Sowjetunion,* **5,** 1934, 503–510.
44. W. Döring, *Metallwirtschaft,* **14,** 1935, 918–919.

45. F. Laves and H. Witte, *Metallwirtschaft*, **14**, 1935, 645–649.
46. K. Anderko, E. J. Klimek, D. W. Levinson, and W. Rostoker, *Trans. ASM*, **49**, 1957, 778–791; discussion, 791–793.
47. J. B. Clark and F. N. Rhines, *Trans. AIME*, **209**, 1957, 425–430; see also discussion in [46].
48. L. L. Wyman and J. J. Park; see discussion in [46].

$\bar{1}.4259$
0.5741

Mg-Zr Magnesium-Zirconium

Mg-rich Alloys. Phase relations proposed by various investigators agree in that the maximum solubility of Zr in liquid Mg [1–7] is lower than that in solid Mg

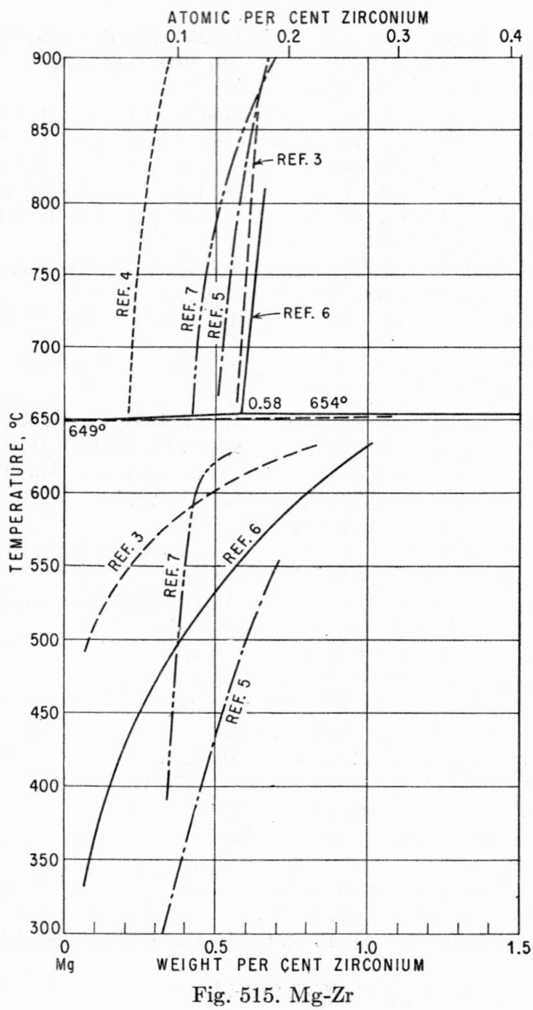

Fig. 515. Mg-Zr

[8, 2, 3, 5–7], which gives rise to a peritectic equilibrium. Figure 515 illustrates the quantitative differences.

Data by [3, 5, 6] show conclusively that the solubility of Zr in liquid Mg is close to 0.6 wt. % Zr at temperatures between 650 and 800°C; the solubility values due to [4] are definitely too low. The solid solubility appears to be best represented by the data of [6]. The solubility at the peritectic temperature is unknown; values suggested by [3, 7] are largely speculative. The peritectic temperature was found to lie about 1.5 [5] or 4–5°C [6, 7] above the melting point of Mg, the latter being reported as 649°C [5, 6], rather than the generally accepted value of 650°C.

Although [7] believed indication was found of the existence of an intermediate phase, presumably Mg_2Zr (65.22 wt. % Zr), it appears to be conclusively established that no intermediate phase occurs [2–6, 9].

As to the effect of Zr on the parameter values of Mg, see [10].

Zr-rich Alloys. It has been concluded from X-ray and metallographic studies that up to approximately 65 wt. (about 87 at.) % Mg [2] or 38 wt. (about 70 at.) % Mg [7] is soluble in α-Zr. However, the evidence presented is not too convincing; in fact, there is considerable doubt [9] whether true alloys were investigated. [9] have attempted to prepare Zr-rich alloys using various techniques. Good evidence was found that Zr will take at least some Mg into solid solution [9, 11] and that the β-Zr solid solution can be retained under certain undefined conditions [9].

1. G. Siebel; see A. Beck, "Magnesium und seine Legierungen," p. 78, Springer-Verlag OHG, Berlin, 1939 (A. Beck, "The Technology of Magnesium and Its Alloys," Hughes and Co., London, 1940).
2. H. Nowotny, E. Wormnes, and A. Mohrnheim, *Z. Metallkunde*, **32,** 1940, 39–42.
3. F. Sauerwald, *Z. anorg. Chem.*, **255,** 1947, 212–220.
4. G. Siebel, *Z. Metallkunde*, **39,** 1948, 22–27.
5. G. A. Mellor, *J. Inst. Metals*, **77,** 1950, 163–174.
6. J. H. Schaum and H. C. Burnett, *J. Research Natl. Bur. Standards*, **49,** 1952, 155–162.
7. O. A. Carson and D. T. Austin, *Can. Mining J.*, **73,** 1952, 70–75.
8. H. Vosskühler; see A. Beck [1].
9. D. J. McPherson and M. Hansen, unpublished work; Final Report COO-89 (Apr. 14, 1952) for U.S. Atomic Energy Commission, Contract AT(11-1)-49.
10. R. S. Busk, *Trans. AIME*, **188,** 1950, 1460–1464.
11. H. H. Hausner and H. S. Kalish, *U.S. Atomic Energy Commission, Publ.* SEP-38, 1950.

$\bar{1}.7578$
0.2422

Mn-Mo Manganese-Molybdenum

In 1906, [1] isolated several stoichiometric compositions (Mo_2Mn, MoMn, $MoMn_2$, $MoMn_4$, and $MoMn_6$) from Mo-Mn alloys by an alcohol–acetic acid solution. It appears certain that these compositions were chance results, having no bearing on the constitution of the Mo-Mn system.

The results of a thermoanalytical and micrographic analysis [2] of 14 alloys with up to 12.9 wt. (7.8 at.) % Mo are summarized in the partial phase diagram of Fig. 516. The three transformations in solid Mn were observed at 1139, 1083, and 660°C. The authors admit that the last of these values is too low because of supercooling of β-Mn; therefore, in Fig. 516 the lower eutectoid line has been drawn at about 700°C rather than 640°C as in the original diagram (cf. thermal data plotted in Fig. 516). The solubility of Mo in α-Mn appears to be negligible; an alloy with 0.06 wt. (0.03 at.)

% Mo annealed at 600°C was two-phase. The thermal effect at 1070°C has been ascribed by [2] to a polymorphic transformation of a compound of unknown composition which enters into equilibrium with the Mn phases (see below).

A σ phase was detected in the Mo-Mn system by [3]. In later investigations it was established that the σ phase (*a*) exists only above 1115°C [4], (*b*) has a quite narrow range of homogeneity, at about 64 at. % Mn [4], (*c*) has the tetragonal lattice parameters $a = 9.10$ A, $c = 4.74$ A, $c/a = 0.52$ at 63.7 at. (50.1 wt.) % Mn and ordered atomic arrangement [5].

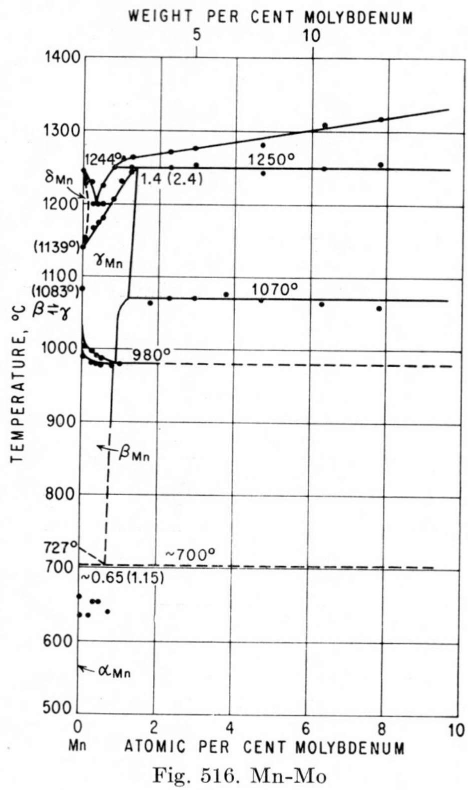

Fig. 516. Mn-Mo

Another intermediate phase was observed by [4, 6] in alloys annealed at 1000°C. [6] found this phase to extend from about 80 to 88 at. % Mn. Since it was not observed by [4] in alloys sintered at 1125°C, it may be stable only up to about 1100°C.

According to [7], no Laves phase exists in the Mo-Mn system.

[8] reported that a 0.33 wt. % Mn alloy was one-phase. According to X-ray and microscopic work by [9], the solubility of Mn in solid Mo changes strongly with the temperature. Whereas about 20 wt. (30 at.) % Mn was found dissolved in alloys quickly cooled (600°C per min) from 1800°C, only about 10 wt. (16 at.) % Mn remained in solution after slow cooling to room temperature.

The lattice parameter of Mo decreases with increasing Mn content [8, 9].

1. G. Arrivaut, *Compt. rend.*, **143,** 1906, 285–287, 464–465.
2. V. N. Eremenko and E. F. Zhel'vis, *Ukrain. Khem. Zhur.*, **16,** 1950, 370–383. The

alloys were melted in corundum crucibles under a slag of CaF_2 + NaF and were analyzed.
3. J. S. Kasper, B. F. Decker, and J. R. Belanger, *J. Appl. Phys.*, **22,** 1951, 361–362.
4. B. F. Decker, R. M. Waterstrat, and J. S. Kasper, *Trans. AIME*, **197,** 1953, 1476.
5. B. F. Decker, R. M. Waterstrat, and J. S. Kasper, *Trans. AIME*, **200,** 1954, 1406–1407.
6. P. Greenfield and P. A. Beck, *Trans. AIME*, **200,** 1954, 253–257.
7. R. P. Elliott, Armour Research Foundation, Chicago, Ill., Technical Report 1, OSR Technical Note OSR-TN-247, August, 1954.
8. Climax Molybdenum Co., First Annual Report, Apr. 1, 1950.
9. E. Pipitz and R. Kieffer, *Z. Metallkunde*, **46,** 1955, 187–194.

0.5934
$\bar{1}$.4066

Mn-N Manganese-Nitrogen

The tentative isobaric phase diagram presented in Fig. 517 was outlined by [1]. It is substantially based on the findings of [2–6] and differs somewhat from the diagram proposed by [6]. [2] first identified the three intermediate phases ϵ, ζ, and η by X-ray analysis and suggested a schematic diagram.

The Range 0–20 At. % N. [7] studied the equilibrium pressure of N_2 over Mn-N preparations with up to 34.8 at. (12 wt.) % N and concluded that they consisted of solid solutions of N in Mn or a nitride. [8] found the solubility at 1030°C as 20.02 at. (6 wt.) % N (γ_{Mn} phase in Fig. 517). The solubility of N in α-Mn was determined as about 0.5 at. (0.13 wt.) %, and that in β-Mn was claimed to be "much greater" [2]. The homogeneity range of β_{Mn} was found to extend to lower temperatures [6].

In the vicinity of 8.1 at. (2.2 wt.) % N, [2] detected a homogeneity range between 650 and 850°C. This f.c. tetragonal phase, designated as δ, decomposes at a temperature between 600 and 400°C into α_{Mn} and ϵ (Mn_4N) [2]. According to [6], it is identical with the f.c. tetragonal γ_{Mn} phase obtained on quenching.

The melting point of Mn appears to be raised by N, since the saturated solid solution of N in γ-Mn melts at approximately 1300°C, giving off N_2 [4, 6]. The temperature decreases as N_2 is rejected [6]. [4] observed that an alloy with 12.8 at. (3.6 wt.) % N was not fully molten at 1260°C. After melting at 1300–1340°C, the manganese contained 0.4–0.5 at. (0.10–0.13 wt.) % N.

The ϵ Phase (Mn_4N). This phase, the only ferromagnetic phase of the system [5], was previously thought to correspond to the composition Mn_7N_2 [22.2 at. (6.8 wt.) % N] [9]. However, [2] proved it to have a narrow homogeneity range between 20.0 and 21.4 at. (5.99 and 6.5 wt.) % N at temperatures below 400°C and to be isostructural with Fe_4N, f.c.c. with superstructure. Since the ϵ phase was found to extend above 400°C to lower nitrogen contents [12.5 at. (3.5 wt.) %], [2] suggested that its phase field might merge with that of the above-mentioned δ phase, which [6] proposed to be identical with γ_{Mn} [10].

Magnetic measurements confirmed that below 800°C ϵ was stable between 20 and 21 at. (6.0–6.35 wt.) % N, and that between 800 and 1200°C a homogeneous phase field extended from 11.50 to 21.25 at. (3.21–6.44 wt.) % N [5]. The Curie temperature of ϵ is 465–475°C [5].

On the basis of X-ray work, [6, 1] proposed that not only the δ phase of [2] (see above) but also the extended phase field above 800°C [2, 5] was identical with γ_{Mn} (Fig. 517).

The existence of ϵ (as well as ζ) was verified by tensimetric studies [3]. At temperatures below 800°C, its homogeneity range was given as 20–20.8 at. (6.0–6.3 wt.) % N. Shifting of the Mn-rich phase boundary of ζ—from 28.0 at. (9 wt.) %

at 540°C to 24.7 at. (7.2 wt.) % N at 750°C—was interpreted as indicating a continuous transition between ε and ζ. Since a continuous transition between the f.c.c. ε phase and the hexagonal ζ phase appears unlikely, a two-phase field was assumed in Fig. 517.

The ζ Phase. The ζ phase was first prepared by [11]. It was found to contain 28.4 at. (9.2 wt.) % N and, therefore, designated as Mn_5N_2 (9.26 wt. % N). Although

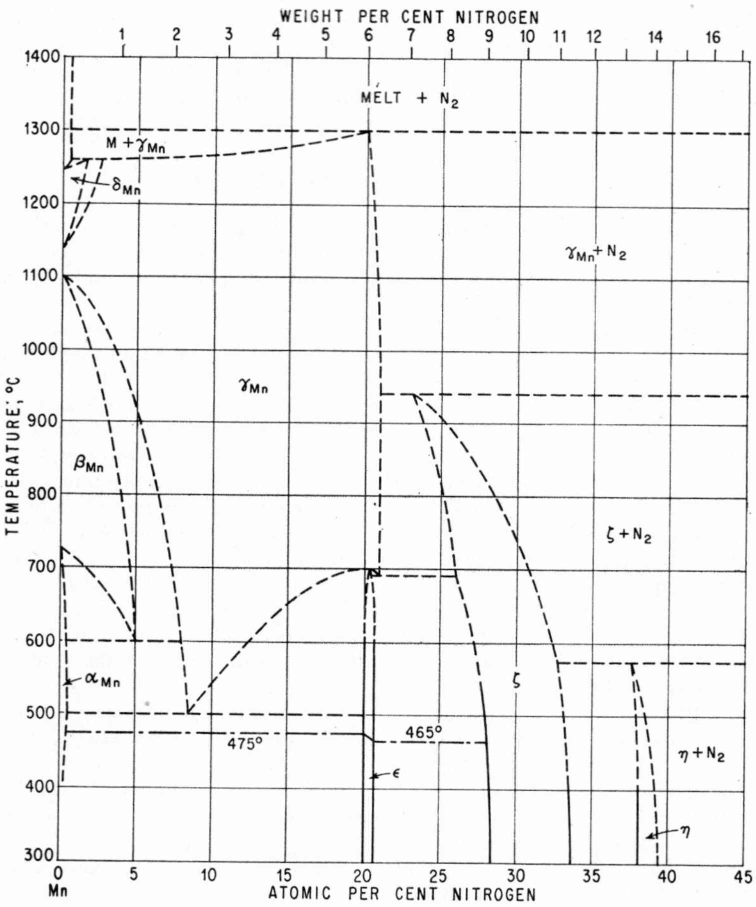

Fig. 517. Mn-N. (Tentative diagram.)

disputed by [12], the existence of an intermediate phase in this range was confirmed by [9] and by magnetic measurements of [13]. The homogeneity range at 400°C was found by X-ray work to be 28.4–34.6 at. (9.2–11.9 wt.) % N; i.e., it includes the compositions Mn_5N_2 and Mn_2N [2]. [6] observed that ζ dissociates at temperatures below 1000°C (Fig. 517).

The η Phase. The η phase was found to exist at 38.15 at. (13.6 wt.) % N [11, 14], 39.0 at. (14.0 wt.) % N [14], and between 38 and 39 at. (13.5–14 wt.) % N [2]. It decomposes already below 600°C into ζ and N_2 [2, 6]; see Fig. 517.

Investigations by [15] were not considered in drawing Fig. 517, since the results apparently do not represent equilibrium conditions.

Crystal Structures. The lattice parameter of α_{Mn} increases from a = 8.912 A (0% N) to a = 8.915 A and that of β_{Mn} from a = 6.302 A to a = 6.318 A [2]. The γ_{Mn} phase—δ phase according to [2]—is an interstitial solid solution of N in f.c.c. γ-Mn [2, 6]. It is f.c. tetragonal after quenching (metastable condition), with an axial ratio of 0.96–0.98 [2, 6]. Lattice constants were reported by [2, 4, 6].

ϵ (Mn_4N) has a f.c.c. superstructure, with a = 3.852–3.868 A [2, 5, 6]; it is isostructural with Fe_4N. ζ (Mn_2N) is h.c.p., with random interstitial distribution of N atoms [2, 5, 6], a = 2.779 A, c = 4.529 A, c/a = 1.630 at 27.2 at. % N and a = 2.834 A, c = 4.537 A, c/a = 1.601 at 35.0 at. % N [2].

η is f.c. tetragonal, with random interstitial distribution of N atoms [2, 6, 15], with a = 4.202 A, c = 4.039 A, c/a = 0.9612 at 37.8 at. % N and a = 4.215 A, c = 4.137 A, c/a = 0.9815 at 38.8 at. % N [2]. [16] reported a f.c.c. phase with approximately 21 wt. (51 at.) % N (!), a = 4.435 A.

1. The diagram is based on a review of the literature by Dr. U. Zwicker and Dr. J. Motz. Their assistance in preparing this system is gratefully acknowledged.
2. G. Hägg, *Z. physik. Chem.*, **B4,** 1929, 346–370; see also **B6,** 1929, 229–230.
3. R. Schenck and A. Kortengräber, *Z. anorg. Chem.*, **210,** 1933, 273–285.
4. R. Ochsenfeld, *Ann. Physik*, **12,** 1932, 353–384.
5. C. Guillaud and J. Wyart, *Rev. mét.*, **45,** 1948, 271–276.
6. U. Zwicker, *Z. Metallkunde*, **42,** 1951, 274–276.
7. I. Shukow, *Zhur. Russ. Fiz.-Khim. Obshchestva*, **40,** 1908, 457–459; see also R. Lorenz and J. Woolcock, *Z. anorg. Chem.*, **176,** 1928, 290–291.
8. G. Valensi, *J. chim. phys.*, **26,** 1929, 152–157, 202–218.
9. E. Wedekind and T. Veit, *Ber. deut. chem. Ges.*, **41,** 1908, 3769–3773; **44,** 1911, 2663.
10. The possibility of a continuous transition of δ into γ_{Mn} and of γ_{Mn} into ϵ was also discussed by [2].
11. O. Prelinger, *Monatsh. Chem.*, **15,** 1894, 391.
12. F. Haber and G. van Oordt, *Z. anorg. Chem.*, **44,** 1905, 370–375.
13. T. Ishiwara, *Science Repts. Tôhoku Univ.*, **5,** 1916, 70–73.
14. G. G. Henderson and J. C. Galletly, *J. Soc. Chem. Ind.*, **27,** 1908, 387–389.
15. L. Duparc, P. Wegener, and C. Cimerman, *Helv. Chim. Acta*, **12,** 1929, 806–817.
16. Z. Nishiyama and R. Iwanaga, *Mem. Inst. Sci. Ind. Research, Osaka Univ.*, **9,** 1952, 74–75; *Chem. Abstr.*, **47,** 1953, 10943.

$\bar{1}.7718$
0.2282

Mn-Nb Manganese-Niobium

The phase $NbMn_2$ (45.82 wt. % Nb) was identified by determination of its crystal structure. It is hexagonal of the $MgZn_2$ (C14) type, a = 4.87_5 A, c = 7.97 A, c/a = 1.635 [1]. There appears to be no σ-type phase in this system [2].

Supplement. [3] determined the melting temperature of $NbMn_2$ as approximately 1480°C and confirmed the $MgZn_2$ structure for this compound. The following lattice parameters were measured: a = 4.879 A, c = 7.902 A, c/a = 1.620. No indications for a polymorphic transformation were found.

1. H. J. Wallbaum, *Z. Krist.*, **103,** 1941, 391–402; *Arch. Eisenhüttenw.*, **14,** 1941, 521–526.
2. P. Greenfield and P. A. Beck, *Trans. AIME*, **200,** 1954, 253–257.
3. R. P. Elliott, Armour Research Foundation, Chicago, Ill., Technical Report 1, OSR Technical Note OSR-TN-247, 1954.

$\bar{1}.9712$
0.0288

Mn-Ni Manganese-Nickel

The phase boundaries of the Mn-Ni system have been well established only for temperatures above 700–800°C. The intention to reinvestigate the low-temperature portion of this system has been recently announced by [1].

Solidification. Thermal data on the solidification of the entire system or of parts of it were reported by [2–8, 1, 9, 10]. Those of [1] are undoubtedly the most reliable ones and were plotted in Fig. 518. [1] found the minimum in the liquidus and solidus curves at 38 at. (39.5 wt.) % Ni, 1018°C; data given by other authors are 38 wt. (36.5 at.) % Ni, 1005°C [7] and 39.8 wt. (38.2 at.) % Ni, 1020°C [10]. Evidence for a peritectic reaction: melt + $\delta_{Mn} \rightleftarrows \gamma_{Mn}$ was first found by [7], and this was corroborated by [8, 1]. However, [9] claimed to have found that the δ-γ transformation has a maximum at about 2.8 at. % Ni, 1180°C, and ends in the nonvariant reaction $\delta \rightleftarrows \gamma$ + melt at 1170°C between 4.5 and 7.5 wt. (4.2 and 7.0 at.) % Ni.

Solid State. The existence of a continuous series of solid solutions at higher temperatures had been suggested by [2], and this could be corroborated by later work [7, 8, 1, 9, 10] (both γ-Mn [11] and Ni have f.c.c. lattices).

Mn-*rich Alloys.* [12] gave X-ray evidence that the $\gamma_{Mn} \rightleftarrows \beta_{Mn}$ transformation is lowered by addition of Ni. In micrographic studies, [7] found a broad ($\beta + \gamma$) field, extending from 4 to 21 at. % Ni below 800°C. The phase diagram by [8], worked out by means of thermal, microscopic, dilatometric, and resistometric analysis, shows the eutectoid reaction β (10.5 at. % Ni) $\rightleftarrows \alpha$ (3.4 at. % Ni) + γ (13 at. % Ni) at about 615°C. [13] found that Mn-rich alloys annealed at 500°C consisted of α and γ, which finding also suggests a eutectoidal decomposition of the β_{Mn} phase. [1] studied the constitution micrographically down to 600°C and observed a narrow ($\beta + \gamma$) field extending from 13.6 to 15.5 at. % Ni at 800°C. The $\alpha_{Mn} \rightleftarrows \beta_{Mn}$ transformation was also shown to be lowered. [9] also found that both the $\alpha_{Mn} \rightleftarrows \beta_{Mn}$ and the $\beta_{Mn} \rightleftarrows \gamma_{Mn}$ transformations are lowered by addition of Ni. Above 800°C their micrographically determined $\gamma/(\beta + \gamma)$ boundary is in fair agreement with that of [1].

In Fig. 518, the ($\beta + \gamma$) field above 800°C is that given by [1]. As to the constitution at lower temperatures, that shown in Fig. 518 is considered to be the best possible estimate from the conflicting evidence available.

Alloys of Intermediate Composition. In the first investigations of the system [2–4], numerous thermal effects were observed around the equiatomic composition between 480 and 970°C, which could be corroborated only in part by later work. The existence of a δ (= MnNi) phase was well established by [3, 4]; he also suggested phases based on the compositions Mn_3Ni_2 and Mn_3Ni_4. [14] claimed to have discovered a f.c.c. intermediate phase corresponding to the latter, but this was not confirmed by [15]. More recent thermal [7, 8, 1, 10], microscopic [7, 8, 1, 10], X-ray [15, 7, 1], resistometric [15, 8], and dilatometric [8, 10] studies have given ample evidence that only one "compound," based on the composition MnNi (51.66 wt. % Ni), exists in the concentration region under discussion. It undergoes, however, a polymorphic transformation [7, 8, 1, 10] at higher temperatures.

The phase equilibria involving the high-temperature modification, (MnNi)(H), were most thoroughly studied by [1] (Fig. 518); the results of [10] are in good agreement with those of [1]. The phase boundaries at lower temperatures have not been reliably established as yet. According to microscopic work by [8] and [10], the (MnNi) phase exists in the range 28–63 at. % (at 100°C) and 36.5–61.5 at. % Ni (at 300°C), with the single-phase region lying at 42–54 at. % (at 100°C) and 46.5–54.5 at. % Ni (at 300°C), respectively. Based on these data and also on those obtained by [7, 1] for 600°C, the low-temperature phase boundaries in Fig. 518 have been drawn tentatively.

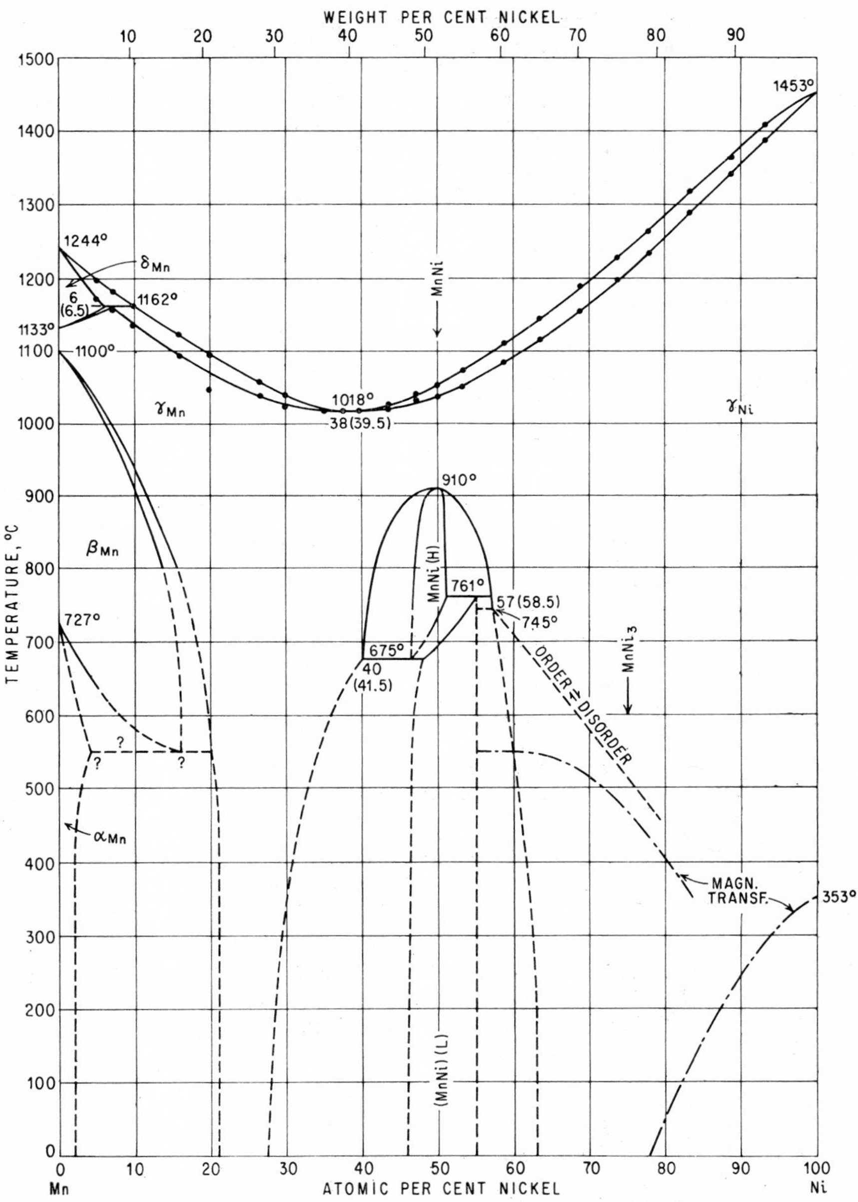

Fig. 518. Mn-Ni. (See also Note Added in Proof.)

The thermo-emf and specific resistance of alloys containing 50–53 at. % Ni have been measured by [16] between room temperature and 900°C.

Ni-*rich Alloys.* [14] found that alloys containing 60–85 wt. % Ni become strongly ferromagnetic after slow cooling or annealing at about 400°C [17]. This was interpreted by [14] as having been caused by the formation of an ordered atomic distribution based on the composition $MnNi_3$ (76.22 wt. % Ni). This ordering reaction has been studied subsequently by numerous authors [18, 15, 19–22, 7, 23–25, 8, 26, 27, 10, 28]. [14] already recognized that the temperatures critical for ordering lie above the Curie temperatures. The variation of the temperature of ordering, and/or of the Curie temperature, with composition has been measured by [14, 7, 10]. The order-disorder curve in Fig. 518 is based largely on thermal and elasticity measurements by [7] and the magnetic-transition curve on the data reported by [14, 7] which are in excellent agreement. At the composition $MnNi_3$, the transitions were found at 510 [20, 7], 520 [22], and 460 [20, 24], 480°C [7], respectively.

The lowering of the Curie point of Ni by addition of Mn has been studied by [14, 29, 19, 30, 7]. The data by [14, 7] lie at somewhat higher temperatures than those by [19, 30]. In Fig. 518 an interpolating curve has been drawn. Below about 85 at. % Ni this curve can, of course, be measured only in quenched alloys.

Crystal Structures. The low-temperature modification of the (MnNi) phase has a "face-centered tetragonal" [14, 15, 7] structure, CuAu ($L1_0$) type [1], with a = 3.732 A, c = 3.537 A, c/a = 0.95 at 52.5 wt. (50.8 at.) % Ni [7]. For the variation of the lattice parameters with the composition, see [15, 7].

Photographs taken from specimens quenched from above the transformation temperature show the lines of a f.c.c. lattice [7, 1]; however, work with a high-temperature camera [1] has shown conclusively that the (MnNi)(H) phase has a b.c.c. structure, with a = 2.974 A at 47.7 at. % Ni, 745°C.

Superstructure lines due to ordering in the $MnNi_3$ alloy could be observed in X-ray [18, 26] as well as neutron-diffraction [25, 34] studies. The superstructure is of the Cu_3Au ($L1_2$) type.

According to X-ray work on alloys quenched from high temperatures, the lattice parameter of the f.c.c. γ phase increases with increasing Mn content [15, 31, 7].

It was found by [15, 1] that Ni expands the lattice parameter of β-Mn. For the variation of the lattice parameters of the metastable tetragonal (γ'_{Mn}) phase with composition, see [32, 7].

Note Added in Proof. Recent paper on the constitution of Mn-rich Mn-Ni alloys: [33].

1. B. R. Coles and W. Hume-Rothery, *J. Inst. Metals*, **80,** 1951, 85–92. Discussion, *J. Inst. Metals*, **80,** 1952, 694–700.
2. S. Zemczuzny, G. Urasow, and A. Rykowskow, *Z. anorg. Chem.*, **57,** 1908, 261–266.
3. A. Dourdine, *J. Russ. Met. Soc.*, **1,** 1912, 11–23, 341–395 (in Russian); *Rev. mét.*, **12,** 1915, 125–133.
4. A. Dourdine, *Rev. mét.*, **29,** 1932, 507–518, 565–573.
5. N. Parravano, *Gazz. chim. ital.*, **42,** 1912, 372.
6. J. M. Paul and G. V. Beard, *J. Phys. Coll. Chem.*, **52,** 1948, 750–753.
7. W. Köster and W. Rauscher, *Z. Metallkunde*, **39,** 1948, 178–184.
8. N. N. Kurnakov and M. Ya. Troneva, *Doklady Akad. Nauk S.S.S.R.*, **68,** 1949, 73–76 (in Russian).
9. V. N. Eremenko and V. I. Skuratovskaya, *Ukrain. Khem. Zhur.*, **18,** 1952, 213–218 (in Russian).
10. V. N. Eremenko and T. D. Shtepa, *Ukrain. Khem. Zhur.*, **18,** 1952, 219–231 (in Russian).

11. U. Zwicker, *Z. Metallkunde,* **42,** 1951, 247, 327–330; Z. S. Basinski and J. W. Christian, *J. Inst. Metals,* **80,** 1952, 659–666.
12. E. Persson and E. Öhman, *Nature,* **124,** 1929, 333–334.
13. U. Zwicker, *Z. Metallkunde,* **42,** 1951, 246–252.
14. S. Kaya and A. Kussmann, *Z. Physik,* **72,** 1931, 293–309.
15. S. Valentiner and G. Becker, *Z. Physik,* **93,** 1935, 795–803.
16. T. Sato, *J. Phys. Soc. Japan,* **6,** 1951, 223–227. *Met. Abstr.,* **19,** 1952, 710.
17. A. Gray, *Phil. Mag.,* **24,** 1912, 1.
18. U. Dehlinger; see A. Kussmann, B. Scharnow, and W. Steinhaus, "Festschrift der Heraeus Vacuumschmelze," p. 319, Hanau (Main), 1933.
19. S. Valentiner, *Z. Physik,* **97,** 1935, 745–757.
20. N. Thompson, *Proc. Phys. Soc. (London),* **52,** 1940, 217–228.
21. S. Kaya and M. Nakayama, *Proc. Phys. Math. Soc., Japan,* **22,** 1940, 126–141. *Met. Abstr.,* **7,** 1940, 347. These authors were not able to define a critical temperature in which ordering sets in and state that the transition occurs over the range 600–300°C.
22. N. Vol'kenstein and A. Komar, *Zhur. Eksptl. i Teoret. Fiz.,* **11,** 1941, 723–724. *Met. Abstr.,* **10,** 1943, 279.
23. A. Komar and I. Portnyagin, *Doklady Akad. Nauk S.S.S.R.,* **60,** 1948, 569–570. *Met. Abstr.,* **20,** 1952, 91.
24. T. Sato, *J. Phys. Soc. Japan,* **3,** 1948, 198–201. *Chem. Abstr.,* **44,** 1950, 4851.
25. C. G. Shull and S. Siegel, *Phys. Rev.,* **75,** 1949, 1008–1010.
26. B. L. Averbach, *J. Appl. Phys.,* **22,** 1951, 1088–1089.
27. L. R. Aronin, *J. Appl. Phys.,* **23,** 1952, 642–643.
28. G. R. Piercy and E. R. Morgan, *Can. J. Physics,* **31,** 1953, 529–536.
29. P. Chevenard (1923); see Dourdine [4].
30. V. Marian, *Ann. phys.,* **7,** 1937, 459–527.
31. L. Vegard, *Skrifter Norske Videnskaps-Akad. Oslo, Mat. Naturv. Kl.,* no. 2, 1947, pp. 56–65.
32. E. Öhman, *Z. physik. Chem.,* **B8,** 1931, 87.
33. A. Hellawell and W. Hume-Rothery, *Phil. Trans. Roy. Soc.,* **A249,** 1957, 417–459; *Titanium Abstr. Bull.,* **2,** 1957, 2592.
34. C. G. Shull and M. K. Wilkinson, *Phys. Rev.,* **97,** 1955, 304–310.

0.2488
$\bar{1}$.7512

Mn-P Manganese-Phosphorus

By means of thermal [1–3], X-ray [4, 5, 2, 3], microscopic [3], and tensimetric investigations [6], it has been conclusively established that the following phosphides exist: Mn_3P (15.82 wt. % P) [5, 3], Mn_2P (21.99 wt. % P) [4, 5, 2, 6, 3], Mn_3P_2 (27.32 wt. % P) [3], MnP (36.06 wt. % P) [1, 4, 5, 2, 6, 3], and MnP_3 (62.85 wt. % P) [6]. Only the most recently published phase diagram presented in Fig. 519 [3] indicates the existence of all these phases, with the exception of the highest phosphide MnP_3, which could be identified only by tensimetric [6] and X-ray work [6]. The phases Mn_4P [2, 6] and Mn_5P_2 [1], shown in previously suggested phase diagrams, could not be confirmed.

Results of thermal analyses [1–3] agree in that the liquidus was found to have a maximum at the composition Mn_5P_2 [1] or Mn_2P [2, 3], but deviate as to the constitution of the alloys lower and higher in P. According to the diagram by [1], the phase Mn_5P_2 (melting point, 1390°C) forms eutectics with Mn and MnP at 9.5 at. % P, 964°C, and 40.5 at. % P, 1095°C, respectively. The thermal analysis by [2] revealed the existence of Mn_2P (melting point, 1326°C) instead of Mn_5P_2 and that of a lower phosphide, formed peritectically at 1085°C, which was erroneously identified as Mn_4P instead of Mn_3P. The Mn-Mn_4P (Mn_3P) eutectic was found at 10 at. % P, 960°C, and the Mn_2P-MnP eutectic at about 40.4 at. % P, 1080°C [2]. No indication of the existence of Mn_3P_2, which forms peritectically at 1090°C and decomposes into Mn_2P and MnP at about 1002°C (Fig. 519), was found by [1, 2]. Nor could the existence of a transformation of Mn_4P (Mn_3P) at 1016°C [2] be confirmed [3]; no roentgenographic evidence for such a transformation could be found [2, 3].

The phosphides in the region 0–50 at. % P are phases of fixed composition [5, 3]. The monophosphide was reported to undergo a transformation at 135°C [7]. MnP_3 is able to dissolve an excess of P [6].

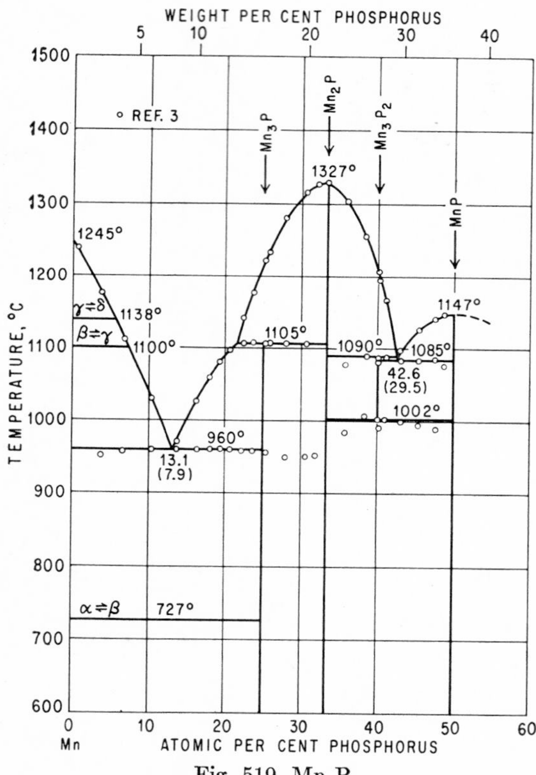

Fig. 519. Mn-P

The existence of Mn_3P [8] and Mn_3P_2 [9–11] had been reported earlier on the basis of inconclusive evidence, however.

Crystal Structures. Mn_3P is b.c. tetragonal with 32 atoms per unit cell and isotypic with Fe_3P, a = 9.178 A, c = 4.608 A, c/a = 0.502 [5]; a = 9.02 A, c = 4.58 A, c/a = 0.508 [3]. Mn_2P is isotypic with Fe_2P (C22 type), a = 6.09 A, c = 3.45_7 A, c/a = 0.568 [4, 5, 3]. MnP is the prototype of the B31 type of structure, a = 5.917 A, b = 5.260 A, c = 3.173 A [4, 5]. The powder photogram of MnP_3 was reported to be relatively simple [6].

Ferromagnetism of Mn-P Alloys. [12] reported Mn_2P to be ferromagnetic with a Curie temperature of about 40°C. [13] prepared alloys with 17–33.7 wt.

(26.6–47.4 at.) % P, by direct synthesis at 1000°C followed by slow cooling, and found their Curie temperature to be 25°C, independent of composition. The saturation magnetization increased with the P content. [14] prepared samples containing 47.1–58.2 at. % P by passing PCl_3 vapor over Mn at temperatures up to 660°C. They found the relative magnetic permeability and the abruptness of the permeability decrease (at about 18°C) to be distinctly greater in MnP than in compositions to either side and concluded that MnP is the only carrier of ferromagnetism in Mn-P alloys above 10°C.

1. S. F. Zemczuzny and N. Efremow, *Z. anorg. Chem.*, **57**, 1908, 241–252.
2. F. Wiechmann, *Z. anorg. Chem.*, **234**, 1937, 130–141.
3. J. Berak and T. Heumann, *Z. Metallkunde*, **41**, 1950, 19–23.
4. K. E. Fylking, *Arkiv Kemi, Mineral. Geol.*, **11B**, 1934, no. 48; see also **17A**, 1943, no. 7.
5. O. Årstad and H. Nowotny, *Z. physik. Chem.*, **B38**, 1937, 356–358.
6. W. Biltz and F. Wiechmann, *Z. anorg. Chem.*, **234**, 1937, 117–129.
7. L. F. Bates, *Phil. Mag.*, **8**, 1930, 714–732.
8. A. Schrötter, *Ber. Wien. Akad.*, **1**, 1849, 305.
9. F. Wöhler and Merkel, *Liebigs Ann.*, **86**, 1853, 371.
10. A. Granger, *Compt. rend.*, **124**, 1897, 190–191.
11. E. Wedekind and T. Veit, *Ber. deut. chem. Ges.*, **40**(2), 1907, 1268–1269.
12. H. Nowotny, *Z. Elektrochem.*, **49**, 1943, 254–260; quoted in [14].
13. C. Guillaud and H. Créveaux, *Compt. rend.*, **224**, 1947, 266–268.
14. K. H. Sweeny and A. B. Scott, *J. Chem. Phys.*, **22**, 1954, 917–921.

$\bar{1}.4234$
0.5766

Mn-Pb Manganese-Lead

The phase diagram in Fig. 520 is based on thermal data [1] of melts prepared using Mn of only 98.7% purity, having a melting point of 1228°C instead of 1245°C [2]. The composition of the Pb-rich melt at the monotectic temperature was not determined, though it was claimed to lie below 90 wt. (70.5 at.) % Pb, since an alloy

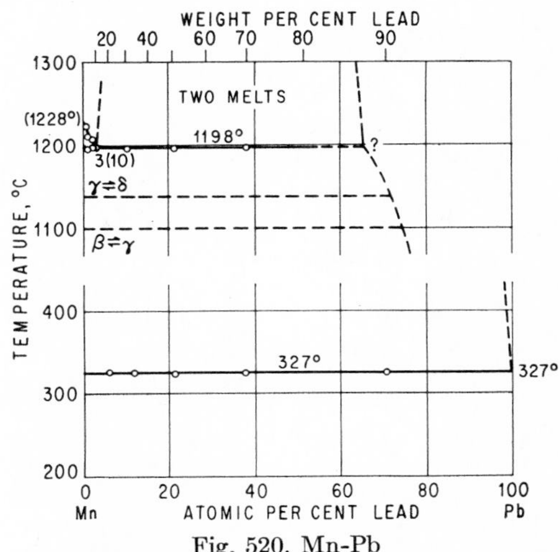

Fig. 520. Mn-Pb

of this composition did not show layer formation. Reinvestigation of the system under appropriate experimental conditions would be desirable.

1. R. S. Williams, *Z. anorg. Chem.*, **55,** 1907, 31–33.
2. Melts were prepared under an atmosphere of nitrogen!

$\bar{1}.7116$
0.2884

Mn-Pd Manganese-Palladium

Solidification. The liquidus and solidus temperatures indicated in Fig. 521 were determined by [1]. Prior to this work, [2] had reported thermal-analysis data

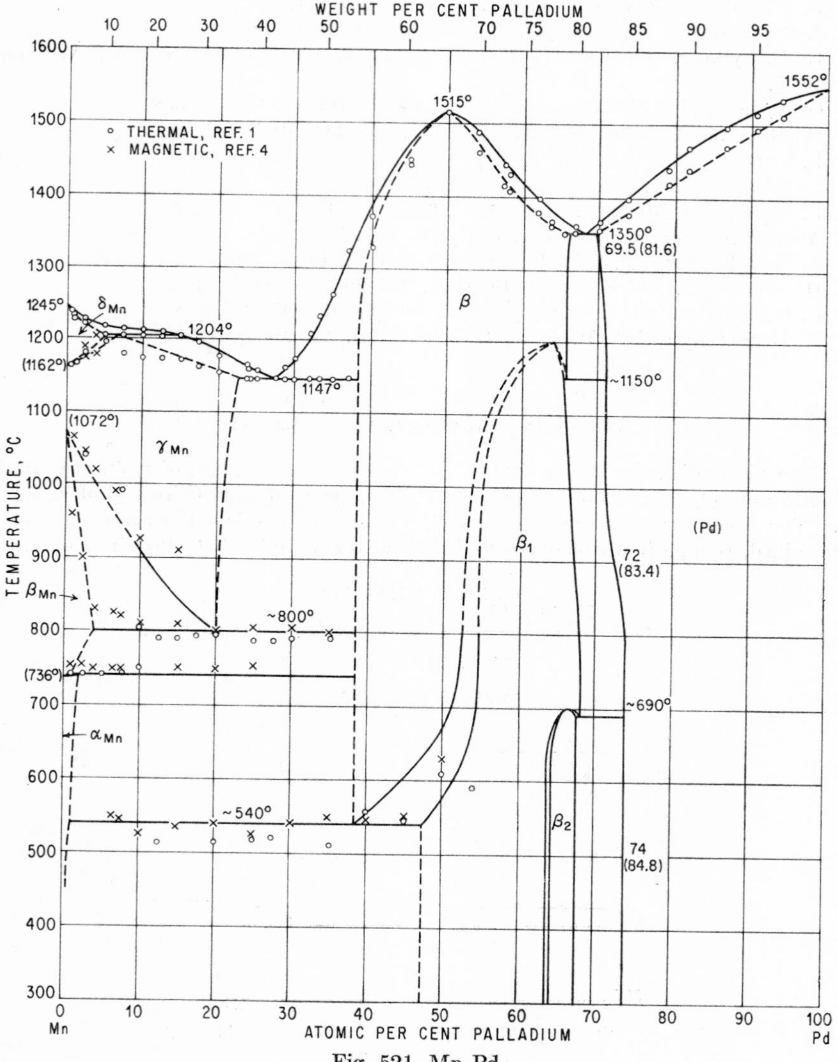

Fig. 521. Mn-Pd

according to which a minimum would exist at approximately 25 at. % Pd, 1110°C. However, the maximum at 50 at. % was not found.

[1] concluded (erroneously) from roentgenographic work that alloys with 50–100 at. % Mn, after quenching from 1200°C, had the f.c.c. structure of Pd. They believed, therefore, that there existed a series of solid solutions covering the range 40–100 at. % Pd. Since this appeared unlikely—in view of the fact that the liquidus had a maximum and a minimum in this composition range—[3] reinvestigated the system by means of micrographic analysis and X-ray investigations after quenching as well as at high temperatures. It was proved that a miscibility gap exists which extends from about 66 to 69.5 at. % Pd at the eutectic temperature, 1350°C (Fig. 521).

Solid State. Phase relations in the range 0–40 at. % Pd were determined by thermal analysis [1] and measurements of magnetic susceptibility vs. temperature [4], which were corroborated by limited microscopic [1, 3] and X-ray data [1, 5, 3]. None of the phase boundaries represents equilibrium conditions.

The phase boundaries in the region 40–100 at. % Pd are based on micrographic [3] and X-ray analysis data [3]. Since the β phase cannot be retained on quenching, the phase boundaries between the β and β_1 fields were outlined by high-temperature X-ray studies. On the other hand, the $\beta_1 \rightarrow \beta_2$ transformation is very sluggish; long-time annealings were necessary to complete this phase change.

Crystal Structures. Lattice constants of all phases after quenching as well as those of (Pd), β, β_1, and β_2, obtained by high-temperature work, were reported by [3]. [5] gave lattice constants of the γ_{Mn} phase after quenching from 1150°C. The c/a ratio of the metastable f.c. tetragonal γ_{Mn} phase (0.95–0.96 at 2.5–4 at. % Pd) was found to approach the value 1 at about 7 at. % Pd, after quenching from 1150°C [5]. However, [3] reported the c/a ratio of the alloy with 11.2 at. % Pd, after quenching from 1100°C, to be 1.016. In quenched alloys with 18 and 25.4 at. % Pd, the γ_{Mn} phase was found to be f.c.c. (see Cu-Mn).

β is b.c.c. of the CsCl (B2) type [3]. β_1 has a tetragonally distorted CsCl type of structure, and β_2 probably has a f.c. tetragonal structure with a complex superstructure [3]. Lattice parameters of the β and β_1 phases at 49 at. % Pd were found as follows: β, a = 3.150 A at 662°C; β_1, a = 2.992 A, c/a = 1.155 at 597°C; β_1, a = 2.890 A, c/a = 1.245 at room temperature. Lattice constants of the β_2 phase are a = 3.817 A, c/a = 1.059 at 66.3 at. % Pd [3].

1. G. Grube, K. Bayer, and H. Bumm, *Z. Elektrochem.*, **42,** 1936, 805–815.
2. A. T. Grigoriev, *Izvest. Inst. Fiz.-Khim. Anal.*, **7,** 1935, 75–87.
3. E. Raub and W. Mahler, *Z. Metallkunde*, **45,** 1954, 430–436.
4. G. Grube and O. Winkler, *Z. Elektrochem.*, **42,** 1936, 815–830.
5. G. Grube and O. Winkler, *Z. Elektrochem.*, **45,** 1939, 784–787.

$\bar{1}$.4493
0.5507

Mn-Pt Manganese-Platinum

[1] concluded from thermal data (Fig. 522) and measurements of the temperature coefficient of electrical resistance of alloys quenched from 1050°C that Mn-Pt melts crystallize to form an uninterrupted series of solid solutions [2]. It is more likely, however, that there is a continuous series of solid solutions between γ-Mn and Pt (rather than δ-Mn and Pt) since both these phases are f.c.c. Indeed, an alloy with 4.8 at. % Pt, after quenching from 1150°C, proved to have the metastable f.c. tetragonal structure (c/a = 0.96) formed on quenching by transformation of the cubic solid solution [3].

The curve of the temperature coefficient of electrical resistance of alloys annealed at 860°C showed peaks at the compositions Mn_4Pt, MnPt, and $MnPt_3$, believed to indicate the formation of ordered phases of these compositions [1]. This is certainly true for the compositions MnPt [1] and $MnPt_3$ [1, 4]; however, as far as the constitution of the Mn-rich alloys is concerned, it is to be expected that it is more complicated than was assumed by [1] on the basis of very scanty evidence.

The existence of an order-disorder transformation in the range of $MnPt_3$ was conclusively established by magnetic and X-ray investigations [4]. The ordered alloys have the structure of the Cu_3Au ($L1_2$) type. Alloys with 59.7 and 63.4 at. %

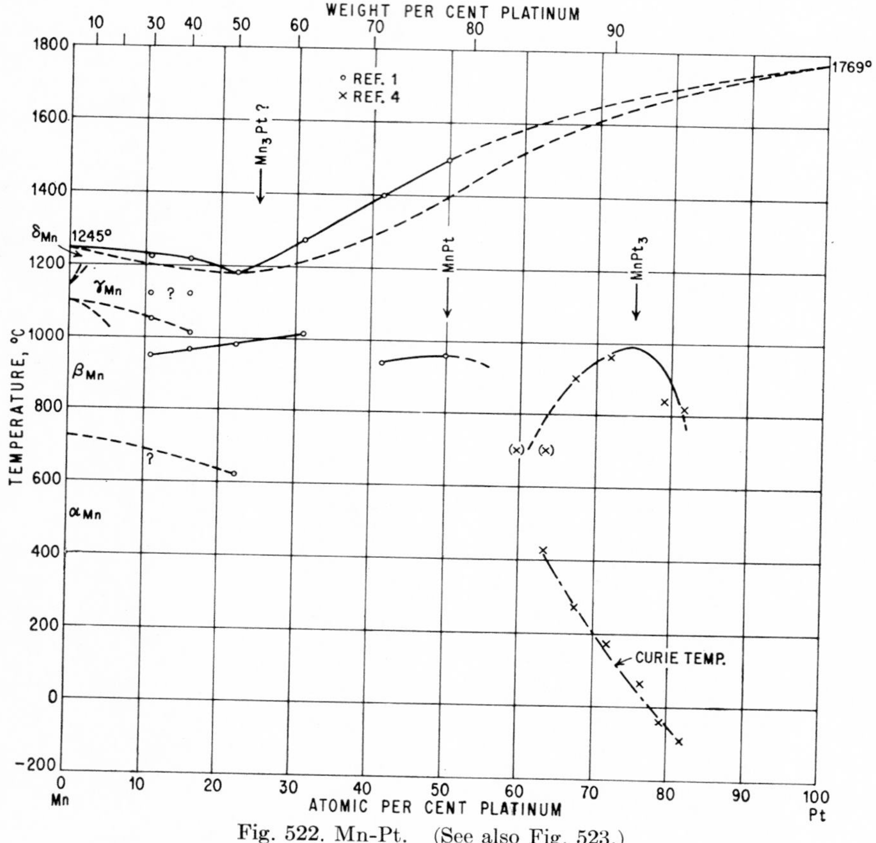

Fig. 522. Mn-Pt. (See also Fig. 523.)

Pt were found to be two-phase. The order-disorder transition temperatures indicated in Fig. 522 were obtained by dilatometric work, and the Curie temperatures by magnetic studies.

Supplement. The assumption that there is a continuous series of solid solutions in the Mn-Pt phase diagram has not been corroborated by [5], who studied the system by X-ray and micrographic analysis and arrived at the phase relations shown in Fig. 523. The liquidus of Fig. 523 is tentative since no cooling curves were taken. The saturation limits of the α_{Mn} and β_{Mn} phases lie below 2.9 at. (10 wt.) % Pt. The γ_{Mn} phase becomes ordered (Cu_3Au type) below 1050°C in the neighborhood of the composition Mn_3Pt. The structure of the β_1 phase is primitive tetragonal, with a =

2.827 A, $c/a = 1.298$ at 50 at. % Pt. There is microscopic evidence that a high-temperature modification (β in Fig. 523), presumably of the CsCl (B2) type, exists. The Cu_3Au-type ordering near the composition $MnPt_3$ could be confirmed; the transformation curve shown in Fig. 523 is that of [4] (see Fig. 522).

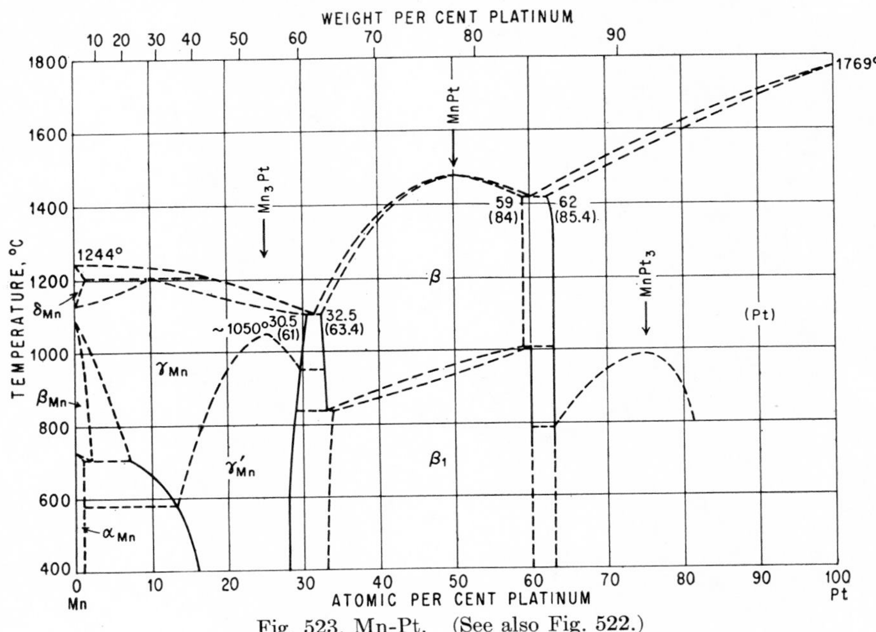

Fig. 523. Mn-Pt. (See also Fig. 522.)

X-ray data of 19 alloys with 2.9–82.9 at. % Pt after various heat-treatments were tabulated by [5].

1. V. A. Nemilov, T. A. Vidusova, and M. N. Pivovarova, *Izvest. Akad. Nauk S.S.S.R.* (*Khim.*), **1937**, 743–752.
2. The aluminothermic Mn used contained 0.35% Si, 0.30% Fe, and 0.29% Al.
3. U. Zwicker, *Z. Metallkunde*, **42**, 1951, 246–253.
4. M. Auwärter and A. Kussmann, *Ann. Physik*, **7**, 1950, 169–172.
5. E. Raub and W. Mahler, *Z. Metallkunde*, **46**, 1955, 282–290.

$\bar{1}.3614$
0.6386

Mn-Pu Manganese-Plutonium

The phase diagram of Fig. 523a has been established by [1] by means of thermal, microscopic, X-ray, densimetric, and microhardness investigations on alloy samples weighing some hundreds of milligrams at the most. "The temperature of the $\delta \rightarrow \epsilon$ polymorphous transformation is 30–40°C lower than the temperature of the corresponding change in the pure metal, a fact that may be associated with a slight solubility of Mn in the solid ϵ-phase of Pu. It is possible that the change occurring at 430°C is $\eta \rightarrow \epsilon$, rather than $\delta \rightarrow \epsilon$ [2]" [1].

The sole intermediate phase $PuMn_2$ is of slightly variable composition and has the f.c.c. $MgCu_2$ (C15) type of structure [1, 3]. The lattice parameter varies between

$a = 7.29$ A (Pu-rich) and $a = 7.26$ A (Mn-rich) [3]. Pu has a slight solubility in β-Mn; the cell parameter was found to vary over the range 6.293–6.317 A [1].

1. S. T. Konobeevsky, Conference of the Academy of Sciences of the U.S.S.R. on the Peaceful Uses of Atomic Energy, Division of Chemical Science, July, 1955; English translation by U.S. Atomic Energy Commission, pp. 210–211, Washington, 1956.
2. In pure Pu six allotropic forms were found to occur: α, 119°C; β, 218°C; γ, 310°C; δ, 450°C; η, 472°C; ε, 640°C (melting point).

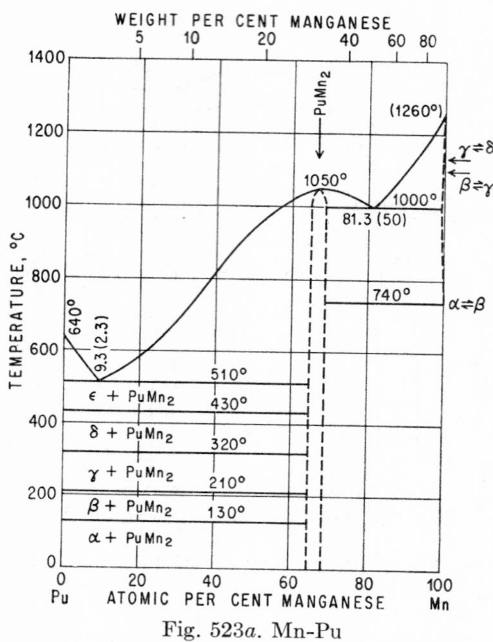

Fig. 523*a*. Mn-Pu

3. A. S. Coffinberry and M. B. Waldron, Review of Physical Metallurgy of Pu, in "Metallurgy and Fuels," Progress in Nuclear Energy, ser. V, vol. 1, Pergamon Press Ltd., London, 1956.

$\bar{1}.7274$
0.2726

Mn-Rh Manganese-Rhodium

An exploratory X-ray and microscopic study [1] of the Mn-Rh system yielded the phase relations outlined in the partial phase diagram of Fig. 524. The γ_{Mn} phase becomes ordered (Cu_3Au type) below 900°C in the neighborhood of the composition Mn_3Rh. The intermediate β phase, homogeneous between about 35 and 55 at. % Rh, has the CsCl (B2) type of structure, with $a = 3.051$ A at 50 at. % Rh.

The saturation limit of the extended solid solution of Mn in Rh was located at 67, 82, and 86 at. % Rh, respectively, at 1300, 1000, and 800°C. X-ray data of 11 alloys with 5.5–80.8 at. % Rh after various heat-treatments were tabulated by [1].

1. E. Raub and W. Mahler, *Z. Metallkunde*, **46**, 1955, 282–290.

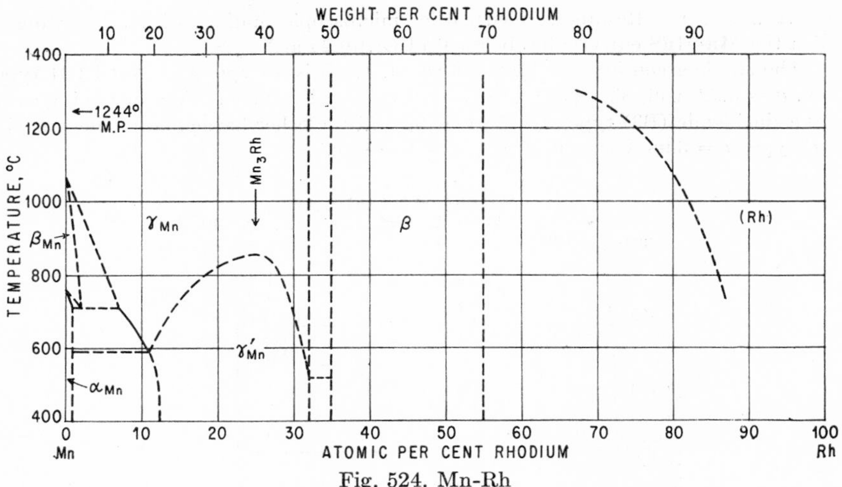

Fig. 524. Mn-Rh

$\bar{1}.7325$
0.2675

Mn-Ru Manganese-Ruthenium

According to an exploratory X-ray and microscopic study by [1] in which the solid-state phase boundaries were outlined (Fig. 525), no intermediate phase occurs in the Mn-Ru system.

1. E. Raub and W. Mahler, *Z. Metallkunde*, **46**, 1955, 282–290.

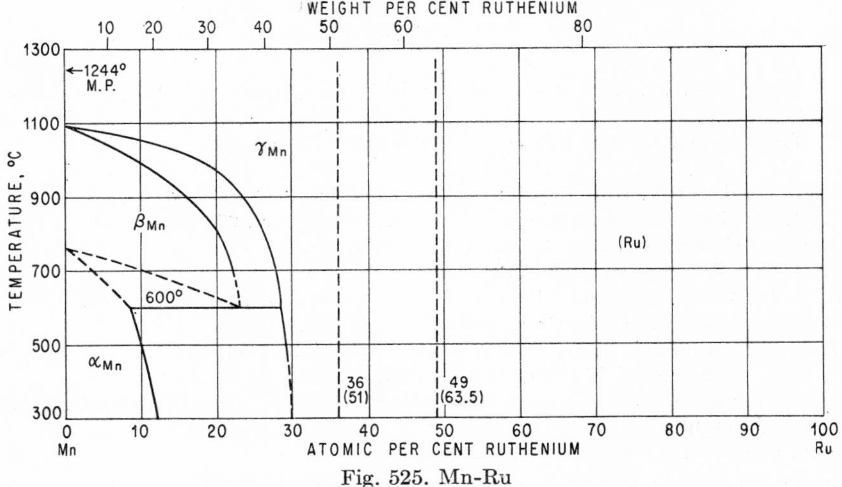

Fig. 525. Mn-Ru

0.2338
$\bar{1}.7662$

Mn-S Manganese-Sulfur

The phase diagram in Fig. 526 is due to [1]. Chemical analysis of the two solidified layers showed the miscibility gap in the liquid state to extend from about 0.3

to 33.2 wt. % S. Results verified earlier microscopic findings of [2], according to which the Mn-MnS eutectic lies below 0.14 wt. (0.24 at.) % S.

The stable green form of MnS (36.86 wt. % S) is isotypic with NaCl (B1 type) [3–8], a = 5.22 A [6, 8]. There are two unstable red forms of MnS, one being cubic of the zincblende (B3) type, a = 5.61 A [6], and the other hexagonal of the wurtzite (B4) type, a = 3.98 A, c = 6.44 A, c/a = 1.618 [6].

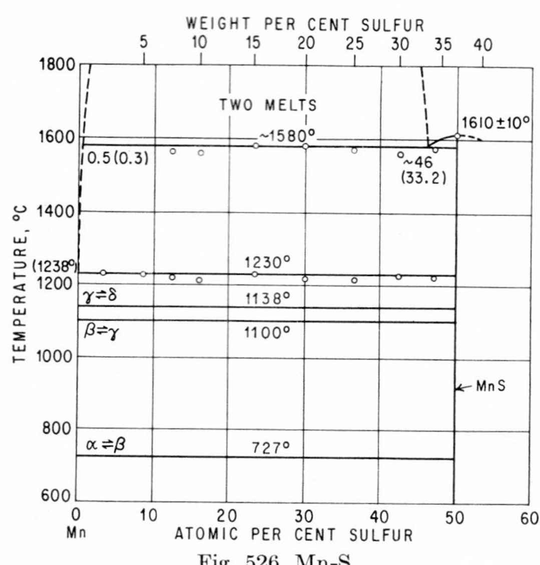

Fig. 526. Mn-S

MnS_2 (53.87 wt. % S) is isotypic with FeS_2 (C2 type) [9, 6, 10, 11], a = 6.108 A [10, 11].

There is no additional sulfide between MnS and MnS_2 [11].

1. R. Vogel and W. Hotop, *Arch. Eisenhüttenw.*, **11**, 1937-1938, 41–54.
2. H. Le Chatelier and M. Ziegler, *Bull. soc. chim. France*, **27**, 1902, 1140.
3. R. W. G. Wyckoff, *Am. J. Sci.*, **2**, 1921, 239–249.
4. H. Ott, *Z. Krist.*, **63**, 1926, 222–235.
5. H. B. Weiser and W. O. Milligan, *J. Phys. Chem.*, **35**, 1931, 2330–2344.
6. H. Schnaase, *Z. physik. Chem.*, **B20**, 1933, 89–117.
7. B. S. Ellefson and N. W. Taylor, *J. Chem. Phys.*, **2**, 1934, 58–64.
8. F. A. Kröger, *Z. Krist.*, **100**, 1939, 543–545.
9. P. P. Ewald and W. Friedrich, *Ann. Physik*, **44**, 1914, 1183–1196; *Physik. Z.*, **15**, 1914, 399–401.
10. F. Offner, *Z. Krist.*, **89**, 1934, 182–184.
11. W. Biltz and F. Wiechmann, *Z. anorg. Chem.*, **228**, 1936, 268–274.

$\bar{1}.6543$
0.3457

Mn-Sb Manganese-Antimony

Solidification temperatures were determined by [1, 2]. Qualitatively, results are in substantial agreement; however, there are quite considerable temperature differences, probably due mainly to the different purities of the manganese used [3]. The temperatures of the two eutectics (Fig. 527) were given as 900 [1] vs. 922°C [2] and 577 [1] vs. 570°C [2] and that of the peritectic reaction as 853 [1] and 872°C [2].

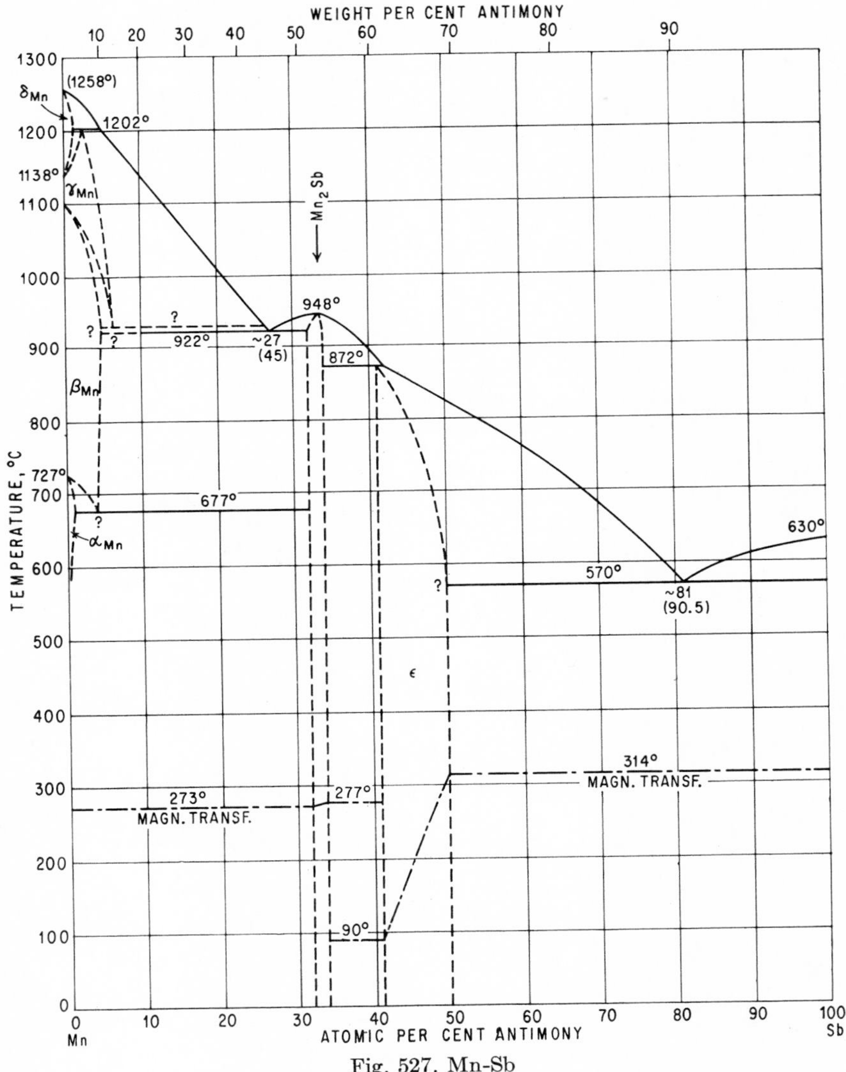

Fig. 527. Mn-Sb

The maximum at the composition Mn_2Sb (52.57 wt. % Sb) was found at 919 [1] and 948°C [2].

The effect of Sb on the transformation temperatures of Mn was studied only in an exploratory fashion. Since the $\gamma \rightleftarrows \delta$ transformation was still unknown, only the $\alpha \rightleftarrows \beta$ and $\beta \rightleftarrows \gamma$ transformations were considered, and these were found at 750 and 1060°C, respectively [2]. In the original diagram, the three-phase equilibrium at 1202°C (Fig. 527) was shown as the peritectic reaction: γ_{Mn} + melt $\rightleftarrows \beta_{Mn}$. However, it very likely corresponds to the reaction δ_{Mn} + melt $\rightleftarrows \gamma_{Mn}$. Since no additional horizontal was detected by [2], Fig. 527 was drawn to show the necessary three-phase equilibrium between β_{Mn}, γ_{Mn}, and melt, just slightly above the eutectic horizontal at 922°C, assuming that the two thermal effects could not be observed separately. Manganese-rich alloys quenched from a temperature above the $\beta \rightleftarrows \gamma$ transformation were found to transform completely to β-Mn on quenching [4].

Whereas [1] concluded the existence of two intermediate phases of variable composition with approximately 31–35 and 39.5–50 at. % Sb, other investigators believed they found microscopic [2] and roentgenographic evidence [5] for an additional phase of fixed composition, i.e., the compound MnSb (68.91 wt. % Sb), which was assumed to be formed peritectically at 809°C [2]. However, X-ray analysis [6, 7] and magnetic investigations [8] have conclusively established that there is no justification for assuming the existence of MnSb besides the NiAs-type phase of variable composition designated as ϵ in Fig. 527.

The intermediate phase richest in Mn is based on the composition Mn_2Sb [1, 2, 9]; its homogeneity range extends from about 32 to 33.3 at. % Sb [8]. The homogeneity range of the ϵ phase was given as about 40–50 at. % Sb [6, 7] and about 42–50 at. % Sb [8].

Both Mn_2Sb and ϵ are ferromagnetic [10–13, 8]. Their Curie temperatures were determined by [1, 12, 13, 8]; those of the last author are shown in Fig. 527.

Crystal Structures. Mn_2Sb is tetragonal and isotypic with Cu_2Sb (C38 type), a = 4.09 A, c = 6.57 A, c/a = 1.606 [9, 14]. The ϵ phase is hexagonal of the NiAs (B8) type [6, 7, 15], a = 4.128 A, c = 5.789 A, c/a = 1.402 at about 50 at. % Sb [15, 16].

Supplement. The lattice constants and the point position parameters of the structure of Mn_2Sb have been refined by [17]. For the former the values a = 4.078 A, c = 6.557 A are reported.

1. R. S. Williams, *Z. anorg. Chem.*, **55,** 1907, 2–7.
2. T. Murakami and A. Hatta, *Science Repts. Tôhoku Univ.*, **22,** 1933, 88–100.
3. 98.7% [1] vs. 99.8% [2].
4. U. Zwicker, *Z. Metallkunde*, **42,** 1951, 251.
5. A. Hatta and A. Osawa, *Nippon Kinzoku Gakkai-Shi*, **2,** 1938, 270–283; *Met. Abstr.*, **5,** 1938, 682.
6. I. Oftedal, *Z. physik. Chem.*, **128,** 1927, 135–153; **132,** 1928, 215; see also A. Westgren, *Metallwirtschaft*, **9,** 1930, 920.
7. H. Pfisterer and K. Schubert, *Z. Metallkunde*, **41,** 1950, 364–365.
8. C. Guillaud, *Ann. phys.*, **4,** 1949, 671–703; *Rev. mét.*, **46,** 1949, 453–456; *J. recherches centre natl. recherche sci., Labs. Bellevue (Paris)*, **1947,** 15–21.
9. F. Halla and H. Nowotny, *Z. physik. Chem.*, **B34,** 1936, 141–144.
10. F. Heusler, *Z. anorg. Chem.*, **17,** 1904, 262.
11. E. Wedekind, *Z. Elektrochem.*, **11,** 1905, 850–851; *Z. physik. Chem.*, **66,** 1909, 614–632.
12. K. Honda, *Ann. Physik*, **32,** 1910, 1017–1023.
13. K. Honda and T. Ishiwara, *Science Repts. Tôhoku Univ.*, **6,** 1917, 9–21.
14. [5] also reported a tetragonal structure but gave lattice constants of a = 8.131 A, c = 12.999 A, c/a = 1.599, 24 atoms per unit cell.
15. B. T. M. Willis and H. P. Rooksby, *Proc. Phys. Soc. (London)*, **67A,** 1954, 290–296.
16. [5] reported the a axis to be twice as large as that given by [6, 7, 15].
17. L. Heaton and N. S. Gingrich, *Acta Cryst.*, **8,** 1955, 207–210.

$\bar{1}.8424$
0.1576

Mn-Se Manganese-Selenium

α-MnSe (58.97 wt. % Se) is isotypic with NaCl (B1 type) [1, 2], a = 5.46 A [1]. Besides this stable modification, two unstable forms were reported to exist, which transform into α-MnSe: β-MnSe, isotypic with zincblende (B3 type), a = 5.83 A; γ-MnSe, isotypic with wurtzite (B4 type), a = 4.13 A, c = 6.73 A, c/a = 1.63 [2].

$MnSe_2$ (74.19 wt. % Se) has the structure of the pyrite (C2) type, a = 6.43 A [3].

As for phase transitions of MnSe at temperatures below room temperature, see [4, 5].

1. E. Broch, *Z. physik. Chem.*, **127**, 1927, 446–454.
2. A. Baroni, *Z. Krist.*, **99**, 1938, 336–339.
3. N. Elliott, *J. Am. Chem. Soc.*, **59**, 1937, 1958–1962.
4. R. Lindsay, *Phys. Rev.*, **84**, 1951, 569–571.
5. M. Murakami, *Bull. Fac. Eng., Hiroshima Univ.*, no. 2, 1953, 67–72; *Chem. Abstr.*, **48**, 1954, 1747.

0.2913
$\bar{1}.7087$

Mn-Si Manganese-Silicon

Partial System Mn-MnSi. Phase relations presented in Fig. 528 are based on thermal and microscopic work [1]. The existence of the three intermediate phases shown was verified by means of X-ray investigations: Mn_3Si (14.56 wt. % Si) by [2, 3], Mn_5Si_3 (23.48 wt. % Si) by [2, 3], and MnSi (33.84 wt. % Si) by [4].

Previously, [5] had concluded from exploratory thermal and metallographic studies that the phases Mn_2Si (instead of Mn_5Si_3) and MnSi existed. The Mn_2Si (Mn_5Si_3)-MnSi eutectic was found as about 31 wt. (46.8 at.) % Si, 1230°C, in good agreement with the findings of [1].

The effect of Si on the $\gamma \rightleftarrows \delta$ transformation of Mn is unknown. The $\beta \rightleftarrows \gamma$ transformation appears to be raised, giving rise to a peritectic equilibrium at 1155°C (Fig. 528). The existence of a wide range of β_{Mn} solid solutions (up to about 14 at. % Si) was confirmed by roentgenographic work [3]; see also [6].

Between 9 and 13 wt. (16.2–22.6 at.) % Si, [1] observed thermal effects at 980°C. The nature of this transformation is not known. It may correspond to the formation of an additional intermediate phase (Mn_5Si?) or be caused by impurities of the manganese used [1]. Some indications of the presence of an unknown phase (or phases) in alloys with 14 and 18 at. % Si, after annealing at 600–800°C, were reported by [3]. It is likely that phase relations in this composition range are more complicated than indicated in Fig. 528.

Partial System MnSi-Si. The thermal data shown in Fig. 528 are due to [5, 7]. Microexamination revealed the existence of an intermediate phase in the vicinity of 45 wt. (61.6 at.) % Si, later identified as $MnSi_2$ (50.56 wt. % Si) [4]. By metallographic studies of a series of arc-melted alloys covering the range 50–100 at. % Si, it has been conclusively established that $MnSi_2$ has a concealed maximum and forms a eutectic with Si [8].

Crystal Structures. The lattice constant of β-Mn decreases by dissolving Si [9, 4, 3]. Mn_3Si is b.c.c. of the A2 type, a = 2.86 A [2, 3]. Mn_5Si_3 is hexagonal with 16 atoms per unit cell ($D8_8$ type), a = 6.912 A, c = 4.812 A, c/a = 0.696 [4, 2, 3]. MnSi is isotypic with FeSi (B20 type), a = 4.557 A [4]; and $MnSi_2$ is tetragonal with 48 atoms per unit cell and a = 5.524 A, c = 17.46 A, c/a = 3.16 [4].

Note Added in Proof. A transformation in the Mn_3Si phase near 600°C is suggested by measurements of heat capacity and thermal dilatation [10].

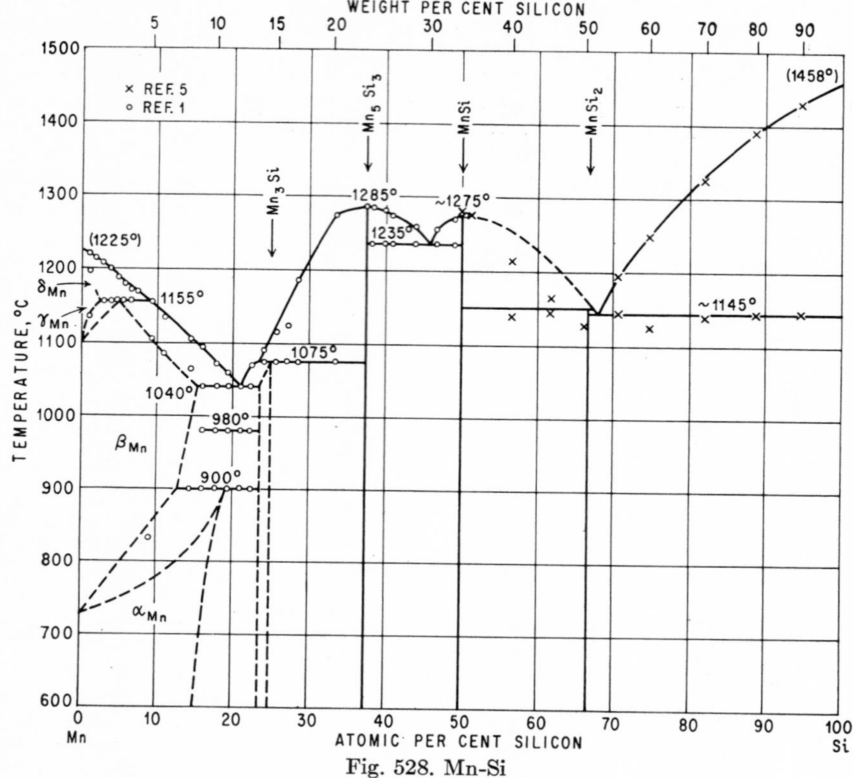

Fig. 528. Mn-Si

1. R. Vogel and H. Bedarff, *Arch. Eisenhüttenw.*, **7**, 1933-1934, 423–425. Alloys were prepared using Mn of 97% purity; alloys were melted under argon and not analyzed.
2. F. Laves, *Z. Krist.*, **89**, 1934, 189–191.
3. K. Åmark, B. Borén, and A. Westgren, *Metallwirtschaft*, **15**, 1936, 835–836; see also [4].
4. B. Borén, *Arkiv Kemi, Mineral. Geol.*, **11A**, 1933, no. 10.
5. F. Doerinckel, *Z. anorg. Chem.*, **50**, 1906, 117–132.
6. U. Zwicker, *Z. Metallkunde*, **42**, 1951, 251.
7. Mn 99.4%, Si 98–98.9%.
8. W. R. Johnson and M. Hansen, unpublished work; AF Technical Report 6383, June, 1951.
9. A. Westgren and G. Phragmén, *Z. Physik*, **33**, 1925, 785.
10. K. N. Davydov and P. V. Gel'd, *Fiz. Metal. i Metalloved.*, **2**, 1956, 192; *Met. Abstr.*, **24**, 1957, 802.

$\bar{1}.6654$
0.3346

Mn-Sn Manganese-Tin

Thermal-analysis data shown in Fig. 529 are due to [1]. The author concluded that three intermediate phases, Mn_4Sn (35.07 wt. % Sn), Mn_2Sn (51.93 wt. % Sn), and probably MnSn (68.36 wt. % Sn), exist. The effect of Sn on the temperatures of transformation of Mn is unknown. Magnetic measurements [2] proved to be essentially in accord with the phase diagram by [1]. Potential measurements in 1 *N* KOH [3] showed a discontinuous change at the composition Mn_3Sn (41.87 wt. % Sn). [4], who electrolyzed manganese into amalgamated tin, claimed to have found the phases Mn_2Sn, MnSn, $MnSn_2$ (81.21 wt. % Sn), and $MnSn_3$ (86.64 wt. % Sn).

Results of roentgenographic analyses [5, 6] substantially agree as to the crystal structure of the three intermediate phases existing, but differ as far as quantitative findings are concerned (see below). The solid solubility of Sn in both α-Mn (at 490–640°C) and β-Mn (at 815°C) was estimated, on the basis of measurements of lattice parameter, to be "rather small" [5]. On the other hand, [6] concluded from parameter measurements that about 17 wt. (8.7 at.) % Sn was soluble in β-Mn [7]. Microscopic examination [1] placed the solubility limit, after annealing at 950°C, at approximately

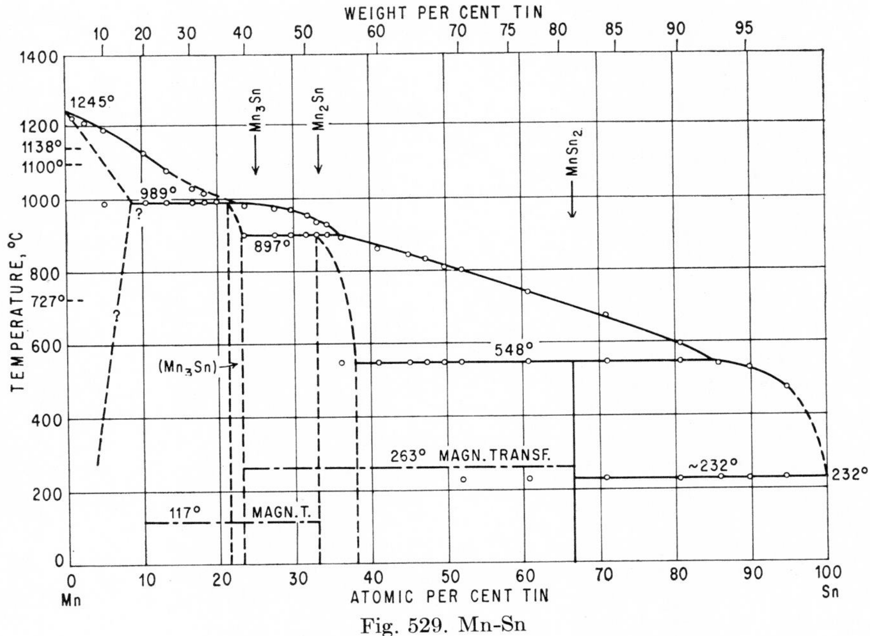

Fig. 529. Mn-Sn

8 wt. (4 at.) % Sn (?). The solid solubility of Mn in Sn appears to be negligibly small [8, 5].

On the basis of crystal-structure determinations, the phase in equilibrium with Mn was concluded to be of the ideal composition Mn_3Sn (25 at. % Sn), but having an estimated homogeneity range of about 23–24.5 at. % Sn [5]. [6] assumed the composition $Mn_{11}Sn_3$ (21.43 at. % Sn), since this alloy proved to be single-phase, and claimed that it had a vary narrow range of homogeneity, indeterminable by X-ray diffraction (see Supplement).

The intermediate phase formed peritectically at 897°C is Mn_2Sn [5, 6]. According to [5], X-ray data indicate that it has the roughly estimated homogeneity range of 35–40 at. % Sn (i.e., the 33.3 at. % alloy would be heterogeneous). However, [6] reported it to be homogeneous between about 32.8 and 36.1 at. % Sn (see Supplement).

The Sn-rich intermediate phase was identified as the compound $MnSn_2$ [5, 6].

[5] suggested it to have a maximum melting point, rather than being formed peritectically (Fig. 529), but [6] stated that the sluggishness of the peritectic reaction at 548°C can result in the formation of eutectic-like structures. Thermal data [1] are definitely in favor of an incongruent melting point.

Crystal Structures. Mn_3Sn is hexagonal of the Ni_3Sn ($D0_{19}$) type, with 8 atoms per unit cell [5, 6]. However, [6] found some indication that the structure may be more complicated. According to [5], the parameters are a = 5.665 A, c = 4.533 A, c/a = 0.800 if saturated with Mn at 490–640°C and a = 5.674 A, c = 4.537 A, c/a = 0.800 if saturated with Sn at 475–635°C. [6] gave a = 5.66 A, c = 4.52 A, $c/a = 0.79_8$.

The structure of Mn_2Sn is of the "filled" NiAs (B8) type. In Fig. 529 its homogeneity range was tentatively assumed as 33–38 at. % Sn. Lattice parameters were reported to be as follows: a = 4.385 A, c = 5.504 A, c/a = 1.255 (saturated with Mn at 475°C), a = 4.351 A, c = 5.497 A, c/a = 1.263 (saturated with Sn at 450°C) [5]; $a = 4.40_1$ A, $c = 5.46_8$ A, $c/a = 1.24_2$ (Mn-side), $a = 4.37_9$ A, $c = 5.48_6$ A, $c/a = 1.25_3$ (Sn-side) [6].

$MnSn_2$ is b.c. tetragonal of the $CuAl_2$ (C16) type [6], with a = 6.659 A, c = 5.436 A, c/a = 0.816 [5] or a = 6.660 A, c = 5.445 A, c/a = 0.817 [6].

Supplement. [9] studied alloys with 20–100 at. % Sn, apparently in the as-cast state, by means of magnetic measurements and X-ray diffraction. He concluded, without having knowledge of the findings of [5, 6], the existence of three intermediate phases with 20–22, 31–37, and 50 at. % Sn. The first two phases are identical with Mn_3Sn and Mn_2Sn, identified by [5, 6]. However, MnSn definitely does not exist since the intermediate phase richest in Sn is $MnSn_2$ [5, 6].

1. R. S. Williams, *Z. anorg. Chem.*, **55,** 1907, 24–31. The Mn used contained 98.71% Mn, 0.64% Fe, and 0.32% SiO_2 (rest?). Alloys were melted in an N_2 atmosphere!
2. K. Honda, *Ann. Physik*, **32,** 1910, 1023–1025.
3. N. Puschin, *Zhur. Russ. Fiz.-Khim. Obshchestva*, **39,** 1907, 869–897; *Chem. Abstr.*, **2,** 1908, 62.
4. A. S. Russell, T. R. Kennedy, and R. P. Lawrence, *J. Chem. Soc.*, **1934,** 1750–1754.
5. O. Nial, Dissertation, University of Stockholm, 1945; *Arkiv Kemi, Mineral. Geol.*, **17B,** 1944, no. 11.
6. H. Nowotny and K. Schubert, *Z. Metallkunde*, **37,** 1946, 17–23.
7. U. Zwicker, *Z. Metallkunde*, **42,** 1951, 251.
8. D. Hanson and E. J. Sandford, *J. Inst. Metals*, **56,** 1935, 196–200.
9. C. Guillaud, *J. recherches centre natl. recherche sci., Labs. Bellevue (Paris)*, **1947,** 15–21.

$\bar{1}.4824$
0.5176

Mn-Ta Manganese-Tantalum

The phase $TaMn_2$ (37.79 wt. % Mn) is hexagonal of the $MgZn_2$ (C14) type [1, 2], a = 4.87 A, c = 7.95 A, c/a = 1.633 [1]; a = 4.864 A, c = 7.947 A, c/a = 1.634 [2]. By X-ray examination of quenched alloys, no polymorphic change could be found between 800 and 1400°C [2]. There appears to be no σ-type phase in this system [3].

1. H. J. Wallbaum, *Z. Krist.*, **103,** 1941, 391–402; *Arch. Eisenhüttenw.*, **14,** 1941, 521–526.
2. R. P. Elliott, Armour Research Foundation, Chicago, Ill., Technical Report 1, OSR Technical Note OSR-TN-247, August, 1954, p. 24.
3. P. Greenfield and P. A. Beck, *Trans. AIME*, **200,** 1954, 253–257.

$\bar{1}.6339$
0.3661

Mn-Te Manganese-Tellurium

X-ray analysis [1] of a series of alloys covering the composition range 33.3–77.8 at. % Te and prepared by powder-metallurgy techniques [2] showed that only the two tellurides MnTe (69.91 wt. % Te) [3, 4] and $MnTe_2$ (82.29 wt. % Te) [5, 6] exist. Both phases have very restricted homogeneity ranges [1].

MnTe is isotypic with NiAs (B8 type), a = 4.132 A, c = 6.712 A, c/a = 1.624 [4]; a = 4.146 A, c = 6.709 A, c/a = 1.618 [1]. $MnTe_2$ has the structure of the FeS_2 (C2) type, a = 6.957 A [5, 6]; a = 6.951 A [1].

Note Added in Proof. The magnetic and electric properties of antiferromagnetic MnTe were investigated by [7]. The results suggest a change of crystal structure of the compound at ~130°C.

1. S. Furberg, *Acta Chem. Scand.*, **7**, 1953, 693–694.
2. R. Ochsenfeld (*Ann. Physik*, **12**, 1932, 354–355) was unable to obtain homogeneous melts.
3. E. Wedekind and T. Veit, *Ber. deut. chem. Ges.*, **44**, 1911, 2667–2668.
4. I. Oftedal, *Z. physik. Chem.*, **128**, 1927, 135–153.
5. I. Oftedal, *Z. physik. Chem.*, **135**, 1928, 291–299.
6. N. Elliott, *J. Am. Chem. Soc.*, **59**, 1937, 1958–1962.
7. E. Uchida, H. Kondoh, and N. Fukuoka, *J. Phys. Soc. Japan*, **11**, 1956, 27–32; *Met. Abstr.*, **23**, 1956, 814.

$\bar{1}.3741$
0.6259

Mn-Th Manganese-Thorium

The following intermediate phases have been identified [1]: (a) $ThMn_2$ (32.12 wt. % Mn), isotypic with $MgZn_2$ (C14 type), a = 5.48 A, c = 8.95 A, c/a = 1.633; (b) Th_6Mn_{23} [79.30 at. (47.57 wt.) % Mn], f.c.c. of a new type, a = 12.523 A; (c) $ThMn_{12}$ [92.31 at. (73.96 wt.) % Mn] with a b.c. tetragonal structure similar to but not identical with that of the corresponding Fe, Co, and Ni compounds, a = 8.74 A, c = 4.95 A, c/a = 0.566. [2] reported that a eutectic occurs at 911°C at a composition greater than 20 at. % Mn and that the addition of 1 at. % Mn lowers the Th lattice constant to a = 5.0746 A.

1. J. V. Florio, R. E. Rundle, and A. I. Snow, *Acta Cryst.*, **5**, 1952, 449–457.
2. H. A. Wilhelm and O. N. Carlson, unpublished information (1946); quoted in H. A. Saller and F. A. Rough, Compilation of U.S. and U.K. Uranium and Thorium Constitutional Diagrams, *U.S. Atomic Energy Comm., Publ.* BMI-1000, June, 1955.

0.0595
$\bar{1}.9405$

Mn-Ti Manganese-Titanium

The principal investigation was carried out by [1], using arc-cast iodide Ti-base alloys. The boundaries of the α and β fields and the composition of the intermediate phases were determined micrographically (Fig. 530). Thermal analysis was used to determine solidification temperatures in the region above 30 at. % Mn.

It was concluded that a phase of the approximate composition TiMn (53.42 wt. % Mn) was formed peritectically at about 1200°C by reaction of melt (about 42 at. % Mn) with a phase nearly corresponding to the composition $TiMn_2$ (69.64 wt. % Mn) (see inset of Fig. 530). This latter phase had been identified earlier [2, 3].

Although there is agreement as to the existence of TiMn, its mode of formation is a controversial matter. In a reinvestigation of the range 40–70 wt. % Mn, no

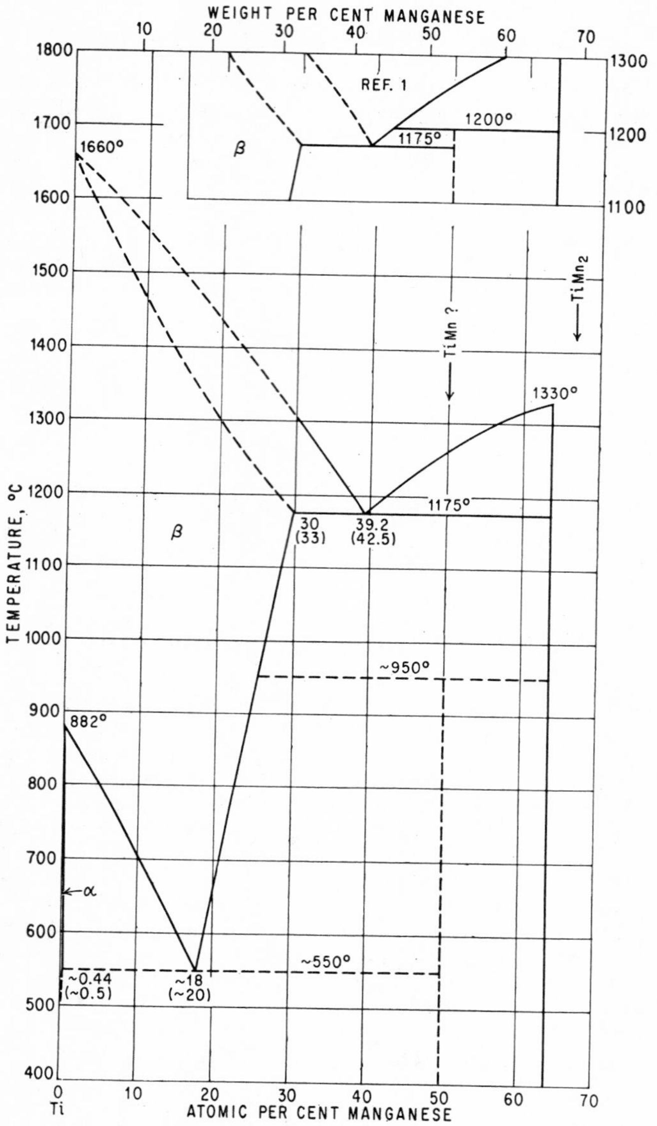

Fig. 530. Mn-Ti. (See also Note Added in Proof.)

evidence for the occurrence of TiMn at higher temperatures, and accordingly its peritectic formation, could be found [4]. Instead, both X-ray and metallographic data revealed that the phase is formed in the solid state by a peritectoid reaction between 900 and 1000°C. The evidence presented was disputed by [5]. On the other hand, however, [6], investigating the Ti-Mn system based on magnesium-reduced Ti, reported the peritectoid formation of TiMn from β and $TiMn_2$ at about 950°C. In a later report, however, these same authors [7] refuted the evidence and left the controversial region open [8]. (See Supplement.)

[9] determined the $\beta/(\alpha + \beta)$ boundary up to about 4 at. % Mn by measuring the hydrogen pressure in equilibrium with alloys containing extremely dilute solutions of hydrogen, as a function of temperature. The curve was found to lie below that reported by [1], which is shown in Fig. 530, about 55°C lower at 4 at. % Mn [10]. For retention of the b.c.c. β structure on quenching, 6.4 wt. (5.6 at.) % Mn is the minimum concentration [1]. The M_s curve for iodide Ti–Mn alloys was determined by [11].

The eutectoid decomposition of β is an extremely sluggish reaction. No conclusive evidence for the decomposition in iodide Ti-base alloys was found, although annealing times of 20 days at 540°C and 60 days at 500°C were used [1]. In alloys based on magnesium-reduced Ti, the eutectoid temperature was found at about 675°C [7, 12].

Crystal Structures. $TiMn_2$ is isotypic with $MgZn_2$ (C14 type) [2, 3, 1, 13], $a = 4.82$ A, $c = 7.89_6$ A, $c/a = 1.638$ [3]; $a = 4.825$ A, $c = 7.917$ A, $c/a = 1.641$ [13]. TiMn was reported to be isotypic with the σ-phase family of structures and, therefore, designated as TiMn, although its composition is not exactly known [14]; lines indexed as a tetragonal structure gave parameters of $a = 8.880$ A, $c = 4.542$ A, $c/a = 0.512$.

[2] reported a phase Ti_2Mn (36.44 wt. % Mn) as being f.c.c., with 96 atoms per unit cell. No evidence for its existence was found by [15, 1, 7]; see especially [15].

Lattice constants of several β solid-solution alloys were determined by [1].

Supplement. In a brief note, [16] have presented data indicating that in the 60–70 wt. % Mn region several intermediate phases exist. Until further evidence is brought out, the diagram in Fig. 530 should be accepted.

According to [17], $TiMn_2$ undergoes no polymorphic transformation in the temperature range 600–1300°C. This author reported the lattice constants $a = 4.817$ A, $c = 7.885$ A, $c/a = 1.637$.

Note Added in Proof. An abstract of a recent paper on the constitution of Mn-rich Ti-Mn alloys [18] reads as follows: "Results are tabulated and a phase diagram is shown for the Ti-Mn system at temperatures above 1100°C and with 50–100 at. % Mn. Ti was soluble in δ-Mn; the liquidus and solidus curves for δ solid solution fell to a eutectic horizontal at 1204°C. The liquid was then in equilibrium with δ and with an intermediate phase, probably $TiMn_3$, formed by a peritectic reaction which occurred at 1230–1250°C. The liquidus rose to a maximum at 1325°C, at which temperature the liquid was in equilibrium with $TiMn_2$. A secondary arrest, at 1181°C, was shown for alloys in the range 50–67 at. % Mn."

1. D. J. Maykuth, H. R. Ogden, and R. I. Jaffee, *Trans. AIME,* **197,** 1953, 225–230.
2. F. Laves and H. J. Wallbaum, *Naturwissenschaften,* **27,** 1939, 674–675.
3. H. J. Wallbaum, *Z. Krist.,* **103,** 1941, 391–402.
4. W. Rostoker, R. P. Elliott, and D. J. McPherson, *Trans. AIME,* **197,** 1953, 1566–1567.
5. D. J. Maykuth, H. R. Ogden, and R. I. Jaffee, *Trans. AIME,* **197,** 1953, 1567–1568.
6. J. W. Holladay, J. G. Kura, and J. H. Jackson, Battelle Memorial Institute, Progress Report 11, Contract AF 33(038)-8544 to Wright Patterson Air Force Base, Oct. 12, 1951.
7. J. W. Holladay, J. G. Kura, and J. H. Jackson, Battelle Memorial Institute, Summary Report on Contract AF 33(038)-8544 to Wright Patterson Air Force Base, Jan. 12, 1952.
8. E. Ence and H. Margolin, *Trans. AIME,* **200,** 1954, 346–348.

9. A. D. McQuillan, *J. Inst. Metals,* **80,** 1951-1952, 363–368.
10. A. D. McQuillan, *J. Inst. Metals,* **82,** 1953-1954, 47–48.
11. P. Duwez, *Trans. ASM,* **45,** 1953, 934–940.
12. P. D. Frost, W. M. Parris, L. L. Hirsch, J. R. Doig, and C. M. Schwartz, *Trans. ASM,* **46,** 1954, 1056–1071; discussion, pp. 1071–1074.
13. B. W. Levinger, R. P. Elliott, and W. Rostoker, Armour Research Foundation, Final Report, Contract DA 11-022-ORD-272 to Watertown Arsenal, February, 1953.
14. R. P. Elliott and W. Rostoker, *Trans. AIME,* **197,** 1953, 1203–1204.
15. W. Rostoker, *Trans. AIME,* **194,** 1952, 209–210.
16. H. Margolin and E. Ence, *Trans. AIME,* **200,** 1954, 1267–1268.
17. R. P. Elliott, Armour Research Foundation, Chicago, Ill., Technical Report 1, OSR Technical Note OSR-TN-247, August, 1954, p. 29.
18. A. Hellawell and W. Hume-Rothery, *Phil. Trans. Roy. Soc.,* **A249,** 1957, 417–459; *Titanium Abstr. Bull.,* **2,** 1957, 2592.

$\bar{1}.4294$
0.5706

Mn-Tl Manganese-Thallium

This system requires reinvestigation under appropriate experimental conditions since the phase diagram shown in Fig. 531 is based on thermal-analysis data of only

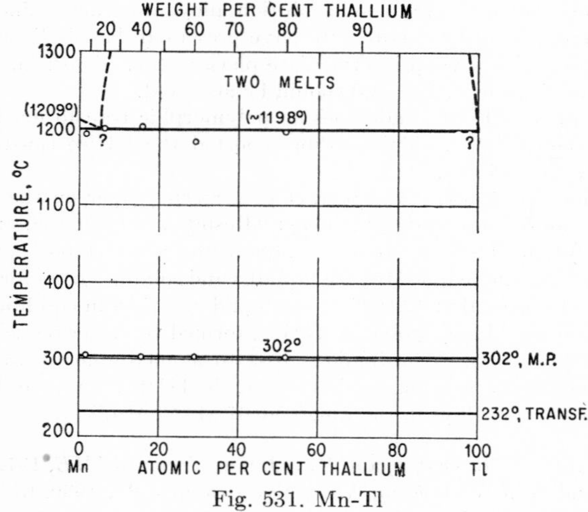

Fig. 531. Mn-Tl

five alloys [1] prepared using Mn of only 98.8% purity with a melting point of 1209°C instead of 1245°C [2].

1. N. Baar, *Z. anorg. Chem.,* **70,** 1911, 358–362.
2. Alloys were prepared in an atmosphere of nitrogen(!).

$\bar{1}.3631$
0.6369

Mn-U Manganese-Uranium

The phase diagram in Fig. 532 is based on thermal, microscopic [1], and roentgenographic studies [2, 3]. The effect of U on the temperature of the polymorphic transformations of Mn was not determined [4]. The maximum solubility of Mn in

γ-U was estimated as less than 1 wt. (4.2 at.) % Mn. The β_U solid solution with about 0.25 wt. (1.1 at.) % Mn decomposes eutectoidally at 626°C.

UMn_2 (68.43 wt. % U) is isotypic with $MgCu_2$ (C15 type), a = 7.147 A [2, 3]. U_6Mn is b.c. tetragonal, with 28 atoms per unit cell, a = 10.29 A, c = 5.24 A, c/a = 0.509 [2].

Note Added in Proof. In contrast to the findings of [1, 4], data by [5] indicate that γ_{Mn} is stabilized to room temperature by small additions of U.

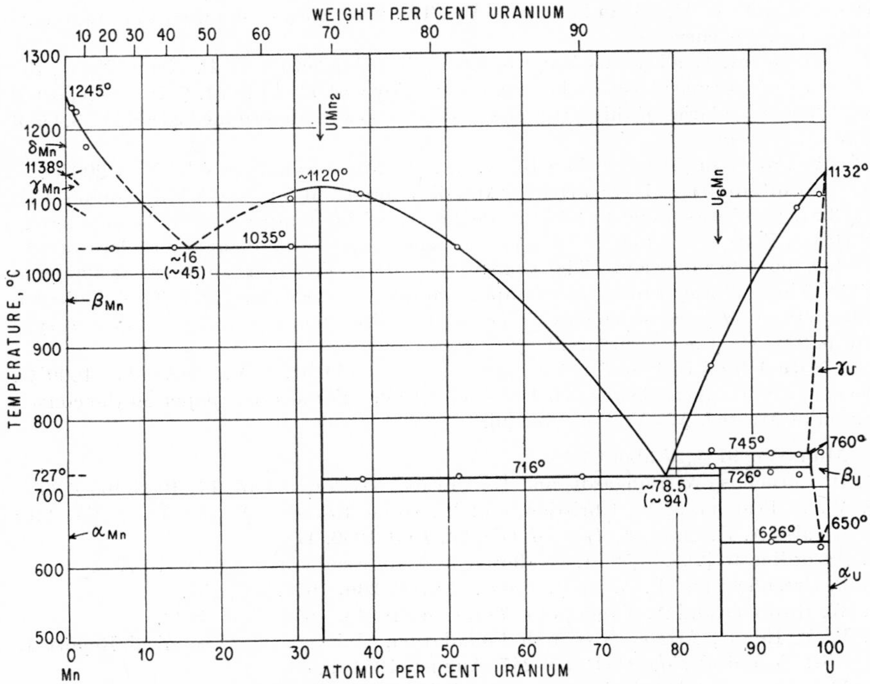

Fig. 532. Mn-U. (See also Note Added in Proof.)

According to [5, 6], β-U can be kept metastable at room temperature by small additions of Mn.

1. H. A. Wilhelm and O. N. Carlson, *Trans. ASM*, **42,** 1950, 1311–1325.
2. N. C. Baenziger, R. E. Rundle, A. I. Snow, and A. S. Wilson, *Acta Cryst.*, **3,** 1950, 34–40.
3. P. J. Bowles, T. S. Hutchison, P. C. L. Pfeil, and M. B. Waldron, AERE-M/R-581, 1950; *Nuclear Sci. Abstr.*, **5,** 1951, 2424.
4. "The region from 90 wt. % Mn over to pure Mn has not been studied thoroughly. Solubility of U in Mn is postulated on the basis of existing data since no eutectic was observed above 95 wt. % either in thermal analysis or in the microstructures. X-ray studies have shown the phases present in a 36 wt. % Mn sample to be UMn_2 plus β-Mn. This indicates a retention of the β-phase due to solid solution of U in β-Mn" [1].
5. Atomic Energy Research Establishment, United Kingdom, unpublished information (May, 1950); quoted in [7].

6. H. A. Saller and F. A. Rough, unpublished information (1949); quoted in [7].
7. H. A. Saller and F. A. Rough, Compilation of U.S. and U.K. Uranium and Thorium Constitutional Diagrams, *U.S. Atomic Energy Comm., Publ.* BMI-1000, June, 1955.

0.0327
$\bar{1}$.9673

Mn-V Manganese-Vanadium

From microscopic and X-ray investigations of alloys prepared using low-grade metals [1], it was concluded that three intermediate phases exist, with compositions in the vicinity of 11, 50, and 85 at. % V. The same alloys were used to outline part of the liquidus curve [2].

The 50 at. % alloy was found to have a b.c.c. structure [3, 4]. In addition, the existence of a σ-type phase in a heterogeneous alloy with about 24 at. % V was reported [3]. The boundaries of this latter phase at 1000°C were determined as about 13.4 and 24.5 at. % V [5].

According to [6], more than 15 wt. (at.) % Mn is soluble in solid V at 900°C.

Supplement. [7] reported: "An alloy prepared at the VMn_2 composition [33.3 at. % V] was found to have an incipient melting temperature of approximately 1370°C. This value may be subject to error because of an apparent loss of Mn during the determination. The as-cast structure and samples annealed at 1200 and 1300°C had a b.c.c. structure. Samples annealed at 1000, 800, and 600°C had two-phase structures, one of which was b.c.c. The other phase . . . was similar to typical patterns of the σ structure."

Note Added in Proof. A σ-phase alloy with 16.6 at. % V, annealed at 1000°C, became ferromagnetic below 106°K (−167°C) [8]. For a recent paper on the constitution of Mn-rich Mn-V alloys, see [9].

1. Mn about 97%, V about 96%.
2. H. Cornelius, W. Bungardt, and E. Schiedt, *Metallwirtschaft,* **17,** 1938, 977–980.
3. W. B. Pearson, J. W. Christian, and W. Hume-Rothery, *Nature,* **167,** 1951, 110; see also A. H. Sully, *J. Inst. Metals,* **80,** 1951-1952, 178.
4. According to [2], the 50 at. % alloy consists of a cubic phase, a = 10.72 A.
5. P. Greenfield and P. A. Beck, *Trans. AIME,* **200,** 1954, 253–257.
6. W. Rostoker and A. Yamamoto, *Trans. ASM,* **46,** 1954, 1136–1163.
7. R. P. Elliott, Armour Research Foundation, Chicago, Ill., Technical Report 1, OSR Technical Note OSR-TN-247, August, 1954, p. 31.
8. M. V. Nevitt and P. A. Beck, *Trans. AIME,* **203,** 1955, 671.
9. A. Hellawell and W. Hume-Rothery, *Phil. Trans. Roy. Soc.,* **A249,** 1957, 417–459; *Titanium Abstr. Bull.,* **2,** 1957, 2592.

$\bar{1}$.4752
0.5248

Mn-W Manganese-Wolfram

W was found to be insoluble in liquid Mn [1, 2]; see also [3, 4].

1. D. Kremer, *Abhandl. Inst. Metallhütt. u. Elektromet. Tech. Hochsch., Aachen,* **1,** 1916, 1–19.
2. U. Zwicker, *Z. Metallkunde,* **42,** 1951, 251.
3. G. Arrivaut, *Compt. rend.,* **143,** 1906, 594–596.
4. C. Sargent, *J. Am. Chem. Soc.,* **22,** 1901, 783.

$\bar{1}$.9244
0.0756

Mn-Zn Manganese-Zinc

The constitution has been established by the investigations of [1] and [2] which cover the ranges 50–100 and 0–50 wt. % Zn, respectively (Fig. 533). The latter

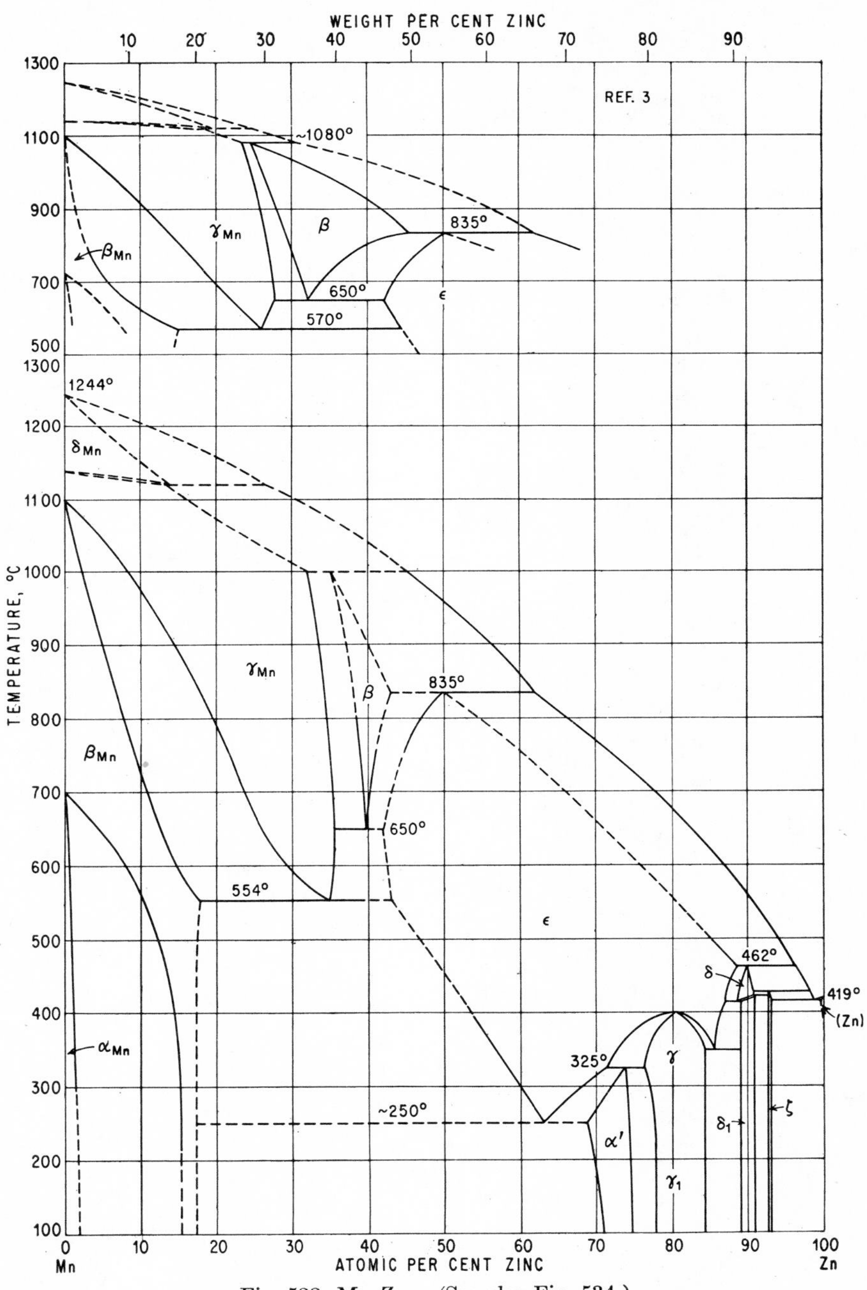

Fig. 533. Mn-Zn. (See also Fig. 534.)

region was also studied by [3]; results are not published as yet (see upper part of Fig. 533). Earlier investigations of Zn-rich alloys [4–7], which yielded contradictory results entirely incompatible with those of [1], are obsolete.

Phase relations in the range 0–50 wt. (45.7 at.) % Zn are due to [2, 3]. They are based on X-ray [2, 3], micrographic [3], and thermal data [2] of alloys prepared using powder-metallurgy methods. Results are in qualitative agreement; they differ mainly as to the extent of the γ_{Mn}- and β-phase fields.

The region above 50 wt. % Zn was the subject of a thorough study by means of thermal, micrographic, and X-ray methods [1]. Figure 534 exhibits the complicated phase relationships in alloys with more than 80 at. % Zn on a larger scale. [8] determined the liquidus curve in the range 2–10 at. % Mn by chemical analysis of samples taken from melts equilibrated at various temperatures. The temperatures lie 10–20°C

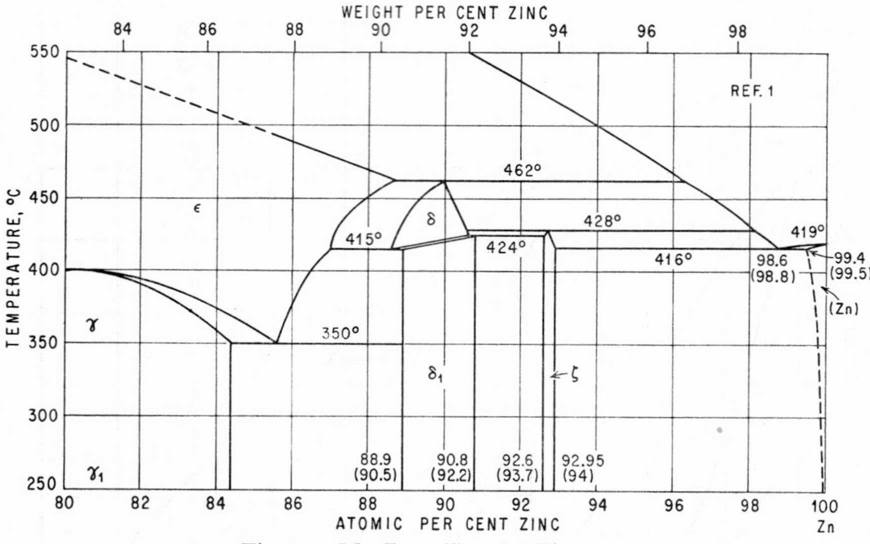

Fig. 534. Mn-Zn. (See also Fig. 533.)

above those found thermally [1] and are to be regarded as equilibrium temperatures [17]. The temperature of the ζ + (Zn) eutectic was reported as 417.25 [9] vs. 416°C according to [1]; its composition was given as 99.2 wt. (99.06 at.) % Zn [11] as compared with 98.8 wt. (98.6 at.) % Zn [1].

The solid solubility of Mn in Zn was found to be 0.50 ± 0.05 wt. (0.58 at.) % [1] or 0.45 wt. (0.53 at.) % at the eutectic temperature and less than 0.02 wt. (at.) % at 200°C [10]; for earlier determinations, see [11].

Crystal Structures. Lattice parameters of the solid solutions based on the α, β, and γ modifications of Mn were reported by [2]. Quenched γ_{Mn} alloys were found to be f.c. tetragonal (metastable) up to 19.5 at. % Zn and f.c.c. at higher Zn contents. Additional γ_{Mn} parameters were given by [12].

The β phase is b.c.c. (A2 type), with a = 3.06 A at 39.7–40.7 at. % Zn [1, 2]. ϵ is h.c.p. (A3 type) [1, 2, 13], with a = 2.73 A, c/a = 1.631 at 45.7 at. % Zn [1, 2] and a = 2.76 A, c/a = 1.609 at 81.5 at. % Zn [2]. The α' phase is f.c.c. with probably ordered distribution (Cu_3Au, $L1_2$ type); a = 3.86 A at 73.9 at. % Zn [1].

The γ phase, which transforms at some lower temperature to γ_1, is isotypic with

γ-brass ($D8_{1-3}$ type), a = 9.16 A at 81.5 at. % Zn. The structure of γ_1 is similar to that of γ, possibly slightly distorted. The ideal composition Mn_5Zn_{21} corresponds to 80.77 at. % Zn and lies within the homogeneity range [1].

The δ_1 phase was reported to be isostructural with the corresponding phase of the Fe-Zn system [1], which is hexagonal, with 550 ± 8 atoms per unit cell. ζ ($MnZn_{13}$ = 92.86 at. % Zn) is monoclinic, with 28 atoms per unit cell, and isostructural with the corresponding Fe-Zn phase [1].

In earlier X-ray investigations of Zn-rich alloys, carried out without knowledge of the complicated phase relations, the structure of the ϵ and γ phases had been recognized [14-16].

Supplement. Ferromagnetism, the carrier of which is presumably the ϵ phase, has been observed by [18].

1. J. Schramm, *Z. Metallkunde*, **32,** 1940, 399-407; see also **30,** 1938, 131-135.
2. E. V. Potter and R. W. Huber, *Trans. ASM*, **41,** 1949, 1001-1022.
3. J. Schramm, unpublished work; private communication.
4. N. Parravano and U. Perret, *Gazz. chim. ital.*, **45**(1), 1915, 1-6.
5. P. Siebe, *Z. anorg. Chem.*, **108,** 1919, 171-173.
6. P. Gieren, Dissertation, Technische Hochschule Berlin, 1919; *Z. Metallkunde*, **11,** 1919, 16-17; **12,** 1920, 141-142.
7. C. L. Ackermann, *Z. Metallkunde*, **19,** 1927, 200-204.
8. G. Edmunds, *Trans. AIME*, **156,** 1944, 263-276.
9. E. Weisse, A. Blumenthal, and H. Hanemann, *Z. Metallkunde*, **34,** 1942, 221.
10. J. L. Rodda, unpublished work; "Metals Handbook," 1948 ed., p. 1229, The American Society for Metals, Cleveland, Ohio.
11. W. M. Peirce, *Trans. AIME*, **68,** 1923, 777-779.
12. U. Zwicker, *Z. Metallkunde*, **42,** 1951, 246-252.
13. K. Moeller, *Z. Metallkunde*, **35,** 1943, 27-28.
14. N. Parravano and V. Montoro, *Mem. accad. ital., Classe sci. fis. mat. nat., Chim.*, **1,** 1930, no. 4 (*Strukturbericht*, **2,** 1928-1932, 711).
15. N. Parravano and V. Caglioti, *Rend. accad. nazl. Lincei*, **14,** 1931, 166-169; *Mem. accad. ital., Classe sci. fis. mat. nat., Chim.*, **3,** 1932, no. 3 (*Strukturbericht*, **2,** 1928-1932, 712).
16. N. Parravano and V. Caglioti, *Ricerca sci.*, **7,** 1936, 223-224.
17. For a note on supercooling of Mn-Zn alloys with a composition close to pure Zn, see J. Schramm, discussion on F. Pawlek, *Z. Metallkunde*, **36,** 1944, 111-112.
18. H. Nowotny and H. Bittner, *Monatsh. Chem.*, **81,** 1950, 898-901.

$\bar{1}.7797$
0.2203

Mn-Zr Manganese-Zirconium

The partial phase diagram in Fig. 535 is the result of thermal and micrographic studies of arc-melted alloys prepared using magnesium-reduced Zr [1]. The solubility of Mn in α Zr was not determined but appears to be very low.

The intermediate phase in equilibrium with Zr was estimated to contain between 40 and 50 wt. % Mn [1]. It was identified previously as $ZrMn_2$ (54.64 wt. % Mn), which is isotypic with $MgZn_2$ (C14 type), a = 5.039 A, c = 8.240 A, c/a = 1.635 [2].

Supplement. [3] confirmed that the structure of $ZrMn_2$ is isotypic with $MgZn_2$, a = 5.019 A, c = 8.249 A, c/a = 1.644. The melting temperature of this compound was found to be 1340°C; however, [3] admits that this value may be subject to error because of the volatilization of Mn during the determination.

1. A. H. Roberson, E. T. Hayes, and V. V. Donaldson, "Zirconium and Zirconium Alloys," pp. 283-291, The American Society for Metals, Cleveland, Ohio, 1953.

2. H. J. Wallbaum, *Z. Krist.*, **103**, 1941, 391–402; *Arch. Eisenhüttenw.*, **14**, 1941, 521–526.
3. R. P. Elliott, Armour Research Foundation, Chicago, Ill., Technical Report 1, OSR Technical Note OSR-TN-247, August, 1954, p. 38.

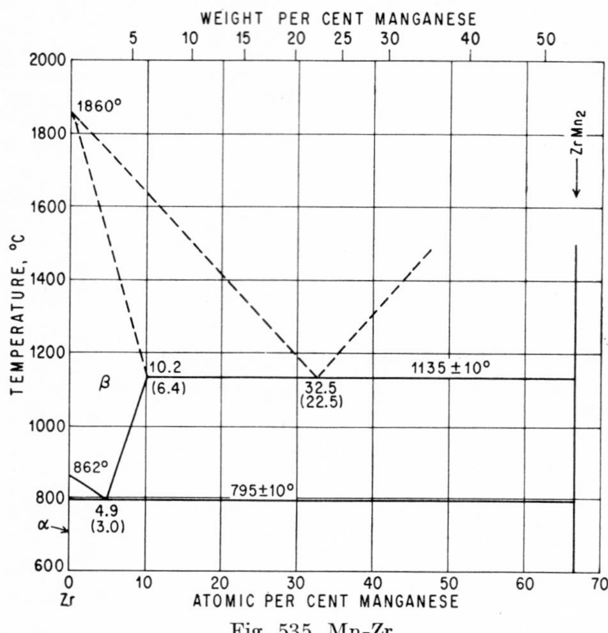

Fig. 535. Mn-Zr

0.8357
$\bar{1}$.1643

Mo-N Molybdenum-Nitrogen

X-ray analysis of preparations, obtained by nitriding Mo powder with pure NH_3 at temperatures between 400 and 725°C, showed the existence of the three intermediate phases indicated in the schematic diagram of Fig. 536 [1]. The solubility of N in Mo in equilibrium with γ is negligibly small. The β phase, stable only above approximately 600°C, is homogeneous at about 28 at. (5.4 wt.) % N. It has a f.c. tetragonal lattice of the Mo atoms, $a = 4.18_8$ A, $c = 4.02_4$ A, $c/a = 0.961$, after quenching from above 850°C. The position of the N atoms is unknown.

The homogeneity range of γ or Mo_2N (6.80 wt. % N) is about 1 at. % wide at temperatures below 800°C. At higher temperatures it appears to be shifted to lower N concentrations. The Mo atoms form an f.c.c. lattice; the N atoms are randomly distributed in the largest interstices. The lattice parameter increases from $a =$ 4.163 A on the Mo side to $a = 4.168$ A on the N side. After quenching from 1000°C, the parameter was found as $a = 4.136$ A at 32.7 at. % N.

The δ phase or MoN (12.74 wt. % N) was previously believed to have a simple hexagonal lattice of the Mo atoms, with $a = 2.866$ A, $c = 2.810$ A, $c/a = 0.980$ [1]. Actually it has a hexagonal superstructure formed by doubling the axes of the substructure: $a = 5.725$ A, $c = 5.608$ A, $c/a = 0.980$, 16 atoms per unit cell [2].

The existence of β and γ was confirmed by [3, 2]. An additional Mo-rich phase of unknown composition, presumably formed below about 850°C, was reported by [3]. The phase Mo_3N_2 (8.87 wt. % N) was tentatively identified [4]; however, no indication of such a phase had been found by [1].

The absorption of N_2 was determined by [5–9]. Results are greatly at variance. The quantity of N_2 absorbed in the temperature range 900–1200°C [6–8] was slight in several cases (and decreased with rise in temperature); in others, however, the absorption was one order of magnitude higher [7]. [9] reported the following absorption values in cc per 100 g: 0.84 at 1200°C, 3.44 at 1600°C, 8.4 at 2000°C, and 16 at 2400°C.

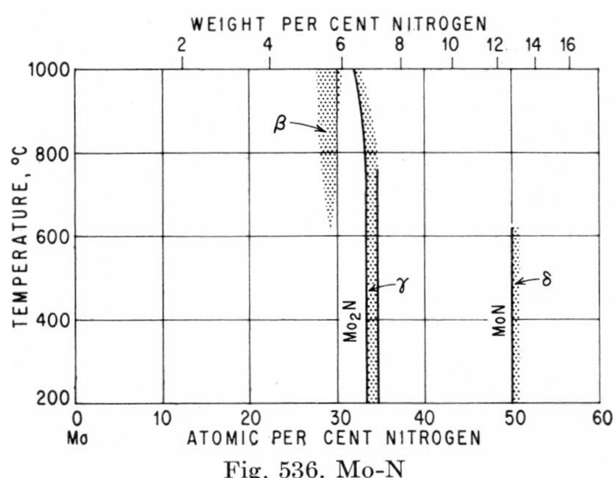

Fig. 536. Mo-N

1. G. Hägg, *Z. physik. Chem.*, **B7**, 1930, 339–356.
2. N. Schönberg, *Acta Chem. Scand.*, **8**, 1954, 204–207.
3. A. Sieverts and G. Zapf, *Z. anorg. Chem.*, **229**, 1936, 161–174.
4. M. Mathis, *Bull. soc. chim. France*, **18**, 1951, 443–451.
5. I. I. Zhukov, *Izvest. Inst. Fiz.-Khim. Anal.*, **3**, 1926, 14–51; *Chem. Abstr.*, **21**, 1927, 3800.
6. E. Martin, *Arch. Eisenhüttenw.*, **3**, 1929-1930, 407–416.
7. A. Sieverts and K. Brüning, *Arch. Eisenhüttenw.*, **7**, 1933-1934, 641–645.
8. A. Sieverts and G. Zapf, *Z. anorg. Chem.*, **229**, 1936, 161–174.
9. F. J. Norton and A. L. Marshall; data published in S. Dushman, "Scientific Foundations of Vacuum Technique," p. 599, John Wiley & Sons, Inc., New York, 1949. See also F. J. Norton and A. L. Marshall, *Trans.* AIME, **156**, 1944, 369–370.

0.0140
$\bar{1}$.9860

Mo-Nb Molybdenum-Niobium

Measurements of the lattice parameter of three alloys with about 25.5, 51, and 76 at. % Nb indicate that Mo and Nb form an uninterrupted series of b.c.c. solid solutions [1].

[2] reported that the diffraction patterns of powder specimens of the composition $NbMo_2$ annealed at 600, 1000, and 1200°C were b.c.c. [3] prepared several alloys by sintering powder compacts in vacuo. Microscopic and X-ray investigations showed all alloys to be one-phase. The lattice parameters of the b.c.c. cell showed a small negative deviation from additivity. Dilatometric measurements did not reveal any transformation up to 1100°C.

1. H. Bückle, *Z. Metallkunde*, **37**, 1946, 53–56.
2. R. P. Elliott, Armour Research Foundation, Chicago, Ill., Technical Report 1, OSR Technical Note OSR-TN-247, August, 1954, p. 20.
3. V. N. Eremenko, *Ukrain. Khim. Zhur.*, **20**, 1954, 227–231; *Met. Abstr.*, **23**, 1956, 636–637.

0.2135
$\bar{1}$.7865

Mo-Ni Molybdenum-Nickel

Solidification. The features of the solidification equilibria, established by the work of [1–3], were verified by [4, 5], using thermal analysis [1, 5] and thermoresistometric [2], thermomagnetometric [4], microscopic [5], and incipient-fusion methods [5] (Fig. 537). Results differ somewhat as regards the temperatures of the three-phase equilibria and compositions of the characteristic points.

The temperature of the peritectic formation of MoNi (62.05 wt. % Mo) was reported as 1340–1345 [1], 1347 ± 7 [2], 1345 [3], 1370 ± 5 [4], and about 1350°C [5],

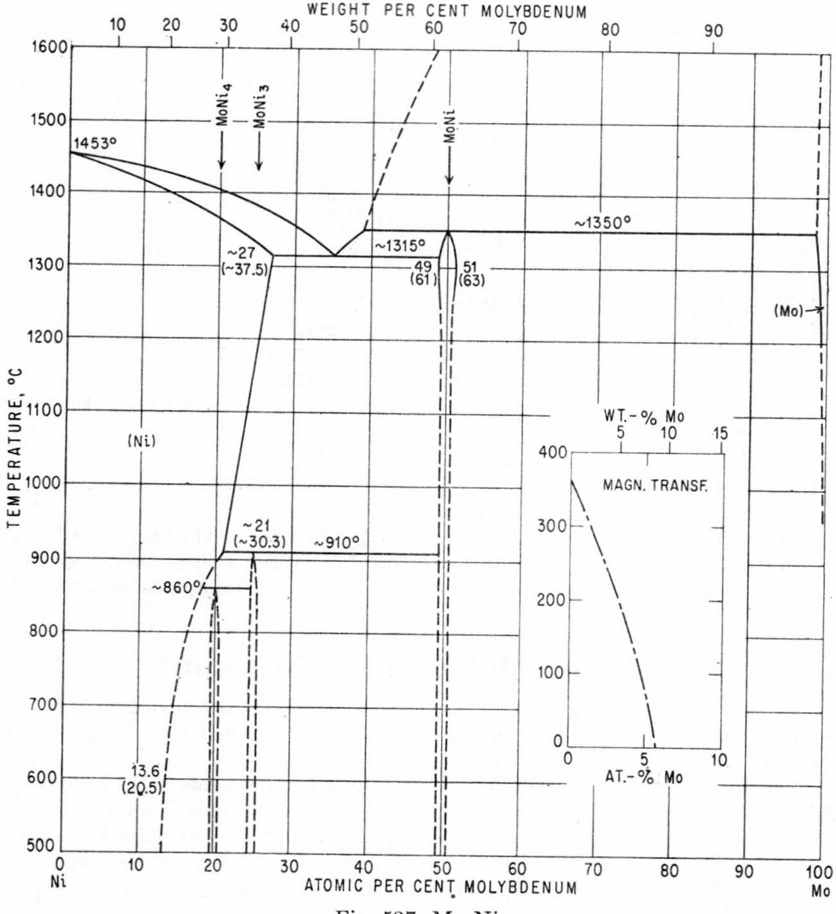

Fig. 537. Mo-Ni

and the composition of the peritectic melt in wt. % Mo (at. % in parentheses) as about 53 (40.8) [1], 50–55 (38–42.8) [3], about 51 (38.9) [4], and about 47 (35.2) % Mo [5]. The eutectic temperature was found as about 1305 [1], 1315 [6], 1303 ± 7 [2], 1300–1310 [3], and 1320°C [4, 5] and the eutectic composition as approximately 49 (37) [1, 3], 46.5 (34.8) [4], and 42.5 (31.2) % Mo [5]. The liquidus temperatures given by [5] are incompatible with those of [1, 3] and the micrographically determined composition of the peritectic melt [4] in that they indicate a shift to lower Mo contents, by 4–7 wt. %. The solidus of the (Ni) phase shown in Fig. 537 is based on measurements of electrical resistance vs. temperature [2].

Solid State. The existence of two three-phase equilibria in the solid state, involving the formation of the intermediate phases of the approximate composition $MoNi_3$ (35.27 wt. % Mo) and $MoNi_4$ (29.01 wt. % Mo), was detected by [2, 3], using measurements of electrical resistance [2] and of magnetic susceptibility vs. temperature [3], as well as X-ray studies. Micrographic work [4] substantiated these findings (Fig. 537). The peritectoid temperatures were reported as 925°C, on heating [3], 890 ± 10°C [4]; and 875 ± 5°C, on heating [2, 3], 840 ± 10°C [4], respectively. The boundaries of the phases $MoNi_3$ and $MoNi_4$ have not been determined accurately as yet.

The solid solubility of Mo in Ni was reported as about 37 wt. (26.5 at.) % at the eutectic temperature [2, 4], 34 wt. (24 at.) % at 1100°C (parametrically) [2], 30 wt. (20.8 at.) % at 890°C [4], and 20.5 wt. (13.6 at.) % Mo at 600°C [4]. According to lattice-parameter measurements [7], the solubility of Ni in Mo is 1.3, 1.2, 1.05, and 0.8 at. (0.8, 0.75, 0.65, and 0.5 wt.) % Ni at 1340, 1315, 1290, and 1200°C, respectively; see also [10]. [4] gave about 0.9 wt. (1.45 at.) % Ni at the peritectic temperature.

Curie temperatures of the (Ni) phase were determined by [6, 8, 9, 2, 3]. The transformation curve in Fig. 537 is a compromise of the data by [8, 9, 2, 3].

Crystal Structures. Lattice parameters of the (Ni) phase were reported by [8, 2] and those of the (Mo) phase by [7]. The phase $MoNi_4$ was reported to have a f.c. tetragonal superstructure, with a = 3.62 A, c = 3.57 A, c/a = 0.986 [2, 3]. In contrast to this, [11] concluded this phase to have an ordered f.c. tetragonal structure, with a = 5.731 A, c = 3.571 A, c/a = 0.623. The ordered structure can be described in terms of a tetragonal cell 2.5 times the volume of the distorted f.c.c. cell of the (Ni) phase. The phase $MoNi_3$ was reported to have a h.c.p. lattice, with $a = 2.54_5$ A, c/a = 1.65 [3]; see also [4].

1. N. Baar, *Z. anorg. Chem.*, **70**, 1911, 353–358.
2. G. Grube and H. Schlecht, *Z. Elektrochem.*, **44**, 1938, 413–422.
3. G. Grube and O. Winkler, *Z. Elektrochem.*, **44**, 1938, 423–428.
4. F. H. Ellinger, *Trans. ASM*, **30**, 1942, 607–637.
5. D. S. Bloom and N. J. Grant, *Trans. AIME*, **200**, 1954, 261–263.
6. Dreibholz, *Z. physik. Chem.*, **108**, 1924, 8–11.
7. J. L. Ham, Climax Molybdenum Co., Arc-cast Molybdenum-base Alloys, First Annual Report, 1950, pp. 99–107.
8. W. Köster and W. Schmidt, *Arch. Eisenhüttenw.*, **8**, 1934-1935, 23–27.
9. V. Marian, *Ann. phys.*, **7**, 1937, 459–527.
10. J. L. Ham, *Trans. Am. Soc. Mech. Eng.*, **73**, 1951, 723–731.
11. D. Harker, *J. Chem. Phys.*, **12**, 1944, 315–317.

$\bar{1}.7028$
0.2972

Mo-Os Molybdenum-Osmium

The following conclusions were drawn [1] from a microscopic and roentgenographic investigation of three alloys with 30, 50, and 70 wt. (17.8, 33.5, and 54.1 at.)

% Os: (*a*) The solid solubility of Os in Mo is slight. (*b*) There are two intermediate phases in this range, namely, Mo_3Os (39.8 wt. % Os), formed in the solid state, and "Mo_5Os_3" (54.32 wt. % Os). The former is cubic of the "β-W" (A15) type, and the latter, according to its powder pattern, has a structure equal or similar to that of "Mo_5Ru_3." Since this phase is a σ phase [2], it is very likely that "Mo_5Os_3" also has a σ-type structure (see Mo-Ru). (*c*) About 40 at. (25.2 wt.) % Mo is soluble in Os at 1200°C. The lattice constants of Os (a = 2.724 A, c/a = 1.585) are increased to a = 2.758 A, c/a = 1.601 [1].

1. E. Raub, *Z. Metallkunde*, **45**, 1954, 23–30.
2. P. Greenfield and P. A. Beck, *Trans. AIME*, **206**, 1956, 265–276.

0.4910
$\bar{1}$.5090

Mo-P Molybdenum-Phosphorus

The existence of the phosphides Mo_3P (9.72 wt. % P) [1–3], MoP (24.40 wt. % P) [4, 5, 2, 3], and MoP_2 (39.23 wt. % P) [6, 2, 3] has been established. [7] reported that Mo and MoP form a eutectic at 12 wt. (29.7 at.) % P and at an estimated temperature of about 1650°C. This would suggest that Mo_3P is formed by a reaction in the solid state, since, if the eutectic would consist of Mo and Mo_3P, its composition would have to lie below 9.7 wt. % P. The melting point of MoP lies above 1700°C [7].

Mo_3P is isotypic with Fe_3P [2, 3], b.c. tetragonal, with 24 atoms per unit cell and a = 9.729 A, c = 4.923 A, c/a = 0.51 [3]. MoP is isotypic with WC, hexagonal with 2 atoms per unit cell, a = 3.23 A, c = 3.20 A, c/a = 0.99 [3]. The X-ray pattern of MoP_2 is similar to that of WP_2 [2].

1. J. L. Andrieux and M. Chêne, *Compt. rend.*, **209**, 1939, 672–674; J. L. Andrieux, *Rev. mét.*, **45**, 1948, 49–59.
2. F. E. Faller, W. Biltz, K. Meisel, and M. Zumbusch, *Z. anorg. Chem.*, **248**, 1941, 209–228.
3. N. Schönberg, *Acta Chem. Scand.*, **8**, 1954, 226–239.
4. F. Wöhler and Rautenberg, *Liebigs Ann.*, **109**, 1859, 374.
5. H. Hartmann and U. Conrad, *Z. anorg. Chem.*, **233**, 1937, 313.
6. E. Heinerth and W. Biltz, *Z. anorg. Chem.*, **198**, 1931, 175–176.
7. R. Vogel and D. Horstmann, *Arch. Eisenhüttenw.*, **24**, 1953, 369–374.

$\bar{1}$.6656
0.3344

Mo-Pb Molybdenum-Lead

[1] prepared alloys with up to 20 wt. % Mo by dissolving Mo in boiling Pb. The primary phase is elemental Mo, which precipitates from the melt at very high temperatures.

1. W. Guertler, *Z. Metallkunde*, **15**, 1923, 152.

$\bar{1}$.9539
0.0461

Mo-Pd Molybdenum-Palladium

Micrographic and X-ray work covering the range 5.5–72.2 at. (5–70 wt.) % Mo [1] revealed that no intermediate phase exists. The solid solubility of Mo in Pd, determined parametrically, decreases from 47.5 at. (44.9 wt.) % at 1200°C to 38 at. (35.5 wt.) % at 850°C and 35.7 at. (33.2 wt.) % Mo at 800°C. The solubility of Pd in Mo appears to be very small.

The lattice constant of the (Pd) phase passes through a minimum at about 11 at. % (a = 3.878 A) and increases to a = 3.903 A at 47.5 at. % Mo.

Supplement. [2] examined, microscopically and by X-ray diffraction, 50 and 75 at. % Mo alloys annealed at 1200°C. They confirm the absence of an intermediate phase but suggest that the solubility of Pd in solid Mo is greater than 25 at. % Pd.

1. E. Raub, *Z. Metallkunde,* **45,** 1954, 23–30.
2. P. Greenfield and P. A. Beck, *Trans. AIME,* **206,** 1956, 265–276.

$\bar{1}.6915$
0.3085

Mo-Pt Molybdenum-Platinum

From microexamination and measurements of hardness of alloys with up to 58 at. (40.5 wt.) % Mo, annealed for 170 hr at 1300°C and cooled, [1] concluded that, up to the maximal Mo content studied, the system comprises a series of solid solutions. Also, [2] reported that alloys with up to 50 at. (33 wt.) % Mo, annealed for 6 days at 1000°C, gave X-ray patterns showing a single f.c.c. phase with lattice constants near that of Pt.

[3] investigated the micro- and X-ray structures of alloys with 5–82.6 at. (2.5–70 wt.) % Mo after annealing at and quenching from temperatures between 800 and 1700°C. The following findings were reported: (*a*) At temperatures above approximately 1000°C, the f.c.c. Pt-base α solid solution extends to about 42 at. (26.3 wt.) % Mo. Alloys with more than about 25 at. (14 wt.) % Mo undergo a transformation to a f.c. tetragonal superstructure α', with $a = 3.895 \pm 0.004$ A, $c/a = 1.005$–1.009, independent of the Mo content. The $\alpha \to \alpha'$ transformation cannot always be suppressed by rapid quenching, even from temperatures of 1400–1700°C. The lattice constant of α reaches a minimum value ($a = 3.909$ A) at 10–20 at. % Mo and increases to a value not exceeding that of Pt at the saturation limit at high temperatures, about 42 at. % Mo. (*b*) There is a hexagonal phase of variable composition ranging from 46 to 72 at. (29.7–55.8 wt.) % Mo, with $a = 2.786$ A, $c/a = 1.611$ at the Pt side and $a = 2.80$ A, $c/a = 1.603$ at the Mo side of the homogeneity range. (*c*) The solubility of Pt in Mo is "very slight."

Supplement. [4] examined four alloys, with 36–72 at. % Mo and annealed at 1200°C, microscopically and by X-ray diffraction. In agreement with [3], an intermediate phase of h.c.p. structure was observed.

1. V. A. Nemilov and N. M. Voronov, *Izvest. Sektora Platiny,* **14,** 1937, 157–162.
2. R. Hultgren and R. I. Jaffee, *J. Appl. Phys.,* **12,** 1941, 501–502.
3. E. Raub, *Z. Metallkunde,* **45,** 1954, 23–30.
4. P. Greenfield and P. A. Beck, *Trans. AIME,* **206,** 1956, 265–276.

$\bar{1}.7118$
0.2882

Mo-Re Molybdenum-Rhenium

The solubility of Re in solid Mo is 42 at. % at 2400°C, falling to 27 at. % at 1000°C [1]. The existence of a σ-type phase has been reported by [2–4, 1]. It is formed at about 2500°C by the peritectic reaction: melt + (Re) $\rightleftarrows \sigma$ [1], and its range of homogeneity extends from 50 to 74 [1], 48 to 67 at. % Re (at 1200°C) [2]. The lattice parameters at 60 at. % Re are $a = 9.588$ A, $c = 4.983$ A, $c/a = 0.5197$ [3]. According to [1], the σ phase is stable only above 1200°C.

At higher Re contents, about 80 at. % [1], another intermediate phase exists, having the α-Mn (A12) type structure [2, 1], with $a = 9.55$ A [2]. According to [1], this phase forms peritectoidally at 1850°C from the σ and (Re) phases. The maximum solubility of Mo in solid Re is 14 at. % [1].

A f.c.c. phase with $a = 3.70$ A, claimed by [5] to be present in a 25 wt. (14.65 at.) % Re alloy annealed at 2000°C, could not be confirmed by [1].

1. A. G. Knapton, *Bull. Inst. Metals*, **3**, 1957, 161.
2. P. Greenfield and P. A. Beck, *Trans. AIME*, **206**, 1956, 265–276.
3. A. G. Knapton, *Bull. Inst. Metals*, **3**, 1955, 21.
4. J. Niemiec and W. Trzebiatowski, *Bull. acad. polon. sci.*, **4**, 1956, 601–603; *Met. Abstr.*, **24**, 1957, 819.
5. C. J. McHargue and H. W. Maynor, *Trans. AIME*, **197**, 1953, 1382.

$\bar{1}.9696$
0.0304

Mo-Rh Molybdenum-Rhodium

[1] studied the X-ray and microstructures of alloys with 2.8–71.5 at. (2.5–70 wt.) % Mo after annealing at and quenching from 1200–1400°C. The saturation limit of Mo in Rh at 1300°C lies at 7 at. (6.6 wt.) % Mo. The only intermediate phase, designated as ϵ, is one of variable composition whose homogeneity range extends from 15 to 52.5 at. (14.3–50.8 wt.) % Mo, at 1400°C. This phase is h.c.p., with a = 2.712 A, c/a = 1.603 at the Rh side and a = 2.755 A, c/a = 1.605 (mean values) at the Mo side of the region. ϵ and Mo form a eutectic which lies somewhat below 71.5 at. (70 wt.) % Mo; therefore, ϵ has a maximum melting point. The solubility of Rh in Mo appears to be negligibly small.

Supplement. [2] examined microscopically and by X-ray diffraction four alloys with 27–75 at. % Mo quenched from 1200°C. In agreement with [1], only one intermediate phase, of h.c.p. structure with a = 2.715 A, c = 4.340 A, c/a = 1.598 at 27 at. % Mo, was observed. Its homogeneity range was found to extend from approximately 27 to over 50 at. % Mo (cf. above).

1. E. Raub, *Z. Metallkunde*, **45**, 1954, 23–30.
2. P. Greenfield and P. A. Beck, *Trans. AIME*, **206**, 1956, 265–276.

$\bar{1}.9747$
0.0253

Mo-Ru Molybdenum-Ruthenium

Roentgenographic and microscopic investigations [1] of alloys with 10.5–71.2 at. (10–70 wt.) % Mo, annealed at and quenched from temperatures between 800 and 1600°C, showed the following: (*a*) The solid solubility of Mo in Ru in the temperature range 800–1600°C is about 35 at. (33.8 wt.) % Mo; that of Ru in Mo seems to be very slight at 1200°C but to increase at higher temperatures. (*b*) An intermediate phase exists which is homogeneous at the composition Mo_5Ru_3 [62.5 at. (61.13 wt.) % Mo] and has a complex powder pattern. This phase must be identical with the σ-type phase found to be present around 60 at. % Mo [2]. It is formed peritectoidally at some temperature between 1200 and 1600°C or higher and appears to have a rather small homogeneity range [1]. The lattice constants of σ are a = 9.54 A, c = 4.95 A, c/a = 0.519 [2].

The lattice constants of Ru are increased to a = 2.746 A, c = 4.402 A, c/a = 1.603 at the saturation limit, 35 at. % Mo [1]. Lattice-spacing data of alloys with 0.99–3.07 at. % Mo were reported by [3].

Supplement. As a result of X-ray diffraction examination of sintered powder compacts, [4] claims that the σ phase lies close to 70 at. % Mo and is stable only above 1200°C. This latter conclusion is apparently influenced by incorrect translation of the paper by [1]; see [5]. As to the composition of the σ phase, the results obtained by [1, 2] on melted alloys appear to be more reliable.

1. E. Raub, *Z. Metallkunde*, **45**, 1954, 23–30.
2. P. Greenfield and P. A. Beck, *Trans. AIME*, **206**, 1956, 265–276.

3. A. Hellawell and W. Hume-Rothery, *Phil. Mag.*, **45**, 1954, 797–806.
4. D. S. Bloom, *Trans. AIME*, **203**, 1955, 420.
5. [1] stated: "On the basis of the microscopic studies, Ru_3Mo_5 cannot exist up to the melting point. It rather forms peritectically [should be peritectoidally] at lower temperature." [4] assumed erroneously that [1] had found evidence for "Ru_3Mo_5" being a high-temperature phase.

0.4760 $\bar{1}.5240$ Mo-S Molybdenum-Sulfur

In the chemical literature the existence of Mo_2S_3, MoS_2, Mo_2S_5, MoS_3, and MoS_4 has been reported [1]. However, work by [2] has shown that, by synthesis, preparations are obtained which contain only the phases Mo and MoS_2. MoS_3 can be prepared only by decomposition of complex thiomolybdates [3].

MoS_2 is hexagonal, with 6 atoms per unit cell (C7 type), $a = 3.15 \pm 0.005$ A, $c = 12.32$ A, $c/a = 3.911$ [4, 5]; see also [6].

Supplement. By thermal decomposition of MoS_2 at temperatures around 1100°C, [7] obtained a sulfide which was identified by X-ray and chemical analysis as Mo_2S_3. [7] stated that "it is certain that the molybdenum sulphide in equilibrium with sulphur vapor and molybdenum metal from 1025 to 1150°C is molybdenum sesquisulphide, Mo_2S_3."

1. "Gmelins Handbuch der anorganischen Chemie," 8th ed., System No. 53, pp. 182–188, Verlag Chemie, G.m.b.H., Berlin, 1935.
2. P. Ehrlich, unpublished work; see *FIAT Rev. Germ. Sci.*, 1939-1946, Inorganic Chemistry, Part II, p. 34.
3. W. Biltz and A. Köcher, *Z. anorg. Chem.*, **248**, 1941, 172–174.
4. R. G. Dickinson and L. Pauling, *J. Am. Chem. Soc.*, **45**, 1923, 1465–1471.
5. O. Hassel, *Z. Krist.*, **61**, 1925, 92–99.
6. A. E. van Arkel, *Rec. trav. chim.*, **45**, 1926, 437–444.
7. C. L. McCabe, *Trans. AIME*, **203**, 1955, 61–63.

0.0846 $\bar{1}.9154$ Mo-Se Molybdenum-Selenium

The existence of Mo_2Se_3, $MoSe_2$, Mo_2Se_5, and $MoSe_3$, prepared by chemical reactions, has been reported [1].

1. E. Wendehorst, *Z. anorg. Chem.*, **173**, 1928, 268–272.

0.5335 $\bar{1}.4665$ Mo-Si Molybdenum-Silicon

The existence of the following intermediate phases has been established by roentgenographic [1–4] and microscopic investigations [5, 3, 4]: Mo_3Si (8.89 wt. % Si) [2, 5, 3, 4], Mo_3Si_2 (16.32 wt. % Si) [2, 5, 3, 4], and $MoSi_2$ (36.92 wt. % Si) [1, 2, 5, 4]. The findings of [3, 5] are based on the study of arc-melted alloys; [2] and [4] used sintered and hot-pressed powder compacts, respectively. The silicides MoSi [6] and Mo_2Si_3 [7], reported earlier, definitely do not occur [5, 4].

The partial diagram for the region up to Mo_3Si_2, presented in Fig. 538, is due to [3] and based on metallographic and melting-point data. Independently, [4] outlined the phase relations of the entire composition range, using the same methods. Their diagram shows the phases Mo_3Si_2 and $MoSi_2$ to have maximum melting points of

2100 ± 50 and 2030 ± 50°C, respectively. However, in contrast to the findings of [3], Mo_3Si was believed to be formed peritectically by the reaction of melt with (Mo) rather than with Mo_3Si_2. Recently, [8] proposed a corrected diagram similar to that of [3].

Whereas the location of the Mo_3Si_2-$MoSi_2$ eutectic is unknown, that of the $MoSi_2$-Si eutectic was found thermally as approximately 98.5 at. (95 wt.) % Si and about 1410°C [4]. According to [5], the latter would lie between 93.5 and 85 wt. (98 and 95.2 at.) % Si, on the basis of microexamination of arc-melted alloys.

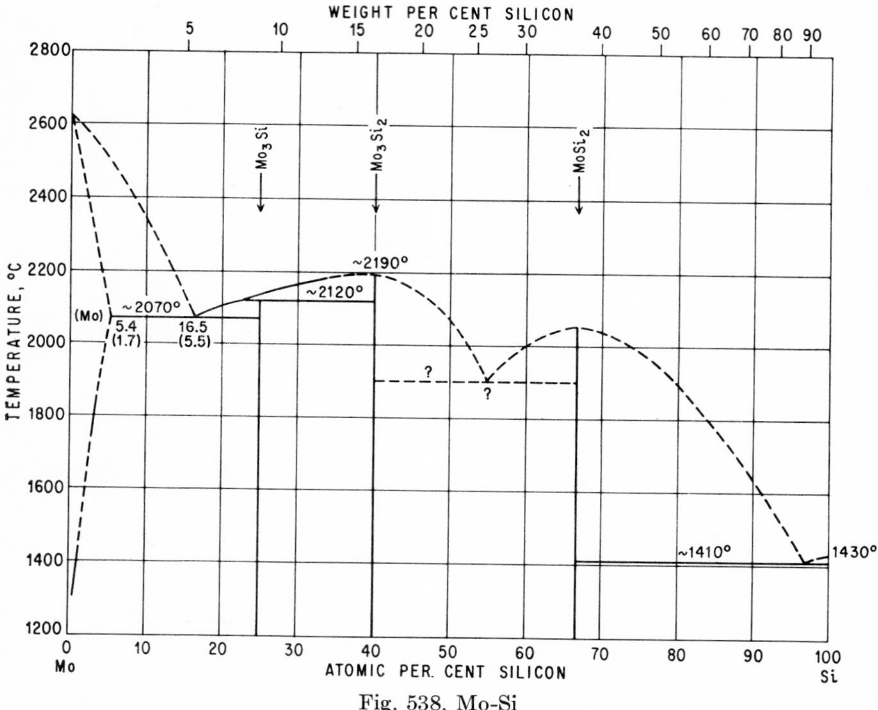

Fig. 538. Mo-Si

The solid solubility of Si in Mo was determined parametrically as 0.9, 1.1, 2.7 (?), and 5.4 at. [0.27, 0.32, 0.80 (?), and 1.65 wt.] % at 1315, 1370, 1425 (?), and 2070°C [3, 9]. [4] gave about 3.35 at. (1 wt.) % Si at 1800°C.

Crystal Structures. The lattice parameter of Mo is reduced to 3.140 A by 3.15 at. % Si [3]. Mo_3Si is isotypic with "β-W" (A15 type), a = 4.89–4.90 A [10, 3, 4]. $MoSi_2$ is tetragonal, with 6 atoms per unit cell (C11b type), $a = 3.20_6$ A, $c = 7.87_7$ A, c/a = 2.457 [1]; $a = 3.20_3$ A, $c = 7.88_7$ A, c/a = 2.462 [4]; $a = 3.20_2$ A, $c = 7.85_2$ A, $c/a = 2.45_2$ [11]. As to the unknown structure of Mo_3Si_2, see [2].

[12] reported that, in rapidly solidified melts containing about 2 wt. % C, a phase was identified which is isotypic with Mn_5Si_3 ($D8_8$ type), $a = 7.28_6$ A, $c = 5.00_2$ A, $c/a = 0.686_5$.

Supplement. [13] indexed the powder pattern of Mo_3Si_2 with a tetragonal unit cell of the dimensions a = 9.66 A, c = 4.99 A, $c/a = 0.51_7$; the cell contains 6 Mo_3Si_2. [14] determined the melting point of $MoSi_2$ as 1950°C.

1. W. H. Zachariasen, *Z. physik. Chem.*, **128**, 1927, 39–48.
2. L. Brewer, A. W. Searcy, D. H. Templeton, and C. H. Dauben, *J. Am. Ceram. Soc.*, **33**, 1950, 291–294; see also [10].
3. Climax Molybdenum Co., Arc-cast Molybdenum-base Alloys, First Annual Report, 1950; Second Annual Report, 1951.
4. R. Kieffer and E. Cerwenka, *Z. Metallkunde*, **43**, 1952, 101–105.
5. W. R. Johnson and M. Hansen, AF Technical Report 6383, June, 1951.
6. E. Wedekind and J. Pintsch, German Patent 294,267 (1913).
7. E. Vigouroux, *Compt. rend.*, **129**, 1899, 1238–1239.
8. H. Nowotny, E. Parthé, R. Kieffer, and F. Benesovsky, *Monatsh. Chem.*, **85**, 1954, 255–272.
9. J. L. Ham, *Trans. ASME*, **73**, 1951, 723–731.
10. D. H. Templeton and C. H. Dauben, *Acta Cryst.*, **3**, 1950, 261–262.
11. H. Nowotny, R. Kieffer, and H. Schachner, *Monatsh. Chem.*, **83**, 1952, 1248.
12. H. Schachner, E. Cerwenka, and H. Nowotny, *Monatsh. Chem.*, **85**, 1954, 245–254.
13. E. Parthé, H. Schachner, and H. Nowotny, *Monatsh. Chem.*, **86**, 1955, 182–185.
14. G. A. Geach and F. O. Jones, *Plansee Proc.*, **1955**, 80–91; *Met. Abstr.*, **24**, 1957, 366.

$\bar{1}.9076$
0.0924

Mo-Sn Molybdenum-Tin

Mo-Sn alloys could not be obtained by reduction of the metal oxides with carbon [1].

1. C. L. Sargent, *J. Am. Chem. Soc.*, **22**, 1900, 783–791.

$\bar{1}.7246$
0.2754

Mo-Ta Molybdenum-Tantalum

X-ray [1, 2] and metallographic [3, 2] investigations have shown that a continuous series of solid solutions exists. Earlier, [4] had reported that the two liquids

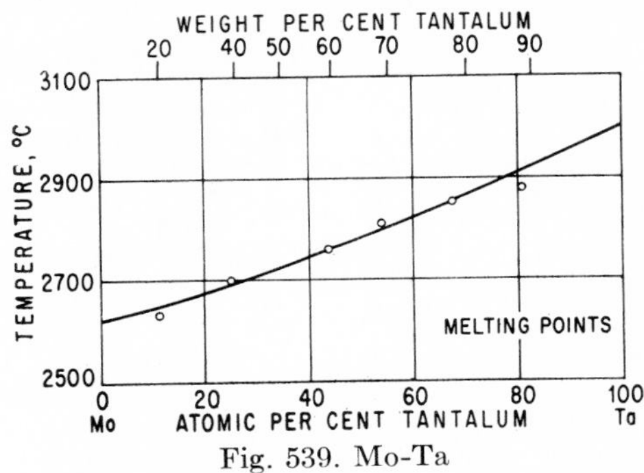

Fig. 539. Mo-Ta

alloy in all proportions. Results of melting-point determinations [2] are indicated in Fig. 539; the data points represent temperatures between the liquidus and solidus, considered to be accurate to about ±50°C.

Lattice-spacing data were reported by [1] and especially [2]. There is a slight deviation from Vegard's rule which reaches a maximum at 62.5 at. % Ta [2]. For parameters of Ta-rich alloys, see also [5].

Supplement. [6] prepared an alloy of the composition $TaMo_2$ to investigate the possibility of a precipitation from the solid phase. "The structure of a specimen held for 13 days at 600°C was identical to filings annealed at 800°C, single-phase b.c.c."

1. H. Bückle, *Z. Metallkunde*, **37**, 1946, 53–56.
2. G. A. Geach and D. Summers-Smith, *J. Inst. Metals*, **80**, 1951-1952, 143–146.
3. W. C. Schumb, S. F. Radtke, and M. B. Bever, *Ind. Eng. Chem.*, **42**, 1950, 826–829; *J. Inst. Metals*, **80**, 1951-1952, 528.
4. W. von Bolton, *Z. Elektrochem.*, **11**, 1905, 51.
5. R. H. Myers, *Metallurgia*, **42**, 1950, 3.
6. R. P. Elliott, Armour Research Foundation, Chicago, Ill., Technical Report 1, OSR Technical Note OSR-TN-247, August, 1954, p. 25.

$\bar{1}.8762$
0.1238

Mo-Te Molybdenum-Tellurium

[1] identified the synthetically prepared phases Mo_2Te_3 (66.61 wt. % Te) and $MoTe_2$ (72.68 wt. % Te) and determined their thermal stability.

1. A. Morette, *Compt. rend.*, **215**, 1942, 86–88; *Ann. chim.* (*Paris*), **19**, 1944, 130–143.

$\bar{1}.6163$
0.3837

Mo-Th Molybdenum-Thorium

X-ray examination of sintered Mo-Th alloys annealed at and slowly cooled from 1000 and 600°C indicated the absence of mutual solid solubilities as well as of intermediate phases at lower temperatures [1].

1. E. Pipitz and R. Kieffer, *Z. Metallkunde*, **46**, 1955, 187–194.

0.3017
$\bar{1}.6983$

Mo-Ti Molybdenum-Titanium

The phase diagram in Fig. 540 was established by micrographic and X-ray analysis and incipient-melting determinations of arc-melted iodide Ti–base alloys [1]. Equilibrium was obtained by annealing for 90–650 hr at eight temperature levels between 855 and 600°C following homogenization for 20–40 hr at 1250°C. Prior to this investigation, a limited amount of information was available [2–6], including evidence for the existence of a complete solid solubility between β-Ti and Mo [3, 4] and an outline of the $\alpha + \beta$ field [6].

Independently of the work by [1], the constitution up to 20 wt. (11 at.) % Mo at temperatures between 980 and 650°C was studied by micrographic analysis of arc-melted alloys, prepared using iodide Ti [7]. Results differ as to the slope of the $\beta/(\alpha + \beta)$ boundary and its position in the lower temperature range. No satisfactory explanation can be offered for this discrepancy; however, the fact that investigations of alloys based on iodide Ti [1] as well as sponge Ti [1, 6, 8] have resulted in quite similar findings is strongly in favor of the work by [1]. Additional support is provided by the $\beta/(\alpha + \beta)$ boundary calculated using an expression for the free energy of a homogeneous phase as a function of temperature [9].

[7] determined the temperatures at which the martensitic-type transformation $\beta \rightarrow \alpha'$ takes place. The minimum concentration for retaining the β structure by quenching is between 10 and 12 wt. (5.3–6.4 at.) % Mo [1, 7, 10].

Lattice parameters of the β phase between 11 and 100 at. % Mo were reported by [1]. The trend of the parameter curve between 72.5 and 100 at. % Mo as given by [5] is in general agreement with these findings.

1. M. Hansen, E. L. Kamen, H. D. Kessler, and D. J. McPherson, *Trans. AIME*, **191**, 1951, 881–888.

2. W. Kroll, *Z. Metallkunde*, **29,** 1937, 189–192.
3. E. I. Larsen, E. F. Swazy, L. S. Busch, and R. H. Freyer, *Metal Progr.*, **55,** 1949, 359–361.
4. B. W. Gonser, *Ind. Eng. Chem.*, **42,** 1950, 222–226.

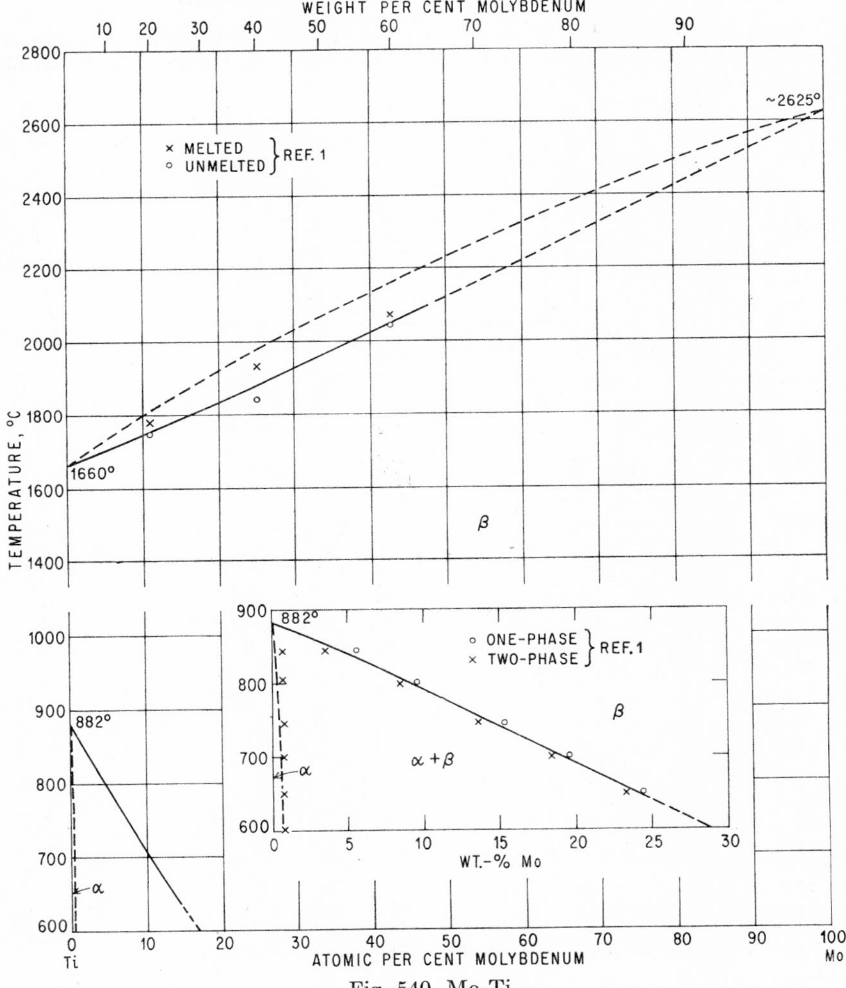

Fig. 540. Mo-Ti

5. J. L. Ham, Climax Molybdenum Co., Arc-cast Molybdenum-base Alloys, First Annual Report, Project NRD 31–331, Apr. 1, 1950.
6. C. M. Craighead, O. W. Simmons, and L. W. Eastwood, *Trans. AIME*, **188,** 1950, 485–513.
7. P. Duwez, *Trans. AIME*, **191,** 1951, 765–771; **194,** 1952, 518.
8. *J. Inst. Metals*, **80,** 1951-1952, 704.
9. W. Rostoker, *Trans. AIME*, **191,** 1951, 1203–1205.

10. D. J. DeLazaro, M. Hansen, R. E. Riley, and W. Rostoker, *Trans. AIME*, **194**, 1952, 265–269.

$\bar{1}.6053$
0.3947

Mo-U Molybdenum-Uranium

The phase relationships (Fig. 541) were established by micrographic, X-ray, and thermal-analysis methods [1]. Alloys were prepared by melting in beryllia crucibles under vacuum.

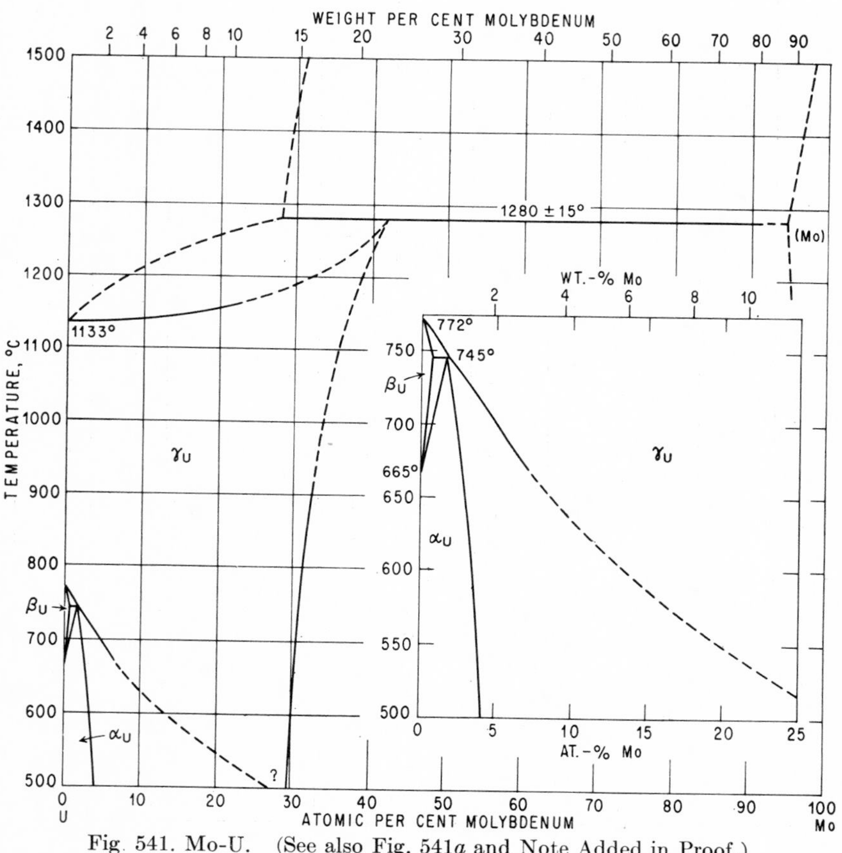

Fig. 541. Mo-U. (See also Fig. 541*a* and Note Added in Proof.)

The maximal solubility of Mo in β-U is about 0.7 at. (0.28 wt.) % at the peritectoid temperature of 745°C; the composition of the α_U phase at this temperature is about 1.7 at. (0.69 wt.) % Mo. The solubility of U in Mo at 1080°C was estimated as 4.5 ± 0.6 at. % U (but see Note Added in Proof).

The slope of the $\gamma/(\alpha + \gamma)$ boundary, determined micrographically down to 600°C [1], suggested that the γ_U phase with 20 at. % Mo was stable at room temperature. However, [2, 3] presented microscopic and X-ray evidence that an alloy with 20.5 at. (9.4 wt.) % Mo was definitely heterogeneous at 450°C. This was confirmed by a reinvestigation [4] which showed that equilibrium had not been attained in the earlier heat-treatments [1]. Arc-melted alloys with 20.5–31 at. % Mo were homogenized and then annealed at 500°C for 14 days. These tests placed the $\gamma/(\alpha + \gamma)$ boundary between 26 and 29 at. % Mo at 500°C [4]. The slope of the corrected (tentative) $\gamma/(\alpha + \gamma)$ boundary and that of the $\gamma/[\gamma + (\text{Mo})]$ boundary (Fig. 541) strongly suggest, however, that γ_U decomposes eutectoidally into α_U + (Mo). Evidence for such a transformation is given in a photomicrograph of an alloy with 23 at. % Mo annealed at 450°C [4]. The temperature of the eutectoid decomposition may well lie above 500°C. See Note Added in Proof.

Lattice parameters of the b.c.c. γ_U phase with 17.3–31.2 at. % Mo were reported by [5], and additional data were given by [1]. The γ_U phase can be retained by quenching alloys containing more than about 9 at. % Mo, but with lower Mo contents decomposition to α by a martensitic type of reaction occurs [1].

The alloy with 20 at. % Mo, after annealing at 450°C, showed, in addition to a strong α-phase pattern, a heavy pattern of another phase, γ', found to correspond to a b.c. tetragonal structure, with a = 3.420 A, c = 3.278 A, c/a = 0.959 [3]. Similar

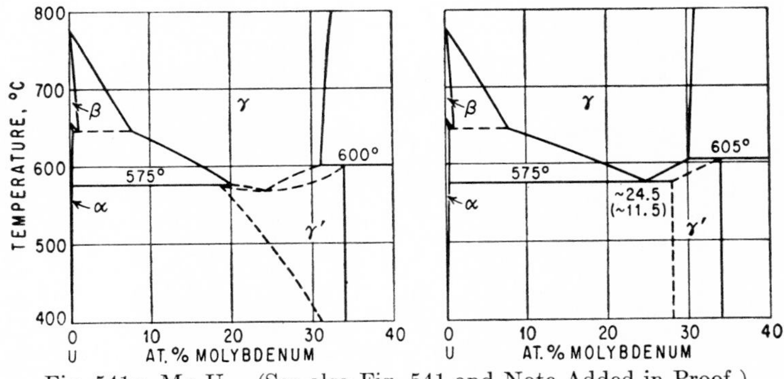

Fig. 541*a*. Mo-U. (See also Fig. 541 and Note Added in Proof.)

data were reported by [4]. It appears likely that the transition $\gamma \rightarrow \gamma'$ is connected with the decomposition of γ, rather than a disorder-order change as proposed by [4].

Note Added in Proof. It has been well established by more recent work that γ-phase alloys containing about 30 at. % Mo transform to an ordered structure, called γ' (or δ, ϵ), on cooling below 600°C. The transformation equilibria are, however, still a matter of controversy.

A group of representatives of several American laboratories have agreed on the phase relations shown in the left partial phase diagram of Fig. 541*a*. It is mainly based on hitherto unpublished information and has been reproduced from the review by [6]. As regards the solubility of Mo in U, the diagram shows the maximum solubility in α- and β-U to be about 0.5 and 1.0 at. (0.2 and 0.4 wt.) % Mo, respectively. Mo lowers, rather than raises [1], the α/β transformation in U. Another version of the γ/γ'-transformation equilibria (Fig. 541*a*, right) has been proposed by [7], on the basis of information by [8] that a eutectoid $\gamma \rightleftarrows \alpha + \gamma'$ occurs at about 11.5 wt. (24.4 at.) % Mo, 575°C, as well as of their own study of transformation kinetics. This version has the advantage of simpler phase relations; however, since [7] observed eutectoid-like structures after heat-treatments in the $\alpha + \gamma$ field, it cannot be considered as finally established.

The crystal structure of the γ' (MoU_2) phase has been shown by [9] to be of the tetragonal $MoSi_2$ (C11b) type, with a = 3.427 A, c = 9.834 A, c/a = 2.871.

According to [10], the solubility of U in solid Mo appears to be a maximum of about 2 at. % at 1285°C.

1. P. C. L. Pfeil, *J. Inst. Metals,* **77,** 1950, 553–570.
2. A. U. Seybolt and R. K. McKechnie, *J. Inst. Metals,* **78,** 1950-1951, 760.
3. C. W. Tucker, *J. Inst. Metals,* **78,** 1950-1951, 760–761.
4. P. C. L. Pfeil, *J. Inst. Metals,* **78,** 1950-1951, 762–763.
5. A. S. Wilson and R. E. Rundle, *Acta Cryst.,* **2,** 1949, 126–127.
6. H. A. Saller and F. A. Rough, Compilation of U.S. and U.K. Uranium and Thorium Constitutional Diagrams, *U.S. Atomic Energy Comm., Publ.* BMI-1000, June, 1955.
7. R. J. Van Thyne and D. J. McPherson, *Trans. ASM,* **49,** 1957, 598–619; discussion, pp. 619–621.
8. E. K. Halteman, private communication to [7].
9. E. K. Halteman, *Acta Cryst.,* **10,** 1957, 166–169.
10. H. A. Saller, F. A. Rough, and D. C. Bennett, unpublished information (March, 1952); quoted in [6].

0.2749
$\bar{1}$.7251

Mo-V Molybdenum-Vanadium

On the basis of metallographic and X-ray studies, it was reported that at least 48 at. (33 wt.) % V was soluble in Mo at 1100°C [1]. Arc-cast alloys with 38.5–100 at. (25–100 wt.) % V were found to consist of heavily cored solid solutions. Annealing at 900°C for 170 hr resulted in the precipitation of small amounts of a second phase of unknown identity, in alloys with 56–94.5 at. (40–90 wt.) % V [2].

These findings permit the conclusion that Mo and V form an uninterrupted series of solid solutions at high temperatures.

The lattice parameter of Mo is reduced to about 3.10 A by 31 at. % V [3].

Supplement. The following alloy specimens yielded diffraction patterns which showed only the lines of a b.c.c. structure: 66.7 at. % V, annealed 13 days at 600°C [4]; ~33.5 and ~66 at. % V, annealed 5 hr at 1000°C and 5 hr at 600°C followed by slow cooling [5]. These results suggest that a continuous series of b.c.c. solid solutions exists also at lower temperatures.

1. J. L. Ham, *Trans. ASME,* **73,** 1951, 723–731.
2. W. Rostoker and A. Yamamoto, *Trans. ASM,* **46,** 1954, 1136–1163.
3. J. L. Ham, Climax Molybdenum Co., Arc-cast Molybdenum-base Alloys, First Annual Report, Project NRD 31-331, Apr. 1, 1950, pp. 60–66.
4. R. P. Elliott, Armour Research Foundation, Chicago, Ill., Technical Report 1, OSR Technical Note OSR-TN-247, August, 1954, p. 18.
5. E. Pipitz and R. Kieffer, *Z. Metallkunde,* **46,** 1955, 187–194, especially 191.

$\bar{1}$.7174
0.2826

Mo-W Molybdenum-Wolfram

By means of melting-point determinations [1–3], microexamination [1-3], X-ray analysis [4–6], and measurements of the temperature coefficient of electrical resistance [3], it has been established that the two metals form a continuous series of solid solutions.

The melting points of sintered samples, designated in Fig. 542 by crosses, were obtained by optical temperature measurements [1]. [2] determined fusion points of wires by the fusion-wattage method and reported the temperatures indicated in

Fig. 542 by circles. These temperatures probably lie somewhere between the solidus and liquidus. The same method was also used by [1]; the scattering of individual determinations for a particular composition varied between 40 and 140°C; however, the fusion-temperature-vs.-composition curve follows the same trend as the data presented in Fig. 542 [7]. This also holds for the melting points reported by [3], which are based on measurements of brightness temperatures. They lie on a nearly straight line (over the composition in at. %) connecting the melting points of Mo and W given as 2420 and 3280°C, respectively [8].

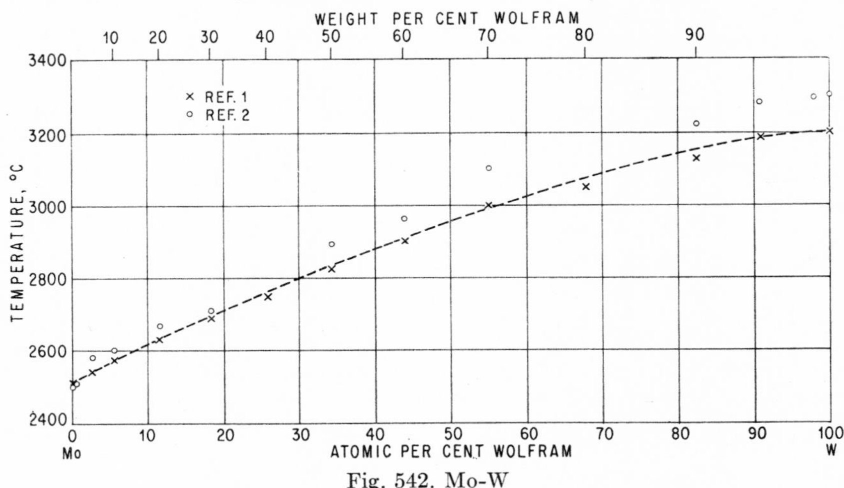

Fig. 542. Mo-W

The lattice parameter of the b.c.c. cell changes additively with composition [5, 6]. [9] reported that an ordered structure had been detected in this system; however, neither reference nor data are given.

1. F. A. Fahrenwald, *Trans. AIME*, **54**, 1917, 570–573, 583–585; **56**, 1917, 612–619.
2. Z. Jeffries, *Trans. AIME*, **56**, 1917, 600–611.
3. W. Geiss and J. A. M. van Liempt, *Z. anorg. Chem.*, **128**, 1923, 355–360.
4. E. C. Bain, *Chem. Met. Eng.*, **28**, 1923, 24.
5. A. E. van Arkel, *Physics*, **4**, 1924, 33–41; **6**, 1926, 64–69; *Z. Krist.*, **67**, 1928, 235–238.
6. H. Bückle, *Z. Metallkunde*, **37**, 1946, 53–56.
7. See contribution to the discussion by J. W. Richards, *Trans. AIME*, **56**, 1917, 618–619.
8. See the remarks by C. J. Smithells, "Tungsten," p. 253, Chapman & Hall, Ltd., London, 1953.
9. S. Kaya and A. Kussmann, *Z. Physik*, **72**, 1931, 306.

0.1666
$\bar{1}.8334$

Mo-Zn Molybdenum-Zinc

It has been reported by [1] that small amounts of Mo dissolve in Zn, the microstructure showing primary crystals of an intermediate phase in the Zn matrix. According to [2], however, Mo and Zn do not alloy at temperatures up to 1350°C; their results suggest strongly that the intermediate phase observed by [1] was actually $FeZn_7$, Fe being present as a contaminant in commercial Mo.

1. W. Guertler, *Z. Metallkunde*, **15**, 1923, 152.
2. W. Köster and H. Schmid, *Z. Metallkunde*, **46**, 1955, 462–463.

0.0220
$\bar{1}$.9780

Mo-Zr Molybdenum-Zirconium

Micrographic and thermal analysis as well as incipient-melting tests of arc-melted alloys based on iodide Zr were used to establish the phase diagram in Fig. 543 [1]. The

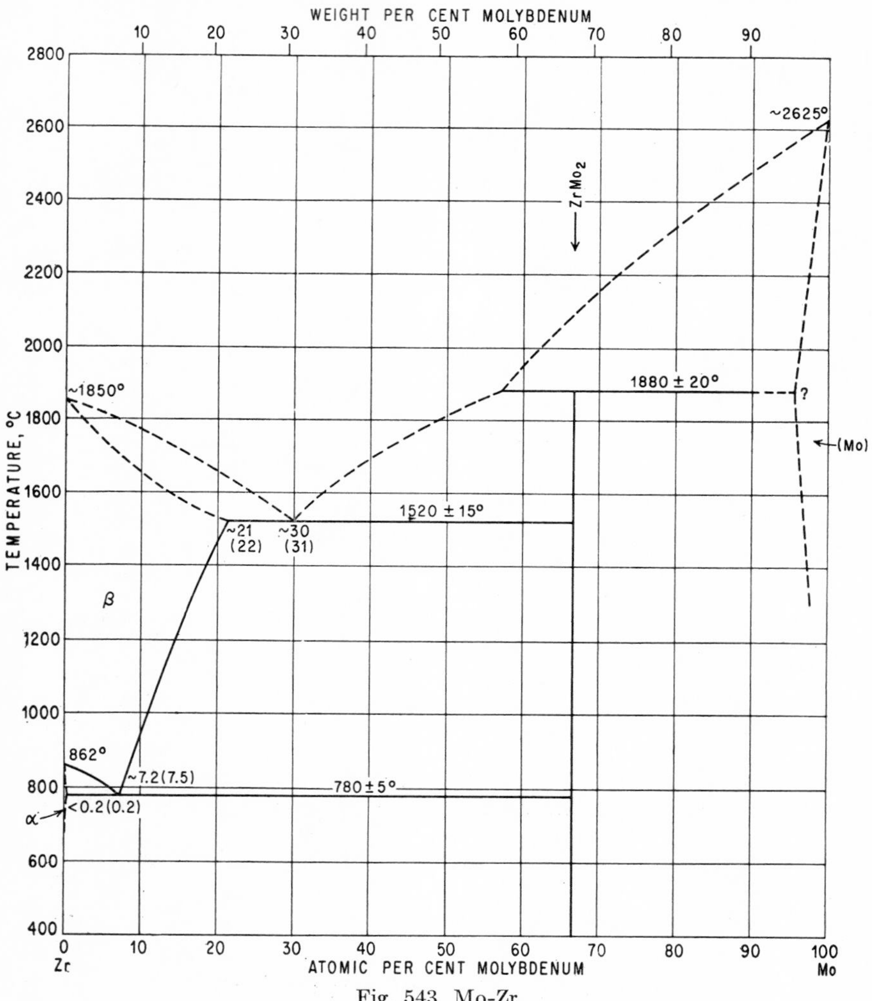

Fig. 543. Mo-Zr

only intermediate phase existing, $ZrMo_2$ (67.78 wt. % Mo), is isotypic with $MgCu_2$ (C15 type), $a = 7.60 \pm 0.05$ A [2, 1]. It was suggested that the phase $ZrMo_3$ (75.94 wt. % Mo), reported by [3] to be cubic with $a = 4.95$ A, was probably $(Mo,Zr)_3O$ [4]. The solid solubility of Zr in Mo was found to be between 3 and 5 at. (wt.) % in the neighborhood of 1100°C [5].

Supplement. The $MgCu_2$-type structure of $ZrMo_2$ has been corroborated by [6, 7]; a = 7.596 A according to [6]. Parametric work by [7] showed the lattice parameter of Mo to be greatly increased by addition of Zr and the limit of solubility in slowly cooled alloys as approximately 7 at. (wt.) % Zr (cf. above).

1. R. F. Domagala, D. J. McPherson, and M. Hansen, *Trans. AIME,* **197,** 1953, 73–79.
2. P. Duwez and C. B. Jordan, *J. Am. Chem. Soc.,* **73,** 1951, 5509.
3. H. J. Wallbaum, *Naturwissenschaften,* **30,** 1942, 149.
4. G. Hägg and N. Schönberg, *Acta Cryst.,* **7,** 1954, 351–352.
5. J. L. Ham, *Trans. ASME,* **73,** 1951, 723–731.
6. R. P. Elliott, Armour Research Foundation, Chicago, Ill., Technical Report 1, OSR Technical Note OSR-TN-247, August, 1954, p. 38.
7. E. Pipitz and R. Kieffer, *Z. Metallkunde,* **46,** 1955, 187–194, especially 190–191.

$\bar{1}.1783$
0.8217

N-Nb Nitrogen-Niobium

Discrepancies as to the composition and crystal structure of Nb-N phases [1–9] have been cleared up by systematic X-ray analysis of oxygen-free preparations with 4.8–50 at. % N [10] (but see Supplement). According to this work, there are (*a*) three different forms (NbN I, NbN II, and NbN III) of the phase based on the composition NbN; (*b*) a phase stable between about 42.9 and 44.1 at. % N; and (*c*) a phase with about 28.6–33.3 at. % N (Nb_2N), already reported by [11].

The formation of NbN I, stable at 50 at. % N, is favored by relatively low nitriding temperatures (1300°C). It is hexagonal, with a = 2.956 A, c = 11.27 A, c/a = 3.815. NbN III with 47.1–48.5 at. % N is the phase previously considered as the only stable phase NbN [1, 2, 4, 5, 7–9] with NaCl-type structure. It is formed at higher nitriding temperatures and by annealing of NbN I at about 1450°C. Its lattice parameter is a = 4.389 A (at 48.5 at. % N) [10]. NbN II, with about 48.7 at. % N, first observed by [3, 11], appears to be stable only under certain conditions of temperature and pressure. It is hexagonal, with a = 2.94 A, c = 5.46 A, c/a = 1.86 [11, 10]; i.e., the c axis is about half that of NbN I.

The phase with about 42.9–44.1 at. % N has a tetragonally distorted NaCl structure [4, 10], with a = 4.386 A, c = 4.312 A, c/a = 0.983 at 42.9 at. % N and a = 4.386 A, c = 4.331 A, c/a = 0.988 at 44.5 at. % N [10]. Nb_2N has a h.c.p. arrangement of the Nb atoms with interstitial N atoms [11], a = 3.052 A, c = 4.957 A, c/a = 1.624 at the low-N side and a = 3.052 A, c = 4.996 A, c/a = 1.637 at the high-N side of the composition range [10]. [12] confirmed that samples with 42–44 at. % N essentially consist of the f.c. tetragonal phase; however, in contrast to [10, 11] it was suggested that the limiting composition at the low-N side had a composition corresponding to Nb_2N. In a joint publication [13], agreement was reached in favor of the findings by [10, 11].

A nitride of the composition Nb_3N_5 (62.5 at. % N), reported earlier [14, 15], could not be confirmed.

The solid solubility of N in Nb was determined by aging experiments, using measurements of internal friction, and found to be about 0.05, 0.04, 0.03, 0.019, and 0.005 wt. % N at 1100, 950, 800, 600, and 300°C, respectively [16]. The melting point of NbN was given as about 2050°C [1]. The kinetics of reaction of Nb with N_2 were studied by [17].

Supplement. The findings of [10] were only partially corroborated by a recent X-ray powder-diffraction study [18] of specimens prepared by nitriding Nb metal powder or Nb hydride with dry oxygen-free ammonia (Fig. 544).

As to the (Nb_2N), or β, phase, [18] confirmed the data given by [10]. However, [18] was unable to prepare the two phases with NaCl and tetragonally distorted NaCl structure in an oxygen-free state. He found, instead, a hexagonal WC-type phase, γ, between the approximate phase limits $NbN_{0.80}$ and $NbN_{0.90}$ (44.4 and 47.4 at. % N) with the unit-cell dimensions varying as a = 2.950–2.958 A, c = 2.772–2.779 A, c/a = 0.940–0.939.

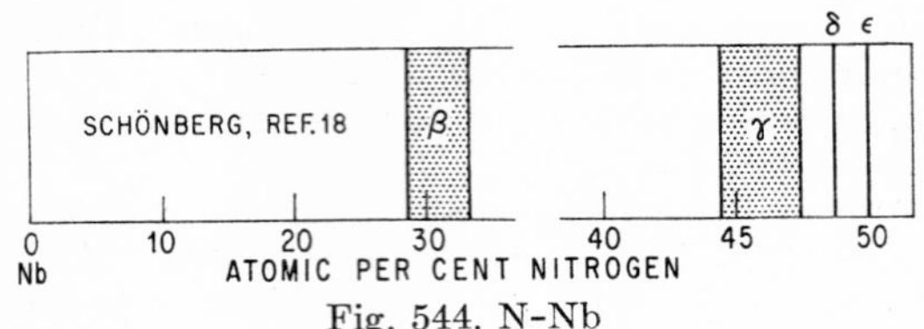

Fig. 544. N-Nb

The δ phase of [18] is identical with the phase NbN II reported by [10]. Unit-cell dimensions according to [18] are a = 2.968 A, c = 5.535 A, c/a = 1.865. Likewise, ϵ is identical with NbN I; a = 2.952 A, c = 11.25 A, c/a = 3.811, according to [18].

1. E. Friederich and L. Sittig, *Z. anorg. Chem.*, **143,** 1925, 308–309.
2. K. Becker and F. Ebert, *Z. Physik,* **31,** 1925, 268–272.
3. J. S. Umanski, *J. Phys. Chem. U.S.S.R.,* **14,** 1940, 332–339.
4. G. Aschermann, E. Friederich, E. Justi, and J. Kramer, *Physik. Z.,* **42,** 1941, 349–360.
5. F. H. Horn and W. T. Ziegler, *J. Am. Chem. Soc.,* **69,** 1947, 2762–2769.
6. R. E. Rundle, *Acta Cryst.,* **1,** 1948, 180–187.
7. G. T. Armstrong, *J. Am. Chem. Soc.,* **71,** 1949, 3583–3587.
8. D. B. Cook, M. W. Zemansky, and H. A. Boorse, *Phys. Rev.,* **79,** 1950, 1021.
9. P. Duwez and F. Odell, *J. Electrochem. Soc.,* **97,** 1950, 299–304.
10. G. Brauer and J. Jander, *Z. anorg. Chem.,* **270,** 1952, 160–178; see also [13].
11. G. Brauer, *Z. Elektrochem.,* **46,** 1940, 397–402.
12. H. Rögener, *Z. Physik,* **132,** 1952, 446–467.
13. G. Brauer, J. Jander, and H. Rögener, *Z. Physik,* **134,** 1953, 432–434.
14. R. D. Hau and E. F. Smith, *J. Am. Chem. Soc.,* **27,** 1905, 1395–1396.
15. W. Muthmann, L. Weiss, and R. Riedelbauch, *Liebigs Ann.,* **355,** 1907, 92.
16. C. Y. Ang and C. Wert, *Trans. AIME,* **197,** 1953, 1032–1036.
17. E. A. Gulbransen and K. F. Andrew, *J. Electrochem. Soc.,* **96,** 1949, 364–376; *Trans. AIME,* **188,** 1950, 586–599.
18. N. Schönberg, *Acta Chem. Scand.,* **8,** 1954, 208–212.

$\bar{2}.9872$
1.0128

N-Nd Nitrogen-Neodymium

NdN (8.85 wt. % N) is cubic of the NaCl (B1) type, a = 5.15 A [1].

1. A. Iandelli and E. Botti, *Atti reale accad. nazl. Lincei, Rend.,* **25,** 1937, 638–640.

$\bar{1}.3778$
0.6222

N-Ni Nitrogen-Nickel

According to [1], N_2 is insoluble in and does not react with Ni at temperatures up to 1400°C. This was confirmed by [2] for temperatures up to 900°C; see also [3].

By heating Ni powder with NH_3 at temperatures between 300 and 1000°C (also under pressure), [4] was unable to obtain preparations with more than 0.32 wt. % N (plus 0.17 wt. % H). Some evidence for a slight solid solubility of N in Ni was found by parameter measurements [4]. [5, 6] obtained preparations with up to 7.5 wt. (25.3 at.) % N by treating Ni with NH_3 at 500–550°C, and [7], using the same process, preparations with up to 9 wt. (29.3 at.) % N. X-ray analysis showed the existence of the nitride Ni_3N (7.37 wt. % N) having a h.c.p. structure of the Ni atoms with interstitial N atoms, a = 2.670 A, c = 4.307 A, c/a = 1.613 [6]. Ni_3N decomposes above 360°C. The solid solubility of N in Ni was estimated as about 0.07 wt. (0.3 at.) % [6]. [8] studied the structure of Ni layers prepared by cathodic atomizing in a nitrogen atmosphere.

Supplement. [9] studied Ni-N preparations by means of X-ray and thermomagnetic methods. It was found that addition of N raises the lattice parameter of Ni (3.516 A) to 3.521 A and lowers its Curie point to 342°C. The hexagonal nitride Ni_3N (a = 2.668 A, c = 4.294 A, c/a = 1.609) decomposed at 190°C on heating in vacuo and at about 450°C on heating in a nitrogen atmosphere.

According to [10], the lattice parameters of Ni_3N are a = 2.6677 ± 5 A, c = 4.3122 ± 5 A, c/a = 1.6165. No range of composition could be detected.

1. A. Sieverts and W. Krumbhaar, *Ber. deut. chem. Ges.*, **43**, 1910, 894.
2. P. Laffitte and P. Grandadam, *Compt. rend.*, **200**, 1935, 1039–1041.
3. "Gmelins Handbuch der anorganischen Chemie," 8th ed., System No. 4, p. 247, Verlag Chemie, G.m.b.H., Berlin, 1936.
4. G. Hägg, *Nova Acta Reg. Soc. Sci. Upsaliensis*, (4)**7**, 1929, 22–23.
5. G. T. Beilby and G. G. Henderson, *J. Chem. Soc.*, **79**, 1901, 1251–1252.
6. R. Juza and W. Sachsze, *Z. anorg. Chem.*, **251**, 1943, 201–212.
7. P. Grandadam, *Ann. Chim.*, **4**, 1935, 83–146.
8. W. Büssem and F. Gross, *Z. Physik*, **87**, 1934, 778–799.
9. R. Bernier, *Ann. chim. (Paris)*, **6**, 1951, 124–131.
10. K. H. Jack, *Acta Cryst.*, **3**, 1950, 393.

$\bar{2}.7716$
1.2284

N-Np Nitrogen-Neptunium

NpN is isotypic with NaCl (B1 type), a = 4.897 A [1].

1. W. H. Zachariasen, *Acta Cryst.*, **2**, 1949, 388–390; see also I. Sheft and S. Fried, *J. Am. Chem. Soc.*, **75**, 1953, 1236–1237.

$\bar{2}.7828$
1.2172

N-Pa Nitrogen-Protactinium

The existence of PaN_2, analogous to UN_2, has been reported [1].

1. P. Sellers, S. Fried, R. Elson, and W. Zachariasen, *U.S. Atomic Energy Comm., Publ.* AECD-3167, 1952.

$\bar{2}.8300$
1.1700

N-Pb Nitrogen-Lead

Nitrogen is insoluble in the following metals (investigated up to the temperature given in parentheses): Pb (600°C), Pd (1400°C), Sb (800°C), Sn (800°C), Tl (600°C), and Zn (600°C) [1]. For Pb, Sb, Sn, and Tl, see also [2].

1. A. Sieverts and W. Krumbhaar, *Ber. deut. chem. Ges.*, **43**, 1910, 894.
2. K. Iwase, *Science Repts. Tôhoku Univ.*, **15**(1), 1926, 557.

$\bar{1}.1182$
0.8818

N-Pd Nitrogen-Palladium

See N-Pb, page 985.

$\bar{2}.9974$
1.0026

N-Pr Nitrogen-Praseodymium

PrN (9.04 wt. % N) is isotypic with NaCl (B1 type), a = 5.16 A [1].

1. A. Iandelli and E. Botti, *Atti reale accad. nazl. Lincei, Rend.*, **25,** 1937, 129–132.

$\bar{2}.7626$
1.2374

N-Pu Nitrogen-Plutonium

The existence of PuN has been established by determination of its crystal structure, which is of the NaCl (B1) type, a = 4.905 A [1]. [2] studied two methods of preparation as well as the chemical properties of PuN. The lattice parameter was found as a = 4.905 ± 2 A.

1. W. H. Zachariasen, *Acta Cryst.*, **2,** 1949, 388–390.
2. F. Brown, H. M. Ockenden, and G. A. Welch, *J. Chem. Soc.*, **1955,** 4196–4201.

$\bar{2}.8762$
1.1238

N-Re Nitrogen-Rhenium

A rhenium nitride of low stability with maximal 30 at. % N (Re_2N?) was prepared by reaction of NH_3 with NH_4ReO_4 (or $ReCl_3$) at 300–350°C and reported to have a f.c.c. structure of Re atoms with interstitial N atoms, a = 3.93 A [1]; see also [2]. The solubility of N in Re is probably very small [1].

The nitride is not formed by reaction of Re with N [1, 3–5].

1. H. Hahn and A. Konrad, *Z. anorg. Chem.*, **264,** 1951, 174–180.
2. W. Geilmann and F. W. Wrigge, *Z. anorg. Chem.*, **199,** 1931, 65–76.
3. C. Agte, H. Alterthum, K. Becker, G. Heyne, and K. Moers, *Z. anorg. Chem.*, **196,** 1931, 129–159.
4. I. Noddack and W. Noddack, "Das Rhenium," p. 33, L. Voss, Leipzig, 1933.
5. W. Trzebiatowski, *Z. anorg. Chem.*, **233,** 1937, 376–384.

$\bar{1}.0609$
0.9391

N-Sb Nitrogen-Antimony

See N-Pb, page 985.

$\bar{1}.4922$
0.5078

N-Sc Nitrogen-Scandium

ScN (23.76 wt. % N) is isotypic with NaCl, a = 4.45 A [1].

1. K. Becker and F. Ebert, *Z. Physik*, **31,** 1925, 268–272.

$\bar{1}.6978$
0.3022

N-Si Nitrogen-Silicon

Si_3N_4 (39.94 wt. % N) is formed by annealing Si in N_2 at 1300–1500°C [1]. It was reported to be isomorphous with Ge_3N_4 and most likely to have an orthorhombic structure, with a = 13.38 A, b = 8.60 A, c = 7.74 A [2]. However, [3] have found Ge_3N_4 to be tetragonal.

1. L. Weiss and T. Engelhardt, *Z. anorg. Chem.*, **65**, 1910, 38–104; H. Funk, *Z. anorg. Chem.*, **133**, 1924, 67–72; E. Friederich and L. Sittig, *Z. anorg. Chem.*, **143**, 1925, 313–314; W. B. Hincke and L. R. Brantley, *J. Am. Chem. Soc.*, **52**, 1930, 48–52.
2. W. C. Leslie, K. G. Carroll, and R. M. Fischer, *Trans. AIME*, **194**, 1952, 204–206; see also discussion, p. 1201.
3. R. Juza and H. Hahn, *Z. anorg. Chem.*, **244**, 1940, 125–132.

$\bar{1}$.0719
0.9281

N-Sn Nitrogen-Tin

See N-Pb, page 985.

The formation of a tin nitride of the probable formula Sn_3N_4 by volatilization of tin in a nitrogen atmosphere has been reported by [1].

1. G. Berraz, *Anales soc. cient. argentina, Sección Santa Fé*, **7**, 1935, 6–7; *Chem. Abstr.*, **31**, 1937, 949.

$\bar{2}$.8890
1.1110

N-Ta Nitrogen-Tantalum

The existence of two nitrides has been established: Ta_2N (3.73 wt. % N) [1–3] and TaN (7.19 wt. % N) [4–9]. The lower nitride possibly is not homogeneous at the ideal composition; see the systems N-Nb and N-V. The nitrides Ta_3N_5 [4] and TaN_2 [10] have not been confirmed; it is very unlikely that they exist. The melting point of TaN was given as 2890 [5] and 3090 ± 50°C [8].

Crystal Structures. The structure of TaN was reported to be hexagonal [presumably of the wurtzite (B4) type], with a = 3.06 A, c = 4.95 A, c/a = 1.62 [11]. However, [1] found that Ta_2N had the structure and parameters (a = 3.07 A, c = 4.96 A, c/a = 1.62) observed by [11] for TaN [12]. Confirming this, [2] noted that the hexagonal structure must be actually that of Ta_2N; and [3] reported supporting evidence, giving the parameters of Ta_2N as a = 3.05 A, c = 4.93 ± 0.01 A, c/a = 1.61–1.62. TaN was reported to have a complex unidentified structure [9]; see also [3].

As regards the kinetics of reaction of Ta with N_2, see [13].

Supplement. Systematic X-ray investigations of the Ta-N system have been carried out by [14] and [15]. Whereas [14] reported the existence of two (possibly three) intermediate phases, [15] observed four.

[14] synthesized their specimens from the elements at 1400°C. The solubility of N in Ta was found to be 4 at. % N at 1000°C but to decrease rather sharply with fall in temperature. Slowly cooled specimens in the low-N region exhibited powder lines of a phase which was not identified (see below). The (Ta_2N) phase was found to be homogeneous within the range $TaN_{0.41-0.50}$ (29.1–33.3 at. % N), with the hexagonal lattice constants a = 3.048 A, c = 4.919 A, c/a = 1.61 at 33.3 at. % N. The structure of the compound TaN was determined as hexagonal, a = 5.1911 A, c = 2.9107 A, c/a = 0.561, 3 TaN per unit cell.

[15] nitrided Ta or Ta hydride with ammonia at temperatures between 700 and 1100°C. The following intermediate phases were identified: (a) β or $TaN_{\sim 0.05}$ (~5 at. % N). B.c.c. sublattice; weak superlattice lines can be indexed by a cubic lattice of tripled lattice parameter, a = 10.11 A. (b) γ or $TaN_{\sim 0.40-0.45}$ (29–31 at. % N). Hexagonal with the unit cell dimensions varying as a = 3.041–3.048 A, c =

4.907–4.918 A, $c/a = 1.614$. Good agreement with the results of [14]. (*c*) δ or $TaN_{\sim 0.80-0.90}$ (44.5–47.5 at. % N). Hexagonal with the unit cell dimensions varying as $a = 2.925$–2.938 A, $c = 2.876$–2.883 A, $c/a = 0.983$–0.981. [14] studied no alloy within this composition range. (*d*) ϵ or TaN. Hexagonal with cell dimensions invariably found as $a = 5.185$ A, $c = 2.908$ A, $c/a = 0.561$; 3 TaN per unit cell, B35 (CoSn) type. The structure is identical with that described by [14] (see above).

Neither [14] nor [15] found nitrides containing more than 50 at. % N.

1. G. Aschermann, E. Friederich, E. Justi, and J. Kramer, *Physik. Z.*, **42,** 1941, 349–360.
2. E. Rundle, *Acta Cryst.*, **1,** 1948, 180–187.
3. P. Chiotti, *J. Am. Ceram. Soc.*, **35,** 1952, 123–130.
4. A. Joly, *Bull. soc. chim. France*, **25,** 1876, 506.
5. E. Friederich and L. Sittig, *Z. anorg. Chem.*, **143,** 1925, 308–309.
6. A. E. van Arkel and J. H. de Boer, *Z. anorg. Chem.*, **148,** 1925, 348.
7. K. Moers, *Z. anorg. Chem.*, **198,** 1931, 243–261, especially 256–257.
8. C. Agte and K. Moers, *Z. anorg. Chem.*, **198,** 1931, 239.
9. F. H. Horn and W. T. Ziegler, *J. Am. Chem. Soc.*, **69,** 1947, 2762–2769.
10. W. Muthmann, L. Weiss, and R. Riedelbauch, *Liebigs Ann.*, **355,** 1907, 92.
11. A. E. van Arkel, *Physica*, **4,** 1924, 286–301.
12. In a reaction product of variable N content, K. Becker and F. Ebert (*Z. Physik*, **31,** 1925, 268–272) suggested a cubic phase (Ta-rich solid solution?) with $a = 3.62$ A to be present, besides a phase of lower symmetry (Ta_2N?).
13. E. A. Gulbransen and K. F. Andrew, *J. Electrochem. Soc.*, **96,** 1949, 364–376; *Trans. AIME*, **188,** 1950, 586–599.
14. G. Brauer and K. H. Zapp, *Z. anorg. Chem.*, **277,** 1954, 129–139. The structure of TaN was already described in *Naturwissenschaften*, **40,** 1953, 604.
15. N. Schönberg, *Acta Chem. Scand.*, **8,** 1954, 199–203.

$\bar{2}.7807$
1.2193

N-Th Nitrogen-Thorium

The existence of two nitrides has been established: ThN (5.69 wt. % N) and Th_2N_3 (8.30 wt. % N). ThN has the NaCl (B1) type of structure [1], with $a = 5.21$ A [2]; its melting point is 2630 ± 50°C [2]. Th_2N_3 is isotypic with La_2O_3 ($D5_2$ type), $a = 3.883$ A, $c = 6.187$ A, $c/a = 1.593$ [3]. [2] gave the lattice constants as $a = 3.88$ A, $c = 6.17$ A, $c/a = 1.59$. [4] reported that an electron-diffraction pattern of a reaction product of Th with N_2 indicated an apparently new phase, believed to be a nitride and having a b.c.c. structure, with $a = 4.55$ A.

Nitride preparations previously identified as Th_3N_4 [57.14 at. (7.45 wt.) % N] [5–7] probably were actually Th_2N_3.

Supplement. The reactions of Th with N_2 were studied by [8] over the temperature range 670–1605°C. Surface films of ThN and Th_2N_3 were observed. From diffusion data, the following limiting solubilities of N in Th were obtained: 0.33, 0.20, 0.11, and 0.05 wt. (5.2, 3.2, 1.8, and 0.8 at.) % N at 1500, 1200, 1000, and 850°C, respectively (taken from a graph).

1. R. E. Rundle, *Acta Cryst.*, **1,** 1948, 180–187.
2. P. Chiotti, *J. Am. Ceram. Soc.*, **35,** 1952, 123–130.
3. W. H. Zachariasen, *Acta Cryst.*, **2,** 1949, 388–390.
4. G. W. Fox et al., *U.S. Atomic Energy Comm., Publ.* ISC-224, 1952.
5. H. Moissan and A. Étard, *Compt. rend.*, **122,** 1896, 573; *Ann. chim. et phys.*, **12,** 1897, 427.

6. C. Matignon and M. Delépine, *Compt. rend.*, **132,** 1901, 37.
7. B. Neumann, C. Kröger, and H. Haebler, *Z. anorg. Chem.*, **207,** 1932, 145–149.
B. Neumann, C. Kröger, and H. Kunz, *Z. anorg. Chem.*, **218,** 1934, 379–401.
8. A. F. Gerds and M. W. Mallett, *J. Electrochem. Soc.*, **101,** 1954, 175–180.

$\bar{1}.4660$
0.5340

N-Ti Nitrogen-Titanium

The phase diagram of Fig. 545 is based on micrographic and incipient-melting work [1]. Alloys with up to 43 at. % N were prepared by arc-melting charges of iodide Ti and TiN. The approximate composition in at. % N (wt. % in parentheses) of the three phases participating in the peritectic reactions at 2020 ± 25°C and 2350 ± 25°C are as follows: 2020°C, melt 4 (1.2), β 6.5 (2), α 12.5 (4); 2350°C melt 15.5 (5), α 20.5 (7), δ 27 (9.8). The boundaries of the ($\alpha + \delta$) field are accurate within ±0.5 wt. (±1.1 at.) %. The composition of the phase formed peritectoidally at about 1050°C was not established since the reaction was not completed after 96 hr at 1000°C. It may be based on the formula Ti_3N (8.88 wt. % N). The melting point of TiN of 2950 ± 50°C was reported by [2]; about 2930°C was given by [3].

Previous Work. X-ray analysis of alloys prepared by sintering mixtures of magnesium-reduced Ti and TiN showed the existence of a wide range of α solid solution [4] up to approximately 18.5–19 at. % N [5]. The high-Ti boundary of the δ (TiN) phase was found at about 29.5 at. (10.9 wt.) % N [5]; see Fig. 545.

The solubility of N in β-Ti at 1000 and 1560°C, based on diffusion data, was given as 0.95 and 4.7 at. % N [6]. These values are slightly higher than those given by [1], the difference being about 0.2 at. % at 1000°C and about 0.7 at. % N at 1400°C, the temperature up to which [1] determined the $\beta/(\alpha + \beta)$ boundary. [7] examined the effect of N, in amounts up to 0.7 wt. (2.3 at.) % N, on the transformation of iodide Ti. Results are in excellent agreement with those of [1].

The composition range of the δ phase extends beyond 50 at. % N. [8] and [9] investigated the crystal structure of preparations with up to 51.27 and 53.7 at. % N, respectively (see below). Whether the latter composition constitutes the upper limit of the (TiN) phase is unknown.

There appears to be no doubt that, besides the ϵ phase, TiN is the only intermediate phase of the system [10, 11, 3, 12, 13] and that a phase of the composition Ti_3N_4 (57.14 at. % N) [14–18] does not exist [10]. The product of the composition Ti_5N_6 (54.55 at. % N) [14] was recognized as TiN [15]. Also, the compound TiN_2 [14] does not exist [15].

As regards the kinetics of reaction of Ti with N_2, see [19–21].

Crystal Structures. [22] determined the effect of up to 0.4 wt. (1.3 at.) % N on the lattice parameters of iodide Ti, and [5, 1] carried the measurements throughout the α solid-solution range. [23] confirmed that N expands the α-Ti lattice.

TiN is isotypic with NaCl. Numerous lattice-parameter values have been reported [24–28, 9, 8, 5, 29–32]. Six investigators found values within the range 4.243 ± 0.002 A [9, 28, 8, 5, 29, 31]. The parameter increases from 4.22 A at the low-N boundary of the δ phase, reaches a maximum (a = 4.243 A) at 50 at. % [5], and decreases to a = 4.221 A at 53.7 at. % N [9]. This proves that the solid solutions on both sides of the stoichiometric composition are of the subtraction type. The ϵ phase was reported to be tetragonal, with a = 4.92 A, c = 5.16 A, c/a = 1.05 [1].

Supplement. Two more values for the lattice parameter of TiN have been published: a = 4.249 ± 2 A [33], a = 4.238 A [34]. The partial pressures p_{Ti} and p_{N_2} above TiN have been measured by [35] in the temperature range 1714–1968°C by means of the Knudsen effusion method.

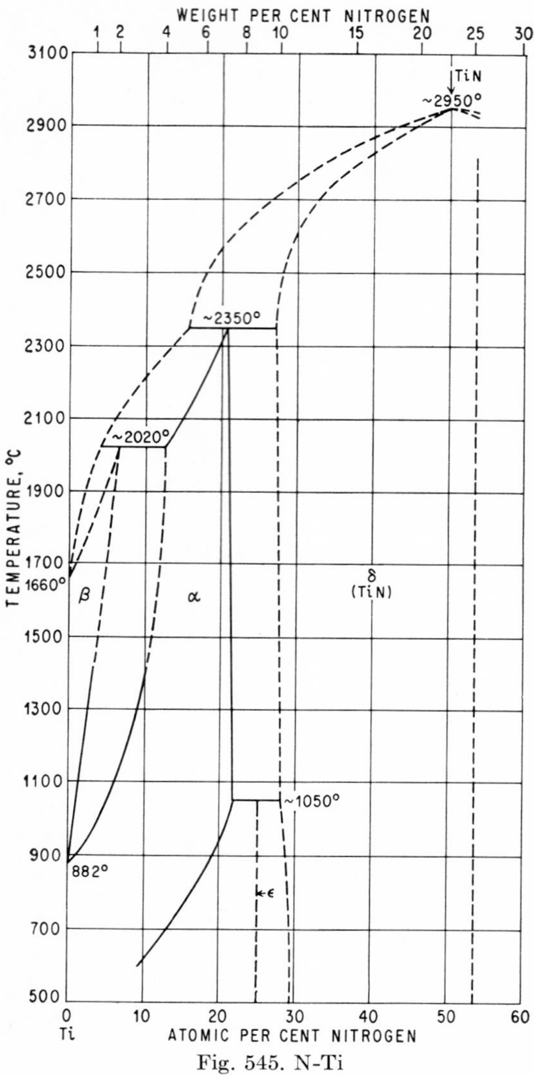

Fig. 545. N-Ti

The diffusion of nitrogen in Ti has been studied by [36]. Specimens reacted above and at 950°C showed two ($\delta + \alpha$) and three ($\delta + \epsilon + \alpha$) surface layers, respectively. However, a triple layer was also observed in specimens reacted at 850°C, below the transformation temperature of Ti. [36] stated, therefore, that "the available data on the Ti-N equilibrium diagram [Fig. 545] below the transformation temperature are insufficient to permit complete interpretation of the surface-layer structures present."

1. A. E. Palty, H. Margolin, and J. P. Nielsen, *Trans. ASM*, **46**, 1954, 312–328.
2. C. Agte and K. Moers, *Z. anorg. Chem.*, **198**, 1931, 239.

3. E. Friederich and L. Sittig, *Z. anorg. Chem.*, **143,** 1925, 297–300.
4. J. D. Fast, *Metallwirtschaft,* **17,** 1938, 641–644.
5. P. Ehrlich, *Z. anorg. Chem.*, **259,** 1949, 1–41.
6. R. Wasilewski, reported by C. A. Papp and M. W. Mallett, Symposium on High Temperature Chemistry, Chicago, 1952; see also M. W. Mallett, E. M. Baroody, H. R. Nelson, and C. A. Papp, *J. Electrochem. Soc.*, **100,** 1953, 103–106.
7. R. I. Jaffee, H. R. Ogden, and D. J. Maykuth, *Trans. AIME,* **188,** 1950, 1261–1266.
8. W. Hume-Rothery, G. V. Raynor, and A. T. Little, *J. Iron Steel Inst.*, **145,** 1942, 133; see also discussion, p. 481.
9. A. Brager, *Acta Physicochim. U.R.S.S.*, **11,** 1939, 617–632.
10. O. Ruff and F. Eisner, *Ber. deut. chem. Ges.*, **41,** 1908, 2250–2264; **42,** 1909, 900.
11. L. Weiss and K. Kaiser, *Z. anorg. Chem.*, **65,** 1910, 393–394.
12. A. E. van Arkel and J. H. de Boer, *Z. anorg. Chem.*, **148,** 1925, 348.
13. H. Moers, *Z. anorg. Chem.*, **198,** 1931, 243–261, especially 256.
14. F. Wöhler, *Liebigs Ann.*, **73,** 1850, 46.
15. C. Friedel and J. Guérin, *Ann. chim. et phys.*, **8,** 1876, 24.
16. E. A. Schneider, *Z. anorg. Chem.*, **8,** 1895, 88–91.
17. H. Geisow, Dissertation, München, 1902.
18. N. Whitehouse, *J. Soc. Chem. Ind.*, **26,** 1907, 738.
19. L. G. Carpenter and F. R. Reavell, *Metallurgia,* **39,** 1948, 63–65.
20. E. A. Gulbransen and K. F. Andrew, *J. Electrochem. Soc.*, **96,** 1949, 364–376; *Trans. AIME,* **185,** 1949, 741–748.
21. L. S. Richardson and N. J. Grant, *Trans. AIME,* **200,** 1954, 69–70.
22. H. T. Clark, *Trans. AIME,* **185,** 1949, 588–589.
23. D. A. Sutcliffe, Royal Aircraft Establishment, Farnborough, Great Britain, Tech. Note Met. 141, March, 1951.
24. A. E. van Arkel, *Physica,* **4,** 1924, 286–301.
25. K. Becker and F. Ebert, *Z. Physik,* **31,** 1925, 268–272.
26. W. Hofmann and A. Schrader, *Arch. Eisenhüttenw.*, **10,** 1936, 65–66.
27. A. Brager, *Acta Physicochim. U.R.S.S.*, **10,** 1939, 593–600; see also [9].
28. W. Dawihl and W. Rix, *Z. anorg. Chem.*, **244,** 1940, 191–197.
29. P. Duwez and F. Odell, *J. Electrochem. Soc.*, **97,** 1950, 299–304.
30. P. Chiotti, *J. Am. Ceram. Soc.*, **35,** 1952, 123–130.
31. A. Münster and K. Sagel, *Z. Elektrochem.*, **57,** 1953, 571–579.
32. H. J. Beattie and F. L. VerSnyder, *Trans. ASM,* **45,** 1953, 397–423.
33. O. Schmitz-Dumont and K. Steinberg, *Naturwissenschaften,* **41,** 1954, 117.
34. N. Schönberg, *Acta Chem. Scand.* **8,** 1954, 215.
35. M. Hoch, D. P. Dingledy, and H. L. Johnston, *J. Am. Chem. Soc.*, **77,** 1955, 304–306.
36. R. J. Wasilewski and G. L. Kehl, *J. Inst. Metals,* **83,** 1954, 94–104.

$\bar{2}.8359$
1.1641

N-Tl Nitrogen-Thallium

See N-Pb, page 985.

$\bar{2}.7697$
1.2303

N-U Nitrogen-Uranium

In a systematic investigation of the system by chemical and X-ray work [1], the following nitrides have been established: UN (5.56 wt. % N), U_2N_3 (8.11 wt. % N), and UN_2 (10.53 wt. % N). The melting point of UN was found as 2650 ± 100°C [2].

UN, the lowest nitride existing, is of singular composition and isotypic with NaCl, a = 4.890 A [1]. U_2N_3 is b.c.c. of the Mn_2O_3 ($D5_3$) type, with a = 10.70 A. It forms solid solutions with N, at least up to $UN_{1.75}$ [63.64 at. (9.34 wt.) % N], whereby its structure, essentially a distorted type of the CaF_2 structure, gradually changes toward the UN_2 structure, which is of the CaF_2 (C1) type. The parameter decreases to 10.60 A at $UN_{1.75}$ [1]. UN_2 can be prepared only at high N_2 pressure (126 atm) and has the ideal CaF_2 structure, a = 5.32 A [1]. Apparently at high pressures U_2N_3 disproportionates into UN and UN_2. The solubility of nitrogen in U appears to be very slight [1]. See Note Added in Proof.

A nitride U_3N_4 [57.14 at. (7.23 wt.) % N], reported in the older literature [3, 4], does not exist; according to later work it was a mixture of UN and U_2N_3. Likewise, U_5N_2 and U_5N_4 [4] were not definite nitrides, as already suggested by [5].

Note Added in Proof. [6] studied the surface reactions of massive uranium with nitrogen in the temperature range 550–900°C at 1 atm nitrogen pressure. In X-ray examinations of the surface reaction layers [6, 7], the patterns of UN and UN_2, but not that of b.c.c. U_2N_3 [1], were observed. Instead, the reaction layer intermediate between the layers of UN and UN_2 yielded a new pattern which was ascribed [7] to a U_2N_3 phase of the hexagonal La_2O_3 ($D5_2$) type, with a = 3.69 A, c = 5.83 A, c/a = 1.58. Th_2N_3 crystallizes in the same type of structure. "Since the X-ray diffraction pattern does not agree with the U_2N_3 pattern of [1], further proof of the N content of the new phase is needed to resolve the difference between U_2N_3 of [1] and the new nitride phase. It is possible, however, that this and the U_2N_3 structure of [1] are polymorphic modifications of the same compound. No study has been made of the thermal stability of the new phase" [7].

No evidence of solution of nitrogen in the U metal phase could be found by [6]. "This is what one would expect since the solubility of N in U is very low (<100 ppm or <0.01 wt. %) over the entire range 550–900°C" [6].

1. R. E. Rundle, N. C. Baenziger, A. S. Wilson, and R. A. McDonald, *J. Am. Chem. Soc.*, **70,** 1948, 99–105. A more detailed account of the investigation is given in R. E. Rundle (and 7 coauthors), *U.S. Atomic Energy Comm., Publ.* AECD-2247, 1948.
2. P. Chiotti, *J. Am. Ceram. Soc.*, **35,** 1952, 123–130.
3. C. Rammelsberg, *Pogg. Ann.*, **55,** 1842, 323; H. Moissan, *Compt. rend.*, **122,** 1896, 276; V. Kohlschütter, *Liebigs Ann.*, **317,** 1901, 166; A. Colani, *Compt. rend.*, **137,** 1903, 383.
4. O. Heusler, *Z. anorg. Chem.*, **154,** 1926, 366–373.
5. R. Lorenz and J. Woolcock, *Z. anorg. Chem.*, **176,** 1928, 289–304.
6. M. W. Mallett and A. F. Gerds, *J. Electrochem. Soc.*, **102,** 1955, 292–296.
7. D. A. Vaughan, *Trans. AIME*, **206,** 1956, 78–79.

$\bar{1}.4392$
0.5608

N-V Nitrogen-Vanadium

The existence of VN (21.56 wt. % N) has long been established [1–11]. It is the nitride richest in N; VN_2, reported earlier [1, 12], does not exist.

A systematic X-ray analysis of the system, covering the range up to 50 at. % N [10], showed the existence of another intermediate phase varying in composition between 27.0 and 30.0 at. % N (9.2 and 10.5 wt. % N), i.e., between the composition V_3N and V_2N, previously suggested as nitrides [12, 13]. Since the corresponding phase in the system N-Nb extends from 28.6 to 33.3 at. % N, it appears more appropriate to designate this phase as V_2N, being stable only with an excess of V, rather than V_3N; see also the system N-Ta.

Also, the phase VN has a homogeneity range extending from 41.5 [10] to at least 51.0 at. % N [9]. Its melting point was given as about 2050°C [3]. The solid solubility of N in V, reported by [10] to be very small, appears to be at least 3.5 at. (1.0 wt.) % N at higher temperatures (after rapid solidification) but less than 3.5 at. % at 900°C [14]. A solubility range is also indicated by parameter measurements [15]; alloys contained up to 0.19 wt. (0.66 at.) % N and 0.04–0.15 wt. % O.

As for the kinetics of reaction of N_2 with V, see [16].

Crystal Structures. V_2N is hexagonal, with lattice constants of a = 2.837 A, c = 4.542 A, c/a = 1.601 and a = 2.841 A, c = 4.550 A, c/a = 1.602 at the high-V and high-N sides of the homogeneity range, respectively [10]. VN is isotypic with NaCl [4, 5, 7–11]. The parameter of the subtraction-type structure increases linearly from a = 4.071 A at 41.5 at. % V to 4.134 A at the stoichiometric composition [10] and decreases to a = 4.113 A at 51.0 at. % N [9]. Parameters of the VN phase have also been reported by [4, 7, 8, 11]; those by [7–10] agree within 4.136 ± 0.002 A.

An arc-melted alloy with about 5 wt. (16 at.) % N, after annealing at 900°C and quenching, showed the same twinned microstructure and b.c. tetragonal crystal structure (a = 2.970 A, c = 3.395 A, c/a = 1.143) as a V-O alloy with 5 wt. % O [14]. As to the nature and origin of the tetragonal phase, see the system O-V (page 1072).

Supplement. [17] reported the lattice parameter a = 4.169 A for $VN_{1.00}$. [18] stated in a brief note that an X-ray investigation of the V-N system confirmed in all respects the data given by [10]. In referring to the tetragonal phase observed by [14], [18] claims that "the preparation methods used by [14] indicate that they obtained oxides instead of nitrides."

1. H. E. Roscoe, *Ann. Pharm. Suppl.*, **6,** 1868, 114; **7,** 1870, 191.
2. N. Whitehouse, *J. Soc. Chem. Ind.*, **26,** 1907, 738–739.
3. E. Friederich and L. Sittig, *Z. anorg. Chem.*, **143,** 1925, 303–304.
4. K. Becker and F. Ebert, *Z. Physik*, **31,** 1925, 268–272.
5. G. Hägg, *Z. physik. Chem.*, **B6,** 1929, 221–232.
6. K. Moers, *Z. anorg. Chem.*, **198,** 1931, 243–261, especially 256.
7. W. Dawihl and W. Rix, *Z. anorg. Chem.*, **244,** 1940, 191–197.
8. A. Brager and V. Epelbaum, *Acta Physicochim. U.R.S.S.*, **13,** 1940, 600–603; see also pp. 595–599.
9. V. Epelbaum and B. Ormont, *Acta Physicochim. U.R.S.S.*, **22,** 1947, 319–330.
10. H. Hahn, *Z. anorg. Chem.*, **258,** 1949, 58–68.
11. P. Duwez and F. Odell, *J. Electrochem. Soc.*, **97,** 1950, 299–304.
12. X. Uhrlaub, *Pogg. Ann.*, **103,** 1858, 134.
13. W. Muthmann, L. Weiss, and R. Riedelbauch, *Liebigs Ann.*, **355,** 1907, 58.
14. W. Rostoker and A. Yamamoto, *Trans. ASM*, **46,** 1954, 1136–1163.
15. S. Beatty, *Trans. AIME*, **194,** 1952, 987–988.
16. E. A. Gulbransen and K. F. Andrew, *J. Electrochem. Soc.*, **97,** 1950, 396–404.
17. N. Schönberg, *Acta Chem. Scand.*, **8,** 1954, 215.
18. N. Schönberg, *Acta Chem. Scand.*, **8,** 1954, 211 (footnote).

$\bar{2}.8818$
1.1182

N-W Nitrogen-Wolfram

According to [1], no reaction takes place between W and N_2 at temperatures up to 1500°C. This was confirmed by [2] for temperatures up to 900°C; however, they found that reaction with NH_3 starts already at 140°C. [3] were not able to form a nitride by reaction of W with NH_3, probably because of the low rate of nitriding.

From X-ray analysis of W-N alloys with up to 18.2 at. (1.67 wt.) % N, prepared by nitriding W powder with NH_3 at 700–800°C, [4] concluded the existence of a nitride of the ideal composition W_2N (3.67 wt. % N), probably stable, however, at a lower N concentration. It has a f.c.c. lattice of W atoms with interstitial N atoms, a = 4.126 A, i.e., is structurally similar to the corresponding phase in the system Mo-N. On the basis of parameter measurements, N appears to be insoluble in W [4] (see Supplement).

[5] reported the existence of another nitride W_2N (called γ phase), formed by nitriding at 825–875°C and having a simple cubic structure with 4 W atoms per cell and a lattice constant varying with the temperature of nitriding from a = 4.130 A (825°C) to 4.122 A (875°C). It was suggested to be closely related to the phase W_2N reported by [4]. The real unit cell possibly is cubic with 256 W atoms (cube edge = $4a$) [5] (see Supplement). The phases W_2N and W_3N_2 were tentatively identified by [6]. The formation of the nitride WN_2 (13.22 wt. % N) by electrically heating a wolfram wire to about 2500°C in a nitrogen atmosphere was reported by [7, 8].

Adsorption isotherms of N on finely divided W powder for temperatures from 400 to 750°C and various pressures were determined by [9]. At saturation, the presence of W_2N was indicated.

Supplement. The data obtained by [10] for the solubility of nitrogen in W at 1 atm pressure are collected in Table 35.

Table 35

°C	Mg/100 g	Wt. % N	At. % N
2400	0.38	0.38×10^{-3}	0.50×10^{-2}
2000	0.11	0.11×10^{-3}	0.14×10^{-2}
1600	0.019	0.19×10^{-4}	0.25×10^{-3}
1200	0.0013	0.13×10^{-5}	0.17×10^{-4}

The γ phase claimed by [5] could not be verified by [11] and was shown [12] to be an oxide-nitride rather than a nitride of W. [11] stated that he could confirm in all respects the results of [4], and [12] corroborated the f.c.c. phase of the approximate composition W_2N. At higher N contents, [11] observed a hexagonal phase containing probably somewhat less than 50 at. % N. This δ or WN phase is isomorphous with WC and has the lattice constants a = 2.893 A, c = 2.826 A, c/a = 0.977; it decomposes in vacuo at 600°C with formation of W_2N.

1. A. Sieverts and E. Bergner, *Ber. deut. chem. Ges.*, **44**, 1911, 2401.
2. P. Laffitte and P. Grandadam, *Compt. rend.*, **200**, 1935, 1039–1041.
3. G. G. Henderson and J. C. Galletly, *J. Chem. Soc. Ind.*, **27**, 1908, 387–389.
4. G. Hägg, *Z. physik. Chem.*, **B7**, 1930, 356–360.
5. R. Kiessling and Y. H. Liu, *Trans. AIME*, **191**, 1951, 639–642.
6. M. Mathis, *Bull. soc. chim. France*, **18**, 1951, 443–451.
7. I. Langmuir, *J. Am. Chem. Soc.*, **35**, 1913, 931–945; *Z. anorg. Chem.*, **85**, 1914, 261–278.
8. C. J. Smithells and H. P. Rooksby, *J. Chem. Soc.*, **1927**, 1882–1888.
9. R. T. Davis, *J. Am. Chem. Soc.*, **68**, 1946, 1395–1402.
10. F. J. Norton and A. L. Marshall, private communication to S. Dushman, "Scientific Foundations of Vacuum Technique," p. 599, John Wiley & Sons, Inc., New York, 1949.
11. N. Schönberg, *Acta Chem. Scand.*, **8**, 1954, 204–207.
12. R. Kiessling and L. Peterson, *Acta Met.*, **2**, 1954, 675–679.

$\bar{1}.3309$
0.6691

N-Zn Nitrogen-Zinc

See N-Pb, page 985. There is no reaction between Zn and N_2 at temperatures up to 400°C [1]. Like [2], [3] were unable to find any measurable solubility of nitrogen in Zn.

1. P. Laffitte and P. Grandadam, *Compt. rend.*, **200,** 1935, 1039–1041.
2. A. Sieverts and W. Krumbhaar, *Ber. deut. chem. Ges.*, **43,** 1910, 894 (see N-Pb).
3. W. Hofmann and J. Maatsch, *Z. Metallkunde*, **47,** 1956, 89–95.

$\bar{1}.1863$
0.8137

N-Zr Nitrogen-Zirconium

The phase diagram (Fig. 546) is due to [1]. It is based on comprehensive micrographic and incipient-melting work and complementary X-ray data. Alloys up to 6 wt. % N were prepared by arc-melting charges of iodide Zr and nitrided Zr sponge; for the higher N range, nitrided Zr sponge was used. The phase boundaries in the region up to about 29.5 at. (6 wt.) % N are considered to be accurate within ±0.2 wt. %. The high-N boundary of the (α + ZrN) field is less accurate, perhaps to about ±0.5 wt. %, since it is based on limited X-ray results. This boundary indicates that the homogeneity range of the ZrN phase is much smaller than that of the TiN phase. However, the width of the two-phase field (Fig. 546) is in agreement with X-ray photograms obtained by [2], according to which an alloy with about 42.5 at. % N consisted of ZrN and α.

The solubility of N in β-Zr up to about 1600°C was determined from diffusion data [3, 4]. The solubility in high-Hf Zr [3] coincides within 0.1 wt. % with the $\beta/(\alpha + \beta)$ boundary of [1], whereas the solubility in low-Hf Zr [4] is somewhat higher than the solubility according to [1], the solubility at 1600°C being about 0.55 wt. (3.4 at.) % according to [1] and about 0.85 wt. (5.4 at.) % according to [4]. Qualitative data as to the solubility of N in α-Zr and the effect of N on the transformation point of Zr have been reported by [5]. The melting point of ZrN was given as 2930 [6] and 2980 ± 50°C [7].

The reaction of Zr with N_2 was studied by [8].

ZrN is the only intermediate phase in the system [6, 9–13]. The nitrides Zr_3N_2 [14], Zr_3N_4 [15], Zr_2N_3 [16, 17], and Zr_3N_8 [16] do not exist.

ZrN is isotypic with NaCl; lattice parameters have been reported by [18–21, 2]. The parameter a = 4.575 A [20, 21] appears to be the most accurate.

Supplement. [22] found the lattice constant of the (ZrN) phase to vary between 4.537 and 4.562 A; he stated that "the high value of a = 4.63 A for zirconium nitride given by some authors is most likely due to the presence of oxygen in the lattice." The vapor pressure above ZrN between 1963 and 2193°C was measured by [23] using the Knudsen effusion method.

1. R. F. Domagala, D. J. McPherson, and M. Hansen, *Trans. AIME*, **206,** 1956, 98–105.
2. P. Chiotti, *J. Am. Ceram. Soc.*, **35,** 1952, 123–130.
3. M. W. Mallett, E. M. Baroody, H. R. Nelson, and C. A. Papp, *J. Electrochem. Soc.*, **100,** 1953, 103–106.
4. M. W. Mallett, J. Belle, and B. B. Cleland, *J. Electrochem. Soc.*, **101,** 1954, 1–5.
5. J. H. de Boer and J. D. Fast, *Rec. trav. chim.*, **55,** 1936, 459–467; J. D. Fast, *Metallwirtschaft*, **17,** 1938, 641–644.
6. E. Friederich and L. Sittig, *Z. anorg. Chem.*, **143,** 1925, 300–303.

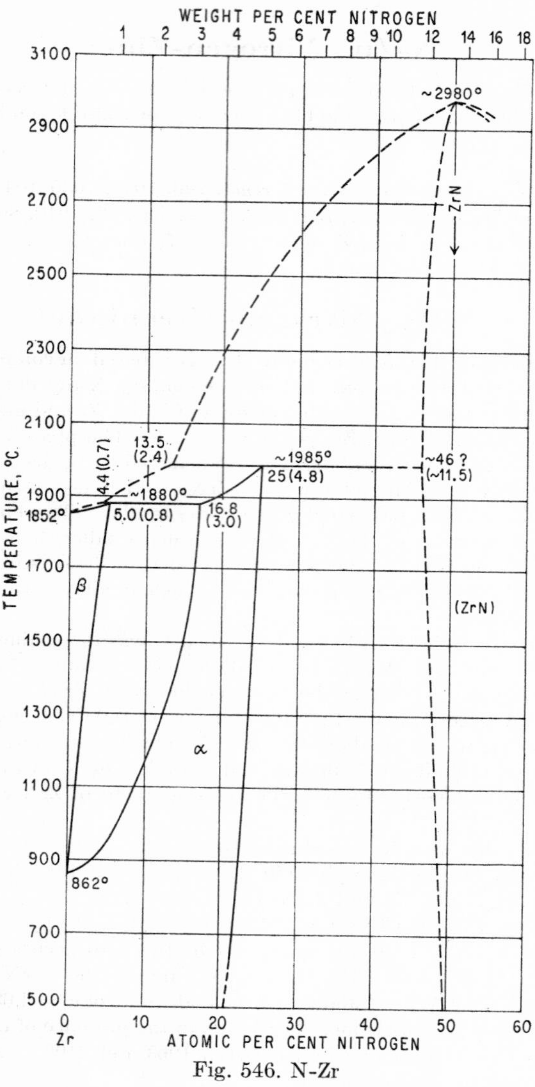

Fig. 546. N-Zr

7. C. Agte and K. Moers, *Z. anorg. Chem.*, **198,** 1931, 239.
8. E. A. Gulbransen, *Trans. AIME*, **185,** 1949, 515–525; *J. Electrochem. Soc.*, **96,** 1949, 364–376.
9. A. E. van Arkel and J. H. de Boer, *Z. anorg. Chem.*, **148,** 1925, 347–348.
10. J. H. de Boer and J. D. Fast, *Z. anorg. Chem.*, **153,** 1926, 7.
11. G. Hägg, *Z. physik. Chem.*, **B6,** 1929, 221–232.
12. K. Moers, *Z. anorg. Chem.*, **198,** 1931, 243–261, especially 255–256.
13. P. Clausing, *Z. anorg. Chem.*, **208,** 1932, 401–419.
14. E. Wedekind, *Liebigs Ann.*, **395,** 1913, 177, 180.

15. P. Bruère and E. Chauvenet, *Compt. rend.*, **167**, 1918, 203.
16. J. M. Mathews, *J. Am. Chem. Soc.*, **20**, 1898, 843–846.
17. E. Wedekind, *Z. anorg. Chem.*, **45**, 1905, 292–293.
18. A. E. van Arkel, *Physica*, **4**, 1924, 286–301.
19. K. Becker and F. Ebert, *Z. Physik*, **31**, 1925, 268–272.
20. C. E. Ransley and H. P. Rooksby, *J. Iron Steel Inst.*, **145**, 1942, 140.
21. P. Duwez and F. Odell, *J. Electrochem. Soc.*, **97**, 1950, 299–304.
22. N. Schönberg, *Acta Chem. Scand.*, **8**, 1954, 627 (footnote).
23. M. Hoch, D. P. Dingledy, and H. L. Johnston, *J. Am. Chem. Soc.*, **77**, 1955, 304–306.

$\bar{1}.0453$
0.9547

Na-Pb Sodium-Lead

Solidification. The liquidus was determined completely by [1–3] and partially by [4–9], all using thermal analysis. The results of [2], with the improvements made by [6] between 25 and 50 at. % Pb, were confirmed from 50 to 100 and 0 to 100 at. % Pb by [5] and [3], respectively, and form the basis for the liquidus of Fig. 547. The compositions reported for the two eutectics richest in Pb do not agree too well: 62.5 [1], 62.7 [2], and 60.5 [8]; and 80 [1], 78.9 [2], ~82 [7], and 79.0 [8] at. % Pb; for Fig. 547 the data 62 and 80 at. % Pb were chosen.

A maximum depression of the solidification point of Na of 0.2°C by 1.23 wt. (0.14 at.) % Pb was reported by [10] as compared with 0.3°C by 0.38 wt. (0.04 at.) % Pb by [11].

Some solidus data due to [6, 12] are plotted in Fig. 547.

Intermediate Phases. For the early progress on the knowledge of the number and the formulas of the intermediate Na-Pb phases, involving the papers of [13, 14, 1, 2, 5, 6], reference is made to [14a].

As to the intermediate phase richest in Na, the formula Na_4Pb was assigned by [2, 13], and [2] concluded from his thermal data that a certain amount of Pb is taken into solid solution. X-ray investigations by [15, 16] gave contradictory results. [15] indexed their powder pattern by a f.c.c. cell with $a = 13.30 \pm 0.035$ A and suggested the formula $Na_{31}Pb_8$ (γ-brass type) rather than Na_4Pb; on the other hand, [16] found by single-crystal work that "$Na_{15}Pb_4$" [21.05 at. (70.60 wt.) % Pb] (Fig. 547) has the b.c.c. $Cu_{15}Si_4$ ($D8_6$) type of structure, with $a = 13.32$ A. "It appears possible that there are actually two phases of different structure with approximate composition Na_4Pb" [17]. (But see Note Added in Proof.)

No structural data for the intermediate phases Na_5Pb_2 [28.57 at. (78.28 wt.) % Pb] [1, 6, 9] and Na_2Pb (81.83 wt. % Pb) [6] are available. According to the thermal investigation of [6], Na_5Pb_2 has a certain range of homogeneity (Fig. 547). (See Supplement and Note Added in Proof.)

The compound NaPb (90.01 wt. % Pb) [13, 2, 6] has a homogeneity range extending from about 48 to 50 at. % Pb [6]. A careful X-ray analysis by [17] gave a tetragonal structure, with $a = 10.580 \pm 5$ A, $c = 17.746 \pm 15$ A, $c/a = 1.677$, and 64 atoms per unit cell [18].

The intermediate phase richest in Pb (known as β phase) has the cubic Cu_3Au ($L1_2$) type of structure [19, 20]; its homogeneity range, however, does not include the ideal composition $NaPb_3$. The position and extent of its phase field was investigated by thermal [2, 21], resistometric [21, 12, 8], hardness [21, 8], and roentgenographic [19, 20] analysis. However, the results obtained differ considerably [22]; for Fig. 547 the data reported by [12], which are in fair agreement with those of [19], were used: 67.5 and 73 at. % Pb at the eutectic temperatures, 67.5 and 72.3 at. % at room temperature. The lattice parameter increases with increasing Pb content: 4.882–4.893 A [19], 4.871–4.880 A [20].

Emf measurements were carried out by [23] (90–95.5 wt. % Pb) and [24] (0–100 % Pb); for details see the original papers.

Solutions of Na-Pb alloys in liquid NH_3 and/or the preparation of solid alloys using these solutions were studied by [14, 25–27, 19].

Solubility of Na in Solid Pb. The lattice parameter of Pb decreases as Na goes into solid solution [26, 19, 28–30] although the metallic radius of Na is larger than that of Pb. The solubility was investigated by many methods: thermally [2, 21], microscopically [31, 32, 29], by measurements of hardness [33, 34, 21, 12, 28, 8, 29], roentgenographically [19, 28], by measurements of density [33], and resistometrically [34, 21, 12]. The data obtained are very contradictory, lying between 4.1 and 18 at.

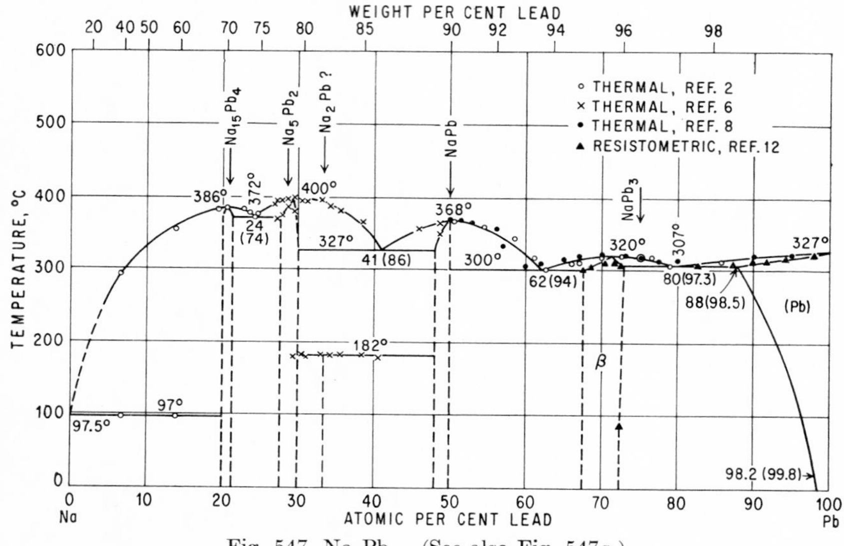

Fig. 547. Na-Pb. (See also Fig. 547*a*.)

% Na for the maximal solubility at the eutectic temperature and between 1.8 and 6.8 at. % Na for room temperature. Since [30] showed—by using extremely high quenching rates—that Pb may take into supersaturated solution up to 24 at. (3.4 wt.) % Na, this would account for some of the discrepancies in the above results. The data 11.1 [28], 5.3 [29], and 1.8 [32, 29] at. (1.37, 0.62, and 0.2 wt.) % Na at 290, 160, and 20°C, respectively, were chosen for Fig. 547.

According to [35], additions of Na considerably delay the recrystallization of Pb.

Supplement. In a preliminary note [36] on a thermal and microscopic reinvestigation of the region in the vicinity of the composition Na_2Pb, it is stated that this "study disproves the existence of a compound at the composition Na_2Pb and demonstrates the existence of a hitherto unrecognized open-maximum compound at the Na_9Pb_4 [30.77 at. % Pb] composition. Also, the neighboring compound Na_5Pb_2 was shown to be a peritectic rather than an open-maximum compound." Since no detailed data were available at the time of writing, the above findings were not considered in Fig. 547.

Note Added in Proof. A detailed account of the investigation by [36] has been published recently [37]. The revised partial phase diagram is shown in Fig. 547*a*.

As regards the 190 (or 182)°C thermal arrest, [6] had found the halting time to be greatest at the composition Na_2Pb and concluded therefrom the existence of a compound of this composition. The thermal and microscopic findings of [37] definitely disprove the existence of Na_2Pb. [37] suggest that the 190°C thermal effects represent a lattice transformation or an order-disorder change in the compound Na_9Pb_4. Surprisingly, no effects could be observed within the (Na_5Pb_2 + Na_9Pb_4) two-phase field.

In a powder X-ray examination [38] of alloys containing between 13.4 and 21.9 at. % Pb and heat-treated in various ways, the pattern of b.c.c. $Na_{15}Pb_4$ reported by [16],

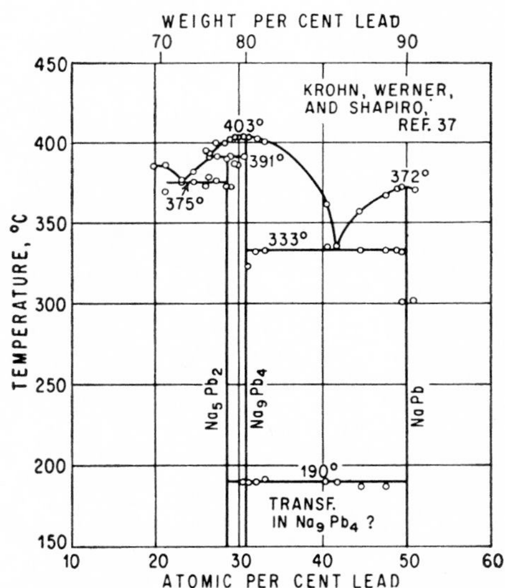

Fig. 547*a*. Na-Pb. (See also Fig. 547.)

but not that of f.c.c. "$Na_{31}Pb_8$" claimed by [15], could be found. "It appears overwhelmingly probable that the data of [15] apply not to an alloy of approximate composition Na_4Pb but to some other substance or substances as yet unidentified" [38].

1. N. S. Kurnakov (and A. N. Kusnetzov), *Z. anorg. Chem.*, **23,** 1900, 455–462.
2. C. H. Mathewson, *Z. anorg. Chem.*, **50,** 1906, 172–180. The alloys were melted in glass tubes under hydrogen and their compositions checked by numerous chemical analyses.
3. H. Siegert, quoted by [12]. A thermal analysis confirmed previous results of [2, 6]; no detailed data are given.
4. C. T. Heycock and F. H. Neville, *J. Chem. Soc.*, **61,** 1892, 904.
5. J. Goebel, *Z. Metallkunde*, **14,** 1922, 425–432; *Z. anorg. Chem.*, **106,** 1919, 211–212.
6. G. Calingaert and W. J. Boesch, *J. Am. Chem. Soc.*, **45,** 1923, 1901–1904.
7. Thermal data between 75 and 100 at. % Pb by Ageev and Shoikhet, published in [28].
8. N. S. Kurnakov, S. A. Pogodin, E. S. Shpichinetskij, and V. S. Zorin, *Izvest. Sektora Fiz.-Khim. Anal.*, **13,** 1940, 233–250 (in Russian).
9. I. T. Krohn and H. Shapiro (1952); thermal analysis shows the existence of Na_5Pb_2 and Na_9Pb_4; quoted by [17] in advance of publication (see Supplement).

10. G. Tammann, *Z. physik. Chem.*, **3**, 1889, 447.
11. C. T. Heycock and F. H. Neville, *J. Chem. Soc.*, **55**, 1889, 675.
12. H. Klaiber, *Z. Elektrochem.*, **42**, 1936, 258–264.
13. W. H. Greene and W. H. Wahl, *J. Franklin Inst.*, **130**, 1890, 483–484.
14. A. Joannis, *Compt. rend.*, **114**, 1892, 585–586; *Ann. chim. et phys.*, (8)**7**, 1906, 79.
14a. M. Hansen, "Der Aufbau der Zweistofflegierungen," Springer-Verlag OHG, Berlin, 1936.
15. C. W. Stillwell and W. K. Robinson, *J. Am. Chem. Soc.*, **55**, 1933, 127–129.
16. E. Zintl and A. Harder, *Z. physik. Chem.*, **B34**, 1936, 238–254.
17. R. E. Marsh and D. P. Shoemaker, *Acta Cryst.*, **6**, 1953, 197–205.
18. These results corroborate unpublished data of L. S. Ramsdell. See [17].
19. E. Zintl and A. Harder, *Z. physik. Chem.*, **154**, 1931, 58–91.
20. E. S. Makarov and Z. V. Popova, *Izvest. Akad. Nauk S.S.S.R. (Khim.)*, **1951**, 377–387.
21. N. S. Kurnakov and S. A. Pogodin, *Izvest. Inst. Fiz.-Khim. Anal.*, **6**, 1933, 275 (in Russian).
22. Data for the extreme lower and upper boundaries are 64.9 and 75 at. % Pb, respectively.
23. F. Haber and M. Sack, *Z. Elektrochem.*, **8**, 1902, 246–248; M. Sack, *Z. anorg. Chem.*, **34**, 1903, 317–331.
24. R. Kremann and P. v. Reininghaus, *Z. Metallkunde*, **12**, 1920, 273–279.
25. C. A. Kraus and H. F. Kurtz, *J. Am. Chem. Soc.*, **47**, 1925, 43–60.
26. E. Zintl, *Naturwissenschaften*, **17**, 1929, 782–783.
27. E. Zintl, J. Goubeau, and W. Dullenkopf, *Z. physik. Chem.*, **A154**, 1931, 37–39.
28. N. V. Ageev and N. Ya. Talyzin, *Izvest. Sektora Fiz.-Khim. Anal.*, **13**, 1940, 251–255 (in Russian).
29. E. Schmid, *Z. Metallkunde*, **35**, 1943, 85–92.
30. G. Falkenhagen and W. Hofmann, *Z. Metallkunde*, **43**, 1952, 69–81.
31. L. Lewin, *Münch. med. Wochschr.*, **65**, 1918, 38–39.
32. E. Schulz, *Metallwirtschaft*, **20**, 1941, 418–421.
33. J. Goebel, *Z. Ver. deut. Ing.*, **63**, 1919, 425; *Z. Metallkunde*, **14**, 1922, 425–432.
34. G. Tammann and H. Rüdiger, *Z. anorg. Chem.*, **192**, 1930, 16–26.
35. W. Hofmann and H. Hanemann, *Z. Metallkunde*, **30**, 1938, 47–49.
36. I. T. Krohn, R. C. Werner, and H. Shapiro, American Chemical Society, Abstract of Papers 126th Meeting, 1954, 5R.
37. I. T. Krohn, R. C. Werner, and H. Shapiro, *J. Am. Chem. Soc.*, **77**, 1955, 2110–2113. For the sake of clarity, not all cooling-curve data have been plotted in Fig. 547*a*.
38. D. P. Shoemaker, N. E. Weston, and J. Rathlev, *J. Am. Chem. Soc.*, **77**, 1955, 4226–4228.

$\bar{1}.3335$
0.6665

Na-Pd Sodium-Palladium

No systematic investigations on the Na-Pd system are available [1]. The solidification point of Na is depressed for maximal 0.4°C by the addition of 2.8 wt. (0.62 at.) % Pd [2].

1. "Gmelins Handbuch der anorganischen Chemie," System No. 68, Part A(5), Verlag Chemie, G.m.b.H., Weinheim/Bergstrasse, Germany, 1951.
2. G. Tammann, *Z. physik. Chem.*, **3**, 1889, 448.

$\bar{1}.0711$
0.9289

Na-Pt Sodium-Platinum

There is no systematic investigation on the system Na-Pt [1]. The freezing point of Na is almost unaffected by addition of Pt up to about 1.5 wt. (0.18 at.) % [2, 3]. Pt is attacked by molten Na and by Na vapor [4–6]. Molten Na attacks Pt noticeably only above 450°C [7, 3]; on cooling, pure Pt separates from the solution [3].

By compression of the metals, a compound NaPt is assumed to form [8]. In the electrolysis of molten Na salts between Pt electrodes, Na separates cathodically with alloy formation [9].

1. "Gmelins Handbuch der anorganischen Chemie," System No. 68, Part A(6), p. 747, Verlag Chemie, G.m.b.H., Weinheim/Bergstrasse, Germany, 1951.
2. G. Tammann, *Z. physik. Chem.*, **3,** 1889, 446.
3. L. Hackspill, *7th Intern. Cong. Appl. Chem., London, 1909,* Sec. II, p. 267; for abstract see [1].
4. J. Dewar and A. Scott, *Chem. News,* **40,** 1879, 294.
5. V. Meyer, *Ber. deut. chem. Ges.*, **13,** 1880, 391.
6. F. Haber and M. Sack, *Z. Elektrochem.*, **8,** 1902, 250.
7. C. T. Heycock and F. H. Neville, *J. Chem. Soc.*, **55,** 1889, 666–676.
8. W. Spring, according to [3], p. 266.
9. A. Brester, *Arch. néerl.*, **1,** 1866, 296; abstract, see [1].

$\bar{1}.4298$
0.5702

Na-Rb Sodium-Rubidium

Thermal analyses of the Na-Rb system were carried out by [1–3]. The phase diagram (Fig. 548) is simple eutectic, and no indications for the formation of an intermediate phase or of solid solutions were found; a roentgenographic analysis at low temperatures [3] corroborated these results.

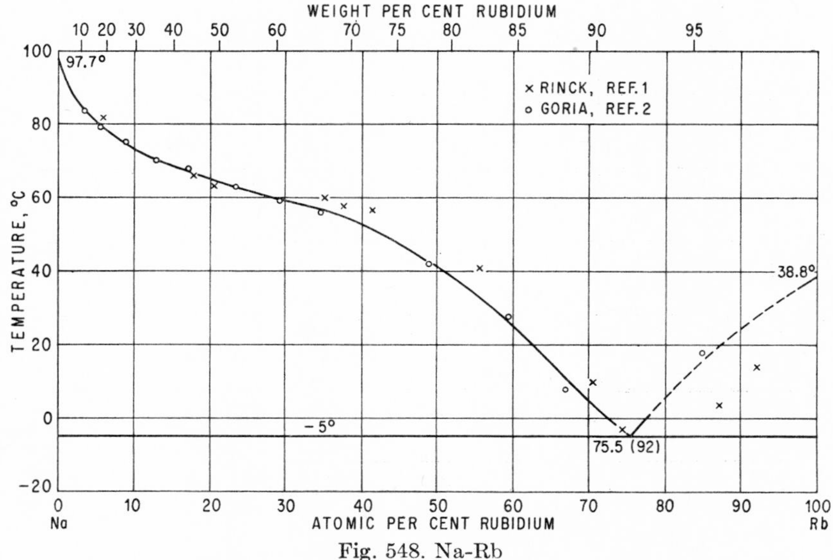

Fig. 548. Na-Rb

The data given for the eutectic are in good agreement: −4.5°C, 75 at. (91.8 wt.) % Rb [1]; −5.2°C, 92.0 wt. (75.6 at.) % Rb [2].

1. E. Rinck, *Compt. rend.*, **197,** 1933, 1404–1406.
2. C. Goria, *Gazz. chim. ital.*, **65,** 1935, 865–870.
3. B. Böhm and W. Klemm, *Z. anorg. Chem.*, **243,** 1939, 69–85. The thermal data (up to 70 at. % Rb) are plotted only in a very small diagram. They are said to corroborate completely the results of previous authors.

$\bar{1}.2762$
0.7238

Na-Sb Sodium-Antimony

The constitution of Na-Sb alloys was studied thermally by [1] (Fig. 549). According to this investigation, which gave no special attention to solid solubilities, there exist the two compounds Na_3Sb (63.83 wt. % Sb) and NaSb (84.11 wt. % Sb). Na_3Sb had already been prepared by [2, 3], and [4] had found that the melting point

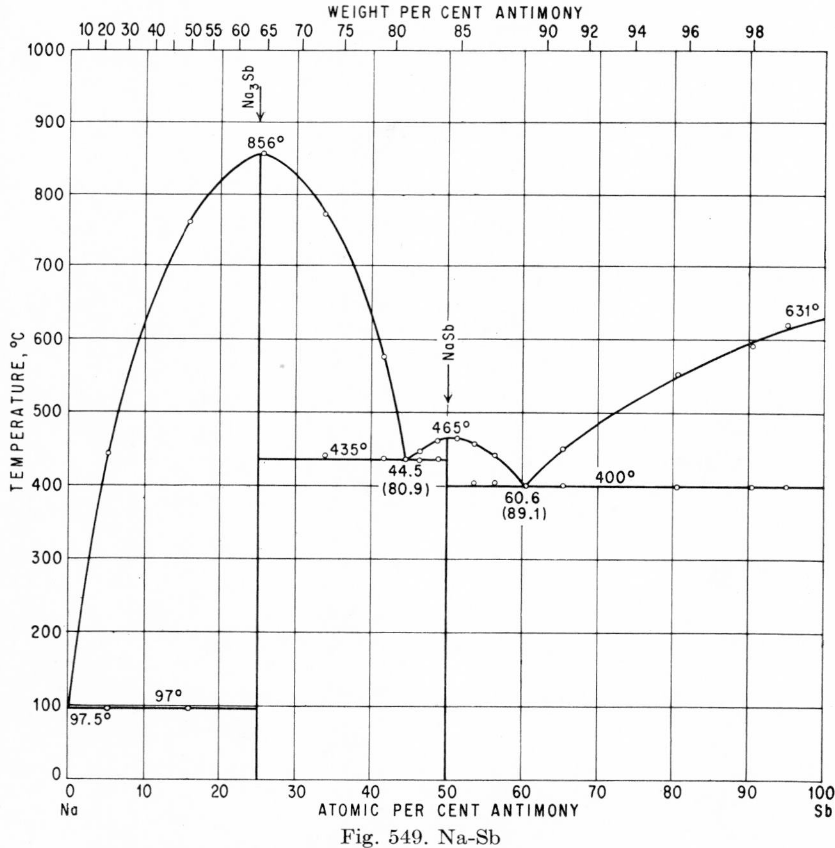

Fig. 549. Na-Sb

of Na is raised by small additions of Sb. The diagram of [1] in the range 50–100 at. % Sb was corroborated roentgenographically by [5].

For experiments with liquid NH_3 solutions, see [2, 6–8]; on complete removal of the ammonia, only the metallic phases of Fig. 549 are obtained. (See also [5].) Emf measurements between 70 and 100 wt. % Sb were carried out by [9] and calorimetric measurements by [10].

Crystal Structures. Na_3Sb was shown [11] to have the hexagonal Na_3As ($D0_{18}$) type of structure, with a = 5.366 A, c = 9.515 A, c/a = 1.773 (24.5 at. % Sb). NaSb has a monoclinic unit cell containing 16 atoms [5].

1. C. H. Mathewson, *Z. anorg. Chem.*, **50,** 1906, 192–195. Na_3Sb was melted in an iron crucible; all other alloys were melted in glass tubes in a hydrogen atmosphere. Some alloys were analyzed for loss of Na during preparation, and all compositions were corrected.
2. A. Joannis, *Compt. rend.*, **114,** 1892, 587.
3. P. Lebeau, *Compt. rend.*, **130,** 1900, 502.
4. G. Tammann, *Z. physik. Chem.*, **3,** 1889, 446.
5. E. Zintl and W. Dullenkopf, *Z. physik. Chem.*, **B16,** 1932, 183–194.
6. E. B. Peck, *J. Am. Chem. Soc.*, **40,** 1918, 335.
7. C. A. Kraus and H. F. Kurtz, *J. Am. Chem. Soc.*, **47,** 1925, 53–54.
8. E. Zintl, J. Goubeau, and W. Dullenkopf, *Z. physik. Chem.*, **154,** 1931, 1–46, especially 6, 33–35.
9. R. Kremann and E. Pressfreund, *Z. Metallkunde*, **13,** 1921, 27–29.
10. O. Kubaschewski and W. Seith, *Z. Metallkunde*, **30,** 1938, 7–9.
11. E. Zintl and G. Brauer, *Z. physik. Chem.*, **B37,** 1937, 323–352.

$\bar{1}.4643$
0.5357

Na-Se Sodium-Selenium

Figure 550 shows the phase diagram established by [1] using thermal analysis. Liquidus temperatures of the high-melting alloys with 20–40 at. % Se could not be determined since no suitable crucible material was available. The diagram shows the existence of five compounds: Na_2Se (63.19 wt. % Se), Na_2Se_2 (77.44 wt. % Se), Na_2Se_3 (83.74 wt. % Se), Na_2Se_4 (87.29 wt. % Se), and Na_2Se_6 (91.15 wt. % Se). The depression of the freezing point of Se by addition of Na does not exceed 0.2°C.

The compounds Na_2Se, Na_2Se_2, and Na_2Se_4 had been prepared previously [2]. In potentiometric titration of Se with Na in liquid NH_3, [3] observed effects indicating the existence of Na_2Se_x (x = 1, 2, 3, 4, 5, 6); according to Fig. 550, Na_2Se_5 does not form from the melt.

The magnetic properties of Na selenides were studied by [4]. Na_2Se was shown [5, 4] to have the cubic CaF_2 (C1) type of structure, with a = 6.823 A [5]; it does not take Se into solid solution [4].

1. C. H. Mathewson, *J. Am. Chem. Soc.*, **29,** 1907, 867–880. The alloys were melted in glass tubes in a hydrogen atmosphere and analyzed.
2. See chemical handbooks. Cf. also F. W. Bergstrom, *J. Am. Chem. Soc.*, **48,** 1926, 146–151.
3. E. Zintl, J. Goubeau, and W. Dullenkopf, *Z. physik. Chem.*, **154,** 1931, 1–46, especially 28–30.
4. W. Klemm, H. Sodomann, and P. Langmesser, *Z. anorg. Chem.*, **241,** 1939, 281–304.
5. E. Zintl, A. Harder, and B. Dauth, *Z. Elektrochem.*, **40,** 1934, 588–593.

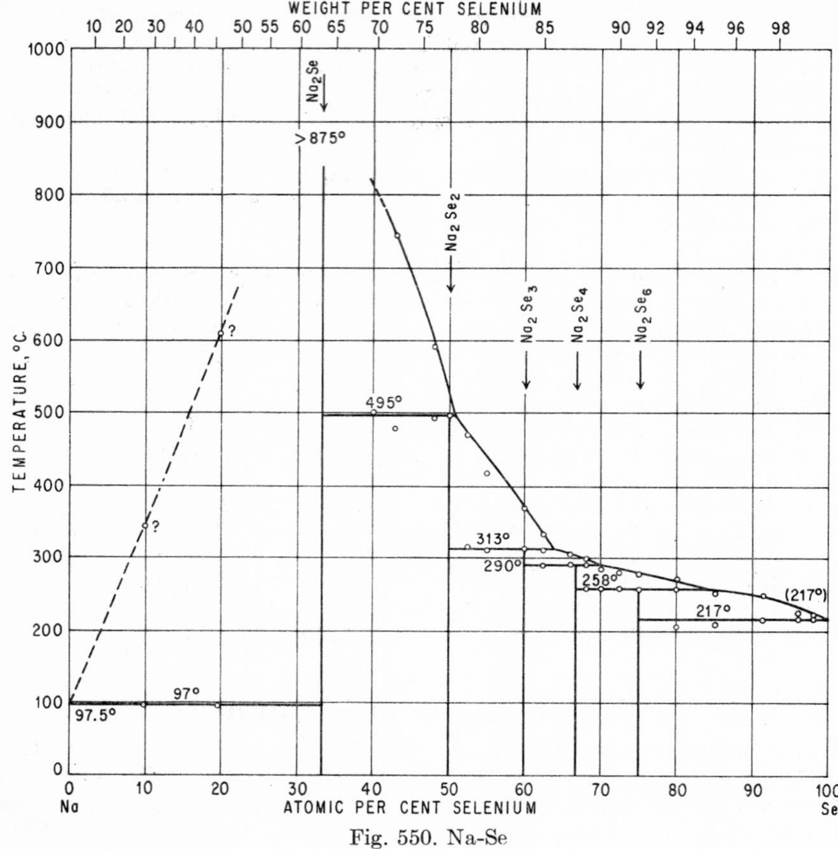

Fig. 550. Na-Se

$\bar{1}.9131$
0.0869

Na-Si Sodium-Silicon

The preparation of Na silicides has been reported by [1–3]. A compound with the probable formula $NaSi_2$ (70.95 wt. % Si) was prepared by reduction of quartz sand (SiO_2) with Na [2]. From powder photographs a tetragonal unit cell with $a = 4.97_5$ kX, $c = 16.7_0$ kX, $c/a = 3.36$, containing 12 atoms, was tentatively proposed.

NaSi (54.98 wt. % Si) was synthesized from the elements [3]. This compound, which dissociates completely when heated to 420°C in vacuum, shows a rather complex powder pattern (representation given).

1. H. Moissan, *Bull. soc. chim.*, **27,** 1902, 1204; *Compt. rend.*, **134,** 1902, 1083.
2. H. Nowotny and E. Scheil, *Z. Metallkunde*, **38,** 1947, 76–80.
3. E. Hohmann, *Z. anorg. Chem.*, **257,** 1948, 113–126.

$\bar{1}.2872$
0.7128

Na-Sn Sodium-Tin

Early Papers. The solubility of Sn in liquid Na is very small [1, 2]; according to [2], the maximal depression of the freezing point of Na by alloying with Sn is 0.07°C by 0.62 wt. (0.12 at.) % Sn. The influence of Na on the freezing point of Sn was investigated by [3], and a depression to 220°C by 0.96 wt. (4.8 at.) % Na observed. The compounds Na_4Sn and Na_2Sn were isolated by [4, 5], respectively; [6] reported the existence of relatively high melting compounds.

The Phase Diagram. As a result of a thermal analysis, [7] reported the existence of the compounds Na_4Sn, Na_2Sn, Na_4Sn_3, NaSn, and $NaSn_2$. Of these, the last three were regarded as undergoing polymorphic transformations. However, [8] found—by means of a careful thermal and microscopic analysis—that the system was even more complex. His phase diagram (Fig. 551) shows the nine compounds Na_4Sn (56.34 wt. % Sn), Na_3Sn (63.24 wt. % Sn), Na_2Sn (72.08 wt. % Sn) [9], Na_4Sn_3 [42.86 at. (79.47 wt.) % Sn], NaSn (83.77 wt. % Sn) which undergoes a polymorphic

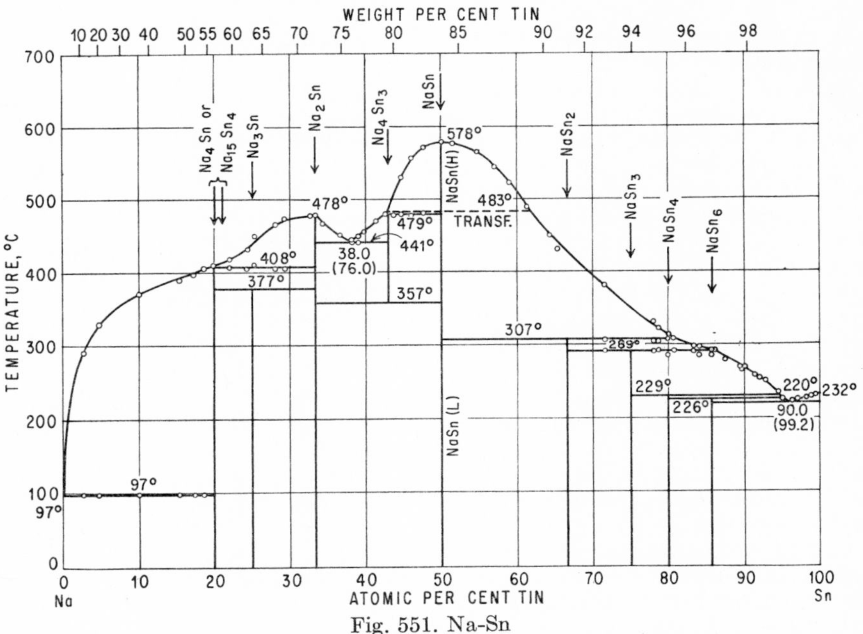

Fig. 551. Na-Sn

transformation at 483°C, $NaSn_2$ (91.17 wt. % Sn), $NaSn_3$ (93.93 wt. % Sn), $NaSn_4$ (95.38 wt. % Sn), and $NaSn_6$ [85.71 at. (96.87 wt.) % Sn]. No evidence was found for the existence of any solid solutions.

Emf measurements at room temperature were carried out by [10, 11]. [10] found breaks at the compositions Na_4Sn, Na_2Sn, NaSn, and $NaSn_2$; above 97 wt. (86 at.) % Sn, nobler protective films were formed. This also holds more or less for the measurements of alloys with 90–94.4 wt. % Sn by [11]. According to [12], Na dissolved in liquid NH_3 reacts with Sn or Sn salts; from color changes of the solution they concluded the existence of Na_4Sn and another compound richer in Sn. On treating a Na-Sn alloy with liquid NH_3, [13] prepared a saltlike compound Na_4Sn_9, stable only in NH_3 solution. Calorimetric measurements were carried out by [14–19]. Papers by [20, 21] deal with volume changes in the formation of Na-Sn compounds.

Crystal Structures. In alloying Sn with excess Na and extracting the free Na by liquid NH_3 at −50°C, the Na-richest compound was prepared and analyzed to be $Na_{15}Sn_4$ [21.05 at. (57.92 wt.) % Sn] rather than Na_4Sn [22]. X-ray analysis revealed an orthorhombic translation group, with $a = 9.81$ A, $b = 22.83$ A, $c = 5.57$ A. From these (preliminary) lattice parameters and the density a unit cell content of 39.7 atoms was computed [compare with $2(15 + 4) = 38$]. The structure was not evaluated; however, it is thought to be related to that of $Na_{15}Pb_4$ [22]. On the basis of the evidence available [8, 22], it is difficult to decide whether the composition of this phase should be expressed by Na_4Sn or $Na_{15}Sn_4$.

Solubility of Na in Solid Sn. No solid solution of Na in Sn could be detected by [8]; a 0.52 at. % Na alloy, after annealing at 200°C for 17 days, showed quite clearly two constituents. On the other hand, [23] claimed that Na is soluble in solid Sn up to 0.48 wt. (2.4 at.) % Na [24].

1. C. T. Heycock and F. H. Neville, *J. Chem. Soc.*, **55,** 1889, 668.
2. G. Tammann, *Z. physik. Chem.*, **3,** 1889, 448.
3. C. T. Heycock and F. H. Neville, *J. Chem. Soc.*, **57,** 1890, 380.
4. P. Lebeau, *Compt. rend.*, **130,** 1900, 502–505.
5. H. Bailey, *Chem. News*, **65,** 1892, 18.
6. N. S. Kurnakov, *Z. anorg. Chem.*, **23,** 1900, 455.
7. C. H. Mathewson, *Z. anorg. Chem.*, **46,** 1905, 94–112.
8. W. Hume-Rothery, *J. Chem. Soc.*, **131,** 1928, 947–963. The alloys were melted and the cooling curves taken in glass tubes in an atmosphere of nitrogen. In nearly all alloys both components were determined analytically.
9. In 1926, W. Hume-Rothery (*J. Inst. Metals*, **35,** 1926, 347–348) reported 470°C as the melting point of Na_2Sn.
10. R. Kremann and J. Gmachl-Pammer, *Z. Metallkunde*, **12,** 1920, 257–262.
11. F. Haber and M. Sack, *Z. Elektrochem.*, **8,** 1902, 248–250; M. Sack, *Z. anorg. Chem.*, **34,** 1903, 331–332.
12. C. A. Kraus and H. F. Kurtz, *J. Am. Chem. Soc.*, **47,** 1925, 51–52.
13. E. Zintl and A. Harder, *Z. physik. Chem.*, **A154,** 1931, 47–57.
14. W. Biltz and W. Holverscheit, *Z. anorg. Chem.*, **140,** 1924, 261.
15. W. Biltz and F. Meyer, *Z. anorg. Chem.*, **176,** 1928, 23.
16. W. Biltz, *Z. Metallkunde*, **29,** 1937, 73.
17. O. Kubaschewski and W. Seith, *Z. Metallkunde*, **30,** 1938, 7.
18. R. L. McKisson and L. A. Bromley, *U.S. Atomic Energy Comm., Publ.* UCRL-689, 1950; *Met. Abstr.*, **18,** 1951, 600.
19. R. L. McKisson and L. A. Bromley, *Trans. AIME*, **194,** 1952, 33–38.
20. F. Löwig, *Ann. Chemie*, **84,** 1852, 308.
21. W. Biltz, *Z. Metallkunde*, **26,** 1934, 230–232.
22. E. Zintl and A. Harder, *Z. physik. Chem.*, **B34,** 1936, 238–254.
23. N. L. Pokrovsky and N. D. Galanina, *Zhur. Fiz. Khim.*, **23,** 1949, 324–331; *Met. Abstr.*, **20,** 1953, 551.
24. Alloys containing 0.01–0.48 wt. % Na were cast and annealed in vacuo in contact with glass plates, and *cast surfaces* were examined microscopically.

$\bar{1}.2558$
0.7442

Na-Te Sodium-Tellurium

This system has been investigated thermally by [1] (0–100 % Te) and [2] (46–100 at. % Te). [1] concluded from their data the existence of (*a*) a miscibility gap in the liquid state between about 63 and 75 at. % Te and (*b*) the compounds Na_2Te [3], Na_3Te_2, and Na_3Te_7. However, the redetermination of the Te-rich part of the phase diagram by means of thermal [2] and resistometric [4] measurements failed to corroborate the miscibility gap, and the two compounds richest in Te were shown to have the formulas NaTe (84.73 wt. % Te) [5] and $NaTe_3$ (94.33 wt. % Te) rather than

Na_3Te_2 and Na_3Te_7. In Fig. 552 the thermal data of [1] were plotted only up to 47 at. % Te, and the Te-rich part of the diagram was plotted according to the results of [2, 4].

The formulas derived from studies of solutions of "polytellurides" in liquid NH_3 [6–9]—Na_2Te, NaTe (Na_2Te_2), Na_2Te_3, and $NaTe_2$ (Na_2Te_4)—do not fully agree

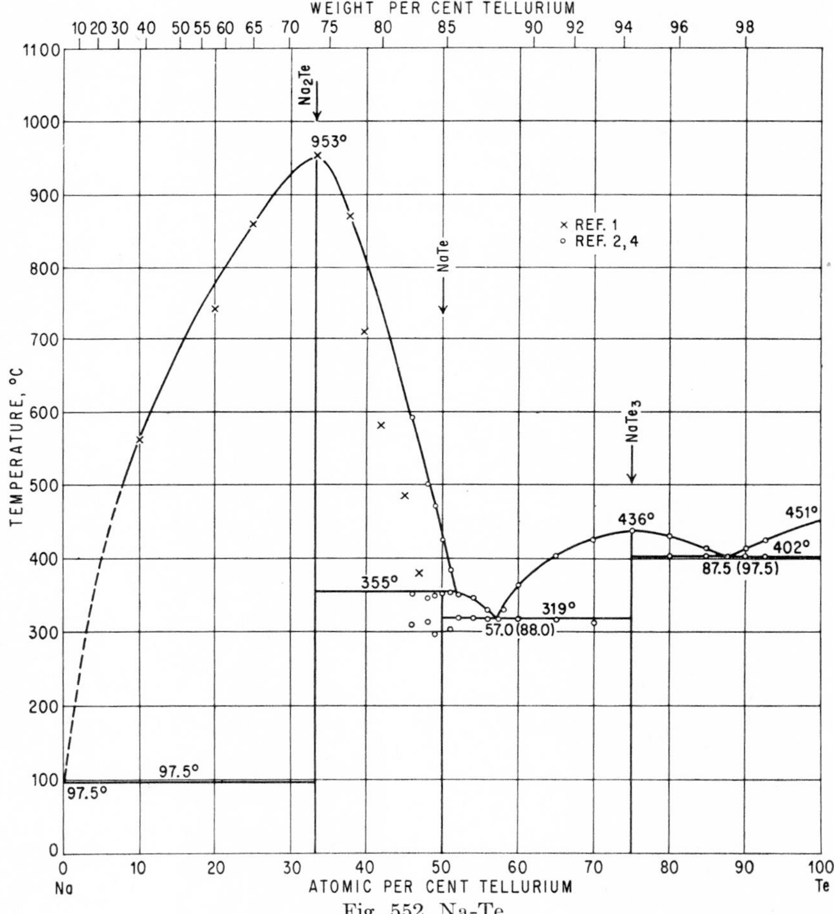

Fig. 552. Na-Te

with those of Fig. 552; however, these findings have no direct bearing on the binary Na-Te system.

Na_2Te was shown [10, 11] to have the cubic CaF_2 (C1) type of structure, with $a = 7.329$ A [10]; it does not take Te into solid solution [11]. The magnetic properties of Na tellurides were studied by [11].

1. G. Pellini and E. Quercigh, *Atti reale accad. Lincei,* (5)**19**(2), 1910, 350–356.
2. C. A. Kraus and S. W. Glass, *J. Phys. Chem.,* **33,** 1929, 995–999.
3. Na_2Te (73.51 wt. % Te) has long been known; see the chemical handbooks.
4. C. A. Kraus and S. W. Glass, *J. Phys. Chem.,* **33,** 1929, 984–994.

5. "An exact determination of the composition of this compound is not possible by the method of thermal analysis, although the general form of the curve indicates a compound of approximately this composition" [2].
6. C. Hugot, *Compt. rend.*, **129**, 1899, 299, 388.
7. C. A. Kraus and C. Y. Chiu, *J. Am. Chem. Soc.*, **44**, 1922, 1999–2008.
8. E. Zintl, J. Goubeau, and W. Dullenkopf, *Z. physik. Chem.*, **154**, 1931, 1–46, especially 30–31.
9. C. A. Kraus and J. A. Ridderhof, *J. Am. Chem. Soc.*, **56**, 1934, 79–86.
10. E. Zintl, A. Harder, and B. Dauth, *Z. Elektrochem.*, **40**, 1934, 588–593.
11. W. Klemm, H. Sodomann, and P. Langmesser, *Z. anorg. Chem.*, **241**, 1939, 281–304.

$\bar{2}.9960$
1.0040

Na-Th Sodium-Thorium

The phase diagram of Fig. 553 is based on thermal-analysis data of [1]. The lengths of the peritectic and eutectic arrests point to the formula Na_4Th (71.61 wt. % Th) for the compound formed at 121°C.

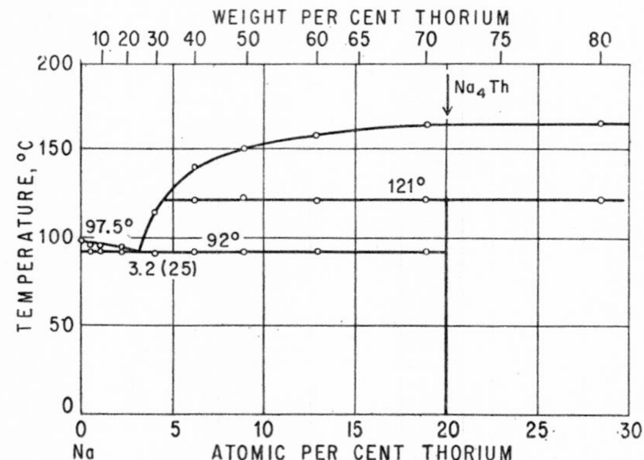

Fig. 553. Na-Th. (The correctness of this diagram was disputed.)

The correctness of the diagram has been disputed [2]. It was observed that Th is not attacked by Na at temperatures of 650–800°C.

1. G. Grube and L. Botzenhardt, *Z. Elektrochem.*, **48**, 1942, 418–425. The alloys were melted in iron crucibles in an argon atmosphere.
2. L. R. Kelman, Argonne National Laboratory; private communication to M. Hansen.

$\bar{1}.0512$
0.9488

Na-Tl Sodium-Thallium

The phase diagram of Fig. 554 was established by means of thermal (56 alloys) and resistometric (15 alloys, in the range 50–100 at. % Tl) measurements [1]. Both cooling and heating curves were taken in the thermal work (solidification equilibria, $\alpha_{Tl} \rightleftarrows \beta_{Tl}$ transformation); solid-state reactions were mainly investigated by temperature-resistivity curves. The liquidus is in good agreement with one previously worked out by thermal analysis [2] and also with individual data in the Na-rich region by

[3, 4]. The formulas of the intermediate phases occurring are Na_6Tl (14.29 at. = 59.71 wt. % Tl), Na_2Tl (81.63 wt. % Tl), NaTl (89.89 wt. % Tl), and $NaTl_2$ (94.67 wt. % Tl). Alloys with 76, 80, and 83 at. % Tl showed age hardening [1].

Further Papers. Emf measurements by [5] (amalgamated Tl/0.1 *N* NaI in pyridine/$Na_{1-x}Tl_x$) appear to have suffered from the formation of nobler protective films; breaks were observed at the compositions NaTl and $\sim NaTl_3$.

Work in NH_3 solution has been carried out by [6, 7]. [7] found by potentiometric titration evidence for NaTl and $NaTl_2$, for the latter also by roentgenographic work.

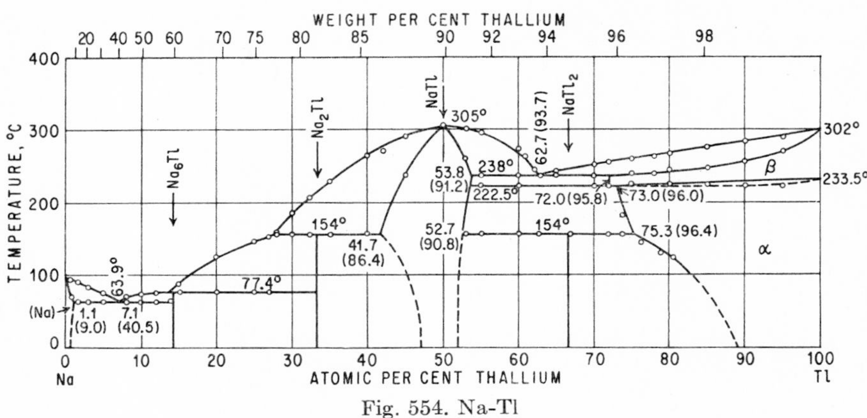

Fig. 554. Na-Tl

Crystal Structure. NaTl has the cubic B32 prototype structure [8], with $a = 7.488 \pm 3$ A (50.04 at. % Tl).

1. G. Grube and A. Schmidt, *Z. Elektrochem.*, **42**, 1936, 201–209. The alloys were melted in iron crucibles in an argon atmosphere. Most of the alloys were analyzed chemically for their Na content and some for both components.
2. N. S. Kurnakov and N. A. Puschin, *Z. anorg. Chem.*, **30**, 1902, 86–101.
3. G. Tammann, *Z. physik. Chem.*, **3**, 1889, 446–447.
4. C. T. Heycock and F. H. Neville, *J. Chem. Soc.*, **55**, 1889, 671.
5. R. Kremann and P. v. Reininghaus, *Z. Metallkunde*, **12**, 1920, 279–282.
6. C. A. Kraus and H. F. Kurtz, *J. Am. Chem. Soc.*, **47**, 1925, 43–60.
7. E. Zintl, J. Goubeau, and W. Dullenkopf, *Z. physik. Chem.*, **A154**, 1931, 1–46, especially 15, 40–43.
8. E. Zintl and W. Dullenkopf, *Z. physik. Chem.*, **B16**, 1932, 195–205.

$\bar{2}.9850$
1.0150

Na-U Sodium-Uranium

Massive U is not appreciably attacked by liquid Na at 500°C even after several days of exposure [1]. From a cryoscopic study, [2] concluded that the solubility of U in liquid Na at 97.8°C probably lies between 0.00 and 0.05 wt. (0 and 0.005 at.) %.

1. F. Foote, *U.S. Atomic Energy Commission, Rept.* CT-2857, 1945, quoted in J. J. Katz and E. Rabinowitch, "The Chemistry of Uranium," Part I, National Nuclear Energy Series, Div. VIII, vol. 5, p. 177, McGraw-Hill Book Company, Inc., New York, 1951.
2. T. B. Douglas, *J. Research Natl. Bur. Standards*, **52**, 1954, 223–226.

$\bar{1}.5462$
0.4538

Na-Zn Sodium-Zinc

The phase diagram of Fig. 555 is based mainly on the thermal and microscopic work of [1]. The diagram is characterized by an extended miscibility gap in the liquid state and the occurrence of one intermediate phase, adjacent to pure Zn. As to the formula of this compound, [1] concluded from his thermal and chemical analyses $NaZn_{11-12}$; and [2], from potentiometric titrations in NH_3 solution, $\sim NaZn_{12}$ [3]. According to more recent roentgenographic work [4–6], the formula is $NaZn_{13}$ [92.86 at. (97.37 wt.) % Zn] (Fig. 555); the compound has a f.c.c. structure [4–6], the $D2_3$ prototype, with a = 12.2836 ± 3 A [6] and 112 atoms per unit cell.

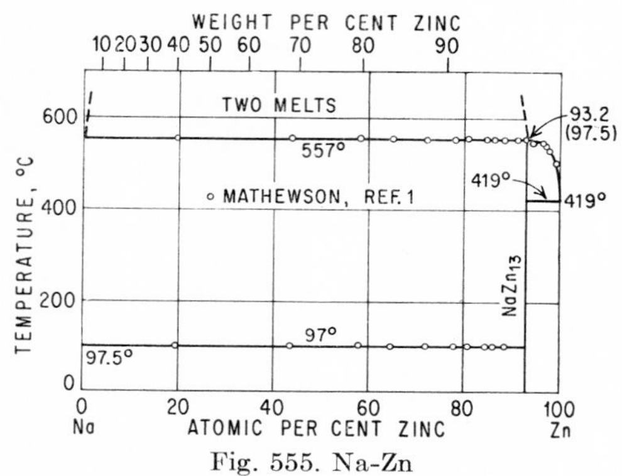

Fig. 555. Na-Zn

Some older observations of [7, 8] on the alloying of Na and Zn and the results of emf measurements by [9] are in essential agreement with the phase diagram of Fig. 555.

1. C. H. Mathewson, *Z. anorg. Chem.*, **48,** 1906, 195–200. References to some older papers may be found there. The alloys were prepared in glass tubes under hydrogen; those containing 96–99.6 wt. % Zn were analyzed.
2. E. Zintl, J. Goubeau, and W. Dullenkopf, *Z. physik. Chem.*, **154,** 1931, 1–46, especially 43.
3. No indication was found for the existence of a compound $NaZn_4$, which C. A. Kraus and H. F. Kurtz (*J. Am. Chem. Soc.*, **47,** 1925, 43) and W. M. Burgess and A. Rose (*J. Am. Chem. Soc.*, **51,** 1929, 2127) claimed to have prepared.
4. J. A. A. Ketelaar, *J. Chem. Phys.*, **5,** 1937, 668.
5. E. Zintl and W. Haucke, *Z. Elektrochem.*, **44,** 1938, 104–111; *Naturwissenschaften*, **25,** 1937, 717.
6. D. P. Shoemaker, R. E. Marsh, F. J. Ewing, and L. Pauling, *Acta Cryst.*, **5,** 1952, 637–644.
7. C. T. Heycock and F. H. Neville, *J. Chem. Soc.*, **55,** 1889, 674.
8. F. Haber and M. Sack, *Z. Elektrochem.*, **8,** 1902, 250.
9. R. Kremann and P. v. Reininghaus, *Z. Metallkunde*, **12,** 1920, 282–285.

0.1995
$\bar{1}.8005$

Nb-Ni Niobium-Nickel

The constitution of Ni-rich Nb-Ni alloys has been investigated in the range 0–37 wt. % Nb by [1] and 0–65 wt. % Nb by [2, 3] by means of thermal [1, 3], microscopic [2, 1, 3], hardness [2, 3], and roentgenographic [1] measurements (Fig. 556).

From pure Ni to the Ni-rich eutectic (24 wt. % Nb, 1265°C [1]; 23.5 wt. % Nb, 1270°C [3]) the liquidus and solidus of [1, 3] are in concord; the intermediate phase $NbNi_3$ (34.54 wt. % Nb), however, melts with an open maximum [3] rather than peritectically [1]. According to measurements of hardness, $NbNi_3$ has a homogeneity range from 32.5 to 36 wt. (23.3–26.2 at.) % Nb at 900°C [3]. A further intermediate phase (γ, presumably NbNi) forms peritectically at 1280°C [3] and seems to undergo a polymorphic transformation (Fig. 556) [4]. The existence of $NbNi_3$ was corroborated microscopically and roentgenographically by [5, 6].

The **solubility of Nb in solid Ni** as a function of temperature has been measured roentgenographically by [1] and microscopically [7] by [3] (Fig. 556). With comparable annealing times (between 5 and 48 hr) the roentgenographic data are appreciably lower and in Fig. 556 were given the greater emphasis. Single (unprecise) values were given by [2, 6].

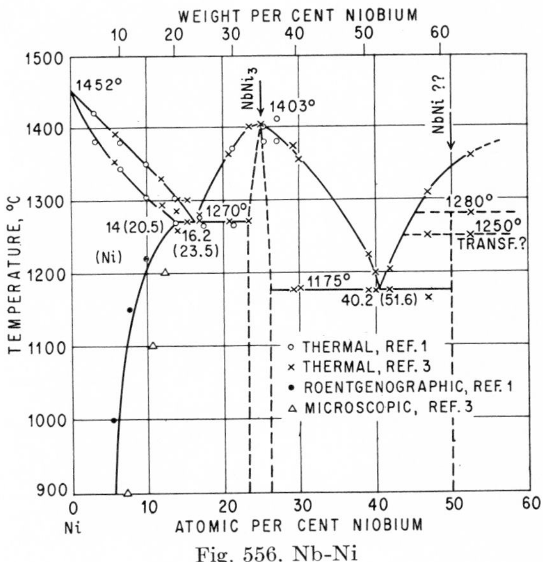

Fig. 556. Nb-Ni

Microscopic work on Nb-rich alloys [6] revealed a solubility of Ni in solid Nb of less than 5 wt. (7.7 at.) %.

Crystal Structure. $NbNi_3$ has a deformed h.c.p. structure with ordered atomic arrangement [8] ($TiCu_3$ type [9]); no lattice parameters are given. Previously, [5] reported an "elementlike" and [10] a h.c.p. structure. According to [11], no σ phase exists in the Nb-Ni system.

Note Added in Proof. [12] reported the lattice parameters of $NbNi_3$ (orthorhombic $TiCu_3$ type [9]) as $a = 5.10_6$ kX, $b = 4.55_6$ kX, $c = 4.25_1$ kX.

1. G. Grube, O. Kubaschewski, and K. Zwiauer, *Z. Elektrochem.*, **45,** 1939, 881–884. The alloys were prepared by reduction of a Nb_2O_5 + Ni mixture by hydrogen.
2. S. A. Pogodin and A. N. Zelikman, *Metallurg,* **1939**(1), 8–14 (in Russian).
3. S. A. Pogodin and A. N. Zelikman, *Compt. rend. acad. sci. U.R.S.S.*, **31,** 1941, 895–897 (in German); *Izvest. Sektora Fiz.-Khim. Anal.*, **16,** 1943, 158–166 (in Russian). The alloys were prepared from the elements (Ni, 99.5; Nb, 98.7 wt. % pure) in alumina crucibles and analyzed.

4. NbNi (?) may be in equilibrium with the Nb phase [3].
5. H. J. Wallbaum, *Arch. Eisenhüttenw.*, **14,** 1940-1941, 521–526.
6. O. Kubaschewski and A. Schneider, *J. Inst. Metals,* **75,** 1948-1949, 403–416, especially 414.
7. The value for 900°C [7.05 at. (10.7 wt.) % Nb] was confirmed by hardness and resistivity data.
8. H. J. Wallbaum, *Naturwissenschaften,* **31,** 1943, 91–92.
9. According to N. Karlsson, *J. Inst. Metals,* **79,** 1951, 391, $TiCu_3$ is orthorhomb (pseudohexagonal).
10. S. Hidekel, in [3].
11. P. Greenfield and P. A. Beck, *Trans. AIME,* **200,** 1954, 253–257.
12. K. Schubert et al., *Naturwissenschaften,* **44,** 1957, 229–230.

0.7639
$\bar{1}.2361$

Nb-O Niobium-Oxygen

Solubility of Oxygen in Solid Nb. [1] estimated from cursory X-ray work the extent of the Nb primary solid solution to be less than 4.76 at. (0.86 wt.) % O. The results of a more detailed roentgenographic and microscopic study in the range

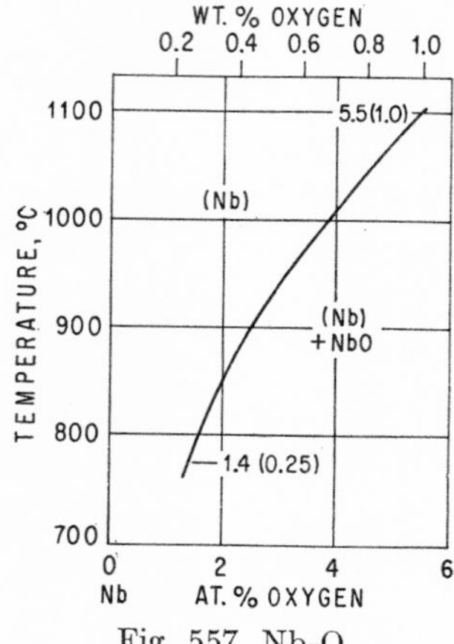

Fig. 557. Nb-O

between 775 and 1100°C [2] may be seen from Fig. 557; the solubilities at the temperatures mentioned are 1.4 at. (0.25 wt.) % and 5.5 at. (1.0 wt.) % O, respectively.

Oxides of Nb. According to [1], whose paper concludes a fruitful discussion [3–6] on the number and formulas of Nb oxides, there exist the three compounds NbO, NbO_2, and Nb_2O_5 [7]. NbO (14.69 wt. % O) has no extensive homogeneity range and possesses a cubic structure [4, 1], with a = 4.211 A and 6 atoms per unit cell (NaCl lattice with ordered vacancies). NbO_2 (25.62 wt. % O) has no homogeneity range of measurable extent. Powder diagrams showed lines in addition to the expected rutile diagram—reported by [8]—and it is assumed [4, 1] that the lattice is not quite the same as that of rutile.

Nb_2O_5 [71.43 at. (30.09 wt.) % O] melts at 1460 ± 5°C [1], and its homogeneity range extends down to the composition $Nb_2O_{4.8}$ [4, 1]. In heating the compound prepared at lower temperatures, two irreversible transformations were observed [1]. The structure of the three polymorphic forms are not yet known; the low-temperature form appears to be isotypic with Ta_2O_5.

The magnetic susceptibilities in the Nb-Nb_2O_5 system have been measured by [9].

The *kinetics* of the reaction of Nb with oxygen have been studied by [10–13] (see also [14]).

Note Added in Proof. According to [15], the structure of the low-temperature form of Nb_2O_5 (called γ) is closely related to that of α-U_3O_8, orthorhombic with $a = 6.19$ A, $b = 3.65$ A, $c = 3.94$ A. It is stated, however, that a few additional weak lines "apparently refer to a larger multiple cell."

1. G. Brauer, *Z. anorg. Chem.*, **248,** 1941, 1–31.
2. A. U. Seybolt, *Trans. AIME*, **200,** 1954, 774–776.
3. G. Grube, O. Kubaschewski, and K. Zwiauer, *Z. Elektrochem.*, **45,** 1939, 885–888.
4. G. Brauer, *Naturwissenschaften*, **28,** 1940, 30.
5. O. Kubaschewski, *Z. Elektrochem.*, **46,** 1940, 284–287.
6. G. Brauer, *Z. Elektrochem.*, **46,** 1940, 397–402.
7. For older (obsolete) literature on Nb oxides, see chemical handbooks or [3, 1].
8. V. M. Goldschmidt, W. Zachariasen, et al., *Norske Videnskaps-Akad. Oslo, Mat.-Naturv. Kl.*, **1,** 1926; abstract, *Strukturbericht*, **1,** 1913-1928, 211.
9. G. Brauer, *Z. anorg. Chem.*, **256,** 1948, 10–14.
10. D. J. McAdam and G. G. Geil, *J. Research Natl. Bur. Standards*, **28,** 1942, 593–635.
11. Technical Bulletin, Fansteel Metallurgical Corp., Chicago, Ill., 1945; quoted from [12, 13].
12. E. A. Gulbransen and K. F. Andrew, *J. Electrochem. Soc.*, **96,** 1949, 364–376.
13. E. A. Gulbransen and K. F. Andrew, *Trans. AIME*, **188,** 1950, 586–599.
14. R. T. Phelps, E. A. Gulbransen, and J. W. Hickman, *Ind. Eng. Chem., Anal. Ed.*, **18,** 1946, 391–400 (Electron diffraction and electron microscope study of oxide films formed on metals and alloys at moderate temperatures).
15. W. T. Holser, *Acta Cryst.*, **9,** 1956, 196.

$\bar{1}.6888$
0.3112

Nb-Os Niobium-Osmium

According to [1], Nb_3Os has the cubic "β-W" (A15) type of structure with $a = 5.121 \pm 2$ A.

1. S. Geller, B. T. Matthias, and R. Goldstein, *J. Am. Chem. Soc.*, **77,** 1955, 1502–1504.

0.4770
$\bar{1}.5230$

Nb-P Niobium-Phosphorus

Nb-monophosphide NbP (25.00 wt. % P) [1–3] and Nb-diphosphide NbP_2 (40.00 wt. % P) [2, 3] have been prepared. According to [2], NbP has a range of homogeneity between about $NbP_{0.8-1.2}$ and is dimorphous; a monotropic transition is assumed.

[3] found two closely related tetragonal structures: $a = 3.32$ A, $c = 5.69$ A, $c/a = 1.71$ (~4 atoms per unit cell, partially ordered, "α-NbP"), and $a = 3.325$ A, $c = 11.38$ A, $c/2a = 1.71$ (8 atoms per unit cell, ordered, "β-NbP"), of the compositions $NbP_{0.95}$ and $NbP_{1.0}$, respectively [4].

1. E. Heinerth and W. Biltz, *Z. anorg. Chem.*, **198,** 1931, 173–174.
2. A. Reinecke, F. Wiechmann, M. Zumbusch, and W. Biltz, *Z. anorg. Chem.*, **249,** 1942, 14–20.
3. N. Schönberg, *Acta Chem. Scand.*, **8,** 1954, 226–239 (in English).
4. The alloys were presumably slowly cooled to room temperature.

$\bar{1}.9399$
0.0601

Nb-Pd Niobium-Palladium

[1] examined seven Nb-Pd alloys varying in composition between 32 and 52 at. % Pd, which had been annealed at and quenched from 1000°C, by microscopic and X-ray methods. Only one intermediate phase was found which exists over a very narrow range of composition at about 40 at. % Pd and possesses the tetragonal σ-phase structure, with $a = 9.89$ A, $c = 5.11$ A, $c/a = 0.52$.

1. P. Greenfield and P. A. Beck, *Trans. AIME,* **206,** 1956, 265–276.

$\bar{1}.6775$
0.3225

Nb-Pt Niobium-Platinum

[1] examined the Nb-Pt system for Laves phases, but found none. [2] studied seven alloys ranging from 25 to 57 at. % Pt, which had been annealed at and quenched from 1000°C, by microscopic and X-ray diffraction methods. Two intermediate phases were found. Nb_3Pt (41.19 wt. % Pt) has the "β-W" (A15) type of structure, with $a = 5.11$ A. Another intermediate phase exists at about 37.5 at. % Pt, having the tetragonal σ-phase structure, $a = 9.89$ A, $c = 5.11$ A, $c/a = 0.52$. [3] reported $a = 5.153 \pm 3$ A for the lattice parameter of "β-W" (A15) type Nb_3Pt.

1. H. J. Wallbaum, *Naturwissenschaften,* **31,** 1943, 91–92.
2. P. Greenfield and P. A. Beck, *Trans. AIME,* **206,** 1956, 265–276.
3. S. Geller, B. T. Matthias, and R. Goldstein, *J. Am. Chem. Soc.,* **77,** 1955, 1502–1504.

$\bar{1}.6978$
0.3022

Nb-Re Niobium-Rhenium

[1] mentioned the existence of a σ phase at the composition NbRe (66.73 wt. % Re). [2] studied eight alloys containing between 35 and 82 at. % Re, which had been annealed at and quenched from 1200 or 1000°C, by microscopic and X-ray methods. Annealing at 1200°C showed that no σ phase exists at this temperature. An alloy containing 63 at. % Re was found to have a structure of the cubic α-Mn (A12) type, with $a = 9.670$ A. At 1000°C, however, a σ phase enters into equilibrium between the b.c.c. Nb-base solid solution and the α-Mn type phase. The σ phase appears to have a very narrow range of composition at about 50 at. % Re, in agreement with [1]; its tetragonal lattice parameters are $a = 9.72$ A, $c = 5.07$ A, $c/a = 0.52$. The α-Mn structure is stable from about 60 up to at least 82 at. % Re. The solid solution of Re in Nb extends to about 40 at. % Re.

Note Added in Proof. [3] reported the existence of the phase "Nb_7Re_{22}" (75.86 at. % Re), having the χ (chi) type (= α-Mn type) of structure, with $a = 9.67_6$ A.

1. P. Duwez, oral discussion, The Institute of Metals Meeting, March, 1952. Quoted by [2].
2. P. Greenfield and P. A. Beck, *Trans. AIME,* **206,** 1956, 265–276.
3. J. Niemiec and W. Trzebiatowski, *Bull. acad. polon. sci.,* **4,** 1956, 601–603 (in English); *Met. Abstr.,* **24,** 1957, 819.

$\bar{1}.9556$
0.0444

Nb-Rh Niobium-Rhodium

[1] examined seven Nb-Rh alloys varying in composition between 24 and 48 at. % Rh, which had been annealed at and quenched from 1000°C, by microscopic and

X-ray methods. Two intermediate phases were found. Nb_3Rh (26.97 wt. % Rh) has the "β-W" (A15) structure, a = 5.115 A. Another intermediate phase was found over a range of about 1 at. % at approximately 40 at. % Rh; it possesses the tetragonal σ-phase structure, a = 9.774 A, c = 5.054 A, c/a = 0.517. The Rh-rich phase in equilibrium with σ may be either another compound or the Rh-rich solid solution.

1. P. Greenfield and P. A. Beck, *Trans. AIME*, **206,** 1956, 265–276.

$\bar{1}.9607$
0.0393

Nb-Ru Niobium-Ruthenium

[1] examined six Nb-Ru alloys varying in composition between 32 and 78 at. % Ru, which had been annealed at and quenched from 1200°C, by microscopic and X-ray methods. The (Nb) primary solid solution extends up to at least 32 at. % Ru. A tetragonal lattice, with a = 3.00 A, c = 3.38 A, c/a = 1.13, was tentatively assigned to an intermediate phase at about 48 and 49 at. % Ru. The X-ray diffraction patterns of 58 and 68 at. % Ru alloys were not interpreted; one or two additional intermediate phases of unidentified structure are apparently present in this composition range. The solubility limit of Nb in solid Ru is less than 22 at. % Nb.

[2] measured the lattice spacings of dilute (up to 2.41 at. %) solid solutions of Nb in Ru; a, c, and c/a were found to increase with increasing Nb content.

1. P. Greenfield and P. A. Beck, *Trans. AIME*, **206,** 1956, 265–276.
2. A. Hellawell and W. Hume-Rothery, *Phil. Mag.*, **45,** 1954, 797–806.

0.4620
$\bar{1}.5380$

Nb-S Niobium-Sulfur

According to work on preparation and tensimetric and qualitative X-ray studies [1], there exist two intermediate phases in the Nb-S system, viz., a phase characterized as a sesquisulfide Nb_2S_3 but having a very broad homogeneity range up to the approximate composition NbS_4, and a monosulfide NbS (25.66 wt. % S) which is homogeneous in the $NbS_{0.5-1.0}$ interval.

These results could not be fully corroborated by more recent X-ray powder work [2, 3]. "Two sulfide phases, NbS and NbS_2 (40.84 wt. % S), with rather small homogeneity ranges were found to exist. Because of the impossibility of preparing NbS in a pure state, the homogeneity limits could not be determined with any certainty but seem to be approximately $NbS_{0.9}$ and $NbS_{1.2}$" [3]. For NbS in equilibrium with Nb, the hexagonal WC structure [2, 3] with a = 3.32 A, c = 3.23 A, c/a = 0.97 [3] was observed, and for the phase with an excess of S, the NiAs (B8) structure [2, 3] with a = 3.32 A, c = 6.46 A, c/a = 1.95 [3]. The transition is most likely a disorder-order transformation of the positions of the S atoms [2]. "The NbS_2 phase does not exist with a greater S content than corresponding to the composition 1:2 and has a narrow homogeneity range—probably smaller than 1 at. %" [3]. It was found to be of the rhombohedral $CdCl_2$ (C19) type, with a = 6.24 A, α = 30.95° [3].

1. W. Biltz and A. Köcher, *Z. anorg. Chem.*, **237,** 1938, 369–380. The paper contains references to some older (obsolete) literature.
2. N. Schönberg, *Acta Met.*, **2,** 1954, 427–432.
3. G. Hägg and N. Schönberg, *Arkiv Kemi*, **7,** 1954, 371–380.

0.0706
$\bar{1}.9294$

Nb-Se Niobium-Selenium

On the preparation of an (undefined) Nb-Se alloy, see [1].

1. W. v. Bolton, *Z. Elektrochem.*, **13,** 1907, 149.

0.5195
$\bar{1}.4805$

Nb-Si Niobium-Silicon

$NbSi_2$ (37.68 wt. % Si) [1–3] has the hexagonal $CrSi_2$ (C40) type of structure [1], with $a = 4.795 \pm 5$ A, $c = 6.589 \pm 5$ A, $c/a = 1.374$. A eutectic ($NbSi_2$ + Si) lies above 73 wt. (90 at.) % Si [3, 4].

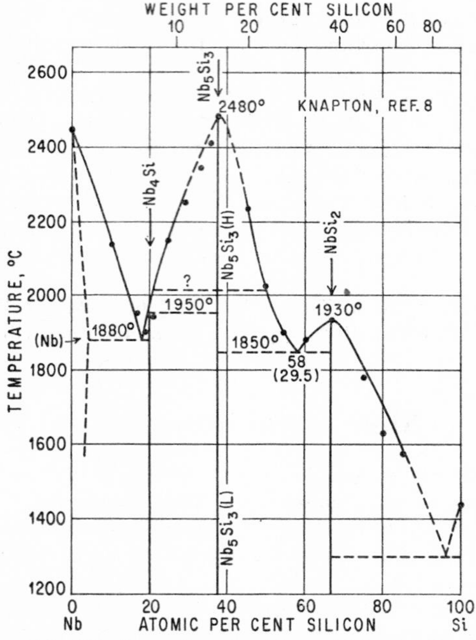

Fig. 558. Nb-Si. (See also Note Added in Proof.)

In a systematic X-ray investigation, [2] found—besides $NbSi_2$—two additional intermediate phases near the composition Nb_2Si (13.13 wt. % Si) and suggested that they represent two modifications of (Nb_2Si) (see Note Added in Proof). More recently the compound Nb_5Si_3 (15.35 wt. % Si)—which certainly corresponds to one of the "Nb_2Si modifications" just mentioned—has been prepared [5, 6] and shown [5] to have the hexagonal Mn_5Si_3 ($D8_8$) type of structure, with $a = 7.53_6$ A, $c = 5.24_8$ A, $c/a = 0.696$ (cf. Supplement). [6] also observed a phase Nb_2Si "apparently formed only at very high temperatures."

The solubility of Si in solid Nb is less than 5 at. (1.6 wt.) % [2].

Supplement. Three additional papers dealing with the constitution of the Nb-Si system have been published recently.

[7] reported the existence of the silicides $NbSi_{0.55\pm0.1}$ (~35.5 at. % Si), Nb_5Si_3 (37.5 at. % Si, having the Mn_5Si_3 type of structure), and $NbSi_2$; they suggest that other silicides may exist in the region between $NbSi_{0.55}$ and Nb_5Si_3. Indications for eutectics between Nb-$NbSi_{0.55}$ and Nb_5Si_3-$NbSi_2$ were found.

[8] established the phase diagram shown in Fig. 558 by means of melting-point determinations, X-ray examination, and metallography. "From metallographic and X-ray examinations, it is evident that Nb_4Si is isomorphous with Ta_4Si and Zr_4Si, but the $D0_{19}$ structure of the Ta_4Si phase reported by [9] has not been detected in alloys arc-melted from components of high purity." The compound Nb_5Si_3 was found to exist in two modifications, with the transformation temperature lying between 1900 and 2100°C. [8] was unable to find the $D8_8$ structure reported by [5] for Nb_5Si_3 and showed that contamination by carbon causes the formation of this type of structure. The terminal solid solubilities were not determined; however, an appreciable solubility of Si in Nb is indicated by a lattice expansion observed in powder compacts sintered at 1300°C.

[10] reported the finding of a tetragonal phase, a = 9.97 A, c = 5.08 A, c/a = 0.51, in the Nb-Si system. Possible formulas are said to be Nb_3Si_2 or Nb_5Si_3.

Note Added in Proof. The crystal structures of the (carbon-free) polymorphic forms of Nb_5Si_3 (or Nb_3Si_2) have been determined. The low-temperature modification is tetragonal and isotypic with Cr_5B_3, with $a = 6.57_0$ A, $c = 11.88_4$ A [11–13]. The high-temperature modification is also tetragonal, isotypic with $(V, Ta, Cr, Mo, W)_5Si_3$, and has the lattice parameters $a = 10.01_8$ A, $c = 5.07_7$ A, c/a = 0.506 [11, 13].

[13] reinvestigated the system by melting-point determinations and microscopic as well as X-ray work on sintered or arc-melted alloys. Alloys with 20–33.3 at. % Si were found to be two-phase ($Nb + Nb_5Si_3$) between 1550 and 1900°C. No indications for the existence of the Nb_4Si phase reported by [8] could be found. The X-ray patterns ascribed by [2] to phases of the composition Nb_2Si could be identified as actually being Nb_5Si_3 patterns. The eutectic ($NbSi_2 + Si$) was found at 92 wt. (97.4 at.) % Si, 1405°C. A revised phase diagram is given.

[14] reported the melting point of Nb_5Si_3 as 2410°C.

1. H. J. Wallbaum, *Z. Metallkunde*, **33,** 1941, 378–381.
2. G. Brauer and W. Scheele, in *FIAT Rev. Ger. Sci.*, Inorganic Chemistry, Part II, p. 103, 1948.
3. W. R. Johnson and M. Hansen, Technical Report 6383, June, 1951, U.S. Air Force, Wright Air Development Center, Wright-Patterson Air Force Base, Dayton, Ohio.
4. There is a very high tendency toward supercooling in the solidification of Si-rich alloys [3].
5. H. Schachner, E. Cerwenka, and H. Nowotny, *Monatsh. Chem.*, **85,** 1954, 245–254.
6. G. F. Hardy and J. K. Hulm, *Phys. Rev.*, **93,** 1954, 1004–1016.
7. L. Brewer and O. Krikorian, *U.S. Atomic Energy Comm., Publ.* UCRL-2544, 1954.
8. A. G. Knapton, *Nature,* **175,** 1955, 730.
9. R. Kieffer, F. Benesowsky, H. Nowotny, and H. Schachner, *Z. Metallkunde,* **44,** 1953, 242.
10. E. Parthé, H. Schachner, and H. Nowotny, *Monatsh. Chem.*, **86,** 1955, 182–185.
11. E. Parthé, H. Nowotny, and H. Schmid, *Monatsh. Chem.*, **86,** 1955, 385–396.
12. E. Parthé, B. Lux, and H. Nowotny, *Monatsh. Chem.*, **86,** 1955, 859–867.
13. R. Kieffer, F. Benesovsky, and H. Schmid, *Z. Metallkunde,* **47,** 1956, 247–253.
14. G. A. Geach and F. O. Jones, *Plansee Proc.*, **1955,** 80–91; *Met. Abstr.*, **24,** 1957, 366.

$\bar{1}.8936$
0.1064

Nb-Sn Niobium-Tin

According to [1], the compound Nb_3Sn (29.87 wt. % Sn) possesses the "β-W" (A15) type of structure, with a = 5.3 A, and seems to be formed peritectically between

1200 and 1550°C. [2] reported the lattice parameter $a = 5.289 \pm 2$ A. Nb_3Sn has the highest superconducting transition temperature known to date: 18.05°K [1, 2].

1. B. T. Matthias, T. H. Geballe, S. Geller, and E. Corenzwit, *Phys. Rev.*, **95**, 1954, 1435.
2. S. Geller, B. T. Matthias, and R. Goldstein, *J. Am. Chem. Soc.*, **77**, 1955, 1502–1504.

$\bar{1}.7107$
0.2893

Nb-Ta Niobium-Tantalum

X-ray powder patterns of three Nb-Ta alloys prepared by sintering at high temperatures revealed the existence of a continuous series of solid solutions. The lattice-parameter-vs.-composition (in at. %) curve is a straight line.

1. H. Bückle, *Z. Metallkunde*, **37**, 1946, 53–56.

$\bar{1}.6024$
0.3976

Nb-Th Niobium-Thorium

On the basis of data obtained from microscopic examination, melting observations, cooling curves, X-ray analyses, measurements of solubility and resistance, the

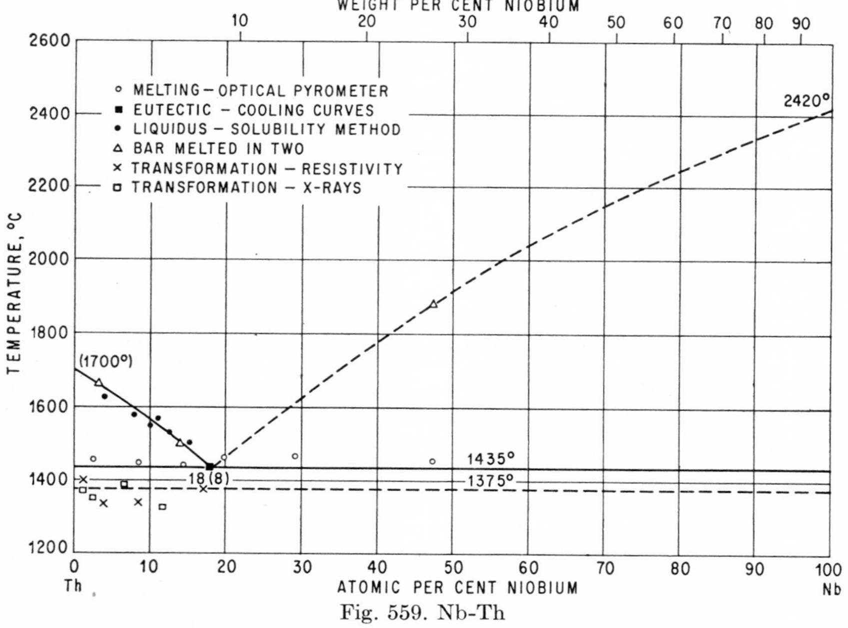

Fig. 559. Nb-Th

phase diagram of Fig. 559 has been proposed by [1]. The Th liquidus data were obtained primarily from solubility measurements (chemical analyses of the formerly liquid phase which had been in equilibrium with Th). The position of the eutectic was determined by thermal, microscopic, and resistometric work.

The allotropic transformation in Th reported by [2] at 1400 ± 25°C—f.c.c. (L) ⇄ b.c.c. (H)—has been corroborated by [1]. High-temperature X-ray studies and electrical-resistivity curves suggested that the addition of Nb lowers the temperature of the Th transformation slightly (1375°C in Fig. 559). However, whether this lowering is actually due to the solubility of Nb in the high-temperature form of Th or to some secondary effect (scavenging action?) is considered [1] to be questionable. According to X-ray parametric work [1], the solubility of Nb in solid Th at lower temperatures is not greater than 0.1 wt. (0.25 at.) %; that of Th in solid Nb is negligible.

1. O. N. Carlson, J. M. Dickinson, H. E. Lunt, and H. A. Wilhelm, *Trans. AIME*, **206,** 1956, 132–136. The alloys were prepared by arc melting under argon from Th metal sponge (>99.9 wt. % pure) and Nb (>99 wt. % pure). Important alloys were analyzed.
2. P. Chiotti, *J. Electrochem. Soc.*, **101,** 1954, 567; quoted from [1].

0.2877
$\bar{1}.7123$

Nb-Ti Niobium-Titanium

The phase diagram given in Fig. 560 was established by micrographic and lattice-parametric analyses and incipient-melting determinations of arc-melted iodide titanium-base alloys [1]. It is based on nominal compositions; weight losses after melting were negligible. The fact that β-Ti and Nb form a continuous series of solid

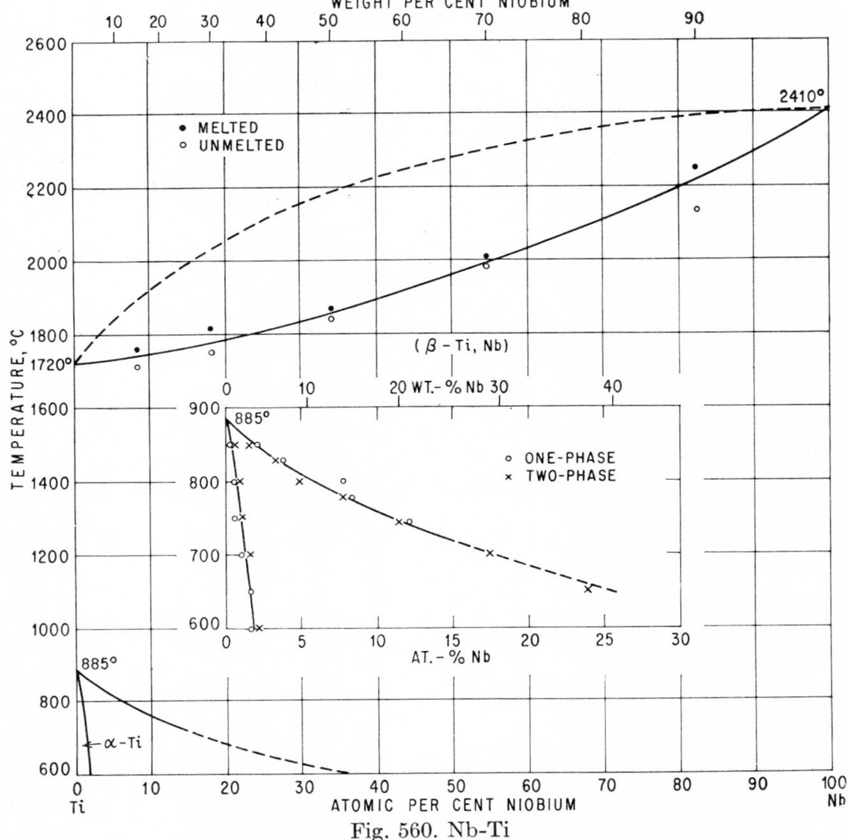

Fig. 560. Nb-Ti

solutions was previously reported [2], as the result of studies of sintered alloys [3]. The $\beta/(\alpha + \beta)$ boundary at 700°C and lower temperatures could not be located precisely because of the presence of some residual coring due to lack of complete homogenization [1]. The $\beta/(\alpha + \beta)$ boundary, calculated [4] by using an expression for the free energy of a homogeneous phase as a function of composition, was found to be in good agreement with the experimentally determined one, if the $\alpha/(\alpha + \beta)$ boundary was redrawn in a new position which also lay within the confines of the brackets imposed by the data points shown in Fig. 560.

The temperature at which the martensitic type of transformation $\beta \rightarrow \alpha'$ takes place in alloys up to 28 wt. % Nb was determined by thermal analysis under high rates of cooling [5]. The lattice-parameter-vs.-composition curve follows a small negative deviation from Vegard's law [1].

For measurements of electrical resistivity of (β-Ti, Nb)-phase alloys, see [6].

1. M. Hansen, E. L. Kamen, H. D. Kessler, and D. J. McPherson, *Trans. AIME*, **191**, 1951, 881–888.
2. B. W. Gonser, *Ind. Eng. Chem.*, **42**, 1950, 222–226.
3. Private communication by R. I. Jaffee, Oct. 30, 1950.
4. W. Rostoker, *Trans. AIME*, **191**, 1951, 1203–1205.
5. P. Duwez, *Trans. ASM*, **45**, 1953, 934–940.
6. S. L. Ames and A. D. McQuillan, *Acta Met.*, **2**, 1954, 831–836.

$\bar{1}.5914$
0.4086

Nb-U Niobium-Uranium

The phase diagram shown in Fig. 560*a* deviates from that of [1] in that it shows the monotectoid reaction $\gamma_U \rightleftarrows \alpha + \gamma_{Nb}$ at 634 [3] rather than about 655°C [1]. [4] located it between 600 and 650°C. No experimental details of the work of [1] are available.

According to the review by [2], "other investigators [5–7] have confirmed the major features of the system. Work by [5] indicates that the $(\gamma_U + \gamma_{Nb})$ region exists, but the boundaries are uncertain, except that the maximum of this region is well below 1000°C. Data by [7] for the same region show its limits to be about 16.5 and 66 at. % Nb at about 645°C, with the maximum at about 825°C. There appears to be fair agreement at the high-U end of the system. [1] interpreted the system as having a peritectoid reaction when α is formed. Work by [5] supports this view, although studies by [6] indicated a eutectoid rather than a peritectoid reaction. In any case, it is agreed that the solubilities in the α and β phase regions are limited."

"If diffusion samples are prepared in this system and are annealed first in the γ region and then in the α temperature range, an intermediate layer is observed (see also [8]). After very long periods of annealing, this layer decomposes. Similar material has been observed during annealing of γ-treated alloy samples. The appearance of this layer may be explained simply as a microstructural phenomenon which occurs at a certain composition and treatment, or it may be explained by the formation and decomposition of a metastable phase. It appears certain, however, that after long-time annealing at α temperatures (500–600°C) there is no intermediate phase in this system [6]" [2].

Transformation kinetics of γ alloys with 10 and 20 wt. % Nb have been studied by [4].

1. B. Sawyer, unpublished information (October, 1946), quoted in [2].
2. H. A. Saller and F. A. Rough, Compilation of U.S. and U.K. Uranium and Thorium Constitutional Diagrams, *U.S. Atomic Energy Comm., Publ.* BMI-1000, June, 1955.
3. A. E. Dwight; see discussion on [4].

4. R. J. Van Thyne and D. J. McPherson, *Trans. ASM*, **49**, 1957, 576–591; discussion, pp. 591–597.
5. B. A. Rogers, unpublished information (1954), quoted in [2].
6. H. A. Saller and F. A. Rough, unpublished information (1952), quoted in [2].
7. Atomic Energy Research Establishment, United Kingdom, unpublished information (1952, 1953), quoted in [2].
8. A. B. McIntosh and K. Q. Bagley, *J. Inst. Metals*, **84**, 1956, 260.

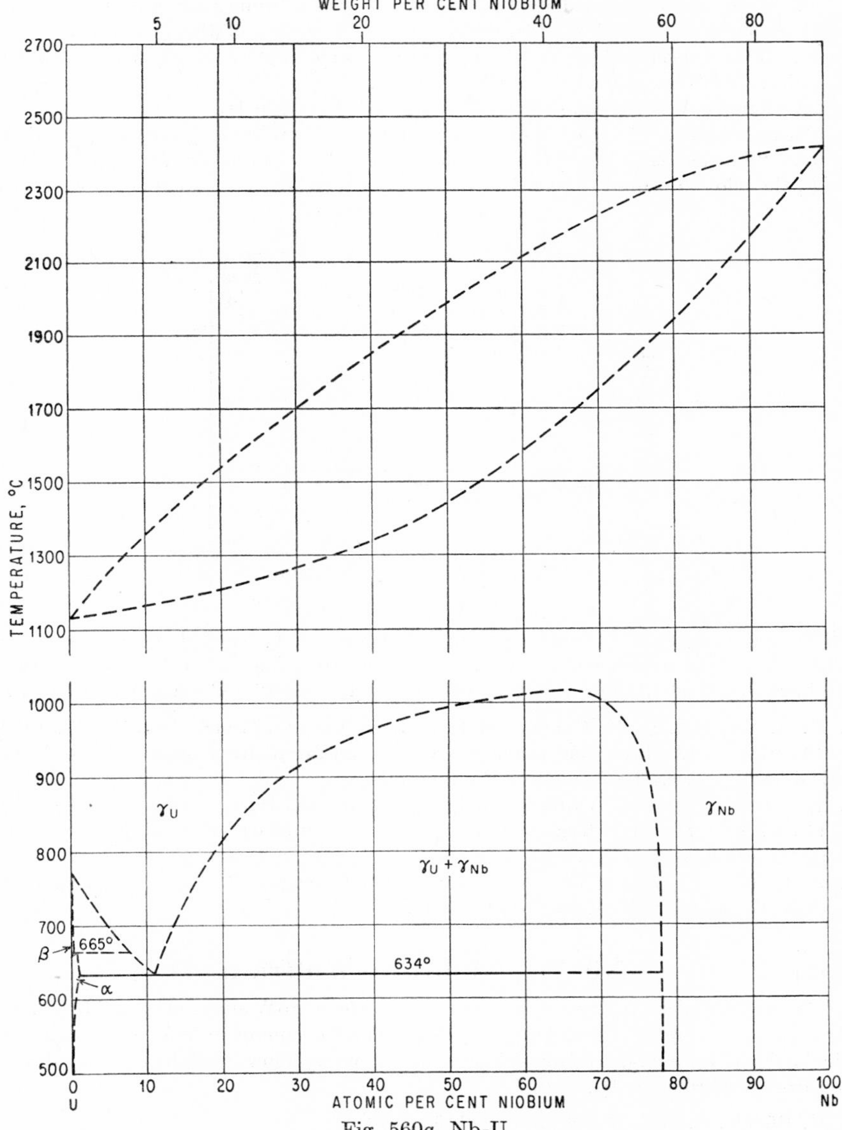

Fig. 560*a*. Nb-U

0.2609
$\bar{1}$.7391

Nb-V Niobium-Vanadium

The results of melting-point observations by [1]—plotted in Fig. 561 on the basis of temperature measurements by an optical pyrometer—and of microscopic and roentgenographic investigations of arc-melted alloys by [2] (11 alloys in the range 0–70 wt. % Nb) and [1] (9 alloys, 10–90 wt. % Nb) indicate the existence of a continuous series of solid solutions.

[2] observed in annealed (900°C, 170 hr) alloys containing 30 and 40 wt. % Nb small amounts of grain-boundary precipitate and suggested the formation of a σ phase. These observations, however, were not corroborated by [1], who concluded that above 650°C no solid-state reactions occur.

A lattice-parameter-vs.-composition graph is found in [1].

Supplement. An alloy of the composition V_2Nb (47.7 wt. % Nb) was examined by [3]. "The as-cast alloy of this composition had limited ductility. Metallographically, the alloy appeared to be principally a solid solution with a peppery precipitate.

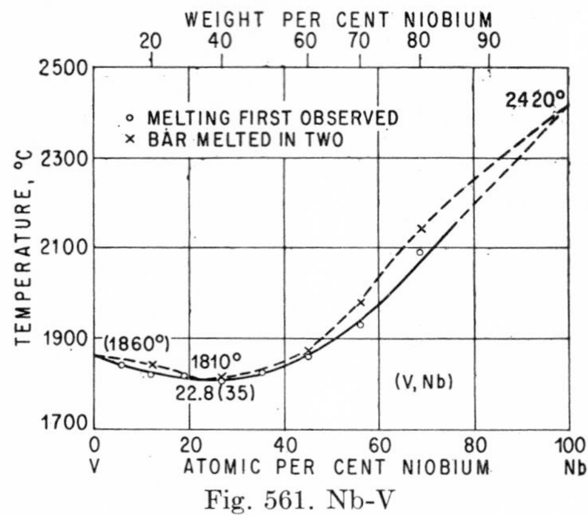

Fig. 561. Nb-V

Diffraction analysis of a sample annealed 24 hr at 800°C showed the patterns of a vanadium solid solution with a faint pattern of Nb. The diffraction pattern of a specimen annealed 13 days at 600°C was identical to the 800°C annealed specimen."

1. H. A. Wilhelm, O. N. Carlson, and J. M. Dickinson, *Trans. AIME,* **200,** 1954, 915–918. The alloys were prepared in an argon arc-melting furnace. No chemical analyses were performed on the final alloys.
2. W. Rostoker and A. Yamamoto, *Trans. ASM,* **46,** 1954, 1136–1163, especially 1155–1158. See also discussion, p. 1165 (P. Greenfield). The alloys were prepared in a helium arc-melting furnace.
3. R. P. Elliott, Armour Research Foundation, Chicago, Ill., Technical Report 1, OSR Technical Note OSR-TN-247, 1954.

$\bar{1}$.7034
0.2966

Nb-W Niobium-Wolfram

X-ray and microhardness measurements on a few Nb-W alloys prepared by sintering at high temperatures revealed the existence of a continuous series of solid solutions. The lattice-parameter-vs.-composition (in at. %) curve shows a slight negative deviation from Vegard's law [1].

1. H. Bückle, *Z. Metallkunde,* **37,** 1946, 53–56.

0.0080
$\bar{1}.9920$

Nb-Zr Niobium-Zirconium

The phase diagram shown in Fig. 562 has been established by means of incipient-melting tests, thermoresistometric and dilatometric measurements, and X-ray work on quenched alloys [1]. Because of the sluggishness of the solid-state reactions shown by all alloys containing more than 5 wt. % Nb, the diagram is based on data taken with rising temperature. No effort was made to determine points on the liquidus, but indications were found for melting intervals being narrow near 20 wt. (at.) % Nb.

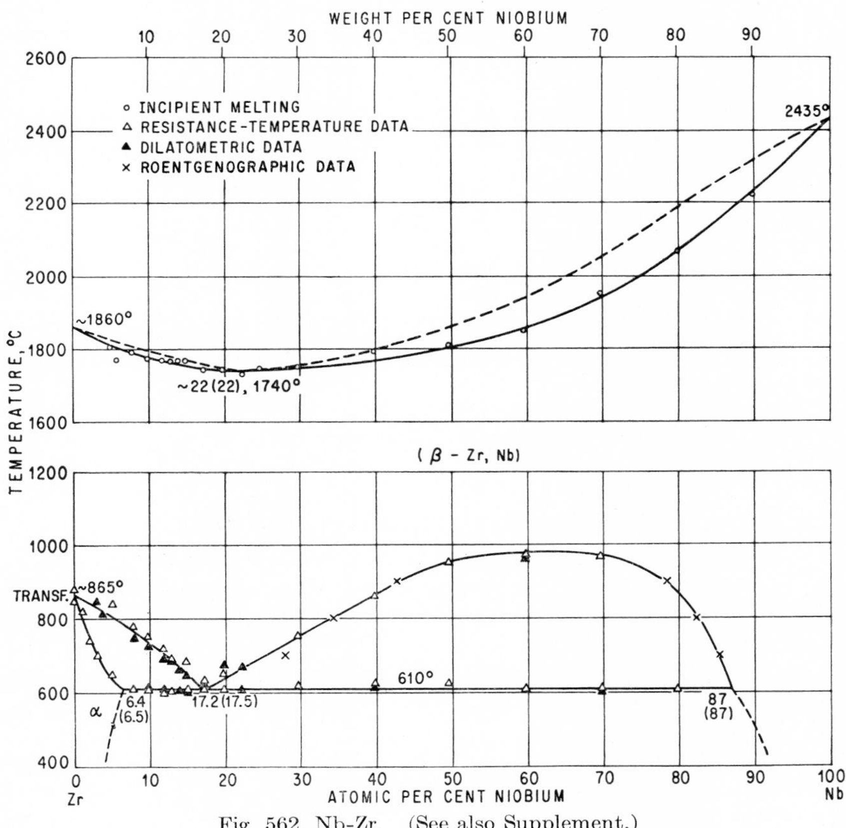

Fig. 562 Nb-Zr. (See also Supplement.)

No phase boundaries were determined below the eutectoid (monotectoid) temperature; however, age-hardening effects could be observed in Zr-rich alloys. The lattice-constant-vs.-composition curve of b.c.c. alloys cooled rapidly from 1100°C is almost a straight line.

Previously, some constitutional work had been carried out by [2–4]. According to [3], less than 0.5 wt. % Nb is soluble in Zr at 800°C. [4] investigated the system up to about 25 wt. % Nb; he suggested a eutectoid at about 625°C and 10 wt. % Nb and estimated that the solubility of Nb in Zr at 625°C was near 6 wt. %.

Supplement. [5] examined metallographically seven alloys containing up to 50.5 wt. % Nb, in the as-cast condition as well as annealed at and quenched from temperatures in the range 600–1600°C. Their observations were interpreted to show that (*a*) the eutectoid temperature is about 800°C, (*b*) a continuous series of solid solutions exists only above 1180°C, and (*c*) the solubility of Nb in α-Zr is less than 5 wt. % Nb. [6] gave additional evidence in support of the phase boundaries as established by [1]. There is agreement on the need for a systematic metallographic investigation under proper experimental conditions.

1. B. A. Rogers and D. F. Atkins, *U.S. Atomic Energy Comm., Publ.* ISC-500, 1954. The alloys were melted in a W-electrode Cu-crucible arc furnace in an argon atmosphere. Also *Trans. AIME,* **203,** 1955, 1034–1041.
2. S. M. Shelton, *U.S. Atomic Energy Comm., Publ.* AF-TR-5932, 1949; *Met. Abstr.,* **21,** 1954, 869.
3. C. R. Simcoe and W. L. Mudge, *U.S. Atomic Energy Comm. Report* WAPD-38, 1951, quoted from [1].
4. E. S. Hodge, *U.S. Atomic Energy Comm. Report* TID-5061, 1952, pp. 461–470, quoted from [1].
5. R. F. Domagala and D. J. McPherson, *Trans. AIME,* **206,** 1956, 619–620 (discussion on [1]).
6. B. A. Rogers and D. F. Atkins, *Trans. AIME,* **206,** 1956, 620 (author's reply).

0.7106
$\bar{1}$.2894

Nd-Si Neodymium-Silicon

1. G. Brauer and H. Haag, *Naturwissenschaften,* **37,** 1950, 210–211 (Crystal structure of disilicides of lanthanides).
2. G. Brauer and H. Haag, *Z. anorg. Chem.,* **267,** 1952, 198–212. (Preparation and crystal structure of disilicides of certain rare-earth metals.)

$\bar{1}$.7825
0.2175

Nd-U Neodymium-Uranium

"Nd is much less soluble in U than is Ce. The solid solubility of Ce in U is reported to be 0.5 wt. % near the melting point of U. Ce is only partially miscible with U in the liquid state. Probably there are no compounds in the Nd-U system [1]" [2].

1. National Physical Laboratory, United Kingdom, unpublished information (1949-1950).
2. H. A. Saller and F. A. Rough, Compilation of U.S. and U.K. Uranium and Thorium Constitutional Diagrams, *U.S. Atomic Energy Comm., Publ.* BMI-1000, June, 1955.

0.5644
$\bar{1}$.4356

Ni-O Nickel-Oxygen

Ni-NiO. *Solidification.* The existence of a eutectic between Ni and NiO near pure Ni was recognized early [1, 2]. According to [3], it is located at 0.24 wt. (0.87 at.) % O (1.1 wt. % NiO), 1438°C (Fig. 563). Solubility data on oxygen in liquid Ni up to about 6 at. % O, measured by [4, 5], are plotted in the diagram. [6], who located the eutectic at 1.1 wt. % NiO also, reported a solubility gap in the liquid state. This is in contradiction to the statement of [3] that NiO "is soluble in molten Ni in all proportions"; nor was it corroborated in the region up to 13.8 wt. (37 at.) % O by [7]. The phase diagram published recently by [8] also shows complete solubility in the liquid state. It is likely, therefore, that the liquidus shown in the figure continues smoothly to the melting point of NiO (21.42 wt. % O) which, according to more recent data [9, 10], lies at about 2000°C.

Solubility of Oxygen in Solid Ni. According to [11], the solubility of oxygen in solid Ni increases with decreasing temperature as follows: 0.012, 0.014, 0.019, and 0.020 wt. (0.044, 0.051, 0.070, and 0.073 at.) % O at 1200, 1000, 800, and 600°C, respectively.

The Compound NiO. The structure of NiO at normal temperatures had long been thought to be of the NaCl (B1) type [12, 13] (examination of NiO films by [14–17]). However, a critical examination of X-ray powder photographs disclosed [18, 19] a very slight rhombohedral distortion of the lattice. Later it was shown by [20] that the rhombohedral room-temperature cell with $a = 2.9518 \pm 5$ A, $\alpha = 60° 4.2'$ becomes cubic (NaCl type) only above 200°C. At 275°C the lattice parameter $a = 4.1946 \pm 5$ A was measured.

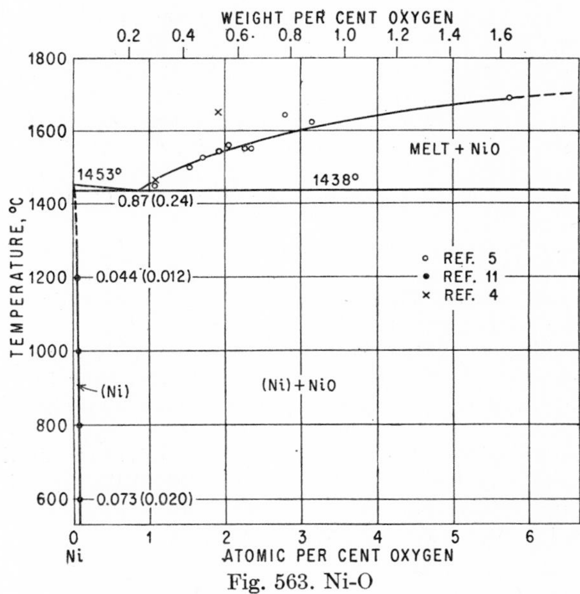

Fig. 563. Ni-O

In a series of paper [21–23, 10] it has been shown that the temperature dependence of several properties exhibits anomalies at about 250°C, because of an antiferromagnetic Curie point.

X-ray [19] and conductometric [10] observations suggest that excess oxygen can be taken into the lattice (vacant Ni sites according to [19]).

Higher Oxides of Ni. See [24, 8] and the chemical literature.

Note Added in Proof. The solubilities of oxygen in liquid Ni determined by [25] agree broadly with those of [5].

1. R. Ruer and K. Kaneko, *Metallurgie*, **9,** 1912, 422.
2. *Natl. Bur. Standards* (*U.S.*), *Circ.* 100, 1921.
3. P. D. Merica and R. G. Waltenberg, *Trans. AIME*, **71,** 1925, 715–716.
4. F. R. Hensel and J. A. Scott, AIME, Preprint, 1932, not available to the author; abstract, *J. Inst. Metals*, **50,** 1932, 511, and *Trans. AIME*, **104,** 1933, 139.

5. H. A. Wriedt and J. Chipman, *Trans. AIME,* **203,** 1955, 477–479.
6. H. Nishimura, M. Morinaga, and T. Ikeda, *Suiyokwai-Shi,* **9,** 1937, 251–255. *Chem. Abstr.,* **31,** 1937, 6599.
7. C. v. Bohlen und Halbach and W. Leitgebel, *Metall u. Erz,* **38,** 1941, 117–123.
8. D. P. Bogatskii, *Zhur. Obshchei Khim.,* **21,** 1951, 3–10 (in Russian); *Chem. Abstr.,* **46,** 1952, 7861. "State diagram of the system Ni-O and physicochemical nature of the solid phases in that system." The paper deals mainly with the preparation of NiO and of higher oxides (Ni_2O_3 and NiO_2) and their reduction-oxidation equilibria.
9. "Metals Handbook," 1948 ed., p. 1231, The American Society for Metals, Cleveland, Ohio.
10. M. Foëx, *Bull. soc. chim. France,* 1952, pp. 373–379.
11. A. U. Seybolt, Dissertation, Yale University, New Haven, Conn., 1936; quoted by E. N. Skinner in [9].
12. See the compilations in *Strukturbericht,* **1,** 1913-1928, 114, 123–124, 268, and **2,** 1928-1932, 222–224.
13. R. W. Cairns and E. Ott, *J. Am. Chem. Soc.,* **55,** 1933, 527–533.
14. B. A. Preston, *Phil. Mag.,* **17,** 1934, 466–470 (Electron diffraction).
15. N. Smith, *J. Am. Chem. Soc.,* **58,** 1936, 173–179 (Electron diffraction).
16. V. I. Arkharov and K. M. Graevsky, *Zhur. Tekh. Fiz.,* **14,** 1944, 132–145; *Met. Abstr.,* **13,** 1946, 165.
17. V. I. Arkharov and G. D. Lomakin, *Zhur. Tekh. Fiz.,* **14,** 1944, 155–161; *Met Abstr.,* **13,** 1946, 165.
18. H. P. Rooksby, *Nature,* **152,** 1943, 304.
19. Y. Shimomura and Z. Nishiyama, *Mem. Inst. Sci. Ind. Research, Osaka Univ.,* **6,** 1948, 30–34; also *Nippon Kinzoku Gakkai-Shi,* **13,** 1949, 2–3 (in English).
20. H. P. Rooksby, *Acta Cryst.,* **1,** 1948, 226.
21. M. Foëx, *Compt. rend.,* **227,** 1948, 193 (Dilatometric study).
22. M. Foëx and Ch. H. la Blanchetais, *Compt. rend.,* **228,** 1949, 1579–1580; also *J. phys. radium,* **12,** 1951, 170–171 (presented by F. Trombe) (Magnetic susceptibility).
23. Ch. H. la Blanchetais, *J. phys. radium,* **12,** 1951, 765–771 (Magnetic study).
24. F. François and M. L. Delwaulle, *Compt. rend.,* **205,** 1937, 282–284.
25. A. M. Samarin and V. P. Fedotov, *Izvest. Akad. Nauk S.S.S.R., Otdel. Tekh. Nauk,* **1956,** 119–125; *Met. Abstr.,* **24,** 1957, 777.

$\bar{1}.4894$
0.5106

Ni-Os Nickel-Osmium

According to microscopic, roentgenographic, magnetic, and hardness investigations of 11 annealed alloys (48 hr at 1200°C, furnace cooling) by [1], only two solid phases—the (Ni) and (Os) primary solid solutions—exist in the Ni-Os system [2]. Ni dissolves about 15 wt. (5 at.) % Os, and Os about 16 wt. (38 at.) % Ni. The solubility of Os in Ni increases with the temperature: a 25 wt. (9.3 at.) % Os alloy after quenching from 1200°C proved to be homogeneous.

The lattice parameter of Ni increases whereas the parameters of the hexagonal Os remain unchanged, with the addition of the second component. The Curie temperature of Ni is lowered to 100°C at the solubility limit.

1. W. Köster and E. Horn, "Festschrift aus Anlass des 100-jährigen Jubiläums der Firma W.C. Heraeus G.m.b.H.," pp. 114–123, Hanau, 1951.
2. The microstructures suggest a peritectic reaction in solidification.

0.2776
$\bar{1}$.7224

Ni-P Nickel-Phosphorus

Work on Preparation. The following compounds of Ni and P were claimed to be prepared by dry or aqueous methods: Ni_4P, Ni_3P, Ni_5P_2, Ni_2P, Ni_3P_2, Ni_6P_5, NiP, Ni_2P_3, NiP_2, NiP_3, and NiP_4 [1]. Obviously not only definite compounds, but phase mixtures also, were analyzed.

Systematic Investigations. Up to 36 at. % P the phase diagram of Fig. 564 is based on the thermal and microscopic study of [2]. His results [3] indicate the existence of Ni_3P (14.96 wt. % P), Ni_5P_2 [28.57 at. (17.43 wt.) % P] which undergoes a polymorphic transformation, and Ni_2P (20.88 wt. % P). The high-temperature form

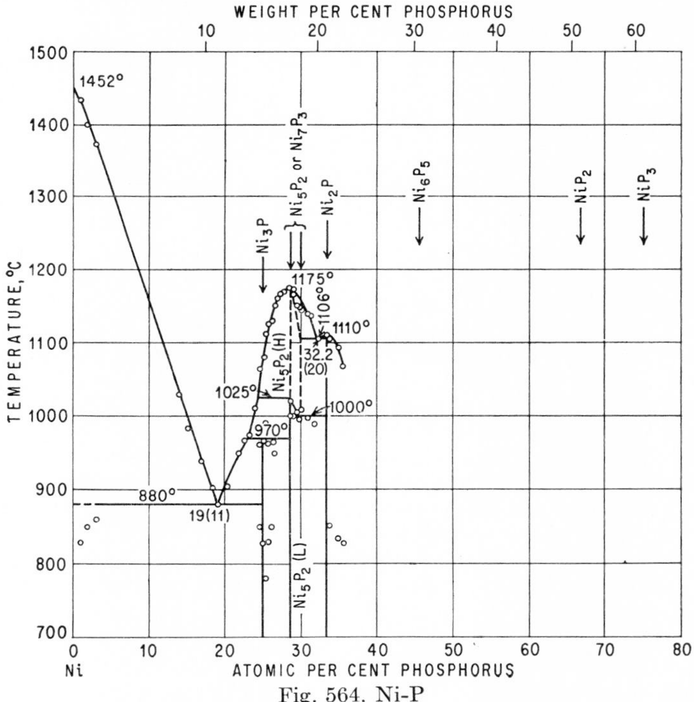

Fig. 564. Ni-P

of Ni_5P_2 is assumed to dissolve some P. These compounds were corroborated by a tensimetric and roentgenographic analysis [4] in which the intermediate phases ~Ni_6P_5 [45.45 at. (30.54 wt.) % P], NiP_2 (51.35 wt. % P), and NiP_3 (61.28 wt. % P) were also found. Representations of powder patterns of Ni_6P_5 and the higher phosphides are given.

Slowly cooled alloys with 15–34 at. % P were roentgenographically investigated by [5]. Ni_3P is isotypic with Fe_3P and Mn_3P; the lattice parameters of the tetragonal unit cell are $a = 8.93_4$ A, $c = 4.39_8$ A, $c/a = 0.4923$. Ni_2P has the hexagonal Fe_2P (C22) type of structure, with $a = 5.86_2$ A, $c = 3.37_2$ A, $c/a = 0.57_5$. Powder patterns of alloys of the compositions Ni_5P_2 and Ni_7P_3 were rather similar (a homogeneity range is suggested), and the stronger reflections of the Ni_7P_3 pattern could be indexed by a cubic cell. A later note [6] stated that "the phase Ni_5P_2 has actually the formula Ni_7P_3" (18.44 wt. % P). More detailed X-ray work is desirable, also with regard to the polymorphic transformation of Ni_5P_2 reported by [1]. According to [5], the homogeneity ranges of Ni, Ni_3P, and Ni_2P are very narrow. The heat of formation has a maximum at the composition Ni_5P_2 [7].

1. Only the more recent papers are listed here: C. Paal and L. Friederici, *Ber. deut. chem. Ges.*, **64B,** 1931, 1766–1776. R. Scholder, A. Apel, and H. L. Haken, *Z. anorg. Chem.*, **232,** 1937, 1–16. M. Chêne, *Ann. chim. (Paris)*, **15,** 1941, 187–285. References and short reviews of older papers may be found in these and also in the following publications: [2], [4], and M. Hansen, "Der Aufbau der Zweistofflegierungen," Springer-Verlag OHG, Berlin, 1936.
2. N. Konstantinow, *Z. anorg. Chem.*, **60,** 1908, 405–415. All alloys were analyzed.
3. Since [2] based his temperature measurements between the melting points of Cu and Ni on a Ni melting point of 1484°C (instead of 1452°C), all thermal data above 1083°C were corrected in Fig. 564. The maximum in the liquidus at the composition Ni_5P_2 is lowered by about 10°C.
4. W. Biltz and M. Heimbrecht, *Z. anorg. Chem.*, **237,** 1938, 132–144.
5. H. Nowotny and E. Henglein, *Z. physik. Chem.*, **B40,** 1938, 281–284.
6. H. Nowotny and E. Henglein, *Monatsh. Chem.*, **79,** 1948, 390 (footnote).
7. F. Weibke and G. Schrag, *Z. Elektrochem.*, **47,** 1941, 222–238.

$\bar{1}.4522$
0.5478

Ni-Pb Nickel-Lead

The constitution of the Ni-Pb system was worked out independently by [1, 2] (Fig. 565). Since [1] based his temperature measurements on a Ni melting point of 1484 instead of 1452°C, his thermal data above 1000°C were corrected graphically; they agree well then with those of [2]. For the extent of the miscibility gap in the liquid state at 1340°C, the data 40–93 wt. % Pb and 28–84 wt. % Pb [3] were interpolated and extrapolated from the length of the thermal arrests by [1, 2], respectively. In Fig. 565 the values 34 (average) and 93 wt. (12.7 and 79 at.) % Pb were chosen. For the end of solidification, [1] found about 323 ± 1°C, and [2], 329 ± 1°C (melting point of Pb, 327°C); in Fig. 565 a peritectic reaction on the Pb side is assumed. (But see Note Added in Proof.)

From the duration of the thermal arrests, [2] concluded a solubility of Pb in solid Ni of about 4 wt. (1.2 at.) %. This agrees with an observation by [4] that a 2 wt. % Pb alloy was one-phase; however, [4] admitted the possibility that Pb for the most part vaporized during alloying. [5] found the solubility of the Pb isotope ThB in solid Ni to be lower than 10^{-6} wt. %.

The solubility of Ni in solid Pb was determined [6] by means of a magnetic method (see Fig. 565, inset); the values found at 327°C and "room temperature" are 0.195 and 0.023 wt. (0.68 and 0.08 at.) % Ni, respectively. The Curie temperature of Ni is raised about 5°C by the addition of Pb [2].

Note Added in Proof. By means of chemical analysis of equilibrated and quenched alloys, [7] found the following solubilities of Ni in liquid Pb, in wt. % Ni (at. % in parentheses): 0.153, 0.264, 0.386, 0.465, 0.850, and 1.425 (0.53, 0.92, 1.35, 1.62, 2.94, and 4.85) at 370, 450, 500, 540, 635, and 727°C, respectively. Cooling curves showed the end of solidification to be at 324°C, which points to a eutectic rather than a peritectic reaction near pure Pb. The eutectic composition is about 0.11 wt. (0.38 at.) % Ni. The solubility of Ni in solid Pb should, therefore, be reinvestigated.

1. A. Portevin, *Rev. mét.*, **4,** 1907, 814–818.
2. G. Voss, *Z. anorg. Chem.*, **57,** 1908, 45–48.

3. "My extrapolation resulted in a higher Pb content, approximately 90 wt. % Pb," quoted from M. Hansen, "Der Aufbau der Zweistofflegierungen," Springer-Verlag OHG, Berlin, 1936.
4. G. Tammann and G. Bandel, *Z. Metallkunde*, **25**, 1933, 156.
5. [4], p. 155.
6. G. Tammann and W. Oelsen, *Z. anorg. Chem.*, **186**, 1930, 266–267.
7. E. Pelzel, *Metall*, **9**, 1955, 692–694.

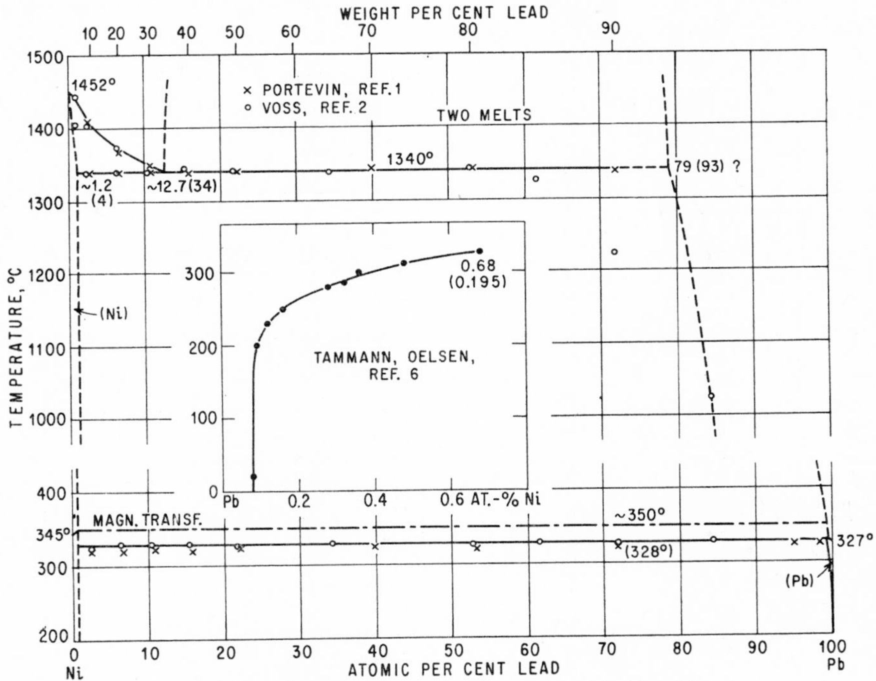

Fig. 565. Ni-Pb. (See also Note Added in Proof.)

$\bar{1}.7404$
0.2596

Ni-Pd Nickel-Palladium

According to thermal analyses on stationary [1] and vibrating [2] melts, roentgenographic work [3] on long-annealed [4] alloys, microscopic [1, 5] observations, and measurements of the temperature coefficient of the electrical resistance [5] and the hardness [5], as well as magnetic investigations [1, 6–9], the two metals form a continuous series of solid solutions. The claim of [2] (for which no detailed evidence was given) that "a series of experiments indicated the existence of a miscibility gap in the solid state" was not corroborated by later work, especially that of [3]. In Fig. 566 liquidus and solidus were drawn according to the data of [2], the curve of the magnetic transformation according to those of [9].

[3] showed that the lattice parameters follow a smooth curve with a positive deviation from Vegard's law. At considerable variance with this, [10] found an irregularly shaped curve but admitted that their (electrolytic) method of preparation "allows the existence of some deviation from the equilibrium state."

Note Added in Proof. [11] determined the constant θ in the Curie-Weiss law of paramagnetism for alloys with 83–100 at. % Pd.

1. F. Heinrich, *Z. anorg. Chem.*, **83,** 1913, 322–327. In the thermal analysis, strong supercooling was observed.
2. W. Fraenkel and A. Stern, *Z. anorg. Chem.*, **166,** 1927, 164–165.
3. R. Hultgren and C. A. Zapffe, *Trans. AIME*, **133,** 1939, 58–68.
4. Alloys were annealed 2 weeks at 600°C, 3 weeks at 400°C, as well as slowly cooled from 600°C over a period of 15 days.

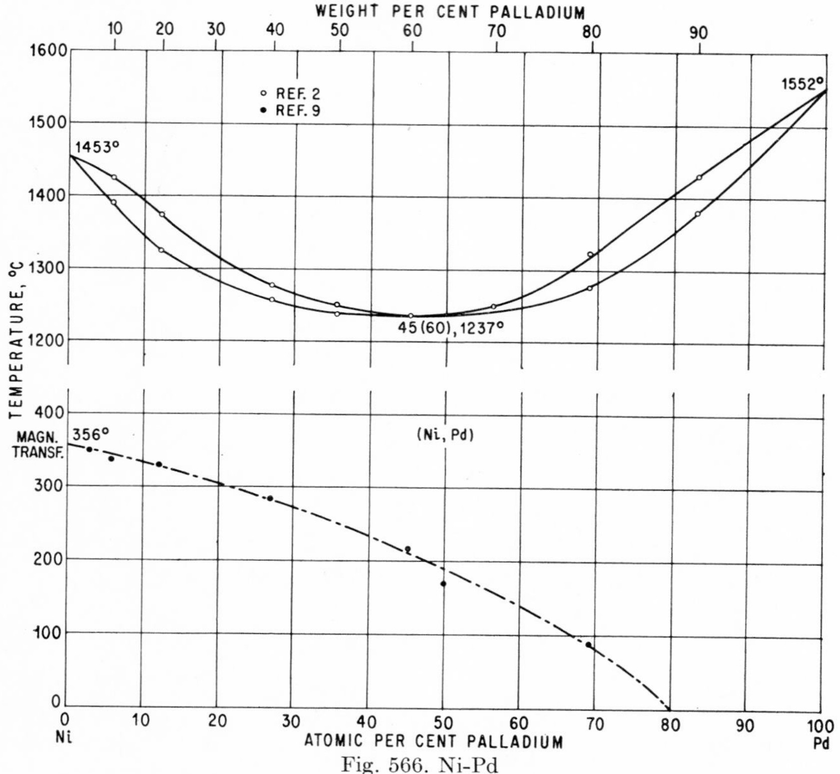

Fig. 566. Ni-Pd

5. A. T. Grigoriev, *Izvest. Inst. Platiny*, **9,** 1932, 13–22 (in Russian).
6. C. Sadron, *Ann. phys.*, **17,** 1932, 416; abstract, "Gmelins Handbuch der anorganischen Chemie," System No. 68 A(5), p. 610, Verlag Chemie, G.m.b.H., Weinheim/Bergstrasse, Germany, 1951.
7. L. Néel, *Ann. phys.*, **18,** 1932, 94.
8. C. Manders, *Ann. phys.*, **5,** 1936, 225.
9. V. Marian, *Ann. phys.*, **7,** 1937, 459–527.
10. Y. D. Kondrashev, I. P. Tverdovsky, and Z. L. Vert, *Doklady Akad. Nauk S.S.S.R.*, **78,** 1951, 729–731 (in Russian).
11. J. Cohen, *Compt. rend.*, **243,** 1956, 1613–1616.

1.6196
0.3804

Ni-Pr Nickel-Praseodymium

The diagram in Fig. 567 is essentially a schematic drawing based on the thermal data of only nine alloys [1]. The compositions were selected on the assumption that the compositions of the intermediate phases in this system are the same as those of the phases in the system Ce-Ni studied in more detail (see Ce-Ni). In the absence of tabulated data, the data points given had to be taken from a smaller diagram and, hence, are not accurate. Whereas the existence of the phases $PrNi_5$ (32.44 wt. % Pr), $PrNi_2$ (54.56 wt. % Pr), PrNi (70.60 wt. % Pr), and Pr_3Ni (87.81 wt. % Pr) appears to be established, the composition of the other two phases given as $PrNi_4$ (37.51 wt. % Pr) and $PrNi_3$ (44.46 wt. % Pr) is uncertain.

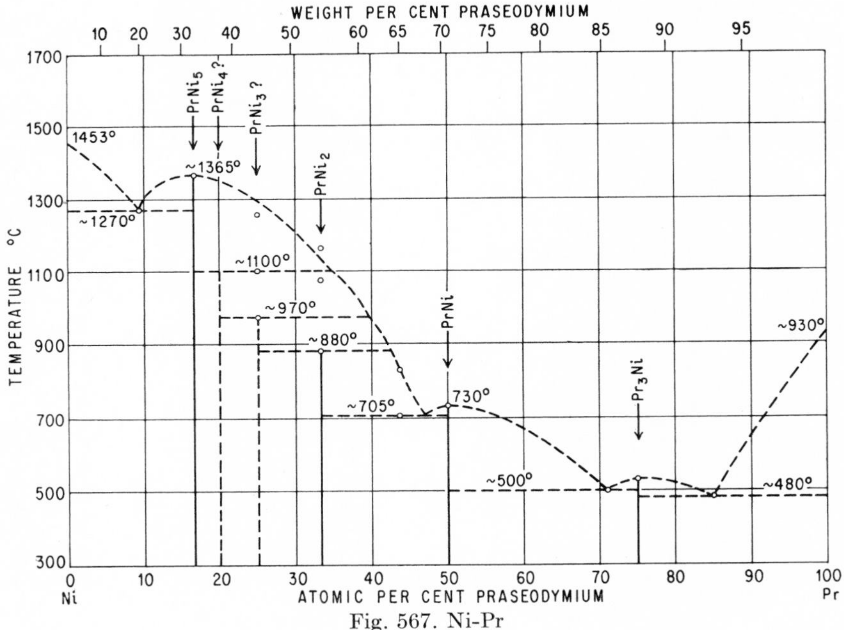

Fig. 567. Ni-Pr

$PrNi_2$ has the $MgCu_2$ (C15) type of structure, $a = 7.20_5$ A [1]. $PrNi_5$ is hexagonal of the $CaCu_5$ type, $a = 4.948$ A, $c = 3.973$ A, $2c/a = 1.606$ [1].

1. R. Vogel (W. Fülling), *Z. Metallkunde*, **38,** 1947, 97–103.

1.4780
0.5220

Ni-Pt Nickel-Platinum

According to thermal, microscopic, and electrical investigations by [1] and roentgenographic measurements by [2, 3], the two metals form a continuous series of solid solutions at higher temperatures. Whether or not a minimum in the liquidus occurs cannot be decided from the solidification data of [1] plotted in Fig. 568.

According to magnetic [4, 2] and X-ray diffraction [2, 5] work, an ordering reaction takes place around the composition Ni_3Pt (52.58 wt. % Pt). This reaction was observed in alloys containing 21–31 at. % Pt [2] and between 500 and 620°C [5].

[6] found the disordering temperature of a 25.07 at. % Pt alloy by means of temperature-resistivity curves to be 580 ± 5°C. The ordered phase has the Cu_3Au ($L1_2$) type of structure [2, 5]. Another superstructure was observed conductometrically and roentgenographically [3] below 645°C in the equiatomic region (Fig. 568). It has the tetragonal CuAu ($L1_0$) type of structure, and its homogeneity range extends from about 40 to 55 at. % Pt at 400°C. The lattice parameters of the 50 at. (76.89 wt.) % Pt alloy were found to be $a = 3.823$ A, $c = 3.589$ A, $c/a = 0.939$.

The variation of the Curie temperature with the composition was measured by [7, 8, 4, 2, 3]. For Fig. 568 the curve measured by [2] on alloys quenched from 900°C was chosen, modified to conform to a Curie temperature of 353°C for pure Ni. Alloys

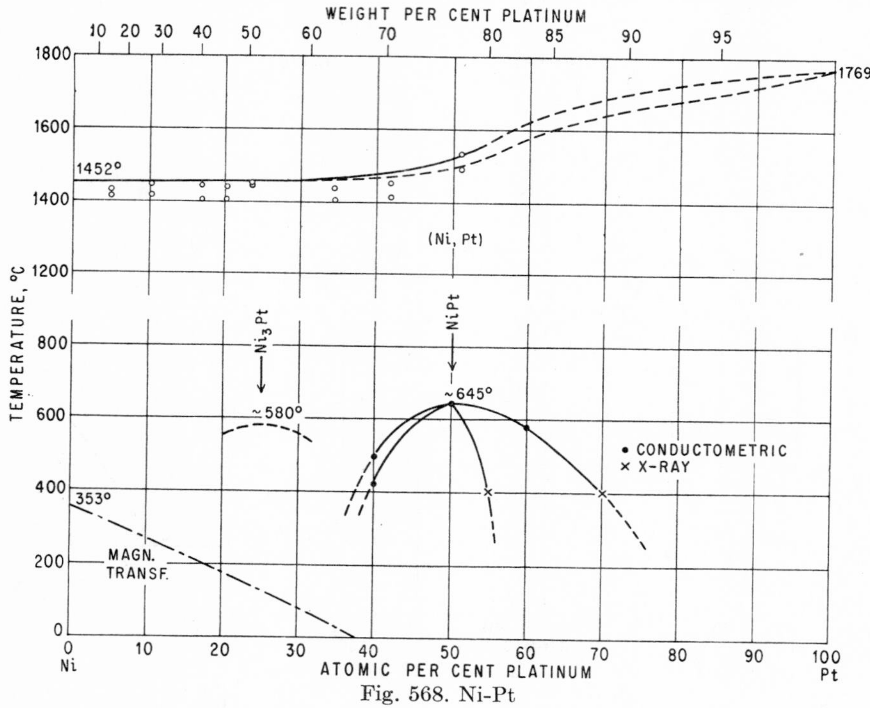

Fig. 568. Ni-Pt

with 21–31 at. % Pt annealed at 450°C showed Curie temperatures which were 20–40°C lower than those of the quenched alloys, which effect is due to the formation of the ordered (Ni_3Pt) phase [2] (see also [4]).

1. N. S. Kurnakow and V. A. Nemilov, *Izvest. Inst. Platiny U.S.S.R.*, **8**, 1931, 17–24; *Z. anorg. Chem.*, **210**, 1933, 13–20.
2. A. Kussmann and H. Nitka, *Physik. Z.*, **39**, 1938, 373–375; *Metallwirtschaft*, **17**, 1938, 657–659.
3. U. Esch and A. Schneider, *Z. Elektrochem.*, **50**, 1944, 268–274.
4. V. Marian, *Ann. phys.*, **7**, 1937, 459–527.
5. A. Kussmann and H. E. v. Steinwehr, *Z. Metallkunde*, **40**, 1949, 263–266.
6. R. A. Oriani and T. S. Jones, *Acta Met.*, **1**, 1953, 243.
7. F. W. Constant, *Phys. Rev.*, **34**, 1929, 1222–1223.
8. C. Manders, *Ann. phys.*, **5**, 1936, 167–231.

Ī.3902
0.6098

Ni-Pu Nickel-Plutonium

The phase diagram of Fig. 568*a* is mainly due to a Russian investigation [1] by several standard methods of alloy samples weighing hundreds of milligrams at the most. Five intermediate phases found by [1] were corroborated by American and Canadian workers [2], who showed that a sixth compound, Pu_2Ni_{17}, occurs. In Fig. 568*a*, Pu_2Ni_{17} has been tentatively assumed to be formed peritectically at about 1240°C.

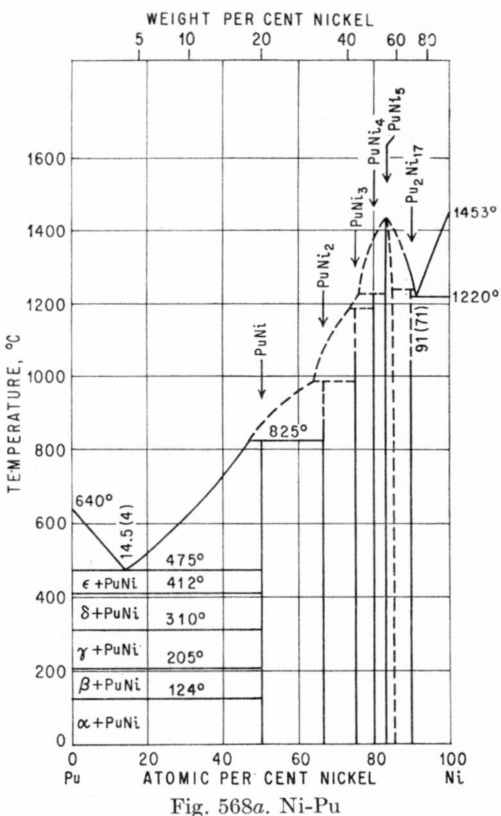

Fig. 568*a*. Ni-Pu

Crystal Structures. $PuNi_2$: f.c.c. $MgCu_2$ (C15) type of structure [1, 2], with $a = 7.141$ A (Pu-rich), $a = 7.115$ A (Ni-rich) [2]. $PuNi_3$: unknown structure, isostructural with $PuCo_3$ [2]. $PuNi_5$: hexagonal $CaZn_5$ type of structure, with $a = 4.872$ A, $c = 3.980$ A (Pu-rich), and $a = 4.861$ A, $c = 3.982$ A (Ni-rich) [2]. Pu_2Ni_{17}: hexagonal Th_2Ni_{17} type of structure, $a = 8.30$ A, $c = 8.00$ A [2].

1. S. T. Konobeevsky, *Conference of the Academy of Sciences of the U.S.S.R. on the Peaceful Uses of Atomic Energy, Division of Chemical Science*, July, 1955; English translation by U.S. Atomic Energy Commission, Washington, 1956.
2. See review by A. S. Coffinberry and M. B. Waldron on the Physical Metallurgy of Plutonium in "Metallurgy and Fuels," Progress in Nuclear Energy, ser. V, vol. 1, Pergamon Press Ltd. London, 1956.

$\bar{1}.4983$
0.5017
Ni-Re Nickel-Rhenium

The constitution of Re-Ni alloys has not been investigated as yet. It is known only that the two metals alloy easily at higher temperatures [1] and that Re-Ni alloys can be electrolytically deposited from aqueous solutions [2–4].

1. K. Moers, *Z. anorg. Chem.*, **196,** 1931, 149.
2. C. G. Fink and P. Deren, *Trans. Am. Electrochem. Soc.*, **66,** 1934, 474.
3. C. B. F. Young, *Metal Ind.* (*N.Y.*), **34,** 1936, 176–177.
4. L. E. Netherton and M. L. Holt, *Metal Finishing*, **48,** 1950, 75; and *J. Electrochem. Soc.*, **98,** 1951, 106–109.

$\bar{1}.7612$
0.2388
Ni-Ru Nickel-Ruthenium

"From the steady course of the magnetic properties in alloys up to about 20 wt. (12 at.) % Ru the existence of a solid solution can be concluded [1]. Thermal treatment does not noticeably influence the magnetic properties [1, 2]" [3].

1. C. Manders, *Ann. phys.*, **5,** 1936, 181.
2. V. Marian, *Ann. phys.*, **7,** 1937, 511.
3. "Gmelins Handbuch der anorganischen Chemie," System No. 68 A (5), p. 541, Verlag Chemie, G.m.b.H., Weinheim/Bergstrasse, Germany, 1951.

0.2625
$\bar{1}.7375$
Ni-S Nickel-Sulfur

The phase diagram of Fig. 569 is based mainly on the following investigations: thermal [1, 2], microscopic [1, 3–5], roentgenographic [6, 7, 2, 5, 8], densimetric [1, 7, 5], tensimetric [7, 9], and H_2/H_2S equilibria [10, 11].

The liquidus was thermally determined by [1] up to 45 at. % S; alloys with higher S content cannot be melted at atmospheric pressure.

The boundaries of the Ni_3S_2(H) phase (26.70 wt. % S), which cannot be retained by quenching, were determined by [1, 11]. The view of [12] that no Ni_3S_2(L) phase exists was not confirmed by later investigations. Much work has been done in the region between 40 and 50 at. % S. The homogeneity range for the Ni_6S_5 [45.45 at. (31.28 wt.) % S] high-temperature phase is taken from [11], and the formula Ni_7S_6 [46.15 at. (31.90 wt.) % S] for the low-temperature phase from [5, 8]. However, as long as the crystal structures of these phases are unknown, the constitution in this region cannot be regarded as finally established.

The value 797°C given by [13] as the melting point of NiS (35.33 wt. % S) is certainly too low. NiS exists in two modifications; the transition was studied by [2] by means of a thermal-difference method (see also [14]). [7, 2] assumed a rather extended solubility of S in NiS(H) which was not corroborated by the work of [8, 11]; for Fig. 569 the data given by [11] were used.

The existence of Ni_3S_4 [57.14 at. (42.14 wt.) % S], long known as a mineral, was established by [9, 8]. From a tensimetric analysis, [7] concluded that NiS_2 (52.22 wt. % S) dissolves sulfur up to about the composition NiS_4. The results of more recent roentgenographic work [8], however, do not agree with such a wide homogeneity range (cf. Crystal Structures).

Crystal Structures. Ni_3S_2(L) is rhombohedral [15, 8, 16], with a = 4.080 A, α = 89°25′ [16]. No structural data of the high-temperature form are available.

Ni_6S_5, stable between 400 and 560°C, is orthorhombic, with a = 11.22 A, b = 16.56 A, c = 3.27 A (very probably 44 atoms per unit cell) [8]. Ni_7S_6 yields a complex powder pattern; "one cannot rule out the possibility that here we meet a degenerate pentlandide structure, and in this case, the ideal composition may be Ni_9S_8" [8].

It was shown by [7, 2] that the two long-known [17–19] modifications of NiS, a hexagonal and a rhombohedral one, have an enantiotropic relationship. For the NiAs (B8) type high-temperature form the parameters a = 3.428–3.420 A, c = 5.340–5.315 A, c/a = 1.558–1.554 (with varying S content) were given by [8]. The formation of vacant Ni positions with increasing S content had already been recognized by [7]. The low-temperature form is rhombohedral, B13 type, with the (hexagonal

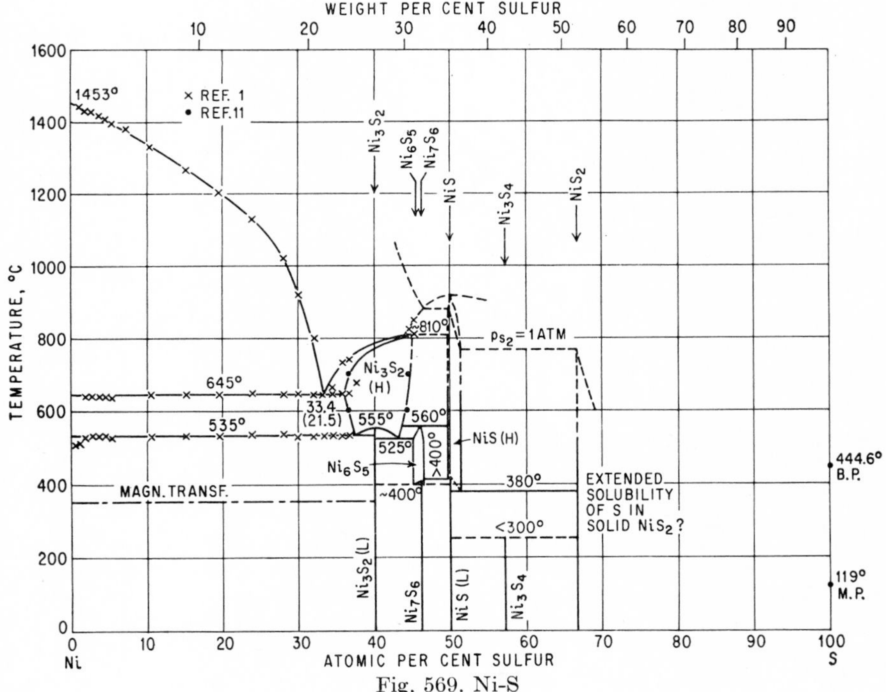

Fig. 569. Ni-S

lattice parameters [8] a = 9.596–9.587 A, c = 3.145 A (with varying S content). The homogeneity range is estimated as less than 0.1 at. % S [8]. Ni_3S_4, formerly ascribed to the spinel ($H1_1$) type [20], now is listed as Co_3S_4 ($D7_2$) type, with a = 9.457 A [8]. NiS_2 has the cubic pyrite (C2) type of structure [21, 8]; the lattice parameters given by [8] for the Ni- and S-rich boundary are a = 5.676 and 5.678 A, respectively.

Solubility of S in Solid Ni. [3] and [4] were able to identify free Ni_3S_2 in Ni microscopically when 0.05 and 0.005 wt. (0.09 and 0.01 at.) % S, respectively, were present. The lattice constant of sulfur-saturated Ni quenched from 650°C or lower was identical with that of pure Ni; when quenched from 850°C, a slight increase in the parameter was observed [8]. Metallographic methods for the identification of sulfides in Ni have been described by [22].

1. K. Bornemann, *Metallurgie*, **5,** 1908, 13–19; **7,** 1910, 667–674. The Ni powder used contained 0.26 Fe and 0.40 SiO_2 (wt. %) as main contaminations. For the sake of clarity some thermal points were omitted in Fig. 569, in particular, those belonging to a nonvariant line at 503°C between Ni_3S_2 and Ni_6S_5 for which no indication was found by later investigators, especially [11].
2. W. Biltz, F. Weibke, P. Ehrlich, and K. Meisel, *Z. anorg. Chem.*, **228,** 1936, 285–296.
3. G. Masing and L. Koch, *Z. Metallkunde,* **19,** 1927, 278.
4. P. D. Merica and R. G. Waltenberg, *Trans. AIME,* **71,** 1925, 709–716; *Natl. Bur. Standards Tech. Paper,* **281,** 1925, 155–182.
5. G. Peyronel and E. Pacilli, *Atti accad. Italia, Rend., Classe sci. fis. mat. e nat.*, **3,** 1941, 278–288.
6. R. Brill and E. Halle, *Angew. Chem.*, **48,** 1935, 787 (Existence of Ni_3S_2).
7. W. Biltz, A. Voigt, and K. Meisel, *Z. anorg. Chem.*, **228,** 1936, 275–285.
8. D. Lundqvist, *Arkiv Kemi, Mineral. Geol.*, **24A**(21), 1947, 1–12; **24A**(23), 1947, 1–7.
9. Y. I. Gerasimov, N. I. Pirtskhalov, and V. V. Stepin, *J. Gen. Chem. (U.S.S.R.)*, **6,** 1936, 1736–1743, quoted from [8].
10. R. Schenck and P. von der Forst, *Z. anorg. Chem.*, **241,** 1939, 145–157. See also R. Schenck and E. Raub, *Z. anorg. Chem.*, **178,** 1929, 234.
11. T. Rosenqvist, *J. Iron Steel Inst.*, **176,** 1954, 37–57.
12. W. Guertler and W. Savelsberg, *Metall u. Erz,* **29,** 1932, 84–90.
13. W. Biltz, *Z. anorg. Chem.*, **59,** 1908, 280.
14. K. Shimomura, *J. Sci. Hiroshima Univ.*, **A16,** 1952, 319–323; *Chem. Abstr.*, **47,** 1953, 9704.
15. A. Westgren, *Z. anorg. Chem.*, **239,** 1938, 82–84.
16. M. A. Peacock, *Univ. Toronto Studies, Geol. Ser.*, no. 51, 1947, pp. 59–69; *Am. Mineralogist,* **32,** 1947, 484. Abstract in *Structure Repts.*, **11,** 1947-1948, 299.
17. N. Alsén, *Geol. Fören. i Stockholm Förh.*, **47,** 1925, 19–72. H. W. V. Willems, *Physica,* **7,** 1927, 203–207.
18. N. H. Kolkmeijer and A. L. T. Moesveld, *Z. Krist.*, **80,** 1931, 91–102.
19. G. R. Levi and A. Baroni, *Z. Krist.*, **92,** 1935, 210–215.
20. G. Menzer, *Z. Krist.*, **64,** 1926, 506–507. W. F. de Jong, *Z. anorg. Chem.*, **161,** 1927, 311–315.
21. W. F. de Jong and H. W. V. Willems, *Z. anorg. Chem.*, **160,** 1927, 185–189.
22. A. M. Hall, *Trans. AIME,* **152,** 1943, 278–284.

$\bar{1}.6831$
0.3169

Ni-Sb Nickel-Antimony

Early Papers. The phase diagram of [1], established by means of thermal and microscopic work, showed the existence of Ni_4Sb, Ni_5Sb_2, NiSb, and of an additional Sb-rich phase. NiSb was also found by [2] in his work on preparation, but [3] suggested from emf measurements Ni_3Sb rather than Ni_4Sb as the formula for the compound richest in Ni.

Phase Diagram. More recently the phase diagram has been carefully reinvestigated, especially by the thermal, electrical, dilatometric, and microscopic work of [4].

The liquidus was determined thermally by [1, 4], the results being in fair agreement. For Fig. 570 the data of [4] were used. The limits of the (Ni) primary solid solution shown in Fig. 570 were established by microscopic work of [4]; the solubility at the eutectic temperature was found to be 16 wt. (8.4 at.) % Sb, with no measurable change down to room temperature. The superlattice phase $Ni_{15}Sb$ (12.15 wt. % Sb) was discovered roentgenographically [5] and confirmed by thermomagnetic [6], dilatometric, and electrical measurements [7].

The constitution shown in Fig. 570 for alloys with 20–30 at. % Sb, involving the phases Ni_3Sb (δ) (40.88 wt. % Sb), Ni_5Sb_2 (β) (45.34 wt. % Sb), and Ni_7Sb_3 (θ) (47.06 wt. % Sb), was taken from the paper of [4]. For δ and θ, also, the formulas $Ni_{13}Sb_4$ [8] and Ni_9Sb_4 [5], respectively, were suggested. Quenching of β yields martensitic microstructures (β') [4]. The (NiSb) phase (67.48 wt. % Sb) has been studied roentgenographically (homogeneity range, 47.7–52.7 at. % Sb) [5], microscopically (43–53 at. % at 1070°C, 46.2–52.3 at. % Sb at "room temperature") [4], and by several standard methods (46.4–54.4 at. % Sb) by [9].

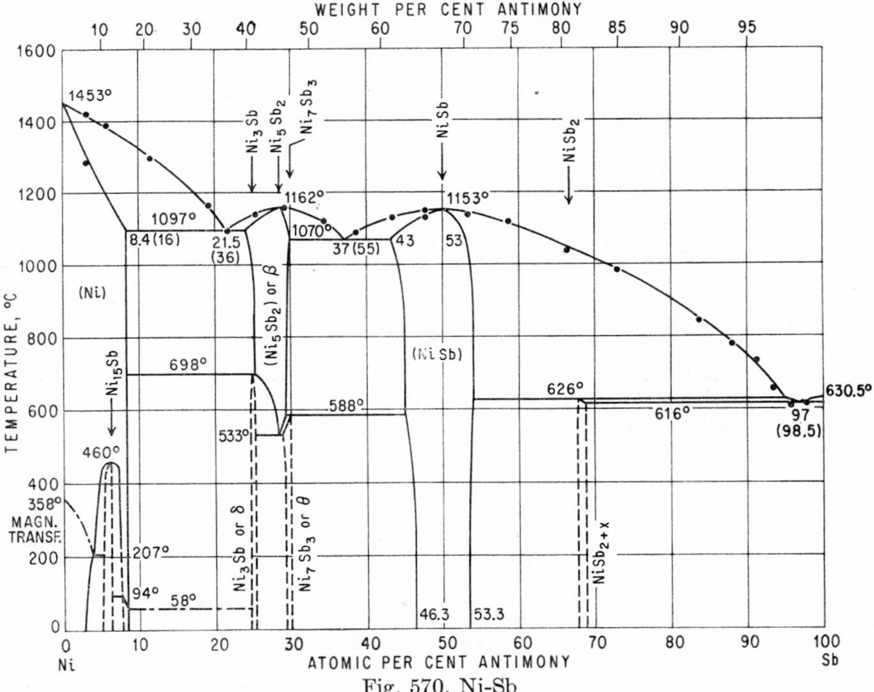

Fig. 570. Ni-Sb

As to the intermediate phase richest in Sb, the formulas Ni_4Sb_5 [1], $NiSb_2$ [8], Ni_5Sb_{11} [5], Ni_2Sb_5 [4], and $NiSb_{2+x}$ (x = 0.1–0.2) [10] (Fig. 570) have been named. Its crystal structure (see below) points to the composition AB_2, but the structure probably has vacant Ni positions [10].

The solubility of Ni in solid Sb was found microscopically [4] to be not more than 0.05 wt. (0.1 at.) % Ni at the eutectic temperature.

Crystal Structures. The lattice constant of the (Ni) phase increases with the concentration of Sb [5]. The cubic superstructure cell of $Ni_{15}Sb$ contains 32 atoms, and the parameter varies with composition from 7.062 to 7.121 kX [5].

As to the δ phase, [8] indexed the "$Ni_{13}Sb_4$" pattern cubic and [5] the "Ni_3Sb" pattern hexagonal (64 atoms per unit cell). For the θ phase (Ni_7Sb_3 in Fig. 570), [5] gave lattice parameters of a tetragonal cell (a = 8.095 kX, c = 11.389 kX, c/a = 1.407) containing 52 atoms ("Ni_9Sb_4"). Tetragonal symmetry was also found for the

(Ni_5Sb_2) high-temperature phase [5, 11]. [8] annealed alloys of the composition Ni_2Sb above and below 580°C; the powder patterns of the quenched alloys were identical and could be indexed by a tetragonal cell, with a = 5.78 kX, c = 6.00 kX, c/a = 1.036.

(NiSb) was shown to have the NiAs (B8) type of structure [12, 8, 9], with a = 3.942 A, c = 5.155 A, c/a = 1.308 (50 at. % Sb) [13]. The Sb-richest compound $(NiSb_{2+x}$ in Fig. 570) has the marcasite (C18) structure [8, 10], probably with vacant Ni positions [10], and the parameters a = 5.180 A, b = 6.314 A, c = 3.838 A [10].

1. K. Lossew, *Z. anorg. Chem.*, **49,** 1906, 58–71.
2. E. Vigouroux, *Compt. rend.*, **147,** 1908, 976–978.
3. N. A. Puschin, *Zhur. Russ. Fiz.-Khim. Obshchestva*, **39,** 1907, 528–566. Abstract in *Chem. Zentr.*, **1907**(2), p. 2028.
4. N. Shibata, *Sci. Rept. Tôhoku Univ.*, **29,** 1941, 697–727 (in English); also *Nippon Kinzoku Gakkai-Shi*, **3,** 1939, 237–249.
5. A. Osawa and N. Shibata, *Nippon Kinzoku Gakkai-Shi*, **4,** 1940, 362–368; *Met. Abstr.*, **8,** 1941, 56.
6. The Curie-temperature-vs.-composition curve of V. Marian, *Ann. phys.*, **7,** 1937, 459–527 (especially 499–502), which is nearly a straight line, was obviously measured on alloys not in a state of equilibrium. As for the Curie temperature of pure Ni, the value 358°C measured by Marian was used in Fig. 570 instead of the value ~380°C found by [7].
7. N. Shibata, *Nippon Kinzoku Gakkai-Shi*, **5,** 1941, 41–45.
8. U. Fürst and F. Halla, *Z. physik. Chem.*, **B40,** 1938, 285–307.
9. N. V. Ageev and E. S. Makarov, *Izvest. Akad. Nauk S.S.S.R., Otdel. Khim. Nauk*, **1943,** 87–98; E. S. Makarov, *Izvest. Sektora Fiz.-Khim. Anal.*, **16,** 1943, 149–157.
10. T. Rosenqvist, *Acta Met.*, **1,** 1953, 761–763.
11. The unit cells proposed by [5] should be looked upon with great reservation, since even established structures (NiSb and $\sim NiSb_2$) were incorrectly indexed.
12. W. F. de Jong, *Physica*, **5,** 1925, 241–243; N. Alsén, *Geol. Fören. i Stockholm Förh.*, **47,** 1925, 19; W.F. de Jong and H. W. V. Willems, *Physica*, **7,** 1927, 74–79; I. Oftedal, *Z. physik. Chem.*, **128,** 1927, 135–158.
13. D. F. Hewitt, *Econ. Geol.*, **43,** 1948, 408–417; *Structure Repts.*, **11,** 1947-1948, 32.

$\bar{1}.8712$
0.1288

Ni-Se Nickel-Selenium

The compound NiSe (57.36 wt. % Se) has long been known; see the chemical literature. From his work on preparation, [1] suggested the existence of the additional compounds Ni_2Se, Ni_2Se_3 or Ni_3Se_4, and $NiSe_2$; however, only $NiSe_2$ (72.90 wt. % Se) has been corroborated as yet.

Two crystallized modifications of NiSe were observed by [2] (cf. Ni-S); one has—as shown before by [3]—the NiAs (B8) type of structure, with a = 3.67 A, c = 5.34 A, c/a = 1.46 [3, 2], and the other the rhombohedral millerite (B13) type of structure, with the (hexagonal) lattice parameters a = 9.86 A, c = 3.19 A, c/a = 0.32. $NiSe_2$ was shown [4, 5] to have the pyrite (C2) type of structure, with a = 5.960 A [5].

The one-phase field between the compositions NiSe and $NiSe_2$ as shown in the partial phase diagram of [6] is not based on experimental investigations and is certainly incorrect (cf. the crystal structures).

1. H. Fonces-Diacon, *Compt. rend.*, **131,** 1900, 556–558.
2. G. R. Levi and A. Baroni, *Z. Krist.*, **92,** 1935, 210–215.

3. N. Alsén, *Geol. Fören. i Stockholm Förh.*, **47**, 1925, 19–72. *Strukturbericht*, **1**, 1913-1928, 138.
4. W. F. de Jong and H. W. V. Willems, *Z. anorg. Chem.*, **170**, 1928, 241–245.
5. S. Tengnér, *Z. anorg. Chem.*, **239**, 1938, 126–132.
6. E. S. Makarov, *Izvest. Akad. Nauk S.S.S.R., Otdel. Khim. Nauk*, (6)**1945**, 569–580.

0.3200
$\bar{1}.6800$

Ni-Si Nickel-Silicon

Early Work. From work on preparation and analyses of residues, the existence of the phases Ni_4Si [1], Ni_2Si [2, 1, 3], and NiSi [3] has been suggested, and from density measurements [3] that of Ni_2Si, Ni_3Si_2, and NiSi. Essential features of the constitution of this system were revealed by the thermal and microscopic work of [4]; the interpretation of their data was later modified by [5].

Liquidus. Thermal analyses were carried out between 0 and 100% Si by [4, 6] and between 0 and 20 wt. % Si by [7, 8]. In Fig. 571 the liquidus of [8] was used in the region 0–20 wt. % Si and that of [6] in the region 20–100 wt. % Si. For a comparison of the results obtained by various investigators, see Table 36 and [9].

Table 36. Characteristic Data of the System Ni-Si (Temperatures in °C, Compositions in Wt. % Si)

	[6]	[7]	[8]
Eutectic (Ni) + β_3	1151°C, 11.5 (21.4 at. % Si)	1156°C, 11.5	1152°C, 11.5
Melting point of Ni_5Si_2 (γ)	1255°C	1266°C	1282°C
Eutectic (γ + Ni_2Si)	1242°C, 17.5 (30.7 at. % Si)	~1250°C, 17.4	1265°C, 17.5
Melting point of Ni_2Si	1285°C	1306°C	1318°C

Intermediate Phases and Their Crystal Structures. The constitution of the solid alloys shown in the figure is mainly based on thermal [6], microscopic [6, 10], X-ray powder [6, 11, 10], and dilatometric [10] work.

An intermediate phase of the approximate composition Ni_3Si (13.76 wt. % Si) had already been found by [4]. Several authors [6, 7, 8] assumed this phase to be stable only above 1100°C [12], but it was clearly shown by [11, 13, 10, 14] that Ni_3Si exists down to room temperature. According to [10], Ni_3Si undergoes two transformations at ~1120 and 1040°C; the three modifications are designated β_3, β_2, and β_1 in the figure. The composition of these phases alters slightly with temperature [10], resulting in eutectoid-like microstructures. According to [11, 13, 10, 14], β_1 has the cubic Cu_3Au ($L1_2$) type of structure, with $a = 3.507 \pm 5$ A [14].

Ni_5Si_2 [28.57 at. (16.07 wt.) % Si], named γ by [6], has, as [10] showed microscopically, also a small range of homogeneity. [11] indexed the powder pattern tentatively with an orthohexagonal cell containing 91 atoms.

The phase boundaries of the δ and θ phases, located at or near the composition Ni_2Si (19.31 wt. % Si), were established by [6, 11]. Based on work of [11] on powder photographs, [15] derived the structures of both δ and θ by means of measurements on monocrystals: δ or Ni_2Si(L), is orthorhombic, with $a = 7.06$ A, $b = 4.99$ A, $c = 3.72$ A, and 12 atoms per unit cell. θ, or (Ni_2Si)(H), is hexagonal, with $a = 3.805$ A, $c = 4.890$ A, $c/a = 1.285$ [for a 24 wt. (39.8 at.) % Si alloy quenched from 964°C], and 6 atoms per unit cell. These parameters differ only slightly from those found by [11, 16].

Ni_3Si_2 (24.19 wt. % Si) is formed by a peritectoid reaction [4, 6]. The powder pattern was indexed by [11] with an orthohexagonal cell containing 45 atoms. [17] reported NiSi (32.37 wt. % Si) to have the cubic FeSi structure. However, [11] and

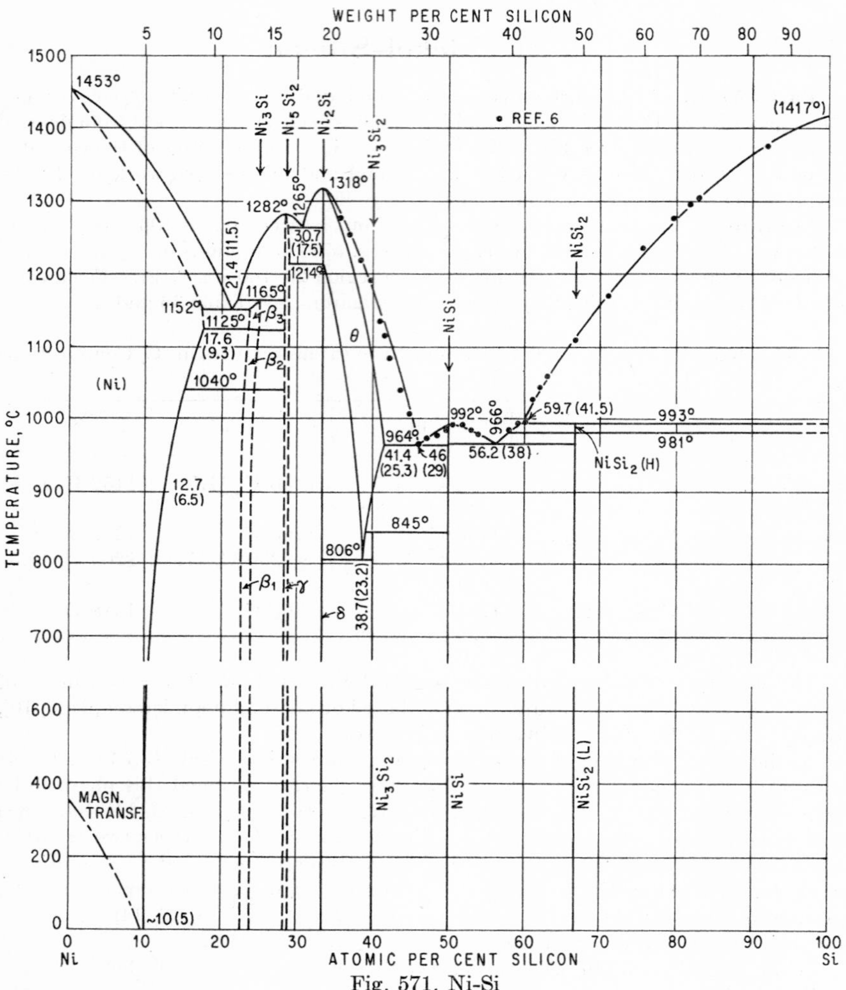

Fig. 571. Ni-Si

[18] showed this compound to be noncubic, and [19], in measurements on monocrystals, found an orthorhombic structure, with $a = 5.62$ A, $b = 5.18$ A, $c = 3.34$ A, and 8 atoms per unit cell [pseudohexagonal, related to B8 type [19]; MnP (B31) type according to [19a]].

At normal temperatures $NiSi_2$ (48.90 wt. % Si) has the CaF_2 (C1) type of structure, with $a = 5.406 \pm 3$ A [16, 18]. According to [11], it is doubtful whether the high-temperature modification of $NiSi_2$ (stable above 981°C [6], cf. [20]) can be quenched. [18] suggested an order-disorder transformation.

The results of a microscopic study of alloys with 38–98 wt. % Si [21] are in keeping with the constitution shown in the figure.

The heat-of-formation–composition curve shows two marked points of inflection corresponding to the compositions Ni_2Si and NiSi [22].

Primary Solid Solutions. (Ni). Data as to the solubility of Si in solid Ni were reported by [5, 23–25, 6, 26, 11, 10, 14]. The values found for "room temperature," by means of microscopic, X-ray, and magnetic studies mainly, lie close to 5 wt. (10 at.) % Si. At higher temperatures, however, the data points are badly scattered; in the temperature range 1100–1150°C, values between 6.5 and 9.3 wt. (12.7 and 17.6 at.) % Si were reported. For Fig. 571 the results of a microscopic, X-ray, and dilatometric study [10] were used [27], indicating a maximal solubility of 9.3 wt. (17.6 at.) % Si at 1120°C and a solubility of 6.5 wt. (12.7 at.) % Si at 900°C.

Lattice parameters of the (Ni) phase were measured by [11], Curie temperatures by [28, 23, 6, 26]; the results of [28, 6, 26] are in good agreement.

(Si). [17] observed a contraction, [11] an expansion of the Si lattice by alloying of Ni. The latter concluded a solubility of Ni in solid Si of 1–2 wt. % at higher temperatures. [18] found that the change in the lattice parameter of Si by alloying Ni does not exceed 0.001 A.

Supplement. [29] stated: "If Toman's [15] description of δ-Ni_2Si is changed from Pbnm to Pmnb, and if the origin is chosen differently, this compound belongs to the C23 [$PbCl_2$] type."

1. E. Vigouroux, *Compt. rend.*, **142,** 1905, 1270–1271.
2. E. Vigouroux, *Compt. rend.*, **121,** 1895, 686.
3. R. Frilley, *Rev. mét.*, **8,** 1911, 484–491.
4. W. Guertler and G. Tammann, *Z. anorg. Chem.*, **49,** 1906, 93–112. The nickel and the silicon used were rather impure.
5. W. Guertler, "Handbuch der Metallographie," vol. I, part 2, pp. 676–689, Verlagsbuchhandlung Gebrüder Borntraeger, Berlin, 1917.
6. K. Iwasé and M. Okamoto, *Science Repts. Tôhoku Imp. Univ., K. Honda Anniv. Vol.*, 1936, pp. 777–792 (in English); also *Tetsu-to-Hagane*, **22,** 1936, 869–875 (in Japanese). The alloys were melted from the elements (Ni 99.9, Si 98.7 wt. % pure) in porcelain tubes under hydrogen.
7. A. C. Forsyth and R. L. Dowdell, *Trans. AIME*, **137,** 1940, 373–387.
8. K. Ruttewit and G. Masing, *Z. Metallkunde*, **32,** 1940, 60.
9. In the region 13–20 wt. % Si the liquidus of [8] lies appreciably higher than that of [6]. According to [8] this is due to the impure Si used by [6]. If this assumption is correct, the liquidus above 20 wt. % Si would lie at higher temperatures than shown in Fig. 571.
10. M. Okamoto, *Nippon Kinzoku Gakkai-Shi*, **2,** 1938, 544–551 (in Japanese; tables and illustrations in English), not available to the author; *Met. Abstr.*, **6,** 1939, 6.
11. A. Osawa and M. Okamoto, *Nippon Kinzoku Gakkai-Shi*, **2,** 1938, 378–388; also *Science Repts. Tôhoku Imp. Univ.*, **27,** 1939, 326–347 (in English).
12. Eutectoidal decomposition at about 1125°C.
13. K. Kusumoto, *Nippon Kinzoku Gakkai-Shi*, **2,** 1938, 617–619 (in Japanese; table and illustrations in English); *Chem. Abstr.*, **33,** 1939, 3244.
14. N. F. Lashko, *Doklady Akad. Nauk S.S.S.R.*, **81,** 1951, 605–607 (in Russian); *Chem. Abstr.*, **46,** 1952, 10080.
15. K. Toman, *Acta Cryst.*, **5,** 1952, 329–331.
16. K. Schubert and H. Pfisterer, *Naturwissenschaften*, **37,** 1950, 112–113.

17. B. Borén, *Arkiv Kemi, Mineral. Geol.*, **11A**, 1933, 22–23.
18. K. Schubert and H. Pfisterer, *Z. Metallkunde*, **41**, 1950, 438.
19. K. Toman, *Acta Cryst.*, **4**, 1951, 462–464.
19a. K. Schubert, private communication (1954).
20. Guertler and Tammann [4] found thermal effects at about 950°C.
21. W. R. Johnson and M. Hansen, AF Technical Report 6383, 1951 (work at Armour Research Foundation, Chicago, Ill.).
22. W. Oelsen and H. O. v. Samson-Himmelstjerna, *Mitt. Kaiser-Wilhelm Inst. Eisenforsch. Düsseldorf*, **18**, 1936, 131–133.
23. O. Dahl and N. Schwartz, *Metallwirtschaft*, **11**, 1932, 277–279.
24. O. Dahl, *Z. Metallkunde*, **24**, 1932, 277–281.
25. B. Blumenthal and M. Hansen, unpublished X-ray work on long-annealed alloys (1932) (1100°C, 7.5 wt. % Si; 800°C, 6.1 wt. % Si).
26. V. Marian, *Ann. phys.*, **7**, 1937, 459–527 (Magnetic measurements).
27. Since only an abstract of this paper was available, a critical examination of the data could not be made.
28. A. Kussmann and B. Scharnow, *Z. Metallkunde*, **23**, 1931, 216.
29. F. Laves, in C. J. Smithells, "Metals Reference Book," p. 221 (footnote), Interscience Publishers, Inc., New York, 1955.

$\bar{1}.6941$
0.3059

Ni-Sn Nickel-Tin

The phases and phase relationships have been studied by means of thermal [1–5], microscopic [6, 2–5], chemical [7], and roentgenographic [8, 9, 4, 5, 10] methods. Results published prior to 1943 were very conflicting. A detailed discussion of the earlier findings and discrepancies appears to be unnecessary, since the literature has been reviewed in a number of publications [11, 4, 5, 10].

It had been recognized early that the three intermediate phases Ni_3Sn (40.27 wt. % Sn) [1, 6, 7, 2, 3], Ni_3Sn_2 (57.42 wt. % Sn) [7, 2, 3], and NiSn (66.91 wt. % Sn) [7, 2] exist. In addition, [3] had claimed the existence of Ni_4Sn (33.58 wt. % Sn), on the basis of inconclusive evidence, however. The latter phase was also assumed by [4], but [5] definitely established by thermal and micrographic studies that it does not occur. Also, the phase NiSn [8] was replaced by a phase richer in Sn, Ni_3Sn_4 (72.95 wt. % Sn) [4, 5, 12, 10].

In a preliminary note, [9] reported that besides the phase later identified as Ni_3Sn_4 there were two phases with very narrow homogeneity ranges at approximately 51 and 54 at. % Sn. The former alloy was reported by [10] to consist of Ni_3Sn_2 and another phase with about 54 at. % Sn, possibly different from Ni_3Sn_4. However, [5] did not find evidence for any phase in this range besides Ni_3Sn_4. (See Note Added in Proof.)

The existence of two miscibility gaps in the liquid state in the composition ranges of about 38–60 and 70–92 at. % Sn had been assumed [3, 4] on the basis of thermal-analysis data only; no evidence for layer formation was found. [5, 10] have definitely established that no liquid immiscibility exists.

The interpretation of the transformation in the solid state in the range of approximately 10–40 at. % Sn differs widely [2–4]. It now appears to be well established that the thermal effects found in the temperature range 760–925°C [2–5] are caused by a transformation of the phase Ni_3Sn [5] rather than by the eutectoid decomposition [3, 4] and peritectoid formation [3] of intermediate phases other than Ni_3Sn.

As to the phase diagram in Fig. 572, largely based on data reported by [5, 10, 13, 4], the following remarks may be made.

The solid solubility of Sn in Ni between 1100 and 500°C was determined using the

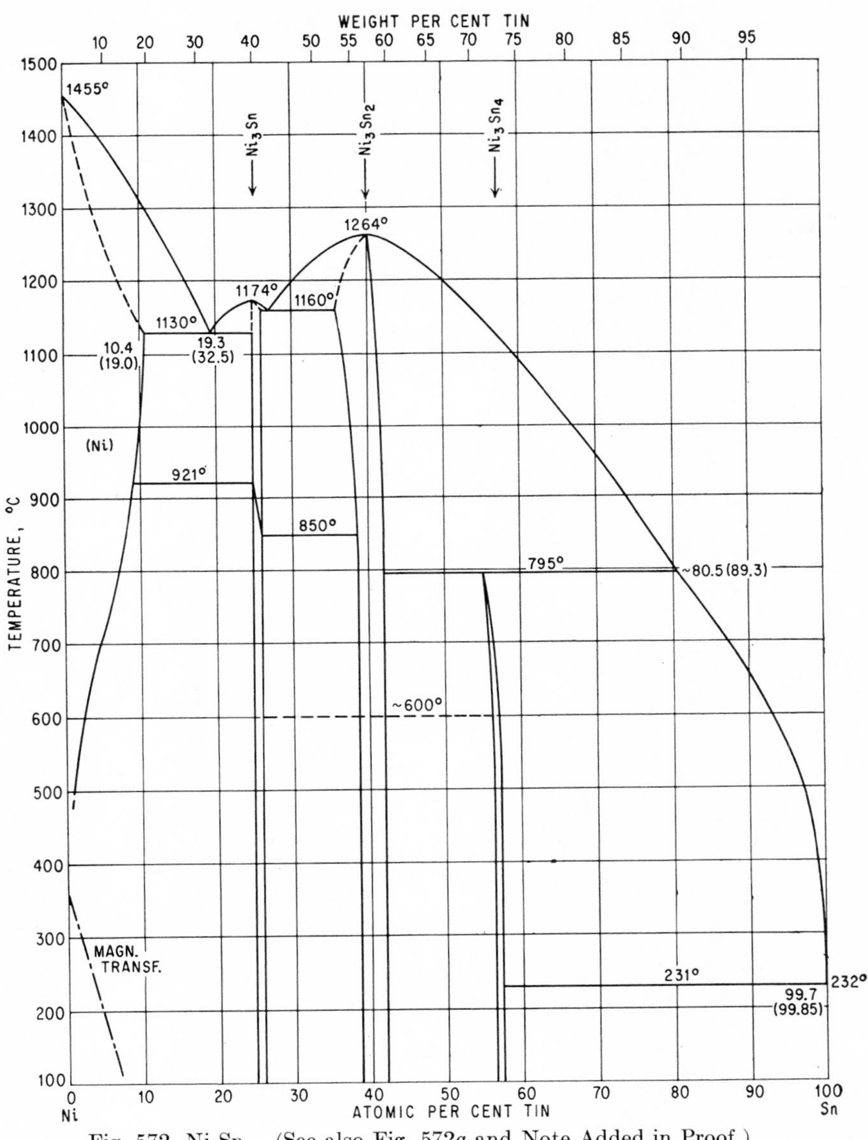

Fig. 572. Ni-Sn. (See also Fig. 572*a* and Note Added in Proof.)

lattice-parametric method [13, 4]. In the interval 1100–800°C, solubilities differ by 0.6–1.3 at. % Sn, those by [13] being higher, at 700°C and below by only 0.2–0.3 at. % Sn, those by [4] being higher. As both works appear to be fully equivalent, the solubility boundary in Fig. 572 was obtained by graphical interpolation of these findings showing a solubility in at. % Sn (wt. % in parentheses) of 10.4 (19.0), 9.9 (18.2), 8.7 (16.1), 6.9 (13.0), 4.7 (9.1), 2.4 (4.7), and 1.0 (1.9) at 1100, 1000, 900, 800, 700, 600, and 500°C, respectively.

Whereas earlier work [2–4] had indicated that Ni_3Sn is formed by a peritectic reaction, [5] showed that it melts without decomposition. The homogeneity range of this phase was reported as 24.8–26.0 at. (40.0–41.5 wt.) % Sn [5]. It undergoes a transformation at 921–850°C [5] which is probably of the order-disorder type [5, 10, 14]; see below.

The boundaries of the Ni_3Sn_2 phase were outlined micrographically [5] and by lattice-parameter measurements [10]. According to [5], it is stable between 36 and about 40.5 at. (53.2 and 58 wt.) % Sn at 1160°C and between 38.6 and 42.1 at. (56 and 59.5 wt.) % Sn at 500°C. [10] reported 38.4 and 42.5 at. (55.8 and 59.9 wt.) % Sn at 1000°C and 39.4–42.9 at. (56.8–60.3 wt.) % Sn at 600°C.

The phase Ni_3Sn_4, detected by [4], changes its composition with temperature [5]. It contains 55.5 at. (71.1 wt.) % Sn at 700–795°C and about 56.3–57.2 at. (72.2–73 wt.) % Sn at 500°C [5]. Results by [10] agree on the whole. [5] found slightly different powder patterns when quenched from temperatures above and below 600°C and concluded that a transformation occurs at about this temperature. No thermal

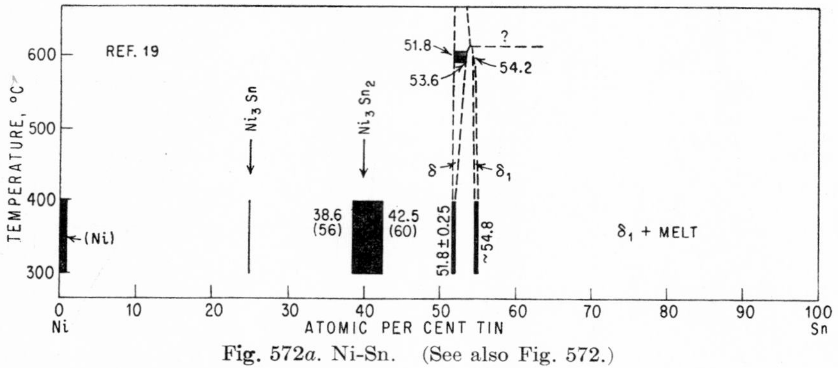

Fig. 572*a*. Ni-Sn. (See also Fig. 572.)

effect could be observed, however. Neither [10] nor [12] were able to verify this transformation. (See Note Added in Proof.)

It is generally agreed [4, 9, 10] that Ni is practically insoluble in Sn. According to [15], there is a eutectic at 99.7 at. (99.85 wt.) % Sn and a temperature 0.8°C below the melting point of Sn. By means of determinations of the solubility of Ni in liquid Sn, [16] determined the liquidus curve between 240 and 550°C and concluded the eutectic to lie at 99.65 at. (99.82 wt.) % Sn and practically the melting point of Sn. According to [4], the eutectic coincides with 100% Sn.

Curie-point measurements on more or less supersaturated Ni-rich alloys were carried out by [17, 18]. The curve shown in Fig. 572 is due to [18] and was measured on alloys quenched from 1000°C. The boundary of the supersaturated primary solid solution was reached at 10.5 at. % Sn (and −32°C).

Crystal Structures. Lattice parameters of the (Ni) phase were reported by [13, 4]. The low-temperature form of Ni_3Sn has the ordered h.c.p. structure of the Mg_3Cd ($D0_{19}$) type, with a = 5.286 A, c = 4.242 A, c/a = 0.803 [14]. Parameters vary from a = 5.293 A, c = 4.243 A, c/a = 0.802 at the Ni-rich side to a = 5.309 A, c = 4.255 A, c/a = 0.802 at the Sn-rich side [10]. The high-temperature form is probably of the h.c.p. Mg (A3) type.

Ni_3Sn_2 is hexagonal of the NiAs (B8) type, partially filled [8, 4, 10]. Lattice dimensions decrease from a = 4.145 A, c = 5.213 A, c/a = 1.258 (saturated with Ni) to a = 4.048 A, c = 5.123 A, c/a = 1.266 (saturated with Sn) [10]. There is indication for the formation of a superlattice of the hexagonal structure below 600°C, with a = 16.47 A, c = 5.188 A [10].

Ni_3Sn_4 is monoclinic, with 14 atoms per unit cell, similar to the CoSn (B35) type of structure, and not polymorphous [12], as suggested by [5]. At higher temperatures, the narrow homogeneity range shifts to higher Ni contents: $a = 12.22_4$ A, $b = 4.06_4$ A, $c = 5.22_5$ A, $\beta = 105°3'$ at 57.3 at. % Sn; $a = 12.31_5$ A, $b = 4.06_4$ A, $c = 5.18_0$ A, $\beta = 103°48'$ at 54.8 at. % Sn [12].

Note Added in Proof. [19] studied the constitution of 39 alloys prepared by thermal decomposition of Ni-Sn amalgams by means of X-ray diffraction. The phases and phase boundaries found (corresponding to equilibrium temperatures of 300–400°C) are shown in Fig. 572*a*. In contrast to Fig. 572, two intermediate phases, called δ and δ_1, were found to occur in the composition range 50–60 at. % Sn. δ is identical with the monoclinic "Ni_3Sn_4" phase of [12]. The X-ray pattern of the δ_1 phase is similar to, but not identical with, that of the δ phase. Heat-treatments at 600°C showed that the ranges of homogeneity of both the δ and δ_1 phases vary with temperature (Fig. 572*a*). The temperature of the (tentatively assumed) peritectic formation of δ_1 was not determined.

1. H. Gautier, *Bull. soc. encour. ind. natl.*, **1,** 1896, 1313; Société d'encouragement pour l'industrie nationale, Paris, Commission des alliages, "Contribution à l'étude des alliages," pp. 112–113, Typ. Chamerot et Renouard, Paris, 1901. Liquidus curve only.
2. L. Guillet, *Rev. mét.*, **4,** 1907, 531–551; *Compt. rend.*, **144,** 1907, 752–753.
3. G. Voss, *Z. anorg. Chem.*, **57,** 1908, 35–45.
4. W. Mikulas, L. Thomassen, and C. Upthegrove, *Trans. AIME*, **124,** 1937, 111–133.
5. T. Heumann, *Z. Metallkunde*, **35,** 1943, 206–211.
6. G. Charpy, *Bull. soc. encour. ind. natl.*, **2,** 1897, 384.
7. E. Vigouroux, *Compt. rend.*, **144,** 1907, 639–641, 712–714, 1351–1353; **145,** 1907, 246–248, 429–431.
8. I. Oftedal, *Z. physik. Chem.*, **132,** 1928, 208–216.
9. E. Fetz and E. R. Jette, *J. Chem. Phys.*, **4,** 1936, 537; see also *Trans. AIME*, **124,** 1937, 133–136.
10. O. Nial, Dissertation, University of Stockholm, 1945; *Svensk Kem. Tidskr.*, **59,** 1947, 172–183.
11. M. Hansen, "Der Aufbau der Zweistofflegierungen," pp. 951–957, Springer-Verlag OHG, Berlin, 1936.
12. H. Nowotny and K. Schubert, *Naturwissenschaften*, **32,** 1944, 76; *Z. Metallkunde*, **37,** 1946, 23–31.
13. E. R. Jette and E. Fetz, *Metallwirtschaft*, **14,** 1935, 165–168.
14. P. Rahlfs, *Metallwirtschaft*, **16,** 1937, 343–345.
15. C. T. Heycock and F. H. Neville, *J. Chem. Soc.*, **57,** 1890, 378.
16. D. Hanson, E. S. Sandford, and H. Stevens, *J. Inst. Metals*, **55,** 1934, 117–119.
17. K. Honda, *Ann. Physik*, **32,** 1910, 1011–1015.
18. V. Marian, *Ann. phys.*, **7,** 1937, 459–527; especially 494–497.
19. F. Lihl and H. Kirnbauer, *Monatsh. Chem.*, **86,** 1955, 745–751; see also F. Lihl, *Z. Metallkunde*, **46,** 1955, 438–439.

$\bar{1}.5112$
0.4888

Ni-Ta Nickel-Tantalum

The following systematic analyses of the Ni-Ta system have been made: thermal and microscopic, up to 80 wt. % Ta [1]; thermal, X-ray, and microscopic, up to 90 wt.

% Ta [2]; thermal and microscopic (also hardness and electrical resistivity), up to 59 wt. % Ta [3].

Liquidus. Up to about 30 at. % Ta the thermal results of [1–3] are in fair agreement, and—with the exception of the melting point of $TaNi_3$ [4]—an interpolated curve was drawn in Fig. 573. It should be mentioned, however, that whereas [2, 3] found a eutectic between Ni and $TaNi_3$ [38 wt. (16.6 at.) % Ta, 1365°C [2]; 37.5 wt. (16.3 at.) % Ta, 1360°C [3]], [1] assumed a minimum in both the liquidus and the solidus at about 35 wt. (14.9 at.) % Ta. Although neither [2] nor [3] gave conclusive evidence for the assumed eutectic, the course of the (Ni) solvus line is in favor of this concept.

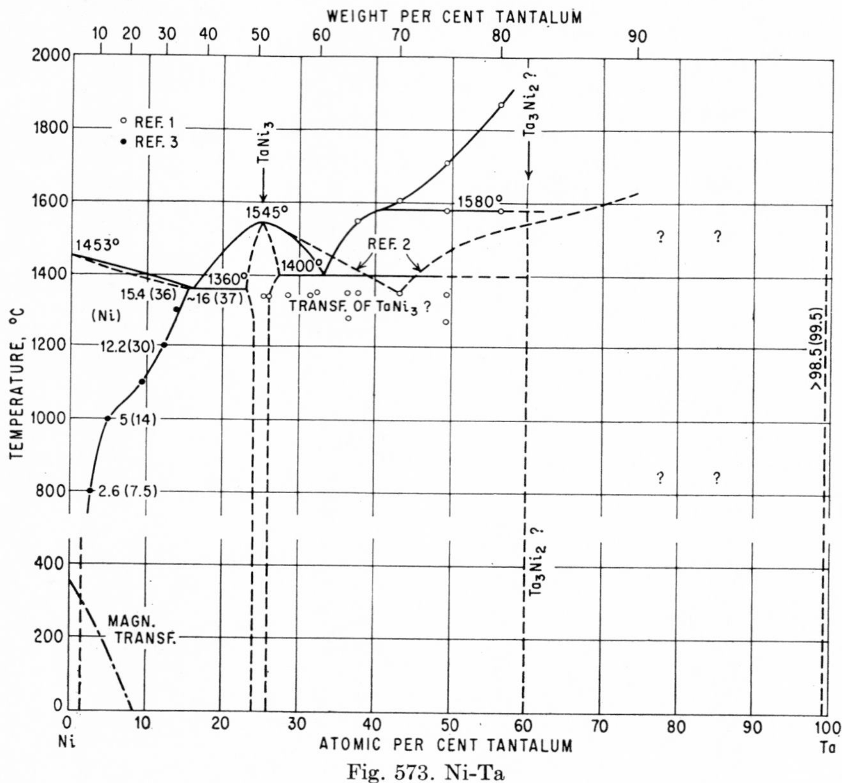

Fig. 573. Ni-Ta

Above 30 at. % Ta the thermal results of [1] and [2] deviate greatly, as can best be seen from the positions found for a second eutectic—60.5 wt. (33.2 at.) % Ta, 1400°C [1]; ~70 wt. (43 at.) % Ta, 1350°C [2]. In Fig. 573 the solid-line liquidus is that of [1] and the dashed-line liquidus that of [2].

Intermediate Phases. $TaNi_3$ (50.67 wt. % Ta) possesses a certain range of homogeneity [1–3, 5] which, however, has not yet been exactly determined. To a second, peritectically formed, compound the formulas TaNi (75.50 wt. % Ta) [1] and Ta_3Ni_2 (82.21 wt. % Ta) [2] (Fig. 573) have been assigned. For the peritectic temperature, the values 1580 and 1525°C were reported by [1] and [2], respectively.

The region between $TaNi_3$ and Ta_3Ni_2 (?) needs more careful constitutional work. Thermal effects observed by [1] in the solid state, especially at 1350°C (see Fig. 573), could be due to a transformation in the ($TaNi_3$) phase or to the peritectoid formation of an additional compound with a composition between TaNi and Ta_3Ni_2. A transformation in the $TaNi_3$ phase at 1350°C was reported by [6] (X-ray work) but denied by [2] (thermal work). The fact that a 62 wt. % Ta alloy (unannealed) is three-phase [1] points to the existence of an additional compound.

"According to unpublished results of [7], between 95 and 100 wt. % Ta and below 1600°C the alloys consist of the Ta-rich solid solution containing less than 0.5 wt. % Ni, and an intermediate phase" [8].

The solid solubility of Ta in Ni as a function of temperature was microscopically determined by [3]. The following solubilities, in wt. %, were found (at. % and annealing periods in parentheses): ~36 (15.4), 33 (13.8, 4 hr), 30 (12.2, 6 hr), 24 (9.3, 48 hr), 14 (5.0, 100 hr), and 7.5 (2.6, 250 hr) at 1360, 1300, 1200, 1100, 1000, and 800°C, respectively (Fig. 573). The value 30 wt. % Ta, 1000°C given by [2] obviously corresponds to a state of supersaturation.

Crystal Structures. The lattice parameter of the (Ni) phase increases with the Ta content [2]. According to [9], the "modification of $TaNi_3$ stable below 1350°C" has a deformed h.c.p. structure with ordered atomic arrangement ($TiCu_3$ type). [10] then showed that the unit cell is orthorhombic (slightly deformed h.c.p.), with $a = 5.114$ A, $b = 4.250$ A, $c = 4.542$ A (25 at. % Ta).

According to [11], no σ phase exists in the Ni-Ta system.

The Curie temperature vs. concentration curve in Fig. 573 was interpolated from data given by [1, 2]. It represents the magnetic-transformation temperatures of the supersaturated solid solution.

1. E. Therkelsen, *Metals & Alloys*, **4,** 1933, 105–108; *Metal Ind.* (*London*), **43,** 1933, 175–178.
2. O. Kubaschewski and H. Speidel, *J. Inst. Metals*, **75,** 1949, 417–430.
3. I. I. Kornilov and E. N. Pylaeva, *Izvest. Sektora Fiz.-Khim. Anal.*, **23,** 1953, 110–117.
4. 1545°C [1], 1540°C [2], and 1504°C [3].
5. I. I. Kornilov and E. N. Pylaeva, *Doklady Akad. Nauk S.S.S.R.*, **91,** 1953, 841–843; *Met. Abstr.*, **21,** 1954, 524. "Separation of Ni_3Ta from alloys of system Ni-Ta."
6. H. J. Wallbaum, *Arch. Eisenhüttenw.*, **14,** 1940-1941, 521–526.
7. W. H. Lenz and W. M. Shafer, unpublished work.
8. Quoted from "Metals Handbook," 1948 ed., p. 1235, The American Society for Metals, Cleveland, Ohio.
9. H. J. Wallbaum, *Naturwissenschaften*, **31,** 1943, 91–92.
10. N. Karlsson, *J. Inst. Metals*, **79,** 1951, 391–405, especially 398–399.
11. P. Greenfield and P. A. Beck, *Trans. AIME*, **200,** 1954, 253–257.

$\bar{1}.6627$
0.3373

Ni-Te Nickel-Tellurium

NiTe (68.50 wt. % Te) was first prepared by [1] and was shown [2–5] to have the NiAs (B8) type of structure, with $a = 3.98$ A, $c = 5.38$ A, $c/a = 1.35$ [5]. [4] detected roentgenographically a possibly continuous transition between the NiAs (B8) type NiTe alloy and the CdI_2 (C6) type $NiTe_2$ (81.30 wt. % Te) alloy, the latter having the parameters [6] $a = 3.869$ A, $c = 5.308$ A, $c/a = 1.372$. This solid solution was also investigated roentgenographically, densimetrically, and magnetically by [5].

For discussions of crystal-chemical relations, see [7–9].

Note Added in Proof. For a recent magnetic study of Ni-Te alloys, see [10].

1. C. Fabre, *Compt. rend.*, **105**, 1887, 277.
2. I. Oftedal, *Z. physik. Chem.*, **128**, 1927, 135–153.
3. W. F. de Jong and H. W. V. Willems, *Physica*, **7**, 1927, 74–79.
4. S. Tengnér, *Z. anorg. Chem.*, **239**, 1938, 126–132; *Naturwissenschaften*, **26**, 1938, 429.
5. W. Klemm and N. Fratini, *Z. anorg. Chem.*, **251**, 1943, 222–232.
6. The lattice parameters found by M. A. Peacock and R. M. Thompson (*Univ. Toronto Studies, Geol. Ser.*, no. 50, 1946, pp. 63–73; *Chem. Abstr.*, **40**, 1946, 6368) are somewhat smaller than those of [4].
7. F. Laves and H. J. Wallbaum, *Z. angew. Mineral.*, **4**, 1941-1942, 17–46.
8. E. S. Makarov, *Izvest. Akad. Nauk S.S.S.R., Otdel. Khim. Nauk*, **1944**, 201–208; *Met. Abstr.*, **12**, 1945, 148.
9. E. S. Makarov, *Izvest. Akad. Nauk S.S.S.R., Otdel. Khim. Nauk*, **1945**, 569–580; *Structure Repts.*, **10**, 1945-1946, 26–29.
10. E. Uchida and H. Kondoh, *J. Phys. Soc. Japan*, **11**, 1956, 21–27; *Met. Abstr.*, **23**, 1956, 814.

$\bar{1}.4028$
0.5972

Ni-Th Nickel-Thorium

[1] have investigated the system by thermal, microscopic, and X-ray powder methods. They found two melting-point maxima (Fig. 574) corresponding to the

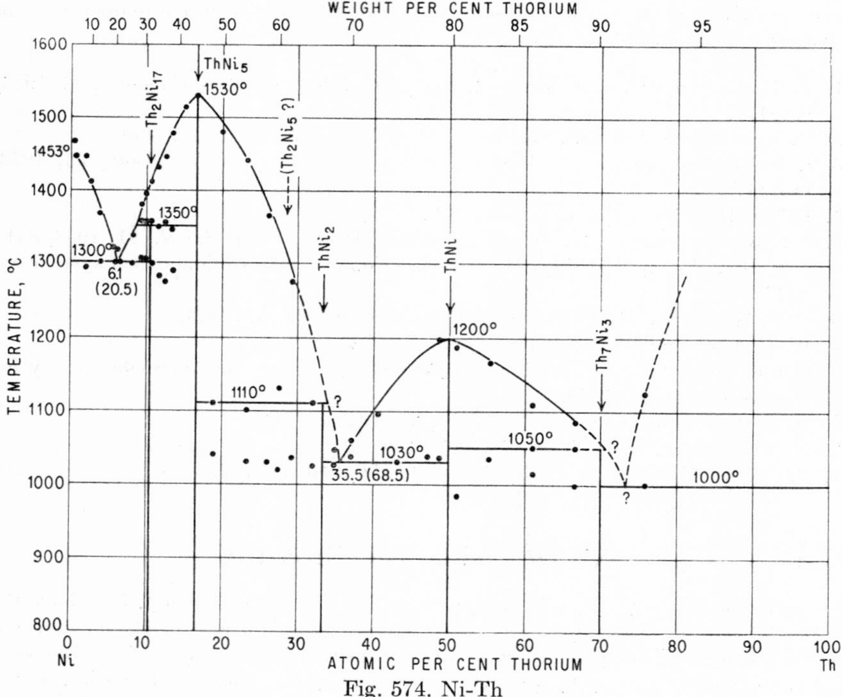

Fig. 574. Ni-Th

compounds $ThNi_5$ [16.67 at. (44.17 wt.) % Th] and ThNi (79.82 wt. % Th) and three peritectically formed compounds to which they assigned the formulas $ThNi_9$ (30.53 wt. % Th), probably Th_2Ni_5 [28.57 at. (61.27 wt.) % Th], and Th_2Ni (88.78 wt. %

Th) [2]. "From the X-ray patterns the existence of an additional phase with a composition between $ThNi_5$ and Th_2Ni_5 has to be assumed; however, the thermal and microscopic work failed to indicate unequivocally such a phase" [1].

The results of a detailed crystallographic study [3, 8] of the intermediate phases occurring at room temperature deviate from those of [1] as far as the compositions of the peritectically formed compounds are concerned. Instead of $ThNi_9$, Th_2Ni_5, and Th_2Ni [1], the formulas Th_2Ni_{17} [10.53 at. (31.77 wt.) % Th], $ThNi_2$ (66.42 wt. % Th), and Th_7Ni_3 (90.22 wt. % Th) were suggested, besides $ThNi_5$ and ThNi. Since these new formulas are well founded on crystal-structure determinations (see below), the phase diagram given by [1] was slightly modified [4] to accommodate the new results (Fig. 574).

Age hardening in Ni-rich alloys was observed by [1] and the "room temperature" solubility of Th in solid Ni estimated to be below 0.1 wt. (0.03 at.) %. Previously the data <0.01 wt. % for "room temperature" and <0.05 wt. % for 1000°C had been given [5]. No solubility of Ni in solid Th was found by [1].

Crystal Structures. Th_2Ni_{17} crystallizes hexagonally, with a = 8.37 A, c = 8.14 A, c/a = 0.972, and 38 atoms per unit cell, in a new structural type [3, 8]. $ThNi_5$ was shown [6, 3, 8] to have the hexagonal $CaCu_5$ type of structure, with a = 4.921 A, c = 3.990 A, c/a = 0.811 [6], and a = 4.97 A, c = 4.01 A, according to [8], who found some evidence that the composition may be variable, with a possible range from $ThNi_5$ to $ThNi_4$. $ThNi_2$ is isotypic with AlB_2 (hexagonal C32) type and has the parameters a = 3.95 A, c = 3.83 A, c/a = 0.970 [3, 8]. ThNi crystallizes orthorhombically, with a = 14.51 A, b = 4.31 A, c = 5.73 A, and 16 atoms per unit cell, in a new structural type [3, 8]. Th_7Ni_3 is hexagonal (isotypic with Th_7Fe_3), with a = 9.86 A, c = 6.23 A, c/a = 0.632, and 20 atoms per unit cell [7, 8].

1. L. Horn and C. Bassermann, *Z. Metallkunde*, **39**, 1948, 272–275; the alloys were melted in Al_2O_3 crucibles in a hydrogen or argon atmosphere; however, Th losses by oxidation could not be fully prevented. All alloys were chemically analyzed, at least for Th. For the sake of clarity not all thermal data were plotted in Fig. 574.
2. Th_2Ni was first claimed to have been prepared by E. Chauvenet, *Bull. acad. belg.*, **1908**, 684.
3. J. V. Florio and R. E. Rundle, *U.S. Atomic Energy Comm., Publ.* ISC-273, 1952; *Met. Abstr.*, **21**, 1954, 526. See also the summary table in J. V. Florio, R. E. Rundle, and A. I. Snow, *Acta Cryst.*, **5**, 1952, 450.
4. The compositions of the peritectic points at 1110 and 1050°C and of the eutectic at 1000°C [91 wt. (72 at.) % Th, according to [1]] were tentatively assumed to lie at somewhat higher Th contents.
5. W. Hessenbruch and L. Horn, *Z. Metallkunde*, **36**, 1944, 145–146.
6. T. Heumann, *Nachr. Akad. Wiss. Göttingen, Math.-Phys. Kl.*, **1**, 1948, 21–26; *Structure Repts.*, **11**, 1947-1948, 59.
7. N. C. Baenziger, *U.S. Atomic Energy Comm., Publ.* AECD-3237, 1948.
8. J. V. Florio, N. C. Baenziger, and R. E. Rundle, *Acta Cryst.*, **9**, 1956, 367–372.

0.0882
$\bar{1}$.9118

Ni-Ti Nickel-Titanium

Ni-rich Alloys. The existence of Ni-rich solid solutions was indicated by measurements of conductivity [1, 2]. The constitution of the alloys in the composition range 70–100 wt. % Ni was disclosed by [3] by means of thermal and micrographic analysis. Their findings showed that a Ni-rich solid solution forms a eutectic with the compound $TiNi_3$ (78.61 wt. % Ni; melting point 1378°C) at 83.8 wt. (80.8 at.) %

Ni and 1287°C (Fig. 575). The solid solubility of Ti in Ni was found to decrease from 10.8 wt. % Ti at 1287°C to 3.3 wt. % Ti at 850°C. By lattice-parameter measurements and microexamination, [4] found higher solubilities, namely, 11 wt. (13.2 at.) % Ti at 1150°C and 8 wt. (9.6 at.) % Ti at 750°C (Fig. 575). Lattice parameters of

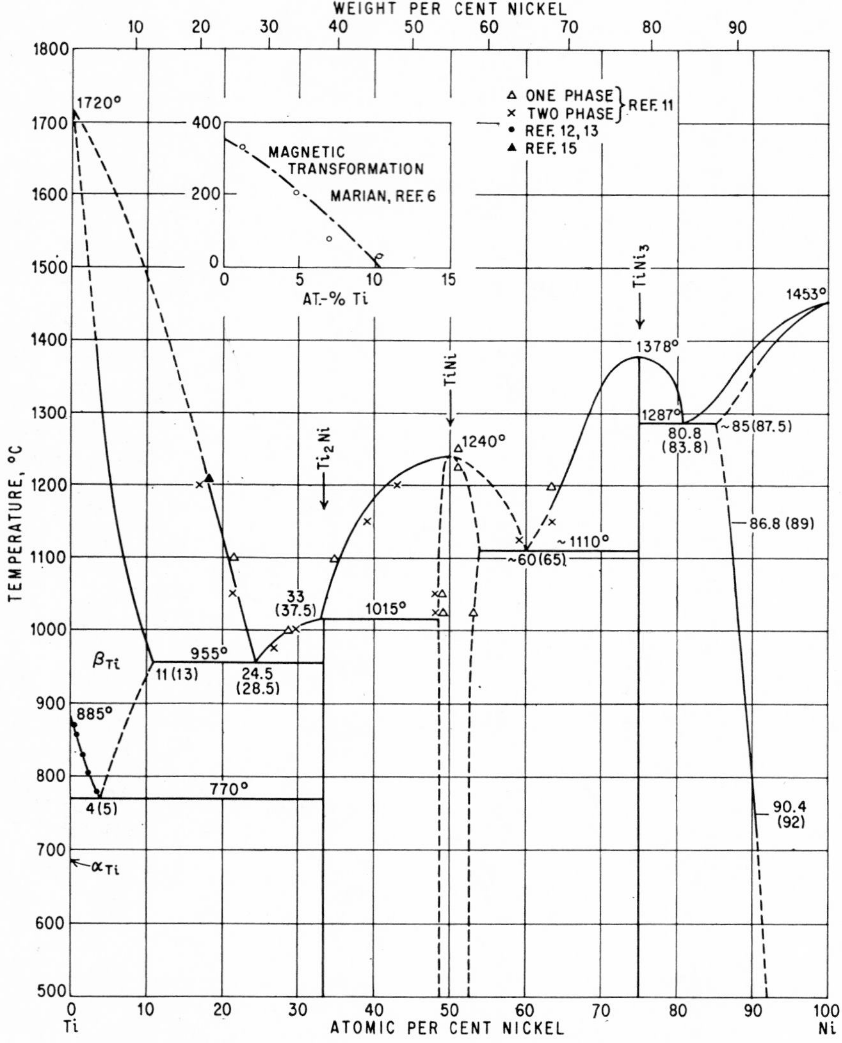

Fig. 575. Ni-Ti. (See also Fig. 575*a* and Note Added in Proof.)

(Ni)-phase alloys were also measured by [5] (up to 3 at. % Ti). The Curie-point temperature vs. composition curve determined by [6] is shown in the inset of Fig. 575.

[7] reported that, besides $TiNi_3$, two additional compounds exist, TiNi (55.06 wt. % Ni) and Ti_2Ni (37.99 wt. % Ni). These phases were identified by X-ray investigations. In 1941, [8] carried out thermal (in the region 50–80 wt. % Ni) and roent-

genographic work and published a phase diagram in which the results of [3] (between 70 and 100 wt. % Ni) were adopted. It was shown that (*a*) the TiNi-$TiNi_3$ eutectic occurs at approximately 66 wt. (61.3 at.) % Ni and 1100°C, (*b*) the phase TiNi is formed by a peritectic reaction at 1270°C (cf. below), and (*c*) TiNi has a certain undetermined range of homogeneity. It should be mentioned that the titanium used by [8] was only 95 wt. % pure and that the alloys were melted in Pythagoras, or corundum, crucibles. Between 0 and 54 wt. % Ni the diagram of [8] was hypothetical.

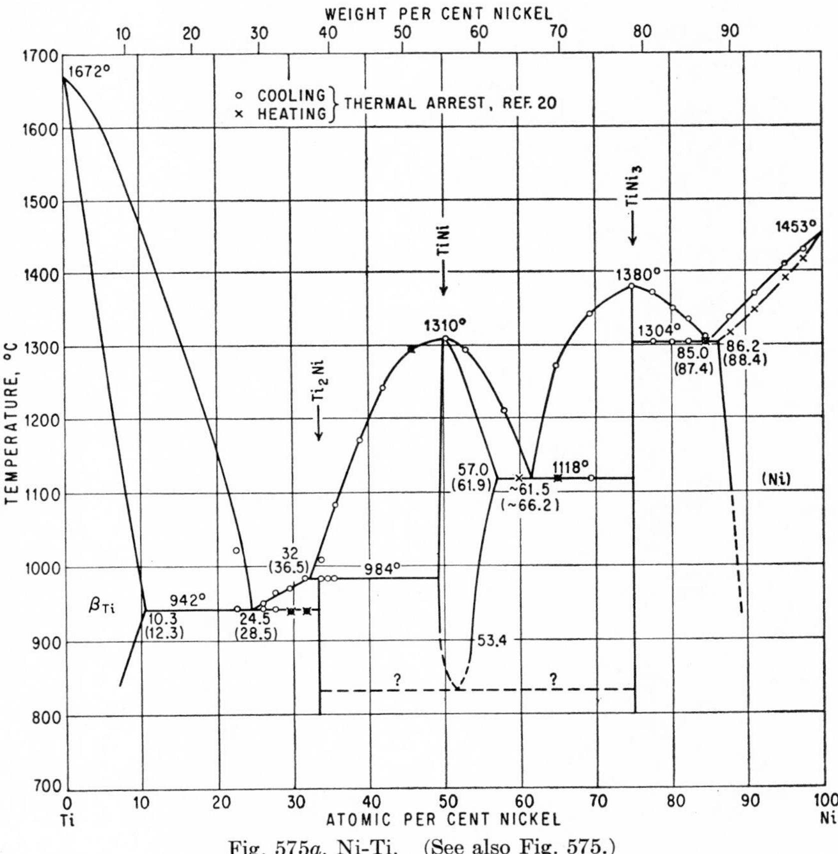

Fig. 575*a*. Ni-Ti. (See also Fig. 575.)

Ti-rich Alloys. The first attempt to reveal the phase relationships in the Ti-rich alloys was made by [9], using Mg-reduced titanium-base alloys prepared by powder-metallurgy methods. Their tentative diagram, as obtained by micrographic analysis, shows the eutectoid at about 7.3 wt. (6.0 at.) % Ni and approximately 765°C; the β_{Ti} phase, with 12 wt. (10 at.) % Ni, forms a eutectic at 33 wt. (28.7 at.) % Ni and about 965°C with an intermediate phase.

According to unpublished work by [10] on Mg-reduced titanium-base alloys, the eutectoid point would be located (by extrapolation) at 7 wt. (5.8 at.) % Ni and 765°C.

[11] have studied the system in the range 0–68 wt. % Ni with iodide titanium–base alloys by metallographic, X-ray, and melting-point methods. Their results are

incorporated in Fig. 575 and indicate that there is a eutectic at about 28.5 wt. (24.5 at.) % Ni, a peritectic melt at approximately 37.5 wt. (33 at.) % Ni, and a maximum melting point at the composition TiNi. The data of [8], which were interpreted to indicate the peritectic formation of the TiNi phase, were shown to be incorrect. The phase TiNi was proved to have a certain range of homogeneity. The composition of the eutectoid was given as 6.2 wt. (5.1 at.) % Ni, and the maximum solubility of Ni in α-Ti as <0.2 wt. %.

[12] determined the $\beta_{Ti}/(\alpha + \beta)$ boundary up to 3.6 at. % Ni by measuring the hydrogen pressure in equilibrium with an extremely dilute solution of hydrogen in Ti-rich (iodide titanium base) alloys as a function of temperature. His boundary (Fig. 575) deviates considerably from those presented by the other investigators [9–11]; however, after this boundary has been rechecked at one composition [13] by both the hydrogen method and microscopic observations on quenched alloys [14], there remains little doubt that it is the correct one. The $\beta_{Ti}/(\beta + Ti_2Ni)$ boundary, investigated by [9–11], should, therefore, be redetermined.

[15] described methods for determining liquidus points of reactive alloys; the point found for an 18.2 at. % Ni alloy is plotted in Fig. 575.

Crystal Structures. Ti_2Ni possesses a f.c.c. translation group [7, 16], with $a = 11.333$ A [16] and approximately 96 atoms per unit cell. [17] suggested the Fe_3W_3C ($E9_2$) type of structure.

The compound TiNi has the CsCl (B2) type of structure [7, 16, 20, 21], with $a = 3.013$ A [20], $a = 3.015$ A [21]. [16] found that TiNi decomposes, upon annealing for 10 days at 800 or 650°C, into Ti_2Ni and $TiNi_3$; however, this could not be corroborated by [11] (48 hr, 750°C). See Note Added in Proof.

The structure of $TiNi_3$ forms the hexagonal $D0_{24}$ prototype [18, 16, 19]. The lattice constants are $a = 5.1010$ A, $c = 8.3067$ A, $c/a = 1.6284$ [19].

Note Added in Proof. The phase boundaries above 900°C of the entire system have been carefully redetermined by [20]. In Fig. 575*a*, the results of thermal analysis are plotted. The Ti-rich portion of the diagram as well as the extent of the (TiNi)-phase field was worked out by microscopical methods. As regards the (Ni) solvus line, which was found by microscopic work to be in good agreement with that proposed by [4] (Fig. 575), [20] assume that "the true solubility curve would show a greater decrease at low temperatures than the straight line in [Fig. 575*a*], although the latter represents the structures found after annealing for 500 hr at 1050°C." Precision lattice parameters of the intermediate phases as well as the (Ni) phase are given in [20]. In corroboration of the results of [16], the TiNi phase was found to decompose into $Ti_2Ni + TiNi_3$ at lower temperatures.

In an X-ray study, [22] found the limiting solubility of Ti in solid Ni to be 10 at. %.

The formation and subsequent decomposition of metastable phases (such as retained β, α') in Ti-Ni alloys containing up to 11 at. % Ni have been studied by [23].

1. O. Laue, *Abhandl. Inst. Metallhütt. u. Elektrochem. Tech. Hochsch., Aachen,* **1,** 1916, 21–37; see also M. Hansen, "Der Aufbau der Zweistofflegierungen," Springer-Verlag OHG, Berlin, 1936.
2. A. M. Hunter and J. W. Bacon, *Trans. Am. Electrochem. Soc.,* **37,** 1920, 520.
3. R. Vogel and H. J. Wallbaum, *Arch. Eisenhüttenw.,* **12,** 1938-1939, 299–304.
4. A. Taylor and R. W. Floyd, *J. Inst. Metals,* **80,** 1952, 577–587.
5. T. H. Hazlett and E. R. Parker, *Trans. ASM,* **46,** 1954, 701–715.
6. V. Marian, *Ann. phys.,* **7,** 1937, 459–527.
7. F. Laves and H. J. Wallbaum, *Naturwissenschaften,* **27,** 1939, 674–675.
8. H. J. Wallbaum, *Arch. Eisenhüttenw.,* **14,** 1940-1941, 521–526.

9. J. R. Long, E. T. Hayes, D. C. Root, and C. E. Armantrout, *U.S. Bur. Mines Rept. Invest.* 4463, 1949; J. R. Long, *Metal Progr.*, **55,** 1949, 364–365.
10. C. M. Craighead, F. Fawn, and L. W. Eastwood, Battelle Memorial Institute, Second Progress Report on Contract AF 33 (038)-3736 to Wright Patterson Air Force Base, 1949.
11. H. Margolin, E. Ence, and J. P. Nielsen, *Trans. AIME,* **197,** 1953, 243–247.
12. A. D. McQuillan, *J. Inst. Metals,* **80,** 1951-1952, 363–368.
13. A. D. McQuillan, *J. Inst. Metals,* **82,** 1953, 47–48.
14. Microstructural evidence is presented showing that delay in quenching of alloy specimens is the most probable cause of disagreement in the reported boundaries.
15. W. Hume-Rothery and D. M. Poole, *J. Inst. Metals,* **82,** 1954, 490–492.
16. P. Duwez and J. L. Taylor, *Trans. AIME,* **188,** 1950, 1173–1176. For a discussion, see *Trans. AIME,* **191,** 1951, 551.
17. W. Rostoker, *Trans. AIME,* **194,** 1952, 209–210.
18. F. Laves and H. J. Wallbaum, *Z. Krist.*, **A101,** 1939, 78–93.
19. A. Taylor and R. W. Floyd, *Acta Cryst.*, **3,** 1950, 285–289.
20. D. M. Poole and W. Hume-Rothery, *J. Inst. Metals,* **83,** 1955, 473–480; discussion, **84,** 1956, 532–535.
21. T. V. Philip and P. A. Beck, *Trans. AIME,* **209,** 1957, 1269–1271.
22. I. I. Kornilov and A. Ya. Snetkov, *Izvest. Akad. Nauk S.S.S.R., Otdel. Tekh. Nauk,* **1955,** 84–88; *Met. Abstr.*, **24,** 1957, 870.
23. D. H. Polonis and J. G. Parr, *Trans. AIME,* **206,** 1956, 531–536; discussion, **209,** 1957, 524–526.

$\bar{1}.4581$
0.5419

Ni-Tl Nickel-Thallium

Figure 576 shows the phase diagram worked out by [1] using thermal analysis and microscopic examination. According to the length of the thermal arrests at 1387°C, the monotectic horizontal extends from 3 wt. (0.9 at.) Tl to practically 100% Tl. The existence of a (Ni) primary solid solution is also indicated by the depression of the magnetic-transformation temperature of Ni on addition of Tl.

1. G. Voss, *Z. anorg. Chem.*, **57,** 1908, 49–52.

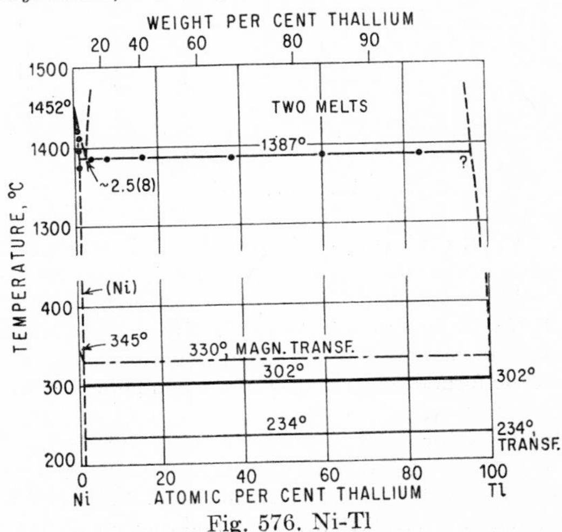

Fig. 576. Ni-Tl

Ī.3918
0.6082

Ni-U Nickel-Uranium

A few U-Ni alloys were prepared and studied by [1, 2]. The entire phase diagram has been investigated more recently by [3, 4]. [3] found—as a result of thermal, optical, and roentgenographic investigations, only the last of which, however, are published as yet—the compounds U_6Ni [14.29 at. (3.95 wt.) % Ni] [5], UNi, UNi_2 (33.02 wt. % Ni), and UNi_5 [83.33 at. (55.20 wt.) % Ni]. [4], by means of thermal, microscopic, and X-ray powder work, established the rather complex phase diagram

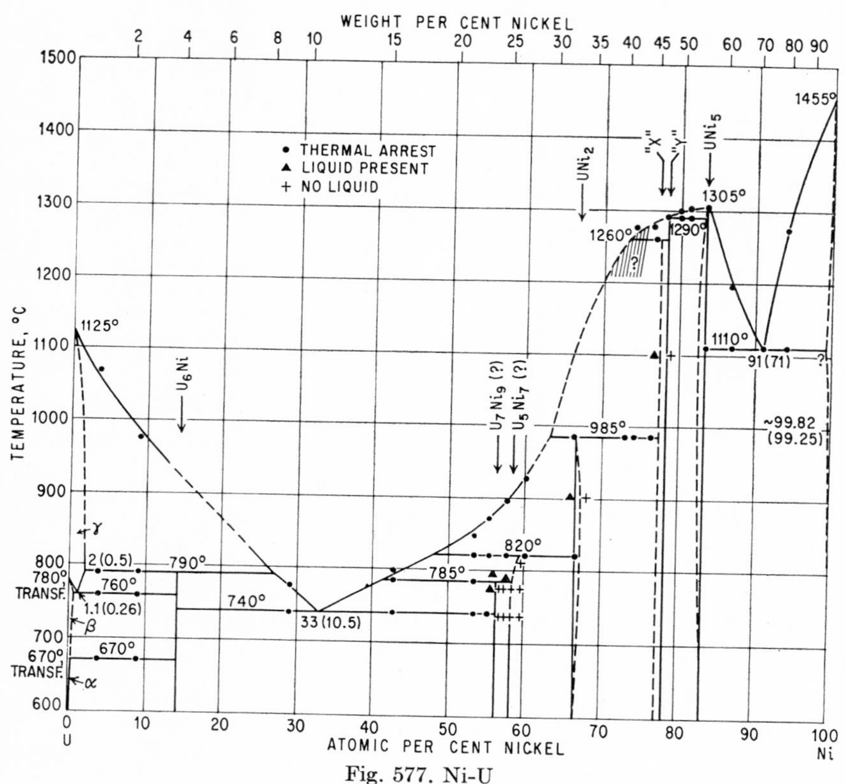

Fig. 577. Ni-U

shown in Fig. 577. U_6Ni, UNi_2, and UNi_5 (but not UNi) were corroborated; and evidence for the existence of four additional intermediate phases [6]—$\sim U_7Ni_9$ [56.25 at. (24.1 wt.) % Ni], $\sim U_5Ni_7$ [58.33 at. (25.65 wt.) % Ni] [7], *X* [~77 at. (45.2 wt.) % Ni; U_2Ni_7?], and *Y* [78–79 at. (46.6–48.1) wt. % Ni]—was obtained. In the shaded area of Fig. 577 very anomalous thermal effects were observed (for details, see the original paper), and the formation equilibria of the compounds *X* and *Y* could not be determined with certainty.

Primary Solid Solutions. The values for the solubility of Ni in U shown in Fig. 577 [nearly 2 at. (0.5 wt.) % Ni at 790°C and decreasing with temperature] are in good agreement with some earlier data [8]. As to the solubility of U in Ni, [4] observed that an alloy containing 0.74 wt. (0.18 at.) % U and quenched from 1000°C consisted of only one phase but, when reheated to 950°C and quenched, showed some precipitated phase.

Crystal Structures. The following results were reported by [3]: U_6Ni, b.c. tetragonal structure, with a = 10.37 ± 4 A, c = 5.21 ± 2 A, c/a = 0.502, and 28 atoms per unit cell (isotypic with U_6[Mn, Fe, Co]); "UNi" (cf. above), structure too complex to be interpreted on the basis of powder data available; UNi_2, hexagonal $MgZn_2$ (C14) type, with a = 4.966 A, c = 8.252 A, c/a = 1.662; UNi_5, f.c.c. [isotypic with $PdBe_5$, ~$MgCu_2$ (C15) type], with a = 6.7830 ± 5 A.

1. P. A. Heller, *Metall u. Erz*, **19**, 1922, 397–398. The alloys obtained were very impure.
2. G. Tammann and G. Bandel, *Arch. Eisenhüttenw.*, **6**, 1932-1933, 296.
3. N. S. Baenziger, R. E. Rundle, A. I. Snow, and A. S. Wilson, *Acta Cryst.*, **3**, 1950, 34–40.
4. J. D. Grogan and R. J. Pleasance (B. E. Williams), *J. Inst. Metals*, **82**, 1953, 141–147. The alloys were prepared, from high-purity metals, in stabilized zirconia crucibles in vacuo in an induction furnace. Most of the alloys were analyzed.
5. U_6Ni was first reported by F. Foote et al., Manhattan Project Report CT-3013, 1945.
6. "The exact ranges of composition and the formulae of these compounds have not been determined; the formulae suggested are adopted for the sake of convenience" [4].
7. It is assumed that the composition of the phase U_5Ni_7 (which is peritectically formed at 820°C) changes toward the Ni-rich area above 785°C (see Fig. 577).
8. A solubility of 1 and 2 at. % Ni in β-U and γ-U, respectively, was reported by Foote [5] (quoted from J. J. Katz and E. Rabinowitch, "The Chemistry of Uranium," Part I, National Nuclear Energy Series, Div. VIII, vol. 5, p. 175, McGraw-Hill Book Company, Inc., New York, 1951).

0.0614
$\bar{1}$.9386

Ni-V Nickel-Vanadium

Early Work. In 1915, [1] investigated Ni-rich alloys and found an extended primary solid solution with solid-state reactions above 20 wt. % V.

Phase Diagram and Crystal Structures. Up to 60 at. % V the phase diagram of Fig. 578 is mainly due to careful thermal, microscopic, and roentgenographic (at both high and normal temperatures) analyses by [2]. The phase boundaries were worked out down to about 650°C and are thought to be accurate to within 1 at. %. For the V-rich part of the diagram only some cursory results of [2, 3, 4] are available (see below).

The VNi_3 (22.44 wt. % V) (θ) phase has the tetragonal $TiAl_3$ (DO_{22}) type of structure [2], with a = 3.542 A, c = 7.213 A, c/a = 2.036 (26.12 at. % V) for the b.c. unit cell. A graph showing the parameter-composition relations can be found in [2].

As to the VNi_2 (30.27 wt. % V) (δ) phase which, like VNi_3, separates out from the (Ni) solid solution, "the evidence as a whole indicates a narrow phase region in the range 33.6–34.1 at. % V" [2]. Its powder lines could be accounted for by a monoclinic distortion of the (Ni) phase, and a b.c. orthorhombic pseudo cell can be ascribed.

The existence of a σ phase in the Ni-V system had been predicted by [5] and was experimentally discovered by [6, 7]. Its Ni-rich boundary was established by [2], who also found that a transition ($\sigma' \rightleftarrows \sigma$) occurs below 800°C (Fig. 578) (see also [8]). The V-rich boundary is not yet exactly known; [9] found for 1200°C the boundary at 68.5 at. % V and [2] (see also [10]) for 1100°C at 74 at. % V. In Fig. 578 an intermediate line was drawn. The structure of the σ phase has not yet been completely elucidated. It is isotypic with "σ phases" of several other binary systems [10], and for literature on theoretical considerations on σ phases in general reference is made to the Cr-Fe system. Tentative unit cells with tetragonal and orthorhombic symmetry were suggested by [6] and [11], respectively. Laue and rotation photographs made by [12, 10] showed that the structure is tetragonal, with $a = 8.966$ A, $c = 4.641$ A, $c/a = 0.5176$ (60 at. % V). For parameter changes with composition, see [2, 10]. The $\sigma' \rightleftarrows \sigma$ transformation may be of the disorder-order type [8] (see Supplement).

A compound V_3Ni (72.25 wt. % V), of the "β-W" (A15) type with $a = 4.71$ A, was discovered by [3, 4]; and it was suggested [3] that it may be formed peritectically at

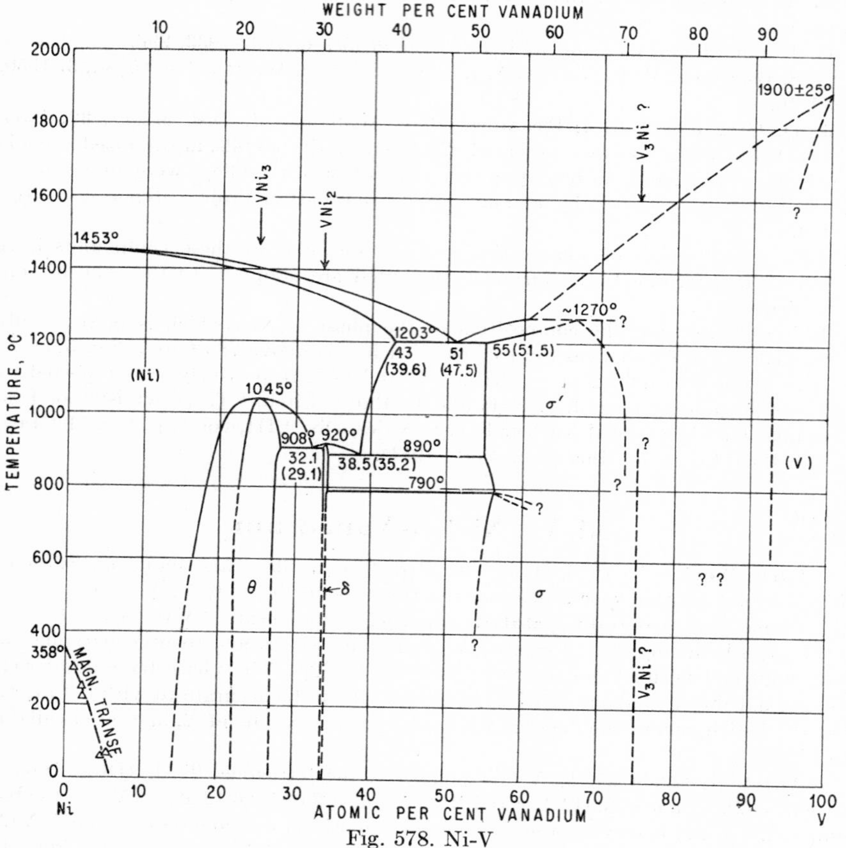

Fig. 578. Ni-V

1380°C. However, [13] pointed out that in the work of [2] a powder photograph of an alloy with about 80 at. % V quenched from 1100°C still showed σ as a main constituent, and suggested that V_3Ni may be formed at a lower temperature [14].

[4] stated that a further "unknown phase is present in alloys containing 15 and 20 wt. % Ni." The solubility limit of Ni in solid V is said to be between 7 and 10 wt. (6.1 and 8.8 at.) % Ni [4].

The Curie temperature of Ni-V alloys has been measured by [1, 15]; the results of [15] are plotted in Fig. 578.

Lattice parameters of the (Ni) primary solid solution were measured by [2].

Supplement. [16] was unable to corroborate the existence of a compound V_3Ni. [17] concluded that there is no reliable X-ray evidence for ordering in the σ phase; "forbidden reflexions" are thought to be more likely due to the Renninger "double reflexion" (*Umweganregung*) effect than to ordering. However, a neutron-diffraction study of alloys with 56.8, 65.0, and 69.9 at. % V annealed at 500°C yielded clear evidence of ordering in the σ phase [18]. According to [19], the σ-phase alloy containing 64.2 at. % V becomes ferromagnetic below 52°K (−221°C).

1. H. Giebelhausen, *Z. anorg. Chem.*, **91,** 1915, 254–256.
2. W. B. Pearson and W. Hume-Rothery, *J. Inst. Metals,* **80,** 1952, 641–652.
3. W. Rostoker, *J. Inst. Metals,* **80,** 1952, 698 (Discussion).
4. W. Rostoker and A. Yamamoto, *Trans. ASM,* **46,** 1954, 1136–1163.
5. P. A. Beck and W. D. Manly, *Trans. AIME,* **185,** 1949, 354.
6. P. Duwez and S. R. Baen, *ASTM Symposium, Spec. Tech. Publ.* 110, 1950, pp. 48–54.
7. W. B. Pearson, J. W. Christian, and W. Hume-Rothery, *Nature,* **167,** 1951, 110.
8. W. B. Pearson, *Nature,* **169,** 1952, 934.
9. P. Greenfield and P. A. Beck, *Trans. AIME,* **200,** 1954, 253–257.
10. W. B. Pearson and J. W. Christian, *Acta Cryst.*, **5,** 1952, 157–162.
11. P. Pietrokowsky and P. Duwez, *Trans. AIME,* **188,** 1950, 1283–1284.
12. W. B. Pearson and J. W. Christian, *Nature,* **169,** 1952, 70–71.
13. W. B. Pearson, discussion on [4], *Trans. ASM,* **46,** 1954, 1166.
14. The formation of V_3Si of the "β-W" type by reaction between V vapor and silicon containers was observed by [2]. Cf. discussion on [4], *Trans. ASM,* **46,** 1954, 1167 (authors' reply).
15. V. Marian, *Ann. phys.*, **7,** 1937, 459–527.
16. P. A. Beck, private communication.
17. J. A. Bland, *Acta Cryst.*, **7,** 1954, 477–478.
18. J. S. Kasper and R. M. Waterstrat, *Acta Cryst.*, **9,** 1956, 289–295.
19. M. V. Nevitt and P. A. Beck, *Trans. AIME,* **203,** 1955, 671.

$\bar{1}.5039$
0.4961

Ni-W Nickel-Wolfram

The thermal, microscopic, and resistometric work on Ni-rich alloys of [1], which was greatly influenced by supercooling, as later studies have shown, has only historical interest today.

Solidification. Thermal analyses were carried out by [2, 3], and a few observations of incipient melting were made by [4]. The results are compared in Table 37 and in Fig. 579. Microscopic methods also were used by [2, 4] in the determination of the eutectic composition and of the maximum solubility. In Fig. 579 the data of [4] were favored. Beyond the eutectic point, the liquidus appears to rise rapidly toward the melting point of W.

Solid State. The solvus of the (Ni) phase shown in Fig. 579 was determined micrographically by [4] on long-annealed alloys down to 600°C. Resistometric data below 950°C were in keeping with this boundary. The somewhat higher values found parametrically by [5] were measured on appreciably shorter heat-treated alloys. The Curie temperature in the (Ni) phase decreases [2, 3, 6] (Fig. 579), the lattice parameter increases [5, 4, 7, 8], with increasing W content. Precipitation hardening was observed by [4, 7]. A microradiographic study of the segregation in a 5 at. % W alloy was made by [9].

Table 37. Characteristics of the System Ni-W (Compositions in Wt. % W)

	Ref. 2	Ref. 3	Ref. 4
Maximum in liquidus	35 (14.7 at. % W), 1525°C	34.2 (WNi_6), 1506°C	35, 1505 ± 5°C
Eutectic	52, 1510°C	47.5, 1493°C	~45, 1495 ± 5°C
Maximum solid solubility of W in Ni at the eutectic temperature	~47	46	40

In the early papers [2, 10, 3] the formation of an intermediate phase WNi_6 out of the (Ni) solid solution at about 940°C had been assumed and the W-saturated (Ni) phase had been thought to decompose eutectoidally at 905 [2] or 920°C [10, 3]. However, [11] found that the Ni lattice persisted at the composition WNi_6; and later X-ray powder [5, 4, 7] and microscopic [4, 7] work established conclusively that the formula of the only intermediate phase, which is formed peritectoidally (at 970 ± 10°C according to [4]), is WNi_4 (43.93 wt. % W) [12].

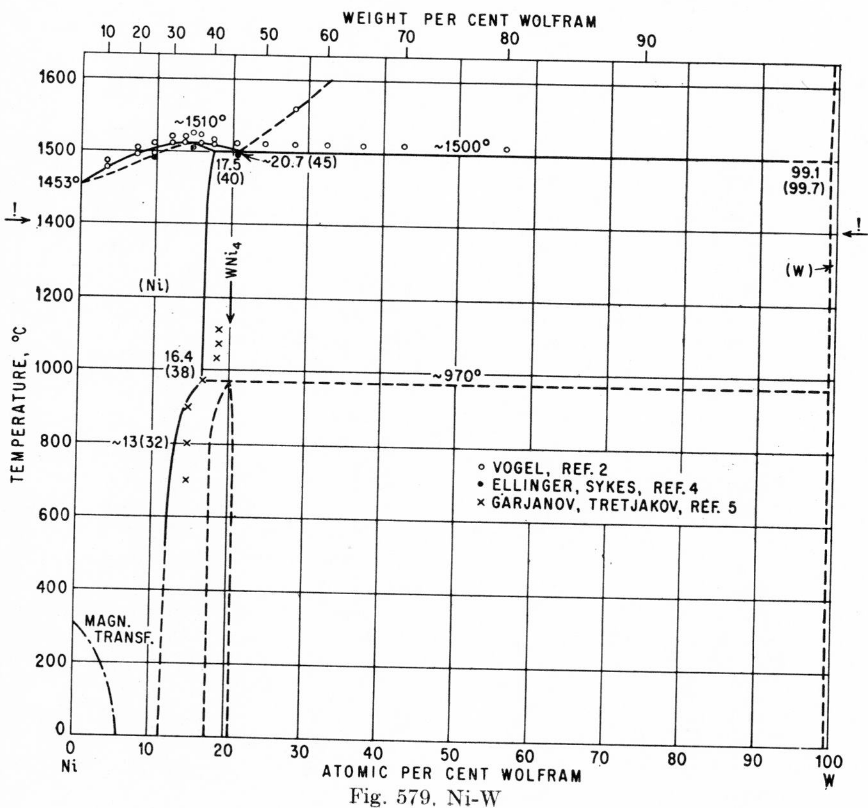

Fig. 579. Ni-W

WNi_4 has a certain range of homogeneity [5, 4, 7]. [4] estimated its extent as about 2 wt. (3 at.) % at 900°C, and [7] suggested the low-W boundary to lie at 17.6 at. % W, or even lower, below 850°C. As to the structure of WNi_4, [5] suggested a cubic one; but [7] showed in a careful study that it is tetragonal b.c. (space group I4/m), with $a = 5.730 \pm 1$ A, $c = 3.553 \pm 1$ A, $c/a = 0.620$, and 10 atoms per unit cell in ordered positions.

Conflicting evidence as to the solubility of Ni in solid W has been reported. From their debyograms, [5] estimated a solubility of 4–5 wt. %, which value is certainly too high. [11, 4] were unable to observe any effect of Ni on the lattice parameter of W. Based on their microscopic work, [4] assumed a solubility of about 0.3 wt. (0.9 at.) % Ni at the eutectic temperature (Fig. 579). The lattice parameter of W saturated with Ni at 1150°C published by [7] is somewhat lower than that of pure W.

1. R. Irmann, *Metall u. Erz*, **12,** 1915, 358–364; the paper contains references to some older literature.
2. R. Vogel, *Z. anorg. Chem.*, **116,** 1921, 231–242. The alloys up to 50 wt. % W were melted without protective atmosphere. The Ni used contained 0.47 wt. % Fe and 0.35 wt. % SiO_2.
3. S. Takeda, *Science Repts. Tôhoku Imp. Univ., K. Honda Anniv. Vol.*, 1936, pp. 864–881 (in English). The paper gives only a short outline of the results but no details of the thermal, dilatometric, roentgenographic, microscopic, and magnetic studies carried out.
4. F. H. Ellinger and W. P. Sykes, *Trans. ASM*, **28,** 1940, 619–643; the alloys were sintered or melted (W 99.9 wt. % pure) in a hydrogen atmosphere.
5. F. K. Garjanov and V. I. Tretjakov, *Zhur. Tekh. Fiz.*, **8,** 1938, 1326–1332 (in Russian).
6. V. Marian, *Ann. phys.*, **7,** 1937, 459–527.
7. E. Epremian and D. Harker, *Trans. AIME*, **185,** 1949, 267–273.
8. T. H. Hazlett and E. R. Parker, *Trans. ASM*, **46,** 1954, 701–715.
9. K. A. Osipov and S. G. Fedotov, *Doklady Akad. Nauk S.S.S.R.*, **78,** 1951, 51–53.
10. K. Winkler and R. Vogel, *Arch. Eisenhüttenw.*, **6,** 1932-1933, 165.
11. K. Becker and F. Ebert, *Z. Physik*, **16,** 1923, 168.
12. H. Bückle (*Recherche aéronaut.*, **24,** 1951, 49–55) was unable to find any intermediate phase. This is certainly due to the fact that he worked with ternary (low-Cr) alloys which were annealed at a relatively low temperature (800°C).

$\bar{1}.9531$
0.0469

Ni-Zn Nickel-Zinc

The phase diagram shown in Fig. 580 is essentially that established by [1–6] in a series of investigations where thermal, microscopic, X-ray, and magnetic methods were used. Except for the (Ni)-phase boundary at lower temperatures (see below), future work may be expected to bring only minor quantitative refinements.

Previously the following systematic investigations had been carried out: thermal and microscopic for 50–100 wt. % Zn [7], 80–100 wt. % Zn [8], and the whole system [9]; thermal, thermoresistometric, dilatometric, magnetic, microscopic, and X-ray for the whole system [10]; X-ray for 47–100 wt. % Zn [11] and the whole system [12]; differential thermal and microscopic for the Zn-rich region [12]. For a detailed discussion of these papers, reference is made to [13]. It had been already observed by [14] that Ni depresses the melting point of Zn only slightly. According to [15], the eutectic point lies between 0.12 and 0.25 wt. (0.13 and 0.28 at.) % Ni.

Intermediate Phases [16] **and Their Crystal Structures** (see Fig. 580). In the equiatomic region two phases β and β_1 occur. The β high-temperature phase has the CsCl (B2) type of structure, with $a = 2.914$ A [2], and cannot be retained by quenching. Incorrect structures were reported by [10, 1, 12].

The β_1 room-temperature phase has the CuAu ($L1_0$) type of structure [1, 11, 12, 2, 4], with $a = 2.7468$ A, $c = 3.1901$ A, $c/a = 1.1614$ [at 50 at. (47.3 wt.) % Ni] [4]. The changes of the lattice parameters with composition [4, 17] and temperature [17] have been measured.

The γ phase is isotypic with γ-brass ($D8_{1-3}$) type [18, 12, 4]. The lattice-parameter-vs.-composition curve goes through a maximum [4] and, at the Zn-rich boundary, indicates the value $a = 8.9168$ A.

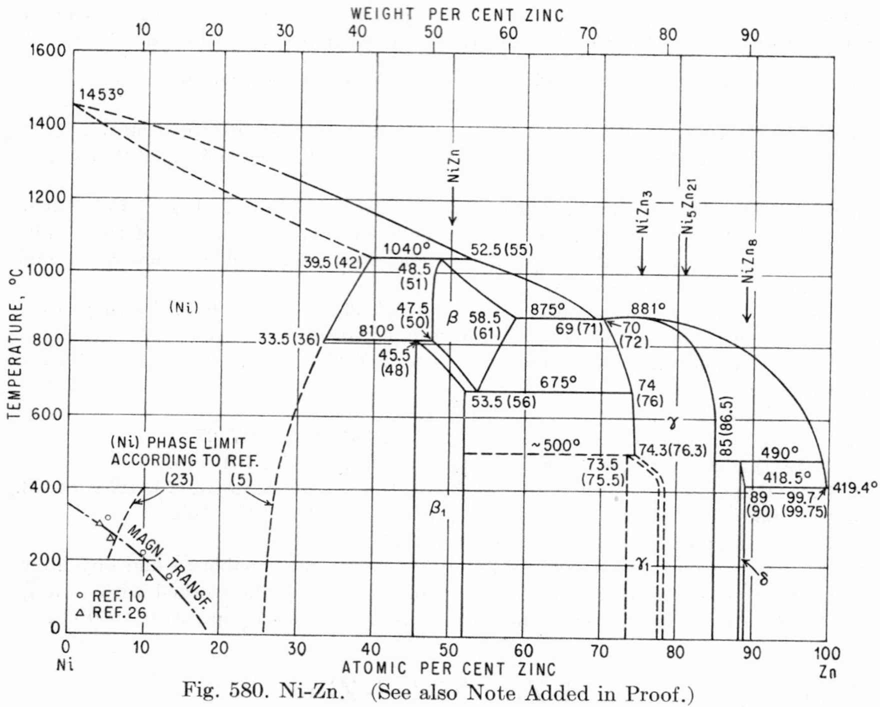

Fig. 580. Ni-Zn. (See also Note Added in Proof.)

γ_1 has a slightly distorted γ-brass structure [18, 4, 19] (line splitting). The structure and the formula of the intermediate phase richest in Zn ($NiZn_8$ or δ in Fig. 580) are not yet conclusively established. The formulas $NiZn_9$ [20], Ni_2Zn_{15} [1], Ni_3Zn_{22} [12], and Ni_4Zn_{31} [3, 5] have been assigned, and its structure was indexed hexagonal by [1, 3] and tetragonal by [12]; according to [4] its structure—like that of the γ_1 phase—is related to the γ-brass type.

Magnetic susceptibilities in the γ-phase region were measured by [21, 19] (see also [6]). Emf measurements were carried out by [22].

Solid Solubility of Zn in Ni. The (Ni) solvus line in Fig. 580 was taken from the diagram published by [5]. The solubility "at room temperature" had been roentgenographically determined on very slowly (for several weeks) cooled alloys to be about 28 wt. (26 at.) % Zn. However, a more recent investigation [23] indicated a much lower solubility; according to age-hardening tests and microscopic work on melted and sintered alloys as well as X-ray measurements on alloys prepared by means of the "amalgam method" [24], the solubility at about 300°C is only 7–8 wt. (6.3–7.2 at.) % Zn (see Fig. 580), the same order of magnitude as in the Fe-Zn system (but see Note Added in Proof). A redetermination of the whole (Ni) solvus line is, therefore, necessary.

The lattice parameter of the (Ni) phase increases [12, 2, 4], its Curie temperature decreases [10, 26] (Fig. 580), with increasing Zn content.

Solid Solubility of Ni in Zn. The solubility of Ni in solid Zn is small. [27] concluded from changes in mechanical properties of Zn-rich alloys and from microscopic work the following solubilities (taken from a graph): 0.03, 0.01, and 0.007 wt. (at.) % Ni at 400, 240, and 100°C, respectively [see also 11, 4].

On crystal chemistry of Ni-Zn intermediate phases, see [28, 29].

Note Added in Proof. In a diagram showing the results of a determination of phase boundaries at 250°C by means of the "amalgam method" [30], the following homogeneity ranges are indicated (in at. % Zn): (Ni), 0–24.5; β_1, 43–50; γ_1, 76–77; γ, 82–86; and δ, ~89 (cf. Fig. 580). As regards the extent of the (Ni) primary solid solution, no mention is made of previous findings (see above) of a much lower solubility of Zn in Ni at lower temperatures.

1. W. Heike, J. Schramm, and O. Vaupel, *Metallwirtschaft*, **11**, 1932, 525–530, 539–542; **12**, 1933, 115–120.
2. W. Heike, J. Schramm, and O. Vaupel, *Metallwirtschaft*, **15**, 1936, 655–662.
3. J. Schramm and O. Vaupel, *Metallwirtschaft*, **15**, 1936, 723–726.
4. J. Schramm, *Z. Metallkunde*, **30**, 1938, 122–130.
5. J. Schramm, *Z. Metallkunde*, **30**, 1938, 131–135.
6. J. Schramm, *Z. Metallkunde*, **30**, 1938, 327–334.
7. V. Tafel, *Metallurgie*, **4**, 1907, 781–785; **5**, 1908, 413–414, 428–430.
8. G. Voss, *Z. anorg. Chem.*, **57**, 1908, 67–69.
9. H. Hafner, Dissertation, Freiberg i. Sachsen, Germany, 1927.
10. K. Tamaru, *Kinzoku-no-Kenkyu*, **9**, 1932, 511–526; also *Science Repts. Tôhoku Imp. Univ.*, **21**, 1932, 344–363.
11. V. Caglioti, *Atti congr. nazl. chim. pura ed appl.*, **4**, 1933, 431–441; abstract in *Chem. Zbl.*, **105**(1), 1934, 1283.
12. K. Tamaru and A. Osawa, *Kinzoku-no-Kenkyu*, **12**, 1935, 131–147; also *Science Repts. Tôhoku Imp. Univ.*, **23**, 1935, 794–815.
13. M. Hansen, "Der Aufbau der Zweistofflegierungen," Springer-Verlag OHG, Berlin, 1936.
14. C. T. Heycock and F. H. Neville, *J. Chem. Soc.*, **71**, 1897, 403.
15. W. M. Peirce, *Trans. AIME*, **68**, 1923, 776–777.
16. In reading the original literature, attention should be paid to the fact that different systems of (Greek) phase designation were used.
17. W. Mahler (and K. Schubert), Thesis W. Mahler, Technische Hochschule, Stuttgart, 1950.
18. W. Ekman, *Z. physik. Chem.*, **B12**, 1931, 69–77. See also A. Westgren, *Z. Metallkunde*, **22**, 1930, 372.
19. H. Nowotny and H. Bittner, *Monatsh. Chem.*, **81**, 1950, 887–906.
20. P. Charrier, *Compt. rend.*, **47**, 1924, 330–333.
21. J. G. Dorfman and S. K. Sidorow, *Zhur. Eksptl. i Teoret. Fiz.*, **9**, 1939, 25; quoted from [19].
22. E. Vigouroux and A. Bourbon, *Bull. soc. chim. France*, **9**, 1911, 873.
23. F. Lihl, *Z. Metallkunde*, **43**, 1952, 310–312.
24. Thermal decomposition of Ni-Zn amalgam. See also [25].

25. A. S. Russell, T. R. Kennedy, and R. P. Lawrence, *J. Chem. Soc.*, **1934**, 1750–1754 ($NiZn$, $NiZn_3$, and $NiZn_{4.2}$ prepared).
26. V. Marian, *Ann. phys*, **7**, 1937, 459–527, especially 483–486; also *J. phys. radium*, **8**, 1937, 313–315.
27. F. Pawlek, *Z. Metallkunde*, **36**, 1944, 105–111.
28. W. Hume-Rothery, J. O. Betterton, and J. Reynolds, *J. Inst. Metals*, **80**, 1952, 609–616.
29. K. Schubert, *Z. Metallkunde*, **43**, 1952, 1–10.
30. F. Lihl, *Z. Metallkunde*, **46**, 1955, 438.

$\bar{1}.8085$
0.1915

Ni-Zr Nickel-Zirconium

Up to 50 at. % Ni the phase diagram of Fig. 581 is due to [1]. Two intermediate phases were recognized in this region and tentatively identified as Zr_2Ni (24.34 wt. %

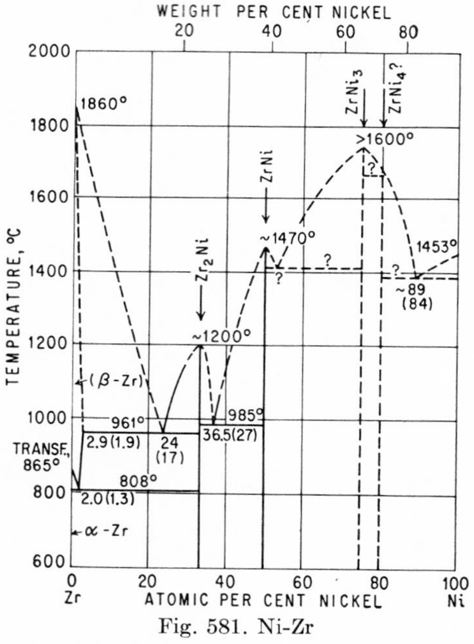

Fig. 581. Ni-Zr

Ni) and ZrNi (39.15 wt. % Ni). Melting points of the eutectics and the compounds, located by microscopic examination, were measured by means of an optical pyrometer. The eutectoid temperature was determined by differential thermal analysis, and the (β-Zr) field was outlined using metallographic techniques [2]. The solubility of Ni in α-Zr was not determined precisely since it was believed to be very small; no shifting of the α-Zr powder lines could be observed in low-Ni alloys. Some preliminary work on Zr-rich alloys by [3–5] is in keeping with the constitution shown in Fig. 581.

Early work on the Zr-Ni system was restricted to Ni-rich alloys. Rapidly cooled alloys containing up to 55 wt. % Zr—melted in vacuo [6]—were microscopically examined by [7]. The phase diagram of Fig. 581 was tentatively completed [8] according to their results [8a]. Both compounds, $ZrNi_3$ (65.87 wt. % Ni) and $ZrNi_4$ (?) (72.02 wt. % Ni), were said to melt above 1600°C. $ZrNi_3$ was corroborated by [9 and said to have an element-like structure. [7] assume that the solubility of Zr in solid Ni is lower than 0.5 wt. (0.3 at.) %. The electrical resistance and its temperature coefficient of Ni-rich alloys were measured by [10]; the results do not allow positive conclusions to be drawn as to the solid solubility in Ni.

1. E. T. Hayes, A. H. Roberson, and O. G. Paasche, *Trans. ASM*, **45,** 1953, 893–900.
2. The (β-Zr) phase could not be retained by any of the quenching methods used.
3. S. M. Shelton, *U.S. Atomic Energy Comm., Publ.* AF-TR-5932, 1949.
4. C. T. Anderson et al., *U.S. Bur. Mines, Rept. Invest.* 4658, March, 1950.
5. F. B. Litton, *Iron Age,* **167**(15), 1951, 112–114.
6. Ni-Zr alloys prepared aluminothermically by K. Metzger, Dissertation, München, Technische Hochschule, 1910; H. S. Cooper, *Chem. & Met. Eng.*, **16,** 1917, 660; *Trans. Am. Electrochem. Soc.*, **43,** 1923, 244; J. W. Marden and M. N. Rich, *U.S. Bur. Mines Bull.* 186, 1921, pp. 107–108, were badly contaminated by Al, Si, and Fe.
7. T. E. Allibone and C. Sykes, *J. Inst. Metals,* **39,** 1928, 179–182. Alundum crucibles were used, and, therefore, the alloys were contaminated with Al and O.
8. M. Hansen, "Der Aufbau der Zweistofflegierungen," p. 969, Springer-Verlag OHG, Berlin, 1936.

8a. In keeping with the results of [7], E. S. Davenport and W. P. Kierman, *J. Inst. Metals,* **39,** 1928, 189 (discussion), reported that an 8.5 wt. % Zr alloy showed primary Ni crystals and eutectic.

9. H. J. Wallbaum, *Arch. Eisenhüttenw.*, **14,** 1940-1941, 521–526.
10. C. Sykes, *J. Inst. Metals,* **41,** 1929, 179–181.

Np-P Neptunium-Phosphorus

1. I. Sheft and S. Fried, *J. Am. Chem. Soc.*, **75,** 1953, 1236–1237; see also *Structure Repts.*, **12,** 1949, 112.

Np-S Neptunium-Sulfur

1. W. H. Zachariasen, *Acta Cryst.*, **2,** 1949, 291–296.
2. W. H. Zachariasen, in G. T. Seaborg, J. J. Katz, and W. M. Manning, (eds.), "The Transuranium Elements," Part II, National Nuclear Energy Series, Div. IV, vol. 14B, pp. 1451–1461, McGraw-Hill Book Company, Inc., New York, 1949. Abstract, *Structure Repts.*, **12,** 1949, 180–181.
3. I. Sheft and S. Fried, *J. Am. Chem. Soc.*, **75,** 1953, 1236–1237; see also *Structure Repts.*, **12,** 1949, 112–113.

Np-Se Neptunium-Selenium

Attempts to prepare compounds of Np with Se and Te were not successful [1].

1. I. Sheft and S. Fried, *J. Am. Chem. Soc.*, **75,** 1953, 1236–1237.

Np-Si Neptunium-Silicon

1. W. H. Zachariasen, *Acta Cryst.*, **2,** 1949, 94–99.
2. I. Sheft and S. Fried, *J. Am. Chem. Soc.*, **75,** 1953, 1236–1237; see also *Structure Repts.*, **12,** 1949, 112.

Np-Te Neptunium-Tellurium

See Np-Se.

$\bar{2}$.8877
1.1123

O-Pb Oxygen-Lead*

The solubility of O in molten Pb at various temperatures was investigated by determining the O content of specimens saturated with O and subsequently quenched or rapidly cooled [1–5]. Results are indicated in Fig. 582. Considering the experimental difficulties (quenching procedure, analytical method), the agreement in the range between 350°C and about 600°C is satisfactory. At higher temperatures, however, results differ considerably. Individual determinations at 1000 and 1200°C (i.e., above the monotectic temperature) yielded solubilities of 0.10 wt. (1.28 at.) % O [6] and 0.135 wt. (1.72 at.) % O [7], respectively, the former value probably being too low. These data suggest that the highest solubilities between 600 and 800°C reported by [4] are more likely than those of [3, 5].

The solubility at 350°C, found to lie between 0.00013 and 0.0006 wt. (0.0017–0.0078 at.) % O, indicates the existence of a eutectic Pb-PbO in this range. At higher temperatures, molten Pb and PbO are immiscible. The monotectic temperature is believed to lie slightly below the melting point of PbO [6], given as 884 ± 8 [8], 879 [9], 884 ± 1 [10], and 890°C [11].

The unstable suboxide Pb_2O, reported by [12], could not be verified by [13–16].

PbO exists in two modifications, a red and a yellow one. The transformation point was found as 587 [9], 585 [17], about 530 [18], 489 [10], and 486–489°C [19]. Although 489°C appears to be widely accepted, the higher temperature should probably be preferred, since the reduction of higher oxides at 550°C yielded the low-temperature form [20]. According to [20, 21], PbO has a very narrow, if any, homogeneity range [22, 23]. Previously, [17] had reported this phase to be homogeneous between 50 and 52.4 at. % O. The lower portion of Fig. 582 shows the phase diagram [6].

Crystal Structures. The low-temperature form of PbO (red) is tetragonal of the B10 type [24, 13, 14, 25, 26, 27], a = 3.9759 A, c = 5.023 A, c/a = 1.263 at 27°C [27]. Yellow PbO is orthorhombic, with 8 atoms per unit cell [28, 14, 26, 29, 20, 30, 27], a = 5.489 A, b = 4.755 A, c = 5.891 A at 27°C [27].

1. H. W. Worner, *J. Inst. Metals*, **66,** 1940, 131–139.
2. K. Grosheim-Krisko, W. Hofmann, and H. Hanemann, *Z. Metallkunde*, **36,** 1944, 91–93.
3. K. Barteld and W. Hofmann, *Erzmetall*, **5,** 1952, 102–105.
4. J. Fischer, H. Bechtel, and E. Schulz, unpublished work, 1952.
5. K. Sano and S. Minowa, *Mem. Fac. Eng. Nagoya Univ.*, **5,** 1953, 80–82.
6. E. Gebhardt and W. Obrowski, *Z. Metallkunde*, **45,** 1954, 332–338.
7. C. W. Dannat and F. D. Richardson, *Metal Ind. (London)*, **83,** 1953, 63–66.
8. Data from various sources, Landolt-Börnstein, "Zahlenwerte und Funktionen," 6th ed., vol. II, Springer-Verlag OHG, Berlin, 1956.
9. F. M. Jaeger and H. C. Germs, *Z. anorg. Chem.*, **119,** 1921, 147–148.
10. E. Cohen and N. W. H. Addink, *Z. physik. Chem.*, **168,** 1934, 188–201.
11. W. Hofmann and J. Kohlmeyer, *Z. Metallkunde*, **45,** 1954, 339–341.

* The assistance of Dr. J. Motz and Dr. K. Löhberg in preparing this system is gratefully acknowledged.

12. A. Ferrari, *Gazz. chim. ital.*, **56,** 1926, 630–637.
13. A. E. van Arkel, *Rec. trav. chim.*, **44,** 1925, 652–654.
14. J. A. Darbyshire, *J. Chem. Soc.*, **1932,** 211–219.
15. M. Le Blanc and E. Eberius, *Z. physik. Chem.*, **160,** 1932, 129–140.
16. R. Fricke and P. Ackermann, *Z. physik. Chem.*, **161,** 1932, 227–230.
17. M. Le Blanc and E. Eberius, *Z. physik. Chem.*, **160,** 1932, 69–100.
18. E. Rencker and M. Bassière, *Compt. rend.*, **202,** 1936, 765–767.

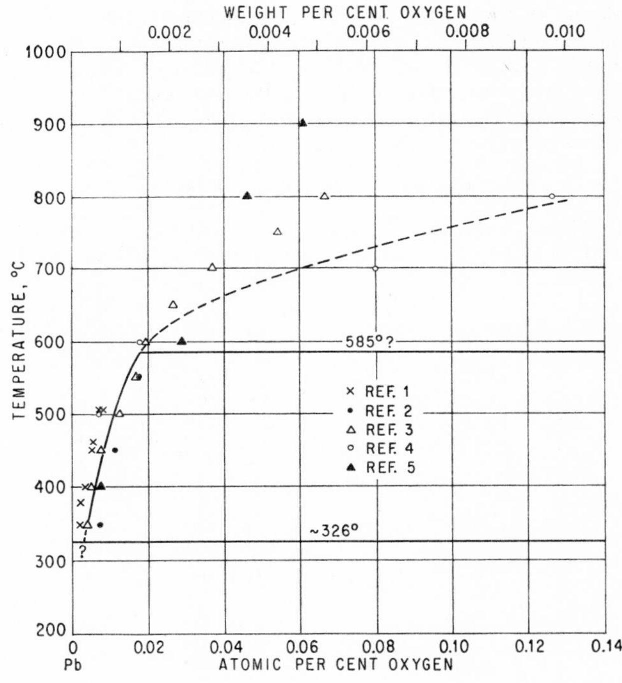

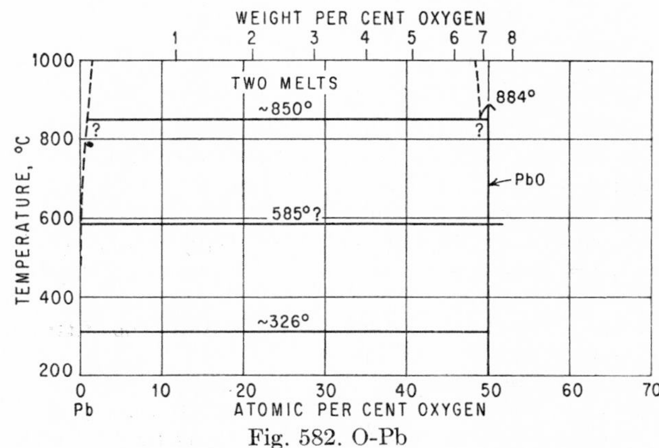

Fig. 582. O-Pb

19. N. Kamayama and T. Fukumoto, *J. Soc. Chem. Ind. Japan*, **49**, 1946, 155–157; *Chem. Abstr.*, **42**, 1948, 6689.
20. A. Byström, *Arkiv Kemi, Mineral. Geol.*, **20A**, 1945, no. 11.
21. F. Gronvold, H. Haraldsen, and J. Vihovde, *Acta Chem. Scand.*, **8**, 1954, 1935.
22. T. Katz, *Ann. chim. (Paris)*, **5**, 1950, 5–65.
23. *Structure Repts.*, **13**, 1950, 170–171.
24. R. G. Dickinson and J. B. Friauf, *J. Am. Chem. Soc.*, **46**, 1924, 2457–2462.
25. W. J. Moore and L. Pauling, *J. Am. Chem. Soc.*, **63**, 1941, 1392–1394.
26. G. L. Clark and R. Rowan, *J. Am. Chem. Soc.*, **63**, 1941, 1302–1305, 1305–1310.
27. H. E. Swanson and R. K. Fuyat, *Natl. Bur. Standards (U.S.), Circ.* 539, II, 1953.
28. F. Halla and F. Pawlek, *Z. physik. Chem.*, **128**, 1927, 49–70.
29. A. Byström, *Arkiv Kemi, Mineral. Geol.*, **17B**, 1943, no. 8.
30. A. Byström, *Arkiv Kemi, Mineral. Geol.*, **25A**, 1947, no. 13.

$\bar{1}.1297$
0.8703

O-Sn Oxygen-Tin

As a result of an investigation [1–3] of the thermal behavior of tin oxides, especially of SnO (11.88 wt. % O), the partial phase diagram of Fig. 583 was proposed

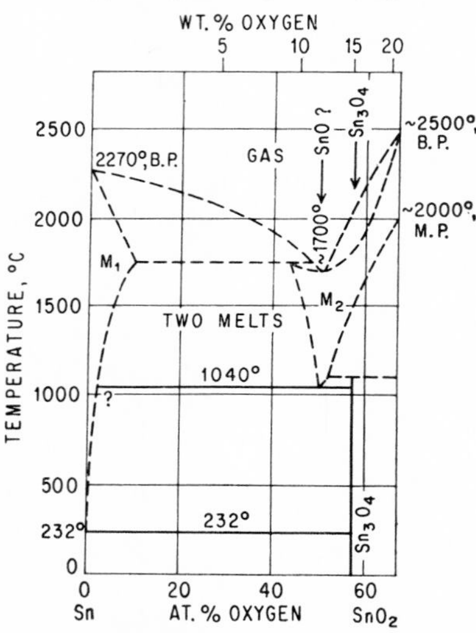

Fig. 583. O-Sn. Schematic diagram for 1 atm pressure. (See also Note Added in Proof.)

by [3]. It is assumed that the long-known compound SnO can be obtained in a metastable state only; above 400°C, SnO decomposes into Sn + Sn_3O_4 (15.24 wt. % O) at a noticeable rate. The high-temperature section of the diagram is hypothetical.

SnO was shown to have the tetragonal PbO (B10) type of structure [4], with $a = 3.804 \pm 6$ A, $c = 4.826 \pm 8$ A, $c/a = 1.269$. The results of [5] were found to be incorrect.

SnO_2 (21.23 wt. % O) has the tetragonal rutile (C4) type of structure [6], with $a = 4.738$ A, $c = 3.188$ A, $c/a = 0.673$ [7].

Note Added in Proof. [8] found the boiling point of SnO under normal pressure to be 1425°C. In Fig. 583 it had been estimated [3] to be about 1700°C.

No evidence of the compound Sn_3O_4 could be found by [9]. They stated: "[3]

give a T-x diagram [Fig. 583] at one atmosphere for the binary system Sn + SnO_2, based on the occurrence of Sn_3O_4 up to about 1040°C and of SnO (liquid) at higher temperatures. In the authors' opinion the latter premises are open to doubt. The present work clearly shows that, at least in the temperature range of 300–1127°C, liquid Sn, solid SnO_2 and gaseous SnO are the only stable phases."

[10] studied the oxidation of Sn under reduced pressures by means of electron diffraction. The patterns of SnO and SnO_2 were observed.

[11] redetermined the lattice constants (a = 4.737 ± 1 A, c = 3.185 ± 1 A) and the atomic parameter of SnO_2.

1. H. Spandau and E. J. Kohlmeyer, *Z. anorg. Chem.*, **254,** 1947, 65–82. This paper contains a review of the extended chemical literature on SnO.
2. H. Spandau, *Angew. Chem.*, **A60,** 1948, 73–74. *Chem. Abstr.*, **44,** 1950, 9228.
3. H. Spandau and E. J. Kohlmeyer, *Z. Metallkunde,* **40,** 1949, 374–376.
4. W. J. Moore and L. Pauling, *J. Am. Chem. Soc.*, **63,** 1941, 1392–1394.
5. C. R. Levi, *Nuovo Cimento,* **1,** 1924, 335–346; G. R. Levi and E. G. Natta, *Nuovo Cimento,* **3,** 1926, 114; *Strukturbericht,* **1,** 1913-1928, 120.
6. *Strukturbericht,* **1,** 1913-1928, 210–211; **2,** 1928-1932, 263–264.
7. H. E. Swanson and E. Tatge, *Natl. Bur. Standards (U.S.), Circ.* 539, 1953, pp. 54–55.
8. H. Spandau and T. Ullrich, *Z. anorg. Chem.*, **274,** 1953, 271–280.
9. J. C. Platteeuw and G. Meyer, The System Tin-Oxygen, *Trans. Faraday Soc.*, **52,** 1956, 1066–1073.
10. J. Trillat, L. Tertian, and M. Plattard, *Compt. rend.*, **240,** 1955, 526–528.
11. W. H. Baur, *Acta Cryst.*, **9,** 1956, 515–520.

$\bar{2}.9467$
1.0533

O-Ta Oxygen-Tantalum

A number of Ta oxides (TaO, Ta_2O_3, TaO_2, Ta_2O_5) had been described in the chemical literature [1]; however, the pentoxide was the only one firmly established. Systematic investigations by means of X-ray powder work were carried out only recently [2–4]. [2] suggested the existence of a high-temperature modification of Ta_2O_5 (transition point 1320 ± 20°C) but were unable to detect any lower oxides. The transition in Ta_2O_5 was corroborated by [3], who also observed a suboxide, probably Ta_2O, having an orthorhombic translation group, with a = 5.29 A, b = 4.92 A, c = 3.05 A, and 6 atoms per unit cell [5].

On the other hand, [4] reported no less than five tantalum oxides: The β phase, with an upper phase limit corresponding to the formula Ta_4O (2.16 wt. % O) and an orthorhombic structure, the lattice parameters of which were found to vary as a = 7.194–7.238 A, b = 3.266–3.273 A, c = 3.204–3.216 A; the γ phase, with the probable composition TaO (8.13 wt. % O) and the NaCl (B1) type of structure, with the a parameter varying between 4.422 and 4.439 A; the δ phase, with the formula TaO_2 (15.03 wt. % O) and the tetragonal rutile (C4) type of structure, with a = 4.709 A, c = 3.065 A, c/a = 0.651; the ϵ phase, yielding a complex powder pattern and existing somewhere within the interval TaO_2-Ta_2O_5; and the ζ phase, Ta_2O_5 [71.43 at. (18.11 wt.) % O], which is isomorphous (as already reported by [2, 3]) with the low-temperature form of Nb_2O_5 and for which no high-temperature modification (cf. above) could be found [6]. Since the intermediate oxides could not be prepared in a pure state and the various methods of preparation used certainly did not exclude the possibility of contamination, the results of [4] should be checked.

Solubility of O in Solid Ta. Expansion of the Ta lattice on addition of oxygen was observed by [7, 8, 3, 4]. [7] found by measurements of electrical resistance that

Ta absorbs up to 20 volumes (1.9 at. %) of oxygen at 1800°K (1527°C) before a compound is formed. According to the parametric work of [3], the solubility at 1050°C is 0.9 wt. (9.3 at.) % O but appears to be considerably higher near 2000°C. [4] estimated the solubility to be about 5 at. % O. Heating above about 2200°C in vacuo results in the expulsion of the dissolved oxygen [7, 8]. See Note Added in Proof.

Kinetics. For the kinetics of the reaction of gaseous oxygen with Ta in various temperature and pressure ranges, see [9–13]. Anodic oxide films were studied by [14, 15]. There is agreement that the oxide film is essentially Ta_2O_5; for some peculiarities, see [12, 15, 13].

Note Added in Proof. Physical, technological, and chemical properties of Ta-O alloys with up to 7.25 at. % O have been studied by [16–18]. Measurements of electrical conductivity [17] and hardness [16] yielded the following limiting solubilities of O in Ta (in at. % O): 3.7, 2.9, 2.3, and about 1.4 at 1500, 1200, 1000, and 750°C, respectively.

According to [19], the low-temperature (β) form of Ta_2O_5 is orthorhombic and isostructural with the low-temperature form of Nb_2O_5 and with α-U_3O_8, the lattice parameters being a = 6.20 A, b = 3.67 A, c = 3.90 A. [19] added, however, that X-ray diffraction patterns "consistently show a few weak lines which apparently refer to a larger multiple cell."

1. Quoted from [2].
2. S. Lagergren and A. Magnéli, *Acta Chem. Scand.*, **6,** 1952, 444–446.
3. R. J. Wasilewski, *J. Am. Chem. Soc.*, **75,** 1953, 1001–1002.
4. N. Schönberg, *Acta Chem. Scand.*, **8,** 1954, 240–245.
5. The debyogram of Ta_2O is very similar to that of "Nb_2O" (Kubaschewski; see Nb-O); however, attention should be given to the fact that Brauer (see Nb-O) has clearly shown that the pattern thought to belong to "Nb_2O" (which does not exist) was actually that of Nb_2N.
6. It may be that the high-temperature form of Ta_2O_5 of [2, 3] is identical with the ϵ phase of [4]. In some papers Ta_2O_5 is said to be orthorhombic, but obviously no structural analysis has been made as yet.

7. M. R. Andrews, *J. Am. Chem. Soc.*, **54,** 1932, 1845–1854.
8. R. H. Myers, *Metallurgia*, **41,** 1950, 301–304.
9. D. J. McAdam and G. G. Geil, *J. Research Natl. Bur. Standards*, **28,** 1942, 593–635.
10. E. A. Gulbransen and K. F. Andrew, *J. Electrochem. Soc.*, **96,** 1949, 364–376.
11. E. A. Gulbransen and K. F. Andrew, *Trans. AIME*, **188,** 1950, 586–599.
12. J. T. Waber, G. E. Sturdy, E. M. Wise, and C. R. Tipton, *J. Electrochem. Soc.*, **99,** 1952, 121–129.
13. R. C. Peterson, W. M. Fassell, and M. E. Wadsworth, *Trans. AIME*, **200,** 1954, 1038–1044.
14. W. G. Burgers, A. Claassen, and J. Zernike, *Z. Physik*, **74,** 1932, 593–603.
15. D. A. Vermilyea, *Acta Met.*, **1,** 1953, 282–294.
16. E. Gebhardt and H. Preisendanz, *Z. Metallkunde*, **46,** 1955, 560–568.
17. E. Gebhardt and H. Koeberle, results of measurements of electrical conductivity given in [16] in advance of publication.
18. E. Gebhardt and H. Seghezzi, *Z. Metallkunde*, **48,** 1957, 430–435.
19. W. T. Holser, *Acta Cryst.*, **9,** 1956, 196.

$\bar{1}$.5238
0.4762

O-Ti Oxygen-Titanium

Ti-TiO Region. The constitution of the system up to 30 wt. % O has been established by [1] (Fig. 584). Micrographic analysis of cast and heat-treated samples, X-ray diffraction analysis, and incipient-melting techniques were used as principal

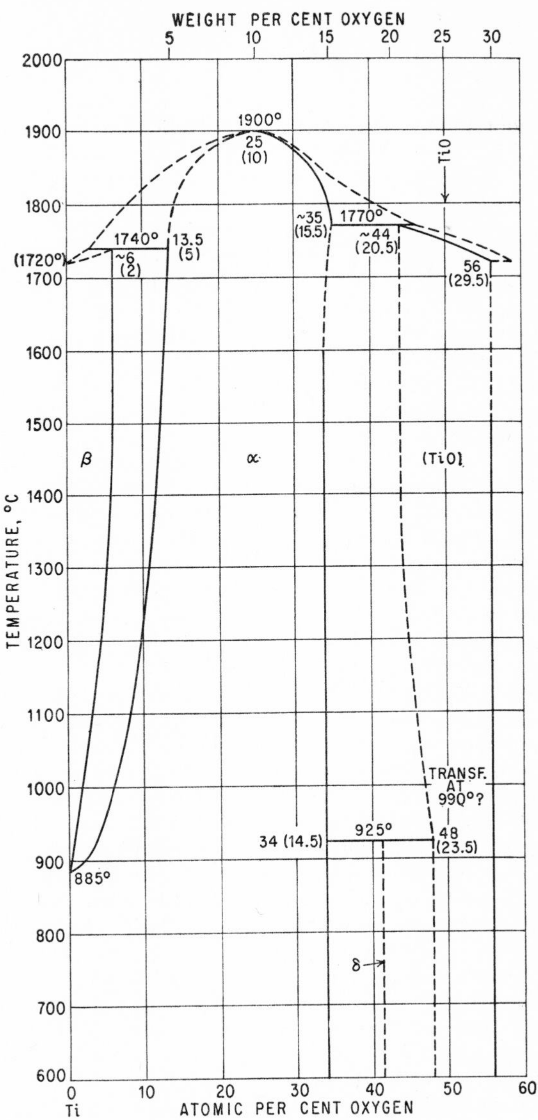

Fig. 584. O-Ti. (See also Note Added in Proof.)

tools. A hitherto unreported intermediate phase, designated δ [~19 wt. (41.3 at.) % O], was found to be formed from a peritectoid reaction at 925°C. The stoichiometric formulas Ti_3O_2 (18.2 wt. % O) or Ti_4O_3 (20.0 wt. % O) most closely approach the composition of the δ phase. The intermediate phase in equilibrium with TiO at oxygen contents higher than 29.5 wt. % was determined to be Ti_2O_3.

Comparisons with Other Literature. The results of an investigation at New York University [2] confirm the diagram of [1] in almost every detail. The same alloy range was studied, iodide Ti, and similar techniques of investigation were employed. The phase boundaries and characteristic temperatures determined superimpose on the diagram of Fig. 584 within the limits of experimental accuracy except in two regions: (*a*) The $\beta/(\beta + \alpha)$ boundary shifts increasingly to higher oxygen contents above about 1200°C, resulting in a peritectic composition at 3 wt. (8.5 at.) % O (instead of 2 wt. % as in Fig. 584) [3]. See Note Added in Proof. (*b*) The $TiO/(TiO + Ti_2O_3)$ boundary is placed at about 30.5 wt. % O (instead of 29.5 wt. %). In the first case, the highest annealing temperature employed in [2] was 1400°C, and the boundary appears to have been extrapolated upward to the peritectic temperature. Accordingly, the work of [1], which employs experimental data above 1400°C, is to be preferred. In regard to the second point, the diagram of [2] shows only a dashed boundary. [4, 5] as well as [6, 7] investigated the phases occurring in the Ti-O system by X-ray diffraction measurements. In the Ti-TiO region the following homogeneity ranges were found: α at 0–30 [4, 5] and 0–33.3 at. % O [6, 7] and (TiO) at 38–55 [4, 5] and ~40–56 at. % O [6, 7]. A higher degree of impurity content in the Ti used by [4, 5] might account for his positioning of the phase boundaries at lower oxygen contents than those indicated in later works.

[8] showed that the introduction of oxygen raises the $\alpha \rightleftarrows \beta$ transformation in Ti, but did not obtain quantitative data. The transformation has been investigated by [9, 10, 1, 2]. [10] employed measurements of thermoelectric power at various temperatures as a principal tool, whereas other investigators used micrographic analysis of annealed and quenched alloys. Iodide Ti was the starting material in each case. A comparison of the data of these studies gives the following picture: The $\beta/(\beta + \alpha)$ phase boundary up to 1 wt. (3 at.) % O is in close agreement among three investigators, [10] showing a greater solubility of oxygen in β for each temperature. [1] and [2] agree closely on the $\alpha/(\alpha + \beta)$ boundary up to 2 wt. (6 at.) % O. [9] show a somewhat narrower, and [10] a considerably narrower, $(\alpha + \beta)$ field (see also [3]). Slight preference must be given to the boundaries presented by [1], because they are connected with experimentally determined boundaries at higher oxygen contents.

Crystal Structures. Measurements of the lattice parameters in the α solid solution were carried out by [5, 11, 7, 1, 2, 12]. The δ phase, reported by [1] to exist at about 19 wt. (41.3 at.) % O at temperatures below 925°C, could not be completely isolated for X-ray studies, even on very long annealing. Alternatively, the extra lines produced by the appearance of the phase were examined as a group. The lines, with one exception, could be indexed for the tetragonal system. The calculated lattice parameters were $a = 5.344$ A, $c = 6.658$ A, $c/a = 1.246$. The phase TiO (25.04 wt. % O) has the f.c.c. NaCl (B1) type of structure [13]. The parameter $a = 4.243$ A was reported by [14]. [15] found the lower value $a = 4.162$ A. The parametric curve for the solid solutions based on TiO was determined by [4, 6, 7, 1]; according to [5, 7], this phase possesses large numbers of vacant lattice sites. [4] considered this phase to have the ionic type of bonding; [16] has recently presented calculations which indicate that the bonding may rather be of the metallic type.

There are some indications—from high-temperature heat content [17] as well as X-ray [18] measurements—that TiO might undergo a structural change at about 990°C [17]; see Note Added in Proof.

Higher Oxides. For a discussion of the higher oxides of titanium (Ti_2O_3, Ti_3O_5, and TiO_2), reference is made to [19–21, 32]. [21] constructed a tentative ("not impossible") phase diagram Ti-TiO_2 on the basis of literature data (which, above 56 at. % O, are unsystematic).

For the *kinetics* of the oxidation of Ti under various conditions of temperature and pressure, see [22–26, 12, 27, 33]; as to TiO_2 layers formed, only the rutile modification was observed, regardless of the temperature of reaction.

Note Added in Proof. Phase boundaries above 700°C and up to 35 wt. % O have been redetermined by [29]. Considering the high temperatures involved and the limited accuracy of oxygen analyses, the agreement between the results of [1] (Fig. 584) and those of [29] is remarkably good. For a comparison, see Fig. 3 of [29] and Table 37*a*. The following remarks may be made: (*a*) There is agreement in the results of [2, 3, 29] that the solubilities of oxygen in β_{Ti} above 1200°C shown in Fig. 584 [1] are too low. (*b*) The form of the solidification equilibria of the (TiO) phase claimed by [29] is rather improbable. Both the liquidus and the solidus are assumed to be nearly isothermal lines (with the peritectic horizontal α + melt $\rightleftarrows$ TiO in an intermediate position) between about 45 and 55 at. % O. (*c*) In contrast to the (likewise improbable) vertical oxygen-rich α-phase boundary of [1] (Fig. 584), [29] found a marked decrease in oxygen solubility below 1000°C.

According to [29], Ti_2O_3 melts at 1800°C with an open maximum.

The transformation in the (TiO) phase has been studied by [29, 28]. According to thermal analysis [29], the transformation temperatures appear to have a maximum (about 980°C) at 50 at. % O. X-ray results of [28] appear to indicate that the low-temperature modification (called β by [29]) of TiO has a distorted NaCl-type structure of low symmetry. [29] observed lines of a tetragonal (other than δ) and of a b.c.c. phase (?).

According to a short abstract, [30] studied the Ti-O system by X rays, microscopy, and measurement of melting point, with emphasis on the oxygen-rich portions of the diagram.

[31] reported the identification of seven discrete intermediate phases in the narrow composition range $TiO_{1.75}$-$TiO_{1.90}$ (63.6–65.5 at. % O).

Table 37*a*. Characteristic Data of the Phase Diagram by [29] [Compositions in At. % O (Wt. % in Parentheses), Temperatures in °C]

Peritectic at ~1720°	Melt [2.1 (0.7)] + α [16.0 (6.0)] $\rightleftarrows$ β [10.3 (3.7)]
Maximum in liquidus at ~1870°	α [23.7 (9.4)]
Peritectic at ~1740°	α [36.7 (16.2)] + melt [56 (30)] $\rightleftarrows$ (TiO) [41.7 (19.3)]
Peritectoid at ~910°	α [28.6 (11.8)] + (TiO) [48.3 (23.8)] $\rightleftarrows$ δ [41.3 (19)]

1. E. S. Bumps, H. D. Kessler, and M. Hansen, *Trans. ASM*, **45**, 1953, 1008–1025; discussion, pp. 1025–1028. Alloys were prepared by arc-melting charges of iodide Ti and high-purity TiO_2. The phase diagram is based on nominal alloy compositions, but a number of chemical analyses were made to check the validity of this presentation.
2. I. Cadoff and A. E. Palty, *U.S. Atomic Energy Comm., Publ.* WAL-401-14-30, 1953.
3. R. J. Wasilewski and G. L. Kehl, *J. Inst. Metals*, **83**, 1954, 94–104, extrapolated data for the $\beta/(\beta + \alpha)$ boundary (from measurements of diffusion at 950–1414°C) which are said to be in good agreement with those of [2]. At 1400°C a solubility of about 6.5 at. % O was found (taken from a graph).
4. P. Ehrlich, *Z. Elektrochem.*, **45**, 1939, 362–372.
5. P. Ehrlich, *Z. anorg. Chem.*, **247**, 1941, 53–64.
6. H. Krainer and K. Konopicky, *Berg-Hüttenmänn. Monatsh.*, **92**, 1947, 166–178.
7. H. Krainer, *Arch. Eisenhüttenw.*, **21**, 1950, 119–127.
8. J. H. de Boer, W. G. Burgers, and J. D. Fast, *Proc. Acad. Amsterdam*, **39**, 1936, 515–519.

9. R. I. Jaffee, H. R. Ogden, and D. J. Maykuth, *Trans. AIME,* **188,** 1950, 1261–1266.
10. A. E. Jenkins and H. W. Worner, *J. Inst. Metals,* **80,** 1951-1952, 157–166.
11. H. T. Clark, *Trans. AIME,* **185,** 1949, 588–589.
12. A. E. Jenkins, *J. Inst. Metals,* **82,** 1954, 213–221.
13. In the discussion on [1], this structure has been erroneously called "body-centered cubic."
14. H. Brakken, *Z. Krist.,* **67,** 1928, 547–549.
15. W. Dawihl and K. Schröter, *Z. anorg. Chem.,* **233,** 1937, 178–183.
16. W. Rostoker, *Trans. AIME,* **194,** 1952, 981–982.
17. B. F. Naylor, *J. Am. Chem. Soc.,* **68,** 1946, 1077–1080.
18. A. E. Jenkins; see discussion on [1].
19. "Gmelins Handbuch der anorganischen Chemie," System No. 41, Verlag Chemie, G.m.b.H., Weinheim/Bergstrasse, Germany, 1951.
20. J. Barksdale, "Titanium," The Ronald Press Company, New York, 1949.
21. R. C. de Vries and R. Roy, *Bull. Am. Ceram. Soc.,* **33,** 1954, 370–372.
22. L. G. Carpenter and F. R. Reavell, *Metallurgia,* **39,** 1948, 63.
23. J. W. Hickman and E. A. Gulbransen, *J. Anal. Chem.,* **20,** 1948, 158–165.
24. E. A. Gulbransen and K. F. Andrew, *Trans. AIME,* **185,** 1949, 741–748.
25. M. H. Davies and C. E. Birchenall, *Trans. AIME,* **191,** 1951, 877–880.
26. P. H. Morton and W. M. Baldwin, *Trans. ASM,* **44,** 1952, 1004–1028.
27. L. S. Richardson and N. J. Grant, *Trans. AIME,* **200,** 1954, 69–70.
28. C. Wang and N. J. Grant, *Trans. AIME,* **206,** 1956, 184–185.
29. T. H. Schofield and A. E. Bacon, *J. Inst. Metals,* **84,** 1955, 47–53.
30. H. Nishimura and H. Kimura, *J. Japan. Inst. Metals,* **20,** 1956, 524–528; *Titanium Abstr. Bull.,* **2,** 1957, 2279.
31. S. Andersson and A. Magnéli, *Naturwissenschaften,* **43,** 1956, 495–496.
32. W. H. Baur, *Acta Cryst.,* **9,** 1956, 515–520.
33. W. Kinna and W. Knorr, *Z. Metallkunde,* **47,** 1956, 594–598.

$\bar{1}.4970$
0.5030

O-V Oxygen-Vanadium

The constitution of the V-O system is a complex one and is far from being fully understood at the present time. In Table 38, a survey of the systematic investigations carried out thus far is given. In the following, only V-rich alloys (up to about 50 at. % O) will be discussed in detail.

Table 38. Investigations of the V-O System

Year	Ref.	Range investigated	Methods used
1932	1	V-VO_2	Microscopic, X-ray
1939	2	V_2O_3-V_2O_5	X-ray, magnetic
1942	3	V-V_2O_3	X-ray
1948	4	V_2O_3-V_2O_5	Electrical conductivity, X-ray
1948	5	VO_2-V_2O_5	X-ray, microscopic
1953	6	V-VO	X-ray, microscopic (mainly)
1953	7	VO-V_2O_5	X-ray
1954	8	VO-$VO_{1.30}$	X-ray
1955	9	V-VO	Microscopic, X-ray, incipient melting

V-VO. For this interval two conflicting tentative phase diagrams have been set forth (see below). [1] (cf. Table 38) observed the phases (V) and (VO); however, no precise data as to the extension of the homogeneity ranges were given [10]. No evidence of the compound V_2O, reported in the early literature, could be found by [11, 1]. A tetragonal phase (now known as β; cf. below) was first observed by [3], who, however, erroneously suggested a continuous transition from the b.c.c. V lattice to the tetragonal one. [3] showed the homogeneity range of the (VO) phase to extend from $VO_{0.9}$ to $VO_{1.3}$ (47.4–56.5 at. % O); their assumption that this phase decomposes at low temperatures was not corroborated by later work. From equilibria

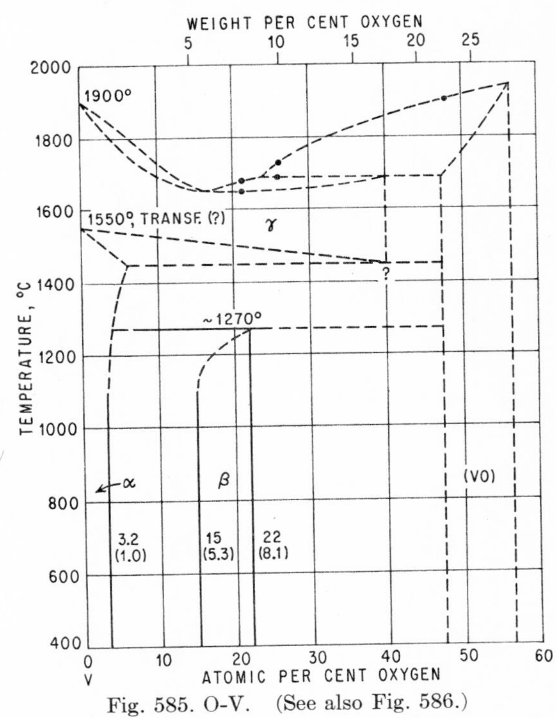

Fig. 585. O-V. (See also Fig. 586.)

measurements (V, O, alkaline-earth metals) and limited microscopic observations, [12] suggested a solubility of O in solid V of 0.25 wt. (0.8 at.) % at 1000°C.

A study by [6], mainly by means of roentgenographic, microscopic, and hydrogen equilibria methods, of the solubility of oxygen in solid V and of the nature of the phase relationship in the V-rich region, particularly at temperatures below 1300°C, resulted in a partial phase diagram (Fig. 585), of which the region above 1270°C is largely speculative. By parametric analyses the solubility of O in solid V was found to be 3.2 at. (1.0 wt.) % and the tetragonal β phase to be homogeneous between 15 and 22 at. (5.3 and 8.1 wt.) % O. [6] stated that, "while the microstructures observed were very interesting and instructive in a qualitative sense, it was not feasible to use metallographic evidence in locating phase boundaries." The homogeneity range of the (VO) phase shown in Fig. 585 had been taken from the work of [3], whereas the constitution above 1300°C is based on a few optical-pyrometer liquidus and solidus points, the eutectoid-like appearance of some regions in microstructures of alloys containing more than 22 at. % O, and indications found by thermo-emf measurements for a phase transformation in pure V at about 1550°C (not corroborated by [13]).

An X-ray powder analysis by [8] confirmed the results on the (V) phase and the β phase ($VO_{0.15}$-$VO_{0.25}$, i.e., 13.0–20.0 at. % O, according to [8]) obtained by [6] and showed the homogeneity range of the (VO) phase to extend between $VO_{0.8}$ and $VO_{1.2}$ (44.5 and 54.5 at. % O) (cf. above).

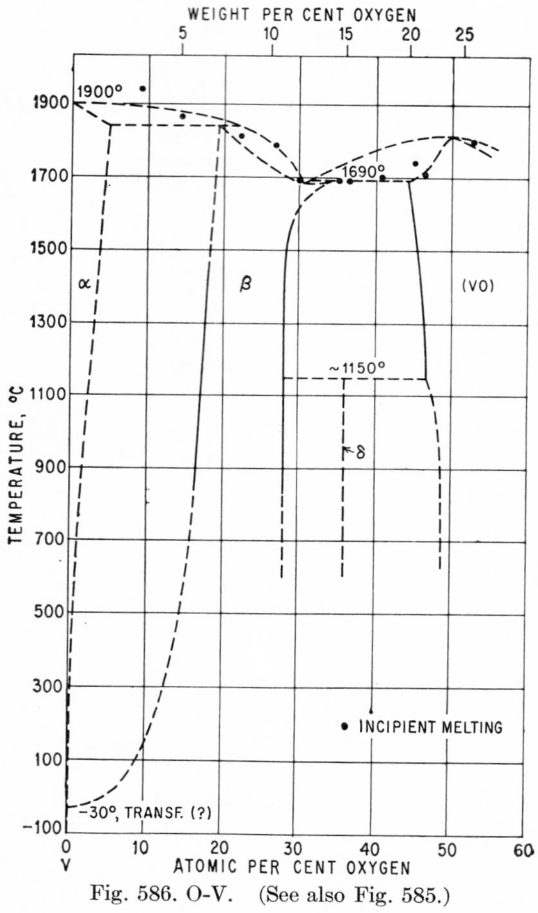

Fig. 586. O-V. (See also Fig. 585.)

[9], however, arrived at a quite different constitution for the V-VO region (Fig. 586). They studied 16 alloys heat-treated at temperatures up to the melting range by microscopic and X-ray diffraction work and found some evidence for a phase transformation in pure V at about −30°C by measurements of electrical resistivity. [9] consider the β phase as a stabilization of a tetragonal low-temperature modification of V (cf. O-Ti, O-Zr). The solubility of oxygen in solid V is assumed to decrease from about 1.6 wt. (5 at.) % O at 1840°C (taken from a graph) to less than 0.25 wt. (0.8 at.) % O at room temperature, and the "continuity of parameters establishes that the β phase at room temperature is homogeneous over the range 2 to 10.3 wt. [6.1 to

26.8 at.] % O." Some evidence for the existence of an additional "δ" phase between β and (VO) [14] was found [9].

In summarizing, the following may be said: Apart from the question of the existence or nonexistence of the δ phase, the two partial phase diagrams show no major disagreements in the temperature range 400–1200°C. As to the solubility of O in solid V, the data of [6] have a better basis than those of [9]. On the other hand, the high-temperature region of the diagram of [6] is largely speculative, whereas that of [9] is based on (limited) experimental data. Whether or not V undergoes an allotropic transformation, and at what temperature, must be decided by further experimental work.

Crystal Structures. The lattice parameter of the V phase increases with increasing oxygen content [6, 8]. The β phase has a b.c. tetragonal structure [3, 6, 15, 8, 9], with a = 2.948 A, c = 3.478 A, c/a = 1.180 at 21.0 at. % O [15]. No predominant X-ray diffraction pattern of the δ phase (reported by [9] only) could be obtained. Tentative indexing suggested a hexagonal lattice.

The (VO) phase has the NaCl (B1) type of structure [1, 3, 7, 8, 9], with a = 4.093 A at 50 at. % O [8]. The lattice parameter increases with increasing O content.

Higher Oxides. A great number of intermediate phases in the VO-V_2O_5 interval has been reported in the literature. Besides the papers already listed in Table 38, the references [16–28, 31, 32] pertain to this subject.

The kinetics of oxidation of V has been investigated by [29, 30].

1. C. H. Mathewson, E. Spire, and C. H. Samans, *Trans. ASST*, **20,** 1932, 357–384.
2. E. Hoschek and W. Klemm, *Z. anorg. Chem*, **242,** 1939, 63–69.
3. W. Klemm and L. Grimm, *Z. anorg. Chem.*, **250,** 1942, 42–55.
4. W. Klemm and P. Pirscher, *Optik*, **3,** 1948, 75–80.
5. F. Aebi, *Helv. Chem. Acta*, **31,** 1948, 8–21.
6. A. U. Seybolt and H. T. Sumsion, *Trans. AIME*, **197,** 1953, 292–299.
7. G. Andersson, *Research*, **6,** 1953, 45S–46S; *Acta Chem. Scand.*, **8,** 1954, 1599–1606.
8. N. Schönberg, *Acta Chem. Scand.*, **8,** 1954, 221–225.
9. W. Rostoker and A. S. Yamamoto, *Trans. ASM*, **47,** 1955, 1002–1017; preliminary results were published in *Trans. ASM*, **46,** 1954, 1136–1163; see also discussion, pp. 1163–1167.
10. Microscopic work suggested that solubility increases with temperature in both boundaries of the heterogeneous V-VO area.
11. J. Koppel and A. Kaufmann, *Z. anorg. Chem.*, **45,** 1905, 352–358.
12. N. P. Allen, O. Kubaschewski, and O. v. Goldbeck, *J. Electrochem. Soc.*, **98,** 1951, 417–424; discussion, **99,** 1952, 274 (The free-energy diagram of the V-O system).
13. J. O. McCalden and P. Duwez, *Trans. AIME*, **200,** 1954, 619–620.
14. "There is insufficient evidence to identify δ as V_2O" [9].
15. C. W. Tucker, A. U. Seybolt, and H. T. Sumsion, *Acta Met.*, **1,** 1953, 390–393.
16. V. M. Goldschmidt, T. Barth, and G. Lunde, *Skrifter Norske Videnskaps.-Akad., Oslo, Mat. Nat. Kl.*, no. 7, 1925. See also *Strukturbericht*, **1,** 1913-1928, 264.
17. V. M. Goldschmidt, *Skrifter Norske Videnskaps.-Akad., Oslo, Mat. Nat. Kl.*, no. 1, 1926; see also *Strukturbericht*, **1,** 1913-1928, 211.
18. W. H. Zachariasen, *Skrifter Oslo*, no. 4, 1928; see also *Strukturbericht*, **2,** 1928-1932, 310.
19. J. A. A. Ketelaar, *Z. Krist.*, **A95,** 1936, 9–27; *Chem. Weekblad*, **33,** 1936, 51–57; *Nature*, **137,** 1936, 316.
20. F. Machatschki, *Naturwissenschaften*, **24,** 1936, 742–743.
21. C. T. Anderson, *J. Am. Chem. Soc.*, **58,** 1936, 564–567.
22. M. Foëx, *Compt. rend.*, **223,** 1946, 1126–1128.

23. O. A. Cook, *J. Am. Chem. Soc.*, **69**, 1947, 331–333.
24. M. Foëx, *Compt. rend.*, **229**, 1949, 880–882.
25. G. Foëx and J. Wucher, *Compt. rend.*, **229**, 1949, 882–884.
26. A. Byström, K. A. Wilhelmi, and O. Brotzen, *Acta Chem. Scand.*, **4**, 1950, 1119–1130.
27. M. Foëx, S. Goldsztaub, R. Wey, J. Jaffray, R. Lyand, and J. Wucher, *J. recherches centre natl. recherche sci., Labs. Bellevue (Paris)*, **4**(21), 237–259.
28. J. Jaffray and A. Dumas, *J. recherches centre natl. recherche sci., Labs. Bellevue (Paris)*, **5**(27), 360–366.
29. D. J. McAdam and G. G. Geil, *J. Research Natl. Bur. Standards*, **28**, 1942, 593.
30. E. A. Gulbransen and K. F. Andrew, *J. Electrochem. Soc.*, **97**, 1950, 396–404.
31. M. S. Archer, D. S. P. Roebuck, and F. J. Whitby, *Nature*, **174**, 1954, 754–755.
32. S. Andersson and A. Magnéli, *Naturwissenschaften*, **43**, 1956, 495–496.

$\bar{2}.9395$
1.0605

O-W Oxygen-Wolfram

In 1933 an extensive review of the then available literature was made [1], and the following excerpts show the conclusions reached: "Nowadays the oxides WO_2, WO_3, and W_2O_5 or W_4O_{11} are considered to be stable compounds. Besides those three oxides, the monoxide WO is known as an intermediate stage in oxidation. Other oxides described in the literature (W_2O_3, W_4O_3, W_5O_9, W_3O_8, and W_5O_{14}) are uncertain. There are data on the dissociation equilibria of the stable oxides (computed from measurements of reduction equilibria with H or CO which do not agree too well with each other) which should not be considered as very reliable." For references and detailed discussions of the older literature, see [1].

This discussion will deal with the more recent literature, which corroborates the oxides WO_2 (14.82 wt. % O), W_4O_{11} [73.33 at. (19.30 wt.) % O], and WO_3 (20.70 wt. % O) mentioned above, in the order of increasing oxygen content.

The *solubility of oxygen in solid* W appears to be negligible. In the roentgenographic examination of alloys with compositions between 0 and 66 at. % O, [2] observed a mixture of WO_2 and metallic W, the lattice parameter of the latter being the same (within the limits of accuracy) as that of pure W.

W_3O (?). According to [3], the existence of "β-W" as a modification of W is contradicted by the low density of the phase and by its decomposition into W and WO_2 at about 700°C. [3] assume that "β-W" is a metallic W oxide with the probable ideal formula W_3O (2.82 wt. % O); the unit cell would then contain 6 W + 2 O atoms distributed at random over the eight positions of the A15 type of structure. See Note Added in Proof.

WO_2, having a very narrow range of homogeneity [2], had been previously [4, 2] thought to be of the tetragonal rutile type; however, [5, 6] have shown it to be actually monoclinic, with a = 5.560 A, b = 4.884 A, c = 5.546 A, β = 118.93°, and 12 atoms per unit cell. The structure is isomorphous with that of MoO_2.

"γ" **Phase.** From X-ray diffraction and electrical-resistivity studies, [2] concluded the existence of a one-phase region between the compositions $WO_{2.65}$ and $WO_{2.76}$. This interval includes the formulas W_4O_{11} (= $WO_{2.75}$) and $W_{18}O_{49}$ (= $WO_{2.72}$). The monoclinic symmetry of this phase was recognized by [7] ("W_4O_{11}"); and [8] elucidated its structure ("$W_{18}O_{49}$"), which has the lattice parameters a = 18.32 A, b = 3.79 A, c = 11.04 A, β = 115°2′, and 67 atoms per unit cell.

"β" **Phase.** [2] concluded another one-phase field between the compositions $WO_{2.88}$ and $WO_{2.92}$. This was corroborated by [5], who found a monoclinic phase with a narrow range of homogeneity at about $WO_{2.90}$. The structure of this phase was determined by [9], who found the lattice parameters a = 12.1 A, b = 3.78 A, c = 23.4 A, β = 95° for "$W_{20}O_{58}$" (= $WO_{2.90}$). The structure is closely related to that of ReO_3 (DO_9 type).

WO_3. Because of its ferroelectric properties, WO_3 has become of great interest in the last few years. (*a*) Room-temperature form: [10] described the structure as being triclinic (pseudomonoclinic) (DO_{10} type); however, the atomic positions given by [10] correspond to an orthorhombic symmetry, as was pointed out in [11]. Later reports on this structure vary. When [12] interpreted the powder pattern of WO_3, he did not find any deviation from a monoclinic quadratic form, whereas [13], by measuring single crystals optically, came to the conclusion that the symmetry was triclinic. [14] stated that the structure was orthorhombic. A renewed study by [15] found the symmetry to be monoclinic, with a = 7.285 A, b = 7.517 A, c = 3.835 A, β = 90.90°; the positions of the W atoms given by [10] were verified with Fourier methods. (*b*) Low-temperature form: According to [16], a polymorphic transformation occurs in WO_3 near −50°C, resulting in a structure with a higher symmetry than that of the room-temperature modification. (*c*) High-temperature form: A polymorphic transformation somewhere between 700 and 750°C into a tetragonal high-temperature modification has been reported by [17–19, 13, 14, 20]. According to [18], its structure, with a = 5.250 ± 2 A, c = 3.915 ± 2 A, c/a = 0.746, and 8 atoms per unit cell, may be described as a distorted structure of the ReO_3 (DO_9) type. [13, 14, 20] give an a axis about $\sqrt{2}$ times greater than that of [18]. According to [13], the triclinic (cf. above) room-temperature modification becomes orthorhombic above 350°C before transforming into the tetragonal form at 735°C.

It appears that the WO_3 room-temperature structure can tolerate a slight deficiency in oxygen [2, 5].

Note Added in Proof. [21] prepared W by several different methods and examined the products by X-ray diffraction. They concluded that "β-W can only be formed by chemical processes, and then only when certain oxides are present or likely to be formed in the system. Our findings appear to offer additional support for the proposal by [3] that the β-form is, in reality, an oxide of W."

[22] measured the thermal expansion of WO_3 from room temperature to 700°C by means of X-ray diffraction. They found that the monoclinic [15] room-temperature cell undergoes a transformation near 330°C to an orthogonal form which is probably orthorhombic.

1. "Gmelins Handbuch der anorganischen Chemie," System No. 54, p. 107, Verlag Chemie, G.m.b.H., Berlin, 1933.
2. O. Glemser and H. Sauer, *Z. anorg. Chem.*, **252,** 1943, 144–159.
3. G. Hägg and N. Schönberg, *Acta Cryst.*, **7,** 1954, 351–352.
4. V. M. Goldschmidt, *Skrifter Akad. Oslo*, **1926,** 17.
5. G. Hägg and A. Magnéli, *Arkiv Kemi, Mineral. Geol.*, **19A**(2), 1944.
6. A. Magnéli, *Arkiv Kemi, Mineral. Geol.*, **24A**(2), 1947.
7. F. Ebert and H. Flasch, *Z. anorg. Chem.*, **226,** 1935, 65–81.
8. A. Magnéli, *Arkiv Kemi*, **1,** 1949, 223–230.
9. A. Magnéli, *Nature*, **165,** 1950, 356–357; also *Arkiv Kemi*, **1,** 1950, 513–523.
10. H. Braekken, *Z. Krist.*, **78,** 1931, 484–488.
11. *Strukturbericht*, **2,** 1928-1932, 32–33.
12. A. Magnéli, *Acta Chem. Scand.*, **3,** 1949, 88–89.
13. J. Wyart and M. Foëx, *Compt. rend.*, **232,** 1951, 2459–2461.
14. R. Ueda and T. Ichinokawa, *Busseiron Kenkyu*, no. 36, 1951, pp. 64–68; *Chem. Abstr.*, **46,** 1952, 4874.
15. G. Andersson, *Acta Chem. Scand.*, **7,** 1953, 154–158.

16. B. T. Matthias and E. A. Wood, *Phys. Rev.*, **84**, 1951, 1255.
17. W. L. Kehl, R. G. Hay, and D. Wahl, *Phys. Rev.*, **82**, 1951, 774.
18. W. L. Kehl, R. G. Hay, and D. Wahl, *J. Appl. Phys.*, **23**, 1952, 212–215.
19. R. Ueda and T. Ichinokawa, *Phys. Rev.*, **82**, 1951, 563–564.
20. M. Foëx and J. Wyart, *Bull. soc. franç. minéral.*, **76**, 1953, 102–109; *Chem. Abstr.*, **47**, 1953, 9092.
21. M. G. Charlton and G. L. Davis, *Nature*, **175**, 1955, 131–132.
22. C. Rosen, E. Banks, and B. Post, *Acta Cryst.*, **9**, 1956, 475–476.

$\bar{1}.2440$
0.7560

O-Zr Oxygen-Zirconium

The constitution of the system Zr-ZrO_2 has been determined by [1]. Arc-melted iodide Zr crystal bar-based alloys were carefully annealed at temperature levels between 600 and 2000°C. Determination of the phase boundaries was accomplished by metallographic evaluation of specimens quenched from the various temperatures. Incipient-melting techniques were used to determine solidus curves, and X-ray diffraction work was employed to study the lattice parameters of the α solid solution and the phase ZrO_2. Figure 587 shows the diagram established by [1], supplemented by results from other sources as far as the transformation in ZrO_2 is concerned (cf. below). In the following, other literature pertaining to the constitution is discussed.

Primary Solid Solutions. [2] showed that the introduction of oxygen raises the $\alpha \rightleftarrows \beta$ transformation in Zr but determined the limits of the two-phase region only approximately. The maximum solubility of oxygen in α-Zr, 29.2 at. % according to [1], had previously [3–5] been reported to be about 40 at. %. Several physical and mechanical properties of low-oxygen alloys were measured by [6] and the lattice parameters of the α solid solution by [4, 7, 1].

Solidus Temperatures. [8] worked out a melting-point–composition diagram by optically measuring the temperature at which wire samples visibly melted. The results do not agree with those of [1] and have to be considered as obsolete.

ZrO_2. In the literature on ZrO_2, several ZrO_2 modifications have been described [9–15, 1]; however, only two, a monoclinic and a tetragonal one—both distorted CaF_2 structures—have been conclusively established as yet.

The room-temperature modification forms the monoclinic C43 prototype of structure [16], with a = 5.17 A, b = 5.26 A, c = 5.30 A; β = 80°10′ [13]. Above about 1000°C [17, 18, 10, 11, 14, 15] the tetragonal form, with a = 5.08 A, c = 5.17 A, $c/a = 1.01_8$ [17, 10], becomes stable. The transformation takes place over a wide range of temperatures [10, 11, 14, 15]. Whereas [17, 18, 10] reported that the tetragonal form can be retained at room temperature, [1] were unable to retain it by quenching. [18, 10] claimed to have produced a third (hexagonal?) polymorphic form of ZrO_2 by prolonged heating above 1900°C. A cubic form of ZrO_2, frequently reported in the early literature, forms only if certain foreign oxides (especially MgO) are present.

According to [8, 1], the homogeneity range of ZrO_2 (25.97 wt. % O) extends down to 63 at. (23 wt.) % O.

For the melting point of ZrO_2 the following values have been reported: 2677 ± 20 [19], 2687 ± 20 [20], 2715 ± 20 [21], and 2710 ± 15°C [22]. In Fig. 587 the average value 2700°C has been used.

Do Lower Oxides Exist? The existence of "suboxides" (ZrO and Zr_2O_3) has been postulated by some authors [23–28]; however, no conclusive evidence has ever been given [29]. [8] and [1] were unable to find any evidence for the existence of ZrO or another suboxide, at least above 700°C. In view of this, the findings of [30], that an order-disorder transition (simple cubic-b.c.c.) occurs in a 50 at. % O alloy at 1970°C (!), are hard to understand. A cubic ZrO phase (a = 4.584–4.620 A) has also been reported recently by [31] in a short note.

The kinetics of the oxidation of Zr has been studied by [32, 5, 33–35, 37, 38].

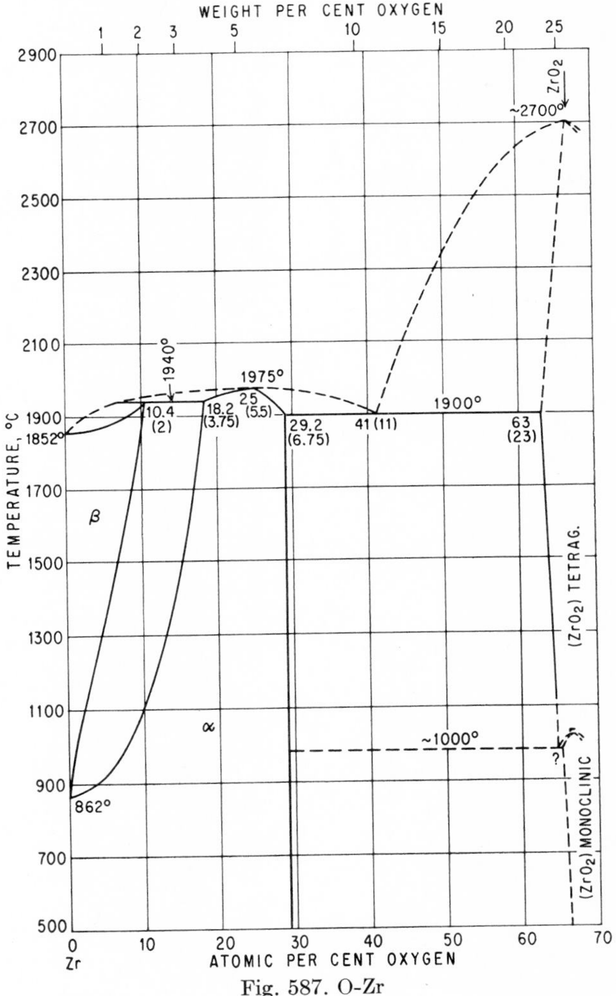

Fig. 587. O-Zr

Reviews of the literature of the system Zr-O and the compound ZrO_2 were made by [36] and [14], respectively.

1. R. F. Domagala and D. J. McPherson, *Trans. AIME*, **200,** 1954, 238–246.
2. J. H. de Boer and J. D. Fast, *Rec. trav. chim.*, **55,** 1936, 459–467.
3. J. D. Fast, *Metallwirtschaft*, **17,** 1938, 641–644.
4. J. H. de Boer and J. D. Fast, *Rec. trav. chim.*, **59,** 1940, 161–167.
5. W. G. Guldner and L. A. Wooten, *J. Electrochem. Soc.*, **93,** 1948, 223–235.

6. R. M. Treco, *Trans. ASM*, **45**, 1953, 872–890.
7. R. M. Treco, *Trans. AIME*, **197**, 1953, 344–348.
8. D. Cubicciotti, *J. Am. Chem. Soc.*, **73**, 1951, 2032–2035.
9. *Strukturbericht*, **1**, 1913-1928, 196, 777–778; **2**, 1928-1932, 265, 267–269; **3**, 1933-1935, 301; **4**, 1936, 119; **7**, 1939, 87–88.
10. W. M. Cohn, *J. Electrochem. Soc.*, **68**, 1935, 65–71.
11. R. F. Geller and P. J. Yavorsky, *J. Research Natl. Bur. Standards*, **35**, 1945, 87–110.
12. J. W. Hickman and E. A. Gulbransen, *Anal. Chem.*, **20**, 1948, 158–165 (Examination of oxide layers by electron diffraction).
13. P. Duwez and F. Odell, *J. Am. Ceram. Soc.*, **33**, 1950, 274–283.
14. A. Dietzel and H. Tober, *Ber. deut. keram. Ges.*, **30**, 1953, 47–61, 71–82.
15. P. Murray and E. B. Allison, *Trans. Brit. Ceram. Soc.*, **53**, 1954, 335–361 (Study of the monoclinic $\rightleftarrows$ tetragonal phase transformation in ZrO_2). *Chem. Abstr.*, **48**, 1954, 11891.
16. St. v. Náray-Szabó, *Z. Krist.*, **A94**, 1936, 414–416.
17. O. Ruff and F. Ebert, *Z. anorg. Chem.*, **180**, 1929, 19–41.
18. W. M. Cohn and S. Tolksdorf, *Z. physik. Chem.*, **B8**, 1929, 331–356.
19. F. Henning, *Naturwissenschaften*, **13**, 1925, 661.
20. P. Clausing, *Z. anorg. Chem.*, **204**, 1932, 33–39.
21. N. Zirnowa, *Z. anorg. Chem.*, **218**, 1934, 193.
22. W. A. Lambertson and F. H. Gunzel, *U.S. Atomic Energy Comm., Publ.* AECD-3465, 1952.
23. E. Wedekind, *Z. anorg. Chem.*, **45**, 1905, 391.
24. L. Weiss and E. Neumann, *Z. anorg. Chem.*, **65**, 1910, 248–278.
25. A. E. van Arkel and J. H. de Boer, *Z. anorg. Chem.*, **148**, 1925, 346.
26. J. Jacob, *Schweiz. mineralog. petrog. Mitt.*, **17**, 1937, 154–163.
27. E. Zintl, W. Morawietz, and E. Gastinger, *Z. anorg. Chem.*, **245**, 1940, 8–11.
28. H. Jakobs, *J. Appl. Phys.*, **17**, 1946, 596–603.
29. Some of the phenomena observed may easily be explained on the basis of the solid solutions shown in Fig. 587.
30. M. Hoch and H. L. Johnston, *Acta Cryst.*, **7**, 1954, 660 (abstract of paper presented at meeting of International Union of Crystallography); see also M. Hoch, W. O. Grooves, and H. L. Johnston, *Am. Chem. Soc., Abstr. of Papers*, **126**, 1954, 42R-43R.
31. N. Schönberg, *Acta Chem. Scand.*, **8**, 1954, 627 (footnote).
32. D. J. McAdam and G. G. Geil, *J. Research Natl. Bur. Standards*, **28**, 1942, 593–635.
33. E. A. Gulbransen and K. F. Andrew, *J. Electrochem. Soc.*, **96**, 1949, 364–376.
34. E. A. Gulbransen and K. F. Andrew, *Trans. AIME*, **185**, 1949, 515–525.
35. D. Cubicciotti, *J. Am. Chem. Soc.*, **72**, 1950, 4138–4141.
36. C. F. P. Bevington, S. L. Martin, and D. H. Mathews, *Proc. Intern. Congr. Pure and Appl. Chem. 11th Congr., London*, **11**, 1947, 3–16.
37. J. Belle and M. W. Mallett, *J. Electrochem. Soc.*, **101**, 1954, 339.
38. E. A. Gulbransen and K. F. Andrew, *Trans. AIME*, **209**, 1957, 394–400.

0.7882
$\bar{1}.2118$

Os-P Osmium-Phosphorus

X-ray analysis has shown that Os forms only one phosphide, OsP_2 (24.57 wt. % P) [1]. The structure was not determined.

1. W. Biltz, H. J. Ehrhorn, and K. Meisel, *Z. anorg. Chem.*, **240**, 1939, 117–128.

$\bar{1}.9887$
0.0113

Os-Pt Osmium-Platinum

This system is characterized by the existence of a miscibility gap between terminal solid solutions, as shown by metallographic examination of alloys with 11, 25, 50, 60, and 75 wt. (about 10.8, 24.5, 49.4, 59.4, and 74.5 at.) % Pt, after annealing for 2 hr at 1200°C. The alloy with 11 wt. % Pt was found to be two-phase, indicating that the solid solubility of Pt in Os is smaller. The limit of solid solubility of Os in Pt lies just below 25 wt. % Os [1].

1. O. Winkler, *Z. Elektrochem.*, **49**, 1943, 221–228.

$\bar{1}.9008$
0.0992

Os-Pu Osmium-Plutonium

The partial phase diagram of Fig. 587*a* has been established by means of metallographic, X-ray, thermal, and dilatometric work on alloy samples weighing some

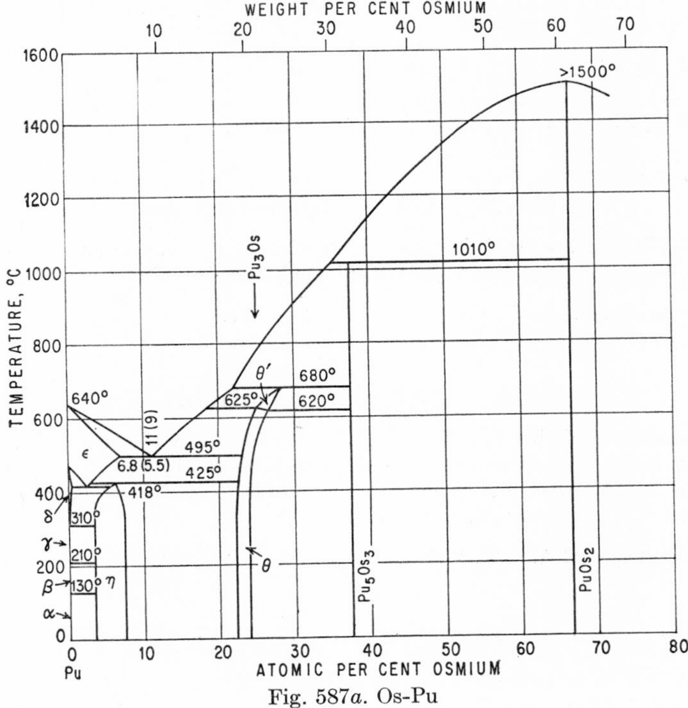

Fig. 587*a*. Os-Pu

hundreds of milligrams at the most [1]. The eutectoid at 418°C contains 1.96 wt. (2.45 at.) % Os. According to the results of thermal analysis, δ-Pu dissolves about 0.3 wt. (0.4 at.) % Os at 418°C. The solubilities of Os in the α, β, and γ phases of Pu were estimated metallographically to be less than 0.24 wt. (0.30 at.) % Os.

$PuOs_2$ (61.4 wt. % Os) is of the hexagonal $MgZn_2$ (C14) type of structure, with $a = 5.337$ A, $c = 8.682$ A, $c/a = 1.627$ [1].

1. S. T. Konobeevsky, *Conference of the Academy of Sciences of the U.S.S.R. on the Peaceful Uses of Atomic Energy, Division of Chemical Science*, July, 1955; English translation by U.S. Atomic Energy Commission, Washington, 1956.

0.0090
$\bar{1}$.9910

Os-Re Osmium-Rhenium

Os and Re, having the same h.c.p. crystal structure and nearly equal lattice parameters, form a continuous series of solid solutions. This was shown by metallographic examination and lattice-spacing measurements of alloys with 25, 50, and 75 wt. % Re [1].

1. O. Winkler, *Z. Elektrochem.*, **49**, 1943, 221–228.

0.7732
$\bar{1}$.2268

Os-S Osmium-Sulfur

According to tensimetric and roentgenographic investigations, only the sulfide OsS_2 (25.22 wt. % S) exists [1]. The compound is isotypic with pyrite (C2 type), $a = 5.65$ A [2], $a = 5.619$ A [3].

1. R. Juza, *Z. anorg. Chem.*, **219**, 1934, 129–140.
2. I. Oftedal, *Z. physik. Chem.*, **135**, 1928, 291–299.
3. K. Meisel, *Z. anorg. Chem.*, **219**, 1934, 141–142.

0.3818
$\bar{1}$.6182

Os-Se Osmium-Selenium

$OsSe_2$ (45.36 wt. % Se) is isotypic with pyrite (C2 type), $a = 5.945$ A [1]. OsSe could not be prepared by direct synthesis; its existence is unlikely.

1. L. Thomassen, *Z. physik. Chem.*, **B2**, 1929, 353–357.

0.8307
$\bar{1}$.1693

Os-Si Osmium-Silicon

Roentgenographic investigations covering the full range of composition revealed the existence of two silicides, with the approximate compositions Os_2Si_3 (18.14 wt. % Si) and $OsSi_2$ (22.80 wt. % Si). Os_2Si_3 was reported to be tetragonal, with $a = 5.58$ A, $c = 5.12$ A, $c/a = 0.919$ [1].

1. J. H. Buddery and A. J. E. Welch, *Nature*, **167**, 1951, 362.

0.2048
$\bar{1}$.7952

Os-Sn Osmium-Tin

Residues isolated from Os-Sn alloys proved to be Os [1]. These findings were confirmed by X-ray studies [2]: "In the powder photographs only the reflections of Os and Sn were found in positions identical with those of the pure metals. Thus no intermediate phase is formed and Os and Sn are practically insoluble in each other."

1. H. St. Claire Deville and H. Debray, *Ann. mines*, **16**, 1859, 12; H. Debray, *Compt. rend.*, **104**, 1889, 1472, 1667–1669.
2. O. Nial, Dissertation, University of Stockholm, 1945; *Svensk Kem. Tidskr.*, **59**, 1947, 172–177.

0.0218
$\bar{1}.9782$

Os-Ta Osmium-Tantalum

According to [1], a σ phase exists in this system having the approximate composition range 65–75 at. % Ta at 1200°C. The tetragonal lattice parameters of the 75 at. % Ta alloy were found as a = 9.934 A, c = 5.189 A.

1. M. V. Nevitt and J. W. Downey, *Trans. AIME*, **209**, 1957, 1072.

0.1733
$\bar{1}.8267$

Os-Te Osmium-Tellurium

$OsTe_2$ (57.30 wt. % Te) is isotypic with pyrite (C2 type), a = 6.382 A [1]. The low melting point of $OsTe_2$ (500–600°C) given by [1] is not correct [2]. There is no lower telluride [1].

1. L. Thomassen, *Z. physik. Chem.*, **B2**, 1929, 351–353.
2. L. Wöhler, K. Ewald, and H. G. Krall, *Ber. deut. chem. Ges.*, **66**, 1933, 1638–1652.

0.5989
$\bar{1}.4011$

Os-Ti Osmium-Titanium

It has been reported that Ti and Os form the compound TiOs (79.88 wt. % Os), which has a crystal structure of the CsCl (B2) type [1], and that phases of the compositions $TiOs_2$ and $TiOs_3$ do not exist [2]. According to [3], the lattice parameter of CsCl-type TiOs is a = 3.07 A.

1. F. Laves and H. J. Wallbaum, *Naturwissenschaften*, **27**, 1939, 674–675.
2. H. J. Wallbaum, *Naturwissenschaften*, **31**, 1943, 91–92.
3. C. B. Jordan, *Trans. AIME*, **203**, 1955, 832–833.

$\bar{1}.9025$
0.0975

Os-U Osmium-Uranium

The compound UOs_2 (61.50 wt. % Os) is of the f.c.c. $MgCu_2$ (C15) type of structure, with a = 7.4974 ± 5 kX (at 24°C) [1].

1. T. J. Heal and G. I. Williams, *Acta Cryst.*, **8**, 1955, 494–498.

0.0146
$\bar{1}.9854$

Os-W Osmium-Wolfram

X-ray analysis (including lattice-parameter measurements) of eight alloys (as-cast condition) covering the range 15–90 wt. (15.4–90.3 at.) % W had the following results: (*a*) About 48.5 at. (47.2 wt.) % W is soluble in Os; (*b*) there is an intermediate phase, probably of the composition W_3Os (74.37 wt. % W); and (*c*) the solid solubility of Os in W is of the order of magnitude of 5 at. % [1]. [2] suggested that W_3Os is a σ phase; this could be corroborated by [3] (a = 9.933 ± 7 A, c/a = 0.515). However, [4] reported the approximate composition range 65–67 at. % W (at 1200°C) and the lattice parameters a = 9.686 A, c = 5.012 A, c/a = 0.517.

1. E. Raub and P. Walter, "Festschrift aus Anlass des 100-jährigen Jubiläums der Firma W. C. Heraeus G.m.b.H.," pp. 124–146, Hanau, 1951.
2. P. A. Beck, private communication to E. Raub.
3. H. Beeskow; see E. Raub, *Z. Metallkunde*, **48**, 1957, 53–54.
4. M. V. Nevitt and J. W. Downey, *Trans. AIME*, **209**, 1957, 1072.

0.4638
$\bar{1}.5362$

Os-Zn Osmium-Zinc

According to [1, 2], Os and Zn do not form a compound since the residue of Zn-rich alloys, after treating with dilute HCl, was found to be pure Os.

1. H. St. Claire Deville and H. Debray, *Compt. rend.*, **94**, 1882, 1557–1560.
2. H. Debray, *Compt. rend.*, **104**, 1887, 1667.

0.3191
$\bar{1}.6809$

Os-Zr Osmium-Zirconium

$ZrOs_2$ (80.66 wt. % Os) is isotypic with $MgZn_2$ (C14 type), $a = 5.189$ A, $c = 8.526$ A, $c/a = 1.643$ [1].

1. H. J. Wallbaum, *Naturwissenschaften*, **30**, 1942, 149.

$\bar{1}.1746$
0.8254

P-Pb Phosphorus-Lead

There is no reliable information as to the behavior of P toward Pb in the molten state. It may be that the compound Pb_3P_2 (9.06 wt. % P), prepared by [1] by means of an aqueous reaction, also exists in Pb-rich alloys. However, [2] was unable to prepare any phosphides either by direct synthesis or by chloride methods.

1. A. Brukl, *Z. anorg. Chem.*, **125**, 1922, 255–256.
2. A. Granger, *Ann. chim. et phys.*, **14**, 1898, 5–90; abstract, *Chem. Zentr.*, **1898**(1), 1262.

$\bar{1}.4628$
0.5372

P-Pd Phosphorus-Palladium

As early as 1849, [1] synthesized the compound PdP_2 (36.73 wt. % P). [2] recognized that the embrittlement of Pd by small additions of P is due to the formation of a low-melting (eutectic) grain-boundary constituent and that the solubility of P in solid Pd is only about 0.01 wt. (0.03 at.) %.

The phase diagram of Fig. 588 was established by thermal, tensimetric, microscopic, and qualitative X-ray investigations [3]. There exist four intermediate phases, β or (Pd_5P) (5.49 wt. % P), γ or (Pd_3P) (8.82 wt. % P), Pd_5P_2 [28.57 at. (10.40 wt.) % P], and PdP_2. Up to 35.6 at. % P, the thermal analysis could be carried out in open crucibles (under nitrogen); at the melting point of PdP_2 ($\sim$1150°C), however, the phosphorus vapor pressure amounted to several atmospheres [4]. In agreement with [2], the solubility of P in solid Pd was found to be negligible. The decreasing solubility of Pd in the β phase was concluded from some microscopic work. The tensimetric and X-ray work indicated a certain solubility of P in PdP_2. A representation of the powder pattern of PdP_2 is given.

1. A. Schrötter, *Ber. Wien. Akad.*, **2**, 1849, 301–303.
2. A. Jedele, *Z. Metallkunde*, **27**, 1935, 271–275.
3. G. Wiehage, F. Weibke, and W. Biltz (with K. Meisel and F. Wiechmann), *Z. anorg. Chem.*, **228**, 1936, 357–371.
4. The equilibrium pressure in the two-phase field (melt + PdP_2) at 955°C is 189 mm Hg (see Fig. 588).

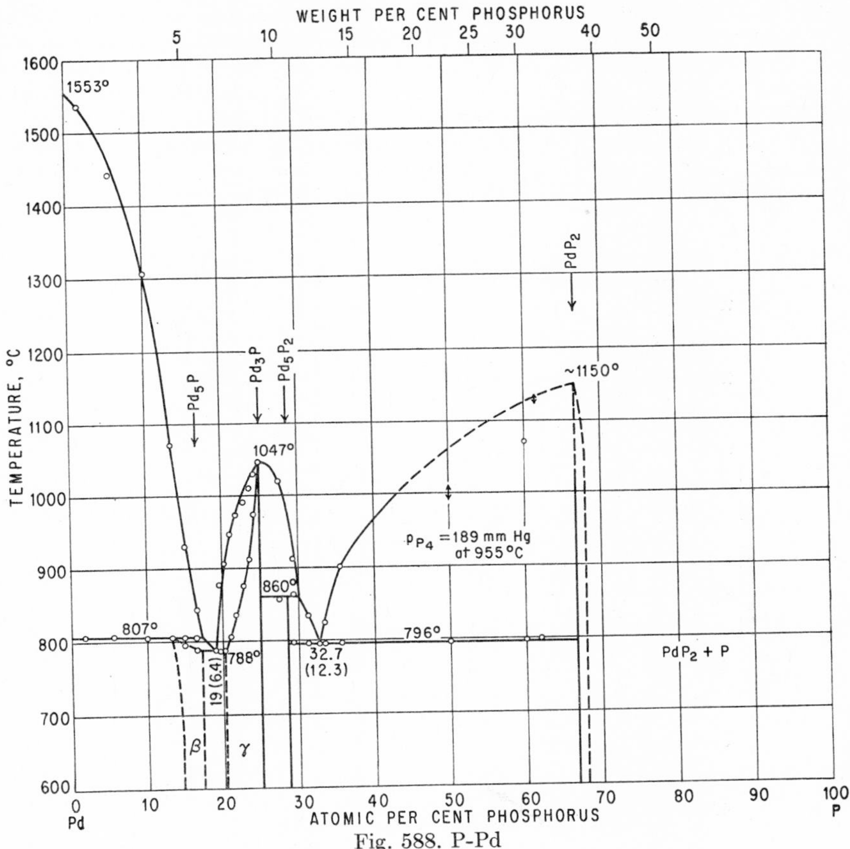

Fig. 588. P-Pd

$\bar{1}.3420$
0.6580

P-Pr Phosphorus-Praseodymium

According to [1], PrP (18.02 wt. % P) has the NaCl (B1) type of structure, with $a = 5.87_2$ A.

1. A. Iandelli and E. Botti, *Atti reale accad. nazl. Lincei Rend.*, **24**, 1936, 459–464; *Strukturbericht*, **5**, 1937, 43.

$\bar{1}.2005$
0.7995

P-Pt Phosphorus-Platinum

In early papers dealing with preparation [1], the "phosphides" Pt_2P, PtP, Pt_3P_5, and PtP_2 were described. In all cases except PtP_2, phase mixtures obviously were analyzed.

The phase diagram of Fig. 589 was established by thermal, microscopic, and qualitative X-ray investigations [2]. It was clearly shown that only two intermediate phases exist, $Pt_{20}P_7$ [25.93 at. (5.26 wt.) % P] and PtP_2 (24.09 wt. % P). The formula for the compound low in P, which is formed peritectically at 590°C, was concluded mainly from microscopic work; a 26 at. % P alloy was homogeneous; alloys containing 25.5 and 25.0 at. % P were two-phase. The extent of the miscibility gap in the liquid state at 683°C was derived from thermal and analytical work; its upper limit was estimated from observations on the viscosity of the melt. The melting point of PtP_2 (which is stable under atmospheric pressure up to 1400°C) lies above 1500°C.

The solubility limit of P in solid Pt was microscopically determined [3] as 0.005 wt. (0.03 at.) % . The embrittlement of Pt by small additions of P is due to the formation of the low-melting eutectic at the grain boundaries [2, 3].

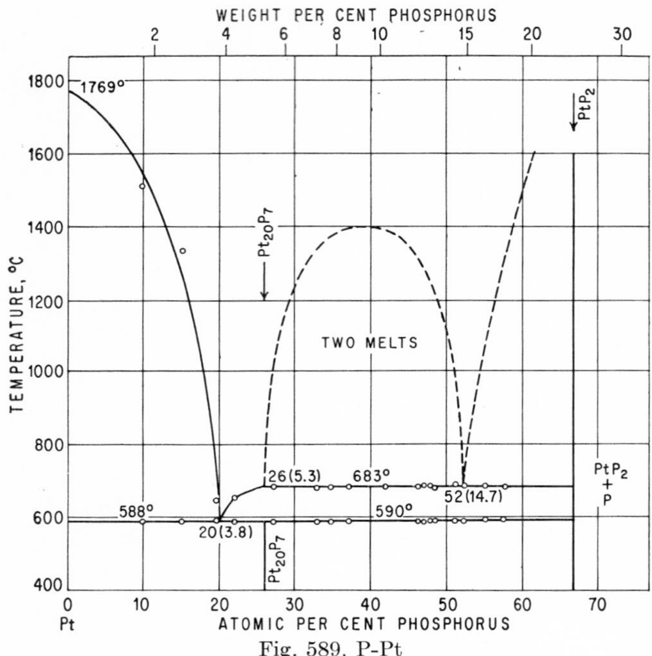

Fig. 589. P-Pt

$Pt_{20}P_7$ has a very complex powder pattern (a representation is given in [2]). According to [4], PtP_2 has the pyrite (C2) type of structure, with $a = 5.69_4$ A.

1. See "Gmelins Handbuch der anorganischen Chemie," System No. 68 C, p. 135, Verlag Chemie, G.m.b.H., Berlin, 1939.
2. W. Biltz, F. Weibke, E. May, and K. Meisel, *Z. anorg. Chem.*, **223,** 1935, 129–143.
3. A. Jedele, *Z. Metallkunde*, **27,** 1935, 271–275.
4. L. Thomassen, *Z. physik. Chem.*, **B4,** 1929, 281–283.

$\bar{1}.1126$
0.8874

P-Pu Phosphorus-Plutonium

PuP (11.5 wt. % P) has been reported to be of the NaCl (B1) type, with $a = 5.644 \pm 4$ A. The compound "appears to undergo decomposition, rather than melting, in the neighborhood of 2000°C" [1].

1. A. E. Gorum, *Acta Cryst.*, **10,** 1957, 143–144.

$\bar{1}.2208$
0.7792

P-Re Phosphorus-Rhenium

According to tensimetric and qualitative X-ray analyses by [1], there exist the four intermediate phases Re_2P (7.68 wt. % P), ReP (14.26 wt. % P), ReP_2 (24.95 wt. % P), and ReP_3 (33.28 wt. % P), all of which yield complex powder patterns. Representations of these are given.

1. H. Haraldsen, *Z. anorg. Chem.*, **221**, 1935, 397–417.

$\bar{1}.4786$
0.5214

P-Rh Phosphorus-Rhodium

The compound Rh_2P (13.08 wt. % P) was shown [1] to have the fluorite (C1) type of structure, with $a = 5.516$ A. The (incomplete) phase diagram of Fig. 590

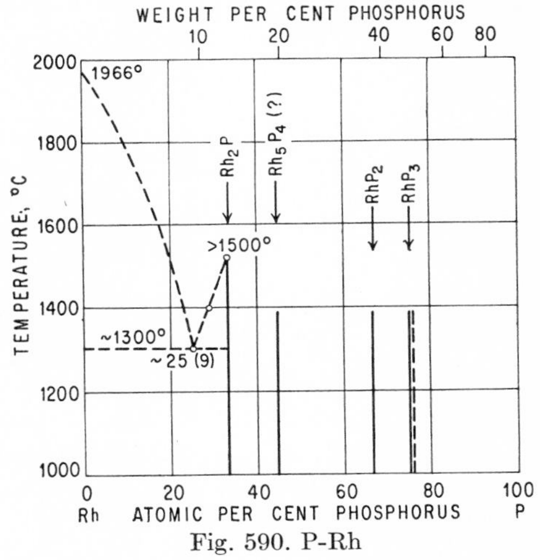

Fig. 590. P-Rh

was worked out by [2] by means of tensimetric and qualitative X-ray studies as well as some melting-point observations. Four intermediate phases exist: Rh_2P, Rh_5P_4 (?) [44.44 at. (19.40 wt.) % P], RhP_2 (37.58 wt. % P), and RhP_3 (47.45 wt. % P); only RhP_3 was found to have a homogeneity range of measurable extent. In the range 35–45 at. % P the temperatures of the beginning of melting were found to lie somewhat below 1400°C. Representations of the rather complex powder patterns of Rh_5P_4, RhP_2, and RhP_3 are given. By means of high-temperature microscopy, [3] found the temperature of the eutectic (Rh + Rh_2P) to be 1254°C (cf. Fig. 590).

1. M. Zumbusch, *Z. anorg. Chem.*, **243**, 1940, 322–329.
2. F. E. Faller, E. F. Strotzer, and W. Biltz, *Z. anorg. Chem.*, **244**, 1940, 317–328.
3. G. Reinacher, *Rev. mét.*, **54**, 1957, 321–336.

$\bar{1}.4837$
0.5163

P-Ru Phosphorus-Ruthenium

According to tensimetric and qualitative X-ray powder work by [1], there exist the three intermediate phases Ru_2P (?) (13.22 wt. % P), RuP (23.35 wt. % P), and RuP_2 (37.85 wt. % P). Representations of the rather complex powder patterns are given. No indications for extended homogeneity ranges were found. By means of high-temperature microscopy, [2] observed melting of the eutectic grain-boundary network in a 0.5 wt. (1.6 at.) % P alloy at 1425°C.

1. W. Biltz, H. J. Ehrhorn, and K. Meisel, *Z. anorg. Chem.*, **240,** 1939, 117–128.
2. G. Reinacher, *Rev. mét.*, **54,** 1957, 321–336.

$\bar{1}.4055$
0.5945

P-Sb Phosphorus-Antimony

Early reports (eighteenth and nineteenth century) on the existence of an antimony phosphide [1] are not conclusive.

According to cursory experiments in sealed quartz tubes [2], the two elements have only a restricted mutual solubility in the molten state.

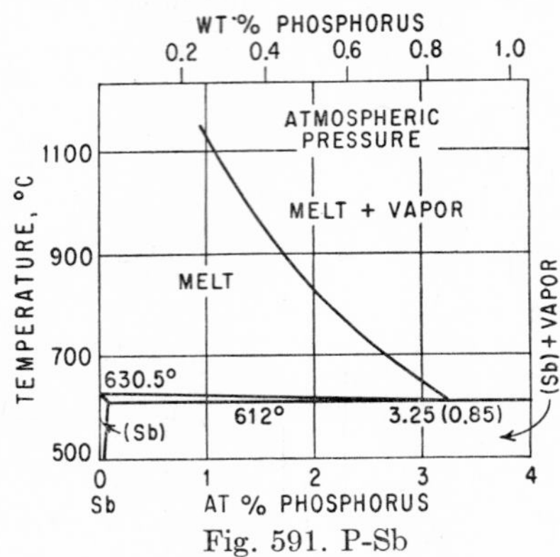

Fig. 591. P-Sb

The phase relations in the Sb-rich region at atmospheric pressure were studied by [3] by means of microscopic, thermal, and chemical analyses (Fig. 591). The maximum solubility of P in liquid Sb amounts to 0.85 wt. (3.25 at.) %; on cooling below 612°C the saturated melt decomposes into a solid and a vapor phase ("second boiling point"). According to [3], P and Sb do not form any compounds with each other.

1. "Gmelins Handbuch der anorganischen Chemie," System No. 18 B, p. 560, Gmelin-Verlag G.m.b.H., Clausthal-Zellerfeld, Germany, 1949.
2. W. Klemm and I. v. Falkowski, *FIAT Rev. Ger. Sci.*, **23,** Inorganic Chemistry, Part I, p. 274, 1949; see also *Z. anorg. Chem.*, **256,** 1948, 343.
3. R. Vogel and D. Horstmann, *Arch. Eisenhüttenw.*, **23,** 1952, 127–128.

$\bar{1}.4166$
0.5834

P-Sn Phosphorus-Tin

The phosphides Sn_5P_2 [1], Sn_2P [2], Sn_3P_2 [3], SnP [4], and SnP_2 [5], concluded in the early literature from inadequate criteria, were not corroborated by later work.

[3] determined the liquidus and solidus temperature of a 3 wt. % P alloy (Fig. 592) and chemically isolated crystals of the intermediate phase richest in Sn, corresponding approximately to the composition Sn_3P_2. This finding excluded the existence of Sn_5P_2 and Sn_2P mentioned above.

[6] found that, at atmospheric pressure, Sn cannot be alloyed with more than

about 13 wt. % P, confirming an old observation by [7]. [6] determined—by means of electrolytical isolation—the composition of the Sn-richest compound as Sn_4P_3 [42.86 at. (16.37 wt.) % P]. Furthermore, SnP_3 (43.91 wt. % P) could be prepared and isolated.

The phase diagram of Fig. 592 was established by [8] by means of thermal, microscopic, and residue analytical investigations. Melts up to 8 wt. (25 at.) % P were thermally analyzed under atmospheric pressure; those with higher P contents had to be prepared and analyzed in sealed glass tubes. Consequently, the alloys were in contact with P vapor of varying and undetermined pressure, and the phase diagram of Fig. 592 corresponds above 25 at. % P to an undefined (and presumably simplified) section through the three-dimensional pressure-temperature-composition diagram.

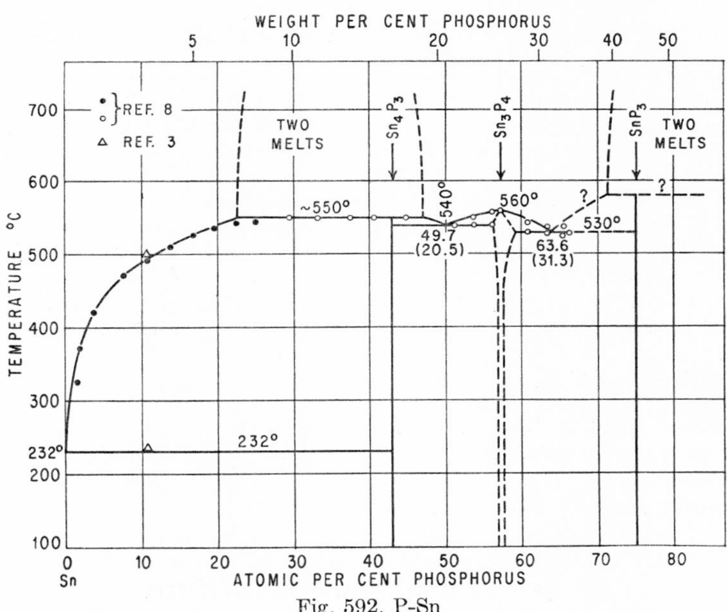

Fig. 592. P-Sn

The following remarks on Fig. 592 should be made: (*a*) Microscopic observations by [8] pointed to the formula Sn_4P_3 for the compound richest in Sn (cf. above). In analyses of numerous residues, however, [8] always found higher Sn contents, from which fact he concluded solid solubility of Sn in Sn_4P_3. It is more probable, however, that elemental Sn had not been completely extracted [9] and, therefore, the questionable homogeneity range is not shown in Fig. 592. (*b*) According to thermal and chemical analytical data, the miscibility gap in the liquid state in the Sn-rich region extends from 7 to 19 wt. (22.5–47 at.) % P. (*c*) [8] detected the intermediate phase Sn_3P_4 [57.14 at. (25.81 wt.) % P] and corroborated the existence of SnP_3. The equilibria of formation of SnP_3 shown in Fig. 592, however, are hypothetical. (*d*) Supercooling occurred in solidification. [8], strangely enough, frequently chose temperatures of greatest supercooling as liquidus points; therefore, the true liquidus temperatures may be expected to lie 10–20°C above those shown in Fig. 592.

1. S. Natanson and G. Vortmann, *Ber. deut. chem. Ges.*, **10,** 1877, 1460.
2. M. Ragg, *Österr. Chem.-Ztg.*, **1,** 1898, 94.

3. J. E. Stead, *J. Soc. Chem. Ind.*, **16,** 1897, 206; see also W. Campbell, *J. Franklin Inst.*, **154,** 1902, 216.
4. A. Schrötter, *Ber. Wien. Akad.*, **2,** 1849, 301; Vigier, *Bull. soc. chim. France*, **2,** 1861, 5; O. Emmerling, *Ber. deut. chem. Ges.*, **12,** 1879, 155; see also [1].
5. O. Emmerling; see [4].
6. P. Jolibois, *Compt. rend.*, **148,** 1909, 636–638.
7. Pelletier, *Ann. chim. et phys.*, **13,** 1792, 120.
8. A. C. Vivian, *J. Inst. Metals*, **23,** 1920, 325–360; discussion, pp. 361–366.
9. The better (electrolytical) isolation method of [6] yielded the composition Sn_4P_3 for the compound.

$\bar{1}.2336$
0.7664

P-Ta Phosphorus-Tantalum

According to work on preparation and tensimetric and qualitative X-ray analyses [1], two phosphides exist: TaP (14.62 wt. % P) and TaP_2 (25.51 wt. % P). No indications for extended homogeneity ranges were found. As for TaP, two modifications were observed and assumed to have a monotropic relationship (cf. the Nb-P system and [2]). Representations of powder patterns are given. It has been suggested [1, 2] that the Nb and Ta phosphides are isomorphous. This was confirmed by [3] as far as the two modifications of the monophosphide are concerned. Therefore, "α-TaP," with probably the approximate composition $TaP_{0.95}$, has a partially ordered tetragonal structure, with $a = 3.320$ A, $c = 5.69$ A, $c/a = 1.71$, and $\sim$4 atoms per unit cell; and "β-TaP," found to be homogeneous at 50 at. % P, has an ordered tetragonal structure, with $a = 3.330$ A, $c = 11.39$ A, $c/2{:}a = 1.71$, and 8 atoms per unit cell.

The existence of TaP_2 was confirmed by [3].

1. M. Zumbusch and W. Biltz, *Z. anorg. Chem.*, **246,** 1941, 35–45.
2. W. Biltz et al., *Z. anorg. Chem.*, **249,** 1942, 20–22 (Comparison of V, Nb, and Ta phosphides).
3. N. Schönberg, *Acta Chem. Scand.*, **8,** 1954, 226–239.

$\bar{1}.3851$
0.6149

P-Te Phosphorus-Tellurium

The compound P_2Te_3 (86.07 wt. % Te), stable in dry air, has been prepared by [1].

1. E. Montignie, *Bull. soc. chim. France*, **9,** 1942, 658–661; *Met. Abstr.*, **13,** 1946, 47.

$\bar{1}.1253$
0.8747

P-Th Phosphorus-Thorium

According to work on preparation and roentgenographic studies, two intermediate phases exist in the Th-P system [1–3]. The compound richest in P is Th_3P_4 [57.14 at. (15.10 wt.) % P], which has a b.c.c. structure [2] ($D7_3$ type), with $a = 8.617 \pm 2$ A.

The composition and structure of a "subphosphide" could not be conclusively established. It has been tentatively called ThP, and its structure may be the NaCl (B1) type, but deficient in P, with $a = 5.830 \pm 2$ A [2, 3].

1. E. F. Strotzer, W. Biltz, and K. Meisel, *Z. anorg. Chem.*, **238,** 1938, 69–80.
2. K. Meisel, *Z. anorg. Chem.*, **240,** 1939, 300–312.
3. M. Zumbusch, *Z. anorg. Chem.*, **245,** 1941, 402–408.

1.8107
0.1893

P-Ti Phosphorus-Titanium

By synthetic preparation from Ti and a large excess of P, [1] obtained the limiting compound $TiP_{0.92}$, a phase previously reported by [2]. The formula TiP would require 39.27 wt. % P. X-ray examinations proved $TiP_{0.92}$ to be single-phase and gave some indication of the existence of a subphosphide, probably Ti_2P (24.44 wt. % P).

More recently X-ray powder work was carried out by [3]. The existence of a "subphosphide" (of unknown composition and structure) was confirmed, and TiP was found to have a hexagonal structure, with a = 3.487 A, c = 11.65 A, c/a = 3.34, and 8 atoms per unit cell (superstructure of the NiAs type). [3] also stated that the existence of a Ti diphosphide was confirmed; however, there appears to be no previous mention of such a phase in the literature.

The lowest per cent addition at which P appeared as an insoluble compound in the microstructure of magnesium-reduced titanium-base alloys, hot-rolled and subsequently annealed for ½ hr at 790°C, was found to be 0.07 wt. (0.1 at.) % P [4].

1. W. Biltz, A. Rink, and F. Wiechmann, *Z. anorg. Chem.*, **238,** 1938, 395–405.
2. J. Gewecke, *Liebigs Ann.*, **361,** 1908, 79–88.
3. N. Schönberg, *Acta Chem. Scand.*, **8,** 1954, 226–239.
4. R. M. Goldhoff, H. L. Shaw, C. M. Craighead, and R. I. Jaffee, *Trans. ASM*, **45,** 1953, 941–965.

1.1806
0.8194

P-Tl Phosphorus-Thallium

There are conflicting reports in the early literature on preparation as to whether or not a thallium phosphide exists [1].

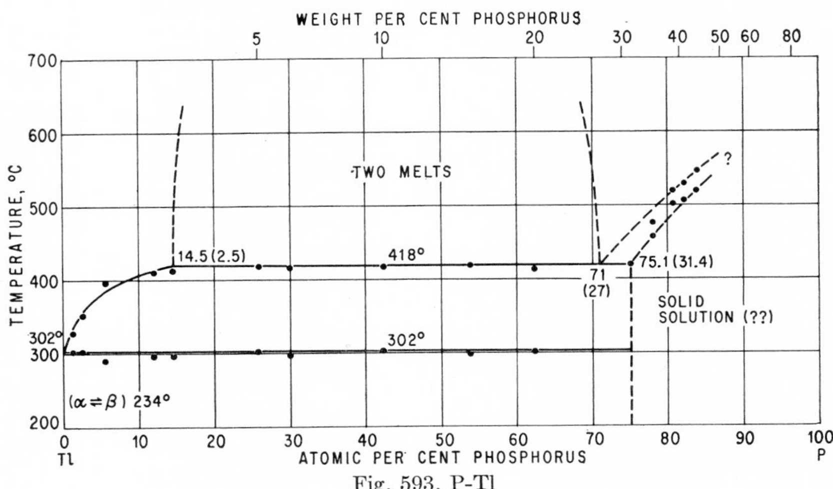

Fig. 593. P-Tl

A thermal analysis [2] of alloys containing up to 84 at. % P was carried out in evacuated and sealed hard-glass tubes and yielded the data plotted in Fig. 593. Consequently, the alloys were in contact with P vapor of varying and undetermined pressure, and the phase diagram is not valid for either atmospheric or even any constant pressure. As to the "solid solution" in the P-rich region, [2] noted that a high pressure is necessary for its formation. In drawing Fig. 593 the diagram given by [2] was slightly modified in order to fit the phase rule. In the original diagram, liquidus and solidus of the solid solution meet at 75.1 at. % P, 418°C. Since three coexisting phases are required at the monotectic temperature and since [2] stated that only alloys containing between about 2.5 and 27 wt. (14.5–71 at.) % P separate into two liquid layers, the liquidus has been drawn as shown in Fig. 593.

In a more recent paper, [3] claimed to have prepared a phosphide Tl_3P (4.81 wt. % P) by alloying Tl and white P in a sealed tube at about 400°C.

1. "Gmelins Handbuch der anorganischen Chemie," System No. 38, pp. 425–426, Verlag Chemie, G.m.b.H., Berlin, 1940.
2. Q. A. Mansuri, *J. Chem. Soc.*, **130,** 1927, 2993–2995.
3. E. Montignie, *Bull. soc. chim. France,* **4,** 1937, 295–296; quoted from [1].

$\bar{1}.1143$
0.8857

P-U Phosphorus-Uranium

The phosphides UP (11.51 wt. % P) [1] and U_3P_4 [57.14 at. (14.78 wt.) % P] [2] are mentioned in the early literature on preparation.

In a systematic tensimetric and X-ray analysis of alloys prepared by direct synthesis, [3] showed three compounds to exist in the U-P system, namely, UP, U_3P_4, and UP_2 (20.65 wt. % P).

UP has the NaCl (B1) type of structure, with a = 5.600 A [4, 5]; U_3P_4 the cubic Th_3P_4 ($D7_3$) type of structure, with $a = 8.21_3$ A [4]; and UP_2 the tetragonal Cu_2Sb (C38) type of structure, with a = 3.800 A, c = 7.762 A, c/a = 2.043 [6].

1. C. Rammelsberg, *Sitzber. kgl. preuss. Akad. Wiss.*, **1872,** 449; quoted from [7].
2. A. Colani, *Ann. chim. et phys.*, **12,** 1907, 59.
3. M. Heimbrecht, M. Zumbusch, and W. Biltz, *Z. anorg. Chem.*, **245,** 1941, 391–401.
4. M. Zumbusch, *Z. anorg. Chem.*, **245,** 1941, 402–408.
5. R. E. Rundle and N. C. Baenziger, *U.S. Atomic Energy Comm., Rept.* CC-1778, 1944; quoted from [7].
6. A. Iandelli, *Atti accad. nazl. Lincei, Rend. Classe sci. fis. mat. e nat.*, **13,** 1952, 144–151; *Chem. Abstr.*, **47,** 1953, 11062.
7. J. J. Katz and E. Rabinowitch, "The Chemistry of Uranium," Part I, National Nuclear Energy Series, Div. VIII, vol. 5, p. 241, McGraw-Hill Book Company, Inc., New York, 1951.

$\bar{1}.7839$
0.2161

P-V Phosphorus-Vanadium

[1] reported the preparation and isolation (as metallic needles) of the two phosphides V_2P and VP by electrolysis of a fused mixture.

According to a systematic investigation by [2] (work on preparation, tensimetric and qualitative X-ray studies), at least three phosphides exist: V_3P, VP (37.81 wt. % P), and VP_2 (54.87 wt. % P). The powder pattern of V_3P showed a great similarity to that of Cr_3P (see Cr-P). For the monophosphide a homogeneity range between about 45 and somewhat more than 50 at. % P was assumed. The powder pattern of the VP_2 phase was very similar to the patterns of NbP_2 and TaP_2 (see Nb-P, P-Ta). Between V_3P and VP, indications of the existence of additional compounds were found.

In a recent X-ray powder analysis of samples with compositions between $VP_{0.2}$

and $VP_{1.2}$, however, [3] found that the only compound existing in the V-VP interval is the VP phase, having a small homogeneity range and the NiAs (B8) type of structure, with $a = 3.18$ A, $c = 6.22$ A, $c/a = 1.96$. The existence of VP_2 was confirmed.

It may be that the "subphosphides" reported by [1, 2] were stabilized by impurities.

1. M. Chêne, *Compt. rend.*, **208,** 1939, 1144–1146; *Ann. chim.*, **15,** 1941, 187–285; *Chem. Abstr.*, **33,** 1939, 4541; **36,** 1942, 2791. See also J. L. Andrieux, *Rev. mét.*, **45,** 1948, 49–59; *Chem. Abstr.*, **42,** 1948, 8089.
2. M. Zumbusch and W. Biltz, *Z. anorg. Chem.*, **249,** 1942, 1–13, 20–22; the investigation suffered from wartime restrictions. Only 3 g of pure V was available. Most of the preparations were badly contaminated by oxygen.
3. N. Schönberg, *Acta Chem. Scand.*, **8,** 1954, 226–239; the alloys were prepared from metal powder and red P in evacuated and sealed silica tubes at about 800°C.

$\bar{1}.2264$
0.7736

P-W Phosphorus-Wolfram

According to tensimetric [1] and X-ray powder analyses [1, 2] of alloys prepared by direct synthesis, only two compounds, WP (14.42 wt. % P) and WP_2 (25.20 wt. % P), exist in the W-P system. The phosphides W_2P [3–5] and W_3P_4 [3] could not be corroborated. WP had been reported previously by [6] and WP_2 by [7, 8].

An unstable and "amorphous" product of the composition W_4P was obtained by [5] by means of electrolysis of phosphate melts

WP has the orthorhombic MnP (B31) type of structure [1, 2], with $a = 6.219$ A, $b = 5.717$ A, $c = 3.238$ A [2]. The powder pattern of WP_2 is similar to that of MoP_2 (representation given) [1].

1. F. E. Faller and W. Biltz (with K. Meisel), *Z. anorg. Chem.*, **248,** 1941, 209–228.
2. N. Schönberg, *Acta Chem. Scand.*, **8,** 1954, 226–239.
3. F. Wöhler and H. Wright, *Liebigs Ann.*, **79,** 1851, 244.
4. H. Hartmann, F. Ebert, and O. Bretschneider, *Z. anorg. Chem.*, **198,** 1931, 121, 125.
5. H. Hartmann and J. Orban, *Z. anorg. Chem.*, **226,** 1936, 257–264.
6. E. Defacqz, *Compt. rend.*, **132,** 1901, 32–35.
7. E. Defacqz, *Compt. rend.*, **130,** 1900, 915–917.
8. E. Heinerth and W. Biltz, *Z. anorg. Chem.*, **198,** 1931, 171–173.

$\bar{1}.6756$
0.3244

P-Zn Phosphorus-Zinc

Two phosphides of zinc are known, namely, Zn_3P_2 (24.00 wt. % P) [1] and ZnP_2 (48.65 wt. % P) [2–5]. A phase diagram has not yet been established; it can be said only that the liquidus rises steeply from the melting point of pure Zn [6].

According to single-crystal work by [5], Zn_3P_2 possesses a tetragonal structure, $D5_9$ (Zn_3P_2) type, with $a = 8.113 \pm 20$ A, $c = 11.47 \pm 3$ A, $c/a = 1.418 \pm 5$. Former structure determinations by [7, 8] proved to be incorrect.

ZnP_2 has a primitive tetragonal lattice, with $a = 5.08$ A, $c = 18.69$ A, $c/a = 3.68$, and 24 atoms per unit cell [5].

1. Long known in the chemical literature.
2. Hvoslef, *Ann. Pharm.*, **100,** 1856, 99.
3. Renault, *Compt. rend.*, **76,** 1873, 283.
4. P. Jolibois, *Compt. rend.*, **147,** 1908, 801–803.
5. M. v. Stackelberg and R. Paulus, *Z. physik. Chem.*, **B28,** 1935, 427–460.

6. R. Vogel and D. Horstmann, *Arch. Eisenhüttenw.*, **24,** 1953, 247–249.
7. L. Passerini, *Gazz. chim. ital.*, **58,** 1928, 655–664; *Strukturbericht,* **2,** 1928-1932, 325.
8. M. v. Stackelberg and R. Paulus, *Z. physik. Chem.*, **B22,** 1933, 305–322.

$\bar{1}.5309$
0.4691

P-Zr Phosphorus-Zirconium

By means of direct syntheses, tensimetric and qualitative X-ray analyses, three Zr phosphides were found by [1], namely, ZrP_2 (40.44 wt. % P) [2], ZrP (25.35 wt. % P), and a "subphosphide," perhaps of the composition Zr_3P.

More recent X-ray powder work [3] confirmed the existence of ZrP_2 and a phosphide of unknown composition and structure within the interval $ZrP_{0.3-0.5}$. "In the vicinity of the composition $ZrP_{0.9}$ a phosphide occurred which had a comparatively small homogeneity range. The phase is of NaCl (B1) type, and the length of the cube edge was found to vary between 5.261 and 5.278 A. The NaCl structure gives ZrP as the ideal formula for this phase. A small fraction of the P positions is evidently empty. It, therefore, seems practical to call the phase ZrP and to denote it by α-ZrP" [3]. A phase called β-Zr by [3] "appeared to be homogeneous at the composition ZrP, and is isomorphous with the TiP phase." It has, therefore, a hexagonal structure, with $a = 3.677$ A, $c = 12.52$ A, $c/a = 3.40$, and 8 atoms per unit cell, which can be considered a superstructure of the NiAs type.

1. E. F. Strotzer, W. Biltz, and K. Meisel, *Z. anorg. Chem.*, **239,** 1938, 216–224.
2. This compound had already been prepared by J. Gewecke, *Liebigs Ann.*, **361,** 1908, 79–88.
3. N. Schönberg, *Acta Chem. Scand.*, **8,** 1954, 226–239.

0.2882
$\bar{1}.7118$

Pb-Pd Lead-Palladium

[1] found the compound Pd_3Pb (39.29 wt. % Pb) by residue analysis. [2] investigated the influence of small additions of Pd on the melting point of Pb. A systematic investigation of the Pd-Pb system by means of thermal and microscopic work was carried out by [3]; his thermal data are plotted in Fig. 594. [3] assumed five intermediate phases: Pd_3Pb and $PdPb_2$ (79.52 wt. % Pb) melting with open maxima; and Pd_2Pb (49.26 wt. % Pb), $\sim Pd_6Pb_5$, and PdPb (66.01 wt. % Pb) formed by peritectic reactions at 830, 596, and 495°C, respectively. Since strong peritectic coring occurred and no heat-treatment was applied to reach equilibrium, the formulas of the peritectic compounds could be derived from the length of the peritectic arrests only. [4] suggested the simple formula Pd_3Pb_2 (56.42 wt. % Pb) for the compound called $\sim Pd_6Pb_5$ by [3]. According to thermal and microscopic evidence, the homogeneity range of Pd_3Pb extends down to about 22 at. (35.3 wt.) % Pb [3]. The solid solubility of Pd in Pb and the temperature dependence of that of Pb in Pd [about 24 wt. (14 at.) % at the eutectic temperature] were not investigated by [3]. Emf measurements on as-cast alloys by [5] revealed the existence of $PdPb_2$ only.

The existence of Pd_3Pb [6], Pd_3Pb_2 [6], PdPb [6], and $PdPb_2$ [7] has been corroborated by X-ray diffraction work (cf. Crystal Structures). In deviation from the phase diagram proposed by [3], the microscopic and roentgenographic work of [6] indicated that (*a*) Pd_3Pb_2—rather than Pd_2Pb—is peritectically formed at 830°C (Fig. 594); (*b*) a compound Pd_2Pb probably does not exist [8]; (*c*) the constitution between Pd_3Pb_2 and PdPb may be more complex, especially at high temperatures, than assumed by [3]. See [9].

Crystal Structures. The lattice parameter of the Pb-saturated (Pd) phase is appreciably higher than that of pure Pd [6]. Pd_3Pb has the cubic Cu_3Au ($L1_2$) type of structure, with a = 4.021 A for an alloy in the ($Pd_3Pb + Pd_3Pb_2$) phase field [6].

The structure of Pd_3Pb_2 is of the (partially filled-in) NiAs (B8) type, and the homogeneity range extends between about 38.5 and 41 at. % Pb [6]. At the Pd-rich side, the lattice parameters a = 4.465 A, c = 5.704 A, c/a = 1.275 were measured.

Single-crystal work revealed a monoclinic translation group for the phase PdPb [6]: a = 7.09 A, b = 8.44 A, c = 5.57 A, β = 71°. There are indications that this phase—analogous to Ni_3Sn_4—exists at a composition somewhat richer in Pb than 50

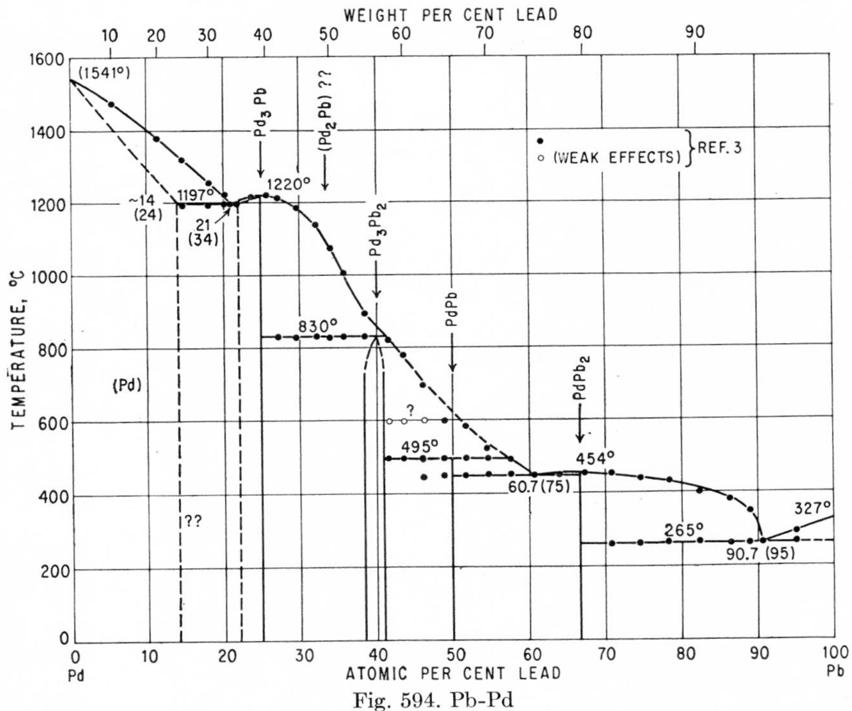

Fig. 594. Pb-Pd

at. % [6]. According to [7], $PbPb_2$ has the tetragonal $CuAl_2$ (C16) type of structure, with a = 6.849 ± 5 A, c = 5.833 ± 5 A, c/a = 0.85.

1. A. Bauer, *Ber. deut. chem. Ges.*, **4**, 1871, 451.
2. C. T. Heycock and F. H. Neville, *J. Chem. Soc.*, **61**, 1892, 906.
3. R. Ruer, *Z. anorg. Chem.*, **52**, 1907, 345–357.
4. M. Hansen, "Der Aufbau der Zweistofflegierungen," p. 979, Springer-Verlag OHG, Berlin, 1936.
5. N. A. Puschin and N. P. Paschsky, *Z. anorg. Chem.*, **62**, 1909, 360–363.
6. H. Nowotny, K. Schubert, and U. Dettinger, *Z. Metallkunde*, **37**, 1946, 137–145.
7. H. J. Wallbaum, *Z. Metallkunde*, **35**, 1943, 218–221.
8. Since the X-ray pattern of a 51.4 wt. (35.3 at.) % Pb alloy quenched from 800°C showed foreign lines, [6] admit that "it is probable that there take place reactions or exist allotropic forms at higher temperatures."

9. "Presumably there exists a phase with a composition between Pd_3Pb_2 and PdPb which, at lower temperatures, decomposes into these two" [6].

$\bar{1}.9942$
0.0058

Pb-Po Lead-Polonium

See [1].

1. W. H. Beamer and C. R. Maxwell, *J. Chem. Phys.*, **14**, 1946, 569; **17**, 1949, 1293–1298; abstract, *Structure Repts.*, **12**, 1949, 121–122.

0.1674
$\bar{1}.8326$

Pb-Pr Lead-Praseodymium

Five melts, three of which had compositions corresponding to the formulas $PrPb_3$ (18.48 wt. % Pr), PrPb (40.48 wt. % Pr), and Pr_2Pb (57.63 wt. % Pr), were used [1]

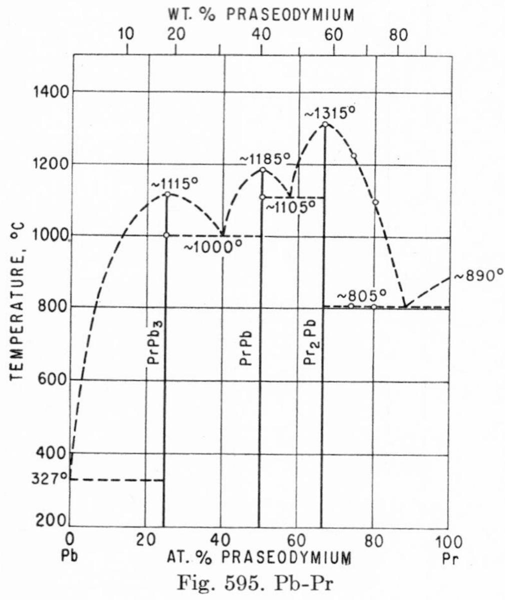

Fig. 595. Pb-Pr

roughly to outline the diagram by thermal analysis (Fig. 595). Results were claimed to prove that lead forms the same compounds with praseodymium as with cerium and lanthanum. Earlier, $PrPb_3$ was identified by determination of its crystal structure: f.c.c. of the Cu_3Au ($L1_2$) type, a = 4.867 A [2]. Its melting point was reported to be 1150°C [2].

1. R. Vogel and T. Heumann, *Z. Metallkunde*, **35**, 1943, 29–42.
2. A. Rossi, *Gazz. chim. ital.*, **64**, 1934, 832–834.

0.0259
$\bar{1}.9741$

Pb-Pt Lead-Platinum

[1] isolated the compound PtPb (51.49 wt. % Pb). [2] investigated the freezing-point depression of Pb by addition of small amounts of Pt.

A systematic thermal and microscopic analysis of this system has been carried out by [3]. Three peritectic horizontals, indicating the formation of three intermediate phases, were found (see the thermal data plotted in Fig. 596). Since strong coring occurred, the formula for only one of these phases, PtPb, could be conclusively established by [3].

As to the intermediate phase richest in Pb, X-ray diffraction work showed it to have the composition $PtPb_4$ (80.94 wt. % Pb) [4–7]. No such agreement exists in the more recent literature as far as the compound richest in Pt is concerned. [5]

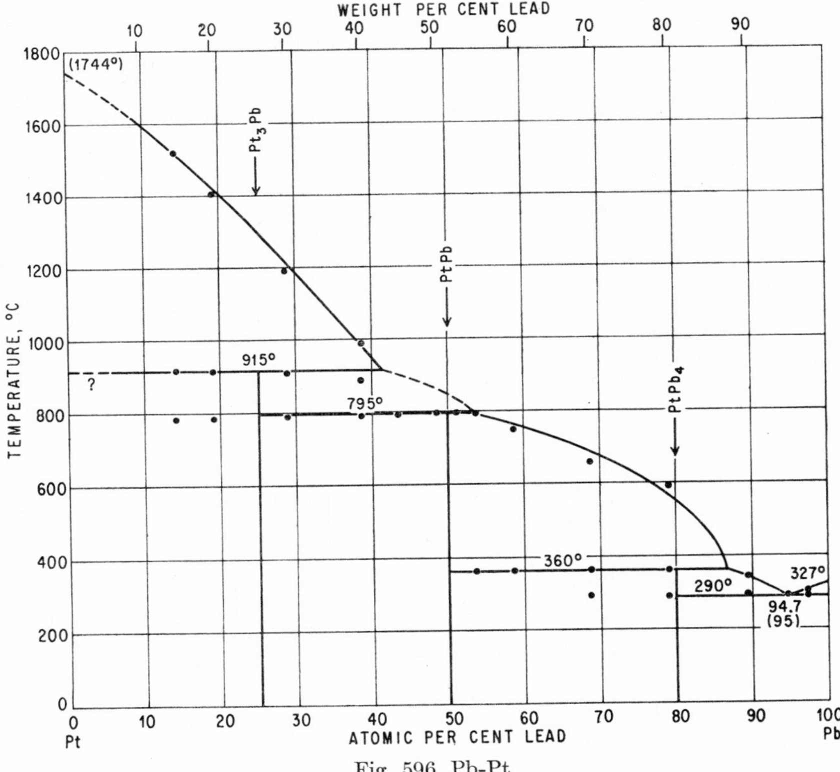

Fig. 596. Pb-Pt

reported the composition Pt_3Pb (26.14 wt. % Pb), whereas [7]—who unfortunately were unaware of the work of [5]—suggested the true formula lying in the $Pt_{5-7}Pb$ interval. For Fig. 596, the formula Pt_3Pb was chosen since it has a good structural and crystallochemical background.

The emf-vs.-composition curve measured by [8] on as-cast (!) alloys shows breaks at the compositions PtPb and $PtPb_2$.

Crystal Structures. According to [5], the intermediate phase richest in Pt, Pt_3Pb, has the cubic Cu_3Au ($L1_2$) type of structure, with $a = 4.053$ A. However, the (primitive cubic) superstructure lines were not visible owing to the closeness in atomic numbers of Pt and Pb. A cubic unit cell has been reported by [7] also.

PtPb possesses the hexagonal NiAs (B8) structure [9, 5], with $a = 4.258$ A, $c = 5.467$ A, $c/a = 1.284$ [5]. [7] reported a hexagonal cell, with $a = 4.24$ A, $c = 5.48$ A.

The structure of $PtPb_4$ has been elucidated by [6]. It is related to the $CuAl_2$ (C16) type; and the tetragonal unit cell, with $a = 6.666 \pm 10$ A, $c = 5.978 \pm 10$ A, $c/a = 0.89_7$, contains 10 atoms. The unit-cell dimensions found by [7] are in fair agreement with those of [6].

1. A. Bauer, *Ber. deut. chem. Ges.*, **3,** 1870, 836; **4,** 1871, 449.
2. C. T. Heycock and F. H. Neville, *J. Chem. Soc.*, **61,** 1892, 909.
3. F. Doerinckel, *Z. anorg. Chem.*, **54,** 1908, 358–365.
4. H. J. Wallbaum, *Z. Metallkunde*, **35,** 1943, 218.
5. H. Nowotny, K. Schubert, and U. Dettinger, *Z. Metallkunde*, **37,** 1946, 137–145.
6. U. Rösler and K. Schubert, *Z. Metallkunde*, **42,** 1951, 395–400.
7. R. Graham, G. C. S. Waghorn, and P. T. Davies, *Acta Cryst.*, **7,** 1954, 634–635 (abstract only).
8. N. A. Puschin and P. N. Lascenko, *Z. anorg. Chem.*, **62,** 1909, 34–39.
9. E. Zintl and H. Kaiser (and A. Harder), *Z. anorg. Chem.*, **211,** 1933, 128.

$\bar{1}.9380$
0.0620

Pb-Pu Lead-Plutonium

The phase diagram of Fig. 596*a* is due to [1]. It was established by work on alloy samples weighing some hundreds of milligrams at the most. There are two

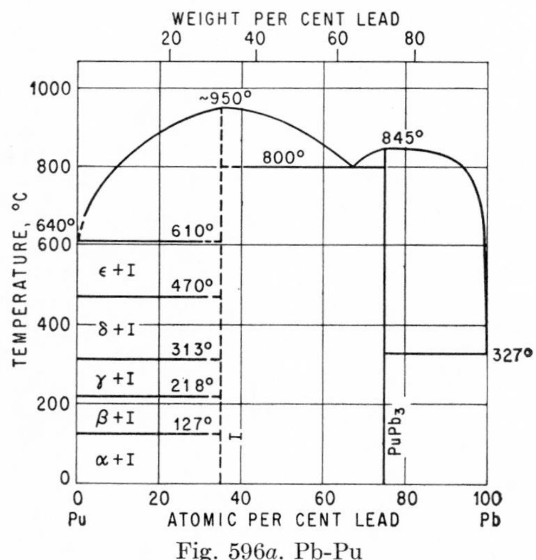

Fig. 596*a*. Pb-Pu

compounds in this system. "The first compound [called I], which is found in the Pb-concentration range of 30–50 at. %, is readily oxidized, and it ignites in the air" [1]. The second compound, $PuPb_3$, has, according to [1], the ordered Cu_3Au ($L1_2$) type of structure, with $a = 4.81$ A. However, [2] were unable to detect the (primitive cubic) superlattice lines ($a = 4.808 \pm 1$ A).

1. S. T. Konobeevsky, *Conference of the Academy of Sciences of the U.S.S.R. on the Peaceful Uses of Atomic Energy, Division of Chemical Science*, July, 1955; English translation by U.S. Atomic Energy Commission, Washington, 1956.

2. Investigators at Los Alamos; see A. S. Coffinberry and M. B. Waldron, Review of the Physical Metallurgy of Plutonium, in "Metallurgy and Fuels," Progress in Nuclear Energy, ser. V, vol. 1, Pergamon Press Ltd., London, 1956.

0.3040
$\bar{1}$.6960

Pb-Rh Lead-Rhodium

The compound Rh_2Pb (50.17 wt. % Pb) has been isolated by [1, 2]. $RhPb_2$ (80.11 wt. % Pb) was found to have the tetragonal $CuAl_2$ (C16) type of structure, with $a = 6.664 \pm 3$ A, $c = 5.865 \pm 3$ A, $c/a = 0.88$ [3].

1. H. Debray, *Compt. rend.*, **90,** 1880, 1195–1199; see also **104,** 1887, 1581.
2. L. Wöhler and L. Metz, *Z. anorg. Chem.*, **149,** 1925, 311.
3. H. J. Wallbaum, *Z. Metallkunde*, **35,** 1943, 218–221.

0.3091
$\bar{1}$.6909

Pb-Ru Lead-Ruthenium

According to [1], Ru and Pb do not form any compounds with each other. In treating Pb-rich alloys with dilute HNO_3 a residue consisting of pure Ru was obtained. It is not certain, however, whether the Ru had been dissolved in the molten lead.

1. H. Debray, *Compt. rend.*, **104,** 1887, 1580, 1667.

0.8104
$\bar{1}$.1896

Pb-S Lead-Sulfur

According to the Pb-PbS phase diagram established by [1] (Fig. 597) using thermal and microscopic analysis, Pb and PbS (13.40 wt. % S) are completely miscible in the liquid state and no subsulfides [2] exist. Since the liquidus data lie nearly on a horizontal line between about 20 and 40 at. % S, a miscibility gap in the liquid state had been suggested by [3]; however, later experiments [4, 5] corroborated the complete miscibility found by [1]. [1] could not determine the liquidus above 50 at. % S since strong evaporation of S occurred. [6] observed a further rise: the S-richest melt investigated contained 14.1 wt. (51.5 at.) % S and solidified at 1130°C.

The data for the melting point of PbS found by [1, 7, 6, 8–11] lie in the interval 1119 ± 16°C.

The solubility of S in solid Pb was found by [12] by means of microscopic and resistometric work to be less than 0.0006 at. (0.0001 wt.) % at 300°C. At this concentration the sulfur appeared as primary PbS phase. PbS has the NaCl (B1) type of structure [13, 14]. Precision lattice-parameter measurements were carried out by [15, 16]; according to [16], $a = 5.9362$ A at 26°C.

1. K. Friedrich and A. Leroux, *Metallurgie*, **2,** 1905, 536–539.
2. The compounds Pb_4S and Pb_2S were mentioned in the early literature. It had been shown already by F. Roessler (*Z. anorg. Chem.*, **9,** 1895, 41–44) that PbS is the compound in equilibrium with pure Pb.
3. W. Guertler, "Handbuch der Metallographie," vol. 1, pp. 993–995, Verlagsbuchhandlung Gebrüder Borntraeger, Berlin, 1912; see also W. Guertler and K. L. Meissner, *Metall u. Erz*, **18,** 1921, 145–152.
4. W. Guertler and G. Landau, *Metall u. Erz*, **31,** 1934, 169–171; these authors also determined 21 liquidus points in sulfurizing molten Pb by means of H_2S at various temperatures. Since the data are badly scattered, they are not plotted in Fig. 597.

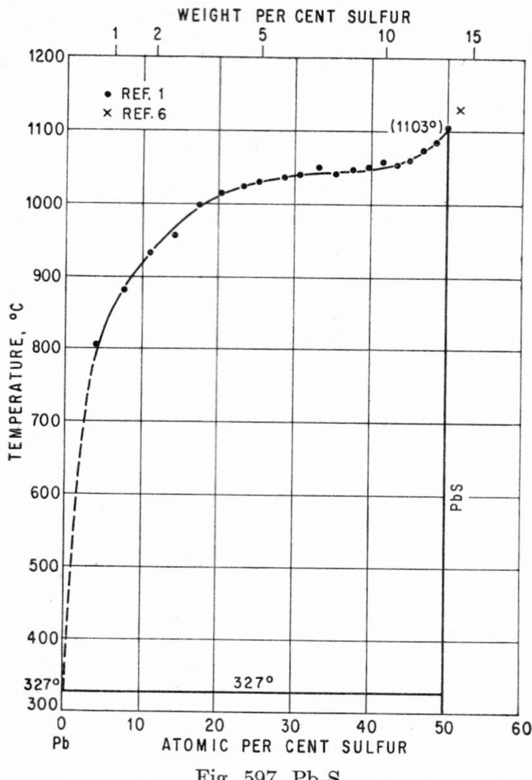

Fig. 597. Pb-S

5. W. Leitgebel and E. Miksch, *Metall u. Erz*, **31,** 1934, 290–293.
6. K. Friedrich, *Metallurgie*, **5,** 1908, 23–27, 51–52.
7. K. Friedrich, *Metallurgie*, **4,** 1907, 481, 672 (1114°C).
8. W. Biltz, *Z. anorg. Chem.*, **59,** 1908, 273 (1112 ± 2°C).
9. W. Truthe, *Z. anorg. Chem.*, **76,** 1912, 163 (1106°C).
10. W. Heike, *Metallurgie*, **9,** 1912, 317 (1106°C).
11. E. J. Kohlmeyer, *Metall u. Erz*, **29,** 1932, 108–109 (1135°C).
12. J. N. Greenwood and H. W. Worner, *J. Inst. Metals*, **65,** 1939, 435–445.
13. W. P. Davey, *Phys. Rev.*, **17,** 1921, 402.
14. *Strukturberichte*, **1, 2, 3.**
15. B. Wasserstein, *Am. Mineralogist*, **36,** 1951, 102–115.
16. H. E. Swanson and R. K. Fuyat, *Natl. Bur. Standards (U.S.), Circ.* 539(II), 1953, p. 18.

0.2309
$\bar{1}$.7691

Pb-Sb Lead-Antimony

The literature on the constitution of the Pb-Sb system was carefully reviewed by [1] in 1951.

Liquidus. The course of the whole liquidus line was first studied by [2]; the simple eutectiferous type of constitution found was confirmed microscopically by

[3–5]. Liquidus data were subsequently reported by [6–10, 53] (whole system) as well as [11–14] (parts of the system or single values). As to the interval 0–16 wt. (0–24.5 at.) % Sb, the thermal data of [14] are the most reliable ones (estimated accuracy of ±1–2°C [1]) and were used for Fig. 598. Since in hypereutectic alloys there occurs strong supercooling, the degree of which depends on the cooling rate, [14] extrapolated his data between 11 and 16 wt. % Sb to infinitely slow cooling. Between 16 and about 75 wt. % Sb the liquidus data published vary as much as 30°C. To account for supercooling it has been suggested by [1] that preference be given to the highest experimental data recorded. Therefore, in the region 45–75 wt. % Sb the data by [6, 10] were used for Fig. 598. The gap between 16 and 45 wt. % Sb has been bridged over by a smooth line; the fact that it lies up to more than 10°C above the highest

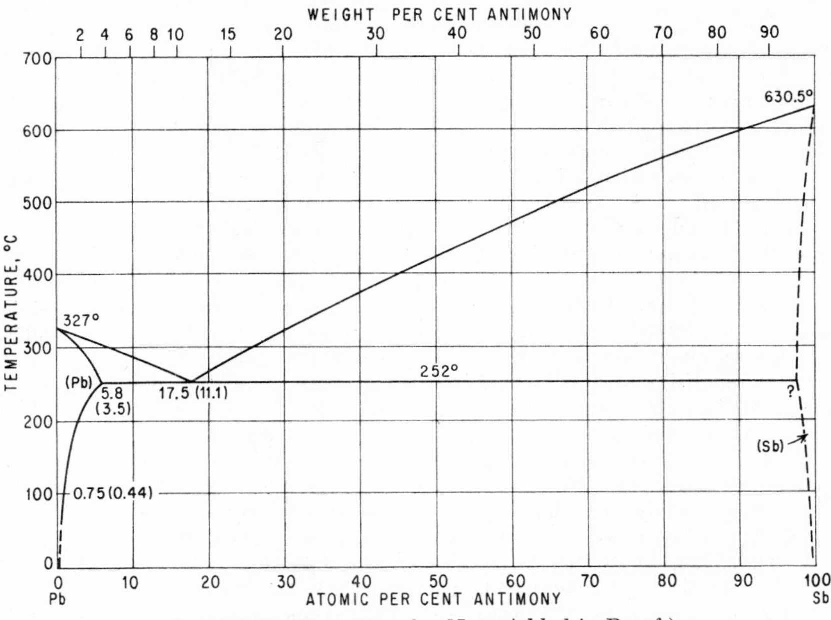

Fig. 598. Pb-Sb. (See also Note Added in Proof.)

experimental data should not be too disturbing in view of the now well-known supercooling effect. In the region 80–100 wt. % Sb, the data by various authors are in good agreement.

As to the composition of the eutectic point, values between 12.1 and 13 wt. % Sb were reported by [2–4, 6, 7, 11, 12, 9, 10, 15–17, 13]. However, from the careful thermal work of [14], supported by chemical analyses of eutectic material by [18, 19], it is clear that the true value is appreciably lower than the previously accepted value of about 13 wt. % Sb: [14] found 11.1 wt. (17.5 at.) % Sb (Fig. 598); [18], 11.4 wt. %; and [19], 11.6 wt. % Sb.

The tendency toward supercooling, mentioned above, also accounts for the low values (below 250°C) reported for the eutectic temperature by many authors [2, 4, 6, 7, 11, 15, 20]. Higher values were found by the following investigators: 250 [9, 13], 250–252 [10], 251.5 ± 0.5 [21], 252.0 [14] (Fig. 598), 254 (22), and 258°C (heating) [12].

Solubility of Sb in Solid Pb. The existence of a (Pb) primary solid solution, for which indications had early been found, was definitely established by [12]. Systematic investigations of the solid-solubility (solvus) curve were made by [23] (resistivity-temperature curves), [24] (conductivity-concentration curves), [25] (X-ray parametric work), and [26] (tensile-strength isotherms). The work of [24, 25] on carefully heat-treated alloys showed that the solubility below 200°C is appreciably lower than the data of [23] indicated. In Fig. 598, weight has been given to the data of [25]; however, as to the solubility at the eutectic temperature, the value 3.5 wt. (5.8 at.) % Sb [20, 21] has been chosen instead of the value 2.94 wt. % Sb given by [25] (for 247°C). Single data points reported by [27, 20, 28, 13] are in fair agreement with the solvus curve of Fig. 598. The results obtained by [12, 29, 21] for the Pb-rich solidus curve are in good agreement. "The early work of [12] employed differential heating-curve methods up to approximately 2 wt. % Sb, and was substantially confirmed by [29], who used annealing and quenching methods over the same composition range. [21], determining the temperature at which sudden fracture occurred on slow heating, obtained results in good agreement with previous work, and extended the curve to the eutectic temperature" [1].

Solubility of Pb in Solid Sb. No accurate data for either the solidus or the solvus curve are available. The solidus has been investigated by [9] by means of resistivity-temperature curves (after annealing for 24 hr at 200–500°C); at 247°C a saturation concentration lying between 3 and 5.5 wt. % Pb was found. Other values reported for the solubility at the eutectic temperature are 2.5 wt. % (thermo-resistometric) [20], 4.5 wt. % Pb (calculated thermodynamically) [13]. [30] were unable to detect any measurable change in the size of the Sb cell in the investigation of "annealed" (no details given) alloys and concluded that the solubility does not exceed 0.5 wt. % Pb. From magnetic measurements on Sb single crystals containing Pb by [31], a solubility exceeding 6 wt. % Pb is to be expected; but there are indications that the single crystals were in a state of supersaturation.

In summarizing, it may be said that the maximum solubility of Pb in Sb very likely lies between 2.5 and 4.5 wt. (1.5 and 2.7 at.) % Pb. A reinvestigation of this part of the phase diagram is very desirable. See Note Added in Proof.

Further Investigations. There had been some discussion as to whether or not an intermediate phase exists in the Pb-Sb system; see, e.g., [12, 27, 32, 33]. Today the occurrence of any such phase can be considered definitely disproved.

Many physical properties of Pb-Sb alloys have been studied: electrical conductivity [34, 27, 24, 10, 16, 35, 28, 36], thermo-emf [37, 10, 35, 36], emf [38–40, 10], magnetic susceptibility [9, 41–43, 31], Hall effect [35], specific heat [44], density [45, 46, 36], and hardness [16, 47]. Only a few of these investigations have more than qualitative bearing on the constitution.

Lattice parameters of the (Pb) solid solution have been measured by [25, 46, 48]. [46] showed that there is no need to assume a molecular constitution of the solid solution as was deduced by [49, 50] by means of thermodynamic analyses.

The boiling-point curve for the Pb-Sb system was determined by [51]. The microsegregation in Pb-Sb alloys has been studied by [52].

Note Added in Proof. [53] studied the system by means of calorimetric measurements ("quantitative thermal analysis"). As regards the solubility of Pb in solid Sb, indications of a retrograde solidus curve were found, the maximum solubility being about 3 wt. (1.8 at.) % Pb at 420°C.

1. G. V. Raynor, Annotated Equilibrium Diagrams, no. 9, The Institute of Metals, London, 1951.
2. Roland-Gosselin, *Bull. soc. encour. ind. natl.*, **1,** 1896, 1307.
3. G Charpy, *Bull. soc. encour. ind. natl.*, **2,** 1897, 394.
4. J. E. Stead, *J. Soc. Chem. Ind.*, **16,** 1897, 200–208, 507

5. W. Campbell, *J. Franklin Inst.*, **154**, 1902, 205–207.
6. W. Gontermann, *Z. anorg. Chem.*, **55**, 1907, 419–425.
7. R. Loebe, *Metallurgie*, **8**, 1911, 8–9.
8. F. Wüst and R. Durrer, "Temperatur-Wärmeinhaltskurven wichtiger Metall-Legierungen," Verlag Verein Deutscher Ingenieure, Berlin, 1921; see V. Fischer, *Z. tech. Physik*, **6**, 1925, 148.
9. H. Endo, *Science Repts. Tôhoku Imp. Univ.*, **14**, 1925, 503–507.
10. W. Broniewski and L. Sliwowski, *Rev. mét.*, **25**, 1928, 397–404.
11. E. Heyn and O. Bauer, "Untersuchungen über Lagermetalle," p. 224, Leonhard Simion, Berlin, 1914 (*Beiheft Verhandl. Ver. Gewerbefl.*, 1914).
12. R. S. Dean, *J. Am. Chem. Soc.*, **45**, 1923, 1683–1688.
13. H. Seltz and B. J. DeWitt, *J. Am. Chem. Soc.*, **61**, 1939, 2594–2597.
14. B. Blumenthal, *Trans. AIME*, **156**, 1944, 240–250; discussion, pp. 250–252.
15. E. Abel, O. Redlich, and J. Adler, *Z. anorg. Chem.*, **174**, 1928, 270.
16. P. Saldau, *J. Inst. Metals*, **41**, 1929, 289–309.
17. F. D. Weaver, *J. Inst. Metals*, **56**, 1935, 212.
18. O. Quadrat and J. Jiriste, *Chim. & ind. (Paris)*, **1934**, 485–489; *Chem. listy*, **29**, 1935, 304–308.
19. W. Hofmann and R. Engel, *Z. Metallkunde*, **44**, 1953, 132–133.
20. M. LeBlanc and H. Schöpel, *Z. Elektrochem.*, **39**, 1933, 695–701.
21. W. S. Pellini and F. N. Rhines, *Trans. AIME*, **152**, 1943, 65–71.
22. H. v. Hofe and H. Hanemann, *Z. Metallkunde*, **32**, 1940, 115.
23. R. S. Dean, W. E. Hudson, and M. F. Fogler, *Ind. Eng. Chem.*, **17**, 1925, 1246–1247.
24. E. E. Schuhmacher and G. M. Bouton, *J. Am. Chem. Soc.*, **49**, 1927, 1667–1675.
25. I. Obinata and E. Schmid, *Metallwirtschaft*, **12**, 1933, 101–103.
26. E. E. Cherkashin and V. N. Tolmachev, *Trudy Inst. Khim. Kharkov Gosudarst. Univ.*, **5**, 1940, 263–271 (in Russian); not available to the author. *Met. Abstr.*, **11**, 1944, 360: "The boundaries of the sol. soln. determined at various temperatures by this method agree well with those given in the literature."
27. R. S. Dean, L. Zickrick, and F. C. Nix, *Trans. AIME*, **73**, 1926, 505–540.
28. E. Kurzyniec, *Bull. intern. acad. polon. sci.*, **8/10A**, 1938, 498–517 (in German).
29. E. E. Schuhmacher and F. C. Nix, *Proc. AIME, Inst. Metals Div.*, **1927**, 195–205.
30. D. Solomon and W. Morris-Jones, *Phil. Mag.*, **10**, 1930, 470–475. In the X-ray diffraction analysis of the Pb-Sb system only the lines of Pb and Sb were observed.
31. S. H. Browne and C. T. Lane, *Phys. Rev.*, **60**, 1941, 895–899.
32. L. O. Howard, *Proc. AIME, Inst. Metals Div.*, **1928**, 369–373 (Hardness measurements).
33. M. G. Raeder and J. Brun, *Z. physik. Chem.*, **133**, 1928, 26–27.
34. A. W. Smith, *J. Franklin Inst.*, **192**, 1921, 101.
35. E. Stephens, *Phil. Mag.*, **5**, 1930, 547–560.
36. V. A. Yurkov, *Doklady Akad. Nauk S.S.S.R.*, **91**, 1953, 891–893; *Met. Abstr.*, **21**, 1954, 608.
37. E. Rudolfi, *Z. anorg. Chem.*, **67**, 1910, 83–85.
38. A. P. Laurie, *J. Chem. Soc.*, **65**, 1894, 1035.
39. N. A. Pushin, *Zhur. Russ. Fiz.-Khim. Obshchestva*, **39**, 1907, 869–897.
40. S. D. Muzaffar, *Trans. Faraday Soc.*, **19**, 1923, 56–58.
41. K. Honda and H. Endo, *J. Inst. Metals*, **37**, 1927, 29–49.
42. F. L. Meara, *Phys. Rev.*, **37**, 1931, 467; see also *Physics*, **2**, 1932, 33–41.
43. Y. Shimizu, *Science Repts. Tôhoku Imp. Univ.*, **21**, 1932, 845–846. Criticism of the susceptibility data by Endo [9].

44. R. Durrer, *Physik. Z.*, **19**, 1918, 86–88.
45. J. Goebel, *Z. Metallkunde*, **14**, 1922, 358–360.
46. N. V. Ageev and I. V. Krotov, *J. Inst. Metals*, **59**, 1936, 301–308.
47. N. Aoki, *Kinzoku-no-Kenkyu*, **11**, 1934, 1–20 (in Japanese); *Met. Abstr.*, **1**, 1934, 172.
48. C. Tyzack and G. V. Raynor, *Acta Cryst.*, **7**, 1954, 505–510.
49. Yap Chu Phay, *Trans. AIME*, **93**, 1931, 185–206.
50. F. H. Jeffery, *Trans. Faraday Soc.*, **28**, 1932, 567–569.
51. W. Leitgebel, *Z. anorg. Chem.*, **202**, 1931, 305–324.
52. A. C. Simon and E. L. Jones, *J. Electrochem. Soc.*, **100**, 1953, 1–10.
53. W. Oelsen, F. Johannsen, and A. Podgornik, *Erzmetall*, **9**, 1956, 459–469.

0.4190
$\bar{1}$.5810

Pb-Se Lead-Selenium

The compound PbSe (27.59 wt. % Se) has long been known; it also occurs naturally (clausthalite). The thermal investigation of the system by [1, 2] showed that PbSe is the only existing intermediate phase. Emf measurements (of rather poor quality, however) by [3] gave the same result.

The thermal analysis by [2] was restricted to the range up to 50 at. % Se; for a comparison with the results of [1], see Fig. 599. For the melting point of PbSe the

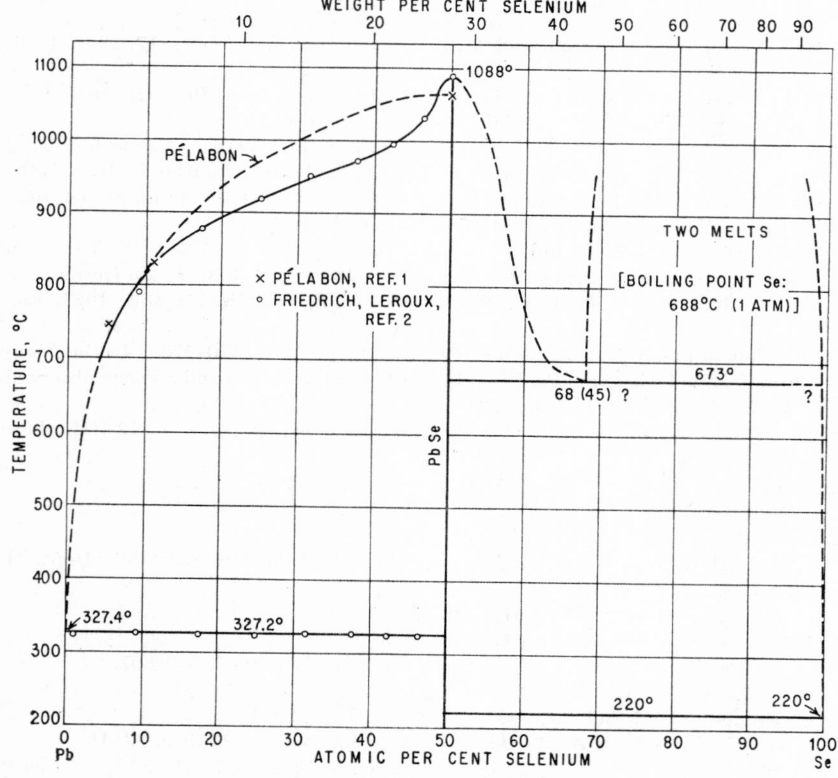

Fig. 599. Pb-Se. (See also Fig. 600.)

values 1065 [1] and 1088°C [2] were reported. As to the liquidus in the range 50–68 at. % Se only scanty data were given by [1]; the dashed curve (Fig. 599) has been taken from his diagram. Above about 45 wt. (68 at.) % Se layer formation was observed by [1]: the upper layer proved to be pure Se; the lower one contained 46.2 wt. % Se. The monotectic horizontal at 673°C was drawn by [1] up to about 88 wt. (95 at.) % Se only; apparently alloys richer in Se could not be investigated because of the volatilization of Se.

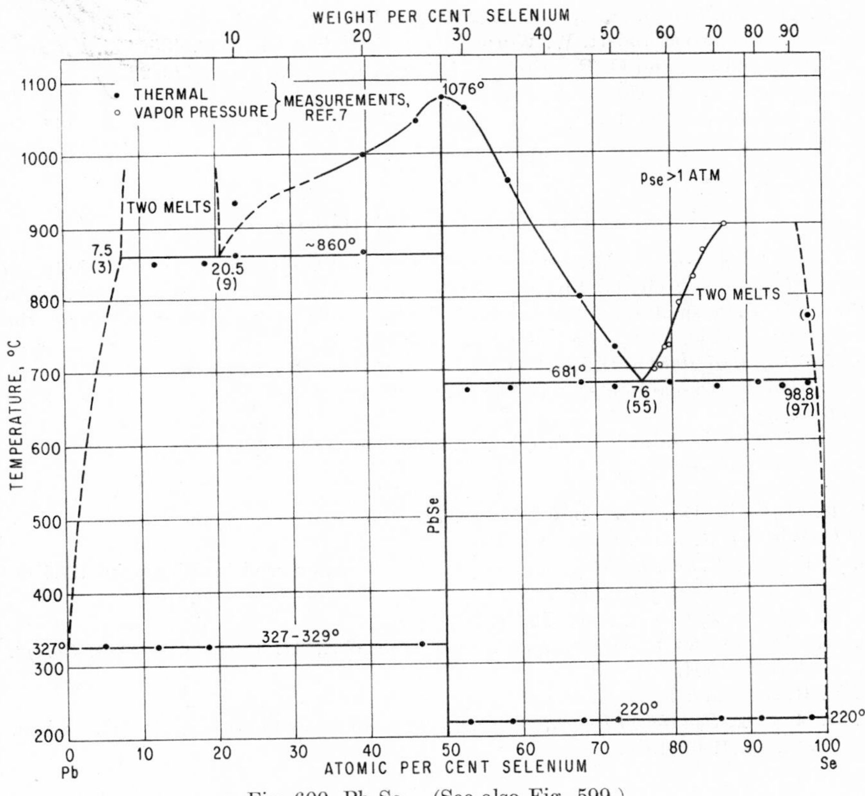

Fig. 600. Pb-Se. (See also Fig. 599.)

A careful microscopic, thermal, and resistometric study of Pb-rich alloys [4] revealed that (*a*) a eutectic exists at 0.005 wt. (0.013 at.) % Se, melting at a temperature 0.2°C below the melting point of Pb; (*b*) the solubility of Se in solid Pb amounts to 0.0015 wt. (0.004 at.) % at 300°C.

According to studies of the semiconducting properties of PbSe, this compound has a very narrow range of homogeneity with the invariant melting point lying at the composition 50.005 at. % Se [5].

PbSe has the NaCl (B1) type of structure, with a = 6.147 ± 5 A [6].

Supplement. The constitution of Pb-Se alloys has been reinvestigated by [7], who used thermal, microscopic, and X-ray analysis as well as vapor-pressure measurements. The phase diagram arrived at by [7] is shown in Fig. 600. It deviates in two main respects from that of Fig. 599 in showing (*a*) a second miscibility gap in the liquid state, between about 7.5 and 20.5 at. % Se, and (*b*) a peritectic rather than a eutectic horizontal as solidus line for Pb-rich alloys. Whereas point (*a*) appears to be fairly well established by [7], this does not hold for point (*b*); the detailed study of [4] indicating a eutectic near pure Pb is considered to be more reliable. The lattice constant of PbSe is given by [7] as $a = 6.128$ A.

1. H. Pélabon, *Compt. rend.*, **144**, 1907, 1159–1161.
2. K. Friedrich and A. Leroux, *Metallurgie*, **5**, 1908, 355–358.
3. H. Pélabon, *Compt. rend.*, **154**, 1912, 1414–1416.
4. J. N. Greenwood and H. W. Worner, *J. Inst. Metals*, **65**, 1939, 435–445.
5. A. E. Goldberg and G. R. Mitchell, *J. Chem. Phys.*, **22**, 1954, 220–222.
6. *Strukturbericht*, **1**, 1913-1928, 125, 137.
7. R. Nozato and K. Igaki, *Bull. Naniwa Univ. (Japan)*, **A3**, 1955, 125–141 (in English).

0.8679
$\bar{1}.1321$

Pb-Si Lead-Silicon

According to early observations by [1, 2] and thermal data (seven alloys between 45 and 99.7 at. % Si) by [3], Pb and Si do not alloy. However, [4, 5] found that molten Pb can dissolve small quantities of Si at higher temperatures. According to [5], the solubility at 1250, 1330, 1400, 1450, and 1550°C is in wt. % (at. % in parentheses) 0.02 (0.15), 0.07 (0.51), 0.15 (1.1), 0.21 (1.53), and 0.78 (5.5), respectively.

From lattice-parameter measurements, [6] concluded that there are no solid solutions.

The eutectic data 44.2 at. (?) % Pb, 398°K (125°C) listed in an abstract [7] must be erroneous.

1. H. St. Claire Deville, *J. prakt. Chem.*, **72**, 1857, 208.
2. C. Winkler, *J. prakt. Chem.*, **91**, 1864, 193.
3. S. Tamaru, *Z. anorg. Chem.*, **61**, 1909, 42–44; the silicon used (melting point 1385°C) contained 6.1 Fe, 1.7 Al (wt. %)!
4. E. Vigouroux, *Compt. rend.*, **123**, 1896, 115.
5. H. Moissan and F. Siemens, *Ber. deut. chem. Ges.*, **37**, 1904, 2086–2089; *Compt. rend.*, **138**, 1904, 657–661.
6. E. R. Jette and E. B. Gebert, *J. Chem. Phys.*, **1**, 1933, 753–755.
7. O. Hájíček, *Hutnické Listy*, **3**, 1948, 265–270 (not available to the author). *Chem. Abstr.*, **43**, 1949, 4935.

0.2420
$\bar{1}.7580$

Pb-Sn Lead-Tin

Early Literature. The literature up to 1909 has been reviewed by [1]; only a few of these old papers are considered in the following review.

Liquidus. More or less complete determinations of the whole liquidus or of parts of it were made by [2–20]. In Fig. 601 weight has been given to the results of [18], which have support from some older [21] and also from more recent [19, 20] data.

Besides many of the authors listed above, [23–27] have also determined the eutectic point (most of these data are listed in [28]). The most reliable data are those of [25, 18]: 183°C (well supported by many investigators) and 38.1 wt. (26.1 at.) % Pb.

The (Sn) Phase. The values reported more recently for the (maximum) solubility of Pb in solid Sn at the eutectic temperature lie within the range 1.5–2.6 wt. % Pb [16–18, 29–31]; for Fig. 601, 2.5 wt. (1.45 at.) % Pb was chosen.

Solidus data were determined by the following methods: resistometric [16, 30], thermal [17], microscopic [18, 30], initial melting [31], and bending test [29]. The results are in substantial agreement, except for those of [29] (debatable method) near the eutectic temperature where a solubility of only 1.6 wt. % Pb is indicated.

Data reported for the solid-solubility (solvus) curve [16, 32, 33] are much at variance with each other. The curve shown in Fig. 601 is essentially based on the measurements of resistivity by [33] and has an intermediate position between those found by [16] (resistometric) and [32] (dilatometric, resistometric). For room-temperature solubility, [33] gave the value 0.4 wt. % Pb; however, tensile strength [34] and (qualitative) X-ray diffraction [35] work indicate that the solubility may be even lower (see also [36]), less than 0.3 wt. (0.17 at.) % Pb according to [34].

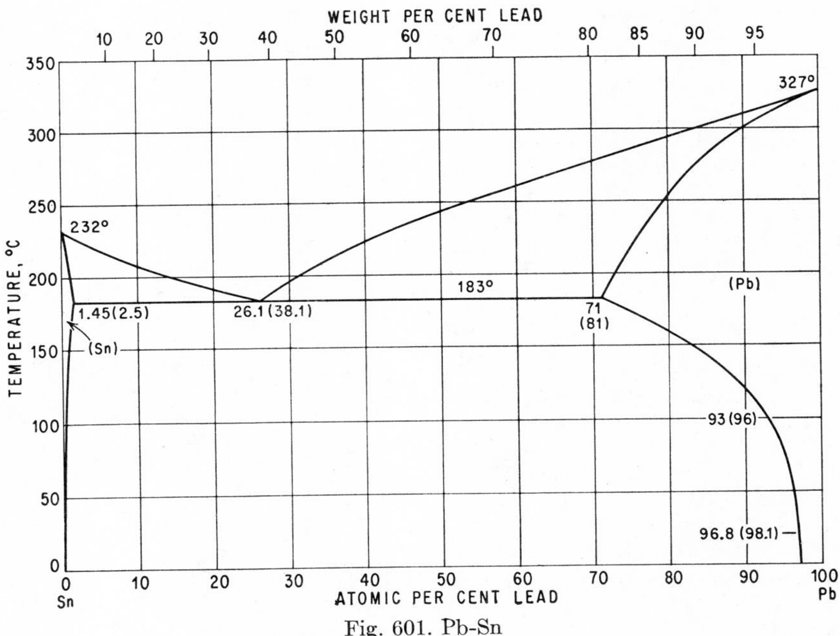

Fig. 601. Pb-Sn

High-temperature X-ray work [37] showed that Pb expands the lattice spacings of Sn.

The (Pb) Phase. *Solidus.* Determinations of the solidus were carried out by [24, 13, 16–19]. The results of [17] (differential thermal), [18] (micrographic), and [19] (resistometric) are in good agreement and were used for Fig. 601.

Solid-solubility (Solvus) Curve. A graphic comparison of the widely differing results of various investigators [10, 38, 39, 16–18, 31, 33] can be found in the papers of [31, 33]. The most consistent results have been obtained by means of resistometric methods [39, 16, 18, 31, 33]. In Fig. 601, weight has been given to the data of [31] (above 110°C) and [33] (100°C and lower), the curve being smoothed out in the junction range. The room-temperature value found by [33] (3.2 at. % Sn) has support from an X-ray value by [40] (3.4 at. % Sn).

As to the maximum solubility of Sn in Pb at the eutectic temperature, the values given by [10, 41, 38, 39, 13, 15–17] lie between 16 and 18 wt. % Sn; [18] found 19.5 and

[31] about 18.2 wt. (28 at.) % Sn. In Fig. 601 the value 19 wt. (29 at.) % Sn was used. The lattice parameter of Pb is lowered by addition of Sn [35, 40, 42].

Further Remarks. The observation of a marked thermal effect at about 150°C by several authors has stimulated an extensive study [43] of the nature of this phenomenon. It was first shown by [41] that neither does an allotropic transformation in the (Pb) phase occur nor is an intermediate phase formed, but that this effect is caused by the greatly decreasing solubility of Sn in the (Pb) phase with decreasing temperature (see also the kinetic investigation by [31]).

The following investigations of physical properties of Sn-Pb alloys have only qualitative bearing on the constitution: electrical resistivity [44], thermo-emf [45], magnetic susceptibility [46–48], emf [49], and hardness [50, 44, 51].

A thermodynamic calculation of the phase diagram, based on specific-heat measurements, was carried out by [52].

1. W. Guertler, *Z. Elektrochem.*, **15**, 1909, 125–129.
2. C. T. Heycock and F. H. Neville, *J. Chem. Soc.*, **55**, 1889, 667; **57**, 1890, 376; **61**, 1892, 908.
3. F. C. Weld, *Am. Chem. J.*, **13**, 1891, 121–122.
4. B. Wiesengrund, *Wied. Ann.*, **52**, 1894, 777; see also *Z. anorg. Chem.*, **53**, 1907, 138–140; **63**, 1909, 210.
5. W. C. Roberts-Austen, *Engineering*, **63**, 1897, 223.
6. G. Charpy, *Bull. soc. encour. ind. natl.*, **2**, 1897, 384; also *Metallographist*, **1**, 1898, 87–106; Société d'encouragement pour l'industrie nationale, Paris, Commision des alliages, "Contribution à l'étude des alliages," pp. 131–132, Typ. Chamerot et Renouard, Paris, 1901. See also *Z. anorg. Chem.*, **53**, 1907, 138–140; **63**, 1909, 210.
7. A. W. Kapp, *Ann. Physik*, **6**, 1901, 754; see also *Z. anorg. Chem.*, **53**, 1907, 138–140; **63**, 1909, 210.
8. N. S. Kurnakov, *Zhur. Russ. Fiz.-Khim. Obshchestva*, **37**, 1905, 579. See *Z. anorg. Chem.*, **60**, 1908, 32 (footnote).
9. A. Stoffel, *Z. anorg. Chem.*, **53**, 1907, 138–140.
10. W. Rosenhain and P. A. Tucker, *Phil. Trans. Roy. Soc.* (*London*), **A209**, 1908, 89; *Proc. Roy. Soc.* (*London*), **A81**, 1909, 331–334.
11. P. N. Degens, *Z. anorg. Chem.*, **63**, 1909, 207–224.
12. P. Müller, *Metallurgie*, **7**, 1910, 765.
13. S. Konno, *Science Repts. Tôhoku Imp. Univ.*, **10**, 1921, 57–74.
14. L. J. Gurevich and J. S. Hromatko, *Trans. AIME*, **64**, 1921, 234.
15. K. Kaneko and A. Araki, *Nippon Kogyo Kwai-Shi*, **41**, 1925, 437–455; abstract, *J. Inst. Metals*, **33**, 1927, 422.
16. F. H. Jeffery, *Trans. Faraday Soc.*, **24**, 1928, 209–211; **26**, 1930, 588–590.
17. K. Honda and H. Abe, *Science Repts. Tôhoku Imp. Univ.*, **19**, 1930, 315–330.
18. D. Stockdale, *J. Inst. Metals*, **49**, 1932, 267–282; see also discussion.
19. R. Hultgren and S. A. Lever, *Trans. AIME*, **185**, 1949, 67–71.
20. H. J. Fisher and A. Phillips, *Trans. AIME*, **200**, 1954, 1062.
21. "Over the range 0–40 wt. % Sn the results of [10, 16, 18] are in good agreement. . . . Agreement between the results of several investigators for the branch 68–100 wt. % Sn is good: 68–88 wt. % Sn [16, 18]; 88–100 wt. % Sn [10, 14, 17, 18]." Quoted from Raynor [22].
22. G. V. Raynor, Annotated Equilibrium Diagrams, no. 6, The Institute of Metals, London, 1947.
23. D. Mazzotto, *Mem. R. Ist. Lombardo*, **16**, 1886, 1.
24. D. Mazzotto, *Nuovo cimento*, **18**, 1909, 180.
25. D. Stockdale, *J. Inst. Metals*, **43**, 1930, 193–211.

26. P. Saldau, *J. Inst. Metals*, **41**, 1929, 292–293; *Z. anorg. Chem.*, **194**, 1930, 291.
27. O. Hájíček, *Hutnické Listy*, **3**, 1948, 265–270 (not available to the author). *Chem. Abstr.*, **43**, 1949, 4935.
28. M. Hansen, "Der Aufbau der Zweistofflegierungen," Springer-Verlag OHG, Berlin, 1936.
29. C. D. Homer and H. Plummer, *J. Inst. Metals*, **64**, 1939, 169–200. See also discussion, pp. 201–208 (especially Stockdale).
30. A. Stockburn, *J. Inst. Metals*, **66**, 1940, 33–38.
31. G. Borelius, F. Larris, and E. Ohlsson, *Arkiv Mat., Astron. Fysik*, **31A**(10), 1944 (in English). See also G. Borelius, *Trans. AIME*, **191**, 1951, 477–484.
32. Y. Matuyama, *Science Repts. Tôhoku Imp. Univ.*, **20**, 1931, 661.
33. E. Kurzyniec and Z. Wojtaszek, *Bull. intern. acad. polon. sci., Classe sci. math. nat.*, **A1951**, 131–146 (in English). The ratios of the resistance at the temperature of boiling N_2 or H_2 to the resistance at the temperature of melting ice was measured on carefully heat-treated alloys.
34. H. S. Kalish and F. J. Dunkerley, *Trans. AIME*, **180**, 1949, 637–656.
35. W. C. Phebus and F. C. Blake, *Phys. Rev.*, **25**, 1925, 107.
36. G. Tammann and G. Bandel, *Z. Metallkunde*, **25**, 1933, 156.
37. J. A. Lee and G. V. Raynor, *Proc. Phys. Soc. (London)*, **B67**, 1954, 737–747.
38. D. Mazzotto, *Mem. accad. sci., Modena*, **10**, 1912. See with Parravano and Scortecci [39]; also *Intern. Z. Metallog.*, **4**, 1913, 286–287.
39. N. Parravano and A. Scortecci, *Gazz. chim. ital.*, **50**(2), 1920, 83–92.
40. I. Obinata and E. Schmid, *Metallwirtschaft*, **12**, 1933, 101–103.
41. D. Mazzotto, *Intern. Z. Metallog.*, **1**, 1911, 289–346.
42. C. Tyzack and G. V. Raynor, *Acta Cryst.*, **7**, 1954, 505–510.
43. For references and a detailed discussion, see Hansen [28].
44. G. Tammann and H. Rüdiger, *Z. anorg. Chem.*, **192**, 1930, 1–44.
45. Batelli; see with W. Broniewski, *Rev. mét.*, **7**, 1910, 355. E. Rudolfi, *Z. anorg. Chem.*, **67**, 1910, 78–80.
46. K. Honda and T. Soné, *Science Repts. Tôhoku Imp. Univ.*, **2**, 1913, 1–14. J. F. Spencer and M. E. John, *Proc. Roy. Soc. (London)*, **A116**, 1927, 61–72; *J. Soc. Chem. Ind.*, **50**, 1931, 38.
47. E. L. Dupuy, *Compt. rend.*, **158**, 1914, 793–794.
48. K. Honda and H. Endo, *J. Inst. Metals*, **37**, 1927, 34–38.
49. N. Puschin, *Zhur. Russ. Fiz.-Khim. Obshchestva*, **39**, 1907, 528. S. D. Muzaffar, *Z. anorg. Chem.*, **126**, 1923, 254–256. M. G. Raeder and D. Efjestad, *Z. physik. Chem.*, **A140**, 1929, 125–126 (Hydrogen overvoltage).
50. S. Ssaposhnikow, *Zhur. Russ. Fiz.-Khim. Obshchestva*, **40**, 1908, 92–95. J. Goebel, *Z. Metallkunde*, **14**, 1922, 362–366. C. di Capua and M. Arnone, *Rend. accad. Lincei*, **33**, 1924, 293–297. A. Mallock, *Nature*, **121**, 1928, 827. W. Schischokin and W. Agejewa, *Z. anorg. Chem.*, **193**, 1930, 241. F. Marmet, Dissertation, Aachen, 1930; abstract, *Physik. Ber.*, **12**, 1931, 40–41.
51. N. Aoki, *Kinzoku-no-Kenkyu*, **11**, 1934, 1–20 (in Japanese); *Met. Abstr.*, **1**, 1934, 172.
52. Ya. E. Geguzin and B. Ya. Pines, *Zhur. Fiz. Khim.*, **25**, 1951, 1228–1238.

0.3738
$\bar{1}.6262$

Pb-Sr Lead-Strontium

The Pb-rich part of the Pb-Sr phase diagram shown in Fig. 602 was established by [1] by means of thermal and microscopic analysis. A slight decrease in the lattice parameter of Pb by addition of Sr indicates a certain solubility of Sr in solid Pb [2].

According to [2], the structure of $SrPb_3$ (12.36 wt. % Sr) corresponds to a slightly tetragonally distorted $Cu_3Au(L1_2)$ type; $a = 4.965 \pm 3$ A, $c = 5.035 \pm 3$ A, $c/a = 1.014$. If it exists at all, the compound SrPb has neither the β-brass nor the NaTl type of structure [3].

1. E. Piwowarsky, *Z. Metallkunde*, **14**, 1922, 300–301. The alloys were prepared under hydrogen.
2. E. Zintl and S. Neumayr, *Z. Elektrochem.*, **39**, 1933, 86–97.
3. E. Zintl and G. Brauer, *Z. physik. Chem.*, **B20**, 1933, 245–271.

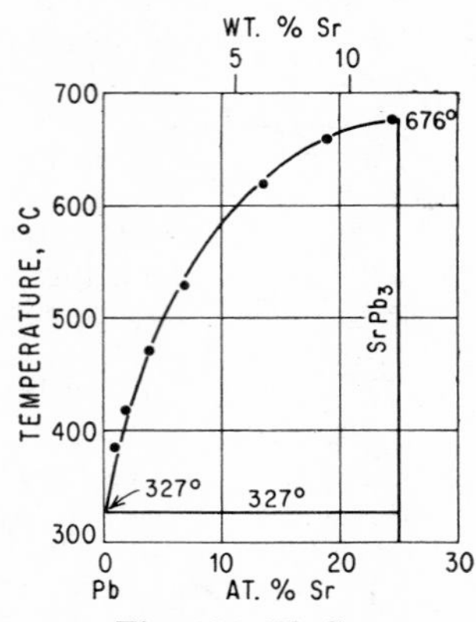

Fig. 602. Pb-Sr

0.2105
$\bar{1}.7895$

Pb-Te Lead-Tellurium

Solidification equilibria of the Pb-Te system were investigated by [1–3] and [4]. For quantitative differences in the results of [1–3], see Fig. 603 and Table 39. As for the composition of the (PbTe + Te) eutectic, the value of [1] is the most reliable one. According to more recent microscopic and thermal work [4], the (Pb + PbTe) eutectic lies at 0.025 wt. (0.04 at.) % Te and at a temperature 0.7°C below the solidification point of pure Pb.

Table 39

	Ref. 1	Ref. 2	Ref. 3
Solidification point of Pb	(322°C)	?	(326°C)
(Pb + PbTe) eutectic	322°C	?	332–334°C
	0% Te		0% Te (!)
Solidification point of PbTe	917°C	860°C	> 904°C
(PbTe + Te) eutectic	400°C	403°C	412°C
	78.5 wt. % Te	85 wt. % Te	~76 wt. % Te
Solidification point of Te	(446°C)	452°C	(441°C)

The intermediate phase PbTe (38.12 wt. % Te), known in mineralogy as altaite, had previously been prepared by [5–7]. Its existence is further corroborated by measurements of emf [8], thermo-emf [9], and magnetic susceptibility [10].

Solid Solutions in the Pb-Te System [11]. *The* (Pb) *Phase.* [4] determined the solubility of Te in solid Pb on carefully annealed alloys to 0.004 wt. (0.0065 at.) % Te at 300°C by means of microexamination and measurements of electrical resistance. [12], who found that Pb-rich alloys can be age-hardened, questioned this low solubility value but admitted that their cast alloys could have been strongly supersaturated. In fact, in quenching alloys from the molten state at extremely high rates, [13] were able to produce supersaturated solid solutions containing up to 0.2 wt. (0.32 at.) % Te. For the great effect of small additions of Te on the properties of Pb, see [14, 12].

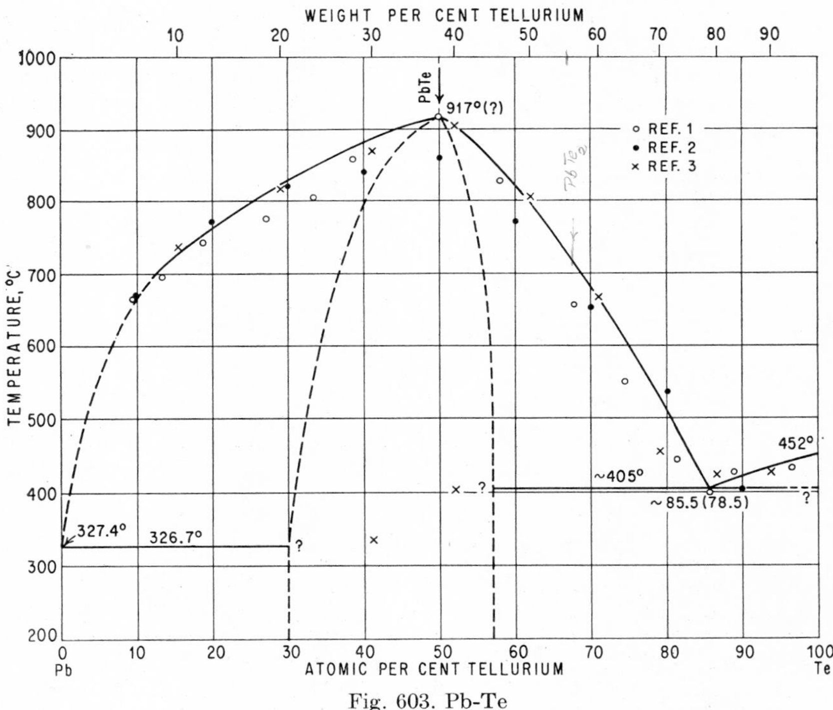

Fig. 603. Pb-Te

(PbTe). The measurements of magnetic susceptibility by [10] showed that PbTe has a wide range of homogeneity, its limits being (at first approximation) 20–25 and 45 wt. (29–35 and 57 at.) % Te (Fig. 603). However, see [11].

The (Te) *Phase.* The magnetic-susceptibility-vs.-composition curve by [10] indicates a great influence of small additions of Pb on the susceptibility of Te, and [10] assumed the solubility to exceed 5 wt. % Pb. However, since no data are given between 90 and 100 wt. % Te, nothing certain can be said about the actual extent of the solid solution. It is, furthermore, improbable that the terminal solid solution is more extended on the Te side than on the Pb side of the phase diagram, since the crystal structure of Te is more complex than that of the f.c.c. Pb. In Fig. 603 no (Te)-phase field has been drawn.

Crystal Structures. Addition of Te lowers the lattice parameter of Pb [13]. PbTe was shown to have the NaCl (B1) type of structure [15, 16], with $a = 6.45$ A (50 at. % Te?) [16].

1. H. Fay and C. B. Gillson, *Am. Chem. J.*, **27,** 1902, 81–95.
2. H. Pélabon, *Ann. chim. et phys.*, **17,** 1909, 557–558. Most of the liquidus temperatures plotted in Fig. 603 were taken from a graph.
3. M. Kimura, *Mem. Coll. Sci. Kyoto Univ.*, **1,** 1915, 149–152 (in German).
4. J. N. Greenwood and H. W. Worner, *J. Inst. Metals,* **65,** 1939, 435–445.
5. J. Margottet, Thèse 422, Paris, 1879.
6. C. Fabre, *Compt. rend.*, **105,** 1887, 277.
7. C. A. Tibbals, *J. Am. Chem. Soc.*, **31,** 1909, 909.
8. N. A. Puschin, *Z. anorg. Chem.*, **56,** 1907, 12–15.
9. W. Haken, *Ann. Physik*, **32,** 1910, 329–330.
10. H. Endo, *Science Repts. Tôhoku Imp. Univ.*, **16,** 1927, 209–211; see also K. Honda and H. Endo, *J. Inst. Metals,* **37,** 1927, 42–43.
11. Fay and Gillson [1] and Kimura [3] concluded from their thermal and microscopic work that neither the components (especially Te) nor PbTe form any solid solutions. Heat-treatments were not carried out by these investigators.
12. W. Hofmann and H. Hanemann, *Z. Metallkunde,* **33,** 1941, 62–63.
13. G. Falkenhagen and W. Hofmann, *Z. Metallkunde,* **43,** 1952, 69–81.
14. W. Singleton and B. Jones, *J. Inst. Metals,* **51,** 1933, 71–92.
15. L. S. Ramsdell, *Am. Mineralogist,* **10,** 1925, 281–304.
16. V. M. Goldschmidt; see *Strukturbericht,* **1,** 1913-1928, 138.

$\bar{1}.9507$
0.0493

Pb-Th Lead-Thorium

Th dissolves in molten Pb and, on solidification, separates out in elementary form [1].

1. See report on DRP (German Patent) No. 146,503, 1900, in *Chem. Zentr.*, **1903**(2), 1156.

0.6361
$\bar{1}.3639$

Pb-Ti Lead-Titanium

[1] have metallographically examined Mg-reduced Ti-base alloys with about 0.5, 1.3, and 2.2 wt. % Pb (which contained 0.3–1 wt. % W) after quenching from various temperatures between 950 and 790°C. Their finding that Pb "raises the β solvus temperatures $\beta/(\alpha + \beta)$ markedly" was not corroborated by later work.

[2] have studied the constitution in the range of composition 37.5–100 wt. % Pb by means of microscopic and X-ray diffraction analysis. Only one intermediate phase was identified, which very likely has the composition Ti_4Pb (51.96 wt. % Pb). The crystal structure, as determined on an alloy with 37.5 wt. % Pb, is of the hexagonal Ni_3Sn ($D0_{19}$) type, with $a = 5.98_5$ A, $c = 4.84_6$ A, $c/a = 0.809$; therefore, this phase apparently has the ideal, although not realizable, composition Ti_3Pb.

More recently the system has been investigated with 23 alloys in the ranges 0–58 wt. % Pb and 500°C to liquidus temperatures [3]. Microscopic and X-ray analysis of cast and annealed specimens and melting-point observations were the principal tools of investigation. The diagram obtained is shown in Fig. 604. The existence of Ti_4Pb was corroborated; however, in deviation from the findings of [2], some evidence for the existence of another intermediate phase (γ), richer in Pb, was found. Of the formulas Ti_3Pb (59.05 wt. % Pb) and Ti_2Pb (68.38 wt. % Pb), the latter is considered the more likely. The retention of β solid solution on quenching was found to occur only in alloys containing 35 wt. % Pb or more.

[3] found the lattice parameters $a = 5.96_2$ A, $c = 4.81_4$ A, $c/a = 0.808$ for Ti_4Pb

in a 54.6 wt. % Pb alloy. Insufficient amounts of γ phase were present in the alloys to permit identification by X-ray methods; however, the patterns proved that γ could not be Pb.

Note Added in Proof. [4] prepared the intermediate phase Ti_4Pb by heating a mixture of molten Pb and powdered Ti.

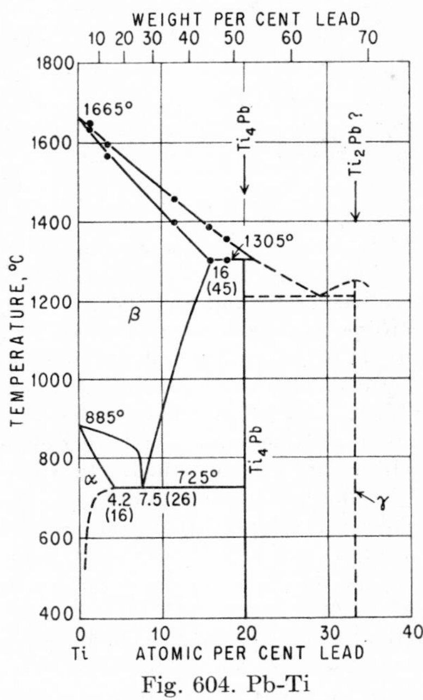

Fig. 604. Pb-Ti

1. C. M. Craighead, O. W. Simmons, and L. W. Eastwood, *Trans. AIME,* **188,** 1950, 485–513.
2. H. Nowotny and J. Pesl, *Monatsh. Chem.,* **82,** 1951, 344–347.
3. P. Farrar and H. Margolin, *Trans. AIME,* **203,** 1955, 101–104. The alloys were prepared by arc melting under argon. Extreme difficulty was encountered in preparing homogeneous alloys without excessive Pb losses because of the low boiling point of Pb as compared with the Ti melting point.
4. B. N. Rabinovich and D. M. Chizhikov, *Izvest. Akad. Nauk S.S.S.R., Otdel. Tekh. Nauk,* **1956,** 114–117; *Met. Abstr.,* **24,** 1957, 806; *Titanium Abstr. Bull.,* **2,** 1956, 2011.

0.0060
$\bar{1}.9940$

Pb-Tl Lead-Thallium

Solidification. The solidification near pure Tl and pure Pb was studied by [1, 2]. It was first investigated by [3] over the whole composition range (Fig. 605). Careful and repeated determinations of the liquidus near its maximum have shown that it lies near 37.5 at. % Pb rather than at a distinctive composition like Tl_2Pb or Tl_3Pb_2 (see also [4]). The maximum is extremely flat. At higher Pb contents liquidus and solidus run close together, the maximal distance being 2.5°C at 60–70 at. % Pb.

Further thermal analyses were carried out by [5] (whole system) and [6] (0–40 wt. % Pb). Apart from the (incorrect) statement by [5] that the maximum in the liquidus corresponds to a compound Tl_2Pb, their results are in substantial agreement with those of [3] (cf. Table 40).

Table 40

	Ref. 3	Ref. 5	Ref. 6
Peritectic temperature, °C.	310	310	311
Extent of the peritectic horizontal, wt. % Pb.	5.5–25	5–24	5.5–23
Composition of β_{Tl} at the peritectic temperature, wt. % Pb.	6.5		

The compositions reported by [3, 5, 6] for the Pb-rich end of the peritectic horizontal (23–25 wt. % Pb) were derived from thermal data only and, therefore, are bound to be much too high (see below and Fig. 605).

Solid Alloys. According to thermal [3] and thermoresistometric [7, 8] studies, the transformation in pure Tl is lowered by addition of Pb (Fig. 605). Contrary to the statement by [9], however, β_{Tl} is not stable at room temperature [10]. An (obviously metastable) f.c.c. form of Tl was observed by [11] in quenched alloys with 5 and 10 wt. % Pb. There is little and conflicting evidence as far as the phase boundaries in the Tl-rich region are concerned. In the following a compilation of the data reported for the limits of the miscibility gap adjacent to the (Tl) phases (α, β) is made (see also Table 40): [8] by resistivity isotherms, 6–20 wt. % Pb at 150°C, 6–22 wt. % Pb at 250°C; [12] by emf measurements, about 3.5–7.5 at. (wt.) % Pb at 250°C; and [10] by X-ray diffraction, <3.1 to about 12.5 at. (wt.) % Pb (slowly cooled alloys).

In Fig. 605 the boundary of the (Pb) phase has been drawn arbitrarily over the whole temperature range at 12.5 at. % Pb. The decrease in solubility of Pb in α-Tl with decreasing temperature (to about 3% at room temperature) shown in Fig. 605 has support from measurements of electrical resistivity by [13]. It is also in accord with the finding of [10] listed above.

Numerous investigations were carried out to clarify the nature of the maximum in the liquidus. X-ray diffraction analyses [14, 15, 12, 10, 16] showed that the f.c.c. Pb lattice persists (with decreasing parameter) down to low Pb contents, 12.5 at. % according to [10]. This refutes the occurrence of any "genuine" compound in the Tl-Pb system. However, possible ordering reactions cannot easily be detected by X-ray work, owing to the near equality in scattering powers of Tl and Pb. [12] noticed that the lattice constants do not vary linearly with composition and concluded—supported by data from emf measurements—that below 45.4 at. % Pb the phase field of ordered (Tl_7Pb) extends. The precision measurements by [10] did not corroborate the discontinuity in slope of the lattice-constant curve at 45.4 at. % Pb reported by [12], but their data show a singularity near 25 at. % Pb. According to their interpretation, the ordered structures Tl_7Pb (12.5 at. % Pb) and Tl_3Pb exist, the latter having the Cu_3Au ($L1_2$) type of structure and the former a cubic unit cell containing 32 atoms (crystallographic type ABC_6; for details see the original paper). "In the intermediate ranges, solid solutions of the types $Tl_6(Tl,Pb)Pb$ and $(Tl,Pb)_3Pb$ exist" [10]. [10] also formulated a thermodynamic treatment that accounts satisfactorily for the existence of a maximum in melting point displaced from the composition Tl_3Pb of the ordered structure [17]. It remains to be established whether any (very narrow) two-phase fields exist in the region around the composition of the maximum in the liquidus.

Investigations other than X-ray gave the following results: No indications for the existence of any distinctive composition were found in measurements of conductivity isotherms [18, 19, 8, 13], of the thermo-emf [19], specific volume [19], hardness [6, 19], emf [20, 21], and hydrogen overvoltage [22], and of the superconductivity and resistivity at low temperatures [23] (see also [24]). Indications for the existence of a

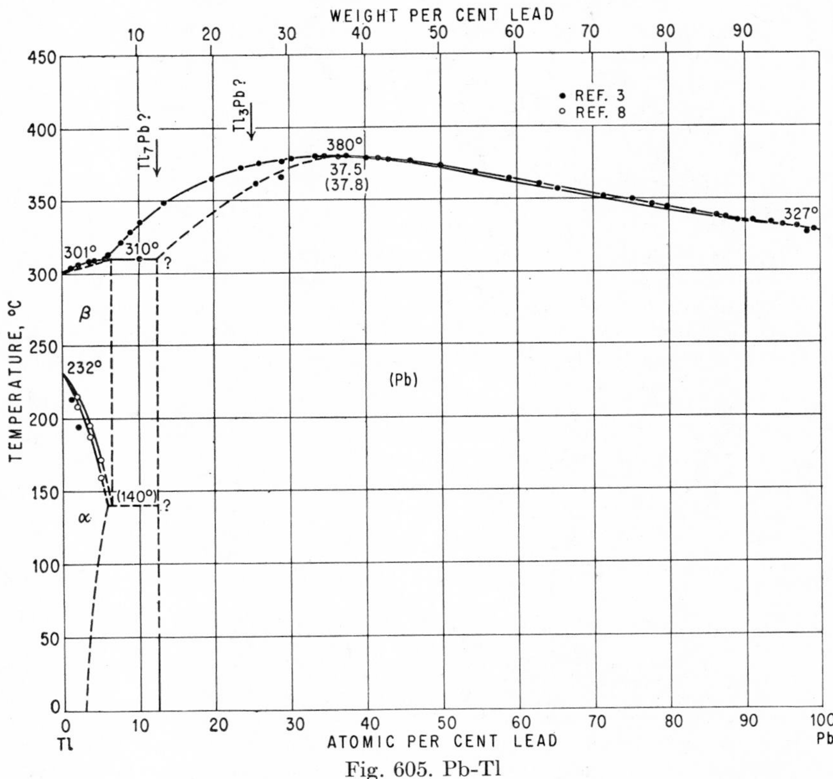

Fig. 605. Pb-Tl

distinctive composition Tl_2Pb were claimed to be found by [25] (magnetic susceptibility), [26] (activity measurements of molten alloys), [27] (investigations on Pb-Tl catalyzers), [9] (X-ray, see [28]), [29] (hardness), and [30] (review).

Boiling points of Tl-Pb alloys were determined by [31].

1. C. T. Heycock and F. H. Neville, *J. Chem. Soc.*, **61,** 1892, 910.
2. C. T. Heycock and F. H. Neville, *J. Chem. Soc.*, **65,** 1894, 35.
3. N. S. Kurnakov and N. A. Puschin, *Zhur. Russ. Fiz.-Khim. Obshchestva*, **38,** 1906, 1146–1167; also *Z. anorg. Chem.*, **52,** 1907, 430–451.
4. N. S. Kurnakov and N. I. Korenev, *Izvest. Inst. Fiz.-Khim. Anal.*, **6,** 1933, 47–68 (in Russian).
5. K. Lewkonja, *Z. anorg. Chem.*, **52,** 1907, 452–456.
6. C. di Capua, *Rend. accad. Lincei*, **23**(2), 1923, 343–346.
7. W. Guertler, "Handbuch der Metallographie," vol. 1, Part 1, pp. 542–543, Verlagsbuchhandlung Gebrüder Borntraeger, Berlin, 1912.
8. W. Guertler and A. Schulze, *Z. physik. Chem.*, **104,** 1923, 269–309.

9. V. M. Goldschmidt, *Z. physik. Chem.*, **133,** 1928, 409–410.
10. Y. Tang and L. Pauling, *Acta Cryst.*, **5,** 1952, 39–44.
11. S. Sekito, *Z. Krist.*, **74,** 1930, 193–195.
12. A. Ölander, *Z. physik. Chem.*, **168,** 1934, 274–282; see also *Z. Metallkunde*, **27,** 1935, 141.
13. G. Tammann and H. Rüdiger, *Z. anorg. Chem.*, **192,** 1930, 35–39.
14. E. McMillan and L. Pauling, *J. Am. Chem. Soc.*, **49,** 1927, 666–669.
15. J. Halla and R. Staufer, *Z. Krist.*, **67,** 1928, 440–454; **68,** 1928, 299–300.
16. C. Tyzack and G. V. Raynor, *Acta Cryst.*, **7,** 1954, 505–510 (Lattice parameters in the range 85–100 at. % Pb).
17. This treatment implies that (partial) ordering persists up to the melting range.
18. N. S. Kurnakov and S. F. Zemczuzny, *Z. anorg. Chem.*, **64,** 1909, 155–162.
19. L. Rolla, *Gazz. chim. ital.*, **45**(1), 1915, 185–191.
20. E. Bekier, *Chem. Polski*, **15,** 1917, 119–131; abstract, *Chem. Zentr.*, **1918(1),** 1001.
21. R. Kremann and A. Lobinger, *Z. Metallkunde*, **12,** 1920, 247–249.
22. M. G. Raeder and D. Efjestad, *Z. physik. Chem.*, **140,** 1929, 126–127.
23. W. Meissner, H. Franz, and H. Westerhoff, *Ann. Physik*, **13,** 1932, 968–979.
24. C. Benedicks, *Z. Metallkunde*, **25,** 1933, 201–202.
25. H. Endo, *Science Repts. Tôhoku Imp. Univ.*, **14,** 1925, 497–498; **16,** 1927, 211–212.
26. J. H. Hildebrand and J. N. Sharma, *J. Am. Chem. Soc.*, **51,** 1929, 469–471.
27. E. Pietsch and F. Seuferling, *Z. Elektrochem.*, **37,** 1931, 660–662. E. Pietsch, *Metallwirtschaft*, **12,** 1933, 223–224.
28. "An alloy $PbTl_2$ was obtained in hexagonal as well as in f.c.c. form; the former seems to be stable at normal temperatures" (?).
29. W. P. Schischokin and W. Ageeva, *Zvet. Metally*, **1932,** 119–136 (in Russian); abstract, *J. Inst. Metals*, **53,** 1933, 552.
30. E. Jänecke, *Z. Metallkunde*, **26,** 1934, 153–155; **27,** 1935, 141.
31. W. Leitgebel, *Z. anorg. Chem.*, **202,** 1931, 305–324.

$\bar{1}.9397$
0.0603

Pb-U Lead-Uranium

[1] showed by X-ray analysis that UPb_3 (72.30 wt. % Pb) is the intermediate phase richest in Pb.

The whole system was studied by [2, 3]. Both used thermal, microscopic, and X-ray powder methods, and [3] additionally made use (for determination of the Pb-rich liquidus) of a hot centrifuge and of measurements of resistivity. The thermal results of [2, 3] are plotted in Fig. 606, which shows two compounds to exist, namely, UPb (46.53 wt. % Pb) and UPb_3. The principal difference between the diagrams arrived at by [2, 3] lies in the kind of solidification equilibria between 0 and 70 at. % Pb: [2] assumed a miscibility gap in the liquid state extending from 1.5 to 60 wt. % Pb at (the synthetic temperature) 1280°C, whereas [3] interpreted their results in terms of a flattened liquidus.

In order to confirm his conclusions concerning the liquid immiscibility, the author of [2] recently gave more details of the experimental findings on which these conclusions were founded [5]. There appears to remain little doubt that a miscibility gap does exist (Fig. 606).

The (U + UPb) eutectic is situated near pure U, at 0.4 wt. (0.5 at.) % Pb according to [2]. Whereas [2] observed no effect of Pb on the allotropic-transformation temperatures of U, [3] found a slight depression, to the extent of about 5°C in each case (Fig. 606). The melting point of UPb_3 has not been definitely established.

According to measurements of resistivity [3], the solubility of U in liquid Pb is 0.02, 0.10, and 0.25 at. (0.02, 0.11, and 0.29 wt.) % at 500, 615, and 720°C, respectively. Data by [5], obtained by a filtration technique, indicate a somewhat lower solubility, in wt. % (at. % in parentheses): 0.002, 0.046, 0.26, and 0.59 (0.002, 0.040, 0.23, and 0.51) at 416, 612, 806, and 1000°C, respectively.

Lattice spacings measured by [3] indicated little solubility of Pb in solid U. As to the UPb_3 and Pb phases, no indications for solid solutions were found roentgenographically [1–3]. Experimental difficulties prevented an X-ray study of UPb [2, 3]. According to neutron-diffraction work [5], UPb is b.c. tetragonal, with a = 11.04 A, c = 10.60 A, c/a = 0.961, and 48 atoms per unit cell.

UPb_3 yields the X-ray pattern of a f.c.c. (A1) lattice [1–3]; though no superstructure lines are visible [4], the (ordered) Cu_3Au ($L1_2$) type of structure is highly probable. As regards the lattice parameter of UPb_3, [5] stated: "[3] give the lattice

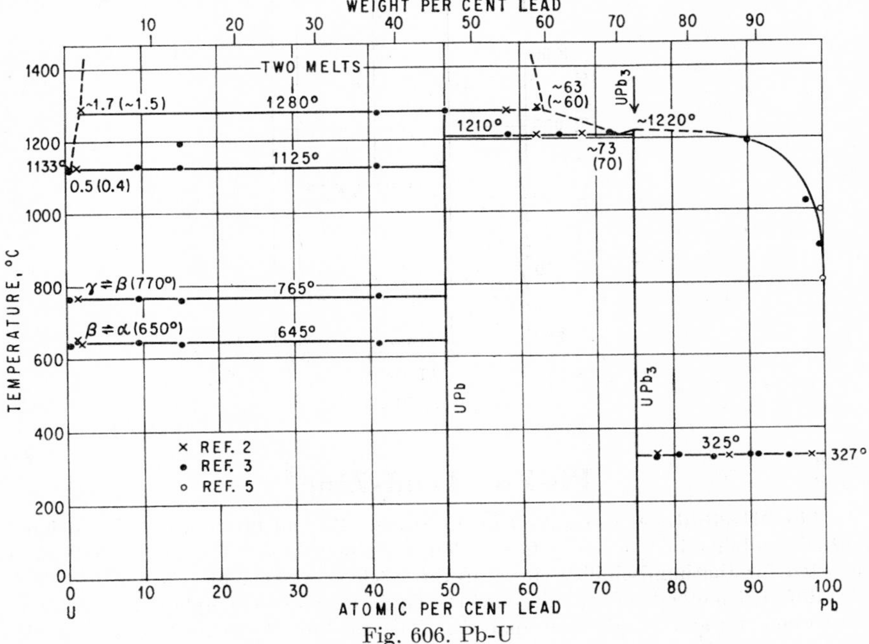

Fig. 606. Pb-U

parameter as a = 4.7834 A. This value disagrees with that reported in an earlier paper by the author [2] (a = 4.791 A). This discrepancy may be due to an oversight by [3] in converting kX units to Angstroms, as that is exactly the ratio of the two values."

1. A. Iandelli and R. Ferro, *Ann. chim. (Rome)*, **42,** 1952, 598–608.
2. R. J. Teitel, *Trans. AIME*, **194,** 1952, 397–400.
3. B. R. Frost and J. T. Maskrey, *J. Inst. Metals*, **82,** 1953, 171–180.
4. The scattering powers of U and Pb are similar.
5. R. J. Teitel, *J. Inst. Metals*, **85,** 1957, 409–412.

0.0518
$\bar{1}$.9482

Pb-W Lead-Wolfram

Solidification temperatures of alloys containing up to 30 at. % W were determined by [1] (Fig. 607). [1] claimed that the primary crystals were pure W (which would mean that no intermediate phases are formed); however, no attempts were made to identify them thoroughly.

According to [2], Pb and W do not alloy.

1. S. Inouye, *Mem. Coll. Sci. Kyoto Univ.*, **4**, 1919, 43–46. The alloys were melted in porcelain tubes under H_2. To dissolve W the molten Pb had to be heated to 1300°C.
2. D. Kremer, *Abhandl. Inst. Metallhütt. u. Elektromet. Tech. Hochsch., Aachen*, **1**, 1917, 11–12.

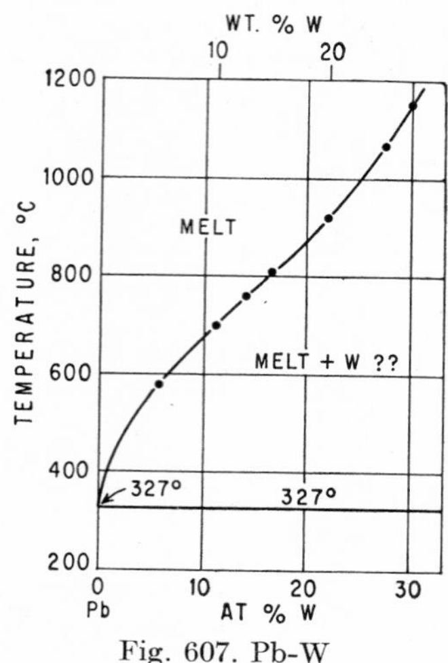

Fig. 607. Pb-W

0.5010
$\bar{1}$.4990

Pb-Zn Lead-Zinc*

The Miscibility Gap in the Liquid State. Zn and Pb show only a restricted mutual solubility in the liquid state at lower temperatures. The more recent data reported for the immiscibility curve are in excellent agreement (cf. Fig. 608): [1], four different methods of separation of the liquid layers and sample taking; [2], emf measurements; and [3], thermal analysis. The data for the critical point (see Fig. 608) were experimentally determined by [3] and also computed by [4] (cf. below); this point lies well below the boiling-point curve for atmospheric pressure measured by [5].

Table 41

Reference	Monotectic temperature, °C	Miscibility gap
Heycock, Neville [12, 13]	418	0.94–? wt. % Pb
Arnemann [14]	418	0.5–96.6 wt. % Pb
Waring et at. [1, 15]	417.8	0.9–98.0 wt. % Pb (0.29–93.9 at. % Pb)
Lumsden [4]	418.0	0.27–94.8 at. % Pb (0.85–98.3 wt. % Pb)

* The assistance of Dr. K. Claus in preparing this system is gratefully acknowledged.

[1] also duplicated the experimental procedures for the determination of the immiscibility curve used by [6, 7] (freezing the melt before sampling) and [8, 9] (sampling the two liquid layers) and showed them to be inadequate, especially those of [6–8]. Electrical-resistivity data by [10] in the range 80–100 wt. % Pb and analytical data by [11] are in fair agreement with the solubility curve of Fig. 608.

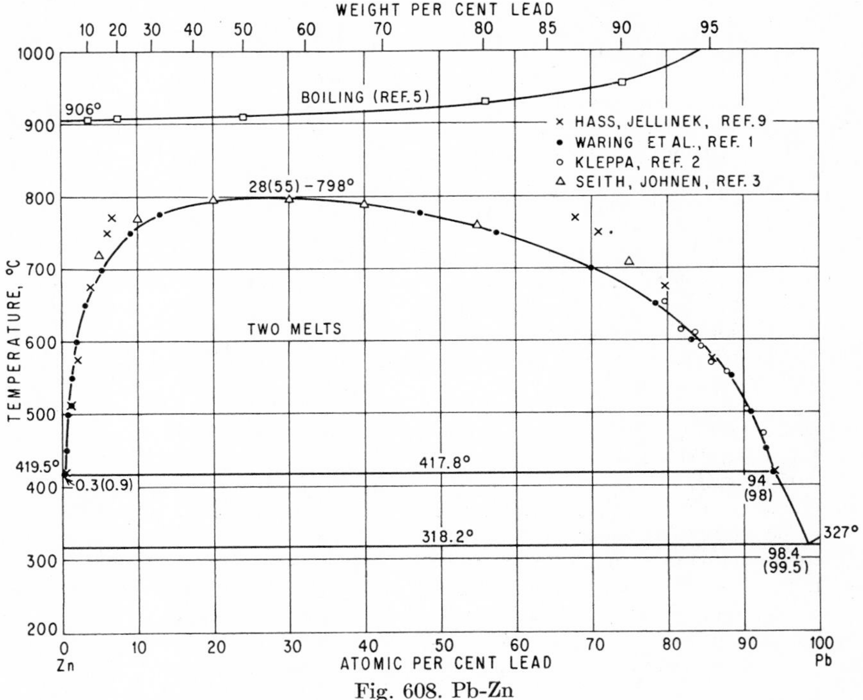

Fig. 608. Pb-Zn

Data reported for the monotectic temperature and the extent of the miscibility gap at this temperature are collected in Table 41.

For Fig. 608 the values 417.8°C, 0.3 and 94 at. (0.9 and 98 wt.) % Pb were chosen.

Eutectic. Data are listed in Table 42.

Table 42

Reference	Method	Eutectic data
Heycock, Neville [12, 13]	Thermal	0.7 wt. % Zn, 318°C
Arnemann [14]	Thermal	1.2 wt. % Zn, 317°C
Konno [16]	Electrical resistivity	316°C
Hodge, Heyer [17]	Thermal, microscopic	0.5 wt. (1.6 at.) % Zn, 318.2°C
Waring et al. [1]	Thermal	318.2°C
Bray [18]	Thermal	0.5 wt. % Zn, 318°C
Lumsden [4]	Computed	1.81 at. (0.58 wt.) % Zn, 318.35°C

The experimental results of [17, 1, 18] are in good agreement and also consistent with the calculated values. Liquidus data in the composition range eutectic–pure Pb were measured by [12, 13, 17, 18]; see also [4].

Solid Solubilities. The solubility of Pb in solid Zn is extremely small. Measurements of electrical resistivity [19, 20] failed to show any solubility, and careful microscopic work by [21] indicated that the solubility is definitely less than 0.0019 wt. % Pb, and probably even less than 0.0002 wt. ($\sim 6 \times 10^{-5}$ at.) % Pb.

According to thermal work [17, 18] the solubility of Zn in solid Pb at the eutectic temperature is about 0.05 wt. (0.16 at.) % Zn. The decrease in the lattice parameter of Pb ($\sim$0.01 A) by addition of Zn observed by [22] is also in favor of a slight solubility. However, measurements of resistivity by [20] on alloys annealed at 295°C failed to indicate any solubility. According to the solidus computed by [4], the solubility at the eutectic temperature is 0.33 at. (0.10 wt.) % Zn; this value "is believed to be the best estimate possible with the available experimental data."

[4] has put forward a four-constant equation, derived from the experimental miscibility gap by successive approximations, and used it to extrapolate experimentally less well-known data. [23] applied his "subregular" solution model to the immiscibility curve in the liquid state. The measurements of emf and density by [24, 25] and [26], respectively, and the microscopic observations of [27] on Zn single crystals containing Pb have no bearing on the constitution of the Zn-Pb system.

1. R. K. Waring, E. A. Anderson, R. D. Springer, and R. L. Wilcox, *Trans. AIME*, **111,** 1934, 254–263.
2. J. Kleppa, *J. Am. Chem. Soc.*, **74,** 1952, 6052–6056.
3. W. Seith and H. Johnen, *Z. Elektrochem.*, **56,** 1952, 140–143. Twenty alloys, sealed in evacuated quartz tubes, were investigated; unfortunately, the results are presented in a small diagram only.
4. J. Lumsden, *Discussions Fadaray Soc.*, **4,** 1948, 60–68. "Thermodynamics of Alloys," pp. 335–340, The Institute of Metals, London, 1952.
5. W. Leitgebel, *Z. anorg. Chem.*, **202,** 1931, 305–324.
6. A. Matthiessen and M. v. Bose, *Proc. Roy. Soc.* (*London*), **11,** 1861, 430.
7. C. R. A. Wright, *J. Soc. Chem. Ind.*, **11,** 1892, 492–494; **13,** 1894, 1014–1017.
8. W. Spring and L. Romanoff, *Z. anorg. Chem.*, **13,** 1896, 29–35.
9. K. Hass and K. Jellinek, *Z. anorg. Chem.*, **212,** 1933, 356–361.
10. P. Müller, *Metallurgie,* **7,** 1910, 739–740; 759–762.
11. W. Seith and G. Helmhold, *Z. Metallkunde,* **42,** 1951, 137–141.
12. C. T. Heycock and F. H. Neville, *J. Chem. Soc.*, **61,** 1892, 905.
13. C. T. Heycock and F. H. Neville, *J. Chem. Soc.*, **71,** 1897, 394, 402.
14. P. T. Arnemann, *Metallurgie,* **7,** 1910, 201–211.
15. "Waring and Springer extrapolated the data of [1] to the monotectic temperature by plotting 1/T against the logarithm of the mol fraction of Pb in the liquid phase. The results alter slightly the interpolated values originally published by Waring, et al [1]." Quoted from "Metals Handbook," 1948 ed., p. 1239, The American Society for Metals, Cleveland, Ohio. The monotectic concentration, 0.9 wt. % Pb, shown in Table 41 is that selected by Waring and Springer.
16. S. Konno, *Science Repts. Tôhoku Imp. Univ.*, **10,** 1921, 57–74.
17. J. M. Hodge and R. H. Heyer, *Metals & Alloys,* **5,** 1931, 297–301, 313.
18. J. L. Bray, *Trans. AIME,* **124,** 1937, 199–207.
19. W. M. Peirce, *Trans. AIME,* **68,** 1923, 768–769.
20. A. Pasternak, *Bull. intern. acad. polon. sci., Classe sci. math. nat.,* **A1951,** 177–192 (in English).

21. J. L. Rodda, *Mining and Met.*, **19,** 1938, 367.
22. G. Venturello and S. Allaria, *Metallurgia ital.*, **36,** 1944, 157–158.
23. H. K. Hardy, *Acta Met.*, **1,** 1953, 610–611.
24. A. P. Laurie, *J. Chem. Soc.*, **55,** 1889, 678–679.
25. R. Kremann and R. Knabel, "Elektrochemische Metallkunde," pp. 283–284, Verlagsbuchhandlung Gebrüder Borntraeger, Berlin, 1921.
26. E. Maey, *Z. physik. Chem.*, **50,** 1905, 215.
27. W. Hofmann, *Z. Metallkunde*, **38,** 1947, 383.

0.3563
$\bar{1}$.6437

Pb-Zr Lead-Zirconium

[1] showed that Zr-Pb alloys can be prepared by means of a sintering process. According to [2], who made an X-ray powder analysis of alloys containing 20–80 at. % Pb, only one intermediate phase exists, namely, Zr_5Pb_3 [37.50 at. (57.68 wt.) % Pb] [3], having the hexagonal Mn_5Si_3 ($D8_8$) type of structure, with a = 8.53 A, c = 5.86 A, c/a = 0.68_7.

1. H. S. Cooper, *Trans. Am. Electrochem. Soc.*, **43,** 1923, 224.
2. H. Nowotny and H. Schachner, *Monatsh. Chem.*, **84,** 1953, 169–180. The alloys were prepared by heating mixtures of the pure metals (Zr: 94.7 wt. %) in sealed "Pythagoras" tubes up to 1600°C.
3. Since no complete alloying could be reached with charges containing 60 at. % or more Zr, the possible existence of a phase analogous to Ti_4Pb could not be ruled out.

$\bar{1}$.7376
0.2624

Pd-Pt Palladium-Platinum

Measurements of the electrical conductivity [1], temperature coefficient of electrical resistance [1], thermal conductivity [2], thermoelectric force [1], and magnetic susceptibility [3, 4] indicate that the two metals form an uninterrupted series of solid solutions. This was verified by thermal [5] and microscopic investigations [5, 6]. Unfortunately, the thermal data [5] could not be obtained. [6] concluded from the cored microstructure of rapidly solidified alloys that the liquidus and solidus rise continuously from the melting point of Pd to that of Pt.

The existence of some kind of transformation between 1400 and 700°C was claimed [6], since the hardness of alloys quenched from 1400°C was higher than after annealing at 600–700°C (especially in the ranges 10–40 and 60–90 at. % Pt). Until more reliable evidence is available, this latter conclusion has to be treated with reserve. X-ray studies are still lacking.

1. W. Geibel, *Z. anorg. Chem.*, **70,** 1911, 242–246.
2. F. A. Schulze, *Physik. Z.*, **12,** 1911, 1028–1031.
3. E. Vogt, *Ann. Physik*, **14,** 1932, 19–26.
4. Y. Shimizu, *Science Repts. Tôhoku Univ.*, **21,** 1932, 838–839.
5. V. A. Nemilov, T. A. Vidusova, A. A. Rudnitsky, and M. M. Putsykina, *Izvest. Sektora Platiny*, **20,** 1947, 176–224; *Met. Abstr.*, **18,** 1951, 652.
6. G. Tammann and H. J. Rocha, "Festschrift zum 50-jährigen Bestehen der Platinschmelze G. Siebert G.m.b.H.," pp. 309–316, Hanau, 1931.

$\bar{1}$.7579
0.2421

Pd-Re Palladium-Rhenium

Alloys, annealed for 3 days at 1000°C and furnace-cooled, proved to be microscopically single-phase up to 7.4 wt. (4.4 at.) % Re and two-phase at 20.4 wt. (12.8 at.)

% Re and above, the second phase being Re (identified by X-ray analysis). The lattice constant of Pd decreases with increasing Re content; values are not reported [1].

1. T. A. Vidusova, *Izvest. Sektora Platiny*, **28**, 1954, 251–255.

0.0157
$\bar{1}.9843$

Pd-Rh Palladium-Rhodium

Metallographic examination of alloys, in the as-cast condition and after annealing at 1200°C, has shown that this system consists of a continuous series of solid solutions, with the liquidus curve rising continuously from the melting point of Pd to that of Rh [1].

1. G. Tammann and H. J. Rocha, "Festschrift zum 50-jährigen Bestehen der Platinschmelze G. Siebert G.m.b.H.," pp. 317–320, Hanau, 1931.

0.0208
$\bar{1}.9792$

Pd-Ru Palladium-Ruthenium

The two metals cannot form an uninterrupted series of solid solutions since they have different crystal structures. [1] reported lattice spacings of the Ru-rich solid solutions with up to 2.61 at. (2.74 wt.) % Pd.

1. A. Hellawell and W. Hume-Rothery, *Phil. Mag.*, **45**, 1954, 797–806.

0.5221
$\bar{1}.4779$

Pd-S Palladium-Sulfur

According to thermal [1], X-ray [1, 2], and tensimetric studies [2], there are three sulfides stable at room temperature: Pd_4S (6.99 wt. % S), PdS (23.10 wt. % S), and PdS_2 (37.54 wt. % S). In addition, one phase with 26–27 at. % S exists which is stable only between 635 and 555°C [1] (Fig. 609). The solid solubility of S in Pd is vanishingly small [1]. Also, Pd_4S and PdS are of fixed composition [1, 2]. From measurements of the sulfide-hydrogen equilibrium at 580°C, [6] apparently concluded the existence of Pd_2S. However, the data more likely point to the high-temperature β phase.

PdS is the prototype of the B34 type of structure, tetragonal with 16 atoms per unit cell [3, 4], with $a = 6.43$ A, $c = 6.63$ A, $c/a = 1.031$ [3]. PdS_2 is not isotypic with pyrite [2], as was suggested by [5]. According to [7], PdS_2 has an orthorhombic structure, with $a = 5.460$ A, $b = 5.541$ A, $c = 7.531$ A, and 12 atoms per unit cell, which is related to the pyrite-type structure.

1. F. Weibke and J. Laar (K. Meisel), *Z. anorg. Chem.*, **224**, 1935, 49–61.
2. W. Biltz and J. Laar (K. Meisel), *Z. anorg. Chem.*, **228**, 1936, 257–267.
3. T. F. Gaskell, *Z. Krist.*, **96**, 1937, 203–213.
4. F. A. Bannister, *Z. Krist.*, **96**, 1937, 201–202; F. A. Bannister and M. H. Hey, *Mineral. Mag.*, **23**(138), 1932-1934, 200.
5. L. Wöhler, K. Ewald, and H. G. Krall, *Ber. deut. chem. Ges.*, **66**, 1933, 1638–1652.
6. P. von der Forst, *FIAT Rev. Ger. Sci.*, 1939-1946, Inorganic Chemistry, Part V, pp. 59–62.
7. F. Grønvold and E. Røst, *Acta Cryst.*, **10**, 1957, 329–331.

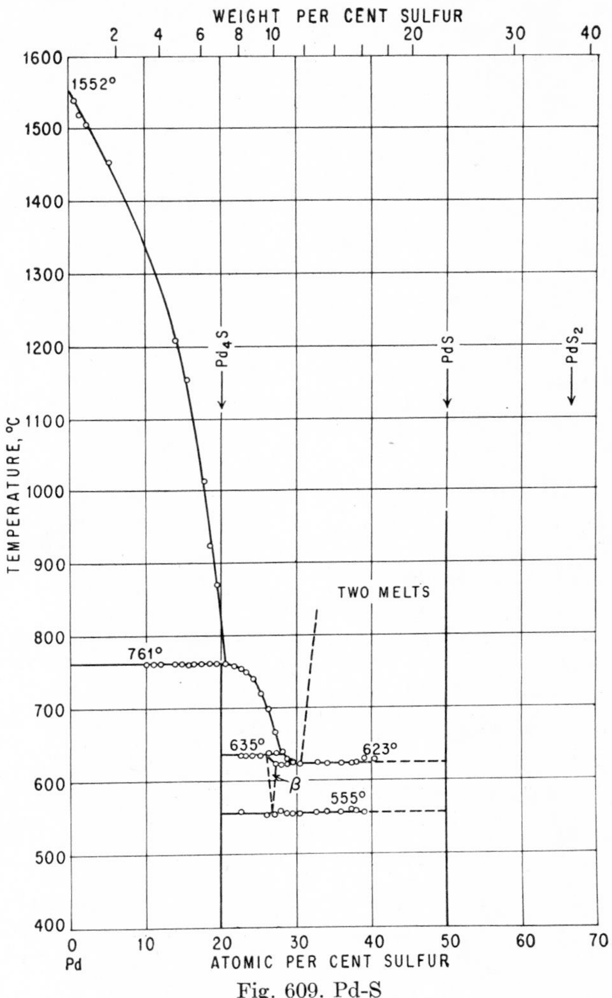

Fig. 609. Pd-S

$\bar{1}.9427$
0.0573

Pd-Sb Palladium-Antimony

Results of two thermal studies [1, 2] are in substantial agreement, although there are differences in the temperatures of the liquidus (Fig. 610) and those of the three-phase equilibria (Table 43). Figure 610 represents a compromise of these data; however, preference was given to the interpretation of [2] as regards the transformations at about 955 and about 550°C. The eutectoid decomposition of the γ phase was confirmed by microexamination [2]. Solubilities in the solid state have not yet been determined. The phase $PdSb_2$ had been identified earlier [3].

Crystal Structures. The γ phase (designated as Pd_5Sb_3) has the structure of the filled NiAs (B8) type, $a = 4.45$ A, $c/a = 1.31$ at 660°C [4]. PdSb was reported

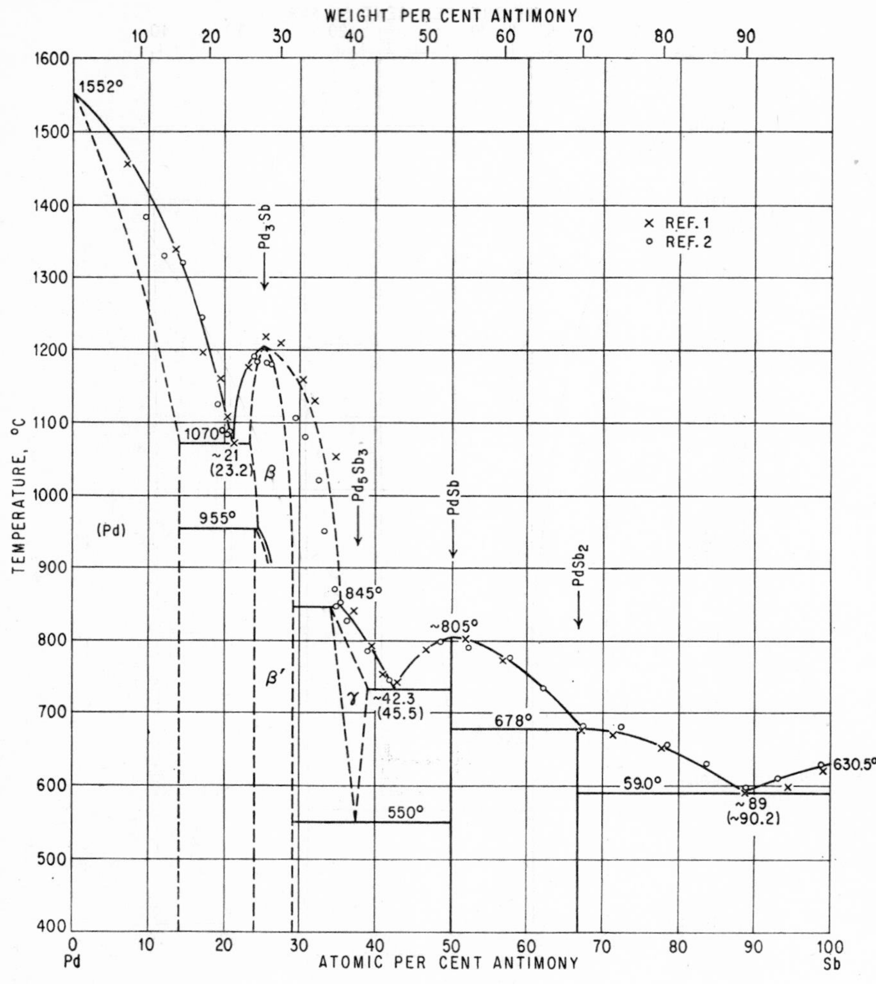

Fig. 610. Pd-Sb

to be isotypic with NiAs (B8 type), $a = 4.078$ A, $c = 5.593$ A, $c/a = 1.371$ [5]. $PdSb_2$ is isotypic with pyrite (C2 type), $a = 6.452$ A [5].

1. W. Sander, *Z. anorg. Chem.*, **75,** 1912, 97–106.
2. A. T. Grigorjew, *Z. anorg. Chem.*, **209,** 1932, 308–320; *Ann. inst. platine*, **1929**(7), 32–44.
3. F. Roessler, *Z. anorg. Chem.*, **9,** 1895, 69–70.
4. K. Schubert, H. Beeskow, et al., *Naturwissenschaften*, **40,** 1953, 269.
5. L. Thomassen, *Z. physik. Chem.*, **135,** 1928, 383–392.

Table 43. Approximate Characteristic Temperatures in °C and Compositions in At. % Sb (Wt. % in Parentheses) of the System Pd-Sb

Equilibrium	Ref. 1	Ref. 2
Melting point of Pd	1541°C	1549°C
$L \rightleftarrows (Pd) + \beta$	1070°C	1069°C
	21.2 (23.5)	20.8 (23)
Melting point of Pd_3Sb	1220°C	1185°C
Transformation $\beta \rightleftarrows \beta'$	950°C	955°C
$L + \beta \rightleftarrows \gamma$	839°C	850°C
$L \rightleftarrows \gamma + PdSb$	733°C	731°C
	41.8 (45)	42.3 (45.5)
Melting point of PdSb	805°C	800°C
$\gamma \rightleftarrows \beta' + PdSb$	530°C	550°C*
	(cooling)	(cooling)
$L + PdSb \rightleftarrows PdSb_2$	677°C	680°C
$L \rightleftarrows PdSb_2 + Sb$	587°C	593°C
	88.8 (90)	89.2 (90.4)

* 560–576°C on heating.

0.1308
$\bar{1}$.8692

Pd-Se Palladium-Selenium

Palladium forms three selenides: Pd_4Se (15.61 wt. % Se) [1], PdSe (42.53 wt. % Se) [2, 1, 3], and $PdSe_2$ (59.68 wt. % Se) [4, 5]. $PdSe_2$ appears to have the structure of the CdI_2 (C6) type [4]. See the system Pd-S.

Note Added in Proof. The following structural results have been reported recently: Pd_9Se_8, primitive cubic, a = 10.64 A [6]; PdSe, tetragonal PdS (B34) type of structure, a = 6.73 A, c = 6.91 A [6]; $PdSe_2$, orthorhombic, deformed pyrite-type structure [7, 6], with a = 5.741 A, b = 5.866 A, c = 7.691 A [7]; a = 5.72 A, b = 5.80 A, c = 7.67 A [6].

1. F. Roessler, *Z. anorg. Chem.*, **9,** 1895, 56–58.
2. H. Rössler, *Liebigs Ann.*, **180,** 1876, 244; quoted from [1].
3. L. Moser and K. Atynski, *Monatsh. Chem.*, **45,** 1924, 235–250.
4. L. Wöhler, K. Ewald, and H. G. Krall, *Ber. deut. chem. Ges.*, **66,** 1933, 1638–1652.
5. L. Thomassen, *Z. physik. Chem.*, **B2,** 1929, 374.
6. K. Schubert et al., *Naturwissenschaften*, **44,** 1957, 229–230.
7. F. Grønvold and E. Røst, *Acta Cryst.*, **10,** 1957, 329–331.

0.5796
$\bar{1}$.4204

Pd-Si Palladium-Silicon

Liquidus temperatures were determined by [1]; however, only the concentrations and temperatures of the characteristic points were reported. They are indicated as crosses in Fig. 611. The existence of the silicides Pd_2Si (11.63 wt. % Si) and PdSi (20.84 wt. % Si) was confirmed by a recent thermal and microscopic investigation [2]; see data points in Fig. 611. Also, PdSi could be isolated from alloys with more than 60 wt. % Si [1].

On cooling alloys with less than 20 wt. % Si, strong thermal effects were observed in the vicinity of 600°C which were interpreted to indicate "the crystallization of a supersaturated solution" [1]. It is possible that this heat evolution—apparently not observed by [2]—is caused by the formation of a phase having the approximate composition Pd_3Si (8.07 wt. % Si) detected by [3], perhaps by reaction of Pd and Pd_2Si. Another more likely possibility is that, instead of Pd_3Si, Pd forms a silicide corresponding to Pt_5Si_2 with 28.6 at. % Si and that this phase—as in the system Pt-Si—crystallizes from the melt. Unfortunately, [2] did not examine alloys between 22.5 and 32 at. % Si and, therefore, may have overlooked this phase (or Pd_3Si). The thermal effect at approximately 600°C [1] may then have been caused by supercooling of the crystallization and/or peritectic formation of Pd_5Si_2 or Pd_3Si (Fig. 611).

Pd_2Si is isotypic with Fe_2P (C22 type), a = 6.49 A, c = 3.43 A, c/a = 0.528 [3]; see also [4]. PdSi is isotypic with MnP (B31 type), a = 6.133 A, b = 5.599 A, c = 3.381 A [5].

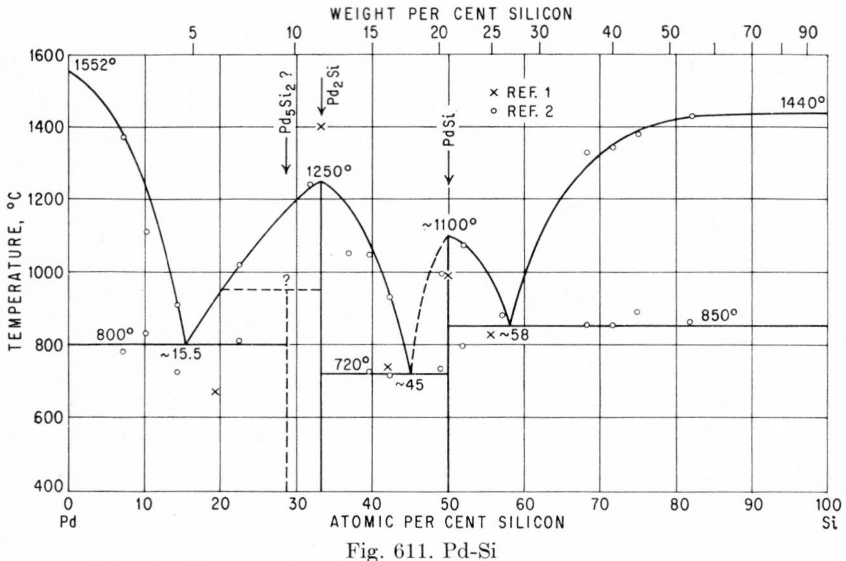

Fig. 611. Pd-Si

Note Added in Proof. By means of high-temperature microscopy, [6] observed that melting of the eutectic grain-boundary network in a 5.4 at. % Si alloy occurs at 798°C (cf. Fig. 611).

1. P. Lebeau and P. Jolibois, *Compt. rend.*, **146,** 1908, 1028–1031.
2. A. T. Grigorev, T. A. Strunina, and A. S. Adamova, *Izvest. Sektora Platiny,* **27,** 1952, 219–222.
3. K. Anderko and K. Schubert, *Z. Metallkunde,* **44,** 1953, 307–312.
4. J. H. Buddery and A. J. E. Welch, *Nature,* **167,** 1951, 362.
5. H. Pfisterer and K. Schubert, *Z. Metallkunde,* **41,** 1950, 358–367.
6. G. Reinacher, *Rev. mét.,* **54.** 1957, 321–336.

$\bar{1}.9537$
0.0463

Pd-Sn Palladium-Tin

The existence of the following intermediate phases was detected by roentgenographic investigations and, for the most part, fully established by determining their crystal structures: Pd_2Sn (35.74 wt. % Sn) [1, 2], "Pd_3Sn_2" (42.58 wt. % Sn) [1, 2], PdSn (52.90 wt. % Sn) [1, 2], $PdSn_2$ (69.01 wt. % Sn) [1–4], and $PdSn_4$ (81.65 wt. % Sn) [2, 5]. The compositions of the alloys investigated are indicated by crosses in the lower part of Fig. 612.

Based on lattice-parameter measurements, the solid solubility of Sn in Pd was estimated as about 9.5 at. (10.5 wt.) % Sn [1] and determined as about 26 at. (28 wt.) % Sn [2]. It is possible that the f.c.c. solid solution transforms into an ordered structure of the composition Pd_3Sn (27.05 wt. % Sn) [2], similar to Pd_3Pb.

The phase Pd_2Sn has a complex powder pattern which was not elucidated (see Note Added in Proof). The phase of variable composition around 40 at. % Sn has a "partially filled" NiAs structure (B8 type). Its homogeneity range was estimated to be more than 3% and maximal 5% [2]; [3] gave it as about 36–41.5 at. % Sn for alloys annealed at 480°C. Although the 40 at. % Sn alloy lies within the homogeneity range, it does not appear justified to designate the phase as Pd_3Sn_2. Numerous lattice parameters of various alloys in different conditions of heat-treatment were reported

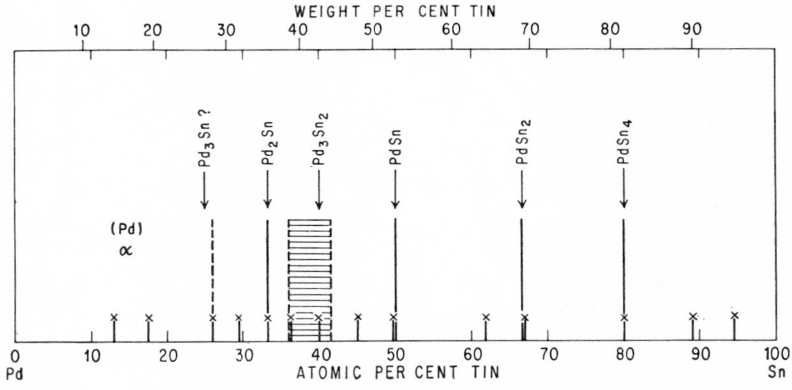

Fig. 612. Pd-Sn. (See also Note Added in Proof.)

[1, 2], indicating that both axes of the hexagonal cell decrease with increasing Sn content. For the alloy with 40 at. % Sn, the following parameters were given: $a = 4.388$ A, $c = 5.645$ A, $c/a = 1.286$ [1]; and $a = 4.399$ A, $c = 5.666$ A, $c/a = 1.288$ [2].

According to [2], PdSn has the structure of the orthorhombic MnP (B31) type, with $a = 3.87$ A, $b = 6.13$ A, $c = 6.32$ A. For a quenched melt of the composition $PdSn_2$, [4] reported a monoclinically deformed B31 type, with $a = 3.93$ A, $b = 6.18$ A, $c = 6.38$ A, $\beta = 88.5°$. [4] apparently assumes that PdSn, having a certain range of homogeneity, becomes monoclinically deformed at its Sn-rich side. This indicates that $PdSn_2$ is formed by a peritectic reaction involving PdSn, which would be prevented by quenching from the molten state.

$PdSn_2$ was reported to be isotypic with $CoGe_2$, orthorhombic (pseudotetragonal), with $a = b = 6.491$ A, $c = 12.179$ A [3]; and [4] described the structure as a tetragonal "polyfluorite" type of structure, $a = 6.546$ A, $c = 24.57$ A. $PdSn_4$, isotypic with $PtSn_4$, is orthorhombic, with 20 atoms per unit cell and $a = 6.40$ A, $b = 6.43$ A, $c = 11.44$ A [5].

[6] investigated the effect of very small additions (up to 0.28 at. %) of Pd on the melting point of Sn and found a eutectic at 0.22 at. % Pd and a temperature 0.6°C below the melting point of Sn. The solid solubility of Pd in Sn is practically nil [1].

Note Added in Proof. [7] reported that (*a*) a CuAu ($L1_0$) type of phase exists, with $a = 4.07$ A, $c/a = 0.91_5$, between 15 and 25 at. % Sn after quenching from 900°C, and (*b*) Pd_2Sn has the orthorhombic Co_2Si (C37) type of structure, with $a = 8.12$ A, $b = 5.65$ A, $c = 4.31$ A.

1. O. Nial, Dissertation, University of Stockholm, 1945; *Svensk Kem. Tidskr.*, **59,** 1947, 172–183.
2. H. Nowotny, K. Schubert, and U. Dettinger, *Z. Metallkunde,* **37,** 1946, 137–145.
3. K. Schubert and H. Pfisterer, *Z. Metallkunde,* **41,** 1950, 433–441.
4. E. Hellner, *Fortschr. Mineral.,* **29/30,** 1951, 59–61.
5. K. Schubert and U. Rösler, *Z. Metallkunde,* **41,** 1950, 298–300.
6. C. T. Heycock and F. H. Neville, *J. Chem. Soc.,* **57,** 1890, 380.
7. K. Schubert et al., *Naturwissenschaften,* **44,** 1957, 229–230.

$\bar{1}.7708$
0.2292

Pd-Ta Palladium-Tantalum

X-ray investigation of three arc-melted alloys with 50, 62, and 72 at. % Ta, annealed at 1000°C and quenched, showed that no intermediate phase exists at this temperature [1].

1. P. Greenfield and P. A. Beck, *Trans. AIME,* **206,** 1956, 265–276.

$\bar{1}.9223$
0.0777

Pd-Te Palladium-Tellurium

PdTe (54.46 wt. % Te) is isotypic with NiAs (B8 type), $a = 4.135$ A, $c = 5.674$ A, $c/a = 1.372$, and $PdTe_2$ (70.52 wt. % Te) has the structure of the CdI_2 (C6) type, $a = 4.036$ A, $c = 5.128$ A, $c/a = 1.271$ [1].

1. L. Thomassen, *Z. physik. Chem.,* **B2,** 1929, 365–367, 375–376.

0.3478
$\bar{1}.6522$

Pd-Ti Palladium-Titanium

According to [1, 2], the following intermediate phases exist: $TiPd_3$ (13.02 wt. % Ti), a phase of the approximate composition Ti_2Pd_3 (23.03 wt. % Ti), and Ti_2Pd (47.31 wt. % Ti). $TiPd_3$ is isomorphous with $TiNi_3$ ($D0_{24}$ type), $a = 2a' = 5.486$ A, $c = 8.976$ A, $c/a = 1.636$ [2]. The structure of Ti_2Pd_3 is unknown, and Ti_2Pd has a structure of the Ti_2Ni type [1]. A phase of the composition TiPd does not occur.

1. F. Laves and H. J. Wallbaum, *Naturwissenschaften,* **27,** 1939, 674–675.
2. H. J. Wallbaum, *Naturwissenschaften,* **31,** 1943, 91–92.

$\bar{1}.7177$
0.2823

Pd-Tl Palladium-Thallium

An alloy of the approximate composition Pd_2Tl was reported to have the structure of the NiAs (B8) type [1].

1. E. Hellner, *Fortschr. Mineral.,* **27,** 1948, 32–33.

$\bar{1}.6514$
0.3486

Pd-U Palladium-Uranium

The phase diagram shown in Fig. 612*a* is the result of a thermal, metallographic, and X-ray diffraction investigation of the Pd-U system by [1]. The solubility of Pd in γ-U was found to be 2.3 wt. (5.0 at.) % at 998°C and less than 0.5 wt. (1.1 at.) % Pd at 760°C. In the α and β forms, the solubility is less than 0.14 wt. (0.3 at.) % Pd at all temperatures down to 625°C.

The position of the (Pd) phase solid-solubility boundary was determined both by microscopical examination and by lattice-parameter measurements. It appears to

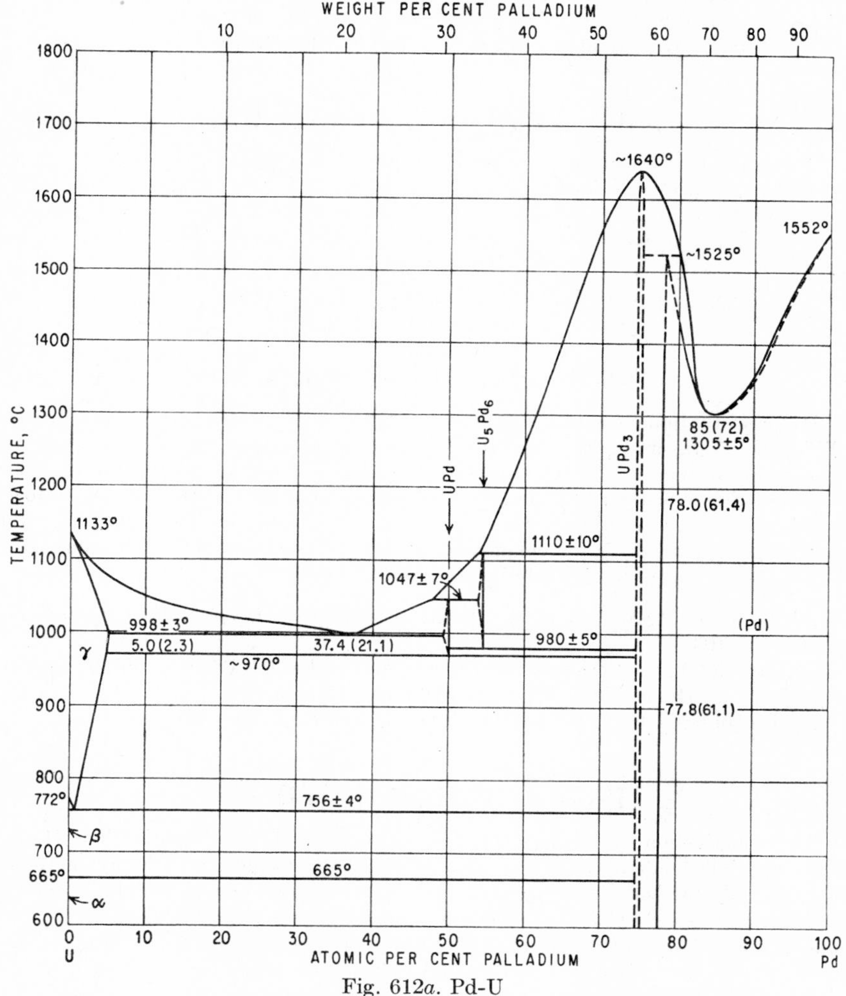

Fig. 612*a*. Pd-U

the reviewer, however, that the wide extent of the (Pd)-phase field assumed in Fig. 612*a* needs corroboration. It might well be that the minimum in the liquidus at 1305°C actually corresponds to a eutectic point separating the f.c.c. (Pd) phase from a likewise f.c.c. intermediate phase existing around 80 at. % Pd. This type of constitution would give a simple explanation of the fact that, according to [2], several physical properties show a discontinuity at approximately 10 at. % U. A narrow two-phase field may have been overlooked by [1] (cf. Au-Mn system).

The compound UPd_3 (57.34 wt. % Pd) is of the hexagonal $TiNi_3$ ($D0_{24}$) type of structure [3, 1], with a = 5.769 ± 1 A, c = 9.640 ± 1 A, c/a = 1.671 (75.05 at. % Pd) [3]. [1] found marked line shifting due to a small solubility range.

[2] measured the paramagnetic susceptibilities of UPd_3 and (Pd) phase alloys.

1. J. A. Catterall, J. D. Grogan, and R. J. Pleasance, *J. Inst. Metals*, **85**, 1956, 63–67.
2. L. F. Bates and S. J. Leach, *Proc. Phys. Soc.*, **B69**, 1956, 997–1005.
3. T. J. Heal and G. I. Williams, *Acta Cryst.*, **8**, 1955, 494–498.

0.3210
$\bar{1}.6790$

Pd-V Palladium-Vanadium

According to X-ray investigation of five arc-melted alloys with 50–76 at. % V, annealed at 1200°C and quenched, it is "probable that there is at least one intermediate phase of unknown structure, although the data obtained are not decisive. The solution of V in Pd is approximately 50 at. %, but the solution of Pd in V is less than 20 at. %" [1].

1. P. Greenfield and P. A. Beck, *Trans. AIME*, **206**, 1956, 265–276.

$\bar{1}.7635$
0.2365

Pd-W Palladium-Wolfram

[1] found that alloys with up to 14.5 at. (22.6 wt.) % W consist of solid solutions of W in Pd. X-ray and microscopic investigations showed that no intermediate phase exists in this system [2, 3]. The saturation limit of the Pd-rich solid solution, after homogenization annealing at 1350–1400°C, lies close to 19.9 at. (30 wt.) % W. However, the lattice constant of Pd does not change appreciably by addition of W. Pd appears to be practically insoluble in solid W [2]. The solidus point of the alloy with 13.3 at. (21 wt.) % W lies slightly below 1400°C.

1. V. A. Nemilov, A. A. Rudnitsky, and R. S. Polyakova, *Izvest. Sektora Platiny*, **23**, 1949, 101–103.
2. E. Raub and P. Walter, "Festschrift aus Anlass des 100-jährigen Jubiläums der Firma W. C. Heraeus G.m.b.H.," pp. 124–146, Hanau, 1951.
3. P. Greenfield and P. A. Beck, *Trans. AIME*, **206**, 1956, 265–276.

0.2127
$\bar{1}.7873$

Pd-Zn Palladium-Zinc*

The constitution of this system was the subject of two investigations [1, 2], carried out independently, both using thermal, micrographic, and X-ray methods (Figs. 613 and 614). Besides certain differences in the location of the solubility limits, the diagram proposed by [2] shows two additional phases with about 60 at. % Zn and about 92.5 at. % Zn. The intermediate phases have been designated differently by [1] and [2]; the denotation in Figs. 613 and 614 is essentially that introduced by [1]. Neither of the two diagrams can be considered an equilibrium diagram, since the phase limits have only been outlined, rather than determined quantitatively.

Constitution below 900°C. The following points may be emphasized: (*a*) The boundary of the α phase in Fig. 613 appears to be more correct, since [1] have shown by lattice-spacing determinations that the solubility of Zn in Pd is about 9 at. % Zn at 530°C and about 18 at. % Zn at 780°C. (*b*) As far as the width of the β_1-phase

* The assistance of Dr. K. Claus and Dr. U. Zwicker in preparing this system is gratefully acknowledged.

field is concerned, the boundary on the Pd side differs by only 1.5 at. %, whereas on the Zn side there is a difference of about 5 at. % at 400–500°C. Micrographic data points given by [1] appear to be in favor of approximately 37 and 56 at. % Zn. (*c*) There is a considerable discrepancy as regards the width of the β'-phase field. According to

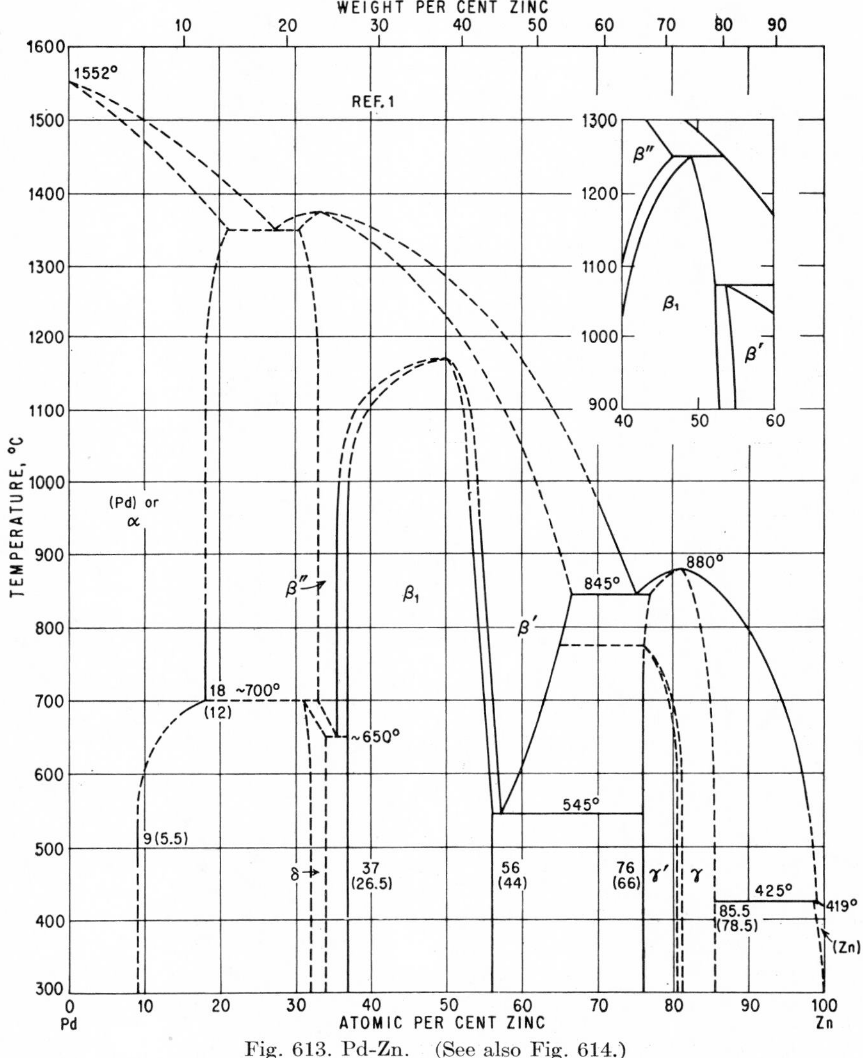

Fig. 613. Pd-Zn. (See also Fig. 614.)

Fig. 614 one would expect a peritectic equilibrium L + $\beta_1 \rightleftarrows \beta'$ between 900 and 1000°C. On the other hand, [1] have found some microscopic evidence for a continuous transition between β'' and β', giving rise to a critical point at 50 at. %, as indicated in Fig. 613. However, this subject has not been clarified fully, and as an alternative the schematic diagram presented in the inset of Fig. 613 was suggested

[1]. (*d*) Nor can a final decision be reached in regard to the existence or nonexistence of the ξ phase stable below 600°C (Fig. 614). (*e*) Both diagrams agree as to the existence of two phases γ' and γ in the range 76–85 at. % Zn. It should be noted, however, that no thermal evidence was found for the presence of a three-phase equilibrium involving the γ' and γ phases. (*f*) The η phase appears to have been overlooked by [1], since [2] reported evidence for the presence of a phase in this region having an axial ratio of $c/a = 1.55$.

Constitution above 900°C. Phase relations in this temperature range (Fig. 613) have been roughly outlined by a limited amount of micrographic work. Because of the high vapor pressure of the alloys, samples had to be heat-treated in sealed evacuated silica bulbs.

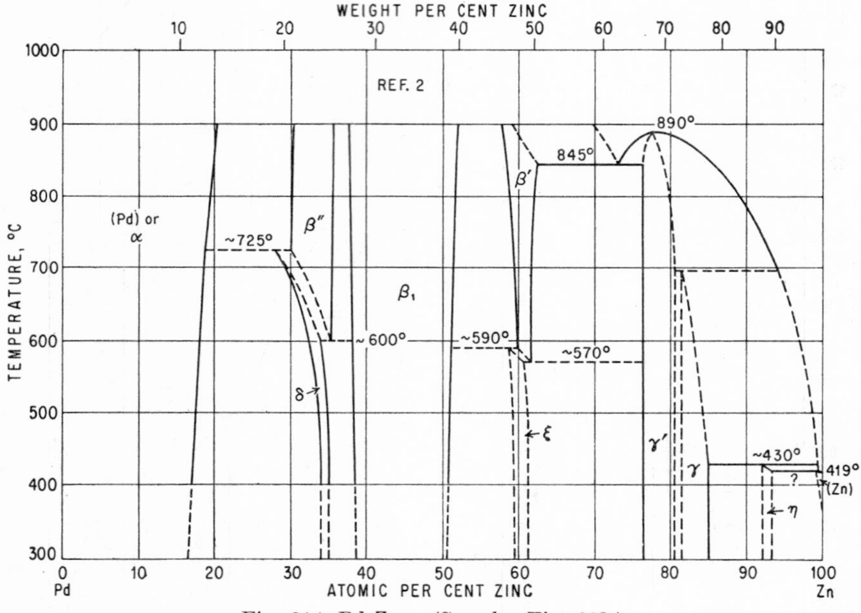

Fig. 614. Pd-Zn. (See also Fig. 613.)

Crystal Structures. The β'' and β' phases have the same structure after quenching, b.c.c. of the CsCl (B2) type [1, 2], with $a = 3.05_5$ A at 34.6 at. % Zn [1] and 32.5 at. % Zn [2], $a = 3.04$ A at 60.5 at. % Zn [1]. The β_1 phase—designated by [2] as δ—is isotypic with CuAu($L1_0$ type) [3, 1, 2], with $a = 4.13$ A, $c/a = 0.810$ at 44.5 at. % Zn and $a = 4.09$ A, $c/a = 0.81_9$ at 55.2 at. % Zn [1].

The structure of the δ phase—ξ according to [2]—is of low symmetry [2], and that of the ξ phase—ξ_1 according to [2]—appears to be somewhat similar to that of γ-brass [2].

The occurrence of a γ-brass type phase which includes the ideal composition Pd_5Zn_{21} [80.8 at. (72.02 wt.) % Zn] was already detected by [4]. The structures of γ' and γ differ slightly: that of γ' appears to be more complicated (perhaps orthorhombic) whereas that of γ is of the ideal γ-brass type [1, 2]. Lattice constants are as follows: $a = 9.07$ A at 83 at. % Zn, $a = 9.02$ A at the Zn side of the γ-phase field [1], $a = 9.11$ A at 81.6 at. % Zn [2]. The η phase is h.c.p. (not A3 type), $c/a = 1.55$ [2].

1. W. Köster and U. Zwicker, "Festschrift aus Anlass des 100-jährigen Jubiläums der Firma W. C. Heraeus G.m.b.H.," pp. 76–90, Hanau, 1951.
2. H. Nowotny, E. Bauer, and A. Stempfl, *Monatsh. Chem.*, **82,** 1951, 1086–1093.
3. H. Nowotny and H. Bittner, *Monatsh. Chem.*, **81,** 1950, 679–680.
4. W. Ekman, *Z. physik. Chem.*, **B12,** 1931, 57–78; A. Westgren, *Z. Metallkunde*, **22,** 1930, 368–373.

0.0681
$\bar{1}.9319$

Pd-Zr Palladium-Zirconium

Preliminary experiments have shown that the lattice of Pd is expanded by Zr [1].

1. A. Hellawell and W. Hume-Rothery, *Phil. Mag.*, **45,** 1954, 797–806.

0.8162
$\bar{1}.1838$

Po-S Polonium-Sulfur

Po_2S_3 has been prepared [1].

1. "Gmelins Handbuch der anorganischen Chemie," System No. 12, p. 135, Verlag Chemie, G.m.b.H., Berlin, 1941.

0.2367
$\bar{1}.7633$

Po-Sb Polonium-Antimony

For solid solubility of Po in Sb, Sn, Te and Zn, see [1].

1. G. Tammann and A. v. Löwis of Menar, *Z. anorg. Chem.*, **205,** 1932, 145–162.

0.2478
$\bar{1}.7522$

Po-Sn Polonium-Tin

See Po-Sb.

0.2163
$\bar{1}.7837$

Po-Te Polonium-Tellurium

See Po-Sb.

0.5068
$\bar{1}.4932$

Po-Zn Polonium-Zinc

See Po-Sb.

0.0635
$\bar{1}.9365$

Pr-Sb Praseodymium-Antimony

According to [1], PrSb (46.35 wt. % Sb) has the NaCl (B1) type of structure, with $a = 6.36_6$ A.

1. A. Iandelli and E. Botti, *Atti reale accad. nazl. Lincei, Rend.*, **25,** 1937, 498–502. *Strukturbericht*, **5,** 1937, 44.

0.7004
$\bar{1}.2996$

Pr-Si Praseodymium-Silicon

$PrSi_2$ (28.50 wt. % Si) has the tetragonal $ThSi_2$ type of structure, with a = 4.148 A, c = 13.67 A, c/a = 3.30 (12 atoms per unit cell) [1].

1. G. Brauer and H. Haag, *Z. anorg. Chem.*, **267**, 1952, 198–212.

0.0745
$\bar{1}.9255$

Pr-Sn Praseodymium-Tin

Only five melts, three of which had compositions corresponding to the formulas Pr_2Sn (29.64 wt. % Sn), Pr_2Sn_3 (55.82 wt. % Sn), and $PrSn_3$ (71.65 wt. % Sn), were used roughly to outline the diagram by thermal analysis (Fig. 615). Results were

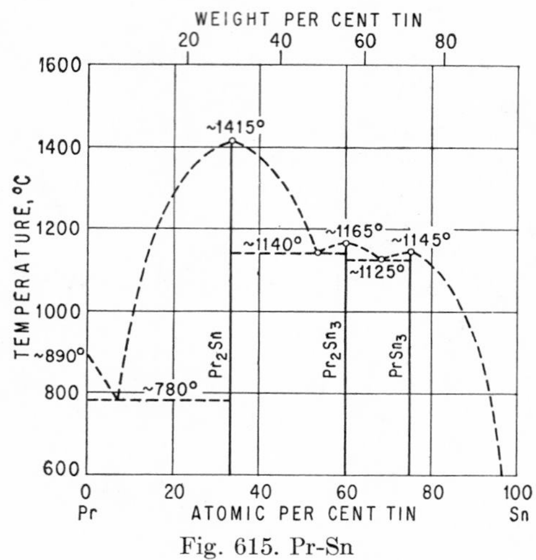

Fig. 615. Pr-Sn

claimed to prove that the same compounds exist as in the systems Ce-Sn and La-Sn [1]. Earlier, $PrSn_3$ was identified by determining its crystal structure: f.c.c. of the Cu_3Au ($L1_2$) type, a = 4.713 A. Its melting point was reported to be 1160°C [2].

1. R. Vogel and T. Heumann, *Z. Metallkunde*, **35**, 1943, 29–42.
2. A. Rossi, *Gazz. chim. ital.*, **64**, 1934, 832–834.

$\bar{1}.8385$
0.1615

Pr-Tl Praseodymium-Thallium

Only three melts, having compositions nearly corresponding to the formulas Pr_2Tl (42.04 wt. % Tl), PrTl (59.19 wt. % Tl), and $PrTl_3$ (81.31 wt. % Tl), were used roughly to outline the diagram by thermal analysis (Fig. 616). This information, corroborated by metallographic study of some additional alloys, was thought to be sufficient to claim that praseodymium forms the same compounds with thallium as cerium and lanthanum [1].

1. R. Vogel and T. Heumann, *Z. Metallkunde*, **35**, 1943, 29–42.

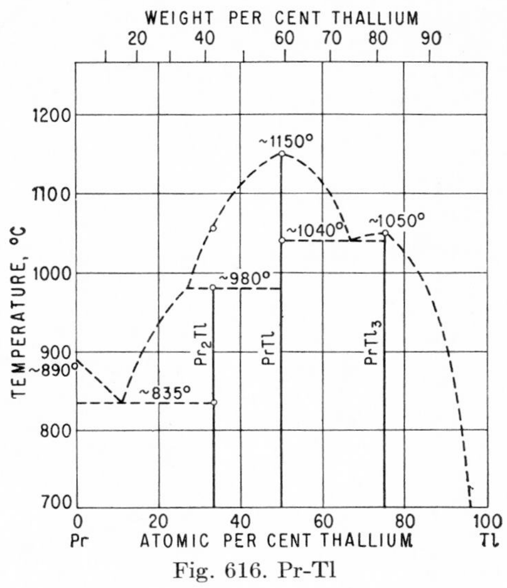

Fig. 616. Pr-Tl

$\bar{1}.7723$
0.2277

Pr-U Praseodymium-Uranium

"The information on the U-Pr system is identical with that for the U-La and U-Nd systems. Pr is much less soluble in U than is Ce. The solid solubility of Ce in U is reported to be 0.5 wt. % near the melting point of uranium, and Ce is only partially miscible with uranium in the liquid state. Probably there are no compounds in the U-Pr system [1]" [2].

1. National Physical Laboratory, United Kingdom, unpublished information (1949-1950).
2. Quoted from H. A. Saller and F. A. Rough, Compilation of U.S. and U.K. Uranium and Thorium Constitutional Diagrams, *U.S. Atomic Energy Comm., Publ.* BMI-1000, June, 1955.

0.3335
$\bar{1}.6665$

Pr-Zn Praseodymium-Zinc

PrZn (31.70 wt. % Zn) has the cubic CsCl (B2) type of structure, with $a = 3.67_8$ A [1]. $PrZn_{11}$ [91.67 at. (83.62 wt.) % Zn] is isotypic with b.c. tetragonal $BaCd_{11}$, with $a = 10.65$ A, $c = 6.85$ A, $c/a = 0.643$ (48 atoms per unit cell) [2].

1. A. Iandelli and E. Botti, *Gazz. chim. ital.*, **67**, 1937, 638–644.
2. M. J. Sanderson and N. C. Baenziger, *Acta Cryst.*, **6**, 1953, 627–731.

0.0203
$\bar{1}.9797$

Pt-Re Platinum-Rhenium

The phase diagram in Fig. 617 is based on lattice-spacing determinations, metallographic investigations, and fusion-point determinations of sintered as well as arc-melted alloys, which had been homogenized and subsequently annealed [1]. Annealing times ranged between 29 days at 600°C and 5–20 hr at 1900°C. Earlier studies had shown that at least 10 [2] or at least 17 wt. % Re [3] was soluble in Pt.

The lattice parameter of the (Pt) phase gradually decreases from a = 3.922 A at 0% Re to a = 3.897 A at the saturation limit (about 42 at. % Re). The c axis of

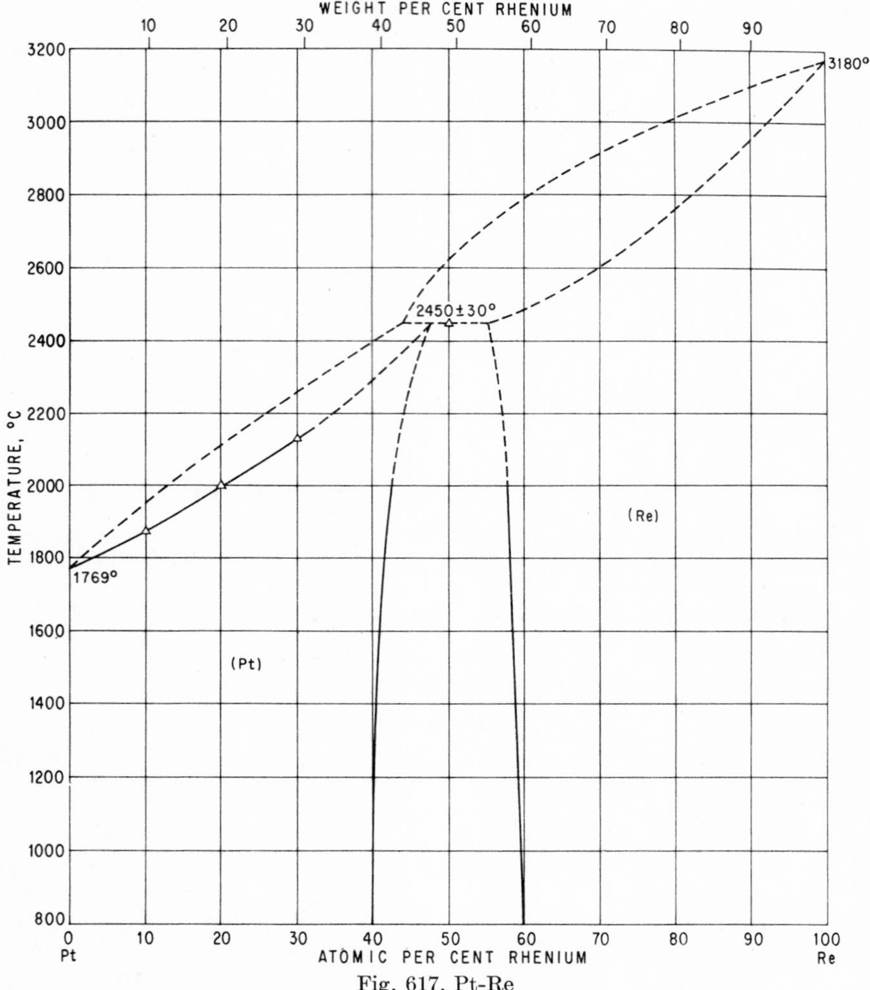

Fig. 617. Pt-Re

the hexagonal Re cell (a = 2.762 A, c = 4.458 A, c/a = 1.614) decreases and the a axis increases to a minimum and maximum value, respectively, in the range 70–80 at. % Re. The cell dimensions of the saturated phase (about 58 at. % Re) are a = 2.766 A, c = 4.431 A, c/a = 1.602 [1].

1. W. Trzebiatowski and J. Berak, *Bull. acad. polon. sci.*, **2,** 1954, 37–40.
2. W. Goedecke, "Festschrift zum 50-jährigen Bestehen der Platinschmelze G. Siebert G.m.b.H.," pp. 79–80, Hanau, 1931.
3. A. A. Rudnicky and R. S. Poliakova, *Izvest. Sektora Platiny*, **27,** 1952, 223 (quoted from [1]).

0.2781
$\bar{1}.7219$

Pt-Rh Platinum-Rhodium

The liquidus points [1] and melting points [2] shown in Fig. 618 indicate that the components form a continuous series of solid solutions. The liquidus temperatures reported by [1] are also given as corrected values, since they are based on a melting point of Rh of 1920°C instead of 1966°C [2]. The fact that the liquidus curve in the range 65–95 at. % Rh [1] runs almost horizontally at the melting point of Rh (actually 10–15°C above this temperature) may be due to insufficient mixing of the melts; i.e., the solidification point of the less dense Rh was determined. The melting points given by [2] probably lie between the solidus and liquidus. Within the accuracy of

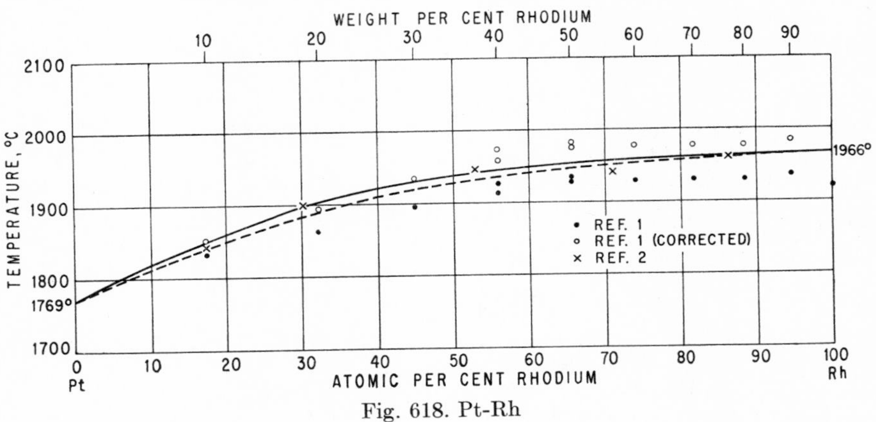

Fig. 618. Pt-Rh

±20°C only one, nearly equal, melting point was found on heating and cooling. This indicates that the solidification interval is quite small.

The existence of an uninterrupted series of solid solutions was confirmed by microscopic examination [1–3], X-ray analysis [4, 2, 5], and measurements of the electrical resistivity [2, 3], temperature coefficient of electrical resistance [2, 3], and hardness [2, 3]. [6] suggested that a maximum in the magnetic-susceptibility-vs.-composition curve at about 25 at. % Rh may indicate an order-disorder type of transformation. However, no superlattice lines in the roentgenograms were found in the positions to be expected theoretically [5].

The dimension of the f.c.c. unit cell changes practically additively with composition in at. % [4, 5].

1. L. Müller, *Ann. Physik*, **7**, 1930, 9–47.
2. J. S. Acken et al., *J. Research Bur. Standards*, **12**, 1934, 249–258.
3. V. A. Nemilov and N. M. Voronov, *Izvest. Inst. Platiny*, **12**, 1935, 27–35; *Z. anorg. Chem.*, **226**, 1936, 185–191.
4. J. Weerts and F. Beck, *Z. Metallkunde*, **24**, 1932, 139–140.
5. H. J. Goldschmidt and T. Land, *J. Iron Steel Inst.*, **155**, 1947, 221–226.
6. E. Hildebrand, *Ann. Physik*, **30**, 1937, 603–607.

0.2832
$\bar{1}.7168$

Pt-Ru Platinum-Ruthenium

Microexamination and measurements of the temperature coefficient of electrical resistance showed the solid solubility of Ru in Pt to exceed 79 at. (66.2 wt.) % Ru [1].

This was verified for the range up to 70 at. % Ru by X-ray work [2]. The lattice constant of Pt decreases from 3.923 to 3.83_7 A at 67.5 at. % Ru [2].

1. V. A. Nemilov and A. A. Rudnizky, *Izvest. Akad. Nauk S.S.S.R.* (*Khim.*), 1937, 33–40.
2. N. V. Ageev and V. G. Kuznetsov, *Izvest. Akad. Nauk S.S.S.R.* (*Khim.*), 1937, 753–755.

0.7845
$\bar{1}.2155$

Pt-S Platinum-Sulfur

According to tensimetric investigations there are only two Pt sulfides, PtS (14.11 wt. % S) and PtS_2 (24.73 wt. % S) [1]. The solid solubility of S in Pt was claimed to be about 0.23 wt. (1.4 at.) % S [2]. The structure of PtS is the prototype of the B17 type of structure, tetragonal with a = 4.93 A, c = 6.13 A, c/a = 1.243, using synthetic samples with 52.5 and 56.0 at. % S [3]. PtS_2 is isotypic with CdI_2 (C6 type), $a = 3.54_4$ A, $c = 5.02_9$ A, $c/a = 1.41_9$ [4].

1. W. Biltz and R. Juza, *Z. anorg. Chem.*, **190,** 1930, 166–173.
2. A. Jedele, *Z. Metallkunde*, **27,** 1935, 273.
3. F. A. Bannister and M. H. Hey, *Mineralog. Mag.*, **23,** 1932, 188–206; *Nature,* **130,** 1932, 142.
4. L. Thomassen, *Z. physik. Chem.*, **B2,** 1929, 371–374.

0.2050
$\bar{1}.7950$

Pt-Sb Platinum-Antimony

The system was the subject of two investigations, using thermal analysis and microexamination of alloys as cooled from the melt [1, 2]. There is substantial agreement as regards the composition range 50–100 at. % Sb, indicating the existence of the phases PtSb (38.41 wt. % Sb) and $PtSb_2$ (55.50 wt. % Sb) [3]. However, interpretation of the thermal data in the region 10–50 at. % Sb differs considerably, as shown by the insets of Fig. 619. [2] suggested that the phases Pt_4Sb (13.49 wt. % Sb) and PtSb undergo transformations at 671 and 660°C, respectively, and denied the formation of an additional phase between Pt_4Sb and PtSb. The Pt_4Sb-PtSb eutectic was placed at 33.7 at. % Sb, 633 ± 8°C. In contrast to this, [1] believed this eutectic to be located at 33.6 at. % Sb, 691 ± 10°C, and the phase Pt_5Sb_2 (19.97 wt. % Sb) to be formed peritectoidally at 637 ± 8°C.

Since no systematic heat-treatments were carried out to check the thermal data, it is not possible to decide strongly in favor of one or the other interpretation. In the main diagram of Fig. 619, preference was given to the data of [2], chiefly because the thermal effects at about 671 and 660°C show only slight scattering and, therefore, appear to indicate two (transformation) horizontals. On the other hand, the corresponding thermal effects reported by [1] scatter irregularly within the wide limits of 678 and 711°C. It should be noted, however, that according to [1] the alloy with 20 wt. % Sb was found to be single-phase, after annealing for 2½ hr at 640°C, indicating the existence of Pt_5Sb_2. At any rate, the phase relations shown are to be considered as tentative.

Crystal Structures. PtSb has the structure of the NiAs (B8) type, a = 4.13 A, $c = 5.48_3$ A, $c/a = 1.32_5$ [4], and $PtSb_2$ is isotypic with pyrite (C2 type), $a = 6.44_1$ A [5].

1. K. Friedrich and A. Leroux, *Metallurgie,* **6,** 1909, 1–3.
2. V. A. Nemilov and N. M. Voronov, *Izvest. Inst. Platiny,* **12,** 1935, 17–25; *Z. anorg. Chem.*, **226,** 1936, 177–184.

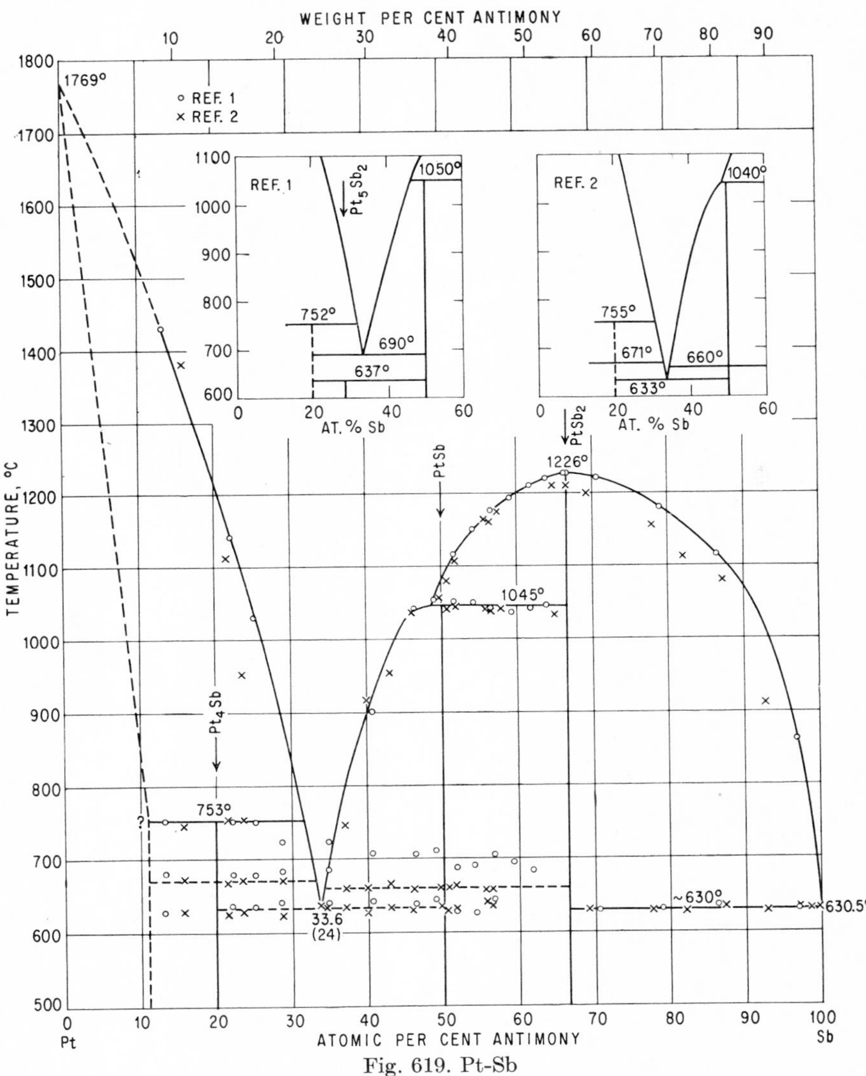

Fig. 619. Pt-Sb

3. $PtSb_2$ was first identified by F. Roessler, *Z. anorg. Chem.*, **9**, 1895, 66–67.
4. L. Thomassen, *Z. physik. Chem.*, **B4**, 1929, 277–287.
5. L. Thomassen, *Z. physik. Chem.*, **B2**, 1929, 349–379.

0.3931
$\bar{1}.6069$

Pt-Se Platinum-Selenium

The existence of PtSe (28.80 wt. % Se), reported by [1], is likely but has not yet been definitely established. $PtSe_2$ (44.72 wt. % Se) is isotypic with CdI_2 (C6 type), $a = 3.73$ A, $c = 5.07$ A, $c/a = 1.35_9$ [2].

1. F. Roessler, *Z. anorg. Chem.*, **9**, 1895, 53–55, 59–60; see also [2].
2. L. Thomassen, *Z. physik. Chem.*, **B2**, 1929, 369–371.

0.8420
$\bar{1}.1580$

Pt-Si Platinum-Silicon

According to thermal and microscopic investigations [1], Pt forms the silicides Pt_5Si_2 (5.44 wt. % Si), Pt_2Si (6.71 wt. % Si), and PtSi (12.58 wt. % Si). Pt_2Si undergoes a polymorphic transformation at approximately 700°C (Fig. 620). PtSi had already been identified earlier [2, 3] by isolation from Si-rich alloys. The silicides

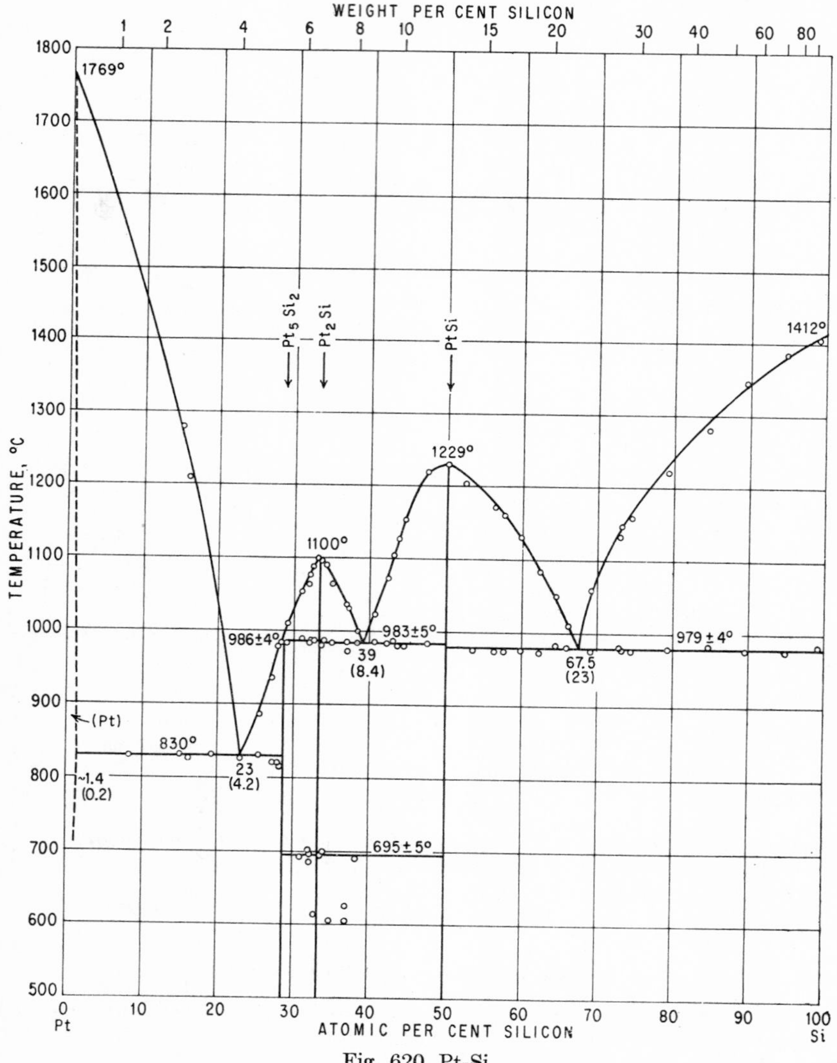

Fig. 620. Pt-Si

Pt_3Si_2, Pt_4Si_3, and $PtSi_{12}$, reported in the older literature [4], do not exist. The solid solubility of Si in Pt, after annealing for 190 hr at 800°C, is about 1.4 at. (0.2 wt.) % Si [1].

Pt_2Si is tetragonal [5, 6] and structurally related to Fe_2B, which has the $CuAl_2$ (C16) type of structure [6]. The lattice constants reported by [5], $a = 2.77_5$ A, $c = 2.95_6$ A, are not the dimensions of the true unit cell, which is larger [6]. PtSi is orthorhombic of the MnP (B31) type, $a = 5.93_2$ A, $b = 5.59_5$ A, $c = 3.60_3$ A [7].

1. N. M. Voronov, *Izvest. Sektora Platiny,* **13,** 1936, 145–166.
2. P. Lebeau and A. Novitzky, *Compt. rend.,* **145,** 1907, 241–243.
3. E. Vigouroux, *Compt. rend.,* **145,** 1907, 376–378.
4. L. Baraduc-Muller, *Rev. mét.,* **7,** 1910, 757–758.
5. J. H. Buddery and A. J. E. Welch, *Nature,* **167,** 1951, 362.
6. K. Anderko and K. Schubert, *Naturwissenschaften,* **39,** 1952, 351; see also *Z. Metallkunde,* **44,** 1953, 307–312.
7. H. Pfisterer and K. Schubert, *Z. Metallkunde,* **41,** 1950, 358–367.

0.2161
$\bar{1}.7839$

Pt-Sn Platinum-Tin

Thermal-analysis data by [1] and [2], shown in Fig. 621, agree in principle although there are some quantitative differences (see Table 44 and data points in Fig. 621). Both authors assumed the existence of the intermediate phases Pt_3Sn (16.85 wt. % Sn), PtSn (37.81 wt. % Sn), and Pt_2Sn_3 (47.70 wt. % Sn), occurring in two modifications with the transformation point at 743 [1] or 746°C [2], and a phase richer in Sn, namely, Pt_3Sn_8 (?) (61.85 wt. % Sn) according to [1] and $PtSn_4$ (70.86 wt. % Sn) according to [2]. Whereas [1] showed Pt_3Sn to be formed by a peritectic reaction at 1365°C, [2] assumed this phase to melt congruently. Later work [3] decided this discrepancy in favor of [2]; a eutectic (Pt)-Pt_3Sn was found microscopically, the composition of which was tentatively located in Fig. 621 at 22.5 at. (15 wt.) % Sn.

Table 44. Melting Points and Temperatures of Three-phase Equilibria in °C on the Basis of Data by [1] and [2]

	Ref. 1	Ref. 2
Pt	1740 (1773)	
$L \rightleftarrows \alpha + Pt_3Sn$	1365	
Pt_3Sn	......	~1406
$L \rightleftarrows Pt_3Sn + PtSn$	1080	1060
PtSn	1281	1324
$L + PtSn \rightleftarrows Pt_2Sn_3$	850	846
$L + Pt_2Sn_3 \rightleftarrows PtSn_2$	743	746
$L + PtSn_2 \rightleftarrows PtSn_4$	539	505
$L \rightleftarrows PtSn_4 + Sn$	232	224

Earlier, [4] had isolated a residue of the approximate composition Pt_2Sn_3 from an alloy with 86 wt. % Sn [5], whereas [6] reported to have isolated $PtSn_4$ crystals from an alloy with 98 wt. % Sn.

The solid solubility of Sn in Pt is unknown. [7] estimated it to be about 8 at. (5 wt.) % Sn at 750–800°C, basing it on the increase of the lattice parameter of Pt from $a = 3.924$ A to $a = 3.93_3$ A for the saturated solid solution at this temperature [3] and on general considerations. [8] arrived at the same estimated composition at 635–1090°C, basing it, however, on the increase of the parameter of Pt to $a = 3.998$ A, erroneously regarded as the parameter of the saturated solid solution (!). As [3] have pointed out, the latter value, obtained by [8] with alloys containing as much as 38.4 and 42.1 at. % Sn (!), is that of the phase Pt_3Sn.

X-ray work showed that the solid solubility of Pt in Sn is negligibly small [8]; [9] gave <0.1 at. % Pt at 200°C, on the basis of indirect evidence.

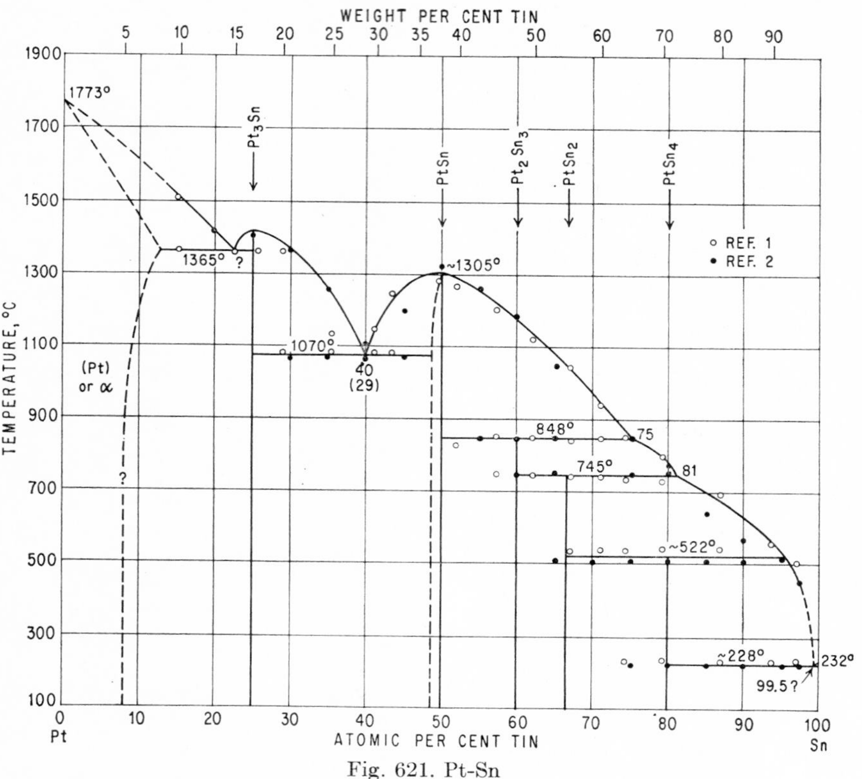

Fig. 621. Pt-Sn

By means of X-ray studies, the following intermediate phases were identified: Pt_3Sn [3], PtSn [10, 8], Pt_2Sn_3 [3], $PtSn_2$ (54.87 wt. % Sn) [11, 8], and $PtSn_4$ [12]. As a consequence, the phase diagrams by [1, 2] had to be modified, as suggested by [3, 7]; see Fig. 621. The temperature of the alleged polymorphic transformation of Pt_2Sn_3 (see above) was assumed as that of the peritectic formation of $PtSn_2$.

Crystal Structures. Pt_3Sn is isotypic with Cu_3Au ($L1_2$ type), $a = 4.01$ A [3, 13]. The structure of PtSn is that of the NiAs (B8) type, $a = 4.111$ A, $c = 5.439$ A, $c/a = 1.323$ [10]; $a = 4.103$–4.115 A, $c = 5.437$–5.441 A, $c/a = 1.325$–1.322 [8]. The structure of Pt_2Sn_3 is hexagonal of a new type, 10 atoms per unit cell, $a = 4.334$ A, $c = 12.960$ A, $c/a = 2.990$ [3]. $PtSn_2$ is isotypic with CaF_2 (C1 type), $a = 6.425$ A [11, 8]. $PtSn_4$ is orthorhombic with 20 atoms per unit cell, and $a = 6.388$ A, $b = 6.419$ A, $c = 11.357$ A [12].

1. F. Doerinckel, *Z. anorg. Chem.*, **54**, 1907, 349–358.
2. N. Podkopajew, *Zhur. Russ. Fiz.-Khim. Obshchestva*, **40**, 1908, 249–260; see abstract with diagram in *Chem. Zentr.*, **1908**(2), 493–494.
3. K. Schubert and H. Pfisterer, *Z. Metallkunde*, **40**, 1949, 405–411.
4. H. St. Claire Deville and H. Debray, *Ann. chim. et phys.*, **56**, 1859, 385, 430.
5. A phase of this composition cannot be isolated from an alloy with 86 wt. % Sn.
6. H. Debray, *Compt. rend.*, **104**, 1887, 1470, 1557.
7. K. Schubert and E. Jahn, *Z. Metallkunde*, **40**, 1949, 399–400.
8. O. Nial, Dissertation, University of Stockholm, 1945; *Svensk Kem. Tidskr.*, **59**, 1947, 172–183.
9. E. Jenckel and L. Roth, *Z. Metallkunde*, **30**, 1938, 135–144.
10. I. Oftedal, *Z. physik. Chem.*, **132**, 1927, 208–216.
11. H. J. Wallbaum, *Z. Metallkunde*, **35**, 1943, 200–201.
12. K. Schubert and U. Rösler, *Z. Metallkunde*, **41**, 1950, 298–300.
13. Nial [8] reported $a = 3.998$ A for the saturated solid solution of Sn in Pt; however, this is the parameter of Pt_3Sn [3].

0.0332
$\bar{1}.9668$

Pt-Ta Platinum-Tantalum

X-ray analysis of eight arc-melted alloys with 57–83.5 at. % Ta, annealed at 1000°C and quenched, showed the existence of a σ phase, with limits between 65.8 and 69.5 at. % Ta at the Pt-rich end and between 81.5 and 83.5 at. % Ta at the Ta-rich end, $a = 9.95$ A, $c = 5.16$ A, $c/a = 0.52$. In alloys with 57–65.8 at. % Ta, σ was found to coexist with another unidentified intermediate phase [1].

1. P. Greenfield and P. A. Beck, *Trans. AIME*, **206**, 1956, 265–276.

0.1847
$\bar{1}.8153$

Pt-Te Platinum-Tellurium

$PtTe_2$ (56.66 wt. % Te), first identified by isolation from an alloy with about 90 wt. % Te [1], was reported to be isotypic with CdI_2 (C6 type), $a = 4.01_8$ A, $c = 5.21_1$ A, $c/a = 1.29_7$ [2]. The existence of PtTe and Pt_2Te was proposed by [1]; however, the homogeneity of these products was not established. Attempts to prepare PtTe by direct synthesis failed; the reaction product consisted mainly of $PtTe_2$ [2].

1. C. Roessler, *Z. anorg. Chem.*, **15**, 1897, 405–411.
2. L. Thomassen, *Z. physik. Chem.*, **B2**, 1929, 365–369.

0.6102
$\bar{1}.3898$

Pt-Ti Platinum-Titanium

The following intermediate phases have been reported to exist: $TiPt_3$ (7.56 wt. % Ti), a phase of the approximate composition Ti_2Pt_3 (14.06 wt. % Ti), Ti_2Pt (32.92 wt. % Ti) [1], and a phase in the vicinity of the composition Ti_3Pt (42.40 wt. % Ti) [2]. TiPt does not occur [1]. $TiPt_3$ has the structure of the Cu_3Au ($L1_2$) type, with $a = 3.898$ A [3]; Ti_2Pt is isotypic with Ti_2Ni [1]; and Ti_3Pt has the "β-W" (A15) type of structure, $a = 5.031$ A [2].

1. F. Laves and H. J. Wallbaum, *Naturwissenschaften*, **27**, 1939, 674–675.
2. P. Duwez and C. B. Jordan, *Acta Cryst.*, **5**, 1952, 213–214.
3. H. J. Wallbaum, *Naturwissenschaften*, **31**, 1943, 91–92.

$\bar{1}.9801$
0.0199

Pt-Tl Platinum-Thallium

The melting point of Tl is lowered to a eutectic point at 98.4 wt. (98.3 at.) % Tl, 291°C [1]; see inset of Fig. 622. [2] reported that (*a*) the liquidus point of the 35 wt. (34 at.) % Tl alloy was 855°C and (*b*) the compound PtTl (51.15 wt. % Tl) existed.

These data make it possible to draw the diagram shown in Fig. 622 [3, 4]. Microexamination confirmed the existence of PtTl and two eutectics [2]. PtTl is isotypic with CoSn (B35 type), $a = 5.616$ A, $c = 4.648$ A, $c/a = 0.828$ [5]. Widening of the

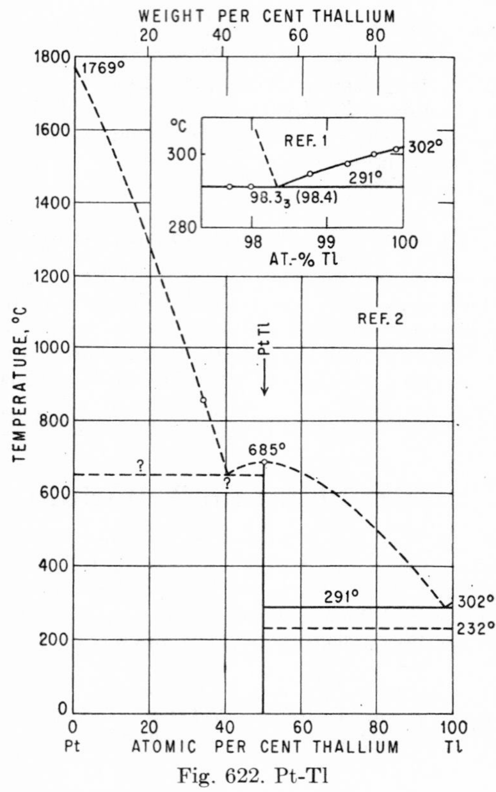

Fig. 622. Pt-Tl

Pt lattice by Tl additions indicates a slight solubility estimated as about 1.5 wt. (1.4 at.) % Tl [5].

1. C. T. Heycock and F. H. Neville, *J. Chem. Soc.*, **65,** 1894, 34.
2. L. Hackspill, *Compt. rend.*, **146,** 1908, 820–822.
3. M. Hansen, "Der Aufbau der Zweistofflegierungen," pp. 1024–1025, Springer-Verlag OHG, Berlin, 1936.
4. The liquidus curve above the boiling point of Tl (approximately 1460°C) cannot be realized under normal pressure.
5. E. Zintl and A. Harder, *Z. Elektrochem.*, **41,** 1935, 767–771; *Strukturbericht*, **6,** 1938, 5.

$\bar{1}.9138$
0.0862

Pt-U Platinum-Uranium

Evidence of retention of β-U by quenching alloys having low platinum contents has been observed [1, 2]. The compound UPt_2 has been reported [3] to have the hexagonal $MgNi_2$ (C36) type of structure; no lattice parameters were given.

UPt_3 is of the hexagonal Mg_3Cd ($D0_{19}$) type of structure, $a = 5.764 \pm 1$A, $c = 4.899 \pm 1$ A, $c/a = 0.851$ [5].

1. H. A. Saller and F. A. Rough, unpublished information (1950); quoted in [4].
2. A. U. Seybolt, unpublished information (1944); quoted in [4].
3. Fulmer Research Institute, United Kingdom, unpublished information (1953); quoted in [4].
4. H. A. Saller and F. A. Rough, Compilation of U.S. and U.K. Uranium and Thorium Constitutional Diagrams, *U.S. Atomic Energy Comm., Publ.* BMI-1000, June, 1955.
5. T. J. Heal and G. I. Williams, *Acta Cryst.*, **8,** 1955, 494–498.

0.5834
$\bar{1}.4166$

Pt-V Platinum-Vanadium

X-ray analysis of four arc-melted alloys with 40, 49, 65, and 75 at. % V, annealed at 1200°C and quenched, revealed the existence of V_3Pt (43.91 wt. % V) having the structure of the so-called β-W (A15) type, $a = 4.808$ A [1]. The solid solubility of V in Pt exceeds 40 at. % V. It appears probable that there occurs between V_3Pt and the Pt-rich solid solution another intermediate phase, but this has not been definitely established [1]. [2] reported that "it appears that V will dissolve at least 25 wt. % of Pt. The lines on the powder photograph of this alloy were broad and therefore no change in lattice constant of the V could be detected."

1. P. Greenfield and P. A. Beck, *Trans. AIME,* **206,** 1956, 265–276.
2. S. Geller, B. T. Matthias, and R. Goldstein, *J. Am. Chem. Soc.*, **77,** 1955, 1502–1504.

0.0259
$\bar{1}.9741$

Pt-W Platinum-Wolfram

The phase diagram in Fig. 623 is based on the work of [1, 2]; it is considered to be a tentative diagram [2]. The liquidus temperatures in the range 0–50 wt. (51.5 at.) % W are due to [1], and the solidus temperatures indicated as crosses in Fig. 623 were determined by [2]. Whereas the solidus points of the alloys with 26.2 and 51.5 at. % W are believed to be nearly correct, those of the W-rich solid solution are certainly too high. On the other hand, the melting point of W (99.9+%) was found as only 3220°C, as compared with the currently agreed value of 3380°C [3].

The solid solubility of W in Pt was found to exceed 50 wt. % W [1], 35 wt. % [4], and 55 wt. % W [5]. Microexamination and qualitative X-ray analysis of sintered alloys, quenched from their fusion temperatures, showed the maximum solubility to be about 62 wt. (63.4 at.) % W [2]. A steep rise of the hardness-vs.-composition curve above 25 at. % W was suggested to indicate the existence of an ordered phase WPt_3 [5]. Also, there is some microscopic evidence for an order-disorder transformation in this range [5]. The limit of the solid solution of Pt in W at the solidus temperature was found to lie between 4 and 6 wt. (3.8–5.7 at.) % Pt [2].

Absence of an intermediate phase was verified by [6] by means of X-ray analysis of an alloy with 70 at. % W.

1. L. Müller, *Ann. Physik*, **7,** 1930, 9–47.
2. R. I. Jaffee and H. P. Nielsen, *Trans. AIME,* **180,** 1949, 603–615.
3. C. J. Smithells, "Tungsten," p. 198, Chapman & Hall, Ltd., London, 1955.
4. R. Hultgren and R. I. Jaffee, *J. Appl. Phys.*, **12,** 1941, 501–502.
5. V. A. Nemilov and A. A. Rudnitskii, *Izvest. Sektora Platiny,* **21,** 1948, 234–238.
6. P. Greenfield and P. A. Beck, *Trans. AIME,* **206,** 1956, 265–276.

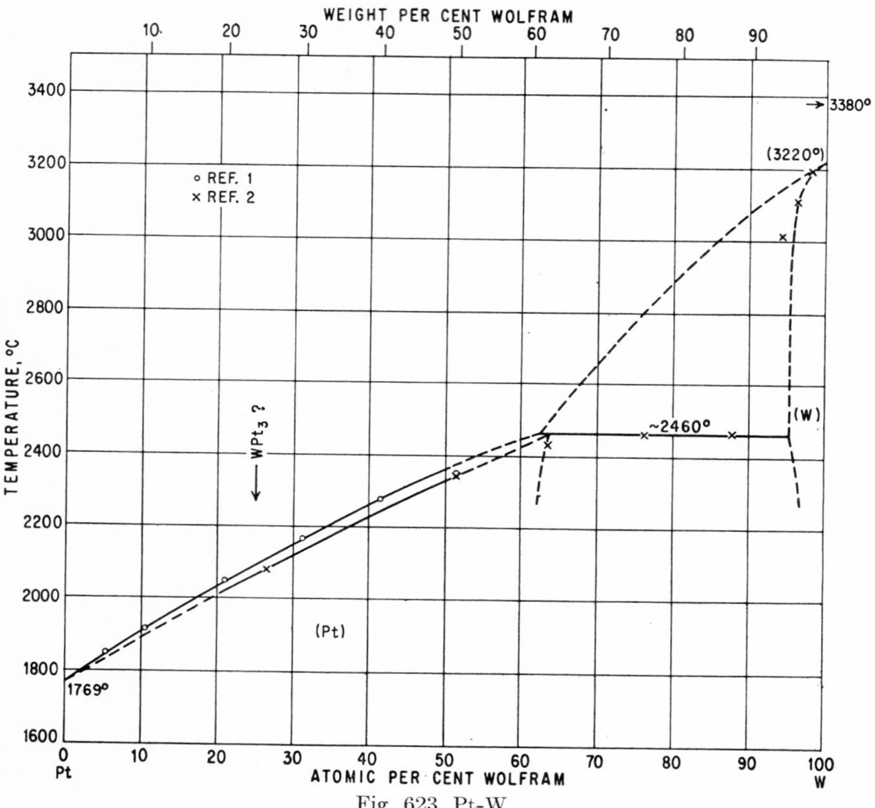

Fig. 623. Pt-W

0.4751
$\bar{1}$.5249

Pt-Zn Platinum-Zinc

The constitution in the temperature range 750–400°C (Fig. 624) was outlined by means of X-ray analysis, microexamination, and magnetic-susceptibility measurements of alloys quenched from 750 and 400°C [1]. Alloys investigated are indicated by arrows. Thermal-analysis data were reported for the range 87.4–96.5 at. % Zn [1, 2].

Evidence was found of the ordered phase Pt_3Zn being isotypic with Cu_3Au ($L1_2$ type), $a = 3.89_3$ A, and having a formation temperature of about 800°C. The disappearance of the superstructure lines after quenching from 500°C is believed to be due to precipitation of ϑ causing interference with the ordering process. The ϑ phase is f.c. tetragonal of the CuAu ($L1_0$) type, with the axial ratio decreasing from about

0.954 on the Pt side to 0.860 on the Zn side [3, 1]; $a = 4.03_7$ A, $c = 3.47_3$ A, c/a = 0.860 at 47.2 at. % Zn [1].

ξ ($PtZn_2$) is probably hexagonal, with $a = 4.11_1$ A, $c = 2.744$ A, $c/a = 0.668$ at 64.7 at. % Zn; it is structurally related to the AlB_2 (C32) type. The structure of ξ_1 is similar to that of ξ but of lower symmetry. γ is isotypic with γ-brass, and the structure of γ_1 is similar but more complicated. γ_1 was already reported by [4] and designated as Pt_5Zn_{21} (80.77 at. % Zn), $a = 18.116$ A [4, 5]. The intermediate phase

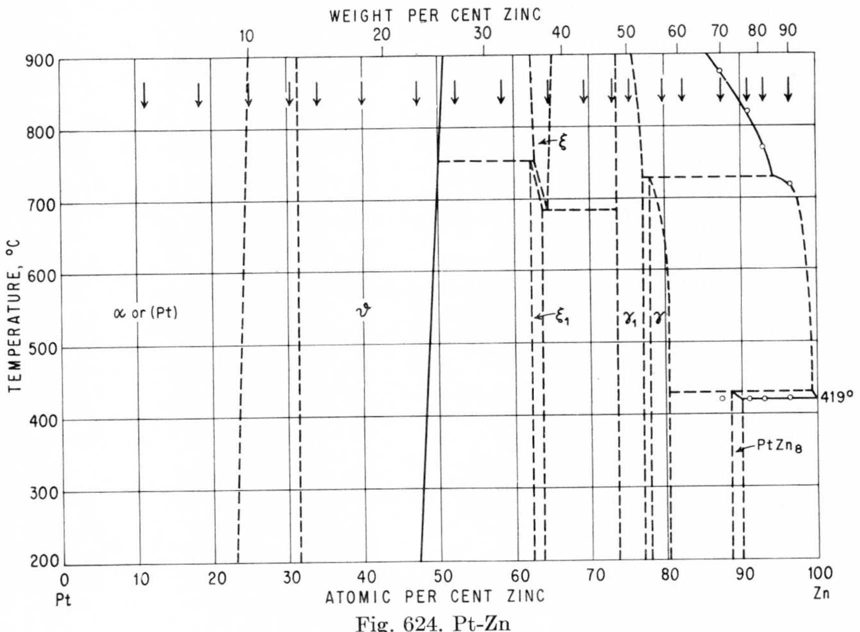

Fig. 624. Pt-Zn

richest in Zn corresponds to the composition $PtZn_8$ (88.89 at. % Zn). It has a very complicated diffraction pattern, indicating that it is of low symmetry.

1. H. Nowotny, E. Bauer, A. Stempfl, and H. Bittner, *Monatsh. Chem.*, **83**, 1952, 221–236.
2. C. T. Heycock and F. H. Neville (*J. Chem. Soc.*, **71**, 1897, 421) investigated the effect of Pt on the melting point of Zn but found only the solidus temperature, at the melting point of Zn.
3. H. Nowotny, E. Bauer, and A. Stempfl, *Monatsh. Chem.*, **81**, 1950, 1164.
4. W. Ekman, *Z. physik. Chem.*, **B12**, 1931, 69–77.
5. W. Hume-Rothery, J. O. Betterton, and J. Reynolds, *J. Inst. Metals*, **80**, 1951-1952, 609–616.

0.3305
$\bar{1}.6695$

Pt-Zr Platinum-Zirconium

$ZrPt_3$ (13.48 wt. % Zr) is isotypic with $TiNi_3$ ($D0_{24}$ type), $a = 5.644$ A, c = 9.225 A, $c/a = 1.634$ [1].

1. H. J. Wallbaum, *Naturwissenschaften*, **31**, 1943, 91–92.

0.8724
$\bar{1}$.1276

Pu-S Plutonium-Sulfur

1. W. H. Zachariasen, *Acta Cryst.*, **1,** 1948, 265–268; **2,** 1949, 57–60, 291–296.
2. W. H. Zachariasen, in G. T. Seaborg, J. J. Katz, and W. M. Manning (eds.), "The Transuranium Elements," Part II, National Nuclear Energy Series, Div. IV, vol. 14B, pp. 1454–1461, McGraw-Hill Book Company, Inc., New York, 1950.
3. B. M. Abraham, N. R. Davidson, and E. F. Westrum, in G. T. Seaborg, J. J. Katz, and W. M. Manning (eds.), "The Transuranium Elements," Part I, National Nuclear Energy Series, Div. IV, vol. 14B, pp. 814–817, McGraw-Hill Book Company, Inc., New York, 1950. (Preparation and properties of some plutonium sulfides and oxysulfides.)

0.9298
$\bar{1}$.0702

Pu-Si Plutonium-Silicon

The existence of the following silicides of Pu has been reported: (*a*) Pu_5Si_3 (?) and Pu_3Si_2 (?); the powder patterns of these phases are still unsolved [1]. (*b*) PuSi, orthorhombic FeB (B27) type of structure, with $a = 5.727 \pm 5$ A, $b = 7.933 \pm 3$ A, $c = 3.847 \pm 1$ A [1]. (*c*) β-$PuSi_2$ or Pu_2Si_3. There is a discrepancy in the results by [2] and [3]. [2] investigated the crystal structure of products having approximately the composition Pu_2Si_3. These samples gave only the Debye lines of a single hexagonal phase ($a = 3.884 \pm 3$ A, $c = 4.082 \pm 3$ A) which they found to be isostructural with β-USi_2, which is of the AlB_2 (C32) type. On the other hand, [3] observed at the composition Pu_2Si_3 a number of weak lines in addition to the Debye lines obtained by [2]. [3] concluded, therefore, that a phase of the composition Pu_2Si_3 has a crystal structure that is a distortion of the AlB_2 structure. (*d*) α-$PuSi_2$; this phase is isotypic with b.c. tetragonal $ThSi_2$ [4]. According to [1], it is of variable composition, the lattice parameters at the Si-rich limit being $a = 3.967 \pm 1$ A, $c = 13.72 \pm 3$ A. Near the Pu-rich limit a splitting of the high-angle lines was observed [1], indicating a change of structure in this composition range.

1. Los Alamos Scientific Laboratory; see [5].
2. O. J. C. Runnalls and R. R. Boucher, *Acta Cryst.*, **8,** 1955, 592.
3. F. H. Ellinger; see [5].
4. W. H. Zachariasen, *Acta Cryst.*, **2,** 1949, 94–99.
5. A. S. Coffinberry and M. B. Waldron, Review of the Physical Metallurgy of Plutonium, in "Metallurgy and Fuels," Progress in Nuclear Energy, ser. V, vol. 1, Pergamon Press Ltd., London, 1956.

0.3040
$\bar{1}$.6960

Pu-Sn Plutonium-Tin

$PuSn_3$ (59.8 wt. % Sn) has the Cu_3Au ($L1_2$) type of structure, with $a = 4.630 \pm 1$ A [1].

1. Los Alamos Scientific Laboratory; see A. S. Coffinberry and M. B. Waldron, Review of the Physical Metallurgy of Pu, in "Metallurgy and Fuels," Progress in Nuclear Energy, ser. V, vol. 1, Pergamon Press Ltd., London, 1956.

0.2725 $\bar{1}.7275$

Pu-Te Plutonium-Tellurium

PuTe (34.8 wt. % Te) has the NaCl (B1) type of structure, with $a = 6.183 \pm 4$ A [1].

1. A. E. Gorum, *Acta Cryst.*, **10**, 1957, 144.

0.0127 $\bar{1}.9873$

Pu-Th Plutonium-Thorium

A phase in the region of $ThPu_2$ (32.7 wt. % Th) has been identified by X-ray and metallographic techniques. Tentative indexing suggests an orthorhombic unit cell, with $a = 9.820$ A, $b = 8.164$ A, $c = 6.681$ A, containing six formula units [1].

1. Atomic Energy Research Establishment, Harwell, England; see A. S. Coffinberry and M. B. Waldron, Review of the Physical Metallurgy of Pu, in "Metallurgy and Fuels," Progress in Nuclear Energy, ser. V, vol. 1, Pergamon Press Ltd., London, 1956.

0.0017 $\bar{1}.9983$

Pu-U Plutonium-Uranium

"The η phase has a homogeneity range that extends from about 2 to about 70 at. % U. Its field lies at higher temperatures in the Pu-U diagram than does that of the ζ phase. The unit cell dimensions ($a = 10.57$ A, $c = 10.76$ A, 52 atoms per unit cell) were determined for a specimen having the composition Pu_3U. The crystal lattice is apparently simple tetragonal. The homogeneity range of the ζ phase extends from about 25 to about 74 at. % U. It is the stable phase at room temperature, where the composition limits are somewhat narrower. The unit cell dimension reported ($a = 10.664 \pm 5$ A, 58 atoms per unit cell) is for the composition PuU. Much better powder patterns have been obtained for this phase than for the η phase, and the ζ unit cell appears to be a cube at room temperature. But it manifests the anomalous behaviour of expanding anisotropically when heated. The lines of its powder pattern split in exact accordance with tetragonal symmetry. For this reason it seems likely that the crystal structure of the ζ phase may be tetragonal with $c/a = 1.000$ at room temperature" [1].

1. Work at Los Alamos Scientific Laboratory, quoted from A. S. Coffinberry and M. B. Waldron, Review of the Physical Metallurgy of Pu, in "Metallurgy and Fuels," Progress in Nuclear Energy, ser. V, vol. 1, Pergamon Press Ltd., London, 1956. (Phase diagram not declassified at the time of writing.)

0.6713 $\bar{1}.3287$

Pu-V Plutonium-Vanadium

The phase diagram of the Pu-V system is of the simple eutectic type (Fig. 624*a*). The addition of V does not appreciably affect the phase-transition temperatures of Pu [1].

1. S. T. Konobeevsky, *Conference of the Academy of Sciences of the U.S.S.R. on the Peaceful Uses of Atomic Energy, Division of Chemical Science*, July, 1955; English translation by U.S. Atomic Energy Commission, Washington, 1956.

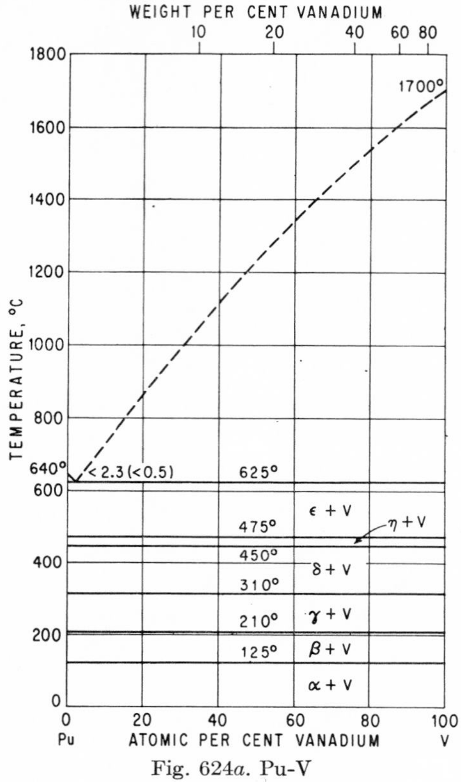

Fig. 624*a*. Pu-V

0.4183
$\bar{1}$.5817

Pu-Zr Plutonium-Zirconium

"A compound near the high-Pu end of the Pu-Zr phase diagram forms by solid-state reaction. It was first detected through discovery of its Debye pattern on X-ray films taken at Los Alamos with heat-treated specimens. It has since been found to form under similar conditions at Harwell. Density measurements have indicated that it is unlikely to be a solid solution of Zr in α-Pu. The Los Alamos workers have conservatively estimated that the value of x (in Pu_xZr) must be at least as great as 3 but is probably greater. Harwell opinion is that 'x' is equal to or greater than 10" [1].

1. Work at Los Alamos Scientific Laboratory and at the Atomic Energy Research Establishment, Harwell, England; quoted from A. S. Coffinberry and M. B. Waldron, Review of the Physical Metallurgy of Pu, in "Metallurgy and Fuels," Progress in Nuclear Energy, ser. V, vol. 1, Pergamon Press Ltd., London, 1956.

$\bar{1}$.8464
0.1536

Rb-Sb Rubidium-Antimony

Photoelectric studies suggest the existence of the compound Rb_3Sb (32.20 wt. % Sb) [1, 2].

1. A. Sommer, *Nature,* **148,** 1941, 468.
2. A. Sommer, *Proc. Phys. Soc.* (*London*), **55,** 1943, 145–154.

0.4833
$\bar{1}.5167$

Rb-Si Rubidium-Silicon

The compound RbSi (24.73 wt. % Si) has been prepared by direct synthesis. On thermal decomposition a substance of the approximate composition $RbSi_8$ [88.89 at. (72.45 wt.) % Si] was obtained. Representations of powder patterns of both phases are given [1].

1. E. Hohmann, *Z. anorg. Chem.*, **257**, 1948, 113–126.

0.2578
$\bar{1}.7422$

Re-Rh Rhenium-Rhodium

It was claimed that Rh-rich alloys with up to at least 10 wt. % Re consist of solid solutions [1].

1. W. Goedecke, "Festschrift zum 50-jährigen Bestehen der Platinschmelze G. Siebert G.m.b.H.," pp. 80–81, Hanau, 1931.

0.7642
$\bar{1}.2358$

Re-S Rhenium-Sulfur

Tensimetric investigations have shown that ReS_2 (25.60 wt. % S) is the only rhenium sulfide in equilibrium with sulfur vapor [1]; see also [2]. It decomposes at temperatures above about 1100°C [1]. Re_2S_7 (37.58 wt. % S) can be obtained only by chemical reactions [1–4]; it decomposes above about 250°C [5].

ReS_2, first believed to have the CdI_2 (C6) type of structure [6], was reported to be isotypic with MoS_2 (C7 type), $a = 3.14$ A, $c = 12.20$ A, $c/a = 3.88_5$ [7]; see also [8].

1. R. Juza and W. Biltz, *Z. Elektrochem.*, **37**, 1931, 498–501.
2. "Gmelins Handbuch der anorganischen Chemie," 8th ed., System No. 70, pp. 117–119, Verlag Chemie, G.m.b.H., Berlin, 1941.
3. H. V. A. Briscoe, P. L. Robinson, and E. M. Stoddart, *J. Chem. Soc.*, **1931**, 1439–1443.
4. W. Geilmann and G. Lange, *Z. anal. Chem.*, **126**, 1943, 321–334; W. Geilmann, *Z. anal. Chem.*, **126**, 1943, 418–426.
5. W. Biltz and F. Weibke, *Z. anorg. Chem.*, **203**, 1931, 3–8.
6. K. Meisel; ref. [1], also *Z. angew. Chem.*, **44**, 1931, 243.
7. J. Lagrenaudie, *J. phys. radium*, **15**, 1954, 299–300.
8. G. Hägg and N. Schönberg, *Arkiv Kemi*, **7**, 1954, 376, footnote.

0.3728
$\bar{1}.6272$

Re-Se Rhenium-Selenium

Re_2Se_7 (59.72 wt. % Se), prepared by wet chemical reaction, decomposes thermally at 325–330°C in vacuum to form $ReSe_2$ (45.88 wt. % Se) [1].

1. H. V. A. Briscoe, P. L. Robinson, and E. M. Stoddart, *J. Chem. Soc.*, **134**, 1931, 1439–1443.

0.8217
$\bar{1}.1783$

Re-Si Rhenium-Silicon

The following silicides have been identified by X-ray analysis of alloys prepared by sintering [1]: Re_3Si (4.79 wt. % Si), ReSi (13.10 wt. % Si), and $ReSi_2$ (23.17 wt. % Si).

$ReSi_2$, already reported earlier, is isotypic with $MoSi_2$ (C11b type), $a = 3.129$ A, $c = 7.674$ A, $c/a = 2.452$ [2]. ReSi is cubic of the FeSi (B20) type [1].

1. A. W. Searcy and R. A. McNees, *J. Am. Chem. Soc.*, **75**, 1953, 1578–1580.
2. H. J. Wallbaum, *Z. Metallkunde*, **33**, 1941, 378–381.

0.1958
$\bar{1}$.8042

Re-Sn Rhenium-Tin

[1] stated that "the powder photographs of the alloys [compositions not given] contained only the reflections of Re and Sn in the same positions as those of the pure metals. Thus no intermediate phase exists and Re and Sn seem to be mutually almost insoluble."

1. O. Nial, Dissertation, University of Stockholm, 1945; *Svensk Kem. Tidskr.*, **59**, 1947, 172–183.

0.0128
$\bar{1}$.9872

Re-Ta Rhenium-Tantalum

X-ray analysis and microexamination of nine alloys, covering the range 25–52 at. % Ta, revealed the existence of the following phases, after annealing for 3–5 days at 1200°C and quenching [1]: (*a*) a phase isotypic with α-Mn (A12 type) from a composition below 25 at. % Ta to about 37 at. % Ta, $a = 9.711$ A; (*b*) a σ-type phase [2] having a narrow homogeneity range at about 41 at. % Ta, with $a = 9.72$ A, $c = 5.07$ A, $c/a = 0.52$; (*c*) a terminal Ta-base solid solution up to 48–52 at. % Re, $a = 3.16$ A at 49 at. % Ta [3]. [4] reported the existence of a χ-type phase (α-Mn or A12 type of structure) of the formula Ta_7Re_{22} (24.14 at. % Ta), with the lattice parameter $a = 9.69_7$ A (cf. above).

1. P. Greenfield and P. A. Beck, *Trans. AIME*, **206**, 1956, 265–276.
2. Already reported by P. Duwez, unpublished work; see [1].
3. Private communication by P. Greenfield and P. A. Beck.
4. J. Niemiec and W. Trzebiatowski, *Bull. acad. polon. sci.*, **4**, 1956, 601–603; *Met. Abstr.*, **24**, 1957, 819.

0.5899
$\bar{1}$.4101

Re-Ti Rhenium-Titanium

The b.c.c. β_{Ti} primary solid solution appears to extend up to nearly 50 at. % Re (at 1200°C), with the lattice parameter (β-Ti: $a = 3.282$ A) decreasing nearly linearly to $a = 3.199$ A at Ti_3Re and to $a = 3.104$ A at nearly 50 at. % Re. No CsCl-type ordering could be found in the TiRe alloy after annealing for 1 week at 600°C [1].

The isolation of a compound of the (almost unvarying) composition Ti_5Re_{24} (82.76 at. % Re) has been reported by [2]. This compound was found to be stable in the investigated temperature range 600–1700°C and to have the α-Mn (A12) type of structure (probably with an ordered atom distribution).

1. T. V. Philip and P. A. Beck, *Trans. AIME*, **209**, 1957, 1269–1271.
2. W. Trzebiatowski and J. Niemiec, *Roczniki Chem.*, **29**, 1955, 277–283; *Titanium Abstr. Bull.*, **2**, 1956, 1512.

0.5631
$\bar{1}$.4369

Re-V Rhenium-Vanadium

X-ray analysis and microscopic examination of four alloys with 40–66 at. % V, annealed for 3–5 days at 1200°C and quenched, showed that there is no intermediate phase in this system. The solid-solution limit of Re in V is about 38 at. % Re, while the solution limit of V in Re is approximately 40 at. % V [1].

1. P. Greenfield and P. A. Beck, *Trans. AIME*, **206**, 1956, 265–276.

0.0056
$\bar{1}$.9944

Re-W Rhenium-Wolfram

[1] determined the fusion points of sintered alloys shown in Fig. 625. The authors claimed that these were solidus temperatures of the system, believed to indicate the existence of a phase of the composition Re_3W_2, which forms eutectics with Re and W at about 33 at. % W, 2820 ± 50°C and 50 at. % W, 2890 ± 50°C.

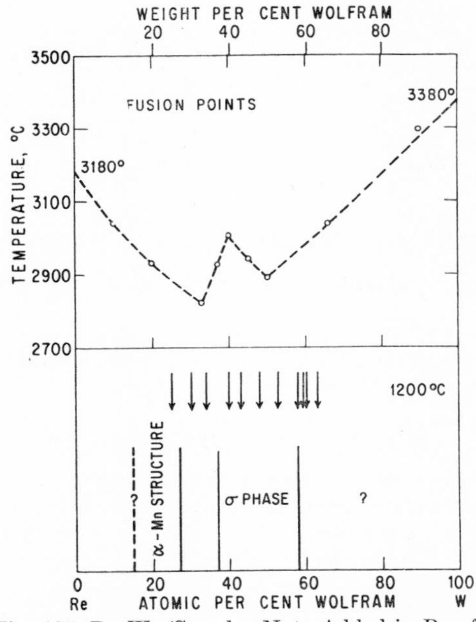

Fig. 625. Re-W. (See also Note Added in Proof.)

However, if this type of constitution were to exist, the solidus would consist of two eutectic horizontals at the temperatures given, provided that these temperatures actually are eutectic temperatures. It is obvious that the fusion points do not represent equilibrium temperatures but lie somewhere in liquid + solid fields. Also, the microstructures of the alloys with 33 and 50 at. % W are incompatible with the occurrence of eutectics of these compositions.

These conclusions [2] are corroborated by the solid-state constitution at 1200°C, as revealed by X-ray analysis of arc-melted alloys whose compositions are indicated by arrows in the lower portion of Fig. 625 [3]. Two intermediate phases of variable composition were identified: (*a*) a phase isotypic with α-Mn (A12 type), ranging from about 27 at. % W to an undetermined lower W composition (a = 9.588 A), and (*b*) a phase with a σ-type structure which ranges from approximately 37 to 58 at. % W; a = 9.55 A, c = 4.98 A, c/a = 0.52. The phase richer in W, coexisting with σ, was not identified.

Note Added in Proof. The finding of a σ-type phase in the Re-W system has also been reported by [4, 5]. According to [4], its homogeneity range extends from approximately 35 to 45 at. % W, and the lattice parameters are a = 9.645 A, c = 5.038 A at 40.5 at. % W.

1. K. Becker and K. Moers, *Metallwirtschaft*, **9**, 1930, 1063–1066.
2. M. Hansen, "Der Aufbau der Zweistofflegierungen," pp. 1028–1030, Springer-Verlag OHG, Berlin, 1936.
3. P. Greenfield and P. A. Beck, *Trans. AIME*, **206**, 1956, 265–276.
4. A. G. Knapton, *Bull. Inst. Metals*, **3**, 1955, 21.
5. J. Niemiec and W. Trzebiatowski, *Bull. acad. polon. sci.*, **4**, 1956, 601–603; *Met. Abstr.*, **24**, 1957, 819.

0.3101
$\bar{1}$.6899

Re-Zr Rhenium-Zirconium

$ZrRe_2$ (80.33 wt. % Re) is isotypic with $MgZn_2$ (C14 type), a = 5.262 A, c = 8.593 A, c/a = 1.633 [1].

1. H. J. Wallbaum, *Naturwissenschaften*, **30**, 1942, 149.

0.0051
$\bar{1}$.9949

Rh-Ru Rhodium-Ruthenium

Lattice constants of Ru-rich solid solutions with 1.93, 3.12, and 5.30 at. % Rh were determined by [1].

1. A. Hellawell and W. Hume-Rothery, *Phil. Mag.*, **45**, 1954, 797–806.

0.5064
$\bar{1}$.4936

Rh-S Rhodium-Sulfur

Tensimetric analysis of the whole system, corroborated by qualitative X-ray studies, has established the existence of the compounds Rh_9S_8, Rh_3S_4, Rh_2S_3, and Rh_2S_5 [1]. No evidence was found for RhS_2, which was reported to be isotypic with pyrite (C2 type) [2]. The powder pattern of Rh_2S_5 is somewhat similar to that of FeS_2 [2].

1. R. Juza, O. Hülsmann, and K. Meisel, *Z. anorg. Chem.*, **225**, 1935, 369–385.
2. L. Thomassen, *Z. physik. Chem.*, **B4**, 1929, 283–285.

$\bar{1}$.9270
0.0730

Rh-Sb Rhodium-Antimony

RhSb (54.20 wt. % Sb) is isotypic with MnP (B31 type), a = 6.333 A, b = 5.952 A, c = 3.876 A [1].

1. H. Pfisterer and K. Schubert, *Z. Metallkunde*, **41**, 1950, 358–367.

0.1150
$\bar{1}$.8850

Rh-Se Rhodium-Selenium

By reaction of $RhCl_3$ with Se, no higher selenide than Rh_2Se_5 is obtained [1, 2]. Attempts to prepare lower selenides by stepwise thermal dissociation of Rh_2Se_5 were unsuccessful [2]. Rh_2Se_5 has a "pseudopyrite" type of structure [2].

Supplement. [3] examined by X-ray analysis Rh-Se alloys prepared by direct synthesis at 900°C and containing 40–71.4 at. % Se. Between about 64 and at least 71.4 at. % Se, the phase ($RhSe_2$) of the cubic pyrite (C2) type of structure is homogeneous, with the *a* lattice constant decreasing from 6.015 to 5.985 A with increasing Se content. [3] also found indications for the existence of another intermediate phase with a composition somewhere between 40 and 60 at. % Se.

The line splitting around line positions of a pyrite type of structure observed by [2] is considered [3] to have been caused by the failure to attain equilibrium in their preparations.

1. L. Wöhler, K. Ewald, and H. G. Krall, *Ber. deut. chem. Ges.*, **66,** 1933, 1638–1652.
2. W. Biltz, F. W. Wrigge, P. Ehrlich, and K. Meisel, *Z. anorg. Chem.*, **233,** 1937, 282–285.
3. S. Geller and B. B. Cetlin, *Acta Cryst.*, **8,** 1955, 272–274.

0.5639
$\bar{1}$.4361

Rh-Si Rhodium-Silicon

Qualitative X-ray analysis revealed the existence of Rh_3Si_2, RhSi, Rh_2Si_3 (or $RhSi_2$), and an unidentified compound lower in silicon than Rh_3Si_2 [1]. RhSi was found to be cubic of the FeSi (B20) type, $a = 4.675$ A [2]. By means of high-temperature microscopy [3] observed that melting of the eutectic grain-boundary network in a 1.5 wt. (5.3 at.) % Si alloy occurs at 1389°C. The solubility of Si in solid Rh appears to be less than 0.5 wt. (1.8 at.) % Si [3].

1. J. H. Buddery and A. J. E. Welch, *Nature,* **167,** 1951, 362.
2. S. Geller and E. A. Wood, *Acta Cryst.*, **7,** 1954, 441–443.
3. G. Reinacher, *Rev. mét.*, **54,** 1957, 321–336.

$\bar{1}$.9380
0.0620

Rh-Sn Rhodium-Tin

The following intermediate phases [1] were identified by roentgenographic analysis: Rh_2Sn (38.58 wt. % Sn) [2], "Rh_3Sn_2" (43.47 wt. % Sn) [3, 4], RhSn (53.56 wt. % Sn) [3, 2], and $RhSn_2$ (69.76 wt. % Sn) [3, 2, 5, 6]. In addition, a phase richer in Sn, very likely $RhSn_4$ (82.19 wt. % Sn) [2], exists. Three eutectics have been found [2]: one of the Rh-base solid solution and Rh_2Sn, RhSn-$RhSn_2$ at about 66 wt. (62.7 at.) % Sn, and $RhSn_4$(?)-Sn. RhSn and $RhSn_4$ are formed by peritectic reactions [2], and this probably holds for "Rh_3Sn_2" also. Rh_2Sn and $RhSn_2$ appear to melt without decomposition. A *schematic* phase diagram is shown in Fig. 626, with the alloys investigated indicated by crosses.

The lattice parameter of Rh ($a = 3.804$ A) is increased to $a = 3.86$ A for the saturated solid solution [2] of unknown composition. The powder pattern of Rh_2Sn has not been elucidated [2] (see Note Added in Proof). The phase designated as "Rh_3Sn_2" has the structure of the partially filled NiAs (B8) type; its homogeneity range is estimated as about 37.6–40.5 at. % Sn. Lattice parameters are $a = 4.340$ A, $c = 5.553$ A, $c/a = 1.279$ if saturated with Rh [4]; and $a = 4.338$ A, $c = 5.544$ A, $c/a = 1.278$ if saturated with Sn [3]. RhSn is cubic of the FeSi (B20) type, $a = 5.131$ A [3, 2]. $RhSn_2$ exists in two modifications, with a transformation point above 500°C [3, 2]. The high-temperature form is tetragonal of the $CuAl_2$ (C16) type, with $a = 6.412$ A, $c = 5.655$ A, $c/a = 0.882$ [3, 5]; the low-temperature form is isotypic with $CoGe_2$, orthorhombic (pseudotetragonal) unit cell, with 7 Rh and 16 Sn atoms and $a = b = 6.332$ A, $c = 11.99$ A [5]. [6] described the structure as tetragonal of a "polyfluorite" type, $a = 6.380$ A, $c = 17.88$ A.

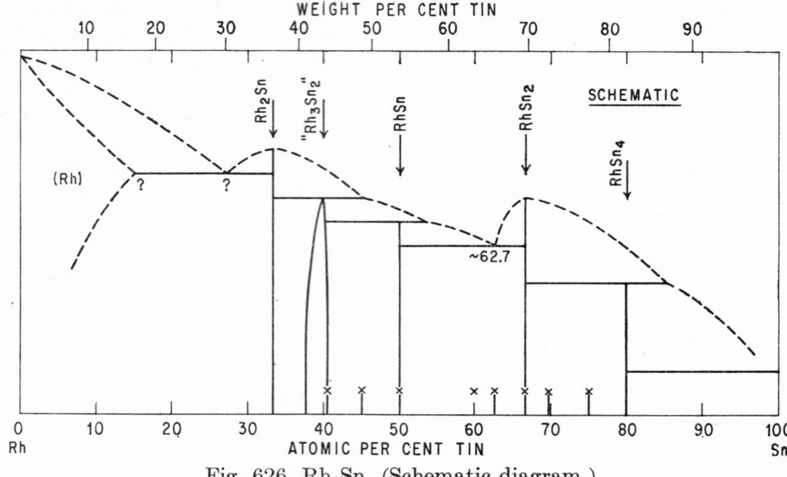

Fig. 626. Rh-Sn. (Schematic diagram.)

Note Added in Proof. Rh_2Sn is of the orthorhombic Co_2Si (improved C37) type of structure, with a = 8.21 A, b = 5.52 A, c = 4.22 A [7].

1. H. St. Claire Deville and H. Debray (*Ann. chim. et phys.*, **56,** 1859, 385) and H. Debray (*Compt. rend.*, **104,** 1887, 1471–1472) claimed to have isolated crystals of RhSn and $RhSn_3$, respectively, from Sn-rich alloys. The former compound cannot be isolated from such alloys, and the alleged phase $RhSn_3$ is more likely $RhSn_4$ [2].
2. K. Schubert, *Z. Naturforsch.*, **2a,** 1947, 120.
3. O. Nial, Dissertation, University of Stockholm, 1945; *Svensk Kem. Tidskr.*, **59,** 1947, 172–183.
4. H. Nowotny, K. Schubert, and U. Dettinger, *Z. Metallkunde*, **37,** 1946, 137–145.
5. K. Schubert and H. Pfisterer, *Z. Metallkunde*, **41,** 1950, 433–441.
6. E. Hellner, *Fortschr. Mineral.*, **29–30,** 1951, 59–61.
7. K. Schubert et al., *Naturwissenschaften*, **44,** 1957, 229–230.

$\bar{1}.7551$
0.2449

Rh-Ta Rhodium-Tantalum

Microscopic and X-ray diffraction work on eight Ta-Rh alloys containing 16.4–81 at. % Rh, annealed at and quenched from 1000°C, was carried out by [1]. Only one intermediate phase was found. It exists over a narrow range at approximately 40 at. % Rh and has the tetragonal σ-phase structure, with a = 9.754 A, c = 5.058 A, c/a = 0.518.

1. P. Greenfield and P. A. Beck, *Trans. AIME*, **206,** 1956, 265–276.

$\bar{1}.9066$
0.0934

Rh-Te Rhodium-Tellurium

[1] prepared $RhTe_2$ by reaction of $RhCl_3$ with Te. [2] prepared a compound in a similar manner but claimed the true formula to be Rh_2Te_5. The results of more recent X-ray work by [3] indicate that there are at least three intermediate phases in this system: RhTe, with the hexagonal NiAs (B8) type of structure, a = 3.99 ± 1 A,

$c = 5.66 \pm 1$ A; $RhTe_2$ (high-temperature form) (quenched from 1200°C), with the hexagonal CdI_2 (C6) type of structure, $a = 3.92 \pm 1$ A, $c = 5.41 \pm 1$ A; and $RhTe_2$ (low-temperature form) (prepared at 900°C), of the cubic pyrite (C2) type of structure, $a = 6.441 \pm 2$ A.

1. L. Wöhler, K. Ewald, and H. G. Krall, *Ber. deut. chem. Ges.*, **66**, 1933, 1638–1652.
2. W. Biltz, F. W. Wrigge, P. Ehrlich, and K. Meisel, *Z. anorg. Chem.*, **233**, 1937, 282–285.
3. S. Geller, *J. Am. Chem. Soc.*, **77**, 1955, 2641–2644.

0.3321
$\bar{1}$.6679

Rh-Ti Rhodium-Titanium

According to [1], phases of the compositions $TiRh_2$ and $TiRh_3$ do not exist.

1. H. J. Wallbaum, *Naturwissenschaften*, **31**, 1943, 91–92.

0.3053
$\bar{1}$.6947

Rh-V Rhodium-Vanadium

[1] studied four V-Rh alloys containing 25, 42, 50 and 60 at. % Rh, annealed at and quenched from 1200°C, by microexamination and X-ray diffraction. Two intermediate phases were found. One of these, V_3Rh (40.24 wt. % Rh), possesses the cubic "β-W" (A15) type of structure, with $a = 4.767$ A. An alloy containing 60 at. % Rh consists of a second intermediate phase with a h.c.p. crystal structure.

1. P. Greenfield and P. A. Beck, *Trans. AIME*, **206**, 1956, 265–276.

$\bar{1}$.7478
0.2522

Rh-W Rhodium-Wolfram

[1] examined the microstructures of 10 alloys varying in composition between 2.9 and 69 at. (5 and 80 wt.) % W and determined the crystal structures of the phases present after annealing at 1600–1650°C. From measurements of the lattice constants the following single-phase fields were derived: (*a*) f.c.c. terminal solid solutions of W in Rh up to about 12.3 at. (20 wt.) % W; (*b*) a h.c.p. phase having a homogeneity range of about 19.5–46 at. (30.3–60.4 wt.) % W; (*c*) b.c.c. terminal solid solutions of Rh in W in the approximate region 98.5–100 at. (99.2–100 wt.) % W. The solubility of W in Rh and in the hexagonal phase decreases with fall in temperature.

Supplement. The results by [2], who investigated three alloys with 19.2, 53, and 75 at. % W, annealed at and quenched from 1200°C, by microexamination and X-ray diffraction, are in keeping with the findings of [1]. [2] reported the following lattice parameters for the h.c.p. phase: $a = 2.708$ A, $c = 4.328$ A, $c/a = 1.598$ (at 19.2 at. % W).

1. E. Raub and P. Walter, "Festschrift aus Anlass des 100-jährigen Jubiläums der Firma W. C. Heraeus G.m.b.H.," pp. 124–146, Hanau, 1951.
2. P. Greenfield and P. A. Beck, *Trans. AIME*, **206**, 1956, 265–276.

0.1970
$\bar{1}$.8030

Rh-Zn Rhodium-Zinc

Exploratory X-ray work showed that there is a phase, very likely of the ideal composition Rh_5Zn_{21} [80.78 at. (72.74 wt.) % Zn], which is isotypic with γ brass [1].

1. A. Westgren and W. Ekman, *Arkiv Kemi, Mineral. Geol.*, **10B,** no. 11, 1930; W. Ekman, *Z. physik. Chem.*, **B12,** 1931, 59–77; A. Westgren, *Z. Metallkunde,* **22,** 1930, 372.

0.5013
$\bar{1}$.4987

Ru-S Ruthenium-Sulfur

Tensimetric and roentgenographic studies showed that Ru forms only one stable sulfide, RuS_2 (38.68 wt. % S) [1]. It is isotypic with pyrite (C2 type), $a = 5.612 \pm 0.003$ A [2]; $a = 5.58 \pm 0.02$ A [3]; $a = 5.60 \pm 0.01$ A [4].

1. R. Juza and W. Meyer, *Z. anorg. Chem.*, **213,** 1933, 273–282.
2. W. F. de Jong and A. Hoog, *Rec. trav. chim.*, **46,** 1927, 173–176.
3. I. Oftedal, *Z. physik. Chem.*, **135,** 1928, 291–299; experimental work prior to [2].
4. F. A. Bannister and M. H. Hey, *Mineralog. Mag.*, **23,** 1928, 188–206; *Am. Mineralogist,* **17,** 1932, 454.

0.1099
$\bar{1}$.8901

Ru-Se Ruthenium-Selenium

$RuSe_2$ (60.83 wt. % Se) is isotypic with pyrite (C2 type), $a = 5.933$ A [1]. In an attempt to prepare RuSe (43.71 wt. % Se) by direct synthesis, a product was obtained which consisted mainly of $RuSe_2$ [1].

1. L. Thomassen, *Z. physik. Chem.*, **B2,** 1929, 359–361.

0.5588
$\bar{1}$.4412

Ru-Si Ruthenium-Silicon

[1] claimed to have isolated crystals of the compound RuSi (21.62 wt. % Si). According to [2], silicides of the approximate stoichiometric compositions Ru_3Si_2 (15.55 wt. % Si), RuSi, and Ru_2Si_3 (29.27 wt. % Si) have been identified by X-ray diffraction. RuSi was found to be isotypic with CsCl (B2 type), $a = 2.90_6$ A. Another phase in the vicinity of RuSi is also cubic, $a = 4.70_9$ A. Ru_2Si_3 (?) was reported to be tetragonal, $a = 5.53$ A, $c = 4.47$ A, $c/a = 0.808$ [2]. The latter phase is possibly identical with $RuSi_2$ (35.58 wt. % Si), reported to have a tetragonal structure of an unknown type [3]. By means of high-temperature microscopy, [4] observed that melting of a eutectic grain-boundary phase in a 3 wt. (10.1 at.) % Si alloy occurs at about 1488°C. The solubility of Si in solid Ru appears to be less than 0.5 wt. %, since an as-cast alloy of this composition is clearly two-phase [4].

1. H. Moissan and W. Manchot, *Ber. deut. chem. Ges.*, **36,** 1903, 2993–2996; *Compt. rend.*, **137,** 1903, 229–232.
2. J. H. Buddery and A. J. E. Welch, *Nature*, **167,** 1951, 362.
3. H. J. Wallbaum, *Naturwissenschaften,* **32,** 1944, 76.
4. G. Reinacher, *Rev. mét.*, **54,** 1957, 321–336.

$\bar{1}$.9329
0.0671

Ru-Sn Ruthenium-Tin

Residues of the approximate composition $RuSn_2$ (70.01 wt. % Sn) [1] or $RuSn_3$ (77.78 wt. % Sn) [2] have been isolated from Sn-rich alloys. It was not proved whether they were homogeneous. According to roentgenographic studies [3], alloys with 70–100 at. % Sn, annealed at 475°C or quenched from 650°C, contained a cubic phase, probably of the composition Ru_3Sn_7 (73.14 wt. % Sn) [4], with a = 9.351 A and 40 atoms per unit cell (new structure type). It is likely that other intermediate phases exist. Ru is practically insoluble in Sn [3].

1. H. St. Claire Deville and H. Debray, *Ann. chim. et phys.*, **56,** 1859, 385.
2. H. Debray, *Compt. rend.*, **104,** 1887, 1470.
3. O. Nial, Dissertation, University of Stockholm, 1945; *Svensk Kem. Tidskr.*, **59,** 1947, 172–183.
4. This composition lies between those claimed by [1, 2].

$\bar{1}.7499$
0.2501

Ru-Ta Ruthenium-Tantalum

[1] examined Ta-Ru alloys containing 38, 50, 55, 64, and 73 at. % Ru, which had been annealed at and quenched from 1200°C, by X-ray and microscopic methods. The following results were obtained: (*a*) the 38 at. % Ru alloy has the CsCl (B2) type of structure (ordered Ta-base b.c.c. solid solution); (*b*) the alloy with about 50 at. % Ru is apparently tetragonal, a = 3.02 A, c = 3.37 A, c/a = 1.11; (*c*) a second intermediate phase with unknown structure exists between 55 and 64 at. % Ru; (*d*) the solubility limit of Ta in solid Ru is less than 27 at. (40 wt.) % Ta.

1. P. Greenfield and P. A. Beck, *Trans. AIME*, **206,** 1956, 265–276.

$\bar{1}.9014$
0.0986

Ru-Te Ruthenium-Tellurium

$RuTe_2$ (71.51 wt. % Te) is isotypic with pyrite (C2 type), a = 6.37 A [1]. In an attempt to prepare RuTe (55.65 wt. % Te) by direct synthesis, a product was obtained which consisted mainly of $RuTe_2$ [1].

1. L. Thomassen, *Z. physik. Chem.*, **B2,** 1929, 259–261.

0.3270
$\bar{1}.6730$

Ru-Ti Ruthenium-Titanium

It has been reported [1, 2] that Ru and Ti form the compound TiRu (32.02 wt. % Ti), which has the crystal structure of the CsCl (B2) type, and that phases of the compositions $TiRu_2$ and $TiRu_3$ do not exist. [3] corroborated the CsCl (B2) type of structure for TiRu and gave the lattice parameter a = 3.06 A.

1. F. Laves and H. J. Wallbaum, *Naturwissenschaften*, **27,** 1939, 674–675.
2. H. J. Wallbaum, *Naturwissenschaften*, **31,** 1943, 91–92.
3. C. B. Jordan, *Trans. AIME*, **203,** 1955, 832–833.

$\bar{1}.6306$
0.3694

Ru-U Ruthenium-Uranium

URu_3 (56.17 wt. % Ru) is of the cubic Cu_3Au ($L1_2$) type of structure, with a = 3.988 ± 2 A [1].

1. T. J. Heal and G. I. Williams, *Acta Cryst.*, **8,** 1955, 494–498.

0.3002
$\bar{1}.6998$

Ru-V Ruthenium-Vanadium

[1] examined V-Ru alloys containing about 36, 50, 71, and 82 at. % Ru, which had been annealed at and quenched from 1200°C, by X-ray and microscopic methods.

The 36 at. % Ru alloy was found to have the CsCl (B2) structure, with unit cell dimensions very similar to those of the V-base b.c.c. solid solutions. At 50 at. % Ru an intermediate phase was found having a tetragonal lattice, $a = 2.96$ A, $c = 3.09$ A, $c/a = 1.04$. Whether this tetragonal phase is present at 1200°C or is a result of transformation on cooling to room temperature is not known. The solubility limit of V in solid Ru is less than 18 at. (10 wt.) % V.

1. P. Greenfield and P. A. Beck, *Trans. AIME*, **206**, 1956, 265–276.

$\bar{1}.7427$
0.2573

Ru-W Ruthenium-Wolfram

Microexamination and X-ray diffraction studies of six alloys with 5.8–56.3 at. (10–70 wt.) % W showed that no intermediate phase occurs. The solid solubility of W in the h.c.p. Ru, after annealing at 1400°C, was found parametrically to be about 36.5 at. (51.2 wt.) % W. The solid solubility of Ru in W is quite restricted and estimated to be 1.5 at. % Ru at the most [1].

Supplement. The absence of intermediate phases in this system has been corroborated by [2]. According to recent work by [3], however, there exists an intermediate phase of the σ type of structure. "It is formed by peritectic reaction of melt with the (W) phase and has about the composition W_3Ru_2 (about 73 wt. % W). The lattice parameters are $a = 9.55$ A, $c/a = 0.52$. It is stable only at temperatures above 1650°C; at lower temperatures it decomposes eutectoidally into (Ru) + (W)" [3].

1. E. Raub and P. Walter, "Festschrift aus Anlass des 100-jährigen Jubiläums der Firma W. C. Heraeus G.m.b.H.," pp. 124–146, Hanau, 1951.
2. P. Greenfield and P. A. Beck, *Trans. AIME*, **206**, 1956, 265–276.
3. W. Obrowski, to be published in *Naturwissenschaften*.

0.0472
$\bar{1}.9528$

Ru-Zr Ruthenium-Zirconium

[1] reported the existence of $ZrRu_2$ (69.04 wt. % Ru), which is isotypic with $MgZn_2$ (C14 type), $a = 5.141$ A, $c = 8.507$ A, $c/a = 1.655$. The lattice spacings of dilute solutions of Zr in solid Ru were measured by [2]. Alloys with 0.35 and 0.50 at. (0.31 and 0.45 wt.) % Zr were one-phase and two-phase, respectively [3].

1. H. J. Wallbaum, *Naturwissenschaften*, **30**, 1942, 149.
2. A. Hellawell and W. Hume-Rothery, *Phil. Mag.*, **45**, 1954, 797–806.
3. The alloys were prepared in an argon arc furnace, partly homogenized (5 min, near the melting point), ground to powder, and then strain-annealed at 1050°C for 12 hr.

$\bar{1}.4205$
0.5795

S-Sb Sulfur-Antimony

Figure 627 shows the phase diagram due to [1]; it is based on the thermal data points indicated. In the S-rich range, the monotectic temperature of 530°C could be observed only up to 80 at. % S; "at higher S contents boiling of S interfered." The solidus temperature in this composition range was found as 110°C, i.e., practically at the melting point of S. The solid solubility of S in Sb was concluded to be about 0.3 at. (0.08 wt.) % S, based on measurements of electrical resistivity [2].

Previously, [3] had reported quite similar characteristic temperatures for the partial system Sb-Sb_2S_3 without giving tabulated data [4]. Between about 1.5 and 21 wt. (5.5–50 at.) % S, the monotectic temperature was found as 615°C and the eutectic temperature as 515–519°C. The Sb-Sb_2S_3 eutectic was placed at about 24.5 wt. (55.2 at.) % S. The melting point of Sb_2S_3 (28.31 wt. % S) was determined as 555°C, as compared with 546 and 554°C, according to [1] and [5], respectively. For additional values ranging between 540 and 550°C, see [6].

Sb_2S_5 (39.70 wt. % S) does not crystallize from the melt; it can be prepared only by wet chemical reactions.

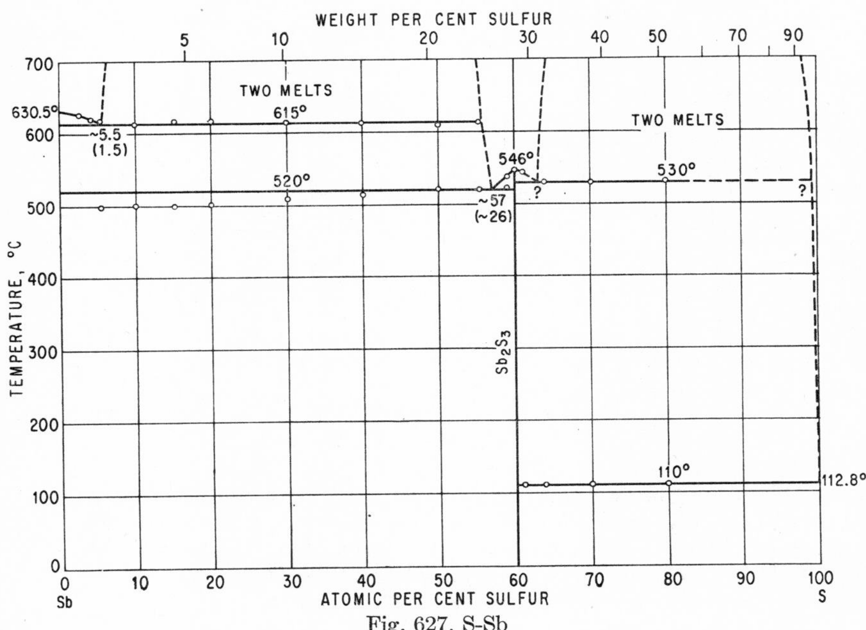

Fig. 627. S-Sb

The structure of Sb_2S_3 is the prototype of the $D5_8$ type of structure [7–10], with a = 11.22 A, b = 11.30 A, c = 3.84 A [10].

1. F. M. Jaeger, *Akad. Amsterdam Verslag*, **20**, 1911-1912, 498; F. M. Jaeger and H. S. van Klooster, *Z. anorg. Chem.*, **78**, 1912, 246–248.
2. J. Olie and H. R. Kruyt, *Akad. Amsterdam Verslag*, **20**, 1911–1912, 69; see also [1].
3. H. Pélabon, *Compt. rend.*, **138**, 1904, 277–279; *Ann. chim. et phys.*, **17**, 1909, 530–535.
4. Work by P. Chrétien and J. Guinchant (*Compt. rend.*, **142**, 1906, 709–711) gave inconclusive results.
5. E. Jensen, *Avhandl. Norske Videnskaps-Akad. Oslo, Mat.-Naturw. Kl.*, no. 2, 1947; *Chem. Abstr.*, **43**, 1949, 3274.
6. "Gmelins Handbuch der anorganischen Chemie," 8th ed., System No. **18** B, p. 513, Gmelin-Verlag G.m.b.H., Clausthal-Zellerfeld, Germany, 1949.
7. C. Gottfried, *Z. Krist.*, **65**, 1927, 428–434.
8. C. Gottfried and E. Lubberger, *Z. Krist.*, **71**, 1929, 257–262.
9. J. G. Albright, *Phys. Rev.*, **37**, 1931, 458.
10. W. Hofmann, *Z. Krist.*, **86**, 1933, 225–245.

Ī.6086
0.3914

S-Se Sulfur-Selenium

The phase diagram shown in Fig. 628 was established by means of thermal and dilatometric investigations [1]. As to the crystal structure and lattice dimensions of the α, γ, and δ phases, see [2], [3], and [4], respectively.

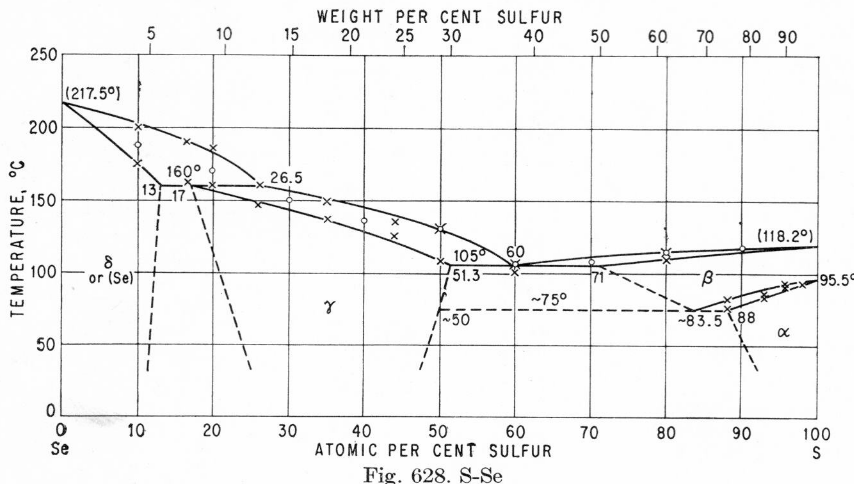

Fig. 628. S-Se

1. W. E. Ringer, *Z. anorg. Chem.*, **32,** 1902, 183–218. In this paper reference is made to previous work. See also L. Losana, *Gazz. chim. ital.*, **53,** 1923, 396–397.
2. F. Halla and F. X. Bosch, *Z. physik. Chem.*, **B10,** 1930, 149–156.
3. F. Halla, E. Mehl, and F. X. Bosch, *Z. physik. Chem.*, **B12,** 1931, 377–386.
4. G. P. Barnard, *Proc. Phys. Soc. (London)*, **47,** 1935, 482.

0.0575
Ī.9425

S-Si Sulfur-Silicon

There are two silicon sulfides: SiS (53.31 wt. % S) and SiS_2 (69.54 wt. % S). Sulfurization of Si always results in the formation of SiS_2 [1], having a melting point of about 1090°C and a boiling point lying between 1100 and 1200°C [1]. SiS, which is prepared by reaction of SiS_2 with Si under vacuum at 850°C [2], has a sublimation point of about 940°C.

The structure of SiS_2 is the prototype of the C42 type of structure; a = 9.57–9.59 A, b = 5.61–5.66 A, c = 5.54–5.55 A [3, 4]. The structure of SiS is not known, and the product may possibly be a disproportionate intimate mixture of SiS_2 and Si [2].

1. E. J. Kohlmeyer and H. W. Retzlaff, *Z. anorg. Chem.*, **261,** 1950, 248–260; additional references are given in this paper. See also H. Gabriel and C. Alvarez-Tostado, *J. Am. Chem. Soc.*, **74,** 1952, 262–264; L. Malatesta, *Gazz. chim. ital.*, **78,** 1948, 702–706.
2. W. C. Schumb and W. J. Bernard, *J. Am. Chem. Soc.*, **77,** 1955, 904–905.
3. A. Zintl and K. Loosen, *Z. physik. Chem.*, **174,** 1935, 301–311.
4. W. Büssem, H. Fischer, and E. Gruner, *Naturwissenschaften*, **23,** 1935, 740.

$\bar{1}.4316$
0.5684

S-Sn Sulfur-Tin

Only the partial system Sn-SnS could be established by means of thermal studies [1, 2], since all the sulfur evaporated from melts with more than about 23.5 wt. (53.2 at.) % S (Fig. 629). The existence of a range of liquid immiscibility, although highly

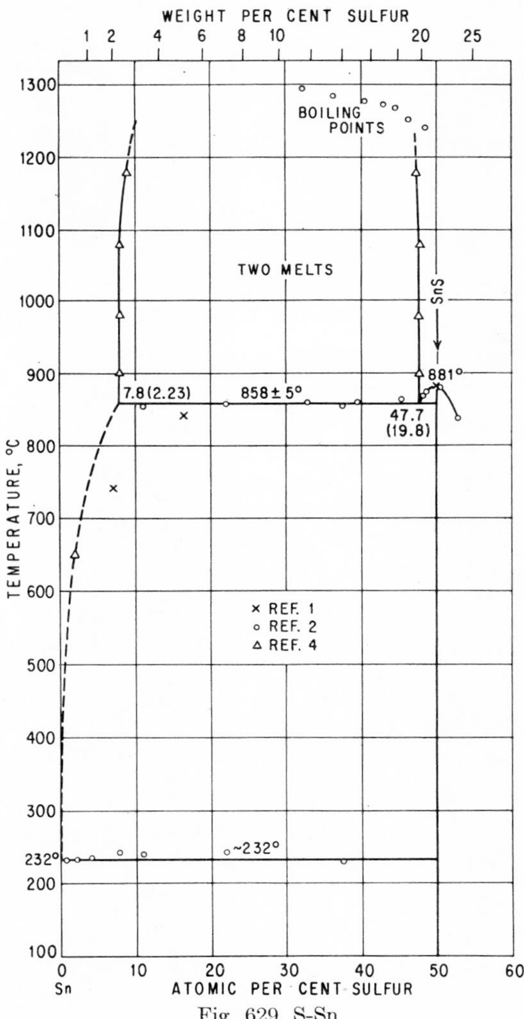

Fig. 629. S-Sn

probable because of the constant liquidus temperature (858 ± 5°C) between about 11 and 45 at. % S and the formation of two layers in this range, was denied by [2] for the following reasons: (*a*) the boiling points determined for the range 32.5–48.6 at. % S were found to change with composition rather than to remain constant; (*b*) no thermal effect was observed for the monotectic reaction between 48 and 50 at. % S.

Nevertheless, a miscibility gap has been assumed to occur by [3]. Its existence was conclusively established by determining the compositions of the conjugate liquid phases in equilibrium at temperatures between 900 and 1180°C [4]. The latter authors also determined the liquidus point of the melt with 0.5 wt. (1.8 at.) % S to lie between 700 and 600°C. The eutectic temperature was confirmed to "lie very close, indeed" to the melting point of Sn [4].

Crystal Structures. The structure of SnS (21.27 wt. % S) is the prototype of the B29 type of structure, a = 3.99 A, b = 4.34 A, c = 11.20 A [5]. According to [6], SnS is isotypic with GeS (B16 type). SnS_2 (35.08 wt. % S) is isotypic with CdI_2, $a = 3.64_6$ A, $c = 5.88_0$ A, $c/a = 1.61_3$ [7].

1. H. Pélabon, *Compt. rend.*, **142,** 1906, 1147–1149; *Ann. chim. et phys.*, **17,** 1909, 526.
2. W. Biltz and W. Mecklenburg, *Z. anorg. Chem.*, **64,** 1909, 226–235.
3. M. Hansen, "Der Aufbau der Zweistofflegierungen," Springer-Verlag OHG, pp. 1034–1035, Berlin, 1936.
4. J. S. Anderson and M. J. Ridge, *Trans. Faraday Soc.*, **39,** 1943, 98–102.
5. W. Hofmann, *Z. Krist.*, **92,** 1935, 161–173; *Fortschr. Mineral.*, **19,** 1935, 30–31.
6. K. Schubert, private communication.
7. I. Oftedal, *Z. physik. Chem.*, **134,** 1928, 301–310.

$\bar{1}.5634$
0.4366

S-Sr Sulfur-Strontium

SrS (26.79 wt. % S) is isotypic with NaCl (B1 type) [1–3], a = 6.008 A [3].

1. S. Holgersson, *Z. anorg. Chem.*, **126,** 1923, 179.
2. V. M. Goldschmidt; see *Strukturbericht,* **1,** 1913-1928, 127.
3. W. Primak, H. Kaufman, and R. Ward, *J. Am. Chem. Soc.*, **70,** 1948, 2043–2046.

$\bar{1}.2486$
0.7514

S-Ta Sulfur-Tantalum

Tensimetric and X-ray investigations [1] were claimed to indicate the existence of TaS (15.06 wt. % S), observed in the range 50–65.5 at. % S, TaS_2 (26.18 wt. % S) [2], TaS_3 (34.72 wt. % S), and a subsulfide (observed in the range 23–50 at. % S) of undetermined composition. X-ray results reported by [3] agree with those given by [1] only in one point, viz., the existence of TaS_2. "No intermediate phase in the Ta-TaS_2 interval was found to exist, and the use of sulfur in great excess only caused the formation of disulfide and free sulfur."

TaS_2 was found by [1] to be hexagonal of the CdI_2 (C6) or MoS_2 (C7) type or possibly another lattice type; $a = 3.40_2$ A, $c = 5.91_4$ A, $c/a = 1.73_8$. According to [3], TaS_2 exists in four different modifications, α-, β-, γ-, and δ-TaS_2, all of them having layer lattices. α-TaS_2 has a homogeneity range of approximately 5 at. %, while no deviations from the stoichiometric composition were found for the other three modifications. Only α-TaS_2 could be prepared in a pure state; the other three modifications could not be entirely separated from each other. Also, it was not possible to transform one phase into another by means of heat-treatment at different temperatures. As an increasing amount of the most complicated δ-TaS_2 modification was obtained at prolonged reaction times, this phase is probably the most stable one. The close similarity between the structures makes it very probable, however, that their energies differ only slightly.

The lattice constants and number Z of formula units TaS_2 of the unit cells were reported as follows [3]: α-TaS_2 (C6 type), a = 3.319 A, c = 6.275 A, c/a = 1.891 at about 61.5 at. % S; a = 3.346 A, c = 5.860 A, c/a = 1.751, Z = 1, at the stoichiometric composition; β-TaS_2 (C27 type), a = 3.32 A, c = 12.30 A, c/a = 3.70, Z = 2; γ-TaS_2 (C19 type), a = 3.32 A, c = 18.29 A, c/a = 5.51, Z = 3; and δ-TaS_2 (new structure type), a = 3.34 A, c = 35.94 A, c/a = 10.75, Z = 6.

1. W. Biltz and A. Köcher, *Z. anorg. Chem.*, **238,** 1938, 81–93.
2. H. Biltz and C. Kircher, *Ber. deut. chem. Ges.*, **43,** 1910, 1636.
3. G. Hägg and N. Schönberg, *Arkiv Kemi*, **7,** 1954, 371–380.

$\bar{1}.5104$
0.4896

S-Tc Sulfur-Technetium

The sulfide Tc_2S_7 was identified by chemical analysis [1].

1. C. L. Rulfs and W. W. Meinke, *J. Am. Chem. Soc.*, **74,** 1952, 235–236.

$\bar{1}.4002$
0.5998

S-Te Sulfur-Tellurium

Solidification data points reported by [1–4] are shown in Fig. 630. Between 0 and 60 at. % S, they lie in a scattering band of up to about 15°C in width. In the range 60–100 at. % S, scattering is considerably larger and reaches a maximal differ-

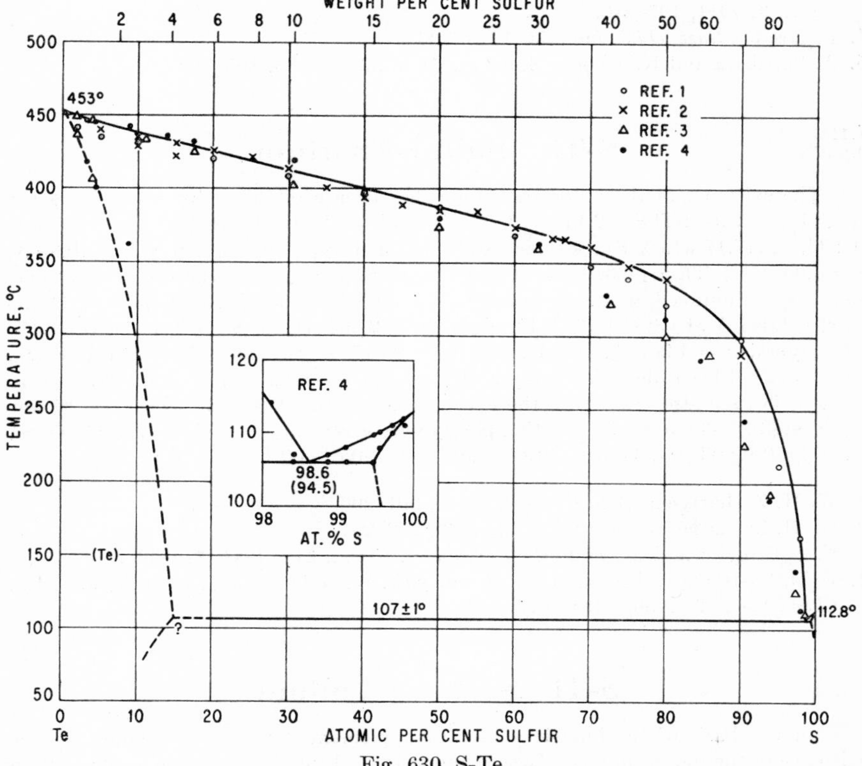

Fig. 630. S-Te

ence of about 70°C at 90 at. % S. In drawing the liquidus curve in Fig. 630, preference was given to the highest temperatures, especially in the region where the liquidus is steep.

The eutectic temperature was reported as 105.5–110.5 [1], 101–110 [2], 107–109 [3], and 106–109°C [4], the mean values being 108 [1], 106 [2], 108.5 [3], and 107°C [4]. The eutectic composition was given as about 99 at. (96.15 wt.) % S [1], 99.5 at. (98 wt.) % S [2], 98.2 at. (93 wt.) % S [3], and 98.6 at. (94.5 wt.) % S [4]. The last value appears to be the most reliable one; see inset of Fig. 630.

The solid solubility of S in Te has not yet been determined accurately. The values reported for the solubility at the eutectic temperature are based only on thermal data: about 26 at. (8 wt.) % S [2], 7.5 at. (2 wt.) % S [3], and 14.5 at. (4 wt.) % S [4]. For some lower temperature, [5] gave 5.9 at. (1.55 wt.) % S, on the basis of measurements of thermo-emf and electrical resistance. The solid solubility of Te in β-S (stable above 95.5°C) was given as approximately 0.5 at. (2 wt.) % Te [3, 4], based on thermal data only.

The α-β transformation point of S (95.5°C) is lowered to about 90°C by 0.3 at. (1.2 wt.) % Te [1].

1. G. Pellini, *Atti reale accad. Lincei,* **18**(1), 1909, 701–706.
2. F. M. Jaeger, *Verslag Akad. Wetenschap. Amsterdam,* **18,** 1910, 606–620; F. M. Jaeger and J. B. Menke, *Z. anorg. Chem.,* **75,** 1912, 241–255.
3. M. Chikashige, *Z. anorg. Chem.,* **72,** 1911, 109–118; *Mem. Coll. Sci. Eng. Kyoto Univ.,* **3,** 1911, 197–210.
4. L. Losana, *Gazz. chim. ital.,* **53,** 1923, 396–410.
5. A. Petrikaln and K. Jacoby, *Z. anorg. Chem.,* **210,** 1933, 195–202.

$\bar{1}.1403$
0.8597

S-Th Sulfur-Thorium

The existence of the following sulfides has been established by determining the crystal structures: ThS (12.14 wt. % S), isotypic with NaCl (B1 type), a = 5.683 A [1]; Th_2S_3 (17.17 wt. % S), isotypic with Sb_2S_3 ($D5_8$ type), a = 10.99 A, b = 10.85 A, c = 3.96 A [1]; Th_7S_{12} [63.16 at. (19.15 wt.) % S], hexagonal structure of new type, 19 atoms per unit cell, with a = 11.063 A, c = 3.991 A at 63.1 at. % S and a = 11.086 A, c = 4.010 A at 63.8 at. % S [2]; and ThS_2 (21.65 wt. % S), isotypic with $PbCl_2$ (C23 type), a = 4.267 A, b = 7.264 A, c = 8.617 A [1]. The phases ThS, Th_2S_3, and ThS_2 had been identified earlier, by means of tensimetric and qualitative X-ray analysis [3]; see also [4]. Also, the existence of Th_3S_7 [70 at. (24.38 wt.) % S] was reported [3]. The following melting points were given by [4]: ThS, >2200°C; Th_2S_3, 1950 ± 50°C; Th_7S_{12}, 1770 ± 30°C; and ThS_2, 1905 ± 30°C.

1. W. H. Zachariasen, *Acta Cryst.,* **2,** 1949, 291–296.
2. W. H. Zachariasen, *Acta Cryst.,* **2,** 1949, 288–291.
3. E. F. Strotzer and M. Zumbusch, *Z. anorg. Chem.,* **247,** 1941, 415–428.
4. E. D. Eastman, L. Brewer, L. A. Bromley, P. W. Gilles, and N. L. Lofgren, *J. Am. Chem. Soc.,* **72,** 1950, 4019–4023.

$\bar{1}.8257$
0.1743

S-Ti Sulfur-Titanium

On the basis of tensimetric analysis, supplemented by X-ray studies [1], the existence of the following phases was suggested: a subsulfide of unknown composition between $TiS_{0.2}$ (16.7 at. % S) and TiS, a monosulfide TiS, a sesquisulfide between $TiS_{1.1}$ (52.4 at. % S) and Ti_2S_3, a disulfide between $TiS_{1.7}$ (63.1 at. % S) and TiS_2, and a trisulfide TiS_3. Ti_2S_3 [2] and TiS_2 [2, 3] had been reported earlier. TiS and Ti_2S_3 were claimed to form solid solutions at higher temperatures, which decompose into TiS and Ti_2S_3 on cooling [1].

Reinvestigation of the system by X-ray methods [4] did not confirm the existence of Ti_2S_3 and TiS_3 but verified, besides TiS and TiS_2, a subsulfide with unknown structure and with the approximate composition Ti_2S. However, [5] maintained the existence of Ti_2S_3, having a structure of low symmetry. The phases Ti_4S_5, Ti_3S_4, and Ti_3S_5, claimed by [6], cannot be regarded as definite compounds. TiS_3 appears to be well established by [1].

The lowest per cent addition at which sulfur appeared as an insoluble compound (Ti_2S) in the microstructure of magnesium-reduced Ti-base alloys, hot-rolled and subsequently annealed for ½ hr at 790°C, was found to be 0.02 wt. % S [7].

Crystal Structures. [8] reported the structure of TiS to be of the NiAs (B8) or a related type. [9] confirmed the NiAs type of structure for TiS, $a = 3.30$ A, $c = 6.44$ A, $c/a = 1.96$. TiS_2 is isotypic with CdI_2 (C6 type), $a = 3.40_4$ A, $c = 5.7_0$ A, $c/a = 1.67$ [3]; $a = 3.40_5$ A, $c = 5.68_4$ A, $c/a = 1.66_9$ [5]. It ranges in composition between about 65 and at least 66.67 at. % S [5]. Lattice dimensions on the Ti side are $a = 3.40_2$ A, $c = 5.66_9$ A, $c/a = 1.66_6$ [5].

Note Added in Proof. By means of X-ray powder and single-crystal work, [10] obtained the following structural results: (*a*) TiS is dimorph. The low-temperature form (prepared at 700°C by [9, 10]) is confirmed to be of the NiAs (B8) type. Annealing at 1000°C results in the transformation to a superstructure of the NiAs type, with the hexagonal lattice parameters $a = 3.417 \pm 5$ A, $c = 26.4 \pm 1$ A, $c/a = 7.7_5$. Even prolonged annealing at 700°C does not retransform it into the NiAs modification. (*b*) The phase stable around the composition Ti_3S_4 (called Ti_2S_3 by previous authors) is hexagonal, with $a = 3.431 \pm 6$ A, $c = 11.44 \pm 1$ A, $c/a = 3.33_6$, and isotypic with TiP and TiAs. (*c*) TiS_3 is monoclinic, with $a = 4.9_9$ A, $b = 3.3_8$ A, $c = 17._6$ A, $\beta = 97.5°$, and 4 formula units per unit cell.

In a paper dealing mainly with the effects of S on the mechanical properties of Ti, [11] reported some observations relating to the phase relationships in Ti-rich Ti-S alloys. Microscopic work showed the solubility of S in α-Ti as well as in β-Ti (at 900°C) to be between 0.009 and 0.017 wt. %. But "there are indications from grain-size studies that S is less soluble in β- than in α-Ti" [11]. Thermal analysis and quenching studies showed no measurable effect of S on the $\alpha \rightleftarrows \beta$ transformation in Ti. As-cast microstructures indicate the existence of a eutectic (β_{Ti} + sulfide) somewhere between 4 and 13.5 wt. (6 and 19 at.) % S. According to [12], the eutectic temperature is 1215°C.

1. W. Biltz, P. Ehrlich, and K. Meisel, *Z. anorg. Chem.*, **234,** 1937, 97–116.
2. "Gmelins Handbuch der anorganischen Chemie," 8th ed., System No. 41, pp. 337–343, Verlag Chemie, G.m.b.H., Weinheim/Bergstrasse, Germany, 1951.
3. I. Oftedal, *Z. physik. Chem.*, **134,** 1928, 301–310.
4. G. Hägg and N. Schönberg, *Arkiv Kemi*, **7,** 1954, 371–380.
5. P. Ehrlich, *Z. anorg. Chem.*, **260,** 1949, 13.
6. M. Picon, *Bull. soc. chim.*, **1,** 1934, 919–926.
7. R. M. Goldhoff, H. L. Shaw, C. M. Craighead, and R. I. Jaffee, *Trans. ASM*, **45,** 1953, 941–971.
8. A. Faessler and M. Goehring, *Naturwissenschaften*, **31,** 1943, 568.
9. N. Schönberg, *Acta Met.*, **2,** 1954, 427–432.
10. H. Hahn and B. Harder, *Z. anorg. Chem.*, **288,** 1956, 241–256.

11. L. W. Berger, D. N. Williams, and R. I. Jaffee, *Trans. ASM*, **49**, 1957, 300–312; discussion, 312–314.
12. R. A. Perkins, see discussion on [11].

$\bar{1}.1956$
0.8044

S-Tl Sulfur-Thallium

The solidification diagram was first roughly outlined by [1]; no tabulated data were given. Results indicated the existence of two miscibility gaps in the liquid state in the monotectic ranges >0–33.3 at. % S (Tl_2S) at 448°C and about 71–100 at. % S at 125°C. In the intermediate region 33.3–46.5 at. % S an additional three-phase equilibrium was found at about 295°C which was suggested by [2] to indicate the peritectic formation of the phase TlS (13.56 wt. % S), rather than Tl_8S_7 [46.67 at. (12.07 wt.) % S] proposed by [1]. Later, TlS was identified by X-ray studies [3, 4] and chemical analysis of a preparation obtained by chemical reaction [5].

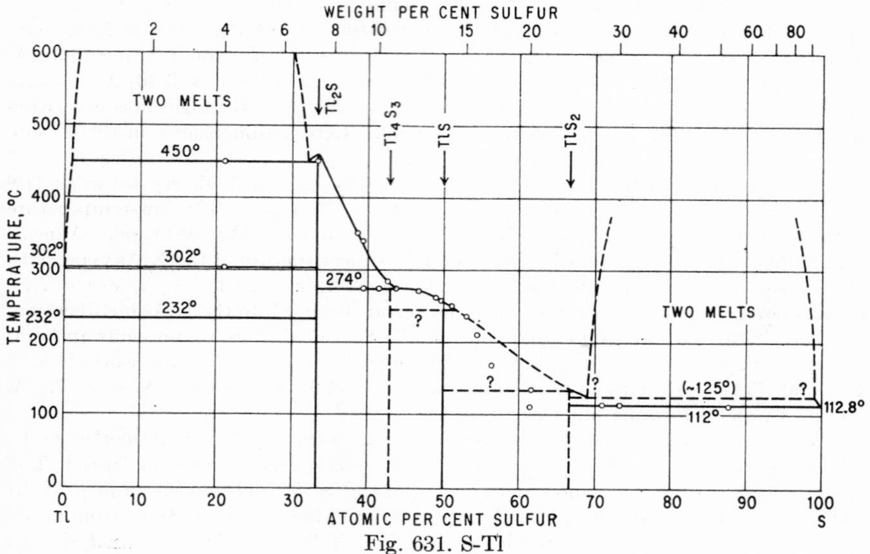

Fig. 631. S-Tl

Figure 631 shows the thermal data points reported by [6]. They were concluded to prove the existence of Tl_2S (7.28 wt. % S) and Tl_8S_7. However, the composition Tl_8S_7 lies definitely outside the range of the peritectic horizontal at 274°C, since the peritectic melt contains only about 44 at. % S. This indicates the composition of the phase formed peritectically at 274°C to be Tl_4S_3 (or perhaps Tl_3S_2). Indeed, a phase of the likely composition Tl_4S_3 [42.86 at. (10.53 wt.) % S] was reported by [4], on the basis of X-ray analysis. No thermal evidence for the crystallization of TlS was found by [6] (Fig. 631).

This holds also for the phase TlS_2 (23.89 wt. % S), identified by X-ray investigation [4]. According to [6], the composition TlS_2 lies well within the range of liquid immiscibility, 61–100 at. % S, whereas according to the older diagram [1] it lies outside this region. The peritectic formation of TlS_2 necessitates an additional horizontal, besides those corresponding to the monotectic equilibrium at about 125 [1] or 132°C [6] and the eutectic crystallization at about the melting point of S. It should be stated, however, that the investigations in this range were rather cursory. The phase relationships between 50 and 70 at. % S shown in Fig. 631 should be regarded as strictly *tentative.*

On the basis of the work of [4], a phase Tl_2S_5 (71.43 at. % S), assumed by [1, 5], is not a definite compound. This composition may lie close to that of the Tl-rich melt at the monotectic temperature.

Crystal Structure. Tl_2S has a hexagonal-rhombohedral structure with 27 formula weights per unit cell, a = 12.22 A, c = 18.21 A, c/a = 1.490 [3]; the structure is very similar to that of the CdI_2 (C6) type; see also [4]. TlS, formerly suggested to be monoclinic [3], is tetragonal and isotypic with TlSe (B37 type), a = 7.79 A, c = 6.80 A, c/a = 0.88_6 [4]. The powder diagrams of the phases of the approximate composition Tl_4S_3 (observed from 39.8 to 42.9 at. % S) and TlS_2 were not indexed [4].

1. H. Pélabon, *Compt. rend.*, **145,** 1907, 118–121; *Ann. chim. et phys.*, **17,** 1909, 526–566.
2. M. Hansen, "Der Aufbau der Zweistofflegierungen," pp. 1037–1038, Springer-Verlag OHG, Berlin, 1936.
3. J. A. A. Ketelaar and E. W. Gorter, *Z. Krist.*, **101,** 1939, 367–375.
4. H. Hahn and W. Klingler, *Z. anorg. Chem.*, **260,** 1949, 110–119.
5. V. Scatturin and E. Frasson, *Ann. chim.* (*Rome*), **43,** 1953, 561–563.
6. A. P. Obukhov and N. S. Bubyreva, *Izvest. Sektora Fiz.-Khim. Anal.*, **17,** 1949, 276–280.

$\bar{1}.1293$
0.8707

S-U Sulfur-Uranium

The sulfides US (11.87 wt. % S) [1], U_2S_3 (16.81 wt. % S) [1], and US_2 (21.22 wt. % S) [2, 3] have been reported in the older chemical literature. More recently, the existence of U_4S_3 [42.86 at. (9.18 wt.) % S], U_2S_3, US_2, and US_3 (28.78 wt. % S) was claimed on the basis of tensimetric and qualitative X-ray analysis [4]; and that of US [5], U_2S_3 [5], and US_2 [6, 7] was definitely established by determination of the crystal structures. As regards U_2S_3 and US_3, see also [8, 7] and [12], respectively.

US is isotypic with NaCl (B1 type), a = 5.484 A [5], and U_2S_3 is isotypic with Sb_2S_3 ($D5_8$ type), a = 10.41 A, b = 10.65 A, c = 3.89 A [5]. US_2 exists in two forms [6, 9, 7]. α-US_2 (or US_2 I), stable above approximately 1350°C [9], is tetragonal of a new structure type, with 10 U atoms per unit cell and a = 10.27 A, c = 6.31 A [6, 7]. The low-temperature modification, β-US_2 (or US_2 II), is isotypic with $PbCl_2$ (C23 type), having parameters of a = 4.23 A, b = 7.09 A, c = 8.47 A according to [6] and a = 4.12 A, b = 7.11 A, c = 8.46 A (or kX?) according to [8]. U_4S_3 [4] seems to be identical with US, since [10] reported the parameter of the cubic structure as a = 5.50_5 A (see above).

The following melting points were reported by [11]: US and U_2S_3, >2000°C; US_2, 1850 ± 100°C.

1. G. Alibegoff, *Liebigs Ann.*, **233,** 1886, 134–135.
2. E. Peligot, *Ann. chim. et phys.*, **5,** 1842, 20.
3. A. Colani, *Compt. rend.*, **137,** 1903, 382; *Ann. chim. et phys.*, **12,** 1907, 80.
4. E. F. Strotzer, O. Schneider, and W. Biltz, *Z. anorg. Chem.*, **243,** 1940, 307–321.
5. W. H. Zachariasen, *Acta Cryst.*, **2,** 1949, 291–296.
6. Unpublished work by W. H. Zachariasen, quoted from J. J. Katz and E. Rabinowitch, "The Chemistry of Uranium," Part I, pp. 330–335, National Nuclear Energy Series, Div. VIII, vol. 5, McGraw-Hill Book Company, Inc., New York, 1951.

7. M. Picon and J. Flahaut, *Compt. rend.*, **237**, 1953, 1160–1162.
8. R. Flatt and W. Hess, *Helv. Chim. Acta*, **21**, 1938, 525–529.
9. M. Picon and J. Flahaut, *Compt. rend.*, **237**, 1953, 808–810.
10. M. Zumbusch, *Z. anorg. Chem.*, **243**, 1940, 322–329.
11. E. D. Eastman, L. Brewer, L. A. Bromley, P. W. Gilles, and N. L. Lofgren, *J. Am. Chem. Soc.*, **72**, 1950, 4019–4023.
12. R. W. M. D'Eye, P. G. Sellman, and J. R. Murray, *J. Chem. Soc.*, **1952**, 2558.

$\bar{1}.7989$
0.2011

S-V Sulfur-Vanadium

The following phases were identified by X-ray and/or tensimetric analysis: VS (38.63 wt. % S) [1, 2], V_2S_3 (48.56 wt. % S) [1–3], and VS_4 (71.57 wt. % S) [1, 3]; see also [4]. VS forms a eutectic with V at about 12 wt. (17.8 at.) % S, 1312°C [5] (see Fig. 632) and is isotypic with NiAs (B8 type), $a = 3.3_5$ A, $c = 5.79_6$ A, $c/a = 1.730$ [1]; $a = 3.36_7$ A, $c = 5.82_5$ A, $c/a = 1.730$ [2].

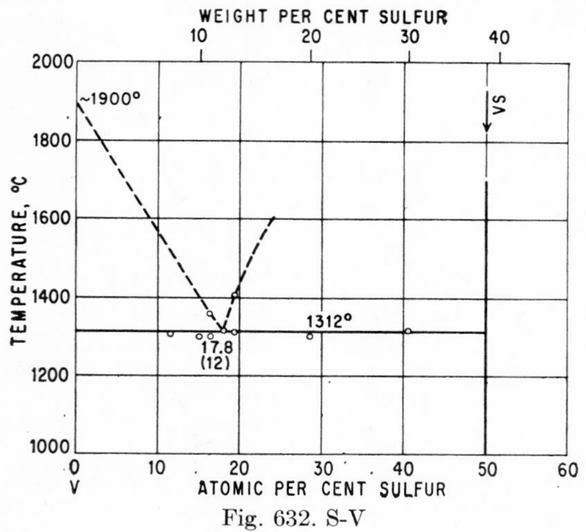

Fig. 632. S-V

1. W. Biltz and A. Köcher, *Z. anorg. Chem.*, **241**, 1939, 324–337.
2. E. Hoschek and W. Klemm, *Z. anorg. Chem.*, **242**, 1939, 60–62.
3. W. Klemm and E. Hoschek, *Z. anorg. Chem.*, **226**, 1936, 359–369.
4. G. Gaudefroy, *Compt. rend.*, **237**, 1953, 1705–1707.
5. R. Vogel and A. Wüstefeld, *Arch. Eisenhüttenw.*, **12**, 1938-1939, 261.

$\bar{1}.2414$
0.7586

S-W Sulfur-Wolfram

X-ray analysis [1, 2] showed that WS_2 (25.86 wt. % S) is the only thermally stable intermediate phase in this system, which can be prepared by direct synthesis; see also [3]. Its homogeneity range was determined as 66.1–66.67 at. % S [2]. WS_3 (34.34 wt. % S), prepared in various ways [1–3], decomposes into WS_2 and S in vacuum at 170°C [1].

WS_2 is isotypic with MoS_2 (C7 type) [4, 1, 2], with $a = 3.151$ A, $c = 12.29$ A,

$c/a = 3.90$ at 66.1 at. % S and $a = 3.146$ A, $c = 12.34$ A, $c/a = 3.92$ at the ideal composition [2].

1. O. Glemser, H. Sauer, and P. König, *Z. anorg. Chem.*, **257**, 1948, 241–246.
2. P. Ehrlich, *Z. anorg. Chem.*, **257**, 1948, 247–253.
3. D. Beischer and G. Oechsel, *FIAT Rev. Ger. Sci.*, Inorganic Chemistry, Part II, p. 33, 1948 (otherwise unpublished).
4. A. E. van Arkel, *Rec. trav. chim.*, **45**, 1926, 437–444.

$\bar{1}.6906$
0.3094

S-Zn Sulfur-Zinc

The solidification of Zn-rich melts (solutions of S in molten Zn and Zn in molten ZnS) has not yet been investigated. It is likely that the phase diagram is analogous to that of the systems Se-Zn and Cd-Se.

According to [1], zincblende melts at 1049°C; however, this value is definitely too low. From the phase diagrams of the systems of ZnS with PbS, Cu_2S, Ag_2S, and FeS, [2] extrapolated the melting point as 1600–1700°C, closer to 1700°C. Under normal pressure, however, ZnS probably will not melt since it volatilizes without decomposition at considerably lower temperatures. [3] reported that zincblende and synthetic ZnS sublime even at a temperature as low as 1200°C. [4] found that ZnS volatilizes at 1000°C to such a degree that small crystals of wurtzite are formed. The sublimation point of pure natural zincblende (cubic) was determined as 1178 ± 2°C and that of synthetic wurtzite (hexagonal), obtained by heating precipitated ZnS in nitrogen at 1700–1800°C, as 1185 ± 6°C [5]. Since these two values are practically identical, [5] concluded that zincblende transforms before or on sublimation into wurtzite; i.e., the hexagonal modification would have to be considered the form stable at high temperatures. This is in agreement with observations of [4], who determined the transformation point zincblende → wurtzite as 1020 ± 5°C. The melting point of wurtzite under a pressure of 100–150 atm was found as 1800–1900°C [6].

Zincblende and wurtzite are the prototypes of the B3 [7, 8] and B4 [16] types of structure, respectively. The parameter of zincblende, both natural and synthetic, was determined by [9–15], the most probable value being $a = 5.4060$ A [14]. The lattice constants of wurtzite, of natural origin and prepared chemically, were reported by [16–20, 14, 15]. The most probable values are $a = 3.820$ A, $c = 6.260$ A, $c/a = 1.639$ [14]. Besides the latter there are three modifications of wurtzite with structures determined by [21]; see also [22].

1. R. Cussak, *Neues Jahr. Mineral.*, **1899**(1), 196.
2. K. Friedrich, *Metallurgie*, **5**, 1908, 114–118.
3. F. O. Doeltz and C. A. Graumann, *Metallurgie*, **3**, 1906, 442–443.
4. E. T. Allen and J. L. Crenshaw, *Z. anorg. Chem.*, **79**, 1912, 127–146, 174–179.
5. W. Biltz, *Z. anorg. Chem.*, **59**, 1908, 277–278.
6. E. Tiede and A. Schleede, *Ber. deut. chem. Ges.*, **53**, 1920, 1719–1720.
7. W. H. Bragg and W. L. Bragg, *Proc. Roy. Soc. (London)*, **A89**, 1913, 248, 277, 468.
8. P. P. Ewald, *Ann. Physik*, **44**, 1914, 257–282.
9. W. Gerlach, *Physik. Z.*, **23**, 1922, 114–120.
10. W. M. Lehmann, *Z. Krist.*, **60**, 1924, 379–413.
11. F. Rinne, *Z. Krist.*, **59**, 1924, 230–248.
12. W. Hartwig, *Sitzber. preuss. Akad. Wiss.*, **10**, 1926, 79–80.
13. V. M. Goldschmidt; see *Strukturbericht*, **1**, 1913-1928, 129.

14. H. E. Swanson and R. K. Fuyat, *Natl. Bur. Standards (U.S.), Circ.* 539, II, 1953, pp. 14, 16.
15. G. Kullerud, *Norsk. Geol. Tidsskr.*, **32,** 1953, 61–147.
16. W. L. Bragg, *Phil. Mag.*, **39,** 1920, 647.
17. G. Aminov, *Z. Krist.*, **58,** 1923, 203–219.
18. F. Ulrich and W. Zachariasen, *Z. Krist.*, **62,** 1925, 260–273.
19. M. L. Fuller, *Phil. Mag.*, **8,** 1929, 658–664.
20. R. M. Thompson, *Am. Mineralogist*, **35,** 1950, 451–455.
21. C. Frondel and C. Palache, *Science*, **107,** 1948, 602.
22. H. Ehrenberg, *Neues Jahr. Mineral., Geol., Paläont. Beil.*, **64,** 1931, 397–422.

$\bar{1}.5460$
0.4540

S-Zr Sulfur-Zirconium

On the basis of tensimetric and X-ray analysis, [1] claimed the existence of the following phases: a subsulfide in the $ZrS_{0.20-0.33}$ interval, a "0.75 type" sulfide in the $ZrS_{0.5-0.9}$ interval (Zr_4S_3?), Zr_2S_3, ZrS_2, and ZrS_3. In contrast to this, [2] identified by X-ray investigation "a subsulfide of undetermined structure and composition besides the ZrS and ZrS_2 phases which both have narrow homogeneity ranges." ZrS is tetragonal and isotypic with AgZr [2], and ZrS_2 has the structure of the CdI_2 (C6) type, $a = 3.69$ A, $c = 5.86$ A, $c/a = 1.58_8$ [3]. ZrS_2 has a melting point of about 1550°C [4]. As regards ZrS_3, see also [5].

1. E. F. Strotzer, W. Biltz, and K. Meisel, *Z. anorg. Chem.*, **242,** 1939, 249–271.
2. G. Hägg and N. Schönberg, *Arkiv Kemi*, **7,** 1954, 371–380.
3. A. E. van Arkel, *Physica*, **4,** 1924, 286–301.
4. R. Vogel and A. Hartung, *Arch. Eisenhüttenw.*, **15,** 1941-1942, 414–415.
5. R. W. M. D'Eye, P. G. Sellman, and J. R. Murray, *J. Chem. Soc.*, **1952,** 2558.

0.1881
$\bar{1}.8119$

Sb-Se Antimony-Selenium

The phase relationships were already substantially established by early investigations, using thermal [1–3] and microscopic methods [4, 5]. The lower portion of Fig. 633 shows the liquidus and solidus curves reported by [2]; no tabulated data were given. The existence of the phase Sb_2Se_3 (49.31 wt. % Se), long since known, was confirmed, and a range of liquid immiscibility, with the monotectic temperature at 566°C, was found. Results claimed by [6, 7] proved to be erroneous [8, 9].

The upper and lower portions of Fig. 633 represent the phase diagrams determined by [8] and [9], respectively, by means of both thermal and microscopic investigations. The diagram of [9] differs from those due to [2] and [8] in that it does not show a miscibility gap in the liquid state between Sb and Sb_2Se_3 [10]. Since the difference in density of the two melts is small, formation of distinct layers takes place only on slow cooling [8]. In addition, temperatures in the diagram of [9] are too low throughout; the melting point of Sb was given as 594 instead of 630.5°C. The diagram in the upper portion of Fig. 633 should be considered the most reliable.

Emf measurements were claimed to indicate, besides Sb_2Se_3, the existence of SbSe [11]. This phase is incompatible with the diagrams of [8] and [9].

Sb_2Se_3 is isotypic with Sb_2S_3 ($D5_8$ type), $a = 11.5_8$ A, $b = 11.6_8$ A, $c = 3.9_8$ A (or kX?) [12]. [13] refined the structure of Sb_2Se_3 and reported the lattice parameters $a = 11.62 \pm 1$ A, $b = 11.77 \pm 1$ A, $c = 3.962 \pm 7$ A.

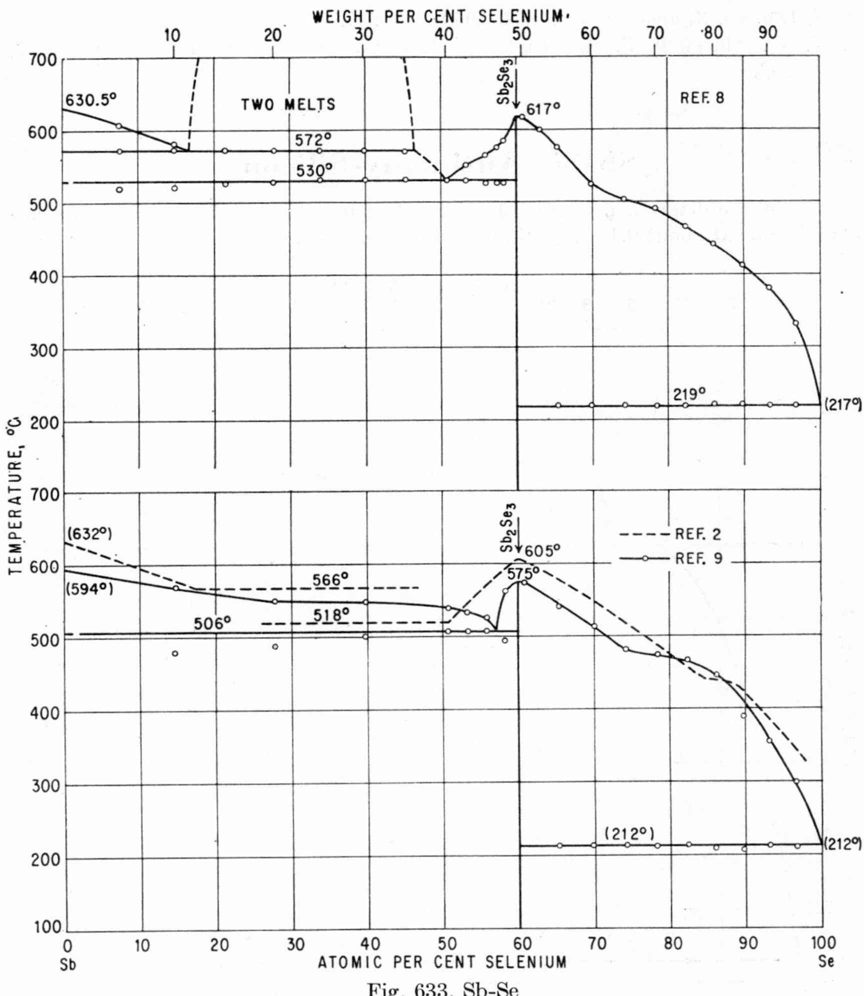

Fig. 633. Sb-Se

1. H. Pélabon, *J. chim. phys.*, **2,** 1904, 321–339.
2. H. Pélabon, *Compt. rend.*, **142,** 1906, 207–210.
3. H. Pélabon, *Ann. chim. et phys.*, **17,** 1909, 535.
4. H. Pélabon, *Compt. rend.*, **153,** 1911, 343–346.
5. H. Pélabon, *Ann. chim. et phys.*, **13,** 1920, 126.
6. P. Chrétien, *Compt. rend.*, **142,** 1906, 1339–1341.
7. P. Chrétien, *Compt. rend.*, **142,** 1906, 1412–1413.
8. N. Parravano, *Gazz. chim. ital.*, **43**(1), 1913, 210–220.
9. M. Chikashige and M. Fujita, *Mem. Coll. Sci., Kyoto Univ.*, **2,** 1917, 233–237.
10. Chikashige and Fujita studied the system without having knowledge of the paper by Parravano.
11. R. Kremann and R. Wittek, *Z. Metallkunde*, **13,** 1921, 90–97.

12. E. Dönges, *Z. anorg. Chem.*, **263,** 1950, 289–291.
13. N. W. Tideswell, F. H. Kruse, and J. D. McCullough, *Acta Cryst.*, **10,** 1957, 99–102.

0.6370
$\bar{1}$.3630

Sb-Si Antimony-Silicon

The phase diagram presented in Fig. 634 is due to [1]. The eutectic was calculated to lie at about 0.1 at. % Si and a temperature 0.4°C below the melting point

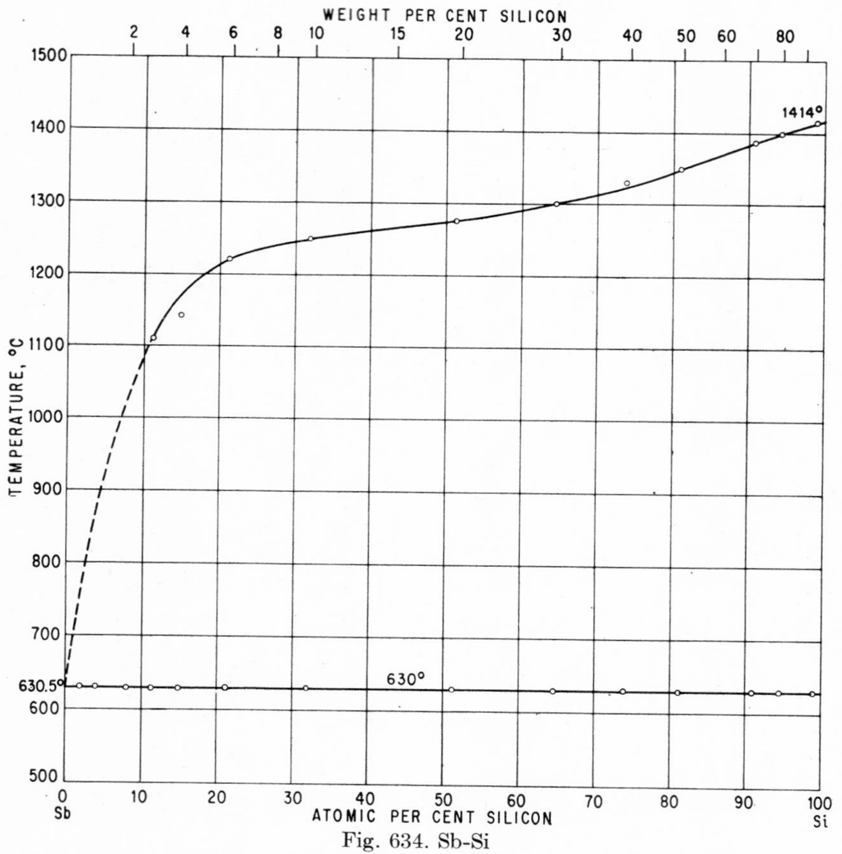

Fig. 634. Sb-Si

of Sb [2]. Lattice-parameter determinations, using alloys with 40, 50, and 80 wt. % Si, showed that Si is insoluble in solid Sb; the solubility of Sb in Si was estimated to be smaller than 0.5 at. % Sb [3].

1. R. S. Williams, *Z. anorg. Chem.*, **55,** 1907, 1921. The silicon used was 98% pure.
2. C. D. Thurmond, *J. Phys. Chem.*, **57,** 1953, 827–830.
3. E. R. Jette and E. B. Gebert, *J. Phys. Chem.*, **1,** 1933, 753–755.

0.0110
$\bar{1}$.9890

Sb-Sn Antimony-Tin

The constitution of this system was the subject of numerous investigations covering the entire composition range [1–10] and parts thereof [11–15]. As a result, the phase relations and boundaries appear to be well established, with the exception of the boundary of the (Sb) phase. The diagram shown in Fig. 635 is generally accepted

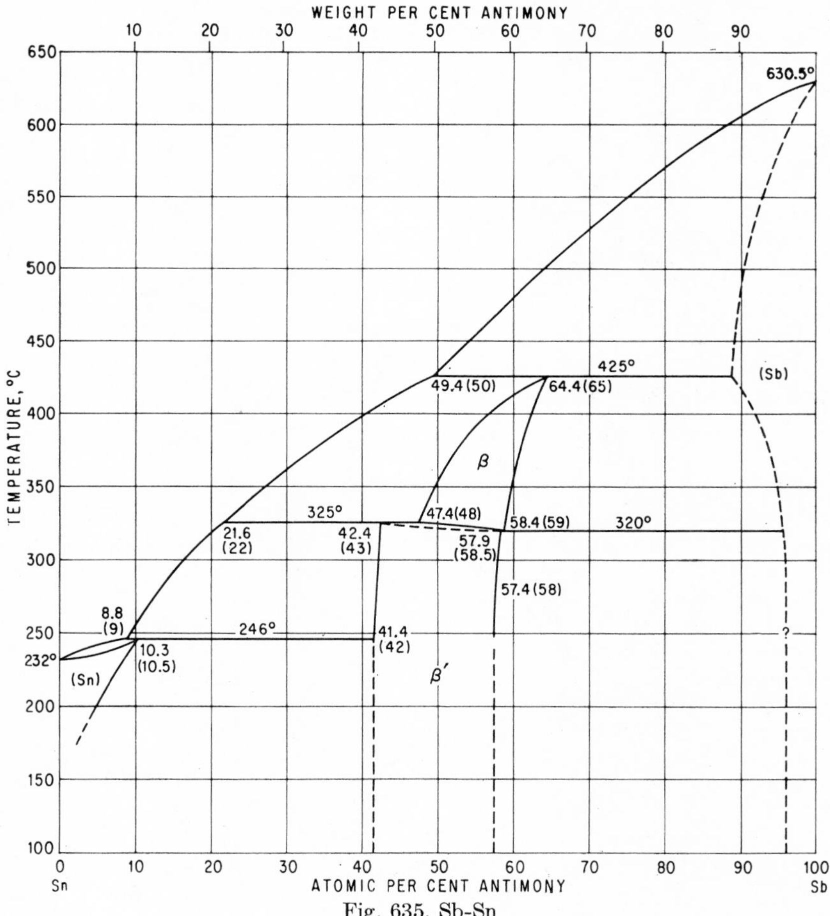

Fig. 635. Sb-Sn

at present. It is chiefly based on the work of [9]. As to the data published prior to 1936, reference is made to the review of [16] which includes the diagrams according to [2–4, 7–9].

Solidification. Liquidus temperatures covering the whole system [2, 4, 7–10] lie in a scattering band of, in the main, only 10°C width. [11, 14] investigated the effect of Sb additions on the melting point of Sn, up to 2.6 and 9.8 at. % Sb, respectively. The temperatures of the three peritectic reactions were found as follows: (*a*) melt + β' ⇄ (Sn), 243 [2–4], 244 [5], 246 [14, 8, 9, 15, 10], and 247°C [7]; (*b*) melt + $\beta \rightleftarrows \beta'$, 310 [2], 319 [3], 320 [8], 325 [9], and 326°C [10]; and (*c*) melt + (Sb) $\rightleftarrows$ β, 420 [4], 422 [5], 423 [8], 425 [7, 9], and 430°C [2, 3, 10]. The compositions of the peritectic melts vary between 8 and 9 wt. (7.8 and 8.8 at.) % Sb, between about 21.5 and 22.5 wt. (21.1 and 22.1 at.) % Sb, and between about 49 and 54 wt. (48.4 and 53.4 at.) % Sb, respectively.

The Intermediate Phase. In the early stage of the study of the constitution, it was believed that the intermediate phase was one of the fixed (or nearly fixed) composition SnSb (50.64 wt. % Sb) [17–19, 4, 5]. Other investigators assumed the existence of two intermediate phases, such as Sn_2Sb and SnSb [20], SnSb and Sn_3Sb_4 (or Sn_4Sb_5) [2], or Sn_3Sb_2 and SnSb [13]. It was shown by [3] and later confirmed by [7, 8, 9, 21, 10] that the β phase has a rather wide homogeneity range which shifts to higher Sb contents as the temperature rises [3, 7, 9, 10]. According to the crystal structure (see below), the ideal composition of β would be SnSb.

[3] first gave conclusive evidence of the existence of a transformation of β. This was verified by [8–10] but overlooked by other investigators [4, 5, 7]. Figure 635 shows the phase boundaries and transformation temperatures of the β phase according to [9]; they are based on thermal, thermoresistometric, and micrographic data. Results of [10] are in good agreement with these. The transformation was suggested to be of the order-disorder type [21].

The Solid Solubility of Sb in Sn. The crystallization of solid solutions of Sb in Sn, first recognized by [12], was confirmed by [17, 18, 2] and all later investigators. The only reliable determination of the solubility boundary is that by [15]. According to this micrographic work, the solubility decreases from 10.5 wt. (10.3 at.) % Sb at 246°C to 7.9 (7.7) at 225°C, and 4 wt. (3.9 at.) % Sb at 190°C. At some lower temperature the solubility decreases to between 2.2 and 2.8 wt. (2.15 and 2.74 at.) % Sb [22].

The Solid Solubility of Sn in Sb was estimated to be of the order of 10 at. % Sn [4, 13, 7, 8, 21]. The solubility boundary outlined by [3] does not represent equilibrium conditions. The same appears to be true for the solidus curve (Fig. 635), which is based on a few thermoresistometric data [9], and the upper portion of the solvus curve (400–425°C), obtained by some cursory microscopic work [9].

Crystal Structures. The structure of the β' phase was reported to be of the NaCl (B1) type [23–28], with a = 6.136 A at 45 at. % Sb and a = 6.127 A at 52 at. % Sb [25]. According to [21], the structure is not cubic but rhombohedral, with a = 6.226 A, α = 89.38° at 52.4 at. % Sb; a = 6.129 A, α = 89.70° at the Sn-rich boundary; and a = 6.150 A, α = 89.18° at the low-Sn boundary. Lattice dimensions of the (Sn) phase were determined by [25, 21] and especially [29], and those of the (Sb) phase by [25, 21].

1. Roland-Gosselin, reported by H. Gautier, *Bull. soc. encour. ind. natl.*, **1,** 1896, 1293–1318; Société d'encouragement pour l'industrie nationale, Paris, Commission des alliages, "Contribution a l'étude des alliages," pp. 93–118, Typ. Chamerot et Renouard, Paris, 1901.
2. W. Reinders, *Z. anorg. Chem.*, **25,** 1900, 113–125.
3. F. E. Gallagher, *J. Phys. Chem.*, **10,** 1906, 93–98.
4. R. S. Williams, *Z. anorg. Chem.*, **55,** 1907, 12–19.
5. R. Loebe, *Metallurgie*, **8,** 1911, 9–10.
6. S. F. Zemczuzny, unpublished work, reported by [13].
7. W. Broniewski and L. Sliwowski, *Rev. mét.*, **25,** 1928, 312–321.
8. M. Tasaki, *Mem. Coll. Sci., Kyoto Univ.*, **12,** 1929, 229–230.
9. K. Iwasé, N. Aoki, and A. Osawa, *Science Repts. Tôhoku Univ.*, **20,** 1931, 353–368; *Kinzoku-no-Kenkyu*, **7,** 1930, 147–160.

10. R. Blondel, *Publs. sci. et tech. ministère Air (France)*, **1936**(89), 10–14; R. Blondel and P. Laffitte, *Compt. rend.*, **200**, 1935, 1472–1474.
11. C. T. Heycock and F. H. Neville, *J. Chem. Soc.*, **57**, 1890, 387.
12. A. Van Bijlert, *Z. physik. Chem.*, **8**, 1891, 357–362.
13. N. Konstantinow and W. Smirnow, *Intern. Z. Metallog.*, **2**, 1912, 152–171.
14. L. J. Gurevich and J. S. Hromatko, *Trans. AIME*, **64**, 1921, 233–235.
15. D. Hanson and W. T. Pell-Walpole, *J. Inst. Metals*, **58**, 1936, 299–308.
16. M. Hansen, "Der Aufbau der Zweistofflegierungen," pp. 1044–1051, Springer-Verlag OHG, Berlin, 1936.
17. J. E. Stead, *J. Soc. Chem. Ind.*, **16**, 1897, 204–206; **17**, 1898, 1111–1112; see also *J. Inst. Metals*, **22**, 1919, 127–130.
18. G. Charpy, *Bull. soc. encour. ind. natl.*, **2**, 1897, 407; Société d'encouragement pour l'industrie nationale, Paris, Commission des alliages, "Contribution à l'étude des alliages," p. 144, Typ. Chamerot et Renouard, Paris, 1901.
19. N. A. Pushin, *Zhur. Fiz.-Khim. Obshchestva*, **39**, 1906, 549.
20. H. Behrens and H. Baucke, *Verslag Koninkl. Akad. Wetenschap., Amsterdam*, **7**, 1898-1899, 58.
21. G. Hägg and A. G. Hybinette, *Phil. Mag.*, **20**, 1935, 913–929.
22. F. J. Dunkerley, H. B. Hunter, and F. G. Stone, *Trans. AIME*, **185**, 1949, 1005–1016.
23. V. M. Goldschmidt, *Skrifter Norske Videnskaps-Akad. Oslo, I., Math.-nat. Kl.*, **1927**, no. 8.
24. A. Osawa, *Nature*, **124**, 1929, 14; see also *Strukturbericht*, **1**, 1913-1928, 601.
25. E. G. Bowen and W. Morris-Jones, *Phil. Mag.*, **12**, 1931, 441–462.
26. H. S. van Klooster and M. D. Debacher, *Metals & Alloys*, **4**, 1933, 23–24.
27. M. v. Schwarz and O. Summa, *Z. Metallkunde*, **25**, 1933, 92–97.
28. G. S. Farnham, *J. Inst. Metals*, **55**, 1934, 69.
29. J. A. Lee and G. V. Raynor, *Proc. Phys. Soc.*, **B67**, 1954, 737–747.

$\bar{1}.9796$
0.0204

Sb-Te Antimony-Tellurium

Results of thermal studies [1–5] are in substantial agreement (Table 45; Fig. 636). Since [1] were unable to detect a eutectic crystallization in the range 0–60 at. % Te, they concluded that Sb and Sb_2Te_3 formed a continuous series of solid solutions. Thermal and microscopic evidence showed this not to be the case [3, 4].

Table 45. Characteristic Temperatures in °C and Compositions in At. % Te (Wt. % in Parentheses) of the System Sb-Te

	Ref. 1	Ref. 2	Ref. 3	Ref. 4	Ref. 5*
Melting point of Sb......	624°C	632°C	627°C	630°C	
Sb-Sb_2Te_3 eutectic.......	~550°C	....	541°C	540°C	(~540°C)
	~29	~29	29	29	~31
	(~30)	(~30)	(30)	(30)	(~32)
Melting point of Sb_2Te_3..	629°C	(~606°C)	~620°C	622°C	(~630°C)
Sb_2Te_3-Te eutectic.......	421°C	425°C	420°C	424°C	(415–420°C)
	~86.5	90	90	89	~87.5
	(~87)	(90.5)	(90.5)	(89.5)	(~88)
Melting point of Te......	446°C	452°C	442°C	442°C	

* No tabulated data were given.

Solubilities in the solid state have not been investigated. There is no doubt, however, that the phase Sb_2Te_3 (61.12 wt. % Te) is of variable composition [4, 5]. According to measurements of the magnetic susceptibility, the homogeneity range may extend from approximately 56 to 63 at. % Te [5]; see also [6]. Also, there is a certain solid solubility of Te in Sb, as indicated by electrical-conductivity [6] and magnetic-susceptibility data [7].

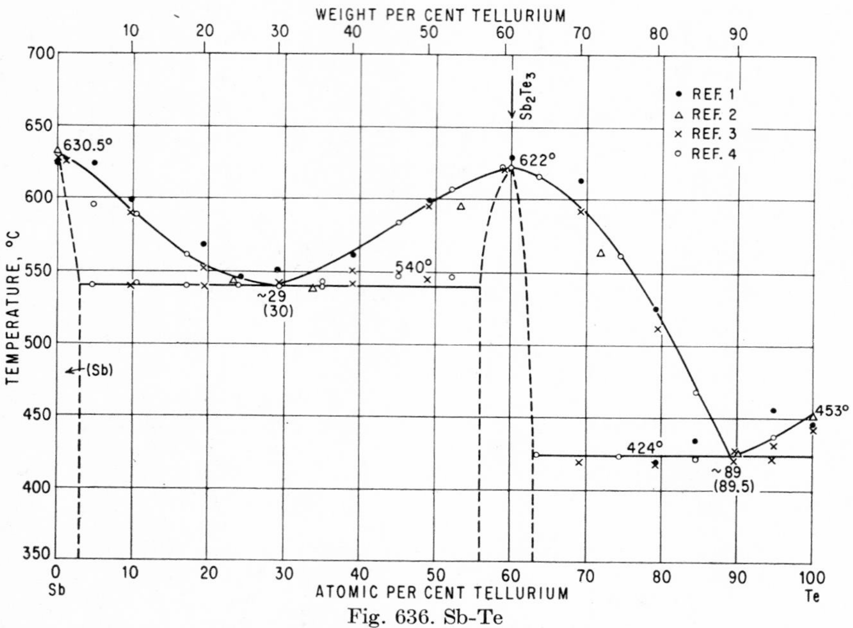

Fig. 636. Sb-Te

Crystal Structure. [8] briefly noted that Sb_2Te_3 is rhombohedral and isotypic with Bi_2Te_3 (C33 type). According to [9], this corresponds to a structure with the hexagonal axes $a = 4.2_5$ A, $c = 6.0_7$ A, $c/a = 1.4_3$. The actual structure, however, is a superstructure, with $a = 4.2_5$ A, $c = 30.4$ A, $c/a = 7.1_5$.

1. H. Fay and H. E. Ashley, *Am. Chem. J.*, **27,** 1902, 95–105.
2. H. Pélabon, *Compt. rend.*, **142,** 1906, 207–210.
3. Y. Kimata, *Mem. Coll. Sci., Kyoto Univ.*, **1,** 1915, 115–118.
4. N. S. Konstantinow and V. I. Smirnow, *Izvest. St. Petersburg Politekh. Inst. Imp. Petra Velik*, **23,** 1915, 713–720.
5. H. Endo, *Science Repts. Tôhoku Univ.*, **16,** 1927, 213–215; see also K. Honda and H. Endo, *J. Inst. Metals*, **37,** 1927, 42.
6. W. Haken, *Ann. Physik*, **32,** 1910, 312–316.
7. K. Honda and T. Soné, *Science Repts. Tôhoku Univ.*, **2,** 1913, 9–10.
8. L. S. Ramsdell, *Am. Mineralogist*, **15,** 1930, 119.
9. E. Dönges, *Z. anorg. Chem.*, **265,** 1951, 56–61.

$\bar{1}.7198$
0.2802

Sb-Th Antimony-Thorium

According to X-ray analysis of alloys prepared by direct synthesis from the elements at temperatures around 1000°C, the following compounds exist in the Th-Sb system [1]: (*a*) ThSb (65.59 wt. % Th), with the NaCl (B1) type of structure, a = 6.318 A; (*b*) Th_3Sb_4 (58.84 wt. % Th), with the b.c.c. Th_3P_4 ($D7_3$) type of structure, a = 9.372 A; and (*c*) $ThSb_2$ (48.80 wt. % Th), with the tetragonal Cu_2Sb (C38) type of structure, a = 4.353 A, c = 9.172 A, c/a = 2.107.

1. R. Ferro, *Acta Cryst.*, **9**, 1956, 817–818.

0.4052
$\bar{1}.5948$

Sb-Ti Antimony-Titanium

The constitution was studied in the range 40–100 wt. (21–100 at.) % Sb, using 35 alloys prepared by powder-metallurgy methods (sintered at 1300–1600°C and

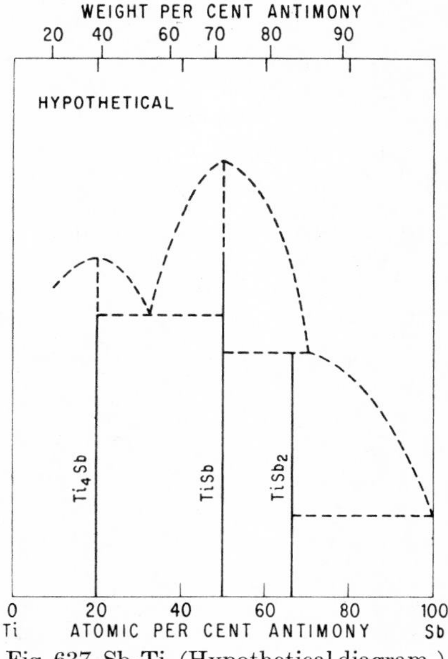

Fig. 637. Sb-Ti. (Hypothetical diagram.)

subsequently annealed for 60 hr at 600°C) [1]. The phases Ti_4Sb (38.86 wt. % Sb), TiSb (71.77 wt. % Sb), and $TiSb_2$ (83.56 wt. % Sb), which have restricted homogeneity ranges, were identified by metallographic and X-ray investigations. Ti_4Sb and TiSb melt congruently (melting points were not determined), whereas $TiSb_2$ is formed by the peritectic reaction TiSb + melt → $TiSb_2$. Ti_4Sb and TiSb form a eutectic. The melting point of Sb is raised to 680°C by addition of 0.5 wt. (1.3 at.) % Ti. Ti appears to be insoluble in solid Sb as the lattice parameters of Sb are not changed by Ti additions. On the basis of the above information, the *hypothetical* phase diagram in Fig. 637 can be drawn.

Rolled samples of arc-cast alloys with 0.06, 0.12, and 0.14 wt. % Sb (prepared using magnesium-reduced Ti) were quenched from various temperatures between 790 and 950°C and examined metallographically [2]. They consisted of β phase at 900°C, mostly α phase at 870°C, and all α at 840°C and lower temperatures.

Crystal Structures. Ti_4Sb has a hexagonal superlattice with a = 5.958 A, c = 4.808 A, c/a = 0.807 [3]. This phase apparently has the ideal, although not realizable, composition Ti_3Sb (45.87 wt. % Sb), because its structure is of the Ni_3Sn ($D0_{19}$) type.

TiSb is isotypic with NiAs (B8 type), $a = 4.07$ A, $c = 6.30_6$ A, $c/a = 1.55$ [1], and $TiSb_2$ has the structure of the $CuAl_2$ (C16) type, $a = 6.66_6$ A, $c = 5.81_7$ A, $c/a = 0.87_3$ [1].

1. H. Nowotny and J. Pesl, *Monatsh. Chem.*, **82,** 1951, 336–343.
2. C. M. Craighead, O. W. Simmons, P. J. Maddex, C. T. Greenidge, and L. W. Eastwood, unpublished work; quoted from WADC Technical Report 53-41, 1953, p. 101.
3. H. Nowotny, R. Funk, and J. Pesl, *Monatsh. Chem.*, **82,** 1951, 513–525.

$\bar{1}.7750$
0.2250

Sb-Tl Antimony-Thallium

The thermal data points in Fig. 638 were reported by [1] and interpreted as indicating (*a*) a eutectic at about 20 wt. (29.56 at.) % Sb, 195°C, consisting of Sb and an α-Tl solid solution with approximately 15 wt. (23 at.) % Sb, and (*b*) the peritectoid formation at 187°C of a phase having the probable composition Tl_3Sb, according to the reaction α-Tl solid solution + Sb → Tl_3Sb [2]. The thermal effects at 226°C in the range 0–19.5 at. % Sb were believed to correspond to the $\alpha \rightleftarrows \beta$ transformation of Tl. Later work [3, 4] confirmed the existence of an intermediate phase but showed the phase equilibria in the Tl-rich alloys to be entirely different. Since a systematic study of these alloys after equilibration is still lacking, the phase relations shown are not yet fully established.

[3] investigated the microstructure and X-ray structure of alloys with 25.0 (I), 28.1 (II), 31.25 (III), and 37.5 at. % Sb (IV), and higher Sb contents (treatment not stated). Alloy I consisted of a solid solution of Sb in β-Tl and alloy II of a mixture of β_{Tl} plus an intermediate phase having the approximate composition $Tl_{11}Sb_5 = \gamma$ (31.25 at. % Sb). Alloy III proved to be single-phase (γ), and alloy IV consisted of γ plus small amounts of Sb.

In contrast to this, metallographic and X-ray studies [4] of alloys covering the range 5–20 wt. (8.1–29.6 at.) % Sb gave results which allowed, in principle, the phase relations exhibited in Fig. 638 to be drawn [5]. The intermediate phase γ (very likely formed peritectically) was found to correspond to the composition Tl_7Sb_2 (22.22 at. % Sb) [6]. Also, it was reported that β-Tl formed a series of solid solutions, apparently stabilized down to ordinary temperature at about 10–11 at. % Sb [7, 4]. It should be noted, however, that β-Tl was believed to be f.c.c. [3, 7, 4, 8] rather than b.c.c. [9].

Recently, [10] fixed the homogeneity ranges of the phases α_{Tl}, β_{Tl}, and γ at 180°C (after annealing for 300–400 hr) by determining the value $r = R_x/R_0$, i.e., the ratio of electrical resistivity at −196°C or −253°C, respectively, to the resistivity at 0°C (R_0) [11]. Thus the phase limits at 180°C, indicated by crosses in Fig. 638, were found: 13.5, 14.6, 15.5, 22.2, and 23.0 at. % Sb.

The eutectic point, found by [1, 4] to lie at 29.6 at. % Sb, was reported by [12] to be located at 31.4 at. % Sb, 195°C. The nature of the transformation at 187°C [1] is still unknown.

Crystal Structures. The γ phase was claimed to be of the CsCl (B2) type [13, 3]. However, [4] proved that [3] observed only the strongest interferences of this phase. They showed that γ has a complicated cubic structure ($L2_2$ type) with 54 atoms per unit cell and $a = 11.61$ A; see also [14]. The f.c.c. structure observed in alloys with 10 [7], 25 [3], 8.1 [4, 8], and 11.2 at. % Sb [4] apparently represents the structure of a metastable phase resulting from the $\beta \rightarrow \alpha$ transformation of the Tl-rich solid solution. β-Tl is b.c.c. [9].

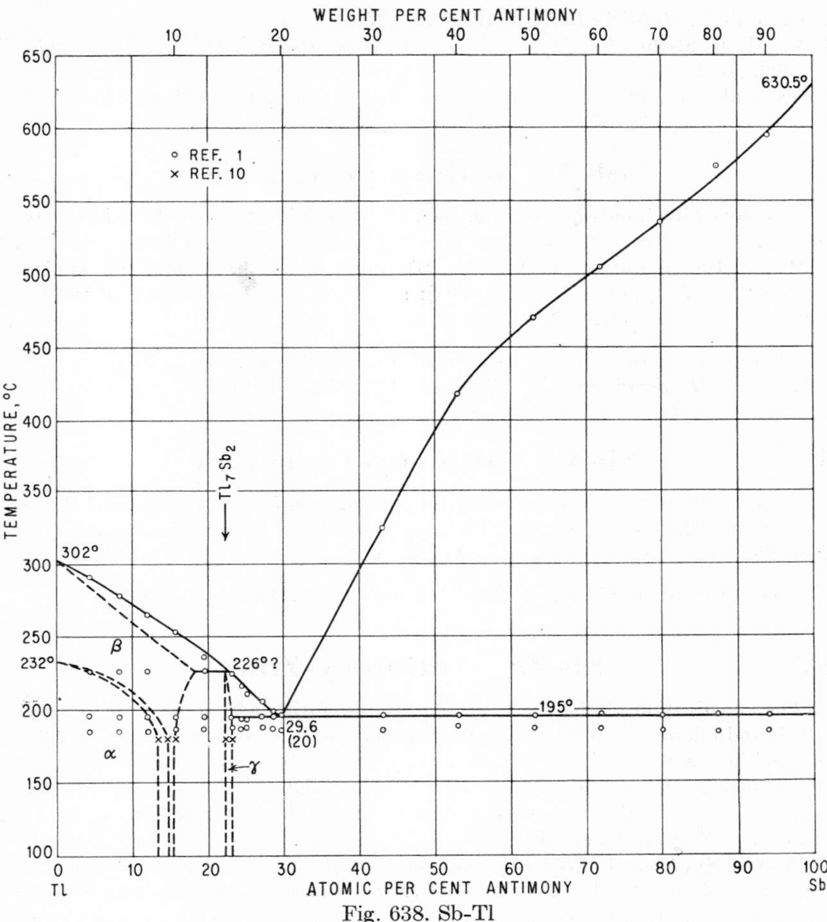

Fig. 638. Sb-Tl

1. R. S. Williams, *Z. anorg. Chem.*, **50,** 1906, 127–132.
2. The faint thermal effects at the eutectic temperature between 0 and 22 at. % Sb are caused by lack of diffusion on solidification.
3. T. Barth, *Z. physik. Chem.*, **127,** 1927, 113–120.
4. E. Persson and A. Westgren, *Z. physik. Chem.*, **136,** 1928, 208–214.
5. M. Hansen, "Der Aufbau der Zweistofflegierungen," pp. 1054–1057, Springer-Verlag OHG, Berlin, 1936.
6. The compositions $Tl_{11}Sb_5$ [3] and Tl_7Sb_2 [4] differ by as much as 9 at. %. The former composition nearly coincides with the eutectic point.
7. V. M. Goldschmidt, *Z. physik. Chem.*, **133,** 1928, 410.
8. S. Sekito, *Z. Krist.*, **74,** 1930, 189.
9. H. Lipson and A. R. Stokes, *Nature,* **148,** 1941, 437.
10. E. Kurzyniec, *Bull. intern. acad. polon. sci., Classe sci. math. nat.,* **A1951,** 159–138.
11. 30 alloys, covering the range 1–35 at. % Sb, were used.

12. O. Hájiček, *Hutnické Listy*, **3**, 1948, 265–270.
13. V. M. Goldschmidt, *Skrifter Norske Videnskaps-Akad. Oslo, I, Math. nat. Kl.*, 1926, no. 2.
14. F. R. Morral and A. Westgren, *Svensk Kem. Tidskr.*, **46**, 1934, 153–156.

$\bar{1}.7088$
0.2912

Sb-U Antimony-Uranium

The following phases were identified by determining their crystal structures: USb (33.84 wt. % Sb), U_3Sb_4 [57.14 at. (40.54 wt.) % Sb], and USb_2 (50.57 wt. % Sb) [1, 2]. USb is isotypic with NaCl (B1 type), a = 6.191 A [1]; U_3Sb_4 is isotypic with Th_3P_4 ($D7_3$ type), a = 9.095 A [1]; and USb_2 has the structure of the Cu_2Sb (C38) type, with a = 4.272 A, c = 8.741 A, c/a = 2.044 [2].

1. R. Ferro, *Atti accad. nazl. Lincei, Rend.*, **13**, 1952, 53–61.
2. R. Ferro, *Atti accad. nazl. Lincei, Rend.*, **13**, 1952, 151–157.

0.3784
$\bar{1}.6216$

Sb-V Antimony-Vanadium

X-ray study of five alloys covering the range 25–66.7 at. % Sb revealed the existence of VSb_2 (82.70 wt. % Sb), besides other unidentified intermediate phases. VSb_2 is isotypic with $CuAl_2$ (C16 type), a = 6.55_5 A, c = 5.63_5 A, c/a = 0.859_6 [1].

1. H. Nowotny, R. Funk, and J. Pesl, *Monatsh. Chem.*, **82**, 1951, 519–522.

0.2701
$\bar{1}.7299$

Sb-Zn Antimony-Zinc

The constitution of the entire system was investigated by [1–3], using thermal analysis and microexamination [4]. Partial diagrams showing the results of [2, 3] are presented in Fig. 639. The phase diagram due to [5], based on a careful reinvestigation by means of thermal, thermoresistometric, dilatometric, and metallographic methods, is now generally accepted (Fig. 640). Individual thermal data for the range 9–97.5 at. % Zn were reported by [6].

Partial System Sb-Zn_3Sb_2. According to [1], the intermediate phases of fixed composition, ZnSb (34.94 wt. % Zn) and Zn_3Sb_2 (44.61 wt. % Zn), have congruent melting points of 544 and 560°C, respectively, and form a eutectic at about 52.6 at. (37.4 wt.) % Zn, 539°C. This was disproved by [2, 3, 5], who established that ZnSb forms peritectically at a temperature found as 537°C on cooling [2] or 546°C on heating [3, 5]. [5] detected that an additional phase existed between ZnSb and Zn_3Sb_2, viz., Zn_4Sb_3 [57.14 at. (41.72 wt.) % Zn]. This phase forms peritectically at 563°C (Fig. 640), only 3°C below the maximum melting point of Zn_3Sb_2. Its existence was confirmed by [7, 8].

A comprehensive thermal analysis by [2] (85 melts) established the existence of a stable and an unstable system. As indicated in Fig. 639, quite different liquidus temperatures (between 32 and 46.3 at. % Zn) and solidus temperatures (between 0 and 59.1 at. % Zn) were found, according to whether or not the solidifying melt was inoculated by adding crystals of ZnSb. Whereas [2] believed the unstable phase below 46.3 at. % Zn to be Zn_3Sb_2, the diagram of [5] suggests this phase to be Zn_4Sb_3 (Fig. 640).

Without inoculation, primary separation of Sb (unstable between 32 and 38 at. % Zn) or Zn_4Sb_3 (unstable between 38 and 46.3 at. % Zn) was followed by crystallization of the unstable eutectic Sb-Zn_4Sb_3 at 482°C. Below this temperature, at 390–

480°C, a sudden rise in temperature (average 40°C, maximum 80°C) occurred, resulting from the tendency to eliminate the unstable equilibrium, immediately followed by a fall in temperature.

With inoculation, the stable phase ZnSb crystallized primarily between 32 and 46.3 at. % Zn, followed by the solidification of the stable Sb-ZnSb eutectic at 505°C. Above 46.3 at. % Zn the primary crystallization of Zn_4Sb_3 was followed by the peritectic reaction: melt + $Zn_4Sb_3 \rightarrow$ ZnSb at 537°C. The formation of the unstable system, therefore, is due to the fact that the primary crystallization of ZnSb (between 32 and 46.3 at. % Zn) as well as the peritectic reaction at 537°C (between 46.3 and about 56 at. % Zn) can be suppressed by supercooling [9].

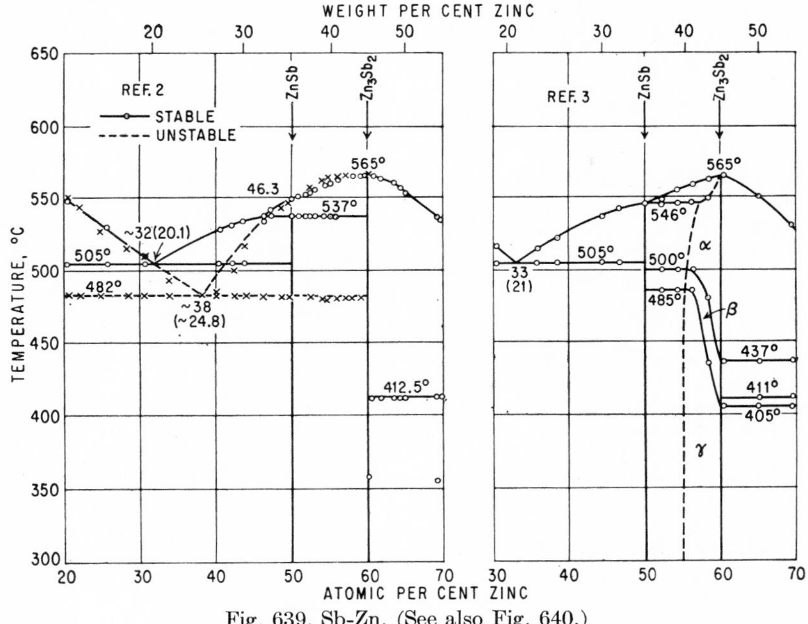

Fig. 639. Sb-Zn. (See also Fig. 640.)

[5] found much smaller differences between stable and unstable liquidus points on cooling; the transformation into the stable phase ZnSb already took place at 520–530°C, with an increase in temperature to 535–540°C. Unstable equilibrium was found by [5] only in alloys with less than 56.4 at. % Zn, whereas [2] believed the unstable eutectic horizontal at 482°C to extend to at least 59.1 at. % Zn. [5] concluded, therefore, that, in alloys above 56.4 at. % Zn, [2] had observed the $\gamma \rightarrow \beta$ transformation which occurs on cooling at 485°C.

The formation of the unstable equilibrium was not observed by [3], since heating curves of annealed specimens were taken. Otherwise, his results are in good agreement with those of [2] (Fig. 639). Two reactions in the solid state were detected. They were interpreted as transformations of the solid solution of Sb in Zn_3Sb_2: $\alpha \rightleftarrows \beta$ between 500 and 437°C and $\beta \rightleftarrows \gamma$ between 485 and 405°C; see, however, Fig. 640.

The work of [5] (Fig. 640) indicated that Zn_4Sb_3 existed in two modifications (β and γ) and Zn_3Sb_2 in three modifications (ϵ, ζ, and η), all of which were established by microscopic investigation. They were found to be of variable composition. In addition, the γ form of Zn_4Sb_3 undergoes a transformation at 527–530°C, without change in microstructure. The same holds for the transformation in ZnSb at about 300°C, found dilatometrically. β-Zn_4Sb_3 has another transformation point in the range −10 to −20°C [8]. A transformation of Zn_3Sb_2 at about 358°C, found by [1, 2] in the whole region Zn_3Sb_2-Zn, could not be observed by [3, 5]. It has been omitted in Fig. 640.

It was claimed by [10], on the basis of X-ray studies and density measurements, that Zn_3Sb_2 (ϵ) decomposed into ZnSb and Zn. Although no other explanation can be given for the evidence reported, Zn_3Sb_2 could decompose only into Zn_4Sb_3 and Zn. Additional work is necessary to verify the conclusion of [10].

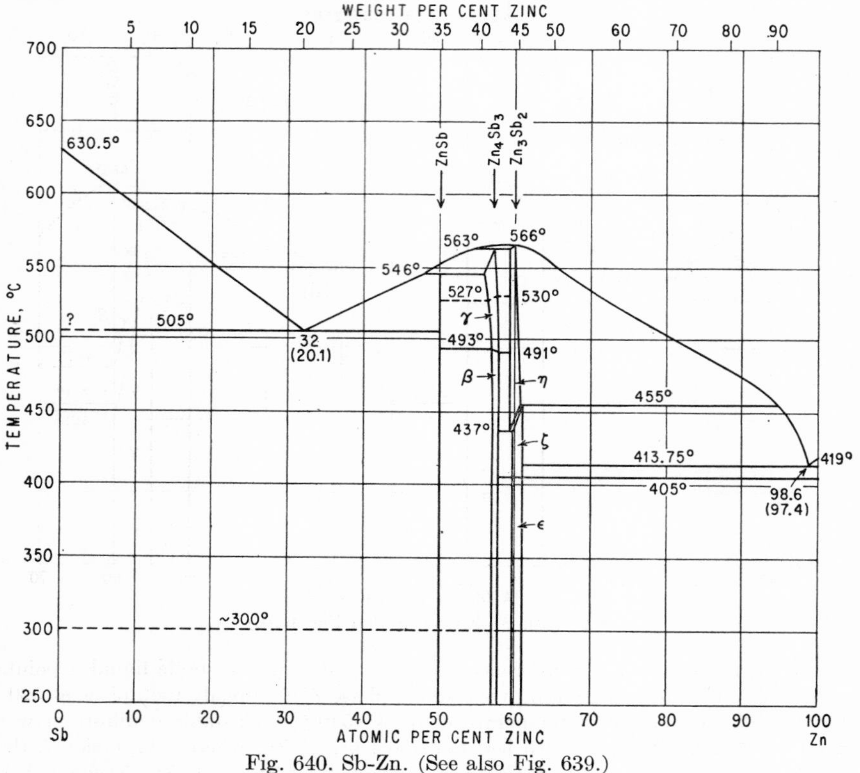

Fig. 640. Sb-Zn. (See also Fig. 639.)

By studies of various physical properties, the existence of ZnSb [11–14] and Zn_3Sb_2 [12, 13] was confirmed. Also, curves of the magnetic susceptibility [13], electrical resistivity [15], and density [16] of liquid alloys show discontinuities at these compositions.

According to measurements of the magnetic susceptibility [12, 13], the solid solubility of Zn in Sb can only be very small. However, [7]—on the basis of emf-measurement data and the assumption that the solid solution can be considered ideal—computed solidus data between 592 and 505°C and arrived at a solubility of 6.7 at. % Zn (!) at the eutectic temperature (505°C).

Partial System Zn_3Sb_2-Zn. The liquidus curves determined by [1, 2, 3, 5] are in substantial agreement, the difference being largest (up to 20°C) in the range 90–97.5

at.% Zn. Also, liquidus points were reported for the ranges 88–100 at. % Zn [17] and 97.5–100 at. % Zn [18].

The eutectic composition was found as 98.8 at. (97.8 wt.) % Zn [18, 2, 17] or 98.6 at. (97.4 wt.) % Zn [19], and the eutectic temperature as 412–413°C [18, 5, 17] or 413.75°C [19].

The boiling points of the melts with 17, 38, 65, and 88 at. % Zn were determined to be 1273, 1090, 973, and 925°C, respectively [20].

Crystal Structure. ZnSb was reported to be orthorhombic, with eight atoms per unit cell and a = 6.18 A, b = 8.29 A, c = 3.95 A [10]. According to [21], however, these results are faulty; the structure may rather be regarded as a strongly deformed diamond structure, with 16 atoms per unit cell and a = 6.218 A, b = 7.741 A, c = 8.115 A [22]. No other representative of this structure is so far known.

1. K. Mönkemeyer, *Z. anorg. Chem.*, **43,** 1905, 182–196.
2. S. F. Zemczuzny, *Z. anorg. Chem.*, **49,** 1906, 384–399; *Zhur. Russ. Fiz.-Khim. Obshchestva*, **38,** 1906, 17–32.
3. B. E. Curry, *J. Phys. Chem.*, **13,** 1909, 589–597.
4. Previously, the liquidus curve had been outlined by Roland-Gosselin; see H. Gautier, *Bull. soc. encour. ind. natl.*, **1,** 1896, 1311; Société d'encouragement pour l'industrie nationale, Paris, Commission des alliages, "Contribution a l'étude des alliages," p. 111, Typ. Chamerot et Renouard, Paris, 1901.
5. T. Takei, *Science Repts. Tôhoku Univ.*, **16,** 1927, 1031–1056.
6. R. Blondel, *Publ. sci. tech. ministère air* (*Paris*), no. 89, 1936; *Met. Abstr.*, **4,** 1937, 87; see also R. Blondel and P. Laffitte, *Compt. rend.*, **200,** 1935, 1472–1474.
7. B. DeWitt and H. Seltz, *J. Am. Chem. Soc.*, **61,** 1939, 3170–3173.
8. H. Bruns and G. Lautz, *Z. Naturforsch.*, **9a,** 1954, 694.
9. Of the early investigators, [4] determined the curves of the unstable system and [1] those of the stable system.
10. F. Halla, H. Nowotny, and H. Tompa, *Z. anorg. Chem.*, **214,** 1933, 197–200.
11. A. W. Smith, *Phys. Rev.*, **32,** 1911, 178.
12. K. Honda and T. Soné, *Science Repts. Tôhoku Univ.*, **2,** 1913, 6–7.
13. H. Endo, *Science Repts. Tôhoku Univ.*, **16,** 1927, 215–218; K. Honda and H. Endo, *J. Inst. Metals*, **37,** 1927, 40.
14. F. L. Meara, *Physics*, **2,** 1932, 33–41; abstracted in *J. Inst. Metals*, **50,** 1932, 354.
15. Y. Matuyama, *Science Repts. Tôhoku Univ.*, **16,** 1927, 447–474.
16. Y. Matuyama, *Science Repts. Tôhoku Univ.*, **18,** 1929, 737–744; see, however, F. Sauerwald, *Z. Metallkunde*, **14,** 1922, 457–458.
17. P. T. Arnemann, *Metallurgie*, **7,** 1910, 205.
18. C. T. Heycock and F. H. Neville, *J. Chem. Soc.*, **71,** 1897, 394, 402.
19. E. Weisse, A. Blumenthal, and H. Hanemann, *Z. Metallkunde*, **34,** 1942, 221.
20. W. Leitgebel, *Z. anorg. Chem.*, **202,** 1931, 305–324.
21. K. E. Almin, *Acta Chem. Scand.*, **2,** 1948, 400–407.
22. A. Ölander, *Z. Krist.*, **91,** 1935, 243–247 (structure of CdSb).

0.1254
$\bar{1}$.8746

Sb-Zr Antimony-Zirconium

The partial phase diagram of Fig. 641 was established by [1] by means of microscopic examination of slowly cooled and quenched specimens, incipient-melting tests, X-ray diffraction work, and differential cooling curves (Zr transformation). The phase boundaries of the α_{Zr} and β_{Zr} phases cannot be considered as accurately established; the microscopic investigation was obviously hindered by the high carbon pickup, and an attempt to determine the solubility of Sb in α-Zr by X-ray parametric work failed because of the oxygen and nitrogen content of the samples.

X-ray powder and single-crystal patterns of Zr_2Sb (40.03 wt. % Sb) were indexed with a hexagonal unit cell (a = 8.4 A, c = 5.6 A, c/a = 0.67) containing 5 molecules (15 atoms).

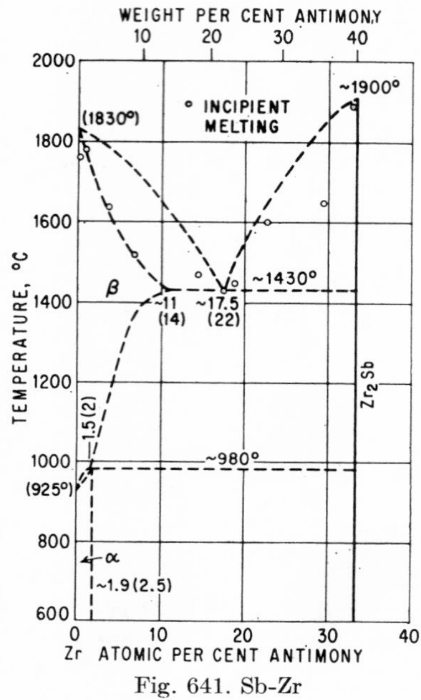

Fig. 641. Sb-Zr

The β_{Zr} phase could not be retained by quenching.

1. R. F. Russi and H. A. Wilhelm, *U.S. Atomic Energy Comm., Publ.* AECD-3610, 1951 (declassified Jan. 18, 1954). Most of the alloys were prepared by diluting a master alloy of 24 wt. % Sb with sponge Zr (containing up to 2 wt. % Hf) in graphite crucibles in vacuo. The carbon pickup was appreciable. The master alloy was made by diffusing Sb into sponge Zr in a sealed stainless-steel container.

0.4489
$\bar{1}$.5511

Se-Si Selenium-Silicon

$SiSe_2$ (84.90 wt. % Se) [1–4] is isotypic with SiS_2 (C42 type), a = 9.76 A, b = 6.03 A, c = 5.76 A [3]. In analogy with SiS, the compound SiSe is likely to exist [3].

1. Sabatier, *Compt. rend.*, **113,** 1891, 132.
2. H. Gabriel and C. Alvarez-Tostado, *J. Am. Chem. Soc.*, **74,** 1952, 262–264.
3. A. Weiss and A. Weiss, *Z. Naturforsch.*, **7b,** 1952, 483–484.
4. A. Weiss and A. Weiss, *Z. Naturforsch.*, **8b,** 1953, 104–105.

$\bar{1}$.8230
0.1770

Se-Sn Selenium-Tin

The approximate course of the liquidus curve was determined by [1]. The data points in Fig. 642 indicate the existence of SnSe (39.95 wt. % Se) and another phase in the range 60–70 at. % Se, suggested to be $SnSe_2$ (57.09 wt. % Se). Since no liquidus points were determined between 9.4 and 50 at. % Se, no indication of a miscibility gap in the liquid state was found.

[2] concluded from thermal-analysis data that, besides SnSe, the phase Sn_2Se_3 (49.95 wt. % Se), rather than $SnSe_2$, occurs, which was proposed to be peritectically formed at about 650°C; see inset of Fig. 642. This would exclude the existence of $SnSe_2$ claimed in former investigations which were based on preparations from chemical reactions. In order to eliminate this discrepancy, the data points given by [2]

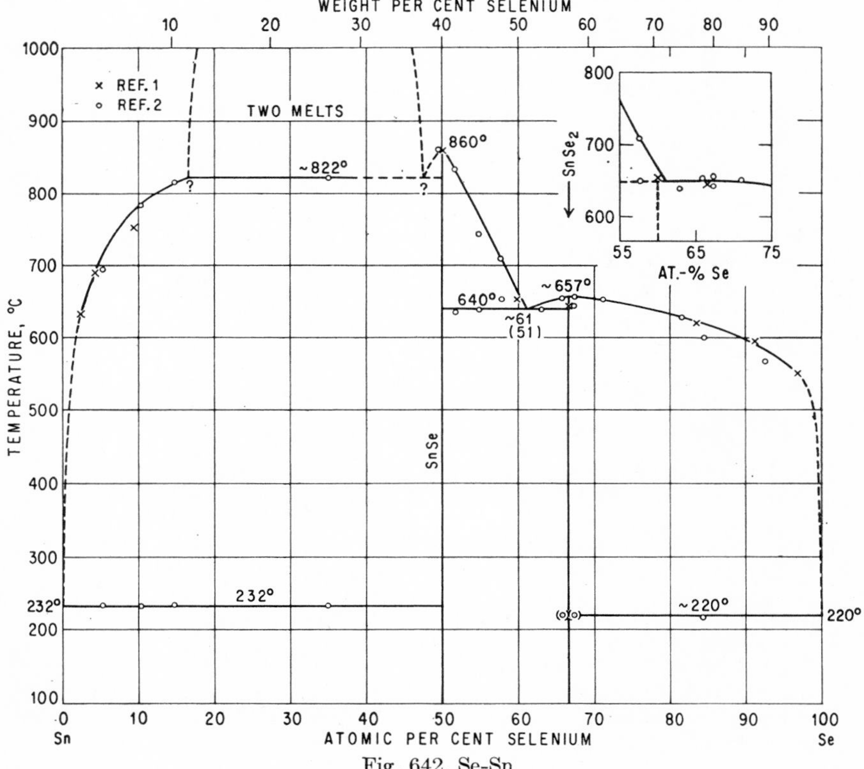

Fig. 642. Se-Sn

were interpreted in the main diagram to indicate a flat liquidus maximum at the composition $SnSe_2$ and a eutectic at about 61 at. % Se, 640°C. This interpretation appears to be more compatible with the data points than that suggested by [2] in the inset of Fig. 642. In addition, a miscibility gap in the liquid state is very likely to exist, although the original diagram of [2] did not show a monotectic horizontal (see also S-Sn).

X-ray investigation of lamellae prepared by deposition from the vapor phase was claimed to indicate the existence of both Sn_2Se_3 and $SnSe_2$, besides SnSe [3]. If this is true, the diagram in the range 50–70 at. % Se would have to be corrected so as to show the phase Sn_2Se_3 formed peritectically at approximately 650°C (as indicated in the inset of Fig. 642) and the phase $SnSe_2$, having a maximal melting point of about 657°C and forming a eutectic with Sn_2Se_3 at about 640°C.

The solubility of Se in Sn at 200°C was reported to be far below 0.05 at. % Se [4].

Crystal Structures. Information as regards the structure of SnSe is conflicting. Whereas [5] reported it to be orthorhombic, with a = 4.33 A, b = 3.98 A, c = 11.18 A (these parameters are identical with those given for SnS!), [3] claimed SnSe to be isotypic with NaCl (B1 type), a = 5.99 A, and thus isostructural with SnTe (see Note Added in Proof). The structure of Sn_2Se_3 was given as tetragonal, with two formula units per unit cell and a = 6.77 A, c = 5.86 A, c/a = 0.866 [3]. The X-ray pattern of $SnSe_2$ could not be elucidated [3]. As to the discrepancy in the existence of Sn_2Se_3 and/or $SnSe_2$, see above.

Note Added in Proof. In a short note, [6] reported SnSe to be isomorphous with GeS and SnS, orthorhombic with a = 4.46 A, b = 4.19 A, c = 11.57 A. The structure may be considered as a distorted NaCl-type structure.

1. H. Pélabon, *Compt. rend.*, **142**, 1906, 1147–1149; *Ann. chim. et phys.*, **17**, 1909, 526–566.
2. W. Biltz and W. Mecklenburg, *Z. anorg. Chem.*, **64**, 1909, 226–235.
3. L. S. Palatnik and V. V. Levitin, *Doklady Akad. Nauk S.S.S.R.*, **96**, 1954, 975–978.
4. E. Jenckel and L. Roth, *Z. Metallkunde*, **30**, 1938, 135–144.
5. Y. Matukuva, T. Yamamoto, and A. Okazaki, *Mem. Fac. Sci., Kyusyu Univ.*, **B1**, 1953, 98–101, quoted from *Chem. Abstr.*, **48**, 1954, 7376.
6. A. Okazaki and I. Ueda, *J. Phys. Soc. Japan*, **11**, 1956, 470; *Met. Abstr.*, **24**, 1956, 31.

$\bar{1}.9548$
0.0452

Se-Sr Selenium-Strontium

SrSe (47.40 wt. % Se) has the NaCl (B1) type of structure [1–3], a = 6.2320 A [3].

1. M. K. Slattery, *Phys. Rev.*, **25**, 1925, 333–337; **21**, 1923, 213; **20**, 1922, 84.
2. V. M. Goldschmidt, *Skrifter Akad. Oslo*, **1926**, no. 8; *Ber. deut. chem. Ges.*, **60**, 1927, 1263.
3. W. Primak, H. Kaufman, and R. Ward, *J. Am. Chem. Soc.*, **70**, 1948, 2043–2046.

$\bar{1}.7915$
0.2085

Se-Te Selenium-Tellurium

Thermal investigations [1, 2] and X-ray analysis [3] have shown that the two elements form an uninterrupted series of solid solutions. In the range 20–100 at. % Te, the liquidus temperatures reported by [1, 2] and presented in Fig. 643 differ by approximately 10–15°C, those of [2] being lower. If the latter, which are based on Se and Te melting points of 197 and 441°C, are corrected, they practically coincide with those given by [1] (Fig. 643). The large deviation of the liquidus temperature between 0 and 20 at. % Te is caused by the fact that these Se-rich melts tend to supercool strongly and to form vitreous rather than crystalline solids. This is the reason why [2] found an actually nonexistent minimum at about 4 at. % Te. The solidification interval is small, about 10°C [1] or less. The much wider range suggested by [2] is certainly incorrect [4].

The variation of the hexagonal unit cell dimensions with composition departs only slightly from linearity [3].

1. G. Pellini and G. Vio, *Atti reale accad. Lincei, Rend.*, **15**, 1906(2), 46–53.
2. Y. Kimata, *Mem. Coll. Sci., Kyoto Univ.*, **1**, 1915, 119–122.
3. E. Grison, *J. Chem. Phys.*, **19**, 1951, 1109–1113.
4. Solidus points were constructed from cooling-curve data.

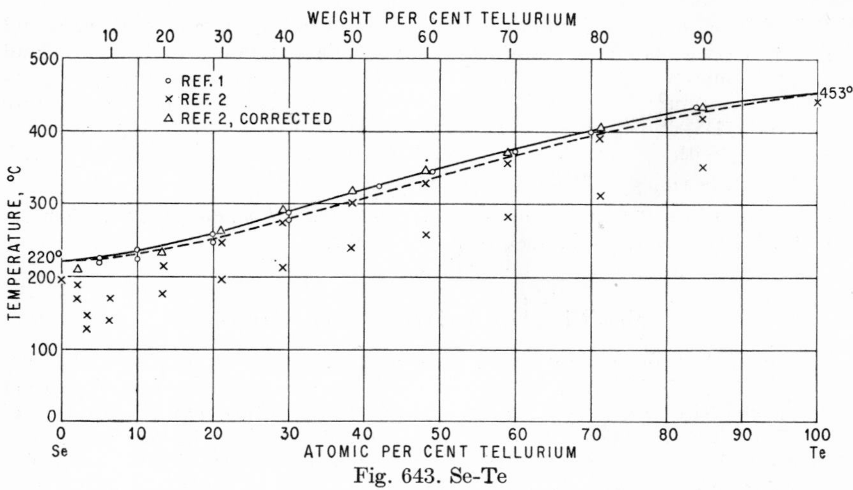

Fig. 643. Se-Te

$\bar{1}.5317$
0.4683

Se-Th Selenium-Thorium

By means of melting-point determinations and metallographic, X-ray, and tensimetric methods, the existence of the phases ThSe (25.38 wt. % Se), Th_2Se_3 (33.79 wt. % Se), Th_7Se_{12} [63.16 at. (36.84 wt.) % Se], $ThSe_2$ (40.49 wt. % Se), and Th_3Se_7 (44.25 wt. % Se) was established [1]. Figure 644 shows the *tentative* phase diagram.

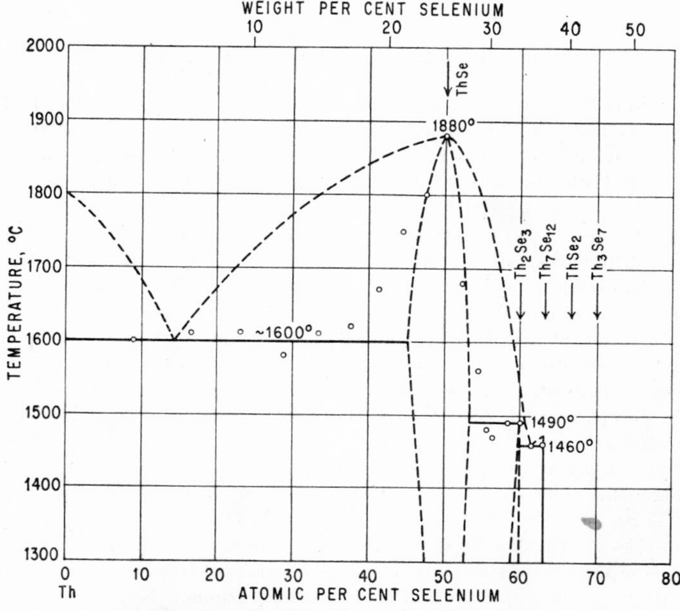

Fig. 644. Se-Th. (Tentative diagram.)

The Th-ThSe eutectic was found to lie between 9 and 16.7 at. % Se. Both ThSe and Th_2Se_3 have homogeneity ranges which extend from about 48.7 to 52.4 at. % Se and from approximately 58.3 at. % Se to the stoichiometric composition, respectively.

ThSe is isotypic with NaCl (B1 type), a = 5.875 A [1]. Th_2Se_3 has the structure of the Sb_2S_3 ($D5_8$) type, with a = 11.34 A, b = 11.57 A, c = 4.27 A [1]. Th_7Se_{12} is hexagonal [1], with a = 11.56_9 A, c = 4.23 A [2], and $ThSe_2$ is probably isotypic with $PbCl_2$ (C23 type) [1], a = 4.420 A, b = 7.610 A, c = 9.064 A [2].

1. R. W. M. D'Eye, P. G. Sellman, and J. R. Murray, *J. Chem. Soc.*, **1952**, **2555–2562.**
2. R. W. M. D'Eye, *J. Chem. Soc.*, **1953**, 1670–1672.

0.2171
$\bar{1}$.7829

Se-Ti Selenium-Titanium

X-ray analysis of preparations varying in composition between TiSe (62.24 wt. % Se) and $TiSe_2$ (76.73 wt. % Se) showed that there is a continuous transition from the hexagonal NiAs (B8) type of structure of TiSe to the hexagonal CdI_2 (C6) type of structure of $TiSe_2$. TiSe: a = 3.566 A, c = 6.232 A, c/a = 1.748. $TiSe_2$: a = 3.548 A, c = 5.998 A, c/a = 1.690 [1]. Earlier, [2] had found for $TiSe_2$: a = 3.540 A, c = 6.007 A, c/a = 1.697. Indications for the existence of a subselenide (Ti_2Se?) were found [1]. There is no further compound between $TiSe_2$ and Se [1].

The lowest per cent addition at which Se appeared as insoluble compound in the microstructure of magnesium-reduced Ti-base alloys, hot-rolled and subsequently annealed for ½ hr at 790°C, was found to be 0.25 wt. % Se [3].

1. P. Ehrlich, *Z. anorg. Chem.*, **260**, 1949, 1–18.
2. I. Oftedal, *Z. physik. Chem.*, **134**, 1928, 301–310.
3. R. M. Goldhoff, H. L. Shaw, C. M. Craighead, and R. I. Jaffee, *Trans. ASM*, **45**, 1953, 941–971.

$\bar{1}$.5870
0.4130

Se-Tl Selenium-Thallium

The liquidus curve outlined by [1] indicated the existence of Tl_2Se (16.19 wt. % Se) and TlSe (27.87 wt. % Se), having melting points of approximately 390 and 338°C, respectively. In addition, two ranges of liquid immiscibility were found to exist, between Tl and Tl_2Se and between about 71 and 100 at. % Se, with monotectic temperatures of about 390 and 195°C, respectively.

The constitution was more thoroughly investigated by [2] and [3]. Apart from temperature differences, results are in substantial agreement (Fig. 645). Both studies established the existence of an additional phase, Tl_2Se_3 (36.69 wt. % Se), which is formed peritectically at about 280 [2] or 274°C [3] and undergoes a polymorphic transformation at about 167 [2] or 192°C [3]. The temperature of the Tl_2Se_3-Se eutectic and that of the Se-rich monotectic were found as 151 [2] or 172°C [3] and 177 [2] or 202°C [3], respectively. Temperatures above the melting point of Se (217°C) given by [2] were corrected, since the melting point of Tl was reported as only 287°C, instead of 302°C.

The liquidus points in the range 25–100 at. % Se, reported by [4], are also presented in Fig. 645. The maxima at 40 and 60 at. % Se correspond to the compositions Tl_3Se_2 and Tl_2Se_3 although the author claimed that they prove the existence of Tl_2Se, rather than Tl_3Se_2. Obviously, the results of this investigation are in error, probably because the composition of the melts investigated differed from the intended compositions in that they are shifted to higher Se concentrations. Nevertheless, they also indicate two phases with congruent melting points.

Tl_2Se and TlSe were confirmed by X-ray investigations, whereas Tl_2Se_3 could not be verified [5, 6]. This discrepancy needs clarification, although the thermal data of [2, 3] leave little doubt that there is an additional phase between TlSe and Se. [7] disputed the suitability of thermal analysis in this case to establish the constitution.

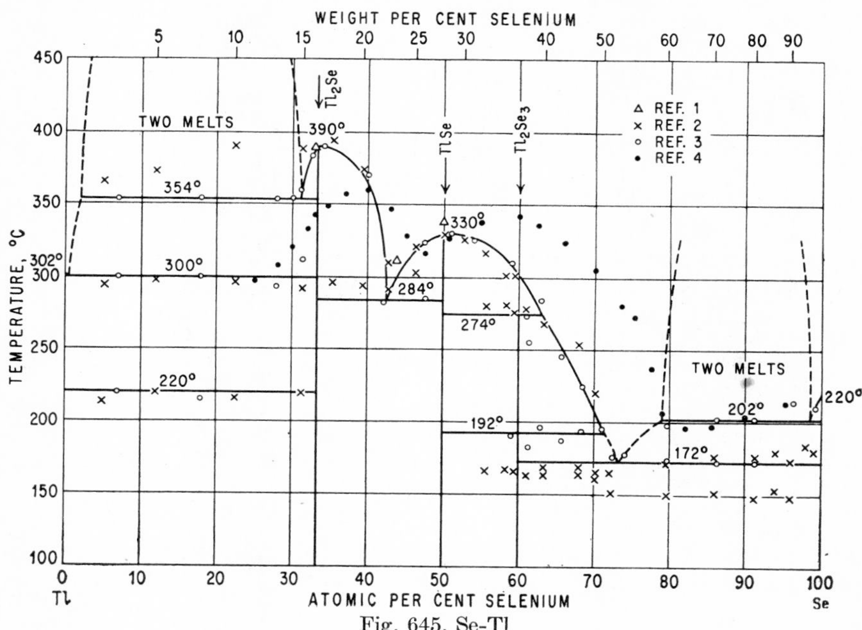

Fig. 645. Se-Tl

Crystal Structures. The structure of TlSe is tetragonal, with 16 atoms per unit cell (B37 type), $a = 8.03_6$ A, $c = 7.01_4$ A, $c/a = 0.873$ [5]. Tl_2Se is not isotypic with Tl_2S [6].

1. H. Pélabon, *Compt. rend.*, **145,** 1907, 118–121; *Ann. chim. et phys.*, **17,** 1909, 526.
2. T. Murakami, *Mem. Coll. Sci., Kyoto Univ.*, **1,** 1915, 153–159.
3. A. P. Obukhov and N. S. Bubyreva, *Izvest. Sektora Fiz.-Khim. Anal.*, **17,** 1949, 276–280.
4. A. Baroni, *Atti reale accad. Lincei, Rend.*, **25,** 1937, 621–626.
5. J. A. A. Ketelaar, W. H. t'Hart, M. Moerel, and D. Polder, *Z. Krist.*, **101,** 1939, 396–405.
6. H. Hahn and W. Klingler, *Z. anorg. Chem.*, **260,** 1949, 110–119.
7. L. Rolla, *Atti reale accad. Lincei, Rend.*, **28,** 1919, 355–359.

$\bar{1}.5207$
0.4793

Se-U Selenium-Uranium

X-ray analysis of U-Se preparations [1] revealed the existence of USe (24.91 wt. % Se) [2], U_2Se_3 (33.22 wt. % Se) [2], USe_2 (39.88 wt. % Se), and a selenide with a higher Se content than USe_2. USe_2 seems to occur in two modifications.

USe is isotypic with NaCl (B1 type), $a = 5.750$ A [1].

1. R. Ferro, *Z. anorg. Chem.*, **275,** 1954, 320–326.
2. A. Colani, *Compt. rend.*, **137,** 1903, 382–383; *Ann. chim. et phys.*, **12,** 1907, 85.

0.1903
$\bar{1}$.8097

Se-V Selenium-Vanadium

The following phases were identified by X-ray studies: VSe (60.78 wt. % Se); a phase observed in the range 55.5–61.5 at. % Se, probably of the composition V_2Se_3 (69.92 wt. % Se); and VSe_2 (75.61 wt. % Se). The latter appears to be stable only with an excess of V, i.e., up to about 65–65.5 at. % Se.

VSe is isotypic with NiAs (B8 type), $a = 3.58_7$ A, $c = 5.98_9$ A, $c/a = 1.67_0$ at 51 at. % Se. V_2Se_3 is definitely not hexagonal but of lower symmetry. VSe_2 is isostructural with CdI_2 (C6 type), with $a = 3.41_0$ A, $c = 5.98_9$ A, $c/a = 1.75_6$ at 61.8 at. % Se and $a = 3.35_5$ A, $c = 6.13_4$ A, $c/a = 1.82_8$ at 66.3 at. % Se [1].

1. H. Hoschek and W. Klemm, *Z. anorg. Chem.*, **242**, 1939, 49–62.

$\bar{1}$.6328
0.3672

Se-W Selenium-Wolfram

The existence of WSe_2 (46.20 wt. % Se) [1, 2] and WSe_3 (56.29 wt. % Se) [1, 3] has been reported. WSe_2 is isostructural with MoS_2 (C7 type), $a = 3.29$ A, $c = 12.97$ A, $c/a = 3.94$ [2].

1. H. Uelsmann, *Liebigs Ann.*, **116**, 1860, 122.
2. O. Glemser, H. Sauer, and P. König, *Z. anorg. Chem.*, **257**, 1948, 241–246.
3. L. Moser and K. Atynski, *Monatsh. Chem.*, **45**, 1925, 241.

$\bar{1}$.6593
0.3407

Se-Yb Selenium-Ytterbium

YbSe (31.33 wt. % Se) is isostructural with NaCl (B1 type), $a = 5.879$ A [1].

1. H. Senff and W. Klemm, *Z. anorg. Chem.*, **242**, 1939, 92–96.

0.0820
$\bar{1}$.9180

Se-Zn Selenium-Zinc

Cooling-curve data and metallographic examination [1] have shown that the components do not mix in the liquid state (Fig. 646). Therefore, the phase ZnSe (54.70 wt. % Se) forms only at the interface of the two layers. The results of the thermal analysis seem to indicate that no compound exists. The quantity of ZnSe formed increases with temperature; at normal pressure, however, the reaction temperature is limited by the boiling point of Se. The melting point of ZnSe lies considerably above 1100°C.

ZnSe is isotypic with zincblende (B3 type) [2–4], with $a = 5.672$ A [4], $a = 5.667$ A [5].

1. M. Chikashige and R. Kurosawa, *Mem. Coll. Sci., Kyoto Univ.*, **2**, 1917, 245–248.
2. W. P. Davey, *Phys. Rev.*, **21**, 1923, 380.
3. W. F. de Jong, *Z. Krist.*, **63**, 1926, 471.
4. W. Zachariasen, *Z. physik. Chem.*, **124**, 1926, 440–444.
5. H. E. Swanson, R. K. Fuyat, and G. M. Ugrienic, *Natl. Bur. Standards Circ. (U.S.)*, 539, III, 1954, p. 23.

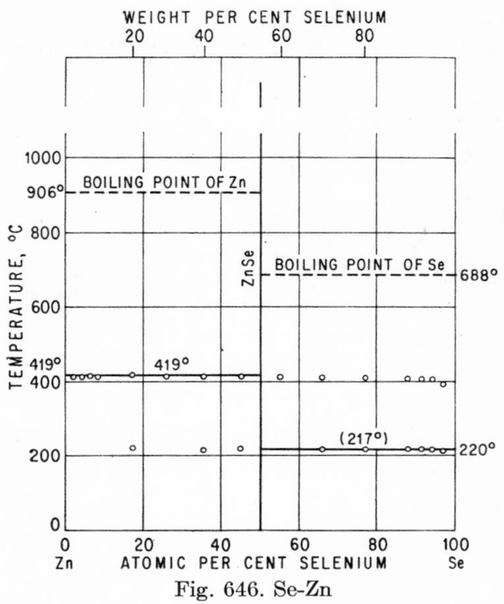

Fig. 646. Se-Zn

$\bar{1}.9373$
0.0627

Se-Zr Selenium-Zirconium

$ZrSe_2$ (63.39 wt. % Se) is isostructural with CdI_2 (C6 type), $a = 3.80$ A, $c = 6.19$ A, $c/a = 1.63$ [1].

1. A. E. van Arkel, *Physica,* **4,** 1924, 286–301; see also F. Hund, *Z. Physik,* **14,** 1925, 833. As to the preparation of $ZrSe_2$, see O. Ruff and R. Wallstein, *Z. anorg. Chem.,* **128,** 1923, 100; A. E. van Arkel and J. H. de Boer, *Z. anorg. Chem.,* **148,** 1925, 348.

$\bar{1}.2712$
0.7288

Si-Sm Silicon-Samarium

$SmSi_2$ (27.19 wt. % Si) has a weakly distorted $ThSi_2$ type of structure with the (pseudo-)tetragonal lattice parameters $a = 4.04_9$ A, $c = 13.36$ A, $c/a = 3.29$ [1].

1. G. Brauer and H. Haag, *Z. anorg. Chem.,* **267,** 1952, 198–212.

$\bar{1}.3741$
0.6259

Si-Sn Silicon-Tin

[1] found that Si-Sn alloys are mixtures of the components. This was corroborated by thermal and microscopic [2] (Fig. 647) as well as X-ray diffraction [3] work. The lattice constants of the elements in the alloys were found to be practically identical with those of the pure elements [3].

Estimates of the eutectic data were made by [4] using a regular solution equation: for the eutectic composition, x_{Si} (atom fraction) $= 10^{-6}$ (10^{-4} at. % Si) was computed; for the eutectic temperature, a value 0.001°C below the melting point of Sn.

1. E. Vigouroux, *Compt. rend.*, **123**, 1896, 116.
2. S. Tamaru, *Z. anorg. Chem.*, **61**, 1909, 40–42. The alloys were melted in porcelain tubes under H_2. Silicon of two different grades of purity was used (see Fig. 647), with Fe and Al as main impurities.
3. E. R. Jette and E. B. Gebert, *J. Chem. Phys.*, **1**, 1933, 753–755.
4. C. D. Thurmond, *J. Phys. Chem.*, **57**, 1953, 827–830.

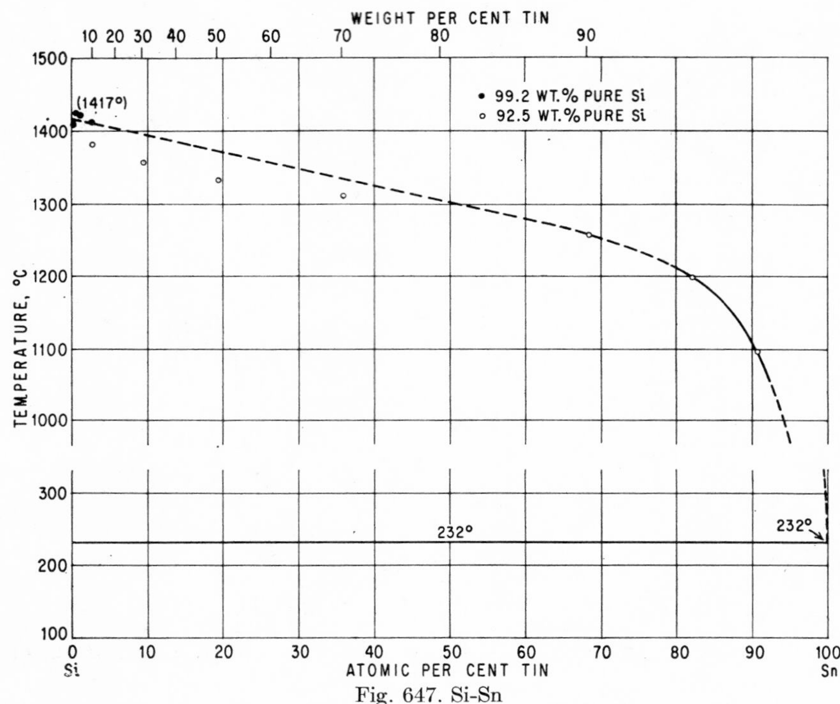

Fig. 647. Si-Sn

Ī.5059
0.4941

Si-Sr Silicon-Strontium

SrSi (24.27 wt. % Si) [1] and $SrSi_2$ (39.06 wt. % Si) [2, 1] have been prepared.

1. L. Wöhler and W. Schuff, *Z. anorg. Chem.*, **209**, 1932, 33–59.
2. C. S. Bradley, *Chem. News*, **82**, 1900, 149.

Ī.1912
0.8088

Si-Ta Silicon-Tantalum

In 1907, [1] prepared the disilicide $TaSi_2$ (23.70 wt. % Si).

Systematic investigations of the constitution of the Ta-Si system on alloys prepared by powder-metallurgical techniques were carried out by [2] (X-ray) and [3, 4] (X-ray, microscopic, and melting points). Their results are in substantial agreement as far as the number and the compositions of the observed intermediate phases are concerned. Besides $TaSi_2$ they reported the following compounds: $TaSi_{0.60} = Ta_5Si_3$ [37.50 at. (8.52 wt.) % Si], $TaSi_{0.4\pm0.1}$, and $TaSi_{0.2\pm0.05}$ [2]; Ta_5Si_3, $Ta_2Si = TaSi_{0.5}$

(7.21 wt. % Si), and $Ta_{4.5\pm0.5}Si$ [3, 4], respectively. In each case the composition brackets express the uncertainty of the formula; no indications for any extended solid-solubility ranges were found by either [2] or [3, 4]. The formula Ta_2Si has support from a crystal-structure determination (see below).

Melting points determined pyrometrically by [4] are plotted in Fig. 648; these authors also carried out a thermal analysis in the range 90–100 wt. % Si. The conclusion of [5]—based on limited microscopic work—that a monosilicide TaSi does exist has no support from the work of [2–4].

Crystal Structures. Addition of Si slightly expands the lattice parameter of Ta; at 1800°C the solubility is less than 0.2 wt. (1.3 at.) % Si [3]. Also, the following

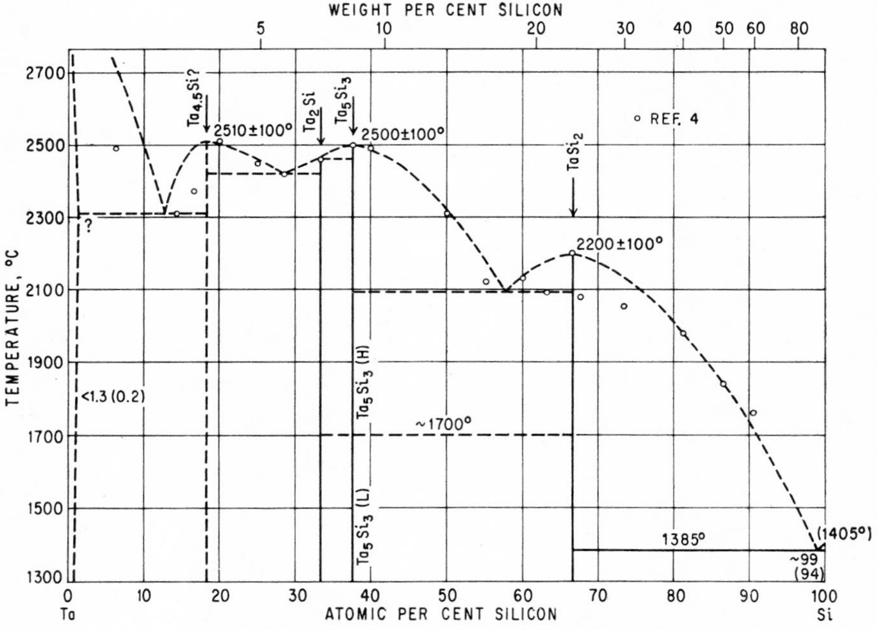

Fig. 648. Si-Ta. (See also Note Added in Proof.)

structures were determined by means of X-ray powder photographs: $\sim Ta_{4.5}Si$, hexagonal Ni_3Sn ($D0_{19}$) type, with $a = 6.10_5$ A, $c = 4.91_9$ A, $c/a = 0.806$ (partially at random atomic distribution); Ta_2Si, tetragonal $CuAl_2$ (C16) type, with $a = 6.15_7$ A, $c = 5.03_9$ A, $c/a = 0.818$; Ta_5Si_3, hexagonal Mn_5Si_3 ($D8_8$) type, with $a = 7.47_4$ A, $c = 5.22_5$ A, $c/a = 0.699$ (but see Note Added in Proof). There is evidence for the existence of a high-temperature modification of Ta_5Si_3 above 1600–1800°C.

According to X-ray powder and single-crystal work by [6], $TaSi_2$ has the hexagonal $CrSi_2$ (C40) type of structure, with $a = 4.783 \pm 5$ A, $c = 6.565 \pm 5$ A, $c/a = 1.373$. This structure was corroborated by [3, 4]. Solid Si dissolves practically no Ta [3].

Note Added in Proof. As has been shown recently [7–9], the occurrence of the $D8_8$ type of structure with Ta_5Si_3 is caused by the presence of small amounts of contaminants such as C, N, and O. The pure binary compound Ta_5Si_3 (or Ta_3Si_2) occurs in two tetragonal modifications, designated as T1 and T2 by [9], which appear to be the high- and low-temperature form, respectively [9]. The T1 form is isotypic with the corresponding V, Nb, Cr, Mo, and W compounds and has the lattice parameters $a = 9.88$ A, $c = 5.06$ A, $c/a = 0.512$ [8]. The T2 form is isotypic with Cr_5B_3, $a = 6.51_6$ A, $c = 11.87_3$ A, $c/a = 1.82_2$ [9].

[10] determined the melting points of Ta_5Si_3 and $TaSi_2$ as 2495 and 2200°C, respectively (cf. Fig. 648).

1. O. Hönigschmid, *Monatsh. Chem.*, **28**, 1907, 1017–1018.
2. L. Brewer, A. W. Searcy, D. H. Templeton, and C. H. Dauben, *J. Am. Ceram. Soc.*, **33**, 1950, 291–294.
3. H. Nowotny, H. Schachner, R. Kieffer, and F. Benesovsky, *Monatsh. Chem.*, **84**, 1953, 1–12. This paper gives a detailed description of the roentgenographic results.
4. R. Kieffer, F. Benesovsky, H. Nowotny, and H. Schachner, *Z. Metallkunde*, **44**, 1953, 242–246.
5. W. R. Johnson and M. Hansen, U.S. Air Force Technical Report 6383, June, 1951.
6. H. J. Wallbaum, *Z. Metallkunde*, **33**, 1941, 378–381.
7. A. G. Knapton, *Nature*, **175**, 1955, 730.
8. E. Parthé, H. Nowotny, and H. Schmid, *Monatsh. Chem.*, **86**, 1955, 385–396.
9. E. Parthé, B. Lux, and H. Nowotny, *Monatsh. Chem.*, **86**, 1955, 859–867.
10. G. A. Geach and F. O. Jones, *Plansee Proc.*, **1955**, 80–91; *Met. Abstr.*, **24**, 1957, 366.

$\bar{1}.3427$
0.6573

Si-Te Silicon-Tellurium

The existence of SiTe (81.96 wt. % Te) and $SiTe_2$ (90.08 wt. % Te) has been reported by [1, 2]. SiTe is said to have a complex cubic structure [2]. $SiTe_2$ appears to exist in two modifications: red crystals of the hexagonal CdI_2 (C6) type, with $a = 4.28 \pm 1$ A, $c = 6.71 \pm 1$ A, $c/a = 1.56_7$; and a fibrous modification occasionally obtained which presumably is isotypic with (orthorhombic) $SiSe_2$ [1, 2].

1. A. Weiss and A. Weiss, *Z. Naturforsch.*, **8b**, 1953, 104 (short communication).
2. A. Weiss and A. Weiss, *Z. anorg. Chem.*, **273**, 1953, 124–128.

$\bar{1}.0828$
0.9172

Si-Th Silicon-Thorium

$ThSi_2$ (19.49 wt. % Si), first reported by [1], was shown [2] to have a b.c. tetragonal structure, with $a = 4.134$ A, $c = 14.375$ A, $c/a = 3.48$, and 12 atoms per unit cell. According to [3], $ThSi_2$ exists in two modifications, the known tetragonal "α" and the hexagonal "β" form. Furthermore, a tetragonal compound, Th_3Si_2 (7.47 wt. % Si), was found [3].

Supplement. According to [4], α-$ThSi_2$ is transformed to β-$ThSi_2$ above 1400°C. Attempts to reverse this transition have failed. β-$ThSi_2$, which melts between 1600 and 1650°C, was shown to be isotypic with β-USi_2 [hexagonal AlB_2 (C32) type] and to have the lattice parameters $a = 3.986 \pm 5$ A, $c = 4.223 \pm 5$ A, $c/a = 1.059$. Prolonged heating at about 1450°C appears to produce ThSi, which is isotypic with USi [orthorhombic FeB (B27) type].

Note Added in Proof. A eutectic at 10 at. % Si and at a temperature above 1300°C has been found [5]. The lattice parameters of tetragonal Th_3Si_2 are $a = 7.841$ A, $c = 4.166$ A [6].

1. O. Hönigschmid, E. Wedekind, and K. Fetzer, *Chem. Ztg.*, **29**, 1905, 1031. O. Hönigschmid, *Monatsh. Chem.*, **28**, 1907, 1017–1018.

2. G. Brauer and A. Mitius, *Z. anorg. Chem.*, **249**, 1942, 325–339.
3. W. H. Zachariasen, to be published; quoted by G. F. Hardy and J. K. Hulm, *Phys. Rev.*, **93**, 1954, 1004–1016, who searched for superconductivity with these phases.
4. E. L. Jacobson (and A. W. Searcy), M.S. thesis, E. L. Jacobson, Purdue University, Lafayette, Ind., June, 1952.
5. F. Foote, unpublished information (March, 1945); quoted in [7].
6. W. H. Zachariasen, unpublished information (1953); quoted in [7].
7. H. A. Saller and F. A. Rough, Compilation of U.S. and U.K. Uranium and Thorium Constitutional Diagrams, *U.S. Atomic Energy Comm., Publ.* BMI-1000, June, 1955.

$\bar{1}.7682$
0.2318

Si-Ti Silicon-Titanium

The phase diagram of Fig. 649 is mainly based on work by [1], who used micrographic, thermal, and X-ray diffraction analysis as well as detection of incipient melting as principal tools of investigation. The statement by [2] that the melting point of Ti is lowered by Si is in accordance with Fig. 649; however, the apparent melting points reported for sintered compacts with 3 and 10 wt. % Si—about 1225°C and about 1100°C, respectively—deviate considerably.

Alloys with 0–5 At. % Si. Micrographic investigations of the solubility of Si in solid Ti on cold-worked iodide Ti-base alloys annealed at and quenched from temperatures between 1100 and 650°C were carried out by [1, 3]. The compositions found for the eutectoid point are ~0.9 wt. (1.5 at.) % Si [1] and 0.65 wt. (1.1 at.) % Si [3]. Since the two investigations were performed under similar experimental conditions, it is difficult to account for the disagreement. In Fig. 649 weight has been given to the results of [3] since this author obtained the greater amount of data. As to the maximum solubility of Si in β-Ti at the eutectic temperature (1330 ± 5°C [1], 1320 ± 6°C [3]), and the solubility in α-Ti (see Fig. 649), the results of [1, 3] are in substantial agreement. Earlier it had been reported [4] that the solubility of Si in α-Ti is less than 0.44 wt. %.

Intermediate Phases. According to [1], there are three compounds in this system: Ti_5Si_3 [37.50 at. (26.03 wt.) % Si], TiSi (36.96 wt. % Si), and $TiSi_2$ (53.98 wt. % Si). The phase coexisting with the α and β solid solutions extends over a range of composition [1, 5]; whereas [1] observed heterogeneous microstructures below 26 (37.5) and above 28 wt. (40 at.) % Si, [5] claimed that an alloy with 33.5 at. % Si lay within the homogeneity range. The crystal structure of this phase is based on the composition Ti_5Si_3 (see below).

As to the peritectically formed compound, [1] gave strong evidence for the composition TiSi. [6], who first claimed the existence of TiSi, did not prove an alloy of this composition to be of single phase. Based on some cursory X-ray work, [5] suggested that several intermediate phases may exist in the region around 50 at. % Si.

The existence of the disilicide $TiSi_2$, first characterized by [7], was established conclusively by [8] by determining its crystal structure. [9] observed an unusual double bend in the resistivity-vs.-temperature curve of this compound above 1200°C.

[1] found no evidence of the existence of the compounds Ti_2Si [10, 6] and Ti_2Si_3 [11] reported earlier.

The solubility of Ti in solid Si is considered [12] to be very small or negligible.

Crystal Structures. Ti_5Si_3 was shown [13] to have the hexagonal Mn_5Si_3 ($D8_8$) type of structure. [5] observed a decrease in unit cell dimensions with increasing Si content. Lattice parameters reported by [13, 5, 14] are in poor agreement; as to those by [13, 14], the exact alloy compositions are not known. Those by [5] are certainly influenced by contaminations (O, Mo). The parameters given by [13] are $a = 7.465 \pm 2$ A, $c = 5.162 \pm 2$ A, $c/a = 0.692$.

X-ray diffraction patterns of the TiSi phase were tabulated by [1]; the structure was not determined.

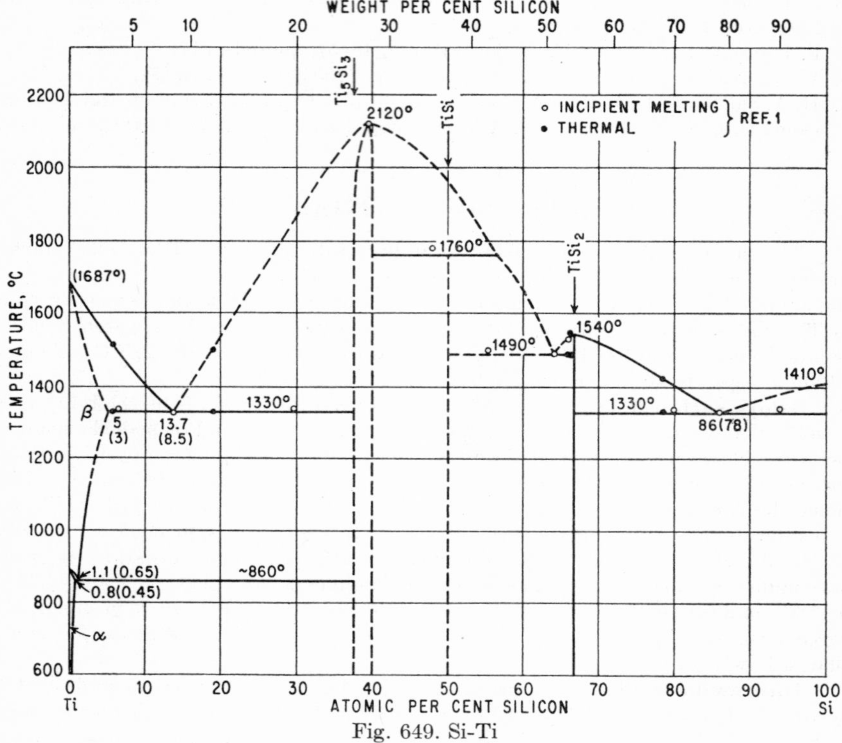

Fig. 649. Si-Ti

The structure of $TiSi_2$ forms the orthorhombic C54 prototype [8]; the lattice parameters are $a = 8.252$ A, $b = 4.783$ A, $c = 8.540$ A [8, 15]. [16] reported, however, that $TiSi_2$ is of the orthorhombic $ZrSi_2$ (C49) type of structure, with $a = 3.62 \pm 1$ A, $b = 13.76 \pm 2$ A, $c = 3.60 \pm 1$ A.

1. M. Hansen, H. D. Kessler, and D. J. McPherson, *Trans. ASM,* **44,** 1952, 518–536. See also discussion, pp. 536–538. Arc-melted alloys are based on magnesium-reduced and iodide Ti, the latter for alloys in the 0–2 wt. % Si range.
2. E. I. Larsen, E. F. Swazy, L. S. Busch, and R. H. Freyer, *Metal Progr.,* **55,** 1949, 359–361.
3. D. A. Sutcliffe, *Rev. mét.,* **51,** 1954, 524–536; also *Metal Treatment,* **21,** 1954, 191–197.
4. C. M. Craighead, O. W. Simmons, and L. W. Eastwood, *Trans. AIME,* **188,** 1950, 497–498.
5. L. Brewer and O. Krikorian, *U.S. Atomic Energy Comm., Publ.* UCRL-2544, 1954.
6. P. P. Alexander, *Metals & Alloys,* **9,** 1938, 179–181.
7. O. Hönigschmid, *Compt. rend.,* **143,** 1906, 224–226.
8. F. Laves and H. J. Wallbaum, *Z. Krist.,* **101,** 1939, 78–93.

9. F. W. Glaser and D. Moskowitz, *Powder Met. Bull.*, **6**, 1953, 178–185.
10. L. Levy, *Compt. rend.*, **121**, 1895, 1148–1150.
11. P. Askenasy and C. Ponnaz, *Z. Elektrochem.*, **14**, 1908, 810–811.
12. [1], discussion, p. 538.
13. P. Pietrokowsky and P. Duwez, *Trans. AIME*, **191**, 1951, 772–773.
14. W. Freundlich, A. Chrétien, and M. Bichara, *Compt. rend.*, **239**, 1954, 1141–1143.
15. H. Nowotny, H. Schroth, R. Kieffer, and F. Benesovsky, *Monatsh. Chem.*, **84**, 1953, 582.
16. P. G. Cotter, J. A. Kohn, and R. A. Potter, *J. Am. Ceram. Soc.*, **39**, 1956, 11–12 (not available to the author); *Met. Abstr.*, **24**, 1957, 569–570.

$\bar{1}.1381$
0.8619

Si-Tl Silicon-Thallium

According to a thermal investigation by [1], Tl and Si do not alloy. The only thermal arrests observed were practically at the melting points of the elements.

1. S. Tamaru, *Z. anorg. Chem.*, **61**, 1909, 44–45. The silicon used contained 6.1 Fe and 1.7 Al (wt. %)!

$\bar{1}.0718$
0.9282

Si-U Silicon-Uranium

In 1908 the aluminothermic preparation of USi_2 (19.09 wt. % Si) was reported by [1]. Systematic thermal, microscopic, and X-ray studies of the U-Si phase diagram have been carried out within the Manhattan Project, and their results have recently become available in abstract form [2]. Six intermediate phases were found to which were tentatively assigned the compositions $U_{10}Si_3$, U_5Si_3, USi, U_2Si_3, USi_2, and USi_3. However, [3] has shown by X-ray structure analyses that some of the above formulas were incorrect: "$U_{10}Si_3$" is actually U_3Si, "U_5Si_3" is U_3Si_2, and "U_2Si_3" is actually a polymorphic form of USi_2. Therefore, the identified compounds are U_3Si (3.78 wt. % Si), U_3Si_2 (7.29 wt. % Si), USi (10.55 wt. % Si), α- and β-USi_2, and USi_3 (26.14 wt. % Si) (see Note Added in Proof).

The thermal data plotted in Fig. 650 were taken from the phase diagram published by [4]. In drawing Fig. 650 the diagram presented by [4] had to be tentatively modified to accommodate the X-ray results of [3]. Since no information on the α-USi_2–β-USi_2 relationship is available, this polymorphism is not indicated in Fig. 650 (see also Crystal Structures).

According to [2], the maximum solubility of Si in γ-U is about 1.75 at. (0.21 wt.) % at 980°C and the solubility in β-U less than 1 at. (0.12 wt.) %. The $\beta \rightleftarrows \gamma$ transformation temperature of U is somewhat increased by Si; that of the $\alpha \rightleftarrows \beta$ transformation appears unchanged at about 665°C.

Crystal Structures. U_3Si has a b.c. tetragonal structure, with $a = 6.029 \pm 2$ A, $c = 8.696 \pm 3$ A, $c/a = 1.442$, and 16 atoms per unit cell, which is related to the Cu_3Au structure [3]. U_3Si_2 has a primitive tetragonal structure, with $a = 7.3299 \pm 4$ A, $c = 3.9004 \pm 5$ A, $c/a = 0.532$, and 10 atoms per unit cell [3]. USi was shown [3] to have the orthorhombic FeB (B27) type of structure, with $a = 5.66 \pm 1$ A, $b = 7.67 \pm 1$ A, $c = 3.91 \pm 1$ A. USi_3 has a cubic structure [2, 5] of the Cu_3Au ($L1_2$) type [6], with $a = 4.0353$ A [6].

As for the compound USi_2, three different crystal structures have been reported. According to [3], an α and a β form exist, having the following structures: α, isotypic with tetragonal $ThSi_2$, $a = 3.98 \pm 3$ A, $c = 13.74 \pm 8$ A, $c/a = 3.45$; β, hexagonal AlB_2 (C32) type of structure, with $a = 3.86 \pm 1$ A, $c = 4.07 \pm 1$ A, $c/a = 1.05$. This modification was also observed by [5]. In the structural analysis of the alloy

USi_2 synthesized from the elements in a menstruum of molten Al, [7] found only diffraction patterns of a cubic structure with $a = 4.053 \pm 3$ A (or kX?) and the atomic distribution 1 U at (000), 0.07 U + 2.14 Si at random at the face centers of the cube. In a later investigation [8], they observed the lines of the cubic and of the tetragonal form, but none of the hexagonal form of USi_2. A preparation USi_3 yielded the

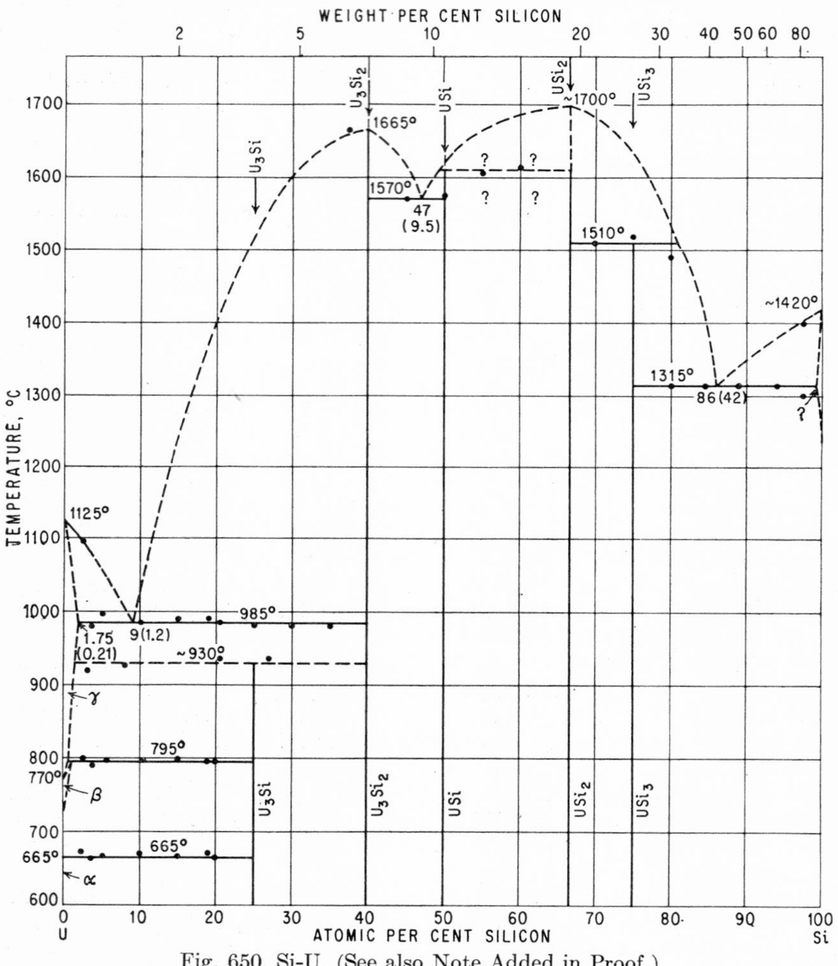

Fig. 650. Si-U. (See also Note Added in Proof.)

pattern of the cubic form. [8] considered the constitution in the USi_2-USi_3 range as not yet fully established.

In summary, it can be said that cubic USi_3 certainly exists. A simple explanation of the cubic phase observed by [7, 8] would be to identify it with USi_3; however, the chemical analytical work by [7] seems to exclude this possibility. Further experimental work is necessary to clarify the polymorphic relations of the USi_2 phases.

Note Added in Proof. Details of the Manhattan Project work on U-Si alloys have been published recently by [9]. Whereas [9] accept the formula U_3Si_2 [3], rather than U_5Si_3, as correct, they maintain that (*a*) the intermediate phase richest in U is homogeneous at about 23 at. % Si rather than at exactly the stoichiometric composition U_3Si, and (*b*) the compound U_2Si_3 presumably exists. Their (modified) phase diagram shows USi and U_2Si_3 to be formed peritectically at 1575 and 1610°C, respectively (cf. Fig. 650).

According to [9], all the U-Si compounds, other than $\sim U_3Si$ (also called ϵ), are "hopelessly brittle."

1. E. Defacqz, *Compt. rend.*, **147,** 1908, 1050–1052. See also *Rev. mét.*, **7,** 1910, 762.
2. Work by A. R. Kaufmann, B. D. Cullity, G. Bitsianes, P. Gordon, M. Cohen, and R. B. Bostian at the Massachusetts Institute of Technology; published in J. J. Katz and E. Rabinowitch, "The Chemistry of Uranium," Part I, National Nuclear Energy Series, Div. VIII, vol. 5, pp. 226–231, McGraw-Hill Book Company, Inc., New York, 1951.
3. W. H. Zachariasen, *Acta Cryst.*, **2,** 1949, 94–99. See also *Acta Cryst.*, **1,** 1948, 265–268.
4. Katz and Rabinowitch; see [2].
5. A. Iandelli and R. Ferro, *Ann. chim. (Rome)*, **42,** 1952, 598–606.
6. B. R. T. Frost and J. T. Maskrey, *J. Inst. Metals*, **82,** 1953, 177–178.
7. G. Brauer and H. Haag, *Z. anorg. Chem.*, **259,** 1949, 197–200.
8. G. Brauer and H. Haag, *Z. anorg. Chem.*, **267,** 1952, 211–212.
9. A. Kaufmann, B. Cullity, and G. Bitsianes, *Trans. AIME*, **209,** 1957, 23–27.

$\bar{1}.7414$
0.2586

Si-V Silicon-Vanadium

Early Papers. [1] and [2] claimed to have prepared the silicides V_2Si (21.61 wt. % Si) and VSi_2 (52.44 wt. % Si); see also [3]. A thermal and microscopic analysis of alloys containing 40–100 wt. % Si was carried out by [4] (Fig. 651). The eutectic halting times indicate a eutectic at or above 95 wt. (97 at.) % Si. The peak of the liquidus and the disappearance of the eutectic arrest point to the existence of a compound VSi_2. Microscopic examination of unetched alloys suggests a solid solution of V in VSi_2 down to at least 40 wt. (55 at.) % Si.

Later Developments. The existence of the intermediate phases V_3Si (15.52 wt. % Si) [5, 6], V_5Si_3 [37.50 at. (24.85 wt.) % Si] [7], and VSi_2 [8] has been established by X-ray structure analyses (see Crystal Structures). Microscopic work in the V-Si system has been carried out by [9, 10]. [9] obtained no reliable results; they based their ternary work (Fe-V-Si) on the assumption that V_2Si and VSi_2 exist. [10] found the 95 wt. % Si alloy to be close to the eutectic composition, which is in agreement with the results of [4]. Primary VSi_2 crystals in a 91 wt. % Si alloy showed coring, which may be taken as evidence that this compound also dissolves Si.

[11] measured 1750°C as the melting point for VSi_2, whereas the thermal data of [4] indicated about 1654°C (Fig. 651).

A cursory microscopic, X-ray diffraction, and incipient-melting study of V-rich alloys (up to 40 wt. % Si) has been made by [12]. The solubility limit of Si in solid V was found to be just less than 5 wt. (8.7 at.) % Si; a 2.5 wt. (4.5 at.) % Si alloy annealed at 900°C showed a rejected phase indicating a decrease in solubility with decreasing temperature. A eutectic between (V) and V_3Si was found at about 7.5 wt. (13 at.) % Si and 1840°C. "Examination of cast structures indicates that V_3Si enters into equilibrium with a phase richer in Si by means of a peritectic reaction. The composition of this phase was not determined, but it is unlikely to be V_2Si as reported in the literature" [12]. A VSi (35.54 wt. % Si) phase is considered more probable [12].

An intermediate phase of unknown structure with the approximate composition V_3Si_2 (26.88 wt. % Si) has been reported by [13]. In summarizing, it can be said that the existence of only V_3Si, V_5Si_3, and VSi_2 has been conclusively established as yet.

Crystal Structures. V_3Si has the "β-W" (A15) type of structure [5, 6], with $a = 4.721 \pm 3$ A [5]. The compound V_5Si_3, prepared by powder-metallurgical techniques, was found to have the hexagonal Mn_5Si_3 ($D8_8$) type, with $a = 7.13_5$ A, $c = 4.84_2$ A, $c/a = 0.678_6$ [7] (but see Note Added in Proof). VSi_2 was shown [8] to have the hexagonal $CrSi_2$ (C40) type of structure, with $a = 4.571 \pm 3$ A, $c = 6.372 \pm 4$ A, $c/a = 1.359$.

Note Added in Proof. According to recent X-ray results [14, 15], the phase V_5Si_3 (or possibly V_3Si_2) has the tetragonal Ni_3P type of structure, with $a = 9.42_9$ A,

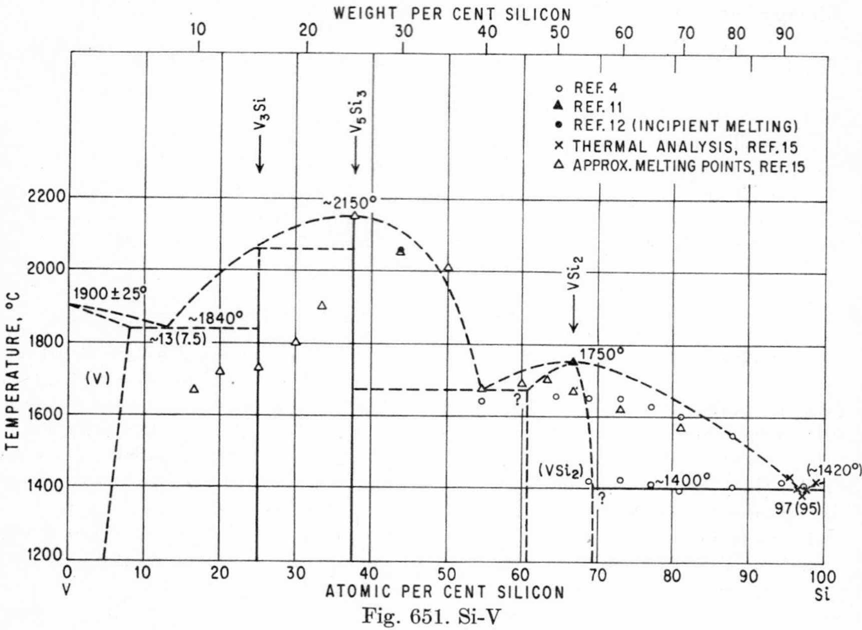

Fig. 651. Si-V

$c = 4.75_7$ A [isotypic with (Nb, Ta, Cr, Mo, W)$_5$Si$_3$]. The previously reported $D8_8$ type of structure corresponds to a ternary phase containing carbon.

[15] studied 24 alloys prepared by pressure sintering or arc melting. The results of thermal analysis and of the determination of approximate melting points by means of a Seger-cone method are plotted in Fig. 651 [16]. In addition, microscopic and X-ray work was carried out. [15] concluded that there are three intermediate phases in the V-Si system, namely, V_3Si, $\sim V_5Si_3$, and VSi_2. The eutectic (VSi_2 + Si) was located at 95 wt. % Si, 1385°C, in good agreement with the findings of [4, 10].

1. H. Moissan and A. Holt, *Compt. rend.*, **135**, 1902, 78–81, 493–497. VSi_2 is said to have the lower melting point.
2. P. Lebeau, *Ann. chim. et phys.*, **1**, 1904, 553.
3. L. Baraduc-Muller, *Rev. mét.*, **7**, 1910, 762–763.
4. H. Giebelhausen, *Z. anorg. Chem.*, **91**, 1915, 251–253.
5. H. J. Wallbaum, *Z. Metallkunde*, **31**, 1939, 362.
6. W. B. Pearson and W. Hume-Rothery, *J. Inst. Metals*, **80**, 1952, 652.

7. H. Schachner, E. Cerwenka, and H. Nowotny, *Monatsh. Chem.*, **85,** 1954, 245–254.
8. H. J. Wallbaum, *Z. Metallkunde,* **33,** 1941, 378–381.
9. R. Vogel and C. Jentsch-Uschinski, *Arch. Eisenhüttenw.*, **13,** 1940, 403–408.
10. W. R. Johnson and M. Hansen, U.S. Air Force Technical Report 6383, June, 1951.
11. E. Cerwenka, Thesis, Technische Hochschule, Graz, 1951. Quoted from P. Schwarzkopf and R. Kieffer, "Refractory Hard Metals," The Macmillan Company, New York, 1953.
12. W. Rostoker and A. Yamamoto, *Trans. ASM,* **46,** 1954, 1136–1163.
13. G. F. Hardy and J. K. Hulm, *Phys. Rev.*, **93,** 1954, 1013–1014.
14. E. Parthé, H. Nowotny, and H. Schmid, *Monatsh. Chem.*, **86,** 1955, 385–396.
15. R. Kieffer, F. Benesovsky, and H. Schmid, *Z. Metallkunde,* **47,** 1956, 247–253; R. Kieffer, H. Schmid, and F. Benesovsky, *Plansee Proc.*, **1955,** 154–156.
16. The melting-point determinations of V-rich alloys were seriously hampered by heavy carbon contamination. In their preliminary phase diagram, [15] accept 1840°C [12] as the temperature of the eutectic ($V + V_3Si$).

$\bar{1}.1839$
0.8161

Si-W Silicon-Wolfram

In the early literature several attempts to prepare W silicides were described [1–7], and the compounds W_2Si_3 [3, 7], WSi_2 [4, 5], and WSi_3 [7] were claimed to exist.

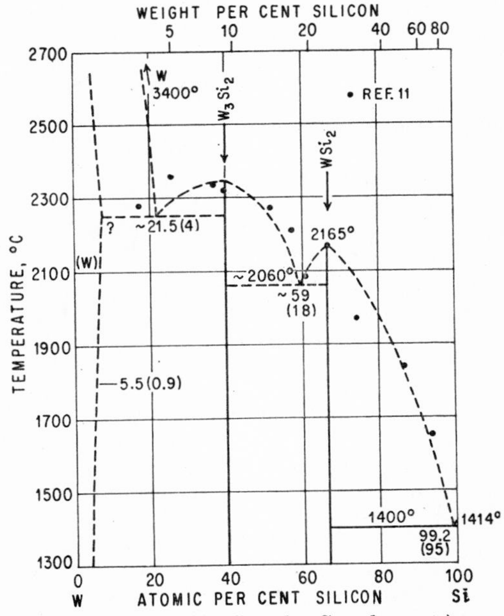

Fig. 652. Si-W. (See also Supplement.)

For details, see [8, 9]. Of these compounds, only WSi_2 (23.40 wt. % Si) was corroborated by more recent work.

Systematic investigations of the constitution of the W-Si system were carried out on sintered alloys by [10] (X-ray diffraction) and [11] (thermal, microscopic, and X-ray diffraction). Both papers agree on the existence of only one additional compound besides WSi_2, which [10] located in the range $WSi_{0.7\pm0.1}$ (37.5–44.4 at. % Si) and [11] showed to have the formula W_3Si_2 [40 at. (9.24 wt.) % Si]. The positions of the powder lines of this phase were the same whether W or WSi_2 was present [10]. [10] did not reach the melting zone of any alloy [12]. [11] made melting-point determinations using the Seger-cone method (Fig. 652) and took cooling curves of alloys containing 85–100 wt. % Si. Only for the Si-rich eutectic could relatively exact determinations be made: 5 wt. (0.8 at.) % W, 1400 ± 5°C. According to microscopic work by [13], this eutectic lies between 4 and 12 wt. % W. For the melting point of WSi_2, the values 2050 and 2165°C have been reported by [14, 11], respectively.

According to parametric work on quenched alloys, the solubility of Si in solid W is about 0.9 wt. (5.5 at.) % at 1800°C [11].

Crystal Structures. WSi_2 was shown [15] to have the tetragonal $MoSi_2$ (C11) type of structure, with $a = 3.21_0$ A, $c = 7.82_9$ A, $c/a = 2.43_9$ [16]. W_3Si_2 yields a complex powder pattern [11] which resembles that of $MoSi_{0.65\pm0.05}$ [10].

Supplement. According to recent work, the phase W_3Si_2 (or possibly W_5Si_3) is isostructural with the corresponding silicides of V, Nb, Ta, Cr, and Mo [17–19] and has a tetragonal structure [20, 18], with $a = 9.645$ A, $c = 4.969$ A, and 4 formula units "W_5Si_3" per unit cell [20]; $a = 9.56$ A, $c = 4.94$ A, $c/a = 0.51_7$ [18].

The W-Si phase diagram has been reinvestigated by [21] by means of pyrometric melting-point determinations as well as chemical, microscopical, and X-ray analyses of numerous alloys prepared from 98–99 wt. % pure W and 95–97 wt. % pure Si by arc melting in hydrogen atmosphere. The characteristic data of their phase diagram are in good agreement with those of Fig. 652: eutectic ($W + W_5Si_3$), ~21 at. % Si, 2210°C; melting point of W_5Si_3, ~2370°C; eutectic ($W_5Si_3 + WSi_2$), ~59.5 at. % Si, 2010°C; melting point of WSi_2, ~2160°C; eutectic (WSi_2 + Si), ~99 at. % Si, 1390°C.

1. H. Moissan, *Compt. rend.*, **123,** 1896, 13.
2. H. N. Warren, *Chem. News*, **78,** 1898, 318–319.
3. E. Vigouroux, *Compt. rend.*, **127,** 1898, 393–395.
4. E. Defacqz, *Compt. rend.*, **144,** 1907, 848–851.
5. O. Hönigschmid, *Monatsh. Chem.*, **28,** 1907, 1017–1028.
6. L. Baraduc-Muller, *Rev. mét.*, **7,** 1910, 761.
7. R. Frilley, *Rev. mét.*, **8,** 1911, 502–510.
8. "Gmelins Handbuch der anorganischen Chemie," System No. 54, pp. 204–205, Verlag Chemie G.m.b.H., Berlin, 1933.
9. P. Schwarzkopf and R. Kieffer, "Refractory Hard Metals," The Macmillan Company, New York, 1953.
10. L. Brewer, A. W. Searcy, D. H. Templeton, and C. H. Dauben, *J. Am. Ceram. Soc.*, **33,** 1950, 291–294.
11. R. Kieffer, F. Benesovsky, and E. Gallistl, *Z. Metallkunde*, **43,** 1952, 284–291.
12. They gave only "lower limits to eutectic temperatures," but not the eutectic temperatures themselves, as has been erroneously quoted in [11].
13. W. R. Johnson and M. Hansen, U.S. Air Force Technical Report 6383, June, 1951.
14. E. Cerwenka, Thesis, Technische Hochschule, Graz, 1951. Quoted by [9].
15. W. Zachariasen, *Z. physik. Chem.*, **128,** 1927, 39–48. *Strukturbericht*, **1,** 1913-1928, 219, 741, 783–784.
16. H. Nowotny, R. Kieffer, and H. Schachner, *Monatsh. Chem.*, **83,** 1952, 1248.
17. A. G. Knapton, *Nature*, **175,** 1955, 730.
18. E. Parthé, H. Schachner, and H. Nowotny, *Monatsh. Chem.*, **86,** 1955, 182–185.
19. E. Parthé, H. Nowotny, and H. Schmid, *Monatsh. Chem.*, **86,** 1955, 385–396.
20. B. Aronsson, *Acta Chem. Scand.*, **9,** 1955, 137–140.
21. R. Blanchard and J. Cueilleron, *Compt. rend.*, **244,** 1957, 1782–1785.

$\bar{1}.4996$
0.5004

Si-Y Silicon-Yttrium

The compound YSi_2 (38.72 wt. % Si) has been prepared by [1]. Its structure is obviously related to that of $ThSi_2$ [1]; a representation of a powder pattern is given.

1. G. Brauer and H. Haag, *Z. anorg. Chem.*, **267**, 1952, 198–212.

$\bar{1}.2104$
0.7896

Si-Yb Silicon-Ytterbium

For a representation of the powder pattern of an ytterbium silicide $YbSi_x$, see [1].

1. G. Brauer and H. Haag, *Z. anorg. Chem.*, **267**, 1952, 198–212.

$\bar{1}.6331$
0.3669

Si-Zn Silicon-Zinc

According to observations by [1–3], Si dissolved in molten Zn at higher temperatures separates out on cooling in elementary form. [4] was unable to alloy the two elements in an electric furnace (vaporization of Zn). [5] determined the solubility of Si in Zn at 600–850°C by quenching saturated melts from these temperatures and analyzing the Si content of the samples. The following solubilities were found, in wt. % (at. % in parentheses): 0.06, 0.15, 0.57, 0.92, and 1.62 (0.14, 0.35, 1.32, 2.12, and 3.7) at 600, 650, 730, 800, and 850°C, respectively.

From lattice-parameter measurements, [6] concluded that no mutual solid solubilities exist.

1. H. St. Claire Deville and H. Caron, *Compt. rend.*, **45**, 1857, 163; **57**, 1863, 740.
2. C. Winkler, *J. prakt. Chem.*, **91**, 1864, 193.
3. E. Baerwind, Thesis, Berlin, 1914. Abstract, *Intern. Z. Metallog.*, **7**, 1915, 213.
4. E. Vigouroux, *Compt. rend.*, **123**, 1896, 115.
5. H. Moissan and F. Siemens, *Ber. deut. chem. Ges.*, **37**, 1904, 2086–2089; *Compt. rend.*, **138**, 1904, 657, 1299.
6. E. R. Jette and E. B. Gebert, *J. Chem. Phys.*, **1**, 1933, 753–755.

$\bar{1}.4885$
0.5115

Si-Zr Silicon-Zirconium

Early Papers. In the early literature the preparation of the compounds $ZrSi_2$ (38.12 wt. % Si) [1, 2] and ZrSi (?) (23.55 wt. % Si) [3] was described [4].

Phase Diagram. The results of cursory work on Zr-rich [5–7] and Si-rich [8] alloys are referred to below. The phase diagram of Fig. 653 is mainly based on the work by [9]; metallography of cast and heat-treated specimens, incipient-melting tests, and thermal analyses were their principal tools. The solubility of Si in solid Zr was found to be very limited in both α [<0.1 wt. (0.3 at.) %] and β [<0.2 wt. (0.65 at.) %] Zr; whether a eutectoid or a peritectoid reaction takes place could not be ascertained. These solubility limits are in substantial agreement with observations by [5, 6]; the discussion remark by [7] that Zr dissolves in excess of 3 wt. % Si at 1000°C was not corroborated. The compositions reported by [5, 6, 9, 10] for the Zr-rich eutectic lie between 2.85 and 3 wt. (8.7 and 9.1 at.) % Si. [11] observed primary compound crystals in an as-cast 8 at. % Si alloy.

[9] claimed to have found the following compounds: Zr_4Si (7.15 wt. % Si), Zr_2Si

(13.34 wt. % Si), Zr_3Si_2, Zr_4Si_3 [42.86 at. (18.77 wt.) % Si], Zr_6Si_5 [45.45 at. (20.42 wt.) % Si], ZrSi, and $ZrSi_2$; only one of them, Zr_6Si_5, displayed an open maximum. At considerable variance with these findings are the results of X-ray and microscopic work and of melting-point determinations on sintered alloys by [12, 13, 10]. These authors corroborated, it is true, the existence of Zr_2Si, ZrSi, and $ZrSi_2$, but found Zr_5Si_3 [37.50 at. (15.59 wt.) % Si] to be the only compound melting with an open maximum and were unable to detect the phases Zr_4Si, Zr_3Si_2, Zr_4Si_3, and Zr_6Si_5 reported by [9].

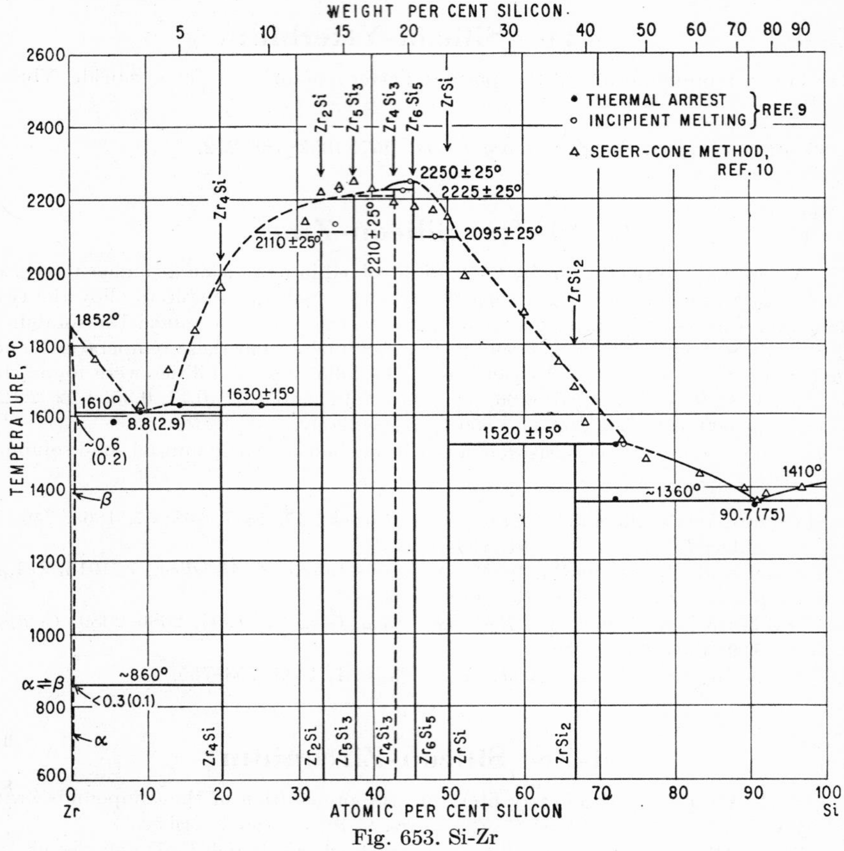

Fig. 653. Si-Zr

Recent X-ray powder analyses by [14] of their own alloys and of those supplied by [9] confirmed—with one exception—the results of [9] as to existing compounds; obviously Zr_5Si_3 should replace the formula Zr_3Si_2. [14] traced the failure of [12, 10] to detect all intermediate phases to an oxygen or carbon content of sintered alloys. Evidence for the existence of Zr_4Si has also been given by [7, 11]; [14] even suggested a high-temperature modification for this compound.

Data reported for the Si-rich eutectic are slightly above 73 wt. % Si (microscopic) [8]; 75 wt. (90.7 at.) % Si, 1355 ± 15°C [9]; and ~75 wt. % Si, 1360°C (thermal analysis [10]). [9] stated that "the solubility of Zr in Si is considerably less than 5 wt. %, but no attempt was made to precisely determine the value."

Crystal Structures. The α-Zr lattice is little affected by addition of Si [12].

Zr_2Si was shown [12, 11, 14] to have the tetragonal $CuAl_2$ (C16) type of structure. The lattice parameters reported vary considerably; those of [11] are the most reliable ones: $a = 6.6120 \pm 4$ A, $c = 5.2943 \pm 3$ A, $c/a = 0.8007$. No change in cell dimensions with composition could be detected by [11].

Zr_5Si_3 has the Mn_5Si_3 ($D8_8$) type of structure [12, 14, 10], with $a = 7.88_6$ A, $c = 5.55_8$ A, $c/a = 0.704_8$ at 37.5 at. % Si [12, 10]. According to [12, 10], this phase has a narrow, but roentgenographically well observable, range of homogeneity.

The powder pattern of ZrSi was indexed by [9] by means of a large hexagonal cell; however, [13] showed this phase to have the orthorhombic FeB (B27) type, with $a = 6.98_2$ A, $b = 3.78_6$ A, $c = 5.30_1$ A.

$ZrSi_2$ has an orthorhombic structure, with $a = 3.72$ A, $b = 14.76$ A, $c = 3.67$ A [13], and 12 atoms per unit cell. This unit cell was already known in 1928 [15]. The atomic-position parameters of the structure evaluated by [16] (C49 type in *Strukturbericht*, **5**, 1937) were criticized by [17] and the parameters redetermined by [18, 13].

Note Added in Proof. [19] found the lattice parameters $a = 3.72 \pm 1$ A, $b = 14.69 \pm 2$ A, $c = 3.66 \pm 1$ A for C49-type $ZrSi_2$.

1. E. Wedekind, *Z. Chem. u. Ind. Kolloide*, **7**, 1900, 249.
2. O. Hönigschmid, *Compt. rend.*, **143**, 1906, 224–226; *Monatsh. Chem.*, **27**, 1907, 1067–1069.
3. E. Wedekind, *Ber. deut. chem. Ges.*, **35**, 1902, 3932.
4. The literature up to 1910 was summarized by L. Baraduc-Muller, *Rev. mét.*, **7**, 1910, 763.
5. S. M. Shelton, *U.S. Atomic Energy Comm., Publ.* AF-TR-5932, 1949. *Met. Abstr.*, **21**, 1954, 869.
6. C. T. Anderson, E. T. Hayes, A. H. Roberson, and W. J. Kroll, *U.S. Bur. Mines, Rept. Invest.* 4658, 1950.
7. A. H. Roberson and E. T. Hayes, *Trans. ASM*, **44**, 1952, 536 (discussion remark).
8. W. R. Johnson and M. Hansen, U.S. Air Force Technical Report 6383, June, 1951.
9. C. E. Lundin, D. J. McPherson, and M. Hansen, *Trans. ASM*, **45**, 1953, 901–914. The alloys were prepared from an iodide Zr crystal bar (99.8 wt. % pure) and high-purity Si (99.99 wt. %) under protective He atmosphere in a nonconsumable-electrode-arc furnace.
10. R. Kieffer, F. Benesovsky, and R. Machenschalk, *Z. Metallkunde*, **45**, 1954, 493–498.
11. P. Pietrokowsky, *Acta Cryst.*, **7**, 1954, 435–438.
12. H. Schachner, H. Nowotny, and R. Machenschalk, *Monatsh. Chem.*, **84**, 1953, 677–685.
13. H. Schachner, H. Nowotny, and H. Kudielka, *Monatsh. Chem.*, **85**, 1954, 1140–1153.
14. L. Brewer and O. Krikorian, *U.S. Atomic Energy Comm., Publ.* UCRL-2544, 1954.
15. H. Seyfarth, *Z. Krist.*, **67**, 1928, 295–328.
16. St. v. Náray-Szabó, *Z. Krist.*, **A97**, 1937, 223–228.
17. G. Brauer and A. Mitius, *Z. anorg. Chem.*, **249**, 1942, 338–339.
18. G. Brauer and H. Haag, unpublished work, quoted by [13].
19. P. G. Cotter, J. A. Kohn, and R. A. Potter, *J. Am. Ceram. Soc.*, **39**, 1956, 11–12; *Met. Abstr.*, **24**, 1957, 569–570.

0.1318
$\bar{1}.8682$

Sn-Sr Tin-Strontium

From thermal and microscopic investigations on Sn-rich Sn-Sr alloys (especially after annealing for several hours at 240°C), [1] concluded the existence of two com-

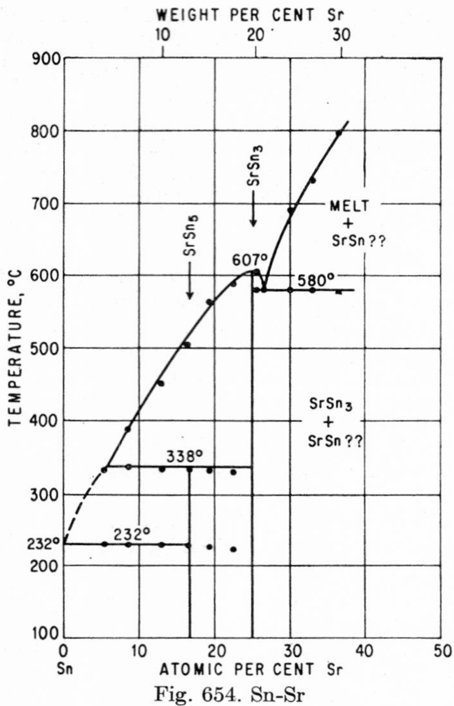

Fig. 654. Sn-Sr

pounds, $SrSn_5$ [16.67 at. (12.87 wt.) % Sr] and $SrSn_3$ (19.75 wt. % Sr). Figure 654 shows the phase diagram established by [1].

1. K. W. Ray, *Ind. Eng. Chem.*, **22,** 1930, 519–522.

$\bar{1}.8171$
0.1829

Sn-Ta Tin-Tantalum

According to [1], Ta_3Sn (17.95 wt. % Sn) has the "β-W" (A15) type of structure, with $a \sim 5.3$ A. The compound seems to be formed by a peritectic reaction between 1200 and 1550°C. [2] reported the lattice parameter $a = 5.276 \pm 1$ A and stated that the structure is probably partially disordered.

1. B. T. Matthias, T. H. Geballe, S. Geller, and E. Corenzwit, *Phys. Rev.*, **95,** 1954, 1435.
2. S. Geller, B. T. Matthias, and R. Goldstein, *J. Am. Chem. Soc.*, **77,** 1955, 1502–1504.

$\bar{1}.9686$
0.0314

Sn-Te Tin-Tellurium

The solidification equilibria of the Sn-Te alloys were investigated by [1] (cursory only), [2–4], and [5] (0–3 wt. % Te only). The liquidus of Fig. 655 is essentially based on the data of [1, 3, 4]; most of those reported by [2] are appreciably too low. A compilation of distinctive temperatures and compositions of the diagram may be found in Table 46 [6].

Table 46

	Ref. 1	Ref. 2	Ref. 3	Ref. 4
End of solidification, 0–50 at. % Te.	. . . .	232°C	232°C	232°C
Melting point of SnTe.	~780°C	769°C	~796°C	781°C
SnTe-Te eutectic.	388°C	399°C	405°C	393°C
	85 wt. % Te	85 wt. % Te	85 wt. % Te	86 wt. % Te
Melting point of Te.	452°C	446°C	455°C	437°C

The compound SnTe (51.81 wt. % Te), the existence of which is evident from the thermal results, had been previously prepared by [7] and found to distill undecomposed. Whether it forms solid solutions with Sn or Te has not been adequately

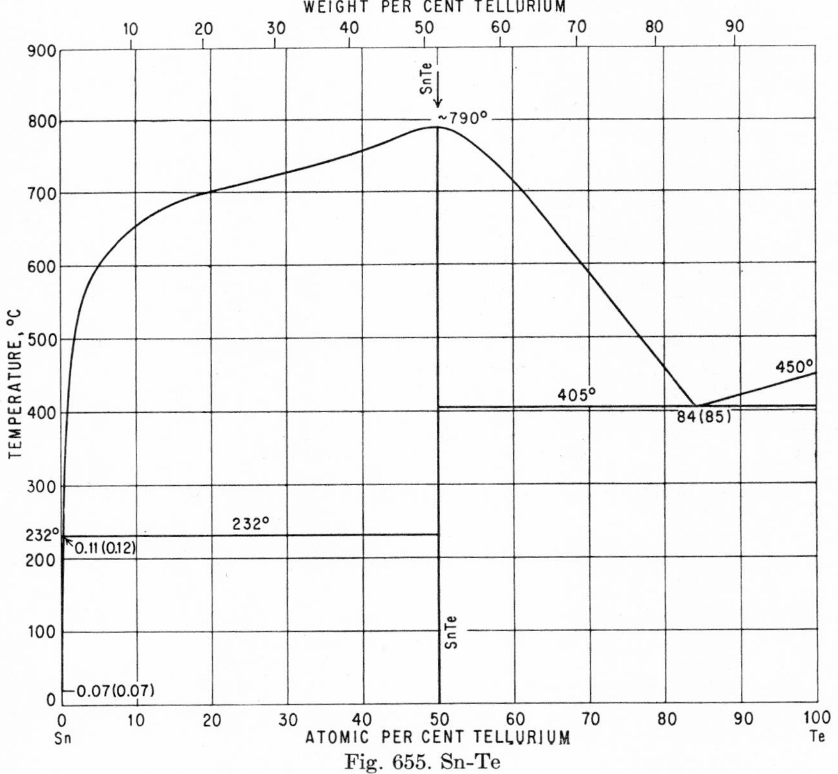

Fig. 655. Sn-Te

studied as yet; however, measurements of magnetic susceptibility [8, 9] exclude the existence of any extended solutions.

Microscopic work [2–4] and measurements of the electrical conductivity and the thermo-emf [10] as well as magnetic susceptibility [8, 9] and of the emf [11] are in general agreement with the phase diagram of Fig. 655. The work of [10, 8, 9] indicates that the solubility of Sn in solid Te is negligible.

The solubility of Te in solid Sn has been studied by [12, 13]. According to microscopic work by [12], the solubility is 0.12 wt. (0.11 at.) % Te at 230°C and 0.07 wt. (at.) % Te at 20°C; these authors also found some evidence for a eutectic near pure Sn. A much lower solubility—~0.025 at. (wt.) % Te at 200°C—has been reported by [13]. In Fig. 655, weight has been given to the data of [12].

Magnetic-susceptibility isotherms of molten Sn-Te alloys have sharp minima at the composition SnTe which indicate the existence of SnTe molecules in the melt [9]. SnTe was shown [14] to have the NaCl (B1) type of structure, with $a = 6.298 \pm 4$ A.

1. H. Pélabon, *Compt. rend.*, **142,** 1906, 1147–1149. *Ann. chim. et phys.*, **17,** 1909, 526.
2. H. Fay, *J. Am. Chem. Soc.*, **29,** 1907, 1265–1268.
3. W. Biltz and W. Mecklenburg, *Z. anorg. Chem.*, **64,** 1909, 226–235.
4. M. Kobayashi, *Z. anorg. Chem.*, **69,** 1911, 6–9.
5. W. T. Pell-Walpole, unpublished results. Quoted from "Metals Handbook," 1948 ed., p. 1240, The American Society for Metals, Cleveland, Ohio.
6. The relatively flat part of the liquidus between about 18 and 40 at. % Te—especially marked in the data of [3]—suggests that there may be in reality a horizontal line (miscibility gap in the liquid state). However, only strong segregation, but not formation of two layers, was reported by [3].
7. A. Ditte, *Compt. rend.*, **96,** 1893, 1792.
8. K. Honda and T. Soné, *Science Repts. Tôhoku Imp. Univ.*, **2,** 1913, 10–11.
9. H. Endo, *Science Repts. Tôhoku Imp. Univ.*, **16,** 1927, 218–220; see also K. Honda and H. Endo, *J. Inst. Metals*, **37,** 1927, 38–41.
10. W. Haken, *Ann. Physik*, **32,** 1910, 316–318.
11. N. Puschin, *Z. anorg. Chem.*, **56,** 1908, 15–17.
12. D. Hanson and W. T. Pell-Walpole, *J. Inst. Metals*, **63,** 1938, 109–122.
13. E. Jenckel and L. Roth, *Z. Metallkunde*, **30,** 1938, 135–144.
14. V. M. Goldschmidt, *Skrifter Norske Videnskaps-Akad. Oslo, Mat. Naturv. Kl.*, no. 8, 1927, pp. 7–156. *Strukturbericht*, **1,** 1913-1928, 137.

$\bar{1}.7087$
0.2913

Sn-Th Tin-Thorium

According to [1], Th dissolves in molten Sn and precipitates on cooling in elementary form.

1. Report on German Patent (DRP) 146,503, 1900, in *Chem. Zentr.*, **1903**(2), 1156.

0.3941
$\bar{1}.6059$

Sn-Ti Tin-Titanium

Preliminary Work. [1] concluded from microscopic work that at least up to 1.7 wt. % Sn is soluble in both Ti modifications. The existence of the intermediate phases Ti_3Sn (45.24 wt. % Sn) [2] and Ti_5Sn_3 [37.50 at. (59.78 wt.) % Sn] [3] was established by crystal-structure determinations (see Crystal Structures).

Phase Diagram. Four papers on the constitution of this system have been published recently. [4] investigated sponge as well as iodide Ti-base alloys with 0–25 at. % Sn mainly in the temperature interval 600–1120°C by means of metallographic and X-ray powder methods. [5] reported on alloys containing 0–12 at. % Sn. The partial phase diagram presented by these authors is to be the subject of a forthcoming publication [5a], and the details of the investigation are not discussed in [5]; however, as [6] pointed out, it is fairly clear that iodide-base alloys were used. [7], using iodide

Ti, determined the entire constitution diagram by means of metallographic, X-ray, incipient-melting, and thermal-analysis methods. Finally, [8] rechecked the influence of Sn on the $\alpha \rightleftarrows \beta$ transformation in Ti using the "hydrogen-pressure method."

Ti-*rich Alloys.* Sn depresses the freezing point of Ti [4, 5, 7] until a eutectic is reached for which the data 17 at. %, 1550°C, and 18.0 at. % Sn, 1605°C, were found by [4] and [7], respectively. The conflicting results obtained by [4, 5, 7, 8] as to the effect of Sn on the transformation in Ti can be seen from Fig. 656. Although the minimum in the $(\alpha + \beta)$ field reported by [4] was not well established (insufficient number of alloys, short annealing times), it could be corroborated—at a slightly higher Sn content—by an independent method [8], and its inclusion in the recommended phase diagram of Fig. 657 appears justified from the evidence available. From this minimum a peritectoid [4, 5] rather than a eutectoid [7] reaction as termination of the $(\alpha + \beta)$ field necessarily ensues.

In the phase diagram given by [7], Ti_3Sn is shown as a strictly stoichiometric compound; however, [4] found some microscopic and roentgenographic indications for a certain range of homogeneity of this phase, at least toward higher Ti contents. See Note Added in Proof.

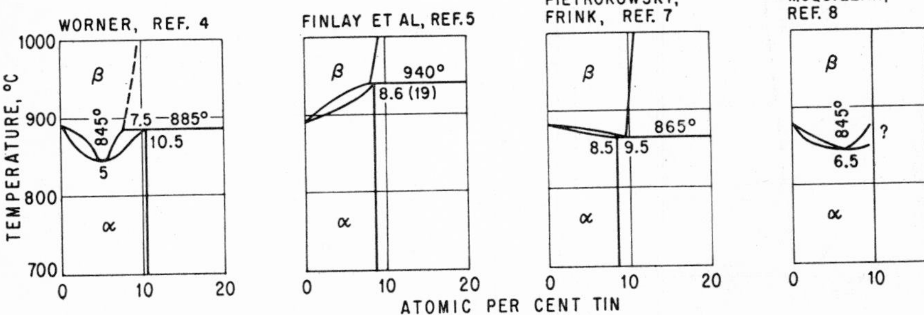

Fig. 656. Sn-Ti. (See also Figs. 657 and 657*a*.)

Rest of the Phase Diagram. The phase relationships for this system beyond Ti_3Sn are due to [7]. Three additional intermediate phases exist (Fig. 657): Ti_2Sn (55.34 wt. % Sn), Ti_5Sn_3, and Ti_6Sn_5 [45.45 at. (67.37 wt.) % Sn]. Thermal analyses carried out by [7] were restricted to Sn-rich alloys; an influence of Ti on the freezing point of Sn could not be conclusively established.

Crystal Structures. The lattice parameters of α-Ti are increased by addition of Sn [4, 7]. Ti_3Sn has the hexagonal Ni_3Sn (DO_{19}) type of structure [2, 4], with $a = 5.916 \pm 4$ A, $c = 4.764 \pm 4$ A, $c/a = 0.805$ [2]. Ti_2Sn was determined [7] to be a filled NiAs (B8) type, with $a = 4.65_3$ A, $c = 5.70_0$ A, $c/a = 1.22$ (at 34.8 at. % Sn). Ti_5Sn_3 has the hexagonal Mn_5Si_3 ($D8_8$) type, with $a = 8.049 \pm 2$ A, $c = 5.454 \pm 2$ A, $c/a = 0.678$ [3]; and Ti_6Sn_5, according to powder as well as single-crystal work by [7], has a hexagonal translation group, with $a = 9.22 \pm 1$ A, $c = 5.69 \pm 1$ A, $c/a = 0.617$, and 22 atoms per unit cell. The lattice parameters of the Sn phase in a $(Ti_6Sn_5 + Sn)$ alloy were essentially the same as those of pure Sn [7].

Note Added in Proof. [9] concluded from a micrographic study of 13 alloy compositions in the range 2–25 at. % Sn that there appears to be a continuous transition from the α_{Ti} primary solid solution to the ordered Ti_3Sn, or γ, phase. The partial phase diagram established by [9] is shown in Fig. 657*a*. The following comments may be made: (*a*) The minimum in the α-β equilibria shown in Fig. 657*a* was accepted from the work of [8]. (*b*) Although very long annealing times were used (up to 7½ months), equilibrium could not be reached in most alloys below 900°C

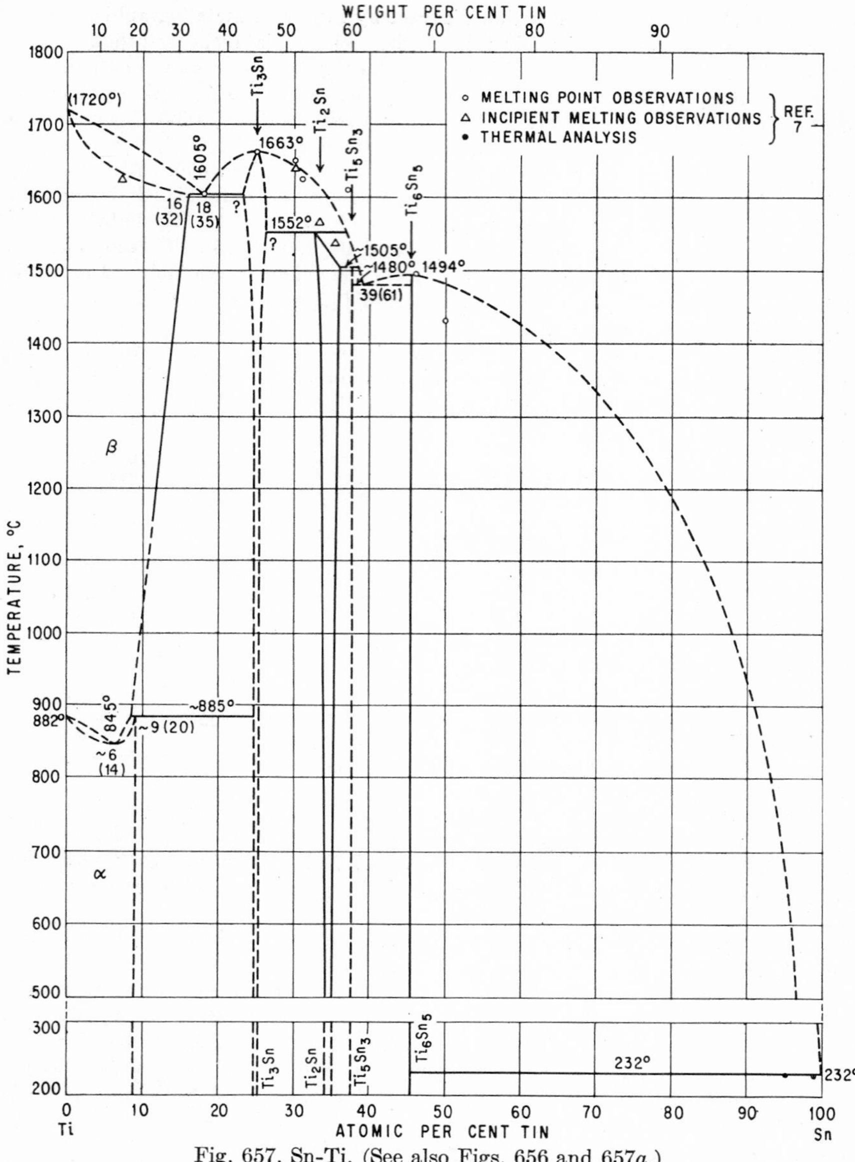

Fig. 657. Sn-Ti. (See also Figs. 656 and 657*a*.)

(see crosses in Fig. 657*a*). Strange to say, no effort was made to enhance the rate of diffusion by working the specimens prior to annealing. (*c*) There is no doubt that Ti_3Sn, or γ, is not a strictly stoichiometric compound, as is pretended by [7], but dissolves appreciable amounts of Ti at lower temperatures. [9] found the 15.3 at. % Sn alloy to be one-phase after annealing for 21 weeks at 880°C. The findings of [7], however, seem to exclude the possibility of a continuous transition $\alpha \rightarrow \gamma$. Thus [7] observed that a 9 at. % Sn alloy, quenched from the β-phase field, then cold-rolled, is clearly two-phase after annealing for 20 days at 788°C [10]. [9] admits that a narrow ($\alpha + \gamma$) two-phase field may exist (see dashed lines in Fig. 657*a*); it appears to the reviewer, however, that the limiting solubility of Sn in α_{Ti} is appreciably lower than is assumed in this figure.

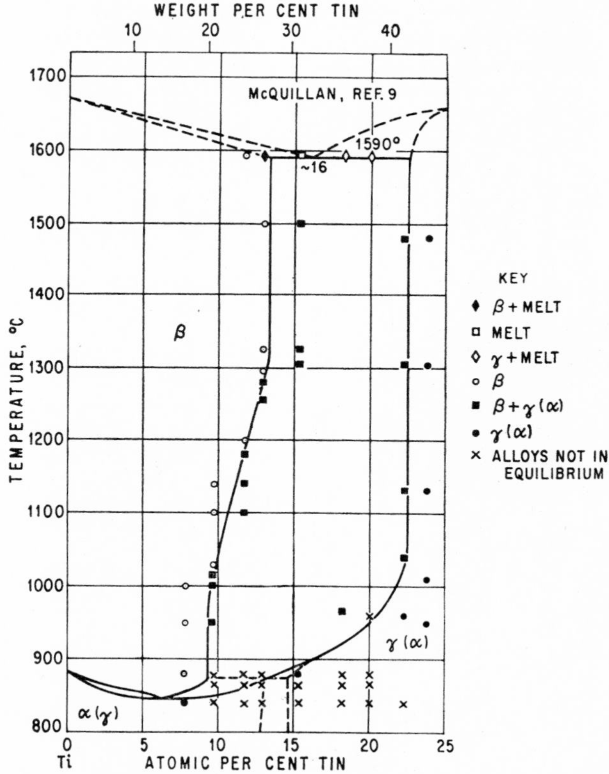

Fig. 657*a*. Sn-Ti. (See also Figs. 656 and 657.)

1. C. M. Craighead, O. W. Simmons, and L. W. Eastwood, *Trans. AIME*, **188**, 1950, 485–513, especially 498.
2. P. Pietrokowsky, *Trans. AIME*, **194**, 1952, 211–212.
3. P. Pietrokowsky and P. Duwez, *Trans. AIME*, **191**, 1951, 772–773.
4. H. W. Worner, *J. Inst. Metals*, **81**, 1953, 521–528.
5. W. L. Finlay, R. I. Jaffee, R. W. Parcel, and R. C. Durstein, *J. Metals*, **6**, 1954, 25.

5a. R. I. Jaffee, H. R. Ogden, D. J. Maykuth, and W. L. Finlay (paper in preparation).

6. D. W. Levinson, D. J. McPherson, and W. Rostoker, Supplement to Constitution of Ti Alloy Systems, WADC Technical Report 54-502, 1954.
7. P. Pietrokowsky and E. P. Frink, Jet Propulsion Laboratory, California Institute of Technology, Progress Report 20-212, March, 1954; also *Trans. ASM*, **49**, 1957, 339–358.
8. A. D. McQuillan, *J. Inst. Metals*, **83**, 1955, 181–184.

9. M. K. McQuillan, *J. Inst. Metals*, **84**, 1956, 307–312; discussion, 532–535.
10. See fig. 11 of [7] (*Trans. ASM* paper).

$\bar{1}.7640$
0.2360

Sn-Tl Tin-Thallium

Solidification. Thermal analyses were carried out by [1] (80–100 wt. % Sn, 16 alloys), [2] (entire system, 50 alloys), and [3] (entire system, 16 alloys). The results of [1, 2] are in perfect agreement, and they were used in drawing the liquidus of Fig. 658. The data by [3] show deviations up to ±6°C from this curve. Of the eutectic data reported—56.5 wt. (69.1 at.) % Sn, 170°C [2]; 57.6 wt. (70.1 at.) % Sn, 166°C [3]—those of [2] are the more reliable ones.

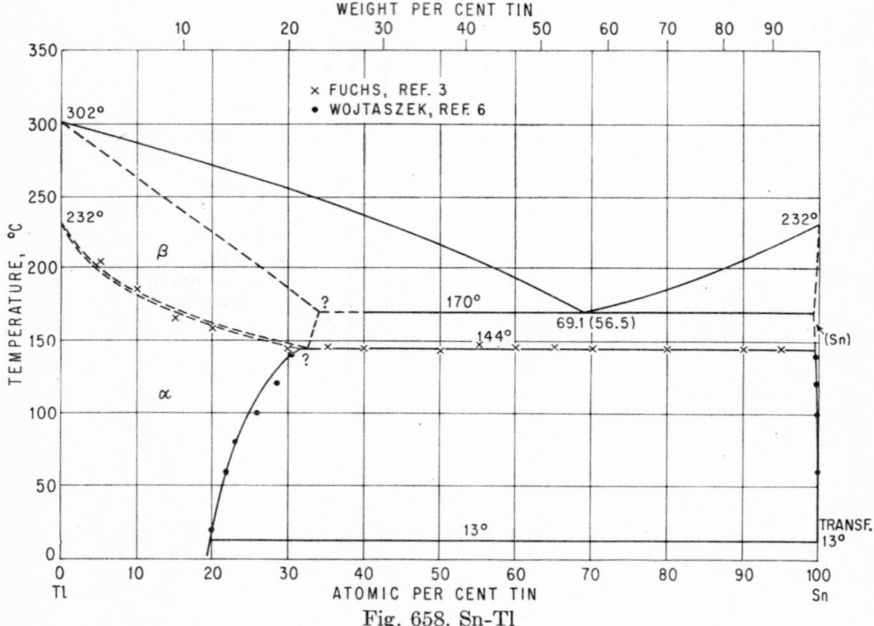

Fig. 658. Sn-Tl

The existence of a primary solid solution of Sn in Tl was concluded by [2] from the anomalous low freezing-point depression, and [3] estimated from thermal data (eutectic arrests) its extent to 17.5 wt. (27 at.) % Sn at the eutectic temperature. This value is certainly too low.

Transformation in the Solid State. The α (h.c.p.) $\rightleftarrows \beta$ (b.c.c.) transformation in pure Tl is lowered by addition of Sn, as the thermal data [3] plotted in Fig. 658 indicate. According to [4], quenched (temperature?) alloys with 10 and 20 wt. % Sn possess f.c.c. lattices (metastable state).

Solid Solubilities. The mutual solid solubilities of Tl and Sn up to 140°C were determined by [6] on carefully annealed alloys by measurements of electrical resistivity [7]. The results, also plotted in Fig. 658, are 30.4, 28.5, 25.9, 23.0, 21.8, and 20.0 at. % Sn in Tl at 140, 120, 100, 80, and 60°C, and room temperature, respectively; 0.35, 0.275, 0.225, 0.175, and 0.125 at. % Tl in Sn at 140, 120, 100, and 60°C, and room temperature, respectively. Solubility data for 140°C (30 at. % Sn in Tl, 0.4 at. % Tl in Sn) had already been measured by this method [8]; they are in excellent agreement with those of [6].

Cursory resistometric measurements by [9] gave no reliable basis for deriving quantitative data for the solid solubility of Sn in Tl.

Further Work. Superconducting-point-vs.-composition curves measured by [10, 11] are more complex than the simple constitution of Fig. 658 suggests, and [12]—on the basis of the curve by [10]—postulated an intermediate phase at about 90–95.5 at. % Sn which is assumed to be peritectically formed at about 170–200°C. Any corroborating experimental evidence is missing. The results of an X-ray analysis by [5], unfortunately carried out on unannealed alloys and therefore of only qualitative value, are in keeping with the constitution of Fig. 658. Measurements of emf [13] and magnetic susceptibility [14] yielded no clear picture of the constitution.

1. C. T. Heycock and F. H. Neville, *J. Chem. Soc.*, **57**, 1890, 379.
2. N. S. Kurnakov and N. A. Puschin, *Z. anorg. Chem.*, **30**, 1902, 101–108.
3. P. Fuchs, *Z. anorg. Chem.*, **107**, 1919, 308–312.
4. S. Sekito, *Z. Krist.*, **74**, 1930, 193–195.
5. H. J. C. Ireton, J. P. Blewett, and J. F. Allen, *Can. J. Research*, **9**, 1933, 415–418.
6. Z. Wojtaszek, *Bull. intern. acad. polon. sci., Classe sci. math. nat.*, **A1952**, 147–157 (in English).
7. Measuring of r values, i.e., ratio of resistivity at temperature of boiling N_2 or H_2 to resistivity at temperature of melting ice.
8. E. Kurzyniec, *Roczniki Chem.*, **18**, 1938, 651–659 (in German, pp. 658–659). Quoted by [6].
9. G. Tammann and H. Rüdiger, *Z. anorg. Chem.*, **192**, 1930, 40–42.
10. W. Meissner, H. Franz, and H. Westerhoff, *Ann. Physik*, **13**, 1932, 510–521; *Metallwirtschaft*, **10**, 1931, 293–294.
11. J. F. Allen, *Phil. Mag.*, **16**, 1933, 1005–1044. Quoted in [5].
12. C. Benedicks, *Z. Metallkunde*, **25**, 1933, 201.
13. R. Kremann and A. Lobinger, *Z. Metallkunde*, **12**, 1920, 251–253.
14. F. Meara, *Physics*, **2**, 1932, 33–41.

$\bar{1}.6977$
0.3023

Sn-U Tin-Uranium

According to [1], U forms at least three compounds with Sn, all being brittle and exceedingly pyrophoric. One of them has been definitely located [1, 2]: USn_3 (59.93 wt. % Sn), Cu_3Au ($L1_2$) type of structure, with a = 4.626 A [2], peritectic [3] melting point 1350°C [3, 2]. The formulas U_5Sn_4 and U_3Sn_5 given for two other compounds are not yet conclusively established [3]. The solubility of Sn in α-U is nil, and that of U in (liquid) Sn at 600°C is <0.02 at. (<0.04 wt.) % [3].

Note Added in Proof. The phase diagram illustrated in Fig. 658*a* has been reproduced from [4]. It is based on the results of metallographic examination, thermal analysis, and X-ray diffraction [5].

1. R. E. Rundle and A. S. Wilson, *Acta Cryst.*, **2**, 1949, 148–150.
2. B. R. T. Frost and J. T. Maskrey, *J. Inst. Metals*, **82**, 1953, 177.
3. J. J. Katz and E. Rabinowitch, "The Chemistry of Uranium," Part I, National Nuclear Energy Series, Div. VIII, vol. 5, pp. 174–179, McGraw-Hill Book Company, Inc., New York, 1951.
4. H. A. Saller and F. A. Rough, Compilation of U.S. and U.K. Uranium and Thorium Constitutional Diagrams, *U.S. Atomic Energy Comm., Publ.* BMI-1000, June, 1955.
5. H. A. Wilhelm and D. A. Treick, unpublished information (September, 1944).

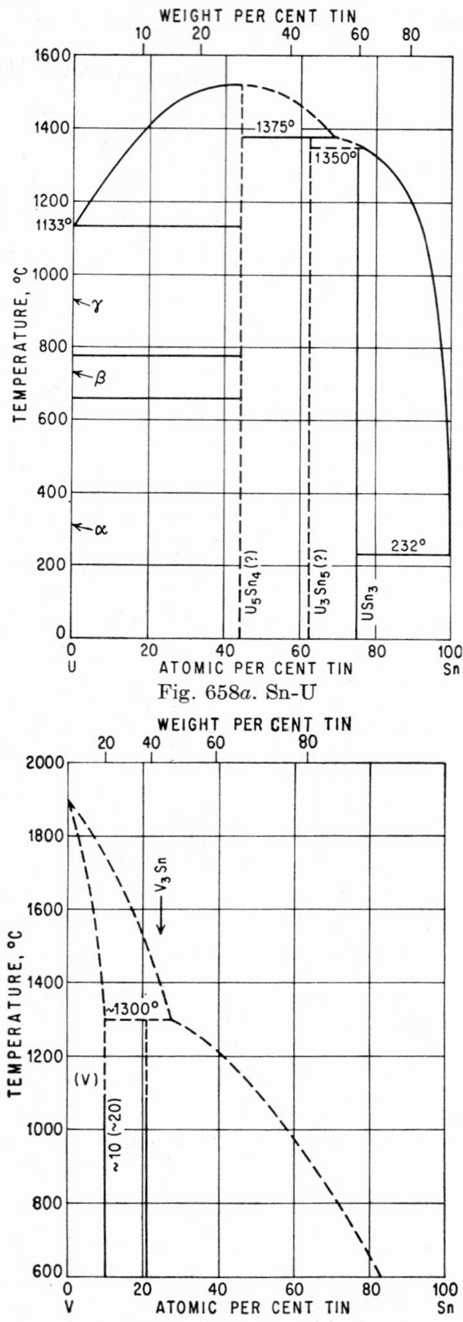

Fig. 658*a*. Sn-U

Fig. 658*b*. Sn-V

0.3673
$\bar{1}$.6327

Sn-V Tin-Vanadium

Examination of alloys containing up to 10 wt. (4.6 at.) % Sn showed a range of solid solution within the limits studied [1]. [2] reported the compound V_3Sn (43.71 wt. % Sn) to have the cubic "β-W" (A15) type of structure, with $a = 4.94 \pm 1$ A.

The tentative phase diagram of Fig. 658*b* was established by [3] by means of microscopic and X-ray analysis of 10 alloys annealed at temperatures up to 1100°C. The solubility of Sn in solid V was found to be about 20 wt. (10 at.) % Sn; the lattice parameter of the b.c.c. cell increases from 3.01 to 3.08 A with increasing Sn content. The V_3Sn phase was observed to be homogeneous around 38.5 wt. (21.2 at.) % Sn rather than at its stoichiometric composition.

1. W. Rostoker and A. Yamamoto, *Trans. ASM*, **46,** 1954, 1136–1163, especially 1157.
2. S. Geller, B. T. Matthias, and R. Goldstein, *J. Am. Chem. Soc.*, **77,** 1955, 1502–1504.
3. W. Köster and K. Haug, *Z. Metallkunde*, **48,** 1957, 327.

$\bar{1}$.8098
0.1902

Sn-W Tin-Wolfram

Several attempts [1–4] to alloy W and Sn were unsuccessful. [4] found that W wire is not dissolved in molten Sn heated to 1680°C.

1. Caron, *Ann. chim. et phys.*, **68,** 1863, 143.
2. C. Sargent, *J. Am. Chem. Soc.*, **22,** 1900, 783.
3. D. Kremer, *Abhandl. Inst. Metallhütt. u. Elektromet. Tech. Hochsch., Aachen,* **1,** 1916(2), 11–12 (Aluminothermic method).
4. E. F. Northrup, *Bull. AIME*, **1919,** 1443; quoted from "Gmelins Handbuch der anorganischen Chemie," System No. 54, p. 104, Verlag Chemie, G.m.b.H., Berlin, 1933.

0.2590
$\bar{1}$.7410

Sn-Zn Tin-Zinc

The determinations of liquidus points by [1–3] were lessened in importance by the more detailed and exact measurements (35 alloys) by [4]. Later determinations of the whole curve by [5] (16 alloys), [6] (13 alloys), [31] (15 alloys) and of parts of it by [7] (alloys with 1 and 9 wt. % Sn), [8] (16 alloys with 96–100 wt. % Sn), [9] (9 alloys), and [10] (alloy with 70 wt. % Sn) are in very good agreement [11] with the liquidus of [4] (Fig. 659). The liquidus was computed by [12] on the basis of emf data and the simplifying assumption that solid solubilities are zero (cf. below).

For the eutectic point, the following data have been reported (in wt. % Sn unless stated otherwise): 91.5, 204°C [1]; ~92.5, 198°C [2]; 90.5, 198°C [4]; 92, 199°C [5]; ~91.7, 199°C [6]; 199°C [7]; 85.5 at. (91.5 wt.) % Sn [13]; 92, 196.7°C [9]; 91, 198°C [31]; and 82.5 at. (89.5 wt.) % Sn, 198°C [14]. For Fig. 659, the values 85 at. (91 wt.) % Sn, 198°C, were chosen. The orientation relationship of the eutectic components was studied by [15] by means of X-ray diffraction.

The absence of an intermediate phase in the system Sn-Zn is clearly indicated by thermal and microscopic investigations [16] as well as measurements of various physical properties as a function of composition. Among others, there are investigations on the following properties: electrical and thermal conductivity [17], thermo-emf [18], magnetic susceptibility [19], emf [20–22, 6], and hardness [23].

Solid Solubilities. The solubility of Sn in solid Zn is very restricted. Microscopic work by [24, 9] failed to indicate any measurable solubility. [25] found the solubility at 400°C (after quenching) to be slightly greater than 0.1 wt. (0.06 at.) %; after slow-cooling the melt to room temperature, a 0.05 wt. (0.03 at.) % Sn alloy was homogeneous, whereas one containing 0.1 wt. % Sn was two-phase. From measurements of tensile strength, [26] concluded the solubility to lie between 0.05 and 0.1 wt. % Sn at the eutectic temperature.

Data reported in the literature for the solubility of Zn in solid Sn at or near the eutectic temperature range from 0 to 7 wt. % Zn. The recent X-ray parametric work by [27] showed clearly that the solubility is at least 0.7 at. (0.4 wt.) %; this eliminates the data 0.2 at. (0.1 wt.) % and 0.325 wt. % reported by [28] (for 180°C) and [29] (for eutectic temperature), respectively. Microscopic work was carried out by [13] and [9]; [13] annealed his alloys for 6 weeks at 185°C and found the solubility to lie between

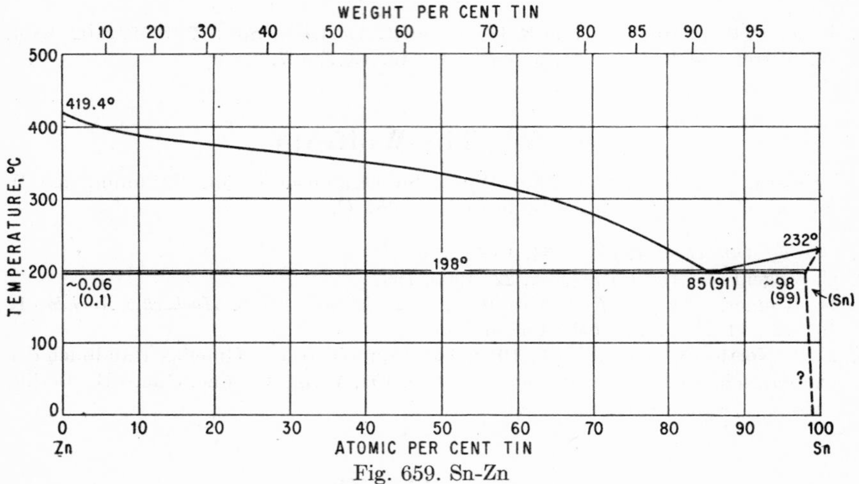

Fig. 659. Sn-Zn

1 and 3 at. (0.5 and 1.7 wt.) % Zn; [9], annealing for 11 days at 150°C, reported the solubility value 2.5 wt. %. Resistometric work by [30] on alloys annealed at 190°C yielded the bracket 1–2 at. (0.5–1.1 wt.) % Zn. For Fig. 659, 2 at. (1 wt.) % Zn has been used as a reasonable estimate.

1. F. Rudberg, *Pogg. Ann.*, **18,** 1830, 240.
2. D. Mazzotto, *Mem. R. Ist. Lombardo*, **16,** 1886, 1; cf. K. Bornemann, *Metallurgie*, **7,** 1910, 90–91.
3. H. Gautier, *Bull. soc. encour. ind. natl.*, **1,** 1896, 1293. Also, Société d'encouragement pour l'industrie nationale, Paris, Commission des alliages, "Contribution à l'étude des alliages," pp. 93–118, Typ. Chamerot et Renouard, Paris, 1901.
4. C. T. Heycock and F. H. Neville, *J. Chem. Soc.*, **57,** 1890, 382; **71,** 1897, 392–393.
5. R. Lorenz and D. Plumbridge, *Z. anorg. Chem.*, **83,** 1913, 228–231. R. Lorenz, *Z. anorg. Chem.*, **85,** 1914, 435–436.
6. E. Crepaz, *Giorn. chim. ind. ed appl.*, **5,** 1923, 115–116.
7. P. Arnemann, *Metallurgie*, **7,** 1910, 205–206.
8. L. J. Gurevich and J. S. Hromatko, *Trans. AIME*, **64,** 1921, 234–235.
9. R. Blondel, *Publ. sci. et tech. ministère air (France)*, no. 89, 1936; see also R. Blondel and P. Laffitte, *Compt. rend.*, **200,** 1935, 1472–1474.

10. H. Johnen, Dissertation, Münster, Germany, 1952.
11. In a special test, Johnen [10] was unable to find any indications for a miscibility gap in the liquid state.
12. F. J. Dunkerley and G. J. Mills, "Thermodynamics in Physical Metallurgy," The American Society for Metals, Cleveland, Ohio, 1950.
13. P. Saldau, *J. Inst. Metals*, **41,** 1929, 289–309.
14. O. Hájíček, *Hutnické Listy*, **3,** 1948, 265–270. Not available to the author; *Chem. Abstr.*, **43,** 1949, 4935.
15. M. Straumanis and N. Brakss, *Z. physik. Chem.*, **B38,** 1937, 140–155. *Strukturbericht*, **5,** 1937, 134.
16. For microscopic studies, see G. Charpy, *Bull. soc. encour. ind. natl.*, **1898,** 670; B. E. Curry, *J. Phys. Chem.*, **13,** 1909, 597–598.
17. F. A. Schulze, *Ann. Physik*, **9,** 1902, 565–567, 583–584. See also A. Schulze, *Z. anorg. Chem.*, **159,** 1927, 330–332.
18. E. Rudolfi, *Z. anorg. Chem.*, **67,** 1910, 72–75.
19. H. Endo, *Science Repts. Tôhoku Imp. Univ.*, **14,** 1925, 488–489.
20. A. P. Laurie, *J. Chem. Soc.*, **55,** 1889, 679.
21. M. Herschkowitsch, *Z. physik. Chem.*, **27,** 1898, 141.
22. P. Fuchs, *Z. anorg. Chem.*, **109,** 1920, 87.
23. W. Schischokin and W. Ageeva, *Z. anorg. Chem.*, **193,** 1930, 237–244.
24. W. M. Peirce, *Trans. AIME*, **68,** 1923, 781.
25. G. Tammann and W. Crone, *Z. anorg. Chem.*, **187,** 1930, 300.
26. G. Tammann and H. J. Rocha, *Z. Metallkunde*, **25,** 1933, 133.
27. J. A. Lee and G. V. Raynor, *Proc. Phys. Soc.* (*London*), **B67,** 1954, 737–747.
28. E. Jenckel and L. Roth, *Z. Metallkunde*, **30,** 1938, 135–144.
29. C. E. Homer and H. Plummer, *J. Inst. Metals*, **64,** 1939, 169–200.
30. E. Kurzyniec, *Bull. intern. acad. polon. sci.*, **8/10A,** 1938, 489–497 (in German).
31. W. Oelsen, *Z. Metallkunde*, **48,** 1957, 1–8 (Calorimetry and Thermodynamics of Sn-Zn Alloys).

0.1144
$\bar{1}.8856$

Sn-Zr Tin-Zirconium

The phase diagram of Fig. 660 is essentially the result of a systematic investigation by [1]; micrographic analysis of as-cast and heat-treated specimens was the principal tool, and thermal analysis and X-ray diffraction served as auxiliary techniques. As for the solubility of Sn in α-Zr and the composition of two intermediate phases, the diagram given by [1] was modified to account for the results of other investigations.

Solubility of Sn in α- and β-Zr. [1] determined the solubility of Sn in solid Zr between 600 and 1472°C micrographically by the "composition-bracket" method (Figs. 660 and 661). The annealing times for the cold-worked Grade 1 Zr alloys were 445 hr at 604°C and 20 hr at 1100°C.

The α_{Zr} boundary was also studied by [2], using the X-ray parametric method between 700 and 900°C (quenched alloys) and a strain-aging method [3] between 300 and 500°C. The annealing times used by [2] for the X-ray wire and powder samples were 1,600 hr at 700°C, up to 1,000 hr at 800°C, and 300 hr at 900°C. The results obtained are plotted in Fig. 661, together with those of [1]; as can be seen, [2] found a less pronounced decrease in solubility with falling temperature. In Fig. 660, weight has been given to the results of [2].

Intermediate Phases. According to [1], there exist three intermediate phases in the Zr-Sn system. The compound richest in Zr was shown [1] to have the singular composition Zr_4Sn (24.55 wt. % Sn); its powder pattern could be tentatively indexed with a f.c. tetragonal unit cell (a = 7.645 A, c = 12.461 A, c/a = 1.63).

The second of the three intermediate phases melts with an open maximum; [1] found the formula Zr_3Sn_2 (40 at. % Sn) to be in closest agreement with the observed single-phase composition. [4] claimed "Zr_5Sn_3" (37.50 at. % Sn) to have the hexagonal Mn_5Si_3 ($D8_8$) type of structure, with a = 8.461 A, c = 5.795 A, c/a = 0.685; he stated: "X-ray diffraction patterns of this [37.5 at. % Sn] alloy in the as-cast condition and after homogenizing at 980°C for 10 days indicated that precautions would be necessary to prevent what appeared to be a solid-state transformation. Attempts to

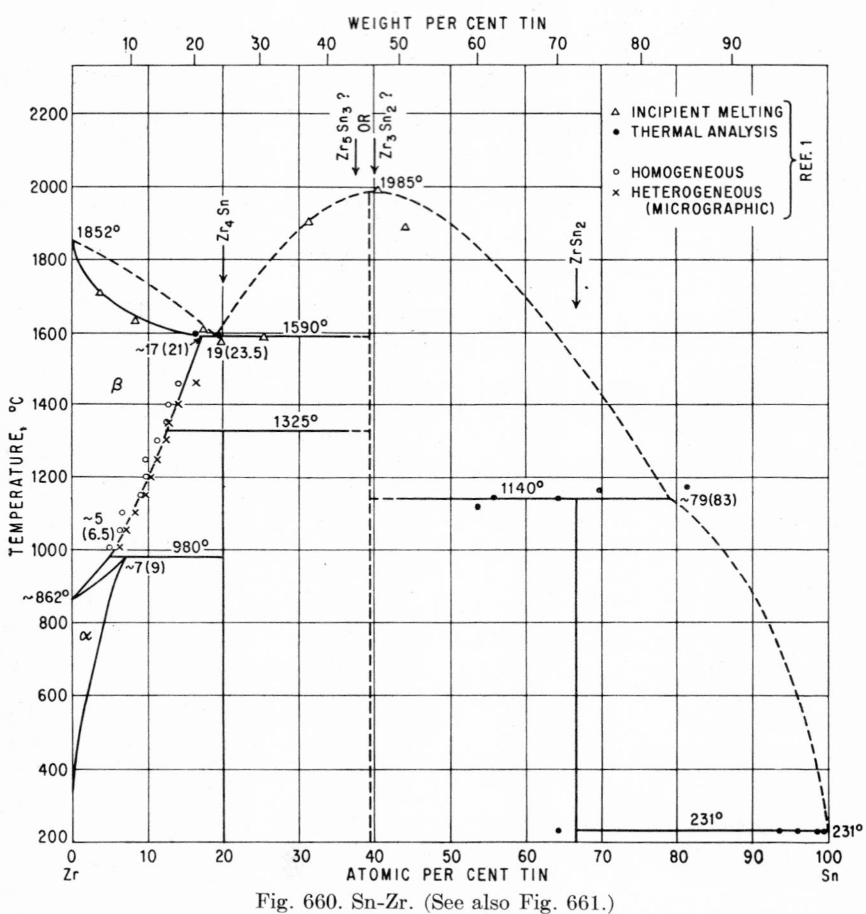

Fig. 660. Sn-Zr. (See also Fig. 661.)

obtain an essentially single phase of the compound Zr_5Sn_3 using quenching apparatus similar to that of Greninger were unsuccessful. X-ray photographs in which the superfluous lines were absent were obtained only by quenching powdered samples into liquid argon. A series of experiments revealed that a single-phase diffraction photograph could not be obtained by quenching the powders from below 1200°C."

[5] reported a compound in the composition interval $ZrSn_{0.55\pm0.05}$ (33.3–37.5 at. % Sn) and suggested a relationship to the $D8_8$ type. Obviously composition and crystal structure of the intermediate phase melting with an open maximum cannot yet be

considered as finally established; perhaps this compound even undergoes a polymorphic transformation.

Chemical analyses of electrolytically isolated primary crystals of the intermediate phase richest in Sn suggested the formula ZrSn [1]; the complex powder pattern was tentatively indexed as orthorhombic [1]. However, [5] reported the existence of the compound $ZrSn_2$ (72.24 wt. % Sn), having the orthorhombic $TiSi_2$ (C54) type of structure, with $a = 9.57_3$ A, $b = 5.64_4$ A, $c = 9.92_7$ A. Obviously, $ZrSn_2$ rather than ZrSn exists.

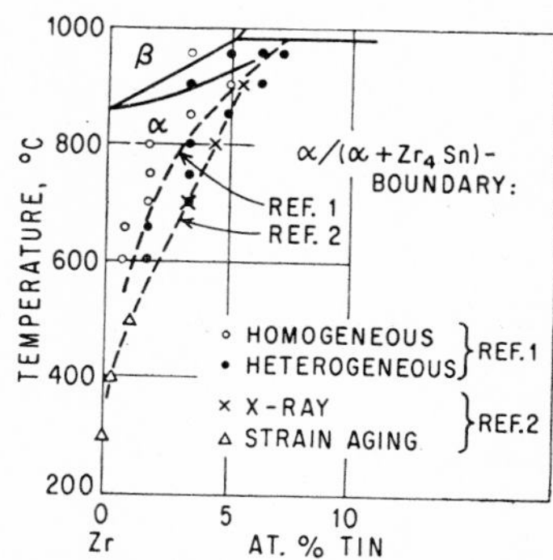

Fig. 661. Sn-Zr. (See also Fig. 660.)

Sn displays no appreciable solid solubility for Zr; whether there is a eutectic or a peritectic near pure Sn could not be determined by [1].

1. D. J. McPherson and M. Hansen, *Trans. ASM*, **45,** 1953, 915–931; discussion, pp. 932–933. The alloys were arc-melted under helium. Westinghouse Grade 3 crystal bar Zr was used for the majority of alloys. A limited number of alloys were prepared with the higher-purity Westinghouse Grade 1 Zr for final refinement of certain phase boundaries.
2. G. R. Speich and S. A. Kulin, "Zirconium and Zirconium Alloys," pp. 197–207, The American Society for Metals, Cleveland, Ohio, 1953. The crystal bar Zr–base alloys were prepared by arc melting.
3. This method involves observation of changes in yield stress of tensile specimens after aging at the test temperature. The limitations of this unusual method are critically discussed in [2].
4. P. Pietrokowsky, unpublished work done at the Jet Propulsion Laboratory, California Institute of Technology, under Contract DA-04-495-ORD-18 with the U.S. Army.
5. H. Nowotny and H. Schachner, *Monatsh. Chem.*, **84,** 1953, 169–180.

$\bar{1}.8368$
0.1632

Sr-Te Strontium-Tellurium

The compound SrTe (59.28 wt. % Te) has the NaCl (B1) type of structure, with $a = 6.660 \pm 6$ A [1].

1. V. M. Goldschmidt, *Skrifter Akad. Oslo*, 1926, no. 8; *Strukturbericht*, **1,** 1913-1928, 135.

$\bar{1}.6322$
0.3678

Sr-Tl Strontium-Thallium

The compound SrTl (69.99 wt. % Tl) has the CsCl (B2) type of structure, with a = 4.032 ± 6 A [1].

1. E. Zintl and G. Brauer, *Z. physik. Chem.*, **B20**, 1933, 245–271.

0.1272
$\bar{1}.8728$

Sr-Zn Strontium-Zinc

The compound $SrZn_{13}$ [92.86 at. (90.66 wt.) % Zn] has the f.c.c. $NaZn_{13}$ ($D2_3$) type of structure, with a = 12.239 ± 5 A [1]. According to X-ray single-crystal work by [2], $SrZn_5$ possesses an orthorhombic structure, with a = 5.32 A, b = 6.72 A, c = 13.15 A, and 24 atoms per unit cell. The structure is related to that of $CaZn_5$.

1. J. A. A. Ketelaar, *J. Chem. Phys.*, **5**, 1937, 668.
2. N. C. Baenziger and J. W. Conant, *Acta Cryst.*, **9**, 1956, 361–364.

$\bar{1}.8917$
0.1083

Ta-Th Tantalum-Thorium

"[1] examined Th and Ta after reaction at 2000°C and found only α-Th, Ta, and ThO_2 by X-ray diffraction examination. On the other hand, a report from [2] indicates the existence of an intermetallic compound in the system. No details of this work are available. [1] reports that the solubility of Th in Ta is quite limited. This report is based on a determination of the lattice coefficients of the Ta solid solution. No reliable data are available on the solubility of Ta in Th" [3].

1. H. W. Wilhelm, unpublished information (1946).
2. Atomic Energy Research Establishment, United Kingdom, unpublished information (1952).
3. Quoted from H. A. Saller and F. A. Rough, Compilation of U.S. and U.K. Uranium and Thorium Constitutional Diagrams, *U.S. Atomic Energy Comm., Publ.* BMI-1000, June, 1955.

0.5770
$\bar{1}.4230$

Ta-Ti Tantalum-Titanium

[1] reported, as a result of studies using powder-metallurgical techniques, that β-Ti and Ta form a continuous series of solid solutions. Later work [2] on arc-melted alloys showed that the α solubility extended to between 2 and 5 wt. (0.5 and 1.4 at.) % Ta at 790°C.

The phase diagram of the whole system was determined independently by [3] and [4] (Figs. 662 and 663). [3] based his work on micrographic analysis of arc-cast alloys, prepared using iodide Ti and Ta of 99.98 wt. % purity and homogenized at temperatures not far below the solidus temperature. Annealing times varied between 24 hr at 850°C and 15 days at 650°C. Melting points (Fig. 662) were determined by optical observation.

[4] used Ta of 99.9+ wt. % purity, iodide Ti as a base for the Ti-rich alloys, and Mg-reduced Ti for the intermediate and high-alloy ranges [5]. Micrographic analysis was the main method of investigation. Both optical observation and incipient-melting techniques were used to determine the melting range. Arc-cast ingots with up to 60 wt. % Ta were hot-worked at temperatures between 750 and 980°C, according to

the Ta content, and homogenization-annealed. Prior to final isothermal annealing, alloys with up to 27 wt. % Ta were given cold reductions of 50%. Annealing times varied between 2 hr at 1000°C and 168 hr at 670–600°C.

The $\beta/(\alpha + \beta)$ curves determined by both investigations are presented in Fig. 663. The largest deviation in temperature for a given composition is about 60°C.

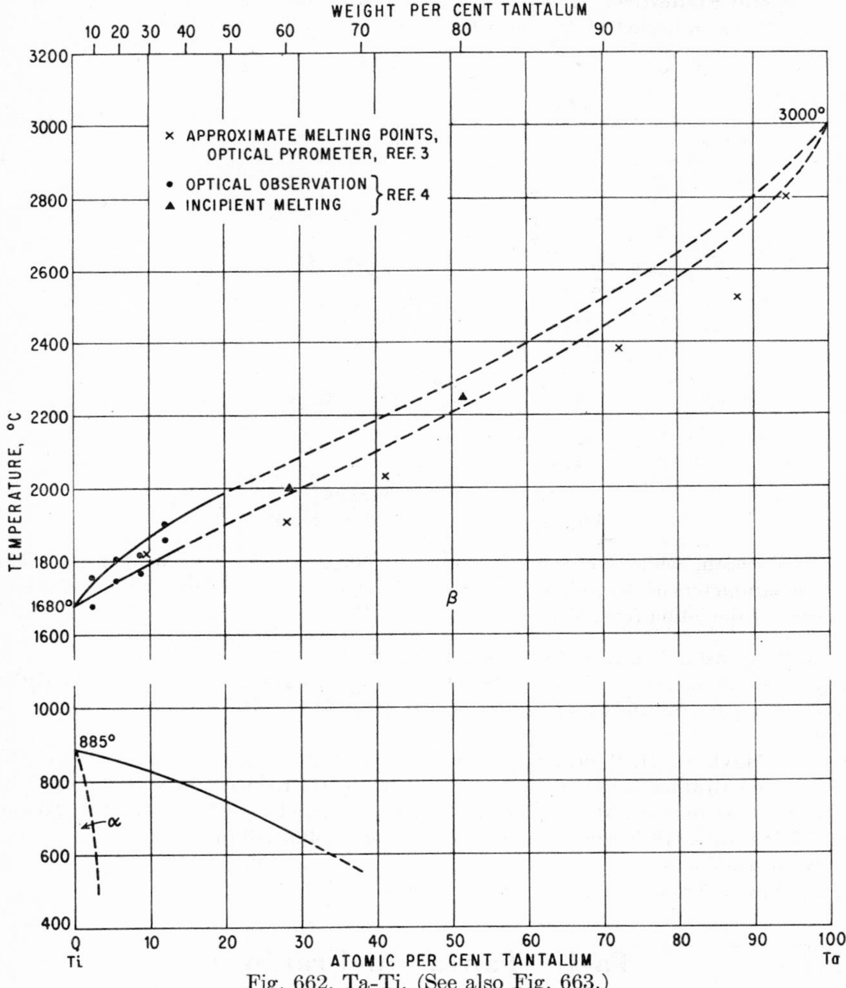

Fig. 662. Ta-Ti. (See also Fig. 663.)

It appears justified to draw an average curve which is correct within ±30°C at about 20 at. % Ta (Fig. 662).

Whereas [4] bracketed the α-phase boundary in the position shown in Fig. 663 (about 3 at. % at 700°C), [3] estimated that the solid solubility of Ta in α-Ti is less than 0.5 at. % at 700°C. On the basis of subsequent metallographic work, this value was later [6] admitted to be too low; and a solubility of about 1.5 at. % at 700°C was suggested instead.

By thermal analysis at cooling rates in the range of 100–10,000°C per sec, [7] determined the temperature at which the martensite transformation $\beta \rightarrow \alpha'$ takes place in alloys up to 18 wt. % Ta. The composition above which the β solid solution is retained on quenching was found to lie between 38 and 45 wt. % Ta [3] or 40 and 50 wt. % Ta [4].

Crystal Structure. The α parameters determined for three alloys with up to 1.49 at. % Ta, annealed 14 days at 530°C, showed that Ta expands the α-Ti lattice in

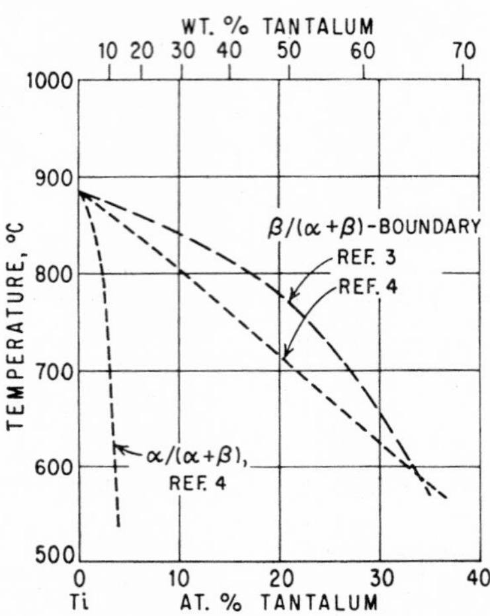

Fig. 663. Ta-Ti. (See also Fig. 662.)

the c direction, while showing no significant effect on the a parameter [4]. The lattice parameters of the β alloys between 10 and 94.3 at. (30 and 98.5 wt.) % Ta show a negative deviation from Vegard's law [3].

1. B. W. Gonser, *Ind. Eng. Chem.*, **42,** 1950, 222–226.
2. Battelle Memorial Institute, AF Technical Report 6218, Part 2, p. 40, June, 1950.
3. D. Summers-Smith, *J. Inst. Metals,* **81,** 1952/53, 73–76; see also discussion, p. 426.
4. D. J. Maykuth, H. R. Ogden, and R. I. Jaffee, *Trans. AIME,* **197,** 1953, 231–237.
5. The constitution of Ti-rich alloys made with Mg-reduced Ti was reported by L. W. Eastwood, R. M. Goldhoff, J. W. Holladay, and J. G. Kura, Battelle Memorial Institute, AF Technical Report 6515, Part I, June, 1951.
6. See discussion on [3].
7. P. Duwez, *Trans. ASM,* **45,** 1953, 934–940.

$\bar{1}.8807$
0.1193

Ta-U Tantalum-Uranium

The main features of the Ta-U phase diagram have been delineated by [1] using arc-melted alloys made from Ta and U of 99.65 and 99.9 wt. % purity, respectively (Fig. 664). The tools of investigation were micrographic analysis of as-melted and heat-treated (up to several hundred hours) alloys, thermal analysis, and X-ray powder work. The liquidus up to 2000°C was determined by chemical analysis of saturated and quenched melts.

No indications for the existence of an intermediate phase were found. Only the order of magnitude of solid-solution limits was established. The maximum solubility

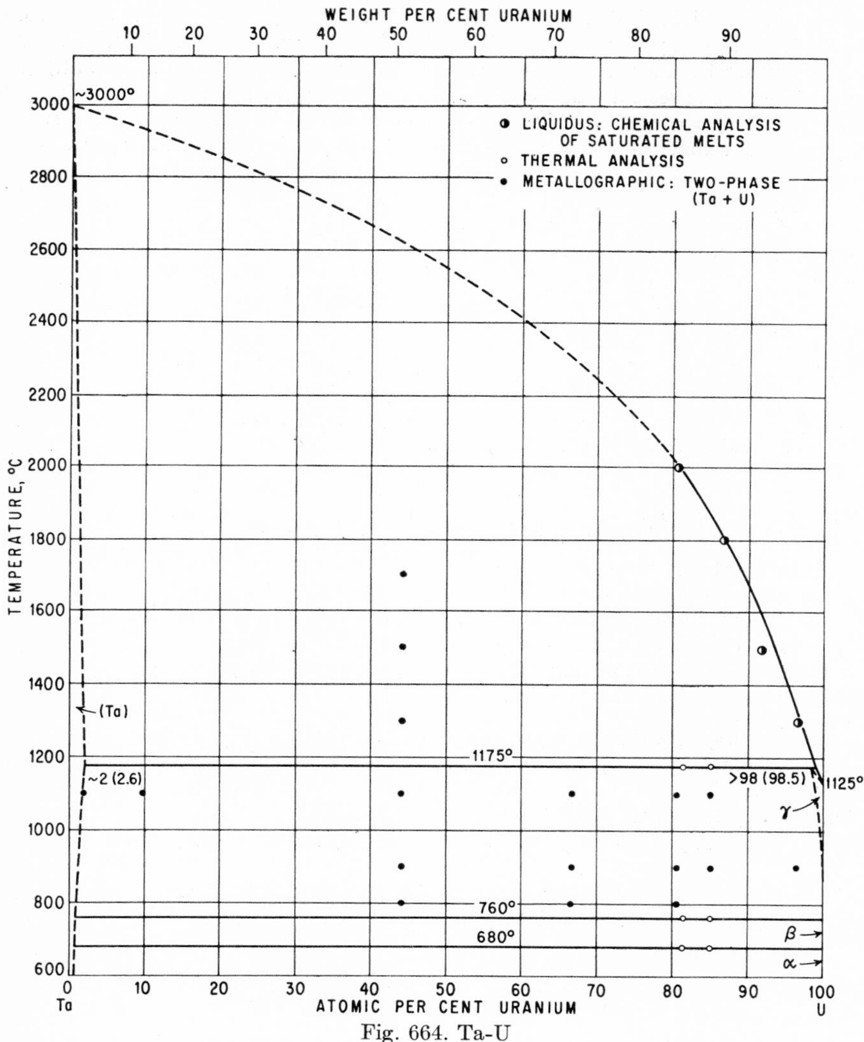

Fig. 664. Ta-U

of U in Ta is about 2 at. (2.6 wt.) %, and that of Ta in γ-U less than about 2 at. (1.5 wt.) %. No effects of Ta on the transformations in U could be observed. The lattice parameter of Ta is slightly increased by addition of U.

1. C. H. Schramm, P. Gordon, and A. R. Kaufmann, *Trans. AIME,* **188,** 1950, 195–204.

0.5502
$\bar{1}$.4498

Ta-V Tantalum-Vanadium

A microscopic and X-ray diffraction analysis of 10 arc-melted V-Ta alloys containing up to 80 wt. (53 at.) % Ta by [1] gave the following results: As-cast alloys showed simple solid-solution structures throughout. Alloys annealed at 900°C for 170 hr showed clearly the rejection of a second phase between 25 and 80 wt. (8.6 and 53 at.) % Ta, with a maximum proportion of rejected phase at the 50 wt. (22 at.) % composition. The second phase was tentatively indexed with a tetragonal unit cell (a = 6.104 kX, c = 8.833 kX), and it was suggested to be a σ phase. However, [2] pointed out that the powder-pattern lines tabulated in [1] do not support this contention.

Supplement. A 33.3 at. % Ta alloy has been examined by [3]: "The diffraction pattern of filings of the as-cast structure annealed for 10 min at 800°C proved to be a principal pattern of b.c.c. vanadium solid solution with a second phase indexed as a member of the σ structure. The lines of this structure, however, did not agree with those reported by [1]. Filings of a specimen annealed for 13 days at 600°C gave a diffraction pattern of two b.c.c. phases, one of vanadium and one of tantalum."

1. W. Rostoker and A. Yamamoto, *Trans. ASM*, **46,** 1954, 1136–1163.
2. P. Greenfield, discussion on [1], *Trans. ASM*, **46,** 1954, 1165.
3. R. P. Elliott, Armour Research Foundation, Chicago, Ill., Technical Report 1, OSR Technical Note OSR-TN-247, 1954.

$\bar{1}.9928$
0.0072

Ta-W Tantalum-Wolfram

In 1905, [1] reported that Ta and W alloy in all proportions. The existence of a continuous series of solid solutions was established by lattice-parameter [2, 3, 4] as well as microhardness [3] measurements. The lattice-parameter-vs.-composition curve shows a slight negative deviation from Vegard's rule [3, 4].

Diffraction patterns of an alloy of the composition TaW_2, annealed at 800 and 600°C, failed to indicate any solid-state reaction [5].

1. W. v. Bolton, *Z. Elektrochem.*, **11,** 1905, 51.
2. C. Agte and K. Becker, *Z. tech. Physik*, **11,** 1930, 107–111; *Physik. Z.*, **32,** 1931, 65–80.
3. H. Bückle, *Z. Metallkunde*, **37,** 1946, 53–56.
4. C. H. Schramm, P. Gordon, and A. R. Kaufmann, *Trans. AIME*, **188,** 1950, 195–204.
5. R. P. Elliott, Armour Research Foundation, Chicago, Ill., Technical Report 1, OSR Technical Note OSR-TN-247, 1954.

0.2973
$\bar{1}.7027$

Ta-Zr Tantalum-Zirconium

Cursory work only has been carried out as yet on the constitution of Zr-Ta alloys. The more reliable experimental data are plotted in Fig. 665. Some evidence for a Zr-rich eutectic was found by [1–3]; according to microscopic observations [1, 2], it lies at about 20 wt. (11 at.) % Ta [4]. The existence of a rather wide solid solubility of Ta in Zr at higher temperatures was recognized by [1, 5, 2, 6, 7], that of a eutectoid by [5, 7]. Microscopic work on alloys annealed at and quenched from 600, 800, and 1000°C (Fig. 665) showed that the eutectoid point lies roughly at 7 wt. (3.5 at.) % Ta, 820°C [7]. No reliable data as to the β-phase limit at higher temperatures are available. The solubility of Ta in α-Zr is less than 2 wt. (1 at.) % at 800°C [7] and less than 1 wt. (0.5 at.) % at "room temperature" [5].

Little is known about the constitution above 20 at. % Ta. According to [3], the solidus temperatures increase rapidly above 45 at. % Ta. An intermediate phase with cubic lattice has been claimed by [5] at about 40 wt. (25 at.) % Ta. According to [3], the general trend of the solidus curve suggests a compound ZrTa (?). [7] observed no one-phase structures other than β in alloys containing 2–80 wt. % Ta, although specific stoichiometric compositions of possible compounds (Zr_2Ta, ZrTa, Zr_2Ta_3, and $ZrTa_2$) were included in the alloys prepared; this means that the system may possess

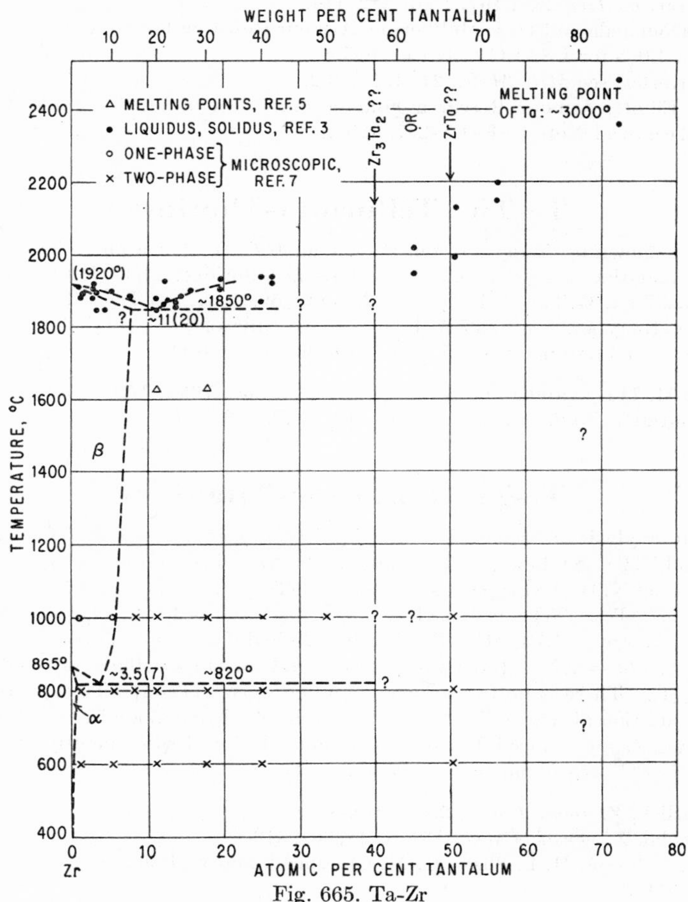

Fig. 665. Ta-Zr

a compound of some other composition (Zr_3Ta_2?) or may have no intermediate phase at all.

According to [8], the Zr-Ta constitutional diagram is being explored.

Supplement. According to [9], an alloy of the composition $ZrTa_2$ annealed 48 hr at 800°C gave a pattern of a mixture of a b.c.c. and a h.c.p. phase. Metallographic investigation of the as-cast ingot indicated a single phase.

1. S. M. Shelton, *U.S. Atomic Energy Comm., Publ.* AF-TR-5932, 1949; *Met. Abstr.*, **21,** 1954, 869.
2. C. T. Anderson, E. T. Hayes, A. H. Roberson, and W. J. Kroll, *U.S. Bur. Mines Rept.* 4658, 1950.

3. J. R. Havill, M.S. thesis, Oregon State College, Corvallis, Ore., 1952.
4. The eutectic structure could not be resolved microscopically. This may account for the fact that some authors reported alloys with up to 30 wt. % Ta as single-phase β.
5. R. E. Kleint, M.S. thesis, Oregon State College, Corvallis, Ore., 1949.
6. F. B. Litton, *Iron Age*, **167,** 1951, 112–114.
7. R. F. Domagala and D. J. McPherson, Armour Research Foundation, Chicago, Ill., Project B068 for USAEC, August, 1955.
8. J. H. Keeler; see *Met. Abstr.*, **21,** 1954, 432.
9. R. P. Elliott, Armour Research Foundation, Chicago, Ill., Technical Report 1, OSR Technical Note OSR-TN-247, 1954.

$\bar{1}.7402$
0.2598

Te-Th Tellurium-Thorium

It was shown by X-ray and tensimetric analysis [1] that a phase Th_3Te [2] does not exist and that Th forms only the following tellurides: ThTe (35.47 wt. % Te), $ThTe_2$ (52.37 wt. % Te), and Th_3Te_8 [72.73 at. (59.45 wt.) % Te]. ThTe is isotypic with CsCl (B2 type), a = 3.827 A [1]. The structures of $ThTe_2$ and Th_3Te_8 have not been determined because of their apparently low symmetry.

1. R. W. M. D'Eye and P. G. Sellman, *J. Chem. Soc.*, **1954,** 3760–3766.
2. E. Montignie, *Bull. soc. chim. France,* **14,** 1947, 748–749.

0.4255
$\bar{1}.5745$

Te-Ti Tellurium-Titanium

X-ray analysis of preparations varying in composition between TiTe (72.71 wt. % Te) and $TiTe_2$ (84.20 wt. % Te) showed that there is a continuous transition from the hexagonal NiAs (B8) type of structure of TiTe to the hexagonal CdI_2 (C6) type of structure of $TiTe_2$. TiTe: $a = 3.84_2$ A, $c = 6.40_2$ A, $c/a = 1.66_7$. $TiTe_2$: $a = 3.76_4$ A, $c = 6.52_6$ A, $c/a = 1.73_4$ [1]. Previously, [2] had found for $TiTe_2$: $a = 3.78_2$ A, $c = 6.5_4$ A, $c/a = 1.73$. Indications were found for the existence of a subtelluride (Ti_2Te?) [1]. There is no further compound between $TiTe_2$ and Te [1]. The lowest per cent addition at which Te appeared as an insoluble compound in the microstructure of magnesium-reduced Ti-base alloys, hot-rolled and subsequently annealed for ½ hr at 790°C, was found to be 0.47 wt. % Te [3].

1. P. Ehrlich, *Z. anorg. Chem.*, **260,** 1949, 1–18.
2. I. Oftedal, *Z. physik. Chem.*, **134,** 1928, 301–310.
3. R. M. Goldhoff, H. L. Shaw, C. M. Craighead, and R. I. Jaffee, *Trans. ASM*, **45,** 1953, 941–971.

$\bar{1}.7954$
0.2046

Te-Tl Tellurium-Thallium

The liquidus curve was determined by [1] (see upper portion of Fig. 666); and complete phase diagrams, based on thermal-analysis data, were established by [2] and [3]. The intermediate phase lowest in Te content [4], which forms a monotectic equilibrium with Tl-rich melt and has a maximal melting point of 442 [1], 430 [2], or 434°C [3], was assumed to be Tl_5Te_3 [37.5 at. (27.25 wt.) % Te] [1, 3] or Tl_3Te_2 (29.39 wt. % Te) [2]. The thermal data points in this range (Fig. 666) would be compatible with both of these compositions; they differ by only 2.5 at. %.

On the other hand, X-ray work [5] gave strong evidence for the existence of a phase of the approximate composition Tl_2Te (23.79 wt. % Te). According to Fig. 666, this composition definitely lies outside the range of high liquidus points, however. The compound Tl_2Te was previously reported by [6, 7], but the homogeneity of these preparations was not substantiated. Determination of the structure of this phase would clarify the discrepancy.

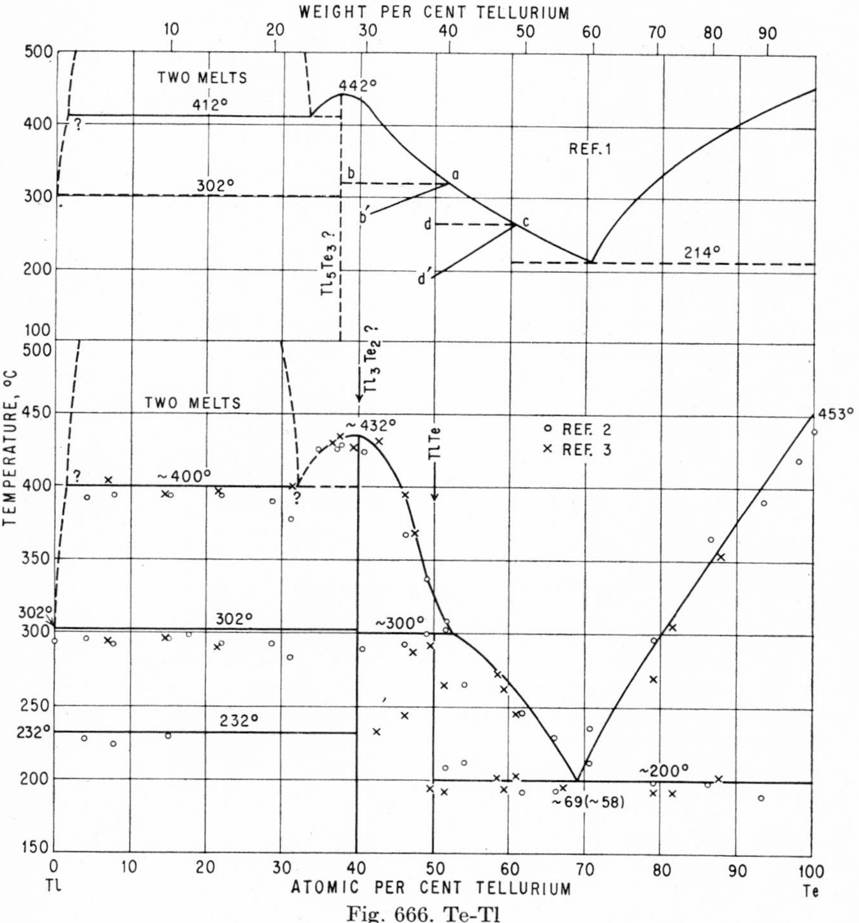

Fig. 666. Te-Tl

The existence of TlTe (38.44 wt. % Te) was established by thermal analysis [2, 3] and X-ray study [5]. It is formed peritectically at about 305 [2] or about 292°C [3]. The TlTe-Te eutectic was given as about 70.5 at. % Te, 214°C [1], about 69.5 at. % Te, 210°C [2], and about 67 at. % Te, 194°C [3].

The diagram in the upper portion of Fig. 666 shows two sloping lines *ab′* and *cd′*. These were interpreted [8] as indicating thermal effects which are shifted, by supercooling phenomena, to temperatures which are increasingly lower with lower Te content [9]. If this holds, there would exist two horizontals *ab* and *cd* at about 320 and 265°C, respectively, that correspond to the peritectic formation of TlTe and possibly Tl_2Te_3. The latter peritectic reaction was not observed by [2, 3], however [10].

This point needs clarification, although [2] reported that the thermal data were corroborated by the microstructures.

Crystal Structures. TlTe probably has a more complex structure than TlS and TlSe. The X-ray diagram of Tl_2Te (?) resembles that of Tl_2Se, but there are some differences in the powder patterns [5].

1. H. Pélabon, *Compt. rend.*, **145,** 1907, 118–121; *Ann. chim. et phys.*, **17,** 1909, 554.
2. M. Chikashige, *Z. anorg. Chem.*, **78,** 1912, 68–74. The temperatures reported are too low throughout.
3. A. P. Obukhov and N. S. Bubyreva, *Izvest. Sektora Fiz.-Khim. Anal.*, **17,** 1949, 276–280.
4. Pélabon believed that the compositions of the monotectic and eutectic melts also corresponded to compounds.
5. H. Hahn and W. Klingler, *Z. anorg. Chem.*, **260,** 1949, 110–119.
6. C. Fabre, *Compt. rend.*, **105,** 1887, 277.
7. A. Brukl, *Monatsh. Chem.*, **45,** 1924, 479.
8. W. Guertler, "Handbuch der Metallographie," vol. 1, Part 1, pp. 936–937, Verlagsbuchhandlung Gebrüder Borntraeger, Berlin, 1912.
9. This was also observed by [2, 3]; see data points in Fig. 666.
10. The thermal data points given by [2] for the crystallization of the TlTe-Te eutectic scatter widely between 189 and 212°C. With the exception of the melts containing about 51.6, 54.2, and 70.6 at. % Te, the eutectic temperature was found at 189–199°C.

    It is possible that the data points of the alloys with 51.6 and 54.2 at. % Te (208 and 211°C) correspond to a peritectic reaction. The eutectic data points of [3] also scatter rather widely; however, there is no indication of a peritectic reaction.

$\bar{1}.7292$
0.2708

Te-U Tellurium-Uranium

The existence of the following phases has been established by X-ray analysis [1]: UTe (34.90 wt. % Te), U_3Te_4 (41.68 wt. % Te), U_2Te_3 (44.57 wt. % Te), and UTe_2 (51.74 wt. % Te). UTe and U_2Te_3 had been reported previously [2]. [3] described the preparation of a telluride of the approximate composition $UTe_{2.2}$. It is uncertain whether this is identical with UTe_2; [1] believes that it is a separate phase.

UTe has the structure of the NaCl (B1) type, with $a = 6.163$ A [1]. U_3Te_4 is isotypic with Th_3P_4 ($D7_3$ type), $a = 9.397$ A [1]. UTe_2 is tetragonal, with $a = 4.006$ A, $c/a = 1.865$; it is not isostructural with the CaC_2 (C11) or Cu_2Sb (C38) types of structure [1].

1. R. Ferro, *Z. anorg. Chem.*, **275,** 1954, 320–326.
2. A. Colani, *Compt. rend.*, **137,** 1903, 383; *Ann. chim. et phys.*, **12,** 1907, 87.
3. E. Montignie, *Bull. soc. chim. France*, **14,** 1947, 748–749.

0.3987
$\bar{1}.6013$

Te-V Tellurium-Vanadium

The existence of VTe (71.47 wt. % Te), first reported by [1], was conclusively established by determining its crystal structure [2]. It is isotypic with NiAs (B8 type), $a = 3.81_3$ A, $c = 6.13_2$ A, $c/a = 1.60_9$ [2].

1. E. Hoschek and W. Klemm, *Z. anorg. Chem.*, **242,** 1939, 59.
2. P. Ehrlich, *Z. anorg. Chem.*, **260,** 1949, 1–18.

$\bar{1}.8412$
0.1588

Te-W Tellurium-Wolfram

WTe_2 (58.12 wt. % Te) was claimed to be identified [1]. [2] prepared seven alloys with compositions ranging from $WTe_{1.50}$ to WTe_{15} by heating the elements in sealed, evacuated quartz tubes at 700–800°C, and studied them by X-ray and thermal analysis. But one compound, WTe_2, was detected having an orthorhombic structure with $a = 3.490 \pm 6$ A, $b = 6.277 \pm 15$ A, $c = 14.07 \pm 2$ A.

1. A. Morette, *Compt. rend.*, **216**, 1943, 566–568; *Ann. chim.* (*Paris*), **19**, 1944, 130–143.
2. O. Knop and H. Haraldsen, *Can. J. Chem.*, **34**, 1956, 1142–1145 (not available to the author); *Chem. Abstr.*, **50**, 1956, 16250 *Met. Abstr.*, **25**, 1957, 17.

$\bar{1}.8677$
0.1323

Te-Yb Tellurium-Ytterbium

YbTe (42.44 wt. % Te) has the NaCl (B1) type of structure, $a = 6.35_3$ A [1].

1. H. Senff and W. Klemm, *Z. anorg. Chem.*, **242**, 1939, 92–96.

0.2904
$\bar{1}.7096$

Te-Zn Tellurium-Zinc

The phase diagram in Fig. 667 is due to [1]. Since Zn evaporated from melts with more than 35 wt. % Zn, the liquidus curve in this range could not be determined under

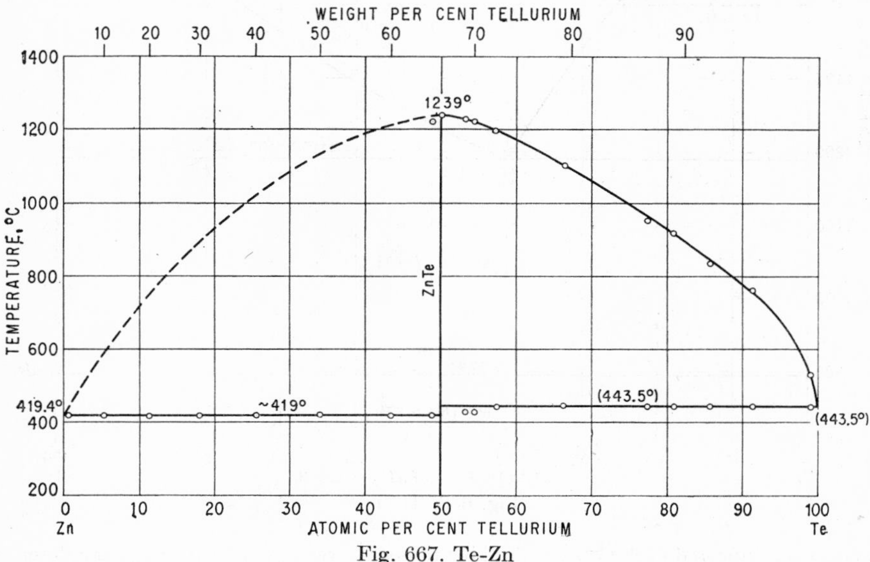

Fig. 667. Te-Zn

normal pressure. If the melts were heated only up to about 900°C, at which temperature large proportions of ZnTe (66.12 wt. % Te) were formed, the solidus temperature could be determined; it is practically identical with the melting point of Zn [1]. ZnTe [2] is isotypic with zincblende (B3 type), $a = 6.10_1$ A [3].

1. M. Kobayashi, *Intern. Z. Metallog.*, **2**, 1912, 65–69; *Mem. Coll. Sci. Eng., Kyoto Univ.*, **3**, 1911, 217–221.
2. "Gmelins Handbuch der anorganischen Chemie," 8th ed., System No. **32**, p. 247, Verlag Chemie, G.m.b.H., Leipzig and Berlin, 1924.
3. W. Zachariasen, *Z. physik. Chem.*, **124**, 1926, 277–284.

0.6854
$\bar{1}.3146$

Th-Ti Thorium-Titanium

The phase diagram (Fig. 668) is based on X-ray, microscopic, thermal, and diffusion studies of arc-melted alloys prepared using a "special high-purity Th sponge"

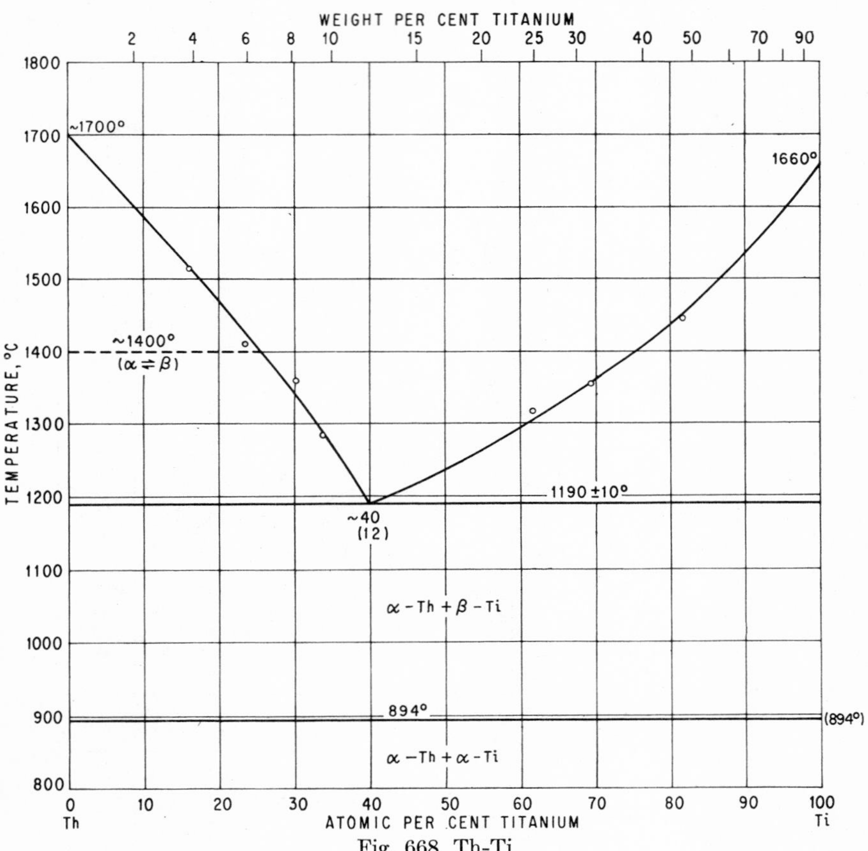

Fig. 668. Th-Ti

and magnesium-reduced Ti [1]. Liquidus temperatures (see data points) were determined by holding melts of approximately eutectic composition in crucibles of Th and Ti at various temperatures and analyzing the portion which was molten and represented the saturated liquid solution. The eutectic temperature was found by the incipient-melting technique. As shown by lattice-spacing measurements of specimens vacuum-annealed at 1050°C for 100 hr, the mutual solid solubility is very small.

The temperature of the $\alpha \rightleftarrows \beta$ transformation of Ti, found as 894°C on cooling, is not affected by Th, indicating that there is also no appreciable solubility of Th in β-Ti.

1. O. N. Carlson, J. M. Dickinson, H. E. Lunt, and H. A. Wilhelm, *Trans. AIME*, **206**, 1956, 132–136.

$\bar{1}.9890$
0.0110

Th-U Thorium-Uranium

The phase diagram shown in Fig. 669 is based on thermal data and solubility studies, information obtained from X-ray examination and microscopic evidence [1].

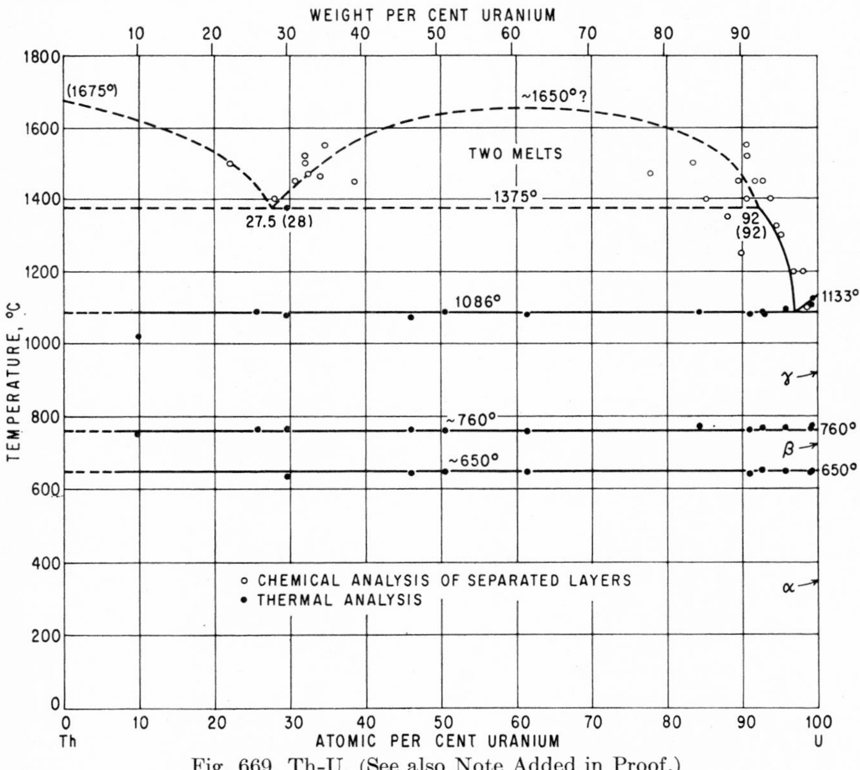

Fig. 669. Th-U. (See also Note Added in Proof.)

The extent of the miscibility gap in the liquid state was determined by chemical analysis of the separated layers. The Th liquidus, which rises sharply from the eutectic, was established in a similar manner [2] since no thermal data could be obtained.

No intermediate phases were found. Thermal, microscopic, and X-ray diffraction work indicated that the mutual solid solubilities are negligible.

Note Added in Proof. According to more recent work, there is an appreciable solubility of U in solid Th. On the basis of hardness curves and metallographic examination, [3] reported the tentative solubility data 5 at. % at 700–800°C and 1 at. % at 600°C.

[4] studied Th-rich alloys by high-temperature X-ray work. The following

results were obtained: (*a*) The $\alpha \rightleftarrows \beta$ transition in Th occurs at 1330 ± 20°C and is slightly lowered by additions of U. (*b*) The lattice parameter of f.c.c. α-Th is lowered by additions of U. (*c*) The solubility of U in α-Th, as determined by parametric work, is, in wt. % U (at. % in parentheses), 2.5, 4.5, and 7.5 (2.4, 4.4, and 7.3) at 950, 1150, and 1250°C. (*d*) A 2 wt. % U alloy showed signs of melting at 1450°C.

[4] suggest that the monotectic temperature is 1330°C rather than 1375°C (cf. Fig. 669), the lower value being equally consistent with the data of [1].

1. O. N. Carlson, *U.S. Atomic Energy Comm., Publ.* AECD-3206 (and ISC-102), 1951.
2. Excess solid Th was immersed in molten U until equilibrium was reached. By chemical analysis the amount of Th dissolved was determined and a point on the liquidus curve obtained.
3. F. A. Rough, unpublished information (1954); quoted in H. A. Saller and F. A. Rough, *U.S. Atomic Energy Comm., Publ.* BMI-1000, June, 1955.
4. W. B. Wilson, A. E. Austin, and C. M. Schwartz, Report BMI-1111, Battelle Memorial Institute, Columbus, Ohio, July, 1956.

0.6586
$\bar{1}$.3414

Th-V Thorium-Vanadium

According to [1], Th and V form a simple eutectiferous system (Fig. 670). No attempt was made to establish the liquidus line. Microscopic studies showed the

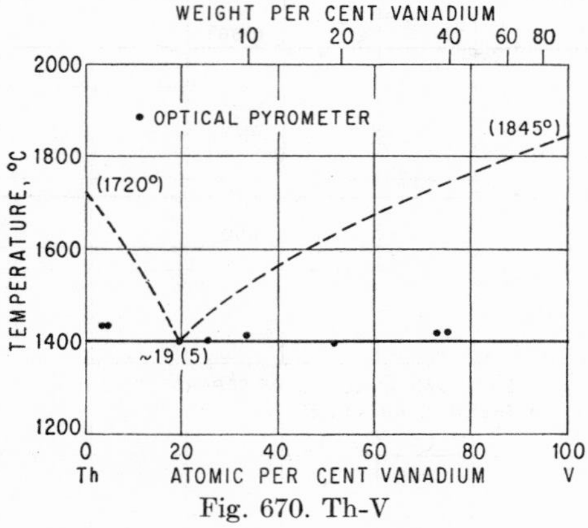

Fig. 670. Th-V

eutectic composition to be approximately 5 wt. (19 at.) % V. Melting (solidus) temperatures were determined by means of an optical pyrometer.

No indications for any extended solid solubilities were found in cursory microscopic and X-ray diffraction work. "High-temperature X-ray analysis as performed by [2] indicated negligible solubility of V in Th even at the eutectic temperature, and also disclosed a transition in Th from f.c.c. to b.c.c. at about 1425°C. Small amounts of V do not alter the temperature of this transformation" [1]. The suggested transformation in Th [3] has not been included in Fig. 670.

1. H. L. Levingston and B. A. Rogers, *U.S. Atomic Energy Comm., Publ.* AECD-3602 (and ISC-340), 1953. The alloys were prepared in a helium arc-melting furnace.
2. P. Chiotti, private communication (to Levingston and Rogers).
3. Further literature on the transformation: P. Chiotti, *J. Electrochem. Soc.*, **101**, 1954, 567–570. J. O. McCaldin and P. Duwez, *Trans. AIME*, **200**, 1954, 619–620.

0.1011
$\bar{1}.8989$

Th-W Thorium-Wolfram

From measurements of the temperature coefficient of the electrical resistance on W filaments containing metallic (?) Th, [1] concluded that W does not take Th into solid solution. The contradictory results by [2], according to which the solubility of Th in W increases with the temperature, are explained by [1] as probably caused by the presence of a third component.

In the X-ray examination of thoriated-tungsten filament, [3] claimed to have found an intermediate phase ("a definite compound of tungsten in thorium") having "the same f.c.c. structure as Th but with the atoms somewhat closer together (a = 4.97 kX, a_{Th} = 5.04 kX)." The evidence given is insufficient.

Measurements of diffusion and electrical resistivity by [4] clearly indicate a certain solubility of Th in solid W.

Note Added in Proof. An account of hitherto unpublished information, given by [5], reads as follows: "[6] reports that there are no compounds in the Th-W system. Data from [7] indicate that there is a eutectic in the system at about 8 at. % W. The eutectic probably occurs at 1475°C, according to [6]. The solubility in Th is less than 1 at. % W [7]."

1. W. Geiss and J. A. M. van Liempt, *Z. anorg. Chem.*, **168**, 1927, 110–111.
2. H. v. Wartenberg, J. Broy, and R. Reinicke, *Z. Elektrochem.*, **29**, 1923, 214–217.
3. A. St. John, *Trans. AIME*, **78**, 1928, 228–237. See also discussion, pp. 237–240.
4. G. R. Fonda, A. H. Young, and A. Walker, *Physics*, **4**, 1933, 1–6.
5. H. A. Saller and F. A. Rough, Compilation of U.S. and U.K. Uranium and Thorium Constitutional Diagrams, *U.S. Atomic Energy Comm., Publ.* BMI-1000, June, 1955.
6. H. A. Wilhelm, unpublished information (1946).
7. Atomic Energy Research Establishment, United Kingdom, unpublished information (1952).

0.5503
$\bar{1}.4497$

Th-Zn Thorium-Zinc

An X-ray diffraction analysis of Th-Zn alloys containing up to 42 wt. (17 at.) % Th has been carried out by [1]. Single-crystal and powder patterns of a 30 wt. % Th alloy, nearly one-phase, indicated this compound to have the hexagonal $CaCu_5$ type of structure, with $a = 5.24_8$ A, c = 4.451 A, c/a = 0.846. Chemical analysis as well as intensity calculations points to the formula $ThZn_9$ (28.29 wt. % Th), better written as $(Th_{0.6}Zn_{0.4})Zn_5$.

X-ray reflections of a second, unidentified compound were observed in 20, 30, and 42 wt. % Th alloys; this compound is thought to be formed by a peritectic reaction with $ThZn_9$. The Zn lattice was unaffected by addition of Th, indicating a negligible, or zero, solubility of Th is solid Zn.

The validity of the formula $ThZn_9$ has been questioned by [2] from geometrical considerations.

Note Added in Proof. Th_2Zn is stable up to 1040°C, at which temperature it decomposes to form Th and Zn [3]. It has the tetragonal $CuAl_2$ (C16) type of structure, with a = 7.60 A, c = 5.64 A [4].

1. H. Nowotny, *Z. Metallkunde*, **37**, 1946, 31–34.
2. T. Heumann, *Nachr. Akad. Wiss. Göttingen Math. Phys. Kl.*, **A2**(1), 1950, 1–6.
3. H. A. Wilhelm, unpublished information (1946); quoted in H. A. Saller and F. A. Rough, *U.S. Atomic Energy Comm., Publ.* BMI-1000, June, 1955.
4. N. C. Baenziger, R. E. Rundle, and A. I. Snow, *Acta Cryst.*, **9**, 1956, 93–94.

0.4056
$\bar{1}$.5944

Th-Zr Thorium-Zirconium

Cursory work by [1, 2] showed that at least 5 wt. % Th is soluble in solid Zr at higher temperatures.

The (incomplete) phase diagram established by [3] is shown in Fig. 671. The liquidus was not determined; solidus data were obtained mainly by the optical-pyrometer method. Reactions in the solid state were studied by differential thermal, micrographic, and X-ray diffraction analyses.

The solubility of Th in α-Zr was found to be negligible; no change in the α-Zr lattice with increasing Th content of the alloys could be observed (but see Note Added in Proof). The limit of solubility of Zr in solid Th in the temperature range up to 700°C is about 5 wt. (12 at.) % Zr; the lattice parameter of Th drops on addition of Zr. The constitution between 800°C and the solidus line could not be conclusively established. Indications were found for a continuous series of solid solutions below the solidus line, which requires a b.c.c. high-temperature form of Th (see Th-V!); however, [3] "does not contend that another allotropic form of Th must exist. His position is that if one were to exist it would help to resolve certain inconsistencies in this system which are difficult to explain on any other basis." [3] further stated, "The exact significance of the 900°C horizontal line is not fully understood at this writing. The thermal data show that the 900°C cooling-curve break is of maximum intensity somewhere in the vicinity of 35 wt. % Zr [65 wt. % Th]. . . . Thus a solubility maximum or pseudo-compound must exist at this composition." For further details

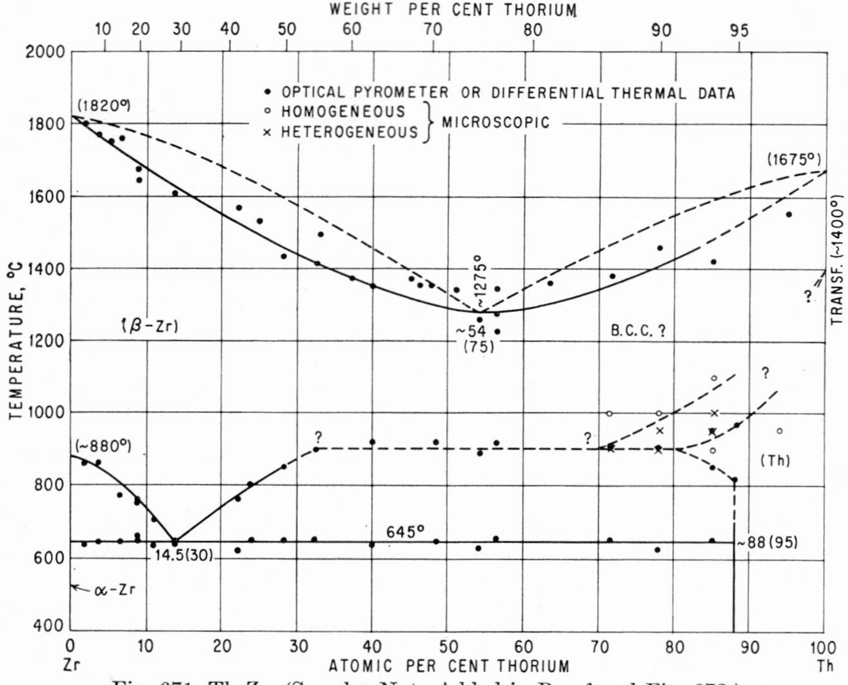

Fig. 671. Th-Zr. (See also Note Added in Proof and Fig. 672.)

of the investigation by [3], especially as to a discussion of a possible peritectic reaction at 1350°C and 60–70 wt. % Th, reference is made to the original paper.

Note Added in Proof. [4] reinvestigated the uncertain regions of the Th-Zr system and arrived at the complete phase diagram of Fig. 672, on which the following comments may be made: (*a*) The solidus was simply redrawn to fit a minimum temperature of 1350°C rather than 1275°C as shown in Fig. 671. (*b*) The boundaries of the ($\alpha_{Th} + \beta_{Th}$) two-phase field were determined by micrographic as well as high-temperature X-ray work. (*c*) The existence of an uninterrupted series of b.c.c.

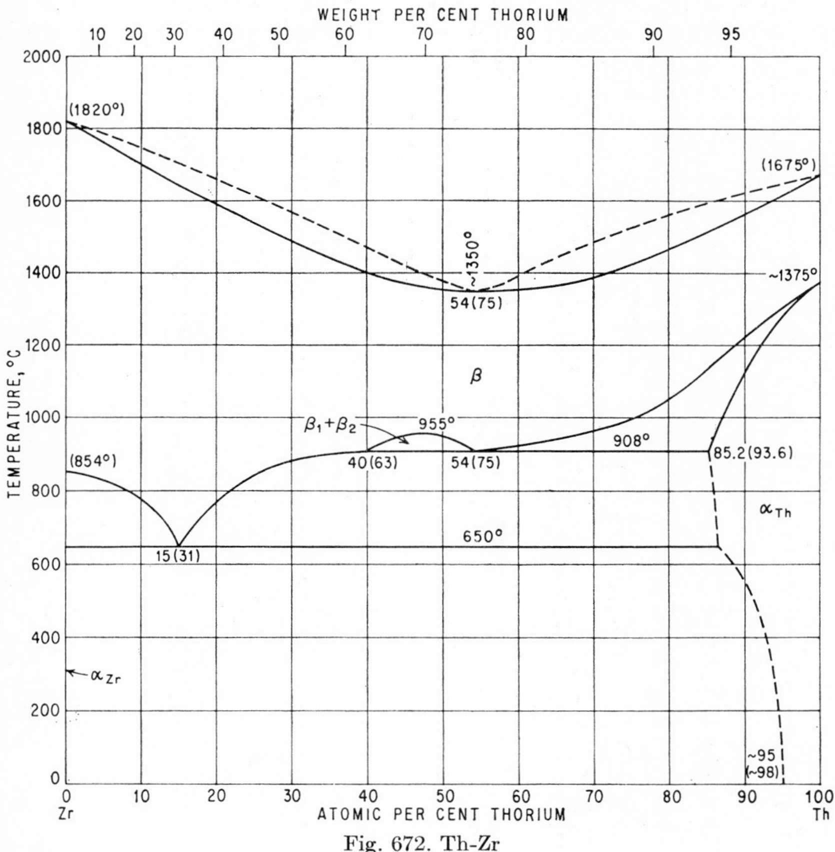

Fig. 672. Th-Zr

solid solutions at higher temperatures was verified by high-temperature X-ray measurements. Vegard's law is obeyed. (*d*) The solid immiscibility region above the 908°C horizontal line could be verified only by high-temperature X-ray work. Microscopic work as well as measurements of electrical resistivity failed to detect this loop. (*e*) The solubility of Zr in solid Th at "room temperature" was estimated from parameter measurements to be about 2 wt. (5 at.) %. Similar work on the zirconium end of the system placed the limit of solid solubility of Th in Zr at "0.5 wt. % Th or less." There is a decrease in the *a* parameter and an increase in the *c* parameter of the h.c.p. cell by the addition of small amounts of Th to α-Zr.

1. S. M. Shelton, *U.S. Atomic Energy Comm., Publ.* AF-TR-5932, 1949; *Met. Abstr.*, **21**, 1954, 869.
2. C. T. Anderson, E. T. Hayes, A. H. Roberson, and W. J. Kroll, *U.S. Bur. Mines Rept. Invest.* 4658, 1950.
3. O. N. Carlson, *U.S. Atomic Energy Comm., Publ.* AECD-3206 (and ISC-102), 1951.
4. E. D. Gibson, B. A. Loomis, and O. N. Carlson, *Trans. ASM*, **50**, 1958, preprint 24.

$\bar{1}.3036$
0.6964

Ti-U Titanium-Uranium

The inset of Fig. 673 shows the tentative partial phase diagram due to [1]. It is based on thermal, micrographic, and X-ray data obtained with alloys prepared by

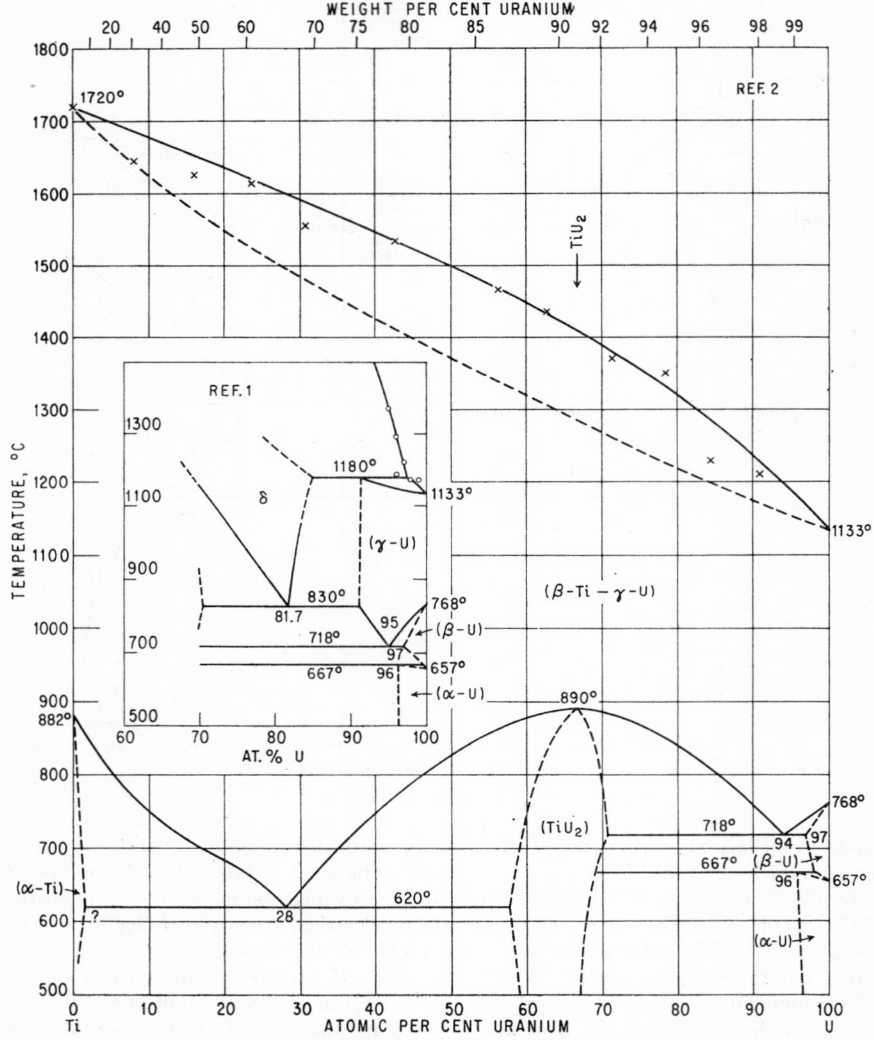

Fig. 673. Ti-U. (See also Note Added in Proof and Fig. 673*a*.)

induction melting in BeO crucibles. The liquidus curve, determined by means of cooling curves, was found to rise steeply from the melting point of U to 1925°C at 30 at. % Ti; only the U-rich portion is presented in the inset. Prior to final heat-treatment, specimens for the microscopic analysis were homogenized for 8 days at temperatures approximately 100°C below the solidus temperature and then furnace-

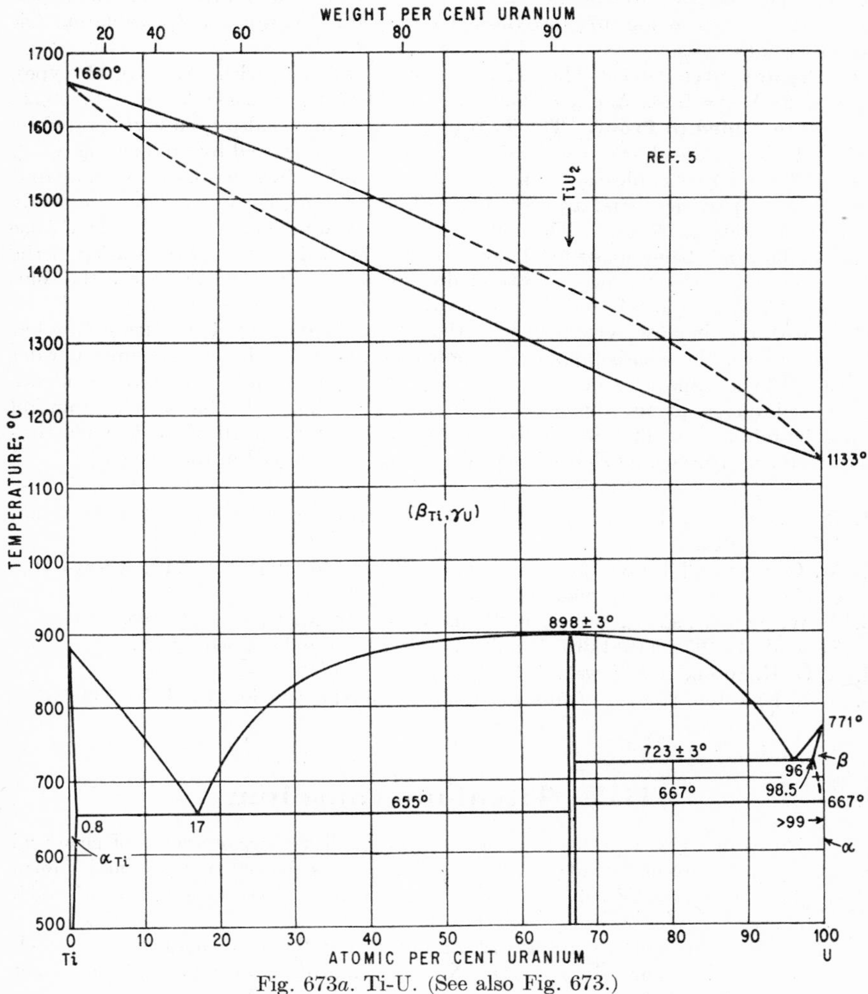

Fig. 673*a*. Ti-U. (See also Fig. 673.)

cooled. The existence of the high-temperature phase designated as δ was concluded from microscopic and X-ray studies.

In contrast to these results, [2] found no evidence for the δ phase. Instead, β-Ti and γ-U were concluded to form a complete series of solid solutions; see main diagram in Fig. 673. Alloys were prepared by arc melting, using iodide Ti and high-purity U. The ingots were hot-worked under protective conditions and homogenized for 7 days at 1095°C and subsequently annealed at and quenched from various temperatures between 540 and 1095°C. The solid-state boundaries were outlined by micrographic analysis, complemented by X-ray work used for phase identification. The solubility limits of Ti in α-U and β-U were not determined; the work of [1] was accepted in this region of the system. The liquidus temperatures indicated in Fig. 673 and the solidus in the range 0–60 at. % U were determined by visual observations [2, 3]. The temperatures of the eutectoids between α-Ti and TiU_2 and between β-U and TiU_2, shown in Fig. 673, are those determined in the work of [3] by dilatometric and thermal means.

Crystal Structure. The TiU_2 phase is isotypic with AlB_2 (C32 type), $a = 4.828$ A, $c = 2.847$ A, $c/a = 0.590$ [4]; $a = 4.817$ A, $c = 2.844$ A, $c/a = 0.590$ [2].

Note Added in Proof. The Ti-U phase diagram has also been investigated by [5], with the results shown in Fig. 673*a*. Alloys were prepared by arc melting, using iodide Ti and two grades of U, namely, bar material "which contained numerous coarse oxide particles" and a purer powder U which was, however, "not completely free from oxide particles." The bulk of the work was carried out on bar U-base alloys, allowance being made for Ti losses due to oxidation. The constitution of the U-rich end of the system was reexamined using the purer material. Since the solubilities of Ti in α- and β-U and the composition of the U-rich eutectoid reported by [1] (Fig. 673) were in close agreement with the results obtained by [5] on the impure bar U-base alloys, the phase boundaries shown in Fig. 673*a*, based on purer powder U-base alloys, appear to be more reliable. Both microscopic and X-ray evidence suggested but a narrow homogeneity range for TiU_2. The b.c.c. solid solution could be retained on quenching only in the composition range 10–30 at. % U; lattice-parameter measurements showed that Vegard's law is obeyed in this range.

1. R. W. Buzzard, R. B. Liss, and D. P. Fickle, *J. Research Natl. Bur. Standards*, **50**, 1953, 209–214.
2. M. C. Udy and F. W. Boulger, *Trans. AIME*, **200**, 1954, 207–210; see also discussion, *Trans. AIME*, **200**, 1954, 1317–1318.
3. A. W. Seybolt, D. E. White, F. W. Boulger, and M. C. Udy, *Reactor Science & Technol.*, **1**, 1951, 118–119. Results of this work are discussed by [2].
4. A. G. Knapton, *Acta Cryst.*, **7**, 1954, 457–458.
5. A. G. Knapton, *J. Inst. Metals*, **83**, 1955, 497–504; discussion, **84**, 1956, 532–535.

$\bar{1}.9732$
0.0268

Ti-V Titanium-Vanadium

Microscopic [1–3] and X-ray analysis [2], as well as measurements of electrical resistivity [1], of arc-melted iodide titanium-base alloys showed that β-Ti and V form a complete series of solid solutions. This was verified by [4], using alloys of much lower purity.

The solidus curve shown in Fig. 674 was outlined by the incipient-melting method, using "only a few compositions" [1]. No individual data were given. The exact composition of the minimum was not determined; it was placed at approximately 30 wt. (28.7 at.) % V and 1620°C [5]. Solidus and liquidus temperatures reported by [4] lie roughly 100°C below those indicated in Fig. 674, with the minimum at approximately 33 wt. (31.6 at.) % V, 1530°C, and melting points of Ti and V were found as 1680 and 1860 ± 20°C, respectively.

After [6] had outlined the $(\alpha + \beta)$ field, its boundaries were more carefully determined by means of micrographic studies [1, 2]; see inset of Fig. 674, with the composition in wt. % V. The solubility of V in α-Ti at 650°C was found to be between 3.0

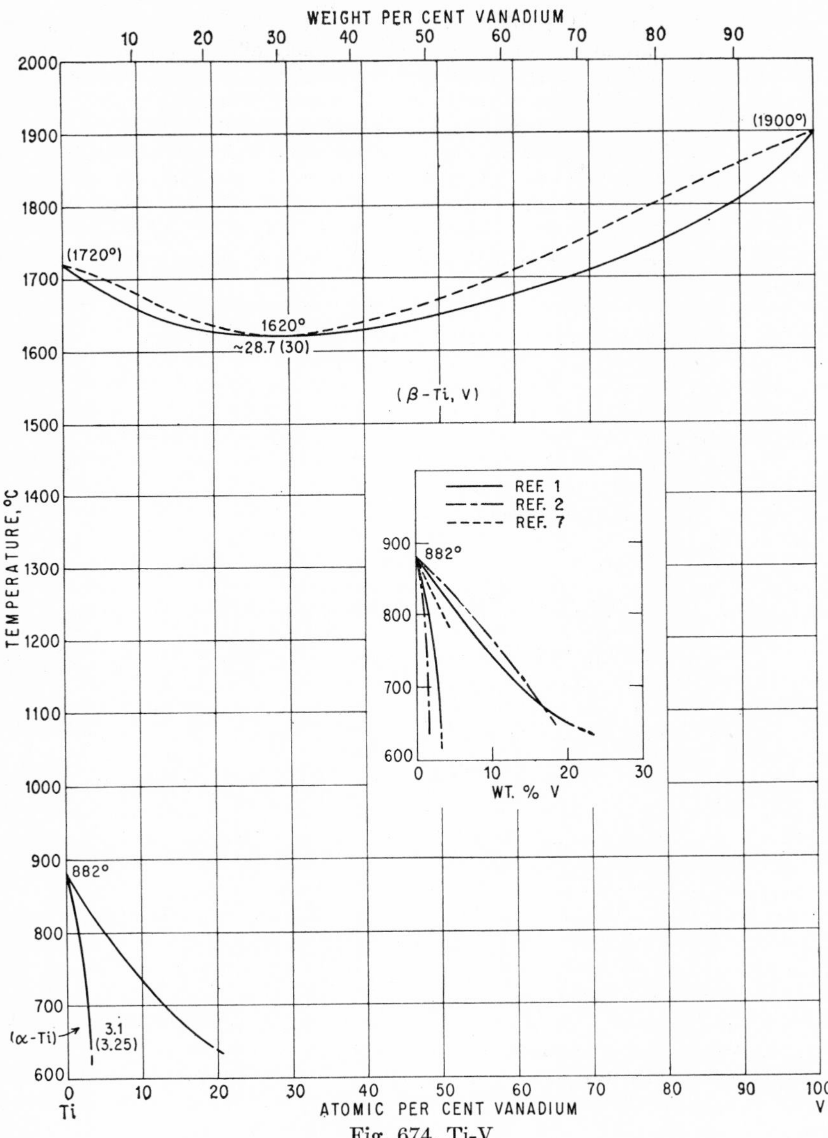

Fig. 674. Ti-V

and 3.5 wt. (2.83–3.30 at.) % V [1], and between 1.0 and 2.0 wt. (0.94–1.88 at.) % V [2], respectively. As the annealing times were of comparable length, no obvious reason can be given for this discrepancy. It may be mentioned, however, that the number of data points used to bracket the $\alpha/(\alpha + \beta)$ boundary in the work by [1] is larger and, in addition, their alloys were severely cold-worked prior to annealing. On the other hand, [2], who reported that "no extensive study was made in the range 0–2 wt. % V," used homogenized as-cast specimens.

[7] determined the $\beta/(\alpha + \beta)$ boundary up to 4.5 at. (4.75 wt.) % V by measuring the hydrogen pressure in equilibrium with an extremely dilute solution of hydrogen in the (iodide Ti base) alloys as a function of temperature. The slope of the curve is markedly steeper in the range of composition tested. It was concluded that the solubility of V in α-Ti at 700°C would lie between 0.5 and 1.0 wt. % V.

Crystal Structure. Between the composition at which the β phase can be retained on quenching, 14.9 wt. % V [2], and V, the lattice parameter changes nearly linearly with composition [1, 2], with a slight negative deviation from Vegard's law [2].

1. H. K. Adenstedt, J. R. Pequinot, and J. M. Raymer, *Trans. ASM*, **44**, **1952**, 990–1003.
2. P. Pietrokowsky and P. Duwez, *Trans. AIME*, **194**, 1952, 627–630.
3. W. Rostoker and A. Yamamoto, *Trans. ASM*, **46**, 1954, 1136–1163.
4. R. M. Powers and H. A. Wilhelm, Ames Laboratory Report ISC-228, U.S. Atomic Energy Commission publication, September, 1952; quoted from WADC Technical Report 54-502, September, 1954, pp. 42–45.
5. The melting point of the vanadium used (99.8%) was found as 1900 ± 25°C.
6. C. M. Craighead, O. W. Simmons, and L. W. Eastwood, *Trans. AIME*, **188**, 1950, 485–513.
7. A. D. McQuillan, *J. Inst. Metals*, **80**, 1951-1952, 363–368.

$\bar{1}.4157$
0.5843

Ti-W Titanium-Wolfram

β-Ti was first thought to form a continuous series of solid solutions with W, as a result of studies using alloys prepared by solid-state diffusion [1]. The work by [2], however, showed that this was not true (Fig. 675). These investigators used iodide Ti as a base for the Ti-rich alloys and magnesium-reduced Ti as a base for the intermediate and high-alloy ranges [3].

Micrographic analysis of the arc-melted alloys (those up to 6 at. % W were hot- and cold-worked prior to annealing) was the main method of investigation. Both optical observation and incipient-melting techniques were used to determine the melting range [4]. The eutectoid temperature was bracketed between 725 and 700°C. The partial diagram in the inset of Fig. 675 is plotted in wt. %.

A number of liquidus points, indicated in Fig. 675, were reported by [5]; no experimental details were given. The composition above which the β solid solution is retained on quenching lies between 20 and 25 wt. (6.1 and 8.0 at.) % W [6].

Reference is made to a forthcoming paper by [7].

1. B. W. Gonser, *Ind. Eng. Chem.*, **42**, 1950, 222–226.
2. D. J. Maykuth, H. R. Ogden, and R. I. Jaffee, *Trans. AIME*, **197**, 1953, 231–237.
3. The constitution of Ti-rich alloys made with magnesium-reduced Ti was determined by L. W. Eastwood, R. M. Goldhoff, J. W. Holladay, and J. G. Kura, Battelle Memorial Institute, AF Technical Report 6515, Part 1, June, 1951.
4. W. Kroll (*Z. Metallkunde*, **29**, 1937, 189–192) reported that 5 wt. % W raises the melting point of Ti.
5. H. Nowotny, E. Parthé, R. Kieffer, and F. Benesovsky, *Z. Metallkunde*, **45**, 1954, 97–99.
6. P. Duwez, *Trans. ASM*, **45**, 1953, 934–940.
7. See preliminary report by W. Trzebiatowski, J. Berak, J. Niemic, and T. Romotowski, *Roczniki Chem.*, **25**, 1951, 516–517; *Chem. Abstr.*, **46**, 1952, 9047.

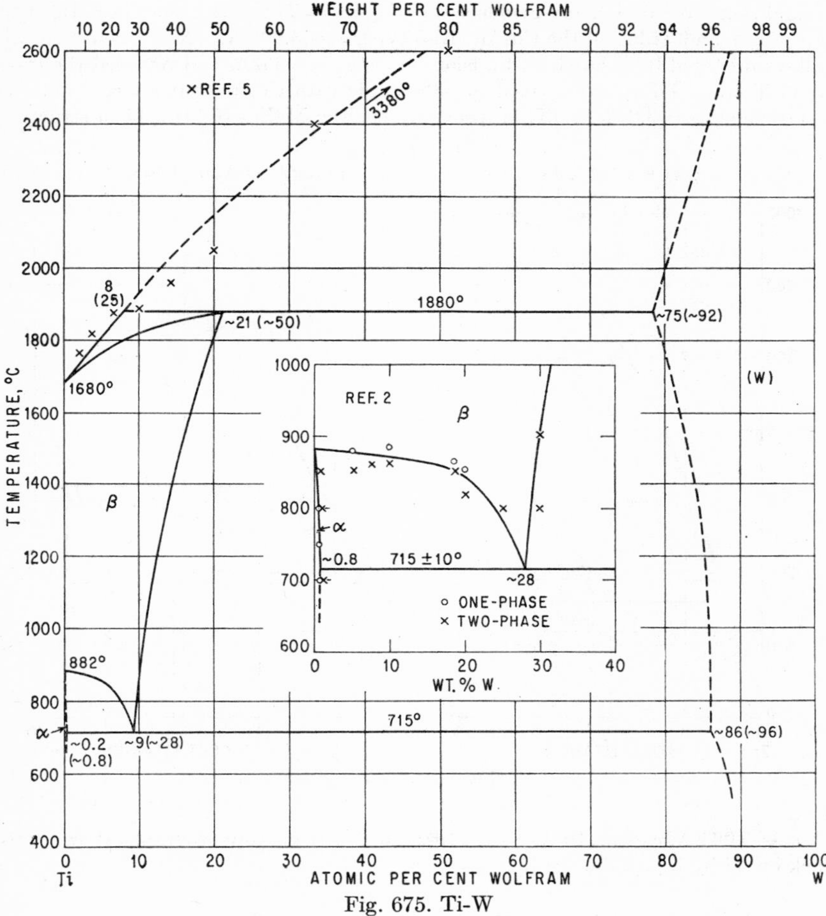

Fig. 675. Ti-W

$\bar{1}.8649$
0.1351

Ti-Zn Titanium-Zinc

The constitution of the Zn-rich alloys, as evidenced by thermal and microscopic studies [1, 2], is presented in Fig. 676. [1] determined the freezing ranges by means of a sensitive thermoanalytical method and found a eutectic at 0.45 wt. % Ti and 418°C. In addition, thermal effects were observed at 460 and 490°C which are probably due to peritectic reactions [3]. No attempt was made to identify the phases formed by these reactions. The solid solubility of Ti in Zn was reported to be smaller than 0.02 wt. % Ti at 400°C.

[2] claim that there is considerable evidence in favor of a eutectic at 0.12 wt. % Ti [4]. The composition of the intermediate phase formed peritectically at 485°C was given as 4.7 wt. % Ti, which corresponds to the formula $TiZn_{15}$ [4.66 wt. (6.25 at.) % Ti]. The composition of the second intermediate phase, 6.9 wt. % Ti, would correspond to the formula $TiZn_{10}$ [6.81 wt. (9.09 at.) % Ti]. The solid solubility of Ti in Zn is between 0.007 and 0.015 wt. % Ti at 300°C.

[5] briefly reported X-ray evidence for the phases TiZn, with the CsCl (B2) type of structure, and $TiZn_3$, of the Cu_3Au ($L1_2$) type, $a = 3.029$ A. X-ray investigations of alloys prepared by solid-state diffusion [6] confirmed $TiZn_3$ and revealed the existence of $TiZn_2$. $TiZn_3$ was verified to be isotypic with Cu_3Au, but a larger unit cell was obtained, $a = 3.932$ A. $TiZn_2$ probably has the $MgZn_2$ (C14) type of structure,

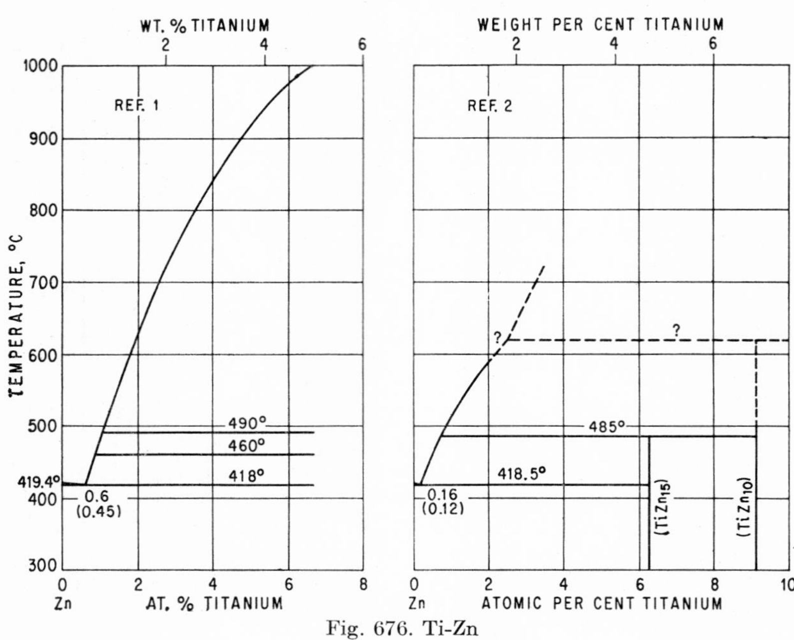

Fig. 676. Ti-Zn

with $a = 5.064$ A, $c = 8.210$ A, $c/a = 1.621$ [6]. For microphotographs showing the attack on Ti by liquid Zn, see [7].

1. E. Gebhardt, *Z. Metallkunde,* **33,** 1941, 355–357.
2. E. A. Anderson, E. J. Boyle, and P. W. Ramsey, *Trans. AIME,* **156,** 1944, 278–286.
3. As the data points determined by thermal analysis are not tabulated, they have been omitted from Fig. 676.
4. No data points are given.
5. F. Laves and H. J. Wallbaum, *Naturwissenschaften,* **27,** 1939, 674–675.
6. P. Pietrokowsky, *Trans. AIME,* **200,** 1954, 219–226.
7. K. Ruttewit and E. Eichmeyer, *Metall,* **11,** 1957, 659–662.

$\bar{1}.7202$
0.2798

Ti-Zr Titanium-Zirconium

Measurements of the electrical resistivity and temperature coefficient of resistance of three alloys with 12.3, 78.9, and 86.6 at. % Zr indicated the existence of a continuous series of solid solutions of the low-temperature modifications of the component metals [1]. Later work on the transformation diagram gave conclusive evidence that

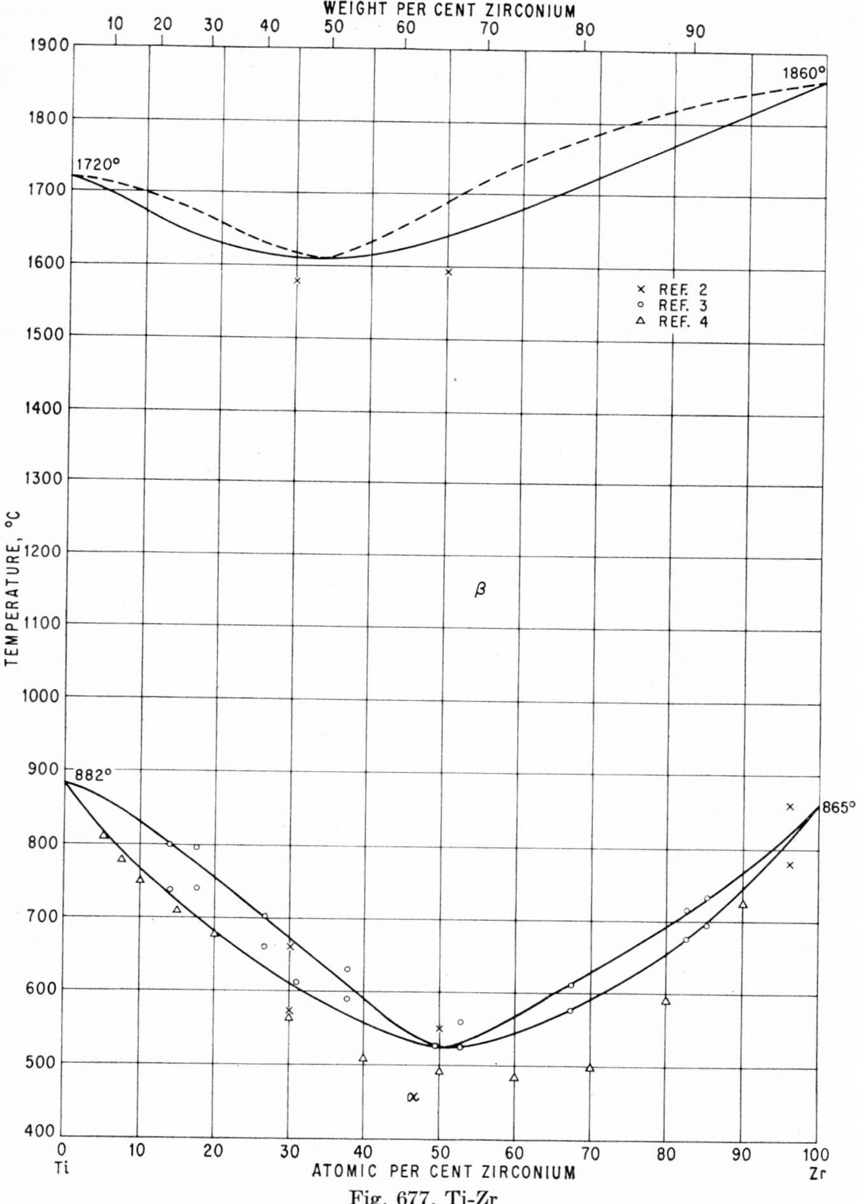
WEIGHT PER CENT ZIRCONIUM
10
20
30
40
50
60
70
80
90
1900
1800
1700
1600
1500
1400
1300
1200
1100
1000
900
800
700
600
500
400
TEMPERATURE, °C
1720°
1860°
× REF. 2
○ REF. 3
△ REF. 4
β
882°
865°
α
0
Ti
10
20
30
40
50
60
70
80
90
100
Zr
ATOMIC PER CENT ZIRCONIUM

Fig. 677. Ti-Zr

there is complete miscibility between both the h.c.p. α and b.c.c. β forms [2–4]. The melting-point curve (Fig. 677) was determined by the incipient-melting method, using four alloys with 27, 55, 76, and 86 wt. % Zr. Results were reproducible within 20°C; individual data were not given [3]. Alloys were prepared using magnesium-reduced Ti and Zr; their carbon content varied between 0.17 and 1.85 wt. % and was 0.8 wt. % on the average. The oxygen content was approximately 0.15 wt. % [3]. Previously, [2] had determined the fusion points of alloys with 30 and 50 at. % Zr [5] (Fig. 677).

Temperatures of the α-β transformation were determined by means of thermoresistometric [2, 3] and thermal studies [4]; see data points in Fig. 677. The alloys were prepared from magnesium-reduced metals melted in carbon crucibles (see above) [3] or iodide Ti and Zr [2, 4], respectively. The transformation points reported by [4] were determined for high rates of cooling, ranging from 100 to 8000°C per sec. They were practically independent of the rate of cooling, indicating that the transformation is probably of the martensitic type. The minimum was placed at approximately 50 at. % Zr, 525°C [3], and close to 60 at. % Zr, 485°C [4]. The boundaries of the $(\alpha + \beta)$ field in alloys with up to 10 wt. (5.5 at.) % Zr were roughly outlined micrographically by [6], using short-time annealed alloys prepared with magnesium-reduced Ti.

Crystal Structure. Lattice parameters of the α phase and β phase (partially retained between 20 and 80 at. % Zr on quenching from 980°C) were determined by [2–4] and [4], respectively.

1. J. H. de Boer and P. Clausing, *Physica*, **10**, 1930, 267–269.
2. J. D. Fast, *Rec. trav. chim.*, **58**, 1939, 973–983.
3. E. T. Hayes, A. H. Roberson, and O. G. Paasche, *U.S. Bur. Mines Rept. Invest.* 4826, November, 1951.
4. P. Duwez, *J. Inst. Metals*, **80**, 1951-1952, 525–527.
5. Alloys were prepared by thermal decomposition of mixtures of the iodides.
6. C. M. Craighead, F. Fawn, and L. W. Eastwood, Battelle Memorial Institute, Second Progress Report on Contract 33 (038)-3736 to Wright-Patterson Air Force Base, Oct. 31, 1949.

$\bar{1}.9338$
0.0662

Tl-U Thallium-Uranium

According to [1], the U-Tl compound richest in Tl is UTl_3 (72.03 wt. % Tl), yielding the reflections of a f.c.c. lattice, with $a = 4.675$ A. Because of the similar scattering power of the U and Tl atoms, it could not be decided whether or not the compound is ordered. [1] suggest an ordered structure (Cu_3Au or $L1_2$ type) because of the appreciable difference in atomic radii between U and Tl.

1. A. Iandelli and R. Ferro, *Ann. chim.* (*Rome*), **42**, 1952, 598–606.

0.4950
$\bar{1}.5050$

Tl-Zn Thallium-Zinc

A thermal and microscopic analysis carried out by [1] showed that the system Zn-Tl is characterized by a miscibility gap in the liquid state which extends from about 2.5 to 95 wt. (0.8–86 at.) % Tl at the monotectic temperature (416–417°C) [2]. The extent of the gap at higher temperatures was measured by [5] by means of differential thermal analysis (Fig. 678).

Solid solubilities appear to be negligible; the influence of Zn on the transformation in Tl has not been studied.

The results of emf measurements by [6] using the cell: Zn/normal solution of $ZnSO_4/Zn_xTl_{1-x}$ are in agreement with the constitution of Fig. 678.

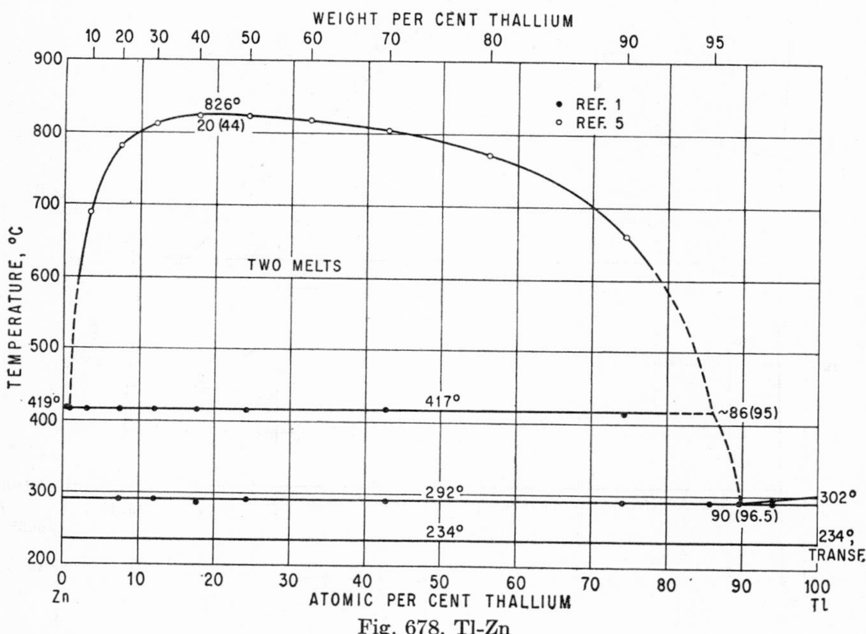

Fig. 678. Tl-Zn

1. A. v. Vegesack, *Z. anorg. Chem.*, **52**, 1907, 30–34.
2. The depression of the freezing point of Zn by addition of Tl was studied by [3] and [4]. According to [3], 2.5 wt. (0.8 at.) % Tl lowers the freezing point to 414°C; [4] investigated two alloys with 0.25 and 0.5 at. % Tl.
3. C. T. Heycock and F. H. Neville, *J. Chem. Soc.*, **71**, 1897, 383.
4. K. Honda and T. Ishigaki, *Science Repts. Tôhoku Imp. Univ.*, **14**, 1925, 228. Quoted from "Gmelins Handbuch der anorganischen Chemie," System No. 38, p. 212, Verlag Chemie, G.m.b.H., Berlin, 1940.
5. H. Johnen, Thesis, University of Münster (Germany), 1952; W. Seith, H. Johnen, and J. Wagner, *Z. Metallkunde*, **46**, 1955, 773–779.
6. R. Kremann and A. Lobinger, *Z. Metallkunde*, **12**, 1920, 246–247.

0.6696
$\bar{1}$.3304

U-V Uranium-Vanadium

The phase diagram in Fig. 679 was established by means of thermal and micrographic studies [1]. Alloys were prepared by arc melting (0–50 at. % U) and induction melting in BeO crucibles (50–100 at. % U). The solubility of V in β-U is about 2 at. (0.4 wt.) % V at 700°C, and that in α-U is less than 1.5 at. (0.3 wt.) % V at 600°C.

1. H. A. Saller and F. A. Rough, *Trans. AIME*, **197**, 1953, 545–548.

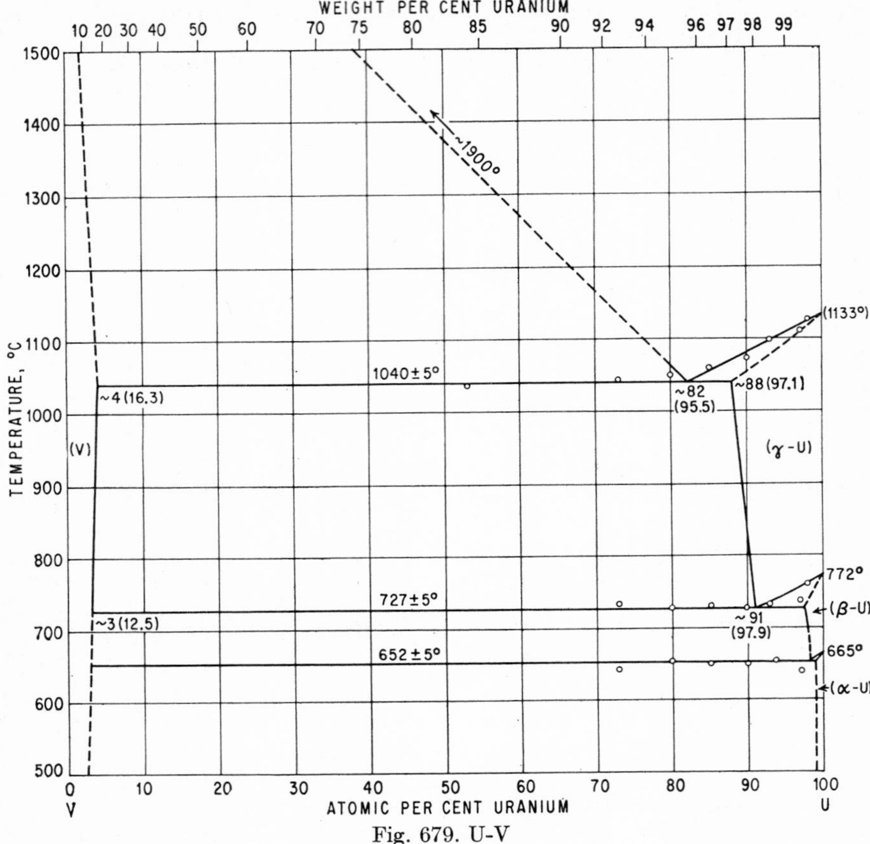

Fig. 679. U-V

0.1121
$\bar{1}.8879$

U-W Uranium-Wolfram

The phase diagram in Fig. 680 was established by thermal, micrographic, and X-ray analysis of arc-melted alloys, as well as by solubility determinations to fix the liquidus curve up to about 2440°C. Heat-treatments in vacuum followed by quenching were carried out in the temperature range 800–2300°C. X-ray evidence "placed the elevated temperature solution limits at probably less than 1 at. % at both ends of the diagram" [1].

Recent work by [2] confirmed the results of [1]. At 1000°C the solubility of W in γ-U was found to be about 0.9 ± 0.3 at. % and that of U in W to be about 0.1 at. %. The small influence of W on the transformation temperatures of U suggests that the solubility of W in γ-U and β-U is very slight.

1. C. H. Schramm, P. Gordon, and A. R. Kaufmann, *Trans. AIME,* **188,** 1950, 195–204.
2. D. Summers-Smith, *J. Inst. Metals,* **83,** 1954-1955, 383–384.

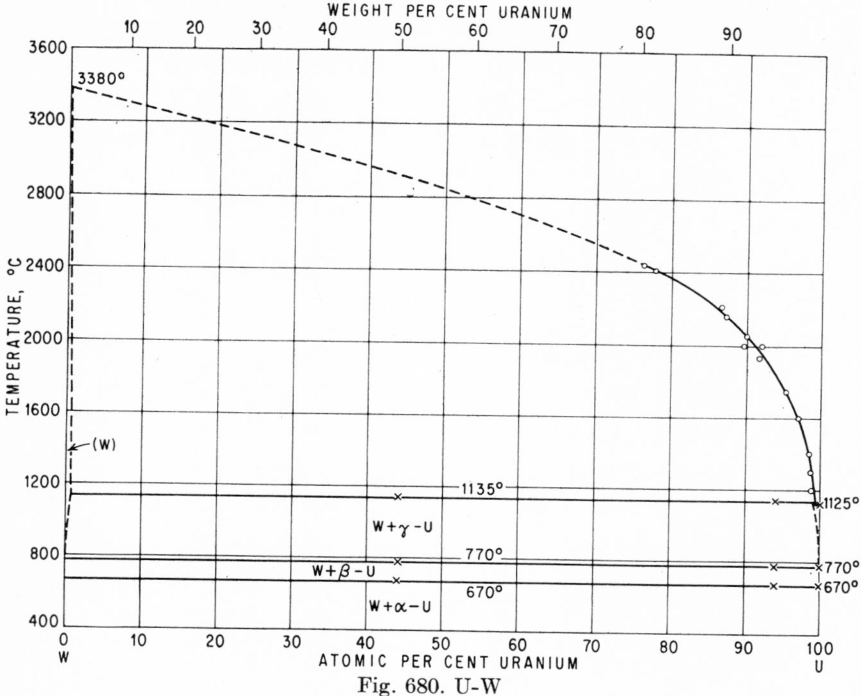

Fig. 680. U-W

0.5613
$\bar{1}$.4387

U-Zn Uranium-Zinc

Figure 680*a* shows the phase diagram at 5 atm pressure (boiling point of Zn under atmospheric pressure: 907°C), constructed by [1] on the basis of thermal analysis of alloys in sealed containers up to 1150°C and on the basis of metallographic, X-ray, and analytical data. The Zn-rich liquidus up to 700°C was determined by means of chemical analyses of samples taken from equilibrated melts. The results are plotted (on a wt. % scale) in the inset of Fig. 680*a*, together with some data of previous investigators [2, 3]. The compound UZn_9 (28.81 wt. % U), first prepared by [4, 5], was shown by [1] to be hexagonal, with $a = 8.99$ A, $c = 8.98$ A. The results of thermal and metallographic analysis indicated little or no solid solubility of Zn in U.

Under an external pressure of 1 atm, UZn_9 decomposes, as indicated by vapor-pressure measurements [1], into Zn vapor and a U-rich liquid at 945 ± 5°C. At 910°C, the compound is in equilibrium with Zn vapor and Zn-rich liquid containing approximately 14.6 wt. (4.5 at.) % U. For a phase diagram at 1 atm pressure, see [1].

1. P. Chiotti, H. H. Klepfer, and K. J. Gill, *Trans. AIME*, **209**, 1957, 51–57.
2. E. E. Hayes and P. Gordon, unpublished work (1948); quoted in [1].
3. A. R. Kaufmann and S. Isserow, unpublished work (1954); quoted in [1].
4. J. Chipman et al., unpublished work (1943); quoted in [1].
5. J. H. Carter et al., unpublished work (1943); quoted in [1].

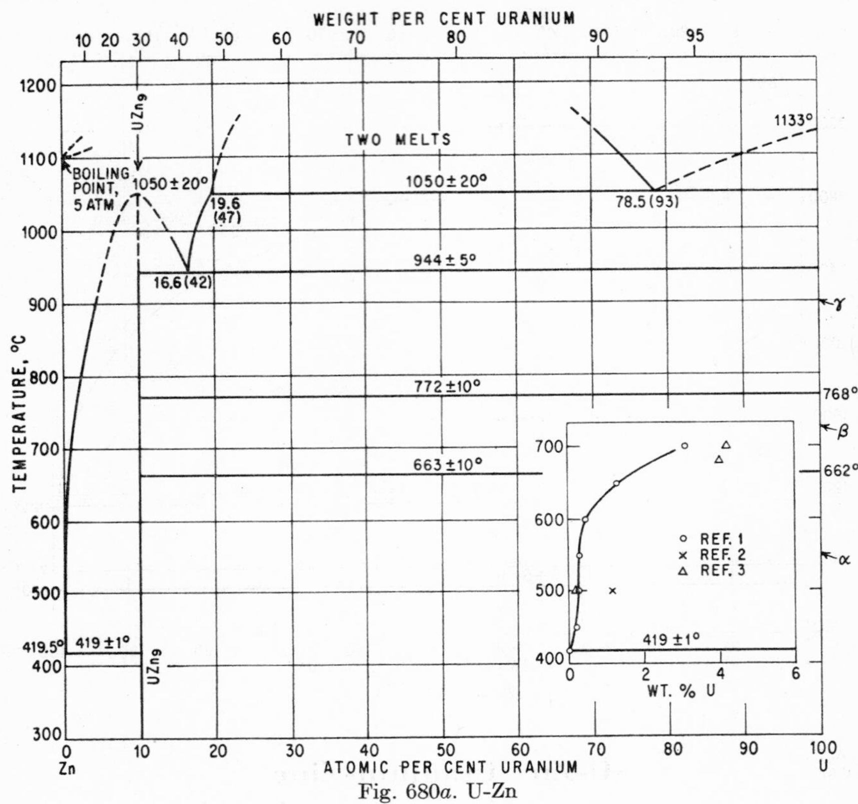

Fig. 680*a*. U-Zn

0.4166
$\bar{1}.5834$

U-Zr Uranium-Zirconium

Research data, published in classified literature, were reviewed by [1]. The following features of the constitution had been recognized: (*a*) increase of the melting point of U by 20 at. % Zr to about 1225°C; (*b*) a continuous series of solid solutions between β-Zr and γ-U; (*c*) a eutectoid at about 90 at. % U, 688°C; (*d*) another eutectoid horizontal at about 600°C, extending across almost the entire system; (*e*) an intermediate phase, ϵ, stable at and below 600°C in the approximate range 20–30 at. % U and having a b.c. tetragonal structure.

More recently, a phase diagram, compiled from various sources, was published [2]; see inset of Fig. 681. It indicates all the salient features mentioned above. The homogeneity range of the ϵ phase, which may exist in more than one modification, is tentatively given as 25–32 at. % U. The phase is believed to decompose on heating at about 600°C into α-U and α-Zr. Crystals of the ϵ phase (Zr_3U) could be isolated from annealed alloys by treating with HCl-ethylacetate [2]. The residue was reported to exhibit "a unique diffraction pattern which originates from a component of the alloy system which is neither α-U nor α-Zr."

The main diagram in Fig. 681 is due to [3]. It is based on thermal, micrographic, dilatometric, and high-temperature X-ray investigations. X-ray examination of alloys with 25–35 at. % U at 599°C revealed both α-U and α-Zr, indicating the absence

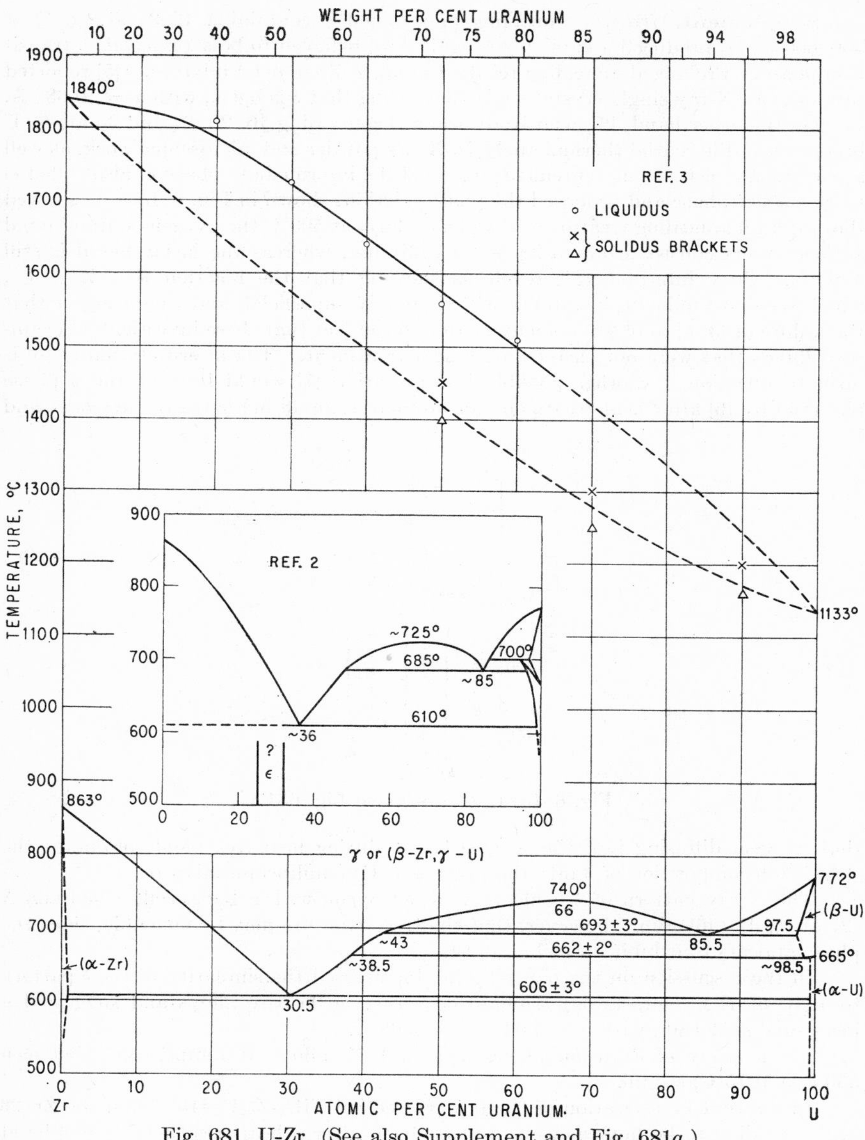

Fig. 681. U-Zr. (See also Supplement and Fig. 681*a*.)

of an intermediate phase. The solid solubility of U in α-Zr was found to be 0.4–1.0 at. % U at 606°C. The maximum solubility of Zr in β-U is 2.5 at. % at 693°C. The β-U solid solution transforms by a eutectoid reaction at 662°C and about 1.5 at. % Zr. The maximum solubility of Zr in α-U at the eutectoid was placed tentatively at about 1 at. % Zr.

Crystal Structures. The parameter of the b.c.c. γ phase varies linearly with composition [3]. The ϵ phase, Zr_3U, reported by [1, 2], was claimed to be b.c. tetragonal [1].

Supplement. By the annealing of retained γ containing 12–34 at. % U at 300–600°C, [4] produced a cubic phase (called δ), believed to be a (*metastable*) transition phase. Prolonged annealing resulted in an (α-Zr + α-U) mixture. [5] reported results of an X-ray single-crystal study, indicating that δ is b.c.c., with a = 10.688 A.

On the other hand, [6], who studied alloys containing 10, 20, 27 and 36 at. % U by means of differential thermal analysis, X-ray powder and microscopic work, as well as electrical-resistance measurements, claimed the intermediate phase δ (also called ϵ) to be a *stable* phase and proposed the phase relations shown in Fig. 681*a*. [6] showed that after an annealing treatment of δ-phase alloys at 500°C the oxygen-contaminated surface layers consist of an (α-Zr + α-U) mixture, whereas the inside metal is still δ-phase. They interpret this result as showing that the reaction $\delta \rightarrow$ (Zr + U), which seems to indicate instability of δ, is purely superficial, and they suggest that the failure of [3, 4] to observe δ may be due to the fact that these investigators examined filings that were not cleaned after heat-treatment. The interpretation of [6] is open to question. Another possible interpretation [7] would be that the δ phase observed by [6] after heat-treatment (of no more than 48 hr) is really *metastable* and

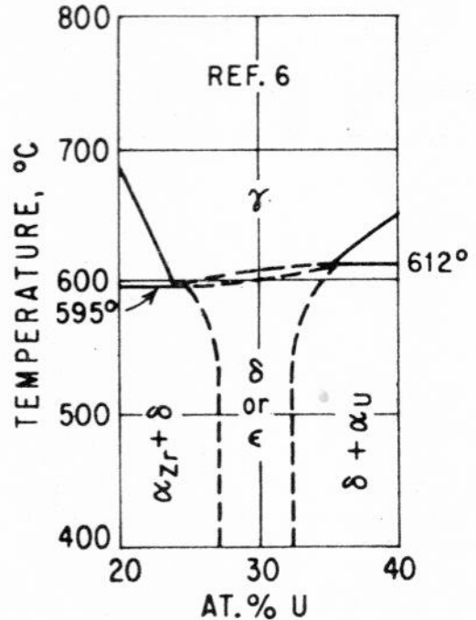

Fig. 681*a*. U-Zr. (See also Fig. 681.)

that oxygen, diffusing into the surface layers during heat-treatment, promotes the sluggish decomposition of δ into the (α-Zr + α-U) equilibrium mixture.

The X-ray pattern of δ could be indexed by [6] with a b.c.c. cell, a = 10.68 A (27 at. % U); [6] admit, however, that the true structure may be noncubic, since the phase appears to exhibit optical activity.

In the discussion on the paper by [6], [8] stressed the similarity of the δ pattern to that of the (metastable) ω phase occurring in Ti-base alloys and suggested a hexagonal structure, with a = 5.03 A, c = 3.08 A.

For a study of diffusion phenomena in U-Zr alloys at temperatures between 550 and 1075°C, see [9].

In a recent investigation of the ternary section TiU_2-Zr_3U, [10] found the Zr–26 at. % U alloy to be homogeneous δ-phase even after a heat-treatment of 670 hr at 510°C. [10] consider the δ phase a stable intermediate phase.

1. O. N. Carlson, *U.S. Atomic Energy Comm., Publ.* AECD-3206 (and ISC-102), 1951.
2. R. P. Larsen, R. S. Shor, H. M. Feder, and D. S. Flikkema, *U.S. Atomic Energy Comm., Publ.* ANL-5135, 1954.
3. D. Summers-Smith, *J. Inst. Metals,* **83,** 1954-1955, 277–282.
4. M. H. Mueller, S. T. Zegler, and P. A. Beck, unpublished information (1955).

5. M. H. Mueller, *Acta Cryst.*, **8**, 1955, 849–850.
6. A. N. Holden and W. E. Seymour, *Trans. AIME*, **206**, 1956, 1312–1316.
7. H. Westphal, Metallgesellschaft A.G., Frankfurt/Main, private communication.
8. J. M. Silcock, *Trans. AIME*, **209**, 1957, 521.
9. Y. Adda, J. Philibert, and H. Faraggi, *Rev. mét.*, **54**, 1957, 597–610.
10. H. A. Saller, F. A. Rough, A. A. Bauer, and J. R. Doig, *Trans. AIME*, **209**, 1957, 878–881.

$\bar{1}.4425$
0.5575

V-W Vanadium-Wolfram

"Considerable difficulty was encountered in getting W into solution. The problem was solved by making a homogeneous master alloy of 50 wt. [21.7 at.] % W by repeated melting. Homogeneous alloys containing up to 10 wt. % W were successfully melted. In this range of composition only V-base solid solutions were observed. The microstructure of the as-cast master alloy showed small amounts of a second phase. The identification of this phase was not pursued. It seems likely that solid solubility of W in V persists to higher than 40 wt. [15.6 at.] %" [1].

1. W. Rostoker and A. Yamamoto, *Trans. ASM*, **46**, 1954, 1136–1163, especially 1159–1160. The alloys were prepared in a helium arc-melting furnace.

$\bar{1}.7470$
0.2530

V-Zr Vanadium-Zirconium

The phase diagram of Fig. 682 has been established by [1] mainly by means of incipient-melting, dilatometric, and resistometric measurements as well as microscopic and X-ray diffraction work.

No attempt was made to determine the liquidus for the system. Incipient-melting (solidus) temperatures were measured by optical pyrometry. Microscopic work located the eutectic composition at about 30 wt. (43.4 at.) % V. Dilatometry and measurement of electrical-resistance changes were relied upon to establish the eutectoid horizontal and the lower boundary of the (β-Zr) region; since the resistance data were a better representation of equilibrium conditions in the alloys, they have been plotted in Fig. 682. The limit of solubility of Zr in V indicated by microscopic examination of quenched specimens was not completely certain because of the presence of foreign inclusions in the alloys; the best estimate which could be made by [1] from the evidence obtained was that V dissolves about 5 wt. (3 at.) % Zr at 600°C.

Previous Work. The compound ZrV_2 (52.76 wt. % V) was first reported by [2]. In a cursory microscopic study of Zr-rich alloys, [3, 4] found that Zr dissolves some V and that a eutectic is formed at about 4.7 wt. % V; from Fig. 682 it is obvious that actually the eutectoid was present. On the basis of microscopic, incipient-melting, and X-ray work on 11 alloys with 20–98 wt. % V, [5] published a tentative phase diagram which showed the same eutectic-peritectic phase relationships as given in Fig. 682; however, the temperatures assigned to these isothermal lines, 1360 and 1740°C (!), respectively, could not be corroborated by [1] (see also discussion on [5]). The solubility limit of Zr in solid V was found to be about 3 wt. (1.7 at.) % by [5].

Crystal Structure. [2] found ZrV_2 to have the hexagonal $MgZn_2$ (C14) type of structure, with $a = 5.288$ A, $c = 8.664$ A, $c/a = 1.639$. This was not confirmed by [6], who reported the cubic $MgCu_2$ (C15) type, with $a = 7.44$ A, instead; "alloys annealed at temperatures from 600 to 1400°C [7] . . . indicated that there was at no temperature an $MgZn_2$-modification as reported by [2]." [1] also was unable to find indications for a polymorphic transformation in the intermediate phase and stated

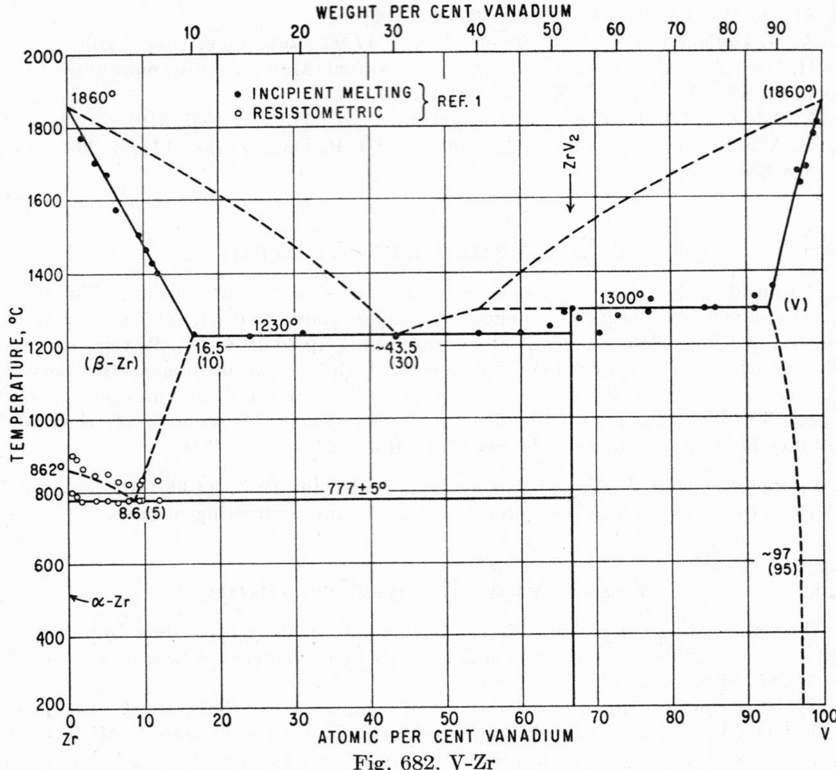

Fig. 682. V-Zr

that "lines were visible which did not seem to be reasonably accounted for by the terminal solid solutions or by the structure assigned to ZrV_2 by [2]."

1. J. T. Williams, *Trans. AIME*, **203,** 1955, 345–350. The alloys were prepared in a helium-arc furnace. Vanadium of about 99.5 wt. % purity and (for all dilatometric and resistance specimens and about two-thirds of the solidus-temperature specimens) iodide Zr were used.
2. H. J. Wallbaum, *Naturwissenschaften*, **30,** 1942, 149.
3. S. M. Shelton, *U.S. Atomic Energy Comm., Publ.* AF-TR-5932, 1949; *Met. Abstr.*, **21,** 1954, 869.
4. C. T. Anderson, E. T. Hayes, A. H. Roberson, and W. J. Kroll, *U.S. Bur. Mines Rept. Invest.* 4658, 1950.
5. W. Rostoker and A. Yamamoto, *Trans. ASM*, **46,** 1954, 1136–1163; discussion, pp. 1163–1167.
6. R. P. Elliott, Armour Research Foundation, Chicago, Ill., Technical Report 1, OSR Technical Note OSR-TN-247, 1954.
7. According to [6], the temperature of the peritectic decomposition of ZrV_2 is 1500°C.

0.4492
$\bar{1}$.5508

W-Zn Wolfram-Zinc

According to [1], W and Zn do not alloy at temperatures up to 1350°C [2]. On the other hand, [3] suggested on the basis of some exploratory work that there exists, as in the system Cr-Zn, a Zn-rich intermediate phase. However, since this phase was obtained in only a very small amount, it might well be that a contaminating element (Fe?), rather than W itself, alloyed with Zn (cf. Mo-Zn).

1. W. Köster and H. Schmid, *Z. Metallkunde*, **46**, 1955, 462–463.
2. Previously, D. Kremer [*Abhandl. Inst. Metallhütt. u. Elektromet. Tech. Hochsch., Aachen*, **1**, 1916(2), 9–10] had reported that molten Zn does not dissolve W.
3. T. Heumann, *Z. Metallkunde*, **39**, 1948, 51–52.

0.3045
$\bar{1}$.6955

W-Zr Wolfram-Zirconium

The phase diagram of the system W-Zr has been established by [1] on the basis of metallographic analysis, incipient-melting data, and X-ray diffraction work. In the discussion on this paper, [2] published original experimental data; the methods of investigation used were, for the most part, very similar to those described in [1], but [2] also used dilatometry to determine the temperature of the $\alpha \rightleftarrows \beta$ transformation and made measurements of the liquidus temperatures of alloys in order to fix the eutectic point. The results of the two independent studies are in unusually good agreement, as can be seen from Table 47 and Fig. 683.

Table 47

	Ref. 1	Ref. 2	Fig. 683
Eutectoid temperature........	850 ± 15°C	865 ± 5°C	860°C
Eutectic....................	18 wt. (10 at.) % W, 1650 ± 15°C	10 at. % W, 1700 ± 50°C	10 at. % W, 1660°C
Peritectic melt..............	50 wt. (33 at.) % W, 2175 ± 25°C	33 at. % W, 2150°C	33 at. % W, 2150°C
Solubility of Zr in solid W at peritectic temperature......	<10 wt. (18 at.) % Zr	6 at. % Zr	6 at. % Zr

According to [1], the solubility of W in α-Zr is considerably less than 0.5 wt. (0.25 at.) % (the alloy lowest in W prepared). No clearly retained (β-Zr) structures were observed.

Cursory microscopic work on Zr-rich alloys had been carried out by [3, 4].

Crystal Structure. ZrW_2 (80.13 wt. % W) has the cubic $MgCu_2$ (C15) type of structure [5, 6], with a = 7.63 A [5], 7.615 A [6]. There was no indication of any polymorphic transformation in this compound [6].

1. R. F. Domagala, D. J. McPherson, and M. Hansen, *Trans. AIME*, **197**, 1953, 73–79. The alloys were prepared in a helium-arc furnace. Iodide Zr (99.8 wt. % pure) and W of 99.9 wt. % purity were used.
2. G. A. Geach and G. F. Slattery, *Trans. AIME*, **197**, 1953, 747–748.
3. S. M. Shelton, *U.S. Atomic Energy Comm., Publ.* AF-TR-5932, 1949. *Met. Abstr.*, **21**, 1954, 869.
4. C. T. Anderson, E. T. Hayes, A. H. Roberson, and W. J. Kroll, *U.S. Bur. Mines Rept. Invest.*, 4658, 1950.
5. A. Claassen and W. G. Burgers, *Z. Krist.*, **A86**, 1933, 100–105.
6. R. P. Elliott, Armour Research Foundation, Chicago, Ill., Technical Report 1, OSR Technical Note OSR-TN-247, 1954.

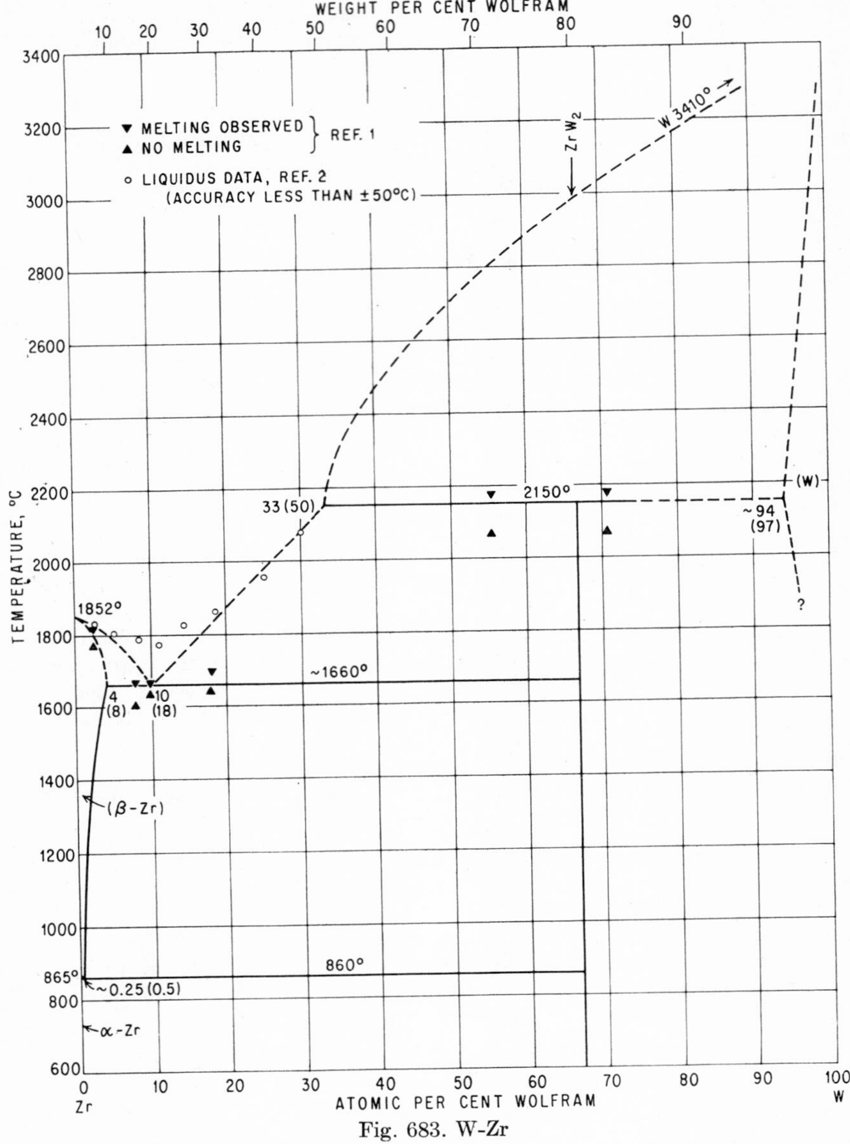

Fig. 683. W-Zr

$\bar{1}.8554$
0.1446

Zn-Zr Zinc-Zirconium

[1] investigated 13 Zn-rich alloys containing up to 2.6 wt. % Zr by means of thermal, microscopic, and X-ray diffraction analyses. A eutectic at about 0.1 wt. (0.07 at.) % Zr, 416°C, and two additional, presumably peritectic, three-phase lines at 545 and 970°C were found (Fig. 684). Composition and crystal structure of the inter-

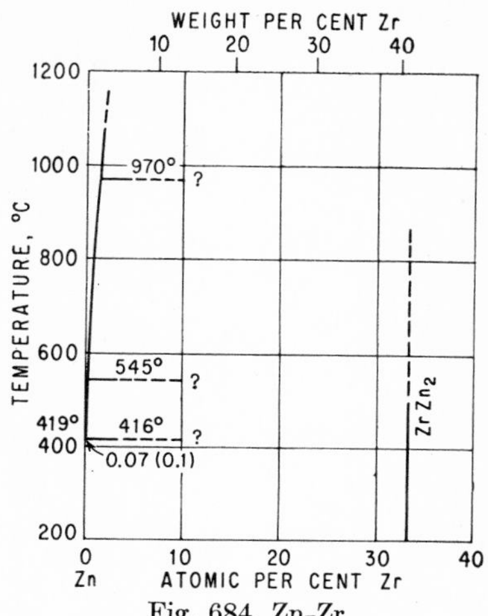

Fig. 684. Zn-Zr

mediate phases were not determined. The solubility of Zr in solid Zn is about 0.02 wt. (0.015 at.) % at 400°C.

[2] detected the intermediate phase $ZrZn_2$ (41.09 wt. % Zr), having the cubic $MgCu_2$ (C15) type of structure, with $a = 7.3958 \pm 3$ A (35 at. % Zr).

1. E. Gebhardt, *Z. Metallkunde*, **33,** 1941, 355–357.
2. P. Pietrokowsky, *Trans. AIME*, **200,** 1954, 219–226.

Appendix

Periodic Table of Elements

(International atomic weights, 1955)

1 H 1.0080																		2 He 4.003
3 Li 6.940	4 Be 9.013												5 B 10.82	6 C 12.011	7 N 14.008	8 O 16.0000	9 F 19.00	10 Ne 20.183
11 Na 22.991	12 Mg 24.32												13 Al 26.98	14 Si 28.09	15 P 30.975	16 S 32.066	17 Cl 35.457	18 Ar 39.944
19 K 39.100	20 Ca 40.08	21 Sc 44.96		22 Ti 47.90	23 V 50.95	24 Cr 52.01	25 Mn 54.94	26 Fe 55.85	27 Co 58.94	28 Ni 58.71	29 Cu 63.54	30 Zn 65.38	31 Ga 69.72	32 Ge 72.60	33 As 74.91	34 Se 78.96	35 Br 79.916	36 Kr 83.80
37 Rb 85.48	38 Sr 87.63	39 Y 88.92		40 Zr 91.22	41 Nb 92.91	42 Mo 95.95	43 Tc (99)	44 Ru 101.1	45 Rh 102.91	46 Pd 106.4	47 Ag 107.880	48 Cd 112.41	49 In 114.82	50 Sn 118.70	51 Sb 121.76	52 Te 127.61	53 I 126.91	54 Xe 131.30
55 Cs 132.91	56 Ba 137.36	57 La 138.92	58 to 71	72 Hf 178.50	73 Ta 180.95	74 W 183.86	75 Re 186.22	76 Os 190.2	77 Ir 192.2	78 Pt 195.09	79 Au 197.0	80 Hg 200.61	81 Tl 204.39	82 Pb 207.21	83 Bi 209.00	84 Po 210	85 At (210)	86 Rn 222
87 Fr (223)	88 Ra 226.05	89 Ac 227	90 to ?															

Lanthanides	58 Ce 140.13	59 Pr 140.92	60 Nd 144.27	61 Pm (145)	62 Sm 150.35	63 Eu 152.0	64 Gd 157.26	65 Tb 158.93	66 Dy 162.51	67 Ho 164.94	68 Er 167.27	69 Tm 168.94	70 Yb 173.04	71 Lu 174.99
Actinides	90 Th 232.05	91 Pa 231	92 U 238.07	93 Np (237)	94 Pu (242)	95 Am (243)	96 Cm (245)	97 Bk (249)	98 Cf (249)	99 Es	100 Fm	101 Md (256)	102 No	

Table A. Some Physical Properties of the Elements

The data of this table were taken mainly from C. J. Smithells, "Metals Reference Book," 2d ed., Interscience Publishers, Inc., New York, 1955, and F. H. Spedding and A. H. Daane, The Preparation and Properties of Rare Earth Metals, in "Metallurgy and Fuels," Progress in Nuclear Energy, ser. V, vol. 1, Pergamon Press Ltd., London, 1956.

The table shows the 1952 atomic weights on which all atomic per cent–weight per cent conversions in this book are based. For 1955 atomic weights, see the Periodic Table of Elements (page 1261).

For transformation points, see Table B.

Symbol	Element	Atomic number	Atomic weight (1952)	Melting point, °C	Boiling point, °C	Density, g/cc at 20°C
Ag......	Silver	47	107.880	960.5	2210	10.49
Al.......	Aluminum	13	26.98	660.1	2480	2.70
As.......	Arsenic	33	74.91	(817)	616 (sublimes)	5.73
Au......	Gold	79	197.2	1063	(2950)	19.3
B.......	Boron	5	10.82	2100–2200	(2550)	2.54
Ba......	Barium	56	137.36	710	(1700)	3.75
Be.......	Beryllium	4	9.013	1284	(2970)	1.85
Bi.......	Bismuth	83	209.00	271.3	(1560)	9.80
C.......	Carbon	6	12.010	(5000)	(5000)	3.52 (diamond) 2.25 (graphite)
Ca......	Calcium	20	40.08	850	1482[a]	1.54
Cb......	Columbium (see Nb, niobium)					
Cd......	Cadmium	48	112.41	320.9	765	8.64
Ce......	Cerium	58	140.13	804 ± 5	3600	6.77
Co......	Cobalt	27	58.94	1495	(3550)	8.92
Cp......	Cassiopeium (see Lu, Lutecium)					
Cr......	Chromium	24	52.01	1875 ± 15[b]	2430	7.14
Cs......	Cesium	55	132.91	29.7	700	1.87
Cu......	Copper	29	63.54	1083	2600	8.96
Dy......	Dysprosium	66	162.46	1475–1500	2600	8.56
Er......	Erbium	68	167.2	1475–1525	(2600)	9.06
Eu......	Europium	63	152.0	(900)	(1700)	5.17
Fe......	Iron	26	55.85	1534	(3070)	7.87
Ga......	Gallium	31	69.72	29.8	(2070)	5.91
Gd......	Gadolinium	64	156.9	1350 ± 20	(3000)	7.87
Ge......	Germanium	32	72.60	937[c]	(2700)	5.32
H.......	Hydrogen	1	1.0080	−259.2	−252.5	
Hf......	Hafnium	72	178.6	2220 ± 30[d]	(5400)	13.1
Hg......	Mercury	80	200.61	−38.87	357	13.55
Ho......	Holmium	67	164.94	1475–1525	(2700)	8.80

Table A. Some Physical Properties of the Elements (*Continued*)

Symbol	Element	Atomic number	Atomic weight (1952)	Melting point, °C	Boiling point, °C	Density, g/cc at 20°C
In.......	Indium	49	114.76	156.4	(1730)	7.31
Ir.......	Iridium	77	193.1	2450	(5300)	22.4
K.......	Potassium	19	39.100	63.6	775	0.86
La......	Lanthanum	57	138.92	920 ± 5	4515	6.16
Li.......	Lithium	3	6.940	180	1370	0.534
Lu......	Lutecium	71	174.99	1650–1750	(3500)	9.85
Mg......	Magnesium	12	24.32	649	1103	1.74
Mn......	Manganese	25	54.93	1243	2097	7.44
Mo......	Molybdenum	42	95.95	2620 ± 10	(4800)	10.2
N.......	Nitrogen	7	14.008	−210.0	−195.8	
Na......	Sodium	11	22.997	97.8	883	0.97
Nb......	Niobium	41	92.91	2468 ± 10[e]	(3300)	8.57
Nd......	Neodymium	60	144.27	1024 ± 5	3300	7.01
Ni.......	Nickel	28	58.69	1453	(3000)	8.9
Np......	Neptunium	93	237	640	....	17.6–19.5
O.......	Oxygen	8	16.0000	−218.8	−183.0	1.426 (solid)
Os.......	Osmium	76	190.2	(2700)	(5500)	22.5
P.......	Phosphorus	15	30.975	44.1 (yellow)	280 (yellow)	2.34 (metallic)
Pa......	Protactinium	91	231	1300 (?)	....	15.37[f]
Pb......	Lead	82	207.21	327.4	1740	11.34
Pd......	Palladium	46	106.7	1552	(3900)	12.02
Po......	Polonium	84	210	246	....	9.4
Pr.......	Praseodymium	59	140.92	935 ± 5	3450	6.77
Pt.......	Platinum	78	195.23	1769	(4500)	21.45
Pu......	Plutonium	94	239	639.5 ± 2[g]	....	19.74[g]
Rb......	Rubidium	37	85.48	38.8	680	1.53
Re......	Rhenium	75	186.31	3180 ± 20[h]	(5900)	21.02[h]
Rh......	Rhodium	45	102.91	1966	(4500)	12.44
Ru......	Ruthenium	44	101.7	(2500)	(4900)	12.2
S........	Sulfur	16	32.066	119.0 (monoclinic)	444.5	2.0–2.1
Sb.......	Antimony	51	121.76	630.5	1640	6.68
Sc.......	Scandium	21	44.96	1550–1600	(2750)	2.995
Se.......	Selenium	34	78.96	217	685	4.82
Si.......	Silicon	14	28.09	1412[i]	(2600)	2.32
Sm......	Samarium	62	150.43	1052 ± 5	(1900)	7.54
Sn.......	Tin	50	118.70	231.9	(2200)	7.30
Sr.......	Strontium	38	87.63	770	1360	2.6
Ta......	Tantalum	73	180.88	3000 ± 50	(5300)	16.6
Tb......	Terbium	65	159.2	1400–1500	(2800)	8.25
Te.......	Tellurium	52	127.61	450	1390	6.24
Th......	Thorium	90	232.12	(1750)	(4000)	11.7

Table A. Some Physical Properties of the Elements (*Continued*)

Symbol	Element	Atomic number	Atomic weight (1952)	Melting point, °C	Boiling point, °C	Density, g/cc at 20°C
Ti.......	Titanium	22	47.90	1668 ± 5	(3500)	4.51
Tl.......	Thallium	81	204.39	303	1460	11.85
Tm......	Thulium	69	169.4	1500–1550	(2400)	9.32
U.......	Uranium	92	238.07	1133	(3820)[e]	19.05
V.......	Vanadium	23	50.95	1900 ± 25	(3350)	6.1
W.......	Wolfram	74	183.92	3380	(6000)	19.3
Y.......	Yttrium	39	88.92	1475–1525	(3500)	4.47
Yb......	Ytterbium	70	173.04	824 ± 5	(1800)	6.96
Zn......	Zinc	30	65.38	419.5	907	7.14
Zr.......	Zirconium	40	91.22	1852	(3600)	6.49

[a] G. W. Thomson and E. Garelis, Ethyl Corporation, 1952.

[b] L. L. Wyman and J. T. Sterling, "Ductile Chromium," p. 180, American Society for Metals, Cleveland, Ohio, 1957.

[c] E. S. Greiner and P. Breidt, *Trans. AIME,* **203,** 1955, 187.

[d] D. K. Deardorff and E. T. Hayes, *Trans. AIME,* **206,** 1956, 509.

[e] T. H. Schofield, *J. Inst. Metals,* **85,** 1957, 372.

[f] P. A. Sellers et al., *J. Am. Chem. Soc.,* **76,** 1954, 5935.

[g] *Nuclear Met.,* no. 1 (AIME), 1955.

[h] C. T. Sims et al., *Trans. AIME,* **203,** 1955, 168.

[i] M. Olette, *Compt. rend.,* **244,** 1957, 1033.

Table B. Structural Data of the Elements

The data of this table were taken mainly from the following sources:

C. S. Barrett, "Structure of Metals," 2d ed., McGraw-Hill Book Company, Inc., New York, 1952.

W. Hume-Rothery and G. V. Raynor, "The Structure of Metals and Alloys," 3d ed., The Institute of Metals, London, 1954.

F. H. Spedding and A. H. Daane, The Preparation and Properties of Rare Earth Metals, in "Metallurgy and Fuels," Progress in Nuclear Energy, ser. V, **vol. 1,** Pergamon Press Ltd., London, 1956 (also *Acta Cryst.*, **9,** 1956, 559; *Trans. AIME*, **209,** 1957, 895).

F. Laves, *Naturwissenschaften*, **25,** 1937, 731–732.

Unless otherwise stated, lattice constants apply to room temperature. The radii of the metalloid atoms given in brackets are those observed in interstitial structures and are not corrected for coordination number 12. All data are given in "true angstrom" units (10^{-8} cm).

Element	Allotropic forms	Structure	Lattice parameters, A			Distance of closest approach, A	Goldschmidt atomic radii (12-fold co-ordination), A	Remarks
			a	b or α	c			
Ag........	....	F.c.c. A1	4.0856	....	....	2.888	1.44	
Al........	....	F.c.c. A1	4.0490	....	....	2.862	1.43	
As........	....	Rhombohedral A7	4.159	53°49′	....	2.51	1.48	
Au........	....	F.c.c. A1	4.0783	....	....	2.884	1.44	
B........	Tetragonal needles		8.75	....	5.04	1.75 }	[0.97]	
	Orthorhombic (?) plates		17.90	8.95	10.15	.... }		
Ba........	....	B.c.c. A2	5.025	....	....	4.35	2.24	
Be........	....	H.c.p. A3	2.2854	....	3.5841	2.225	1.13	
Bi........	....	Rhombohedral A7	4.7356	57°14′	....	3.111	1.82	
C........	Graphite, α	Hexagonal A9	2.4614	....	6.7014	1.42 }		
	Graphite, β	Rhombohedral	2.461	....	10.064	..... }	[0.77]	
	Diamond	Cubic A4	3.568	....	....	1.544 }		
Ca*........	<464°C	F.c.c. A1	5.582	....	....	3.94	1.97	* *J. Electrochem. Soc.*, **103,** 1956, 409 (the widely quoted h.c.p. phase could not be found in a 99.9+ wt. % Ca sample)
	>464°C	B.c.c. A2	4.48 (500°C)					
Cb........	....	....	....	....	....	....	....	See niobium, Nb
Cd........	....	H.c.p. A3	2.9787	....	5.617	2.979	1.52	

Table B. Structural Data of the Elements (*Continued*)

Element	Allotropic forms	Structure	Lattice parameters, A			Distance of closest approach, A	Goldschmidt atomic radii (12-fold co-ordination), A	Remarks
			a	b or α	c			
Ce	<730°C	F.c.c. A1	5.1612	...	...	3.65	1.82	A h.c.p. form appears to be metastable only; see *Phys. Rev.*, **76,** 1949, 301, and *J. Chem. Phys.*, **18,** 1950, 145; for low-temperature transitions see *Acta Cryst.*, **10,** 1957, 832
	>730°C	?						
Co	ε(<400°C)	H.c.p. A3	2.507	...	4.069	2.506	1.25	
	α or γ (>400°C)	F.c.c. A1	3.552	...	...	2.511		
Cp	...	...	...	...	...	...	...	See lutecium, Lu
Cr	...	B.c.c. A2	2.8850*	...	...	2.498	1.28	* *Acta Cryst.*, **8,** 1955, 367
Cs	...	B.c.c. A2	6.067*	...	...	5.25	2.70	* At 78°K
Cu	...	F.c.c. A1	3.6153	...	...	2.556	1.28	
Dy	...	H.c.p. A3	3.590	...	5.6475	3.503	1.77	
Er	...	H.c.p. A3	3.559	...	5.587	3.468	1.76	
Eu	...	B.c.c. A2	4.577*	...	...	3.964	2.04	* *J. Chem. Phys.*, **25,** 1956, 1123
Fe	α (<910°C)	B.c.c. A2	2.8664	...	...	2.481	1.27	
	γ (910–1390°C)	F.c.c. A1	3.656*	...	...	2.585	...	* At 950°C
	δ (>1390°C)	B.c.c. A2	2.94†	...	...	...	...	† At 1425°C
Ga	...	Orthorhombic A11	3.526	4.520	7.660	2.442	1.39	
Gd	...	H.c.p. A3	3.636	...	5.783	3.573	1.80	
Ge	...	Cubic A4	5.658	...	...	2.450	1.39	
H	(Para)	Hexagonal	3.76	...	6.13*	...	[0.46]	* At −271°C
Hf	α (<~1700°C)	H.c.p. A3	3.206	...	5.087	3.15	1.59	
	β (>~1700°C)	B.c.c. A2	3.51					
Hg	...	Rhombohedral A10	2.9925	70°44.6′*	...	3.006	1.55	* At 78°K; *Acta Cryst.*, **10, 1957, 58**
Ho	...	H.c.p. A3	3.5773	...	5.6158	3.486	1.77	
In	...	Tetragonal A6	4.594	...	4.951	3.25	1.57	
Ir	...	F.c.c. A1	3.8389	...	...	2.714	1.36	
K	...	B.c.c. A2	5.344	...	...	4.627	2.38	

La..........	α (<330°C)	H.c.p.	3.770	....	12.159	3.739	1.88	
	β (330–864°C)	F.c.c. A1	5.31					
	γ (>864°C)	?						
Li..........	....	B.c.c. A2	3.5089	....	....	3.039	1.57	
	Cooling to 78°K	H.c.p. A3	3.111	....	5.093			
	Cold-worked at 78°K	F.c.c. A1	4.40					
Lu..........	....	H.c.p. A3	3.503	....	5.551	3.434	1.72	
Mg..........	....	H.c.p. A3	3.2092	....	5.2103	3.196	1.60	
Mn..........	α (<727°C)	Cubic A12	8.912	....	....	2.24	1.30	
	β (727–1095°C)	Cubic A13	6.313	....	....	2.373		
	γ (1095–1133°C)	F.c.c. A1	3.862*	....	....	2.731	....	* At 1095°C
	δ (>1133°C)	B.c.c. A2	3.080†	....	....	....	....	† At 1134°C
Mo..........	....	B.c.c. A2	3.1466	....	....	2.725	1.40	
N..........	α	Cubic	5.67*	....	....	1.06 }	[0.71]	* At −252°C
	β	Hexagonal	4.04	....	6.60†	.... }		† At −234°C
Na..........	....	B.c.c. A2	4.2906	....	....	3.715	1.92	
	<51°K	H.c.p. A3	3.767	....	6.154			
Nb..........	....	B.c.c. A2	3.3007	....	....	2.859	1.47	
Nd..........	<862°C	H.c.p.	3.658	....	11.799	3.63	1.82	
	>862°C	?						
Ni..........	....	F.c.c. A1	3.5238	....	....	2.491	1.25	
Np..........	α (<278°C)	Orthorhombic	4.73	4.90	3.67			
	β (278–540°C)	Tetragonal	4.90	....	3.39			
	γ (>540°C)	B.c.c. A2 (?)	3.53					
O..........	α (−252°C)	Orthorhombic	5.51	3.83	3.45	..... }		
	β (−238°C)	Rhombohedral	6.20	99.1°	....	..... }	[0.60]	
	γ (−225°C)	Cubic	6.84	....	....	..... }		
Os..........	....	H.c.p. A3	2.733	....	4.319	2.675	1.35	
P..........	(Metallic)	Orthorhombic A16	3.32	4.39	10.52	2.18	1.3	
Pa..........	....	Tetragonal	3.925	....	3.238*	3.21	....	* *J. Am. Chem. Soc.*, **76,** 1954, 5935
Pb..........	....	F.c.c. A1	4.9495	....	....	3.499	1.75	
Pd..........	....	F.c.c. A1	3.8902	....	....	2.750	1.37	
Po*..........	α (<75°C)	Simple cubic	3.352	....	....	3.352	1.7_5	* *J. Chem. Phys.*, **17,** 1949, 1293
	β (>75°C)	Rhombohedral	3.366	98°13′				
Pr..........	α (<792°C)	H.c.p.	3.672	....	11.835	3.640	1.83	
	β (>792°C)	F.c.c. A1 (?)						
Pt..........	....	F.c.c. A1	3.9237	....	....	2.775	1.39	

Table B. Structural Data of the Elements (*Continued*)

Element	Allotropic forms	Structure	Lattice parameters, A			Distance of closest approach, A	Goldschmidt atomic radii (12-fold co-ordination), A	Remarks
			a	b or α	c			
Pu*........	α (<122°C)	Monoclinic	a = 6.183, b = 4.824, c = 10.973, γ = 101.8°			....	....	* *Nuclear Met.*, no. 1 (AIME), 1955; *Acta Cryst.*, **8,** 1955, 431; **10,** 1957, 776; *Trans. AIME*, **206,** 1956, 1256
	β (122–206°C)							
	γ (206–319°C)	F.c. orthorhombic	3.1587	5.7682	10.162 (235°)			
	δ (319–451°C)	F.c.c. A1	4.636 (350°)	....	....	3.28	1.64	
	δ (451–480°C)	B.c. tetragonal	3.33	....	4.46 (470°)			
	ϵ (480–m.p.)	B.c.c. A2	3.639 (510°)					
Rb..........	....	B.c.c. A2	5.710	....	....	4.996	2.57	
Re..........	....	H.c.p. A3	2.7609	....	4.4583	2.740	1.37	
Rh..........	....	F.c.c. A1	3.8034	....	....	2.689	1.34	There are indications of a transformation between 1100 and 1200°C
Ru..........	....	H.c.p. A3	2.7038	....	4.2816	2.649	1.34	
S...........	α (yellow)	Orthorhombic A17	10.44	12.84	24.37*	2.04	....	* *Acta Cryst.*, **8,** 1955, 661
	β	Monoclinic						
Sb..........	....	Rhombohedral A7	4.5064	57.1°	....	2.903	1.61	
Sc..........	....	H.c.p. A3	3.3090	....	5.273	3.256	1.64	The widely quoted f.c.c. structure may be due to scandium nitride
Se...........	(Metallic)	Hexagonal A8	4.3640	....	4.9594	2.32	1.6	
Si...........	....	Cubic A4	5.4282	....	....	2.351	1.34	
Sm..........	<917°C	Rhombohedral–h.c.p.	3.621	....	26.25*	3.59	1.80	* *Acta Cryst.*, **7,** 1954, 532
	>917°C	?						
Sn..........	Gray (<13°C)	Cubic A4	6.491	....	....	2.81	1.58	
	White (>13°C)	Tetragonal A5	5.8311	....	3.1817	3.022		

Sr*.........	α (<215°C)	F.c.c. A1	6.085	....	....	4.31	2.15	* *Acta Cryst.*, **6, 1953, 100**
	β (215–605°C)	H.c.p. A3	4.32	....	7.06 (248°C)			
	γ (>605°C)	B.c.c. A2	4.85 (614°C)					
Ta.........	....	B.c.c. A2	3.3026	....	....	2.860	1.47	
Tb.........	....	H.c.p. A3	3.601	....	5.694	3.526	1.78	
Te.........	....	Hexagonal A8	4.4559	....	5.9268	2.87	1.7	
Th.........	α (<1400 ± 25°C)	F.c.c. A1	5.0843*	....	....	3.595	1.80	* *Acta Cryst.*, **9**, 1956, 376
	β (>1400 ± 25°C)	B.c.c. A2	4.12 (1400°)					
Ti.........	α (<882.5°C)	H.c.p. A3	2.9504	....	4.6833	2.89	1.47	
	β (>882.5°C)	B.c.c. A2	3.306*	....	....	....	....	* At 900°C
Tl.........	α (<234°C)	H.c.p. A3	3.4564	....	5.531	3.407	1.71	
	β (>234°C)	B.c.c. A2	3.882*	....	....	....	....	* At 262°C (in quenched specimens a f.c.c. modification has been observed)
Tm.........	....	H.c.p. A3	3.5375	....	5.555	3.448	1.75	
U.........	α (<660°C)	Orthorhombic A20	2.858	5.877	4.955	2.77		
	β (660–775°C)	Tetragonal	10.758	....	5.656*	....	....	* At 720°C
	γ (>775°C)	B.c.c. A2	3.49†	....	....	3.02	1.56	† At 800°C
V.........	....	B.c.c. A2	3.039	....	....	2.632	1.36	
W.........	....	B.c.c. A2	3.1648	....	....	2.739	1.41	
Y.........	....	H.c.p. A3	3.647	....	5.731	3.556	1.80	
Yb.........	<798°C	F.c.c. A1	5.486	....	....	3.879	1.94	
	>798°C	?						
Zn.........	....	H.c.p. A3	2.664	....	4.945	2.664	1.37	
Zr.........	α (<865°C)	H.c.p. A3	3.230	....	5.133	3.17	1.60	
	β (> 865°C)	B.c.c. A2	3.62*	....	....	....	....	* At 867°C

Table C. Crystal-structure Types According to "Strukturbericht"

This table, which is largely based on a compilation by F. Laves in C. J. Smithells, "Metals Reference Book," 2d ed., 1955, gives a short description of structural types collected in the seven volumes of "Strukturbericht" (*Zeitschrift für Kristallographie*, Leipzig) which cover the period 1913 to 1939. Only structural types of elements and binary alloys are listed; corrective or supplementary information due to more recent research has been considered.

For a classification in the "Strukturbericht" notation of structural types discovered since 1939, see the compilation by Laves mentioned above. The "Structure Reports," successors to "Strukturbericht," discontinued the use of the notation for newly detected structural types.

Name of type	Crystal system	Space group	Number of atoms per unit cell	"Strukturbericht" volume and page, or other source
		A Types (Elements)		
A1 (Cu)	Cubic	O_h^5—Fm3m	4	I, 13
A2 (W)	Cubic	O_h^9—Im3m	2	I, 15
A3 (Mg)	Hexagonal	D_{6h}^4—$P6_3$/mmc	2	I, 16
A4 (diamond)	Cubic	O_h^7—Fd3m	8	I, 19
A5 (Sn, metallic)	Tetragonal	D_{4h}^{19}—I4/amd	4	I, 21
A6 (In)	Tetragonal	D_{4h}^{17}—I4/mmm	2	I, 23
A7 (As)	Rhombohedral	D_{3d}^5—$R\bar{3}m$	2	I, 25
A8 (Se)	Hexagonal	D_3^4—$P3_121$	3	I, 27
A9 (graphite)	(α) Hexagonal	D_{6h}^4—$P6_3$/mmc	4	I, 28
	(β) Rhombohedral	D_{3d}^5—$R\bar{3}m$	6	"Metals Reference Book," 1955, p. 212
A10 (Hg)	Rhombohedral	D_{3d}^5—$R\bar{3}m$	1	I, 737
A11 (Ga)	Orthorhombic	D_{2h}^{18}—Abma	8	II, 1; III, 3
A12 (α-Mn)	Cubic	T_d^3—$I\bar{4}3m$	58	II, 2
A13 (β-Mn)	Cubic	O^6—$P4_332$ (?)	20	II, 3; III, 221, 326
A15 (β-W or Cr_3Si)	Cubic	O_h^3—Pm3n	8	II, 6; see *Acta Cryst.*, **7**, 1954, 351
A19 (Po)	Monoclinic	C_2^3—C2	12	IV, 4
A20 (α-U)	Orthorhombic	D_{2h}^{17}—Cmcm	4	VI, 3

Table C. Crystal-structure Types According to "Strukturbericht" (*Continued*)

Name of type	Crystal system	Space group	Number of atoms per unit cell	"Strukturbericht" volume and page, or other source
		B Types (Compounds *XY*)		
B1 (NaCl)	Cubic	O_h^5—Fm3m	8	I, 72
B2 (CsCl)	Cubic	O_h^1—Pm3m	2	I, 74
B3 (ZnS, zincblende)	Cubic	T_d^2—F$\bar{4}$3m	8	I, 76
B4 (ZnS, wurtzite)	Hexagonal	C_{6v}^4—P6_3mc	4	I, 78
B8[a] (α) NiAs	Hexagonal	D_{6h}^4—P6_3/mmc	4	I, 84; "Metals Reference Book," 1955, p. 215
(β) Ni_2In	Hexagonal	D_{6h}^4—P6_3/mmc	6	
B9 (HgS, cinnabar)	Hexagonal	D_3^4—P$3_1$21	6	I, 87
B10 (LiOH, PbO)	Tetragonal	D_{4h}^7—P4/nmm	4	I, 89; *J. Am. Chem. Soc.*, **63,** 1941, 1392
B11 (γ-TiCu)	Tetragonal	D_{4h}^7—P4/nmm	4	I, 94; *J. Inst. Metals,* **79,** 1951, 401
B13 (NiS, millerite)	Rhombohedral	C_{3v}^5—R3m	6	II, 6
B16 (GeS)	Orthorhombic	D_{2h}^{16}—Pbnm	8	II, 8
B17 (PtS, cooperite)	Tetragonal	D_{4h}^9—P4_2/mmc	4	II, 9
B18 (CuS, covellite)	Hexagonal	D_{6h}^4—P6_3/mmc	12	II, 10
B19 (AuCd)	Orthorhombic	D_{2h}^5—Pmcm	4	II, 11
B20 (FeSi)	Cubic	T^4—P$2_1$3	8	II, 13; III, 13
B26 (CuO)	Monoclinic	C_{2h}^6—C2/c	8	III, 11
B27 (FeB)	Orthorhombic	D_{2h}^{16}—Pbnm	8	III, 12
B29 (SnS)[b]	Orthorhombic	D_{2h}^{16}—Pmcn	8	III, 14
B31 (MnP)	Orthorhombic	D_{2h}^{16}—Pcmn	8	III, 17
B32 (NaTl)	Cubic	O_h^7—Fd3m	16	III, 19
B34 (PdS)	Tetragonal	C_{4h}^2—P4_2/m	16	V, 3
B35 (CoSn)	Hexagonal	D_{6h}^1—P6/mmm	6	VI, 4
B37 (TlSe)	Tetragonal	D_{4h}^{18}—I4/mcm	16	VII, 6

Table C. Crystal-structure Types According to "Strukturbericht" (*Continued*)

Name of type	Crystal system	Space group	Number of atoms per unit cell	"Strukturbericht" volume and page, or other source
C Types (Compounds XY_2)				
C1 (CaF_2)	Cubic	O_h^5—Fm3m	12	I, 148
C2 (FeS_2, pyrites)	Cubic	T_h^6—Pa3	12	I, 150
C3 (Cu_2O)	Cubic	O_h^4—Pn3m	6	I, 153
C4 (TiO_2, rutile)	Tetragonal	D_{4h}^{14}—$P4_2/mnm$	6	I, 155
C5 (TiO_2, anatase)	Tetragonal	D_{4h}^{19}—$I4_1/amd$	12	I, 158
C6 (CdI_2)[c]	Hexagonal	D_{3d}^3—$P\bar{3}m1$	3	I, 161
C7 (MoS_2)	Hexagonal	D_{6h}^4—$P6_3/mmc$	6	I, 164
C11a (CaC_2)	Tetragonal	D_{4h}^{17}—I4/mmm	6	I, 740
C11b ($MoSi_2$)	Tetragonal	D_{4h}^{17}—I4/mmm	6	I, 740
C12 ($CaSi_2$)	Rhombohedral	D_{3d}^5—$R\bar{3}m$	6	I, 175
C14 ($MgZn_2$)	Hexagonal	D_{6h}^4—$P6_3/mmc$	12	I, 180
C15 ($MgCu_2$)	Cubic	O_h^7—Fd3m	24	I, 490
C16 ($CuAl_2$)	Tetragonal	D_{4h}^{18}—I4/mcm	12	I, 491
C18 (FeS_2, marcasite)	Orthorhombic	D_{2h}^{12}—Pnnm	6	I, 495; V, 52
C21 (TiO_2, brookite)	Orthorhombic	D_{2h}^{15}—Pbca	24	II, 14
C22 (Fe_2P)	Hexagonal	D_3^2—P321	9	II, 15
C23 ($PbCl_2$)	Orthorhombic	D_{2h}^{16}—Pmnb	12	II, 16
C29 (SrH_2)	Orthorhombic	D_{2h}^{16}—Pnma	12	III, 24
C32 (AlB_2)	Hexagonal	D_{6h}^1—P6/mmm	3	III, 28
C33 (Bi_2Te_2S, tetradymite)	Rhombohedral	D_{3d}^5—$R\bar{3}m$	5	III, 28
C34 ($AuTe_2$, calaverite)	Monoclinic	C_{2h}^3—C2/m	6	III, 30
C36 ($MgNi_2$)	Hexagonal	D_{6h}^4—$P6_3/mmc$	24	III, 31
C37 (Co_2Si)	Orthorhombic	D_{2h}^{16}—Pnam	12	*Acta Cryst.*, **8**, 1955, 83 (cf. III, 32)

C38 (Cu_2Sb)	Tetragonal	D^7_{4h}—P4/nmm	6	III, 33
C40 ($CrSi_2$)	Hexagonal	D^4_6—$P6_222$	9	III, 35
C42 (SiS_2)	Orthorhombic	D^{26}_{2h}—Icma	12	III, 37
C43 (ZrO_2)	Monoclinic	C^5_{2h}—$P2_1/c$	12	IV, 9
C44 (GeS_2)	Orthorhombic	C^{19}_{2v}—Fdd2	72	IV, 11
C46 ($AuTe_2$, krennerite)	Orthorhombic	C^4_{2v}—Pma2	24	IV, 15
C49 ($ZrSi_2$)	Orthorhombic	D^{17}_{2h}—Cmcm	12	V, 5; see also *Z. anorg. Chem.*, **249**, 1942, 338, and *Monatsh. Chem.*, **85**, 1954, 1140
C52 (TeO_2)	Orthorhombic	D^{15}_{2h}—Pcab	24	VII, 8
C54 ($TiSi_2$)	Orthorhombic	D^{24}_{2h}—Fddd	24	VII, 12

D Types (Compounds X_mY_n)

$D0_2$ ($CoAs_3$)	Cubic	T^5_h—Im3	32	I, 232
$D0_3$ (BiF_3 or Li_3Bi)	Cubic	O^5_h—Fm3m	16	II, 22
$D0_9$ (ReO_3 or Cu_3N)	Cubic	O^1_h—Pm3m	4	II, 31
$D0_{11}$ (Fe_3C)	Orthorhombic	D^{16}_{2h}—Pbnm	16	II, 33
$D0_{17}$ (BaS_3)	Orthorhombic	D^3_2—$P2_12_12$	16	IV, 18
$D0_{18}$ (Na_3As)	Hexagonal	D^4_{6h}—$P6_3/mmc$	8	V, 6
$D0_{19}$ (Mg_3Cd or Ni_3Sn)	Hexagonal	D^4_{6h}—$P6_3/mmc$	8	V, 7
$D0_{20}$ ($NiAl_3$)[d]	Orthorhombic	D^{16}_{2h}—Pnma	16	V, 8
$D0_{21}$ (Cu_3P)	Hexagonal	D^4_{3d}—$P\bar{3}c1$	24	II, 7; see also VII, 99
$D0_{22}$ ($TiAl_3$)	Tetragonal	D^{17}_{4h}—I4/mmm	8	VII, 13
$D0_{23}$ ($ZrAl_3$)	Tetragonal	D^{17}_{4h}—I4/mmm	16	VII, 14
$D0_{24}$ ($TiNi_3$)	Hexagonal	D^4_{6h}—$P6_3/mmc$	16	VII, 14
$D1_3$ ($BaAl_4$)	Tetragonal	D^{17}_{4h}—I4/mmm	10	III, 45
$D2_1$ (CaB_6)	Cubic	O^1_h—Pm3m	7	II, 37
$D2_3$ ($NaZn_{13}$)	Cubic	O^6_h—Fm3c	112	VI, 8
$D5_1$ (α-Al_2O_3, corundum)	Rhombohedral	D^6_{3d}—$R\bar{3}c$	10	I, 240
$D5_2$ (La_2O_3)	Hexagonal	D^3_{3d}—$P\bar{3}m1$	5	I, 744
$D5_3$ (Mn_2O_3)	Cubic	T^7_h—Ia3	80	II, 38

Table C. Crystal-structure Types According to "Strukturbericht" (*Continued*)

Name of type	Crystal system	Space group	Number of atoms per unit cell	"Strukturbericht" volume and page, or other source
$D5_4$ or $D6_1$ (Sb_2O_3, senarmontite)	Cubic	O_h^7—Fd3m	80	I, 245
$D5_8$ (Sb_2S_3)	Orthorhombic	D_{2h}^{16}—Pbnm	20	III, 49
$D5_9$ (Zn_3P_2)	Tetragonal	D_{4h}^{15}—$P4_2$/nmc	40	III, 51
$D5_{10}$ (Cr_3C_2)	Orthorhombic	D_{2h}^{16}—Pbnm	20	III, 53
$D5_{11}$ (Sb_2O_3, valentinite)	Orthorhombic	D_{2h}^{10}—Pccn	20	IV, 20
$D5_{12}$ (β-Bi_2O_3)	Tetragonal	D_{2d}^7—$P\bar{4}2b$	40	V, 9
$D5_{13}$ (Ni_2Al_3)	Hexagonal	D_{3d}^3—$P\bar{3}m1$	5	V, 10
$D7_1$ (Al_4C_3)	Rhombohedral	D_{3d}^5—$R\bar{3}m$	7	III, 56
$D7_2$ (Co_3S_4)	Cubic	O_h^7—Fd3m	56	VI, 9
$D7_3$ (Th_3P_4)	Cubic	T_d^6—$I\bar{4}3d$	28	VII, 15
$D8_1$ (Fe_3Zn_{10})	Cubic	O_h^9—Im3m	52	I, 497
$D8_2$ (Cu_5Zn_8)	Cubic	T_d^3—$I\bar{4}3m$	52	I, 497; VII, 198
$D8_3$ (Cu_9Al_4)	Cubic	T_d^1—$P\bar{4}3m$	52	III, 57
$D8_4$ ($Cr_{23}C_6$)	Cubic	O_h^5—Fm3m	116	III, 59
$D8_5$ (W_6Fe_7)	Rhombohedral	D_{3d}^5—$R\bar{3}m$	13	III, 61
$D8_6$ ($Cu_{15}Si_4$)	Cubic	T_d^6—$I\bar{4}3d$	76	III, 62
$D8_7$ (V_2O_5)	Orthorhombic	D_{2h}^{13}—Pmmn	14	IV, 22; *Acta Chem. Scand.*, **4**, 1950, 1119
$D8_8$ (Mn_5Si_3)	Hexagonal	D_{6h}^3—$P6_3$/mcm	16	IV, 24
$D8_9$ (Co_9S_8)	Cubic	O_h^5—Fm3m	68	IV, 26
$D8_{10}$ (Cr_5Al_8)	Rhombohedral	C_{3v}^5—R3m	26	V, 11
$D8_{11}$ (Co_2Al_5)	Hexagonal	D_{6h}^4—$P6_3$/mmc	28	VI, 11

Table C. Crystal-structure Types According to "Strukturbericht" (*Continued*)

Name of type	Crystal system	Space group	Number of atoms per unit cell	"Strukturbericht" volume and page, or other source
		Other Types		
$H1_1$ (Al_2MgO_4, spinel)[e]	Cubic	O_h^7—Fd3m	56	I, 350; IV, 171
$L1_0$ (CuAu)[f]	Tetragonal	D_{4h}^1—P4/mmm	4	I, 484
$L1_1$ (PtCu)[f]	Rhombohedral	D_{3d}^5—R$\bar{3}$m	32	I, 485
$L1_2$ (Cu_3Au)[f]	Cubic	O_h^1—Pm3m	4	I, 486
$L2_2$ (Tl_7Sb_2)	Cubic	O_h^9—Im3m	54	III, 175
$L'1_0$ (Fe_4N)	Cubic	O_h^1—Pm3m or O_h^5—Fm3m	5	I, 487

[a] Between the main types α and β there exist a number of intermediate arrangements due to the variation of the stoichiometric formulas. Similarly, there is virtually a continuous change from the B8 type to the C6 type.

[b] According to *Naturwissenschaften*, **41,** 1954, 448, identical with the B16 (GeS) type.

[c] There is virtually a continuous change from this type to the B8 type.

[d] Another choice of coordinates and of origin shows that the structure is very similar to the $D0_{11}$ (Fe_3C) type.

[e] In some cases lattice sites may be vacant, for example, γ-Al_2O_3 or In_2S_3.

[f] Superstructure of A1 (Cu) type.

Table D. Temperature Conversions*

The middle column of figures (in boldface type) contains the reading (°F or °C) to be converted. If converting from degrees Fahrenheit to degrees centigrade, read the centigrade equivalent in the column headed "°C." If converting from degrees centigrade to degrees Fahrenheit, read the Fahrenheit equivalent in the column headed "°F."

°F		°C	°F		°C	°F		°C
....	**−458**	−272.22	....	**−368**	−222.22	....	**−278**	−172.22
....	**−456**	−271.11	....	**−366**	−221.11	....	**−276**	−171.11
....	**−454**	−270.00	....	**−364**	−220.00	....	**−274**	−170.00
....	**−452**	−268.89	....	**−362**	−218.89	−457.6	**−272**	−168.89
....	**−450**	−267.78	....	**−360**	−217.78	−454.0	**−270**	−167.78
....	**−448**	−266.67	....	**−358**	−216.67	−450.4	**−268**	−166.67
....	**−446**	−265.56	....	**−356**	−215.56	−446.8	**−266**	−165.56
....	**−444**	−264.44	....	**−354**	−214.44	−443.2	**−264**	−164.44
....	**−442**	−263.33	....	**−352**	−213.33	−439.6	**−262**	−163.33
....	**−440**	−262.22	....	**−350**	−212.22	−436.0	**−260**	−162.22
....	**−438**	−261.11	....	**−348**	−211.11	−432.4	**−258**	−161.11
....	**−436**	−260.00	....	**−346**	−210.00	−428.8	**−256**	−160.00
....	**−434**	−258.89	....	**−344**	−208.89	−425.2	**−254**	−158.89
....	**−432**	−257.78	....	**−342**	−207.78	−421.6	**−252**	−157.78
....	**−430**	−256.67	....	**−340**	−206.67	−418.0	**−250**	−156.67
....	**−428**	−255.56	....	**−338**	−205.56	−414.4	**−248**	−155.56
....	**−426**	−254.44	....	**−336**	−204.44	−410.8	**−246**	−154.44
....	**−424**	−253.33	....	**−334**	−203.33	−407.2	**−244**	−153.33
....	**−422**	−252.22	....	**−332**	−202.22	−403.6	**−242**	−152.22
....	**−420**	−251.11	....	**−330**	−201.11	−400.0	**−240**	−151.11
....	**−418**	−250.00	....	**−328**	−200.00	−396.4	**−238**	−150.00
....	**−416**	−248.89	....	**−326**	−198.89	−392.8	**−236**	−148.89
....	**−414**	−247.78	....	**−324**	−197.78	−389.2	**−234**	−147.78
....	**−412**	−246.67	....	**−322**	−196.67	−385.6	**−232**	−146.67
....	**−410**	−245.56	....	**−320**	−195.56	−382.0	**−230**	−145.56
....	**−408**	−244.44	....	**−318**	−194.44	−378.4	**−228**	−144.44
....	**−406**	−243.33	....	**−316**	−193.33	−374.8	**−226**	−143.33
....	**−404**	−242.22	....	**−314**	−192.22	−371.2	**−224**	−142.22
....	**−402**	−241.11	....	**−312**	−191.11	−367.6	**−222**	−141.11
....	**−400**	−240.00	....	**−310**	−190.00	−364.0	**−220**	−140.00
....	**−398**	−238.89	....	**−308**	−188.89	−360.4	**−218**	−138.89
....	**−396**	−237.78	....	**−306**	−187.78	−356.8	**−216**	−137.78
....	**−394**	−236.67	....	**−304**	−186.67	−353.2	**−214**	−136.67
....	**−392**	−235.56	....	**−302**	−185.56	−349.6	**−212**	−135.56
....	**−390**	−234.44	....	**−300**	−184.44	−346.0	**−210**	−134.44
....	**−388**	−233.33	....	**−298**	−183.33	−342.4	**−208**	−133.33
....	**−386**	−232.22	....	**−296**	−182.22	−338.8	**−206**	−132.22
....	**−384**	−231.11	....	**−294**	−181.11	−335.2	**−204**	−131.11
....	**−382**	−230.00	....	**−292**	−180.00	−331.6	**−202**	−130.00
....	**−380**	−228.89	....	**−290**	−178.89	−328.0	**−200**	−128.89
....	**−378**	−227.78	....	**−288**	−177.78	−324.4	**−198**	−127.78
....	**−376**	−226.67	....	**−286**	−176.67	−320.8	**−196**	−126.67
....	**−374**	−225.56	....	**−284**	−175.56	−317.2	**−194**	−125.56
....	**−372**	−224.44	....	**−282**	−174.44	−313.6	**−192**	−124.44
....	**−370**	−223.33	....	**−280**	−173.33	−310.0	**−190**	−123.33

Table D. Temperature Conversions* (*Continued*)

°F		°C	°F		°C	°F		°C
−306.4	**−188**	−122.22	−126.4	**−88**	−66.67	+53.6	**+12**	−11.11
−302.8	**−186**	−121.11	−122.8	**−86**	−65.56	+57.2	**+14**	−10.00
−299.2	**−184**	−120.00	−119.2	**−84**	−64.44	+60.8	**+16**	−8.89
−295.6	**−182**	−118.89	−115.6	**−82**	−63.33	+64.4	**+18**	−7.78
−292.0	**−180**	−117.78	−112.0	**−80**	−62.22	+68.0	**+20**	−6.67
−288.4	**−178**	−116.67	−108.4	**−78**	−61.11	+71.6	**+22**	−5.56
−284.8	**−176**	−115.56	−104.8	**−76**	−60.00	+75.2	**+24**	−4.44
−281.2	**−174**	−114.44	−101.2	**−74**	−58.89	+78.8	**+26**	−3.33
−277.6	**−172**	−113.33	−97.6	**−72**	−57.78	+82.4	**+28**	−2.22
−274.0	**−170**	−112.22	−94.0	**−70**	−56.67	+86.0	**+30**	−1.11
−270.4	**−168**	−111.11	−90.4	**−68**	−55.56	+89.6	**+32**	0.00
−266.8	**−166**	−110.00	−86.8	**−66**	−54.44	+93.2	**+34**	+1.11
−263.2	**−164**	−108.89	−83.2	**−64**	−53.33	+96.8	**+36**	+2.22
−259.6	**−162**	−107.78	−79.6	**−62**	−52.22	+100.4	**+38**	+3.33
−256.0	**−160**	−106.67	−76.0	**−60**	−51.11	+104.0	**+40**	+4.44
−252.4	**−158**	−105.56	−72.4	**−58**	−50.00	107.6	**42**	5.56
−248.8	**−156**	−104.44	−68.8	**−56**	−48.89	111.2	**44**	6.67
−245.2	**−154**	−103.33	−65.2	**−54**	−47.78	114.8	**46**	7.78
−241.6	**−152**	−102.22	−61.6	**−52**	−46.67	118.4	**48**	8.89
−238.0	**−150**	−101.11	−58.0	**−50**	−45.56	122.0	**50**	10.00
−234.4	**−148**	−100.00	−54.4	**−48**	−44.44	125.6	**52**	11.11
−230.8	**−146**	−98.89	−50.8	**−46**	−43.33	129.2	**54**	12.22
−227.2	**−144**	−97.78	−47.2	**−44**	−42.22	132.8	**56**	13.33
−223.6	**−142**	−96.67	−43.6	**−42**	−41.11	136.4	**58**	14.44
−220.0	**−140**	−95.56	−40.0	**−40**	−40.00	140.0	**60**	15.56
−216.4	**−138**	−94.44	−36.4	**−38**	−38.89	143.6	**62**	16.67
−212.8	**−136**	−93.33	−32.8	**−36**	−37.78	147.2	**64**	17.78
−209.2	**−134**	−92.22	−29.2	**−34**	−36.67	150.8	**66**	18.89
−205.6	**−132**	−91.11	−25.6	**−32**	−35.56	154.4	**68**	20.00
−202.0	**−130**	−90.00	−22.0	**−30**	−34.44	158.0	**70**	21.11
−198.4	**−128**	−88.89	−18.4	**−28**	−33.33	161.6	**72**	22.22
−194.8	**−126**	−87.78	−14.8	**−26**	−32.22	165.2	**74**	23.33
−191.2	**−124**	−86.67	−11.2	**−24**	−31.11	168.8	**76**	24.44
−187.6	**−122**	−85.56	−7.6	**−22**	−30.00	172.4	**73**	25.56
−184.0	**−120**	−84.44	−4.0	**−20**	−28.89	176.0	**80**	26.67
−180.4	**−118**	−83.33	−0.4	**−18**	−27.78	179.6	**82**	27.78
−176.8	**−116**	−82.22	+3.2	**−16**	−26.67	183.2	**84**	28.89
−173.2	**−114**	−81.11	+6.8	**−14**	−25.56	186.8	**86**	30.00
−169.6	**−112**	−80.00	+10.4	**−12**	−24.44	190.4	**88**	31.11
−166.0	**−110**	−78.89	+14.0	**−10**	−23.33	194.0	**90**	32.22
−162.4	**−108**	−77.78	+17.6	**−8**	−22.22	197.6	**92**	33.33
−158.8	**−106**	−76.67	+21.2	**−6**	−21.11	201.2	**94**	34.44
−155.2	**−104**	−75.56	+24.8	**−4**	−20.00	204.8	**96**	35.56
−151.6	**−102**	−74.44	+28.4	**−2**	−18.89	208.4	**98**	36.67
−148.0	**−100**	−73.33	+32.0	**0**	−17.78	212.0	**100**	37.78
−144.4	**−98**	−72.22	+35.6	**+2**	−16.67	215.6	**102**	38.89
−140.8	**−96**	−71.11	+39.2	**+4**	−15.56	219.2	**104**	40.00
−137.2	**−94**	−70.00	+42.8	**+6**	−14.44	222.8	**106**	41.11
−133.6	**−92**	−68.89	+46.4	**+8**	−13.33	226.4	**108**	42.22
−130.0	**−90**	−67.78	+50.0	**+10**	−12.22	230.0	**110**	43.33

Table D. Temperature Conversions* (*Continued*)

°F		°C	°F		°C	°F		°C
233.6	**112**	44.44	413.6	**212**	100.00	593.6	**312**	155.56
237.2	**114**	45.56	417.2	**214**	101.11	597.2	**314**	156.67
240.8	**116**	46.67	420.8	**216**	102.22	600.8	**316**	157.78
244.4	**118**	47.78	424.4	**218**	103.33	604.4	**318**	158.89
248.0	**120**	48.89	428.0	**220**	104.44	608.0	**320**	160.00
251.6	**122**	50.00	431.6	**222**	105.56	611.6	**322**	161.11
255.2	**124**	51.11	435.2	**224**	106.67	615.2	**324**	162.22
258.8	**126**	52.22	438.8	**226**	107.78	618.8	**326**	163.33
262.4	**128**	53.33	442.4	**228**	108.89	622.4	**328**	164.44
266.0	**130**	54.44	446.0	**230**	110.00	626.0	**330**	165.56
269.6	**132**	55.56	449.6	**232**	111.11	629.6	**332**	166.67
273.2	**134**	56.67	453.2	**234**	112.22	633.2	**334**	167.78
276.8	**136**	57.78	456.8	**236**	113.33	636.8	**336**	168.89
280.4	**138**	58.89	460.4	**238**	114.44	640.4	**338**	170.00
284.0	**140**	60.00	464.0	**240**	115.56	644.0	**340**	171.11
287.6	**142**	61.11	467.6	**242**	116.67	647.6	**342**	172.22
291.2	**144**	62.22	471.2	**244**	117.78	651.2	**344**	173.33
294.8	**146**	63.33	474.8	**246**	118.89	654.8	**346**	174.44
298.4	**148**	64.44	478.4	**248**	120.00	658.4	**348**	175.56
302.0	**150**	65.56	482.0	**250**	121.11	662.0	**350**	176.67
305.6	**152**	66.67	485.6	**252**	122.22	665.6	**352**	177.78
309.2	**154**	67.78	489.2	**254**	123.33	669.2	**354**	178.89
312.8	**156**	68.89	492.8	**256**	124.44	672.8	**356**	180.00
316.4	**158**	70.00	496.4	**258**	125.56	676.4	**358**	181.11
320.0	**160**	71.11	500.0	**260**	126.67	680.0	**360**	182.22
323.6	**162**	72.22	503.6	**262**	127.78	683.6	**362**	183.33
327.2	**164**	73.33	507.2	**264**	128.89	687.2	**364**	184.44
330.8	**166**	74.44	510.8	**266**	130.00	690.8	**366**	185.56
334.4	**168**	75.56	514.4	**268**	131.11	694.4	**368**	186.67
338.0	**170**	76.67	518.0	**270**	132.22	698.0	**370**	187.78
341.6	**172**	77.78	521.6	**272**	133.33	701.6	**372**	188.89
345.2	**174**	78.89	525.2	**274**	134.44	705.2	**374**	190.00
348.8	**176**	80.00	528.8	**276**	135.56	708.8	**376**	191.11
352.4	**178**	81.11	532.4	**278**	136.67	712.4	**378**	192.22
356.0	**180**	82.22	536.0	**280**	137.78	716.0	**380**	193.33
359.6	**182**	83.33	539.6	**282**	138.89	719.6	**382**	194.44
363.2	**184**	84.44	543.2	**284**	140.00	723.2	**384**	195.56
366.8	**186**	85.56	546.8	**286**	141.11	726.8	**386**	196.67
370.4	**188**	86.67	550.4	**288**	142.22	730.4	**388**	197.78
374.0	**190**	87.78	554.0	**290**	143.33	734.0	**390**	198.89
377.6	**192**	88.89	557.6	**292**	144.44	737.6	**392**	200.00
381.2	**194**	90.00	561.2	**294**	145.56	741.2	**394**	201.11
384.8	**196**	91.11	564.8	**296**	146.67	744.8	**396**	202.22
388.4	**198**	92.22	568.4	**298**	147.78	748.4	**398**	203.33
392.0	**200**	93.33	572.0	**300**	148.89	752.0	**400**	204.44
395.6	**202**	94.44	575.6	**302**	150.00	755.6	**402**	205.56
399.2	**204**	95.56	579.2	**304**	151.11	759.2	**404**	206.67
402.8	**206**	96.67	582.8	**306**	152.22	762.8	**406**	207.78
406.4	**208**	97.78	586.4	**308**	153.33	766.4	**408**	208.89
410.0	**210**	98.89	590.0	**310**	154.44	770.0	**410**	210.00

Table D. Temperature Conversions* (*Continued*)

°F		°C	°F		°C	°F		°C
773.6	**412**	211.11	953.6	**512**	266.67	1580.0	**860**	460.00
777.2	**414**	212.22	957.2	**514**	267.78	1598.0	**870**	465.56
780.8	**416**	213.33	960.8	**516**	268.89	1616.0	**880**	471.11
784.4	**418**	214.44	964.4	**518**	270.00	1634.0	**890**	476.67
788.0	**420**	215.56	968.0	**520**	271.11	1652.0	**900**	482.22
791.6	**422**	216.67	971.6	**522**	272.22	1670.0	**910**	487.78
795.2	**424**	217.78	975.2	**524**	273.33	1688.0	**920**	493.33
798.8	**426**	218.89	978.8	**526**	274.44	1706.0	**930**	498.89
802.4	**428**	220.00	982.4	**528**	275.56	1724.0	**940**	504.44
806.0	**430**	221.11	986.0	**530**	276.67	1742.0	**950**	510.00
809.6	**432**	222.22	989.6	**532**	277.78	1760.0	**960**	515.56
813.2	**434**	223.33	993.2	**534**	278.89	1778.0	**970**	521.11
816.8	**436**	224.44	996.8	**536**	280.00	1796.0	**980**	526.67
820.4	**438**	225.56	1000.4	**538**	281.11	1814.0	**990**	532.22
824.0	**440**	226.67	1004.0	**540**	282.22	1832.0	**1000**	537.78
827.6	**442**	227.78	1007.6	**542**	283.33	1850.0	**1010**	543.33
831.2	**444**	228.89	1011.2	**544**	284.44	1868.0	**1020**	548.89
834.8	**446**	230.00	1014.8	**546**	285.56	1886.0	**1030**	554.44
838.4	**448**	231.11	1018.4	**548**	286.67	1904.0	**1040**	560.00
842.0	**450**	232.22	1022.0	**550**	287.78	1922.0	**1050**	565.56
845.6	**452**	233.33	1040.0	**560**	293.33	1940.0	**1060**	571.11
849.2	**454**	234.44	1058.0	**570**	298.89	1958.0	**1070**	576.67
852.8	**456**	235.56	1076.0	**580**	304.44	1976.0	**1080**	582.22
856.4	**458**	236.67	1094.0	**590**	310.00	1994.0	**1090**	587.78
860.0	**460**	237.78	1112.0	**600**	315.56	2012.0	**1100**	593.33
863.6	**462**	238.89	1130.0	**610**	321.11	2030.0	**1110**	598.89
867.2	**464**	240.00	1148.0	**620**	326.67	2048.0	**1120**	604.44
870.8	**466**	241.11	1166.0	**630**	332.22	2066.0	**1130**	610.00
874.4	**468**	242.22	1184.0	**640**	337.78	2084.0	**1140**	615.56
878.0	**470**	243.33	1202.0	**650**	343.33	2102.0	**1150**	621.11
881.6	**472**	244.44	1220.0	**660**	348.89	2120.0	**1160**	626.67
885.2	**474**	245.56	1238.0	**670**	354.44	2138.0	**1170**	632.22
888.8	**476**	246.67	1256.0	**630**	360.00	2156.0	**1180**	637.78
892.4	**478**	247.78	1274.0	**690**	365.56	2174.0	**1190**	643.33
896.0	**480**	248.89	1292.0	**700**	371.11	2192.0	**1200**	648.89
899.6	**482**	250.00	1310.0	**710**	376.67	2210.0	**1210**	654.44
903.2	**484**	251.11	1328.0	**720**	382.22	2228.0	**1220**	660.00
906.8	**486**	252.22	1346.0	**730**	387.78	2246.0	**1230**	665.56
910.4	**488**	253.33	1364.0	**740**	393.33	2264.0	**1240**	671.11
914.0	**490**	254.44	1382.0	**750**	398.89	2282.0	**1250**	676.67
917.6	**492**	255.56	1400.0	**760**	404.44	2300.0	**1260**	682.22
921.2	**494**	256.67	1418.0	**770**	410.00	2318.0	**1270**	687.78
924.8	**496**	257.78	1436.0	**780**	415.56	2336.0	**1280**	693.33
928.4	**498**	258.89	1454.0	**790**	421.11	2354.0	**1290**	698.89
932.0	**500**	260.00	1472.0	**800**	426.67	2372.0	**1300**	704.44
935.6	**502**	261.11	1490.0	**810**	432.22	2390.0	**1310**	710.00
939.2	**504**	262.22	1508.0	**820**	437.76	2408.0	**1320**	715.56
942.8	**506**	263.33	1526.0	**830**	443.33	2426.0	**1330**	721.11
946.4	**508**	264.44	1544.0	**840**	448.89	2444.0	**1340**	726.67
950.0	**510**	265.56	1562.0	**850**	454.44	2462.0	**1350**	732.22

Table D. Temperature Conversions* (*Continued*)

°F		°C	°F		°C	°F		°C
2480.0	**1360**	737.78	3380.0	**1860**	1015.6	4280.0	**2360**	1293.3
2498.0	**1370**	743.33	3398.0	**1870**	1021.1	4298.0	**2370**	1298.9
2516.0	**1380**	748.89	3416.0	**1880**	1026.7	4316.0	**2380**	1304.4
2534.0	**1390**	754.44	3434.0	**1890**	1032.2	4334.0	**2390**	1310.0
2552.0	**1400**	760.00	3452.0	**1900**	1037.8	4352.0	**2400**	1315.6
2570.0	**1410**	765.56	3470.0	**1910**	1043.3	4370.0	**2410**	1321.1
2588.0	**1420**	771.11	3488.0	**1920**	1048.9	4388.0	**2420**	1326.7
2606.0	**1430**	776.67	3506.0	**1930**	1054.4	4406.0	**2430**	1332.2
2624.0	**1440**	782.22	3524.0	**1940**	1060.0	4424.0	**2440**	1337.8
2642.0	**1450**	787.78	3542.0	**1950**	1065.6	4442.0	**2450**	1343.3
2660.0	**1460**	793.33	3560.0	**1960**	1071.1	4460.0	**2460**	1348.9
2678.0	**1470**	798.89	3578.0	**1970**	1076.7	4478.0	**2470**	1354.4
2696.0	**1480**	804.44	3596.0	**1980**	1082.2	4496.0	**2480**	1360.0
2714.0	**1490**	810.00	3614.0	**1990**	1087.8	4514.0	**2490**	1365.6
2732.0	**1500**	815.56	3632.0	**2000**	1093.3	4532.0	**2500**	1371.1
2750.0	**1510**	821.11	3650.0	**2010**	1098.9	4550.0	**2510**	1376.7
2768.0	**1520**	826.67	3668.0	**2020**	1104.4	4568.0	**2520**	1382.2
2786.0	**1530**	832.22	3686.0	**2030**	1110.0	4586.0	**2530**	1387.8
2804.0	**1540**	837.78	3704.0	**2040**	1115.6	4604.0	**2540**	1393.3
2822.0	**1550**	843.33	3722.0	**2050**	1121.1	4622.0	**2550**	1398.9
2840.0	**1560**	848.89	3740.0	**2060**	1126.7	4640.0	**2560**	1404.4
2858.0	**1570**	854.44	3758.0	**2070**	1132.2	4658.0	**2570**	1410.0
2876.0	**1580**	860.00	3776.0	**2080**	1137.8	4676.0	**2580**	1415.6
2894.0	**1590**	865.56	3794.0	**2090**	1143.3	4694.0	**2590**	1421.1
2912.0	**1600**	871.11	3812.0	**2100**	1148.9	4712.0	**2600**	1426.7
2930.0	**1610**	876.67	3830.0	**2110**	1154.4	4730.0	**2610**	1432.2
2948.0	**1620**	882.22	3848.0	**2120**	1160.0	4748.0	**2620**	1437.8
2966.0	**1630**	887.78	3866.0	**2130**	1165.6	4766.0	**2630**	1443.3
2984.0	**1640**	893.33	3884.0	**2140**	1171.1	4784.0	**2640**	1448.9
3002.0	**1650**	898.89	3902.0	**2150**	1176.7	4802.0	**2650**	1454.4
3020.0	**1660**	904.44	3920.0	**2160**	1182.2	4820.0	**2660**	1460.0
3038.0	**1670**	910.00	3938.0	**2170**	1187.8	4838.0	**2670**	1465.6
3056.0	**1680**	915.56	3956.0	**2180**	1193.3	4856.0	**2680**	1471.1
3074.0	**1690**	921.11	3974.0	**2190**	1198.9	4874.0	**2690**	1476.7
3092.0	**1700**	926.67	3992.0	**2200**	1204.4	4892.0	**2700**	1482.2
3110.0	**1710**	932.22	4010.0	**2210**	1210.0	4910.0	**2710**	1487.8
3128.0	**1720**	937.78	4028.0	**2220**	1215.6	4928.0	**2720**	1493.3
3146.0	**1730**	943.33	4046.0	**2230**	1221.1	4946.0	**2730**	1498.9
3164.0	**1740**	948.89	4064.0	**2240**	1226.7	4964.0	**2740**	1504.4
3182.0	**1750**	954.44	4082.0	**2250**	1232.2	4982.0	**2750**	1510.0
3200.0	**1760**	960.00	4100.0	**2260**	1237.8	5000.0	**2760**	1515.6
3218.0	**1770**	965.56	4118.0	**2270**	1243.3	5018.0	**2770**	1521.1
3236.0	**1780**	971.11	4136.0	**2280**	1248.9	5036.0	**2780**	1526.7
3254.0	**1790**	976.67	4154.0	**2290**	1254.4	5054.0	**2790**	1532.2
3272.0	**1800**	982.22	4172.0	**2300**	1260.0	5072.0	**2800**	1537.8
3290.0	**1810**	987.78	4190.0	**2310**	1265.6	5090.0	**2810**	1543.3
3308.0	**1820**	993.33	4208.0	**2320**	1271.1	5108.0	**2820**	1548.9
3326.0	**1830**	998.89	4226.0	**2330**	1276.7	5126.0	**2830**	1554.4
3344.0	**1840**	1004.4	4244.0	**2340**	1282.2	5144.0	**2840**	1560.0
3362.0	**1850**	1010.0	4262.0	**2350**	1287.8	5162.0	**2850**	1565.6

Table D. Temperature Conversions* (*Continued*)

°F		°C	°F		°C	°F		°C
5180.0	**2860**	1571.1	5702.0	**3150**	1732.2	7952.0	**4400**	2426.6
5198.0	**2870**	1576.7	5792.0	**3200**	1760.0	8042.0	**4450**	2454.4
5216.0	**2880**	1582.2	5882.0	**3250**	1787.7	8132.0	**4500**	2482.2
5234.0	**2890**	1587.8	5972.0	**3300**	1815.5	8222.0	**4550**	2510.0
5252.0	**2900**	1593.3	6062.0	**3350**	1843.3	8312.0	**4600**	2537.7
5270.0	**2910**	1598.9	6152.0	**3400**	1871.1	8402.0	**4650**	2565.5
5288.0	**2920**	1604.4	6242.0	**3450**	1898.8	8492.0	**4700**	2593.3
5306.0	**2930**	1610.0	6332.0	**3500**	1926.6	8582.0	**4750**	2621.1
5324.0	**2940**	1615.6	6422.0	**3550**	1954.4	8672.0	**4800**	2648.8
5342.0	**2950**	1621.1	6512.0	**3600**	1982.2	8762.0	**4850**	2676.6
5360.0	**2960**	1626.7	6602.0	**3650**	2010.0	8852.0	**4900**	2704.4
5378.0	**2970**	1632.2	6692.0	**3700**	2037.7	8942.0	**4950**	2732.2
5396.0	**2980**	1637.8	6782.0	**3750**	2065.5	9032.0	**5000**	2760.0
5414.0	**2990**	1643.3	6872.0	**3800**	2093.3	9122.0	**5050**	2787.7
5432.0	**3000**	1648.9	6962.0	**3850**	2121.1	9212.0	**5100**	2815.5
5450.0	**3010**	1654.4	7052.0	**3900**	2148.8	9302.0	**5150**	2843.3
5468.0	**3020**	1660.0	7142.0	**3950**	2176.6	9392.0	**5200**	2871.1
5486.0	**3030**	1665.6	7232.0	**4000**	2204.4	9482.0	**5250**	2898.8
5504.0	**3040**	1671.1	7322.0	**4050**	2232.2	9572.0	**5300**	2926.6
5522.0	**3050**	1676.7	7412.0	**4100**	2260.0	9662.0	**5350**	2954.4
5540.0	**3060**	1682.2	7502.0	**4150**	2287.7	9752.0	**5400**	2982.2
5558.0	**3070**	1687.8	7592.0	**4200**	2315.5	9842.0	**5450**	3010.0
5576.0	**3080**	1693.3	7682.0	**4250**	2343.3	9932.0	**5500**	3037.7
5594.0	**3090**	1698.9	7772.0	**4300**	2371.1	10,022.0	**5550**	3065.5
5612.0	**3100**	1704.4	7862.0	**4350**	2398.8	10,112.0	**5600**	3093.3

* From "ASM Metals Handbook," 1948 ed., by permission of the American Society for Metals.

Table E. Interconversion of Atomic and Weight Percentages

Values of $\log \frac{x}{100 - x} + 10$

$x = 0\text{–}4.99$

x	0	0.01	0.02	0.03	0.04	0.05	0.06	0.07	0.08	0.09
0.0	$-\infty$	6.0000	3011	4772	6022	6992	7784	8454	9034	9546
.1	7.0004	7.0419	0797	1145	1467	1767	2048	2312	2560	2796
.2	3019	3231	3434	3627	3812	3990	4161	4325	4484	4637
.3	4784	4927	5065	5200	5330	5456	5579	5698	5814	5928
.4	6038	6146	6251	6353	6454	6552	6648	6742	6833	6923
0.5	7.7012	7.7098	7183	7266	7347	7428	7506	7584	7660	7734
.6	7808	7800	7951	8021	8090	8157	8224	8290	8355	8419
.7	8482	8544	8605	8665	8725	8783	8841	8899	8955	9011
.8	9066	9120	9174	9227	9279	9331	9382	9433	9483	9534
.9	9582	9630	9678	9725	9772	9819	9865	9910	9955	8.0000
1.0	8.0044	8.0087	0130	0173	0216	0258	0299	0340	0371	0422
.1	0462	0502	0541	0580	0619	0657	0695	0733	0770	0808
.2	0844	0881	0917	0953	0988	1024	1059	1094	1128	1162
.3	1196	1230	1263	1297	1330	1362	1395	1427	1459	1491
.4	1522	1554	1585	1616	1647	1677	1707	1738	1767	1797
1.5	8.1826	8.1856	1885	1914	1943	1971	2000	2028	2056	2084
.6	2111	2139	2166	2193	2220	2247	2274	2300	2327	2353
.7	2379	2405	2431	2456	2482	2507	2532	2557	2582	2607
.8	2632	2656	2680	2705	2729	2753	2777	2800	2824	2848
.9	2871	2894	2917	2940	2963	2986	3009	3031	3054	3076
2.0	8.3098	8.3120	3142	3164	3186	3208	3229	3250	3272	3293
.1	3314	3335	3356	3377	3398	3419	3439	3460	3480	3501
.2	3521	3541	3561	3581	3601	3621	3640	3660	3680	3699
.3	3718	3738	3757	3776	3795	3814	3833	3852	3870	3889
.4	3908	3926	3945	3963	3981	3999	4018	4036	4054	4072
2.5	8.4089	4107	4125	4142	4160	4178	4195	4212	4230	4247
.6	4264	4281	4298	4315	4332	4349	4366	4383	4399	4416
.7	4432	4449	4466	4482	4498	4514	4531	4547	4563	4579
.8	4595	4611	4627	4643	4658	4674	4690	4705	4721	4736
.9	4752	4767	4782	4798	4813	4828	4843	4858	4874	4888
3.0	8.4904	4918	4933	4948	4963	4978	4992	5007	5021	5036
.1	5050	5065	5079	5094	5108	5122	5136	5150	5165	5179
.2	5193	5207	5221	5235	5248	5262	5276	5290	5304	5317
.3	5331	5344	5358	5372	5385	5398	5402	5425	5438	5452
.4	5465	5478	5491	5504	5518	5531	5544	5557	5570	5583

Table E. Interconversion of Atomic and Weight Percentages (*Continued*)

x = 0–4.99

x	0	0.01	0.02	0.03	0.04	0.05	0.06	0.07	0.08	0.09
3.5	8.5595	5608	5621	5634	5646	5659	5672	5685	5697	5710
.6	5722	5735	5747	5760	5772	5784	5797	5809	5821	5834
.7	5846	5858	5870	5882	5894	5906	5918	5930	5942	5954
.8	5966	5978	5990	6002	6013	6025	6037	6048	6060	6072
.9	6083	6095	6107	6118	6130	6141	6152	6164	6175	6186
4.0	8.6198	6209	6220	6232	6243	6254	6265	6276	6288	6299
.1	6310	6321	6332	6343	6354	6365	6375	6386	6397	6408
.2	6419	6430	6440	6451	6462	6472	6483	6494	6504	6515
.3	6526	6536	6547	6557	6568	6578	6588	6599	6609	6620
.4	6630	6640	6650	6661	6671	6681	6691	6702	6712	6722
4.5	8.6732	6742	6752	6762	6772	6782	6792	6802	6812	6822
.6	6832	6842	6852	6862	6872	6881	6891	6901	6911	6920
.7	6930	6940	6949	6959	6969	6978	6988	6998	7007	7017
.8	7026	7036	7045	7054	7064	7073	7083	7092	7102	7111
.9	7120	7130	7139	7148	7157	7167	7176	7185	7194	7203

x = 5.0–98.9

x	0	0.1	0.2	0.3	0.4	0.5	0.6	0.7	0.8	0.9
5	8.7212	7303	7392	7479	7565	7649	7732	7814	7894	7973
6	8050	8127	8202	8276	8349	8421	8492	8562	8631	8699
7	8766	8832	8898	8962	9026	9089	9151	9213	9274	9334
8	9393	9452	9510	9567	9624	9680	9736	9791	9845	9899
9	9952	9.0005	0057	0109	0160	0211	0261	0311	0360	0409
10	9.0458	0506	0553	0600	0647	0694	0740	0785	0831	0876
11	0920	0964	1008	1052	1095	1138	1180	1222	1264	1306
12	1347	1388	1429	1469	1509	1549	1589	1628	1667	1706
13	1744	1783	1821	1858	1896	1933	1970	2007	2044	2080
14	2116	2152	2188	2224	2259	2294	2329	2364	2398	2433
15	9.2467	2501	2534	2568	2602	2635	2668	2701	2734	2766
16	2798	2831	2863	2895	2926	2958	2989	3021	3052	3083
17	3114	3145	3175	3205	3236	3266	3296	3326	3356	3385
18	3415	3444	3473	3502	3531	3560	3589	3618	3646	3674
19	3703	3731	3759	3787	3815	3842	3870	3898	3925	3952
20	9.3979	4007	4034	4060	4087	4114	4141	4167	4193	4220
21	4246	4272	4298	4324	4350	4376	4401	4427	4453	4478
22	4503	4529	4554	4579	4604	4629	4654	4679	4703	4728
23	4752	4777	4801	4826	4850	4874	4898	4922	4946	4970
24	4994	5018	5042	5065	5089	5112	5136	5159	5182	5206

Table E. Interconversion of Atomic and Weight Percentages (*Continued*)
x = 5.0–98.9

x	0	0.1	0.2	0.3	0.4	0.5	0.6	0.7	0.8	0.9
25	9.5229	5252	5275	5298	5321	5344	5367	5389	5412	5435
26	5457	5480	5502	5525	5547	5570	5592	5614	5636	5658
27	5680	5702	5724	5746	5768	5790	5812	5833	5855	5877
28	5898	5920	5941	5963	5984	6005	6027	6048	6069	6090
29	6111	6132	6154	6175	6196	6216	6237	6258	6279	6300
30	9.6320	6341	6362	6382	6403	6423	6444	6464	6484	6505
31	6525	6545	6566	6586	6606	6626	6646	6666	6686	6706
32	6726	6746	6766	6786	6806	6826	6846	6865	6885	6905
33	6924	6944	6964	6983	7003	7022	7042	7061	7081	7100
34	7119	7139	7158	7177	7197	7216	7235	7254	7273	7293
35	9.7312	7331	7350	7369	7388	7407	7426	7445	7463	7482
36	7501	7520	7539	7558	7576	7595	7614	7633	7651	7670
37	7689	7707	7726	7744	7763	7781	7800	7819	7837	7856
38	7874	7892	7911	7929	7948	7966	7984	8003	8021	8039
39	8057	8076	8094	8112	8130	8148	8167	8185	8203	8221
40	9.8239	8257	8275	8293	8311	8329	8347	8365	8383	8401
41	8419	8437	8455	8473	8491	8509	8527	8545	8563	8580
42	8598	8616	8634	8652	8670	8637	8705	8723	8741	8758
43	8776	8794	8811	8829	8847	8864	8882	8900	8917	8935
44	8953	8970	8988	9005	9023	9041	9058	9076	9093	9111
45	9.9129	9146	9164	9181	9199	9216	9234	9251	9269	9286
46	9304	9321	9339	9356	9374	9391	9409	9426	9443	9461
47	9478	9496	9513	9531	9548	9565	9583	9600	9618	9635
48	9652	9670	9687	9705	9722	9739	9757	9774	9792	9809
49	9826	9844	9861	9878	9896	9913	9931	9948	9965	9983
50	10.0000	0017	0035	0052	0070	0087	0104	0122	0139	0156
51	0174	0191	0209	0226	0243	0261	0278	0295	0313	0330
52	0348	0365	0382	0400	0417	0435	0452	0470	0487	0504
53	0522	0539	0557	0574	0592	0609	0626	0644	0661	0679
54	0696	0714	0731	0749	0766	0784	0801	0819	0836	0854
55	10.0872	0889	0907	0924	0942	0959	0977	0995	1012	1030
56	1047	1065	1083	1100	1118	1136	1153	1171	1189	1206
57	1224	1242	1260	1277	1295	1313	1331	1348	1366	1384
58	1402	1420	1437	1455	1473	1491	1509	1527	1545	1563
59	1581	1599	1617	1635	1653	1671	1689	1707	1725	1743
60	10.1761	1779	1797	1815	1833	1852	1870	1888	1906	1924
61	1943	1961	1979	1998	2016	2034	2053	2071	2089	2108
62	2126	2145	2163	2182	2200	2219	2237	2256	2274	2293
63	2311	2330	2349	2367	2386	2405	2424	2442	2461	2480
64	2499	2518	2537	2555	2574	2593	2612	2631	2650	2669

Table E. Interconversion of Atomic and Weight Percentages (*Continued*)
$x = 5.0\text{–}98.9$

x	0	0.1	0.2	0.3	0.4	0.5	0.6	0.7	0.8	0.9
65	10.2688	2708	2727	2746	2765	2784	2803	2823	2842	2861
66	2881	2900	2919	2939	2958	2978	2997	3017	3036	3056
67	3076	3095	3115	3135	3154	3174	3194	3214	3234	3254
68	3274	3294	3314	3334	3354	3374	3394	3414	3434	3455
69	3475	3495	3516	3536	3556	3577	3597	3618	3639	3659
70	10.3680	3701	3721	3742	3763	3784	3805	3826	3847	3868
71	3889	3910	3931	3952	3973	3995	4016	4037	4059	4080
72	4102	4123	4145	4167	4188	4210	4232	4254	4276	4298
73	4320	4342	4364	4386	4408	4430	4453	4475	4498	4520
74	4543	4565	4588	4611	4633	4656	4679	4702	4725	4748
75	10.4771	4794	4818	4841	4864	4888	4911	4935	4959	4982
76	5006	5030	5054	5078	5102	5126	5150	5174	5199	5223
77	5248	5272	5297	5322	5346	5371	5396	5421	5446	5472
78	5497	5522	5548	5573	5599	5624	5650	5676	5702	5728
79	5754	5780	5807	5833	5860	5886	5913	5940	5967	5994
80	10.6021	6048	6075	6103	6130	6158	6185	6213	6241	6269
81	6297	6326	6354	6383	6411	6440	6469	6498	6527	6556
82	6585	6615	6645	6674	6704	6734	6764	6795	6825	6856
83	6886	6917	6948	6979	7011	7042	7074	7105	7137	7169
84	7202	7234	7267	7299	7332	7365	7398	7432	7466	7499
85	10.7533	7567	7602	7636	7671	7706	7741	7776	7812	7848
86	7884	7920	7956	7993	8030	8067	8104	8142	8180	8218
87	8256	8294	8333	8372	8411	8451	8491	8531	8571	8612
88	8653	8694	8736	8778	8820	8862	8905	8948	8992	9036
89	9080	9124	9169	9215	9260	9306	9353	9400	9447	9494
90	10.9542	9591	9640	9689	9739	9789	9840	9891	9943	9995
91	11.0048	0101	0155	0210	0265	0320	0376	0433	0490	0548
92	0607	0666	0726	0787	0849	0911	0974	1038	1102	1168
93	1234	1301	1369	1438	1508	1579	1651	1742	1798	1873
94	1950	2027	2106	2186	2268	2351	2435	2521	2608	2697
95	11.2788	2880	2974	3070	3168	3268	3370	3474	3581	3690
96	3802	3917	4034	4154	4278	4405	4535	4669	4807	4950
97	5096	5248	5405	5568	5736	5911	6092	6282	6479	6686
98	6902	7129	7368	7621	7889	8174	8478	8804	9156	9538

Table E. Interconversion of Atomic and Weight Percentages (*Continued*)
x = 99.0–99.99

x	0	0.01	0.02	0.03	0.04	0.05	0.06	0.07	0.08	0.09
99.0	11.9956	12.0000	0045	0090	0135	0181	0228	0275	0322	0370
99.1	12.0418	0466	0517	0567	0618	0669	0721	0773	0826	0880
99.2	0934	0989	1045	1101	1159	1217	1275	1335	1395	1456
99.3	1518	1581	1645	1710	1776	1843	1910	1979	2049	2120
99.4	2192	2266	2340	2416	2494	2572	2653	2734	2817	2902
99.5	2988	3077	3167	3258	3352	3448	3546	3647	3749	3854
99.6	3962	4072	4186	4302	4421	4544	4670	4800	4935	5073
99.7	5216	5363	5516	5675	5839	6010	6188	6373	6566	6769
99.8	6981	7204	7440	7688	7952	8233	8533	8855	9203	9581
99.9	9996	13.0454	0966	1546	2216	3008	3978	5228	6989	14.0000

Add value of log (A/B) when converting from atomic to weight percentage.
Subtract value of log (A/B) when converting from weight to atomic percentage.

Index of Elements

Systems with a diagram are indicated by an asterisk(*)